MICROPROCESSOR SUPPORT CHIPS SOURCEBOOK

MICROPROCESSOR SUPPORT CHIPS SOURCEBOOK

ALAN CLEMENTS

McGRAW-HILL BOOK COMPANY

London · New York · St Louis · San Francisco · Auckland · Bogotá
Caracas · Hamburg · Lisbon · Madrid · Mexico · Milan · Montreal
New Delhi · Panama · Paris · San Juan · São Paulo · Singapore
Sydney · Tokyo · Toronto

Published by McGRAW-HILL Book Company Europe
Shoppenhangers Road, Maidenhead, Berkshire, SL62QL, England
Telephone 0628 23432   Fax 0628 770224

British Library Cataloguing in publication Data

Clements, Alan, *1948–*
  Microprocessor support chips sourcebook.
  I. Title
  6231.3916

  ISBN 0-07-707463-7

Library of Congress Cataloguing in publication Data

Clements, Alan, *1948–*
  Microprocessor support chips sourcebook / Alan Clements.
     p.   cm.
  Includes bibliographical references and index.
  1. Microprocessors.   2. Computer interfaces.   3. System design.
I. Title
TK7895.M5C57   1991
004.16—dc20   91–28246
                                                         CIP

Copyright © 1991 McGraw-Hill International (UK) Limited. All rights reserved. No part of this publication may be reproduced, stored in a retrieval system, or transmitted, in any form or by any means, electronic, mechanical, photocopying, recording, or otherwise, without the prior permission of the authors.

234 CL 932

Printed and bound in Great Britain by Clays Ltd, St. Ives, plc.

Contents

Preface	xi

Chapter 1 POWER SUPPLIES

Theory and applications of the MC34063 and µA 78S40 switching regulator control circuits
Motorola Inc., Application note AN920
By Jade Alberkrack — 2

TCA5600 Universal microprocessor power supply controller
Motorola Inc., Data sheet — 39

LM2984C Microprocessor power supply controller
National Semiconductor Corp., Data sheet — 51

MC34164/MC33164 Micropower undervoltage sensing circuits
Motorola Inc., Data sheet — 65

Chapter 2 BUSES

Transmission line concepts
National Semiconductor Corp., FAST Applications Handbook — 75

Summary of electrical characteristics of some well known digital interface standards
National Semiconductor Corp., Application Note AN-216
By Don Tarver — 126

Transmission line drivers and receivers for EIA standards RS-422 and RS-423
National Semiconductor Corp., Application Note AN-214
By John Abbott — 139

Transceivers and repeaters meeting the EIA RS-485 interface standard
National Semiconductor Corp., Application Note AN-409
By Sivakumar Sivasothy — 149

Chapter 3 DIGITAL LOGIC AND SYSTEMS DESIGN

Texas instruments logic families – technology and characteristics 157
Texas Instruments, Inc.

State machine design 217
Monolithic Memories

Testability 246
Monolithic Memories

Designer's guide to high performance low-power Schottky logic 268
AMD, by David A. Laws and Roy J. Levy

A metastability primer 282
Philips components, Application Note AN219 by Charles Dike

Arbitration in shared resource systems 286
Philips Components, Application Note AN216

Multiple μP interfacing with FAST ICs 291
Philips Components, Application Note AN207

PHD16N8-5 Programmable high-speed decoder logic ($16 \times 16 \times 8$) 304
Philips Components, Data sheet

Chapter 4 MEMORY SYSTEMS

MCM6206 32K × 8-bit fast static random access memory 313
Motorola data sheet

HM628128 131072-word × 8–bit high-speed CMOS static RAM 319
Hitachi data sheet

Memory design for the low power microsystem environment 329
Intel coporation, March 1985, Application note AP-238
by Dennis Knudson

Am27H010 1 Megabit (131,072 × 8-bit) high speed CMOS EPROM 360
Advanced Micro Devices, Data Sheet

27210 IM (641k × 16) Word-wide EPROM 378
Intel Corp., Data sheet

27C203 Fast pipelined 256K (16k × 16) EPROM 388
Intel Corp., Data sheet

CONTENTS

28F020 2048K (256K × 8) CMOS flash memory 404
Intel Corp., Data sheet

Guide to FLASH memory reprogramming 432
Intel Corp., Application Note AP-325
By Saul Zales

ETOX flash memory reliability data summary 456
Intel Corp., Reliability report RR-60

E/M28HC256 256K High speed EPROM 493
SEEQ Technology, Inc., Data sheet

E^2ROM interfacing 503
SEEQ Technology, Incorporated

MCM 514100 4M × 1 CMOS dynamic RAM 520
Motorola data sheet

Page, nibble, and static column modes: high-speed, serial-access options on 1 M-bit+ DRAMs 538
Motorola Application Note AN986

MCM 514100 4M × 1 CMOS dynamic memory controller (DMC) 542
Motorola Inc., Data sheet

DP8420A/21A/22A microCMOS programmable 256K/1M/4M dynamic RAM controller/drivers 562
National Semiconductor Data sheet

Interfacing the DP8420A/21A/22A to the 68000/008/010 634
National Semiconductor Corp., Application note AN-538
By Joe Tate and Rusty Meier

High-speed cache directory optimizes throuput of new high-end microprocessors 650
Texas Instruments, Inc.
By Nicholas Efthymiou and Loren Schiele

SN74ACT21562. SN74ACT2154 2K × 8 cache address comparators 661
Texas Instruments Inc., Data sheet

Cache tag RAM chips simplify cache memory design 674
Integrated Device Technology Inc., Application note AN-07
By David C. Wyland

Error detection and correction using SN54/74LS630 or SN54/74LS631 683
Texas Instruments, Inc., Application report CA-201
By Dale Hunt and Thomas J. Tyson

16-bit parallel error detection and correction circuits 695
Texas Instruments, Inc., Data sheet

The IDT FourPort RAM facilitates multiprocessor designs 701
Integrated Device Technology Inc., Application note AN-43
By Robert Stodiek

Introdution to IDT's FourPort RAM 707
Integrated Device Technology Inc., Application note AN-45
By John R. Mick

Understanding the IDT7201/7202 FIFO 720
Integrated Device Technology Inc., Application note AN-01
By Michael J. Miller

Dual-port RAMs simplify communication in computer systems 728
Integrated Device Technology Inc., Application note AN-02
By David C. Wyland

IDT7201 CMOS parallel first-in/first-out FIFO 741
Integrated Device Technology Inc.

Chapter 5 PERIPHERALS

SCN68562 Dual universal communications controller (DUSCC) 755
Signetics data sheet

VL82C106 PC/AT combo I/O chip 803
VLSI Technology, Inc., Data sheet

DP8490 Enhanced asynchronous SCSI interface (EASI) 839
National Semiconductor Corp., Data sheet

A SCSI printer controller using either the DP8490 EASI or DP5380 879
ASI and users guide
National Semiconductor Corp., Application note AN-563
By Andrew M. Davidson

MM58274C Microprocessor compatible real time clock 899
National Semiconductor Corp., Data sheet

The MM58274C adds reliable real-time keeping to any microprocessor 912
system
National Semiconductor Corp., Application note AN-365
By Peter K. Thomson

CONTENTS

DP8473 Floppy disk controller PLUS 2 National Semiconductor Corp.	928
Floppy disk data separator design guide for the DP8473 National Semiconductor Corp., Application note AN-505 By B. Lutz, P. Melloni, L. Wakeman	954
Design guide for DP8473 in a PC-AT National Semiconductor Corp., Application note AN-631 By Robert Lutz	983
8005 Advanced ethernet data link controller SEEQ Technology Incorporated, Data sheet	993
8020 MCC Manchester code converter SEEQ Technology Incorporated, Data sheet	1033
8005 Advanced EDLC user's guide SEEQ Technology Incorporated, Applications brief 7	1047

Preface

The Microcomputer Peripheral Sourcebook offers a selection of the best tutorial material written by semiconductor manufacturers together with data sheets describing a wide range of microprocessor peripherals. I have defined the term 'peripheral' to include many of the components found in a microcomputer (e.g., memory devices and buses).

Engineers and academics employ two sources of information when they design microprocessor systems: primary sources and secondary sources. Secondary sources are the microprocessor textbooks written by academics and engineers. The advantage of using a secondary source of information is that the writer had carried out a lot of work for you. He or she has collated a large amount of primary information, digested it and summarized it in a readable form. The disadvantage is that the writer has decided what should and should not be included and has interpreted the source material in his or her own way (which may not always be correct).

Primary sources of information are produced by semiconductor manufacturers to describe and define their products. This book is composed of primary information sources and can be used as a reference when either designing or studying microprocessor systems.

We have not attempted to provide little more than a catalogue of device types. This book covers a series of important topics in microprocessor systems design; it is not microprocessor-specific because it deals with the design of the whole system. It is not concerned with how the microprocessor itself works or is programmed, and its contents are both broader and deeper than existing texts. This is partially because details of microprocessors themselves are omitted (which leaves much more space for peripherals) and basic introductory material is not included.

We look at many of the important design issues relevant to anyone involved in microprocessor systems design and include bread and butter topics often neglected in other texts (such as the power supply). In particular, we concentrate on topics fundamental to the design of all microprocessor systems (e.g., the memory, and bus technology). Texts that deal with specific microprocessors devote so much time to the basics that they have insufficient space to cover allied topics in microprocessor systems design. The designer is expected to learn all about these 'missing topics' when he or she is engaged in professional activity (after they have graduated).

By taking some of the best papers, articles and application notes from leading semiconductor manufacturers, the Microcomputer Peripheral Sourcebook is a comprehensive and authoritative source book with a strongly tutorial nature. it is not simply a design cookbook, as that would be too specific and would rapidly become out of date. Engineers need to know the underlying design principles of any circuit or system. Since the material is supplied by major component manufacturers, it is highly practical. Another use of this book is to give the student an idea of how components are actually used in practice.

The editor of a book covering a range of microprocessor interface components and peripherals devices has to decide whether to aim for depth or for breadth. We have chosen depth over breadth. Simply providing a list of semiconductor devices along with their pinouts and other random parameters is of little use. We have selected fewer devices and have provided extensive application notes where appropriate. Some chapters include very little information about actual devices and cover the basic principles involved in the design of part of a microcomputer (e.g., buses). Indeed, this book is intended to be as much a design reference book as a component reference book.

As we have already said, the material selected for this book is not based on any specific microprocessor but is relevant to almost all devices. For once, both Intel and Motorola designers will be able to read it. Moreover, it will be applicable to both 8-bit designers and 16/32-bit designers.

We cover many vital practical topics that are frequently skipped over in conventional course books. This omission arises because some of the material is not considered as 'mainstream' and does not fit into existing academic courses.

Chapter one begins at the beginning with the microprocessor's power supply. Power supplies are not as glamorous as the other parts of a microcomputer and developments in power supply technology are often neglected. We first look at the switching power supply and then describe components that sit between the microprocessor and the power supply (e.g., the under-voltage detector).

The various sub-units that make up a microcomputer communicate with each other by means of serial and parallel buses. Chapter two describes the basic principles underlying the design of buses. For example, we look at the transmission line behaviour of buses. Because microprocessors are becoming faster, the traditional back-plane bus can no longer be regarded as no more that a 'length of wire that just moves signals from A to B'.

Chapter three covers topics of interest to digital designers. Although LSI and VLSI chips perform most of the computation, storage, and control functions in a microcomputer, standard MSI and LSI logic elements are needed to glue together all these large chips. In this chapter we discuss basic topics like the characteristics of logic families and also some more esoteric topics like metastability.

The largest chapter in this book is chapter four that deals with memory components. A few years ago, there were just a couple of types of basic semiconductor memory element. Today, the systems designer has a much wider range of memory components to choose from. In chapter four we provide sample data sheets for some of the most important categories of memory component.

Memory systems design involves much more than selecting the appropriate storage device. Chapter four also covers some special-purpose devices such as multiport memories, FIFOs, cache memories and error detecting memories.

Chapter five describes some of the peripheral chips that interface a microprocessor to the outside world or to peripherals such as disk drives. We have not attempted to be comprehensive and include a little material about many devices. Instead we have concentrated on some of the most important chips (the serial interface, the SCSI controller, the disk drive controller, and the local area network controller).

Chapter 1 Power supplies

There are two ways of dealing with a microcomputer's power supply. The first is to buy a ready-made power supply and the second is to build your own. You might wish to construct your own power supply for reasons of economics, or because no existing supply conforms to your requirements.

A few years ago most microcomputer power supplies employed the linear technology of the transformer, diode bridge and regulator. Today, the need for high currents and compact supplies has made the switching regulator a popular choice with designers. The first paper in this chapter is an application note that describes how a switching regulator operates and how it can be used to construct a power supply to your own specifications.

The second data sheet describes a power supply controller that has been designed specifically for battery and automotive applications. Besides general purpose power supply controllers we include the data sheet of a device that performs a whole series of functions in a microprocessor system. For example, it provides three independent output supplies and even resets the microprocessor during the power-up phase.

Semiconductor manufacturers have designed several chips that sit between a conventional power supply and a microprocessor and carry out a monitoring function. We have included the data sheet of one simple three-terminal device that simply monitors the output of the power supply and resets the processor if the voltage drops below a preset limit. This device can be used to force an orderly shut down before the loss of power becomes so serious that the system can no longer operate.

MOTOROLA SEMICONDUCTOR APPLICATION NOTE

Order this document by AN920/D Rev. 2

AN920
Rev. 2

Theory and Applications of the MC34063 and µA78S40 Switching Regulator Control Circuits

Prepared by
Jade Alberkrack

This paper describes in detail the principle of operation of the MC34063 and µA78S40 switching regulator subsystems. Several converter design examples and numerous applications circuits with test data are included.

INTRODUCTION

The MC34063 and µA78S40 are monolithic switching regulator subsystems intended for use as dc to dc converters. These devices represent a significant advancement in the ease of implementing highly efficient and yet simple switching power supplies. The use of switching regulators is becoming more pronounced over that of linear regulators because the size reductions in new equipment designs require greater conversion efficiency. Another major advantage of the switching regulator is that it has increased application flexibility of output voltage. The output can be less than, greater than, or of opposite polarity to that of the input voltage.

PRINCIPAL OF OPERATION

In order to understand the difference in operation between linear and switching regulators we must compare the block diagrams of the two step-down regulators shown in Figure 1. The linear regulator consists of a stable reference, a high gain error amplifier, and a variable resistance series-pass element. The error amplifier monitors the output voltage level, compares it to the reference and generates a linear control signal that varies between two extremes, saturation and cutoff. This signal is used to vary the resistance of the series-pass element in a corrective fashion in order to maintain a constant output voltage under varying input voltage and output load conditions.

The switching regulator consists of a stable reference and a high gain error amplifier identical to that of the linear regulator. This system differs in that a free running oscillator and a gated latch have been added. The error amplifier again monitors the output voltage, compares it to the reference level and generates a control signal. If the output voltage is below nominal, the control signal will go to a high state and turn on the gate, thus allowing the oscillator clock pulses to drive the series-pass element alternately from cutoff to saturation. This will continue until the output voltage is pumped up slightly above its nominal value. At this time, the control signal will go low and turn off the gate, terminating any further switching of the series-pass element. The output voltage will eventually decrease to below nominal due to the presence of an external load, and will initiate the switching process again. The increase in conversion efficiency is primarily due to the operation of the series-pass element only in the saturated or cutoff state. The voltage drop across the element, when saturated, is small as is the dissipation. When in cutoff, the current through the element and likewise the power dissipation are also small. There are other variations of switching control. The most common are the fixed frequency pulse width modulator and the fixed on-time variable off-time types, where the on-off switching is uninterrupted and regulation is achieved by duty cycle control. Generally speaking, the example given in Figure 1b does apply to MC34063 and µA78S40.

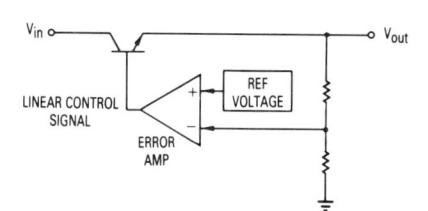

a. Linear Regulator

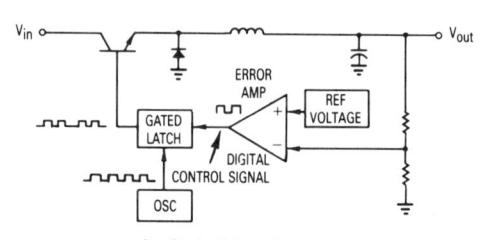

b. Switching Regulator

Figure 1. Step-Down Regulators

©MOTOROLA INC., 1989

GENERAL DESCRIPTION

The MC34063 series is a monolithic control circuit containing all the active functions required for dc to dc converters. This device contains an internal temperature compensated reference, comparator, controlled duty cycle oscillator with an active peak current limit circuit, driver, and a high current output switch. This series was specifically designed to be incorporated in step-up, step-down and voltage-inverting converter applications. These functions are contained in an 8 pin dual in-line package shown in Figure 2a.

The µA78S40 is identical to the MC34063 with the addition of an on-board power catch diode, and an uncommitted operational amplifier. This device is in a 16 pin dual in-line package which allows the reference and the non-inverting input of the comparator to be pinned out. These additional features greatly enhance the flexibility of this part and allow the implementation of more sophisticated applications. These may include series-pass regulation of the main output or of a derived second output voltage, a tracking regulator configuration or even a second switching regulator.

FUNCTIONAL DESCRIPTION

The oscillator is composed of a current source and sink which charges and discharges the external timing capacitor C_T between an upper and lower preset threshold. The typical charge and discharge currents are 35 µA and 200 µA respectively, yielding about a 1 to 6 ratio. Thus the ramp-up period is 6 times longer than that of the ramp-down as shown in Figure 3. The upper threshold is equal to the internal reference voltage of 1.25 volts and the lower is approximately equal to 0.75 V. The oscillator runs continuously at a rate controlled by the selected value of C_T.

During the ramp-up portion of the cycle, a Logic '1' is present at the 'A' input of the AND gate. If the output voltage of the switching regulator is below nominal, a Logic '1' will also be present at the 'B' input. This condition will set the latch and cause the 'Q' output to go to a Logic '1', enabling the driver and output switch to conduct. When the oscillator reaches its upper threshold, C_T will start to discharge and Logic '0' will be present at the 'A' input of the AND gate. This logic level is also connected to an inverter whose output presents a Logic '1' to the reset input of the latch. This condition will cause 'Q' to go low, disabling the driver and output switch. A logic truth table of these functional blocks is shown in Figure 4.

The output of the comparator can set the latch only during the ramp-up of C_T and can initiate a partial or full on-cycle of output switch conduction. Once the comparator has set the latch, it cannot reset it. The latch will remain set until C_T begins ramping down. Thus the comparator can initiate output switch conduction, but **cannot terminate it** and the latch is always reset when C_T **begins** ramping down. The comparator's output will be at a Logic '0' when the output voltage of the switching regulator is above nominal. Under these conditions, the comparator's output can inhibit a portion of the output switch on-cycle, a complete cycle, a complete cycle plus a portion of one cycle, multiple cycles, or multiple cycles plus a portion of one cycle.

Current limiting is accomplished by monitoring the voltage drop across an external sense resistor placed in series with V_{CC} and the output switch. The voltage drop developed across this resistor is monitored by the I_{pk} Sense pin. When this voltage becomes greater than 330 mV, the current limit circuitry provides an additional

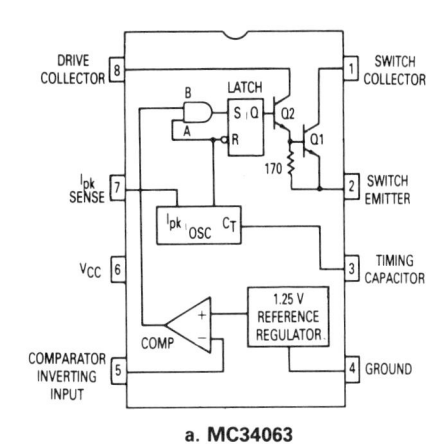

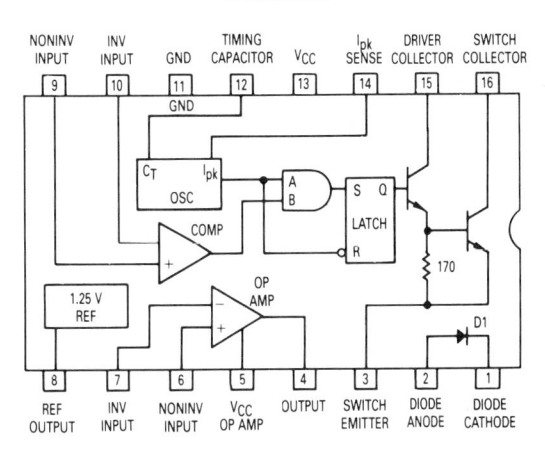

a. MC34063

b. µA78S40

Figure 2. Functional Block Diagrams

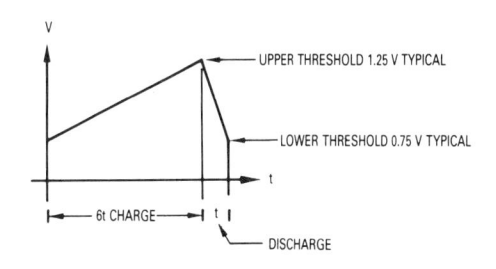

Figure 3. C_T Voltage Waveform

MOTOROLA AN920

Active Condition of Timing Capacitor C_T	AND Gate Inputs A	AND Gate Inputs B	Latch Inputs S	Latch Inputs R	Output Switch	Comments on State of Output Switch
Begins Ramp-Up	⤴	0	0	⤵	0	Switching Regulator's Output is ≥ nominal ('B' = 0).
Begins Ramp-Down	⤵	0	0	⤴	0	No change since 'B' was 0 before C_T Ramp-Down.
Ramping Down	0	⤴	0	1	0	No change even though switching regulators output < nominal. Output switch cannot be initiated during C_T Ramp-Down.
Ramping Down	0	⤵	0	1	0	No change since output switch conduction was terminated when 'A' went to 0.
Ramping Up	1	⤴	⤴	0	⤴	Switching regulator's output went < nominal during C_T Ramp-Up ('B' → 1). Partial on-cycle for output switch.
Ramping Up	1	⤵	⤵	0	1	Switching Regulators output went ≥ nominal ('B' → 0) during C_T Ramp-Up. No change since 'B' cannot reset latch.
Begins Ramp-Up	⤴	1	⤴	⤵	⤴	Complete on-cycle since 'B' was 1 before C_T started Ramp-Up.
Begins Ramp-Down	⤵	1	⤵	⤴	⤵	Output switch conduction is always terminated whenever C_T is Ramping Down.

Figure 4. Logic Truth Table of Functional Blocks

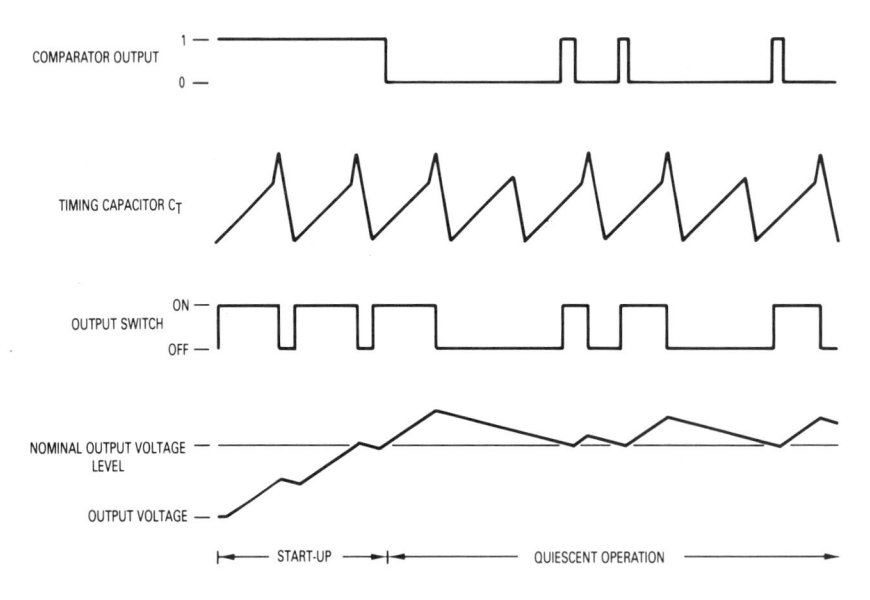

Figure 5. Typical Operating Waveforms

current path to charge the timing capacitor C_T. This causes it to rapidly reach the upper oscillator threshold, thereby shortening the time of output switch conduction and thus reducing the amount of energy stored in the inductor. This can be observed as an increase in the slope of the charging portion of the C_T voltage waveform as shown in Figure 5. Operation of the switching regulator in an overload or shorted condition will cause a very short but finite time of output conduction followed by either a normal or extended off-time interval provided by the oscillator ramp-down time of C_T. The extended interval is the result of charging C_T beyond the upper oscillator threshold by overdriving the current limit sense input. This can be caused by operating the switching regulator with a severely overloaded or shorted output or having the input voltage grossly above the nominal design value.

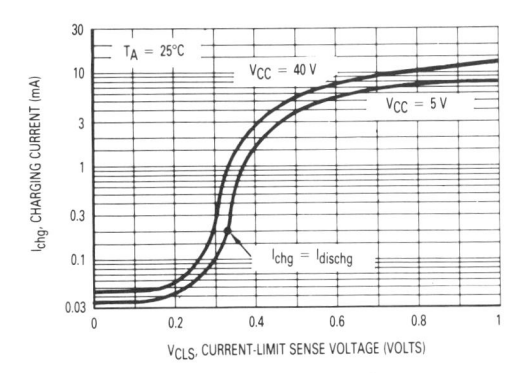

Figure 6. Timing Capacitor Charge Current versus Current-Limit Sense Voltage

Under extreme conditions, the voltage across C_T will approach V_{CC} and can cause a relatively long off-time. This action may be considered a feature since it will reduce the power dissipation of the output switch considerably. This feature may be disabled on the µA78S40 only, by connecting a small signal PNP transistor as a clamp. The emitter is connected to C_T, the base to the reference output, and the collector to ground. This will limit the maximum charge voltage across C_T to less than 2.0 volts. With the use of current limiting, saturation of the storage inductor may be prevented as well as achieving a soft start-up.

In practice the current limit circuit will somewhat modify the charging slope and peak amplitude of C_T each time the output switch is required to conduct. This is because the threshold voltage of the current limit sense circuit exhibits a "soft" voltage turn-on characteristic and has a turn-off time delay that causes some overshoot. The 330 mV threshold is defined where the charge and discharge currents are of equal value with V_{CC} = 5.0 V, as shown in Figure 6. The current limit sense circuit can be disabled by connecting the I_{pk} Sense pin to V_{CC}.

To aid in system design flexibility, the driver collector, output switch collector and emitter are pinned out separately. This allows the designer the option of driving the output switch transistor into saturation with a selected forced gain or driving it near saturation when connected as a Darlington. The output switch has a typical current gain of 70 at 1.0 amp and is designed to switch a maximum of 40 volts collector-to-emitter, with up to 1.5 amps peak collector current.

The µA78S40 has the additional features of an on-chip uncommitted operational amplifier and catch diode. The op amp is a high gain single supply type with an input common-mode voltage range that includes ground. The output is capable of sourcing up to 150 mA and sinking 35 mA. A separate V_{CC} pin is provided in order to reduce the integrated circuit standby current and is useful in low power applications if the operational amplifier is not incorporated into the main switching system. The catch diode is constructed from a lateral PNP transistor and is capable of blocking up to 40 volts and will conduct currents up to 1.5 amps. There is, however, a "catch" when using it.

Because the integrated circuit substrate is common with the internal and external circuitry ground, the cathode of the diode cannot be operated much below ground or forward biasing of the substrate will result. This totally eliminates the diode from being used in the basic voltage inverting configuration as in Figure 15, since the substrate, pin 11, is common to ground. The diode can be considered for use only in **low power** converter applications where the total system component count must be held to a minimum. The substrate current will be about 10 percent of the catch diode current in the step-up configuration and about 20 percent in the step-down and voltage-inverting in which pin 11 is common to the negative output. System efficiency will suffer when using this diode and the package dissipation limits must be observed.

STEP-DOWN SWITCHING REGULATOR OPERATION

Shown in Figure 7a is the basic step-down switching regulator. Transistor Q1 interrupts the input voltage and provides a variable duty cycle squarewave to a simple LC filter. The filter averages the squarewaves producing a dc output voltage that can be set to any level less than the input by controlling the percent conduction time of Q1 to that of the total switching cycle time. Thus,

$$V_{out} = V_{in}\left(\% \, t_{on}\right) \text{ or } V_{out} = V_{in}\left(\frac{t_{on}}{t_{on}+t_{off}}\right)$$

The MC34063/µA78S40 achieves regulation by varying the on-time and the total switching cycle time. An explanation of the step-down converter operation is as follows: Assume that the transistor Q1 is off, the inductor current I_L is zero, and the output voltage V_{out} is at its nominal value. The output voltage across capacitor C_O will eventually decay below nominal because it is the only component supply current into the external load R_L. This voltage deficiency is monitored by the switching control circuit and causes it to drive Q1 into saturation. The inductor current will start to flow from V_{in} through Q1 and, C_O in parallel with R_L, and rise at a rate of $\Delta I/\Delta T = V/L$. The voltage across the inductor is equal to $V_{in} - V_{sat} - V_{out}$ and the peak current at any instant is:

$$I_L = \left(\frac{V_{in} - V_{sat} - V_{out}}{L}\right) t$$

At the end of the on-time, Q1 is turned off. As the magnetic field in the inductor starts to collapse, it generates a reverse voltage that forward biases D1 and, the peak current will decay at a rate of $\Delta I/\Delta T = V/L$ as energy is supplied to C_O and R_L. The voltage across the inductor during this period is equal to $V_{out} + V_F$ of D1 and, the current at any instant is:

$$I_L = I_{L(pk)} - \left(\frac{V_{out} + V_F}{L}\right) t$$

Assume that during quiescent operation the average output voltage is constant and that the system is operating in the discontinuous mode. Then $I_{L(peak)}$ attained during t_{on} must decay to zero during t_{off} and a ratio of t_{on} to t_{off} can be determined.

MOTOROLA AN920

$$\left(\frac{V_{in} - V_{sat} - V_{out}}{L}\right) t_{on} = \left(\frac{V_{out} + V_F}{L}\right) t_{off}$$

$$\therefore \frac{t_{on}}{t_{off}} = \frac{V_{out} + V_F}{V_{in} - V_{sat} - V_{out}}$$

Note that the volt-time product of t_{on} must be equal that of t_{off} and the inductance value is not of concern when determining their ratio. If the output voltage is to remain constant, the average current into the inductor must be equal to the output current for a complete cycle. The peak inductor current with respect to output current is:

$$\left(\frac{I_{L(pk)}}{2}\right) t_{on} + \left(\frac{I_{L(pk)}}{2}\right) t_{off} = \left(I_{out} t_{on}\right) + \left(I_{out} t_{off}\right)$$

$$\frac{I_{L(pk)}(t_{on} + t_{off})}{2} = I_{out}(t_{on} + t_{off})$$

$$\therefore I_{L(pk)} = 2 I_{out}$$

The peak inductor current is also equal to the peak switch current $I_{pk(switch)}$ since the two are in series. The on-time t_{on} is the maximum possible switch conduction time. It is equal to the time required for C_T to ramp up from its lower to upper threshold. The required value for C_T can be determined by using the minimum oscillator charging current and the typical value for the oscillator voltage swing both taken from the data sheet electrical characteristics table.

$$C_T = I_{chg(min)} \left(\frac{\Delta t}{\Delta V}\right)$$
$$= 20 \times 10^{-6} \left(\frac{t_{on}}{0.5}\right)$$
$$= 4.0 \times 10^{-5} t_{on}$$

a. Step-Down $V_{out} \leq V_{in}$

b. Step-Up $V_{out} \geq V_{in}$

c. Voltage-Inverting $|V_{out}| \leq \geq V_{in}$

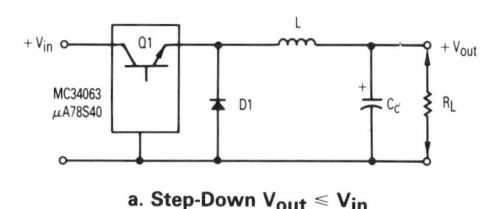

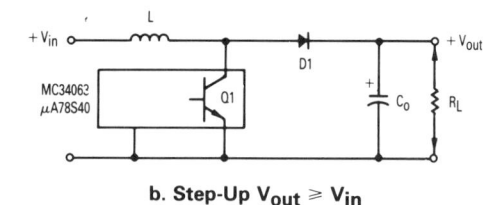

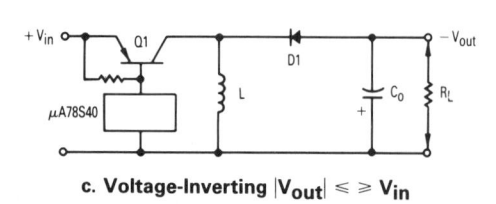

Figure 7. Basic Switching Regulator Configurations

The off-time t_{off}, is the time that diode D1 is in conduction and it is determined by the time required for the inductor current to return to zero. The off-time is **not** related to the ramp-down time of C_T. The cycle time of the LC network is equal to $t_{on(max)} + t_{off}$ and the minimum operating frequency is:

$$f_{min} = \frac{1}{t_{on(max)} + t_{off}}$$

A minimum value of inductance can now be calculated for L. The known quantities are the voltage across the inductor and the required peak current for the selected switch conduction time.

$$L_{(min)} = \frac{V_{in} - V_{sat} - V_{out}}{I_{pk(switch)}} t_{on}$$

This minimum value of inductance was calculated by assuming the onset of continuous conduction operation with a fixed input voltage, maximum output current, and a minimum charge-current oscillator.

The net charge per cycle delivered to the output filter capacitor C_o, must be zero, $Q+ = Q-$, if the output voltage is to remain constant. The ripple voltage can be calculated from the known values of on-time, off-time, peak inductor current, and output capacitor value.

$$V_{ripple(p-p)} = \left(\frac{1}{C_o}\right) \int_0^{t_1} i \, t \, dt + \left(\frac{1}{C_o}\right) \int_{t_1}^{t_2} i' \, t \, dt$$

Where $i' \, t = \frac{\frac{1}{2}I_{pk}t}{t_{on}/2}$ and $i' \, t = \frac{\frac{1}{2}I_{pk}t}{t_{off}/2}$

$$= \frac{1}{C_o} \left| \frac{I_{pk}}{t_{on}} \frac{t^2}{2} \right|_0^{t_1} + \frac{1}{C_o} \left| \frac{I_{pk}}{t_{off}} \frac{t^2}{2} \right|_{t_1}^{t_2}$$

And $t_1 = \frac{t_{on}}{2}$ and $t_2 - t_1 = \frac{t_{off}}{2}$

Substituting for t_1 and $t_2 - t_1$ yields:

$$= \frac{1}{C_o} \frac{I_{pk}}{t_{on}} \frac{(t_{on}/2)^2}{2} + \frac{1}{C_o} \frac{I_{pk}}{t_{off}} \frac{(t_{off}/2)^2}{2}$$

$$= \frac{I_{pk}(t_{on} + t_{off})}{8 C_o}$$

A graphical derivation of the peak-to-peak ripple voltage can be obtained from the capacitor current and voltage waveforms in Figure 8.

The calculations shown account for the ripple voltage contributed by the ripple current into an ideal capacitor. In practice, the calculated value will need to be increased due to the internal equivalent series resistance ESR of the capacitor. The additional ripple voltage will be equal to $I_{pk}(ESR)$. Increasing the value of the filter capacitor will reduce the output ripple voltage. However, a point of diminishing return will be reached because the comparator requires a finite voltage difference across its inputs to control the latch. This voltage difference to completely change the latch states is about 1.5 mV and the minimum achievable ripple at the output will be the feedback divider ratio multiplied by 1.5 mV or:

$$V_{ripple(p-p)min} = \frac{V_{out}}{V_{ref}}(1.5 \times 10^{-3})$$

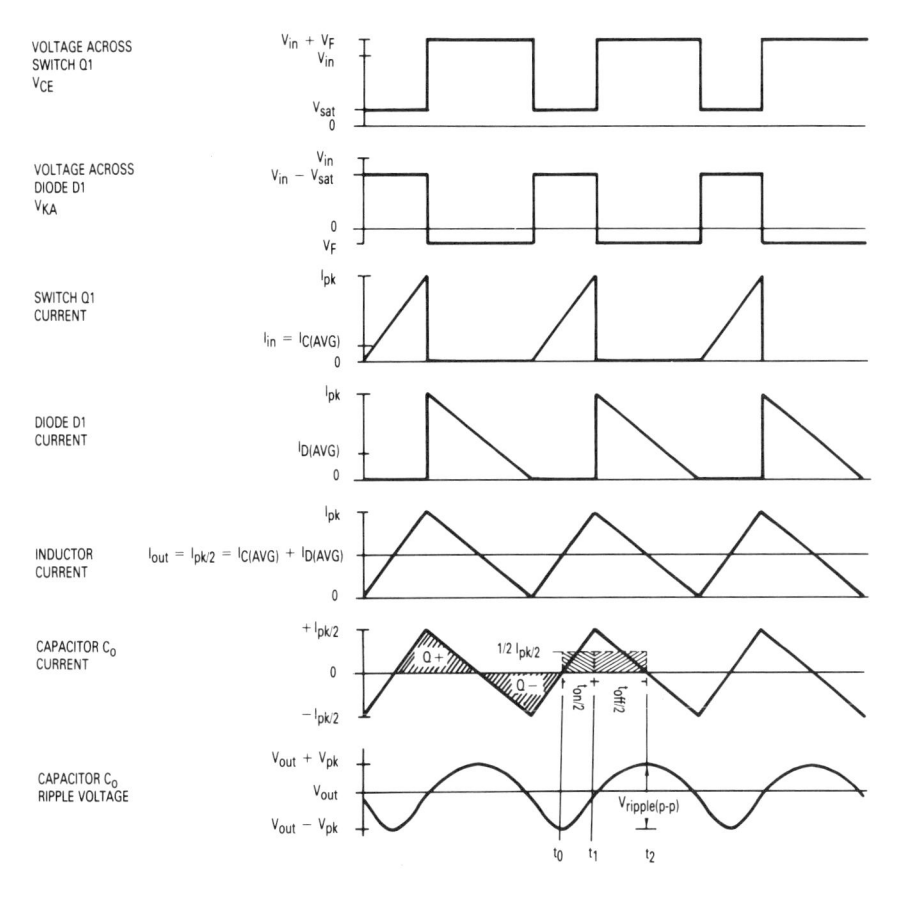

Figure 8. Step-Down Switching Regulator Waveforms

This problem becomes more apparent in a step-up converter with a high output voltage. Figures 12 and 13 show two different ripple reduction techniques. The first uses the μA78S40 operational amplifier to drive the comparator in the feedback loop. The second technique uses a zener diode to level shift the output down to the reference voltage.

Step-Down Switching Regulator Design Example

A schematic of the basic step-down regulator is shown in Figure 9. The μA78S40 was chosen in order to implement a minimum component system, however, the MC34063 with an external catch diode can also be used. The frequency chosen is a compromise between switching losses and inductor size. There will be a further discussion of this and other design considerations later. Given are the following conditions:

V_{out} = 5.0 V
I_{out} = 50 mA
f_{min} = 50 kHz
$V_{in(min)}$ = 24 V − 10% or 21.6 V
$V_{ripple(p-p)}$ = 0.5% V_{out} or 25 mVp-p

1. Determine the ratio of switch conduction t_{on} versus diode conduction t_{off} time.

$$\frac{t_{on}}{t_{off}} = \frac{V_{out} + V_F}{V_{in(min)} - V_{sat} - V_{out}}$$

$$= \frac{5.0 + 0.8}{21.6 - 0.8 - 5.0}$$

$$= 0.37$$

2. The cycle time of the LC network is equal to $t_{on(max)} + t_{off}$.

$$t_{on(max)} + t_{off} = \frac{1}{f_{min}}$$

$$= \frac{1}{50 \times 10^3}$$

$$= 20 \ \mu s \text{ per cycle}$$

3. Next calculate t_{on} and t_{off} from the ratio of t_{on}/t_{off} in #1 and the sum of $t_{on} + t_{off}$ in #2. By using substitution and some algebraic gymnastics, an equation can be written for t_{off} in terms of t_{on}/t_{off} and $t_{on} + t_{off}$.

The equation is:

$$t_{off} = \frac{t_{on(max)} + t_{off}}{\frac{t_{on}}{t_{off}} + 1}$$

$$= \frac{20 \times 10^{-6}}{0.37 + 1}$$

$$= 14.6 \ \mu s$$

Since $t_{on(max)} + t_{off} = 20 \ \mu s$
$t_{on(max)} = 20 \ \mu s - 14.6 \ \mu s$
$= 5.4 \ \mu s$

Note that the ratio of $t_{on}/(t_{on} + t_{off})$ does not exceed the maximum of 6/7 or 0.857. This maximum is defined by the 6:1 ratio of charge-to-discharge current of timing capacitor C_T (refer to Figure 3).

4. The maximum on-time, $t_{on(max)}$, is set by selecting a value for C_T.

$$C_T = 4.0 \times 10^{-5} \ t_{on}$$
$$= 4.0 \times 10^{-5} \ (5.4 \times 10^{-6})$$
$$= 216 \ pF$$

Use a standard 220 pF capacitor.

5. The peak switch current is:

$$I_{pk(switch)} = 2 \ I_{out}$$
$$= 2 \ (50 \times 10^{-3})$$
$$= 100 \ mA$$

6. With knowledge of the peak switch current and maximum on time, a minimum value of inductance can be calculated.

$$L_{(min)} = \left(\frac{V_{in(min)} - V_{sat} - V_{out}}{I_{pk(switch)}} \right) t_{on(max)}$$

$$= \left(\frac{21.6 - 0.8 - 5.0}{100 \times 10^{-3}} \right) 5.4 \times 10^{-6}$$

$$= 853 \ \mu H$$

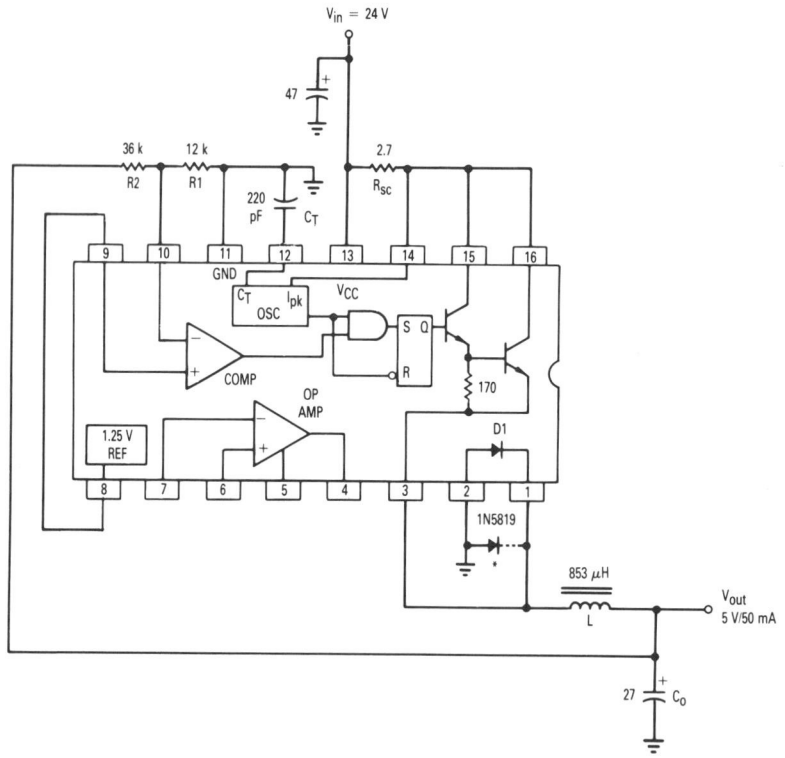

Test	Conditions	Results
Line Regulation	V_{in} = 18 to 30 V, I_{out} = 50 mA	Δ = 16 mV or ± 0.16%
Load Regulation	V_{in} = 24 V, I_{out} = 25 to 50 mA	Δ = 28 mV or ± 0.28%
Output Ripple	V_{in} = 21.6 V, I_{out} = 50 mA	24 mV_{p-p}
Short Circuit Current	V_{in} = 24 V, R_L = 0.1 Ω	105 mA
Efficiency, Internal Diode	V_{in} = 24 V, I_{out} = 50 mA	45.3%
Efficiency, External Diode*	V_{in} = 24 V, I_{out} = 50 mA	72.6%

Figure 9. Step-Down Design Example

7. A value for the current limit resistor R_{sc} can be determined by using the current level of $I_{pk(switch)}$ when V_{in} = 24 V.

$$I'_{pk(switch)} = \left(\frac{V_{in} - V_{sat} - V_{out}}{L_{(min)}}\right) t_{on(max)}$$

$$= \left(\frac{24 - 0.8 - 5.0}{853 \times 10^{-6}}\right) 5.4 \times 10^{-6}$$

$$= 115 \text{ mA}$$

$$R_{sc} = \frac{0.33}{I'_{pk(switch)}}$$

$$= \frac{0.33}{115 \times 10^{-3}}$$

$$= 2.86 \, \Omega \text{ use } 2.7 \, \Omega$$

This value may have to be adjusted downward to compensate for conversion losses and any increase in $I_{pk(switch)}$ current if V_{in} varies upward. Do not set R_{sc} to exceed the maximum $I_{pk(switch)}$ limit of 1.5 A when using the internal switch transistor.

8. A minimum value for an ideal output filter capacitor can now be obtained.

$$C_o = \frac{I_{pk(switch)} (t_{on} + t_{off})}{8 \, V_{ripple \, (p-p)}}$$

$$= \frac{0.1 \, (20 \times 10^{-6})}{8 \, (25 \times 10^{-3})}$$

$$= 10 \, \mu F$$

Ideally this would satisfy the design goal, however, even a solid tantalum capacitor of this value will have a typical ESR (equivalent series resistance) of 0.3 Ω which will contribute 30 mV of ripple. The ripple components are not in phase, but can be assumed to be for a conservative design. In satisfying the example shown, a 27 μF tantalum with an ESR of 0.1 Ω was selected. The ripple voltage should be kept to a low value since it will directly affect the system line and load regulation.

9. The nominal output voltage is programmed by the R1, R2 resistor divider. The output voltage is:

$$V_{out} = 1.25 \left(\frac{R2}{R1} + 1\right)$$

The divider current can go as low as 100 μA without affecting system performance. In selecting a minimum current divider R1 is equal to:

$$R1 = \frac{1.25}{100 \times 10^{-6}}$$

$$= 12,500 \, \Omega$$

Rearranging the above equation so that R2 can be solved yields:

$$R2 = R1 \left(\frac{V_{out}}{1.25} - 1\right)$$

If a standard 5% tolerance 12 k resistor is chosen for R1, R2 will also be a standard value.

$$R2 = 12 \times 10^3 \left(\frac{5.0}{1.25} - 1\right)$$

$$= 36 \text{ k}$$

Using the above derivation, the design is optimized to meet the assumed conditions. At $V_{in(min)}$, operation is at the onset of continuous mode and the output current capability will be greater than 50 mA. At $V_{in(nom)}$ i.e.: 24 V, the current limit will activate slightly above the rated I_{out} of 50 mA.

STEP-UP SWITCHING REGULATOR OPERATION

The basic step-up switching regulator is shown in Figure 7b and the waveform is in Figure 10. Energy is stored in the inductor during the time that transistor Q1 is in the 'on' state. Upon turn-off, the energy is transferred in series with V_{in} to the output filter capacitor and load. This configuration allows the output voltage to be set to any value greater than that of the input by the following relationship:

$$V_{out} = V_{in} \left(\frac{t_{on}}{t_{off}}\right) + V_{in} \text{ or } V_{out} = V_{in} \left(\frac{t_{on}}{t_{off}} + 1\right)$$

An explanation of the step-up converter's operation is as follows: Initially, assume that transistor Q1 is off, the inductor current is zero, and the output voltage is at its nominal value. At this time, load current is being supplied only by C_o and it will eventually fall below nominal. This deficiency will be sensed by the control circuit and it will initiate an on-cycle, driving Q1 into saturation. Current will start to flow from V_{in} through the inductor and Q1 and rise at a rate of $\Delta I / \Delta T = V/L$. The voltage across the inductor is equal to $V_{in} - V_{sat}$ and the peak current is:

$$I_L = \left(\frac{V_{in} - V_{sat}}{L}\right) t$$

When the on-time is completed, Q1 will turn off and the magnetic field in the inductor will start to collapse generating a reverse voltage that forward biases D1, supplying energy to C_o and R_L. The inductor current will decay at a rate of $\Delta I / \Delta T = V/L$ and the voltage across it is equal to $V_{out} + V_F - V_{in}$. The current at any instant is:

$$I_L = I_{L(pk)} - \left(\frac{V_{out} + V_F - V_{in}}{L}\right) t$$

Assuming that the system is operating in the discontinuous mode, the current through the inductor will reach zero after the t_{off} period is completed. Then $I_{L(pk)}$ attained during t_{on} must decay to zero during t_{off} and a ratio of t_{on} to t_{off} can be written.

$$\left(\frac{V_{in} - V_{sat}}{L}\right) t_{on} = \left(\frac{V_{out} + V_F - V_{in}}{L}\right) t_{off}$$

$$\therefore \frac{t_{on}}{t_{off}} = \frac{V_{out} + V_F - V_{in}}{V_{in} - V_{sat}}$$

Note again, that the volt-time product of t_{on} must be equal to that of t_{off} and the inductance value does not affect this relationship.

The inductor current charges the output filter capacitor through diode D1 only during t_{off}. If the output voltage is to remain constant, the net charge per cycle delivered to the output filter capacitor must be zero, $Q+ = Q-$.

$$I_{chg} \, t_{off} = I_{dischg} \, t_{on}$$

Figure 10 shows the step-up switching regulator waveforms. By observing the capacitor current and making

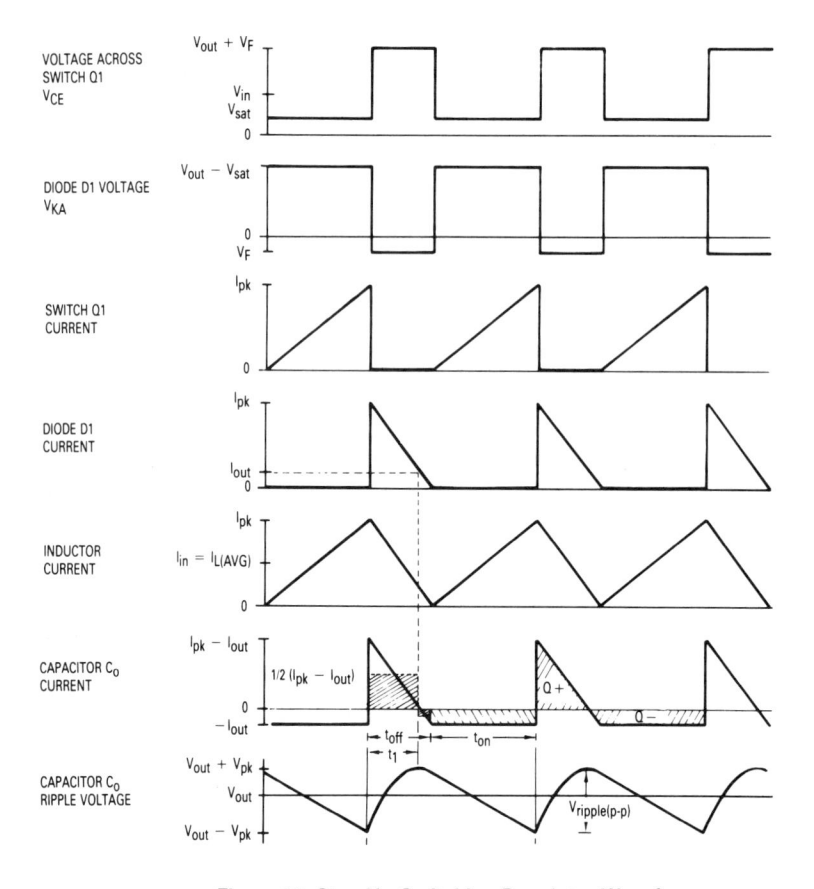

Figure 10. Step-Up Switching Regulator Waveforms

some substitutions in the above statement, a formula for peak inductor current can be obtained.

$$\left(\frac{I_{L(pk)}}{2}\right) t_{off} = I_{out} (t_{on} + t_{off})$$

$$I_{L(pk)} = 2 I_{out} \left(\frac{t_{on}}{t_{off}} + 1\right)$$

The peak inductor current is also equal to the peak switch current, since the two are in series.

With knowledge of the voltage across the inductor during t_{on} and the required peak current for the selected switch conduction time, a minimum inductance value can be determined.

$$L_{(min)} = \left(\frac{V_{in} - V_{sat}}{I_{pk(switch)}}\right) t_{on(max)}$$

The ripple voltage can be calculated from the known values of on-time, off-time, peak inductor current, output current and output capacitor value. Referring to the capacitor current waveforms in Figure 10, t_1 is defined as the capacitor charging interval. Solving for t_1 in known terms yields:

$$\frac{I_{pk} - I_{out}}{t_1} = \frac{I_{pk}}{t_{off}}$$

$$\therefore t_1 = \left(\frac{I_{pk} - I_{out}}{I_{pk}}\right) t_{off}$$

And the current during t_1 can be written:

$$I = \left(\frac{I_{pk} - I_{out}}{t_1}\right) t$$

The ripple voltage is:

$$V_{ripple(p-p)} = \left(\frac{1}{C_o}\right) \int_0^{t_1} \frac{I_{pk} - I_{out}}{t_1} \, t \, dt$$

$$= \frac{1}{C_o} \left. \frac{I_{pk} - I_{out}}{t_1} \frac{t^2}{2} \right|_0^{t_1}$$

$$= \frac{1}{C_o} \frac{(I_{pk} - I_{out})}{2} t_1$$

Substituting for t_1 yields:

$$= \frac{1}{C_o} \frac{(I_{pk} - I_{out})}{2} \frac{(I_{pk} - I_{out})}{I_{pk}} t_{off}$$

$$= \frac{(I_{pk} - I_{out})^2 \, t_{off}}{2 I_{pk} C_o}$$

A simplified formula that will give an error of less than 5% for a voltage step-up greater than 3 with an ideal capacitor is shown:

$$V_{ripple(p-p)} \approx \left(\frac{I_{out}}{C_o}\right) t_{on}$$

This neglects a small portion of the total Q− area. The area neglected is equal to:

$$A = (t_{off} - t_1) \frac{I_{out}}{2}$$

Step-Up Switching Regulator Design Example

The basic step-up regulator schematic is shown in Figure 11. The μA78S40 again was chosen in order to implement a minimum component system. The following conditions are given:

V_{out} = 28 V
I_{out} = 50 mA
f_{min} = 50 kHz
$V_{in(min)}$ = 9.0 V − 25% or 6.75 V
$V_{ripple(p-p)}$ = 0.5% V_{out} or 140 mVp-p

1. Determine the ratio of switch conduction t_{on} versus diode conduction t_{off} time.

$$\frac{t_{on}}{t_{off}} = \frac{V_{out} + V_F = V_{in(min)}}{V_{in(min)} - V_{sat}}$$

$$= \frac{28 + 0.8 - 6.75}{6.75 - 0.3}$$

$$= 3.42$$

2. The cycle time of the LC network is equal to $t_{on(max)} + t_{off}$.

$$t_{on(max)} + t_{off} = \frac{1}{f_{min}}$$

$$= \frac{1}{50 \times 10^3}$$

$$= 20 \ \mu s \text{ per cycle}$$

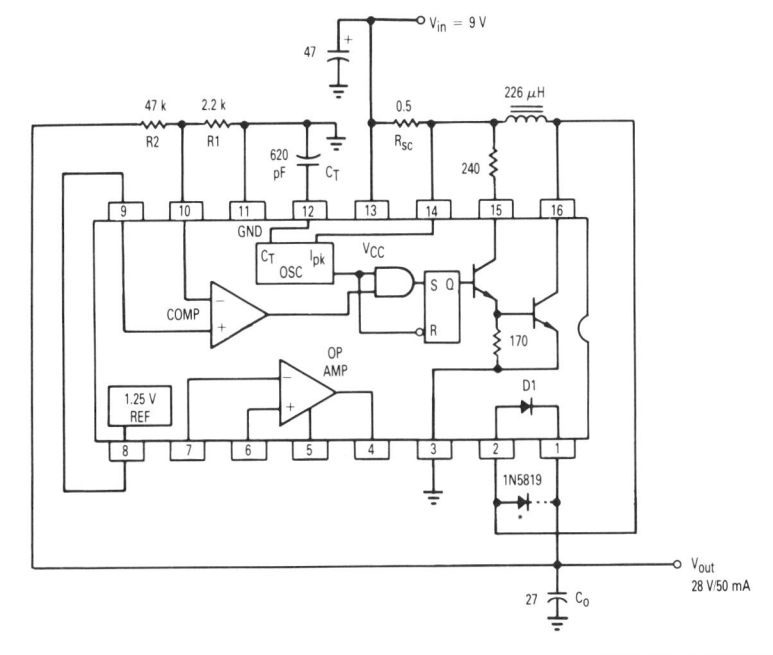

Test	Conditions	Results
Line Regulation	V_{in} = 6.0 to 12 V, I_{out} = 50 mA	Δ = 120 mV or ± 0.21%
Load Regulation	V_{in} = 9.0 V, I_{out} = 25 to 50 mA	Δ = 50 mV or ± 0.09%
Output Ripple	V_{in} = 6.75 V, I_{out} = 50 mA	90 mV$_{p-p}$
Efficiency, Internal Diode	V_{in} = 9.0 V, I_{out} = 50 mA	62.2%
Efficiency, External Diode*	V_{in} = 9.0 V, I_{out} = 50 mA	74.2%

Figure 11. Step-Up Design Example

3. Next calculate t_{on} and t_{off} from the ratio of t_{on}/t_{off} in #1 and the sum of $t_{on} + t_{off}$ in #2.

$$t_{off} = \frac{20 \times 10^{-6}}{3.42 + 1}$$
$$= 4.5 \, \mu s$$
$$t_{on} = 20 \, \mu s - 4.5 \, \mu s$$
$$= 15.5 \, \mu s$$

Note that the ratio of $t_{on}/(t_{on} + t_{off})$ does not exceed the maximum of 0.857.

4. The maximum on-time, $t_{on(max)}$, is set by selecting a value for C_T.

$$C_T = 4.0 \times 10^{-5} \, t_{on}$$
$$= 4.0 \times 10^{-5} \, (15.5 \times 10^{-6})$$
$$= 620 \, pF$$

5. The peak switch current is:

$$I_{pk(switch)} = 2 \, I_{out} \left(\frac{t_{on}}{t_{off}} + 1 \right)$$
$$= 2 \, (50 \times 10^{-3}) \, (3.42 + 1)$$
$$= 442 \, mA$$

6. A minimum value of inductance can be calculated since the maximum on-time and peak switch current are known.

$$L_{(min)} = \left(\frac{V_{in(min)} - V_{sat}}{I_{pk(switch)}} \right) t_{on}$$
$$= \left(\frac{6.75 - 0.3}{442 \times 10^{-3}} \right) 15.5 \times 10^{-6}$$
$$= 226 \, \mu H$$

7. A value for the current limit resistor, R_{SC}, can now be determined by using the current level of $I_{pk(switch)}$ when $V_{in} = 9.0$ V.

$$I'_{pk(switch)} = \left(\frac{V_{in} - V_{sat}}{L_{(min)}} \right) t_{on(max)}$$
$$= \left(\frac{9.0 - 0.3}{226 \times 10^{-6}} \right) 15.5 \times 10^{-6}$$
$$= 597 \, mA$$

$$R_{sc} = \frac{0.33}{I'_{pk(switch)}}$$
$$= \frac{0.33}{597 \times 10^{-3}}$$
$$= 0.55 \, \Omega \qquad \text{use } 0.5 \, \Omega$$

Note that current limiting in this basic step-up configuration will **only** protect the switch transistor from overcurrent due to inductor saturation. If the output is severely overloaded or shorted D1, L, or R_{sc} may be destroyed since they form a direct path from V_{in} to V_{out}. Protection may be achieved by current limiting V_{in} or replacing the inductor with 1:1 turns ratio transformer.

8. An approximate value for an ideal output filter capacitor is:

$$C_o \approx \frac{I_{out}}{V_{ripple(p-p)}} t_{on}$$
$$\approx \frac{50 \times 10^{-3}}{140 \times 10^{-3}} 15.5 \times 10^{-6}$$
$$\approx 5.5 \, \mu F$$

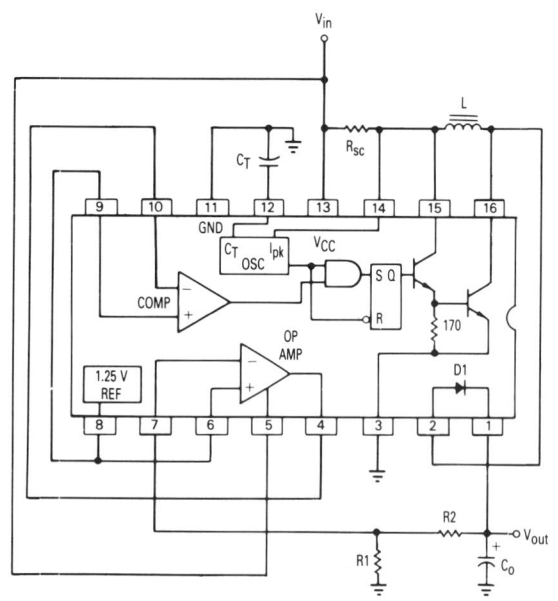

Op Amp is used in feedback loop to drive the comparator.

Figure 12. µA78S40 Ripple Reduction Technique

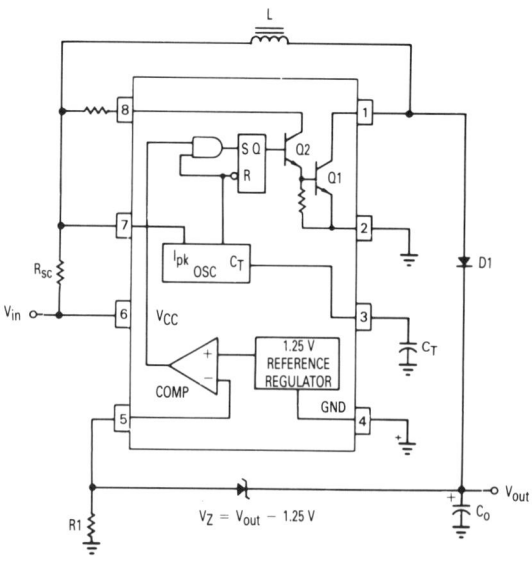

Zener is used to level shift output down to 1.25 V reference.

Figure 13. MC34063 Ripple Reduction Technique

The ripple contribution due to the gain of the comparator:

$$V_{ripple(p-p)} = \frac{V_{out}}{V_{ref}} \, 1.5 \times 10^{-3}$$

$$= \frac{28}{1.25} \, 1.5 \times 10^{-3}$$

$$= 33.6 \text{ mV}$$

A 27 µF tantalum capacitor with an ESR of 0.10 Ω was again chosen. The ripple voltage due to the capacitance value is 28.7 mV and 44.2 mV due to ESR. This yields a total ripple voltage of:

$$E_{ripple(p-p)} = \frac{V_{out}}{V_{ref}} \, 1.5 \times 10^{-3} + \frac{I_{out}}{C_o} t_{on} + I_{pk}\, ESR$$

$$= 33.6 \text{ mV} + 28.7 \text{ mV} + 44.2 \text{ mV}$$

$$= 107 \text{ mV}$$

9. The nominal output voltage is programmed by the R1, R2 divider.

$$V_{out} = 1.25 \left(1 + \frac{R2}{R1}\right)$$

A standard 5% tolerance, 2.2 k resistor was selected for R1 so that the divider current is about 500 µA.

$$R1 = \frac{1.25}{500 \times 10^{-6}}$$

$$= 2500 \text{ Ω use 2.2 k}$$

$$\text{Then } R2 = R1 \left(\frac{V_{out}}{1.25} - 1\right)$$

$$= 2200 \left(\frac{28}{1.25} - 1\right)$$

$$= 47{,}080 \text{ Ω use 47 k}$$

10. In this design example, the output switch transistor is driven into saturation with a forced gain of 20 at an input voltage of 7.0 V. The required base drive is:

$$I_B = \frac{I_{pk(switch)}}{B_f}$$

$$= \frac{442 \times 10^{-3}}{20}$$

$$= 22.1 \text{ mA}$$

The current required to drive the internal 170 Ω base-emitter resistor is:

$$I_{170\,\Omega} = \frac{V_{BE(switch)}}{170}$$

$$= \frac{0.7}{170}$$

$$= 4.1 \text{ mA}$$

The driver collector current is equal to sum of 22.1 mA + 4.1 mA = 26.2 mA. Allow 0.3 V for driver saturation and 0.2 V for the drop across R_{sc} (0.5 × 442 mA I_{pk}).
Then the driver collector resistor is equal to:

$$R_{driver} = \frac{V_{in} - V_{sat(driver)} - V_{RSC}}{I_B + I_{170\,\Omega}}$$

$$= \frac{7.0 - 0.3 - 0.2}{(22.1 + 4.1) \times 10^{-3}}$$

$$= 248 \text{ Ω use 240 Ω}$$

VOLTAGE-INVERTING SWITCHING REGULATOR OPERATION

The basic voltage-inverting switching regulator is shown in Figure 7c and the operating waveforms are in Figure 14. Energy is stored in the inductor during the conduction time of Q1. Upon turn-off, the energy is transferred to the output filter capacitor and load. Notice that in this configuration the output voltage is derived only from the inductor. This allows the magnitude of the output to be set to any value. It may be less than, equal to, or greater than that of the input and is set by the following relationship:

$$V_{out} = V_{in} \left(\frac{t_{on}}{t_{off}}\right)$$

The voltage-inverting converter operates almost identically to that of the step-up previously discussed. The voltage across the inductor during t_{on} is $V_{in} - V_{sat}$ but during t_{off}, the voltage is equal to the negative magnitude of $V_{out} + V_F$. Remembering that the volt-time product of t_{on} must be equal to that of t_{off}, a ratio of t_{on} to t_{off} can be determined.

$$(V_{in} - V_{sat})\, t_{on} = (|V_{out}| + V_F)\, t_{off}$$

$$\therefore \frac{t_{on}}{t_{off}} = \frac{|V_{out}| + V_F}{V_{in} - V_{sat}}$$

The derivations and the formulas for $I_{pk(switch)}$, $L_{(min)}$, and C_O are the same as that of the step-up converter.

Voltage-Inverting Switching Regulator Design Example

A circuit diagram of the basic voltage-inverting regulator is shown in Figure 15.
The µA78S40 was selected for this design since it has the reference and both comparator inputs pinned out. The following operating conditions are given:

$V_{out} = -15$ V
$I_{out} = 500$ mA
$f_{min} = 50$ kHz
$V_{in(min)} = 15$ V -10% or 13.5 V
$V_{ripple(p-p)} = 0.4\% \, V_{out}$ or 60 mVp-p

1. Determine the ratio of switch conduction t_{on} versus diode conductions t_{off} time.

$$\frac{t_{on}}{t_{off}} = \frac{|V_{out}| + V_F}{V_{in} - V_{sat}}$$

$$= \frac{15 + 0.8}{13.5 - 0.8}$$

$$= 1.24$$

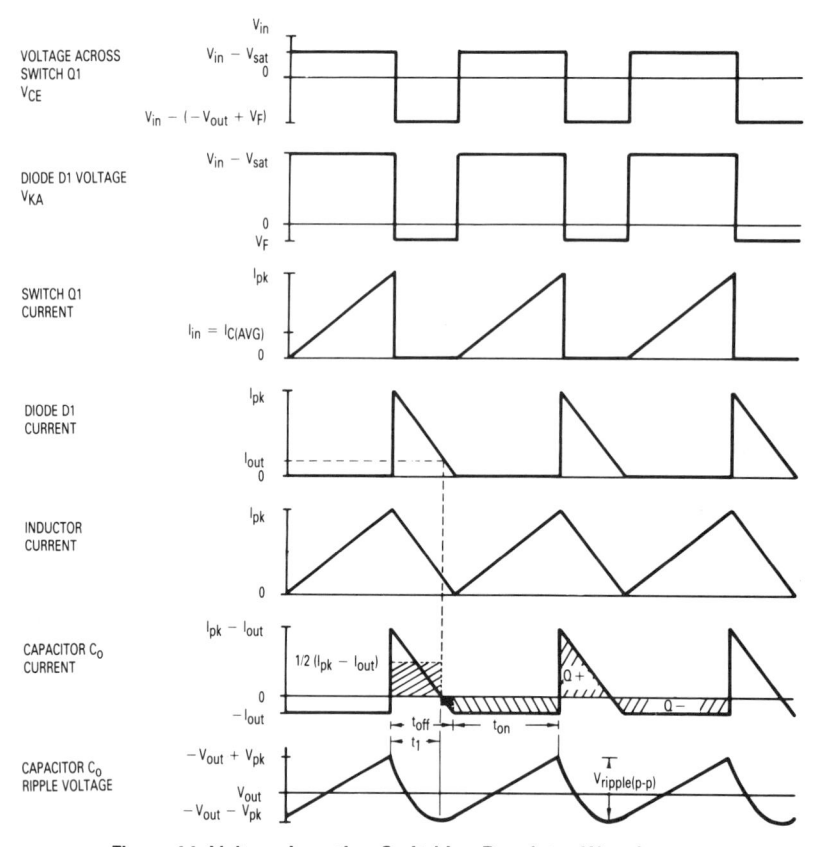

Figure 14. Voltage-Inverting Switching Regulator Waveforms

2. The cycle time of the LC network is equal to $t_{on(max)} + t_{off}$.

$$t_{on(max)} + t_{off} = \frac{1}{f_{min}}$$
$$= \frac{1}{50 \times 10^3}$$
$$= 20 \ \mu s$$

3. Calculate t_{on} and t_{off} from the ratio of t_{on}/t_{off} in #1 and the sum of $t_{on} + t_{off}$ in #2.

$$t_{off} = \frac{20 \times 10^{-6}}{1.24 + 1}$$
$$= 8.9 \ \mu s$$
$$t_{on} = 20 \ \mu s - 8.9 \ \mu s$$
$$= 11.1 \ \mu s$$

Note again that the ratio of $t_{on}/(t_{on} + t_{off})$ does not exceed the maximum of 0.857.

4. A value of C_T must be selected in order to set $t_{on(max)}$.

$$C_T = 4.0 \times 10^{-5} \ t_{on}$$
$$= 4.0 \times 10^{-5} \ (11.1 \times 10^{-6})$$
$$= 444 \ pF \qquad \text{use } 430 \ pF$$

5. The peak switch current is:

$$I_{pk(switch)} = 2 \ I_{out} \left(\frac{t_{on}}{t_{off}} + 1\right)$$
$$= 2 \ (500 \times 10^{-3}) \ (1.24 + 1)$$
$$= 2.24 \ A$$

6. The minimum required inductance value is:

$$L_{(min)} = \left(\frac{V_{in(min)} - V_{sat}}{I_{pk(switch)}}\right) t_{on}$$
$$= \left(\frac{13.5 - 0.8}{2.24}\right) 11.1 \times 10^{-6}$$
$$= 66.5 \ \mu H$$

7. The current-limit resistor value was selected by determining the level of $I_{pk(switch)}$ for $V_{in} = 16.5$ V.

$$I'_{pk(switch)} = \left(\frac{V_{in} - V_{sat}}{L_{(min)}}\right) t_{on}$$
$$= \left(\frac{16.5 - 0.8}{66.5 \times 10^{-6}}\right) 11.1 \times 10^{-6}$$
$$= 2.62 \ A$$

$$R_{sc} = \frac{0.33}{I'_{pk(switch)}}$$
$$= \frac{0.33}{2.62}$$
$$= 0.13 \ \Omega \quad \text{use } 0.12 \ \Omega$$

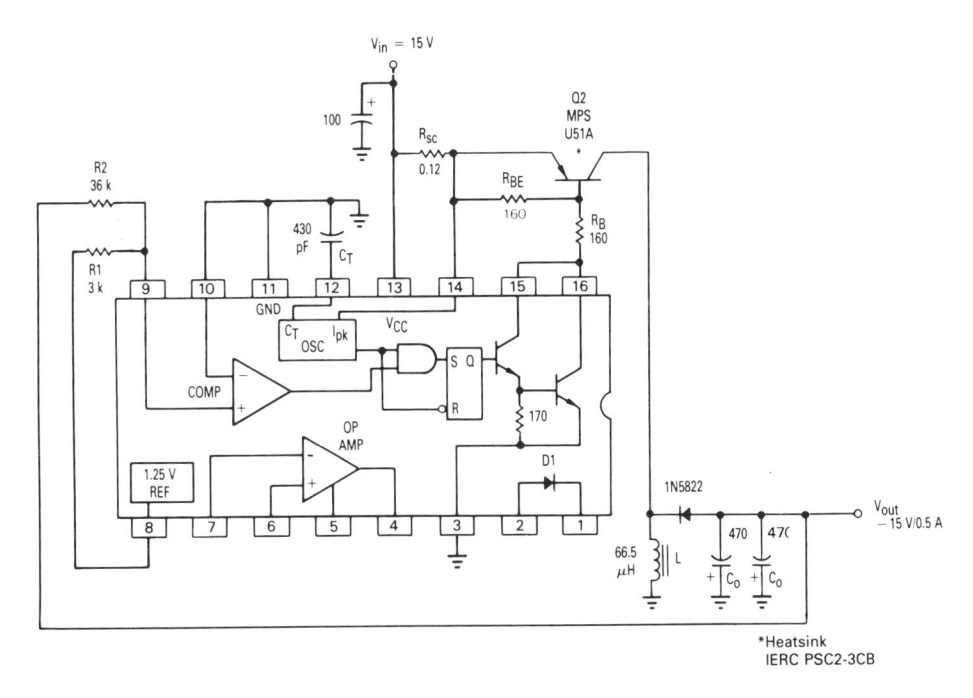

Test	Conditions	Results
Line Regulation	V_{in} = 12 to 16 V, I_{out} = 0.5 A	Δ = 3.0 mV or ±0.01%
Load Regulation	V_{in} = 15 V, I_{out} = 0.1 to 0.5 A	Δ = 27 mV or ± 0.09%
Output Ripple	V_{in} = 13.5 V, I_{out} = 0.5 A	35 mV$_{p-p}$
Short Circuit Current	V_{in} = 15 V, R_L = 0.1 Ω	2.5 A
Efficiency	V_{in} = 15 V, I_{out} = 0.5 A	80.6%

Figure 15. Voltage-Inverting Design Example

8. An approximate value for an ideal output filter capacitor is:

$$C_o \approx \left(\frac{I_{out}}{V_{ripple(p-p)}}\right) t_{on}$$

$$\approx \frac{0.5}{60 \times 10^{-3}} 11.1 \times 10^{-6}$$

$$\approx 92.5 \ \mu F$$

The ripple contribution due to the gain of the comparator is:

$$V_{ripple(p-p)} = \frac{|V_{out}|}{V_{ref}} 1.5 \times 10^{-3}$$

$$= \frac{15}{1.25} 1.5 \times 10^{-3}$$

$$= 18 \ mV$$

For a given level of ripple, the ESR of the output filter capacitor becomes the dominant factor in choosing a value for capacitance. Therefore two 470 μF capacitors with an ESR of 0.020 Ω each was chosen. The ripple voltage due to the capacitance value is 5.9 mV and 22.4 mV due to ESR. This yields a total ripple voltage of:

$$E_{ripple(p-p)} = \frac{|V_{out}|}{V_{ref}} 1.5 \times 10^{-3} + \frac{I_{out}}{C_o} t_{on} + I_{pk} \ ESR$$

$$= 18 \ mV + 5.9 \ mV + 22.4 \ mV$$

$$= 46.3 \ mV$$

9. The nominal output voltage is programmed by the R1, R2 divider. Note that with a negative output voltage, the inverting input of the comparator is referenced to ground. Therefore, the voltage at the junction of R1, R2 and the noninverting input must also be at ground potential when V_{out} is in regulation. The magnitude of V_{out} is:

$$|V_{out}| = 1.25 \frac{R2}{R1}$$

A divider current of about 400 μA was desired for this example.

$$R1 = \frac{1.25}{400 \times 10^{-6}}$$

$$= 3{,}125\ \Omega\ \text{use}\ 3.0\ \text{k}$$

$$\text{Then}\ R2 = \frac{|V_{out}|}{1.25} R1$$

$$= \frac{15}{1.25}\ 3.0 \times 10^3$$

$$= 36\ \text{k}$$

10. Output switch transistor Q2 is driven into a soft saturation with a forced gain of 35 at an input voltage of 13.5 V in order to enhance the turn-off switching time. The required base drive is:

$$I_B = \frac{I_{pk(switch)}}{B_f}$$

$$= \frac{2.24}{35}$$

$$= 64\ \text{mA}$$

The value for the base-emitter turn-off resistor R_{BE} is determined by:

$$R_{BE} = \frac{10\ B_f}{I_{pk(switch)}}$$

$$= \frac{10\ (35)}{2.24}$$

$$= 156.3\ \Omega\ \text{use}\ 160\ \Omega$$

The additional base current required due to R_{BE} is:

$$I_{R_{BE}} = \frac{V_{BE}\ (Q2)}{R_{BE}}$$

$$= \frac{0.8}{160}$$

$$= 5.0\ \text{mA}$$

Then I_B (Q2) is equal to the sum of 64 mA + 5.0 mA = 69 mA. Allow 0.8 V for the IC driver saturation and 0.3 V for the drop across R_{sc} (0.12 × 2.24 A I_{pk}).

Then the base driver resistor is equal to:

$$R_B = \frac{V_{in(min)} - V_{sat\ (IC)} - V_{RSC} - V_{BE}\ (Q2)}{I_B + I_{160\ \Omega}}$$

$$= \frac{13.5 - 0.8 - 0.3 - 1.0}{(64 + 5) \times 10^{-3}}$$

$$= 165.2\ \Omega\ \text{use}\ 160\ \Omega$$

STEP UP/DOWN SWITCHING REGULATOR OPERATION

When designing at the board level it sometimes becomes necessary to generate a constant output voltage that is less than that of the battery. The step-down circuit shown in Figure 16a will perform this function efficiently. However, as the battery discharges, its terminal voltage will eventually fall below the desired output, and in order to utilize the remaining battery energy the step-up circuit shown in Figure 16b will be required.

General Applications

By combining circuits a and b a unique step-up/down configuration can be created (Figure 17) which still employs a *simple inductor* for the voltage transformation. Energy is stored in the inductor during the time that transistors Q1 and Q2 are in the 'on' state. Upon turn-off, the energy is transferred to the output filter capacitor and load forward biasing diodes D1 and D2. Note that during t_{on} this circuit is identical to the basic step-up, but during t_{off} the output voltage is derived only from the inductor and is with respect to ground instead of V_{in}. This allows the output voltage to be set to any value, thus it may be less than, equal to, or greater than that of the input. Current limit protection cannot be employed in the basic step-up circuit. If the output is severely overloaded or shorted L or D2 may be destroyed since they form a direct path from V_{in} to V_{out}. The step-up/down configuration allows the control circuit to implement current limiting because Q1 is now in series with V_{out}, as is in the step-down circuit.

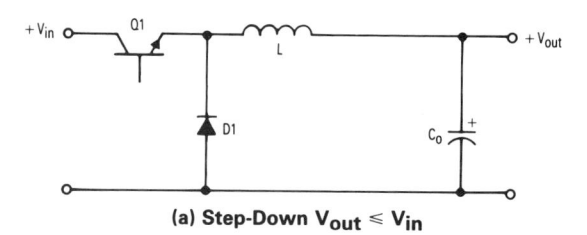

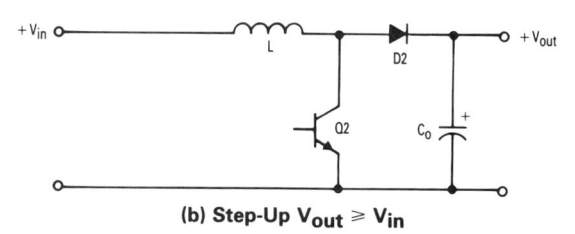

(a) Step-Down $V_{out} \leq V_{in}$

(b) Step-Up $V_{out} \geq V_{in}$

Figure 16. Basic Switching Regulator Configurations

Step-Up/Down $V_{out} \leq\geq V_{in}$

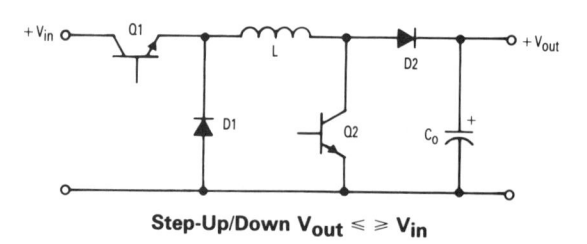

Figure 17. Combined Configuration

Step Up/Down Switching Regulator Design Example

A complete step-up/down switching regulator design example is shown in Figure 18. An external switch transistor was used to perform the function of Q2. This regulator was designed to operate from a standard 12 V battery pack with the following conditions:

V_{in} = 7.5 to 14.5 V $\qquad V_{out}$ = 10 V
f_{min} = 50 kHz $\qquad I_{out}$ = 120 mA
$V_{ripple(p-p)}$ = 1% V_{out} or 100 mV_{p-p}

The following design procedure is provided so that the user can select proper component values for his specific converter application.

1) Determine the ratio of switch conduction t_{on} versus diode conduction t_{off} time.

$$\frac{t_{on}}{t_{off}} = \frac{V_{out} + V_{FD1} + V_{FD2}}{V_{in(min)} - V_{satQ1} - V_{satQ2}}$$

$$= \frac{10 + 0.6 + 0.6}{7.5 - 0.8 - 0.8}$$

$$= 1.9$$

2) The cycle time of the LC network is equal to $t_{on(max)} + t_{off}$.

$$t_{on(max)} + t_{off} = \frac{1}{f_{min}}$$

$$= \frac{1}{50 \times 10^3}$$

$$= 20 \ \mu s \text{ per cycle}$$

3) Next calculate t_{on} and t_{off} from the ratio of t_{on}/t_{off} in #1 and the sum of $t_{on(max)} + t_{off}$ in #2.

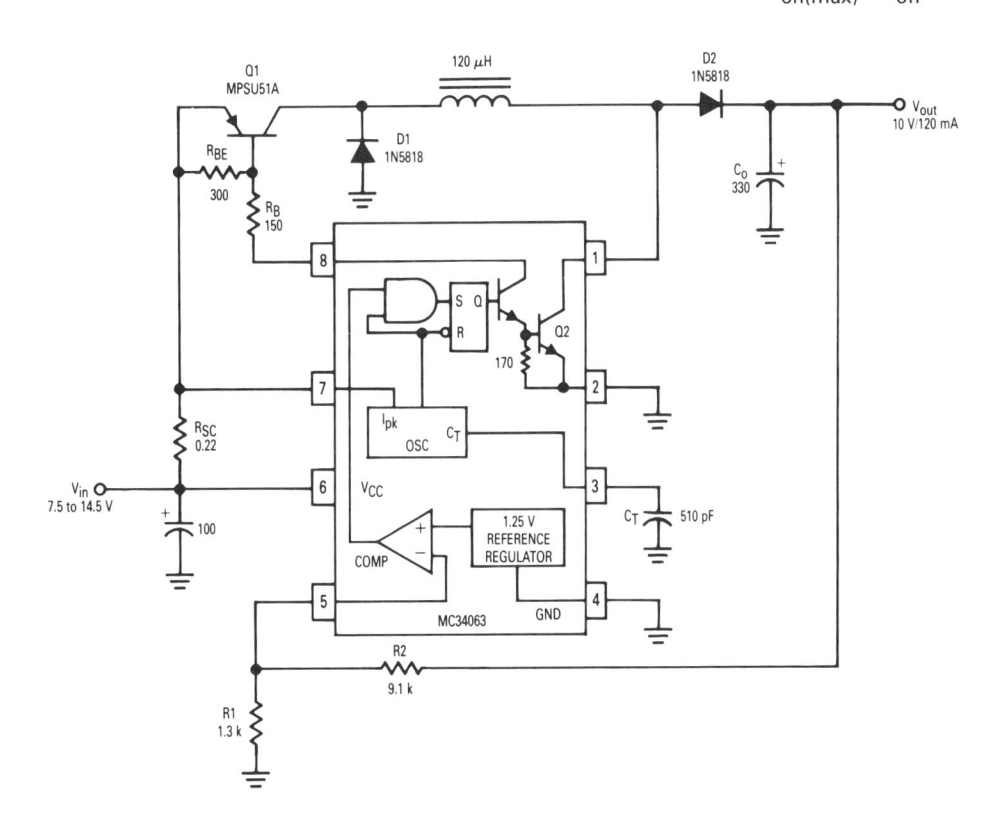

Test	Conditions	Results
Line Regulation	V_{in} = 7.5 to 14.5 V, I_{out} = 120 mA	Δ = 22 mV or ± 0.11%
Load Regulation	V_{in} = 12.6 V, I_{out} = 10 to 120 mA	Δ = 3.0 mV or ± 0.015%
Output Ripple	V_{in} = 12.6 V, I_{out} = 120 mA	95 mV_{p-p}
Short Circuit Current	V_{in} = 12.6 V, R_L = 0.1 Ω	1.54 A
Efficiency	V_{in} = 7.5 to 14.5 V, I_{out} = 120 mA	74%

Figure 18. Step-Up/Down Switching Regulator Design Example

$$t_{off} = \frac{t_{on(max)} + t_{off}}{\frac{t_{on}}{t_{off}} + 1}$$

$$= \frac{20 \times 10^{-6}}{1.9 + 1}$$

$$= 6.9 \ \mu s$$

$$t_{on} = 20 \ \mu s - 6.9 \ \mu s$$

$$= 13.1 \ \mu s$$

4) The maximum on-time is set by selecting a value for C_T.

$$C_T = 4 \times 10^{-5} \ t_{on(max)}$$
$$= 4 \times 10^{-5} \ (13.1 \times 10^{-6})$$
$$= 524 \ pF$$

Use a standard 510 pF capacitor.

5) The peak switch current is:

$$I_{pk(switch)} = 2I_{out}\left(\frac{t_{on}}{t_{off}} + 1\right)$$

$$= 2 \ (120 \times 10^{-3}) \ (1.9 + 1)$$

$$= 696 \ mA$$

6) A minimum value of inductance can now be calculated since the maximum on-time and peak switch current are known.

$$L_{min} = \left(\frac{V_{in(min)} - V_{satQ1} - V_{satQ2}}{I_{pk(switch)}}\right) t_{on}$$

$$= \left(\frac{7.5 - 0.8 - 0.8}{696 \times 10^{-3}}\right) 13.1 \times 10^{-6}$$

$$= 111 \ \mu H$$

A 120 μH inductor was selected for $L_{(min)}$.

7) A value for the current limit resistor, R_{SC}, can be determined by using the current limit level of $I_{pk(switch)}$ when V_{in} = 14.5 V.

$$I'_{pk(switch)} = \left(\frac{V_{in} - V_{satQ1} - V_{satQ2}}{L_{(min)}}\right) t_{on(max)}$$

$$= \left(\frac{14.5 - 0.8 - 0.8}{120 \times 10^{-6}}\right) 13.1 \times 10^{-6}$$

$$= 1.41 \ A$$

$$R_{SC} = \frac{0.33}{I'_{pk(switch)}}$$

$$= \frac{0.33}{1.41}$$

$$= 0.23 \ \Omega$$

Use a standard 0.22 Ω resistor.

8) A minimum value for an *ideal* output filter capacitor is:

$$C_o \approx \left(\frac{I_{out}}{V_{ripple(p-p)}}\right) t_{on}$$

$$\approx \left(\frac{120 \times 10^{-3}}{100 \times 10^{-3}}\right) 13.1 \times 10^{-6}$$

$$\approx 15.7 \ \mu F$$

Ideally this would satisfy the design goal, however, even a solid tantalum capacitor of this value will have a typical ESR (equivalent series resistance) of 0.3 Ω which will contribute an additional 209 mV of ripple. Also there is a ripple component due to the gain of the comparator equal to:

$$V_{ripple(p-p)} = \left(\frac{V_{out}}{V_{Ref}}\right) 1.5 \times 10^{-3}$$

$$= \left(\frac{10}{1.25}\right) 1.5 \times 10^{-3}$$

$$= 12 \ mV$$

The ripple components are not in phase, but can be assumed to be for a conservative design. From the above it becomes apparent that ESR is the dominant factor in the selection of an output filter capacitor. A 330 μF with an ESR of 0.12 Ω was selected to satisfy this design example by the following:

$$ESR \approx \frac{V_{ripple(p-p)} - \left(\frac{I_{out}}{C_o}\right) t_{on} - \left(\frac{V_{out}}{V_{Ref}}\right) 1.5 \times 10^{-3}}{I_{pk(switch)}}$$

9) The nominal output voltage is programmed by the R1, R2 resistor divider.

$$R2 = R1 \left(\frac{V_{out}}{V_{Ref}} - 1\right)$$

$$= R1 \left(\frac{10}{1.25} - 1\right)$$

$$= 7 \ R1$$

If 1.3 k is chosen for R1, then R2 would be 9.1 k, both being standard resistor values.

10) Transistor Q1 is driven into saturation with a forced gain of approximately 20 at an input voltage of 7.5 V. The required base drive is:

$$I_B = \frac{I_{pk(switch)}}{\beta_F}$$

$$= \frac{696 \times 10^{-3}}{20}$$

$$= 35 \ mA$$

The value for the base-emitter turn-off resistor R_{BE} is determined by:

$$R_{BE} = \frac{10 \ \beta_F}{I_{pk(switch)}}$$

$$= \frac{10 \ (20)}{696 \times 10^{-3}}$$

$$= 287 \ \Omega$$

A standard 300 Ω resistor was selected.

The additional base current required due to R_{BE} is:

$$I_{RBE} = \frac{V_{BEQ1}}{R_{BE}}$$

$$= \frac{0.8}{300}$$

$$= 3 \ mA$$

AN920

The base drive resistor for Q1 is equal to:

$$R_B = \frac{V_{in(min)} - V_{sat(driver)} - V_{RSC} - V_{BEQ1}}{I_B + I_{RBE}}$$

$$= \frac{7.5 - 0.8 - 0.15 - 0.8}{(35 + 3) \times 10^{-3}}$$

$$= 151 \, \Omega$$

A standard 150 Ω resistor was used.

The circuit performance data shows excellent line and load regulation. There is some loss in conversion efficiency over the basic step-up or step-down circuits due to the added switch transistor and diode 'on' losses. However this unique converter demonstrates that with a simple inductor, a step-up/down converter with current limiting can be constructed.

DESIGN CONSIDERATIONS

As previously stated, the design equations for L_{min} were based upon the assumption that the switching regulator is operating on the onset of continuous conduction with a fixed input voltage, maximum output load current, and a minimum charge-current oscillator. Typically the oscillator charge-current will be greater than the specific minimum of 20 microamps, thus t_{on} will be somewhat shorter and the actual LC operating frequency will be greater than predicted.

Also note that the voltage drop developed across the current-limit resistor R_{sc} was not accounted for in the t_{on}/t_{off} and $L_{(min)}$ design formulas. This voltage drop must be considered when designing high current converters that operate with an input voltage of less than 5.0 volts.

When checking the initial switcher operation with an oscilloscope, there will be some concern of circuit instability due to the apparent random switching of the output. The oscilloscope will be difficult to synchronize. This is **not** a problem. It is a normal operating characteristic of this type of switching regulator and is caused by the asynchronous operation of the comparator to that of the oscillator. The oscilloscope may be synchronized by varying the input voltage or load current slightly from the design nominals.

High frequency circuit layout techniques are imperative with switching regulators. To minimize EMI, all high current loops should be kept as short as possible using heavy copper runs. The low current signal and high current switch and output grounds should return on separate paths back to the input filter capacitor. The R1, R2 output voltage divider should be located as close to the IC as possible to eliminate any noise pick-up into the feedback loop. The circuit diagrams were purposely drawn in a manner to depict this.

All circuits used molypermalloy power toroid cores for the magnetics where only the inductance value is given. The number of turns, wire and core size information is not given since no attempt was made to optimize their design. Inductor and transformer design information may be obtained from the magnetic core and assembly companies listed on the switching regulator component source table.

In some circuit designs, mainly step-up and voltage-inverting, a ratio of $t_{on}/(t_{on} + t_{off})$ greater than 0.857 may be required. This can be obtained by the addition of the ratio extender circuit shown in Figure 19. This circuit uses germanium components and is temperature sensitive. A negative temperature coefficient timing capacitor will help reduce this sensitivity. Figure 20 shows the output switch on and off time versus C_T with and without the ratio extender circuit. Notice that without the circuit, the ratio of $t_{on}/(t_{on} + t_{off})$ is limited to 0.857 only for values of C_T greater than 2.0 nF. With the circuit, the ratio is variable depending upon the value chosen for C_T since t_{off} is now nearly a constant. Current limiting must be used on all step-up and voltage-inverting designs using the ratio extender circuit. This will allow the inductor time to reset between cycles of overcurrent during initial power-up of the switcher. When the output filter capacitor reaches its nominal voltage, the voltage feedback loop will control regulation.

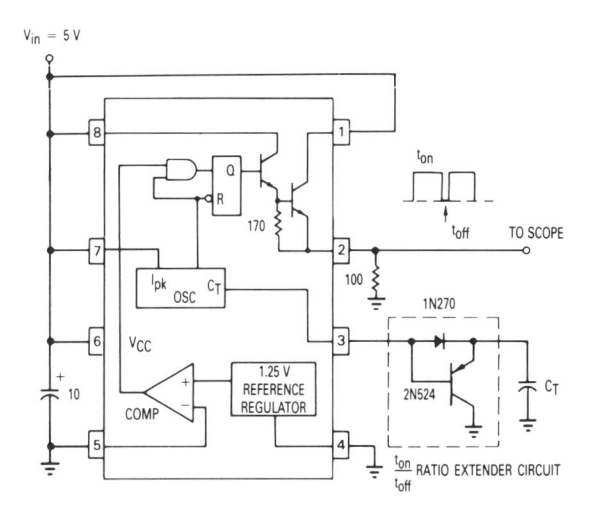

Figure 19. Output Switch On-Off Time Test Circuit

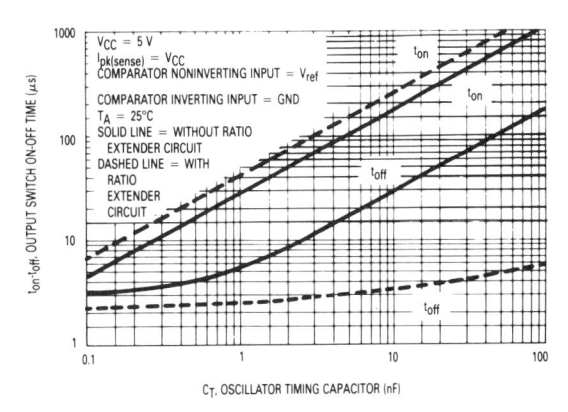

Figure 20. Output Switch On-Off Time versus Oscillator Timing Capacitor

APPLICATIONS SECTION

Listed below is an index of all the converter circuits shown in this application note. They are categorized into three major groups based upon the main output configuration. Each of these circuits was constructed and tested, and a performance table is included.

Index of Converter Circuits

	Main Output Configuration	Input V	Output 1 V/mA	Output 2 V/mA	Output 3 V/mA	Figure No.
STEP-DOWN						
µA78S40	Low Power with Minimum Components	24	5/50	—	—	9
MC34063	Medium Power	36	12/750	—	—	21
MC34063	Buffered Switch and Second Output	28	5/5000	12/300	—	22
µA78S40	Linear Pass from Main Output	33	24/500	15/50	—	23
µA78S40	Buffered Switch and Buffered Linear Pass from Main Output	28	15/3000	12/1000	—	24
µA78S40	Negative Input and Negative Output	−28	−12/500	—	—	25
STEP-UP						
µA78S40	Low Power with Minimum Components	9.0	28/50	—	—	11
MC34063	Medium Power	12	36/225	—	—	26
MC34063	High Voltage, Low Power	4.5	190/5.0	—	—	27
µA78S40	High Voltage, Medium Power Photoflash	4.5	334/45	—	—	28
µA78S40	Linear Pass from Main Output	2.5	9/100	6/30	—	29
µA78S40	Buffered Linear Pass from Main Output EE PROM Programmer	4.5	See Circuit	—	—	30
µA78S40	Buffered Switch and Buffered Linear Pass from Main Output	4.5	15/1000	12/500	−12/50	31
µA78S40	Dual Switcher, Step-Up and Step Down with Buffered Switch	12	28/250	5/250	—	32
STEP-UP/DOWN						
MC34063	Medium Power Step-Up/Down	7.5 to 14.5	10/120	—	—	18
VOLTAGE-INVERTING						
MC34063	Low Power	5	−12/100	—	—	33
µA78S40	Medium Power with Buffered Switch	15	−15/500	—	—	15
µA78S40	High Voltage, High Power with Buffered Switch	28	−120/850	—	—	34
µA78S40	42 Watt Off-Line Flyback Switcher	115 Vac	5/4000	12/700	−12/700	35
µA78S40	Tracking Regulator with Buffered Switch and Buffered Linear Pass from Input	15	−12/500	12/500	—	37

AN920 MOTOROLA

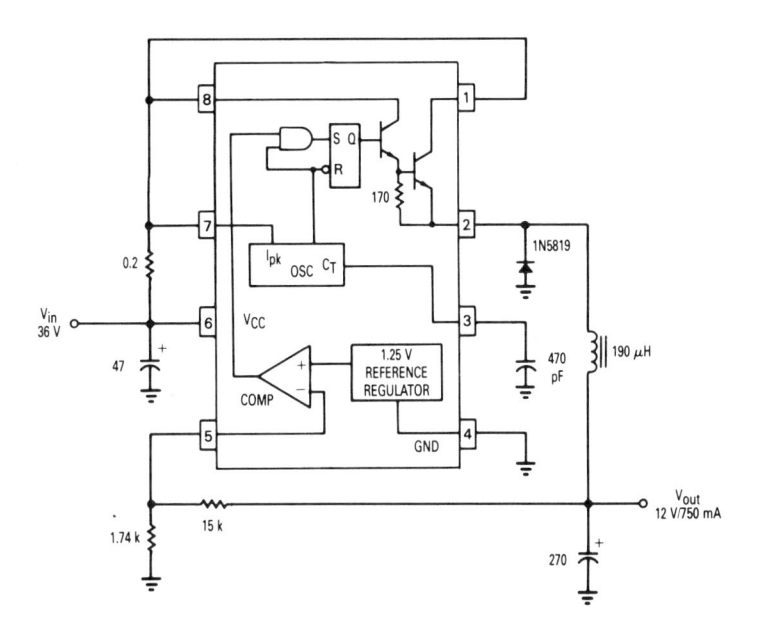

Test	Conditions	Results
Line Regulation	V_{in} = 20 to 40 V, I_{out} = 750 mA	Δ = 15 mV or ± 0.063%
Load Regulation	V_{in} = 36 V, I_{out} = 100 to 750 mA	Δ = 40 mV or ± 0.17%
Output Ripple	V_{in} = 36 V, I_{out} = 750 mA	60 mV$_{p-p}$
Short Circuit Current	V_{in} = 36 V, R_L = 0.1 Ω	1.6 A
Efficiency	V_{in} = 36 V, I_{out} = 750 mA	89.5%

A maximum power transfer of 9.0 watts is possible from an 8 pin dual-in-line package with V_{in} = 36 V and V_{out} = 12 V.

Figure 21. Step-Down

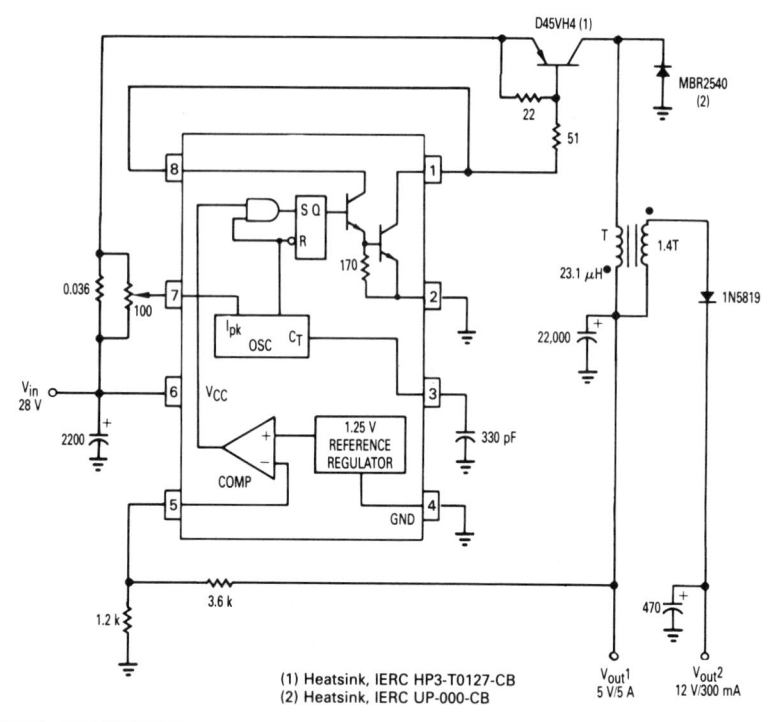

Test		Conditions	Results
Line Regulation	$V_{out}1$	V_{in} = 20 to 30 V, $I_{out}1$ = 5.0 A, $I_{out}2$ = 300 mA	Δ = 9.0 mV or ±0.09%
Load Regulation	$V_{out}1$	V_{in} = 28 V, $I_{out}1$ = 1.0 to 5.0 A, $I_{out}2$ = 300 mA	Δ = 20 mV or ± 0.2%
Output Ripple	$V_{out}1$	V_{in} = 28 V, $I_{out}1$ = 5.0 A, $I_{out}2$ = 300 mA	60 mV$_{p-p}$
Short Circuit Current	$V_{out}1$	V_{in} = 28 V, R_L = 0.1 Ω	11.4 A
Line Regulation	$V_{out}2$	V_{in} = 20 to 30 V, $I_{out}1$ = 5.0 A, $I_{out}2$ = 300 mA	Δ = 72 mV or ±0.3%
Load Regulation	$V_{out}2$	V_{in} = 20 V, $I_{out}2$ = 100 to 300 mA, $I_{out}1$ = 5.0 A	Δ = 12 mV or ± 0.05%
Output Ripple	$V_{out}2$	V_{in} = 28 V, $I_{out}1$ = 5.0 A, $I_{out}2$ = 300 mA	25 mV$_{p-p}$
Short Circuit Current	$V_{out}2$	V_{in} = 28 V, R_L = 0.1 Ω	11.25 A
Efficiency		V_{in} = 28 V, $I_{out}1$ = 5.0 A, $I_{out}2$ = 300 mA	80%

A second output can be easily derived by winding a secondary on the main output inductor and phasing it so that energy is delivered to $V_{out}2$ during t_{off}. The second output power should not exceed 25% of the main output. The 100 Ω potentiometer is used to divide down the voltage across the 0.036 Ω resistor and thus fine tune the current limit.

Figure 22. Step-Down with Buffered Switch and Second Output

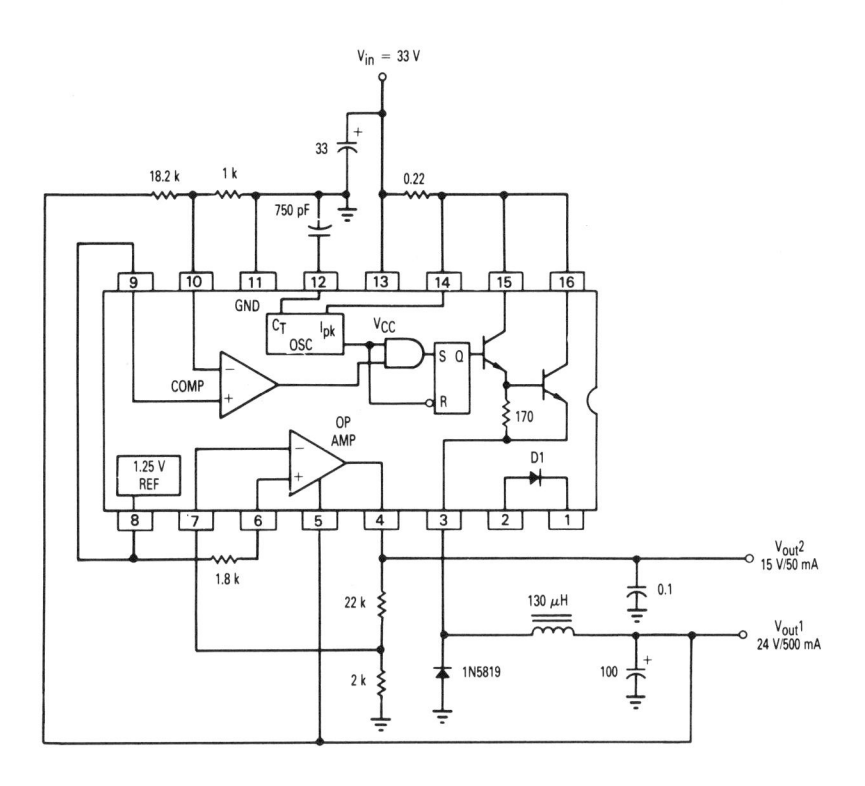

Test		Conditions	Results
Line Regulation	$V_{out}1$	V_{in} = 30 to 36 V, $I_{out}1$ = 500 mA, $I_{out}2$ = 50 mA	Δ = 30 mV or ± 0.63%
Load Regulation	$V_{out}1$	V_{in} = 33 V, $I_{out}1$ = 100 to 500 mA, $I_{out}2$ = 50 mA	Δ = 70 mV or ± 0.15%
Output Ripple	$V_{out}1$	V_{in} = 33 V, $I_{out}1$ = 500 mA, $I_{out}2$ = 50 mA	80 mV$_{p-p}$
Short Circuit Current	$V_{out}1$	V_{in} = 33 V, R_L = 0.1 Ω	2.5 A
Line Regulation	$V_{out}2$	V_{in} = 30 to 36 V, $I_{out}1$ = 500 mA, $I_{out}2$ = 50 mA	Δ = 20 mV or 0.067%
Load Regulation	$V_{out}2$	V_{in} = 33 V, $I_{out}2$ = 0 to 50 mA, $I_{out}1$ = 500 A	Δ = 60 mV or ± 0.2%
Output Ripple	$V_{out}2$	V_{in} = 33 V, $I_{out}1$ = 500 mA, $I_{out}2$ = 50 mA	70 mV$_{p-p}$
Short Circuit Current	$V_{out}2$	V_{in} = 33 V, R_L = 0.1 Ω	90 mA
Efficiency		V_{in} = 33 V, $I_{out}1$ = 500 mA, $I_{out}2$ = 50 mA	88.2%

Figure 23. Step-Down with Linear Pass from Main Output

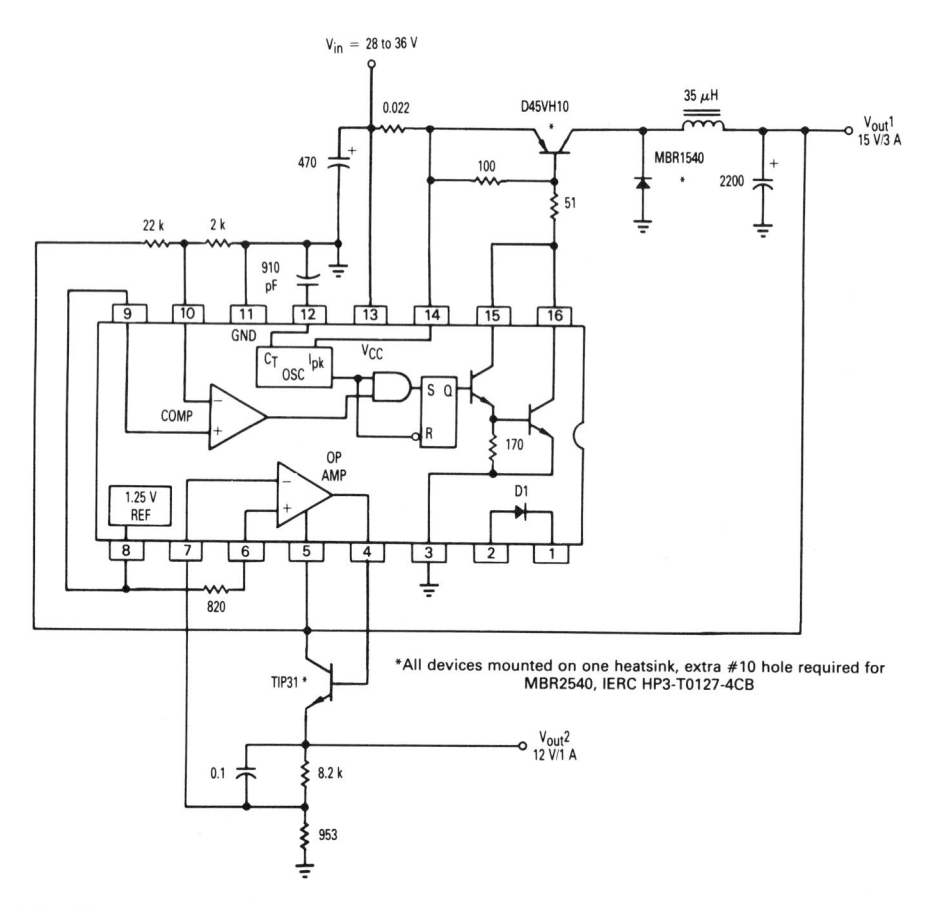

Test		Conditions	Results
Line Regulation	$V_{out}1$	V_{in} = 28 to 36 V, $I_{out}1$ = 3.0 A, $I_{out}2$ = 1.0 A	Δ = 13 mV or ± 0.043%
Load Regulation	$V_{out}1$	V_{in} = 36 V, $I_{out}1$ = 1.0 to 4.0 A, $I_{out}2$ = 1.0 A	Δ = 21 mV or ± 0.07%
Output Ripple	$V_{out}1$	V_{in} = 36 V, $I_{out}1$ = 3.0 A, $I_{out}2$ = 1.0 A	120 mV$_{p-p}$
Short Circuit Current	$V_{out}1$	V_{in} = 36 V, R_L = 0.1 Ω	12.6 A
Line Regulation	$V_{out}2$	V_{in} = 28 to 36 V, $I_{out}1$ = 3.0 A, $I_{out}2$ = 1.0 A	Δ = 2.0 mV or ± 0.008%
Load Regulation	$V_{out}2$	V_{in} = 36 V, $I_{out}2$ = 0 to 1.5 A	Δ = 2.0 mV or ± 0.008%
Output Ripple	$V_{out}2$	V_{in} = 36 V, $I_{out}1$ = 3.0 A, $I_{out}2$ = 1.0 A	25 mV$_{p-p}$
Short Circuit Current	$V_{out}2$	V_{in} = 36 V, R_L = 0.1 Ω	3.6 A
Efficiency		V_{in} = 36 V, $I_{out}1$ = 3.0 A, $I_{out}2$ = 1.0 A	78.5%

Figure 24. Step-Down with Buffered Switch and Buffered Linear Pass from Main Output

AN920   MOTOROLA

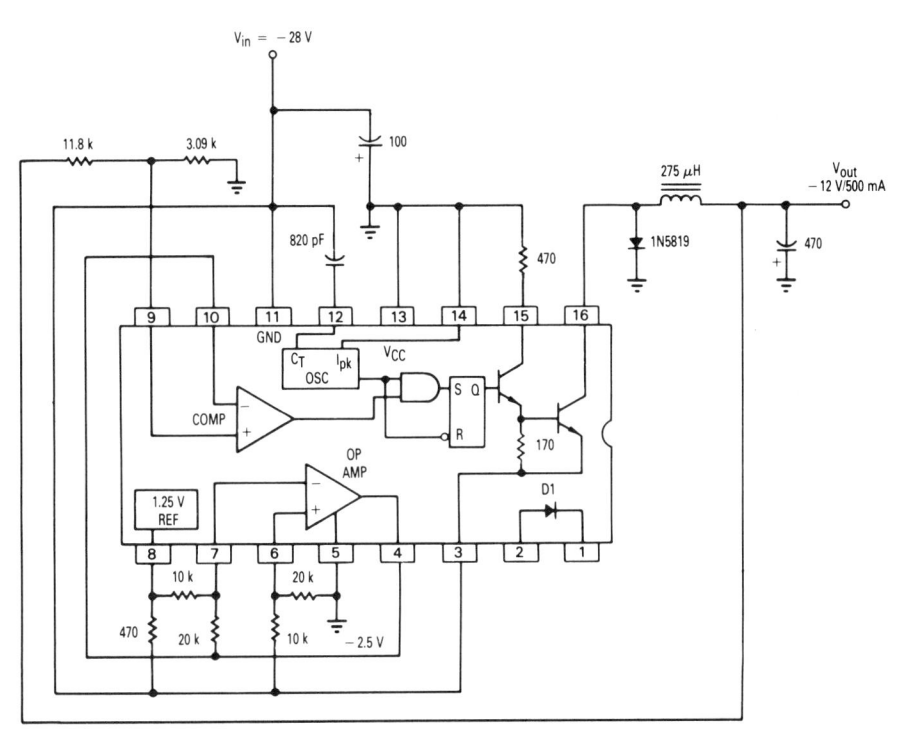

Test	Conditions	Results
Line Regulation	V_{in} = –22 to –28 V, I_{out} = 500 mA	Δ 25 mV or ± 0.104%
Load Regulation	V_{in} = –28 V, I_{out} = 100 to 500 mA	Δ = 10 mV or ± 0.042%
Output Ripple	V_{in} = –28 V, I_{out} = 500 mA	130 mV$_{p-p}$
Efficiency	V_{in} = –28 V, I_{out} = 500 mA	85.5%

In this step-down circuit, the output switch must be connected in series with the negative input, causing the internal 1.25 V reference to be with respect to $-V_{in}$. A second reference of –2.5 V with respect to ground is generated by the Op Amp. Note that the 10 k and 20 k resistors must be matched pairs for good line regulation and that no provision is made for output short-circuit protection.

Figure 25. Step-Down with Negative Input and Negative Output

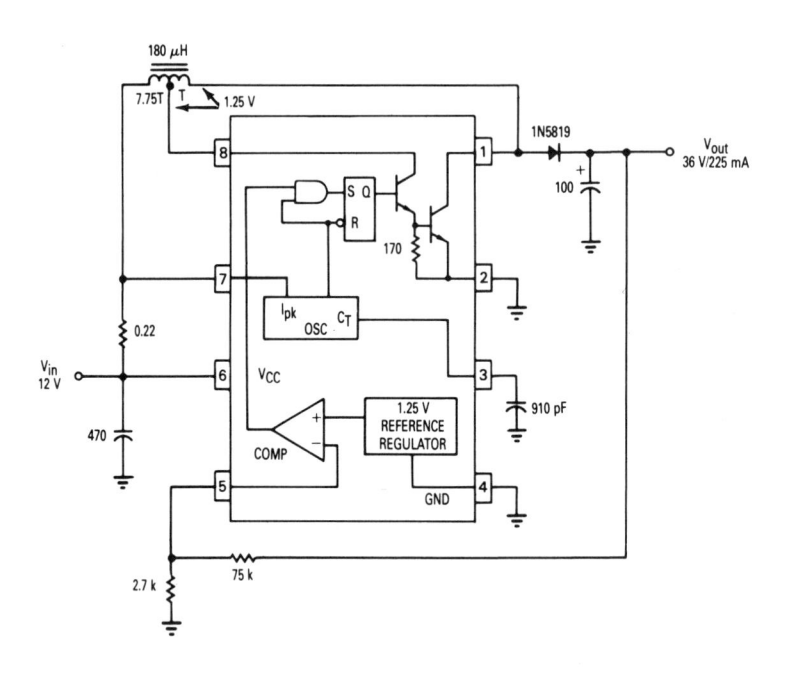

Test	Conditions	Results
Line Regulation	V_{in} = 11 to 15 V, I_{out} = 225 mA	Δ = 20 mV or ± 0.028%
Load Regulation	V_{in} = 12 V, I_{out} = 50 to 225 mA	Δ = 30 mV or ± 0.042%
Output Ripple	V_{in} = 12 V, I_{out} = 225 mA	100 mV$_{p-p}$
Efficiency	V_{in} = 12 V, I_{out} = 225 mA	90.4%

A maximum power transfer of 8.1 watts is possible with V_{in} = 12 V and V_{out} = 36 V. The high efficiency is partially due to the use of the tapped inductor. The tap point is set for a voltage differential of 1.25 V. The range of V_{in} is somewhat limited when using this method.

Figure 26. Step-Up

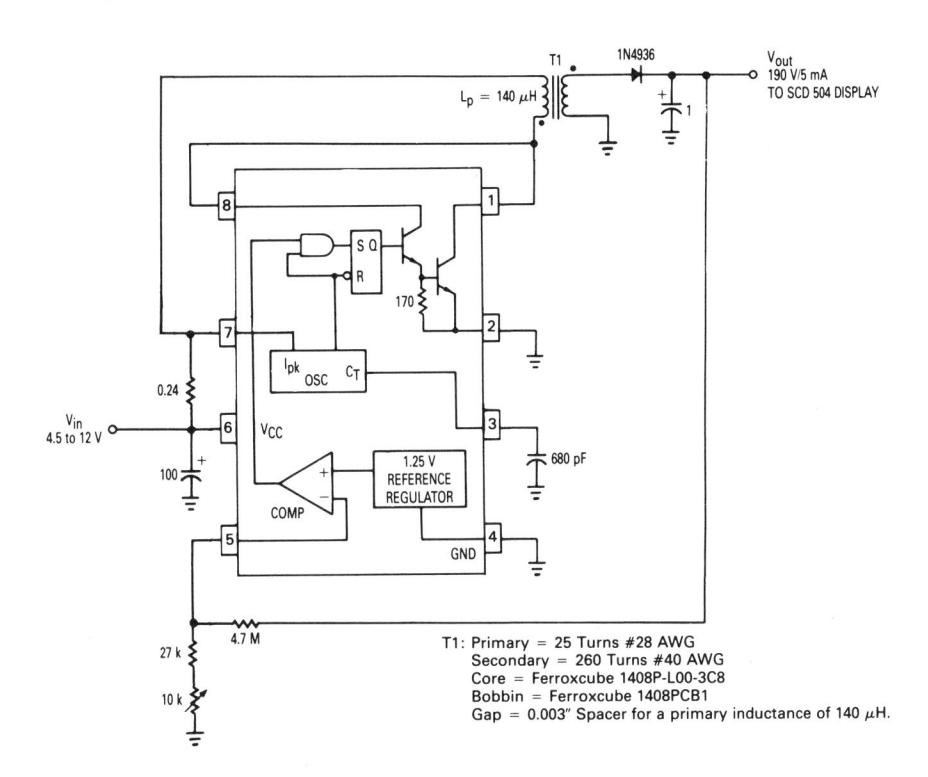

T1: Primary = 25 Turns #28 AWG
Secondary = 260 Turns #40 AWG
Core = Ferroxcube 1408P-L00-3C8
Bobbin = Ferroxcube 1408PCB1
Gap = 0.003" Spacer for a primary inductance of 140 µH.

Test	Conditions	Results
Line Regulation	V_{in} = 4.5 to 12 V, I_{out} = 5.0 mA	Δ = 2.3 V or ± 0.61%
Load Regulation	V_{in} = 5.0 V, I_{out} = 1.0 to 6.0 mA	Δ = 1.4 V or ± 0.37%
Output Ripple	V_{in} = 5.0 V, I_{out} = 5.0 mA	250 mV$_{p-p}$
Short Circuit Current	V_{in} = 5.0 V, R_L = 0.1 Ω	113 mA
Efficiency	V_{in} = 5.0 V, I_{out} = 5.0 mA	68%

This circuit was designed to power the Motorola Solid Ceramic Displays from a V_{in} of 4.5 to 12 V. The design calculations are based on a step-up converter with an input of 4.5 V and a 24 V output rated at 45 mA. The 24 V level is the maximum step-up allowed by the oscillator ratio of $t_{on}/(t_{on} + t_{off})$. The 45 mA current level was chosen so that the transformer primary power level is about 10% greater than that required by the load. The maximum V_{in} of 12 V is determined by the sum of the flyback and leakage inductance voltages present at the collector of the output switch during turn-off must not exceed 40 V.

Figure 27. High-Voltage, Low Power Step-Up for Solid Ceramic Display

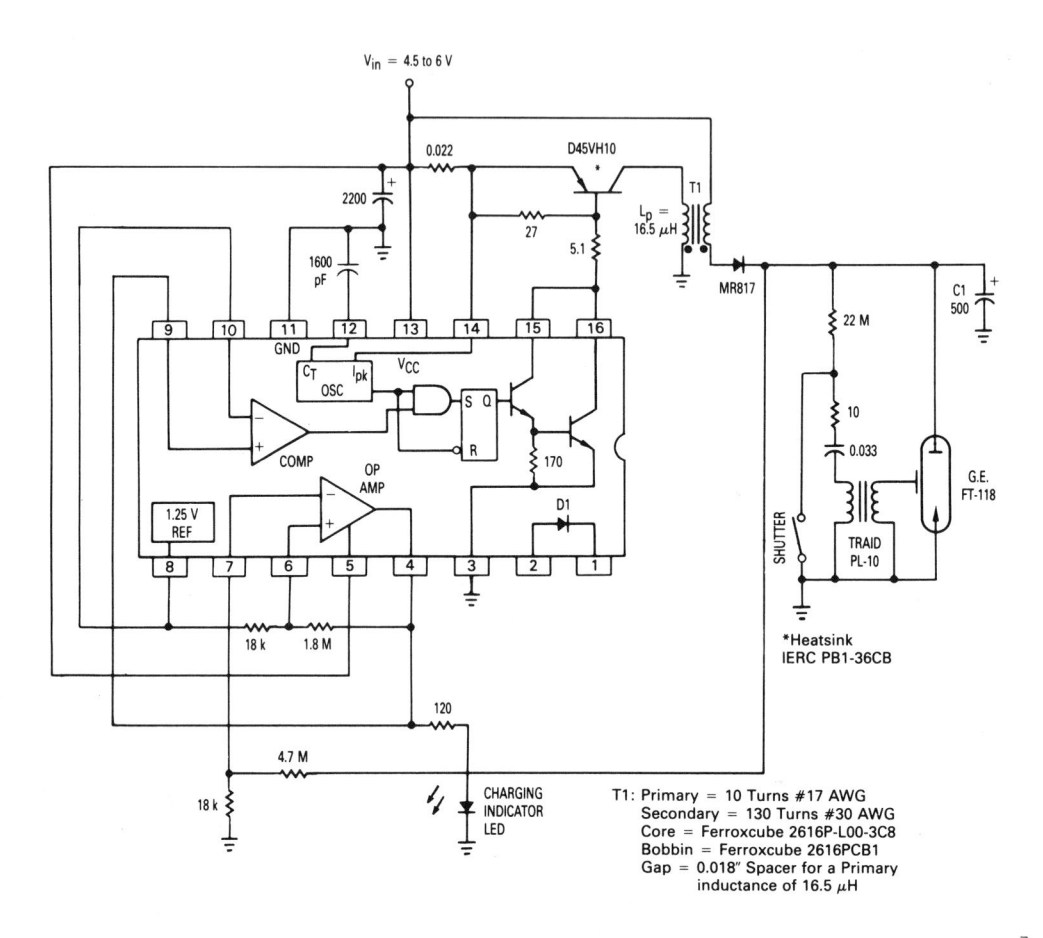

With V_{in} of 6.0 V, this step-up converter will charge capacitor C1 from 0 to 334 V in 4.7 seconds. The switching operation will cease until C1 bleeds down to 323 V. The charging time between flashes is 4.0 seconds. The output current at 334 V is 45 mA.

Figure 28. High-Voltage Step-Up with Buffered Switch for Photoflash Applications

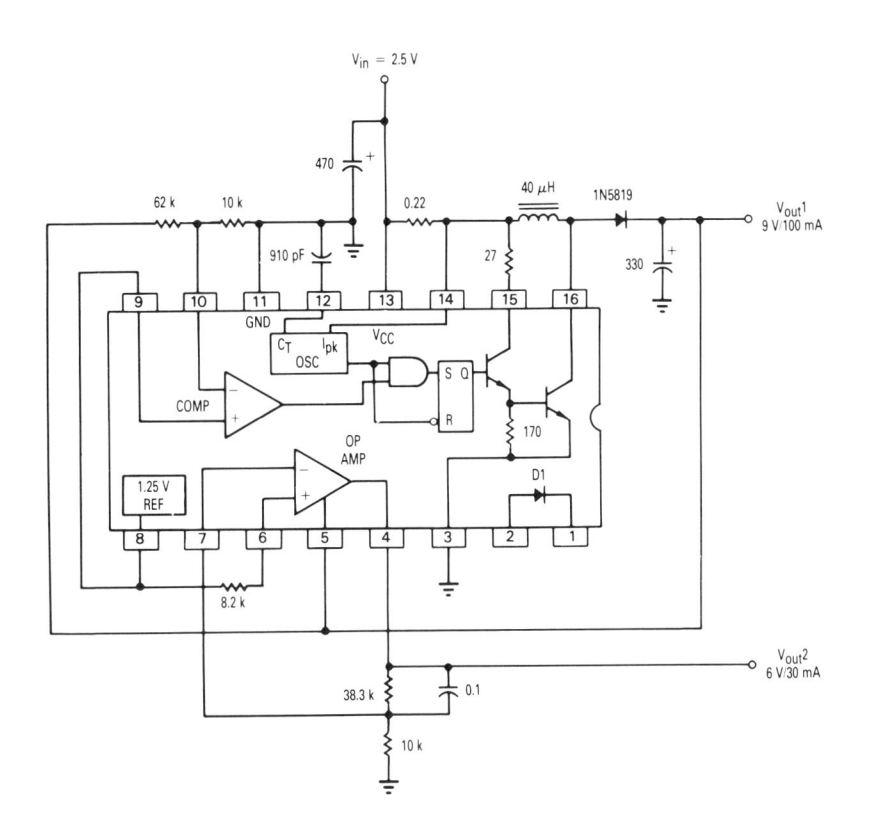

Test		Conditions	Results
Line Regulation	$V_{out}1$	V_{in} = 2.5 to 3.5 V, $I_{out}1$ = 100 mA, $I_{out}2$ = 30 mA	Δ = 20 mV or ± 0.11%
Load Regulation	$V_{out}1$	V_{in} = 2.5 V, $I_{out}1$ = 25 to 100 mA, $I_{out}2$ = 30 mA	Δ = 20 mV or ± 0.11%
Output Ripple	$V_{out}1$	V_{in} = 2.5 V, $I_{out}1$ = 100 mA, $I_{out}2$ = 30 mA	60 mV$_{p-p}$
Line Regulation	$V_{out}2$	V_{in} = 2.5 to 3.5 V, $I_{out}1$ = 100 mA, $I_{out}2$ = 30 mA	Δ = 1.0 mV or ± 0.0083%
Load Regulation	$V_{out}2$	V_{in} = 2.5 V, $I_{out}2$ = 0 to 50 mA, $I_{out}1$ = 100 mA	Δ = 1.0 mV or ± 0.0083%
Output Ripple	$V_{out}2$	V_{in} = 2.5 V, $I_{out}1$ = 100 mA, $I_{out}2$ = 30 mA	5.0 mV$_{p-p}$
Short Circuit Current	$V_{out}2$	V_{in} = 2.5 V, R_L = 0.1 Ω	150 mA
Efficiency		V_{in} = 3.0 V, $I_{out}1$ = 100 mA, $I_{out}2$ = 30 mA	68.3%

Figure 29. Step-Up with Linear Pass from Main Output

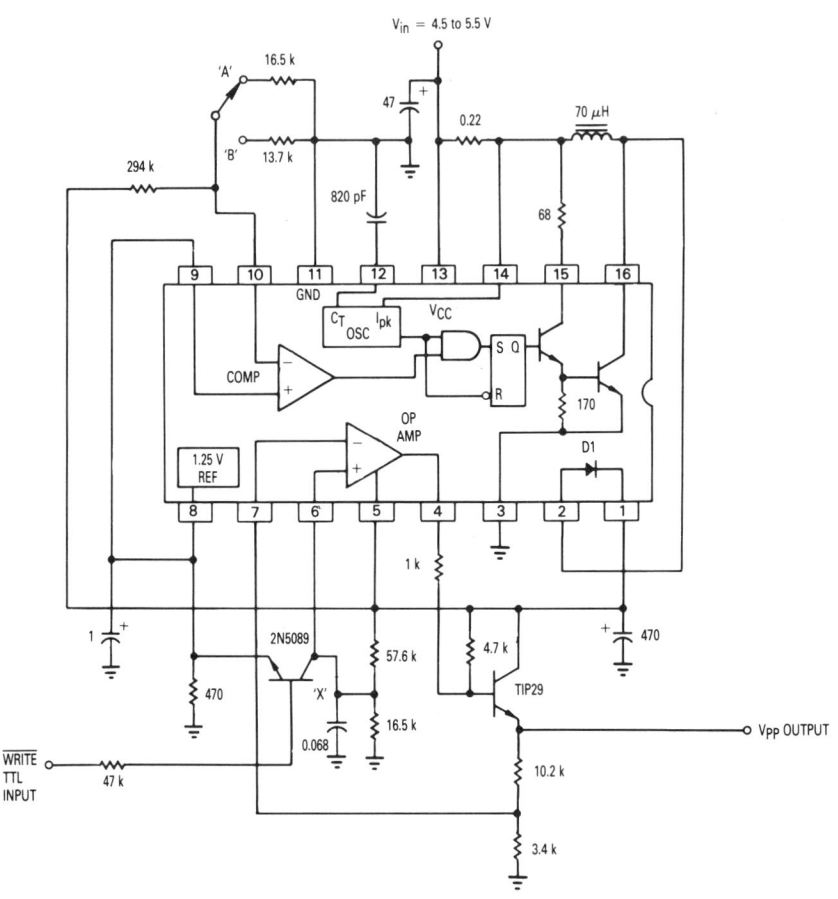

Switch Position	WRITE	Voltage @ Pins 1 & 5	Voltage @ X	Voltage @ V$_{PP}$
A	< 1.5	23.52	5.24	20.95
A	> 2.25	23.52	1.28	5.12
B	< 1.5	28.07	6.25	25.01
B	> 2.25	28.07	1.28	5.12

All values are in volts. V$_{ref}$ = 1.245 V.
Contributed by Steve Hageman of Calex Mgf. Co. Inc.

Used in conjunction with two transistors, the μA78S40 can generate the required Vpp voltage of 21 or 25 volts needed to program and erase EEPROMs from a single 5.0 volt supply. A step-up converter provides a selectable regulated voltage at Pins 1 and 5. This voltage is used to generate a second reference at point 'X' and to power the linear regulator consisting of the internal op amp and a TIP29 transistor. When the WRITE input is less than 1.5 V, the 2N5089 transistor is OFF, allowing the voltage at 'X' to rise exponentially with an approximate time constant of 600 μs as required by some EEPROMs. The linear regulator amplifies the voltage at 'X' by four, generating the required Vpp output voltage for the byte-erase write cycle. When the WRITE input is greater than 2.25 V, the 2N5089 turns ON clamping point 'X' to the internal reference level of 1.245 volts. The Vpp output will not be at approximately 5.1 volts or 4.0 (1.245 + V$_{sat}$ 2N5089). The μA78S40 reference can only source current, therefore a reference pre bias of 470 Ω is used. The Vpp output is short-circuit protected and can supply a current of 100 mA at 21 V or 75 mA at 25 V over an input range of 4.5 to 5.5 V.

Figure 30. Step-Up with Buffered Linear Pass from Main Output for Programming EEPROMs

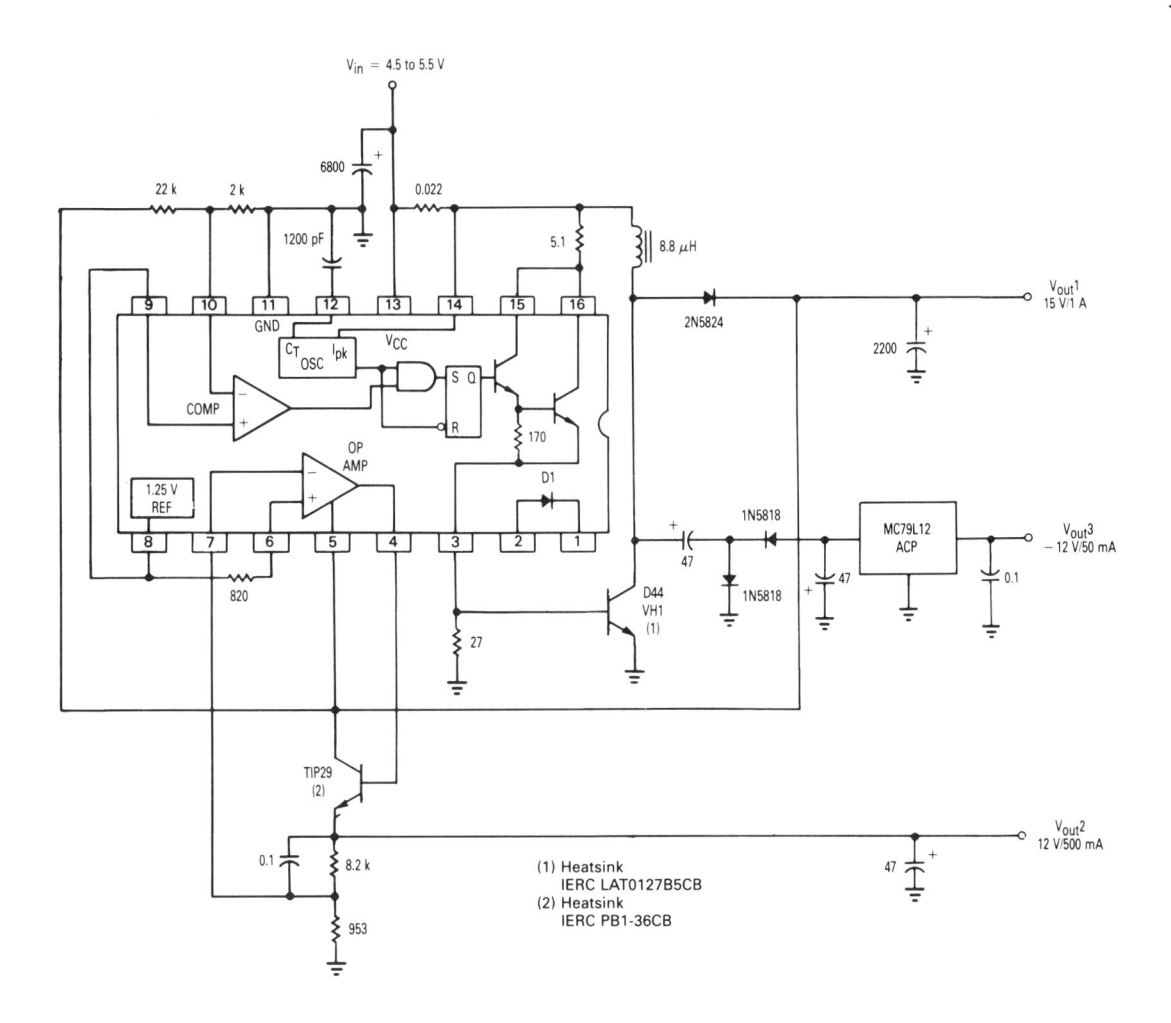

Test		Conditions	Results
Line Regulation	$V_{out}1$	V_{in} = 4.5 to 5.5 V	Δ = 18 mV or ± 0.06%
Load Regulation	$V_{out}1$	V_{in} = 5.0 V, $I_{out}1$ = 0.25 to 1.0 A	Δ = 25 mV or ± 0.083%
Output Ripple	$V_{out}1$	V_{in} = 5.0 V	75 mV$_{p-p}$
Line Regulation	$V_{out}2$	V_{in} = 4.5 to 5.5 V	Δ = 3.0 mV or ± 0.013%
Load Regulation	$V_{out}2$	V_{in} = 5.0 V, $I_{out}2$ = 100 to 500 mA	Δ = 5.0 mV or ± 0.021%
Output Ripple	$V_{out}2$	V_{in} = 5.0 V	20 mV$_{p-p}$
Short Circuit Current	$V_{out}2$	V_{in} = 5.0 V, R_L = 0.1 Ω	2.7 A
Line Regulation	$V_{out}3$	V_{in} = 4.5 to 5.5 V	Δ = 2.0 mV or ± 0.008%
Load Regulation	$V_{out}3$	V_{in} = 5.0 V, $I_{out}3$ = 0 to 50 mA	Δ = 29 mV or ± 0.12%
Output Ripple	$V_{out}3$	V_{in} = 5.0 V	15 mV$_{p-p}$
Short Circuit Current	$V_{out}3$	V_{in} = 5.0 V, R_L = 0.1 Ω	130 mA
Efficiency		V_{in} = 5.0 V	71.8%

All outputs are at nominal load current unless otherwise noted.

Figure 31. Step-Up with Buffered Switch and Buffered Linear Pass from Main Output

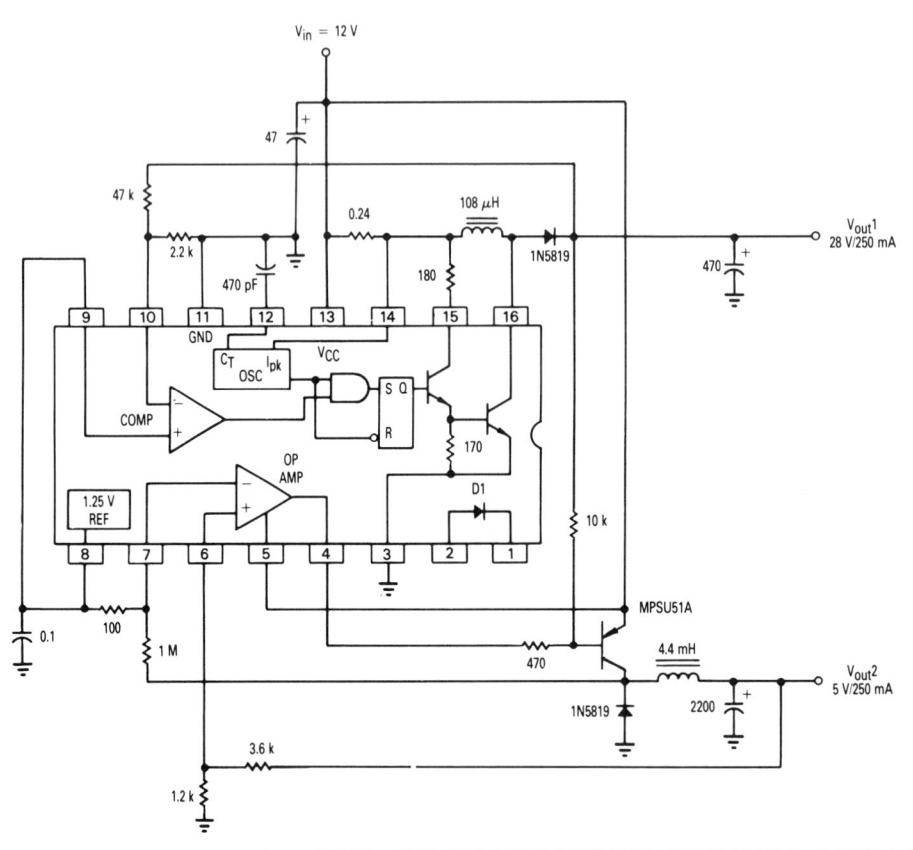

Test		Conditions	Results
Line Regulation	$V_{out}1$	V_{in} = 9.0 to 15 V, $I_{out}1$ = 250 mA, $I_{out}2$ = 250 mA	Δ = 30 mV or ± 0.054%
Load Regulation	$V_{out}1$	V_{in} = 12 V, $I_{out}1$ = 100 to 300 mA, $I_{out}2$ = 250 mA	Δ = 20 mV or ± 0.036%
Output Ripple	$V_{out}1$	V_{in} = 12 V, $I_{out}1$ = 250 mA, $I_{out}2$ = 250 mA	35 mV$_{p-p}$
Short Circuit Current	$V_{out}1$	V_{in} = 12 V, R_L = 0.1 Ω	1.7 A
Line Regulation	$V_{out}2$	V_{in} = 9.0 to 15 V, $I_{out}1$ = 250 mA, $I_{out}2$ = 250 mA	Δ = 4.0 mV or ± 0.04%
Load Regulation	$V_{out}2$	V_{in} = 12 V, $I_{out}2$ = 100 to 300 mA, $I_{out}1$ = 250 mA	Δ = 18 mV or ± 0.18%
Output Ripple	$V_{out}2$	V_{in} = 12 V, $I_{out}1$ = 250 mA, $I_{out}2$ = 250 mA	70 mV$_{p-p}$
Efficiency		V_{in} = 12 V, $I_{out}1$ = 250 mA, $I_{out}2$ = 250 mA	81.8%

This circuit shows a method of using the µA78S40 to construct two independent converters. Output 1 uses the typical step-up circuit configuration while Output 2 makes the use of the op amp connected with positive feedback to create a free running step-down converter. The op amp slew rate limits the maximum switching frequency at rated load to less than 2.0 kHz.

Figure 32. Dual Switcher, Step-Up and Step-Down with Buffered Switch

$$|V_{out}| = 1.25 \left(1 + \frac{R2}{R1}\right)$$

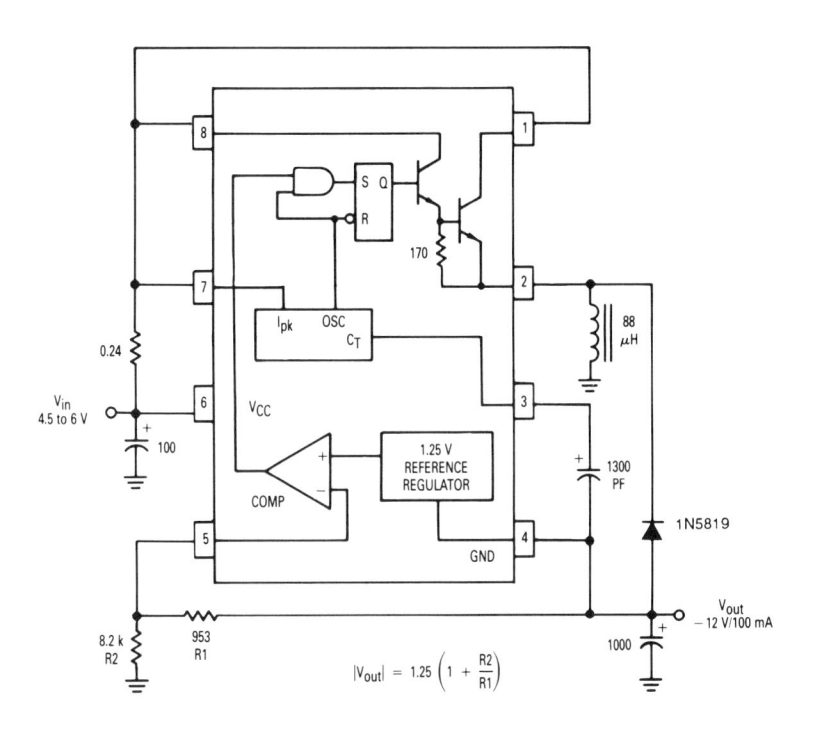

Test	Conditions	Results
Line Regulation	V_{in} = 4.5 to 5.0 V, I_{out} = 100 mA	Δ = 2.0 mV or ± 0.008%
Load Regulation	V_{in} = 5.0 V, I_{out} = 10 to 100 mA	Δ = 10 mV or ± 0.042%
Output Ripple	V_{in} = 5.0 V, I_{out} = 100 mA	35 mV$_{p-p}$
Short Circuit Current	V_{in} = 5.0 V, R_L = 0.1 Ω	1.4 A
Efficiency	V_{in} = 5.0 V, I_{out} = 100 mA	60%

The above circuit shows a method of using the MC34063 to construct a low power voltage-inverting converter. Note that the integrated circuit ground, pin 4, is connected directly to the negative output, thus allowing the internally connected comparator and reference to function properly for output voltage control. With this configuration, the sum of V_{in} + V_{out} + V_F must not exceed 40 V. The conversion efficiency is modest since the output switch is connected as a Darlington and its on-voltage is a large portion of the minimum operating input voltage. A 12% improvement can be realized with the addition of an external PNP saturated switch when connected in a similar manner to that shown in Figure 15.

Figure 33. Low Power Voltage-Inverting

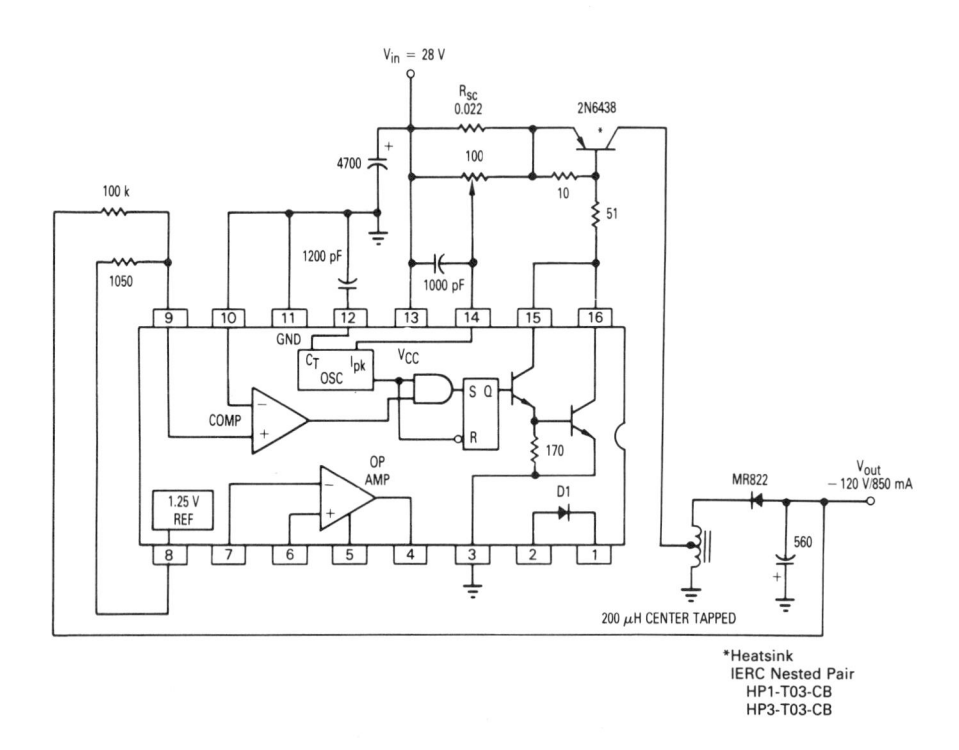

Test	Conditions	Results
Line Regulation	V_{in} = 24 to 28 V, I_{out} = 850 mA	Δ = 100 mV or ± 0.042%
Load Regulation	V_{in} = 28 V, I_{out} = 100 to 850 mA	Δ = 70 mV or ± 0.029%
Output Ripple	V_{in} = 28 V, I_{out} = 850 mA	450 mV$_{p-p}$
Short Circuit Current	V_{in} = 28 V, R_L = 0.1 Ω	6.4 A
Efficiency	V_{in} = 28 V, I_{out} = 850 mA	81.8%

This high power voltage-inverting circuit makes use of a center tapped inductor to step-up the magnitude of the output. Without the tap, the output switch transistor would need a V_{CE} breakdown greater than 148 V at the start of t_{off}; the maximum rating of this device is 120 V. All calculations are done for the typical voltage-inverting converter with an input of 28 V and an output of −120 V. The inductor value will be 50 μH or 200 μH center tapped for the value of C_T used. The 1000 pF capacitor is used to filter the spikes generated by the high switching current flowing through the wiring and R_{sc} inductance.

Figure 34. High Power Voltage-Inverting with Buffered Switch

An economical 42 watt off-line flyback switcher is shown in Figure 35. In this circuit the μA78S40 is connected to operate as a fixed frequency pulse width modulator. The oscillator sawtooth waveform is connected to the noninverting input of the comparator and a preset voltage of 685 mV, derived from the reference is connected to the inverting input. The preset voltage reduces the maximum percent on-time of the output switch from a nominal of 85.7% to about 45%. The maximum must be less than 50% when an equal turns ratio of primary to clamp winding is used. Output regulation and isolation is achieved by the use of the TL431 as an output reference and comparator, and a 4N35 optocoupler. As the 5.0 V output reaches its nominal level, the TL431 will start to conduct current through the LED in the 4N35. This in turn will cause the optoreceiver transistor to turn-on, raising the voltage at Pin 10 which will cause a reduction in percent on-time of the output switch.

The peak drain current at 42 W output is 2.0 A. As the output loading is increased, the MPS6515 will activate the $I_{pk(sense)}$ pin and shorten t_{on} on a cycle by cycle basis. If an output is shorted the $I_{pk(sense)}$ circuit will cause C_T to charge beyond the upper oscillator trip point and the oscillator frequency will decrease. This action will result in a lower average power dissipation for the output switching transistor.

Each output has a series inductor and a second shunt filter capacitor forming a Pi filter. This is used to reduce the level of high frequency ripple and spikes. Care must be taken with the layout of grounds in the Pi filter network. Each input and output filter capacitor must have separate ground returns to the transformer as shown on the circuit diagram. A complete printed circuit board with component layout is shown in Figure 36.

The μA78S40 may be used in any of the previously shown circuit designs as a fixed frequency pulse width modulator, however consideration must be given to the proper selection of the feedback loop elements in order to insure circuit stability.

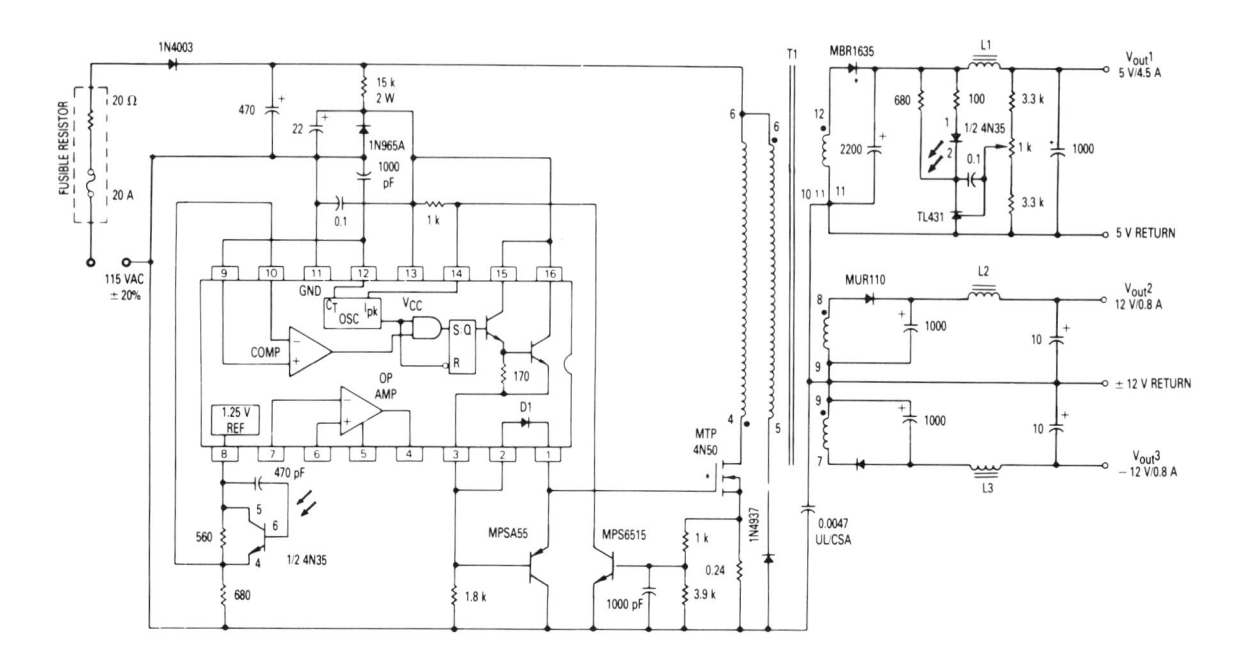

Test		Conditions	Results
Ling Regulation	$V_{out}1$	V_{in} = 92 to 138 Vac	Δ = 1.0 mV or ± 0.01%
Load Regulation	$V_{out}1$	V_{in} = 115 Vac, $I_{out}1$ = 1.0 to 4.5 A	Δ = 3.0 mV or ± 0.03%
Output Ripple	$V_{out}1$	V_{in} = 115 Vac	40 mV$_{p-p}$
Short Circuit Current	$V_{out}1$	V_{in} = 115 Vac, R_L = 0.1 Ω	19.2 A
Line Regulation	$V_{out}2$ or 3	V_{in} = 92 to 138 Vac	Δ = 10 mV or ± 0.04%
Load Regulation	$V_{out}2$ or 3	V_{in} = 115 Vac, $I_{out}2$ or 3 = 0.25 to 0.8 A	Δ = 384 mV or ± 1.6%
Output Ripple	$V_{out}2$ or 3	V_{in} = 115 Vac	80 mV$_{p-p}$
Short Circuit Current	$V_{out}2$ or 3	V_{in} = 115 Vac, R_L = 0.1 Ω	10.8 A
Efficiency		V_{in} = 115 Vac	75.7%

All outputs are at nominal load current unless otherwise noted.

*Heatsink
Thermalloy 6072B-MT
T1 — Primary:
  Pins 4 and 6 = 72 Turns #24 AWG   Bifilar Wound
  Pins 5 and 6 = 72 Turns #26 AWG
Secondary 5.0 V:
  6 Turns (two strands) #18 AWG Bifilar Wound
Secondary ± 12 V:
  14 Turns #23 AWG Bifilar Wound
Core and Bobbin:
  Coilcraft PT3995
Gap: 0.030" Spacer in each leg for a primary inductance of 550 μH.
  Primary to primary leakage inductance must be less than 30 μH.
L1 — Coilcraft Z7156:
  Remove one layer for final inductance of 4.5 μH.
L2, L3 — Coilcraft Z7157:
  25 μH at 1.0 A

Figure 35. 42 Watt Off-Line Flyback Switcher with Primary Power Limiting

Component Layout — Bottom View

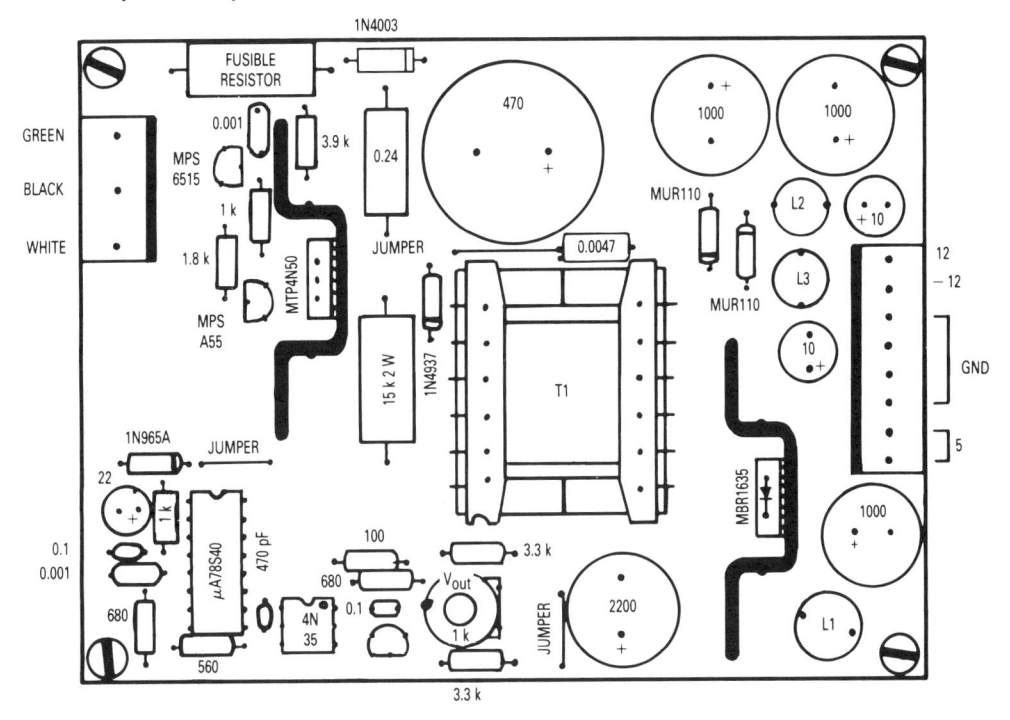

Printed Circuit Board Negative — Bottom View

Figure 36. 42 Watt Off-Line

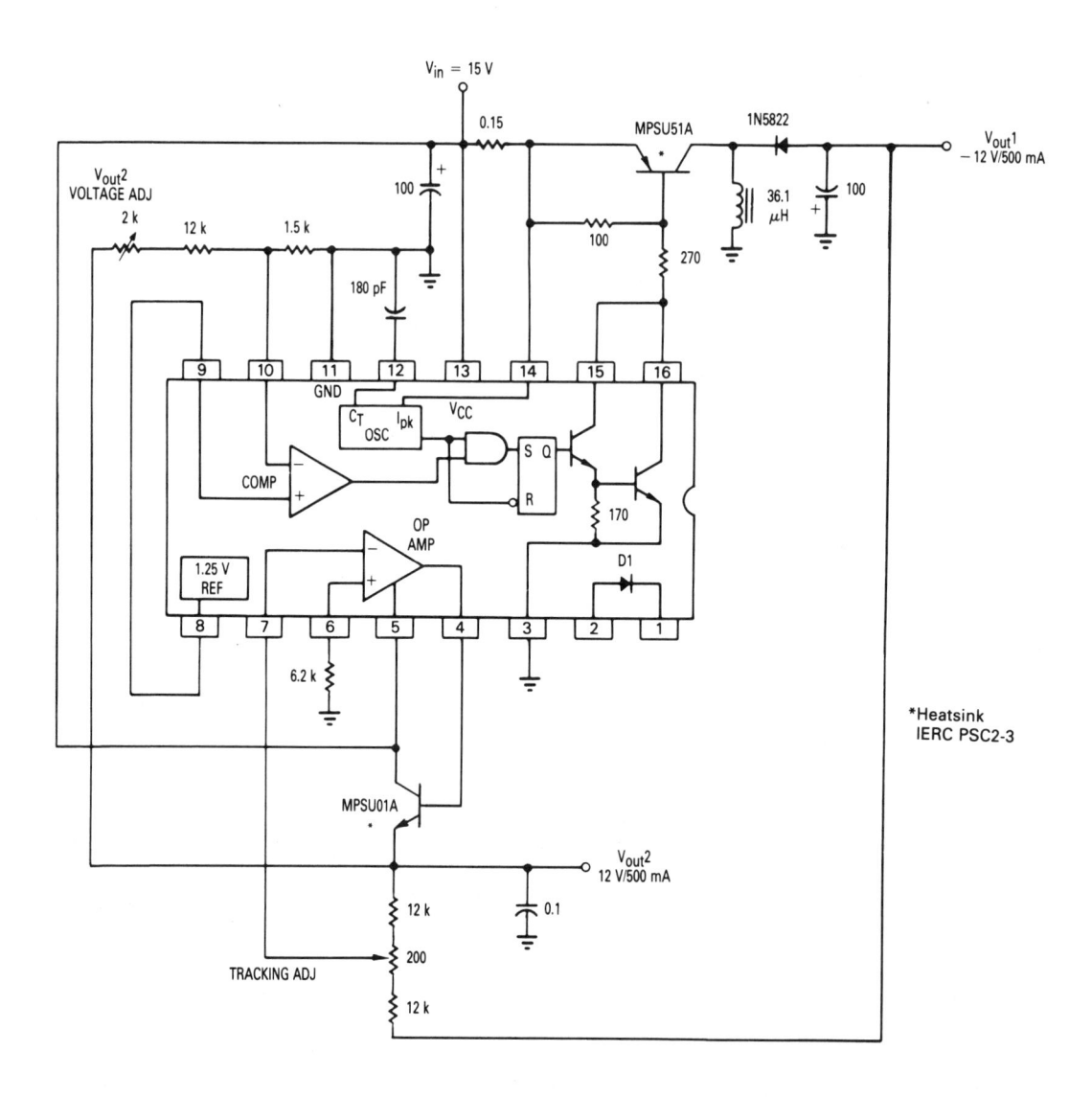

Test		Conditions	Results
Line Regulation	$V_{out}1$	V_{in} = 14.5 to 18 V, $I_{out}1$ = 500 mA, $I_{out}2$ = 500 mA	Δ = 10 mV or ± 0.042%
Load Regulation	$V_{out}1$	V_{in} = 15 V, $I_{out}1$ = 100 to 500 mA, $I_{out}2$ = 500 mA	Δ = 2.0 mV or ± 0.008%
Output Ripple	$V_{out}1$	V_{in} = 15 V, $I_{out}1$ = 500 mA, $I_{out}2$ = 500 mA	125 mV$_{p-p}$
Line Regulation	$V_{out}2$	V_{in} = 14.5 to 18 V, $I_{out}1$ = 500 mA, $I_{out}2$ = 500 mA	Δ = 10 mV or ± 0.042%
Load Regulation	$V_{out}2$	V_{in} = 15 V, $I_{out}2$ = 100 to 500 mA, $I_{out}1$ = 500 mA	Δ = 5.0 mV or ± 0.021%
Output Ripple	$V_{out}2$	V_{in} = 15 V, $I_{out}1$ = 500 mA, $I_{out}2$ = 500 mA	140 mV$_{p-p}$
Efficiency		V_{in} = 15 V, $I_{out}1$ = 500 mA, $I_{out}2$ = 500 mA	77.2%

This tracking regulator provides a ± 12 V output from a single 15 V input. The negative output is generated by a voltage-inverting converter while the positive is a linear pass regulator taken from the input. The ± 12 V outputs are monitored by the op amp in a corrective fashion so that the voltage at the center of the divider is zero ± V_{IO}. The op amp is connected as a unity gain inverter when $|V_{out}1| = |V_{out}2|$.

Figure 37. Tracking Regulator, Voltage-Inverting with Buffered Switch and Buffered Linear Pass from Input

MOTOROLA SEMICONDUCTOR TECHNICAL DATA

TCA5600
TCF5600

Advance Information

UNIVERSAL MICROPROCESSOR POWER SUPPLY/CONTROLLER

UNIVERSAL MICROPROCESSOR POWER SUPPLY CONTROLLER

SILICON MONOLITHIC INTEGRATED CIRCUITS

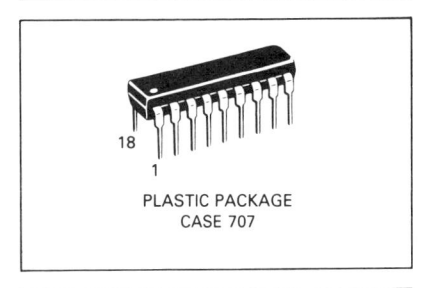

The TCA5600 is a versatile power supply control circuit for microprocessor based systems and mainly intended for automotive applications and battery powered instruments. To cover a wide range of applications, the device offers high circuit flexibility with minimum of external components.

Functions included in this IC are a temperature compensated voltage reference, on chip dc/dc converter, programmable and remote controlled voltage regulator, fixed 5.0 V supply voltage regulator with external PNP power device, undervoltage detection circuit, power-on RESET delay and watchdog feature for safe and hazard free microprocessor operations.

- 6.0 to 30 V Operation Range
- 2.5 V Reference Voltage Accessible for Other Tasks
- Fixed 5.0 V ± 4% Microprocessor Supply Regulator Including Current Limitation, Overvoltage Protection and Undervoltage Monitor
- Programmable 6.0 to 30 V Voltage Regulator Exhibiting High Peak Current (150 mA), Current Limiting and Thermal Protection
- Two Remote Inputs to Select the Regulator's Operation Mode: OFF, 5.0 V, 5.0 V Standby and Programmable Output Voltage
- Self Contained dc/dc Converter Fully Controlled By the Programmable Regulator to Guarantee Safe Operation Under All Working Conditions
- Programmable Power-On RESET Delay
- Watchdog Select Input
- Negative Edge Triggered Watchdog Input
- Low Current Consumption in the V_{CC1} Standby Mode
- All Digital Control Ports are TTL- and MOS-Compatible

APPLICATIONS INCLUDE
- Microprocessor Systems with E^2PROMs
- High Voltage Crystal and Plasma Displays
- Decentralized Power Supplies in Computer and Telecommunication Systems

PLASTIC PACKAGE
CASE 707

PIN CONNECTIONS

RESET	1	18	WDS
V_{out1} Sense	2	17	Delay
V_{CC1}	3	16	I_{out1} Sense
WDI	4	15	Base Drive
V_{ref}	5	14	V_{CC2}
INH1	6	13	Gnd
V_{out2} Prog	7	12	Current Sense
V_{out2} Output	8	11	INH2
Converter Output	9	10	Converter Input

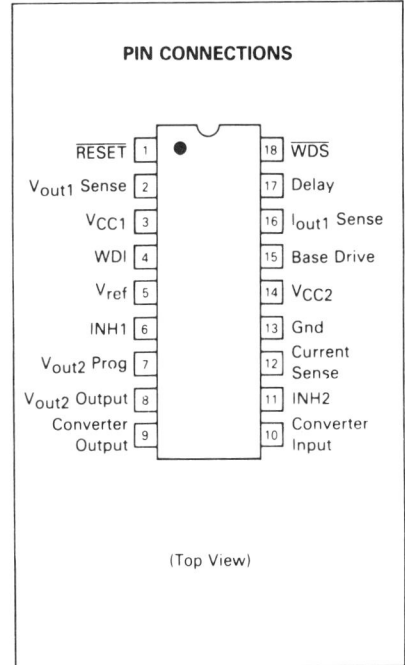

(Top View)

RECOMMENDED OPERATION CONDITIONS

Characteristic	Symbol	Min	Max	Unit
Power Supply Voltage	V_{CC1}	5.0	30	V
	V_{CC2}	5.5	30	
Collector Current	I_C	—	800	mA
Output Voltage	V_{out2}	6.0	30	V
Reference Source Current	I_{ref}	0	2.0	mA

ORDERING INFORMATION

Device	Operating Junction Temperature Range	Package
TCA5600	0 to +125°C	Plastic DIP
TCF5600	−40 to +150°C	Plastic DIP

This document contains information on a new product. Specifications and information herein are subject to change without notice.

MAXIMUM RATINGS (T_A = +25°C unless otherwise noted, Note 1)

Rating	Symbol	Value	Unit
Power Supply Voltage (Pin 3, 14)	V_{CC1}, V_{CC2}	35	Vdc
Base Drive Current (Pin 15)	I_B	20	mA
Collector Current (Pin 10)	I_C	1.0	A
Forward Rectifier Current (Pin 10–Pin 9)	I_F	1.0	A
Logic Inputs INH1, INH2, $\overline{WDS}$ (Pin 6, 11, 18)	V_{INP}	−0.3 V to V_{CC1}	Vdc
Logic Input Current WDI (Pin 4)	I_{WDI}	±0.5	mA
Output Sink Current $\overline{RESET}$ (Pin 1)	I_{RES}	10	mA
Analog Inputs (Pin 2) (Pin 7)	— —	−0.3 to 10 −0.3 to 5.0	V
Reference Source Current (Pin 5)	I_{ref}	5.0	mA
Power Dissipation (Note 2) T_A = +75°C TCA5600 T_A = +85°C TCF5600	P_D	500 650	mW
Thermal Resistance (Junction to Air)	$R_{\theta JA}$	100	°C/W
Operating Temperature Range TCA5600 TCF5600	T_A	0 to +75 −40 to +85	°C
Operating Junction Temperature TCA5600 TCF5600	T_J	+125 +150	°C
Storage Temperature Range	T_{stg}	−65 to +150	°C

NOTES:
1. Values beyond which damage may occur.
2. Derate at 10 mW/°C for junction temperature above +75°C (TCA5600).
   Derate at 10 mW/°C for junction temperature above +85°C (TCF5600).

FIGURE 1 — FUNCTIONAL BLOCK DIAGRAM

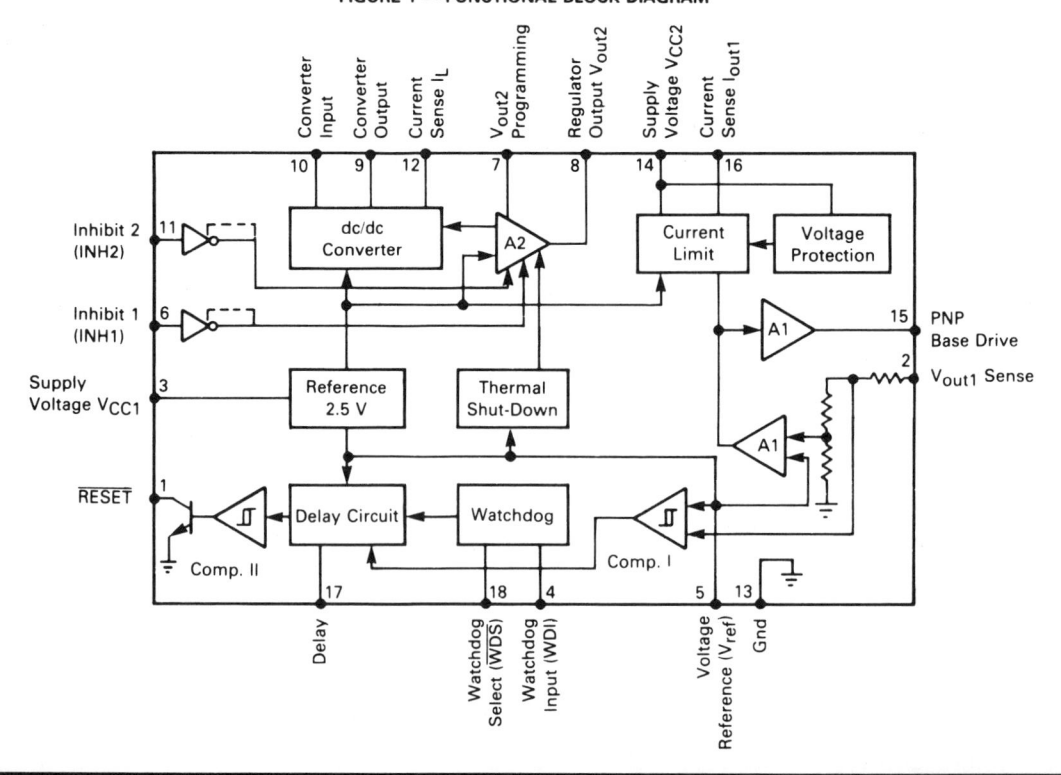

ELECTRICAL CHARACTERISTICS (V_{CC1} = V_{CC2} = 12 V; T_J = 25°C; I_{ref} = 0; I_{out1} = 0 (Note 3); R_{SC} = 0.5 Ω; INH1 = "High"; INH2 = "High"; $\overline{WDS}$ = "High"; I_{out2} = 0 (Note 4); if not otherwise specified)

Characteristic	Figure	Symbol	Min	Typ	Max	Unit
REFERENCE SECTION						
Nominal Reference Voltage	1	$V_{ref\ nom}$	2.42	2.5	2.58	V
Reference Voltage I_{ref} = 0.5 mA, $T_{low} \leq T_J \leq T_{high}$ (Note 5), 6.0 V $\leq V_{CC1} \leq$ 18 V		V_{ref}	2.4	—	2.6	V
Line Regulation (6.0 V $\leq V_{CC2} \leq$ 18 V)		Reg$_{line}$	—	2.0	15	mV
Average Temperature Coefficient $T_{low} \leq T_J \leq T_{high}$ (Note 5)	2	$\dfrac{\Delta V_{ref}}{\Delta T_J}$	—	—	+/−0.5	mV/°C
Ripple Rejection Ratio f = 1.0 kHz, V_{sin} = 1.0 V_{pp}	3	RR	60	70	—	dB
Output Impedance 0 $\leq I_{ref} \leq$ 2.0 mA		Z_O	—	1.0	—	Ohm
Standby Current Consumption V_{CC2} = Open	4	I_{CC1}	—	3.0	—	mA

NOTES:
3. The external PNP power transistor satisfies the following minimum specifications:
   $h_{FE} \geq 60$ at I_C = 500 mA and V_{CE} = 5.0 V; $V_{CE(sat)} \leq$ 300 mV at I_B = 10 mA and I_C = 300 mA
4. Regulator V_{out2} programmed for nominal 24 V output by means of R4, R5 (see Figure 1)
5. T_{low} = 0°C for TCA5600; T_{low} = −40°C for TCF5600.
   T_{high} = 125°C for TCA5600; T_{high} = 150°C for TCF5600.

5.0 V MICROPROCESSOR VOLTAGE REGULATOR SECTION

Characteristic	Figure	Symbol	Min	Typ	Max	Unit
Nominal Output Voltage		$V_{out1(nom)}$	4.8	5.0	5.2	V
Output Voltage 5.0 mA $\leq I_{out1} \leq$ 300 mA, $T_{low} \leq T_J \leq T_{high}$ (Note 5) 6.0 V $\leq V_{CC2} \leq$ 18 V	5 6	V_{out1}	4.75	—	5.25	V
Line Regulation (6.0 V $\leq V_{CC2} \leq$ 18 V)		Reg$_{line}$	—	10	50	mV
Load Regulation (5.0 mA $\leq I_{out1} \leq$ 300 mA)		Reg$_{load}$	—	20	100	mV
Base Current Drive (V_{CC2} = 6.0 V, V_{15} = 4.0 V)		I_B	10	15	—	mA
Ripple Rejection Ratio f = 1.0 kHz, V_{sin} = 1.0 V_{pp}	3	RR	50	65	—	dB
Undervoltage Detection Level (R_{SC} = 5.0 Ω)	7	V_{low}	4.5	0.93 × V_{out1}	—	V
Current Limitation Threshold (R_{SC} = 5.0 Ω)		V_{RSC}	210	250	290	mV
Average Temperature Coefficient $T_{low} \leq T_J \leq T_{high}$ (Note 5)		$\dfrac{\Delta V_{out1}}{\Delta T_J}$	—	—	±1.0	mV/°C

TCA5600 • TCF5600   MOTOROLA

Characteristic	Figure	Symbol	Min	Typ	Max	Unit
PROGRAMMABLE VOLTAGE REGULATOR SECTION (Note 6)						
Nominal Output Voltage		$V_{out2(nom)}$	23	24	25	V
Output Voltage 1.0 mA ≤ I_{out2} ≤ 100 mA, T_{low} ≤ T_J ≤ T_{high} (Notes 5, 7)	8	V_{out2}	22.8	—	25.2	V
Load Regulation 1.0 mA ≤ I_{out2} ≤ 100 mA (Note 7)		Reg$_{load}$	—	40	200	mV
DC Output Current		I_{out2}	100	—	—	mA
Peak Output Current (Internally Limited)		$I_{out2\,p}$	150	200	—	mA
Ripple Rejection Ratio f = 20 kHz, V = 0.4 V_{pp}		RR	45	55	—	dB
Output Voltage (Fixed 5.0 V) 1.0 mA ≤ I_{out2} ≤ 20 mA, T_{low} ≤ T_J ≤ T_{high}, INH1 = "High" (Note 5)		$V_{out2(5.0\,V)}$	4.75	—	5.25	V
OFF State Output Impedance (INH2 = "Low")		R_{out1}	—	10	—	kΩ
Average Temperature Coefficient T_{low} ≤ T_J ≤ T_{high} (Note 5)		$\Delta V_{out2} / \Delta T_J V_{out2}$	—	—	±0.25	mV/°C V

NOTES:
6. V_9 = 28 V, INH1 = "Low" for this Electrical Characteristic section unless otherwise specified.
7. Pulse tested t_p ≤ 300 μs

Characteristic	Figure	Symbol	Min	Typ	Max	Unit
DC/DC CONVERTER SECTION						
Collector Current Detection Level "High" R_C = 10 k "Low"	9	$V_{12(H)}$ $V_{12(L)}$	350 —	400 50	450 —	mV
Collector Saturation Voltage I_C = 600 mA (Note 7)	10	$V_{CE(sat)}$	—	—	1.6	V
Rectifier Forward Voltage Drop I_F = 600 mA (Note 7)	11	V_F	—	—	1.4	V
WATCHDOG AND $\overline{RESET}$ CIRCUIT SECTION						
Threshold Voltage "High" (static) "Low"		$V_{C5(H)}$ $V_{C5(L)}$	— —	2.5 1.0	— —	V
Current Source T_{low} ≤ T_J ≤ T_{high} (Note 5) Power-Up $\overline{RESET}$ Watchdog Time Out Watchdog $\overline{RESET}$		I_{C5}	−1.8 — —	−2.5 5×I_{C5} −50×I_{C5}	−3.2 — —	μA
Watchdog Input Voltage Swing		V_{WDI}	—	—	±5.5	V
Watchdog Input Impedance		r_i	12	15	—	kΩ
Watchdog Reset Pulse Width (C8 = 1.0 nF) (Note 9)		t_p	—	—	10	μs
DIGITAL PORTS: $\overline{WDS}$, INH 1, INH 2, $\overline{RESET}$ (Note 8)						
Input Voltage Range		V_{INP}	—	—	−0.3 to V_{CC1}	V
Input HIGH Current 2.0 V ≤ V_{IH} ≤ 5.5 V 5.5 V ≤ V_{IH} ≤ V_{CC1}		I_{IH}		— —	100 150	μA
Input LOW Current −0.3 V ≤ V_{IL} ≤ 0.8 V for INH1, INH2, −0.3 V ≤ V_{IL} ≤ 0.4 V for $\overline{WDS}$		I_{IL}		—	−100	μA
Leakage Current Immunity (INH2, High "Z" State)	12	I_Z	±20	—	—	μA
Output LOW Voltage $\overline{RESET}$ (I_{OL} = 6.0 mA)		V_{OL}	—	—	0.4	V
Output HIGH Current $\overline{RESET}$ (V_{OH} = 5.5 V)		V_{OH}	—	—	20	μA

NOTES:
8. Temperature range T_{low} ≤ T_J ≤ T_{high} applies to this Electrical Characteristics section.
9. For test purposes, a negative pulse is applied to Pin 4 (−2.5 V ≥ V_4 ≥ −5.5 V).

MOTOROLA   TCA5600 • TCF5600

TYPICAL CHARACTERISTICS

FIGURE 1 — REFERENCE VOLTAGE versus SUPPLY VOLTAGE

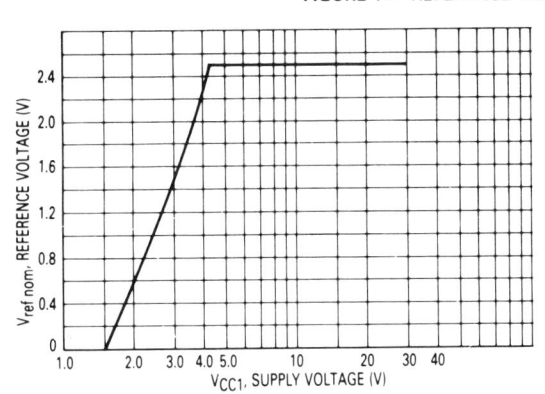

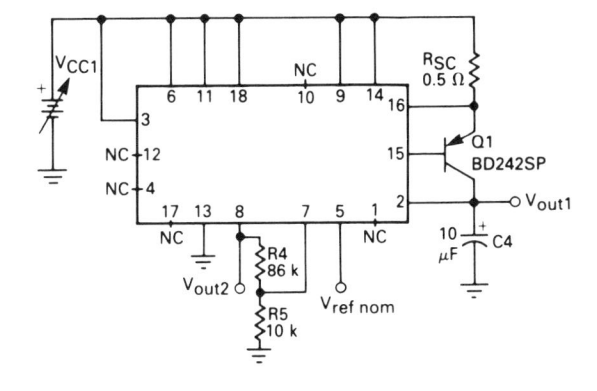

FIGURE 2 — REFERENCE STABILITY versus TEMPERATURE

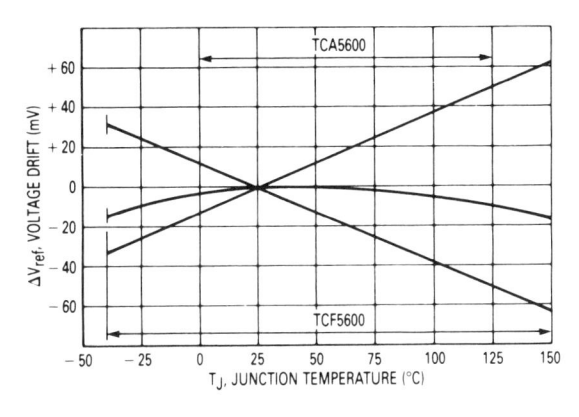

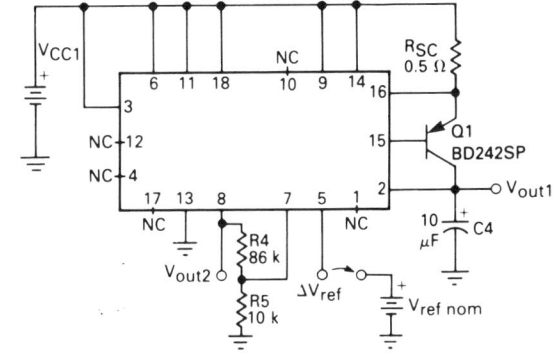

FIGURE 3 — RIPPLE REJECTION versus FREQUENCY

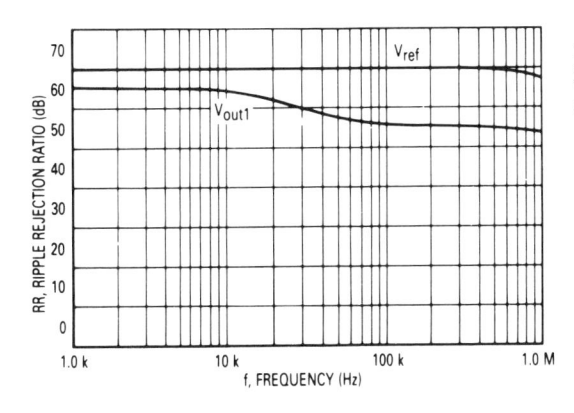

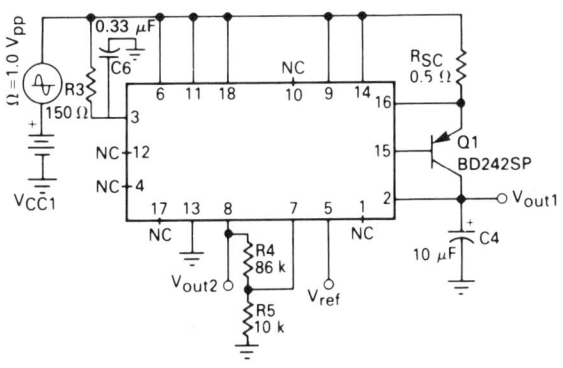

TCA5600 • TCF5600

MOTOROLA

FIGURE 4 — STAND-BY CURRENT versus SUPPLY VOLTAGE

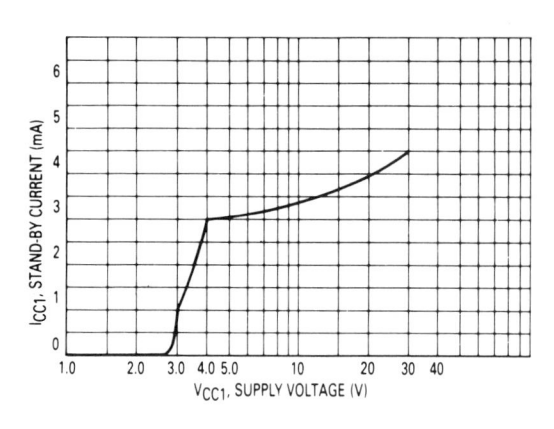

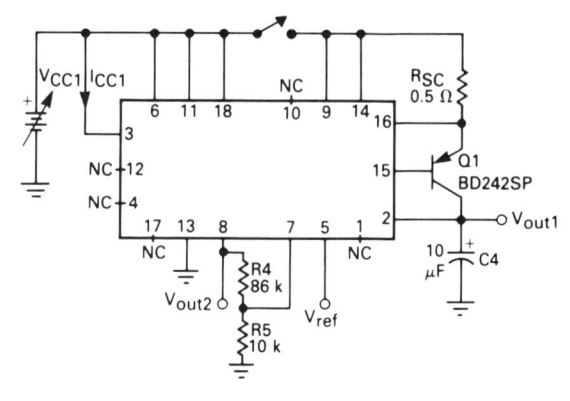

FIGURE 5 — POWER-UP BEHAVIOR OF THE 5.0 V REGULATOR

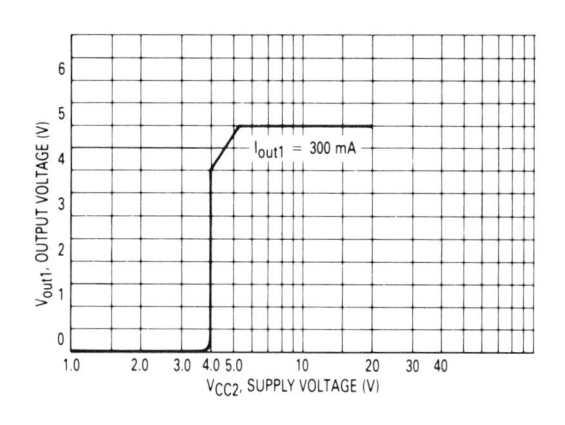

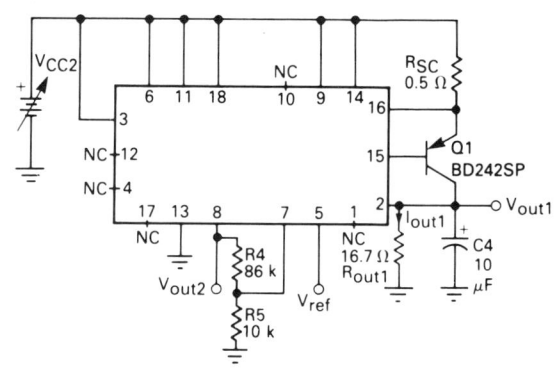

FIGURE 6 — FOLDBACK CHARACTERISTICS OF THE 5.0 V REGULATOR

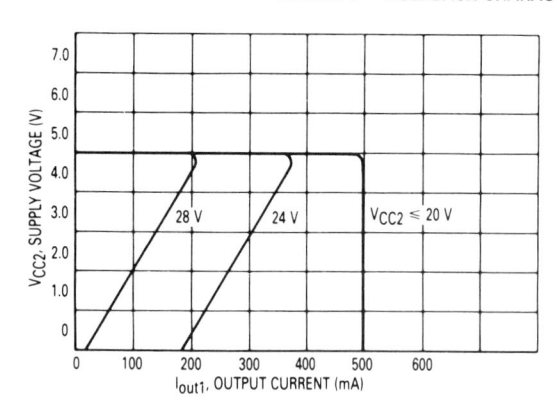

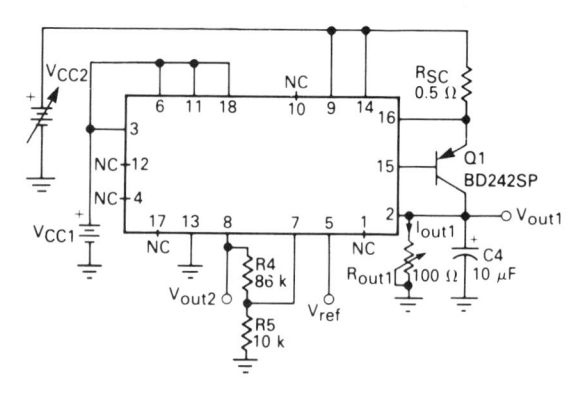

MOTOROLA   TCA5600 • TCF5600

FIGURE 7 — UNDERVOLTAGE LOCKOUT CHARACTERISTICS

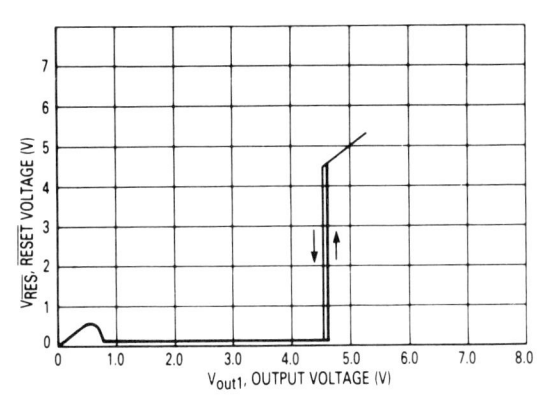

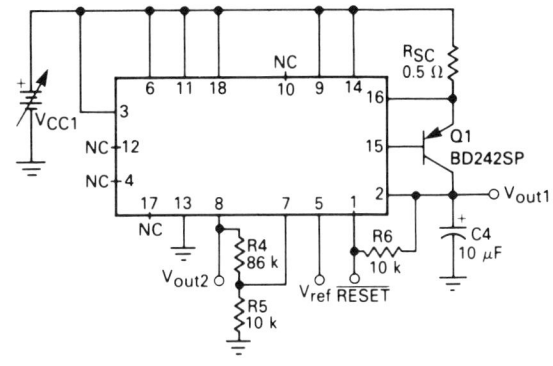

FIGURE 8 — OUTPUT CURRENT CAPABILITY OF THE PROGRAMMING REGULATOR

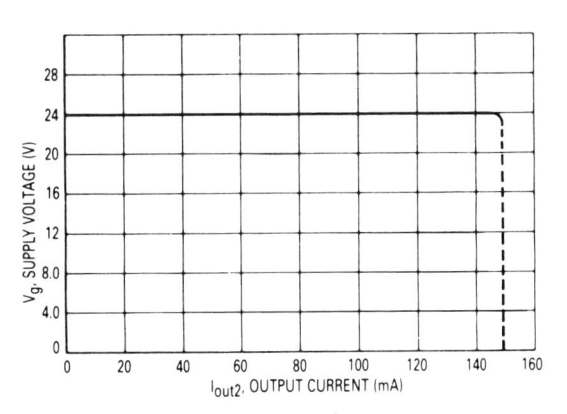

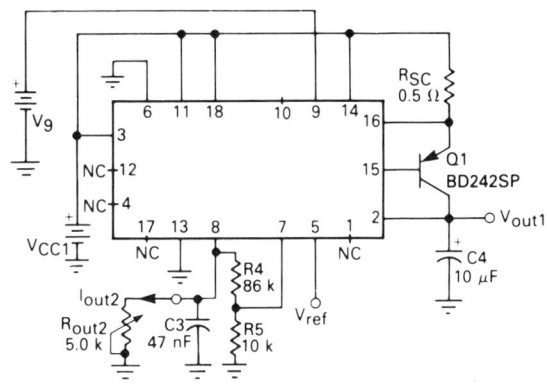

FIGURE 9 — COLLECTOR CURRENT DETECTION LEVEL

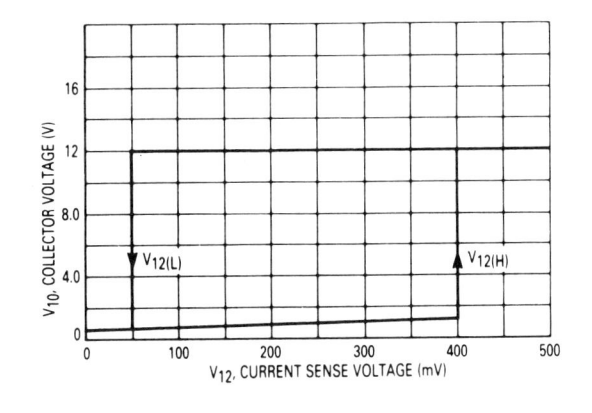

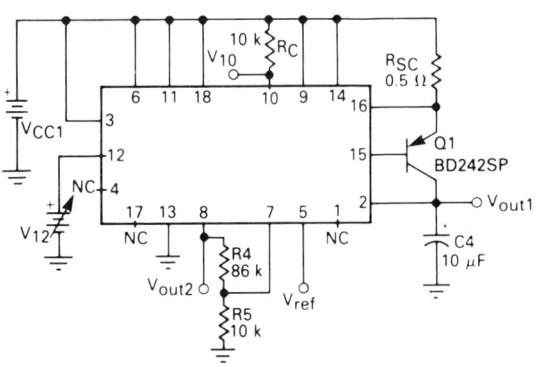

TCA5600 • TCF5600   MOTOROLA

FIGURE 10 — POWER SWITCH CHARACTERISTICS

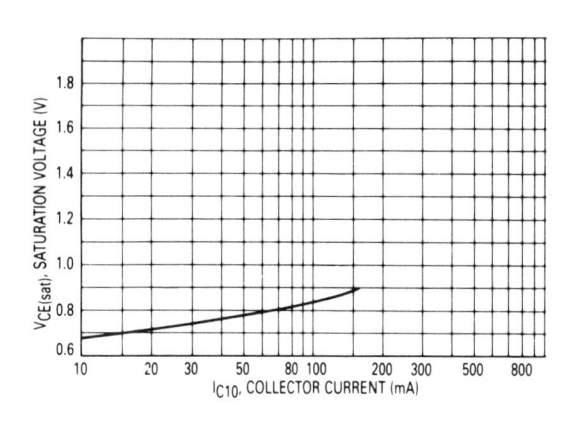

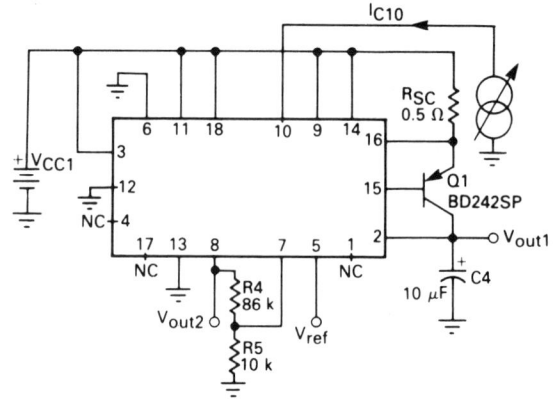

FIGURE 11 — RECTIFIER CHARACTERISTICS

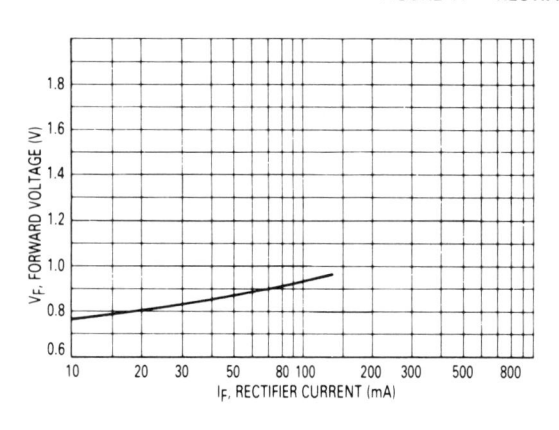

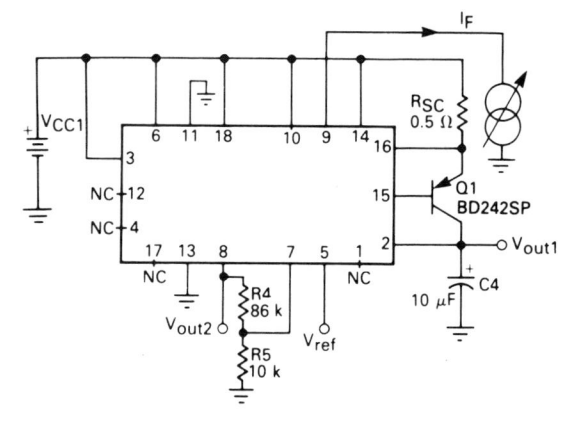

FIGURE 12 — INH 2 LEAKAGE CURRENT IMMUNITY

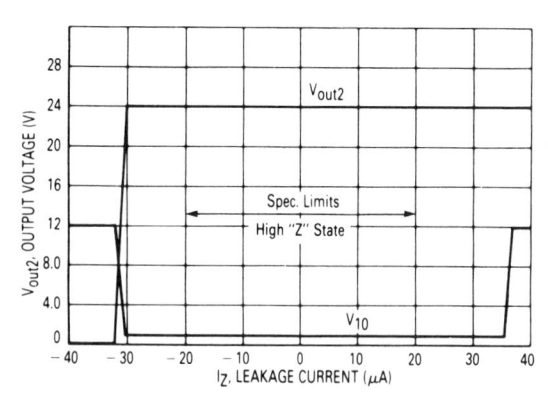

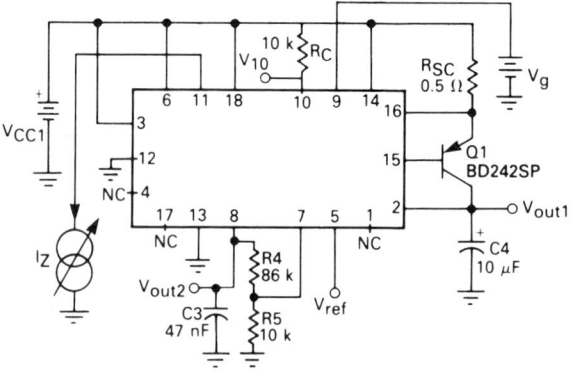

MOTOROLA   TCA5600 • TCF5600

APPLICATIONS INFORMATION
(See Figure 18)

1. VOLTAGE REFERENCE V_{ref}

The voltage reference V_{ref} is based upon a highly stable bandgap voltage reference and is accessible on Pin 5 for additional tasks. This circuit part has its own supply connection on Pin 3 and is therefore able to operate in standby mode. The RC network R3, C6 improves the ripple rejection on both regulators.

2. DC/DC CONVERTER

The dc/dc converter performs according to the fly back principle and does not need a time base circuit. The maximum coil current is well defined by means of the current sensing resistor R1 under all working conditions (start-up phase, circuit overload, wide supply voltage range and extreme load current change). Figure 13 shows the simplified converter schematic:

FIGURE 13 — SIMPLIFIED CONVERTER SCHEMATIC

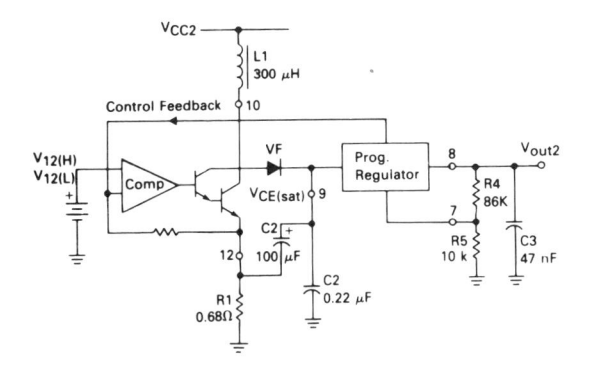

A simplified method on "how to calculate the coil inductance" is given below. The operation point at min. supply voltage (V_{CC2}) and max. output current (I_{out2}) for a fixed output voltage (V_{out2}) determines the coil data. Figure 14 shows the typical voltage and current wave forms on the coil L1 (coil losses neglected).

The equations (1) and (2) yield the respective coil voltage $V_L -$ and $V_L +$ (see Figure 14):

$$V_L + = V_{out2} + \Delta V_{(Pin\ 9\ -\ Pin\ 8)} + V_F - V_{CC2} \quad (1)$$
$$V_L - = V_{CC2} - V_{CE(sat)} - V_{12(H)} \quad (2)$$

($\Delta V_{(Pin\ 9\ -\ Pin\ 8)}$: input/output voltage drop of the regulator, 2.5 V typical)

(V_F, $V_{CE(sat)}$, $V_{12(H)}$: see electrical characteristics)

FIGURE 14 — VOLTAGE AND CURRENT WAVEFORM ON THE COIL (not to scale)

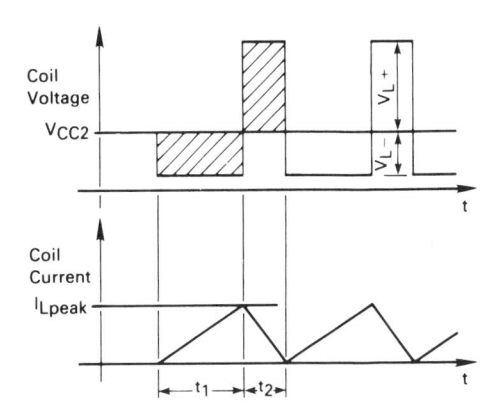

The time ratio α for the charging time to dumping time is defined by equation (3):

$$\alpha = \frac{t_1}{t_2} = \frac{V_L +}{V_L -} \quad (3)$$

The coil charging time t_1 is found using equation (4):

$$t_1 = \frac{1}{(1 + \frac{1}{\alpha}) \cdot f} \quad (4)$$

(f : min. oscillation frequency which should be chosen above the audio frequency band (e.g. 20 kHz))

Knowing the dc output current I_{out2} of the programmable regulator, the peak coil current $I_{L(peak)}$ can now be calculated:

$$I_{L(peak)} = 2 \cdot I_{out2} \cdot (1 + \alpha) \quad (5)$$

The coil inductance L1 of the nonsaturated coil is given by equation (6):

$$L1 = \frac{t_1}{I_{L(peak)}} \cdot V_L - \quad (6)$$

The formula (6a) yields the current sensing resister R1 for a defined peak coil current $I_{L(peak)}$:

$$R1 = \frac{V_{12(H)}}{I_{L(peak)}} \quad (6a)$$

TCA5600 • TCF5600

MOTOROLA

In order to limit the by-pass current through capacitor C7 during the energy dumping phase the value C2>>C7 should be implemented.

For all other operation conditions, the feedback signal from the programmable voltage regulator controls the activity of the converter.

3. PROGRAMMABLE VOLTAGE REGULATOR

This series voltage regulator is programmable by the voltage divider R4, R5 for a nominal output voltage 6.0 V ≤ V_{out2} ≤ 30 V.

$$R4 = \frac{(V_{out2} - V_{ref\,nom}) \cdot R5}{V_{ref\,nom}} \quad (7)$$

(R5 = 10 k, $V_{ref\,nom}$ = 2.5 V)

Current limitation and thermal shutdown capability are standard features of this regulator. The voltage drop $\Delta V_{(Pin\,9\,-\,Pin\,8)}$ across the series pass transistor generates the feedback signal to control the dc/dc converter (see Figure 13).

4. CONTROL INPUTS INH1, INH2

The dc/dc converter and/or the regulator V_{out2} are remote controllable through the TTL, MOS compatible inhibit inputs INH1 and INH2 where the latter is a 3-level detector (Logic "0", high impedance "Z", Logic "1"). Both inputs are setup to provide the following truth table:

FIGURE 15 — INH1, INH2 TRUTH TABLE

Mode	INH1	INH2	V_{out2}	dc/dc
1	0	0	OFF	INT
2	0	High "Z"	V_{out2}	ON
3	0	1	V_{out2}	INT
4	1	0	OFF	INT
5	1	High "Z"	5.0 V	ON
6	1	1	5.0 V	INT

INT: Intermittent operation of the converter means that the converter operates only if $V_{CC2} < V_{out2}$.
ON: The converter loads the storage capacitor C2 to its full charge (V_9 = 33 V), allowing fast response time of the regulator V_{out2} when addressed by the control software.
OFF: High impedance (internal resistor 10 k to ground)

Figure 16 represents a typical timing diagram for an E²PROM programming sequence in a microprocessor based system. The high "Z" state enables the dc/dc converter to ramp during t₃ to the voltage V_9 at Pin 9 to a high level before the write cycle takes place in the memory.

5. MICROPROCESSOR SUPLY REGULATOR

Together with an external PNP power transistor (Q1), a 5.0 V supply exhibiting low voltage drop is obtained to power microprocessor systems and auxilliary circuits. Using a power Darlington with adequate heat sink in the output stage boosts the output current I_{out1} above 1 amp.

FIGURE 16 — TYPICAL E²PROM PROGRAMMING SEQUENCE (not to scale)

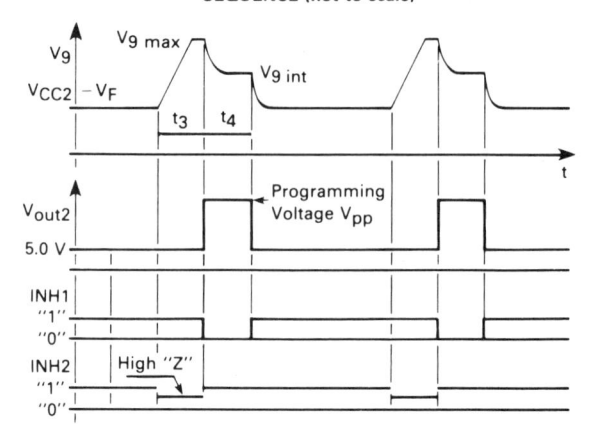

The current limitation circuit measures the emitter current of Q1 by means of the sensing resistor R_{SC}.

$$R_{SC} = \frac{V_{RSC}}{I_E} \quad (8)$$

(I_E: emitter current of Q1)
(V_{RSC}: threshold voltage (see electrical characteristics))

The voltage protection circuit performs a fold-back characteristic above a nominal operating voltage V_{CC2} ≥ 18 V.

6. DELAY AND WATCHDOG CIRCUIT

The under voltage monitor supervises the power supply V_{out1} and releases the delay circuit $\overline{RESET}$ as soon as the regulator output reaches the microprocessor operating range (e.g. V_{LOW} ≥ 0.93 · $V_{out1(nom)}$). The $\overline{RESET}$ output has an open-collector and may be connected in a "wired-OR" configuration.

The watchdog circuit consists of a retriggerable monostable with a negative edge sensitive control input WDI. The watchdog feature may be disabled by means of the watchdog select input $\overline{WDS}$ driven to a "1". Figure 17 displays the typical $\overline{RESET}$ timing diagram.

The commuted current source I_{C5} on Pin 17, threshold voltage $V_{C5(L)}$, $V_{C5(H)}$ and an external capacitor C5 define the $\overline{RESET}$ delay and the watchdog timing. The relationship of the timing signals are indicated by the equations (9) to (11).

$\overline{RESET}$ delay: $\quad t_d = \dfrac{C5 \cdot V_{C5(H)}}{|I_{C5}|} \quad (9)$

Watchdog time-out: $\quad t_{wd} = \dfrac{C5 \cdot (V_{C5(H)} - V_{C5(L)})}{5 \cdot I_{C5}} \quad (10)$

Watchdog $\overline{RESET}$: $\quad t_r = \dfrac{C5 \cdot (V_{C5(H)} - V_{C5(L)})}{50 \cdot |I_{C5}|} \quad (11)$

(I_{C5}, $V_{C5(H)}$, $V_{C5(L)}$: see electrical characteristics.)

FIGURE 17 — TYPICAL $\overline{\text{RESET}}$ TIMING DIAGRAM
(not to scale)

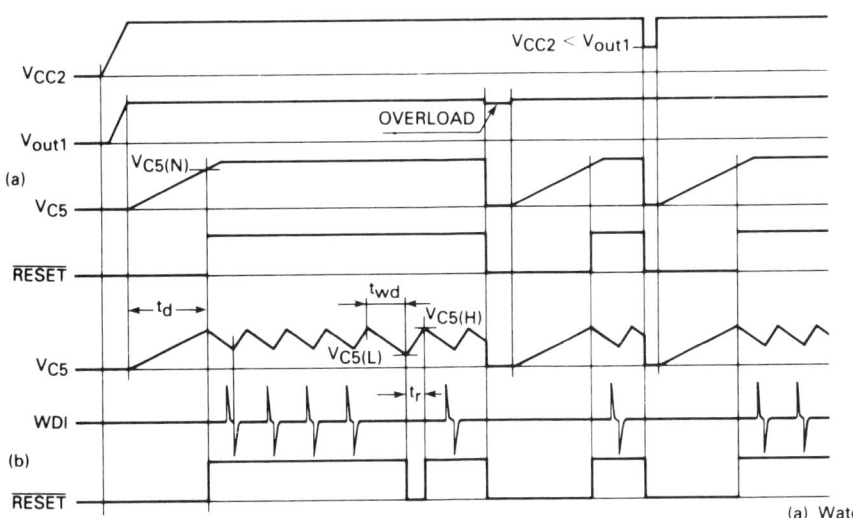

(a) Watchdog inhibited, $\overline{\text{WDS}}$ = "1"
(b) Watchdog operational, $\overline{\text{WDS}}$ = "0"

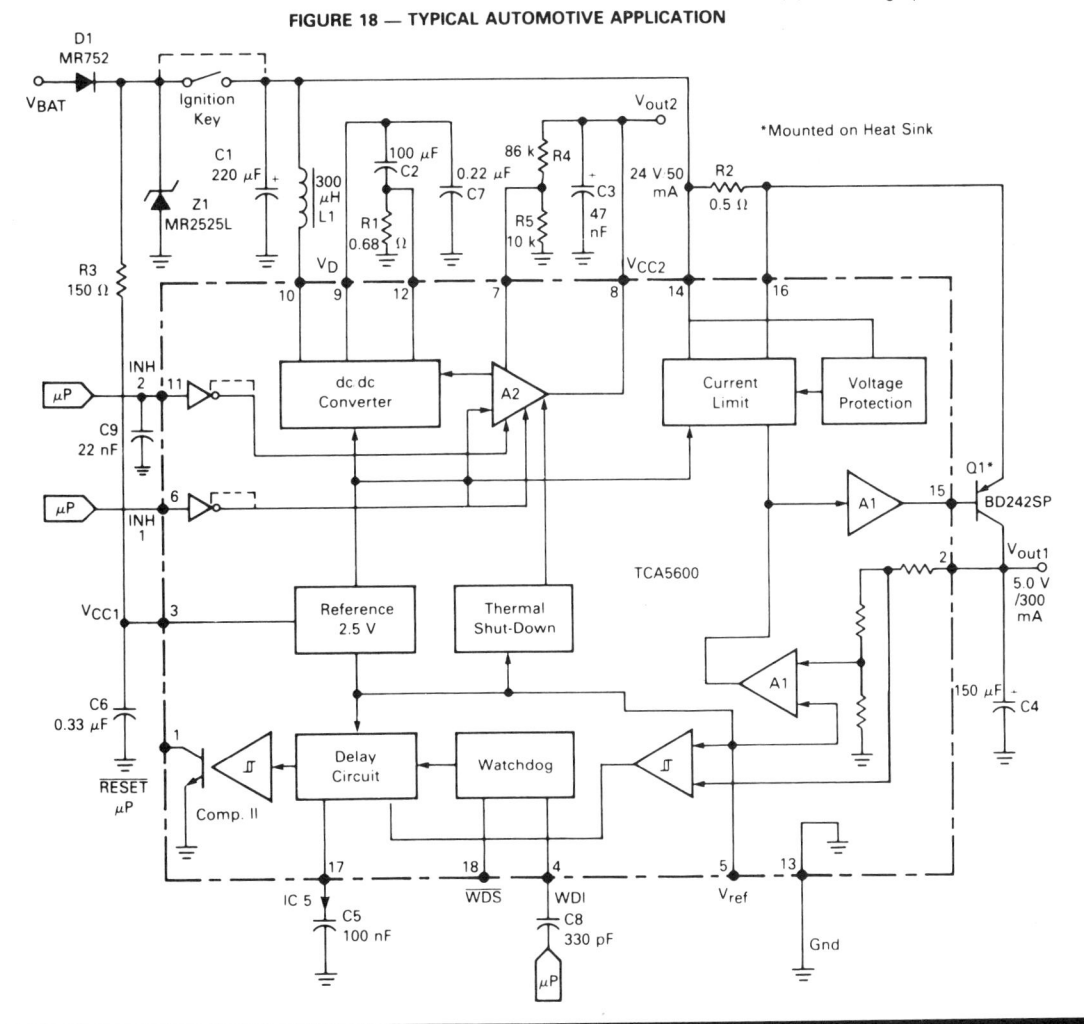

FIGURE 18 — TYPICAL AUTOMOTIVE APPLICATION

*Mounted on Heat Sink

TCA5600 • TCF5600 MOTOROLA

OUTLINE DIMENSIONS

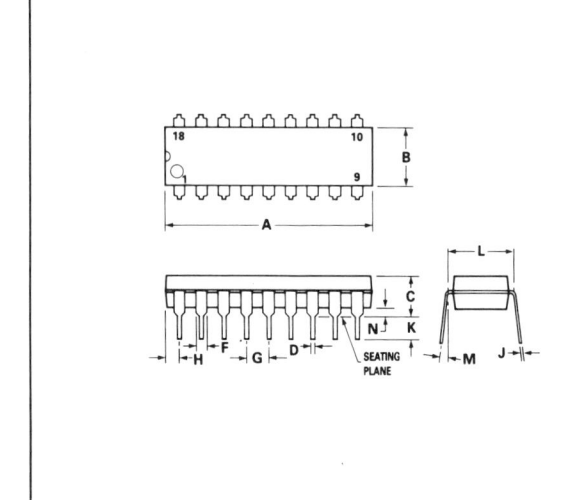

NOTES:
1. POSITIONAL TOLERANCE OF LEADS (D), SHALL BE WITHIN 0.25mm(0.010) AT MAXIMUM MATERIAL CONDITION, IN RELATION TO SEATING PLANE AND EACH OTHER.
2. DIMENSION L TO CENTER OF LEADS WHEN FORMED PARALLEL.
3. DIMENSION B DOES NOT INCLUDE MOLD FLASH.

DIM	MILLIMETERS MIN	MILLIMETERS MAX	INCHES MIN	INCHES MAX
A	22.22	23.24	0.875	0.915
B	6.10	6.60	0.240	0.260
C	3.56	4.57	0.140	0.180
D	0.36	0.56	0.014	0.022
F	1.27	1.78	0.050	0.070
G	2.54 BSC		0.100 BSC	
H	1.02	1.52	0.040	0.060
J	0.20	0.30	0.008	0.012
K	2.92	3.43	0.115	0.135
L	7.62 BSC		0.300 BSC	
M	0°	15°	0°	15°
N	0.51	1.02	0.020	0.040

PLASTIC PACKAGE
CASE 707-02

Motorola reserves the right to make changes without further notice to any products herein to improve reliability, function or design. Motorola does not assume any liability arising out of the application or use of any product or circuit described herein; neither does it convey any license under its patent rights nor the rights of others. Motorola products are not authorized for use as components in life support devices or systems intended for surgical implant into the body or intended to support or sustain life. Buyer agrees to notify Motorola of any such intended end use whereupon Motorola shall determine availability and suitability of its product or products for the use intended. Motorola and Ⓜ are registered trademarks of Motorola, Inc. Motorola, Inc. is an Equal Employment Opportunity/Affirmative Action Employer.

Literature Distribution Centers:
USA: Motorola Literature Distribution; P.O. Box 20912; Phoenix, Arizona 85036.
EUROPE: Motorola Ltd.; European Literature Center; 88 Tanners Drive, Blakelands, Milton Keynes, MK14 5BP, England.
ASIA PACIFIC: Motorola Semiconductors H.K. Ltd.; P.O. Box 80300; Cheung Sha Wan Post Office; Kowloon Hong Kong.
JAPAN: Nippon Motorola Ltd.; 3-20-1 Minamiazabu, Minato-ku, Tokyo 106 Japan.

Ⓜ MOTOROLA

LM2984C Microprocessor Power Supply System

General Description

The LM2984C positive voltage regulator features three independent and tracking outputs capable of delivering the power for logic circuits, peripheral sensors and standby memory in a typical microprocessor system. The LM2984C includes circuitry which monitors both its own high-current output and also an external μP. If any error conditions are sensed in either, a reset error flag is set and maintained until the malfunction terminates. Since these functions are included in the same package with the three regulators, a great saving in board space can be realized in the typical microprocessor system. The LM2984C also features very low dropout voltages on each of its three regulator outputs (0.6V at the rated output current). Furthermore, the quiescent current can be reduced to 1 mA in the standby mode.

Designed also for vehicular applications, the LM2984C and all regulated circuitry are protected from reverse battery installations or 2-battery jumps. Familiar regulator features such as short circuit and thermal overload protection are also provided. Fixed outputs of 5V are available in the plastic TO-220 power package.

Features

- Three low dropout tracking regulators
- Output current in excess of 500 mA
- Low quiescent current standby regulator
- Microprocessor malfunction RESET flag
- Delayed RESET on power-up
- Accurate pretrimmed 5V outputs
- Reverse battery protection
- Overvoltage protection
- Reverse transient protection
- Short circuit protection
- Internal thermal overload protection
- ON/OFF switch for high current outputs
- 100% electrical burn-in in thermal limit

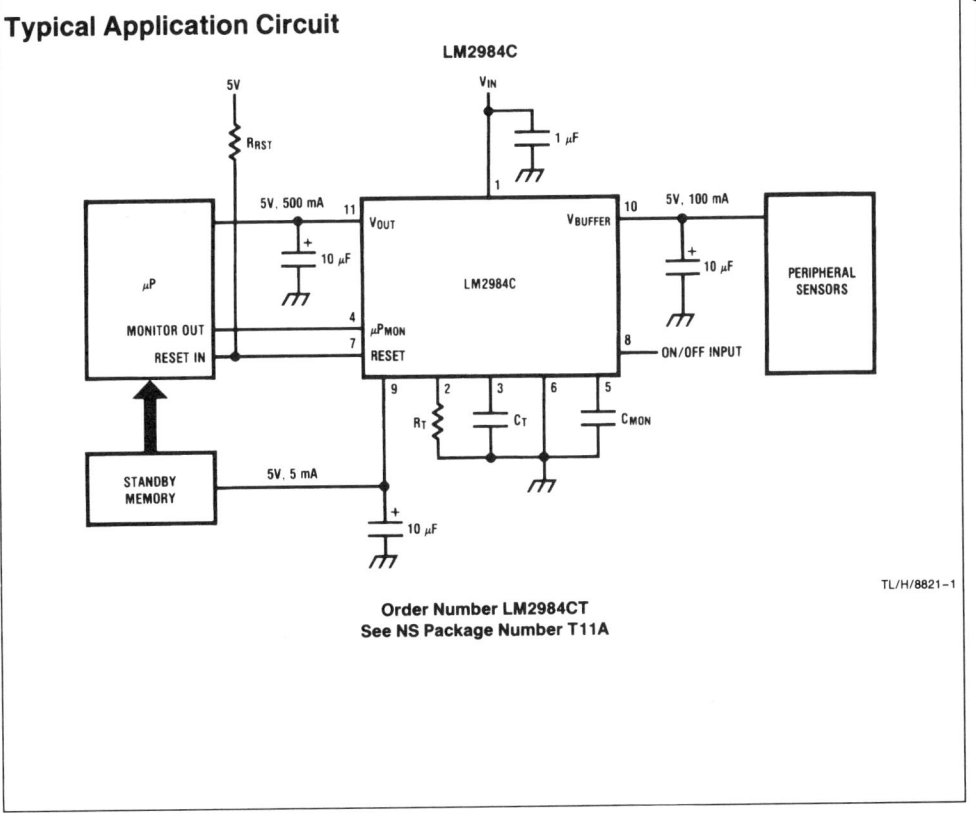

Typical Application Circuit

Order Number LM2984CT
See NS Package Number T11A

Absolute Maximum Ratings

Specifications for Military/Aerospace products are not contained in this datasheet. Refer to the associated reliability electrical test specifications document.

Input Voltage
 Survival Voltage (<100 ms) 35V
 Operational Voltage 26V

Internal Power Dissipation Internally Limited
Operating Temperature Range (T_A) 0°C to +125°C
Maximum Junction Temperature (Note 1) 150°C
Storage Temperature Range −65°C to +150°C
Lead Temperature (Soldering, 10 sec.) 230°C
ESD rating is to be determined.

Electrical Characteristics

V_{IN} = 14V, I_{OUT} = 5 mA, C_{OUT} = 10 μF, T_j = 25°C unless otherwise indicated

Parameter	Conditions	Typical	Tested Limit (Note 2)	Design Limit (Note 3)	Units
V_{OUT} (Pin 11)					
Output Voltage	5 mA ≤ I_o ≤ 500 mA 6V ≤ V_{IN} ≤ 26V	5.00	4.85 5.15	4.75 5.25	V_{min} V_{max}
Line Regulation	9V ≤ V_{IN} ≤ 16V	2	25		mV$_{max}$
	7V ≤ V_{IN} ≤ 26V	5	50		mV$_{max}$
Load Regulation	5 mA ≤ I_{OUT} ≤ 500 mA	12	50		mV$_{max}$
Output Impedance	250 mA$_{dc}$ and 10 mA$_{rms}$, f_o = 120 Hz	24			mΩ
Quiescent Current	I_{OUT} = 500 mA	38	100		mA$_{max}$
	I_{OUT} = 250 mA	14	50		mA$_{max}$
Output Noise Voltage	10 Hz–100 kHz, I_{OUT} = 100 mA	100			μV
Long Term Stability		20			mV/1000 hr
Ripple Rejection	f_o = 120 Hz	70	60		dB$_{min}$
Dropout Voltage	I_{OUT} = 500 mA	0.53	0.80	1.00	V_{max}
	I_{OUT} = 250 mA	0.28	0.50	0.60	V_{max}
Current Limit		0.92	0.75		A$_{min}$
Maximum Operational Input Voltage	Continuous DC	32	26	26	V_{min}
Maximum Line Transient	V_{OUT} ≤ 6V, R_{OUT} = 100Ω	45	35	35	V_{min}
Reverse Polarity Input Voltage DC	V_{OUT} ≥ −0.6V, R_{OUT} = 100Ω	−30	−15	−15	V_{min}
Reverse Polarity Input Voltage Transient	T ≤ 100 ms, R_{OUT} = 100Ω	−55	−35	−35	V_{min}

Electrical Characteristics

$V_{IN} = 14V$, $I_{buf} = 5$ mA, $C_{buf} = 10$ μF, $T_j = 25°C$ unless otherwise indicated

Parameter	Conditions	Typical	Tested Limit (Note 2)	Design Limit (Note 3)	Units
V_buffer (Pin 10)					
Output Voltage	5 mA ≤ I_o ≤ 100 mA 6V ≤ V_{IN} ≤ 26V	5.00	4.85 5.15	4.75 5.25	V_{min} V_{max}
Line Regulation	9V ≤ V_{IN} ≤ 16V	2	25		mV_{max}
	7V ≤ V_{IN} ≤ 26V	5	50		mV_{max}
Load Regulation	5 mA ≤ I_{buf} ≤ 100 mA	15	50		mV_{max}
Output Impedance	50 mA_{dc} and 10 mA_{rms}	200			mΩ
Quiescent Current	I_{buf} = 100 mA	8.0	15.0		mA_{max}
Output Noise Voltage	10 Hz–100 kHz, I_{OUT} = 100 mA	100			μV
Long Term Stability		20			mV/1000 hr
Ripple Rejection	f_o = 120 Hz	70	60		dB_{min}
Dropout Voltage	I_{buf} = 100 mA	0.35	0.50	0.60	V_{max}
Current Limit		0.23	0.15		A_{min}
Maximum Operational Input Voltage	Continuous DC	32	26	26	V_{min}
Maximum Line Transient	V_{buf} ≤ 6V, R_{buf} = 100Ω	45	35	35	V_{min}
Reverse Polarity Input Voltage DC	V_{buf} ≥ −0.6V, R_{buf} = 100Ω	−30	−15	−15	V_{min}
Reverse Polarity Input Voltage Transient	T ≤ 100 ms, R_{buf} = 100Ω	−55	−35	−35	V_{min}

Electrical Characteristics

$V_{IN} = 14V$, $I_{stby} = 1$ mA, $C_{stby} = 10$ μF, $T_j = 25°C$ unless otherwise indicated

Parameter	Conditions	Typical	Tested Limit (Note 2)	Design Limit (Note 3)	Units
V_standby (Pin 9)					
Output Voltage	1 mA ≤ I_o ≤ 7.5 mA 6V ≤ V_{IN} ≤ 26V	5.00	4.85 5.15	4.75 5.25	V_{min} V_{max}
Line Regulation	9V ≤ V_{IN} ≤ 16V	2	25		mV_{max}
	7V ≤ V_{IN} ≤ 26V	5	50		mV_{max}
Load Regulation	0.5 mA ≤ I_{stby} ≤ 7.5 mA	6	50		mV_{max}
Output Impedance	5 mA_{dc} and 1 mA_{rms}, f_o = 120 Hz	0.9			Ω
Quiescent Current	I_{stby} = 7.5 mA	1.2	2.0		mA_{max}
	I_{stby} = 2 mA	0.9	1.5		mA_{max}

Electrical Characteristics (Continued)

V_{IN} = 14V, I_{stby} = 1 mA, C_{stby} = 10 µF, T_j = 25°C unless otherwise indicated

Parameter	Conditions	Typical	Tested Limit (Note 2)	Design Limit (Note 3)	Units
$V_{standby}$ (Continued)					
Output Noise Voltage	10 Hz–100 kHz, I_{stby} = 1 mA	100			µV
Long Term Stability		20			mV/1000 hr
Ripple Rejection	f_o = 120 Hz	70	60		dB_{min}
Dropout Voltage	I_{stby} = 1 mA	0.26	0.50	0.50	V_{max}
Dropout Voltage	I_{stby} = 7.5 mA	0.38	0.60	0.70	V_{max}
Current Limit		15	12		mA_{min}
Maximum Operational Input Voltage	4.5V ≤ V_{stby} ≤ 6V, R_{stby} = 1000Ω	45	35	35	V_{min}
Maximum Line Transient	V_{stby} ≤ 6V, R_{stby} = 1000Ω	45	35	35	V_{min}
Reverse Polarity Input Voltage DC	V_{stby} ≥ −0.6V, R_{stby} = 1000Ω	−30	−15	−15	V_{min}
Reverse Polarity Input Voltage Transient	T ≤ 100 ms, R_{stby} = 1000Ω	−55	−35	−35	V_{min}

Electrical Characteristics

V_{IN} = 14V, T_j = 25°C, C_{OUT} = 10 µF, C_{buf} = 10 µF, C_{stby} = 10 µF unless otherwise specified

Parameter	Conditions	Typical	Tested Limit (Note 2)	Design Limit (Note 3)	Units
Tracking and Isolation					
Tracking V_{OUT}–V_{stby}	I_{OUT} ≤ 500 mA, I_{buf} = 5 mA, I_{stby} ≤ 7.5 mA	±30	±100		mV_{max}
Tracking V_{buf}–V_{stby}	I_{OUT} = 5 mA, I_{buf} ≤ 100 mA, I_{stby} ≤ 7.5 mA	±30	±100		mV_{max}
Tracking V_{OUT}–V_{buf}	I_{OUT} ≤ 500 mA, I_{buf} ≤ 100 mA, I_{stby} = 1 mA	±30	±100		mV_{max}
Isolation* V_{buf} from V_{OUT}	R_{OUT} = 1Ω, I_{buf} ≤ 100 mA	5.00	4.50 5.50		V_{min} V_{max}
Isolation* V_{stby} from V_{OUT}	R_{OUT} = 1Ω, I_{stby} ≤ 7.5 mA	5.00	4.50 5.50		V_{min} V_{max}
Isolation* V_{OUT} from V_{buf}	R_{buf} = 1Ω, I_{OUT} ≤ 500 mA	5.00	4.50 5.50		V_{min} V_{max}
Isolation* V_{stby} from V_{buf}	R_{buf} = 1Ω, I_{stby} ≤ 7.5 mA	5.00	4.50 5.50		V_{min} V_{max}

*Isolation refers to the ability of the specified output to remain within the tested limits when the other output is shorted to ground.

Electrical Characteristics

V_{IN} = 14V, I_{OUT} = 5 mA, I_{buf} = 5 mA, I_{stby} = 5 mA, R_t = 130k, C_t = 0.33 μF, C_{mon} = 0.47 μF, T_j = 25°C unless otherwise specified

Parameter	Conditions	Typical	Tested Limit (Note 2)	Design Limit (Note 3)	Units
Computer Monitor/Reset Functions					
I_{reset} Low	V_{IN} = 4V, V_{rst} = 0.4V	5	2	1	mA_{min}
V_{reset} Low	V_{IN} = 4V, I_{rst} = 1 mA	0.10	0.40		V_{max}
R_t voltage	(Pin 2)	1.22	1.15		V_{min}
		1.22	1.30		V_{max}
Power On Reset Delay	$V\mu P_{mon}$ = 5V (T_{dly} = 1.2 $R_t C_t$)	50	45		ms_{min}
		50	55		ms_{max}
V_{OUT} Low Reset Threshold	(Note 4)	4.00	3.60		V_{min}
		4.00	4.40		V_{max}
V_{OUT} High Reset Threshold	(Note 4)	5.50	5.25		V_{min}
		5.50	6.00		V_{max}
Reset Output Leakage	$V\mu P_{mon}$ = 5V, V_{rst} = 12V	0.01	1		μA_{max}
μP_{mon} Input Current (Pin 4)	$V\mu P_{mon}$ = 2.4V	7.5	25		μA_{max}
	$V\mu P_{mon}$ = 0.4V	0.01	10		μA_{max}
μP_{mon} Input Threshold Voltage		1.22	0.80	0.80	V_{min}
		1.22	2.00	2.00	V_{max}
μP Monitor Reset Oscillator Period	$V\mu P_{mon}$ = 0V (T_{window} = 0.82 $R_t C_{mon}$)	50	45		ms_{min}
		50	55		ms_{max}
μP Monitor Reset Oscillator Pulse Width	$V\mu P_{mon}$ = 0V ($RESET_{pw}$ = 2000 C_{mon})	1.0	0.7	0.5	ms_{min}
		1.0	1.3	2.0	ms_{max}
Minimum μP Monitor Input Pulse Width	(Note 5)	2			μs_{max}
Reset Fall Time	R_{rst} = 10k, V_{rst} = 5V, C_{rst} ≤ 10 pF	0.20	1.00		μs_{max}
Reset Rise Time	R_{rst} = 10k, V_{rst} = 5V, C_{rst} ≤ 10 pF	0.60	1.00		μs_{max}
On/Off Switch Input Current (Pin 8)	V_{ON} = 2.4V	7.5	25		μA_{max}
	V_{ON} = 0.4V	0.01	10		μA_{max}
On/Off Switch Input Threshold Voltage		1.22	0.80	0.80	V_{min}
		1.22	2.00	2.00	V_{max}

Note 1: Thermal resistance without a heatsink for junction-to-case temperature is 3°C/W. Thermal resistance case-to-ambient is 40°C/W.

Note 2: Tested Limits are guaranteed and 100% production tested.

Note 3: Design Limits are guaranteed (but not 100% production tested) over the indicated temperature and supply voltage range. These limits are not used to calculate outgoing quality levels.

Note 4: An internal comparator detects when the main regulator output (V_{OUT}) drops below 4.0V or rises above 5.5V. If either condition exists at the output, the Reset Error Flag is held low until the error condition has terminated. The Reset Error Flag is then allowed to go high again after a delay set by R_t and C_t. (See Applications Section.)

Note 5: This parameter is a measure of how short a pulse can be detected at the μP Monitor Input. This parameter is primarily influenced by the value of C_{mon}. (See Typical Performance Characteristics and Applications Section.)

Block Diagram

Pin Description

Pin No.	Pin Name	Comments
1	V_{IN}	Positive supply input voltage
2	R_t	Sets internal timing currents
3	C_t	Sets power-up reset delay timing
4	μP_{mon}	Microcomputer monitor input
5	C_{mon}	Sets μC monitor timing
6	Ground	Regulator ground
7	Reset	Reset error flag output
8	ON/OFF	Enables/disables high current regulators
9	$V_{standby}$	Standby regulator output (7.5 mA)
10	V_{buffer}	Buffer regulator output (100 mA)
11	V_{OUT}	Main regulator output (500 mA)

External Components

Component	Typical Value	Component Range	Comments
C_{IN}	1 μF	0.47 μF–10 μF	Required if device is located far from power supply filter.
R_t	130k	24k–1.2M	Sets internal timing currents.
C_t	0.33 μF	0.033 μF–3.3 μF	Sets power-up reset delay.
C_{tc}	0.01 μF	0.001 μF–0.1 μF	Establishes time constant of AC coupled computer monitor.
R_{tc}	10k	1k–100k	Establishes time constant of AC coupled computer monitor. (See applications section.)
C_{mon}	0.47 μF	0.047 μF–4.7 μF	Sets time window for computer monitor. Also determines period and pulse width of computer malfunction reset. (See applications section.)
R_{rst}	10k	5k–100k	Load for open collector reset output. Determined by computer reset input requirements.
C_{stby}	10 μF	10 μF–no bound	A 10 μF is required for stability but larger values can be used to maintain regulation during transient conditions.
C_{buf}	10 μF	10 μF–no bound	A 10 μF is required for stability but larger values can be used to maintain regulation during transient conditions.
C_{OUT}	10 μF	10 μF–no bound	A 10 μF is required for stability but larger values can be used to maintain regulation during transient conditions.

Typical Circuit Waveforms

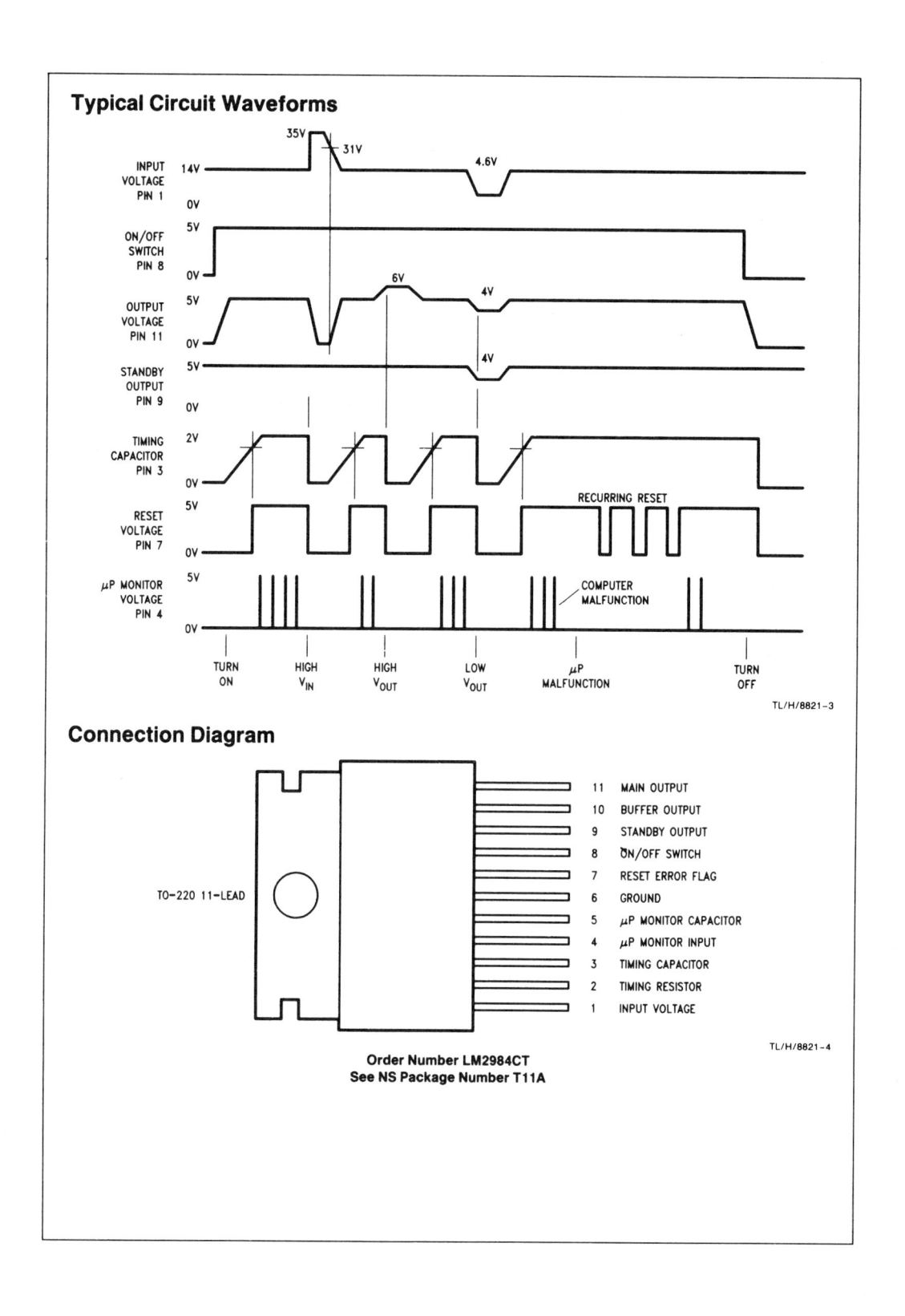

Connection Diagram

TO-220 11-LEAD

- 11 MAIN OUTPUT
- 10 BUFFER OUTPUT
- 9 STANDBY OUTPUT
- 8 ON/OFF SWITCH
- 7 RESET ERROR FLAG
- 6 GROUND
- 5 μP MONITOR CAPACITOR
- 4 μP MONITOR INPUT
- 3 TIMING CAPACITOR
- 2 TIMING RESISTOR
- 1 INPUT VOLTAGE

Order Number LM2984CT
See NS Package Number T11A

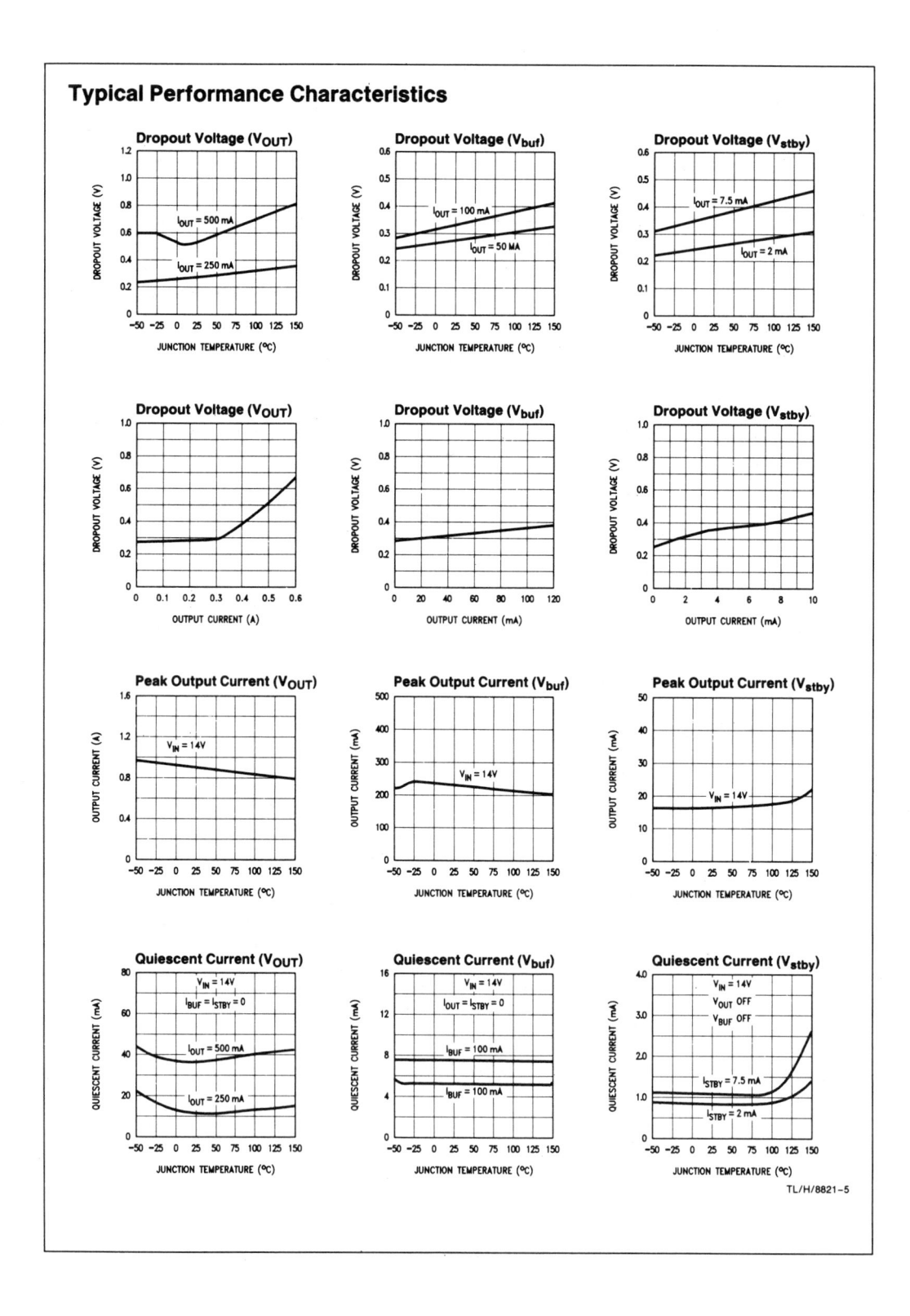

Typical Performance Characteristics

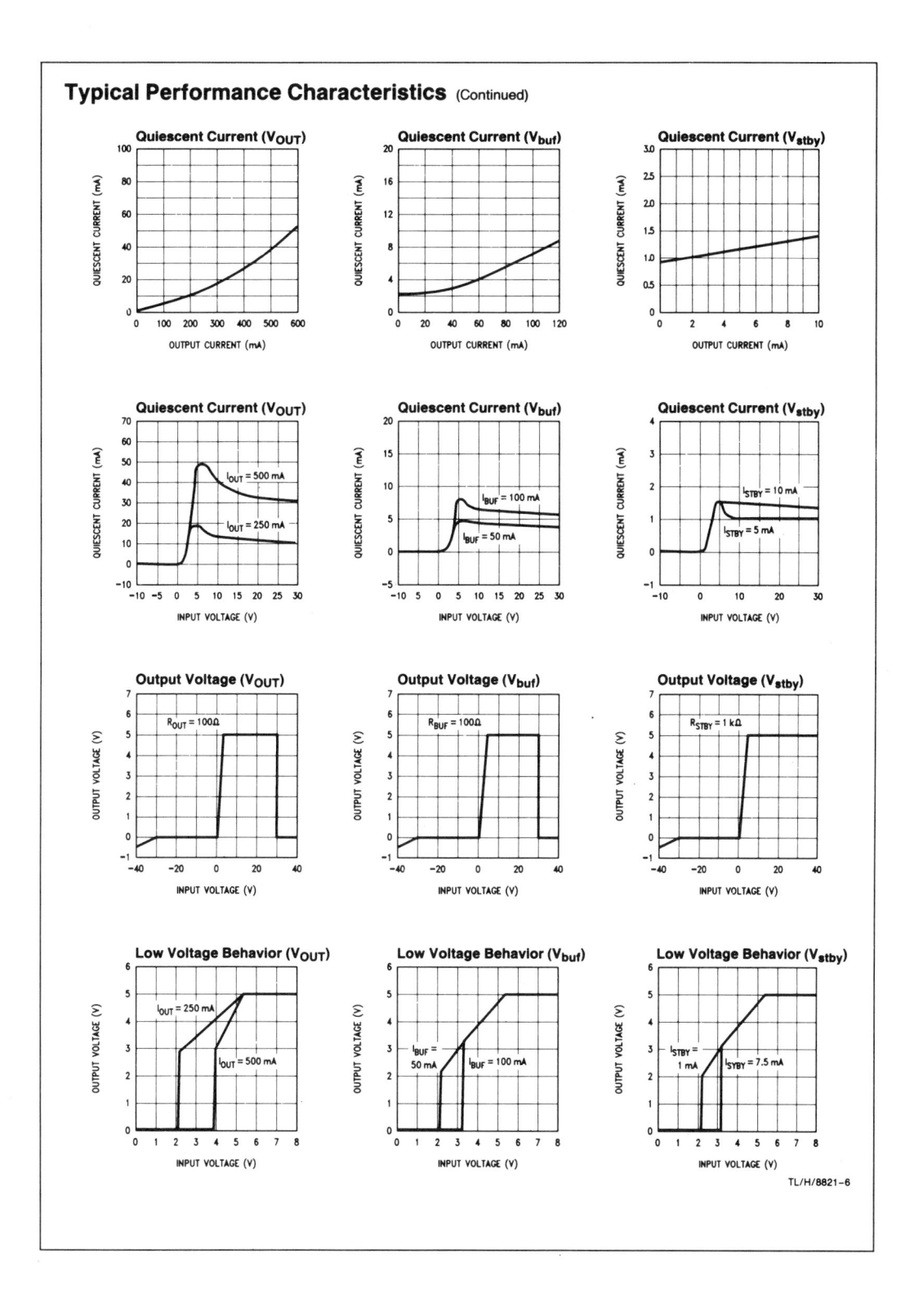

Typical Performance Characteristics (Continued)

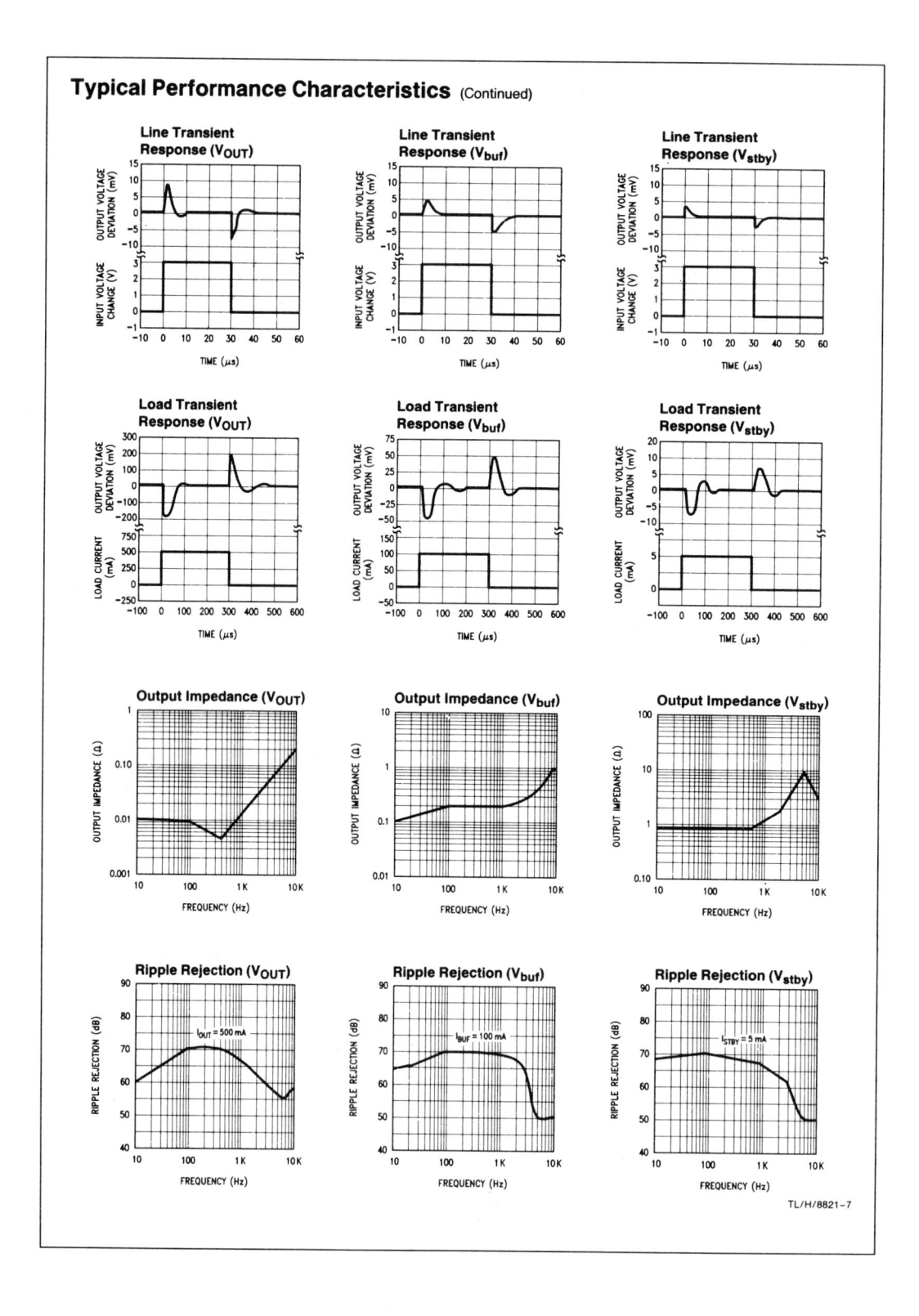

Typical Performance Characteristics (Continued)

Typical Performance Characteristics (Continued)

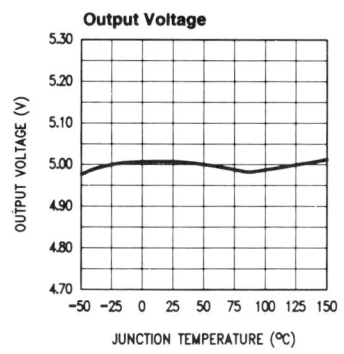

Output Voltage

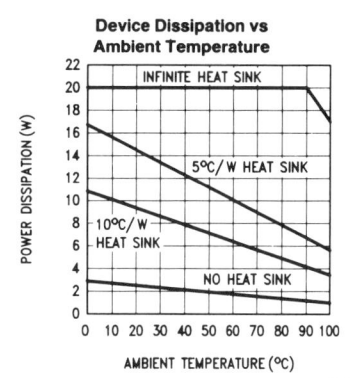

Device Dissipation vs Ambient Temperature

Application Hints

OUTPUT CAPACITORS

The LM2984C output capacitors are required for stability. Without them, the regulator outputs will oscillate, sometimes by many volts. Though the 10 μF shown are the minimum recommended values, actual size and type may vary depending upon the application load and temperature range. Capacitor effective series resistance (ESR) also affects the IC stability. Since ESR varies from one brand to the next, some bench work may be required to determine the minimum capacitor value to use in production. Worst case is usually determined at the minimum ambient temperature and the maximum load expected.

Output capacitors can be increased in size to any desired value above the minimum. One possible purpose of this would be to maintain the output voltages during brief conditions of negative input transients that might be characteristic of a particular system.

Capacitors must also be rated at all ambient temperatures expected in the system. Many aluminum type electrolytics will freeze at temperatures less than $-30°C$, reducing their effective capacitance to zero. To maintain regulator stability down to $-40°C$, capacitors rated at that temperature (such as tantalums) must be used.

Each output **must** be terminated by a capacitor, even if it is not used.

STANDBY OUTPUT

The standby output is intended for use in systems requiring standby memory circuits. While the high current regulator outputs are controlled with the ON/OFF pin described later, the standby output remains on under all conditions as long as sufficient input voltage is supplied to the IC. Thus, memory and other circuits powered by this output remain unaffected by positive line transients, thermal shutdown, etc.

The standby regulator circuit is designed so that the quiescent current to the IC is very low (<1.5 mA) when the other regulator outputs are off.

The capacitor on the output of this regulator can be increased without bound. This will help maintain the output voltage during negative input transients and will also help to reduce the noise on all three outputs. Because the other two track the standby output: therefore any noise reduction here will also reduce the other two noise voltages.

BUFFER OUTPUT

The buffer output is designed to drive peripheral sensor circuitry in a μP system. It will track the standby and main regulator within a few millivolts in normal operation. Therefore, a peripheral sensor can be powered off this supply and have the same operating voltage as the μP system. This is important if a ratiometric sensor system is being used.

The buffer output can be short circuited while the other two outputs are in normal operation. This protects the μP system from disruption of power when a sensor wire, etc. is temporarily shorted to ground, i.e. only the sensor signal would be interrupted, while the μP and memory circuits would remain operational.

The buffer output is similar to the main output in that it is controlled by the ON/OFF switch in order to save power in the standby mode. It is also fault protected against overvoltage and thermal overload. If the input voltage rises above approximately 30V (e.g. load dump), this output will automatically shut down. This protects the internal circuitry and enables the IC to survive higher voltage transients than would otherwise be expected. Thermal shutdown is necessary since this output is one of the dominant sources of power dissipation in the IC.

MAIN OUTPUT

The main output is designed to power relatively large loads, i.e. approximately 500 mA. It is therefore also protected against overvoltage and thermal overload.

This output will track the other two within a few millivolts in normal operation. It can therefore be used as a reference voltage for any signal derived from circuitry powered off the standby or buffer outputs. This is important in a ratiometric sensor system or any system requiring accurate matching of power supply voltages.

ON/OFF SWITCH

The ON/OFF switch controls the main output and the buffer output. The threshold voltage is compatible with most logic families and has about 20 mV of hysteresis to insure 'clean' switching from the standby mode to the active mode and vice versa. This pin can be tied to the input voltage through a 10 kΩ resistor if the regulator is to be powered continuously.

Application Hints (Continued)

POWER DOWN OVERRIDE

Another possible approach is to use a diode in series with the ON/OFF signal and another in series with the main output in order to maintain power for some period of time after the ON/OFF signal has been removed (see *Figure 1*). When the ON/OFF switch is initially pulled high through diode D1, the main output will turn on and supply power through diode D2 to the ON/OFF switch effectively latching the main output. An open collector transistor Q1 is connected to the ON/OFF pin along with the two diodes and forces the regulators off after a period of time determined by the μP. In this way, the μP can override a power down command and store data, do housekeeping, etc. before reverting back to the standby mode.

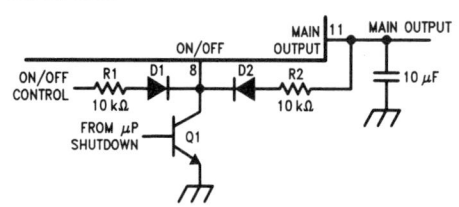

FIGURE 1. Power Down Override

RESET OUTPUT

This output is an open collector NPN transistor which is forced low whenever an error condition is present at the main output or when a μP error is sensed (see μP Monitor section). If the main output voltage drops below 4V or rises above 5.5V, the RESET output is forced low and held low for a period of time set by two external components, R_t and C_t. There is a slight amount of hysteresis in these two threshold voltages so that the RESET output has a fast rise and fall time compatible with the requirements of most μP RESET inputs.

DELAYED RESET

Resistor R_t and capacitor C_t set the period of time that the RESET output is held low after a main output error condition has been sensed. The delay is given by the formula:

$$T_{dly} = 1.2 \, R_t C_t \text{ (seconds)}$$

The delayed RESET will be initiated any time the main output is outside the 4V to 5.5V window, i.e. during power-up, short circuit, overvoltage, low line, thermal shutdown or power-down. The μP is therefore RESET whenever the output voltage is out of regulation. (It is important to note that a RESET is only initiated when the main output is in error. The buffer and standby outputs are not directly monitored for error conditions.)

μP MONITOR RESET

There are two distinct and independent error monitoring systems in the LM2984C. The one described above monitors the main regulator output and initiates a delayed RESET whenever this output is in error. The other error monitoring system is the μP watchdog. These two systems are OR'd together internally and both force the RESET output low when either type of error occurs.

This watchdog circuitry continuously monitors a pin on the μP that generates a positive going pulse during normal operation. The period of this pulse is typically on the order of milliseconds and the pulse width is typically on the order of 10's of microseconds. If this pulse ever disappears, the watchdog circuitry will time out and a RESET low will be sent to the μP. The time out period is determined by two external components, R_t and C_{mon}, according to the formula:

$$T_{window} = 0.82 \, R_t C_{mon} \text{ (seconds)}$$

The width of the RESET pulse is set by C_{mon} and an internal resistor according to the following:

$$RESET_{pw} = 2000 \, C_{mon} \text{ (seconds)}$$

A square wave signal can also be monitored for errors by filtering the C_{mon} input such that only the positive edges of the signal are detected. *Figure 2* is a schematic diagram of a typical circuit used to differentiate the input signal. Resistor R_{tc} and capacitor C_{tc} pass only the rising edge of the square wave and create a short positive pulse suitable for the μP monitor input. If the incoming signal continues in a high state or in a low state for too long a period of time, a RESET low will be generated.

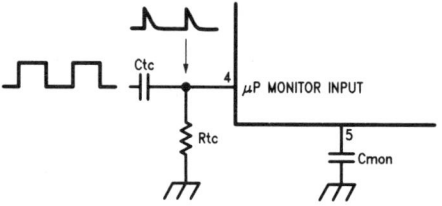

FIGURE 2. Monitoring Square Wave μP Signals

The threshold voltage and input characteristics of this pin are compatible with nearly all logic families.

There is a limit on the width of a pulse that can be reliably detected by the watchdog circuit. This is due to the output resistance of the transistor which discharges C_{mon} when a high state is detected at the input. The minimum detectable pulse width can be determined by the following formula:

$$PW_{min} = 20 \, C_{mon} \text{ (seconds)}$$

Equivalent Schematic Diagram

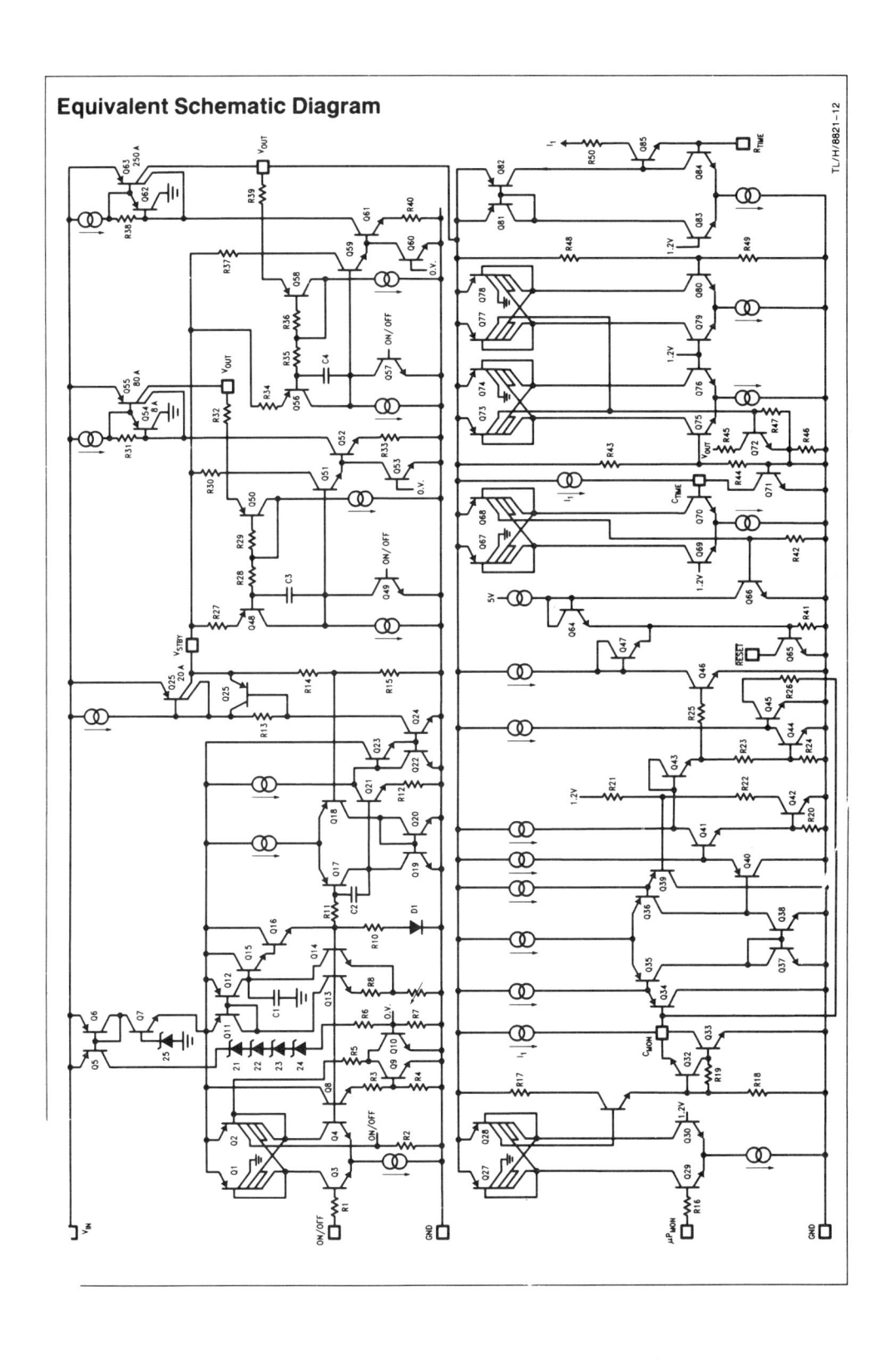

LM2984C Microprocessor Power Supply System

Physical Dimensions inches (millimeters)

Lit. # 108032-1

Order Number LM2984CT
NS Package Number T11A

T11A (REV B)

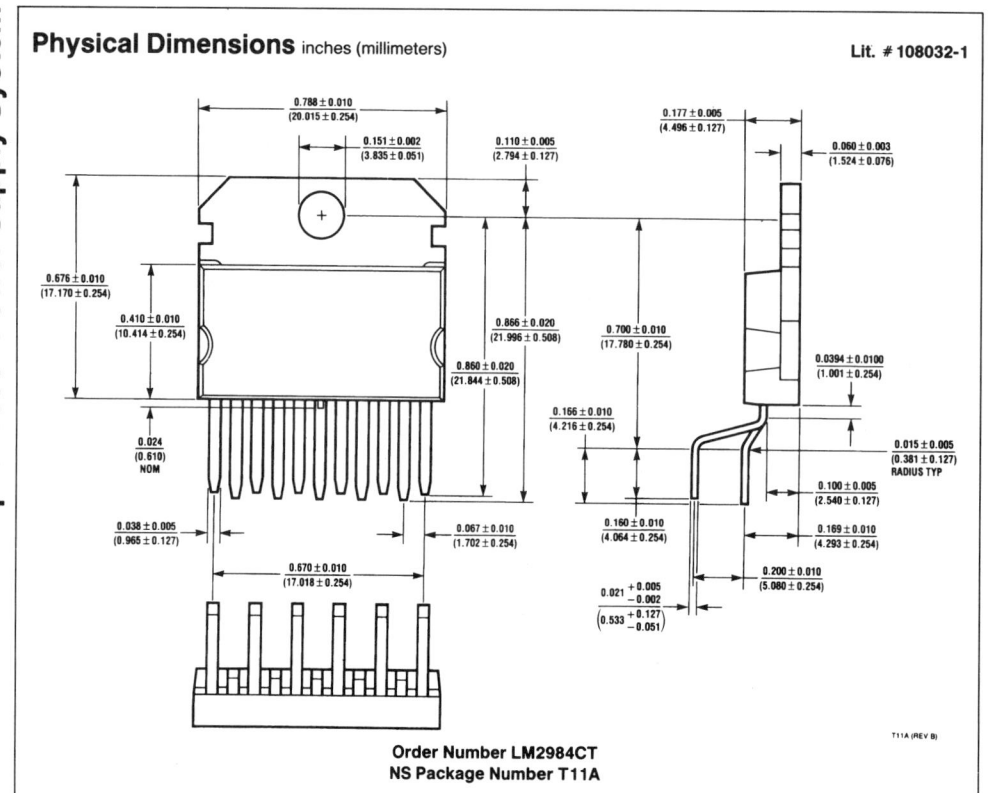

LIFE SUPPORT POLICY

NATIONAL'S PRODUCTS ARE NOT AUTHORIZED FOR USE AS CRITICAL COMPONENTS IN LIFE SUPPORT DEVICES OR SYSTEMS WITHOUT THE EXPRESS WRITTEN APPROVAL OF THE PRESIDENT OF NATIONAL SEMICONDUCTOR CORPORATION. As used herein:

1. Life support devices or systems are devices or systems which, (a) are intended for surgical implant into the body, or (b) support or sustain life, and whose failure to perform, when properly used in accordance with instructions for use provided in the labeling, can be reasonably expected to result in a significant injury to the user.

2. A critical component is any component of a life support device or system whose failure to perform can be reasonably expected to cause the failure of the life support device or system, or to affect its safety or effectiveness.

National Semiconductor (UK) Ltd.	National Semiconductor GmbH	National Semiconductor France S.A.	National Semiconductor (UK) Ltd.	National Semiconductor S.p.A.	National Semiconductor AB
The Maple, Kembrey Park Swindon, Wiltshire SN2 6UT United Kingdom Tel. (07 93) 61 41 41 Telex 444 674	Industriestraße 10 D-8080 Fürstenfeldbruck Tel. (0 81 41) 103-0 Telex 527 649	Expansion 10000 28, rue de la Redoute F-92260 Fontenay-aux-Roses Tel. (1) 46 60 81 40 Telex 250 956	301 Harpur Centre Horne Lane Bedford, MK40 1TR Tel. (02 34) 27 00 27 Telex 826 209	Via Solferino, 19 I-20121 Milano Tel. (02) 6 59 61 46 Telex 332 835	Box 2016 Stensätravägen 13 S-12702 Skärholmen Tel. (08) 97 01 90 Telex 10 731

National does not assume any responsibility for use of any circuitry described, no circuit patent licenses are implied and National reserves the right at any time without notice to change said circuitry and specifications.

MOTOROLA SEMICONDUCTOR TECHNICAL DATA

MC34164 / MC33164

Order this data sheet by MC34164/D

MICROPOWER UNDERVOLTAGE SENSING CIRCUITS

SILICON MONOLITHIC INTEGRATED CIRCUIT

Advance Information

MICROPOWER UNDERVOLTAGE SENSING CIRCUITS

The MC34164 series are undervoltage sensing circuits specifically designed for use as reset controllers in portable microprocessor based systems where extended battery life is required. These devices offer the designer an economical solution for low voltage detection with a single external resistor. The MC34164 series features a bandgap reference, a comparator with precise thresholds and built-in hysteresis to prevent erratic reset operation, an open collector reset output capable of sinking in excess of 6.0 mA, and guaranteed operation down to 1.0 V input with extremely low standby current. These devices are packaged in 3-pin TO-226AA and 8-pin surface mount packages.

Applications include direct monitoring of the 3.0 or 5.0 Volt MPU/logic power supply used in appliance, automotive, consumer, and industrial equipment.

- Temperature Compensated Reference
- Monitors 3.0 Volt (MC34164-3) or 5.0 Volt (MC34164-5) Power Supplies
- Precise Comparator Thresholds Guaranteed Over Temperature
- Comparator Hysteresis Prevents Erratic Reset
- Reset Output Capable of Sinking in Excess of 6.0 mA
- Internal Clamp Diode for Discharging Delay Capacitor
- Guaranteed Reset Operation with 1.0 V Input
- Extremely Low Standby Current: As Low as 9.0 µA
- Economical TO-226AA and Surface Mount Packages

P SUFFIX
PLASTIC PACKAGE
CASE 29
(TO-226AA)

Pin 1. Reset
2. Input
3. Ground

D SUFFIX
PLASTIC PACKAGE
CASE 751
(SO-8)

Pin 1. Reset
2. Input
3. N.C.
4. Ground
5. N.C.
6. N.C.
7. N.C.
8. N.C.

REPRESENTATIVE BLOCK DIAGRAM

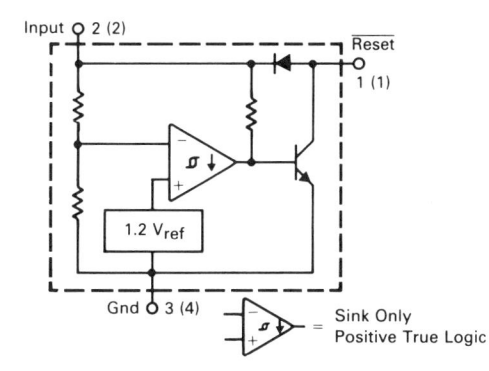

Input 2 (2)
Reset 1 (1)
1.2 V$_{ref}$
Gnd 3 (4)

= Sink Only Positive True Logic

Pin numbers adjacent to terminals are for the 3-pin TO-226AA package.
Pin numbers in parenthesis are for the D suffix SO-8 package.

ORDERING INFORMATION

Device	Temperature Range	Package
MC34164D-3	0°C to +70°C	SO-8
MC34164D-5		
MC34164P-3		TO-226AA
MC34164P-5		
MC33164D-3	−40°C to +85°C	SO-8
MC33164D-5		
MC33164P-3		TO-226AA
MC33164P-5		

This document contains information on a new product. Specifications and information herein are subject to change without notice.

©MOTOROLA INC.; 1990   ADI1727

MAXIMUM RATINGS

Rating	Symbol	Value	Unit
Power Input Supply Voltage	V_{in}	−1.0 to 12	V
Reset Output Voltage	V_O	−1.0 to 12	V
Reset Output Sink Current	I_{Sink}	Internally Limited	mA
Clamp Diode Forward Current, Pin 1 to 2 (Note 1)	I_F	100	mA

Power Dissipation and Thermal Characteristics
P Suffix, Plastic Package
  Maximum Power Dissipation @ T_A = 25°C
  Thermal Resistance Junction to Air
D Suffix, Plastic Package
  Maximum Power Dissipation @ T_A = 25°C
  Thermal Resistance Junction to Air
	
	P_D
	$R_{\theta JA}$
	
	P_D
	$R_{\theta JA}$
		
		700
		178
		
		700
		178
			
			mW
			°C/W
			
			mW
			°C/W

Operating Junction Temperature	T_J	+150	°C

Operating Ambient Temperature Range
  MC34164 Series
  MC33164 Series
	T_A	
		0 to +70
		−40 to +85
			°C

Storage Temperature Range	T_{stg}	−65 to +150	°C

MC34164-3, MC33164-3 SERIES
ELECTRICAL CHARACTERISTICS (For typical values T_A = 25°C, for min/max values T_A is the operating ambient temperature range that applies, see notes 2 & 3).

Characteristic	Symbol	Min	Typ	Max	Unit
COMPARATOR					

Threshold Voltage
  High State Output (V_{in} Increasing)
  Low State Output (V_{in} Decreasing)
  Hysteresis (I_{Sink} = 100 μA)
	
	V_{IH}
	V_{IL}
	V_H
		
		2.55
		2.55
		0.03
			
			2.71
			2.65
			0.06
				
				2.80
				2.80
				—
					V

RESET OUTPUT					

Output Sink Saturation
  (V_{in} = 2.4 V, I_{Sink} = 1.0 mA)
  (V_{in} = 1.0 V, I_{Sink} = 0.25 mA)
	V_{OL}	
		—
		—
			
			0.14
			0.1
				
				0.4
				0.3
					V

Output Sink Current (V_{in}, Reset = 2.4 V)	I_{Sink}	6.0	12	30	mA

Output Off-State Leakage
  (V_{in}, Reset = 3.0 V)
  (V_{in}, Reset = 10 V)
	$I_{R(leak)}$	
		—
		—
			
			0.02
			0.02
				
				0.5
				1.0
					μA

Clamp Diode Forward Voltage, Pin 1 to 2 (I_F = 5.0 mA)	V_F	0.6	0.9	1.2	V
TOTAL DEVICE					
Operating Input Voltage Range	V_{in}	1.0 to 10	—	—	V

Quiescent Input Current
  V_{in} = 3.0 V
  V_{in} = 6.0 V
	I_{in}	
		—
		—
			
			9.0
			24
				
				15
				40
					μA

NOTES:
1. Maximum Package power dissipation limits must be observed.
2. Low duty cycle pulse techniques are used during test to maintain junction temperature as close to ambient as possible.
3. T_{low} =   0°C for MC34164 T_{high} = +70°C for MC34164
          = −40°C for MC33164   = +85°C for MC33164

MC34164-5, MC33164-5 SERIES

ELECTRICAL CHARACTERISTICS (For typical values T_A = 25°C, for min/max values T_A is the operating ambient temperature range that applies, see notes 2 & 3).

Characteristic	Symbol	Min	Typ	Max	Unit
COMPARATOR					
Threshold Voltage					V
High State Output (V_{in} Increasing)	V_{IH}	4.15	4.33	4.45	
Low State Output (V_{in} Decreasing)	V_{IL}	4.15	4.27	4.45	
Hysteresis (I_{Sink} = 100 μA)	V_H	0.02	0.09	—	
RESET OUTPUT					
Output Sink Saturation	V_{OL}				V
(V_{in} = 4.0 V, I_{Sink} = 1.0 mA)		—	0.14	0.4	
(V_{in} = 1.0 V, I_{Sink} = 0.25 mA)		—	0.1	0.3	
Output Sink Current (V_{in}, $\overline{Reset}$ = 4.0 V)	I_{Sink}	7.0	20	50	mA
Output Off-State Leakage	$I_{R(leak)}$				μA
(V_{in}, $\overline{Reset}$ = 5.0 V)		—	0.02	0.5	
(V_{in}, $\overline{Reset}$ = 10 V)		—	0.02	2.0	
Clamp Diode Forward Voltage, Pin 1 to 2 (I_F = 5.0 mA)	V_F	0.6	0.9	1.2	V
TOTAL DEVICE					
Operating Input Voltage Range	V_{in}	1.0 to 10	—	—	V
Quiescent Input Current	I_{in}				μA
V_{in} = 5.0 V		—	12	20	
V_{in} = 10 V		—	32	50	

NOTES:
2. Low duty cycle pulse techniques are used during test to maintain junction temperature as close to ambient as possible.
3. T_{low} = 0°C for MC34164 T_{high} = +70°C for MC34164
   = −40°C for MC33164   = +85°C for MC33164

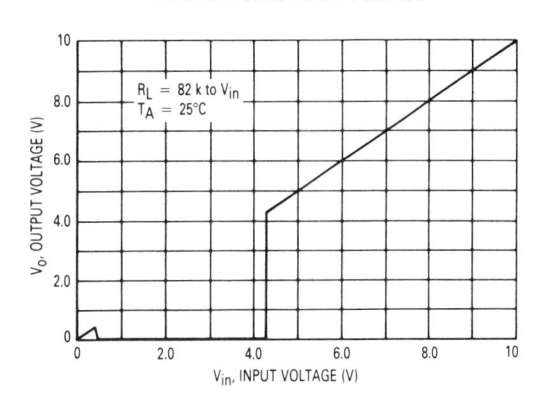

FIGURE 1 — MC3X164-3 RESET OUTPUT VOLTAGE versus INPUT VOLTAGE

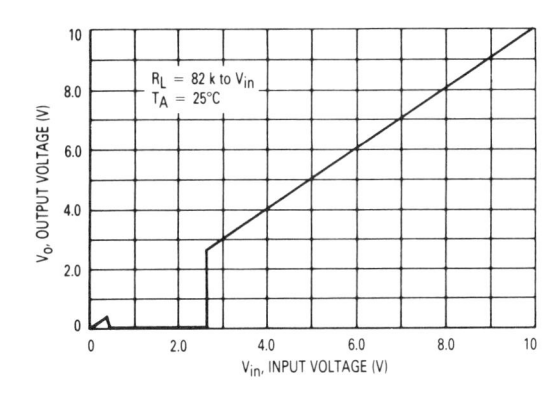

FIGURE 2 — MC3X164-5 RESET OUTPUT VOLTAGE versus INPUT VOLTAGE

FIGURE 3 — MC3X164-3 $\overline{\text{RESET}}$ OUTPUT VOLTAGE versus INPUT VOLTAGE

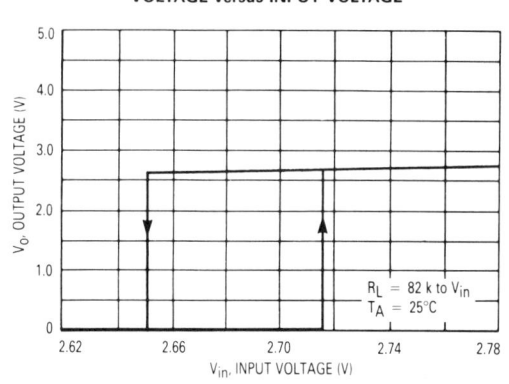

FIGURE 4 — MC3X164-5 $\overline{\text{RESET}}$ OUTPUT VOLTAGE versus INPUT VOLTAGE

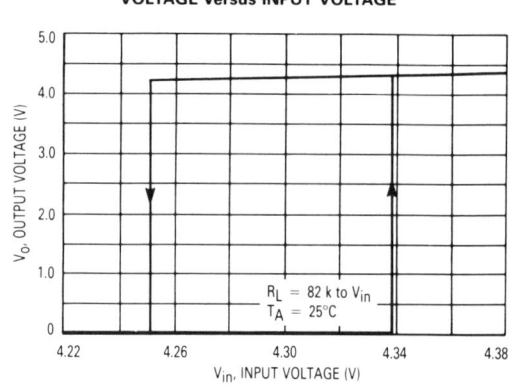

FIGURE 5 — MC3X164-3 COMPARATOR THRESHOLD VOLTAGE versus TEMPERATURE

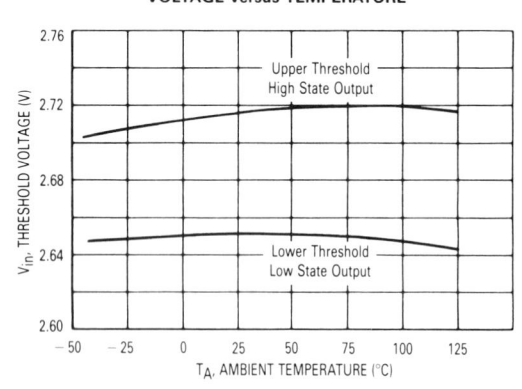

FIGURE 6 — MC3X164-5 COMPARATOR THRESHOLD VOLTAGE versus TEMPERATURE

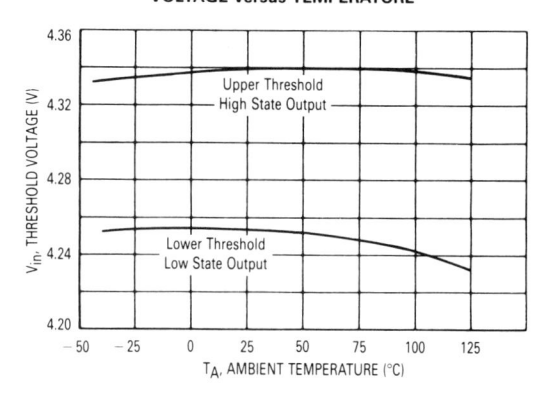

FIGURE 7 — MC3X164-3 INPUT CURRENT versus INPUT VOLTAGE

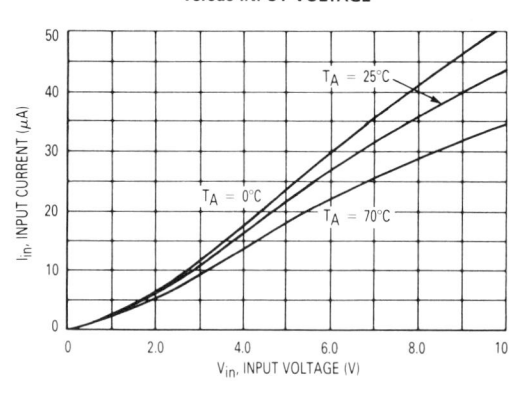

FIGURE 8 — MC3X164-5 INPUT CURRENT versus INPUT VOLTAGE

MOTOROLA

MC34164 • MC33164 SERIES

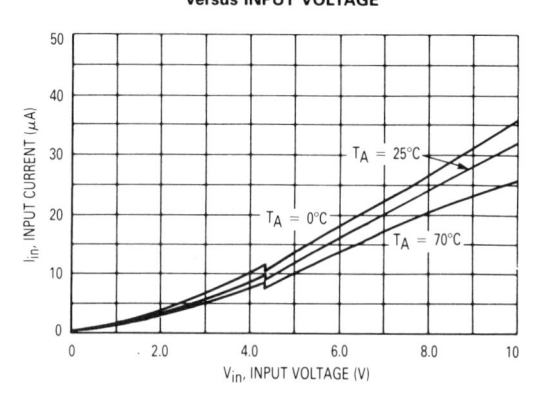

FIGURE 9 — MC3X164-3 RESET OUTPUT SATURATION versus SINK CURRENT

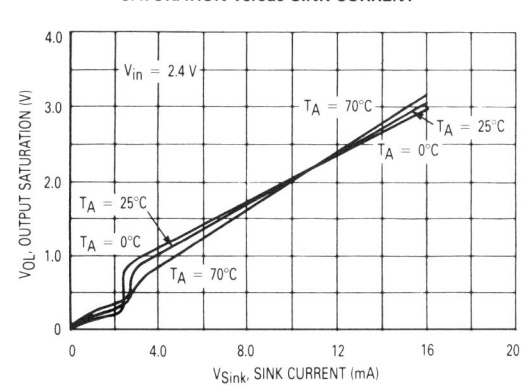

FIGURE 10 — MC3X164-5 RESET OUTPUT SATURATION versus SINK CURRENT

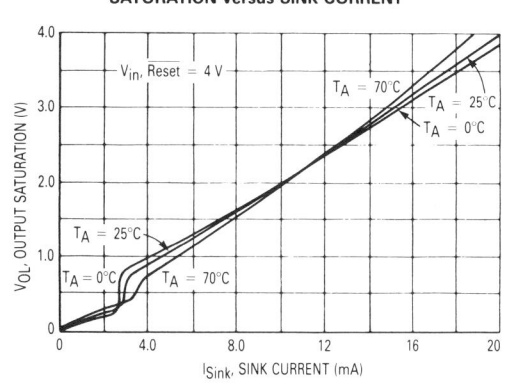

FIGURE 11 — CLAMP DIODE FORWARD CURRENT versus VOLTAGE

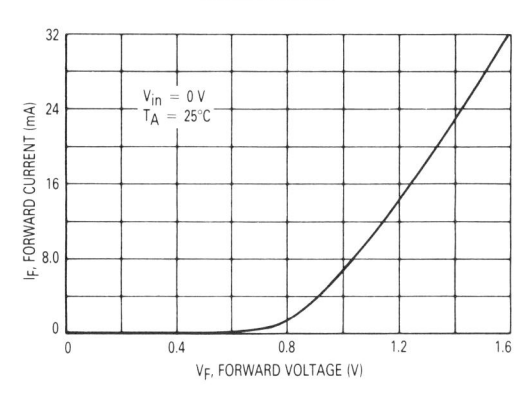

FIGURE 12 — RESET DELAY TIME (MC3X164-5 shown)

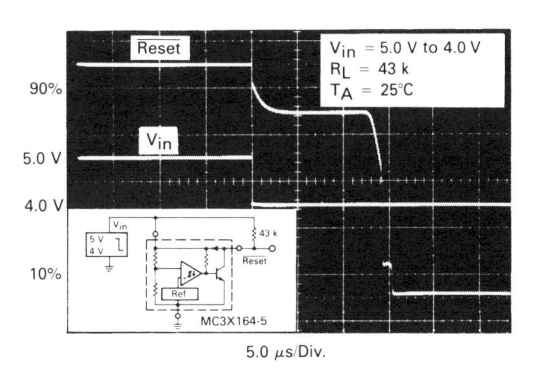

FIGURE 13 — LOW VOLTAGE MICROPROCESSOR RESET

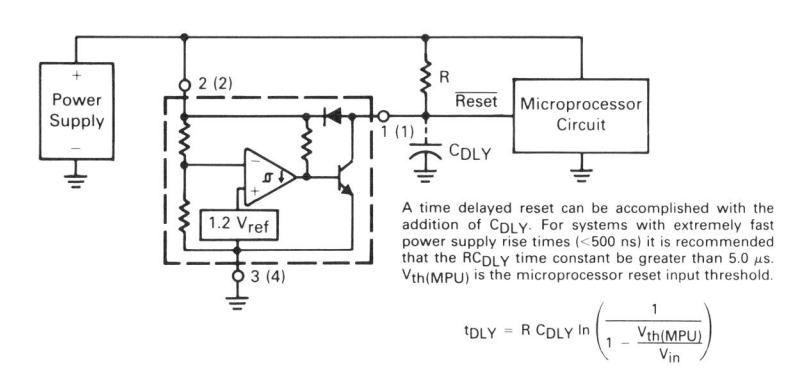

A time delayed reset can be accomplished with the addition of C_{DLY}. For systems with extremely fast power supply rise times (<500 ns) it is recommended that the RC_{DLY} time constant be greater than 5.0 µs. $V_{th(MPU)}$ is the microprocessor reset input threshold.

$$t_{DLY} = R\, C_{DLY} \ln\left(\frac{1}{1 - \dfrac{V_{th(MPU)}}{V_{in}}}\right)$$

FIGURE 14 — LOW VOLTAGE MICROPROCESSOR RESET WITH ADDITIONAL HYSTERESIS
(MC3X164-5 shown)

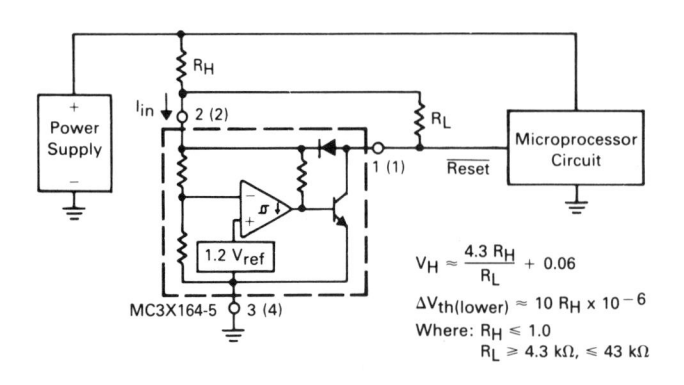

TEST DATA			
V_H (mV)	ΔV_{th} (mV)	R_H (Ω)	R_L (kΩ)
60	0	0	43
103	1.0	100	10
123	1.0	100	6.8
160	1.0	100	4.3
155	2.2	220	10
199	2.2	220	6.8
280	2.2	220	4.3
262	4.7	470	10
306	4.7	470	8.2
357	4.7	470	6.8
421	4.7	470	5.6
530	4.7	470	4.3

$$V_H \approx \frac{4.3\,R_H}{R_L} + 0.06$$

$$\Delta V_{th(lower)} \approx 10\,R_H \times 10^{-6}$$

Where: $R_H \leq 1.0$
$R_L \geq 4.3\,k\Omega, \leq 43\,k\Omega$

Comparator hysteresis can be increased with the addition of resistor R_H. The hysteresis equation has been simplified and does not account for the change of input current I_{in} as V_{CC} crosses the comparator threshold (Figure 8). An increase of the lower threshold $\Delta V_{th(lower)}$ will be observed due to I_{in} which is typically 10 μA at 4.3 V. The equations are accurate to ±10% with R_H less than 1.0 kΩ and R_L between 4.3 kΩ and 43 kΩ.

FIGURE 15 — VOLTAGE MONITOR

FIGURE 16 — SOLAR POWERED BATTERY CHARGER

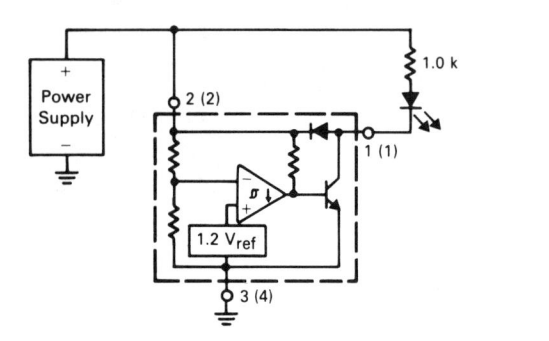

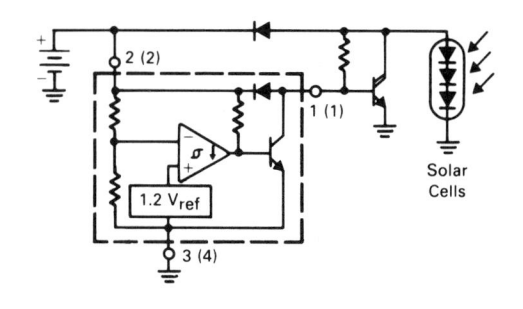

FIGURE 17 — MOSFET LOW VOLTAGE GATE DRIVE PROTECTION USING THE MC3X164-5

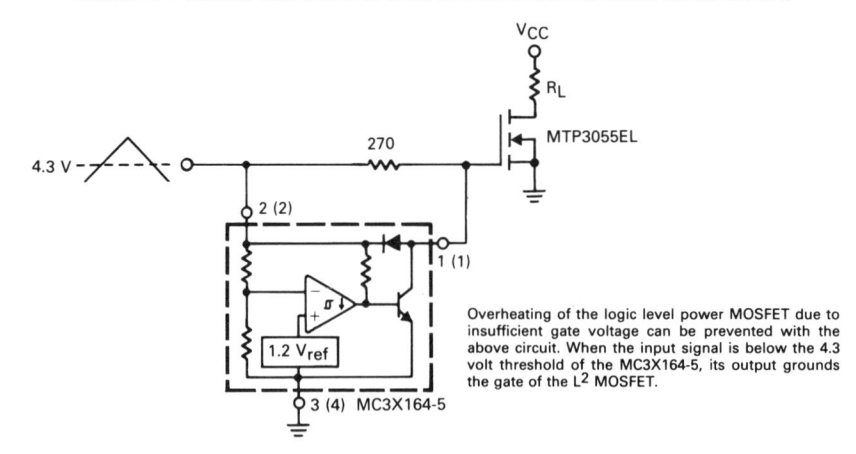

Overheating of the logic level power MOSFET due to insufficient gate voltage can be prevented with the above circuit. When the input signal is below the 4.3 volt threshold of the MC3X164-5, its output grounds the gate of the L² MOSFET.

OUTLINE DIMENSIONS

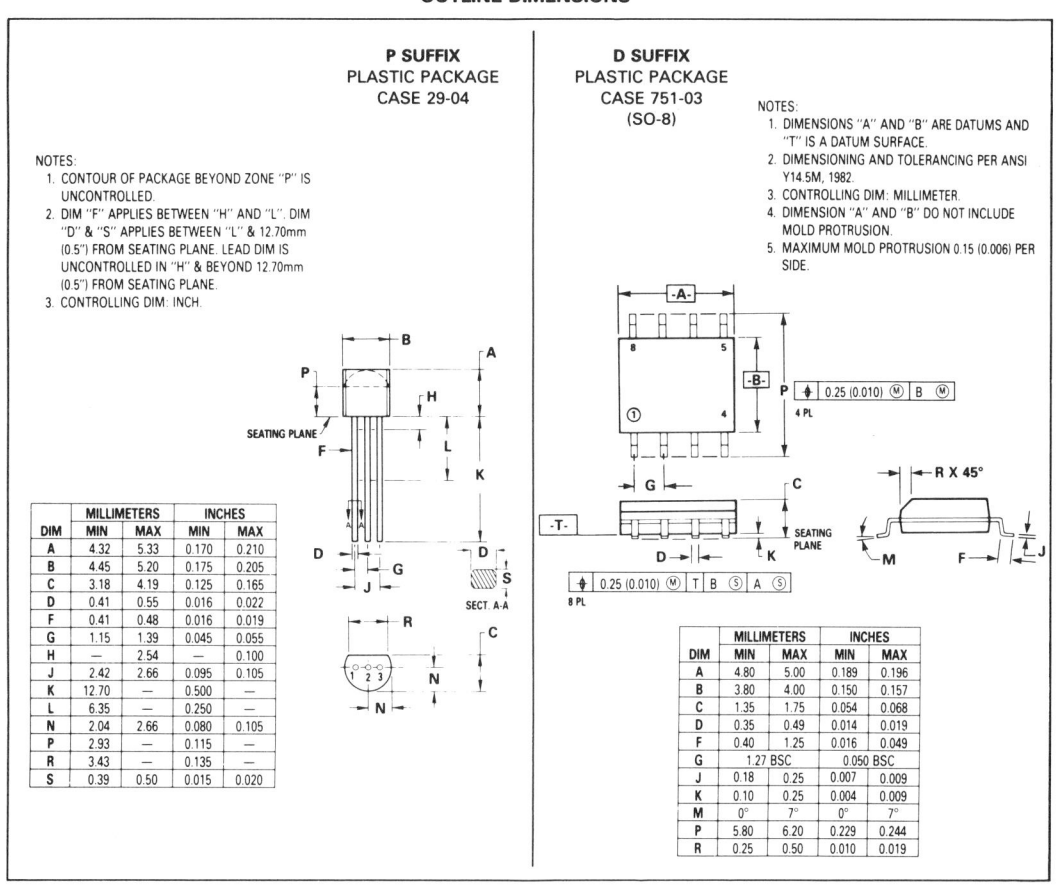

P SUFFIX
PLASTIC PACKAGE
CASE 29-04

NOTES:
1. CONTOUR OF PACKAGE BEYOND ZONE "P" IS UNCONTROLLED.
2. DIM "F" APPLIES BETWEEN "H" AND "L". DIM "D" & "S" APPLIES BETWEEN "L" & 12.70mm (0.5") FROM SEATING PLANE. LEAD DIM IS UNCONTROLLED IN "H" & BEYOND 12.70mm (0.5") FROM SEATING PLANE.
3. CONTROLLING DIM: INCH.

DIM	MILLIMETERS MIN	MILLIMETERS MAX	INCHES MIN	INCHES MAX
A	4.32	5.33	0.170	0.210
B	4.45	5.20	0.175	0.205
C	3.18	4.19	0.125	0.165
D	0.41	0.55	0.016	0.022
F	0.41	0.48	0.016	0.019
G	1.15	1.39	0.045	0.055
H	—	2.54	—	0.100
J	2.42	2.66	0.095	0.105
K	12.70	—	0.500	—
L	6.35	—	0.250	—
N	2.04	2.66	0.080	0.105
P	2.93	—	0.115	—
R	3.43	—	0.135	—
S	0.39	0.50	0.015	0.020

D SUFFIX
PLASTIC PACKAGE
CASE 751-03
(SO-8)

NOTES:
1. DIMENSIONS "A" AND "B" ARE DATUMS AND "T" IS A DATUM SURFACE.
2. DIMENSIONING AND TOLERANCING PER ANSI Y14.5M, 1982.
3. CONTROLLING DIM: MILLIMETER.
4. DIMENSION "A" AND "B" DO NOT INCLUDE MOLD PROTRUSION.
5. MAXIMUM MOLD PROTRUSION 0.15 (0.006) PER SIDE.

DIM	MILLIMETERS MIN	MILLIMETERS MAX	INCHES MIN	INCHES MAX
A	4.80	5.00	0.189	0.196
B	3.80	4.00	0.150	0.157
C	1.35	1.75	0.054	0.068
D	0.35	0.49	0.014	0.019
F	0.40	1.25	0.016	0.049
G	1.27 BSC		0.050 BSC	
J	0.18	0.25	0.007	0.009
K	0.10	0.25	0.004	0.009
M	0°	7°	0°	7°
P	5.80	6.20	0.229	0.244
R	0.25	0.50	0.010	0.019

Motorola reserves the right to make changes without further notice to any products herein to improve reliability, function or design. Motorola does not assume any liability arising out of the application or use of any product or circuit described herein; neither does it convey any license under its patent rights nor the rights of others. Motorola products are not authorized for use as components in life support devices or systems intended for surgical implant into the body or intended to support or sustain life. Buyer agrees to notify Motorola of any such intended end use whereupon Motorola shall determine availability and suitability of its product or products for the use intended. Motorola and Ⓜ are registered trademarks of Motorola, Inc. Motorola, Inc. is an Equal Employment Opportunity/Affirmative Action Employer.

Chapter 2 Buses

You probably could say that buses are the simplest part of a computer conceptually, but the most complex part in practice. Although buses simply move signals from one place to another, the electrical characteristics of both the signals and the bus have to be taken into account by the systems designer.

We begin this chapter with a tutorial on the electrical characteristics of computer buses and the components used to make buses. A second tutorial deals with the design of a video system for reduced EMI (electromagnetic interference) and complements the tutorial.

Having looked at the characteristics of the high speed parallel bus, we look at some of the drivers and receivers used to implement serial buses.

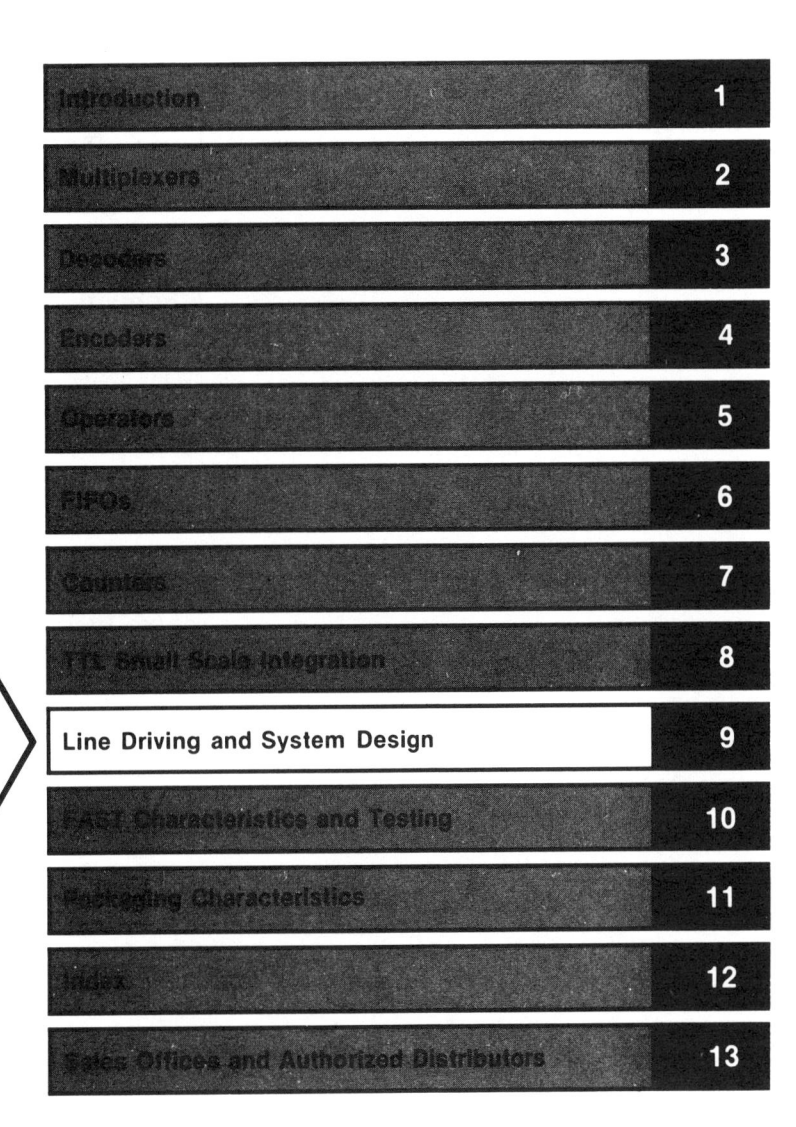

Introduction	1
Multiplexers	2
Decoders	3
Encoders	4
Operators	5
FIFOs	6
Counters	7
TTL Small Scale Integration	8
Line Driving and System Design	**9**
FAST Characteristics and Testing	10
Packaging Characteristics	11
Index	12
Sales Offices and Authorized Distributors	13

National Semiconductor

Line Driving and System Design

Introduction

Successful high-speed system design is dependent on careful system timing design and good board layout. The pitfalls are many and varied, and this section addresses some of those problem areas and simplifies the design requirements. All systems must interconnect signals either by short lines on printed circuit board, long lines on a backplane, twisted pair cables, or coaxial cables, etc. At high frequency, all of these mediums must be treated as transmission lines. Two properties of transmission lines, characteristic impedance (Z_O) and propagation delay (t_{PD}), are of concern. Transmission lines store energy, the magnitude of which is dependent on line length, impedance, applied voltage and source impedance. This stored energy must be dissipated by the terminating device and is also available to be coupled in other circuits by crosstalk. The effects of termination on line reflection and crosstalk are discussed, as well as good board layout practices.

Transmission Line Concepts
Line Driving
Decoupling
Design Considerations
    Ground—An Essential Link
    Crosstalk
The Capacitor

Transmission Line Concepts

The interactions between wiring and circuitry in high-speed systems are more easily determined by treating the interconnections as transmission lines. A brief review of basic concepts is presented and simplified methods of analysis are used to examine situations commonly encountered in digital systems. Since the principles and methods apply to any type of logic circuit, normalized pulse amplitudes are used in sample waveforms and calculations.

Simplifying Assumptions

For the great majority of interconnections in digital systems, the resistance of the conductors is much less than the input and output resistance of the circuits. Similarly, the insulating materials have very good dielectric properties. These circumstances allow such factors as attenuation, phase distortion and bandwidth limitations to be ignored. With these simplifications, interconnections can be dealt with in terms of characteristic impedance and propagation delay.

Characteristic Impedance

The two conductors that interconnect a pair of circuits have distributed series inductance and distributed capacitance between them, and thus constitute a transmission line. For any length in which these distributed parameters are constant, the pair of conductors have a characteristic impedance Z_O. Whereas quiescent conditions on the line are determined by the circuits and terminations, Z_O is the ratio of transient voltage to transient current passing by a point on the line when a signal change or other electrical disturbance occurs. The relationship between transient voltage, transient current, characteristic impedance, and the distributed parameters is expressed as follows:

(E9-1) $$\frac{V}{I} = Z_O = \sqrt{\frac{L_O}{C_O}}$$

where L_O = inductance per unit length, and C_O = capacitance per unit length. Z_O is in ohms, L_O in henries, and C_O in farads.

Propagation Velocity

Propagation velocity (ν) and its reciprocal, delay per unit length (δ), can also be expressed in terms of L_O and C_O. A consistent set of units is nanoseconds, microhenries and picofarads, with a common unit of length.

(E9-2) $$\nu = \frac{1}{\sqrt{L_O C_O}} \qquad \delta = \sqrt{L_O C_O}$$

Equations 9-1 and 9-2 provide a convenient means of determining the L_O and C_O of a line when delay, length and impedance are known. For a length l and delay T, δ is the ratio T/l. To determine L_O and C_O, combine Equations 9-1 and 9-2.

(E9-3) $$L_O = \delta Z_O$$

(E9-4) $$C_O = \frac{\delta}{Z_O}$$

More formal treatments of transmission line characteristics, including loss effects, are available from many sources.

Termination and Reflection

A transmission line with a terminating resistor is shown in Figure 9-1. As indicated, a positive step function voltage travels from left to right. To keep track of reflection polarities, it is convenient to consider the lower conductor as the voltage reference and to think in terms of current flow in the top conductor only. The generator is assumed to have zero internal impedance. The initial current I_1 is determined by V_1 and Z_O.

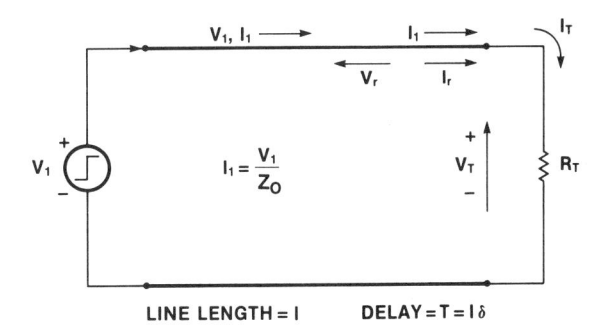

Figure 9-1

If the terminating resistor matches the line impedance, the ratio of voltage to current traveling along the line is matched by the ratio of voltage to current which must, by Ohm's law, always prevail at R_T. From the viewpoint of the voltage step generator, no adjustment of output current is ever required; the situation is as though the transmission line never existed and R_T had been connected directly across the terminals of the generator.

From the R_T viewpoint, the only thing the line did was delay the arrival of the voltage step by the amount of time T.

When R_T is not equal to Z_O, the initial current starting down the line is still determined by V_1 and Z_O but the final steady state current, after all reflections have died out, is determined by V_1 and R_T (ohmic resistance of the line is assumed to be negligible). The ratio of voltage to current in the initial wave is not equal to the ratio of voltage to current demanded by R_T. Therefore, at the instant the initial wave arrives at R_T, another voltage and current wave must be generated so that Ohm's law is satisfied at the line-load interface. This reflected wave, indicated by V_r and I_r in Figure 9-1, starts to return toward the generator. Applying Kirchoff's laws to the end of the line at the instant the initial wave arrives results in the following:

$$I_1 + I_r = I_T = \text{current into } R_T$$

(E9-5)

Since only one voltage can exist at the end of the line at this instant of time, the following is true:

$$V_1 + V_r = V_T$$

thus $$I_T = \frac{V_T}{R_T} = \frac{V_1 + V_r}{R_T}$$

also $$I_1 = \frac{V_1}{Z_O} \text{ and } I_r = -\frac{V_r}{Z_O}$$

(E9-6)

with the minus sign indicating that V_r is moving toward the generator.

Combining the foregoing relationships algebraically and solving for V_r yields a simplified expression in terms of V_1, Z_O and R_T.

$$\frac{V_1}{Z_O} - \frac{V_r}{Z_O} = \frac{V_1 + V_r}{R_T} = \frac{V_1}{R_T} + \frac{V_r}{R_T}$$

(E9-7)

$$V_1 \left(\frac{1}{Z_O} - \frac{1}{R_T}\right) = V_r \left(\frac{1}{R_T} + \frac{1}{Z_O}\right)$$

$$V_r = V_1 \left(\frac{R_T - Z_O}{R_T + Z_O}\right) = \rho_L V_1$$

The term in parentheses is called the coefficient of reflection (ρ_L). With R_T ranging between zero (shorted line) and infinity (open line), the coefficient ranges between -1 and $+1$ respectively. The subscript L indicates that ρ_L refers to the coefficient at the load end of the line.

Equation 9-7 expresses the amount of voltage sent back down the line, and since

$$V_T = V_1 + V_r$$

then $$V_T = V_1 (1 + \rho_L)$$

(E9-8)

V_T can also be determined from an expression which does not require the preliminary step of calculating ρ_L. Manipulating $(1 + \rho_L)$ results in

$$1 + \rho_L = 1 + \frac{R_T - Z_O}{R_T + Z_O} = 2\left(\frac{R_T}{R_T + Z_O}\right)$$

Substituting in Equation 9-8 gives

$$V_T = 2\left(\frac{R_T}{R_T + Z_O}\right) V_1$$

(E9-9)

The foregoing has the same form as a simple voltage divider involving a generator V_1 with internal impedance Z_O driving a load R_T, except that the amplitude of V_T is doubled.

The arrow indicating the direction of V_r in Figure 9-1 correctly indicates the V_r direction of travel, but the direction of I_r flow depends on the V_r polarity. If V_r is positive, I_r flows toward the generator, opposing I_1. This relationship between the polarity of V_r and the direction of I_r can be deduced by noting in Equation 9-7 that if V_r is positive it is because R_T is greater than Z_O. In turn, this means that the initial current I_1 is larger than the final quiescent current, dictated by V_1 and R_T. Hence I_r must oppose I_1 to reduce the line current to the final quiescent value. Similar reasoning shows that if V_r is negative, I_r flows in the same direction as I_1.

It is sometimes easier to determine the effect of V_r on line conditions by thinking of it as an independent voltage generator in series with R_T. With this concept, the direction of I_r is immediately apparent; its magnitude, however, is the ratio of V_r to Z_O, i.e., R_T is already accounted for in the magnitude of V_r. The relationships between incident and reflected signals are represented in Figure 9-2 for both cases of mismatch between R_T and Z_O.

Figure 9-2 Reflections for $R_T \neq Z_O$

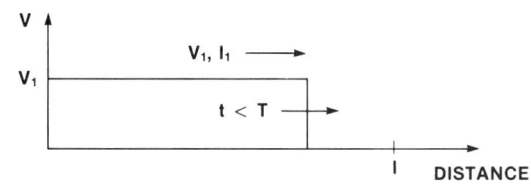

a. Incident Wave

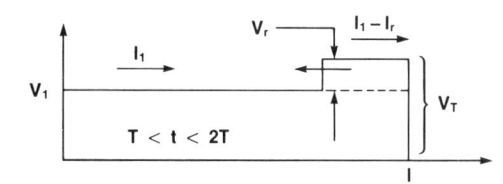

b. Reflected Wave for $R_T > Z_O$

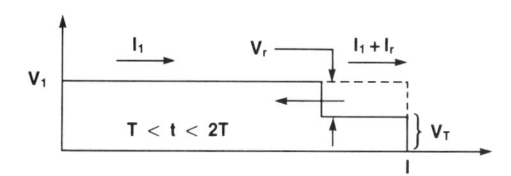

c. Reflected Wave for $R_T > Z_O$

The incident wave is shown in Figure 9-2a, before it has reached the end of the line. In Figure 9-2b, a positive V_r is returning to the generator. To the left of V_r the current is still I_1, flowing to the right, while to the right of V_r the net current in the line is the difference between I_1 and I_r. In Figure 9-2c, the reflection coefficient is negative, producing a negative V_r. This, in turn, causes an increase in the amount of current flowing to the right behind the V_r wave.

Source Impedance, Multiple Reflections

When a reflected voltage arrives back at the source (generator), the reflection coefficient at the source determines the response to V_r. The coefficient of reflection at the source is governed by Z_O and the source resistance R_S.

$$\rho_S = \frac{R_S - Z_O}{R_S + Z_O}$$

(E9-10)

If the source impedance matches the line impedance, a reflected voltage arriving at the source is not reflected back toward the load end. Voltage and current on the line are stable with the following values.

$$V_T = V_1 + V_r, \text{ and } I_T = I_1 - I_r$$

(E9-11)

If neither source impedance nor terminating impedance matches Z_O, multiple reflections occur; the voltage at each end of the line comes closer to the final steady state value with each succeeding reflection. An example of a line mismatched on both ends is shown in Figure 9-3. The source is a step function of 1V amplitude occurring at time t_0. The initial value of V_1 starting down the line is 0.75V due to the voltage divider action of Z_O and R_S. The time scale in the photograph shows that the line delay is approximately 6ns. Since neither end of the line is terminated in its characteristic impedance, multiple reflections occur.

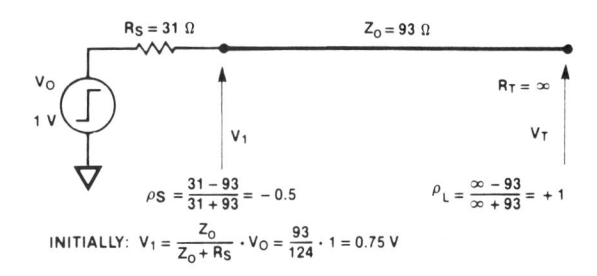

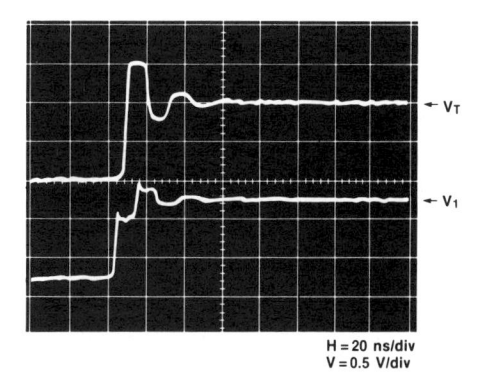

$\rho_S = \frac{31 - 93}{31 + 93} = -0.5 \qquad \rho_L = \frac{\infty - 93}{\infty + 93} = +1$

INITIALLY: $V_1 = \frac{Z_O}{Z_O + R_S} \cdot V_O = \frac{93}{124} \cdot 1 = 0.75 \text{ V}$

H = 20 ns/div
V = 0.5 V/div

Figure 9-3 Multiple Reflections Due to Mismatch at Load and Source

The amplitude and persistence of the ringing shown in Figure 9-3 become greater with increasing mismatch between the line impedance and source and load impedances. Reducing R_S (Figure 9-3) to 13Ω increases ρ_S to -0.75, and the effects are illustrated in Figure 9-4. The initial value of V_T is 1.8V with a reflection of 0.9V from the open end. When this reflection reaches the source, a reflection of $(0.9) \times (-0.75V)$ starts back toward the open end. Thus, the second increment of voltage arriving at the open end is negative-going. In turn, a negative-going reflection of $(0.9) \times (-0.75V)$ starts back toward the source. This negative increment is again multiplied by $-0.75V$ at the source and returned toward the open end. It can be deduced that the difference in amplitude between the first two positive peaks observed at the open end is

$$V_T - V'_T = (1 + \rho_L)V_1 - (1 + \rho_L)V_1 \rho_L^2 \rho_S^2$$

$$= (1 + \rho_L)V_1 (1 - \rho_L^2 \rho_S^2).$$

(E9-12)

The factor $(1 - \rho_L^2 \rho_S^2)$ is similar to the damping factor associated with lumped constant circuitry. It expresses the attenuation of successive positive or negative peaks of ringing.

Lattice Diagram

In the presence of multiple reflections, keeping track of the incremental waves on the line and the net voltage at the ends becomes a bookkeeping chore. A convenient and systematic method of indicating the conditions which combines magnitude, polarity and time utilizes a graphic construction called a lattice diagram. A lattice diagram for the line conditions of Figure 9-3 is shown in Figure 9-5.

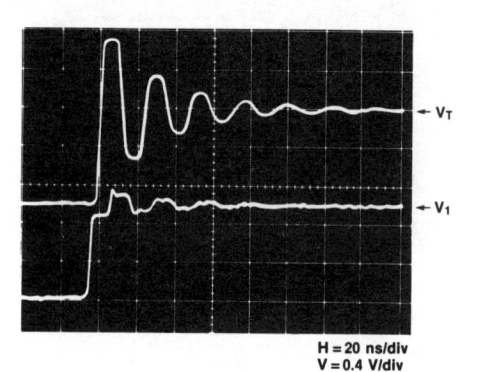

H = 20 ns/div
V = 0.4 V/div

Figure 9-4 Extended Ringing when R_S of Figure 9-3 is Reduced to 13Ω

The vertical lines symbolize discontinuity points, in this case the ends of the line. A time scale is marked off on each line in increments of 2T, starting at t_0 for V_1 and T for V_T. The diagonal lines indicate the incremental voltages traveling between the ends of the line; solid lines are used for positive voltages and dashed lines for negative. It is helpful to write the reflection and transmission multipliers ρ and $(1 + \rho)$ at each vertical line, and to tabulate the incremental and net voltages in columns alongside the vertical lines. Both the lattice diagram and the waveform photograph show that V_1 and V_T asymptotically approach 1V, as they must with a 1V source driving an open-ended line.

Shorted Line

The open-ended line in Figure 9-3 has a reflection coefficient of +1 and the successive reflections tend toward the steady state conditions of zero line current and a line voltage equal to the source voltage. In contrast, a shorted line has a reflection coefficient of -1 and successive reflections must cause the line conditions to approach the steady state conditions of zero voltage and a line current determined by the source voltage and resistance.

Shorted line conditions are shown in Figure 9-6a with the reflection coefficient at the source end of the line also negative. A negative coefficient at both ends of the line means that any voltage approaching either end of the line is reflected in the opposite polarity. Figure 9-6b shows the response to an input step-function with a duration much longer than the line delay. The initial voltage starting down the line is about $+0.75$ V, which is inverted at the shorted end and returned toward the source as -0.75 V. Arriving back at the source end of the line, this voltage is multiplied by $(1 + \rho_S)$, causing a $-0.37V$ net change in V_1. Concurrently, a reflected voltage of $+0.37V$ ($-0.75V$ times ρ_S of -0.5) starts back toward the shorted end of the line. The voltage at V_1 is reduced by 50% with each successive round trip of reflections, thus leading to the final condition of zero volts on the line.

When the duration of the input pulse is less than the delay of the line, the reflections observed at the source end of the line constitute a train of negative pulses, as shown in Figure 9-6c. The amplitude decreases by 50% with each successive occurrence as it did in Figure 9-6b.

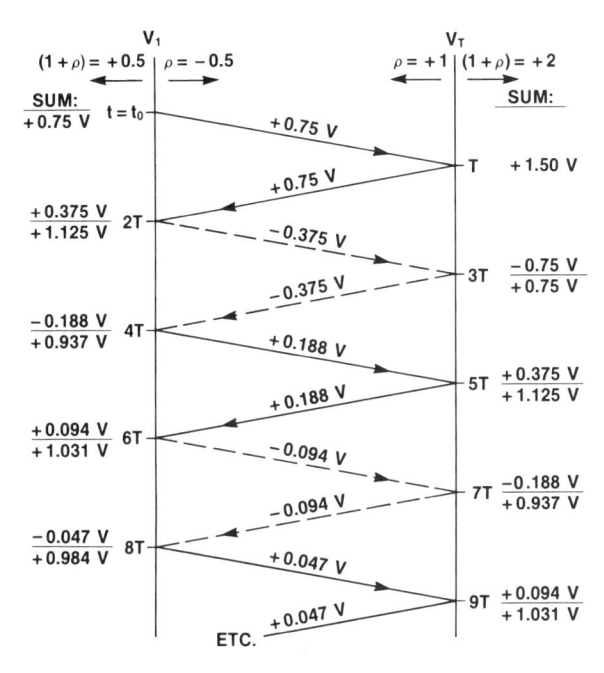

Figure 9-5 Lattice Diagram for the Circuit of Figure 9-3

Series Termination

Driving an open-ended line through a source resistance equal to the line impedance is called series termination. It is particularly useful when transmitting signals which originate on a PC board and travel through the backplane to another board, with the attendant discontinuities, since reflections coming back to the source are absorbed and ringing thereby controlled. Figure 9-7 shows a 93Ω line driven from a 1V generator through a source impedance of 93Ω. The photograph illustrates that the amplitude of the initial signal sent down the line is only half of the generator voltage, while the voltage at the open end of the line is doubled to full amplitude $(1 + \rho_L = 2)$. The reflected voltage arriving back at the source raises V_1 to the full amplitude of the generator signal. Since the reflection coefficient at the source is zero, no further changes occur and the line voltage is equal to the generator voltage. Because the initial signal on the line is only half the normal signal swing, the loads must be connected at or near the end of the line to avoid receiving a 2-step input signal.

A TTL output driving a series-terminated line is severely limited in its fanout capabilities due to the IR drop associated with the collective I_{IL} drops of the inputs being driven. For most TTL families other than FAST it should not be considered since either the input currents are so high (TTL, S, H) or the input threshold is very low (LS). In either case the noise margins are severely degraded to the point where the circuit becomes unusable. In FAST, however, the I_{IL} of 0.6mA, if sunk through a resistor of 93Ω used as a series terminating resistor, will reduce the low level noise margin 55.8mV for each standard FAST input driven.

Figure 9-6 Reflections of Long and Short Pulses on a Shorted Line

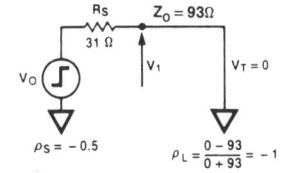

a. Reflection Coefficients for Shorted Line

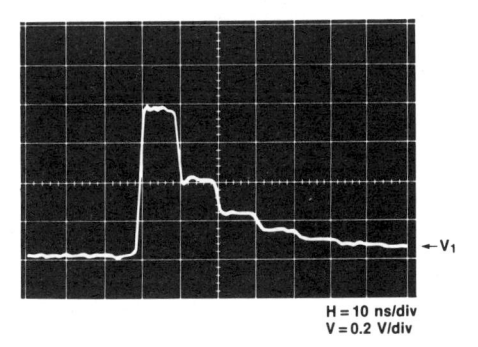

b. Input Pulse Duration >> Line Delay

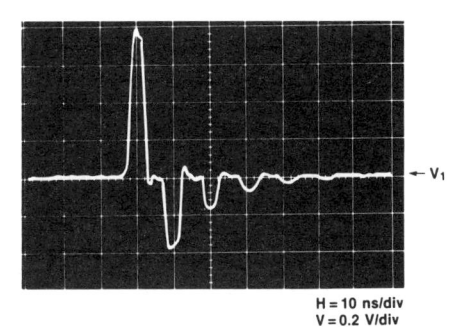

c. Input Pulse Duration < Line Delay

Figure 9-7 Series-Terminated Line Waveforms

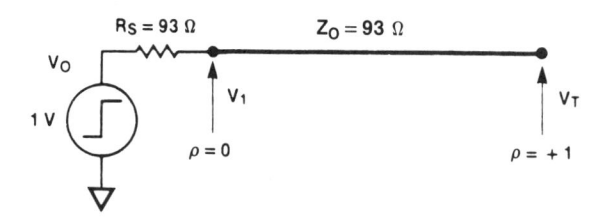

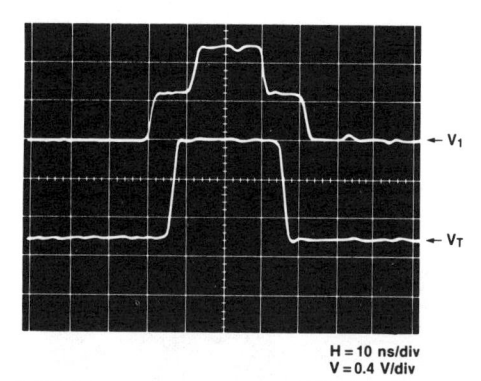

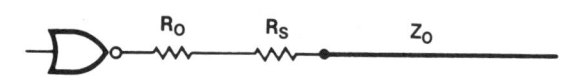

Figure 9-9 Extra Delay with Termination Capacitance

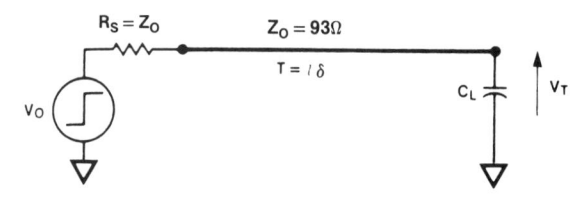

a. Series-Terminated Line with Load Capacitance

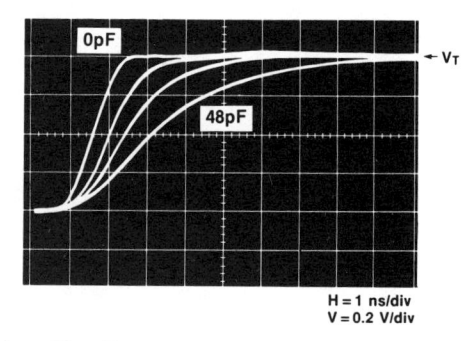

Figure 9-8 TTL Element Driving a Series-Terminated Line

b. Output Rise Time Increase with Increasing Load Capacitance

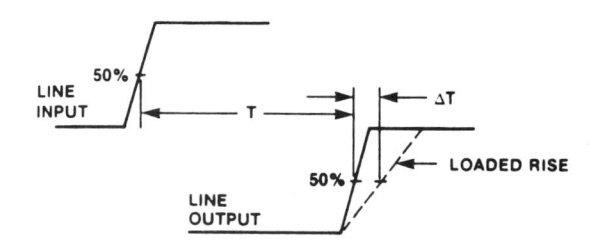

c. Extra Delay ΔT Due to Rise Time Increase

Extra Delay with Termination Capacitance

Designers should consider the effect of the load capacitance at the end of the line when using series termination. Figure 9-9 shows how the output waveform changes with increasing load capacitance. Figure 9-9b shows the effect of load capacitances of 0, 12, 24, 48pF. With no load, the delay between the 50% points of the input and output is just the line delay T. A capacitive load at the end of the line causes an extra delay ΔT due to the increase in rise time of the output signal. The midpoint of the signal swing is a good approximation of the FAST threshold since $V_{OL} = 0.5V$ and $V_{OH} = 2.5V$ and the actual input switching threshold of FAST is 1.5V at 25°C.

The increase in propagation delay can be calculated by using a ramp approximation for the incident voltage and characterizing the circuit as a fixed impedance in series with the load capacitance, as shown in Figure 9-10. One general solution serves both series and parallel termination cases by using an impedance Z' and a time constant, τ, defined in Figure 9-10a and 9-10b. Calculated and observed increases in delay time to the 50% point show close agreement when τ is less than half the ramp time. At large ratios of τ/a (where a = ramp time), measured delays exceed calculated values by approximately 7%. Figure 9-11, based on measured values, shows the increase in delay to the 50% point as a function of the Z'C time constant, both normalized to the 10% to 90% rise time of the input signal. As an example of using the graph, consider a 100Ω series-terminated line with 30pF load capacitance at the end of the line. The 3ns rise time assumed is typical of FAST in an actual line driving application. From Figure 9-10a, Z' is equal to 100Ω; the ratio Z'C t_r is 1. From the graph, the ratio $\Delta T/t_r$ is 0.8. Thus the increase in the delay to the 50% point of the output waveform is 0.8 t_r, or 2.4ns, which is then added to the no-load line delay T to determine the total delay.

Had the 100Ω line in the foregoing example been parallel rather than series terminated at the end of the line, Z' would be 50Ω. The added delay would be only 1.35ns with the same 30pF loading at the end. The added delay would be only 0.75ns if the line were 50Ω and parallel-terminated. The various trade-offs involving type of termination, line impedance, and loading are important considerations for critical delay paths.

Figure 9-10 Determining the Effect of End-of-Line Capacitance

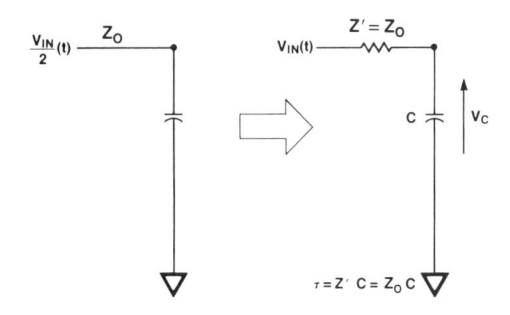

Figure 9-11 Increase in 50% Point Delay Due to Capacitive Loading at the End of the Line, Normalized to t_r

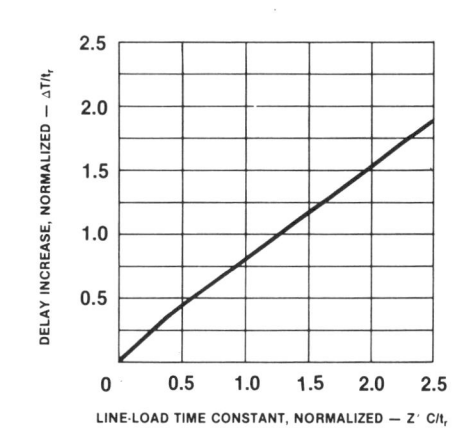

a. Thevenin Equivalent for Series-Terminated Case

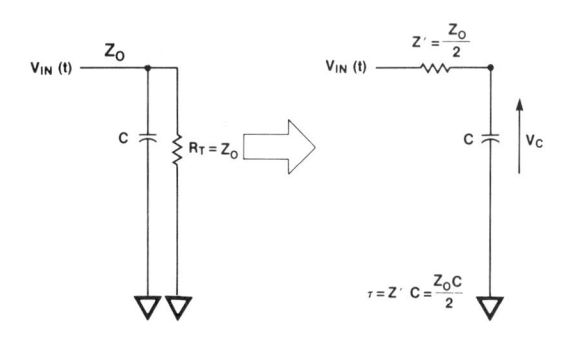

b. Thevenin Equivalent for Parallel-Terminated Case

Distributed Loading Effects on Line Characteristics

When capacitive loads such as TTL inputs are connected along a transmission line, each one causes a reflection with a polarity opposite to that of the incident wave. Reflections from two adjacent loads tend to overlap if the time required for the incident wave to travel from one load to the next is equal to or less than the signal rise time. Figure 9-12a illustrates an arrangement for observing the effects of capacitive loading, while Figure 9-12b shows an incident wave followed by reflections from two capacitive loads. The two capacitors causing the reflections are separated by a distance requiring a travel time of 1ns. The two reflections return to the source 2ns apart, since it takes 1ns longer for the incident wave to reach the second capacitor and an additional 1ns for the second reflection to travel back to the source. In the upper trace of Figure 9-12b, the input signal rise time is 1ns and there are two distinct reflections, although the trailing edge of the first overlaps the leading edge of the second. The input rise time is longer in the middle trace, causing a greater overlap. In the lower trace, the 2ns input rise time causes the two reflections to merge and appear as a single reflection which is relatively constant (at $\approx -10\%$) for half its duration. This is about the same reflection that would occur if the 93Ω line had a middle section with an impedance reduced to 75Ω.

With a number of capacitors distributed all along the line of Figure 9-12a, the combined reflections modify the observed input waveform as shown in the top trace of Figure 9-12c. The reflections persist for a time equal to the 2-way line delay (15ns), after which the line voltage attains its final value. The waveform suggests a line terminated with a resistance greater than its characteristic impedance ($R_T > Z_O$). This analogy is strengthened by observing the effect of reducing R_T from 93Ω to 75Ω which leads to the middle waveform of Figure 9-12c. Note that the final (steady state) value of the line voltage is reduced by about the same amount as that caused by the capacitive reflections. In the lower trace of Figure 9-12c the source resistance R_S is reduced from 93Ω to 75Ω, restoring both the initial and final line voltage values to the same amplitude as the final value in the upper trace. From the standpoint of providing a desired signal voltage on the line and impedance matching at either end, the effect of distributed capacitive loading can be treated as a reduction in line impedance.

Figure 9-12 Capacitive Reflections and Effects on Line Characteristics

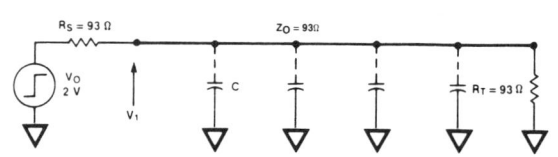

a. Arrangement for Observing Capacitive Loading Effects

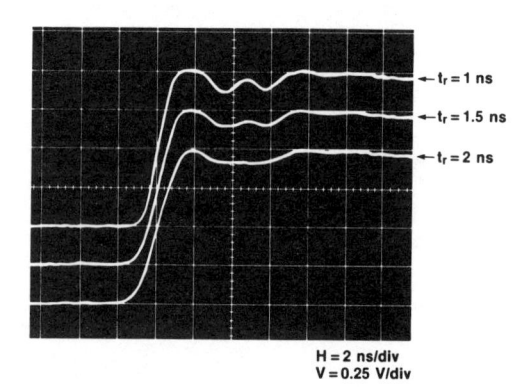

H = 2 ns/div
V = 0.25 V/div

b. Capacitive Reflections Merging as Rise Time Increases

H = 5 ns/div
V = 0.25 V/div

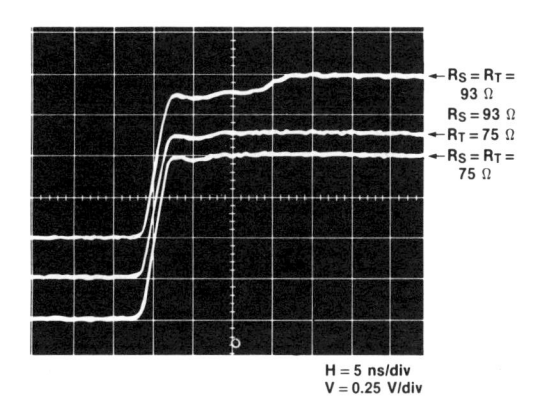

← $R_S = R_T = 93\ \Omega$
$R_S = 93\ \Omega$
← $R_T = 75\ \Omega$
← $R_S = R_T = 75\ \Omega$

c. Matching the Altered Impedance of a Capacitively Loaded Line

The reduced line impedance can be calculated by considering the load capacitance C_L as an increase in the intrinsic line capacitance C_O along that portion of the line where the loads are connected. Denoting this length of line as l, the distributed value C_D of the load capacitance is as follows:

$$C_D = \frac{C_L}{l}$$

C_D is then added to C_O in Equation 9-1 to determine the reduced line impedance Z_O.

(E9-13)
$$Z_O' = \sqrt{\frac{L_O}{C_O + C_D}} = \sqrt{\frac{L_O}{C_O\left(1 + \frac{C_D}{C_O}\right)}}$$

$$Z_O' + \frac{\sqrt{\frac{L_O}{C_O}}}{\sqrt{1 + \frac{C_D}{C_O}}} = \frac{Z_O}{\sqrt{1 + \frac{C_D}{C_O}}}$$

In the example of Figure 9-12c, the total load capacitance (lC_O) is 60pF. Note that the ratio C_D/C_O is the same as C_L/lC_O. The calculated value of the reduced impedance is thus

(E9-14)
$$Z_O' = \frac{93}{\sqrt{1 + \frac{33}{60}}} = \frac{93}{\sqrt{1.55}} = 75\Omega$$

This correlates with the results observed in Figure 9-12c when R_T and R_S are reduced to 75Ω.

The distributed load capacitance also increases the line delay, which can be calculated from Equation 9-2.

$$\delta' = \sqrt{L_O(C_O + C_D)} = \sqrt{L_O C_O}\sqrt{1 + \frac{C_D}{C_O}} = \delta\sqrt{1 + \frac{C_D}{C_O}}$$

(E9-15)

The line used in the example of Figure 9-12c has an intrinsic delay of 6ns and a loaded delay of 7.5ns which checks with Equation 9-15.

$$l\delta' = l\delta\sqrt{1.55} = 6\sqrt{1.55} = 7.5\ ns$$

(E9-16)

Equation 9-15 can be used to predict the delay for a given line and load. The ratio C_D/C_O (hence the loading effect) can be minimized for a given loading by using a line with a high intrinsic capacitance C_O.

A plot of Z' and δ' for a 50Ω line as a function of C_D is shown in Figure 9-13. This figure illustrates that relatively modest amounts of load capacitance will add appreciably to the propagation delay of a line. In addition, the characteristic impedance is reduced significantly.

Figure 9-13 Capacitive Loading Effects on Line Delay and Impedance

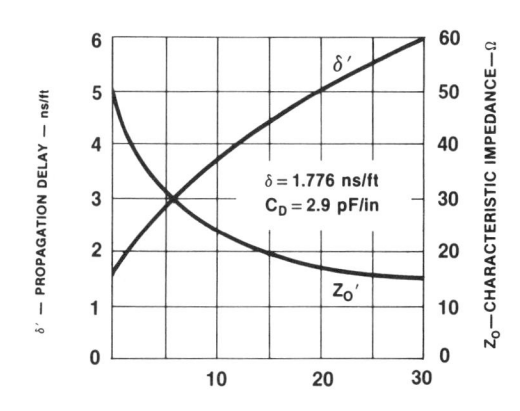

$\delta = 1.776$ ns/ft
$C_D = 2.9$ pF/in

δ' — PROPAGATION DELAY — ns/ft
Z_O — CHARACTERISTIC IMPEDANCE — Ω
C_D — DISTRIBUTED CAPACITANCE — pF/in

9-12

Worst case reflections from a capacitively loaded section of transmission line can be accurately predicted by using the modified impedance of Equation 9-9. When a signal originates on an unloaded section of line, the effective reflection coefficient is as follows:

(E9-17) $$\rho = \frac{Z_O' - Z_O}{Z_O' + Z_O}$$

Mismatched Lines

Reflections occur not only from mismatched load and source impedances but also from changes in line impedance. These changes could be caused by bends in coaxial cable, unshielded twisted-pair in contact with metal, or mismatch between PC board traces and backplane wiring. With the coax or twisted-pair, line impedance changes run about 5 to 10% and reflections are usually no problem since the percent reflection is roughly half the percent change in impedance. However, between PC board and backplane wiring, the mismatch can be 2 or 3 to 1. This is illustrated in Figure 9-14 and analyzed in the lattice diagram of Figure 9-15. Line 1 is driven in the series-terminated mode so that reflections coming back to the source are absorbed.

The reflection and transmission at the point where impedances differ are determined by treating the downstream line as though it were a terminating resistor. For the example of Figure 9-14, the reflection coefficient at the intersection of lines 1 and 2 for a signal traveling to the right is as follows:

(E9-18) $$\rho_{12} = \frac{Z_2 - Z_1}{Z_2 + Z_1} = \frac{93 - 50}{143} = +0.3$$

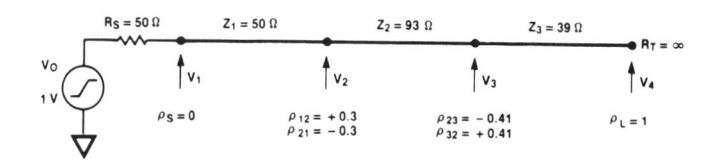

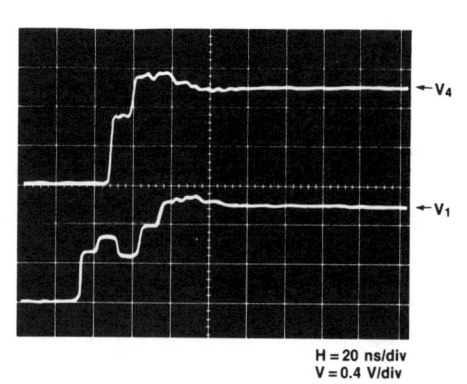

Figure 9-14 Reflections from Mismatched Lines

Thus the signal reflected back toward the source and the signal continuing along line 2 are, respectively, as follows:

$$V_{1r} = \rho_{12} V_1 = +0.3 V_1$$

$$V_2 = (1 + \rho_{12}) V_1 = +1.3 V_1$$

(E9-19)

At the intersection of lines 2 and 3, the reflection coefficient for signals traveling to the right is determined by treating Z_3 as a terminating resistor.

$$\rho_{23} = \frac{Z_3 - Z_2}{Z_3 + Z_2} = \frac{39 - 93}{132} = -0.41$$

(E9-20)

When V_2 arrives at this point, the reflected and transmitted signals are as follows:

$$V_{2r} = \rho_{23} V_2 = -0.41 V_2$$
$$= (-0.41)(1.3) V_1)$$
$$= -0.53 V_1$$

(E9-21a)

$$V_3 = (1 + \rho_{23}) V_2 = 0.59 V_2$$
$$= (0.59)(1.3) V_1$$
$$= 0.77 V_1$$

(E9-21b)

Voltage V_3 is doubled in magnitude when it arrives at the open-ended output, since ρ_L is $+1$. This effectively cancels the voltage divider action between R_S and Z_1.

$$V_4 = (1 + \rho_L) V_3 = (1 + \rho_L)(1 + \rho_{23}) V_2$$
$$= (1 + \rho_L)(1 + \rho_{23})(1 + \rho_{12}) V_1$$
$$= (1 + \rho_L)(1 + \rho_{23})(1 + \rho_{12}) \frac{V_O}{2}$$
$$V_4 = (1 + \rho_{23})(1 + \rho_{12}) V_O$$

(E9-22)

Thus, Equation 9-22 is the general expression for the initial step of output voltage for three lines when the input is series-terminated and the output is open-ended. Note that the reflection coefficients at the intersections of lines 1 and 2 and lines 2 and 3 in Figure 9-15 have reversed signs for signals traveling to the left. Thus the voltage reflected from the open output and the signal

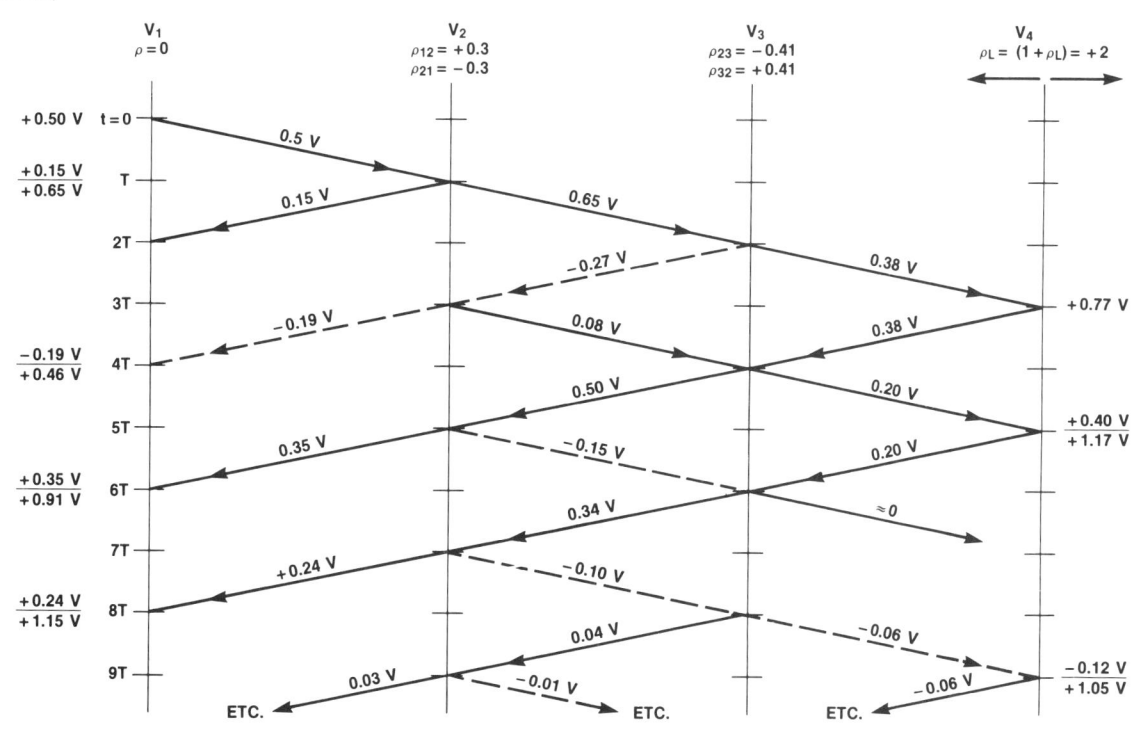

Figure 9-15 Lattice Diagram for the Circuit of Figure 9-14

reflecting back and forth on line 2 both contribute additional increments of output voltage in the same polarity as V_O. Lines 2 and 3 have the same delay time; therefore, the two aforementioned increments arrive at the output simultaneously at time 5T on the lattice diagram (Figure 9-15).

In the general case of series lines with different delay times, the vertical lines on the lattice diagram should be spaced apart in the ratio of the respective delays. Figure 9-16 shows this for a hypothetical case with delay ratios 1:2:3. For a sequence of transmission lines with the highest impedance line in the middle, at least three output voltage increments with the same polarity as V_O occur before one can occur of opposite polarity. On the other hand, if the middle line has the lowest impedance, the polarity of the second increment of output voltage is the opposite of V_O. The third increment of output voltage has the opposite polarity, for the time delay ratios of Figure 9-16.

When transmitting logic signals, it is important that the initial step of line output voltage pass through the threshold region of the receiving circuit, and that the next two increments of output voltage augment the initial step. Thus in a series-terminated sequence of three mismatched lines, the middle line should have the highest impedance.

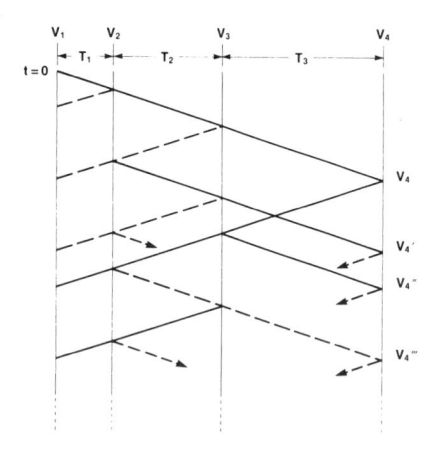

Figure 9-16 Lattice Diagram for Three Lines with Delay Ratios 1:2:3

Rise Time Versus Line Delay

When the 2-way line delay is less than the rise time of the input wave, any reflections generated at the end of the line are returned to the source before the input transition is completed. Assuming that the generator has a finite source resistance, the reflected wave adds algebraically to the input wave while it is still in transition, thereby changing the shape of the input. This effect is illustrated in Figure 9-17, which shows input and output voltages for several comparative values of rise time and line delay.

In Figure 9-17b where the rise time is much shorter than the line delay, V_1 rises to an initial value of 1V. At time T later, V_T rises to 0.5 V, i.e., $1 + \rho_L = 0.5$. The negative reflection arrives back at the source at time 2T, causing a net change of -0.4V, i.e., $(1 + \rho_S)(-0.5) = -0.4$.

The negative coefficient at the source changes the polarity of the other 0.1V of the reflection and returns it to the end of the line, causing V_T to go positive by another 50 mV at time 3T. The remaining 50 mV is inverted and reflected back to the source, where its effect is barely distinguishable as a small negative change at time 4T.

In Figure 9-17c, the input rise time (0 to 100%) is increased to such an extent that the input ramp ends just as the negative reflection arrives back at the source end. Thus the input rise time is equal to 2T.

The input rise time is increased to 4T in Figure 9-17d, with the negative reflection causing a noticeable change in input slope at about its midpoint. This change in slope is more visible in the double exposure photo of Figure 9-17e, which shows V_1 (t_r still set for 4T) with and without the negative reflection. The reflection was eliminated by terminating the line in its characteristic impedance.

The net input voltage at any particular time is determined by adding the reflection to the otherwise unaffected input. It must be remembered that the reflection arriving back at the input at a given time is proportional to the input voltage at a time 2T earlier. The value of V_1 in Figure 9-17d can be calculated by starting with the 1V input ramp.

$$V_1 = \frac{1}{t_r} \cdot t \text{ for } 0 \leq t \leq 4T$$
$$= 1 \text{ V} \quad \text{for } t \geq 4T$$

(E9-23)

Figure 9-17 Line Voltages for Various Ratios of Rise Time to Line Delay

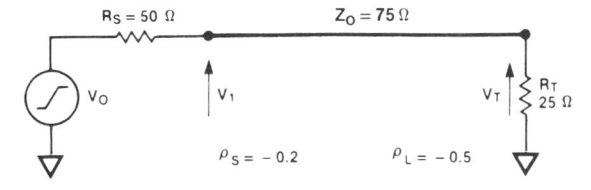

a. Test Arrangement for Rise Time Analysis

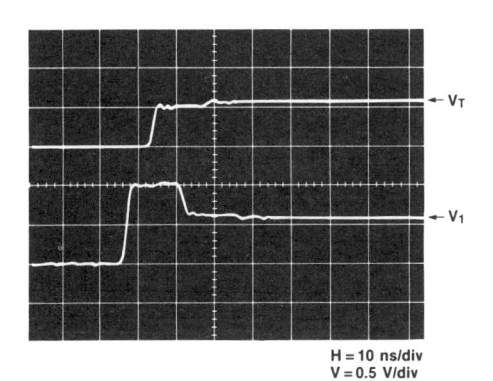

b. Line Voltages for $t_r << T$

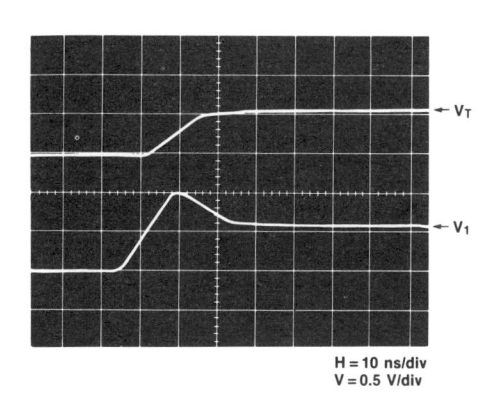

c. Line Voltages for $t_r = 2T$

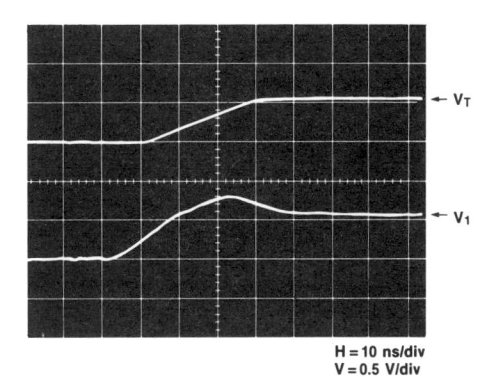

d. Line Voltages for $t_r = 4T$

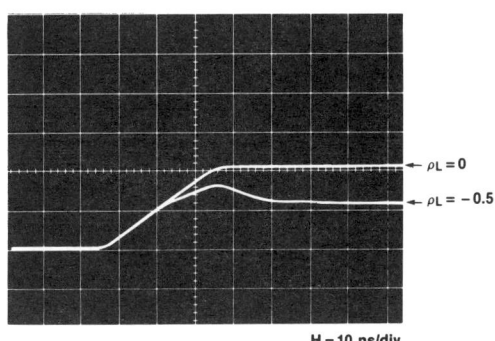

e. Input Voltage With and Without Reflection

The reflection from the end of the line is

$$V_r = \frac{\rho_L(t-2T)}{t_r};$$
(E9-24)

the portion of the reflection that appears at the input is

$$V'_r = \frac{(1+\rho_S)\,\rho_L\,(t-2T)}{t_r};$$
(E9-25)

the net value of the input voltage is the sum.

$$V'_1 = \frac{t}{t_r} + \frac{(1+\rho_S)\,\rho_L\,(t-2T)}{t_r}$$
(E9-26)

The peak value of the input voltage in Figure 9-17d is determined by substituting values and letting t equal 4T.

$$V'_1 = \frac{(0.8)\,(-0.5)\,(4T-2T)}{t_r}$$
(E9-27)
$$= 1 - .04\,(0.5) = 0.8\ V$$

After this peak point, the input ramp is no longer increasing but the reflection is still arriving. Hence the net value of the input voltage decreases. In this example, the later reflections are too small to be detected and the input voltage is thus stable after time 6T. For the general case of repeated reflections, the net voltage $V_{1(t)}$ seen at the driven end of the line can be expressed as follows, where the signal caused by the generator is $V_{1(t)}$:

$$V'_{1(t)} = V_{1(t)}$$
$$\text{for } 0 < t < 2T$$

$$V'_{1(t)} = V_{1(t)} + (1+\rho_S)\, \rho_L\, V_{1(t-2T)}$$
$$\text{for } 2T < t < 4T$$

$$V'_{1(t)} = V_{1(t)} + (1+\rho_S)\, \rho_L\, V_{1(t-2T)}$$
$$+ (1+\rho_S)\, \rho_S \rho_L^2\, V_{1(t-4T)}$$
$$\text{for } 4T < t < 6T$$

$$V'_{1(t)} = V_{1(t)} + (1+\rho_S)\, \rho_L\, V_{1(t-2T)}$$
$$+ (1+\rho_S)\, \rho_S \rho_L^2\, V_{1(t-4T)}$$
$$+ (1+\rho_S)\, \rho_S^2 \rho_L^3\, V_{1(t-6T)}$$
$$\text{for } 6T < t < 8T, \text{ etc.}$$

(E9-28)

The voltage at the output end of the line is expressed in a similar manner.

$$V_{T(t)} = 0$$
$$\text{for } 0 < t < T$$

$$V_{T(t)} = (1+\rho_L)\, V_{1(t-T)}$$
$$\text{for } T < t < 3T$$

$$V_{T(t)} = (1+\rho_L)\, V_{1(t-T)}$$
$$+ (1+\rho_L)\, \rho_S \rho_L\, V_{1(t-3T)}$$
$$\text{for } 3T < t < 5T$$

$$V_{T(t)} = (1+\rho_L)\, V_{1(t-T)}$$
$$+ (1+\rho_L)\, \rho_S \rho_L\, V_{1(t-3T)}$$
$$+ (1+\rho_L)\, \rho_S^2 \rho_L^2\, V_{1(t-5T)}$$
$$\text{for } 5T < t < 7T, \text{ etc.}$$

(E9-29)

Ringing

Multiple reflections occur on a transmission line when neither the signal source impedance nor the termination (load) impedance matches the line impedance. When the source reflection coefficient ρ_S and the load reflection coefficient ρ_L are of opposite polarity, the reflections alternate in polarity. This causes the signal voltage to oscillate about the final steady state value, commonly recognized as ringing.

When the signal rise time is long compared to the line delay, the signal shape is distorted because the individual reflections overlap in time. The basic relationships among rise time, line delay, overshoot and undershoot are shown in a simplified diagram, Figure 9-18. The incident wave is a ramp of amplitude B and rise duration A. The reflection coefficient at the open-ended line output is +1 and the source reflection coefficient is assumed to be −0.8, i.e., $R_O = Z_O/9$.

Figure 9-18b shows the individual reflections treated separately. Rise time A is assumed to be three times the line delay T. The time scale reference is the line output and the first increment of output voltage V_O rises to 2B in the time interval A. Simultaneously, a positive reflection (not shown) of amplitude B is generated and travels to the source, whereupon it is multiplied by −0.8 and returns toward the end of the line. This negative-going ramp starts at time 2T (twice the line delay) and doubles to −1.6B at time 2T + A.

Figure 9-18 Basic Relationships Involved in Ringing

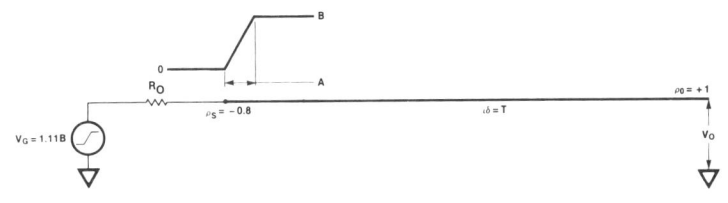

a. Ramp Generator Driving Open-Ended Line

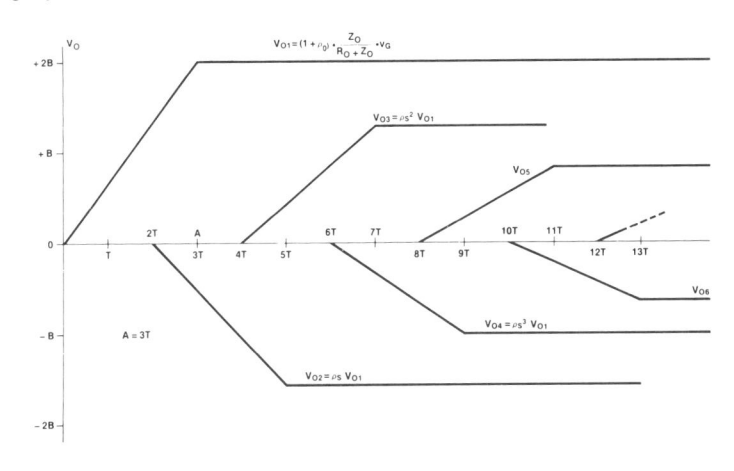

b. Increments of Output Voltage Treated Individually

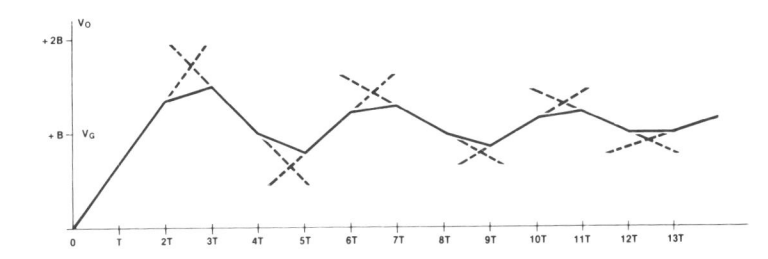

c. Net Output Signal Determined by Superposition

The negative-going increment also generates a reflection of amplitude −0.8B which makes the round trip to the source and back, appearing at time 4T as a positive ramp rising to +1.28B at time 4T + A. The process of reflection and re-reflection continues, and each successive increment changes in polarity and has an amplitude of 80% of the preceding increment.

In Figure 9-18c, the output increments are added algebraically by superposition. The starting point of each increment is shifted upward to a voltage value equal to the algebraic sum of the quiescent levels of all the preceding increments (i.e., 0, 2B, 0.4B, 1.68B, etc.). For time intervals when two ramps occur simultaneously, the two linear functions add to produce a third ramp that prevails during the overlap time of the two increments.

It is apparent from the geometric relationships, that if the ramp time A is less than twice the line delay, the first output increment has time to rise to the full 2B amplitude and the second increment reduces the net

9-18

output voltage to 0.4B. Conversely, if the line delay is very short compared to the ramp time, the excursions about the final value V_G are small.

Figure 9-18c shows that the peak of each excursion is reached when the earlier of the two constituents ramps reaches its maximum value, with the result that the first peak occurs at time A. This is because the earlier ramp has a greater slope (absolute value) than the one that follows.

Actual waveforms such as produced by TTL do not have a constant slope and do not start and stop as abruptly as the ramp used in the example of Figure 9-18. Predicting the time at which the peaks of overshoot and undershoot occur is not as simple as with ramp excitation. A more rigorous treatment is required, including an expression for the driving waveform which closely simulates its actual shape. In the general case, a peak occurs when the sum of the slopes of the individual signal increment is zero.

Summary

The foregoing discussions are by no means an exhaustive treatment of transmission line characteristics. Rather, they are intended to focus attention on the general methods used to determine the interactions between high-speed logic circuits and their interconnections. Considering an interconnection in terms of distributed rather than lumped inductance and capacitance leads to the line impedance concept, i.e., mismatch between this characteristic impedance and the terminations causes reflections and ringing.

Series termination provides a means of absorbing reflections when it is likely that discontinuities and/or line impedance changes will be encountered. A disadvantage is that the incident wave is only one-half the signal swing, which limits load placement to the end of the line. TTL input capacitance increases the rise time at the end of the line, thus increasing the effective delay. With parallel termination, i.e., at the end of the line, loads can be distributed along the line. TTL input capacitance modifies the line characteristics and should be taken into account when determining line delay.

Line Driving

All interconnects, such as coaxial cable, defined impedance transmission lines and feeders, can be considered as transmission lines, whereas printed circuit traces and hook-up wire tend to be ignored as transmission lines. With any high-speed logic family, all interconnects should be considered as transmission lines, and evaluated as such to see if termination is required. Of the many properties of transmission lines, two are of major interest to us: Z_O (the effective equivalent resistive value that causes zero reflection) and t_{PD} (propagation delay down the transmission line). Both of these parameters are geometry dependent. Here are some common configurations:

Printed Circuit Configurations

h = dielectric thickness
c = trace thickness
L = trace length
K = dielectric thickness between ground planes
b = trace width
ϵ_r = dielectric constant

Figure 9-19 Micro Stripline

(E9-30) $$Z_O = \frac{87}{\sqrt{\epsilon_r + 1.41}} \ln\left(\frac{5.98h}{0.8\,b+c}\right) \; \Omega$$

$$t_{PD} = 1.017 \sqrt{0.475\,\epsilon_r + 0.67} \; \text{ns/ft}.$$

Figure 9-20 Stripline

$$Z_O = \frac{60}{\sqrt{\epsilon_r}} \ln\left(\frac{4K}{0.67\,\pi\,b\left(0.8 + \frac{c}{b}\right)}\right) \; \Omega$$

(E9-31) $t_{PD} = 1.017 \sqrt{\epsilon_r}$ ns/ft.

Figure 9-21 Side by Side

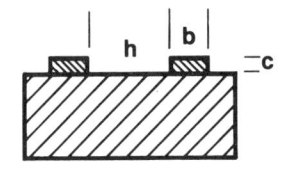

$$Z_O = \frac{120}{\sqrt{\epsilon_r}} \ln\left(\frac{\pi h}{b+c}\right) \ \Omega$$

(E9-32) $\quad t_{PD} = 1.017\sqrt{0.475\,\epsilon_r + 0.67}\ \text{ns/ft}$

Figure 9-22 Flat Parallel Conductors for $b>>h$ and $h>>c$

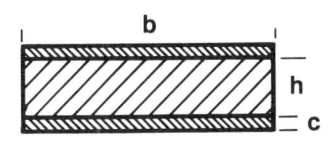

$$Z_O = \frac{377}{\sqrt{\epsilon_r}} \ln\left(\frac{h}{b}\right) \ \Omega$$

(E9-33) $\quad t_{PD} = 1.017\sqrt{0.475\,\epsilon_r + 0.67}\ \text{ns/ft}$

Figure 9-23 Wiring

a. Wire over Ground Plane

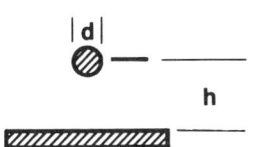

(E9-34) $\quad Z_O = \frac{60}{\sqrt{\epsilon_r}} \ln\left(\frac{4h}{d}\right) \ \Omega$

b. Coaxial Cable

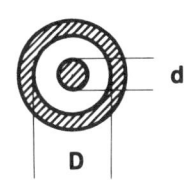

(E9-35) $\quad Z_O = \frac{60}{\sqrt{\epsilon_r}} \ln\left(\frac{D}{d}\right) \ \Omega$

Figure 9-24 Twisted Pair or Ribbon Cable

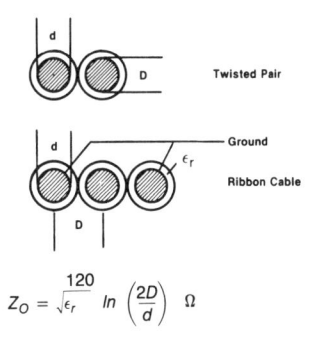

(E9-36) $\quad Z_O = \frac{120}{\sqrt{\epsilon_r}} \ln\left(\frac{2D}{d}\right) \ \Omega$

All of the above rely on the complex relationship

$$Z_O = \sqrt{\frac{R_O + j\omega L_O}{G_O + j\omega C_O}} \ \Omega$$

(E9-37)

and can be simplified to

$$Z_O = \sqrt{\frac{L_O}{C_O}}$$

if we assume

$$G_O \simeq R_O = 0$$

Note that Z_O is real, not complex, appears resistive and is not a function of length.

Also,

$$t_{PD} = \sqrt{L_O C_O}$$

(E9-38)

The inductance of PC trace can be determined by the formula

$$L_O = \left[0.0051\ \ln\left(\frac{4h}{\alpha}\right)\right] + .00127\ \mu H/\text{inch}$$

(E9-39) $\quad$ where $\alpha = \sqrt{\frac{4\,bc}{\pi}}$

For power and ground planes in a multilayer board, the capacitance of the plane can be calculated by the formula for parallel plates separated by a dielectric:

$$C = 0.2212\,\frac{\epsilon_r A}{h}\ \text{pF}$$

(E9-40)

where A = surface area of one plate.

The above formula (E9-40) cannot be used to calculate PC trace capacitance. This must either be measured or an appropriate value may be taken from the following curves.

The impedance of striplines and microstriplines can be found quickly from the following curves. For characteristics of cables, refer to manufacturers' data.

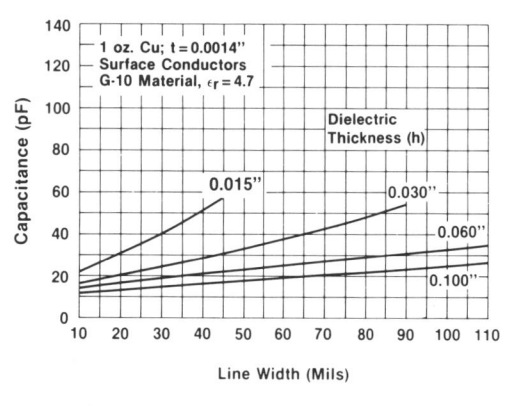

Figure 9-25 Capacitance of Microstriplines

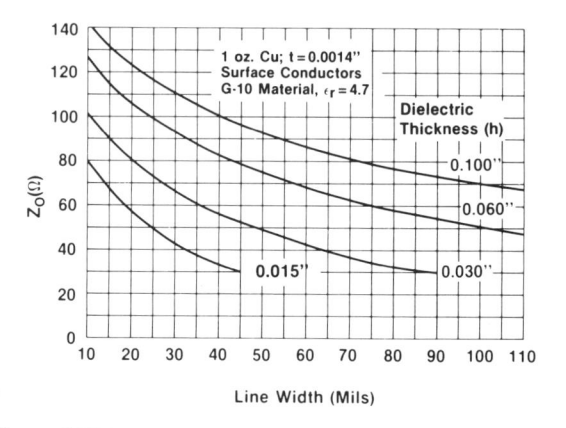

Figure 9-27 Impedance of Microstriplines

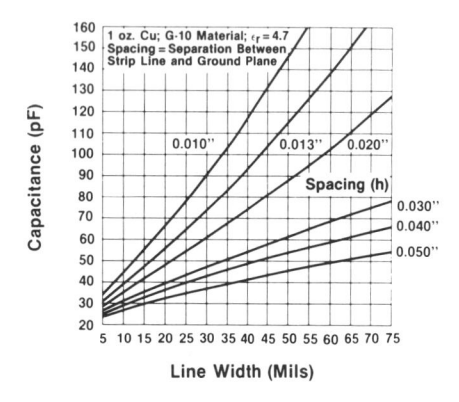

Figure 9-26 Capacitance of Striplines

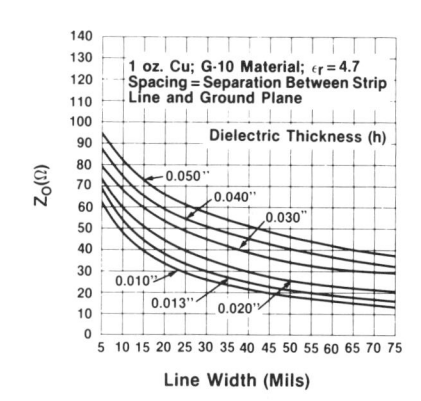

Figure 9-28 Impedance of Striplines

Table 9-1
Relative Dielectric Constants of Various Materials

Material	ϵ_r
Air	1.0
Polyethylene foam	1.6
Cellular polyethylene	1.8
Teflon	2.1
Polyethylene	2.3
Polystyrene	2.5
Nylon	3.0
Silicon rubber	3.1
Polyvinylchloride (PVC)	3.5
Epoxy resin	3.6
Delrin	3.7
Epoxy glass	4.7
Mylar	5.0
Polyurethane	7.0

All the above information on impedance and propagation delays are for the circuit interconnect only. The actual impedance and propagation delays will differ from this by the loading effects of gate input and output capacitances, and by any connectors that may be in line. The effective impedance and propagation delay can be determined from the following formula:

$$Z_O' = \dfrac{Z_O}{\sqrt{1+\left(\dfrac{C_L}{C_O}\right)}}\ \Omega$$

$$t_{PD} = \sqrt{L_O C_O}\ \therefore\ t_{PD}' = t_{PD}\sqrt{1+\left(\dfrac{C_L}{C_O}\right)}$$

(E9-41)

where C_L is the total of all additional loading.

The results of these formulas will frequently give effective impedances of less than half Z_O, and interconnect propagation delays greater than the driving device propagation delays, thus becoming the predominant delay.

Driving Transmission Lines

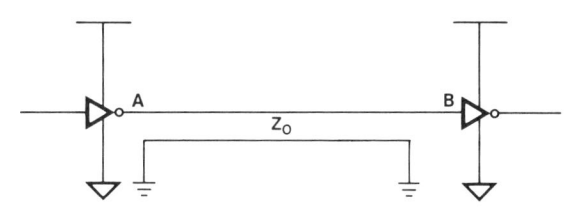

Figure 9-29
1. Unterminated

The maximum length for an unterminated line can be determined by

$$l_{max} = \dfrac{t_r}{2t_{PD}'}\quad \text{For FAST, } t_r = 3\ ns$$

$$\therefore\ l_{max} = 10 \text{ inches for trace on GI0 epoxy glass P.C.}$$

(E9-42)

The voltage wave propagated down the transmission line (V step) is the full output drive of the device into Z_O'. Reflections will not be a problem if $l \leq l_{max}$. Lines longer than l_{max} will be subject to ringing and reflections and will drive the inputs and outputs below ground.

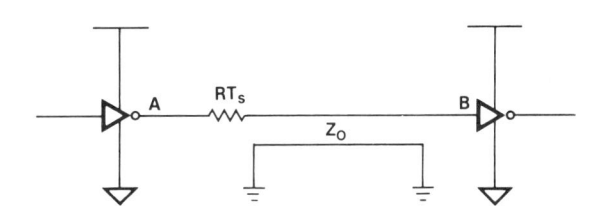

Figure 90-30
2. Series-Terminated

$RT_S = Z_O$

Series termination has limited use in TTL interconnect schemes due to the voltage drop across RT_S in the LOW state, reducing noise margins at the receiver. Series termination is the ideal termination for highly capacitive memory arrays whose DC loadings are minimal. RT_S values of 10 to 50Ω are normally found in these applications.

3. Parallel-Terminated

Four possibilities for parallel termination exist:

A. Z_O' to V_{CC}. This will consume current from V_{CC} when output is LOW;

B. Z_O' to GND. This will consume current from V_{CC} when output is HIGH;

C. Thevenin equivalent termination. This will consume half the current of A and B from the output stage, but will have reduced noise margins, and consume current from V_{CC} with outputs HIGH or LOW. If used on a 3-state bus, this will set the quiescent line voltage to half.

D. AC Termination. An RC termination to GND, $R + X_C = Z_O$, X_C to be less than 2τ of Z_O' at

$$f = \frac{1}{2t_r}$$

(E9-43)

This consumes no DC current with outputs in either state. If this is used on a 3-state bus, then the quiescent voltage on the line can be established at V_{CC} or GND by a high value pull up (down) resistor to the appropriate supply rail.

Parallel-Terminated

A. RT to V_{CC}
  RT = Z_O'

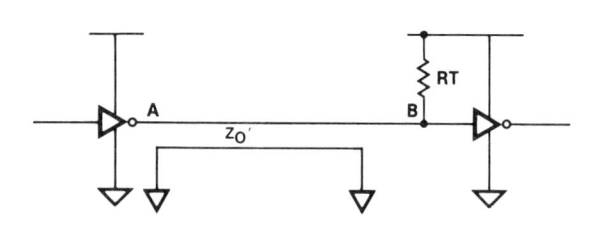

B. RT to GND
  RT = Z_O'

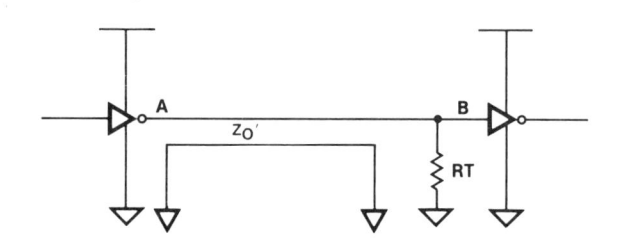

C. Thevenin Termination
  RT = $2Z_O'$

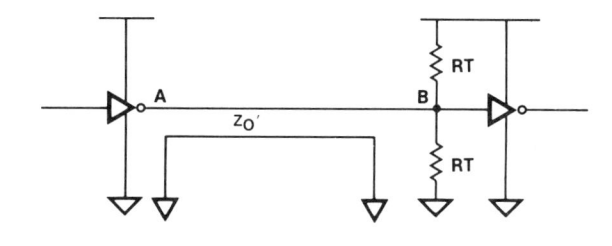

D. AC Termination to GND
  RT + X_{CT} = Z_O'

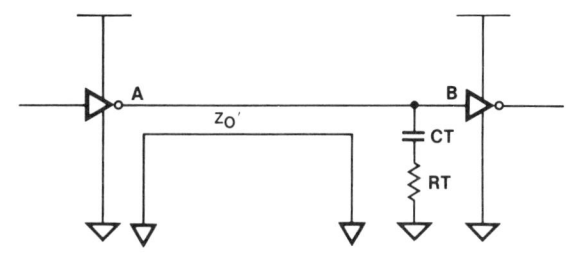

Figure 9-31

Decoupling

Typical Dynamic Impedance of Unbypassed V_{CC} Runs

I_{CC} Drain Due to Line Driving

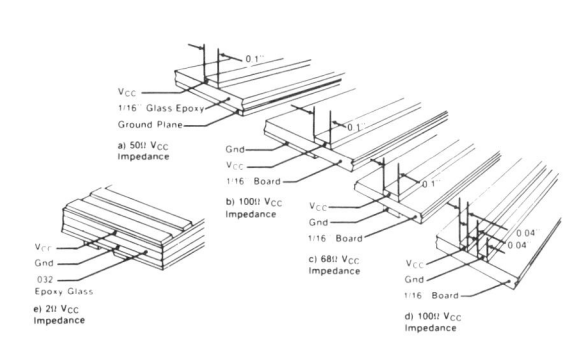

Figure 9-32

This diagram shows several schemes for power and ground distribution on logic boards. Figure 9-32 is a cross-section, with a, b, and c showing a 0.1 inch wide V_{CC} bus and ground on the opposite side. Figure 9-32d shows side-by-side V_{CC} and ground strips, each 0.04 inch wide. Figure 9-32e shows a four layer board with embedded power and ground planes.

In Figure 9-32a, the dynamic impedance of V_{CC} with respect to ground is 50Ω, even though the V_{CC} trace width is generous and there is a complete ground plane. In Figure 9-32b, the ground plane stops just below the edge of the V_{CC} bus and the dynamic impedance doubles to 100Ω. In Figure 9-32c, the ground bus is also 0.1 inch wide and runs along under the V_{CC} bus and exhibits a dynamic impedance of about 68Ω. In Figure 9-32d, the trace widths and spacing are such that the traces can run under a DIP, between two rows of pins. The impedance of the power and ground planes in Figure 9-32e is typically less than 2Ω.

These typical dynamic impedances point out why a sudden current demand due to an IC output switching can cause a momentary reduction in V_{CC}, unless a bypass capacitor is located near the IC.

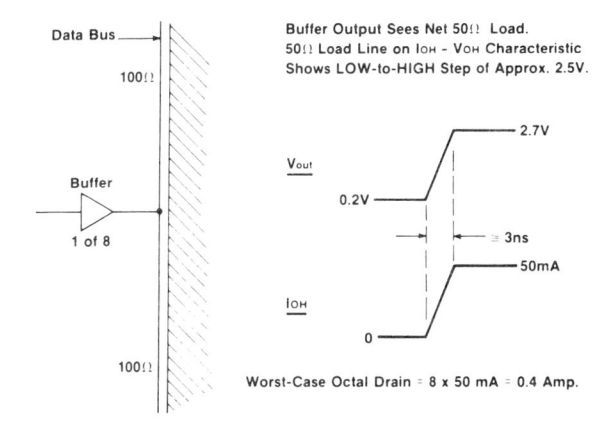

Figure 9-33

This diagram illustrates the sudden demand for current from V_{CC} when a buffer output forces a LOW-to-HIGH transition into the midpoint of a data bus. The sketch shows a wire-over-ground transmission line, but it could also be twisted pair, flat cable or PC interconnect.

The buffer output effectively sees two 100Ω lines in parallel and thus a 50Ω load. For this value of load impedance, the buffer output will force an initial LOW-to-HIGH transition from 0.2V to 2.7V in about 3ns. This net charge of 2.5V into a 50 load causes an output-HIGH current change of 50mA.

If all eight outputs of an octal buffer switch simultaneously, in this application the current demand on V_{CC} would be 0.4 Amp. Clearly, a nearby V_{CC} bypass capacitor is needed to accommodate this demand.

V$_{CC}$ Bypass Capacitor for Octal Driver

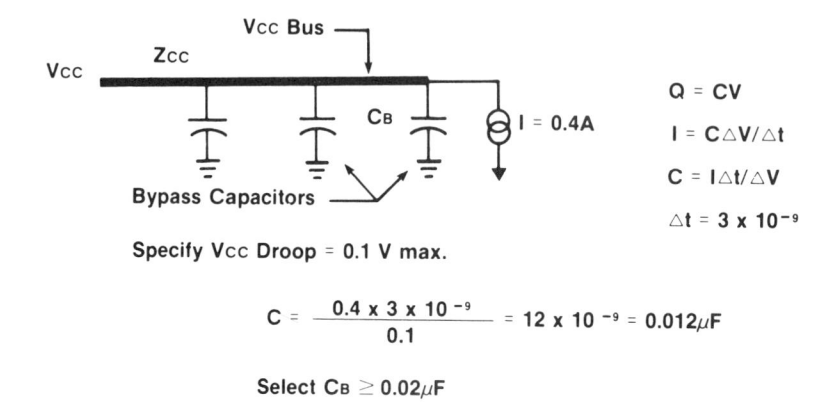

$$Q = CV$$
$$I = C\triangle V/\triangle t$$
$$C = I\triangle t/\triangle V$$
$$\triangle t = 3 \times 10^{-9}$$

Specify V$_{CC}$ Droop = 0.1 V max.

$$C = \frac{0.4 \times 3 \times 10^{-9}}{0.1} = 12 \times 10^{-9} = 0.012 \mu F$$

Select C$_B \geq 0.02 \mu F$

Place one bypass capacitor near each buffer package. Distribute other bypass capacitors evenly throughout the logic, one capacitor per two packages.

Figure 9-34

A V$_{CC}$ bus with bypass capacitors connected periodically along its length is shown above. Also shown is a current source representing the current demand of the buffer in the preceding application.

The equations illustrate an approximation method of estimating the size of a bypass capacitor based on the current demand, the drop in V$_{CC}$ that can be tolerated and the length of time that the capacitor must supply the charge. While the current demand is known, the other two parameters must be chosen. A V$_{CC}$ droop of 0.1V will not cause any appreciable change in performance, while a time duration of 3ns is long enough for other nearby bypass capacitors to help supply charge. If the current demand continues over a long period of time, charge must be supplied by a very large capacitor on the board. This is the reason for the recommendation that a large capacitor be located where V$_{CC}$ comes onto a board. If the buffers are also located near the connector end of the board, the large capacitor helps supply charge sooner.

Design Considerations

Ground—An Essential Link

With the advent of Fairchild Advanced Schottky Technology (FAST) with considerably faster edge rates and switching times, proper grounding practice has become of primary concern in printed circuit layout. Poor circuit grounding layout techniques may result in crosstalk and slowed switching rates. This reduces overall circuit performance and may necessitate costly redesign. Also when FAST chips are substituted for standard TTL-designed printed circuit boards, faster edge rates can cause noise problems. The source of these problems can be sorted into three categories:

1. V_{CC} droop due to faster load capacitance charging;
2. Coupling via ground paths adjacent to both signal sources and loads; and
3. Crosstalk caused by parallel signal paths.

V_{CC} droop can be remedied with better or more bypassing to ground. The rule here is to place $0.01\mu F$ capacitors from V_{CC} to ground for every two FAST circuits used, as near the IC as possible. The other two problems are not as easily corrected, because PC boards may already be manufactured and utilized. In this case, simply replacing TTL circuits with FAST compatible circuits is not always as easy as it may seem, especially on two-sided boards. In this situation IC placement is critical at high speeds. Also when designing high density circuit layout, a ground-plane layer is imperative to provide both a sufficiently low inductance current return path and to provide electromagnetic and electrostatic shielding thus preventing noise problem 2 and reducing, by a large degree, noise problem 3.

Illustrations

Two-Sided PC Board Layout

When considering the two-sided PC board, more than one ground trace is often found in a parallel or non-parallel configuration. For this illustration parallel traces tied together at one end are shown. This arrangement is referred to as a ground comb. The ground comb is placed on one side of the PC board while the signal traces are on the other side, thus the two-sided circuit board.

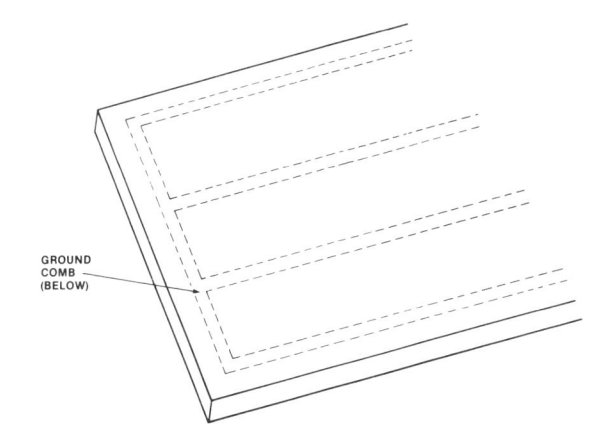

Figure 9-35

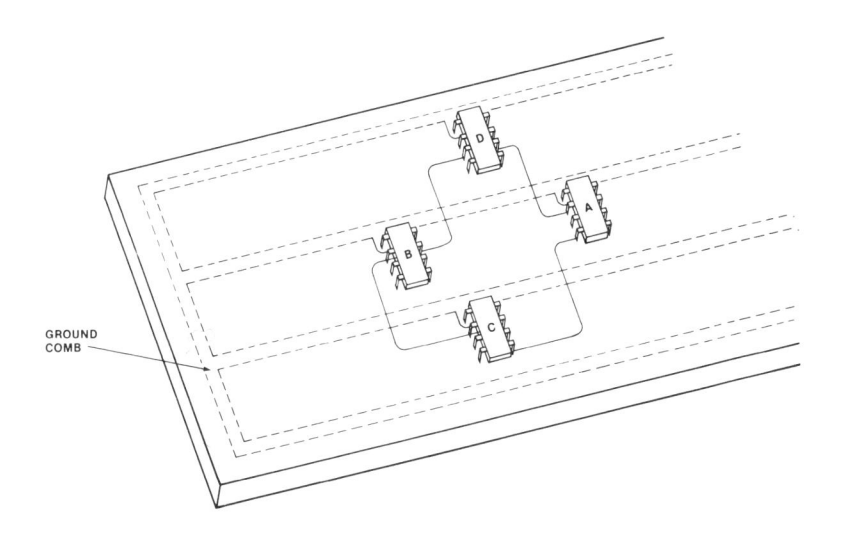

Figure 9-36

Figure 9-36 illustrates how noise is generated even though there is no apparent means of crosstalk between the circuits. If package A has an output which drives package D input and package B output drives package C input, there is no apparent path for crosstalk since mutual signal traces are remotely located. What is significant, and must be emphasized here, is that circuit packages A and B accept their ground link from the same trace. Hence, circuit A may well couple noise to circuit B via the common or shared portion of the trace. This is especially true at high switching speeds.

Ground Trace Coupling

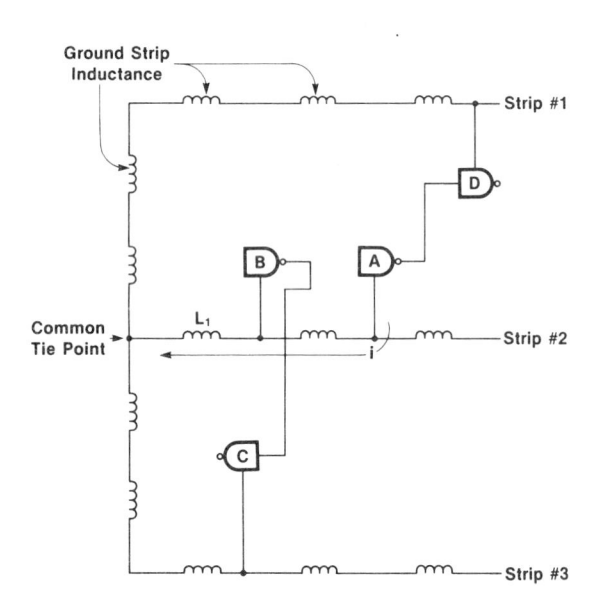

Figure 9-37

Ground trace noise coupling is illustrated by a model circuit in Figure 9-37. With the ground comb configuration, the ground strips may be shown to contain distributed inductance, as is indeed the case. Referring to the above illustration we can see that if we switch gate A from HIGH to LOW, the current for the transition is drawn from ground strip number two. Current flows in the direction indicated by the arrow to the common tie point. It can be seen that gate B shares ground strip number two with gate A from the point where gate B is grounded back to the common tie point. This length is represented by L_1. When A switches states there is a current transient which occurs on the ground strip in the positive direction. This current spike is caused by the ground strip inductance and it is "felt" by gate B. If gate B is in a LOW state (V_{OL}) the spike will appear on the output since gate B's V_{OL} level is with reference to ground. Thus if gate B's ground reference rises momentarily V_{OL} will also rise. Consequently, if gate B is output to another gate (C in the illustration) problems may arise.

Problem

System faults occur if the sum of V_{OL} quiescent level plus current spike amplitude reaches the threshold region of gate C. From this it can be seen that erroneous switching may be transmitted throughout the system. In the illustration the glitch at gate B's output is given by the following formula:

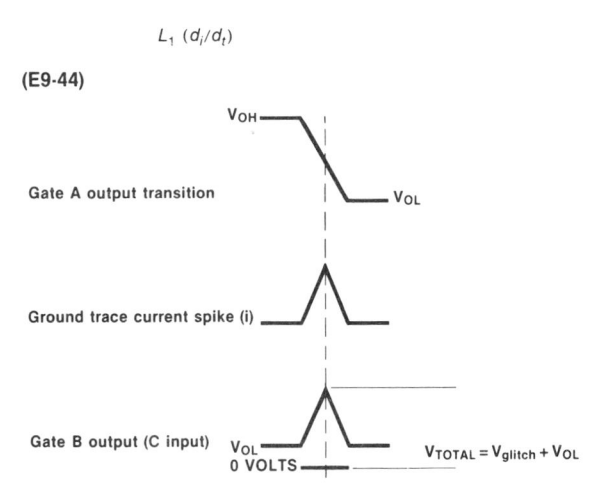

$$L_1 \, (d_i/d_t)$$

(E9-44)

Figure 9-38

Solution

The following sketch (Figure 9-39) shows one method of effectively reducing ground path length when using the ground comb layout. By using topside traces to tie the underside ground comb together we can reduce ground strip distributed inductances. Therefore, current transients are significantly smaller or nonexistent in amplitude. In the application of these topside strips, care need not be exercised in their spacing or arrangement. Parallelism is not of paramount importance either. Another advantage is evident: if one or more of these strips is placed between topside signal traces, crosstalk can be eliminated between those traces.

9-27

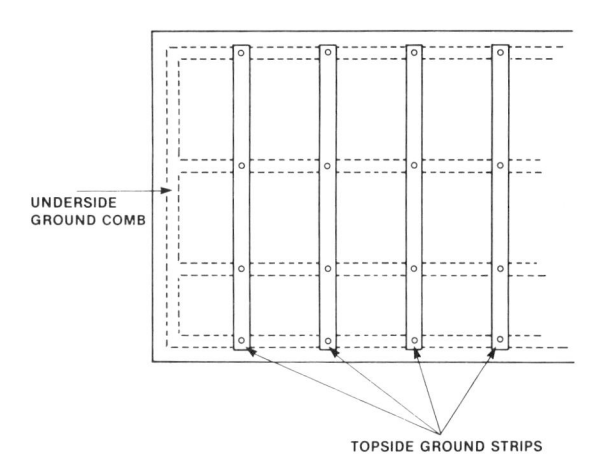

Figure 9-39

Bus Driver Packages

An area which warrants special consideration is bus driver/buffer package placement. Here we refer to products such as the 'F240, 'F241, 'F244, 'F540, 'F827 and 'F828. These units have a minimum of eight outputs. A problem may arise if all eight outputs happen to switch from HIGH to LOW or vice versa at the same time. In this case the chance for a large current transient on the ground circuit is apparent. This is possible even on short runs of ground strip (1 to 2 inches). Here, extra care is advised and it is suggested that buffer/driver groups driving backplanes be segregated to one area in the circuit. This area should have its own ground reference. Ideally it should be a ground plane configuration or contain minimal or negligible length ground trace connections.

General-Purpose Boards (Breadboard)

It is important when breadboarding, creating prototype circuits for evaluation or making special function generators, to use optimum techniques for connecting V_{CC} and ground. Breadboard-type selection is of certain consequence here and should be attended to wisely.

The best choice, when designing with high-speed logic, is board material which has power and ground already connected to circuit trace grids. Boards may offer the designer the option of using IC sockets although these are not recommended for high-speed applications. Socket layout is convenient and may be necessary when special or one-of-a-kind circuits are utilized in initial circuit arrangements. However, when designing with FAST products, consider that sockets increase total lead inductance and interlead capacitance, thus circuit performance may be adversely affected. If boards without ground and power grids must be used or if non-standard pin connections must be accommodated, the use of copper strips or braid is recommended. Copper strip is readily available as shim stock while a brand of solder wick can be used for braid material. Please note here that jumper wires must be avoided because of wire inductance. Wire inductance, like stripline inductance, will slow rise and fall times. Jumpers also promote crosstalk coupling.

Noise Decoupling

As stated earlier under "Ground—An Essential Link", it was noted that the common rule of thumb is to decouple every other FAST package with $0.01\mu F$ capacitors. This is fine for most gates in the majority of applications. However, with buffer/driver packages, decoupling should occur at each package since the possibility of all outputs switching coincidentally exists and can cause large loads on V_{CC}.

An alternative to standard decoupled power traces on the two-sided P.C. board is a product called Q/PAC*. Q/PAC is a low impedance, high capacitance power distribution system which uses wide V_{CC} and ground conductors in close proximity with ceramic insulators to effectively represent integral decoupling capacitors. Packages are available in varying lengths and pinout spacings.

* Rogers Corporation
Q/PAC Division, 5750 East McKellips Road
Mesa, AZ 85205 Telephone: 602-830-3370

Crosstalk

Crosstalk is an interference effect of an active signal line on an inactive signal line in close proximity to the active line. There are two forms of crosstalk that are of concern in system design: forward and reverse crosstalk. The causes of both both kinds are similar, but the effects are significantly different. Four possible crosstalk conditions can exist at the inactive receiver: (1) a positive pulse on a LOW, (2) a negative pulse on a LOW, (3) a positive pulse on a HIGH, (4) a negative pulse on a HIGH. Of the four previously mentioned conditions (1) and (4) are of major concern in logic systems, and (2) and (3) are less problematic. Crosstalk is caused by a number of interrelated factors which fall into two groups: mutual impedance and velocity difference.

Mutual impedance is caused by the mutual inductance and mutual capacitance distributed along two signal lines in close proximity. The electrical effects are akin to transformer action with well defined polarities. The induced crosstalk voltage pulse is of opposite polarity to the inducing pulse. Figure 9-40 shows the schematic representation.

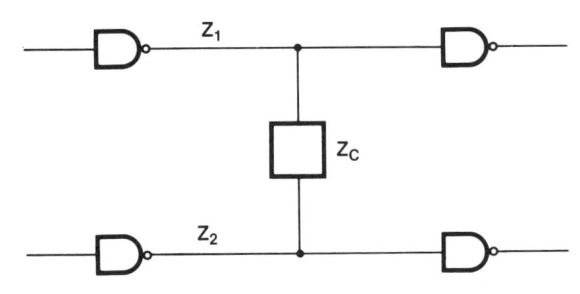

Figure 9-40

Here Z_1 and Z_2 represent the adjacent signal line impedances, and Z_C is the mutual impedance coupling the two signal lines. An equivalent circuit is shown below.

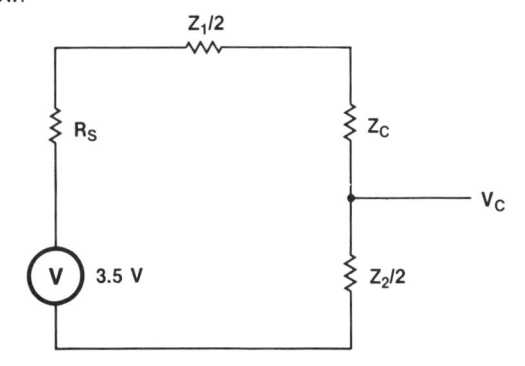

Figure 9-41

R_S is the effective source resistance; for V_{OH}, $R_S = 33\Omega$ and V_{OL}, $R_S = 3\Omega$. These are the typical FAST gate sink and source resistances. V_C is the crosstalk voltage and should be adjusted for polarity. The crosstalk voltage can be calculated with the following simplified formula:

$$V_C = \frac{Z_2/2}{R_S + Z_C + Z_1/2 + Z_2/2} \times V_{OUT}$$

(E9-45)

Velocity differences are caused when a signal propagates along a conductive medium that is in contact with substances of different dielectric constants, i.e., epoxy glass and air in printed circuit board applications. The different dielectric constants of the materials cause the wave propagating at the epoxy glass interface to be travelling slower than the wave at the air interface. This has the effect of generating a pulse that will couple electrostatically into the adjacent signal line and add to the pulse caused by mutual impedance coupling. The velocity difference pulse will have the same rise time as the signal on the active line and its duration will be twice the difference between the arrival of the wave front in air and the wave front in epoxy glass.

Forward Crosstalk

Forward crosstalk is the effect when the active driver and the driver on the non-active line are at the same end: the wave front propagates toward the active and non-active receiver simultaneously. Forward crosstalk is classically attributed almost entirely to velocity differences, but in practice it is a mixture of both velocity difference and mutual impedance effects.

Reverse Crosstalk

Reverse or backward crosstalk is the effect when the active driver and the non-active receiver are at the same end of the signal lines: the wave front propagates toward the active receiver and the non-active driver simultaneously. Reverse crosstalk is due entirely to mutual impedance effects. Forward and reverse crosstalk tests have been performed on both parallel circuit board traces and ribbon cable.

Crosstalk on PC Trace

Crosstalk on printed circuit traces exhibits both velocity difference and mutual impedance effects. This can be seen clearly in Figure 9-42. The jig, two 50Ω parallel traces, 34 inches long and 0.100 inches apart, was characterized using a 5V 3ns rise time signal from a 50Ω source and all traces terminated in 50Ω. Figures 9-43 through 9-50 show the effects of forward and reverse crosstalk on terminated and unterminated cases using the jig of Figure 9-42. All of the cases show no approach to the logic threshold on this test jig; other circuit configuarations and impedances may not act in a similar fashion and crosstalk avoidance procedures may have to be taken.

Figure 9-42 34-Inch, 50 Ω Crosstalk Jig

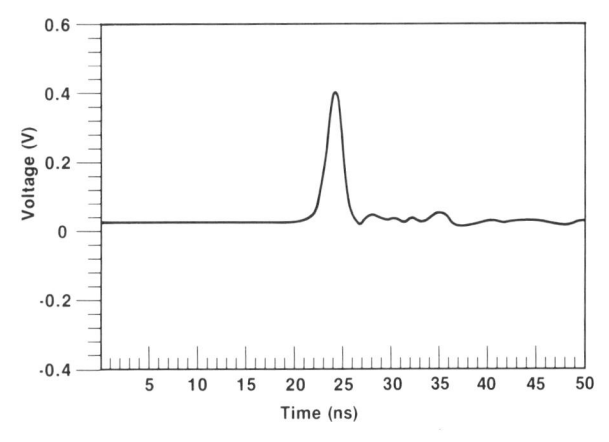

a. Velocity Difference

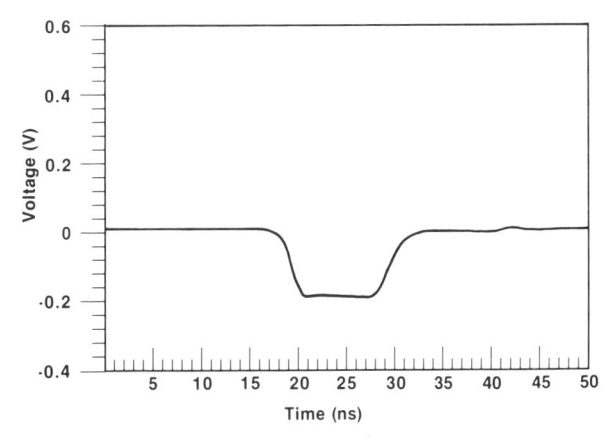

b. Mutual Impedance

Figure 9-43 Reverse PC Board Crosstalk Through 34-Inch, 0.100 Trace Unterminated

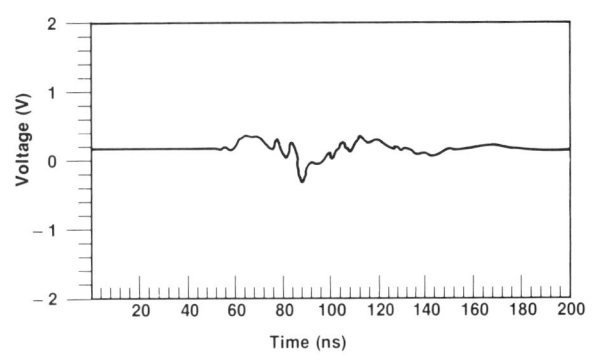

a. Adjacent Receiver

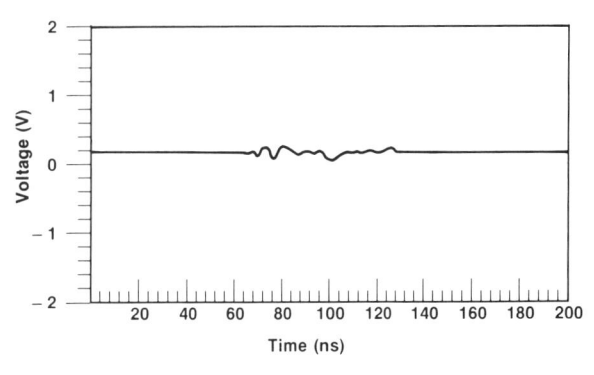

b. Adjacent Driver

Figure 9-43 (continued)

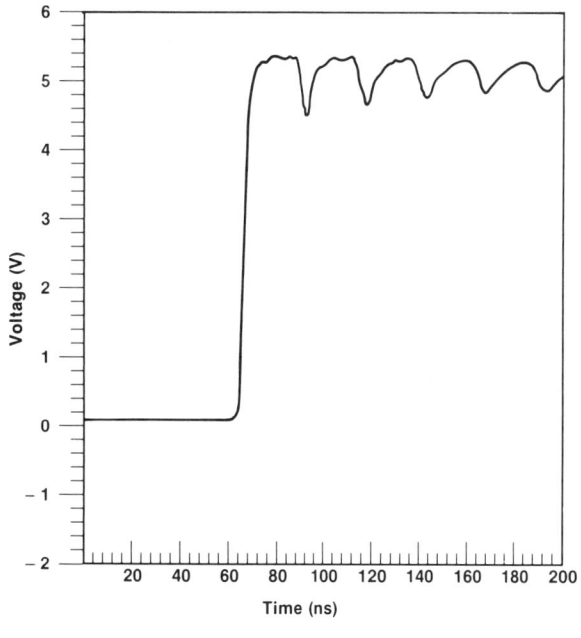

Figure 9-44 Forward PC Board Crosstalk Through 34-Inch Trace Unterminated

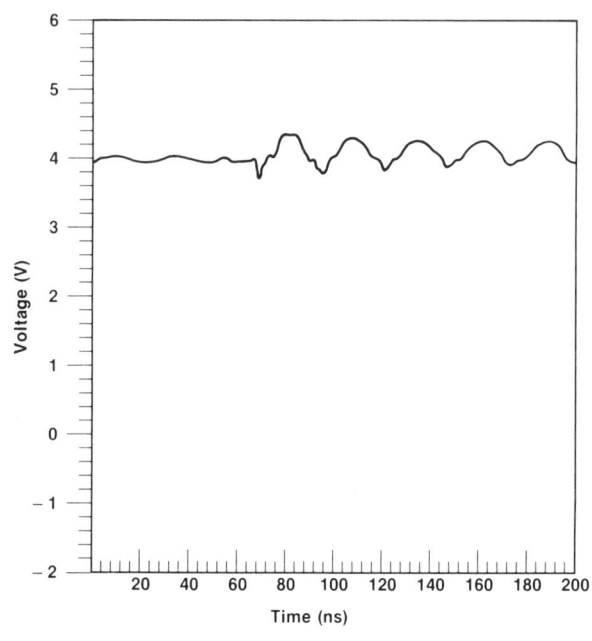

c. Active Receiver

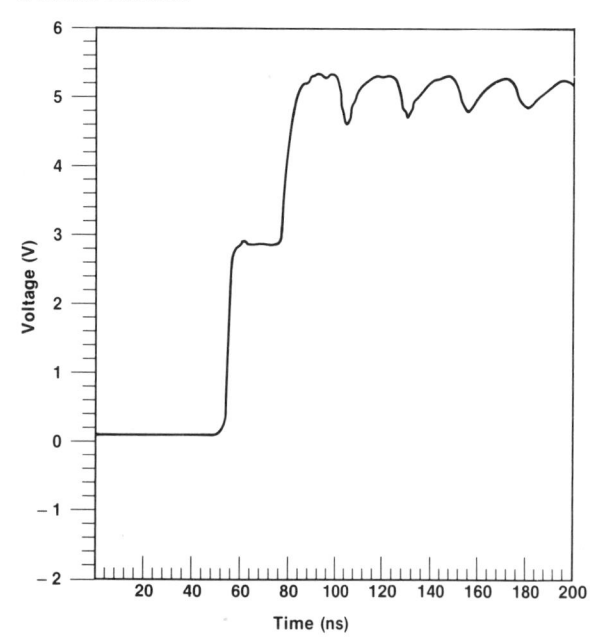

a. Adjacent Receiver

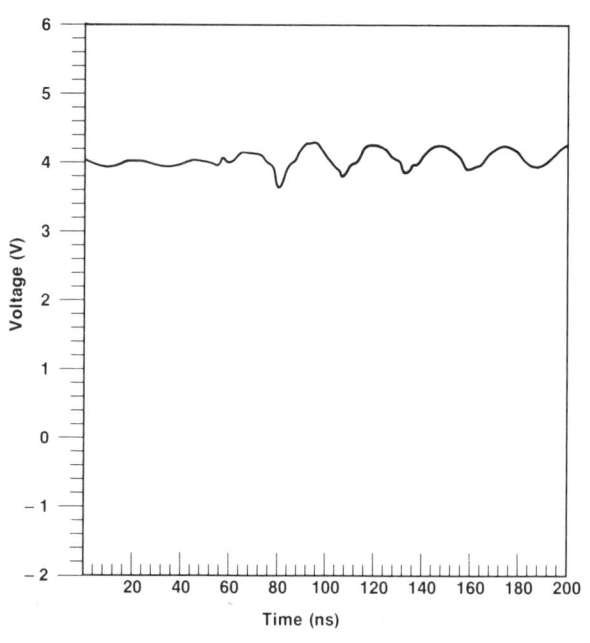

d. Active Driver

b. Adjacent Driver

Figure 9-44 (continued)

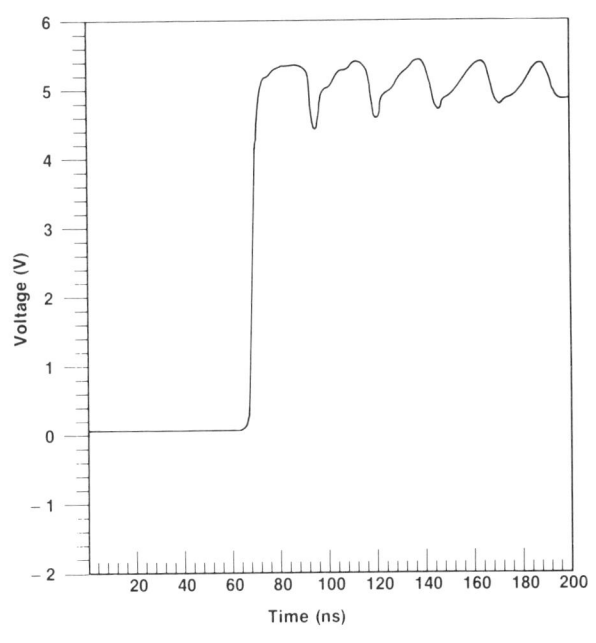

Figure 9-45 PC Board Crosstalk Through 36-Inch, 0.100 Trace, Forward with No Termination

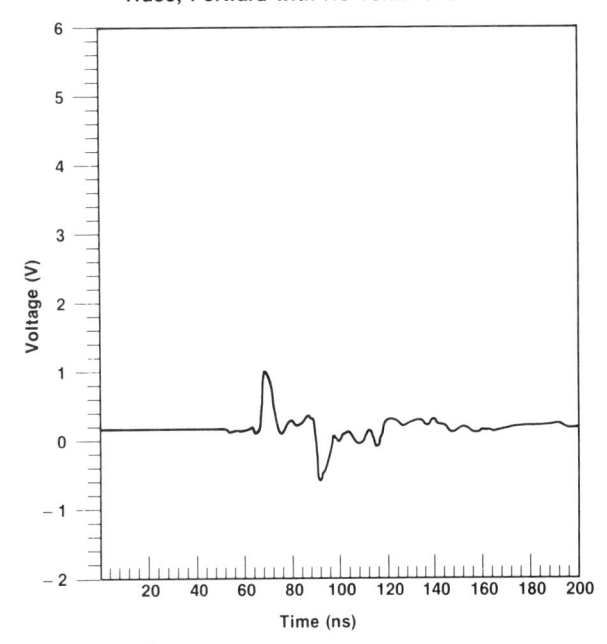

c. Active Receiver

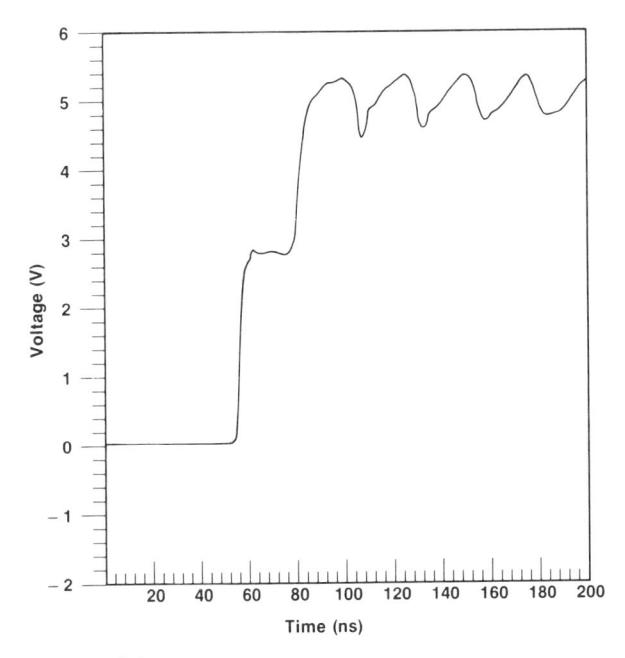

a. Adjacent Receiver

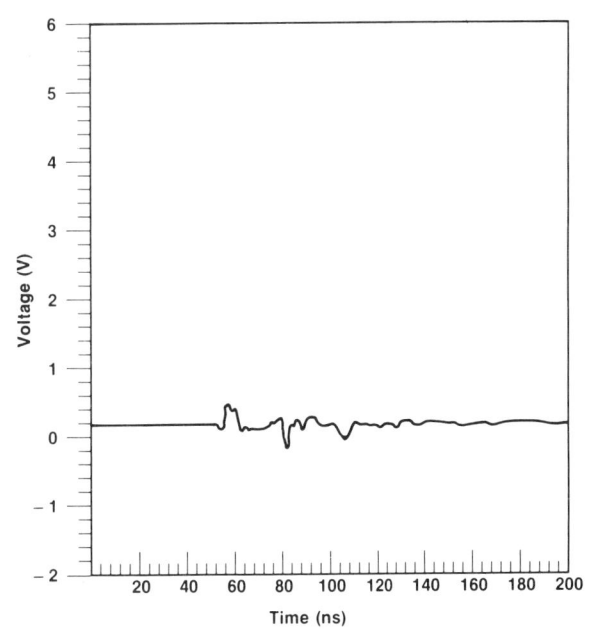

d. Active Driver

b. Adjacent Driver

Figure 9-45 (continued)

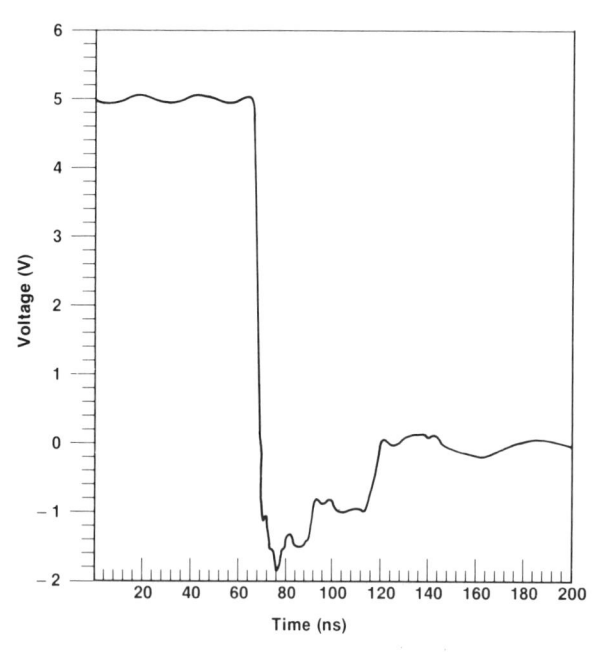

c. Active Receiver

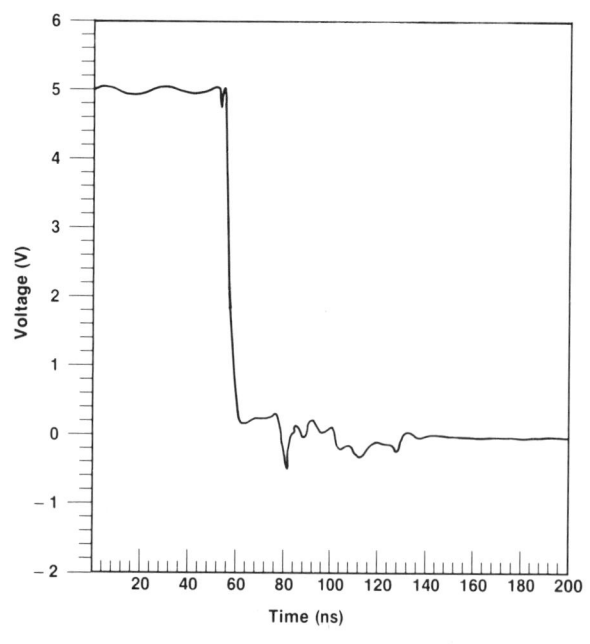

d. Active Driver

Figure 9-46 Reverse PC Board Crosstalk Through 34-Inch, 0.100 Trace Unterminated

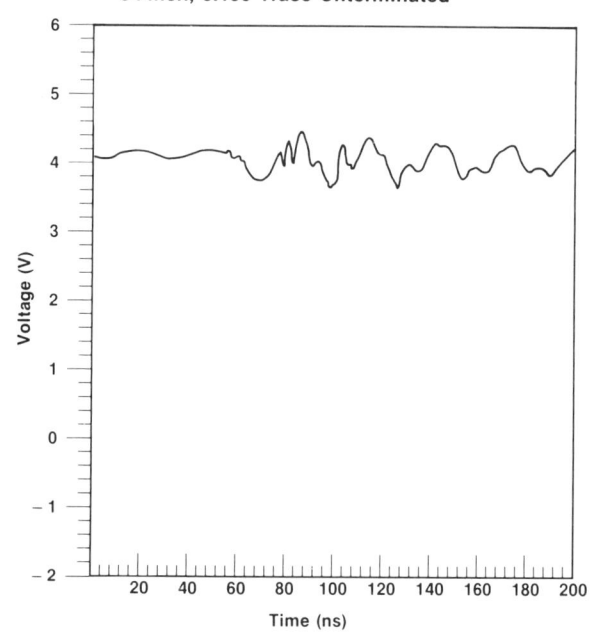

a. Adjacent Receiver

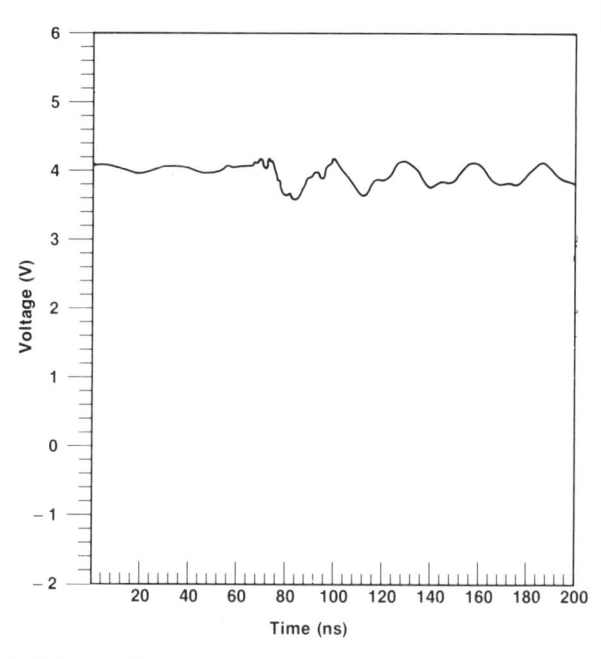

b. Adjacent Driver

9-33

Figure 9-46 (continued)

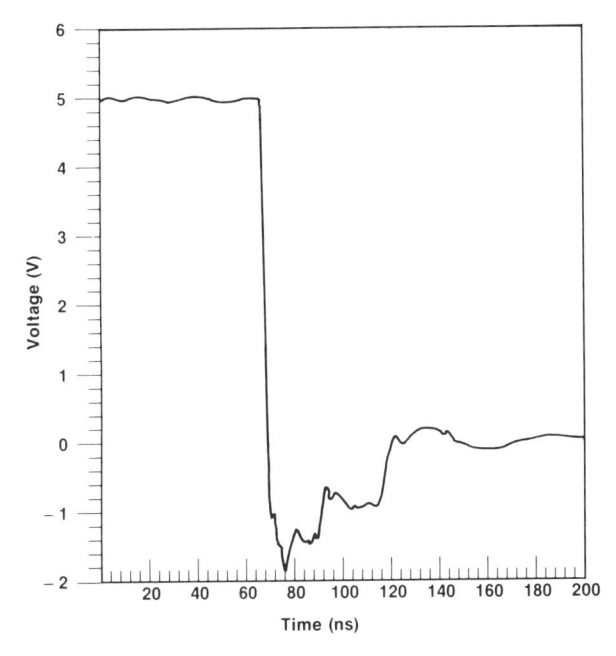

c. Active Receiver

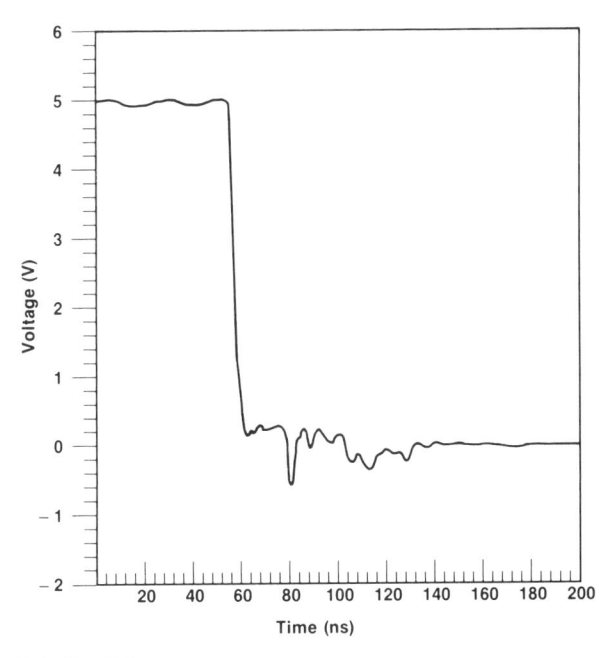

d. Active Driver

Figure 9-47 Reverse PC Board Crosstalk Through 34-Inch, 0.100 Trace Terminated

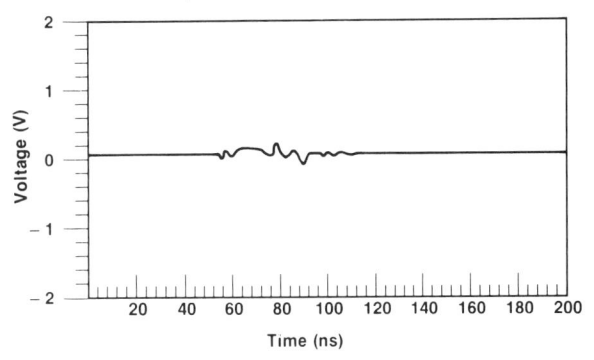

a. Adjacent Receiver

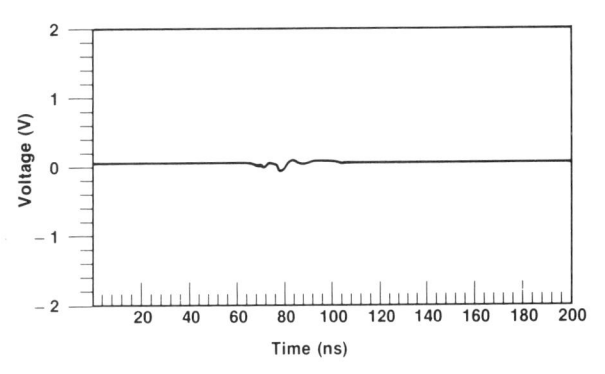

b. Adjacent Driver

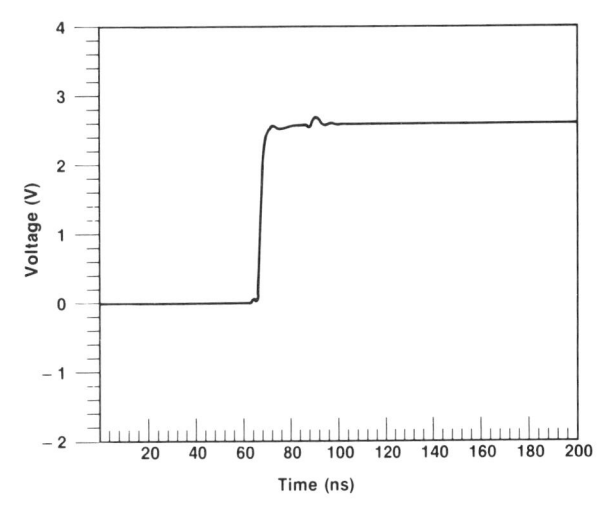

c. Active Receiver

Figure 9-47 (continued)

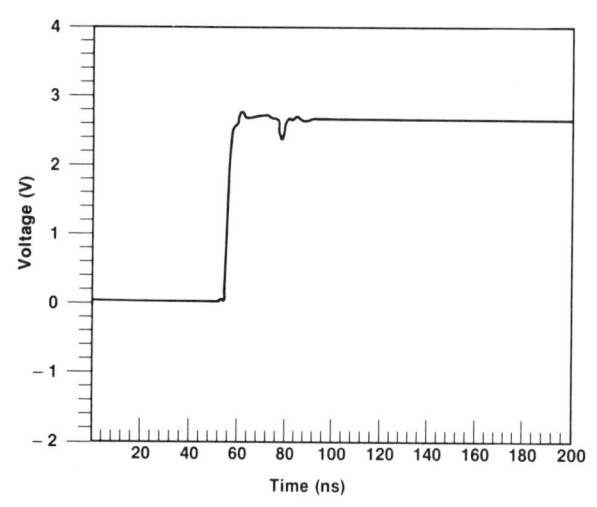

d. Active Driver

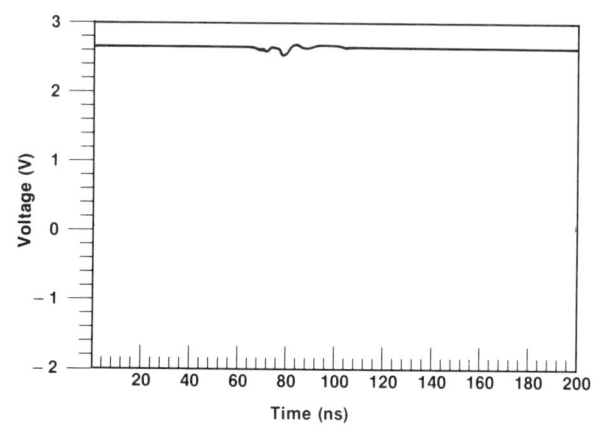

b. Adjacent Driver

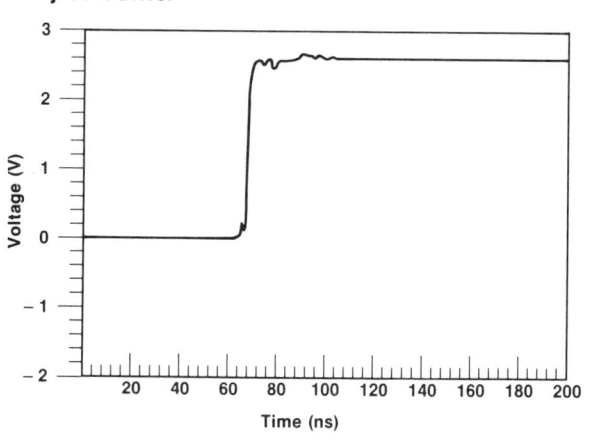

c. Active Receiver

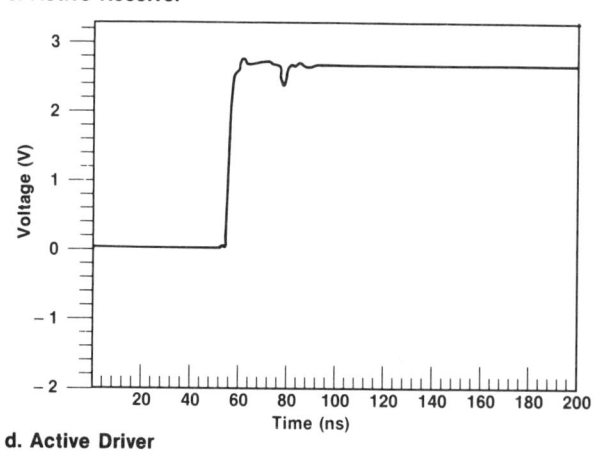

Figure 9-48 Forward PC Board Crosstalk Through 34-Inch Trace Terminated

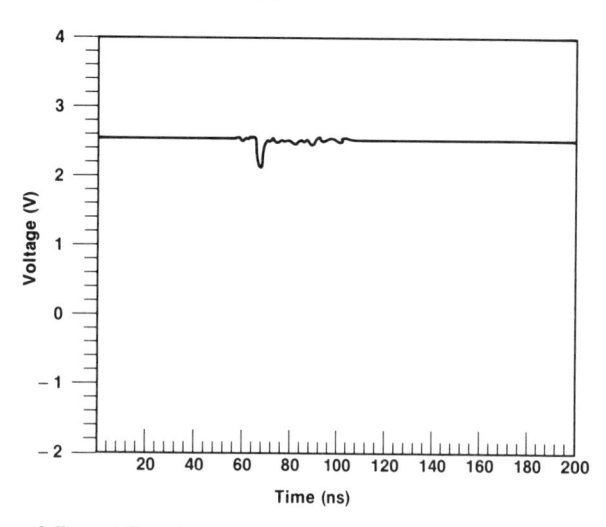

a. Adjacent Receiver

d. Active Driver

9-35

Figure 9-49 Forward PC Board Crosstalk Through 34-Inch Trace with Termination

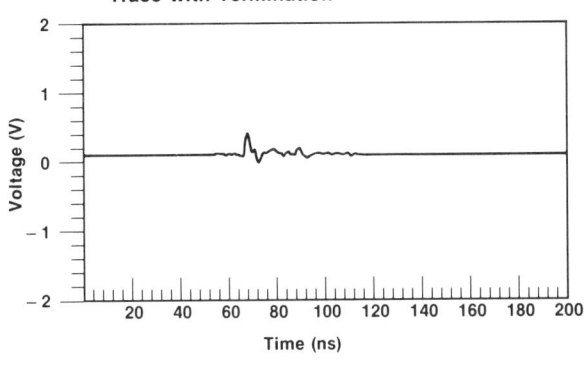

a. Adjacent Receiver

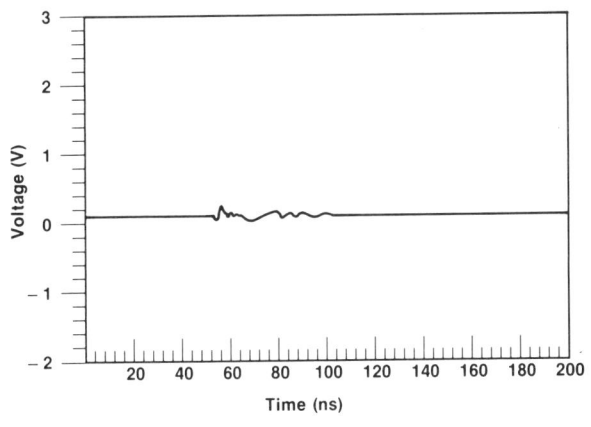

b. Adjacent Driver

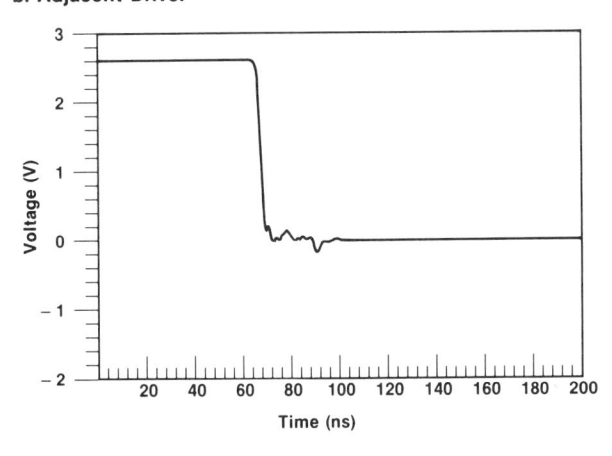

c. Active Receiver

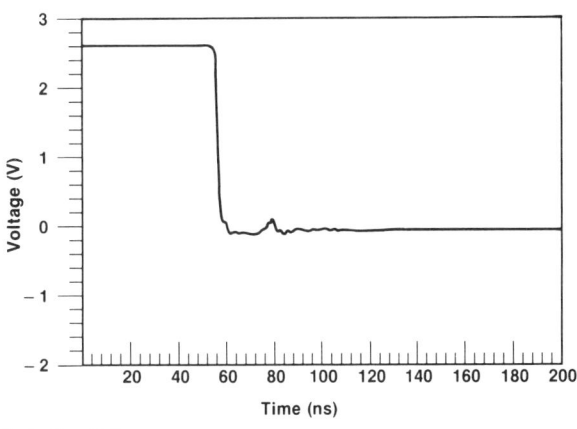

d. Active Driver

Figure 9-50 Reverse PC Board Crosstalk Through 34-Inch, 0.100 Trace Terminated

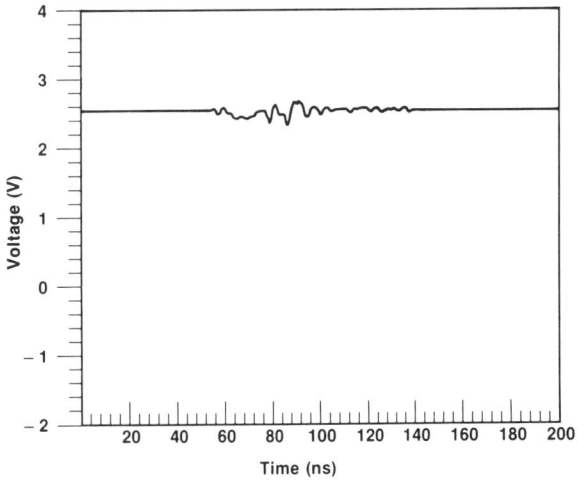

a. Adjacent Receiver

Figure 9-50 (continued)

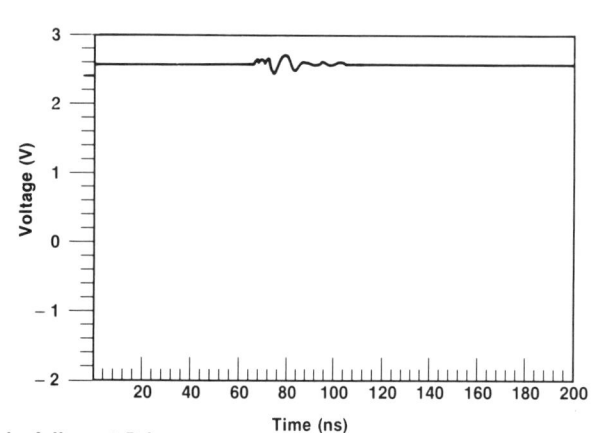

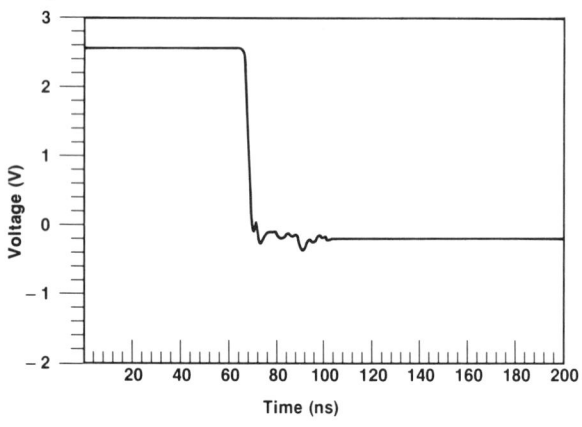

b. Adjacent Driver

c. Active Receiver

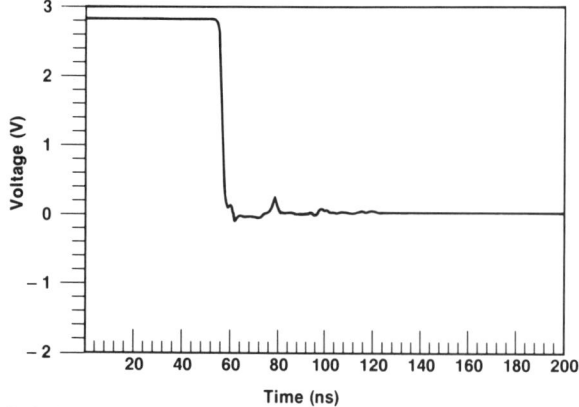

d. Active Driver

Crosstalk on Ribbon Cable

Crosstalk on ribbon cable shows no velocity difference effects—because the cable insulation is a homogeneous medium, all effects are due to mutual impedance. The results of tests on three foot sections of 160Ω ribbon cable are shown in Figures 9-51 through 9-58. From these it can be seen that the unterminated lines exhibit large amounts of ringing due to unterminated energy being transferred between lines. Note also that when the adjacent line is in a HIGH state a charge pump effect occurs, forcing the HIGH output above the V_{CC} supply and into a high impedance state with the output structure turned off and the input exhibiting only leakage currents. This high impedance state causes the current that has been induced into the line to reflect fro both ends and induce crosstalk back into the active line. This action will continue until damped by circuit resistance and leakages. The charge pump effect will leave the adjacent line at around 7V. If this line is then switched low, twice the normal energy is required to switch the line, thus almost doubling the crosstalk generated in the previous case. The terminated lines show the true magnitude of the crosstalk. Note that when the adjacent line is in the LOW state, the crosstalk will cause the driver output to turn off until clamped by the diode in the output structure.

Figure 9-51 Forward Crosstalk Using FAST and 3-Foot Ribbon Cable, Unterminated

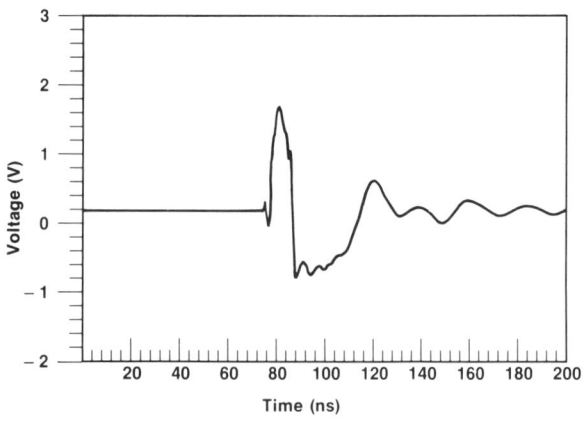

a. Adjacent Receiver

Figure 9-51 (continued)

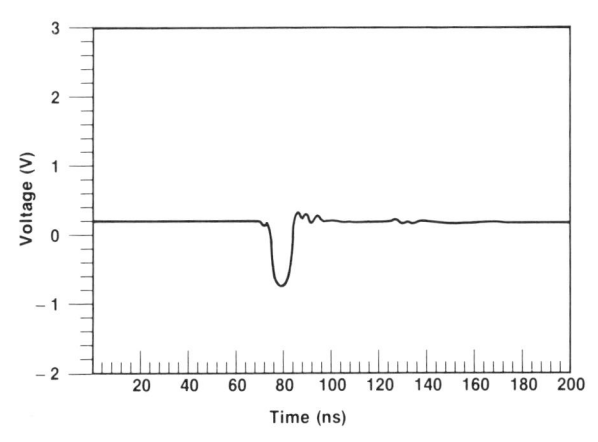

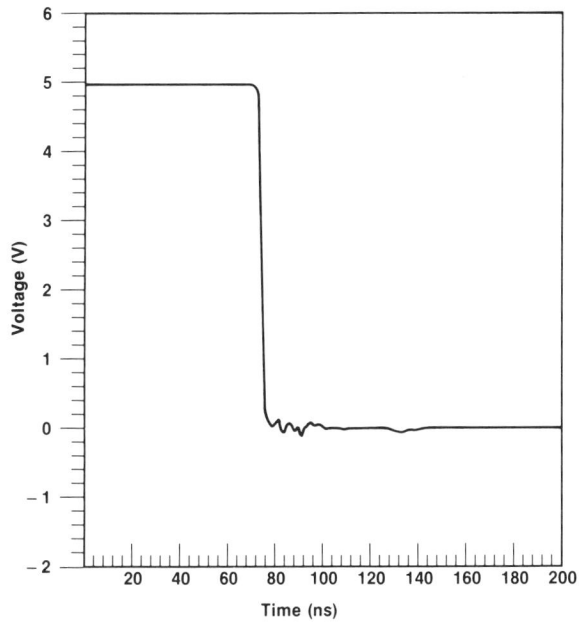

b. Adjacent Driver

c. Active Driver

Figure 9-52 Forward Crosstalk Using FAST and 3-Foot, 2-Conductor Ribbon Cable, Unterminated

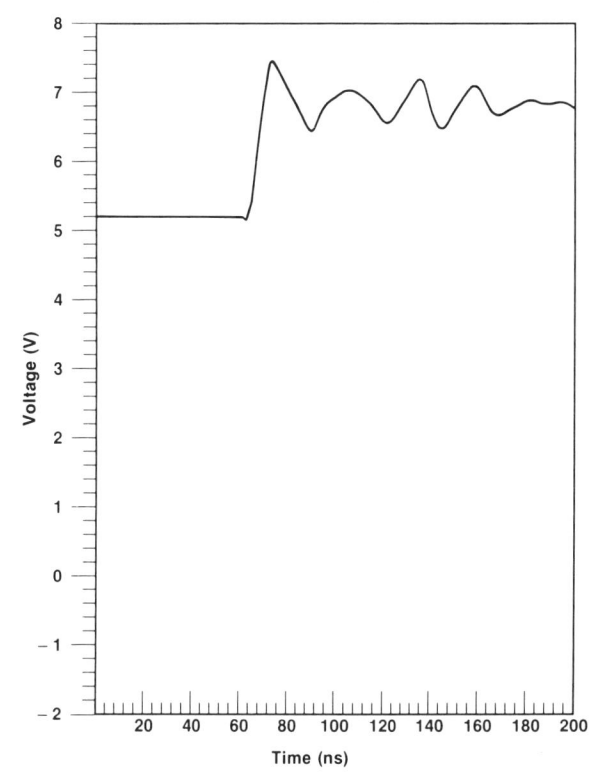

a. Adjacent Receiver

Figure 9-52 (continued)

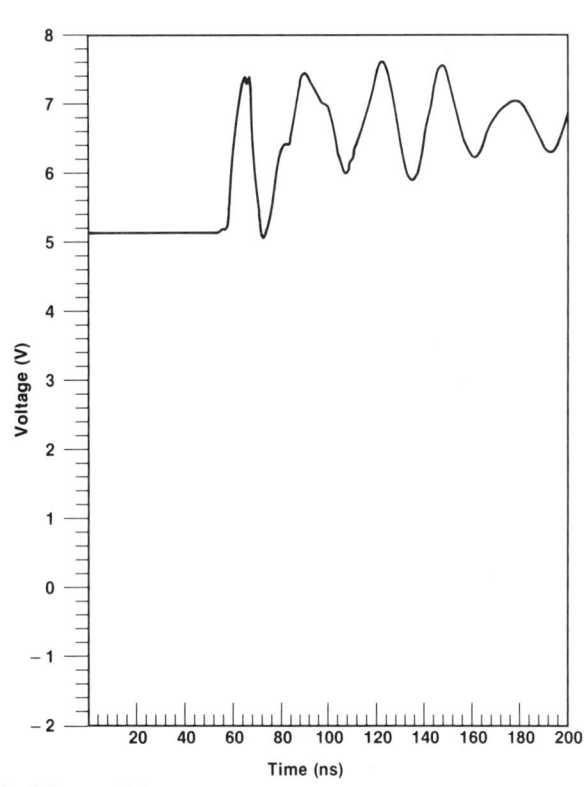

b. Adjacent Driver

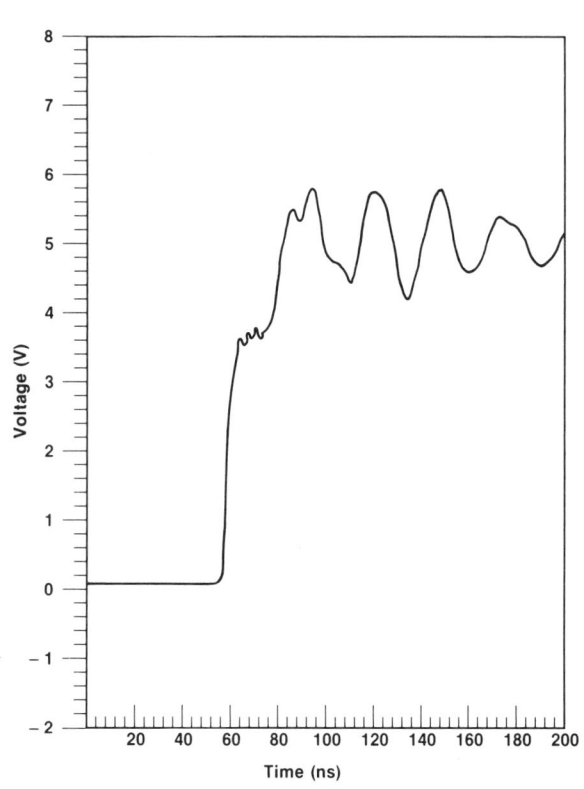

c. Active Driver

Figure 9-53 Reverse Crosstalk Using FAST and 3-Foot, 2-Conductor Ribbon Cable, Unterminated

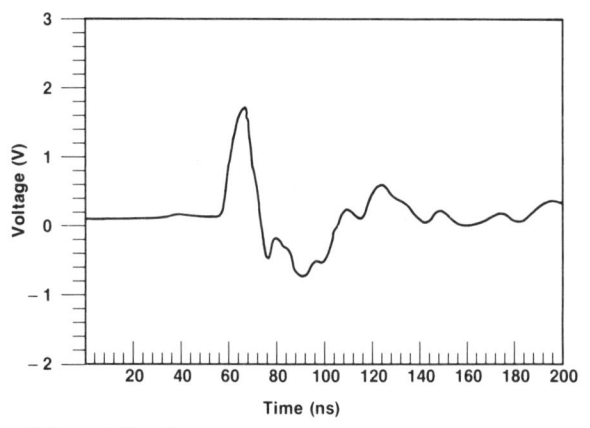

a. Adjacent Receiver

9-39

Figure 9-53 (continued)

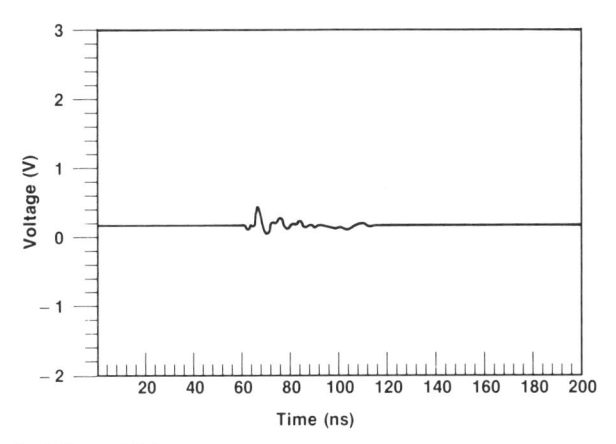

b. Adjacent Driver

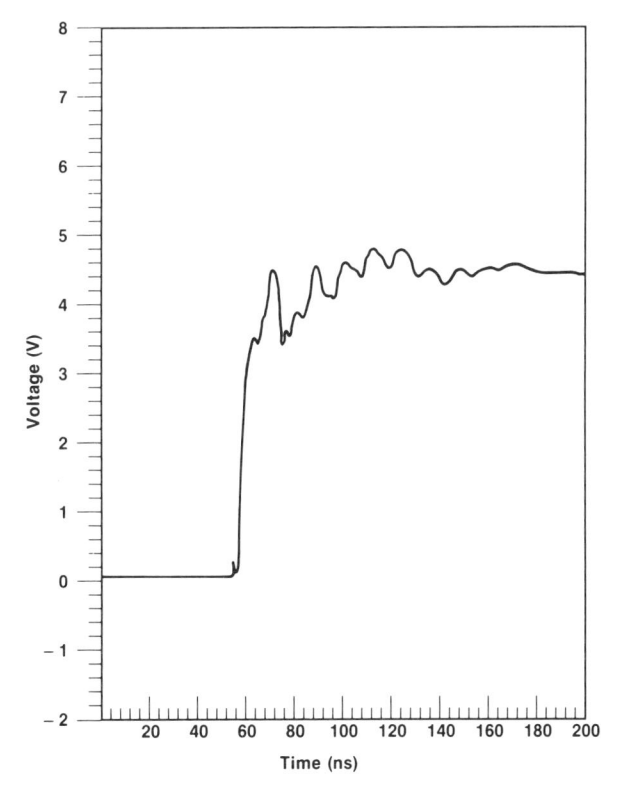

Figure 9-54 Reverse Crosstalk Using FAST and 3-Foot, 2-Conductor Ribbon Cable, Unterminated

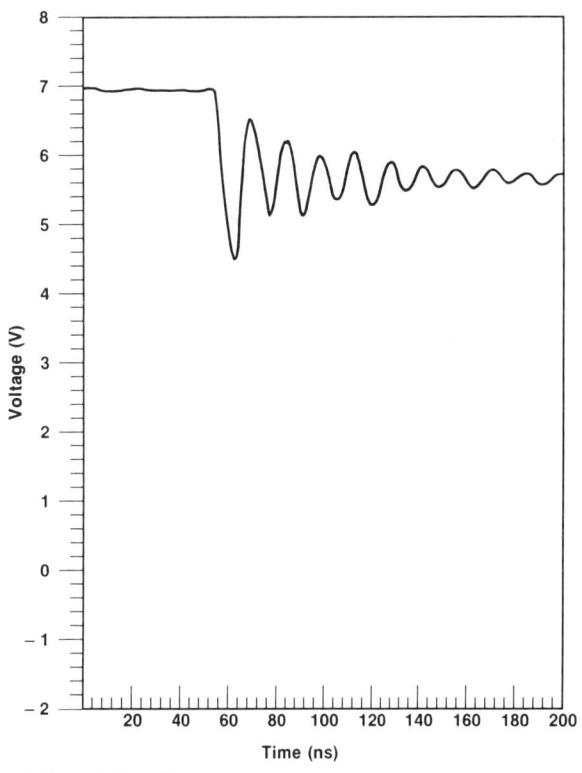

a. Adjacent Receiver

c. Active Driver

Figure 9-54 (continued)

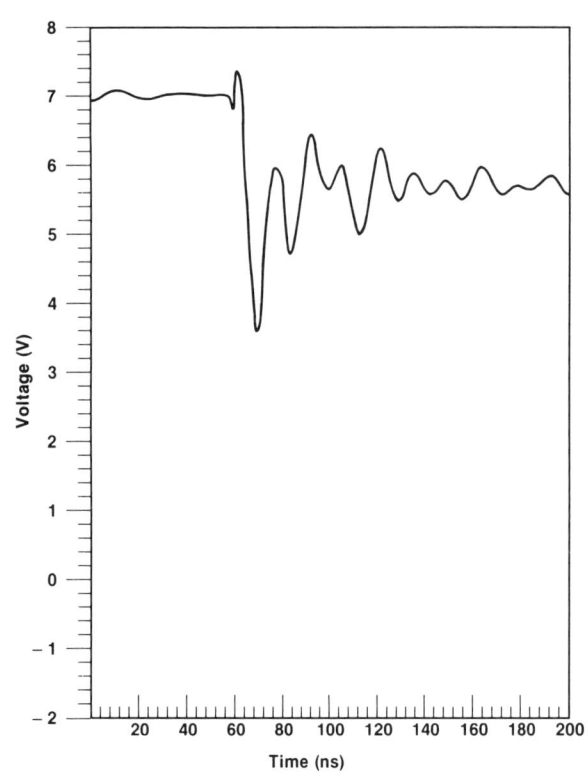

b. Adjacent Driver

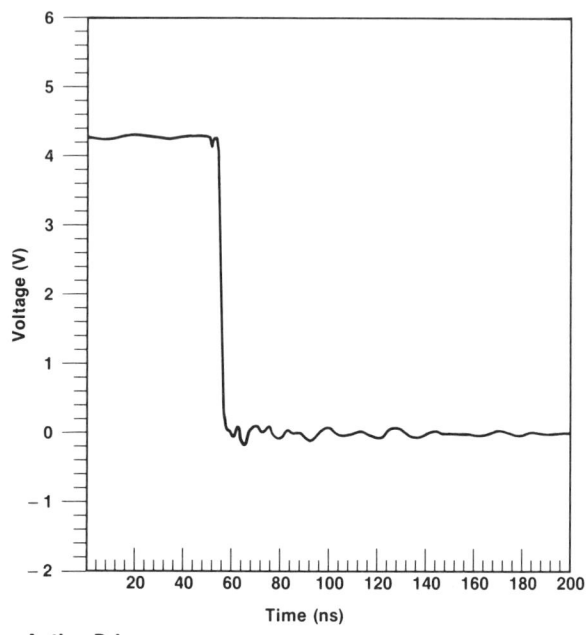

c. Active Driver

Figure 9-55 Forward Crosstalk Using FAST and 3-Foot, 2-Conductor Ribbon Cable, Terminated

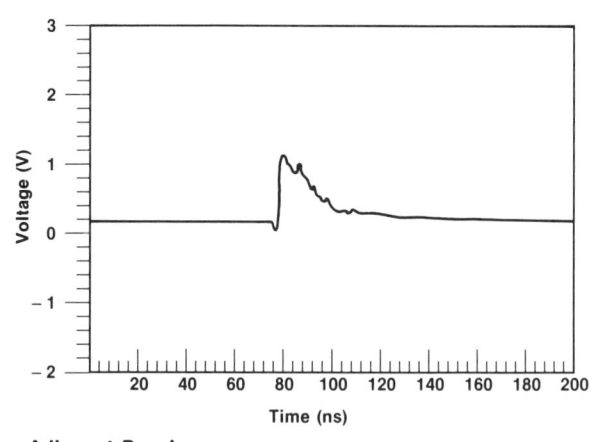

a. Adjacent Receiver

9-41

Figure 9-55 (continued)

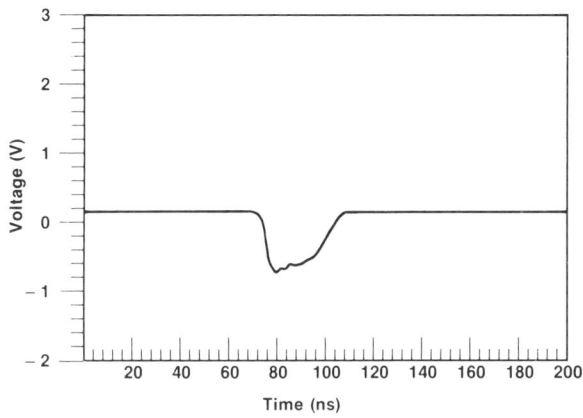

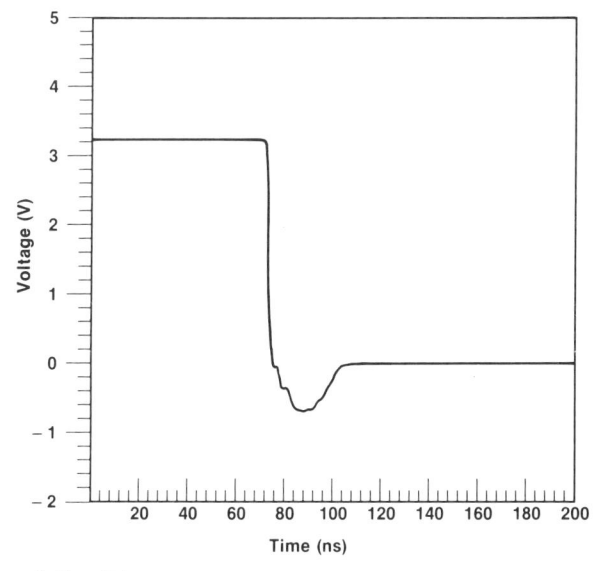

b. Adjacent Driver

c. Active Driver

Figure 9-56 Reverse Crosstalk Using FAST and 3-Foot, 2-Conductor Ribbon Cable, Terminated

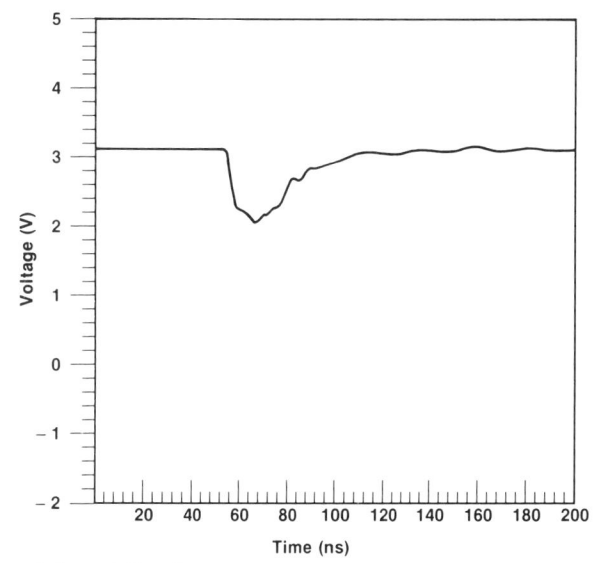

a. Adjacent Receiver

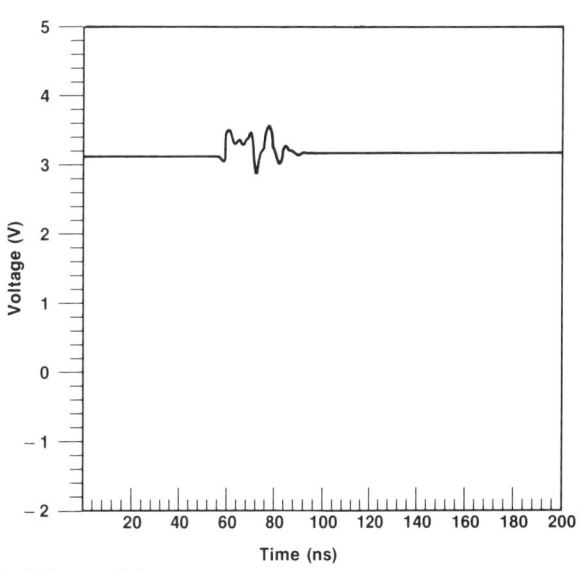

b. Adjacent Driver

Figure 9-56 (continued)

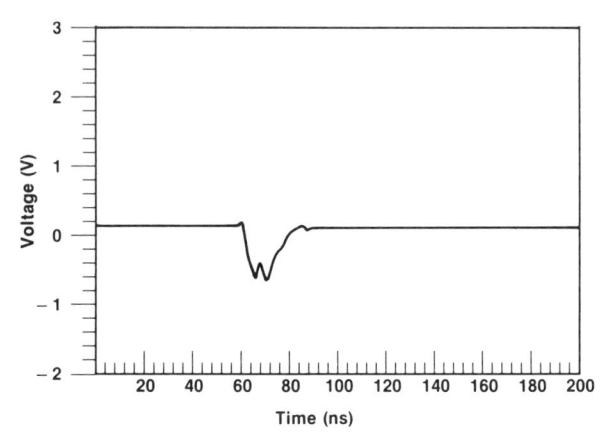

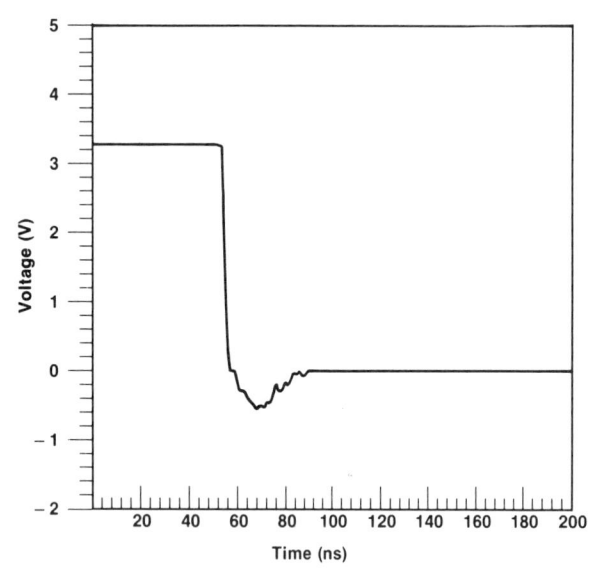

c. Active Driver

b. Adjacent Driver

Figure 9-57 Reverse Crosstalk Using FAST and 3-Foot, 2-Conductor Ribbon Cable, Terminated

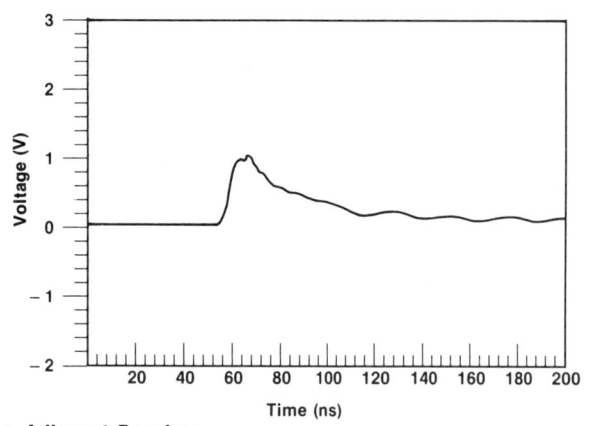

a. Adjacent Receiver

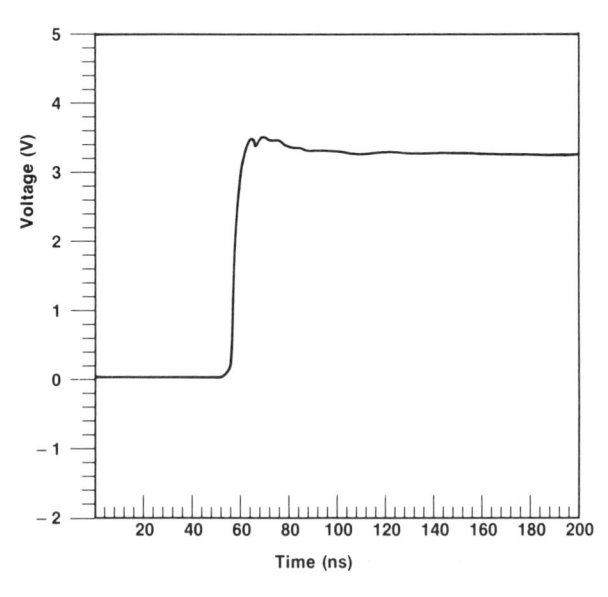

c. Active Driver

Figure 9-58 Forward Crosstalk Using FAST and 3-Foot, 2-Conductor Ribbon Cable, Terminated

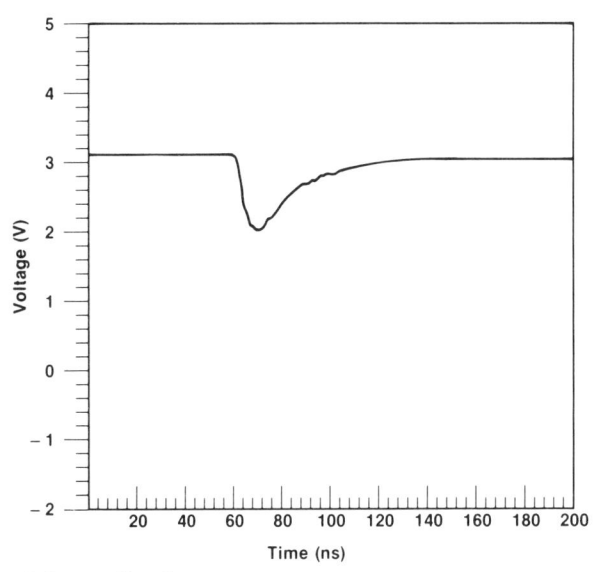

a. Adjacent Receiver

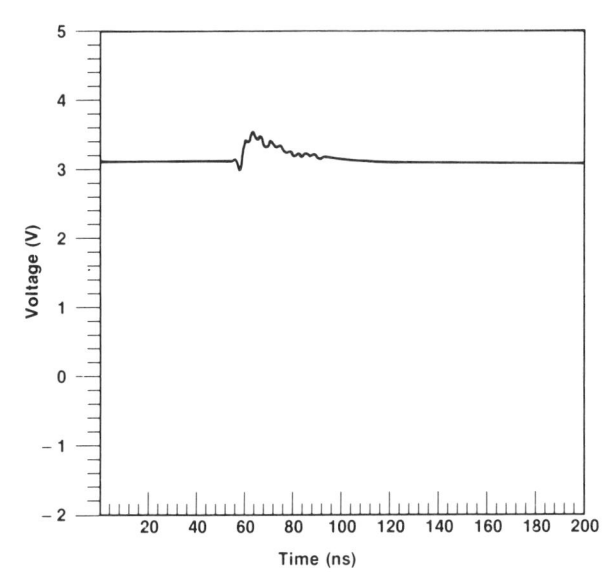

b. Adjacent Driver

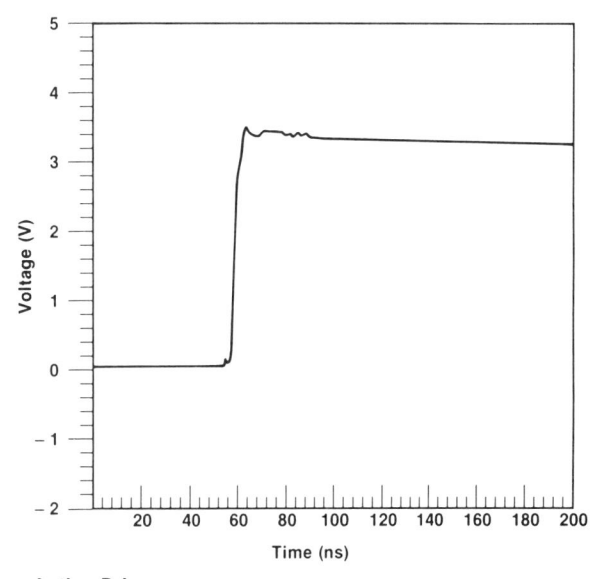

c. Active Driver

Recommendations

In order to minimize crosstalk it is necessary to consider the causes during the design of systems. Some preventative measures are as follows:

1. always use maximum allowable spacing between signal lines;

2. minimize spacing between signal lines and ground lines;

3. run ground strips alongside either the cross-talker or the cross-listener and between the two when possible;

4. in backplane and wire-wrap applications use twisted pair for sensitive functions such as clocks, asynchronous set or clear, asynchronous parallel load (especially leading to LS inputs); and

5. for ribbon or flat cabling make every other conductor a ground line.

In the case where systems or boards are already built and problems are encountered, some temporary or quick fixes may be utilized. They are:

1. with printed circuit boards, glue a source of ground, either a wire or a copper strip, alongside the cross-talker or cross-listener—preferably between them;

2. for the backplane or wire-wrap situation, spiral a ground wire around the talker to confine its electromagnetic field or around the listener in order to shield it, or do both;

3. try the split-resistor termination on the offending line (Figure 9-41);

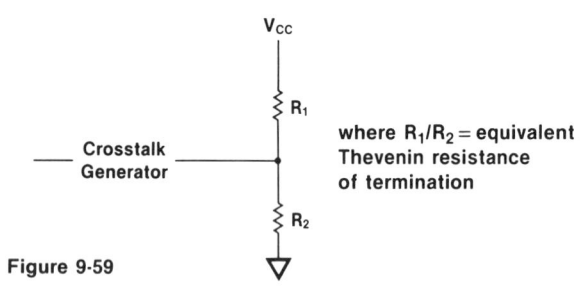

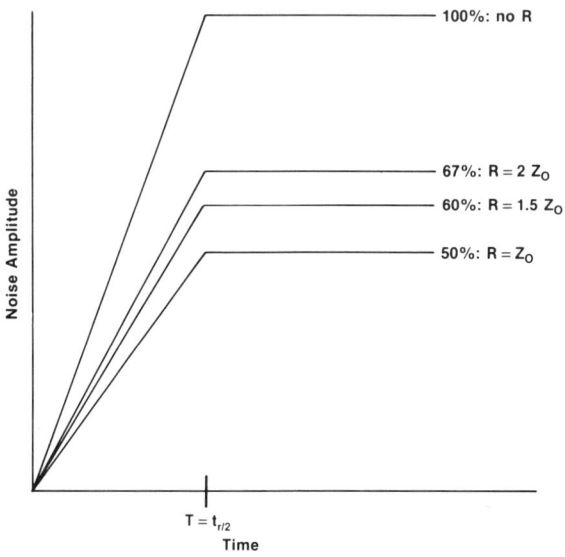

Figure 9-59

where R_1/R_2 = equivalent Thevenin resistance of termination

Figure 9-61

100%: no R
67%: $R = 2 Z_O$
60%: $R = 1.5 Z_O$
50%: $R = Z_O$

$T = t_{r/2}$

4. cut the offending crosstalk trace from the PC board and replace it with a wire. In this method reverse and forward crosstalk can be lessened. The line in this case may be lengthened, thereby increasing propagation delays, but a rerouting of the generating signal line may eliminate the crosstalk.

Termination can be used to reduce the effects of crosstalk. It can be seen here that a little termination is better than no termination.

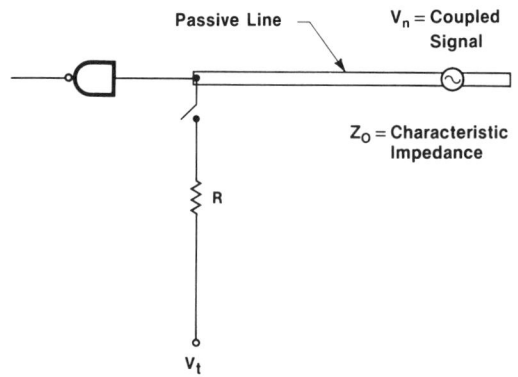

Figure 9-60

V_n = Coupled Signal
Z_O = Characteristic Impedance

Summary

Trace proximity and coupled trace length are the two main factors which affect the amount of reverse crosstalk that occurs. Therefore, if coupled length is long, noise will be at a maximum. For short lengths, noise may appear only as a short spike which can cause difficulties and even system failures.

When two lines do not run between the same points but are in proximity over part of their length, signal propagation time (line delay) along this coupled length is T. If T is long compared to the rise of the signal on the active line, the crosstalk pulse has time to develop its full amplitude. The trailing edge of the noise pulse is caused by the reflection from the driven end of the passive line. When T is half the rise time, the reflection from the driven end of the passive line arrives and begins to pull the noise pulse down just as it reaches full amplitude. Any value of T less than half the rise time of the active signal will cause a reflection to arrive and oppose the noise pulse voltage before it can reach full amplitude. The noise will therefore be lower in amplitude.

9-45

The Capacitor

General Information

A capacitor is a component which is capable of storing electrical energy. It consists of conductive plates (electrodes) separated by insulating material which is called the dielectric. A typical formula for determining capacitance is:

(E9-45)
$$C = \frac{0.224\ KA}{t}$$

C = Capacitance (farads)
K = Dielectric constant (Vacuum = 1)
A = Area in square inches
t = Separation between plates in inches (thickness of dielectric)
0.224 = Conversion constant (0.0884 for metric system in cm)

Capacitance—The standard unit of capacitance is the farad. A capacitor has a capacitance of 1 farad when 1 coulomb charges it to 1 volt. One farad is a very large unit and most capacitors have values in the micro (10^{-6}), nano (10^{-9}), or pico farad (10^{-12}) level.

Dielectric Constant—In the formula for capacitance given above, the dielectric constant of a vacuum is arbitrarily chosen as the number 1. Dielectric constants of other materials are then compared to the dielectric constant of a vacuum. Dielectric constants of some typical materials are as follows:

Material	
Ruby Mica	7
Glass	10
Ceramic (class 1)	5-450
Ceramic (class 2)	200-12,000
Paper	2.5
Mylar	3
Polystyrene	2.6
Polycarbonate	3
Aluminum Oxide	7
Tantalum Oxide	11

Dielectric Thickness—Capacitance is indirectly proportional to the separation between electrodes. Lower voltage requirements mean thinner dielectrics and greater capacitance per volume.

Area—Capacitance is directly proportional to the area of the electrodes. Since the other variables in the equation are usually set by the performance desired, area is the easiest parameter to modify to obtain a specific capacitance within a material group.

Energy which can be stored in a capacitor is given by the formula:

(E9-46)
$$E = 1/2\ CV^2$$

E = Energy in joules (watts-sec)
V = Applied voltage
C = Capacitance in farads

A capacitor is a reactive component which reacts against a change in potential across it. This is shown by the equation for the linear charge of a capacitor:

(E9-47)
$$I_{ideal} = C\ \frac{dV}{dt}$$

where
I = Current
C = Capacitance
dV/dt = Slope of voltage transition across capacitor

Thus an infinite current would be required to instantly change the potential across a capacitor, and the amount of current a capacitor can "sink" is given by the above equation.

A capacitor, as a practical device, exhibits not only capacitance but also resistance and inductance. A simplified schematic for the equivalent circuit is:

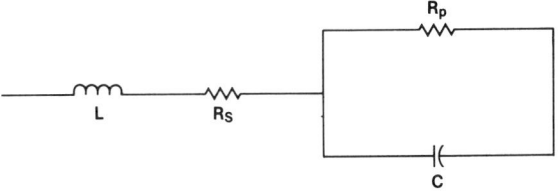

C = Capacitance   L = Inductance
R_S = Series resistance R_p = Parallel resistance

Figure 9-62

All the factors shown above are important in the application of capacitors. The inductance determines the usefulness of the capacitor at high frequency, the parallel resistance affects performance in timing and coupling circuits (normally expressed as Insulation Resistance) and the series resistance is a measure of the loss in the capacitor and is a major factor in Power Factor and/or Dissipation Factor.

Since the insulation resistance (R_p) is normally very high, the total impedance of a capacitor is:

$$Z = \sqrt{R_S^2 + (X_C - X_L)^2}$$

(E9-48)

where
- Z = Total impedance
- R_S = Series resistance
- X_C = Capacitive reactance = $1/2\pi fc$
- X_L = Inductive reactance = $2\pi fL$

The variation of a capacitor's impedance with frequency determines its effectiveness in many applications.

Power Factor and Dissipation Factor are often confused since they are both measures of the loss in a capacitor under AC application and are often almost identical in value. In a "perfect" capacitor the current in the capacitor will lead the voltage by 90°.

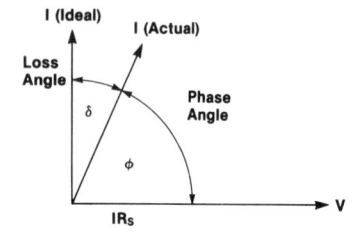

Figure 9-63

In practice the current leads the voltage by some other phase angle due to the series resistance R_s. The complement of this angle is called the loss angle and:
Power Factor (PF) = cos ϕ or sine δ
Dissipation Factor (DF) = tan δ

For small values of δ the tan and sine are essentially equal which has led to the common interchangeability of the two terms in the industry.

The term ESR or Equivalent Series Resistance combines all losses, both series and parallel, in a capacitor at a given frequency so that the equivalent circuit is reduced to a simple R-C series connection.

Figure 9-64

$$\text{Dissipation Factor} = \frac{ESR}{X_c} = (2\pi fc)(ESR)$$

(E9-49)

The DF/PF of a capacitor tells what percent of the apparent power input will turn to heat in the capacitor. The watts loss is:

$$\text{Watts Loss} = (2\pi fcE^2)(DF)$$

(E9-50)

Very low values of dissipation factor are expressed as their reciprocal for convenience. These are called the "Q," or Quality factor of capacitors.

Insulation Resistance is the resistance measured across the terminals of a capacitor and consists principally of the parallel resistance R_p shown in the equivalent circuit. As capacitance values and hence the area of dielectric increases, the IR decreases. The product (C x IR or RC) is often specified in ohm farads or commonly megohm microfarads.

Dielectric Strength is an expression of the ability of a material to withstand an electrical stress. Although dielectric strength is ordinarily expressed in volts, it is actually dependent on the thickness of the dielectric and thus is also more generically a function of volts/mil.

Other specialized factors which may be of interest to the user, especially in high voltage applications, are corona and dielectric absorption.

The phenomenon of Dielectric Absorption is exhibited in the following manner: charging current from a steady unidirectional source continues to flow at a gradually decreasing rate into a capacitor of negligible series resistance for some time after the almost instantaneous charge is completed. A steady value proportional to the capacitor parallel resistance is finally reached. The additional charge apparently is absorbed by the dielectric. Conversely, a capacitor does not discharge instantaneously upon application of a short circuit, but drains gradually after the capacitance proper has been discharged. It is common practice to measure the dielectric absorption by determining the "reappearing voltage" which develops across a capacitor at some point in time after it has been fully discharged under short circuit conditions.

Corona is the ionization of air or other vapors which causes them to conduct current. It is especially prevalent in high voltage units but can occur with low voltages as well where high voltage gradients occur. The energy discharged degrades the performance of the capacitor and can in time cause catastrophic failures.

The usual characteristics that are specified for a capacitor include Capacitance, Dissipation Factor or ESR, Insulation Resistance or Leakage Current and Dielectric Strength. The electrical and environmental parameters that are of most interest with respect to these four basic measurements are temperature, voltage and test frequency. The reference temperature for most capacitor measurements is 25°C. Voltage is dependent on the rating applied by the manufacturer and the test frequency typically depends on the class of product.

As the ambient temperature changes, the dielectric constant and hence the capacitance of many capacitors changes. In general, when the dielectric constant is lower, materials tend to change capacitance less with temperature or with relatively predictable changes that are linear with temperature. High dielectric constant materials tend to have capacitance changes that are non-linear and expressed as percent capacitance change over a temperature range. Increasing temperature usually reduces Insulation Resistance, increases Leakage Current and Power Factor/Dissipation Factor and reduces the voltage rating of the part. Some ceramic capacitors actually exhibit a decrease in DF with increasing temperature. Conversely, reducing temperature normally improves most characteristics.

The effects of applied voltage on capacitors are a prime consideration in use. Capacitance and other parameter changes occur under both AC and DC applied voltages. Even those cases where voltage application does not change the parametric characteristics of a capacitor, the level of voltage applied will determine the life expectancy of the capacitor.

Frequency is the third factor which is of great concern in the application of capacitors. This is an area that is often overlooked by designers. Earlier an equivalent circuit was given for a capacitor. Inductance which is caused by the leads and the electrodes was depicted. As the frequency applied to the capacitor increases it eventually passes through self-resonance and becomes inductive with gradually increasing impedance. Even though a capacitor is beyond the self-resonant point it still blocks DC and has a low impedance and thus is useful in bypass, coupling and many other applications. Care should be taken in feedback, tuning, phase shift and such applications.

Ceramic Capacitors

Ceramic capacitors are the most widely used capacitors. They come in an extremely wide range of mechanical configurations and electrical characteristics. The common mechanical variations are discs, tubulars, feed throughs and monolithics. A 0.01μF disc is about 1/2 inch in diameter while the 0.01μF monolithic chip capacitor is only 0.050" x 0.075" x 0.030". Electrically, ceramic capacitors are broken into two classes. Class 1 ceramic dielectrics are also called temperature compensating ceramics and feature zero TC and other predictable and relatively linear TC bodies. The insulation resistance is high, the losses are low and the parts are essentially unaffected by voltage or frequency and are usually used for tuned circuits, timing applications and other precision circuits.

Where Class 1 ceramics are completely predictable, Class 2 general purpose ceramics are full of surprises for the unsuspecting engineer. Not only does capacitance change with temperature but the "high K" units which are so enticingly small in size may lose 90% of their room temperature capacitance at −55°C. Further care must be exercised when voltage is applied, particularly with monolithic capacitors with their thin dielectrics. AC voltage caused the capacitance to increase and DC voltage causes a capacitance loss. A change in frequency also changes capacitance and DF.

The fact that more ceramic Class 2 capacitors are used than all other types combined proves that the variability of characteristics not only can be overcome by wise selection but can in many cases be an advantage. Considerably more detailed information is given below and a number of articles and booklets are also available on this subject.

The ceramic capacitor is defined as a capacitor manufactured from metallic oxides, sintered at a high temperature. As a general rule, the electrical ceramics used in capacitors are based on complex titanate compounds, principally barium titanate, rare earth titanates, calcium titanates, sodium titanate, etc. Occasionally other materials, such as lead niobiate, may be used.

From a mechanical point of view, ceramic capacitors are manufactured by two basic techniques. One method involves pressing or extruding the ceramic material, firing (sintering) the ceramic and subsequently applying electrodes (typically with silver materials) which are fired onto the ceramic at lower temperatures after the maturation of the ceramic. This is the method employed in the fabrication of single layer devices. The most common form of single layer capacitors is disc capacitors with radial leads or tubular capacitors which are available with axial leads, radial leads or in feed-through form with both bolt and eyelet types being common. There are specialized versions of pressed ceramic capacitors, such as high voltage cartwheels and double cup high voltage units. These and other types may be considered as jumbo size disc pressed units.

The second method of fabricating ceramic capacitors evolved in recent years as a result of the demand for lower voltages and smaller sizes consistent with the advent of semiconductor usage. The miniaturization in the ceramic capacitor area was made possible through the manufacture of monolithic types of ceramic capacitors. These capacitors are manufactured by mixing the ceramic powder in an organic binder (slurry) and casting it by one technique or another into thin layers typically ranging from about 3 mils in thickness down to 1 mil or thinner.

Metal electrodes are deposited onto the green ceramic layers which are then stacked to form a laminated structure. The metal electrodes are arranged so that their terminations alternate from one edge of the capacitor to another. Upon sintering at high temperature the part becomes a monolithic block which can provide extremely high capacitance values in small mechanical volumes. Figure 9-65 shows a pictorial view of a monolithic ceramic capacitor.

While pressed and extruded ceramic capacitors are in general low cost and provide limited capacitance values, monolithic units are typically smaller in size, feature excellent high frequency characteristics because of the small size and provide considerably higher capacitance values with low voltage ratings.

Ceramic capacitors are available in a tremendous variety of characteristics. Electronic Industries Association (EIA) and the military have established categories to help divide the basic characteristics into more easily specified classes. The basic industry specifications for ceramic capacitors is EIA specification RS-198 and as noted in the general section it specifies temperature compensating capacitors as Class 1 capacitors. These are specified by the military under specification MIL-C-20. General purpose capacitors with non-linear temperature coefficients are called Class 2 capacitors by EIA and are specified by the military under MIL-C-11015 and MIL-C-39014. EIA specifications further include a Class 3 category which is defined as reduced titanates.

Class 1 or temperature compensating capacitors are usually made from mixtures of titanates where barium titanate is normally not a major part of the mix. They have predictable temperature coefficients and in general do not have any aging characteristic. Thus they operate in a manner similar to mica capacitors except for the TC which is controllable. Normally the TCs of Class 1 capacitors are deemed to run between P100 and N750. Class 1 extended temperature compensating capacitors are also manufactured in TCs from negative 1400 through negative 5600, however, these may start developing a slight aging characteristic and voltage susceptibility.

Most TC formulations are available in pressed and extruded construction while only NPO (zero TC) is provided by most manufacturers in monolithic construction. NPO ceramics in monolithic capacitors are available in high enough values to cover most applications requiring extreme stability. With the exception of some NPO capacitors, almost all temperature compensating capacitors have a TC curve which is a true curve and not a straight line. The TC tends to become more negative at the cold end than it is from the 25°C reference to +85°C. Both EIA specification RS-198 and military specification MIL-C-20 contain information about curvature. This information is contained in Table 9-2. These charts are based on industry accepted standard TC values.

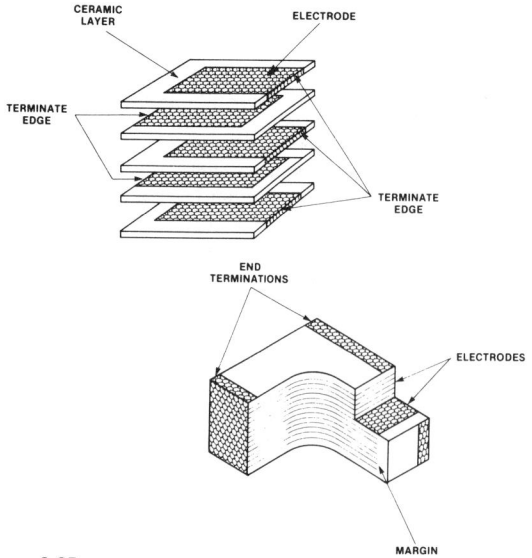

Figure 9-65

Table 9-2

Capacitance in pf	TC TOLERANCES [1]									
	NPO	N030	N080	N150	N220	N330	N470	N750	N1500	N2200
	−55°C to +25°C IN PPM/°C									
10 and over	+30 −75	+30 −80	+30 −90	+30 −105	+30 −120	+60 −180	+60 −210	+120 −340	+250 −670	+500 −1100
	+25°C to +85°C IN PPM/°C									
10 and over	±30	±30	±30	±30	±30	±60	±60	±120	±250	±500
Closest MIL−C−20D Equivalent	CG	HG	LG	PG	RG	SH	TH	UJ	NONE	NONE
EIA Desig.	C0G	S1G	U1G	P2G	R2G	S2H	T2H	U2J	P3K	R3L

[1] Table 9-2 indicates the tolerance available on specific temperature characteristics. It may be noted that limits are established on the basis of measurements at +25°C and +85°C and that TC becomes more negative at low temperature. Wider tolerances are required on low capacitance values because of the effects of stray capacitance.

General purpose ceramic capacitors are called Class 2 capacitors and have become extremely popular because of the high capacitance values available in very small size. Class 2 capacitors are "ferro electric" and vary in capacitance value under the influence of the environmental and electrical operating conditions. Class 2 capacitors are affected by temperature, voltage (both AC and DC), frequency and time. Temperature effects for Class 2 are exhibited as non-linear capacitance changes with temperature.

In specifying capacitance change with temperature, EIA expresses capacitance change over an operating temperature range by a 3-symbol code. The first symbol represents the cold temperature end of the range, the second represents the upper limit of the operating range and a third symbol represents the capacitance change allowed over the operating temperature range. Table 9-3 provides a detailed explanation of the EIA system. As an example, a capacitor with a characteristic X7R would change ±15% over the temperature range −55°C to +125°C and is often identical to military characteristics BX. Parts with characteristics are also sometimes called "K1200" but most manufacturers now use higher dielectric constant than 1200 so the term is now taken to mean only X7R and is commonly called semi-stable material.

Table 9-3

MIL-C-11015D CODE		EIA CODE Percent Capacity Change Over Temperature Range	
Symbol	Temperature Range	RS198	Temperature Range
A	−55°C to +85°C	X7	−55°C to +125°C
B	−55°C to +125°C	X5	−55°C to +85°C
C	−55°C to +150°C	Y5	−30°C to +85°C
		Z5	+10°C to +85°C

Symbol	Cap. Change Zero Volts	Cap. Change Rated Volts	Code	Per Cent Capacity Change
R	+15%, −15%	+15%, −40%	D	±3.3%
			E	±4.7%
W	+22%, −56%	+22%, −66%	F	±7.5%
			P	±10%
X	+15%, −15%	+15%, −25%	R	±15%
			S	±22%
Y	+30%, −70%	+30%, −80%	T	+22% −33%
			U	+22% −56%
Z	+20%, −20%	+20%, −30%	V	+22% −82%

Temperature characteristic is specified by combining range and change symbols, for example BR or AW. Specification slash sheets indicate the characteristic applicable to a given style of capacitor.

Example—A capacitor is desired with the capacitance value at 25°C to increase no more than 7.5% or decrease no more than 7.5% from −30°C to +85°C. EIA Code wil be Y5F.

A Z5U temperature characteristic is also extremely popular. It allows a capacitance change of +22% to −56% over the temperature range of +10°C to +85°C, and is usually made with materials with a dielectric constant in the range of 5000 to 10,000.

Effects of Voltage

Whereas variations in temperature affect all of the parameters of ceramic capacitors, voltage basically affects only the capacitance and dissipation factor. The application of DC voltage reduces both the capacitance and dissipation factor while the application of an AC voltage within a reasonable range tends to increase both capacitance and dissipation factor readings. If a high enough AC voltage is applied, eventually it will reduce capacitance just as a DC voltage will. However, the application of this high an AC voltage is normally not encountered.

Since the magnitude of the effect is dependent on the thickness of the dielectric versus the voltage applied (volts per mil) the curve is based on percent of rated voltage in order to give a basic idea of the order of magnitude of the changes in question. Figure 9-66 shows the effects of AC voltage.

These are of major significance in some applications but are perhaps of even more significance when it comes to measuring the capacitance value and dissipation factor of the capacitors. Capacitor specifications specify the AC voltage at which to measure (normally 1 VAC) and application of the wrong voltage can cause spurious readings. Figure 9-67 gives

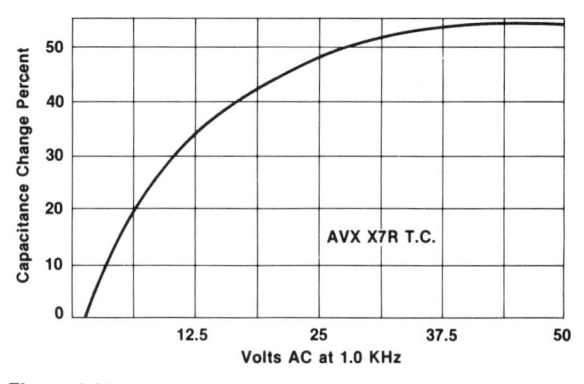

Figure 9-66

the voltage coefficient of dissipation factor for AC based on 1000 cycles reading and 1 kilohertz readings. Applications of different frequencies will affect the percentage changes versus voltages.

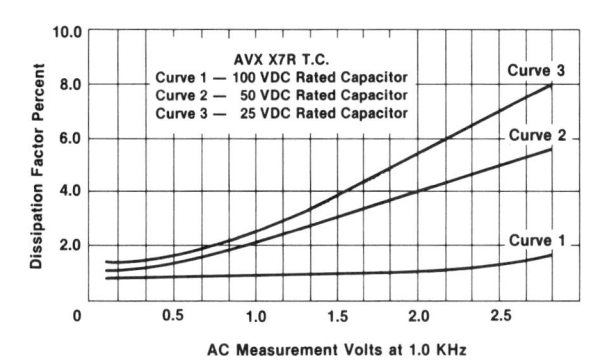

Figure 9-67

The effect of the application of DC voltage is once again dependent on the thickness of the dielectric (volts per mil) and is shown in a similar manner in Figure 9-68. As will be noted in general, the voltage coefficient is more pronounced for higher K dielectrics. These figures are shown for room temperature conditions. Of considerable interest to the user is a combination characteristic known as voltage temperature limit which shows the effects of rated voltage over the operating temperature range. Figure 9-69 shows a capacitor of military specification type BX.

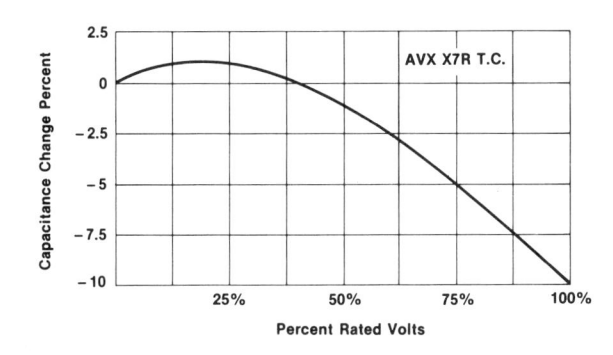

Figure 9-68

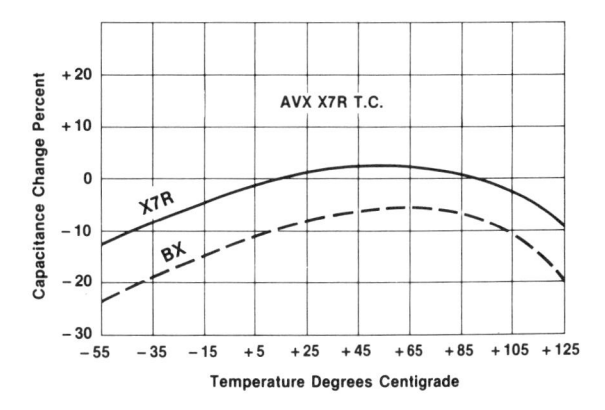

Figure 9-69

Effects of Frequency

Frequency affects capacitance and dissipation factor as is the case with voltage. Curves of capacitance change and dissipation factor change with normal type ceramics are shown in Figure 9-70 and 9-71.

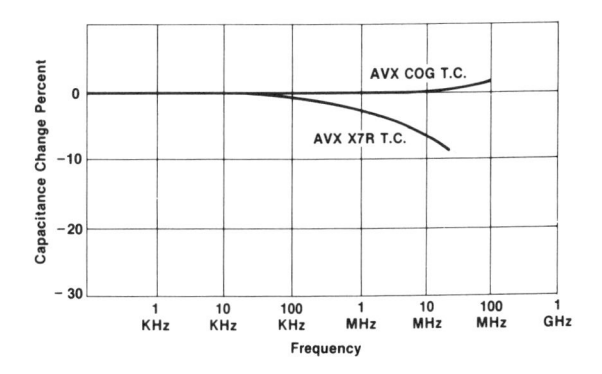

Figure 9-70

Variation of impedance with frequency is an important consideration for decoupling capacitor applications. Lead length, lead configuration and body size all affect the impedance level as well as the ceramic formulation variations.

Special ceramic materials are also made for use at extremely high frequencies.

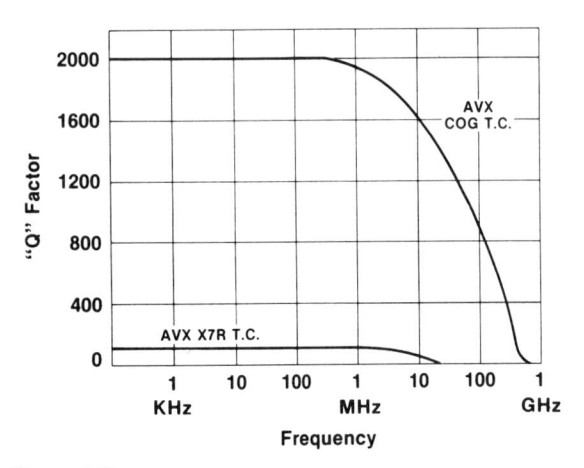

Figure 9-71

Effects of Time
Class 2 ceramic capacitors change capacitance and dissipation factor with time as well as temperature, voltage and frequency. This change with time is known as aging. Aging is caused by a gradual re-alignment of the crystalline structure of the ceramic and produces an exponential loss in capacitance and decrease in dissipation factor versus time. A curve of typical aging rate of semistable ceramic is shown in Figure 9-72 and a table is given showing the aging rates of various dielectrics.

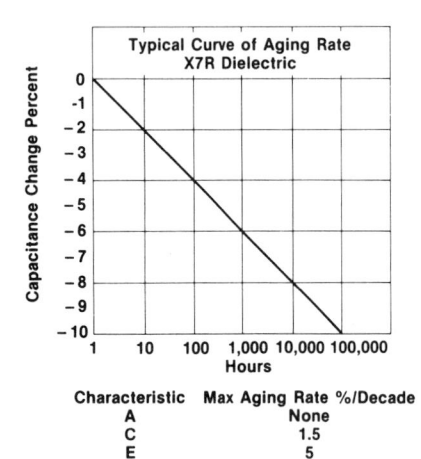

Characteristic	Max Aging Rate %/Decade
A	None
C	1.5
E	5

Figure 9-72

If a ceramic capacitor that has been sitting on the shelf for a period of time is heated above its Curie point (125°C for 4 hours or 150°C for ½ hour will suffice), the part will de-age and return to its initial capacitance and dissipation factor readings. Because the capacitance changes rapidly, immediately after de-aging, the basic capacitance measurements are normally referred to a time period sometime after the de-aging process. Various manufacturers use different time bases but the most popular one is one day or twenty-four hours after "last heat." Aging as noted is expressed in terms of percent per decade with a split in the industry as to whether it should be day decades or hour decades. This is of concern to the user only because the industry only guarantees the capacitor to be within tolerance for two decades after receipt which means a difference between a thousand hours (42 days) depending on the system employed. Permanent change in the aging curve can be caused by the application of voltage and other stresses and if this is of importance to the designer, he should consult the manufacturer for further details. The possible changes in capacitance due to de-aging by heating the unit explain why capacitance changes are allowed after test, such as temperature cycling, moisture resistance, etc., in mil specs. The application of high voltages such as dielectric withstanding voltages also tends to de-age capacitors and once again explains why re-reading of capacitance after 12 or 24 hours is allowed in military specs after dielectric strength tests have been performed.

Effects of Mechanical Stress
Ceramic capacitors exhibit some low level piezoelectric reactions under mechanical stress. As a general statement, the piezoelectric output is higher, the higher the dielectric constant of the ceramic. It therefore is often wise to investigate this effect before using high K dielectrics as coupling capacitors in extremely low level applications. Normally for this type of application semistable (X7R, BX, K1200, etc.) material works considerably better.

Reliability
Historically ceramic capacitors have been considered one of the most reliable types of capacitors in use today. The inherent reliability of the ceramic capacitor can be improved by power aging (burn-in) and other types of screening intended to remove early failures. In testing of ceramic capacitors, fairly high dielectric strength tests are applied which will remove units with any gross mechanical defects such as large voids in the dielectric or other mechanical weaknesses in the basic dielectric. Power aging or burn-in not only will

supplement the detection of this type of failure, but also, if properly applied, tend to eliminate early wear-out modes of failure. Assuming that gross mechanical defects and early wear-out types of failures have been reasonably eliminated by dielectric strength test and burn-in, the effects of temperature and voltage can be considered. The approximate formula for the reliability of a ceramic capacitor is:

(E9-51) $$\frac{L_O}{L_t} = \left(\frac{V_t}{V_O}\right)^X \left(\frac{T_t}{T_O}\right)^Y$$

L_o = Operating life
L_t = Test life
V_t = Test voltage
V_o = Operating voltage
T_t = Test temperature and
T_o = Operating temperature in °C
X, Y = See test

Historically in the ceramic capacitor business the exponent X has always been considered as 3. Considerable work has been done to either verify or disprove the "cube law" for ceramic capacitors, and although it has been shown that it is not always accurate, it has been demonstrated to be a reasonable approximation. The exponent Y for temperature effects typically tends to run about 8. Taking some examples, one can see readily that lowering the voltage and the temperature dramatically improves the anticipated performance of ceramic capacitors, pointing out once again the importance of careful design. One factor as far as reliability that is usually not considered is the impedance of the circuit versus the impedance of the test conditions. All standard ceramic test requirements at the present time specify that currents be limited to 50 milliamps or less. With the advent of semi-conducting devices, however, many applications are operating at considerably higher current availabilities in very low impedance circuits. Because of the differences between test conditions and operating conditions failures can sometimes occur in components that would perform reliably under the high impedance test conditions. Quite often consultation with the manufacturer and special screening are desirable for this type of application.

Reprinted with permission of AVX Capacitors.

Summary of Electrical Characteristics of Some Well Known Digital Interface Standards

National Semiconductor Corp.
Application Note 216
Don Tarver

FOREWORD

Not the least of the problems associated with the design or use of data processing equipment is the problem of providing for or, actually interconnecting the differing types and models of equipment to form specific processing systems.

The magnitude of the problem becomes apparent when one realizes that every aspect of the electrical, mechanical and architectural format must be specified. The most common of the basic decisions confronting the engineer include:

- Type of logic (negative or positive)
- Threshold levels
- Noise immunity
- Form of transmission
  - Balanced/unbalanced, terminated/unterminated
  - Unidirectional/bidirectional, simplex/multiplexed
- Type of transmission line
- Connector type and pin out
- Bit or byte oriented
- Baud rate

If each make and/or model of equipment presented a unique interface at its I/O ports, "interface" engineering would become a major expenditure associated with the use of data processing equipment.

Fortunately, this is not the case as various interested or cognizant groups have analyzed specific recurring interface areas and recommended "official" standards around which common I/O ports could be structured. Also, the I/O specifications of some equipment with widespread popularity such as the IBM 360/370 computer and DEC minicomputer have become "defacto" standards because of the desire to provide/use equipment which interconnect to them.

Compliance with either the "official" or "defacto" standards on the part of equipment manufacturers is voluntary. However, it is obvious that much can be gained and little lost by providing equipment that offers either the "official" or "defacto" standard I/O ports.

As can be imagined, the entire subject of interface in data processing systems is complicated and confusing, particularly to those not intimately involved in the day-to-day aspects of interface engineering or management. However, at the component level the questions simplify to knowing what standards apply and what circuits or components are available to meet the standards.

This application note summarizes the important electrical characteristics of the most commonly accepted interface standards and offers recommendations on how to use National Semiconductor integrated circuits to meet those standards.

1.0 INTRODUCTION

The interface standards covered in this application note are listed in Table I. The body of the text expands upon the scope and application of each listed standard and summarizes important electrical parameters.

Table II summarizes the National Semiconductor IC's applicable to each standard.

TABLE I. Common Line Driver/Receiver Interface Standards Summary

Interface Area	Application	Standard	Origin	Comments
Data Communications Equipment (DCE*) to Data Terminal Equipment (DTE)	U.S.A. Industrial	RS-232C	EIA	Unbalanced, Short Lines
		RS-422	EIA	Balanced, Long Lines
		RS-423	EIA	Unbalanced, RS-232 Up-Grade
		RS-449	EIA	System Standard Covering Use of RS-422, RS-423
		RS-485	EIA	Balanced, Long Line Multipoint
	International	CCITT Vol. VIII V. 24	International Telephone and Telegraph Consultative Committee	Similar to RS-232
		CCITT No. 97 X. 26		Similar to RS-423
		CCITT No. 97 X. 27		Similar to RS-422
	U.S.A. Military	MIL-STD-188C	D.O.D.	Unbalanced, Short Lines
		MIL-STD-188-114	D.O.D.	Similar to RS-422, RS-423
		MIL-STD-1397 (NTDS–Slow)	Navy	42k bits/sec.
		MIL-STD-1397 (NTDS–Fast)	Navy	250k bits/sec
	U.S. Government, Non-Military	FED-STD-1020	GSA	Identical to RS-423
		FED-STD-1030	GSA	Identical to RS-422

TABLE I. Common Line Driver/Receiver Interface Standards Summary (Continued)

Interface Area	Application	Standard	Origin	Comments
Computer to Peripheral	IBM 360/370	System 360/370 Channel I/O	IBM	Unbalanced Bus
	DEC Mini-Computer	DEC Unibus®	DEC	Unbalanced Bus
Instrument to Computer	Nuclear Instrumentation	CAMAC (IEEE Std. 583-1975)	NIM (AEC)	DTL/TTL Logic Levels
	Laboratory Instrumentation	488	IEEE	Unbalanced Bus
Microprocessor to Interface Devices	Microprocessor Circuits	Microbus™	National Semiconductor	Short Line; 8-Bit Parallel, Digital Transmission
Facsimile Equipment to DTE	Facsimile Transmission	RS-357	EIA	Incorporates RS-232
Automatic Calling Equipment to DTE	Impulse Dialing and Multi-Tone Keying	RS-366	EIA	Incorporates RS-232
Numerically Controlled Equipment to DTE	Numerically Controlled Equipment	RS-408	EIA	Short Lines (<4 Ft.)

* Changed to "Data Circuit-Terminating Equipment"

TABLE II. Line Driver/Receiver Integrated Circuit Selection Guide for Digital Interface Standards

Standard Designation	Line Driver 0°C to −70°C	Line Driver −55°C to +125°C	Line Receiver 0° to +70°C	Line Receiver −55°C to +125°C
U.S. INDUSTRIAL STANDARDS				
RS-232C	DS1488 DS75150	Not Applicable Not Applicable	DS1489 (A) DS75154	Not Applicable Not Applicable
RS-357	See RS-232C			
RS-366	See RS-232C			
RS-408	DS75453 DS75454	DS55454 DS55454	DS7820A DS75115	DS7820A DS55115
RS-422	DS3691 DS26LS31C DS3487	DS1691A DS26LS31M DS3587	DS88LS120 DS26LS32C DS3486 DS26LS33C DS88C20 DS88C120	DS78LS120 DS26LS32M DS26LS33M DS78C20 DS78C120
RS-423	DS3691 DS3692	DS1691A DS1692	DS88LS120 DS88C20 DS88C120	DS78LS120 DS78C20 DS78C120
RS-449	See RS-422, RS-423			
RS-485 Transceivers	DS3695 DS3696 DS3697 DS3698 DS75176A		DS3695 DS3696 DS3697 DS3698 DS75176A	
IEEE 488	DS3666 DS75160A DS75161A DS75162A		DS3666 DS75160A DS75161A DS75162A	
CAMAC	See RS-232C, RS-422, RS-423 or IEEE 488			
IBM 360/370 I/O Port	DS75123	Not Applicable	DS75124	Not Applicable

TABLE II. Line Driver/Receiver Integrated Circuit Selection Guide for Digital Interface Standards (Continued)

Standard Designation	Part Number			
	Line Driver		Line Receiver	
	0°C to +70°C	−55°C to +125°C	0° to +70°C	−55°C to +125°C
DEC Unibus®	DS36147 DS8641 Transceiver	DS16147 DS7641 Transceiver	DS8640 DS8641 Transceiver	DS7640 DS7641 Transceiver
Microbus™	DS3628 DP8228 DP8216 DP8212 DP8340B Transceiver	DS1628 DP8228M DP8216M DP8212M	DP8304B Transceiver	
GOVERNMENT STANDARDS				
MIL-STD-188C	DS3692	DS1692	DS88LS120	DS78LS120
MIL-STD-188-114	DS3692	DS1692	DS88LS120	DS78LS120
FED-STD-1020	See RS-423			
FED-STD-1030	See RS-422			
MIL-STD-1397 (NTDS–Slow)	Use Discrete Components and/or Comparators			
MIL-STD-1397 (NTDS–Fast)	Use Discrete Components and/or Comparators			
INTERNATIONAL STANDARDS (CCITT)				
1969 White Book Vol. VIII, V. 24	See RS-232C			
Circular No. 97, X. 26	See RS-422			
Circular No. 97, X. 27	See RS-423			

2.0 (DTE) (DCE)
Data terminal equipment (DTE) to data communications equipment (DCE) interface standards

2.1 Application
The DTE/DCE standards cover the electrical, mechanical and functional interface between or among terminals (i.e., teletypewriters, CRT's etc.) and communications equipment (i.e., modems, cryptographics sets, etc.).

2.2 U.S. Industrial DTE/DCE Standards

2.2.1 EIA RS-232
RS-232C is the oldest and most widely known DTE/DCE interface standard. Viewed by many as a complete standard, it provides for one-way/non-reversible, single ended (unbalanced) non terminated line, serial digital data transmission. *Figure 1* shown below illustrates a typical application. See Table III for Specification Summary.

Important features are:
* Positive logic (±5V min to ±15V max)
* Fault protection
* Slew-rate control
* 50 feet recommended cable length
* 20k bits per second data rate

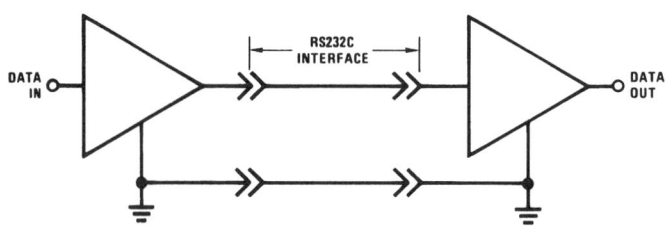

FIGURE 1. EIA RS-232C Application

2.2.2 EIA RS-422, RS-423 and RS-485

In a move to upgrade system capabilities by using state-of-the-art devices and technology the EIA in 1975, introduced two new specifications covering RS-422 balanced and RS-423 unbalanced data transmission. Both of these standards offered major advantages over the popular RS-232C interface. Understanding the advantages of the balanced interface RS-422, the EIA introduced in 1983 the RS-485 Multipoint Systems standard that eliminates several limitations of RS-422.

2.2.2.1 RS-423

RS-423 closely resembles RS-232C in that it, too specifies a one-way/non-reversible, data transmission. Several key advantages of the standard include a 100k Baud data rate at 30 feet and a balanced receiver offering an input voltage common mode (VCM) of ±7V. As shown in *Figure 2* the receiver input is referenced to the driver ground permitting ground differences between the driver and receiver. See Table IV for Specification Summary.

Important features are:

* Positive logic (±4V min to ±6V max)
* Fault protected driver outputs
* Controlled Slew-rate reduces crosstalk and reflections
* 30 feet maximum cable length at 100k Baud
* Differential receiver with ±7V VCM and ±200 mV sensitivity

2.2.2.2 RS-422

RS-422 provides for balanced data tranmission with unidirectional/non-reversible, terminated or non-terminated transmission lines. Several key advantages offered by this standard include the differential receiver defined in RS-423, a differential driver and data rates as high as 10M Baud at 40 feet. *Figure 3* shows a typical interconnect application. See Table V for Specification Summary.

2.2.2.3 RS-485

RS-485 standards accommodates the requirements on a balanced transmission line used in party-line circuit configurations. This standard is similar to RS-422 and is considered to be an extension permitting multipoint applications where multiple drivers and receivers share the same line in data transmission.

Several key characteristics of the standard that differentiate it from RS-422 are; the expanded common mode range of both the driver and receiver, (VCM range +12 to −7V), and characteristics that permit 32 drivers and receivers on the line. *Figure 4* shows a typical party-line application. Note that the transmission line which is intended to be 120Ω twisted pair is terminated at both ends. See Table VI for Specifications Summary.

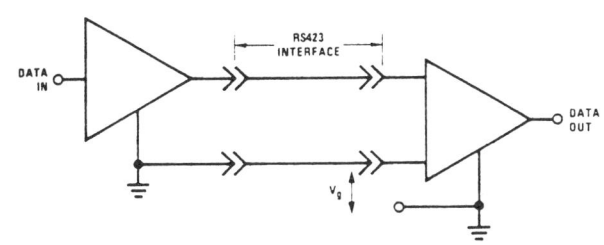

FIGURE 2. EIA RS-423 Application

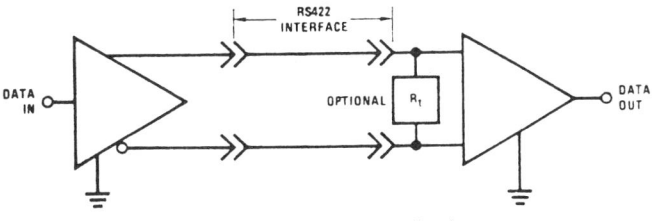

FIGURE 3. EIA RS-222 Application

Note: The termination resistor is defined as optional by RS-422. However this termination resistor is highly recommended to reduce the possibility of line reflections caused by mis-matched impedance between the cable and the driver.

D—Driver
R—Receiver
T—Transceiver

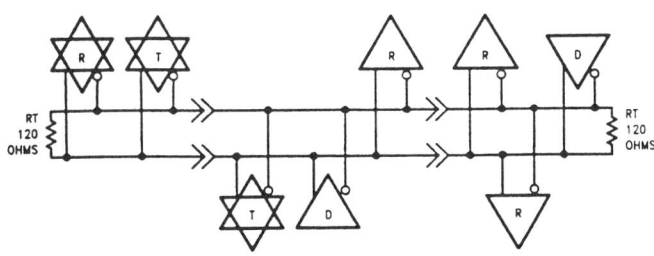

FIGURE 4. A Typical RS-485 Party-Line Configuration

TABLE III. EIA RS-232C Specification Summary

Symbol	Parameter	Conditions	EIA RS-232C Min	EIA RS-232C Typ	EIA RS-232C Max	Units
V_{OH}	Driver Output Voltage Open				25	V
V_{OL}	Circuit		−25			V
V_{OH}	Driver Output Voltage Loaded	$3\,k\Omega \leq R_L \leq 7\,k\Omega$	5		15	V
V_{OL}	Output		−15		−5	V
R_O	Driver Output Resistance Power Off	$-2V \leq V_O \leq 2V$			300	Ω
I_{OS}	Driver Output Short-Circuit Current		−500		500	mA
	Driver Output Slew Rate					
	All Interchange Circuits				30	V/μs
	Control Circuits		6			V/ms
	Rate and Timing Circuits		6			V/ms
		% of Unit Interval	4			%
R_{IN}	Receiver Input Resistance	$3V \leq V_{IN} \leq 25V$	3000		7000	Ω
	Receiver Open Circuit Input Bias Voltage		−2		2	V
	Receiver Input Threshold					
	Output = MARK					
			−3			V
	Output = SPACE				3	V

TABLE IV. EIA RS-423 Specification Summary

Symbol	Parameter	Conditions	EIA RS-423 Min	EIA RS-423 Typ	EIA RS-423 Max	Units
V_O / $\overline{V_O}$	Driver Unloaded Output Voltage		4 / −4		6 / −6	V / V
V_T / $\overline{V_T}$	Driver Loaded Output Voltage	$R_L = 450\,\Omega$	3.6 / −3.6			V / V
R_S	Driver Output Resistance				50	Ω
I_{OS}	Driver Output Short-Circuit Current	$V_O = 0V$			±150	mA
	Driver-Output Rise and Fall Time	Baud Rate ≤ 1k Baud			300	μs
		Baud Rate ≥ 1k Baud			30	% Unit Interval
I_{OX}	Driver Power OFF Current	$V_O = \pm6V$			±100	μA
V_{TH}	Receiver Sensitivity	$V_{CM} \leq \pm7V$			±200	mV
V_{CM}	Receiver Common-Mode Range				±10	V
R_{IN}	Receiver Input Resistance		4000			Ω
	Receiver Common-Mode Input Offset				±3	V

TABLE V. EIA RS-422 Specification Summary

Symbol	Parameter	Conditions	EIA RS-422 Min	EIA RS-422 Typ	EIA RS-422 Max	Units
V_O / $\overline{V_O}$	Driver Unloaded Output Voltage				6 / −6	V / V
V_T / $\overline{V_T}$	Driver Loaded Output Voltage	$R_T = 100\,\Omega$	2 / −2			V / V
R_S	Driver Output Resistance	Per Output			50	Ω
I_{OS}	Driver Output Short-Circuit Current	$V_O = 0V$			150	mA
	Driver Output Rise Time				10	% Unit Interval
I_{OX}	Driver Power OFF Current	$-0.25V \leq V_O \leq 6V$			±100	μA
V_{TH}	Receiver Sensitivity	$V_{CM} = \pm7V$			200	mV
V_{CM}	Receiver Common-Mode Voltage		−12		12	V
	Receiver Input Offset		±3			V
R_{IN}	Receiver Input Resistance		4000			Ω

TABLE VI. EIA RS-485 Specification Summary

Symbol	Parameter	Conditions	EIA RS-485 Min	Typ	Max	Units
V_O / $\overline{V_O}$	Driver Unloaded Output Voltage					V
V_T / $\overline{V_T}$	Driver Loaded Output Voltage	$R_T = 100\Omega$ RS-422	2 / −2			V / V
		$R_T = 54\Omega$, $C_L = 50$ ps RS-485	1.5 / −1.5			V / V
I_{OS}	Driver Output Short-Circuit Current	$V_O = \pm 12V$ $V_O = -7V$			250 / −250	mA / mA
V_{OS}	Driver Common Mode Output Voltage				3	V
$V_{OS} - \overline{V_{OS}}$	Difference in Common Mode Offset				0.2	V
V_{TH}	Receiver Sensitivity	$-7V \leq V_{CM} \leq +12V$			200	mV
V_{CM}	Receiver Common Mode Voltage		−7		+12	V
R_{IN}	Receiver Input Resistance		12k			Ω

2.3 International Standards

2.3.1 CCITT 1969 White Book Vol. VIII, V.24. This standard is identical to RS-232C.

2.3.2 CCITT circular No. 97 Com SPA/13, X. 26. This standard is similar to RS-422 with the exception that the receiver sensitivity at the specified maximum common-mode voltage (±7V) shall be ±300 mV vs ±200 mV for RS-422.

2.3.3 CCITT circular No. 97 Com SPA/13, X. 27. This standard is similar to RS-423 with 2 exceptions:

a) The receiver sensitivity is as specified in paragraph X.26, and

b) The driver output voltage is specified at a load resistance of 3.9 kΩ.

2.4 U.S. Military Standards

2.4.1 MIL-STD-188C (Low Level)

The military equivalent to RS-232C is MIL-STD-188C. Devices intended for RS-232C can be applied to MIL-STD-188C by use of external wave shaping components on the driver end and input resistance and threshold tailoring on the receiver end.

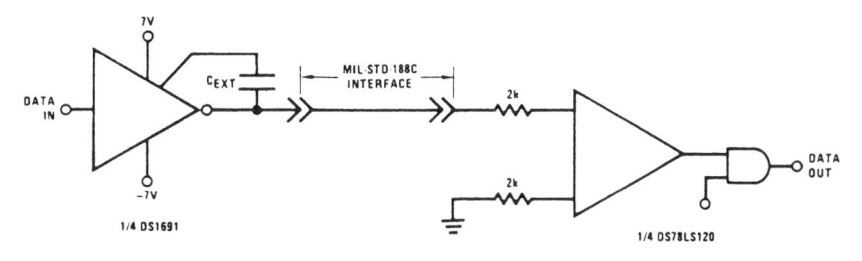

FIGURE 5. MIL-STD-188C Application

TABLE VII. MIL-STD-188C Specification Summary

Symbol	Parameter	Conditions	MIL-STD-188C Low Level Limits Min	Typ	Max	Units
V_{OL}	Driver Output Voltage Open Circuit	(Note 1)	5		7	V
V_{OL}			−7		−5	V
R_O	Driver Output Resistance Power ON	$I_{OUT} \leq 10$ mA			100	Ω
I_{CS}	Driver Output Short-Circuit Current		−100		100	mA
	Driver Output Slew Rate All Interchange Circuits Control Circuits Rate and Timing Circuits	(Note 2)	5		15	% IU
R_N	Receiver Input Resistance	Mode Rate ≤ 200k Baud	6			Ω
	Receiver Input Threshold Output = MARK Output = SPACE	(Note 3)	−100		100	μA / μA

Note 1: Ripple < 0.5%, V_{OH}, V_{OL} matched to within 10% of each other.
Note 2: Waveshaping required on driver output such that the signal rise or fall time is 5% to 15% of the unit interval at the applicable modulation rate.
Note 3: Balance between marking and spacing (threshold) currents actually required shall be within 10% of each other.

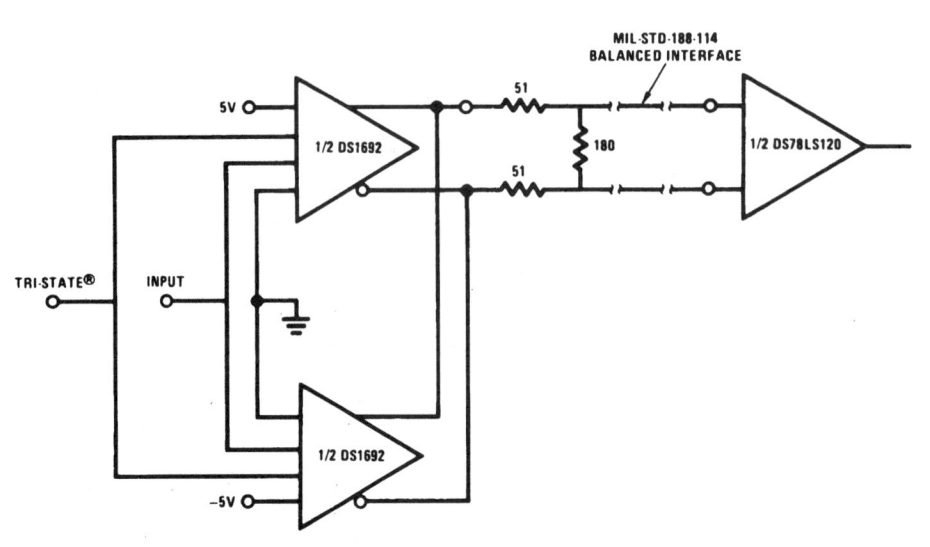

FIGURE 6. MIL-STD-188-114 (Balanced Applications)

2.4.2 MIL-STD-188-114 Balanced

This standard is similar to RS-422 with the exception that the driver offset voltage level is limited to ±0.4V vs ±3V allowed in RS-422.

2.4.3 MIL-STD-188-114 Unbalanced.

This standard is similar to RS-423 with the exception that loaded circuit driver output voltage at $R_L = 450\Omega$ must be 90% of the open circuit output voltage vs ±2V at $R_S = 100\Omega$ for RS-422.

2.4.4 MIL-STD-1397 (Slow and Fast)

2.5 FED-STD-1020/1030

U.S. Government (non-military) standards FED-STD-1020 and 1030 are identical without exception to EIA RS-423 and RS-422, respectively.

3.0 COMPUTER TO PERIPHERAL INTERFACE STANDARDS

To date, the only standards dealing with the interface between processors and other equipment are the "defacto" standards in the form of specifications issued by IBM and DEC covering the models 360/370 I/O ports and the Unibus, respectively.

3.1 GA-22-6974-0

IBM specification GA-22-6974-0 covers the electrical characteristics, the format of information and the control sequences of the data transmitted between 360/370's and up to 10 I/O ports.

The interface is an unbalanced bus using 95Ω, terminated, coax cables. Devices connected to the bus should feature short-circuit protection, hysteresis in the receivers, and open-emitter drivers. Careful attention should be paid to line lengths and quality in order to limit cable noise to less than 400 mV.

TABLE VIII. MIL-STD-1397 Specification Summary

Symbol	Parameter	Conditions	Comparison Limits (MIL-STD) 1397 (Slow)	Comparison Limits (MIL-STD) 1397 (Fast)	Units
	Data Transmission Rate		42	250	k Bits/Sec
V_{OH} V_{OL}	Driver Output Voltage		±1.5 −10 to −15.5	0 −3	V V
I_{OH} I_{OL}	Driver Output Current		≥ −4 1		mA mA
R_S	Driver Power OFF Impedance		≥ 100		kΩ
V_{IH} V_{IL}	Receiver Input Voltage	Fail-Safe Open Circuit	≤ 4.5 ≥ −7.5	≤ −1.1 ≥ −1.9	V V

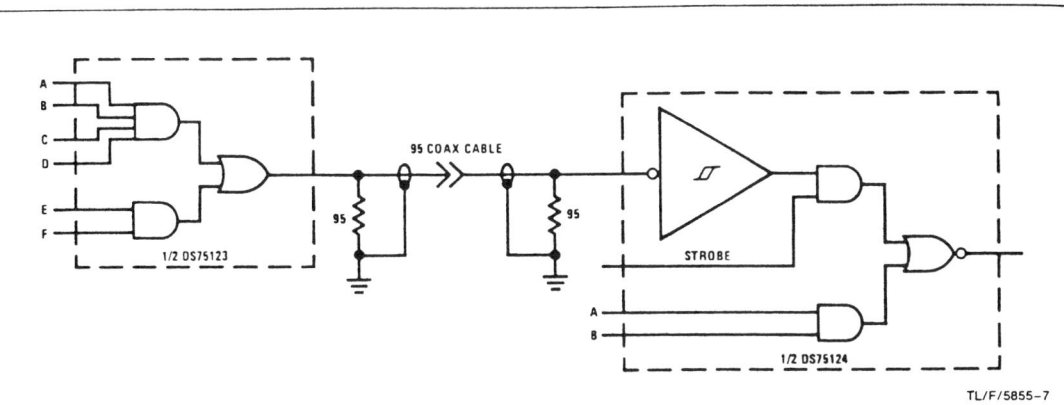

FIGURE 7. IBM 360/370 I/O Application

TABLE IX. IBM 360/370 Specification Summary

Symbol	Parameter	Conditions	IBM 360/370 Min	Typ	Max	Units
V_{OH}	Driver Output Voltage	I_{OH} = 123 mA			7	V
V_{OH}		I_{OH} = 30 µA			5.85	V
V_{OH}		I_{OH} = 59.3 mA	3.11			V
V_{OL}		I_{OL} = −240 µA			0.15	V
V_{IH}	Receiver Input Threshold				1.7	V
V_{IL}	Voltage		0.7			V
I_{IH}	Receiver Input Current	V_{IN} = 3.11V			−0.42	mA
I_{IL}		V_{IN} = 0.15V	0.24			mA
V_{IN}	Receiver Input Voltage Range Power ON		−0.15		7	V
V_{IN}	Power OFF		−0.15		6	V
V_{IN}	Power ON		−0.15		7	V
V_{IN}	Power OFF		−0.15		6	V
R_{IN}	Receiver Input Impedance	$0.15V \leq V_{IN} \leq 3.9V$	7400			Ω
I_{IN}	Receiver Input Current	V_{IN} = 0.15V			240	µA
Z_O	CABLE Impedance		83		101	Ω
R_O	CABLE Termination Line Length (Specified as Noise on Signal and Ground Lines)	$P_D \geq 390$ mW	90		100 400	Ω mV

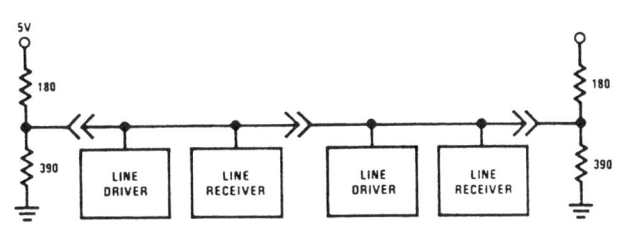

FIGURE 8. DEC Unibus Application

TABLE X. DEC Unibus Specification Summary

Symbol	Parameter	Conditions	DEC Unibus Min	DEC Unibus Typ	DEC Unibus Max	Units
V_{OL}	Driver Output Voltage	I_{OL} = 50 mA			0.7	V
V_O		Absolute Maximum			7	V
V_{IH}	Receiver Input Voltage		1.7			V
V_{IL}					1.3	V
I_{IH}	Receiver Input Current	V_{IN} = 4V			100	µA
I_{IL}		V_{IN} = 4V Power OFF			100	µA

3.2 DEC UNIBUS

Another example of an unofficial industry standard is the interface to a number of DEC minicomputers. This interface, configured as a 120Ω double-terminated data bus is given the name Unibus. Devices connected to the bus should feature hysteresis in the receivers and open-collector driver outputs. Cable noise should be held to less than 600 mV.

4.0 INSTRUMENTATION TO COMPUTER INTERFACE STANDARDS

4.1 INTRODUCTION

The problem of linking instrumentation to processors to handle real-time test and measurement problems was largely a custom interface problem. Each combination of instruments demanded unique interfaces, thus inhibiting the wide spread usage of small processors to day-to-day test, measurement and control applications.

Two groups addressed the problem for specific environments. The results are:

a) IEEE 488 bus standard based upon proposals made by HP, and

b) The CAMAC system pioneered by the nuclear physics community.

4.2 IEEE 488

IEEE 488 covers the functional, mechanical and electrical interface between laboratory instrumentation (i.e. signal generators, DPM's, counters, etc.) and processors such as programmable calculators and minicomputers. Equipment with IEEE 488 I/O ports can be readily daisy chained in any combination of up to 15 equipments (including processor) spanning distances of up to 60 feet. 16 lines (3 handshake, 5 control and 8 data lines) are required.

4.3 CAMAC

The CAMAC system is the result of efforts by those in the nuclear physics community to standardize the interface between laboratory instruments and computers before the introduction of IEEE 488.

It allows either serial or parallel interconnection of instruments via a "crate" controller.

The electrical requirements of the interfaces are compatible with DTL and TTL logic levels.

5.0 MICROPROCESSOR SYSTEMS INTERFACE STANDARDS

5.1 Microprocessor Systems

Microprocessor systems are bus organized systems with two types of bus requirements:

a) Minimal system: for data transfer over short distances (usually on 1 PC board), and,

b) Expanded system: for data transfer to extend the memory or computational capabilities of the system.

5.2 Minimal Systems and Microbus

Microbus considers the interface between MOS/LSI microprocessors and interfacing devices in close physical proximity which communicate over 8-bit parallel unified bus systems. It specifies both the functional and electrical characteristics of the interface and is modeled after the 8060, 8080 and 8090 families of microprocessors as shown in *Figures 10, 11* and *12*.

The electrical characteristics of Microbus are shown in Table XII.

TABLE XI. IEEE 488 Specification Summary

Symbol	Parameter	Conditions	IEEE 488 Min	IEEE 488 Typ	IEEE 488 Max	Units
V_{OH}	Driver Output Voltage	I_{OH} = −5.2 mA	2.4			V
V_{OL}		I_{OL} = 48 mA			0.4	V
I_{OZ}	Driver Output Current TRI-STATE®	V_O = 2.4V			±40	µA
I_{OH}	Open Collector	V_O = 5.25V			250	µA
V_{IH}	Receiver Input Voltage	0.4V Hysteresis Recommended	2.0			V
V_{IL}					0.8	V
I_{IH}	Receiver Input Current	V_{IN} = 2.4V			40	µA
I_{IL}		V_{IN} = 0.4V			−1.6	mA
	Receiver Clamp Current	V_{IN} = −1.5V			12	mA
R_{L1}	Termination Resistor	V_{CC} = 5V (±5%)	2850		3150	
R_{L2}		V = Gnd	5890		6510	

TABLE XII. Microbus Electrical Specification Summary

Symbol	Parameter	Driver	Receiver Standard	Receiver Hysteresis (Recommended)	Units
V_{OL}	Output Voltage (At 1.6 mA)	≤ 0.4V			
V_{OH}	(At −100 μA)	≥ 2.4V			
V_{IL}	Input Voltage		0.8	0.6	V
V_{IH}			2.0	2.0	V
	Internal Capacitive Load at 25°C	15	10	10	pF
t_r	Rise Time (Maximum)	100			ns
t_f	Fall Time (Maximum)	100			ns

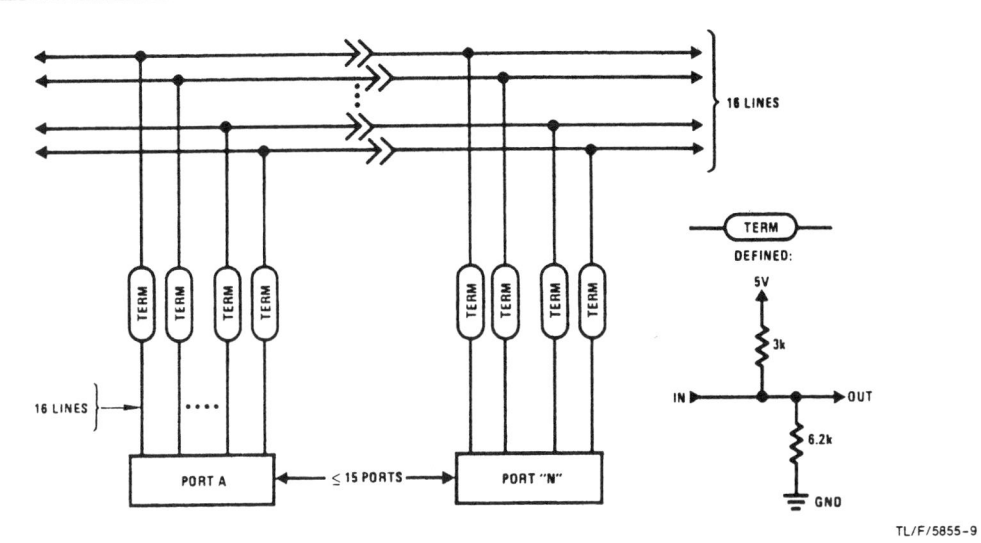

FIGURE 9. IEEE 488 Application

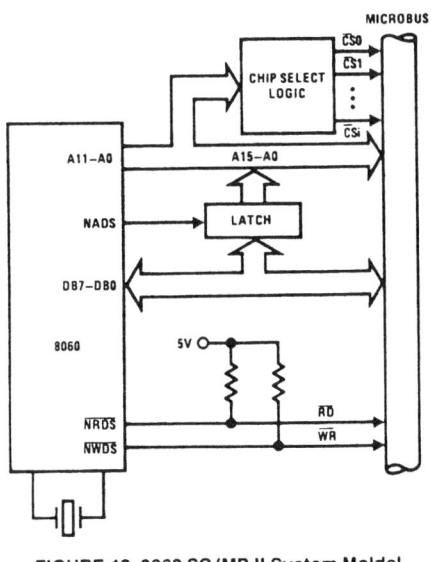

FIGURE 10. 8060 SC/MP II System Moldel

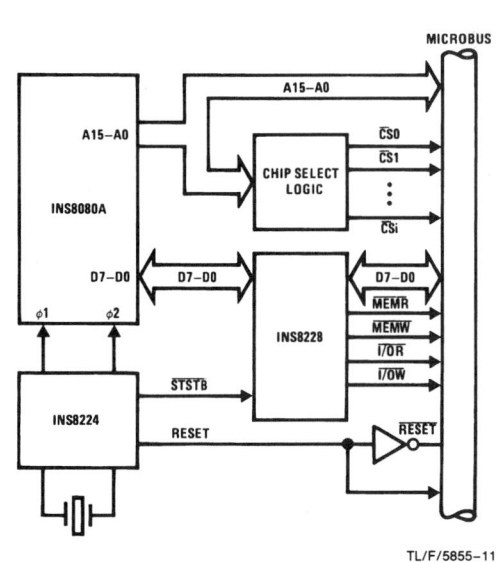

FIGURE 11. 8080 System Model for the Basic Microbus Interface

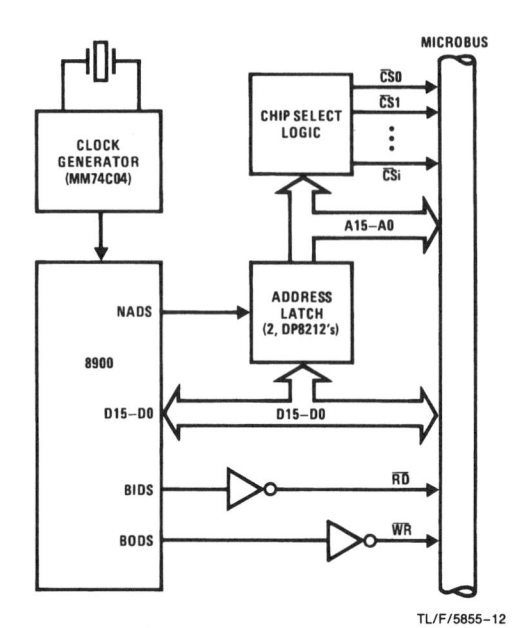

FIGURE 12. 8900 System Model

5.3 Expanded Microprocessor System Interfaces

Since the outputs of most microprocessor devices are limited to a loading of one relative to a TTL load, expanded system will require buffers on both their address and data lines.

To date, no formal standards exist which govern this interface. However, "defacto" standards are emerging in the form of the specifications for "recommended devices" which are mentioned in the data sheets and application notes for the widely sourced microprocessor devices. Here, the answer to the question of how to provide a "standard" interface is simplified to that of proper usage of recommended devices.

Table XIII summarizes the important electrical characteristics of recommended bus drivers for expanded microprocessor systems.

6.0 OTHER INTERFACE STANDARDS

Some other commonly occurring interfaces which have become standardized are:

a) Interface between facsimile terminals and voice frequency communication terminals,

b) Interface between terminals and automatic calling equipment used for data communications, and

c) Interface between numerically controlled equipment and data terminals.

TABLE XIII. Recommended Specification of Bus Drivers for Expanded Microprocessor Systems

Symbol	Parameter	Conditions	Min	Typ	Max	Units
V_{IH}			2			V
V_{IL}					0.8	V
V_{OH}	Driver Output Voltage	$I_{OH} = -10$ mA	2.4			V
V_{OL}		$I_{OL} = 48$ mA			0.5	V
I_{OS}	Short-Circuit Current	$V_{CC} = 5.25$V			−150	mA
C_L	Bus Drive Capability		300			pF

6.1 EIA RS-357

RS-357 defines the electrical, functional and mechanical characteristics of the interface between analog facsimile equipment to be used for telephone data transmission and the data sets used for controlling/transmitting the data.

Figure 13 summarizes the functional and electrical characteristics of RS-357.

6.2 EIA RS-366

RS-366 defines the electrical, functional and mechanical characteristics of the interface between automatic calling equipment for data communications and data terminal equipment.

The electrical characteristics are encompassed by RS-232C.

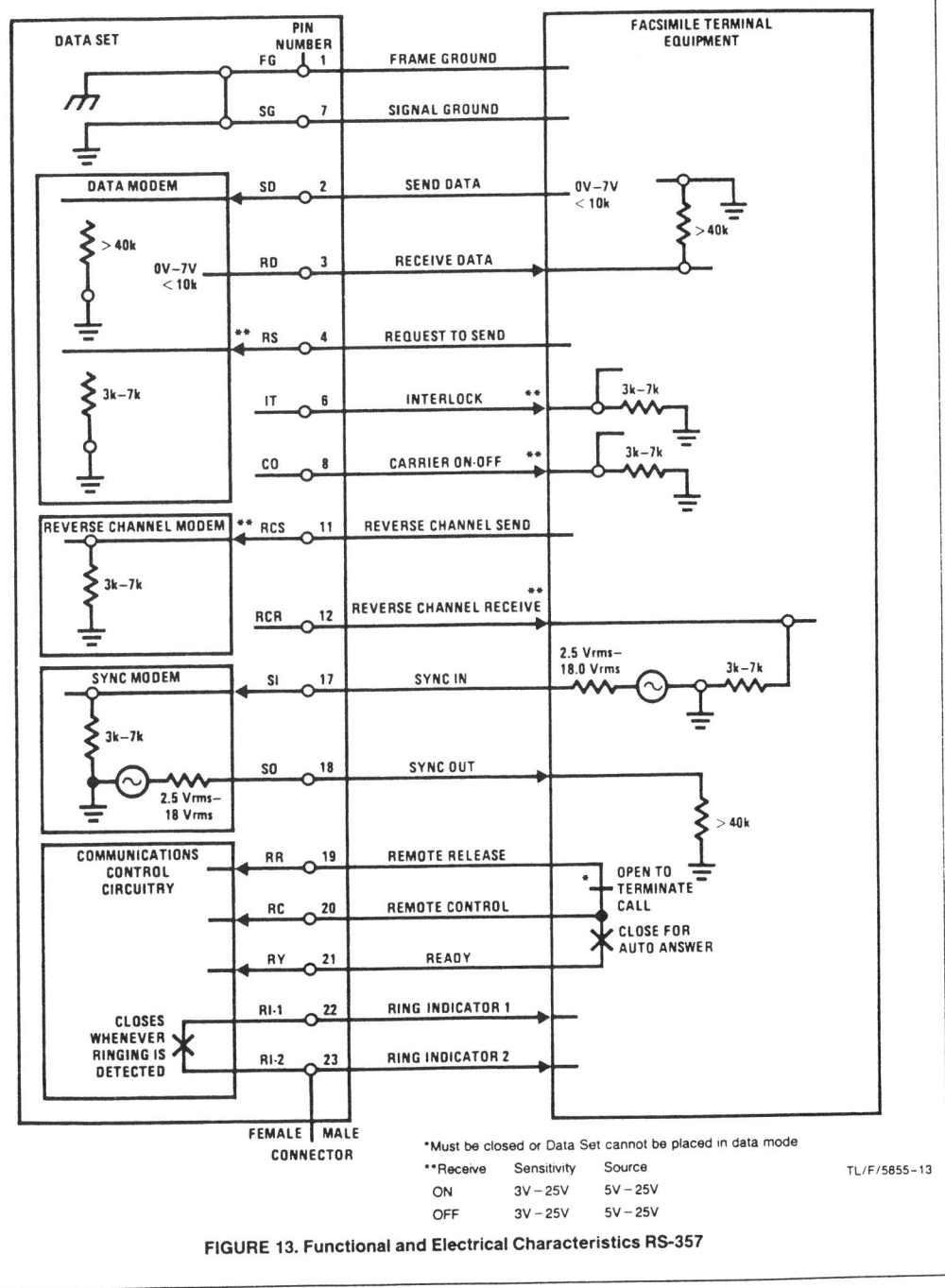

FIGURE 13. Functional and Electrical Characteristics RS-357

6.3 EIA RS-408

RS-408 recommends the standardization of the 2 interfaces shown in *Figure 14*.

The electrical characteristics of NCE to DTE interface are, in summary, those of conventional TTL drivers (series 7400) with:

$V_{OL} \leq 0.4V$ at $I_{OL} = 48$ mA

$V_{OH} \geq 2.4V$ at $I_{OH} \leq -1.2$ mA, and

$C_L \leq 2000$ pF.

Short circuit protection should be provided.

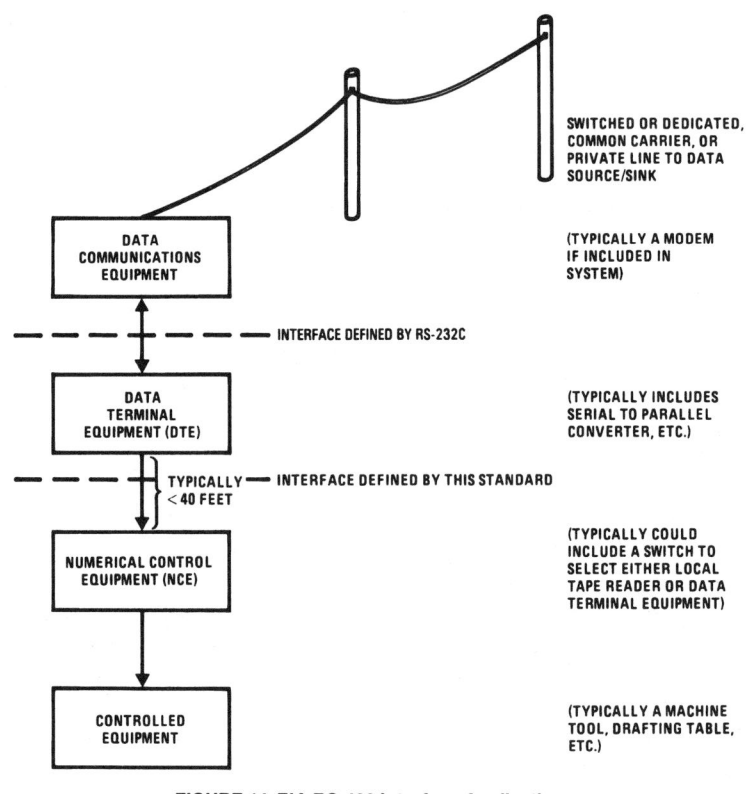

FIGURE 14. EIA RS-408 Interface Applications

Transmission Line Drivers and Receivers for EIA Standards RS-422 and RS-423

National Semiconductor Corp.
Application Note 214
John Abbott

With the advent of the microprocessor, logic designs have become both sophisticated and modular in concept. Frequently the modules making up the system are very closely coupled on a single printed circuit board or cardfile. In a majority of these cases a standard bus transceiver will be adequate. However because of the distributed intelligence ability of the microprocessor, it is becoming common practice for the peripheral circuits to be physically separated from the host processor with data communications being handled over cables (e.g. plant environmental control or security system). And often these cables are measured in hundreds or thousands of feet as opposed to inches on a backplane. At this point the component wavelengths of the digital signals may become shorter than the electrical length of the cable and consequently must be treated as transmission lines. Further, these signals are exposed to electrical noise sources which may require greater noise immunity than the single chassis system.

It is the object of this application note to underscore the more important design requirements for balanced and unbalanced transmission lines, and to show that National's DS1691 driver and DS78LS120 receiver meet or exceed all of those requirements.

THE REQUIREMENTS

The requirements for transmission lines and noise immunity have been adequately recognized by National Semiconductor's application note AN-108 and EIA standards RS-422 (balanced) and RS-423 (unbalanced). A summary review of these notes will show that the controlling factors in a voltage digital interface are:

1) The cable length
2) The modulation rate
3) The characteristic of the interconnection cable
4) The rise time of the signal

RS-422 and RS-423 contain several useful guidelines relative to the choice of balanced circuits versus unbalanced circuits. *Figures 1a* and *1b* are the digital interface for balanced *(1a)* and unbalanced *(1b)* circuits.

Even though the unbalanced interface circuit is intended for use at lower modulation rates than the balanced circuit, its use is not recommended where the following conditions exist:

1) The interconnecting cable is exposed to noise sources which may cause a voltage sufficient to indicate a change of binary state at the load.
2) It is necessary to minimize interference with other signals, such as data versus clock.
3) The interconnecting cable is too long electrically for unbalanced operation *(Figure 2)*.

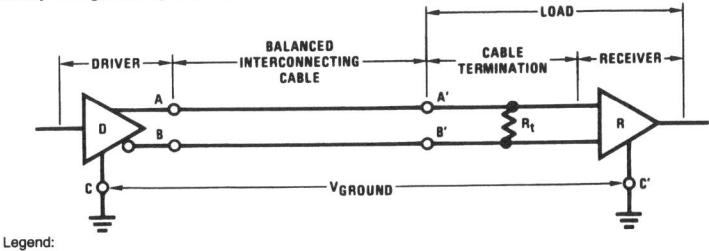

Legend:
R_t = Optional cable termination resistance/receiver input impedance.
V_{GROUND} = Ground potential difference
A, B = Driver interface
A', B' = Load interface
C = Driver circuit ground
C' = Load circuit ground

FIGURE 1a. RS-422 Balanced Digital Interface Circuit

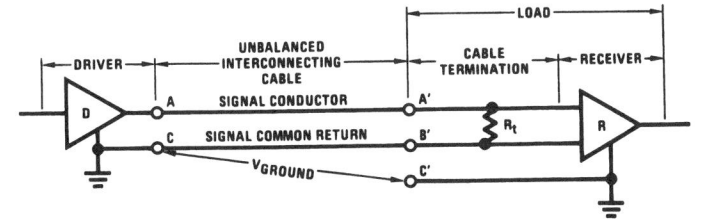

Legend:
R_t = Transmission line termination and/or receiver input impedance
V_{GROUND} = Ground potential difference
A, C = Driver interface
A', B' = Load interface
C = Driver circuit ground
C' = Load circuit ground

FIGURE 1b. RS-423 Unbalanced Digital Interface Circuit

CABLE LENGTH

While there is no maximum cable length specified, guidelines are given with respect to conservative operating distances as a function of modulation rate. *Figure 2* is a composite of the guidelines provided by RS-422 and RS-423 for data modulation versus cable length. The data is for 24 AWG twisted pair cable terminated for worst case (due to IR drop) in a 100Ω load, with rise and fall times equal to or less than one half unit interval at the applied modulation rate.

The maximum cable length between driver and load is a function of the baud rate. But it is influenced by:

1) A maximum common noise range of ±7 volts

   A) The amount of common-mode noise

   Difference of driver and receiver ground potential plus driver offset voltage and coupled peak random noise.

   B) Ground potential differences between driver and load.

   C) Cable balance

   Differential noise caused by imbalance between the signal conductor and the common return (ground)

2) Cable termination

   At rates above 200 kilobaud or where the rise time is 4 times the one way propagation delay time of the cable (RS-422 Sec 7.1.2)

3) Tolerable signal distortion

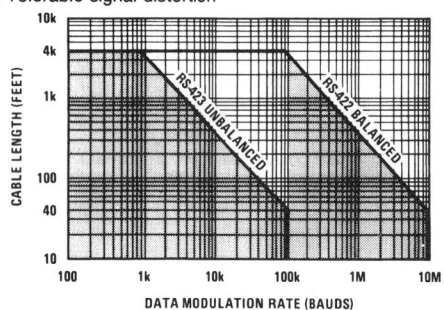

FIGURE 2. Data Modulation Rate vs Cable Length

MODULATION RATE

Section 3 of RS-422 and RS-423 states that the unbalanced voltage interface will normally be utilized on data, timing or control circuits where the modulation rate on these circuits is below 100 kilobauds, and balanced voltage digital interface on circuits up to 10 megabauds. The voltage digital interface devices meeting the electrical characteristics of this standard need not meet the entire modulation range specified. They may be designed to operate over narrower ranges to more economically satisfy specific applications, particularly at the lower modulation rates.

As pointed out in AN-108, the duty cycle of the transmitted signal contributes to the distortion. The effect is the result of rise time. Due to delay and attenuation caused by the cable, it is possible due to AC averaging of the signal, to be unable to reach one binary level before it is changed to another. If the duty cycle is ½ (50%) and the receiver threshold is midway between logic levels, the distortion is small. However, if the duty cycle were ⅛ (12.5%) the signal would be considerably distorted.

CHARACTERISTICS

Driver Unbalanced (RS-423)

The unbalanced driver characteristics as specified by RS-423 Sec 4.1 are as follows:

1) A driver circuit should be a low impedance (50Ω or less) unbalanced voltage source that will produce a voltage applied to the interconnecting cable in the range of 4V to 6V.

2) With a test load of 450Ω connected between the driver output terminal and the driver circuit ground, the magnitude of the voltage (V_T) measured between the driver output and the driver circuit ground shall not be less than 90% of the magnitude for either binary state.

3) During transitions of the driver output between alternating binary states, the signal measured across a 450Ω test load connected between the driver output and circuit ground should be such that the voltage monotonically changes between 0.1 and 0.9 of V_{SS}. Thereafter, the signal shall not vary more than 10% of V_{SS} from the steady state value, until the next binary transition occurs, and at no time shall the instantaneous magnitude of V_T and $\overline{V_T}$ exceed 6V, nor be less than 4V. V_{SS} is defined as the voltage difference between the 2 steady state values of the driver output.

Driver Balanced (RS-422)

The balanced driver characteristics as specified by RS-422 Sec 4.1 are as follows:

1) A driver circuit should result in a low impedance (100Ω or less) balanced voltage source that will produce a differential voltage applied to the interconnecting cable in the range of 2V to 6V.

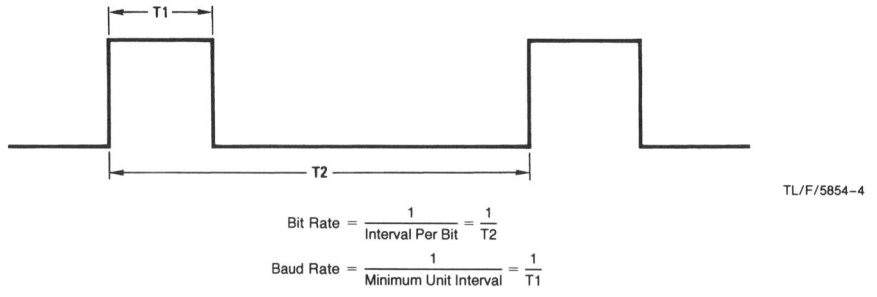

$$\text{Bit Rate} = \frac{1}{\text{Interval Per Bit}} = \frac{1}{T2}$$

$$\text{Baud Rate} = \frac{1}{\text{Minimum Unit Interval}} = \frac{1}{T1}$$

FIGURE 3a. Definition of Baud Rate

1/2 DUTY CYCLE DATA

1/2 DUTY CYCLE LINE RESPONSE V_{TH}

TL/F/5854-5

1/8 DUTY CYCLE DATA

1/8 DUTY CYCLE LINE RESPONSE V_{TH}

TL/F/5854-6

FIGURE 3b. Signal Distortion Due to Duty Cycle

$V_{SS} = |V_t - \overline{V_t}|$
V_{SS} = Difference in steady state voltages

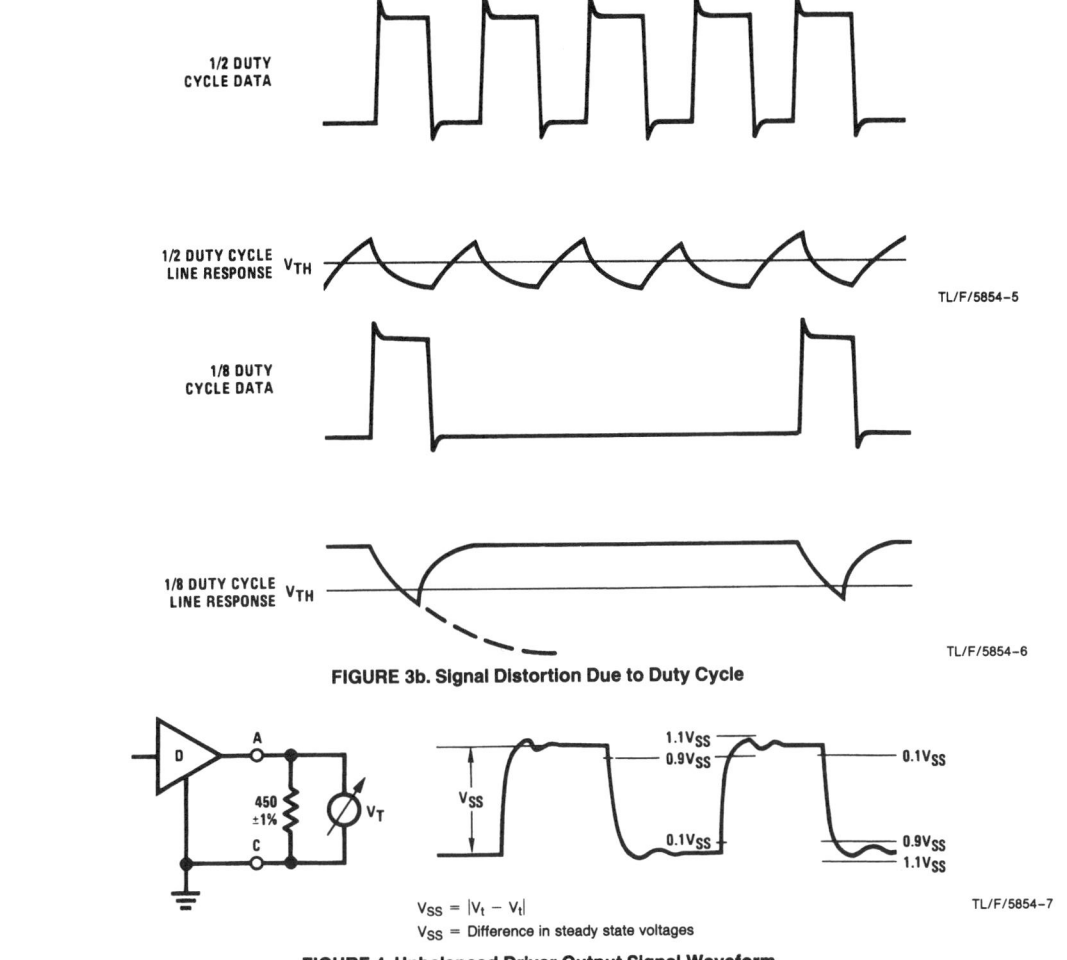

TL/F/5854-7

FIGURE 4. Unbalanced Driver Output Signal Waveform

2) With a test load of 2 resistors, 50Ω each, connected in series between the driver output terminals, the magnitude of the differential voltage (VT) measured between the 2 output terminals shall not be less than either 2.0V or 50% of the magnitude of V_O, whichever is greater. For the opposite binary state the polarity of VT shall be reversed ($\overline{VT}$). The magnitude of the difference in the magnitude of VT and $\overline{VT}$ shall be less than 0.4V. The magnitude of the driver offset voltage (V_{OS}) measured between the center point of the test load and driver circuit ground shall not be greater than 3.0V. The magnitude of the difference in the magnitude of V_{OS} for one binary state and $\overline{V_{OS}}$ for the opposing binary state shall be less than 0.4V.

3) During transitions of the driver output between alternating binary states, the differential signal measured across a 100Ω test load connected between the driver output terminals shall be such that the voltage monotonically changes between 0.1 and 0.9 of V_{SS} within 0.1 of the unit interval or 20 ns, whichever is greater. Thereafter the signal voltage shall not vary more than 10% of V_{SS} from the steady state value, until the next binary transition occurs, and at no time shall the instantaneous magnitude of VT or $\overline{VT}$ exceed 6V, nor less than 2V.

Interconnecting Cable

The characteristics of the interconnecting cable should result in a transmission line with a characteristic impedance in the general range of 100Ω to frequencies greater than 100 kHz, and a DC series loop resistance not exceeding 240Ω. The cable may be composed of twisted or untwisted pair (flat cable) possessing the characteristics specified in RS-422 Sec 4.3 as follows:

1) Conductor size of the 2 wires shall be 24 AWG or larger with wire resistance not to exceed 30Ω per 1000 feet per conductor.

2) Mutual pair capacitance between 1 wire in the pair to the other shall not exceed 20 pF per foot.

3) Stray capacitance between 1 wire in the pair with all other wires connected to ground, shall not exceed 40 pF per foot.

FIGURE 5. Balanced Driver Output Signal Waveform

t_b = Time duration of the unit interval at the applicable modulation rate.
$t_r \leq 0.1\ t_b$ when $t_b \geq 200$ ns
$t_r \leq 20$ ns when $t_b < 200$ ns
V_{SS} = Difference in steady state voltages
$V_{SS} = |V_t - V_{\bar{t}}|$

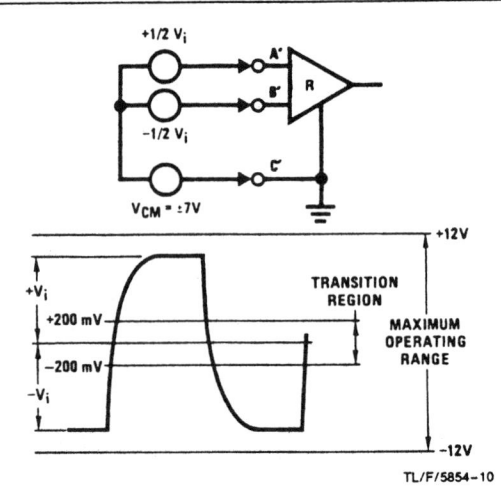

FIGURE 6. Receiver Input Sensitivity Measurement

Note: Designers of terminating hardware should be aware that slow signal transitions with superimposed noise present may give rise to instability or oscillations in the receiving device, and therefore appropriate techniques should be implemented to prevent such behavior. For example, adequate hysteresis and response control may be incorporated into the receiver to prevent such conditions.

Receiver

The load characteristics are identical for both balanced (RS-422) and unbalanced (RS-423) circuits. Each consists of a receiver and optional termination resistance as shown in *Figure 1*. The electrical characteristics single receiver without termination or optional fail-safe provisions are specified in RS-422/423 Sec 4.2 as follows:

1) Over an entire common-mode voltage range of −7V to +7V, the receiver shall not require a differential input voltage of more than 200 mV to correctly assume the intended binary state. The common-mode voltage (V_{CM}) is defined as the algebraic mean of the 2 voltages appearing at the receiver input terminals with respect to the receiver circuit ground. Reversing the polarity of V_T shall cause the receiver to assume the opposite binary state. This allows for operations where there are ground differences caused by IR drop and noise of up to ±7V.

2) To maintain correct operation for differential input signal voltages ranging between 200 mV and 6V in magnitude.

3) The maximum voltage present between either receiver input terminal and receiver circuit ground shall not exceed 10V (3V signal plus 7V common-mode) in magnitude nor cause the receiver to operationally fail. Additionally, the receiver shall tolerate a maximum differential signal of 12V applied across its input terminals without being damaged.

4) The total load including up to 10 receivers shall not have a resistance greater than 90Ω for balanced, and 400Ω for unbalanced at its input points and shall not require a differential input voltage of greater than 200 mV for all receivers to assume the correct binary state.

5) Fail-safe operation per RS-423 Sec 4.2.5 states that other standards and specifications using the electrical characteristics of the unbalanced interface circuit may require that specific interchange leads be made fail-safe to certain fault conditions. Where fail-safe operation is required by such referencing standards and specifications, a provisions shall be incorporated in the load to provide a steady binary condition (either "1" or "0") to protect against certain fault conditions (open or shorted cable).

The designer should be aware that in circuits employing pull-up resistors, the resistors used become part of the termination.

SIGNAL RISE TIME

The signal rise time is a high frequency component which causes interference (near end cross-talk) to be coupled to adjacent channels in the interconnecting cable. The near-end crosstalk is a function of both rise time and cable length, and in considering wave shaping, both should be considered. Since in the balanced voltage digital interface the output is complementary, there is practically no cross-talk coupled and therefore wave shaping is limited to unbalanced circuits.

Per RS-423 Sec 4.1.6, the rise time of the signal should be controlled so that the signal has reached 90% of V_{SS} between 10% and 30% of the unit interval at the maximum modulation rate. Below 1 kilobaud the time to reach 90% V_{SS} shall be between 100 μs and 300 μs. If a driver is to operate over a range of modulation rates and employ a fixed amount of wave shaping which meets the specification for the maximum modulation rate of the operating range, the wave shaping is considered adequate for all lesser modulation rates.

However a major cause of distortion is the effect the transmission line has on the rise time of the transmitted signal. *Figure 7* shows the effect of line attenuation and delay to a voltage step as it progresses down the cable. The increase of the rise time with distance will have a considerable effect on the distortion at the receiver. Therefore in fixing the amount of wave shaping employed, caution should be taken not to use more than the minimum required.

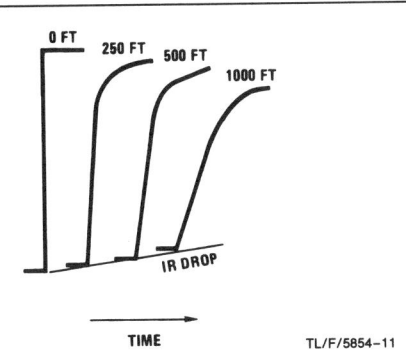

FIGURE 7. Signal Rise Time on Transmission Line vs Line Length

DS1691A, DS78LS120

The Driver

The DS1691A/DS3691 are low power Schottky TTL line drivers designed to meet the above listed requirements of EIA standard RS-422 and RS-423. They feature 4 buffered outputs with high source and sink current capability with internal short circuit protection. The DS1691/DS3691 employ a mode selection pin which allows the circuit to become either a pair of balanced drivers *(Figure 8)* or 4 independent unbalanced drivers *(Figure 9)*. When configured for unbalanced operation *(Figure 10)* a rise time control pin allows the use of an external capacitor to control rise time for suppression of near end cross-talk to adjacent channels in the interconnect cable. *Figure 11* is the typical rise time vs external capacitor used for wave shaping.

The DS3691 configured for RS-422 is connected $V_{CC} = 5V$ $V_{EE} = 0V$, and configured for RS-423 connected $V_{CC} = 5V$ $V_{EE} = -5V$. For applications outside RS-422 conditions and for greater cable lengths the DS1691/DS3691 may be connected with a V_{CC} of 5 volts and V_{EE} of -5 volts. This will create an output which is symmetrical about ground, similar to Mil Standard 188-114.

When configured as balanced drivers *(Figure 8)*, each of the drivers is equipped with an independent TRI-STATE® control pin. By use of this pin it is possible to force the driver into its high impedance mode for applications using party line techniques.

If the common-mode voltage, between driver 1 and all other drivers in the circuit, is small then several line drivers (and receivers) may be incorporated into the system. However, if the common-mode voltage exceeds the TRI-STATE common-mode range of any driver, then the signal will become attenuated by that driver to the extent the common-mode voltage exceeds its common-mode range (see *Figure 12*, top waveform).

It is important then to select a driver with a common-mode range equal to or larger than the common-mode voltage requirement of the system. In the case of RS-422 and RS-423 the minimum common-mode range would be $\pm 7V$. The DS1692/DS3692 driver is tested to a common-mode range of $\pm 10V$ and will operate within the requirements of such a system (see *Figure 12*, bottom waveform).

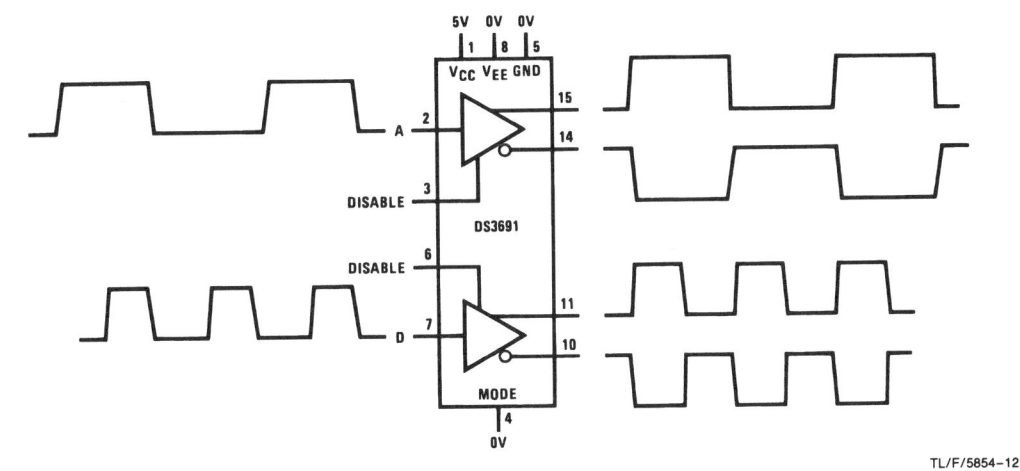

FIGURE 8. DS3691 Connected for Balanced Mode Operation

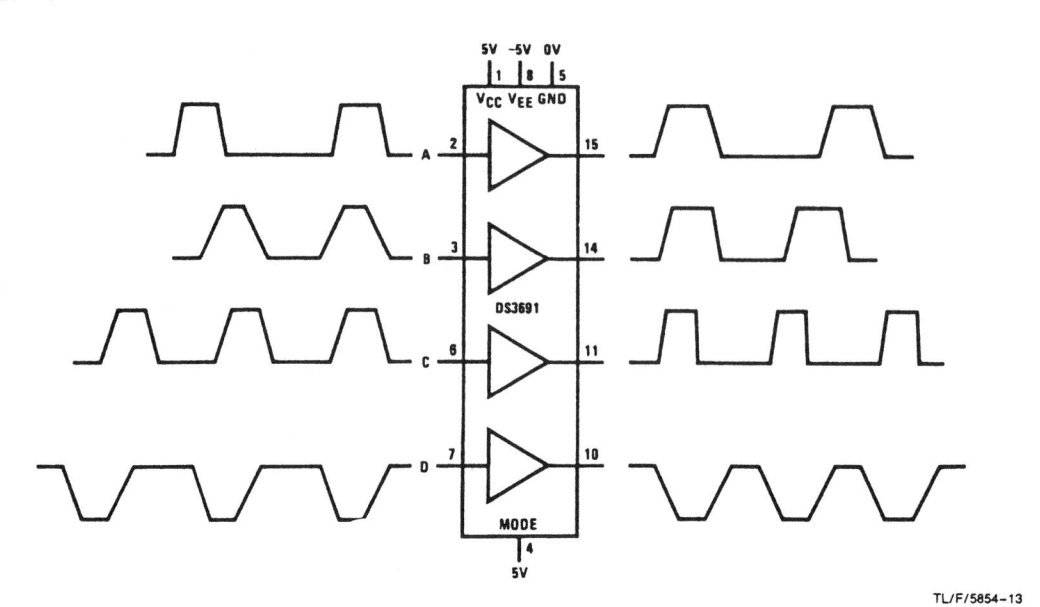

FIGURE 9. DS3691 Connected for Unbalanced Mode Operation

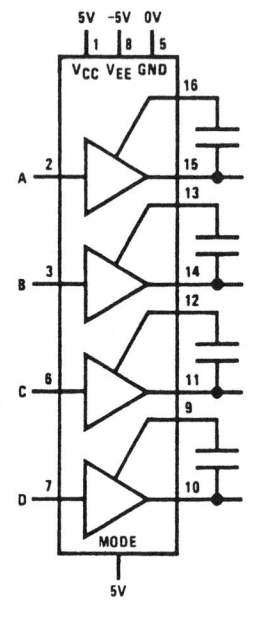

FIGURE 10. Using an External Capacitor to Control Rise Time of DS3691

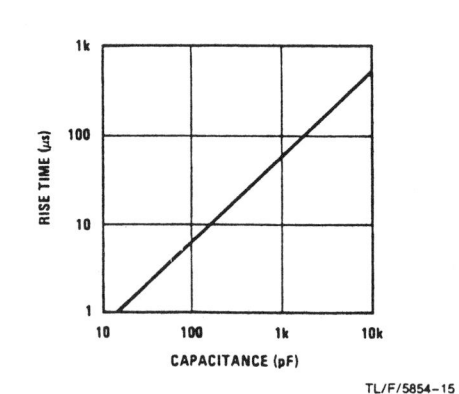

FIGURE 11. DS3691 Rise Time vs External Capacitor

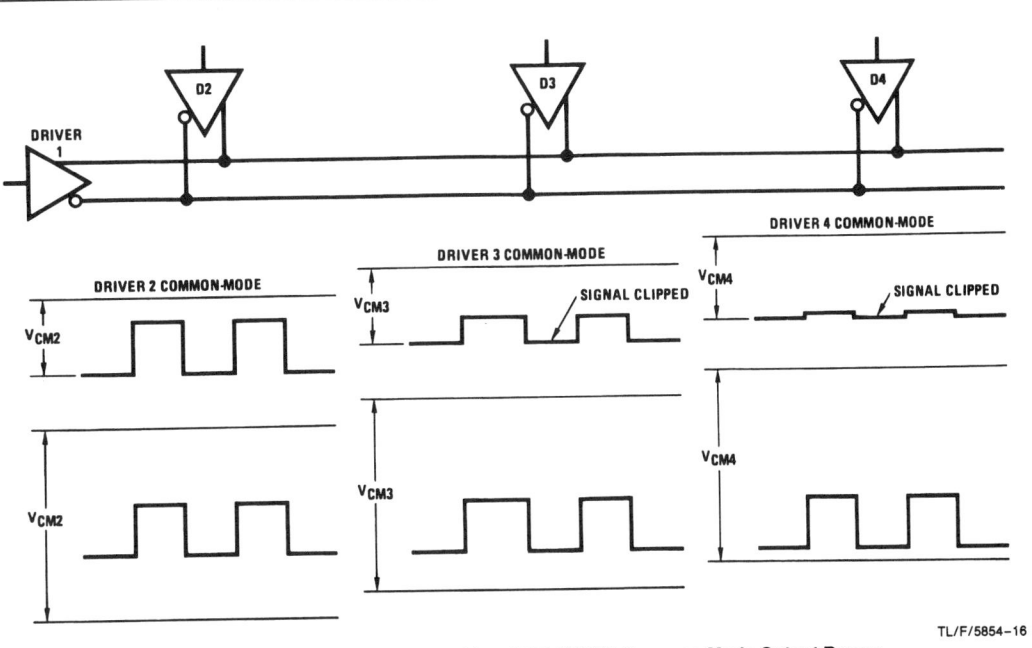

FIGURE 12. Comparison of Drivers without TRI-STATE Common-Mode Output Range (top waveforms) to DS3691 (bottom waveforms)

Top View

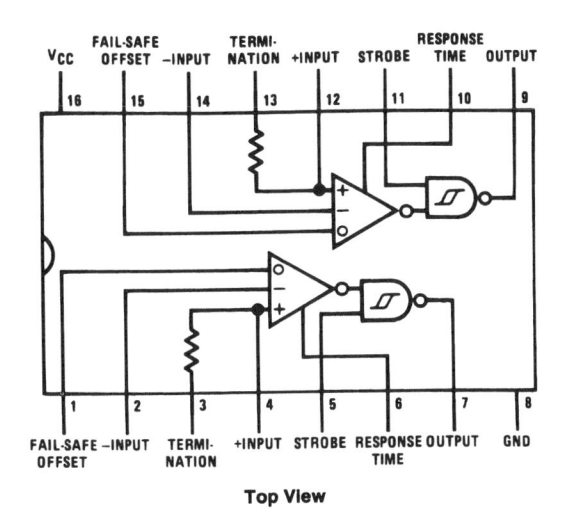

FIGURE 13. DS78LS120/DS88LS120 Dual Differential Line Receiver

DS78LS120/DS88LS120

The Receiver

The DS78LS120/DS88LS120 are high peformance, dual differential TTL compatible line receivers which meet or exceed the above listed requirements for both balanced and unbalanced voltage digital interface.

The line receiver will discriminate a ±200 millivolt input signal over a full common-mode range of ±10 volts and a ±300 millivolt signal over a full common-mode range of ±15 volts.

The DS78LS120/DS88LS120 include response control for applications where controlled rise and fall times and/or high frequency noise rejection are desirable. Switching noise which may occur on the input signal can be eliminated by the 50 mV (referred to input) of hysteresis built into the output gate *(Figure 14)*. The DS78LS120/DS88LS120 makes use of a response control pin for the addition of an external capacitor, which will not affect the line termination impedance of the interconnect cable. Noise pulse width rejection versus the value of the response control capacitor is shown in *Figure 15*. The combination of the filter followed by hysteresis will optimize performance in a worst case noise environment. The DS78C120/DS88C120 is identical in performance to the DS78LS120/DS88LS120, except it's compatible with CMOS logic gates.

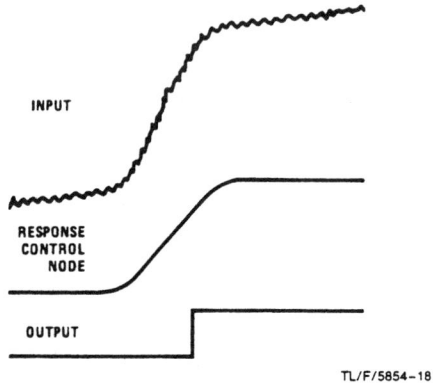

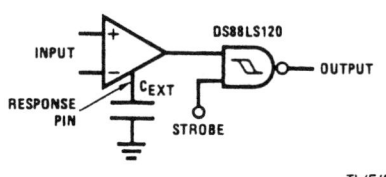

FIGURE 14. Application of DS88LS120 Receiver Response Control and Hysteresis

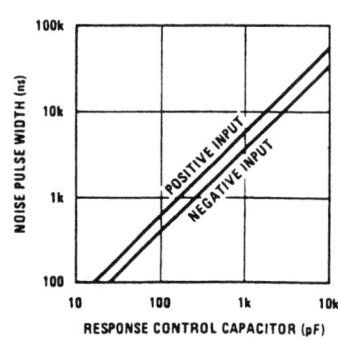

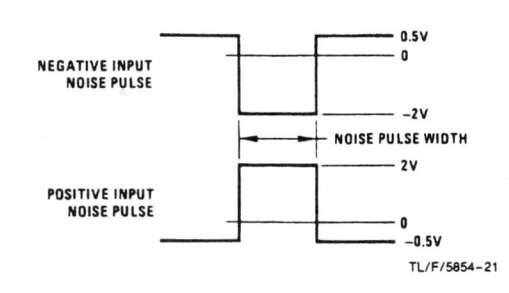

FIGURE 15. Noise Pulse Width vs Response Control Capacitor

AN-214

FAIL-SAFE OPERATION

Communication systems require elements of a system to detect the loss of signals in the transmission lines. And it is desirable to have the system shut-down in a fail-safe mode if the transmission line is open or short. To facilitate the detection of input opens or shorts, the DS78LS120/DS88LS120 incorporates an input threshold voltage offset. This feature will force the line receiver to a specific logic state if presence of either fault condition exists.

The receiver input threshold is ± 200 mV and an input signal greater than ± 200 mV insures the receiver will be in a specific logic state. When the offset control input is connected to a $V_{CC} = 5V$, the input thresholds are offset from 200 mV to 700 mV, referred to the non-inverting input, or -200 mV to -700 mV, referred to the inverting input. Therefore, if the input is open or short, the input will remain in a specific state (see *Figure 16*).

It is recommended that the receiver be terminated in 500Ω or less to insure it will detect an open circuit in the presence of noise.

For unbalanced operation, the receiver would be in an indeterminate logic state if the offset control input was open. Connecting the offset to +5V, offsets the receiver threshold 0.45V. The output is forced to a logic zero state if the input is open or short.

For balanced operation with inputs short or open, receiver C will be in an indeterminate logic state. Receivers A and B will be in a logic zero state allowing the NOR gate to detect the short or open fault condition. The "strobe" input will disable the A and B receivers and therefore may be used to "sample" the fail-safe detector (see *Figure 17*).

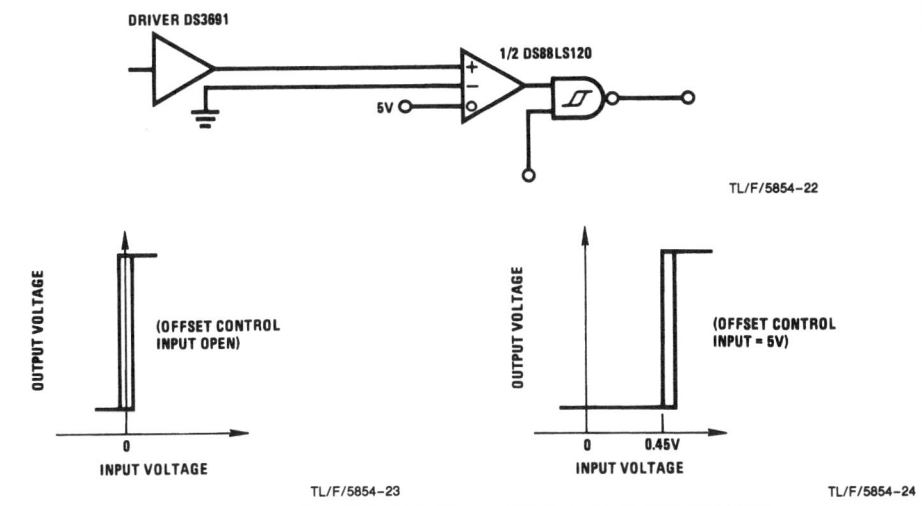

FIGURE 16. Fail-Safe Using the DS88LS120 Threshold Offset for Unbalanced Lines

FIGURE 17. Fail-Safe Using the DS88LS120 Threshold Offset for Balanced Lines

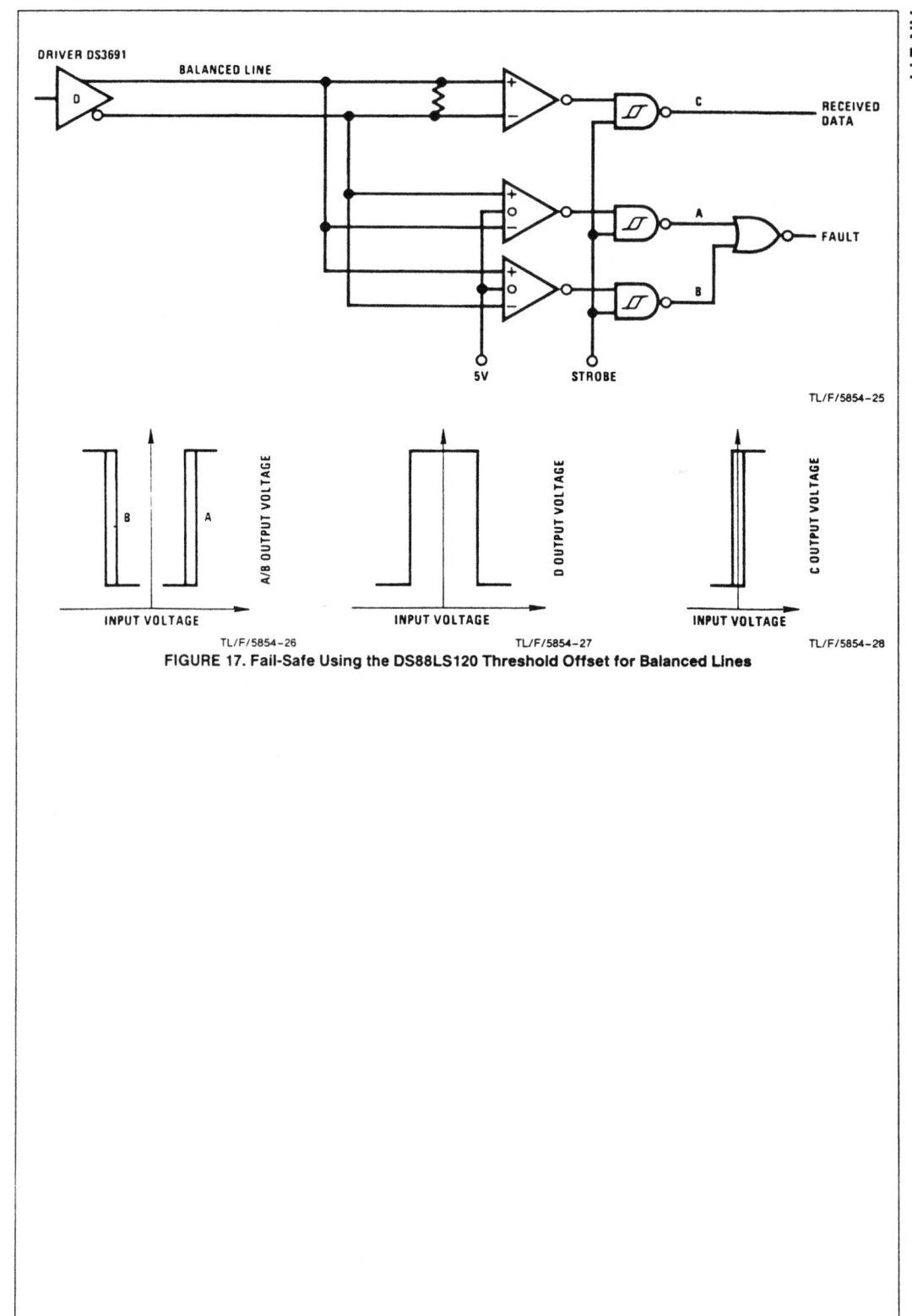

Transceivers and Repeaters Meeting the EIA RS-485 Interface Standard

National Semiconductor Corp.
Application Note 409
Sivakumar Sivasothy

INTRODUCTION

The Electronics Industries Association (EIA), in 1983, approved a new balanced transmission standard called RS-485. The EIA RS-485 standard addresses the problem of data transmission, where a balanced transmission line is used in a party-line configuration. It is similar in many respects to the popular EIA RS-422 standard; in fact RS-485 may be considered the outcome of expanding the scope of RS-422 to allow multipoint—multiple drivers and receivers sharing the same line—data transmission. The RS-485 standard, like the RS-422 standard, specifies only the electrical characteristics of the driver and the receiver to be used at the line interface; it does not specify or recommend any protocol. The protocol is left to the user.

The EIA RS-485 standard has found widespread acceptance and usage since its ratification. Users are now able to configure inexpensive local area networks and multi-drop communication links using twisted pair wire and the protocol of their choice. They also have the flexibility to match cable quality, signalling rate and distance to the specific application and thus obtain the best tradeoff between cost and performance. The acceptance of the RS-485 standard is also reflected by the fact that other standards refer to it when specifying multipoint data links. The ANSI (American National Standards Institute) standards IPI (Intelligent Peripheral Interface) and SCSI (Small Computer Systems Interface) have used the RS-485 standard as the basis for their voltage mode differential interface class. The IPI standard specifies the interface between disc drive controllers and host adapters and requires a data rate of 2.5 megabaud over a 50 meters NRZ data link. The SCSI standard specifies the interface between personal computers, disc drives and printers at data rates up to a maximum of 4 megabaud over 25 meters.

It is not possible to use standard gate structures and meet the requirements of RS-485. The modifications necessary to comply with the DC requirements of the standard, tend to exact a heavy toll on speed and other AC characteristics like skew. However, it is possible to vastly improve the ac performance by employing special design techniques. The DS3695 family of chips made by National Semiconductor meets all the requirements of EIA RS-485, and still provides ac performance comparable with most existing RS-422 devices. The chip set consists of four devices; they are the DS3695/DS3696 transceivers and the DS3697/DS3698 repeaters. National's RS-485 devices incorporate several features in addition to those specified by the RS-485 standard. These features provide greater versatility, easier use and much superior performance. This article discusses the requirements of a multi-point system, and the way in which RS-485 addresses these requirements. It also explains the characteristics necessary and desirable in the multi-point drivers and receivers, so that these may provide high performance and comply with generally accepted precepts of data transmission practice.

WHY RS-485?

Until the introduction of the RS-485 standard, the RS-422 standard was the most widely accepted interface standard for balanced data transmission. The RS-422 drivers and re-

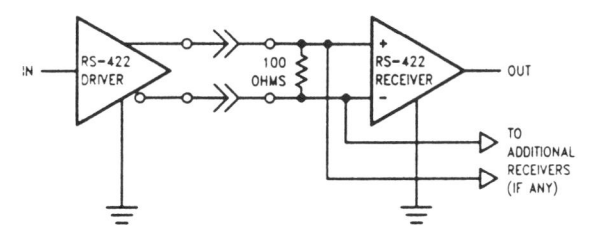

FIGURE 1a. An RS-422 Configuration

FIGURE 1b. A Typical RS-485 Party-Line Configuration

ceivers were intended for use in the configuration shown in *Figure 1a*. The driver is at one end of the line; the termination resistor (equal to 100Ω) and up to 10 receivers reside at the other end of the line. This approach works well in simplex (unidirectional) data transmission applications, but creates problems when data has to be transmitted back and forth between several pieces of equipment. If several Data Terminal Equipments (DTEs) have to communicate with one another over long distances using RS-422 links, two such balanced lines have to be established between each pair of DTEs. The hardware cost associated with such a solution would normally be unacceptable.

A party line is the most economical solution to the above problem. RS-422 hardware could conceivably be used to implement a party line if the driver is provided with TRI-STATE® capability, but such an implementation would be subjected to severe restrictions because of inadequacies in the electrical characteristics of the driver. The biggest problem is caused by ground voltage differences. The common mode voltage on a balanced line is established by the enabled driver. The common mode voltage at the receiver is the sum of the driver offset voltage and the ground voltage difference between the driver and the receiver. In simplex systems only the receiver need have a wide common mode range. Receiver designs that provide a wide common mode range are fairly straightforward. In a party-line network several hundred feet long, in which each piece of equipment is earthed at a local ac outlet, the ground voltage difference between two DTEs could be as much as a few volts. In such a case both the receiver and the driver must have a wide common mode range. Most RS-422 drivers are not designed to remain in the high impedance state over a wide enough common mode range, to make them immune to even small ground drops.

Classical line drivers are vulnerable to ground drops because of their output stage designs. A typical output stage is shown in *Figure 2a*. Two such stages driven by complementary input signals, may be used to provide the complementary outputs of a differential line driver. Transistors Q1 and Q4 form a Darlington pull up for the totem pole output stage; Q2 is the pull down transistor. The phase splitter Q3 switches current between the upper and lower transistors to obtain the desired output state. DSUB is the diode formed by the collector of Q2 and the grounded substrate of the integrated circuit. The output in *Figure 2a* can be put into the high impedance state by pulling down the bases of transistors Q3 and Q4. Unfortunately, the high impedance state cannot be maintained if the output is pulled above the power supply voltage or below ground voltage. In party-line applications, where ground voltage differences of a few volts will be common, it is essential that the drivers be able to hold the high impedance state while their outputs are taken above V_{CC} and below ground.

The output in *Figure 2a* can be taken high until the emitter-base junction of Q1 breaks down. Thereafter, the output will be clamped to a zener voltage plus a base-collector diode voltage above V_{CC}; V_{CC} could be zero if the device is powered off. If the output is taken below ground, it will cause the substrate diode, DSUB, associated with Q2 to turn on and clamp the output voltage at a diode drop below ground. If a disabled driver turns on and clamps the line, the signal put out by the active driver will get clipped and distorted. It is also possible for ground drops to cause dangerously large substrate currents to flow and damage the devices as illustrated in *Figure 2b*. *Figure 2b* depicts two drivers A and B; it shows the pull down transistors (Q2A and Q2B) and their associated substrate diodes (DSUB-A and DSUB-B) for the two drivers A and B. Here driver A is ON in the low output state; driver B is disabled, and therefore, should neither source nor sink current. The ground of driver A is 3 volts lower than that of driver B. Consequently, the substrate diode DSUB-B sees a forward bias voltage of about 2.7V (the collector-emitter voltage of Q2A will be about 0.3V), which causes hundreds of milliamperes of current to flow out of it.

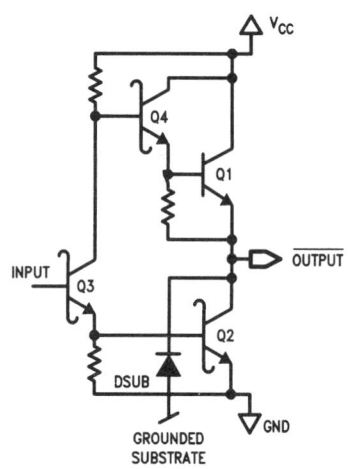

FIGURE 2a. Driver Output Stage (not RS-485)

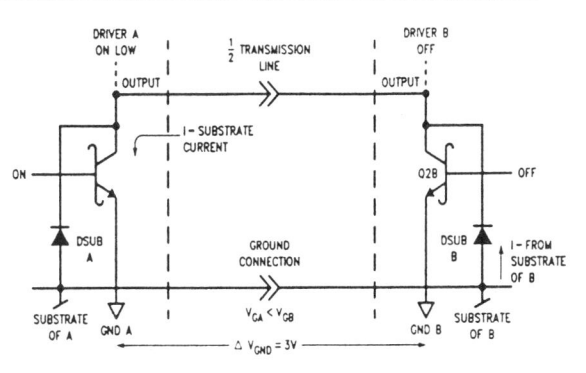

FIGURE 2b. Two DCEs Separated by a Ground Drop

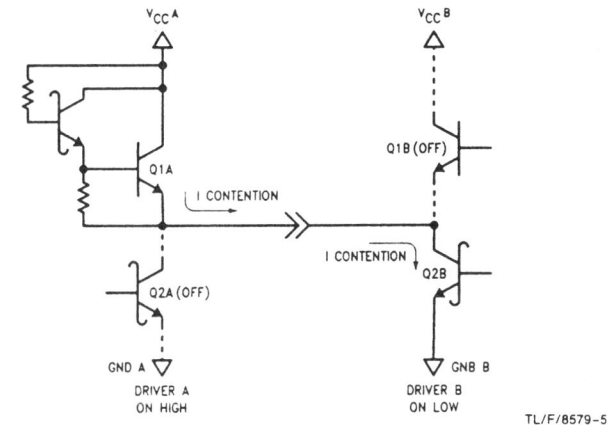

FIGURE 2c. Bus Contention

Another problem is line contention, i.e. two drivers being 'ON' simultaneously. Even if the protocol does not allow two drivers to be on at the same time, such a contingency could arise as a result of a fault condition. A line contention situation, where two drivers are on at the same time, is illustrated in *Figure 2c*. Here, drivers A and B are 'ON' simultaneously; driver A is trying to force a high level on the line whereas driver B is trying to force a low level. Transistors Q1A and Q2B are 'ON' while transistors Q2A and Q1B are 'OFF'. As a result, a large current is sourced by Q1A and sunk by Q2B; the magnitude of this current is limited only by the parasitic resistances of the two devices and the line. The problem is compounded by any ground drop that may exist between the two contending drivers. This large contention current can cause damage to one or both of the contending drivers. Most RS-422 drivers are not designed to handle line contention.

A multi-point driver should also be capable of providing more drive than a RS-422 driver. The RS-422 driver is only required to drive one 100Ω termination resistor, and ten receivers each with an input impedance no smaller than 4 kΩ. A party-line, however, would have to be terminated at both ends; it should also be able to drive more devices to be useful and economical.

Because of the above limitations, it is quite impractical to use RS-22 hardware to interconnect systems on a party-line. Clearly, a new standard had to be generated to meet the more stringent hardware requirements of muti-point data links.

THE RS-485 STANDARD

The RS-485 standard specifies the electrical characteristics of drivers and receivers that could be used to implement a balanced multi-point transmission line (party-line). A data exchange network using these devices will operate properly in the presence of reasonable ground drops, withstand line contention situations and carry 32 or more drivers and receivers on the line. The intended transmission medium is a 120Ω twisted pair line terminated at both ends in its characteristic impedance. The drivers and receivers can be distributed between the termination resistors as shown in *Figure 1b*.

The effects of ground voltage differences are mitigated by expanding the common mode voltage (V_{CM}) range of the driver and the receiver to $-7V < V_{CM} < +12V$. A driver forced into the high impedance state, should be able to have its output taken to any voltage in the common mode range and still remain in the high impedance state, whether powered on or powered off. The receiver should respond properly to a 200 mV differential signal super-imposed on any common mode voltage in this range. With a 5V power supply, the common mode voltage range specified by RS-485 has a 7V spread from either supply terminal. The system will therefore perform properly in the presence of ground drops and longitudinally coupled extraneous noise, provided that the sum of these is less than 7 volts.

The output drive capability of the driver and the input impedance of the receiver are increased to accommodate two termination resistors and several devices (drivers, receivers and transceivers) on the line. The RS-485 standard defines a 'unit load' so that the load presented to the line by each device can be expressed in terms of unit loads (a 12 kΩ resistor, with one end tied to any voltage between ground and $V_{CC}/2$, will satisfy the requirements of a unit load). It was anticipated that most manufacturers would design their drivers and receivers such that the combined load of one receiver and one disabled driver would be less than one unit load. This would require the RS-485 receiver to have three times the input resistance of a RS-422 receiver. The required receiver sensitivity is ±200 mV—the same as for RS-422. The driver is required to provide at least 1.5V across its outputs when tied to a terminated line populated with 32 transceivers. Although this output voltage is smaller than the 2.0V specified for RS-422, a careful design of the driver, with special regard to ac performance, can allow the user to operate a multi-point network at data rates and distances comparable to RS-422.

RS-485 has additional specifications to guarantee device safety in the event of line contention or short circuits. An enabled driver whose output is directly shorted to any voltage in the common mode range, is required to limit its current output to ±250 mA. Even with such a current limit, it is possible for a device to dissipate as much as 3 Watts (if the device draws 250 mA while shorted to 12 volts). Power dissipation of such a magnitude will damage most ICs; therefore, the standard requires that manufacturers include some additional safeguard(s) to protect the devices in such situations.

The ±250 mA current limit also serves another purpose. If a contending driver is abruptly turned off, a voltage transient, of magnitude $I_C Z/2$, is reflected along the line as the line discharges its stored energy (I_C is the contention current and Z is the characteristic impedance of the line). This voltage transient must be small enough to avoid breaking down the output transistors of the drivers on the line. If the contention current is limited to 250 mA, the magnitude of this voltage transient, on a 120Ω line, is limited to 15V, a value that is a good compromise between transistor breakdown voltage and speed.

AC PERFORMANCE

To achieve reliable transmission at high data rates over long distances, the driver should have optimum ac characteristics. The response should be fast and the output transients sharp and symmetrical.

(1) **Propagation Delay:** The propagation delay through the driver should be small compared to the bit interval so that the data stream does not encounter a bottle-neck at the driver. If the propagation delay is comparable to the bit interval, the driver will not have time to reach the full voltage swing it is capable of. In lines a few hundred feet long, the line delay would impose greater limits on data throughput than the driver propagation delay. However, a fast driver would be desirable for short haul networks such as those in automobile vehicles or disc drives; in the latter case high data throughput would be essential. Driver propagation delays less than 20 ns would be very good for a wide range of applications.

(2) **Transition Time:** For distortion free data transmission, the signal at the farthest receiver must have rise and fall times much smaller than the bit interval. Signal distortion results from driver imbalance, receiver threshold offset and skew. RS-485 limits the DC imbalance in the driver output to ±0.2V i.e., 13% of worst-case signal amplitude. Usually, the greatest distortion is caused by offset in the receiver threshold. In a long line in which a 1.5V driver output signal amplitude is attenuated by the loop resistance to about 0.4V, a 200 mV offset in the receiver threshold can cause severe pulse width distortion if the rise time is comparable to the bit interval. For lines longer than about five hundred feet, the rise time would be dominated by the line and not the driver. In short-haul networks, the transient response of the driver can significantly affect signal distortion; a faster transient creates less distortion and hence permits a smaller bit interval and a higher baud rate. A rise time less than 20 ns will be a good target spec., for it will permit a baud rate of 10 Meg over 50' of standard twisted pair wire with less than 5% distortion.

The driver should provide the above risetime and propagation delay numbers while driving a reasonable capacitance, say 100 pF from each output, in addition to the maximum resistive load of 54Ω. A properly terminated transmission line appears purely resistive to the driver. Most manufacturers take this into account and specify their driver delays with 15 pF loads. However, if any disabled transceivers are situated close to the driver (such that the round trip delay is less than the rise time), the input capacitances of these transceivers will appear as lumped circuit loads to the driver. The driver output rise time will then be affected by all other devices in such close proximity. In the case of high speed short-haul networks, where rise time and propagation delay are critical, several devices could be clustered in a short span. In such an instance, specifying propagation delays with 15 pF loads is quite meaningless. A 100 pF capacitive load is more reasonable; even if we allocate a generous 20 pF per transceiver, it allows up to six transceivers to be clustered together in an eight foot span (the eight foot span is the approximate round trip distance travelled by the wavefront in one rise time of 20 ns).

(3) **Skew:** The ideal differential driver will have the following waveform characteristics: the propagation delay times from the input to the high and low output states will be equal; the rise and fall times of the complementary outputs will be equal and the output waveforms will be perfectly symmetrical.

If the propagation delay to the low output state is different from the propagation delay to the high output state, there is said to be 'propagation skew' between output states. If a square wave input is fed into a driver with such skew, the output will be distorted in that it will no longer have a 50% duty cycle.

If the mid-points of the waveforms from the two complementary driver outputs are not identical, there is said to be SKEW between the complementary outputs. This type of skew is undesirable because it impairs the noise immunity of the system and increases the amount of electromagnetic emission.

Figure 3a shows the differential signal from a driver that has no skew. *Figure 3b* shows the case when there is 80 ns of skew. The first signal makes its transition uniformly and passes rapidly through 0V. The second waveform flattens out for tens of nanoseconds near 0V. Unfortunately, this flat region occurs near the receiver threshold. A common mode noise spike hitting the inputs of a slightly unbalanced receiver would create a small differential noise pulse at the receiver inputs. If this noise

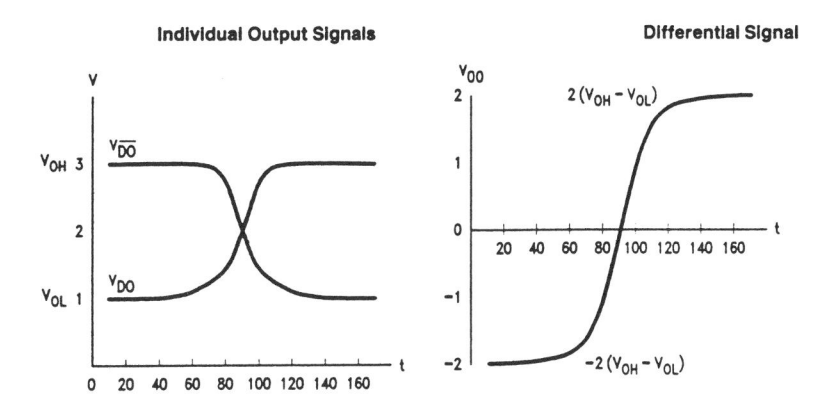

FIGURE 3a. Transients with no Skew

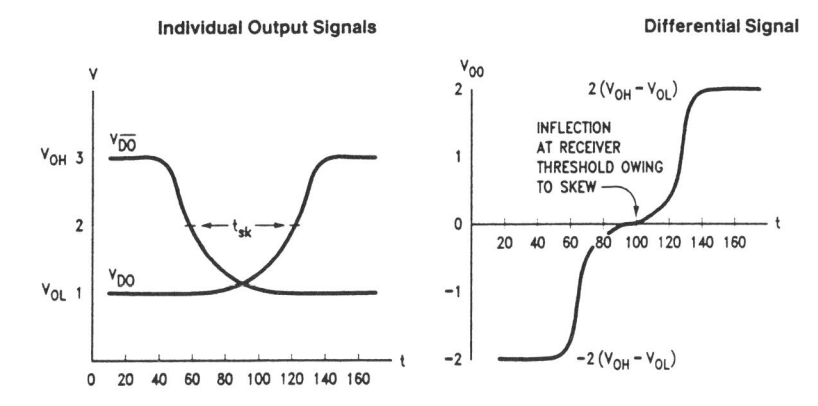

FIGURE 3b. Skewed Transients

pulse occurs when the driver transition is flat near 0V, there will be a glitch at the receiver output. A glitch could also occur if a line reflection reaches the receiver input when the driver transition is temporarily flat. Skew is insidious in that it can cause erroneous outputs to occur at random. It can also increase the amount of electromagnetic interference (EMI) generated by the transmission system. If the complementary outputs are perfectly symmetrical, and the twisted pair medium is perfectly balanced, the radiation from one wire is cancelled exactly by the radiation from the other wire. If there is skew between the outputs, there will be net radiation proportional to the skew.

(4) **Balance:** The impedance seen looking into each of the complementary inputs of the transceiver should be identical. If there is any imbalance at these nodes, the common mode rejection will be degraded. Any DC imbalance, due to a mismatch in the receiver input resistances, will manifest itself as an offset in the receiver threshold, and can be easily detected during testing. AC imbalance is more difficult to detect, but it can hurt noise immunity at high frequencies. A sharp common mode noise spike striking an unbalanced receiver will cause a spurious differential signal. If the receiver is fast enough (as it is bound to be in most cases), it will respond to this noise signal. It is best to keep the imbalance below 4 pF. This number is reasonable to achieve; in addition, the combined imbalance of 32 transceivers will still provide sufficient immunity from h.f. interference.

DESIGN CONSIDERATIONS

The driver poses the greatest design challenge. Its speed, drive and common mode voltage requirements are best met using a bipolar process. National Semiconductor uses an established Schottky process with a 5μ deep epitaxial layer. NPN transistors are fabricated with LVCEO values greater than 15V to satisfy the breakdown requirements. It will be seen that lateral PNP transistors are crucial to the driver. The 5μ EPI process provides adequate lateral PNP transistors, and NPN transistors of sufficient speed.

Figure 4 shows the driver output circuit used by National. It is a standard totem pole output circuit modified to provide a common mode range that exceeds the supply limits. If the driver output is to be taken to $-7V$ while the driver is in TRI-STATE, precautions must be taken to prevent the substrate diodes from turning on. This is achieved in the lower output transistor Q1 by including Schottky diode S1 in series. The only way to isolate the upper half of the totem pole from the substrate is by using a lateral PNP transistor. Lateral PNP transistors are, however, notoriously slow; the trick therefore is not to use the PNP transistor in the switching path. In the circuit shown, the PNP transistor is a current source which feeds NPN transistor Q2 and therefore, does not participate in the switching function. This allows National's driver to have 15 ns propagation delays and 10 ns rise timers. A Darlington stage cannot be used instead of Q2 because it would reduce the voltage swing below the 1.5V specification. Consequently, the rise time is bound to be significantly larger than the fall time, resulting in a large skew. National's driver uses a patented circuit with a plurality of discharge paths, to slow down the falltime so that it matches the rise time, and to keep the two transition times on track over temperature. This keeps the skew small (2 ns typical at 25°C) over the entire operating temperature range. The symmetry of the complementary outputs of National's DS3695 driver can be seen from the photographs in *Figure 5*. The lateral PNP transistor which has been kept out of the switching path has nevertheless got to be turned on or off when the driver is respectively enabled or disabled. Another patented circuit is used to hasten turn-on and turn-off of the lateral PNP transistors so that these switch in 25 ns instead of in 100 ns. Consequently, the driver can be enabled or disabled in 35 ns.

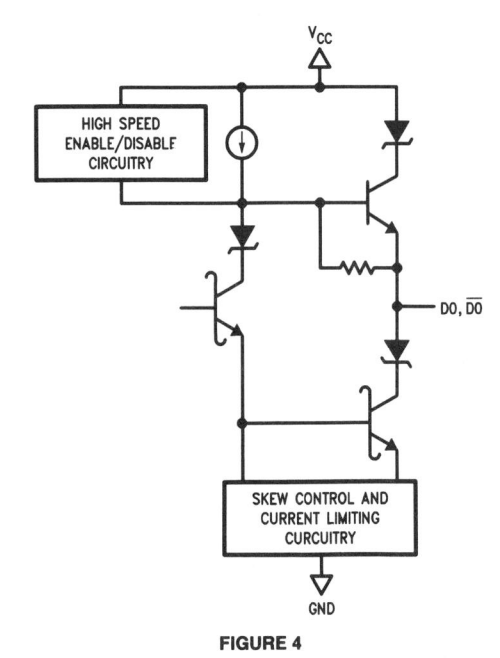

FIGURE 4

Complementary Outputs
of National's RS-485 Driver

Differential Output
of National's RS-485 Driver

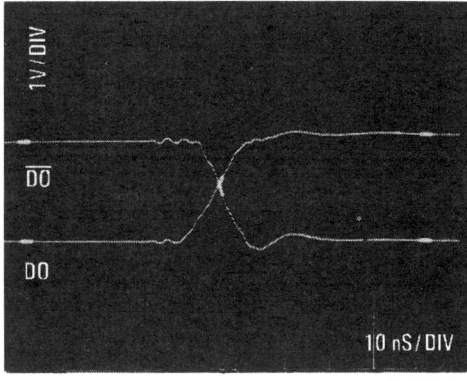

FIGURE 5

The devices must be protected in fault conditions and contention situations. One way of doing this is by sensing current and voltage to determine power, and then if necessary, turning the device off or limiting its output current to prevent damage. This method has the advantage of fast detection of a fault and rapid recovery from one. However, too many contingencies have to be accounted for; the corresponding circuitry will increase the die size and the cost beyond what would be acceptable in many low cost applications. National preferred the simpler and inherently more reliable thermal shutdown protection scheme. Here, the device is disabled when the die temperature exceeds a certain value. This method is somewhat slower (order of milliseconds), but fast enough to protect the part. A fault would usually result from a breakdown in network protocol or from a hardware failure. In either case it is immaterial how long the device takes to shut down or recover as long as it stays undamaged. It would be useful to be notified of the occurrence of a fault in any particular channel, so that remedial action may be taken. Two of National's devices, the DS3696 receiver and the DS3698 repeater, provide a fault reporting pin which can flag the processor or drive an alarm LED in the event of a fault. National also decided to make its devices as single transceivers housed in 8 pin mini DIP packages. If thermal shutdown protection is employed, it is pointless to have dual or quad versions because a faulty channel will shut down a good one. Since most RS-485 applications will employ single channel serial data, the 8 pin package will give optimum flexibility, size and economy.

The receiver has 70 mV (typical) hysteresis for improved noise immunity. Hysteresis can contribute some distortion, especially in short lines, if the rise and fall times are different. However, this is more than adequately compensted for by the noise immunity it provides with long lines where rise times are slow. The matched rise and fall times with National's drivers assure low pulse width distortion even at short distances and high data rates.

Chapter 3 Digital logic and systems design

In this chapter we look at some of the logic that interfaces the various large scale chips used to construct a microcomputer. We have not attempted to provide a primer in either digital logic or semiconductor technology, but have concentrated on the themes that are likely to be of particular interest to the designer.

The first part of this chapter looks at the characteristics of typical logic elements. In particular we examine how logic elements behave electrically and how they are interfaced to other logic elements. Some of the topics covered are the input/output parameters of logic elements and the effects of simultaneous switching in chips with multiple outputs.

The introduction of programmable logic has forced a revolution in the way in which sequential circuits are designed. The middle of this chapter provides a tutorial on the design of synchronous state machines.

In recent years there has been a shift of emphasis in the world of digital design. Once, all the engineer had to concentrate on was getting the system to work. Today, designing a system for testability is as important as designing a working system, since debugging a faulty system is so expensive. We provide a tutorial on the principles of testability.

The final part of this chapter looks at several topics of interest to the systems designer. The first is metastability, an often neglected problem that can cause systems to fail inexplicably. The second is the design of the arbiters that are needed to share resources in a multiprocessor system. The third topic is the use of MSI logic elements in the design of multiprocessor systems, and the final topic is the programmable high-speed address decoder

Texas Instruments Logic Families

– Technology and Characteristics –

IMPORTANT NOTICE

Texas Instruments (TI) reserves the right to make changes in the devices or the device specifications identified in this publication without notice. TI advises its customers to obtain the latest version of device specifications to verify, before placing orders, that the information being relied upon by the customer is current.

TI warrants performance of its semiconductor products, including SNJ and SMJ devices, to current specifications in accordance with TI's standard warranty. Testing and other quality control techniques are utilized to the extent TI deems such testing necessary to support this warranty. Unless mandated by government requirements, specific testing of all parameters of each device is not necessarily performed.

In the absence of written agreement to the contrary, TI assumes no liability for TI applications assistance, customer's product design, or infringement of patents or copyrights of third parties by or arising from use of semiconductor devices described herein. Nor does TI warrant or represent that any license, either express or implied, is granted under any patent right, copyright, or other intellectual property right of TI covering or relating to any combination, machine, or process in which such semiconductor devices might be or are used.

Copyright © 1988, Texas Instruments Incorporated

1. Introduction

The spectrum of logic families offered by Texas Instruments has been greatly extended with the emergence of new technologies with improved performance features (Figure 1-1). There are the bipolar TTL, LS, S, ALS, AS and F families, the HC(T) and AC(T) CMOS families, as well as the BCT series, which combines the advantages of bipolar and CMOS circuit technologies in a unique BiCMOS process.

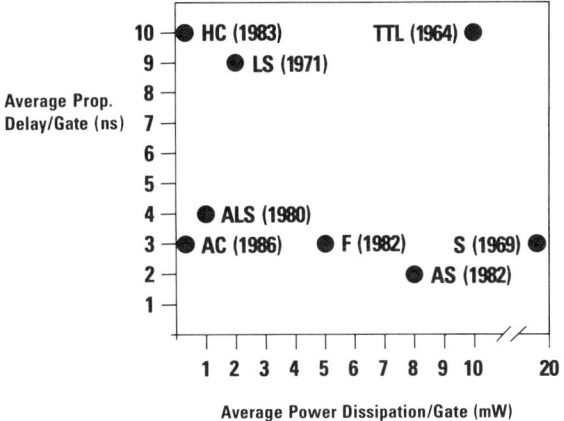

There are no gate functions in the BCT family. The average power consumption of bus interface circuits in this family is approx. 50 % in the active and 10 % in the disabled state, referred to the F family (for the same propagation delay).

Figure 1-1. Spectrum of TI logic families

The following report illustrates the basic characteristics of all logic families mentioned in a direct comparison. The designer is also given guidelines for deciding what logic family will be most successful in a particular application and advised of the points that must be specially watched out for.

Because of its special characteristics and advantages, there will be a separate report dealing with the innovative BiCMOS technology. See page 3-123.

2. Bipolar Logic Families

2.1 Technology

In (Low Power) Schottky the starting material is doped by a diffusion process. This method does not permit precise control of the doping depth; furthermore, sidewall diffusions are unavoidable, these delimiting the manufacturable geometries downwards (Figure 2-1a).

The transition from the diffusion to the ion-implantation process represented a major step towards reduced geometries with improved characteristics, and this is used for the ALS, AS and F families. The sidewall diffusion effect is largely avoided by the socalled IMPACT process (Ion Implanted Composed Mask Technology). Better control of the doping depth can also be achieved. Together with other improvements in the production process, significantly smaller geometries are possible than with diffusion (Figure 2-1b). These smaller geometries lead in turn to smaller capacitances, thus markedly reducing power dissipation.

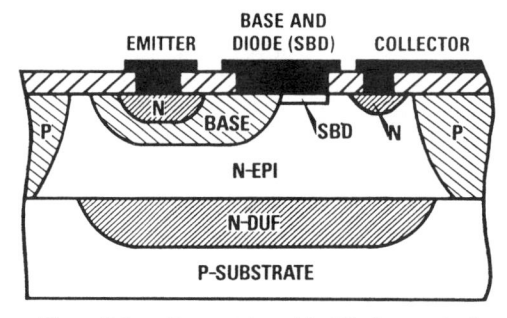

Figure 2-1a. Geometries with diffusion method

Figure 2-1b. Geometries with ion-implantation method

2.2 Inner Circuit

Essential changes were made in ALS/AS and F circuits compared to LS/S. An additional p-n-p transistor at the input (Figure 2-2) reduces the input currents I_{IL} by a factor of four. But the outputs are still capable of sinking or sourcing at least as much current.

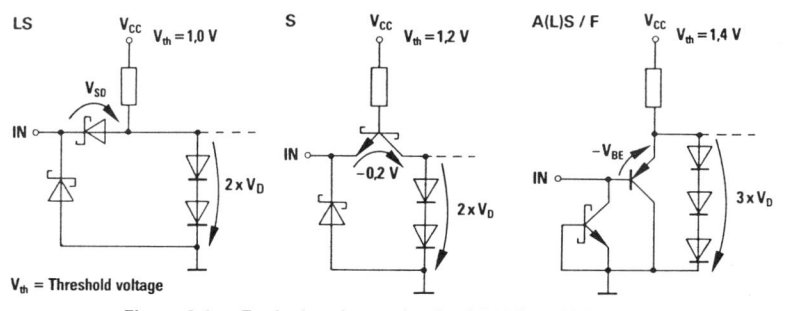

V_{th} = Threshold voltage

Figure 2-2. Equivalent input circuit of S/AS and LS/ALS/F

The additional input stage raises the switching threshold to approx. 1.4 V, thus improving the noise margin at Low level.

Both inputs and outputs exhibit improved ESD protection circuits. An npn transistor is connected in parallel to the usual Schottky clamping diode (Figures 2-3a through e). The Schottky transistor that is formed in this way has a different characteristic and thus reduces the input power when ESD occurs, offering greater protection against destruction of the device.

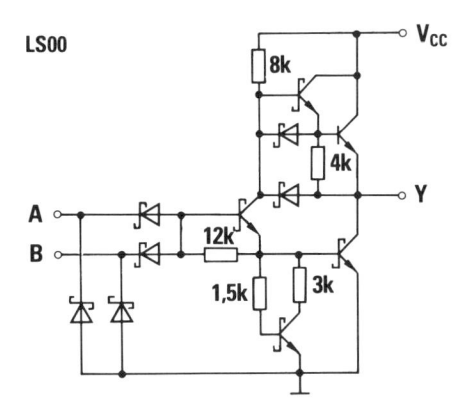

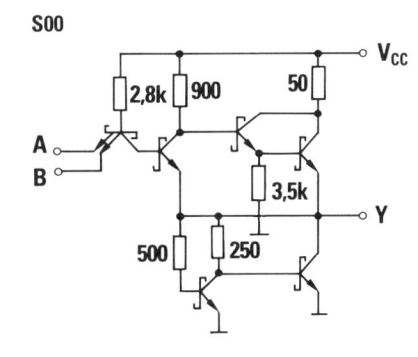

Figure 2-3a.   Typical inner circuit of LS gate

Figure 2-3b.   Typical inner circuit of S gate

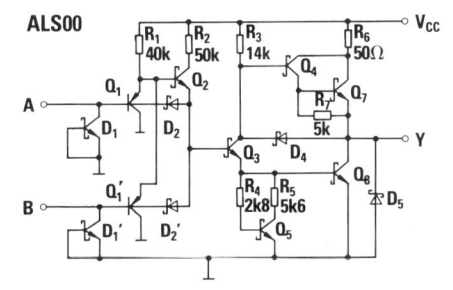

Figure 2-3c.   Typical inner circuit of ALS gate

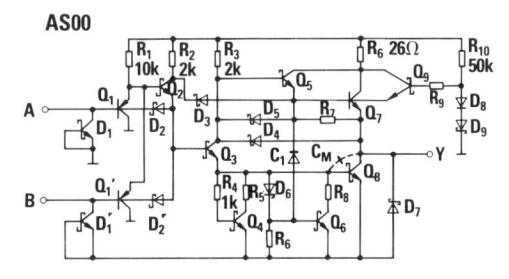

Figure 2-3d.   Typical inner circuit of AS gate

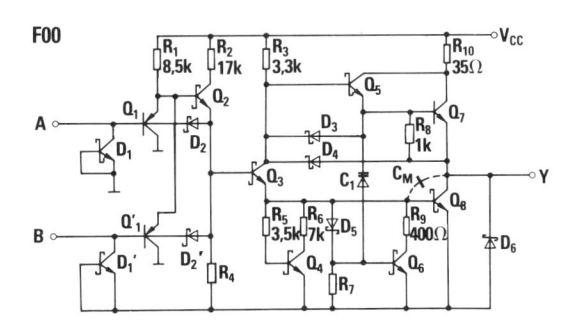

Figure 2-3e.   Typical inner circuit of F gate

Designer's Information

3-11

The inner circuit of AS and F is very similar to ALS. Except for the value of resistor R_1, the input stage with Q_1 and Q_2 is identical. The entire circuitry is designed with lower resistance to achieve high switching speeds. A further reduction of switching time is produced by the following additions to the circuit:

– AC Miller killer

Because of the large Miller capacitance, the lower output transistor (common-emitter circuit) of conventional TTL families exhibits unfavourable turn-off response. During the Low-to-High transition both transistors are briefly turned on and sink a high current. This "contamination" of the supply lines with current spikes (Figure 2-4) can degrade the noise immunity of the entire circuit (see page 3-44).

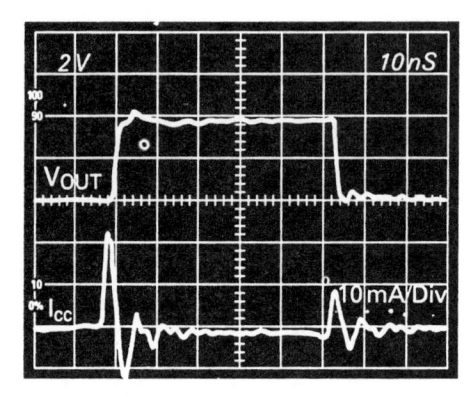

Figure 2-4. Current spikes for AS devices

In the AS and F families the Miller effect is reduced by what is called an AC "Miller killer" (Figure 2-5). When the output goes from Low to High, the emitter potential of Q_7 will rise. Current then flows briefly to the base of Q_6 across diode C_1, which acts as a capacitance. This current turns on the Miller capacitance and discharges it. The turn-off behaviour of Q_8 is improved and thus the current spikes are narrower (Figure 2-4).

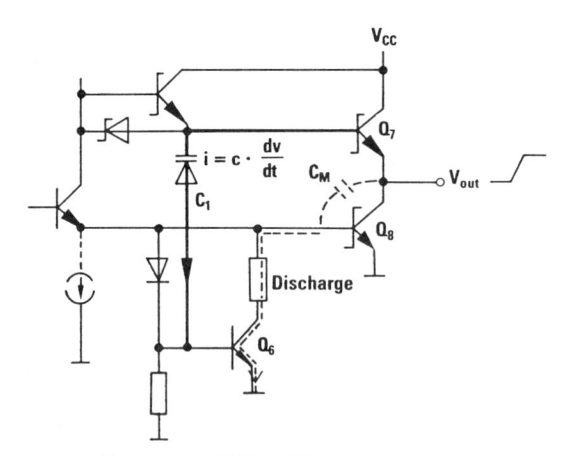

Figure 2-5. "Miller killer" circuit

- **"Zero-voltage" regulator at output**

The output of a gate may be drawn below ground by capacitive coupling of two adjacent gates (Figure 2-6). When the output of gate 2 goes from High to Low, it draws the output of gate 1 below ground. A negative output voltage at the instant of switching may also be the result of (negative) line reflections (Figure 2-7). The switching speed is reduced and malfunctions of flip-flops and counters are possible. If the output voltage drops below 0 V, the voltage regulator will switch the upper output transistor Q_7 with Q_9. Q_7 will drive current into the output and thus quickly reduce the negative potential.

The zero-voltage regulator is implemented in the majority of AS circuits, and also in a number of ALS and F devices.

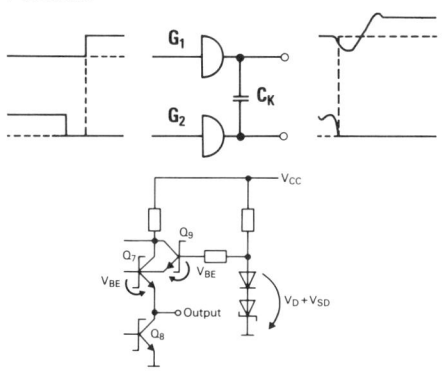

Figure 2-6. Capacitive coupling of gates: voltage regulator

Figure 2-7. Line reflections on gate output

- **DC Miller killer**

On the line the output of an inactive bus driver appears as a large capacitance. This is also a Miller capacitance, which reduces the speed of the bus system.

The lower transistor of the output stage can be turned on briefly across the mutual capacitance C_m. If the level of the bus goes High, a displacement current will flow across C_m to the base of Q_2. A gate/transistor combination is used to avoid this (Figure 2-8). In the high-impedance state Q_3 turns on, tightly connecting the base of Q_2 to ground.

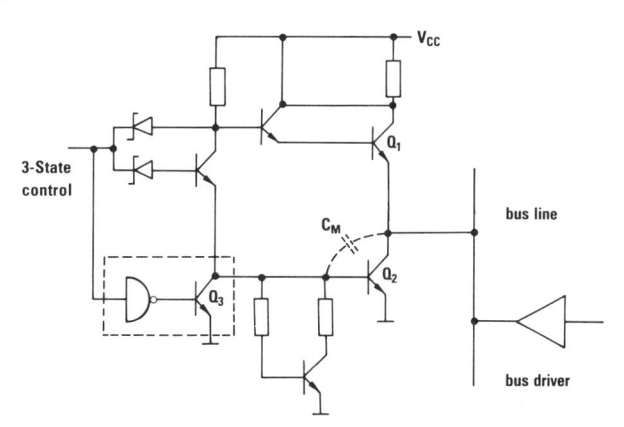

Figure 2-8. Output stage of bus driver

Designer's Information

3-13

— Feedback circuit for tristate devices

In the case of tristate devices a feedback diode, as is used for circuits with totem-pole outputs, would draw the output to ground in the high-impedance state. For this reason a transistor Q_{10} is integrated into the feedback path that draws its base current across R_{10} (Figure 2-9). In the high-impedance state the base circuit will be interrupted and Q_{10} turns off. Thus no current can flow at the output.

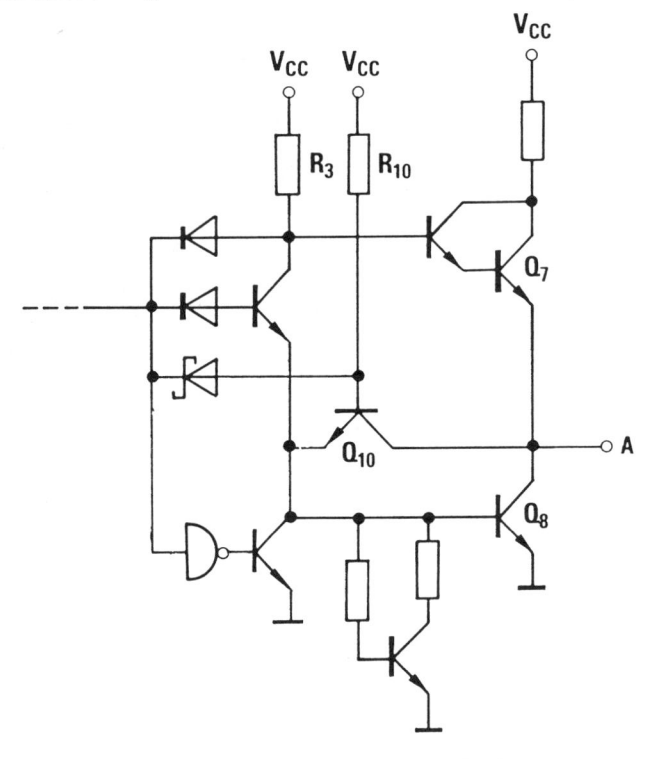

Figure 2-9. Tristate output circuit

2.3 Input Characteristics

Knowledge of input and output characteristics is necessary if full use is to be made of TTL devices. This is particularly true when a circuit interfaces with another device that is not in the same TTL series. In addition, knowledge of the voltage and current relationships of all elements is important for proper design.

Any device that is used to drive a TTL gate must source and sink current. By convention, current flowing towards a device input terminal is designated as positive, and current flowing out of a device input terminal is designated as negative. Low-level input current is negative because it flows out of the input terminal. High-level input current is positive because it flows into the input terminal.

Figure 2-10 shows typical input characteristics (input current vs input voltage) for negative input voltages (−1.2 V to 0 V). The input current rapidly increases below approx. −0.6 V and so negative undershoot on the input is reduced very quickly.

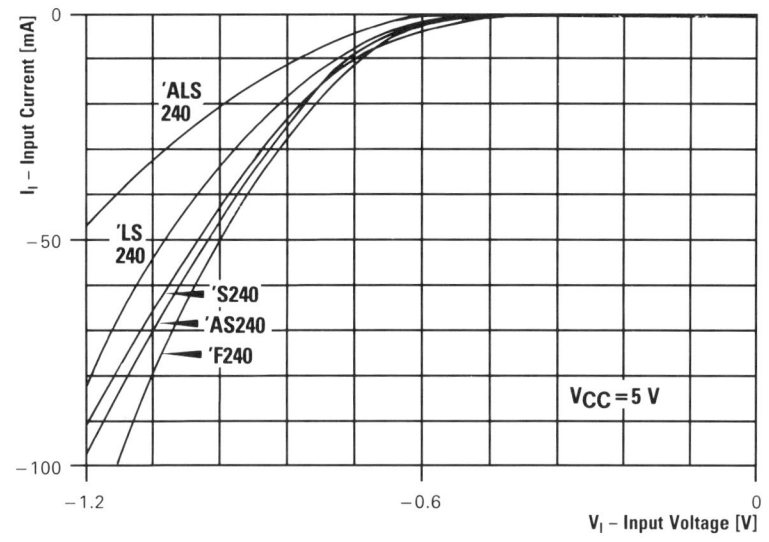

Figure 2-10. Typical '240 input characteristics for $V_I < 0$ V

Figure 2-11 shows the corresponding characteristics for input voltages above 0 V (I_I with different scale). The input currents at Low level (0.4 V) are less than 0.5 mA for the '240. Above approx. 1.8 V the input current is less than 1 μA (max. 20 μA) under normal operating conditions.

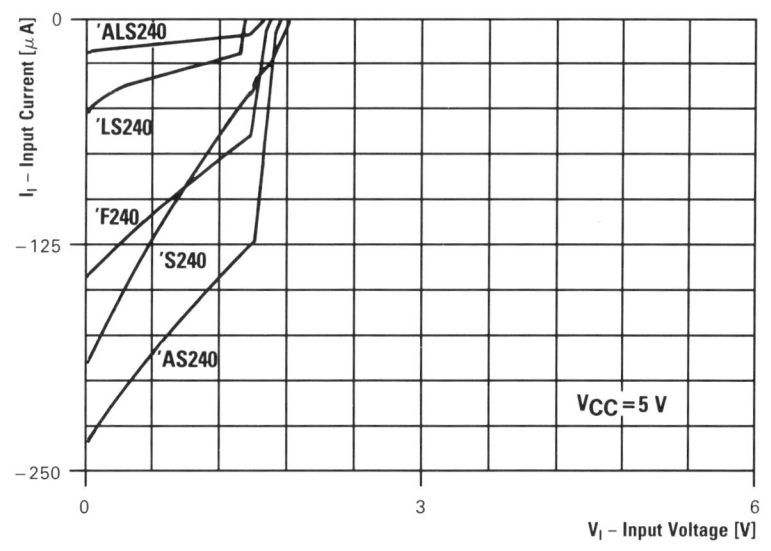

Figure 2-11. Typical '240 input characteristics for $V_I > 0$ V

2.4 Output Characteristics

Most of the bus interface circuits dealt with in this data book feature significantly higher driving capabilities than gates of the same families. This results in higher fan-out (see section 2.6).

Figures 2-12 and 2-13 show the Low and High output characteristics of bipolar '240 circuits. Comparison shows that the driving capability at Low level is generelly higher than at High level.

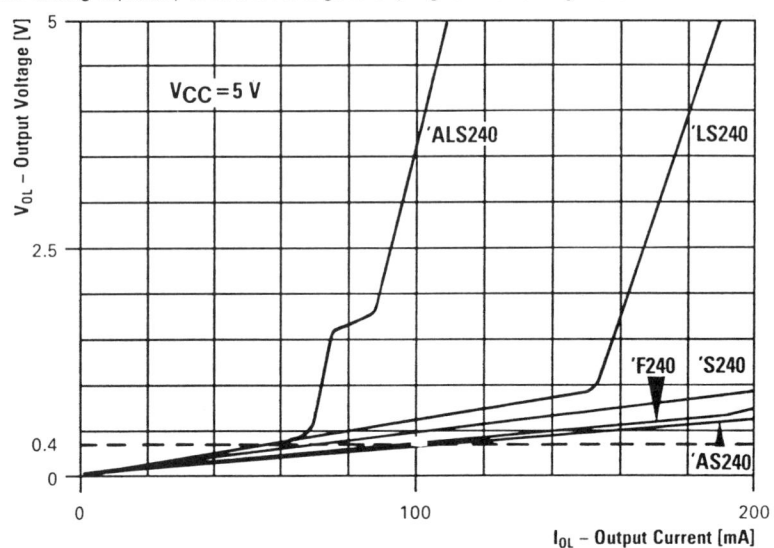

Figure 2-12. Typical '240 output characteristics at Low level

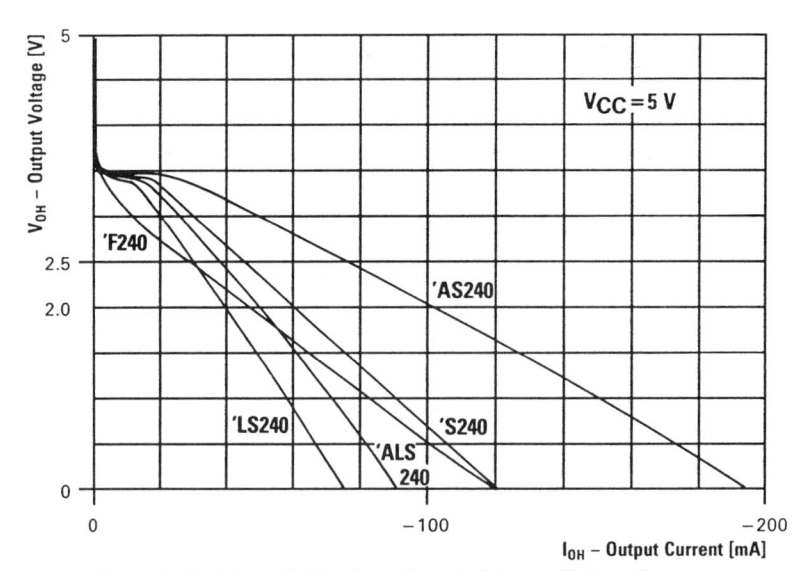

Figure 2-13. Typical '240 output characteristics at High level

2.5 Noise Margin

Because of the improved inner circuit (see Figure 2-3) the input switching threshold V_{th} is higher for ALS, AS and F devices than for LS circuits. Thus the noise margin, especially at Low level*, is higher in the newer families (see Table 2-1).

Table 2-1. Noise margin at Low level

Type	V_{th} [V]	$V_{th} - V_{OL}$ [V]	S_L [%]
LS240	1.1	0.7	23
ALS240	1.4	1.15	39
S240	1.3	0.75	26
AS240	1.5	1.19	39
F240	1.4	0.98	34

V_{th} = threshold voltage

$V_{th} - V_{OL}$ = typical noise margin at Low level

$$S_L = \frac{V_{th} - V_{OL}}{V_{OH} - V_{OL}} \times 100\,\%$$

2.6 Minimum Edge Steepness

A minimum edge steepness must be maintained for all logic circuits. If this minimum value is undercut, unwanted oscillations will occur at the output (Figure 2-14). Such oscillations are critical for edge-triggered storage devices (counters, flip-flops, etc.).

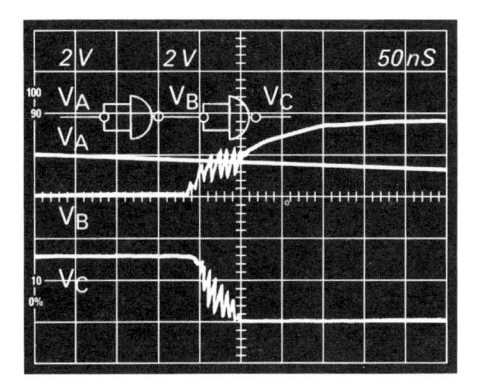

Figure 2-14. Oscillating AS00 output

The following minimum edge steepnesses are required:

LS	50 ns/V
S	50 ns/V
ALS	15 ns/V
AS	8 ns/V
F	8 ns/V

* Noise on the output is higher for Low level than for High level because matching to line impedances is worse in that case.

3. CMOS Logic Families
3.1 Technology
3.1.1 Self-aligning Silicon-gate Process

Just as in the older metal-gate CMOS process (74CXX, 4000 series), the starting material for the HC family is n-substrate with p-wells. But further doping is made by ion implantation instead of diffusion. Thus better control of doping density and the precision of structures can be achieved. The silicon-gate process that is used reduces structure sizes from 5 to 3 μm compared to metal-gate CMOS.

- Metal-gate process: Gate metallization requires separate masking, which is performed after drain and source diffusion. Overlaps of the drain and source regions as a result of tolerances in mask alignment cannot be avoided (Figure 3-1).

- Silicon-gate process: Here the first step is to create the polysilicon gates and field oxide regions. Both of them serve as a mask in the subsequent drain/source implantation process. This is a self-aligning process. The drain/source overlaps of the gates are slight and thus parasitic capacitances are smaller by a factor of seven (Figure 3-2).

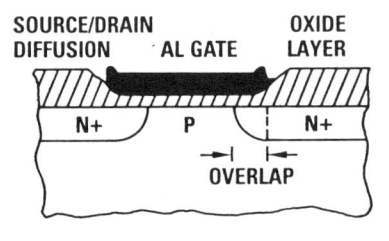

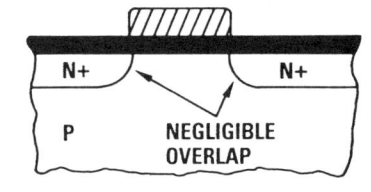

Polysilicon gate as mask for drain/source implantation

Figure 3-1. Metal gate

Figure 3-2. Silicon gate

Additional benefits of silicon-gate technology:

- The finer structures reduce the size of parasitic diodes in the drain and source regions (n-p junctions and p-n junctions). Then the associated capacitances are also smaller and switching speeds increase.
- The use of polysilicon gates improves the driving characteristics of the MOSFETs.
- The silicon-gate process permits thinner gate-oxide layers. The gate capacitance C_g increases.
- The structural precision of the process permits smaller gate lengths.
- Ion implantation reduces the threshold voltage V_{th} (the voltage required to build up the inverse charge in the MOSFET).
- With the polysilicon layer above the field oxide the chip has an additional contact level that provides for higher packing density.

3.1.2 EPIC™ 1 μm CMOS Process

The Enhanced Performance Implanted CMOS (EPIC™) 1 μm process that is used for the Advanced CMOS Logic family from Texas Instruments was derived from the technology used to develop the 1-megabit DRAM. Both high performance and high reliability are designed into the EPIC™ process. The design process is shown in Figure 3-3. Some of the features of the process are as follows:

- 1.0 μm gate length for sub-nanosecond on-chip propagation delays

- Silicide gate-source-drain to reduce internal interconnect resistance, allowing further speed enhancements

- Sidewall oxidation between gate-source and gate-drain to reduce internal capacitance, allowing further speed enhancements
- Epitaxial substrate layer for latch-up suppression
- Copper-doped aluminum metal to protect against electromigration, providing reliability enhancement
- Twin-well structure for high packing density, providing a vehicle for future LSI/VLSI product development.

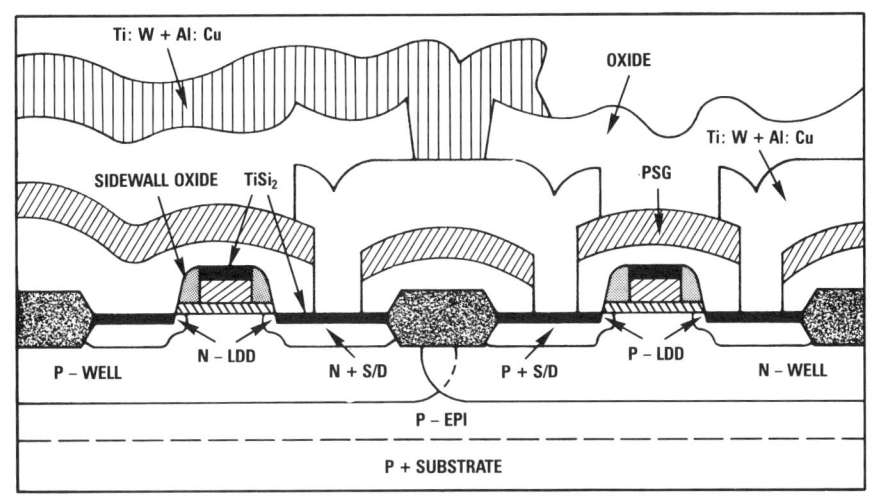

Figure 3-3. EPIC™ 1 μm CMOS process

3.2 Inner Circuit

3.2.1 HCMOS Devices

Each output of an HCMOS device is buffered by two inverter stages*.

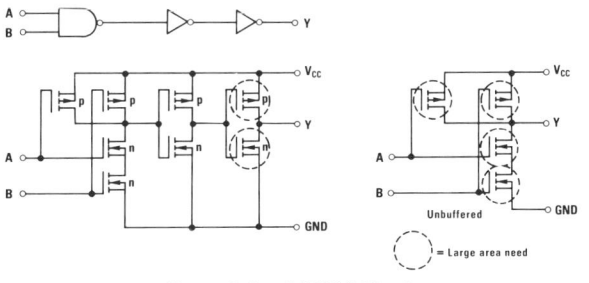

Figure 3-4. 74HC00 Circuit

Benefits of buffered inner circuit:
- Instead of the four large output transistors of the unbuffered version, only two are required. This reduces space needs and increases switching speeds.
- The higher gain of three stages improves noise immunity (approximately ideal transfer characteristic).

* The 74HCU04 inverter is an exception. This unbuffered version is mainly used in oscillator circuits.

Protection Circuitry

Electrostatic discharge (ESD) and latch-up are two traditional causes of CMOS device failure. In order to protect HCMOS devices from ESD and latch-up, additional circuitry has been implemented on the inputs and outputs.

ESD occurs when a buildup of static charges on one surface arcs through a dielectric to another surface that has the opposite charge. The end effect is the ESD causes a short between the two surfaces. These damaged devices (walking-wounded) may still pass normal data sheet tests but will eventually fail. The unique input protection circuitry designed by Texas Instruments provides immunity to typically 4500 V on the inputs and 3000 V on the outputs, which exceeds MIL-STD-883B, Method 3015, requirements for ESD protection (2000 V, 1.5 kΩ, 100 pF).

Figure 3-5 shows the circuitry implemented to provide protection for the input gates against ESD. The diode is forward biased for input voltages greater than V_{CC} + 0.5 V. The two transistors and resistor (actually one transistor diffused across a resistor) act as a resistor-diode network against negative-going transients. As illustrated in Figure 3-6, the ESD protection for the output consists of an additional diffused diode D_3 from the output to V_{CC}. The other diodes, D_1 and D_2, are parasitics.

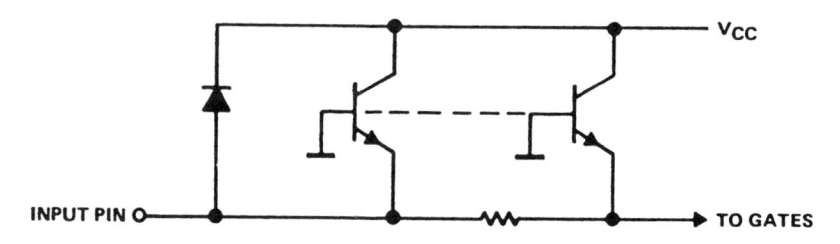

Figure 3-5. ESD input protection circuitry

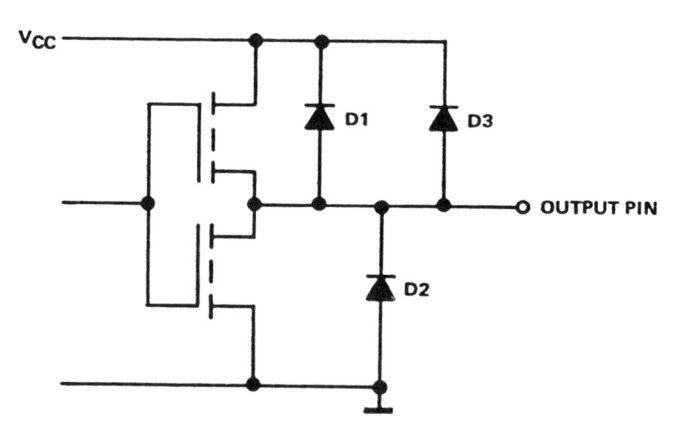

Figure 3-6. ESD output protection circuitry; D_1 and D_2 are parasitic diodes

Internal to nearly all CMOS devices are two parasitic bipolar transistors: one p-n-p and one n-p-n. Figure 3-7 shows the cross-section of a typical CMOS inverter with the parasitic bipolar transistors. Note that, as shown in Figure 3-8, these parasitic bipolar transistors are naturally configured as a thyristor or SCR. These transistors conduct when one more of the p-n junctions become forward biased. When this happens, each parasitic transistor supplies the necessary base current for the other to remain in saturation. This is known as the latch-up condition and could possibly destroy the device if the supply current is not limited.

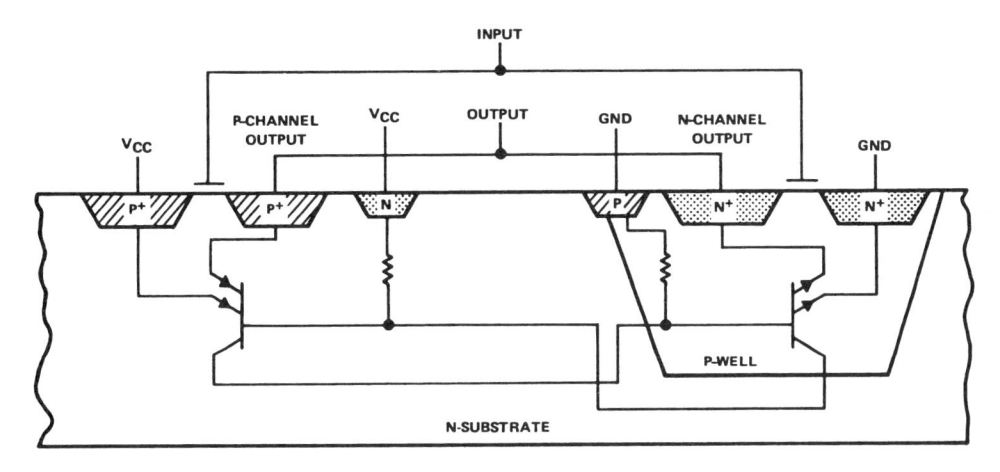

Figure 3-7. Parasitic bipolar transistors in CMOS

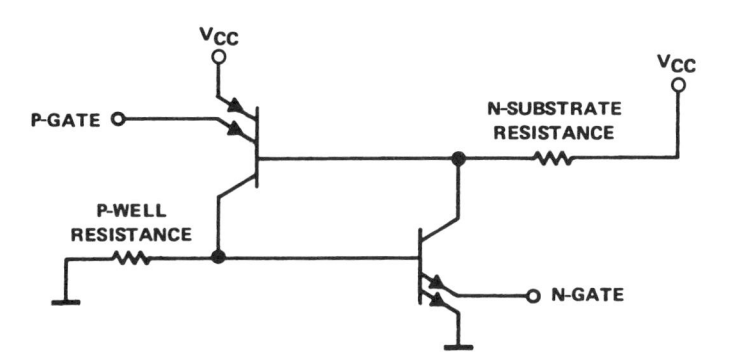

Figure 3-8. Schematic of parasitic SCR: p-gate and n-gate electrodes are connected together

A conventional thyristor is fired (turned on) by applying a voltage to the base of the n-p-n transistor, but the parasitic CMOS thyristor is fired by applying a voltage to the emitter of either transistor. One emitter of the p-n-p transistor is connected to an emitter of the n-p-n transistor, which is also the output of the CMOS gate. The other two emitters of the p-n-p and n-p-n transistors are connected to V_{CC} and ground, respectively. Therefore, to trigger the thyristor there must be a voltage greater than $V_{CC} + 0.5$ V or less than -0.5 V and there has to be sufficient current to cause the latch-up condition.

Latch-up cannot be completely eliminated! The alternative is to impede the thyristor from triggering. Texas Instruments has improved the circuit design by adding four additional diffusions or guard rings alternately connected to V_{CC} and ground as shown in Figure 3-9. The guard rings provide isolation between the device pins and any p-n junction that is not isolated by a transistor gate. All internal p-n junctions are separated by two guard rings. Tests have shown effective latch-up protection ranges from 450 mA to greater than 1 A at 25 °C, and typically greater than 250 mA at 125 °C.

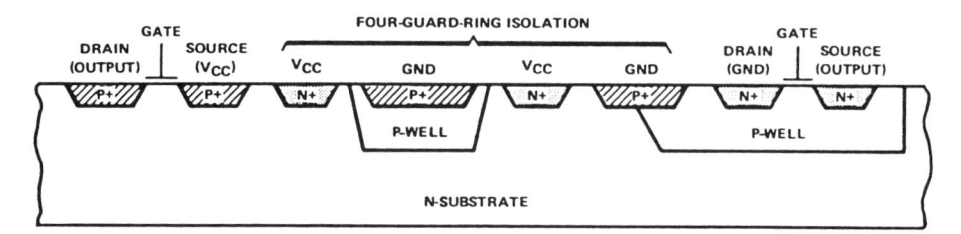

Figure 3-9. Unique latch-up suppression utilizes guard rings to virtually eliminate latch-up

3.2.2 ACL (Advanced CMOS Logic) Devices

For CMOS devices with typical gate delay times of 3 ns, as offered by TI with AC and ACT components, CMOS techniques like those used in HCMOS circuits are no longer practicable. Bipolar circuits have already shown that the design effort significantly increases with the required speed. This is also true for the CMOS process. The internal gate delay time in a circuit manufactured in this technology is about 400 ps. CMOS, therefore, is ideally suited to producing extremely fast VLSI circuits like multipliers of signal processors. As long as only signals on the chip are processed and only loads of a few 100 fF have to be driven, the benefits of ACL technology can be fully utilized.

Outside the circuit these speeds can no longer be achieved with a logic family which, like ACL, shows 5 V signal swing.

As shown in Figure 3-4, a simple NAND gate like the SN74HC00 consists of three stages: an input stage also providing the logic function in this case, an inverting buffer stage, and an output stage that has to supply the current required for driving the connected circuit. Each of these stages has a specific delay time, that of the output stage being of special importance due to the large transistors required for the high output current.

Furthermore, arbitrary edge rates cannot be allowed at the output of a circuit. Slope rates that can be controlled at system level are of the order of 2 ns. But 50 % of the edge rate always represents additional delay. 2 ns rise time corresponds to a 1 ns delay caused by the signal edge. If the delay times of the input and output stages are added to this, the total propagation delay of the circuit becomes approx. 3 ns. This illustrates that at speeds of this kind the limits of high-speed logic families will be reached, especially if 5 V signal swing is used.

A number of special measures have to be taken in the circuit design of ACL devices to ensure their reliable operation in a system. These measures begin with the circuit of the input stage. Every inverter, i.e. the input stage of a CMOS device too, represents an amplifier. In high-speed logic families these amplifiers have an upper frequency limit of some hundred MHz. As long as a defined Low or High level is applied to the input of such inverters, conditions will be stable. During signal transition, however, the amplifier will be linear for a certain time, depending on the signal edge rate. Then an oscillator, which disturbs the working of the circuit, can be formed by capacitive feedback in the circuit or capacitive and/or inductive feedback across the supply lines. For this reason ACL circuits use an input stage with hysteresis which prevents the stage from oscillating, even when rise times are some tens of ns.

Figure 3-10 shows the circuit of an ACL input stage. The familiar inverter (Q_1, Q_2) is followed by a feedback stage consisting of inverter G and transistor Q_3. With the input of the circuit at Low level, transistor Q_2 turns off, while Q_1 and Q_3 conduct. When the input voltage increases, transistor Q_1 conducts onwards from a certain voltage and the voltage on the drain terminals of Q_1 and Q_2 begins to drop. Simultaneously, inverter G responds, and the voltage at its output will switch rapidly from High to Low because of the voltage amplification of this stage, thus turning off transistor Q_3. This immediately changes the resistance resulting from the parallel connection of the two transistors Q_1 and Q_3. The voltage at the output of the input stage changes correspondingly. The same process runs in reverse if the input voltage changes from High to Low. Thus, this input stage has a Schmitt trigger characteristic, the hysteresis being about 100 mV.

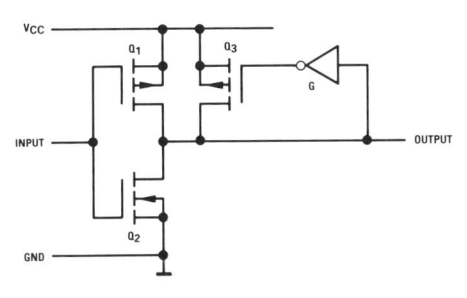

Figure 3-10.   54/74AC input circuit

Protection Circuitry

The input circuitry of the ACL family also incorporates several components for input electrostatic discharge (ESD) protection. These components also function to clamp input voltages greater than V_{CC} or less than GND. The circuitry has an absolute maximum DC input diode current rating of ± 20 mA, which must not be exceeded. Figure 3-11 shows the circuitry used for ESD protection.

The diffused p-n-p transistor cell is formed by placing a diffused p^+ resistor in an n-tank on a p^- epi layer. As for the HCMOS devices, this structure forms a diffused resistor, which is standard in ESD protection circuits for CMOS technologies, as well as a diffused p-n-p transistor. The diffused p-n-p transistor interacts with the SCR input protection circuit to absorb ESD energy. This interaction forms the primary ESD protection circuit for an ACL input. The NMOS transistor Q_m is used to guarantee a low voltage at the internal gate input terminal as well as to provide a secondary level of ESD protection. The resistance of the emitter of the diffused transistor is used to limit the current flow through the NMOS transistor Q_m.

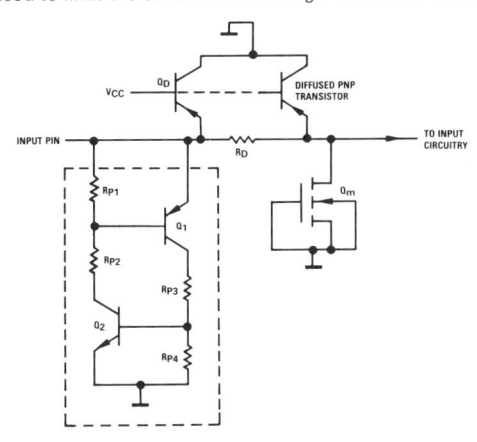

Figure 3-11.   ESD protection circuit

As described in the HCMOS protection circuitry, parasitic thyristors are also found in ACL devices. This means that latch-up protection again has to be provided to prevent damage to the device in the event of voltages greater than $V_{CC} + 0.5\,V$.

One of the advantages of the EPIC™ process used in ACL devices is its low-resistance epitaxial substrate layer. Because of the lower path resistance compared to other CMOS processes, assuming the same current, voltage drops are less critical, and the risk of latch-up is reduced. Together with additional guard rings (two rings each at the input and output) the latch-up protection up to input currents of typically 500 mA is significantly improved compared to HCMOS.

3.3 Input Characteristics

Figure 3-12 shows typical input characteristics for 'HC240 and 'AC11240 devices. Both are similar to bipolar families for $V_I < 0\,V$ (significant current increase for $V_I <$ approx. 0.6 V; see Figure 2-10).

For input voltages greater than approx. $V_{CC} + 0.7\,V$ the HC device in particular shows a strong increase in the input current, which could destroy the circuit if the input voltage were increased further. Input voltages outside the recommended range (0 V ... V_{CC} for HC, 0 V ... $V_{CC} + 0.5\,V$ for AC) should be avoided. If such voltages occur, the input clamp current must be limited to its specified maximum value.

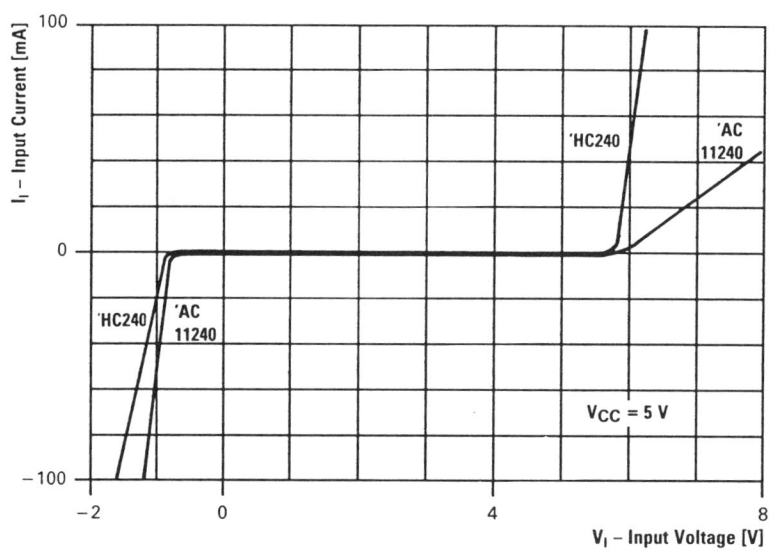

Figure 3-12. Typical 'HC240 and 'AC11240 input characteristics

3.4 Output Characteristics

The input stages of HC and AC devices are dimensioned to reach treshold voltages of about 50 % of V_{CC}. (This is not true for HCT and ACT circuits; see page 3-123).

Since output characteristics are nearly symmetrical in the Low and High states (see Figures 3-13, 3-14), the specified driving capability for both states is the same, too (typ. 6 mA for HC, 24 mA for AC).

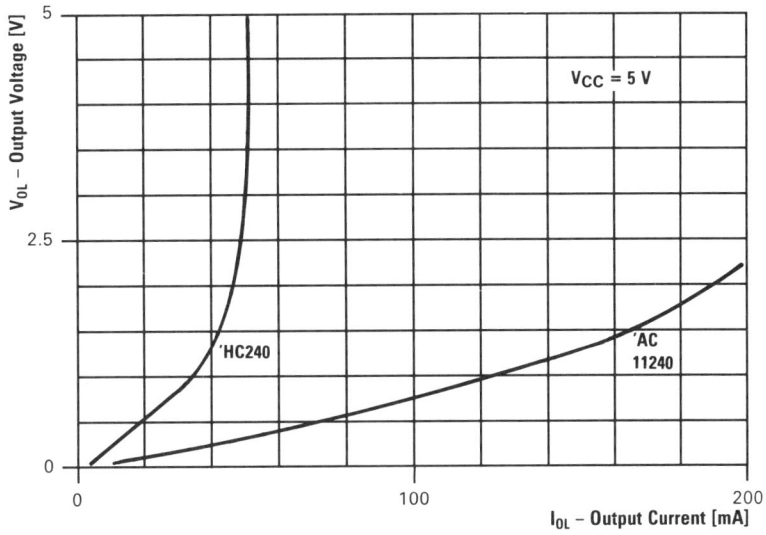

Figure 3-13. Typical 'HC240 and 'AC11240 output characteristics at Low level

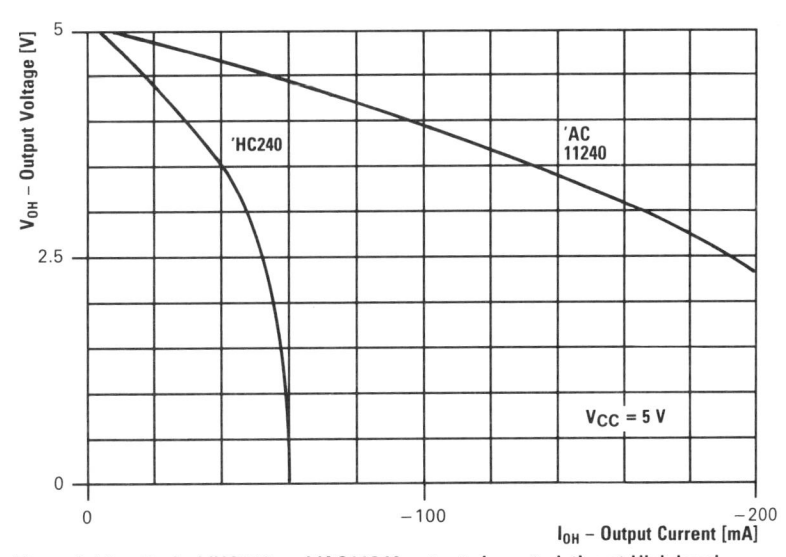

Figure 3-14. Typical 'HC240 and 'AC11240 output characteristics at High level

3.5 Noise Margin

The threshold voltage V_{th} is typically 50 % of V_{CC} in HC and AC circuits and the output voltage swing of such devices almost corresponds to V_{CC}, so the noise margins S_L and S_H at Low and High logic levels are the same. The following relationship applies for $V_{CC} = 5\,V$:

$$S = \frac{V_{th} - V_{OL}}{V_{OH} - V_{OL}} \times 100\,\% = \frac{V_{OH} - V_{th}}{V_{OH} - V_{OL}} \times 100\,\% = \frac{2.4\,V}{4.8\,V} \times 100\,\% = 50\,\%$$

Thus the noise margin is higher in CMOS circuits than in all other bipolar families.

3.6 Minimum Edge Rate

In CMOS circuits, too, there can be unwanted oscillation on the outputs if the minimum edge rate of the input signals is not maintained (see also section 2.6). The following minimum edge rates are required:

$V_{CC}\,[V]$	t_s [ns/V] HC	t_s [ns/V] AC
2.0	625	10
4.5	110	10
6.0	80	10

4. Comparison of Features of All Texas Instruments Logic Families

4.1 Fan-out

Table 4-1 shows the typical fan-out (at Low level) of all logic families available from Texas Instruments. Here the '240 has been taken as an example. In all cases the fan-out at High level is the same or higher.

The table shows that practically any number of CMOS inputs can be driven with CMOS circuits as a result of their very low input current (typ. 1 μA).

The designer has to consider that no direct driving of CMOS inputs from bipolar outputs is possible because of the different logic levels. Here either HCT/ACT circuits or appropriate circuit design measures may be taken for level matching (see page 3-123). The latter solution will significantly reduce fan-out however.

Table 4-1. Typical '240 fan-out in all logic families

To From*	'LS 240	'S 240	'ALS 240	'AS 240	'F 240	'HC 240	'AC 11240	'BCT 240
'LS240	120	60	240	48	24	24000	24000	24
'S240	320	160	640	128	64	64000	64000	64
'ALS240	120	60	240	48	24	24000	24000	24
'AS240	320	160	640	128	64	64000	64000	64
'F240	320	160	640	128	64	64000	64000	64
'HC240	30	15	60	12	6	6000	6000	6
'AC11240	120	60	240	48	24	24000	24000	24
'BCT240	320	160	640	128	64	64000	64000	64

* Fan-out doubles if '240-1 versions are used as drivers.

4.2 Power Consumption

The power consumption of a logic circuit is caused by
- the quiescent or non-switching power,
- the dynamic or switching power.

The quiescent power consumption is the product of supply voltage and the total supply current, i.e.

$P_q = V_{CC} \times I_{CC\,(tot)}$ (outputs w/o load)

For HC and AC devices $I_{CC\,(tot)}$ is the value of I_{CC} specified in device data sheets plus any increase in I_{CC} due to input signal voltage levels. For most applications the increase in I_{CC} due to input signal voltage level can be ignored whenever the input is driven by another CMOS device, unless the input voltage approaches the $V_{IL\,max}$ or $V_{IH\,min}$ limit. Input voltages less than 1 V or greater than $V_{CC} - 1$ V do not cause a significant increase in $I_{CC\,(tot)}$.

The dynamic power consumption is made up of two components
1. P_l = power consumption by the load at the circuit output
2. P_{sw} = power consumption by the internal current peak when the output switches.

The power consumption caused by the capacitive load on the output will be explained taking an 'ALS00 as an example (Figure 2-3c). When the output goes from Low to High, load capacitor C_L is charged via the upper output transistor Q_7. The charge is:

$Q = C_L \times V_{OH}$

When the output switches from High to Low, the capacitor is discharged again via the lower transistor Q_8. Thus every transition from High to Low causes the charge Q to be conducted to ground.

The average current consumption is then calculated as follows:

$$I_{av} = \frac{C_L \times (V_{OH} - V_{OL})}{T} = C_L \times (V_{OH} - V_{OL}) \times F$$

The current is thus directly proportional to the switching frequency. This means that the power consumption of the circuit is also directly proportional to the switching frequency.

Power consumption caused by the device switching logic states occurs because transistors fabricated on integrated circuits are not ideal. Parasitic capacitances exist on the chip; they must be charged and discharged. Additionally, there are short periods of time during switching when both the n-channel and p-channel transistors are partially on, resulting in current spiking between V_{CC} and ground. It is impossible to differentiate between power used to charge parasitic capacitances and current spiking. Therefore, it is necessary to develop a method that will allow the user to estimate the dynamic power consumption in any given system. This is accomplished by specifying a value called power dissipation capacitance or C_{pd}. C_{pd} allows a user to calculate the dynamic power consumption of a device using the formula:

$$P_d = [C_{pd} \times (V_{OH} - V_{OL}) \times V_{CC} \times F_i] + \Sigma\, [C_L \times (V_{OH} - V_{OL}) \times V_{CC} \times F_o]$$

where:

C_L = load capacitance on each output
C_{pd} = power-dissipation capacitance of device (specified in data sheet)
F_i = input switching frequency
F_o = output switching frequency

For CMOS circuits one can approximate $V_{OH} = V_{CC}$ and $V_{OL} = 0\,V$, giving

$$P_d = [C_{pd} \times V_{CC}^2 \times F_i] + \Sigma\, [C_L \times V_{CC}^2 \times F_o]$$

No C_{pd} values are specified for bipolar circuits; they have to be determined empirically.

In determining C_{pd} one must consider that influences on current spikes will also have an effect on C_{pd}. Parameters that have a significant effect on C_{pd}, especially in CMOS devices, are input rise and fall times. When these transition times become longer, the time interval during which both input transistors are partially on will increase accordingly. This leads to increased switching current in the circuit and thus to an increase of C_{pd}. Figure 4-1a and 4-1b illustrate the relationships in the case of HC00 and AC11000.

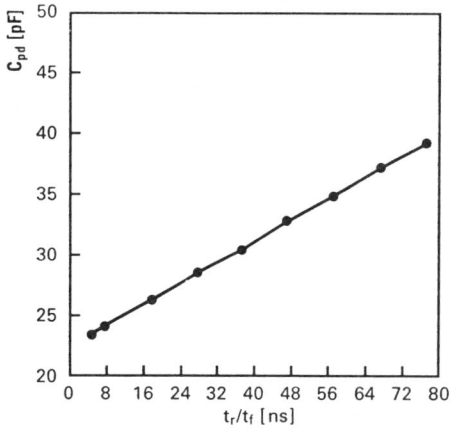

Figure 4-1a. C_{pd} vs input rise and fall times for 'HC00

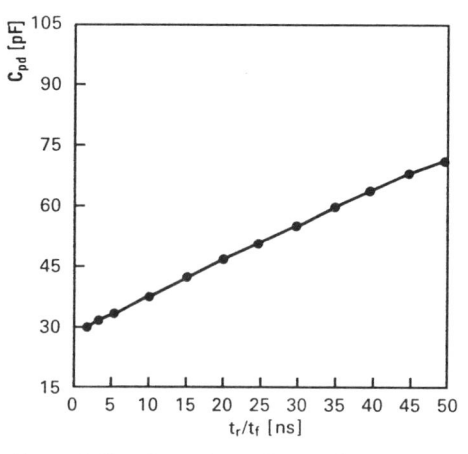

Figure 4-1b. C_{pd} vs input rise and fall times for 'AC11000

Other influences which are smaller but still have to be considered are supply voltage variations (V_{CC}) and device temperature.

Consequently the overall power consumption of a circuit is:

$P_{tot} = P_q + P_d = P_q + P_l + P_{sw}$
$= V_{CC} \times I_{CC\,(tot)} + C_{pd} \times (V_{OH} - V_{OL}) \times V_{CC} \times F_i + \Sigma\,[C_L \times (V_{OH} - V_{OL}) \times V_{CC} \times F_o]$

Figures 4-2a and 4-2b show typical diagrams for power consumption vs frequency. A '00 NAND gate is taken as an example.

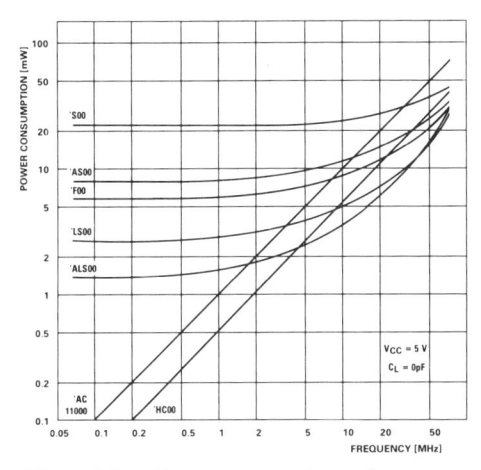

Figure 4-2a. Power consumption vs frequency ($C_L = 0$ pF for all outputs)

Figure 4-2b. Power consumption vs frequency ($C_L = 50$ pF for all outputs)

An analysis of the (measured) curves permits the following conclusions:

- The quiescent power component (P_q) of CMOS circuits is smaller by orders of magnitude. Thus these devices are highly suitable in applications like battery-powered equipment.

- Frequency-dependent internal losses cannot be neglected in any logic family (Figure 4-2a).

- A crossover frequency is given for each CMOS family in comparison with their bipolar equivalents, onwards from which the power dissipation of the CMOS circuit is greater than that of the bipolar device. Even though the relations between the product families that can be read from the diagrams are basically applicable to all other functions (drivers, flip-flops, etc.) it may not be assumed that crossover frequencies always have the same values.

- Crossover frequencies are lower for loaded outputs than in an unloaded state, i.e. the load-dependent components P_l and P_{sw} of the total power consumption are proportionally higher for CMOS circuits. This is due to the higher voltage swing $V_{OH} - V_{OL}$ on the output (typ. 3.1 V for LS, min. 4.8 V for HC and AC at $V_{CC} = 5$ V).

When conclusions are drawn from the measurements reproduced in Figures 4-2a and 4-2b, it is always necessary to consider that the findings from a direct comparison of devices cannot be transferred to the system level. Thus, for example, the effects of the power consumption in a quiescent state may not be ignored for a definitive comparison at system level.

Another problem of the test method is the output load of the individual circuits. Comparison of a bipolar and a CMOS device for the same output load says very little because the bipolar device has to source substantially higher output currents within a system of such devices than the CMOS device, whose connected inputs only sink negligibly small currents.

Because of these differences a direct comparison of devices is not valid for systems. Estimating the power consumption of a system, all factors like the duty cycle of the individual circuits, switching frequencies of functional groups, etc. have to be considered.

4.3 Propagation Delay

Figure 1-1 in the introduction has already shown a comparison of typical propagation delays for gate circuits of all families. For examining the relations with bus drivers circuits, the (maximum) delays of '240s in different technologies are compared in Table 4-2.

Table 4-2. Delays of '240 devices in different logic families

	'LS 240	'S 240	'ALS 240	'AS 240	'F 240	'HC 240	'HCT 240	'AC 11240	'ACT 11240	'BCT 240	Unit
t_{pd} max See note A	18	7	9	6.5	8	23.5	30	8.4	10.6	5.6	ns
t_{en} max See note B	30	15	18	9	10	36	42	9.2	12.5	11.1	ns
t_{dis} max See note C	25	15	12	9.5	9.5	36	42	7.7	10.8	10.6	ns
C_L See note D	45	50	50	50	50	50	50	50	50	50	pF
R_L See note D	667	90				1000	1000				Ω
R_1, R_2 See note D			500	500	500			500	500	500	Ω

Notes:

A Unless otherwise stated in the data sheet, t_{pd} is the worst case value of t_{PLH} and t_{PHL}.
B Unless otherwise stated in the data sheet, t_{en} is the worst case value of t_{PZH} and t_{PZL}.
C Unless otherwise stated in the data sheet, t_{dis} is the worst case value of t_{PHZ} and t_{PLZ}.
D See section 1 for parameter measurement information.

Comparing the delay values given in the table above it is necessary to consider that the test circuitry used to determine the values for the individual families is different. Nevertheless, these test circuits still represent comparable load conditions for the individual devices.

Comparison of the delays in Table 4-2 shows that the speed relationships of bus drivers are analogous to those of gates. The speed of the new BCT drivers is comparable to that of F devices (with reduced power consumption of about 50 % in the active state and 90 % in tristate).

Propagation delays of logic circuits depend on the load capacitance just like power consumption does. As a rule of thumb it can be said that propagation delays are about three times higher at C_L = 500 pF compared to C_L = 50 pF.

Texas Instruments Logic Families

– Design Considerations –

IMPORTANT NOTICE

Texas Instruments (TI) reserves the right to make changes in the devices or the device specifications identified in this publication without notice. TI advises its customers to obtain the latest version of device specifications to verify, before placing orders, that the information being relied upon by the customer is current.

TI warrants performance of its semiconductor products, including SNJ and SMJ devices, to current specifications in accordance with TI's standard warranty. Testing and other quality control techniques are utilized to the extent TI deems such testing necessary to support this warranty. Unless mandated by government requirements, specific testing of all parameters of each device is not necessarily performed.

In the absence of written agreement to the contrary, TI assumes no liability for TI applications assistance, customer's product design, or infringement of patents or copyrights of third parties by or arising from use of semiconductor devices described herein. Nor does TI warrant or represent that any license, either express or implied, is granted under any patent right, copyright, or other intellectual property right of TI covering or relating to any combination, machine, or process in which such semiconductor devices might be or are used.

Copyright © 1988, Texas Instruments Incorporated

1. Introduction

The logic families offered by Texas Instruments cover a wide spectrum of applications in terms of power consumption and propagation delay times. However, distinct parameters and internal properties of the individual logic family as well as different requirements with regard to the specific application mean that the designer must know and consider special design rules. These rules can vary very much for the different technologies. The following is meant to give some assistance by explaining a couple of application-specific problems and the possibilities for their solution.

2. Power Supply Considerations

Power supply regulation

Power supply regulation cannot be treated as if it were an independent characteristic of the device involved. Power supply regulation, along with temperature range, affects noise margins, fan-out, switching speed, and several other parameters. The characteristics most affected are noise margin and fan-out. When these two parameters are within the specified limits, the power supply regulation will normally be within specified limits. However, on a device where auxiliary parameters are more critically specified, more restrictive power supply regulation is normally required. When power supply regulation is slightly outside the specified limits, the device may still operate satisfactorily. However, if high ambient noise levels and extreme temperatures are encountered, failures may occur.

Application of a supply voltage above the absolute maximum rating will result in damage to the circuit.

Since power dissipation in the package is directly related to supply voltage, the maximum recommended supply voltage for TTL devices is specified at 5.5 V. This provides an adequate margin to ensure that functional capability and long-term reliability are not jeopardized.

High-level output voltage is almost directly proportional to supply voltage (i.e. a drop in supply voltage causes a drop in high-level output voltage and an increase in supply voltage results in an increase in high-level output voltage).

Since high-level output voltage is directly related to supply voltage, the output current of the device is also directly related. The output current value is established by choosing output conditions to produce a current that is approximately one half of the true short-circuit current.

It is advantageous to regulate or clamp the maximum supply voltage at the maximum recommended level including noise ripple and spikes.

Supply voltage ripple

Ripple in the supply voltage is generally considered a part of the supply voltage regulation. However, when combined with other effects (e.g. slow rise times), ripple voltage is more significant.

The effect of ripple voltage V_R can appear on either the supply voltage V_{CC} or the ground supply GND. When ripple appears on the supply voltage, it causes modulation of the input signal. The extent of the effect depends upon circuit parameters and source impedance.

Calculation of ripple voltage will be discussed using ALS/AS devices:

The turning on of transistor Q_8 shown in Figures 2-3c and 2-3d on page 3-11 is controlled by the voltage at the base of transistor Q_2 with respect to ground according to the formula:

$$V_B = V_{BE} \text{ of } Q_2 + V_{BE} \text{ of } Q_3 + V_{BE} \text{ of } Q_8$$

When ripple voltage is modulated onto the input voltage, the amplitude depends on the source impedance (Figure 2-1). The amplitude can be determined by the following equation:

$$\triangle V_R = V_R \left(\frac{R_1/\beta}{R_1/\beta + R_2} \right) = V_R \left(\frac{R_1}{R_1 + \beta R_2} \right) \quad (1)$$

where R_1 = source impedance
β = gain of transistor Q_1.

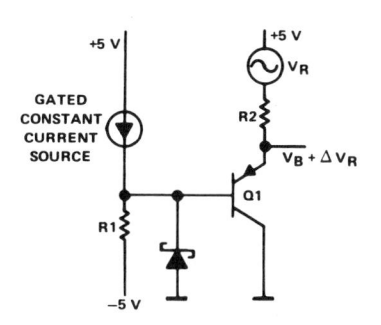

Figure 2-1. Effect of source impedance on input noise

Ripple voltage has the effect of adding extra pulses to the input signal (Figure 2-2). When ripple voltage appears in the ground supply, the threshold voltage is modulated and extra pulses occur (Figure 2-3).

Although decreasing the source impedance will reduce the effects of ripple voltage, it cannot be eliminated entirely because the emitter-base junction has an apparent resistance of approximately 30 Ω. Because of cancellation between the driving gate and the driven gate, low-frequency ripple is not a problem.

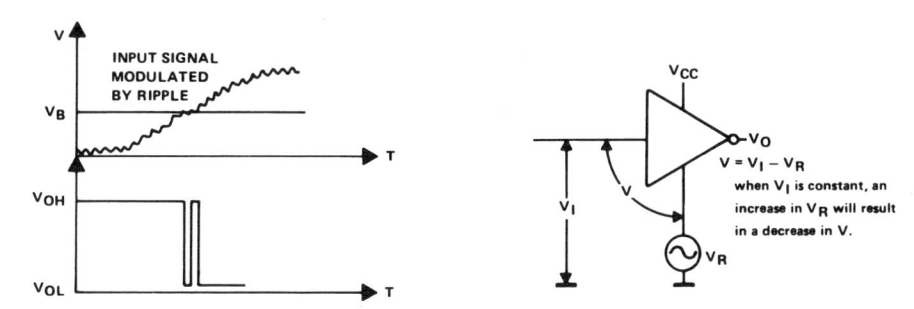

Figure 2-2. Spurious output produced by supply voltage ripple

Figure 2-3. Effect of ground noise on noise margin

Powering up/down sequence

To avoid any possible damage and reliability problems when applying power, the following steps should be followed:

1. Connect ground
2. Connect V_{CC}
3. Connect the input signal

When powering down a high-speed CMOS device, follow the above steps in reverse order.

Connecting unused inputs

Unused inputs should be tied to V_{CC} or ground to prevent the input from floating. If they are left to float, the power consumption of the device will increase and system reliability may be jeopardized. Some devices have common I/O pins. When it is necessary to tie one of these I/O pins to V_{CC} or ground, a 10 kΩ resistor should be used. The direct connection of these pins to V_{CC} or ground could destroy the device.

3. Multiple V$_{CC}$ Supplies or Partial Power-down

CMOS devices offer a designer many desirable features, the most important one being a low power consumption. However, in some systems a designer will find that even the low power consumption of CMOS is insufficient to meet his power supply constraints. Therefore, some designers will utilize partial system power-down or multiple V$_{CC}$ supplies to meet their system power requirements.

Whenever a system incorporates the use of multiple V$_{CC}$ supplies or partial power-down, the designer must take into account several important device parameters if he is using High-speed CMOS (HC) or Advanced CMOS (ACL) devices. This is necessary to avoid excessive power dissipation and prevent damage to a device that could lead to a degradation in the reliability of the device. These parameters are the continuous input and output diode currents (I$_{IK}$ and I$_{OK}$) and the continuous output current (I$_O$). I$_{IK}$ and I$_{OK}$ refer to the continuous current that is flowing through the input and output electrostatic discharge (ESD) protection circuits (Figure 3-1 shows functionally equivalent schematics of the ESD structures for HC and ACL devices).

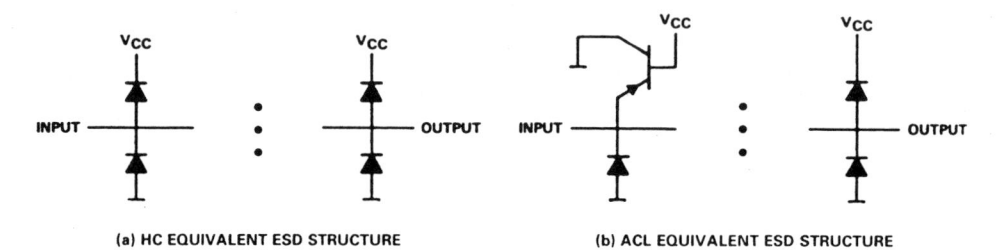

(a) HC EQUIVALENT ESD STRUCTURE   **(b) ACL EQUIVALENT ESD STRUCTURE**

Figure 3-1. Simplified ESD structures for HC and ACL devices

I$_O$ is the continuous current flowing through one of the two output transistors. Table 1 shows the absolute maximum ratings for I$_{IK}$, I$_{OK}$, and I$_O$ for both HC and ACL devices, as listed in device data sheets.

Table 1. Absolute maximum values for I$_O$, I$_{IK}$ and I$_{OK}$

PARAMETER	ABSOLUTE MAXIMUM	
	HIGH-SPEED CMOS (HC)	ADVANCED CMOS (ACL)
I$_O$	± 25 mA (standard) ± 35 mA (high-current)	± 50 mA
I$_{IK}$	± 20 mA	± 20 mA
I$_{OK}$	± 20 mA	± 20 mA

To understand how I$_{IK}$, I$_{OK}$, and I$_O$ can affect a system design, consider an example of a partial system power-down. Figure 3-2 illustrates a partial power-down situation where a device powered with V$_{CC}$ = 5 V is driving a device without power applied. The input voltage to the non-powered device exceeds V$_{CC}$ by more than the threshold voltage (0.6 to 0.8 V), causing the ESD protection structure to conduct whenever the output of the driver is in a High state. Therefore, the driving device will power-up the receiving device and any other device sharing the same V$_{CC}$ line. If no current limiting is provided, the maximum I$_O$ of the driving device and the maximum I$_{IK}$ of the receiving device could be exceeded.

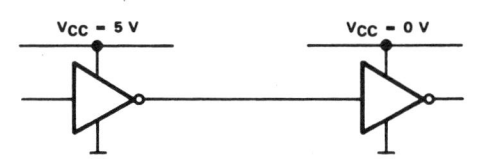

Figure 3-2. Example of partial system power-down

Several methods are available to protect the driving and receiving devices during partial system power-down. If the driving device has three-state outputs, then placing the outputs in a high-impedance state will provide the best solution. However, if this is not a viable option, then some method of current limiting must be provided. Figure 3-3 shows several methods that can be used, with current-limiting series resistors being the simplest. The value of the resistor is chosen to limit the current into the receiving device to less than 20 mA. The major drawback to using a current-limiting resistor is power dissipation. Another drawback is the effect that the resistor has on the input transition time at the receiving device during normal system operation. If the total capacitance of the interconnections and the receiving devices is high (i.e. a high-capacitance bus), then a current-limiting resistor will increase the input transition time. A system designer will have to ensure that the addition of the resistor will not increase the input transition time above the maximum input transition time of the receiving device.

(a) RESISTOR CURRENT LIMITING   (b) RESISTOR-DIODE CURRENT LIMITING

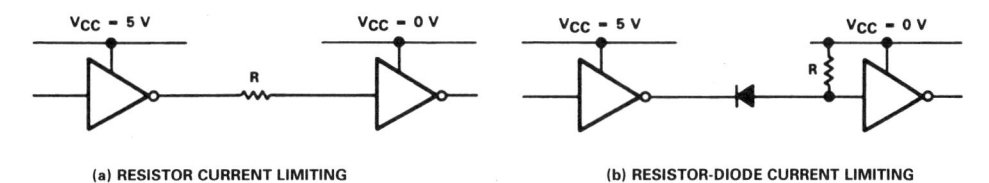

Figure 3-3. Current limiting for a partial system power-down

A second method of current limiting shown in Figure 3-3 involves the use of a pull-up resistor and a diode. The advantage of this method is that it allows for the use of a large resistor, thereby holding power dissipation to a minimum. The disadvantage of this method is that it requires the use of additional components and results in a higher value of V_{IL} at the receiving device.

A second example of how a partial power-down can cause unwanted operation is the case of two drivers connected to the same bus with one device powered down, as shown in Figure 3-4. In this case, the first bus driver will attempt to power-up the second bus driver and any other devices sharing the same V_{CC} line through the output ESD structure of the unpowered device.

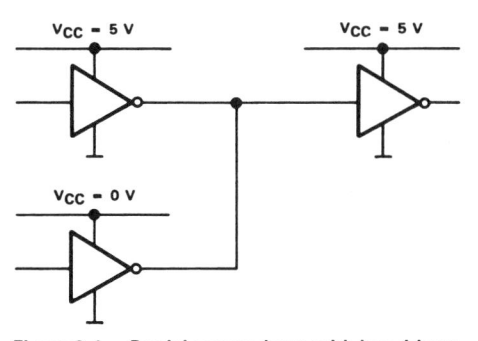

Figure 3-4. Partial power-down with bus drivers

Designer's Information

Several methods are available to solve this type of problem. One method is simply to use a current-limiting resistor as outlined above. Another solution is to isolate the unpowered driver from the V_{CC} line by means of a diode between the power pin and the V_{CC} supply. If the unpowered device is a transceiver, then pull-up or pull-down resistors are required on the output control inputs to disable the outputs. Not disabling the transceiver outputs would allow the transceiver to power-up the unpowered devices that are driven by its outputs. Whenever an isolating diode is used, the V_{CC} at the driver will always be a diode forward drop below the voltage of the supply, resulting in a degradation of V_{OH}. Figure 3-5 illustrates these circuit solutions.

(a) CURRENT-LIMITING RESISTOR

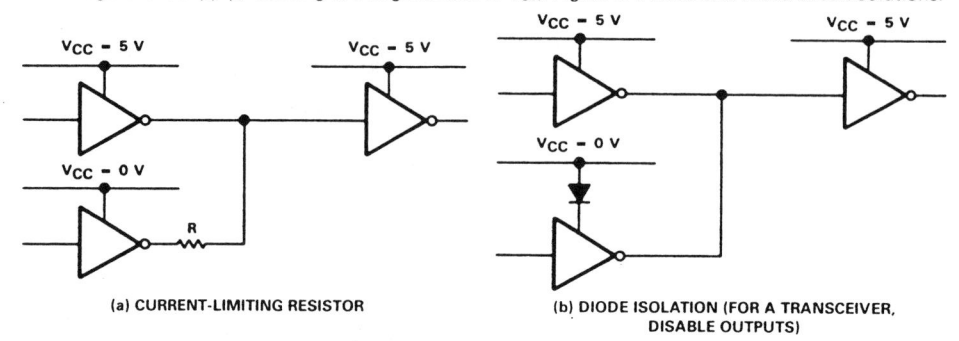

(b) DIODE ISOLATION (FOR A TRANSCEIVER, DISABLE OUTPUTS)

Figure 3-5. Current limiting for bus drivers during partial power-down

Note that for Texas Instruments' new BiCMOS (BCT) devices, no supplementary circuitry is required for partial power-down. Whenever the supply voltage (V_{CC}) falls below a voltage level of 3.5 V, the device's outputs go into tristate (high impedance), remaining there until V_{CC} is above this level again.

Another example of a system that could require current limiting protection is one that uses multiple V_{CC} supplies, or provides each card with its own on-board voltage regulator. If the V_{CC} supplies of two connection devices differ by more than 0.5 Vdc, then a current limiting scheme should be considered if the driving device is a CMOS device and is connected to the higher V_{CC}. This is necessary because V_{OH} of a CMOS device will be the same as V_{CC} whenever the I_{OH} requirement is very small. Therefore the input ESD protection diode could conduct if the V_{CC} of the driver (or V_{OH}) exceeds the V_{CC} of the receiver by more than 0.5 Vdc. It should be pointed out that it is the resulting current flow that causes the degradation of the diode, not the voltage. Note: This applies only to supplies that vary by more than 0.5 Vdc. Dynamic switching currents could cause transient voltage spiking on V_{CC} lines such that a 0.5 V difference between supplies could easily exist. These transients will not cause a problem as long as their duration is short (less than 20 ns).

Partial system power-down offers a designer a convenient method to save on system power consumption. However, when a partial power-down scheme is used, a designer must take steps to ensure that no damage occurs to devices and to avoid excessive power dissipation. He must also take similar precautions when using multiple V_{CC} supplies if the supplies of two connected devices differ by more than 0.5 Vdc.

4. Simultaneous Switching

4.1 Issue of Simultaneous Switching

Simultaneous switching of multiple outputs occurs in most applications. Previously, simultaneous switching noise was not a major issue in CMOS/TTL logic families. This was due to slower rise and fall times of the older logic families. However, with the emergence of advanced logic families – 74HC, 74ALS, 74 F, 74 AS, and 74AC/ACT – analysis of integrated-circuit performance during simultaneous switching conditions becomes necessary.

The advanced CMOS technology inherently causes fast edge rates as compared to bipolar TTL with the same speed performance range. Coupled with the rail-to-rail voltage swings inherent to CMOS logic, the susceptibility to large switching noise spikes becomes greater. Therefore, it is necessary to evaluate the contributing causes of noise during the simultaneous switching of multiple outputs.

The device physics of the actual package play a fundamental role in the voltage noise spike. The major effect on a high-speed logic device is the induced voltage on the GND and V_{CC} pins caused by the transient currents resulting from switching capacitive loads.

$$V_L(t) = -L_p C_L \frac{d^2 V_O(t)}{dt^2} \quad (2)$$

where:

V_L = voltage transient on the ground pin
L_p = self inductance of the ground pin
C_L = load capacitance
$d^2 V$ = change in the slope of the transition edge
dt = change in the time of the transition edge

To reduce the amplitude of the voltage spike sufficiently to minimize the simultaneous switching noise effect, it is necessary to address two of the components in equation (2): (a) dv/dt and (b) L_p.

The dv/dt component of equation (2) is the change in the edge rates (rise/fall times) that is technology driven. The technology utilized determines the typical performance levels as well as the edge rates. All advanced CMOS manufacturers have similar output characteristics. Therefore, all advanced CMOS logic products will exhibit similar edge rates and, as a result, will be affected by the noise problems associated with fast edges and package inductance. Reducing the dv/dt component of equation (2) would increase the device propagation delay and/or reduce the output drive. This is not a reasonable solution due to the performance and transmission line driving capabilities needed by the family for high-frequency operations.

The other factor, the L_p component of equation (2), affects the noise generated. The amount of L_p component, or package inductance, is dependent upon lead lengths as well as the location of V_{CC} and GND pins in the package. By reducing the package inductance, the voltage spike is reduced without any adverse affects to device performance.

Noise spikes resulting from simultaneous switching of multiple outputs depend upon specific application variables. It is impractical for advanced CMOS vendors to specify and test every application condition likely to cause noise, mainly because simultaneous switching effects cannot be accurately tested on production IC testers available. Some of the ways to minimize the effects of simultaneous switching are:

- Implement all circuits in surface mounted technology (SMT)
- Add series damping resistors to outputs
- Use fewer outputs per package (e.g. four out of eight outputs).

None of these options are ideal. SMT will still experience noise levels above the TTL logic threshold level with end-pin V_{CC} and GND. Series damping resistors slow the propagation delay times as well as reduce the output drive current capability and increase the component count. Utilizing fewer outputs limits application usage and decreases logic density dramatically.

4.2 The TI Simultaneous Switching Solution

To decrease the magnitude of voltage noise generated in high-speed advanced CMOS logic during simultaneous switching of multiple outputs, TI has adopted a new pinout for its EPIC™ ACL family. The pinouts for both AC/ACT in all package types will incorporate center V_{CC} and GND pin(s). The center-pin configuration helps to reduce the effective package inductance that is directly related to the voltage noise spike caused during simultaneous switching of multiple outputs. Additional ground pins and supply pins have been added to the center-pin configuration to further reduce the magnitude of the voltage spikes. To further aid the designer in the implementation of ACL, Texas Instruments will utilize a new flow-through pinout architecture which will simplify circuit board layout. Inputs will surround V_{CC} pins, outputs will surround GND pins and, where possible, control pins will be strategically located at the ends of the package.

Use of the new center-pin packages for EPIC™ ACL instead of end-pin packaging allows the designer to minimize simultaneous switching noise effects without having to adopt extreme care and extensive engineering efforts to ensure reliable system performance.

As discussed, the largest contributing factor to the amount of package inductance is the length from the V_{CC} pin or GND pin to the die. Traditional pinouts place V_{CC} and GND pins at opposite ends of the package, resulting in the maximum possible inductance through the lead frame. By relocating the V_{CC} and GND pin locations, the inductance can be significantly reduced. Figures 4-1 and 4-2 plot the pin inductance associated with each pin for both a 14-pin and 24-pin device in the plastic DIP and small outline (SO) packages. As expected, the lowest inductance occurs at the pin locations with the shortest lead length to the internal die. These pins are located at the center of the package. By placing V_{CC} and GND pins at the center of the package, the lead length is reduced from the die to the external package. This results in a significant reduction of the L_p factor of equation (1), directly decreasing the voltage noise spike.

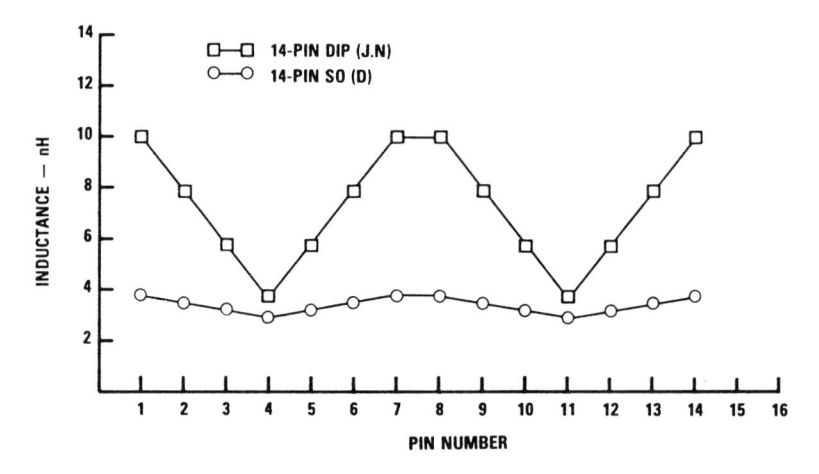

Figure 4-1. Pin inductance for a 14-pin package, end-pin configuration

In order to offer a consistent switching performance across the EPIC™ ACL family, Texas Instruments continued its inductance analysis to determine if the number of V_{CC} and GND pins per package affects the amount of package inductance. It was found that the number of outputs is directly related to the noise amplitude. In functions with three or more outputs, TI added more V_{CC} and GND pins in order to offer an advanced CMOS family which would exhibit similar noise performance across the entire family breadth. The additional V_{CC} and GND pins accomplish this for TI.

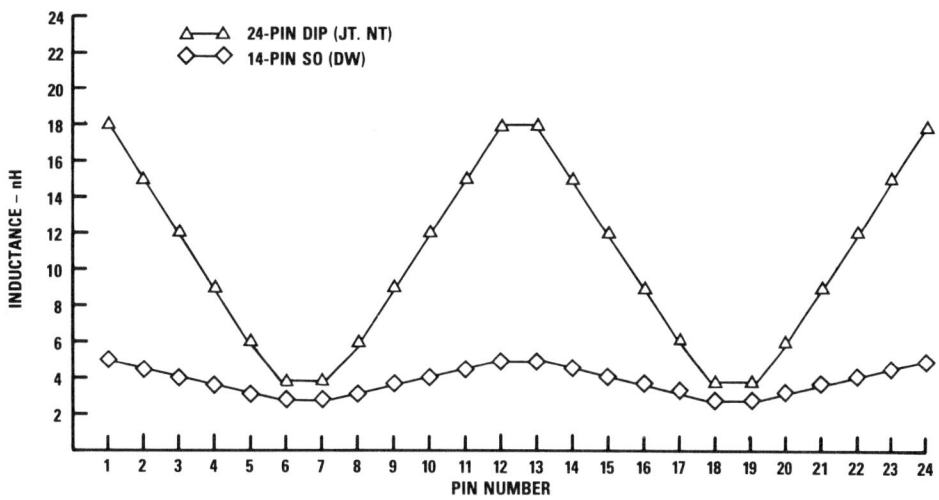

Figure 4-2. Pin inductance for a 24-pin package, end-pin configuration

As indicated in Table 2, the amplitude of the noise voltage is further reduced by increasing the number of V_{CC} and GND pins offered per package. A 20-pin DIP package with conventional end-pin pinout exhibits a significant level of inductance as compared to a 20-pin DIP with multiple V_{CC} and GND pins in the center of the package (13.7 nH vs 1.7 nH for V_{CC} and 1.1 nH for GND). With the SO (DW) package in Table 2, the package inductance is further reduced because the lead length is inherently shorter from V_{CC} or GND. Coupled with multiple V_{CC} and GND pins, the package inductance is reduced further (4.2 nH vs 1.2 nH for V_{CC} and 0.7 nH for GND).

Table 2. Inductance vs package configuration

Type package	Number of pins	End-pin configuration L_p (nH)	Center-pin configuration L_p (nH)
DIP (J, N)	14	10.2	3.2
DIP (J, N)	16	10.5	3.3 1.7[*]
DIP (J, N)	20	13.7	3.4 1.7/1.1[†]
DIP (JT, NT)	24	18.1	1.9/1.2[†]
DIP (J, N)	28	21.0	2.1/1.3[†]
SO (D)	14	3.8	2.6
SO (D)	16	4.3	2.4 1.2[*]
SO (DW)	20	4.2	2.4 1.2/0.7[†]
SO (DW)	24	4.9	1.3/0.8[†]
SO (DW)	28	5.4	1.4/0.9[†]

L_p is the effective power pin inductance.

Notes: [*] Center-pin configuration with two V_{CC} and two GND pins
 [†] Center-pin configuration with two V_{CC} and four GND pins

To optimize simultaneous switching performance of EPIC™ ACLs, TI determined the number of V_{CC} and GND pins to be employed per package. Using the new pinout configuration, the number of pins is directly related to the maximum number of outputs that can be simultaneously switched and to the package inductance that can be tolerated with the associated loads and edge rates. For DIPs and SMT packages, Table 3 indicates the number and locations of V_{CC} and GND pins per package. The use of centrally located V_{CC} and GND pins will enable designers to use EPIC™ ACL products in simultaneous switching applications with minimal noise effects.

Table 3. Number and location of V_{CC}/GND pins for ACL package

Package pin no.	No. outputs switching	Ground pin(s)	V_{CC} pin(s)
14-pin	1 or 2	Pin 4	Pin 11
16-pin	1 or 2	Pin 4	Pin 12
16-pin	3 or 4	Pin 4, 5	Pin 12, 13
20-pin	1 or 2	Pin 5	Pin 15
20-pin	3 or more	Pin 4, 5, 6, 7	Pin 15, 16
24-pin	3 or more	Pin 5, 6, 7, 8	Pin 18, 19
28-pin	3 or more	Pin 6, 7, 8, 9	Pin 21, 22

4.3 Evaluation Results

The measured data shown in Figures 4-3 through 4-5 support the need for V_{CC} and GND pins to be located at the center of the package. The device utilized for this data was a '244 function, an octal buffer/line driver. The data were measured under typical system conditions, which consisted of a 5.5-V supply voltage with a 50-pF capacitor load operating at a temperature of 25°C. The test circuit used is shown in Figure 4-6. The test performed demonstrates an excessive level of voltage produced during the condition of the output held Low while seven outputs are simultaneously switched from High to Low. The resulting instability can impact system integrity if not controlled. Similar results occur for Low-to-High switching with one output held High.

Advanced CMOS '244 devices of two other manufacturers, Figures 4-3 and 4-4, were put through the same testing procedure, and a noise spike of greater than 2.0 V was generated. A 2.0-V spike reduced the effective noise margin by almost 90% which is unacceptable to many system designers because it violates the V_{IL} maximum specifications.

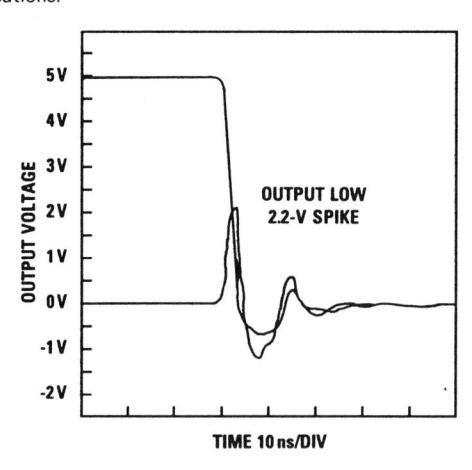

Figure 4-3. Vendor A Advanced CMOS '244 20-pin plastic DIP, 1 V_{CC}/1 GND, end-pin-package

Figure 4-5 shows that the package chosen by TI for the advanced CMOS '244 function, 74AC11244, reduces the noise to a usable level. To gather these data, the test condition was simulated for the 74AC11244 device. Use of the center-pin package controls the level of voltage noise produced, thus enabling the system designer to design with noise levels equal to or less than levels of advanced bipolar families currently in use. For this reason, TI has chosen to offer the EPIC™ ACL product line with center-pin V_{CC} and GND pins. In conjunction with TI, Signetics/Philips is also offering an advanced CMOS logic family with center-pin pinouts.

See Texas Instruments' Advanced CMOS Logic Designer's Handbook for further information on the simultaneous switching phenomenon.

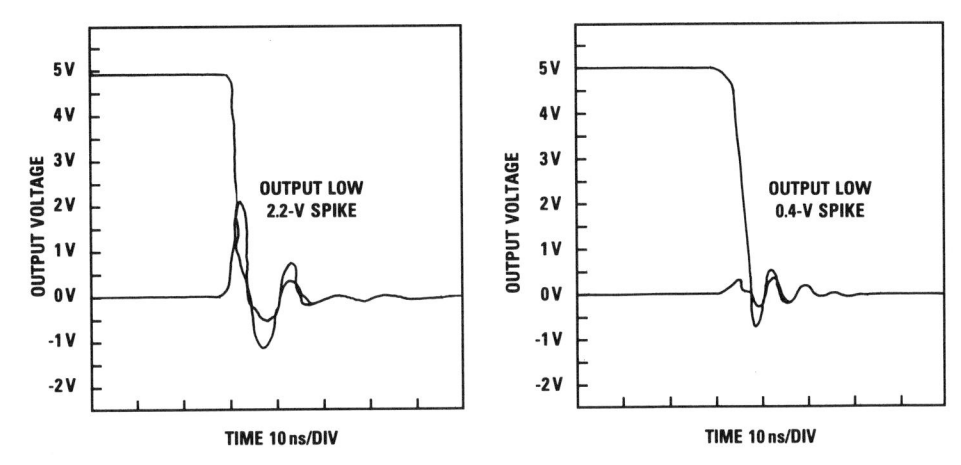

Note: Ground looping was not taken into account.

Figure 4-4. Vendor B Advanced CMOS '244 20-pin plastic DIP, 1 V_{CC}/1 GND, end-pin package

Figure 4-5. Texas Instruments 'AC11244 24-pin plastic DIP, 2 V_{CC}/4 GND, center-pin package

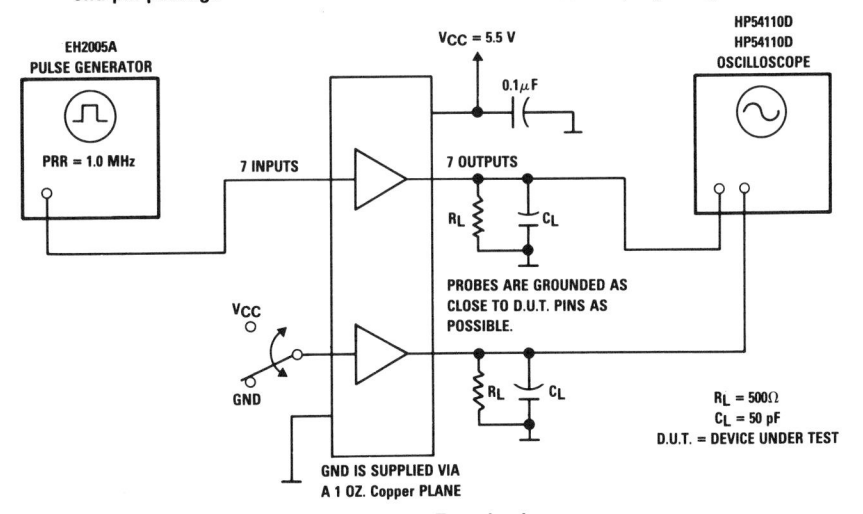

Figure 4-6. Test circuitry

$R_L = 500\Omega$
$C_L = 50$ pF
D.U.T. = DEVICE UNDER TEST

Designer's Information

5. Cross Talk

When currents and voltages are impressed on a connecting line in a system, it is impossible for adjacent lines to remain unaffected. Static and magnetic fields interact and opposing ground currents flow, creating linking magnetic fields. These cross-coupling effects are lumped together and called cross talk.

Figure 5-1 shows an equivalent circuit for a practical type of transmission line. The sending and the receiving line are coupled via the coupling impedance Z_C, representing the mutual inductances and capacitances of the two lines. Resistance coupling can be neglected in most cases.

A voltage source represented by V_{G3} and its internal resistance R_{S3} impresses a pulse on the sending line. The receiving line is at either High or Low level. The extent to which cross talk will occur depends on the type of lines used and their relationship to each other.

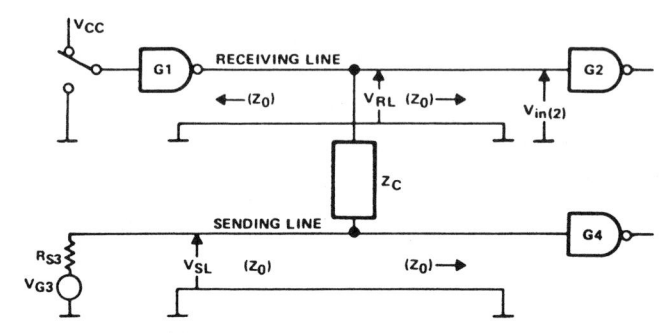

Figure 5-1. Coupling impedances involved in cross talk

The voltage impressed on the sending line is determined by the equation:

$$V_{SL} = \frac{V_{G3} Z_0}{R_{S3} + Z_0} \quad (3)$$

where

V_{G3} = open-circuit logic voltage swing generated by gate G3
R_{S3} = output impedance of gate G3
Z_0 = line impedance
V_{SL} = voltage impressed on sending line

As the voltage impressed on the sending line propagates farther along the line, it can be represented as voltage source V_{SL} (Figure 5-2). V_{SL} is then coupled to the receiving line via the coupling capacitance, where the impedance looking into the line is line impedance in both directions. Therefore the equation becomes

$$V_{RL} = V_{SL} \frac{\frac{Z_0}{2}}{(1.5 Z_0 + Z_C)} \quad (4)$$

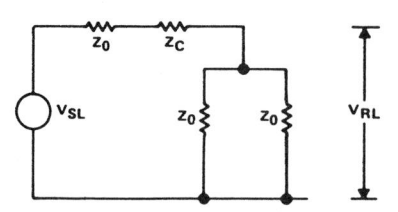

Figure 5-2. Equivalent cross talk network

The voltage impressed on the receiving line (V_{RL}) then propagates along the receiving line to gate G2, which can be considered as an open circuit, and voltage doubling occurs. Therefore:

$$V_{in(2)} = 2\, V_{RL} = V_{G3} \left(\frac{1}{1.5 + \frac{Z_C}{Z_0}}\right) \left(\frac{Z_0}{R_{S3} + Z_0}\right) \quad (5)$$

In the switching period, the transistor has a very low output impedance. Then $R_{S3} \ll Z_0$ and $V_{in(2)}$ can be simplified to the following:

$$V_{in(2)} = V_{G3} \left(\frac{1}{1.5 + \frac{Z_C}{Z_0}}\right) \quad (6)$$

The term $V_{in(2)}/V_{G3}$ can be defined as the cross talk coupling constant.

The worst case for signal line cross talk occurs when sending and receiving lines are close together but widely seperated from a ground return path. The lines then have a high characteristic impedance and a low coupling impedance.

For example, if we assume a coupling impedance of 50 pF at 150 MHz with a line impedance of approximately 200 Ω, then:

$$\frac{V_{in(2)}}{V_{G3}} = 0.62$$

This level is unsatisfactory because none of the very high-speed logic circuits has a guaranteed noise margin greater than one-third of the logic swing. Such potential cross talk can be avoided by not using the close spacing of conductors.

It should be noted that the coupling impedance model as shown in Figure 5-1 is not applicable for the calculation of travelling waves on the lines. In the above calculation, the input and output impedances of the line drivers are ignored. In addition, an equivalent circuit to be used for signal shape predictions has to consider that the coupling impedance is distributed along the line.

A more detailed model regarding these facts is shown in Figures 5-3 and 5-4. Considering the coupling impedance to be distributed, the transmission directions of the adjacent lines have to be taken into account if the input and output impedances of the drivers are different.

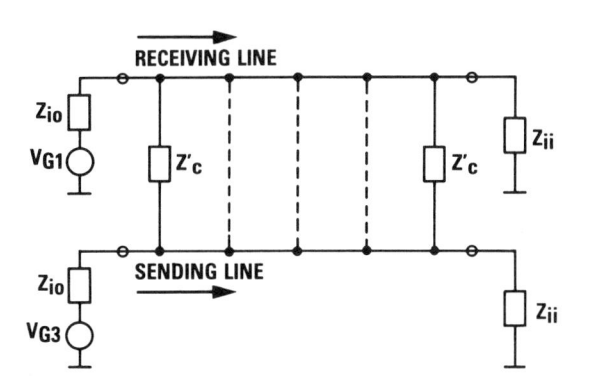

Figure 5-3. Cross talk circuit model – parallel transmission

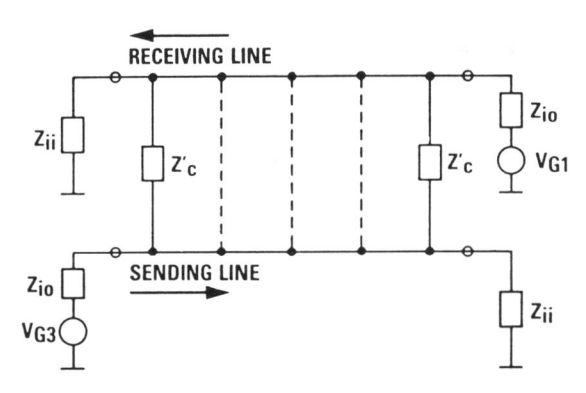

Figure 5-4. Cross talk circuit model – antiparallel transmission

Usually the output resistance of bus interface circuits is in the range of 20–50 Ω, while input resistances can reach more than 100 kΩ. This leads to different cross talk behaviour for parallel and antiparallel lines. If the sending and the receiving lines have the same transmission direction, the cross talk signal voltage coupled to the receiving line is damped by the low output impedance of the line driver immediately. In case of antiparallel transmission lines, the voltage impressed on the receiving line at the line start is being reflected at the driver input (high-ohmic). Here, damping occurs at the end of the receiving line, causing the coupled surges to propagate along the line.

As Figure 5-5 demonstrates, this leads to significantly higher cross talk noise in the latter case, requiring special layout considerations to prevent undefined logic states on the receiving line. Adjacent lines with the same transmission direction usually do not cause cross talk problems. This is valid for all logic families offered by Texas Instruments.

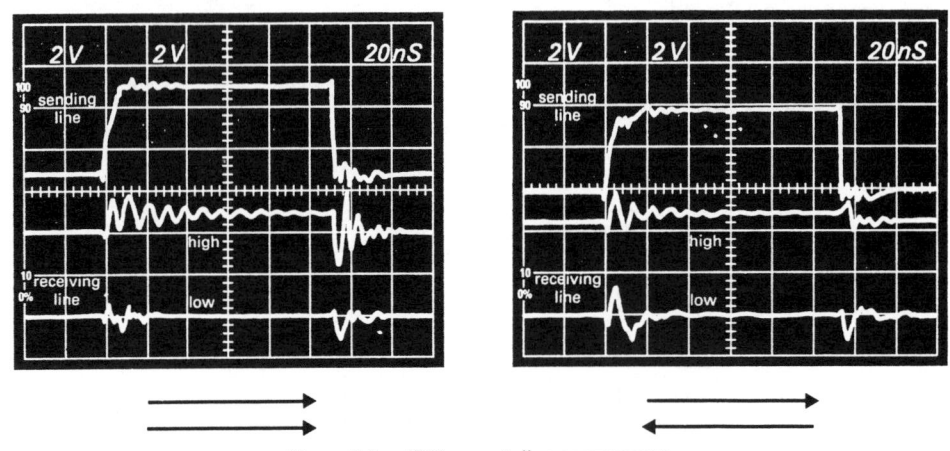

**Figure 5-5. 'F00 cross talk measurement
No ground plane, no shielding line**

The measurements shown in Figure 5-5 (as well as those in the following figures) were taken with 25-cm-long lines on a PCB (line width = 0.5 mm, line distance = 0.5 mm).

Measurements for several actions to reduce cross talk noise are shown in Figures 5-6 through 5-11. All measurements were taken with antiparallel transmission lines.

Figure 5-6:   No ground plane; no shielding line
Figure 5-7:   No ground plane; shielding line grounded at both line ends
Figure 5-8:   Ground plane; no shielding line
Figure 5-9:   Ground plane; no shielding line; line distance increased to 1.5 mm
Figure 5-10: Ground plane; shielding line grounded at one line end
Figure 5-11: Ground plane; shielding line grounded at both line ends

Because cross talk noise may reach its highest levels for Advanced CMOS circuits, which is due to their fast transition times, all the measurements shown in Figures 5-6 through 5-11 were taken with 'AC11000 devices.

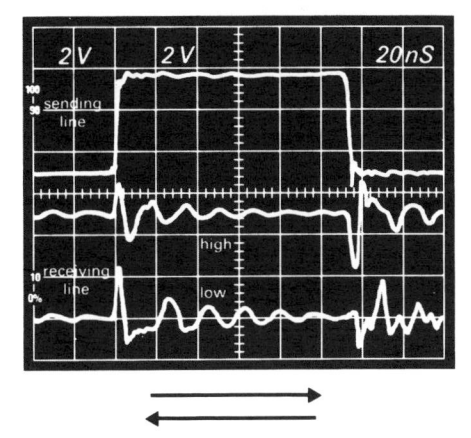

Figure 5-6. AC11000 cross talk measurement
No ground plane; no shielding line

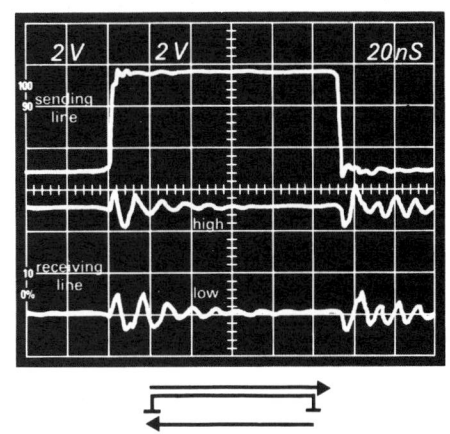

Figure 5-7. AC11000 cross talk measurement
No ground plane; shielding line grounded at both line ends

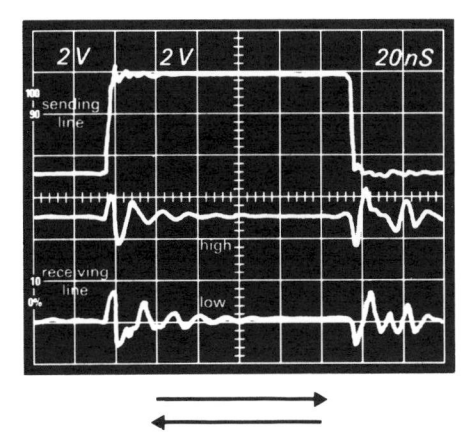

Figure 5-8. AC11000 cross talk measurement
Ground plane; no shielding line

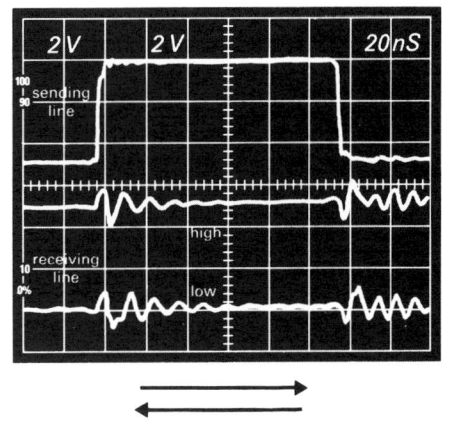

Figure 5-9. AC11000 cross talk measurement
Ground plane; no shielding line;
line distance = 1.5 mm

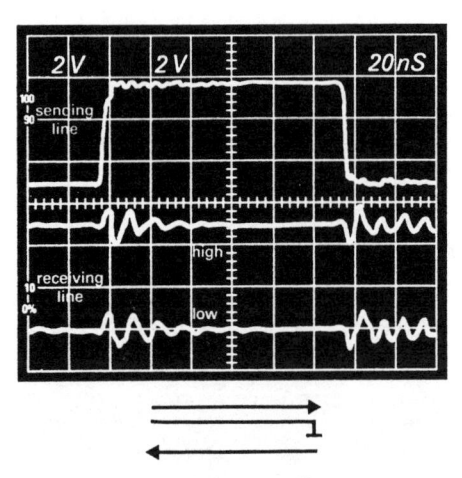

Figure 5-10. AC11000 cross talk measurement Ground plane; shielding line grounded at one line end

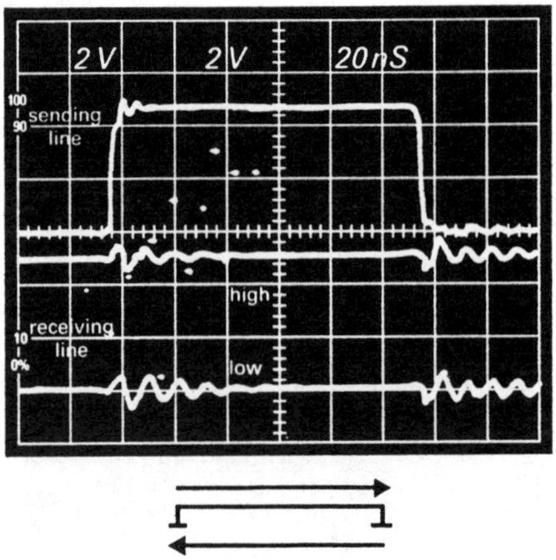

Figure 5-11. AC11000 cross talk measurement Ground plane; shielding line grounded at both line ends

The following can be deducted from the measurement results:

- Cross talk noise is higher when the receiving line level is Low. In specific constellations, unwanted switching will occur. For example, the maximum noise voltage in the Low state of Figure 5-6 is about 2.5 V, which will be accepted as High level.
- A ground plane reduces the cross talk noise amplitude by about 50%.
- Increasing the line distance from 0.5 mm to 1.5 mm reduces cross talk noise by about 30%.
- Adding a shielding line between the transmission lines, grounded at both line ends – (thus also increasing the line distance to 1.5 mm) – reduces cross talk noise by about 60%.
- If this shielding line is only grounded at one line end, the reduction is about 45% (grounding at line start, not shown in Figures 5-6 through 5-11) or 50% (grounding at line end).

6. Metastable Characteristics

Familiar to every system designer is the problem of synchronizing two digital signals operating at different frequencies. This problem is typically solved by synchronizing one of the signals to the local clock through a flip-flop. However, this solution presents an awkward trade-off; the designer cannot guarantee that the setup and hold time specifications associated with the flip-flop will not be violated. The reaction of a flip-flop under this potential metastable condition influences the overall system reliability. This section gives the system designer a better understanding of metastable characteristics.

When the setup or hold time of the flip-flop is violated, the device output response is uncertain. Presently, there is no circuit that can guarantee its reliable operation under this condition. The metastable state is defined as that time period when the output of a logic device is not at a logic level 1 (V_{OUT} greater than 2 V for TTL logic or than 70 % of V_{CC} for CMOS), nor at a logic level 0 (V_{OUT} less than 0.8 V for TTL logic or than 30 % of V_{CC} for CMOS), but instead is between these values. Since the input data are changing while it is being clocked, the system designer does not care if the flip-flop goes to either a High or Low logic level, as long as the output does not hang up in the metastable region. The metastable characteristics for a particular flip-flop will determine how long the device stays in the metastable region. This concept is illustrated in the timing diagram of Figure 6-1.

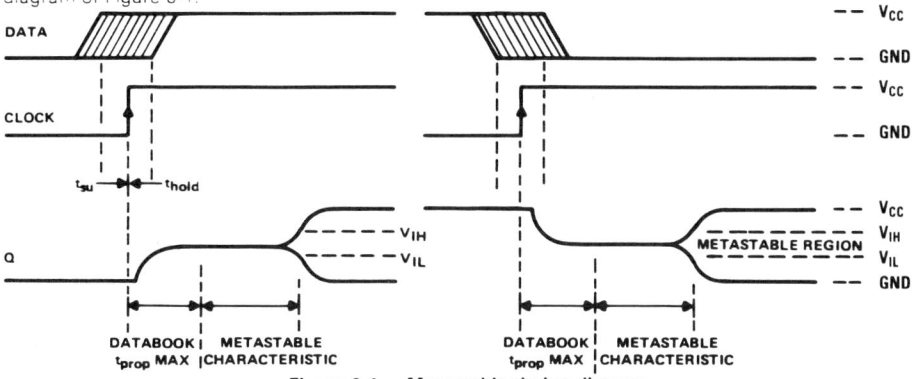

Figure 6-1. Metastable timing diagram

Evaluating the metastable characteristics for a particular flip-flop is not an easy task. The number of times that the output hangs up in the metastable region is extremely small compared to the total number of clock transitions that occur. In addition, the amount of time that the output actually spends in the metastable region varies, depending on the technology in which the flip-flop is being implemented.

From a system-engineering standpoint, a designer cannot use the specified data sheet maximum for propagation delay when using the flip-flop as a data synchronizer. Instead, one needs to know how long to wait after the specified data sheet maximum before using the data to guarantee reliable system operation. Specifically, the designer needs to know for a certain Δt, the wait time between the clocking of the flip-flop and the time when the output Q will be considered valid, what will be the expected Mean Time Between Failures (MTBF).

Conventional test equipment is not designed to measure these parameters. A special test circuit has been developed for measuring MTBF and Δt. By formulating representative values for these two parameters through experimental evaluation, the system designer can make a rational decision about what type of flip-flop to use and how long to wait before using the data.

The circuit in Figure 6-2 can be used to evaluate a flip-flop MTBF for a specific Δt. Two AD9685 voltage comparators are used to detect when the Q output of the device under test (D.U.T.) is in the metastable stage. This is accomplished by comparing the output to both V_{IL} and V_{IH}. When the output voltage from the D.U.T. is between V_{IL} and V_{IH} (in the metastable region), the comparators will be in opposite states. The outputs of the AD9685s are clocked (CBUS.CLK2) into an MC10131 at a specific Δt after the active edge of the D.U.T. clock (CBUS.CLK1). Then the outputs of the MC10131 are exclusive-ORed through an 'ALS86 and clocked (CBUS.CLK3) into an 'AS74 flip-flop.

To maximize the possibility of forcing the D.U.T. into a metastable state, the input signal must occur within a window defined by the setup and hold specifications of the D.U.T. – the "jitter" window. The width of the jitter window should not exceed the specified setup time plus hold time for the device.

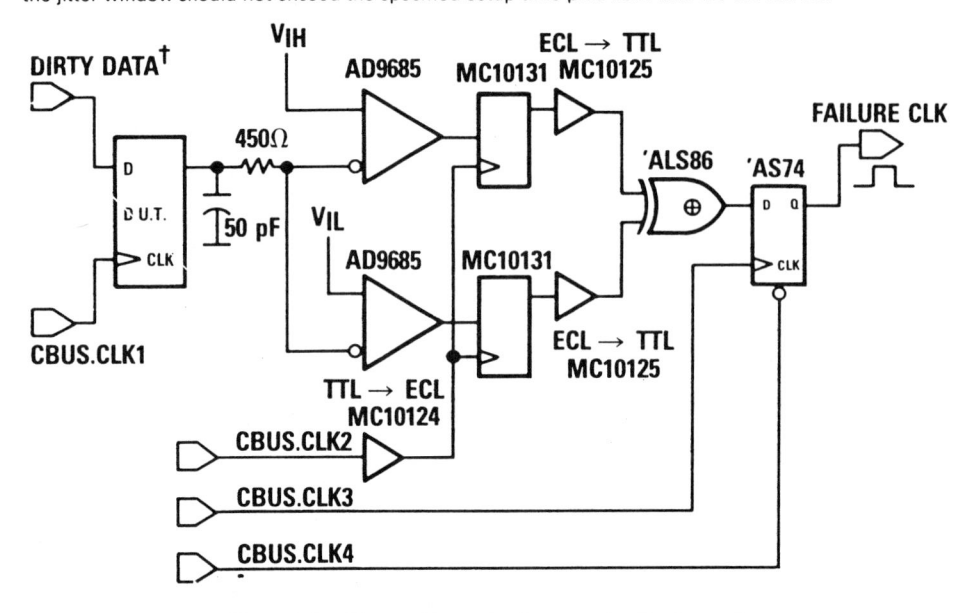

Figure 6-2. Metastable evaluation test circuit schematic

The worst-case condition occurs when the input data always violate the data setup and hold times. This relationship is shown in the timing diagram in Figure 6-3. Any other relationship of CBUS.CLK1 and DIRTY DATA[†] will reduce the chances of setting up a metastable state in the flip-flop. Therefore, the worst-case condition for a given input data frequency will be one half the D.U.T. clock rate and in phase with the clock. By using this test circuit, the MTBF can be determined for several different $\triangle t$. Plotting this data on a semilog coordinate system shows the metastable characteristics of the flip-flop at the input data frequency that was used.

Two factors considerably affect the accuracy of the results produced by the test circuit described above: the centering of the input data jitter around the input clock and the propagation delay of the AD9685 voltage comparators. If the jitter is not centered around the input clock, the probability of entering the metastable state is reduced, producing conditions which are not worst case.

The propagation delay of the voltage comparators adds a delay between the output of the D.U.T. and the clocking of the data into the MC10131s after $\triangle t$. This causes a problem if the output of the D.U.T. comes out of the metastable state, but the comparator outputs do not switch until after CBUS.CLK2 has clocked. The test circuit will record a failure when, in reality, the D.U.T. came out of the metastable region before $\triangle t$. In addition, there must be allowance for line propagation delays. The test results take these additional delays into account by subtracting their total delay from the $\triangle t$ that was recorded. The delays for each part of the test circuit are shown in Figure 6-4.

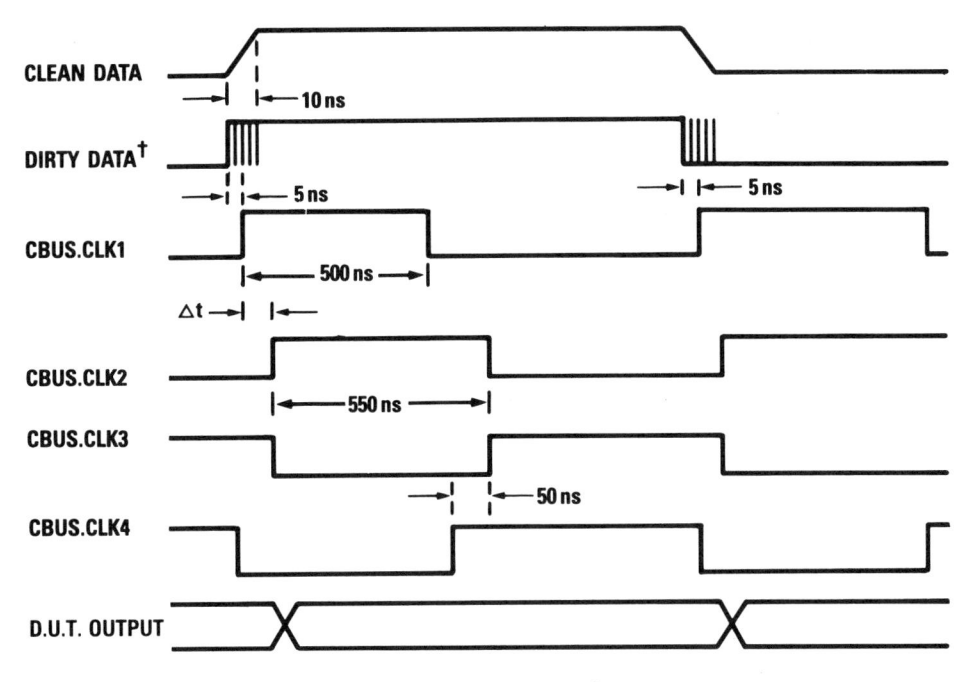

Figure 6-3. Metastable timing diagram †Also may be known as jitter.

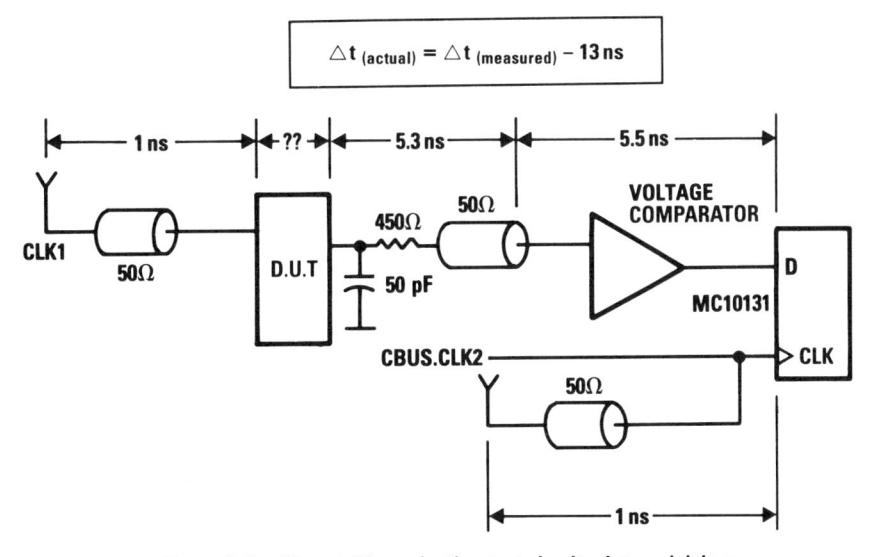

$$\triangle t_{(actual)} = \triangle t_{(measured)} - 13\,\text{ns}$$

Figure 6-4. Metastable evaluation test circuit – internal delays

From the metastability graph, we can develop an MTBF equation as follows. If we assume a linear function on the semilog plot, the equation for the MTBF line is:

$$\log(MTBF) = a\Delta t + b$$

If the log base l is converted to that of natural logs, the equation becomes:

$$\ln(MTBF) = a\Delta t + b$$

where a and b are new constants. Solving for MTBF yields:

$$MTBF = e^{(a\Delta t + b)}$$

or

$$MTBF = Be^{A\Delta t}$$

or

$$\frac{1}{MTBF} = Be^{-A\Delta t}$$

As the exponent Δt gets very small, the MTBF approaches its minimum B. The minimum MTBF occurs when a fail is present at every active edge of the D.U.T. CLK. Thus, the constant B is a function of the frequency of the D.U.T. CLK, f_{cp}, that is, CBUS.CLK1, as well as the frequency of the D.U.T. DATA, f_{data}, the jitter window, and the alignment of the two signals. If the D.U.T. is clocked at 1 MHz, the minimum MTBF will be 1 μs. Here it is assumed that the conditions for failure are present at every active edge of the CLK. That is, every edge of DIRTY DATA is aligned with every active edge of CBUS.CLK. Refer to Figure 6-3.

In general, the presence of a jitter window will not ensure a failure. A failure will only occur when this environment is present and the D.U.T. is sensitive to the environment (a device- or device-technology-dependent parameter). Thus the constant B can be modelled as follows:

$$B = C1 \times f_{data} \times f_{cp} \quad [\text{fails/second}]$$

where C1 is a function of the device technology of the D.U.T., f_{data} is the frequency at which the jitter window is present, and f_{cp} is the frequency at which the D.U.T. is sampling data.

Finally, the constant A corresponds to the slope of the MTBF graph. It is a function of the technology as well. If we replace A with C2, the final metastability equation becomes:

$$\frac{1}{MTBF} = C1 \times f_{data} \times f_{cp} \times e^{-C2\Delta t}$$

As stated earlier, the worst-case condition for the test circuit shown in Figure 6-2 occurs when the data setup and hold time is always violated. This occurs when the input data frequency is 0.5 times the D.U.T. clock frequency. Therefore, the worst-case metastability equation can be written as:

$$\frac{1}{MTBF} = \tfrac{1}{2} f_{cp}^2 \times C1\, e^{(-C2\Delta t)}$$

The constants C1 and C2 describe the metastable characteristics of the device. From the experimental data graphs in Figure 6-5, these constants can be determined for each device family. As an example, the constants are solved below for the AC11074.

C2 is defined by the slope of the line. Picking two data points off the graph yields the following result. Multiplying by 2.302 converts the base of the logs from 10 to e.

$$C2 = \frac{\log 10^6 - \log 10^0}{15.63 - 10.95}(2.302) = \frac{6}{4.68}(2.302) = 2.95 \text{ sec}^{-1}$$

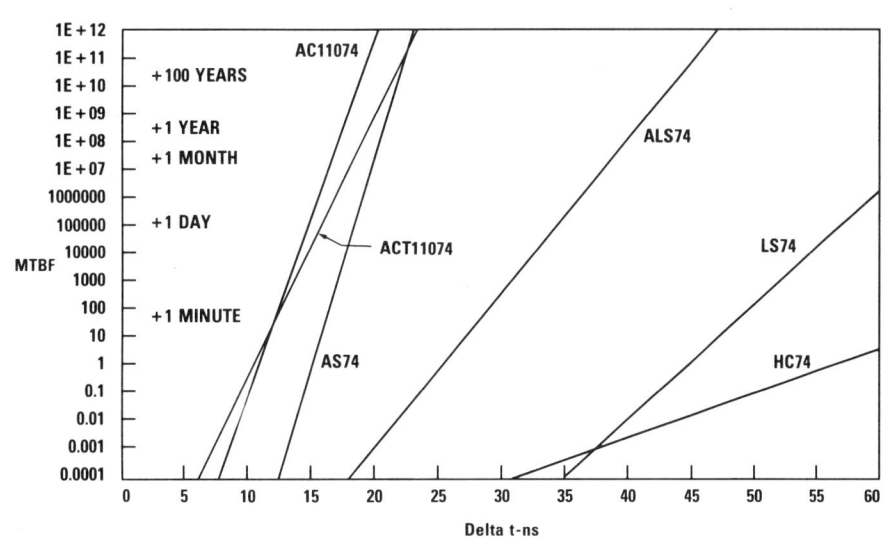

Figure 6-5. '74 metastability performance

By inserting C2 into the equation and using a data point off the graph, C1 can be computed as follows:

$$\frac{1}{\text{MTBF}} = \tfrac{1}{2} f_{cp}^2 \times C1 \; e^{(-2.95 \triangle t)}$$

$$\frac{1}{1} = \tfrac{1}{2} (10^6)^2 \times C1 \; e^{(-2.95 \times 10.95)}$$

$$C1 = 2.13 \times 10^2$$

Inserting C1 and C2 into the equation yields the metastable equation for the AC11074:

$$\frac{1}{\text{MTBF}} = \tfrac{1}{2} f_{cp}^2 \times 2.13 \times 10^2 \; e^{(-2.95 \triangle t)}$$

Given this worst-case equation, the system designer can determine the metastable characteristics when using other input clock frequencies.

Also, for any given clock frequency, the designer can determine how long a wait state must be allotted to ensure reliable system performance under a condition conducive to metastability. When the designer considers how this test under constant metastable conditions relates to the metastable potential of the system being designed, an adequate MTFB can be determined for the particular system being developed. A designer can use this MTBF and, consulting either the metastability graph or the metastability equation, determine the needed $\triangle t$.

Metastable-resistant devices

With the new 'AS3074 through 'AS4374 circuits, Texas Instruments presents a series of metastable-resistant devices. They are specially designed for data synchronization applications where the normal setup and hold time specifications will frequently be violated. Its improved metastable characteristics are achieved through TI-patented circuitry that drastically reduces the probability of entering the metastable state and greatly reduces the recovery time typically associated with a flip-flop coming out of the metastable region.

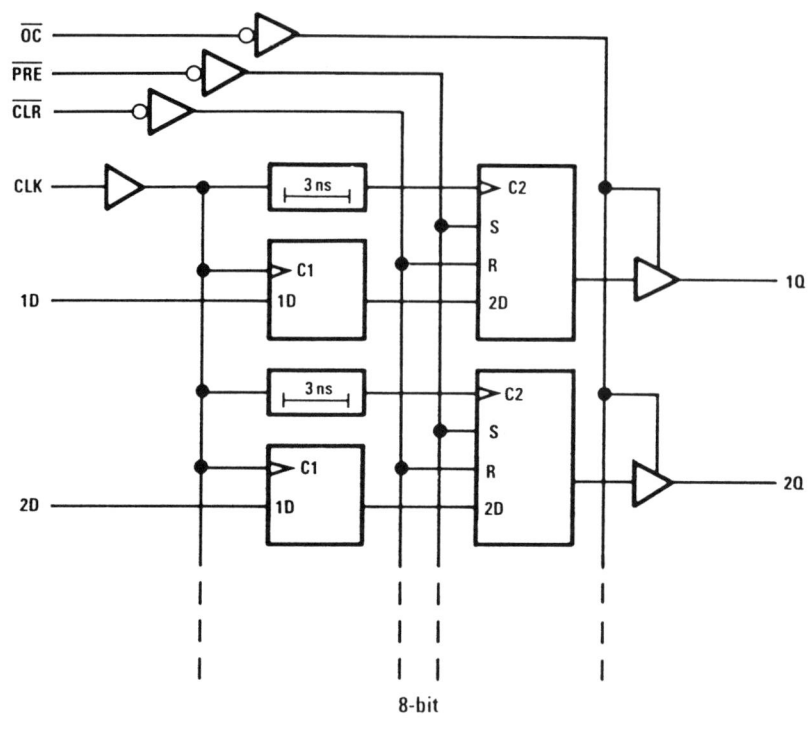

Figure 6-6. Logic diagram of 'AS3674

Figure 6-6 shows the dual-rank architecture employed in the 'AS3674. The second level of flip-flops act as a filter to exponentially decrease the probability of the output entering the metastable region.

Summary

The metastable characteristics of a flip-flop used for data synchronization can greatly affect system reliability. Based on the information presented in this section, the system designer can make a rational decision about what type of flip-flop to use and what its metastable characteristics will be. The selection of what type of flip-flop to use must be based on the speed of the application. As a general rule, the faster the flip-flop, the better its metastable characteristics.

The graphs shown and equations derived represent a reasonable assumption about the metastable characteristics. However, it is strongly reccommended that, when using flip-flops as data synchronizers, an adequate amount of guardband be allowed beyond the characteristics shown before sampling the output.

TTL Compatible CMOS Devices

3
Designer's Information

IMPORTANT NOTICE

Texas Instruments (TI) reserves the right to make changes in the devices or the device specifications identified in this publication without notice. TI advises its customers to obtain the latest version of device specifications to verify, before placing orders, that the information being relied upon by the customer is current.

TI warrants performance of its semiconductor products, including SNJ and SMJ devices, to current specifications in accordance with TI's standard warranty. Testing and other quality control techniques are utilized to the extent TI deems such testing necessary to support this warranty. Unless mandated by government requirements, specific testing of all parameters of each device is not necessarily performed.

In the absence of written agreement to the contrary, TI assumes no liability for TI applications assistance, customer's product design, or infringement of patents or copyrights of third parties by or arising from use of semiconductor devices described herein. Nor does TI warrant or represent that any license, either express or implied, is granted under any patent right, copyright, or other intellectual property right of TI covering or relating to any combination, machine, or process in which such semiconductor devices might be or are used.

Copyright © 1988, Texas Instruments Incorporated

1. Introduction

To simplify the interconnection of TTL circuits with High-speed and Advanced CMOS devices, Texas Instruments introduced the HCT and ACT circuits as subgroups to these logic families. The input circuits of the HCT and ACT devices were modified so that their input threshold voltages are identical to those of the TTL devices. The output characteristics of these devices correspond to the respective family (HC and AC) however.

2. TTL/CMOS Interface

When TTL and HC or AC circuits are interconnected, there is an incompatibility between the output voltage of the TTL circuits and the input voltage of the CMOS device. This particularly concerns the TTL High output voltage V_{OH} and the CMOS High input voltage V_{IH}. There are three ways of resolving this conflict. Firstly, the interface between TTL and HC or AC circuits can be designed with HCT or ACT devices. As already mentioned, their input voltage levels are compatible with TTL circuits. The second alternative is to provide pull-up resistors at the outputs of the TTL circuits to produce a sufficient High output voltage. Finally, special level converters may be used.

The use of HCT or ACT devices is by far the simplest solution as they were specially designed for such applications. There is no need for discrete components like pull-up resistors, and CMOS advantages like low power consumption are benefited from.

If pull-up resistors are used to match TTL output signals to the input voltages required by HC or AC circuits, the designer must calculate the optimal resistance for the application. The lowest resistance is determined by the maximum current I_{OL} the TTL circuit can supply at the output in Low state (V_{OL}):

$$R_{p\,min} = \frac{V_{CC\,max} - V_{OL\,min}}{I_{OL} + n \times I_{IL}}$$

where n is the number of CMOS inputs to be driven and I_{IL} is their input current. As the latter only amounts to a few nA, it can be neglected in all calculations.

More effort is necessary when calculating the upper limit for the resistance. First, adequate High level V_{IH} must be ensured:

$$R_{p\,max} = \frac{V_{CC} - V_{IH\,min}}{n \times I_{IH}}$$

In this case, too, the input current of CMOS devices is negligible, so very large values will be obtained.

What is more important in calculating the maximum permissible resistance, however, is ensuring that the maximum admissible rise time at the input of the CMOS circuit is not exceeded. Thus the resistance is calculated by the following formula:

$$V_{IH} = V_{CC}(1 - e^{(-t/R_p \times C)})$$

where C is the total load capacitance in the circuit. This is composed of the output capacitance of the driving gate, the total input capacitances of the gates to be driven, and the line capacitance. The actual value is calculated by solving the equation for R_p:

$$R_{p\,max} = \frac{-t}{C \times \ln(1 - V_{IH}/V_{CC})}$$

Shorter rise times will result in lower-rated resistors and thus in higher power consumption. The above calculation is based on the assumption that the driving gate has an open-collector output. It is better, however, if a gate with a totem-pole output is used instead. In this case the gate output will ensure that the voltage increases to $V_{OH\,min,\,TTL}$ (e.g. 2.7 V for ALS) within a rise time of $t_{or,\,TTL}$ (typically a few ns). The pull-up resistor then only has to pull the level to $V_{IH\,min,\,CMOS}$ (e.g. 3.5 V with $V_{CC} = 5$ V for HC) within the required time. The calculation of $R_{p\,max}$ with a required rise time t is then as follows:

$$R_{p\,max} = \frac{-(t - t_{or,\,TTL})}{C \times \ln\left(1 - \dfrac{V_{IH\,min,\,CMOS} - V_{OH\,min,\,TTL}}{V_{CC,\,TTL} - V_{OH\,min,\,TTL}}\right)}$$

It can be seen that the upper limiting value of the resistance is primarily dictated by the required rise time. The higher the resistance, the longer the rise times will be and thus the propagation delays, too. As illustrated above, reducing the resistance will increase speed but also power dissipation.

The third method of matching TTL signals to HC or AC circuits is to use special level converters. This method is a disadvantage in that the level converter itself has no logic function, while component outlay and space requirements will increase.

From a designer's point of view, using HCT or ACT circuits to match TTL signal levels to HC or AC devices is by far the simplest and most efficient method. Such devices contain the necessary level converters plus additional logic functions in a single circuit. Furthermore, the designer is not compelled to trade off between signal rise time and thus system speed and power consumption of the stages as when using pull-up resistors.

3. Operating Voltages of TTL-compatible CMOS Circuits

HCT and ACT circuits have a limited operating-voltage range due to the fact that they have to work with TTL voltage levels. The inner circuit with the exception of the input stage is equivalent to that of HC or AC circuits, so these devices would also operate in a voltage range of 2 to 6 V (7 V for ACT). Apart from the fact that HCT and ACT devices are only specified for a supply voltage of 4.5 to 5.5 V, operation at lower voltages produces serious disadvantages. The noise margin – especially at Low level – is reduced with decreasing supply voltage, and the devices are no longer TTL-compatible outside of this range, this being their primary feature in the first place.

4. Noise Margins

The noise margin of a logic family is a very important criterion in system design. This has two components – the noise margin at High level and the noise margin at Low level – and each has to be considered separately. The noise margin at High level is the voltage difference between the guaranteed output voltage V_{OH} of the driving gate and the guaranteed input voltage V_{IH} of the driven gate. Accordingly the noise margin at Low level can be defined as the voltage difference between the guaranteed output voltage V_{OL} and the input voltage V_{IL}. Figure 4-1 illustrates these relations.

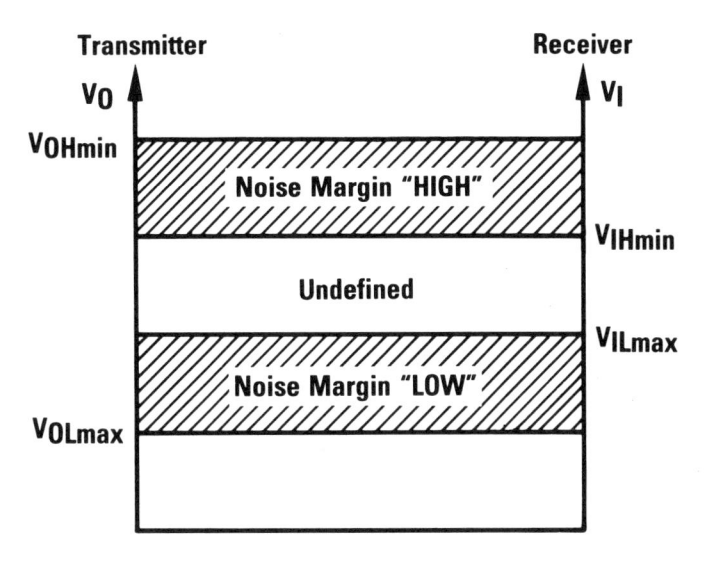

Figure 4-1. Noise margin

It is desirable that both noise margins should be as large as possible and that the undefined region between them be as narrow as possible. If the noise margin is not large enough in a particular application, interference from an internal or external source can falsify a signal and make it fall within the undefined region. Internal noise is caused by inductively or ohmicly conditioned voltage drops or by inductive and capacitive coupling with other signal lines, the latter usually being the most critical case. Figure 4-2 illustrates the voltage relations for HC, HCT, and TTL circuits, and Figure 4-3 those for AC, ACT, and TTL circuits.

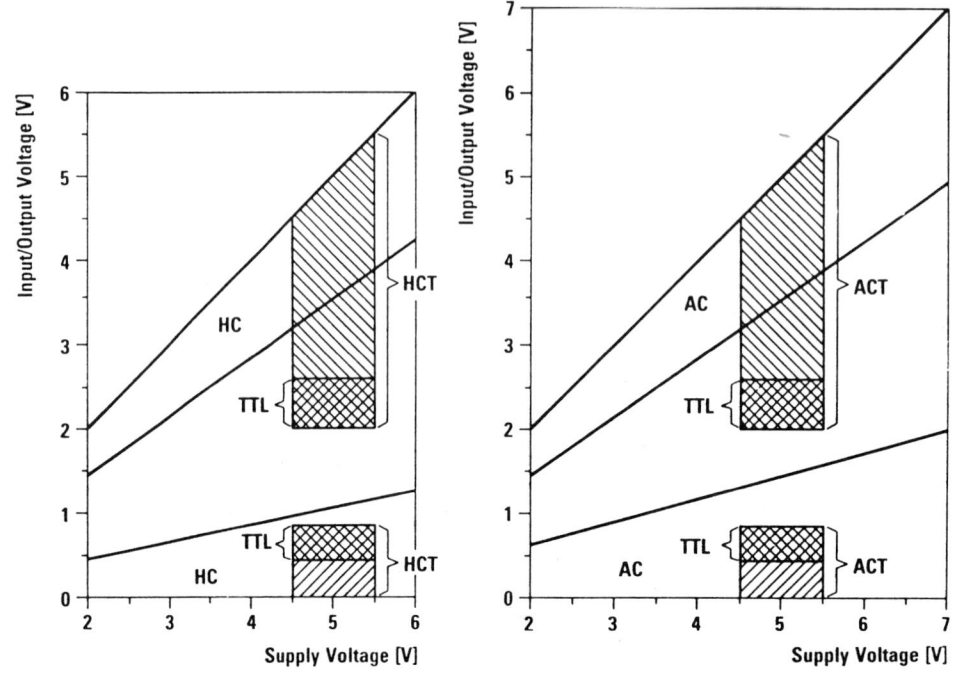

Figure 4-2. Guaranteed noise margins for HC, HCT, and TTL circuits

Figure 4-3. Guaranteed noise margins for AC, ACT, and TTL circuits

A certain percentage will always be coupled from the noisy line to the disturbed line, so it is not the absolute noise margin (in volts) that is of interest but instead the quotient of absolute noise margin and signal voltage swing. Thus the percentage High and Low noise margins S are calculated as follows:

$$S_H = \frac{V_{OH} - V_{IH\,min}}{V_{OH} - V_{OL}} \times 100\,\%$$

$$S_L = \frac{V_{IL\,max} - V_{OL}}{V_{OH} - V_{OL}} \times 100\,\%$$

To obtain realistic values, the voltages guaranteed in the data sheets for V_{OH} and V_{OL} should not be used to calculate the signal swing $V_{OH} - V_{OL}$. These would give the impression of a smaller signal swing and thus a better noise margin than is really the case. Instead the voltage value for Low and High level that the device produces in normal operation must be used. The following table lists the different values for HC, HCT, AC, ACT, and TTL devices and the resulting noise margins. To achieve comparable results, the following calculation is based on a supply voltage level of 5 V.

Table 1. Voltage levels and noise margins

	HC	HCT	AC	ACT	TTL
$V_{OH\,typ}$	4.9 V	4.9 V	4.9 V	4.9 V	3.4 V
$V_{OL\,typ}$	0.1 V	0.1 V	0.1 V	0.1 V	0.3 V
$V_{OH\,typ} - V_{OL\,typ}$	4.8 V	4.8 V	4.8 V	4.8 V	3.1 V
$V_{IH\,min}$	3.5 V	2.0 V	3.5 V	2.0 V	2.0 V
$V_{IL\,max}$	1.0 V	0.8 V	1.5 V	0.8 V	0.8 V
$V_{OH\,typ} - V_{IH\,min}$	1.4 V	2.9 V	1.4 V	2.9 V	0.7 V
$V_{IL\,max} - V_{OL\,typ}$	0.9 V	0.7 V	1.4 V	0.7 V	0.4 V
S_H	29.1 %	60.4 %	29.1 %	60.4 %	22.5 %
S_L	18.7 %	14.6 %	29.1 %	14.6 %	12.9 %

The noise margin S_L at Low level is obviously the more critical value in all three logic families, ranging from 29.1 % (AC) to 12.9 % (TTL). AC and HC devices exhibit significantly better performance than bipolar logic circuits. In practice, however, the extent to which individual devices are capable of attenuating the noise impressed on a line is what is really important. Here, their output impedance plays a primary role.

5. Power Consumption of HCT Circuits

The threshold voltage of a CMOS circuit is determined by the geometries of the input-stage transistors. These transistors are designed to sink the same input current at the required threshold voltage. The resulting voltage at the stage output is equivalent to half the supply voltage V_{CC}. For an HCMOS circuit the channel width of the p-channel transistor of the input stage is approximately twice that of an n-channel transistor. The purpose of this is that both transistors should have the same transmission characteristic and that the threshold voltage of the input stage should be 50 % of the supply voltage V_{CC}. This part of the circuit has been modified in HCT devices so that the n-channel transistor is about seven times wider than the p-channel transistor (Figure 5-1). This shifts the threshold voltage so that it is 1.5 V for a supply voltage of 5 V.

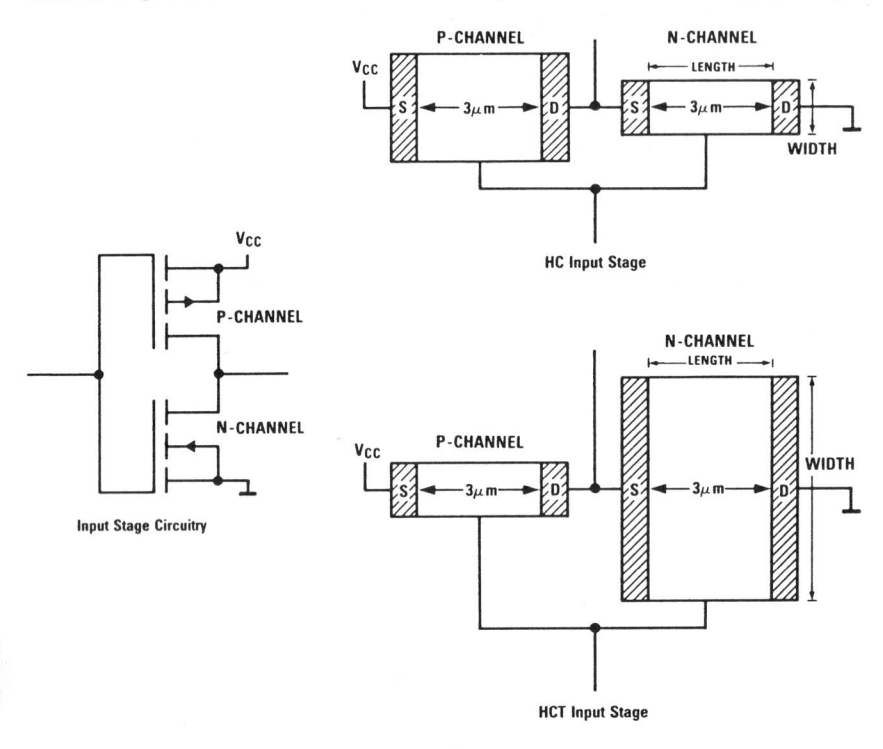

Figure 5-1. Input stage structure of HC and HCT circuits

Some compromises are necessary in HCT circuits to achieve the required parameters. Random reduction of the p-channel transistor is impossible without reducing the circuit speed at the same time because of the decrease in drain current. Therefore the n-channel transistor has to be enlarged to shift the threshold voltage accordingly.

For the input stages of ACT circuits, the dimensioning is also altered compared to the geometry of AC input stages. In addition, the change of the threshold value is produced by integrating another n-channel transistor Q_4 (Figure 5-2).

Both input stages (AC and ACT) are designed to provide approx. 100 mV of hysteresis to increase their noise margins and help ensure that the devices will be free of oscillation with input transitions of ± 10 ns/V. In each case hysteresis is provided via inverter G and PMOS transistor Q_3.

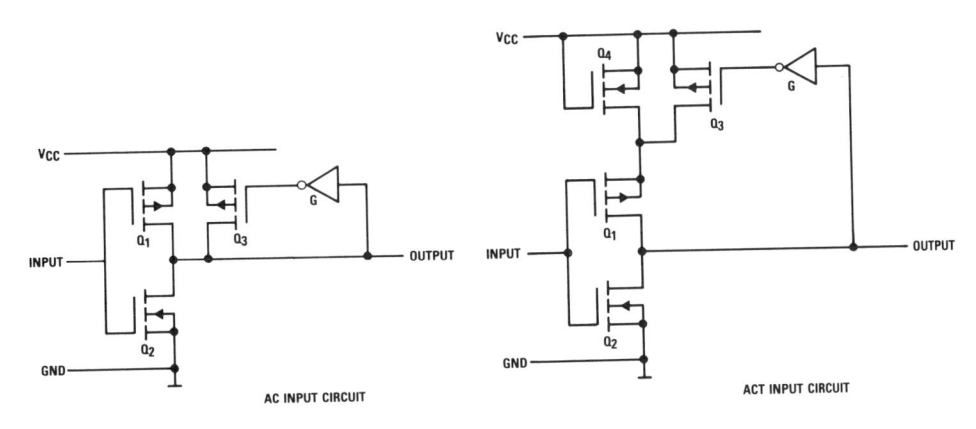

Figure 5-2. Input stages of AC and ACT circuits

However, the dependence of supply current on input voltage levels holds true for HCT and ACT devices. If an input of an HCT or ACT device is driven by a TTL device, it is possible that the HCT or ACT device will use additional supply current whenever the input is High. This increase in supply current will occur whenever the V_{OH} of the TTL driver is low enough to cause both the p-channel and n-channel transistors, in the input of the HCT or ACT device, to be partially on. The increase in supply current will not occur when the TTL device drives the input of the HCT or ACT device Low, because the V_{OL} of the TTL driver is low enough to ensure that the n-channel transistor is fully off. The increase in supply current that can occur when an input is driven by a TTL device is called $\triangle I_{CC}$. This is specified as a maximum value in HCT and ACT data sheets on a per input basis.

The quiescent power consumed by CMOS devices is given by:

$$P_q = V_{CC} \times I_{CC\,(tot)}$$

For HC and AC devices $I_{CC\,(tot)}$ is the value of I_{CC} specified in device data sheets plus any increase in I_{CC} due to input signal voltage levels. For most applications, the increase in I_{CC} due to input signal voltage level can be ignored whenever the input is driven by another CMOS device, unless the input voltage approaches the $V_{IL\,max}$ or $V_{IH\,min}$ limit. Input voltages less than 1 V or greater than $V_{CC} - 1$ V do not cause a significant increase in $I_{CC\,(tot)}$.

For HCT and ACT devices, $I_{CC\,(tot)}$ includes the increase in I_{CC} due to inputs being driven by TTL devices. It is calculated as:

$$I_{CC\,(tot)} = [N \times (\triangle I_{CC}) \times dc] + I_{CC}$$

where:

- N = number of inputs driven by TTL device
- dc = duty cycle
- $\triangle I_{CC}$ = increase in supply current (specified in data sheet)
- I_{CC} = quiescent supply current (specified in data sheet)

6. Delay Times

Another restriction in the use of HCT or ACT circuits results from their greater propagation delays. Although these circuits contain no more stages than HC or AC devices, the time for charge reversal at the output of the first stage is longer because of the smaller p-channel transistor and the higher capacitance of the n-channel transistor. This leads to an increase in propagation delay of approx. 1 to 2 ns for HCT compared to HC circuits, and of typically 2 ns for ACT compared to AC devices.

State Machine Design

Introduction

State machine designs are widely used for sequential control logic, which forms the core of many digital systems. State machines are required in a variety of applications covering a broad range of performance and complexity; low-level control of microprocessor-to-VLSI-peripheral interfaces, bus arbitration and timing generation in conventional microprocessors, custom bit-slice microprocessors, data encryption and decryption and transmission protocols are but a few examples.

Typically, the details of control logic are the last to be settled in the design cycle, since they are continuously affected by changing system requirements and feature enhancements. Programmable logic is a forgiving solution for control logic design because it allows easy modifications to be made without disturbing PC board layout. Its flexibility provides an escape valve that permits design changes without impacting time-to-market.

A majority of registered PAL device applications are sequential control designs where state machine design techniques are employed. As technology advances, new high-speed and high-functionality devices are being introduced which simplify the task of state machine design. A broad spectrum of different functionality-and-performance solutions is available for state machine design. In this discussion we will examine the functions performed by state machines, their implementation on various devices, and their selection. Finally, we will implement a state machine design and go through all of the stages involved in a design tutorial.

What Is a State Machine?

A state machine is a digital device which traverses through a predetermined sequence of states in an orderly fashion. A state is a set of values measured at different parts of the circuit. A simple state machine can consist of PAL-device-based combinatorial logic, output registers, and buried (state) registers. The state in such a sequencer is determined by the values stored in the buried and/or output registers.

A general form of a state machine can be depicted as a device shown in Figure 1. In addition to the device inputs and outputs, a state machine consists of two essential elements: combinatorial logic and memory (registers). This is similar to the registered counter designs discussed on page 2-66, which are essentially simple state machines. The memory is used to store the state of the machine. The combinatorial logic can be viewed as two distinct functional blocks: the next state decoder and the output decoder (Figure 2). The next state decoder determines the next state of the state machine while the output decoder generates the actual outputs. Although they perform two distinct functions, these are usually combined into one combinatorial logic array as in Figure 1.

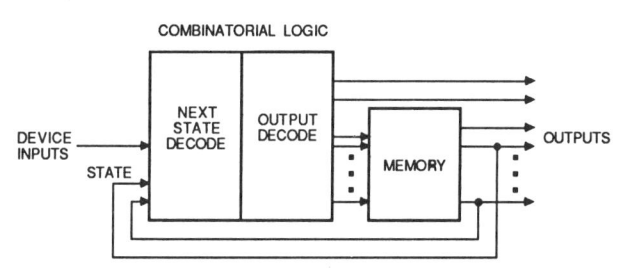

Figure 1. Block Diagram of a Simple State Machine

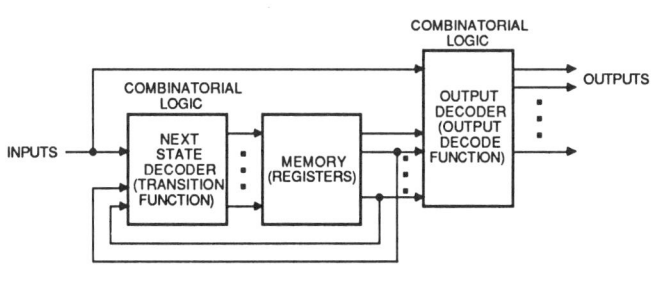

Figure 2. State Machine, with Separate Output & Next State Decoders

State Machine Design

The basic operation of a state machine is twofold:

1. It traverses through a sequence of states, where the next state is determined by next state decoder, depending upon the present state and input conditions.

2. It provides sequences of output signals based upon state transitions. The outputs are generated by the output decoder based upon present state and input conditions.

Using input signals for deciding the next state is also known as branching. In addition to branching, complex sequencers provide the capability of repeating sequences (looping) and subroutines. The transitions from one state to another are called *control sequencing* and the logic required for deciding the next states is called the *transition function* (Figure 2).

The use of input signals in the decision-making process for *output generation* determines the type of a state machine. There are two widely-known types of state machines: Mealy and Moore (Figure 3). Moore state machine outputs are a function of the present state only. In the more general Mealy-type state machine, the outputs are functions of both the state and the input signals. The logic required is known as the *output function*. For either type, the control sequencing depends upon both states and input signals.

Most practical state machines are synchronous sequential circuits which rely on clock signals to trigger the state transitions. A single clock is connected to all of the state and output edge-triggered flip-flops, which allows a state change to occur on the rising edge of the clock. Asynchronous state machines are also possible, which utilize the propagation delay in combinatorial logic for the memory function of the state machine. Such machines are highly susceptible to hazards, hard to design and are seldom used. In our discussion we will focus solely on sequential state machines.

State Machine Applications

State machines are used in a number of system control applications. A sampling of a few of the applications, and how state machines are applied, is described below.

As sequencers for digital signal processing (DSP) applications, state machines offer speed and sufficient functionality without the overkill of complex microprocessors. For simple algorithms, such as those involved in performing a Fast Fourier Transform (FFT), a state machine can control the set of vectors that are multiplied and added in the process. For complex DSP operations, a programmable DSP may be better. On the other hand, the programmable DSP solution is not likely to be as fast as the dedicated hardware approach.

Consider the case of a video controller. It generates addresses for scanning purposes, using counters with various sequences and lengths. But instead of implementing these as actual counters, the sequences involved can be "unlocked" and implemented, instead, as state machine transitions. And there is an advantage beyond mere economy of parts. A count can be set or initiated, then left to take care of itself, freeing the microprocessor for other operations.

In peripheral control the simple state machine approach can be very efficient. Consider the case of run-length-limited (RLL) code. Both encoding and decoding can be translated into state machines, which examine the serial data stream as it is read, and generate the output data.

Industrial control and robotics offer further areas where simple control functions are required. Such tasks as mechanical positioning of a robot arm, simple decision making, and calculation of a trigonometric function, usually do not require the high-power solution of microprocessors with stacks and pointers. Rather, what is required is a device that is capable of storing a limited number of states and allows simple branching upon conditions.

a. Mealy State Machines

b. Moore State Machines

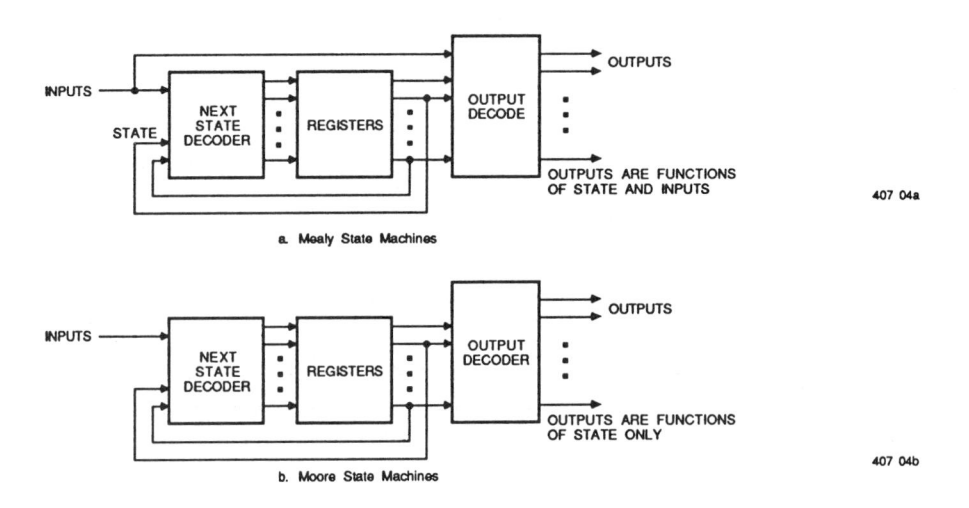

Figure 3. The Two Standard State Machine Models

State Machine Design

Data encryption and decryption present similar problems to those encountered in encoding and decoding for mass media, only here it is desirable to make the scheme not so obvious. A programmable state machine device with a security fuse is ideal for this because memory is internally programmed and cannot be accessed by someone tampering with the system.

Functions Performed

All of the system design functions performed by controllers can be categorized as one of the following state machine functions:

- Arbitration
- Event monitoring
- Multiple condition testing
- Timing delays
- Control signal generation

Later we will take a design example and illustrate how these functions can be used when designing a state machine.

State Machine Theory

Let us take a brief look at the underlying theory for all sequential logic systems, the *finite state machine* (FSM), or simply state machine.

Those parts of digital systems whose outputs depend on their past inputs as well as their current ones can be modeled as finite state machines. The "history" of the machine is summed up in the value of its internal state. When a new input is presented to the FSM, an output is generated which depends on this input and the present state of the FSM, and the machine is caused to move into a new state, referred to as the next state. This new state also depends on both the input and present state. The structure of an FSM is shown pictorially in Figure 2. The internal state is stored in a block labelled "memory". As discussed earlier, two combinatorial functions are required: the transition function, which generates the value of the next state, and the output function, which generates the state machine output.

State Diagram Representation

The behavior of an FSM may be specified in graphical form as shown in Figure 4. This is called a state diagram, or state transition diagram. Each bubble represents a state, and each arrow represents a transition between states. Inputs which cause the transitions are shown next to each transition arrow.

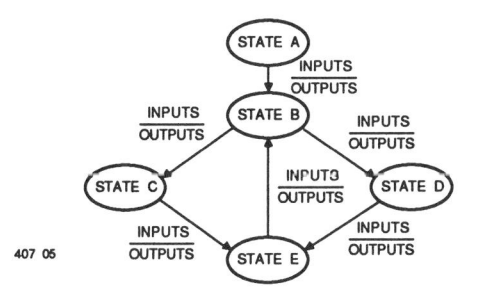

Figure 4. State Machine Representation

Control sequencing is represented in the state transition diagram as shown in Figure 5. Direct control sequencing requires an unconditional transition from state A to state B. Similarly conditional control sequencing shows a conditional transition from state C to either state D or state E, depending upon input signal I1.

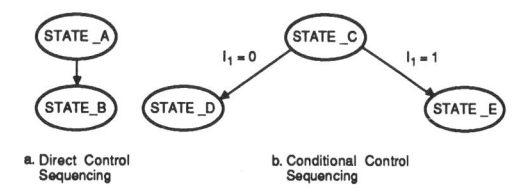

Figure 5. Control Sequencing

For Moore machines the output generation is represented by assigning outputs with states (bubbles) as shown in Figure 6. Similarly, for Mealy machines conditional output generation is represented by assigning outputs to transitions (arrows), as was shown in Figure 4. More detail on Mealy and Moore output generation is given later.

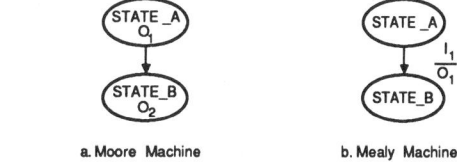

Figure 6. Output Generation

For this notation, there is a specification uncertainty as to which signals are outputs or inputs, as they both occur on the drawing next to the arrow in which they are active. This is usually resolved by separating the input and output signals names with a line (Figures 4 & 6). Sometimes an auxiliary pin list detailing the logic polarity and input or output designations is also used.

State transition diagrams can be made more compact by writing on the transitions not the input values which cause the transition, as in Figure 4, but a Boolean expression defining the input combination or combinations which cause this transition. For example, in Figure 7, some transitions have been shown for a machine with inputs "START", "X1" and "X2". In the transition between states 1 and 2, the inputs X1 and X2 are ignored (that is, they are "don't cares") and thus do not appear on the diagram. This saves space and makes the function more obvious.

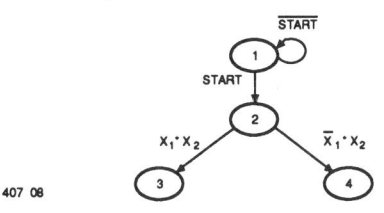

Figure 7. State Transition Diagram with Mnemonics

2-103

State Machine Design

There can be a problem with this method if one is careless. The state transitions in Figure 8 show what can happen. There are three input combinations, {I0, I1, I2, I3} = {1011}, {1101} and {1111}, which make both (/I0*I2 + I3) and (I0*I1 + I0*I2) true. Since a transition to two next states is impossible, this is an error in the specification. It must either be guaranteed that these input combinations never occur, or the transition conditions must be modified. In this example, changing (I0*I1 + I0*I2) to (I0*I1 + I0*I2)*/I3 would solve the problem.

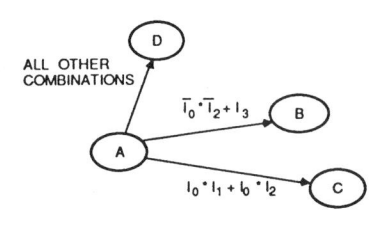

Figure 8. State Diagram with Conflicting Branch Conditions

State Transition Table Representation

A second method for state machine representation is the tabular form known as the state transition table, which has the format shown in Figure 9. Along the top are listed all of the possible input bit combinations and internal states. Each row gives the next state and the next output; the table thus specifies the transition and output functions. This type of table, however, is not suitable for specifying practical machines in which there is a large number of inputs, since each input combination defines a row of the table. With 10 inputs for example, there would have to be 1024 rows! A modified version of this table is often used directly for programmable logic sequencer (PLS) device design.

PRESENT STATE	INPUTS	NEXT STATE	OUTPUTS GENERATED
S_0-S_n	I_0-I_m	S_0-S_n	O_0-O_p

Figure 9. A State Transition Table

Flowcharts

Another popular notation is based on flowcharts. In this notation, states are represented by rectangular boxes, and alternative state transitions are determined by strings of diamond-shaped boxes. The elements may have multiple entry points, but in general have only one exit. The state name is written as the first entry in the rectangular state box. Any Moore outputs present are written next in the state box, with a caret ("^") following those that are unregistered. The state code assignment, if it is known, is written next to the upper right corner of the state box. Decision boxes are diamond or hexagonal shaped boxes containing either an input signal or a logic expression. Two exits labelled "0" and "1" lead to either another decision box, a state box, or a Mealy output. The rounded oval is used for Mealy machine outputs. Once again, a caret ("^") follows those outputs that are unregistered. All of the boxes may need to be expanded to accommodate a number of output signals or a larger expression.

The use of these symbols is shown in Figure 10. Each path through the decision boxes from one state to another defines a particular combination or set of combinations of the input variables. A path does not have to include all input variables; thus it accommodates "don't cares". These decision trees take more space than the expressions would, but in many practical cases, state machine controllers only test a small subset of the input variables in each state and the trees are quite manageable. Also, the chain of decisions often mirrors the designer's way of thinking about the actions of the controller. It is important to note that these tests are not performed sequentially in the FSM; all are performed in parallel by the FSM's state transition logic.

A benefit of this method of specifying transitions is that the problem of Figure 8 can be avoided. Such a conflict would be impossible as one path cannot diverge to define paths to two states.

When there is no danger of conflicts due to multiple next states being defined, this flowchart notation can be compacted by allowing more complex decisions. Expressions can be tested, as shown in Figure 11a, or multiple branches can extend from a decoding box, as in Figure 11b. In the second case it is convenient to group the set of binary inputs into a vector, and branch on different values of this vector.

The three methods of state machine representation—state diagrams, state tables, and flowcharts—are all equivalent and interchangeable, since they all describe the same hardware structure. Each style has its own particular advantages. Although most popular, the state transition diagrams are more complex for problems where state transitions depend on many inputs, since the transition conditions are written directly on the transition arrows. Although cumbersome, the state tables allow the designer tight control over signal logic. Flowcharts are convenient for small problems where there are not more than about ten states and where up to two or three inputs or input expressions are tested in each state. For larger problems, they can become ungainly.

Once a state machine is defined, it must be implemented on a device. Software packages are then used to implement the design on a device. The task is to convert the state machine description into transition and output functions. Software packages also account for device-specific architectural variations and limitations, to provide a uniform user interface.

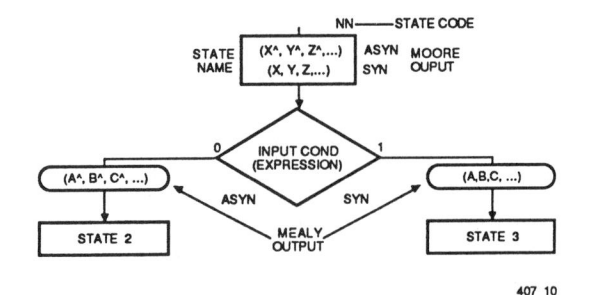

Figure 10. Flowchart Notation

State Machine Design

Some software packages accept all of the three different state machine representations directly as design inputs. However, the most prevalent design methodology is to convert the three state machine design representations to a simple textual representation. Textual representations are accepted by most software packages although the syntax varies. The PALASM 2 software package offers one such simple and easy-to-use state machine textual representation. The task of converting from a state transition diagram and flowchart representation to PALASM 2 software state machine syntax is demonstrated in a design tutorial on page 2-122.

Since the most common of all state machine representations is the state transition diagram representation, we will use it in all subsequent discussions. Transition table and flowchart representation implementations will be very similar.

State Machine Types: Mealy & Moore

With the state machine representation clarified, we can now return to the generic sequencer model of Figure 1, which has been labelled (Figure 12) to show the present state (PS), next state (NS) and output (OB, OA). This will illustrate how Mealy and Moore machines are implemented with most sequencer devices which provide a single combinatorial logic array for both next state and output decode functions. There are four ways of using the sequencer, two of which implement Moore machines and two Mealy. First, let us look at the Mealy forms.

The standard Mealy form is shown in Figure 13, where the signals are labelled as in Figure 12 to indicate which registers and outputs are used. The register outputs PS are fed back into the array and define the present state. The combinatorial logic implements the transition function, which produces the next state flip-flop inputs NS, and the output function, which produces the machine output OB. This is the asynchronous Mealy form.

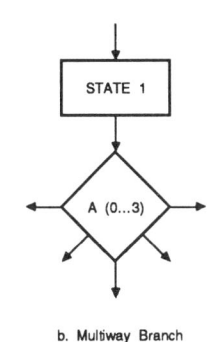

a. Testing Expressions

b. Multiway Branch

Figure 11. Using Flowcharts

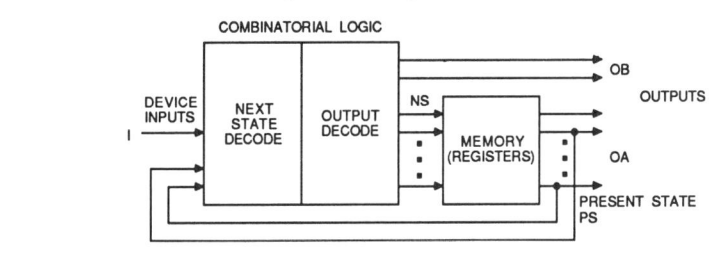

Figure 12. Generic Model of an FSM

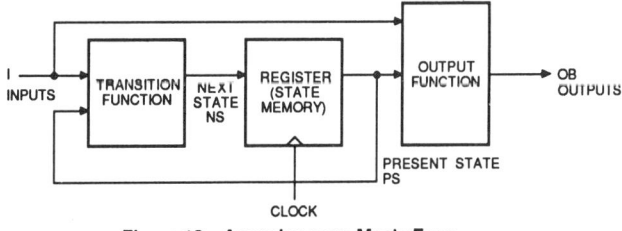

Figure 13. Asynchronous Mealy Form

State Machine Design

An alternative Mealy form is shown in Figure 14. Here the outputs are passed through an extra output register (OA) and thus do not respond immediately to input changes. This is the synchronous Mealy form.

The standard Moore form is given in Figure 15. Here the outputs OB depend only on the present state PS. This is the asynchronous Moore form. The synchronous Moore form is shown in Figure 16. In this case the combinatorial logic can be assumed to be the unity function. The outputs (OB) can be generated directly along with the present state (PS). Although these forms have been described separately, a single sequencer is able to realize a machine which combines them, provided that the required paths exist in the device.

In the synchronous Moore form, the outputs occur in the state in which they are named in the state transition diagram. Similarly in the asynchronous Mealy and Moore forms the outputs occur in the state in which they are named, although delayed a little by the propagation delay of the output decoder. This is because they are combinatorial functions of the state (and inputs in the Mealy case).

However, the synchronous Mealy machine is different. Here an output does not appear in the state in which it is named, since it goes into another register first. It appears when the machine is in the next state, and is thus delayed by one clock cycle. The state diagram in Figure 17 illustrates all of the possibilities on a state transition diagram.

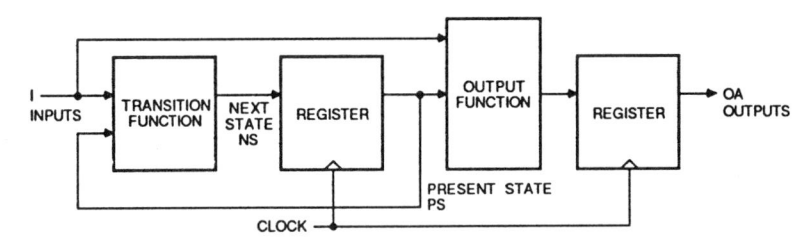

Figure 14. Synchronous Mealy Form

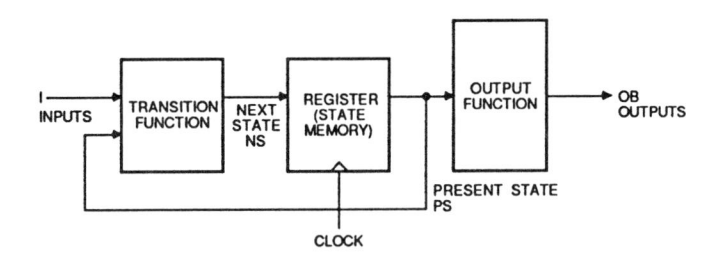

Figure 15. Asynchronous Moore Form

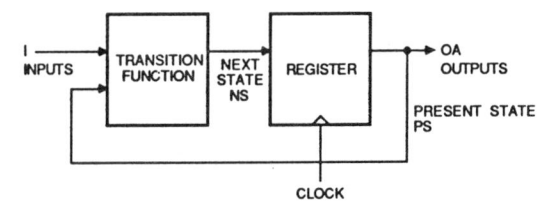

Figure 16. Synchronous Moore Form

State Machine Design

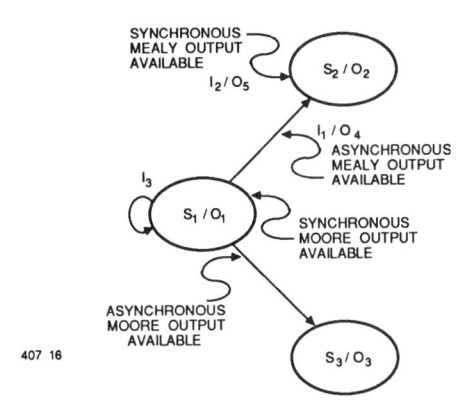

Figure 17. State Diagram Labelling for Different Output Types

As a matter of notation, Moore outputs are often placed within the state bubble and Mealy outputs are placed next to the path or arrow which activates them.

The relationship of Mealy and Moore, synchronous and asynchronous outputs to the states is shown in Figure 18.

Device Selection Considerations

Architecturally, the state machine devices can be divided into three categories:

- Logic-based devices
- Memory-based devices
- Instruction-based devices

Logic-based devices include the PAL and PLS devices. These devices use the sum-of-products logic array to implement the transition and output functions. The memory-based devices like PROSE implement the transition and output functions using a PROM or RAM array. Conceptually this is no different from the logic-based devices, since the memory can be viewed as special logic. The instruction-based sequencers (Am29PL141) offer hardwired instructions and fixed logic blocks to provide the transition function logic with enhanced capabilities like subroutines. Functionally all three types of devices work similarly performing two basic functions: control sequencing and output generation.

There are three major criteria for selecting the correct state machine device for a design:

- Number of inputs/outputs
  - I/O flexibility
  - Number of output registers
- Speed
- Intelligence/functionality
  - Number of product terms
  - Type of flip-flops
  - Number of state registers
  - Number of PROM locations (memory-based sequencers)

Number of I/Os

The number of inputs, outputs and I/O pins determines the signals which can be sampled or generated by a state machine. Figure 19 lists the devices offered for state machine designs, and shows the number of inputs, outputs and I/O pins available on each device.

Timing and Speed

The timing considerations for sequencer design are similar to those for registered logic design (page 2-64). A system clock

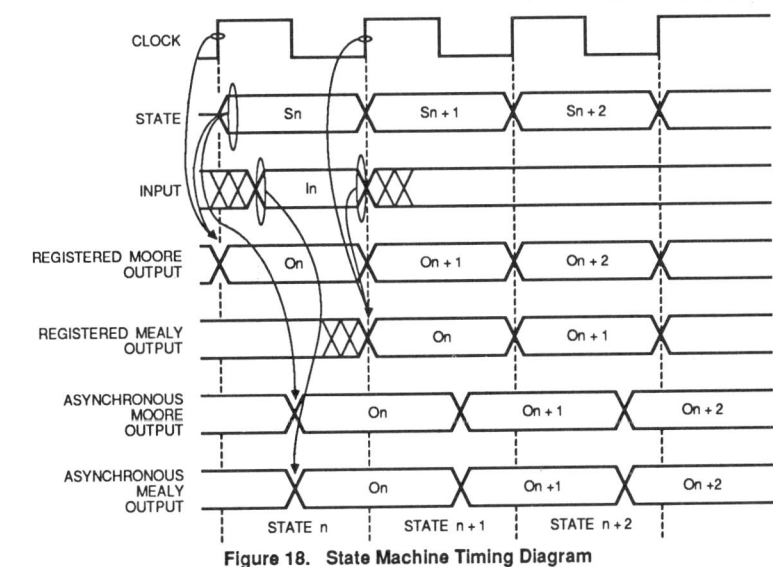

Figure 18. State Machine Timing Diagram

State Machine Design

DEVICE	TOTAL NUMBER OF PINS	DEDICATED INPUT PINS	DEDICATED REGISTERED OUTPUT PINS	I/O PINS	LOGICAL PRODUCT TERMS/ OUTPUT	SPEED GRADES FREQUENCY IN MHz f_{MAX}
TTL						
16R8	20	8	8	-	8	55.5, 37, 25, 16
16R6	20	8	6	2	8	55.5, 37, 25, 16
16R4	20	8	4	4	8	55.5, 37, 25, 16
16RP8	20	8	8	-	8	22.2
16RP6	20	8	6	2	8	22.2
16RP4	20	8	4	4	8	22.2
16RA8	20	8	0-8	8-0	4	20
16X4	20	8	4	4	8	14
23S8	20	9	4	4	8-12	33.3, 28.5
20R8	24	12	8	-	8	37, 25, 16
20R6	24	12	6	2	8	37, 25, 16
20R4	24	12	4	4	8	37, 25, 16
20X10	24	10	10	-	4	22.2
20X8	24	10	8	2	4	22.2
20X4	24	10	4	6	4	22.2
20RP10	24	10	10	-	8	37, 25
20RP8	24	10	8	2	8	37, 25
20RP6	24	10	6	4	8	37, 25
20RP4	24	10	4	6	8	37, 25
20XRP10	24	10	10	-	8	30, 22.2, 14
20XRP8	24	10	8	2	8	30, 22.2, 14
20XRP6	24	10	6	4	8	30, 22.2, 14
20XRP4	24	10	4	6	8	30, 22.2, 14
20RS10	24	10	10	-	8-16	19.2
20RS8	24	10	8	2	8-16	19.2
20RS4	24	10	4	6	8-16	19.2
20RA10	24	10	0-10	10-0	4	33, 20
22RX8	24	14	0-8	8-0	8	28.5
22V10	24	12	0-10	10-0	8-16	40, 28.5, 18
32VX10	24	12	0-10	10-0	8-16	25, 22.2
32R16	40	16	16	-	8-16	16
PLS167	24	14	6	-	Total 48	33
PLS168	24	12	8	-	Total 48	33
PLS105	28	16	8	-	Total 48	37
PMS14R21	24	8	8	-	128 states	25
29PL141	28	8	16	-	64 states	20
CMOS						
C16R8	20	8	8	-	8	28.5
C16R6	20	8	6	2	8	28.5
C16R4	20	8	4	4	8	28.5
C20R8	24	12	8	-	8	20, 15.3
C20R6	24	12	6	2	8	20, 15.3
C20R4	24	12	4	4	8	20, 15.3

Figure 19. Devices for State Machine Design

State Machine Design

DEVICE	TOTAL NUMBER OF PINS	DEDICATED INPUT PINS	DEDICATED REGISTERED OUTPUT PINS	I/O PINS	LOGICAL PRODUCT TERMS/ OUTPUT	SPEED GRADES FREQUENCY IN MHz f_{MAX}
C22V10	24	12	0-10	10-0	8-16	33.3, 20
C29M16	24	5	-	16	8-16	20, 15
C29MA16	24	5	-	16	4-12	20, 15
ECL						
10H20EV/EG8	24	12	0-8	8-0	8-12	125
10020EV/EG8	24	12	0-8	8-0	8-12	125

Figure 19. Devices for State Machine Design (Cont'd.)

cycle forms the basic kernel for evaluating control function behavior. For the most part, all input and output functions are specified in relationship to the positive edge. Registered outputs are available after a period of time t_{CLK}, the clock-to-output propagation delay. Asynchronous outputs require an additional propagation delay (t_{PD}) before they are valid.

For the circuit to operate reliably, all of the flip-flop inputs must be stable at the flip-flop no later than the minimum set-up time (t_{su}) of the flip-flops before the next active clock edge. If one of the inputs changes after this threshold, then the next state or synchronous output could be stored incorrectly; the circuit may even malfunction. To avoid this, the clock period (t_p) must be greater than the sum of the set-up time of the flip-flops and the clock to output time ($t_{su} + t_{CLK}$). This determines the minimum clock period and hence the maximum clock frequency, f_{MAX}, of the circuit. Metastability and erroneous system operation may occur if these specifications are violated.

The timing relationships are shown in Figure 20. In each cycle there are two regions: the stable region, when all signals are steady, and the transition region, when the machine is changing state and signals are unstable. The active clock edge causes the flip-flops to load the value of the new state which has been set up at their inputs. At a time after this, the present state and output flip-flop outputs will start to change to their new values. After a time has elapsed, the slowest flip-flop output will be stable at its new value. Ignoring input changes for the moment, the changes in the state register cause the combinatorial logic to start generating new values for the asynchronous outputs and the inputs to the flip-flops. If the propagation delay of the logic is t_{PD}, then the stable period will start at a time equal to the sum of the maximum values of t_{CLK} and t_{PD}.

Figure 19 also shows the maximum operating frequency of the devices. ECL-based PAL sequencers can implement simple state machines from 60 MHz to above 100 MHz. Conventional TTL PAL sequencers can now reach speeds of 55 MHz. The new PLS devices operate at 37 MHz. Finally, PROSE sequencers provide 25 MHz operation, and the Am29PL141 can run at 20 MHz. The design engineer usually selects the simplest device that provides the desired level of performance.

Asynchronous Inputs

The timing of the inputs to a synchronous state machine are often beyond the control of the designer and may be random, such as sensor or keyboard inputs, or they may come from another

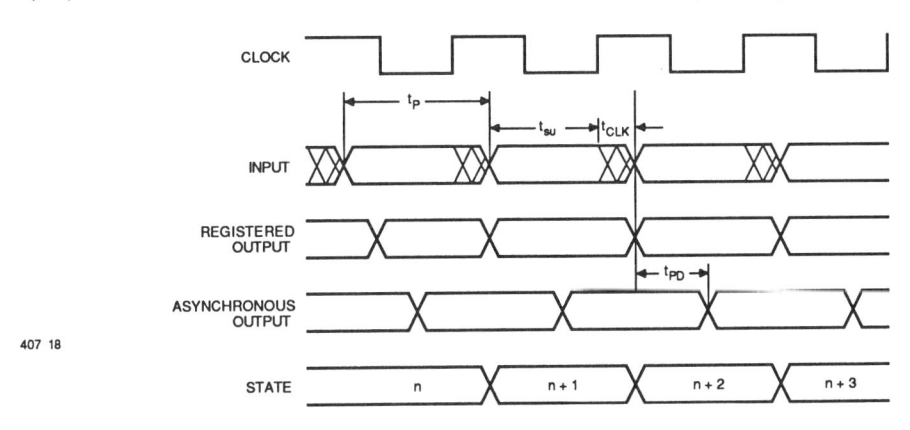

Figure 20. Timing Diagram for Maximum Operating Frequency

State Machine Design

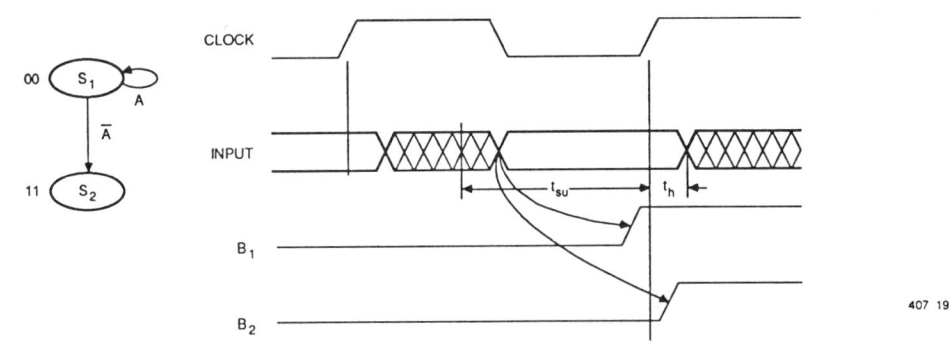

Figure 21. Asynchronous Input Causing Race

synchronous system that has an unrelated clock. In either case no assumptions can be made about the times when inputs can or cannot arrive. This fact causes reliability problems that cannot be completely eliminated, but only reduced to acceptable levels.

Figure 21 shows two possible transitions from state "S1" (code 00) either back to itself, or to state "S2" (code 11). Which transition is taken depends on input variable "A" which is asynchronous to the clock. The transition function logic for both state bits B1 and B2 includes this input. The input A can appear in any part of the clock cycle. For the flip-flops to function correctly, the logic for B1 and B2 must stabilize correctly before the clock. The input should be stable in a window t_{su} (setup time) before the clock and t_h (hold time) after the clock. If the input changes within this window, both the flip-flops may not switch, causing the sequence to jump to states 01 or 10, which are both undefined transitions. This type of erroneous behavior is called an input race.

A solution to this problem is to change the state assignment so that only one state variable depends on the asynchronous input. Thus the 11 code must be changed to 01 or 10. Now, with only one unsynchronized flip-flop input, either the input occurs in time to cause the transition, or it does not, in which case no transition occurs. In the case of a late input, the machine will respond to it one cycle later, provided that the input is of sufficient duration.

There is still the possibility of an input change violating the setup time of the internal flip-flop, driving it into a metastable state. This can produce system failures which can be minimized, but never eliminated (see Metastability, page 3-164). The same problem arises when outputs depend on an asynchronous input.

Very little can be done to handle asynchronous inputs without severely constraining the design of the state machine. The only way to have complete freedom in the use of inputs is to convert them into synchronous inputs. This can be done by allocating a flip-flop to each input as shown in Figure 22. These synchronizing flip-flops are clocked by the sequencer clock, and may even be the sequencer's own internal flip-flops. This method is not foolproof, but significantly reduces the chance of metastability occurring.

Functionality

The functionality of different devices is difficult to compare since different device architectures are available. The number of registers in a device determines the number of state combinations possible. However, all of the possible state combinations are not necessarily usable, since other device constraints may be reached. The number of registers do give an idea of the functionality achievable in a device. Other functionality measures include the number of product terms, type of flip-flop, or the number of PROM locations in a memory-based sequencer. One device may be stronger than another in one of these measures, but overall may be less useful due to other shortcomings. Choosing the best device involves both skill and experience.

A designer has a complete spectrum of devices with different architectures to choose from for a state machine design. These range from the PAL16R8D family of very high speed devices to the new PROSE PMS14R21 and instruction-based Am29PL141 high-functionality devices. The spectrum is completed by mid-range PAL devices including the PAL23S8, PAL22RX8A and PAL32VX10/A, with architectural features specifically designed for state machine designs, and the PLS devices designed exclusively for state machine implementation.

In order to give an idea of device functionality, we will consider each of the architecture options available to the designer and evaluate its functionality.

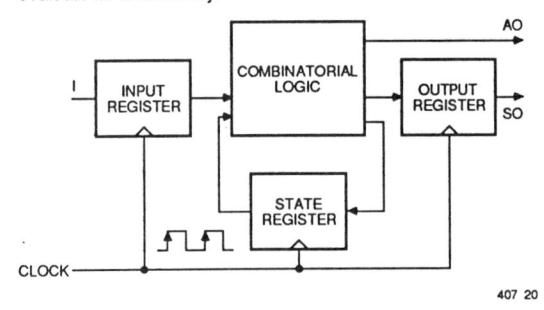

Figure 22. Input Synchronizing Register

State Machine Design

PAL Devices as Sequencers

A vast majority of state machine designs are implemented with PAL devices. Early versions of software required the user to manually write the sum-of-products Boolean equations for using PAL devices. Second generation software allows one to specify the design in "state machine syntax," and handles the translation to sum-of-products logic automatically. PAL devices implement the output and transition functions in sum-of-products form via a user-programmable AND array and a fixed OR array.

PAL devices deliver the fastest speed of any sequencer and are ideally suited for simple control applications characterized by few input and output signals interacting within a dedicated controller in a sequential manner. The number of flip-flops in a PAL device range from 8 to 12, which offer potentially more than one thousand state values. Since some of the flip-flops are used for outputs, and the number of product terms is limited, the usable number of states is reduced drastically. Generally, up to about 35 states can be utilized.

PAL Device Flip-flops

PAL device based sequencers implement small state machine designs, which have a relatively large number of output transitions. Since the output registers change with most state transitions, they can be used simultaneously as state registers, once the state values are carefully selected. Most PAL devices are used for small state machines, and efficiently share the same register for output and state functions. High-functionality PAL device based sequencers provide dedicated buried state registers when sharing is difficult.

As a state machine traverses from one state to another, every output either makes a transition (changes logic level) or holds (stays at the same logic level). Small state machine designs require relatively more transitions and fewer holds. As designs get larger, state machines statistically require relatively fewer transitions and more holds.

Most PAL devices provide D-type output registers. D-type flip-flops use up product terms only for active transitions from logic LOW to HIGH level, and for holds for logic HIGH level only. J-K, S-R, and T-type flip-flops use up product terms for both LOW-to-HIGH and HIGH-to-LOW transitions, but eliminate hold terms. Thus D-type flip-flops are more efficient for small state machine designs. Some high-end PAL devices offer the capability of configuring the flip-flops as J-K, S-R or T-types, which are more efficient for large state machine designs since they require no hold terms.

Many examples of PAL-device-based sequencers can be found in system time base functions, special counters, interrupt controllers, and certain types of video display hardware.

PAL devices are produced in a variety of technologies for multiple applications, and provide a broad range of speed-power options. PAL devices which can be used for sequencer designs are listed in Figure 23. We will consider the following PAL devices in detail.

PAL10H/10020EV/EG8
PAL16R8D family
PAL23S8
PAL22RX8
PAL32VX10

DEVICE	TOTAL NUMBER OF PINS	DEDICATED INPUT PINS	DEDICATED REGISTERED OUTPUT PINS	I/O PINS	LOGICAL PRODUCT TERMS/ OUTPUT	SPEED GRADES FREQUENCY IN MHz f_{MAX}
TTL						
16R8	20	8	8	-	8	55.5, 37, 25, 16
16R6	20	8	6	2	8	55.5, 37, 25, 16
16R4	20	8	4	4	8	55.5, 37, 25, 16
16RP8	20	8	8	-	8	22.2
16RP6	20	8	6	2	8	22.2
16RP4	20	8	4	4	8	22.2
16RA8	20	8	0-8	8-0	4	20
16X4	20	8	4	4	8	14
23S8	20	9	4	4	8-12	33.3, 28.5
20R8	24	12	8	-	8	37, 25, 16
20R6	24	12	6	2	8	37, 25, 16
20R4	24	12	4	4	8	37, 25, 16
20X10	24	10	10	-	4	22.2
20X8	24	10	8	2	4	22.2
20X4	24	10	4	6	4	22.2

Figure 23. Table of PAL Devices for Sequencer Applications

State Machine Design

DEVICE	TOTAL NUMBER OF PINS	DEDICATED INPUT PINS	DEDICATED REGISTERED OUTPUT PINS	I/O PINS	LOGICAL PRODUCT TERMS/ OUTPUT	SPEED GRADES FREQUENCY IN MHz f_{MAX}
20RP10	24	10	10	-	8	37, 25
20RP8	24	10	8	2	8	37, 25
20RP6	24	10	6	4	8	37, 25
20RP4	24	10	4	6	8	37, 25
20XRP10	24	10	10	-	8	30, 22.2, 14
20XRP8	24	10	8	2	8	30, 22.2, 14
20XRP6	24	10	6	4	8	30, 22.2, 14
20XRP4	24	10	4	6	8	30, 22.2, 14
20RS10	24	10	10	-	8-16	19.2
20RS8	24	10	8	2	8-16	19.2
20RS4	24	10	4	6	8-16	19.2
20XRP10	24	10	10	-	8	30, 22.2, 14
20XRP8	24	10	8	2	8	30, 22.2, 14
20XRP6	24	10	6	4	8	30, 22.2, 14
20XRP4	24	10	4	6	8	30, 22.2, 14
20RS10	24	10	10	-	8-16	19.2
20RS8	24	10	8	2	8-16	19.2
20RS4	24	10	4	6	8-16	19.2
20RA10	24	10	0-10	10-0	4	33, 20
22RX8	24	14	0-8	8-0	8	28.5
22V10	24	12	0-10	10-0	8-16	40, 28.5, 18
32VX10	24	12	0-10	10-0	8-16	25, 22.2
32R16	40	16	16	-	8-16	16
CMOS						
C16R8	20	9	8	-	8	28.5
C16R6	20	9	6	2	8	28.5
C16R4	20	9	4	4	8	28.5
C20R8	24	13	8	-	8	20, 15.3
C20R6	24	13	6	2	8	20, 15.3
C20R4	24	13	4	4	8	20, 15.3
C22V10	24	12	0-10	10-0	8-16	33.3, 20
C29M16	24	5	-	16	8-16	20, 15
C29MA16	24	5	-	16	4-12	20, 15
ECL						
10H20EV/EG8	24	12	0-8	8-0	8-12	125
10020EV/EG8	24	12	0-8	8-0	8-12	125

Figure 23. Table of PAL Devices for Sequencer Applications (Cont'd.)

State Machine Design

The ECL PAL10H/10020EV/EG8

At 125 MHz, the ECL PAL10H/10020EV/EG8 provides the highest speed for a state machine design. The PAL10H/10020EV/EG8 has eight outputs and 20 inputs. Half of the outputs have 8 product terms and half have 12 product terms. All of the 20EV8 outputs use D-type flip-flops, while the 20EG8 outputs are transparent latches. Two architectural fuses per output control the polarity of the logic and bypass the flip-flop for combinatorial asynchronous outputs. Two global product terms are present to Set and Reset all flip-flops. Maximum frequency of operation as a state machine is 125 MHz.

Figure 24 details the macrocell for the PAL10H/10020EV8.

The PAL16R8D Family

This is the high-speed end (55 MHz) of the TTL state machine design solutions. PAL16R8D family devices are available in three different versions (PAL16R8/6/4) with varying numbers of registered and combinatorial outputs and I/O pins for convenient design fitting. The PAL16R8 family provides 64 product terms (eight per output).

The software design options available for the PAL16R8D family are the traditional Boolean equations and the state machine syntax provided by PALASM 2 software.

The PAL23S8

At 33.3 MHz and in a 20-pin DIP, this device (Figure 25) provides the highest performance-to-board-space ratio. Designed for high-functionality state machines, this device offers eight output and six dedicated buried registers. It also offers four output macrocells with register bypass for asynchronous outputs. It has nine dedicated device inputs and four I/O pins giving a total of 23 array inputs. For improved state machine initialization, it also offers dedicated synchronous Preset and asynchronous Reset product terms. Pin-compatible with the PAL16R8 family, this device offers varied distribution of 145 product terms for large designs and optimal fit.

The software design options include Boolean equations and state machine syntax.

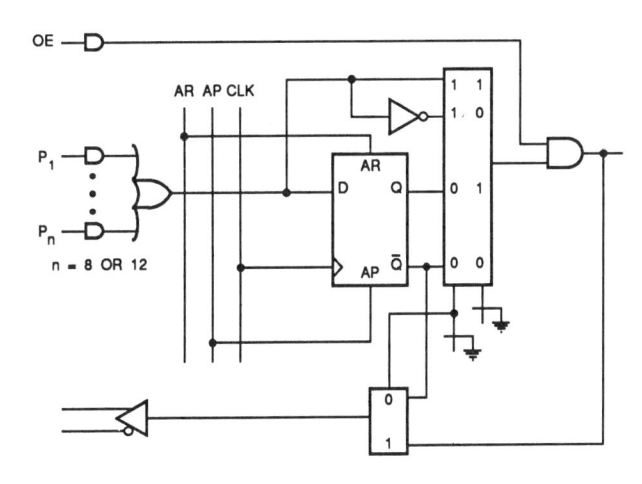

Figure 24. The PAL10H/10020EV8 Output Macrocell

State Machine Design

The PAL22RX8A

The 24-pin PAL22RX8A (Figure 26) provides eight configurable input/output macrocells with register bypass for a user-programmable number of registered or combinatorial outputs and I/O pins. It also provides dedicated asynchronous reset and preset product terms for state machine initialization. The PAL22RX8A provides a product-term-controlled XOR gate with its sum-of-products logic. This allows users to configure individual macrocells with D, J-K, S-R or T-type flip-flops.

The requirements of transition and hold terms are dependent upon state selection, which varies for different applications. The PAL22RX8A provides both D-type and J-K type flip-flops, allowing the designer to select the one requiring fewer product terms.

The PAL22V10

The PAL22V10 provides output macros similar to the PAL22RX8A. It provides both programmable polarity and register bypass. In addition, it has varied product term distribution.

The PAL22V10 has an advantage of a large number of product terms. It provides a varied distribution of eight to sixteen product terms per output. It also provides ten output macrocells, allowing for larger designs.

The PAL32VX10/A

The PAL32VX10/A (Figure 27) provides high functionality and density for PAL device designs. The PAL32VX10/A has configurable flip-flops and initialization product terms for state machine designs. It offers unprecedented design density because of its 10 input/output macros and a large number (8-16/output) of product terms.

The PAL32VX10/A also provides dual feedback paths from each output macrocell. This allows the macrocell register to be used as a buried state register in case the I/O pin is used as a dedicated input. The PAL32VX10/A is designed for large state machine designs which have relatively few output transitions and require separate output and buried registers. The PAL32VX10/A provides up to 10 buried registers for state machine design without any I/O penalty. The PAL32VX10/A also allows trading-off output registers for state registers and vice-versa. This optimizes the use of device resources.

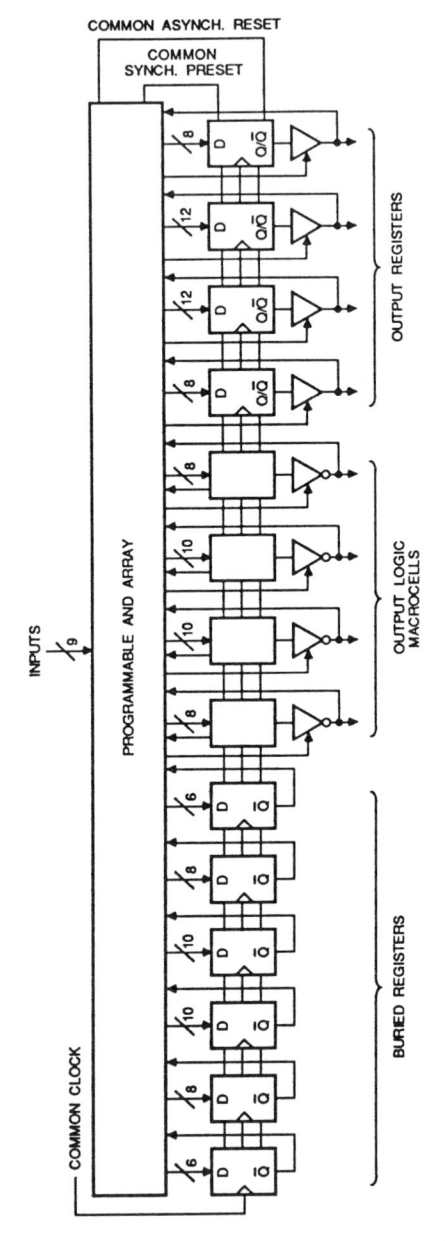

Figure 25. Block Diagram of PAL23S8

State Machine Design

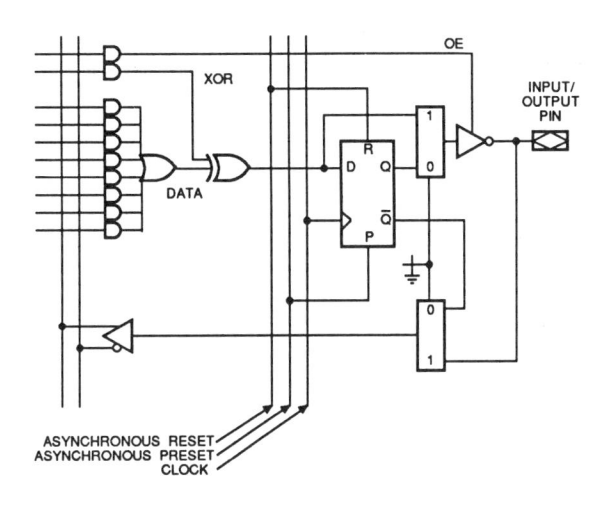

Figure 26. PAL22RX8A Output Macrocell

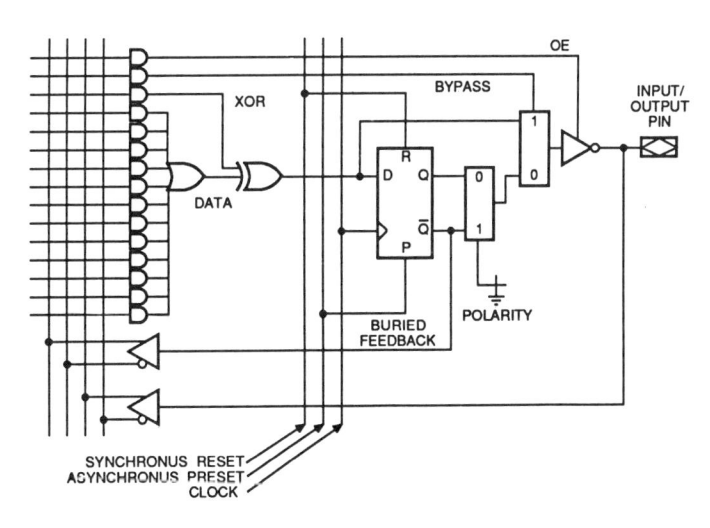

Figure 27. PAL32VX10/A Output Macrocell

State Machine Design

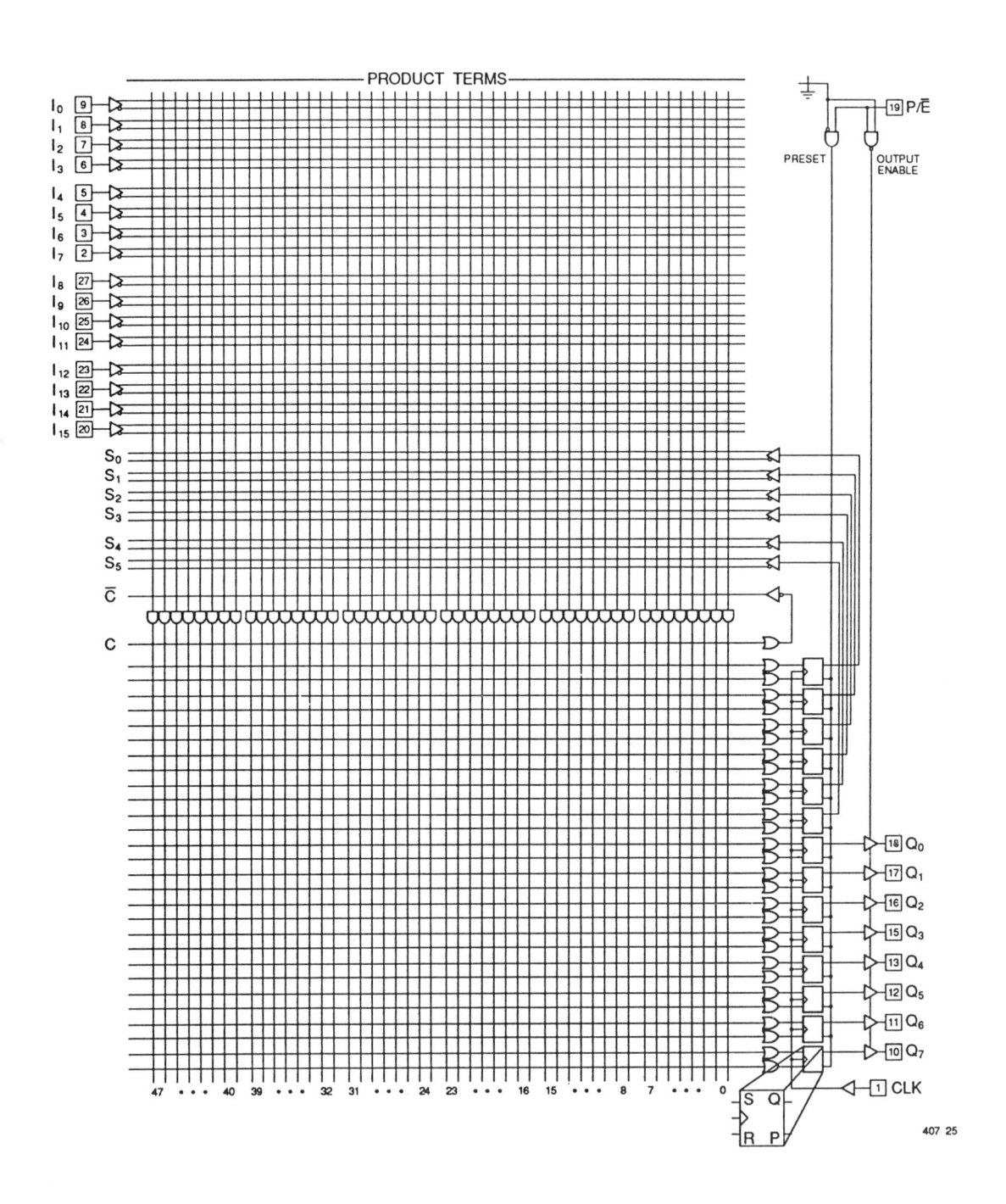

Figure 28. Architecture of a PLS105 Device

State Machine Design

Programmable Logic Sequencers (PLS)

Another alternative for control logic is the PLS (Programmable Logic Sequencer) device (Figure 28). PLS devices have both programmable AND and programmable OR arrays. Forty-eight product terms are available, which can be shared between transition functions and output functions for all of the outputs, as dictated by the application needs. PLS devices also offer S-R flip-flops with buried registers for control sequencing. The flip-flops have a common clock line. The S-R type flip-flops are similar to the J-K type flip-flops discussed earlier, and require no hold terms. Consequently S-R type flip-flops are very efficient for large state machine designs. PLS devices can function as either as Moore machines or Mealy machines with a register on the output path.

With the improvements in design tools, it is possible to specify designs directly from the state diagram. However, one format which has been used often in the past is the PLS table. This table is primarily used where the design is entered manually. The table is also useful for learning about and experimenting with sequencer designs. Often, it is possible to manipulate sequencer programs using the table in ways which are not possible with some of the more rigid CAD languages.

The table has provision for input and output signal pin allocation and naming, and a line for each of the 48 device product terms. The entries used in each line are shown in Figure 29.

VALUE	AND ARRAY TERMS	OR ARRAY TERMS
	(input, present state)	(next state, output)
H	true input/state bit	Set Flip-flop
L	complement	Reset Flip-flop
-	don't care	no change

Figure 29. Symbols Used in a PLS Table

The table defines the output function and transition function. Both functions depend on the *present state* and *inputs*. Each possible link path between two states uses a product term and becomes a line in the PLS table. The collection of all link path lines is a complete definition of the combinatorial logic required. The following example illustrates the relationship between a state diagram and the PLS table.

Let us look at an arbitrary function as shown in Figure 30. The inputs are A,B,C, and D; the state names are S0, S1, S2 and S3; and the outputs are W, X, Y and Z. To construct the PLS table, a state assignment is necessary. Selecting an arbitrary binary code for each state:

S0 = 00
S1 = 01
S2 = 10
S3 = 11

With the two state variables P and Q, the PLS table can be derived, and is shown in Figure 31.

PATH #	INPUTS	PRESENT STATE	NEXT STATE	OUTPUTS
	A B C D	P Q	P Q	W X Y Z
1	H - - -	L L	H L	L H L L
2	L - - -	L L	L H	H L L L
3	- H H -	L H	H H	L L L L
4	- L - -	L H	L L	L L L L
5	- H L -	L H	L L	L L L L
6	- H - -	H L	H H	L L L L
7	- L - -	H L	L L	L L L L
8	- - - H	H H	L L	L L L H
9	- - - L	H H	L L	L L H L

Figure 31. Example PLS Table

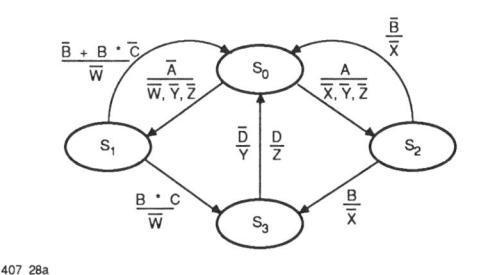

Figure 30. Example State Machine

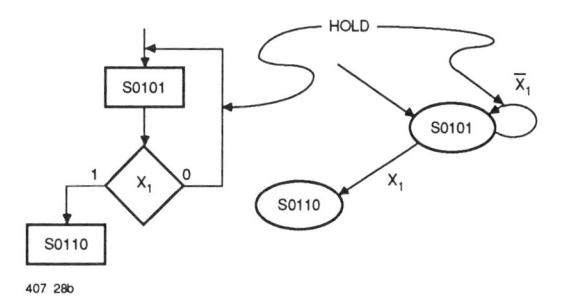

Figure 32. "HOLD" Transitions

State Machine Design

This is now a complete description of the combinational logic of the state machine. Every line in the table is a product term which is active when conditions for a transition to another state are satisfied. For this reason, the product terms in a PLS device are also known as *transition terms*. As the circuit moves from state to state, new transition terms become active as others become inactive.

A comment may be appended to each term to aid documentation of the design.

"Hold" Terms

The flip-flops used in PLS devices are the S-R type. As discussed earlier, S-R flip-flops do not require any product terms for holds and require product terms only for state transition or output generation (Figure 32).

This may be seen in the two lines of the PLS table corresponding to the two link paths shown, in Figure 33.

INPUT X1	PRESENT STATE	NEXT STATE
L	L H L H	- - - -
H	L H L H	- - H L

Figure 33. Hold Transitions in a PLS Table

The first path, corresponding to the first line in Figure 33, is a transition back to the same state. No change is thus required in any of the state variables. '-' entries can thus be used. The first line has only '-' entries in the OR Array part of the table; the transition term is thus not connected and this line can be eliminated altogether. This is a logic minimization which can be done immediately when converting the flowchart or state diagram into a PLS table. The second path is to a different state, "0110". This does require a term since some state variables have to change. Also, note that "hold" state transitions are only free if no output changes are required.

Complement Array Term

An important addition to the PLS device's programmable AND-OR array is the complement array term (Figure 34). This is an extra OR line which is first complemented and then fed back as an additional input into the AND array. Figure 35 shows the three possible configurations of the array at each transition term. The symbols shown are used in an extra truth table column for configuring the complement array.

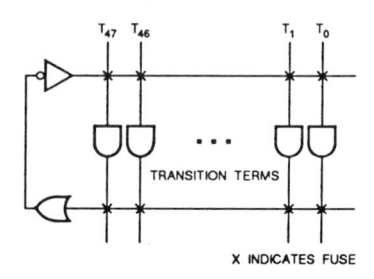

Figure 34. Connections of the Complement Array Term

Proper usage of the complement array can lead to considerable savings in product terms. This can be seen in the typical use of the complement array in the implementation of default or "ELSE" transitions. Figure 36 shows a set of four transitions from state SA, three of which are enabled by one of three combinations of inputs A, B, and C, causing transitions to states S0, S1 and S2. All other combinations cause a transition to state SX.

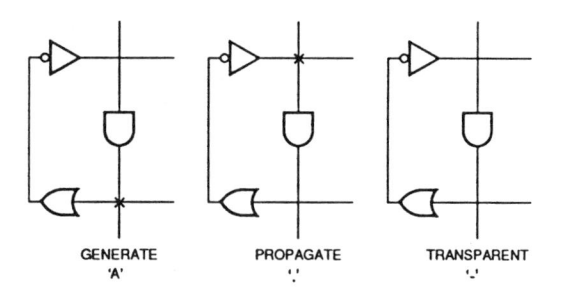

Figure 35. Complement Array Configurations

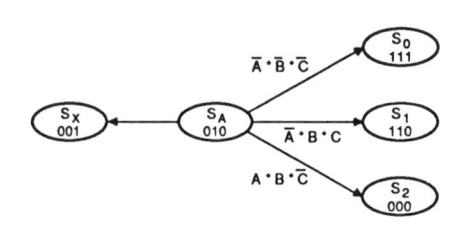

$\bar{A}$	$\bar{B}$	$\bar{C}$	GO TO
$\bar{A}$	$\bar{B}$	$\bar{C}$	S_0
$\bar{A}$	$\bar{B}$	C	S_X
$\bar{A}$	B	$\bar{C}$	S_X
$\bar{A}$	B	C	S_1
A	$\bar{B}$	$\bar{C}$	S_X
A	$\bar{B}$	C	S_X
A	B	$\bar{C}$	S_2
A	B	C	S_X

Figure 36. Transitions with a Default

State Machine Design

Three product terms are required for transition to states S0, S1 and S2 which encode the three different conditions. For the default transition to state SX, the remaining five condition combinations of A, B and C would have to be encoded, requiring five more product terms (Figure 37). But these terms can be thought of as "not the transitions to S0 or S1 or S2". By using the complement array (which provides this NOT signal), only one more product term is required for the default transition (Figure 38).

The last five terms have been compressed into one. If none of the three defined input combinations is true in state SA, the inputs to the complement array will all be zeros, and the complement array output will be a one. This activates the fourth transition term, causing a transition to SX (Figure 39).

The same complement array term can be used in other states too, since the propagated signal is ANDed with the present state code. Therefore all defaults can be realized with one term per state. PLS devices can be an effective solution in state- and branch-intensive applications, typically found in a variety of protocol controllers, waveform generators, and sequence detectors. In such applications, complex arrangements of multiple branches may be present. Multi-step control functions may include many loops of dozens of steps to count packets of data or input events as they occur.

PLS devices can be characterized by the number of inputs, number of product terms, number of flip-flops and number of outputs. The MMI PLS family currently has three members; these are listed in Figure 40.

Software for PLS devices can usually accept input descriptions in three formats: truth tables, Boolean equations and state machine syntax. The first two require the user to work out the explicit logic assignment for all terms and code the detailed fuse states programmed or unprogrammed. The new state machine syntax allows design in a high-level syntax, and is easier to use.

Cn	A	B	C	PRESENT STATE	NEXT STATE	COMMENT
-	L	L	L	L H L	H - H	to S0
-	L	H	H	L H L	H - -	to S1
-	H	H	L	L H L	- L -	to S2
-	L	L	H	L H L	- L H	path to SX
-	L	H	L	L H L	- L H	
-	H	L	L	L H L	- L H	
-	H	L	H	L H L	- L H	
-	H	H	H	L H L	- L H	

Figure 37. Product Terms Required for Transition to State SX

Cn	A	B	C	PRESENT STATE	NEXT STATE	COMMENT
A	L	L	L	LHL	H-H	
A	L	H	H	LHL	H--	
A	H	H	L	LHL	-L-	
•	-	-	-	LHL	-LH	path to state SX

Figure 38. Default Transition Using Complement Array Term

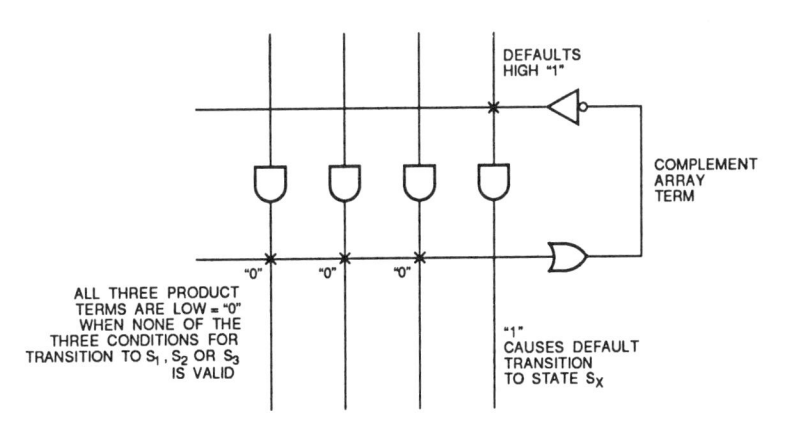

Figure 39. Transitions Using Complement Array Term

State Machine Design

DEVICE	TOTAL NUMBER OF PINS	DEDICATED INPUT PINS	DEDICATED REGISTERED OUTPUT PINS	I/O PINS	LOGICAL PRODUCT TERMS/ OUTPUT	SPEED GRADES FREQUENCY IN MHz f_{MAX}
TTL						
PLS167	24	14	6	-	Total 48	33
PLS168	24	12	8	-	Total 48	33
PLS105	28	16	8	-	Total 48	37

Figure 40. Table of PLS Devices

PROSE Sequencer (PMS14R21)

The PROSE (PMS14R21 PROgrammable SEquencer) is a revolutionary architecture optimized for very high functionality state machine designs. It combines a PROM and a PAL array on a single device, utilizing the efficiencies of both for state machine design.

The PROSE is a high-speed, 14-input, 8-output state machine. It consists of a 128x21 PROM array preceded by a 14H2 PAL array (see Figure 41). The PAL array is efficient for performing logic functions on a large number of input conditions, while the PROM array is optimal for implementing a large number of product terms (decision branches) and states. The combination allows a very efficient implementation of a state machine.

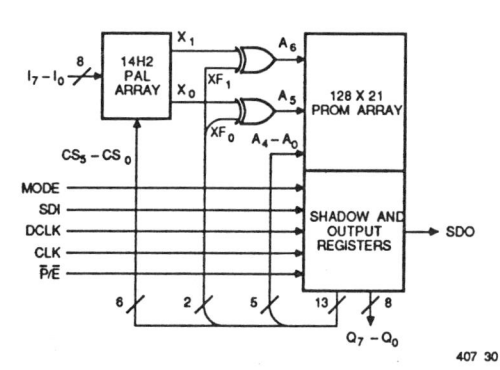

Figure 41. PROSE PMS14R21 Device Architecture

PROM Operation

The PROM consists of 128, 21-bit word locations. Each of these locations in the PROM can be viewed as a state. Eight bits Q(7-0) at each location define the outputs. The other 13 bits are used as feedback: six as condition select bits CS(5-0) to the PAL array, and seven to the PROM array itself for determining the next location to be addressed. Five of these seven constitute the low-order address bits A(4-0) for the next state. The remaining 2 bits are XF(1-0), inputs to the exclusive-OR gates which determine address bits A6 and A5 in conjunction with PAL output signals X1 and X0. Effectively, the seven-bit address of every PROM location is generated by seven bits of the PROM data and constitutes the present state. The next PROM location is the one whose address is stored as data in the present location. Eight of the PROM data bits are used as outputs.

PAL Array Operation

The PAL14H2 has 14 complementary inputs and 2 active-HIGH outputs (X1 and X0). Each of the 2 outputs is a sum of eight product terms. Eight of the inputs, I(7-0), are from an external source, and are used for encoding conditions for branch selection. Six inputs, CS(5-0), are from internal feedback from the PROM, and are used for selection of the conditions that determine branching when decisions are involved in state transitions.

Just as PROMs are efficient for direct control sequencing, PAL arrays are very efficient for encoding input conditions. The function of the PAL14H2 in the PROSE device is to encode all eight test input conditions along with the state information. Based upon these test conditions being true it inverts the polarity of the two most significant PROM address bits. This causes the PROM to address a different location (state), which is equivalent to conditional branching. The PALASM 2 design software selects these branch locations automatically and isolates the designer from low-level details.

With 128 PROM locations, the PROSE device can be used for large state machines. It is very effective in single-thread control functions characterized by long sequences of events generated by one "trigger" event. There may be a single repetitive decision evaluated many times or a long chain of decisions evaluated for sequential inputs. The output control stream may be quite wide and complex depending on the system. Generally, only narrow branches (two or four-way) are needed. Typical applications occur in controllers for error detection, numerical processing, and data encryption and decryption. The PROM output structure can hold any pattern of eight-bit values to be read out as needed. No trade-off has to be made with branching complexity to provide this output flexibility. Byte or word-oriented expansion can be made by arranging multiple devices in parallel.

PROSE state machines are fully supported by PALASM 2 software. A complete design example of a PROSE-based state machine is included on page 2-122.

State Machine Design

Fuse Programmable Controller (Am29PL141)

The Am29PL141 is a high-performance, single-chip, fuse-programmable controller (FPC). This chip is designed to allow the implementation of complex state machines and controllers by programming the appropriate sequence of microinstructions in the on-chip microprogram memory.

Large state machines require the capability of nested routines. Nested state machines are control sequences that may be invoked by one or more control functions, but when invoked, suspend the "calling" control sequence. Following completion of the low-level routine, the higher-level "calling" sequence resumes where it left off. The nested task may have modified part of the machine environment or the data processed by the system. The Am29PL141 provides stacks which allow nested subroutines.

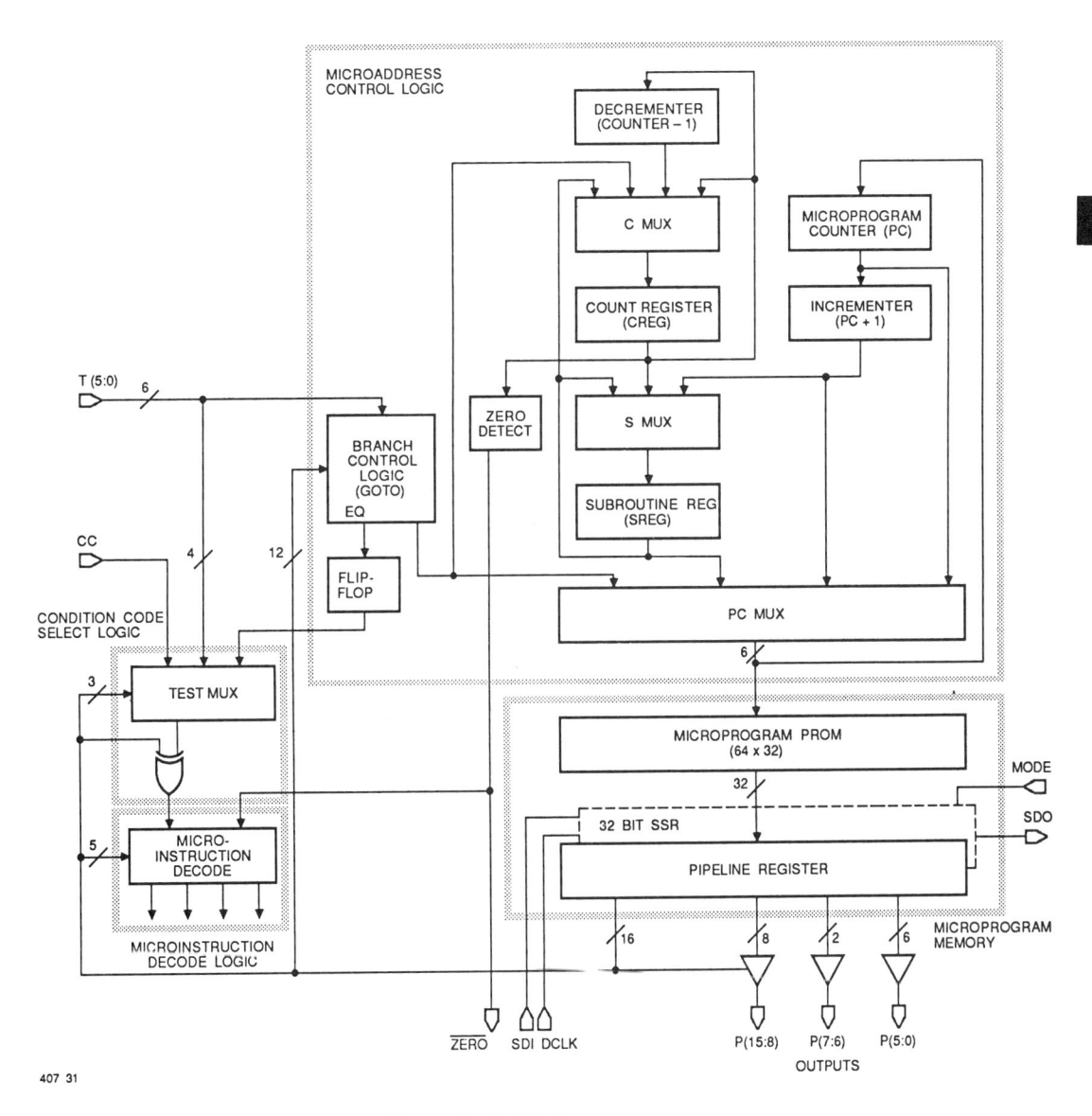

Figure 42. Architecture of Am29PL141

State Machine Design

With its on-chip intelligent microprogram address sequencer (Figure 42), high-speed 64 x 32-bit PROM-based microprogram memory, on-chip pipeline register, and an on-chip diagnostics serial shadow (SSR) register, this chip offers two major advantages: fast operation, and testability. This device is available with a 20-MHz clock rate (50-ns cycle time) in a 28-pin dual-in-line package.

A microprogram address sequencer is the heart of the FPC, and performs all control sequencing functions. The Am29PL141 has 29 high-level microinstructions which include jumps, loops, subroutine calls, and multiway branching. These microinstructions can be conditionally executed based on the test inputs. The output generation function is similar to that of the PROSE device, where output data is stored in PROM locations. A block diagram of the FPC is shown in Figure 42.

The FPC consists of five main logic blocks: the microaddress control logic, condition code selection logic, branch control logic, microinstruction decode logic, and the microprogram memory.

The microaddress control logic generates the proper sequences of addresses for the microinstructions in the microprogram memory. A PLA microinstruction decoder is used to decode the opcode from the microinstruction. This microinstruction decoder also generates all necessary control signals for executing each microinstruction. Depending on the microinstruction, the microaddress can be generated from a number of different sources: branch control, microprogram counter (PC+1), subroutine register, or the microprogram counter (PC).

A six-bit stack is available for looping and subroutine calls. When the loop counter is not being used for counting purposes, it can be configured with the stack register to implement a two-deep stack for nested loops and nested subroutine calls. The condition code selection logic consists of an 8:1 test multiplexer. One of eight test conditions can be selected on which to base the conditional instructions. The polarity of the selected condition code input is controlled by the POL bit in the microword.

The branch control logic is used for generating multiway branch addresses from the external inputs T(5–0). This logic also performs the comparison between the inputs and the pipeline data field for executing the COMPARE instruction. An EQ signal is generated as an output of the COMPARE instruction and can be used for a condition code test in the next clock cycle.

With its 64-location PROM, the Am29PL141 provides up to 64 states for state machine designs. It offers the most sophisticated control sequencing circuitry, which allows multiple branching and nested subroutines. This can enhance the functionality of the device even beyond 64 states.

The Am29PL141 is supported by an assembler (ASM14X) which allows the user to write instructions in a high-level language syntax. The assembler then converts the high-level statements into the specific device microinstructions and PROM data. The JEDEC device output file is also generated for programming the device PROM.

State Machine Design Tutorial

The following discussion is a tutorial on state machine design methodology. In this tutorial we will use the PROSE PMS14R21 device and explain the state machine design process. The PROSE PMS14R21 is supported by the state machine syntax of PALASM 2 design software. This software syntax is identical to and can be used for programming other PAL and PLS devices.

There are four stages in a state machine design. The first requires conceptualizing the design and selecting the correct device; second converting it to its state diagram representation; third converting the state diagram to its state machine syntax text file; and fourth assembling the file and programming the device. To explain all the stages in the design process we will take a general design example and then go through the stages of the design process. As an alternative design methodology, we will also examine the flowchart state representation in the design example.

Conceptualizing the Design

This is the first step in a state machine design process. In this step a complete functional description of the design is formalized along with the requisite truth tables and/or timing diagrams. We will consider a simple traffic signal controller example for illustrative purposes.

To begin with, let us visualize the scene of a traffic intersection (Figure 43), simplified to two one-way streets.

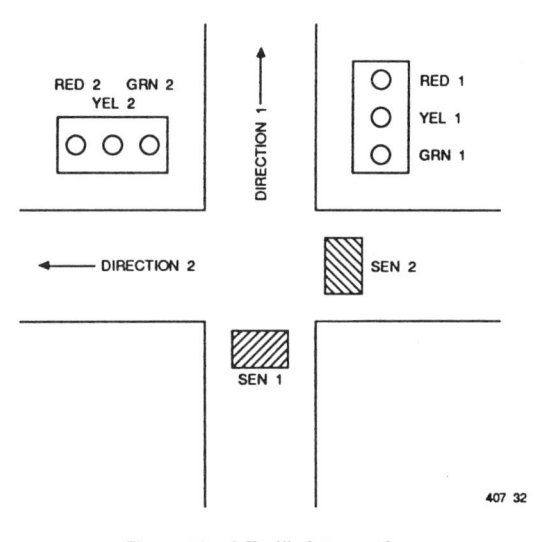

Figure 43. A Traffic Intersection

State Machine Design

The traffic intersection shows two one-way streets: one in direction 1 and the other in direction 2. Each direction has a signal consisting of red, yellow, and green lamps. These lamps are activated with appropriately-named active-HIGH signals (RED_1, YEL_1, GRN_1, RED_2, YEL_2, GRN_2). The signals are generated by the state machine controller. Also, each direction has a sensor which provides an active-HIGH signal (SEN_1, SEN_2) that indicates the presence of a vehicle. The controller has to manage this intersection, with the sensors as inputs and the lamps as outputs.

The assertion of SEN_1 or SEN_2 signals a request for a green light for traffic in the corresponding direction. Once SEN_1 is HIGH and SEN_2 is LOW, indicating traffic in direction 1, the GRN_1 light should be on at all times. Similarly, when SEN_2 is HIGH and SEN_1 is LOW, the GRN_2 light should be on at all times. When SEN_1 and SEN_2 are both HIGH, indicating traffic in both directions, the traffic controller should cycle, allowing equal periods of green signals (GRN_1 and GRN_2) for both directions. This cycling is also done when SEN_1 and SEN_2 are both LOW, indicating no traffic.

SEN_2	SEN_1	
L	H	Allow traffic in direction 1: RED_2, GRN_1 = H
H	L	Allow traffic in direction 2: RED_1, GRN_2 = H
H	H	Cycle with equal durations in both directions
L	L	Cycle with equal durations in both directions

Figure 44. Table for Traffic Flow Direction

Whenever the signals change from direction 1 to direction 2, the appropriate yellow light (YEL_1 or YEL_2) is turned on for the transition duration. This allows time for the traffic in the intersection to pass, before allowing traffic in the other direction. The timing diagram is shown in Figure 45. When both sensors indicate the presence of traffic (SEN_1, SEN_2 = H), extra time is allowed for traffic to pass, before the signals change the direction. The extra time is also allowed when there is no traffic in both directions (SEN_1, SEN_2 = L). The timing diagram for this is shown in Figure 46. Finally, the traffic signal controller shown includes the system clock (CLK), and an initialize or reset signal (INIT). INIT drives the controller to an initial state.

Before we proceed to the next step, we must select a device which will implement this design.

In the previous section we found that the basic selection of a state machine controller device depends upon the speed and I/O considerations. The PROSE PMS14R21 has eight inputs and eight outputs in addition to the initialization input signal PRESET, which are sufficient for this design. Another selection criterion is the intelligence of the PROSE device. The device should be able to implement all of the states required by the design. The PROSE device is the only 24-pin SKINNYDIP device which has the capacity for up to 128 states. Later we will see that these are sufficient for implementing our design.

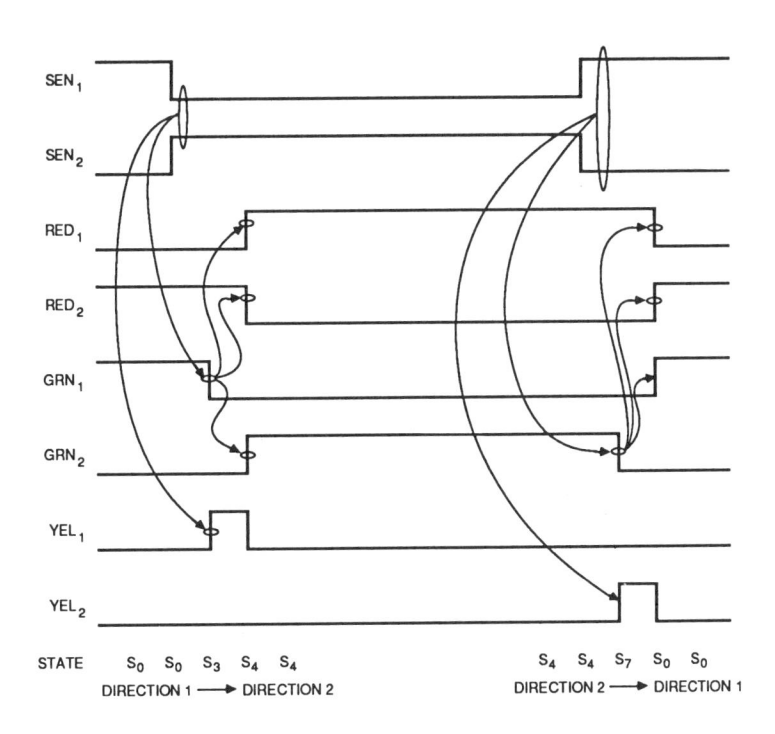

Figure 45. Fast Traffic Flow Transition

State Machine Design

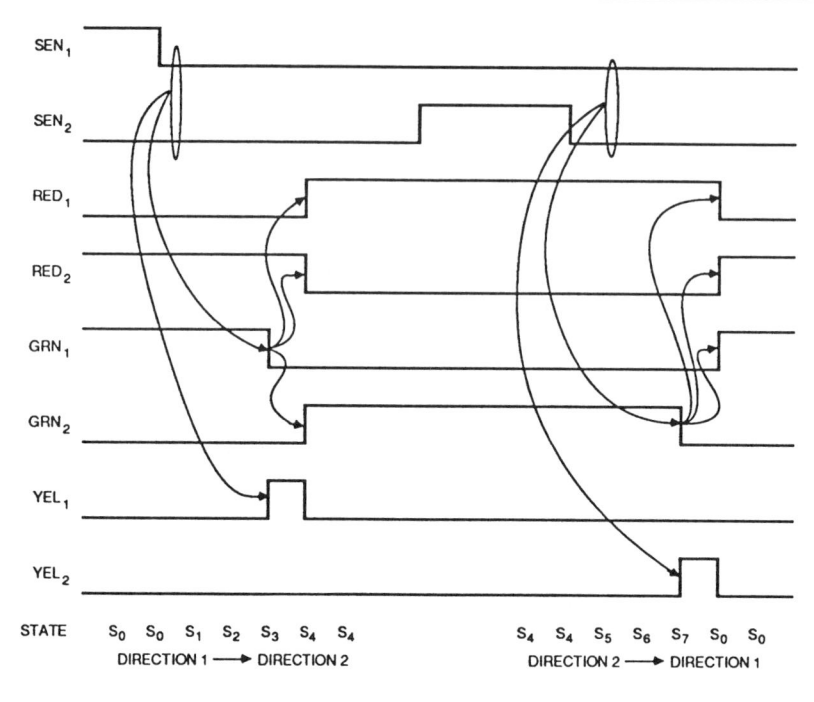

Figure 46. Slow Traffic Flow Transition

State Machine Representation

Once the design problem is defined, the next stage is to convert it to its state machine representation. State machines can be represented by bubble-and-arrow state diagrams (Figure 4). Each bubble represents a state, with each arrow representing a transition. A second method of state machine representation is with flow charts, details of which are explained on page 2-104. State diagrams are by far the most popular representation. Flowcharts are included for completeness; you need not worry if you are not comfortable with them.

The transition from the present state to the next state is dependent upon present state and input conditions. Similarly, outputs can be generated based upon the present state (called a Moore type state machine), or present state and input conditions (called a Mealy type state machine). Our design example is a Moore type state machine. The PALASM 2 syntax supports both Moore and Mealy type state machines.

The first step for constructing a state diagram is to define states. All of the specific events in a design are assigned a state. These include: change of output signals, response to a change in input signals, or time delays. For example, in Figure 45, different states (S_0, S_3 and S_4) have been assigned where the outputs change. Often states with the same outputs can be merged into one unique state, as shown in the case of state S_0. Once the states are defined as bubbles, the transition from one state to another (called control sequencing) and outputs generated for each state (output generation) are defined. The task is to use the timing diagram along with the functional description of the design to develop control sequencing and output generation functions for state representations. The state diagram is an improvement over the timing diagram since it also provide additional branch decision information.

Each of the functions performed by a controller can be viewed as one of the following operations. We will show how each of these operations determine the control sequencing and the output generation for the state machine, and their state diagram and flowchart representations.

- Arbitration
- Multiple condition testing
- Event monitoring
- Control signal generation
- Timing delays

Arbitration is the decision-making process. It is inadequately represented in timing diagrams with arrows (Figures 45 and 46). The table in Figure 44 represents the functional requirement for arbitration. Based upon the sensor inputs, the state machine stays in the same state S_0, when SEN_1=H and SEN_2=L (for allowing traffic in direction 1) or transitions either to state S_3 (SEN_1=L, SEN_2=H), or to state S_1 (SEN_1, SEN_2 are both HIGH or LOW). This is represented in the state diagram and flowchart as the transition from S_0 (Figure 47).

S_0 can also be thought of as an *event monitoring* state where the controller waits for the command from the sensors. Based upon the two bits SEN_0 and SEN_1, the state machine moves to different next states. This is also an example of *multiple condition testing*.

2-124

State Machine Design

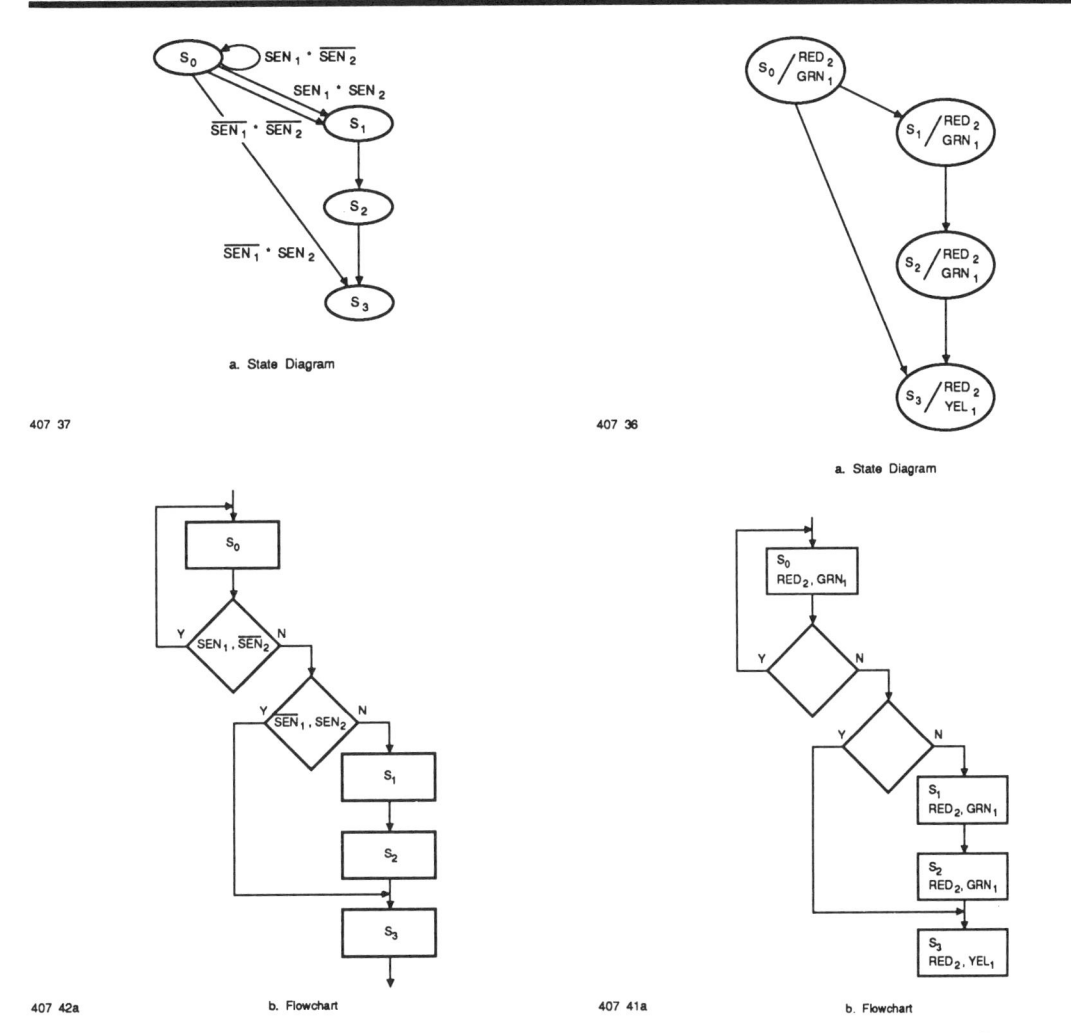

Figure 47. Conditional State Transitions

Figure 48. Output Assignment to States

An example of *control signal generation* (Figure 48) is signal RED_2 which is used to turn on the red light for direction 1. This control signal generation example requires assigning output signal name(s) (RED_2) to the state(s) (S_0, S_1, S_2 and S_3) where it is asserted (see Figure 46). For simplicity, we are only considering the outputs which are asserted. Other outputs in their default states must be accounted for in the final state machine representation.

The signal GRN_1 is kept asserted for two extra clock cycles (Figure 46) to allow the traffic to pass when changing the traffic pattern from direction 2 to direction 1. The state diagram representation (Figure 48) requires assigning unconditional transition (control sequencing) from one state to another, for a number of states (S_1, S_2 and S_3), depending upon the required *time delay*. It also requires assigning the signal name (GRN_1) to all three states,

which results in signal assertion for the extra two clock cycles.

Once the design has been converted into a state diagram (Figure 49) or flowchart (Figure 50) representing all of its timing and functional requirements, the next stage involves conversion to state machine syntax for its textual representation.

State Machine Syntax

The PALASM 2 programming software design file (also called the PAL device Design Specification or PDS) allows both conventional Boolean equations and state machine design constructs. It also allows both Mealy and Moore types of designs with extensive logic minimization capabilities and easy menu-driven operation. Other programming software packages which provide similar design capabilities are also available from various vendors.

State Machine Design

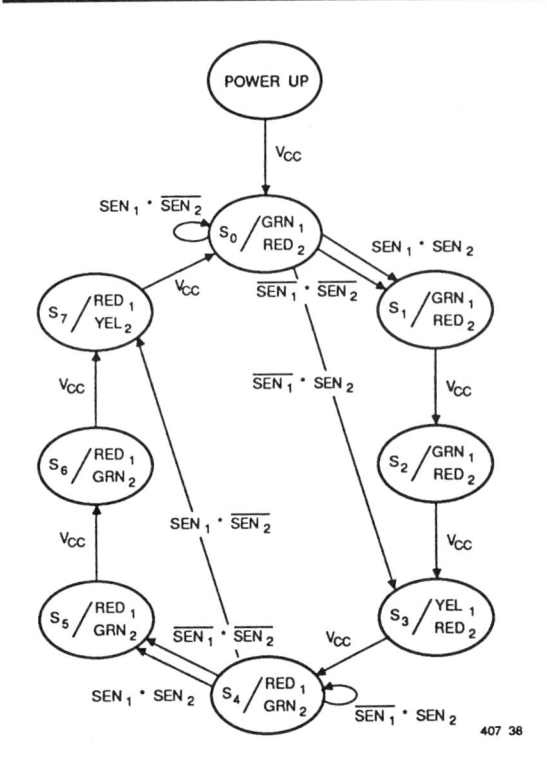

Figure 49. State Diagram—Traffic Signal Controller

A PALASM 2 state machine design file consists of four sections:

- Declaration section
- State section
- Condition section
- Simulation section (optional)

Declaration Section

As illustrated in Figure 51, this section stores the basic information about the device type and pin names. It is also used to store some bookkeeping information such as the company name and design revision numbers.

```
TITLE      TRAFFIC CONTROLLER
PATTERN    STATE MACHINE
REVISION   1
AUTHOR     J. ENGINEER
COMPANY    MONOLITHIC MEMORIES
DATE       JANUARY 30, 1987

CHIP  S_MACHINE PMS14R21
      CLOCK DCLOCK SEN1 SEN2 I2 I3 I4
      I5 I6 I7 SDI GND
      RESET SDO RED1 YEL1 GRN1 RED2
      YEL2 GRN2 O1 O0 MODE VCC
```

Figure 51. Declaration Section

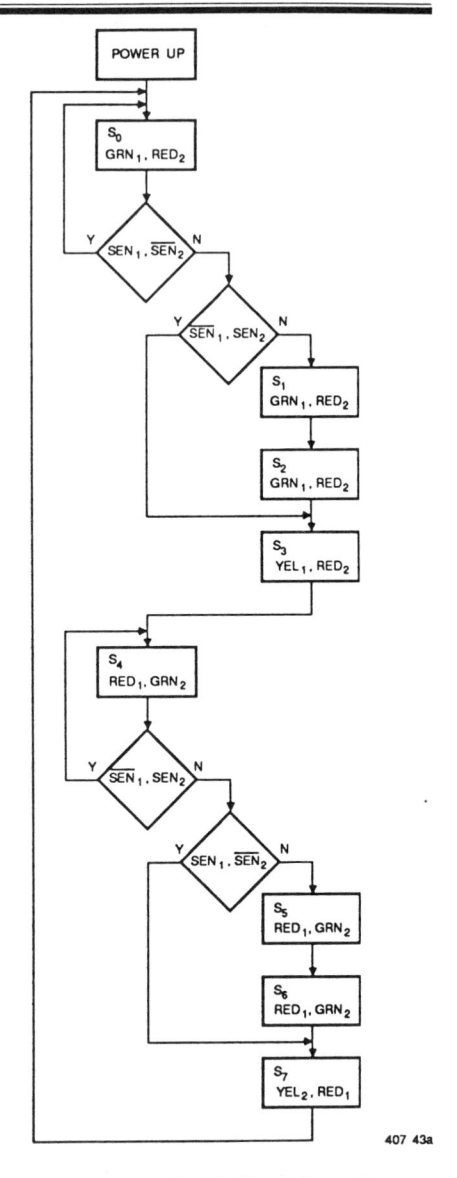

Figure 50. Flowchart—Traffic Signal Controller

State Section

The state section contains all the control sequencing and output generation information. The state section also stores all the state initialization values and default parameters. As illustrated in Figure 52, it initially defines the type of state machine being designed with keywords MOORE_MACHINE or MEALY_MACHINE. For the PROSE device it also selects the use of pin 13 as either MASTER_RESET or OUTPUT_ENABLE.

State Machine Design

It also defines the default output logic values as HIGH or LOW with the keyword DEFAULT_OUTPUT. Once an output is assigned a default logic value, it is explicitly mentioned only when asserted; otherwise it is assumed to be in its default logic value. This usually simplifies the task of drawing a state diagram (see Figure 49). Similarly, the keyword OUTPUT_HOLD defines the outputs which retain their previous values unless explicitly mentioned in a new state. The last two keywords also help isolate the user from the implementation details of J-K or D-type flip-flops when using PAL-device-based state machines. The keyword OUTPUT_HOLD is not used in this design.

```
STATE

    MOORE_MACHINE
    MASTER_RESET
    DEFAULT_OUTPUT /RED1/YEL1/GRN1/RED2/YEL2/GRN2
    POWER_UP := VCC -> S0
```

Figure 52. State Section

Once the default parameters are defined, all of the state transitions are represented. Only state names are used, since the state values are assigned automatically by the software. Every state transition (Figures 53 and 54) can depend upon the present state for direct control sequencing (state_x := VCC -> state_y), or the present state and input conditions for conditional control sequencing (state_x := Condition_name -> state_y). The syntax has a direct relationship to the state diagrams. For conditional control sequencing, the syntax also allows a default state when none of the conditions of transitions to other states are satisfied.

The output generation can also be represented easily. Outputs are specified from each present state. These outputs can be Moore type (state.OUTF := output_name) or Mealy type (state.OUTF := Condition_name -> output_name), depending upon whether or not conditions are used to generate outputs.

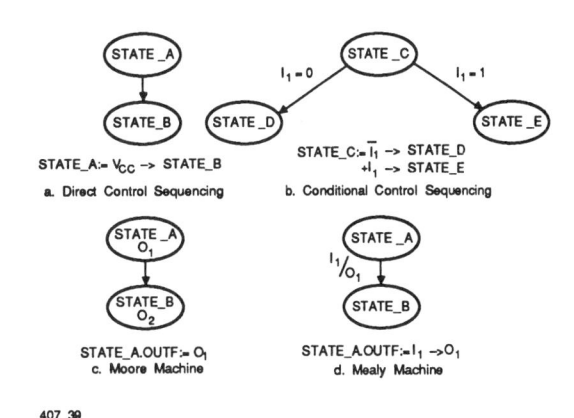

Figure 53. Direct Relationship Between State Diagram and State Syntax

Condition Section

This is the third section of the PALASM 2 state machine file structure. This section assigns names to all of the input conditions used by the state section for both control sequencing and output generation.

Design Example Syntax

All of the state machine tasks represented by the state diagram can be converted easily to textual format. The transition from state S_1 to S_2 and S_3 (Figure 55) illustrates how outputs RED_2 and GRN_1 are generated. These represent control signal generation and timing delay tasks of a state machine. Similarly, the arbitration function can be converted to textual format, showing the

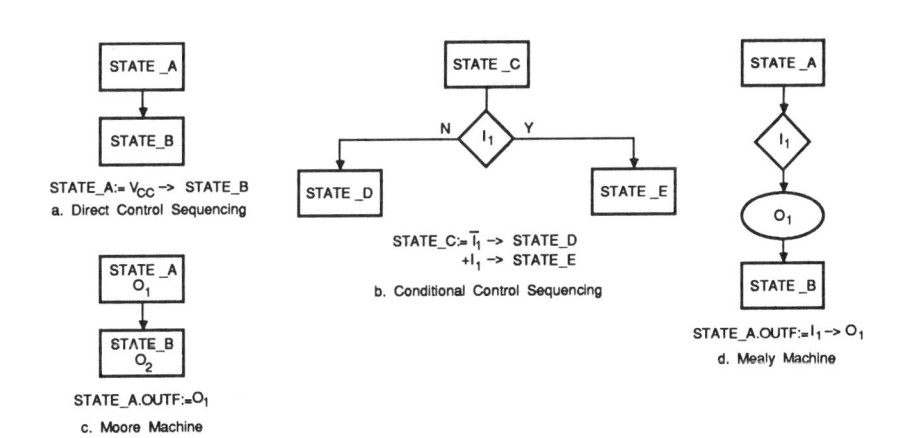

Figure 54. Direct Relationship Between Flowchart and State/Condition Syntax

State Machine Design

transition from S_o to various other states (Figure 56). This is also a good example of multiple condition testing and event monitoring.

```
S0.OUTF := RED2 * GRN1
S1.OUTF := RED2 * GRN1
S2.OUTF := RED2 * GRN1
S3.OUTF := RED2 * YEL1
```

Figure 55. Output Generation Representation

```
S0 :=     C3 -> S1
      +   C0 -> S1
      +   C1 -> S3
          +-> S0
```

Figure 56. Control Sequencing Representation

Simulation Section

The simulation section is similar to that used for conventional Boolean equation simulation. Please see page 4-137 for details.

Assembly and Programming

Once the state diagram is converted to its textual constructs, the file (Figure 57) can be easily assembled by using PALASM 2 software. The software then generates the device programming JEDEC format file. This JEDEC format file can be downloaded to a device hardware programmer provided by various vendors. For detail please see page 4-169 of the PALASM 2 software section.

Traffic Signal Controller

```
TITLE     TRAFFIC CONTROLLER
PATTERN   STATE MACHINE
REVISION  1
AUTHOR    J. ENGINEER
COMPANY   MONOLITHIC MEMORIES
DATE      JANUARY 30, 1987

CHIP    S_MACHIN PMS14R21
        CLOCK DCLOCK SEN1 SEN2 I2 I3 I4
        I5 I6 I7 SDI GND
        RESET SDO RED1 YEL1 GRN1 RED2
        YEL2 GRN2 O1 O0 MODE VCC

STATE

  MOORE_MACHINE         ; Defined as a Moore Machine
  MASTER_RESET          ; Initialization
  DEFAULT_OUTPUT /RED1/YEL1/GRN1/RED2/YEL2/GRN2
  POWER_UP:=VCC -> S0   ; Power up state defined

  S0 := C3 -> S1        ; Traffic in direction 1
  + C0 -> S1            ; Slow signal change
  + C1 -> S3            ; Fast signal change
  +-> S0                ; Otherwise stay in state S0
  S1 := VCC -> S2       ; Time delay
  S2 := VCC -> S3       ; Time delay
  S3 := VCC -> S4       ; Time for yellow
  S4 := C3 -> S5        ; Traffic in direction 2
  + C0 -> S5            ; Slow signal change
  + C2 -> S7            ; Fast signal change
  +-> S4                ; Otherwise stay in state S0
  S5 := VCC -> S6       ; Time delay
  S6 := VCC -> S7       ; Time delay
  S7 := VCC -> S0       ; Time for yellow
```

Figure 57. The Design File

State Machine Design

```
S0.OUTF := GRN1 * RED2      ; Allow direction 1 traffic
S1.OUTF := GRN1 * RED2
S2.OUTF := GRN1 * RED2
S3.OUTF := YEL1 * RED2      ; Change from direction 1 to 2
S4.OUTF := RED1 * GRN2      ; Allow direction 2 traffic
S5.OUTF := RED1 * GRN2
S6.OUTF := RED1 * GRN2
S7.OUTF := RED1 * YEL2      ; Change from direction 2 to 1

CONDITIONS

C0 = /SEN1 * /SEN2          ; Condition for no traffic
C1 = /SEN1 * SEN2           ; Condition for direction 2
C2 = SEN1 * /SEN2           ; Condition for direction 1
C3 = SEN1 * SEN2            ; Condition for traffic in both directions

SIMULATION

TRACE_ON CLOCK SEN1 SEN2 RED1 YEL1 GRN1 RED2 YEL2 GRN2

SETF RESET /CLOCK
CLOCKF CLOCK                ;STATE TRANSITION ONLY ON 1ST CLOCK
CLOCKF CLOCK                ;LIGHTS CHANGE ON 2ND CLOCK
CHECK /RED1 /YEL1 GRN1 /YEL2 /GRN2 RED2
SETF /SEN1 /SEN2
CLOCKF CLOCK
CLOCKF CLOCK
CHECK /RED1 /YEL1 GRN1 RED2 /YEL2 /GRN2
CLOCKF CLOCK
CHECK /RED1 YEL1 /GRN1 RED2 /YEL2 /GRN2
CLOCKF CLOCK
CHECK RED1 /YEL1 /GRN1 /RED2 /YEL2 GRN2
CLOCKF CLOCK
CHECK RED1 GRN2
CLOCKF CLOCK
CHECK RED1 YEL2
CLOCKF CLOCK
CHECK /RED1 /YEL1 GRN1 RED2 /YEL2 /GRN2
SETF /SEN1 SEN2
CLOCKF CLOCK
CHECK /RED1 /YEL1 GRN1 RED2 /YEL2 /GRN2
CLOCKF CLOCK
CLOCKF CLOCK
SETF SEN1 /SEN2
CLOCKF CLOCK
CLOCKF CLOCK
CHECK YEL2 RED1
CLOCKF CLOCK
CHECK GRN1 RED2

TRACE_OFF
```

Figure 57. The Design File (Cont'd.)

Testability

Introduction

With digital logic design, it is all too easy to design a circuit which merely implements a specified function. When production starts it is suddenly found that the circuit cannot be tested, or perhaps that tests cannot be performed economically. Dealing with this situation can, at the very least, have a negative impact on the introduction of the system into the marketplace.

Potential headache can be avoided by taking test issues into consideration during the initial design. Instead of just designing a circuit which implements a specified function, which is the bare minimum that must be accomplished, that function needs to be implemented in a manner which can be tested.

The purpose of this section is to establish the notion of testability and its importance, and then to provide ways of avoiding the most common untestable circuits. The issues will be discussed primarily in the context of logic design in PLD's, although they are also relevant for general logic design.

After testability has been discussed for general circuits, some specific testability circuitry on the PROSE device will be discussed. Finally, test vectors will be reviewed. Various kinds of vectors are mentioned, and the general tools available for vector generation will be summarized.

Defining Testability — A Qualitative Look

A completely testable design is one in which any and all device faults can be systematically detected.

First note that the issue is one of devices, not designs. The design itself must work as specified; that is the main job of the design engineer. Once the design is implemented in a device, the issue is how to test the device to make sure that the design has been correctly implemented. Throughout this paper, then, it will be assumed that a particular design works as is; we will just be addressing its testability.

The easiest and most effective means of testing a circuit is through a systematic series of tests. A random set of tests may also do well, but does not yield much information regarding the testability of a circuit itself. No number of random (or systematic) vectors can test an inherently untestable circuit.

In order to be able to perform a systematic test sequence, every part of the circuit under test must be accessible, so that it can be controlled. Only then can each node be forced high or low as needed. This is essentially a requirement of complete *controllability* of the circuit.

In order to be able to detect faults every part of the circuit must also be visible to the outside world, so that the results of each test can be observed. In this manner, each node can be inspected to determine its logic level. This requires complete *observability*.

These are, of course, the age-old issues of controllability and observability, which are as important for digital logic circuits as they are for so many other kinds of systems. If any portion of a circuit is uncontrollable or unobservable, then the testability of the entire circuit is compromised.

Figure 1 shows a couple of completely untestable circuits. The integrity of the top input in Figure 1a can never be verified. No matter whether it is shorted to ground, to V_{cc}, or whether it is functioning correctly, the output will be the same. That is to say, any faults on the top input cannot be observed at the output.

The circuit in Figure 1a would appear pretty useless as is. It is possible, however, that instead of being directly grounded, the second input may be driven by some distant signal, possibly on a different PC board, which happens to be a a logic low. If you cannot bring this line to a logic high, then it might as well be grounded.

The circuit in Figure 1b essentially has no input. This circuit can be thought of as a latch, but there is no way to change its logic state. Therefore, it is completely uncontrollable.

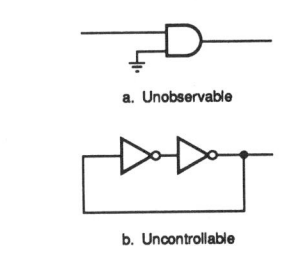

a. Unobservable

b. Uncontrollable

Figure 1. Untestable Circuits

Quantifying Testability

In theory, if we want to quantify the testability of a given circuit, we might first attempt to make a list of all possible things that could go wrong with a circuit (no matter how unlikely), and then verify that all such "faults" can be tested, in all combinations and permutations. But for a circuit of any significance whatsoever, it will rapidly become apparent that this is not a practical solution.

Testability

What we need instead is a measure which can give an empirically reliable indication of the testability of a circuit, or of the quality of a given set of tests. There are several different such measures, but the most popular of these is the *single stuck-at faults* model.

There are several ways of analyzing circuits for single stuck-at faults. For very large circuits, various *testability analysis* schemes have been developed. However, for smaller circuits, especially of the size that would be put into a PLD, the more common method uses simulation.

Simulating Single Stuck-At Faults

A given circuit is first simulated. The quality of the simulation is important; the more complete the simulation the better. A thorough simulation can then serve as a benchmark test sequence later. In this way, the fault simulation procedure also allows us to measure the quality of a given simulation, or set of tests, in addition to the testability of the circuit.

The results of the simulation are recorded. Next, one node in the circuit is modeled with a *"stuck-at" fault* — either *stuck-at-one* (SA1) or *stuck-at-zero* (SA0), as shown in Figure 2. The circuit is now resimulated. If the simulation results of the modified circuit are different from the simulation results of the good circuit, then the fault was detected. If not, then we have a faulty circuit which appears to operate correctly.

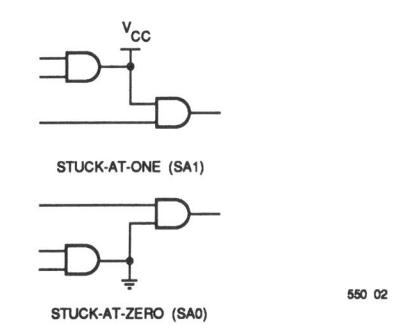

Figure 2. Single "Stuck-At" Faults

This procedure is repeated for each node, one node at a time (hence the name "single" stuck-at faults). The nodes are modeled with both SA1 and SA0 faults, so that for N nodes, we will have 2N simulations. If of those 2N simulations, D of them produced simulation results different from those of the original circuit, then we say that this simulation tested this circuit with a test coverage of D/(2N)*100%. Whereas this specifically tests only for single faults, experience shows that it is also a good test for multiple stuck-at faults.

Undetected Faults

Why are some of the faults not detected? For simple combinatorial logic, there are two basic reasons: either the simulation was not complete enough to find the fault, or the circuit itself cannot be tested for the fault. So when an undetected fault is located, the first step taken is to add vectors to the simulations which will exercise the node being tested. By doing this, we gradually improve the quality of the simulation, and thus the quality of the test sequence that we can use in production.

It is possible that certain nodes will have undetectable faults for which no new vectors can be added. These are the result of an untestable design. It is the joint job of the test and design engineers to generate a test sequence that is as complete as possible. It is the design engineer's responsibility to provide a circuit which is testable. If both of these responsibilities are carried out, the result will be a testable circuit which can be tested with an exhaustive test sequence. This will yield the highest quality system. Note, however, that the overall responsibility is shared between the design and test engineers.

Needless to say, this process of analyzing the testability of a circuit is not done all by hand; software aids are used. There are many different kinds of programs that run on many different kinds of systems, ranging from PCs to workstations to mainframes. Some of them are standalone programs; others are integrated into larger overall environments. Their specific capabilities also vary, but in general, they can simulate a given circuit with a given set of vectors; analyze the test coverage that the vectors provide for the circuit; and generate new tests, either from scratch or by improving on the coverage of a few manually generated "seed" vectors. Most can also point out potential problems areas of a circuit, such as race conditions and logic hazards.

Finally, one frequently asked question is "So what if there is a fault that can never be detected. Who cares?" Theoretically, this question is not unreasonable. However, most companies will not feel comfortable telling a customer "We only tested half of the system, but if anything goes wrong with the other half, you'll never notice it." In addition, as will be seen, many untestable circuits occur as a result of poor design practices.

Testability issues for sequential circuits have implications far beyond the test bed. Indeed, failure to take these issues into account can greatly affect the normal performance of a system. The key for state machines is controllability. The challenge is to make all elements of the circuit controllable, both for testing and for general functionality.

Designing Testable Combinatorial Circuits

All of the previous procedures dealt mostly with the ways in which existing circuits are treated. However, if a finished circuit is found to be untestable, then it must be redesigned for testability. An easier approach is to design for testability from the beginning. Unfortunately there is no direct recipe for a testable design. There are, however, many common ways of making a circuit untestable. Most of this section is devoted to pointing out such problems.

The simplest kind of problem is *redundant logic*. Figure 3a shows one such circuit. It has a purely redundant product term. If the output of either of the product terms is stuck low, for any reason, then as long as the other product term is good, the fault will never be visible at the output.

This may initially look like a benefit, since we have what we could call a "primary" circuit with a "backup." One can cover up some of the failures of the other (but not all failures). If this kind of

Testability

redundancy is truly desired, this is not the way to achieve it. When you ship out this circuit, you do not know if you really have a working primary and backup. The primary may already be malfunctioning; since it was never tested, you will never know. If you want useful, reliable redundancy, test circuitry must be added, as in Figure 3b, so that each part of the circuit can be independently tested.

a. A Purely Redundant Circuit

b. Testable Redundant Circuit

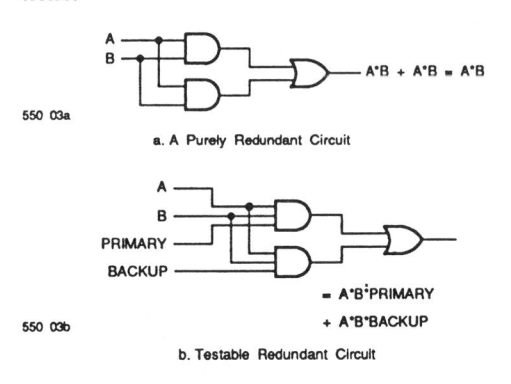

Figure 3. Making Redundancy Testable

Figure 4 shows another redundant circuit. Although the product terms are not identical, the larger AND gate is really redundant. Any stuck-low faults at the output of this gate are not detectable.

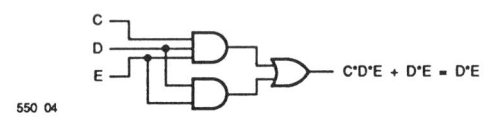

Figure 4. Circuit with a Redundant 3-Input AND Gate

Reconvergent Fanout

Redundant logic is a special case of what is called *reconvergent fanout*. This is a term that refers to circuits that have inputs splitting up, going through independent logic paths, and then reconverging to form a single output, as shown in Figure 5. When this happens, it is very easy to introduce untestable nodes. It may not be easy to identify where such nodes are.

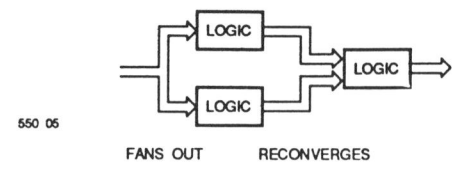

Figure 5. Reconvergent Fanout

Figure 6 is an example of a reconvergent circuit. The inputs are shared between two different product terms, which are eventually summed. This circuit appears harmless enough, but it turns out that the node indicated by "SA1" cannot be tested for a stuck-at-one condition. In other words, there is no way that we can guarantee that that node is operating correctly.

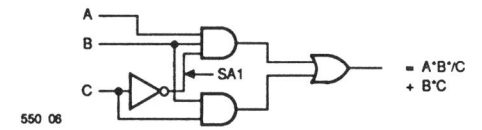

Figure 6. A Reconvergent Circuit with an Untestable Node

It is worth analyzing this circuit a bit more closely. This will give some insight into the kinds of analyses that are necessary when evaluating circuits and generating tests, and into the ways in which untestable nodes are created.

If we wish to prove that the node in question is not stuck high, then we must force it low and prove that we were successful in doing so. Thus we have two requirements: forcing the node low, and seeing the logic low on the output — controlling and observing the node.

First we raise input C high to force the node to a logic low condition, as in Figure 7a. This satisfies our controllability requirement. Next we need to provide a way to propagate this logic low to the output (Figure 7b). This is referred to as *sensitizing a path* to the output. The first step is to get the logic low past the AND gate. But if either input A or B is low, then the output of the AND gate will be low regardless of the node being tested. Thus we must force both A and B to a logic high, so that if there is a low on the output of the AND gate, we will know for sure that it came from the node we are testing. This is shown in Figure 7c.

a. Controllability: Forcing the Node Low

b. Observability: Sensitizing a Path to the Output

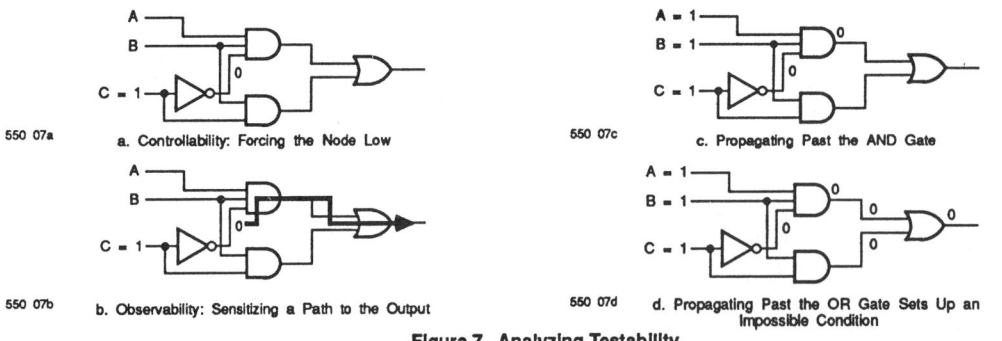

c. Propagating Past the AND Gate

d. Propagating Past the OR Gate Sets Up an Impossible Condition

Figure 7. Analyzing Testability

Testability

Next we wish to get the logic low through the OR gate to the output. To do this, we must insure that the second OR input is always low; if it is high, then the output of the OR gate will be high regardless of the node being tested. If we can keep the lower OR input low, then if the node we are testing was sucessfully forced into a low condition, then the output will be low. Otherwise the output will be high. This can be seen in Figure 7d.

How do we keep the lower OR input low? By making the output of the lower AND gate low, which can be done by setting one of its inputs low. However, we have already required that all of the inputs be high. Thus we have required a set of conditions that cannot be met. One of three things will result:

1. The lower AND gate has both inputs high, and therefore keeps the lower OR input high. In this case, we may have been successful in forcing the node under test low, but we cannot see it at the output.

2. We bring input B low, allowing the lower OR input to go low. However, now the output of the upper AND gate will always be low. So we will see a low at the output, but we cannot be sure exactly where the low came from.

3. We bring input C low, allowing the lower OR input to go low. However, now we are no longer forcing the node under test low.

So we can either force the node low, but cannot see the low at the output; or, we can see a low at the output but cannot be sure of its source; or, we cannot force the node itself low. In any case, we will never be able to guarantee that the node under test is not stuck high.

Note that the two "independent logic blocks" which generate the signals that eventually reconverge are testable by themselves; they are just AND gates. It is only when we hook them together via the OR gate that the overall circuit becomes untestable. Thus *the testability of individual portions of a circuit does not guarantee that the entire circuit will be testable when the testable pieces are all connected.*

We can minimize this circuit using the following steps:

$A^*B^*\overline{C} + B^*C = A^*B^*\overline{C} + B^*C + A^*B^*B$ (by consensus)
$= A^*B^*\overline{C} + B^*C + A^*B$
$= A^*B + B^*C$

Thus the node we were trying to test is really not needed in the logic. The resultant circuit is shown in Figure 8, and is completely testable.

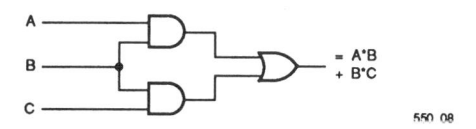

Figure 8. The Minimized Circuit Is Testable

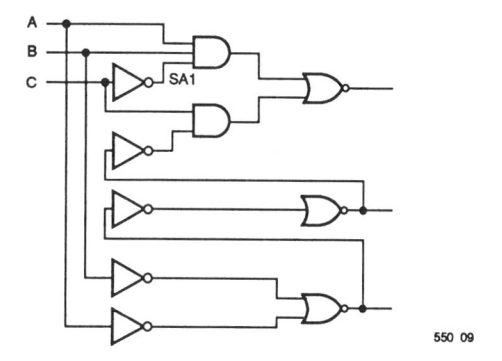

Figure 9. A Messy Reconvergent Circuit

Not all reconvergent circuits are so simple. Figure 9 shows a more complicated reconvergent circuit. Here some signals have to travel through several levels of logic to reach their final destination. This introduces considerable skew into the circuit, and will produce glitches on the outputs during certain transitions. In addition to this, there is again a stuck-at-one fault that cannot be tested.

Circuits like this can result from the design iteration process, as a designer tries to debug a circuit. By adding this and that, eventually the circuit works. But it is a mess, has poor timing characteristics, and is untestable. A little analysis of the logic itself shows that:

the bottom output is
$\overline{(\overline{A} + \overline{B})} = A^*B$

thus the middle output is
$\overline{(A^*B)} = \overline{A} + \overline{B}$

which makes the top output

$\overline{(A^*B^*\overline{C} + C^*(\overline{\overline{A} + \overline{B}}))} = \overline{(A^*B^*\overline{C} + A^*B^*C)}$
$= \overline{(A^*B)}$
$= \overline{A} + \overline{B}$

That is, the top two outputs are actually the same, and the third output is just the inverse of the top two. As convoluted as the original circuit looks, the logic itself is actually trivial. So if three outputs are really needed for some reason, we can generate them independently, as in Figure 10a. If only two outputs are needed, it is even easier. Figures 10b and 10c show two possibilities.

These circuits are much easier to understand, their timing characteristics are better, and they are completely testable.

Testability

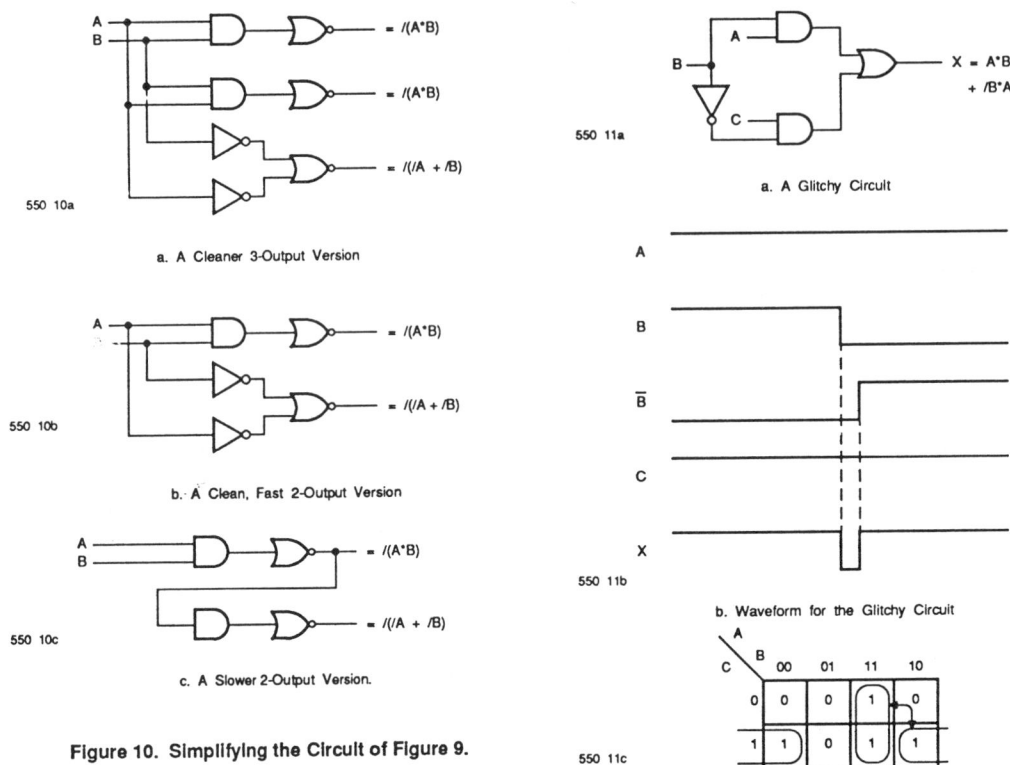

Figure 10. Simplifying the Circuit of Figure 9.

a. A Cleaner 3-Output Version

b. A Clean, Fast 2-Output Version

c. A Slower 2-Output Version.

The Importance of Minimization

The common factor behind all of the untestable circuits we have examined is the fact that all of them were not minimal. By minimizing the logic, we made the circuits testable. This is true in general: *UNMINIMIZED LOGIC CANNOT BE FULLY TESTED*.

Very often, especially when designing with PLDs, an attempt is made to minimize logic only to the point where it fits into a particular PLD. Any further minimization is considered an academic waste of time. This is a grave misconception. Getting rid of all extra product terms, and eliminating all extra literals on the remaining product terms has real value. Failing to do so will result in untestable nodes in the circuit.

Minimizing is not always enjoyable, since hand techniques are usually too tedious, and Karnaugh maps are essentially useless for more than four or five inputs. However, computers have long been used to minimize logic. In particular, PALASM® software (version 2.22 and later) has a minimization routine which can minimize logic automatically before assembly.

Logic Hazards

One occasional side effect of minimization can be the introduction of *glitches* into a circuit. Figure 11a shows such a "glitchy" circuit. The waveform in Figure 11b shows that under steady-state conditions, as long as inputs A and C are high, the output is high

Figure 11. Examining a Glitchy Circuit

a. A Glitchy Circuit

b. Waveform for the Glitchy Circuit

c. "Gap" in the Karnaugh Map Indicates a Logic Hazard

regardless of B. However, as B changes from high to low, causing the top product term to shut off and the bottom one to turn on, the inverter adds a bit of delay to the path that will turn on the lower product term. Thus the top term may shut off before the bottom one gets a chance to turn on. In this case, we have two logic low signals going into the OR gate, giving a low on the output. As soon as the lower product term turns on, the output goes back high, but not before the appearance of the high-low-high glitch.

Figure 11c shows the Karnaugh map for this circuit. It is minimal, but there are two product terms which do not overlap; they are "adjacent" in one location. These represent the two AND gates in the circuit diagram. The arrows indicate the troublesome transition: when A and C are high, and when B changes from high to low or the reverse. We can intuitively think of this as a "gap" between the two adjacent product terms, in which a glitch may occur.

Note that glitching is not a certainty. It is called a *hazard* because in certain situation, given certain timing situations, there is a chance that a glitch will occur.

Note also that the glitch is not really caused by the minimization process itself, but is caused by these "gaps" in the Karnaugh map. Unminimized logic with such gaps may also be glitchy.

Testability

A PROM is a good example of such a circuit. PROMs can be used to implement any logic function of their inputs. However, regardless of the function, it is implemented in a completely unminimized fashion, using complete minterms. So even a function as simple as the one in Figure 12 (which could be implemented using a single product term, grouping all 1's into a single cell) is implemented with each 1 in its own cell. Thus there is a gap between every cell, meaning that every transition is a potential glitch. PROMs are notoriously glitchy, and it is for this reason that the output of a PROM is actually undefined until its access time has elapsed.

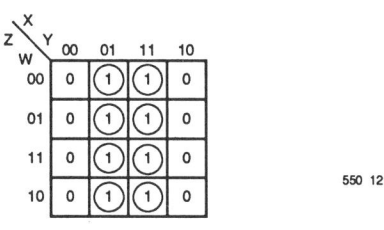

Figure 12. In a PROM, Every Transition Can Glitch

If we go back to the Karnaugh map in Figure 11c, we see that we can eliminate the gap — and the glitch — by adding a product term which overlaps both existing product terms and covers the gap. This is shown in Figure 13a, with the resultant circuit shown in Figure 13b.

a. A Redundant Product Term Can Eliminate the Glitch

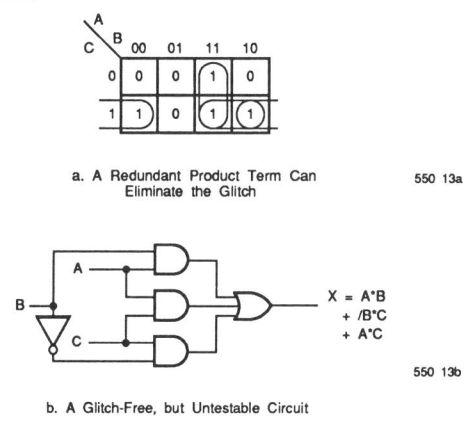

b. A Glitch-Free, but Untestable Circuit

Figure 13. Eliminating Glitches

This circuit is no longer glitchy. Unfortunately, it is also no longer testable, since we have added in a redundant product term which cannot be tested (try it yourself). In order to have a circuit which is both testable and glitch-free, we must add a test input to the circuit which we can use to shut off the outside gates, isolating the middle gate for testing (Figure 14a). When the circuit is operating normally, the extra input is kept at a logic high condition, where it does not interfere with the basic logic function.

The Karnaugh map for this circuit is shown in Figure 14b. Note that all product terms overlap, but now the circuit is minimal. The size of the Karnaugh map has doubled, since we added another input. But if we isolate just that portion which corresponds to the test input being high, which is the normal operating mode (see Figure 14c), it looks exactly like the map of Figure 13a. Of course we should expect this, since we do not want the addition of a test circuit to affect the basic function.

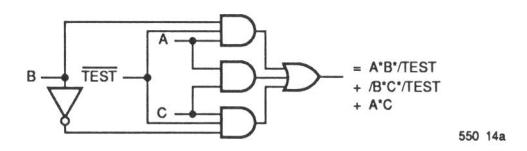

a. A Testable, Glitch-Free Circuit

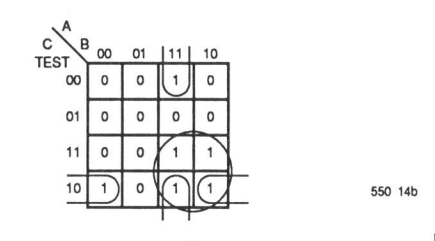

b. Karnaugh Map

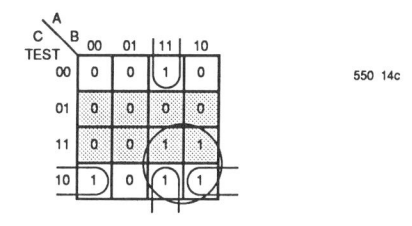

c. Karnaugh Map Showing Non-Test-Mode Portion

Figure 14. Making a Glitch-Free Circuit Testable

Thus, in general, these types of glitches can be eliminated first by adding some redundant logic to get rid of the gaps in the Karnaugh map, and then by adding a test input to make the circuit testable.

Testability

Designing Testable Sequential Circuits

The design of sequential circuits involves considerations above and beyond those required for simple combinatorial circuits. Latches and oscillators are circuits which appear combinatorial, but which use feedback to introduce sequential properties. State machines use flip-flops and feedback to generate what can be complex sequential circuits.

Feedback

Whereas combinatorial circuits depend only on the conditions of present inputs, *sequential* circuits depend on both present conditions and past behavior to determine future behavior. This is made possible primarily by *feedback*. Feedback takes an output signal and routes it back for use as an input to the same circuit, as shown in Figure 15. We now have a situation where an output depends on itself; this can introduce new testability problems.

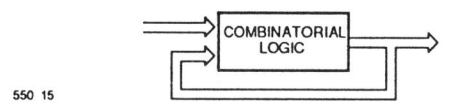

Figure 15. Logic with Feedback

Most sequential circuits (under varying circumstances also called *state machines*, *finite state machines*, and *sequencers*) make use of *flip-flops* as memory elements. These memory elements serve to remember a past condition (called a *state*) so that a future decision can be made based on it. This state is then fed back as in input. With PLDs, the flip-flops and combinatorial logic are contained within a single device, as shown in Figure 16.

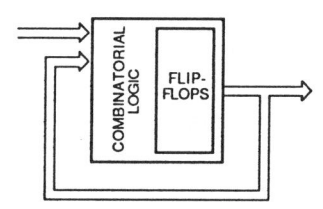

Figure 16. Structure of a Sequential PLD

Of course, the effects of feedback may have to be considered even when there are no flip-flops. The circuit in Figure 15 has feedback, but has no flip-flops. Such a circuit will either function as a *latch* or as an *oscillator*, as will be seen.

Before we look into the special needs of circuits with feedback, bear in mind that all of the testability criteria discussed for combinatorial logic still hold. The blocks of combinatorial logic shown in Figures 15 and 16 must be testable by themselves. What we will discuss here are issues which must be considered in addition to the issues involving combinatorial logic.

Latches

A combinatorial logic circuit which uses positive feedback is a latch. The simplest possible latch is shown in Figure 17a. The output is fed back as an input in its TRUE form. This means, of course, that the output will stay at its present level; hence the name "latch".

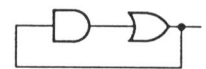

a. Completely Uncontrollable

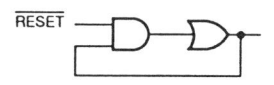

b. Cannot Set Output HIGH

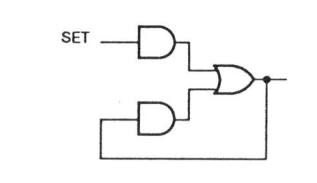

c. Cannot Reset Output LOW

Figure 17. Uncontrollable Latches

The circuit as shown is clearly not useful, since it will always remain in its power-up state. If another input is added, as in Figure 17b, a HIGH output could be made to go LOW by setting the $\overline{\text{RESET}}$ input LOW. However, once the output goes LOW, there is no way to make it go HIGH again. Likewise, the circuit could be modified as in Figure 17c. Now a LOW output can be made HIGH by setting the SET input HIGH. However, once HIGH, the output can never be made to go back LOW.

Controllable latches

For a latch to be useful, it must be completely controllable. The previous latches cannot be completely controlled. In order for a latch to be controllable, it must have both SET and RESET controls, as shown in Figure 18.

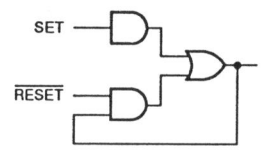

Figure 18. A Controllable Latch

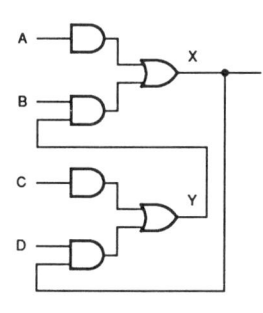

Testability

In PLDs, a latch can be detected by simplifying the logic for each function. If an output is a function of itself in TRUE form, then it is a latch. To be controllable,

- product terms containing the feedback should have at least one other direct input in the product (providing RESET control)

- there should be at least one product term with no feedback (providing SET control).

The circuit in Figure 19a provides an example. At first it is not immediately obvious that the circuit is a latch, but when the logic is simplified, we see that indeed it is. It is controllable since it has both SET and RESET controls. If the logic were shown in Figures 19b or 19c, the latch would be uncontrollable under some circumstances.

Latch hazards

The circuit of Figure 18 can be generalized to have several inputs on both the set and reset controls. Such a circuit is shown in Figure 20. In this case, we have two inputs on the set AND gate. If the two set inputs A and B change from 0 and 1 to 1 and 0, respectively, then there will be a glitch or a false latch at the output if both inputs were 1 at sometime during the transition (Figure 20). For this transition, it is important to make sure that the 1-0 transition be made before the 0-1 transition to avoid anomalous output behavior. Merely delaying one input will not help, since it will delay both rising and falling transitions.

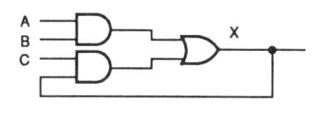

$X = A$
  $+ B^*Y$

$ = A$
  $+ B^*(C + D^*X)$

$ = A$
  $+ B^*C$] SET
  $+ B^*D^*X$
  └──┘
  RESET

a. Latch with SET and RESET

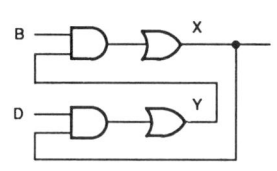

$X = B^*Y$
  $+ B^*D^*X$
  └──┘
  RESET

b. Latch with RESET Only

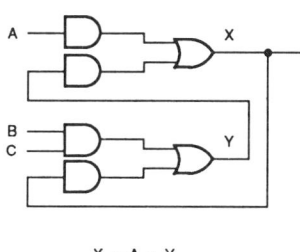

$X = A + Y$
  $ = A$
  $+ B^*C$] SET
  $+ X$

c. Latch with SET Only

Figure 19. More Complex Latches

a. Circuit

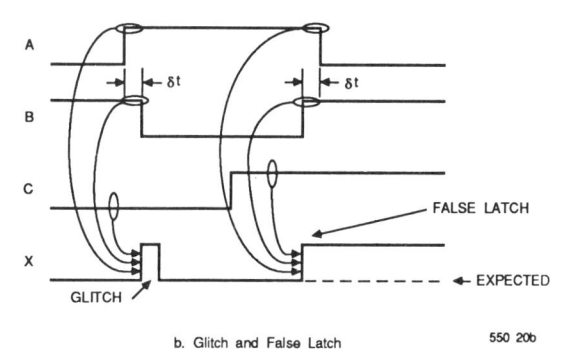

b. Glitch and False Latch

Figure 20. A Latch with More Complex SET Logic

The simplest solution to this problem is the use of an edge-triggered flip-flop to synchronize the signals. This will eliminate any such glitches. If a flip-flop cannot be used, it is possible to delay reaction to a "11" condition to make sure that such a condition is not transitory. A circuit which accomplishes this is

Testability

shown in Figure 21a. This is relatively efficient in that only one delay circuit is required regardless of the number of inputs used on the set control (within the limits of the size of the AND gate). It will require an extra output on a PAL device.

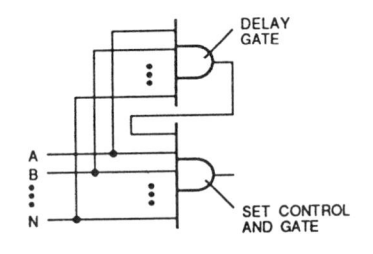

Because we have introduced redundancy, the circuit must be modified to be testable. If the circuit is implemented in a combinatorial PAL device, then programmable three-state can be used to test the circuit, as shown in figure 21b. By enabling output X, the redundant circuit can be observed without regard to Y. Then, to test Y, output X is disabled and then the pin is used as an input to drive the circuitry for Y directly. This provides a simple means of testing the circuit, but it only works if pin X can be measured and driven. The complete circuit is shown in figure 22a.

If node X is not so accessible, then additional circuitry and test inputs must be added. In the worst case, if node X is completely inaccessible, the resulting testable circuit is shown in figure 22b.

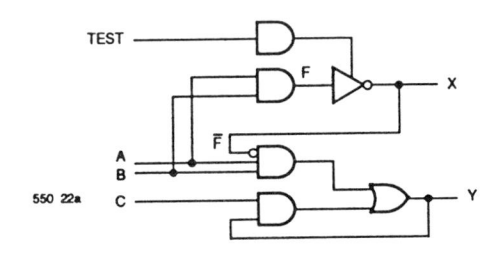

a. Circuit Which Delays "11...1" signals

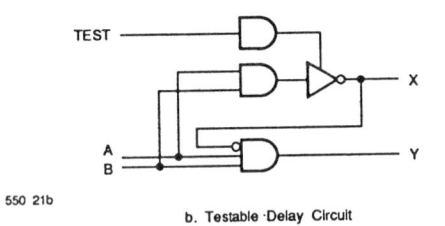

b. Testable Delay Circuit

Figure 21. Delay Circuit

This delay circuit will delay the effect of a "11" input by an extra propagation delay. However, it also provides a window of one propagation delay which will screen out any transitory "11" conditions that occur within that window. This allows up to one propagation delay's worth of skew between inputs during a transition from "01" to "10".

a. Complete Latch Circuit

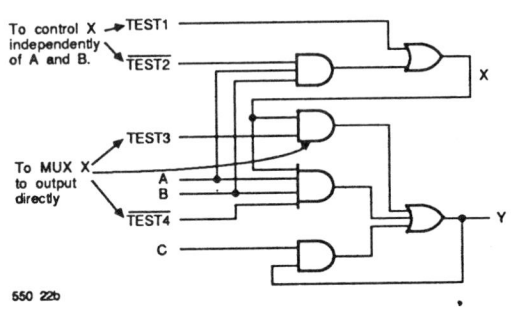

b. Circuit if Node X is Completely Inaccessible

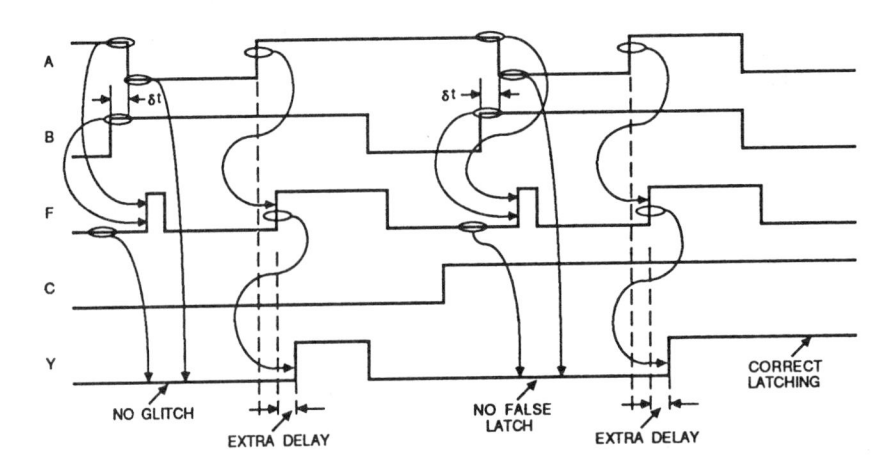

c. Latch Circuit Behavior

Figure 22. A Testable Glitch-Free Latch

Testability

Note that although the three-state capability is not needed, the circuit requires two extra gates, and, worst of all, four test inputs.

Figure 22c shows the behavior of either of the testable glitch-free latches.

Transparent latches

Many designers like to use PLDs to design standard D-type "transparent" latches. A D-type latch is a very simple circuit, shown in basic form in Figure 23a. As it turns out, however, this is a glitchy circuit of the type discussed on page 550-5 above. The problem is compounded in this case, since, given the right timing, the glitch can actually be latched; the glitching problem is no longer transitory. If this type of circuit is desired, it must be designed to be both glitch-free and testable; the resultant circuit is shown in Figure 23b.

OUT = GATE*DATA
    + /GATE*OUT

a. Glitchy

OUT = GATE*DATA*/TEST
    + /GATE*OUT*/TEST
    + DATA*OUT

b. Glitch-Free and Testable

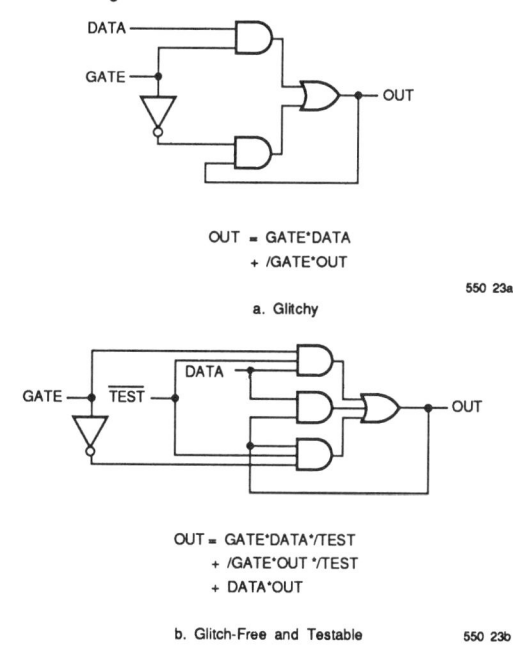

Figure 23. D-Type Transparent Latches

Oscillators

Circuits whose outputs are fed back in TRUE form are latches. If the outputs are fed back in COMPLEMENT form, then the circuit is an oscillator. A simple oscillator circuit is shown in Figure 24.

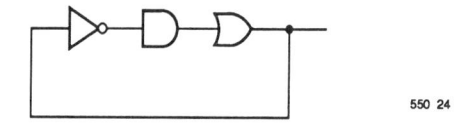

Figure 24. A Simple Oscillator

Latches are very often useful in circuits; oscillators rarely are. Crystals and other specialized oscillators are useful when it is necessary to generate a clock signal, for example. Trying to build an oscillator out of standard logic or PLDs will not yield a very predictable, accurate oscillator; where these circuits occur, it is usually by accident.

An oscillatory circuit may not always be obvious. It also may not oscillate all of the time. The oscillator shown in Figure 24 is uncontrollable; it always oscillates. However, just as we can design controllable latches, we can also design controllable oscillators (on purpose or by accident). This means that there may be an oscillator hidden in the circuit which will sometimes oscillate and sometimes be stable. Such a circuit is shown in Figure 25a.

X = A*B*Z
  = A*B*D*E
  + A*B*C*/X

Y = C*/X
  = /A*C
  + /B*C
  + C*/D*/Y
  + C*E*/Y

Z = D*E
  + Y
  = D*E
  + /A*C
  + /B*C
  + C*/Z

a. Complete Circuit

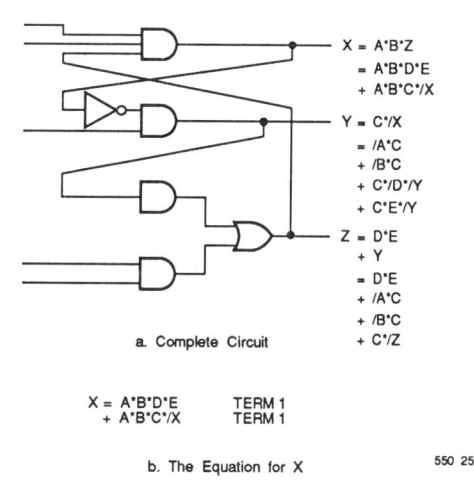

X = A*B*D*E   TERM 1
  + A*B*C*/X   TERM 1

b. The Equation for X

Figure 25. A Conditional Oscillator

Detecting oscillators

The oscillator in the circuit is not obvious. But if we simplify the logic completely, we can see that output X depends on /X; output Y depends on /Y; and output Z depends on /Z. Since the outputs are fed back to themselves in COMPLEMENT form, the circuit constitutes an oscillator.

This circuit will sometimes be stable. If we examine the logic function determining X, we see that it has two product terms, shown in Figure 25b. Term 1 is independent of /X; term 2 is dependent on /X. If inputs A, B, D, and E are all TRUE, then term 1 becomes TRUE, and the output stays HIGH regardless of the status of the rest of the circuit. It is thus stable. However, if signals D and/or E are LOW, then term 1 will be FALSE. If, at the same time, input C is HIGH, then, as long as the output X is LOW, term 2 will be TRUE, making the output HIGH (which makes the product term FALSE, which makes the output LOW, etc.). That is, the circuit oscillates.

In this manner, we can identify the conditions under which a conditional oscillator will oscillate. The mere presence of an oscillator is usually an indication that the circuit needs to be changed. It may be that the circuit only oscillates under conditions that could never possibly exist. One must be very certain of the impossibility of such a condition, however, if a conditional oscillator is to be tolerated. In addition, a thorough test sequence will usually expose a circuit to conditions that it may never encounter in a real system. Thus oscillators may interfere with the test process even if they do not disrupt the system.

Testability

Designing Testable State Machines

State machines have their own set of controllability issues. These essentially boil down to the concepts of *initialization* and *illegal states*.

State machine initialization

The nature of a state machine is that there is a well-defined sequence of states through which the machine will traverse as it operates. This implies the existence of a "first" state. Of course, these initial states vary from design to design. One obvious problem is the fact that many flip-flops — especially older varieties — do not power up in a predictable state.

Power-up initialization

Flip-flops that truly power up into a random state must be initialized explicitly. Lately, however, flip-flops have become available which have "power-up reset". This allows the flip-flops to power up into a predictable state every time. This is helpful when the power-up state also happens to be the initial state. But even if it is not the initial state, a predictable initialization sequence can bring the state machine into its start-up state.

Unfortunately, such initialization schemes rely on the ability of the device to initialize itself when being powered up. If the system needs to be re-initialized, it will have to be completely turned off and then turned on again. Anyone who has had to turn off a computer in order to reboot will know that this is not an elegant way of re-initializing. By building initialization into the design, a means of performing a "warm boot" is provided. It is for this reason that initialization must be considered along with all other aspects of the design.

Some devices, such as the PAL32VX10/A, the PAL22RX8A, and other PAL devices, have mechanisms specifically designed for initializing a state machine. These are usually in the form of global set and reset product terms. By programming the conditions for initialization onto such terms, the device can be re-initialized at any time. Other devices, like the PMS14R21/A and the PLS devices, have pins which can be dedicated as preset pins.

Including initialization in a design

Some of the simpler devices do not have specific provisions for initialization. However, the need is still present in these devices; here the initialization should be included in the design. This is a very simple process; it can be added in after all of the other design details have been worked out. Adding initialization will use up one input pin and potentially one product term on some outputs; this can affect the choice of device for the design.

To provide initialization in an otherwise complete design when Boolean equations are being used:

- determine the start-up state
- assign each bit as being initialized active or inactive, based on the desired start-up state
- if a bit is to be initialized inactive, add "/INIT" to every product term for that bit.
- if a bit is to be initialized active, add one product term consisting solely of "INIT"

Here we have assumed that the initialization pin has been called "INIT". "Active" would mean HIGH for an active high device; LOW for an active low device. "Inactive" is just the reverse.

The equation in Figure 26a can be initialized inactive as shown in Figure 26b, or active as shown in Figure 26c. Initialization is accomplished by asserting the INIT pin and clocking once. This "cookbook" approach is very reliable.

$$Q0 := Q1 * Q2$$
$$+ Q2 * /Q3$$

a. Uninitializable

$$Q0 := Q1 * Q2 * /INIT$$
$$+ Q2 * /Q3 * /INIT$$

b. Initialized Inactive

$$Q0 := Q1 * Q2$$
$$+ Q2 * /Q3$$
$$+ INIT$$

c. Initialized Active

Figure 26. Designing in Initialization

PALASM software also makes it possible to design state machines with a special syntax which essentially allows the state diagram to be transferred directly into a design file. For devices which have no dedicated initialization features, the initialization branches should be explicitly built into the state diagram. The software then performs the remainder of the processing needed.

Testability

Illegal states

A state machine is formed by using a set of flip-flops to remember states, and assigning a code to each state. Since there are 2^n different codes that can be assigned to a group of n flip-flops, there is a good chance that some codes may not be used. For example, if a state machine is to have 6 states, 2 flip-flops will not be sufficient; 3 are needed. But 3 flip-flops allow 8 states, which will result in 2 unused states (see Figure 27).

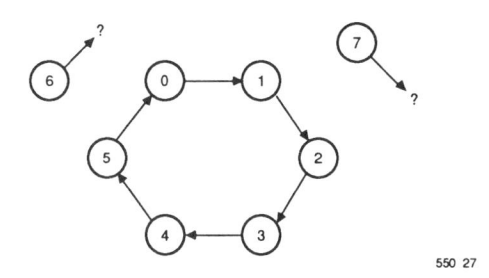

Figure 27. Illegal States

Assuming that the state machine has been designed correctly, there is no reason why these extra states should ever be entered; therefore they are called "illegal" states. Unfortunately, situations do occur, thanks to noise and other unpredictable occurrences, which result in the state machine being in an illegal state. When this happens, the immediate need is to return to a normal sequence of states: *there must be a predictable means of getting from any illegal states into a legal state.*

Illegal state recovery is a controllability issue which actually affects functionality more than it affects testability. But the concepts used for functionality and testing are so closely related that it is worth treating here.

Recovering from illegal states

There are three basic ways of getting out of an illegal state:

- re-initialize
- make sure that one can continue clocking until the machine recovers
- design the machine such that the start-up state is reached from any illegal state in one clock cycle, independent of any conditional inputs

Of course, re-initializing will take the machine back into its start-up state from any state, legal or illegal (Figure 28). The disadvantage here is that outside control is needed to force initialization.

Very often, a path will exist which eventually takes the state machine back into a normal sequence (Figure 29). These paths are not usually designed in; they just happen to be there. In fact, if D-type flip-flops are used, it is surprisingly difficult to get a "closed" set of illegal states (that is, a set such that once one of the illegal states is entered, the machine will forever remain in

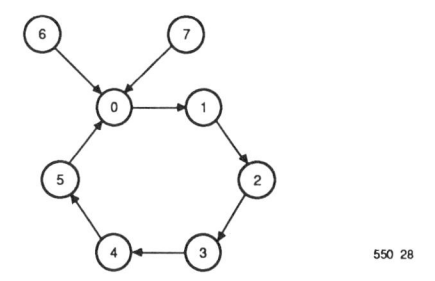

Figure 28. Using Initialization to Recover

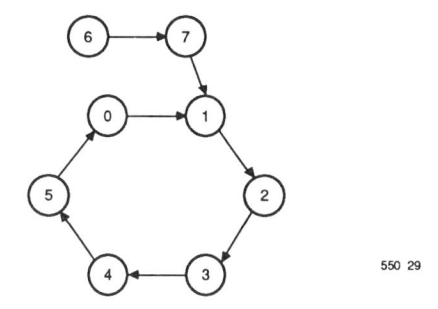

Figure 29. Cycling Back to a Legal State

illegal states) by accident. In most cases, there will be a path which eventually leads back to a legal state. In these cases, merely clocking enough times will cause the machine to recover.

The drawback here is that one does not know ahead of time how many clock cycles will be needed. This necessitates some built-in way of knowing just when a legal state has been re-entered. And once that state has been reached, further cycling may be needed to get to a point where operation can resume.

Designing-in one-step recovery

The most predictable way of dealing with illegal states is to provide a one-step path back to a legal state. Depending on the state desired, more or less work may be involved to do this. For PAL devices, we can consider three cases:

- all illegal states go to state 00...0
- all illegal states go to one state other than 00...0
- each illegal state goes to some legal state

The cause of poor illegal state recovery can be illustrated conceptually with Karnaugh maps (although realistically, Karnaugh maps are often not used). When calculating the equations for a particular bit, it is tempting to use Don't Care cells from the Karnaugh map (Figure 30) to simplify the logic. The success of illegal state recovery depends on how these Don't Care cells are treated.

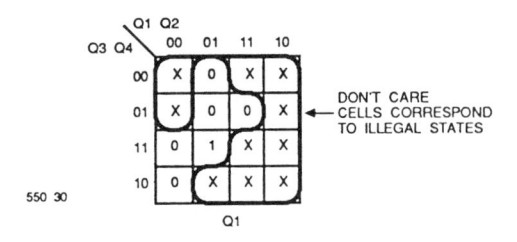

Testability

Figure 30. Illegal State

Recovering into state 00...0

This is the simplest case; it is illustrated in Figure 31. It is accomplished by not using any illegal states to generate the logic for any of the bits. Since most PAL devices have only D-type flip-flops, a bit will go HIGH only as a result of legal states. Any illegal states will cause all bits to be LOW.

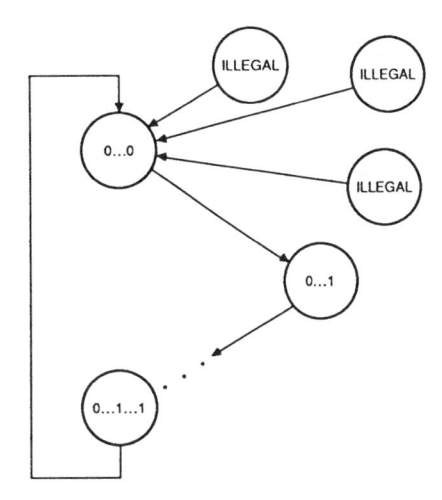

a. State Diagram

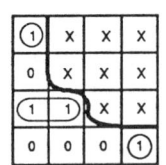

b. Karnaugh Map

Figure 31. Recovering to State 0...0

This procedure does not work when J-K or T-type flip-flops are used. In fact, it is deadly. Whereas a D-type flip-flop defaults to LOW, J-K and T-type flip-flops hold their present state as a default. Thus if illegal states are not considered in the transfer functions, an illegal state will cause the state machine to be locked up in that state.

Recovering into one fixed state

This case is shown in Figure 32a. The procedure can be illustrated conceptually with a Karnaugh map. It must first be decided which legal state will be entered, and the resultant value of each

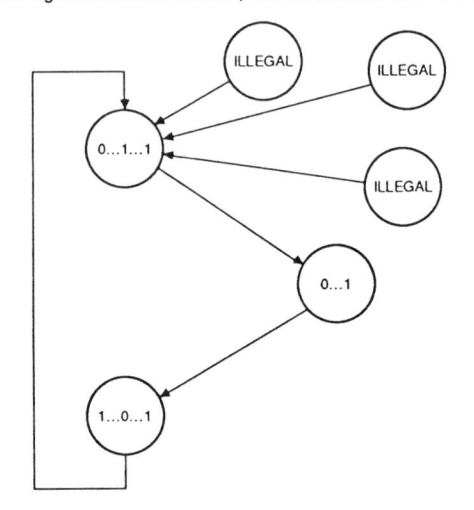

a. State Diagram

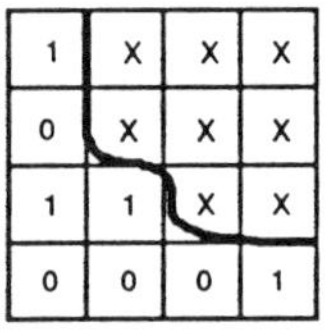

b. Karnaugh Map for Bit Qn

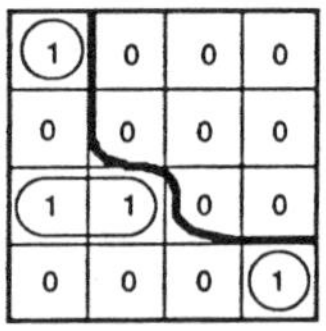

c. Bit Qn Recovers to 0

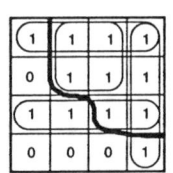

d. Bit Qn Recovers to 1

Figure 32. Recovering to a State Other Than 0...0

Testability

state bit. The Don't Care cells for each bit are then filled with the corresponding next state bit value; if the next state for a bit is to be 1, then Don't Care cells are filled with 1's for that bit's Karnaugh map; the procedure for a 0-bit is analogous. The equations are now taken by including either all Don't Care cells if filled with 1's, or none of them if filled with 0's. This procedure is illustrated in Figures 32b, c, and d.

When Karnaugh maps are not used, the same result can be obtained by explicitly considering all illegal states. When calculating the Boolean equations for:

- a bit that will be 0 after recovery, *no* illegal states should be included.

- a bit that will be 1 after recovery, *all* illegal states should be included.

When J-K flip-flops are used, then the transfer function for either J or K — but not both — will include all illegal states.

- If a bit is to be HIGH after recovery, *J* should account for all illegal states; *K* should account for none.

- If a bit is to be LOW after recovery, *K* should account for all illegal states; *J* should account for none.

This must be done explicitly for J-K flip-flops even if state 0...0 is the recovery state.

When T-type flip-flops are used, there is no easy way out; any recovery must be explicitly designed-in as part of the original function.

Recovering Into Any Legal State

The third case allows one to fill in the Don't Care cells of a Karnaugh map in such a way that some legal next state is always reached in one clock cycle, but such that the 1's and 0's are placed to keep the logic functions simple. This is shown in Figure 33. The disadvantage here is that since different illegal states result in a different legal state, some additional cycling may be required to allow operation to resume.

When Karnaugh maps are not used, this can be implemented more simply by explicitly including the illegal states as part of the complete state diagram. This is especially simple if the state machine input format for PALASM software is being used.

Default transitions

The PMS14R21/A and PLS devices have default branching mechanisms. When PALASM state machine input is used, it is possible to specify a DEFAULT BRANCH. This means that when in any state, if none of the branching conditions are satisfied, some user-definable state is automatically reached. This can be used as a way of recovering from illegal states.

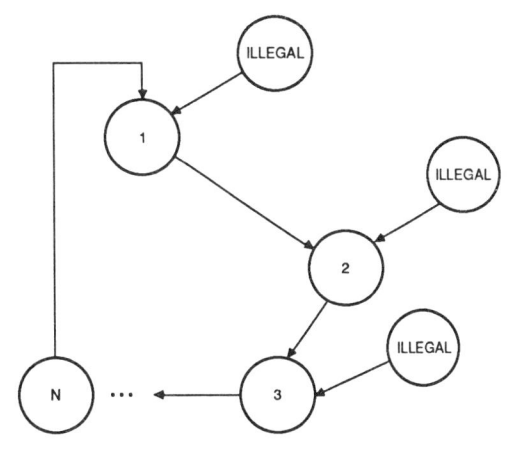

a. State Diagram

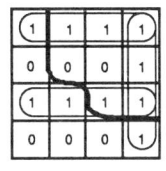

b. Karnaugh Map

Figure 33. Recovery Such That Logic Functions Are As Simple As Possible

In PLS devices, the complement array can serve as a way of recovering from illegal states. In a design, only legal branches are defined. When in an illegal state, since no legal branch is active, the complement array is activated, allowing for some default state to be reached.

Testing illegal state recovery

One of the difficulties of designing illegal state recovery into a circuit is the fact that it is difficult to test. Because the state is illegal, it is impossible to force the circuit into such a state. The use of register preload circumvents this problem. With preload, any state — legal or illegal — can be loaded into the register. If an illegal state is loaded, then the circuit can be tested to verify that correct recovery does indeed occur.

The use of preload must be considered carefully with devices having programmable asynchronous set and reset features. If these are driven by feedback from an output, then situations can occur where preloading one state immediately causes a set or reset to the opposite state (Figure 34). There are two alternatives: either avoid preloading such states, or include a control input in the set and/or reset product terms which can disable the feature when testing.

Testability

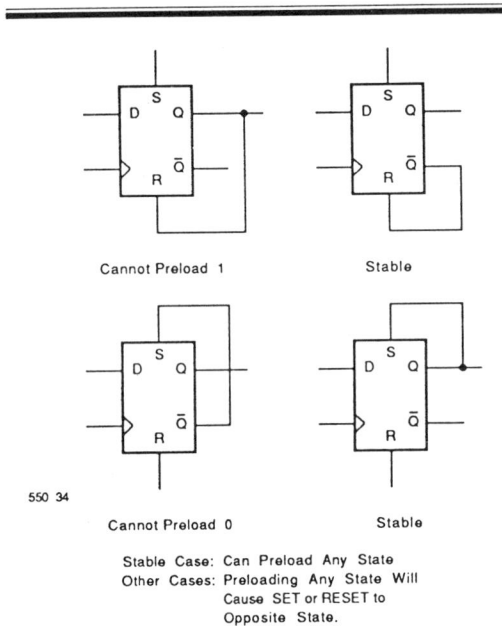

Stable Case: Can Preload Any State
Other Cases: Preloading Any State Will Cause SET or RESET to Opposite State.

Figure 34. Preloading Registers with SET and RESET

Providing for bed-of-nails testing

Most state machine PLDs are equipped with an enable pin for disabling the outputs. This is a key feature when the circuit board is to be tested in a bed-of-nails tester. When the devices driven by by the PLD are tested, it is recommended that the PLD be disabled so that there is no output level contention. Since the enable pin is usually grounded to keep outputs permanently enabled, it can instead be made available for use during testing.

Note that for combinatorial devices, there is generally no output enable pin. The disabling feature is instead implemented through a product term. Designing the part such that the outputs can be disabled during bed-of-nails testing is also encouraged for these combinatorial designs.

Testability

Designing for Testability With the PROSE™ Device

Today's more complex circuits and systems are becoming prohibitively expensive to test using standard methods. Diagnostics-On-Chip™, or DOC™, is a test feature provided in several of Monolithic Memories' devices as a means of increasing testability at the system, board, and chip levels. DOC is especially useful in the PROSE™ device (PMS14R21). It is even used by device programmers to configure the two programmable arrays inside the device (Figure 35).

DOC Architecture

Testability consists of two basic elements: controllability and observability. In a sequential (registered) system, these two elements are lost when a register is not directly accessible. In Figure 36a, the first register is not observable and the last register is not controllable. Figure 36b shows that the addition of a scan path through each register, as in the DOC method, provides the direct access for controllability and observability, which ensures complete testability.

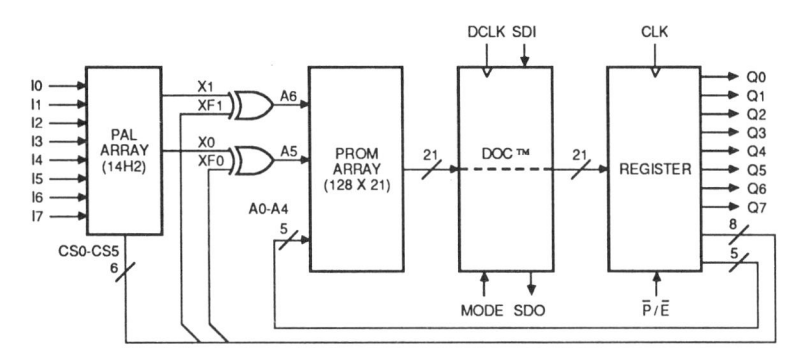

Figure 35. PROSE Block Diagram

a. Standard Sequential Logic

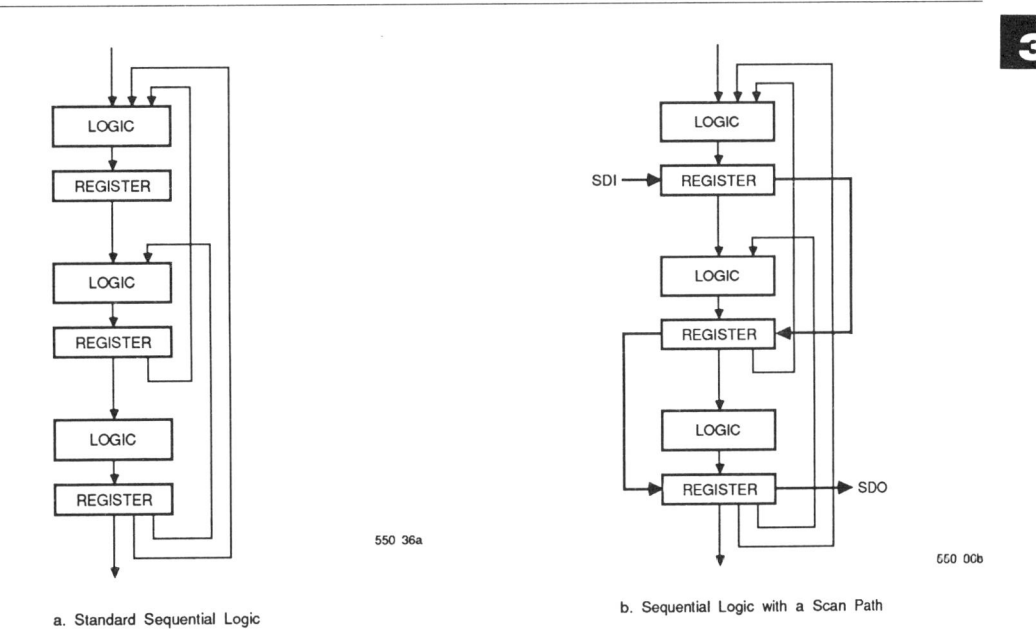

b. Sequential Logic with a Scan Path

Figure 36. Testability Can Be Increased by Providing Direct Access to All Registers

Testability

The heart of the DOC circuitry is the shadow register (see Figure 37). The shadow register is a serial/parallel register, equivalent in length to the pipeline register. It is called a shadow register because it is invisible to the device during normal operation. It is clocked by its own clock input, DCLK.

In normal mode (MODE input is LOW), the shadow register operates as a serial shift register (see Figure 38). The Serial Data Input is SDI, and the Serial Data Output is SDO. The pipeline register can operate at the same time while MODE is LOW.

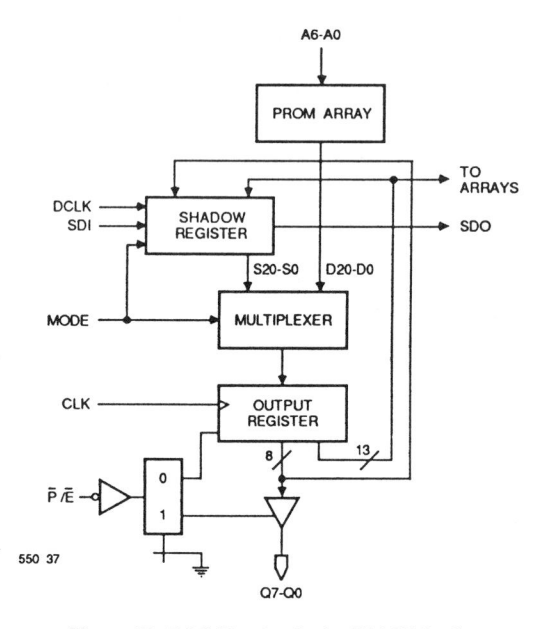

Figure 37. DOC Circuitry in the PROSE Device

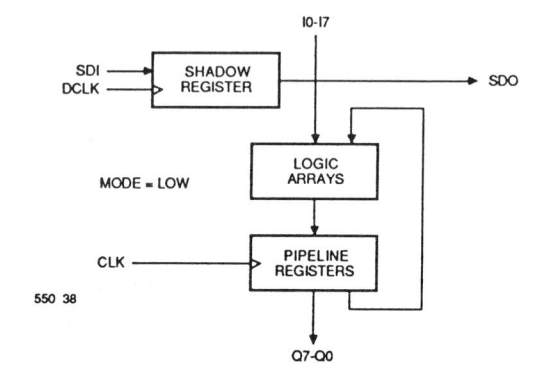

Figure 38. The Shadow Register Operates as an Independent Shift Register when MODE is LOW

In diagnostic mode (MODE is HIGH), the shadow register operates as a parallel register (see Figure 39). It can be parallel loaded from or to the pipeline register by clocking the receiving register. A swap can be performed by clocking both at the same time.

Note that the PROSE device differs from other DOC family members in that the shadow register is loadable only from the output register. Other devices also allow loading of the data on the output pins, if the device is connected to a bus. This does not affect the device-level testability if the outputs do not connect to a bus, of if the bus is otherwise observable.

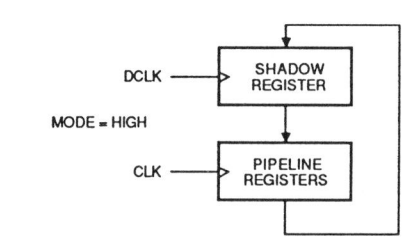

Figure 39. The Shadow Register Can Parallel Transfer its Contents to and from the Pipeline Register when MODE is HIGH

Testability

INPUTS				OUTPUTS			OPERATION
MODE	SDI	CLK	DCLK	Q20-Q0	Q20-Q0	SDO	
L	X	↑	*	Qn ← PROM	HOLD	S20	Load output register from PROM array
L	X	*	↑	HOLD	Sn ← Sn-1 S0 ← SDI	S20	Shift shadow register data
L	X	↑	↑	Qn PROM	Sn ← Sn-1 S0 ← SDI	S20	Load output register from PROM array while shifting shadow register data
H	X	↑	*	Qn ← Sn	HOLD	SDI	Load output register from shadow register
H	L	*	↑	HOLD	Sn ← Qn	SDI	Load shadow register from output register
H	L	↑	↑	Qn ← Sn	Sn ← Qn	SDI	Swap output and shadow registers
H	H	*	↑	HOLD	HOLD	SDI	No operation†

* Clock must be steady or failling.
† Reserved operaton for 74S818 8-Bit Diagnostic Register.

Figure 40. Diagnostics Function Table

All of the functions of the DOC circuitry are described in the function table (Figure 40).

DOC allows access to all twenty-one pipeline flip-flops in the PROSE device through the serial path, requiring only four additional pins. These four pins can be controlled directly, or can be connected to other DOC circuits in series. A series connection of several DOC circuits allows the same four signals to address an unlimited number of flip-flops on a board or system (Figure 41).

A typical test would be performed as follows:

1. Test vector shifted into shadow register(s)
2. Test vector parallel transferred to pipeline register(s)
3. Device/system clocked desired number of times to run test
4. Test results parallel transferred to shadow register(s)
5. Test results shifted out of shadow register(s)

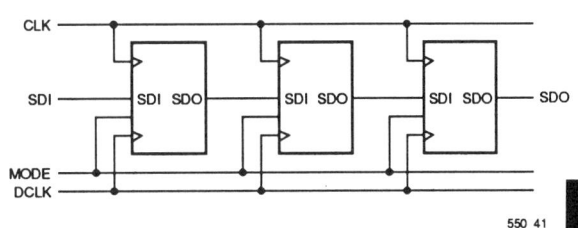

Figure 41. Example Architecture for Use of System-Level Diagnostics

Note that while shifting, the system can return to normal operation. In addition, while test results are being shifted out, a new test vector can be shifted in.

The swap function allows the state of the pipeline register(s) to be restored once the test is complete. When the test vector is parallel transferred to the pipeline register, the pipeline register contents are transferred to the shadow register at the same time for

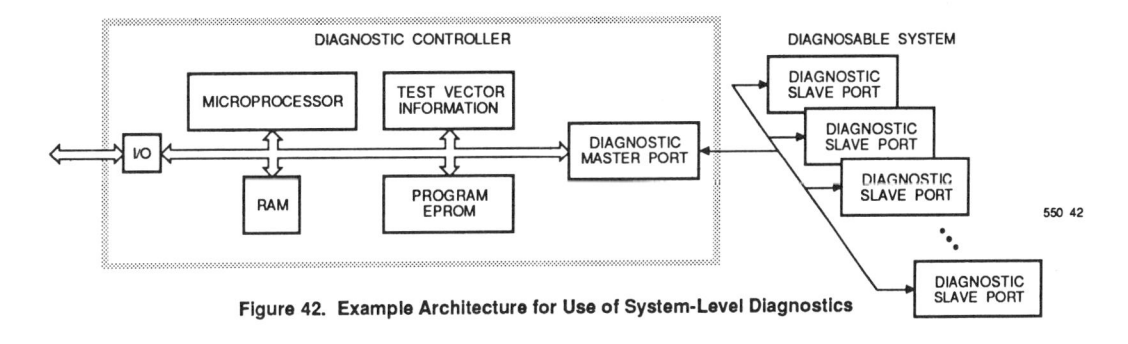

Figure 42. Example Architecture for Use of System-Level Diagnostics

Testability

storage. Later, when the test results are transferred to the shadow register, the original pipeline register information (stored in the shadow register) is transferred back to the pipeline register.

System-Level Testing

At the system level, DOC provides the ability for a diagnostic controller to monitor the interior status of a system. The diagnostic controller could control several scan loops, selecting the loops required for the test needed (Figure 42).

However, many key products, such as microprocessors, are not available with the DOC function and cannot be part of the scan path. This limits the use of DOC for full system-level testing to selected manufacturers who can use this additional testability as an enhanceement to a larger system-level testability strategy.

In addition, little support is available for writing the test vectors that can be run through the DOC scan path. The software that is available is expensive and runs on large computers only. This is another factor that limits the use of DOC on a system level. On the board or chip levels, however, test vectors are much easier to generate and can even be found by running vectors through a known good unit.

A complete system-level test would require that most of the devices in the system shown in Figure 43 incorporate the DOC circuitry. Other devices in the DOC family are shown in Figure 43. Devices with circuitry equivalent to the DOC format are available from several other suppliers as well. Also, many gate array and standard cell manufacturers offer standard functions similar to the DOC scan path and can easily be included in custom designs.

Board-Level Testing

DOC in the PROSE device is especially useful at the board, or functional, level. The PROSE device will usually form the heart of a function, such as a peripheral controller. In addition, it often will serve to off-load the main Central Processing Unit (CPU) and be partially controlled by the CPU. DOC allows direct control of the PROSE device, bypassing a difficult-to-control CPU and taking command of whatever function the PROSE device performs (Figure 44). Here, on-board diagnostics can be easily done with the PROSE device. The alternative is to dedicate edge-connector signals to the DOC path.

The DOC circuitry provides access to the PROSE device's pipeline register. This can be used to set the outputs to a given state, in order to test the effect on the devices surrounding the PROSE device. Or, the pipeline register can be set to a given state and then left to run freely, to verify functionality. If combined with control of the device inputs, the sequencer can be stepped through a number of states, to test the response of the surrounding logic. This is especially useful for bed-of-nails board-level testing; the PROSE device can be tested completely without having to be backdriven.

Device-Level Testing

On the device level, the DOC circuitry effectively provides a Preload function for the register. Instead of loading the register from the outputs, as with standard PAL® devices, the register is preloaded from the shadow register. A standard type of preload is not possible on the PROSE device because thirteen of the flip-flops are buried (Figure 45). Preload is necessary for testing the device functionality, since the buried flip-flops must be set to a known condition before the device can be tested.

PART NUMBER	DESCRIPTION
PMS14R21/A	128-state sequencer
Am29PL141	64-state sequencer
Am27S65/A	4-K Diagnostic PROM
Am27S75/A	8-K Diagnostic PROM
Am27S85/A	16-K Diagnostic PROM
Am9151	4-K Diagnostic Static RAM
74S818	8-bit register
Am29818	8-bit register

Figure 43. DOC Products Family

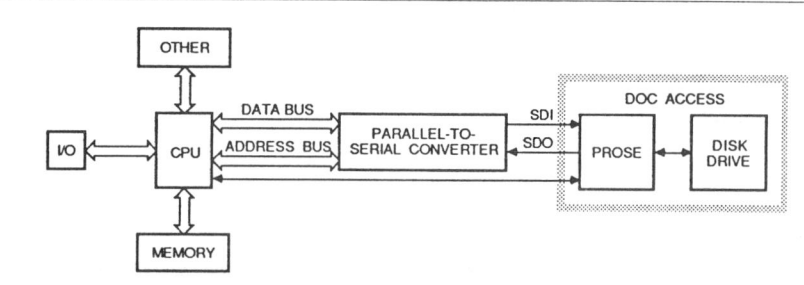

Figure 44. DOC Allows Direct Access to Peripheral Elements In a System, Bypassing the CPU

Testability

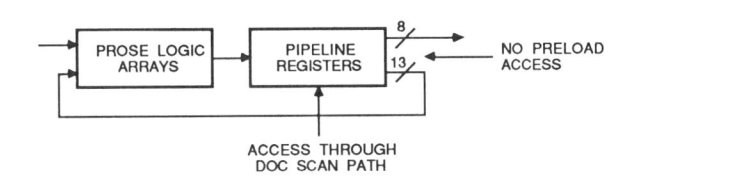

Figure 45. DOC Effectively Adds a Preload Function to Buried Flip-Flops

The DOC circuitry allows more than just a Preload equivalent, however. It also allows observation of the pipeline register, which contains all of the state information. Thus, an individual state transition may be tested by preloading the desired state, setting the inputs, clocking the device, and then observing the resulting state in the pipeline register. State transitions which do not result in a change in outputs are thus easily tested (Figure 46).

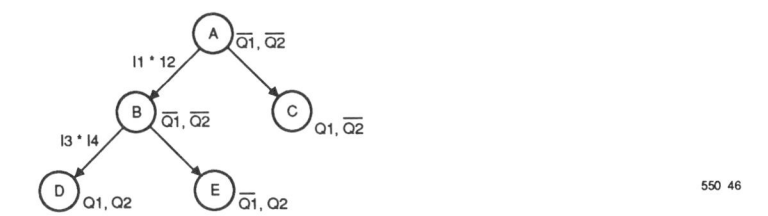

Figure 46. DOC Allows Transitions that Do Not Result in Changes to the Outputs to Be Verified. In this Case, the Transition from A to B Does Not Change the Outputs (Q1 and Q2), but the Internal Feedback Changes (the Condition Select Signals Examine I3 and I4 Instead of I1 and I2). Buried Flip-Flop Observability Is Required to Verify the Transition

Testability

Using Test Vectors

Digital systems are generally tested by applying a sequence of test vectors. A test vector is a group of signals which are applied (forced) and measured (sensed) on a device or a board. The vector thus defines all inputs and expected outputs for a given test. As we have noted, the sequence of tests performed greatly affects the quality of the overall tests, as measured by the fault coverage.

In general, we can talk in terms of three kinds of vectors. *Simulation* (or application) vectors, *functional* test vectors and *signature* test vectors.

Simulation vectors are generated during the design process. Their main purpose is to help the designer verify that the design has been correctly implemented. They represent the way in which the circuit was intended to operate. When PALASM software (or almost any other PLD design software package) is used, simulation may be performed prior to programming a device. The software simulates the operation of the circuit, and then generates vectors from the simulation, adding the vectors to the JEDEC file. These vectors can then be used for testing by programmers that have the capability of performing functional tests.

While simulation vectors may be adequate for verifying that the design is operating as expected, they generally do not provide very extensive test coverage. For this reason, we distinguish functional test vectors from simulation vectors.

It is very difficult to generate a complete set of functional test vectors by hand; computer programs are generally used instead. The simulation vectors are often used as a basis for generating a more comprehensive set of functional test vectors; in this capacity, the simulation vectors serve as *seed* vectors. There are many programs which perform this function although many of the programs require larger computers and take a long time to run. Monolithic Memories also generates functional test vectors for patterns that are used in ProPAL and HAL devices. This is discussed more fully on page 3-106.

More recently, programs which run on the IBM PC-compatible computers have been developed to generate vectors for use in testing PLDs. Most well-known among these are PLDtest™ from Data I/O Corp., and TestPLA™ from Structured Design. These programs use the programming information in the JEDEC file to generate tests.

On most patterns, they can generate test sequences of high quality. If complex internal feedback is used in a particular design, then some manual test generation may still be needed to improve the test coverage. Both of these programs support the use of register preload for initializing states; the TestPLA package can also generate tests for devices which do not have the preload feature.

While functional vectors provide more extensive tests, they may not exercise the circuit in the manner in which it was meant to be used. Thus, for example, a conditional oscillator in a circuit (as discussed above) may not be a problem during simulation, since the conditions causing oscillation are not thought to be possible by the designer. However, the functional vectors will take all situations (some of which may not be physically possible) into account in the tests. Thus more subtle design problems may become apparent when functional test vectors are generated.

Signature vectors are random vectors which are first applied to a device which is known to be good in order to generate a "signature". This same set of vectors is then applied to a device of unknown quality; if the same signature results, the device is said to be good; if a different signature results, then the device is assumed to be faulty.

Signature vectors can vary greatly in the quality of testing they can provide. Since they are generated with no knowledge of the circuit being tested, many more vectors must be used to perform a good test. The quality of the test depends on the circuit being tested, the number of vectors used, the speed with which the tests are applied, and the algorithm used to generate the vectors. The tester must also be able to apply a preload sequence to devices that have registers; otherwise two devices may power up into two different states. In that case, both devices will generate different signatures even if both are good devices.

Quality signature testing can be very cost effective, since no advance knowledge of a device pattern is needed. This reduces the amount of resources that must be dedicated to test vector generation. Signature testing options are discussed more fully on page 3-106.

The different types of vectors are summarized in Table 1 below.

TYPE OF VECTOR	PURPOSE	GENERATED BY:
Simulation (Application)	Used for verifying whether or not a design will operate as expected when implemented.	Sequence defined by the design engineer, usually by hand. Actual vectors generated by design software, placed in the JEDEC file.
Functional	Used for verifying that a device is operating correctly.	Usually generated by a computer program such as PLDtest or TestPLA. The simulation vectors can be used as seed vectors
Signature	Used for verifying that a device is operating correctly without functional vectors.	The tester generates the test sequence during the test.

Table 1. Test vectors

Testability

Summary

The time to start considering ways of testing a circuit is before the circuit has been designed. The key to testability lies in the way the circuit is implemented.

Basic combinatorial logic can be made completely testable simply by minimizing logic. It is not even necessary to analyze the circuit for redundancy or reconvergent fanout; automatically minimizing all logic will eliminate any occurrences.

Where a sequential circuit is generated from simple feedback paths in the logic, the circuit must be analyzed as a combinatorial circuit. All combinatorial logic must be included to determine whether the circuit is a latch or an oscillator. If a latch is desired, it should be completely controllable. If an oscillator is found, it is probably not desired, and will generally indicate a mistake in the design. If a conditional oscillator is to be tolerated, one must be sure that the oscillation conditions can never occur, and that the test procedure will not cause oscillation.

In general, combinatorial circuits should be analyzed completely for the presence of latches and oscillators (wanted or unwanted). This can be done by simplifying each combinatorial logic block to see whether any signal ultimately depends on itself.

When the sequential nature of a circuit is derived through the use of flip-flops to generate a state machine, the two key issues are initialization and illegal state recovery. A combination of device features and careful circuit design will yield circuits that can behave predictably even in unexpected situations.

DOC is a testability feature that is useful in the PROSE™ device; it may be used on multiple levels: system, board, and chip. While the system-level uses may be restricted by the limited availability of support products, the board or functional-level uses are exceptionally handy when the PROSE device acts as a local controller. And on the device level, the DOC circuitry provides a means of accessing the buried flip-flops within the device for functional testing.

It is important to analyze the testability of a circuit before committing it too far. Thus any changes can be made early on. In particular, if the test analysis software points out any logic hazards in your circuit, you can easily remedy them by modifying the design.

These simple steps, taken early in the design phase, can help avoid later redesigns, and ultimately provide a higher quality system.

Finally, the ultimate test quality depends also on the quality of the test sequence used for production, functional test vectors and high quality signature tests will provide you with the highest confidence in the quality of your system.

DESIGNER'S GUIDE TO HIGH PERFORMANCE LOW-POWER SCHOTTKY LOGIC

By David A. Laws and Roy J. Levy.

1.

THE NEW STANDARD LOGIC

Low-power Schottky TTL integrated circuits are now firmly established as the standard logic configuration for new high performance system designs. They have essentially entirely replaced standard "gold-doped" TTL devices in all applications. In addition, they have relegated the other logic families to specialized needs where the ultimate in high speed (ECL) or low power for battery operated operation (CMOS) is mandatory.

This wide acceptance has been achieved because LS offered all of the important features of the earlier TTL families with two significant advantages:

- LS circuits provide performance equal to that of standard TTL at between 20% and 50% of the power requirements. As a result, considerable system cost savings have been made in bulky power supplies and fans.
- LS technology allows more complex designs to be fabricated on a given die size. A far wider selection of systems oriented MSI and LSI functions have therefore been developed in the LS family.

Additional factors in their popularity is that the devices are implemented with the same technology, and are therefore totally compatible with the LSI bit-slice processors and supporting memories which today form the heart of most new high speed designs. Users of LS devices have been able to exploit these features to improve the performance and enhance the functional capability of their systems. In many cases this has been achieved at a lower total cost.

Advanced Micro Devices is a leading supplier of low-power Schottky MSI and LSI devices. Two basic families of product are offered:

AM54/74LS Series

- Typical tpd 10ns/gate at 2mW
- Typical Register fmax = 40MHz

Pin for pin and electrical alternate source devices to the standard performance LS logic family.

AM25LS Series

- Typical tpd 5ns/gate at 2mW
- Typical Register fmax = 65MHz

Advanced Micro Devices' proprietary high performance LS logic family. This includes both original designs and enhanced specification versions of the AM54/74LS devices. Improvements include twice the fan-out over the military temperature range, higher noise margin and faster switching speeds.

The AM25LS improved performance devices are offered by Raytheon Semiconductor and identified by 25LS part numbers. Equivalent Fairchild and Motorola 9LS functions will come close to meeting AM25LS switching speeds on certain products.

The AM25LS proprietary designs have been carefully chosen to improve operation and reduce the cost of building high performance digital systems. A good example is the set of AM25LS14, 15 and 22 digital signal processing elements. Fairchild, Motorola and Texas Instruments have announced plans to alternate source many of the new Advanced Micro Devices' designs.

Both the Am25LS and the Am54LS/74LS families can be freely intermixed. Together with the Am2900 series of bipolar microprocessor functions they will satisfy most of the design requirements of today's advanced systems.

THE SCHOTTKY DIODE STRUCTURE

The major components of switching delays in digital integrated circuits are listed in Figure 1. One of the most significant of these is the storage time constant of a transistor driven into saturation Ts. Standard TTL circuits minimize this parameter with a process technique known as gold doping. This increases the rate of recombination of charge stored in the base region.

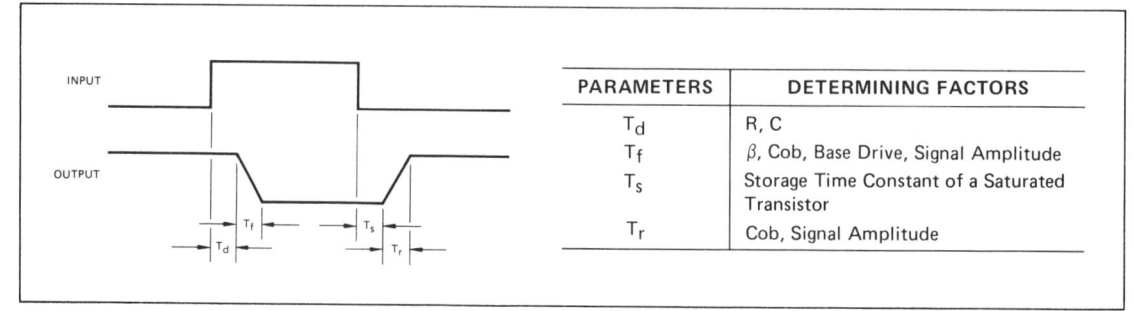

PARAMETERS	DETERMINING FACTORS
T_d	R, C
T_f	β, Cob, Base Drive, Signal Amplitude
T_s	Storage Time Constant of a Saturated Transistor
T_r	Cob, Signal Amplitude

Figure 1. Major Causes of Propagation Delay.

Designer's Guide

The desired result of improved speed is achieved. Unfortunately it also reduces available design β at low temperatures and is marginally effective when hot. This results in lowered performance over the full military temperature range.

The development of the Schottky diode provides a more effective solution. A feature of the Schottky diode is its lower forward voltage at a given current level compared to a diffused (P-N) diode of the same area, Figure 2. Connecting a Schottky diode between the base and collector of a transistor, Figure 3, will shunt excess base current drive from the base to the collector, once the collector drops to a low enough voltage to forward bias the Schottky. This prevents the build up of stored charge and eliminates the Ts component of the delay.

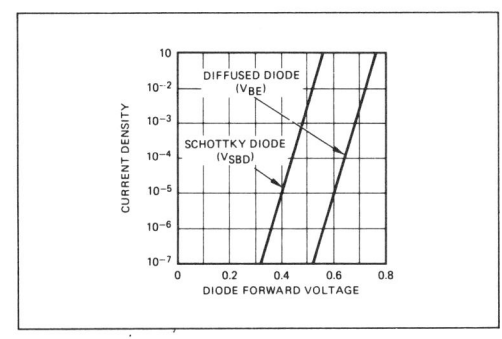

Figure 2. Comparison of V_F for Schottky and Diffused Diodes.

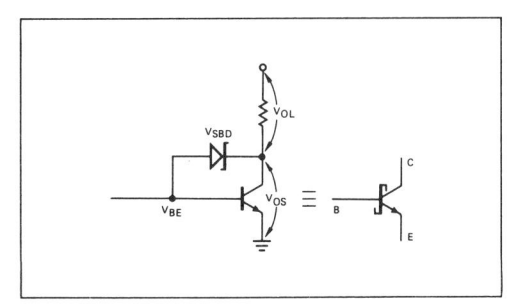

Figure 3. Schottky Clamped Transistor and its Conventional Circuit Symbol.

A Schottky diode is formed at a metal to semiconductor junction when the semiconductor doping is at the level normally found in the collector region of TTL devices. A Schottky-clamped transistor is constructed by extending the metal contact for the base region over the collector as shown in Figure 4. The same metallization structure forms a simple ohmic contact at the base, collector and emitter contact windows because of the higher doping levels in the silicon at these locations.

The selection of the forward voltage drop across the Schottky diode, V_{SBD}, is a compromise between a high value to insure a minimum V_{OL} but low enough to prevent charge storage in the base. Platinum silicide Schottky diodes provide this optimum voltage drop. Platinum is deposited and platinum-silicide is formed by sintering and annealing. As aluminum has a high affinity for silicon, in order to prevent the aluminum interconnect metallization from diffusing through the platinum material, with resulting lower V_{SBD}, a barrier of tungsten-titanium is evaporated after the platinum and before the aluminum metallization. This structure has been extensively evaluated and proven to have excellent reliability characteristics. It is now widely employed in the manufacture of Schottky devices. Reliability data is available from Advanced Micro Devices on request.

CHARACTERISTICS OF SCHOTTKY DEVICES

The primary reason for the development of Schottky devices was to improve AC (switching) performance and the first integrated circuits to employ this technique offered propagation delays as fast as 3ns. However, their fast rise and fall times and high power requirements have restricted their application to highest performance systems. More recently it was realized that the technique could be used to decrease the charging current required to achieve the 10ns speed specification of standard TTL gates. This insures considerably lower operating power requirements. The resulting family of devices are known as Low-Power Schottky (LS) circuits.

While the low current characteristics of LS devices are extremely important, other features of Schottky devices have contributed significantly to improved overall performance;

- Improved yield can be obtained to higher β specifications which reduces the variation of a.c. performance at low temperatures.
- Elimination of the marginal effect of gold doping at high temperature improves switching speed at the upper end.
- PNP transistors with useful β can now be fabricated. Since they reduce input load current requirements, they can be employed on inputs where loading is critical.

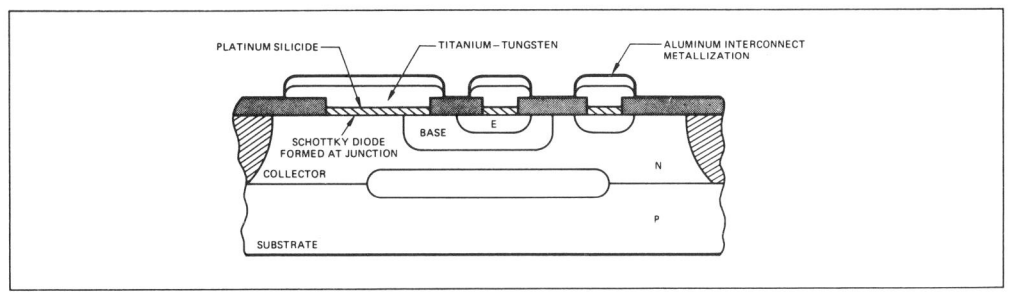

Figure 4. Schottky Diode Clamped Transistor Structure.

Designer's Guide

- The shallow epitaxial layers employed (around 3.5μm) considerably reduce on chip capacitance and series resistance. This is a significant contributor to improved speed performance at low-power.
- Other improvements in general circuit design flexibility include improved control over internal waveform amplitudes, lower junction leakage currents and location of parasitic capacitances at low impedance nodes.

LOW-POWER SCHOTTKY FAMILIES

The first application of the low power technology to a commercially available product was to redesign the most popular elements of the standard, gold-doped 54/74 TTL family in LS. This provided a set of functions pin-for-pin and speed compatible with the earlier TTL parts, but requiring as little as 20% of the power. The basic gate design for a 54LS/74LS element is shown in Figure 5. This offers a typical propagation delay of 10ns at 2mW power dissipation. Similar improvements have been made in power requirements for flip-flops and MSI functions.

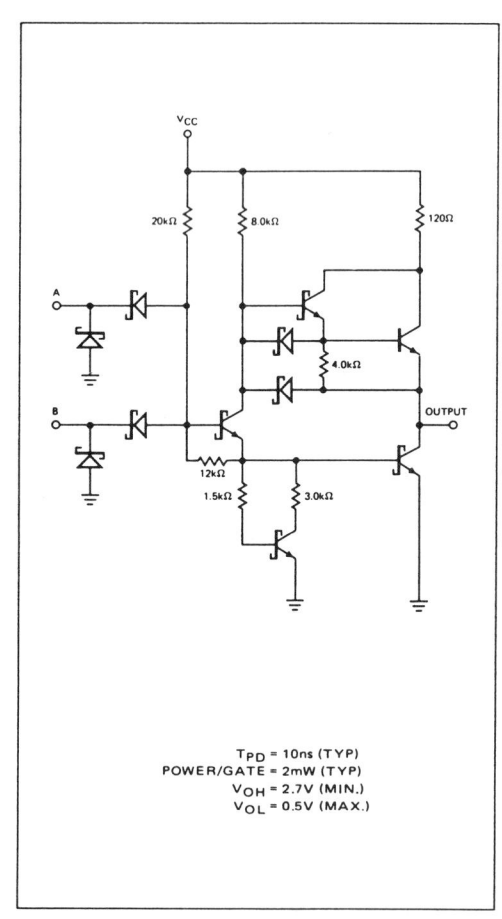

T_{PD} = 10ns (TYP)
POWER/GATE = 2mW (TYP)
V_{OH} = 2.7V (MIN.)
V_{OL} = 0.5V (MAX.)

Figure 5. Low-Power Schottky "74LS" TTL Gate.

This LS family offers many advantages to the system designers over the older standard TTL functions.

- Lower supply currents permit the use of smaller, lower cost power supplies.
- Reduced power dissipation generates less heat and simplifies cooling needs and allows increased board packing density.
- Lower on-chip operating temperatures decrease IC failure rates, thus improving system reliability.
- Lower operating currents reduce output spiking, leading to a decrease in noise generation and associated system problems.
- As the input load current requirements of Low-Power Schottky are only 25% of standard TTL, the new circuits are easier to interface with MOS elements, such as memories and microprocessors.
- Provided input and output loading rules are obeyed, as the functions and pin-outs are identical to those of the earlier TTL families, it is easy to upgrade existing systems.

In addition, no retraining of personnel is necessary before proceeding with a new design using these improved circuits as most engineers are already familiar with the logic functions and capabilities of TTL.

Later improvements in process technology and design techniques have led to what is essentially a second generation of LS devices. Generally described as high-performance LS, these products maintain the same power requirements as 54LS/74LS but offer such improvements as:

- Up to 50% faster speed
- Improved DC noise margin (50mV at full drive)
- Twice the fan-out over the military temperature range

The Advanced Micro Devices' Am25LS Family combines all these high-performance features into products which are direct replacements for the equivalent Am54LS/74LS MSI functions.

INCREASED FUNCTIONAL COMPLEXITY

As devices are operating at lower current levels, smaller area geometries can be employed. Thus, an LS design can often be produced on a smaller die than the equivalent standard TTL function. Further, the recent development of composite and self-aligning masking techniques allows even further reductions in device geometry sizes. These in turn result in faster speeds and the ability to manufacture more complex die.

Lower power dissipation also allows considerably more components to be incorporated onto a single chip without exceeding the recommended chip operating temperature.

The ability to produce large die at economical prices has improved the functional capability and variety of elements available in the LS family compared to standard TTL. Thus, LS technology is being used to implement many high-performance LSI functions in memory, interface and microprocessor, as well as logic families.

An important feature of all LS families is the new 20-pin Dual In-Line Package. This configuration fills the need for a package having the number of terminals necessary to accommodate the more complex products possible with LS, without the physical and cost disadvantages of the older 24-pin outline.

Designer's Guide

The 20-pin DIP has the same 300-mil center to center spacing between rows of pins as the popular 16-pin package. It therefore occupies about one third the board space of the 24-pin DIP with only a minor trade-off in functional capability. For both user and manufacturer this package is also considerably easier and lower cost to handle and test.

The 20-pin DIP is supplied in molded epoxy and hermetic ceramic versions. An hermetic ceramic flat pack is also available for military temperature range devices.

Functionally the 20-pin configuration is optimum for building octal functions. Eight input lines, eight output lines, power supply and ground, leaves two pins available for control signals. Eight-bit devices are ideal for interfacing with popular eight-bit fixed instruction set MOS microprocessors. They are also useful in micro-programmable machines using bit slice processors implemented in multiples of eight-bits. An octal register device in a 20-pin package can reduce count by 50% over the two quad, or even more wasteful, two hex elements frequently used today.

A significant proportion of new Advanced Micro Devices' LS products introduced recently are in the 20-pin package.

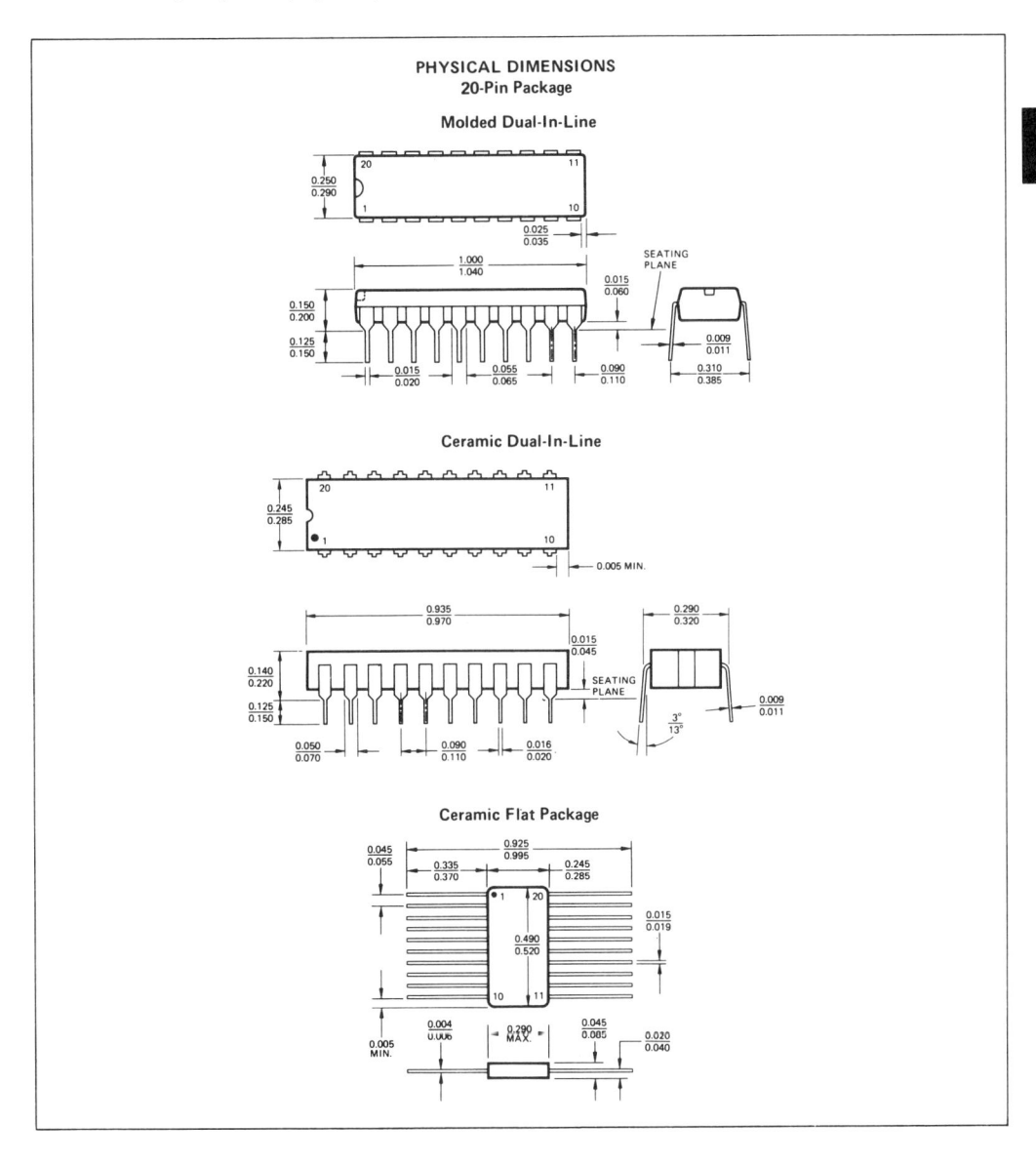

PHYSICAL DIMENSIONS
20-Pin Package

Molded Dual-In-Line

Ceramic Dual-In-Line

Ceramic Flat Package

Designer's Guide

2. D.C. Circuit Characteristics

CIRCUIT CONFIGURATIONS

The basic circuit design configuration of a Low-Power Schottky gate is similar to that of the original standard TTL elements. However, certain refinements have been made to optimize device performance when fabricated with the LS process.

In order to analyze the circuit configuration, Table 1 shows terms used in describing Advanced Micro Devices' LS circuits:

TABLE 1
D.C. CIRCUIT PARAMETER DEFINITIONS

- I_{IL} The current out of an input at a specified LOW voltage.
- I_{IH} The current into an input at a specified HIGH voltage.
- I_{OL} The current into an output when in the LOW state.
- I_{OH} The current out of an output when in the HIGH state (pull-up circuit only).
- I_{SC} The current out of an output in the HIGH state when shorted to ground. (Also called I_{OS})
- V_{CC} The range of supply voltage over which the device is guaranteed to operate.
- V_{IL} The guaranteed maximum input voltage that will be recognized by the device as a logic LOW.
- V_{IH} The guaranteed minimum input voltage that will be recognized by the device as a logic HIGH.
- V_{OL} The maximum guaranteed logic LOW voltage at the output terminal while sinking the specified load current I_{OL}.
- V_{OH} The minimum guaranteed logic HIGH voltage at the output terminal when sourcing the specified source current I_{OH}.

Both the input and output structures of the LS devices themselves have evolved through a number of configurations as designers have attempted to optimize circuit performance.

Depending on the function of the device any one of four commonly used inputs may be employed. The significant characteristics of each of these configurations are summarized in Figure 6.

The first LS designs used the familiar multi-emitter TTL input of Figure 6a. However because of low breakdown voltage and slow speed it is now used only where the geometry offers a significant advantage in circuit mask layout.

The second and still most widely used structure is the simple DTL style input of Figure 6b. This is the fastest version and it has good input breakdown voltage. In output functions having only a single gate delay between input and output, such as a three-state enable input, the low threshold of the DTL configuration causes the output node to be at a sufficiently low voltage to risk leakage problems at high temperature. The input of Figure 6c raises the threshold by one diode to overcome this problem (Figure 7). However because it is slower and uses more silicon area, its use is limited to special situations. A PNP input, Figure 6d, insures low d.c. loading for devices with common input/output pins such as the Am25LS23. However it is slow and has low breakdown voltage, comparable to the multi-emitter TTL structure.

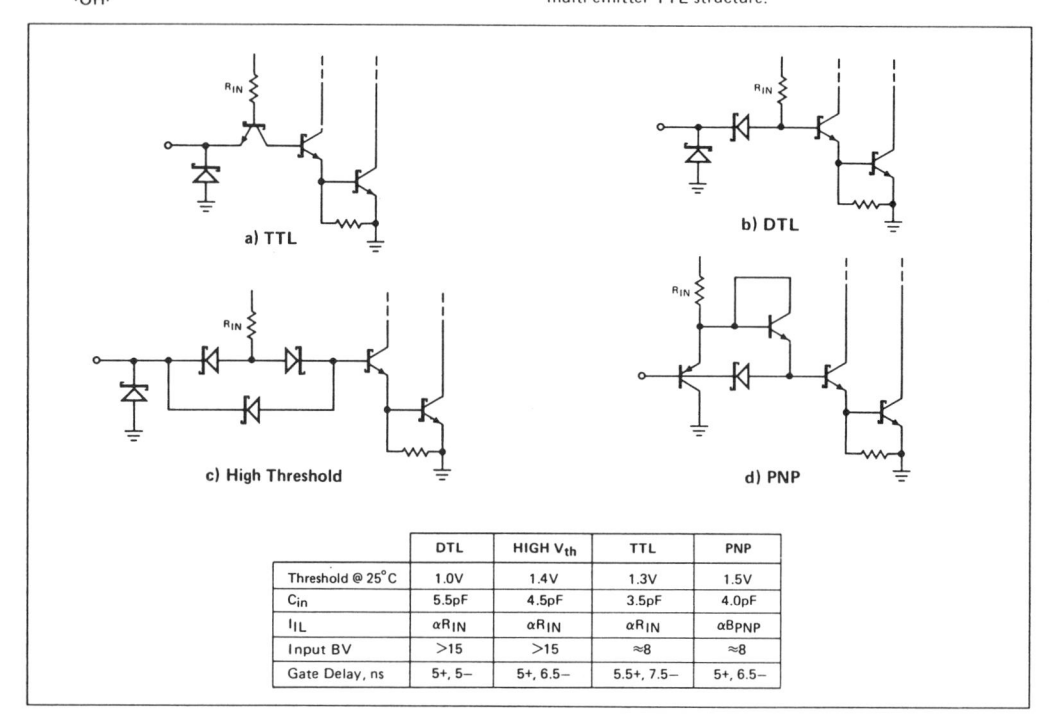

a) TTL

b) DTL

c) High Threshold

d) PNP

	DTL	HIGH V_{th}	TTL	PNP
Threshold @ 25°C	1.0V	1.4V	1.3V	1.5V
C_{in}	5.5pF	4.5pF	3.5pF	4.0pF
I_{IL}	αR_{IN}	αR_{IN}	αR_{IN}	αB_{PNP}
Input BV	>15	>15	≈8	≈8
Gate Delay, ns	5+, 5–	5+, 6.5–	5.5+, 7.5–	5+, 6.5–

Figure 6. Low-Power Schottky Input Configurations.

Designer's Guide

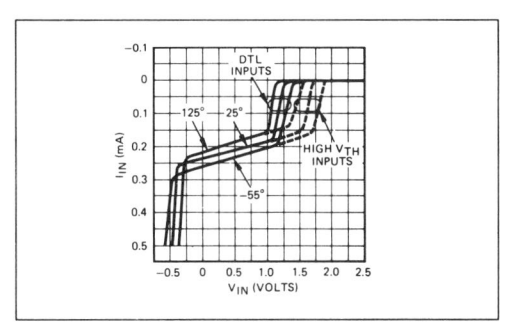

Figure 7. LS Input Characteristics for DTL and High Threshold Inputs.

Figure 8 compares the early LS output configuration with the design most frequently used today. The change was made to provide clamping of positive ringing and to allow the higher I_{SC} currents now specified (see section 3). The typical V_{OH} versus I_{OH} curves of Figure 9 are similar for both versions.

Figure 9. Typical V_{OH} Versus I_{OH} for Low-Power Schottky.

This example displays an I_{SC} of approximately 35mA. Note that both of these designs include the "squaring" network (R_3, R_4 and Q_5) at the base of the output pulldown transistor, Q_4, which was not included on standard TTL families. The result of this is a sharp transition of V_{OUT} with V_{IN} shown in Figure 10 for a simple gate function.

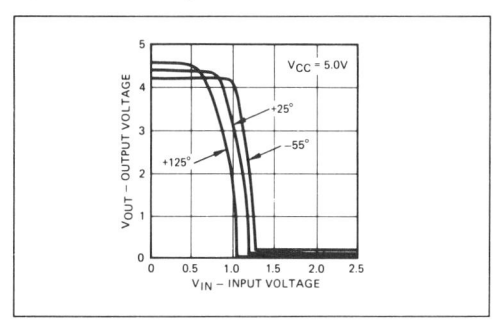

Figure 10. Typical Output Versus Input Voltage Characteristic.

The typical V_{OL} versus I_{OL} output characteristics of LS devices are shown in Figure 11. Most 74LS functions are specified at V_{OL} = 0.4V at I_{OL} = 4mA and 0.5V at 8mA. Am25LS are specified at 0.45V for I_{OL} = 8mA. Some newer designs are being guaranteed at I_{OL} of 12mA and 24mA. This curve indicates that lack of β at low temperature will not permit existing designs to be guaranteed to these higher values without severe yield loss.

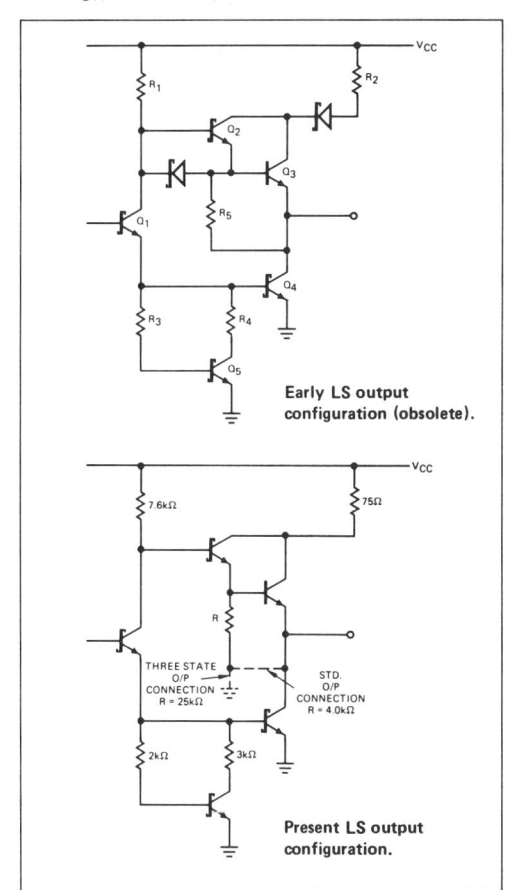

Figure 8. Low-Power Schottky Output Configurations.

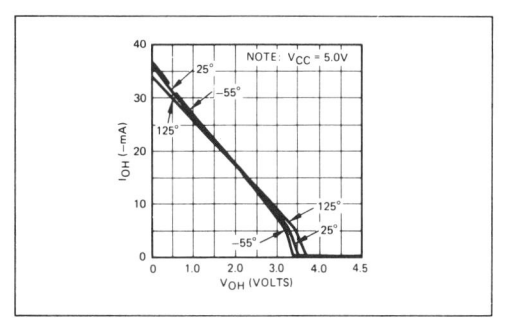

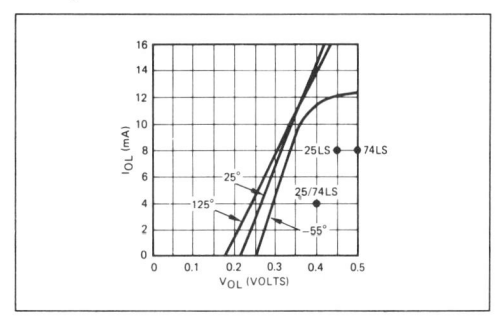

Figure 11. Typical LS V_{OL} Versus I_{OL} Characteristics.

Designer's Guide

TABLE 2
COMPARISON OF TTL DC PARAMETERS

Parameters	54LS/74LS LOW-POWER SCHOTTKY Conditions		Min.	Typ.	Max.	25LS LOW-POWER SCHOTTKY Conditions		Min.	Typ.	Max.	Units
V_{OL}	I_{OL} = 4.0mA				0.4	I_{OL} = 4.0mA				0.4	V
	I_{OL} = 8.0mA (COM'L Only)				0.5	I_{OL} = 8.0mA (MIL, COM'L)				0.45	
V_{OH}	I_{OH} = −400μA	MIL	2.5	3.4		I_{OH} = −440μA	MIL	2.5	3.4		V
		COM'L	2.7	3.4			COM'L	2.7	3.4		
V_{IL}	Logic LOW	MIL			0.7	Logic LOW	MIL			0.7	V
		COM'L			0.8		COM'L			0.8	
V_{IH}	Logic HIGH		2.0			Logic HIGH		2.0			V
I_{IL}	V_{IN} = 0.4V				−0.36	V_{IN} = 0.4V				−0.36	mA
I_{IH}	V_{IN} = 2.7V				20	V_{IN} = 2.7V				20	μA

Parameter	54S/74S AND 25S SCHOTTKY TTL Condition		Min.	Typ.	Max.	STANDARD TTL Condition	Min.	Typ.	Max.	Units
V_{OL}	I_{OL} = 20mA			0.3	0.5	I_{OL} = 16mA		0.2	0.4	Volts
V_{OH}	I_{OH} = −1.0mA	MIL	2.5	3.4		I_{OH} = −300μA	2.4	3.4		Volts
		COM'L	2.7	3.4						
V_{IL}	Logic LOW				0.8	Logic LOW			0.8	Volts
V_{IH}	Logic HIGH		2.0			Logic HIGH	2.0			Volts
I_{IL}	V_{IN} = 0.5V				−2.0	V_{IN} = 0.4V			−1.6	mA
I_{IH}	V_{IN} = 2.7V				50	V_{IN} = 2.4V			40	μA

INPUT/OUTPUT LEVELS

The input thresholds and output logic levels of LS circuits have been designed as far as possible to be compatible with those of standard TTL. Table 2 shows the guaranteed d.c. parameters of the Am54/74LS and second generation Am25LS families. Input current requirements (I_{IH}, I_{IL}) and therefore output drive needs (I_{OH}, I_{OL}) are significantly reduced over standard TTL.

A one unit load input current at logic HIGH, I_{IH}, for Am54LS/74LS is 20μA, compared with 40μA for Am54/74 standard TTL. Similarly at logic LOW, I_{IL} is reduced to −0.36mA from −1.6mA.

Corresponding reductions in the output drive requirements are I_{OL} = 4mA vs. 16mA at V_{OL} = 0.4V and I_{OH} = −400μA compared to 800μA.

FAN-OUT CAPABILITY

The fan-out capability of a logic family indicates the number of inputs which can be driven by a single output. It is defined as the maximum output drive current divided by the input current available.

Logic HIGH Fan-out = I_{OH}/I_{IH}
Logic LOW fan-out = I_{OL}/I_{IL}

Table 3 shows the fan-out capabilities of typical functions from the three families. The lower current operating levels of LS devices allow them to be specified at a logic LOW fan-out over the commercial range of more than twice that of standard TTL (22 vs. 10). The Am25LS family allows this advantage to be extended to the military range.

D.C. NOISE MARGIN

The D.C. noise margins of a digital system are defined from Figure 12 as follows:
  Logic HIGH Noise Margin = $V_{OH1} - V_{IH2}$
  Logic LOW Noise Margin = $V_{IL2} - V_{OL1}$

These parameters for LS devices are shown in Table 2. LS has a minimum logic HIGH output voltage of V_{OH} = 2.5V for military and 2.7V for the commercial temperature range. For standard TTL, V_{OH} is 2.4V. V_{IH} is 2.0V for both families.

Table 3 compares the guaranteed noise margin values for the standard TTL and LS devices. LS devices offer improved margin over standard TTL in the logic HIGH state, which is the most critical with regard to noise generation. At a similar fan-out, 10 for standard TTL and 11 for LS, noise margins in the LOW state are the same over the commercial range.

Designer's Guide

TABLE 3
FAN-OUT AND NOISE MARGIN
COMPARISON OF TTL AND LS FAMILIES.

a) LOGIC "HIGH" STATE

FAMILY	INPUT CURRENT I_{IH}	OUTPUT CURRENT I_{OH}	FAN-OUT MILITARY	FAN-OUT COMMERCIAL	NOISE MARGIN MILITARY	NOISE MARGIN COMMERCIAL
54/74	40µA	−800µA	20	20	400mV	400mV
54LS/74LS	20µA	−400µA	20	20	500mV	700mV
25LS	20µA	−440µA	22	22	500mV	700mV

b) LOGIC "LOW" STATE

FAMILY	INPUT CURRENT I_{IL}	OUTPUT CURRENT I_{OL}	FAN-OUT MILITARY	FAN-OUT COMMERCIAL	NOISE MARGIN MILITARY	NOISE MARGIN COMMERCIAL
54/74	−1.6mA	16mA	10	10	400mV	400mV
54LS/74LS	−0.36mA	4mA	11	11	300mV	400mV
		8mA	No Spec.	22	No Spec.	300mV
25LS	−0.36mA	4mA	11	11	300mV	400mV
		8mA	22	22	250mV	350mV

Military LS devices have a 100mV lower noise margin in the LOW state than standard TTL. In most systems, this does not present a problem as the lower power supply currents being switched with LS generally result in lower system noise generation.

The logic levels guaranteed over the operating temperature ranges are of course worst case. Figures 13 and 14 show the typical values to be considerably better than these.

Am25LS D.C. FEATURES

The D.C. advantages offered by second generation Am25LS over 54/74LS devices can be seen from Table 3 as:

1. In the logic LOW state at a fan-out of 22 (8mA), Am25LS has 50mV greater noise margin (350mV vs. 300mV).

2. Am25LS products are guaranteed at a fan-out of 22 (8mA) over the military range. Am54LS is specified at fan-out of 10 (4mA) only.

3. Am25LS offers a symmetrical fan-out of 22 in both logic HIGH and logic LOW states, allowing full use of the logic LOW drive capability.

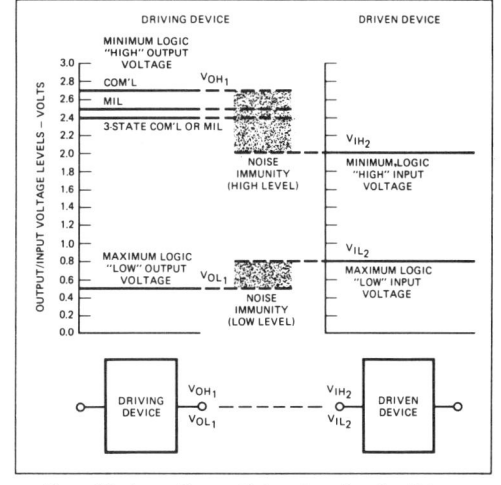

Figure 12. Input/Output Voltage Interface Conditions.

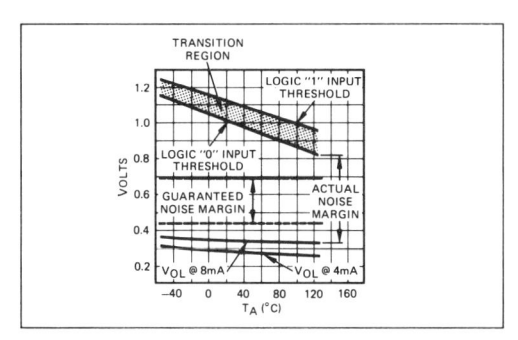

Figure 13. LS Logic "0" Noise Margin.

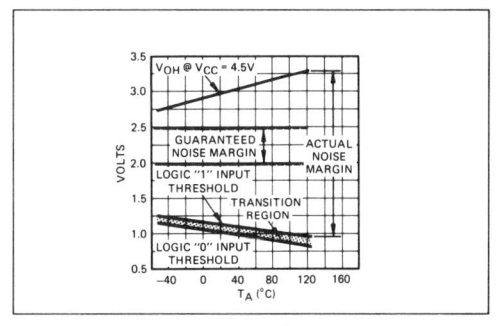

Figure 14. LS Logic "1" Noise Margin.

Designer's Guide

3. A.C. Characteristics

INTRODUCTION

Many Low-Power Schottky functions have been designed specifically to replace standard TTL elements in existing system designs. Their A.C. performance characteristics usually meet or exceed the limits of the earlier devices. The switching terms which are used on data sheets to describe the A.C. performance of these designs are summarized in Table 4. The more important parameters are discussed in detail in this section.

TABLE 4
DEFINITION OF SWITCHING TERMS

(All switching times are measured at the 1.3V logic level unless otherwise noted.)

- f_{MAX} The highest operating clock frequency.
- t_{PLH} The propagation delay time from an input change to an output LOW-to-HIGH transition.
- t_{PHL} The propagation delay time from an input change to an output HIGH-to-LOW transition.
- t_{PW} Pulse width. The time between the leading and trailing edges of a pulse, measured at the 50% points.
- t_r Rise time. The time required for a signal to change from 10% to 90% of its measured values.
- t_f Fall time. The time required for a signal to change from 90% to 10% of its measured values.
- t_s Set-up time. The time interval for which a signal must be applied and maintained at one input terminal before an active transition occurs at another input terminal.
- t_h Hold time. The time interval for which a signal must be retained at one input after an active transition occurs at another input terminal.
- t_R Release time. The time interval for which a signal may be indeterminant at one input terminal before an active transition occurs at another input terminal. (The release time falls within the set-up time interval and is specified by some manufacturers as a negative hold time).
- t_{HZ} also t_{PHZ} HIGH to disable. The delay time from a control input change to the three-state output HIGH-level to high-impedance transition (measured at 0.5V change).
- t_{LZ} also t_{PLZ} LOW to disable. The delay time from a control input change to the three-state output LOW-level to high-impedance transition (measured at 0.5V change).
- t_{ZH} also t_{PZH} Enable HIGH. The delay time from a control input change to the three-state output high-impedance to HIGH-level transition.
- t_{ZL} also t_{PZL} Enable LOW. The delay time from a control input change to the three-state output high-impedance to LOW-level transition.

PROPAGATION DELAYS

The standard designations for delays through combinatorial logic networks are t_{PHL} and t_{PLH}. A delay from an input change to an output going LOW is called t_{PHL}, while t_{PLH} is the delay from an input change to an output going HIGH.

Figure 15 shows a typical waveform with the output changing during the interval indicated by the diagonal, sloping line. Note that all switching times shown are measured at the 1.3 volt logic level.

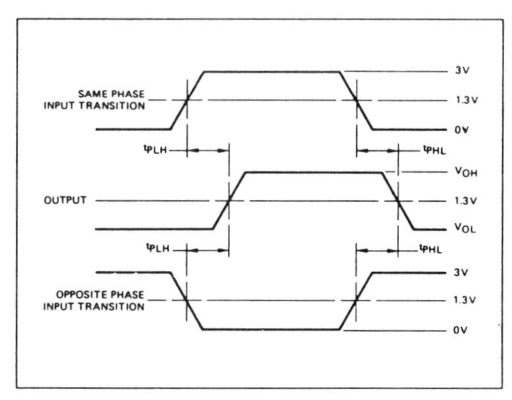

Figure 15. Propagation Delay.

Typical values for a single gate propagation delay t_{PHL} in Low-Power Schottky functions are 8–10ns into a 15pF load. Higher performance LS families, such as Am25LS, exhibit delays in the 4–6ns range. These propagation delays will increase by 2–4ns at an output loading of 50pF or approximately 0.1ns per pF. See Figure 16.

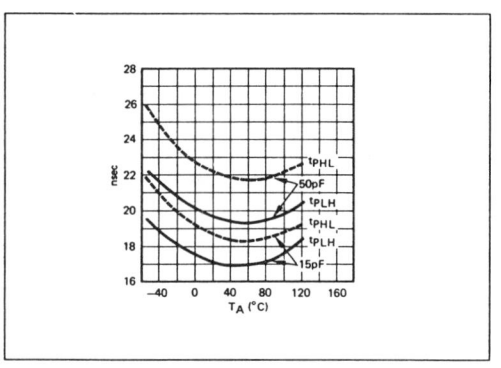

Figure 16. Am25LS138 Typical Propagation Delays Address to Output (3 Levels).

Table 5 shows the worst case delays through typical two and three deep gate MSI functions such as multiplexers and decoders. Speed improvements attainable with the Am25LS higher performance LS devices at this level of complexity are shown to be in the range of 20 to 40%. Guaranteed delays into 50pF loads are being specified on all new Am25LS data sheets. See Table 8.

2-14

Designer's Guide

TABLE 5
COMPARISON OF AC PARAMETERS (T_A = +25°C)

LS138 3-Line to 8-Line Decoder/Demultiplexer

Parameters	Description	Test Conditions	Am25LS138 Min.	Am25LS138 Max.	Am54LS138 / Am74LS138 Min.	Am54LS138 / Am74LS138 Max.	Units
t_{PLH}	Two Level Delay Select to Output			15		20	ns
t_{PHL}				21		41	
t_{PLH}	Three Level Delay Select to Output	V_{CC} = 5.0V, R_L = 2kΩ, C_L = 15pF		23		27	ns
t_{PHL}				27		39	
t_{PLH}	G2A or G2B to Output			15		18	ns
t_{PHL}				23		32	
t_{PLH}	G1 to Output			18		26	ns
t_{PHL}				27		38	

LS158 Quadruple 2-Line to 1-Line Data Selectors/Multiplexers

Parameters	Description	Test Conditions	Am25LS158 Min.	Am25LS158 Max.	Am54LS158 / Am74LS158 Min.	Am54LS158 / Am74LS158 Max.	Units
t_{PLH}	Data to Output			9		12	ns
t_{PHL}				11		12	
t_{PLH}	Strobe to Output	V_{CC} = 5.0V, R_L = 2kΩ, C_L = 15pF		12		17	ns
t_{PHL}				17		18	
t_{PLH}	Select to Output			20		20	ns
t_{PHL}				21		24	

EDGE RATES

The rise and fall times of Low-Power Schottky devices are similar to those of standard TTL. Into a 50pF load fall time, t_f, is typically 6–8ns, while rise time, t_r, is in the 9–12ns range. A.C. parameters are measured at $t_f \leqslant$ 6ns and $t_r \leqslant$ 15ns.

As with standard TTL, careful P.C. board layout rules should be employed to avoid problems which can occur at these relatively fast edge rates. In particular, precautions should be taken to insure that transmission line effects do not cause false switching or ringing and oscillation problems on lines longer than 18 inches. See Section 4 for more information.

SEQUENTIAL DEVICES

Set-up time, t_s, hold time, t_h, and release time, t_R, are the most important parameters for specifying sequential elements such as latches, flip-flops and registers.

For these synchronous devices, inputs must be stable for a certain period of time before the clock or enable pulse. This interval is the region in time during which devices are "sampling" their inputs. As an example, consider a latch with a D input and an active LOW clock. The latch will store the information present on its input just before the clock goes HIGH. The question is, how long does the input level have to be present and stable before the clock goes HIGH? A particular device will "sample" its input at some exact instant, but in a group of devices some are slower than others. The result is an interval of some time called set-up time during which all devices, fast or slow, will "sample" their inputs.

All devices exhibit a hold time. That is a period of time after the clock or enable pulse transition during which the data cannot be changed without loss of input intelligence. This hold time occurs after the clock goes HIGH. Figure 17 shows the input requirements and definitions for data entry. Release time is negative hold time or the time period prior to the clock input after which the data can be released. Typical examples of LS characteristics and the improvements attainable with high performance Am25LS sequential devices are shown in Table 6.

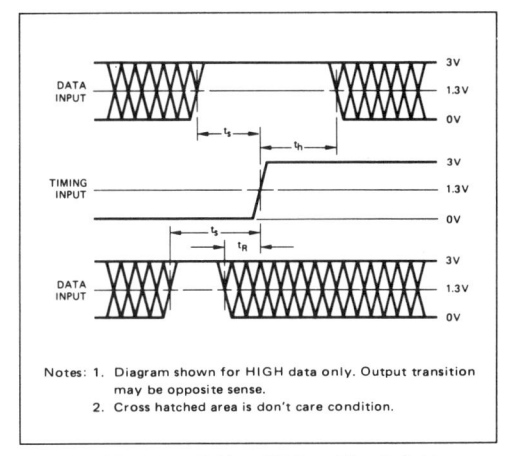

Notes: 1. Diagram shown for HIGH data only. Output transition may be opposite sense.
2. Cross hatched area is don't care condition.

Figure 17. Set-up, Hold, and Release Time Definitions.

f_{MAX}.

A frequently misunderstood parameter on data sheets is maximum clock frequency f_{MAX}. This was defined by the early TTL manufacturers as the maximum toggle frequency which can be attained by the device under ideal conditions with no constraints on t_r, t_f, pulse width, or duty cycle. Although f_{MAX} as specified cannot usually be attained in an operating system, it is a relatively easy parameter to test and provides a convenient measure of comparative performance between different devices. For instance, Table 6 shows the Am54/74LS174 at f_{MAX} = 30MHz (min.) while the high-performance Am25LS is specified at 40MHz (min.). Actual toggle frequency in a system must be determined from the specific signal conditions presented to the device.

2-15

Designer's Guide

TABLE 6
SWITCHING CHARACTERISTICS ($T_A = 25°C$)

LS174/LS175 Hex/Quadruple D-Type Flip-Flops with Clear

Parameters	Description		Test Conditions	Am25LS174 Am25LS175 Min.	Am25LS174 Am25LS175 Max.	Am54LS174 Am54LS175 Am74LS174 Am74LS175 Min.	Am54LS174 Am54LS175 Am74LS174 Am74LS175 Max.	Units
t_{PLH}	Clock to Output				23		30	ns
t_{PHL}					22		35	
t_{PLH}	Clear to $\overline{Q}$ Output, LS175 only				25		25	ns
t_{PHL}	Clear to Output				35		35	
t_{pw}	Pulse Width	Clock	$V_{CC} = 5.0V$, $R_L = 2k\Omega$, $C_L = 15pF$	17		20		ns
		Clear		20		20		
t_s	Data Set-up Time			20		20		ns
t_s	Set-up Time Clear Recovery (in-active) to Clock			20		25		ns
t_h	Data Hold Time			5		5		ns
f_{MAX}	Maximum Clock Frequency			40		30		MHz

EFFECTS OF TEMPERATURE AND POWER SUPPLY VARIATIONS

Standard TTL devices exhibit severe degradation in A.C. performance towards the recommended limits of the operating temperature and power supply voltage ranges.

At elevated temperature and/or high V_{CC} levels, charge storage begins to slow down A.C. response. At the other extreme, low temperature and/or low V_{CC}, the loss of β causes a similar problem. These combined effects can cause more than 50% degradation in performance over the full military temperature and power supply extremes.

As noted in Section 1, Low-Power Schottky technology reduces the impact of both of these effects on performance. β degradation at cold temperatures is far less severe and Schottky clamping largely eliminates the effects of charge storage at high temperature.

General guidelines for variation in the AC response over temperature and power supply variations are not easy to specify. Typical measured variations for a combinatorial and a sequential device are shown in Figures 16 and 18.

The system's designer would like a factor which will allow his system to meet specification with minimum design overkill. However, the component engineer often requires maximum delays to be guaranteed. For system design guidelines, the AC derating factors of Table 7 may be useful.

It must be emphasized that the values of Table 7 are typical. However as it is unlikely that any given system will contain all worst case devices they will usually yield a fairly safe prediction of the system performance which can be achieved.

Individual components will of course be slower than these typical numbers. These must be reflected on procurement specifications. A general rule of thumb would be to double the system design guidelines of Table 7. New Am25LS specifications are now being published with worst case parameters guaranteed over the operating power supply and temperature ranges, as well as at a realistic system load condition of 50pF. A typical example of this format is shown in Table 8.

SHORT CIRCUIT OUTPUT CURRENT

To improve performance, in 1975 TI lowered the short-circuit current limiting resistor value. This increased the I_{SC} (I_{OS}) range from -6 to -42mA up to -30 to -130mA. The overall

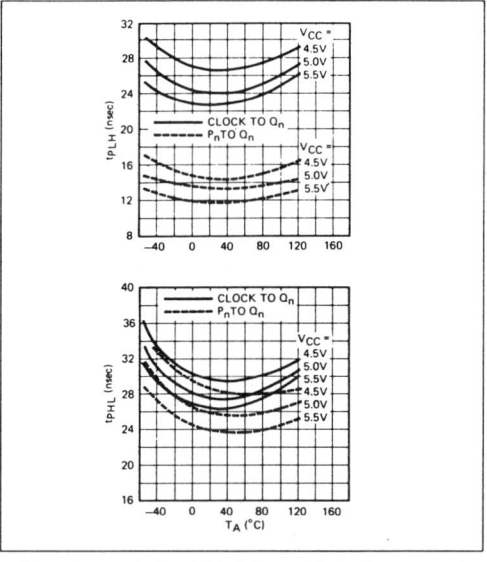

Figure 18. Typical A.C. Variations with Temperature and Power Supply for Am25LS193 Counter.

TABLE 7
GUIDELINES FOR TYPICAL VARIATION OF A.C. PARAMETERS WITH COMBINED TEMPERATURE AND V_{CC} VARIATION

Temperature Range	V_{CC} Variation (Nominal 5V)	AC Derating Factor System	AC Derating Factor Component
COM'L, 0°C to +70°C	None	5%	10%
COM'L, 0°C to +70°C	±0.25V	15%	30%
MIL, −55°C to +125°C	None	15%	30%
MIL, −55°C to +125°C	±0.5V	25%	50%

Designer's Guide

4. Design Guidelines

POWER SUPPLY CONSIDERATIONS

The recommended power supply voltage (V_{CC}) for all TTL circuits, including LS, is +5V. Commercial temperature range devices, designated 74LS or in the case of Am25LS with the suffix C, are specified with a ±5% supply tolerance (±250mV) over the ambient range 0°C to 70°C. Military range parts, designated 54LS or in the case of Am25LS with the suffix M, are guaranteed with a ±10% supply tolerance (±500mV) over an ambient temperature range of −55°C to +125°C. The power supply should be well regulated with a ripple less than 5% and with regulation better than 5%. Even though LS devices generate significantly smaller power supply spikes when switching than standard TTL, on-board regulation is still preferable to isolate this noise to one board.

A low-inductance transmission line power distribution bus with good RF decoupling is necessary for large systems. On all boards, ceramic decoupling capacitors of 0.01µF to 0.1µF should be used at least one for every five packages, and one for every one-shot (monostable), line driver and line receiver package. In addition, a larger tantalum capacitor of 20µF to 100µF should be included on each card. On boards containing a large number of packages, a low impedance ground system is essential. The ground can either be a bus or a ground which is incorporated with the V_{CC} supply to form a transmission line power system. Separate power transmission systems can be attached to the board to provide this same feature without the cost of a multi-layer PC card.

UNUSED INPUTS

An unused input to an AND or NAND gate should not be left floating as it can act as an antenna for noise. On devices with storage, such as latches, registers and counters, it is particularly important to terminate unused inputs (MR, PE, PL, CP) properly since a noise spike on these inputs might change the contents of the memory. This technique optimizes switching speed as the distributed capacitance associated with the floating input, bond wire and package leads is eliminated. To terminate, the input should be held between 2.4V and the maximum input voltage. One method of achieving this is to connect the unused input to V_{CC}. Most LS inputs have a breakdown voltage >7V and require no series resistor. Devices specified with a maximum 5.5 volt breakdown should use a 1kΩ to 10kΩ current limiting series resistor to protect against V_{CC} transients. Another method is to connect the unused input to the output of an unused gate that is forced HIGH. Do not connect an unused input to another input of the same NAND or AND function. Although recommended for standard TTL, with LS this increases the input coupling capacitance and reduces A.C. noise immunity.

TRANSMISSION LINE EFFECTS

The relatively fast rise and fall times of Low-Power Schottky TTL (5 to 15ns) can cause transmission line effects with interconnections as short as 18 inches. With one TTL device driving another and the driver switching from LOW to HIGH, if the propagation delay of the interconnection is long compared to the signal rise time, the arrangement can behave like a transmission line driven by a generator with a non-linear output.

The initial voltage step at the output, just after the driver has switched, propagates down the line and reflects at the end. In the typical case where the line is open ended or terminated in an impedance greater than its characteristics impedance (Z_{OL}), the reflected wave arrives back at the source and increases V_{OUT}. If the total round-trip delay is longer than the rise time of the driving signal, a staircase response results at the driver output and along the line. If one of the driven devices is connected close to the driver, the initial output voltage (V_{OUT}) seen by it might not exceed V_{IH}. The state of the input is undetermined until after the round trip of the transmission line, thus slowing down the response of the system.

The longest interconnection that should be used with LS devices without incurring problems due to line effects is in the 10–12 inch range.

With longer interconnections, transmission line techniques should be used for maximum speed. Good system operation can be obtained by designing around 100 ohm lines. A 0.026 inch (0.65mm) trace on a 0.062 inch epoxy-glass board (E_r = 4.7) with a ground plane on the other side represents a 100Ω line. 28 to 30 gauge wire (0.25 to 0.30mm) twisted pair line has a characteristic impedance of 100 to 115Ω.

LINE DRIVING AND RECEIVING

For lines longer than 2 feet, twisted pairs of coaxial cable should be used. The characteristic impedance or the transmission media should be approximately 120Ω such as twisted pairs of #26 wire or 100Ω coax. A possible choice is cables with a characteristic impedance R_0 of 100Ω such as ribbon cable or flat cable with controlled impedance. Resistive pull-ups at the receiving end can be used to increase noise margin. Where reflection effects are unacceptable, the line must be terminated in its characteristic impedance. A method shown in Figure 19

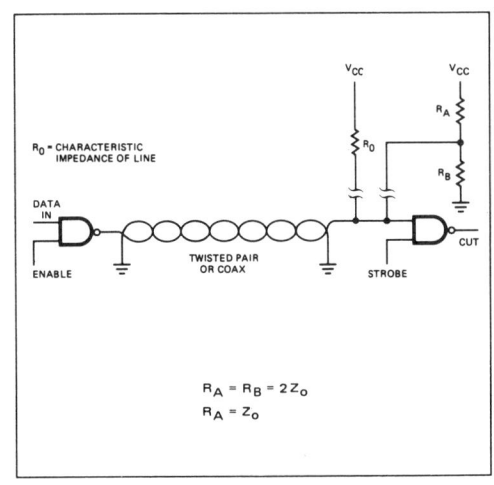

$R_A = R_B = 2Z_o$
$R_A = Z_o$

Figure 19. LS Driving Twisted Pair.

delay when driving very large capacitive loads (>150pF) was reduced somewhat as a result. However, the inherent circuit performance still dominates in normal applications such that the Am25LS and other high performance families remain faster even when driving large capacitive loads.

As an attempt to offer standardized specifications, most manufacturers, including Advanced Micro Devices, Fairchild, Motorola, Raytheon, and Signetics, also lowered their short-circuit current limiting resistor values on new designs to provide a typical I_{SC} of −60mA. Most manufacturers now specify −15 to −100mA to accommodate both old and new circuits. The maximum value of −100mA was chosen, as −130mA was felt to be too high for a noise sensitive system design. The Am25LS high performance family is specified even tighter, with the maximum I_{SC} limited to −85mA.

Early in 1977 TI changed their data sheets yet again to specify I_{SC} from −20mA to −100mA on regular outputs and −30mA to −130mA on three-state outputs.

TABLE 8
Am25LS2513 THREE-STATE PRIORITY ENCODER
A.C. SPECIFICATION FORMAT FOR V_{CC} AND TEMPERATURE EXTREMES AND 50pF LOAD CONDITION

SWITCHING CHARACTERISTICS
(T_A = +25°C, V_{CC} = 5.0V)

Parameters	Description	Test Conditions	Min.	Typ.	Max.	Units
t_{PLH}	$\overline{I}_i$ to An (In-phase)			17	25	ns
t_{PHL}				17	25	
t_{PLH}	$\overline{I}_i$ to An (Out-phase)			11	17	ns
t_{PHL}				12	18	
t_{PLH}	$\overline{I}_i$ to $\overline{EO}$			7.0	11	ns
t_{PHL}				24	36	
t_{PLH}	$\overline{EI}$ to $\overline{EO}$	C_L = 15pF		11	17	ns
t_{PHL}		R_L = 2.0kΩ		23	34	
t_{PLH}	$\overline{EI}$ to An			12	18	ns
t_{PHL}				14	21	
t_{ZH}	G_1 or G_2 to An			23	40	ns
t_{ZL}				20	37	
t_{ZH}	$\overline{G}_3, \overline{G}_4, \overline{G}_5$ to An			20	30	ns
t_{ZL}				18	27	
t_{HZ}	G_1 or G_2 to An	C_L = 5.0pF		17	27	ns
t_{LZ}		R_L = 2.0kΩ		19	28	
t_{HZ}	$\overline{G}_3, \overline{G}_4, \overline{G}_5$ to An			16	24	ns
t_{LZ}				18	27	

SWITCHING CHARACTERISTICS OVER OPERATING RANGE

			Am25LS COM'L		Am25LS MIL		
			T_A = 0°C to +70°C V_{CC} = 5.0V ±5%		T_A = −55°C to +125°C V_{CC} = 5.0V ±10%		
Parameters	Description	Test Conditions	Min.	Max.	Min.	Max.	Units
t_{PLH}	$\overline{I}_i$ to An (In-phase)			31		37	ns
t_{PHL}				30		34	
t_{PLH}	$\overline{I}_i$ to An (Out-phase)			22		27	ns
t_{PHL}				22		25	
t_{PLH}	$\overline{I}_i$ to $\overline{EO}$			15		18	ns
t_{PHL}				48		60	
t_{PLH}	$\overline{EI}$ to $\overline{EO}$	C_L = 50pF		19		21	ns
t_{PHL}		R_L = 2.0kΩ		46		57	
t_{PLH}	$\overline{EI}$ to An			22		25	ns
t_{PHL}				27		32	
t_{ZH}	G_1 or G_2 to An			42		49	ns
t_{ZL}				43		49	
t_{ZH}	$\overline{G}_3, \overline{G}_4, \overline{G}_5$ to An			36		43	ns
t_{ZL}				35		43	
t_{HZ}	G_1 or G_2 to An	C_L = 5.0pF		34		40	ns
t_{LZ}		R_L = 2.0kΩ		34		40	
t_{HZ}	$\overline{G}_3, \overline{G}_4, \overline{G}_5$ to An			30		35	ns
t_{LZ}				31		35	

has the output of the line tied to V_{CC} through a resistor equivalent to the characteristic impedance of the line. As the output impedance of the LS driver is low and must sink the current through it, in addition to the current from the inputs being driven, a useful technique is to terminate the line in a voltage divider with two resistors, each twice the line impedance. This reduces the extra sink current by 50%. Where the line exceeds five feet in length it is preferable to dedicate gates solely to line driving.

For additional noise immunity when driving long lines, a differential line driver and line receiver may be used. These dedicated line interface circuits drive a twisted pair of wires differentially, permit easy termination of lines and provide excellent common mode noise rejection.

The Am26LS31 driver and Am26LS32 and Am26LS33 are quad differential line drivers and receivers satisfying the interface requirements of EIA RS-422 and 423 as well as military applications, Figure 20. They are designed to operate off the standard 5V power supplies of the LS logic devices. More applications information on line termination techniques is provided on the above mentioned device data sheets.

CROSS-TALK AND RINGING

These two problems may be experienced with all forms of high speed digital logic. Crosstalk is the coupling of energy from one circuit to another via real or parasitic capacitance and inductance. Ringing is the possible rebound of the signal into the input threshold region (0.8 – 2.0V) following a HIGH-to-LOW level change. When a driver switches from a HIGH-to-LOW state the output voltage should fall below the threshold value. However, a line having a very low characteristic impedance does not allow transistor Q5 in the NAND gate example to saturate, and the resulting output voltage may not be low enough to switch an adjacent device until two or more line delay times. The low current levels at which LS devices operate, coupled with the low output impedance in both HIGH and LOW Logic states, minimize crosstalk effects. Input clamp diodes provided on all LS devices are extremely effective in reducing ringing phenomenon.

Care should be taken to insure that signals with falling edges faster than 2.5-3ns/volt are not coupled into the input of an LS function. Even though the signal may not pass into the threshold region, if the pulse edge is fast enough, sufficient energy may be capacitively coupled into a sequential device to cause it to change state: High speed Schottky elements in a test setup can exceed this limit. However in an active system, the edges will generally be slowed sufficiently to eliminate any problem.

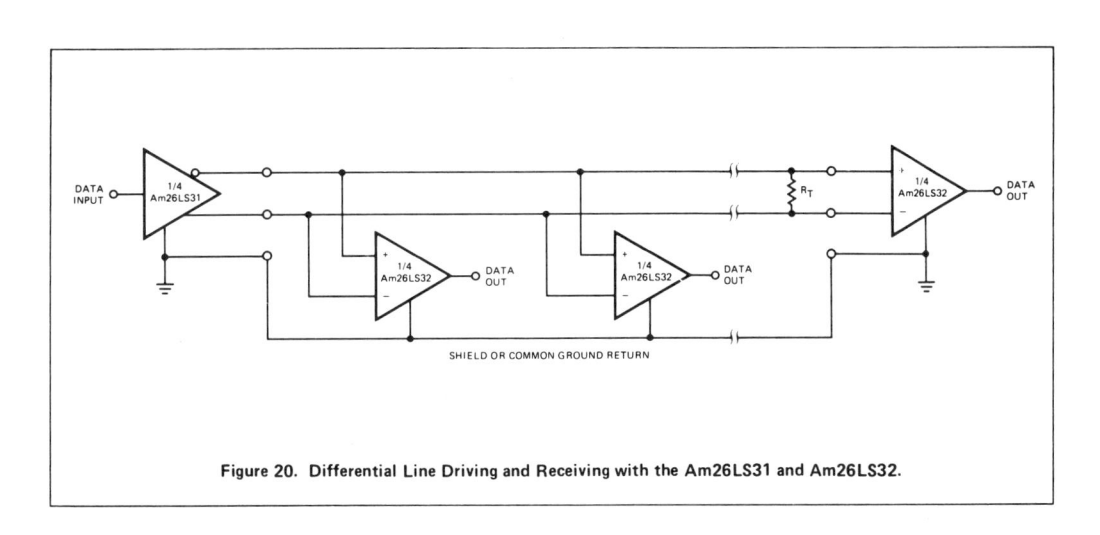

Figure 20. Differential Line Driving and Receiving with the Am26LS31 and Am26LS32.

Philips Components

AN219
A Metastability Primer

Application Note September 1989

Standard Products

Author: Charles Dike

INTRODUCTION

When using a latch or flip-flop in normal circumstances (i.e. when the device's setup and hold times are not being violated) the outputs will respond to a latch enable or clock pulse within some specified time. These are the propagation delays found in the data sheets. If, however, the setup and hold times are violated so that the data input is not a clear one or zero, there is a finite chance that the flip-flop will not immediately latch a high or low but get caught half way in between. This is the metastable state and it is manifested in a bi-stable device by the outputs glitching, going into an undefined state somewhere between a high and low, oscillating, or by the output transition being delayed for an indeterminable time.

Once the flip-flop has entered the metastable state, the probability that it will still be metastable some time later has been shown to be an exponentially decreasing function. Because of this property, a designer can simply wait for some added time after the specified propagation delay before sampling the flip-flop output so that he can be assured that the likelihood of metastable failure is remote enough to be tolerable. On the other hand one consequence of this is that there is some probability (albeit vanishingly small) that the device will remain in a metastable state forever. The designer needs to know the characteristics of metastability so that he can determine how long he must wait to achieve his design goals.

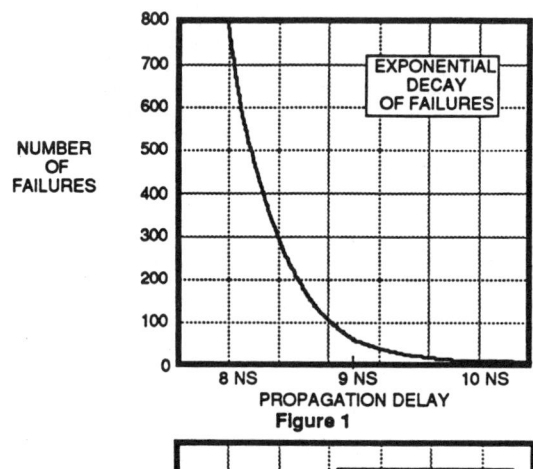

THE CHARACTERISTICS OF METASTABILITY

In order to define the metastability characteristics of a device three things must be known: first, what is the likelihood that the device will enter a metastable state? This propensity is defined by the parameter 'T_0'. Second, once the device is in a metastable state how long would it be expected to remain in that state? This parameter is tau (τ) and is simply the exponential time constant of the decay rate of the metastability. It is sometimes called the metastability time constant. The final parameter is the measured propagation delay of the device. Commonly, the typical propagation delays found in the data book are used for this and it is designated 'h' in the equations (although most designers are familiar with this value as Tpd). Now let's see how tau and T_0 are determined by measurements.

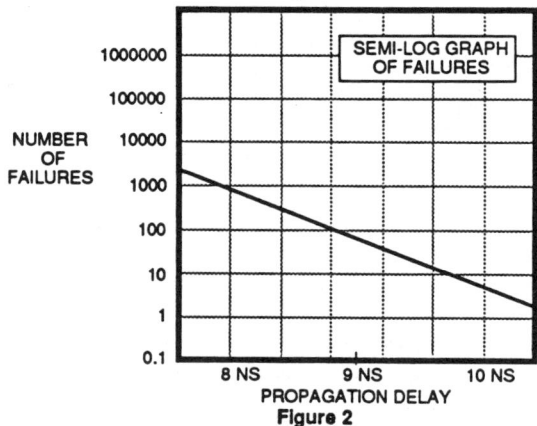

A TEST METHOD

Suppose we wanted to measure the metastability characteristics of a fictitious edge-triggered D-type flip-flop and we had a test system that would count each time the flip-flop is found in a metastable state at some time after a clocking edge. The first thing we would like to know about the flip-flop would be the h or typical propagation delay. We could measure the delay or look it up in the data book (of course, measuring the actual delay would allow more precise results). This fictitious flip-flop has an h of 7 ns. In this test we decide to use a clock frequency of 10 MHz. This frequency is primarily a function of the test systems ability to assimilate the information. The data will run at 5 MHz asynchronously to the clock and with a varying period. This frequency was

Figure 1 — EXPONENTIAL DECAY OF FAILURES (Number of Failures vs Propagation Delay, 8 NS to 10 NS)

Figure 2 — SEMI-LOG GRAPH OF FAILURES (Number of Failures vs Propagation Delay, 8 NS to 10 NS)

A Metastability Primer

Application Note

AN219

chosen because at two transitions per cycle the data signal produces 10 million points each second where it is possible for the flip-flop to go into a metastable state, an average of one point for each clock pulse. An important point about the characteristic of the data signal in relation to the clock is that the data transitions must have an equal probability of occurring anywhere within the clock period or the results could be skewed. In other words, we need to have a uniform distribution of random data transitions (high and low) relative to the clocking edge.

The first measurement we take is to determine the number of times the device is still in a metastable state 8 ns after the clock edge. With this device there are 792 failures after 1 billion clock cycles. Changing the time to 9 ns we measure 65 failures after another 1 billion cycles. Because metastability resolves as an exponentially decaying function the two points define the exponential curve and they can be plotted as shown in Figure 1. An equivalent plot can be made using a semilog scale as in Figure 2. The slope of the line drawn through the two points represents tau. With these two points the tau can be determined by equation (1):

$$(1) \quad \tau = \frac{t_2 - t_1}{\ln(N_1 / N_2)}$$

where N_1 and N_2 are the number of failures at times t_1 and t_2, respectively.

Working thru the numbers gives us a tau of 0.40 ns. Tau of this order is representative of the FAST line of flip-flops.

Earlier we stated that T_0 is an indicator of the likelihood that the device will enter a metastable state. Now we will attempt to explain it. At 9 ns after the clock we observed 65 failures in 1 billion clock cycles. Since the data transits on average once per clock cycle and the period of this clock is 100 ns, from equation (2) we can say that there appears to be an aperture about 0.0065 picoseconds wide at the input of the device that allows metastability to occur for 9 or more nanoseconds. Another way of explaining the same thing would be to suppose that if 1 billion data

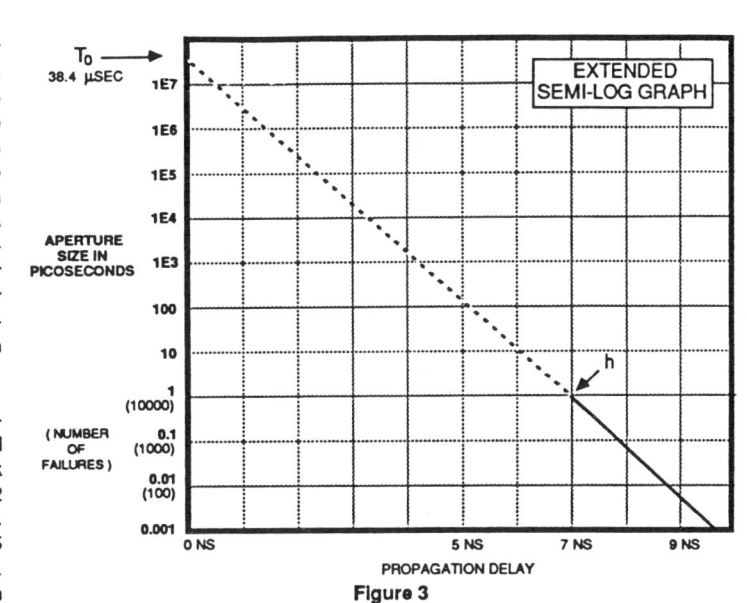

Figure 3

transitions were uniformly and randomly distributed over a clock period of 100 ns: you would expect 65 of these transitions to cause the outputs to go into a metastable state and remain there for at least 9 ns.

$$(2) \quad T_9 = \frac{N_9 P_C}{N_{C9}}$$

Where N_{C9} is the number of clocking events at 9 ns (in this instance, 1 billion), P_C is the period of the clock, and N_9 is the number of failures recorded at 9 ns.

By the same reasoning the window at 8 ns appears to be 0.0792 picoseconds wide. It seems to have grown because there are, of course, more failures after 8 ns than after 9 ns. This aperture has been normalized by researchers to indicate the effective size of the aperture at the clock edge, or time zero. Unfortunately the normalization process tends to obscure the interpretation of T_0. T_0 can be calculated using equation (3). Figure 3 is an extension of Figure 2 and shows the relationship of T_0, h, and tau.

$$(3) \quad T_0 = T_8 e^{\left(\frac{8ns}{\tau}\right)}$$

or equivalently,

$$T_0 = T_9 e^{\left(\frac{9ns}{\tau}\right)}$$

In this case T_0 is 38.4 microseconds and this value is again typical of the FAST line of products.

Figure 3 is an extension of Figure 2 and gives a graphic indication of T_0. The number of failures plots on the same scale as the aperture size but the number of failures is dependent on the number of clock cycles used in the test (we always used 1 billion in this paper) and the ratio of data transitions to clock pulses (1:1 in this paper). On the other hand, the aperture size is independent of these things.

MTBF

Having determined the T_0 and tau of the flip-flop, calculating the mean time between failures (MTBF) is simple. Suppose a designer wants to use the flip-flop for synchronizing asynchronous data that is arriving at 10 MHz, he has a clock frequency of 25 MHz, and has decided that he would like to sample the output of the flip-flop 15 ns after the clock edge. He simply plugs his numbers into equation (4).

$$(4) \quad MTBF = \frac{e^{\left(\frac{t'-8ns}{\tau}\right)}}{T_0 f_C f_i}$$

Application Note

A Metastability Primer

AN219

In this formula f_c is the frequency of the clock, f_i is the average input event frequency, and t' is the time after the clock pulse that the output is sampled (of course t'>h). In this situation the f_i will be twice the data frequency because input events consist of both low and high data transitions. For the numbers above the MTBF is one million seconds or about one failure every 11.6 days. If the designer would have tried to sample the data after only 10 ns the MTBF would have been 3.8 seconds.

Metastability literature can be very confusing because several companies use different nomenclature and often the fundamental parameters are obscured by scale factors, so it is important that the user understand MTBF. Let's try a thought experiment to determine the correct MTBF formula. We know the size of the aperture at 8 ns so we need to know how often that window will occur. This is supplied by the clock period. This gives a ratio of window size to clock period and gives us the likelihood of a transition within the clock period causing a metastable state that lasts beyond the 8 ns point. Now we need to know the number of input events per clock period to determine the MTBF at 8 ns. This is supplied by the average input event period and produces the equation below where P_c and P_i are the periods of the clock and input events, respectively.

$$(5) \quad MTBF = \frac{1}{T_8 \frac{1}{P_c}\frac{1}{P_i}} = \frac{1}{T_8 f_c f_i}$$

This gives the MTBF for 8 ns, but how can the formula be developed to handle other times? It has been stated in this paper that the rate of decay of metastable events is an exponential function with a time constant of tau. Using this information gives the equation below where t' is the time after the clock pulse that the output is sampled.

$$(6) \quad MTBF = \frac{e^{\left(\frac{t'-8ns}{\tau}\right)}}{T_8 f_c f_i} = \frac{e^{\left(\frac{t'}{\tau}\right)}}{T_8 e^{\left(\frac{8ns}{\tau}\right)} f_c f_i}$$

$$= \frac{e^{\left(\frac{t'}{\tau}\right)}}{T_0 f_c f_i}$$

A point should be made here about MTBF. This is the mean time between failures and as such does not indicate the average time between failures. In fact, in this situation, the MTBF is the time before which there is a 63.2% probability that a failure would have occurred. Suppose a device has an MTBF of one million seconds like the example above; because the MTBF is an exponential function there is a 9.5% probability that a failure will occur in the first 1.16 days of operation. This might cause the user to feel that the device is failing more than expected. The user would find that 50% of his failures would occur within 8 days. Figure 4 gives a visual interpretation of this idea: time constant one represents one million seconds in this case.

RECENT DEVELOPMENTS

The quest for better metastability characteristics in flip-flops has recently resulted in the development of flip-flops with taus significantly less than 0.40 ns. Perhaps the most notable of these is the Signetics 74F50XXX series with typical taus of 135 ps. The specifications of these new products can cause confusion among the uninitiated because the typical T_0 on these devices is 9.8 million seconds or about 113 days. This is an example of how the normalization process obscures the interpretation of T_0. In the newest products the taus have decreased faster than the normal propagation delays primarily due to speed limitations of the outputs.

Using the example above and calculating T_7 from equation (3) we see that the window at h is 0.965 ps. Now let's assume that we have a device with the same size window (0.965 ps) at h and an h of 7 ns. The difference between this device and the previous example is that this device has a tau of 150 ps. Clearly, if the device has the same h and the same size of window at h but a smaller tau, the device is better. But let's calculate the T_0.

$$T_0 = T_7 e^{\left(\frac{7ns}{\tau}\right)}$$

T_0=178 million seconds!

Comparing the T_0 of any two devices does not show which device is superior. However, one can expect that the device with the lower tau is superior in all but the most peculiar circumstances.

SUMMARY

This paper is intended to introduce the reader to the terms he will be dealing with regarding metastability and it is hoped that this introduction will help him to digest the more in-depth papers that he will be reading. Signetics uses the parameters described by Thomas Chaney of Washington University in St. Louis, Missouri because they are fundamental and the better metastability papers generally use these parameters. For further reading on the subject, the article "Metastable behavior in digital systems" by Lindsay Kleeman and Antonio Cantoni published in *IEEE Design & Test of Computers* December of 1987, is recommended.

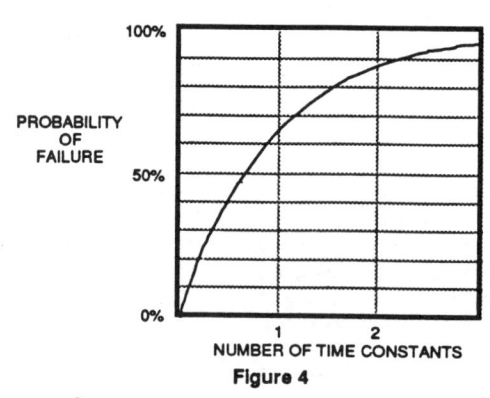

Figure 4

DEFINITIONS

Data Sheet Identification	Product Status	Definition
Objective Specification	**Formative or In Design**	This data sheet contains the design target or goal specifications for product development. Specifications may change in any manner without notice.
Preliminary Specification	**Preproduction Product**	This data sheet contains preliminary data and supplementary data will be published at a later date. Signetics reserves the right to make changes at any time without notice in order to improve design and supply the best possible product.
Product Specification	**Full Production**	This data sheet contains Final Specifications. Signetics reserves the right to make changes at any time without notice in order to improve design and supply the best possible product.

Signetics reserves the right to make changes, without notice, in the products, including circuits, standard cells, and/or software, described or contained herein in order to improve design and/or performance. Signetics assumes no responsibility or liability for the use of any of these products, conveys no license or title under any patent, copyright, or mask work right to these products, and makes no representations or warranties that these products are free from patent, copyright, or mask work right infringement, unless otherwise specified. Applications that are described herein for any of these products are for illustrative purposes only. Signetics makes no representation or warranty that such applications will be suitable for the specified use without further testing or modification.

Philips Components Ltd., Mullard House, Torrington Place, London, WC1E 7HD. Tel: 01-580 6633 Telex: 264341 Fax: 01-636 0394
M89-1389/CC

PHILIPS

AN216
Arbitration In Shared Resource Systems

Application Note

INTRODUCTION

The need for more powerful and faster systems gave birth to multiprocessing and multitasking systems. But to achieve this, cost and reliability were not to be sacrificed. To reduce cost it is vital to share resources, but to do so requires reliable means of arbitration. In a multiprocessing system, a single bus may be shared between various processors or intelligent peripherals. The resources shared by processors (Figure 1) are generally termed as global resources and those shared between the local processor and the peripherals (Figure 2) are typically known as local resources. Whether local or global, there always exists a protocol that will connect and disconnect various devices to and from the shared resources. Various bus architectures in existence today have different ways of doing this.

No matter what the protocol of a specific bus, there is always a method which dictates how arbitration shall be performed between two or more devices. Some systems employ synchronous arbitration and some use an asynchronous approach. The third option is not to use arbitration at all, but instead to employ time-multiplexing. This is used mainly in data communications by dividing the common media into various time slots. Each processor (station) is assigned a predetermined time for using the media. If the station does not need to use the media during its assigned time-slot, it may pass control to the next station. This obviously results in an inefficient use of the bus bandwidth.

Synchronous and asynchronous arbitration have their advantages and disadvantages, and are both used in system designs. Some applications may even use a combination of the two. Generally, synchronous arbitration is used in systems where the designer can take the time to synchronize signals with the master clock. In synchronous arbitration the request is sampled on a clock edge, and therefore if it is asserted close to, but after the sampling clock edge, it will not be recognized

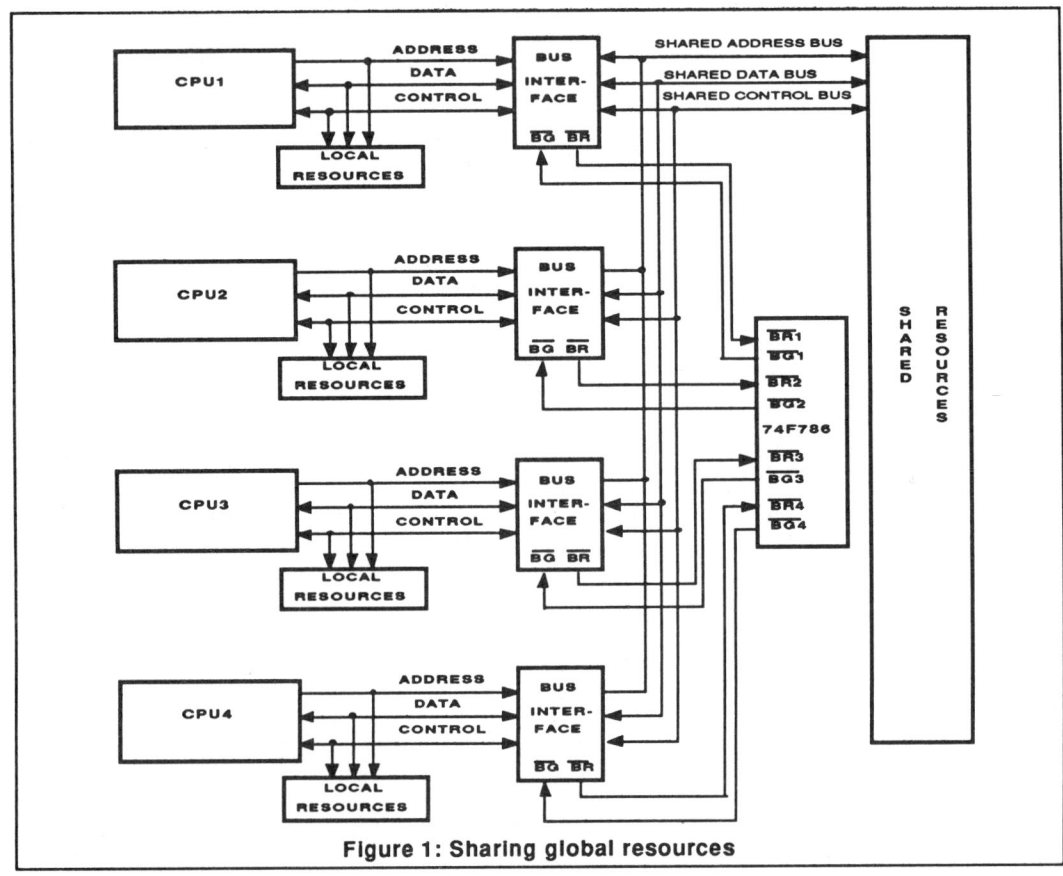

Figure 1: Sharing global resources

July 18, 1988

Application Note

Arbitration In Shared Resource Systems

AN216

until after a whole clock cycle. Todays applications, where speeds are being pushed to their limits may not find that an optimal solution. Therefore more and more designers tend towards asynchronous arbitration because it is much faster on the average. Since applications vary drastically from one to another, some may be better served by first-come-first-serve arbitration, some with fixed priority and some with dynamic priority.

In a first-come-first-served scheme as the name implies, the request to be asserted first is selected first. All other requests made after the first are queued in their respective order of assertion. After the current request is serviced, the request asserted second will be selected and so on. If the request just serviced is asserted again, before all other active requests are serviced, it will be placed at the end of the queue. In a fixed priority method all inputs have a hard-wired priority and cannot be changed. In a dynamic priority assignment the user can change the priority depending upon the system needs. For example, processors performing vital tasks may be placed at a higher priority as compared to processors doing background tasks.

Arbitration, whether synchronous or asynchronous, always brings up the question of "metastability". A hard fact that relates itself all the way back to the beginning of the history of electronics. In its simplest definition it is the state of a flip-flop that is neither a logic"1" or a logic"0", and is a result of violations of its set-up and hold times. This condition must be allowed and dealt with in arbitration and synchronization designs.

Metastability

Various publications have talked about this subject and given recommendations for reducing but not completely eliminating this potential problem. Briefly the suggestions consist of using very fast flip-flops (with very small set-up and hold times), using multiple flip-flops and delay lines and designing of metastable-hardened flip-flops. Please note that a metastable-hardened flip-flop does not necessarily mean that it will never enter a metastable state, but rather it is a flip-flop that is highly optimized to be used in applications where the system designer can not guarantee the minimum set-up and hold times specified by the manufacturer. Since, as of today, the design of a metastable free flip-flop is not practically possible, the next best thing that could be done is design of a flip-flop with significantly reduced set-up and hold times and reduced propagation delays. This will ensure reduced probability of being in a metastable state. Since we still will have some probability of not meeting the minimum set-up and hold times and potentially being in a metastable state, another requirement to be im-

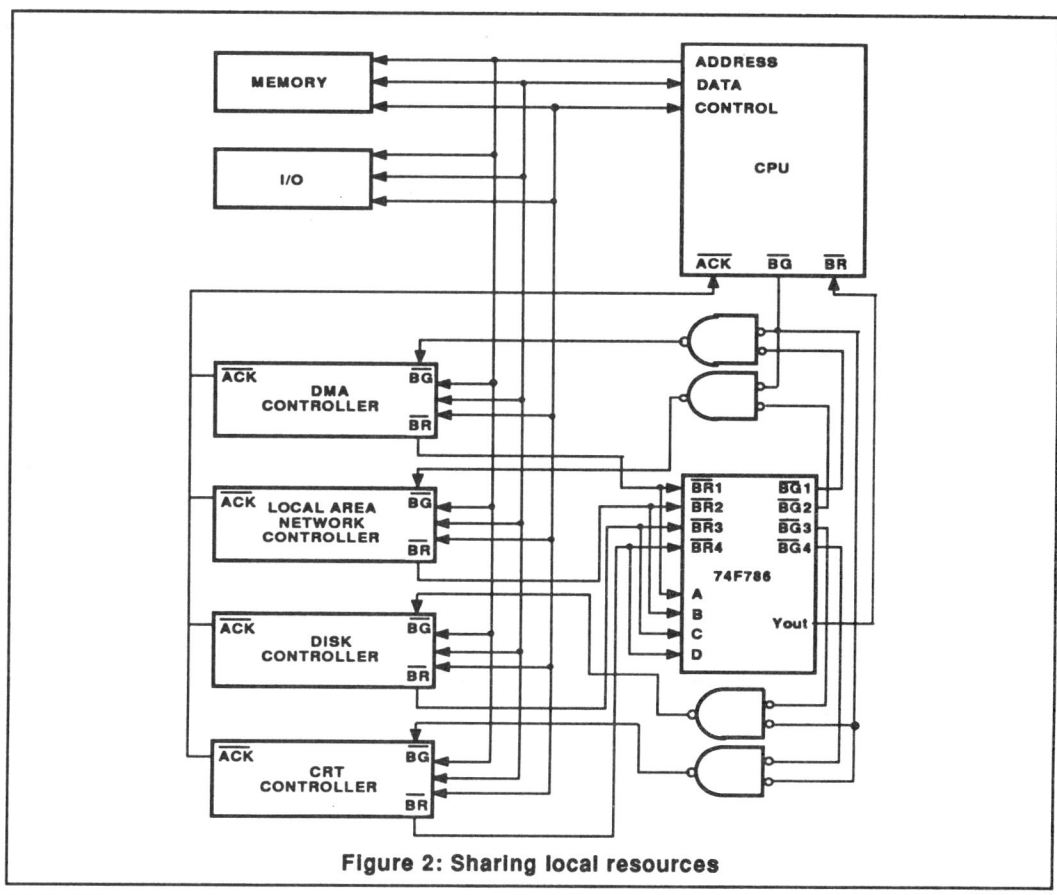

Figure 2: Sharing local resources

July 18, 1988

Arbitration In Shared Resource Systems

Application Note

AN216

posed on this flip-flop would be to hold its previous state and not to propagate this invalid state to its outputs until it has decided to settle in a "0" or a "1" state. By doing so it could be guaranteed that the outputs of a flip-flop will never be in an undetermined state even though the flip-flop may internally be in a metastable state. The penalty that the user would expect to pay in such a design will be a propagation delay that can extend beyond the maximum specified in the data sheet.

74F786- 4-Input Asynchronous Arbiter

The key consideration when arbitrating for shared resources is that access may not be granted to more than one device at a given time. If this could be guaranteed, it would improve reliability. This application note describes a product from Signetics, which guarantees against simultaneous grants and does so at very high speeds. The Signetics 74F786 (Figure 3) is a general purpose asynchronous bus arbiter designed to address the needs for real-time applications, where arbitration is desired between multiple devices sharing common resources. The design goal was to provide for a device, the outputs of which could be guaranteed against logic hazards (glitches), metastability and that no more than one output could be active at a given time. The arbiter has four Bus Request ($\overline{BR}_n$) inputs which allow arbitration between two to four asynchronous inputs. The priority

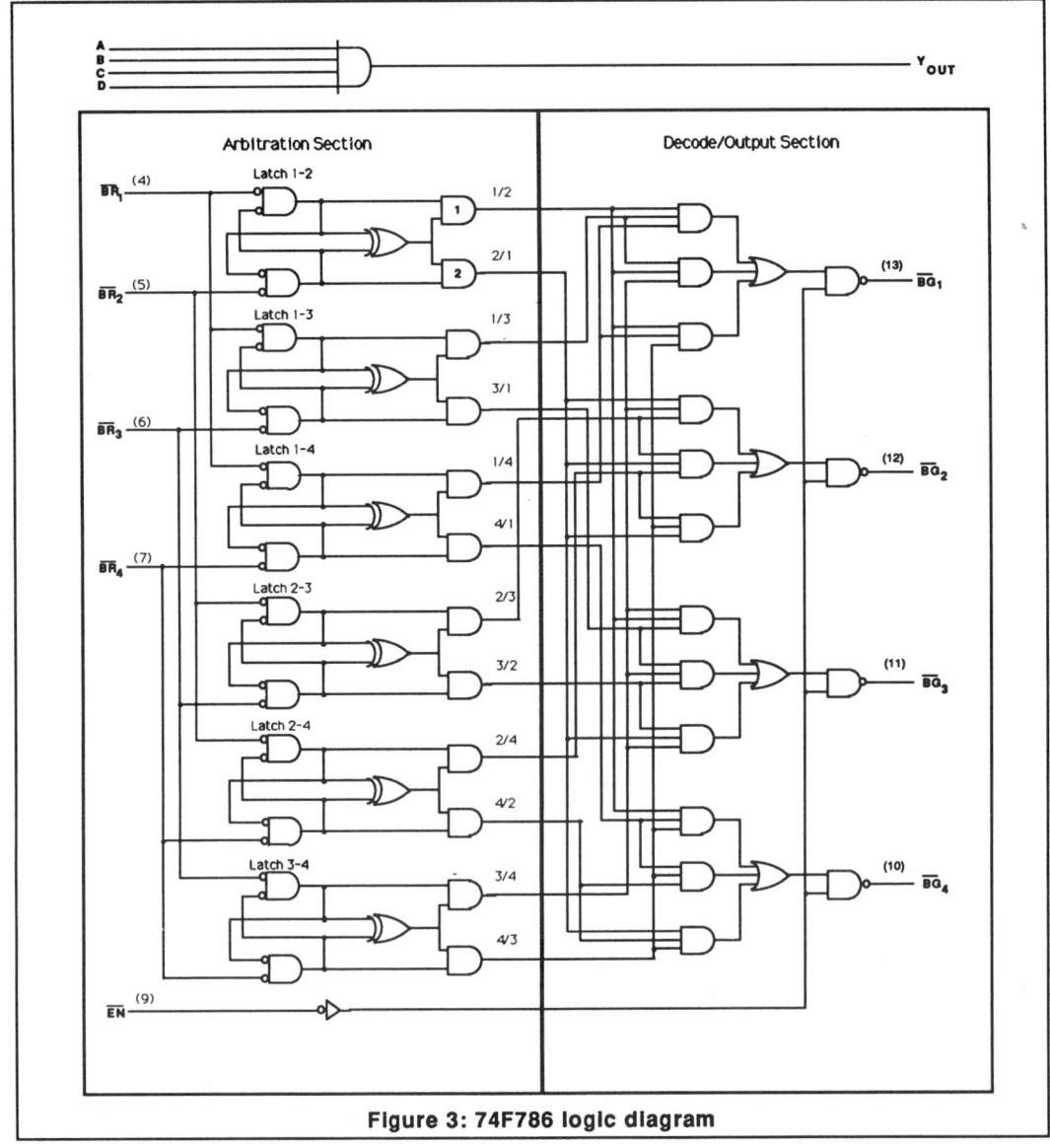

Figure 3: 74F786 logic diagram

July 18, 1988

Arbitration In Shared Resource Systems

Application Note

AN216

is determined on a first-come-first-served basis. Corresponding to each input is a separate Bus Grant ($\overline{BG}_n$) output which indicates which one of the request inputs is served by the arbiter at a given time. All these outputs are enabled by a common enable ($\overline{EN}$) input. Also included on-chip, is a general purpose four-input AND gate which may be used to generate a bus request signal (Figures 2) or as an independent AND gate.

Since the Bus Request inputs have no inherent priority, the arbiter assigns priority to the incoming requests as they are received. Therefore, the first request asserted will have the highest priority. When a Bus Request is received, its corresponding Bus Grant becomes active, provided $\overline{EN}$ is LOW, and no other Bus Grant is active. Typically, a Bus Grant is selected in 6.6 nsec from the time of assertion of a request input. If additional Bus Requests are made after the first request goes LOW, they are queued in their respective order. When the first request is removed, the arbiter services the request with the next highest priority, based upon a first-come-first-served algorithm.

Metastable-Free Outputs

The 74F786 logic diagram (Figure 3) consists of two sections: the arbitration section and the decoding/output section. Within the arbitration section lie six independent 2-input arbiters each of which arbitrates between the two Bus Request ($\overline{BR}_n$) inputs connected to that specific arbiter. Each 2-input arbiter is comprised of two cross-coupled NOR gates, an EX-OR gate and two AND gates. The cross-coupled NOR gates are designed so that they are securely latched when a schottky diode voltage difference appears between the outputs of these NOR gates. The EX-OR gate is designed so that its output will remain LOW until there is at least 1Vbe difference between its inputs. This creates a noise-margin of 1Vbe (base to emitter voltage)-1Vsky (schottky voltage) ≈ 0.3 Volts and assures that the output of the EX-OR will not go HIGH until after the two NOR gates have resolved any contention problems. This guarantees that neither of the outputs of a 2-input arbiter can be in a metastable state, and also that both outputs cannot be high simultaneously. As is clear from Figure 3, the first 2-input arbiter is responsible for deciding between the $\overline{BR}_1$ and $\overline{BR}_2$ inputs. Since both AND gate outputs cannot be high at the same time, the other three possible configurations are; First, AND gate1 is HIGH indicating that $\overline{BR}_1$ arrived at the latch before $\overline{BR}_2$ (designated 1/2); second, AND gate2 is HIGH indicating $\overline{BR}_2$ arrived before $\overline{BR}_1$ (designated 2/1) and third both AND gates are LOW indicating that neither $\overline{BR}_1$ nor $\overline{BR}_2$ has been latched.

Glitch-Free Outputs

The decode section of the 'F786 is responsible for insuring that the outputs do not glitch or produce a logic hazard. While there are three possible Karnaugh mappings, to produce an optimum decode section with a minimum number of transistors and balanced propagation times, the mapping in Table 1 was chosen. Solving Table 1 for $\overline{BG}_1$-$\overline{BG}_4$ yields the following equations:

$\overline{BG}_1 = 1/2 \cdot 1/3 \cdot 1/4 + 1/2 \cdot 1/3 \cdot 3/4 + 1/2 \cdot 1/4 \cdot 4/3$
$\overline{BG}_2 = 2/1 \cdot 2/3 \cdot 2/4 + 2/1 \cdot 2/3 \cdot 3/4 + 2/1 \cdot 2/4 \cdot 4/3$
$\overline{BG}_3 = 3/1 \cdot 3/2 \cdot 3/4 + 1/2 \cdot 3/1 \cdot 3/4 + 2/1 \cdot 3/2 \cdot 3/4$
$\overline{BG}_4 = 4/1 \cdot 4/2 \cdot 4/3 + 1/2 \cdot 4/1 \cdot 4/3 + 2/1 \cdot 4/2 \cdot 4/3$

To see if a glitch can occur let's take the worst possible case, that is, let $\overline{BR}_1$ beat $\overline{BR}_2$, 2 beat 3, 3 beat 4 and 4 beat 1 (a possible situation when all inputs are asserted simultaneously). Also, let's have the outputs of the arbitration section switch sequentially. Initially, all the variables in the equations are false (remember, the outputs of the arbitration section have three possible states). First, when 1/2 goes true 2/1 must remain false. This eliminates several terms from playing a role in deciding which output becomes active. In fact, $\overline{BG}_2$ has been removed from the list and is no longer a contender. At this point, while all the outputs are high (inactive) we have decided that $\overline{BG}_2$ will remain inactive. This leaves us with the following equations.

$\overline{BG}_1 = 1/2 \cdot 1/3 \cdot 1/4 + 1/2 \cdot 1/3 \cdot 3/4 + 1/2 \cdot 1/4 \cdot 4/3$
$\overline{BG}_3 = 3/1 \cdot 3/2 \cdot 3/4 + 1/2 \cdot 3/1 \cdot 3/4 + \cancel{2/1 \cdot 3/2 \cdot 3/4}$
$\overline{BG}_4 = 4/1 \cdot 4/2 \cdot 4/3 + 1/2 \cdot 4/1 \cdot 4/3 + \cancel{2/1 \cdot 4/2 \cdot 4/3}$

Similarly when 2/3 goes true 3/2 must remain false, which further eliminates a term from thus set of 3 equations.

$\overline{BG}_1 = 1/2 \cdot 1/3 \cdot 1/4 + 1/2 \cdot 1/3 \cdot 3/4 + 1/2 \cdot 1/4 \cdot 4/3$
$\overline{BG}_3 = \cancel{3/1 \cdot 3/2 \cdot 3/4} + 1/2 \cdot 3/1 \cdot 3/4 + \cancel{2/1 \cdot 3/2 \cdot 3/4}$
$\overline{BG}_4 = 4/1 \cdot 4/2 \cdot 4/3 + 1/2 \cdot 4/1 \cdot 4/3 + \cancel{2/1 \cdot 4/2 \cdot 4/3}$

Now when 3/4 goes true 4/3 must remain false. This eliminates $\overline{BG}_4$ from the contending list and the contest now is between $\overline{BG}_1$ and $\overline{BG}_3$ as indicated from the following equations.

$\overline{BG}_1 = 1/2 \cdot 1/3 \cdot 1/4 + 1/2 \cdot 1/3 \cdot 3/4 + \cancel{1/2 \cdot 1/4 \cdot 4/3}$
$\overline{BG}_3 = \cancel{3/1 \cdot 3/2 \cdot 3/4} + 1/2 \cdot 3/1 \cdot 3/4 + \cancel{2/1 \cdot 3/2 \cdot 3/4}$

When 4/1 goes true 1/4 must remain false.

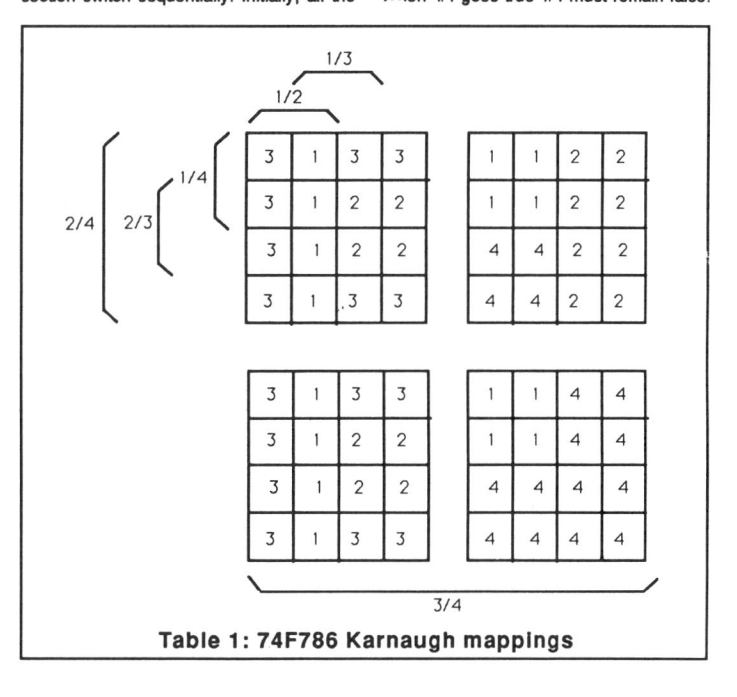

Table 1: 74F786 Karnaugh mappings

July 18, 1988

Application Note

Arbitration In Shared Resource Systems

AN216

Still no decision has been made and is dependent on the two 2-4 and 1-3 latches not taken into account yet. In this case the 2-4 latch status is a don't care, so the outcome of the 1-3 latch dictates the Bus Request granted.

$\overline{BG}_1$ = 1/2. 1/3. 3/4
$\overline{BG}_3$ = 1/2. 3/1. 3/4

If the 1-3 latch settles in the 1/3 state $\overline{BR}_1$ gets the grant, and with 3/1 remaining false, $\overline{BG}_3$ will remain inactive. Similarly if the 1-3 latch goes to the 3/1 state $\overline{BR}_3$ gets the grant, and with 1/3 remaining false $\overline{BG}_1$ will remain inactive.

Notice that the Bus Grant was given in this case without regard to the 2-4 latch. In fact, a quick review shows that neither the 2-3 latch nor the 1-4 latch played a role in making the decision. Each grant is dependent on the state of three latches. By the nature of the encoding logic, as the three activating latches are switched, three outputs are forced to remain in an inactive state. This insures a glitch-free output.

Let's assume that in the example above, the 1-3 latch goes to the 1/3 state and hence $\overline{BG}_1$ is asserted. At this time the other five latches in the circuit will be in 1/2, 4/1, 2/3, 2/4 and 3/4 states. If at this point $\overline{BR}_3$ is removed, then latch 3-4 changes from 3/4 to 4/3 and hence $\overline{BR}_4$ steals the grant (with 1/2. 4/1. 4/3). This concludes that if three or more requests are asserted precisely at the same time, and one of them is removed prior to being serviced, it may cause premature termination of the present grant and assertion of another grant. Therefore, when using three or more Bus Requests it is not advised to remove a request before being serviced. On the other hand, arbitration between two requests does not have this restriction. The user if necessary, may decide to remove an ungranted request at his discretion.

Extended Propagation Delays

Since the outputs of the six 2-input arbiters can not display a metastable condition, the Bus Grant outputs can not display a metastable condition because the decoding/output section does not have any storage element to go metastable. Even though the Bus Grant outputs can't go metastable, the cross-coupled NOR gates can. To determine the metastability characteristics of these NOR gates, the 'F786 was evaluated by Mr. Thomas J. Chaney of Washington University in St. Louis, Missouri, who is considered to be a leading expert in this field. Table 2 gives of the 19 devices supplied to him, the test results from the fastest, the slowest and a typical package. In order to determine the Mean Time Between Package Unresolved (MTBPU) with the relative arrival times of the two input signal transitions uniformly distributed, the following formula is used:

MTBPU= [exp(t'/τ)]/[T$_0$(Input 1 rate)(Input 2 rate)].

Where:

t'= Time given to resolve contention between inputs after they are asserted and τ and T$_0$ are device parameters derived from tests and can most nearly be defined as:

τ= A function of the rate at which a latch in a metastable state resolves that condition.

& T$_0$ = A function of the measurement of the propensity of a latch to enter a metastable state. T$_0$ is also a very strong function of the normal propagation delay of the device.

Solving for t', the resolving time measured from the arrival of the first input, and setting up the equation so the value of T$_0$ in Table 2 (given in nsec.) can be substituted directly is:

t'=(τ)ln[(T$_0$)(3E14)].

The implication of the above equation is that, even though typical propagation delay through the arbiter is about 6.6nsec, contention between inputs may extend this time significantly and can be calculated from Table 2.

Package	Latch	Output Measured	τ (nsec)	T$_0$ (nsec)	h (nsec)	t' for 1 failure/century (Inputs at 10E6hz)
FASTEST	1-2	13	.38	175E2	6.6	16.6
	1-3	13	.39	79E2	6.6	16.4
	1-4	13	.39	69E2	6.6	16.4
	2-3	12	.38	109E2	6.6	16.1
	2-4	12	.39	68E2	6.6	16.5
	3-4	11	.38	181E2	6.6	16.3
SLOWEST	1-2	13	.44	34E2	6.6	18.1
	1-3	13	.44	17E2	6.6	18.0
	1-4	13	.43	26E2	6.6	17.8
	2-3	12	.44	16E2	6.6	17.9
	2-4	12	.46	8E2	6.6	18.5
	3-4	11	.44	29E2	6.6	18.2
TYPICAL	1-2	13	.41	56E2	6.6	17.3
	1-3	13	.42	24E2	6.6	17.2
	1-4	13	.43	17E2	6.6	17.5
	2-3	12	.43	18E2	6.6	17.4
	2-4	12	.39	72E2	6.6	16.6
	3-4	11	.41	49E2	6.6	17.2

Where h= typical propagation delay through the device.

Table 2: 74F786 test results for all latches for three packages. All tests with V$_{cc}$=5.0vdc and at room temperature

July 18, 1988

AN207
Multiple μP Interfacing With FAST ICs

Application Note

FAST Products

INTRODUCTION

As microprocessor costs continue to decrease and the demands on product performance continue to increase, designers are increasingly turning to multiple microprocessor systems to meet the performance challenge. The introduction of many "peripheral controller" type processors has made this choice even more attractive. This application note addresses typical problems associated with interfacing multiple microprocessors, and illustrates the use of Philips Interface Circuits in solving these problems.

A multi-processor system contains two or more processors communicating through parallel ports, multi-port memories, serial data links, and/or shared buses. The most popular multi-processor architectures are "loosely coupled" systems. In loosely coupled systems each processor operates asynchronously with the other processors, usually performing a separate function. Communication is not continuous, and occurs only when necessary.

A special application for multiple microprocessor systems is in redundant systems. As the price of microprocessors dropped, it became economically feasible to achieve greatly increased reliability by employing several processors operating in parallel, performing identical functions. After each operation a vote is taken on the result. If there is disagreement, a fault has been detected, and appropriate corrective action can be taken. Appropriate action might be switching in a third processor, repeating the process, or activating an error sequence and/or an alarm.

In the typical loosely coupled multiple processor system of Figure 1, a main processor "delegates" processing work to four other processors. A keyboard scanner microprocessor scans the keyboard continuously, debounces key closures, performs code conversions, and transmits key codes to the main processor in a format that it can easily assimilate. A separate arithmetic processor accepts parameters from the main processor, performs arithmetic calculations, and provides the results for the main processor to read when it is not busy with other tasks. The display controller accepts data and commands from the main processor, then displays and manipulates data on CRT or other displays. The display controller refreshes the display and supports graphic displays without tying up the main processor. The print spooler is a separate processor that accepts files to be printed from the main processor using high-speed data transfers. Then the print spooler stores and feeds data to the printer at the printer's lower data rate, freeing the main processor for other chores. Each processor module contains its own "local" ROM, RAM, or I/O, so that it performs its task independently, and communicates with other processors only when necessary. As a result, the system as a whole operates closer to its maximum speed.

Some of the advantages of multiple microprocessor systems are:

- Each processor performs a relatively independent task.
  - Design is easily split among team members.
  - Testing is easily performed on a modular level.
  - Modules can be added or modified without affecting other modules.
- Multi-processing allows distributed processing where modules may be physically separated from the main system.

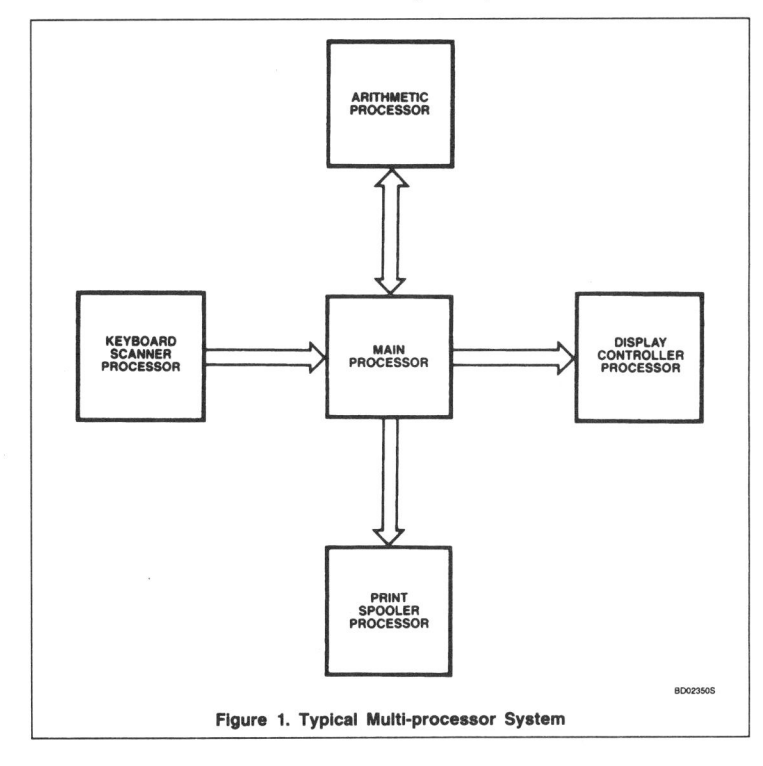

Figure 1. Typical Multi-processor System

June 1987

FAST Products

Application Note

Multiple µP Interfacing With FAST ICs

AN207

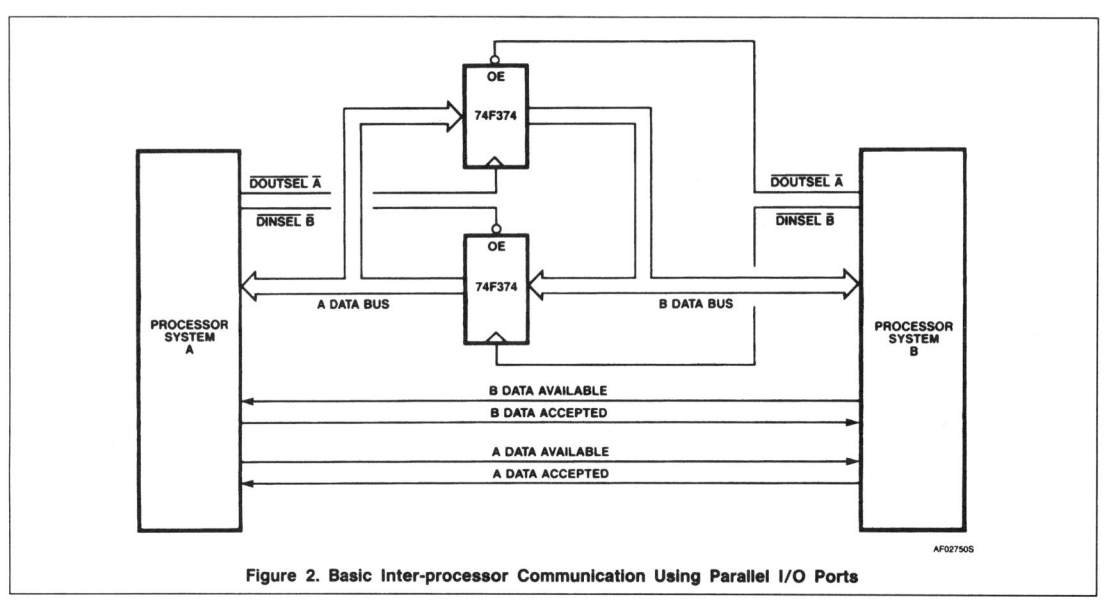

Figure 2. Basic Inter-processor Communication Using Parallel I/O Ports

Figure 3. Handshake for Parallel Port Communication

- Parallel processing greatly increases system performance and throughput.
- Hardware cost is less than single-processor systems with similar performance.
- Reliability can be increased easily by redundant processing.

The following application examples illustrate the use of Philips FAST Interface Circuits in multiple-processor systems.

PARALLEL I/O PORT COMMUNICATIONS

Figure 2 illustrates how parallel I/O ports using Philips FAST Interface devices are used to accomplish simple 2-processor communications. Two 74F374 octal 3-State registers are used to implement bi-directional parallel data communication. Each 74F374 acts as output port to one processor and input port to the other. The handshake lines are needed when the processors operate asynchronously to ensure that data has been received before new data is transmitted. A handshake timing protocol (Figure 3) implemented in software acts as a traffic cop to assure valid data communications. The transmitting processor starts the handshake by setting Data Available to indicate that data is valid. The receiving processor sets Data Accepted to indicate data has been read. The transmitter then resets Data Available allowing the receiver to reset Data Accepted. The transmitter will not send new data until Data Accepted is reset.

COMMUNICATIONS VIA MULTI-PORT MEMORY

Figure 4 illustrates the logic required for two processors to communicate through a multi-port memory. The RAM is accessible from both processor A and processor B via 74F157 multiplexers used to select one processor's bus at a time. Multi-byte messages and data blocks may be written into the memory by one processor and read out by the other at a later time. No byte-by-byte handshake is required. The multi-port memory provides increased system performance at somewhat higher cost compared to a parallel port technique. Because of the use of multi-port memories in microprocessor systems, these systems can become quite complex. Another application note in this series covers interfacing to multi-port memories in greater depth.

June 1987

7-41

FAST Products

Application Note

Multiple µP Interfacing With FAST ICs

AN207

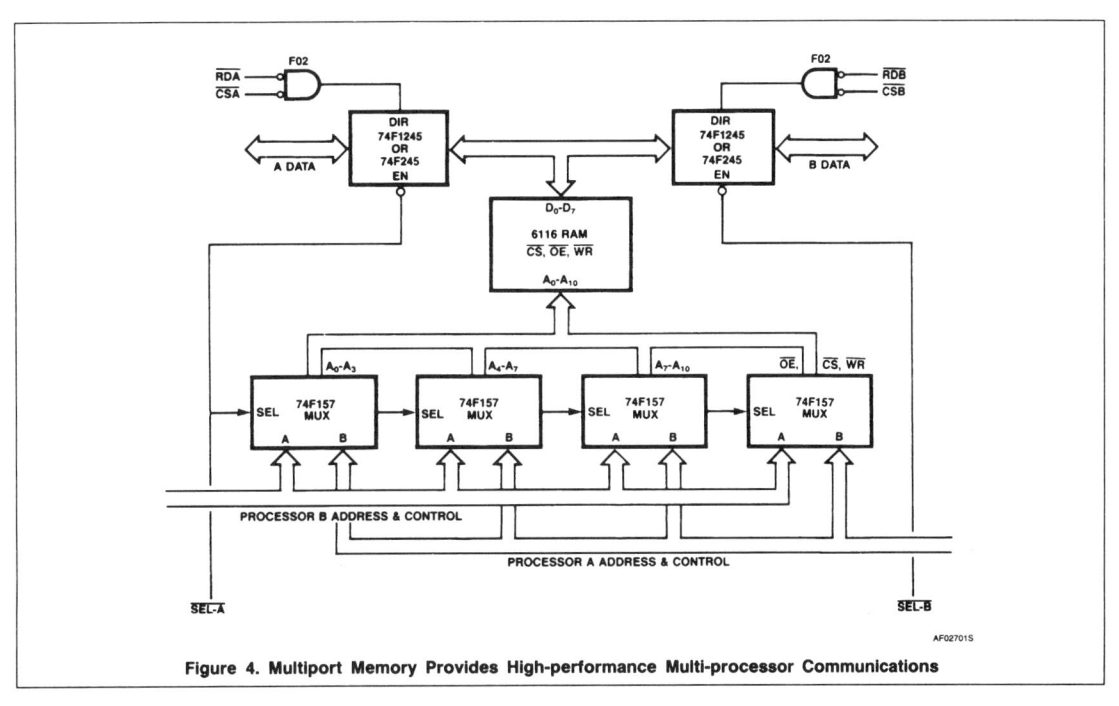

Figure 4. Multiport Memory Provides High-performance Multi-processor Communications

SERIAL COMMUNICATIONS
Although serial communications between multiple processors is slower than the parallel methods examined above, it is usually less expensive and very useful for communicating with remote units. Serial communications via RS-232 or RS-422 links can provide reliable communications over great distances. Implementation of serial communications is simplified by the availability of Universal Asynchronous Receiver Transmitter (UART) devices and well established standards for circuit interfaces and protocols. Figure 5 illustrates local/remote processor communication using Philips SC2681 UART devices. In many cases additional interface lines are required for handshaking.

SHARED BUS ARCHITECTURE
One of the most powerful multiple processor architectures uses the popular shared bus concept. In Figure 6, each processor has its own local bus with some combination of RAM, ROM, and I/O available locally. The shared bus permits use of "global resources" such as global memory and global I/O which are accessible to all processors on the shared bus. Common interfaces such as printer ports do not have to be implemented for each processor, and may be connected to the shared bus. Multiple processors communicate indirectly with one another through the global RAM. This technique provides highest throughput when interconnecting more than two processors. It also reduces cost through sharing of global resources.

Any processor permitted to drive the system address, data, and control buses is known as a "master." Processors not having this capability are "slaves." A useful attribute of shared bus systems is the ability to add whole new functions by connecting a new master to the bus. Figure 7 illustrates a typical shared system bus interface using Philips Interface circuits. Three 74F244 octal 3-State buffers are used to drive the 24 bit system address bus (16 bits in some cases). Two 74F245 octal bidirectional 3-State buffers are used to drive the 16 bit data bus (8 bits in some cases). In addition, half a 74F244 is used to drive the system command bus, composed of the signals $\overline{IORD}$, $\overline{IOWR}$, $\overline{MEMRD}$, and $\overline{MEMWR}$.

Multiple local processors may request use of the shared bus by setting $\overline{BUS\ REQUEST}$ active and waiting for the arbitration logic to assert $\overline{BUS\ GRANT}$. The arbitration logic indicates to the local processor when it may access the shared bus after a request has been made. This is necessary to prevent more than one local processor from accessing the system bus at the same time, resulting in bus contention and possible system failure.

ARBITRATION
Contention by several processors for use of shared resources can create sticky timing problems unless care is exercised in the design of appropriate arbitration logic to resolve timing conflicts. Schemes for bus arbitration vary in speed, cost, and flexibility and involve parallel, serial, transparent, pseudo-transparent, polled, and flag operations.

Parallel Priority Resolution
Parallel priority resolution is most useful in systems with 4 or more masters, where its speed outweighs the disadvantage of the additional hardware. A scheme for system bus arbitration using parallel priority resolution is illustrated in Figure 8.

A master's priority is determined by using a 74F148 priority encoder. Each master's arbitration logic generates a $\overline{REQ}$ to the priority encoder. When there is contention, the master whose $\overline{REQ}$ is connected to the highest priority input will be granted access.

A 74F138 is used to decode the encoder outputs to generate the $\overline{EI}$ (enable input) to the arbitration logic of the master which has been granted access. $\overline{CLEAR}$ is used to remove all masters from the bus during reset or when an error condition is present. $\overline{ARB\ CLOCK}$ is used to synchronize all bus arbitration inputs and outputs to prevent race conditions and to facilitate a standard interface

June 1987

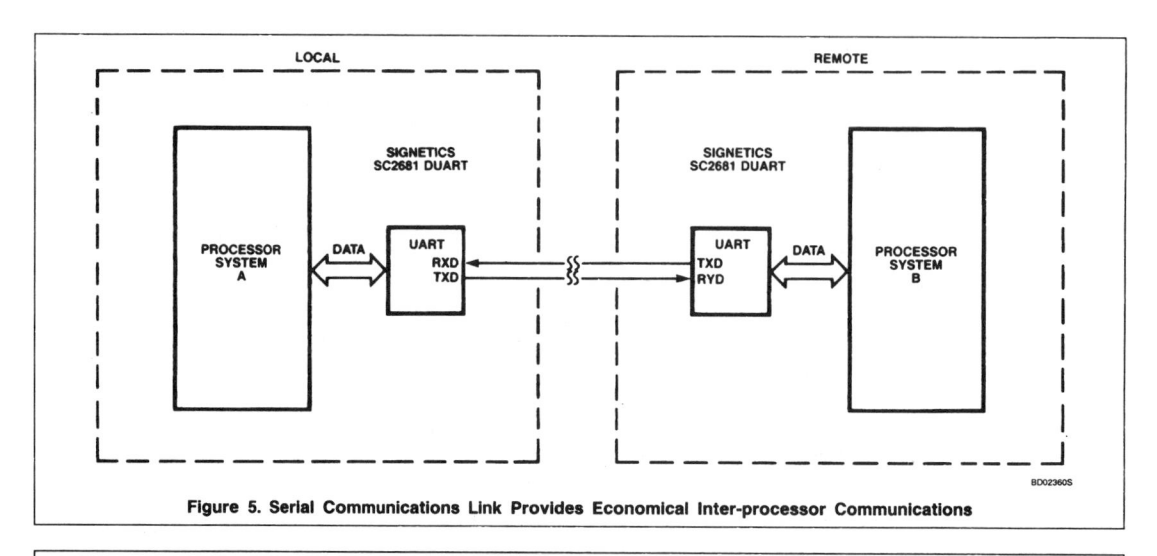

FAST Products

Application Note

Multiple µP Interfacing With FAST ICs

AN207

Figure 5. Serial Communications Link Provides Economical Inter-processor Communications

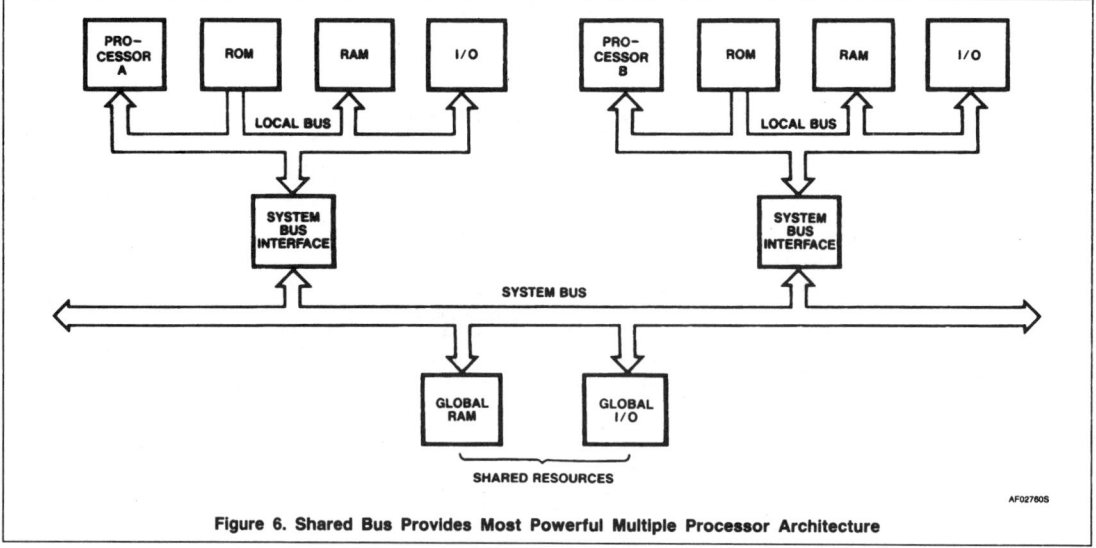

Figure 6. Shared Bus Provides Most Powerful Multiple Processor Architecture

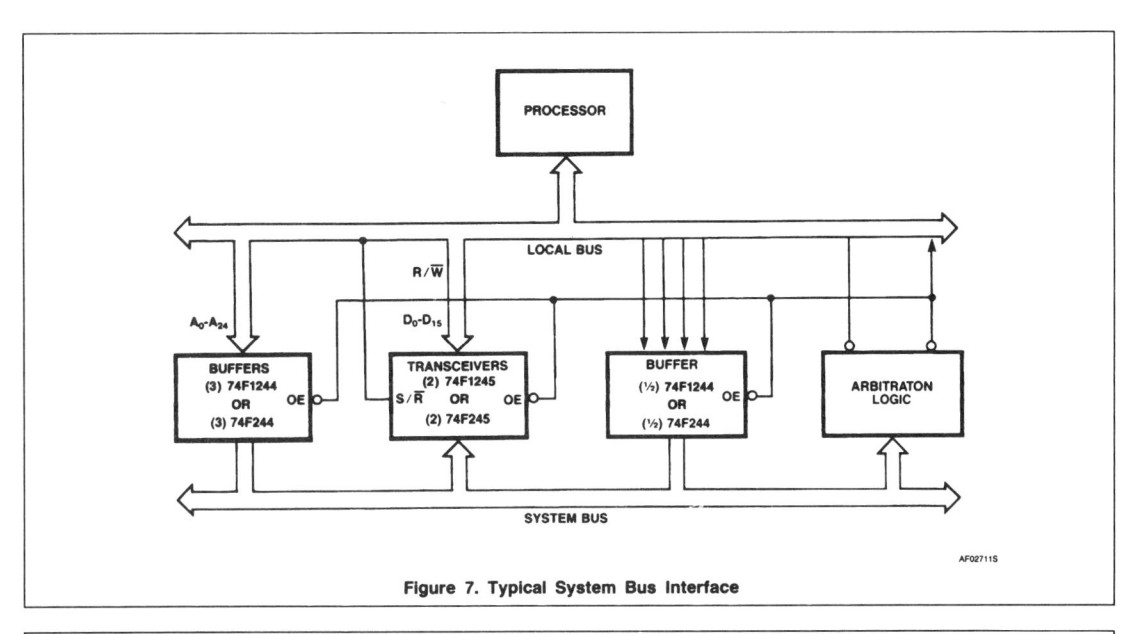

FAST Products

Application Note

Multiple µP Interfacing With FAST ICs

AN207

Figure 7. Typical System Bus Interface

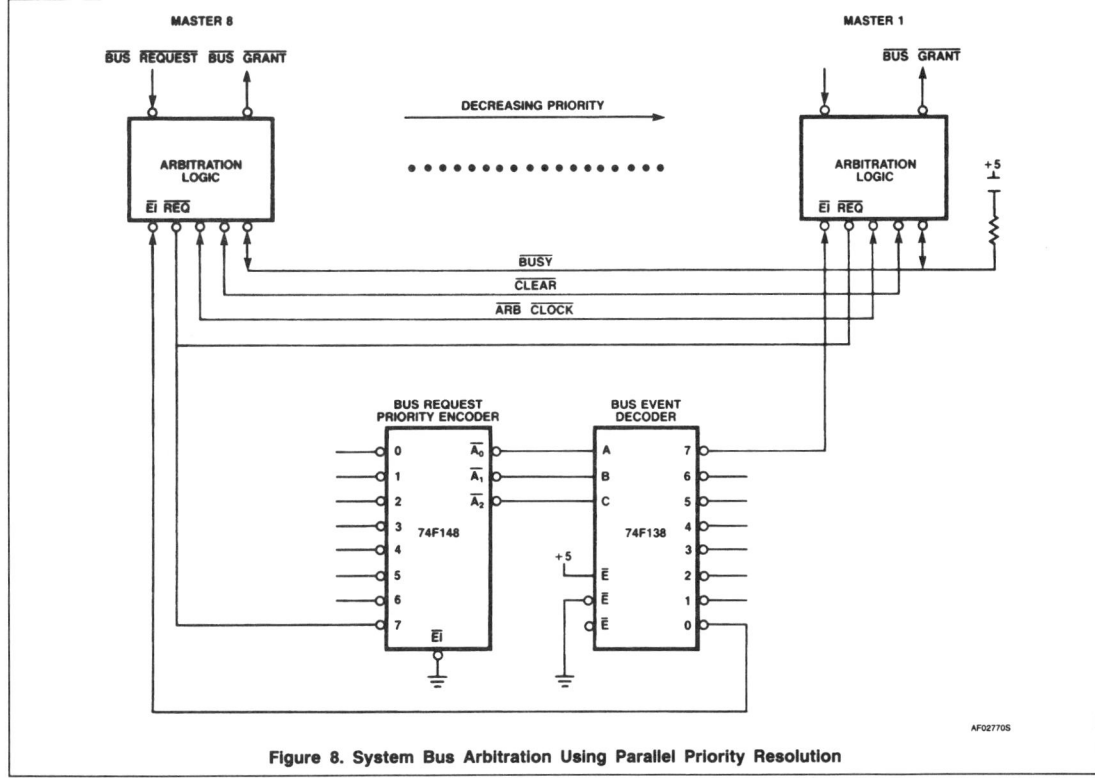

Figure 8. System Bus Arbitration Using Parallel Priority Resolution

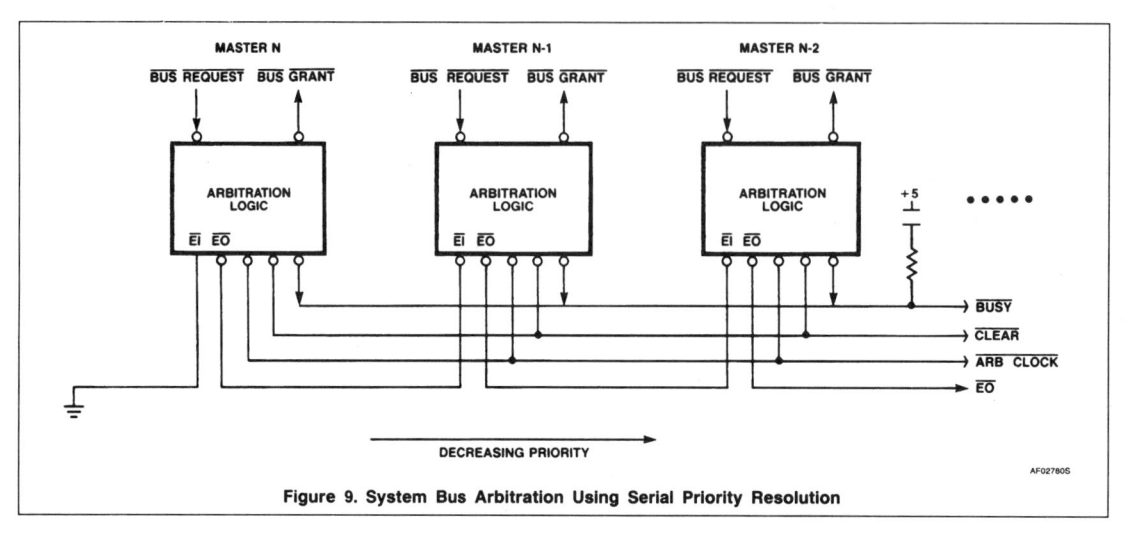

FAST Products

Application Note

Multiple µP Interfacing With FAST ICs

AN207

Figure 9. System Bus Arbitration Using Serial Priority Resolution

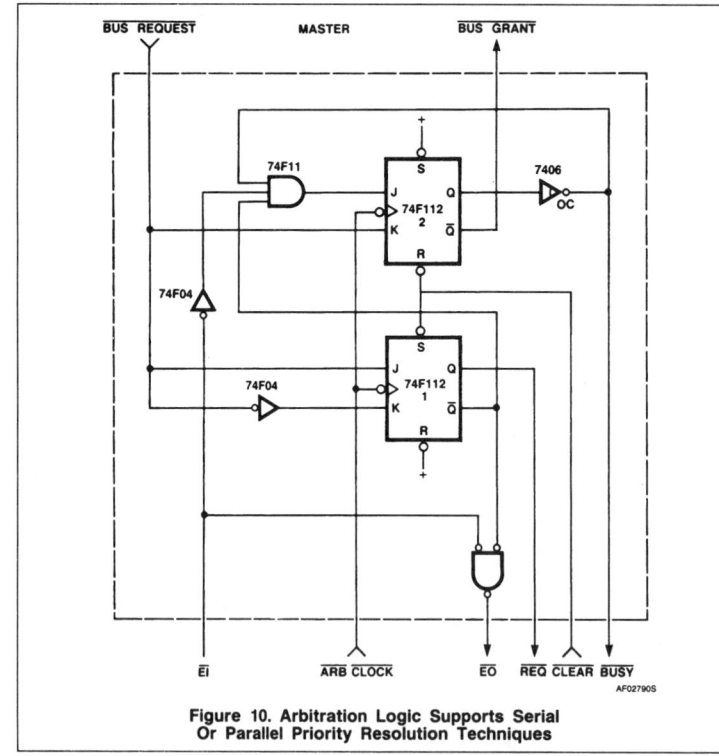

Figure 10. Arbitration Logic Supports Serial Or Parallel Priority Resolution Techniques

design. $\overline{\text{BUSY}}$ is generated by the master currently accessing the bus to indicate that the bus is in use. Even after a master has been granted access by the priority resolution, it must still wait for the current master to vacate the bus, i.e., $\overline{\text{BUSY}}$ going inactive. The arbitration logic generates a $\overline{\text{BUS GRANT}}$ to a master when $\overline{\text{EI}}$ is asserted and $\overline{\text{BUSY}}$ is not.

Serial Priority Resolution

Serial priority resolution eliminates the need for encoder/decoder hardware at the expense of speed. In Figure 9 a master's priority is determined by its physical location in a daisy chain configuration. A master negates its $\overline{\text{EO}}$ (enable output) when its $\overline{\text{EI}}$ (enable input) is negated or when it wants to access the bus. This negates $\overline{\text{EO}}$ for all masters further down the line *to go inactive*. If a master requests the bus, and no higher priority master is requesting the bus, as indicated by $\overline{\text{EI}}$ being asserted, the master may access the bus when the current master is finished. The ARB clock rate is limited to the speed at which the daisy chain signals can propagate through all masters.

Arbitration Logic

Arbitration logic suitable for either parallel or serial priority resolution is illustrated in Figure 10. The logic shown synchronizes a master's $\overline{\text{BUS REQUEST}}$ input to $\overline{\text{ARB CLOCK}}$ using flip-flop 1, asserting $\overline{\text{REQ}}$ and negating $\overline{\text{EO}}$. If $\overline{\text{EI}}$ is asserted and $\overline{\text{BUSY}}$ is not, the master may access the bus on the next falling edge of $\overline{\text{ARB CLOCK}}$. This arbitration is provided by flip-flop 2. $\overline{\text{BUS GRANT}}$ and $\overline{\text{BUSY}}$ are asserted. When the access is complete, the master negates $\overline{\text{BUS REQUEST}}$ inactive. On the falling edge of $\overline{\text{ARB CLOCK}}$, $\overline{\text{REQ}}$ negated and, if $\overline{\text{EI}}$ is asserted $\overline{\text{EO}}$ is asserted. On the next falling edge $\overline{\text{BUSY}}$ and $\overline{\text{BUS GRANT}}$ are negated. The timing diagram for this sequence is illustrated in Figure 11. Note that a master must wait for the current master to complete a transfer and negate $\overline{\text{BUSY}}$ before it may access the bus.

June 1987

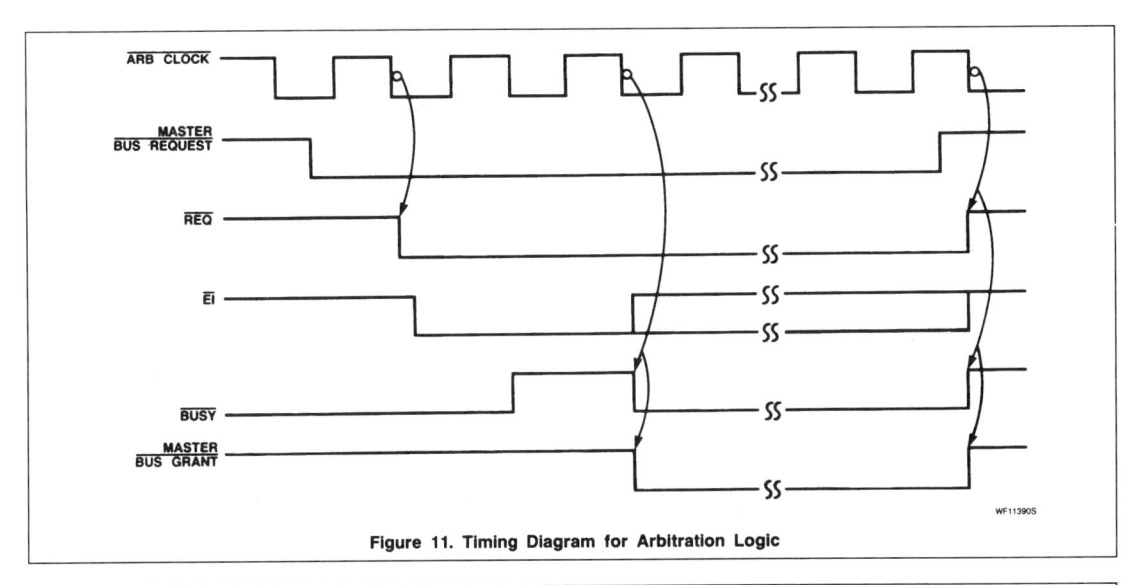

FAST Products

Application Note

Multiple µP Interfacing With FAST ICs

AN207

Figure 11. Timing Diagram for Arbitration Logic

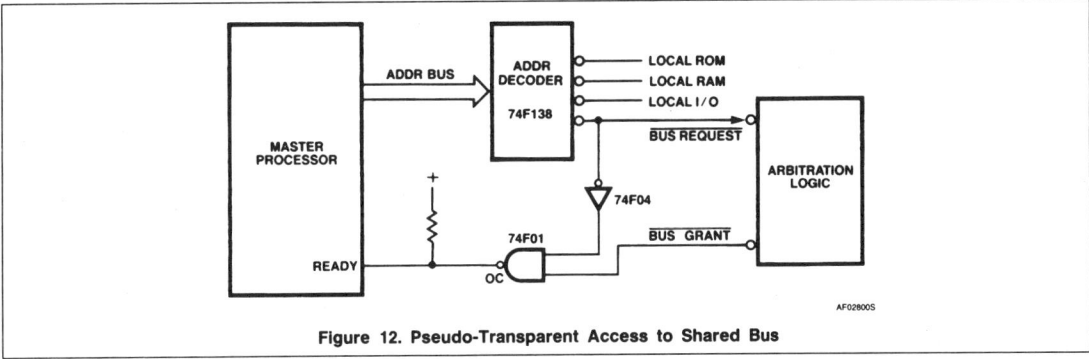

Figure 12. Pseudo-Transparent Access to Shared Bus

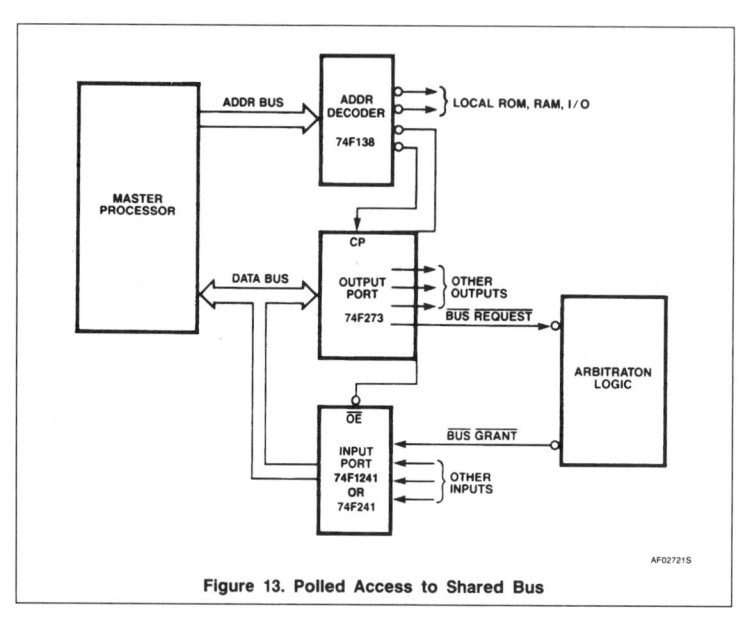

FAST Products

Application Note

Multiple μP Interfacing With FAST ICs

AN207

Figure 13. Polled Access to Shared Bus

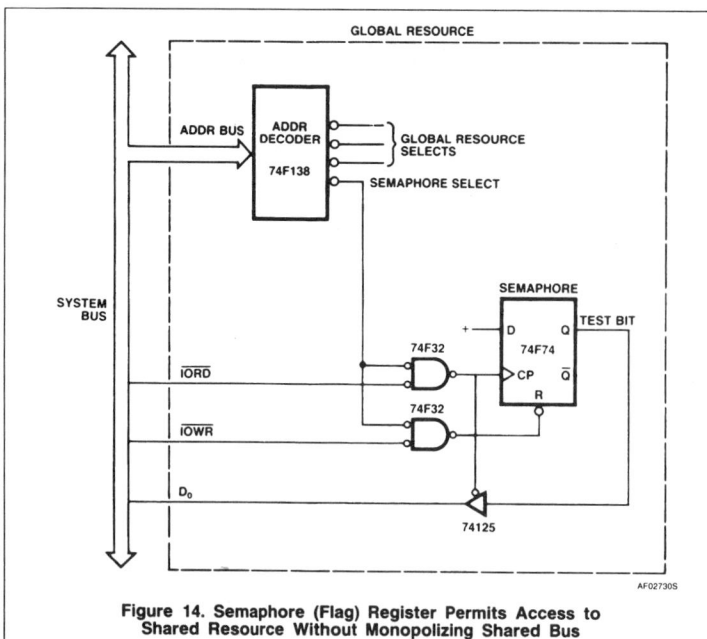

Figure 14. Semaphore (Flag) Register Permits Access to Shared Resource Without Monopolizing Shared Bus

arbitration logic asserts $\overline{\text{BUS GRANT}}$. Then READY is asserted and the shared bus cycle occurs. The processor is unaware of arbitration and unaware that the bus is shared. With this technique, a watchdog timer should be used to ensure that the processor doesn't "hang up" if faulty bus operation prevents access. Access occurs one cycle at a time, preventing any one master from "hogging" the bus.

Polled Access to Shared Bus

The logic in Figure 13 uses an output port to request access to the bus, and polls an input port to determine when access has been granted. Once access is granted, the master retains the bus until it negates the $\overline{\text{BUS REQUEST}}$ output port bit. Large block moves may occur without fear of another master changing the data as with cycle-by-cycle arbitration. However, this approach greatly slows down the response time of the system, because of the waiting while each master performs. All other masters must wait, even if they do not require the use of the same shared resource.

Semaphore (Flag) Arbitration

The logic of Figure 14 improves on the polled access technique by permitting access to a shared resource when that resource is available. A master first reads the semaphore register associated with the resource it wishes to access. The master may not access the resource unless the semaphore bit is false. When the semaphore bit is false, reading the register automatically sets the bit true. When the master reads a false semaphore, it may then access the resource. All other masters reading the semaphore will see it set and will not access the resource. The master may access the resource until it is no longer needed. By writing to the semaphore register, it is automatically reset, allowing other masters to access the resource. Only the one resource, not the entire shared bus, is monopolized by one master at a time. The hardware performs a function similar to a software read-modify-write operation.

The timing for the semaphore operation is shown in Figure 15. If the semaphore bit is false and the register is read, the bit is set true at the end of the read cycle (rising edge of $\overline{\text{IORD}}$). The semaphore bit is reset by doing a "dummy" write to the semaphore register. The bit is set false at the beginning of the cycle ($\overline{\text{IOWR}}$ going low).

Pseudo-Transparent Priority Resolution

The logic of Figure 12 uses "cycle stealing" to permit single byte transfers with pseudo-transparent arbitration. When the address decoder determines that a master requires access to shared bus, it asserts $\overline{\text{BUS REQUEST}}$. The processor's READY line is held negated, "freezing" the processor until the

INTERFACING THE MC68000 TO THE MULTIBUS™*

One of the best examples of a multi-processor shared bus is the MULTIBUS. One of the

*MULTIBUS is a trademark of Intel Corporation.

June 1987

FAST Products

Application Note

Multiple μP Interfacing With FAST ICs

AN207

most popular 16 bit processors in new designs today is the MC68000. Yet, to our knowledge, there are currently (mid-83) no LSI MULTIBUS arbiter ICs available to allow a designer to easily interface the two. There are arbiter ICs available, but they were designed for other processors and are cumbersome and limited in performance when interfaced to the 68000.

The following is the design for a 68000 MULTIBUS interface. The design supports serial or parallel arbitration and performs with a 10MHz bus clock. Operation is similar to the example described previously. Tables 1 and 2 define the MC68000 bus control signals and the MULTIBUS arbitration signals. The timing diagram for MC68000 read and write cycles is shown in Figure 16.

Figure 17 illustrates the control circuitry for the MC68000 to MULTIBUS interface. The master initiates a MULTIBUS transfer by asserting $\overline{\text{MULTIREQ}}$ active. This is usually the output of address decode circuitry. $\overline{\text{AS}}$ clears the request at the end of the transfer. Flip-flops 1, 2, and 3 sample and synchronize the bus request to the falling edge of $\overline{\text{BCLK}}$. Since $\overline{\text{MULTIREQ}}$ is asynchronous to $\overline{\text{BCLK}}$, flip-flop 2 serves as a synchronizer and is clocked on the rising edge of $\overline{\text{BCLK}}$. All inputs to the arbiter are thus synchronous so that race conditions at flip-flop inputs are avoided.

If the bus is not in use ($\overline{\text{BUSY}}$ is not asserted), and no higher priority master requests the bus ($\overline{\text{BRPN}}$ is asserted), the master is granted access on the next falling edge of $\overline{\text{BCLK}}$. Flip-flop 4 provides this function. If these conditions are not satisfied, $\overline{\text{DTACK}}$ is used to force the CPU to wait. Once the master is granted access, it sets $\overline{\text{BUSY}}$ active to indicate that the bus is in use. $\overline{\text{BUSEN}}$ (bus enable) also becomes active and gates the master's address, data, and control buses onto the MULTIBUS. One half cycle later, on the rising edge of $\overline{\text{BCLK}}$, flip-flop 5 sets CMDEN (Command Enable) active. This allows RD or WR strobes to be asserted on the MULTIBUS. This delay is necessary because the MULTIBUS requires data and address valid 50ns before read or write commands. DS is used to generate the read or write strobes.

The MULTIBUS transfer is completed when $\overline{\text{XACK}}$ is asserted terminating the 68000 cycle by asserting $\overline{\text{DTACK}}$. The master maintains control of the MULTIBUS until another master requests access, as indicated by asserted $\overline{\text{CBRQ}}$. If the current master is not performing a MULTIBUS transfer, it loses the bus on the next falling edge of $\overline{\text{BCLK}}$. CMDEN, $\overline{\text{BUSEN}}$, and $\overline{\text{BUSY}}$ are negated. Flip-flop 4 provides this function.

Table 1. MC68000 Bus Control Signals.
(Refer To The Signetics 68000 Microprocessor Data Sheet For More Information.)

CLK	Clock. Time reference for 68000 microprocessor bus control.
$\overline{\text{AS}}$	Address Strobe. Indicates that address on address bus is valid.
$\overline{\text{UDS}}$, $\overline{\text{LDS}}$	Upper and Lower Data Strobe. Indicates that the processor is reading from or writing to the upper data byte ($D_7 - D_{15}$) and/or the lower data byte ($D_0 - D_7$).
R/$\overline{\text{W}}$	Read/Write. Indicates whether the current bus cycle is a read or a write cycle.
$\overline{\text{DTAK}}$	Data Transfer Acknowledge. Input to the 68000 indicating that the data transfer can be completed, on the high to low transition.
$\overline{\text{BCLK}}$	Bus Clock. All arbitration signals listed below must be synchronized to the negative edge of this clock. It is independent of any processor clock.
$\overline{\text{BPRN}}$	Bus Priority In. Indicates that no higher priority master is requesting the bus. Similar to $\overline{\text{EI}}$ in previous examples.
$\overline{\text{BPRO}}$	Bus Priority Out. Used in serial priority resolution circuits. Similar to $\overline{\text{EO}}$ in previous examples.
$\overline{\text{BUSY}}$	Bus Busy. Driven by current bus master to indicate that the bus is in use.
$\overline{\text{BREQ}}$	Bus Request. Used in parallel priority resolution circuits. Similar to $\overline{\text{REQ}}$ in previous examples.
$\overline{\text{CBRQ}}$	Common Bus Request. Driven by all potential bus masters requesting bus. Used to save time by allowing the present bus master to avoid arbitration after each cycle if no other requests are active.
$\overline{\text{XACK}}$	Transfer Acknowledge. Indicates that the MULTIBUS data transfer is completed on high to low transition.

The logic that interfaces the MC68000 to the MULTIBUS is shown in Figure 18. 74F533 inverting octal 3-State latches are used to gate the 20 bit address and 16 bits of data onto the MULTIBUS. Note that the data and address bus is negative true. 74F240 octal 3-State inverting buffers are used to gate 16 bits of data onto and off of the MULTIBUS. Data direction is determined by the MC68000's R/$\overline{\text{W}}$ line. A 74F139, 2 to 4 decoder is used to decode I/O and RD/WR to generate the 4 MULTIBUS commands. I/O is the output of address decode circuitry which decodes I/O addresses. A 74F244 is used to gate the commands onto the MULTIBUS.

Signetics FAST logic family is used in this design to increase speed and bus drive capability while minimizing MULTIBUS loading.

REDUNDANT MICROPROCESSORS ENHANCE RELIABILITY

Figure 19 illustrates how two 6809E microprocessors are used in a parallel redundancy scheme to prevent faulty operation from damaging external systems. Two systems with identical processors, RAM, ROM, and I/O are first synchronized. After synchronization, their data buses are compared every cycle. If the data on the two buses is different, an error has occurred and the system shuts down.

A common clock is used to drive the 6809E processor in each system so that a timing reference is established. Upon reset, both processors execute a sync instruction and the critical output circuits are turned off. When both processors have executed the sync instruction, as indicated by BA = 0 and BS = 1, the START button is used to interrupt the processors and they begin program execution in synchronism. The critical outputs are also turned on. On the falling edge of E, the data buses of the two systems are compared using the 74F521 octal comparator. If the data does not match, at least one system is operating incorrectly. The 74F74 flip-flop latches the error condition and turns off the critical outputs.

A similar technique should be used on outputs to ensure that an output goes active only when the output of both systems goes active.

June 1987

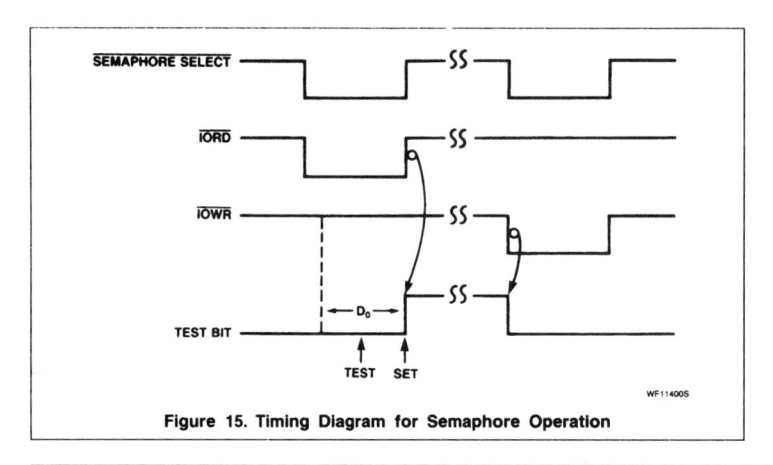

FAST Products

Application Note

Multiple μP Interfacing With FAST ICs

AN207

Figure 15. Timing Diagram for Semaphore Operation

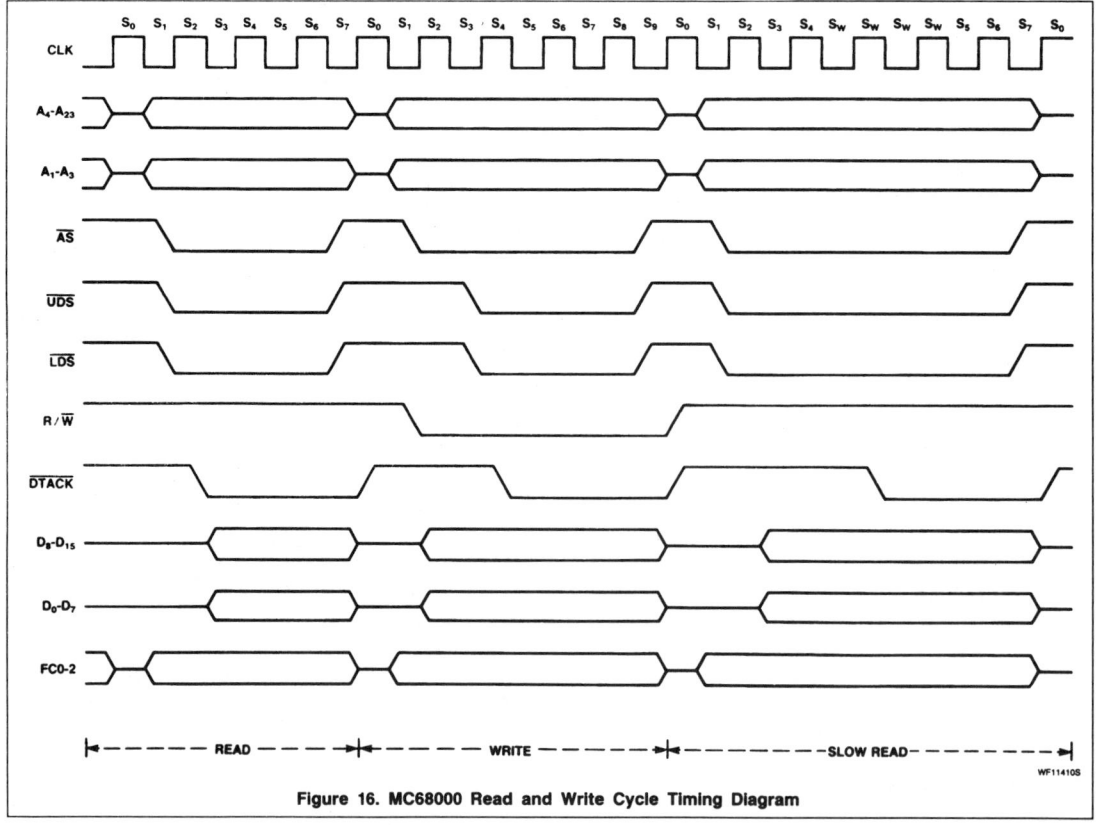

Figure 16. MC68000 Read and Write Cycle Timing Diagram

FAST Products

Application Note

Multiple µP Interfacing With FAST ICs

AN207

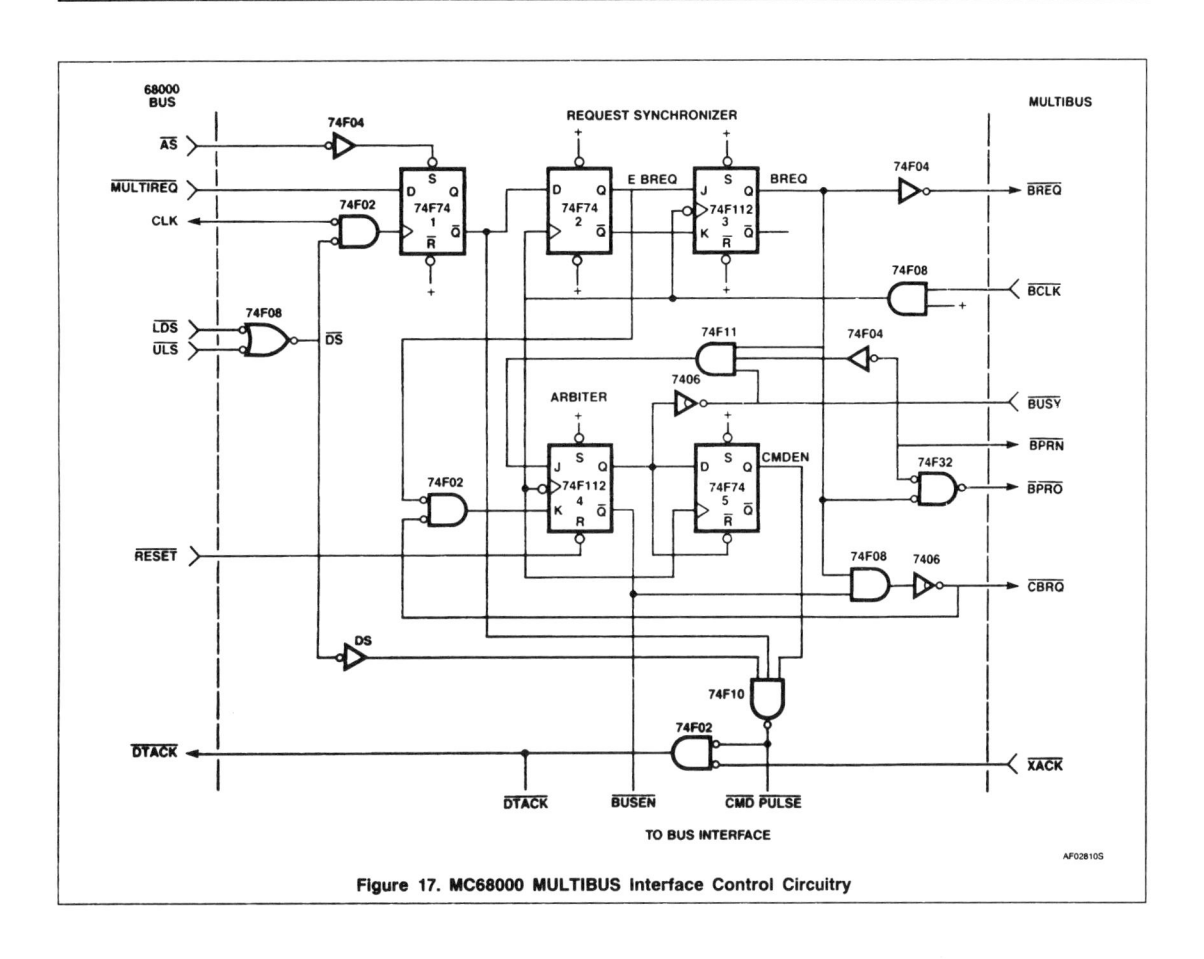

Figure 17. MC68000 MULTIBUS Interface Control Circuitry

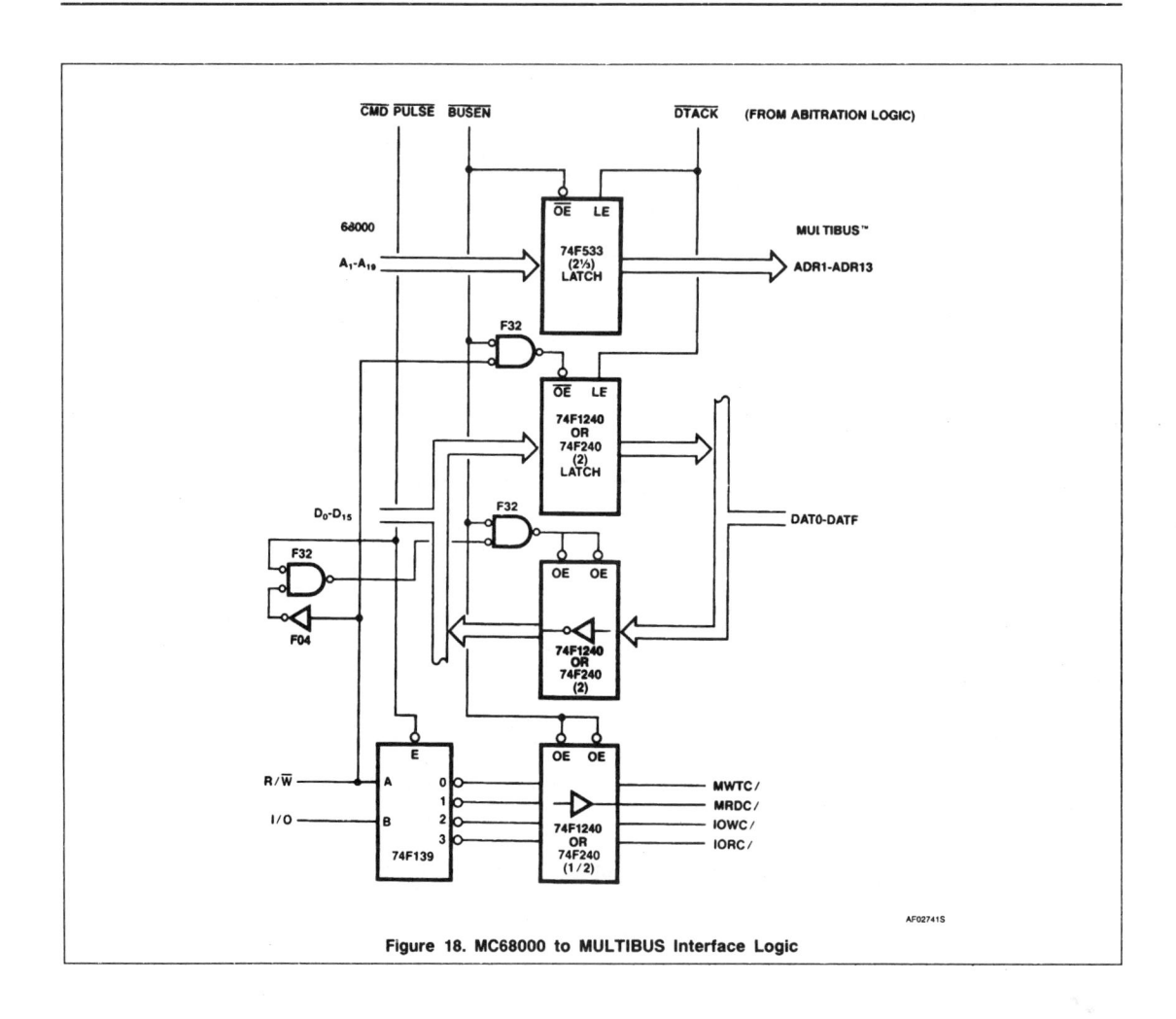

FAST Products

Application Note

Multiple μP Interfacing With FAST ICs

AN207

Figure 18. MC68000 to MULTIBUS Interface Logic

FAST Products

Application Note

Multiple µP Interfacing With FAST ICs

AN207

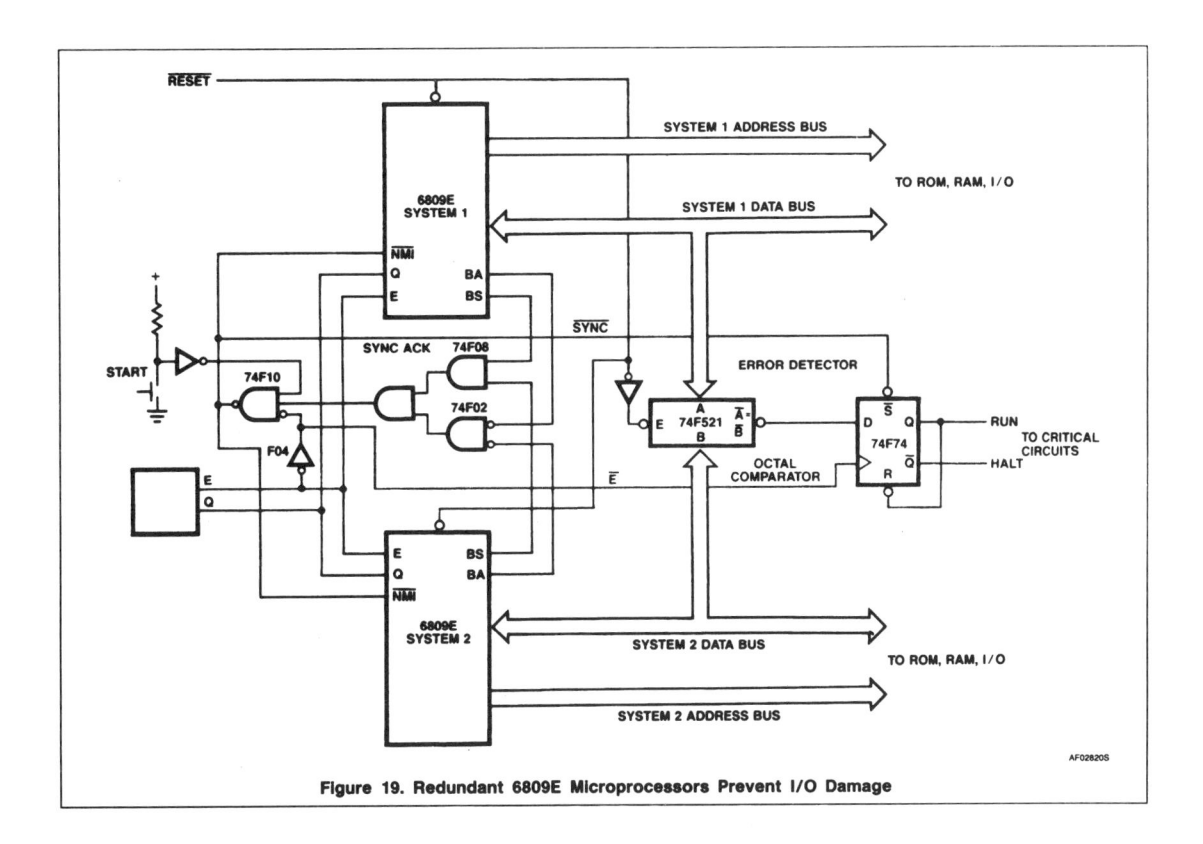

Figure 19. Redundant 6809E Microprocessors Prevent I/O Damage

Philips Components

PHD16N8–5
Programmable High-Speed Decoder Logic (16 × 16 × 8)

Preliminary Specification　　　　　October 1989

DESCRIPTION
The PHD16N8-5 is an ultra fast Programmable High-speed Decoder featuring a 5ns maximum propagation delay. The architecture has been optimized using Philips Components-Signetics state–of–the–art bipolar oxide isolation process coupled with titanium-tungsten fuses to achieve superior speed in any design.

The PHD16N8-5 is a single level logic element comprised of 10 fixed inputs, 8 AND gates, and 8 outputs of which 6 are bidirectional. This gives the device the ability to have as many as 16 inputs. Individual 3-State control of all outputs is also provided.

The device is field-programmable, enabling the user to quickly generate custom patterns using standard programming equipment. Proprietary designs can be protected by programming the security fuse.

The AMAZE software package from Philips Components-Signetics supports easy design entry for the PHD16N8-5 as well as other PLD devices.

Order codes are listed in the pages following.

FEATURES
- Ideal for high speed system decoding
- Super high speed at 5ns t_{PD}
- 10 dedicated inputs
- 8 outputs
  - 6 bidirectional I/O
  - 2 dedicated outputs
- Security fuse to prevent duplication of proprietary designs.
- Individual 3-State control of all outputs
- Field-programmable on industry standard programmers
- Available in 20-pin Plastic DIP and 20-Pin PLCC

APPLICATIONS
- High speed memory decoders
- High speed code detectors
- Random logic
- Peripheral selectors
- Machine state decoders
- Footprint compatible to 16L8
- Fuse/Footprint compatible to TIBPAD

PIN CONFIGURATIONS

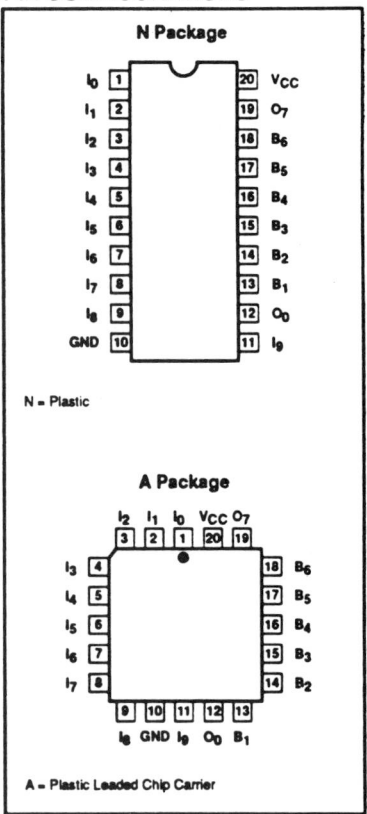

N Package

I_0	1	20	V_{CC}
I_1	2	19	O_7
I_2	3	18	B_6
I_3	4	17	B_5
I_4	5	16	B_4
I_5	6	15	B_3
I_6	7	14	B_2
I_7	8	13	B_1
I_8	9	12	O_0
GND	10	11	I_9

N = Plastic

A Package

Pins (top, left-to-right): I_2(3), I_1(2), I_0(1), V_{CC}(20), O_7(19)
Right side (top-to-bottom): B_6(18), B_5(17), B_4(16), B_3(15), B_2(14)
Bottom (left-to-right): I_8(9), GND(10), I_9(11), O_0(12), B_1(13)
Left side (top-to-bottom): I_3(4), I_4(5), I_5(6), I_6(7), I_7(8)

A = Plastic Leaded Chip Carrier

Preliminary Specification

Programmable High-Speed Decoder Logic (16 × 16 × 8)

PHD16N8-5

LOGIC DIAGRAM

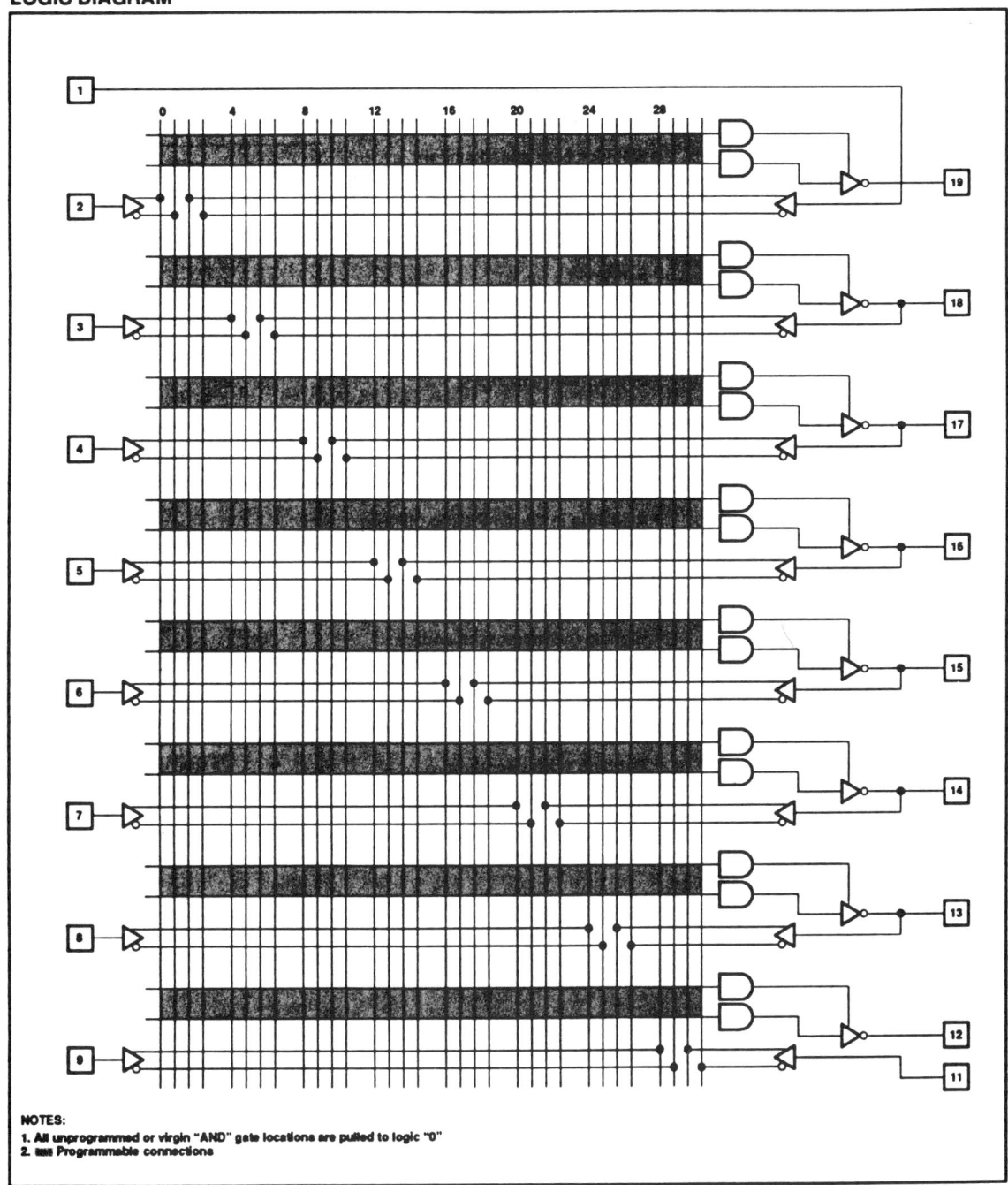

NOTES:
1. All unprogrammed or virgin "AND" gate locations are pulled to logic "0"
2. ■ Programmable connections

Preliminary Specification

Programmable High-Speed Decoder Logic (16 × 16 × 8)

PHD16N8–5

FUNCTIONAL DIAGRAM

[Functional diagram showing inputs I_0 through I_n feeding buffers into an AND-OR array with outputs $B_1 - B_6$ and O_0, O_7]

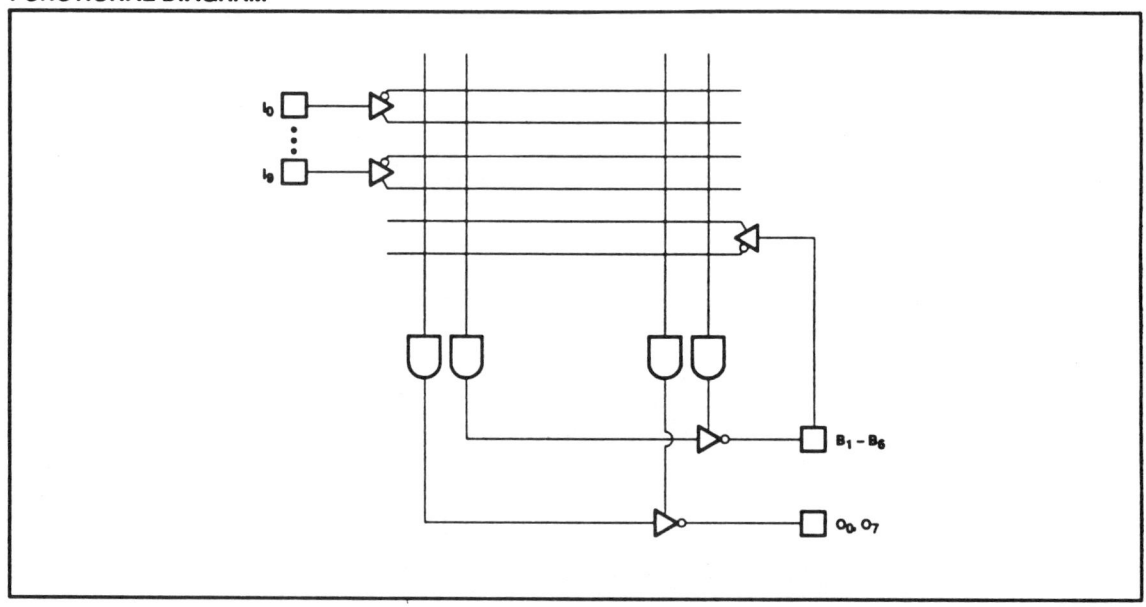

ORDERING INFORMATION

DESCRIPTION	ORDER CODE
20–Pin Plastic Dual In Line Package; (300mil–wide)	PHD16N8–5N
20–Pin Plastic Leaded Chip Carrier; (350mil square)	PHD16N8–5A

THERMAL RATINGS

TEMPERATURE	
Maximum junction	150°C
Maximum ambient	75°C
Allowable thermal rise ambient to junction	75°C

ABSOLUTE MAXIMUM RATINGS[1]

SYMBOL	PARAMETER	Min	Max	UNIT
V_{CC}	Supply voltage	–0.5	+7	V_{DC}
V_{IN}	Input voltage	–0.5	+5.5	V_{DC}
V_{OUT}	Output voltage		+5.5	V_{DC}
I_{IN}	Input currents	–30	+30	mA
I_{OUT}	Output currents		+100	mA
T_A	Operating temperature range	0	+75	°C
T_{STG}	Storage temperature range	–65	+150	°C

NOTES:
1. Stresses above those listed may cause malfunction or permanent damage to the device. This is a stress rating only. Functional operation at these or any other condition above those indicated in the operational and programming specification of the device is not implied.

OPERATING RANGES

SYMBOL	PARAMETER	Min	Max	UNIT
V_{CC}	Supply voltage	+4.75	+5.25	V_{DC}
T_A	Operating free-air temperature	0	+75	°C

Preliminary Specification

Programmable High-Speed Decoder Logic (16 × 16 × 8)

PHD16N8–5

DC ELECTRICAL CHARACTERISTICS $0°C \leq T_A \leq +75°C$, $4.75 \leq V_{CC} \leq 5.25V$

SYMBOL	PARAMETER	TEST CONDITIONS	Min	Typ[1]	Max	UNIT
Input voltage[2]						
V_{IL}	Low	V_{CC} = MIN			0.8	V
V_{IH}	High	V_{CC} = MAX	2.0			V
V_{IC}	Clamp	V_{CC} = MIN, I_{IN} = –18mA		–0.8	–1.5	V
Output voltage						
V_{OL}	Low	V_{CC} = MIN, V_{IN} = V_{IH} or V_{IL} I_{OL} = +24mA			0.5	V
V_{OH}	High	I_{OH} = –3.2mA	2.4			V
Input current						
I_{IL}	Low	V_{CC} = MAX, V_{IN} = +0.40V		–20	–250	μA
I_{IH}	High	V_{IN} = +2.7V			25	μA
I_I	High	V_{IN} = V_{CC} = $V_{CC\,MAX}$			100	μA
Output current						
I_{OZH}	Output leakage [3]	V_{CC} = MAX, V_{OUT} = +2.7V			100	μA
I_{OZL}	Output leakage [3]	V_{OUT} = +0.40V			–100	μA
I_{OS}	Short circuit [4]	V_{OUT} = 0V	–30		–90	mA
I_{CC}	V_{CC} supply current	V_{CC} = MAX		115	180	mA
Capacitance[5]						
C_{IN}	Input	V_{CC} = +5V, V_{IN} = 2.0V @ f = 1MHz		8		pF
C_{OUT}	I/O (B)	V_{OUT} = 2.0V @ f = 1MHz		8		pF

NOTES:
1. Typical limits are at V_{CC} = 5.0V and T_A = +25°C.
2. These are absolute values with respect to device ground and all overshoots due to system or tester noise are included.
3. Leakage current for bidirectional pins is the worst case of I_{IL} and I_{OZL} or I_{IH} and I_{OZH}.
4. Not more than one output should be tested at a time. Duration of the short circuit should not be more than one second.
5. These parameters are not 100% tested, but are periodically sampled.

Programmable High-Speed Decoder Logic (16 × 16 × 8)

PHD16N8-5

Preliminary Specification

AC ELECTRICAL CHARACTERISTICS $0°C \leq T_A \leq +75°C$, $4.75 \leq V_{CC} \leq 5.25V$, $R_1 = 200\Omega$, $R_2 = 390\Omega$

SYMBOL	PARAMETER	FROM	TO	TEST CONDITIONS	LIMITS Min	LIMITS Max	UNIT
t_{PD}[1]	Propagation delay	(I, B) ±	Output ±	C_L = 50pF		5	ns
t_{OE}[2]	Output Enable	(I, B) ±	Output enable	C_L = 50pF		10	ns
t_{OD}[2]	Output Disable	(I, B) ±	Input disable	C_L = 5pF		10	ns

NOTES:
1. t_{PD} is tested with switch S_1 closed and C_L = 50pF.
2. For 3-State output; output enable times are tested with C_L = 50pF to the 1.5V level, and S_1 is open for high-impedance to High tests and closed for high-impedance to Low tests. Output disable times are tested with C_L = 5pF. High-to-High impedance tests are made to an output voltage of $V_T = V_{OH} - 0.5V$ with S_1 open, and Low-to-High impedance tests are made to the $V_T = V_{OL} + 0.5V$ level with S_1 closed.

VIRGIN STATE
A factory shipped virgin device contains all fusible links open, such that:
1. All outputs are disabled.
2. All p-terms are disabled in the AND array.

TIMING DEFINITIONS

SYMBOL	PARAMETER
t_{PD}	Input to output propagation delay.
t_{OD}	Input to Output Disable (3-State) delay (Output Disable).
t_{OE}	Input to Output Enable delay (Output Enable).

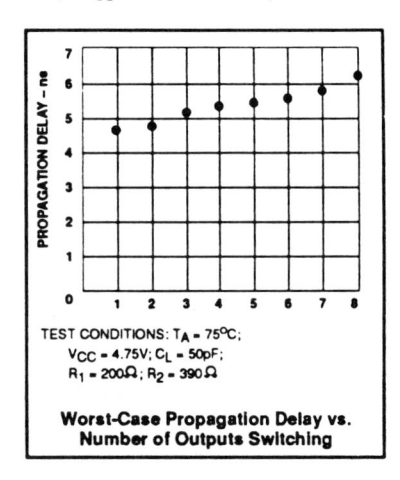

Worst-Case Propagation Delay vs. Number of Outputs Switching

TEST CONDITIONS: T_A = 75°C; V_{CC} = 4.75V; C_L = 50pF; R_1 = 200Ω; R_2 = 390Ω.

TIMING DIAGRAM

WAVEFORM	INPUTS	OUTPUTS
(steady)	MUST BE STEADY	WILL BE STEADY
(changing)	DON'T CARE; ANY CHANGE PERMITTED	CHANGING; STATE UNKNOWN
(3-state)	DOES NOT APPLY	CENTER LINE IS HIGH IMPEDANCE "OFF" STATE

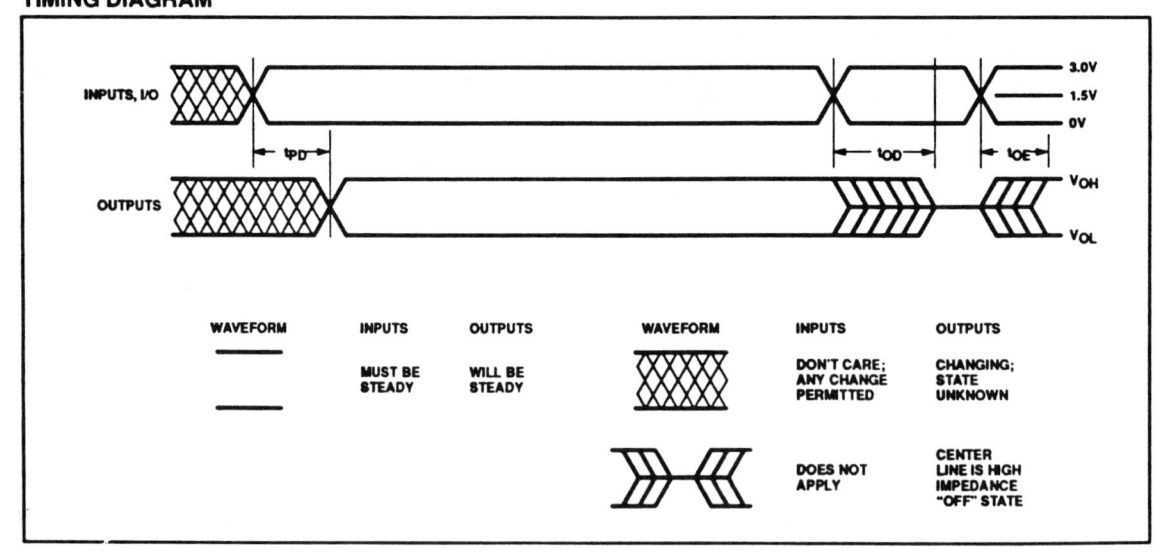

Preliminary Specification

Programmable High-Speed Decoder Logic (16 × 16 × 8)

PHD16N8–5

AC TEST LOAD CIRCUIT

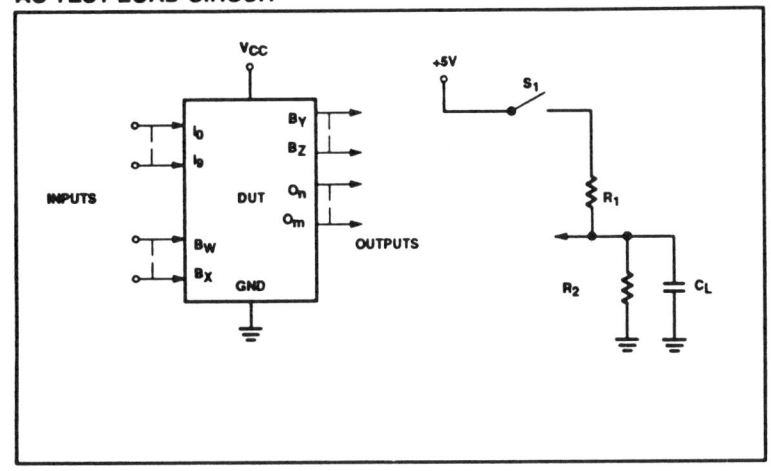

VOLTAGE WAVEFORMS

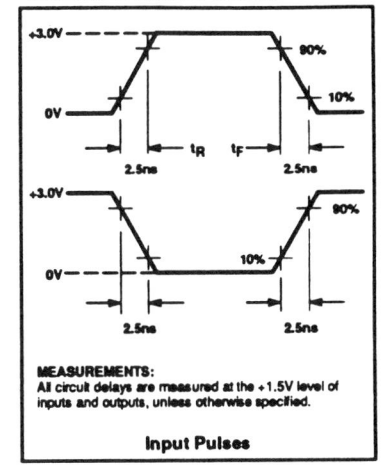

MEASUREMENTS:
All circuit delays are measured at the +1.5V level of inputs and outputs, unless otherwise specified.

Input Pulses

LOGIC PROGRAMMING

PHD16N8–5 logic designs can be generated using any commercially available, JEDEC standard design software.

PHD16N8–5 designs can also be generated using the program table format, detailed on the following page. This program table entry (PTE) format is supported on the Signetics AMAZE PLD design software.

To implement the desired logic functions, each logic variable (I, B, P and D) from the logic equations is assigned a symbol. TRUE (High), COMPLEMENT (Low), DON'T CARE and INACTIVE symbols are defined below.

"AND" ARRAY – (I, B)

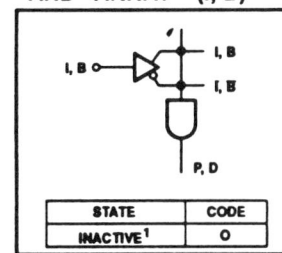

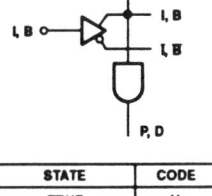

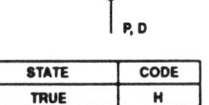

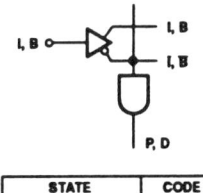

STATE	CODE
INACTIVE[1]	0

STATE	CODE
TRUE	H

STATE	CODE
COMPLEMENT	L

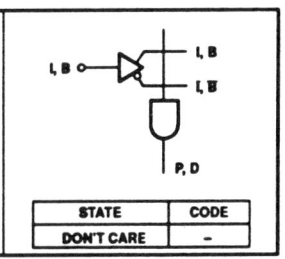

STATE	CODE
DON'T CARE	–

NOTE:
1. This is the initial state.

Preliminary Specification

Programmable High-Speed Decoder Logic (16 × 16 × 8)

PHD16N8–5

PROGRAM TABLE

OR (FIXED)		
DIRECTION	D	
ACTIVE OUTPUT		A
NOT USED		

AND		
INACTIVE	0	
I, B	H	
Ī, B̄	L	
DON'T CARE	—	I, B(Ī)

AND / OR (FIXED) Array

TERM	INPUT (I) 9 8 7 6 5 4 3 2 1 0	INPUTS (B) 6 5 4 3 2 1	OUTPUTS (B, O) 7 6 5 4 3 2 1 0
0			D
1			A
2			D
3			A
4			D
5			A
6			D
7			A
8			D
9			A
10			D
11			A
12			D
13			A
14			D
15			A
PIN	11 9 8 7 6 5 4 3 2 1	18 17 16 15 14 13	19 18 17 16 15 14 13 12

VARIABLE NAME

CUSTOMER NAME
PURCHASE ORDER #
SIGNETICS DEVICE #
CUSTOMER SYMBOLIZED PART # CF(XXXX)
TOTAL NUMBER OF PARTS
PROGRAM TABLE #   REV____ DATE____

NOTES:
1. The PHD16N8-5 is shipped with all links intact.
2. Unused I and B bits in the AND array exist as INACTIVE in the virgin state.
3. All p–terms are inactive until programmed otherwise.
4. Data cannot be entered into the OR array field due to the fixed nature of the device architecture.

Preliminary Specification

Programmable High-Speed
Decoder Logic (16 × 16 × 8)

PHD16N8-5

DECODING 1/2 MEG STATIC MEMORY

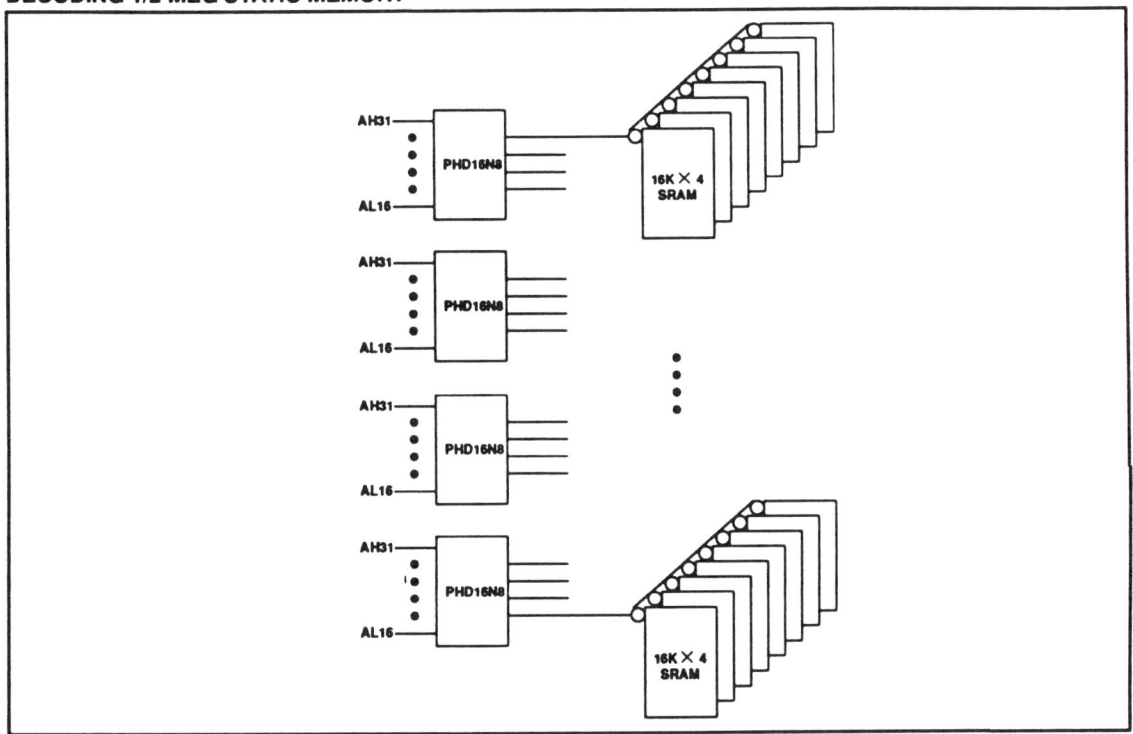

DEFINITIONS

Data Sheet Identification	Product Status	Definition
Objective Specification	Formative or In Design	This data sheet contains the design target or goal specifications for product development. Specifications may change in any manner without notice.
Preliminary Specification	Preproduction Product	This data sheet contains preliminary data and supplementary data will be published at a later date. Signetics reserves the right to make changes at any time without notice in order to improve design and supply the best possible product.
Product Specification	Full Production	This data sheet contains Final Specifications. Signetics reserves the right to make changes at any time without notice in order to improve design and supply the best possible product.

Signetics reserves the right to make changes, without notice, in the products, including circuits, standard cells, and/or software, described or contained herein in order to improve design and/or performance. Signetics assumes no responsibility or liability for the use of any of these products, conveys no license or title under any patent, copyright, or mask work right to these products, and makes no representations or warranties that these products are free from patent, copyright, or mask work right infringement, unless otherwise specified. Applications that are described herein for any of these products are for illustrative purposes only. Signetics makes no representation or warranty that such applications will be suitable for the specified use without further testing or modification.

Philips Components Ltd., Mullard House, Torrington Place, London, WC1E 7HD. Tel: 01-580 6633 Telex: 264341 Fax: 01-636 0394
M90-1038/CC

PHILIPS

Chapter 4 Memory systems

As a microcomputer's memory system often dominates both its price and its performance, it seems fitting that this chapter should dominate this book. We attempt to do two things. The first is to introduce the range of components available to the designer of memory systems. The second is to look at some of the more specialized aspects of memory systems design such as error correcting memories and dynamic memory controllers.

The first part of this chapter provides data sheets for various modern memory devices. In each case, we provide the data sheet of a typical high-performance specimen. We begin with conventional high speed static memory components (a 32K-byte and a 128K-byte device) and a tutorial on the design of low power memory systems. Next we look at read-only memories the EPROM, flash EPROM, and EEPROM. In this section we also provide information on the applications of some of these devices and look at special memory devices (like the fast pipelined EPROM and the 16-bit wide EPROM).

No chapter on modern memories would be complete without a section on dynamic memory, DRAM. Here we provide the data sheet of a modern 4M bit chip and a short application note on some of the special operating modes of a DRAM (e.g., the page mode and static column modes). Although you can design your own interface to the DRAM, many engineers employ a commercial DRAM controller chip. In this section we include two data sheets, one on a modest controller that does little more than multiplex addresses. The other is a complete and highly versatile DRAM controller system. A short application note gives you an idea of some of the calculations that you must perform if you use this very sophisticated DRAM controller in a typical high-performance microcomputer.

Two interesting types of memory component are the FIFO (first-in-first-out) memory used to buffer the transfer of data between two systems, and the multi-port RAM used to provide a mechanism for coupling microprocessors in a multiprocessor system. We provide data sheets and application notes for both these devices.

DRAM suffers from occasional random errors due to the effect of radiation within the die's encapsulation. Errors can be corrected by storing extra bits along with each data word and then using these bits to detect and correct single (or even multiple) errors. We have included an application report that describes the operation of an error detecting and correcting memory. Advances in processor technology have forced memory technology to keep pace. Unfortunately, low-cost 80 ns DRAM is often just not fast enough for processors running at frequencies of 30 MHz and above. Cache memory can be included in microcomputers to improve the performance of high speed microprocessors. We provide applications reports for cache memory and look at one of the chips that can be used to design a cache system.

MOTOROLA
SEMICONDUCTOR
TECHNICAL DATA

Order this data sheet
by MCM6206/D

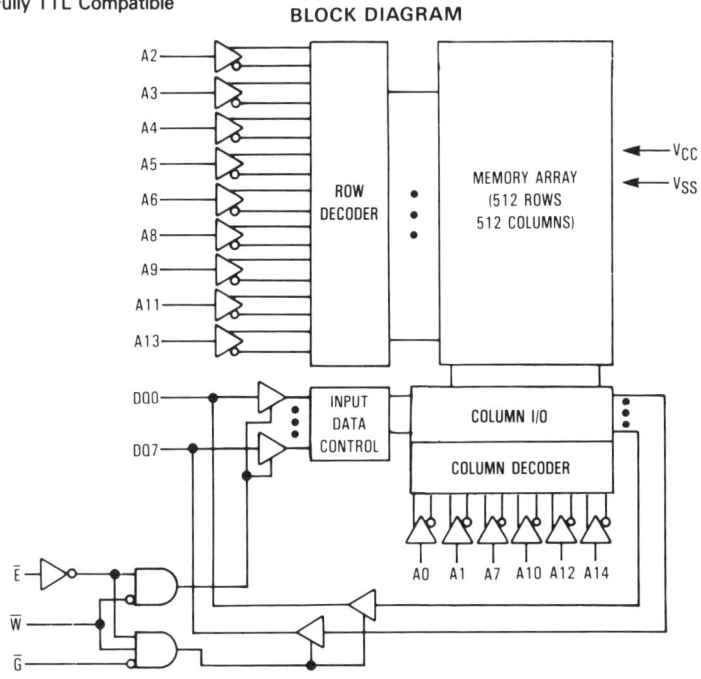

MCM6206

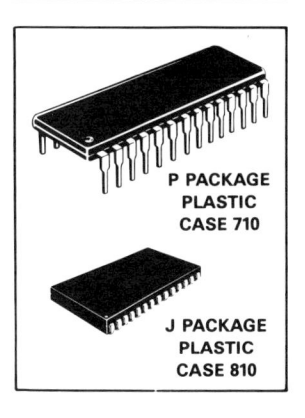

Advance Information
32K × 8 Bit Fast Static Random Access Memory

The MCM6206 is a 262,144 bit static random access memory organized as 32,768 words of 8 bits, fabricated using Motorola's high-performance silicon-gate CMOS technology. Static design eliminates the need for external clocks or timing strobes, while CMOS circuitry reduces power consumption and provides for greater reliability.

Chip enable ($\overline{E}$) controls the power-down feature. It is not a clock but rather a chip control that affects power consumption. In less than a cycle time after $\overline{E}$ goes high, the part automatically reduces its power requirements and remains in this low-power standby mode as long as $\overline{E}$ remains high. This feature provides significant system-level power savings. Another control feature, output enable ($\overline{G}$) allows access to the memory contents as fast as 15 ns (MCM6206-35).

The MCM6206 is packaged in a 600 mil, 28 pin plastic dual-in-line package or a 28 lead 400 mil plastic SOJ package with the JEDEC standard pinout.

- Single 5 V Supply, ±10%
- Fully Static—No Clock or Timing Strobes Necessary
- Fast Access Time—35 or 45 ns (Maximum)
- Low Power Dissipation
- Two Chip Controls; $\overline{E}$ for Automatic Power Down
  $\overline{G}$ for Fast Access to Data
- Three State Outputs
- Fully TTL Compatible

**P PACKAGE
PLASTIC
CASE 710**

**J PACKAGE
PLASTIC
CASE 810**

PIN ASSIGNMENT

A14	1 •	28	V_{CC}
A12	2	27	$\overline{W}$
A7	3	26	A13
A6	4	25	A8
A5	5	24	A9
A4	6	23	A11
A3	7	22	$\overline{G}$
A2	8	21	A10
A1	9	20	$\overline{E}$
A0	10	19	DQ7
DQ0	11	18	DQ6
DQ1	12	17	DQ5
DQ2	13	16	DQ4
V_{SS}	14	15	DQ3

BLOCK DIAGRAM

PIN NAMES

A0-A14	Address
$\overline{W}$	Write Enable
$\overline{E}$	Chip Enable
$\overline{G}$	Output Enable
DQ0-DQ7	Data Input/Output
V_{CC}	+5 V Power Supply
V_{SS}	Ground

This document contains information on a new product. Specifications and information herein are subject to change without notice.

MOTOROLA

©MOTOROLA INC., 1989

ADI1501R1

TRUTH TABLE

$\overline{E}$	$\overline{G}$	$\overline{W}$	Mode	Supply Current	I/O Pin
H	X	X	Not Selected	I_{SB}	High Z
L	H	H	Output Disabled	I_{CC}	High Z
L	L	H	Read	I_{CC}	D_{out}
L	X	L	Write	I_{CC}	D_{in}

X — Don't Care

This device contains circuitry to protect the inputs against damage due to high static voltages or electric fields; however, it is advised that normal precautions be taken to avoid application of any voltage higher than maximum rated voltages to this high-impedance circuit.

ABSOLUTE MAXIMUM RATINGS (See Note)

Rating	Symbol	Value	Unit
Power Supply Voltage	V_{CC}	−0.5 to +7.0	V
Voltage Relative to V_{SS} for Any Pin Except V_{CC}	V_{in}, V_{out}	−0.5 to V_{CC} + 0.5	V
Output Current (per I/O)	I_{out}	±20	mA
Power Dissipation (T_A = 25°C)	P_D	1.0	W
Temperature Under Bias	T_{bias}	−10 to +85	°C
Operating Temperature	T_A	0 to +70	°C
Storage Temperature — Plastic	T_{stg}	−55 to +125	°C

NOTE: Permanent device damage may occur if ABSOLUTE MAXIMUM RATINGS are exceeded. Functional operation should be restricted to RECOMMENDED OPERATING CONDITIONS. Exposure to higher than recommended voltages for extended periods of time could affect device reliability.

DC OPERATING CONDITIONS AND CHARACTERISTICS
(V_{CC} = 5.0 ±10%, T_A = 0 to 70°C, Unless Otherwise Noted)

RECOMMENDED OPERATING CONDITIONS

Parameter	Symbol	Min	Typ	Max	Unit
Supply Voltage (Operating Voltage Range)	V_{CC}	4.5	5.0	5.5	V
Input High Voltage	V_{IH}	2.2	—	V_{CC} + 0.3	V
Input Low Voltage	V_{IL}	−0.3*	—	0.8	V

*V_{IL} (min) = −0.3 V dc; V_{IL} (min) = −3.0 V ac (pulse width ≤ 20 ns)

DC CHARACTERISTICS

Parameter		Symbol	Min	Max	Unit
Input Leakage Current (All Inputs, V_{in} = 0 to V_{CC})		$I_{lkg(I)}$	—	±1.0	µA
Output Leakage Current ($\overline{E}$ = V_{IH}, or $\overline{G}$ = V_{IH}, V_{out} = 0 to 5.5 V)		$I_{lkg(O)}$	—	±1.0	µA
Power Supply Current ($\overline{E}$ = V_{IL}, I_{out} = 0)	(t_{AVAV} = 35 ns)	I_{CC}	—	120	mA
	(t_{AVAV} = 45 ns)	I_{CC}	—	110	mA
Standby Current ($\overline{E}$ = V_{IH}) (TTL Levels)		I_{SB1}	—	20	mA
Standby Current ($\overline{E}$ ≥ V_{CC} − 0.2 V) (CMOS Levels)		I_{SB2}	—	15	mA
Output Low Voltage (I_{OL} = 8.0 mA)		V_{OL}	—	0.4	V
Output High Voltage (I_{OH} = −4.0 mA)		V_{OH}	2.4	—	V

CAPACITANCE (f = 1.0 MHz, T_A = 25°C, periodically sampled and not 100% tested.)

Characteristic	Symbol	Max	Unit
Input Capacitance	C_{in}	6	pF
I/O Capacitance	$C_{I/O}$	8	pF

MOTOROLA · MCM6206

AC OPERATING CONDITIONS AND CHARACTERISTICS
(V_{CC} = 5 V ± 10%, T_A = 0 to 70°C, Unless Otherwise Noted)

Input Pulse Levels . . . . 0 to 3.0 V
Input Rise/Fall Time . . . . 5 ns
Input Timing Measurement Reference Levels . . . . 1.5 V

Output Timing Measurement Reference Levels . . . . 1.5 V
Output Load. . . . . See Figure 1

READ CYCLE 1 & 2 (See Note 1)

Parameter	Symbol	Alt Symbol	MCM6206-35 Min	MCM6206-35 Max	MCM6206-45 Min	MCM6206-45 Max	Unit	Notes
Read Cycle Time	t_{AVAV}	t_{RC}	35	—	45	—	ns	—
Address Access Time	t_{AVQV}	t_{AA}	—	35	—	45	ns	—
$\overline{E}$ Access Time	t_{ELQV}	t_{AC}	—	35	—	45	ns	—
$\overline{G}$ Access Time	t_{GLQV}	t_{OE}	—	15	—	20	ns	—
Enable Low to Enable High	t_{ELEH}	t_{CW}	35	—	45	—	ns	—
Output Hold from Address Change	t_{AXQX}	t_{OH}	5	—	5	—	ns	2
Chip Enable to Output Low-Z	t_{ELQX}	t_{CLZ}	10	—	10	—	ns	2, 3
Output Enable to Output Low-Z	t_{GLQX}	t_{OLZ}	0	—	0	—	ns	2, 3
Chip Enable to Output High-Z	t_{EHQZ}	t_{CHZ}	0	20	0	20	ns	2, 3
Output Enable to Output High-Z	t_{GHQZ}	t_{OHZ}	0	20	0	20	ns	2, 3

NOTES:
1. $\overline{W}$ is high at all times for read cycles.
2. All high-Z and low-Z parameters are considered in a high or low impedance state when the output has made a 500 mV transition from the previous steady state voltage.
3. These parameters are periodically sampled and not 100% tested.

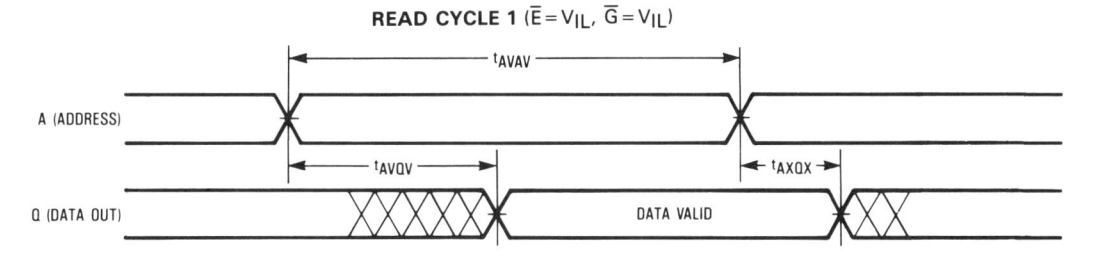

READ CYCLE 1 ($\overline{E} = V_{IL}$, $\overline{G} = V_{IL}$)

READ CYCLE 2

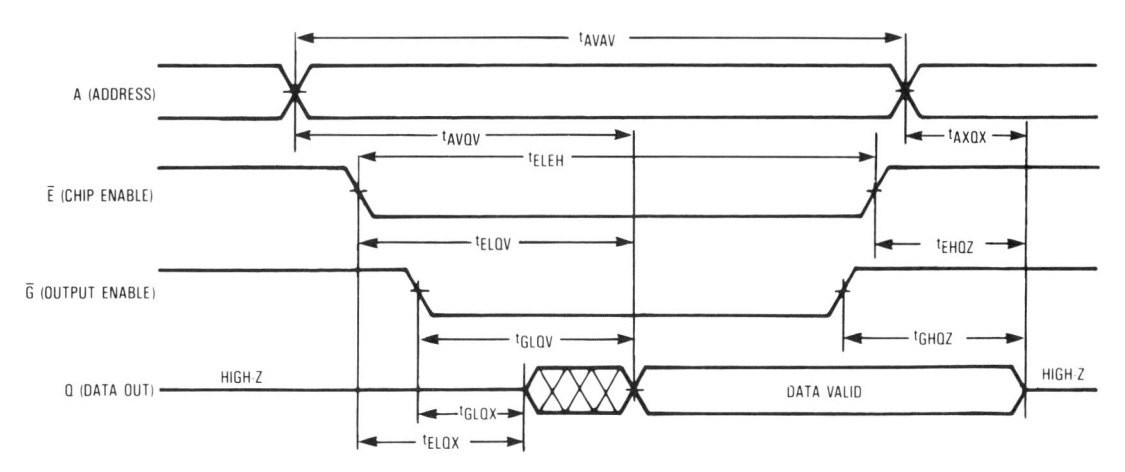

MCM6206　　　　　　　　　　　　　　　　　　　　　　　　　　　　　　　　MOTOROLA

WRITE CYCLE 1 & 2 (See Note 1)

Parameter	Symbol	Alt Symbol	MCM6206-35 Min	MCM6206-35 Max	MCM6206-45 Min	MCM6206-45 Max	Unit	Notes
Write Cycle Time	t_{AVAV}	t_{WC}	35	—	45	—	ns	—
Address Setup to Write Low Address Setup to Enable Low	t_{AVWL} t_{AVEL}	t_{AS}	0	—	0	—	ns	2
Address Valid to Write High Address Valid to Enable High	t_{AVWH} t_{AVEH}	t_{AW}	25	—	35	—	ns	—
Data Valid to Write High Data Valid to Enable High	t_{DVWH} t_{DVEH}	t_{DW}	15	—	20	—	ns	—
Data Hold From Write High Data Hold From Enable High	t_{WHDX} t_{EHDX}	t_{DH}	0	—	0	—	ns	—
Write Recovery Time Enable Recovery Time	t_{WHAX} t_{EHAX}	t_{WR}	0	—	0	—	ns	2
Chip Enable to End of Write Enable Low to Enable High	t_{ELWH} t_{ELEH}	t_{CW}	25	—	35	—	ns	1
Write Pulse Width	t_{WLWH}	t_{WP}	25	—	30	—	ns	3
Write Low to Output High-Z	t_{WLQZ}	t_{WHZ}	0	20	0	20	ns	4, 5
Write High to Output Low-Z	t_{WHQX}	t_{WLZ}	5	—	5	—	ns	4, 5

NOTES:
1. A write cycle starts at the latest transition of a low $\overline{E}$ or low $\overline{W}$. A write cycle ends at the earliest transition of a high $\overline{E}$ or high $\overline{W}$.
2. $\overline{W}$ must be high during all address transitions.
3. If $\overline{G}$ is enabled, allow an additional 15 ns t_{WLWH} to avoid bus contention.
4. All high-Z and low-Z parameters are considered in a high or low impedance state when the output has made a 500 mV transition from the previous steady state voltage.
5. These parameters are periodically sampled and not 100% tested.

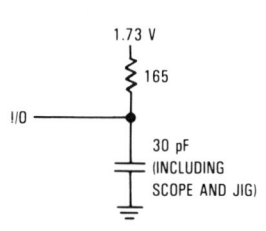

Figure 1. Test Load

WRITE CYCLE 1 ($\overline{W}$ Controlled)

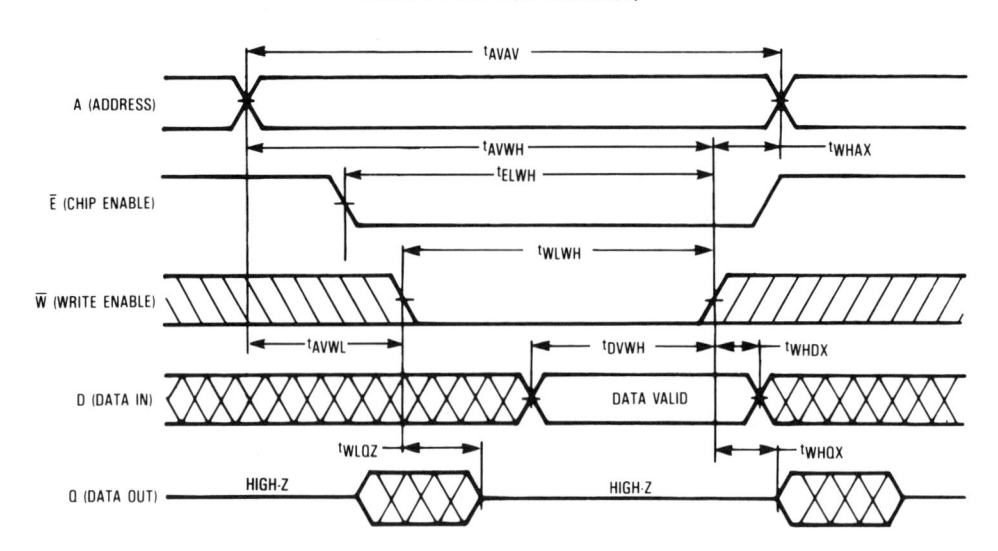

WRITE CYCLE 2 ($\overline{E}$ Controlled)

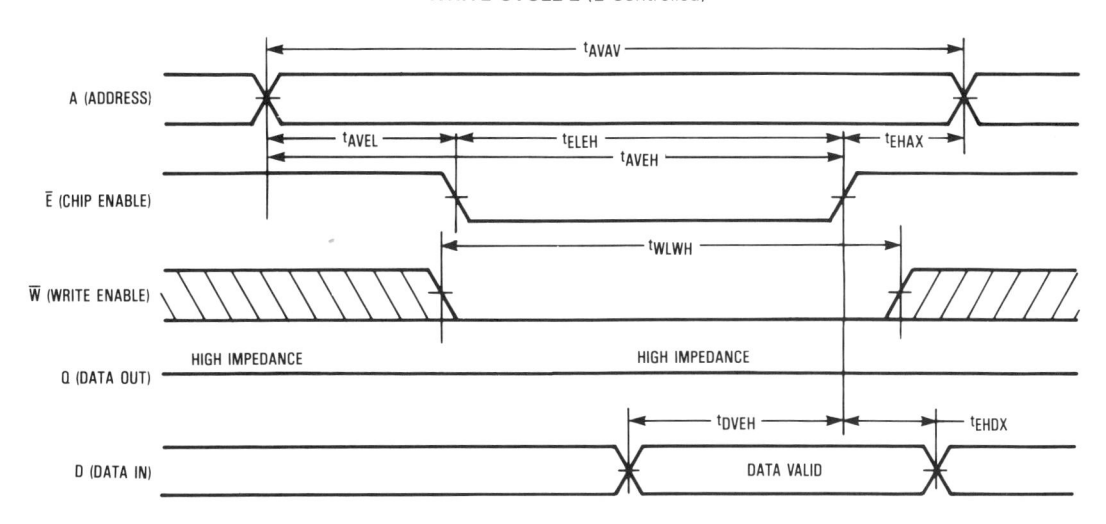

ORDERING INFORMATION
(Order by Full Part Number)

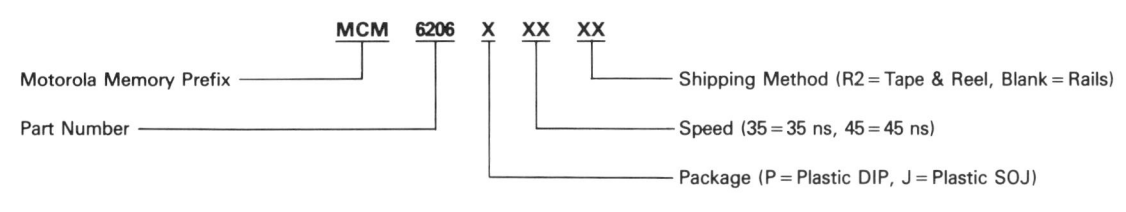

```
MCM 6206 X XX XX
```

- Motorola Memory Prefix — MCM
- Part Number — 6206
- Package (P = Plastic DIP, J = Plastic SOJ)
- Speed (35 = 35 ns, 45 = 45 ns)
- Shipping Method (R2 = Tape & Reel, Blank = Rails)

Full Part Numbers — MCM6206P35   MCM6206P45
MCM6206J35   MCM6206J45
MCM6206J35R2 MCM6206J45R2

PACKAGE DIMENSIONS

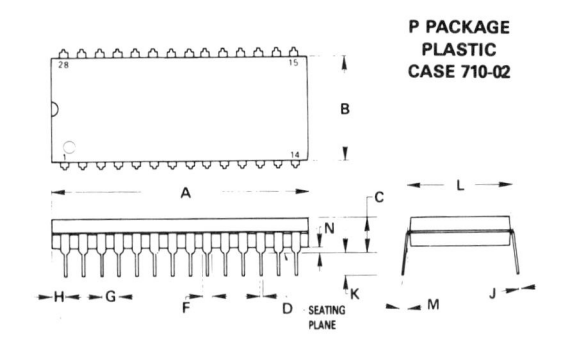

P PACKAGE
PLASTIC
CASE 710-02

DIM	MILLIMETERS MIN	MILLIMETERS MAX	INCHES MIN	INCHES MAX
A	36.45	37.21	1.435	1.465
B	13.72	14.22	0.540	0.560
C	3.94	5.08	0.155	0.200
D	0.36	0.56	0.014	0.022
F	1.02	1.52	0.040	0.060
G	2.54 BSC		0.100 BSC	
H	1.65	2.16	0.065	0.085
J	0.20	0.38	0.008	0.015
K	2.92	3.43	0.115	0.135
L	15.24 BSC		0.600 BSC	
M	0°	15°	0°	15°
N	0.51	1.02	0.020	0.040

NOTES:
1. POSITIONAL TOLERANCE OF LEADS (D), SHALL BE WITHIN 0.25mm(0.010) AT MAXIMUM MATERIAL CONDITION, IN RELATION TO SEATING PLANE AND EACH OTHER.
2. DIMENSION L TO CENTER OF LEADS WHEN FORMED PARALLEL.
3. DIMENSION B DOES NOT INCLUDE MOLD FLASH.

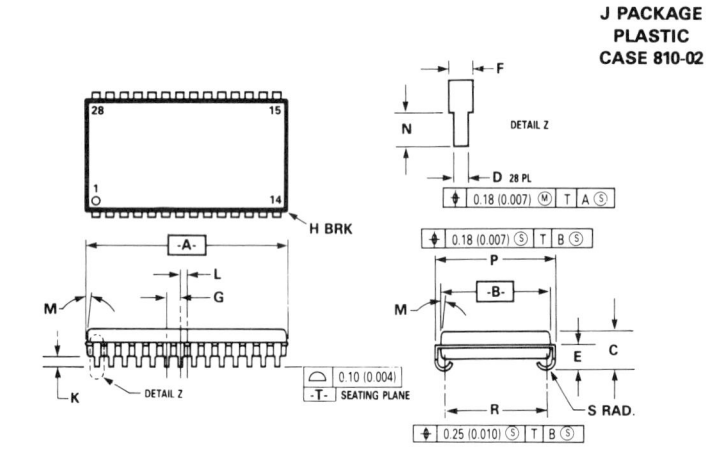

J PACKAGE
PLASTIC
CASE 810-02

DIM	MILLIMETERS MIN	MILLIMETERS MAX	INCHES MIN	INCHES MAX
A	18.29	18.54	0.720	0.730
B	10.04	10.28	0.395	0.405
C	3.26	3.75	0.128	0.148
D	0.39	0.50	0.015	0.020
E	2.24	2.48	0.088	0.098
F	0.67	0.81	0.026	0.032
G	1.27 BSC		0.050 BSC	
H	—	0.50	—	0.020
K	0.89	1.14	0.035	0.045
L	0.64 BSC		0.025 BSC	
M	0°	5°	0°	5°
N	0.89	1.14	0.035	0.045
P	11.05	11.30	0.435	0.445
R	9.15	9.65	0.360	0.380
S	0.77	1.01	0.030	0.040

NOTES:
1. DIMENSIONING AND TOLERANCING PER ANSI Y14.5M, 1982.
2. DIMENSION "A" AND "B" DO NOT INCLUDE MOLD PROTRUSION. MOLD PROTRUSION SHALL NOT EXCEED 0.15 (0.006) PER SIDE.
3. CONTROLLING DIMENSION: INCH.
4. DIM "R" TO BE DETERMINED AT DATUM -T-.

Motorola reserves the right to make changes without further notice to any products herein to improve reliability, function or design. Motorola does not assume any liability arising out of the application or use of any product or circuit described herein; neither does it convey any license under its patent rights nor the rights of others. Motorola products are not authorized for use as components in life support devices or systems intended for surgical implant into the body or intended to support or sustain life. Buyer agrees to notify Motorola of any such intended end use whereupon Motorola shall determine availability and suitability of its product or products for the use intended. Motorola and Ⓜ are registered trademarks of Motorola, Inc. Motorola, Inc. is an Equal Employment Opportunity/Affirmative Action Employer.

Literature Distribution Centers:
USA: Motorola Literature Distribution; P.O. Box 20912; Phoenix, Arizona 85036.
EUROPE: Motorola Ltd.; European Literature Center; 88 Tanners Drive, Blakelands, Milton Keynes, MK14 5BP, England.
ASIA PACIFIC: Motorola Semiconductors H.K. Ltd.; P.O. Box 80300; Cheung Sha Wan Post Office; Kowloon Hong Kong.
JAPAN: Nippon Motorola Ltd.; 3-20-1 Minamiazabu, Minato-ku, Tokyo 106 Japan.

Ⓜ MOTOROLA

MCM6206

HM628128 Series

131072-Word × 8-Bit High Speed CMOS Static RAM

The Hitachi HM628128 is a CMOS static RAM organized 128-kword x 8-bit. It realizes higher density, higher performance and low power consumption by employing 0.8 μm Hi-CMOS process technology.

It offers low power standby power dissipation; therefore, it is suitable for battery back-up systems. The device, packaged in a 525 mil SOP (460-mil body SOP) or a 600-mil plastic DIP, is available for high density mounting.

Features
- High speed: Fast access time 70/85/100/120 ns (max.)
- Low power
  Standby: 10 μW (typ) (L-/L-SL version)
  Operation: 75 mW (typ)
- Single 5 V supply
- Completely static memory
  No clock or timing strobe required
- Equal access and cycle times
- Common data input and output: Three state output
- Directly TTL compatible: All inputs and outputs
- Capability of battery back up operation (L-/L-SL version)
  2 chip selection for battery back up

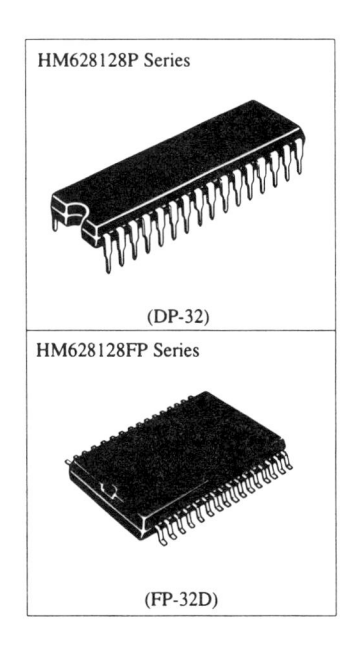

HM628128P Series

(DP-32)

HM628128FP Series

(FP-32D)

Pin Description

Pin Name	Function
A0 – A16	Address
I/O0 – I/O7	Input/output
$\overline{CS1}$	Chip select 1
CS2	Chip select 2
$\overline{WE}$	Write enable
$\overline{OE}$	Output enable
NC	No connection
V__CC_	Power supply
V__SS_	Ground

Pin Arrangement

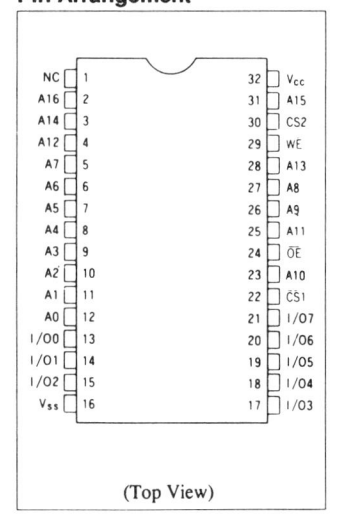

```
         NC  [ 1    32 ] Vcc
        A16  [ 2    31 ] A15
        A14  [ 3    30 ] CS2
        A12  [ 4    29 ] WE
         A7  [ 5    28 ] A13
         A6  [ 6    27 ] A8
         A5  [ 7    26 ] A9
         A4  [ 8    25 ] A11
         A3  [ 9    24 ] OE
         A2  [ 10   23 ] A10
         A1  [ 11   22 ] CS1
         A0  [ 12   21 ] I/O7
        I/O0 [ 13   20 ] I/O6
        I/O1 [ 14   19 ] I/O5
        I/O2 [ 15   18 ] I/O4
         Vss [ 16   17 ] I/O3
```

(Top View)

⦿ HITACHI

HM628128 Series

Ordering Information

Type No.	Access Time	Package	Type No.	Access Time	Package
HM628128P-7	70 ns		HM628128FP-7	70 ns	
HM628128P-8	85 ns		HM628128FP-8	85 ns	
HM628128P-10	100 ns		HM628128FP-10	100 ns	
HM628128P-12	120 ns		HM628128FP-12	120 ns	
HM628128LP-7	70 ns	600 mil 32-pin plastic DIP (DP-32)	HM628128LFP-7	70 ns	525 mil 32-pin plastic DIP (FP-32D)
HM628128LP-8	85 ns		HM628128LFP-8	85 ns	
HM628128LP-10	100 ns		HM628128LFP-10	100 ns	
HM628128LP-12	120 ns		HM628128LFP-12	120 ns	
HM628128LP-7SL	70 ns		HM628128LFP-7SL	70 ns	
HM628128LP-8SL	85 ns		HM628128LFP-8SL	85 ns	
HM628128LP-10SL	100 ns		HM628128LFP-10SL	100 ns	
HM628128LP-12SL	120 ns		HM628128LFP-12SL	120 ns	

Block Diagram

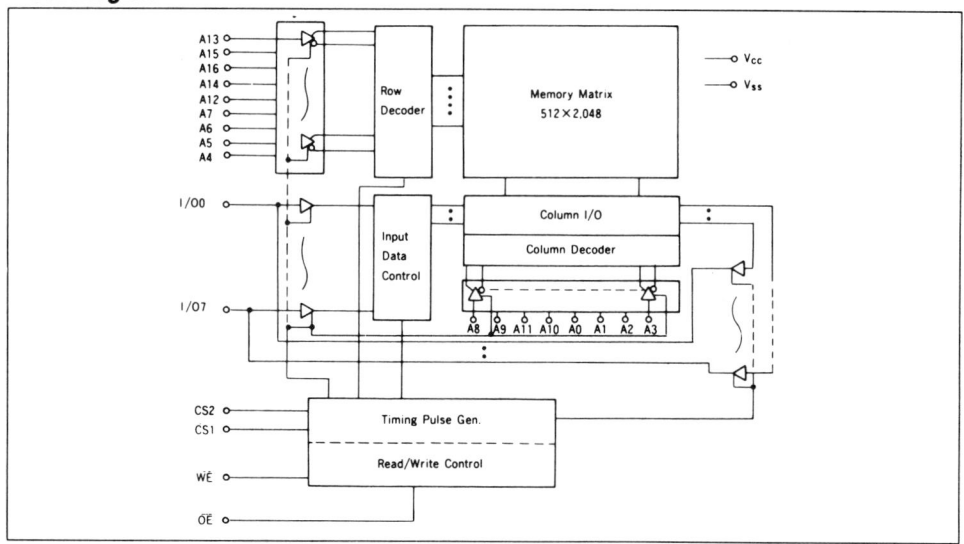

Function Table

$\overline{WE}$	$\overline{CS1}$	CS2	$\overline{OE}$	Mode	Vcc Current	Dout Pin	Ref. Cycle
×	H	×	×	Not selected	I_{SB}, I_{SB1}	High-Z	
×	×	L	×		I_{SB}, I_{SB1}	High-Z	
H	L	H	H	Output disable	I_{CC}	High-Z	
H	L	H	L	Read	I_{CC}	Dout	Read cycle
L	L	H	H	Write	I_{CC}	Din	Write cycle (1)
L	L	H	L	Write	I_{CC}	Din	Write cycle (2)

Note: × : H or L

HITACHI

HM628128 Series

Absolute Maximum Ratings

Item	Symbol	Value	Unit
Voltage on any pin relative to Vss	V_T	−0.5[*1] to +7.0	V
Power dissipation	P_T	1.0	W
Operating temperature	T_{opr}	0 to +70	°C
Storage temperature	T_{stg}	−55 to +125	°C
Storage temperature under bias	T_{bias}	−10 to +85	°C

Note: *1. −3.0 V for pulse half-width ≤ 30 ns

Recommended DC Operating Conditions (Ta = 0 to +70°C)

Item	Symbol	Min	Typ	Max	Unit	Note
Supply voltage	V_{CC}	4.5	5.0	5.5	V	
	V_{SS}	0	0	0	V	
Input high (logic 1) voltage	V_{IH}	2.2	—	6.0	V	
Input low (logic 0) voltage	V_{IL}	−0.3[*1]	—	0.8	V	

Note: *1. −3.0 V for pulse half-width ≤ 30 ns

DC Characteristics (Ta = 0 to +70°C, VCC = 5 V ± 10%, VSS = 0 V)

Item	Symbol	Min	Typ[*1]	Max	Unit	Test Conditions
Input leakage current	$|I_{LI}|$	—	—	2	μA	Vin = Vss to Vcc
Output leakage current	$|I_{LO}|$	—	—	2	μA	$\overline{CS1}$ = V_{IH} or CS2 = V_{IL} or $\overline{OE}$ = V_{IH} or $\overline{WE}$ = V_{IL}, $V_{I/O}$ = Vss to Vcc
Operating power supply current: DC	I_{CC}	—	15	35	mA	$\overline{CS1}$ = V_{IL}, CS2 = V_{IH}, others = V_{IH}/V_{IL}, $I_{I/O}$ = 0 mA
Operating power supply current	I_{CC1}	—	45	70	mA	Min cycle, duty = 100%, $\overline{CS1}$ = V_{IL}, CS2 = V_{IH}, others = V_{IH}/V_{IL}, $I_{I/O}$ = 0 mA
	I_{CC2}	—	15	30	mA	Cycle time = 1 μs, duty = 100%, $I_{I/O}$ = 0 mA, $\overline{CS1}$ ≤ 0.2 V, CS2 ≥ Vcc − 0.2 V, V_{IH} ≥ Vcc − 0.2V, V_{IL} ≤ 0.2V
Standby power supply current: DC	I_{SB}	—	1	3	mA	$\overline{CS1}$ = V_{IH}, CS2 = V_{IH} or CS2 = V_{IL}
Standby power supply current (1): DC	I_{SB1}	—	0.02	2	mA	Vin ≥ 0 V
		—	2[*2]	100[*2]	μA	$\overline{CS1}$ ≥ Vcc − 0.2 V, CS2 ≥ Vcc − 0.2 V or 0 V ≤ CS2 ≤ 0.2 V
		—	2[*3]	50[*3]	μA	
Output low voltage	V_{OL}	—	—	0.4	V	I_{OL} = 2.1 mA
Output high voltage	V_{OH}	2.4	—	—	V	I_{OH} = −1.0 mA

Notes: *1. Typical values are at Vcc = 5.0 V, Ta = +25°C and specified loading.
*2. This characteristics is guaranteed only for L-version.
*3. This characteristics is guaranteed only for L-SL version.

HITACHI

HM628128 Series

Capacitance (Ta = 25°C, f = 1.0 MHz)

Item	Symbol	Min	Typ	Max	Unit	Test Conditions
Input capacitance	Cin	—	—	8	pF	Vin = 0 V
Input/output capacitance	C$_{I/O}$	—	—	10	pF	V$_{I/O}$ = 0 V

Note: This parameter is sampled and not 100% tested.

AC Characteristics (Ta = 0 to +70°C, Vcc = 5 V ± 10%, unless otherwise noted)

Test Conditions
- Input pulse levels: 0.8 V to 2.4 V
- Input rise and fall times : 5 ns
- Input and output timing reference levels: 1.5 V
- Output load: 1 TTL Gate and CL (100pF) (Including scope & jig)

Read Cycle

Item	Symbol	HM628128-7 Min	HM628128-7 Max	HM628128-8 Min	HM628128-8 Max	HM628128-10 Min	HM628128-10 Max	HM628128-12 Min	HM628128-12 Max	Unit	Note
Read cycle time	t$_{RC}$	70	—	85	—	100	—	120	—	ns	
Address access time	t$_{AA}$	—	70	—	85	—	100	—	120	ns	
Chip selection ($\overline{CS1}$) to output valid	t$_{CO1}$	—	70	—	85	—	100	—	120	ns	
Chip selection (CS2) to output valid	t$_{CO2}$	—	70	—	85	—	100	—	120	ns	
Output enable ($\overline{OE}$) to output valid	t$_{OE}$	—	35	—	45	—	50	—	60	ns	
Chip selection ($\overline{CS1}$) to output in low-Z	t$_{LZ1}$	10	—	10	—	10	—	10	—	ns	*1, *2, *3
Chip selection (CS2) to output in low-Z	t$_{LZ2}$	10	—	10	—	10	—	10	—	ns	*1, *2, *3
Output enable ($\overline{OE}$) to output in low-Z	t$_{OLZ}$	5	—	5	—	5	—	5	—	ns	*1, *2, *3
Chip deselection ($\overline{CS1}$) to output in high-Z	t$_{HZ1}$	0	25	0	30	0	35	0	45	ns	*1, *2, *3
Chip deselection (CS2) to output in high-Z	t$_{HZ2}$	0	25	0	30	0	35	0	45	ns	*1, *2, *3
Output disable ($\overline{OE}$) to output in high-Z	t$_{OHZ}$	0	25	0	30	0	35	0	45	ns	*1, *2, *3
Output hold from address change	t$_{OH}$	10	—	10	—	10	—	10	—	ns	

HITACHI

HM628128 Series

Read Timing Waveform*4

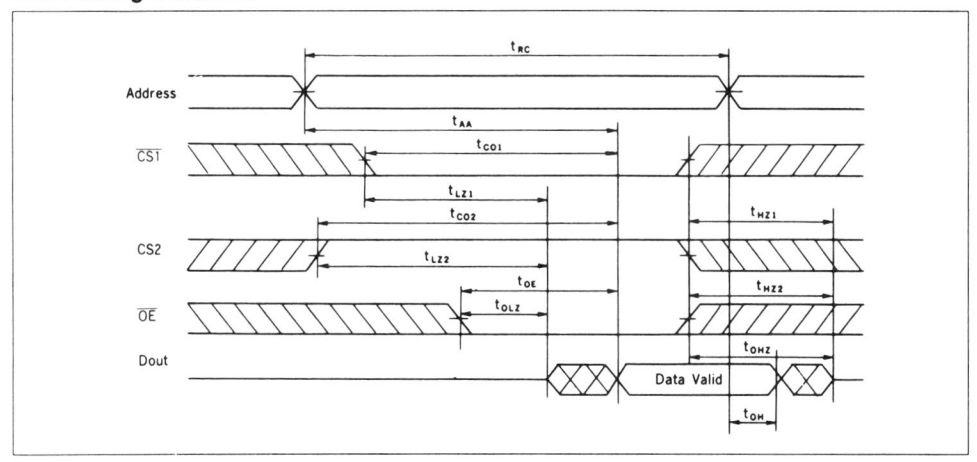

Notes: *1. t_HZ and t_OHZ are defined as the time at which the outputs achieve the open circuit conditions and are not referenced to output voltage levels.
*2. At any given temperature and voltage condition, t_HZ max is less than t_LZ min both for a given device and from device to device.
*3. This parameter is sampled and not 100% tested.
*4. $\overline{\text{WE}}$ is high for read cycle.

Write Cycle

Item	Symbol	HM628128-7 Min	HM628128-7 Max	HM628128-8 Min	HM628128-8 Max	HM628128-10 Min	HM628128-10 Max	HM628128-12 Min	HM628128-12 Max	Unit	Note
Write cycle time	tWC	70	—	85	—	100	—	120	—	ns	
Chip selection to end of write	tCW	60	—	75	—	80	—	85	—	ns	
Address setup time	tAS	0	—	0	—	0	—	0	—	ns	
Address valid to end of write	tAW	60	—	75	—	80	—	85	—	ns	
Write pulse width	tWP	50	—	55	—	60	—	70	—	ns	
Write recovery time	tWR	5	—	5	—	5	—	10	—	ns	
		10	—	10	—	10	—	15	—	ns	*11
Write to output in high-Z	tWHZ	0	25	0	30	0	35	0	40	ns	*10
Data to write time overlap	tDW	30	—	35	—	40	—	45	—	ns	
Write hold from write time	tDH	0	—	0	—	0	—	0	—	ns	
Output active from end of write	tOW	5	—	5	—	5	—	5	—	ns	*10

258

@ HITACHI

HM628128 Series

Write Timing Waveform (1) ($\overline{OE}$ Clock)

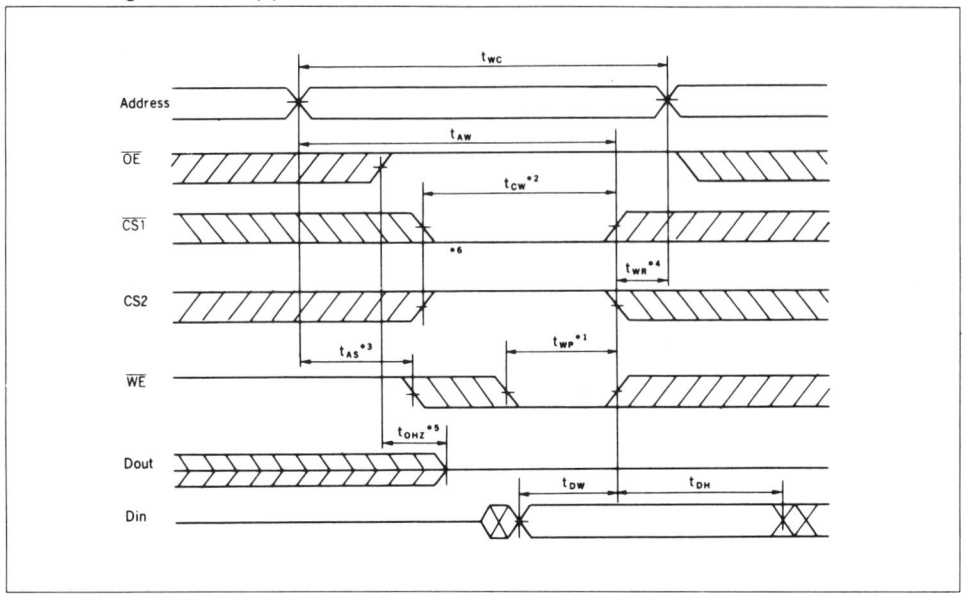

Write Timing Waveform (2) ($\overline{OE}$ Low Fix)

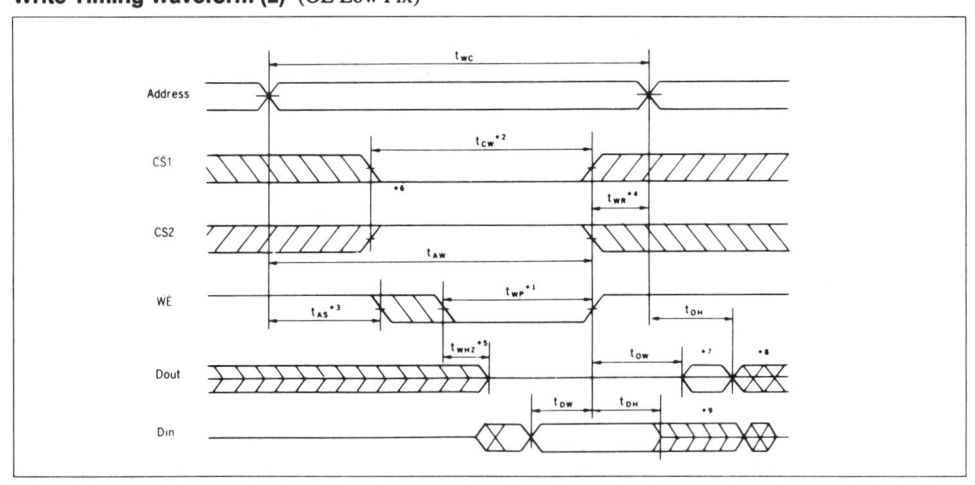

Notes: *1. A write occurs during the overlap of a low $\overline{CS1}$, a high CS2 and a low $\overline{WE}$. A write begins at the latest transition among $\overline{CS1}$ going low, CS2 going high and $\overline{WE}$ going low. A write ends at the earliest transition among $\overline{CS1}$ going high, CS2 going low and $\overline{WE}$ going high. t_{WP} is measured from the beginning of write to the end of write.
*2. t_{CW} is measured from the later of $\overline{CS1}$ going low or CS2 going high to the end of write.
*3. t_{AS} is measured from the address valid to the beginning of write.
*4. t_{WR} is measured from the earliest of $\overline{CS1}$ or $\overline{WE}$ going high or CS2 going low to the end of write cycle.
*5. During this period, I/O pins are in the output state; therefore, the input signals of the opposite phase to the outputs must not be applied.

HITACHI

HM628128 Series

- *6. If $\overline{CS1}$ goes low simultaneously with $\overline{WE}$ going low or after $\overline{WE}$ going low, the outputs remain in high impedance state.
- *7. Dout is the same phase of the latest written data in this write cycle.
- *8. Dout is the read data of next address.
- *9. If $\overline{CS1}$ is low and CS2 is high during this period, I/O pins are in the output state. Therefore, the input signals of the opposite phase to the outputs must not be applied to them.
- *10. This parameter is sampled and not 100% tested.
- *11. This value is measured from CS2 going low to the end of write cycle.

Low Vcc Data Retention Characteristics (Ta = 0 to +70°C)
(This characteristics is guaranteed only for L-and L-SL version.)

Item	Symbol	Min	Typ	Max	Unit	Test Conditions*3
Vcc for data retention	V_{DR}	2.0	—	—	V	$\overline{CS1} \geq V_{CC} - 0.2$ V, $CS2 \geq V_{CC} - 0.2$ V or 0 V $\leq CS2 \leq 0.2$ V, $Vin \geq 0$ V
Data retention current		—	1	50*1	μA	$V_{CC} = 3.0$ V, $Vin \geq 0$ V, $\overline{CS1} \geq V_{CC} - 0.2$ V, $CS2 \geq V_{CC} - 0.2$ V or 0 V $\leq CS2 \leq 0.2$ V
		—	1	15*2	μA	
Chip deselect to data retention time	tCDR	0	—	—	ns	See Retention Waveform
Operation recovery time	tR	5	—	—	ms	

Low Vcc Data Retention Timing Waveform (1) ($\overline{CS1}$ Controlled)

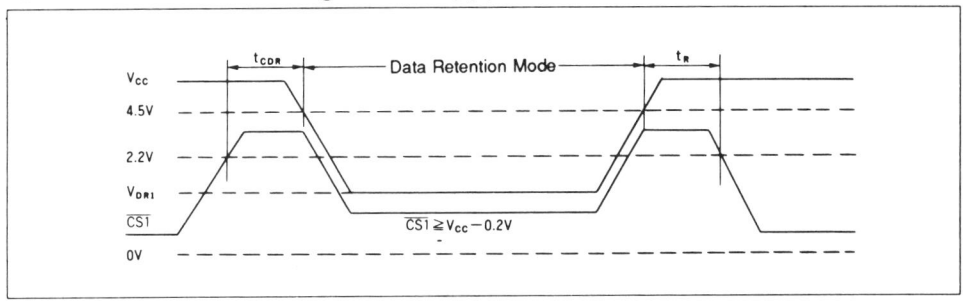

Low Vcc Data Retention Timing Waveform (2) (CS2 Controlled)

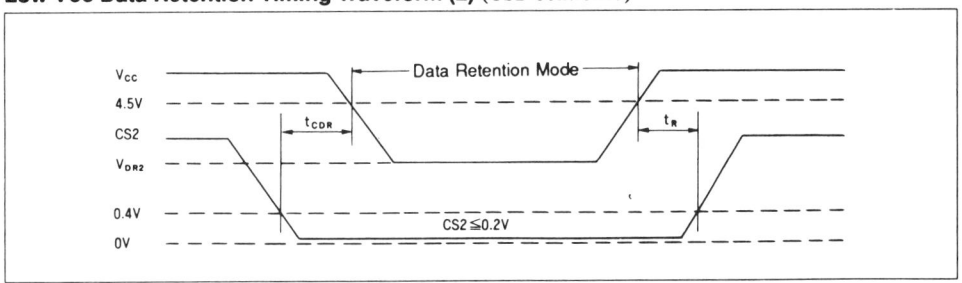

Notes: *1. 20 μA max at Ta = 0 to 40°C.
*2. 3 μA max at Ta = 0 to 40°C.
*3. CS2 controls address buffer, $\overline{WE}$ buffer, $\overline{CS1}$ buffer and $\overline{OE}$ buffer and Din buffer. If CS2 controls data retention mode, Vin levels (address, $\overline{WE}$, $\overline{OE}$, $\overline{CS1}$, I/O) can be in the high impedance state. If $\overline{CS1}$ controls data retention mode, CS2 must be CS2 $\geq V_{CC} - 0.2$ V or 0 V $\leq CS2 \leq 0.2$ V. The other input levels (address, $\overline{WE}$, $\overline{OE}$, I/O) can be in the high impedance state.

HM628128 Series

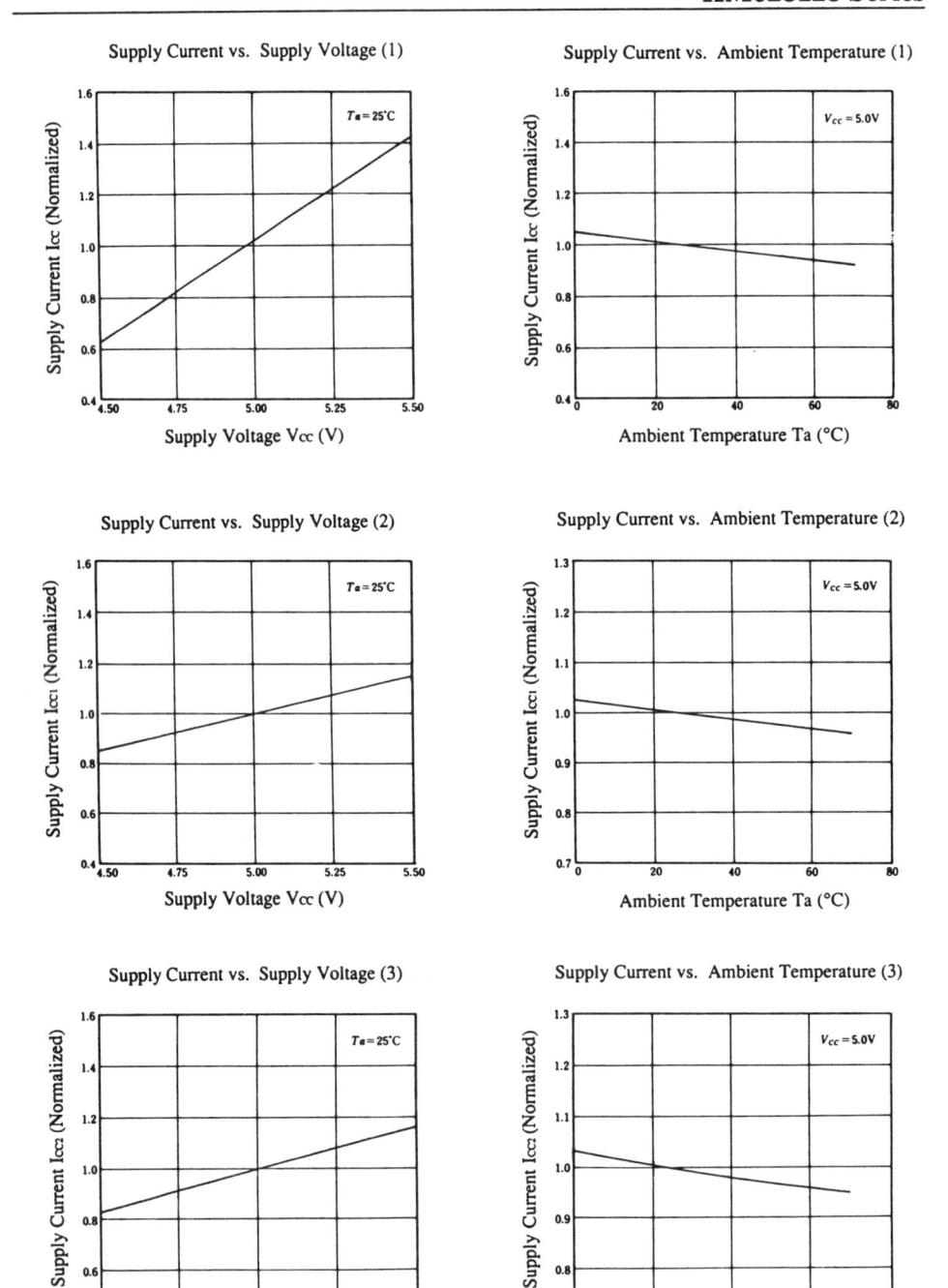

Supply Current vs. Supply Voltage (1) ($T_a = 25°C$)

Supply Current vs. Ambient Temperature (1) ($V_{cc} = 5.0V$)

Supply Current vs. Supply Voltage (2) ($T_a = 25°C$)

Supply Current vs. Ambient Temperature (2) ($V_{cc} = 5.0V$)

Supply Current vs. Supply Voltage (3) ($T_a = 25°C$)

Supply Current vs. Ambient Temperature (3) ($V_{cc} = 5.0V$)

HM628128 Series

Access Time vs. Supply Voltage
$T_a = 25°C$

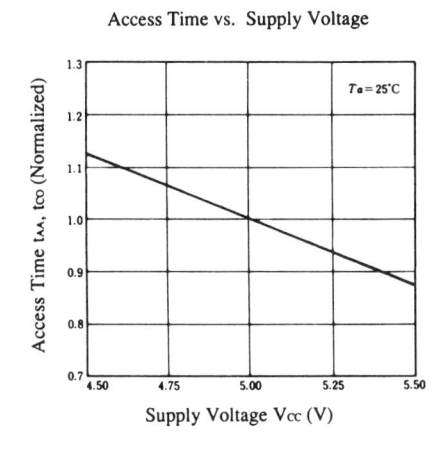

Access Time vs. Ambient Temperature
$V_{cc} = 5.0V$

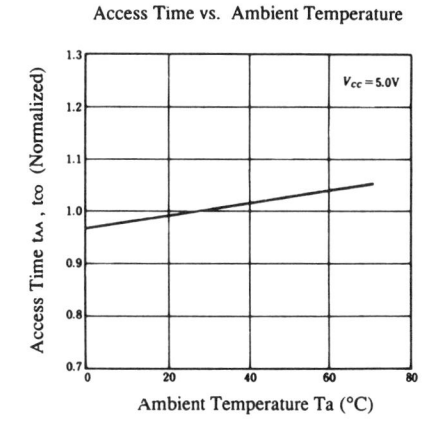

Standby Current vs. Supply Voltage
$T_a = 25°C$

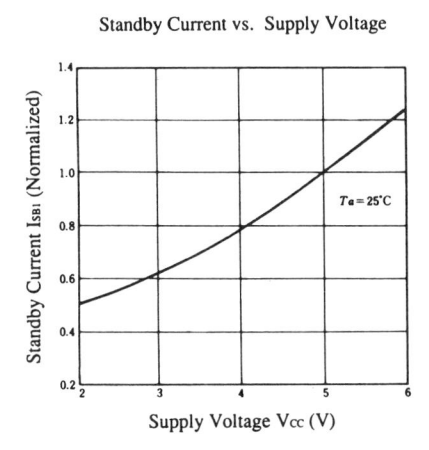

Standby Current vs. Ambient Temperature
$V_{cc} = 5.0V$

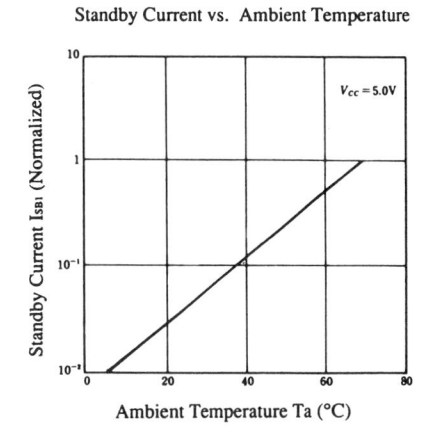

Supply Current vs. Frequency (Read)
$T_a = 25°C$
$V_{cc} = 5.0V$

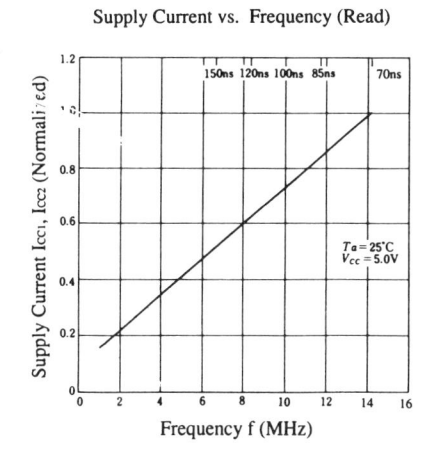

Supply Current vs. Frequency (Write)
$\overline{WE}$ Duty = 80%
$T_a = 25°C$
$V_{cc} = 5.0V$

@HITACHI

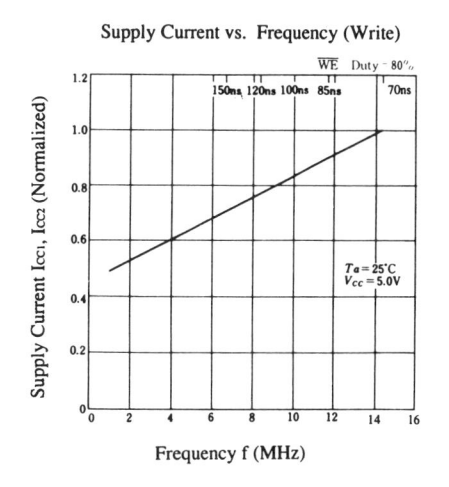

HM628128 Series

Input Low Voltage vs. Supply Voltage

Input High Voltage vs. Supply Voltage

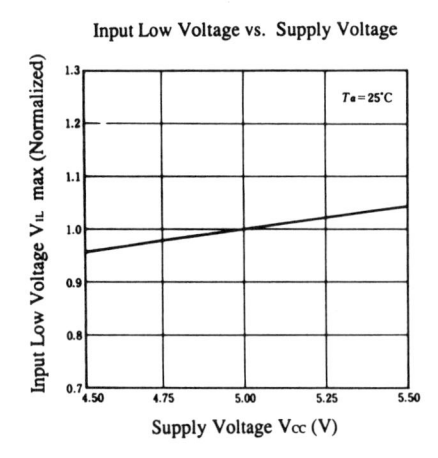

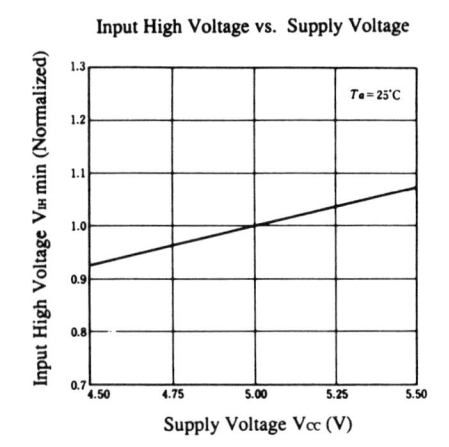

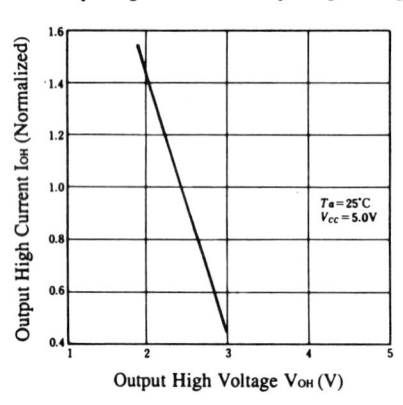

Output High Current vs. Output High Voltage

Output Low Current vs. Output Low Voltage

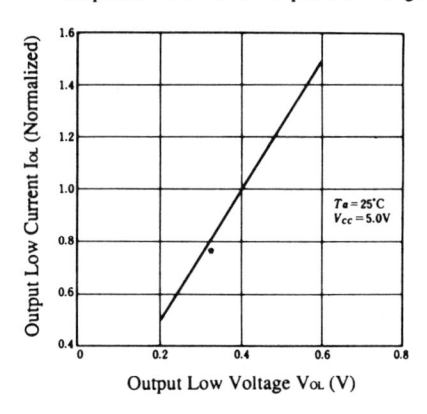

Access Time vs. Load Capacitance

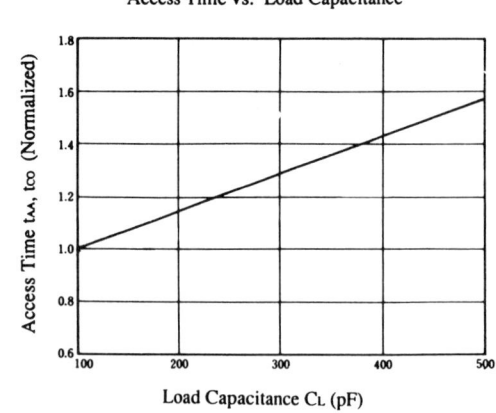

… # intel

APPLICATION NOTE

AP-238

March 1985

Memory Design For The Low Power Microsystem Environment

DENNIS KNUDSON
MEMORY COMPONENTS

© Intel Corporation, 1985

Order Number: 280103-001

AP-238

1.0 INTRODUCTION

1.1 The Low Power Environment

The low power environment, as discussed in this document, refers to microsystems that have low power as a major feature of the design. The purpose here is to provide some guidelines for designing such systems—particularly the memory subsystems. The topics covered include descriptions and trade-offs of various types of memory components, system design examples and discussions of batteries and power switching circuitry.

1.2 Elements

Generally, the low power environment has two basic elements: active and standby power. With respect to memories, another element—the data retention power—is equally important. As discussed in this document, active power is simply the power required during an access to memory. Standby power is that required while the memory is between accesses. Data retention power is the minimum power required to maintain data in the memory.

1.3 Applications

Low power applications in general are divided into those using AC line power and those using batteries either as primary power or as backup power for memory maintenance and/or memory protection. Active or primary power for systems is fairly simple and can be provided by either the AC line or batteries. Back-up power subsystems consists of circuitry supported by batteries only during the absence of regular power. It includes the system memory, batteries and power switching circuits. When in the backup power mode, the batteries provide power to the memory and the support circuits necessary to maintain the stored information. When power fails, the power switching circuits must shut down all of the system except the memory and its support circuits. In so doing, they must provide a clean, glitch-free and orderly transition from line power to battery power. When line power returns, they must transfer control from the battery powered circuits back to line powered circuits, again in an orderly, glitch-free manner.

In the backup power applications, only those components required to maintain volatile memory data when power is removed, or those required to protect the data when power fails need to be low power devices. Some examples are systems requiring fast, non-volatile storage (RAMs) or power failure protection for on-line memory. They include, but are not limited to battery powered systems. For example, when used for storage such as solid state disks, CHMOS DRAMs allow the user to operate directly out of mass memory but they require battery backup to maintain the data whenever the system is powered down.

In battery powered systems or subsystems, all components are battery supported and as such must consume little power. Such systems must be able to run on each set of batteries for an acceptable period defined by the application. When using re-chargeable batteries, the system should run on each charge for the longest likely continuous use in the specific application and recharge time should be acceptable for the probable frequency of use. Obviously, the more power drawn by the components of the system, the bigger the battery required or the shorter the operating time. Above all, a battery powered system design should avoid components or sub-systems that will require special cooling accommodations (i.e. fans).

1.4 Design Considerations

When designing low power memory systems with standard TTL drivers, standby power can be as little as two percent of the active power. To achieve the low standby power possible with CMOS memories, pull-up resistors must be used on all critical control lines and the TTL must be powered down. This requires special considerations. A more effective method is to use CMOS logic.

Newer CMOS products are ideal for these types of applications because they provide the speed required when the system is operating, and dissipate very little heat in the process when compared to NMOS products. When in backup or data retention mode, they require about two tenths of one percent of their active power to maintain information. CMOS drivers for the memory system reduce standby power to its lowest level—lower, typically by a factor of 40, than TTL standby power consumption. The implications to the system designer are lower cooling requirements and smaller power supplies of batteries[1].

2.0 MEMORY OPTIONS FOR LOW POWER OR NON-VOLATILE STORAGE

In this section we will look at various types of memory suitable for low power applications. Typically, memories used in this type of system are CMOS RAMs for their low power and high speed features. Bubble memories are useful here for mass storage if the application requires long term storage without power. For relatively short term storage, power consumption for the bubble can be reduced by power switching[2]. E²PROMs and CMOS EPROMs, also non-volatile, can be used for small, permanent or semi-permanent software storage such as firmware and operating system kernels.

NVRAMs would be useful for buffer storage on a communications data link or in small, fast local memory requiring data retention without power. In 1972, Signetics introduced the 25120 write only memory (WOM) but no lasting applications have developed so it will not be covered here.

2.1 Magnetic Bubble Memory

The bubble memory is a rugged, non-volatile, serial-in/serial-out storage medium. Both the one megabit (7110) and the four megabit (7114) operate at 50 KHz. Briefly, the MBM operates by rotating an external magnetic field to propel a cylindrical magnetic domain (bubble) through a film or magnetic material. Externally, it handles data in a serial fashion similar to magnetic disc or drum storage systems but, without the noise or moving parts, it is faster, more reliable and uses much less power.

Magnetic bubble memory is smaller than any other memory storage system and cost/bit is second only to the dynamic RAM. It is the slowest and has the highest operating and standby power requirements of all solid state memories but it can retain data indefinitely when power is removed.

A complete bubble memory kit consists of the bubble memory, controller, current pulse generator, formatter, coil pre-driver and drive transistors. The kit requires two voltages and dissipates about 1.5 watts in standby and 3.5 watts active per bubble component. Rugged and truly non-volatile, it has many applications including robotics, oil exploration, aircraft navigation and test equipment.

2.2 EPROMs

Non-volatile read-only-memories, EPROMS are erased under ultraviolet light and then re-programmed electrically. They are programmed by injecting "hot" electrons (typically created by a 12.5 volt Vpp) into the floating (isolated) gate of the transistor. Capacitively coupled to the select gate of the transistor, this charge adds to the select gate charge, altering the threshold voltage of the transistor. They are erased by ultraviolet light which induces enough energy into the floating gate to cause the excess electrons to overcome the energy barrier between the floating gate and the insulator surrounding it. Thus, the electrons are dissipated into the select gate and substrate of the transistor.

EPROMs must be removed from the circuit to be erased and re-programmed. In addition, they must first be completely erased before they can be re-programmed. Typical erasure times are 15 to 20 minutes.

Program time is typically 5 minutes (27256 with Int$_e$ligent™ programming) and access times of less than 250 nS are available. Through they do not have the flexibility of reading and writing data in real time, they are close to the dynamic RAM in density and they do not lose data when power is removed.

Some typical applications of EPROMs are firmware, operating system kernels, computer boot programs, communications, portable instruments, office equipment and commercial appliances. CHMOS EPROMs are faster and, with 25% of the active and 0.4% of the standby current of their NMOS counterparts, are well suited for storing initialization programs in low power systems.

2.3 E²PROMs

Electrically erasable (E²) PROMs use the same "floating gate" technology as EPROMs. The difference is that the charge is added or removed via a tunnel oxide in a phenomenon known as Fowler-Nordheim tunneling. At an energy level referred to as the "forbidden" band, the electrons are able to penetrate the oxide without ever reaching the energy level that standard mechanics would predict as being necessary to overcome the barrier[3]. This allows them to be erased and programmed IN-CIRCUIT and it permits individual byte programmability.

Active power is equivalent to that of static RAM but standby power is much higher, second only to magnetic bubble memory. It has the highest cost/bit and lowest density of the solid state memories except for Non-Volatile RAMs. Non-volatile and in-circuit reprogrammable, it is versatile in control functions and in data acquistion and communication systems requiring frequent system reconfiguration.

2.4 NVRAMs

Each cell of the Non-Volatile RAM consists of two memory cells, one static RAM and the other E²PROM. Consequently, it is the least dense of all solid state memories. It functions as a static RAM during operation but provides non-volatile storage of data when powered off. A signal from the power supply indicating that power is being removed (or lost) causes the NVRAM to move the data from the SRAM cell into the E²PROM cell. Thus, data that can be operated on at microprocessor speeds is automatically saved in non-volatile memory, combining the flexibility of SRAM with the long-term, power free storage of E²PROM. Power and cost/bit are roughly equivalent to E²PROMs.

2.5 SRAMs

Static Random Access Memories latch data into either four transistor (4T, for cost) or six transistor (6T, for low power) memory cells. Consequently, they do not need to be refreshed and access time is, essentially, cycle time. These features have historically contributed to the speed advantage of SRAMs over other memory types as well as to simple interface design. However, they also contribute to the density and power disadvantages of the SRAM. As data are gated out by addresses, they do not require clocks but may use one of two signals to control system interface. A chip enable ($\overline{CE}$) may be used to provide a standby mode with reduced power and an output enable ($\overline{OE}$) may be provided to release the data bus.

SRAMs are always one density generation behind dynamic RAMs because of the number of transistors per cell. The new CMOS SRAMs actually enjoy a power advantage over NMOS dynamic RAMs but the advantage is lost once again to dynamic RAMs on the CMOS process. Like dynamic RAMs, SRAMs are volatile memory, losing data when power is lost.

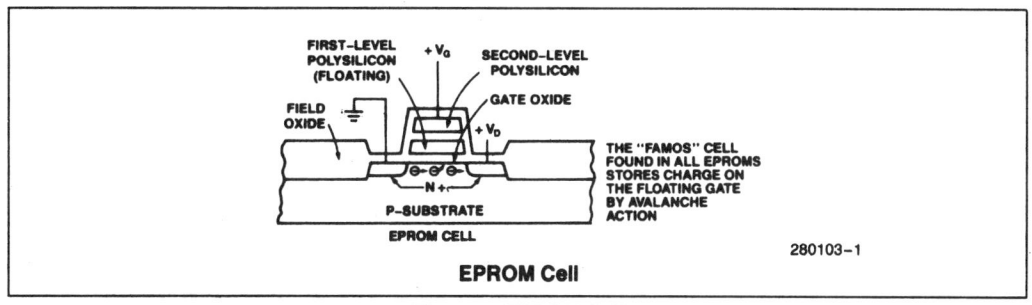

Figure 1. EPROM Cell

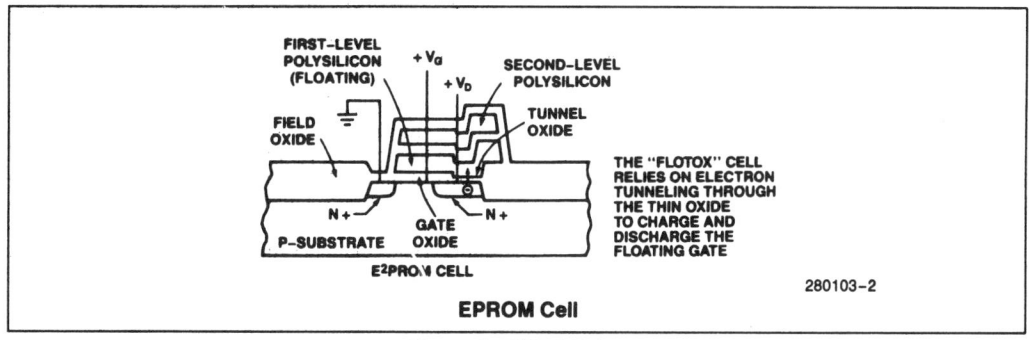

Figure 2. E²PROM Cell

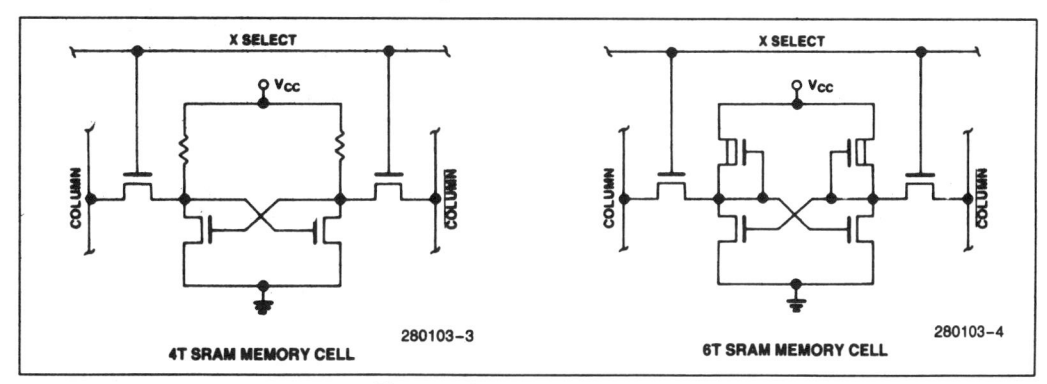

Figure 3. SRAM 4T and 6T Cells

intel AP-238

MEMORY COST vs DATA RETENTION CURRENT/TIME (64K BIT EQUIVALENT)

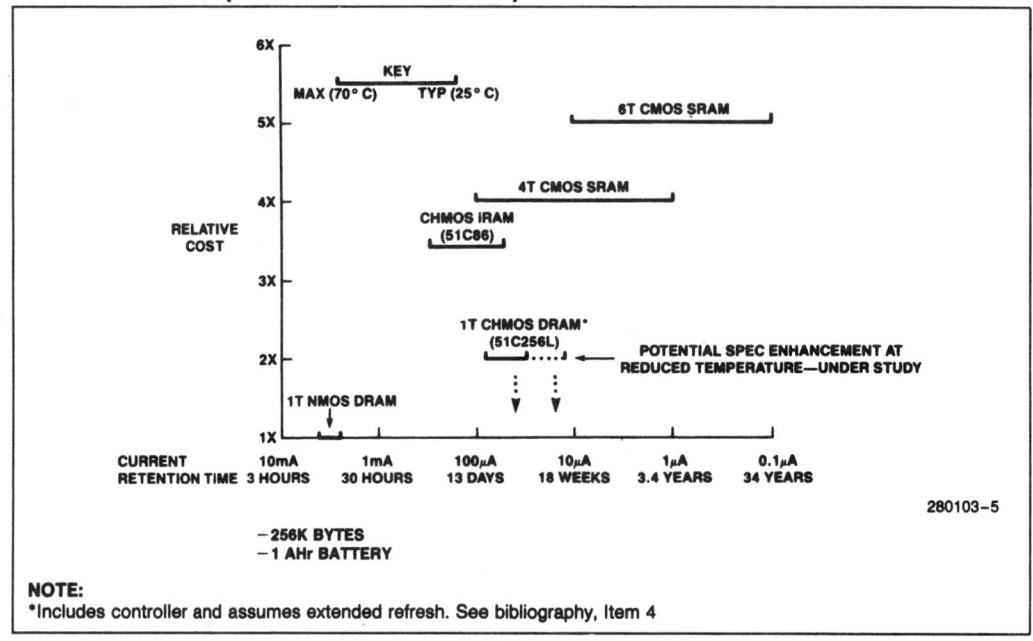

[Chart 1: Relative cost vs current retention time. Key: MAX (70°C), TYP (25°C). Technologies shown: 6T CMOS SRAM (~4.5–5.5X), 4T CMOS SRAM (~3.5–4.5X), CHMOS IRAM (51C86) (~3–4X), 1T CHMOS DRAM* (51C256L) (~2X) with potential spec enhancement at reduced temperature—under study, 1T NMOS DRAM (~1X). Current/retention time axis: 10mA/3 hours, 1mA/30 hours, 100μA/13 days, 10μA/18 weeks, 1μA/3.4 years, 0.1μA/34 years. —256K BYTES, —1 AHr BATTERY]

NOTE:
*Includes controller and assumes extended refresh. See bibliography, Item 4

Chart 1

Some typical applications of SRAMs are in cache memories, bit slice processors, local microprocessor memory (less than 64K), video graphics and (with CMOS SRAMs) battery backup and battery powered systems.

2.6 IRAMs

The integrated RAM works like an SRAM but costs less. It is a dynamic RAM with an on-chip refresh circuitry and it is generally available in two versions, synchronously or asynchronously refreshing. The synchronous version has an on-chip refresh address generator and a "Refresh" input that allows the user to control the refresh occurrence.

The asynchronous version has a timer, refresh address generator, access arbiter and a "Ready" output. When a refresh request occurs coincident with an access request the arbiter resolves which will take precedence. Whenever an access request is held up by refresh, "Ready" goes low, causing the processor to insert WAIT states. With these features, the iRAM provides SRAM characteristics at lower cost per bit and lower power requirements (within like technologies). The iRAM is a volatile memory, particularly well suited to microcontroller memory (up to 64K bytes) and local memory (up to 128K bytes) for microprocessors.

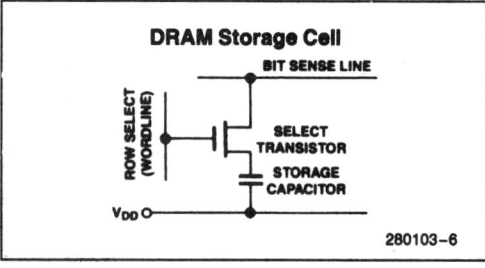

[Figure 4: DRAM Storage Cell — shows row select (wordline) connected to select transistor gate, bit sense line connected to drain, storage capacitor between source and V_{DD}]

Figure 4. DRAM Storage Cell

2.7 DRAMs

The Dynamic RAM stores data on a capacitor in a single transistor cell. This allows the highest density and lowest power per bit of all random access memories, but, because of charge leakage from the storage capacitor, they have to be refreshed periodically. DRAMs are volatile memory and, as such, lose data whenever power is removed or lost.

The primary DRAM application has been, historically, large main memories (greater than 64K) for mainframe, mini and personal computers. The introduction of CHMOS DRAMs has further reduced the power requirements[4] and added new high bandwidth fea-

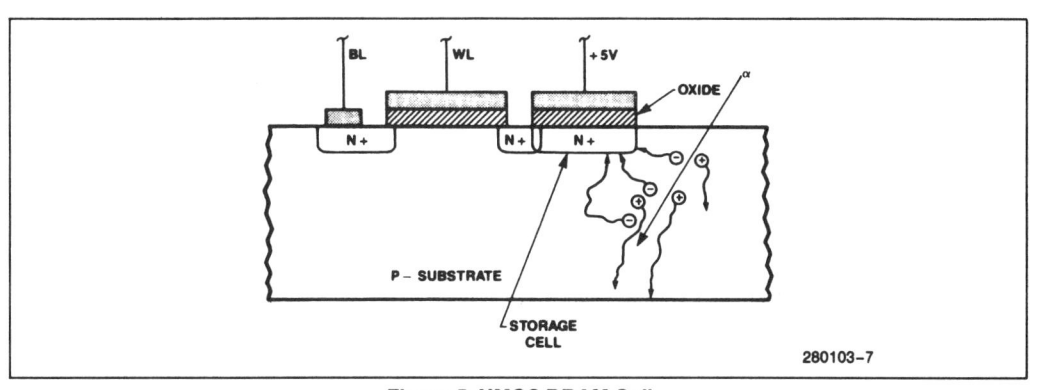

Figure 5. NMOS DRAM Cell

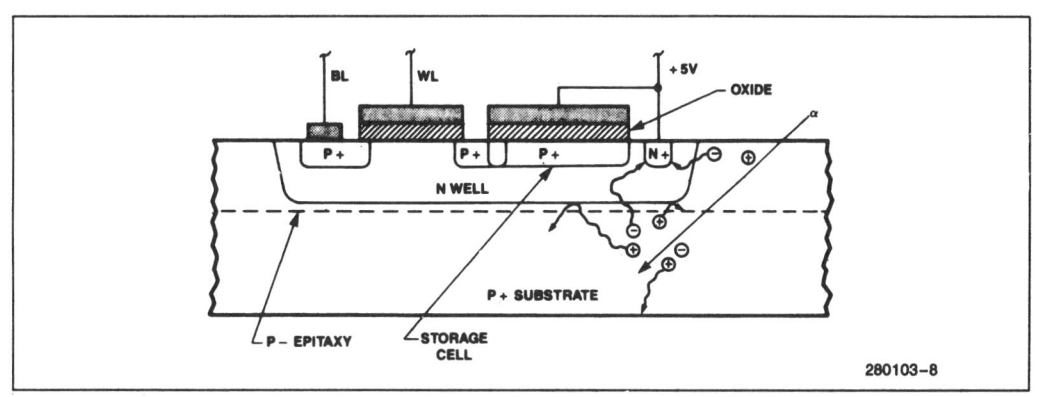

Figure 6. CMOS DRAM Cell

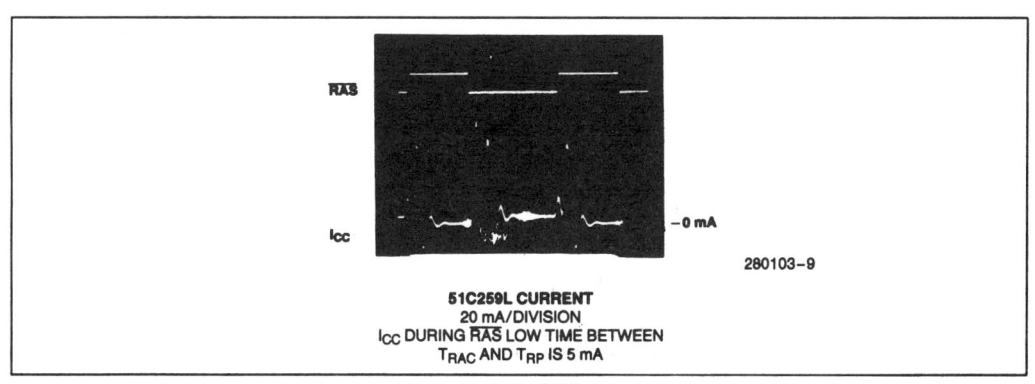

51C259L CURRENT
20 mA/DIVISION
I_{CC} DURING RAS LOW TIME BETWEEN
T_{RAC} AND T_{RP} IS 5 mA

Figure 7. 51C259 Current

tures such as Ripplemode™ and Static Column Decode[5]. The features of CHMOS allow DRAMs to move into non-volatile solid state disks (which will permit the user to execute directly from mass storage), battery powered (portable) computers, battery back-up systems and traditionsl SRAM speed applications such as video graphics[6].

DRAMs on the CHMOS process reduce the standby current by about a factor of forty. The data retention power requirement of the DRAM is further reduced by the extended refresh feature. When operating in the data retention mode, refresh can be extended to 32 ms for the 51C256L, 64 ms for the 51C64L.

Because a small capacitor is the storage mechanism for DRAMs, they have been susceptible to "soft errors". When an alpha particle strikes the die, it generates ions which collect at the capacitor, changing the stored charge[7]. The cell is not damaged but the information it contains can be changed (Figure 5). If it is, a "soft" error has occurred. The CMOS process has changed this by allowing the storage capacitor to be placed in an "n" well which absorbs the ions generated by the stray particle[8] (Figure 6). Consequently, 64K CHMOS DRAMs, at a 1 μS cycle rate, have soft error rates of 10 fits*. At the 256K level, the soft error rate will be less than 400 fits, and typically 40 fits. Furthermore, since the soft error rate is directly related to cycle rate, during the low power data retention mode (i.e. 32 ms refresh) the soft error rate goes to:

1 μs = 40 fits, then for the 256K devices the extended refresh soft error rate is:

32 ms/256 refresh rows = 125 μs

1 μs/125 μs −40 fits = **0.32 fits** per device

*failures in time = number of failures in 10[9] device hours (or 114 years).

3.0 SYSTEM DESIGN USING CHMOS DRAMs

3.1 Active Low Power Design Description

CHMOS DRAMs permit the building of larger and lower powered systems than have previously been possible. As an illustration, a low power, general purpose microcomputer system design will be discussed using all CMOS products. An 80C88 operating in max mode has 256K bytes of DRAM (expandable to 960 Kb) with a DMA controller to handle refresh during normal operation. A standby mode of operation is provided to reduce the power usage during periods of no activity.

During standby the memory refresh is extended to 32 ms which reduces the data retention power.

To minimize the transient effects of current, it is recommended that a 0.1 μF bypass capacitor be provided for each memory device. High speed (low inductance) capacitors are preferred. Most CMOS devices require less than 0.1 μF. The actual decoupling capacitance required can be calculated for specific devices using the formula:

$$C_{BYPASS} = di \cdot dt/dV$$

where; di = change in current
dt = spike duration or 1 μs*, whichever is faster
dV = allowable voltage droop

For example, Figure 7 shows the current requirement for the 51C259. The peak current is 95.63mA so the decoupling requirement is:

$$C = 96 \text{ mA} \times 40 \text{ ns}/100 \text{ mV} = 0.04 \text{ } \mu F$$

In this case 0.1 μF for every other device would be sufficient.

*NOTE:
Assume 1 μs respond time of bulk capacitance, typical of tantalum.

3.1.1 SYSTEM DESIGN EXAMPLE

Referring to Figure 8, during normal ooeration the 80C88 controls the bus through the 82C88 bus controller. During interrupt requests the 82C59A interrupt controller takes control of the bus long enough to send a vector address to the microprocessor, and to refresh the DRAM array, the DMA controller requests the bus and then executes a single (distributed) refresh cycle to all four banks of memory. The DRAM $\overline{RAS}$ signals and the EPROM $\overline{CE}$ are decoded by the higher order addresses A16, A17, A18 and A19 in a 4-to-16 line decoder. The decoder divides the address field into 64K blocks, convenient for the particular DRAM components used here. One output of the decoder is used as Chip Enable for the 16K bytes of EPROM shown here at address FC000H to FFFFFH. The EPROM is located here because the system reset vectors the 80C88 to address FFFF0H to access initialization routines. The EPROM can be expanded downward in the address field to 64K bytes beginning at F0000H, or further still by using another of the 4-to-16 line decoder outputs. Four outputs from the decoder are used as $\overline{RAS}$'s each to a bank of DRAM, totalling 256K bytes at addresses 00000 to 3FFFFH. Using the other outputs to provide $\overline{RAS}$ to eleven more banks would increase the total DRAM by 704K, to 960K bytes.

The reset initialization routines are located in EPROM and, for this system, will include programming routines

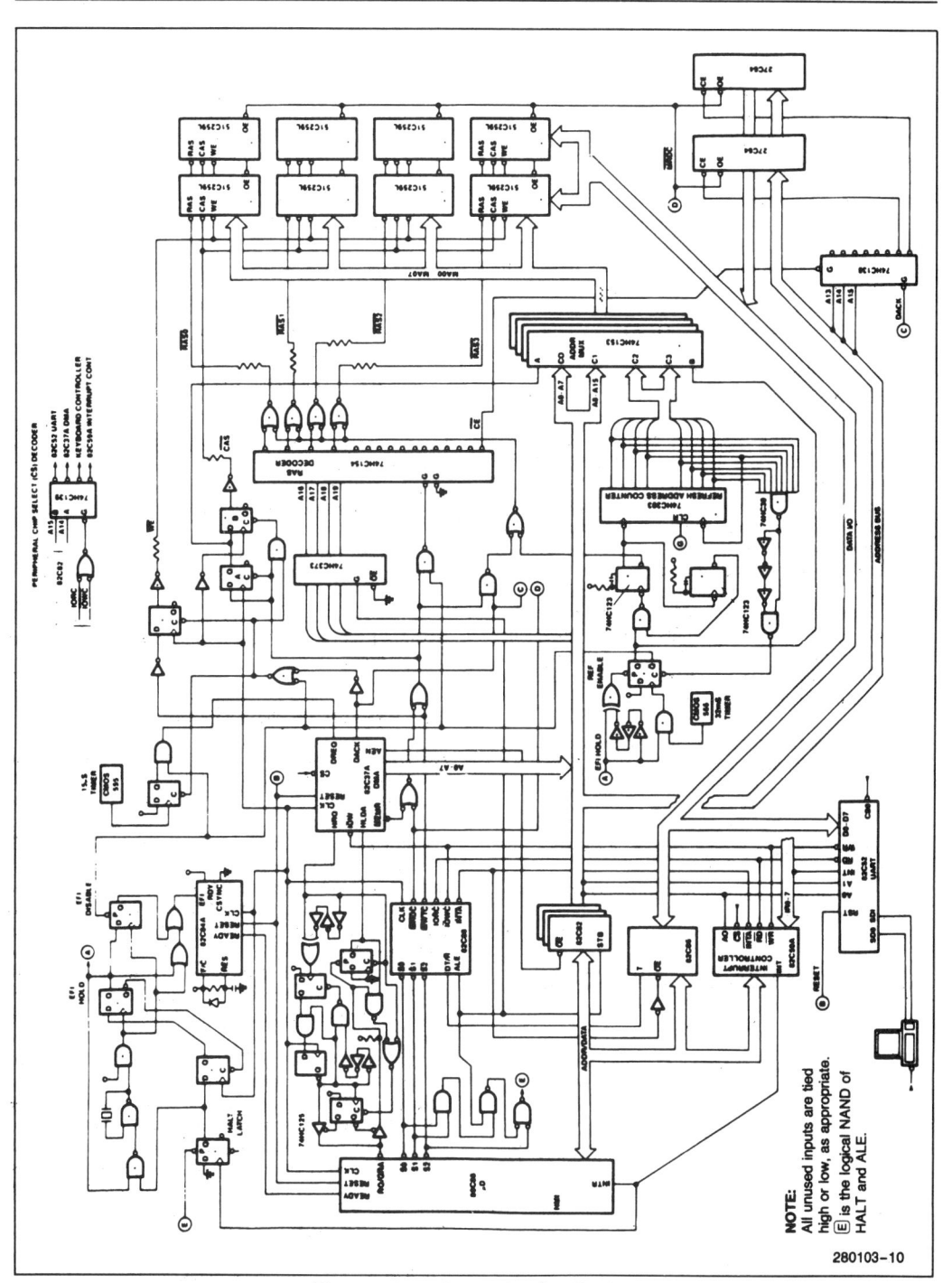

Figure 8. Low Power System Design

intel AP-238

for the DMA controller, the interrupt controller and the UART. Following initialization, the DMA will begin its distributed refresh routine and the operating system software can be loaded through one of the I/O ports.

Write enable ($\overline{WE}$) to the DRAM is generated by the memory write command ($\overline{MWTC}$) from the 82C88 bus controller and the system clock. Column address strobe ($\overline{CAS}$) is generated by $\overline{MWTC}$ or $\overline{MRDC}$ (memory read command) and the system clock. The rising edge of the clock sets the latch (A) which switches the address multiplexer from the row address to the column address. The falling edge of the clock then sets the second latch (B) which sends $\overline{CAS}$ to the memory. During the DMA refresh cycles, the acknowledge signal (DACK) holds CAS and $\overline{WE}$ high and clears the DMA request (DRQ). Note that the $\overline{CAS}$ and $\overline{WE}$ latches are buffered to the array. This is done because reflections from the line stubbing of the memory array back into the outputs of the latches could cause the latches to toggle, glitching the signals. Also, series terminators (~33Ω) are shown in the $\overline{RAS}$, $\overline{CAS}$, $\overline{WE}$ and address lines. These terminators improve the impedance match of the driver to the line, reducing reflections. For best results, they should be located electrically less than one inch from the driver output.

3.1.2 MEMORY

The 51C259L devices used here are low power 256k bit memory devices organized as 64K x 4. This organization helps keep active power low by enabling only two devices during a memory access. If 256K×1 devices were used, all eight would be turned on, drawing about four times the current.

The EPROMs shown are CHMOS 8K x 8's, containing a boot loader and programming instructions for the DMA, the interrupt controller and the UART. The EPROM capacity could be increased to 64K bytes using the rest of the outputs of the 74HC138 decoder.

In max mode, the 80C88 provides two common I/O request/grant ($\overline{RQ/GRA}$) lines to co-processors for bus access, one of which is used here for the DMA (Figure 9). This pin requires a low-active pulse in to request the bus, then sends a low-active pulse out to release (grant) the bus. The processor then waits for another low-active pulse in to indicate that the co-processor is through with the bus. The DMA has separate Hold Request (HRQ) and Hold Acknowledge (HLDA) pins which use active-high levels. Consequently, it is necessary to convert both the high-going and low-going edges of HRQ to low-active pulses. In the period between these pulses, the $\overline{RQ/GRA}$ pin must be released for the 80C88 output "Grant" pulse. The Grant pulse presets a latch for the DMA Hold Acknowledge and returns the $\overline{RQ/GRA}$ pin to the input circuit to wait for the end-of-hold pulse.

3.1.3 STANDBY MODE

One nice feature of the CMOS process is that the system can be put into a very low power standby mode by bringing the system clock to a complete stop. To do this, the external frequency in (EFI) on the 82C84A is used and the EFI clock is stopped, in this case, by a HALT instruction from the 80C88 (Figure 10). The HALT instruction first stops the EFI clock without any glitches and it then stops the oscillator itself, both at a high level. The timing through the 82C84A at this frequency stops the clock output low. Any interrupt (such as a key pressed) will cause the system to exit the standby mode and return to normal operation. The interrupt clears the halt latch which allows the oscillator to begin again. Because the oscillator was stopped at a high level, it will begin by transitioning low. Then, on the high transition, it will clock the EFI hold latch thus clearing it. The following low transition will get through the first stage OR gate. When the burst refresh initiated by clearing the EFI hold latch (see below) is completed, the oscillator will clock the EFI disable latch low and then the oscillator will be enabled through to the EFI input of the 82C84A. The burst

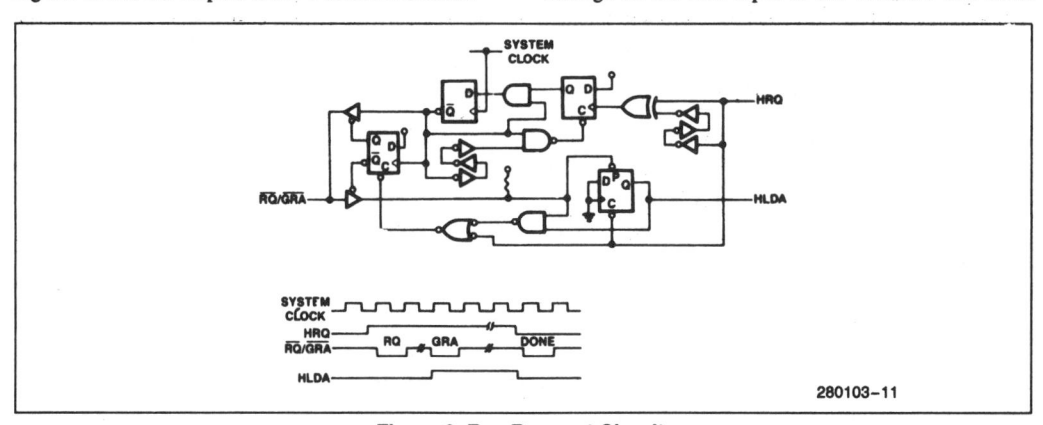

Figure 9. Bus Request Circuit

intel
AP-238

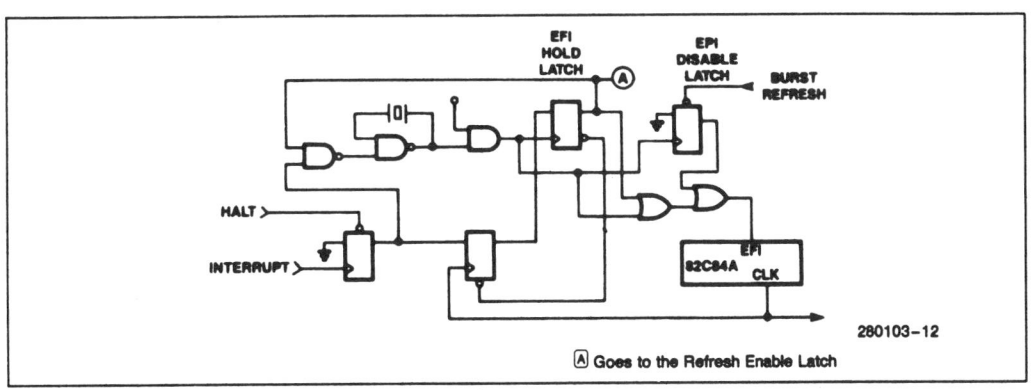

Figure 10. Clock Circuit

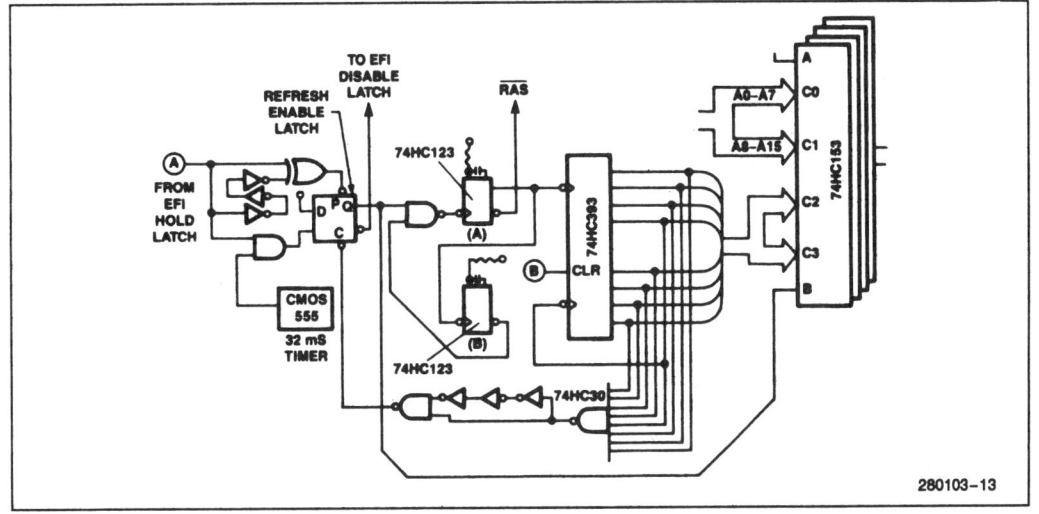

Figure 11. Refresh Circuit

refresh will take about (250 ns × 256 rows =) 64 µs, allowing the oscillator over 900 oscillations to stabilize. If the memory banks are refreshed one at a time (see below), it will take (250 ns × 256 rows × 4 banks =) 256 µs, allowing 3,840 oscillations.

During the standby mode, it is necessary to continue to provide refresh cycles to the DRAM, but it can be reduced to the extended refresh period (32 ms in the case of the 51C259L's used here). In order for the refresh period to be guaranteed, the memory must receive a burst refresh on entering and again on exiting the standby mode. To do this, low-active pulses, generated off both the rising and falling edges of the EFI hold latch, preset the refresh enable latch (Figure 11). During standby, the refresh enable latch is clocked every 32 ms by the 555 timer. Whenever the latch is enabled (Q = 1), the two multivibrators (74HC123's) run freely (but synchronized to each other). One (A) produces $\overline{RAS}$ and the other (B) guarantees t_{RP}. The RAS is gated to all four $\overline{RAS}$ drivers, refreshing all four banks of memory at once. The trailing edge of $\overline{RAS}$ clocks the counter which provides the refresh addresses. The output of the address counter is ANDed to clear the refresh enable latch when the last address has been refreshed. When the counter is clocked from FF (HEX) to 00, a pulse is generated which clears the refresh enable latch.

While refreshing all four banks at once, as is shown here, does not use any more average power, refreshing one row at a time would reduce the peak power requirement during standby by a factor of four (see Figure 14 and the Data Retention design for power calculations, later in this section).

When exiting the standby mode, it is necessary to keep the 80C88 and the DMA off the bus until the last burst refresh has been completed. To do this, (Figure 10) the refresh enable latch presets the EFI disable latch in the clock hold circuitry, blocking the oscillator from the EFI input of the 82C84A and holding the system clock off. At the same time, it disables the DREQ input to the DMA and, for additional security, holds $\overline{CAS}$ and $\overline{WE}$ off. When the burst refresh is complete, the oscillator clears the EFI disable latch, releasing the oscillator to the 82C84A.

3.1.4 POWER ANALYSIS

The system shown (Figure 8) contains ten VLSI devices[9], ten memory devices, thirty logic devices and two timers. To calculate the power consumption, some assumptions are made:

— where not given, typical values are 40% of the maximum value (rule of thumb)
— some values are estimates as final data sheets are not yet available.
— typical values are used as they should average out over a system.
— during normal operation, the memory receives a 15 μs distributed refresh
— The 80C88 can access the memory at 5 MHz/4 = 1.25 MHz max.(one bus cycle takes four T-states)
— The DRAM memory is capable of operating at (1/245 ns =) 4.082 MHz.
— System duty cycle is 2% (most of the system time is spent waiting for instructions such as keyboard entries)
— The speed/power curve is approximately linear

74HC logic current requirement (@25°C):
  536 μA (logic) +144 μA (timers) = 0.68 mA

Eight 51C259L-15 DRAMs, 100% (4.082 MHz) duty cycle, I_{CC} (typ):
— operating = 45 mA each
— standby = 0.01 mA each (CMOS drivers)

Two 27C64 CMOS EPROM, I_{CC} (typ):
— operating = 20 mA × 0.4 = 8 mA each
— standby = 100 μA × 0.4 = 0.04 mA each

During constant DRAM memory accesses, memory activity is:

active: 1.25 MHz/4.082 MHz = 30.6%
refresh: 245 ns/15 μs = 1.7%
standby: 100% − (30.6 + 1.7) = 67.7%

Memory Current is:
active: [2(45 mA) + 6(0.01 mA)] 30.6% = 27.56 mA
refresh: [8(45 mA)] 1.7% = 6.12 mA
standby: [8(0.01 mA)] 67.7% = 0.054 mA
continuous memory access total current = 33.73 mA
At 2% system duty cycle, memory current is:
(I_{oper} × 2%) + I_{ref} + [I_{stby} × (100% − (2 + 1.7)%)] =
(27.56 mA × 2%) + 6.12 mA + [8(0.01 mA)] 96.3% =
6.597 mA

EPROM Current is:
continuous active   16.0 mA
continuous standby   0.08 mA

Table 1 shows the typical current required for the VLSI:

Table 1

Device	Active	Standby
80C88	20.0 mA	0.2 mA
82C84A	6.0 mA	6.0 mA
82C37A	0.4 mA	0.004 mA
82C88	2.0 mA	0.004 mA
82C82 (3 each)	1.5 mA	0.012 mA
82C86	0.5 mA	0.004 mA
82C59A	0.4 mA	0.004 mA
82C52	1.0 mA (est)	0.004 mA
TOTAL	**31.8 mA**	**6.232 mA**

Rounding the DMA duty cycle up to 2% (from 1.7%) for simplicity, the current requirement for the VLSI at a 2% duty cycle is:

(31.8 mA × 2%) + (6.232 mA) × 98% = 6.743 mA

Therefore, total current for the system during normal operation at the 2% duty cycle (DRAM accesses only) is:

logic + DRAM + EPROM + VLSI
0.68 mA + 6.597 mA + 0.08 mA + 6.743 mA = **14.1 mA**

During a RESET operation, for a period of less than 1 ms, current will be:

  DRAM stby   + refresh + EPROM + logic +
[8(0.01 mA)]98.3% + 6.12 mA + 16 mA + 0.68 mA +

μP   +   clk   +   DMA   +   bus   +   addr   +
20 mA + 6 mA + 0.0107 mA + 2.0 mA + 1.5 mA

  data   +   int   +   UART   =
+0.5 mA + 0.004 mA + 0.004 mA = **52.9 mA**

intel

AP-238

Power Supply

Using the formula: $C = di \cdot dt/dV$;

$(52.9 \text{ mA} - 14.1 \text{ mA})1 \text{ ms}/200 \text{ mV} = 194 \text{ }\mu\text{F}$

200 μF of storage capacitance will support V_{CC} during reset to no more than 200 mV of droop. This means the power supply need only be sized for normal operation. Normal operation should not exceed 70% of the power supply capacity (ye olde rule of thumb) so:

$0.7 \text{ I}_{ps} = 14.1 \text{ mA}$
$\text{I}_{ps} = 20.14 \text{ mA}$

a 20 mA supply would suffice and, incidentally, would reduce the maximum droop voltage during reset to:

$(52.9 \text{ mA} - 20 \text{ mA})1 \text{ ms}/200 \text{ }\mu\text{F} = 164.5 \text{ mV}$

3.2 Data Retention Design

A system with data retention can, but need not necessarily be a low power design. The main prerequisite is that that portion of the system to be supported by the battery in the absence of line power must be low power. The elements to be discussed here include:

— saving information in use by the processor
— ensuring refresh for the DRAM
— protecting the memory from false signals:
  a. during back-up
  b. at power up
— minimizing power dissipation during back-up

3.2.1 SAVING INFORMATION

When power fails, the processor will "lose its place", or forget the operations it was actively executing. To prevent this, one can, when the power sense signal occurs, interrupt the processor using the non-maskable interrupt. The NMI would send the processor to a routine causing it to store, in a fixed area of memory, the pointers, stack counter and program counter and then to set a flag. The area of memory used can be anywhere, but it should either be dedicated to this task or it should be a rarely used section such as the bottom of the stack. When power returns, following system initialization, the processor checks the flag and, if set retrieves the saved information and continues from where it was interrupted.

3.2.2 SYSTEM DESIGN EXAMPLE

For the sake of simplicity, we will use the low power design already discussed, modifying it for battery back-up. In Figure 12, power fail circuitry is ORed with the EFI hold signal into the data retention refresh circuitry. A binary counter enabled by 'power sense' counts $\overline{\text{RAS}}$ pulses to the memory, allowing, in this case, fifteen memory accesses (including refresh) to store pointers, etc., and set a flag. It then starts the backup refresh circuitry. The maximum time before starting data retention refreshing would occur when all accesses counted are normal refresh cycles:

$15 \text{ }\mu\text{s} \times 15 \text{ cycles} = 225 \text{ }\mu\text{s}$

This period is easily supported by the capacity of the power supply and V_{DD} grid, but it does limit the number of memory accesses and the time in which to do

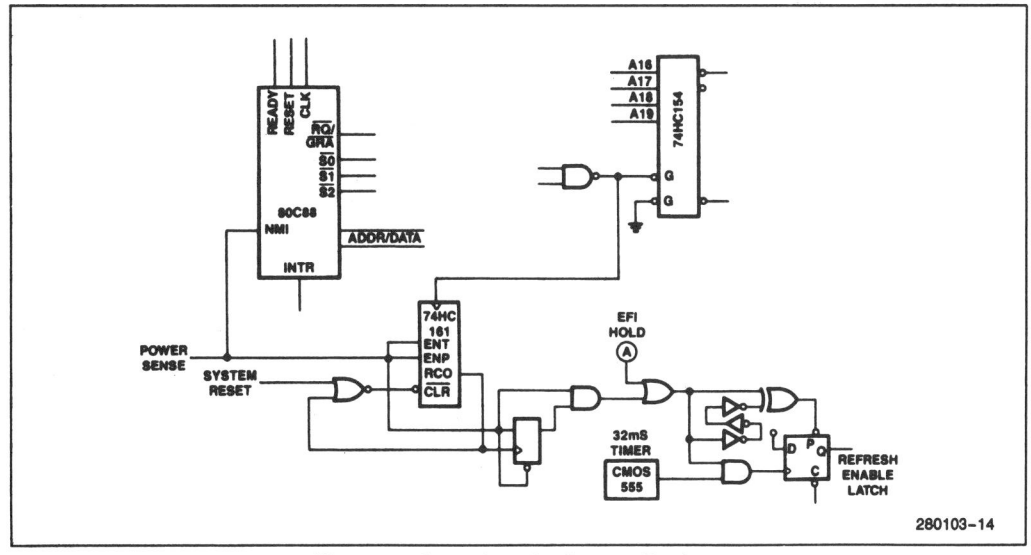

Figure 12. Provisions for Battery Back-up

intel AP-238

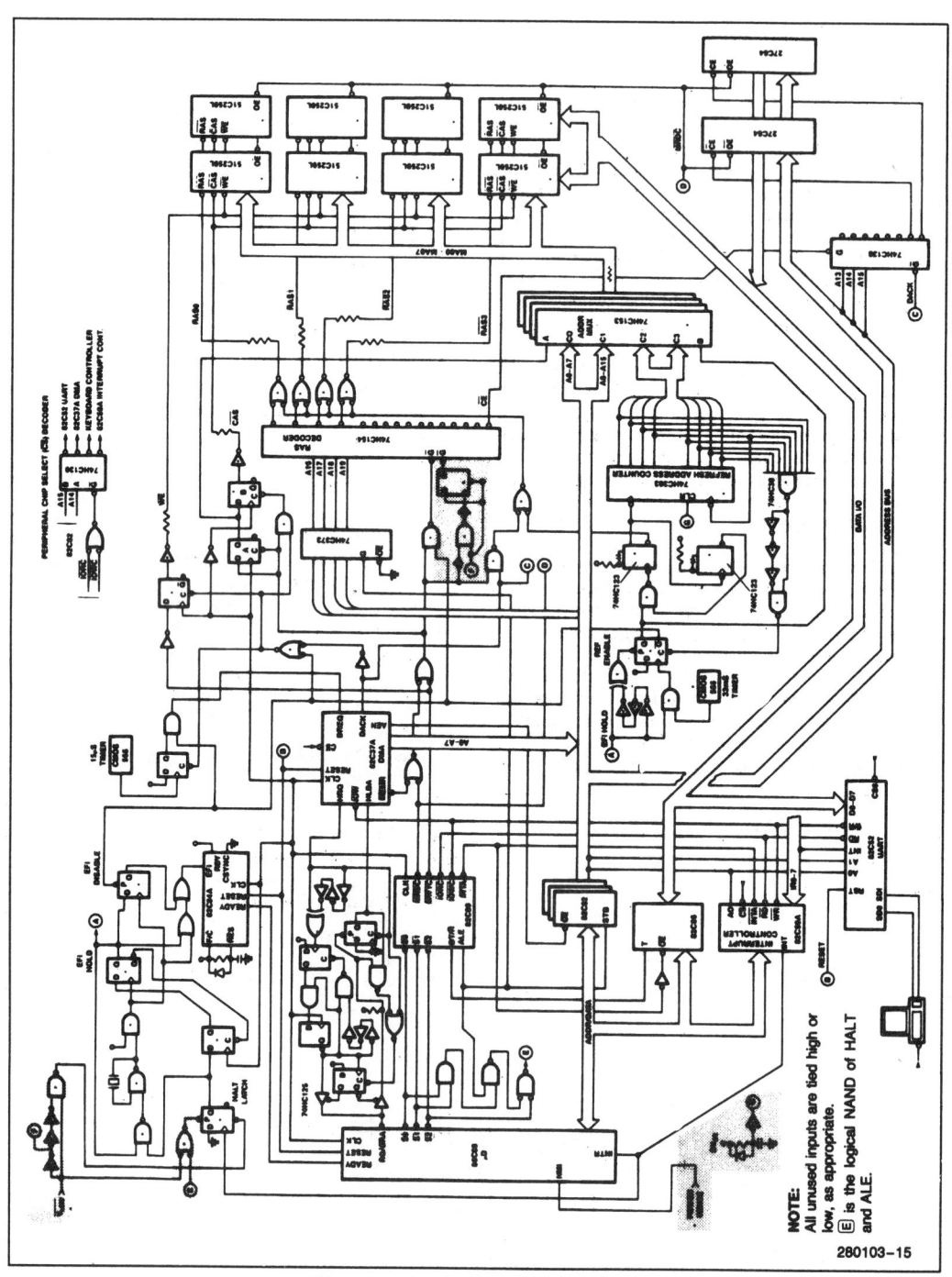

Figure 13. Battery Back-up Design

AP-238

them. Another approach (Figure 13) would be to send power sense to the NMI of the processor, allow the processor unlimited memory accesses and have it issue a HALT instruction when it has finished the necessary sequences. The liability of this method is that the user must determine how many accesses to allow. If too many are permitted, the processor could still be trying to access the memory when the voltage drops below specification. To insure this cannot happen, a low voltage detector is needed to override the processor (Section 5, Power Switching Circuits, Figures 18 and 27). Then, when the V_{low} circuit detects power reduced to the pre-determined low threshold and switches (goes low), it stops the clock using the same circuitry as the HALT instruction. At the same time, it is necessary to stop memory access immediately without glitching $\overline{RAS}$ so V_{low} is ANDed with $\overline{RAS}$ "off" to disable the $\overline{RAS}$ decoder. When power returns and V_{low} switches back high it clears the halt latch, restarting the oscillator. One additional change required is moving the data retention refresh address counter from the system reset to a battery powered reset. The rest of the circuit works as discussed previously.

If a battery with relatively high internal impedance is used, peak power in the data retention circuits would be a concern. In this case, a circuit such as that shown in Figure 14 could be used. The $\overline{RAS}$ pulse generated in the data retention circuit goes to a 2-to-4 line decoder and it goes through a divide-by-four circuit before toggling the 74HC393 address counter. This means four $\overline{RAS}$ pulses will be generated for each address and decoded to the memory banks, refreshing them one at a time. While this does not change the average power used for refreshing the memory, it does reduce the peak power requirement as follows (excluding support circuitry current):

I_{peak} 1 = 45 mA × 8 devices = 360 mA
I_{peak} 2 = (45 mA × 2 devices) + (0.01 mA × 6 devices) = 90.06 mA

3.2.3 MINIMIZE POWER CONSUMPTION

For battery back-up designs, the goal always to be kept in mind is to minimize the amount of power required, at least in the back-up circuitry. To that end, the following general guidelines exist:

- Use ALL CMOS components.
- When possible, use XN memory devices rather than X1. Fewer devices active mean less active power.
- Do not float (leave open or three-state) CMOS inputs. if they are unused, be sure they are tied high or low, as appropriate.
- Reduce the clock rate. This can be done by switching to extended refresh. The design shown here has the data retention refresh timer set to 32mS. Whenever it fires, the memory is refreshed in the burst mode.
- Minimize $\overline{RAS}$ low time.
- Drive all inputs to the rail (CMOS levels). This can be done by using CMOS drivers (preferred) or by using pull-up resistors on the outputs of TTL drivers.

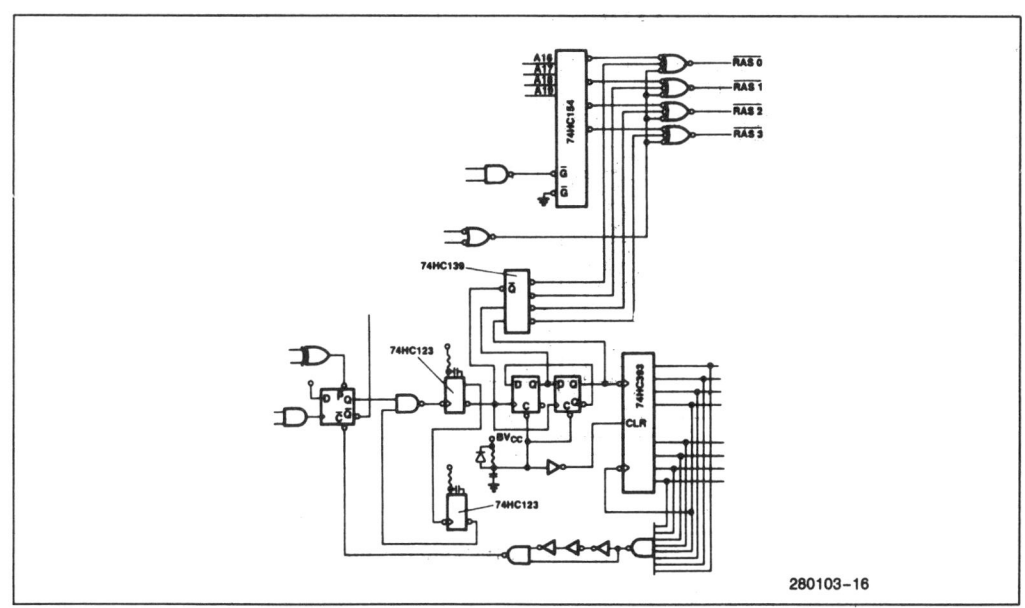

Figure 14. Distributed Refresh Reduces Peak Current

- Minimize the number of pull-up resistors by using CMOS drivers wherever possible.
- Ensure bypass capacitors have low inductance paths across the device (Figure 15).
- Use good design practices. Reduce the load capacitances by keeping related sections physically close.

Control input transitions. When too slow (≥ 25 ns), excess current will flow. When too fast (<5 ns), the increased frequency increases the current requirement. When possible, distribute the power used over time to control peak currents. Figure 14 alternates the refresh to each bank of memory, reducing the peak power requirement.

(For more design guidelines, see #10 in the bibliography)

3.2.4 POWER ANALYSIS

For this example, the power requirements of just the battery back-up section will be analyzed. As shown in Figure 13, those elements identified by an asterisk (*) would be powered by the battery in data retention mode. To determine battery size, typical values may be used, but the application, particularly the ambient temperature should also be considered. The lesser values will be used here.

The current requirements are:

Memory (/dev.): Logic (total, TYP.)
typical operating = 45 mA   464 µA
maximum operating = 65 mA
typical standby = 0.01 mA
maximum standby = 0.1 mA

When supported by the battery, the memory is refreshed every 32 ms. Refresh cycle time is 250 ns and each device has 256 rows to be refreshed (the cycle time can be extended a little to provide margin for component tolerances but extend t_{RP} because as $\overline{RAS}$ low time is extended, power increases). Total current required will be determined by the amount of time spent in each of refresh and standby.

256 rows × 250 ns = 64 µs to refresh
64 µs/32 ms × 100 = 0.2%
8(45 mA) × 0.002 = 720 µA (active)
8(0.01 mA) × 0.998 = 79.84 µA (standby)
720 µA + 79.84 µA = 799.8 µA (averaged over refresh period)

Therefore, the total average current drawn by the battery back-up section of this design, at 25°C, is:

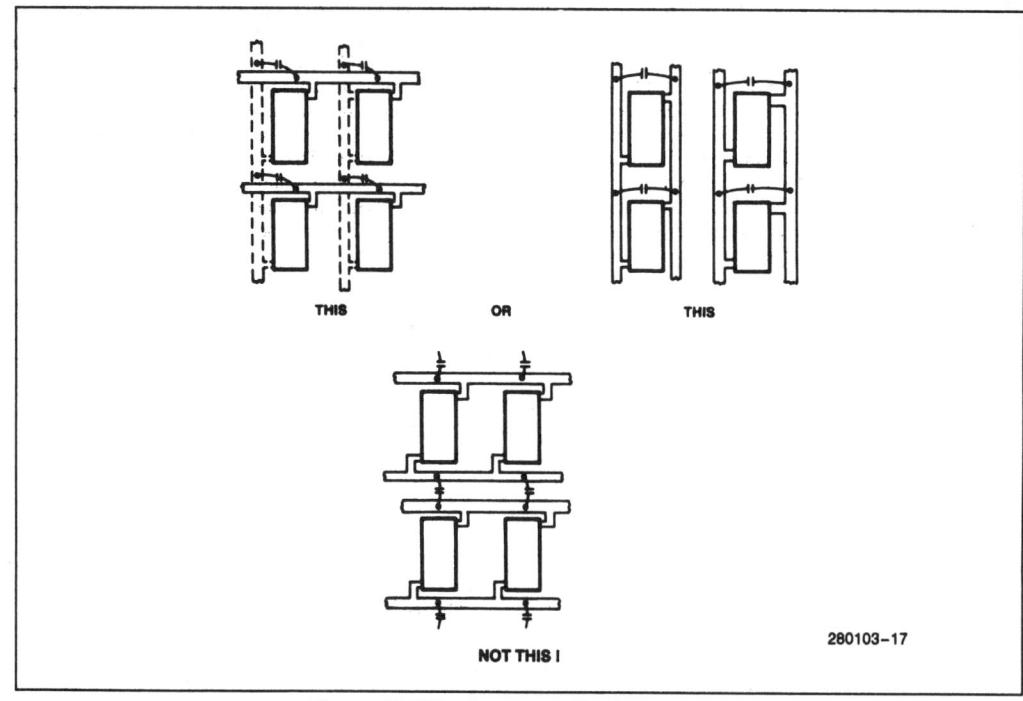

799.8 µA + 464 µA = 1.264 mA

Figure 15. Effective Capacitive Decoupling

AP-238

With this information, a reasonable choice of battery type can be made (see Section 4, Batteries).

4.0 BATTERIES

4.1 Battery Types

Batteries are divided into two basic types, primary and secondary. Primary batteries cannot, reasonably, be recharged. Secondary batteries are those that can be recharged. (For a comprehensive reference on batteries, see #12 in the bibliography.)

Primary batteries are created around a chemical reaction that, for all practical purposes, cannot be reversed. Consequently, when their charge is depleted they must be replaced. Some common examples are the zinc chloride, silver oxide and lithium batteries. They come in various sizes and most are in one of two styles, cylindrical or coin shaped. There are many electrochemical systems under the primary heading, each varying from the others in some unit of measure such as voltage, energy, energy density or cost. These variations can make them more or less suited for various applications (see chart 2)[11].

Secondary batteries use a reversible chemical reaction. Some common examples of secondary batteries are the lead acid car battery and the nickel-cadmium (NiCad) used in many consumer electronics products such as portable hand tools, shavers and photoflash equipment. Lead acid batteries have a low energy density, meaning that, compared to other systems, they have less capacity for their size and weight. NiCads have better energy density and longer cycle life (number of times they can be re-charged) but they have more potential problems with charging and discharging. Secondary batteries are measured similarly to the primaries but with the additions of cycle life and charge rate.

Briefly, here are some pros and cons of various batteries.

4.1.1 PRIMARY BATTERIES:

These have much better charge retention (shelf life) than secondaries and the cost/energy ratio is typically less (when the cost of the secondary is not distributed over cycle life).

4.1.1.1 Carbon Zinc

The common LeClanche battery has a one to three year shelf life. With a 1.5V/cell output, it would take four in series to achieve six volts. They have limited capacity for size (low energy density), a sloping discharge (output voltage decreases continuously over the period of the discharge), low cost and excellent availability in a wide variety of package sizes and capacities.

4.1.1.2 Alkaline Manganese

Similar to the carbon zinc, the alkaline battery has a two to five year shelf life, 1.5V/cell, limited capacity for size and a sloping discharge. It has higher energy density but higher cost than LeClanche and it is widely available in a variety of sizes and capacities.

4.1.1.3 Silver Oxide

This battery has a two to five year shelf life. It has a high energy density but, with 1.5V/cell, it would require four in series of a boost circuit* to achieve six volts. It has a flat discharge curve, relatively high cost and limited package configurations.

4.1.1.4 Mercuric Oxide

Similar to the silver oxide, the mercuric oxide battery has only 1.35V/cell. It has a flat discharge curve, higher cost and a potential ecological problem with the disposal of spent cells.

4.1.1.5 Lithium Manganese

This is only one of several lithium systems. It has an eight to ten year shelf life, 3V/cell and high energy density. It has relatively high internal impedance (as do most lithium systems), so applications requiring greater than a 10 mA drain will affect the voltage output of the battery. Some batteries are designed for higher rate service such as the DL2/3A from Duracell which can handle up to 50 mA drain without a severe impact on the voltage level. These are vented to prevent explosion in the event of abusive use or shorting.

*NOTE:
Texas Instruments makes several boost components, such as RM4193 and TL499C.

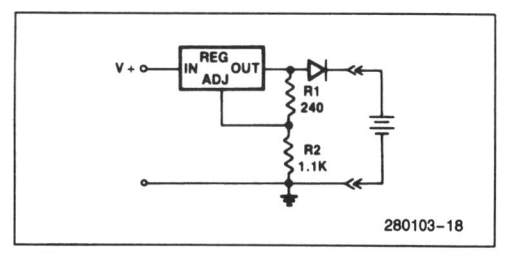

Figure 16. Simple Constant Voltage Charger

4.1.2 SECONDARY BATTERIES:

Because of self-discharge, these batteries have a relatively short shelf life but, with some means of charge

intel   AP-238

Chart 2. Eight Principal Battery Systems and Average Characteristics

Usual Name*	Carbon Zinc*	Zinc Chloride Super Heavy Duty*	Alkaline-Manganese Dioxide (MnO$_2$)	Mercuric Oxide*
Electrochemical System	Zinc-Manganese Dioxide (often called Leclanche)	Zinc-Manganese Dioxide	Zinc-Alkaline Manganese Dioxide	Zinc-Mercuric Oxide
Voltage Per Cell	1.5	1.5	1.5	1.35
Negative Electrode	Zinc	Zinc	Zinc	Zinc
Positive Electrode	Manganese Dioxide	Manganese Dioxide	Manganese Dioxide	Mercuric Oxide
Electrolyte	Aqueous solution of ammonium chloride and zinc chloride	Aqueous solution of zinc chloride (may contain some ammonium chloride)	Aqueous solution of potassium hydroxide	Aqueous solution of potassium hydroxide or sodium hydroxide
Type	Primary	Primary	Primary	Primary
Rechargeability	Not Recommended	Not Recommended	Not Recommended	Not Recommended
Number of Cycles				
Input if Rechargeable				
Overall Equations of Reaction	$2MnO_2 + NH_4Cl + Zn \rightarrow ZnCl_2 + 2NH_3 + H_2O + Mn_2O_3$	$8MnO_2 + 4Zn + ZnCl_2 + 9H_2O \rightarrow 8MnOOH + ZnCl_2 \cdot 4ZnO \cdot 5H_2O$	$2Zn + 2KOH + 3MnO_2 \rightarrow 2ZnO + 2KOH + Mn_3O_4$	$Zn + HgO + KOH \rightarrow ZnO + Hg + KOH$
Typical Commercial Service Capacities	60 mAh to 30 Ah	Several hundred mAh to 18 Ah	30 mAh to 45 Ah	45 mAh to 14 Ah
Energy Density (Commercial) Watt-hour/Lb	20	40	20–45	50
Energy Density (Commercial) Watt-hour/Cubic Inch	2	3	2–5	8
Practical Current Drain Rates Pulse	Yes	Yes	Yes	Yes
Practical Current Drain Rates High (More Than 50 mA)	100 mA/square inch of zinc area ("D" cell)	150 mA/square inch of zinc area ("D" cell)	200 mA/square inch of separator area ("D" cell)	No
Practical Current Drain Rates Low (Less Than 50 mA)	Yes	Yes	Yes	Yes
Discharge Curve (Shape)	Sloping	Sloping	Sloping	Flat
Temperature Range: Storage	−40°F to 120°F (−40°C to 48.9°C)	−40°F to 120°F (−40°C to 48.9°C)	−40°F to 120°F (−40°C to 48.9°C)	−40°F to 140°F (−40°C to 60°C)
Temperature Range: Operating	20°F to 130°F (−6°C to 54.4°C)	0°F to 130°F (−17.8°C to 54.4°C)	−20°F to 130°F (−28.9°C to 54.4°C)	14°F to 130°F (−10°C to 54.4°C)
Effect of Temperature on Service Capacity	Poor low temperature	Good low temperature relative to carbon-zinc	Good low temperature	Good high temperature, low temperature depends upon construction
Impedance	Mode	Low	Very low	Low
Leakage	Low???Under Abusive conditions	Low	Rare	Some salting
Gassing	Medium	Higher than carbon-zinc	Low	Very low
Reliability (Lack of Duds: 95% Confidence Level)	99% at 2 years	99% at 2 years	99% at 2 years	99% at 2 years
Shock resistance	Fair to Good	Good	Good	Good
Cost Initial	Low	Low to Medium	Medium Plus	High
Cost Operating	Low	Low to Medium	Medium to high at high power requirements	High

*NOTE:
Portions of this chart reprinted here courtesy of Union Carbide Corp. Eveready.

intel   AP-238

Chart 2. Eight Principal Battery Systems and Average Characteristics (Continued)

Usual Name*	Silver Oxide*	Nickel-Cadmium*	Lithium	Lead Acid
Electrochemical System	Zinc-Silver Oxide	Nickel-Cadmium	Lithium-Manganese Dioxide	Lead
Voltage Per Cell	1.5 (monovalent)	1.2	3	2
Negative Electrode	Zinc	Cadmium	Lithium	Lead
Positive Electrode	Monovalent Silver Oxide	Nickelic Hydroxide	Manganese Dioxide	Lead Dioxide
Electrolyte	Aqueous solution of potassium hydroxide or sodium hydroxide	Aqueous solution of potassium hydroxide	Non-aqueous mix propylene carbonate dimethoxy ethane	Sulfuric Acid
Type	Primary	Secondary	Primary	Secondary
Rechargeability	Not Recommended	Yes	Not Recommended	Yes
Number of Cycles		300 to 1,000		
Input if Rechargeable		Minimum of 140% of energy withdrawn		160% of energy withdrawn
Overall Equations of Reaction	$Zn + Ag_2O + KOH \rightarrow ZnO + 2Ag + KOH$	$Cd + 2NiOOH + KOH + 2H_2O \rightleftarrows Cd(OH)_2 + 2Ni(OH)_2 + KOH$	$2Li + 2MnO_2 \rightarrow Li_2O + Mn_2O_3$	$PbO_2 + Pb + 2H_2SO_4 \rightleftarrows 2PbSO_4 + 2H_2O$
Typical Commercial Service Capacities	15 mAh to 210 mAh	150 mAh to 4 Ah	35 mAh to 1100 mAh	1.5 Ah to 400 Ah
Energy Density (Commercial) Watt-hour/Lb	50	12–16	100	10–100
Energy Density (Commercial) Watt-hour/Cubic Inch	8	1.2–1.5	6.6	0.8–7
Practical Current Drain Rates Pulse	Yes	Yes	Yes	Yes
Practical Current Drain Rates High (More Than 50 mA)	No	8.10A ≈ 1 amp/sq. in. of electrode	Cylindrical yes coin no	Yes
Practical Current Drain Rates Low (Less Than 50 mA)	Yes	Yes	Yes	Yes
Discharge Curve (Shape)	Flat	Flat	Flat	Sloping
Temperature Range: Storage	−40°F to 140°F (−40°C to 60°C)	−40°F to 140°F (−40°C to 60°C)	−60°C to +75°C	−60°C to 30°C
Temperature Range: Operating	14°F to 130°F (−10°C to 54.4°C)	Discharge: −4°F to 113°F (−20° to 45°C) Charge CH Types: 32°F to 113°F (0°C to 45°C) Charge CF Type: 60°F to 113°F (15.6°C to 45°C)	−10°C to +70°C	−40°C to 60°C
Effect of Temperature on Service Capacity	Low temperature depends upon construction	Very good at low temperature	Good high temperature Low temperature depends upon construction	Low temperature depends upon construction
Impedance	Low	Very low	Medium	Low
Leakage	Some salting	No	Rare	Low to Medium. Some salting
Gassing	Very low	Low	Low	Medium
Reliability (Lack of Duds: 95% Confidence Level)	99% at 2 years	99% at 2 years		
Shock resistance	Good	Good	Good	Fair to Good
Cost Initial	High	High	High	Medium
Cost Operating	High	Low	High	Very Low

*NOTE:
Portions of this chart reprinted here courtesy of Union Carbide Corp. Eveready.

maintenance, shelf life can be extended virtually indefinitely. For the most part, they have a better high rate performance than the primaries, an exception being lithium cells using soluble cathode materials which have comparable performance (i.e. lithium/sulfur dioxide, lithium/thionyl chloride). They can generate gasses when charging or discharging at excessively high rates. For brevity, and because they are generally the preferred design for this application, only the sealed versions will be considered here.

4.1.2.1 Lead Acid

With low energy density, these batteries are generally packaged three or six cells in series for six or twelve volts and in larger sizes for much higher capacities (i.e. 400 Amp hours). Fully charged, they contain strong sulfuric acid and shorting the outputs may cause an explosion. They self-discharge, forming lead sulfate ($PbSO_4$) crystals which can insulate the plate, reducing capacity. Lead filaments can form between the plates when charging and can short out cells. Excessive overcharging may cause venting. They can deliver large current surges and they have a sloping discharge curve which eases the task of sensing when the battery is low. They are normally charged with a constant voltage but can be charged with constant current if care is taken to avoid overcharging. They cannot be stored in a discharged state. They have a lower cycle life than NiCads but they cost less and they have no "memory" effect or thermal runaway problems. Charging circuits are discussed in more detail in the next section.

4.1.2.2 Nickel Cadmium

These batteries have low energy density and 1.2V/cell. If used in series, they should have closely matched discharge rates or the faster discharged cell can be driven to cell reversal (destroyed) when the battery is allowed to reach total discharge. NiCads are charged with constant current at a rate of 10% of capacity (0.1C) for 14 to 16 hours (standard cell) or 30% (0.3C) in 3 to 5 hours for quick charge cells. If constant voltage is used, care must be taken to limit the charge current during the final 25% of charging to prevent thermal runaway. If charged at less than a 10% rate, cell capacity will fade and a sustained overcharge will cause a lack of capacity on the first discharge. When partially discharged to the same level several times and then recharged, they develop a "memory" effect and will, thereafter, discharge no further than that level. It is a temporary effect in that forcing a deep discharge will erase the memory. They cost more than lead-acid but are more rugged, typically have a longer cycle life and they have a long shelf life in any state of charge. Normal discharge rate is fairly flat. Charging circuits are discussed in more detail in the next section.

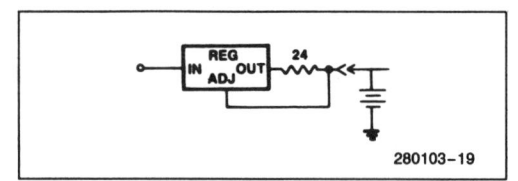

Figure 17. Simple Constant Current Charger

4.1.3 OTHER BATTERY SYSTEMS

This is by no means an exhaustive description of battery systems. Many other primaries and some additional secondaries are available and more are in development. For example, there are a number of primary lithium systems such as lithium/thionyl chloride, lithium/copper oxide, lithium/silver chromate, and so on. A silver oxide secondary is available with the highest energy and power density and best charge retention of any commercially available secondary. Its high cost and low cycle life, however, have held it to limited applications. Many more secondaries are in development, including lithium, nickel, zinc, aluminum, sodium and calcium systems[12]. The concern of this document, however, was for those systems available today and was further narrowed for this application by energy density, availability, voltage/cell, cost, and so on. The LeClanche was included in this description primarily as a commonly known system for reference.

4.2 Key Parameters For Battery Selection

In the process of selecting a particular battery, first consider the application in which it will be used. That will usually encompass many of the characteristics that will identify the battery of choice. For example, in what environment (temperature, vibration, etc.) will the battery exist? What voltage is required. And so on. If necessary, applications engineers from the battery manufacturers can help in the selection process, but here are the main parameters to be considered, in no particular order. The application will generally prioritize them.

4.2.1 VOLTAGE:

What is the voltage range within which the application will be operating?

How low a voltage can be tolerated?

This will show either that a sloping discharge curve is acceptable or it may indicate a requirement for a battery with a flat discharge curve.

AP-238

4.2.2 ENERGY:

What will be the discharge rate and for what period of time?

This will define the ampere-hour capacity required of the battery.

4.2.3 ENERGY DENSITY:

How much space will be available (or perhaps weight tolerable) in the area where the battery is to be located?

It may be possible to use multiple cells in series to achieve, say, six volts, but more than two cells may be two cumbersome. It is worth noting here that re-chargeable batteries require less service than primaries which have to be replaced periodically. So, particularly in stationary systems, they could be located with the system power supply, permitting larger sizes for greater capacity.

4.2.4 ENVIRONMENT:

What are the temperatures, standby and operating, to which the battery will be subjected?

Will it have to withstand vibration?

This will define how rugged the battery must be.

4.2.5 SERVICE LIFE:

How long must the batteries last, of at least, what is an acceptable period?

This will help define the capacity required of the battery. In the case of the re-chargeables, it can also define the cycle life of the battery.

4.2.6 COST:

Of course! What is acceptable?

Higher cost for a flat discharge rate or, perhaps, for higher service?

Higher initial but lower operating (or the reverse) cost?

Depending on the priority of this issue, it may be the deciding factor between a preferred battery and one that is merely acceptable.

Other factors that may be considered include shelf life, duty cycle, safety, reliability and availability.

4.3 Choosing a Battery

Of course, the issues being considered here are low power systems and systems with battery backup. A hypothetical example of each is examined to illustrate the battery selection process.

4.3.1 EXAMPLES

The first application is a low power system. It is portable, so it must be kept both small and light. The voltage will be in the range of six volts to four-and-a-half volts, using not more than two batteries in series that together do not exceed one half inch in height. Peak current drain is 8mA and duty cycle is 1%. A sloping discharge curve will give an early "battery low" indication which will permit continued operation for a reasonable period to allow time for purchase of a replacement. So size, voltage and discharge curve are the first considerations. Within those parameters we will look for greatest capacity and lowest cost. Referring back to the chart of battery comparisons (Chart 2), the carbon zinc and alkaline manganese have limited capacity for size (energy density) and silver oxide and mercuric oxide would require four cells or a boost circuit. The lithium/manganese dioxide has relatively high energy density, sufficient voltage to achieve six volts with just two cells and an acceptable discharge curve. Both initial and operating costs are high but expense is not a high priority (in this example). Current drain is low enough that the internal impedance of the battery should not be a problem. In the re-chargeables, lead-acid has a low energy density (too big and too heavy). The NiCad has slightly better energy density, it will require four cells and a charging circuit, but has low operating cost. Therefore, it will increase the size of the system (small size was a high priority) but keep operating costs down (cost was a low priority).

The battery of choice for this system is the lithium.

The second application is battery backup. A small computer, stationary, has a four megabyte CHMOS DRAM memory (sixty-four 51C259L's) organized as 2M × 16 and serving as both main memory and as mass storage. Using the 64K × 4 51C2259L means that only four devices will be active at a time, keeping both operating and (if banks are refreshed one at a time) refresh currents low. Since the system is not portable and the battery is for backup, it can be located with the primary power supply so size is not as restricted as in the previous example. This will permit a larger, higher capacity battery to be used for the relatively large memory, but it will probably be less accessible. This implies a battery that rarely needs servicing. Since the system normally draws power from the AC line, a recharging circuit can be added to the system quite easily. Therefore, a rechargeable battery will be used, but which one? Size was not an issue so cost can be. Nickel cadmium batteries are more expensive than lead acid and require more care in charging and discharging. So the battery of choice for this system is the lead acid.

3-80

Table 2

System	# = 6V	Capacity	Size (Inches)	Typical Life Cycle @ 14mA
Alkaline	4 each	1600 mAh	1.97×2.25×0.56	114.3 hours
Silver Oxide	1 each	150 nAh	0.99×0.51×0.51	10.7 hours
Mercuric Oxide	1 each	500 mAh	1.77×0.68×0.68	35.7 hours
Lithium Mang.	1 each	1200 mAh	1.41×1.37×0.76	85.7 hours
Lead Acid	1 each	6000 mAh	3.6×4.5×1.9	428.6 hours
NiCad (quick chg)	5 each	1200 mAh	1.7×4.5×0.9	85.7 hours

4.3.2 BATTERY SELECTION FOR THE LOW POWER SYSTEM

The low power system discussed in Section 3 could be powered by a secondary battery instead of the power supply. It would have to support about 14 mA during normal operation and a random pulsed high rate of 53 mA (I_{reset}). As discussed in Section 3, storage capacitance could help during the high rate, but a battery with some high rate capacity could reduce the size of the capacitance required. What remains is to determine an acceptable tradeoff between life cycle, size and cost. Beginning with primaries, Table 2 has a few randomly selected examples of battery systems to further illustrate the concept.

Either the NiCad or the lead acid would work well in this particular design. The final choice would depend on other requirements of the application such as cycle life (NiCad) or cost (lead acid).

4.3.3 BATTERY SELECTION FOR BATTERY BACK-UP SYSTEM

The battery back-up system discussed in Section 3 required 1.26 mA average drain at 6 Vdc. Most line power failures are very short, so assuming AC powers the memory at all times except failures and the longest failure might be four hours, the minimum battery capacity is:

1.26 mA × 4 = 5.04 mAh

Referring to Chart 2, it can be seen that any of the battery types can easily support this requirement. Primary batteries in this system will require a battery low indicator (see Section 5, Power Switching Circuits) and regular servicing. Since AC line power is normally used, this is an excellent application for secondary batteries with a recharging circuit. As above, if the system is to be mobile, smaller NiCads might be used. If stationary or if cost is a significant factor, less expensive lead-acid might be preferred.

If the system is used eight hours a day, five days a week and the batteries are required to support the memory whenever normal power is off, the minimum capacity would have to support the weekend, so:

1.26 mA × 63 hours = 79.4 mAh

With the addition of off-hours during the week·

1.26 mA × 64 hours = 80.6 mAh

Total: 79.4 + 80.6 = 160 mAh per week

Referring above to Table 2, the highest capacity primary would last:

1600 mAh/160 mAh per week = 10 weeks.

The secondaries would last to the extent of their cycle life, typically from one to four years at five cycles per week.

AP-238

5.0 POWER SWITCHING CIRCUITS

A battery back-up system required some means to smoothly enter and exit the battery back-up mode of operation. It must, in an orderly fashion, stop the system and switch the volatile memory over to battery power before the line power is lost. It must also return the memory to line power when that power is restored and it must do it without loss of the data stored and with an orderly transfer of the bus back to the now-active system. In addition, it may be necessary to sense when the battery is in a low state of charge and flag the operator. Where appropriate, it must provide battery charging circuits.

Briefly, these are the elements to be covered in this section:

a. Power sensing
b. Switching to back-up power
c. Sensing battery low
d. Recharging

5.1 Power Sensing

The power sensing circuit is connected to the front-end of the power supply. Essentially, it must provide a look-ahead signal to tell the system that the line power is going down. The system can then invoke a predetermined series of operations that take it to back-up power quickly and in an orderly manner. The sensing circuit is powered at all times by the battery.

5.1.1 µWATTS CMOS LOW POWER DETECTOR

The first circuit shown here is a microwatt CMOS low power detector. It provides two outputs. The first tells the system that power is going down early enough that the system has time to store pointers, stack counter and program counter and shift over to battery power. The second output tells the system to stop all activities, power is gone.

5.1.2 DIFFERENTIAL COMPARATOR FOR POWER FAIL DETECTION

The second circuit uses a comparator that relies on the ripple out of the bridge rectifier. When power starts to go down, the ripple becomes more pronounced. As it fluctuates below the trigger level of the circuit, the comparator output goes high, signalling the beginning of a power loss. As the ripple swings high again (momentarily shutting the comparator off), it provides a little more boost to the regulator, keeping power up a little longer. Because of the relatively low frequency of the ripple (60 × 2 = 120 cps), this provides at least (1/120 =) 8 ms of continued power. Time for the system to shut itself down in orderly sequence

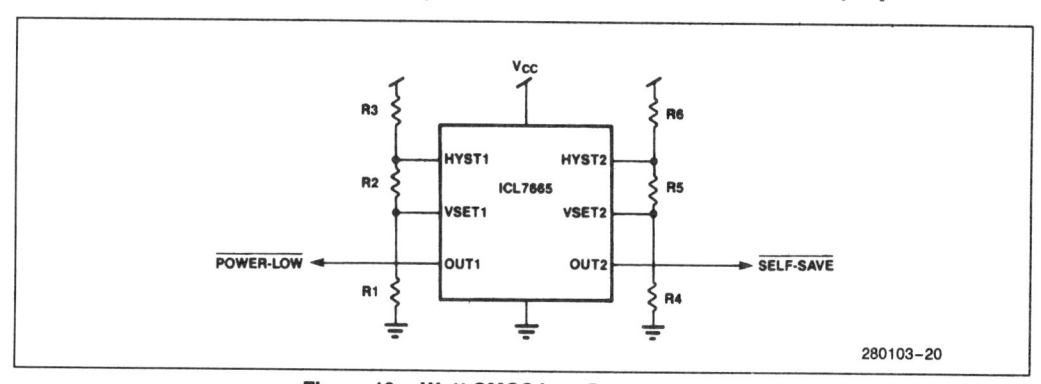

Figure 18. µWatt CMOS Low Power Detector

5.2 Power Switching Circuits

5.2.1 SIMPLE SWITCHING CIRCUIT

The first requirement of a power switching circuit is that the power for the portion of the system what will be supported by the battery when power fails (CMOS) be isolated from the power for the rest of the system (TTL).

This can be done quite simply with a low-leakage diode as shown in the simple diode switching circuit, below. A second diode is installed in the TTL line to keep the normal voltage to both circuits at the same level. V_{CC} must be higher to allow for the diode drop. The lithium battery shown also has a diode to prevent charging of the primary cell. The NiCad has a resistor both to allow charging and to limit the charge current. Battery power to the system will be lower by the voltage drop across the resistor. Relays could be used, but a capacitor would have to support the CMOS circuits during switching and is not recommended.

5.2.2 DUAL VOLTAGE SWITCH

The second switching circuit shown uses transistor isolation. This is a dual voltage circuit, providing normal voltage when the system is operating, and a lower, support voltage on battery power. The voltage drop across the transistor Q2 is small enough (100 to 200mV) that it is unnecessary to match the TTL and CMOS V_{CC} as was done in the diode circuit. During normal operation, Q1 is turned on and the power supply powers the circuit. At the same time, the battery is charged by the base current of Q1. When power goes down, Q2 turns on, Q1 turns off and the battery powers the circuit.

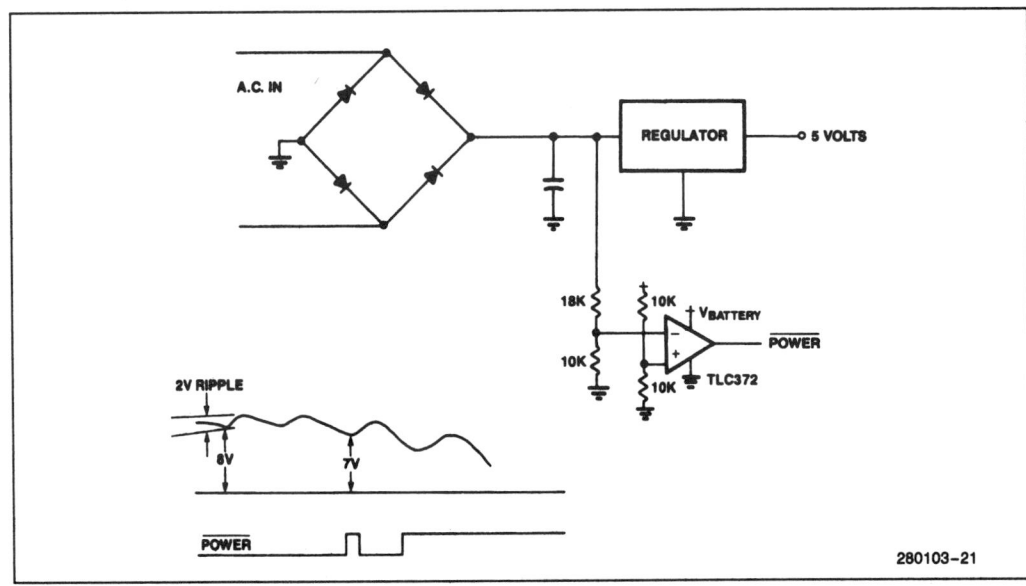

Figure 19. Power Failure Detection

5.2.3 SINGLE-BUS SWITCH

The next switching circuit is a single voltage transistor circuit for a system that has all circuits backed up by the battery. Q2 and its bias network are a regulating circuit that establishes the voltage at which the circuit switches from normal to battery power. As power goes down, that voltage turns off Q2 which, in turn, shuts off Q1 and the battery takes over. The battery is charged by constant voltage, and R_c in the NiCad circuit establishes the maximum charging current for the battery.

5.2.4 DUAL-BUS SWITCH

The last switching circuit uses diode isolation for the two circuits and a transistor switch for the battery. Normally biased off by the unregulated voltage, as power goes down, the base voltage drops, turning on the transistor and the battery powers the circuit without a voltage drop across a resistor. During normal operation, the battery is charged by constant voltage through the 140Ω resistor and 1N914 diode. The zener diode protects the system from over-voltage when the battery is not in the circuit.

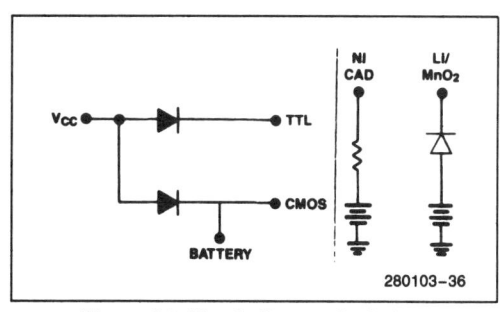

Figure 20. Simple Battery Isolation

Note that the battery charging accommodation in Figure 23 does not provide any protection from overcharging the battery. The others (Figures 20, 21 and 22) are somewhat protected by the lower input voltage used. Overcharging is addressed in the two stage battery charging circuits later in this section.

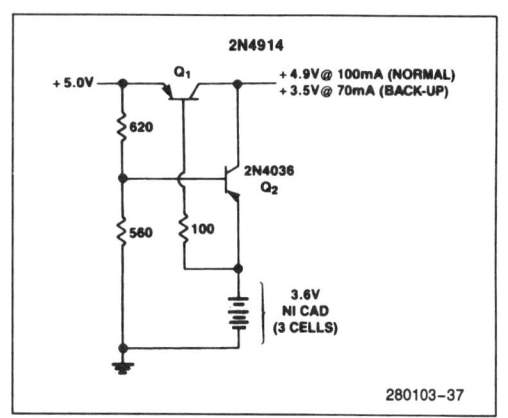

Figure 21. Dual Voltage Circuit

5.3 Protection From False Signals

5.3.1 RESET CIRCUITS

In a battery back-up system, care should be taken to protect the memory from false signals, during both back-up and power-up, which could alter the data that is to be preserved. To avoid any possibility of this happening, write enable ($\overline{WE}$) is held high during back-up and the normal chip select ($\overline{CS}$ or $\overline{RAS}$) is gated with the system power-up reset. For additional insurance, gate $\overline{WE}$ with the reset as well. The simplest reset circuit (Figure 25) is the resistor/capacitor, but it can briefly cause oscillations out of the associated logic circuits as it moves slowly through the threshold level. Putting a Schmitt triggered device on it improves the situation (Figure 24). The three transistor reset circuit (Figure 26) prevents oscillations by providing a more normal risetime to the logic circuits following the delay. When power first comes up, the RC time constant on its base keeps Q1 off which, in turn, keeps Q2 and Q3 off, holding $\overline{RESET}$ low. When the charge on the capacitor rises to threshold, Q1 turns on, turning on Q2 and Q3 and pulling $\overline{RESET}$ high.

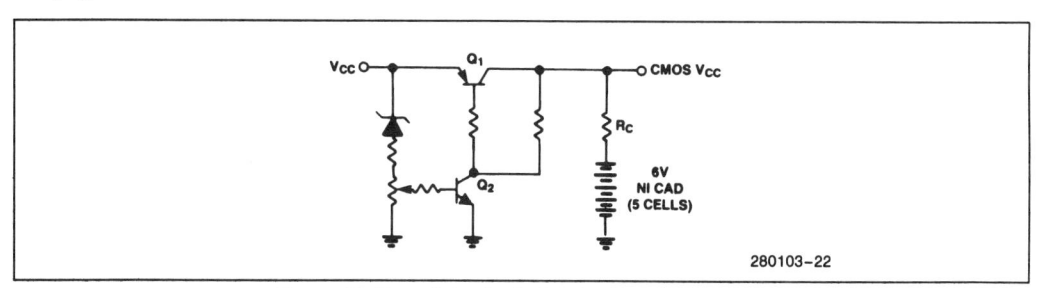

Figure 22. Transistor Isolation for Battery Back-up

intel AP-238

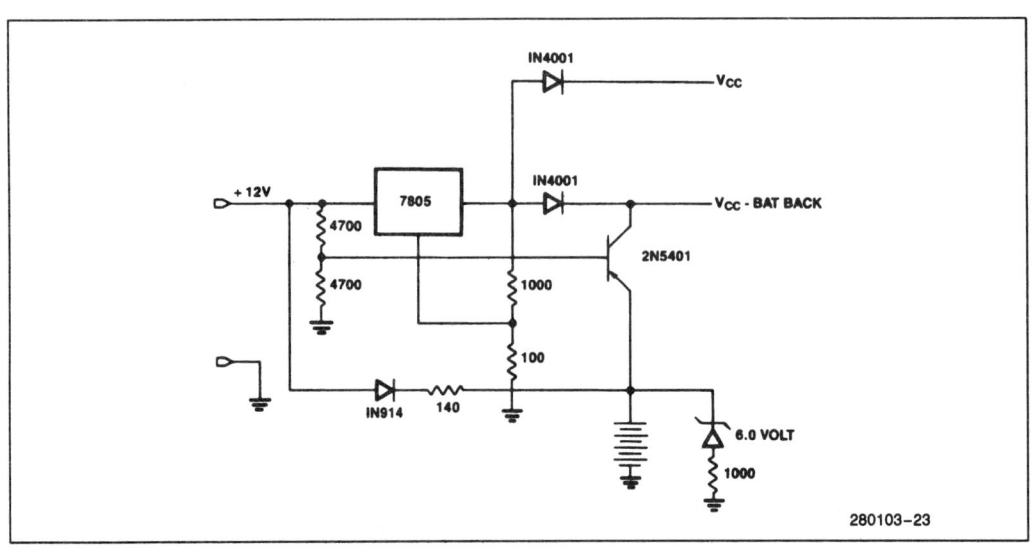

Figure 23. Transistor Switch for Battery

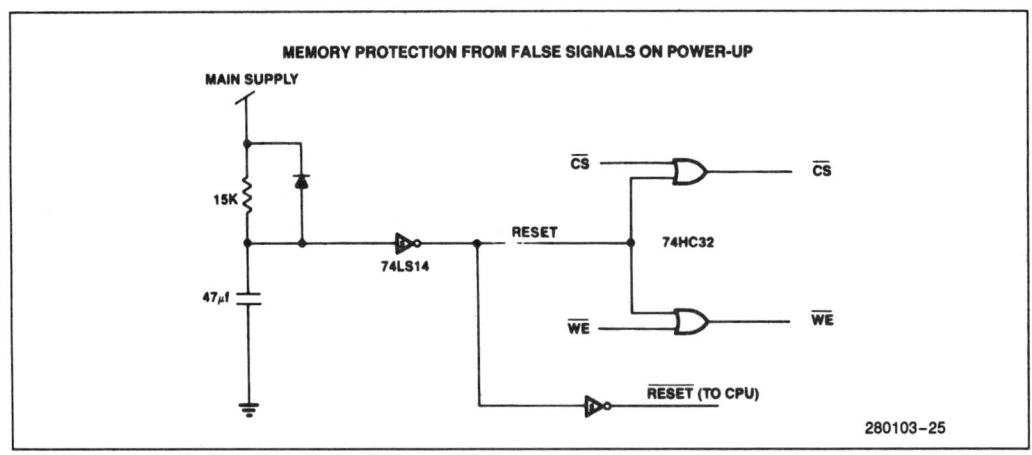

Figure 25. Schmitt Trigger Isolated Reset

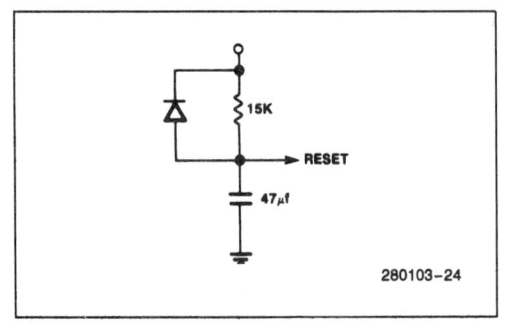

Figure 24. Simple Reset Circuit

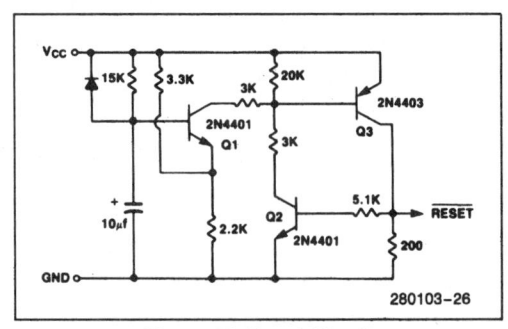

Figure 26. Reset Circuit Protects From Oscillation

AP-238

5.3.2 VOLTAGE LOW DETECTOR

If, when a power failure sequence is initiated, the processor is allowed an extensive series of memory accesses, care must be taken to block it from continuing when the voltage drops below the memory components specification. A low-voltage detector will provide a V_{low} signal to override circuits. The circuit for the detector could use a voltage comparator, an op amp configured as a voltage detector or, as shown in Figure 27, a low voltage detector [13]. Connected to system power, the external resistor network allows the user to define the trip voltage.

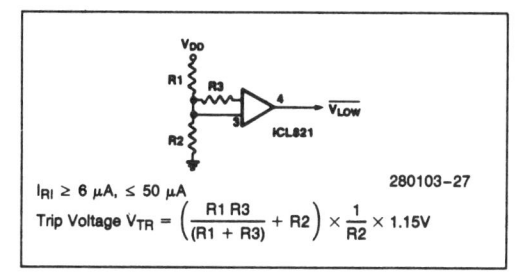

$I_{R1} \geq 6 \mu A, \leq 50 \mu A$

Trip Voltage $V_{TR} = \left(\frac{R1\,R3}{(R1 + R3)} + R2 \right) \times \frac{1}{R2} \times 1.15V$

Figure 27. Voltage Low Detector

5.3.3 SENSING BATTERY LOW

An indicator may be needed to flag the operator when the battery is in a low state of charge, particularly in a battery powered system or in a back-up system using primary batteries. One problem to be considered, however, is that the circuit will be a power drain on the battery. Two approaches are shown.

In Figure 28, the LED is on when the battery is charged, goes off when it reaches a state of discharge determined by resistors R1 and R2. Using a low power LED such as the HLMP-7000 from Hewlett Packard, the circuit draws about 2.03 mA from a charged battery.

In Figure 29, the LED turns on when the battery is low (determined by R1 and R2). It stays on until the battery reaches a predetermined (R3,R4) cutoff level, then shuts off to protect secondary batteries from overdischarge damage (not needed for primaries). This circuit draws about 430 μA from a charged battery, 2.4 mA from a low battery and ~200 μA from a battery at or below the cutoff level. Flashing the LED would extend the battery warning period.

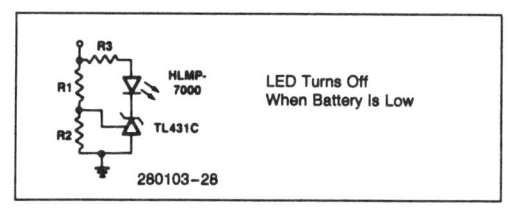

Figure 28. Battery Low Indicator #1

Another concern with detecting a low state of charge is with batteries having a flat rate of discharge [14] (see Battery Comparison Chart 2 in Section 4). This is a positive feature in every other aspect, but the state of charge is difficult to detect until the knee of the discharge curve is reached. At that point, there is very little capacity left, allowing virtually no margin of time to replace primary cells.

The LM431C shunt regulator has a very sharp turn-on characteristic, permitting V_{ref} to be set for a narrow voltage window. However, a circuit using a differential comparator such as the LinCMOS* TLC372C could provide a more precise setting to detect the state of charge before the knee.

5.4 Two Stage Battery Charging Circuits

For systems using secondary batteries, a charging circuit is needed. Section 4 on batteries gave two simple circuits, one for constant voltage (lead-acid), the other for constant current (NiCad) charging. It should be noted here that lead-acid batteries can be charged with a constant current if care is taken to avoid overcharging. Since overcharging is a waste of power and can be damaging to the battery, it should be avoided anyway.

At the same time, a continuing trickle charge is necessary to overcome the self-discharge characteristic of secondaries. This implies a dual potential (voltage) or split rate (current) charger. The switching circuits discussed earlier incorporated simple charging circuits with little protection from overcharging. The circuits to be discussed here provide that protection through two stage charging. The main charging circuit is typically set for a charge rate of 0.1 C (10% of capacity).

When the battery reaches full charge, the circuit switches off, leaving a trickle charge circuit enabled, providing a continuous charge at a 0.01C rate. Temperature affects the charge acceptance of the battery so, if it is to be re-charged at higher rates or in higher than normal room temperatures, temperature compensation must be incorporated into the charger.

NOTE:
Because batteries come in various chemical systems and capacities, the charging circuit MUST be tailored to the specific battery used.

5.4.1 FAST CONSTANT VOLTAGE

The first circuit is a fast constant voltage charger. Referring to Figure 31, the current is limited initially to about 1.5 amps by the internal current limit of the

*LinCMOS is a trademark of Texas Instruments

intel
AP-238

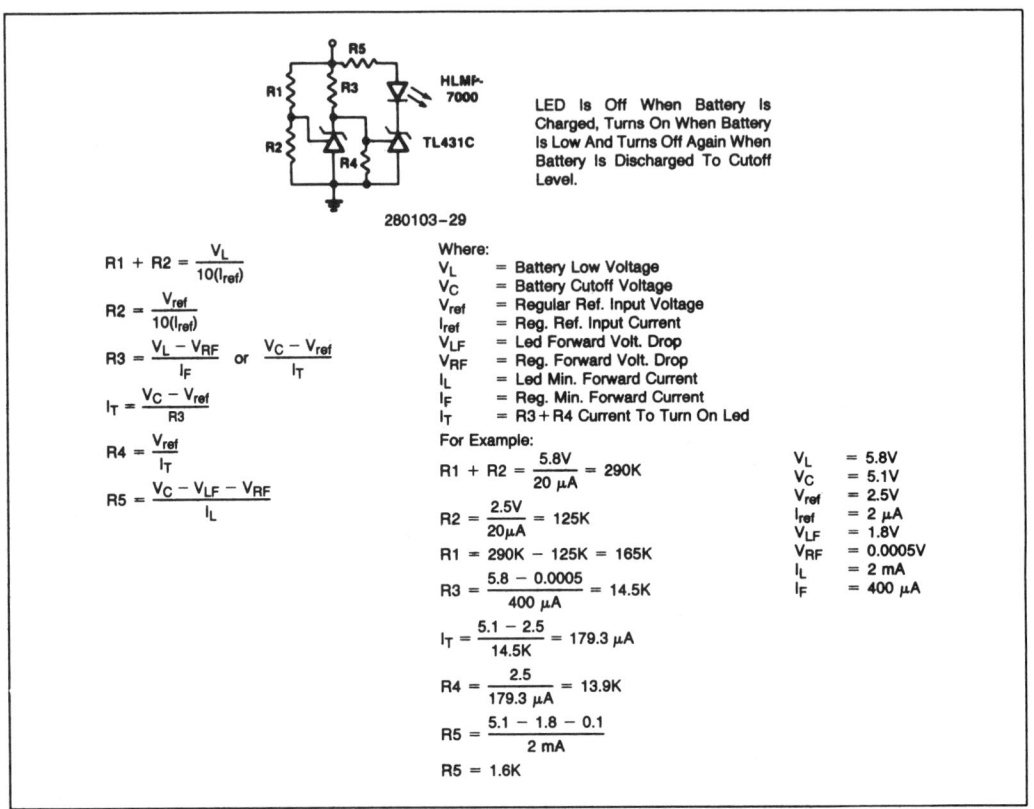

LED Is Off When Battery Is Charged, Turns On When Battery Is Low And Turns Off Again When Battery Is Discharged To Cutoff Level.

280103-29

$$R1 + R2 = \frac{V_L}{10(I_{ref})}$$

$$R2 = \frac{V_{ref}}{10(I_{ref})}$$

$$R3 = \frac{V_L - V_{RF}}{I_F} \text{ or } \frac{V_C - V_{ref}}{I_T}$$

$$I_T = \frac{V_C - V_{ref}}{R3}$$

$$R4 = \frac{V_{ref}}{I_T}$$

$$R5 = \frac{V_C - V_{LF} - V_{RF}}{I_L}$$

Where:
- V_L = Battery Low Voltage
- V_C = Battery Cutoff Voltage
- V_{ref} = Regular Ref. Input Voltage
- I_{ref} = Reg. Ref. Input Current
- V_{LF} = Led Forward Volt. Drop
- V_{RF} = Reg. Forward Volt. Drop
- I_L = Led Min. Forward Current
- I_F = Reg. Min. Forward Current
- I_T = R3 + R4 Current To Turn On Led

For Example:

$$R1 + R2 = \frac{5.8V}{20\,\mu A} = 290K$$

$$R2 = \frac{2.5V}{20\,\mu A} = 125K$$

$$R1 = 290K - 125K = 165K$$

$$R3 = \frac{5.8 - 0.0005}{400\,\mu A} = 14.5K$$

$$I_T = \frac{5.1 - 2.5}{14.5K} = 179.3\,\mu A$$

$$R4 = \frac{2.5}{179.3\,\mu A} = 13.9K$$

$$R5 = \frac{5.1 - 1.8 - 0.1}{2\,mA}$$

$$R5 = 1.6K$$

- V_L = 5.8V
- V_C = 5.1V
- V_{ref} = 2.5V
- I_{ref} = 2 μA
- V_{LF} = 1.8V
- V_{RF} = 0.0005V
- I_L = 2 mA
- I_F = 400 μA

Figure 29. Battery Low Indicator #2

Discharge Curves of Lithium Manganese Dioxide Cells (LiMnO$_2$)

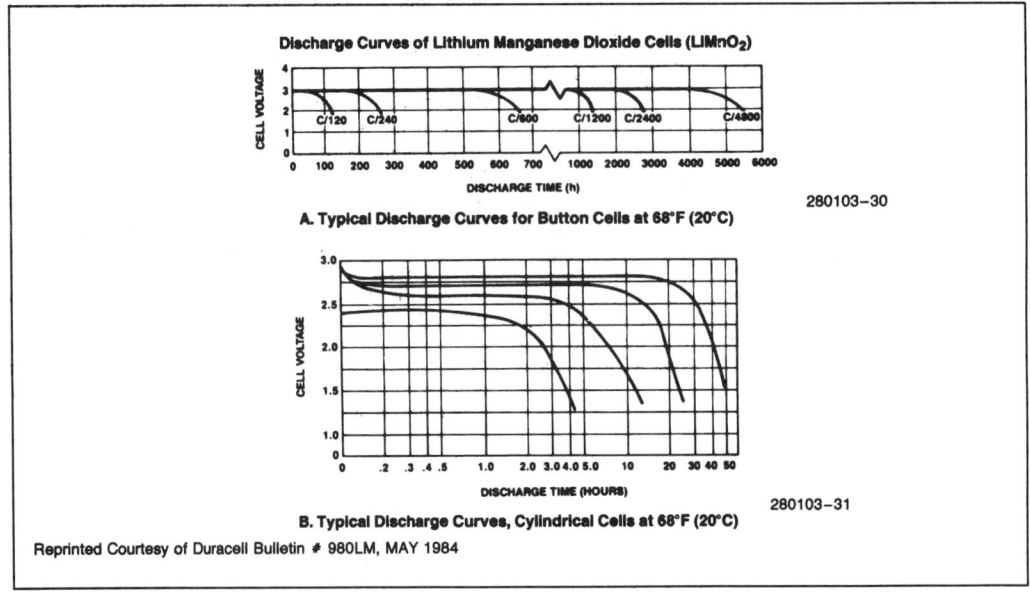

A. Typical Discharge Curves for Button Cells at 68°F (20°C)

280103-30

B. Typical Discharge Curves, Cylindrical Cells at 68°F (20°C)

280103-31

Reprinted Courtesy of Duracell Bulletin # 980LM, MAY 1984

Figure 30. Discharge Curves of Lithium Manganese Dioxide Cells

3-87

LM317. As the charge on the battery increases, the current to the battery (through R6) decreases. When the voltage across R6 decreases below the voltage across R2, the outout of the op amp goes low, reducing the effective value of R1 and lowering the output of the voltage regulator. This terminates the voltage regulator charging of the battery and leaves the trickle charge resistor R7 to maintain the battery at the 0.01C rate. In addition, when the output of the op amp goes low, it turns on the LED, indicating the battery has reached full charge. A momentary contact switch is provided to initiate a charge cycle on a partially discharged battery. The diodes on the positive terminal of the battery prevent the battery from being discharged through the charger circuits when the line power is off.

5.4.2 CURRENT LIMITED CONSTANT VOLTAGE

The second circuit is a current limited constant voltage charger. The charge current is set by resistor R3. When the voltage across the battery reaches (for lead acid, $2.35 \times 3 =$) 7.05 volts indicating full charge, the shunt regulator turns on which, through the transistor, pulls the LM317 'adjustment' pin low shutting down the voltage regulator. This leaves the trickle charge resistor to maintain the battery.

5.4.3 CONSTANT CURRENT

The final circuit is a constant current charger. R3 adjusts the range wherein the voltage regulator turns on and off, R9 and R12 adjust the point at which the op amp switches. The charge current is set by R1. At 5.0V the op amp switches low (adjusted by R3 and R12), turning off Q2 and the shunt regulator and turning on Q1. The charge current is then determined by the differential between the 'output' and 'adjustment' pins of the voltage regulator, developed across R1 (and the transistor). In this case:

$R1 = (1.2V - V_{CE})/0.1C =$

$R1 = (1.2V - 0.1V/50 \text{ mA} = 22\Omega$

When the battery reaches full charge (for NiCad, 1.4V/cell $\times$ 5 cells = 7.0V) the op amp switches high (set by the zener diode and R9), turning Q2 and the shunt regulator on and Q1 off. This pulls the adjustment pin of the voltage regulator low, terminating the charging of the battery by the regulator and leaving R2 to trickle

charge the battery at the 0.01C rate of (500 mAh $\times$ 0.01 =) 5 mA.

$R2 = [12V - (V_{batt} + V_{diode})]/0.01C =$

$R2 = [12V - (7.0V + 0.8V)]/5 \text{ mA} = 840\Omega$

6.0 SUMMARY

The new availability of a wide variety of fast CMOS products can lead to many new, low power applications such as lap or portable computers and high-speed solid state disks. The designer will have to consider a number of issues unique to this low power environment. For example, CMOS products have a much greater change from standby to active current than NMOS products so, while they expand the options (low power, high bandwidth, etc.) available to the design engineer, care must be taken to power them properly. This means proper decoupling and possibly more use of four layer printed circuit boards with power and ground inner layers. (Figure 34)

Many new products are in development (such as an 82C85 clock generator with the ability to stop the clock and an 82C08 CHMOS DRAM controller that supports extended refresh) that will help the designer take advantage of the features of the CMOS products and (s)he will have to stay abreast of them.

Systems to be operated with the limited power of a battery must be designed with careful attention to keeping active power low and making maximum use of the low power features of CMOS, features like stopping the clock and extended refresh. At the same time, means must be provided to notify the user when the battery needs recharging (or replacing).

As power is reduced, battery powered systems will become more common and the designer must become more familiar with the chemical systems available. We think rechargeable batteries will be the more prevalent in this application so knowledge of methods both for detecting batteries in a low state of charge and for recharging will be necessary. Battery backup is a little more complex in that special care must be taken to protect the memory when switching from one power source to another.

With these new low power products, new and much more powerful computing products can be developed. The potential applications are legion.

intel
AP-238

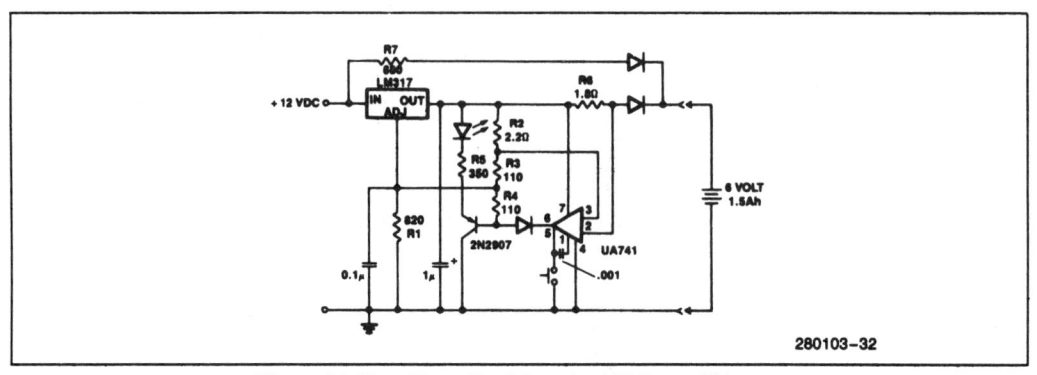

Figure 31. Two Stage Fast Constant Voltage Charger

Current Limited Constant Voltage Charger

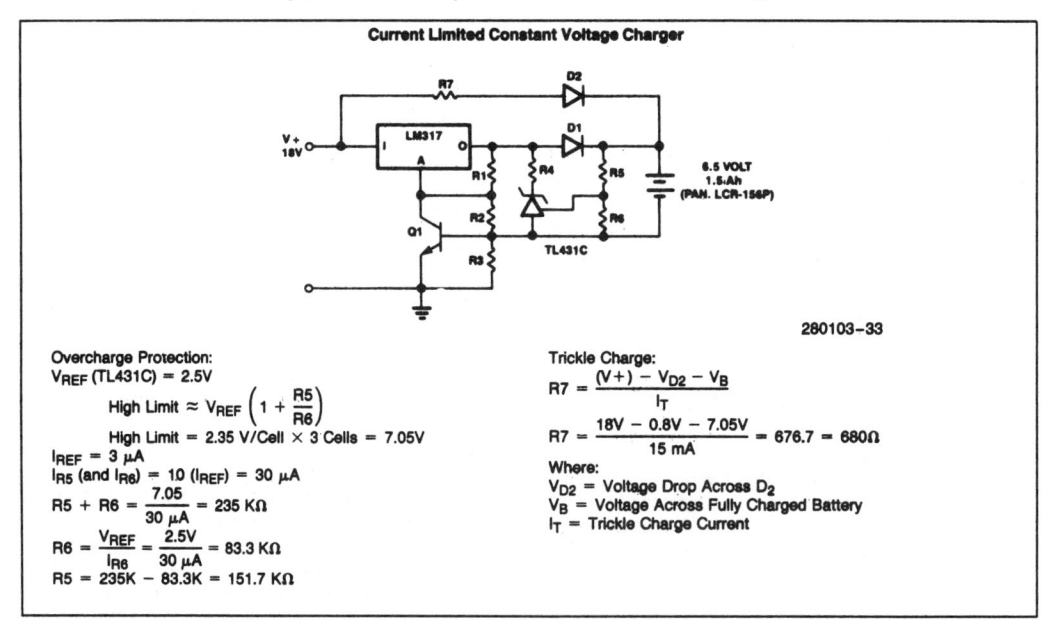

Overcharge Protection:
V_{REF} (TL431C) = 2.5V

High Limit $\approx V_{REF} \left(1 + \dfrac{R5}{R6}\right)$

High Limit = 2.35 V/Cell × 3 Cells = 7.05V

I_{REF} = 3 μA

I_{R5} (and I_{R6}) = 10 (I_{REF}) = 30 μA

$R5 + R6 = \dfrac{7.05}{30 \, \mu A} = 235 \, K\Omega$

$R6 = \dfrac{V_{REF}}{I_{R6}} = \dfrac{2.5V}{30 \, \mu A} = 83.3 \, K\Omega$

$R5 = 235K - 83.3K = 151.7 \, K\Omega$

Trickle Charge:

$R7 = \dfrac{(V+) - V_{D2} - V_B}{I_T}$

$R7 = \dfrac{18V - 0.8V - 7.05V}{15 \, mA} = 676.7 = 680\Omega$

Where:
V_{D2} = Voltage Drop Across D_2
V_B = Voltage Across Fully Charged Battery
I_T = Trickle Charge Current

Figure 32. Two Stage Current Limited Constant Voltage Charger

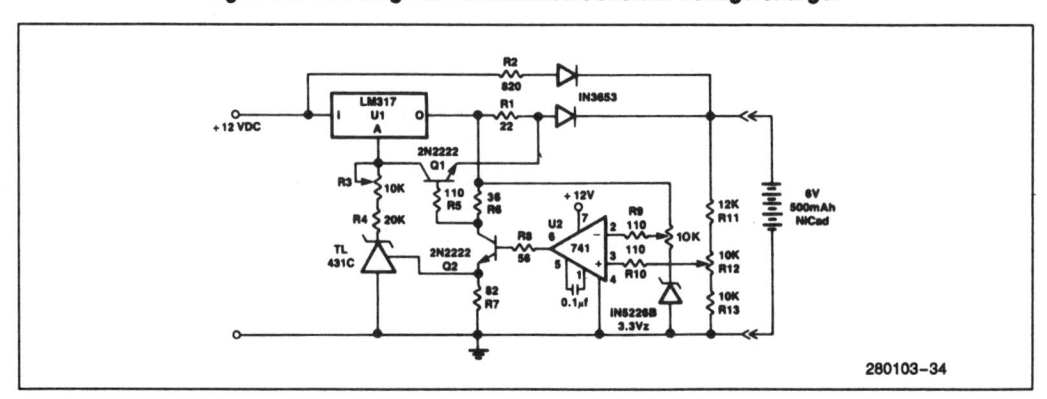

Figure 33. Two Stage Constant Current Charger

3-89

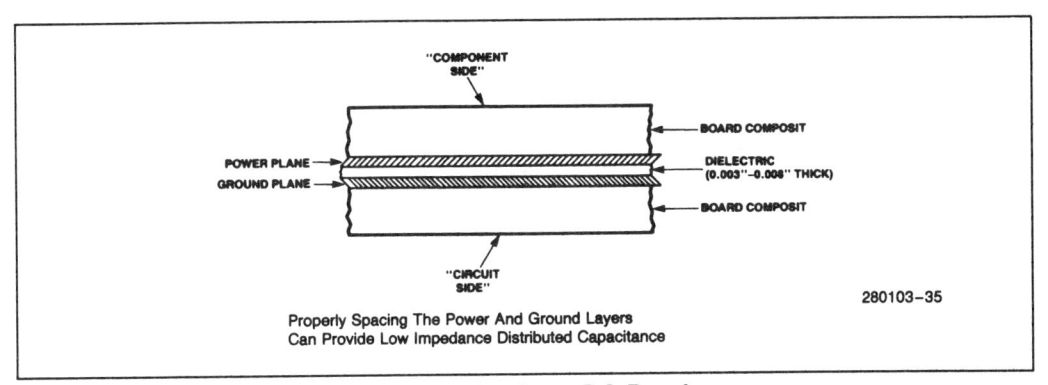

Figure 34. Four Layer P.C. Board

BIBLIOGRAPHY

1. "System Implications of CMOS Dynamic RAMS" Joseph P. Altnether; Intel Corp. IEEE, November, 1983, #230902-001

2. "Using CMOS Minimizes Bubble Memory Power Consumption" Peggy M. Lammer, Ulmont Smith, Jr.; Intel Corp. AP164, November 1983

3. "EEPROM Technology Update" Reginald G. Huff III, Alex Goldberger; Exel Microelectronics Integrated Circuits, September, 1984

4. "Low Power With CHMOS DRAMS" John Fallin; Intel Corp. AP171, July, 1984

5. "Static Column Architecture In CHMOS Dynamic RAMS -A Graphics Memory Solution" William H. Righter; Intel Corp. WESCON 1983, Electronics Conventions Inc., Nov. 1983, #230903-001

6. "CHMOS DRAMS In Graphics Applications" John Fallin; Intel Corp. AP172, July 1984

7. "Alpha Particle Induced Soft Errors In Dynamic RAMs" T.C. May, H.C. Woods; Intel Corp. IEEE Trans. Electron Devices, Vol ED-26, N0 1, pp 2-9, Jan. 1979

8. "CMOS 256K RAM With Wideband Output Stands By On Microwatts" A. Mohsen, R. Kung, J. Schutz, P. Madland, C. Simonsen, E. Hamdy, K. Yu; Intel Corp. Electronics, June 1984

9. "CHMOS Data Catalogue" Intel Corp., 1985

   "CMOS Digital Data Book" Harris Corp., 1984

10. "Designer's guide to: High-Speed CMOS" Larry Wakeman, Roger Kozlowski EDN, April 19, May 2, May 17, June 14, 1984

11. "Eveready Battery Engineering Data" Vol. III Union Carbide Corp., published 1984

12. "Handbook of Batteries and Fuel Cells" Edited by David Linden Published by McGraw-Hill, 1984

13. "Hot Ideas In CMOS" Intersil, 1983

14. "Lithium Battery Product Bulletin" Bulletin #980LM Duracell, Inc. May 1984

Preliminary

Am27H010

1 Megabit (131,072 x 8-Bit) High Speed CMOS EPROM

Advanced Micro Devices

DISTINCTIVE CHARACTERISTICS

- **Industry's fastest**
  - -45ns 1 megabit CMOS EPROM
- **Pin compatible with Am27C010**
- **High speed Flashrite™ programming**
  - Typically less than 30 seconds
- **Versions available in industrial and military temperature ranges**
- **± 10% power supply tolerance available**

GENERAL DESCRIPTION

The Am27H010 is an ultra-high speed 1 megabit CMOS UV EPROM. It utilizes the standard JEDEC pinout making it functionally compatible with the Am27C010, but with significantly faster access capability. This superior random access capability results from a focused high-speed design implemented with AMD's advanced CMOS process technology. This offers users bipolar speeds with higher density, lower cost and proven reliability.

This device is ideal for use with the fastest processors. At 45ns, the Am27H010 completely eliminates performance-draining wait states without using bank-interleaving and caching techniques. Designers may take full advantage of high speed digital signal processors and microprocessors by allowing code to be executed at full speed directly out of EPROM. Typical applications include laser printers, switching networks, graphics, workstations and digital signal processing.

The Am27H010 supports AMD's Flashrite programming algorithm which allows the entire chip to be programmed in typically less than 30 seconds.

It is available in DIP as well as surface mount packages and is offered in commercial, industrial, and extended temperature ranges.

BLOCK DIAGRAM

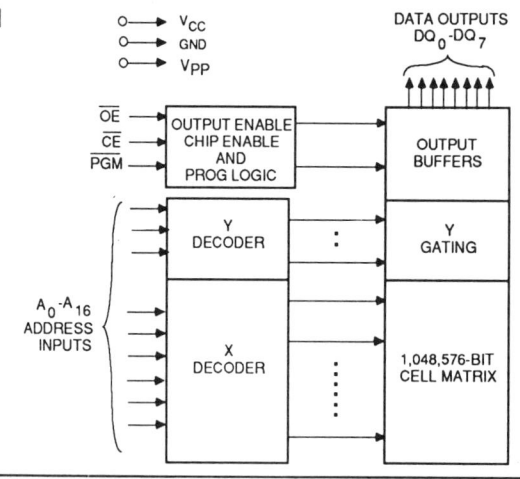

12750-001A

PRODUCT SELECTOR GUIDE

Family Part No.	Am27H010			
Ordering Part No:				
V_{CC} ± 5%	–45V05	–	–	–90V05
V_{CC} ± 10%	–45	–55	–70	–90
Max. Access Time (ns)	45	55	70	90
$\overline{CE}$ ($\overline{E}$) Access Time (ns)	45	55	70	90
$\overline{OE}$ ($\overline{G}$) Access Time (ns)	20	25	35	40

Flashrite is a trademark of Advanced Micro Devices

Publication# 12750 Rev. A Amendment /0

CONNECTION DIAGRAMS
Top View

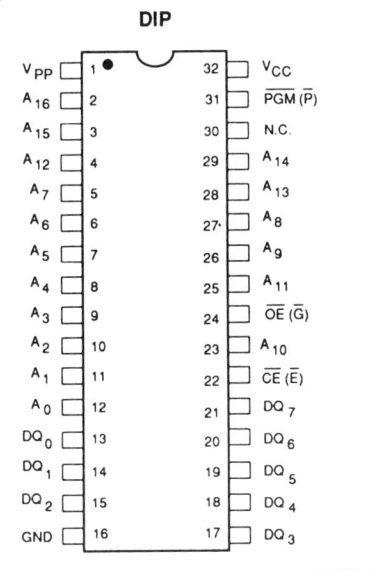

DIP

Pin	Signal		Pin	Signal
1	V_{PP}		32	V_{CC}
2	A_{16}		31	$\overline{PGM}$ ($\overline{P}$)
3	A_{15}		30	N.C.
4	A_{12}		29	A_{14}
5	A_7		28	A_{13}
6	A_6		27	A_8
7	A_5		26	A_9
8	A_4		25	A_{11}
9	A_3		24	$\overline{OE}$ ($\overline{G}$)
10	A_2		23	A_{10}
11	A_1		22	$\overline{CE}$ ($\overline{E}$)
12	A_0		21	DQ_7
13	DQ_0		20	DQ_6
14	DQ_1		19	DQ_5
15	DQ_2		18	DQ_4
16	GND		17	DQ_3

12750-002A

LCC*

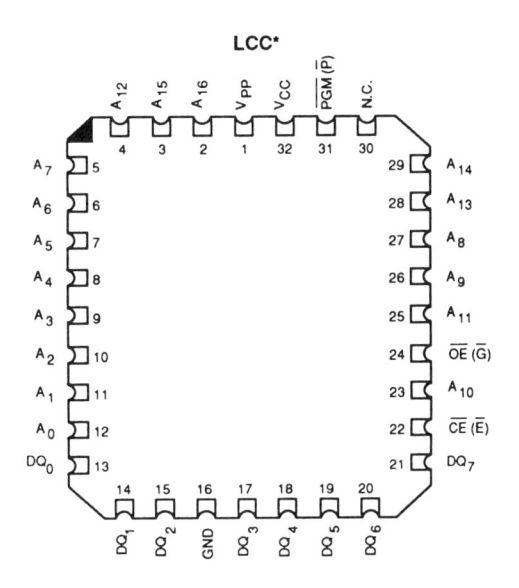

12750-003A

Note:
JEDEC nomenclature is in parentheses.

* Also available in 32-pin rectangular plastic leaded chip

PIN DESCRIPTION

A_0–A_{16}	=	Address Inputs
$\overline{CE}$ ($\overline{E}$)	=	Chip Enable Input
DQ_0–DQ_7	=	Data Inputs/Outputs
$\overline{OE}$ ($\overline{G}$)	=	Output Enable Input
$\overline{PGM}$ ($\overline{P}$)	=	Program Enable Input
V_{CC}	=	V_{CC} Supply Voltage
V_{PP}	=	Program Supply Voltage
GND	=	Ground
NC	=	No Internal Connect

LOGIC SYMBOL

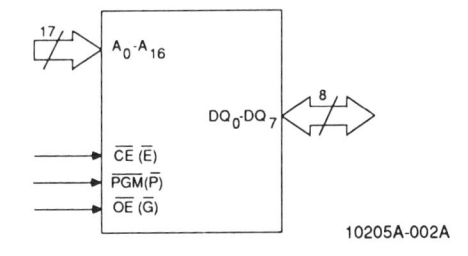

10205A-002A

Am27H010

ORDERING INFORMATION
Standard Information

AMD standard products are available in several packages and operating ranges. The order number (Valid Combination) is formed by a combination of:
a. Device Number
b. Speed Option
c. Package Type
d. Temperature Range
e. Optional Processing

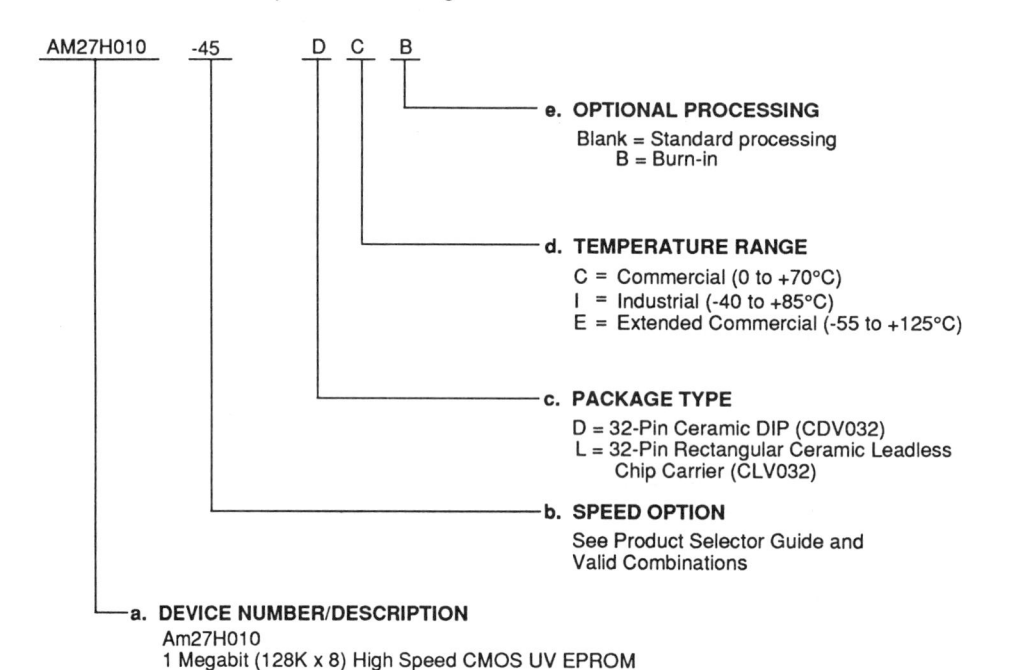

```
AM27H010  -45  D C B
```

e. OPTIONAL PROCESSING
Blank = Standard processing
B = Burn-in

d. TEMPERATURE RANGE
C = Commercial (0 to +70°C)
I = Industrial (−40 to +85°C)
E = Extended Commercial (−55 to +125°C)

c. PACKAGE TYPE
D = 32-Pin Ceramic DIP (CDV032)
L = 32-Pin Rectangular Ceramic Leadless Chip Carrier (CLV032)

b. SPEED OPTION
See Product Selector Guide and Valid Combinations

a. DEVICE NUMBER/DESCRIPTION
Am27H010
1 Megabit (128K x 8) High Speed CMOS UV EPROM

Valid Combinations	
AM27H010-45 AM27H010-45V05 AM27H010-90V05	DC, DCB, DI, DIB, LC, LI, LCB, LIB
AM27H010-55 AM27H010-70 AM27H010-90	DC, DCB, DE, DEB, DI, DIB, LC, LCB, LI, LIB, LE, LEB

Valid Combinations

Valid Combinations list configurations planned to be supported in volume for this device. Consult the local AMD sales office to confirm availability of specific valid combinations, to check on newly released combinations, and to obtain additional data on AMD's standard military grade products.

ORDERING INFORMATION (Cont'd.)
OTP Products (Preliminary)

AMD standard products are available in several packages and operating ranges. The order number (Valid Combination) is formed by a combination of:
    a. Device Number
    b. Speed Option
    c. Package Type
    d. Temperature Range
    e. Optional Processing

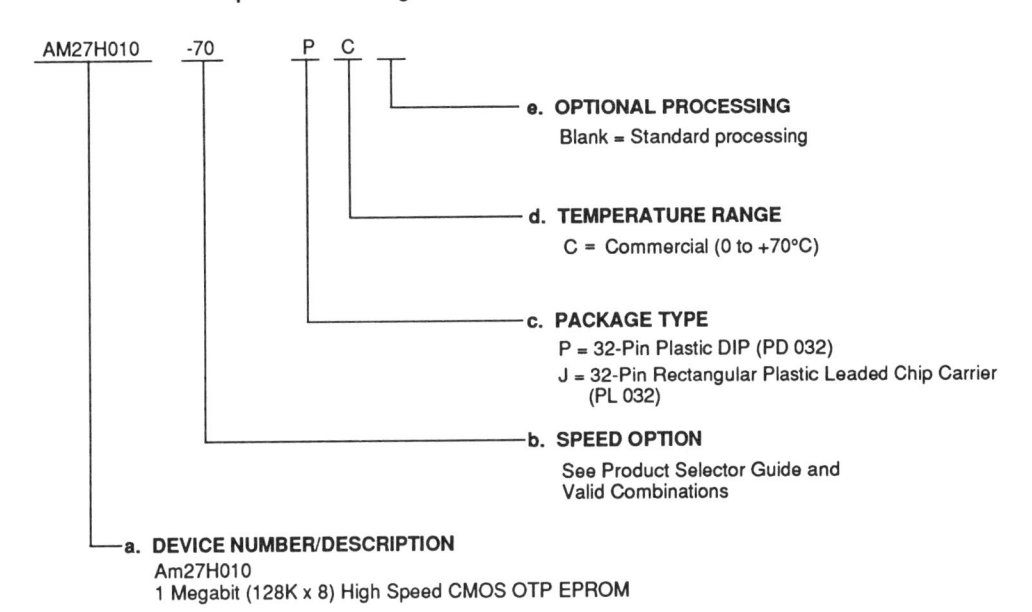

AM27H010   -70   P C

e. OPTIONAL PROCESSING
Blank = Standard processing

d. TEMPERATURE RANGE
C = Commercial (0 to +70°C)

c. PACKAGE TYPE
P = 32-Pin Plastic DIP (PD 032)
J = 32-Pin Rectangular Plastic Leaded Chip Carrier (PL 032)

b. SPEED OPTION
See Product Selector Guide and Valid Combinations

a. DEVICE NUMBER/DESCRIPTION
Am27H010
1 Megabit (128K x 8) High Speed CMOS OTP EPROM

Valid Combinations	
AM27H010-55	
AM27H010-70	PC, JC
AM27H010-90	
AM27H010-90V05	

Valid Combinations

Valid Combinations list configurations planned to be supported in volume for this device. Consult the local AMD sales office to confirm availability of specific valid combinations, to check on newly released combinations, and to obtain additional data on AMD's standard military grade products.

ORDERING INFORMATION (Cont'd.)
APL Products

AMD products for Aerospace and Defense applications are available in several packages and operating ranges. APL (Approved Products List) products are fully compliant with MIL-STD-883C requirements. The order number (Valid Combination) is formed by a combination of:
- a. Device Number
- b. Speed Option
- c. Device Class
- d. Package Type
- e. Lead Finish

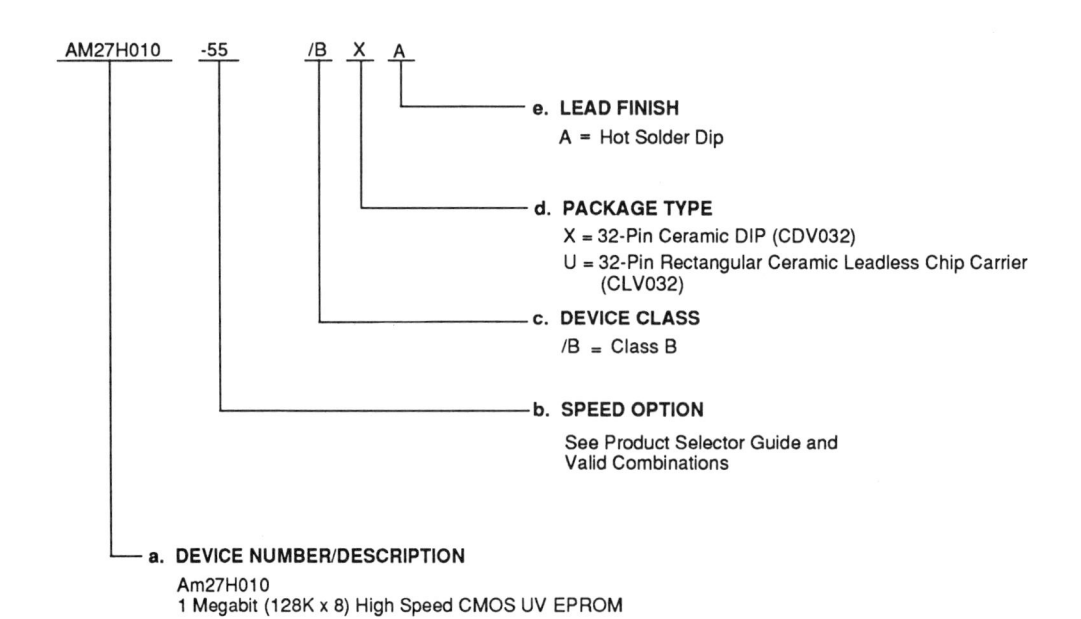

```
AM27H010  -55  /B  X  A
```

e. LEAD FINISH
A = Hot Solder Dip

d. PACKAGE TYPE
X = 32-Pin Ceramic DIP (CDV032)
U = 32-Pin Rectangular Ceramic Leadless Chip Carrier (CLV032)

c. DEVICE CLASS
/B = Class B

b. SPEED OPTION
See Product Selector Guide and Valid Combinations

a. DEVICE NUMBER/DESCRIPTION
Am27H010
1 Megabit (128K x 8) High Speed CMOS UV EPROM

Valid Combinations	
AM27H010-55	/BXA, /BUA
AM27H010-70	
AM27H010-90	

Valid Combinations
Valid Combinations list configurations planned to be supported in volume for this device. Consult the local AMD sales office to confirm availability of specific valid combinations or to check for newly released valid combinations.

Group A Tests
Group A tests consist of Subgroups 1, 2, 3, 7, 8, 9, 10, 11.

FUNCTIONAL DESCRIPTION
Erasing the Am27H010

In order to clear all locations of their programmed contents, it is necessary to expose the Am27H010 to an ultraviolet light source. A dosage of 15 W seconds/cm^2 is required to completely erase an Am27H010. This dosage can be obtained by exposure to an ultraviolet lamp—wavelength of 2537 Angstroms (Å)—with intensity of 12,000 µW/cm^2 for 15 to 20 minutes. The Am27H010 should be directly under and about one inch from the source and all filters should be removed from the UV light source prior to erasure.

It is important to note that the Am27H010, and similar devices, will erase with light sources having wavelengths shorter than 4000 Å. Although erasure times will be much longer than with UV sources at 2537 Å, nevertheless the exposure to fluorescent light and sunlight will eventually erase the Am27H010 and exposure to them should be prevented to realize maximum system reliability. If used in such an environment, the package window should be covered by an opaque label or substance.

Programming the Am27H010

Upon delivery, or after each erasure, the Am27H010 has all 1,048,576 bits in the "ONE", or HIGH state. "ZEROs" are loaded into the Am27H010 through the procedure of programming.

The programming mode is entered when 12.75 ± 0.25 V is applied to the V_{PP} pin, $\overline{CE}$ and $\overline{PGM}$ is at V_{IL}, and $\overline{OE}$ is at V_{IH}. For programming, the data to be programmed is applied 8 bits in parallel to the data output pins.

The Flashrite programming algorithm reduces programming time by using initial 100 µs pulses followed by a byte verification to determine whether the byte has been successfully programmed. If the data does not verify, an additional pulse is applied for a maximum of 25 pulses. This process is repeated while sequencing through each address of the EPROM.

The Flashrite programming algorithm programs and verifies at V_{CC} = 6.25 V and V_{PP} = 12.75 V. After the final address is completed, all bytes are compared to the original data with V_{CC} = V_{PP} = 5.25 V.

Program Inhibit

Programming of multiple Am27H010s in parallel with different data is also easily accomplished. Except for $\overline{CE}$, all like inputs of the parallel Am27H010 may be common. A TTL low-level program pulse applied to an Am27H010 $\overline{CE}$ input with V_{PP} = 12.75 ± 0.25 V, $\overline{PGM}$ is LOW, and $\overline{OE}$ HIGH will program that Am27H010. A high-level $\overline{CE}$ input inhibits the other Am27H010 from being programmed.

Program Verify

A verify should be performed on the programmed bits to determine that they were correctly programmed. The verify should be performed with $\overline{OE}$ and $\overline{CE}$ at V_{IL}, $\overline{PGM}$ at V_{IH}, and V_{PP} between 12.5 V to 13.0 V.

Auto Select Mode

The auto select mode allows the reading out of a binary code from an EPROM that will identify its manufacturer and type. This mode is intended for use by programming equipment for the purpose of automatically matching the device to be programmed with its corresponding programming algorithm. This mode is functional in the 25°C ± 5°C ambient temperature range that is required when programming the Am27H010.

To activate this mode, the programming equipment must force 12.0 ± 0.5 V on address line A_9 of the Am27H010. Two identifier bytes may then be sequenced from the device outputs by toggling address line A_0 from V_{IL} to V_{IH}. All other address lines must be held at V_{IL} during auto select mode.

Byte 0 ($A_0 = V_{IL}$) represents the manufacturer code, and byte 1 ($A_0 = V_{IH}$), the device identifier code. For the Am27H010, these two identifier bytes are given in the Mode Select table. All identifiers for manufacturer and device codes will possess odd parity, with the MSB (DQ_7) defined as the parity bit.

Read Mode

The Am27H010 has two control functions, both of which must be logically satisfied in order to obtain data at the outputs. Chip Enable ($\overline{CE}$) is the power control and should be used for device selection. Assuming that addresses are stable, address access time (t_{ACC}) is equal to the delay from $\overline{CE}$ to output (t_{CE}). Output Enable ($\overline{OE}$) is the output control and should be used to gate data to the output pins, independent of device selection. Data is available at the outputs t_{OE} after the falling edge of $\overline{OE}$, assuming that $\overline{CE}$ has been LOW and addresses have been stable for at least $t_{ACC} - t_{OE}$.

Standby Mode

The Am27H010 has a standby mode which reduces the maximum V_{CC} current to 50% of the active current. It is placed in standby mode when $\overline{CE}$ is at V_{IH}. The amount of current drawn in standby mode depends on the frequency and the number of address pins switching. The Am27H010 is specified with 50% of the address lines toggling at 10 MHz. A reduction of the frequency or quantity of address lines toggling will significantly reduce the actual standby current.

Output OR-Tieing

To accommodate multiple memory connections, a two-line control function is provided to allow for:

1. Low memory power dissipation, and
2. Assurance that output bus contention will not occur.

It is recommended that $\overline{CE}$ be decoded and used as the primary device-selecting function, while $\overline{OE}$ be made a common connection to all devices in the array and connected to the READ line from the system control bus. This assures that all deselected memory devices are in their low-power standby mode and that the output pins are only active when data is desired from a particular memory device.

System Applications

During the switch between active and standby conditions, transient current peaks are produced on the rising and falling edges of Chip Enable. The magnitude of these transient current peaks is dependent on the output capacitance loading of the device. At a minimum, a 0.1-µF ceramic capacitor (high frequency, low inherent inductance) should be used on each device between V_{CC} and GND to minimize transient effects. In addition, to overcome the voltage drop caused by the inductive effects of the printed circuit board traces on EPROM arrays, a 4.7-µF bulk electrolytic capacitor should be used between V_{CC} and GND for each eight devices. The location of the capacitor should be close to where the power supply is connected to the array.

MODE SELECT TABLE

Mode	Pins	$\overline{CE}$	$\overline{OE}$	$\overline{PGM}$	A_0	A_9	V_{PP}	Outputs
Read		V_{IL}	V_{IL}	X	X	X	V_{IH}	D_{OUT}
Output Disable		V_{IL}	V_{IH}	X	X	X	V_{IH}	Hi-Z
Standby		V_{IH}	X	X	X	X	V_{IH}	Hi-Z
Program		V_{IL}	V_{IH}	V_{IL}	X	X	V_{PP}	D_{IN}
Program Verify		V_{IL}	V_{IL}	V_{IH}	X	X	V_{PP}	D_{OUT}
Program Inhibit		V_{IH}	X	X	X	X	V_{PP}	Hi-Z
Auto Select	Manufacturer Code	V_{IL}	V_{IL}	X	V_{IL}	V_H	V_{CC}	01H
(Note 3 & 5)	Device Code	V_{IL}	V_{IL}	X	V_{IH}	V_H	V_{CC}	0EH

Notes:
1. $V_H = 12.0 V \pm 0.5 V$
2. X = Either V_{IH} or V_{IL}
3. $A_1 - A_8 = A_{10} - A_{16} = V_{IL}$
4. See DC Programming Characteristics for V_{PP} voltage during programming.
5. The Am27H010 uses the same Flashrite algorithm during program as the Am27C010.

ABSOLUTE MAXIMUM RATINGS

Storage Temperature
  OTP product   −65 to 125°C
  All other products   −65 to 150°C
Ambient Temperature
  with Power Applied   −55 to +125°C
Voltage with Respect to Ground:
  All pins except A_9, V_{PP}, and
  V_{CC} −0.6 to V_{CC} +0.5 V
  A_9 and V_{PP} −0.6 to 13.5
  V_{CC} −0.6 to 7.0 V

Stresses above those listed under "Absolute Maximum Ratings" may cause permanent damage to the device. This is a stress rating only; functional operation of the device at these or any other conditions above those indicated in the operational sections of this specification is not implied. Exposure of the device to absolute maximum rating conditions for extended periods may affect device reliability.

Notes:
1. Minimum DC voltage on input or I/O is −0.5 V. During transitions, the inputs may undershoot GND to −2.0V for periods of up to 10 ns. Maximum DC voltage on input and I/O is V_{CC} +0.5 V which may overshoot to V_{CC} +2.0 V for periods up to 20 ns.
2. For A_9 and V_{PP} the minimum DC input is −0.5 V. During transitions, A_9 and V_{PP} may undershoot GND to −2.0 V for periods of up to 10 ns. A_9 and V_{PP} must not exceed 13.5 V for any period of time.

OPERATING RANGES

Commercial (C) Devices
  Case Temperature (T_C)   0 to +70°C
Industrial (I) Devices
  Case Temperature (T_C)   −40 to +85°C
Extended Commercial (E) Devices
  Case Temperature (T_C)   −55 to +125°C
Military (M) Devices
  Case Temperature (T_C)   −55 to +125°C
Supply Read Voltages:
  V_{CC} for Am27H010-XXV05   +4.75 to +5.25 V
  V_{CC} for Am27H010-XX   +4.50 to +5.50 V

Operating ranges define those limits between which the functionality of the device is guaranteed.

DC CHARACTERISTICS over operating range unless otherwise specified. (Notes 1, 4, 5, & 8) (for APL Products, Group A, Subgroups 1, 2, 3, 7, and 8 are tested unless otherwise noted)

Parameter Symbol	Parameter Description	Test Conditions		Min.	Max.	Unit
V_{OH}	Output HIGH Voltage	I_{OH} = −4 mA		2.4		V
V_{OL}	Output LOW Voltage	I_{OL} = 12 mA			0.45	V
V_{IH}	Input HIGH Voltage (Note 9)			2.0	V_{CC} + 0.5	V
V_{IL}	Input LOW Voltage (Note 9)			−0.3	+0.8	V
I_{LI}	Input Load Current	V_{IN} = 0 V to + V_{CC}			1.0	µA
I_{LO}	Output Leakage Current	V_{OUT} = 0 V to + V_{CC}			10	µA
I_{CC1}	V_{CC} Active Current (Note 5)	$\overline{CE}$ = V_{IL}, f = 10 MHz I_{OUT} = 0 mA (Open Outputs)	C Devices		50	mA
			I/E/M Devices		60	
I_{CC2}	V_{CC} Standby Current	$\overline{CE}$ = V_{IH}	C Devices		25	mA
			I/E/M Devices		35	
I_{PP1}	V_{PP} Current During Read (Note 6)	$\overline{CE}$ = $\overline{OE}$ = V_{IL}, V_{PP} = V_{CC}			100	µA

PRELIMINARY

DC CHARACTERISTICS (Cont.)

Capacitance (Notes 2, 3, and 7)

Parameter Symbol	Parameter Description	Test Conditions	CDV032 Typ.	CDV032 Max.	CLV032 Typ.	CLV032 Max.	Unit
C_{IN1}	Address Input Capacitance	$V_{IN} = 0$ V	10	12	6	9	pF
C_{IN2}	$\overline{OE}$ Input Capacitance	$V_{IN} = 0$	10	12	7	9	pF
C_{IN3}	$\overline{CE}$ Input Capacitance	$V_{IN} = 0$ V	10	10	7	9	pF
C_{OUT}	Output Capacitance	$V_{OUT} = 0$ V	8	12	6	9	pF

Notes:
1. V_{CC} must be applied simultaneously or before V_{PP}, and removed simultaneously or after V_{PP}.
2. Typical values are for nominal supply voltages.
3. This parameter is only sampled, not 100% tested.
4. **Caution:** the Am27H010 must not be removed from (or inserted into) a socket when V_{CC} or V_{PP} is applied.
5. I_{CC1} is tested with $\overline{OE} = V_{IH}$ to simulate open outputs.
6. Maximum active power usage is the sum of I_{CC} and I_{PP}.
7. $T_A = +25°C$, $f = 1$ MHz.
8. Minimum DC Input Voltage is –0.5 V. During transitions, the inputs may undershoot to –2.0 V for periods less than 10 ns. Maximum DC Voltage on output pins is V_{CC} +0.5 V which may overshoot to V_{CC} +2.0 V for periods less than 10 ns.
9. Tested under static DC conditions.

SWITCHING CHARACTERISTICS over operating range unless otherwise specified (Notes 1, 3, & 4) (for APL Products, Group A, Subgroups 9, 10, and 11 are specified unless otherwise noted)

Parameter Symbols		Parameter Description	Test Conditions		Am27H010				Unit
JEDEC	Standard				−45V05, −45	−55	−70	−90V05, −90	
t_{AVQV}	t_{ACC}	Address to Output Delay	$\overline{CE} = \overline{OE} = V_{IL}$, $C_L = C_{L1}$	Min.	–	–	–	–	ns
				Max.	45	55	70	90	
t_{ELQV}	t_{CE}	Chip Enable to Output Delay	$\overline{OE} = V_{IL}$, $C_L = C_{L1}$	Min.	–	–	–	–	ns
				Max.	45	55	70	90	
t_{GLQV}	t_{OE}	Output Enable to Output Delay	$\overline{CE} = V_{IL}$, $C_L = C_{L1}$	Min.	–	–	–	–	ns
				Max.	20	25	35	40	
t_{EHQZ}, t_{GHQZ}	t_{DF} (Note 2)	Chip Enable HIGH or Output Enable HIGH, Whichever Comes First, to Output Float	$C_L = C_{L2}$	Min.	0	0	0	0	ns
				Max.	20	25	35	40	
t_{AXQX}	t_{OH}	Output Hold from Addresses, $\overline{CE}$, or $\overline{OE}$, Whichever Occured First		Min.	0	0	0	0	ns
				Max.	–	–	–	–	

Notes:
1. V_{CC} must be applied simultaneously or before V_{PP}, and removed simultaneously or after V_{PP}.
2. This parameter is only sampled, not 100% tested.
3. **Caution:** The Am27H10 must not be removed from (or inserted into) a socket or board when V_{PP} or V_{CC} is applied.
4. Output Load: 1 TTL gate and $C = C_L$
   Input Rise and Fall Times: 5 ns
   Input Pulse Levels: 0 to 3 V
   Timing Measurement Reference Level – 1.5 V for inputs and outputs

SWITCHING TEST CIRCUIT

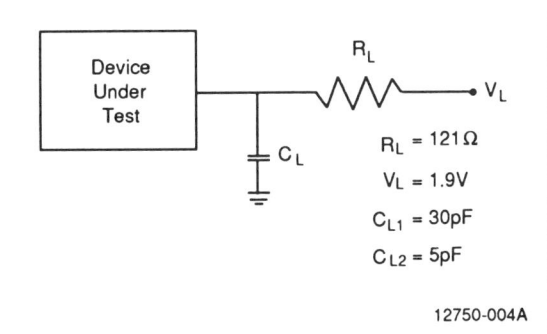

Device Under Test — R_L — V_L
C_L
$R_L = 121\Omega$
$V_L = 1.9V$
$C_{L1} = 30pF$
$C_{L2} = 5pF$

12750-004A

SWITCHING TEST WAVEFORM

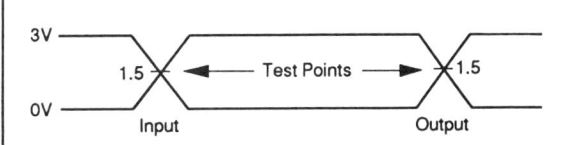

3V — 1.5 ← Test Points → 1.5
0V — Input / Output

AC Testing: Inputs are driven at 3.0 V for a logic "1" and 0 V for a logic "0". Input pulse rise and fall times are ≤5 ns.

12750-005A

KEY TO SWITCHING WAVEFORMS

WAVEFORM	INPUTS	OUTPUTS
 ————	Must be Steady	Will be Steady
 /////	May Change from H to L	Will be Changing from H to L
 \\\\\\	May Change from L to H	Will be Changing from L to H
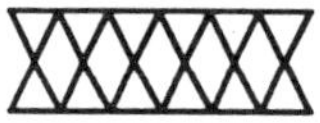 XXXXX	Don't Care; Any Change Permitted	Changing, State Unknown
 >>><<<	Does Not Apply	Center Line is High-Impedance "Off" State

KS000010

SWITCHING WAVEFORMS

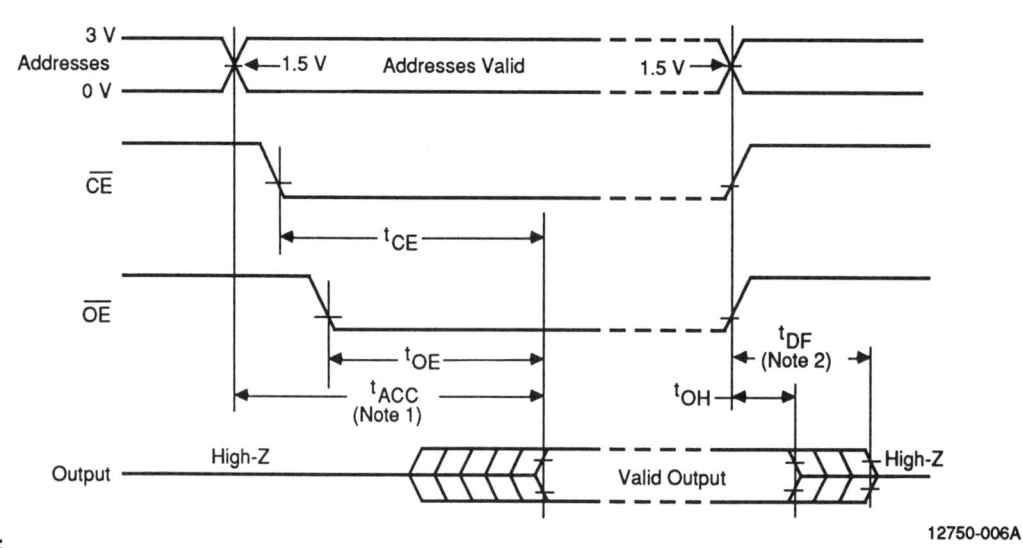

12750-006A

Notes:
1. $\overline{OE}$ may be delayed up to t_{ACC} - t_{OE} after the falling edge of $\overline{CE}$ without impact on t_{ACC}.
2. t_{DF} is specified from $\overline{OE}$ or $\overline{CE}$, whichever occurs first.

Am27H010

FLASHRITE PROGRAMMING FLOW CHART

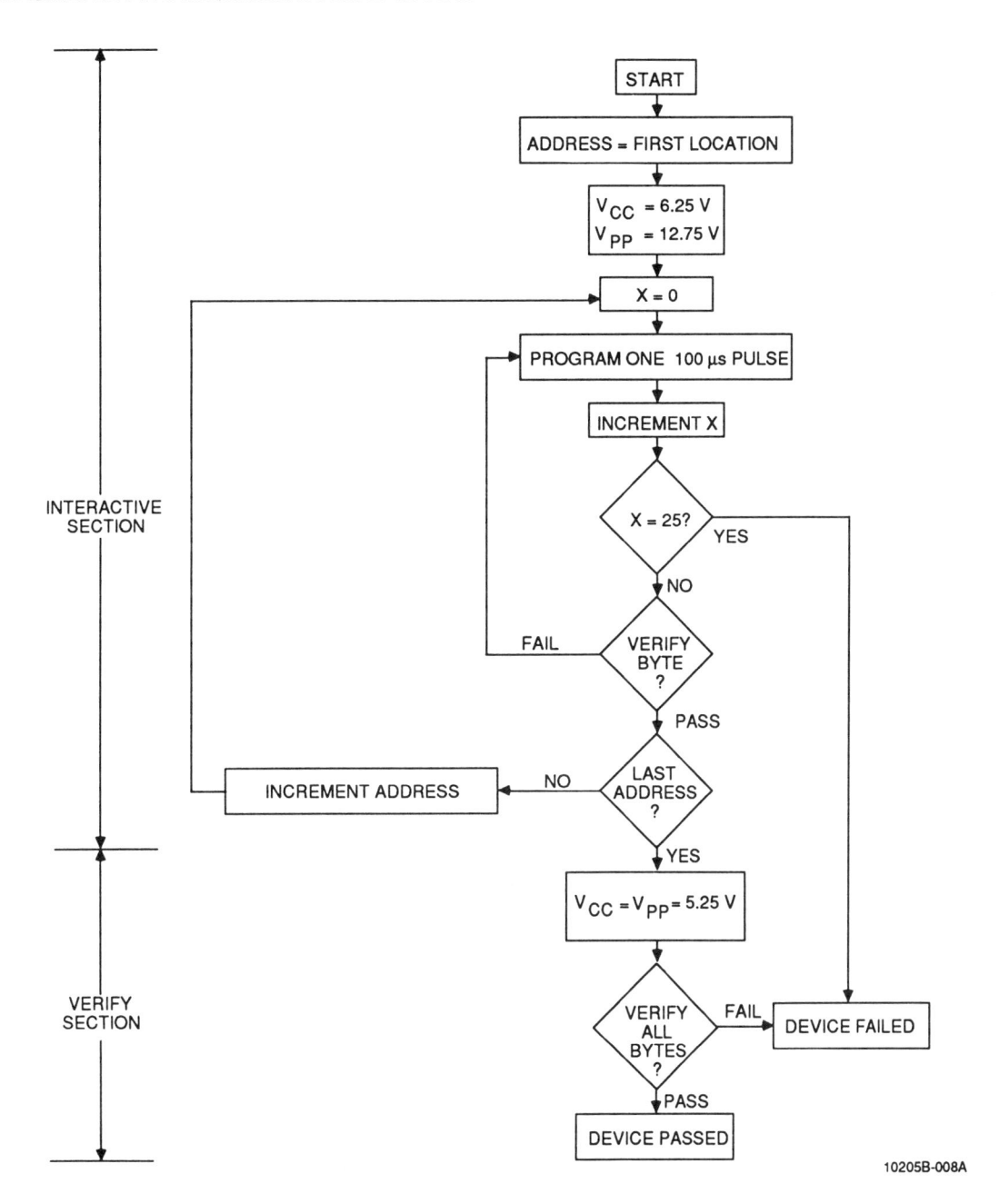

```
                                    START
                                      │
                                      ▼
                          ADDRESS = FIRST LOCATION
                                      │
                                      ▼
                               V_CC = 6.25 V
                               V_PP = 12.75 V
                                      │
                                      ▼
              ┌──────────────────▶  X = 0
              │                       │
              │                       ▼
              │      ┌──▶ PROGRAM ONE 100 µs PULSE
              │      │                │
              │      │                ▼
              │      │          INCREMENT X
              │      │                │
              │      │                ▼
              │      │            ╱ X = 25? ╲──YES──┐
              │      │            ╲         ╱       │
              │      │                │ NO          │
              │      │                ▼             │
              │      │          ╱ VERIFY ╲          │
              │      └──FAIL──  ╲ BYTE ? ╱          │
              │                      │              │
              │                      │ PASS         │
              │                      ▼              │
              │                ╱  LAST  ╲           │
              │   INCREMENT ◀─NO ADDRESS            │
              │    ADDRESS      ╲   ?  ╱            │
              │                      │              │
              │                      │ YES          │
              │                      ▼              │
              │              V_CC = V_PP = 5.25 V   │
              │                      │              │
              │                      ▼              │
              │                ╱ VERIFY ╲           │
              │                ╲  ALL   ╱──FAIL──▶ DEVICE FAILED
              │                ╱ BYTES ╲            ▲
              │                ╲   ?   ╱            │
              │                      │              │
              │                      │ PASS         │
              │                      ▼              │
              │              DEVICE PASSED          │
              │                                     │
              └─────────────────────────────────────┘
```

INTERACTIVE SECTION

VERIFY SECTION

10205B-008A

DC PROGRAMMING CHARACTERISTICS ($T_A = +25°C \pm 5°C$) (Notes 1, 2, & 3)

Parameter Symbol	Parameter Description	Test Conditions	Min.	Max.	Unit
I_{LI}	Input Current (All Inputs)	$V_{IN} = V_{IL}$ or V_{IH}		10.0	µA
V_{IL}	Input LOW Level (All Inputs)		−0.3	0.8	V
V_{IH}	Input HIGH Level		2.0	$V_{CC} + 0.5$	V
V_{OL}	Output LOW Voltage During Verify	$I_{OL} = 12$ mA		0.45	V
V_{OH}	Output HIGH Voltage During Verify	$I_{OH} = -4$ mA	2.4		V
V_H	A_9 Auto Select Voltage		11.5	12.5	V
I_{CC3}	V_{CC} Supply Current (Program & Verify)			50	mA
I_{PP2}	V_{PP} Supply Current (Program)	$\overline{CE} = V_{IL}$, $\overline{OE} = V_{IH}$		30	mA
V_{CC1}	Supply Voltage		6.00	6.50	V
V_{PP1}	Programming Voltage		12.5	13.0	V

SWITCHING PROGRAMMING CHARACTERISTICS ($T_A = +25°C \pm 5°C$) (Notes 1, 2, and 3)

Parameter Symbols JEDEC	Standard	Parameter Description	Min.	Max.	Unit
t_{AVEL}	t_{AS}	Address Setup Time	2		µs
t_{DZGL}	t_{OES}	$\overline{OE}$ Setup Time	2		µs
t_{DVEL}	t_{DS}	Data Setup Time	2		µs
t_{GHAX}	t_{AH}	Address Hold Time	0		µs
t_{EHDX}	t_{DH}	Data Hold Time	2		µs
t_{GHQZ}	t_{DFP}	Output Enable to Output Float Delay	0	130	ns
t_{VPS}	t_{VPS}	V_{PP} Setup Time	2		µs
t_{ELEH1}	t_{PW}	$\overline{PGM}$ Program Pulse Width	95	105	µs
t_{VCS}	t_{VCS}	V_{CC} Setup Time	2		µs
t_{ELPL}	t_{CES}	$\overline{CE}$ Setup Time	2		µs
t_{GLQV}	t_{OE}	Data Valid from $\overline{OE}$		150	ns

Notes:
1. V_{CC} must be applied simultaneously or before V_{PP}, and removed simultaneously or after V_{PP}.
2. When programming the Am27H010, a 0.1-µF capacitor is required across V_{PP} and ground to suppress spurious voltage transients which may damage the device.
3. Programming characteristics are sampled but not 100% tested at worst-case condtions.

FLASHRITE PROGRAMMING ALGORITHM WAVEFORM
(Notes 1 and 2)

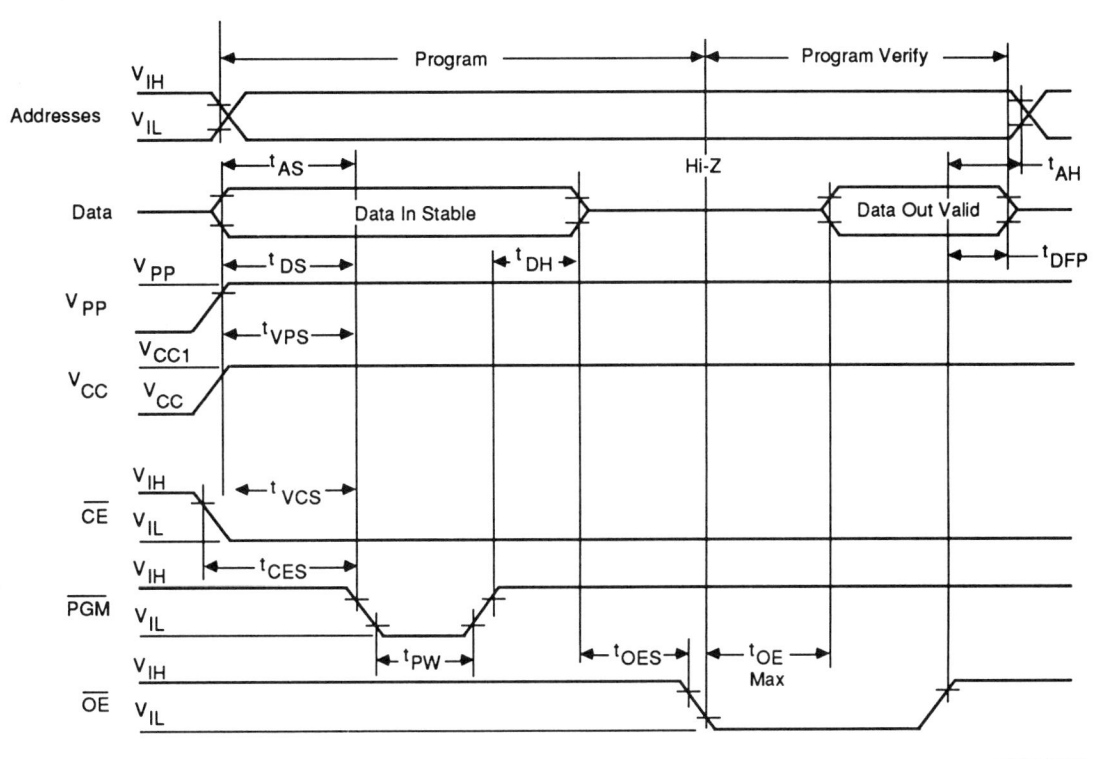

Notes:
1. The input timing reference level is 0.8 V for a V_{IL} and 2V for a V_{IH}.
2. t_{OE} and t_{DFP} are characteristics of the device but must be accomodated by the programmer.

PHYSICAL DIMENSIONS*
CDV032

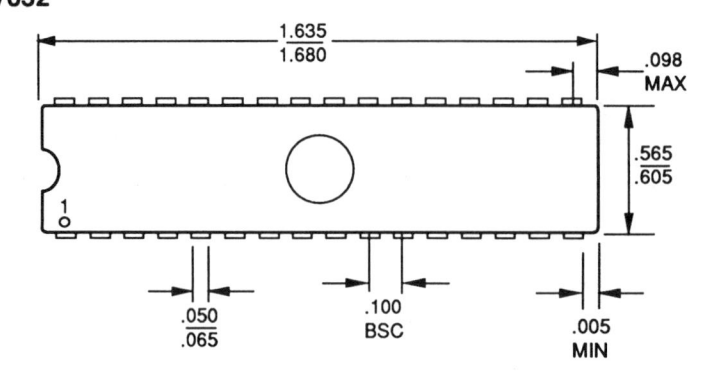

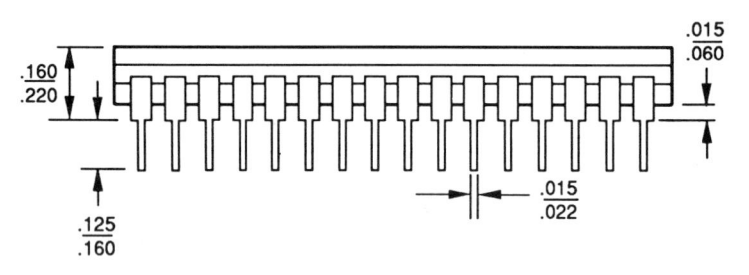

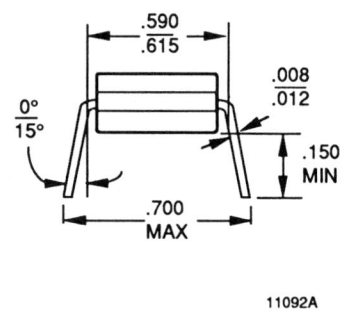

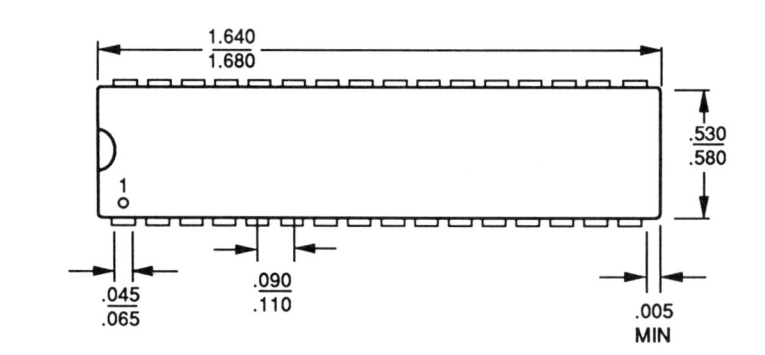

PD 032

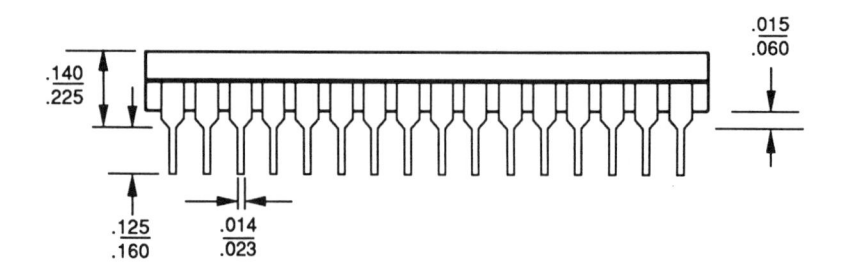

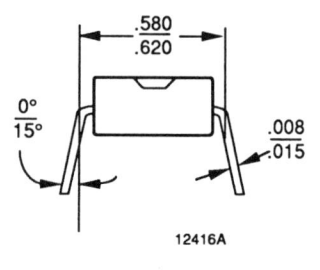

* For reference only. All dimensions are measured in inches. BSC is an ANSI standard for Basic Space Centering.

PHYSICAL DIMENSIONS (Cont'd.)
CLV032

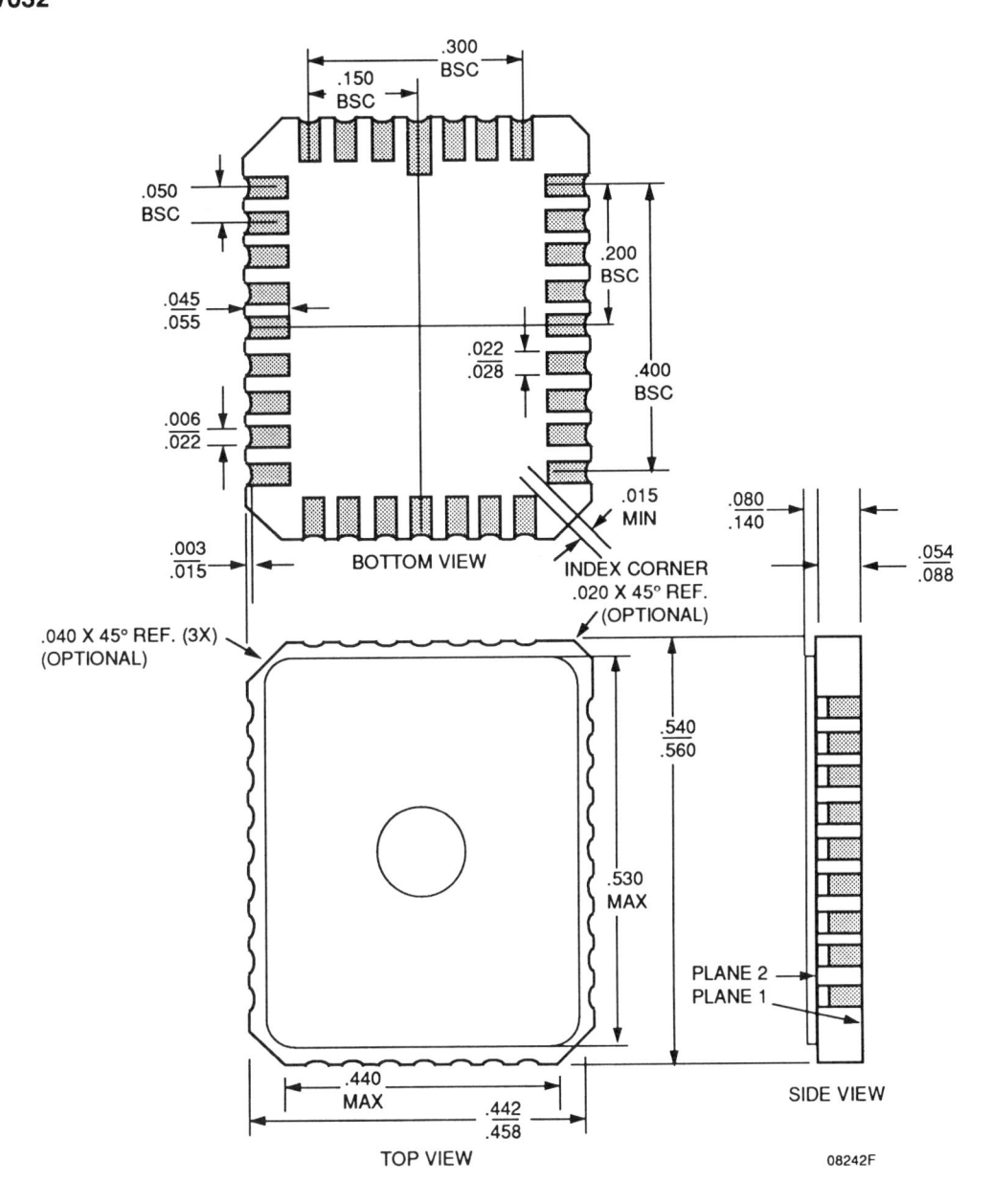

PHYSICAL DIMENSIONS (Cont'd.)
PL 032

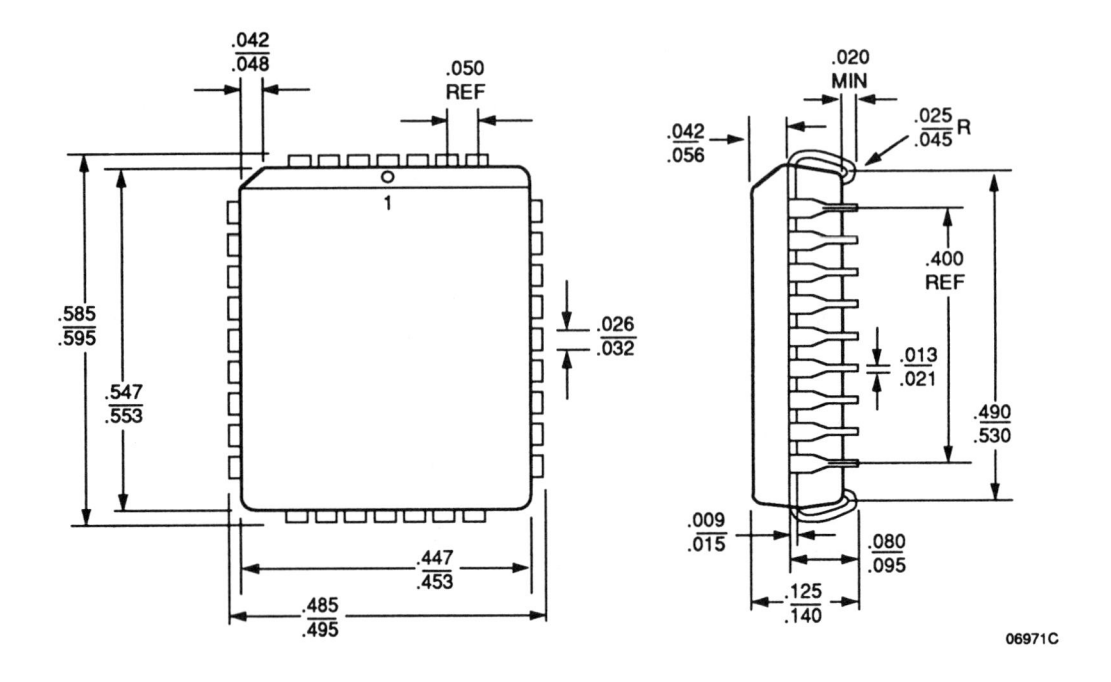

Sales Offices

North American

ALABAMA	(205) 882-9122
ARIZONA	(602) 242-4400
CALIFORNIA,	
Culver City	(213) 645-1524
Newport Beach	(714) 752-6262
Roseville	(916) 786-6700
San Diego	(619) 560-7030
San Jose	(408) 452-0500
Woodland Hills	(818) 992-4155
CANADA, Ontario,	
Kanata	(613) 592-0060
Willowdale	(416) 224-5193
COLORADO	(303) 741-2900
CONNECTICUT	(203) 264-7800
FLORIDA,	
Clearwater	(813) 530-9971
Ft. Lauderdale	(305) 776-2001
Orlando (Casselberry)	(407) 830-8100
GEORGIA	(404) 449-7920
ILLINOIS,	
Chicago (Itasca)	(312) 773-4422
Naperville	(312) 505-9517
KANSAS	(913) 451-3115
MARYLAND	(301) 796-9310
MASSACHUSETTS	(617) 273-3970
MICHIGAN	(313) 347-1522
MINNESOTA	(612) 938-0001
NEW JERSEY,	
Cherry Hill	(609) 662-2900
Parsippany	(201) 299-0002
NEW YORK,	
Liverpool	(315) 457-5400
Poughkeepsie	(914) 471-8180
Rochester	(716) 272-9020
NORTH CAROLINA	(919) 878-8111
OHIO,	
Columbus (Westerville)	(614) 891-6455
OREGON	(503) 245-0080
PENNSYLVANIA	(215) 398-8006
SOUTH CAROLINA	(803) 772-6760
TEXAS,	
Austin	(512) 346-7830
Dallas	(214) 934-9099
Houston	(713) 785-9001
UTAH	(801) 264-2900

International

BELGIUM, Bruxelles	TEL	(02) 771-91-42
	FAX	(02) 762-37-12
	TLX	846-61028
FRANCE, Paris	TEL	(1) 49-75-10-10
	FAX	(1) 49-75-10-13
	TLX	263282F
WEST GERMANY,		
Hannover area	TEL	(0511) 736085
	FAX	(0511) 721254
	TLX	922850
München	TEL	(089) 4114-0
	FAX	(089) 406490
	TLX	523883
Stuttgart	TEL	(0711) 62 33 77
	FAX	(0711) 625187
	TLX	721882
HONG KONG,	TEL	852-5-8654525
Wanchai	FAX	852-5-8654335
	TLX	67955AMDAPHX
ITALY, Milan	TEL	(02) 3390541
		(02) 3533241
	FAX	(02) 3498000
	TLX	843-315286
JAPAN,		
Kanagawa	TEL	462-47-2911
	FAX	462-47-1729
Tokyo	TEL	(03) 346-7550
	FAX	(03) 342-5196
	TLX	J24064AMDTKOJ
Osaka	TEL	06-243-3250
	FAX	06-243-3253

International (Continued)

KOREA, Seoul	TEL	822-784-0030
	FAX	822-784-8014
LATIN AMERICA,		
Ft. Lauderdale	TEL	(305) 484-8600
	FAX	(305) 485-9736
	TLX	5109554261 AMDFTL
NORWAY, Hovik	TEL	(03) 010156
	FAX	(02) 591959
	TLX	79079HBCN
SINGAPORE	TEL	65-3481188
	FAX	65-3480161
	TLX	55650 AMDMMI
SWEDEN,		
Stockholm	TEL	(08) 733 03 50
(Sundbyberg)	FAX	(08) 733 22 85
	TLX	11602
TAIWAN	TEL	886-2-7213393
	FAX	886-2-7723422
	TLX	886-2-7122066
UNITED KINGDOM,		
Manchester area	TEL	(0925) 828008
(Warrington)	FAX	(0925) 827693
	TLX	851-628524
London area	TEL	(0483) 740440
(Woking)	FAX	(0483) 756196
	TLX	851-859103

North American Representatives

CANADA	
Burnaby, B.C.	
DAVETEK MARKETING	(604) 430-3680
Calgary, Alberta	
DAVETEK MARKETING	(403) 291-4984
Kanata, Ontario	
VITEL ELECTRONICS	(613) 592-0060
Mississauga, Ontario	
VITEL ELECTRONICS	(416) 676-9720
Lachine, Quebec	
VITEL ELECTRONICS	(514) 636-5951
IDAHO	
INTERMOUNTAIN TECH MKTG, INC	(208) 888-6071
ILLINOIS	
HEARTLAND TECH MKTG, INC	(312) 577-9222
INDIANA	
Huntington - ELECTRONIC MARKETING CONSULTANTS, INC	(317) 921-3450
Indianapolis - ELECTRONIC MARKETING CONSULTANTS, INC	(317) 921-3450
IOWA	
LORENZ SALES	(319) 377-4666
KANSAS	
Merriam – LORENZ SALES	(913) 384-6556
Wichita – LORENZ SALES	(316) 721-0500
KENTUCKY	
ELECTRONIC MARKETING CONSULTANTS, INC	(317) 921-3452
MICHIGAN	
Birmingham - MIKE RAICK ASSOCIATES	(313) 644-5040
Holland – COM-TEK SALES, INC	(616) 392-7100
Novi – COM-TEK SALES, INC	(313) 344-1409
MISSOURI	
LORENZ SALES	(314) 997-4558
NEBRASKA	
LORENZ SALES	(402) 475-4660
NEW MEXICO	
THORSON DESERT STATES	(505) 293-8555
NEW YORK	
East Syracuse – NYCOM, INC	(315) 437-8343
Woodbury – COMPONENT CONSULTANTS, INC	(516) 364-8020
OHIO	
Centerville – DOLFUSS ROOT & CO	(513) 433-6776
Columbus – DOLFUSS ROOT & CO	(614) 885-4844
Strongsville – DOLFUSS ROOT & CO	(216) 238-0300
PENNSYLVANIA	
DOLFUSS ROOT & CO	(412) 221-4420
PUERTO RICO	
COMP REP ASSOC, INC	(809) 746-6550
UTAH, R² MARKETING	(801) 595-0631
WASHINGTON	
ELECTRA TECHNICAL SALES	(206) 821-7442
WISCONSIN	
HEARTLAND TECH MKTG, INC	(414) 792-0920

Advanced Micro Devices reserves the right to make changes in its product without notice in order to improve design or performance characteristics. The performance characteristics listed in this document are guaranteed by specific tests, guard banding, design and other practices common to the industry. For specific testing details, contact your local AMD sales representative. The company assumes no responsibility for the use of any circuits described herein.

Advanced Micro Devices, Inc. 901 Thompson Place, P.O. Box 3453, Sunnyvale, CA 94088, USA
Tel: (408) 732-2400 • TWX: 910-339-9280 • TELEX: 34-6306 • TOLL FREE: (800) 538-8450
APPLICATIONS HOTLINE TOLL FREE: (800) 222-9323 • (408) 749-5703

© 1990 Advanced Micro Devices, Inc.
2/2/90
WCP-11M-3/90-0 Printed in USA

intel

27210
1M (64K x 16) WORD-WIDE EPROM

PRELIMINARY

- **High-Performance HMOS* II-E**
  — 150 ns Access Time
  — Low 150 mA Active Power
- **Complete Upgrade Capability**
  — PGM "Don't Care" Status Allows Wiring in Higher Order Addresses
- **Fast Programming**
  — Quick-Pulse Programming™ Algorithm—8 Seconds Typical
- **New Word-Wide Pinout**
  — Clean, "Flow-Through" Architecture
- **Standard EPROM Features**
  — TTL Compatibility
  — Two Line Control
  — int$_e$ligent Identifier™ For Automated Programming
- **40-Pin DIP and Compact 44-Lead PLCC Packaging**
  (See Packaging Spec., Order #231369)

The Intel 27210 is a 5V only, 1,048,576-bit, Erasable Programmable Read Only Memory. It is organized as 64K-words of 16 bits each. It defines a new-clean memory architecture, oriented toward high-performance 16-bit and 32-bit CPUs, which simplifies circuit layout and offers a pin-compatible growth path to higher densities.

The 27210's unique circuit design provides for no-hardware-change upgrades to 4M-bits in the future. Since the $\overline{PGM}$ pin is a "don't care" state during read mode, direct connections to higher order addresses, A16 and A17, can be made without affecting the device's read operation. The 27210 will also be offered[1] in One-Time Programmable 40-pin plastic DIP and 44-lead PLCC—with the same 4M-bit upgrade path.

The 27210 provides the highest density and performance available to 16-bit and 32-bit microprocessors. Its by-16 organization makes it an ideal single-chip firmware solution in most microprocessor applications. The 27210's large capacity is sufficient for storage of operating system kernels in addition to standard bootstrap and diagnostic code. Direct execution of operating system software is made possible by the 27210's fast 150 ns access time, which yields no-WAIT-state operation in such high-performance CPUs as the 10 MHz 80286.

The 27210 is part of a three-product megabit EPROM family. Other family members are the 27010 and 27011. These two products have byte-wide organizations geared toward simple upgrades from lower densities. The 27010 is organized as 128K x 8 in a 32-pin DIP package which is pin-compatible with JEDEC-standard 28-pin 512K EPROMs. The 8 x 16K x 8 27011 utilizes page addressing, allowing "drop in" replacement of the 512 K-bit 27513 and continued no-hardware-change upgrades to 32 M-bits in the same JEDEC-compatible 28-pin site.

The 27210 shares several features with standard JEDEC EPROMs, including two-line output control for simplified interfacing and the int$_e$ligent Identifier™ feature for automated programming. It can also be programmed rapidly using Intel's Quick-Pulse Programming™ Algorithm, typically within 8 seconds.

The 27210 is manufactured using a scaled verison of Intel's advanced HMOS* II-E process which assures highest reliability and manufacturability.

*HMOS is a patented process of Intel Corporation.

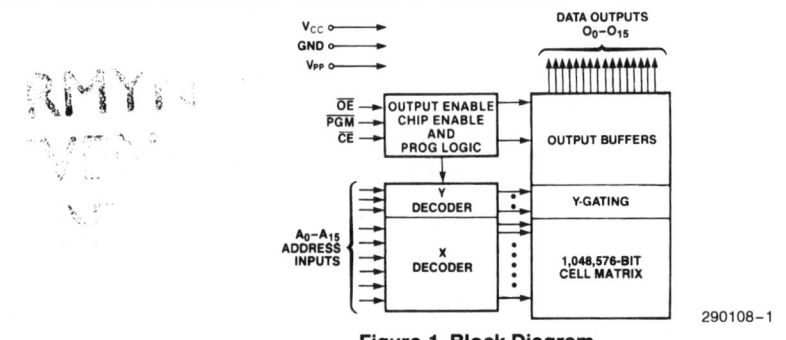

Figure 1. Block Diagram

290108-1

Intel Corporation assumes no responsibility for the use of any circuitry other than circuitry embodied in an Intel product. No other circuit patent licenses are implied. Information contained herein supersedes previously published specifications on these devices from Intel. **November 1986**
© Intel Corporation, 1986
Order Number: 290108-002

intel 27210

PRELIMINARY

Pin Names

A_0–A_{17}	ADDRESSES
$\overline{CE}$	CHIP ENABLE
$\overline{OE}$	OUTPUT ENABLE
O_0–O_{15}	OUTPUTS
$\overline{PGM}$	PROGRAM
N.C.	NO INTERNAL CONNECT
D.U.	DON'T USE

DIP Pin Configuration (27210 P27210, 40-pin):

4M	2M	27210 pin		27210 pin	2M	4M
V_{PP}	V_{PP}	V_{PP}	1 — 40	V_{CC}	V_{CC}	V_{CC}
$\overline{CE}$	$\overline{CE}$	$\overline{CE}$	2 — 39	$\overline{PGM}$	$\overline{PGM}$	A_{17}
O_{15}	O_{15}	O_{15}	3 — 38	NC	A_{16}	A_{16}
O_{14}	O_{14}	O_{14}	4 — 37	A_{15}	A_{15}	A_{15}
O_{13}	O_{13}	O_{13}	5 — 36	A_{14}	A_{14}	A_{14}
O_{12}	O_{12}	O_{12}	6 — 35	A_{13}	A_{13}	A_{13}
O_{11}	O_{11}	O_{11}	7 — 34	A_{12}	A_{12}	A_{12}
O_{10}	O_{10}	O_{10}	8 — 33	A_{11}	A_{11}	A_{11}
O_9	O_9	O_9	9 — 32	A_{10}	A_{10}	A_{10}
O_8	O_8	O_8	10 — 31	A_9	A_9	A_9
GND	GND	GND	11 — 30	GND	GND	GND
O_7	O_7	O_7	12 — 29	A_8	A_8	A_8
O_6	O_6	O_6	13 — 28	A_7	A_7	A_7
O_5	O_5	O_5	14 — 27	A_6	A_6	A_6
O_4	O_4	O_4	15 — 26	A_5	A_5	A_5
O_3	O_3	O_3	16 — 25	A_4	A_4	A_4
O_2	O_2	O_2	17 — 24	A_3	A_3	A_3
O_1	O_1	O_1	18 — 23	A_2	A_2	A_2
O_0	O_0	O_0	19 — 22	A_1	A_1	A_1
$\overline{OE}$	$\overline{OE}$	$\overline{OE}$	20 — 21	A_0	A_0	A_0

NOTE: Compatible Higher Density Word Wide EPROM Pin Configurations are Shown in the Blocks Adjacent to the 27210 Pins

Figure 2. Cerdip/Plastic(P) DIP Pin Configurations

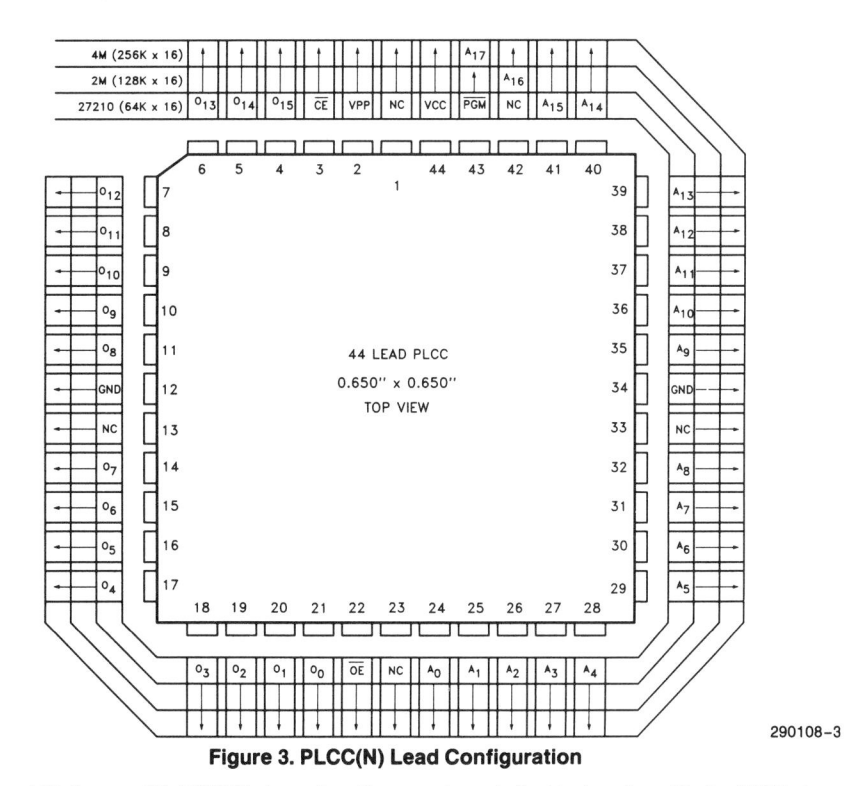

PLCC Lead Configuration (44 LEAD PLCC, 0.650" × 0.650", TOP VIEW):

Top edge pins (6–44,43,42,41,40 / 1,2,3,4,5):
- 27210 (64K × 16): O_{13}, O_{14}, O_{15}, $\overline{CE}$, VPP, NC, VCC, $\overline{PGM}$, NC, A_{15}, A_{14}
- 2M (128K × 16): A_{16}
- 4M (256K × 16): A_{17}

Left side (pins 7–17): O_{12}, O_{11}, O_{10}, O_9, O_8, GND, NC, O_7, O_6, O_5, O_4

Bottom (pins 18–28): O_3, O_2, O_1, O_0, $\overline{OE}$, NC, A_0, A_1, A_2, A_3, A_4

Right side (pins 29–39): A_5, A_6, A_7, A_8, NC, GND, A_9, A_{10}, A_{11}, A_{12}, A_{13}

Figure 3. PLCC(N) Lead Configuration

NOTE:
1. Intel "Universal Site" compatible EPROM pin configurations are shown in the blocks adjacent to the 27210 pins.

intel 27210 PRELIMINARY

EXTENDED TEMPERATURE (EXPRESS) EPROMS

The Intel EXPRESS EPROM family is a series of electrically programmable read only memories which have received additional processing to enhance product characteristics. EXPRESS processing is available for several densities of EPROM, allowing the choice of appropriate memory size to match system applications. EXPRESS EPROM products are available with 168 ±8 hour, 125°C dynamic burn-in using Intel's standard bias configuration. This process exceeds or meets most industry specifications of burn-in. The standard EXPRESS EPROM operating temperature range is 0°C to 70°C. Extended operating temperature range (−40°C to +85°C) EXPRESS products are available. Like all Intel EPROMs, the EXPRESS EPROM family is inspected to 0.1% electrical AQL. This may allow the user to reduce or eliminate incoming inspection testing.

EXPRESS EPROM PRODUCT FAMILY

PRODUCT DEFINITIONS

Type	Operating Temperature	Burn-in 125°C (hr)
Q	0°C to +70°C	168 ±8
T	−40°C to +85°C	None
L	−40°C to +85°C	168 ±8

EXPRESS OPTIONS

27210 Versions

| Packaging Options ||
Speed Versions	Cerdip
-170/05	Q
-200/05	T, L, Q
-250/05	T, L, Q
-170/10	T, L, Q
-200/10	T, L, Q
-250/10	T, L, Q

READ OPERATION

D.C. CHARACTERISTICS

Electrical Parameters of Express EPROM Products are identical to standard EPROM parameters except for:

Symbol	Parameter	T27210, L27210 Min	T27210, L27210 Max	Test Conditions
I_{SB}	V_{CC} Standby Current (mA)		50	$\overline{CE} = V_{IH}, \overline{OE} = V_{IL}$
I_{CC1}[1]	V_{CC} Active Current (mA)		170	$\overline{OE} = \overline{CE} = V_{IL}$
	V_{CC} Active Current at High Temperature (mA)		150	$\overline{OE} = \overline{CE} = V_{IL}, V_{PP} = V_{CC}$, $T_{Ambient} = 85°C$

NOTE:
1. The maximum current value is with Outputs O_0 to O_{15} unloaded.

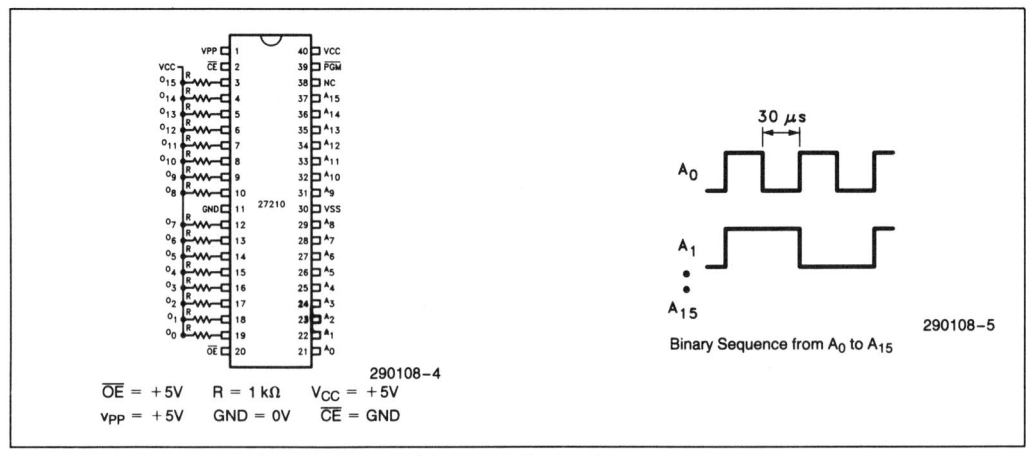

$\overline{OE} = +5V$ $R = 1 k\Omega$ $V_{CC} = +5V$
$V_{PP} = +5V$ $GND = 0V$ $\overline{CE} = GND$

Binary Sequence from A_0 to A_{15}

Burn-In Bias and Timing Diagrams

intel 27210

PRELIMINARY

ABSOLUTE MAXIMUM RATINGS*

Operating Temperature During
Read .... 0°C to +70°C
Temperature Under Bias .... −10°C to +80°C
Storage Temperature .... −65°C to +125°C
All Input or Output Voltages with
Respect to Ground .... −0.6V to +6.25V
Voltage on A_9 with
Respect to Ground .... −0.6V to +13.0V
V_{PP} Supply Voltage with Respect to
Ground During Programming .... −0.6V to +14V
V_{CC} Supply Voltage
with Respect to Ground .... −0.6V to +7.0V

*Notice: Stresses above those listed under "Absolute Maximum Ratings" may cause permanent damage to the device. This is a stress rating only and functional operation of the device at these or any other conditions above those indicated in the operational sections of this specification is not implied. Exposure to absolute maximum rating conditions for extended periods may affect device reliability.

NOTICE: Specifications contained within the following tables are subject to change.

READ OPERATION

D.C. CHARACTERISTICS 0°C ≤ T_A ≤ +70°C

Symbol	Parameter	Min	Typ[3]	Max	Units	Conditions
I_{LI}	Input Load Current			1	µA	V_{IN} = 5.5V
I_{LO}	Output Leakage Current			1	µA	V_{OUT} = 5.5V
I_{PP1}[2]	V_{PP} Load Current Read			1	µA	V_{PP} = 5.5V
I_{SB}	V_{CC} Current Standby			40	mA	$\overline{CE}$ = V_{IH}
I_{CC1}[2]	V_{CC} Current Active			150	mA	$\overline{CE}$ = $\overline{OE}$ = V_{IL}
V_{IL}	Input Low Voltage	−0.1		+0.8	V	
V_{IH}	Input High Voltage	2.0		V_{CC}+1	V	
V_{OL}	Output Low Voltage			0.45	V	I_{OL} = 2.1 mA
V_{OH}	Output High Voltage	2.4			V	I_{OH} = −400 µA

A.C. CHARACTERISTICS 0°C ≤ T_A ≤ +70°C

Versions[5]	V_{CC} ±5%	27210-150/05		27210-170/05 P27210-170/05 N27210-170/05		27210-200/05 P27210-200/05 N27210-200/05		27210-250/05 P27210-250/05 N27210-250/05		Unit
	V_{CC} ±10%			27210-170/10 N27210-170/10		27210-200/10 P27210-200/10 N27210-200/10		27210-250/10 P27210-250/10 N27210-250/10		
Symbol	Characteristics	Min	Max	Min	Max	Min	Max	Min	Max	
t_{ACC}	Address to Output Delay		150		170		200		250	ns
t_{CE}	$\overline{CE}$ to Output Delay		150		170		200		250	ns
t_{OE}	$\overline{OE}$ to Output Delay		65		70		75		100	ns
t_{DF}[4]	$\overline{OE}$ High to Output Float	0	50	0	55	0	60	0	60	ns
t_{OH}	Output Hold from Addresses $\overline{CE}$ or $\overline{OE}$ Whichever Occurred First	0		0		0		0		ns

NOTES:
1. V_{CC} must be applied simultaneously or before V_{PP} and removed simultaneously or after V_{PP}.
2. The maximum current value is with Outputs O_0 to O_{15} unloaded.
3. Typical values are for T_A = 25°C and nominal supply voltages.
4. This parameter is only sampled and is not 100% tested. Output Float is defined as the point where data is no longer driven—see timing diagram.
5. Packaging options: No prefix = Cerdip; Plastic DIP = P; PLCC = N.
6. Both GND pins should be connected to the ground trace on circuit boards.

intel 27210   PRELIMINARY

CAPACITANCE[2] $T_A = 25°C$, $f = 1MHz$

Symbol	Parameter	Typ[1]	Max	Unit	Conditions
C_{IN}	Input Capacitance	4	6	pF	$V_{IN} = 0V$
C_{OUT}	Output Capacitance	8	12	pF	$V_{OUT} = 0V$
C_{VPP}	V_{PP} Input Capacitance		25	pF	$V_{PP} = 0V$

A.C. TESTING INPUT/OUTPUT WAVEFORM

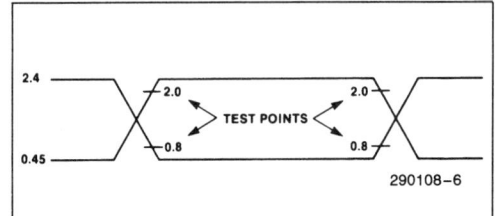

A.C. testing inputs are driven at 2.4V for a Logic "1" and 0.45V for a Logic "0". Timing measurements are made at 2.0V for a Logic "1" and 0.8V for a Logic "0."

A.C. TESTING LOAD CIRCUIT

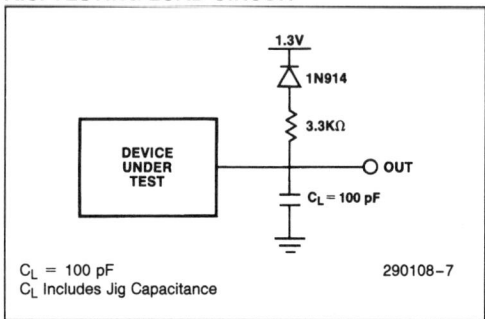

$C_L = 100$ pF
C_L Includes Jig Capacitance

A.C. WAVEFORMS

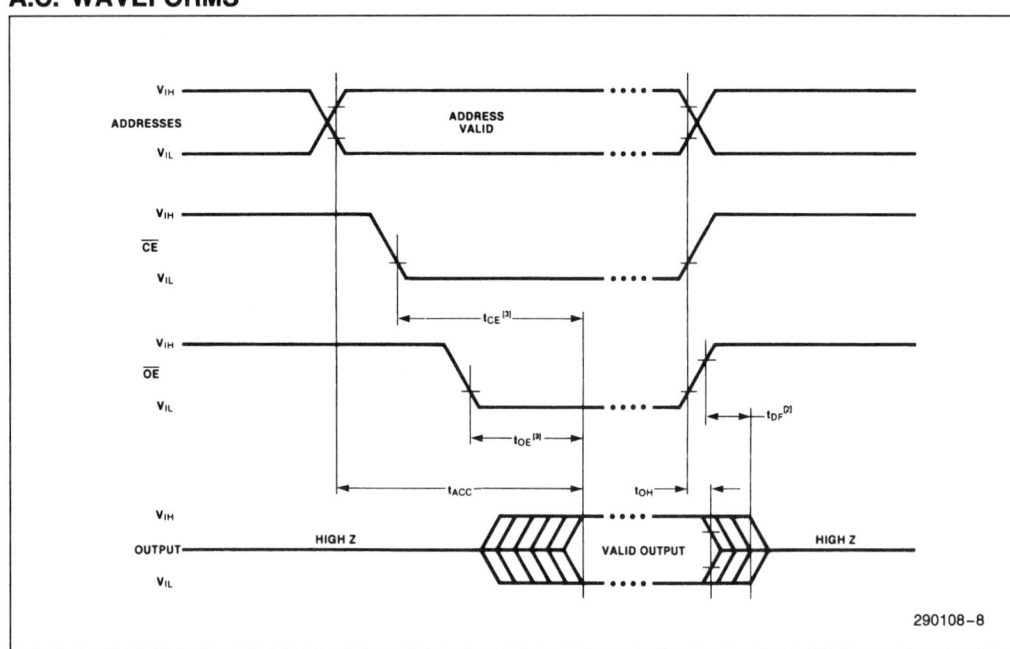

NOTES:
1. Typical values are for $T_A = 25°C$ and nominal supply voltages.
2. This parameter is only sampled and is not 100% tested.
3. $\overline{OE}$ may be delayed up to $t_{CE} - t_{OE}$ after the falling edge of $\overline{CE}$ without impact on t_{CE}.

intel 27210 PRELIMINARY

DEVICE OPERATION

The modes of operation of the 27210 are listed in Table 1. A single 5V power supply is required in the read mode. All inputs are TTL levels except for V_{PP} and 12V on A_9 for int$_e$ligent Identifier.

Table 1. Modes Selection

Mode \ Pins	$\overline{CE}$	$\overline{OE}$	$\overline{PGM}$	A_9	A_0	V_{PP}	V_{CC}	Outputs
Read	V_{IL}	V_{IL}	X	X(1)	X	X	5.0V	D_{OUT}
Output Disable	V_{IL}	V_{IH}	X	X	X	X	5.0V	High Z
Standby	V_{IH}	X	X	X	X	X	5.0V	High Z
Programming	V_{IL}	V_{IH}	V_{IL}	X	X	(Note 4)	(Note 4)	D_{IN}
Program Verify	V_{IL}	V_{IL}	V_{IH}	X	X	(Note 4)	(Note 4)	D_{OUT}
Program Inhibit	V_{IH}	X	X	X	X	(Note 4)	(Note 4)	High Z
int$_e$ligent Identifier Manufacturer(3)	V_{IL}	V_{IL}	X	V_H(2)	V_{IL}	V_{CC}	5.0V	0089 H
int$_e$ligent Identifier Device(3)	V_{IL}	V_{IL}	X	V_H(2)	V_{IH}	V_{CC}	5.0V	00FFH

NOTES:
1. X can be V_{IL} or V_{IH}
2. V_H = 12.0V ±0.5V
3. A_1–A_8, A_{10}–A_{15} = V_{IL}
4. See Table 2 for V_{CC} and V_{PP} voltages.

Read Mode

The 27210 has two control functions, both of which must be logically active in order to obtain data at the outputs. Chip Enable ($\overline{CE}$) is the power control and should be used for device selection. Output Enable ($\overline{OE}$) is the output control and should be used to gate data from the output pins, independent of device selection. Assuming that addresses are stable, the address access time (t_{ACC}) is equal to the delay from $\overline{CE}$ to output (t_{CE}). Data is available at the outputs after a delay of t_{OE} from the falling edge of $\overline{OE}$, assuming that $\overline{CE}$ has been low and addresses have been stable for at least t_{ACC}-t_{OE}.

Standby Mode

EPROMs can be placed in standby mode which reduces the maximum current of the device by applying a TTL-high signal to the $\overline{CE}$ input. When in standby mode, the outputs are in a high impedance state, independent of the $\overline{OE}$ input.

Two Line Output Control

Because EPROMs are usually used in larger memory arrays, Intel has provided 2 control lines which accommodate this multiple memory connection. The two control lines allow for:

a) the lowest possible memory power dissipation, and

b) complete assurance that output bus contention will not occur

To use these two control lines most efficiently, $\overline{CE}$ should be decoded and used as the primary device selecting function, while $\overline{OE}$ should be made a common connection to all devices in the array and connected to the $\overline{READ}$ line from the system control bus. This assures that all deselected memory devices are in their low power standby mode and that the output pins are active only when data is desired from a particular memory device.

SYSTEM CONSIDERATIONS

The power switching characteristics of EPROMs require careful decoupling of the devices. The supply current, I_{CC}, has three segments that are of interest to the system designer—the standby current level, the active current level, and the transient current peaks that are produced by the falling and rising edges of Chip Enable. The magnitude of these transient current peaks is dependent on the output capacitive and inductive loading of the device. The associated transient voltage peaks can be suppressed by complying with Intel's Two-Line Control, and by properly selected decoupling capacitors. It is recommended that a 0.1 μF ceramic capacitor be used on every device between V_{CC} and GND. This should be a high frequency capacitor for low inherent inductance and should be placed as close to the device as possible. In addition, a 4.7 μF bulk electrolytic capacitor should be used between V_{CC} and GND for every eight devices. The bulk capacitor should be located near where the power supply is connected to the array. The purpose of the bulk capacitor is to overcome the voltage droop caused by the inductive effect of PC board-traces.

PROGRAMMING MODES

Caution: Exceeding 14V on V_{PP} will permanently damage the device.

Initially, and after each erasure, all bits of the EPROM are in the "1" state. Data is introduced by selectively programming "0s" into the desired bit locations. Although only "0s" will be programmed, both "1s" and "0s" can be present in the data word. The only way to change a "0" to a "1" is by ultraviolet light erasure (Cerdip EPROMs).

The device is in the programming mode when V_{PP} is raised to its programming voltage (See Table 2) and $\overline{CE}$ and $\overline{PGM}$ are both at TTL low. The data to be programmed is applied 16 bits in parallel to the data output pins. The levels required for the address and data inputs are TTL.

Program Inhibit

Programming of multiple EPROMS in parallel with different data is easily accomplished by using the Program Inhibit mode. A high-level $\overline{CE}$ or $\overline{PGM}$ input inhibits the other devices from being programmed.

Except for $\overline{CE}$, all like inputs (including $\overline{OE}$) of the parallel EPROMs may be common. A TTL low-level pulse applied to the $\overline{PGM}$ input with V_{PP} at its programming voltage and $\overline{CE}$ at TTL-Low will program the selected device.

Program Verify

A verify should be performed on the programmed bits to determine that they have been correctly programmed. The verify is performed with $\overline{OE}$ at V_{IL}, $\overline{CE}$ at V_{IL}, $\overline{PGM}$ at V_{IH} and V_{PP} and V_{CC} at their programming voltages.

int_eligent Identifier™ Mode

The int_eligent Identifier Mode allows the reading out of a binary code from an EPROM that will identify its manufacturer and type. This mode is intended for use by programming equipment for the purpose of automatically matching the device to be programmed with its corresponding programming algorithm. This mode is functional in the 25°C ±5°C ambient temperature range that is required when programming the device.

To activate this mode, the programming equipment must force 11.5V to 12.5V on address line A9 of the EPROM. Two identifier bytes may then be sequenced from the device outputs by toggling address line A0 from V_{IL} to V_{IH}. All other address lines must be held at V_{IL} during the int_eligent Identifier Mode.

Byte 0 (A0 = V_{IL}) represents the manufacturer code and byte 1 (A0 = V_{IH}) the device identifier code. These two identifier bytes are given in Table 1.

INTEL EPROM PROGRAMMING SUPPORT TOOLS

Intel offers a full line of EPROM Programmers providing state-of-the-art programming for Intel programmable devices. The modular architecture of Intel's EPROM programmers allows you to add new support as it becomes available, with very low cost add-ons. For example, even the earliest users of the iUP-FAST 27/K module may take advantage of Intel's new Quick-Pulse Programming Algorithm, the fastest in the industry.

Intel EPROM programmers may be controlled from a host computer using Intel's PROM Programming software (iPPS). iPPS makes programming easy for a growing list of industry standard hosts, including the IBM PC, XT, AT and PCDOS compatibles, Intellec Development Systems, Intel's iPDS Personal Development System, and the Intel Network Development System (iNDS-II). Stand-alone operation is also available, including device previewing, editing, programming, and download of programming data from any source over an RS232C port.

For further details consult the EPROM Programming section of the Development Systems Handbook.

ERASURE CHARACTERISTICS (FOR CERDIP EPROMS)

The erasure characteristics are such that erasure begins to occur upon exposure to light with wavelengths shorter than approximately 4000 Angstroms (Å). It should be noted that sunlight and certain types of fluorescent lamps have wavelengths in the 3000-4000Å range. Data shows that constant exposure to room level fluorescent lighting could erase the EPROM in approximately 3 years, while it would take approximately 1 week to cause erasure when exposed to direct sunlight. If the device is to be exposed to these types of lighting conditions for extended periods of time, opaque labels should be placed over the window to prevent unintentional erasure.

The recommended erasure procedure is exposure to shortwave ultraviolet light which has a wavelength of 2537 Angstroms (Å). The integrated dose (i.e., UV intensity × exposure time) for erasure should be a minimum of 15 Wsec/cm^2. The erasure time with this dosage is approximately 15 to 20 minutes using an ultraviolet lamp with a 12000 μW/cm^2 power rating. The EPROM should be placed within 1 inch of the lamp tubes during erasure. The maximum integrated dose an EPROM can be exposed to without damage is 7258 Wsec/cm^2 (1 week @ 12000 μW/cm^2). Exposure of the device to high intensity UV light for longer periods may cause permanent damage.

intel®   27210   PRELIMINARY

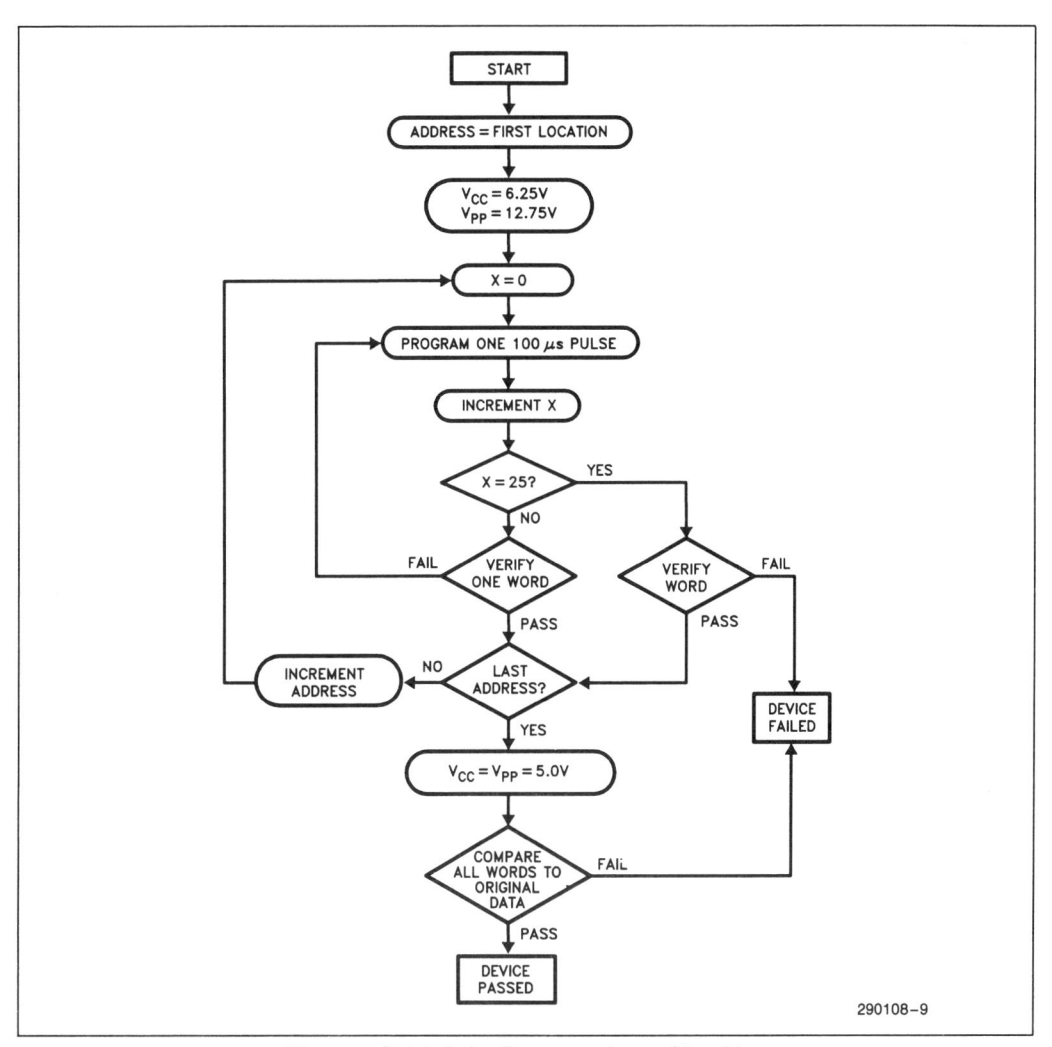

Figure 4. Quick-Pulse Programming™ Algorithm

Quick-Pulse Programming™ Algorithm

Intel's 27210 EPROMs can be programmed using the Quick-Pulse Programming Algorithm, developed by Intel to substantially reduce the throughput time in the production programming environment. This algorithm allows these devices to be programmed in under eight seconds, almost a hundred fold improvement over previous algorithms. Actual programming time is a function of the PROM programmer being used.

The Quick-Pulse Programming Algorithm uses initial pulses of 100 microseconds followed by a word verification to determine when the addressed word has been successfully programmed. Up to 25 100 μs pulses per word are provided before a failure is recognized. A flow chart of the Quick-Pulse Programming Algorithm is shown in Figure 4.

For the Quick-Pulse Programming Algorithm, the entire sequence of programming pulses and word verifications is performed at V_{CC} = 6.25V and V_{PP} at 12.75V. When programming of the EPROM has been completed, all data words should be compared to the original data with V_{CC} = V_{PP} = 5.0V.

27210 PRELIMINARY

D.C. PROGRAMMING CHARACTERISTICS $T_A = 25°C \pm 5°C$

Table 2

Symbol	Parameter	Limits Min	Limits Max	Unit	Test Conditions (Note 1)
I_{LI}	Input Leakage Current (All Inputs)		10	μA	$V_{IN} = 6V$
V_{IL}	Input Low Level (All Inputs)	−0.1	0.8	V	
V_{IH}	Input High Level	2.0	$V_{CC}+1$	V	
V_{OL}	Output Low Voltage During Verify		0.45	V	$I_{OL} = 2.1$ mA
V_{OH}	Output High Voltage During Verify	2.4		V	$I_{OH} = -400$ μA
I_{CC2}[3]	V_{CC} Supply Current (Program & Verify)		160	mA	$\overline{CE} = \overline{PGM} = V_{IL}$
I_{PP2}	V_{PP} Supply Current (Program)		50	mA	$\overline{CE} = \overline{PGM} = V_{IL}$
V_{ID}	A_9 int$_e$ligent Identifier Voltage	11.5	12.5	V	$V_{CC} = 5V$
V_{PP}	Quick-Pulse Programming Algorithm	12.5	13.0	V	
V_{CC}	Quick-Pulse Programming Algorithm	6.0	6.5	V	

A.C. PROGRAMMING CHARACTERISTICS

$T_A = 25°C \pm 5°C$ (See Table 2 for V_{CC} and V_{PP} voltages.)

Symbol	Parameter	Min	Typ	Max	Unit	Conditions* (Note 1)
t_{AS}	Address Setup Time	2			μs	
t_{OES}	$\overline{OE}$ Setup Time	2			μs	
t_{DS}	Data Setup Time	2			μs	
t_{AH}	Address Hold Time	0			μs	
t_{DH}	Data Hold Time	2			μs	
t_{DFP}	$\overline{OE}$ High to Output Float Delay	0		130	ns	(Note 2)
t_{VPS}	V_{PP} Setup Time	2			μs	
t_{VCS}	V_{CC} Setup Time	2			μs	
t_{CES}	$\overline{CE}$ Setup Time	2			μs	
t_{PW}	$\overline{PGM}$ Initial Program Pulse Width	95	100	105	μs	Quick-Pulse Programming
t_{OE}	Data Valid from $\overline{OE}$			150	ns	

*A.C. CONDITIONS OF TEST

Input Rise and Fall Times (10% to 90%) . . . . . . 20 ns
Input Pulse Levels . . . . 0.45V to 2.4V
Input Timing Reference Level . . . . . . 0.8V and 2.0V
Output Timing Reference Level . . . . . . 0.8V and 2.0V

NOTES:
1. V_{CC} must be applied simultaneously or before V_{PP} and removed simultaneously or after V_{PP}.
2. This parameter is only sampled and is not 100% tested. Output Float is defined as the point where data is no longer driven—see timing diagram.
3. The maximum current value is with outputs O_0-O_{15} unloaded.

intel 27210 PRELIMINARY

PROGRAMMING WAVEFORMS

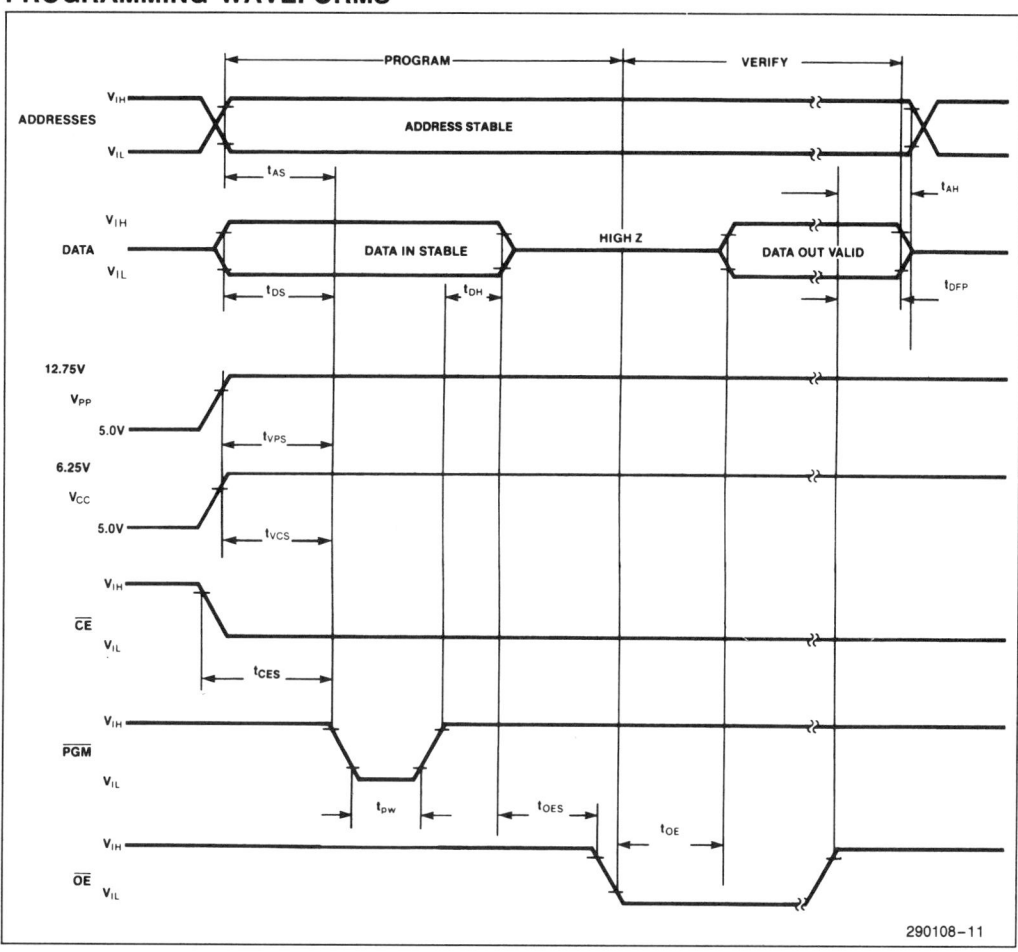

NOTES:
1. The Input Timing Reference Level is 0.8V for V_{IL} and 2V for a V_{IH}.
2. t_{OE} and t_{DFP} are characteristics of the device but must be accommodated by the programmer.
3. When programming the 27210, a 0.1 µF capacitor is required across V_{PP} and ground to suppress spurious voltage transients which can damage the device.

intel® 27C203
FAST PIPELINED 256K (16K x 16) EPROM

PRELIMINARY

- **Pipelined Interface**
  - Clock for Data Latching
  - Optional Synchronous Chip Select
- **Quick-Pulse Programming™**
  - 4 Second Throughput for Automated Manufacturing
- **Very High Speed**
  - Supports up to 20MHz Pipelined 80386/376's at Zero Wait-States
- **Excellent Drive Capability**
  - 4 mA Source/16 mA Sink Current Handles Large Fanout
- **System Initialize Feature**
  - Initialization Register Forces Startup Vector for State Machine Applications
- **High-Performance/Low Power CMOS**
  - High Density Memory With 100 mA I_{CC} Maximum
- **Versatile Package Options**
  - Standard DIP
  - Compact Surface-Mount Cerquad
  - Compact Surface-Mount PLCC**

The Intel 27C203 is a high performance 262,144-bit erasable programmable read-only memory organized as 16K words of 16 bits each. Its density and word-wide configuration, combined with its pipelined bus interface, provides a high integration firmware solution for today's speed-critical applications.

The 27C203 supports the pipelined bus architectures of the Intel 80376 and 80386 microprocessors. Pipelining relaxes memory interface requirements by utilizing an overlapping "early address," effectively stretching the read operation an additional bus cycle. Thus, much higher bus bandwidths are achievable than with a standard interface. An 80386/376 design employing the 27C203 can accommodate system clock rates up to 20 MHz at zero wait-states, while utilizing slower, less expensive interface logic in 16 MHz versions. The 256K bit capacity is well suited for high-end embedded control applications. The 27C203's simple interface and X16 organization help minimize overall chip count.

State machine designers can also use the 27C203 as a synchronous logic element, especially where the designer needs a pipelined "data-flow-through" transfer function. For example, it can be applied to weighting or summing functions in high frequency filters using digital signal processing techniques. To simplify system design and debugging, the 27C203 contains a hardware-controlled initial condition vector through the INIT function pin.

The 27C203 is available in three package versions. The standard 40-pin Dual-In-Line Package (DIP) provides for conventional device handling and socketing. The 44-lead OTP™ (One-Time-Programmable) Plastic Leaded Chip Carrier (PLCC) allows lowest cost, automated, surface-mount manufacturing. The 44-lead Cerquad Package allows reprogramming within the same compact dimensions as the PLCC package.

The 27C203 is manufactured on Intel's advanced CHMOS* III-E, a process optimized for high performance.

The 27C203 will be replaced by the 27C203C mid 1990. The 27C203C will be functionally identical to the 27C203 in the read mode. The 27C203C's memory array will use a 1-transistor cell, versus the 27C203's 2-transistor cell, and will program with a standard one-pass algorithm.

*CHMOS is a patented process of Intel Corporation.   **PLCC package availability TBD

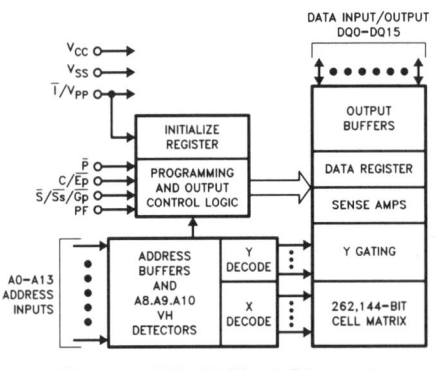

Figure 1. 27C203 Block Diagram

290176-1

Intel Corporation assumes no responsibility for the use of any circuitry other than circuitry embodied in an Intel product. No other circuit patent licenses are implied. Information contained herein supersedes previously published specifications on these devices from Intel.   **October 1989**
© Intel Corporation, 1989   Order Number: 290176-002

intel

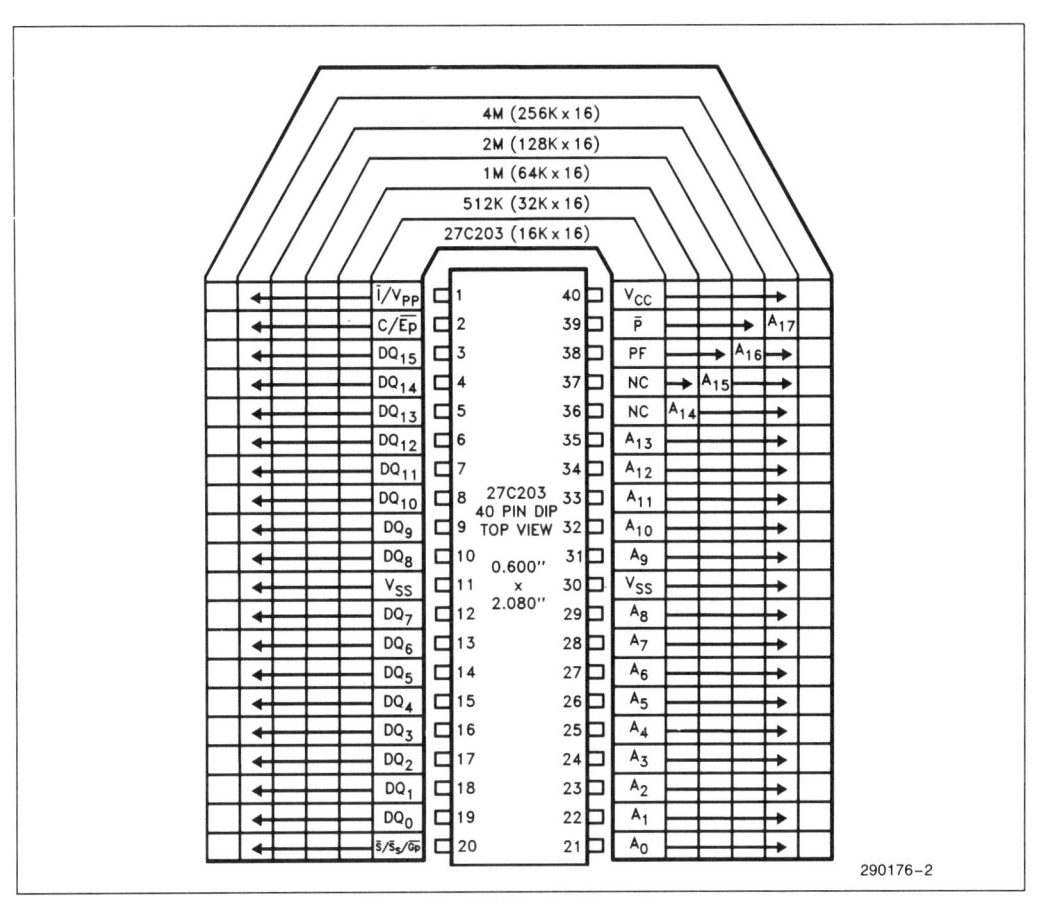

27C203

PRELIMINARY

Figure 2a. DIP Pin Configuration

27C203 Pin Names

A0–A13	Address
DQ0–DQ15	Data Inputs/Outputs
C/$\overline{Ep}$	Clock/Program Chip Enable
$\overline{S}/\overline{Ss}/\overline{Gp}$	Chip Select/Program Output Enable
$\overline{P}$	Programming Enable
PF	Program Function
$\overline{I}$/V$_{PP}$	Initialize Register Enable/V$_{PP}$
NC	No Connect

NOTE:
1. Each 27C203 V$_{CC}$ and V$_{SS}$ pin specified must be tied individually to their respective power supplies. See System Design Considerations.

intel 27C203 PRELIMINARY

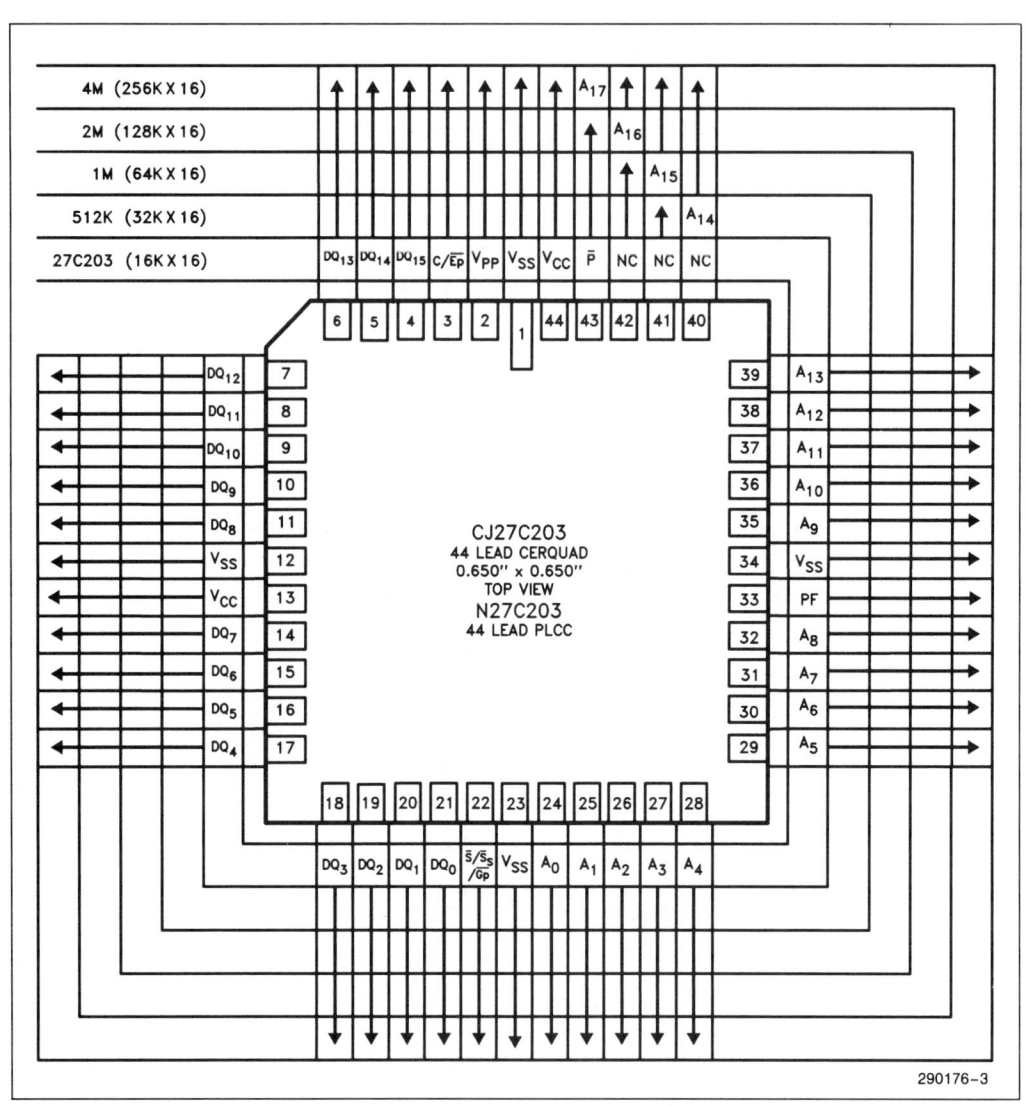

Figure 2b. Cerquad/PLCC Lead Configuration

intel   27C203   PRELIMINARY

ABSOLUTE MAXIMUM RATINGS*

Operating temperature during read ... 0°C to +70°C
Temperature under bias .... -55°C to +125°C
Storage temperature .... -65°C to +125°C
All input or output voltages
   with respect to ground .... -2V to +7V[1]
Voltage on A9 with
   with respect to ground .... -2V to +14V[1]
V_{PP} supply voltage with respect
   to ground during programming .. -2V to +14V[1]
V_{CC} supply voltage with
   respect to ground .... -2V to +7V[1]

*Notice: Stresses above those listed under "Absolute Maximum Ratings" may cause permanent damage to the device. This is a stress rating only and functional operation of the device at these or any other conditions above those indicated in the operational sections of this specification is not implied. Exposure to absolute maximum rating conditions for extended periods may affect device reliability.

NOTICE: Specifications contained within the following tables are subject to change.

EXTENDED TEMPERATURE (EXPRESS) EPROMS

The Intel EXPRESS EPROM family receives additional processing to enhance product characteristics. EXPRESS EPROMs are available with 168 +/- 8 hour, 125°C dynamic burn-in using Intel's standard bias configuration. This process meets or exceeds most industry burn-in specifications. The standard EXPRESS EPROM operating temperature range is 0°C to +70°C. Extended operating temperature range (-40°C to +85°C) EXPRESS products are also available. Like all Intel EPROMs, the EXPRESS EPROM family is inspected to 0.1% electrical AQL. This allows reduction or elimination of incoming testing.

EXPRESS OPTIONS

Versions

Speed Versions	Packaging Options	
	DIP	Cerquad
-45V05	Q, T, L	Q, T, L
-45V10	Q, T, L	Q, T, L
-55V05	Q, T, L	Q, T, L
-55V10	Q, T, L	Q, T, L

EXPRESS EPROM PRODUCT FAMILY

Product Definitions

Type	Operating Temperature	Burn-in 125°C (hr)
Q	0°C to +70°C	168 ± 8
T	-40°C to +85°C	NONE
L	-40°C to +85°C	168 ± 8

NOTE:
1. Minimum D.C. input voltage is -0.5V. During transitions the inputs may undershoot to -2.0V for periods less than 20 ns. Maximum D.C. voltage on output pins is V_{CC} + 0.5V which may overshoot to V_{CC} +2V for periods less than 20 ns.

intel 27C203 PRELIMINARY

D.C. CHARACTERISTICS

Symbol	Parameter	Min	Typ	Max	Units	Conditions
I_{LI}	Input Load Current			10	μA	$V_{IN} = 5.5V$
I_{LO}	Output Leakage Current			10	μA	$V_{OUT} = 5.5V$
I_{PP1}	V_{PP} Load Current Read			10	μA	$V_{PP} = 5.5V$
I_{CC}[1]	V_{CC} Current Active		60	100	mA	
V_{IL}	Input Low Voltage	−0.1		+0.8	V	
V_{IH}	Input High Voltage	2.0		V_{CC} + 1	V	
V_{OL}	Output Low Voltage			.45	V	I_{OL} = 16 mA
V_{OH}	Output High Voltage	2.4			V	I_{OH} = −4mA

A.C. CHARACTERISTICS

Versions (2)	$V_{CC} \pm 5\%$ (5.0 ± 0.25V)	C27C203-45V05 CJ27C203-45V05		C27C203-55V05 CJ27C203-55V05		Units
	$V_{CC} \pm 10\%$ (5.0 ± 0.50V)	C27C203-45V10 CJ27C203-45V10		C27C203-55V10 CJ27C203-55V10		
Symbol	Characteristics	Min.	Max.	Min.	Max.	
t_{AVCH}	Address Valid to Clock	45		55		ns
t_{CHAX}	Address Hold from Clock	0		0		ns
t_{CHQV}[3]	Clock to Valid Output		40		50	ns
t_{CHQZ}[4]	Clock to Output Tri-state for Synchronous Chip Deselect, S_S		15		15	ns
t_{CHCL}, t_{CLCH}	Clock Pulse Width	25		25		ns
t_{SVCH}	Setup Time to Clock for Synchronous Chip Select, $\overline{S_S}$.	10		10		ns
t_{CHSX}	Hold Time for Sync. Select	10		10		ns
t_{CHQX}[4]	Data Hold Time From Clock	0		0		ns
t_{SHQX}[4]	Data Hold Time From Select	0		0		ns
t_{SLQV}[3]	Valid Output Delay from Asynchronous Chip Select, $\overline{S}$.		35		40	ns
t_{SHQZ}[4]	Output Tri-state from Asynchronous Chip Deselect, $\overline{S}$.		10		10	ns
t_{ILQV}[3]	Initialize Register Valid Output Delay		35		40	ns
t_{ILIH}	Initialize Pulse Width	20		25		ns
t_{IHCH}	Recovery Time from Initialize to Clock	5		5		ns

NOTES:
1. V_{CC} current assumes no output loading, i.e., $I_{OH} = I_{OL}$ 0mA
2. Packaging options: C = Ceramic Side-Brazed, CJ = Cerquad.
3. A derating factor of 6 ns/100 pF should be used with output loading greater than 30 pF. This derating factor is only sampled and not 100% tested.
4. These parameters are only sampled and not 100% tested.

intel 27C203 PRELIMINARY

A.C. WAVEFORMS—PIPELINED BUS APPLICATION (80376/80386)

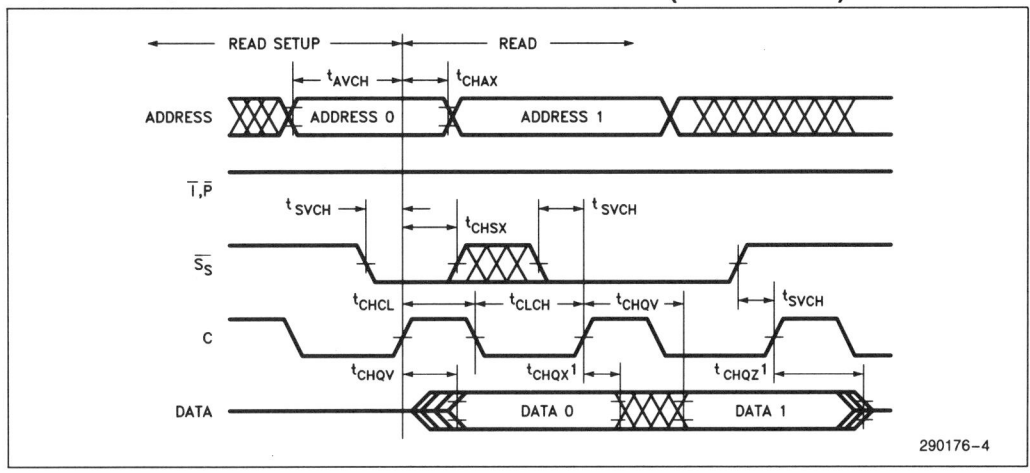

A.C. WAVEFORMS—STATE MACHINE APPLICATION (ASYNCHRONOUS CHIP SELECT)

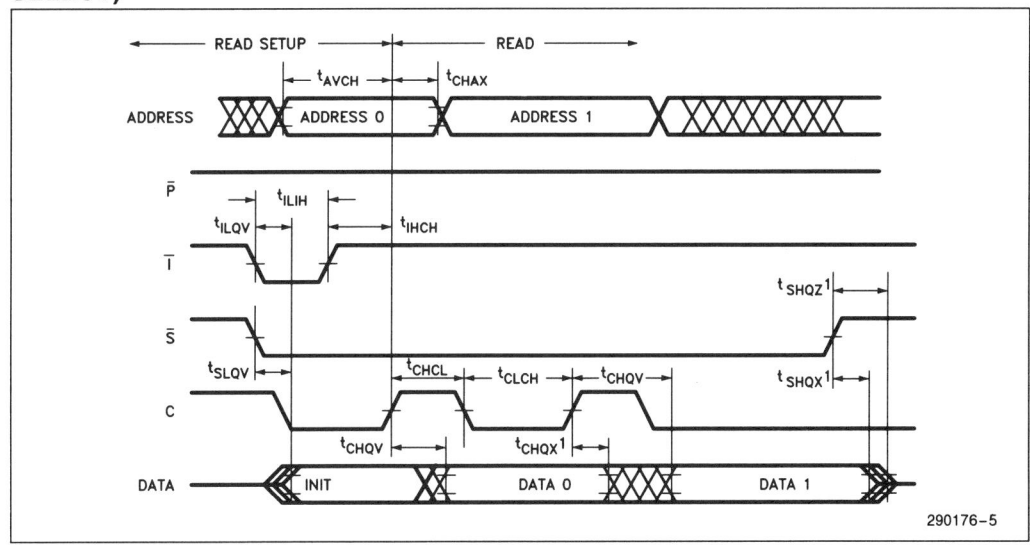

NOTE:
1. These parameters are only sampled and not 100% tested.

CAPACITANCE[1] $T_A = 25°C, f = 1$ MHz

Symbol	Parameter	Typ	Max	Unit	Conditions
C_{IN}	Input Capacitance	12	15	pF	$V_{IN} = 0V$
C_{OUT}	Output Capacitance	12	15	pF	$V_{OUT} = 0V$
C_{VPP}	V_{PP} Input Capacitance		75	pF	$V_{PP} = 0V$

intel® 27C203

PRELIMINARY

A.C. TESTING INPUT/OUTPUT WAVEFORM

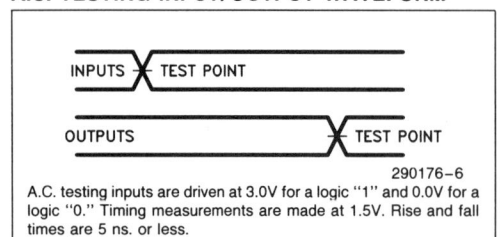

```
INPUTS  ── TEST POINT ──
OUTPUTS ──────── TEST POINT ──
```
290176-6

A.C. testing inputs are driven at 3.0V for a logic "1" and 0.0V for a logic "0." Timing measurements are made at 1.5V. Rise and fall times are 5 ns. or less.

A.C. TESTING LOAD CURRENT

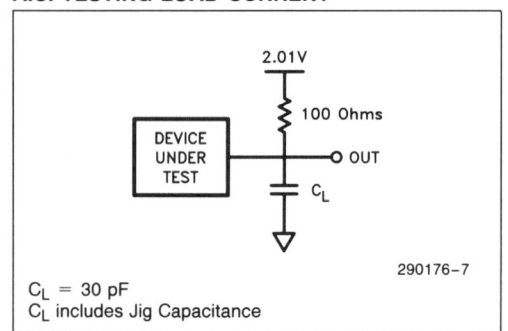

2.01V, 100 Ohms, DEVICE UNDER TEST, OUT, C_L

290176-7

C_L = 30 pF
C_L includes Jig Capacitance

Table 1. Mode Table

Mode	Pins	$\overline{S}/\overline{S_s}/\overline{G_p}$	$C/\overline{E_p}$	pF	A10	A9	A8	A0	$\overline{P}$	$\overline{I}/V_{PP}$	V_{CC}	DQ
Read Setup	Asynchronous Chip Select	X	╱	V_{IH}	A10	A9	A8	A0	V_{IH}	V_{IH}	V_{CC}	DQ Prior
	Synchronous Chip Select	V_{IL}	╱	V_{IH}	A10	A9	A8	A0	V_{IH}	V_{IH}	V_{CC}	DQ Prior
Read	Asynchronous Chip Select	V_{IL}	X	X	X	X	X	X	V_{IH}	V_{IH}	V_{CC}	Q_{OUT}
	Synchronous Chip Select	X	╱	X	X	X	X	X	V_{IH}	V_{IH}	V_{CC}	Q_{OUT}
Initialize Register Read	Asynchronous Chip Select	V_{IL}	X	X	X	X	X	X	V_{IH}	V_{IL}	V_{CC}	Q_{OUT}
	Synchronous Chip Select	V_{IL}	╱	X	X	X	X	X	V_{IH}	V_{IL}	V_{CC}	Q_{OUT}
Deselect	Asynchronous Chip Select	V_{IH}	X	X	X	X	X	X	V_{IH}	V_{IH}	V_{CC}	High Z
	Synchronous Chip Select	V_{IH}	╱	X	X	X	X	X	V_{IH}	V_{IH}	V_{CC}	High Z
Int$_e$ligent Identifier	Manufacturer	V_{IL}	V_{IL}	X	X	V_H	X	V_{IL}	V_{IH}	V_{IH}	V_{CC}	Q_{OUT} = 0089H
	Device	V_{IL}	V_{IL}	X	X	V_H	X	V_{IH}	V_{IH}	V_{IH}	V_{CC}	Q_{OUT} = 66F6H
Blank Check	Ones	V_{IL}	V_{IL}	V_{IH}	A10	A9	A8	A0	V_{IH}	V_{PP}	(Note 2)	Zeros
	Zeros	V_{IL}	V_{IL}	V_{IL}	A10	A9	A8	A0	V_{IH}	V_{PP}	(Note 2)	Ones
Program	Ones	V_{IH}	V_{IL}	V_{IH}	A10	A9	A8	A0	V_{IL}	V_{PP}	(Note 2)	D_{IN}
	Zeros	V_{IH}	V_{IL}	V_{IL}	A10	A9	A8	A0	V_{IL}	V_{PP}	(Note 2)	D_{IN}

7

27C203

PRELIMINARY

Table 1. Mode Table (Continued)

Pins Mode		$\overline{S}/\overline{S_s}/\overline{G_p}$	$C/\overline{E_p}$	pF	A10	A9	A8	A0	$\overline{P}$	$\overline{I}/V_{PP}$	V_{CC}	DQ
Program Verify	Ones	V_{IL}	V_{IL}	V_{IH}	A10	A9	A8	A0	V_{IH}	V_{PP}	(Note 2)	Ones
	Zeros	V_{IL}	V_{IL}	V_{IL}	A10	A9	A8	A0	V_{IH}	V_{PP}	(Note 2)	Zeros
Synchronous Chip Select Program		V_{IH}	V_{IL}	X	X	X	V_H	X	V_{IL}	V_{PP}	(Note 2)	High Z
Initialize Register Program		V_{IH}	V_{IL}	X	V_H	X	X	X	V_{IL}	V_{PP}	(Note 2)	D_{IN}
Initialize Register Read		V_{IL}	V_{IL}	X	X	X	X	X	V_{IH}	V_{IL}	(Note 2)	Q_{OUT}
Program Inhibit		X	V_{IH}	X	X	X	X	X	X	V_{PP}	(Note 2)	High Z

NOTES:
1. Refer to A.C. Test Points for V_{IH} and V_{IL} levels. A V_{IH} to V_{IL} transition is represented by ↘, and a V_{IL} to V_{IH} transition is represented by ↗. X represents either V_{IH} or V_{IL}. See D.C. programming characteristics for V_H and V_{PP}.
2. V_{CC} = 6.25V ±0.25V during programming.

READ MODE

The 27C203 read operation is synchronous. Data registers on the output buffers (Figure 3) allow high-bandwidth, pipelined read operations without the additional components that would normally be required. By using these data registers along with a programmable initial word feature, state machine designs can be implemented without the use of external latches.

The chip select (output enable) can be programmed to be synchronous ($\overline{Ss}$) or left in the default asynchronous state ($\overline{S}$) (Figure 3). The chip select is programmed to be synchronous in pipelined bus applications so the data output triggers off the rising edge of the clock (C) signal. In state machine applications, the chip select can be programmed or left unprogrammed, depending on the needs of the design. Pipelined and state machine applications are detailed below.

PIPELINED BUS APPLICATION (80376 OR 80386)

The 27C203 pipelined EPROM interfaces with the 80376/80386 pipelined bus when programmed in the synchronous chip select mode. Figure 4 shows a 376™ embedded processor with the 27C203 in a simple embedded system.

Valid address information must be stable for the minimum address setup time (t_{AVCH}) before the data can be shifted to the outputs. The data is loaded in the data registers on the rising edge of the 27C203's C. The same C edge latches the synchronous chip select state into the chip select register. To activate the next output, the $\overline{Ss}$ input must be low before the C rising edge. Following the rising edge of the C, the outputs become valid (t_{CHQV}).

To deselect the device in the next output state, the $\overline{Ss}$ must be high for at least the synchronous select setup time (t_{SVCH}) before the C rising edge. Follow-

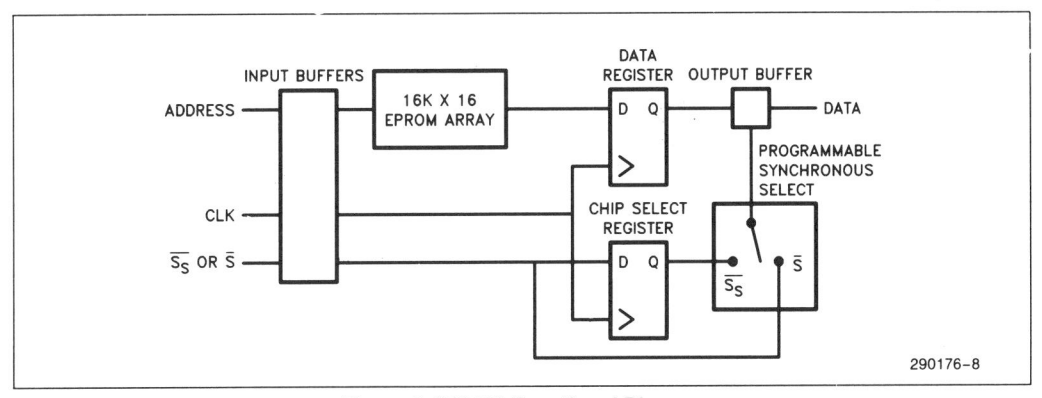

Figure 3. 27C203 Functional Diagram

intel 27C203 PRELIMINARY

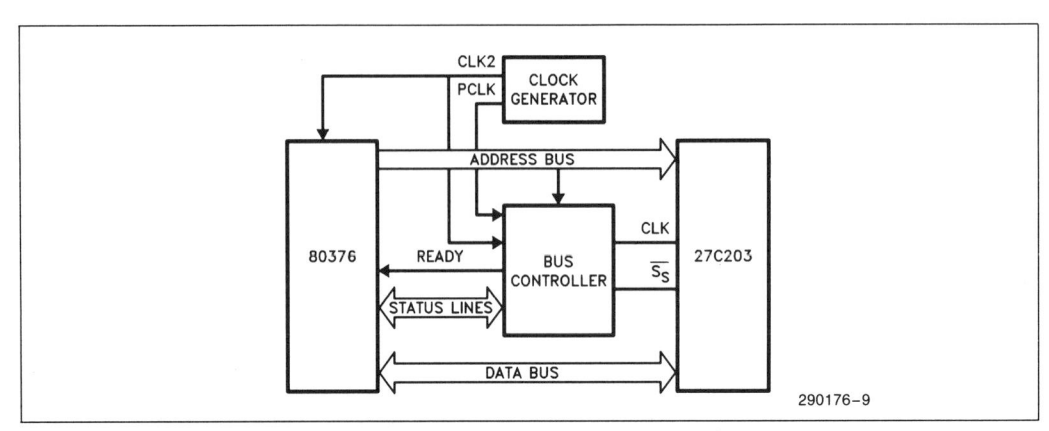

Figure 4. 27C203 in Simple 80376 Embedded System

ing the rising edge of the C, the outputs become tri-stated (t_{CHQZ}).

Buffering may be needed to avoid bus contention when different types of pipelined memories are bussed together. For common 27C203's with synchronous chip selects, the chip deselect times are sufficiently fast to minimize active output overlap. The synchronization of one device's selection with another's deselection is accomplished through common clock inputs (C must be common with multiple synchronously selected devices).

STATE MACHINE APPLICATION

A programmable initial word feature, combined with the data registers, optimize the 27C203 for state machine designs. The 16-bit initial word is active (t_{ILQV}) when the Initialize Register Enable pin ($\overline{I}$) is brought low for a minimum pulse width (t_{ILIH}). A recovery time (t_{IHCH}) is required before the next C rising edge.

The chip select can be programmed to be either synchronous ($\overline{Ss}$) or left in the default asynchronous state ($\overline{S}$), depending on the needs of the application. With asynchronous operation, data is read when chip select ($\overline{S}$) is brought low. The registered data will then become valid in the chip select access time (t_{SLQV}). To deselect the device, $\overline{S}$ must be raised high and the outputs will become tri-stated (t_{SHQZ}).

SYSTEM DESIGN CONSIDERATIONS

The 27C203 is a high performance memory with exceptionally strong output drive capability for fast output response. Input levels are sensitive to power supply stability. Special considerations must be given to the large supply current swings associated with switching 16 powerful output drivers. Minimizing power supply inductance is critical. Wide, short traces with low inductance must be connected from the circuit board's ground and V_{CC} plane to each device. V_{CC} transients must be suppressed with high-frequency 0.1 μF capacitors. Where the solid ground plane of a multilayer board is not available, bulk electrolytic capacitors (typically 4.7μF) should decouple the V_{CC} and the ground supplies for each group of four 27C203 devices.

The 30pF load capacitance specified in the A.C. Test Conditions is not a system design limitation. The 27C203 has the output drive capability to handle the capacitive loading of large memory arrays. However, access times must be appropriately derated for loading in excess of 30pF. t_{CHQV}, t_{SLQV} and t_{ILQV} should be derated 6ns/100pF with output loading greater than 30pF.

PROGRAMMING MODE

Caution: Exceeding 14V on V_{PP} will permanently damage the device.

To minimize data access time delays, the 27C203 utilizes a 2-transistor cell with differential sensing. The 2-transistor cells are partitioned into "ONES" and "ZEROS" arrays. Programming is done in two passes; once in the "ONES" half of the memory array and once in the "ZEROS" half. The same data word must be presented to the device in both passes. A program function (PF) pin is provided to control this partitioning (see Figure 5).

When V_{PP} and V_{CC} are raised to their programming voltage (Table 2), the 27C203 becomes asynchronous and the programming mode is entered. Chip Select ($\overline{S}/\overline{Ss}$) becomes a programming output enable ($\overline{Gp}$), and Clock (C) becomes a chip enable ($\overline{Ep}$). The programming pin ($\overline{P}$) is brought to the TTL-low level to activate the EPROM programming pulse while the programming output enable ($\overline{Gp}$) is held at a TTL-high level.

Outputs are active whenever $\overline{Gp}$ and $\overline{Ep}$ are low. $\overline{Ep}$ does not need to be toggled as it does while serving the read mode Clock function. This facilitates gang programming by allowing $\overline{Gp}$ to be a common verify control line while $\overline{Ep}$ controls individual device selection, eliminating bus contention.

PROGRAM INHIBIT

Programming of multiple EPROMs in parallel with different data is easily accomplished by using the Program Inhibit mode. A high-level $\overline{P}$ or $\overline{Ep}$ prevents devices from being programmed.

Except for $\overline{Ep}$, all like inputs of the parallel EPROMs may be common. A TTL low-level pulse applied to the $\overline{P}$ input, with V_{PP} at programming voltage, $\overline{Ep}$ at a TTL-low level and $\overline{G}_p$ at V_{IH} will program the selected device.

PROGRAM VERIFY/BLANK CHECK

The program verify mode is activated when the programming output enable pin ($\overline{Gp}$) is at a TTL-low level, V_{PP} and V_{CC} are at their programming levels, and the program control line ($\overline{P}$) is at a TTL-high level. Blank Check individually confirms unprogrammed status of both arrays (**a blank check cannot be done with a normal read operation with the 2-transistor cell**). Program verify is used to check programmed status. The elevated V_{CC} voltage used during Program Verify ensures high programming margins and long-term data retention with maximum noise immunity.

The program/verify cycle begins with the "ONES" array; the Program Function pin (PF) is held at a TTL-high level. The "ONES" programming is completed when the desired "ONES" bits are verified changed from their unprogrammed state ("ZEROS")

Programming is not complete, however, until the "ZEROS" array program/verify cycle is also finished. The "ZEROS" cycle begins when PF is brought to a TTL-low level. The "ZEROS" programming is completed when the desired "ZEROS" bits are verified changed from their unprogrammed state ("ONES").

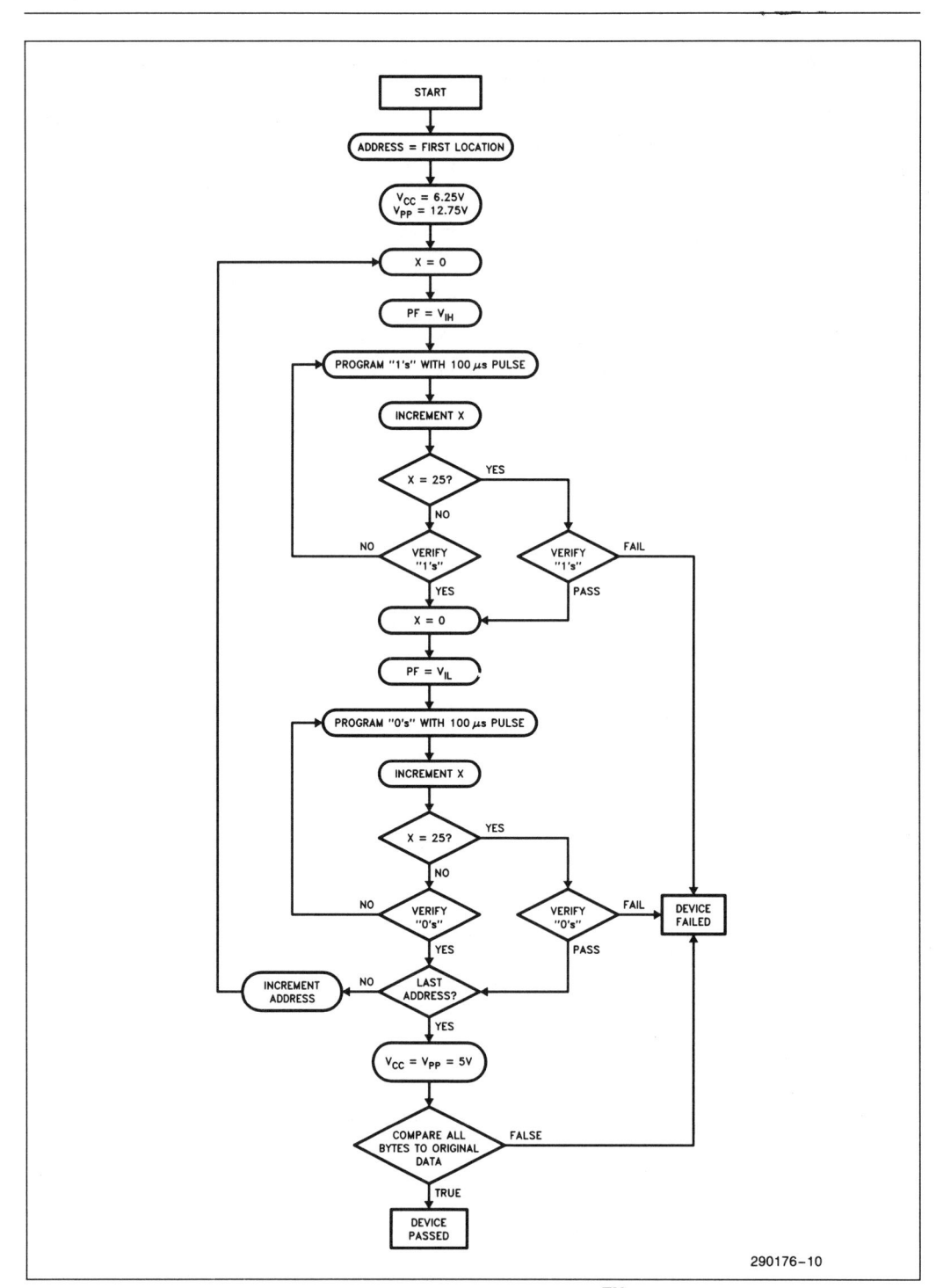

Figure 5. 27C203 Quick-Pulse Programming™ Algorithm

Quick-Pulse Programming™ Operations

The Quick-Pulse Programming algorithm is used to program Intel's 27C203. Developed to substantially reduce production programming throughput time, this algorithm can program a 27C203 in under four seconds. Actual programming time depends on the PROM programmer used.

The Quick-Pulse Programming algorithm uses a 100 microsecond initial-pulse followed by a word verification to determine when the addressed word is correctly programmed. The algorithm terminates if 25 100 μs pulses fail to program a word. This is repeated for both the "ONES" and "ZEROS" programming. Figure 5 shows the 27C203 Quick-Pulse Programming algorithm flowchart.

The entire program-pulse/word-verify sequence is performed with V_{CC} = 6.25V and V_{PP} = 12.75V. When programming is complete, all words should be compared to the original data in a standard read mode with V_{CC} = 5.0V.

int$_e$ligent Identifier™ Mode

The int$_e$ligent Identifier Mode will determine an EPROM's manufacturer and device type. Programming equipment can automatically match a device with its proper programming algorithm.

The Identifier mode is activated when A9 is at high voltage (V_H). A0 determines the information being accessed. The first address (A0 = V_{IL}) accesses the manufacturer code and the second (A0 = V_{IH}) accesses the device code.

SPECIAL FEATURE PROGRAMMING

The Chip Select synchronous state ($\overline{Ss}$) and the initialize register contents are both UV-erasable and user-programmable. Programming operations for these special 27C203 features are described below and waveforms are shown in Figure 7.

SYNCHRONOUS CHIP SELECT PROGRAMMING

The chip select pin's default (erased) state is asynchronous ($\overline{S}$). The synchronous state ($\overline{Ss}$) is programmed by raising A8 to V_H (Figure 7) for twenty five 100 microsecond program pulses. Program verify is accomplished with a functionality check; when $\overline{Ss}$ is programmed, the state of the outputs, valid or tri-state, can only be changed with a clock (C) rising edge.

INITIALIZE REGISTER CONTENTS PROGRAMMING

When the initialize register enable pin is brought low, the initialize register contents override the output generated by normal read oprations. The contents are all "ZEROS" in the erased state. To program the register, A10 is raised to V_H (Figure 7). The remaining setup and programming parameters are the same as the program "ONES" mode, with the data inputs set to the desired initial-condition-vector data. The initialize register read mode or the initialize register verify mode allow programming verification.

Erasure Characteristics (for UV-Window Packages)

Light with wavelengths shorter than 4000 Angstroms (Å) causes EPROM erasure. Sunlight and some florescent lamps have wavelengths in the 3000Å–4000Å range. Constant exposure to room-level florescent light can erase an EPROM in about three years while direct sunlight erasure can occur within one week. Covering windowed EPROMs with opaque labels prevents such unintended erasure.

The recommended erasure procedure is to expose the EPROM to a 2537Å ultraviolet (UV) source for a minimum integrated dose (intensity x exposure time) of 15 W-sec/cm^2. The EPROM should be placed within one inch of the light source and will erase in 15-20 minutes with a typical 12000 μW/cm^2 lamp.

Overexposure to high-intensity UV light can cause permanent device damage. The maximum integrated dose allowed is 7258 W-sec/cm^2 (approximately 1 week at 12000 μW/cm).

intel 27C203 PRELIMINARY

D.C. Programming Characteristics $T_A = 25°C \pm 5°C$

Table 2

Symbol	Parameter	Limits Min	Limits Max	Units	Conditions
I_{IL}	Input Low Load Current		10	μA	$V_{IN} = 0V$
I_{IH}	Input High Load Current		10	μA	$V_{IN} = 6V$
I_{OZL}	Tri-state Low Leakage		10	μA	$V_{OUT} = 0V$
I_{OZH}	Tri-state High Leakage		15	μA	$V_{OUT} = 6V$
I_{CC} (1)	V_{CC} Current Active		100	mA	$\overline{Ep} = \overline{P} = V_{IL}$
V_{IL}	Input Low Voltage	−0.1	+0.8	V	
V_{IH}	Input High Voltage	2.0	$V_{CC} + 1$	V	
V_{OL}	Verify Output Low Volt.		0.45	V	$I_{OL} = 16$ mA
V_{OH}	Verify Output High Volt.	2.4		V	$I_{OH} = -4$ mA
I_{PP}	V_{PP} Supply Current		50	mA	$\overline{Ep} = \overline{P} = V_{IL}$
V_H	Input High Voltage	11.5	12.5	V	$V_{CC} = 5V$
V_{PP}	V_{PP} Supply Voltage	12.5	13.0	V	
V_{CC}	V_{CC} Supply Voltage	6.0	6.5	V	

A.C. Programming Characteristics $T_A = 25°C \pm 5°C$

Symbol	Parameter	Limits Min	Limits Max	Units	Conditions
t_{AVPL}	Address Setup Time	1		μs	
t_{DZGL}	$\overline{Gp}$ Setup Time	1		μs	
t_{DVPL}	Data Setup Time			μs	
t_{FVPL}	PF Setup Time	1		μs	
t_{GHAX}	Address Hold Time	0		μs	
*t_{PHDX}	Data Hold Time	1		μs	
t_{GHQX}	Output Deselect Time	0	50	ns	(Note 2)
t_{VPS}	V_{PP} Setup time	1		μs	
t_{VCS}	V_{CC} Setup Time	1		μs	
t_{ELPL}	$\overline{Ep}$ Setup Time	1		μs	
t_{AHPL}	Special Functions Address Setup Time	1		μs	A8 or A10 = V_H
t_{PLPH}	Program ($\overline{P}$) Pulse Width	95	105	μs	Quick-Pulse™ Programming
t_{GLQV}	Data Valid from $\overline{OE}$		50	μs	
t_{AVQV}	Intelligent Identifier Valid from Address Valid		100	ns	$A_9 = V_H$ Other A's = V_{IL}
t_{ILQV}	Initialize Register Verify from $\overline{I}$ Low		100	ns	$A_{10} = V_H$
t_{PLPH2}	Synchronous Chip Select Program Pulse Width	95	105	μs	$A_8 = V_H$

NOTES:
1. The maximum current value is with outpus DQ0–DQ15 unloaded.
2. This parameter is only sampled and is not 100% tested. Output float is defined as the point where data is no longer driven. See timing diagram.
3. V_{CC} must be applied simultaneously or before V_{PP} and removed simultaneously or after V_{PP}.

*AC Conditions of Test
Input Rise and Fall Times (10% to 90%) . . . . 20 ns
Input Pulse Levels . . . . 0.45V to 2.4V
Input Timing Reference Level . . . . 0.8V to 2.0V
Output Timing Reference Level . . . . 0.8V to 2.0V

intel 27C203 PRELIMINARY

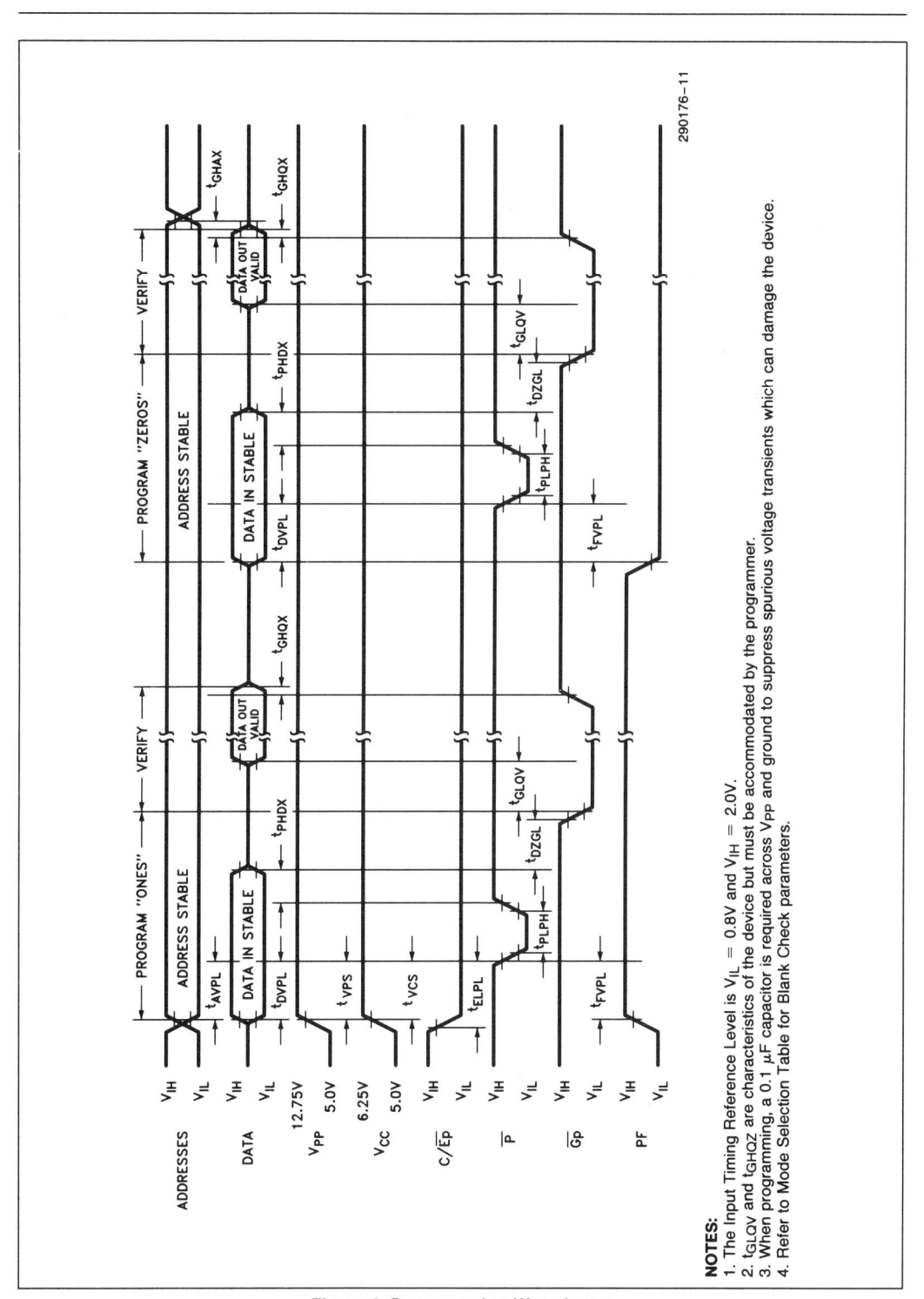

Figure 6. Programming Waveforms

NOTES:
1. The Input Timing Reference Level is $V_{IL} = 0.8V$ and $V_{IH} = 2.0V$.
2. t_{GLQV} and t_{GHQZ} are characteristics of the device but must be accommodated by the programmer.
3. When programming, a 0.1 μF capacitor is required across Vpp and ground to suppress spurious voltage transients which can damage the device.
4. Refer to Mode Selection Table for Blank Check parameters.

intel

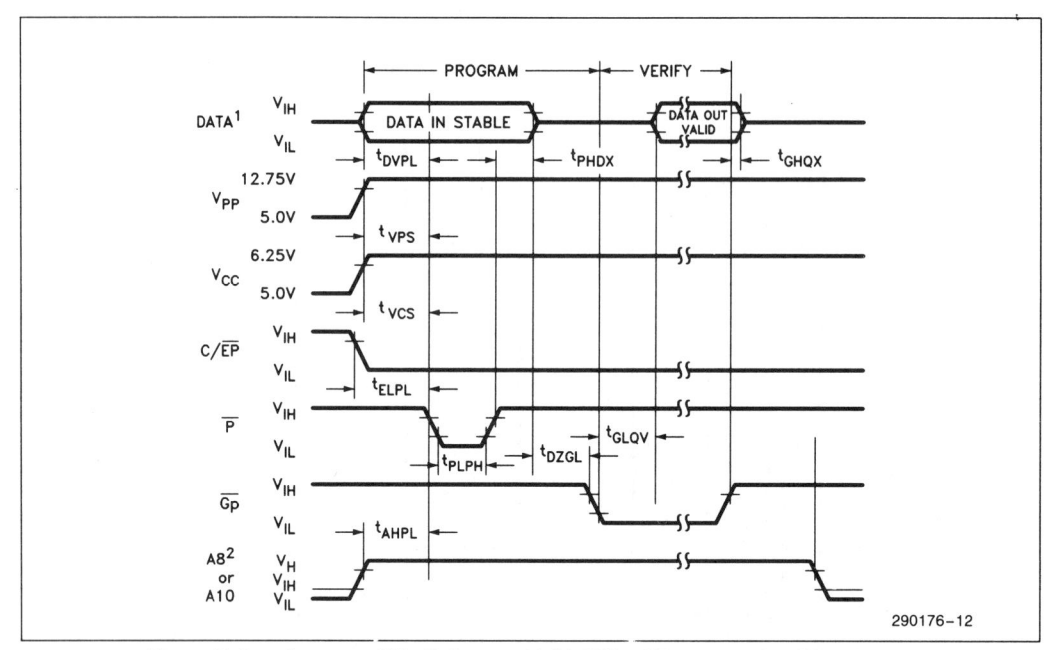

27C203

PRELIMINARY

Figure 7. Synchronous Chip Select and Initial Word Programming Waveforms

NOTES:
1. Data waveform applies to initial word programming function only.
2. Waveform applies to A8 when programming the Synchronous Chip Select function. A10 programs the Initialize register.

EUROPEAN SALES OFFICES

DENMARK
Intel
Glentevej 61, 3rd Floor
2400 Copenhagen NV
Tel (45) (31) 19 80 33
Telex 19567

FINLAND
Intel
Ruosilantie 2
00390 Helsinki
Tel (358) 0 544 644
Telex 123332

FRANCE
Intel
1, rue Edison - BP 303
78054 St Quentin-en-Yvelines Cedex
Tel (33) (1) 30 57 70 00
Telex 699016

NETHERLANDS
Intel
Postbus 84130
3099 CC Rotterdam
Tel (31) 10 407 11 11
Telex 22283

ISRAEL
Intel
Atidim Industrial Park-Neve Sharet
P.O. Box 43202
Tel-Aviv 61430
Tel (972) 03-498080
Telex 371215

ITALY
Intel
Milanofiori Palazzo E
20090 Assago
Milano
Tel (39) (02) 89200950
Telex 341286

NORWAY
Intel
Hvamveien 4 - PO Box 92
2013 Skjetten
Tel (47) (6) 842 420
Telex 78018

SPAIN
Intel
Zurbaran, 28
28010 Madrid
Tel (34) 308 25 52
Telex 46880

SWEDEN
Intel
Dalvagen 24
171 36 Solna
Tel (46) 8 734 01 00
Telex 12261

SWITZERLAND
Intel
Zuerichstrasse
8185 Winkel-Rueti bei Zuerich
Tel (41) 01/860 62 62
Telex 825977

U.K.
Intel
Pipers Way
Swindon, Wilts SN3 1RJ
Tel (44) (0793) 696000
Telex 444447/8

WEST GERMANY
Intel
Dornacher Strasse 1
8016 Feldkirchen bei Muenchen
Tel (49) 089/90992-0
Telex 5-23177

Intel
Hohenzollern Strasse 5
3000 Hannover 1
Tel (49) 0511/344081
Telex 9-23625

Intel
Abraham Lincoln Strasse 16-18
6200 Wiesbaden
Tel (49) 06121/7605-0
Telex 4-186183

Intel
Zettachring 10A
7000 Stuttgart 80
Tel (49) 0711/7287-280
Telex 7-254826

EUROPEAN DISTRIBUTORS/REPRESENTATIVES

AUSTRIA
Bacher Electronics GmbH
Rotenmuehlgasse 26
1120 Wien
Tel (43) (0222) 83 56 46
Telex 31532

BELGIUM
Inelco Belgium S A
Av. des Croix de Guerre 94
1120 Bruxelles
Oorlogskruisenlaan, 94
1120 Brussel
Tel (32) (02) 216 01 60
Telex 64475 or 22090

DENMARK
ITT-Multikomponent
Naverland 29
2600 Glostrup
Tel (45) (0) 2 45 66 45
Telex 33 355

FINLAND
OY Fintronic AB
Melkonkatu 24A
00210 Helsinki
Tel (358) (0) 6926022
Telex 124224

FRANCE
Almex
Zone industrielle d'Antony
48, rue de l'Aubépine
BP 102
92164 Antony cedex
Tel (33) (1) 46 66 21 12
Telex 250067

Jermyn-Generim
60, rue des Gémeaux
Silic 580
94653 Rungis cedex
Tel (33) (1) 49 78 49 78
Telex 206967

Métrologie
Tour d'Asnières
4, av. Laurent-Cely
92606 Asnières Cedex
Tel (33) (1) 47 90 62 40
Telex 611448

Tekelec-Airtronic
Cité des Bruyères
Rue Carle-Vernet — BP 2
92310 Sèvres
Tel (33) (1) 45 34 75 35
Telex 204552

IRELAND
Micro Marketing Ltd
Glenageary Office Park
Glenageary
Co Dublin
Tel (21)(353) (01) 85 63 25
Telex 31584

ISRAEL
Eastronics Ltd
11 Rozanis St
P.O.B 39300
Tel-Aviv 61392
Tel (972) 03-475151
Telex 33638

ITALY
Intesi
Divisione ITT Industries GmbH
Viale Milanofiori
Palazzo E/5
20090 Assago (MI)
Tel (39) 02/824701
Telex 311351

Lasi Elettronica S.p.A
V. le Fulvio Testi, 126
20092 Cinisello Balsamo (MI)
Tel (39) 02/2440012
Telex 352040

ITT Multicomponents
Viale Milanofiori E/5
20090 Assago (MI)
Tel (39) 02/824701
Telex 311351

Silverstar
Via Dei Gracchi 20
20146 Milano
Tel (39) 02/49961
Telex 332189

NETHERLANDS
Koning en Hartman Elektrotechniek B V
Energieweg 1
2627 AP Delft
Tel (31) (01) 15/609906
Telex 38250

NORWAY
Nordisk Elektronikk (Norge) A/S
Postboks 123
Smedsvingen 4
1364 Hvalstad
Tel (47) (02) 84 62 10
Telex 77546

PORTUGAL
ATD Portugal LDA
Rua Dos Lusiados, 5
5 Sala B
1300 Lisboa
Tel (35) (1) 648 091

Ditram
Avenida Miguel Bombarda, 133
1000 Lisboa
Tel (35) (1) 734 884
Telex 14182

SPAIN
ATD Electronica, S A
Plaza Ciudad de Viena, 6
28040 Madrid
Tel (34) (1) 234 40 00
Telex 42754

ITT-SESA
Calle Miguel Angel, 21 3
28010 Madrid
Tel (34) (1) 419 54 00
Telex 27461

Metrologia Iberica, S A
Ctra. de Fuencarral, n 80
28100 Alcobendas (Madrid)
Tel (34) (1) 653 86 11

SWEDEN
Nordisk Elektronik AB
Huvudstagatan 1
Box 1409
171 27 Solna
Tel (46) 08-734 97 70
Telex 105 47

SWITZERLAND
Industrade A G
Hertistrasse 31
8304 Wallisellen
Tel (41) (01) 8328111
Telex 56788

TURKEY
EMPA Electronic
Lindwurmstrasse 95A
8000 Muenchen 2
Tel (49) 089/53 80 570
Telex 528573

UNITED KINGDOM
Accent Electronic Components Ltd
Jubilee House
Jubilee Road
Letchworth
Herts SG6 1TL
Tel (44) (0462) 686666
Telex 826293

Bytech-Comway Systems
3 The Western Centre
Western Road
Bracknell RG12 1RW
Tel (44) (0344) 55333
Telex 847201

Jermyn
Vestry Estate
Otford Road
Sevenoaks
Kent TN14 5EU
Tel (44) (0732) 450144
Telex 95142

MMD
Unit 8 Southview Park
Caversham
Reading
Berkshire RG4 0AF
Tel (44) (0734) 481666
Telex 846669

Rapid Systems
Rapid House
Denmark Street
High Wycombe
Buckinghamshire HP11 2ER
Tel (44) (0494) 450244
Telex 837931

Rapid Silicon
Rapid House
Denmark Street
High Wycombe
Buckinghamshire HP11 2ER
Tel (44) (0494) 442266
Telex 837931

WEST GERMANY
Electronic 2000 AG
Stahlgruberring 12
8000 Muenchen 82
Tel (49) 089/42001-0
Telex 522561

ITT Multikomponent GmbH
Postfach 1265
Bahnhofstrasse 44
7141 Moeglingen
Tel (49) 07141/4879
Telex 7264472

Jermyn GmbH
Im Dachsstueck 9
6250 Limburg
Tel (49) 06431/508-0
Telex 415257-0

Metrologie GmbH
Meglingerstrasse 49
8000 Muenchen 71
Tel (49) 089/78042-0
Telex 5213189

Proelectron Vertriebs GmbH
Max Planck Strasse 1-3
6072 Dreieich
Tel (49) 06103/30434 3
Telex 417903

YUGOSLAVIA
H R Microelectronics Corp
2005 de la Cruz Blvd, Ste 223
Santa Clara, CA 95050
USA
Tel (1) (408) 988-0286
Telex 387452

Rapido Electronic Components, S P A
Via C Beccaria, 8
34133 Trieste
Italia
Tel (39) 040/360655
Telex 460461

Eurolit 05/89

intel

ADVANCE INFORMATION

28F020
2048K (256K x 8) CMOS FLASH MEMORY

- **Flash Electrical Chip-Erase**
  — 2 Second Typical Chip-Erase
- **Quick-Pulse Programming™ Algorithm**
  — 10 μs Typical Byte-Program
  — 4 Second Chip-Program
- **10,000 Erase/Program Cycles Minimum**
- **12.0V $\pm$5% V_{PP}**
- **High-Performance Read**
  — 150 ns Maximum Access Time
- **CMOS Low Power Consumption**
  — 10 mA Typical Active Current
  — 50 μA Typical Standby Current
  — 0 Watts Data Retention Power
- **Command Register Architecture for Microprocessor/Microcontroller Compatible Write Interface**
- **Noise Immunity Features**
  — $\pm$10% V_{CC} Tolerance
  — Maximum Latch-Up Immunity through EPI Processing
- **ETOX™ II Nonvolatile Flash Technology**
  — EPROM-Compatible Process Base
  — High-Volume Manufacturing Experience
- **JEDEC-Standard Pinouts**
  — 32-Pin Plastic Dip
  — 32-Lead PLCC
  — 32-Lead TSOP
  (See Packaging Spec., Order #231369)
- **Integrated Program/Erase Stop Timer**

Intel's 28F020 CMOS flash memory offers the most cost-effective and reliable alternative for read/write random access nonvolatile memory. The 28F020 adds electrical chip-erasure and reprogramming to familiar EPROM technology. Memory contents can be rewritten: in a test socket; in a PROM-programmer socket; on-board during subassembly test; in-system during final test; and in-system after-sale. The 28F020 increases memory flexibility, while contributing to time- and cost-savings.

The 28F020 is a 2048-kilobit nonvolatile memory organized as 262,144 bytes of 8 bits. Intel's 28F020 is offered in 32-pin plastic DIP, 32-lead PLCC, and 32-lead TSOP packages. Pin assignments conform to JEDEC standards for byte-wide EPROMs.

Extended erase and program cycling capability is designed into Intel's ETOX™ II (EPROM Tunnel Oxide) process technology. Advanced oxide processing, an optimized tunneling structure, and lower electric field combine to extend reliable cycling beyond that of traditional EEPROMs. With the 12.0V V_{PP} supply, the 28F020 performs a minimum of 10,000 erase and program cycles well within the time limits of the Quick-Pulse Programming™ and Quick-Erase™ algorithms.

Intel's 28F020 employs advanced CMOS circuitry for systems requiring high-performance access speeds, low power consumption, and immunity to noise. Its 150 nanosecond access time provides no-WAIT-state performance for a wide range of microprocessors and microcontrollers. Maximum standby current of 100 μA translates into power savings when the device is deselected. Finally, the highest degree of latch-up protection is achieved through Intel's unique EPI processing. Prevention of latch-up is provided for stresses up to 100 mA on address and data pins, from $-1V$ to $V_{CC} + 1V$.

With Intel's ETOX II process base, the 28F020 levers years of EPROM experience to yield the highest levels of quality, reliability, and cost-effectiveness.

Intel Corporation assumes no responsibility for the use of any circuitry other than circuitry embodied in an Intel product. No other circuit patent licenses are implied. Information contained herein supersedes previously published specifications on these devices from Intel. **May 1990**
© INTEL CORPORATION, 1990
Order Number: 290245-002

intel 28F020   ADVANCE INFORMATION

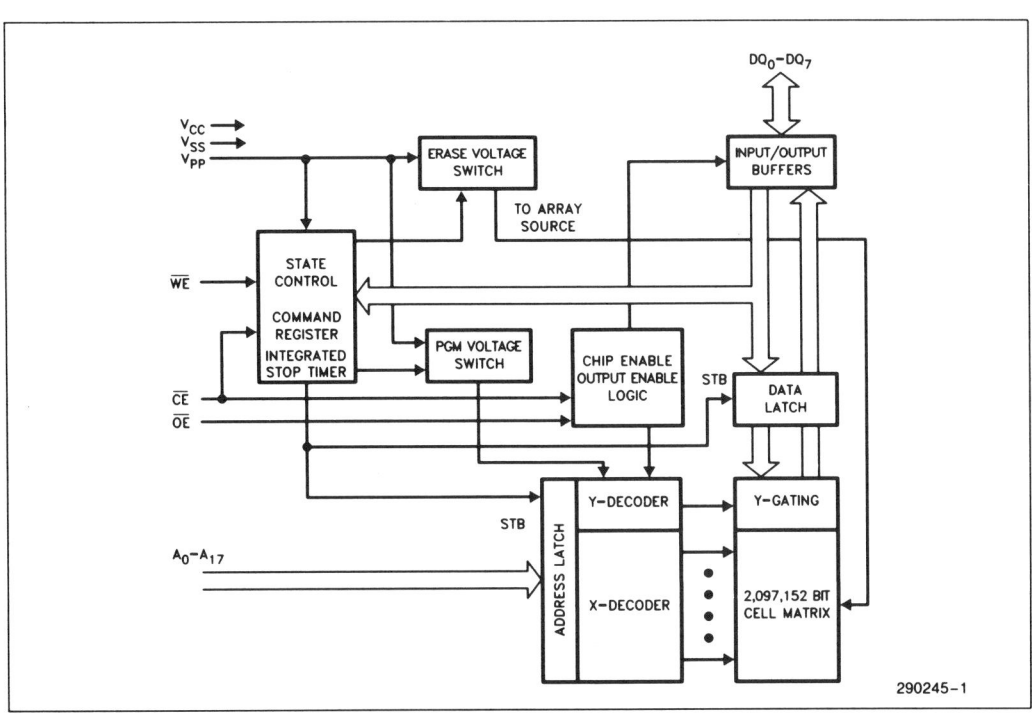

Figure 1. 28F020 Block Diagram

Table 1. Pin Description

Symbol	Type	Name and Function
A_0–A_{17}	INPUT	**ADDRESS INPUTS** for memory addresses. Addresses are internally latched during a write cycle.
DQ_0–DQ_7	INPUT/OUTPUT	**DATA INPUT/OUTPUT:** Inputs data during memory write cycles; outputs data during memory read cycles. The data pins are active high and float to tri-state OFF when the chip is deselected or the outputs are disabled. Data is internally latched during a write cycle.
$\overline{CE}$	INPUT	**CHIP ENABLE:** Activates the device's control logic, input buffers, decoders and sense amplifiers. $\overline{CE}$ is active low; $\overline{CE}$ high deselects the memory device and reduces power consumption to standby levels.
$\overline{OE}$	INPUT	**OUTPUT ENABLE:** Gates the devices output through the data buffers during a read cycle. $\overline{OE}$ is active low.
$\overline{WE}$	INPUT	**WRITE ENABLE:** Controls writes to the control register and the array. Write enable is active low. Addresses are latched on the falling edge and data is latched on the rising edge of the $\overline{WE}$ pulse. **Note:** With $V_{PP} \leq 6.5V$, memory contents cannot be altered.
V_{PP}		**ERASE/PROGRAM POWER SUPPLY** for writing the command register, erasing the entire array, or programming bytes in the array.
V_{CC}		**DEVICE POWER SUPPLY** (5V ±10%)
V_{SS}		**GROUND**

intel

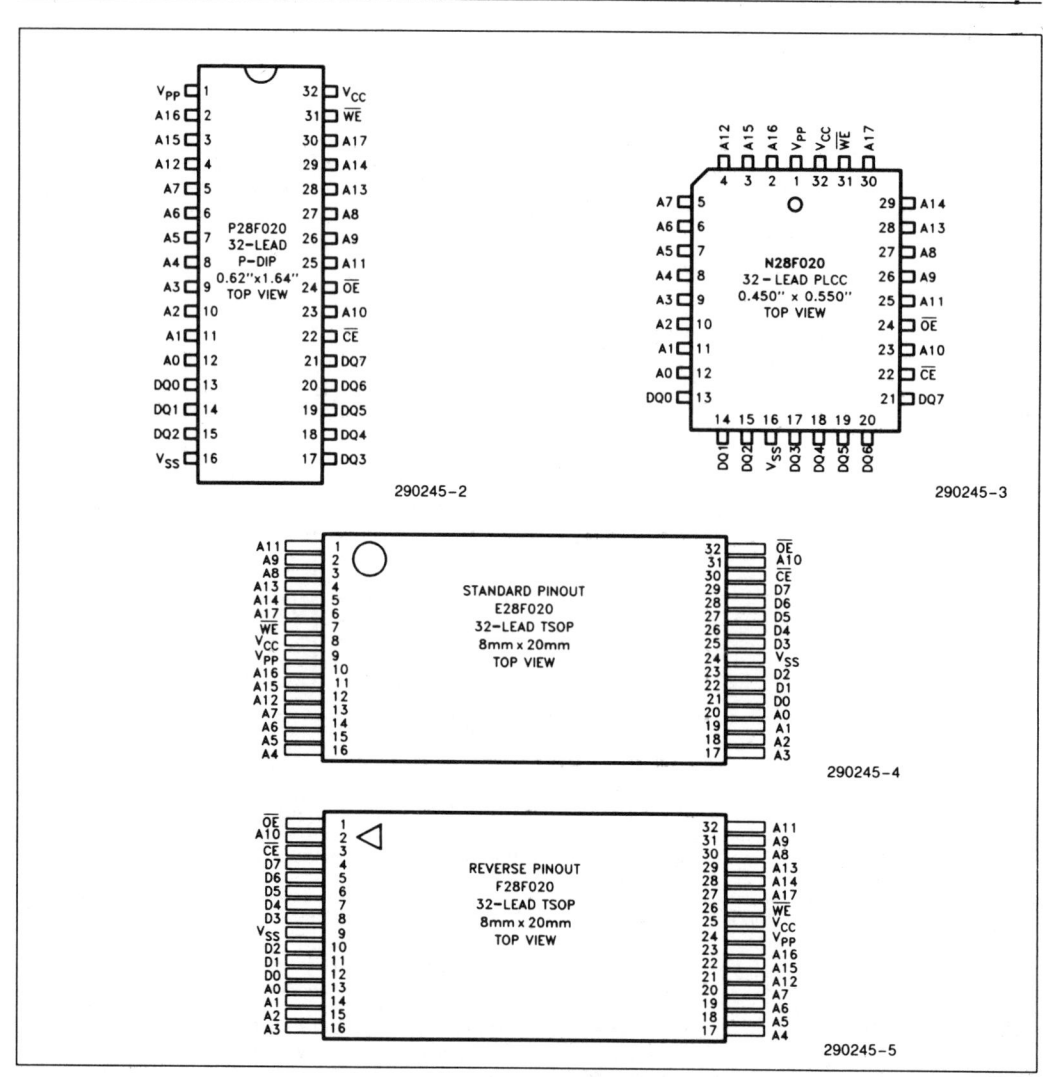

28F020 ADVANCE INFORMATION

Figure 2. 28F020 Pin Configurations

APPLICATIONS

The 28F020 flash memory provides nonvolatility along with the capability to typically perform over 100,000 electrical chip-erasure/reprogram cycles. These features make the 28F020 an innovative alternative to disk, EEPROM, and battery-backed static RAM. Where periodic updates of code and datatables are required, the 28F020's reprogrammability and nonvolatility make it the obvious and ideal replacement for EPROM.

Primary applications and operating systems stored in flash eliminate the slow disk-to-DRAM download process. This results in dramatic enhancement of performance and substantial reduction of power consumption — a consideration particularly important in portable equipment. Flash memory increases flexibility with electrical chip erasure and in-system update capability of operating systems and application code. With updatable BIOS, system manufacturers can easily accommodate last-minute changes as revisions are made.

In diskless workstations and terminals, network traffic reduces to a minimum and systems are instant-on. Reliability exceeds that of electromechanical media. Often in these environments, power interruptions force extended re-boot periods for all networked terminals. This mishap is no longer an issue if boot code, operating systems, communication protocols and primary applications are flash-resident in each terminal.

For embedded systems that rely on dynamic RAM/disk for main system memory or nonvolatile backup storage, the 28F020 flash memory offers a solid state alternative in a minimal form factor. The 28F020 provides higher performance, lower power consumption, instant-on capability, and allows an "execute in place" memory hierarchy for code and data table reading. Additionally, the flash memory is more rugged and reliable in harsh environments where extreme temperatures and shock can cause disk-based systems to fail.

The need for code updates pervades all phases of a system's life — from prototyping to system manufacture to after-sale service. The electrical chip-erasure and reprogramming ability of the 28F020 allows in-circuit alterability; this eliminates unnecessary handling and less-reliable socketed connections, while adding greater test, manufacture, and update flexibility.

Material and labor costs associated with code changes increases at higher levels of system integration — the most costly being code updates after sale. Code "bugs", or the desire to augment system functionality, prompt after-sale code updates. Field revisions to EPROM-based code requires the removal of EPROM components or entire boards. With the 28F020, code updates are implemented locally via an edge-connector, or remotely over a communcations link.

For systems currently using a high-density static RAM/battery configuration for data accumulation, flash memory's inherent nonvolatility eliminates the need for battery backup. The concern for battery failure no longer exists, an important consideration for portable equipment and medical instruments, both requiring continuous performance. In addition, flash memory offers a considerable cost advantage over static RAM.

Flash memory's electrical chip erasure, byte programmability and complete nonvolatility fit well with data accumulation and recording needs. Electrical chip-erasure gives the designer a "blank slate" in which to log or record data. Data can be periodically off-loaded for analysis and the flash memory erased producing a new "blank slate".

A high degree of on-chip feature integration simplifies memory-to-processor interfacing. Figure 3 depicts two 28F020s tied to the 80C186 system bus. The 28F020's architecture minimizes interface circuitry needed for complete in-circuit updates of memory contents.

The outstanding feature of the TSOP (Thin Small Outline Package) is the 1.2 mm thickness. With standard and reverse pin configurations, TSOP reduces the number of board layers and overall volume necessary to layout multiple 28F020s. TSOP is particularly suited for portable equipment and applications requiring large amounts of flash memory. Figure 4 illustrates the TSOP Serpentine layout.

With cost-effective in-system reprogramming, extended cycling capability, and true nonvolatility, the 28F020 offers advantages to the alternatives: EPROMs, EEPROMs, battery backed static RAM, or disk. EPROM-compatible read specifications, straight-forward interfacing, and in-circuit alterability offers designers unlimited flexibility to meet the high standards of today's designs.

intel 28F020 ADVANCE INFORMATION

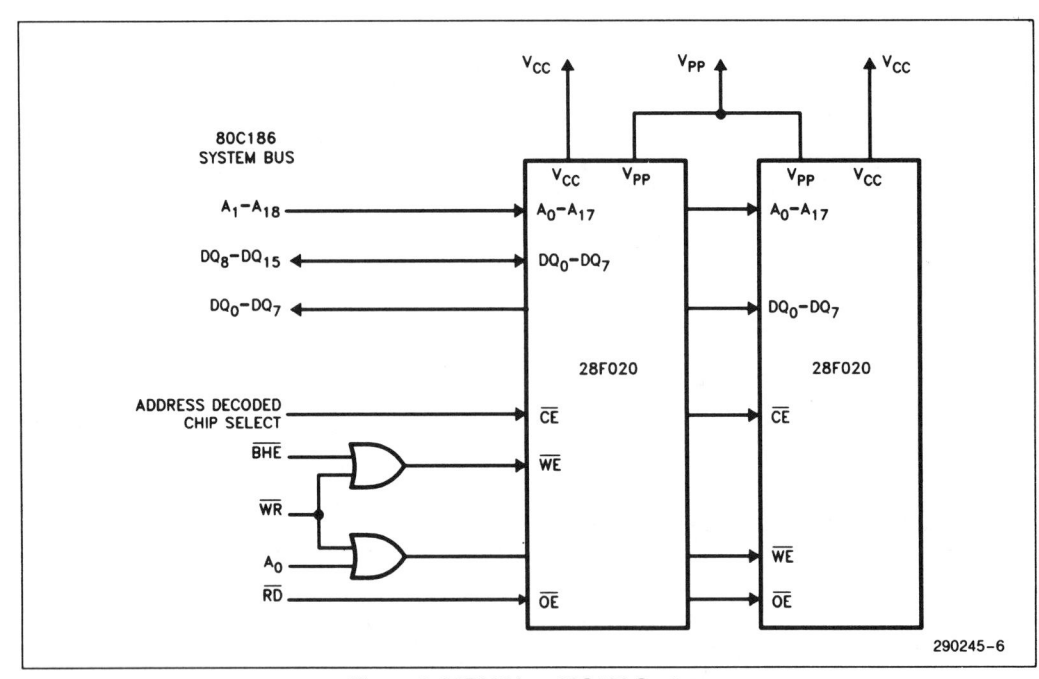

Figure 3. 28F020 in a 80C186 System

PRINCIPLES OF OPERATION

Flash-memory augments EPROM functionality with in-circuit electrical erasure and reprogramming. The 28F020 introduces a command register to manage this new functionality. The command register allows for: 100% TTL-level control inputs; fixed power supplies during erasure and programming; and maximum EPROM compatibility.

In the absence of high voltage on the V_{PP} pin, the 28F020 is a read-only memory. Manipulation of the external memory-control pins yields the standard EPROM read, standby, output disable, and int_eligent Identifier™ operations.

The same EPROM read, standby, and output disable operations are available when high voltage is applied to the V_{PP} pin. In addition, high voltage on V_{PP} enables erasure and programming of the device. All functions associated with altering memory contents—int_eligent Identifier, erase, erase verify, program, and program verify—are accessed via the command register.

Commands are written to the register using standard microprocessor write timings. Register contents serve as input to an internal state-machine which controls the erase and programming circuitry. Write cycles also internally latch addresses and data needed for programming or erase operations. With the appropriate command written to the register, standard microprocessor read timings output array data, access the int_eligent Identifier codes, or output data for erase and program verification.

Integrated Stop Timer

Successive command write cycles define the durations of program and erase operations; specifically, the program or erase time durations are normally terminated by associated program or erase verify commands. An integrated stop timer provides simplified timing control over these operations; thus eliminating the need for maximum program/erase timing specifications. Programming and erase pulse durations are minimums only. When the stop timer terminates a program or erase operation, the device enters an inactive state and remains inactive until receiving the appropriate verify or reset command.

Write Protection

The command register is only active when V_{PP} is at high voltage. Depending upon the application, the system designer may choose to make the V_{PP} power supply switchable—available only when memory updates are desired. When $V_{PP} = V_{PPL}$, the contents of the register default to the read command, making the 28F020 a read-only memory. In this mode, the memory contents cannot be altered.

intel

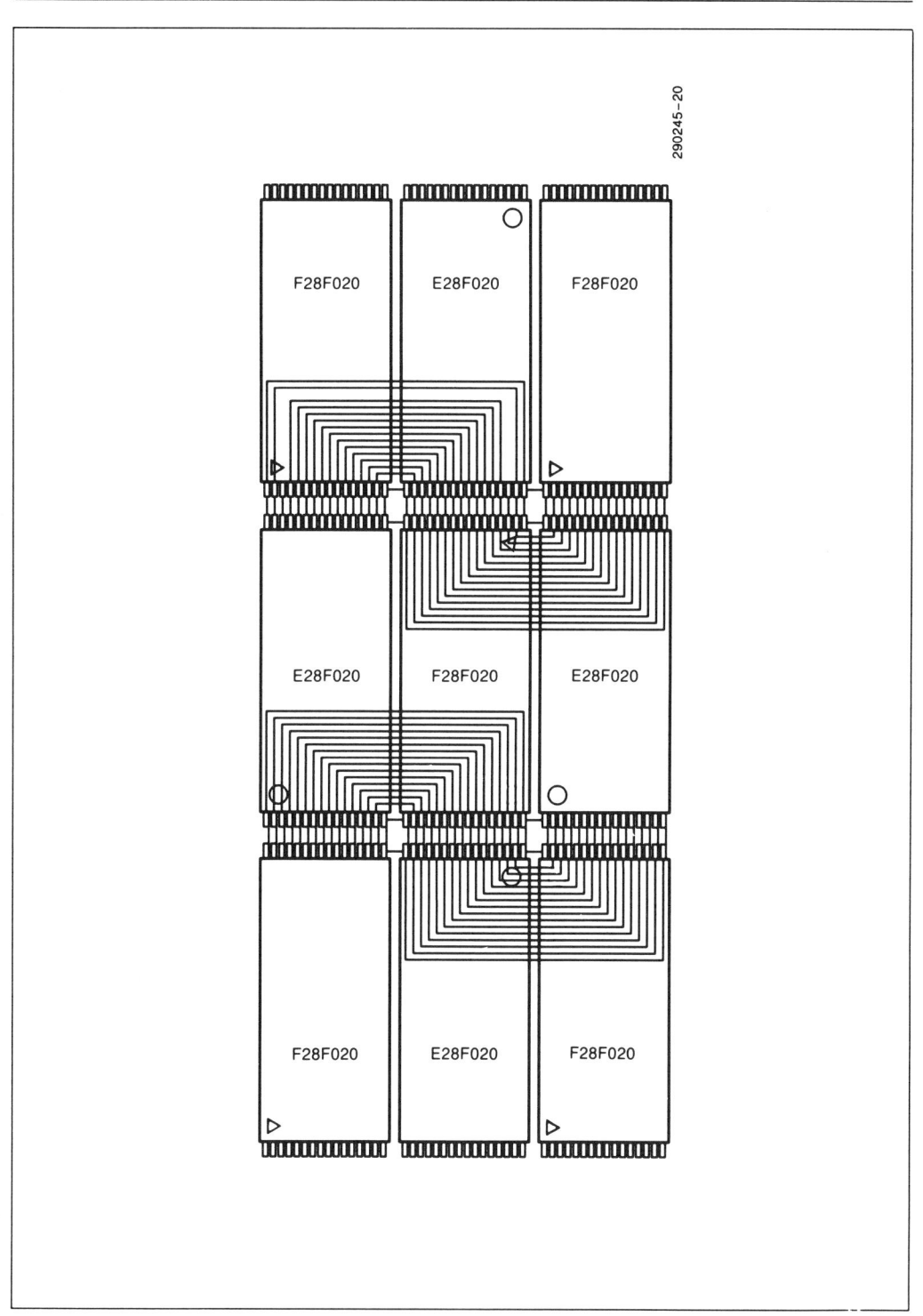

28F020 ADVANCE INFORMATION

Figure 4. TSOP Serpentine Layout

Table 2. 28F020 Bus Operations

Pins Operation		V_{PP} [1]	A_0	A_9	$\overline{CE}$	$\overline{OE}$	$\overline{WE}$	DQ_0-DQ_7
READ-ONLY	Read	V_{PPL}	A_0	A_9	V_{IL}	V_{IL}	V_{IH}	Data Out
	Output Disable	V_{PPL}	X	X	V_{IL}	V_{IH}	V_{IH}	Tri-State
	Standby	V_{PPL}	X	X	V_{IH}	X	X	Tri-State
	int$_e$ligent Identifier™ (Mfr)[2]	V_{PPL}	V_{IL}	V_{ID}[3]	V_{IL}	V_{IL}	V_{IH}	Data = 89H
	int$_e$ligent Identifier™ (Device)[2]	V_{PPL}	V_{IH}	V_{ID}[3]	V_{IL}	V_{IL}	V_{IH}	Data = BDH
READ/WRITE	Read	V_{PPH}	A_0	A_9	V_{IL}	V_{IL}	V_{IH}	Data Out[4]
	Output Disable	V_{PPH}	X	X	V_{IL}	V_{IH}	V_{IH}	Tri-State
	Standby[5]	V_{PPH}	X	X	V_{IH}	X	X	Tri-State
	Write	V_{PPH}	A_0	A_9	V_{IL}	V_{IH}	V_{IL}	Data In[6]

NOTES:
1. Refer to D.C. Characteristics. When $V_{PP} = V_{PPL}$ memory contents can be read but not written or erased.
2. Manufacturer and device codes may also be accessed via a command register write sequence. Refer to Table 3. All other addresses low.
3. V_{ID} is the int$_e$ligent Identifier high voltage. Refer to DC Characteristics.
4. Read operations with $V_{PP} = V_{PPH}$ may access array data or the int$_e$ligent Identifier™ codes.
5. With V_{PP} at high voltage, the standby current equals $I_{CC} + I_{PP}$ (standby).
6. Refer to Table 3 for valid Data-In during a write operation.
7. X can be V_{IL} or V_{IH}.

Or, the system designer may choose to "hardwire" V_{PP}, making the high voltage supply constantly available. In this case, all Command Register functions are inhibited whenever V_{CC} is below the write lockout voltage V_{LKO}. (See Power Up/Down Protection.) The 28F020 is designed to accommodate either design practice, and to encourage optimization of the processor-memory interface.

BUS OPERATIONS

Read

The 28F020 has two control functions, both of which must be logically active, to obtain data at the outputs. Chip-Enable ($\overline{CE}$) is the power control and should be used for device selection. Output-Enable ($\overline{OE}$) is the output control and should be used to gate data from the output pins, independent of device selection. Refer to A.C. read timing waveforms.

When V_{PP} is high (V_{PPH}), the read operation can be used to access array data, to output the int$_e$ligent Identifier™ codes, and to access data for program/erase verification. When V_{PP} is low (V_{PPL}), the read operation can **only** access the array data.

Output Disable

With Output-Enable at a logic-high level (V_{IH}), output from the device is disabled. Output pins are placed in a high-impedance state.

Standby

With Chip-Enable at a logic-high level, the standby operation disables most of the 28F020's circuitry and substantially reduces device power consumption. The outputs are placed in a high-impedance state, independent of the Output-Enable signal. If the 28F020 is deselected during erasure, programming, or program/erase verification, the device draws active current until the operation is terminated.

int$_e$ligent Identifier™ Operation

The int$_e$ligent Identifier operation outputs the manufacturer code (89H) and device code (BDH). Programming equipment automatically matches the device with its proper erase and programming algorithms.

intel 28F020 ADVANCE INFORMATION

With Chip-Enable and Output-Enable at a logic low level, raising A9 to high voltage V_{ID} (see DC Characteristics) activates the operation. Data read from locations 0000H and 0001H represent the manufacturer's code and the device code, respectively.

The manufacturer- and device-codes can also be read via the command register, for instances where the 28F020 is erased and reprogrammed in the target system. Following a write of 90H to the command register, a read from address location 0000H outputs the manufacturer code (89H). A read from address 0001H outputs the device code (BDH).

Write

Device erasure and programming are accomplished via the command register, when high voltage is applied to the V_{PP} pin. The contents of the register serve as input to the internal state-machine. The state-machine outputs dictate the function of the device.

The command register itself does not occupy an addressable memory location. The register is a latch used to store the command, along with address and data information needed to execute the command.

The command register is written by bringing Write-Enable to a logic-low level (V_{IL}), while Chip-Enable is low. Addresses are latched on the falling edge of Write-Enable, while data is latched on the rising edge of the Write-Enable pulse. Standard microprocessor write timings are used.

Refer to A.C. Write Characteristics and the Erase/Programming Waveforms for specific timing parameters.

COMMAND DEFINITIONS

When low voltage is applied to the V_{PP} pin, the contents of the command register default to 00H, enabling read-only operations.

Placing high voltage on the V_{PP} pin enables read/write operations. Device operations are selected by writing specific data patterns into the command register. Table 3 defines these 28F020 register commands.

Table 3. Command Definitions

Command	Bus Cycles Req'd	First Bus Cycle Operation[1]	First Bus Cycle Address[2]	First Bus Cycle Data[3]	Second Bus Cycle Operation[1]	Second Bus Cycle Address[2]	Second Bus Cycle Data[3]
Read Memory	1	Write	X	00H			
Read int$_e$ligent Identifier™ Codes[4]	3	Write	X	90H	Read	(4)	(4)
Set-up Erase/Erase[5]	2	Write	X	20H	Write	X	20H
Erase Verify[5]	2	Write	EA	A0H	Read	X	EVD
Set-up Program/Program[6]	2	Write	X	40H	Write	PA	PD
Program Verify[6]	2	Write	X	C0H	Read	X	PVD
Reset[7]	2	Write	X	FFH	Write	X	FFH

NOTES:
1. Bus operations are defined in Table 2.
2. IA = Identifier address: 00H for manufacturer code, 01H for device code.
   EA = Address of memory location to be read during erase verify.
   PA = Address of memory location to be programmed.
   Addresses are latched on the falling edge of the Write-Enable pulse.
3. ID = Data read from location IA during device identification (Mfr = 89H, Device = BDH).
   EVD = Data read from location EA during erase verify.
   PD = Data to be programmed at location PA. Data is latched on the rising edge of Write-Enable.
   PVD = Data read from location PA during program verify. PA is latched on the Program command.
4. Following the Read int$_e$ligent ID command, two read operations access manufacturer and device codes.
5. Figure 6 illustrates the Quick-Erase™ Algorithm.
6. Figure 5 illustrates the Quick-Pulse Programming™ Algorithm.
7. The second bus cycle must be followed by the desired command register write.

intel 28F020 ADVANCE INFORMATION

Read Command

While V_{PP} is high, for erasure and programming, memory contents can be accessed via the read command. The read operation is initiated by writing 00H into the command register. Microprocessor read cycles retrieve array data. The device remains enabled for reads until the command register contents are altered.

The default contents of the register upon V_{PP} power-up is 00H. This default value ensures that no spurious alteration of memory contents occurs during the V_{PP} power transition. Where the V_{PP} supply is hard-wired to the 28F020, the device powers-up and remains enabled for reads until the command-register contents are changed. Refer to the A.C. Read Characteristics and Waveforms for specific timing parameters.

int_eligent Identifier™ Command

Flash-memories are intended for use in applications where the local CPU alters memory contents. As such, manufacturer- and device-codes must be accessible while the device resides in the target system. PROM programmers typically access signature codes by raising A9 to a high voltage. However, multiplexing high voltage onto address lines is not a desired system-design practice.

The 28F020 contains an int_eligent Identifier operation to supplement traditional PROM-programming methodology. The operation is initiated by writing 90H into the command register. Following the command write, a read cycle from address 0000H retrieves the manufacturer code of 89H. A read cycle from address 0001H returns the device code of BDH. To terminate the operation, it is necessary to write another valid command into the register.

Set-up Erase/Erase Commands

Set-up Erase is a command-only operation that stages the device for electrical erasure of all bytes in the array. The set-up erase operation is performed by writing 20H to the command register.

To commence chip-erasure, the erase command (20H) must again be written to the register. The erase operation begins with the rising edge of the Write-Enable pulse and terminates with the rising edge of the next Write-Enable pulse (i.e., Erase-Verify Command).

This two-step sequence of set-up followed by execution ensures that memory contents are not accidentally erased. Also, chip-erasure can only occur when high voltage is applied to the V_{PP} pin. In the absence of this high voltage, memory contents are protected against erasure. Refer to A.C. Erase Characteristics and Waveforms for specific timing parameters.

Erase-Verify Command

The erase command erases all bytes of the array in parallel. After each erase operation, all bytes must be verified. The erase verify operation is initiated by writing A0H into the command register. The address for the byte to be verified must be supplied as it is latched on the falling edge of the Write-Enable pulse. The register write terminates the erase operation with the rising edge of its Write-Enable pulse.

The 28F020 applies an internally-generated margin voltage to the addressed byte. Reading FFH from the addressed byte indicates that all bits in the byte are erased.

The erase-verify command must be written to the command register prior to each byte verification to latch its address. The process continues for each byte in the array until a byte does not return FFH data, or the last address is accessed.

In the case where the data read is not FFH, another erase operation is performed. (Refer to Set-up Erase/Erase). Verification then resumes from the address of the last-verified byte. Once all bytes in the array have been verified, the erase step is complete. The device can be programmed. At this point, the verify operation is terminated by writing a valid command (e.g. Program Set-up) to the command register. Figure 6, the Quick-Erase™ algorithm, illustrates how commands and bus operations are combined to perform electrical erasure of the 28F020. Refer to A.C. Erase Characteristics and Waveforms for specific timing parameters.

Set-up Program/Program Commands

Set-up program is a command-only operation that stages the device for byte programming. Writing 40H into the command register performs the set-up operation.

Once the program set-up operation is performed, the next Write-Enable pulse causes a transition to an active programming operation. Addresses are internally latched on the falling edge of the Write-Enable pulse. Data is internally latched on the rising edge of the Write-Enable pulse. The rising edge of Write-Enable also begins the programming operation. The programming operation terminates with the next rising edge of Write-Enable, used to write the program-verify command. Refer to A.C. Programming Characteristics and Waveforms for specific timing parameters.

9

intel 28F020 ADVANCE INFORMATION

Program-Verify Command

The 28F020 is programmed on a byte-by-byte basis. Byte programming may occur sequentially or at random. Following each programming operation, the byte just programmed must be verified.

The program-verify operation is initiated by writing C0H into the command register. The register write terminates the programming operation with the rising edge of its Write-Enable pulse. The program-verify operation stages the device for verification of the byte last programmed. No new address information is latched.

The 28F020 applies an internally-generated margin voltage to the byte. A microprocessor read cycle outputs the data. A successful comparison between the programmed byte and true data means that the byte is successfully programmed. Programming then proceeds to the next desired byte location. Figure 5, the 28F020 Quick-Pulse Programming™ algorithm, illustrates how commands are combined with bus operations to perform byte programming. Refer to A.C. Programming Characteristics and Waveforms for specific timing parameters.

Reset Command

A reset command is provided as a means to safely abort the erase- or program-command sequences. Following either set-up command (erase or program) with two consecutive writes of FFH will safely abort the operation. Memory contents will not be altered. A valid command must then be written to place the device in the desired state.

EXTENDED ERASE/PROGRAM CYCLING

EEPROM cycling failures have always concerned users. The high electrical field required by thin oxide EEPROMs for tunneling can literally tear apart the oxide at defect regions. To combat this, some suppliers have implemented redundancy schemes, reducing cycling failures to insignificant levels. However, redundancy requires that cell size be doubled—an expensive solution.

Intel has designed extended cycling capability into its ETOX II flash memory technology. Resulting improvements in cycling reliability come without increasing memory cell size or complexity. First, an advanced tunnel oxide increases the charge carrying ability ten-fold. Second, the oxide area per cell subjected to the tunneling electric field is one-tenth that of common EEPROMs, minimizing the probability of oxide defects in the region. Finally, the peak electric field during erasure is approximately 2 MV/cm lower than EEPROM. The lower electric field greatly reduces oxide stress and the probability of failure—increasing time to wearout by a factor of 100,000,000.

The 28F020 is specified for a minimum of 10,000 program/erase cycles. The device is programmed and erased using Intel's Quick-Pulse Programming™ and Quick-Erase™ algorithms. Intel's algorithmic approach uses a series of operations (pulses), along with byte verification, to completely and reliably erase and program the device.

For further information, see Reliability Report RR-60.

QUICK-PULSE PROGRAMMING™ ALGORITHM

The Quick-Pulse Programming algorithm uses programming operations of 10 μs duration. Each operation is followed by a byte verification to determine when the addressed byte has been successfully programmed. The algorithm allows for up to 25 programming operations per byte, although most bytes verify on the first or second operation. The entire sequence of programming and byte verification is performed with V_{PP} at high voltage. Figure 5 illustrates the Quick-Pulse Programming algorithm.

QUICK-ERASE™ ALGORITHM

Intel's Quick-Erase algorithm yields fast and reliable electrical erasure of memory contents. The algorithm employs a closed-loop flow, similar to the Quick-Pulse Programming™ algorithm, to simultaneously remove charge from all bits in the array.

Erasure begins with a read of memory contents. The 28F020 is erased when shipped from the factory. Reading FFH data from the device would immediately be followed by device programming.

For devices being erased and reprogrammed, uniform and reliable erasure is ensured by first programming all bits in the device to their charged state (Data = 00H). This is accomplished, using the Quick-Pulse Programming algorithm, in approximately four seconds.

Erase execution then continues with an initial erase operation. Erase verification (data = FFH) begins at address 0000H and continues through the array to the last address, or until data other than FFH is encountered. With each erase operation, an increasing number of bytes verify to the erased state. Erase efficiency may be improved by storing the address of the last byte verified in a register. Following the next erase operation, verification starts at that stored address location. Erasure typically occurs in two seconds. Figure 6 illustrates the Quick-Erase algorithm.

intel   28F020   ADVANCE INFORMATION

Bus Operation	Command	Comments
Standby		Wait for V$_{PP}$ Ramp to V$_{PPH}$(1)
		Initialize Pulse-Count
Write	Set-up Program	Data = 40H
Write	Program	Valid Address/Data
Standby		Duration of Program Operation (t$_{WHWH1}$)
Write	Program[2] Verify	Data = C0H; Stops Program Operation[3]
Standby		t$_{WHGL}$
Read		Read Byte to Verify Programming
Standby		Compare Data Output to Data Expected
Write	Read	Data = 00H, Resets the Register for Read Operations
Standby		Wait for V$_{PP}$ Ramp to V$_{PPL}$(1)

290245-7

NOTES:
1. See D.C. Characteristics for the value of V$_{PPH}$ and V$_{PPL}$.
2. Program Verify is only performed after byte programming. A final read/compare may be performed (optional) after the register is written with the Read command.
3. Refer to principles of operation.
4. **CAUTION: The algorithm MUST BE FOLLOWED to ensure proper and reliable operation of the device.**

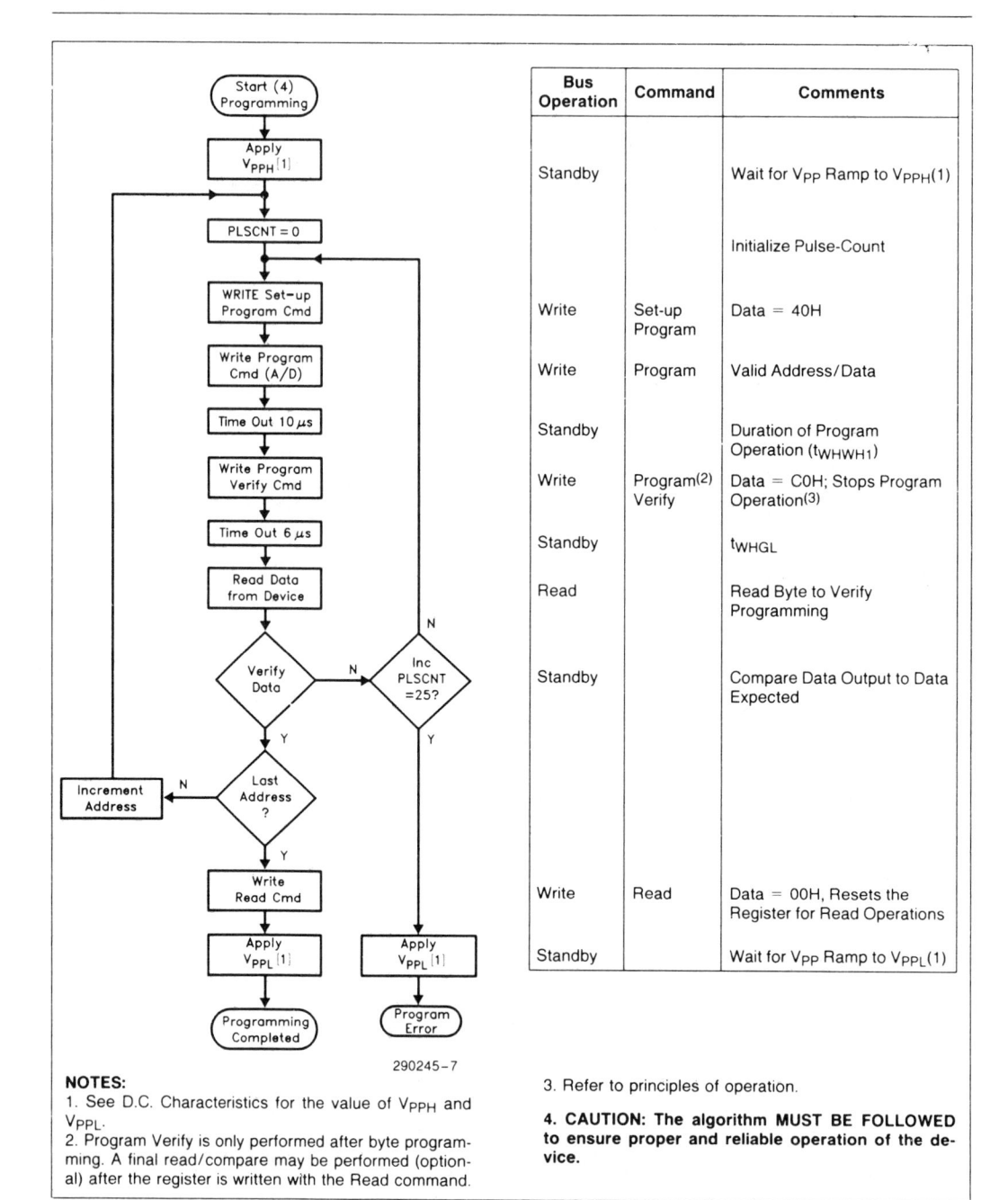

Figure 5. 28F020 Quick-Pulse Programming™ Algorithm

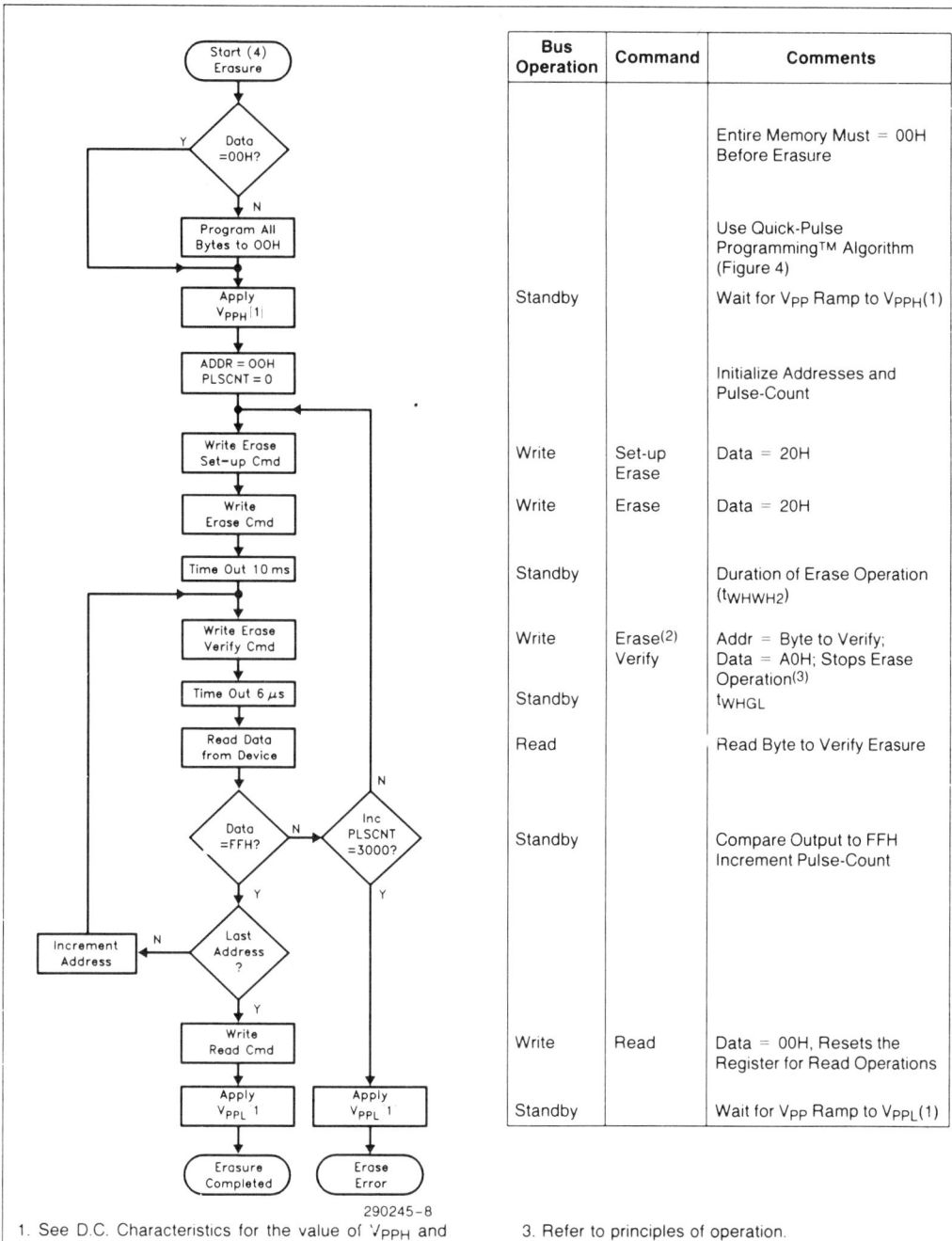

intel 28F020　ADVANCE INFORMATION

Bus Operation	Command	Comments
		Entire Memory Must = 00H Before Erasure
		Use Quick-Pulse Programming™ Algorithm (Figure 4)
Standby		Wait for V$_{PP}$ Ramp to V$_{PPH}$(1)
		Initialize Addresses and Pulse-Count
Write	Set-up Erase	Data = 20H
Write	Erase	Data = 20H
Standby		Duration of Erase Operation (t$_{WHWH2}$)
Write	Erase(2) Verify	Addr = Byte to Verify; Data = A0H; Stops Erase Operation(3)
Standby		t$_{WHGL}$
Read		Read Byte to Verify Erasure
Standby		Compare Output to FFH Increment Pulse-Count
Write	Read	Data = 00H, Resets the Register for Read Operations
Standby		Wait for V$_{PP}$ Ramp to V$_{PPL}$(1)

1. See D.C. Characteristics for the value of V$_{PPH}$ and V$_{PPL}$.
2. Erase Verify is performed only after chip-erasure. A final read/compare may be performed (optional) after the register is written with the read command.
3. Refer to principles of operation.
4. **CAUTION: The algorithm MUST BE FOLLOWED to ensure proper and reliable operation of the device.**

Figure 6. 28F020 Quick-Erase™ Algorithm

28F020

ADVANCE INFORMATION

DESIGN CONSIDERATIONS

Two-Line Output Control

Flash-memories are often used in larger memory arrays. Intel provides two read-control inputs to accommodate multiple memory connections. Two-line control provides for:

a. the lowest possible memory power dissipation and,
b. complete assurance that output bus contention will not occur.

To efficiently use these two control inputs, an address-decoder output should drive chip-enable, while the system's read signal controls all flash-memories and other parallel memories. This assures that only enabled memory devices have active outputs, while deselected devices maintain the low power standby condition.

Power Supply Decoupling

Flash-memory power-switching characteristics require careful device decoupling. System designers are interested in three supply current (I_{CC}) issues—standby, active, and transient current peaks produced by falling and rising edges of chip-enable. The capacitive and inductive loads on the device outputs determine the magnitudes of these peaks.

Two-line control and proper decoupling capacitor selection will suppress transient voltage peaks. Each device should have a 0.1 μF ceramic capacitor connected between V_{CC} and V_{SS}, and between V_{PP} and V_{SS}.

Place the high-frequency, low-inherent-inductance capacitors as close as possible to the devices. Also, for every eight devices, a 4.7 μF electrolytic capacitor should be placed at the array's power supply connection, between V_{CC} and V_{SS}. The bulk capacitor will overcome voltage slumps caused by printed-circuit-board trace inductance, and will supply charge to the smaller capacitors as needed.

V_{PP} Trace on Printed Circuit Boards

Programming flash-memories, while they reside in the target system, requires that the printed circuit board designer pay attention to the V_{PP} power supply trace. The V_{PP} pin supplies the memory cell current for programming. Use similar trace widths and layout considerations given the V_{CC} power bus. Adequate V_{PP} supply traces and decoupling will decrease V_{PP} voltage spikes and overshoots.

Power Up/Down Protection

The 28F020 is designed to offer protection against accidental erasure or programming during power transitions. Upon power-up, the 28F020 is indifferent as to which power supply, V_{PP} or V_{CC}, powers up first. **Power supply sequencing is not required.** Internal circuitry in the 28F020 ensures that the command register is reset to the read mode on power up.

A system designer must guard against active writes for V_{CC} voltages above V_{LKO} when V_{PP} is active. Since both $\overline{WE}$ and $\overline{CE}$ must be low for a command write, driving either to V_{IH} will inhibit writes. The control register architecture provides an added level of protection since alteration of memory contents only occurs after successful completion of the two-step command sequences.

28F020 Power Dissipation

When designing portable systems, designers must consider battery power consumption not only during device operation, but also for data retention during system idle time. Flash nonvolatility increases the usable battery life of your system because the 28F020 does not consume any power to retain code or data when the system is off. Table 4 illustrates the power dissipated when updating the 28F020.

Table 4. 28F020 Typical Update Power Dissipation

Operation	Power Dissipation (Watt-Seconds)
Array Program/Program Verify[1]	0.34
Array Erase/Erase Verify[2]	0.37
One Complete Cycle[3]	1.05

NOTES:
1. Formula to calculate typical Program/Program Verify Power = [V_{PP} · # Bytes · typical # Prog Pulse (t_{WHWH1} · I_{PP2} typical + t_{WHGL} · I_{PP4} typical)] + [V_{CC} · # Bytes · typical # Prog Pulses (t_{WHWH1} · I_{CC2} typical + t_{WHGL} · I_{CC4} typical)].
2. Formula to calculate typical Erase/Erase Verify Power = [V_{PP} (I_{PP3} typical · t_{ERASE} typical + I_{PP5} typical + t_{WHGL} · # Bytes)] + [V_{CC} (I_{CC3} typical · t_{ERASE} typical + I_{CC5} typical · t_{WHGL} · # Bytes)].
3. One Complete Cycle = Array Preprogram + Array Erase + Program.

intel 28F020 ADVANCE INFORMATION

ABSOLUTE MAXIMUM RATINGS*

Operating Temperature
   During Read .... 0°C to +70°C[1]
   During Erase/Program .... 0°C to +70°C
Temperature Under Bias .... −10°C to +80°C
Storage Temperature .... −65°C to +125°C
Voltage on Any Pin with
   Respect to Ground .... −2.0V to +7.0V[2]
Voltage on Pin A_9 with
   Respect to Ground .... −2.0V to +13.5V[2,3]
V_{PP} Supply Voltage with
   Respect to Ground
   During Erase/Program .... −2.0V to +14.0V[2,3]
V_{CC} Supply Voltage with
   Respect to Ground .... −2.0V to +7.0V[2]
Output Short Circuit Current .... 100 mA[4]

*Notice: Stresses above those listed under "Absolute Maximum Ratings" may cause permanent damage to the device. This is a stress rating only and functional operation of the device at these or any other conditions above those indicated in the operational sections of this specification is not implied. Exposure to absolute maximum rating conditions for extended periods may affect device reliability.

NOTICE: Specifications contained within the following tables are subject to change.

NOTES:
1. Operating temperature is for commercial product defined by this specification.
2. Minimum D.C. input voltage is −0.5V. During transitions, inputs may undershoot to −2.0V for periods less than 20 ns. Maximum D.C. voltage on output pins is V_{CC} + 0.5V, which may overshoot to V_{CC} + 2.0V for periods less than 20 ns.
3. Maximum D.C. voltage on A_9 or V_{PP} may overshoot to +14.0V for periods less than 20 ns.
4. Output shorted for no more than one second. No more than one output shorted at a time.

OPERATING CONDITIONS

Symbol	Parameter	Limits Min	Limits Max	Unit	Comments
T_A	Operating Temperature	0	70	°C	For Read-Only and Read/Write Operations
V_{CC}	V_{CC} Supply Voltage	4.50	5.50	V	

D.C. CHARACTERISTICS—TTL/NMOS COMPATIBLE

Symbol	Parameter	Min	Typical	Max	Unit	Test Conditions
I_{LI}[1]	Input Leakage Current			±1.0	µA	V_{CC} = V_{CC} Max, V_{IN} = V_{CC} or V_{SS}
I_{LO}[1]	Output Leakage Current			±10	µA	V_{CC} = V_{CC} Max, V_{OUT} = V_{CC} or V_{SS}
I_{CCS}[1]	V_{CC} Standby Current			1.0	mA	V_{CC} = V_{CC} Max, $\overline{CE}$ = V_{IH}
I_{CC1}[1]	V_{CC} Active Read Current		10	30	mA	V_{CC} = V_{CC} Max, $\overline{CE}$ = V_{IL}, f = 6 MHz, I_{OUT} = 0 mA
I_{CC2}[1]	V_{CC} Programming Current		1.0	10	mA	Programming in Progress
I_{CC3}[1]	V_{CC} Erase Current		5.0	15	mA	Erasure in Progress
I_{CC4}[1]	V_{CC} Program Verify Current		5.0	15	mA	V_{PP} = V_{PPH}, Program Verify in Progress
I_{CC5}[1]	V_{CC} Erase Verify Current		5.0	15	mA	V_{PP} = V_{PPH}, Erase Verify in Progress
I_{PPS}[1]	V_{PP} Leakage Current			±10	µA	V_{PP} ≤ V_{CC}

intel 28F020 ADVANCE INFORMATION

D.C. CHARACTERISTICS—TTL/NMOS COMPATIBLE (Continued)

Symbol	Parameter	Limits Min	Limits Typical	Limits Max	Unit	Test Conditions
$I_{PP1}{}^{(1)}$	V_{PP} Read Current or Standby Current		90	200	μA	$V_{PP} > V_{CC}$
			±90	±200		$V_{PP} \leq V_{CC}$
$I_{PP2}{}^{(1)}$	V_{PP} Programming Current		8	30	mA	$V_{PP} = V_{PPH}$, Programming in Progress
$I_{PP3}{}^{(1)}$	V_{PP} Erase Current		10	30	mA	$V_{PP} = V_{PPH}$
$I_{PP4}{}^{(1)}$	V_{PP} Program Verify Current		2.0	5.0	mA	$V_{PP} = V_{PPH}$, Program Verify in Progress
$I_{PP5}{}^{(1)}$	V_{PP} Erase Verify Current		2.0	5.0	mA	$V_{PP} = V_{PPH}$, Erase Verify in Progress
V_{IL}	Input Low Voltage	−0.5		0.8	V	
V_{IH}	Input High Voltage	2.0		$V_{CC} + 0.5$	V	
V_{OL}	Output Low Voltage			0.45	V	$I_{OL} = 5.8$ mA, $V_{CC} = V_{CC}$ Min
V_{OH1}	Output High Voltage	2.4			V	$I_{OH} = -2.5$ mA, $V_{CC} = V_{CC}$ Min
V_{ID}	A9 int$_e$ligent Identifer™ Voltage	11.50		13.00	V	
$I_{ID}{}^{(1)}$	A9 int$_e$ligent Identifer™ Current		90	200	μA	$A_9 = V_{ID}$
V_{PPL}	V_{PP} during Read-Only Operations	0.00		6.5	V	NOTE: Erase/Program are Inhibited when $V_{PP} = V_{PPL}$
V_{PPH}	V_{PP} during Read/Write Operations	11.40		12.60	V	
V_{LKO}	V_{CC} Erase/Write Lock Voltage	2.5			V	

D.C. CHARACTERISTICS—CMOS COMPATIBLE

Symbol	Parameter	Limits Min	Limits Typical	Limits Max	Unit	Test Conditions
$I_{LI}{}^{(1)}$	Input Leakage Current			±1.0	μA	$V_{CC} = V_{CC}$ Max, $V_{IN} = V_{CC}$ or V_{SS}
$I_{LO}{}^{(1)}$	Output Leakage Current			±10	μA	$V_{CC} = V_{CC}$ Max, $V_{OUT} = V_{CC}$ or V_{SS}
$I_{CCS}{}^{(1)}$	V_{CC} Standby Current		50	100	μA	$V_{CC} = V_{CC}$ Max, $\overline{CE} = V_{CC} \pm 0.2V$
$I_{CC1}{}^{(1)}$	V_{CC} Active Read Current		10	30	mA	$V_{CC} = V_{CC}$ Max, $\overline{CE} = V_{IL}$, $f = 6$ MHz, $I_{OUT} = 0$ mA
$I_{CC2}{}^{(1)}$	V_{CC} Programming Current		1.0	10	mA	Programming in Progress
$I_{CC3}{}^{(1)}$	V_{CC} Erase Current		5.0	15	mA	Erasure in Progress
$I_{CC4}{}^{(1)}$	V_{CC} Program Verify Current		5.0	15	mA	$V_{PP} = V_{PPH}$, Program Verify in Progress
$I_{CC5}{}^{(1)}$	V_{CC} Erase Verify Current		5.0	15	mA	$V_{PP} = V_{PPH}$, Erase Verify in Progress
$I_{PPS}{}^{(1)}$	V_{PP} Leakage Current			±10	μA	$V_{PP} \leq V_{CC}$

15

intel 28F020 ADVANCE INFORMATION

D.C. CHARACTERISTICS—CMOS COMPATIBLE (Continued)

Symbol	Parameter	Limits Min	Typical	Unit Max	Test Conditions
I_{PP1} (1)	V_{PP} Read Current or Standby Current	90	200	µA	$V_{PP} > V_{CC}$
		±90	±200		$V_{PP} \leq V_{CC}$
I_{PP2} (1)	V_{PP} Programming Current		8	30	mA	$V_{PP} = V_{PPH}$, Programming in Progress
I_{PP3} (1)	V_{PP} Erase Current		10	30	mA	$V_{PP} = V_{PPH}$, Erasure in Progress
I_{PP4} (1)	V_{PP} Program Verify Current		2.0	5.0	mA	$V_{PP} = V_{PPH}$, Program Verify in Progress
I_{PP5} (1)	V_{PP} Erase Verify Current		2.0	5.0	mA	$V_{PP} = V_{PPH}$, Erase Verify in Progress
V_{IL}	Input Low Voltage	−0.5		0.8	V	
V_{IH}	Input High Voltage	0.7 V_{CC}		V_{CC} + 0.5	V	
V_{OL}	Output Low Voltage			0.45	V	$I_{OL} = 5.8$ mA, $V_{CC} = V_{CC}$ Min
V_{OH1}	Output High Voltage	0.85 V_{CC}			V	$I_{OH} = -2.5$ mA, $V_{CC} = V_{CC}$ Min
V_{OH2}		$V_{CC} - 0.4$				$I_{OH} = -100$ µA, $V_{CC} = V_{CC}$ Min
V_{ID}	A9 inteligent Identifer™ Voltage	11.50		13.00	V	
I_{ID} (1)	A9 inteligent Identifer™ Current		90	200	µA	A9 = V_{ID}
V_{PPL}	V_{PP} during Read-Only Operations	0.00		6.5	V	NOTE: Erase/Programs are Inhibited when $V_{PP} = V_{PPL}$
V_{PPH}	V_{PP} during Read/Write Operations	11.40		12.60	V	
V_{LKO}	V_{CC} Erase/Write Lock Voltage	2.5			V	

CAPACITANCE(2) $T_A = 25°C$, f = 1.0 MHz

Symbol	Parameter	Limits Min	Max	Unit	Conditions
C_{IN}	Address/Control Capacitance		6	pF	$V_{IN} = 0V$
C_{OUT}	Output Capacitance		12	pF	$V_{OUT} = 0V$

NOTES for D.C. Characteristics and Capacitance:
1. All currents are in RMS unless otherwise noted. Typical values at $V_{CC} = 5.0V$, $V_{PP} = 12.0V$, T = 25°C. These currents are valid for all product versions (packages and speeds).
2. Sampled, not 100% tested.

intel 28F020 ADVANCE INFORMATION

A.C. TESTING INPUT/OUTPUT WAVEFORM

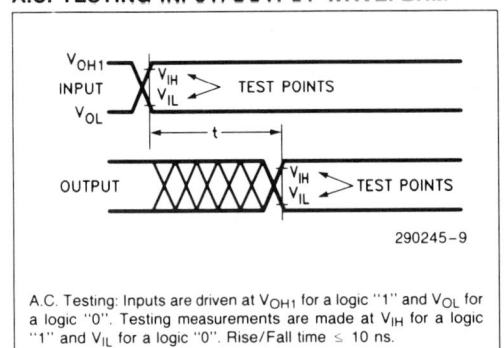

A.C. Testing: Inputs are driven at V_{OH1} for a logic "1" and V_{OL} for a logic "0". Testing measurements are made at V_{IH} for a logic "1" and V_{IL} for a logic "0". Rise/Fall time ≤ 10 ns.

A.C. TESTING LOAD CIRCUIT

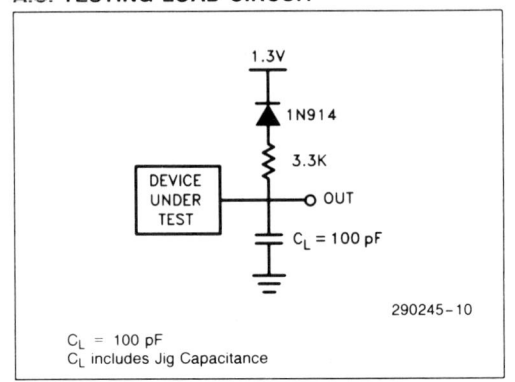

C_L = 100 pF
C_L includes Jig Capacitance

A.C. TEST CONDITIONS

Input Rise and Fall Times (10% to 90%) . . . . . . 10 ns
Input Pulse Levels . . . . V_{OL} and V_{OH1}
Input Timing Reference Level . . . . V_{IL} and V_{IH}
Output Timing Reference Level . . . . V_{IL} and V_{IH}

A.C. CHARACTERISTICS—Read-Only Operations

Versions		28F020-150		28F020-200		Unit
Symbol	Characteristic	Min	Max	Min	Max	
t_{AVAV}/t_{RC}	Read Cycle Time	150		200		ns
t_{ELQV}/t_{CE}	Chip Enable Access Time		150		200	ns
t_{AVQV}/t_{ACC}	Address Access Time		150		200	ns
t_{GLQV}/t_{OE}	Output Enable Access Time		50		55	ns
t_{ELQX}/t_{LZ}	Chip Enable to Output in Low Z	0		0		ns
t_{EHQZ}	Chip Disable to Output in High Z		55		55	ns
t_{GLQX}/t_{OLZ}	Output Enable to Output in Low Z	0		0		ns
t_{GHQZ}/t_{DF}	Output Disable to Output in High Z		35		40	ns
t_{OH}	Output Hold from Address, $\overline{CE}$, or $\overline{OE}$ Change[1]	0		0		ns
t_{WHGL}	Write Recovery Time before Read	6		6		μs

NOTES:
1. Whichever occurs first.
2. Rise/Fall Time ≤ 10 ns.

intel

28F020

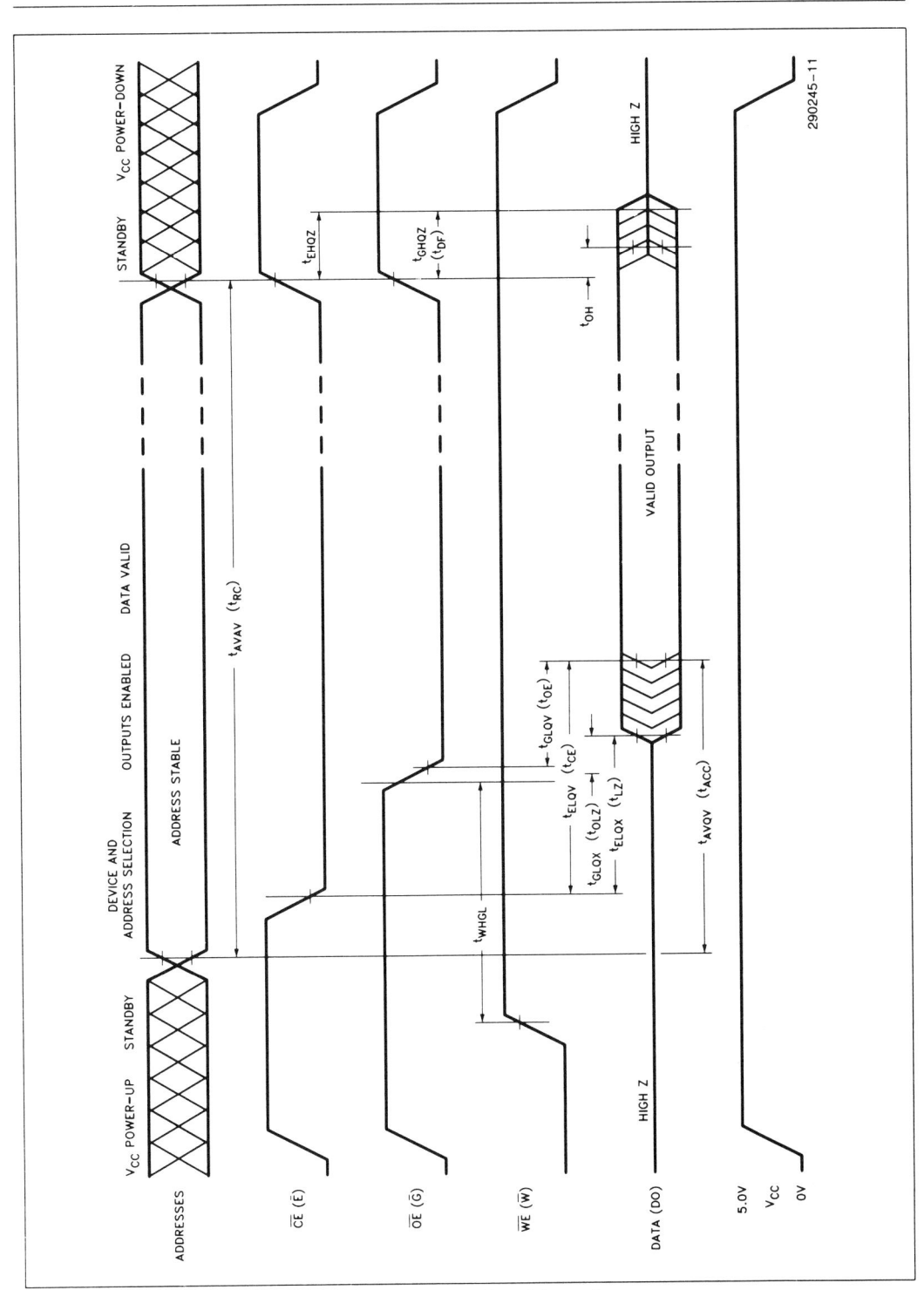

ADVANCE INFORMATION

A.C. Waveforms for Read Operations

290245–11

intel 28F020 ADVANCE INFORMATION

A.C. CHARACTERISTICS—Write/Erase/Program Operations[1,2]

Versions		28F020-150		28F020-200		Unit
Symbol	Characteristic	Min	Max	Min	Max	
t_{AVAV}/t_{WC}	Write Cycle Time	150		200		ns
t_{AVWL}/t_{AS}	Address Set-Up Time	0		0		ns
t_{WLAX}/t_{AH}	Address Hold Time	60		60		ns
t_{DVWH}/t_{DS}	Data Set-Up Time	50		50		ns
t_{WHDX}/t_{DH}	Data Hold Time	10		10		ns
t_{WHGL}	Write Recovery Time before Read	6		6		μs
t_{GHWL}	Read Recovery Time before Write	0		0		μs
t_{ELWL}/t_{CS}	Chip Enable Set-Up Time before Write	20		20		ns
t_{WHEH}/t_{CH}	Chip Enable Hold Time	0		0		ns
t_{WLWH}/t_{WP}	Write Pulse Width[2]	50		50		ns
t_{WHWL}/t_{WPH}	Write Pulse Width High	20		20		ns
t_{WHWH1}	Duration of Programming Operation	10	(3)	10	(3)	μs
t_{WHWH2}	Duration of Erase Operation	9.5	(3)	9.5	(3)	ms
t_{VPEL}	V_{PP} Set-Up Time to Chip Enable Low	100		100		ns

NOTES:
1. Read timing characteristics during read/write operations are the same as during read-only operations. Refer to A.C. Characteristics for Read-Only Operations.
2. Rise/Fall time ≤ 10 ns.
3. The integrated stop timer terminates the programming/erase operations, thereby eliminating the need for a maximum specification.

ERASE AND PROGRAMMING PERFORMANCE

Parameter	Limits						Unit
	28F020-150			28F020-200			
	Min	Typ	Max	Min	Typ	Max	
Chip Erase Time[3,4]		2(1)	30		2(1)	30	Sec
Chip Program Time[4]		4(1)	25(2)		4(1)	25(2)	Sec
Erase/Program Cycles[5]	10,000	100,000		10,000	100,000		Cycles

NOTES:
1. 25°C, 12.0V V_{PP}, 10,000 Cycles.
2. Minimum byte programming time excluding system overhead is 16 μsec (10 μsec program + 6 μsec write recovery), while maximum is 400 μsec/byte (16 μsec x 25 loops allowed by algorithm). Max chip programming time is specified lower than the worst case allowed by the programming algorithm since most bytes program significantly faster than the worst case byte.
3. Excludes 00H Programming prior to Erasure.
4. Excludes System-Level Overhead.
5. Refer to RR-60 "ETOX™ II Flash Memory Reliability Data Summary" for typical cycling data and failure rate calculations.

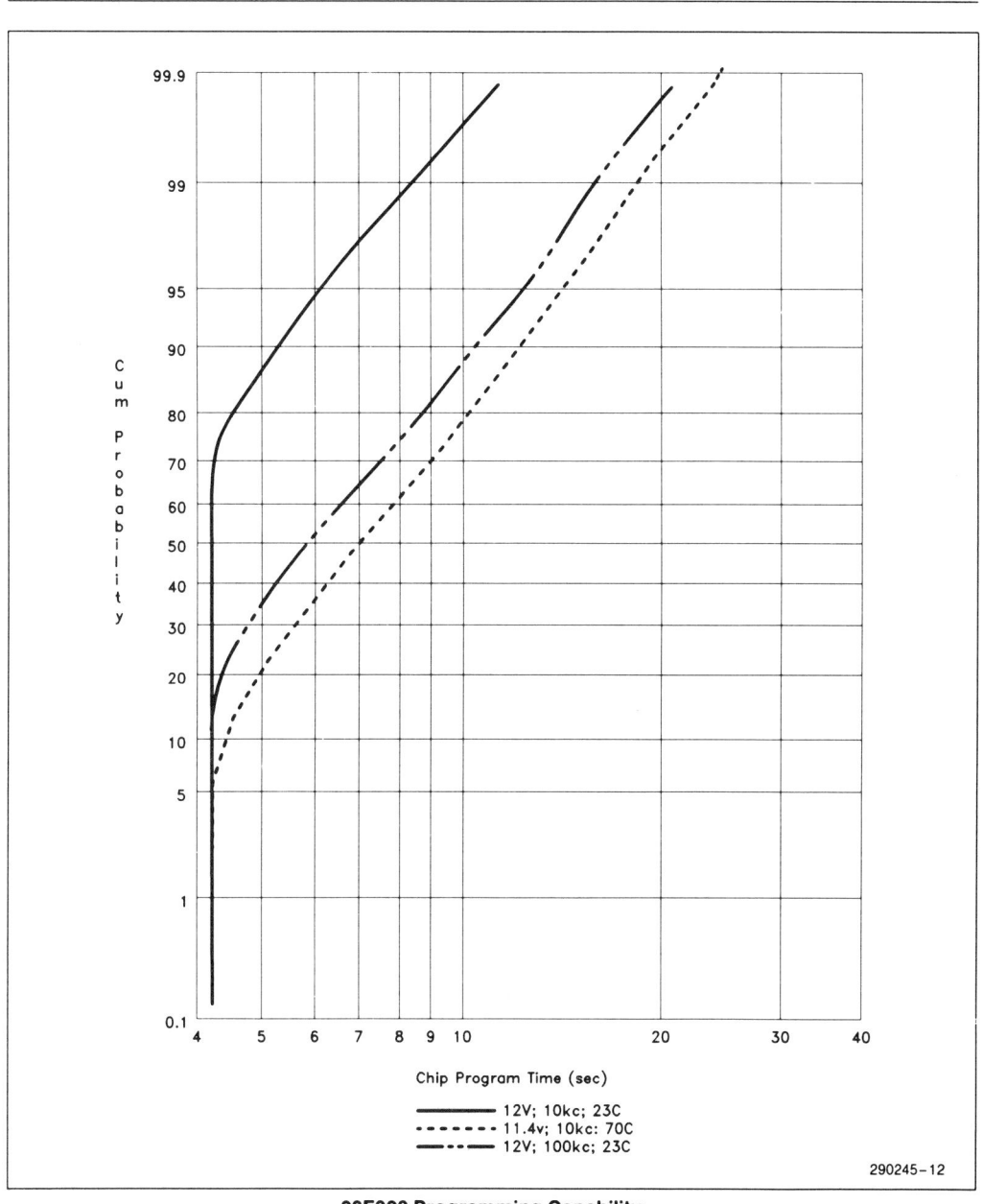

28F020 Programming Capability

intel® 28F020 ADVANCE INFORMATION

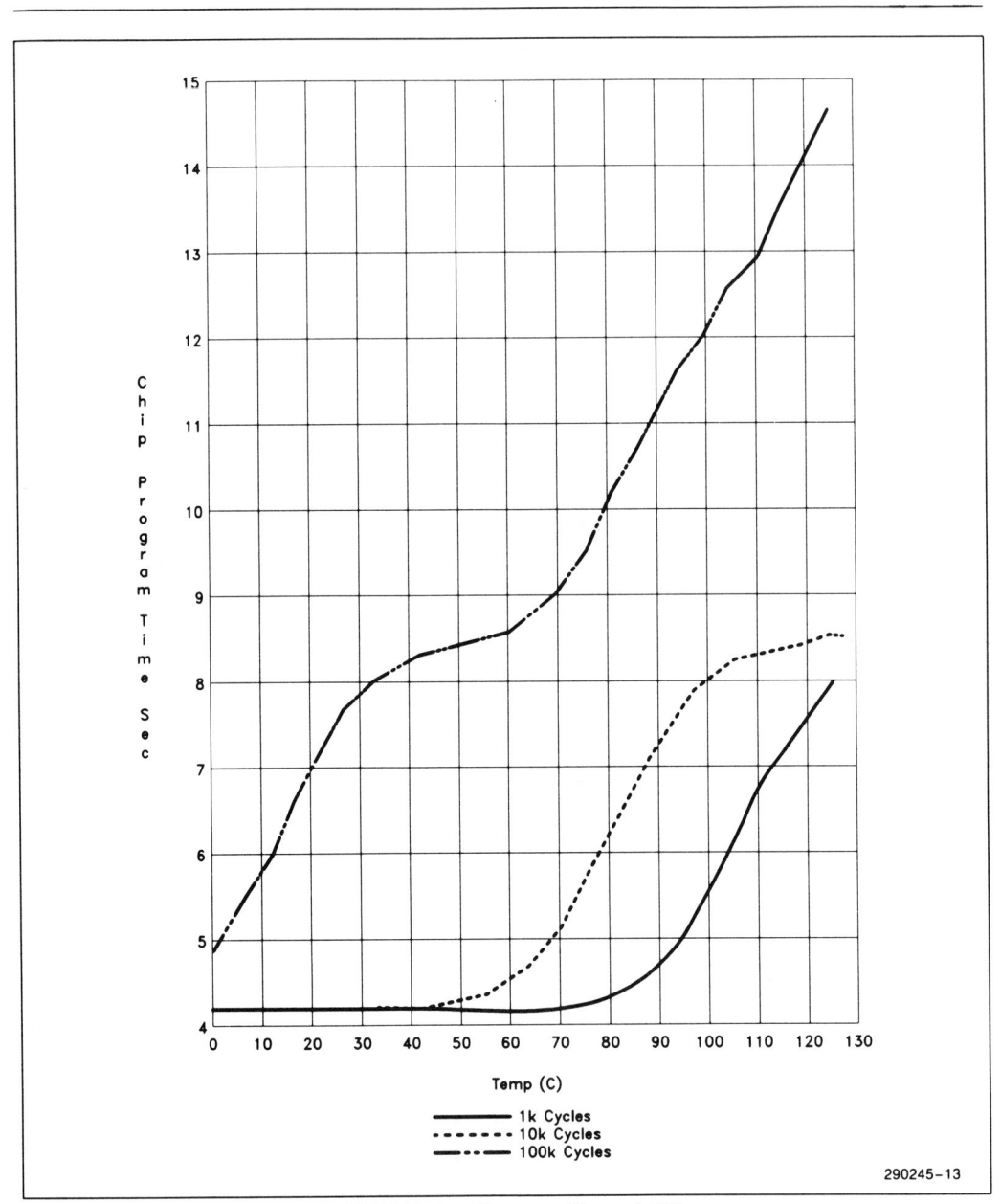

28F020 Typical Program Time at 12V

290245-13

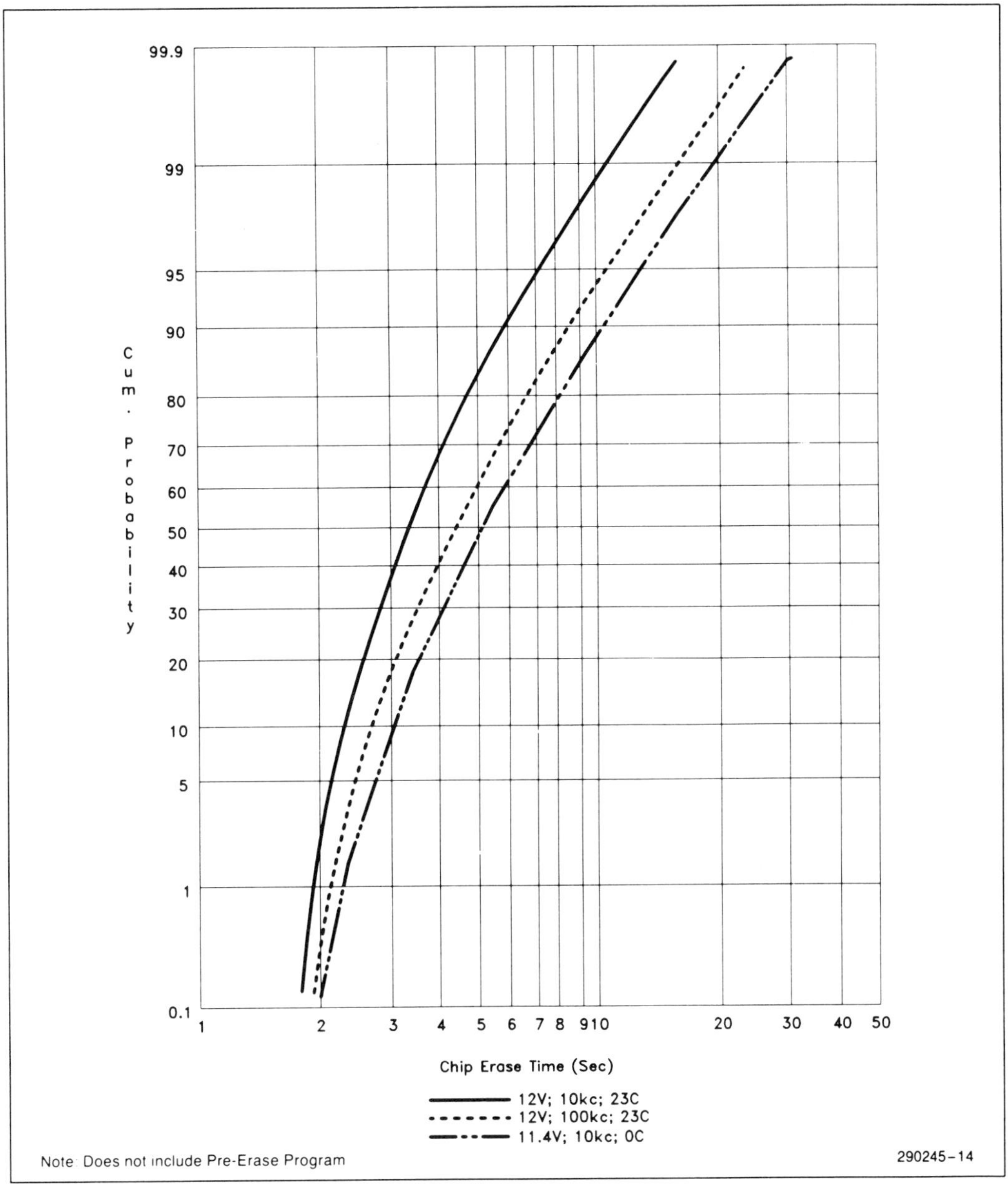
99.9
99
95
90
80
70
60
50
40
30
20
10
5
1
0.1
Cum. Probability
1
2
3
4
5
6
7
8
910
20
30
40
50
Chip Erase Time (Sec)
12V; 10kc; 23C
12V; 100kc; 23C
11.4V; 10kc; 0C
Note: Does not include Pre-Erase Program
290245-14

28F020 Erase Capability

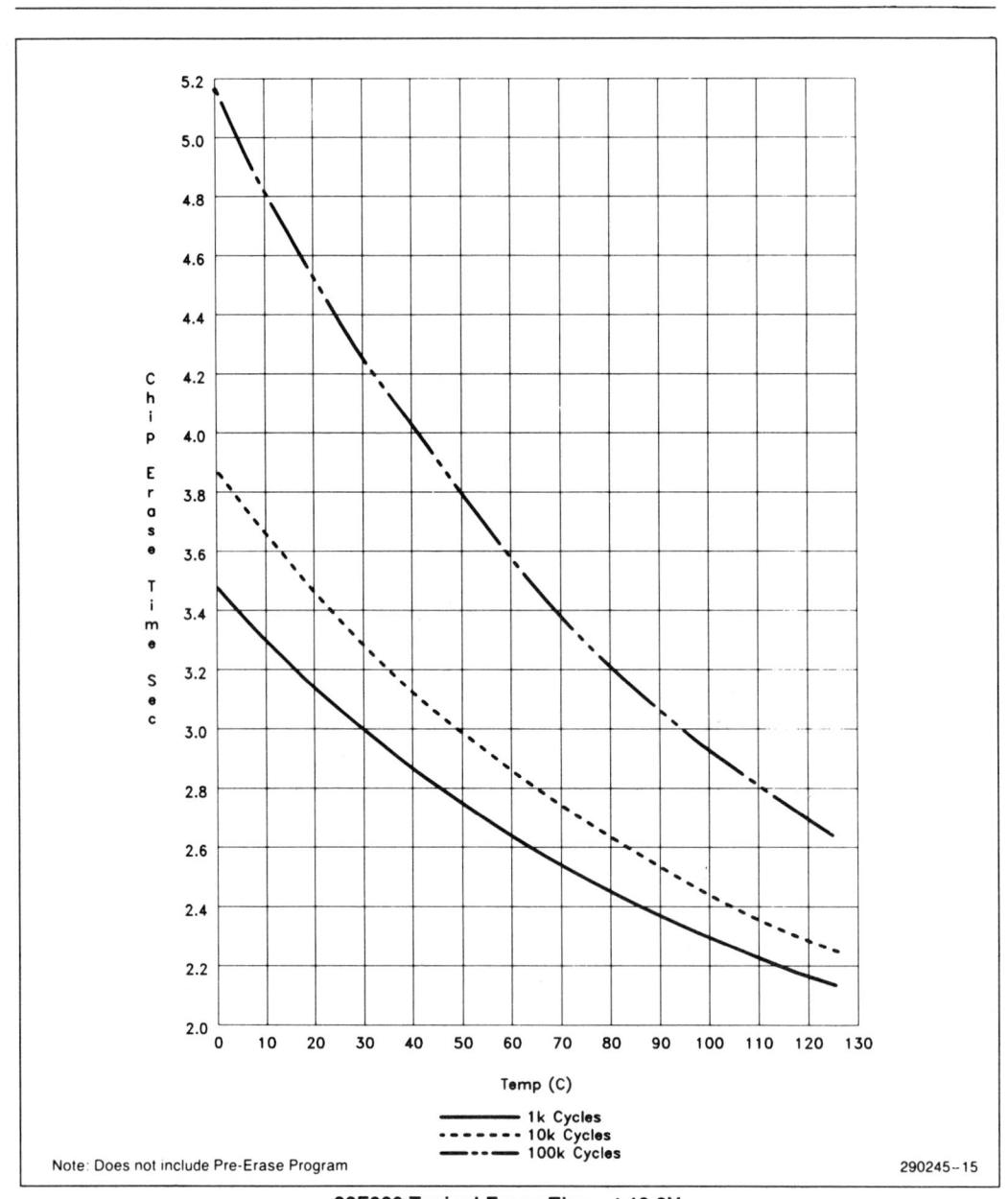

28F020 Typical Erase Time at 12.0V

intel 28F020 ADVANCE INFORMATION

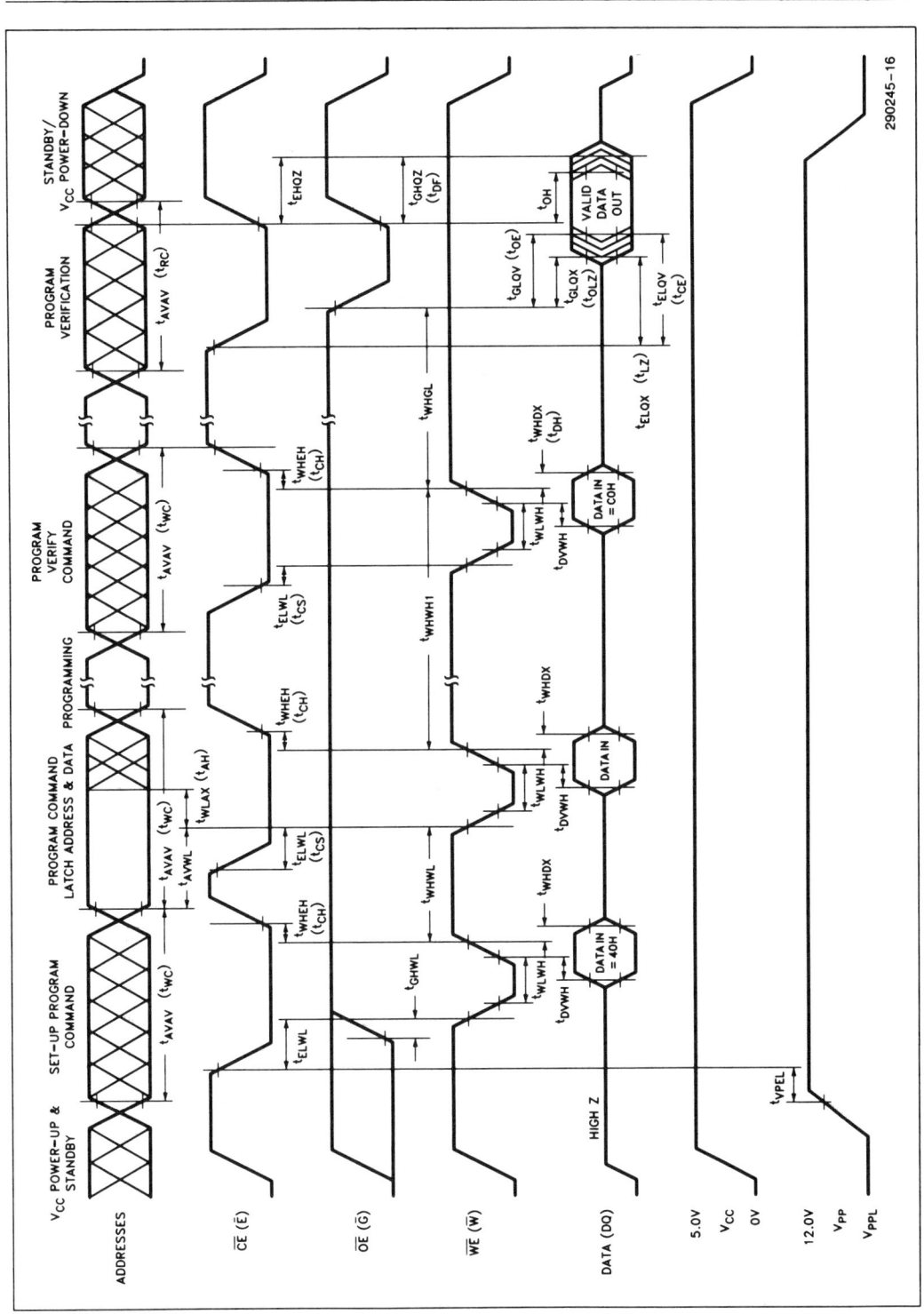

A.C. Waveforms for Programming Operations

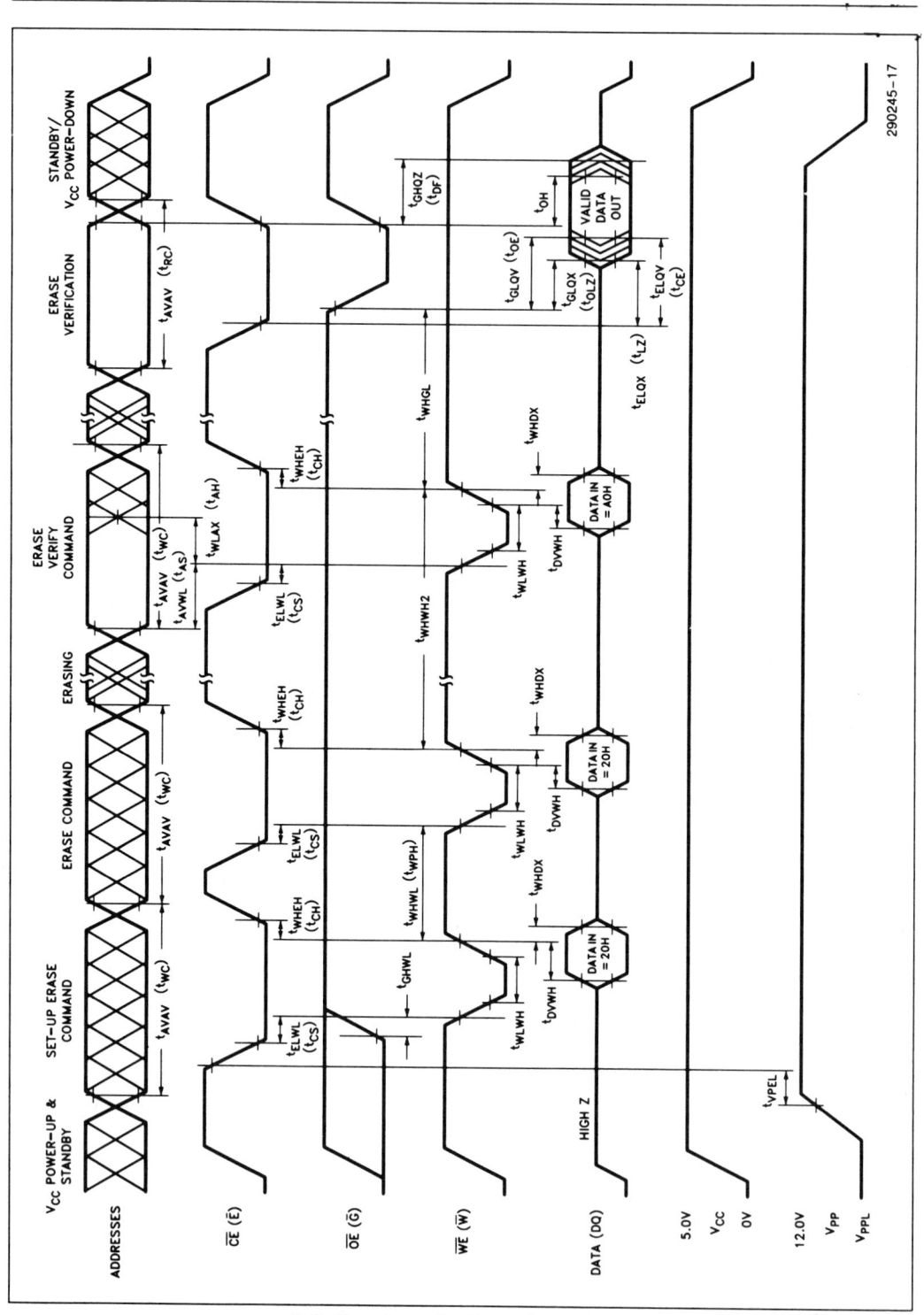

A.C. Waveforms for Erase Operations

intel 28F020 ADVANCE INFORMATION

ALTERNATIVE CE-CONTROLLED WRITES

Versions		28F020-150		28F020-200		Unit
Symbol	Characteristic	Min	Max	Min	Max	
t_{AVAV}	Write Cycle Time	150		200		ns
t_{AVEL}	Address Set-Up Time	0		0		ns
t_{ELAX}	Address Hold Time	80		80		ns
t_{DVEH}	Data Set-Up Time	50		50		ns
t_{EHDX}	Data Hold Time	10		10		ns
t_{EHGL}	Write Recovery Time before Read	6		6		μs
t_{GHEL}	Read Recovery Time before Write	0		0		μs
t_{WLEL}	Write Enable Set-Up Time before Chip Enable	0		0		ns
t_{EHWH}	Write Enable Hold Time	0		0		ns
t_{ELEH}	Write Pulse Width[1]	70		70		ns
t_{EHEL}	Write Pulse Width High	20		20		ns
t_{PEL}	V_{PP} Set-Up Time to Chip Enable Low	100		100		ns

NOTE:
1. Chip-Enable Controlled Writes: Write operations are driven by the valid combination of Chip-Enable and Write-Enable. In systems where Chip-Enable defines the write pulse width (within a longer Write-Enable timing waveform) all set-up, hold and inactive Write-Enable times should be measured relative to the Chip-Enable waveform.

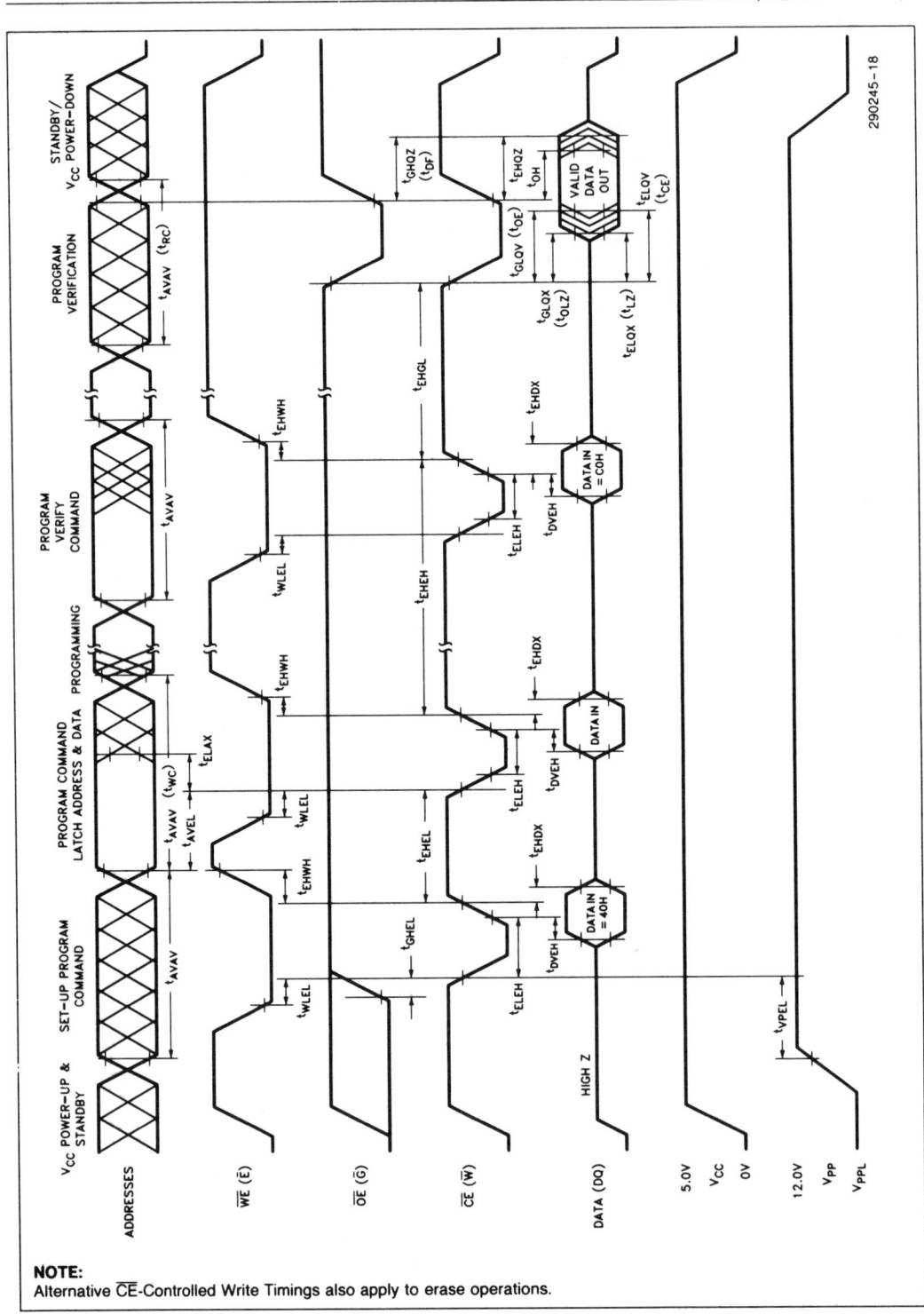

NOTE:
Alternative CE-Controlled Write Timings also apply to erase operations.

Alternate A.C. Waveforms for Programming Operations

intel 28F020 ADVANCE INFORMATION

ORDERING INFORMATION

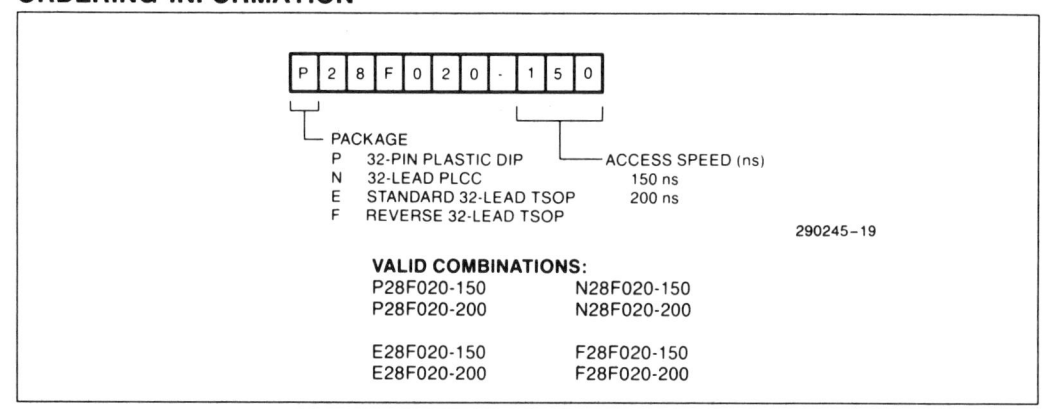

```
P 2 8 F 0 2 0 - 1 5 0
```

- PACKAGE
  - P 32-PIN PLASTIC DIP
  - N 32-LEAD PLCC
  - E STANDARD 32-LEAD TSOP
  - F REVERSE 32-LEAD TSOP

- ACCESS SPEED (ns)
  - 150 ns
  - 200 ns

290245-19

VALID COMBINATIONS:

P28F020-150	N28F020-150
P28F020-200	N28F020-200
E28F020-150	F28F020-150
E28F020-200	F28F020-200

ADDITIONAL INFORMATION

		Order Number
ER-20,	"ETOX™ II Flash Memory Technology"	294005
ER-24,	"The Intel 28F020 Flash Memory"	294008
RR-60,	"ETOX™ II Flash Memory Reliability Data Summary"	293002
AP-316,	"Using Flash Memory for In-System Reprogrammable Nonvolatile Storage"	292046
AP-325	"Guide to Flash Memory Reprogramming"	292059

Printed in England by Carlton Barclay/2K/0790
Memory Components

intel

APPLICATION NOTE

AP-325

May 1989

Guide to Flash Memory Reprogramming

TECHNICAL LITERATURE
supplied by

Jermyn Distribution
Vestry Estate, Sevenoaks, Kent TN14 5EU.
Tel: (0732) 450144 · Telex: 95142 · Fax: (0732) 451251

SAUL ZALES
APPLICATIONS ENGINEERING

© Intel Corporation, 1989

Order Number: 292059-001

Intel Corporation makes no warranty for the use of its products and assumes no responsibility for any errors which may appear in this document nor does it make a commitment to update the information contained herein.

Intel retains the right to make changes to these specifications at any time, without notice.

Contact your local sales office to obtain the latest specifications before placing your order.

The following are trademarks of Intel Corporation and may only be used to identify Intel Products:

> 376, 386, 387, 486, 4-SITE, Above, ACE51, ACE96, ACE186, ACE196, ACE960, BITBUS, COMMputer, CREDIT, Data Pipeline, ETOX, Genius, î, i486, i860, ICE, iCEL, ICEVIEW, iCS, iDBP, iDIS, I²ICE, iLBX, iMDDX, iMMX, Inboard, Insite, Intel, int$_e$l, Intel386, int$_e$lBOS, Intel Certified, Intelevision, int$_e$ligent Identifier, int$_e$ligent Programming, Intellec, Intellink, iOSP, iPDS, iPSC, iRMK, iRMX, iSBC, iSBX, iSDM, iSXM, Library Manager, MAPNET, MCS, Megachassis, MICROMAINFRAME, MULTIBUS, MULTICHANNEL, MULTIMODULE, MultiSERVER, ONCE, OpenNET, OTP, PROMPT, Promware, QUEST, QueX, Quick-Erase, Quick-Pulse Programming, Ripplemode, RMX/80, RUPI, Seamless, SLD, SugarCube, UPI, and VLSiCEL, and the combination of ICE, iCS, iRMX, iSBC, iSBX, iSXM, MCS, or UPI and a numerical suffix.

MDS is an ordering code only and is not used as a product name or trademark. MDS® is a registered trademark of Mohawk Data Sciences Corporation.

*MULTIBUS is a patented Intel bus.

CHMOS and HMOS are patented processes of Intel Corp.

Intel Corporation and Intel's FASTPATH are not affiliated with Kinetics, a division of Excelan, Inc. or its FASTPATH trademark or products.

Additional copies of this manual or other Intel literature may be obtained from:

> Intel Corporation
> Literature Sales
> P.O. Box 58130
> Santa Clara, CA 95052-8130

GUIDE TO FLASH MEMORY REPROGRAMMING

CONTENTS PAGE

INTRODUCTION TO REPROGRAMMING 1
You Are in Control 1

FUNDAMENTALS OF FLASH OPERATION 1
Adaptive vs. Brute Force Algorithms 1
Moving Charge & Other Factors You Should Know 2

ERASURE—THE GOLDEN RULE 5
Margin for Error 5
Most Common Development Issues 6
Device Initialization and Reset 6
The Erase Algorithm Interpreted 7
The Program Algorithm Illuminated 10
Ramifications of the Golden Rule 10

DEBUGGING YOUR CODE AND OTHER TIPS ON TESTING 11
Software Drivers Save You Time 11
Timers, Test Loops and Assembly Level Programming 11
Programming—The Key to Proper Erasure 11
16- and 32-Bit Systems 12
Logic Analyzers and In-Circuit Emulators 12
Testing Your Software—One More Time 12
Watchdog Timer Debug Circuit 13

TROUBLE SHOOTING GUIDE 14
Determining the Root Cause 14

AP-325

INTRODUCTION TO REPROGRAMMING

You Are in Control

Rewriting any type of memory requires hardware or software control. Traditional EEPROM designers combined all control functions into each chip's periphery. This provided a highly functional chip but at a high price. On the other hand, DRAM designers provided a bulk memory with little integrated peripheral circuitry. Each system designer then accommodated the DRAM with external refresh signals and learned quickly that failure to refresh yielded non-functioning memory boards. Initially, software drivers controlled DRAM refresh; today controllers provide the same function.

Similarly, early disk drives required every user to write software to manipulate drive head movement. Failure to follow drive specifications and algorithms caused irreversible head crashes. Leading-edge engineers faced these challenges and triumphed, as evidenced by the sophisticated systems available today.

Today, software directs flash memory reprogramming. One sends a control signal to a device to begin and end programming or erasure. It is a simple process, however care must be taken. If algorithms are not properly followed, a device may be rendered inoperable. This document discusses proper software and debug technique, which yields dependable flash memory operation.

FUNDAMENTALS OF FLASH MEMORY OPERATION

Adaptive vs Brute Force Algorithms

Many designers use EPROMs regularly. Few consider the programming algorithms because the PROM programmer vendors take care of that function.

Two types of algorithms are in use today:
- Adaptive Algorithms
- Brute Force Algorithms

Adaptive algorithms such as Intel's Quick-Pulse Programming™ and Quick-Erase™ algorithms reduce programming time. A feedback mechanism recognizes when each byte has been programmed sufficiently. You may ask how is the point of sufficiency determined? One simply adds the net effects of V_{CC} and temperature variations, and superimposes on those factors the normal EPROM charge leakage to obtain the answer. The next question is how can these factors be checked?

NOTE:
EPROM and EEPROM charge leakage occurs over a very long time—typically 100 years. Reliability papers often discuss charge leakage in terms of the memory's data retention characteristic.

If you look at EPROM programming algorithms, you will notice that V_{CC} is elevated during programming. The elevated V_{CC} acts as the feedback mechanism for the adaptive algorithm. Reading the device and checking for program completion is called verification, or margining. (One is checking the margin to V_{CC} fluctuations.) For example, if the part can be verified at 6.25V, then it can withstand the fluctuations and normal charge leakage.

During the past few years most major EPROM manufacturers have converted to adaptive algorithms. The algorithm loops back and programs a byte again if the first program operation does not verify at the elevated voltage.

Brute force algorithms simply program each byte multiple times, typically with long program durations. This type of algorithm has no in-system margin verification. That is they *assume but never verify* program margin to the typical environment effects.

Many flash memories that specify a brute force algorithm may fail to retain data for 10 years. Additionally, they may not read the data correctly even at specified V_{CC} and temperature extremes.

Intel's flash memory program and erase algorithms are both adaptive. They offer margin verification without requiring users to elevate V_{CC} in-system. When issued a command to program verify, the memory's command register logic taps an internally-generated elevated V_{CC} from the user-supplied external V_{PP} (12V). This is why it is essential that you provide the specified V_{PP} voltage and follow the given adaptive algorithms. Intel's adaptive algorithms, combined with the command register architecture, assures reliable code and data storage and dependable system operation.

Figure 1 shows an example of an adaptive byte programming algorithm. Appendix A compares the algorithm in Figure 1 to a brute force approach.

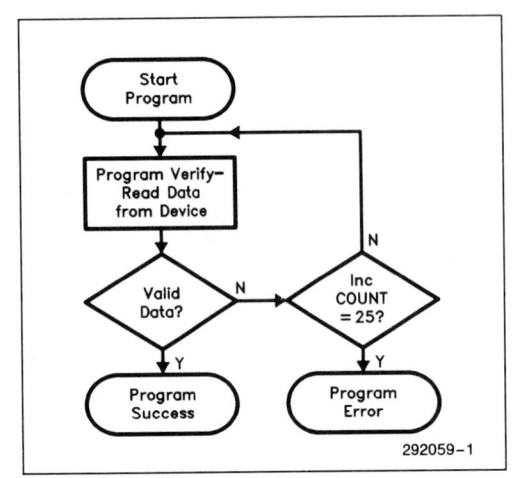

Figure 1. The flow chart shows the fundamental nature of an adaptive algorithm. Based on the outcome of program verification, the flow may loop back for another program operation.

Moving Charge and Other Factors You Should Know

This section discusses the mechanics of flash memory programming. For most system designers, transistor-level discussions were last heard in college. We may recall that DRAM consists of a storage capacitor and a transistor. We remember this clearly because failure to refresh that capacitor causes systems to malfunction. **In like fashion, one should understand the fundamentals of flash memory reprogramming. The understanding will enable error-free memory operation and reliable system performance.**

In simplest terms, each data bit equates to a memory cell. Intel's flash memory uses one transistor per cell with the smallest possible architecture. This delivers the lowest cost per bit and highest capacity, levering system software (rather than bulky, complex cells) for reprogramming control.

Figure 2 shows a simplified cross section of Intel's flash memory transistor. Note the structure; the cell is a stacked gate MOS transistor. An isolated floating gate stores the memory charge. The floating gate consists of a layer of (conductive) polysilicon surrounded by (non-conductive) oxide layers.

On a DRAM cell, each transistor connects to a capacitor which stores the memory charge. The major difference between flash memory and DRAM derives from their cell structure. The DRAM cell loses its charge if not refreshed within a few milliseconds. On the other hand, the flash memory floating gate maintains its charge for typically 100 years. The structure is isolated and insulated by the field and gate oxides—hence the name "floating" gate.

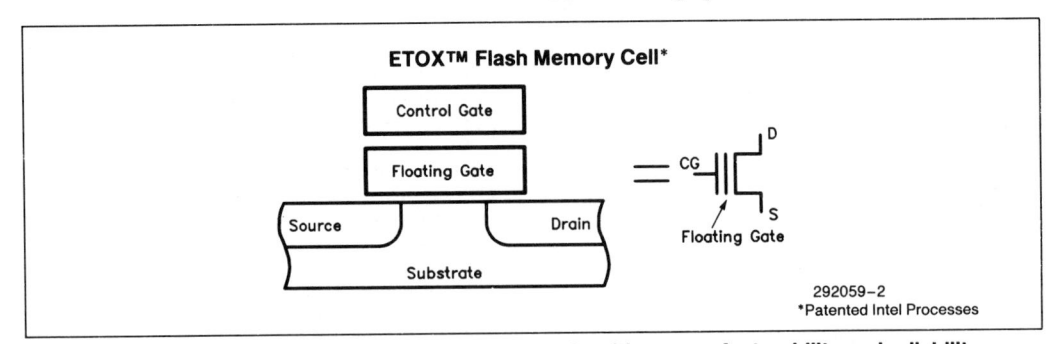

Figure 2. Simplicity of design assures increasing densities, manufacturability and reliability. These are the attributes that drive mainstream memories.

> **CONTRARY TO INTUITION**
> **CHARGE = DATA "0"**
> **NOT DATA "1"**
>
> **PROGRAMMING:**
> ADDS CHARGE TO FLOATING GATE
> → DATA = 0
>
> **ERASURE:**
> REMOVES SOME CHARGE FROM FLOATING GATE
> → DATA = 1
>
> **PROGRAMMING DATA WITH MIXED 0s and 1s:**
> → ONLY DATA "0" BITS GET CHARGED
> → DATA "1" BITS REMAIN UNCHARGED

Changing the memory contents is simple. Figure 3 shows two memory cells—one being erased and one being programmed. Erasure removes charge from all bits simultaneously. Programming adds charge to selected bits. During erasure, not all charge is removed. The erase verify operation tells the system when enough charge has been removed. At that point, the flash memory behaves like a U.V.-erased EPROM.

Removing too much charge by erasing too long renders the memory unprogrammable. Excessive erasure lowers the cell threshold to the point where the transistor is always on and always reads data "1". (Recall that the cell threshold, V_t, determines when the transistor turns on or off.) You must control the erase timing within the algorithm specifications.

A second erase consideration relates to the first. Prior to erasing the chip, you must blanket program all bytes to data 00h, regardless of the previous data. This step equalizes the charge on all transistors.

If you skip this step and proceed directly to erasure, an interesting thing happens. Consider a typical byte programmed with data 0AAh (1010 1010b). While programming this data, bits with data "1" remain erased (charge removed), and bits with data "0" are programmed (charged added). Following programming, normal read operations sense whether a memory transistor has more or less charge and drives the outputs accordingly.

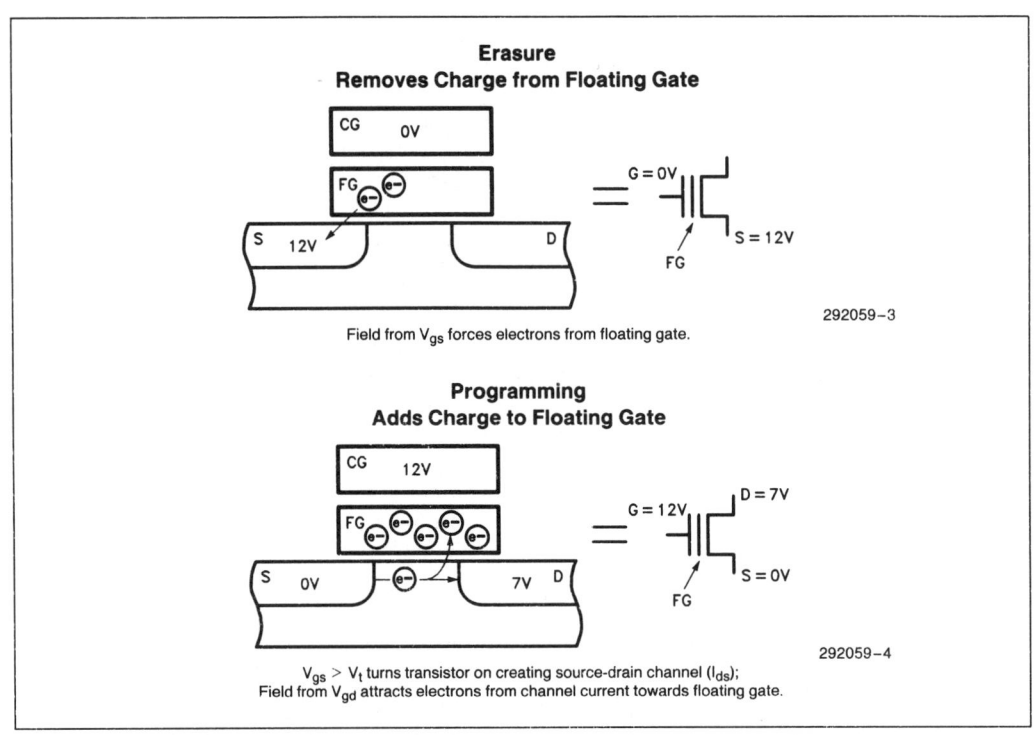

Figure 3. Flash memory cells during erasure and programming. Note the movement of charge on and off the floating gate. The charge adjusts the cell threshold, which tells the outputs whether a bit (transistor) is on or off.

AP-325

Erasure then removes charge from all bits. The bits that have had charge added (data "0") have some quantity of charge removed; bits with less charge (data "1") have charge removed as well. This is akin to excessively erasing the data "1" bits. Pre-programming all bits to data "0" equalizes the charge which allows for controlled, uniform erasure of all bits in the device (i.e., all 1,048,576 bits in a 28F010).

The sections entitled "Margin for Error" and the "Erase Algorithm Interpreted, Program all Bytes to 00h" discuss this concept in greater detail.

ERASURE—THE GOLDEN RULE

Erasure removes charge from all memory cells in parallel. This lowers the cells' threshold voltages from the programmed level (6.5V) below V_{CC} to the erased level (3.2V) **The device continues erasing until told to stop by the verify command.**

Margin for Error

Allowing erasure to continue too long depletes the charge in floating gates. So you ask—how long is too long? Figure 4 shows the margin for error of a typical device. Following the algorithm would have stopped erasure after 1 second. Cell depletion occurred after 10 seconds giving a 10x margin for error. This 10x margin exists if the erased cell erases in 1 second or 10 seconds (i.e., within the algorithm limits). This chart shows one typical example where the device happened to take 1 second to erase.

Flash memory has generous margin for error over the stopping point defined by the algorithm. The stopping point is defined as the point when all bytes in the chip verify to FFh data. The erase operation duration (Twhwh2) is specified at 10 ms ± 500 μs. Five hundred microseconds offers substantial allowance for system latency during erasure and even for slop in the timer generation. Processors or controllers can execute many lines of code in 500 μs, and the margin for error simply adds another guardband.

Proper software and system design will never rely on the additional margin for error. Remember, you control the program code and system operation during erasure. Once you have fully debugged your driver code, the issue of software control disappears entirely.

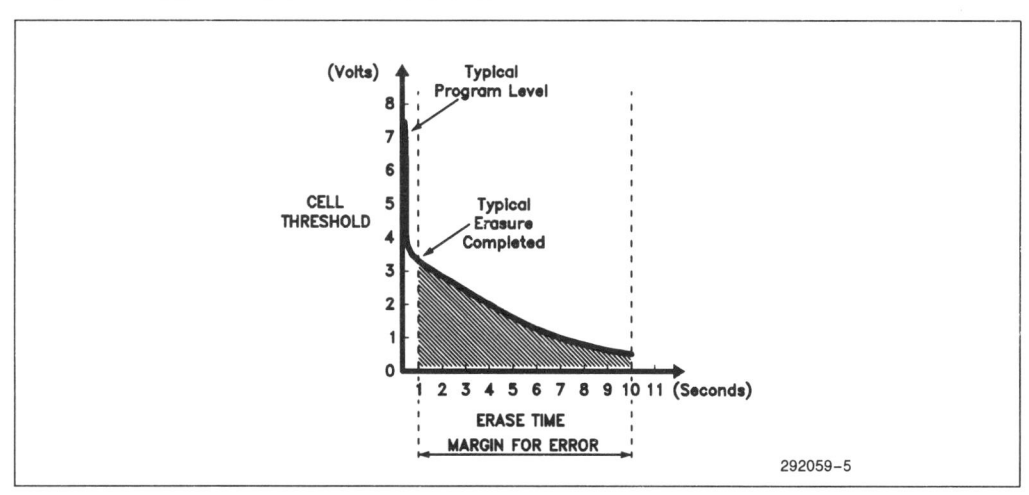

Figure 4. The logarithmic-decaying nature of erasure allows for 10x error in erase time before a device becomes inoperable. Remember, each device has its own erase time, thus the use of an adaptive algorithm.

Most Common Development Issues

Having covered the fundamentals of flash memory reprogramming, let's move on to the system's hardware and software perspectives. The following list of questions might have occurred to you...

- You have defined a system power supply with regulated 5V and 12V outputs (V_{CC} and V_{PP}). Due to the smaller capacitive load on the V_{PP} supply, V_{PP} powers up much faster than the V_{CC} supply. Will this affect the device?
- How does the flash command register architecture reset?
- Suppose your code sends a signal to start erasure, and never tells it to stop?
- Suppose your software delay timers are not calibrated. Instead of stopping erasure after ten milliseconds, the code issues the stop command after 10 seconds?
- Suppose your 6 μs timer used between the erase and erase verify modes is only 2 μs?
- Suppose you decide to skip the first erase operation (program all bytes to zero) because the device is already programmed with data?
- Suppose you are programming and erasing devices in a 16- or 32-bit system?

The answer to all these questions can be found in the following sections. The questions and the reasons all relate to the discussion of how the cell works.

Device Initialization and Reset

Many logic devices which contain command or control registers also have a reset pin. This pin serves two purposes: it resets the device's internal logic; and it synchronizes the device's clock to the system clock.

Intel's flash memory command register and reprogramming circuitry **reset to the read mode** by three means:

1. raising or lowering V_{PP} with $V_{CC} = 5V$;
2. raising V_{CC} with $V_{PP} = 12V$;
3. issuing the reset command twice in succession.

NOTE:
Method 3 stops erasure or programming as well as resets the chip.

A few cases require closer consideration.

Case 1. The System Controls V_{PP} with a Switch

Assuming V_{CC} is stable with V_{PP} switches on, then the command register defaults to the read mode. No power-on reset is required.

Designers might opt to include the V_{PP} switch for either (or both) of two reasons. The first reason is power minimization. Depending on the technology used, a voltage regulator or pump's efficiency can range from 40%–85%. Switching off the V_{PP} supply minimizes system power consumption. See Appendix B for an example V_{PP} generation circuit with ON/OFF control capability.

The second reason is absolute data protection. This feature is not available to 5V-only EEPROM because the reprogramming voltages are generated internally. On that class of memory device, logic glitches can spuriously change data during system power up or power down. Flash memory's 12V power requirement offers absolute control over these concerns; with V_{PP} below V_{CC} + 2V, data protection is guaranteed. Internally, the electric fields are simply too weak to spuriously write data.

Case 2. V_{PP} Powers Up before V_{CC}

Systems with V_{PP} hardwired to a regulated transformer might encounter this case. Typically, V_{CC} will charge many more bypass capacitors than V_{PP}. V_{CC} will therefore power up much more slowly.

The flash memory power-down ($V_{CC} = 0V$) default state blocks V_{PP} from disturbing the array. These conditions hold while V_{CC} is below ~2V. Once V_{CC} rises above ~2V, the internal logic kicks in and resets the device to the read mode. (This is analogous to the internal V_{PP} reset condition described in Case 1.)

Should the three control pins glitch during the power-up phase ($\overline{CE}$low, $\overline{WE}$low, and $\overline{OE}$high), then the command register acts to filter the data. The command port will only react to the correct command sequence.

Designers might opt to hardwire V_{PP} for a number of reasons. The first reason is cost minimization. A regulated 12V secondary from a transformer is commonly available. Adding a switch or a power supply sequencer adds cost and complexity. The second reason involves consideration of the end application. Using the flash memory as a read/write memory requires optimization for the write cycle. Powering V_{PP} on before each write would waste considerable time.

intel AP-325

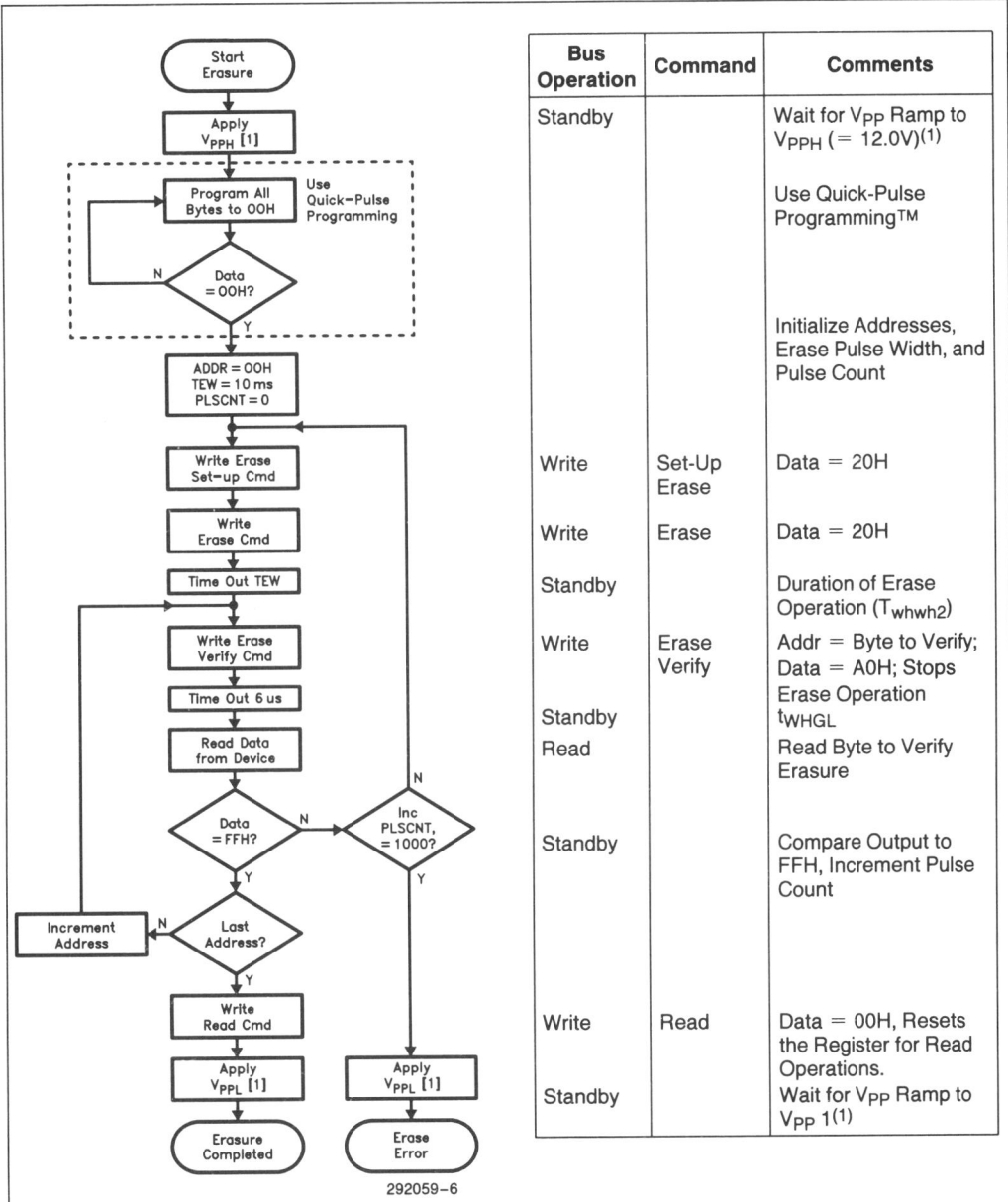

Bus Operation	Command	Comments
Standby		Wait for V_{PP} Ramp to V_{PPH} (= 12.0V)[1]
		Use Quick-Pulse Programming™
		Initialize Addresses, Erase Pulse Width, and Pulse Count
Write	Set-Up Erase	Data = 20H
Write	Erase	Data = 20H
Standby		Duration of Erase Operation (T_{whwh2})
Write	Erase Verify	Addr = Byte to Verify; Data = A0H; Stops Erase Operation
Standby		t_{WHGL}
Read		Read Byte to Verify Erasure
Standby		Compare Output to FFH, Increment Pulse Count
Write	Read	Data = 00H, Resets the Register for Read Operations.
Standby		Wait for V_{PP} Ramp to V_{PP} 1[1]

292059-6

NOTES:
1. The V_{PP} power supply can be hard-wired to the device or switchable. When V_{PP} is switched, V_{PP} 1 may be ground, no-connect with a resistor tied to ground, or less than V_{CC} + 2.0V.
2. Erase verify is performed after chip-erasure. A final read/compare may be performed (optional) after the register is written with the read command.
3. CAUTION: The algorithm MUST BE FOLLOWED to ensure proper and reliable operation of the device.

Figure 5. Quick-Erase™ Algorithm for the 28F010 and 28F512

Case 3. Warm Resets

Warm resets, where the system maintains power while rebooting, requires closer inspection. Consider the situation where the system is reprogramming the flash memory and a hardware or software reset occurs.

The boot software would not realize that programming or erasure is ongoing and would not know to stop the reprogramming operation. Therefore safeguard against this condition with one of two means: 1) ensure that control logic switches V_{PP} off during reset; or 2) reset the flash memory before resetting the processor. For a software reset, simply add the flash memory reset command to the interrupt sequence. For hardware resets, wire the reset switch to the interrupt controller instead of directly to the reset input. Hardware resets would then execute the software interrupt sequence.

The Erase Algorithm Interpreted

The following section offers a block by block explanation of the Quick-Erase™ algorithm shown in Figure 5. Understanding the reasons behind a function will enable you to appreciate the importance of following the algorithm explicitly. Deviations will negatively affect the part's performance and should not be attempted. Note: the effect may not be immediately apparent.

Apply V_{PPH} (Optional, see Discussion on Device Initialization)

Switch on the local V_{PP} supply prior to erasure and programming. The time required for V_{PP} to reach its steady state $12 \pm 0.6V$ depends on the capacitive load and the impedance of the printed circuit board trace. If you measure this delay on a wire-wrapped prototype system, remember that temperature, printed circuit traces and the board's layout change the load seen by the V_{PP} generator. Allow V_{PP} sufficient time to ramp before proceeding with the next step.

Program All Bytes to 00h → Data = 00h?

Prior to erasure, blanket program all addresses in the flash memory to 00h (charge state), **regardless of the previous data**. Verify that each address equals 00h before proceeding to the next address. If you use only part of the memory array, you still need to pre-program the **entire** array for erasure. An example where this is an issue is using a 512K in a 256K socket. A second example is a system where internal microcontroller memory overlaps the external flash memory space.

Programming data 00h equalizes the charge on every bit in the array. This is necessary because erasure removes charge from all cells regardless of their previous state.

For example, reconsider the byte containing AAh data (1010 1010b). If you skip the pre-program step, then during erasure when the data "0" bits get charge removed, the previously erased bits (data "1") lose additional charge. This drives the cell threshold a little lower. The next time you erase the chip and change the code, the threshold will drop to 2.8V.

If the memory transistor is not pre-programmed to data "0" before the next erasure, then its threshold will drop on successive reprogramming cycles (denoted by E3, E4, etc. in Figure 6). Repeated violations of the blanket programming requirement drives the threshold to the point where the transistor is stuck on (data = "1").

Variable Initialization

Initialize two variables and a constant: ADDR (address), PLSCNT (pulse count), and TEW (erase pulse width). The pulse count increments from 0 to a maximum of 1000 erase tries. The erase pulse width remains constant at 10 ms. The address increments from the flash memory starting address to the ending address during verification.

AP-325

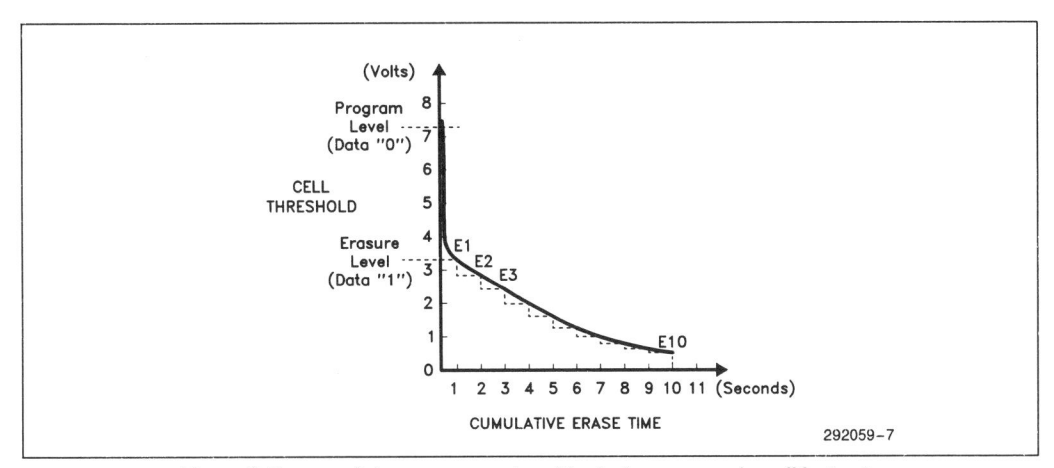

Figure 6. Successful erasure requires blanket programming all bytes to the data "0" level first. This prevents threshold decline on successive erase cycles (E2, E3, etc.). Very low thresholds cause the chip to malfunction.

Write Erase Set-Up Command

Write the erase set-up command (20h) to any flash memory address. This prepares the selected device for erasure, but does not activate the process. A second erase command (20h) is required. Any other data written to the flash memory between the set-up and erase commands will abort the sequence. Once the process is started, it will not stop until told to do so. The correct stop erasure command is Erase Verify (A0h). However, any command including the Reset command is an illegal sequence and will stop erasure as well.

Write Erase Command

The erase command starts the erase process. Internally, the device switches the voltages on all memory cell drains, gates, and sources to the erase configuration.

Time Out Tew (10 ms)

Start your software or hardware timer. Until commanded to verify, the flash memory continues the erase process. Therefore, **assign a high priority to the timer interrupt**. If a higher priority interrupt occurs, stop the erase process and switch contexts (store all variables, registers, etc.). This will allow reentry into the erase procedure in a controlled fashion.

Write Erase Verify Command

Write the erase verify command (A0h) to the flash memory at the address given by the ADDR variable.

The erase verify command performs many tasks. Internally, the device stops erasure and latches the given address for verification. Additionally, the command changes the voltages on the memory cell drains, gates and sources to the erase verify configuration.

Time Out 6 μs

This time out accounts for the internal slew rate of switching the memory array from the erase to the erase verify configuration. Do not attempt to read from the device before 6 μs has passed; the device will appear to still be programmed. This is because you have not allowed sufficient time for the memory to change configurations. Your code will then interpret this as a need for extra erase operations, and will continue erasing the device.

NOTE:
6 μs is a minimum specification. You can use the 10 μs timer developed for the programming algorithm.

intel

AP-325

Read Data from Device
Read the data at the address given by the ADDR variable. This should be the same address driven with the erase verify command.

Data = FFh?
Compare the output data at address ADDR to FFh. If the data equals FFh, then that address has been erased. Continue verification until the last address has been verified or until the maximum erase pulse count (1000) has been reached. Typically, most devices will fully erase within 50–100 erase loops.

Last Address → Increment Address
Check the ADDR variable to see if the last address has been verified. If not, increment the ADDR variable and re-write the erase verify command. Remember to write the erase verify command to address ADDR, since the verify command latches the address. Also, if your system has 64K byte segment boundaries, be sure to increment the base pointer every 64K byte addresses.

Write Read Command
After full chip verification, write the read command (00h) to switch the device to read mode. If you plan to reprogram the device immediately, this step is not necessary.

Apply V_{PPL} (Optional)
Switch V_{PP} off. With V_{PP} left on, the command register offers data protection by requiring a precise sequence to initiate programming or erasure. However, V_{PP} controls overall command register operation. Turning V_{PP} off disables the command register, thus providing absolute data protection. Without the high voltage, the reprogramming mechanisms cannot occur and the component becomes a read only memory (ROM).

Abort/Reset
Whenever a system error condition occurs (reset or reboot), write the Reset command (FFh) to each flash memory twice in succession. This is a good initialization practice in systems leaving V_{PP} at 12V. The processor would be unaware if prior to the reset, it had been in the middle of erasure, and this sequence aborts erasure.

The Program Algorithm Illuminated

The full algorithm will not be interpreted here, although a few items should be noted. You can find a conceptual version of the Quick-Pulse Programming™ algorithm in Chapter 2, Figure 1, and the complete flow chart in Appendix C.

First, similar to erasure, a two-write sequence starts programming. The first write is the Program Set-Up Command which primes the chip for programming. The chip then latches the address and data to be programmed on the second write. You can abort programming by writing the Reset Command twice in succession instead of the data to be programmed.

Second, the device continues programming until commanded to stop by the Program Verify Command. Similar to the Erase Verify Command, this command performs a couple of functions. Internally, it halts programming and latches the given address for verification. Additionally, the command changes the voltages on the memory array's drains, gates, and sources to the program verify configuration.

The cell programming mechanism is self-limiting. However, do not assume that programming all twenty-five 10 µs operations in one pass is the best way to attain reliable operation. To a certain degree, programming stresses the memory cell. The stress is considerably lower than that applied to EEPROM (2 MV/cm lower to be specific). But why stress a component without cause? The adaptive Quick-Pulse Programming algorithm with its fast program operations, minimizes all stresses and affords the greatest reliability.

Finally, after writing the Program Verify Command to the device, wait a minimum of 6 µs before reading the device. The time out accounts for the internal slew rate of switching the memory array from the program to program verify configuration. Do not attempt to read from the device before 6 µs has passed; the device may appear unprogrammed. Your code will then interpret this as a need for extra program operations. It may needlessly reach the 25 operation limit, even though the byte most probably programmed on the first pass.

Ramifications of the Golden Rule

Always follow the erase command with an erase verify command to stop erasure. Interrupt-driven systems must give high priority to servicing the reprogramming timer interrupts. Systems that reset upon a watchdog time-out must reset the flash memory device before rebooting. (See discussion on device initialization.) Likewise any non-maskable interrupt should software- or hardware-reset the flash memory before performing a context switch.

Use an oscilloscope to calibrate all time delays before attempting erasure. The delay modules include the 6 µs, 10 µs, and 10 ms timer routines.

Blanket program all addresses to 00h data before erasing. Verify correct implementation of the programming algorithm with a PROM programmer before attempting erasure. (Chapter 4 explains how this can be done.)

AP-325

16- and 32-bit systems require special attention. Each flash device has its own erase characteristic. Do not assume that if the low byte of a data word is not erased, then the high byte must not be either.

Always follow the listed guidelines and take care while developing your code to eliminate the erase control issue. Consider it similar to implementing any control function. Once the code is debugged and stable, the issue goes away.

You might ask, is it not possible to control erasure through hardware? The alternative to software control is integrated hardware control or an external controller. Either choice adds cost and complexity to the memory solution. Intel's ETOX™ flash memory offers the most reliable, dense, manufacturable, and fastest read/write nonvolatile memory. Other EEPROM approaches which integrate hardware control have drawbacks of multiple transistors per memory cell. This property negatively affects all those attributes offered by Intel's ETOX flash memory.

DEBUGGING YOUR CODE AND OTHER TIPS ON TESTING

As with any software checkout, a few simple principles enable complete flash memory algorithm debug. The following sections offer some hints to make your job easier.

Software Drivers Save You Time

Intel saves you time by offering various processor-family flash memory drivers. You simply edit the files to suit your system. Then assemble the driver, link, and locate it, and you are ready for debug.

These drivers offer the framework for successful flash memory reprogramming, and require some customization to fit your particular system. If your processor's driver is not available, you may use the available driver software as an example. One caution in advance: The drivers have been written in assembly language to give the most speed- and memory- efficient code. However, most people prefer high-level programming.

High-level programming can be used for everything except software-timer generation. Compilers may give different routines with different object code on each compilation. Therefore, the timers must be either hardware-based or coded in assembly language. Software timers also present some risk if there is a frequency upgrade change on the controlling processor. Regenerate and check your timer routines whenever the system clock rate changes.

Timers, Test Loops and Assembly Level Programming

Timing circuits or software play the most crucial role in flash memory reprogramming. Good timers precisely control their function; sloppy timers produce faults. An example of a sloppy timer is one produced by a compiler. Each time high-level languages recompile, the low-level object coding may change. Thus, a timing loop may be 10 ms one compilation, and much longer or shorter the next time.

You can check your timing method with the following simple technique. Develop test loops which call the various timers' routines. For example, implement the 6 μs, 10 μs, and 10 ms timers used with the 28F512 and 28F010. If you have a spare port or peripheral output, use it to trigger an oscilloscope. Follow the trigger call with the timer routines. If you do not have a spare port, write to the flash memory address space before and after each timer call. You can trigger the oscilloscope off of the flash memory $\overline{CE}$ signal. Remember to power-down the system and remove the flash memory before attempting these methods.

Once the timer code has been verified, you can link and locate it to a higher-level erase/program algorithm implementation.

Programming—The Key to Proper Erasure

Earlier sections described the importance of programming 00's prior to erasure. This procedure equalizes the charge on all memory cells; following this step all bits erase in unison.

A conclusive debug technique can check your programming software. Simply use your software to program zeroes into the flash memory and then verify this step using a conventional PROM programmer. Load the PROM programmer's buffer with all zeroes and compare the buffer to the flash memory. If your programmer does not service the flash memory, call the company for the latest software upgrade. Alternatively, one can easily rig the 512K flash memory to look like a 512K EPROM. Simply jumper V_{CC} on a 32-pin socket to a few pins. Note that the 27512 EPROM and 28F512 flash memory have different int$_e$ligent Identifier™ codes. Override the identifier code check to use this method. See Figure 7 for socket details.

Some microcontrollers have limited address space or internal memory that masks certain external address space. Even if you do not use sections of the flash memory, you must still access these sections to program zeroes before erasure. Map and decode port bits to access unused address space, and verify that all bytes are programmed to zero before proceeding with erasure.

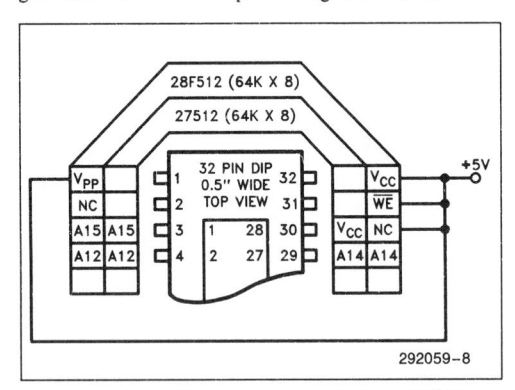

Figure 7. A 28F512 can be read in a PROM programmer as a 27512 by jumpering the appropriate pins to V_{CC}. The same method applies to the 28F010.

16- and 32-Bit Systems— Achieving Optimum Reprogramming Throughput

Erasing flash memory in 16- and 32-bit systems requires special consideration. One could implement the program and erase algorithms in a byte-wise fashion, but this is time-consuming. Alternately, one can treat the multiple flash memory as a data word, and gain optimum performance.

The primary consideration with the latter approach is that one device may program or erase faster than the other(s). Subsequent programming or erasure of a slower device compromises the functionality of the faster device by subjecting it to the slow-device timing.

Consider an example of erasure in a 16-bit system. After 10 passes through the erase operations, both devices verify through address 07C3h. Then at address 07C5h, the processor reads data word 83FFh.

Since erase data is FFFFh, a few bits in the upper byte/device have not erased. The natural flow of the algorithm would dictate another erase operation. But what about the lower device? Could it be completely erased?

Of course it could be; every device erases at a different rate and the algorithm has only checked up to address 07C5h.

You **can** take advantage of the data rate of wider data buses by utilizing the command register, the reset command and an analysis of erasure. Each erase operation is 10 ms. Each byte verification takes 6 μs. Therefore, erasure takes three orders of magnitude longer than verification. Optimization for erasure yields the optimum performance because verification is a second order effect.

Let us reconsider the previous example. At address 07C5h, the data word does not verify. On the next erase operation edit the erase (and erase verify) command word such that only the high byte gets the erase command, and the low byte gets a reset command. (i.e., change the command word from 2020h to 20FFh).

See Application Note AP-316 for a detailed flow chart for this approach. Note this document is based on the 28F256; however, most concepts carry through to the higher density devices. (Literature Order number 292046-001).

Logic Analyzers and In-Circuit Emulators

Many programmers use logic analyzers and in-circuit emulators to debug code. These approaches are fine for flash memory algorithm debug if certain conditions are met.

1. Check timing routines with an oscilloscope; there is no alternative.

2. Know your code and set breakpoints intelligently. One designer had the bad luck of throwing in a breakpoint on the line immediately following an erase command. Because of the breakpoint, the device started erasing and never received the verify command. He hit the system reset and stopped erasure after a few minutes of contemplating what to do next; it was too late

3. Single step through your reprogramming code, if and only if, the flash memory device is removed from the system.

Testing Your Software—One More Time

Some flash memories specify 100 erase/program cycles. This is a minimum specification; Intel flash memories cycle 10,000–100,000 times. With this in mind, feel free to check and recheck reprogramming operations. There is **no** reliability risk in doing so.

AP-325

One confidence-raising test is similar to that done on systems: stress the system/software by executing test code numerous times consecutively. Set the reprogramming drivers in a loop, and let them run 20–40 times. On each consecutive pass, use a constant data pattern such as 0AAh. This tests the reprogramming code from a quasi-static perspective. Missing is the true system environment. In the true system environment, multiple inputs compete with the flash memory for interrupt priority. Also, RF noise from motors can cause spikes and glitching on V_{PP} or V_{CC}. Additionally, fully loaded systems or partially loaded systems might have different V_{PP} response characteristics or noise levels. Signals that look clean in the lab, might not be all that clean in the true operating environment. Therefore, flash reprogramming tests should be done in the true system environment as a final test.

Watchdog Timer Debug Circuit

This section describes a simple tool you can build for debugging your code. An EPLD watches the flash memory data bus and control signals for the erase sequence—erase set-up, erase, and erase verify commands. Once the CPU initiates erasure, the debug tool starts a 15 ms timer. Should the timer count down prior to receiving the erase verify command, then the circuit switches V_{PP} off.

This tool does not check for the other items discussed in the **Tips on Testing** section; you must still check those yourself.

Figure 8 shows the circuit schematic. The EPLD source code and the name of an Intel EPLD applications engineer is located in Appendix D.

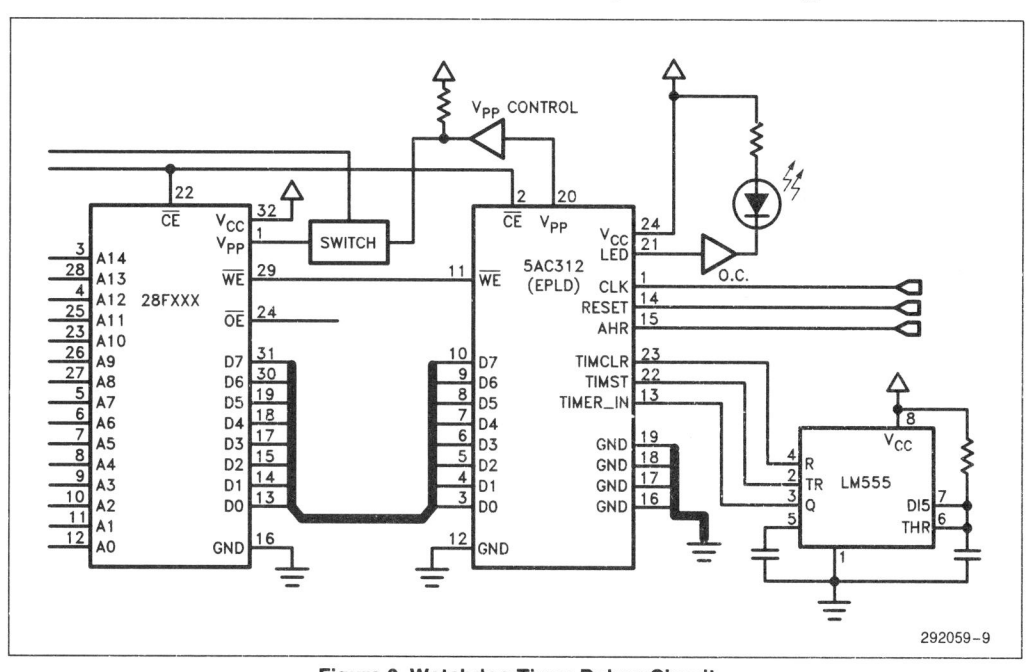

Figure 8. Watchdog Timer Debug Circuit

intel® AP-325

TROUBLE SHOOTING GUIDE

Determining the Root Cause—Software Error vs Device Damage

The three major indications of a flash memory problem are labeled in the following section. The subsequent paragraphs define potential root causes to investigate.

I. **The Device Does Not Program**

Did it program before?

A. **No.**

1. Trigger your oscilloscope on $\overline{CE}$ while probing V_{PP}. Verify that V_{PP} has reached a steady-state 12V when the device is first written.
2. Set the time-base to 10 μs/division (the duration of the program operation). Trigger on $\overline{CE}$ and probe $\overline{WE}$ (look at both traces). Check the duration of the 10 μs program operation time delay. Also, check the duration of the 6 μs delay between writing the program verify command and read.
3. Look for ringing on V_{PP} when V_{PP} has been switched on. Over-voltage stress on V_{PP} (ringing with amplitude greater than 13V) will destroy V_{PP}'s silicon structure.
4. Power the system down and back up. Look for destructive glitches on V_{CC} or V_{PP} (greater than 7V and 13V respectively).
5. Verify erasure and programming on a PROM programmer (if available). Fill the programmer buffer first with 00h data and program the buffer to the flash memory. Then erase the device, and repeat with AAh data. Repeat the last step with 55h data. This sequence fully exercises the array, the input buffers and the output buffers. If all tests pass then check for a hardware or software error.

B. **Yes.**

1. Have you done anything that may have ESD zapped the devices (i.e., touched the devices while not being grounded, re-wired the protoboard with the components socketed, etc.)? If yes, check part as outlined in section 1.A.5.
2. Have you attempted erasure? If yes, verify your algorithm as outlined in Chapter 4. Also, implement the in-system int$_e$ligent Identifier™ mode. If the device outputs an incorrect code, then either an output has been zapped or the golden rule has been violated. Section 1.A.5 describes a method of checking for ESD damage.

II. **The Device Does not Erase**

Did it erase before?

A. **No.**

Follow steps 1–5 outlined in section I. When performing step 2, adjust the oscilloscope time base to 10 ms/div.

B. **Yes.**

Has the board design, clock rate or software changed? System clock rates directly affect the accuracy of software timers. See AP-316 for a discussion on software timing versus clock rate.

III. **The Device Erases Spuriously**

Exercise all system functions while monitoring the flash memory chip selects. Verify that I/O mapped addresses or logic are not accidentally selecting the flash memory. For example, the space bar character sent from a keyboard controller happens to be 20h. If the flash memory is accidentally selected while this data is on the bus, then erasure will commence on the following cycle when the condition occurs again.

intel® AP-325

APPENDIX A

TWO APPROACHES TO ALGORITHMS

Adaptive

[Flowchart: Start Program → Program Verify-Read Data from Device → Valid Data? → Y: Program Success; N: Inc COUNT = 25? → N: loop back; Y: Program Error]

292059-10

Brute Force

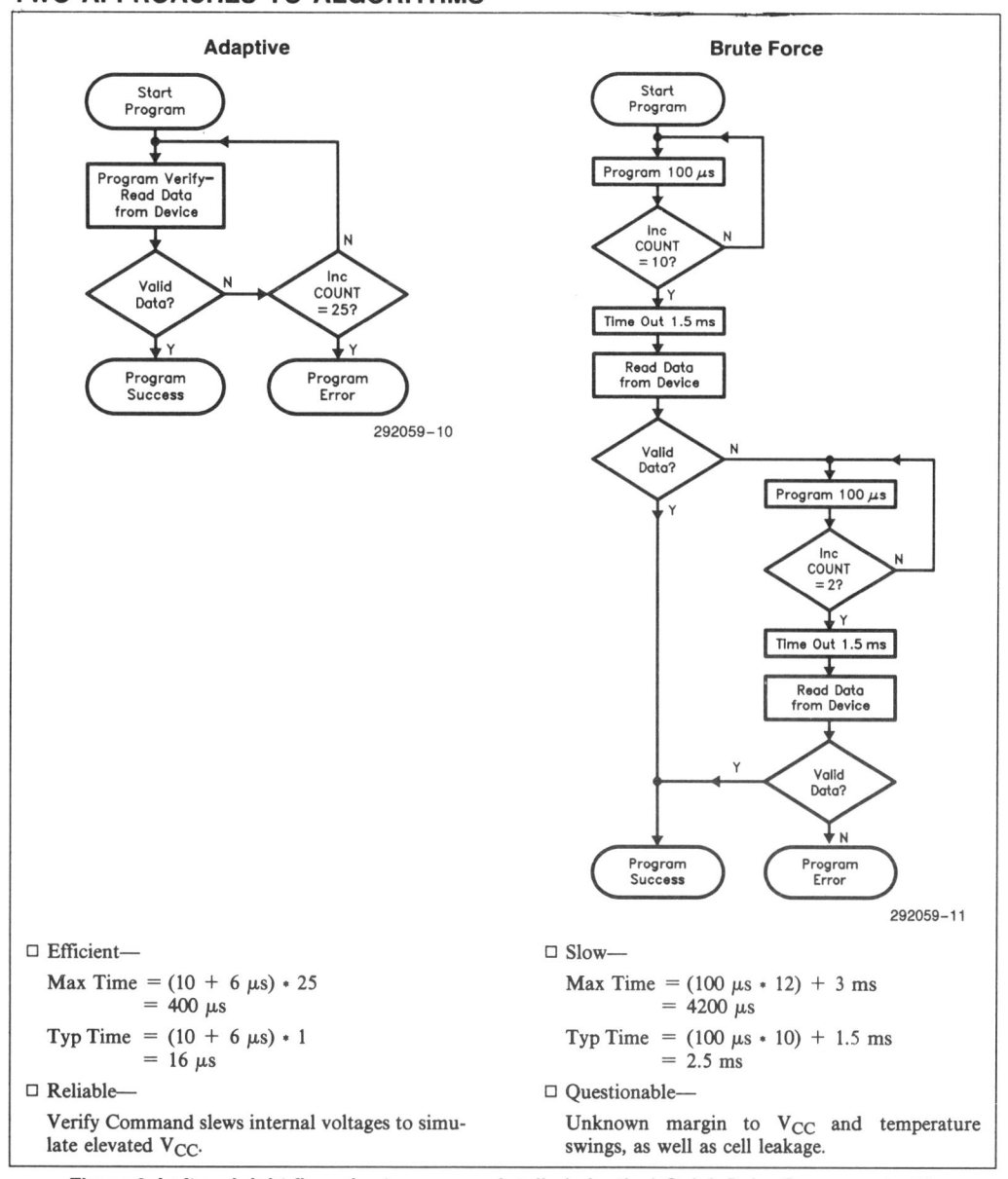

[Flowchart: Start Program → Program 100 μs → Inc COUNT = 10? → N: loop; Y: Time Out 1.5 ms → Read Data from Device → Valid Data? → Y: continues; N: Program 100 μs → Inc COUNT = 2? → N: loop; Y: Time Out 1.5 ms → Read Data from Device → Valid Data? → Y: Program Success; N: Program Error]

292059-11

- ☐ Efficient—
  Max Time $= (10 + 6\ \mu s) \cdot 25$
  $= 400\ \mu s$
  Typ Time $= (10 + 6\ \mu s) \cdot 1$
  $= 16\ \mu s$
- ☐ Reliable—
  Verify Command slews internal voltages to simulate elevated V_{CC}.

- ☐ Slow—
  Max Time $= (100\ \mu s \cdot 12) + 3\ ms$
  $= 4200\ \mu s$
  Typ Time $= (100\ \mu s \cdot 10) + 1.5\ ms$
  $= 2.5\ ms$
- ☐ Questionable—
  Unknown margin to V_{CC} and temperature swings, as well as cell leakage.

Figure 9. Left and right flow charts compare Intel's (adaptive) Quick-Pulse Programming™ algorithm and another company's (brute force) approach to flash memory programming.

APPENDIX B

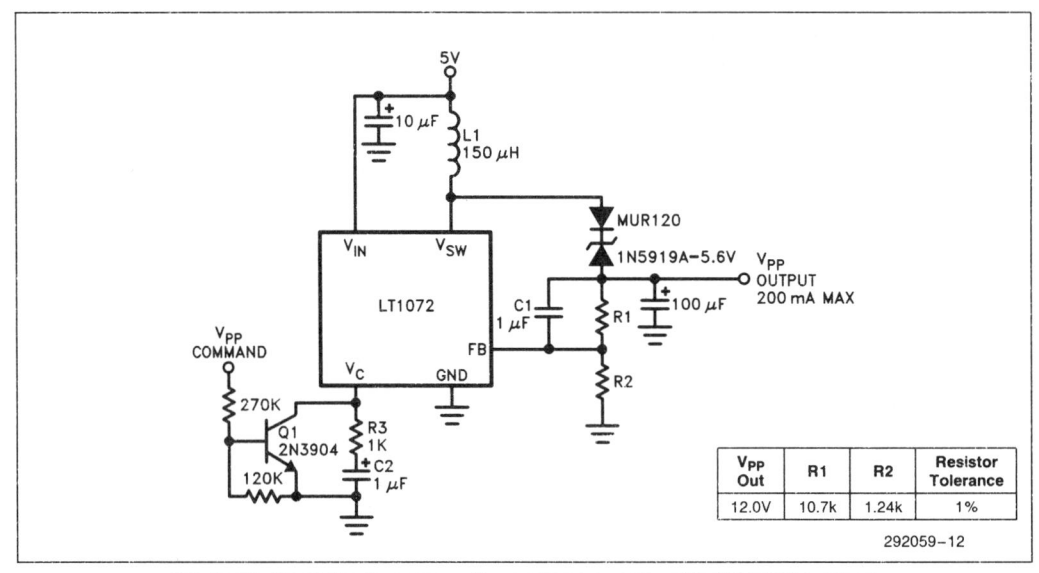

Figure 10. Basic flash memory V$_{PP}$ voltage supply with ON/OFF control.
When V$_{PP}$ COMMAND goes low, the Linear Technology LT1072 switching regulator produces 12V.
This circuit is just one example of a V$_{PP}$ supply.

APPENDIX C
QUICK-PULSE PROGRAMMING™ ALGORITHM

Bus Operation	Command	Comments
Standby		Wait for V_{PP} Ramp to V_{PPh} (= 12.0V)[1]
		Initialize Pulse-Count
Write	Set-Up Program	Data = 40H
Write	Program	Valid Address/Data
Standby		Duration of Program Operation (t_{WHWH1})
Write	Program[2] Verify	Data = C0H; Stops Program Operation
Standby		t_{WHGL}
Read		Read Byte to Verify Programming
Standby		Compare Data Output to Data Expected
Write	Read	Data = 00H, Resets the Register for Read Operations.
Standby		Wait for V_{PP} Ramp to V_{PPl} [1]

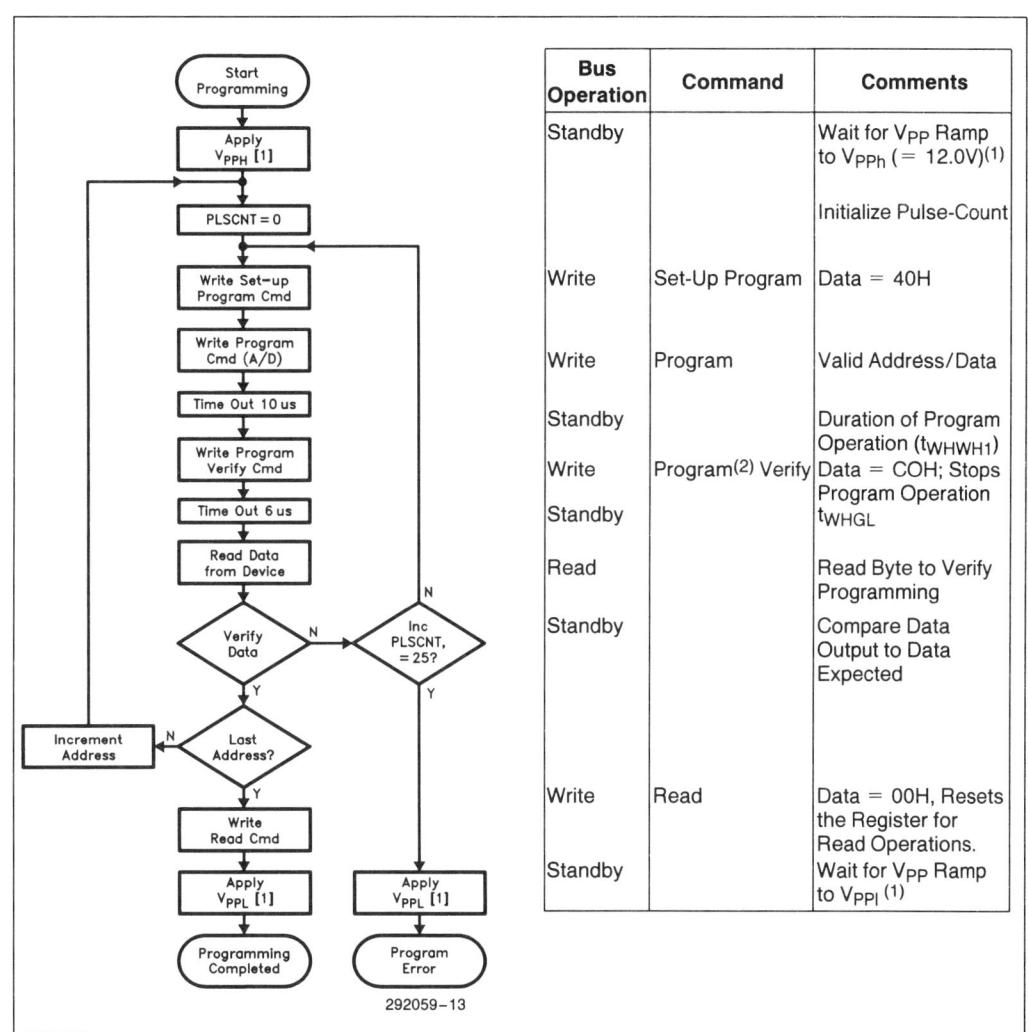

292059-13

NOTES:
1. The V_{PP} power supply can be hard-wired to the device or switchable. When V_{PP} is switched, V_{PP} l may be ground, no-connect with a resistor tied to ground, or less than V_{CC} + 2.0V.
2. Program Verify is only performed after byte programming. A final read/compare may be performed (optional) after the register is written with the Read command.
3. CAUTION: The algorithm MUST BE FOLLOWED to ensure proper and reliable operation of the device.

APPENDIX D
WATCHDOG TIMER CIRCUIT

EPLD Source Code and Applications Contact Person

```
Thom Bowns_PLFG Applications
Intel
January 5, 1989
EPLD HOTLINE: 1-800-323-EPLD
Rev. 008
5AC312
Watchdog timer to cut Vpp from FLASH if erase cycle too long.
OPTIONS: TURBO = ON
PART: 5AC312

INPUTS:     CLK, D0@3, D1@4, D2@5, D3@6, D4@7, D5@8, D6@9, D7@10,
            nCE@2, nWE@11, TIMER_IN@13, RESET@14, AHR@15

OUTPUTS:    TIMCLR@23, TIMST@22, LED@21, VPP@20

NETWORK:
            CLK = INP (CLK)
            D0 = INP (D0)
            D1 = INP (D1)
            D2 = INP (D2)
            D3 = INP (D3)
            D4 = INP (D4)
            D5 = INP (D5)
            D6 = INP (D6)
            D7 = INP (D7)
            nCE = INP (nCE)
            nWE = INP (nWE)
            TIMER_IN = INP (TIMER_IN)
            RESET = INP (RESET)
            AHR = INP (AHR)           % RESET active high if AHR is high %
            COND1 = NOCF (C1d)        % Conditions 1-4 are routed        %
            COND2 = NOCF (C2d)        % through combinatorial feedbacks  %
            COND3 = NOCF (C3d)        % to reduce product term count.    %
            COND4 = NOCF (C4d)        %                                  %
            CLR = NOCF (CLRd)

EQUATIONS:
            CLRd = RESET * AHR + !RESET * !AHR;
            TIMEOUT = /TIMER_IN;
            C1d = (/nCE * /nWE * 20H);      % Write 20 %
            C2d = (/nCE * /nWE * A0H);      % Write A0 %
            C3d = (/nCE * /nWE * /20H);     % Write other than 20 %
            C4d = (/nCE * /nWE * /A0H);     % Write other than A0 %
            20H = /D7 * /D6 * /D5 * /D4 * /D3 * /D2 * /D1 * /D0;
            A0H = /D7 * /D6 * /D5 * /D4 * /D3 * /D2 * /D1 * /D0;
```

AP-325

```
MATCHING: WATCHDOG
CLOCK:    CLK

STATES:    [ VPP   LED   TIMST   TIMCLR   XSB ]
  START    [  0     0      0        0       0  ]
   S1      [  1     0      0        0       0  ]
   S2      [  1     0      1        0       0  ]
   S3      [  1     0      1        1       0  ]
   S4      [  1     0      0        1       0  ]
   S5      [  1     0      1        1       1  ]
   S6      [  1     0      1        0       1  ]
   S7      [  0     1      1        0       0  ]

           % TRANSITION STATEMENTS %

START: S1                          % From power up, go to S1 right away %
S1:    IF COND1    THEN S2         % If write 20, go to next state      %
S2:    IF /COND1   THEN S3         % Until not write 20, hold           %
       IF CLR      THEN S1
S3:    IF COND1    THEN S4         % If another write 20, start timer   %
       IF COND3    THEN S7         % If write other than 20, error      %
       IF CLR      THEN S1
S4:    S5                          % Trigger timer then go to S5 loop   %
S5:    IF COND2    THEN S6         % If write A0, stop timer            %
       IF TIMEOUT  THEN S7         % If timer times out,
                                     go to error state                  %
       IF COND4    THEN S7         % If write other than A0, error      %
S6:    S1                          % Stop timer and go back to S1       %
S7:    IF CLR      THEN S1         % Error state. wait for a RESET.     %

END$
```

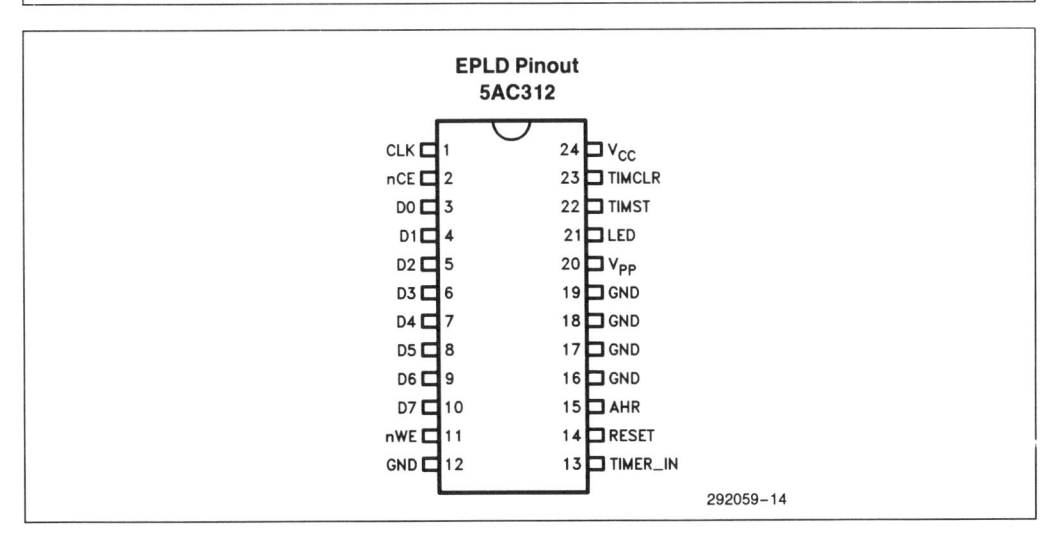

**EPLD Pinout
5AC312**

```
CLK   □ 1        24 □ Vcc
nCE   □ 2        23 □ TIMCLR
D0    □ 3        22 □ TIMST
D1    □ 4        21 □ LED
D2    □ 5        20 □ VPP
D3    □ 6        19 □ GND
D4    □ 7        18 □ GND
D5    □ 8        17 □ GND
D6    □ 9        16 □ GND
D7    □ 10       15 □ AHR
nWE   □ 11       14 □ RESET
GND   □ 12       13 □ TIMER_IN
```

292059-14

APPENDIX E

Checklist: Most Common Mistakes that May Lead to Excessive Erasure

- not programming all bytes to 00 data prior to erasure;
- not observing the 6 μs set-up times between programming or erasure and verification;
- attempting to program before V_{PP} is at 12V (Capacitive Load)
- not latching the erase verify address with the erase verify command, or changing the address on the subsequent read cycle;
- an erase operation longer than the specified duration in the data sheets.

Chapters three and four discuss the correct methods of developing and debugging code to diminish the possibility of making these mistakes.

EUROPEAN SALES OFFICES

DENMARK

Intel
Glentevej 61, 3rd Floor
2400 Copenhagen NV
Tel: (45) (31) 19 80 33
Telex: 19567

FINLAND

Intel
Ruosilantie 2
00390 Helsinki
Tel: (358) 0 544 644
Telex: 123332

FRANCE

Intel
1, rue Edison - BP 303
78054 St Quentin-en-Yvelines Cedex
Tel: (33) (1) 30 57 70 00
Telex: 699016

NETHERLANDS

Intel
Postbus 84130
3099 CC Rotterdam
Tel: (31) 10 407 11 11
Telex: 22283

ISRAEL

Intel
Atidim Industrial Park-Neve Sharet
P.O. Box 43202
Tel Aviv 61430
Tel: (972) 03-498080
Telex: 371215

ITALY

Intel
Milanofiori Palazzo E
20090 Assago
Milano
Tel: (39) (02) 89200950
Telex: 341286

NORWAY

Intel
Hvamveien 4 - PO Box 92
2013 Skjetten
Tel: (47) (6) 842 420
Telex: 78018

SPAIN

Intel
Zurbaran, 28
28010 Madrid
Tel: (34) 308 25 52
Telex: 46880

SWEDEN

Intel
Dalvagen 24
171 36 Solna
Tel: (46) 8 734 01 00
Telex: 12261

SWITZERLAND

Intel
Zuerichstrasse
8185 Winkel-Rueti bei Zuerich
Tel: (41) 01/860 62 62
Telex: 825977

U.K.

Intel
Pipers Way
Swindon, Wilts SN3 1RJ
Tel: (44) (0793) 696000
Telex: 444447/8

WEST GERMANY

Intel
Dornacher Strasse 1
8016 Feldkirchen bei Muenchen
Tel: (49)/089/90992-0
Telex: 5-23177

Intel
Hohenzollern Strasse 5
3000 Hannover 1
Tel: (49) 0511/344081
Telex: 9-23625

Intel
Abraham Lincoln Strasse 16-18
6200 Wiesbaden
Tel: (49) 06121/7605-0
Telex: 4-186183

Intel
Zettachring 10A
7000 Stuttgart 80
Tel: (49) 0711/7287-280
Telex: 7-254826

EUROPEAN DISTRIBUTORS/REPRESENTATIVES

AUSTRIA

Bacher Electronics G m b H
Rotenmuehlgasse 26
1120 Wien
Tel: (43) (0222) 83 56 46
Telex: 31532

BELGIUM

Inelco Belgium S A
Av des Croix de Guerre 94
1120 Bruxelles
Oorlogskruisenlaan, 94
1120 Brussel
Tel: (32) (02) 216 01 60
Telex: 64475 or 22090

DENMARK

ITT-Multikomponent
Naverland 29
2600 Glostrup
Tel: (45) (0) 2 45 66 45
Telex: 33 355

FINLAND

OY Fintronic AB
Melkonkatu 24A
00210 Helsinki
Tel: (358) (0) 6926022
Telex: 124224

FRANCE

Almex
Zone industrielle d'Antony
48, rue de l'Aubépine
BP 102
92164 Antony cedex
Tel: (33) (1) 46 66 21 12
Telex: 250067

Jermyn-Generim
60, rue des Gémeaux
Silic 580
94653 Rungis cedex
Tel: (33) (1) 49 78 49 78
Telex: 260967

Métrologie
Tour d'Asnières
4, av. Laurent-Cely
92606 Asnières Cedex
Tel: (33) (1) 47 90 62 40
Telex: 611448

Tekelec-Airtronic
Cité des Bruyères
Rue Carle-Vernet — BP 2
92310 Sèvres
Tel: (33) (1) 45 34 75 35
Telex: 204552

IRELAND

Micro Marketing Ltd
Glenageary Office Park
Glenageary
Co Dublin
Tel: (21) (353) (01) 85 63 25
Telex: 31584

ISRAEL

Eastronics Ltd
11 Rozanis St
P O B 39300
Tel-Aviv 61392
Tel: (972) 03-475151
Telex: 33638

ITALY

Intesi
Divisione ITT Industries GmbH
Viale Milanofiori
Palazzo E/5
20090 Assago (MI)
Tel: (39) 02/824701
Telex: 311351

Lasi Elettronica S p A
V le Fulvio Testi, 126
20092 Cinisello Balsamo (MI)
Tel: (39) 02/2440012
Telex: 352040

ITT Multicomponents
Viale Milanofiori E/5
20090 Assago (MI)
Tel: (39) 02/824701
Telex: 311351

Silverstar
Via Dei Gracchi 20
20146 Milano
Tel: (39) 02/49961
Telex: 332189

NETHERLANDS

Koning en Hartman Elektrotechniek B V
Energieweg 1
2627 AP Delft
Tel: (31) (01) 15/609906
Telex: 38250

NORWAY

Nordisk Elektronikk (Norge) A/S
Postboks 123
Smedsvingen 4
1364 Hvalstad
Tel: (47) (02) 84 62 10
Telex: 77546

PORTUGAL

ATD Portugal LDA
Rua Dos Lusiados, 5
5 Sala B
1300 Lisboa
Tel: (35) (1) 648 091

Ditram
Avenida Miguel Bombarda, 133
1000 Lisboa
Tel: (35) (1) 734 884
Telex: 14182

SPAIN

ATD Electronica, S A
Plaza Ciudad de Viena, 6
28040 Madrid
Tel: (34) (1) 234 40 00
Telex: 42754

ITT-SESA
Calle Miguel Angel, 21-3
28010 Madrid
Tel: (34) (1) 419 54 00
Telex: 27461

Metrologia Iberica, S A
Ctra. de Fuencarral, n 80
28100 Alcobendas (Madrid)
Tel: (34) (1) 653 86 11

SWEDEN

Nordisk Elektronik AB
Huvudstagatan 1
Box 1409
171 27 Solna
Tel: (46) 08-734 97 70
Telex: 105 47

SWITZERLAND

Industrade A G
Hertistrasse 31
8304 Wallisellen
Tel: (41) (01) 8328111
Telex: 56788

TURKEY

EMPA Electronic
Lindwurmstrasse 95A
8000 Muenchen 2
Tel: (49) 089/53 80 570
Telex: 528573

UNITED KINGDOM

Accent Electronic Components Ltd
Jubilee House
Jubilee Road
Letchworth
Herts SG6 1TL
Tel: (44) (0462) 686666
Telex: 826293

Bytech-Comway Systems
3 The Western Centre
Western Road
Bracknell RG12 1RW
Tel: (44) (0344) 55333
Telex: 847201

Jermyn
Vestry Estate
Otford Road
Sevenoaks
Kent TN14 5EU
Tel: (44) (0732) 450144
Telex: 95142

MMD
Unit 8 Southview Park
Caversham
Reading
Berkshire RG4 0AF
Tel: (44) (0734) 481666
Telex: 846669

Rapid Systems
Rapid House
Denmark Street
High Wycombe
Buckinghamshire HP11 2ER
Tel: (44) (0494) 450244
Telex: 837931

Rapid Silicon
Rapid House
Denmark Street
High Wycombe
Buckinghamshire HP11 2ER
Tel: (44) (0494) 442266
Telex: 837931

WEST GERMANY

Electronic 2000 AG
Stahlgruberring 12
8000 Muenchen 82
Tel: (49) 089/42001-0
Telex: 522561

ITT Multikomponent GmbH
Postfach 1265
Bahnhofstrasse 44
7141 Moeglingen
Tel: (49) 07141/4879
Telex: 7264472

Jermyn GmbH
Im Dachstueck 9
6250 Limburg
Tel: (49) 06431/508-0
Telex: 415257-0

Metrologie GmbH
Meglingerstrasse 49
8000 Muenchen 71
Tel: (49) 089/78042-0
Telex: 5213189

Proelectron Vertriebs GmbH
Max Planck Strasse 1-3
6072 Dreieich
Tel: (49) 06103/30434 3
Telex: 417903

YUGOSLAVIA

H.R. Microelectronics Corp
2005 de la Cruz Blvd., Ste 223
Santa Clara, CA 95050
USA
Tel: (1) (408) 988-0286
Telex: 387452

Rapido Electronic Components, S P A
Via C Beccaria, 8
34133 Trieste,
Italia
Tel: (39) 040/360555
Telex: 460461

Eurolit 05/89

intel® RELIABILITY REPORT

RR-60

November 1989

ETOX™ Flash Memory Reliability Data Summary

© Intel Corporation, 1989

Order Number: 293002-006

Intel Corporation makes no warranty for the use of its products and assumes no responsibility for any errors which may appear in this document nor does it make a commitment to update the information contained herein.

Intel retains the right to make changes to these specifications at any time, without notice.

Contact your local sales office to obtain the latest specifications before placing your order.

The following are trademarks of Intel Corporation and may only be used to identify Intel Products:

376, 386, 387, 486, 4-SITE, Above, ACE51, ACE96, ACE186, ACE196, ACE960, BITBUS, COMMputer, CREDIT, Data Pipeline, DVI, ETOX, FaxBACK, Genius, i, î, i486, i750, i860, ICE, iCEL, ICEVIEW, iCS, iDBP, iDIS, I2ICE, iLBX, iMDDX, iMMX, Inboard, Insite, Intel, int$_e$l, Intel386, int$_e$lBOS, Intel Certified, Intelevision, int$_e$ligent Identifier, int$_e$ligent Programming, Intellec, Intellink, iOSP, iPAT, iPDS, iPSC, iRMK, iRMX, iSBC, iSBX, iSDM, iSXM, Library Manager, MAPNET, MCS, Megachassis, MICROMAINFRAME, MULTIBUS, MULTICHANNEL, MULTIMODULE, MultiSERVER, ONCE, OpenNET, OTP, PRO750, PROMPT, Promware, QUEST, QueX, Quick-Erase, Quick-Pulse Programming, Ripplemode, RMX/80, RUPI, Seamless, SLD, SugarCube, ToolTALK, UPI, Visual Edge, VLSiCEL, and ZapCode, and the combination of ICE, iCS, iRMX, iSBC, iSBX, iSXM, MCS, or UPI and a numerical suffix.

MDS is an ordering code only and is not used as a product name or trademark. MDS® is a registered trademark of Mohawk Data Sciences Corporation.

*MULTIBUS is a patented Intel bus.

CHMOS and HMOS are patented processes of Intel Corp.

Intel Corporation and Intel's FASTPATH are not affiliated with Kinetics, a division of Excelan, Inc. or its FASTPATH trademark or products.

Additional copies of this manual or other Intel literature may be obtained from:

Intel Corporation
Literature Sales
P.O. Box 7641
Mt. Prospect, IL 60056-7641

ETOX™ FLASH MEMORY RELIABILITY DATA SUMMARY

CONTENTS PAGE

THE IMPORTANCE OF RELIABILITY 1
Quality ≠ Reliability 1
Monitor Program 1

ETOX™ FLASH MEMORY TECHNOLOGY OVERVIEW 1
Similarities with EPROM 2
Differences from EPROM 2
Erase/Write Cycling 3

ETOX FLASH MEMORY RELIABILITY TESTING 3

FAILURE RATE CALCULATIONS 4

RELIABILITY DATA SUMMARY 6

27F64 7

28F256 8

28F512 10

28F010 12

PLASTIC RELIABILITY DATA SUMMARY 14
Introduction 14
Plastic Package Characteristics 14
Electrical Characteristics 14
Reliability/Quality Stresses 14

QUALITY/RELIABILITY STANDARDS 15

N28F256 16

APPENDIX A. FAILURE RATE CALCULATIONS FOR 60% UPPER CONFIDENCE LEVEL A-1

APPENDIX B. FLASH MEMORY BIT MAPS AND DIE PHOTOGRAPHS B-1

RR-60

THE IMPORTANCE OF RELIABILITY

Reliability of the non-volatile memories in your end product is critical to your total system reliability. The use of Intel flash memories can make a difference. Reliability is not just tested, but designed into each component Intel manufactures.

Quality ≠ Reliability

A quality component is one that meets your specification when received and tested. A reliable component continues to meet your specification even years after you have shipped your product.

CONSIDER QUALITY VS. RELIABILITY

The true cost of any component involves more than just the purchase price. The true component cost encompasses the initial purchase price, cost of rework during system production, and the cost of field repairs due to component failures. "Rework" costs during system production are incurred prior to shipment of your end product, and are a function of the quality of the component you purchase.

Repair costs incurred in the field after end product shipments, are a function of the reliability of the components. In addition to the increasing real cost of a system field service call, there is the intangible cost of a poor reliability reputation to the end use of your product. These costs depend upon the reliability of the components you purchase. Thus, reliability may impact costs during the system lifetime more than the initial quality of the components!

In-circuit reprogrammability of flash memories enables the addition of production line testing and system level screening. This capability, along with the inherent reliability of Intel flash components, provides your systems with significant reliability enhancements. Soldering the flash memory directly to the board enhances contact integrity. Since flash memories do not have to be removed for reprogramming, reliability risk due to handling is eliminated upon device installation. In addition, single socket testing reduces component handling during incoming inspection.

Monitor Program

Reliability is designed into each component Intel manufactures. From the moment the design is put to paper, stringent reliability standards must be met at each step for a product to bear the Intel name.

Designing-in reliability, however, is only the beginning. Ongoing tests must be conducted to ensure that the original reliability specifications remain as valid in volume production as they were when the device was first qualified.

Intel's Reliability Monitor Program, devised to measure and control device reliability in production, is available to our customers. The Monitor Program subjects all of Intel's technologies to a 48 hour dynamic burn-in at 125°C (with a portion of these devices continued for a 1000 hour lifetest) and provides answers about device reliability that are not generally available from limited testing programs. When test rejects are encountered, failure analysis is performed on each failed part. Isolating the fault and determining the failure mechanism is a critical part of the Monitor Program.

The primary objective is to deliver reliable, quality devices. Actions that Intel takes to meet this objective may include a process or design change, or added reliability screen. Each decision is made with our customers in mind so that they receive the parts—and the performance—that they ordered by specifying Intel. Reliability qualification assures that all new production material meets Intel's reliability standards. The Reliability Monitor Program ensures that these high standards are continually maintained over the duration of a device's life. This reliability improves the lifetime reputation of your product, reducing the required number of field service calls.

ETOX™ FLASH MEMORY TECHNOLOGY OVERVIEW

Intel's ETOX™ and ETOX II (EPROM tunnel oxide) flash memory technologies* consist of a non-volatile memory cell that electrically erases in bulk array form. Derived from Intel's CHMOS** II-E EPROM technology, ETOX flash memory technology combines the EPROM program mechanism with the E^2PROM erase mechanism. The memory cell is composed of a single transistor with a floating gate for charge storage, like the conventional EPROM. The primary difference between flash memory and EPROM cells is the flash memory cell's thinner gate oxide, which enables the electrical erase capability. This report compares and contrasts ETOX technology and EPROM reliability, describes Intel's flash reliability testing methodology, and summarizes the reliability data of Intel's flash memories.

---

*Intel's ETOX flash memory process has patents pending.
**CHMOS is a patented process of Intel Corporation.

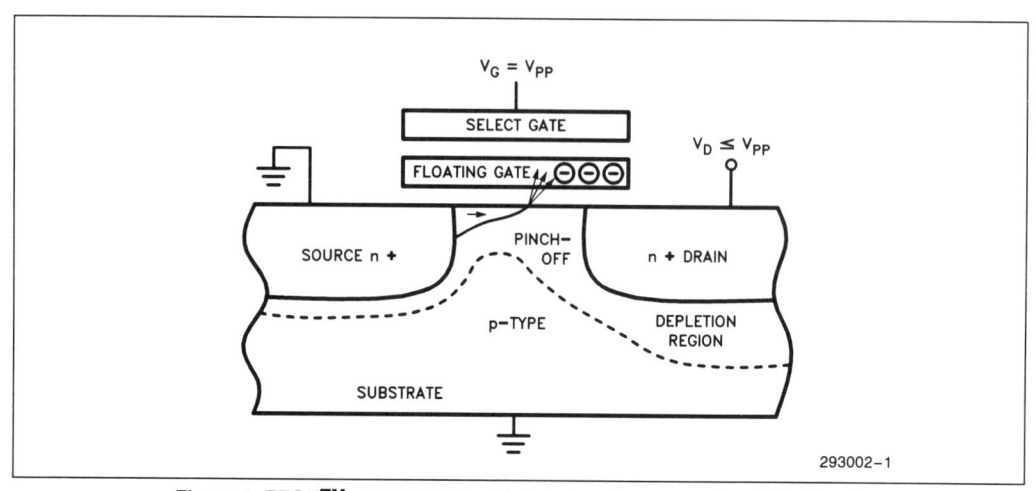

Figure 1. ETOX™ Flash Memory Cell during Programming (Side View)

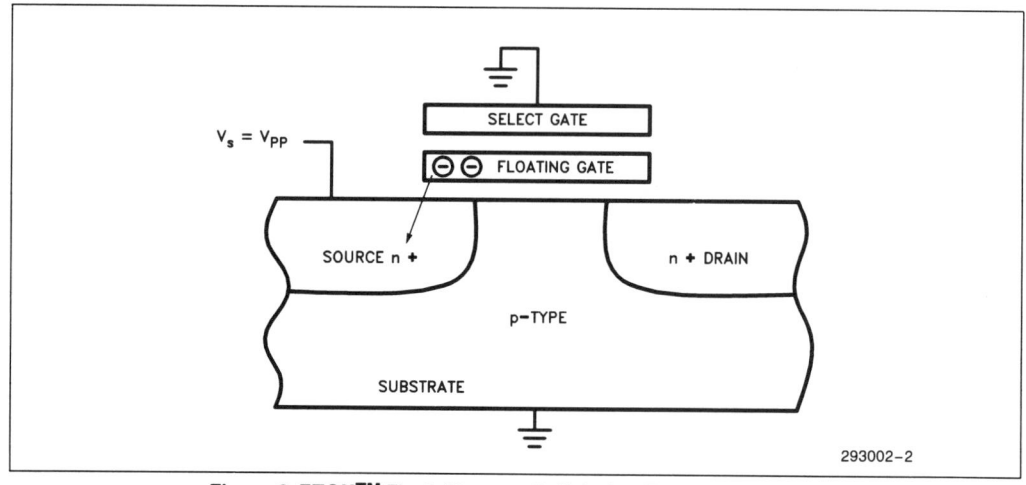

Figure 2. ETOX™ Flash Memory Cell during Erase (Side View)

Similarities with EPROM

When in program mode, a flash memory behaves exactly like a conventional EPROM. A high drain voltage generates "HOT" electrons that are swept across the channel. High voltage on the control gate attracts these free electrons across the lower gate oxide into the floating gate, where they are trapped. See Figure 1. Thus, ETOX flash memory cells exhibit the same reliability characteristics as conventional EPROMs during program mode even with a thinner oxide. When in read mode, a flash memory behaves just like an EPROM.

Differences from EPROM

With respect to functionality, the major difference between flash memory technology and EPROM technology lies with the erase mechanism. For EPROM cells, ultraviolet light neutralizes the charge on the floating gate, thus erasing the cell. For ETOX flash memory cells, an electric field across the lower gate oxide pulls electrons off the floating gate to the source region, thus erasing the cell. See Figure 2. This erase mechanism is an E^2PROM adaptation using "Fowler-Nordheim"[1] tunneling. The electric field during erase is the only new stress compared to EPROM that may impact overall reliability.

Erase/Write Cycling

Failure mechanisms traditionally associated with cycling electrically erasable memories include charge loss due to defective bits, destructive oxide breakdown, and electron trapup. ETOX flash memory technology minimizes these failure mechanisms by improvements in process technology, reducing the electric field stressing the gate oxide, and using efficient erase/write algorithms to control programming and erasure.

OXIDE QUALITY

Thin oxides used in tunnelling have been a reliability concern for electrically erasable memories. The quality of the ETOX tunnel oxide is approximately 10 times better than that of other tunnel oxide approaches. This breakthrough in tunnel oxide quality results from explicit process improvements and through the implicit advantages of the ETOX flash cell approach.

OXIDE BREAKDOWN

Oxide breakdown, due to erase/write cycling, has also been a major reliability concern for thin oxide tunnelling. ETOX technology addresses this concern by reducing the amount of stress placed on the tunnel oxide during programming and erasure. First, erasing the flash cell involves tunnelling only through the gate/source overlap, thus reducing the area under stress. This, coupled with the improvement in oxide quality, lowers the probability of an oxide defect. Secondly, the flash cell is erased using a lower-voltage erase pulse, resulting in lower stress on the tunnel oxide. This lower electric field across the tunnel oxide (10MV/cm versus 12MV/cm) yields a theoretical wear out time 10^8 times longer than other E^2PROM approaches.

ELECTRON TRAPUP

The phenomenon of electron trapup, the gradual reduction of electron mobility through the tunnel oxide, results in increasing program and erase times as cycling occurs. The program and erase algorithms must apply more pulses to add charge to or bleed charge off the floating gate to ensure data retention and integrity. This is seen as a failure to program or erase within the algorithm's allowed time and not as a hard failure. The Quick-Pulse Programming™ and Quick-Erase™ algorithms maintain an efficient program and erase time for the specified number of cycles listed in the flash memory data sheets.

ETOX™ FLASH MEMORY RELIABILITY TESTING

Intel flash memories undergo comprehensive testing to insure electrical reliability. This testing is done at qualification and during ongoing monitor checks.

Information on flash memory reliability testing procedures follows.

High Temperature 5.25V Dynamic Lifetest—This test is used to accelerate failure mechanisms by operating the devices at an elevated temperature of 125°C. During the test, the memory is sequentially addressed and the outputs are exercised, but not monitored or loaded. A checkerboard data pattern is used to simulate random patterns expected during actual use. Results of lifetesting have been summarized along with the failure analysis.

In order to best determine long-term failure rate, all devices used for lifetesting are subjected to standard INTEL testing. The 48 hour burn-in results are an indication of infant mortality and are not included in the failure rate calculation. (See Figure 3 for typical burn-in bias and timing diagrams.)

High Temperature High Voltage Dynamic Lifetest—This test is used to accelerate oxide breakdown failures. The test setup is identical to the one used for the dynamic lifetest except V_{CC} is increased. The acceleration factor due to this test can be found in Table 2. This data plus the standard dynamic lifetest data are used to calculate the 0.3 eV failure rate (See Figure 4 for typical bias and timing diagram).

Data Retention Bake—This test is used to accelerate charge loss from the floating gate. The test is performed by subjecting devices containing a 98% programmed pattern to a 250°C bake with no applied bias. In addition to data retention, this test is used to detect mechanical reliability problems (e.g., bond integrity) and process instability.

Temperature Cycle—This test consists of cycling the temperature of the chamber housing the subject devices from −65°C to +150°C and back. One thousand cycles are performed with a complete cycle taking 20 minutes. This test is to detect mechanical reliability problems and microcracks.

Low Temperature Lifetest—This test is performed at −10°C to detect the effects of hot electron injection into the gate oxide as well as package related failures (e.g., metal corrosion, etc.).[4]

intel RR-60

ESD Testing—This test is performed to validate the product's tolerance to Electro Static Discharge damage. All products incorporate ESD protection networks on appropriate pins.

Two types of tests are performed. First, all devices are tested using Mil STD 883 test criteria. In addition, a charged device test is performed to further validate protection occurring during mechanical handling.

Erase/Write Cycling (ETOX™ Flash Memories)—This test consists of repeatedly programming the device to an all 00H pattern and then erasing to all 0FFH data. Worst case voltage levels are used to maximize charge transfer to and from the floating gates. Cycling is used to ensure devices meet reprogrammability requirements as well as precondition for other reliability stresses.

Failure Rate Calculations

Failure rate calculations are given for each relevant activation energy. Failure rate calculations are made using the appropriate energy [2,3,4,5] and the Arrhenius Plot as shown in Figure 5*. The total equivalent device hours at a given temperature can be determined. The failure rate is then calculated by dividing the number of failures by the equivalent device hours and is expressed as a %/1000 hours. To arrive at a confidence level associated failure rate, the failure rate is adjusted by a factor related to the number of device hours using a chi-square distribution. A conservative estimate of the failure rate is obtained by including zero failures at 0.3 eV. Devices submitted to stresses other than lifetest received a 168 hour lifetest prior to stressing.

NOTE:
The activation energies for various failure mechanisms are listed in Table 1. The methodology for calculating failure rates is detailed in Appendix A.

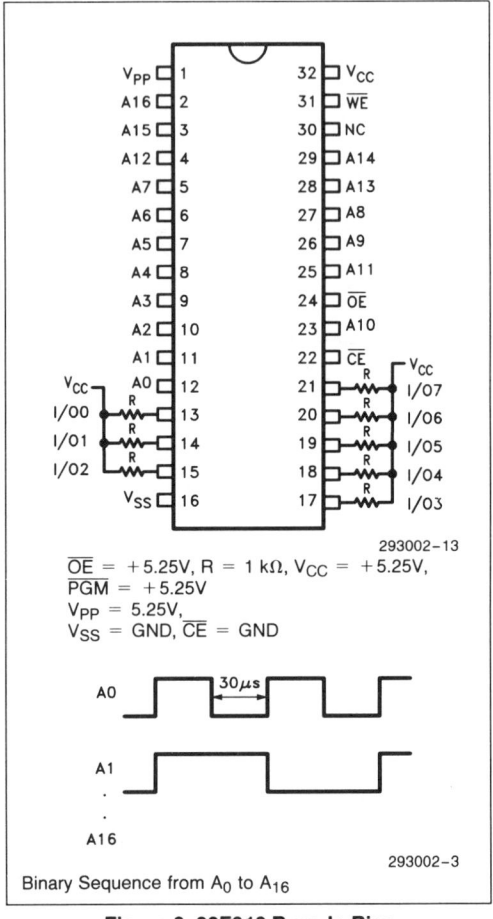

$\overline{OE}$ = +5.25V, R = 1 kΩ, V_{CC} = +5.25V,
$\overline{PGM}$ = +5.25V
V_{PP} = 5.25V,
V_{SS} = GND, $\overline{CE}$ = GND

Binary Sequence from A_0 to A_{16}

Figure 3. 28F010 Burn-In Bias and Timing Diagrams

intel® RR-60

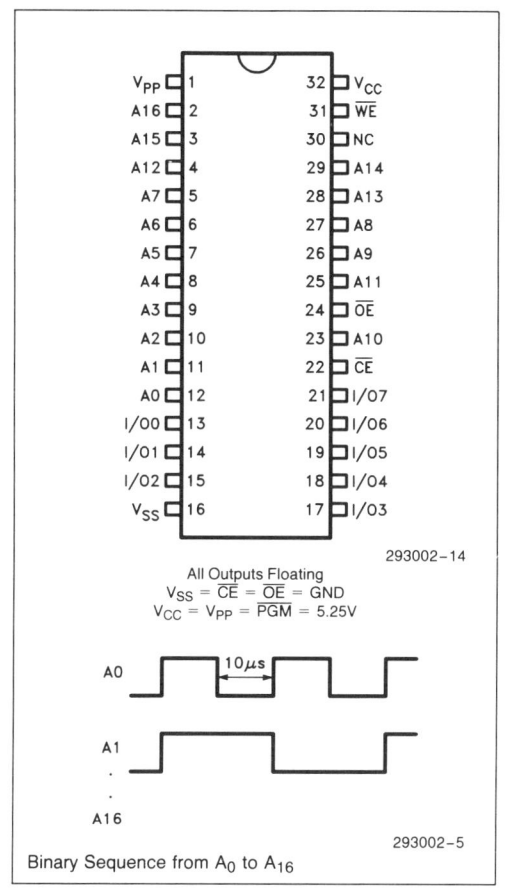

Figure 4. 28F010 Lifetest Bias and Timing Diagram

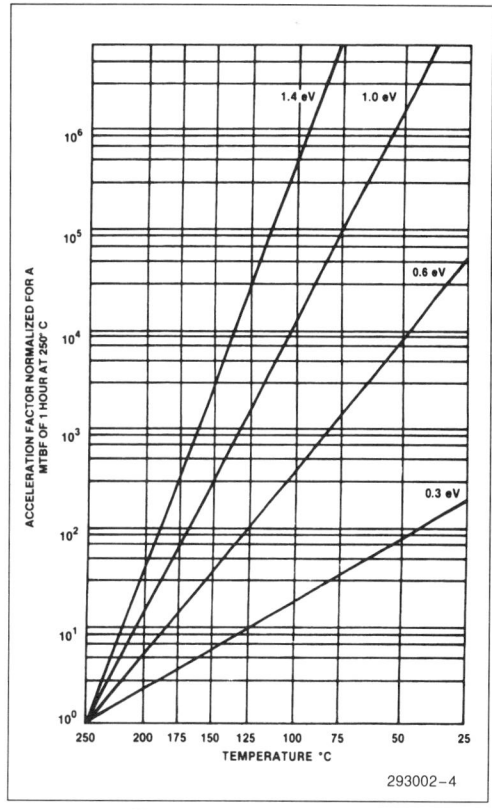

Figure 5. Arrhenius Plot

Table 1. Failure Mechanism Activation Energies Relevant to ETOX Flash Memories

Failure Mode	eV
Oxide	0.3
SBCL/SBCG/MBCL/MBCG	0.6
Contamination	1.0
Speed Degradation	0.3-1.0
Intrinsic Charge Loss	1.4
Contact Spiking	0.8

A typical lifetest bias and timing diagram is shown in Figure 4.

RR-60

Type	Supply Voltage (Volts)	Oxide Thickness (Å)	Operating Stress (MV/CM)	Acceleration Factor at __% Over Stress				
				10%	20%	30%	50%	100%
CHMOS IIE	5	400	1.25	7.5	55	422	3162	5.6E+5
CHMOS III E	5	235	2.13	3.7	13.4	49.1	658	4.3E+5
ETOX™	5	400	1.25	7.5	55	422	3162	5.6E+8
ETOX II	5	235	2.13	3.7	13.4	49.1	658	4.3E+5

ASSUMES:
1. No bias generators
2. Depletion loads
3. Failure rate calculations use the appropriate acceleration factor for stress voltage and maximum operating voltage (conservative).
4. See reference 7 for VAF determination.

Table 2. Time-Dependent Oxide Failure Acceleration

RELIABILITY DATA SUMMARY

The following data is an accumulation of recent qualification and monitor program results. Failure rate calculation methods listed in Appendix A were used to arrive at the tabularized failure rates.

In reviewing the reliability data as presented, questions may arise as to why lot sizes often decrease from one test to another without a corresponding number of identified failures. This is due to a variety of factors. Many tests require smaller sample sizes and as a result all parts from a previous test do not necessarily flow through to a succeeding test.

In addition, various parts are pulled from a sample lot when mechanical or handler failures occur. These "failures" are not a result of the specific test just completed but are nonetheless removed from the sample lot size and are not included in any failure rate calculation. It can also happen that a particular test is done incorrectly through human error or faulty test equipment and "invalid" failures are put aside for retesting at a later date, decreasing the lot size for a succeeding test. If these parts are found to be truly defective, they are treated as failures and listed. If they test out properly, they are removed from any calculation data base.

References

1. M. Lenzlinger. E. H. Snow, "Fowler-Nordheim tunneling into thermally grown SiO_2", Journal of Applied Physics, Vol. 40 (1969), p. 278.
2. S. Rosenberg, D. Crook, B. Euzent, "16th Annual Proceedings of the International Reliability Physics Symposium", pp. 19–25, 1978.
3. S. Rosenberg, B. Euzent, "HMOS Reliability" Reliability Report RR-18, Intel Corporation, 1979.
4. R. M. Alexander, "Calculating Failure Rates from Stress Data, April 1984 International Reliability Physics Symposium.
5. "EPROM Reliability DATA Summary" Reliability Report RR-35, Intel Corporation, 1985.
6. "E^2PROM and NVRAM Reliability DATA Summary" Reliability Report RR-59B, Intel Corporation, 1986.
7. E.S. Anolick, G.R. Nelson, "Low Field Time Dependent Dielectric Integrity", 1979 International Reliability Physics Symposium, pp. 8–12.

NOTE:
The methodology for calculating failure rates is detailed in Appendix A.

27F64

The Intel 27F64 (CERDIP) is a 64K Electrically Bulk-Erasable Flash Memory.

Number of Bits:	65,536
Organization:	8,192 × 8
Pin Out:	28-pin CERDIP
Die Size:	115 × 132 mils
Process:	ETOX™ Flash Memory
Cell Size:	6.0 × 6.0 μM
Programming Voltage:	12.75 VOLTS EXTERNAL
Technology:	CMOS

Table 1. Reliability Data Summary

Year	Burn-In	125°C Dynamic Lifetest				7.0V Dynamic Lifetest			
	48 Hours	168 Hrs	500 Hrs	1K Hours	2K Hrs	48 Hrs	168 Hrs	500 Hrs	1K Hrs
1988	0/3615	1/3602	1/992	0/990	0/990	0/432	0/432	0/432	0/432
Total	0/3615	1/3602	1/992	0/990	0/990	0/432	0/432	0/432	0/432
		A	A						

Table 2. Additional Qualification Tests

Year	Program/Erase Cycling		250°C Data Retention Bake					
			48 Hours		168 Hours		500 Hours	
	100	20K	Noncycled	Cycled	Noncycled	Cycled	Noncycled	Cycled
1988	0/1396	0/100	0/125	0/390	0/125	2/390	1/125	1/388
Total	0/1396	0/100	0/125	0/390	0/125	2/390	1/125	1/388
						A	A	A

Failure Analysis: A—Single bit charge loss

Table 3. 27F64 Failure Rate Prediction

125°C Actual Device Hours	Ea eV	Equivalent Hours		# Fail	Failure Rate %/1K Hours (60% U.C.L.)	
		55°C	70°C		55°C	70°C
2.25E + 06	0.3 BI + ELT	1.33E + 07	8.40E + 06	0	—	—
4.32E + 05	0.3 HVELT VAF*	1.08E + 09	6.80E + 08	0	0.00008	0.00013
2.69E + 06	0.6 BI + ELT + HVELT	9.20E + 07	3.60E + 07	2	0.0045	0.0110
2.69E + 06	1.0 BI + ELT + HVELT	9.60E + 08	2.00E + 07	0	0.0009	0.0046
			Combined Failure Rate:		0.0054	0.0157
			FITs:		54	157

*VAF (Voltage Acceleration Factor) for HVELT = 422

NOTE:
125°C Dynamic Lifetest and 7.0V Dynamic Lifetest samples each contain a split between units which saw 100 p/e cycles before stress and those which did not.

intel

RR-60

28F256

The Intel 28F256 is a 256-kilobit bulk-erasable flash memory.

Number of Bits:	262,144	Process:	ETOX™ Flash Memory
Organization:	32,768 × 8	Technology:	CMOS
Pin Out:	32-pin CERDIP	Cell Size:	6.0 × 6.0 µM
Die Size:	181 × 203 mils	Programming Voltage Options:	(P1) 12.0V +/−5%

Table 1. Reliability Data Summary

Year	Burn-In	125°C Dynamic Lifetest					7.25V Dynamic Lifetest				
	48 Hours	168 Hrs	500 Hrs	1K Hours	2K Hrs	48 Hrs	168 Hrs	500 Hrs	1K Hrs	2K Hrs
1988	1/17622	1/17620	1/945	0/941	0/430	0/458	0/456	0/456	0/152	0/95
1989	0/5428	0/5426	0/432	0/432	0/332	0/466	0/466	0/215	0/215	—
Total	1/23050	1/23046	1/1377	0/1373	0/752	0/924	0/922	0/671	0/367	0/95
	A	B	C							

Table 2. Additional Qualification Tests

Year	Program/Erase Cycling	250°C Data Retention Bake				
		48 Hours		168 Hours		500 Hours	
	100	Noncycled	Cycled	Noncycled	Cycled	Noncycled	Cycled
1988	0/1867	0/735	1/437	0/733	2/436	1/585	3/143
1989	0/800	0/900	0/500	1/874	0/500	5/873	1/497
Total	0/2667	0/1635	1/937	1/1607	3/936	6/1458	4/640
			D	D	D	D	E

Year	Temperature Cycling			Thermal Shock		
	200 Cycles	500 Cycles	1K Cycles	50 Cycles	200 Cycles	500 Cycles
1988	0/233	0/223	0/223	0/224	0/224	0/224
1989	0/175	0/175	0/175	0/125	0/125	0/125
Total	0/398	0/398	0/398	0/349	0/349	0/349

NOTE:
The 250°C Data Retention Bake samples labeled "Noncycled" received no program/erase cycling prior to Bake. "Cycled" units first saw 100 p/e cycles.

intel® RR-60

28F256 Failure Rate Prediction

125°C Actual Device Hours	Ea (eV)	Equivalent Hours 55°C	Equivalent Hours 70°C	# Fail	Fail Rate %/1K Hours 55°C	Fail Rate %/1K Hours 70°C
4.66×10^6	0.3 BI	2.74×10^7	1.78×10^7	0		
6.56×10^5	0.3 × VAF	2.12×10^8	1.38×10^8	0		
		TOTAL 0.3 eV Failures =		0	0.0001	0.0001
4.66×10^6	0.6 BI	1.62×10^8	6.81×10^7	2		
6.56×10^5	0.6 HVELT	2.27×10^7	9.58×10^6	0		
		TOTAL 0.6 eV Failures =		2	0.0017	0.0040
4.66×10^6	1.0 BI	1.72×10^9	4.07×10^8	0		
6.56×10^5	1.0 HVELT	2.42×10^8	5.73×10^7	0		
		TOTAL 1.0 eV Failures =		0	0.0000	0.0002
			Combined Failure Rate:		0.0018	0.0043
			FITs:		18	43

θ_{JA} = 79°C/W
V_{CC} = 5.25V
I_{CC} @55 = 18 mA
I_{CC} @70 = 17 mA
I_{CC} @125 = 16 mA

Temp with θ_{JA}
T(55) = 335.6K
T(70) = 350.2K
T(125) = 404.7K
T(250) = 523.1K
k = 8.62E−05 eV/K

Thermal Accel. Factors
	55°C	70°C
BI/ELT 0.3	5.886	3.821
Accel. 0.6	34.65	14.60
Factors: 1.0	368.3	87.24

Voltage Accel. Factor (VAF) for HVELT on this process is = 422.0

Failure Analysis:
A. 1-Single bit charge gain (pass. defect)
B. 1-Single bit charge gain (metal defect)
C. 1-Multiple bit charge loss (pass. defect)
D. 1-Single bit charge loss
E. 3-Single bit charge loss
   1-Open bond wire

28F512

The Intel 28F512 is a 512-Kbit bulk-erasable flash memory.

Number of Bits:	524,288	Process:	ETOX II Flash Memory
Organization:	65,536 × 8	Technology:	CMOS
Pin Out:	32-pin CERDIP/PLCC	Cell Size:	3.8 × 4.0 μM
Die Size:	227 × 181 mils	Programming Voltage Options:	12.0V ±5%

Table 1. Reliability Data Summary

Year	Burn-In	125°C Dynamic Lifetest			7.0V Dynamic Lifetest			
	48 Hours	168 Hrs	500 Hrs	1K Hours	48 Hrs	168 Hrs	500 Hrs	1K Hrs
1989	0/100	1/100	0/99	1/99	0/300	0/299	0/298	1/220
Total	0/100	1/100	0/99	1/99	0/300	0/299	1/298	1/220
		A		B			C	D

Table 2. Additional Qualification Tests

Year	Program/Erase Cycling	250°C Data Retention Bake					
		48 Hours		168 Hours		500 Hours	
	10K	Noncycled	Cycled	Noncycled	Cycled	Noncycled	Cycled
1989	2/706	1/63	0/62	0/62	0/62	0/62	0/62
Total	2/706	1/63	0/62	0/62	0/62	0/62	0/62
	D	E					

Year	Temperature Cycling			Thermal Shock		
	200 Cycles	500 Cycles	1K Cycles	50 Cycles	200 Cycles	500 Cycles
1989	0/80	0/80	0/80	0/79	0/79	0/79
Total	0/80	0/80	0/80	0/79	0/79	0/79

NOTE:
250°C Data Retention Bake "Cycled" units received 10,000 program/erase cycles prior to Bake. 125°C Dynamic Lifetest and 7.0V Dynamic Lifetest samples contain a mix of cycled and uncycled material.

intel

RR-60

28F512 Failure Rate Prediction

125°C Actual Device Hours	Ea (eV)	Equivalent Hours 55°C	Equivalent Hours 70°C	# Fail	Fail Rate %/1K Hours 55°C	Fail Rate %/1K Hours 70°C
9.44×10^4	0.3 BI	5.65×10^5	3.68×10^5	0		
2.59×10^5	$0.3 \times$ VAF	1.43×10^8	9.46×10^7	0		
		TOTAL 0.3 eV Failures =		0	0.0006	0.0010
9.44×10^4	0.6 BI	3.39×10^6	1.42×10^6	1		
2.59×10^5	0.6 HVELT	9.29×10^6	3.91×10^6	0		
		TOTAL 0.6 eV Failures =		1	0.0159	0.0386
9.44×10^4	0.8 BI	1.12×10^7	3.52×10^6	0		
2.59×10^5	0.8 HVELT	3.07×10^7	9.66×10^6	2		
		TOTAL 0.8 eV Failures =		2	0.0074	0.0242
9.44×10^4	1.0 BI	3.68×10^7	8.71×10^6	1		
2.59×10^5	1.0 HVELT	1.01×10^8	2.39×10^7	0		
		TOTAL 1.0 eV Failures =		1	0.0015	0.0060
		Combined Failure Rate:			0.0254	0.0702
		FITs:			254	702

θ_{JA} = 59°C/W
V_{CC} = 5.25V
I_{CC} @55 = 13 mA
I_{CC} @70 = 12 mA
I_{CC} @125 = 8 mA

Temp with θ_{JA}
T(55) = 332.1K
T(70) = 346.8K
T(125) = 400.6K
T(250) = 523.1K
k = 8.62E−05 eV/K

		Thermal Accel. Factors	
		55°C	70°C
BI/ELT	0.3		3.845
Accel.	0.6	35.93	14.87
	0.8	118.6	36.29
Factors:	1.0	391.2	89.09

Voltage Accel. Factor (VAF)
for HVELT on this process is = 93.3

A. Multiple bit charge loss (contamination)
B. Single bit charge loss
C. 1-V_{MIN} due to single leaky col
D. 1-Basic function due to single leaky col
E. 1-Adjacent column failure due to metal stringer
   1 = ISB Failure Analysis pending assumed valid
F. 1-Multiple bit charge loss

intel

RR-60

28F010

The Intel 28F010 is a 1024-Kbit bulk-erasable flash memory.

Number of Bits:	1,048,576	Process:	ETOX[TM] II Flash Memory
Organization:	2 (512 × 1024)	Technology:	CMOS
Pin Out:	32-pin CERDIP	Cell Size:	3.8 × 4.0 μM
Die Size:	225 × 265 mils	Programming Voltage	12.0V ±5%

Table 1. Reliability Data Summary

Year	Burn-In	125°C Dynamic Lifetest				7.0V Dynamic Lifetest				
	48 Hours	168 Hrs	500 Hrs	1K Hours	2K Hrs	48 Hrs	168 Hrs	500 Hrs	1K Hrs	2K Hrs
1989	0/530	0/500	1/500	1/499	1/98	0/698	1/697	1/697	2/695	1/96
Total	0/530	0/500	1/500	1/499	1/98	0/698	1/697	1/697	2/695	1/96
			A	B	A		B	B	C	B

Table 2. Additional Qualification Tests

Year	Program/Erase Cycling		250°C Data Retention Bake					
			48 Hours		168 Hours		500 Hours	
	10K	100K	Noncycled	Cycled	Noncycled	Cylced	Noncycled	Cycled
1989	13/2169	3/48	0/306	0/307	0/306	0/307	0/85	0/120
Total	13/2169	3/48	0/306	0/307	0/306	0/307	0/85	0/120
	D	E						

Year	Temperature Cycling			Thermal Shock		
	200 Cycles	500 Cycles	1K Cycles	50 Cycles	200 Cycles	500 Cycles
1989	0/369	0/369	0/369	0/318	0/318	0/318
Total	0/369	0/369	0/369	0/318	0/318	0/318

NOTE:
250°C Data Retention Bake "Cycled" units received 10,000 program/erase cycles prior to Bake. 125°C Dynamic Lifetest and 7.0V Dynamic Lifetest samples contain a mix of cycled and uncycled material.

intel RR-60

28F010 Failure Rate Prediction

125°C Actual Device Hours	Ea (eV)	Equivalent Hours 55°C	Equivalent Hours 70°C	# Fail	Fail Rate %/1K Hours 55°C	Fail Rate %/1K Hours 70°C
5.735×10^5	0.3 BI	3.52×10^6	2.24×10^6	0		
7.920×10^5	$0.3 \times$ VAF	4.56×10^8	2.91×10^8	1		
		TOTAL 0.3 eV Failures =		1	0.0004	0.0007
5.735×10^5	0.6 BI	2.16×10^7	8.77×10^6	1		
7.920×10^5	0.6 HVELT	2.98×10^7	1.21×10^7	3		
		TOTAL 0.6 eV Failures =		4	0.0103	0.0248
5.735×10^5	0.8 BI			2		
7.920×10^5	0.8 HVELT			1		
		TOTAL 0.8 eV Failures =		3	0.0025	0.0080
			Combined Failure Rate:		0.0132	0.0335
			FITs:		132	335

$\theta_{JA} = 46°C/W$
$V_{CC} = 5.25V$
$I_{CC} @55 = 15$ mA
$I_{CC} @70 = 13$ mA
$I_{CC} @125 = 11$ mA

Temp with θ_{JA}
T(55) = 331.7K
T(70) = 346.2K
T(125) = 400.8K
T(250) = 523.1K
$k = 8.62E-05$ eV/K

Thermal Accel. Factors

		55°C	70°C
BI/ELT	0.3	6.095	3.925
Accel.	0.6	37.14	15.41
	0.8	123.9	38.35
Factors:	1.0	413.5	95.42

Voltage Accel. Factor (VAF) for HVELT on this process is = 93.3

A. Speed degrade (spiked contact)
B. Single bit charge loss
C. 1-Column failure (spiked contact)
   1-Speed degrade (analysis pending; 0.3 eV)
D. 7-Dual column (spiked contact)
   3-Column failure (spiked contact)
   2-lsb (analysis pending)
   1-Adjacent row (analysis pending)

E. 2-Single bit programming push-out
   1-Single bit failure (oxide defect)*

*NOTE:
Current production testing guarantees programming push-out screens for 10K cycles. 100K cycle screens are currently under development.

Plastic Reliability Data Summary

INTRODUCTION

The following information is written to provide users with the description and reliability summary of Intel's plastic flash product PLCC packages. It includes brief test descriptions, a description of plastic packaging compounds and the reliability data obtained during the qualification and subsequent product monitors of the N28F256. Qualification results for the N28F512 and N28F010 will be available in **Dec 89**.

PLASTIC PACKAGE CHARACTERISTICS

The plastic package is composed of flame retardant plastic/epoxy which meets the rating requirements of US94V0 $\frac{1}{8}$" minimum. The die attach incorporates a silver-filled adhesive die attach on a silver spot plated leadframe. Bonding is accomplished through gold thermal compression bonding and lead finish is either tin plated or 60/40 solder dipped tin/lead.

ELECTRICAL CHARACTERISTICS

Because of the electrical erase capabilities of Flash memories, parts may be programmed, 100% tested and erased in plastic packages.

Flash memories in plastic are tested to the same electrical/parametric levels as their counterparts in CERDIP. The characteristics include input/output voltage levels, speeds, leakage, and power requirement characteristics over the full commercial temperature operating range of 0°C–70°C. Performance capabilities are identical to that of CERDIP product.

RELIABILITY/QUALITY STRESSES

High Temperature 5.25V 125°C Dynamic Lifetest (HTDL)—This test is used to accelerate failure mechanisms by operating the devices at an elevated temperature of 125°C. During the test, the memory is sequentially addressed and outputs are exercised but not monitored or loaded. A checkerboard data pattern is typically used to simulate random patterns expected during actual use. Results of lifetesting have been summarized along with failure analysis. In order to best determine long-term failure rates, all devices used for lifetesting are subjected to a standard Intel screening. The 48-hour burn-in results measure infant mortality and are not included in the failure rate calculations.

High Temperature Extended Lifetest (HTELT)—This test is also performed at 125°C but uses a smaller sample size. The parts are kept in the full active mode for the duration of the test with outputs driven. The test is intended to evaluate the long-term reliability of the product.

High Voltage Extended Lifetest (HVELT)—This test is used to accelerate oxide breakdown failures. The test is set up identical to the one used for dynamic lifetest except for V_{CC} and V_{PP} which are raised to 6.5V. The voltage acceleration factor for this configuration can be found in Table II.

Data Retention Bake—This test is used to accelerate charge loss from the floating gate. The test is performed by subjecting devices containing a 98%+ program pattern to a 140°C bake with no applied bias. In addition to data retention, this test can also be used to detect mechanical reliability problems such as bond integrity or process instabilities.

85/85 TEST

During the 85°C/85% relative humidity test, the devices are subjected to a high temperature, high humidity environment. The object of the test is to accelerate failure mechanisms through an electrolytic process. This is accomplished through a combination of moisture penetration of the plastic, voltage potentials and contamination which, if present, would combine with the moisture to act as an electrolyte. See Figure 6 for typical 85/85 Bias Diagram.

Steam

Steam stressing performed at 121°C, 2 atm. accelerates moisture penetration through the plastic package material to the surface of the die. The objective of this test is to accelerate failures of the device as a result of moisture on the die surface. Corrosion, as typically seen in plastic encapsulated devices, is a very minor contributor to the Flash failure mechanisms. Due to the floating gate storage cell composition, Flash memories have a distinctive failure mode which requires special considerations and solutions.

The floating gate itself is a highly phosphorous doped structure on which electrons are stored, thus creating the non-volatile memory cell. Passivation defects or marginalities can allow moisture penetration to a single Flash cell causing oxide deterioration, thus showing up as a charge loss failure. This becomes the predominant failure mode for Flash product, opposed to corrosion which historically has been the dominant plastic mode of failure. Intel has developed a proprietary, multi-layer passivation which has successfully solved this problem.

intel® RR-60

QUALITY/RELIABILITY STANDARDS

The table below contains Intel's current requirements for qualification for plastic Flash memories. The failure rate criteria has been established based on a survey of major customers world-wide. Intel consistently meets or exceeds these requirements.

HTDL 48-Hr	HTELT 168/500-Hr	140°C Bake 48/168/500-Hr	HVELT 48/168-Hr	Steam 96/168-Hr	85/85 500/1K
<.05%	\<200 FITs Combined Failure Rate			<2%/<5%	<1% Cum.

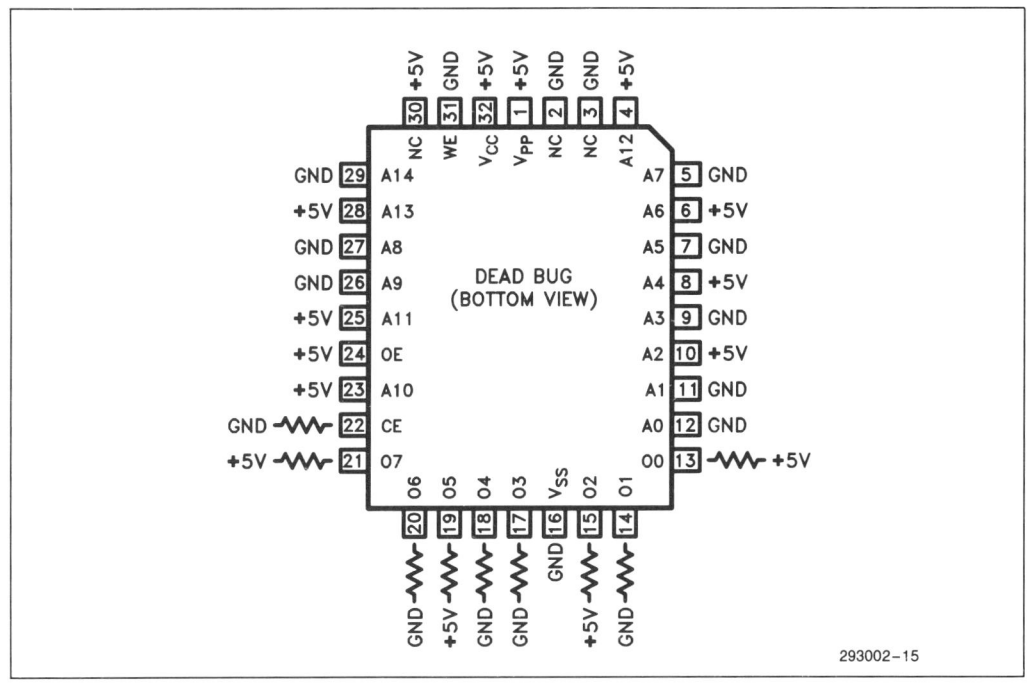

Figure 6. 85°C 85% RH ELT Configuration Diagram

intel

RR-60

N28F256

The N28F256 products are functionally identical to the D28F256 except that they are housed in a PLCC (N) package.

Table 1. Reliability Data Summary

Year	Burn-In	125°C Dynamic Lifetest					7.0V Dynamic Lifetest		
	48 Hours	168 Hrs	500 Hrs	1K Hours	2K Hrs	48 Hrs	168 Hrs	500 Hrs
1988	0/125	0/125	0/125	0/125	0/125	0/100	0/100	0/100
1989	—	—	—	—	—	—	—	—
Total	0/125	0/125	0/125	0/125	0/125	0/100	0/100	0/100

Table 2. Additional Qualification Tests

Year	140°C Data Retention Bake		
	48 Hours	168 Hours	500 Hours
1988	0/100	0/100	0/100
1989	—	—	—
Total	0/100	0/100	0/100

Year	Temperature Cycling			Thermal Shock		
	200 Cycles	500 Cycles	1K Cycles	50 Cycles	200 Cycles	500 Cycles
1988	0/50	0/50	0/50	0/50	0/50	0/50
1989	0/124	0/100	0/98	0/124	0/124	0/124
Total	0/174	0/150	0/148	0/174	0/174	0/174

Year	85°C/85% RH				Steam	
	168 Hrs	500 Hrs	1K Hrs	2K Hrs	168 Hrs	336 Hrs
1988	0/200	0/200	0/200	0/200	1/200	0/197
1989	0/497	0/497	0/497	0/371	0/497	0/493
Total	0/697	0/697	0/697	0/571	1/697	0/690
					A	

NOTE:
PLCC Monitor program is designed to monitor process/package environmental performance. Monitoring of electrical performance is accomplished via CERDIP 28F256 monitors.

RR-60

N28F256 Failure Rate Prediction

125°C Actual Device Hours	Activation Energy (eV)	Equivalent Hours 55°C	Equivalent Hours 70°C	# Fail	Failure Rate %/1K Hours (60% U.C.L.) 55°C	Failure Rate %/1K Hours (60% U.C.L.) 70°C
2.50E + 05	0.3 eV ELT	1.62E + 06	1.02E + 06	0		
5.00E + 04	0.3 eV HVLT × VAF	1.78E + 07	1.12E + 07	0		
	TOTAL 0.3 eV	1.94E + 07	1.22E + 07	0	0.00472	0.00750
2.50E + 05	0.6 eV ELT	1.05E + 07	4.13E + 06	0		
5.00E + 04	0.6 eV HVLT	2.09E + 06	8.27E + 05	0		
	TOTAL 0.6 eV	1.25E + 07	4.96E + 06	0	0.00729	0.01845
2.50E + 05	1.0 eV ELT	1.26E + 08	2.68E + 07	0		
5.00E + 04	1.0 eV HVLT	2.52E + 07	5.36E + 06	0		
	TOTAL 1.0 eV	1.51E + 08	3.22E + 07	0	0.00061	0.00284
			Combined Failure Rate:		0.01261	0.02879
			FITs:		126.1	287.9

VAF = Voltage Acceleration Factor of 55

Failure Analysis:
A. Input leakage; no physical failure analysis performed.

intel
RR-60

APPENDIX A
FAILURE RATE CALCULATIONS FOR 60% UPPER CONFIDENCE LEVEL

Step 1. Accumulate data from 48 hours of burn-in through lifetest of each lot. (Note: 48-hour burn-in results measure infant mortality and are not included in the failure rate calculation.)

Step 2. Determine the failure rate mechanism for each failure and assign an activation energy (E_A) corresponding to each failure mechanism. (See Table 1 below.)

Table 1. Failure Mechanisms Activation Energies Relevent to ETOX™ Flash Memories

Failure Mode	Activation Energy
Defective Big Charge Gain/Loss	0.6 eV
Oxide Breakdown	0.3 eV
Silicon Defects	0.3 eV
Contamination	1.0 eV–1.2 eV
Intrinsic Charge Loss	1.4 eV

Step 3. Calculate the total number of device hours from 48 hours of burn-in through lifetest.

Example: 125°C Burn-In/Lifetest and a 2 lot sample

$$\frac{\text{\# failures}}{\text{total \# devices}}$$

	48 Hours	168 Hours	500 Hours	1K Hours	2K Hours
Lot #1	0/1000	1/1000	0/999	0/998	0/994
Lot #2	0/221	0/201	1/201	1/100	0/99
Totals	0/1221	1/1201	1/1200	1/1098	0/1093

Device Hours = (Number of Devices) (Number of Hours)

Total Device Hours = 1201 (168 hrs − 48 hrs) + 1200 (500 hrs − 168 hrs)

$\quad\quad$ + 1098 (1000 hrs − 500 hrs) + 1093 (2000 hrs − 1000 hrs)

$\quad$ = 1201 (120 hrs) + 1200 (332) + 1098 (500 hrs)

$\quad\quad$ + 1093 (1000 hrs)

$\quad$ = 2.185 × 10^6 Device Hours

intel® RR-60

Step 4. Use E_A tables to find the equivalent device hours at a desired temperature for each activation energy (failure mechanism), or use the Arrhenius relation.

$$R = A \exp\left[\frac{-E_A}{KT}\right]$$

$K = 8.617 \times 10^{-5}$ eV/°K (Boltzman's Constant)
A = proportionality constant
R = mean rate to failure
E_A = activation energy
T = Temperature in Kelvin

$$\frac{R_1}{R_2} = \frac{A_1 \exp\left[\frac{-E_A}{KT_1}\right]}{A_2 \exp\left[\frac{-E_A}{KT_2}\right]} = \exp\left[\frac{E_A}{K}\left(\frac{1}{T_2} - \frac{1}{T_1}\right)\right]$$

Where $A_1 = A_2 = A$ for the same failure mechanism (i.e., same E_A)

Where R_1 and R_2 are rates for a normal operating temperature and an elevated temperature respectively

$$R_1 = R_2 \times \exp\left[\frac{E_A}{K}\left(\frac{1}{T_2} - \frac{1}{T_1}\right)\right]$$

However, since rate (R) has the units $\left(\frac{1}{\text{time}}\right)$, we can think in terms of time to one failure or MTBF.

Thus,

$$R_1 = \frac{1}{t_1} \quad \text{where } t_1 = \text{MTBF at same temperature } T_1$$

and

$$R_2 = \frac{1}{t_2} \quad \text{where } t_2 = \text{MTBF at same temperature } T_2$$

Thus the Arrhenius Relation becomes:

$$\frac{1}{t_1} = \frac{1}{t_2} \times \exp\left[\frac{E_A}{K}\left(\frac{1}{T_2} - \frac{1}{T_1}\right)\right]$$

or

$$t_1 = \exp\left[\frac{E_A}{K}\left(\frac{1}{T_1} - \frac{1}{T_2}\right)\right] \times t_2$$

We then define the Acceleration Factor as:

$$\text{A.F.} = \frac{t_1}{t_2} = \exp\left[\frac{E_A}{K}\left(\frac{1}{T_1} - \frac{1}{T_2}\right)\right]$$

For example: For $E_A = 0.6$ eV, $T_2 = 398°$K, $T_1 = 328°$K

$$t_1 = 41.7 \, t_2$$

Therefore, one hour at 125°C is the equivalent to 41.7 hours at 55°C for a failure mechanism of activation energy $E_A = 0.6$ eV. Then 41.7 is the thermal acceleration factor for time.

intel® RR-60

NOTE:

The Arrhenius Plot is simply ln (Acceleration Factor) vs. 1/Temperature normalized for an MTBF (t_2) of one hour at 250°C (T_2). This plot can also be used to determine the acceleration factor between two temperatures (other than 250°C).

For example: For a 0.3 eV failure at 125°C, the acceleration factor is 8.1 relative to a 0.3 eV failure at 250°C. For a 0.3 eV failure at 25°C, the acceleration factor is 152 relative to 250°C. Therefore, the acceleration factor between 125°C and 25°C is:

$$\text{A.F.} = \frac{t_1}{t_2} = \frac{152}{8.1} = 18.7$$

Step 5. Organize the burn-in/lifetest data by E_A, Total Device Hours at the burn-in/lifetest temperature T_2, Thermal Acceleration Factors for each failure mechanism (E_A), Number of Failures for each failure mechanism, and the calculated equivalent device hours at the desired operating temperature T_1.

NOTE:

The rise in junction temperature due to the thermal resistivity of the package (θ_{JA}) must be added to the desired and actual burn-in/lifetest temperatures.

$$T_{test} = T_J + T_A = \theta_{JA} \, (IV \, @ \, T_A) + T_A$$

E_A (eV)	Total Device Hrs @ T_2	Acceleration Factors	# Fail	Equivalent Hours @ T_1
0.3	T.D.H.	X	N_1	X (T.D.H.)
0.6	T.D.H.	Y	N_2	Y (T.D.H.)
1.0	T.D.H.	Z	N_3	Z (T.D.H.)

The failure rates for individual failure mechanisms and the total combined failure rate can be predicted using the data table and the following formula:

$$\% \text{ fail/1K hrs.} = \frac{\chi^2 (n, \alpha)}{2T} (10^5)$$

Where $\chi^2 (n, \alpha)$ is the value of the chi squared distribution for n degrees of freedom and confidence level of α. T is the total equivalent device hours at T_1. The total combined rate is just the sum of the individual failure rates for each failure mechanism.

For a 60% UCL, the above formula converts to the following:

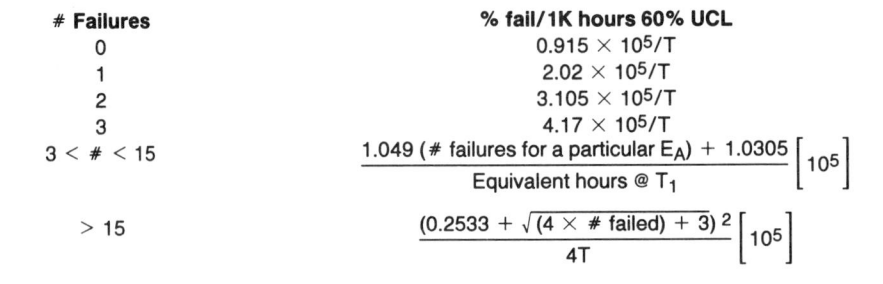

# Failures	% fail/1K hours 60% UCL
0	$0.915 \times 10^5/T$
1	$2.02 \times 10^5/T$
2	$3.105 \times 10^5/T$
3	$4.17 \times 10^5/T$
3 < # < 15	$\frac{1.049 \, (\# \text{ failures for a particular } E_A) + 1.0305}{\text{Equivalent hours @ } T_1} \left[10^5 \right]$
> 15	$\frac{(0.2533 + \sqrt{(4 \times \# \text{ failed}) + 3})^2}{4T} \left[10^5 \right]$

A-3

intel

RR-60

Example 1:

Assume for this example, that I_{CC} active is 57 mA at T_A = 125°C and I_{CC} active is 60 mA at T_A = 55°C.

Also assume that θ_{JA} = 35°C/W.

Then,

T_2 = (35°C/W) (57 mA) (5V) + 125°C

$\approx$ 135°C = 408°K

T_1 = (35°C/W) (60 mA) (5V) + 55°C

$\approx$ 65°C = 338°K

E_A (eV)	Actual Device Hours @ 125°C	Acceleration for 135°C to 65°C	Equivalent Hours at 55°C	# Fail	55°C % Fail/ 1K Hrs
0.3	2.185 × 10⁶	5.85	1.278 × 10⁷	0	0.0081
0.6	2.185 × 10⁶	34.18	7.468 × 10⁷	2	0.0042
1.0	2.185 × 10⁶	359.93	7.864 × 10⁸	1	0.0003
			Total Combined Failure Rate =		0.0126
				=	126 FITs

Example 2:

Assume that an additional lot of 800 flash devices is burned in using a 6.5V lifetest. Using Table 2 below, a voltage acceleration factor of 55 results from a 20% overstress (5.5V to 6.5V).

	48 Hours	168 Hours	500 Hours
Lot #3	0/800	1/800	0/799

Device Hours = 800 (48 hrs − 0 hrs) + 800 (168 hrs − 48 hrs) + 799 (500 hrs − 168 hrs)

= 3.997 × 10⁵

Table 2. Time-Dependent Oxide Failure Accelerations

Type	Supply Voltage (Volts)	Oxide Thickness (Å)	Operating Stress (MV/cm)	Acceleration Factor at __% Over Stress			
				10%	20%	50%	100%
HMOS* E	5	700	0.714	3.2	10	320	99,500
HMOS* IIE	5	400	1.25	7.5	55	23,700	5.6 × 10⁸
ETOX™	5	400	1.25	7.5	55	23,700	5.6 × 10⁸

ASSUMES:
1. No Bias Generators
2. Depletion Loads

intel   RR-60

Since this voltage accelerated stress is used to predict an oxide breakdown failure rate, the 5.5V burn-in/lifetest 55°C hours for $E_A = 0.3$ eV are added to the 6.5V burn-in/lifetest 55°C equivalent hours as follows:

125°C Burn-In/Lifetest	E_A (eV)	Actual Device Hours @ 125°C	Acceleration Factors for 135°C to 65°C	Equivalent Hours @ 55°C
5.5V	0.3	2.185×10^6	5.85	1.278×10^7
6.5V	0.3	3.997×10^5	(5.85×55)	1.286×10^8
		Total Equivalent $E_A = 0.3$ eV Device Hours =		1.414×10^8

The following failure rate predictions include the total equivalent 55°C, $E_A = 0.3$ eV device hours found above:

E_A (eV)	Actual Device Hours @ 125°C	Acceleration Factors for 135°C to 65°C	Equivalent Hours @ 55°C	# Fail	55°C % Fail/ 1K Hrs
0.3	2.185×10^6	5.85	—	—	—
0.3 + 55[1]	3.997×10^5	(5.85×55)	1.414×10^8	1	0.0015
0.6	2.185×10^6	34.18	7.468×10^7	2	0.0042
1.0	2.185×10^6	359.93	7.864×10^8	1	0.0003
		Total Combined Failure Rate =			0.0060
				=	60 FITs

NOTE:
1. The notation 0.3 + 55 is used to show that 6.5V and 5.5V burn-in lifetest equivalent hours have been combined.

APPENDIX B
FLASH MEMORY BIT MAPS AND DIE PHOTOGRAPHS

intel® RR-60

COLUMNS (I/O block 0)

A4	A3	A2	A1	A0	DECIMAL	BITLINE
0	0	0	0	0	0	Y0
1	0	0	0	0	16	Y1
1	0	0	0	1	17	Y2
0	0	0	0	1	1	Y3
0	0	0	1	0	2	Y4
1	0	0	1	0	18	Y5
1	0	0	1	1	19	Y6
0	0	0	1	1	3	Y7
0	0	1	0	0	4	Y8
1	0	1	0	0	20	Y9
1	0	1	0	1	21	Y10
0	0	1	0	1	5	Y11
0	0	1	1	0	6	Y12
1	0	1	1	0	22	Y13
1	0	1	1	1	23	Y14
0	0	1	1	1	7	Y15
...					...	...
1	1	1	1	1	31	Y30
0	1	1	1	1	15	Y31

I/O block 1: y32 to y63
2: y64 to y95
.
.
.
7: y223 to y225

ROWS

A12	A11	A10	A9	A8	A7	A6	A5	DECIMAL	WORDLINE
0	0	0	0	0	0	0	0	0	X0
0	0	0	0	0	0	0	1	1	X1
0	0	0	0	0	0	1	0	2	X2
0	0	0	0	0	0	1	1	3	X3
0	0	0	0	0	1	0	0	4	X4
0	0	0	0	0	1	0	1	5	X5
					...			...	...
1	1	1	1	1	1	1	1	255	X255

ARRAY ORGANIZATION

ROW SELECTS | I/O₀ | I/O₁ | I/O₂ | I/O₃ | I/O₄ | I/O₅ | I/O₆ | I/O₇

COLUMN SELECTS

BITMAP FOR ONE OUTPUT
<— 32 columns —>

x255
x254
x253
.
.
.
x2
x1
x0

I/O₀ — 256 rows

y0 y1 y2 ... y31

27F64 Bit Map

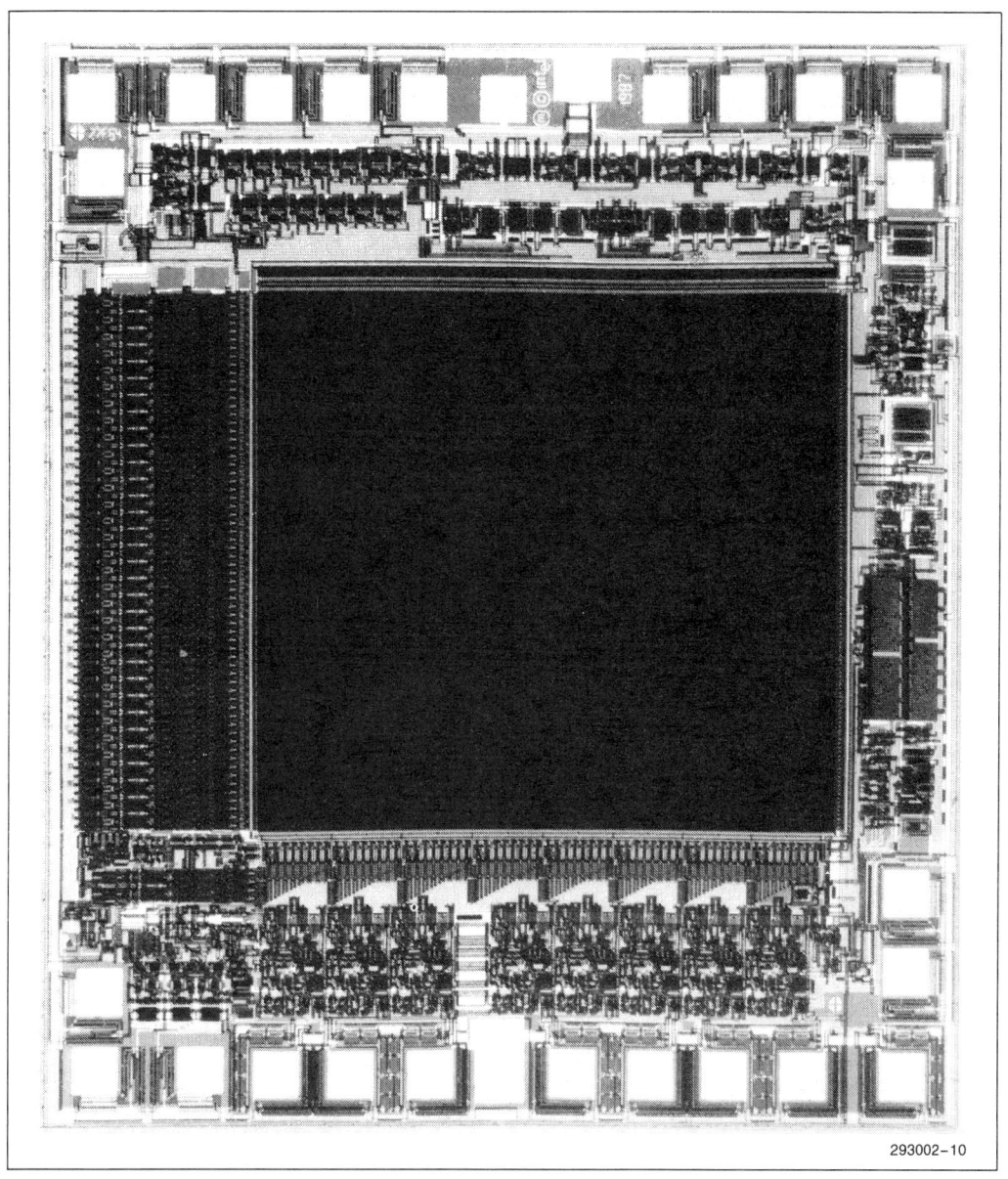

27F64 Die Photograph

intel® RR-60

Columns are numbered 0 through 255 beginning with the column nearest the X-decoder.

Outputs are grouped as follows:

Quadrant Decoding		Column Decoding			LEFT HALF ARRAY O0 O1 O2 O3		RIGHT HALF ARRAY O4 O5 O6 O7	
A4	A3	A2	A1	A0	A10	O0/O7	O1/O6	O2/O5	O3/O4
0	0	0	0	0	0	BL192	BL128	BL64	BL0
0	1	0	0	0	0	BL193	BL129	BL65	BL1
0	1	0	0	0	1	BL194	BL130	BL66	BL2
0	0	0	0	0	1	BL195	BL131	BL67	BL3
0	0	0	0	1	0	BL196	BL132	BL68	BL4
0	1	0	0	1	0	BL197	BL133	BL69	BL5
0	1	0	0	1	1	BL198	BL134	BL70	BL6
0	0	0	0	1	1	BL199	BL135	BL71	BL7
*	.	.	.	.	.	.	.	.	.
*	.	.	.	.	.	.	.	.	.
*	.	.	.	.	.	.	.	.	.
1	0	1	1	0	0	BL248	BL184	BL120	BL56
1	1	1	1	0	0	BL249	BL185	BL121	BL57
1	1	1	1	0	1	BL250	BL186	BL122	BL58
1	0	1	1	0	1	BL251	BL187	BL123	BL59
1	0	1	1	1	0	BL252	BL188	BL124	BL60
1	1	1	1	1	0	BL253	BL189	BL125	BL61
1	1	1	1	1	1	BL254	BL190	BL126	BL62
1	0	1	1	1	1	BL255	BL191	BL127	BL63

28F256 Bitline Decoding

intel
RR-60

X-DECODING: Wordlines are numbered 0 through 511 beginning at the top of the array

WL	A14	A13	A12	A7	A6	A5	A11	A9	A8
WL0	0	0	0	0	0	0	0	0	0
WL1	0	0	0	0	0	0	0	0	1
WL2	0	0	0	0	0	0	0	1	0
WL3	0	0	0	0	0	0	0	1	1
.	.	.	.	.	.	.	.	.	.
WL508	1	1	1	1	1	1	1	0	0
WL509	1	1	1	1	1	1	1	0	1
WL510	1	1	1	1	1	1	1	1	0
WL511	1	1	1	1	1	1	1	1	1

28F256 Wordline Decoding

ARRAY ORGANIZATION

ROW SELECTS

I/O_0 | I/O_1 | I/O_2 | I/O_3 | I/O_4 | I/O_5 | I/O_6 | I/O_7

COLUMN SELECTS

BITMAP FOR ONE OUTPUT

x511
x510
x509
x508
.
.
.
x3
x2
x1
x0

y0 y1 y2 ... y64

28F256 Bit Map

B-5

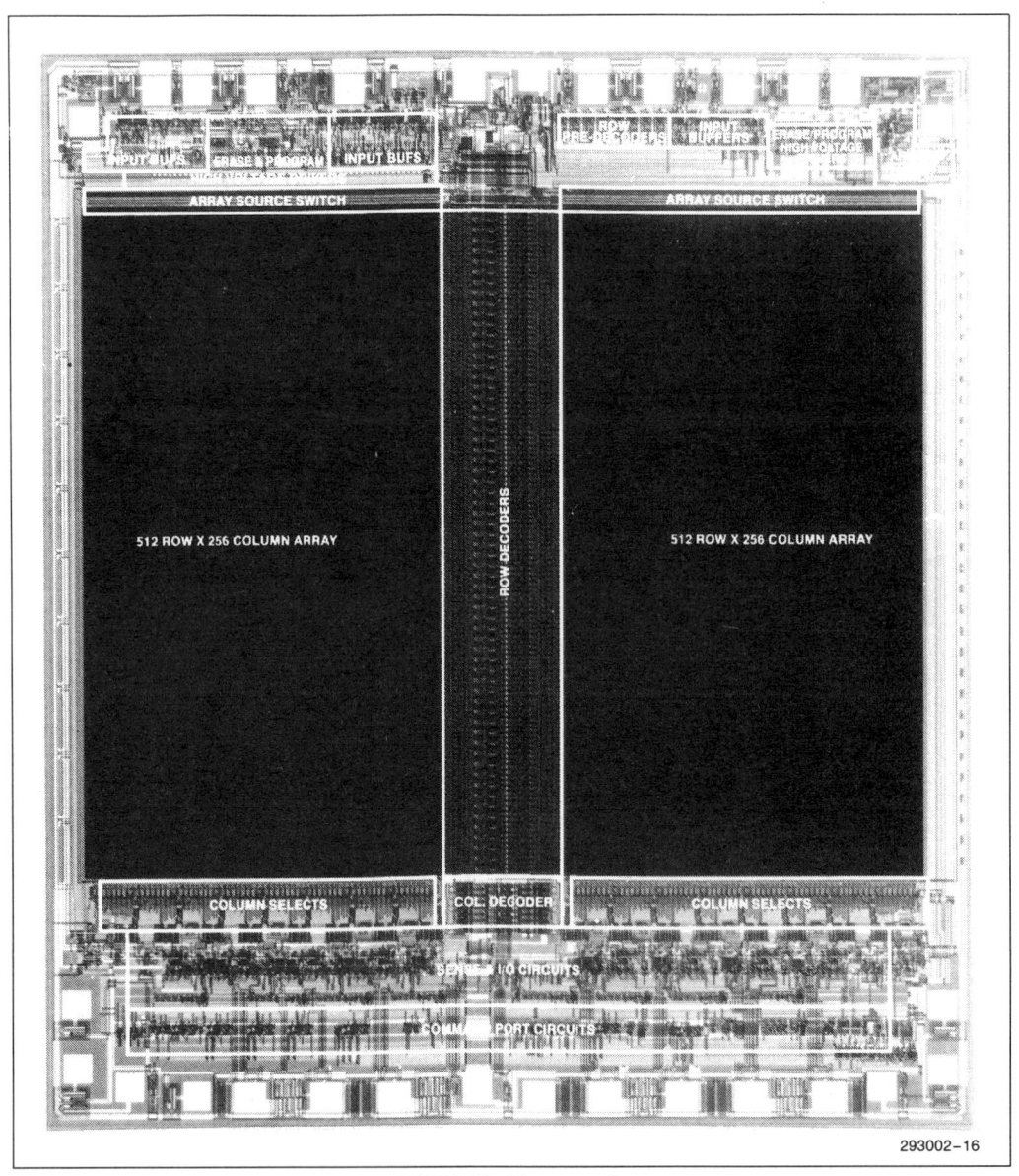

28F256 Die Photograph

intel

RR-60

28F512 (C) Bitline Decoding

Address							Bitlines			
A14	A15	A10	A2	A1	A0	A3	IO0/7	IO1/6	IO2/5	IO3/4
0	0	0	0	0	0	0	BL384	BL256	BL128	BL0
0	0	0	0	0	0	1	BL385	BL257	BL129	BL1
0	0	0	0	0	1	0	BL386	BL258	BL130	BL2
0	0	0	0	0	1	1	BL387	BL259	BL131	BL3
0	0	0	0	1	0	0	BL388	BL260	BL132	BL4
0	0	0	0	1	0	1	BL389	BL261	BL133	BL5
0	0	0	0	1	1	0	BL390	BL262	BL134	BL6
0	0	0	0	1	1	1	BL391	BL263	BL135	BL7
...	...	...	...	...	...	...	...	...	...	...
1	1	1	1	1	0	0	BL508	BL380	BL252	BL124
1	1	1	1	1	0	1	BL509	BL381	BL253	BL125
1	1	1	1	1	1	0	BL510	BL382	BL254	BL126
1	1	1	1	1	1	1	BL511	BL383	BL255	BL127

28F512 (C) Wordline Decoding

X Address									Row
A12	A7	A6	A5	A4	A13	A11	A9	A8	WL
0	0	0	0	0	0	0	0	0	XL0
0	0	0	0	0	0	0	0	1	XL1
0	0	0	0	0	0	0	1	0	XL2
0	0	0	0	0	0	0	1	1	XL3
0	0	0	0	0	0	1	0	0	XL4
0	0	0	0	0	0	1	0	1	XL5
0	0	0	0	0	0	1	1	0	XL6
0	0	0	0	0	0	1	1	1	XL7
0	0	0	0	0	1	0	0	0	XL8
0	0	0	0	0	1	0	0	1	XL9
0	0	0	0	0	1	0	1	0	XL10
0	0	0	0	0	1	0	1	1	XL11
0	0	0	0	0	1	1	0	0	XL12
0	0	0	0	0	1	1	0	1	XL13
0	0	0	0	0	1	1	1	0	XL14
0	0	0	0	0	1	1	1	1	XL15
0	0	0	0	1	1	1	1	1	XL16
0	0	0	0	1	1	1	1	0	XL17
0	0	0	0	1	1	1	0	1	XL18
0	0	0	0	1	1	1	0	0	XL19
0	0	0	0	1	1	0	1	1	XL20
0	0	0	0	1	1	0	1	0	XL21
0	0	0	0	1	1	0	0	1	XL22
0	0	0	0	1	1	0	0	0	XL23
0	0	0	0	1	0	1	1	1	XL24
0	0	0	0	1	0	1	1	0	XL25
0	0	0	0	1	0	1	0	1	XL26
0	0	0	0	1	0	1	0	0	XL27
0	0	0	0	1	0	0	1	1	XL28
0	0	0	0	1	0	0	1	0	XL29
0	0	0	0	1	0	0	0	1	XL30
0	0	0	0	1	0	0	0	0	XL31

intel® RR-60

28F512 (C) Wordline Decoding (Continued)

| X Address |||||||||| Row |
A12	A7	A6	A5	A4	A13	A11	A9	A8	WL
0	0	0	1	0	0	0	0	0	XL32
.	.	.	.	.	.	.	.	.	. . .
0	0	0	1	0	1	1	1	1	XL47
0	0	0	1	1	1	1	1	1	XL48
.	.	.	.	.	.	.	.	.	. . .
0	0	0	1	1	0	0	0	0	XL63
0	0	1	0	0	0	0	0	0	XL64
.	.	.	.	.	.	.	.	.	. . .
0	0	1	0	0	1	1	1	1	XL79
0	0	1	0	1	1	1	1	1	XL80
.	.	.	.	.	.	.	.	.	. . .
0	0	1	0	1	0	0	0	0	XL95
1	1	1	1	0	0	0	0	0	XL480
.	.	.	.	.	.	.	.	.	. . .
1	1	1	1	0	1	1	1	1	XL495
1	1	1	1	1	1	1	1	1	XL496
.	.	.	.	.	.	.	.	.	. . .
1	1	1	1	1	0	0	0	0	XL511

28F512 Die Photo

28F010 (A, B, C) Bitline Decoding

Address							Bitlines			
A16	A15	A10	A2	A1	A0	A3	IO0/7	IO1/6	IO2/5	IO3/4
0	0	0	0	0	0	0	BL384	BL256	BL128	BL0
0	0	0	0	0	0	1	BL385	BL257	BL129	BL1
0	0	0	0	0	1	0	BL386	BL258	BL130	BL2
0	0	0	0	0	1	1	BL387	BL259	BL131	BL3
0	0	0	0	1	0	0	BL388	BL260	BL132	BL4
0	0	0	0	1	0	1	BL389	BL261	BL133	BL5
0	0	0	0	1	1	0	BL390	BL262	BL134	BL6
0	0	0	0	1	1	1	BL391	BL263	BL135	BL7
...	...	...	...	...	...	...	...	...	...	...
1	1	1	1	1	0	0	BL508	BL380	BL252	BL124
1	1	1	1	1	0	1	BL509	BL381	BL253	BL125
1	1	1	1	1	1	0	BL510	BL382	BL254	BL126
1	1	1	1	1	1	1	BL511	BL383	BL255	BL127

28F010 (A, B, C) Wordline Decoding

X Address										Row
A14	A12	A7	A6	A5	A4	A13	A11	A9	A8	WL
0	0	0	0	0	0	0	0	0	0	XL0
0	0	0	0	0	0	0	0	0	1	XL1
0	0	0	0	0	0	0	0	1	0	XL2
0	0	0	0	0	0	0	0	1	1	XL3
0	0	0	0	0	0	0	1	0	0	XL4
0	0	0	0	0	0	0	1	0	1	XL5
0	0	0	0	0	0	0	1	1	0	XL6
0	0	0	0	0	0	0	1	1	1	XL7
0	0	0	0	0	0	1	0	0	0	XL8
0	0	0	0	0	0	1	0	0	1	XL9
0	0	0	0	0	0	1	0	1	0	XL10
0	0	0	0	0	0	1	0	1	1	XL11
0	0	0	0	0	0	1	1	0	0	XL12
0	0	0	0	0	0	1	1	0	1	XL13
0	0	0	0	0	0	1	1	1	0	XL14
0	0	0	0	0	0	1	1	1	1	XL15
0	0	0	0	0	1	1	1	1	1	XL16
0	0	0	0	0	1	1	1	1	0	XL17
0	0	0	0	0	1	1	1	0	1	XL18
0	0	0	0	0	1	1	1	0	0	XL19
0	0	0	0	0	1	1	0	1	1	XL20
0	0	0	0	0	1	1	0	1	0	XL21
0	0	0	0	0	1	1	0	0	1	XL22
0	0	0	0	0	1	1	0	0	0	XL23
0	0	0	0	0	1	0	1	1	1	XL24
0	0	0	0	0	1	0	1	1	0	XL25
0	0	0	0	0	1	0	1	0	1	XL26
0	0	0	0	0	1	0	1	0	0	XL27
0	0	0	0	0	1	0	0	1	1	XL28
0	0	0	0	0	1	0	0	1	0	XL29
0	0	0	0	0	1	0	0	0	1	XL30
0	0	0	0	0	1	0	0	0	0	XL31

28F010 (A, B, C) Wordline Decoding (Continued)

X Address										Row
A14	A12	A7	A6	A5	A4	A13	A11	A9	A8	WL
0	0	0	0	1	0	0	0	0	0	XL32
.	.	.	.	.	.	.	.	.	.	...
0	0	0	0	1	0	1	1	1	1	XL47
0	0	0	0	1	1	1	1	1	1	XL48
.	.	.	.	.	.	.	.	.	.	...
0	0	0	0	1	1	0	0	0	0	XL63
0	0	0	1	0	0	0	0	0	0	XL64
.	.	.	.	.	.	.	.	.	.	...
0	0	0	1	0	0	1	1	1	1	XL79
0	0	0	1	0	1	1	1	1	1	XL80
.	.	.	.	.	.	.	.	.	.	...
0	0	0	1	0	1	0	0	0	0	XL95
1	1	1	1	1	0	0	0	0	0	XL992
.	.	.	.	.	.	.	.	.	.	...
1	1	1	1	1	0	1	1	1	1	XL1007
1	1	1	1	1	1	1	1	1	1	XL1008
.	.	.	.	.	.	.	.	.	.	...
1	1	1	1	1	1	0	0	0	0	XL1023

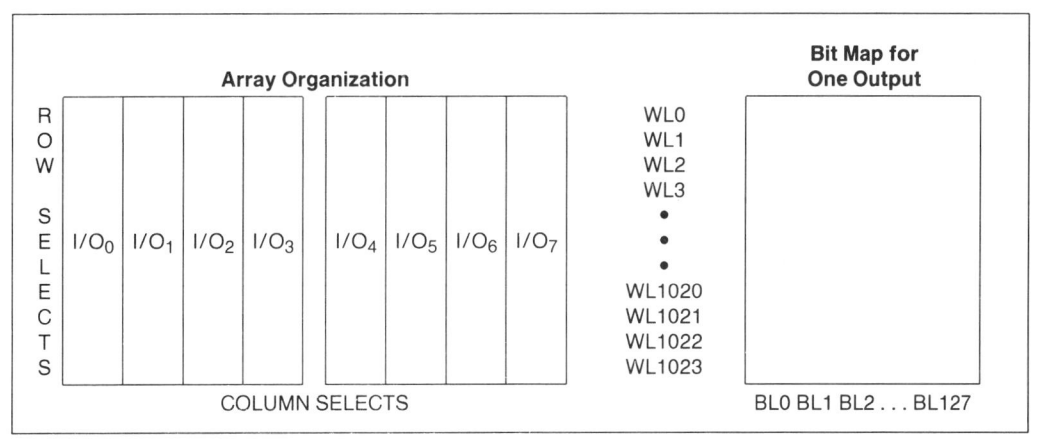

Array Organization

ROW SELECTS: I/O0 | I/O1 | I/O2 | I/O3 | I/O4 | I/O5 | I/O6 | I/O7

COLUMN SELECTS

WL0
WL1
WL2
WL3
•
•
•
WL1020
WL1021
WL1022
WL1023

Bit Map for One Output

BL0 BL1 BL2 ... BL127

28F010 Bit Map

intel®   RR-60

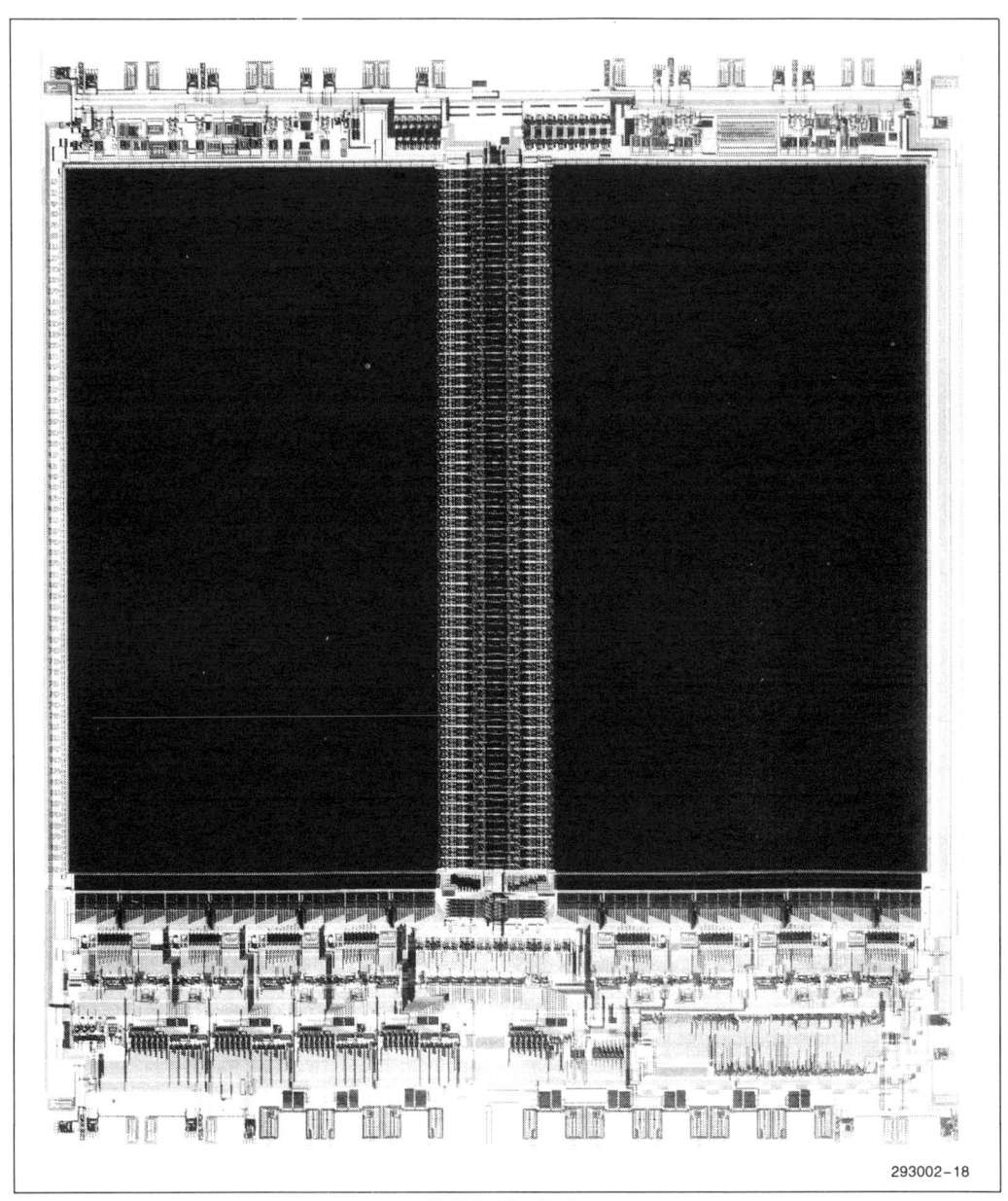

28F010 Die Photo

INTEL CORPORATION, 3065 Bowers Ave., Santa Clara, CA 95051; Tel. (408) 765-8080

INTEL CORPORATION (U.K.) Ltd., Swindon, United Kingdom; Tel. (0793) 696 000

INTEL JAPAN k.k., Ibaraki-ken; Tel. 029747-8511

seeq

E/M28HC256
256K High Speed EEPROM

PRELIMINARY DATA SHEET August 1989

FEATURES

- **Military and Extended Temperature Range**
  - -55° C to +125° C Operation (Military)
  - -40° C to +85° C Operation (Extended)
- **High Speed**
  - 90 nsec Maximum Access Time
- **Low Power CMOS Technology**
  - 80 mA Active Current
  - 50 mA Standby Current
- **Fast Write Cycle Times**
  - 64 Byte Page Write Operation
  - 5 ms Typical Byte/Page Write Time
  - 80 μsec Average Byte Write Time
- **On-Chip Timer**
  - Automatic Erase before Write
- **End of Write Detection**
  - $\overline{DATA}$ Polling
  - Toggle Bit
- **Software Accessible Control Register**
  - Disable Software Protection Mode
  - Chip Erase
  - Disable Automatic Erase before Write
- **Data Protection**
  - Hardware: Power Up/Down Protection Circuitry
  - JEDEC Approved Software Write Protection

- **High Endurance**
  - 10,000 Cycles/Byte
  - 10 Year Data Retention
- **5V +/- 10% Power Supply**
- **CMOS & TTL Compatible I/O**
- **Packages**
  - 28 Pin DIP, 32 Pad LCC & 28 Lead Flatpack

Pin Configuration

DUAL-IN-LINE, FLAT PACK TOP VIEW

A_{14}	1	28	V_{CC}
A_{12}	2	27	$\overline{WE}$
A_7	3	26	A_{13}
A_6	4	25	A_8
A_5	5	24	A_9
A_4	6	23	A_{11}
A_3	7	22	$\overline{OE}$
A_2	8	21	A_{10}
A_1	9	20	$\overline{CE}$
A_0	10	19	I/O_7
I/O_0	11	18	I/O_6
I/O_1	12	17	I/O_5
I/O_2	13	16	I/O_4
GND	14	15	I/O_3

LEADLESS CHIP CARRIER TOP VIEW

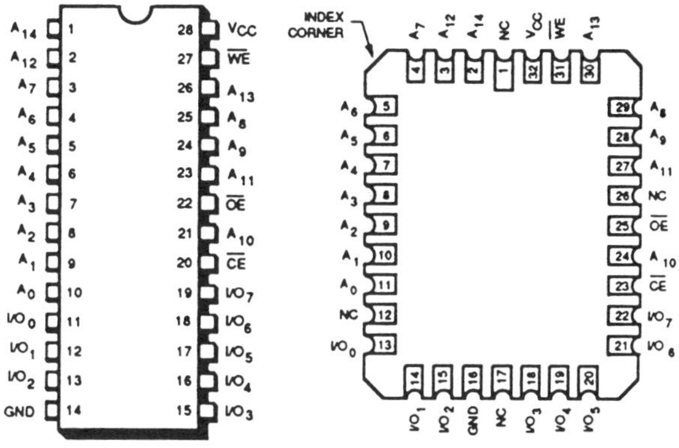

Block Diagram

Pin Names

A_0-A_5	ADDRESSES – COLUMN
A_6-A_{14}	ADDRESSES – ROW
$\overline{CE}$	CHIP ENABLE
$\overline{OE}$	OUTPUT ENABLE
$\overline{WE}$	WRITE ENABLE
I/O_{0-7}	DATA INPUT (WRITE)/DATA OUTPUT (READ)

seeq Technology, Incorporated
MD400082/-

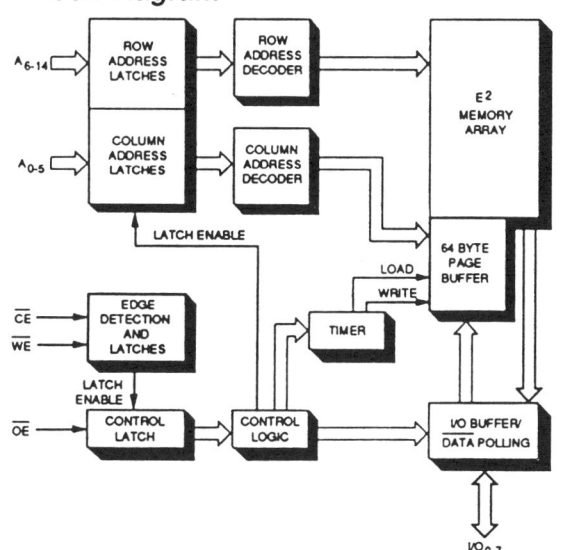

E/M28HC256
PRELIMINARY DATA SHEET

DESCRIPTION

The SEEQ 28HC256 is a high performance 5V only, 32K x 8 Electrically Erasable Programmable Read Only Memory (EEPROM). It is manufactured using SEEQ's advanced 1.0 micron CMOS process and is available in 28 pin Cerdip, 32 pad Leadless Chip Carrier (LCC), and 28 Lead Ceramic Flatpack. The 28HC256 is ideal for high speed applications which require low power consumption, non-volatility, and in-system reprogrammability. The endurance, the number of times which a byte may be written, is specified at 10,000 cycles per byte minimum.

The 90 ns maximum access time meets or exceeds the requirements of most of today's high performance microprocessors. To allow the system designer maximum flexibility, the following features have been added to the device. The 28HC256 has an internal timer which automatically times out the write time. The on-chip timer, along with the high speed input latches, frees the microprocessor for other tasks during the write time. The 28HC256's write cycle time is 5 msec typical. An automatic erase is performed before each write. The $\overline{DATA}$ Polling/Toggle Bit feature can be used to determine the end of a write cycle. A built-in control register allows a software controlled chip erase as well as the ability to disable the autoerase feature. This permits the user to effectively shorten the write time by half. Once the write cycle has been completed, data can be read in a maximum of 90 nsec. All inputs are CMOS/TTL for both write and read modes. Data retention is specified to be greater than 10 years.

DEVICE OPERATION

Operational Modes

There are five operational modes (see Table 1) and, except for the hardware chip erase mode, only CMOS/TTL inputs are required. A write cycle can only be initiated under the conditions shown. Any other conditions for $\overline{CE}$, $\overline{OE}$, and $\overline{WE}$ will inhibit writing and the I/O lines will either be in a high impedance state or have data, depending on the state of the aforementioned three input lines.

Reads

A read is accomplished by presenting the addresses of the desired byte to the address inputs. Once the address is stable, $\overline{CE}$ is brought to a CMOS/TTL low in order to enable the chip. The $\overline{WE}$ pin must be at a CMOS/TTL high during the entire read cycle. The ouput drivers are made active by bringing output enable, $\overline{OE}$, to a CMOS/TTL low. During read, the addresses, $\overline{CE}$, $\overline{OE}$, and I/O latches are transparent.

Writes

To write into a particular location, addresses must be valid and a CMOS/TTL low is applied to the write enable, $\overline{WE}$, pin of a selected ($\overline{CE}$ low) device. This combined with the output enable, $\overline{OE}$, being high, initiates a write cycle. During a byte write cycle, all inputs except data are latched on the falling edge of $\overline{WE}$ or $\overline{CE}$, whichever one occurred last. Write enable needs to be at a CMOS/TTL low only for the specified t_{wp} time. Data is latched on the rising edge of $\overline{WE}$ or $\overline{CE}$, whichever one occurred first. An automatic erase is performed before data is written. Automatic erase before write can be disabled to shorten the write cycle time.

The 28HC256 can write both bytes or blocks of up to 64 bytes. The write mode is discussed below.

Write Cycle Control Pins

For system design simplification, the 28HC256 is designed such that either the $\overline{CE}$ or $\overline{WE}$ pin can be used to initiate a write cycle. The device uses the latest high-to-low transition of either $\overline{CE}$ or $\overline{WE}$ signal to latch addresses and the earliest low-to-high transition to latch the data. Address and $\overline{OE}$ set up and hold are with respect to the later of $\overline{CE}$ or $\overline{WE}$; data set up and hold is with respect to the earlier of $\overline{WE}$ or $\overline{CE}$.

To simplify the following discussion, the $\overline{WE}$ pin is used as the control pin throughout the rest of this document.

Write Mode

One to 64 bytes of data can be loaded randomly into the 28HC256. Addresses A6-A14 select the page address and must remain the same throughout the page load cycle. These addresses are latched on the falling edge of the $\overline{WE}$ signal (assuming $\overline{WE}$ controlled write cycle).

The column addresses, A0-A5, which are used to write into different locations of the page, are latched every time a new write is initiated. These addresses along with $\overline{OE}$

Table 1 Mode Selection

Mode	$\overline{CE}$	$\overline{OE}$	$\overline{WE}$	I/O
Read	V_{IL}	V_{IL}	V_{IH}	D_{OUT}
Standby	V_{IH}	X	X	High Z
Write	V_{IL}	V_{IH}	V_{IL}	D_{IN}
Write Inhibit	X	X	V_{IH}	High Z/D_{OUT}
	V_{IH}	X	X	High Z
	X	V_{IL}	X	High Z/D_{OUT}
	V_{IL}	V_{IL}	V_{IL}	No Operation (High Z)
Chip Erase	V_{IL}	V_{H}	V_{IL}	X

X: Any CMOS/TTL level
V_{H}: 12V ± 10%

E/M28HC256
PRELIMINARY DATA SHEET

state (high) are latched on the falling edge of $\overline{WE}$ signal. For proper write initiation and latching, the $\overline{WE}$ pin has to stay low for a minimum of t_{wp} ns. Data is latched on the rising edge of $\overline{WE}$, allowing easy microprocessor interface.

Upon a low to high $\overline{WE}$ transition, the 28HC256 latches data and starts the internal page load timer. The timer is reset on the falling edge of $\overline{WE}$ signal if another write is initiated before the timer has timed out. The timer stays reset while the $\overline{WE}$ pin is kept low. If no additional write cycles have been initiated in (t_{BLC}) after the last $\overline{WE}$ low to high transition, the part terminates the page load cycle and starts the internal write. During this time, which takes a maximum of t_{wc} the device ignores any additional load attempts. The part can now be read to determine the end of write cycle ($\overline{DATA}$ Polling/Toggle Bit). A 80 μs average byte write time can be achieved if the page is fully utilized. The write time can be further optimized to 40 μs average by disabling automatic erase before write.

Extended Page Load

In order to take advantage of the page mode's faster average byte write time, data must be loaded within the page load cycle time ($t_{BLC\,max}$). Since some applications may not be able to sustain transfers at this minimum rate, the 28HC256 permits an extended page load cycle. To do this, the write cycle must be "stretched" by maintaining $\overline{WE}$ low, assuming a write enable controlled cycle and leaving all other control inputs ($\overline{CE}$, $\overline{OE}$) in the proper page load cycle state. Since the page load timer is reset on the falling edge of $\overline{WE}$, keeping this signal low will prevent the page load cycle timer from beginning. In a $\overline{CE}$ controlled write the same is true, with $\overline{CE}$ holding the timer reset instead of $\overline{WE}$.

$\overline{DATA}$ Polling
I/O7 $\overline{DATA}$ Polling

The 28HC256 has a maximum write cycle time of t_{wc}. However, a write will typically be completed in less than the specified maximum cycle time. $\overline{DATA}$ polling is a method of minimizing write times by determining the actual end point of a write cycle. If a read is performed to any address while the 28HC256 is still writing, the device will present the ones-complement of data bit I/O7. When the 28HC256 has completed its write cycle, a read from the last address written will result in valid data. Thus, software can simply read from the part until the last data byte written is read correctly. A $\overline{DATA}$ polling read should not be initiated until a minimum of t_{LP} nanoseconds after the last byte is written. $\overline{DATA}$ polling attempted during the middle of a page load cycle will present a ones-complement of the most recent data bit I/O7 loaded into the page. Timing for a $\overline{DATA}$ polling read is the same as a normal read once the t_{LP} specification has been met.

I/O6 Toggle Bit Polling

In addition to the polling method described above, the 28HC256 provides I/O6 Toggle Bit to determine the end of the internal write cycle. While the internal write cycle is in progress, I/O6 toggles from 1 to 0 and 0 to 1 on sequential polling reads. When the internal write cycle is complete the toggling stops and the 28HC256 is ready for additional read or write operations. This feature is particularly useful when writing to multiple devices simultaneously.

Hardware Chip Erase

Certain applications may require all bytes to be erased simultaneously. This can be achieved by clearing one byte at a time, however, this would require a clock cycle for each byte or page clear. The high voltage chip erase function completes this task with a single clock cycle, thus reducing the total erase time considerably. Please refer to the Hardware Chip Erase waveforms for timing specifics.

Write Data Protection
Hardware Feature

There is internal circuitry to minimize a false write during Vcc power up or down. This circuitry prevents writing under any one of the following conditions:

1) V_{cc} is less than V_{wr}
2) A high to low Write Enable ($\overline{WE}$) transition has not occurred when the V_{cc} supply is between V_{wl} and V_{cc} with $\overline{CE}$ low and $\overline{OE}$ high.

Writing will also be inhibited when $\overline{WE}$, $\overline{CE}$, or $\overline{OE}$ are in logical states other than that specified for a byte write in the Mode Selection Table.

Software Write Protect (SWP)

The 28HC256 has the ability to enable and disable write operations under software control by accessing an internal control register. Software control of write operations can reduce the probability of inadvertant writes resulting from power up, power down, or momentary power disturbances. The 28HC256 is shipped with the software write protect mode deactivated (default power-up mode) to provide compatibility with parts not having this mode. The software write protection mode is set by performing a page write operation (using page mode write timing) using specific addresses and data.

Set Software Write Protect

A three step write sequence shown below in TABLE 2 is used to set the protect mode. Page mode write timing is to be used. A violation of this sequence or the time-out of the page timer (t_{BLC}) will abort the set protection mode (see note). Reads attempted during the access sequence will

Seeq Technology, Incorporated

E/M28HC256
PRELIMINARY DATA SHEET

be assumed to be a $\overline{DATA}$ polling read and result in the device presenting a ones complement of the last data bit I/O7 written.

Protected Write Operation
Once the software protect mode is set, the software algorithm shown in TABLE 3 must be used for every byte write or page write cycle. The write operation uses the same three sequential steps shown in TABLE 2 to unlock the write protection for each byte/page write. The first three bytes unlock write protection while the fourth and successive bytes if any are written into the device.

Only single byte or page loads can be performed. After completion of internal write cycle, the device returns to the protected mode. The access sequence shown in TABLE 3 must be repeated to write an additional byte or page.

Disable Software Write Protection
The software protection can be disabled by following the six step sequence shown in TABLE 4. The device will be reconfigured to hardware protect mode only after this sequence. Page mode write timing is to be used. A violation of this sequence or the time-out of the page timer (t_{BLC}) will abort the reset protection mode. Reads attempted during the access sequence will be assumed to be $\overline{DATA}$ polling read.

SOFTWARE CONTROLLED SPECIAL FUNCTIONS
Chip erase and disable autoerase functions are accessed using the six step sequence shown in TABLES 5 & 6. The six step access sequence need not be followed by a non-volatile write cycle. The features are available for use both in the protected and unprotected (standard) modes. Page mode write timing is to be used. A violation of this sequence or the time-out of the page timer (t_{BLC}) will abort the access sequence and undesired writes could occur if the part is not software protected. Reads attempted during the access sequence will be assumed to be $\overline{DATA}$ polling read.

Software Chip Erase
5 V only software chip erase is performed by executing the six step access sequence shown in TABLE 5. Control data word 10 hex should be written to the secondary control register. $\overline{DATA}$ polling can be done during chip erase to determine the completion of chip erase. The six step write need not be followed by a byte or page data load. At the end of the six step access sequence, the device begins and completes chip erase internally. Chip erase command can only be issued with the autoerase before write function enabled.

Disable Autoerase
This command disables the automatic erase before write cycle and is used typically after a chip erase operation to reduce the programming time of the device. The six step write sequence shown in TABLE 6 is used to perform the operation. Control data word 40 hex should be written to the secondary control register on the sixth step. At the end of the six step sequence autoerase before write is disabled for the current byte or page write sequence. At end of the internal byte or page write cycle automatic erase before write is re-enabled. Autoerase before write is always enabled on power-up/reset (default).

TABLE 2 Set Software Write Protect Operation Sequence

Step	Mode	Address A14-A0	Data I/O 7-0	Comment
1	Write	5555 Hex	AA Hex	Dummy write.
2	Write	2AAA Hex	55 Hex	Dummy write.
3	Write	5555 Hex	A0 Hex	Dummy write. SWP state activated.
4-67	Write	Address	Data	Write data to address. Byte or Page write.

NOTE: SWP protected state will be activated at the end of write even if a byte or page data load is NOT attempted after the three step access sequence. In such a case, after the three step access sequence AND t_{BLC} timeout, SWP bit is set by performing a non-volatile write cycle. The SWP non-volatile bit is set for protected mode operation during the first access sequence to the part. Once the SWP non-volatile bit is set, subsequent writes require the 3 step sequence to enable byte or page writes. Undesired writes could occur as a result of first access sequence violation while attempting to set SWP.

seeq Technology, Incorporated

E/M28HC256
PRELIMINARY DATA SHEET

TABLE 3 Protected Mode Write Operation Sequence

Step	Mode	Address A14-A0	Data I/O 7-0	Comment
1	Write	5555 Hex	AA Hex	Dummy write.
2	Write	2AAA Hex	55 Hex	Dummy write.
3	Write	5555 Hex	A0 Hex	Dummy write. Enable byte/page writes.
4-67	Write	Address	Data	Write data to address. Byte or page write.

TABLE 4 Disable Protected Mode Operation Sequence

Step	Mode	Address A14-A0	Data I/O 7-0	Comment
1	Write	5555 Hex	AA Hex	Dummy write.
2	Write	2AAA Hex	55 Hex	Dummy write.
3	Write	5555 Hex	80 Hex	Dummy write.
4	Write	5555 Hex	AA Hex	Dummy write.
5	Write	2AAA Hex	55 Hex	Dummy write.
6	Write	5555 Hex	20 Hex	SWP state deactivated.
7-70	Write	Address	Data	Write data to address. Byte or Page Write.

NOTE: The SWP protected mode will be reset at the end of the write even if the six step access sequence is not followed by a byte or page data load. An internal non-volatile write cycle is performed to reset SWP bit after the six step access sequence AND t_{BLC} timeout.

TABLE 5 Chip Erase Operation Sequence

Step	Mode	Address A14-A0	Data I/O 7-0	Comment
1	Write	5555 Hex	AA Hex	Dummy write.
2	Write	2AAA Hex	55 Hex	Dummy write.
3	Write	5555 Hex	80 Hex	Dummy write register.
4	Write	5555 Hex	AA Hex	Dummy write.
5	Write	2AAA Hex	55 Hex	Dummy write.
6	Write	5555 Hex	10 Hex	Chip Erase

TABLE 6 Disable Autoerase Operation Sequence

Step	Mode	Address A14-A0	Data I/O 7-0	Comment
1	Write	5555 Hex	AA Hex	Dummy write.
2	Write	2AAA Hex	55 Hex	Dummy write.
3	Write	5555 Hex	80 Hex	Dummy write register.
4	Write	5555 Hex	AA Hex	Dummy write.
5	Write	2AAA Hex	55 Hex	Dummy write.
6	Write	5555 Hex	40 Hex	Disable Autoerase
7-70	Write	Address	Data	Load Page

seeq Technology, Incorporated

E/M28HC256
PRELIMINARY DATA SHEET

Absolute Maximum Stress Range*

Temperature
  Storage .... −65°C to +150°C
  Under Bias .... −65°C to +135°C

D.C. Voltage applied to all Inputs or Outputs
  with respect to ground .... +7.0 V to −1.2 V

*COMMENT: Stresses above those listed under "Absolute Maximum Ratings" may cause permanent damage to the device. This is a stress rating only and functional operation of the device at these or any other conditions above those indicated in the operational sections of this specification is not implied. Exposure to absolute maximum rating conditions for extended periods may affect device reliability.

Recommended Operating Conditions

	M28HC256	E28HC256
Temperature Range	(Case) −55°C to +125°C	(Ambient) −40°C to +85°C
V_{CC} Power Supply	5 V ± 10%	5 V ± 10%

Endurance and Data Retention

Symbol	Parameter	Value	Units	Condition
N	Minimum Endurance	10,000	Cycles/Byte	MIL-STD 883 Test Method 1033
T_{DR}	Data Retention	>10	Years	MIL-STD 883 Test Method 1008

DC Characteristics Read Operation (Over operating temperature and V_{CC} range, unless otherwise specified)

Symbol	Parameter	Min.	Max.	Units	Test Condition
I_{CC}	Active V_{CC} Current		80	mA	$\overline{CE}$ = $\overline{OE}$ = V_{IL}; All I/O open; Other Inputs = V_{CC} Max. Min. read or write cycle time
I_{SB1}	Standby V_{CC} Current (TTL Inputs)		50	mA	$\overline{CE}$ = V_{IH}, $\overline{OE}$ = V_{IL}; All I/O Open; Other Inputs = V_{IL} to V_{IH}
I_{SB2}	Standby V_{CC} Current (CMOS Inputs)		50	mA	$\overline{CE}$ = V_{CC} −0.3 Other Inputs = V_{IL} to V_{IH} All I/O Open
I_{IL} [2]	Input Leakage Current		1	µA	V_{IN} = V_{CC} Max.
I_{OL} [3]	Output Leakage Current		10	µA	V_{OUT} = V_{CC} Max.
V_{IL}	Input Low Voltage	−1.0	0.8	V	
V_{IH}	Input High Voltage	2.0	V_{CC} + 1	V	
V_{OL}	Output Low Voltage		0.4	V	I_{OL} = 8 mA
V_{OH}	Output High Voltage	2.4		V	I_{OH} = −2 mA
V_{WI}	Write Inhibit Voltage	3.8		V	

NOTES:
1. This parameter is measured only for the initial qualification and after process or design changes which may affect it.
2. Inputs only. Does not include I/O.
3. For I/O only.

seeq Technology, Incorporated

E/M28HC256
PRELIMINARY DATA SHEET

Capacitance [1] $T_A = 25°C$, f = 1 MHz

Symbol	Parameter	Max.	Conditions
C_{IN}	Input Capacitance	6 pF	V_{IN} = OV
C_{OUT}	Data (I/O) Capacitance	12 pF	$V_{I/O}$ = OV

E.S.D. Characteristics

Symbol	Parameter	Value	Test Conditions
V_{ZAP} [1]	E.S.D. Tolerance	>2000 V.	MIL-STD 883 Test Method 3015

A.C. Test Conditions

Output Load: 1 TTL gate and C_L = 100 pF
Input Rise and Fall Times: < 10 ns
Input Pulse Levels: 0.45 V to 2.4 V
Timing Measurement Reference Level:
  Inputs 0.8 V and 2 V
  Outputs 0.8 V and 2 V

AC Characteristics

Read Operation (Over operating temperature and V_{CC} range, unless otherwise specified)

Symbol	Parameter	E/M28HC256-90 Min.	E/M28HC256-90 Max.	E/M28HC256-100 Min.	E/M28HC256-100 Max.	E/M28HC256-120 Min.	E/M28HC256-120 Max.	Units	Test Conditions
t_{RC}	Read Cycle Time	90		100		120		ns	$\overline{CE} = \overline{OE} = V_{IL}$
t_{CE}	Chip Enable Access Time		90		100		120	ns	$\overline{OE} = V_{IL}$
t_{AA}	Address Access Time		90		100		120	ns	$\overline{CE} = \overline{OE} = V_{IL}$
t_{OE}	Output Enable Access Time		40		45		50	ns	$\overline{CE} = V_{IL}$
t_{DF}	Output or Chip Enable High to output in Hi-Z	0	40	0	45	0	50	ns	$\overline{CE} = V_{IL}$
t_{OH}	Output Hold from Address Change, Chip Enable, or Output Enable, whichever occurs first	0		0		0		ns	$\overline{CE} = \overline{OE} = V_{IL}$

Read /$\overline{DATA}$ Polling Cycle

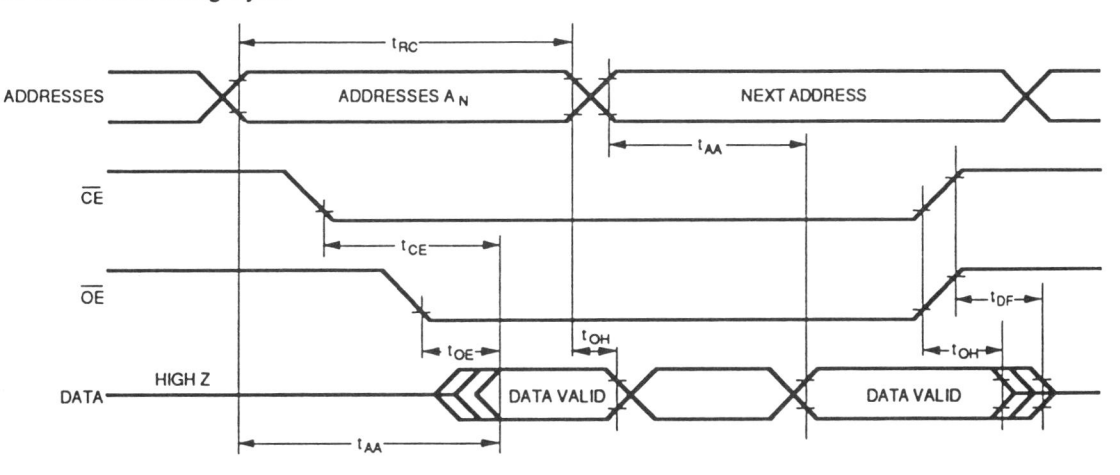

NOTES:
1. This parameter is measured only for the initial qualification and after process or design changes which may affect it.

seeq Technology, Incorporated

E/M28HC256
PRELIMINARY DATA SHEET

AC Characteristics

Write Operation (Over the operating temperature and V_{cc} range, unless otherwise specified)

Symbol	Parameter	E/M28HC256-90 Min.	E/M28HC256-90 Max.	E/M28HC256-100 Min.	E/M28HC256-100 Max.	E/M28HC256-120 Min.	E/M28HC256-120 Max.	Units
t_{WC}	Write Cycle Time		10		10		10	ms
t_{AS}	Address Set-up Time	0		0		0		ns
t_{AH}	Address Hold Time (see note 1)	50		50		50		ns
t_{CS}	Write Set-up Time	0		0		0		ns
t_{CH}	Write Hold Time	0		0		0		ns
t_{CW}	$\overline{CE}$ Pulse Width (note 2)	50		50		50		ns
t_{OES}	$\overline{OE}$ High Set-up Time	0		0		0		ns
t_{OEH}	$\overline{OE}$ High Hold Time	0		0		0		ns
t_{WP}	$\overline{WE}$ Pulse Width (note 2)	50		50		50		ns
t_{DS}	Data Set-up Time	40		40		40		ns
t_{DH}	Data Hold Time	0		0		0		ns
t_{BLC}	Byte Load Timer Cycle (Page Mode Only) (note 3)	0.2	200	0.2	200	0.2	200	µs
t_{LP}	Last Byte Loaded to $\overline{DATA}$ Polling Output		90		100		120	ns

Byte Write Timing

$\overline{WE}$ CONTROLLED WRITE CYCLE

$\overline{CE}$ CONTROLLED WRITE CYCLE

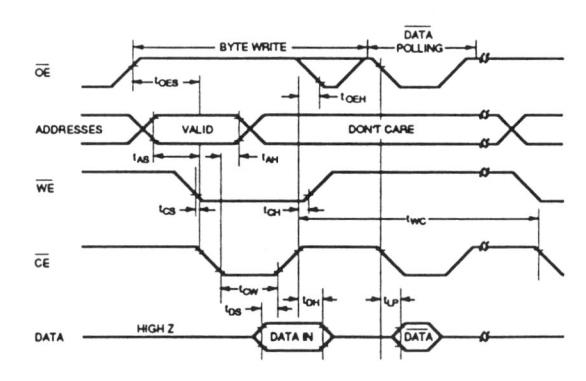

NOTES:
1. Address hold time is with respect to the falling edge of the control signal $\overline{WE}$ or $\overline{CE}$.
2. $\overline{WE}$ and $\overline{CE}$ are noise protected. Less than a 10 nsec write pulse will not activate a write cycle.
3. t_{BLC} min. is the minimum time before the next byte can be loaded. t_{BLC} max. is the minimum time the byte load timer waits before initiating internal write cycle.

seeq Technology, Incorporated

E/M28HC256
PRELIMINARY DATA SHEET

Page Write Timing / $\overline{WE}$ Controlled

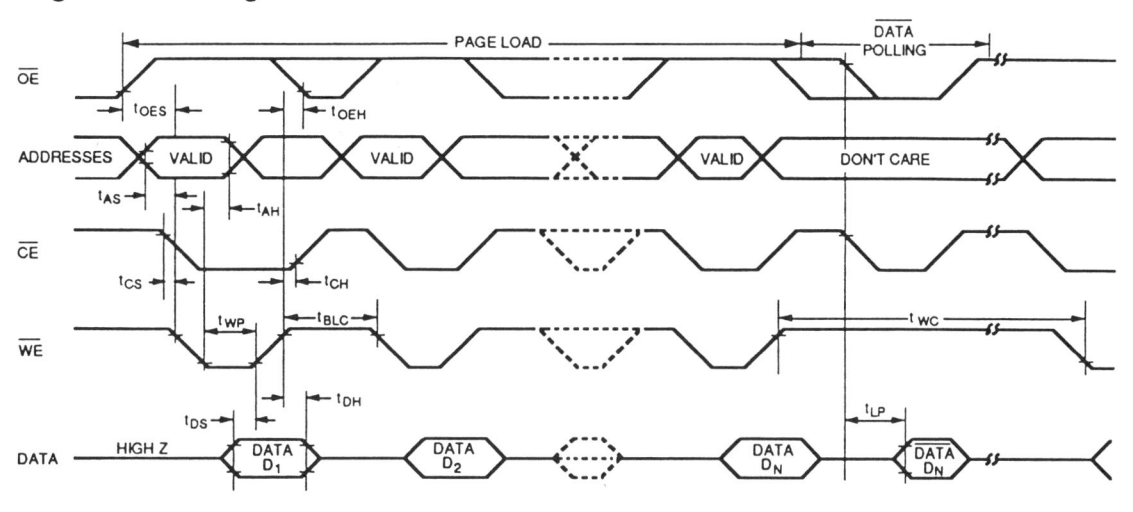

Hardware Chip Erase

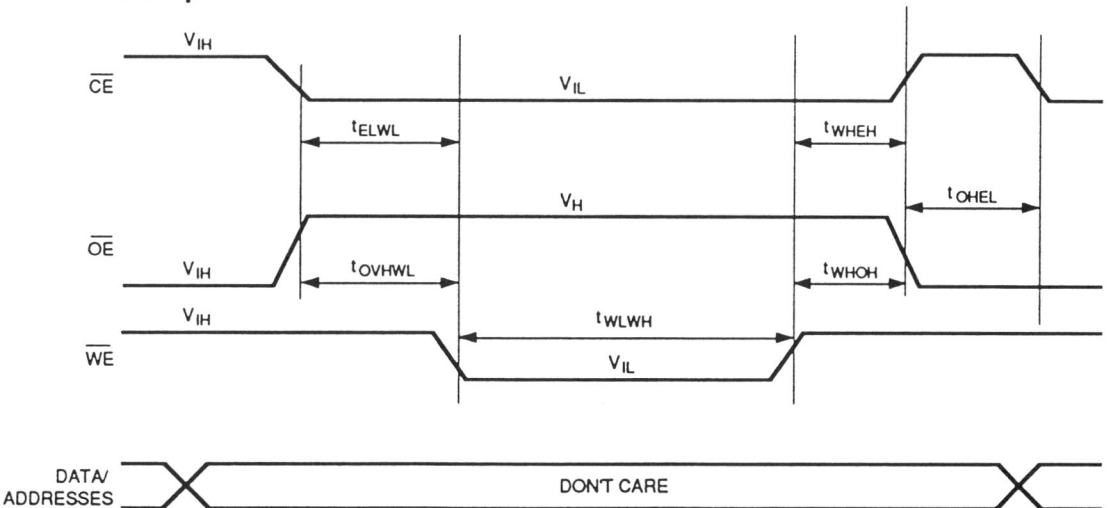

E/M28HC256
PRELIMINARY DATA SHEET

Hardware Chip Erase

Parameter	Description	Min.	Max.	Units
t_{ELWL}	Chip Enable Setup Time	5		μs
t_{OVHEL}	Output Enable Setup Time	5		μs
t_{WLWH}	Write Enable Pulse Width	10		ms
t_{WHEH}	Chip Enable Hold Time	5		μs
t_{WHOH}	Output Enable Hold Time	5		μs
t_{OHEL}	Erase Recovery Time		50	ms
V_H	High Voltage	10.8	13.2	V

Ordering Information

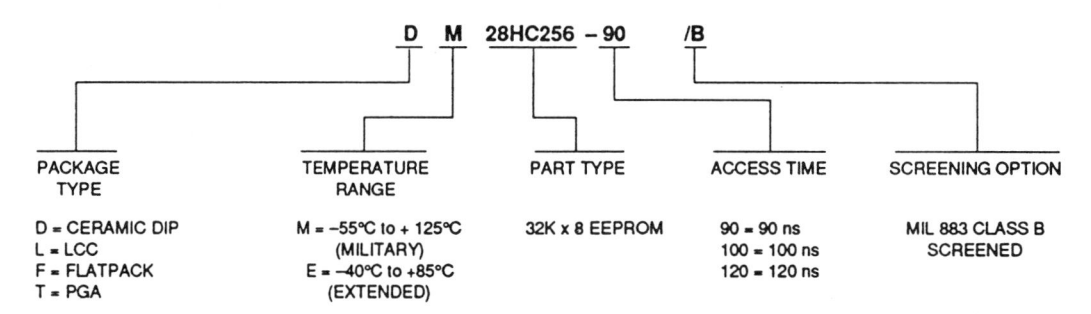

D M 28HC256 – 90 /B

PACKAGE TYPE
D = CERAMIC DIP
L = LCC
F = FLATPACK
T = PGA

TEMPERATURE RANGE
M = –55°C to +125°C (MILITARY)
E = –40°C to +85°C (EXTENDED)

PART TYPE
32K x 8 EEPROM

ACCESS TIME
90 = 90 ns
100 = 100 ns
120 = 120 ns

SCREENING OPTION
MIL 883 CLASS B SCREENED

seeq *Technology, Incorporated*

E²ROM Interfacing

Introduction

The continuing rapid evolution in semiconductor E²ROM memory device technology offers the system designer an ever-increasing choice of function and capability. With these increasing choices for E²ROM devices, however, comes the problem of standardization (or lack thereof) concerning such specifications as endurance, timing characteristics, interface requirements, ad infinitum. Today, there are two popular types of commercially available E²ROM devices.

Both of these types of devices have the JEDEC-approved pinout shown in Figure 1, including the multi-functional pin 1, but differ in the timing of the control interface. The first E²ROM type, the latched type device, such as SEEQ's 52B33 latches the addresses, control, and data inputs on the falling edge of WRITE ENABLE (WE). For this type device, the WE input must remain active low for the duration of the write cycle. The second type of E²ROM, the timer-type device, latches addresses, data, and control signals on the rising edge of WRITE ENABLE or the rising edge of CHIP ENABLE (CE). For the timer device, such as SEEQ's 2864 the WE input need not be held low for the entire write cycle. The primary difference between the latched and timer devices is the control timing required to interface to the microprocessor. Each of these types of devices has advantages depending on system performance and configuration requirements.

When the designer attempts to use the advantages of both in the same system, a problem is encountered.

One of the most frustrating problems facing a system designer is the design of an E²ROM/microprocessor interface that will allow compatible operation of timer and latched type E²ROM devices in the microprocessor-based system. The purpose of this application note is to give examples of cost-effective designs of E²ROM/microprocessor interfaces, which allow the use of both timer and latched E²ROM devices in the system with no changes required to either the controlling software or the hardware. With the interfaces shown in this application note, it is possible to operate with BOTH latched and timer devices simultaneously in the system if the device access times are compatible.

The microprocessor interfaces described in this application note are for the 8085, 8086, 8088, Z80, and 71840. Software examples are provided for the Z80 and 71840 processors. By extension, the Z80 code is easily transportable to 808X processors. In most cases, the hardware required for compatibility consists of only two additional standard (14-pin) TTL packages.

It is hoped that these example interfaces will assist the system designer in implementing E²ROMs in his system. By no means are these special cases presented to limit the system designer, but to provide a starting point for his design. The interface circuits presented are for the family of E²ROM devices (16K, 32K, and 64K). Other extensions of the ideas presented may permit lower power, lower cost, or optimization of other parameters deemed more important.

The body of this application note consists of two sections. First, the Basic Operation section gives the theory of operation of all of the interfaces and should be read to familiarize oneself with those factors common to all of the microprocessor interfaces. Second, the Microprocessor Interface section details the design of the TTL interface required for the given microprocessor.

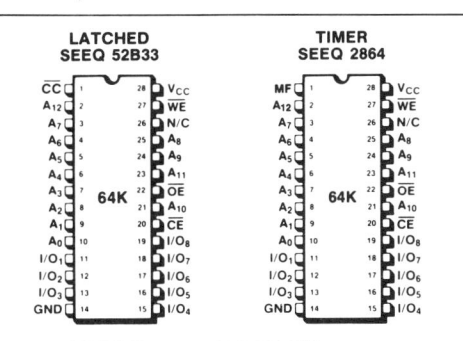

Figure 1. JEDEC Pinout — 64K E²ROMs

seeq *Technology, Incorporated*

Basic Operation

Each of the E²ROM microprocessor interfaces described in the next section integrates hardware and software to achieve compatibility between latched and timer E²ROM devices. Naturally, both hardware and software are processor-dependent. However, the write cycle used is basically the same for all the examples shown.

For compatibility between the latched and timer E²ROM devices, the interface provides control waveforms that have timing compatible with both, since the major difference between latched and timer E²ROM devices is the timing of the write control interface to the microprocessor (see Introduction). The basic waveforms for latched and timer E²ROMs are shown in Figures 2a and 2b, respectively. The latched type E²ROM device acquires data on the leading edge of $\overline{\text{WRITE ENABLE}}$

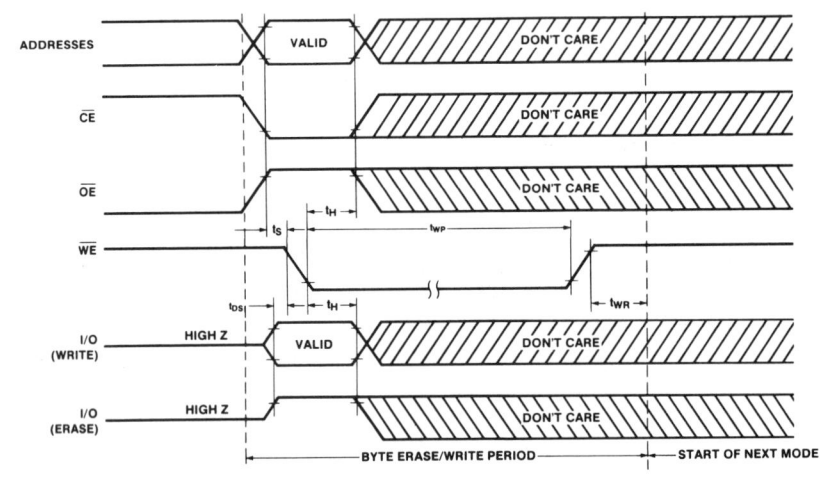

Figure 2a. Latched E²ROM Write Cycle

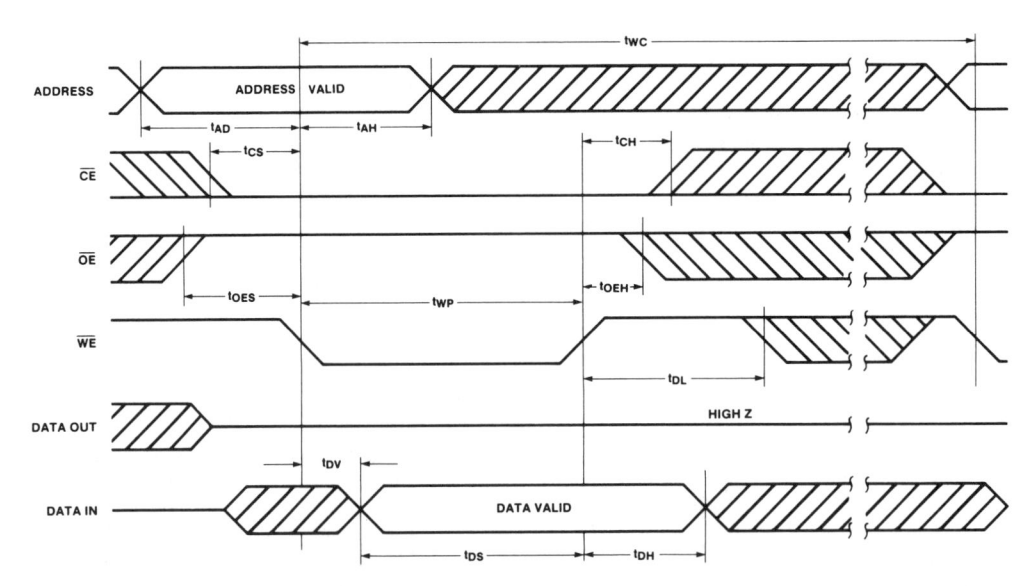

Figure 2b. Timer E²ROM Write Cycle

seeq Technology, Incorporated

($\overline{WE}$). The timer type device acquires data on either the trailing edge of $\overline{WE}$ or the trailing edge of $\overline{CHIP}$ $\overline{ENABLE}$ ($\overline{CE}$). Interface compatibility is achieved between the latched and timer devices by strobing the data, control, and addresses on the leading edge of the Write Enable pulse for the latched device and then by strobing the data on the trailing edge of $\overline{CHIP}$ $\overline{ENABLE}$ for the timer device (see Figure 3). By using this technique, the hardware interface is greatly simplified.

The software part of an E²ROM interface is very simple, but very important. A read operation for both latched and timer E²ROM devices is accomplished by a straightforward issuance of a microprocessor Read command at a particular address (see Figure 4). A write operation, however, involves a more complex process.

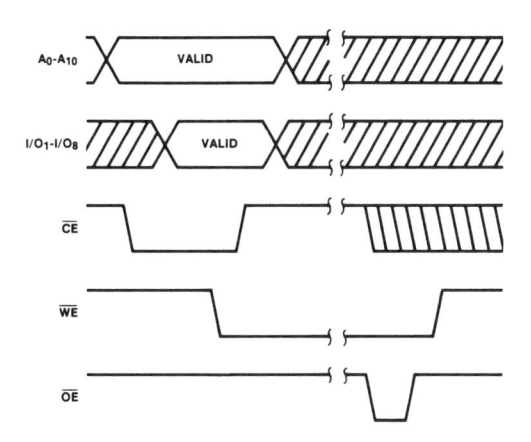

The flow chart for writing to the E²ROM is the same for all microprocessors and is shown in Figure 5. After a Write command is issued, time is required to allow proper writing to the storage cell of the E²ROM device. A Read command is then issued to terminate the write operation. Note that this Read command is not to be used to actually read the E²ROM device, but is inserted to reset the logic circuits used to drive the $\overline{WE}$ input of the E²ROM device.

Between initiation and termination of a write cycle, the interface uses some timing mechanism to assure proper write conditions to the E²ROM and to know when the E²ROM is available for another read/write cycle. The duration of the timeout (t_{WP}) depends upon the type of E²ROM used. For all types, t_{WP} should fall between the minimum and maximum specifications of all E²ROMs for which the application is designed. The latched type of device requires less write time than does the timer type device.

The implementation of this timing can be accomplished in either hardware or software. In hardware timing, a timer can interrupt the processor at regular intervals, or at the end of the desired write time (t_{WP}). In software timing, the processor simply counts down, waiting for the desired t_{WP}. For ease of general implementation, the given examples utilize software timing (see Figure 5). The tradeoffs, however, between software and hardware timing comprise an involved topic. The system designer must make this decision, considering such factors as processor throughput, board space, and expense.

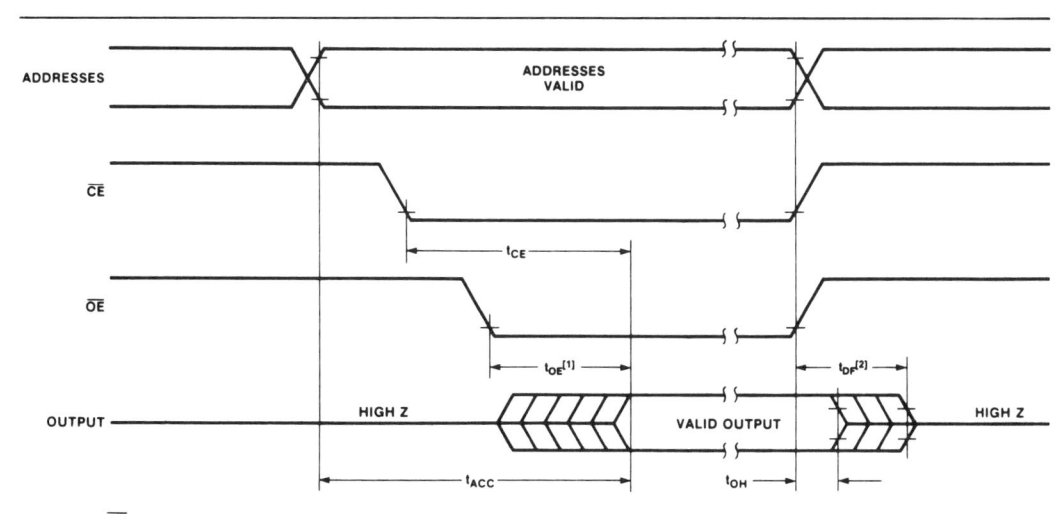

Figure 3. Latched/Timer Compatible E²ROM Write Cycle

Notes:
1. $\overline{OE}$ may be delayed up to t_{ACC} — t_{OE} after the falling edge of $\overline{CE}$ without impact on t_{ACC}.
2. t_{DF} is specified from $\overline{OE}$ or $\overline{CE}$, whichever occurs first.
3. This parameter is periodically sampled.

Figure 4. E²ROM Read Cycle

seeq *Technology, Incorporated*

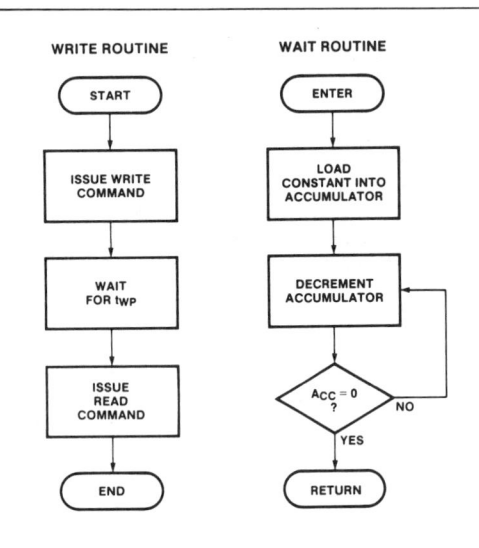

Figure 5. Software Flowchart — E²ROM Write Cycle

After the cycle described by Figure 5 is complete, the E²ROM device is available to be accessed for another Read or Write command. Often, another read will be performed in order to verify the written data. With the solution proposed, this subsequent read cycle will have normal timing, and all required write recovery parameters will be satisfied.

The general description provided above applies to most of the processors shown in the specific examples below. For more detailed information, the reader should refer to the schematic, waveforms, and software that apply to a specific processor.

Microprocessor Interfaces

8085 Interface

The schematic for the 8085 interface to a timer or latched E²ROM device is shown in Figure 6. This interface consists of one each of a 74LS02 and 74LS74 type package and allows the system designer to use the $\overline{WR}$ signal from the 8085 to initiate the write cycle to the E²ROM device. The design permits use of either a timer

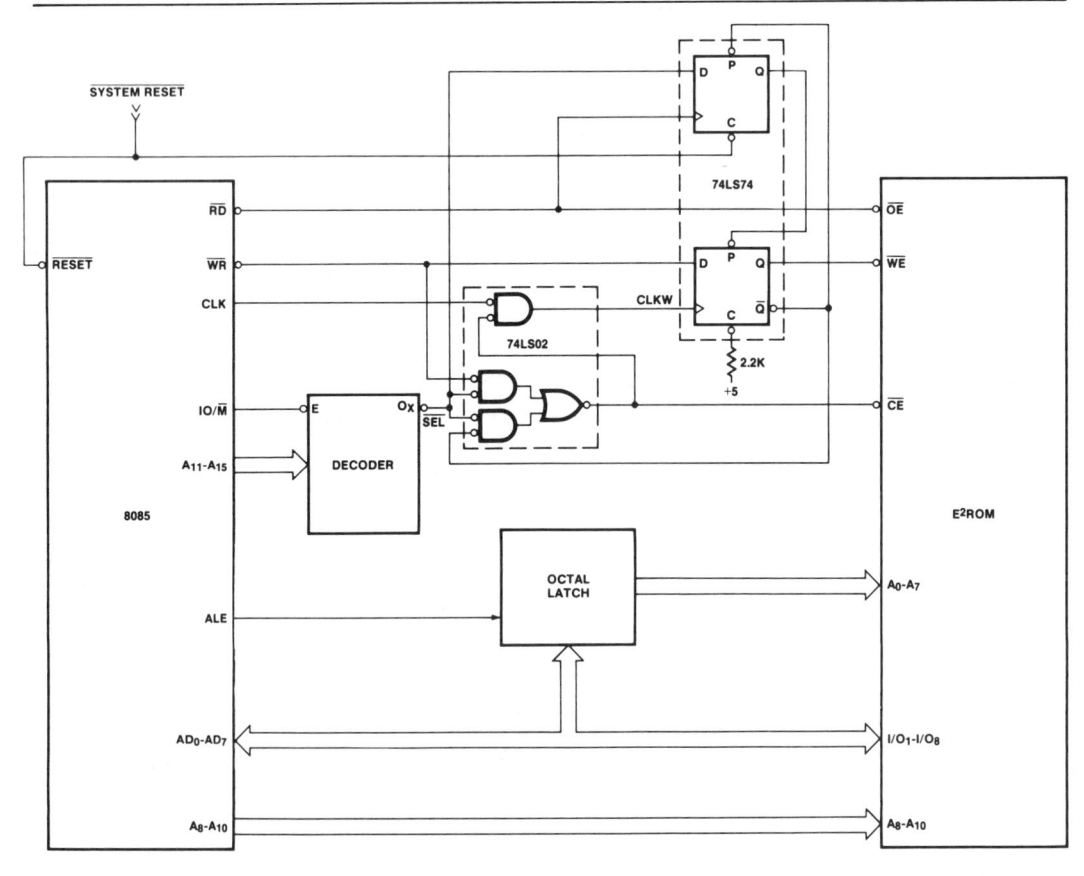

Figure 6. 8085/E²ROM Interface

seeq Technology, Incorporated

5-47

OR a latched E²ROM device with no change required to the controlling software or hardware. The following discussion of the operation of the 8085 interface relies on the 8085 timing diagram summary for read and write cycles shown in Figures 7a and 7b respectively.

Initiating a write cycle requires the software control routine as charted in Figure 5. Should the reader desire a specific example, the Z80 code (see Figure 12) is transportable to the 8085.

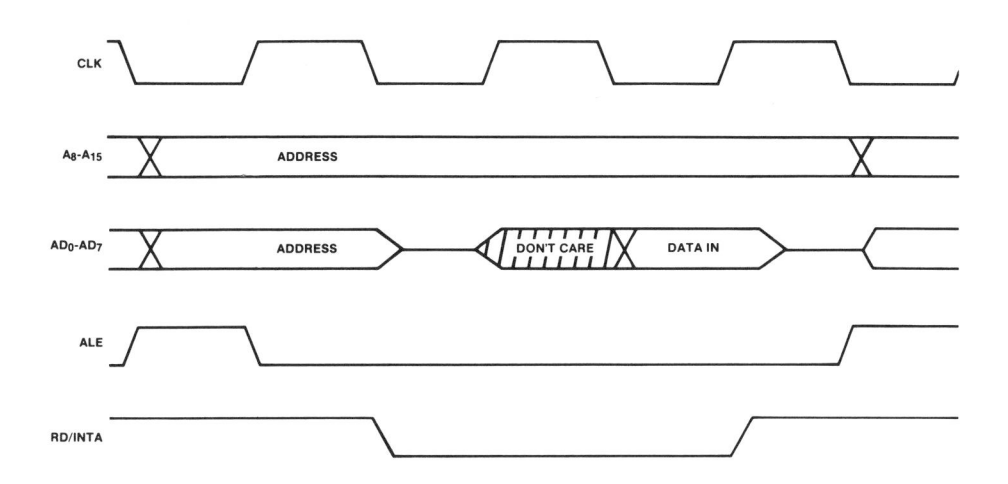

Figure 7a. 8085 Read Timing Summary

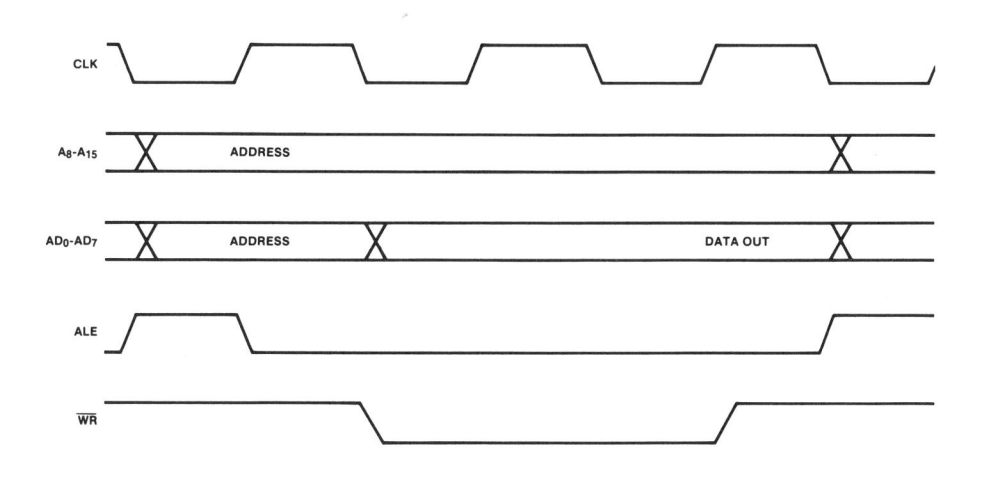

Figure 7b. 8085 Write Timing Summary

seeq *Technology, Incorporated*

The basic write operation waveforms for this interface are shown in Figure 8. The write cycle begins with the addresses becoming valid and being decoded to drive $\overline{\text{SELECT}}$ active low, in order to drive the $\overline{\text{CHIP ENABLE}}$ ($\overline{\text{CE}}$) active low at the E^2ROM device pin (selecting the desired device) (see (A) in Figure 6). An active low level on $\overline{\text{WR}}$ from the 8085 (indicating a write cycle initiation) allows the $\overline{\text{WRITE ENABLE}}$ latch of the interface to be clocked by the next falling edge of the 8085 clock output (CLK) (see (B)). Addresses, data, and control inputs to the latched type E^2ROM are latched in at the falling edge of $\overline{\text{WRITE ENABLE}}$ ($\overline{\text{WE}}$) — shown as (B) in Figure 8. For the timer type E^2ROM device, however, data is latched on the rising edge of $\overline{\text{CHIP ENABLE}}$ ($\overline{\text{CE}}$) — shown as (C) in Figure 8. Note that $\overline{\text{CE}}$ is held active low for a relatively short period of time, while $\overline{\text{WRITE ENABLE}}$ ($\overline{\text{WE}}$) is held low for the entire write time of the E^2ROM device. In this manner, the waveforms shown in Figure 3 are produced, providing signals compatible with both the latched and timer type devices.

To end the write cycle, the 8085 issues a Read command to the E^2ROM device. This read cycle enables the Write Reset latch which in turn presets the $\overline{\text{WRITE ENABLE}}$ latch (shown in Figure 6). The preset to the $\overline{\text{WE}}$ latch brings $\overline{\text{WE}}$ to V_{IH} (see (D) in Figure 8). As indicated in Figure 8, this read cycle does not produce valid data from the E^2ROM. This read cycle is used merely to terminate the write cycle.

The latched and timer devices respond identically in a read cycle. The 8085 read cycle, shown in Figure 7a, produces the read cycle waveforms shown in Figure 4.

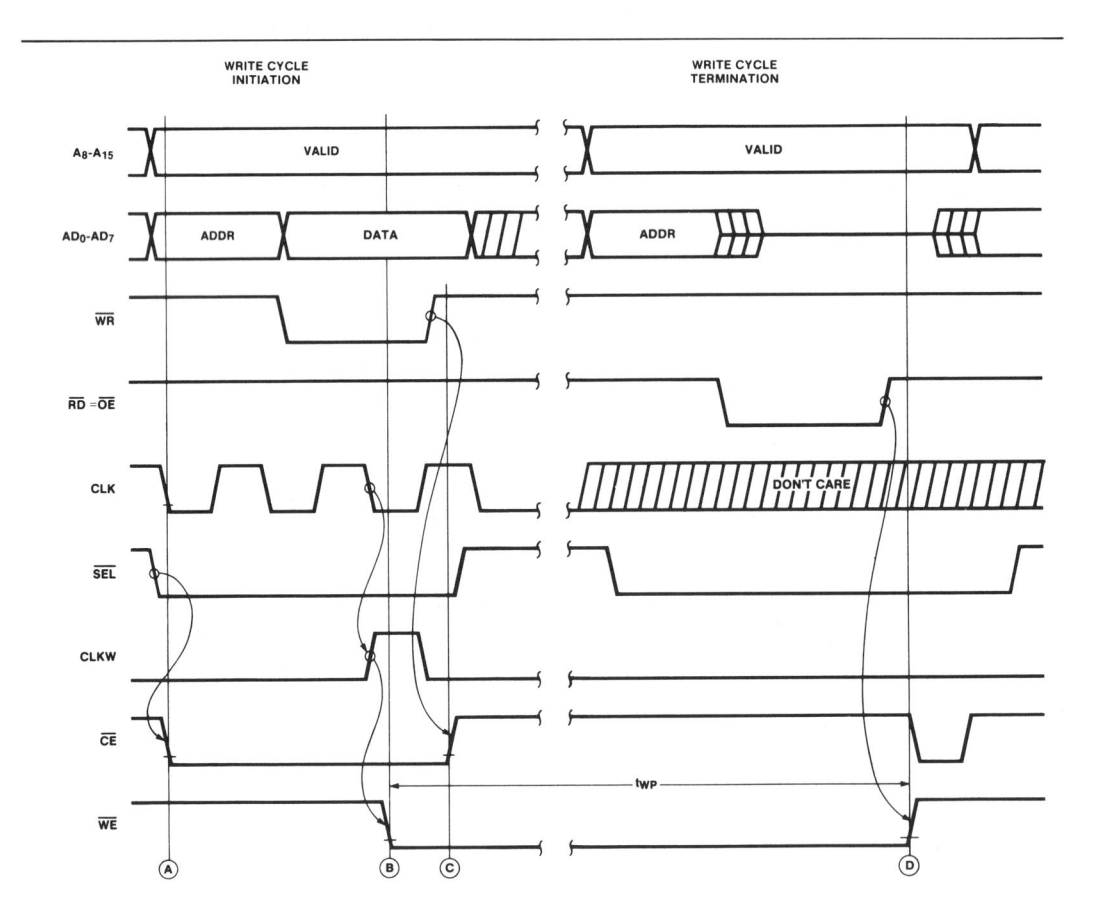

*A₀-A₇: ADDRRESS SIGNALS MULTIPLEXED WITH DATA SIGNALS MUST BE DEMULTIPLEXED USING OCTAL LATCHES.

Figure 8. Timing Diagram — 8085/E^2ROM Interface

seeq Technology, Incorporated

Z80 Interface

A sample interface is shown for a Z80 processor (see Figure 9). The timing diagram for write cycle waveforms at this interface is also shown (see Figure 10). The basic circuit is very similar to the 8085 interface, with the differences based on the fact that the Z80 has data valid at both edges of $\overline{WR}$ (see Figure 11). This simplified timing allows a more simple interface. The CLK output from the processor is not necessary, and $\overline{WR}$ alone provides timing for the write cycle initiation.

The operation of the circuit is otherwise very similar to the 8085 interface. After addresses are brought valid on the address bus, they are decoded to drive $\overline{SEL}$ active low, which drives $\overline{CE}$ active low at the E^{2}ROM device pin (see Figure 9, and (A) in Figure 10). At the falling edge of $\overline{WR}$ (when this device is selected), the $\overline{WE}$ latch is clocked, bringing $\overline{WE}$ active low (see (B) in Figure 10). At this time, the latched type device latches address, data, and control signals, while the timer type device latches address and control signals. At the falling edge of $\overline{WR}$, the gating circuitry brings $\overline{CE}$ high, latching data for the timer type part (see (C) in Figure 10). Within a normal processor cycle, a write cycle has been initiated with timing in accordance with the general approach of Figure 3. Even with additional buffers which may be common in a bus oriented system, this interface can be used with a Z80, Z80A, or Z80B operating with no wait states at up to 6 MHz clock frequency. The individual system designer, of course, must check his own application to ensure satisfaction of applicable setup and hold requirements in the specific system for which the application is intended.

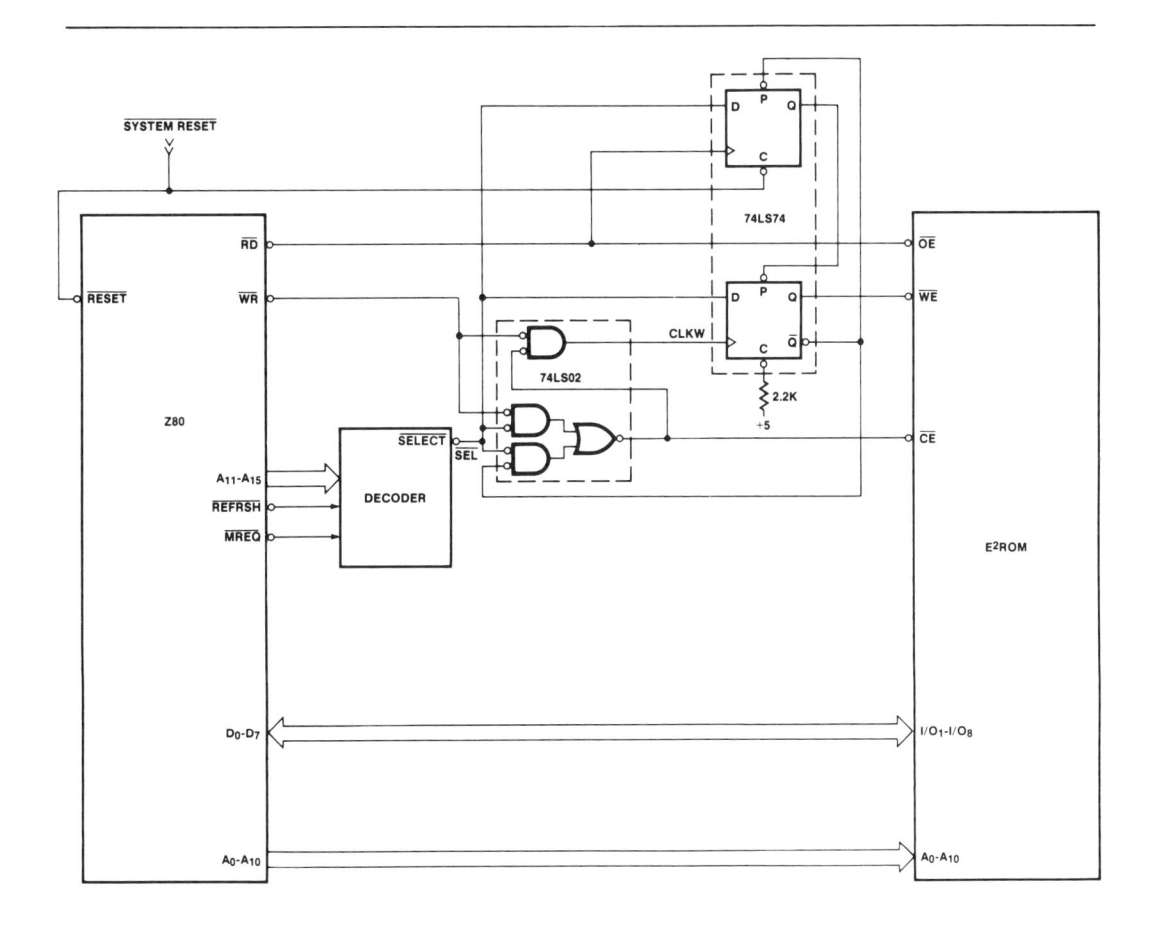

Figure 9. Z80/E^{2}ROM Interface

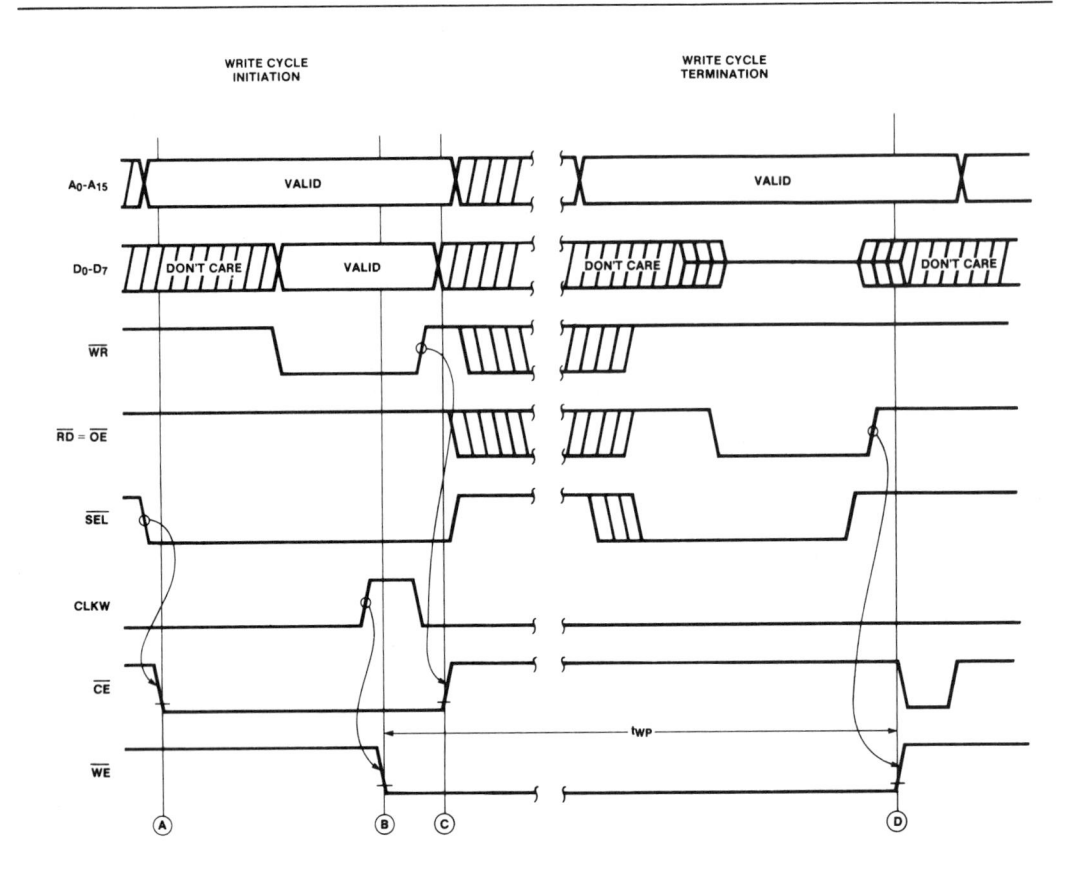

Figure 10. Timing Diagram — E²ROM Interface (Write Cycle)

The termination of a write cycle is very straightforward. As shown in the Basic Operation section (see Figure 5), a read operation to the E²ROM terminates the write cycle, but does not provide valid data. For the interface operation in write cycle termination, the reader should refer to Figure 10. The addresses are brought valid on the address bus, and are decoded to drive $\overline{SEL}$ active low (see (A) in Figure 10). The gating circuitry, however, inhibits $\overline{CE}$, and $\overline{CE}$ remains at V_{IH}. At the rising edge of $\overline{RD}$, the flip-flop receives a positive edge trigger, and clocks in the $\overline{SEL}$ signal to preset the $\overline{WE}$ latch. At this point, $\overline{WE}$ is brought high (see (D) in Figure 10), terminating the write cycle. For the remainder of this processor bus cycle, $\overline{CE}$ becomes valid for a short while. However, $\overline{RD}$ is no longer active low, and no valid data is read in this bus cycle. There is no problem with t_{WR} since the write recovery time occurs during the remaining part of this bus cycle.

Frequently, one may wish to read again from the device, in order to verify data written. This read will be a normal read, following the general waveforms of Figure 4. In a read operation, the interface drives $\overline{CE}$ active low to select the device, and $\overline{RD}$ enables the output from the E²ROM device.

seeq *Technology, Incorporated*

5-51

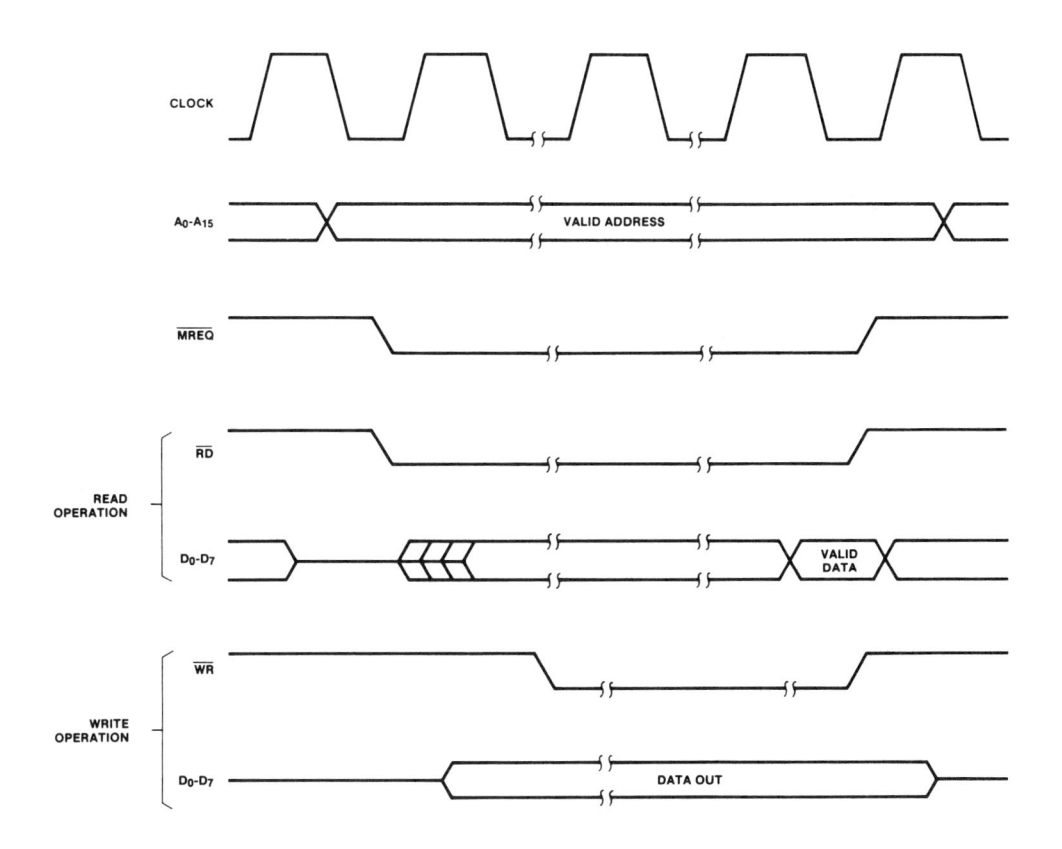

Figure 11. Z80 Read and Write Cycle

```
                            EEWRZ80.1
        LOC   OBJ CODE M STMT SOURCE STATEMENT                        ASM 5.9
                        175
                        176   ;----------------------------------------
                        177   ; Z80 EEROM Write routine.
                        178   ; Incorporates auto-erase and timing
                        179   ; in software.
                        180   ; Accepts: address to be written: Reg DE
                        181   ;          Data     to be written: Reg B
                        182   ; Uses: A, B, D, E Destroys: A
                        183   ;----------------------------------------
                        184
        009B  3EFF      185   EEWR:   LD      A,0FFH     ; FF for erasure.
        009D  12        186           LD      (DE), A    ; BEGIN ERASE
        009E  CDAE00    187           CALL    WaitTwp
        00A1  1A        188           LD      A, (DE)    ; END ERASE
                        189
        00A2  78        190           LD      A,B        ; Data to be written
        00A3  12        191           LD      (DE), A    ; BEGIN WRITE
        00A4  CDAE00    192           CALL    WaitTwp
        00A7  1A        193           LD      A, (DE)    ; Read to end Write
        00A8  1A        194           LD      A, (DE)    ; Read to Verify
        00A9  B8        195           CP      B          ; Check Verification
        00AA  C2C800    196           JP      NZ, ERR1
        00AD  C9        197           RET
                        198
                        199
                        200   ;----------------------------------------
                        201   ; Wait routine for EEROM Byte/ Erase
                        202   ; Uses: Registers A, B,C
                        203   ; Destroys: A,C
                        204   ;----------------------------------------
        00AE  78        205   WaitTwp:LD      A,B
                        206   ; Store B reg in TMP1
        00AF  3202C0    207           LD      (TMP1),A
                        208
                        209   ; Set timing constant for Twp.
                        210   ; This 16-bit constant is loaded
                        211   ; into Registers BC, and depends
                        212   ; on the speed of the CPU clock.
        00B2  3E07      213           LD      A, 07
        00B4  47        214           LD      B,A
        00B5  3E06      215           LD      A, 06
        00B7  4F        216           LD      C,A
                        217
                        218   ; The following loop performs the wait,
                        219   ; by decrementing BC until the 16-bit
                        220   ; number contained in BC equals zero.
                        221
        00B8  3E00      222           LD      A, 00H
        00BA  0B        223   More:   DEC     BC
        00BB  B8        224           CP      B
        00BC  C2BA00    225           JP      NZ, More
        00BF  B9        226           CP      C
        00C0  C2BA00    227           JP      NZ, More
        00C3  3A02C0    228   DUN:    LD      A, (TMP1) ; Restore B Reg
        00C6  47        229           LD      B,A
        00C7  C9        230           RET
```

Figure 12. Z80 E²ROM Erase/Write Routine

8088 Interface

An example interface is shown between an 8088 (operating in minimum mode) and a 16K E^{2}ROM (see Figure 14). The reader may note that this is almost identical to the 8085 E^{2}ROM interface (see Figure 6), with only minor differences. First, the NOR gates used cannot be a standard TTL or LSTTL device, but must be a CMOS or other high impedance input, so that the CLK signal is not loaded. The CLK signal, as output by the 8284, is used as the clock input to the 8088. The V_{OH} level on this signal can fall below specification as a result of a TTL load. A CMOS NOR package, such as a 74C02 or similar device, eliminates this problem. Since the 74LS74 operates from bussed control and data lines, its requirements are not so stringent, and a 74LS74 will work fine in most applications.

The operation of this circuit is almost identical to the operation of the 8085 interface, as a comparison of the timing diagrams will show (see Figures 7b and 15). Because these processors share similar bus timing, the signals differ only in magnitudes of setup and hold times. All required setup and hold times should be confirmed to the satisfaction of the system designer.

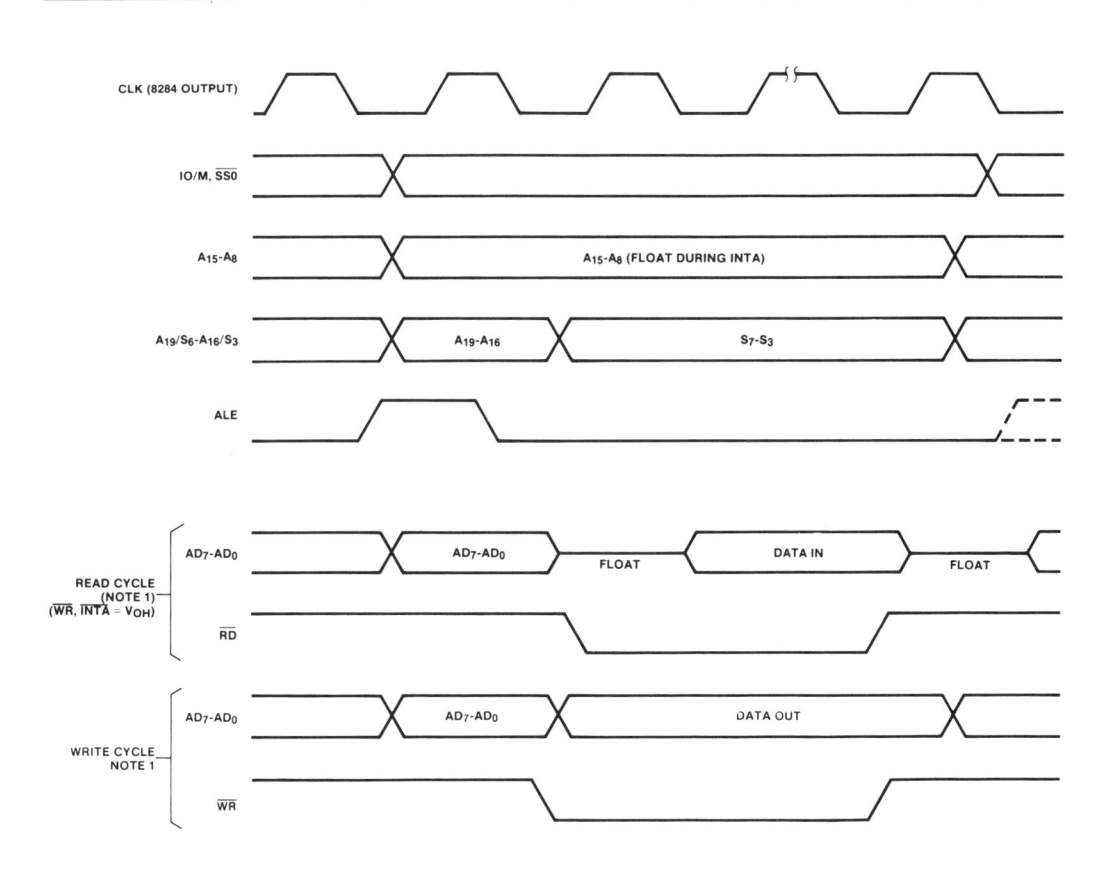

Figure 13. 8088/8086 Bus Timing — Minimum Mode

seeq Technology, Incorporated

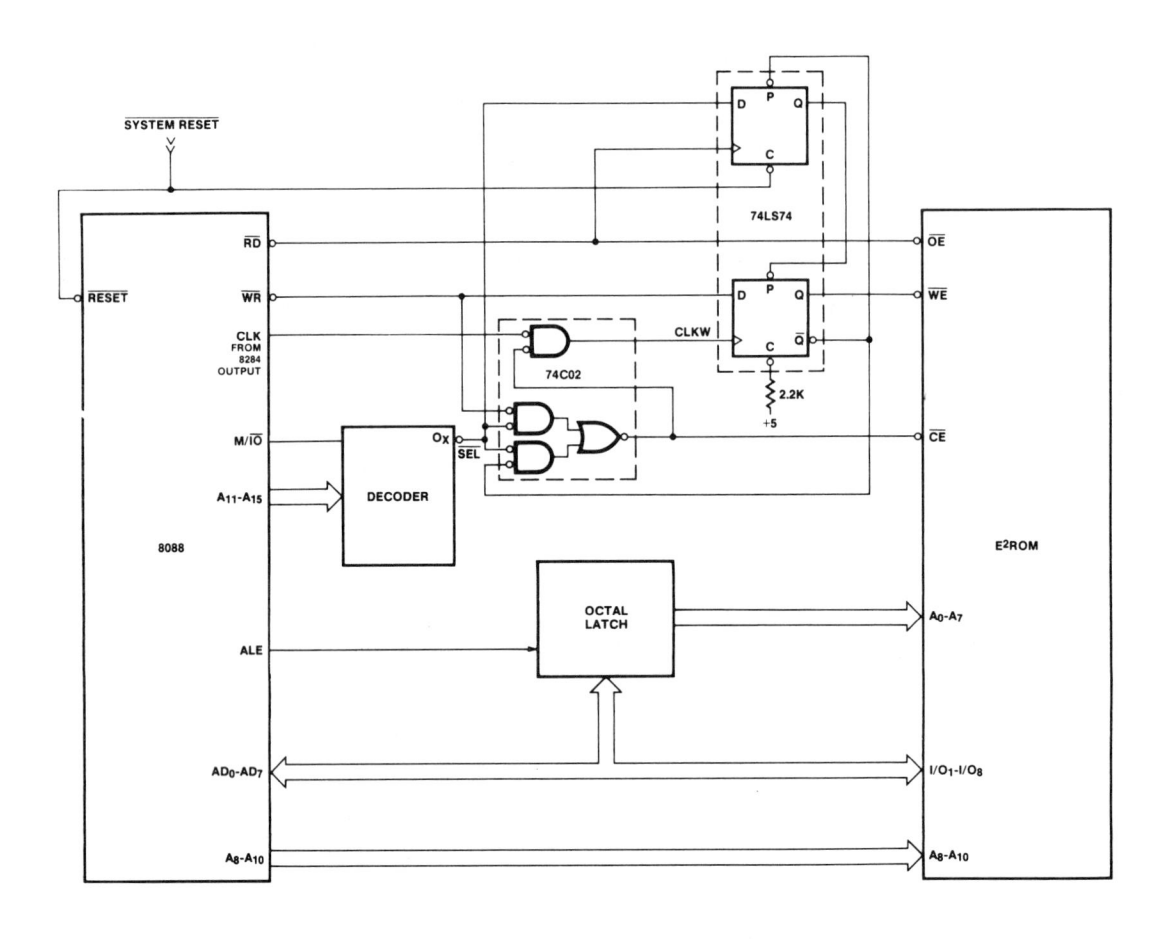

Figure 14. E²ROM Interface — 8088 (Minimum Mode)

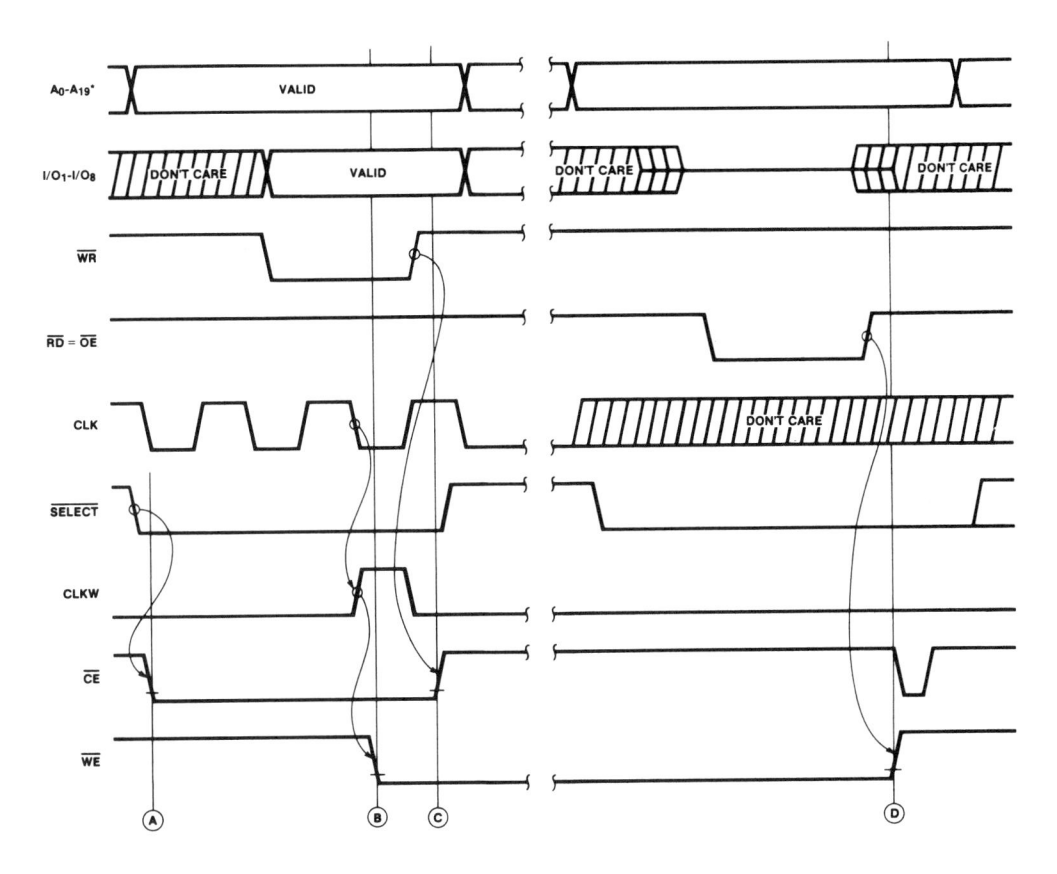

Figure 15. Timing Diagram — 8088/8086 E²ROM Interface

*A₀-A₁₉: ADDRRESS SIGNALS MULTIPLEXED WITH STATUS AND DATA SIGNALS MUST BE DEMULTIPLEXED USING OCTAL LATCHES.

8086 Interface

A sample E²ROM interface shown for the 8086 (see Figure 16) compares very closely in layout and operation to that for the 8088 (see Figure 14). The 8086 interface accounts for the 16-bit 8086 data bus by latching both bytes of address and implementing a pair of devices to read and write an entire word at a time. E²ROM interface control signals are identical to those for the 8088 interface (see Figure 15).

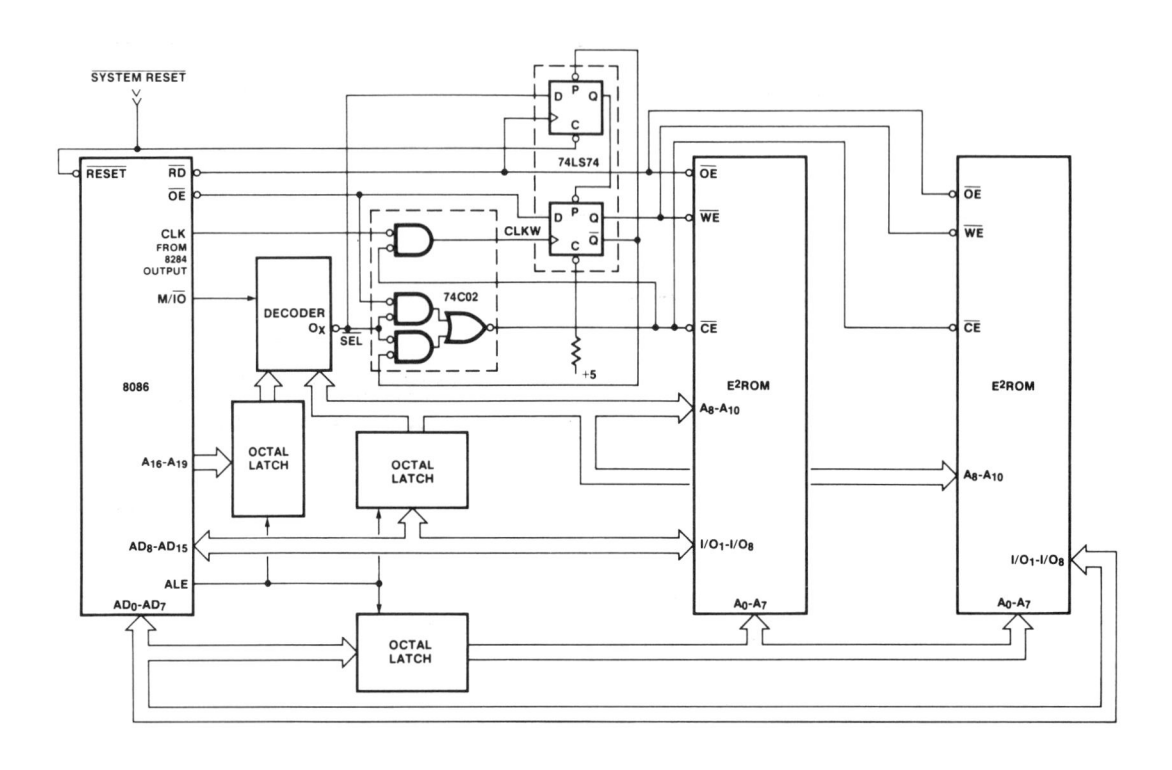

Figure 16. E^{2}ROM Interface — 8086 (Minimum Mode)

71840 (Z8) Interface

An example E^{2}ROM interface is presented for the 71840 (see Figure 17). SEEQ's 71840 is a single-chip microcomputer, with 4K x 8 of **EPROM** and 244 bytes of RAM, which is otherwise compatible with the Z8. Using the architecture and code of the Z8, the 71840 has a logical instruction set, pipelined execution, and a high degree of flexibility, while providing the additional features of Silicon Security™ and EPROM programming. The interface shown includes the octal latch, used to demultiplex the eight address/data bits for the E^{2}ROM. The rest of the circuit acts to produce the control signals shown in Figure 3.

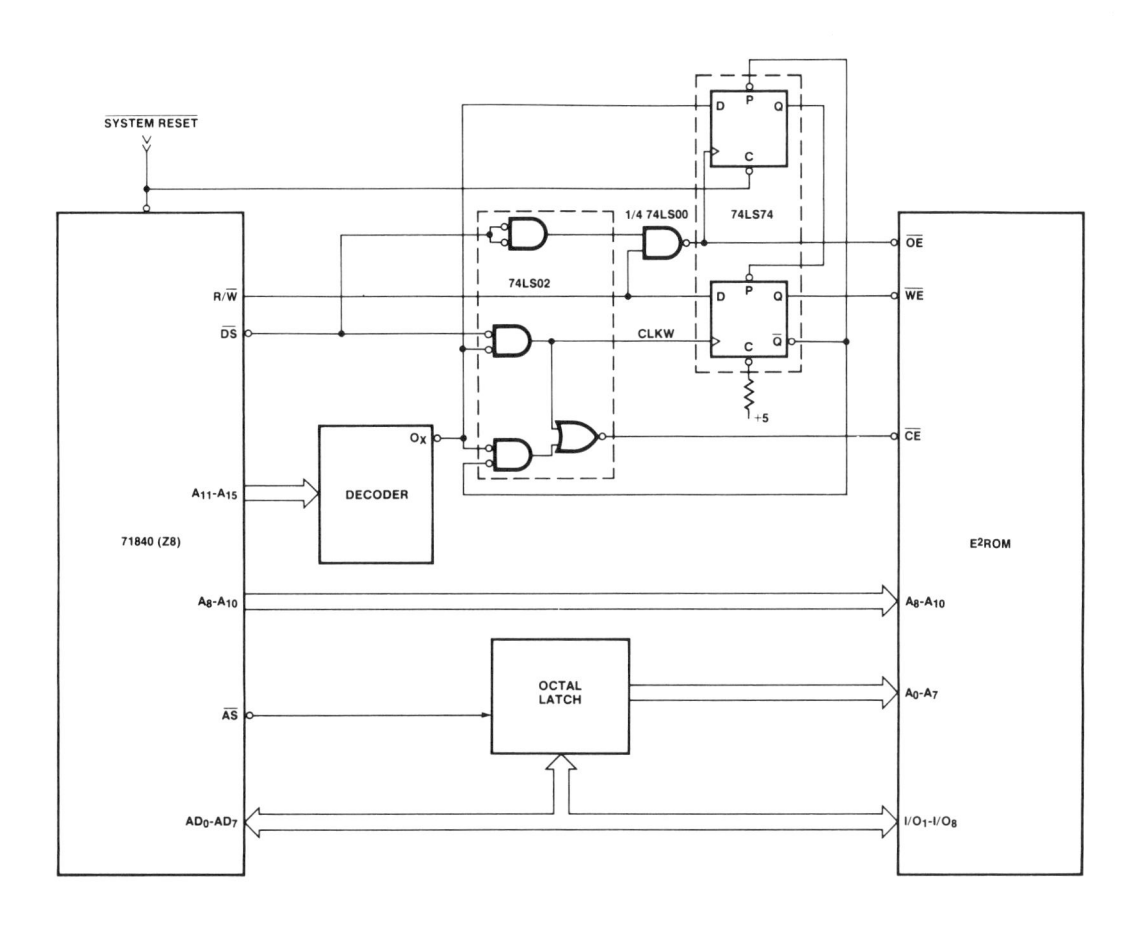

Figure 17. 71840/E²ROM Interface

The operation of the remaining circuitry in write cycle initiation is very simple. The following explanation refers to the timing diagram presented in Figure 18. To initiate the write cycle, the 71840 issues a Write command to the E^{2}ROM device. After addresses become valid, the decoder brings $\overline{\text{SEL}}$ active low (see Ⓐ in Figure 18). Subsequently, the falling edge of $\overline{\text{DATA STROBE}}$ ($\overline{\text{DS}}$) clocks R/$\overline{\text{W}}$ into the $\overline{\text{WE}}$ flip-flop, initiating a write cycle by bringing $\overline{\text{WE}}$ active low (see Ⓑ in Figure 18). At this time, the latched E^{2}ROM latches address, data, and control signals.

At the trailing (rising) edge of $\overline{\text{DS}}$, the gating circuitry inhibits $\overline{\text{CE}}$, and $\overline{\text{CE}}$ is brought to V$_{IH}$ (see Ⓒ in Figure 18). At this point, the timer E^{2}ROM latches data and initiates its write cycle. Both devices have begun the write cycle; now the system is able to time out the write cycle, in order to complete storage of charge within the E^{2}ROM cell.

The termination of the write cycle occurs with equal simplicity (see Figure 18). When the processor reads from the E^{2}ROM, the rising edge of $\overline{\text{DS}}$ causes the $\overline{\text{WE}}$ flip-flop to be preset. This brings $\overline{\text{WE}}$ high (see Ⓓ in Figure 18), ending the write cycle.

An example software driver routine is provided for the 71840 (see Figure 19). This routine will handle initiation, timing, and termination of a write cycle, as well as automatic erasure of the byte to be programmed.

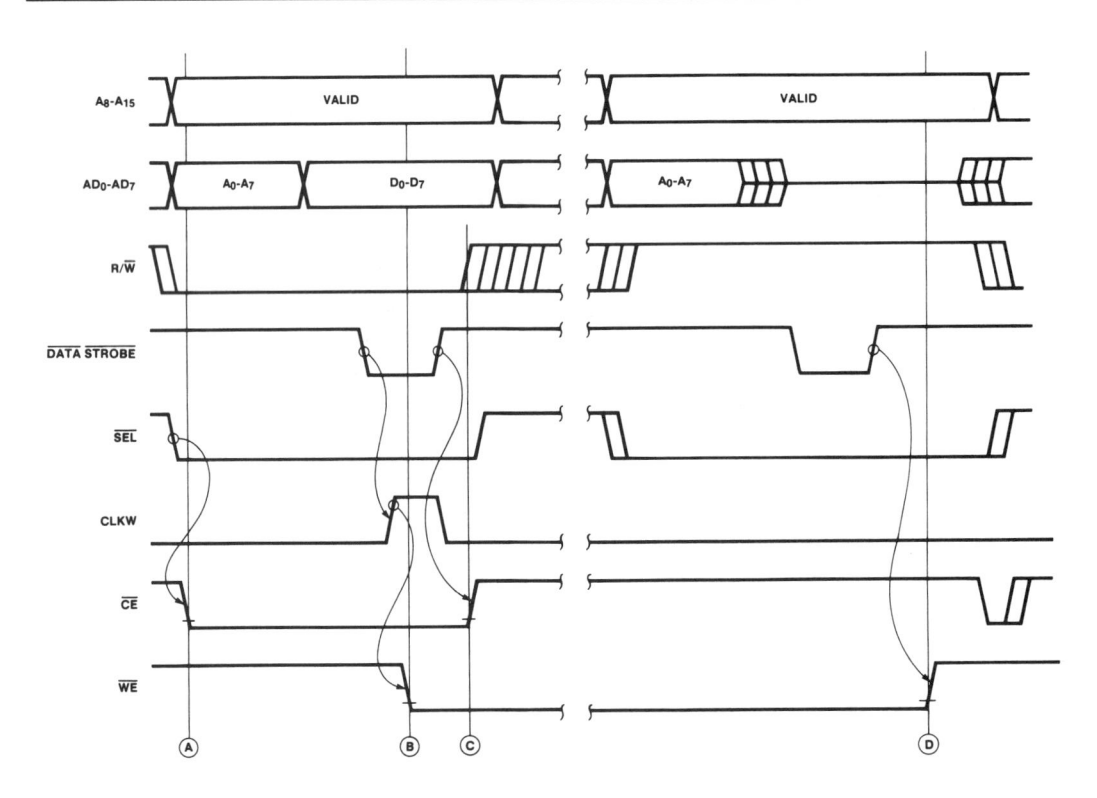

*A$_0$-A$_7$: ADDRRESS SIGNALS MULTIPLEXED WITH DATA SIGNALS MUST BE DEMULTIPLEXED USING OCTAL LATCHES.

Figure 18. Timing Diagram — 71840/E^{2}ROM Interface

```
                              186  //------------------------
                              187  // The following is a general routine for writing
                              188  // data contained in the working register
                              189  // DataReg to an EEROM in
                              190  // the location pointed to by the working register
                              191  // pair AdReg. This EEROM is assumed to be in the
                              192  // external data memory of Z8.
                              193  // Write FF to erase byte.
P 0060  7C  FF                194  EEWR:   LD      OutReg, #%FF
P 0062  92  70                195          LDE     @AdReg, OutReg
P 0064  D6  0071              196          CALL    WaitWP          // Wait for Twp
P 0067  82  80                197          LDE     NowReg, @AdReg  // Turn off WE
                              198  // Now, write the data to the part.
P 0069  92  90                199          LDE     @AdReg, DataReg
P 006B  D6  0071              200          CALL    WaitWP          // Wait for Twp
P 006E  82  80                201          LDE     NowReg, @AdReg  // turn off WE.
                              202
P 0070  AF                    203  FinWr:  RET                     //return from routine
                              204  // End of EEPROM Write Routine
                              205  //----------------------------------------------
                              206
                              207  // Timing routines
P 0071  EC  0A                208  WaitWP: LD      RLoop2, #Twp   // # of ms to wait
                              209                                  // 10-> wait 10 mS.
                              210                                  // 1 -> Wait 1 mS.
                              211
P 0073  D6  007E              212  WPLoop: CALL    Wait1ms
P 0076  00  EE                213          DEC     RLoop2
P 0078  6D  007D              214          JP      Z, DunWP
P 007B  8B  F6                215          JR      WPLoop
P 007D  AF                    216  DunWP:  RET                     // Done with Twp.
                              217
                              218  // Basic 1 msec timing routine-
                              219  // adjust for microprocessor crystal freq.
                              220  // The value of Hex58 (Dec88) works with
                              221  // a Z8 with a 6.144 MHz xtal.
                              222  // Use %6A for 7.3728 MHz xtal. Elimination
                              223  // of NOP, or xtal substitution, will
                              224  // require recalibration.
P 007E  FC  6A                225  Wait1ms: LD     RLoop3, #%6A
                              226
P 0080  FF                    227  Tim1p:  NOP
P 0081  00  EF                228          DEC     RLoop3
P 0083  6D  0088              229          JP      Z, Dun1ms
P 0086  8B  F8                230          JR      Tim1p
                              231
P 0088  AF                    232  Dun1ms: RET                     // Done with wait
                              233
                              234  //End of EEROM Timing Routines
                              235  //------------------------------------
```

Figure 19. 71840 E²ROM Erase/Write Routine

Conclusion

The development in E²ROM memory is continuing at an ever increasing pace. Recent strides in E²ROM cost reduction, access time, and availability have made non-volatile memory suitable for more applications than ever before. It is the purpose of this application note to contribute to this evolution in semiconductor memory by assisting the system designer in the task of E²ROM implementation. Armed with basic hardware and software examples of working E²ROM applications, the designer can more easily complete a feasible E²ROM design, using the flexible, cost-effective devices currently offered.

Acknowledgements

The author wishes to thank J. Oliphant and D. Reynolds for their contributions to this application note.

Silicon Security is a trademark of
SEEQ Technology, Inc.
Z8, Z80 are trademarks of Zilog, Inc.

MOTOROLA
SEMICONDUCTOR
TECHNICAL DATA

Order this data sheet
by MCM514100/D

Advance Information
4M × 1 CMOS Dynamic RAM
Page Mode

MCM514100
MCM51L4100

J PACKAGE
PLASTIC
SMALL OUTLINE
CASE 822A

Z PACKAGE
PLASTIC
ZIG-ZAG IN-LINE
CASE 836

The MCM514100 is a 0.8μ CMOS high-speed, dynamic random access memory. It is organized as 4,194,304 one-bit words and fabricated with CMOS silicon-gate process technology. Advanced circuit design and fine line processing provide high performance, improved reliability, and low cost.

The MCM514100 requires only 11 address lines; row and column address inputs are multiplexed. The device is packaged in a standard 350-mil-wide J-lead small outline package, and a 100-mil zig-zag in-line package (ZIP).

- Three-State Data Output
- Common I/O with Early Write
- Fast Page Mode
- Test Mode
- TTL-Compatible Inputs and Output
- $\overline{RAS}$ Only Refresh
- $\overline{CAS}$ Before $\overline{RAS}$ Refresh
- Hidden Refresh
- 1024 Cycle Refresh: MCM514100 = 16 ms
  MCM51L4100 = 128 ms
- Unlatched Data Out at Cycle End Allows Two Dimensional Chip Selection
- Fast Access Time (t_{RAC}):
  MCM514100-80 and MCM51L4100-80 = 80 ns (Max)
  MCM514100-10 and MCM51L4100-10 = 100 ns (Max)
- Low Active Power Dissipation:
  MCM514100-80 and MCM51L4100-80 = 550 mW (Max)
  MCM514100-10 and MCM51L4100-10 = 468 mW (Max)
- Low Standby Power Dissipation:
  MCM514100 and MCM51L4100 = 11 mW (Max, TTL Levels)
  MCM514100 = 5.5 mW (Max, CMOS Levels)
  MCM51L4100 = 2.2 mW (Max, CMOS Levels)

PIN NAMES

A0–A10	Address Input
D	Data Input
Q	Data Output
$\overline{W}$	Read/Write Enable
$\overline{RAS}$	Row Address Strobe
$\overline{CAS}$	Column Address Strobe
V_{CC}	Power (+5 V)
V_{SS}	Ground
NC	No Connection

PIN ASSIGNMENT

SMALL OUTLINE

D	1	26	V_{SS}
$\overline{W}$	2	25	Q
$\overline{RAS}$	3	24	$\overline{CAS}$
NC	4	23	NC
A10	5	22	A9
A0	9	18	A8
A1	10	17	A7
A2	11	16	A6
A3	12	15	A5
V_{CC}	13	14	A4

ZIG-ZAG IN-LINE

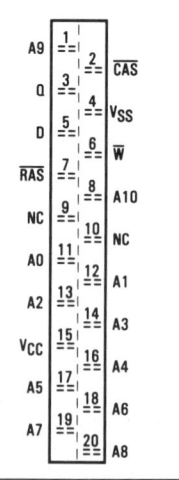

A9	1	
	2	$\overline{CAS}$
Q	3	
	4	V_{SS}
D	5	
	6	$\overline{W}$
$\overline{RAS}$	7	
	8	A10
NC	9	
	10	NC
A0	11	
	12	A1
A2	13	
	14	A3
V_{CC}	15	
	16	A4
A5	17	
	18	A6
A7	19	
	20	A8

This document contains information on a new product. Specifications and information herein are subject to change without notice.

©MOTOROLA INC., 1989

MOTOROLA

ADI1547

BLOCK DIAGRAM

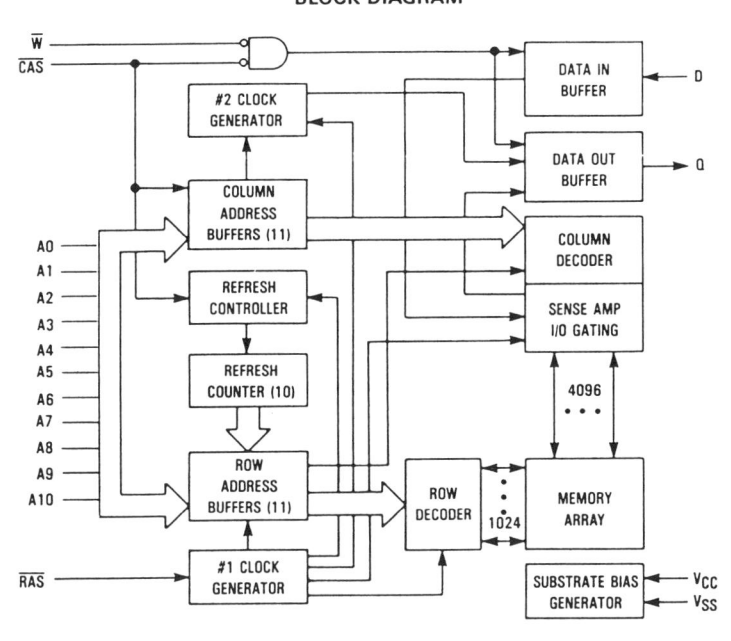

ABSOLUTE MAXIMUM RATINGS (See Note)

Rating	Symbol	Value	Unit
Power Supply Voltage	V_{CC}	−1 to +7	V
Voltage Relative to V_{SS} for Any Pin Except V_{CC}	V_{in}, V_{out}	−1 to +7	V
Data Out Current	I_{out}	50	mA
Power Dissipation	P_D	600	mW
Operating Temperature Range	T_A	0 to +70	°C
Storage Temperature Range	T_{stg}	−55 to +150	°C

NOTE: Permanent device damage may occur if ABSOLUTE MAXIMUM RATINGS are exceeded. Functional operation should be restricted to RECOMMENDED OPERATING CONDITIONS. Exposure to higher than recommended voltages for extended periods of time could affect device reliability.

This device contains circuitry to protect the inputs against damage due to high static voltages or electric fields; however, it is advised that normal precautions be taken to avoid application of any voltage higher than maximum rated voltages to this high-impedance circuit.

DC OPERATING CONDITIONS AND CHARACTERISTICS
(V_{CC} = 5.0 V ± 10%, T_A = 0 to 70°C, Unless Otherwise Noted)

RECOMMENDED OPERATING CONDITIONS

Parameter	Symbol	Min	Typ	Max	Unit	Notes
Supply Voltage (Operating Voltage Range)	V_{CC}	4.5	5.0	5.5	V	1
	V_{SS}	0	0	0	V	1
Logic High Voltage, All Inputs	V_{IH}	2.4	—	6.5	V	1
Logic Low Voltage, All Inputs	V_{IL}	−1.0	—	0.8	V	1

DC CHARACTERISTICS

Characteristic	Symbol	Min	Max	Unit	Notes
V_{CC} Power Supply Current MCM514100-80 and MCM51L4100-80, t_{RC} = 150 ns MCM514100-10 and MCM51L4100-10, t_{RC} = 180 ns	I_{CC1}	— —	100 85	mA	2
V_{CC} Power Supply Current (Standby) ($\overline{RAS}$ = $\overline{CAS}$ = V_{IH})	I_{CC2}	—	2.0	mA	
V_{CC} Power Supply Current During $\overline{RAS}$ only Refresh Cycles ($\overline{CAS}$ = V_{IH}) MCM514100-80 and MCM51L4100-80, t_{RC} = 150 ns MCM514100-10 and MCM51L4100-10, t_{RC} = 180 ns	I_{CC3}	— —	100 85	mA	2
V_{CC} Power Supply Current During Fast Page Mode Cycle ($\overline{RAS}$ = V_{IL}) MCM514100-80 and MCM51L4100-80, t_{PC} = 50 ns MCM514100-10 and MCM51L4100-10, t_{PC} = 60 ns	I_{CC4}	— —	60 50	mA	2, 4
V_{CC} Power Supply Current (Standby) ($\overline{RAS}$ = $\overline{CAS}$ = V_{CC} − 0.2 V) MCM514100 MCM51L4100	I_{CC5}	— —	1.0 400	mA μA	
V_{CC} Power Supply Current During $\overline{CAS}$ Before $\overline{RAS}$ Refresh Cycle MCM514100-80 and MCM51L4100-80, t_{RC} = 150 ns MCM514100-10 and MCM51L4100-10, t_{RC} = 180 ns	I_{CC6}	— —	100 85	mA	2
V_{CC} Power Supply Current, Battery Backup Mode — MCM51L4100 only (t_{RC} = 125 μs; t_{RAS} = 1 μs; $\overline{CAS}$ = $\overline{CAS}$ Before $\overline{RAS}$ Cycle or 0.2 V; A0-A10, $\overline{W}$, D = V_{CC} − 0.2 V or 0.2 V)	I_{CC7}	—	500	μA	
Input Leakage Current (0 V ≤ V_{in} ≤ 6.5 V)	$I_{lkg(I)}$	−10	10	μA	
Output Leakage Current ($\overline{CAS}$ = V_{IH}, 0 V ≤ V_{out} ≤ 5.5 V)	$I_{lkg(O)}$	−10	10	μA	
Output High Voltage (I_{OH} = −5 mA)	V_{OH}	2.4	—	V	
Output Low Voltage (I_{OL} = 4.2 mA)	V_{OL}	—	0.4	V	

CAPACITANCE (f = 1.0 MHz, T_A = 25°C, V_{CC} = 5 V, Periodically Sampled Rather Than 100% Tested)

Parameter		Symbol	Max	Unit	Notes
Input Capacitance	A0-A10, D	C_{in}	5	pF	3
	$\overline{RAS}$, $\overline{CAS}$, $\overline{W}$		7	pF	3
Output Capacitance ($\overline{CAS}$ = V_{IH} to Disable Output)	Q	C_{out}	7	pF	3

NOTES:
1. All voltages referenced to V_{SS}.
2. Current is a function of cycle rate and output loading; maximum current is measured at the fastest cycle rate with the output open.
3. Capacitance measured with a Boonton Meter or effective capacitance calculated from the equation: C = I Δt/ΔV.
4. Measured with one address transition per page mode cycle.

AC OPERATING CONDITIONS AND CHARACTERISTICS
(V_{CC} = 5.0 V ± 10%, T_A = 0 to 70°C, Unless Otherwise Noted)

READ, WRITE, AND READ-WRITE CYCLES (See Notes 1, 2, 3, and 4)

Parameter	Symbol Standard	Symbol Alternate	MCM514100-80 MCM51L4100-80 Min	MCM514100-80 MCM51L4100-80 Max	MCM514100-10 MCM51L4100-10 Min	MCM514100-10 MCM51L4100-10 Max	Unit	Notes
Random Read or Write Cycle Time	t_{RELREL}	t_{RC}	150	—	180	—	ns	5
Read-Write Cycle Time	t_{RELREL}	t_{RWC}	175	—	210	—	ns	5
Page Mode Cycle Time	t_{CELCEL}	t_{PC}	50	—	60	—	ns	
Page Mode Read-Write Cycle Time	t_{CELCEL}	t_{PRWC}	75	—	90	—	ns	
Access Time from $\overline{RAS}$	t_{RELQV}	t_{RAC}	—	80	—	100	ns	6, 7
Access Time from $\overline{CAS}$	t_{CELQV}	t_{CAC}	—	20	—	25	ns	6, 8
Access Time from Column Address	t_{AVQV}	t_{AA}	—	40	—	50	ns	6, 9
Access Time from Precharge $\overline{CAS}$	t_{CEHQV}	t_{CPA}	—	45	—	55	ns	6
$\overline{CAS}$ to Output in Low-Z	t_{CELQX}	t_{CLZ}	0	—	0	—	ns	6
Output Buffer and Turn-Off Delay	t_{CEHQZ}	t_{OFF}	0	20	0	20	ns	10
Transition Time (Rise and Fall)	t_T	t_T	3	50	3	50	ns	
$\overline{RAS}$ Precharge Time	t_{REHREL}	t_{RP}	60	—	70	—	ns	
$\overline{RAS}$ Pulse Width	t_{RELREH}	t_{RAS}	80	10,000	100	10,000	ns	
$\overline{RAS}$ Pulse Width (Fast Page Mode)	t_{RELREH}	t_{RASP}	80	200,000	100	200,000	ns	
$\overline{RAS}$ Hold Time	t_{CELREH}	t_{RSH}	20	—	25	—	ns	
$\overline{CAS}$ Hold Time	t_{RELCEH}	t_{CSH}	80	—	100	—	ns	
$\overline{CAS}$ Pulse Width	t_{CELCEH}	t_{CAS}	20	10,000	25	10,000	ns	
$\overline{RAS}$ to $\overline{CAS}$ Delay Time	t_{RELCEL}	t_{RCD}	20	60	25	75	ns	11
$\overline{RAS}$ to Column Address Delay Time	t_{RELAV}	t_{RAD}	15	40	20	50	ns	12
$\overline{CAS}$ to $\overline{RAS}$ Precharge Time	t_{CEHREL}	t_{CRP}	5	—	10	—	ns	
$\overline{CAS}$ Precharge Time	t_{CEHCEL}	t_{CP}	10	—	10	—	ns	
Row Address Setup Time	t_{AVREL}	t_{ASR}	0	—	0	—	ns	
Row Address Hold Time	t_{RELAX}	t_{RAH}	10	—	15	—	ns	
Column Address Setup Time	t_{AVCEL}	t_{ASC}	0	—	0	—	ns	
Column Address Hold Time	t_{CELAX}	t_{CAH}	15	—	20	—	ns	
Column Address Hold Time Referenced to $\overline{RAS}$	t_{RELAX}	t_{AR}	60	—	75	—	ns	
Column Address to $\overline{RAS}$ Lead Time	t_{AVREH}	t_{RAL}	40	—	50	—	ns	

(continued)

NOTES:
1. V_{IH} min and V_{IL} max are reference levels for measuring timing of input signals. Transition times are measured between V_{IH} and V_{IL}.
2. An initial pause of 200 μs is required after power-up followed by 8 $\overline{RAS}$ cycles before proper device operation is guaranteed.
3. The transition time specification applies for all input signals. In addition to meeting the transition rate specification, all input signals must transition between V_{IH} and V_{IL} (or between V_{IL} and V_{IH}) in a monotonic manner.
4. AC measurements t_T = 5.0 ns.
5. The specifications for t_{RC} (min) and t_{RWC} (min) are used only to indicate cycle time at which proper operation over the full temperature range (0°C ≤ T_A ≤ 70°C) is assured.
6. Measured with a current load equivalent to 2 TTL (−200 μA, +4 mA) loads and 100 pF with the data output trip points set at V_{OH} = 2.0 V and V_{OL} = 0.8 V.
7. Assumes that t_{RCD} ≤ t_{RCD} (max).
8. Assumes that t_{RCD} ≥ t_{RCD} (max).
9. Assumes that t_{RAD} ≥ t_{RAD} (max).
10. t_{OFF} (max) defines the time at which the output achieves the open circuit condition and is not referenced to output voltage levels.
11. Operation within the t_{RCD} (max) limit ensures that t_{RAC} (max) can be met. t_{RCD} (max) is specified as a reference point only; if t_{RCD} is greater than the specified t_{RCD} (max) limit, then access time is controlled exclusively by t_{CAC}.
12. Operation within the t_{RAD} (max) limit ensures that t_{RAC} (max) can be met. t_{RAD} (max) is specified as a reference point only; if t_{RAD} is greater than the specified t_{RAD} (max), then access time is controlled exclusively by t_{AA}.

READ, WRITE, AND READ-WRITE CYCLES (Continued)

Parameter	Symbol Standard	Symbol Alternate	MCM514100-80 MCM51L4100-80 Min	MCM514100-80 MCM51L4100-80 Max	MCM514100-10 MCM51L4100-10 Min	MCM514100-10 MCM51L4100-10 Max	Unit	Notes
Read Command Setup Time	t_{WHCEL}	t_{RCS}	0	—	0	—	ns	
Read Command Hold Time Referenced to $\overline{CAS}$	t_{CEHWX}	t_{RCH}	0	—	0	—	ns	13
Read Command Hold Time Referenced to $\overline{RAS}$	t_{REHWX}	t_{RRH}	0	—	0	—	ns	13
Write Command Hold Time Referenced to $\overline{CAS}$	t_{CELWH}	t_{WCH}	15	—	20	—	ns	
Write Command Hold Time Referenced to $\overline{RAS}$	t_{RELWH}	t_{WCR}	60	—	75	—	ns	
Write Command Pulse Width	t_{WLWH}	t_{WP}	15	—	20	—	ns	
Write Command to $\overline{RAS}$ Lead Time	t_{WLREH}	t_{RWL}	20	—	25	—	ns	
Write Command to $\overline{CAS}$ Lead Time	t_{WLCEH}	t_{CWL}	20	—	25	—	ns	
Data in Setup Time	t_{DVCEL}	t_{DS}	0	—	0	—	ns	14
Data in Hold Time	t_{CELDX}	t_{DH}	15	—	20	—	ns	14
Data in Hold Time Referenced to $\overline{RAS}$	t_{RELDX}	t_{DHR}	60	—	75	—	ns	
Refresh Period MCM514100 MCM51L4100	t_{RVRV}	t_{RFSH}	— —	16 128	— —	16 128	ms	
Write Command Setup Time	t_{WLCEL}	t_{WCS}	0	—	0	—	ns	15
$\overline{CAS}$ to Write Delay	t_{CELWL}	t_{CWD}	20	—	25	—	ns	15
$\overline{RAS}$ to Write Delay	t_{RELWL}	t_{RWD}	80	—	100	—	ns	15
Column Address to Write Delay Time	t_{AVWL}	t_{AWD}	40	—	50	—	ns	15
$\overline{CAS}$ Precharge to Write Delay Time (Page Mode)	t_{CEHWL}	t_{CPWD}	45	—	55	—	ns	15
$\overline{CAS}$ Setup Time for $\overline{CAS}$ Before $\overline{RAS}$ Refresh	t_{RELCEL}	t_{CSR}	5	—	10	—	ns	
$\overline{CAS}$ Hold Time for $\overline{CAS}$ Before $\overline{RAS}$ Refresh	t_{RELCEH}	t_{CHR}	15	—	20	—	ns	
$\overline{RAS}$ Precharge to $\overline{CAS}$ Active Time	t_{REHCEL}	t_{RPC}	0	—	0	—	ns	
$\overline{CAS}$ Precharge Time for $\overline{CAS}$ Before $\overline{RAS}$ Counter Test	t_{CEHCEL}	t_{CPT}	40	—	50	—	ns	
Write Command Set Up Time (Test Mode)	t_{WLREL}	t_{WTS}	10	—	10	—	ns	
Write Command Hold Time (Test Mode)	t_{RELWH}	t_{WTH}	10	—	10	—	ns	
Write to $\overline{RAS}$ Precharge Time ($\overline{CAS}$ Before $\overline{RAS}$ Refresh)	t_{WHREL}	t_{WRP}	10	—	10	—	ns	
Write to $\overline{RAS}$ Hold Time ($\overline{CAS}$ Before $\overline{RAS}$ Refresh)	t_{RELWL}	t_{WRH}	10	—	10	—	ns	

NOTES:
13. Either t_{RRH} or t_{RCH} must be satisfied for a read cycle.
14. These parameters are referenced to $\overline{CAS}$ leading edge in early write cycles and to $\overline{W}$ leading edge in read-write cycles.
15. t_{WCS}, t_{RWD}, t_{CWD}, t_{AWD} and t_{CPWD} are not restrictive operating parameters. They are included in the data sheet as electrical characteristics only; if $t_{WCS} \geq t_{WCS}$ (min), the cycle is an early write cycle and the data out pin will remain open circuit (high impedance) throughout the entire cycle; if $t_{CWD} \geq t_{CWD}$ (min), $t_{RWD} \geq t_{RWD}$ (min), $t_{AWD} \geq t_{AWD}$ (min), and $t_{CPWD} \geq t_{CPWD}$ (min) (page mode), the cycle is a read-write cycle and the data out will contain data read from the selected cell. If neither of these sets of conditions is satisfied, the condition of the data out (at access time) is indeterminate.

READ CYCLE

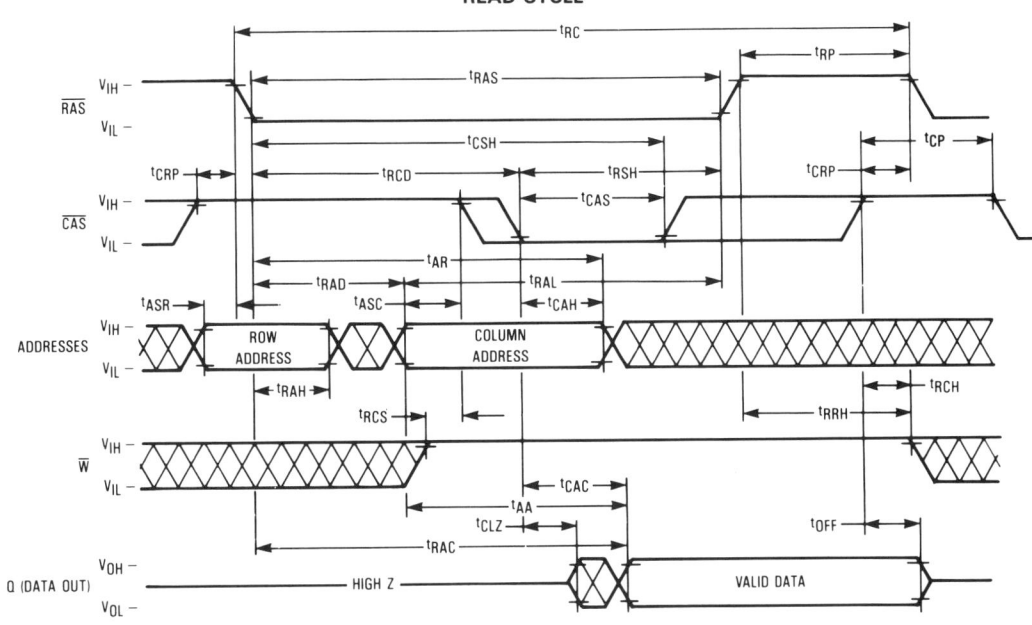

EARLY WRITE CYCLE

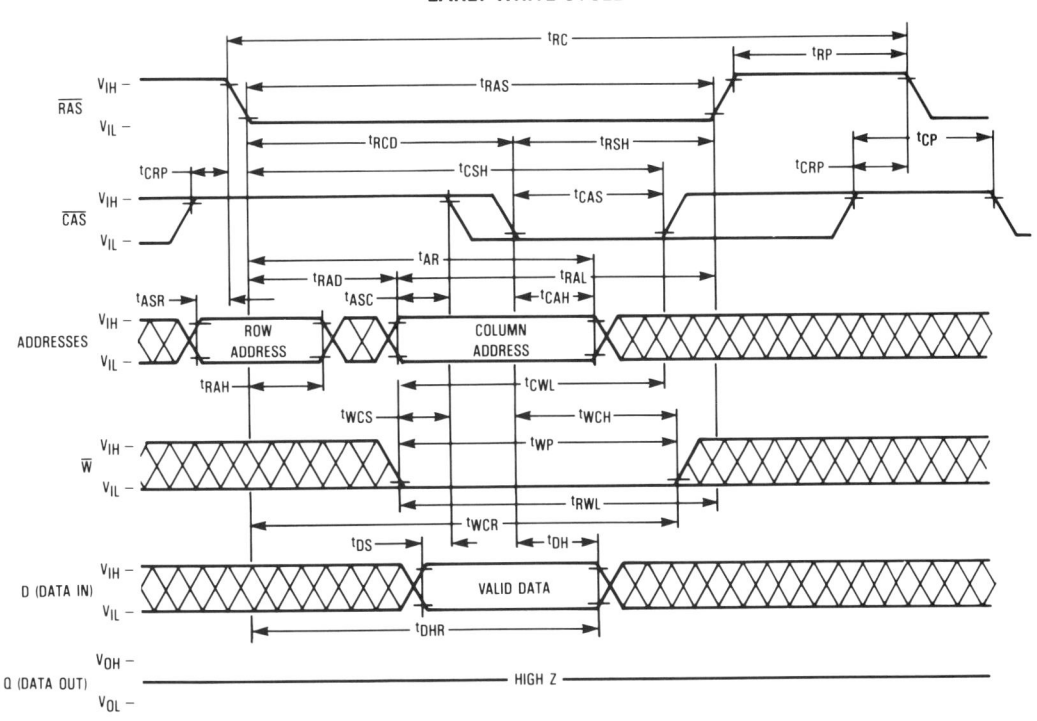

MCM514100 • MCM51L4100

READ-WRITE CYCLE

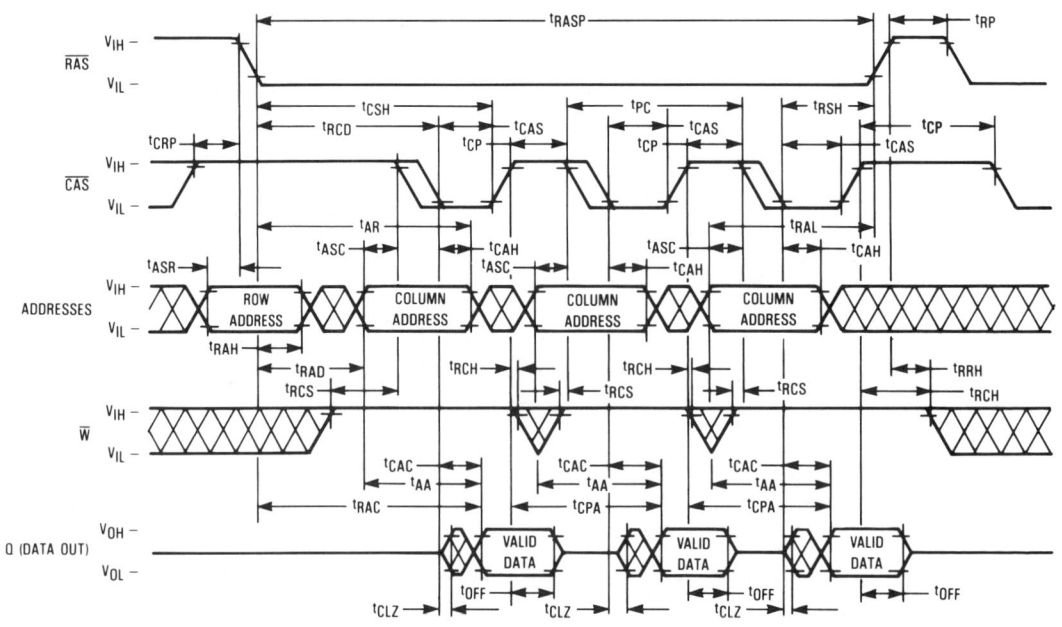

FAST PAGE MODE READ CYCLE

FAST PAGE MODE EARLY WRITE CYCLE

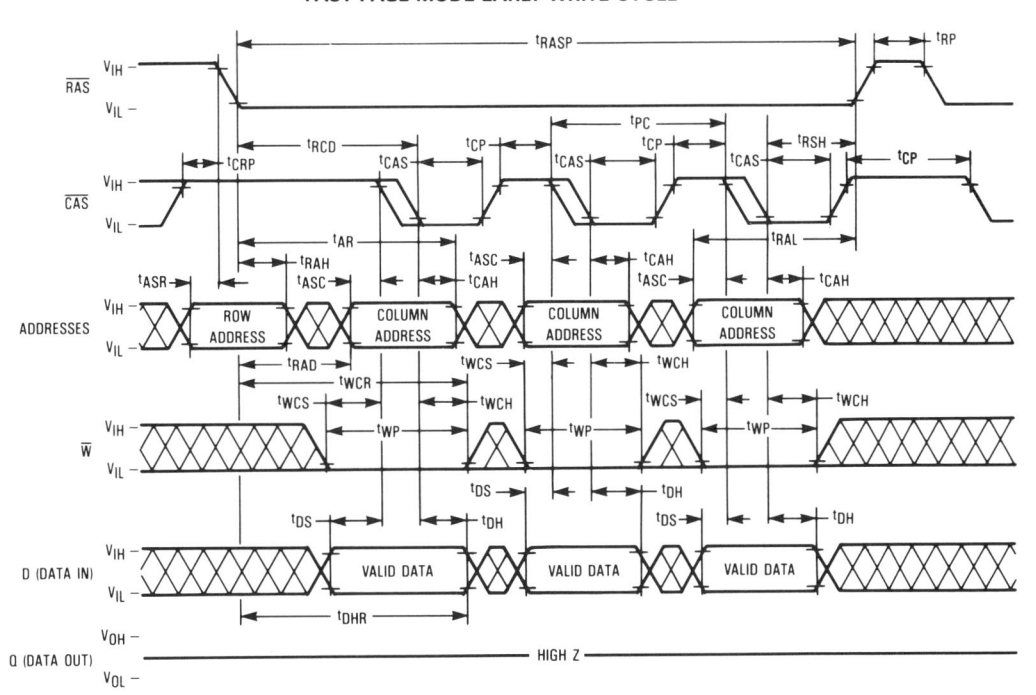

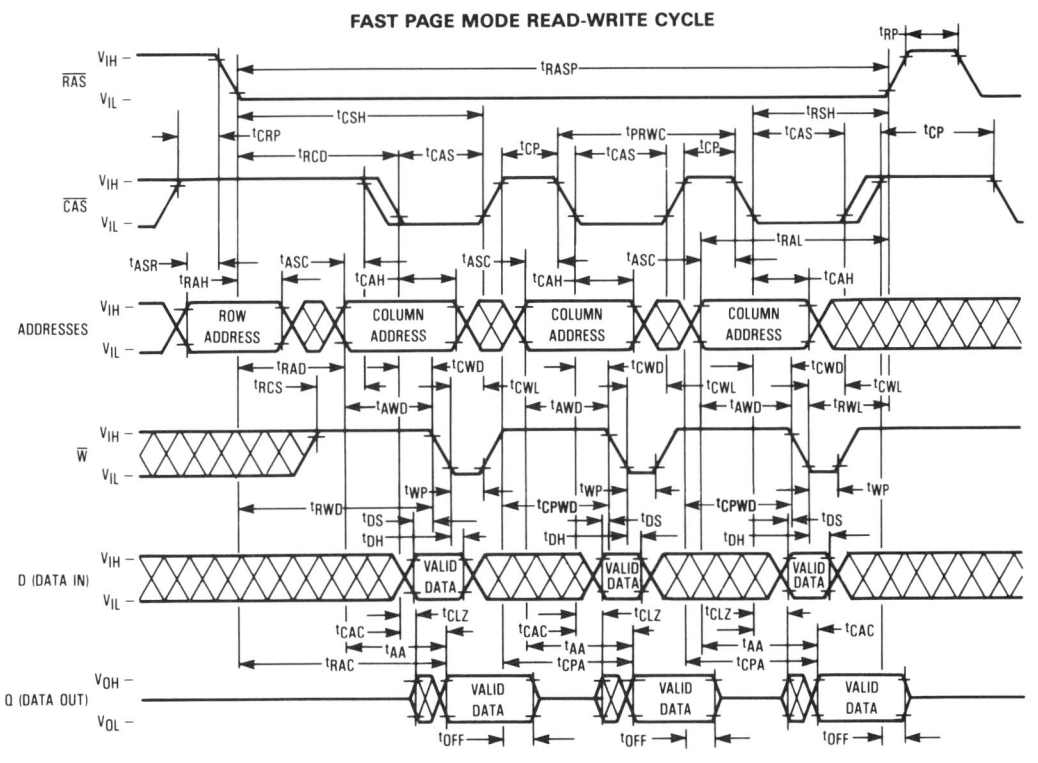

FAST PAGE MODE READ-WRITE CYCLE

RAS ONLY REFRESH CYCLE
($\overline{W}$ and A10 are Don't Care)

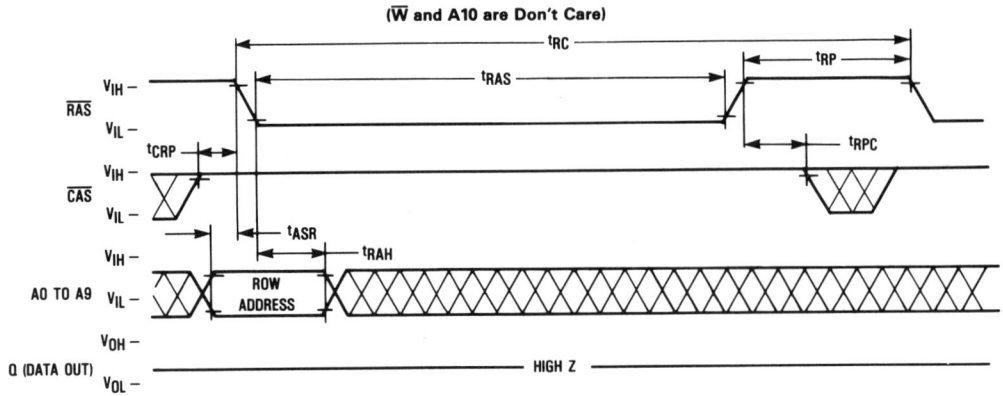

CAS BEFORE RAS REFRESH CYCLE
(A0 to A10 are Don't Care)

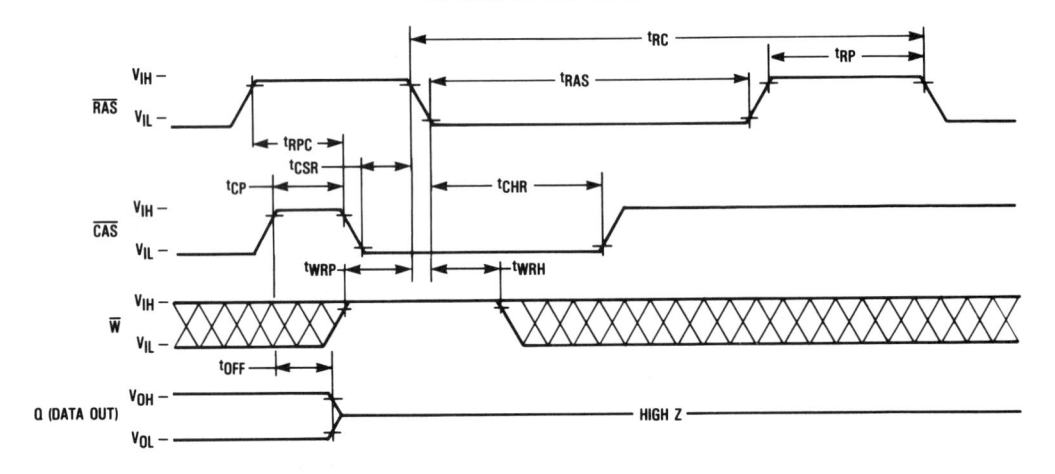

HIDDEN REFRESH CYCLE (READ)

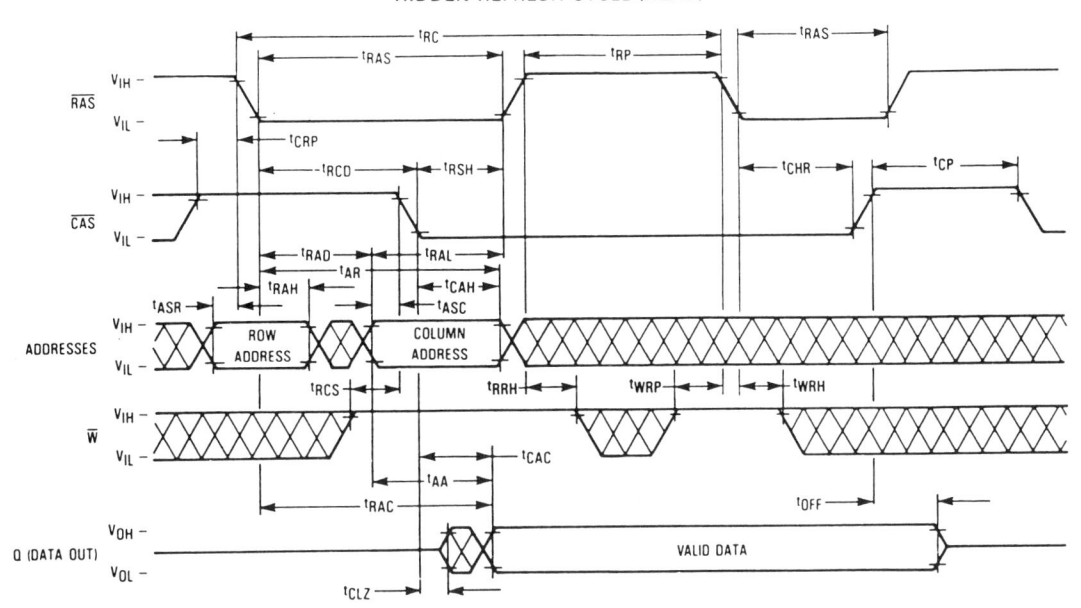

HIDDEN REFRESH CYCLE (EARLY WRITE)

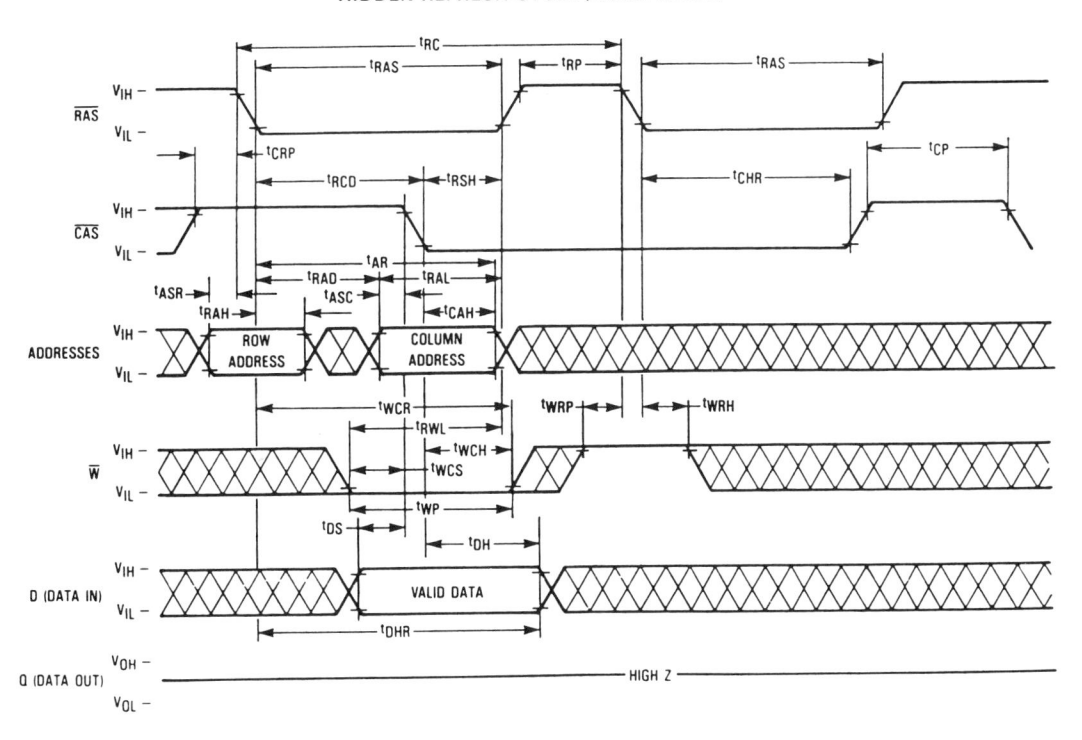

MCM514100 • MCM51L4100

CAS BEFORE RAS REFRESH COUNTER TEST CYCLE

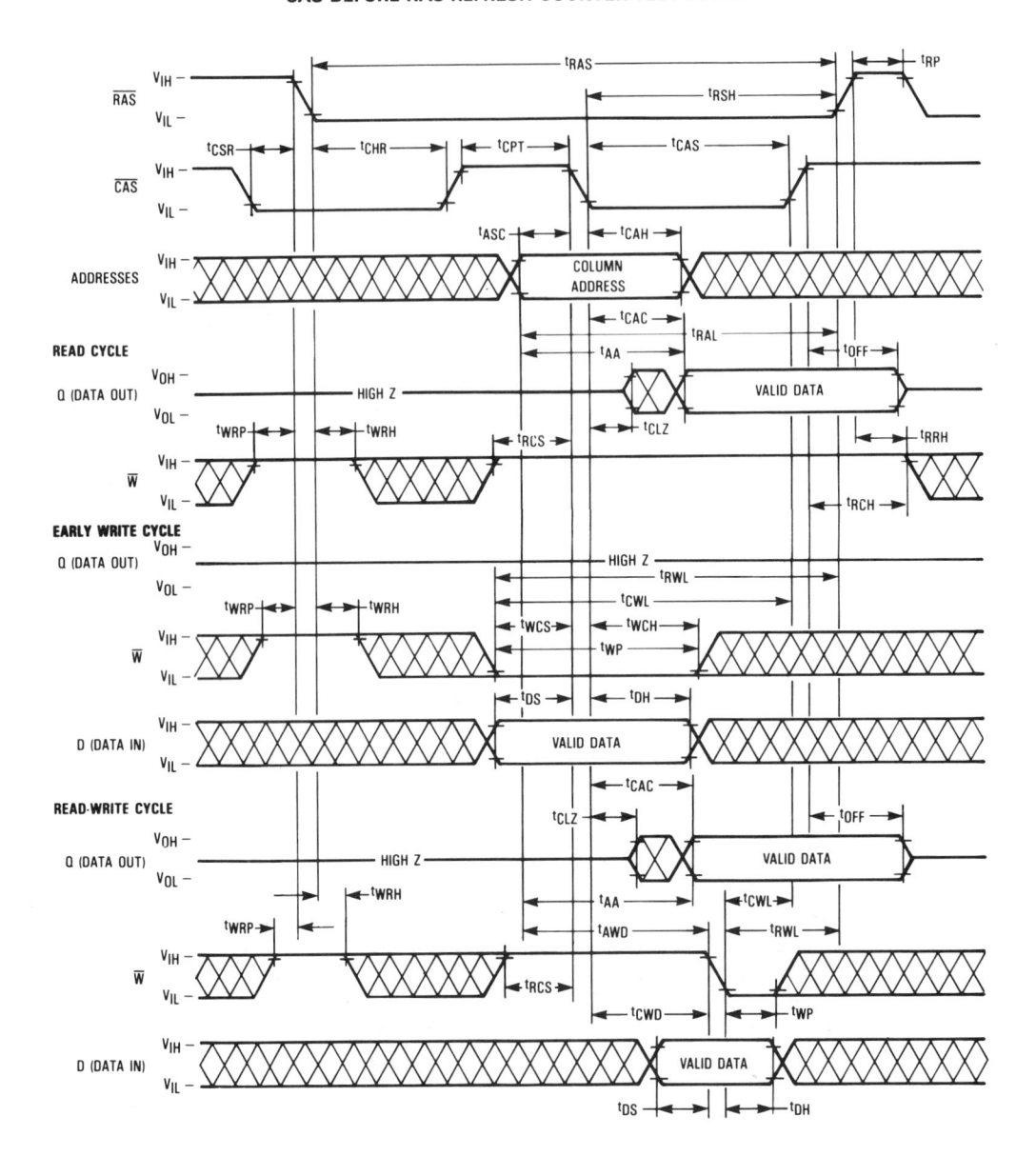

DEVICE INITIALIZATION

On power-up an initial pause of 200 microseconds is required for the internal substrate generator to establish the correct bias voltage. This must be followed by a minimum of eight active cycles of the row address strobe (clock) to initialize all dynamic nodes within the RAM. During an extended inactive state (greater than 16 milliseconds with the device powered up), a wake up sequence of eight active cycles is necessary to assure proper operation.

ADDRESSING THE RAM

The eleven address pins on the device are time multiplexed at the beginning of a memory cycle by two clocks, row address strobe ($\overline{RAS}$) and column address strobe ($\overline{CAS}$), into two separate 11-bit address fields. A total of twenty two address bits, eleven rows and eleven columns, will decode one of the 4,194,304 bit locations in the device. $\overline{RAS}$ active transition is followed by $\overline{CAS}$ active transition (active = V_{IL}, t_{RCD} minimum) for all read or write cycles. The delay between $\overline{RAS}$ and $\overline{CAS}$ active transitions, referred to as the **multiplex window**, gives a system designer flexibility in setting up the external addresses into the RAM.

The external $\overline{CAS}$ signal is ignored until an internal $\overline{RAS}$ signal is available. This "gate" feature on the external $\overline{CAS}$ clock enables the internal $\overline{CAS}$ line as soon as the row address hold time (t_{RAH}) specification is met (and defines t_{RCD} minimum). The multiplex window can be used to absorb skew delays in switching the address bus from row to column addresses and in generating the $\overline{CAS}$ clock.

There are three other variations in addressing the 4M RAM: **$\overline{RAS}$ only refresh cycle**, **$\overline{CAS}$ before $\overline{RAS}$ refresh cycle**, and **page mode**. All three are discussed in separate sections that follow.

READ CYCLE

The DRAM may be read with four different cycles: "normal" random read cycle, page mode read cycle, read-write cycle, and page mode read-write cycle. The normal read cycle is outlined here, while the other cycles are discussed in separate sections.

The normal read cycle begins as described in **ADDRESSING THE RAM**, with $\overline{RAS}$ and $\overline{CAS}$ active transitions latching the desired bit location. The write ($\overline{W}$) input level must be high (V_{IH}), t_{RCS} (minimum) before the $\overline{CAS}$ active transition, to enable read mode.

Both the $\overline{RAS}$ and $\overline{CAS}$ clocks trigger a sequence of events which are controlled by several delayed internal clocks. The internal clocks are linked in such a manner that the read access time of the device is independent of the address multiplex window. However, $\overline{CAS}$ must be active before or at t_{RCD} maximum to guarantee valid data out (Q) at t_{RAC} (access time from $\overline{RAS}$ active transition). If the t_{RCD} maximum is exceeded, read access time is determined by the $\overline{CAS}$ clock active transition (t_{CAC}).

The $\overline{RAS}$ and $\overline{CAS}$ clocks must remain active for a minimum time of t_{RAS} and t_{CAS} respectively, to complete the read cycle. $\overline{W}$ must remain high throughout the cycle, and for time t_{RRH} or t_{RCH} after $\overline{RAS}$ or $\overline{CAS}$ inactive transition, respectively, to maintain the data at that bit location. Once $\overline{RAS}$ transitions to inactive, it must remain inactive for a minimum time of t_{RP} to precharge the internal device circuitry for the next active cycle. Q is valid, but not latched, as long as the $\overline{CAS}$ clock is active. When the $\overline{CAS}$ clock transitions to inactive, the output will switch to High Z (three-state).

WRITE CYCLE

The user can write to the DRAM with any of four cycles; early write, late write, page mode early write, and page mode read-write. Early and late write modes are discussed here, while page mode write operations are covered in another section.

A write cycle begins as described in **ADDRESSING THE RAM**. Write mode is enabled by the transition of $\overline{W}$ to active (V_{IL}). Early and late write modes are distinguished by the active transition of $\overline{W}$, with respect to $\overline{CAS}$. Minimum active time t_{RAS} and t_{CAS}, and precharge time t_{RP} apply to write mode, as in the read mode.

An early write cycle is characterized by $\overline{W}$ active transition at minimum time t_{WCS} before $\overline{CAS}$ active transition. Data in (D) is referenced to $\overline{CAS}$ in an early write cycle. $\overline{RAS}$ and $\overline{CAS}$ clocks must stay active for t_{RWL} and t_{CWL}, respectively, after the start of the early write operation to complete the cycle.

Q remains in three-state condition throughout an early write cycle because $\overline{W}$ active transition precedes or coincides with $\overline{CAS}$ active transition, keeping data-out buffers disabled. This feature can be utilized on systems with a common I/O bus, provided all writes are performed with early write cycles, to prevent bus contention.

A late write cycle occurs when $\overline{W}$ active transition is made after $\overline{CAS}$ active transition. $\overline{W}$ active transition could be delayed for almost 10 microseconds after $\overline{CAS}$ active transition, ($t_{RCD} + t_{CWD} + t_{RWL} + 2t_T) \leq t_{RAS}$, if other timing minimums (t_{RCD}, t_{RWL} and t_T) are maintained. D is referenced to $\overline{W}$ active transition in a late write cycle. Output buffers are enabled by $\overline{CAS}$ active transition but Q may be indeterminate — see note 15 of AC operating conditions table. $\overline{RAS}$ and $\overline{CAS}$ must remain active for t_{RWL} and t_{CWL}, respectively, after $\overline{W}$ active transition to complete the write cycle.

READ-WRITE CYCLE

A read-write cycle performs a read and then a write at the same address, during the same cycle. This cycle is basically a late write cycle, as discussed in the **WRITE CYCLE** section, except $\overline{W}$ must remain high for t_{CWD} minimum after the $\overline{CAS}$ active transition, to guarantee valid Q before writing the bit.

PAGE MODE CYCLES

Page mode allows fast successive data operations at all 2048 column locations on a selected row of the 4M dynamic RAM. Read access time in page mode (t_{CAC}) is typically half the regular $\overline{RAS}$ clock access time, t_{RAC}. Page mode operation consists of keeping $\overline{RAS}$ active while toggling $\overline{CAS}$ between V_{IH} and V_{IL}. The row is latched by $\overline{RAS}$ active transition, while each $\overline{CAS}$ active transition allows selection of a new column location on the row.

A page mode cycle is initiated by a normal read, write, or read-write cycle, as described in prior sections. Once the timing requirements for the first cycle are met, $\overline{CAS}$ transitions to inactive for minimum of t_{CP}, while $\overline{RAS}$ remains low (V_{IL}). The second $\overline{CAS}$ active transition while $\overline{RAS}$ is low initiates the first page mode cycle (t_{PC} or t_{PRWC}). Either a read, write, or read-write operation can be performed in a page mode cycle, subject to the same conditions as in normal operation (previously described). These operations can be intermixed in consecutive page mode cycles and performed in any order. The maximum number of consecutive page mode cycles is limited by t_{RASP}. Page mode operation is ended when $\overline{RAS}$ transitions to inactive, coincident with or following $\overline{CAS}$ inactive transition.

REFRESH CYCLES

The dynamic RAM design is based on capacitor charge storage for each bit in the array. This charge will tend to degrade with time and temperature. Each bit must be periodically **refreshed** (recharged) to maintain the correct bit state. Bits in the MCM514100 require refresh every 16 milliseconds, while refresh time for the MCM51L4100 is 128 milliseconds.

This is accomplished by cycling through the 1024 row addresses in sequence within the specified refresh time. All the bits on a row are refreshed simultaneously when the row is addressed. Distributed refresh implies a row refresh every 15.6 microseconds for the MCM514100, and 124.8 microseconds for the MCM51L4100. Burst refresh, a refresh of all 1024 rows consecutively, must be performed every 16 milliseconds on the MCM514100 and 128 milliseconds on the MCM51L4100.

A normal read, write, or read-write operation to the RAM will refresh all the bits (4096) associated with the particular row decoded. Three other methods of refresh, $\overline{RAS}$-**only refresh**, $\overline{CAS}$ **before** $\overline{RAS}$ **refresh**, and **hidden refresh** are available on this device for greater system flexibility.

$\overline{RAS}$-Only Refresh

$\overline{RAS}$-only refresh consists of $\overline{RAS}$ transition to active, latching the row address to be refreshed, while $\overline{CAS}$ remains high (V_{IH}) throughout the cycle. An external counter is employed to ensure all rows are refreshed within the specified limit.

$\overline{CAS}$ Before $\overline{RAS}$ Refresh

$\overline{CAS}$ before $\overline{RAS}$ refresh is enabled by bringing $\overline{CAS}$ active before $\overline{RAS}$. This clock order activates an internal refresh counter that generates the row address to be refreshed. External address lines are ignored during the automatic refresh cycle. The output buffer remains at the same state it was in during the previous cycle (hidden refresh). $\overline{W}$ must be inactive for time t_{WRP} before and time t_{WRH} after $\overline{RAS}$ active transition to prevent switching the device into a **test mode cycle**.

Hidden Refresh

Hidden refresh allows refresh cycles to occur while maintaining valid data at the output pin. Holding $\overline{CAS}$ active the end of a read or write cycle, while $\overline{RAS}$ cycles inactive for t_{RP} and back to active, starts the hidden refresh. This is essentially the execution of a $\overline{CAS}$ before $\overline{RAS}$ refresh from a cycle in progress (see Figure 1.) $\overline{W}$ is subject to the same conditions with respect to $\overline{RAS}$ active transition (to prevent test mode cycle) as in $\overline{CAS}$ before $\overline{RAS}$ refresh.

$\overline{CAS}$ BEFORE $\overline{RAS}$ REFRESH COUNTER TEST

The internal refresh counter of this device can be tested with a $\overline{CAS}$ **before** $\overline{RAS}$ **refresh counter test**. This test is performed with a read-write operation. During the test, the internal refresh counter generates the row address, while the external address supplies the column address. The entire array is refreshed after 1024 cycles, as indicated by the check data written in each row. See $\overline{CAS}$ **before** $\overline{RAS}$ **refresh counter test cycle** timing diagram.

The test can be performed after a minimum of **8** $\overline{CAS}$ **before** $\overline{RAS}$ initialization cycles. Test procedure:

1. Write "0"s into all memory cells with normal write mode.
2. Select a column address, read "0" out and write "1" into the cell by performing the $\overline{CAS}$ **before** $\overline{RAS}$ **refresh counter test, read-write cycle**. Repeat this operation 1024 times.
3. Read the "1"s which were written in step 2 in normal read mode.
4. Using the same starting column address as in step 2, read "1" out and write "0" into the cell by performing the $\overline{CAS}$ **before** $\overline{RAS}$ **refresh counter test, read-write cycle**. Repeat this operation 1024 times.
5. Read "0"s which were written in step 4 in normal read mode.
6. Repeat steps 1 to 5 using complement data.

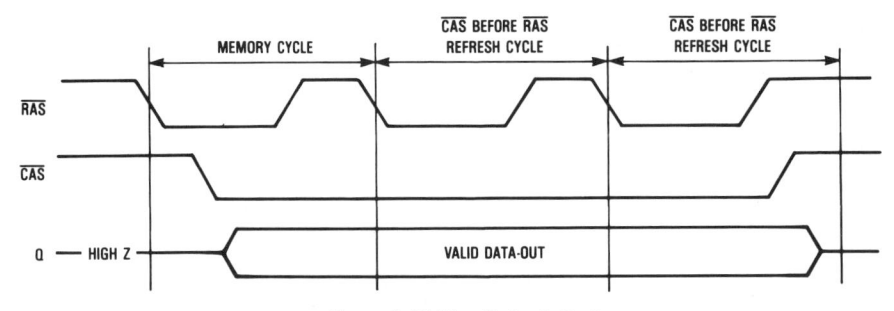

Figure 1. Hidden Refresh Cycle

TEST MODE

The internal organization of this device (512K × 8) allows it to be tested as if it were a 512K × 1 DRAM. Nineteen of the twenty two addresses are used when operating the device in test mode. Row address A10, and column addresses A0 and A10 are ignored by the device in test mode. A test mode cycle reads and/or writes data to a bit in each of the eight 512K blocks (B0–B7) in parallel. External data out is determined by the internal test mode logic of the device. See truth table and test mode block diagram following.

Test mode is enabled by performing a **test mode cycle** (see test mode timing diagram and parameter specifications table). Test mode is disabled by a $\overline{RAS}$ only refresh cycle or $\overline{CAS}$ before $\overline{RAS}$ refresh cycle. The test mode performs refresh with the internal refresh counter like a $\overline{CAS}$ before $\overline{RAS}$ refresh.

Test Mode Truth Table

D	B0	B1	B2	B3	B4	B5	B6	B7	Q
0	0	0	0	0	0	0	0	0	1
1	1	1	1	1	1	1	1	1	1
—	\multicolumn{8}{c}{Any Other}	0

TEST MODE
AC OPERATING CONDITIONS AND CHARACTERISTICS
(V_{CC} = 5.0 V ± 10%, T_A = 0 to 70°C, Unless Otherwise Noted)

READ, WRITE, AND READ-MODIFY-WRITE CYCLES (See Notes 1, 2, 3, and 4)

Parameter	Symbol Standard	Symbol Alternate	MCM514100-80 MCM51L4100-80 Min	Max	MCM514100-10 MCM51L4100-10 Min	Max	Unit	Notes
Random Read or Write Cycle Time	tRELREL	tRC	155	—	185	—	ns	5
Read-Write Cycle Time	tRELREL	tRWC	180	—	215	—	ns	5
Page Mode Cycle Time	tCELCEL	tPC	55	—	65	—	ns	
Page Mode Read-Write Cycle Time	tCELCEL	tPRWC	80	—	95	—	ns	
Access Time from $\overline{RAS}$	tRELQV	tRAC	—	85	—	105	ns	6, 7
Access Time from $\overline{CAS}$	tCELQV	tCAC	—	25	—	30	ns	6, 8
Access Time from Column Address	tAVQV	tAA	—	45	—	55	ns	6, 9
Access Time from Precharge $\overline{CAS}$	tCEHQV	tCPA	—	50	—	60	ns	6
$\overline{RAS}$ Pulse Width	tRELREH	tRAS	85	10,000	105	10,000	ns	
$\overline{RAS}$ Pulse Width (Fast Page Mode)	tRELREH	tRASP	85	200,000	105	200,000	ns	
$\overline{RAS}$ Hold Time	tCELREH	tRSH	25	—	30	—	ns	
$\overline{CAS}$ Hold Time	tRELCEH	tCSH	85	—	105	—	ns	
$\overline{CAS}$ Pulse Width	tCELCEH	tCAS	25	10,000	30	10,000	ns	
Column Address to $\overline{RAS}$ Lead Time	tAVREH	tRAL	45	—	55	—	ns	
$\overline{CAS}$ to Write Delay	tCELWL	tCWD	25	—	30	—	ns	10
$\overline{RAS}$ to Write Delay	tRELWL	tRWD	85	—	105	—	ns	10
Column Address to Write Delay Time	tAVWL	tAWD	45	—	55	—	ns	10
$\overline{CAS}$ Precharge to Write Delay Time (Page Mode)	tCEHWL	tCPWD	50	—	60	—	ns	10

NOTES:
1. V_{IH} min and V_{IL} max are reference levels for measuring timing of input signals. Transition times are measured between V_{IH} and V_{IL}.
2. An initial pause of 200 μs is required after power-up followed by 8 $\overline{RAS}$ cycles before proper device operation is guaranteed.
3. The transition time specification applies for all input signals. In addition to meeting the transition rate specification, all input signals must transition between V_{IH} and V_{IL} (or between V_{IL} and V_{IH}) in a monotonic manner.
4. AC measurements tT = 5.0 ns.
5. The specifications for tRC (min) and tRWC (min) are used only to indicate cycle time at which proper operation over the full temperature range (0°C ≤ T_A ≤ 70°C) is assured.
6. Measured with a current load equivalent to 2 TTL (−200 μA, +4 mA) loads and 100 pF with the data output trip points set at V_{OH} = 2.0 V and V_{OL} = 0.8 V.
7. Assumes that tRCD ≤ tRCD (max).
8. Assumes that tRCD ≥ tRCD (max).
9. Assumes that tRAD ≥ tRAD (max).
10. tWCS, tRWD, tCWD, tAWD and tCFWD are not restrictive operating parameters. They are included in the data sheet as electrical characteristics only; if tWCS ≥ tWCS (min), the cycle is an early write cycle and the data out pin will remain open circuit (high impedance) throughout the entire cycle; if tCWD ≥ tCWD (min), tRWD ≥ tRWD (min), tAWD ≥ tAWD (min), and tCPWD ≥ tCPWD (min) (page mode), the cycle is a read-write cycle and the data out will contain data read from the selected cell. If neither of these sets of conditions is satisfied, the condition of the data out (at access time) is indeterminate.

TEST MODE BLOCK DIAGRAM

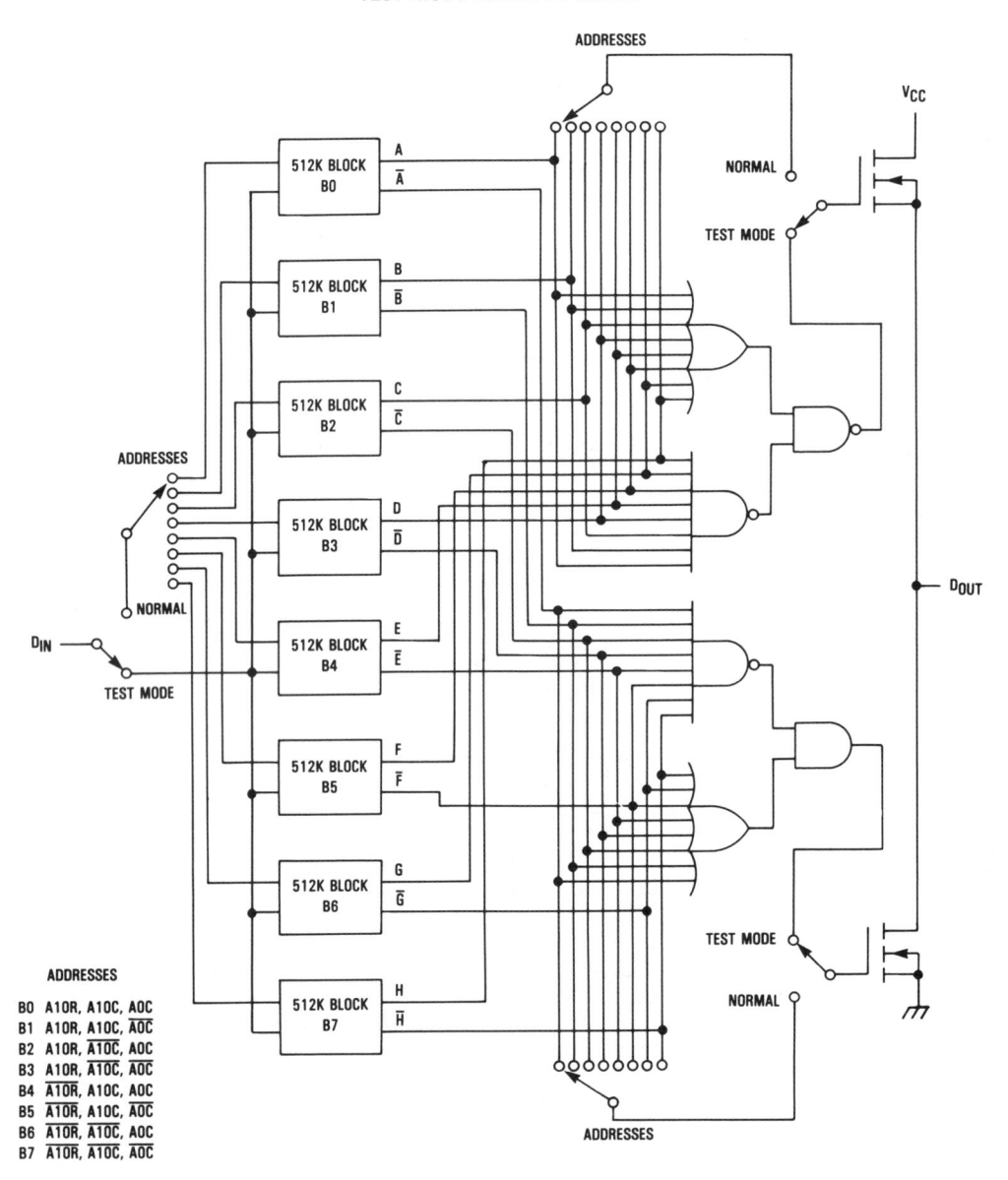

ADDRESSES
B0 A10R, A10C, A0C
B1 A10R, A10C, $\overline{A0C}$
B2 A10R, $\overline{A10C}$, A0C
B3 A10R, $\overline{A10C}$, $\overline{A0C}$
B4 $\overline{A10R}$, A10C, A0C
B5 $\overline{A10R}$, A10C, $\overline{A0C}$
B6 $\overline{A10R}$, $\overline{A10C}$, A0C
B7 $\overline{A10R}$, $\overline{A10C}$, $\overline{A0C}$

MCM514100 • MCM51L4100

MOTOROLA

PACKAGE DIMENSIONS

J PACKAGE
PLASTIC
CASE 822A-01

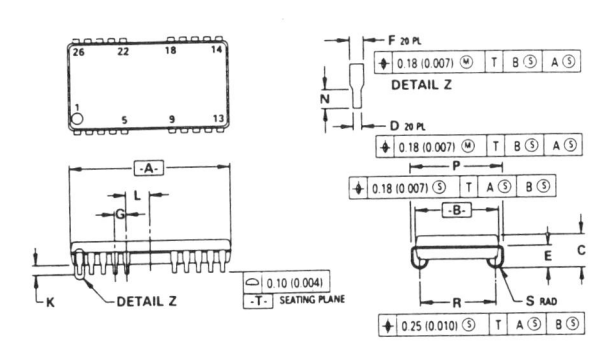

	MILLIMETERS		INCHES	
DIM	MIN	MAX	MIN	MAX
A	17.02	17.27	0.670	0.680
B	8.77	9.01	0.345	0.355
C	3.26	3.75	0.128	0.148
D	0.41	0.50	0.016	0.020
E	2.24	2.48	0.088	0.098
F	0.67	0.81	0.026	0.032
G	1.27 BSC		0.050 BSC	
K	0.64	—	0.025	—
L	2.54 BSC		0.100 BSC	
N	0.89	1.14	0.035	0.045
P	9.66	9.90	0.380	0.390
R	7.88	8.25	0.310	0.325
S	0.77	1.01	0.030	0.040

NOTES:
1. DIMENSIONING AND TOLERANCING PER ANSI Y14.5M, 1982.
2. CONTROLLING DIMENSION: INCH.
3. DIMENSION A & B DO NOT INCLUDE MOLD PROTRUSION. MOLD PROTRUSION SHALL NOT EXCEED 0.15 (0.006) PER SIDE.
4. DIMENSION A & B INCLUDE MOLD MISMATCH AND ARE DETERMINED AT THE PARTING LINE.
5. DIM R TO BE DETERMINED AT DATUM -T-.
6. FOR LEAD IDENTIFICATION PURPOSES, PIN POSITIONS 6, 7, 8, 19, 20, & 21 ARE NOT USED.

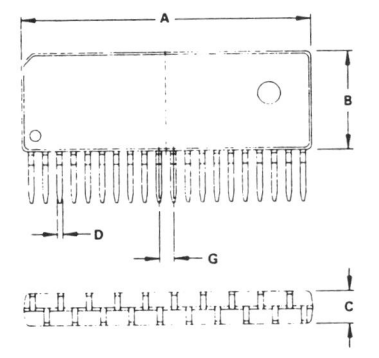

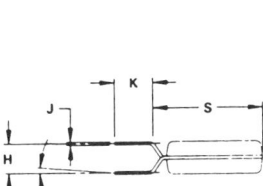

Z PACKAGE
ZIG-ZAG IN-LINE
CASE 836-01

	MILLIMETERS		INCHES	
DIM	MIN	MAX	MIN	MAX
A	25.70	25.77	1.012	1.014
B	8.64	8.73	0.340	0.344
C	2.80	2.90	0.111	0.114
D	0.45	0.55	0.018	0.021
G	1.17	1.37	0.047	0.053
H	2.44	2.64	0.097	0.103
J	0.272	0.278	0.0107	0.0109
K	3.38	3.42	0.133	0.134
M	0°	4°	0°	4°
S	9.83	9.89	0.387	0.389

NOTES
1. DIMENSIONING AND TOLERANCING PER ANSI Y14.5M, 1982
2. CONTROLLING DIMENSION: INCH

Literature Distribution Centers:
USA: Motorola Literature Distribution; P.O. Box 20912; Phoenix, Arizona 85036.
EUROPE: Motorola Ltd.; European Literature Center; 88 Tanners Drive, Blakelands, Milton Keynes, MK14 5BP, England.
ASIA PACIFIC: Motorola Semiconductors H.K. Ltd.; P.O. Box 80300; Cheung Sha Wan Post Office; Kowloon Hong Kong.
JAPAN: Nippon Motorola Ltd.; 3-20-1 Minamiazabu, Minato-ku, Tokyo 106 Japan.

MOTOROLA

MCM514100 • MCM51L4100

TEST MODE CYCLE
(D and A0 to A10 are Don't Care)

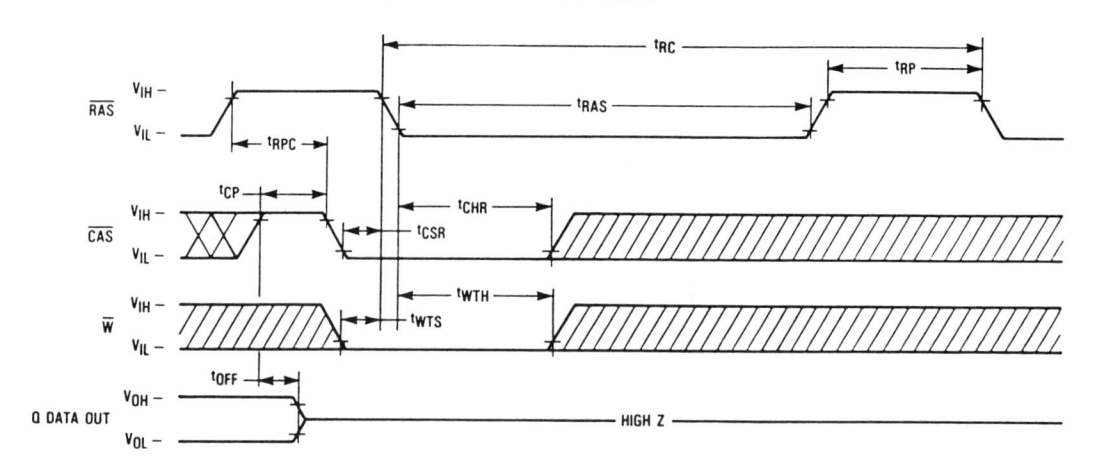

ORDERING INFORMATION
(Order by Full Part Number)

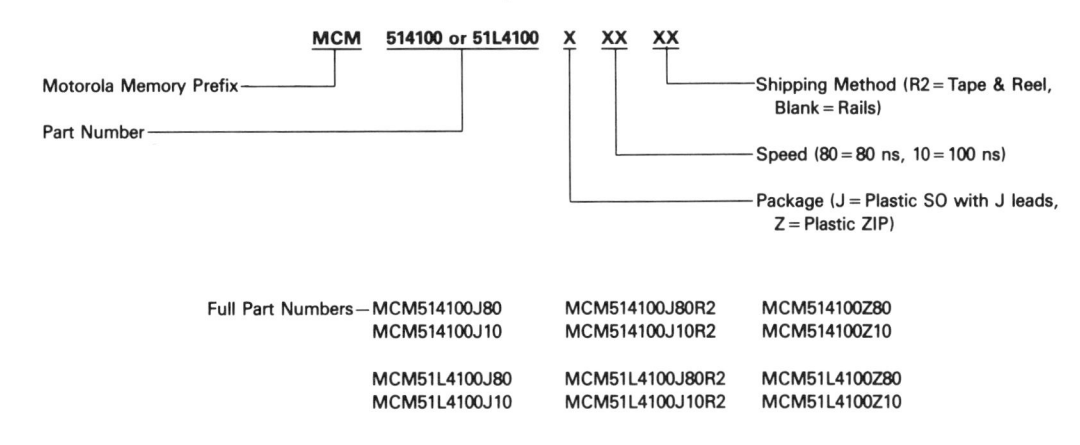

```
MCM  514100 or 51L4100  X  XX  XX
 │         │             │   │   │
 │         │             │   │   └─ Shipping Method (R2 = Tape & Reel,
 │         │             │   │                       Blank = Rails)
 │         │             │   └──── Speed (80 = 80 ns, 10 = 100 ns)
 │         │             └──────── Package (J = Plastic SO with J leads,
 │         │                                Z = Plastic ZIP)
 │         └── Part Number
 └── Motorola Memory Prefix
```

Full Part Numbers — MCM514100J80   MCM514100J80R2   MCM514100Z80
                    MCM514100J10   MCM514100J10R2   MCM514100Z10

                    MCM51L4100J80   MCM51L4100J80R2   MCM51L4100Z80
                    MCM51L4100J10   MCM51L4100J10R2   MCM51L4100Z10

Motorola reserves the right to make changes without further notice to any products herein to improve reliability, function or design. Motorola does not assume any liability arising out of the application or use of any product or circuit described herein; neither does it convey any license under its patent rights nor the rights of others. Motorola products are not authorized for use as components in life support devices or systems intended for surgical implant into the body or intended to support or sustain life. Buyer agrees to notify Motorola of any such intended end use whereupon Motorola shall determine availability and suitability of its product or products for the use intended. Motorola and Ⓜ are registered trademarks of Motorola, Inc. Motorola, Inc. is an Equal Employment Opportunity/Affirmative Action Employer.

MOTOROLA SEMICONDUCTOR APPLICATION NOTE

Order this document by AN986/D

AN986

Page, Nibble, and Static Column Modes: High-Speed, Serial-Access Options on 1M-Bit+DRAMs

The 1M-bit and higher density DRAMs offered by Motorola, in addition to operating in a standard mode at advertised access times, have special operating modes that will significantly decrease access time. These are page, nibble, and static column modes. All three modes are available in the 1M × 1 configuration; page and static column modes are also available on the 256K × 4 configuration. Read, write, and read-write operations can be mixed and performed in any order while these devices are operating in either random or special mode.

The comments that follow refer specifically to successive read operations for page, nibble, and static column modes on the 1M × 1 device. The read operation is chosen for sake of simplicity in illustrating these special operating modes. However, decreased access times will occur for all operations, performed in any order, when the device is operated in any of these modes. General operating comments apply to the 256K × 4 device as well.

All of these special operating modes are useful in applications that require high-speed serial access. Typical examples include video bit map graphics monitors or RAM disks. Page mode is the standard, available since the days of the 16K × 1 DRAM. Static column is the latest mode to be made available on DRAMs, and nibble mode first appeared somewhere in between. Page and static column offer the same column location access, but operate somewhat differently. Nibble is unlike either of the other modes, but faster than both in its niche. All modes are initiated after a standard read or write is performed.

Page and static column modes allow access to any of 1024 column locations on a specific row, while nibble allows access to a maximum of four bits. The location of the first bit in nibble mode determines the other bits to be accessed. Nibble mode allows the fastest access of the three devices (t_{NCAC}), all other parameters held equal, at about 1/4 the standard (t_{RAC}) rate. Page and static column access times (t_{CAC}, t_{AA}) are, respectively, about 1/3 and 1/2 the standard rate.

Cycle time is a better indicator of relative speed improvement, since it measures the minimum time between any two successive reads. Cycle time is approximately 1/4 for nibble and 1/3 for page and static column modes, with respect to a

Table 1. Operating Characteristic Comparison

Parameter		Page	Nibble	Static Column	Random
Access Time (ns)*	t_{CAC}	25	—	—	—
	t_{NCAC}	—	20	—	—
	t_{AA}	—	—	45	—
	t_{RAC}	—	—	—	85
Cycle Time (ns)*	t_{PC}	50	—	—	—
	t_{NC}	—	40	—	—
	t_{SC}	—	—	50	—
	t_{RC}	—	—	—	165
Accessible Bits		1024	4	1024	All
Order of Accessible Bits		Random	Fixed	Random	Random
Conditions	$\overline{RAS}$	Active	Active	Active	Cycle
	$\overline{CAS}$ or $\overline{CS}$**	Cycle	Cycle	Active	Cycle
	Addresses	Cycle	N/A	Cycle	Cycle
	Outputs	Cycle	Cycle	Active	Cycle
Time to Read 4 Bits (ns)*		235	205	235	660
Time to Read 1024 Unique Bits (ns)*		51,235	70,400	51,235	168,960

Values for a 1M × 1 85-ns device. *$\overline{CS}$ on Static Column.

- Page: 4 bit read = $t_{RAC} + 3t_{PC}$
  1024 bit read = $t_{RAC} + 1023t_{PC}$
- Nibble: 4 bit read = $t_{RAC} + 3t_{NC}$
  1024 bit read = $256 \cdot (t_{RAC} + 3t_{NC} + t_{RP})$
- Static Column: 4 bit read = $t_{RAC} + 3t_{SC}$
  1024 bit read = $t_{RAC} + 1023t_{SC}$
- Random: 4 bit read = $4t_{RC}$
  1024 bit read = $1024t_{RC}$

MOTOROLA

©MOTOROLA INC., 1987

random cycle time of 165 nanoseconds. When operated in these high-speed modes, users will typically access most or all of the bits available to that mode, once the mode has been initiated. Thus the best measure of speed for nibble mode is the rate at which four bits are read, while the rate at which 1024 bits are read is the best measure of page or static column mode. When the actual operating conditions are considered, as described elsewhere, the difference between t_{CAC}, t_{NCAC}, and t_{AA} measurements hold relatively little significance.

Page mode is slightly more difficult to interface in a system than static column mode due to extra $\overline{CAS}$ pulses that are required in page mode. Static column generates less noise than page mode, because output buffers and $\overline{CS}$ are always active in this mode. Noise transients, generated every time $\overline{CAS}$ is cycled from inactive to active, are thus eliminated in the static column mode.

PAGE MODE

Page mode allows faster access to any of the 1024 column locations on a given row, typically at one third the standard (t_{RAC}) rate for randomly-performed operations. Page mode consists of cycling the $\overline{CAS}$ clock from active (low) to inactive (high) and back, and providing a column address, while holding the $\overline{RAS}$ clock active (low). A new column location can be accessed with each $\overline{CAS}$ cycle (t_{PC}).

Page mode is initiated with a standard read or write operation. Row address is latched by the $\overline{RAS}$ clock transition to active, followed by column address and $\overline{CAS}$ clock active. Performing a $\overline{CAS}$ cycle (t_{PC}) and supplying a column address while $\overline{RAS}$ clock remains active constitutes the first page mode cycle. Subsequent page mode cycles can be performed as long as $\overline{RAS}$ clock is active. The first access (data valid) occurs at the standard rate (t_{RAC}). All of the read operations in page mode following the initial operation are measured at the faster rate (t_{CAC}), provided all other timing minimums are maintained (see Figure 1a). Page mode cycle time determines how fast successive bits are read (see Figure 1b).

NIBBLE MODE

Nibble mode allows serial access to two, three, or four bits of data at a much higher rate than random operations (t_{RAC}). Nibble mode consists of cycling the $\overline{CAS}$ clock while holding the $\overline{RAS}$ clock active, like page mode. Internal row and column address counters increment at each $\overline{CAS}$ cycle, thus no external column addresses are required (unlike page or static column modes). After cycling $\overline{CAS}$ three times in nibble mode, the address sequence repeats and the same four bits are accessed again, in serial order, upon subsequent cycles of $\overline{CAS}$:

00, 01, 10, 11, 00, 01, 10, 11, . . .

Nibble mode operation is initiated with a standard read or write cycle. Row address is latched by $\overline{RAS}$ clock transition to active, followed by column addresses and $\overline{CAS}$ clock. Performing a $\overline{CAS}$ cycle (t_{NC}) while $\overline{RAS}$ clock remains active constitutes the first nibble mode cycle. Subsequent nibble mode cycles can be performed as long as the $\overline{RAS}$ clock is held active. The first access (data out) occurs at the standard rate (t_{RAC}). All of the read operations in nibble mode following the initial operation are measured at the faster rate (t_{NCAC}), provided all other timing minimums are maintained (see Figure 2a). Nibble mode cycle time determines how fast successive bits are read (see Figure 2b).

STATIC COLUMN MODE

This mode is useful in applications that require less noise than page mode. Output buffers are always on when the device is in this mode and $\overline{CS}$ clock is not cycled, resulting in fewer transients and simpler operation. It allows faster access to any of the 1024 column addresses on a given row, typically at half the standard (t_{RAC}) rate for randomly performed operations. Static column consists of changing column addresses while holding the $\overline{RAS}$ and $\overline{CS}$ clocks active. A new column location can be accessed with each static column cycle (t_{SC}).

Static column mode operation is initiated with a standard read or write cycle. Row address is latched by $\overline{RAS}$ clock transition to active, followed by column addresses and $\overline{CS}$ clock. Performing an address cycle (t_{SC}) while $\overline{RAS}$ and $\overline{CS}$ clocks remain active constitutes the first static column cycle. Subsequent static column cycles can be performed as long as the $\overline{RAS}$ and $\overline{CS}$ clocks are held active. The first access (data out) occurs at the standard (t_{RAC}) rate. All of the read operations in static column following the initial operation are measured at the faster rate (t_{AA}), provided all other timing minimums are maintained (see Figure 3a). Static column cycle time determines how fast successive bits are read (see Figure 3b).

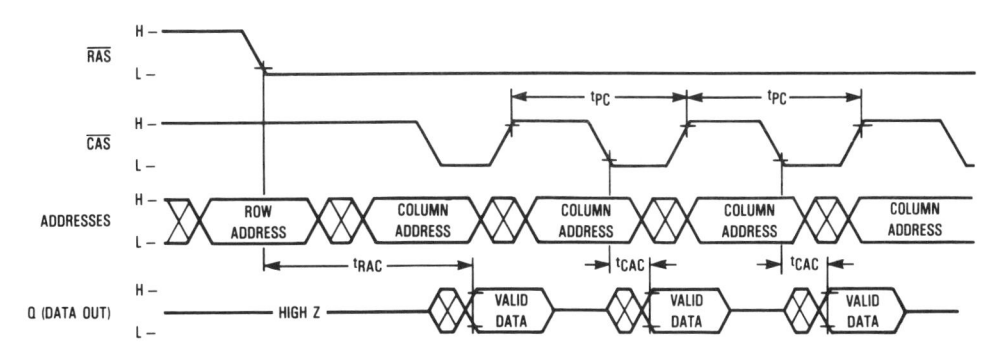

Figure 1a. Page Mode Read Cycle

540

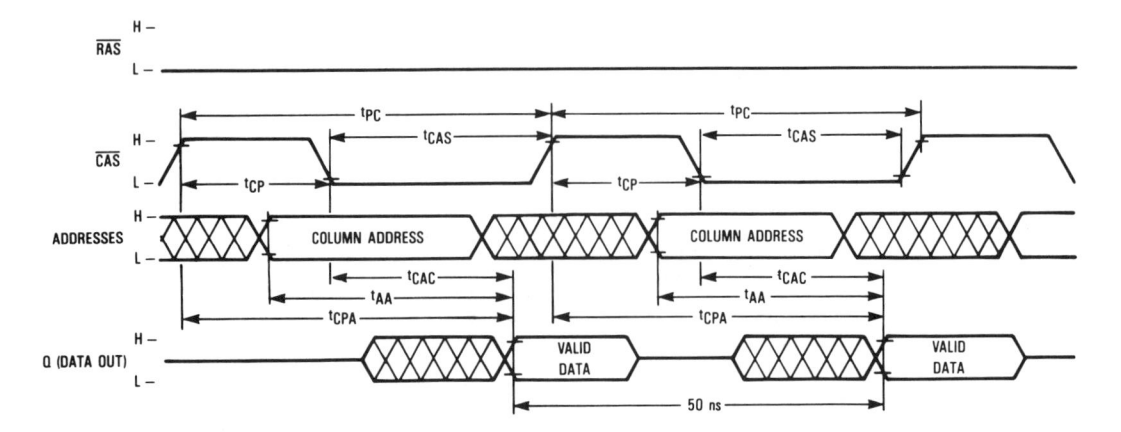

Figure 1b. Page Mode Cycle Minimum Timing

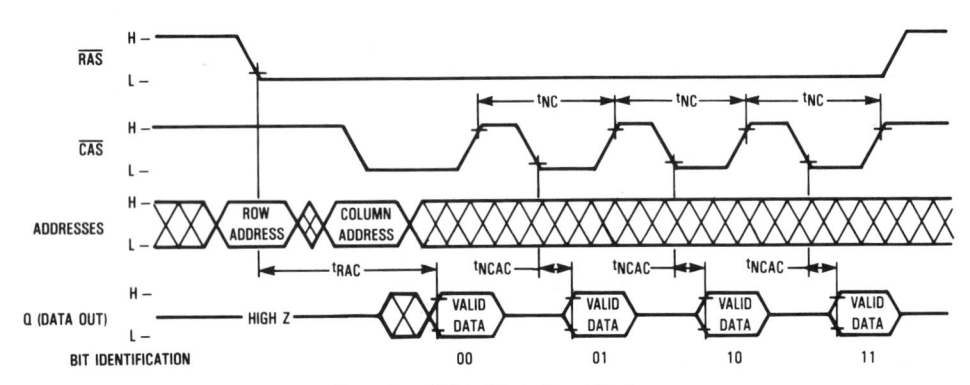

Figure 2a. Nibble Mode Read Cycle

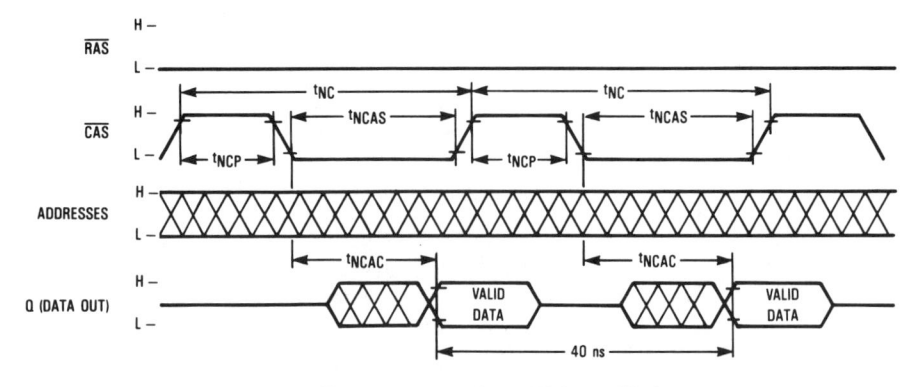

Figure 2b. Nibble Mode Cycle Minimum Timing

AN986

MOTOROLA

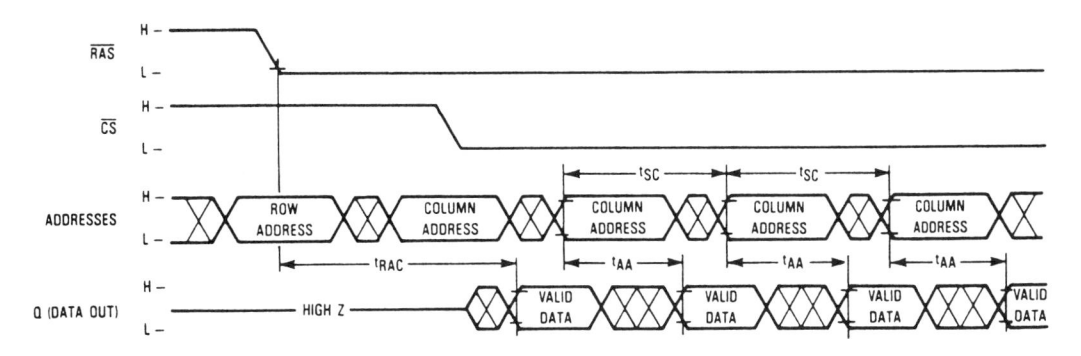

Figure 3a. Static Column Mode Read Cycle

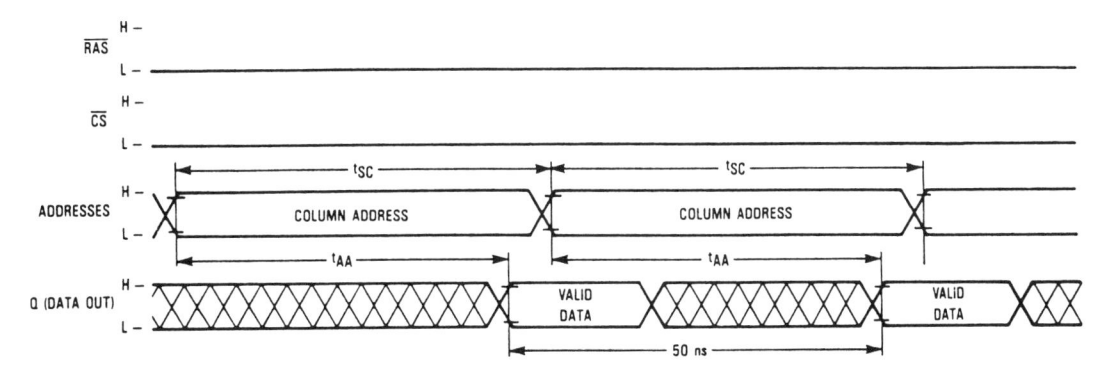

Figure 3b. Static Column Mode Cycle Minimum Timing

Motorola reserves the right to make changes without further notice to any products herein to improve reliability, function or design. Motorola does not assume any liability arising out of the application or use of any product or circuit described herein; neither does it convey any license under its patent rights nor the rights of others. Motorola and Ⓜ are registered trademarks of Motorola, Inc.

MOTOROLA LTD. *Semiconductor Products Group*

COLVILLES ROAD, KELVIN ESTATE, EAST KILBRIDE, GLASGOW, G75 0TG, SCOTLAND

MOTOROLA
SEMICONDUCTOR
TECHNICAL DATA

Order this data sheet
by MC74F29368/D

Advance Information
1 Megabit Dynamic Memory Controller (DMC)

MC74F29368

The MC74F29368 Dynamic Memory Controller (DMC) is designed to be an integral part of today's high performance memory systems. The DMC functions as the address controller between any processor and Dynamic memory array. Using its two 10-bit address latches, the DMC holds the Row and Column addresses for any DRAM up to 1 megabit. These latches and the two 10-bit Row/Column Refresh address counters feed into a 10-bit, 4-input MUX, for output to the Dynamic RAM address lines. Each of the four $\overline{RAS}_n$ and $\overline{CAS}_n$ outputs is selected by the outputs of the Bank Select latch, which is driven by the two high-order address bits.

The MC74F29368 has two basic modes of operation, read/write and refresh. In the refresh mode, the Row and Column counters cycle through the refresh addresses. If memory scrubbing is not being implemented, only the Row counter is used, generating up to 1024 addresses to refresh a 1024-cycle-refresh 1-Mbit DRAM. When memory scrubbing is being performed, both the Row and Column counters are utilized to perform read-modify-write cycles. In this mode, all $\overline{RAS}_n$ outputs will be active, while only $\overline{CAS}_n$ output is active at a time.

- Provides Control for 16K, 64K, and 256K and 1-Megabit Dynamic RAMs
- Outputs Directly Drive up to 88 DRAMs, with a Guaranteed Worst-Case Limit on the Undershoot
- Highest-Order Two Address Bits Select One of Four Banks of RAMs
- Separate Output Enable for Multi-Channel Access to Memory
- Supports Scrubbing Operations and Other Specialty Access Modes
- Upgradable from F2968A 256K DRAM
- Samples Available now in 68-Pin PLCC

LOGIC DIAGRAM

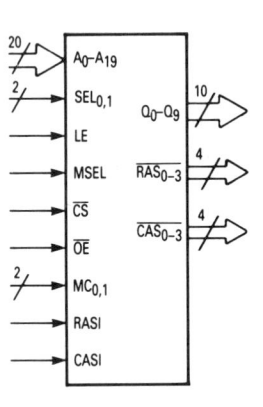

BLOCK DIAGRAM

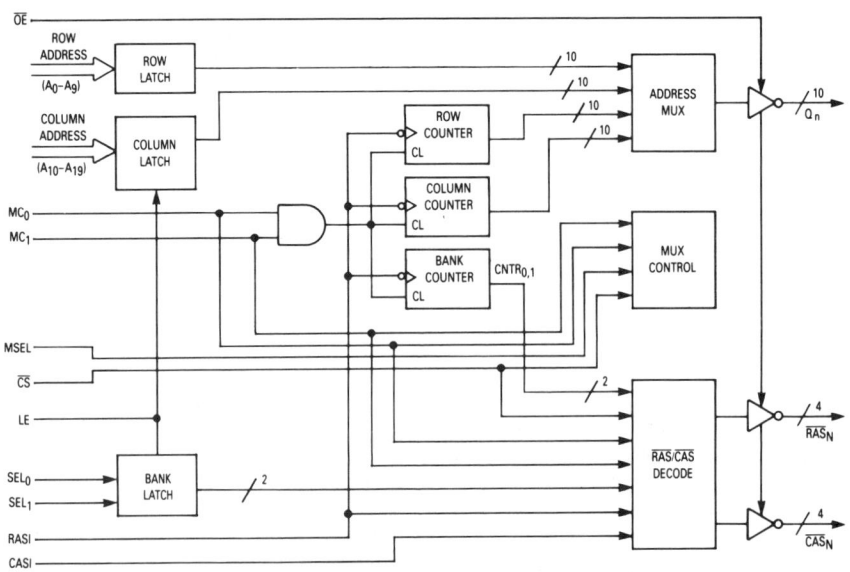

This document contains information on a product under development. Motorola reserves the right to change or discontinue this product without notice.

©MOTOROLA INC., 1989

ADI1492

CONNECTIONS DIAGRAMS

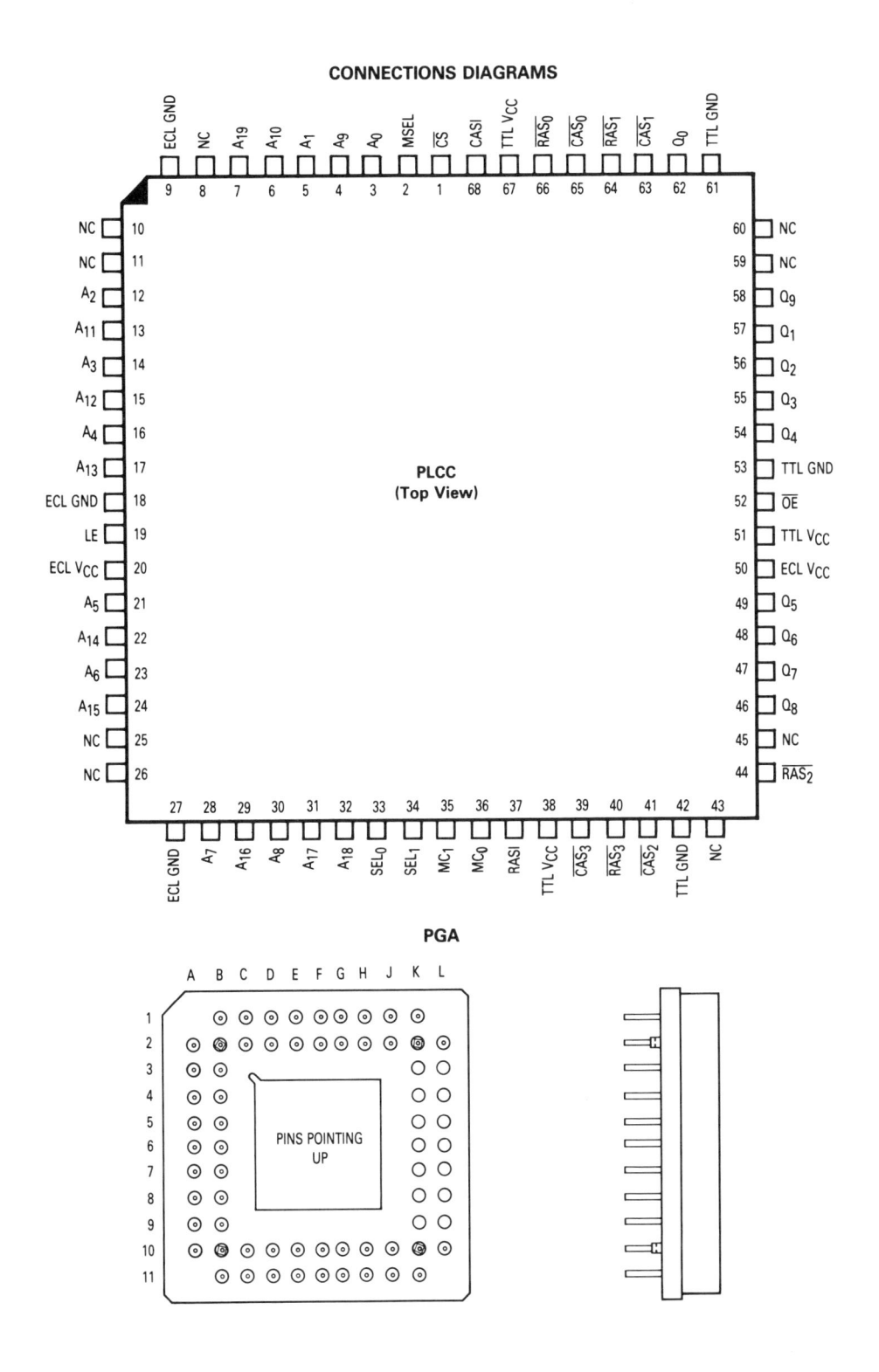

PLCC (Top View)

Pin	Signal
1	$\overline{CS}$
2	MSEL
3	A_0
4	A_9
5	A_1
6	A_{10}
7	A_{19}
8	NC
9	ECL GND
10	NC
11	NC
12	A_2
13	A_{11}
14	A_3
15	A_{12}
16	A_4
17	A_{13}
18	ECL GND
19	LE
20	ECL V_{CC}
21	A_5
22	A_{14}
23	A_6
24	A_{15}
25	NC
26	NC
27	ECL GND
28	A_7
29	A_{16}
30	A_8
31	A_{17}
32	A_{18}
33	SEL_0
34	SEL_1
35	MC_1
36	MC_0
37	RASI
38	TTL V_{CC}
39	$\overline{CAS_3}$
40	$\overline{RAS_3}$
41	$\overline{CAS_2}$
42	TTL GND
43	NC
44	$\overline{RAS_2}$
45	NC
46	Q_8
47	Q_7
48	Q_6
49	Q_5
50	ECL V_{CC}
51	TTL V_{CC}
52	$\overline{OE}$
53	TTL GND
54	Q_4
55	Q_3
56	Q_2
57	Q_1
58	Q_9
59	NC
60	NC
61	TTL GND
62	Q_0
63	$\overline{CAS_1}$
64	$\overline{RAS_1}$
65	$\overline{CAS_0}$
66	$\overline{RAS_0}$
67	TTL V_{CC}
68	CASI

PGA

PINS POINTING UP

PIN DESIGNATIONS

| (SORTED BY PIN NAME) |||| (SORTED BY PIN NUMBER) ||||
PIN NAME	PIN NO.	PIN NAME	PIN NO.	PIN NO.	PIN NAME	PIN NO.	PIN NAME
A_0	E-2	NC	A-10	A-2	GND	G-1	V_{CC}
A_1	D-1	NC	B-2	A-3	A_2	G-2	$\overline{RAS_0}$
A_2	A-3	NC	B-3	A-4	A_3	G-10	SEL_1
A_3	A-4	NC	B-10	A-5	A_4	G-11	SEL_0
A_4	A-5	NC	B-11	A-6	GND	H-1	$\overline{CAS_0}$
A_5	A-8	NC	C-1	A-7	LE	H-2	$\overline{RAS_1}$
A_6	A-9	NC	C-11	A-8	A_5	H-10	MC_0
A_7	D-10	NC	D-11	A-9	A_6	H-11	MC_1
A_8	E-11	NC	K-2	A-10	NC	J-1	$\overline{CAS_1}$
A_9	D-1	$\overline{OE}$	K-5	B-1	A_{19}	J-2	Q_0
A_{10}	C-2	Q_0	J-2	B-2	NC	J-10	V_{CC}
A_{11}	B-4	Q_1	K-3	B-3	NC	J-11	RASI
A_{12}	B-5	Q_2	L-3	B-4	A_{11}	K-1	GND
A_{13}	B-6	Q_3	K-4	B-5	A_{12}	K-2	NC
A_{14}	B-8	Q_4	L-4	B-6	A_{13}	K-3	Q_1
A_{15}	B-9	Q_5	L-7	B-7	V_{CC}	K-4	Q_3
A_{16}	E-10	Q_6	K-7	B-8	A_{14}	K-5	$\overline{OE}$
A_{17}	F-11	Q_7	L-8	B-9	A_{15}	K-6	V_{CC}
A_{18}	F-10	Q_8	K-8	B-10	NC	K-7	Q_6
A_{19}	B-1	Q_9	L-2	B-11	NC	K-8	Q_8
CASI	F-1	RASI	J-11	C-1	NC	K-9	GND
$\overline{CAS_0}$	H-1	$\overline{RAS_0}$	G-2	C-2	A_{10}	K-10	$\overline{RAS_3}$
$\overline{CAS_1}$	J-1	$\overline{RAS_1}$	H-2	C-10	GND	K-11	$\overline{CAS_3}$
$\overline{CAS_2}$	L-10	$\overline{RAS_2}$	L-9	C-11	NC	L-2	Q_9
$\overline{CAS_3}$	K-11	$\overline{RAS_3}$	K-10	D-1	A_9	L-3	Q_2
CS	F-2	SEL_0	G-11	D-2	A_1	L-4	Q_4
GND	A-2	SEL_1	G-10	D-10	A_7	L-5	GND
GND	A-6	V_{CC}	L-6	D-11	NC	L-6	V_{CC}
GND	C-10	V_{CC}	B-7	E-1	MSEL	L-7	Q_5
GND	K-1	V_{CC}	J-10	E-2	A_0	L-8	Q_7
GND	L-5	V_{CC}	G-1	E-10	A_{16}	L-9	$\overline{RAS_2}$
GND	K-9	V_{CC}	K-6	E-11	A_8	L-10	$\overline{CAS_2}$
LE	A-7			F-1	CASI		
MC_0	H-10			F-2	CS		
MC_1	H-11			F-10	A_{18}		
MSEL	E-1			F-11	A_{17}		

PIN DESCRIPTION

A$_0$–A$_{19}$ Address Inputs (Input (20))

A$_0$–A$_9$ are latched in as the 10-bit Row Address for the RAM. These inputs drive Q$_0$–Q$_9$ when the Am29368 is in the Read/Write mode and MSEL is LOW. A$_{10}$–A$_{19}$ are latched in as the Column Address, and will drive Q$_0$–Q$_9$ when MSEL is HIGH and the DMC is in the Read/Write mode. The addresses are latched with the Latch Enable (LE) signal.

$\overline{CAS}_{0-3}$ Column Address Strobe (Output (4))

During normal Read/Write cycles the two select bits (SEL$_0$, SEL$_1$) determine which $\overline{CAS}_n$ output will go active following CASI going HIGH. When memory scrubbing is performed, only the $\overline{CAS}_n$ signal selected by CNTR$_0$ and CNTR$_1$ will be active (see $\overline{CAS}$ Output Function Table). For non-scrubbing cycles, all four $\overline{CAS}_n$ outputs remain HIGH.

CASI Column Address Strobe (Input (1))

This input going active will cause the selected $\overline{CAS}_n$ output to be forced LOW.

$\overline{CS}$ Chip Select (Input (1))

This active-LOW input is used to select the DMC. When $\overline{CS}$ is active, the F29368 operates normally in all four modes. When $\overline{CS}$ goes HIGH, the device will not enter the Read/Write mode. This allows more than one F29368 DMC to control multiple memory banks, thus providing an easy method for expanding the memory size.

LE Latch Enable (Input (1))

This active-HIGH input causes the Row, Column, and Bank Select latches to become transparent, allowing the latches to accept new input data. A LOW input on LE latches the input data, assuming it meets the setup and hold time requirements.

MC$_{0-1}$ Mode Control (Input (2))

These inputs are used to specify which of the four operating modes the DMC should be using. The description of the four operating modes is given in Table 1.

MSEL Multiplexer Select (Input (1))

This input determines whether the Row or Column Address will be sent to the memory address inputs. When MSEL is HIGH the Column Address is selected, while the Row Address is selected when MSEL is LOW. The address may come from either the address latch or refresh address counter depending on MC$_{0,1}$.

$\overline{OE}$ Output Enable (Input (1))

This active-LOW input enables/disables the output signals. When $\overline{OE}$ is HIGH, the outputs of the DMC enter the high-impedance state.

Q$_{0-9}$ Address Outputs (Outputs (10))

These address outputs will feed the DRAM address inputs, and provide drive for memory systems up to 500 picofarads in capacitance.

$\overline{RAS}_{0-3}$ Row Address Strobe (Output (4))

Each one of the Row Address Strobe outputs provides a $\overline{RAS}_n$ signal to one of the four banks of dynamic memory. Each will go LOW only when selected by SEL$_0$ and SEL$_1$ and only after RASI goes HIGH. All four go LOW in response to RASI in either of the Refresh modes.

RASI Row Address Strobe (Input (1))

During normal memory cycles, the decoded $\overline{RAS}_n$ output ($\overline{RAS}_0$, $\overline{RAS}_1$, $\overline{RAS}_2$, or $\overline{RAS}_3$) is forced LOW after receipt of RASI. In either Refresh mode, all four $\overline{RAS}_n$ outputs will go LOW following RASI going HIGH.

SEL$_{0-1}$ Bank Select (Input (2))

These two inputs are normally the two higher-order address bits, and are used in the Read/Write mode to select which bank of memory will be receiving the $\overline{RAS}_n$ and $\overline{CAS}_n$ signals after RASI and CASI go HIGH.

FUNCTIONAL DESCRIPTION

Architecture

The F29368 provides all the required data and refresh addresses needed by the dynamic RAM memory. In normal operation, the Row and Column addresses are multiplexed to the dynamic RAM by using MSEL, with the corresponding $\overline{RAS}_n$ and $\overline{CAS}_n$ signals activated to strobe the addresses into the RAM. High capacitance drivers on the outputs allow the DMC to drive four banks of 16-bit words, including a 6-bit checkword, for a total of 88 DRAMs.

MODE CONTROL FUNCTION TABLE

MC$_1$	MC$_0$	Operating Mode
0	0	Refresh without Scrubbing. Refresh cycles are performed with only the Row Counter being used to generate addresses. In this mode, all four $\overline{RAS}_n$ outputs are active while the four $\overline{CAS}_n$ signals are kept HIGH.
0	1	Refresh with Scrubbing/Initialize. During this mode, refresh cycles are done with both the Row and Column counters generating the addresses. MSEL is used to select between the Row and Column counter. All four $\overline{RAS}_n$ go active in response to RASI, while only one $\overline{CAS}_n$ output goes LOW in response to CASI. The Bank Counter keeps track of which $\overline{CAS}_n$ output will go active. This mode is also used on system power-up so that the memory can be written with a known data pattern.
1	0	Read/Write. This mode is used to perform Read/Write cycles. Both the Row and Column addresses are latched and multiplexed to the address output lines using MSEL. SEL$_0$ and SEL$_1$ are decoded to determine which $\overline{RAS}_n$ and $\overline{CAS}_n$ will be active.
1	1	Clear Refresh Counter. This mode will clear the three refresh counters (Row, Column, and Bank) on the HIGH-to-LOW transition of RASI, putting them at the start of the refresh sequence. In this mode, all four $\overline{RAS}_n$ are driven LOW upon receipt of RASI so that DRAM wake-up cycles may be performed.

ADDRESS OUTPUT FUNCTION TABLE

$\overline{CS}$	MC$_1$	MC$_0$	MSEL	Mode	MUX Output
0	0	0	X	Refresh without Scrubbing	Row Counter Address
0	0	1	1	Refresh with Scrubbing	Column Counter Address
0	0	1	0	Refresh with Scrubbing	Row Counter Address
0	1	0	1	Read/Write	Column Address Latch
0	1	0	0	Read/Write	Row Address Latch
0	1	1	X	Clear Refresh Counter	Zero
1	0	0	X	Refresh without Scrubbing	Row Counter Address
1	0	1	1	Refresh with Scrubbing	Column Counter Address
1	0	1	0	Refresh with Scrubbing	Row Counter Address
1	1	0	X	Read/Write	Zero
1	1	1	X	Clear Refresh Counter	Zero

$\overline{RAS}$ OUTPUT FUNCTION TABLE

RASI	$\overline{CS}$	MC$_1$	MC$_0$	SEL$_1$	SEL$_0$	Mode	$\overline{RAS}_0$	$\overline{RAS}_1$	$\overline{RAS}_2$	$\overline{RAS}_3$
0	X	X	X	X	X	X	1	1	1	1
1	0	0	0	X	X	Refresh without Scrubbing	0	0	0	0
1	0	0	1	X	X	Refresh with Scrubbing	0	0	0	0
1	0	1	0	0	0	Read/Write	0	1	1	1
1	0	1	0	0	1	Read/Write	1	0	1	1
1	0	1	0	1	0	Read/Write	1	1	0	1
1	0	1	0	1	1	Read/Write	1	1	1	0
1	0	1	1	X	X	Clear Refresh Counter	0	0	0	0
1	1	0	0	X	X	Refresh without Scrubbing	0	0	0	0
1	1	0	1	X	X	Refresh with Scrubbing	0	0	0	0
1	1	1	0	X	X	Read/Write	1	1	1	1
1	1	1	1	X	X	Clear Refresh Counter	0	0	0	0

$\overline{CAS}$ OUTPUT FUNCTION TABLE

Inputs							Internal		Outputs			
CASI	$\overline{CS}$	MC_1	MC_0	SEL_1	SEL_0	$CNTR_1$	$CNTR_0$	$\overline{CAS}_0$	$\overline{CAS}_1$	$\overline{CAS}_2$	$\overline{CAS}_3$
1	0	0	0	X	X	X	X	1	1	1	1
1	0	0	1	X	X	0	0	0	1	1	1
1	0	0	1	X	X	0	1	1	0	1	1
1	0	0	1	X	X	1	0	1	1	0	1
1	0	0	1	X	X	1	1	1	1	1	0
1	0	1	0	0	0	X	X	0	1	1	1
1	0	1	0	0	1	X	X	1	0	1	1
1	0	1	0	1	0	X	X	1	1	0	1
1	0	1	0	1	1	X	X	1	1	1	0
1	0	1	1	X	X	X	X	1	1	1	1
1	1	0	0	X	X	X	X	1	1	1	1
1	1	0	1	X	X	0	0	0	1	1	1
1	1	0	1	X	X	0	1	1	0	1	1
1	1	0	1	X	X	1	0	1	1	0	1
1	1	0	1	X	X	1	1	1	1	1	0
1	1	1	0	X	X	X	X	1	1	1	1
1	1	1	1	X	X	X	X	1	1	1	1
0	X	X	X	X	X	X	X	1	1	1	1

Input Latches
For those systems where addresses and data are multiplexed onto a single bus, the DMC has latches to hold the address information. The twenty input latches (Row, Column, and Bank Select) are transparent when Latch Enable (LE) is HIGH and will latch the input data meeting setup and hold time requirements when LE goes LOW. For systems where the processor has separate address and data buses, LE may be permanently enabled HIGH.

Refresh Counters
The two 10-bit refresh counters make it possible to support 128, 256, 512, and 1024 line refresh. External control over which type of refresh is to be performed allows the user maximum flexibility when choosing the refreshing scheme. Transparent (hidden), burst, synchronous or asynchronous refresh modes are all possible.

The refresh counters are advanced at the HIGH-to-LOW transition of RASI. This assures a stable counter output for the next refresh cycle.

Refresh with Error Correction
The F29368 makes it possible to correct single-bit errors in parallel with performing dynamic RAM refresh cycles. This "scrubbing" of memory can be done periodically as a background routine when the memory is not being used by the processor. In a memory scrubbing cycle ($MC_{1,0}$ = 01), the Row Address is strobed into all four banks with all four $\overline{RAS}_n$ outputs going LOW.

The Column Address is strobed into a single bank with the activated $\overline{CAS}_n$ output being selected by the Bank Counter. This type of cycle is used to simultaneously refresh the addressed row in all banks and read and correct (if necessary) one word in memory; thereby reducing the overhead associated with Error Detection and Correction. When doing refresh with memory scrubbing, both the Row and Column counters are multiplexed to the dynamic RAM address lines by using MSEL. Using the Refresh with Memory Scrubbing mode implies the presence of an error correcting facility such as the F2960A EDAC unit. When doing refresh without scrubbing, all four $\overline{RAS}_n$ still go LOW but the $\overline{CAS}_n$ outputs are all driven HIGH so as not to activate the output lines of the memory.

Decoupling
Due to the high switching speeds and high drive capability of the F29368, it is necessary to decouple the device for proper operation. 1 μF multilayer ceramic capacitors are recommended for decoupling (see Figures 1a & 1b). It is important to mount the capacitors as close as possible to the power pins (V_{CC}, GND) to minimize lead inductance and noise. A ground plane is recommended as are provisions for separating TTL & ECL planes.

It is strongly recommended that the F29368 be directly surface mounted whenever possible. Should a PLCC or PGA socket be required, a one-time-insertion-only socket with minimal lead lengths is necessary for proper device functioning.

V$_{ONP}$

The guaranteed maximum undershoot voltage of the F29368 is −1.5 volts. V$_{ONP}$ is measured with respect to ground (see Figure 1a). Note that the ground of the capacitive load must be the same as for the V$_{CC}$ pin(s). As loading increases, V$_{ONP}$ will approach zero.

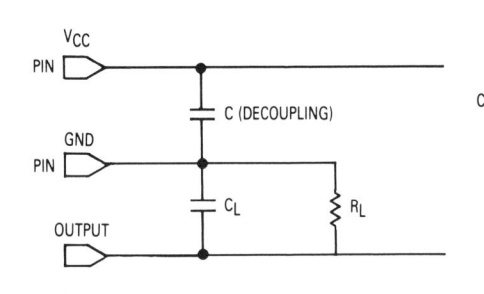

Figure 1a. V$_{ONP}$ with Respect to Ground

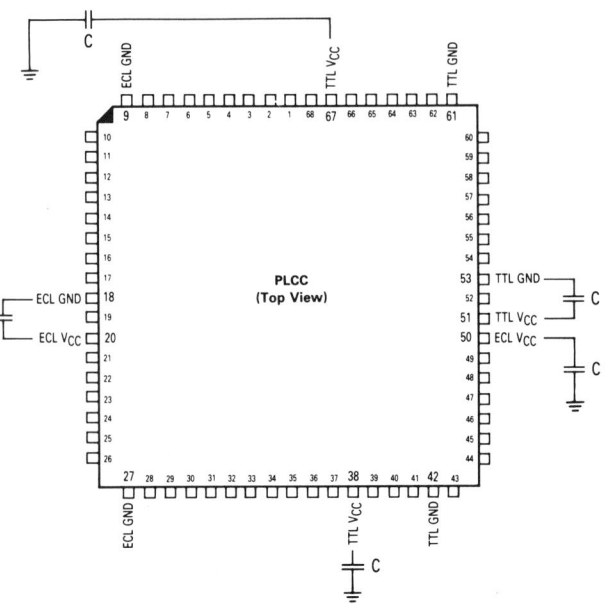

Figure 1b. Decoupling

The RAM Driver symmetrical output design offers significant improvement over a standard Schottky output by providing a balanced drive output impedance (≈25 Ω both HIGH and LOW), and by pulling up to MOS V$_{OH}$ levels. External resistors, not required with the RAM Driver, protect standard Schottky drivers from error causing undershoot but also slow the output rise by adding to the internal R.

The RAM Driver is optimized to drive LOW at maximum speed based on safe undershoot control and to drive HIGH with a symmetrical speed characteristic. This is an optimum approach because the dominant RAM loading characteristic is input capacitance.

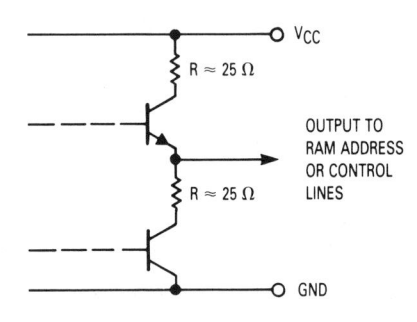

Typical Output Driver

APPLICATIONS

Timing Control

To obtain optimum performance and maximum design flexibility, the timing and control logic for the memory system has been kept a separate function. For systems implementing Error Detection and Correction, the F2969 Memory Timing Controller (MTC) provides all the necessary control signals for the F29368, F2961A/62A EDAC Bus Buffers, and the F2960A EDAC unit (See Figure 2a). Systems not using EDAC, can use the F2970 MTC to provide the control for the F29368 (See Figure 2b). Both the F2969 and F2970 Memory Timing Controllers use a delay line to provide the most accurate timing reference from which the control signals are derived.

Memory Expansion

With a 10-bit address path, the F29368 can control up to one megaword memory when using 1M dynamic RAMs. If a larger memory size is desired, the DMC's chip select ($\overline{CS}$) makes it easy to double the memory size by using two F29368s. Memory can be increased in four megaword increments by adding another DMC unit. A sixteen-megaword memory system implementing EDAC is shown in Figure 3.

549

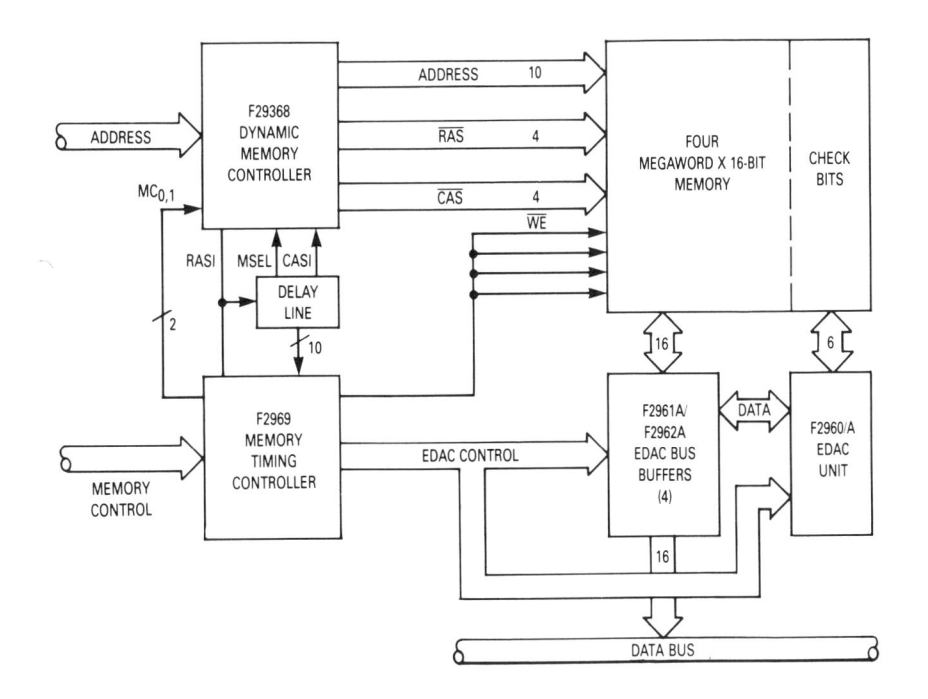

Figure 2a. Four Megaword Dynamic Memory with Error Detection and Correction

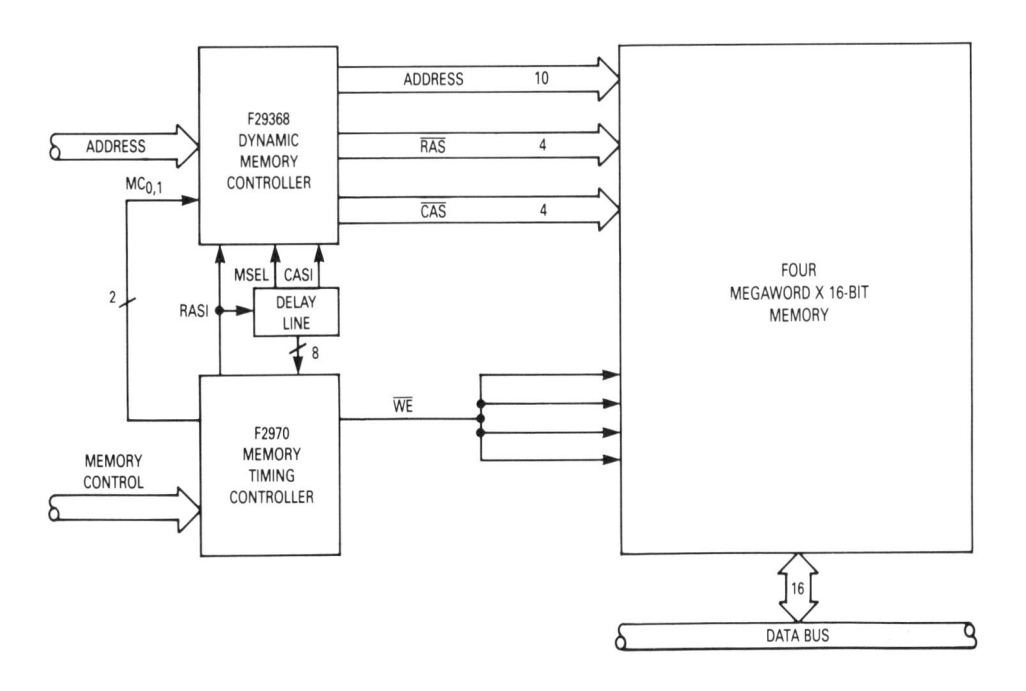

Figure 2b. Four Megaword Dynamic Memory

MOTOROLA
8

MC74F29368

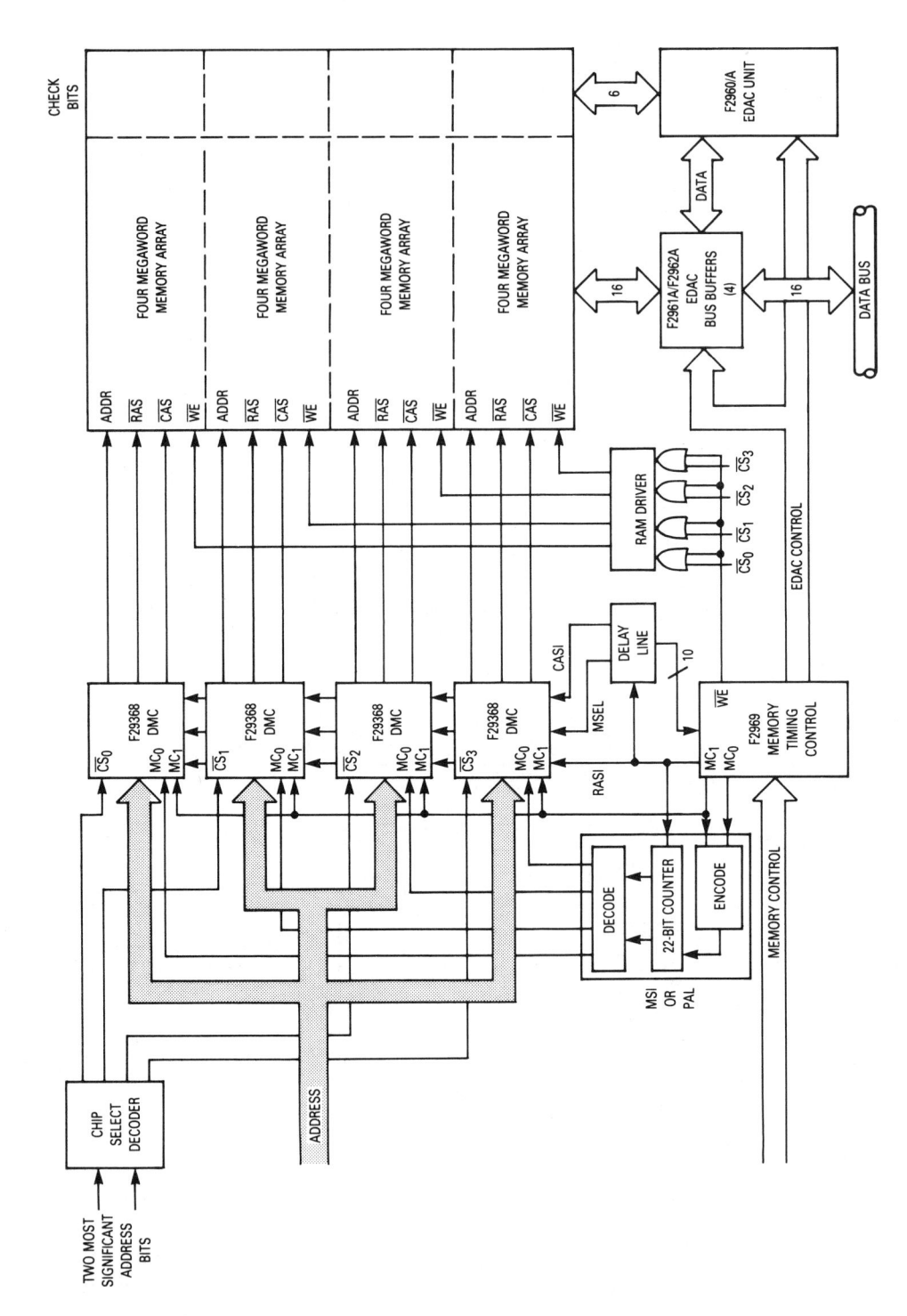

Figure 3. Sixteen Megaword Error Correcting Memory

ABSOLUTE MAXIMUM RATINGS

Storage Temperature	$-65°C$ to $+150°C$
Ambient Temperature with Power Applied	$-55°C$ to $+125°C$
Supply Voltage	-0.5 V to $+7$ V
DC Input Voltage	-0.5 V to $+5.5$ V
DC Input Current	-30 mA to $+5$ mA

GUARANTEED OPERATING RANGES

Commercial

T_A (Ambient)	0 to $+70°C$
V_{CC}	5 V to $\pm 10\%$
Min	4.5 V
Max	5.5 V

Stresses above those listed under MAXIMUM RATINGS may cause permanent device failure. Functionality at or above those limits is not implied. Exposure to absolute maximum ratings for extended periods may affect device reliability.

DC CHARACTERISTICS (over operating ranges unless otherwise specified)

Symbol	Parameter	Test Conditions (Note 1)		Min	Typ	Max	Units
V_{OH}	Output HIGH Voltage	V_{CC} = Min, V_{IN} = V_{IH} or V_{IL}, I_{OH} = -1 mA	COMM	2.7			Volts
			MIL	2.5			
V_{OL}	Output LOW Voltage	V_{CC} = Min, V_{IN} = V_{IH} or V_{IL}	I_{OL} = 1 mA			0.5	Volts
			I_{OL} = 12 mA			0.8	
V_{IH}	Input HIGH Level	Guaranteed input logical-HIGH voltage for all inputs		2			Volts
V_{IL}	Input LOW Level	Guaranteed input logical-LOW voltage for all inputs				0.8	Volts
V_I	Input Clamp Voltage	V_{CC} = Min, I_{IN} = -18 mA				-1.2	Volts
I_{IL}	Input LOW Current	V_{CC} = Max, V_{IN} = 0.4 V				-400	μA
I_{IH}	Input HIGH Current	V_{CC} = Max, V_{IN} = 2.4 V				20	μA
I_I	Input HIGH Current	V_{CC} = Max, V_{IN} = 5.5 V				100	μA
I_{OZH}	Off-State Current	V_O = 2.4 V				50	μA
I_{OZL}	Off-State Current	V_O = 0.4 V				-50	μA
I_{OL}	Output Sink Current	V_{OL} = 2 V		45			mA
I_{SC}	Output Short-Circuit Current	V_{CC} = Max (Note 1)		-60	-170	-275	mA
I_{CC}	Power Supply Current	V_{CC} = Max	25°C, 5 V		260		mA
			0°C to $+70°C$			325	
			$-55°C$ to 125°C*			370	

Note: 1. Not more than one output should be shorted at a time. Duration of the short-circuit test should not exceed one second.
*PGA only.

SWITCHING CHARACTERISTICS (over operating range unless otherwise specified)

Light Capacitive Loading (Small System)

No.	Parameter	Description	Test Conditions	Min	Max	Units
ADDRESS/RASI/CASI/LE LINES (Note 1)						
1	t_{PD}	A_n to Q_n	C_L = 50 pF	3	20	ns
2	t_{PD}	MSEL to Q_n	C_L = 50 pF	3	20	ns
3	t_{PD}	MC_n to Q_n	C_L = 50 pF	5	24	ns
4	t_{PD}	LE to Q_n	C_L = 50 pF	5	25	ns
5	t_{PD}	$\overline{CS}$ to Q_n	C_L = 50 pF		23	ns
6	t_S	A_n/SEL_n to LE	C_L = 50 pF	5		ns
7	t_H	A_n/SEL_n to LE	C_L = 50 pF	5		ns
8	t_H	MC_1 to RASI	C_L = 50 pF	5		ns
9	t_S	$\overline{CS}$ to RASI	C_L = 50 pF	6		ns
10	t_S	SEL_n to RASI	C_L = 50 pF	8		ns
11	t_{PWL}	RASI, CASI	C_L = 50 pF	20		ns
12	t_{PWH}	RASI, CASI	C_L = 50 pF	20		ns
$\overline{RAS_n}$/$\overline{CAS_n}$ LINES (Notes 1 and 3)						
13	t_{PD}	RASI to $\overline{RAS_n}$	C_L = 50 pF	3	18	ns
14	t_{PD}	CASI to $\overline{CAS_n}$	C_L = 50 pF	3	17	ns
15	t_{PD}	LE to $\overline{RAS_n}$	C_L = 50 pF		25	ns
16	t_{PD}	LE to $\overline{CAS_n}$	C_L = 50 pF		24	ns
17	t_{PD}	MC_n to $\overline{RAS_n}$	C_L = 50 pF	3	21	ns
18	t_{PD}	MC_n to $\overline{CAS_n}$	C_L = 50 pF	3	19	ns
19	t_{PD}	$\overline{CS}$ to $\overline{RAS_n}$	C_L = 50 pF		20	ns
20	t_{PD}	$\overline{CS}$ to $\overline{CAS_n}$	C_L = 50 pF		19	ns
21	t_{PD}	SEL_n to $\overline{RAS_n}$	C_L = 50 pF		20	ns
22	t_{PD}	SEL_n to $\overline{CAS_n}$	C_L = 50 pF		18	ns
23	t_{SKEW}	[t_{PD} (RASI to $\overline{RAS_n}$) − t_{PD} (A_n to Q_n)] (MC_n = 10)	C_L = 200 pF	−4	11	ns
			C_L = 350 pF	2	11	
24	t_{SKEW}	[t_{PD} (RASI to $\overline{RAS_n}$) − t_{PD} (MC_n to Q_n)] (MC_n = 00, 01)	C_L = 200 pF	−5	11	ns
			C_L = 350 pF	0	11	
25	t_{SKEW}	[t_{PD} (MSEL to $\overline{Q_n}$) − t_{PD} (RASI to $\overline{RAS_n}$)]	C_L = 200 pF	−15	0	ns
			C_L = 350 pF	−15	−6	
26	t_{SKEW}	[t_{PD} (CASI to $\overline{CAS_n}$) − (MSEL to Q_n)]	C_L = 200 pF	−3	11	ns
			C_L = 350 pF	2	11	
THREE-STATE OUTPUTS/UNDERSHOOT (Note 4)						
27	t_{PLZ}	Output Disable time from LOW, HIGH	C_L = 50 pF	S = 1		16	ns
28	t_{PHZ} (Note 2)			S = 2		16	
29	t_{PZL}	Output Enable Time from LOW, HIGH	C_L = 50 pF	S = 1		17	ns
30	t_{PZH} (Note 2)			S = 2		25	
31	V_{ONP}	Output Undershoot Voltage (Note 5)	C_L = 50 pF		−1.5	V

Notes:
1. Reference Figures 4 and 6 apply to all parameters except 7, 8, 9 and 10.
2. Times are not tested, but are guaranteed by characterization data.
3. C_L = 200 pF loading corresponds to 4 banks, 22 bits (16 data bits + 6 check bits).
   C_L = 350 pF loading corresponds to 4 banks, 39 bits (32 data bits + 7 check bits).
4. See Reference Figures 5 & 7.
5. V_{ONP} is not production tested but is guaranteed by characterization data. Limit specified is for all outputs switching simultaneously with minimum specified loading. As loading increases, V_{ONP} will approach zero.

MC74F29368

SWITCHING CHARACTERISTICS — continued (over operating range unless otherwise specified)

Heavy Capacitive Loading (Large Systems)

No.	Parameter	Description	Test Conditions	Min	Max	Units
ADDRESS/RASI/CASI/LE LINES (Note 1)							
1	t_{PD}	A_n to Q_n	C_L = 500 pF	12	40	ns
2	t_{PD}	MSEL to Q_n	C_L = 500 pF	12	42	ns
3	t_{PD}	MC_n to Q_n	C_L = 500 pF	12	44	ns
4	t_{PD}	LE to Q_n	C_L = 500 pF	12	46	ns
5	t_{PD}	$\overline{CS}$ to Q_n	C_L = 500 pF		45	ns
6	t_S	A_n/SEL_n to LE	C_L = 500 pF	5		ns
7	t_H	A_n/SEL_n to LE	C_L = 500 pF	5		ns
8	t_H	MC_1 to RASI	C_L = 500 pF	5		ns
9	t_S	$\overline{CS}$ to RASI	C_L = 500 pF	6		ns
10	t_S	SEL_n to RASI	C_L = 500 pF	8		ns
11	t_{PWL}	RASI, CASI	C_L = 500 pF	20		ns
12	t_{PWH}	RASI, CASI	C_L = 500 pF	20		ns
$\overline{RAS_n}/\overline{CAS_n}$ LINES (Notes 1 and 3)							
13	t_{PD}	RASI to $\overline{RAS_n}$	C_L = 200 pF	10	25	ns
			C_L = 350 pF	11	33	
14	t_{PD}	CASI to $\overline{CAS_n}$	C_L = 200 pF	10	24	ns
			C_L = 350 pF	11	31	
15	t_{PD}	LE to $\overline{RAS_n}$	C_L = 200 pF		31	ns
			C_L = 350 pF		38	
16	t_{PD}	LE to $\overline{CAS_n}$	C_L = 200 pF		30	ns
			C_L = 350 pF		37	
17	t_{PD}	MC_n to $\overline{RAS_n}$	C_L = 200 pF	10	27	ns
			C_L = 350 pF	11	34	
18	t_{PD}	MC_n to $\overline{CAS_n}$	C_L = 200 pF	10	29	ns
			C_L = 350 pF	11	37	
19	t_{PD}	$\overline{CS}$ to $\overline{RAS_n}$	C_L = 200 pF		25	ns
			C_L = 350 pF		32	
20	t_{PD}	$\overline{CS}$ to $\overline{CAS_n}$	C_L = 200 pF		24	ns
			C_L = 350 pF		31	
21	t_{PD}	SEL_n to $\overline{RAS_n}$	C_L = 200 pF		26	ns
			C_L = 350 pF		34	
22	t_{PD}	SEL_n to $\overline{CAS_n}$	C_L = 200 pF		25	ns
			C_L = 350 pF		33	
23	t_{SKEW}	[t_{PD} (RASI to $\overline{RAS_n}$) − t_{PD} (A_n to Q_n)] (MC_n = 10)	C_L = 200 pF	−20	−3	ns
			C_L = 350 pF	−15	−3	
24	t_{SKEW}	[t_{PD} (RASI to $\overline{RAS_n}$) − t_{PD} (MC_n to Q_n)] (MC_n = 00, 01)	C_L = 200 pF	−20	−1	ns
			C_L = 350 pF	−16	−1	
25	t_{SKEW}	[t_{PD} (MSEL to $\overline{Q_n}$) − t_{PD} (RASI to $\overline{RAS_n}$)]	C_L = 200 pF	−6	18	ns
			C_L = 350 pF	−6	11	
26	t_{SKEW}	[t_{PD} (CASI to $\overline{CAS_n}$) − (MSEL to Q_n)]	C_L = 200 pF	−19	−5	ns
			C_L = 350 pF	−13	−5	

Notes: 1. Reference Figures 4 and 6 apply to all parameters except 7, 8, 9 and 10.
2. Times are not tested, but are guaranteed by characterization data.
3. C_L = 200 pF loading corresponds to 4 banks, 22 bits (16 data bits + 6 check bits).
   C_L = 350 pF loading corresponds to 4 banks, 39 bits (32 data bits + 7 check bits).
4. See Reference Figures 5 & 7.
5. V_{ONP} is not production tested but is guaranteed by characterization data. Limit specified is for all outputs switching simultaneously with minimum specified loading. As loading increases, V_{ONP} will approach zero.

SWITCHING TEST CIRCUITS

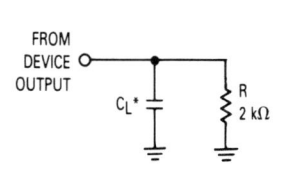

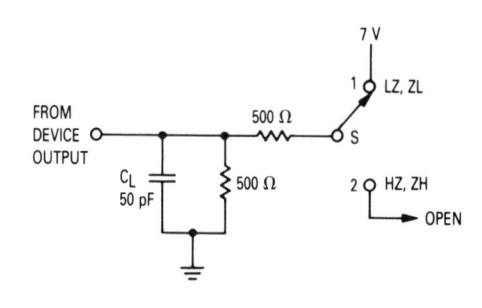

Figure 4. Capacitive Load Switching

*t_{pd} specified at C_L = 50, 200, 350, and 500pF.

Figure 5. Three-State Enable/Disable

SWITCHING TEST WAVEFORMS

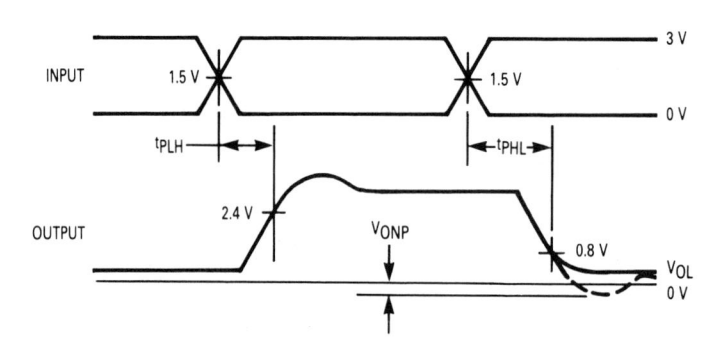

Figure 6. Output Drivers Levels

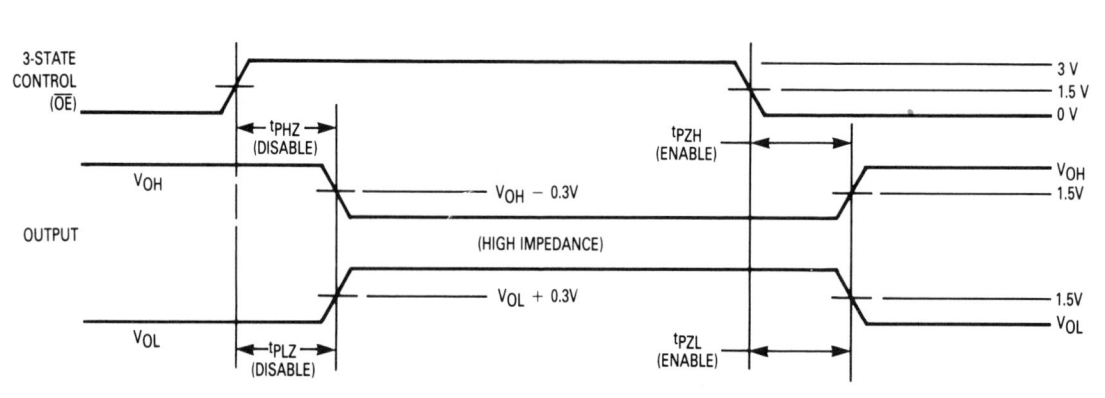

Note: Decoupling is needed for all AC tests

Figure 7. Three-State Control Levels

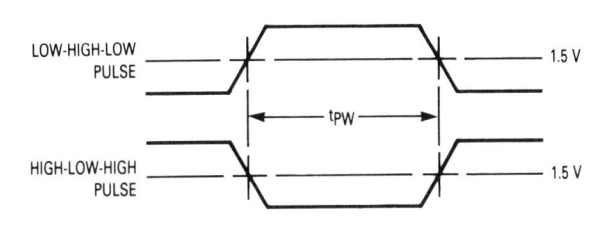

Pulse Width

GENERAL TEST NOTES

Incoming test procedures on this device should be carefully planned, taking into account the complexity and power levels of the part. The following notes may be useful.

1. Insure the part is adequately decoupled at the test head. Large changes in V_{CC} current as the device switches may cause erroneous function failures due to V_{CC} changes.
2. Do not leave inputs floating during any tests, as they may start to oscillate at high frequency.
3. Do not attempt to perform threshold tests at high speed. Following an input transition, ground current may change by as much as 400 mA in 5–8 ns. Inductance in the ground cable may allow the ground pin at the device to rise by 100's of millivolts momentarily.
4. Use extreme care in defining input levels for AC tests. Many inputs may be changed at once, so there will be significant noise at the device pins and they may not actually reach V_{IL} or V_{IH} until the noise has settled. Recommended levels are $V_{IL} \leq 0$ V and $V_{IH} \geq 4$ V for AC tests.
5. Automatic tester hardware and handler add additional round trip A.C. delay to test measurements. Actual propagation delay testing may incorporate a correlation factor to negate the additional delay.

KEY TO SWITCHING WAVEFORMS

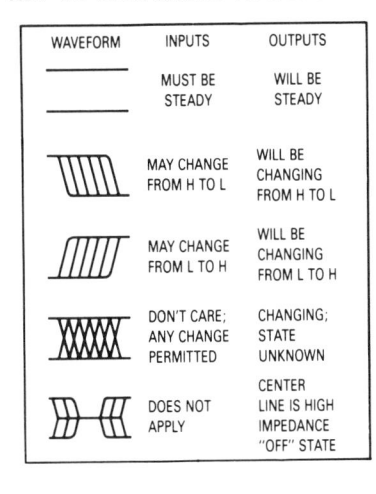

WAVEFORM	INPUTS	OUTPUTS
———	MUST BE STEADY	WILL BE STEADY
(rising hatch)	MAY CHANGE FROM H TO L	WILL BE CHANGING FROM H TO L
(falling hatch)	MAY CHANGE FROM L TO H	WILL BE CHANGING FROM L TO H
(cross-hatch)	DON'T CARE; ANY CHANGE PERMITTED	CHANGING; STATE UNKNOWN
(center line)	DOES NOT APPLY	CENTER LINE IS HIGH IMPEDANCE "OFF" STATE

SWITCHING TEST WAVEFORMS

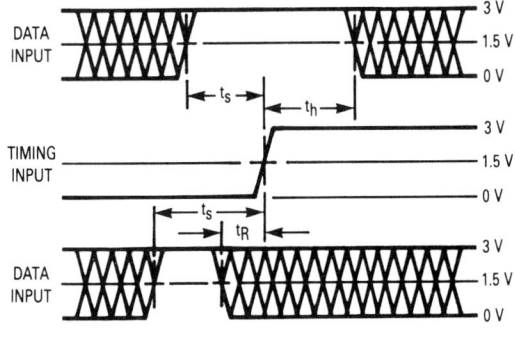

Setup, Hold, and Release Times

Notes: 1. Diagram shown for HIGH data only. Output transition may be opposite sense.
2. Cross-hatched are "don't care" condition.

SWITCHING WAVEFORMS

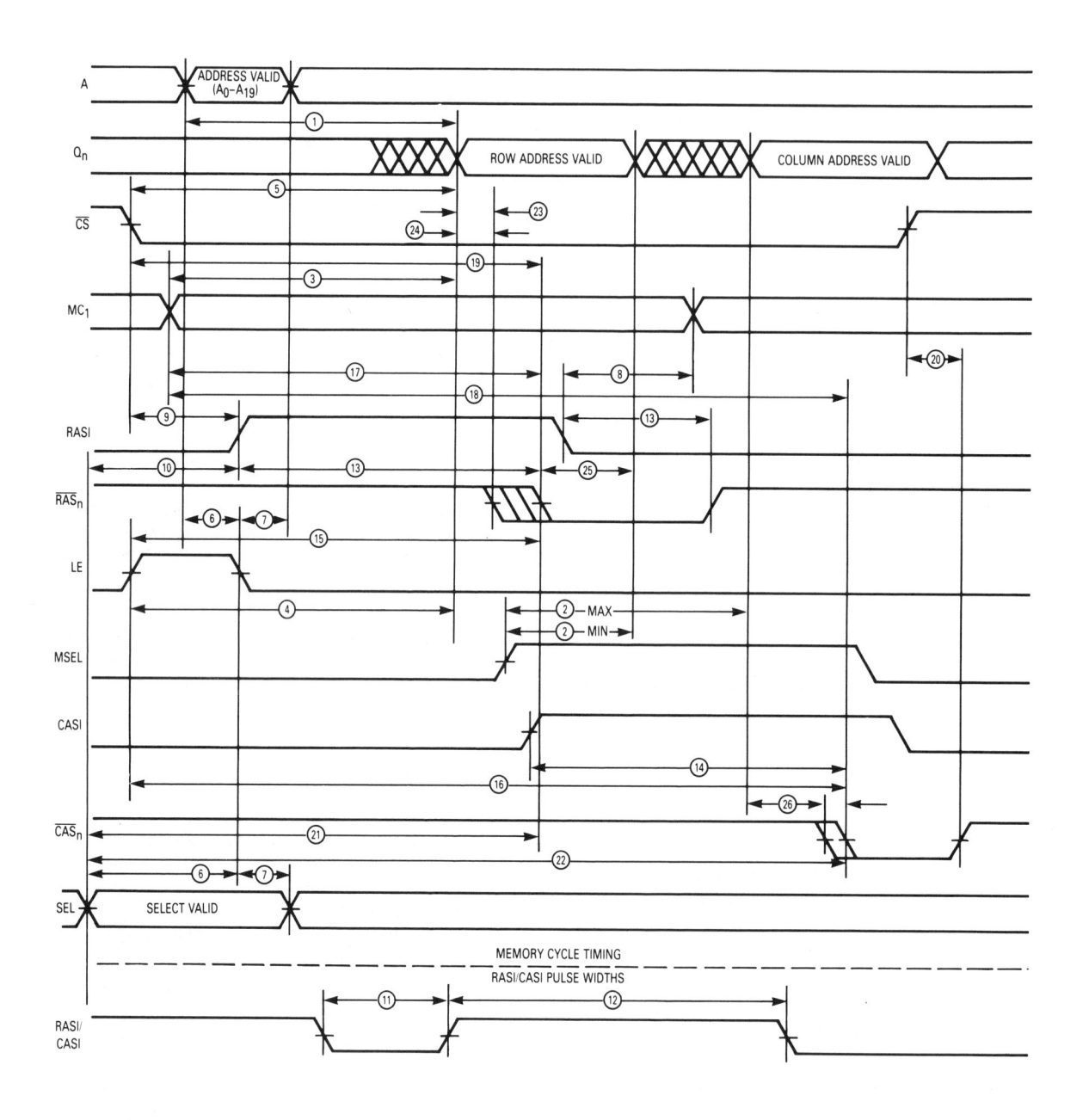

F29368 Dynamic Memory Controller Timing

MEMORY CYCLE TIMING

The relationship between DMC specifications and system timing requirements are shown in Figure 8. T_1, T_2 and T_3 represent the minimum timing requirements at the DMC inputs to guarantee that RAM timing requirements are met and that maximum system performance is achieved.

The minimum requirement for T_1, T_2 and T_3 are as follows:

T_1 Min = t_{ASR} + t_{23}
T_2 Min = t_{RAH} + t_{25}
T_3 Min = T_2 + t_{26} + t_{ASC}

See RAM data sheet for applicable values for t_{RAH}, t_{ASC} and t_{ASR}.

(23) GUARANTEED MAX DIFFERENCE BETWEEN FASTEST RASI TO $\overline{RAS}_n$ DELAY AND THE SLOWEST A_n TO Q_n DELAY ON ANY SINGLE DEVICE

(25) GUARANTEED MAX DIFFERENCE BETWEEN FASTEST MSEL TO Q_n DELAY AND THE SLOWEST RASI TO RAS_n DELAY ON ANY SINGLE DEVICE

(26) GUARANTEED MAX DIFFERENCE BETWEEN FASTEST CASI TO $\overline{CAS}_n$ DELAY AND THE SLOWEST MSEL TO Q_n DELAY ON ANY SINGLE DEVICE

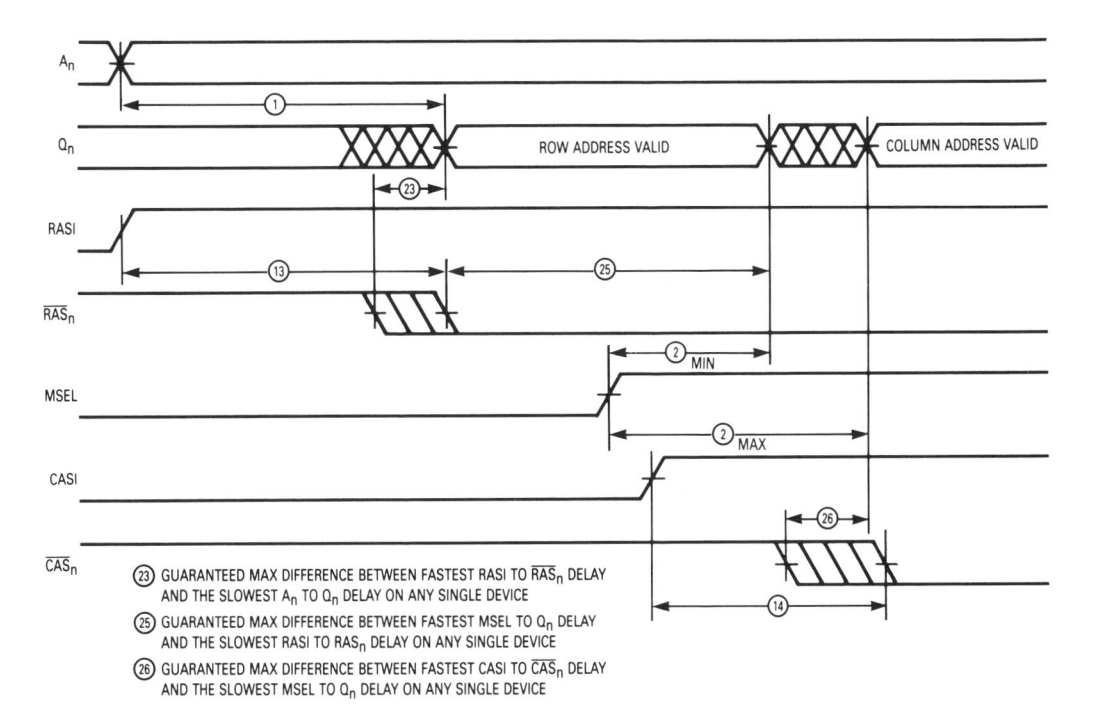

Figure 8. Memory Cycle Timing
a. Specifications Applicable to Memory Cycle Timing (MC_n = 1, 0)

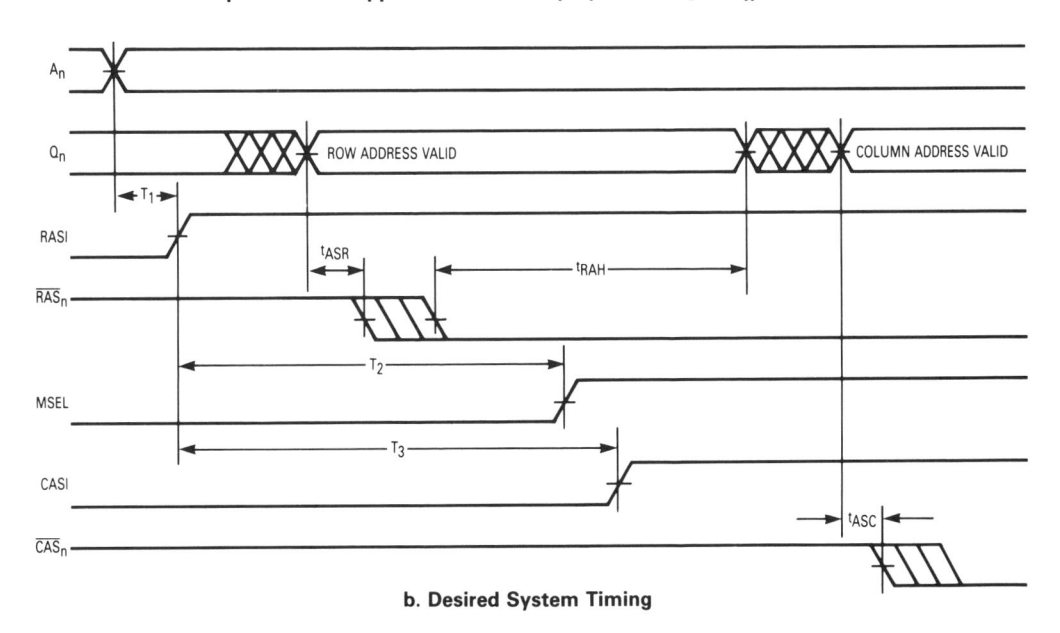

b. Desired System Timing

MOTOROLA
16

MC74F29368

REFRESH CYCLE TIMING

The timing relationships for refresh are shown in Figure 9.
T_4 minimum is calculated as follows:

$T_4 \text{ Min} = t_{ASR} + t_{24}$

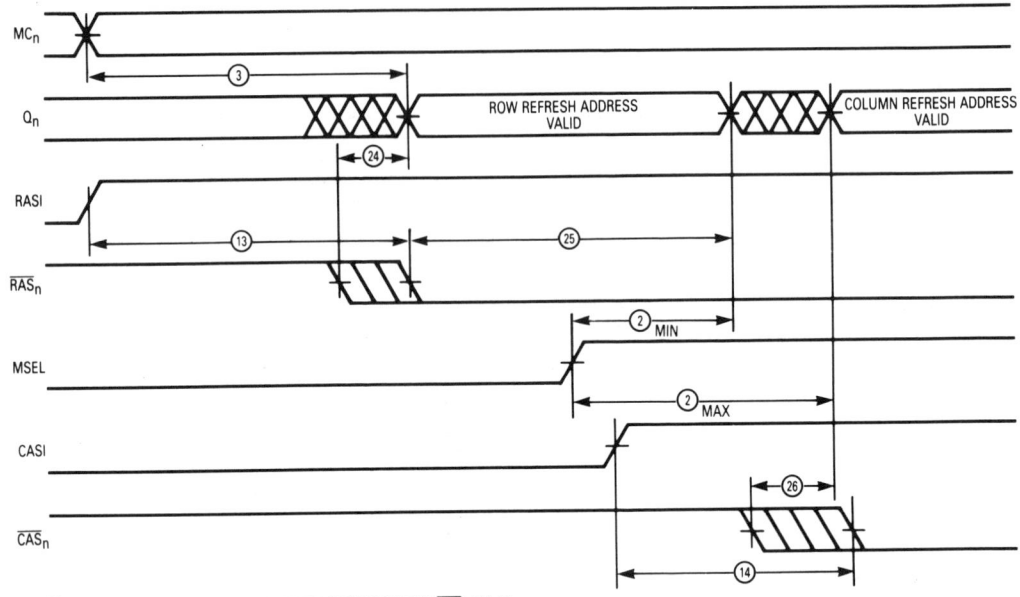

㉔ GUARANTEED MAX DIFFERENCE BETWEEN FASTEST RASI TO $\overline{RAS}_n$ DELAY AND THE SLOWEST MC_n TO Q_n DELAY ON ANY SINGLE DEVICE

㉕ GUARANTEED MAX DIFFERENCE BETWEEN FASTEST MSEL TO Q_n DELAY AND THE SLOWEST RASI TO $\overline{RAS}_n$ DELAY ON ANY SINGLE DEVICE

㉖ GUARANTEED MAX DIFFERENCE BETWEEN FASTEST CASI TO $\overline{CAS}_n$ DELAY AND THE SLOWEST MSEL TO Q_n DELAY ON ANY SINGLE DEVICE

Figure 9. Refresh Cycle Timing
a. Specifications Applicable to Refresh Cycle Timing (MC_n = 00, 01)

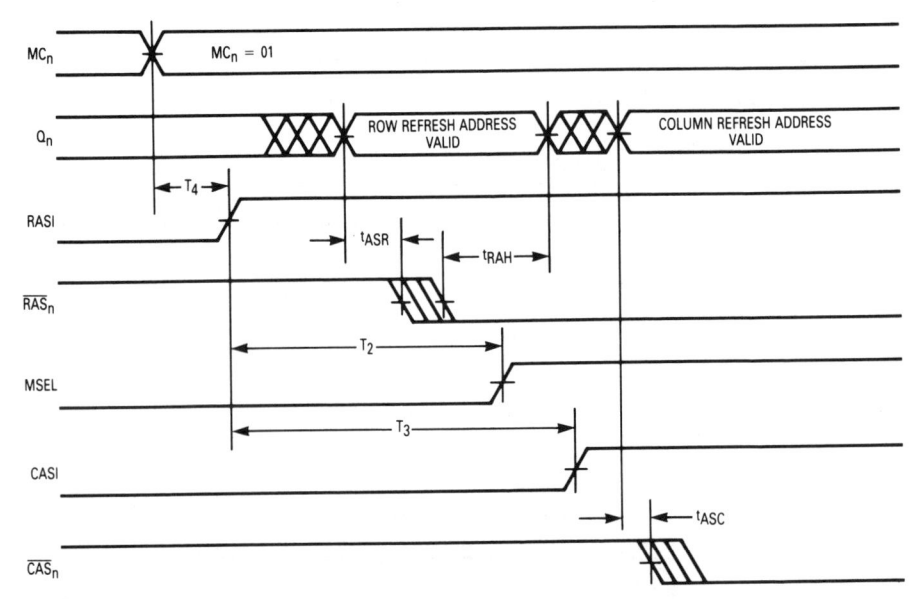

b. Desired Timing: Refresh with Scrubbing

REFRESH CYCLE TIMING

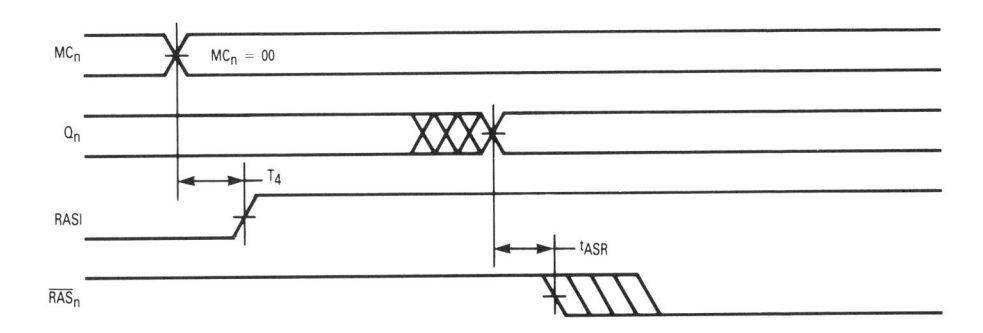

Figure 9. Refresh Cycle Timing (Cont'd.)
c. Desired Timing: Refresh without Scrubbing

TYPICAL PERFORMANCE CURVES

NANOSECONDS VERSUS PICOFARADS

The Switching Characteristics Tables specify the minimum and maximum propagation delays for effective capacitive loads of 50 pF and 500 pF on Address Lines and 50 pF, 200 pF, and 350 pF on $\overline{RAS_n}$ and $\overline{CAS_n}$ Lines. The upper limits represent the maximum calculated load for the following conditions:

Address Lines: 500 pF = 16 data bits + 6 check bits, four banks

500 pF = 32 data bits + 7 check bits, two banks

$\overline{RAS_n}/\overline{CAS_n}$: 200 pF = 16 data bits + 6 check bits, four banks

350 pF = 32 data bits + 7 check bits, four banks

A more comprehensive analysis of loading is given in the table below.

16-Bit Systems (plus 6 Check Bits for Error Detection and Correction)

DMC Output	No. of DRAMs	Line Drive	Total Required Drive for Indicated Banks 1	2	4	Specified Data Sheet Load
Q_n (Address)		5 pF	100 pF	220 pF	440 pF	500 pF
$\overline{RAS_n}$	16 + 6 = 22	8 pF	176 pF	176 pF	176 pF	200 pF
$\overline{CAS_n}$		8 pF	176 pF	176 pF	176 pF	200 pF

32-Bit Systems (plus 7 Check Bits for Error Detection and Correction)

DMC Output	No. of DRAMs	Line Drive	Total Required Drive for Indicated Banks 1	2	4	Specified Data Sheet Load
Q_n (Address)		5 pF	195 pF	390 pF	780 pF	500 pF
$\overline{RAS_n}$	32 + 7 = 39	8 pF	312 pF	312 pF	312 pF	350 pF
$\overline{CAS_n}$		8 pF	312 pF	312 pF	312 pF	350 pF

INPUT/OUTPUT CIRCUIT DIAGRAM

Driven Input

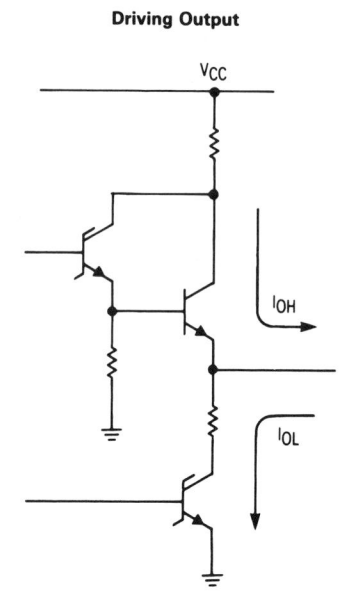

Driving Output

Motorola reserves the right to make changes without further notice to any products herein to improve reliability, function or design. Motorola does not assume any liability arising out of the application or use of any product or circuit described herein; neither does it convey any license under its patent rights nor the rights of others. Motorola products are not authorized for use as components in life support devices or systems intended for surgical implant into the body or intended to support or sustain life. Buyer agrees to notify Motorola of any such intended end use whereupon Motorola shall determine availability and suitability of its product or products for the use intended. Motorola and (M) are registered trademarks of Motorola, Inc. Motorola, Inc. is an Equal Employment Opportunity/Affirmative Action Employer.

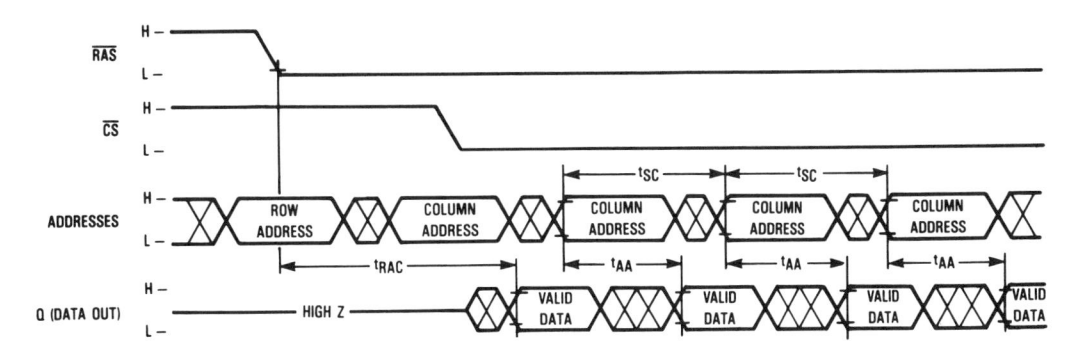

Figure 3a. Static Column Mode Read Cycle

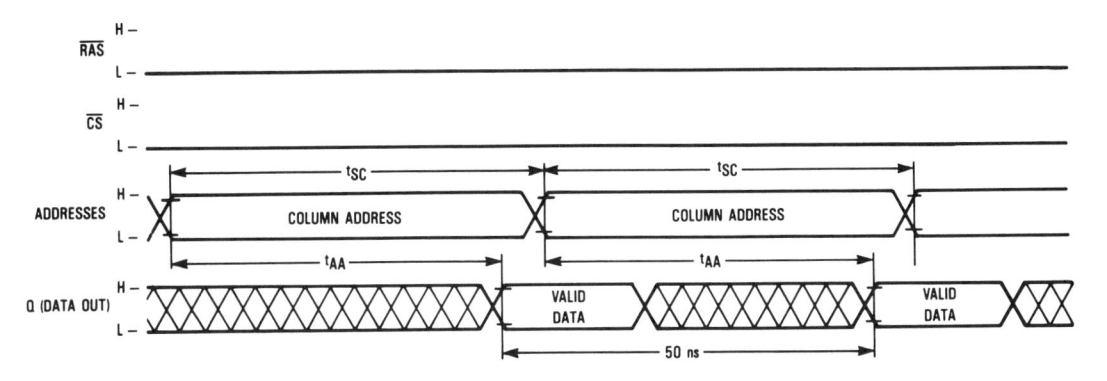

Figure 3b. Static Column Mode Cycle Minimum Timing

Motorola reserves the right to make changes without further notice to any products herein to improve reliability, function or design. Motorola does not assume any liability arising out of the application or use of any product or circuit described herein; neither does it convey any license under its patent rights nor the rights of others. Motorola and Ⓜ are registered trademarks of Motorola, Inc.

MOTOROLA LTD. *Semiconductor Products Group*
COLVILLES ROAD, KELVIN ESTATE, EAST KILBRIDE, GLASGOW, G75 0TG, SCOTLAND

National Semiconductor

DP8420A/21A/22A microCMOS Programmable 256k/1M/4M Dynamic RAM Controller/Drivers

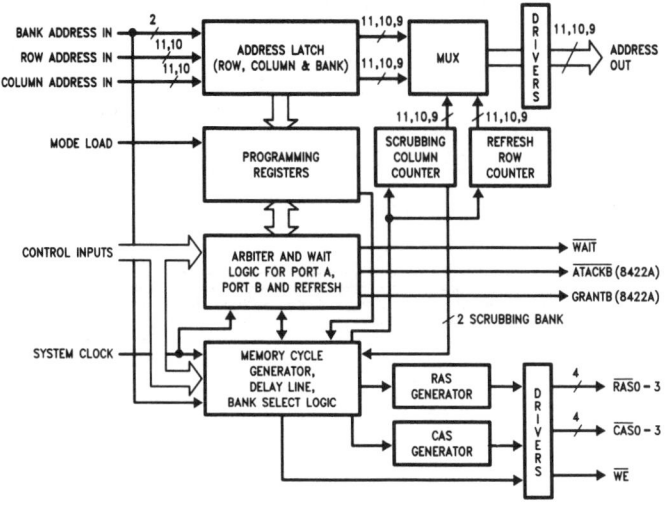

General Description

The DP8420A/21A/22A dynamic RAM controllers provide a low cost, single chip interface between dynamic RAM and all 8-, 16- and 32-bit systems. The DP8420A/21A/22A generate all the required access control signal timing for DRAMs. An on-chip refresh request clock is used to automatically refresh the DRAM array. Refreshes and accesses are arbitrated on chip. If necessary, a $\overline{\text{WAIT}}$ or $\overline{\text{DTACK}}$ output inserts wait states into system access cycles, including burst mode accesses. $\overline{\text{RAS}}$ low time during refreshes and $\overline{\text{RAS}}$ precharge time after refreshes and back to back accesses are guaranteed through the insertion of wait states. Separate on-chip precharge counters for each $\overline{\text{RAS}}$ output can be used for memory interleaving to avoid delayed back to back accesses because of precharge. An additional feature of the DP8422A is two access ports to simplify dual accessing. Arbitration among these ports and refresh is done on chip.

Features

- On chip high precision delay line to guarantee critical DRAM access timing parameters
- microCMOS process for low power
- High capacitance drivers for $\overline{\text{RAS}}$, $\overline{\text{CAS}}$, $\overline{\text{WE}}$ and DRAM address on chip
- On chip support for nibble, page and static column DRAMs
- Byte enable signals on chip allow byte writing in a word size up to 32 bits with no external logic
- Selection of controller speeds: 20 MHz and 25 MHz
- On board Port A/Port B (DP8422A only)/refresh arbitration logic
- Direct interface to all major microprocessors (application notes available)
- 4 $\overline{\text{RAS}}$ and 4 $\overline{\text{CAS}}$ drivers (the $\overline{\text{RAS}}$ and $\overline{\text{CAS}}$ configuration is programmable)

Control	# of Pins (PLCC)	# of Address Outputs	Largest DRAM Possible	Direct Drive Memory Capacity	Access Ports Available
DP8420A	68	9	256 kbit	4 Mbytes	Single Access Port
DP8421A	68	10	1 Mbit	16 Mbytes	Single Access Port
DP8422A	84	11	4 Mbit	64 Mbytes	Dual Access Ports (A and B)

Block Diagram

FIGURE 1

Table of Contents

1.0 INTRODUCTION

2.0 SIGNAL DESCRIPTIONS
2.1 Address, R/W and Programming Signals
2.2 DRAM Control Signals
2.3 Refresh Signals
2.4 Port A Access Signals
2.5 Port B Access Signals (DP8422A)
2.6 Common Dual Port Signals (DP8422A)
2.7 Power Signals and Capacitor Input
2.8 Clock Inputs

3.0 PORT A ACCESS MODES
3.1 Access Mode 0
3.2 Access Mode 1
3.3 Read-Modify-Write Access Cycles

4.0 REFRESH OPTIONS
4.1 Refresh Control Modes
    4.1.1 Automatic Internal Refresh
    4.1.2 Externally Controlled/Burst Refresh
    4.1.3 Refresh Request/Acknowledge
4.2 Refresh Cycle Types
    4.2.1 Conventional Refresh
    4.2.2 Staggered Refresh™
    4.2.3 Error Scrubbing Refresh
4.3 Extending Refresh
4.4 Clearing the Refresh Address Counter
4.5 Clearing the Refresh Request Clock

5.0 PORT A WAIT STATE SUPPORT
5.1 $\overline{WAIT}$ Type Output
    5.1.1 $\overline{WAIT}$ During Single Accesses
    5.1.2 $\overline{WAIT}$ During Page/Burst Accesses
5.2 $\overline{DTACK}$ Type Output
    5.2.1 $\overline{DTACK}$ During Single Accesses
    5.2.2 $\overline{DTACK}$ During Page/Burst Accesses
5.3 Dynamically Increasing the Number of Wait States
5.4 Guaranteeing $\overline{RAS}$ Low Time and $\overline{RAS}$ Precharge Time

6.0 ADDITIONAL ACCESS SUPPORT FEATURES
6.1 Address Latches and Column Increment
6.2 Address Pipelining
6.3 Delay $\overline{CAS}$ During Write Accesses

7.0 $\overline{RAS}$ AND $\overline{CAS}$ CONFIGURATION MODES
7.1 Byte Writing
7.2 Memory Interleaving
7.3 Address Pipelining
7.4 Error Scrubbing
7.5 Page/Burst Mode

8.0 PROGRAMMING AND RESETTING
8.1 External Reset
8.2 Programming Methods
    8.2.1 Mode Load Only Programming
    8.2.2 Chip Selected Access Programming
8.3 Definition of Programming Bits

9.0 TEST MODE

10.0 DRAM CRITICAL TIMING OPTIONS
10.1 Programming Values of t_{RAH} and t_{ASC}
10.2 Calculation of t_{RAH} and t_{ASC}

11.0 DUAL ACCESSING (DP8422A)
11.1 Port B Access Mode
11.2 Port B Wait State Support
11.3 Common Port A and Port B Dual Port Functions
    11.3.1 GRANTB Output
    11.3.2 $\overline{LOCK}$ Input

12.0 ABSOLUTE MAXIMUM RATINGS

13.0 DC ELECTRICAL CHARACTERISTICS

14.0 AC TIMING PARAMETERS

15.0 FUNCTIONAL DIFFERENCES BETWEEN THE DP8420A/21A/22A AND THE DP8420/21/22

16.0 DP8420A/21A/22A USER HINTS

1.0 Introduction

The DP8420A/21A/22A are CMOS Dynamic RAM controllers that incorporate many advanced features including the capabilities of address latches, refresh counter, refresh clock, row, column and refresh address multiplexor, delay line, refresh/access arbitration logic and high capacitive drivers. The programmable system interface allows any manufacturer's microprocessor or bus to directly interface via the DP8420A/21A/22A to DRAM arrays up to 64 Mbytes in size.

After power up, the DP8420A/21A/22A must first be programmed before accessing the DRAM. The chip is programmed through the address bus.

There are two methods of programming the chip. The first method, mode load only, is accomplished by asserting the signal mode load, $\overline{ML}$. A valid programming selection is presented on the row, column, bank and $\overline{ECAS}$ inputs, then $\overline{ML}$ is negated. When $\overline{ML}$ is negated, the chip is programmed with the valid programming bits on the address bus.

The second method, chip selected access, is accomplished by asserting $\overline{ML}$ and performing a chip selected access. When $\overline{CS}$ and $\overline{AREQ}$ are asserted for the access, the chip is programmed. During this programming access, the programming bits affecting the wait logic become effective immediately, allowing the access to terminate. After the access, $\overline{ML}$ is negated and the rest of the programming bits take effect.

Once the DP8420A/21A/22A has been programmed, a 60 ms initialization period is entered. During this time, the DP8420A/21A/22A controllers perform refreshes to the DRAM array so further DRAM warm up cycles are unnecessary.

The DP8420A/21A/22A can now be used to access the DRAM. There are two modes of accessing with the controller. The two modes are Mode 0, which initiates $\overline{RAS}$ synchronously, and Mode 1, which initiates $\overline{RAS}$ asynchronously.

To access the DRAM using Mode 0, the signal ALE is asserted along with $\overline{CS}$ to ensure a valid VRAM access. ALE asserting sets an internal latch and only needs to be pulsed and not held throughout the entire access. On the next rising clock edge, $\overline{RAS}$ will be asserted for that access. The DP8420A/21A/22A will place the row address on the DRAM address bus, guarantee the programmed value of row address hold time of the DRAM, place the column address on the DRAM address bus, guarantee the programmed value of column address setup time and assert $\overline{CAS}$. $\overline{AREQ}$ can be asserted anytime after the clock edge which starts the access $\overline{RAS}$. $\overline{RAS}$ and $\overline{CAS}$ will extend until $\overline{AREQ}$ is negated.

The other access mode, Mode 1, is asynchronous to the clock. When $\overline{ADS}$ is asserted, $\overline{RAS}$ is asserted. The DP8420A/21A/22A will place the row address on the DRAM address bus, guarantee the programmed value of row address hold time, place the column address on the DRAM address bus, guarantee the programmed value of column address setup time and assert $\overline{CAS}$. $\overline{AREQ}$ can be tied to $\overline{ADS}$ or can be asserted after $\overline{ADS}$ is asserted. $\overline{AREQ}$ negated will terminate the access.

The DP8420A/21A/22A have greatly expanded refresh capabilities compared to other DRAM controllers. There are three modes of refreshing available. These modes are internal automatic refreshing, externally controlled/burst refreshing, and refresh request/acknowledge refreshing. Any of these modes can be used together or separately to achieve the desired results.

When using internal automatic refreshing, the DP8420A/21A/22A will generate an internal refresh request from the refresh request clock. The DP8420A/21A/22A will arbitrate between the refresh requests and accesses. Assuming an access is not currently in progress, the DP8420A/21A/22A will assert the signal $\overline{RFIP}$. On the next positive clock edge, refreshing will begin. If an access had been in progress, the refresh will begin after the access has terminated.

To use externally controlled/burst refresh, the user disables the internal refresh request by asserting the input $\overline{DISRFRSH}$. A refresh can now be externally requested by asserting the input $\overline{RFSH}$. The DP8420A/21A/22A will arbitrate between the external refresh request and accesses. Assuming an access is not currently in progress, the DP8420A/21A/22A will assert the output $\overline{RFIP}$. On the next positive clock edge, refreshing will begin. If an access had been in progress, the refresh would take place after the access has terminated.

With refresh request/acknowledge mode, the DP8420A/21A/22A broadcasts the internal refresh request to the system through the $\overline{RFRQ}$ output pin. External circuitry can determine when to refresh the DRAM through the $\overline{RFSH}$ input.

The controllers have three types of refreshing available: conventional, staggered and error scrubbing. Any refresh control mode can be used with any type of refresh. In a conventional refresh, all of the $\overline{RAS}$ outputs will be asserted and negated at once. In a staggered refresh, the $\overline{RAS}$ outputs will be asserted one positive clock edge apart. Error scrubbing is the same as conventional refresh except that a $\overline{CAS}$ will be asserted during a refresh allowing the system to run that data through an EDAC chip and write it back to memory, if a single bit error has occurred. The refreshes can be extended with the EXTEND REFRESH input, EXTNDRF.

The DP8420A/21A/22A have wait support available as $\overline{DTACK}$ or $\overline{WAIT}$. Both are programmable. $\overline{DTACK}$, Data Transfer ACKnowledge, is useful for processors whose wait signal is active high. $\overline{WAIT}$ is useful for processors whose wait signal is active low. The user can choose either at programming. These signals are used by the on-chip arbitor to insert wait states to guarantee the arbitration between accesses and refreshes or precharge. Both signals are independent of the access mode chosen.

$\overline{DTACK}$ will assert a programmed number of clock edges from the event that starts the access $\overline{RAS}$. $\overline{DTACK}$ will be negated, when the access is terminated, by $\overline{AREQ}$ being negated. $\overline{DTACK}$ can also be programmed to toggle with the $\overline{ECAS}$ inputs during burst/page mode accesses.

$\overline{WAIT}$ is asserted during the start of the access (ALE and $\overline{CS}$, or $\overline{ADS}$ and $\overline{CS}$) and will negate a number of clock edges from the event that starts the access $\overline{RAS}$. After $\overline{WAIT}$ is negated, it will stay negated until the next access. $\overline{WAIT}$ can also be programmed to toggle with $\overline{ECAS}$ inputs during a burst/page mode access.

Both signals can be dynamically delayed further through the $\overline{WAITIN}$ signal to the DP8420A/21A/22A.

The DP8420A/21A/22A have address latches, used to latch the bank, row and column address inputs. Once the address is latched, a column increment feature can be used to increment the column address. The address latches can also be programmed to be fall through.

1.0 Introduction (Continued)

The $\overline{RAS}$ and $\overline{CAS}$ drivers can be configured to drive a one, two or four bank memory array up to 32 bits in width. The $\overline{ECAS}$ signals can then be used to select one of four $\overline{CAS}$ drivers for byte writing with no external logic.

When configuring the DP8420A/21A/22A for more than one bank, memory interleaving can be used. By tying the low order address bits to the bank select lines, B0 and B1, sequential back to back accesses will not be delayed since the DP8420A/21A/22A have separate precharge counters per bank. The DP8420A/21A/22A are capable of performing address pipelining. In address pipelining, the DP8420A/21A/22A guarantee the column address hold time and switch the internal multiplexor to place the row address on the address bus. At this time, another memory access to another bank can be initiated.

The DP8422A has all the features previously mentioned. Unlike the DP8420A/21A, the DP8422A has a second port to allow a second CPU to access the memory array. This port, Port B, has two control signals to allow a CPU to access the DRAM array. These signals are access request for Port B, $\overline{AREQB}$, and Advanced Transfer ACKnowledge for Port B, $\overline{ATACKB}$. Two other signals are used by both Port A and Port B for dual accessing purposes. The signals are lock, $\overline{LOCK}$ and grant Port B, GRANTB. All arbitration for the two ports and refresh is done on-chip by the DP8422A through the insertion of wait states. Since the DP8422A has only one input address bus, the address lines have to be multiplexed externally. The signal GRANTB can be used for this purpose since it is asserted when Port B has access to the DRAM array and negated when Port A has access to the DRAM array. Once a port has access to the array, the other port can be "locked out" by asserting the input $\overline{LOCK}$. $\overline{AREQB}$, when asserted, is used by Port B to request an access. $\overline{ATACKB}$, when asserted, signifies that access $\overline{RAS}$ has been asserted for the requested Port B access. By using $\overline{ATACKB}$, the user can generate an appropriate $\overline{WAIT}$ or $\overline{DTACK}$ like signal for the Port B CPU.

The following explains the terminology used in this data sheet. The terms negated and asserted are used. Asserted refers to a "true" signal. Thus, "$\overline{ECAS0}$ asserted" means the $\overline{ECAS0}$ input is at a logic 0. The term "COLINC asserted" means the COLINC input is at a logic 1. The term negated refers to a "false" signal. Thus, "$\overline{ECAS0}$ negated" means the $\overline{ECAS0}$ input is at a logic 1. The term "COLINC negated" means the input COLINC is at a logic 0. The table shown below clarifies this terminology.

Signal	Action	Logic Level
Active High	Asserted	High
Active High	Negated	Low
Active Low	Asserted	Low
Active Low	Negated	High

Connection Diagrams

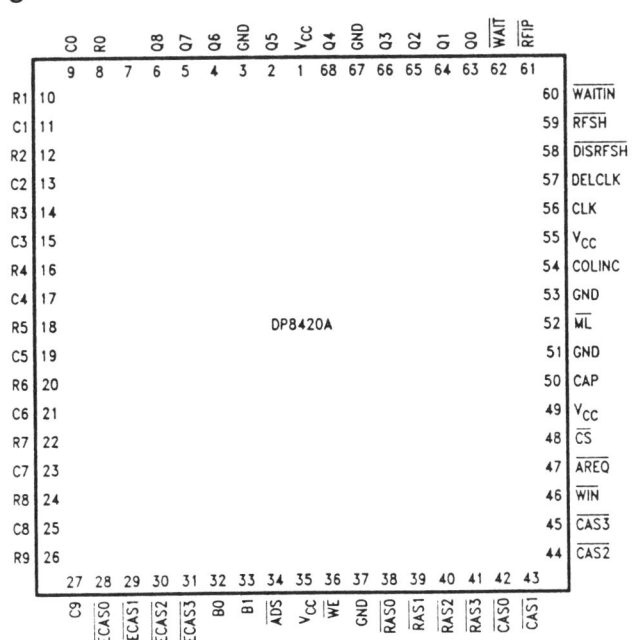

Top View

FIGURE 2

Order Number DP8420AV-20 or DP8420AV-25
See NS Package Number V68A

Connection Diagrams (Continued)

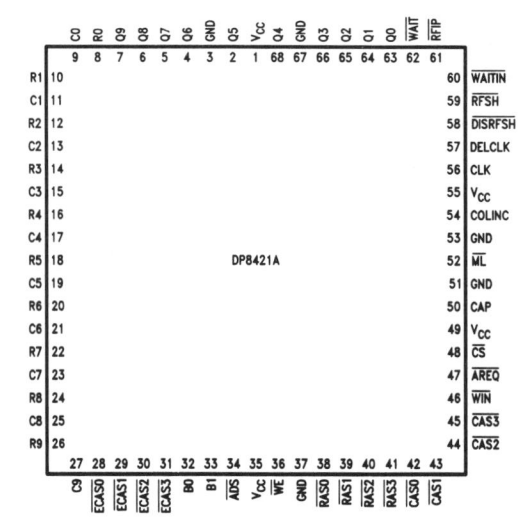

FIGURE 3 — DP8421A (Top View)

Pins (top, left to right): C0, R0, Q9, Q8, Q7, Q6, GND, Q5, V_CC, Q4, GND, Q3, Q2, Q1, Q0, WAIT, RFIP (pins 9, 8, 7, 6, 5, 4, 3, 2, 1, 68, 67, 66, 65, 64, 63, 62, 61)

Left side (top to bottom):
- R1 — 10
- C1 — 11
- R2 — 12
- C2 — 13
- R3 — 14
- C3 — 15
- R4 — 16
- C4 — 17
- R5 — 18
- C5 — 19
- R6 — 20
- C6 — 21
- R7 — 22
- C7 — 23
- R8 — 24
- C8 — 25
- R9 — 26

Right side (top to bottom):
- 60 — WAITIN
- 59 — RFSH
- 58 — DISRFSH
- 57 — DELCLK
- 56 — CLK
- 55 — V_CC
- 54 — COLINC
- 53 — GND
- 52 — ML
- 51 — GND
- 50 — CAP
- 49 — V_CC
- 48 — CS
- 47 — AREQ
- 46 — WIN
- 45 — CAS3
- 44 — CAS2

Bottom (left to right, pins 27–43): C9, ECAS0, ECAS1, ECAS2, ECAS3, B0, B1, ADS, V_CC, WE, GND, RAS0, RAS1, RAS2, RAS3, CAS0, CAS1

TL/F/8588-3

Top View
FIGURE 3

Order Number DP8421AV-20 or DP8421AV-25
See NS Package Number V68A

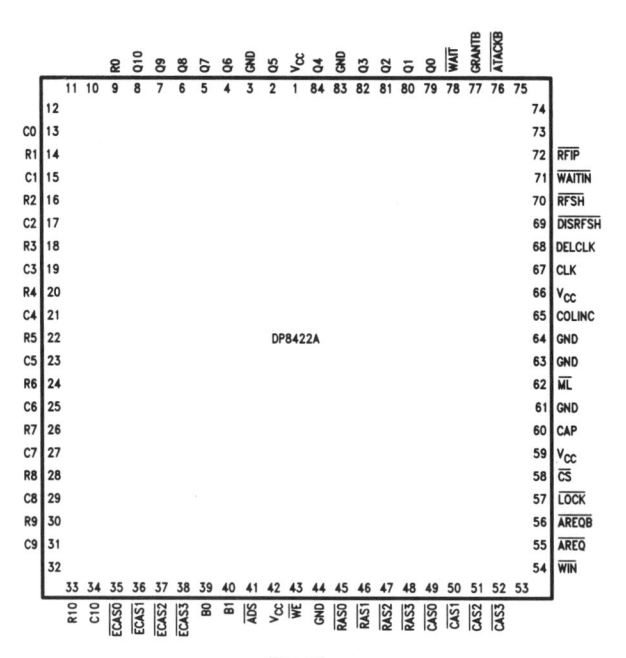

FIGURE 4 — DP8422A (Top View)

Pins (top, left to right): R0, Q10, Q9, Q8, Q7, Q6, GND, Q5, V_CC, Q4, GND, Q3, Q2, Q1, Q0, WAIT, GRANTB, ATACKB (pins 11, 10, 9, 8, 7, 6, 5, 4, 3, 2, 1, 84, 83, 82, 81, 80, 79, 78, 77, 76, 75)

Left side (top to bottom):
- 12
- C0 — 13
- R1 — 14
- C1 — 15
- R2 — 16
- C2 — 17
- R3 — 18
- C3 — 19
- R4 — 20
- C4 — 21
- R5 — 22
- C5 — 23
- R6 — 24
- C6 — 25
- R7 — 26
- C7 — 27
- R8 — 28
- C8 — 29
- R9 — 30
- C9 — 31
- 32

Right side (top to bottom):
- 74
- 73
- 72 — RFIP
- 71 — WAITIN
- 70 — RFSH
- 69 — DISRFSH
- 68 — DELCLK
- 67 — CLK
- 66 — V_CC
- 65 — COLINC
- 64 — GND
- 63 — GND
- 62 — ML
- 61 — GND
- 60 — CAP
- 59 — V_CC
- 58 — CS
- 57 — LOCK
- 56 — AREQB
- 55 — AREQ
- 54 — WIN

Bottom (left to right, pins 33–53): R10, C10, ECAS0, ECAS1, ECAS2, ECAS3, B0, B1, ADS, V_CC, WE, GND, RAS0, RAS1, RAS2, RAS3, CAS0, CAS1, CAS2, CAS3

TL/F/8588-2

Top View
FIGURE 4

Order Number DP8422AV-20 or DP8422AV-25
See NS Package Number V84A

2.0 Signal Descriptions

Pin Name	Device (If not Applicable to All)	Input/Output	Description
2.1 ADDRESS, R/W AND PROGRAMMING SIGNALS			
R0–10 R0–9	DP8422A DP8420A/21A	I I	**ROW ADDRESS:** These inputs are used to specify the row address during an access to the DRAM. They are also used to program the chip when $\overline{ML}$ is asserted (except R10).
C0–10 C0–9	DP8422A DP8420A/21A	I I	**COLUMN ADDRESS:** These inputs are used to specify the column address during an access to the DRAM. They are also used to program the chip when $\overline{ML}$ is asserted (except C10).
B0, B1		I	**BANK SELECT:** Depending on programming, these inputs are used to select a group of $\overline{RAS}$ and $\overline{CAS}$ outputs to assert during an access. They are also used to program the chip when $\overline{ML}$ is asserted.
$\overline{ECAS}$0–3		I	**ENABLE $\overline{CAS}$:** These inputs are used to enable a single or group of $\overline{CAS}$ outputs when asserted. In combination with the B0, B1 and the programming bits, these inputs select which $\overline{CAS}$ output or $\overline{CAS}$ outputs will assert during an access. The $\overline{ECAS}$ signals can also be used to toggle a group of $\overline{CAS}$ outputs for page/nibble mode accesses. They also can be used for byte write operations. If $\overline{ECAS}$0 is negated during programming, continuing to assert the $\overline{ECAS}$0 while negating $\overline{AREQ}$ or $\overline{AREQB}$ during an access, will cause the $\overline{CAS}$ outputs to be extended while the $\overline{RAS}$ outputs are negated (the $\overline{ECAS}$n inputs have no effect during scrubbing refreshes).
$\overline{WIN}$		I	**WRITE ENABLE IN:** This input is used to signify a write operation to the DRAM. If $\overline{ECAS}$0 is asserted during programming, the $\overline{WE}$ output will follow this input. This input asserted will also cause $\overline{CAS}$ to delay to the next positive clock edge if address bit C9 is asserted during programming.
COLINC ($\overline{EXTNDRF}$)		I I	**COLUMN INCREMENT:** When the address latches are used, and $\overline{RFIP}$ is negated, this input functions as COLINC. Asserting this signal causes the column address to be incremented by one. When $\overline{RFIP}$ is asserted, this signal is used to extend the refresh cycle by any number of periods of CLK until it is negated.
$\overline{ML}$		I	**MODE LOAD:** This input signal, when low, enables the internal programming register that stores the programming information.
2.2 DRAM CONTROL SIGNALS			
Q0–10 Q0–9 Q0–8	DP8422A DP8421A DP8421A	O O O	**DRAM ADDRESS:** These outputs are the multiplexed output of the R0–9, 10 and C0–9, 10 and form the DRAM address bus. These outputs contain the refresh address whenever $\overline{RFIP}$ is asserted. They contain high capacitive drivers with 20Ω series damping resistors.
$\overline{RAS}$0–3		O	**ROW ADDRESS STROBES:** These outputs are asserted to latch the row address contained on the outputs Q0–8, 9, 10 into the DRAM. When $\overline{RFIP}$ is asserted, the $\overline{RAS}$ outputs are used to latch the refresh row address contained on the Q0–8, 9, 10 outputs in the DRAM. These outputs contain high capacitive drivers with 20Ω series damping resistors.
$\overline{CAS}$0–3		O	**COLUMN ADDRESS STROBES:** These outputs are asserted to latch the column address contained on the outputs Q0–8, 9, 10 into the DRAM. These outputs have high capacitive drivers with 20Ω series damping resistors.
$\overline{WE}$ ($\overline{RFRQ}$)		O O	**WRITE ENABLE or REFRESH REQUEST:** This output asserted specifies a write operation to the DRAM. When negated, this output specifies a read operation to the DRAM. When the DP8420A/21A/22A is programmed in interleave mode or when $\overline{ECAS}$0 is negated during programming, this output will function as $\overline{RFRQ}$. When asserted, this pin specifies that 13 μs or 15 μs have passed. If $\overline{DISRFSH}$ is negated, the DP8420A/21A/22A will perform an internal refresh as soon as possible. If $\overline{DISRFRSH}$ is asserted, $\overline{RFRQ}$ can be used to externally request a refresh through the input $\overline{RFSH}$. This output has a high capacitive driver and a 20Ω series damping resistor.

2.0 Signal Descriptions (Continued)

Pin Name	Device (If not Applicable to All)	Input/ Output	Description
2.3 REFRESH SIGNALS			
RFIP		O	**REFRESH IN PROGRESS:** This output is asserted prior to a refresh cycle and is negated when all the $\overline{RAS}$ outputs are negated for that refresh.
$\overline{RFSH}$		I	**REFRESH:** This input asserted with $\overline{DISRFSH}$ already asserted will request a refresh. If this input is continually asserted, the DP8420A/21A/22A will perform refresh cycles in a burst refresh fashion until the input is negated. If $\overline{RFSH}$ is asserted with $\overline{DISRFSH}$ negated, the internal refresh address counter is cleared (useful for burst refreshes).
$\overline{DISRFSH}$		I	**DISABLE REFRESH:** This input is used to disable internal refreshes and must be asserted when using $\overline{RFSH}$ for externally requested refreshes.
2.4 PORT A ACCESS			
$\overline{ADS}$ (ALE)		I / I	**ADDRESS STROBE** or **ADDRESS LATCH ENABLE:** Depending on programming, this input can function as $\overline{ADS}$ or ALE. In mode 0, the input functions as ALE and when asserted along with $\overline{CS}$ causes an internal latch to be set. Once this latch is set an access will start from the positive clock edge of CLK as soon as possible. In Mode 1, the input functions as $\overline{ADS}$ and when asserted along with $\overline{CS}$, causes the access $\overline{RAS}$ to assert if no other event is taking place. If an event is taking place, $\overline{RAS}$ will be asserted from the positive edge of CLK as soon as possible. In both cases, the low going edge of this signal latches the bank, row and column address if programmed to do so.
$\overline{CS}$		I	**CHIP SELECT:** This input signal must be asserted to enable a Port A access.
$\overline{AREQ}$		I	**ACCESS REQUEST:** This input signal in Mode 0 must be asserted some time after the first positive clock edge after ALE has been asserted. When this signal is negated, $\overline{RAS}$ is negated for the access. In Mode 1, this signal must be asserted before $\overline{ADS}$ can be negated. When this signal is negated, $\overline{RAS}$ is negated for the access.
$\overline{WAIT}$ ($\overline{DTACK}$)		O / O	**WAIT** or **DTACK:** This output can be programmed to insert wait states into a CPU access cycle. With R7 negated during programming, the output will function as a $\overline{WAIT}$ type output. In this case, the output will be active low to signal a wait condition. With R7 asserted during programming, the output will function as $\overline{DTACK}$. In this case, the output will be negated to signify a wait condition and will be asserted to signify the access has taken place. Each of these signals can be delayed by a number of positive clock edges or negative clock levels of CLK to increase the microprocessor's access cycle through the insertion of wait states.
$\overline{WAITIN}$		I	**WAIT INCREASE:** This input can be used to dynamically increase the number of positive clock edges of CLK until $\overline{DTACK}$ will be asserted or $\overline{WAIT}$ will be negated during a DRAM access.

2.0 Signal Descriptions (Continued)

Pin Name	Device (If not Applicable to All)	Input/Output	Description
2.5 PORT B ACCESS SIGNALS			
$\overline{AREQB}$	DP8422A only	I	**PORT B ACCESS REQUEST:** This input asserted will latch the row, column and bank address if programmed, and requests an access to take place for Port B. If the access can take place, $\overline{RAS}$ will assert immediately. If the access has to be delayed, $\overline{RAS}$ will assert as soon as possible from a positive edge of CLK.
$\overline{ATACKB}$	DP8422A only	O	**ADVANCED TRANSFER ACKNOWLEDGE PORT B:** This output is asserted when the access $\overline{RAS}$ is asserted for a Port B access. This signal can be used to generate the appropriate $\overline{DTACK}$ or $\overline{WAIT}$ type signal for Port B's CPU or bus.
2.6 COMMON DUAL PORT SIGNALS			
GRANTB	DP8422A only	O	**GRANT B:** This output indicates which port is currently granted access to the DRAM array. When GRANTB is asserted, Port B has access to the array. When GRANTB is negated, Port A has access to the DRAM array. This signal is used to multiplex the signals R0–8, 9, 10; C0–8, 9, 10; B0–1; $\overline{WIN}$; $\overline{LOCK}$ and $\overline{ECAS0-3}$ to the DP8422A when using dual accessing.
$\overline{LOCK}$	DP8422A only	I	**LOCK:** This input can be used by the currently granted port to "lock out" the other port from the DRAM array by inserting wait states into the locked out port's access cycle until $\overline{LOCK}$ is negated.
2.7 POWER SIGNALS AND CAPACITOR INPUT			
V_{CC}		I	**POWER:** Supply Voltage.
GND		I	**GROUND:** Supply Voltage Reference.
CAP		I	**CAPACITOR:** This input is used by the internal PLL for stabilization. The value of the ceramic capacitor should be 0.1 μF and should be connected between this input and ground.
2.8 CLOCK INPUTS There are two clock inputs to the DP8420A/21A/22A, CLK and DELCLK. These two clocks may both be tied to the same clock input, or they may be two separate clocks, running at different frequencies, asynchronous to each other.			
CLK		I	**SYSTEM CLOCK:** This input may be in the range of 0 Hz up to 25 MHz. This input is generally a constant frequency but it may be controlled externally to change frequencies or perhaps be stopped for some arbitrary period of time. This input provides the clock to the internal state machine that arbitrates between accesses and refreshes. This clock's positive edges and negative levels are used to extend the $\overline{WAIT}$ ($\overline{DTACK}$) signals. Ths clock is also used as the reference for the $\overline{RAS}$ precharge time and $\overline{RAS}$ low time during refresh. All Port A and Port B accesses are assumed to be synchronous to the system clock CLK.
DELCLK		I	**DELAY LINE CLOCK:** The clock input DELCLK, may be in the range of 6 MHz to 20 MHz and should be a multiple of 2 (i.e., 6, 8, 10, 12, 14, 16, 18, 20 MHz) to have the DP8420A/21A/22A switching characteristics hold. If DELCLK is not one of the above frequencies the accuracy of the internal delay line will suffer. This is because the phase locked loop that generates the delay line assumes an input clock frequency of a multiple of 2 MHz. For example, if the DELCLK input is at 7 MHz and we choose a divide by 3 (program bits C0–2) this will produce 2.333 MHz which is 16.667% off of 2 MHz. Therefore, the DP8420A/21A/22A delay line would produce delays that are shorter (faster delays) than what is intended. If divide by 4 was chosen the delay line would be longer (slower delays) than intended (1.75 MHz instead of 2 MHz). (See Section 10 for more information.) This clock is also divided to create the internal refresh clock.

3.0 Port A Access Modes

The DP8420A/21A/22A have two general purpose access modes. With one of these modes, any microprocessor can be interfaced to DRAM. A Port A access to DRAM is initiated by two input signals: $\overline{ADS}$ (ALE) and $\overline{CS}$. The access is always terminated by one signal: $\overline{AREQ}$. These input signals should be synchronous to the input clock, CLK. One of these access modes is selected at programming through the B1 input signal. In both modes, once an access has been requested by $\overline{CS}$ and $\overline{ADS}$ (ALE), the DP8422A will guarantee the following:

The DP8420A/21A/22A will have the row address valid to the DRAMs' address bus, Q0–8, 9, 10 given that the row address setup time to the DP8420A/21A/22A was met;

The DP8420A/21A/22A will bring the appropriate $\overline{RAS}$ or $\overline{RAS}$s low;

The DP8420A/21A/22A will guarantee the minimum row address hold time, before switching the internal multiplexor to place the column address on the DRAM address bus, Q0–8, 9, 10;

The DP8420A/21A/22A will guarantee the minimum column address setup time before asserting the appropriate $\overline{CAS}$ or $\overline{CAS}$s;

The DP8420A/21A/22A will hold the column address valid the minimum specified column address hold time in address pipelining mode and will hold the column address valid the remainder of the access in non-pipelining mode.

3.1 ACCESS MODE 0

Access Mode 0, shown in *Figure 5a*, is selected by negating the input B1 during programming. This access mode allows accesses to DRAM to always be initiated from the positive edge of the system input clock, CLK. To initiate a Mode 0 access, ALE is pulsed high and $\overline{CS}$ is asserted. Pulsing ALE high and asserting $\overline{CS}$, sets an internal latch which requests an access. If the precharge time from the last access or DRAM refresh had been met and a refresh of DRAM or a Port B access was not in progress, the $\overline{RAS}$ or group of $\overline{RAS}$s would be initiated from the first positive edge of CLK. If a DRAM refresh is in progress or precharge time is required, the controller will wait until these events have taken place and assert $\overline{RAS}$ on the next positive edge of CLK.

Sometime after the first positive edge of CLK after ALE and $\overline{CS}$ have been asserted, the input $\overline{AREQ}$ must be asserted. In single port applications, once $\overline{AREQ}$ has been asserted, $\overline{CS}$ can be negated. Once $\overline{AREQ}$ is negated, $\overline{RAS}$ and $\overline{DTACK}$, if programmed, will be negated. If $\overline{ECAS0}$ is asserted during programming, $\overline{CAS}$ will be negated with $\overline{AREQ}$. If $\overline{ECAS0}$ was negated during programming, a single $\overline{CAS}$ or group of $\overline{CAS}$s will continue to be asserted after $\overline{RAS}$ has been negated given that the appropriate $\overline{ECAS}$s inputs were

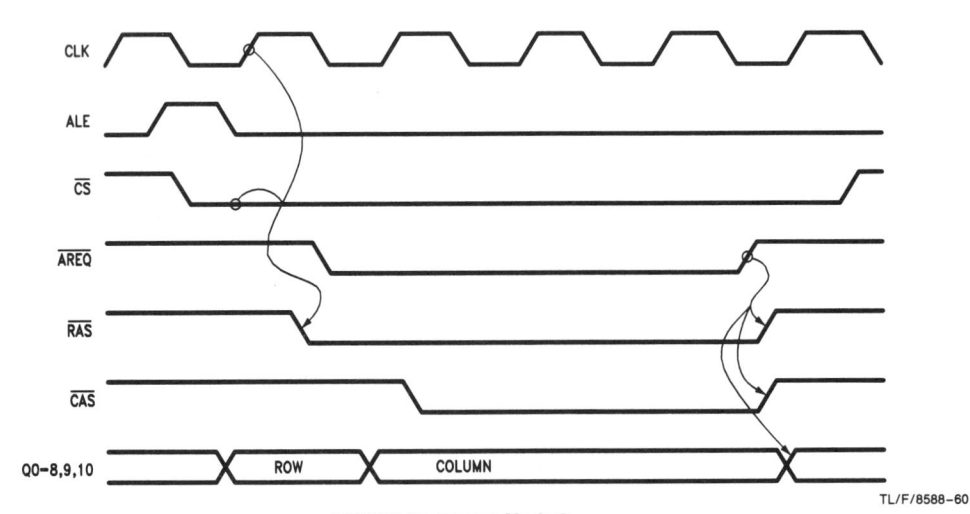

FIGURE 5a. Access Mode 0

3.0 Port A Access Modes (Continued)

asserted as shown in *Figure 5b*. This allows the DRAM to have data present on the data out bus while gaining $\overline{RAS}$ precharge time. ALE can stay asserted several periods of CLK. However, ALE must be negated before or during the period of CLK in which $\overline{AREQ}$ is negated.

When performing address pipelining, the ALE input cannot be asserted to start another access until $\overline{AREQ}$ has been asserted for at least one clock period of CLK for the present access.

3.2 ACCESS MODE 1

Access Mode 1, shown in *Figure 6a*, is selected by asserting the input B1 during programming. This mode allows accesses, which are not delayed by precharge, Port B access or refresh, to start immediately from the access request input, $\overline{ADS}$. To initiate a Mode 1 access, $\overline{CS}$ is asserted followed by $\overline{ADS}$ asserted. If the programmed precharge time from the last access or DRAM refresh had been met and a refresh of the DRAM or Port B access to the DRAM was not in progress, the $\overline{RAS}$ or group of $\overline{RAS}$s selected by programming and the bank select inputs would be asserted from $\overline{ADS}$ being asserted. If a DRAM refresh or Port B access is in progress or precharge time is required, the controller will wait until these events have taken place and assert $\overline{RAS}$ or the group of $\overline{RAS}$s from the next positive edge of CLK.

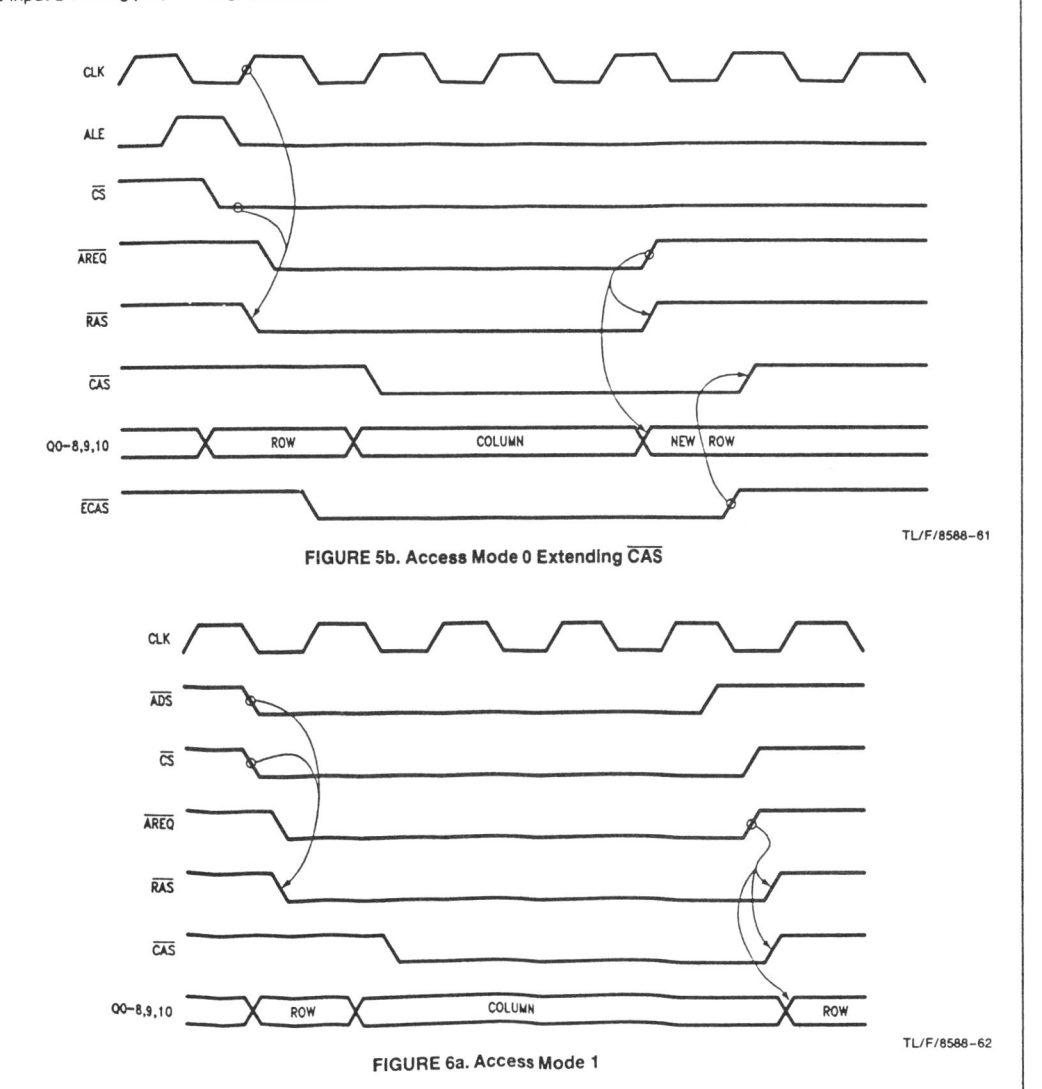

FIGURE 5b. Access Mode 0 Extending $\overline{CAS}$

FIGURE 6a. Access Mode 1

3.0 Port A Access Modes (Continued)

When $\overline{ADS}$ is asserted or sometime after, $\overline{AREQ}$ must be asserted. At this time, $\overline{ADS}$ can be negated and $\overline{AREQ}$ will continue the access. Once $\overline{AREQ}$ is negated, $\overline{RAS}$ and $\overline{DTACK}$, if programmed, will be negated. If $\overline{ECAS0}$ was asserted during programming, $\overline{CAS}$ will be negated with $\overline{AREQ}$. If $\overline{ECAS0}$ was negated during programming, a single $\overline{CAS}$ or group of $\overline{CAS}$s will continue to be asserted after $\overline{RAS}$ has been negated given that the appropriate $\overline{ECAS}$ inputs were asserted as shown in *Figure 6b*. This allows a DRAM to have data present on the data out bus while gaining $\overline{RAS}$ precharge time. $\overline{ADS}$ can continue to be asserted after $\overline{AREQ}$ has been asserted and negated, however a new access would not be started until $\overline{ADS}$ is negated and asserted again. $\overline{ADS}$ and $\overline{AREQ}$ can be tied together in applications not using address pipelining.

If address pipelining is programmed, it is possible for $\overline{ADS}$ to be negated after $\overline{AREQ}$ is asserted. Once $\overline{AREQ}$ is asserted, $\overline{ADS}$ can be asserted again to initiate a new access.

3.3 READ-MODIFY-WRITE CYCLES WITH EITHER ACCESS MODE

There are 2 methods by which this chip can be used to do read-modify-write access cycles. The first method involves doing a late write access where the $\overline{WIN}$ input is asserted some delay after $\overline{CAS}$ is asserted. The second method involves doing a page mode read access followed by a page mode write access with $\overline{RAS}$ held low (see *Figure 5*).

$\overline{CAS}$n must be toggled using the $\overline{ECAS}$n inputs and $\overline{WIN}$ has to be changed from negated to asserted (read to write) while $\overline{CAS}$ is negated. This method is better than changing $\overline{WIN}$ from negated to asserted in a late write access because here a problem may arise with DATA IN and DATA OUT being valid at the same time. This may result in a data line trying to drive two different levels simultaneously. The page mode method of a read-modify-write access allows the user to have transceivers in the system because the data in (read data) is guaranteed to be high impedance during the time the data out (write data) is valid.

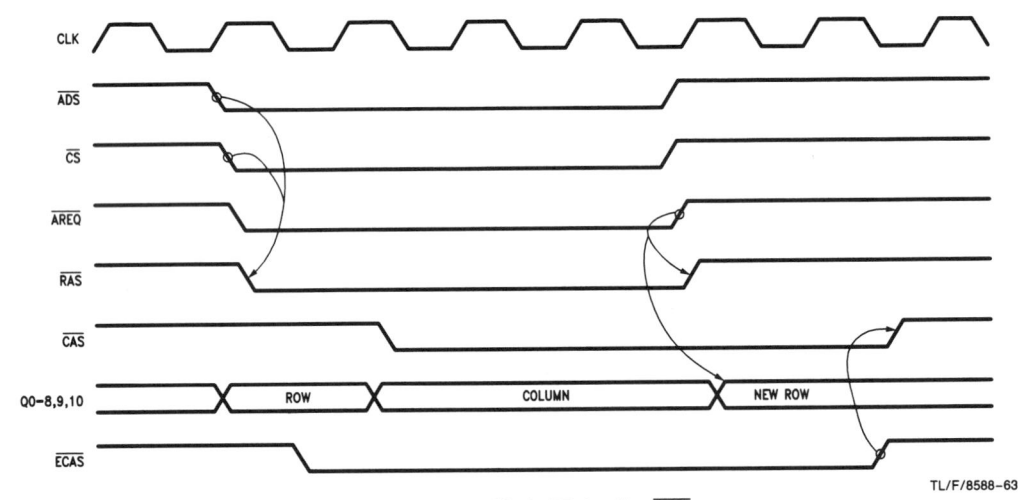

FIGURE 6b. Access Mode 1 Extending $\overline{CAS}$

3.0 Port A Access Modes (Continued)

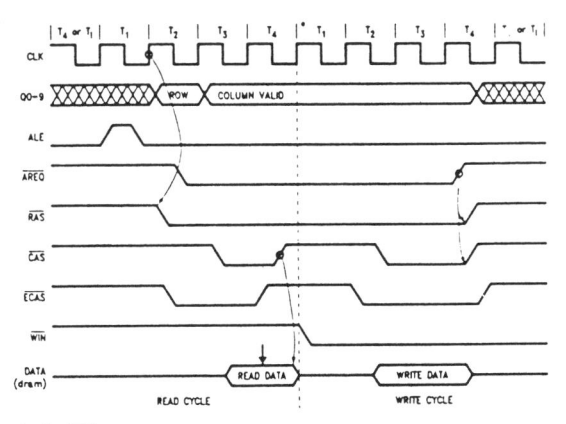

*There may be idle states inserted here by the CPU.

FIGURE 6c. Read-Modify-Write Access Cycle

4.0 Refresh Options

The DP8420A/21A/22A support a wide variety of refresh control mode options including automatic internally controlled refresh, externally controlled/burst refresh, refresh request/acknowledge and any combination of the above. With each of the control modes above, different types of refreshes can be performed. These different types include all $\overline{RAS}$ refresh, staggered refresh and error scrubbing during all $\overline{RAS}$ refresh.

There are three inputs, EXTNDRF, $\overline{RFSH}$ and $\overline{DISRFSH}$, and two outputs, $\overline{RFIP}$ and $\overline{RFRQ}$, associated with refresh. There are also ten programming bits; R0–1, R9, C0–6 and $\overline{ECAS0}$ used to program the various types of refreshing.

The two inputs, $\overline{RFSH}$ and $\overline{DISRFSH}$, are used in the externally controlled/burst refresh mode and the refresh request/acknowledge mode. The output $\overline{RFRQ}$ is used in the refresh requset/acknowledge mode. The input EXTNDRF and the output $\overline{RFIP}$ are used in all refresh modes. Asserting the input EXTNDRF, extends the refresh cycle single or multiple integral clock periods of CLK. The output $\overline{RFIP}$ is asserted one period of CLK before the first refresh $\overline{RAS}$ is asserted. If an access is currently in progress, $\overline{RFIP}$ will be asserted up to one period of CLK before the first refresh $\overline{RAS}$, once $\overline{AREQ}$ or $\overline{AREQB}$ is negated for the access (see *Figure 7a*).

The DP8420A/21A/22A will increment the refresh address counter automatically, independent of the refresh mode used. The refresh address counter will be incremented once all the refresh $\overline{RAS}$s have been negated.

In every combination of refresh control mode and refresh type, the DP8420A/21A/22A is programmed to keep $\overline{RAS}$ asserted a number of CLK periods. The values of $\overline{RAS}$ low time during refresh are programmed with the programming bits R0 and R1.

4.1 REFRESH CONTROL MODES

There are three different modes of refresh control. Any of these modes can be used in combination or singularly to produce the desired refresh results. The three different modes of control are: automatic internal refresh, external/burst refresh and refresh request/acknowledge.

4.1.1. Automatic Internal Refresh

The DP8420A/21A/22A have an internal refresh clock. The period of the refresh clock is generated from the programming bits C0–3. Every period of the refresh clock, an internal refresh request is generated. As long as a DRAM access is not currently in progress and precharge time has been met, the internal refresh request will generate an automatic internal refresh. If a DRAM access is in progress, the DP8420A/21A/22A on-chip arbitration logic will wait until the access is finished before performing the refresh. The refresh/access arbitration logic can insert a refresh cycle between two address pipelined accesses. However, the refresh arbitration logic can not interrupt an access cycle to perform a refresh. To enable automatic internally controlled refreshes, the input $\overline{DISRFSH}$ must be negated.

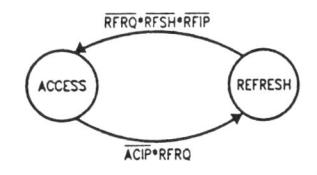

Explanation of Terms

RFRQ = ReFresh ReQuest internal to the DP8420A/21A/22A. RFRQ has the ability to hold off a pending access.

RFSH = Externally requested ReFreSH

RFIP = ReFresh In Progress

ACIP = Port A or Port B (DP8422A only) ACcess In Progress. This means that either RAS is low for an access or is in the process of transitioning low for an access.

FIGURE 7a. DP8420A/21A/22A Access/Refresh Arbitration State Program

4.1.2 Externally Controlled/Burst Refresh

To use externally controlled/burst refresh, the user must disable the automatic internally controlled refreshes by asserting the input $\overline{DISRFSH}$. The user is responsible for generating the refresh request by asserting the input $\overline{RFSH}$. Pulsing $\overline{RFSH}$ low, sets an internal latch, that is used to

4.0 Refresh Options (Continued)

produce the internal refresh request. The refresh cycle will take place on the next positive edge of CLK as shown in *Figure 7b*. If an access to DRAM is in progress or precharge time for the last access has not been met, the refresh will be delayed. Since pulsing $\overline{\text{RFSH}}$ low sets a latch, the user does not have to keep $\overline{\text{RFSH}}$ low until the refresh starts. When the last refresh $\overline{\text{RAS}}$ negates, the internal refresh request latch is cleared.

By keeping $\overline{\text{RFSH}}$ asserted past the positive edge of CLK which ends the refresh cycle as shown in *Figure 8*, the user will perform another refresh cycle. Using this technique, the user can perform a burst refresh consisting of any number of refresh cycles. Each refresh cycle during a burst refresh will meet the refresh $\overline{\text{RAS}}$ low time and the $\overline{\text{RAS}}$ precharge time (programming bits R0–1).

If the user desires to burst refresh the entire DRAM (all row addresses) he could generate an end of count signal (burst refresh finished) by looking at one of the DP8420A/21A/22A high address outputs (Q7, Q8, Q9 or Q10) and the $\overline{\text{RFIP}}$ output. The Qn outputs function as a decode of how many row addresses have been refreshed (Q7 = 128 refreshes, Q8 = 256 refreshes, Q9 = 512 refreshes, Q10 = 1024 refreshes).

4.1.3 Refresh Request/Acknowledge

The DP8420A/21A/22A can be programmed to output internal refresh requests. When the user programs $\overline{\text{ECAS0}}$ negated during programming and/or address pipelining mode, the $\overline{\text{WE}}$ output functions as $\overline{\text{RFRQ}}$. $\overline{\text{RFRQ}}$ will be asserted by one of two events, either the internal refresh clock has expired which signals that another refresh is needed, or by the signal $\overline{\text{RFSH}}$ being pulsed low requesting an external refresh. $\overline{\text{RFRQ}}$ will be asserted from a positive edge of CLK. *Figure 9a* shows an example of an external refresh being requested while $\overline{\text{RFRQ}}$ is negated. When $\overline{\text{RFRQ}}$ is asserted from the expiration of the internal refresh clock signaling a new refresh is needed, it will stay asserted until the $\overline{\text{RFSH}}$ is pulsed low with $\overline{\text{DISRFSH}}$ asserted. This will cause an externally requested/burst refresh to take place. If $\overline{\text{DISRFSH}}$ is negated, an automatic internal refresh will take place as shown in *Figure 9b*.

$\overline{\text{RFRQ}}$ will go high and then assert if additional periods of the internal refresh clock have expired and neither an externally controlled refresh nor an automatically controlled internal refresh have taken place as shown in *Figure 9c*. If a time critical event, or long access like page/static column mode access can not be interrupted, $\overline{\text{RFRQ}}$ pulsing high can be used to increment a counter. The counter can be used to perform a burst refresh of the number of refreshes missed (through the $\overline{\text{RFSH}}$ input).

4.2 REFRESH CYCLE TYPES

Three different types of refresh cycles are available for use. The three different types are mutually exclusive and can be used with any of the three modes of refresh control. The three different refresh cycle types are: all $\overline{\text{RAS}}$ refresh, staggered $\overline{\text{RAS}}$ refresh and error scrubbing during all $\overline{\text{RAS}}$ refresh. In all refresh cycle types, the $\overline{\text{RAS}}$ precharge time is guaranteed: between the previous access $\overline{\text{RAS}}$ ending and the refresh $\overline{\text{RAS0}}$ starting; between refresh $\overline{\text{RAS3}}$ ending and access $\overline{\text{RAS}}$ beginning; between burst refresh $\overline{\text{RAS}}$s.

4.2.1 Conventional $\overline{\text{RAS}}$ Refresh

A conventional refresh cycle causes $\overline{\text{RAS}}$0–3 to all assert from the first positive edge of CLK after $\overline{\text{RFIP}}$ is asserted as shown in *Figure 10*. $\overline{\text{RAS}}$0–3 will stay asserted until the number of positive edges of CLK programmed have passed. On the last positive edge, $\overline{\text{RAS}}$0–3, and $\overline{\text{RFIP}}$ will be negated. This type of refresh cycle is programmed by negating address bit R9 during programming.

4.2.2 Staggered $\overline{\text{RAS}}$ Refresh

A staggered refresh staggers each $\overline{\text{RAS}}$ or group of $\overline{\text{RAS}}$s by a positive edge of CLK as shown in *Figure 11*. The number of $\overline{\text{RAS}}$s, which will be asserted on each positive edge of CLK, is determined by the $\overline{\text{RAS}}$, $\overline{\text{CAS}}$ configuration mode programming bits C4–C6. If single $\overline{\text{RAS}}$ outputs are selected during programming, then each $\overline{\text{RAS}}$ will assert on successive positive edges of CLK. If two $\overline{\text{RAS}}$ outputs are selected during programming then $\overline{\text{RAS0}}$ and $\overline{\text{RAS1}}$ will assert on the first positive edge of CLK after $\overline{\text{RFIP}}$ is asserted. $\overline{\text{RAS2}}$ and $\overline{\text{RAS3}}$ will assert on the second positive edge of CLK after $\overline{\text{RFIP}}$ is asserted. If all $\overline{\text{RAS}}$ outputs were selected during programming, all $\overline{\text{RAS}}$ outputs would assert on the first positive edge of CLK after $\overline{\text{RFIP}}$ is asserted. Each $\overline{\text{RAS}}$ or group of $\overline{\text{RAS}}$s will meet the programmed $\overline{\text{RAS}}$ low time and then negate.

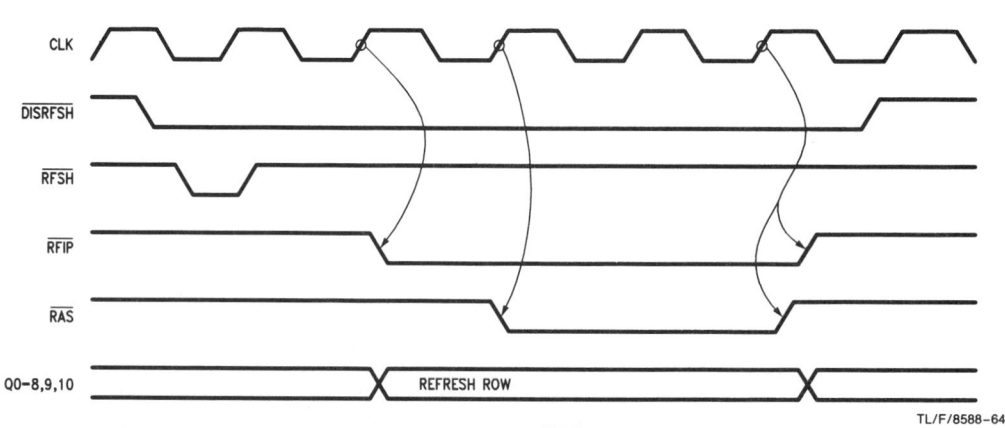

FIGURE 7b. Single External Refreshes (2 Periods of $\overline{\text{RAS}}$ Low during Refresh Programmed)

4.0 Refresh Options (Continued)

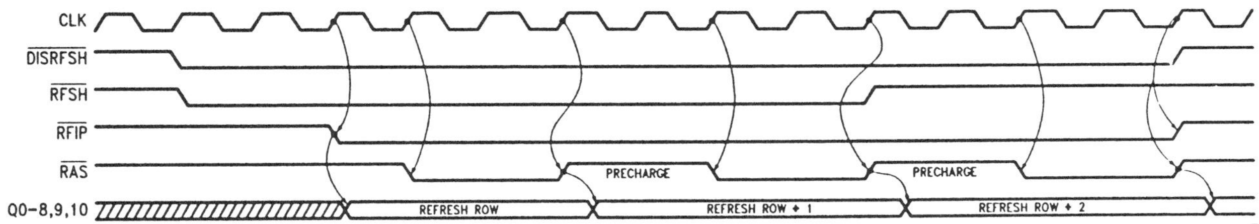

FIGURE 8. External Burst Refresh (2 Periods of RAS Precharge, 2 Periods of Refresh RAS Low during Refresh Programmed)

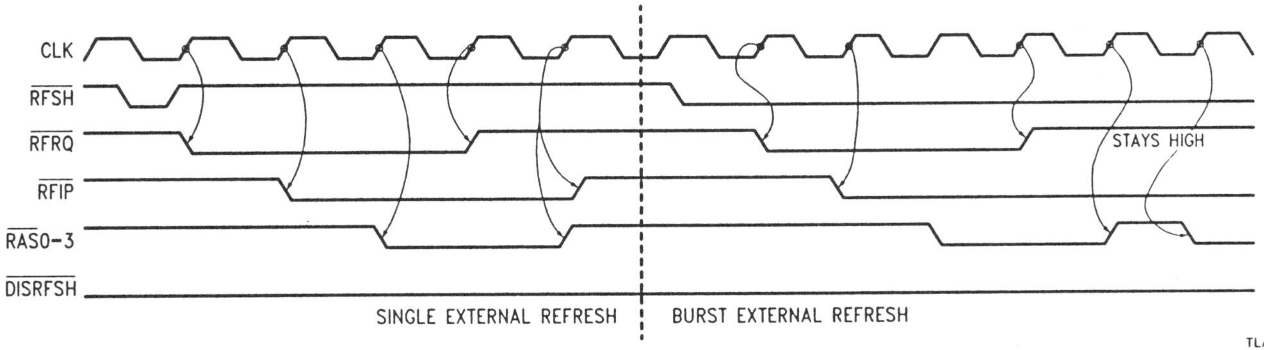

FIGURE 9a. Externally Controlled Single and Burst Refresh with Refresh Request (RFRQ) Output (2 Periods of RAS Low during Refresh Programmed)

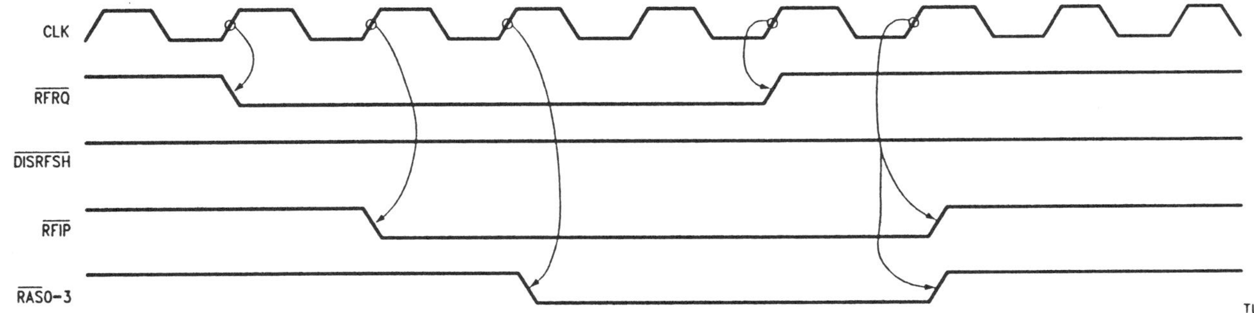

FIGURE 9b. Automatic Internal Refresh with Refresh Request (3T of RAS low during refresh programmed).

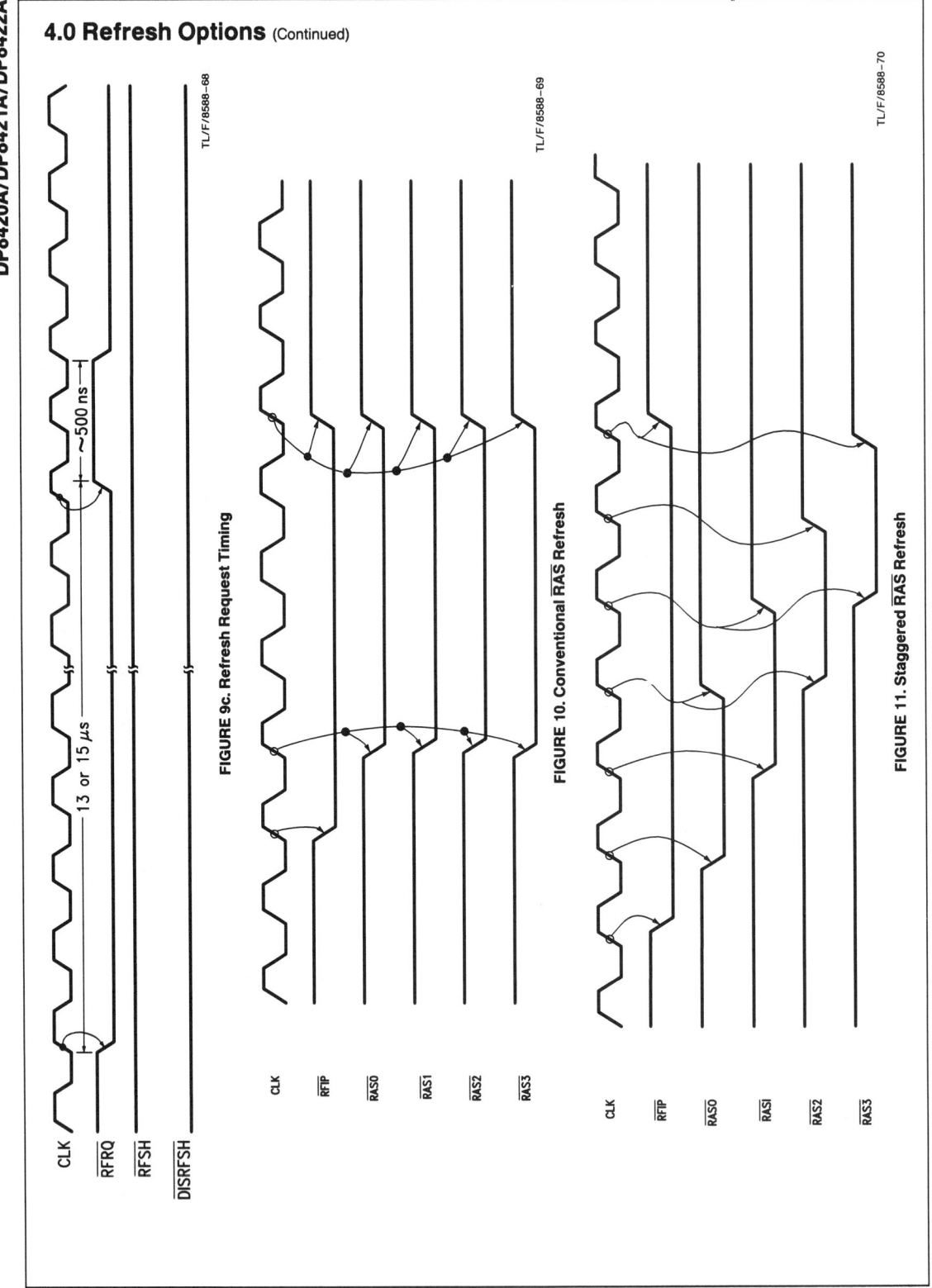

4.0 Refresh Options (Continued)

FIGURE 9c. Refresh Request Timing

FIGURE 10. Conventional RAS Refresh

FIGURE 11. Staggered RAS Refresh

4.0 Refresh Options (Continued)
4.2.3 Error Scrubbing during Refresh

The DP8420A/21A/22A support error scrubbing during all $\overline{RAS}$ DRAM refreshes. Error scrubbing during refresh is selected through bits C4–C6 with bit R9 negated during programming. Error scrubbing can not be used with staggered refresh (see Section 9.0). Error scrubbing during refresh allows a $\overline{CAS}$ or group of $\overline{CAS}$s to assert during the all $\overline{RAS}$ refresh as shown in *Figure 12*. This allows data to be read from the DRAM array and passed through an Error Detection And Correction Chip, EDAC. If the EDAC determines that the data contains a single bit error and corrects that error, the refresh cycle can be extended with the input extend refresh, EXTNDRF, and a read-modify-write operation can be performed by asserting $\overline{WE}$. It is the responsibility of the designer to ensure that $\overline{WE}$ is negated. The DP8422A has a 24-bit internal refresh address counter that contains the 11 row, 11 column and 2 bank addresses. The DP8420A/21A have a 22-bit internal refresh address counter that contains the 10 row, 10 column and 2 bank addresses. These counters are configured as bank, column, row with the row address as the least significant bits. The bank counter bits are then used with the programming selection to determine which $\overline{CAS}$ or group of $\overline{CAS}$s will assert during a refresh.

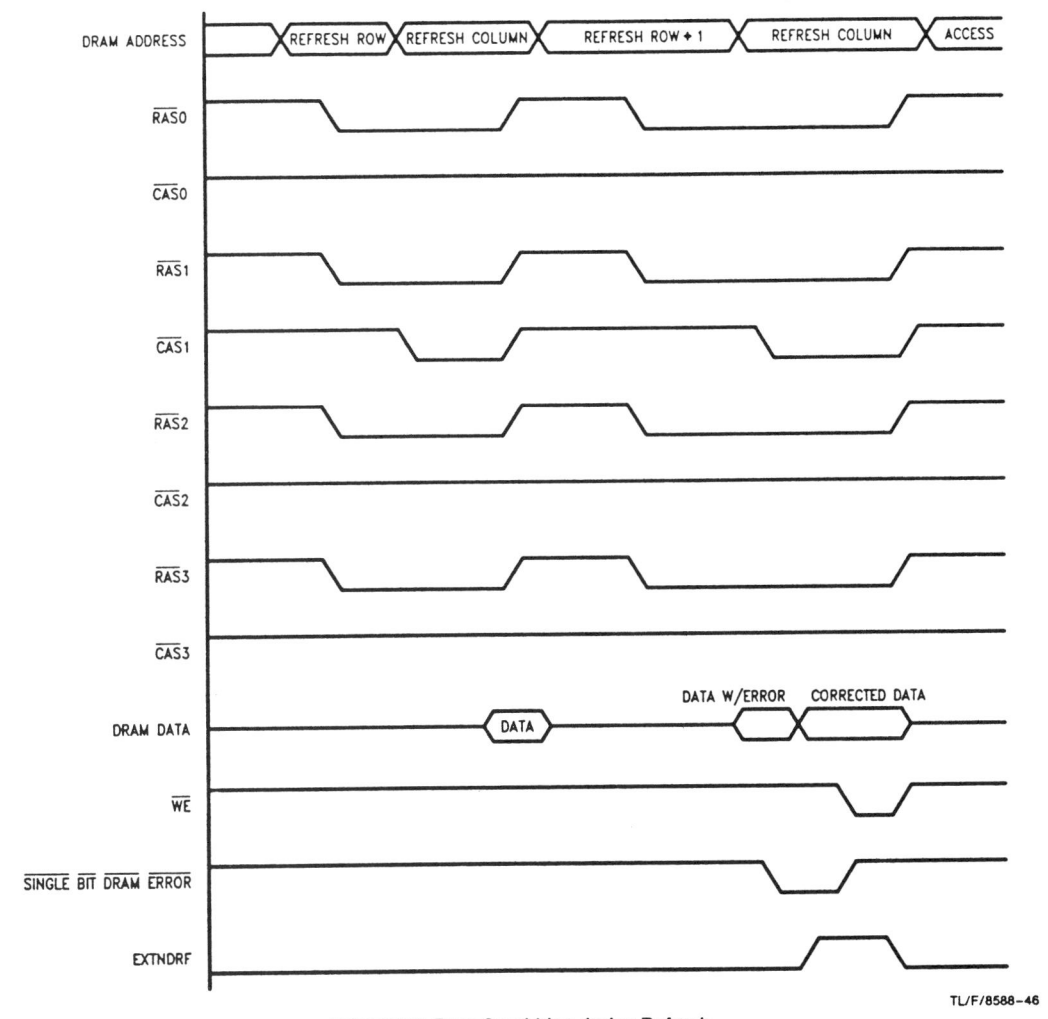

FIGURE 12. Error Scrubbing during Refresh

4.0 Refresh Options (Continued)

4.3 EXTENDING REFRESH

The programmed number of periods of CLK that refresh $\overline{RAS}$s are asserted can be extended by one or multiple periods of CLK. Only the all $\overline{RAS}$ (with or without error scrubbing) type of refresh can be extended. To extend a refresh cycle, the input extend refresh, EXTNDRF, must be asserted before the positive edge of CLK that would have negated all the $\overline{RAS}$ outputs during the refresh cycle and after the positive edge of CLK which starts all $\overline{RAS}$ outputs during the refresh as shown in *Figure 13*. This will extend the refresh to the next positive edge of CLK and EXTNDRF will be sampled again. The refresh cycle will continue until EXTNDRF is sampled low on a positive edge of CLK.

4.4 CLEARING THE REFRESH ADDRESS COUNTER

The refresh address counter can be cleared by asserting $\overline{RFSH}$ while $\overline{DISRFSH}$ is negated as shown in *Figure 14a*. This can be used prior to a burst refresh of the entire memory array. By asserting $\overline{RFSH}$ one period of CLK before $\overline{DISRFSH}$ is asserted and then keeping both inputs asserted, the DP8420A/21A/22A will clear the refresh address counter and then perform refresh cycles separated by the programmed value of precharge as shown in *Figure 14b*. An end-of-count signal can be generated from the Q DRAM address outputs of the DP8420A/21A/22A and used to negate $\overline{RFSH}$.

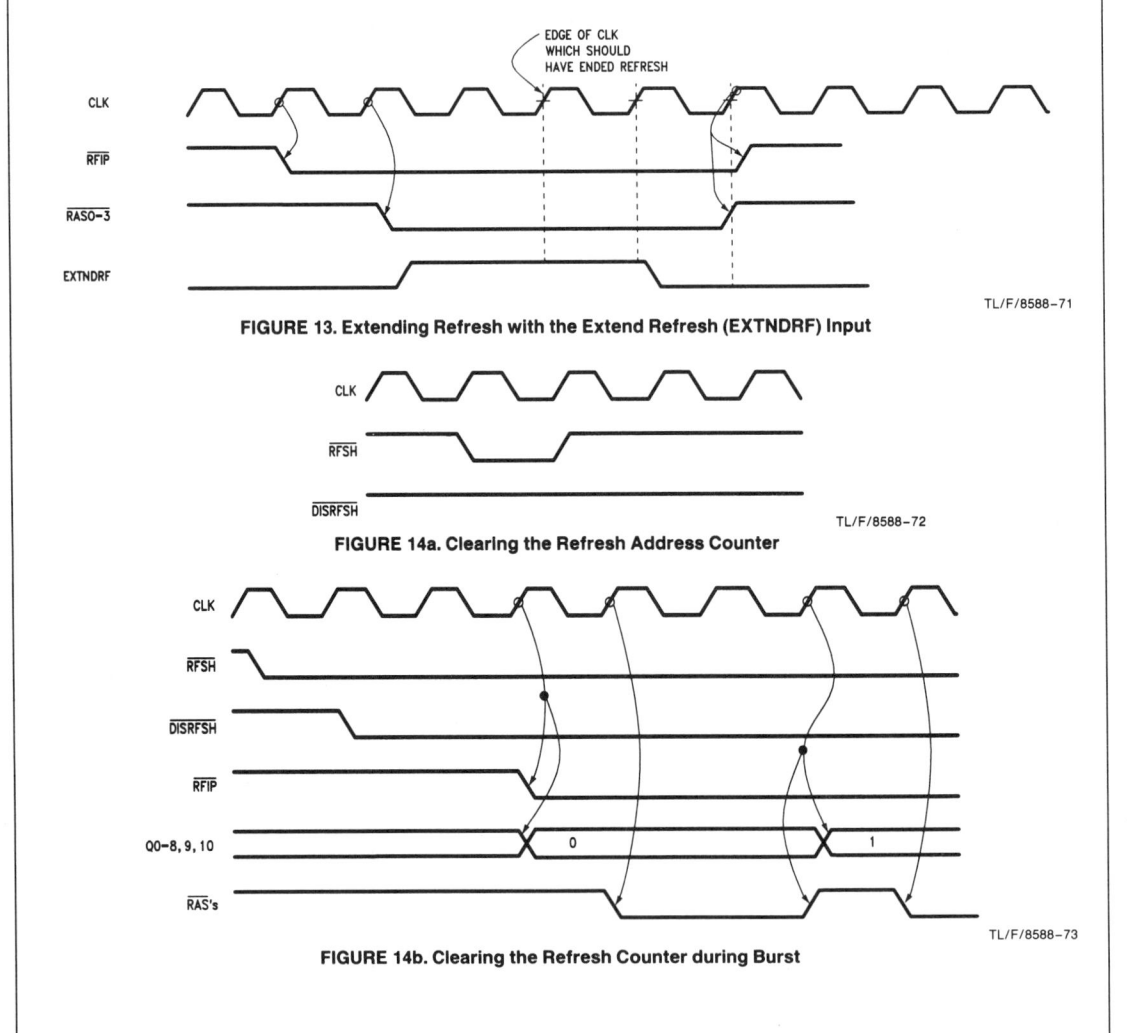

FIGURE 13. Extending Refresh with the Extend Refresh (EXTNDRF) Input

FIGURE 14a. Clearing the Refresh Address Counter

FIGURE 14b. Clearing the Refresh Counter during Burst

4.0 Refresh Options (Continued)

4.5 CLEARING THE REFRESH REQUEST CLOCK

The refresh request clock can be cleared by negating $\overline{\text{DISRFSH}}$ and asserting $\overline{\text{RFSH}}$ for 500 ns, one period of the internal 2 MHz clock as shown in *Figure 15*. By clearing the refresh request clock, the user is guaranteed that an internal refresh request will not be generated for approximately 15 μs, one refresh clock period, from the time $\overline{\text{RFSH}}$ is negated. This action will also clear the refresh address counter.

5.0 Port A Wait State Support

Wait states allow a CPU's access cycle to be increased by one or multiple CPU clock periods. By increasing the CPU's access cycle, all signals associated with that access cycle are extended. The wait or ready input is named differently by CPU manufacturers. However, any CPU's wait or ready input is compatible with either the $\overline{\text{WAIT}}$ or $\overline{\text{DTACK}}$ output of the DP8420A/21A/22A. The CPU samples a wait or ready line to determine if another clock period should be inserted into the access cycle. If another clock period is inserted, the CPU will continue to sample the input every CPU clock period until the input signal changes polarity, allowing the CPU access cycle to terminate. The user determines whether to program $\overline{\text{WAIT}}$ or $\overline{\text{DTACK}}$ (R7) and which value to select for $\overline{\text{WAIT}}$ or $\overline{\text{DTACK}}$ (R2, R3) depending upon the CPU used and where the CPU samples its wait input during an access cycle.

The decision to terminate the CPU access cycle is directly affected by the speed of the DRAMs used. The system designer must ensure that the data from the DRAMs will be present for the CPU to sample or that the data has been written to the DRAM before allowing the CPU access cycle to terminate.

The insertion of wait states also allows a CPU's access cycle to be extended until the DRAM access has taken place. The DP8420A/21A/22A insert wait states into CPU access cycles due to; guaranteeing precharge time, refresh currently in progress, user programmed wait states, the $\overline{\text{WAITIN}}$ signal being asserted and $\overline{\text{GRANTB}}$ not being valid (DP8422A only). If one of these events is taking place and the CPU starts an access, the DP8420A/21A/22A will insert wait states into the access cycle, thereby increasing the length of the CPU's access. Once the event has been completed, the DP8420A/21A/22A will allow the access to take place and stop inserting wait states.

There are six programming bits, R2–R7; an input, $\overline{\text{WAITIN}}$; and an output that functions as $\overline{\text{WAIT}}$ or $\overline{\text{DTACK}}$.

5.1 WAIT TYPE OUTPUT

With the R7 address bit negated during programming, the user selects the $\overline{\text{WAIT}}$ output. As long as $\overline{\text{WAIT}}$ is sampled asserted by the CPU, wait states (extra clock periods) are inserted into the current access cycle as shown in *Figure 16*. Once $\overline{\text{WAIT}}$ is sampled negated, the access cycle is completed by the CPU. $\overline{\text{WAIT}}$ is asserted at the beginning of a chip selected access and is programmed to negate a number of positive edges and/or negative levels of CLK from the event that starts the access. $\overline{\text{WAIT}}$ can also be programmed to function in page/burst mode applications. Once $\overline{\text{WAIT}}$ is negated during an access, and the $\overline{\text{ECAS}}$ inputs are negated with $\overline{\text{AREQ}}$ asserted, $\overline{\text{WAIT}}$ can be programmed to toggle, following the $\overline{\text{ECAS}}$ inputs. Once $\overline{\text{AREQ}}$ is negated, ending the access, $\overline{\text{WAIT}}$ will stay negated until the next chip selected access.

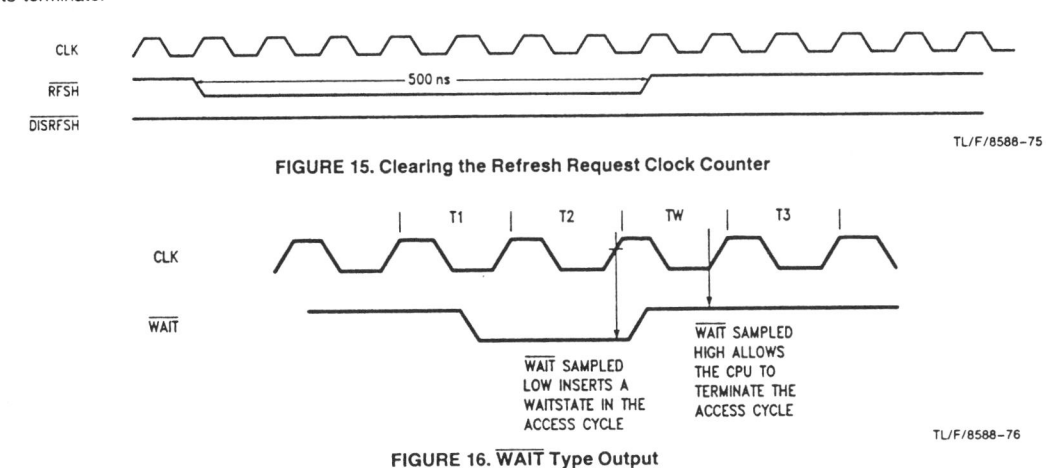

FIGURE 15. Clearing the Refresh Request Clock Counter

FIGURE 16. $\overline{\text{WAIT}}$ Type Output

5.0 Port A Wait State Support (Continued)

5.1.1. Wait during Single Accesses

$\overline{WAIT}$ can be programmed to delay a number of positive edges and/or negative levels of CLK. These options are programmed through address bits R2 and R3 at programming time. The user is given four options described below.

0T during non delayed and delayed acceses: $\overline{WAIT}$ will stay negated during a non-delayed access as shown in Figures 17a and 17c. During an access that is delayed, $\overline{WAIT}$ will assert at the start of the access ($\overline{CS}$ and ALE or $\overline{ADS}$) and negate from the positive edge of $\overline{CLK}$ that starts $\overline{RAS}$ for that access as shown in Figures 17b and 17d.

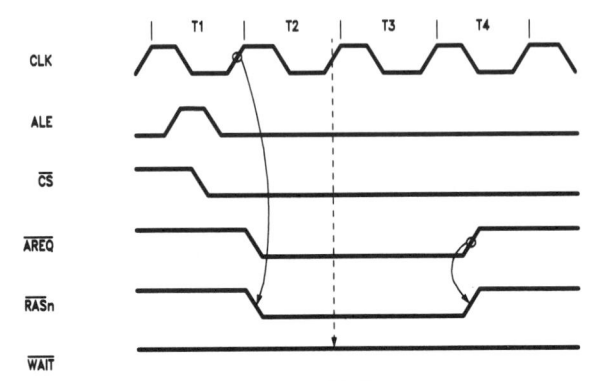

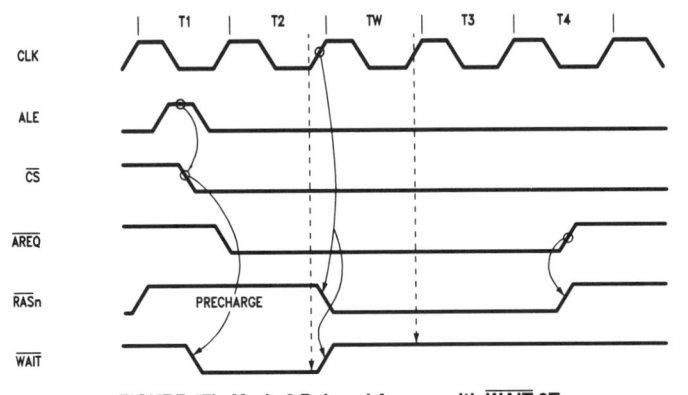

FIGURE 17a. Mode 0 Non-Delayed Access with $\overline{WAIT}$ 0T ($\overline{WAIT}$ is Sampled at the End of the "T2" Clock State)

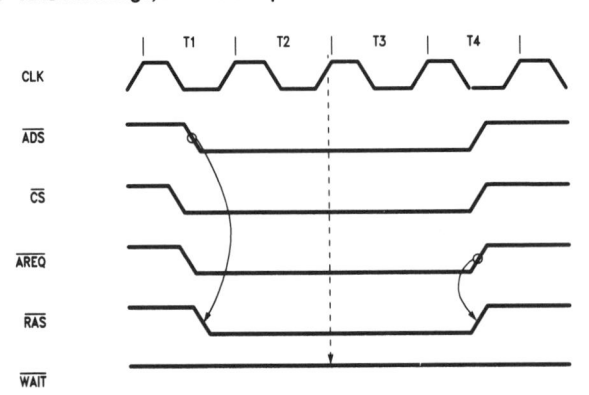

FIGURE 17b. Mode 0 Delayed Access with $\overline{WAIT}$ 0T
("2T" $\overline{RAS}$ Precharge, $\overline{WAIT}$ is Sampled at the End of the "T2" Clock State)

FIGURE 17c. Mode 1 Non-Delayed Access with $\overline{WAIT}$ 0T ($\overline{WAIT}$ is Sampled at the End of the "T2" Clock State)

5.0 Port A Wait State Support (Continued)

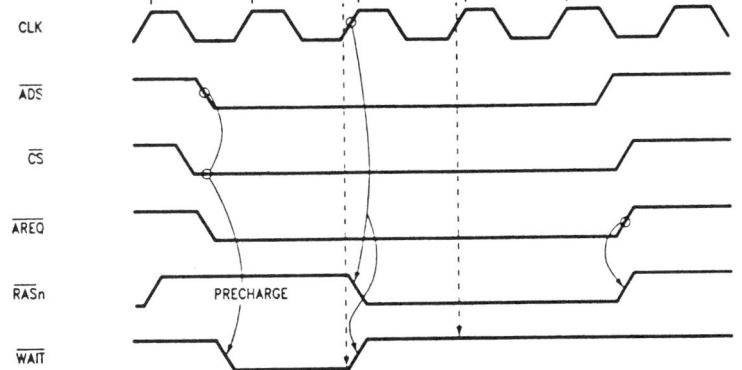

FIGURE 17d. Mode 1 Delayed Access with $\overline{\text{WAIT}}$ 0T ($\overline{\text{WAIT}}$ is Sampled at the End of the "T2" Clock State)

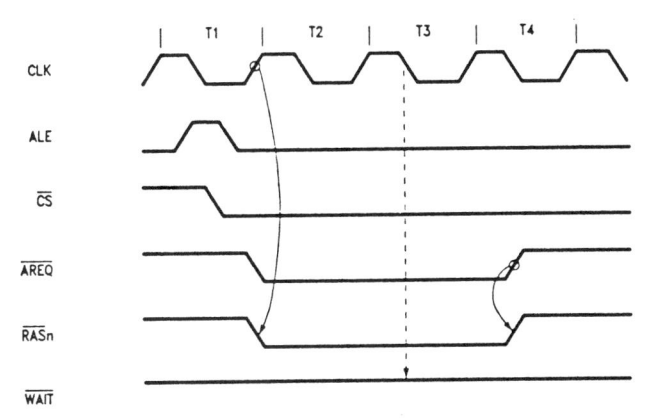

FIGURE 18a. Mode 0 Non-Delayed Access with $\overline{\text{WAIT}}$ 0T ($\overline{\text{WAIT}}$ Is Sampled at the "T3" Falling Clock Edge)

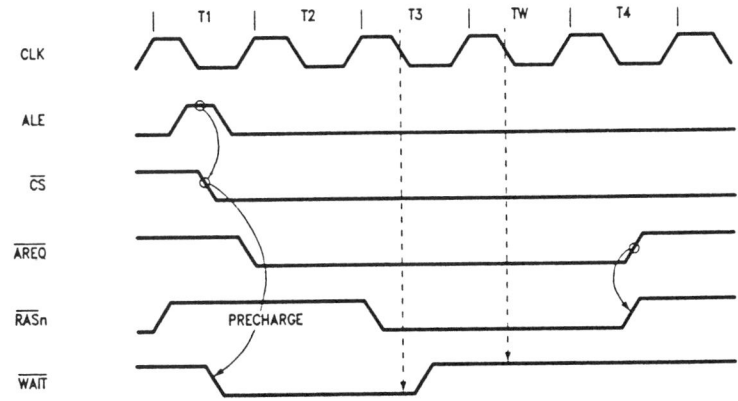

FIGURE 18b. Mode 0 Delayed Access with $\overline{\text{WAIT}}$ ½T ($\overline{\text{WAIT}}$ is Sampled at the "T3" Falling Clock Edge)

0T during non-delayed accesses and ½T during delayed accesses: $\overline{\text{WAIT}}$ will stay negated during a non-delayed access as shown in *Figures 18a* and *18c*. During an access that is delayed, $\overline{\text{WAIT}}$ will assert at the start of the access ($\overline{\text{CS}}$ and ALE or $\overline{\text{ADS}}$) and negate on the negative level of CLK after the positive edge of CLK that asserted $\overline{\text{RAS}}$ for that access as shown in *Figures 18b* and *18d*.

5.0 Port A Wait State Support (Continued)

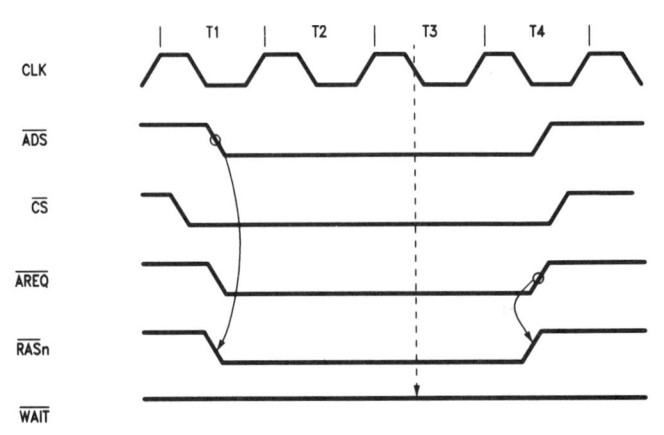

FIGURE 18c. Mode 1 Non-Delayed Access with $\overline{\text{WAIT}}$ 0T ($\overline{\text{WAIT}}$ is Sampled at the "T3" Falling Clock Edge)

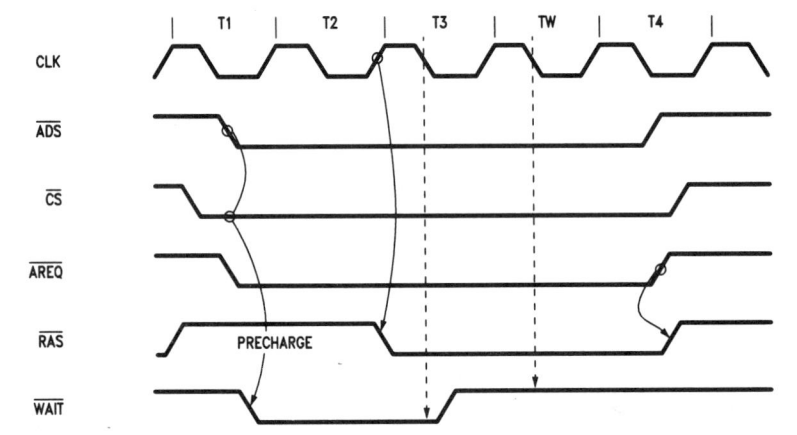

FIGURE 18d. Mode 1 Delayed Access with $\overline{\text{WAIT}}$ ½T ($\overline{\text{WAIT}}$ is Sampled at the "T3" Falling Clock Edge)

5.0 Port A Wait State Support (Continued)

½T during non-delayed and delayed accesses: if mode 0 is used, $\overline{WAIT}$ will assert when ALE is asserted and $\overline{CS}$ is asserted. $\overline{WAIT}$ will then negate on the negative level of CLK after the positive edge of CLK that asserts $\overline{RAS}$ for the access as shown in *Figure 19a*. In Mode 1, $\overline{WAIT}$ will assert from $\overline{CS}$ asserted and $\overline{ADS}$ asserted. $\overline{WAIT}$ will then negate on the negative level of CLK after $\overline{RAS}$ has been asserted for the access as shown in *Figure 19c*. During delayed accesses in both modes, $\overline{WAIT}$ will assert at the start of the access and negate on the negative level of CLK after the positive edge of CLK that started $\overline{RAS}$ for that access as shown in *Figures 19b* and *19d*.

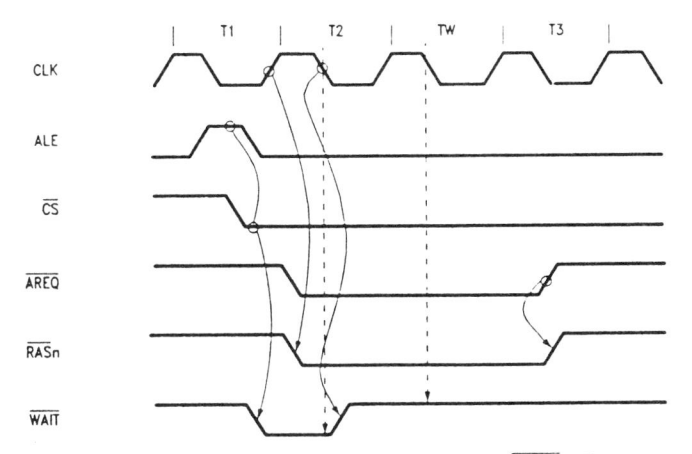

FIGURE 19a. Mode 0 Non-Delayed Access with $\overline{WAIT}$ ½T
($\overline{WAIT}$ is Sampled at the "T2" Falling Clock Edge)

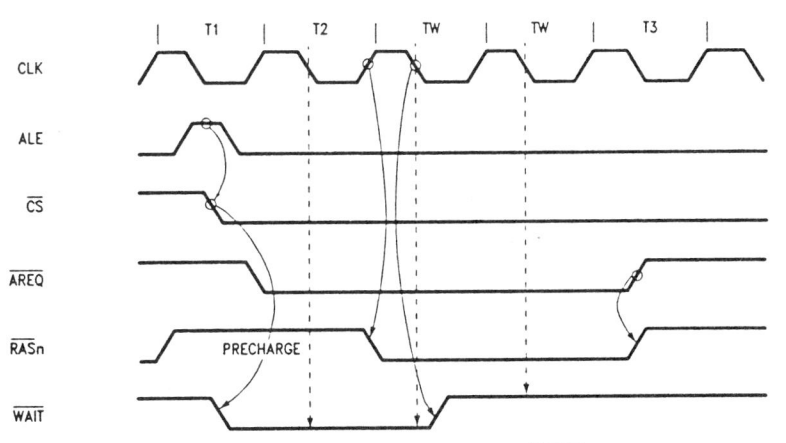

FIGURE 19b. Mode 0 Delayed Access with $\overline{WAIT}$ ½T
($\overline{WAIT}$ is Sampled at the "T2" Falling Clock Edge)

5.0 Port A Wait State Support (Continued)

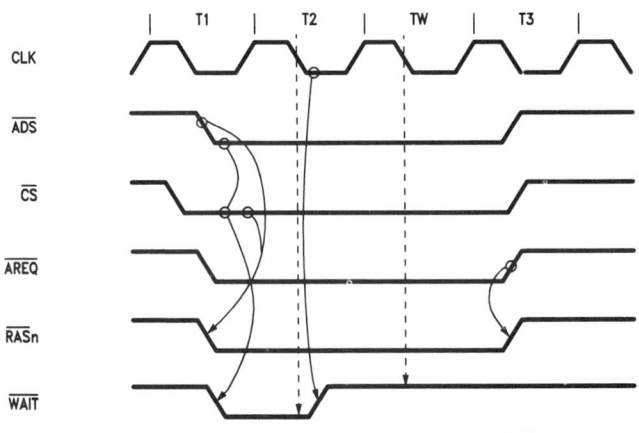

FIGURE 19c. MODE 1 Non-Delayed Access with $\overline{\text{WAIT}}$ ½T
($\overline{\text{WAIT}}$ is Sampled at the "T2" Falling Clock Edge)

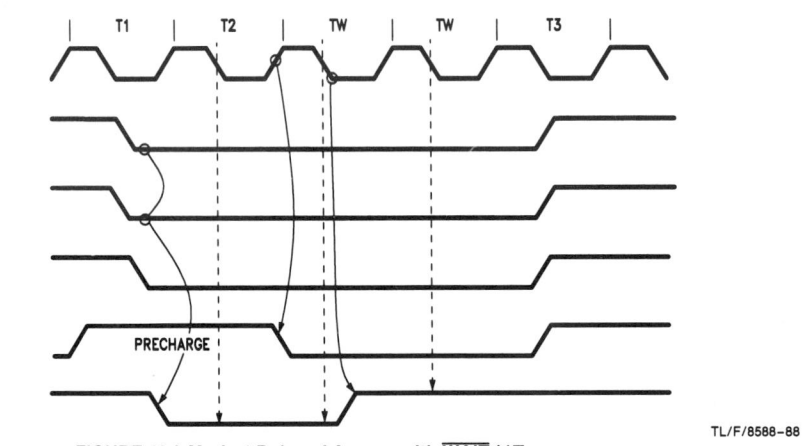

FIGURE 19d. Mode 1 Delayed Access with $\overline{\text{WAIT}}$ ½T
($\overline{\text{WAIT}}$ is Sampled at the "T2" Falling Clock Edge)

5.0 Port A Wait State Support (Continued)

1T during non-delayed and delayed accesses. In Mode 0, $\overline{\text{WAIT}}$ will assert from ALE asserted and $\overline{\text{CS}}$ asserted. $\overline{\text{WAIT}}$ will negate from the next positive edge of CLK that asserts $\overline{\text{RAS}}$ for the access as shown in *Figure 20a*. In Mode 1, $\overline{\text{WAIT}}$ will assert from $\overline{\text{ADS}}$ asserted and $\overline{\text{CS}}$ asserted. $\overline{\text{WAIT}}$ will negate from the first positive edge of CLK after $\overline{\text{ADS}}$ and $\overline{\text{CS}}$ have been asserted as shown in *Figure 20c*. During delayed accesses in both modes, $\overline{\text{WAIT}}$ will assert at the beginning of the access and will negate on the first positive edge of CLK after the positive edge of CLK that starts $\overline{\text{RAS}}$ for the access as shown in *Figures 20b* and *20d*.

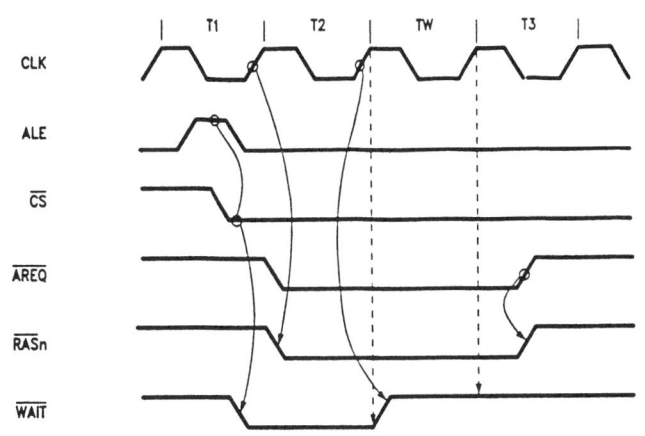

FIGURE 20a. Mode 0 Non-Delayed Access with $\overline{\text{WAIT}}$ 1T ($\overline{\text{WAIT}}$ is Sampled at the End of the "T2" Clock State)

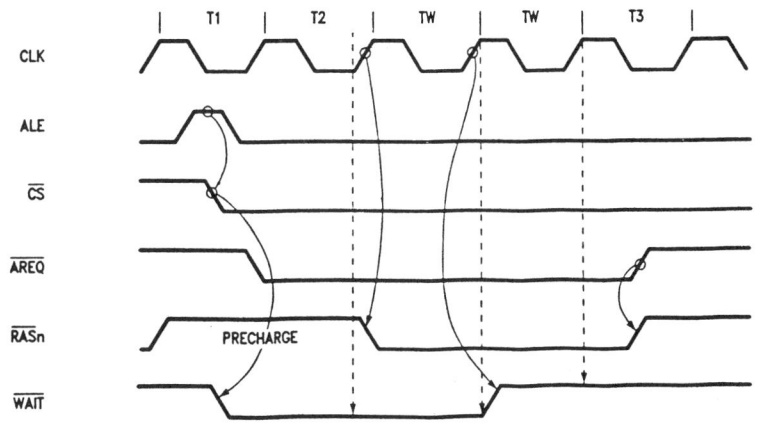

FIGURE 20b. Mode 0 Delayed Access with $\overline{\text{WAIT}}$ 1T ($\overline{\text{WAIT}}$ is Sampled at the End of the "T2" Clock State)

5.0 Port A Wait State Support (Continued)

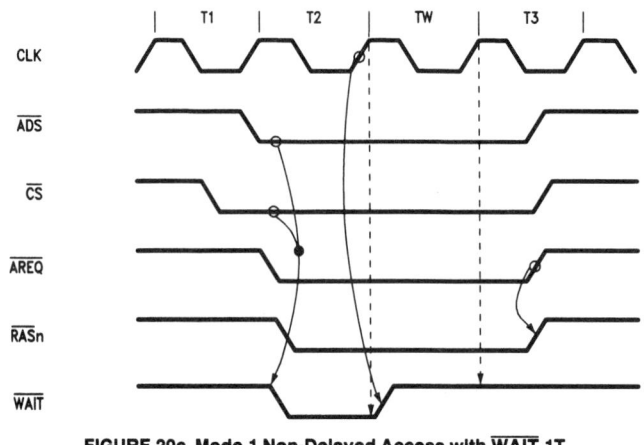

**FIGURE 20c. Mode 1 Non-Delayed Access with WAIT 1T
(WAIT is Sampled at the End of the "T2" Clock State)**

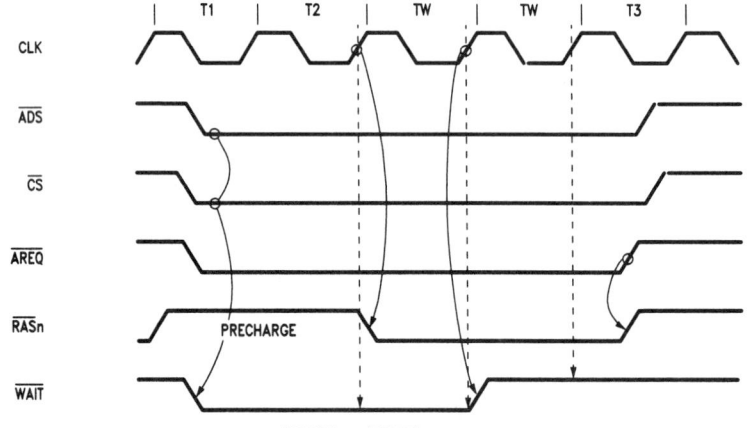

FIGURE 20d. Mode 1 Delayed Access with WAIT 1T (WAIT is Sampled at the End of the "T2" Clock State)

When ending WAIT from a negative level of CLK; if RAS is asserted while CLK is high then WAIT will negate from the negative edge of CLK; if RAS is asserted while CLK is low then WAIT will negate from RAS asserting. When ending WAIT from a positive edge of CLK in Mode 0, the user can think of the positive edge of CLK that starts RAS as 0T and the next positive edge of CLK as 1T. When ending WAIT from a positive edge of CLK in Mode 1, the positive edge of CLK that ADS is setup to can be thought of as 1T in a non-delayed access. In a delayed access, the positive edge of CLK that starts RAS can be thought of as 0T and the next positive edge as 1T.

5.1.2 Wait during Page Burst Accesses

WAIT can be programmed to function differently during page/burst types of accesses. During a page/burst access, the ECAS inputs will be asserted then negated while AREQ is asserted. Through address bits R4 and R5, WAIT can be programmed to assert and negate during this type of access. The user is given four programming options described below.

No Wait States: In this case, WAIT will remain negated even if the ECAS inputs are toggled as shown in *Figure 21*.

5.0 Port A Wait State Support (Continued)

FIGURE 21. No Wait States during Burst ($\overline{WAIT}$ is Sampled at the End of the "T3" Clock State)

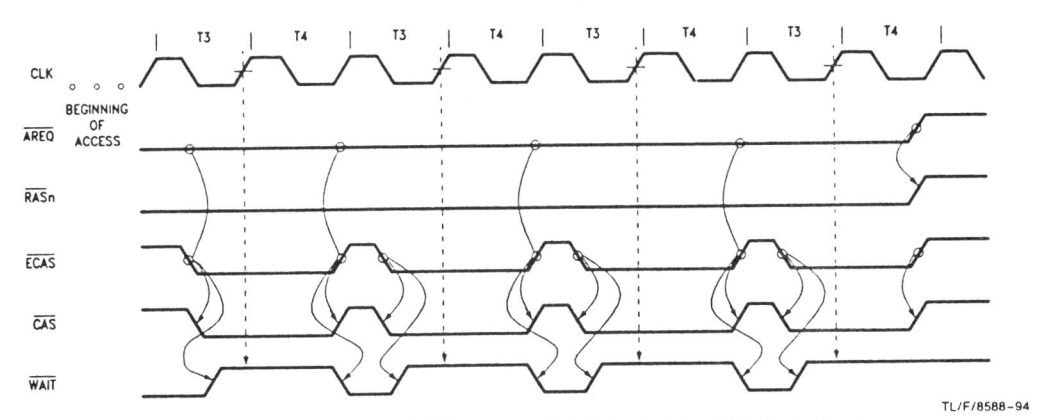

FIGURE 22. 0T during Burst ($\overline{WAIT}$ is Sampled at the End of the "T3" Clock State)

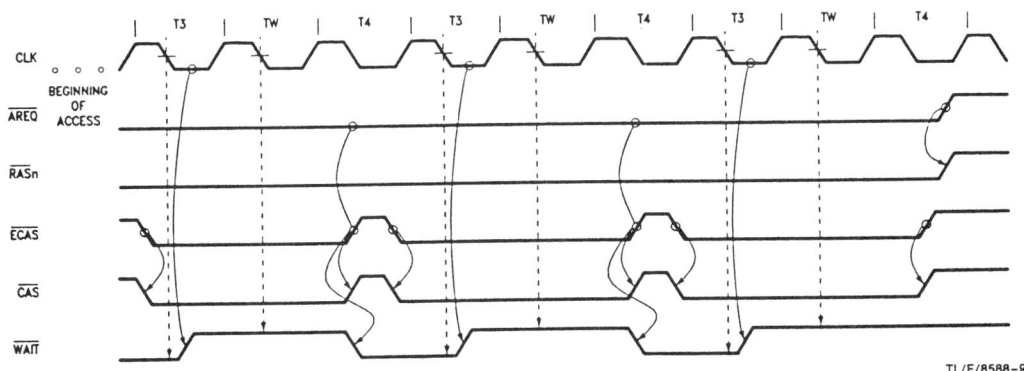

FIGURE 23. ½T during Burst Access ($\overline{WAIT}$ is Sampled at the "T3" Falling Clock Edge)

0T: $\overline{WAIT}$ will be asserted when the $\overline{ECAS}$ inputs are negated with $\overline{AREQ}$ remaining asserted. When a single or group of $\overline{ECAS}$ inputs are asserted, $\overline{WAIT}$ will be negated as shown in *Figure 22*.

½T: $\overline{WAIT}$ will be asserted when the $\overline{ECAS}$ inputs are negated with $\overline{AREQ}$ remaining asserted. When a single or group of $\overline{ECAS}$ inputs are asserted again, $\overline{WAIT}$ will be negated from the first negative level of CLK after a single $\overline{ECAS}$ or group of $\overline{ECAS}$s are asserted as shown in *Figure 23*.

5.0 Port A Wait State Support (Continued)

1T: $\overline{\text{WAIT}}$ will be asserted when the $\overline{\text{ECAS}}$ inputs are negated with $\overline{\text{AREQ}}$ remaining asserted. When a single or group of $\overline{\text{ECAS}}$ inputs are asserted again, $\overline{\text{WAIT}}$ will be negated from the first positive edge of CLK after a single $\overline{\text{ECAS}}$ or group of $\overline{\text{ECAS}}$s are asserted as shown in *Figure 24*.

When ending $\overline{\text{WAIT}}$ from a negative level of CLK; if the $\overline{\text{ECAS}}$s are asserted while CLK is high then $\overline{\text{WAIT}}$ will negate from the negative edge of CLK, if the $\overline{\text{ECAS}}$s are asserted while CLK is low then $\overline{\text{WAIT}}$ will negate from the $\overline{\text{ECAS}}$s asserting. When ending $\overline{\text{WAIT}}$ from a positive edge of CLK, the positive edge of CLK that $\overline{\text{ECAS}}$ is setup to can be thought of as 1T.

5.2 $\overline{\text{DTACK}}$ TYPE OUTPUT

With the R7 address bit asserted during programming, the user selects the $\overline{\text{DTACK}}$ type output. As long as $\overline{\text{DTACK}}$ is sampled negated by the CPU, wait states are inserted into the current access cycle as shown in *Figure 25*. Once $\overline{\text{DTACK}}$ is sampled asserted, the access cycle is completed by the CPU. $\overline{\text{DTACK}}$, which is normally negated, is programmed to assert a number of positive edges and/or negative levels from the event that starts $\overline{\text{RAS}}$ for the access. $\overline{\text{DTACK}}$ can also be programmed to function during page/burst mode accesses. Once $\overline{\text{DTACK}}$ is asserted and the $\overline{\text{ECAS}}$ inputs are negated with $\overline{\text{AREQ}}$ asserted, $\overline{\text{DTACK}}$ can be programmed to negate and assert from the $\overline{\text{ECAS}}$ inputs toggling to perform a page/burst mode operation. Once $\overline{\text{AREQ}}$ is negated, ending the access, $\overline{\text{DTACK}}$ will be negated and stays negated until the next chip selected access.

5.2.1 $\overline{\text{DTACK}}$ during Single Accesses

$\overline{\text{DTACK}}$ can be programmed to delay a number of positive edges and/or negative levels of CLK. These options are programmed through address bits R2 and R3 at programming time. The user is given four options described by the following.

0T during non-delayed accesses and delayed accesses: in Mode 0, $\overline{\text{DTACK}}$ will assert from the positive edge of CLK which starts $\overline{\text{RAS}}$ as shown in *Figure 26a*. In Mode 1, $\overline{\text{DTACK}}$ will assert from $\overline{\text{ADS}}$ and $\overline{\text{CS}}$ as shown in *Figure 26c*. During delayed accesses in both modes, $\overline{\text{DTACK}}$ will assert from the positive edge of CLK which starts $\overline{\text{RAS}}$ for the access as shown in *Figure 26b* and *26d*.

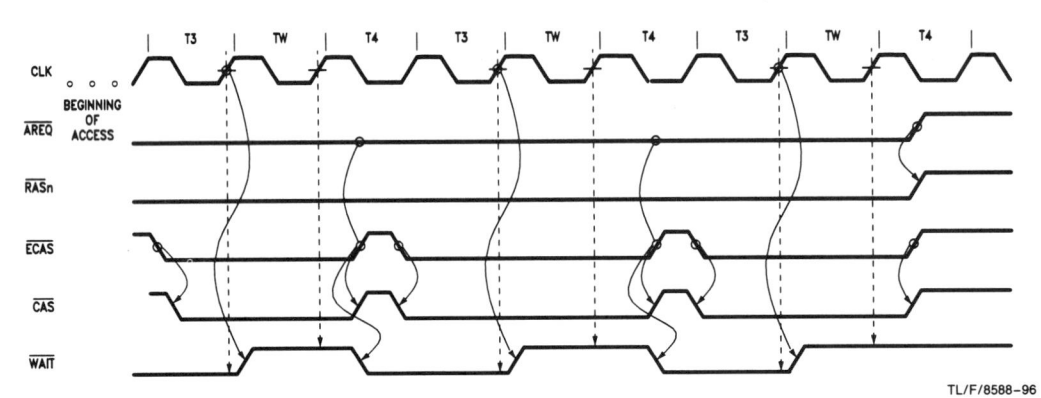

FIGURE 24. 1T during Burst Access ($\overline{\text{WAIT}}$ is Sampled at the End of the "T3" Clock State)

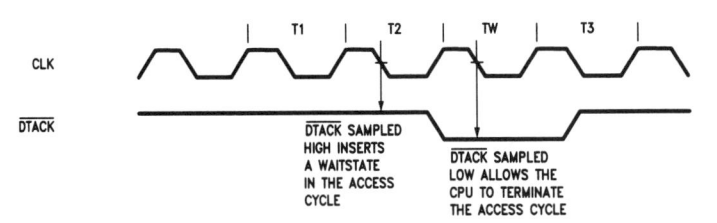

FIGURE 25. $\overline{\text{DTACK}}$ Type Output

5.0 Port A Wait State Support (Continued)

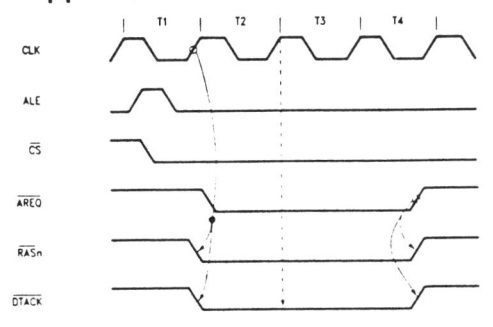

FIGURE 26a. Mode 0 Non-Delayed Access with $\overline{DTACK}$ 0T ($\overline{DTACK}$ is Sampled at the End of the "T2" Clock State)

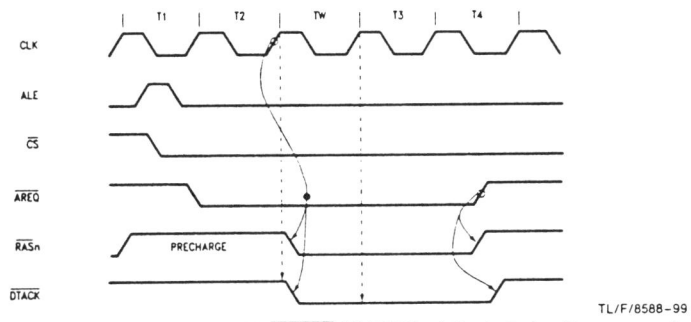

FIGURE 26b. Mode 0 Delayed Access with $\overline{DTACK}$ 0T (2T Clock Periods Are Programmed for $\overline{RAS}$ Precharge, $\overline{DTACK}$ is Sampled at the End of the "T2" Clock State)

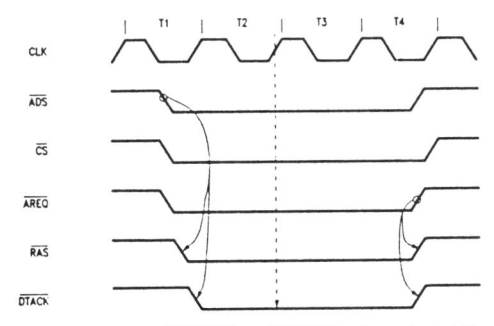

FIGURE 26c. Mode 1 Non-Delayed Access with $\overline{DTACK}$ 0T ($\overline{DTACK}$ is Sampled at the End of the "T2" Clock State)

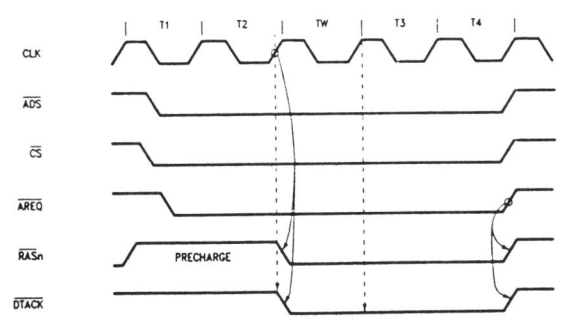

FIGURE 26d. Mode 1 Delayed Access with $\overline{DTACK}$ 0T ($\overline{DTACK}$ is Sampled at the End of the "T2" Clock State)

5.0 Port A Wait State Support (Continued)

½T during non-delayed and delayed accesses: In Mode 0, $\overline{\text{DTACK}}$ will assert on the negative level of CLK after the positive edge of CLK which starts $\overline{\text{RAS}}$ as shown in *Figure 27a*. In Mode 1, $\overline{\text{DTACK}}$ will assert from the negative level of CLK after $\overline{\text{ADS}}$ has been asserted given that $\overline{\text{RAS}}$ is asserted as shown in *Figures 27c* and *27d*. During delayed accesses in both modes, $\overline{\text{DTACK}}$ will assert from the negative level of CLK after the positive edge of CLK which starts $\overline{\text{RAS}}$ for the access as shown in *Figures 27b* and *27e*.

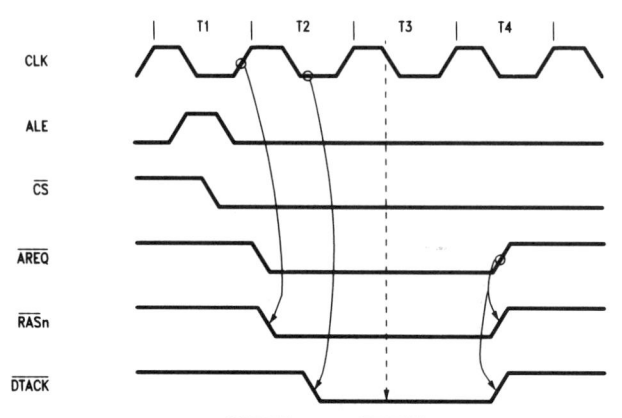

FIGURE 27a. Mode 0 Non-Delayed Access with $\overline{\text{DTACK}}$ of ½T ($\overline{\text{DTACK}}$ is Sampled at the "T3" Falling Clock Edge)

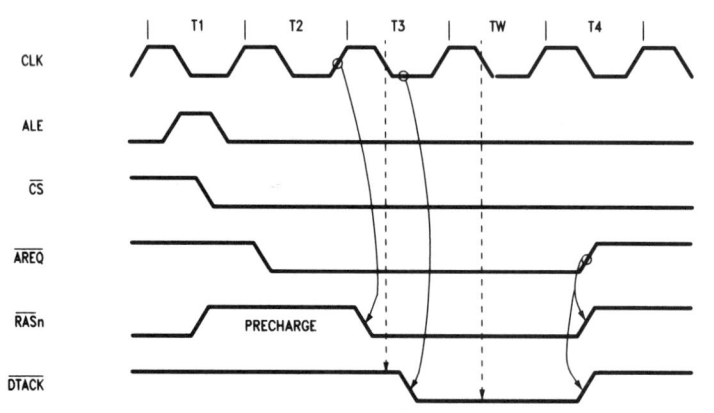

FIGURE 27b. Mode 0 Delayed Access with $\overline{\text{DTACK}}$ of ½T ($\overline{\text{DTACK}}$ is Sampled at the "T3" Falling Clock Edge)

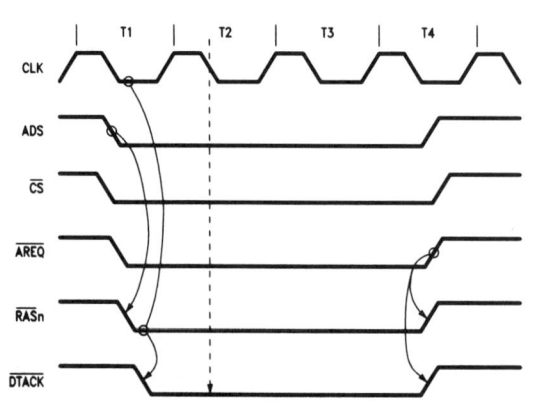

FIGURE 27c. Mode 1 Non-Delayed Access with $\overline{\text{DTACK}}$ of ½T ($\overline{\text{DTACK}}$ is Sampled at the "T2" Falling Clock Edge)

5.0 Port A Wait State Support (Continued)

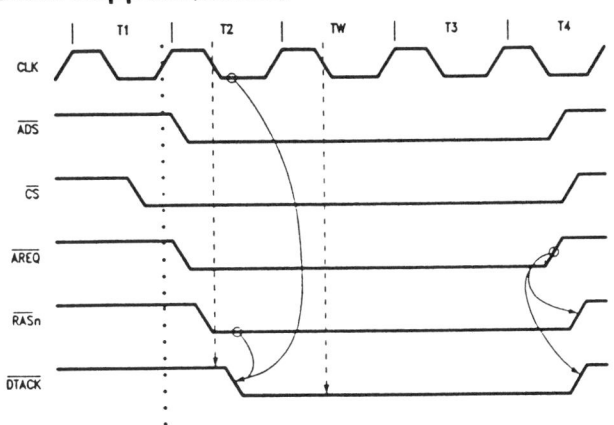

FIGURE 27d. Mode 1 Non-Delayed Access with $\overline{\text{DTACK}}$ of ½T ($\overline{\text{DTACK}}$ is Sampled at the "T2" Falling Clock Edge)

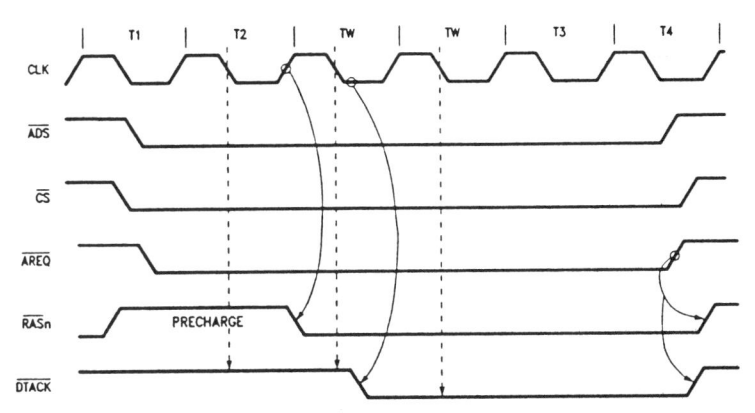

FIGURE 27e. Mode 1 Delayed Access with $\overline{\text{DTACK}}$ of ½T ($\overline{\text{DTACK}}$ is Sampled at the "T2" Falling Clock Edge)

1T during delayed and non-delayed accesses: In Mode 0, $\overline{\text{DTACK}}$ will assert from the first positive edge of CLK after the positive edge of CLK which starts $\overline{\text{RAS}}$ for the access as shown in *Figure 28a*. In Mode 1, $\overline{\text{DTACK}}$ will assert from the positive edge CLK after $\overline{\text{ADS}}$ and $\overline{\text{CS}}$ are asserted as shown in *Figures 28c* and *28d*. During delayed accesses in both modes, $\overline{\text{DTACK}}$ will assert from the first positive edge of CLK after the positive edge of CLK which starts $\overline{\text{RAS}}$ for the access as shown in *Figures 28b* and *28e*.

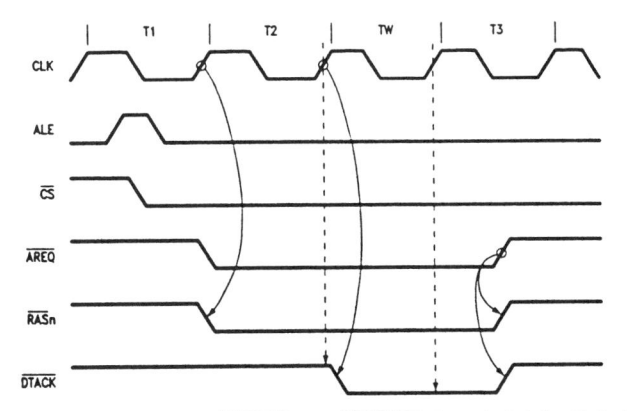

FIGURE 28a. Mode 0 Non-Delayed Access with $\overline{\text{DTACK}}$ of 1T ($\overline{\text{DTACK}}$ is Sampled at the End of the "T2" Clock State)

5.0 Port A Wait State Support (Continued)

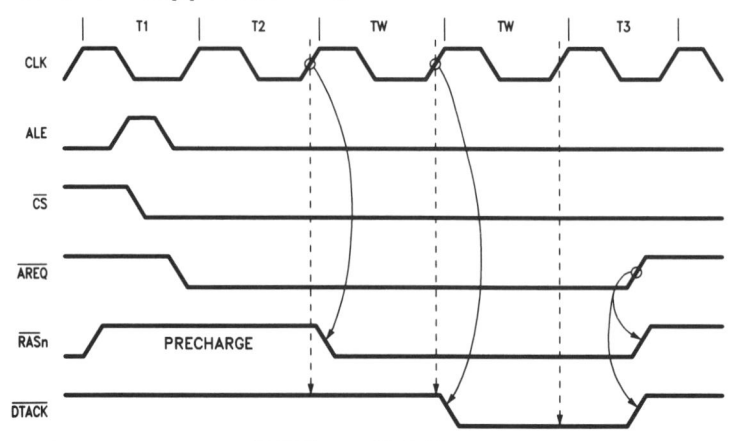

FIGURE 28b. Mode 0 Delayed Access with $\overline{\text{DTACK}}$ of 1T ($\overline{\text{DTACK}}$ is Sampled at the End of the "T2" Clock State)

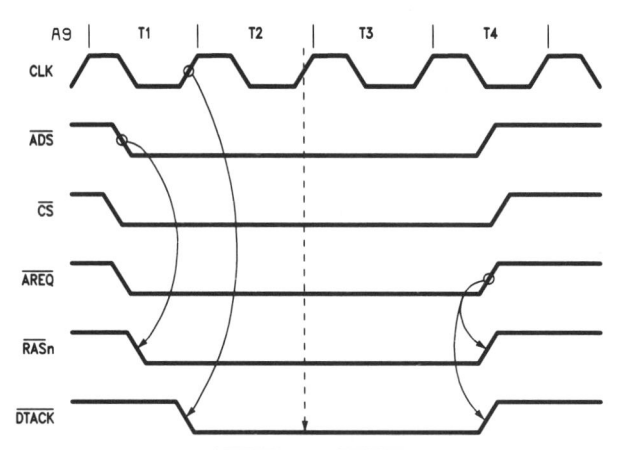

FIGURE 28c. Mode 1 Non-Delayed Access with $\overline{\text{DTACK}}$ of 1T ($\overline{\text{DTACK}}$ is Sampled at the End of the "T2" Clock State)

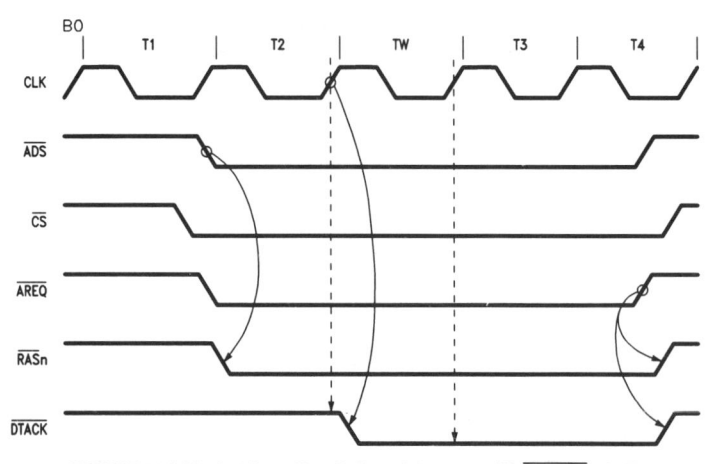

FIGURE 28d. Mode 1 Late Non-Delayed Access with $\overline{\text{DTACK}}$ of 1T ($\overline{\text{DTACK}}$ is Sampled at the End of the "T2" Clock State)

5.0 Port A Wait State Support (Continued)

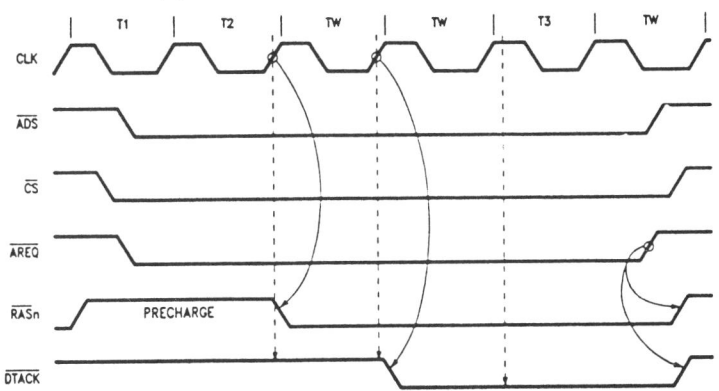

FIGURE 28e. Mode 1 Delayed Access with $\overline{DTACK}$ of 1T ($\overline{DTACK}$ is Sampled at the End of the "T2" Clock State)

1½T during delayed and non-delayed accesses: In Mode 0, $\overline{DTACK}$ will assert from the negative level after the first positive edge of CLK after the positive edge of CLK which starts $\overline{RAS}$ for the access as shown in *Figure 29a*. In Mode 1, $\overline{DTACK}$ will assert from the negative level after the first positive edge of CLK after $\overline{ADS}$ and $\overline{CS}$ are asserted as shown in *Figures 29c* and *29d*. During delayed accesses in both modes, $\overline{DTACK}$ will assert from the negative level after the first positive edge of CLK after the positive edge of CLK which starts $\overline{RAS}$ for the access as shown in *Figures 29b* and *29e*.

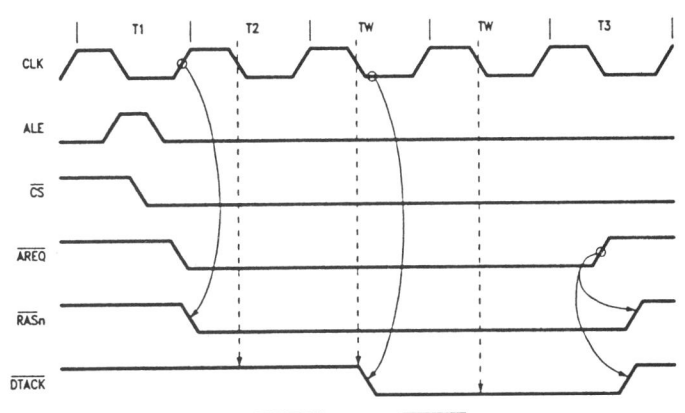

FIGURE 29a. Mode 0 Non-Delayed Access with $\overline{DTACK}$ of 1½T ($\overline{DTACK}$ is Sampled at the "T2" Falling Clock Edge)

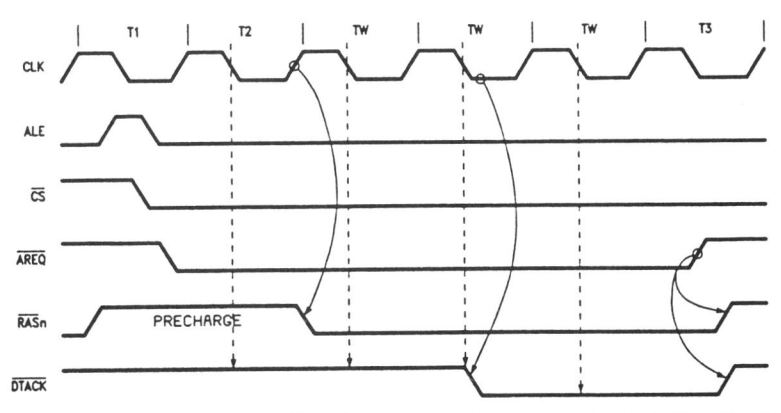

FIGURE 29b. Mode 0 Delayed Access with $\overline{DTACK}$ of 1½T ($\overline{DTACK}$ is Sampled at the "T2" Falling Clock Edge)

5.0 Port A Wait State Support (Continued)

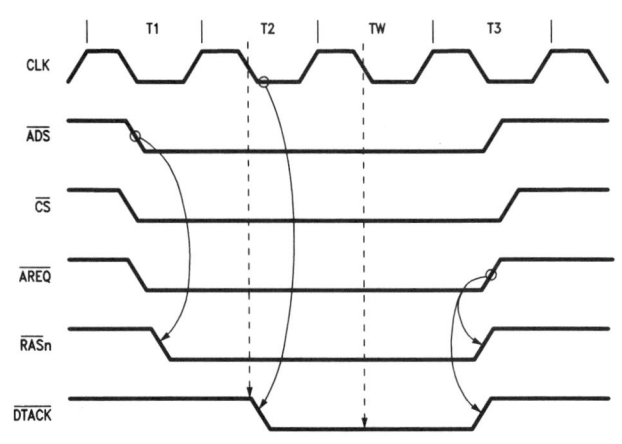

FIGURE 29c. Mode 1 Non-Delayed Access with $\overline{\text{DTACK}}$ of 1½T ($\overline{\text{DTACK}}$ is Sampled at the "T2" Falling Clock Edge)

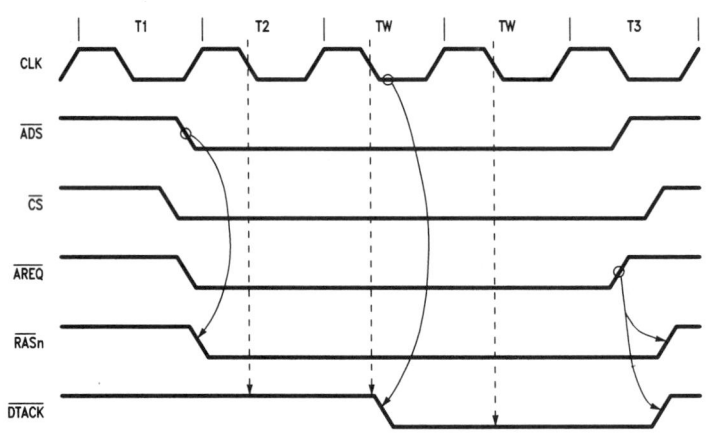

FIGURE 29d. Mode 1 Non-Delayed Access with $\overline{\text{DTACK}}$ of 1½T ($\overline{\text{DTACK}}$ is Sampled at the "T2" Falling Clock Edge)

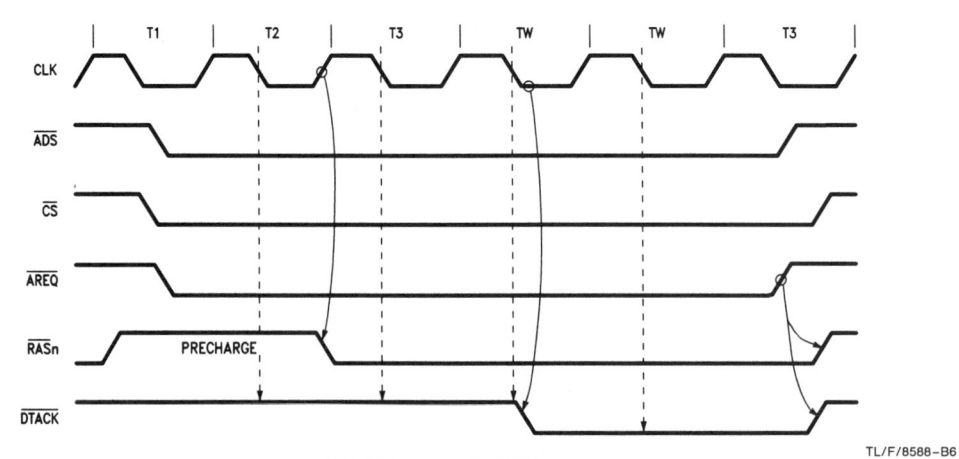

FIGURE 29e. Mode 1 Delayed Access with $\overline{\text{DTACK}}$ of 1½T ($\overline{\text{DTACK}}$ is Sampled at the "T2" Falling Clock Edge)

5.0 Port A Wait State Support (Continued)

When starting $\overline{\text{DTACK}}$ from a negative level of CLK; if $\overline{\text{RAS}}$ is asserted while CLK is high then $\overline{\text{DTACK}}$ will assert from the negative edge of CLK, if $\overline{\text{RAS}}$ is asserted while CLK is low, then $\overline{\text{DTACK}}$ will assert from $\overline{\text{RAS}}$ asserting. When starting $\overline{\text{DTACK}}$ from a positive edge of CLK in Mode 0, the positive edge of CLK that starts $\overline{\text{RAS}}$ can be thought of as 0T. In Mode 1 during non-delayed accesses, the positive edge of CLK that $\overline{\text{ADS}}$ is setup to can be thought of as 1T. During delayed accesses, the positive edge of CLK that starts $\overline{\text{RAS}}$ can be thought of as 0T and the next positive edge of CLK as 1T.

5.2.2 $\overline{\text{DTACK}}$ during Page/Burst Accesses

$\overline{\text{DTACK}}$ can be programmed to function differently during page/burst types of accesses. During a page/burst access, the $\overline{\text{ECAS}}$ inputs will be asserted then negated while $\overline{\text{AREQ}}$ remains asserted. Through address bits R4 and R5, $\overline{\text{DTACK}}$ can be programmed to negate and assert during this type of access. The user is given four programming options described below.

No Wait States: In this case, $\overline{\text{DTACK}}$ will remain asserted even if the $\overline{\text{ECAS}}$ inputs are negated with $\overline{\text{AREQ}}$ asserted as shown in *Figure 30*.

0T: $\overline{\text{DTACK}}$ will be negated when the $\overline{\text{ECAS}}$ inputs are negated with $\overline{\text{AREQ}}$ asserted. When a single or group of $\overline{\text{ECAS}}$ inputs are asserted again, $\overline{\text{DTACK}}$ will be asserted as shown in *Figure 31*.

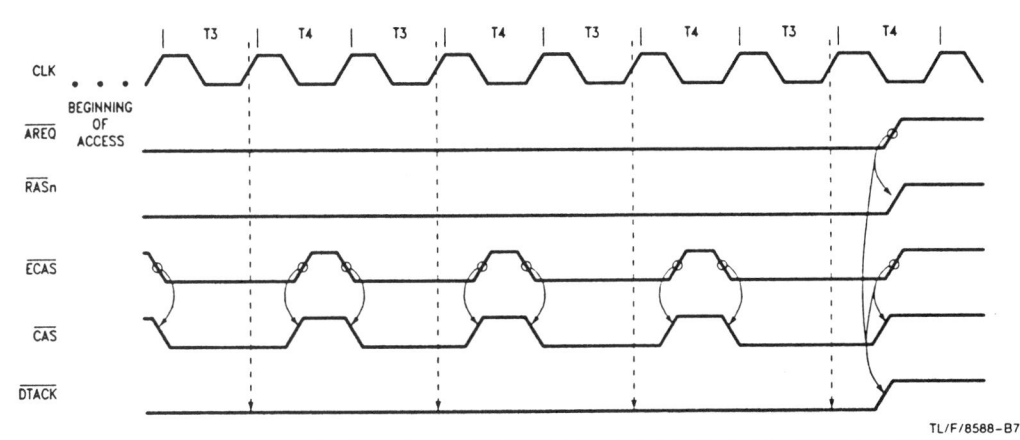

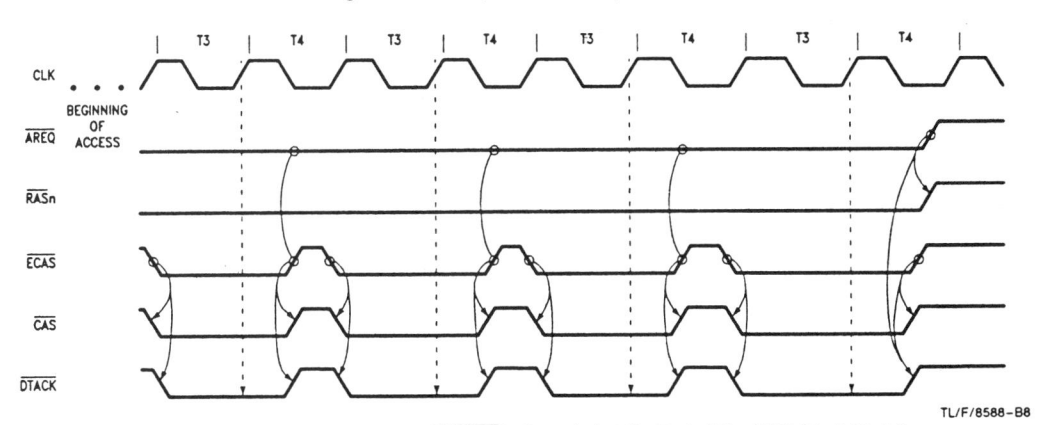

FIGURE 30. No Wait States during Burst Access ($\overline{\text{DTACK}}$ is Sampled at the End of the "T3" Clock State)

FIGURE 31. 0T during Burst Access ($\overline{\text{DTACK}}$ is Sampled at the End of the "T3" Clock State)

5.0 Port A Wait State Support (Continued)

½T: $\overline{\text{DTACK}}$ will be negated when the $\overline{\text{ECAS}}$ inputs are negated with $\overline{\text{AREQ}}$ asserted. When a single or group of $\overline{\text{ECAS}}$ inputs are asserted again, $\overline{\text{DTACK}}$ will be asserted from the first negative level of CLK after the single or group of $\overline{\text{ECAS}}$s are asserted as shown in *Figure 32*.

1T: $\overline{\text{DTACK}}$ will be negated when the $\overline{\text{ECAS}}$ inputs are negated with $\overline{\text{AREQ}}$ asserted. When a single or group of $\overline{\text{ECAS}}$ inputs are asserted again, $\overline{\text{DTACK}}$ will be asserted from the first positive edge of CLK after the single or group of $\overline{\text{ECAS}}$s are asserted as shown in *Figure 33*.

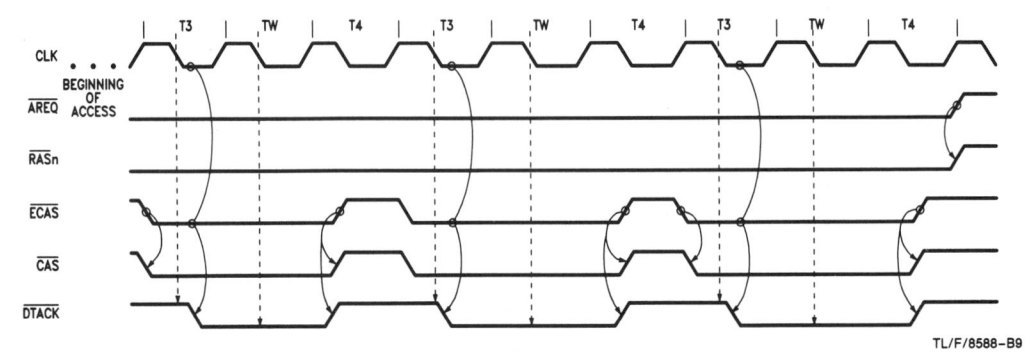

FIGURE 32. ½T during Burst Access ($\overline{\text{DTACK}}$ is Sampled at the "T3" Falling Clock Edge)

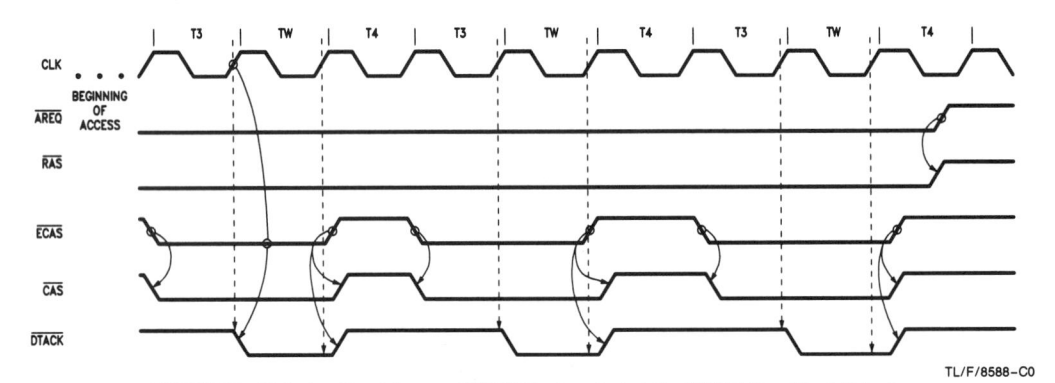

FIGURE 33. 1T during Burst Access ($\overline{\text{DTACK}}$ is Sampled at the "T3" Falling Clock Edge)

5.0 Port A Wait State Support (Continued)

When starting $\overline{\text{DTACK}}$ from a negative level of CLK; if the $\overline{\text{ECAS}}$s are asserted while CLK is high then $\overline{\text{DTACK}}$ will assert from the negative edge of CLK, if the $\overline{\text{ECAS}}$s are asserted while CLK is low then $\overline{\text{DTACK}}$ will assert from the $\overline{\text{ECAS}}$s asserting. When starting $\overline{\text{DTACK}}$ from a positive edge of CLK, the positive edge of CLK that $\overline{\text{ECAS}}$ is setup to can be thought of as 1T.

5.3 DYNAMICALLY INCREASING THE NUMBER OF WAIT STATES

The user can increase the number of positive edges of CLK before $\overline{\text{DTACK}}$ is asserted or $\overline{\text{WAIT}}$ is negated. With the input $\overline{\text{WAITIN}}$ asserted, the user can delay $\overline{\text{DTACK}}$ asserting or $\overline{\text{WAIT}}$ negating either one or two more positive edges of CLK. The number of edges is programmed through address bit R6. If the user is increasing the number of positive edges in a delay that contains a negative level, the positive edges will be met before the negative level. For example if the user programmed $\overline{\text{DTACK}}$ of ½T, asserting $\overline{\text{WAITIN}}$, programmed as 2T, would increase the number of positive edges resulting in $\overline{\text{DTACK}}$ of 2½T as shown in *Figure 34a*. Similarly, $\overline{\text{WAITIN}}$ can increase the number of positive edges in a page/burst access. $\overline{\text{WAITIN}}$ can be permanently asserted in systems requiring an increased number of wait states. $\overline{\text{WAITIN}}$ can also be asserted and negated, depending on the type of access. As an example, a user could invert the $\overline{\text{WRITE}}$ line from the CPU and connect the output to $\overline{\text{WAITIN}}$. This could be used to perform write accesses with 1 wait state and read accesses with 2 wait states as shown in *Figure 34b*.

5.4 GUARANTEEING $\overline{\text{RAS}}$ LOW TIME AND $\overline{\text{RAS}}$ PRECHARGE TIME

The DP8420A/21A/22A will guarantee $\overline{\text{RAS}}$ precharge time between accesses; between refreshes; and between access and refreshes. The programming bits R0 and R1 are used to program combinations of $\overline{\text{RAS}}$ precharge time and $\overline{\text{RAS}}$ low time referenced by positive edges of CLK. $\overline{\text{RAS}}$ low time is programmed for refreshes only. During an access, the system designer guarantees the time $\overline{\text{RAS}}$ is asserted through the DP8420A/21A/22A wait logic. Since inserting wait states into an access increases the length of the CPU signals which are used to create $\overline{\text{ADS}}$ or ALE and $\overline{\text{AREQ}}$, the time that $\overline{\text{RAS}}$ is asserted can be guaranteed.

Precharge time is also guaranteed by the DP8420A/21A/22A. Each $\overline{\text{RAS}}$ output has a separate positive edge of CLK counter. $\overline{\text{AREQ}}$ is negated setup to a positive edge of CLK to terminate the access. That positive edge is 1T. The next positive edge is 2T. $\overline{\text{RAS}}$ will not be asserted until the programmed number of positive edges of CLK have passed as shown in *Figures 35, 37a,* and *37b*. Once the programmed precharge time has been met, $\overline{\text{RAS}}$ will be asserted from the positive edge of CLK. However, since there is a precharge counter per $\overline{\text{RAS}}$, an access using another $\overline{\text{RAS}}$ will not be delayed. Precharge time before a refresh is always referenced from the access $\overline{\text{RAS}}$ negating before $\overline{\text{RAS0}}$ for the refresh asserting. After a refresh, precharge time is referenced from $\overline{\text{RAS3}}$ negating, for the refresh, to the access $\overline{\text{RAS}}$ asserting.

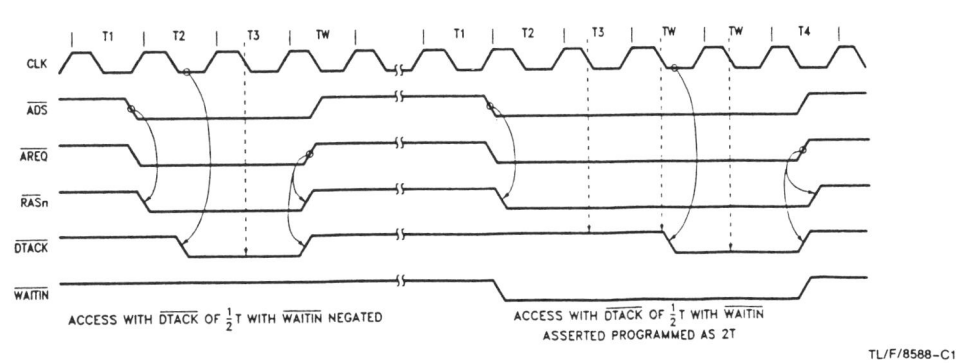

FIGURE 34a. $\overline{\text{WAITIN}}$ Example ($\overline{\text{DTACK}}$ is Sampled at the "T3" Falling Clock Edge)

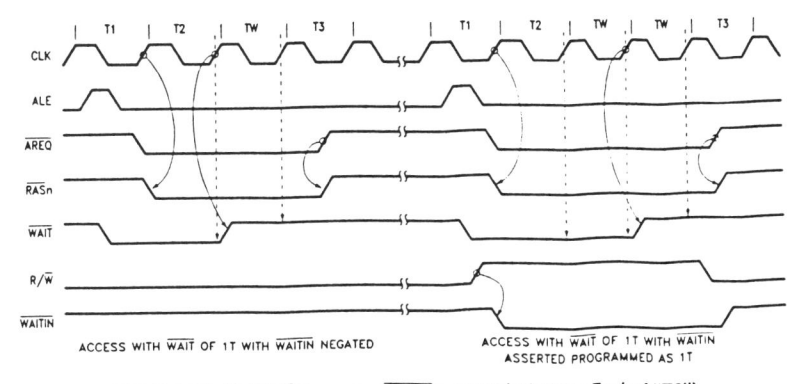

FIGURE 34b. $\overline{\text{WAITIN}}$ Example ($\overline{\text{WAIT}}$ is Sampled at the End of "T2")

5.0 Port A Wait State Support (Continued)

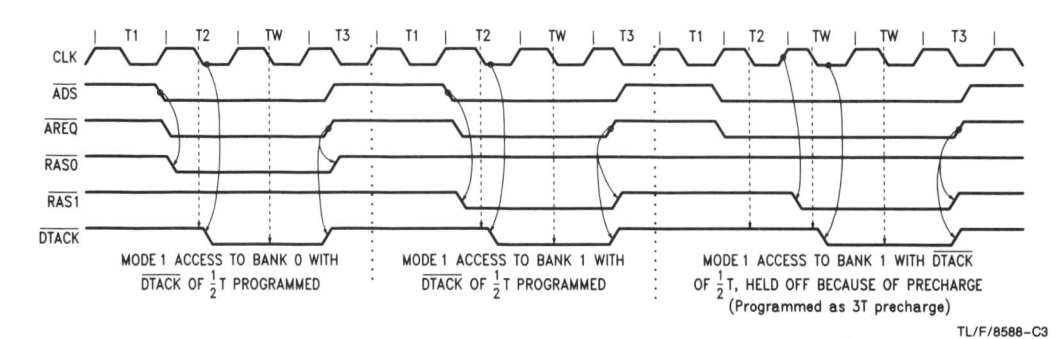

FIGURE 35. Guaranteeing $\overline{RAS}$ Precharge ($\overline{DTACK}$ is Sampled at the "T2" Falling Clock Edge)

6.0 Additional Access Support Features

To support the different modes of accessing, the DP8420A/21A/22A have multiple access features. These features allow the user to take advantage of CPU or DRAM functions. These additional features include: address latches and column increment for page/burst mode support; address pipelining to allow a new access to start to a different bank of DRAM after $\overline{CAS}$ has been asserted and the column address hold time has been met; and delay $\overline{CAS}$, to allow the user with a multiplexed bus to ensure valid data is present before $\overline{CAS}$ is asserted.

6.1 ADDRESS LATCHES AND COLUMN INCREMENT

The address latches can be programmed, through programming bit B0, to either latch the address or remain permanently in fall-through mode. If the address latches are used to latch the address, the rising edge of ALE in Mode 0 places the latches in fall-through. Once ALE is negated, the address present on the row, column and bank inputs is latched. In Mode 1, the address latches are in fall-through mode until $\overline{ADS}$ is asserted. $\overline{ADS}$ asserted latches the address.

Once the address is latched, the column address can be incremented with the input COLINC. With COLINC asserted, the column address is incremented. If COLINC is asserted with all of the bits of the column address asserted, the column address will return to zero. COLINC can be used for sequential accesses of static column DRAMs. COLINC can also be used with the $\overline{ECAS}$ inputs to support sequential accesses to page mode DRAMs as shown in *Figure 36*. COLINC should only be asserted when the signal $\overline{RFIP}$ is negated during an access since this input functions as extend refresh when $\overline{RFIP}$ is asserted. COLINC must be low (negated) when the address is being latched ($\overline{ADS}$ falling edge in Mode 1).

The address latches function differently with the DP8422A. The DP8422A will latch the address of the currently granted port. If Port A is currently granted, the address will be latched as described in Section 6.1. If Port A is not granted, and requests an access, the address will be latched on the first or second positive edge of CLK after GRANTB has been negated depending on the programming bits R0, R1. For Port B, if GRANTB is asserted, the address will be latched with $\overline{AREQB}$ asserted. If GRANTB is negated, the address will latch on the first or second positive edge of CLK after GRANTB is asserted depending on the programming bits R0, R1.

6.0 Additional Access Support Features (Continued)

6.2 ADDRESS PIPELINING

Address pipelining is the overlapping of accesses to different banks of DRAM. If the majority of successive accesses are to a different bank, the accesses can be overlapped. Because of this overlapping, the cycle time of the DRAM accesses are greatly reduced. The DP8420A/21A/22A can be programmed to allow a new row address to be placed on the DRAM address bus after the column address hold time has been met. At this time, a new access can be initiated with $\overline{ADS}$ or ALE, depending on the access mode, while $\overline{AREQ}$ is used to sustain the current access. The DP8422A supports address pipelining for Port A only. This mode can not be used with page, static column or nibble modes of operations because the DRAM column address is switched back to the row address after $\overline{CAS}$ is asserted. This mode is programmed through address bit R8 (see *Figures 37a* and *37b*). In this mode, the output $\overline{WE}$ always functions as $\overline{RFRQ}$.

During address pipelining in Mode 0, shown in *Figure 37c*, ALE cannot be pulsed high to start another access until $\overline{AREQ}$ has been asserted for the previous access for at least one period of CLK. $\overline{DTACK}$, if programmed, will be negated once $\overline{AREQ}$ is negated. $\overline{WAIT}$, if programmed to insert wait states, will be asserted once ALE and $\overline{CS}$ are asserted.

In Mode 1, shown in *Figure 37d*, $\overline{ADS}$ can be negated once $\overline{AREQ}$ is asserted. After meeting the minimum negated pulse width for $\overline{ADS}$, $\overline{ADS}$ can again be asserted to start a new access. $\overline{DTACK}$, if programmed, will be negated once $\overline{AREQ}$ is negated. $\overline{WAIT}$, if programmed, will be asserted once $\overline{ADS}$ is asserted.

In either mode with either type of wait programmed, the DP8420A/21A/22A will still delay the access for precharge if sequential accesses are to the same bank or if a refresh takes place.

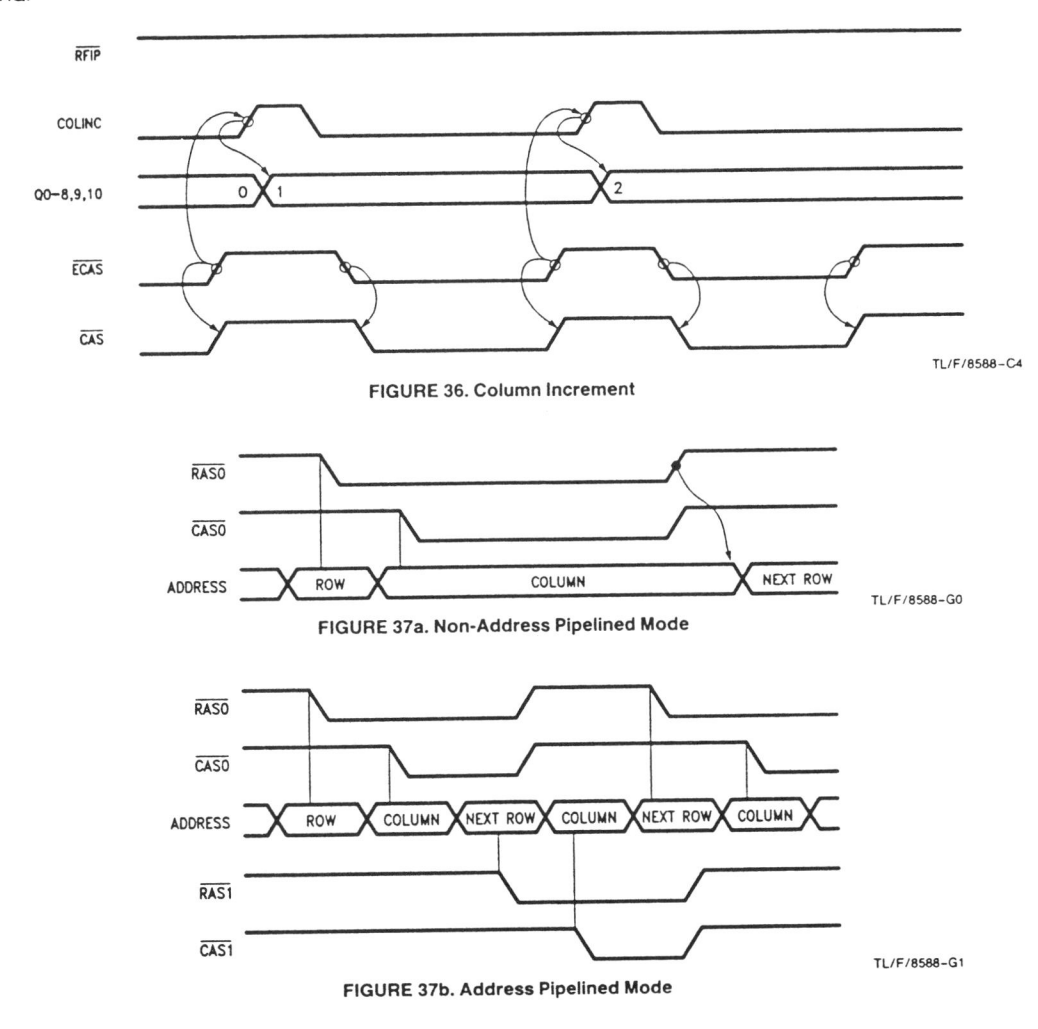

FIGURE 36. Column Increment

FIGURE 37a. Non-Address Pipelined Mode

FIGURE 37b. Address Pipelined Mode

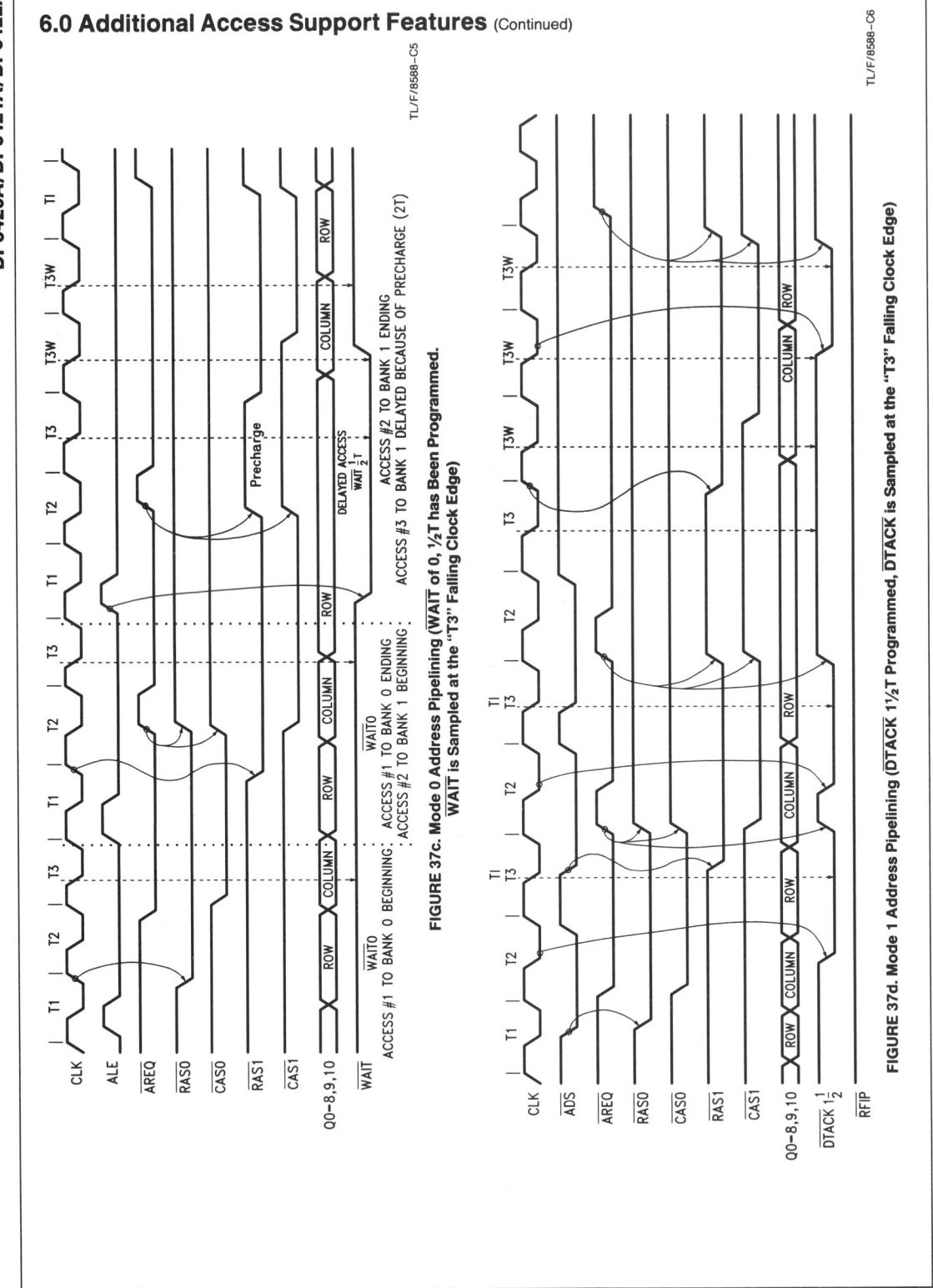

6.0 Additional Access Support Features (Continued)

FIGURE 37c. Mode 0 Address Pipelining (WAIT of 0, 1/2T has Been Programmed. WAIT is Sampled at the "T3" Falling Clock Edge)

FIGURE 37d. Mode 1 Address Pipelining (DTACK 1 1/2T Programmed, DTACK is Sampled at the "T3" Falling Clock Edge)

6.0 Additional Access Support Features (Continued)

6.3 DELAY $\overline{CAS}$ DURING WRITE ACCESSES

Address bit C9 asserted during programming will cause $\overline{CAS}$ to be delayed until the first positive edge of CLK after $\overline{RAS}$ is asserted when the input $\overline{WIN}$ is asserted. Delaying $\overline{CAS}$ during write accesses ensures that the data to be written to DRAM will be setup to $\overline{CAS}$ asserting as shown in *Figures 38a* and *38b*. If the possibility exists that data still may not be present after the first positive edge of CLK, $\overline{CAS}$ can be delayed further with the $\overline{ECAS}$ inputs. If address bit C9 is negated during programming, read and write accesses will be treated the same (with regard to $\overline{CAS}$).

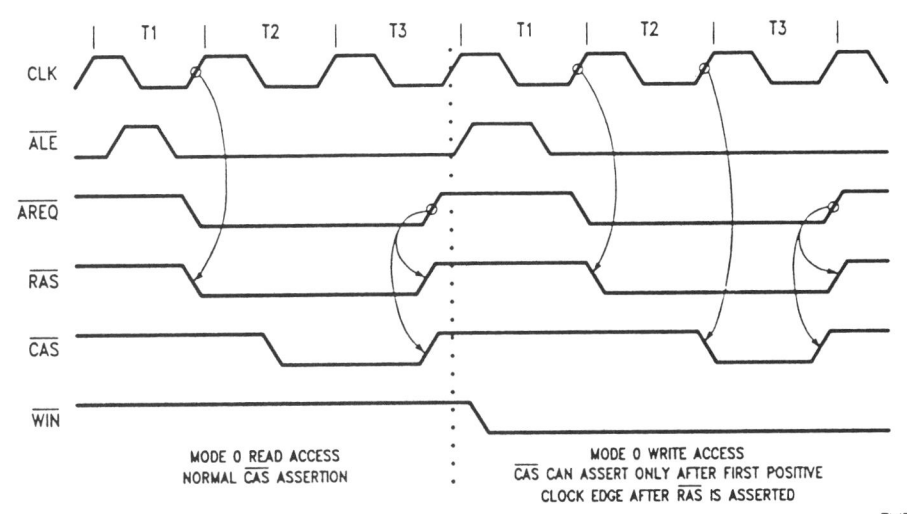

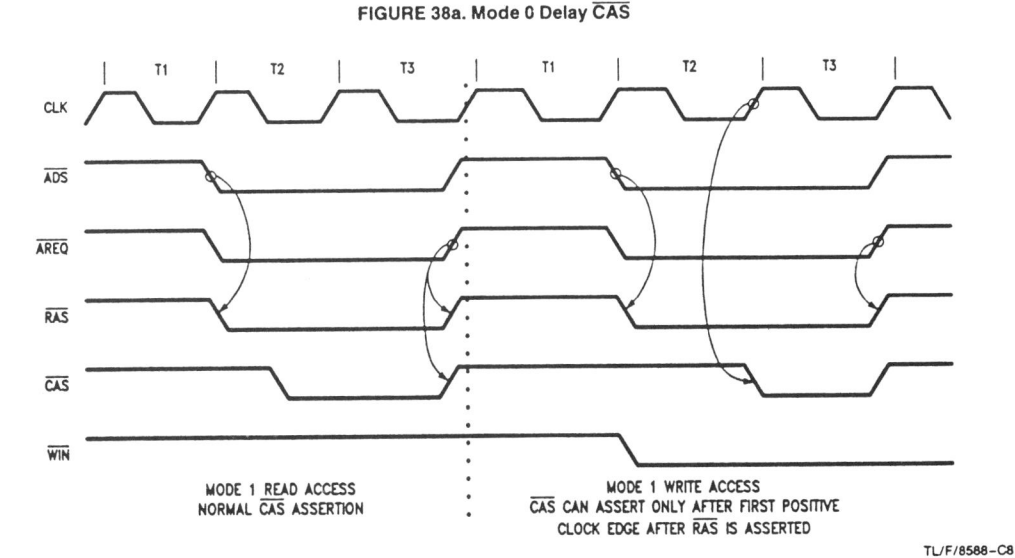

FIGURE 38a. Mode 0 Delay $\overline{CAS}$

FIGURE 38b. Mode 1 Delay $\overline{CAS}$

7.0 $\overline{RAS}$ and $\overline{CAS}$ Configuration Modes

The DP8420A/21A/22A allow the user to configure the DRAM array to contain one, two or four banks of DRAM. Depending on the functions used, certain considerations must be used when determining how to set up the DRAM array. Programming address bits C4, C5 and C6 along with bank selects, B0–1, and $\overline{CAS}$ enables, $\overline{ECAS}$0–3, determine which $\overline{RAS}$ or group of $\overline{RAS}$s and which $\overline{CAS}$ or group of $\overline{CAS}$s will be asserted during an access. Different memory schemes are described. The DP8420A/21A/22A is specified driving a heavy load of 72 DRAMs, representing four banks of DRAM with 16-bit words and 2 parity bits. The DP8420A/21A/22A can drive more than 72 DRAMs, but the AC timing must be increased. Since the $\overline{RAS}$ and $\overline{CAS}$ outputs are configurable, all $\overline{RAS}$ and $\overline{CAS}$ outputs should be used for the maximum amount of drive.

7.1 BYTE WRITING

By selecting a configuration in which all $\overline{CAS}$ outputs are selected during an access, the $\overline{ECAS}$ inputs enable a single or group of $\overline{CAS}$ outputs to select a byte (or bytes) in a word size of up to 32 bits. In this case, the $\overline{RAS}$ outputs are used to select which of up to 4 banks is to be used as shown in *Figures 39a* and *39b*. In systems with a word size of 16 bits, the byte enables can be gated with a high order address bit to produce four byte enables which gives an equivalent to 8 banks of 16-bit words as shown in *Figure 39d*. If less memory is required, each $\overline{CAS}$ should be used to drive each nibble in the 16-bit word as shown in *Figure 39c*.

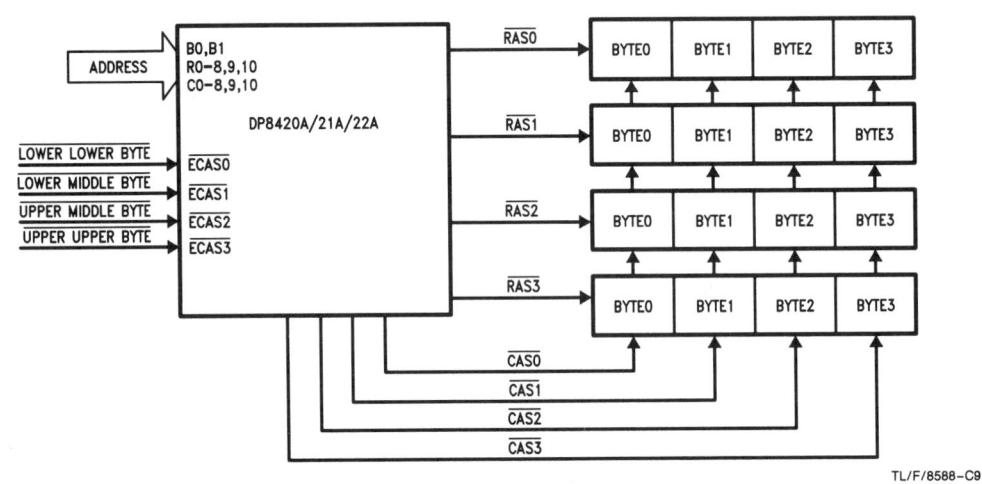

FIGURE 39a. DRAM Array Setup for 32-Bit System (C6, C5, C4 = 1, 1, 0 during Programming)

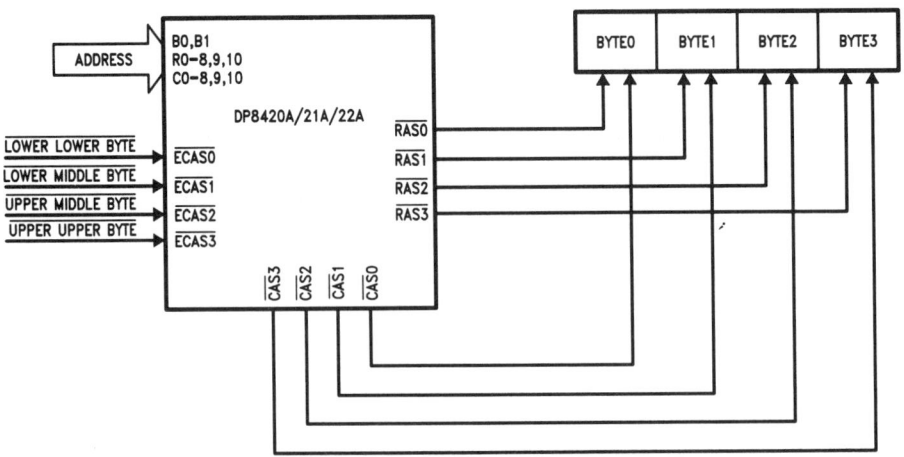

FIGURE 39b. DRAM Array Setup for 32-Bit, 1 Bank System (C6, C5, C4 = 0, 0, 0 Allowing Error Scrubbing or C6, C5, C4 = 0, 1, 1 No Error Scrubbing during Programming)

7.0 $\overline{RAS}$ and $\overline{CAS}$ Configuration Modes (Continued)

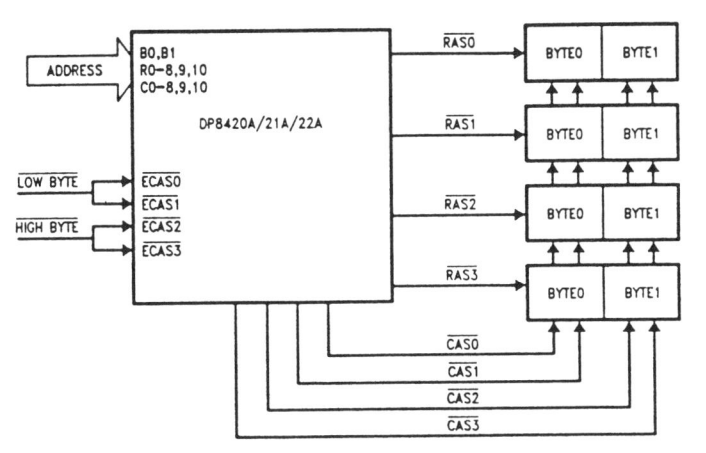

FIGURE 39c. DRAM Array Setup for 16-Bit System (C6, C5, C4 = 1, 1, 0 during Programming)

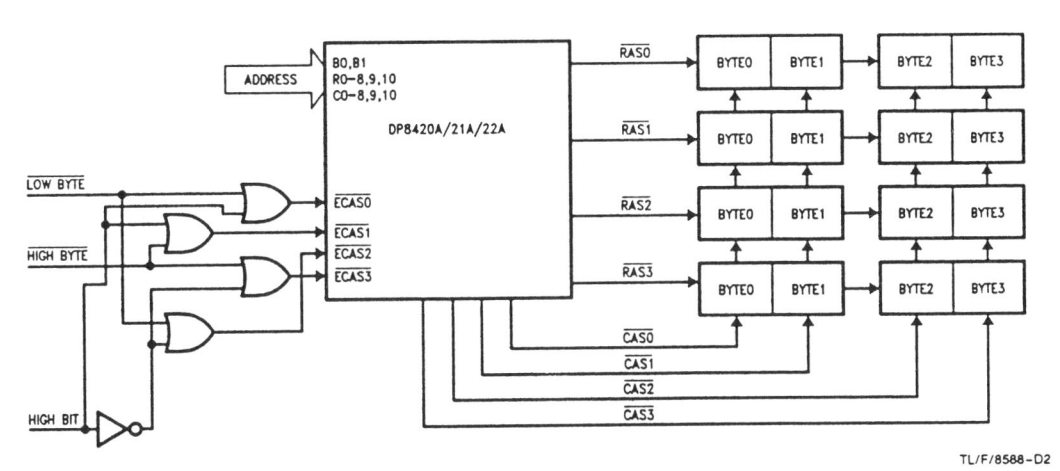

FIGURE 39d. 8 Bank DRAM Array for 16-Bit System (C6, C5, C4 = 1, 1, 0 during Programming)

7.0 RAS and CAS Configuration Modes (Continued)

7.2 MEMORY INTERLEAVING

Memory interleaving allows the cycle time of DRAMs to be reduced by having sequential accesses to different memory banks. Since the DP8420A/21A/22A have separate precharge counters per bank, sequential accesses will not be delayed if the accessed banks use different RAS outputs. To ensure different RAS outputs will be used, a mode is selected where either one or two RAS outputs will be asserted during an access. The bank select or selects, B0 and B1, are then tied to the least significant address bits, causing a different group of RASs to assert during each sequential access as shown in *Figure 40*. In this figure there should be at least one clock period of all RAS's negated between different RAS's being asserted to avoid the condition of a CAS before RAS refresh cycle.

7.3 ADDRESS PIPELINING

Address pipelining allows several access RASs to be asserted at once. Because RASs can overlap, each bank requires either a mode where one RAS and one CAS are used per bank as shown in *Figure 41a* or where two RASs and two CASs are used per bank as shown in *Figure 41b*. Byte writing can be accomplished in a 16-bit word system if two RASs and two CASs are used per bank. In other systems, WEs (or external gating on the CAS outputs) must be used to perform byte writing. If WEs are used separate data in and data out buffers must be used. If the array is not layed out this way, a CAS to a bank can be low before RAS, which will cause a refresh of the DRAM, not an access. To take full advantage of address pipelining, memory interleaving is used. To memory interleave, the least significant address bits should be tied to the bank select inputs to ensure that all "back to back" sequential accesses are not delayed, since different memory banks are accessed.

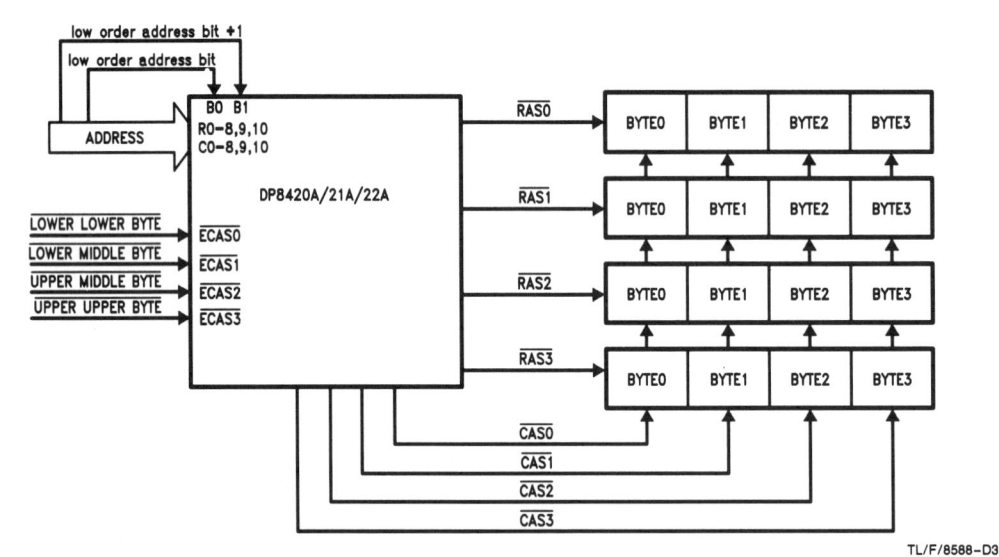

FIGURE 40. Memory Interleaving (C6, C5, C4 = 1, 1, 0 during Programming)

7.0 $\overline{RAS}$ and $\overline{CAS}$ Configuration Modes (Continued)

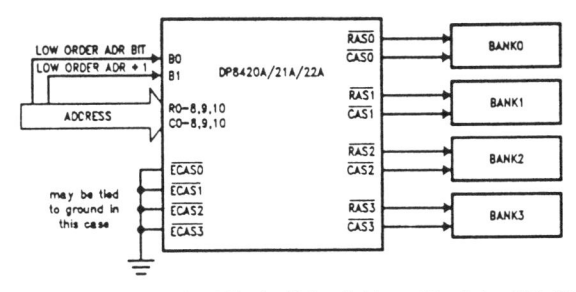

FIGURE 41a. DRAM Array Setup for 4 Banks Using Address Pipelining (C6, C5, C4 = 1, 1, 1 or C6, C5, C4 = 0, 1, 0 (Also Allowing Error Scrubbing) during Programming)

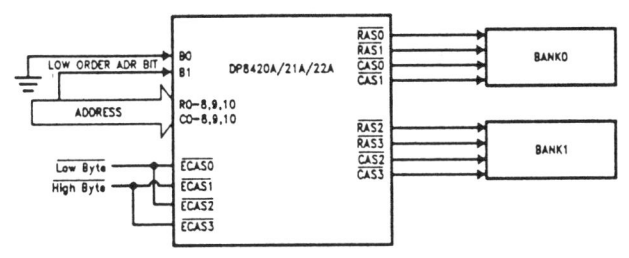

FIGURE 41b. DRAM Array Setup for Address Pipelining with 2 Banks (C6, C5, C4 = 1, 0, 1 or C6, C5, C4 = 0, 0, 1 (Also Allowing Error Scrubbing) during Programming)

7.4 ERROR SCRUBBING

In error scrubbing during refresh, the user selects one, two or four $\overline{RAS}$ and $\overline{CAS}$ outputs per bank. When performing error detection and correction, memory is always accessed as words. Since the $\overline{CAS}$ signals are not used to select individual bytes, the $\overline{ECAS}$ inputs can be tied low as shown in *Figures 42a* and *42b*.

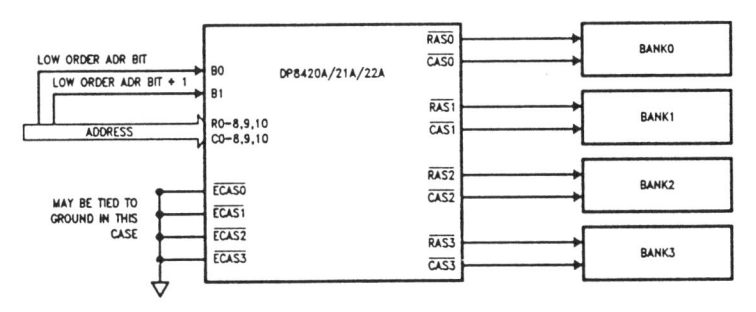

FIGURE 42a. DRAM Array Setup for 4 Banks Using Error Scrubbing (C6, C5, C4 = 0, 1, 0 during Programming)

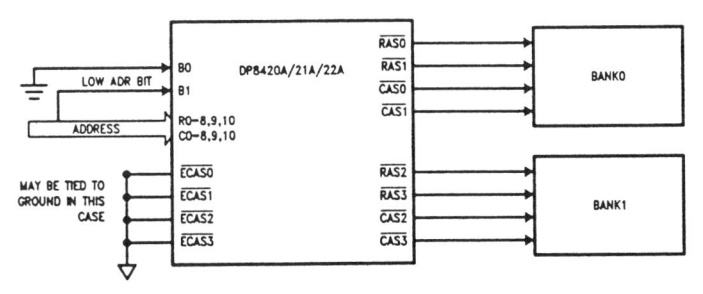

FIGURE 42b. DRAM Array Setup for Error Scrubbing with 2 Banks (C6, C5, C4 = 0, 0, 1 during Programming)

7.0 RAS and CAS Configuration Modes (Continued)

7.5 PAGE/BURST MODE

In a static column, page or burst mode system, the least significant bits must be tied to the column address in order to ensure that the page/burst accesses are to sequential memory addresses, as shown in *Figure 43*. In a nibble mode system, the least significant bits must be tied to the highest column and row address bits in order to ensure that sequential address bits are the "nibble" bits for nibble mode accesses *(Figure 43)*. The ECAS inputs may then be toggled with the DP8420A/21A/22A's address latches in fall-through mode, while AREQ is asserted. The ECAS inputs can also be used to select individual bytes. When using nibble mode DRAMS, the third and fourth address bits can be tied to the bank select inputs to perform memory interleaving. In page or static column modes, the two address bits after the page size can be tied to the bank select inputs to select a new bank if the page size is exceeded.

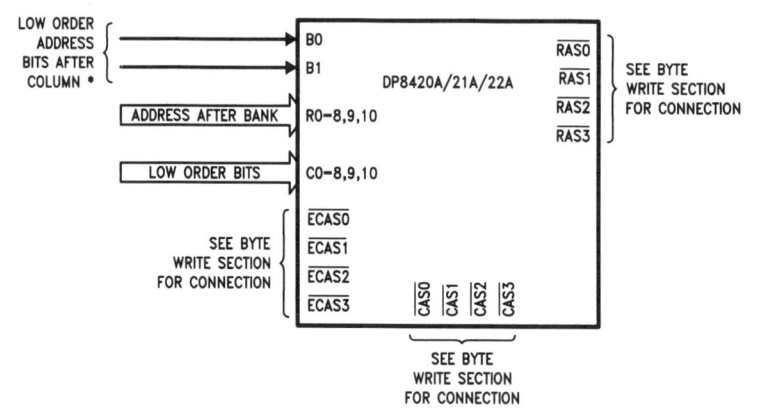

*See table below for row, column & bank address bit map. A0,A1 are used for byte addressing in this example.

Addresses	Nibble Mode*	Page Mode/Static Column Mode Page Size			
		256 Bits/Page	512 Bits/Page	1024 Bits/Page	2048 Bits/Page
Column Address	C9,R9 = A2,A3 C0–8 = X	C0–7 = A2–9 C8–10 = X	C0–8 = A2–10 C9,10 = X	C0–9 = A2–11 C10 = X	C0–10 = A2–12
Row Address	X	X	X	X	X
B0 B1	A4 A5	A10 A11	A11 A12	A12 A13	A13 A14

Assume that the least significant address bits are used for byte addressing. Given a 32-bit system A0,A1 would be used for byte addressing.

X = DON'T CARE, the user can do as he pleases.

*Nibble mode values for R and C assume a system using 1 Mbit DRAMs.

FIGURE 43. Page, Static Column, Nibble Mode System

8.0 Programming and Resetting

The DP8420A/21A/22A must be programmed by one of two possible programming sequences before it can be used. At power up, the DP8420A/21A/22A programming bits are in an undefined state. All internal latches and flip-flops are cleared. After programming, the DP8420A/21A/22A enters a 60 ms initialization period. During this initialization period, the DP8420A/21A/22A performs refreshes about every 15 μs; this makes further DRAM warmup cycles unnecessary. The chip can be programmed as many times as the user wishes. After the first programming, the 60 ms initialization period will not be entered into unless the chip is reset. During the 60 ms initialization period, $\overline{RFIP}$ is asserted. The actual initialization time period is given by the following formula:

T = 4096*(Clock Divisor Select)
     *(Refresh Clock Fine Tune)
     /(DELCK Frequency)

8.1 EXTERNAL RESET

At power up, all internal latches and flip-flops are cleared. The power up state can again be entered by asserting $\overline{ML}$ and $\overline{DISRFSH}$ for 16 positive edges of CLK. After resetting if the user negates $\overline{DISRFSH}$ before negating $\overline{ML}$ as shown in *Figure 44a*, $\overline{ML}$ negated will program the chip. If $\overline{ML}$ is negated before or at the same time as $\overline{DISRFSH}$ as shown in *Figure 44b*, the chip will not be programmed. After the chip is programmed, the 60 ms initialization period will be entered into if this is the first programming after power up or reset.

It is recommended that the user perform a hardware reset of the DP8420A/21A/22A before programming and using the chip.

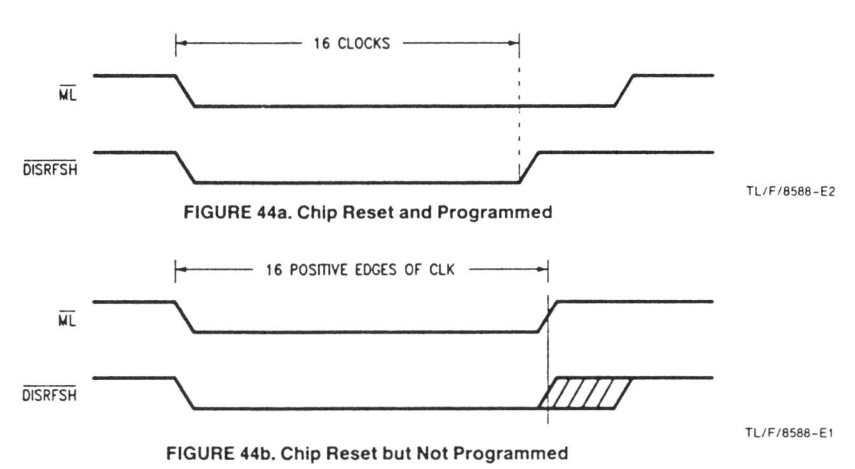

FIGURE 44a. Chip Reset and Programmed

FIGURE 44b. Chip Reset but Not Programmed

8.0 Programming and Resetting (Continued)

8.2 PROGRAMMING METHODS

The DP8420A/21A/22A must be programmed by one of two possible programming sequences before it can be used.

8.2.1 MODE LOAD ONLY PROGRAMMING

MODE LOAD, $\overline{ML}$, asserted enables an internal 23-bit programmable register. To use this method, the user asserts $\overline{ML}$, enabling the internal programming register. After $\overline{ML}$ is asserted, a valid programming selection is placed on the address bus (and $\overline{ECAS0}$), then $\overline{ML}$ is negated. When $\overline{ML}$ is negated, the value on the address bus (and $\overline{ECAS0}$) is latched into the internal programming register and the DP8420A/21A/22A is programmed, as shown in *Figure 45a*. After $\overline{ML}$ is negated, the DP8420A/21A/22A will enter the 60 ms initialization period only if this is the first programming after power up or reset.

Using this method, a set of transceivers on the address bus can be put at TRI-STATE® by the system reset signal. A combination of pull-up and pull-down resistors can be used on the address inputs of the DP8420A/21A/22A to select the programming values, as shown in *Figure 45b*.

8.2.2 CHIP SELECTED ACCESS PROGRAMMING

The chip can also be programmed by asserting $\overline{ML}$ and performing a chip selected access. $\overline{ADS}$ (or ALE) is disabled internally until after programming. To program the chip using this method, $\overline{ML}$ is asserted. After $\overline{ML}$ is asserted, $\overline{CS}$ is asserted and a valid programming selection is placed on the address bus. When $\overline{AREQ}$ is asserted, the chip is programmed with the programming selection on the address bus. After $\overline{AREQ}$ is negated, $\overline{ML}$ can be negated as shown in *Figure 46a*.

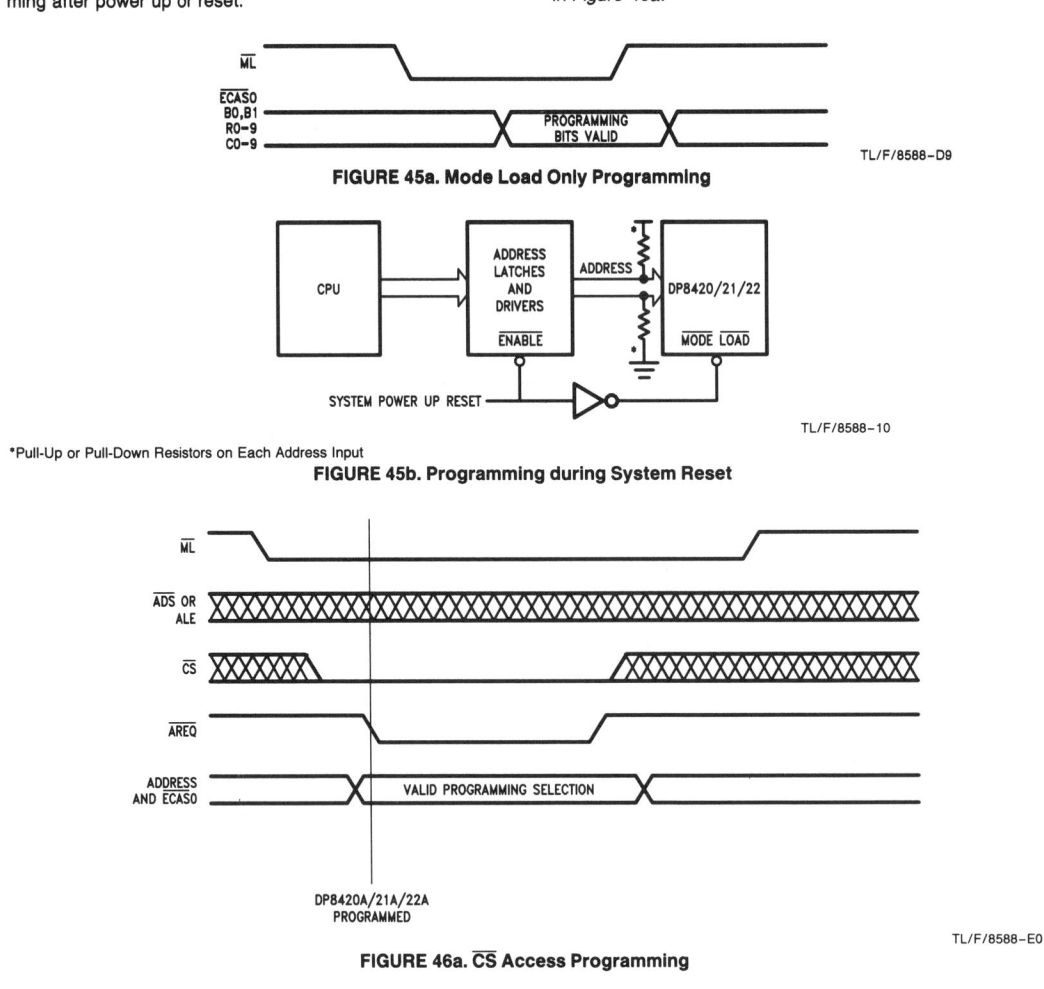

FIGURE 45a. Mode Load Only Programming

*Pull-Up or Pull-Down Resistors on Each Address Input

FIGURE 45b. Programming during System Reset

FIGURE 46a. $\overline{CS}$ Access Programming

6.0 Programming and Resetting (Continued)

Using this method, various programming schemes can be used. For example if extra upper address bits are available, an unused high order address bit can be tied to the signal $\overline{ML}$. Using this method, one need only write to a page of memory, thus asserting the high order bit and in turn programming the chip as shown in *Figure 46b*.

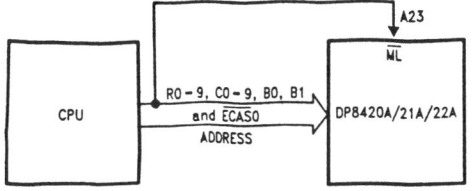

FIGURE 46b. Programming the DP8420A/21A/22A through the Address Bus Only

An I/O port can also be used to assert $\overline{ML}$. After $\overline{ML}$ is asserted, a chip selected access can be performed to program the chip. After the chip selected access, $\overline{ML}$ can be negated through the I/O port as shown in *Figure 46c*.

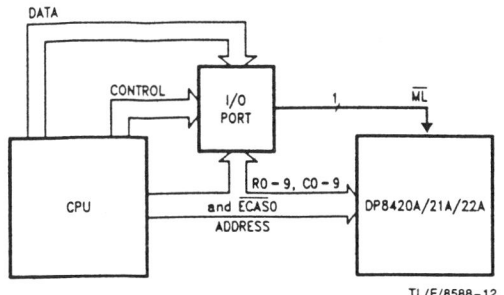

FIGURE 46c. Programming the DP8420A/21A/22A through the Address Bus and an I/O Port

Another simple way the chip can be programmed is the first write after system reset. This method requires only a flip-flop and an OR gate as shown in *Figure 46d*. At reset, the flip-flop is preset, which pulls the $\overline{Q}$ output low. Since $\overline{WR}$ is negated, $\overline{ML}$ is not enabled. The first write access is used to program the chip. When $\overline{WR}$ is asserted, $\overline{ML}$ is asserted. $\overline{WR}$ negated clocks the flip-flop, negates $\overline{ML}$, and programs the DP8420A/21A/22A with the address and $\overline{ECAS0}$ available at that time. $\overline{CS}$ does not need to be asserted using this method.

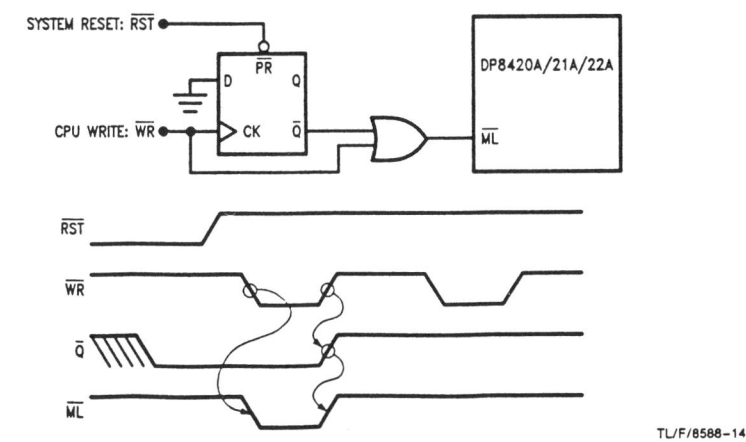

FIGURE 46d. Programming the DP8420A/21A/22A on the First CPU Write after Power Up

8.0 Programming and Resetting (Continued)

8.3 PROGRAMMING BIT DEFINITIONS

Symbol	Description
ECAS0	**Extend $\overline{CAS}$/Refresh Request Select**
0	The $\overline{CAS}$n outputs will be negated with the $\overline{RAS}$n outputs when $\overline{AREQ}$ (or $\overline{AREQB}$, DP8422A only) is negated. The $\overline{WE}$ output pin will function as write enable.
1	The $\overline{CAS}$n outputs will be negated, during an access (Port A (or Port B, DP8422A only)) when their corresponding $\overline{ECAS}$n inputs are negated. This feature allows the $\overline{CAS}$ outputs to be extended beyond the $\overline{RAS}$ outputs negating. Scrubbing refreshes are NOT affected. During scrubbing refreshes the $\overline{CAS}$ outputs will negate along with the $\overline{RAS}$ outputs regardless of the state of the $\overline{ECAS}$ inputs. The $\overline{WE}$ output will function as ReFresh ReQuest ($\overline{RFRQ}$) when this mode is programmed.
B1	**Access Mode Select**
0	**ACCESS MODE 0:** ALE pulsing high sets an internal latch. On the next positive edge of CLK, the access ($\overline{RAS}$) will start. $\overline{AREQ}$ will terminate the access.
1	**ACCESS MODE 1:** $\overline{ADS}$ asserted starts the access ($\overline{RAS}$) immediately. $\overline{AREQ}$ will terminate the access.
B0	**Address Latch Mode**
0	$\overline{ADS}$ or ALE asserted for Port A or $\overline{AREQB}$ asserted for Port B with the appropriate GRANT latch the input row, column and bank address.
1	The row, column and bank latches are fall through.
C9	**Delay $\overline{CAS}$ during WRITE Accesses**
0	$\overline{CAS}$ is treated the same for both READ and WRITE accesses.
1	During WRITE accesses, $\overline{CAS}$ will be asserted by the event that occurs last: $\overline{CAS}$ asserted by the internal delay line or $\overline{CAS}$ asserted on the positive edge of CLK after $\overline{RAS}$ is asserted.
C8	**Row Address Hold Time**
0	Row Address Hold Time = 25 ns minimum
1	Row Address Hold Time = 15 ns minimum
C7	**Column Address Setup Time**
0	Column Address Setup Time = 10 ns minimum
1	Column Address Setup Time = 0 ns minimum
C6, C5, C4	**$\overline{RAS}$ and $\overline{CAS}$ Configuration Modes/Error Scrubbing during Refresh**
0, 0, 0	$\overline{RAS}$0-3 and $\overline{CAS}$0-3 are all selected during an access. $\overline{ECAS}$n must be asserted for $\overline{CAS}$n to be asserted. B0 and B1 are not used during an access. Error scrubbing during refresh.
0, 0, 1	$\overline{RAS}$ and $\overline{CAS}$ pairs are selected during an access by B1. $\overline{ECAS}$n must be asserted for $\overline{CAS}$n to be asserted. B1 = 0 during an access selects $\overline{RAS}$0-1 and $\overline{CAS}$0-1. B1 = 1 during an access selects $\overline{RAS}$2-3 and $\overline{CAS}$2-3. B0 is not used during an Access. Error scrubbing during refresh.
0, 1, 0	RAS and CAS singles are selected during an access by B0-1. $\overline{ECAS}$n must be asserted for $\overline{CAS}$n to be asserted. B1 = 0, B0 = 0 during an access selects $\overline{RAS}$0 and $\overline{CAS}$0. B1 = 0, B0 = 1 during an access selects $\overline{RAS}$1 and $\overline{CAS}$1. B1 = 1, B0 = 0 during an access selects $\overline{RAS}$2 and $\overline{CAS}$2. B1 = 1, B0 = 1 during an access selects $\overline{RAS}$3 and $\overline{CAS}$3. Error scrubbing during refresh.
0, 1, 1	$\overline{RAS}$0-3 and $\overline{CAS}$0-3 are all selected during an access. $\overline{ECAS}$n must be asserted for $\overline{CAS}$n to be asserted. B1, B0 are not used during an access. No error scrubbing. ($\overline{RAS}$ only refreshing)
1, 0, 0	$\overline{RAS}$ pairs are selected by B1. $\overline{CAS}$0-3 are all selected. $\overline{ECAS}$n must be asserted for $\overline{CAS}$n to be asserted. B1 = 0 during an access selects $\overline{RAS}$0-1 and $\overline{CAS}$0-3. B1 = 1 during an access selects $\overline{RAS}$2-3 and $\overline{CAS}$0-3. B0 is not used during an access. No error scrubbing.

8.0 Programming and Resetting (Continued)

8.3 PROGRAMMING BIT DEFINITIONS (Continued)

Symbol	Description
C6, C5, C4	$\overline{RAS}$ and $\overline{CAS}$ Configuration Modes (Continued)
1, 0, 1	$\overline{RAS}$ and $\overline{CAS}$ pairs are selected by B1. $\overline{ECAS}$n must be asserted for $\overline{CAS}$n to be asserted. B1 = 0 during an access selects $\overline{RAS}$0-1 and $\overline{CAS}$0-1. B1 = 1 during an access selects $\overline{RAS}$2-3 and $\overline{CAS}$2-3. B0 is not used during an access. No error scrubbing.
1, 1, 0	$\overline{RAS}$ singles are selected by B0-1. $\overline{CAS}$0-3 are all selected. $\overline{ECAS}$n must be asserted for $\overline{CAS}$n to be asserted. B1 = 0, B0 = 0 during an access selects $\overline{RAS}$0 and $\overline{CAS}$0-3. B1 = 0, B0 = 1 during an access selects $\overline{RAS}$1 and $\overline{CAS}$0-3. B1 = 1, B0 = 0 during an access selects $\overline{RAS}$2 and $\overline{CAS}$0-3. B1 = 1, B0 = 1 during an access selects $\overline{RAS}$3 and $\overline{CAS}$0-3. No error scrubbing.
1, 1, 1	$\overline{RAS}$ and $\overline{CAS}$ singles are selected by B0, 1. $\overline{ECAS}$n must be asserted for $\overline{CAS}$n to be asserted. B1 = 0, B0 = 0 during an access selects $\overline{RAS}$0 and $\overline{CAS}$0. B1 = 0, B0 = 1 during an access selects $\overline{RAS}$1 and $\overline{CAS}$1. B1 = 1, B0 = 0 during an access selects $\overline{RAS}$2 and $\overline{CAS}$2. B1 = 1, B0 = 1 during an access selects $\overline{RAS}$3 and $\overline{CAS}$3. No error scrubbing.
C3	Refresh Clock Fine Tune Divisor
0	Divide delay line/refresh clock further by 30 (If DELCLK/Refresh Clock Clock Divisor = 2 MHz = 15 μs refresh period).
1	Divide delay line/refresh clock further by 26 (If DELCLK/Refresh Clock Clock Divisor = 2 MHz = 13 μs refresh period).
C2, C1, C0	Delay Line/Refresh Clock Divisor Select
0, 0, 0	Divide DELCLK by 10 to get as close to 2 MHz as possible.
0, 0, 1	Divide DELCLK by 9 to get as close to 2 MHz as possible.
0, 1, 0	Divide DELCLK by 8 to get as close to 2 MHz as possible.
0, 1, 1	Divide DELCLK by 7 to get as close to 2 MHz as possible.
1, 0, 0	Divide DELCLK by 6 to get as close to 2 MHz a possible.
1, 0, 1	Divide DELCLK by 5 to get as close to 2 MHz as possible.
1, 1, 0	Divide DELCLK by 4 to get as close to 2 MHz as possible.
1, 1, 1	Divide DELCLK by 3 to get as close to 2 MHz as possible.
R9	Refresh Mode Select
0	$\overline{RAS}$0-3 will all assert and negate at the same time during a refresh.
1	Staggered Refresh. $\overline{RAS}$ outputs during refresh are separated by one positive clock edge. Depending on the configuration mode chosen, either one or two $\overline{RAS}$s will be asserted.
R8	Address Pipelining Select
0	Address pipelining is selected. The DRAM controller will switch the DRAM column address back to the row address after guaranteeing the column address hold time.
1	Non-address pipelining is selected. The DRAM controller will hold the column address on the DRAM address bus until the access $\overline{RAS}$s are negated.
R7	$\overline{WAIT}$ or $\overline{DTACK}$ Select
0	$\overline{WAIT}$ type output is selected.
1	$\overline{DTACK}$ (Data Transfer ACKnowledge) type output is selected.
R6	Add Wait States to the Current Access if $\overline{WAITIN}$ is Low
0	$\overline{WAIT}$ or $\overline{DTACK}$ will be delayed by one additional positive edge of CLK.
1	$\overline{WAIT}$ or $\overline{DTACK}$ will be delayed by two additional positive edges of CLK.

8.0 Programming and Resetting (Continued)

8.3 PROGRAMMING BIT DEFINITIONS (Continued)

Symbol	Description
R5, R4	$\overline{\text{WAIT}}/\overline{\text{DTACK}}$ during Burst (See Section 5.1.2 or 5.2.2)
0, 0	NO WAIT STATES; If R7 = 0 during programming, $\overline{\text{WAIT}}$ will remain negated during burst portion of access. If R7 = 1 programming, $\overline{\text{DTACK}}$ will remain asserted during burst portion of access.
0, 1	1T; If R7 = 0 during programming, $\overline{\text{WAIT}}$ will assert when the $\overline{\text{ECAS}}$ inputs are negated with $\overline{\text{AREQ}}$ asserted. $\overline{\text{WAIT}}$ will negate from the positive edge of CLK after the $\overline{\text{ECAS}}$s have been asserted. If R7 = 1 during programming, $\overline{\text{DTACK}}$ will negate when the $\overline{\text{ECAS}}$ inputs are negated with $\overline{\text{AREQ}}$ asserted. $\overline{\text{DTACK}}$ will assert from the positive edge of CLK after the $\overline{\text{ECAS}}$s have been asserted.
1, 0	½T; If R7 = 0 during programming, $\overline{\text{WAIT}}$ will assert when the $\overline{\text{ECAS}}$ inputs are negated with $\overline{\text{AREQ}}$ asserted. $\overline{\text{WAIT}}$ will negate on the negative level of CLK after the $\overline{\text{ECAS}}$s have been asserted. If R7 = 1 during programming, $\overline{\text{DTACK}}$ will negate when the $\overline{\text{ECAS}}$ inputs are negated with $\overline{\text{AREQ}}$ asserted. $\overline{\text{DTACK}}$ will assert from the negative level of CLK after the $\overline{\text{ECAS}}$s have been asserted.
1, 1	0T; If R7 = 0 during programming, $\overline{\text{WAIT}}$ will assert when the $\overline{\text{ECAS}}$ inputs are negated. $\overline{\text{WAIT}}$ will negate when the $\overline{\text{ECAS}}$ inputs are asserted. If R7 = 1 during programming, $\overline{\text{DTACK}}$ will negate when the $\overline{\text{ECAS}}$ inputs are negated. $\overline{\text{DTACK}}$ will assert when the $\overline{\text{ECAS}}$ inputs are asserted.
R3, R2	$\overline{\text{WAIT}}/\overline{\text{DTACK}}$ Delay Times (See Section 5.1.1 or 5.2.1)
0, 0	NO WAIT STATES; If R7 = 0 during programming, $\overline{\text{WAIT}}$ will remain high during non-delayed accesses. $\overline{\text{WAIT}}$ will negate when $\overline{\text{RAS}}$ is negated during delayed accesses. NO WAIT STATES; If R7 = 1 during programming, $\overline{\text{DTACK}}$ will be asserted when $\overline{\text{RAS}}$ is asserted.
0, 1	½T; If R7 = 0 during programming, $\overline{\text{WAIT}}$ will negate on the negative level of CLK, after the access $\overline{\text{RAS}}$. 1T; If R7 = 1 during programming, $\overline{\text{DTACK}}$ will be asserted on the positive edge of CLK after the access $\overline{\text{RAS}}$.
1, 0	NO WAIT STATES, ½T; If R7 = 0 during programming, $\overline{\text{WAIT}}$ will remain high during non-delayed accesses. $\overline{\text{WAIT}}$ will negate on the negative level of CLK, after the access $\overline{\text{RAS}}$, during delayed accesses. ½T; If R7 = 1 during programming, $\overline{\text{DTACK}}$ will be asserted on the negative level of CLK after the access $\overline{\text{RAS}}$.
1, 1	1T; If R7 = 0 during programming, $\overline{\text{WAIT}}$ will negate on the positive edge of CLK after the access $\overline{\text{RAS}}$. 1½T; If R7 = 1 during programming, $\overline{\text{DTACK}}$ will be asserted on the negative level of CLK after the positive edge of CLK after the access $\overline{\text{RAS}}$.

8.0 Programming and Resetting (Continued)

8.3 PROGRAMMING BIT DEFINITIONS (Continued)

Symbol	Description
R1, R0	$\overline{RAS}$ Low and $\overline{RAS}$ Precharge Time
0, 0	$\overline{RAS}$ asserted during refresh = 2 positive edges of CLK. $\overline{RAS}$ precharge time = 1 positive edge of CLK. $\overline{RAS}$ will start from the first positive edge of CLK after GRANTB transitions (DP8422A).
0, 1	$\overline{RAS}$ asserted during refresh = 3 positive edges of CLK. $\overline{RAS}$ precharge time = 2 positive edges of CLK. $\overline{RAS}$ will start from the second positive edge of CLK after GRANTB transitions (DP8422A).
1, 0	$\overline{RAS}$ asserted during refresh = 2 positive edges of CLK. $\overline{RAS}$ precharge time = 2 positive edges of CLK. $\overline{RAS}$ will start from the first positive edge of CLK after GRANTB transitions (DP8422A).
1, 1	$\overline{RAS}$ asserted during refresh = 4 positive edges of CLK. $\overline{RAS}$ precharge time = 3 positive edges of CLK. $\overline{RAS}$ will start from the second positive edge of CLK after GRANTB transitions (DP8422A).

9.0 Test Mode

Staggered refresh in combination with the error scrubbing mode places the DP8420A/21A/22A in test mode. In this mode, the 24-bit refresh counter is divided into a 13-bit and 11-bit counter. During refreshes both counters are incremented to reduce test time.

10.0 DRAM Critical Timing Parameters

The two critical timing parameters, shown in *Figure 47*, that must be met when controlling the access timing to a DRAM are the row address hold time, tRAH, and the column address setup time, tASC. Since the DP8420A/21A/22A contain a precise internal delay line, the values of these parameters can be selected at programming time. These values will also increase and decrease if DELCLK varies from 2 MHz.

10.1 PROGRAMMABLE VALUES OF tRAH AND tASC

The DP8420A/21A/22A allow the values of tRAH and tASC to be selected at programming time. For each parameter, two choices can be selected. tRAH, the row address hold time, is measured from $\overline{RAS}$ asserted to the row address starting to change to the column address. The two choices for tRAH are 15 ns and 25 ns, programmable through address bit C8.

tASC, the column address setup time, is measured from the column address valid to $\overline{CAS}$ asserted. The two choices for tASC are 0 ns and 10 ns, programmable through address bit C7.

10.2 CALCULATION OF tRAH AND tASC

There are two clock inputs to the DP8420A/21A/22A. These two clocks, DELCLK and CLK can either be tied together to the same clock or be tied to different clocks running asynchronously at different frequencies.

The clock input, DELCLK, controls the internal delay line and refresh request clock. DELCLK should be a multiple of 2 MHz. If DELCLK is not a multiple of 2 MHz, tRAH and tASC will change. The new values of tRAH and tASC can be calculated by the following formulas:

If tRAH was programmed to equal 15 ns then tRAH = 30*(((DELCLK Divisor)* 2 MHz/(DELCLK Frequency)) − 1) + 15 ns.

If tRAH was programmed to equal 25 ns then tRAH = 30*(((DELCLK Divisor)* 2 MHz/(DELCLK Frequency)) − 1) + 25 ns.

If tASC was programmed to equal 0 ns then tASC = 15*((DELCLK Divisor)* 2 MHz/(DELCLK Frequency)) − 15 ns.

If tASC was programmed to equal 10 ns then tASC = 25*((DELCLK Divisor)* 2 MHz/(DELCLK Frequency)) − 15 ns.

Since the values of tRAH and tASC are increased or decreased, the time to $\overline{CAS}$ asserted will also increase or decrease. These parameters can be adjusted by the following formula:

Delay to $\overline{CAS}$ = Actual Spec. + Actual tRAH − Programmed tRAH + Actual tASC − Programmed tASC.

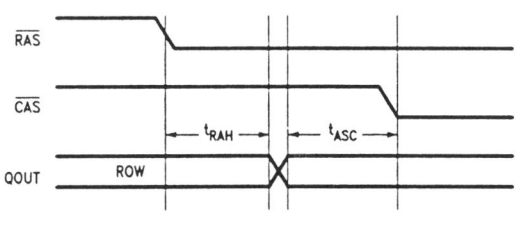

FIGURE 47. tRAH and tASC

11.0 Dual Accessing Functions (DP8422A)

The DP8422A has all the functions previously described. In addition to those features, the DP8422A also has the capabilities to arbitrate among refresh, Port A and a second port, Port B. This allows two CPUs to access a common DRAM array. DRAM refresh has the highest priority followed by the currently granted port. The ungranted port has the lowest priority. The last granted port will continue to stay granted even after the access has terminated, until an access request is received from the ungranted port (see *Figure 48a*). The dual access configuration assumes that both Port A and Port B are synchronous to the system clock. If they are not synchronous to the system clock they should be externally synchronized (Ex. By running the access requests through several Flip-Flops, see *Figure 50a*).

11.1 PORT B ACCESS MODES (DP8422A)

Port B accesses are initiated from a single input, $\overline{AREQB}$. When $\overline{AREQB}$ is asserted, an access request is generated. If GRANTB is asserted and a refresh is not taking place or precharge time is not required, $\overline{RAS}$ will be asserted when $\overline{AREQB}$ is asserted. Once $\overline{AREQB}$ is asserted, it must stay asserted until the access is over. $\overline{AREQB}$ negated, negates $\overline{RAS}$ as shown in *Figure 48b*. Note that if $\overline{ECAS0} = 1$ during programming the $\overline{CAS}$ outputs may be held asserted (beyond $\overline{RASn}$ negating) by continuing to assert the appropriate $\overline{ECASn}$ inputs (the same as Port A accesses). If Port B is not granted, the access will begin on the first or second positive edge of CLK after GRANTB is asserted (See R0, R1 programming bit definitions) as shown in *Figure 48c*, assuming that Port A is not accessing the DRAM ($\overline{CS}$, $\overline{ADS}$/ALE and $\overline{AREQ}$) and $\overline{RAS}$ precharge for the particular bank has completed. It is important to note that for GRANTB to transition to Port B, Port A must **not** be requesting an access at a rising clock edge (or locked) and Port B must be requesting an access at that rising clock edge. Port A can request an access through $\overline{CS}$ and $\overline{ADS}$/ALE or $\overline{CS}$ and $\overline{AREQ}$. Therefore during an interleaved access where $\overline{CS}$ and $\overline{ADS}$/ALE become asserted before $\overline{AREQ}$ from the previous access is negated, Port A will retain GRANTB = 0 whether $\overline{AREQB}$ is asserted or not.

Since there is no chip select for Port B, $\overline{AREQB}$ must incorporate this signal. This mode of accessing is similar to Mode 1 accessing for Port A.

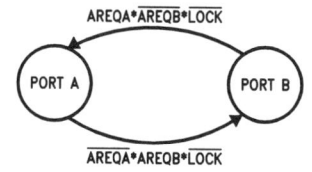

Explanation of Terms

AREQA = Chip Selected access request from Port A
AREQB = Chip Selected access request from Port B
LOCK = Externally controlled LOCKing of the Port that is currently GRANTed.

FIGURE 48a. DP8422A PORT A/PORT B ARBITRATION STATE DIAGRAM. This arbitration may take place during the "ACCESS" or "REFRESH" state (see *Figure 7a*).

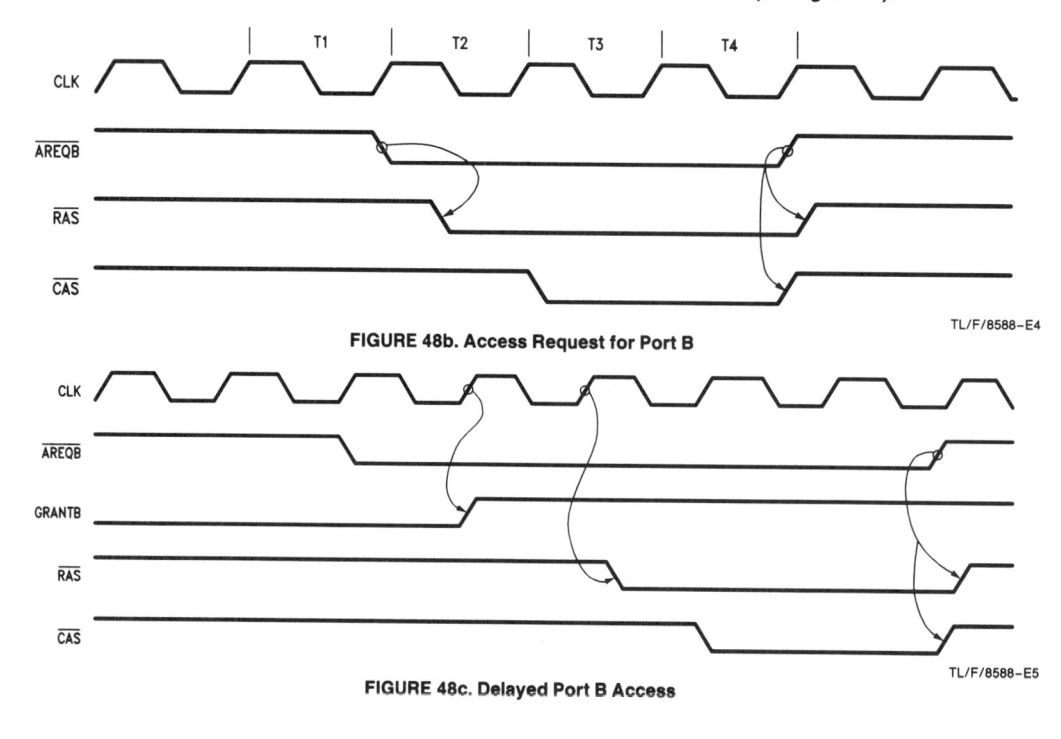

FIGURE 48b. Access Request for Port B

FIGURE 48c. Delayed Port B Access

11.0 Dual Accessing Functions (DP8422A) (Continued)

11.2 PORT B WAIT STATE SUPPORT (DP8422A)

Advanced transfer acknowledge for Port B, $\overline{ATACKB}$, is used for wait state support for Port B. This output will be asserted when $\overline{RAS}$ for the Port B access is asserted, as shown in *Figures 49a* and *49b*. Once asserted, this output will stay asserted until $\overline{AREQB}$ is negated. With external logic, $\overline{ATACKB}$ can be made to interface to any CPU's wait input as shown in *Figure 49c*.

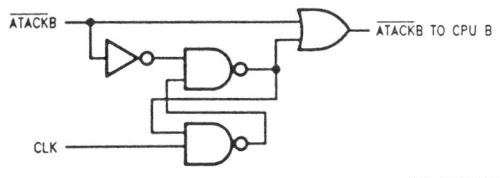

A) Extend $\overline{ATACK}$ to ½T (½ Clock) after $\overline{RAS}$ goes low.

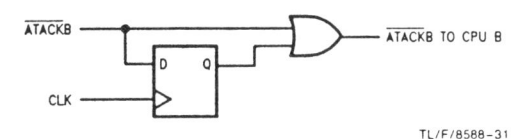

B) Extend $\overline{ATACK}$ to 1T after $\overline{RAS}$ goes low.

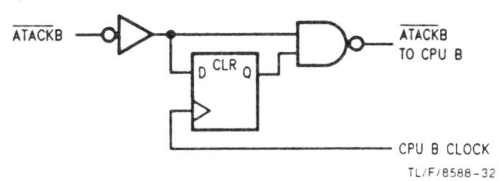

C) Synchronize $\overline{ATACKB}$ to CPU B Clock. This is useful if CPU B runs asynchronous to the DP8422.

FIGURE 49c. Modifying Wait Logic for Port B

11.3 COMMON PORT A AND PORT B DUAL PORT FUNCTIONS

An input, $\overline{LOCK}$, and an output, GRANTB, add additional functionality to the dual port arbitration logic. $\overline{LOCK}$ allows Port A or Port B to lock out the other port from the DRAM. When a Port is locked out of the DRAM, wait states will be inserted into its access cycle until it is allowed to access memory. GRANTB is used to multiplex the input control signals and addresses to the DP8422A.

11.3.1 GRANTB Output

The output GRANTB determines which port has current access to the DRAM array. GRANTB asserted signifies Port B has access. GRANTB negated signifies Port A has access to the DRAM array.

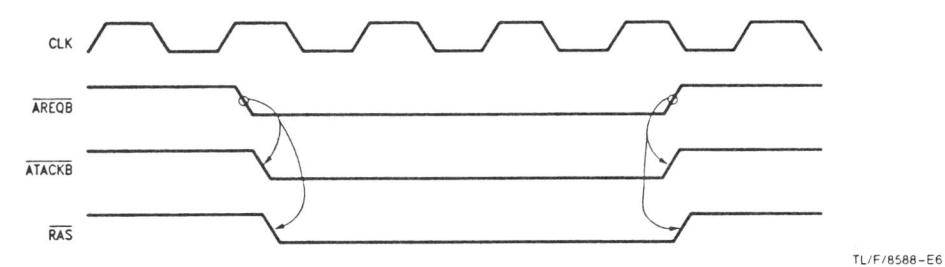

FIGURE 49a. Non-Delayed Port B Access

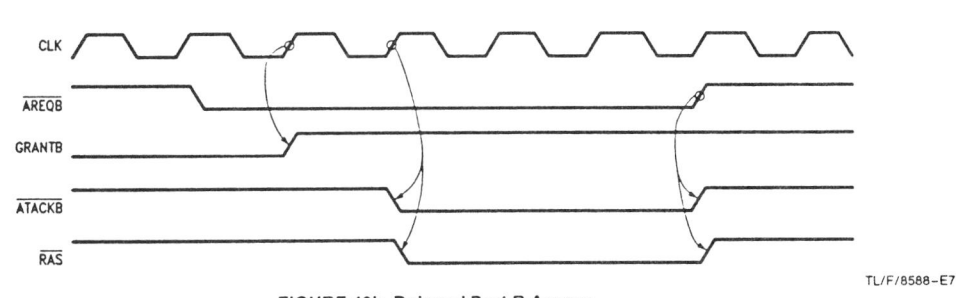

FIGURE 49b. Delayed Port B Access

11.0 Dual Accessing Functions (DP8422A) (Continued)

Since the DP8422A has only one set of address inputs, the signal is used, with the addition of buffers, to allow the currently granted port's addresses to reach the DP8422A. The signals which need to be buffered are R0–10, C0–10, B0–1, $\overline{ECAS}$0–3, $\overline{WE}$, and $\overline{LOCK}$. All other inputs are not common and do not have to be buffered as shown in Figure 50a. If a Port, which is not currently granted, tries to access the DRAM array, the GRANTB output will transition from a rising clock edge from $\overline{AREQ}$ or $\overline{AREQB}$ negating and will preceed the $\overline{RAS}$ for the access by one or two clock periods. GRANTB will then stay in this state until the other port requests an access and the currently granted port is not accessing the DRAM as shown in Figure 50b.

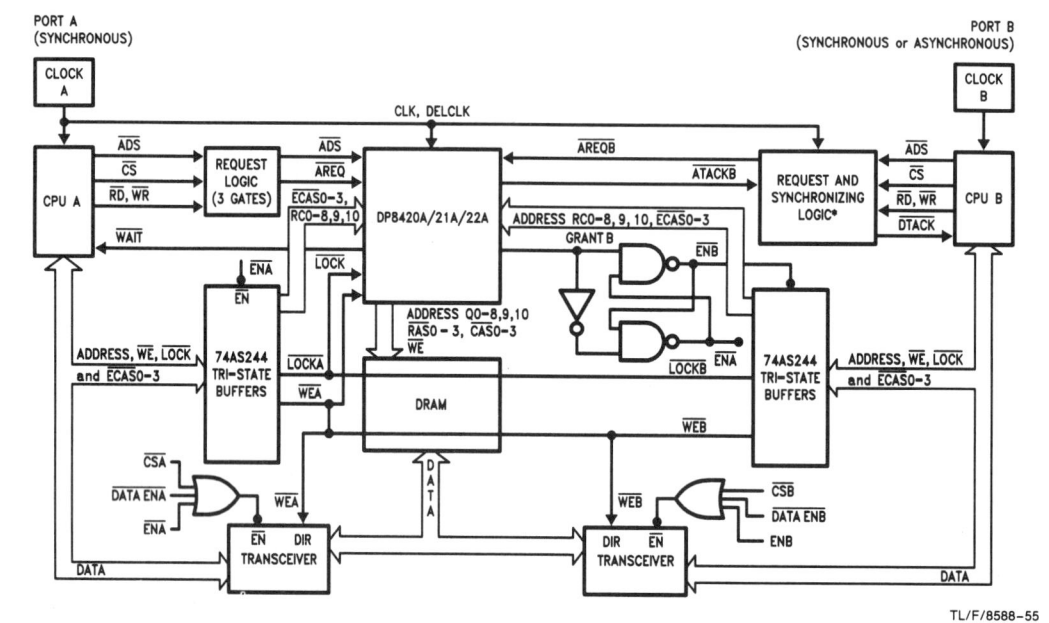

*If Port B is synchronous the Request Synchronizing logic will not be required.

FIGURE 50a. Dual Accessing with the DP8422A (System Block Diagram)

11.0 Dual Accessing Functions (DP8422A) (Continued)

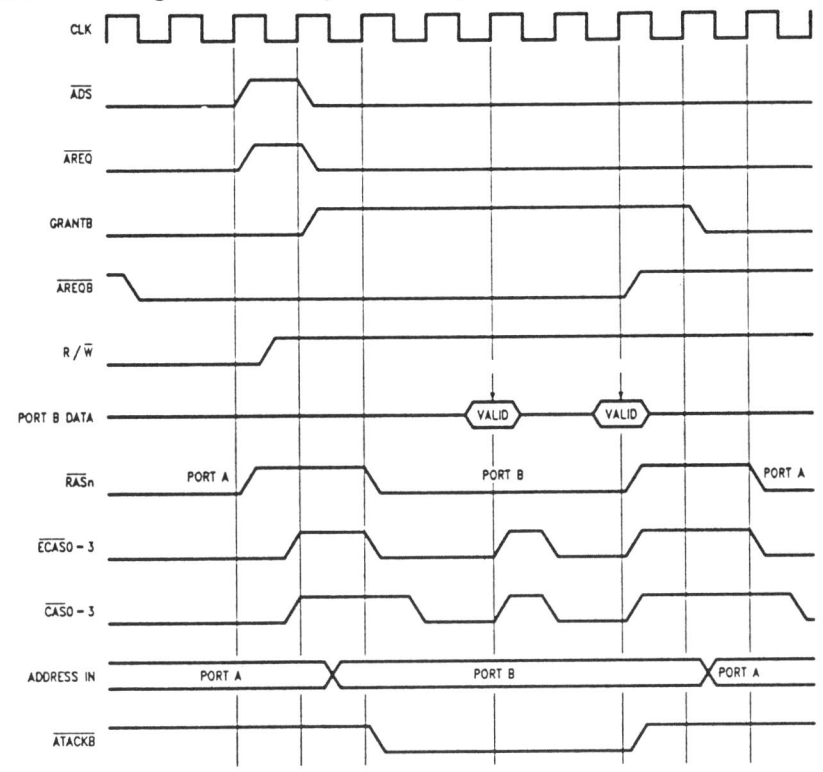

FIGURE 50b. Wait States during a Port B Access

11.3.2 LOCK Input

When the LOCK input is asserted, the currently granted port can "lock out" the other port through the insertion of wait states to that port's access cycle. LOCK does not disable refreshes, it only keeps GRANTB in the same state even if the other port requests an access, as shown in *Figure 51a*. LOCK can be used by either port.

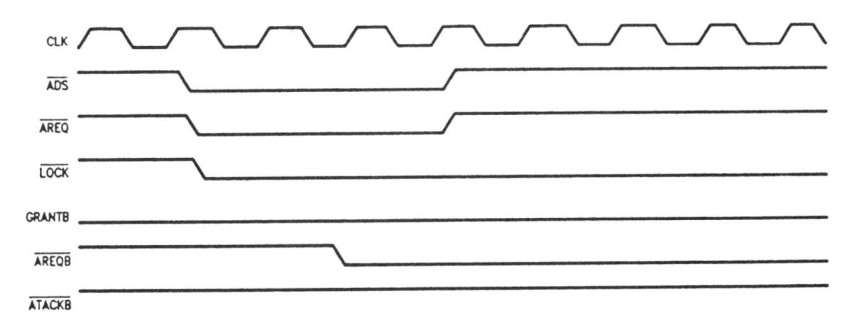

FIGURE 51. LOCK Function

12.0 Absolute Maximum Ratings (Note 1)

If Military/Aerospace specified devices are required, please contact the National Semiconductor Sales Office/Distributors for availability and specifications.

Temperature under Bias .... 0°C to +70°C
Storage Temperature .... −65°C to +150°C
All Input or Output Voltage
with Respect to GND .... −0.5V to +7V
Power Dissipation @ 20 MHz 0.5W
ESD Rating to be determined.

13.0 DC Electrical Characteristics $T_A = 0°C$ to $+70°C$, $V_{CC} = 5V \pm 10\%$, GND = 0V

Symbol	Parameter	Conditions	Min	Typ	Max	Units
V_{IH}	Logical 1 Input Voltage	Tested with a Limited Functional Pattern	2.0		$V_{CC} + 0.5$	V
V_{IL}	Logical 0 Input Voltage	Tested with a Limited Functional Pattern	−0.5		0.8	V
V_{OH1}	Q and $\overline{WE}$ Outputs	$I_{OH} = -10$ mA	$V_{CC} - 1.0$			V
V_{OL1}	Q and $\overline{WE}$ Outputs	$I_{OL} = 10$ mA			0.5	V
V_{OH2}	All Outputs except Qs, $\overline{WE}$	$I_{OH} = -3$ mA	$V_{CC} - 1.0$			V
V_{OL2}	All Outputs except Qs, $\overline{WE}$	$I_{OL} = 3$ mA			0.5	V
I_{IN}	Input Leakage Current	$V_{IN} = V_{CC}$ or GND	−10		10	µA
$I_{IL\ ML}$	$\overline{ML}$ Input Current (Low)	$V_{IN} = $ GND			200	µA
I_{CC1}	Standby Current	CLK at 8 MHz ($V_{IN} = V_{CC}$ or GND)		6	15	mA
I_{CC1}	Standby Current	CLK at 20 MHz ($V_{IN} = V_{CC}$ or GND)		8	17	mA
I_{CC1}	Standby Current	CLK at 25 MHz ($V_{IN} = V_{CC}$ or GND)		10	20	mA
I_{CC2}	Supply Current	CLK at 8 MHz (Inputs Active) ($I_{LOAD} = 0$) ($V_{IN} = V_{CC}$ or GND)		20	40	mA
I_{CC2}	Supply Current	CLK at 20 MHz (Inputs Active) ($I_{LOAD} = 0$) ($V_{IN} = V_{CC}$ or GND)		40	75	mA
I_{CC2}	Supply Current	CLK at 25 MHz (Inputs Active) ($I_{LOAD} = 0$) ($V_{IN} = V_{CC}$ or GND)		50	95	mA
C_{IN}*	Input Capacitance	f_{IN} at 1 MHz			10	pF

*Note: C_{IN} is not 100% tested.

14.0 AC Timing Parameters: DP8420A/DP8421A/DP8422A

Two speed selections are given, the DP8420A/21A/22A-20 and the DP8420A/21A/22A-25. The differences between the two parts are the maximum operating frequencies of the input CLKs and the maximum delay specifications. Low frequency applications may use the "−25" part to gain improved timing.

The AC timing parameters are grouped into sectional numbers as shown below. These numbers also refer to the timing diagrams.

1−36	Common parameters to all modes of operation
50−56	Difference parameters used to calculate: $\overline{RAS}$ low time, $\overline{RAS}$ precharge time, $\overline{CAS}$ high time and $\overline{CAS}$ low time
100−121	Common dual access parameters used for Port B accesses and inputs and outputs used only in dual accessing
200−212	Refresh parameters
300−315	Mode 0 access parameters used in both single and dual access applications
400−416	Mode 1 access parameters used in both single and dual access applications
450−455	Special Mode 1 access parameters which supersede the 400−416 parameters when dual accessing
500−506	Programming parameters

Unless otherwise stated $V_{CC} = 5.0V \pm 10\%$, $0 < T_A < 70°C$, the output load capacitance is typical for 4 banks of 18 DRAMs per bank, including trace capacitance (see Note 2).

Two different loads are specified:

$C_L = 50$ pF loads on all outputs except
$C_L = 150$ pF loads on Q0−8, 9, 10 and $\overline{WE}$; or

$C_H = 50$ pF loads on all outputs except
$C_H = 125$ pF loads on $\overline{RAS}$0−3 and $\overline{CAS}$0−3 and
$C_H = 380$ pF loads on Q0−8, 9, 10 and $\overline{WE}$.

14.0 AC Timing Parameters: DP8420A/DP8421A/DP8422A (Continued)

Number	Symbol	Common Parameter Description	8420A/21A/22A-20 C_L Min	8420A/21A/22A-20 C_L Max	8420A/21A/22A-20 C_H Min	8420A/21A/22A-20 C_H Max	8420A/21A/22A-25 C_L Min	8420A/21A/22A-25 C_L Max	8420A/21A/22A-25 C_H Min	8420A/21A/22A-25 C_H Max
1	fCLK	CLK Frequency	0	20	0	20	0	25	0	25
2	tCLKP	CLK Period	50		50		40		40	
3, 4	tCLKPW	CLK Pulse Width	15		15		12		12	
5	fDCLK	DELCLK Frequency	5	20	5	20	5	20	5	20
6	tDCLKP	DELCLK Period	50	200	50	200	50	200	50	200
7, 8	tDCLKPW	DELCLK Pulse Width	15		15		12		12	
9a	tPRASCAS0	$\overline{RAS}$ Asserted to $\overline{CAS}$ Asserted (tRAH = 15 ns, tASC = 0 ns)	30		30		30		30	
9b	tPRASCAS1	$\overline{RAS}$ Asserted to $\overline{CAS}$ Asserted (tRAH = 15 ns, tASC = 10 ns)	40		40		40		40	
9c	tPRASCAS2	$\overline{RAS}$ Asserted to $\overline{CAS}$ Asserted (tRAH = 25 ns, tASC = 0 ns)	40		40		40		40	
9d	tPRASCAS3	$\overline{RAS}$ Asserted to $\overline{CAS}$ Asserted (tRAH = 25 ns, tASC = 10 ns)	50		50		50		50	
10a	tRAH	Row Address Hold Time (tRAH = 15)	15		15		15		15	
10b	tRAH	Row Address Hold Time (tRAH = 25)	25		25		25		25	
11a	tASC	Column Address Setup Time (tASC = 0)	0		0		0		0	
11b	tASC	Column Address Setup Time (tASC = 10)	10		10		10		10	
12	tPCKRAS	CLK High to $\overline{RAS}$ Asserted following Precharge		27		32		22		26
13	tPARQRAS	$\overline{AREQ}$ Negated to $\overline{RAS}$ Negated		38		43		31		35
14	tPENCL	$\overline{ECAS}$0–3 Asserted to $\overline{CAS}$ Asserted		23		31		20		27
15	tPENCH	$\overline{ECAS}$0–3 Negated to $\overline{CAS}$ Negated		25		33		20		27
16	tPARQCAS	$\overline{AREQ}$ Negated to $\overline{CAS}$ Negated		60		68		47		54
17	tPCLKWH	CLK to $\overline{WAIT}$ Negated		39		39		31		31
18	tPCLKDL0	CLK to $\overline{DTACK}$ Asserted (Programmed as $\overline{DTACK}$ of 1/2, 1, 1½ or if $\overline{WAITIN}$ is Asserted)		33		33		28		28
19	tPEWL	$\overline{ECAS}$ Negated to $\overline{WAIT}$ Asserted during a Burst Access		44		44		36		36
20	tSECK	$\overline{ECAS}$ Asserted Setup to CLK High to Recognize the Rising Edge of CLK during a Burst Access	24		24		19		19	

14.0 AC Timing Parameters: DP8420A/DP8421A/DP8422A (Continued)

Unless otherwise stated $V_{CC} = 5.0V \pm 10\%$, $0°C < T_A < 70°C$, the output load capacitance is typical for 4 banks of 18 DRAMs per bank, including trace capacitance (see Note 2).

Two different loads are specified:
C_L = 50 pF loads on all outputs except
C_L = 150 pF loads on Q0–8, 9, 10 and $\overline{WE}$; or

C_H = 50 pF loads on all outputs except
C_H = 125 pF loads on $\overline{RAS}$0–3 and $\overline{CAS}$0–3 and
C_H = 380 pF loads on Q0–8, 9, 10 and $\overline{WE}$.

Number	Symbol	Common Parameter Description	8420A/21A/22A-20 C_L Min	8420A/21A/22A-20 C_L Max	8420A/21A/22A-20 C_H Min	8420A/21A/22A-20 C_H Max	8420A/21A/22A-25 C_L Min	8420A/21A/22A-25 C_L Max	8420A/21A/22A-25 C_H Min	8420A/21A/22A-25 C_H Max
21	tPEDL	$\overline{ECAS}$ Asserted to $\overline{DTACK}$ Asserted during a Burst Access (Programmed as $\overline{DTACK0}$)		48		48		38		38
22	tPEDH	$\overline{ECAS}$ Negated to $\overline{DTACK}$ Negated during a Burst Access		49		49		38		38
23	tSWCK	$\overline{WAITIN}$ Asserted Setup to CLK	5		5		5		5	
24	tPWINWEH	$\overline{WIN}$ Asserted to $\overline{WE}$ Asserted		34		44		27		37
25	tPWINWEL	$\overline{WIN}$ Negated to $\overline{WE}$ Negated		34		44		27		37
26	tPAQ	Row, Column Address Valid to Q0–8, 9, 10 Valid		29		38		26		35
27	tPCINCQ	COLINC Asserted to Q0–8, 9, 10 Incremented		34		43		30		39
28	tSCINEN	COLINC Asserted Setup to $\overline{ECAS}$ Asserted to Ensure tASC = 0 ns	18		19		17		19	
29a	tSARQCK1	$\overline{AREQ}$, $\overline{AREQB}$ Negated Setup to CLK High with 1 Period of Precharge	46		46		37		37	
29b	tSARQCK2	$\overline{AREQ}$, $\overline{AREQB}$ Negated Setup to CLK High with >1 Period of Precharge Programmed	19		19		15		15	
30	tPAREQDH	$\overline{AREQ}$ Negated to $\overline{DTACK}$ Negated		34		34		27		27
31	tPCKCAS	CLK High to $\overline{CAS}$ Asserted when Delayed by $\overline{WIN}$		31		39		25		32
32	tSCADEN	Column Address Setup to $\overline{ECAS}$ Asserted to Guarantee tASC = 0	14		15		14		16	
33	tWCINC	COLINC Pulse Width	20		20		20		20	
34a	tPCKCL0	CLK High to $\overline{CAS}$ Asserted following Precharge (tRAH = 15 ns, tASC = 0 ns)		81		89		72		79
34b	tPCKCL1	CLK High to $\overline{CAS}$ Asserted following Precharge (tRAH = 15 ns, tASC = 10 ns)		91		99		82		89
34c	tPCKCL2	CLK High to $\overline{CAS}$ Asserted following Precharge (tRAH = 25 ns, tASC = 0 ns)		91		99		82		89
34d	tPCKCL3	CLK High to $\overline{CAS}$ Asserted following Precharge (tRAH = 25 ns, tASC = 10 ns)		101		109		92		99
35	tCAH	Column Address Hold Time (Interleave Mode Only)	32		32		32		32	
36	tPCQR	$\overline{CAS}$ Asserted to Row Address Valid (Interleave Mode Only)		90		90		90		90

14.0 AC Timing Parameters: DP8420A/DP8421A/DP8422A (Continued)

Unless otherwise stated V_{CC} = 5.0V ±10%, 0°C < T_A < 70°C, the output load capacitance is typical for 4 banks of 18 DRAMs per bank, including trace capacitance (see Note 2).

Two different loads are specified:
C_L = 50 pF loads on all outputs except
C_L = 150 pF loads on Q0-8, 9, 10 and $\overline{WE}$; or

C_H = 50 pF loads on all outputs except
C_H = 125 pF loads on $\overline{RAS}$0-3 and $\overline{CAS}$0-3 and
C_H = 380 pF loads on Q0-8, 9, 10 and $\overline{WE}$.

Number	Symbol	Difference Parameter Description	8420A/21A/22A-20 C_L Min	8420A/21A/22A-20 C_L Max	8420A/21A/22A-20 C_H Min	8420A/21A/22A-20 C_H Max	8420A/21A/22A-25 C_L Min	8420A/21A/22A-25 C_L Max	8420A/21A/22A-25 C_H Min	8420A/21A/22A-25 C_H Max
50	tD1	($\overline{AREQ}$ or $\overline{AREQB}$ Negated to $\overline{RAS}$ Negated) Minus (CLK High to $\overline{RAS}$ Asserted)		16		16		14		14
51	tD2	(CLK High to Refresh $\overline{RAS}$ Negated) Minus (CLK High to $\overline{RAS}$ Asserted)		13		13		11		11
52	tD3a	($\overline{ADS}$ Asserted to $\overline{RAS}$ Asserted (Mode 1)) Minus ($\overline{AREQ}$ Negated to $\overline{RAS}$ Negated)		4		4		4		4
53	tD3b	(CLK High to $\overline{RAS}$ Asserted (Mode 0)) Minus ($\overline{AREQ}$ Negated to $\overline{RAS}$ Negated)		4		4		4		4
54	tD4	($\overline{ECAS}$ Asserted to $\overline{CAS}$ Asserted) Minus ($\overline{ECAS}$ Negated to $\overline{CAS}$ Negated)	−7	7	−7	7	−7	7	−7	7
55	tD5	(CLK to Refresh $\overline{RAS}$ Asserted) Minus (CLK to Refresh $\overline{RAS}$ Negated)		6		6		6		6
56	tD6	($\overline{AREQ}$ Negated to $\overline{RAS}$ Negated) Minus ($\overline{ADS}$ Asserted to $\overline{RAS}$ Asserted ((Mode 1))		12		12		10		10

14.0 AC Timing Parameters: DP8420A/DP8421A/DP8422A (Continued)

Unless otherwise stated V_{CC} = 5.0V ±10%, 0°C < T_A < 70°C, the output load capacitance is typical for 4 banks of 18 DRAMs per bank, including trace capacitance (see Note 2).

Two different loads are specified:
C_L = 50 pF loads on all outputs except
C_L = 150 pF loads on Q0–8, 9, 10 and $\overline{WE}$; or

C_H = 50 pF loads on all outputs except
C_H = 125 pF loads on $\overline{RAS}$0–3 and $\overline{CAS}$0–3 and
C_H = 380 pF loads on Q0–8, 9, 10 and $\overline{WE}$.

Number	Symbol	Common Dual Access Parameter Description	8420A/21A/22A-20 C_L Min	8420A/21A/22A-20 C_L Max	8420A/21A/22A-20 C_H Min	8420A/21A/22A-20 C_H Max	8420A/21A/22A-25 C_L Min	8420A/21A/22A-25 C_L Max	8420A/21A/22A-25 C_H Min	8420A/21A/22A-25 C_H Max
100	tHCKARQB	$\overline{AREQB}$ Negated Held from CLK High	3		3		3		3	
101	tSARQBCK	$\overline{AREQB}$ Asserted Setup to CLK High	8		8		7		7	
102	tPAQBRASL	$\overline{AREQB}$ Asserted to $\overline{RAS}$ Asserted		43		48		37		41
103	tPAQBRASH	$\overline{AREQB}$ Negated to $\overline{RAS}$ Negated		41		46		32		36
105	tPCKRASG	CLK High to $\overline{RAS}$ Asserted for Pending Port B Access		55		60		44		48
106	tPAQBATKBL	$\overline{AREQB}$ Asserted to $\overline{ATACKB}$ Asserted		57		57		45		45
107	tPCKATKB	CLK High to $\overline{ATACKB}$ Asserted for Pending Access		67		67		51		51
108	tPCKGH	CLK High to GRANTB Asserted		40		40		32		32
109	tPCKGL	CLK High to GRANTB Negated		35		35		29		29
110	tSADDCKG	Row Address Setup to CLK High That Asserts $\overline{RAS}$ following a GRANTB Change to Ensure tASR = 0 ns for Port B	11		15		11		16	
111	tSLOCKCK	$\overline{LOCK}$ Asserted Setup to CLK Low to Lock Current Port	5		5		5		5	
112	tPAQATKBH	$\overline{AREQ}$ Negated to $\overline{ATACKB}$ Negated		26		26		21		21
113	tPAQBCASH	$\overline{AREQB}$ Negated to $\overline{CAS}$ Negated		59		67		47		54
114	tSADAQB	Address Valid Setup to $\overline{AREQB}$ Asserted	7		11		7		12	
116	tHCKARQG	$\overline{AREQ}$ Negated Held from CLK High	5		5		5		5	
117	tWAQB	$\overline{AREQB}$ High Pulse Width to Guarantee tASR = 0 ns	31		35		26		31	
118a	tPAQBCAS0	$\overline{AREQB}$ Asserted to $\overline{CAS}$ Asserted (tRAH = 15 ns, tASC = 0 ns)		103		111		87		94
118b	tPAQBCAS1	$\overline{AREQB}$ Asserted to $\overline{CAS}$ Asserted (tRAH = 15 ns, tASC = 10 ns)		113		121		97		104
118c	tPAQBCAS2	$\overline{AREQB}$ Asserted to $\overline{CAS}$ Asserted (tRAH = 25 ns, tASC = 0 ns)		113		121		97		104
118d	tPAQBCAS3	$\overline{AREQB}$ Asserted to $\overline{CAS}$ Asserted (tRAH = 25 ns, tASC = 10 ns)		123		131		107		114
120a	tPCKCASG0	CLK High to $\overline{CAS}$ Asserted for Pending Port B Access (tRAH = 15 ns, tASC = 0 ns)		113		121		96		103
120b	tPCKCASG1	CLK High to $\overline{CAS}$ Asserted for Pending Port B Access (tRAH = 15 ns, tASC = 10 ns)		123		131		106		113
120c	tPCKCASG2	CLK High to $\overline{CAS}$ Asserted for Pending Port B Access (tRAH = 25 ns, tASC = 0 ns)		123		131		106		113
120d	tPCKCASG3	CLK High to $\overline{CAS}$ Asserted for Pending Port B Access (tRAH = 25 ns, tASC = 10 ns)		133		141		116		123
121	tSBADDCKG	Bank Address Valid Setup to CLK High That Starts $\overline{RAS}$ for Pending Port B Access	10		10		10		10	

14.0 AC Timing Parameters: DP8420A/DP8421A/DP8422A (Continued)

Unless otherwise stated V_{CC} = 5.0V ±10%, 0°C < T_A < 70°C, the output load capacitance is typical for 4 banks of 18 DRAMs per bank, including trace capacitance (see Note 2).

Two different loads are specified:
C_L = 50 pF loads on all outputs except
C_L = 150 pF loads on Q0-8, 9, 10 and $\overline{WE}$; or

C_H = 50 pF loads on all outputs except
C_H = 125 pF loads on $\overline{RAS}$0-3 and $\overline{CAS}$0-3 and
C_H = 380 pF loads on Q0-8, 9, 10 and $\overline{WE}$.

Number	Symbol	Refresh Parameter Description	8420A/21A/22A-20 C_L Min	8420A/21A/22A-20 C_L Max	8420A/21A/22A-20 C_H Min	8420A/21A/22A-20 C_H Max	8420A/21A/22A-25 C_L Min	8420A/21A/22A-25 C_L Max	8420A/21A/22A-25 C_H Min	8420A/21A/22A-25 C_H Max
200	tSRFCK	$\overline{RFSH}$ Asserted Setup to CLK High	27		27		22		22	
201	tSDRFCK	$\overline{DISRFSH}$ Asserted Setup to CLK High	28		28		22		22	
202	tSXRFCK	EXTENDRF Setup to CLK High	15		15		12		12	
204	tPCKRFL	CLK High to $\overline{RFIP}$ Asserted		39		39		31		31
205	tPARQRF	$\overline{AREQ}$ Negated to $\overline{RFIP}$ Asserted		62		62		50		50
206	tPCKRFH	CLK High to $\overline{RFIP}$ Negated		65		65		51		51
207	tPCKRFRASH	CLK High to Refresh $\overline{RAS}$ Negated		35		40		29		33
208	tPCKRFRASL	CLK High to Refresh $\overline{RAS}$ Asserted		28		33		23		27
209a	tPCKCL0	CLK High to $\overline{CAS}$ Asserted during Error Scrubbing (tRAH = 15 ns, tASC = 0 ns)		82		90		73		80
209b	tPCKCL1	CLK High to $\overline{CAS}$ Asserted during Error Scrubbing (tRAH = 15 ns, tASC = 10 ns)		92		100		83		90
209c	tPCKCL2	CLK High to $\overline{CAS}$ Asserted during Error Scrubbing (tRAH = 25 ns, tASC = 0 ns)		92		100		83		90
209d	tPCKCL3	CLK High to $\overline{CAS}$ Asserted during Error Scrubbing (tRAH = 25 ns, tASC = 10 ns)		102		110		83		100
210	tWRFSH	$\overline{RFSH}$ Pulse Width	15		15		15		15	
211	tPCKRQL	CLK High to $\overline{RFRQ}$ Asserted		46		46		40		40
212	tPCKRQH	CLK High to $\overline{RFRQ}$ Negated		50		50		40		40

14.0 AC Timing Parameters: DP8420A/DP8421A/DP8422A (Continued)

Unless otherwise stated V_{CC} = 5.0V ±10%, 0°C < T_A < 70°C, the output load capacitance is typical for 4 banks of 18 DRAMs per bank, including trace capacitance (see Note 2).

Two different loads are specified:
C_L = 50 pF loads on all outputs except
C_L = 150 pF loads on Q0–8, 9, 10 and $\overline{WE}$; or

C_H = 50 pF loads on all outputs except
C_H = 125 pF loads on $\overline{RAS}$0–3 and $\overline{CAS}$0–3 and
C_H = 380 pF loads on Q0–8, 9, 10 and $\overline{WE}$

Number	Symbol	Mode 0 Access Parameter Description	8420A/21A/22A-20 C_L Min	8420A/21A/22A-20 C_L Max	8420A/21A/22A-20 C_H Min	8420A/21A/22A-20 C_H Max	8420A/21A/22A-25 C_L Min	8420A/21A/22A-25 C_L Max	8420A/21A/22A-25 C_H Min	8420A/21A/22A-25 C_H Max
300	tSCSCK	$\overline{CS}$ Asserted to CLK High	14		14		13		13	
301a	tSALECKNL	ALE Asserted Setup to CLK High Not Using On-Chip Latches or if Using On-Chip Latches and B0, B1, Are Constant, Only 1 Bank	16		16		15		15	
301b	tSALECKL	ALE Asserted Setup to CLK High, if Using On-Chip Latches if B0, B1 Can Change, More Than One Bank	29		29		29		29	
302	tWALE	ALE Pulse Width	18		18		13		13	
303	tSBADDCK	Bank Address Valid Setup to CLK High	20		20		18		18	
304	tSADDCK	Row, Column Valid Setup to CLK High to Guarantee tASR = 0 ns	11		15		11		16	
305	tHASRCB	Row, Column, Bank Address Held from ALE Negated (Using On-Chip Latches)	10		10		8		8	
306	tSRCBAS	Row, Column, Bank Address Setup to ALE Negated (Using On-Chip Latches)	3		3		2		2	
307	tPCKRL	CLK High to $\overline{RAS}$ Asserted		27		32		22		26
308a	tPCKCL0	CLK High to $\overline{CAS}$ Asserted (tRAH = 15 ns, tASC = 0 ns)		81		89		72		79
308b	tPCKCL1	CLK High to $\overline{CAS}$ Asserted (tRAH = 15 ns, tASC = 10 ns)		91		99		82		89
308c	tPCKCL2	CLK High to $\overline{CAS}$ Asserted (tRAH = 25 ns, tASC = 0 ns)		91		99		82		89
308d	tPCKCL3	CLK High to $\overline{CAS}$ Asserted (tRAH = 25 ns, tASC = 10 ns)		101		109		92		99
309	tHCKALE	ALE Negated Hold from CLK High	0		0		0		0	
310	tSWINCK	$\overline{WIN}$ Asserted Setup to CLK High to Guarantee $\overline{CAS}$ is Delayed	−21		−21		−16		−16	
311	tPCSWL	$\overline{CS}$ Asserted to $\overline{WAIT}$ Asserted		26		26		22		22
312	tPCSWH	$\overline{CS}$ Negated to $\overline{WAIT}$ Negated		30		30		25		25
313	tPCLKDL1	CLK High to $\overline{DTACK}$ Asserted (Programmed as $\overline{DTACK0}$)		40		40		32		32
314	tPALEWL	ALE Asserted to $\overline{WAIT}$ Asserted ($\overline{CS}$ is Already Asserted)		35		35		29		29
315		$\overline{AREQ}$ Negated to CLK High That Starts Access $\overline{RAS}$ to Guarantee tASR = 0 ns (Non-Interleaved Mode Only)	41		45		34		39	
316	tPCKCV0	CLK High to Column Address Valid (tRAH = 15 ns, tASC = 0 ns)		78		87		66		75

14.0 AC Timing Parameters: DP8420A/DP8421A/DP8422A (Continued)

Unless otherwise stated V_{CC} = 5.0V ±10%, 0°C < T_A < 70°C, the output load capacitance is typical for 4 banks of 18 DRAMs per bank, including trace capacitance (see Note 2).

Two different loads are specified:
C_L = 50 pF loads on all outputs except
C_L = 150 pF loads on Q0–8, 9, 10 and $\overline{WE}$; or

C_H = 50 pF loads on all outputs except
C_H = 125 pF loads on $\overline{RAS}$0–3 and $\overline{CAS}$0–3 and
C_H = 380 pF loads on Q0–8, 9, 10 and $\overline{WE}$.

Number	Symbol	Mode 1 Access Parameter Description	8420A/21A/22A-20 C_L Min	Max	C_H Min	Max	8420A/21A/22A-25 C_L Min	Max	C_H Min	Max
400a	tSADSCK1	$\overline{ADS}$ Asserted Setup to CLK High	15		15		13		13	
400b	tSADSCKW	$\overline{ADS}$ Asserted Setup to CLK (to Guarantee Correct $\overline{WAIT}$ or $\overline{DTACK}$ Output; Doesn't Apply for $\overline{DTACK}$0)	31		31		25		25	
401	tSCSADS	$\overline{CS}$ Setup to $\overline{ADS}$ Asserted	6		6		5		5	
402	tPADSRL	$\overline{ADS}$ Asserted to $\overline{RAS}$ Asserted		30		35		25		29
403a	tPADSCL0	$\overline{ADS}$ Asserted to $\overline{CAS}$ Asserted (tRAH = 15 ns, tASC = 0 ns)		86		94		75		82
403b	tPADSCL1	$\overline{ADS}$ Asserted to $\overline{CAS}$ Asserted (tRAH = 15 ns, tASC = 10 ns)		96		104		85		92
403c	tPADSCL2	$\overline{ADS}$ Asserted to $\overline{CAS}$ Asserted (tRAH = 25 ns, tASC = 0 ns)		96		104		85		92
403d	tPADSCL3	$\overline{ADS}$ Asserted to $\overline{CAS}$ Asserted (tRAH = 25 ns, tASC = 10 ns)		106		114		95		102
404	tSADDADS	Row Address Valid Setup to $\overline{ADS}$ Asserted to Guarantee tASR = 0 ns	9		13		9		14	
405	tHCKADS	$\overline{ADS}$ Negated Held from CLK High	0		0		0		0	
406	tSWADS	$\overline{WAITIN}$ Asserted Setup to $\overline{ADS}$ Asserted to Guarantee $\overline{DTACK}$0 Is Delayed	0		0		0		0	
407	tSBADAS	Bank Address Setup to $\overline{ADS}$ Asserted	11		11		11		11	
408	tHASRCB	Row, Column, Bank Address Held from $\overline{ADS}$ Asserted (Using On-Chip Latches)	10		10		10		10	
409	tSRCBAS	Row, Column, Bank Address Setup to $\overline{ADS}$ Asserted (Using On-Chip Latches)	3		3		2		2	
410	tWADSH	$\overline{ADS}$ Negated Pulse Width	12		16		12		17	
411	tPADSD	$\overline{ADS}$ Asserted to $\overline{DTACK}$ Asserted (Programmed as $\overline{DTACK}$0)		43		43		35		35
412	tSWINADS	$\overline{WIN}$ Asserted Setup to $\overline{ADS}$ Asserted (to Guarantee $\overline{CAS}$ Delayed during Writes Accesses)	−10		−10		−10		−10	
413	tPADSWL0	$\overline{ADS}$ Asserted to $\overline{WAIT}$ Asserted (Programmed as $\overline{WAIT}$0, Delayed Access)		35		35		29		29
414	tPADSWL1	$\overline{ADS}$ Asserted to $\overline{WAIT}$ Asserted (Programmed $\overline{WAIT}$ 1/2 or 1)		35		35		29		29
415	tPCLKDL1	CLK High to $\overline{DTACK}$ Asserted (Programmed as $\overline{DTACK}$0, Delayed Access)		40		40		32		32
416		$\overline{AREQ}$ Negated to $\overline{ADS}$ Asserted to Guarantee tASR = 0 ns (Non Interleaved Mode Only)	38		42		31		36	
417	tPADSCV0	$\overline{ADS}$ Asserted to Column Address Valid (tRAH = 15 ns, tASC = 0 ns)		83		92		69		78

1-155

14.0 AC Timing Parameters: DP8420A/DP8421A/DP8422A (Continued)

Unless otherwise stated V_{CC} = 5.0V ±10%, 0°C < T_A < 70°C, the output load capacitance is typical for 4 banks of 18 DRAMs per bank, including trace capacitance (see Note 2).

Two different loads are specified:
C_L = 50 pF loads on all outputs except
C_L = 150 pF loads on Q0–8, 9, 10 and $\overline{WE}$; or

C_H = 50 pF loads on all outputs except
C_H = 125 pF loads on $\overline{RAS}$0–3 and $\overline{CAS}$0–3 and
C_H = 380 pF loads on Q0–8, 9, 10 and $\overline{WE}$.

Number	Symbol	Mode 1 Dual Access Parameter Description	8420A/21A/22A-20 C_L Min	Max	C_H Min	Max	8420A/21A/22A-25 C_L Min	Max	C_H Min	Max
450	tSADDCKG	Row Address Setup to CLK High That Asserts $\overline{RAS}$ following a GRANTB Port Change to Ensure tASR = 0 ns	11		15		11		16	
451	tPCKRASG	CLK High to $\overline{RAS}$ Asserted for Pending Access		48		53		38		42
452	tPCLKDL2	CLK to $\overline{DTACK}$ Asserted for Delayed Accesses (Programmed as $\overline{DTACK0}$)		53		53		43		43
453a	tPCKCASG0	CLK High to $\overline{CAS}$ Asserted for Pending Access (tRAH = 15 ns, tASC = 0 ns)		101		109		86		93
453b	tPCKCASG1	CLK High to $\overline{CAS}$ Asserted for Pending Access (tRAH = 15 ns, tASC = 10 ns)		111		119		96		103
453c	tPCKCASG2	CLK High to $\overline{CAS}$ Asserted for Pending Access (tRAH = 25 ns, tASC = 0 ns)		111		119		96		103
453d	tPCKCASG3	CLK High to $\overline{CAS}$ Asserted for Pending Access (tRAH = 25 ns, tASC = 10 ns)		121		129		106		113
454	tSBADDCKG	Bank Address Valid Setp to CLK High That Asserts $\overline{RAS}$ for Pending Access	5		5		4		4	
455	tSADSCK0	$\overline{ADS}$ Asserted Setup to CLK High	12		12		11		11	

Number	Symbol	Programming Parameter Description	8420A/21A/22A-20 C_L Min	Max	C_H Min	Max	8420A/21A/22A-25 C_L Min	Max	C_H Min	Max
500	tHMLADD	Mode Address Held from $\overline{ML}$ Negated	8		8		7		7	
501	tSADDML	Mode Address Setup to $\overline{ML}$ Negated	6		6		6		6	
502	tWML	$\overline{ML}$ Pulse Width	15		15		15		15	
503	tSADAQML	Mode Address Setup to $\overline{AREQ}$ Asserted	0		0		0		0	
504	tHADAQML	Mode Address Held from $\overline{AREQ}$ Asserted	51		51		38		38	
505	tSCSARQ	$\overline{CS}$ Asserted Setup to $\overline{AREQ}$ Asserted	6		6		6		6	
506	tSMLARQ	$\overline{ML}$ Asserted Setup to $\overline{AREQ}$ Asserted	10		10		10		10	

Note 1: "Absolute Maximum Ratings" are the values beyond which the safety of the device cannot be guaranteed. They are not meant to imply that the device should be operated at these limits. The table of "Electrical Characteristics" provides conditions for actual device operation.

Note 2: Input pulse 0V to 3V; tR = tF = 2.5 ns. Input reference point on AC measurements is 1.5V. Output reference points are 2.4V for High and 0.8V for Low.

Note 3: AC Production testing is done at 50 pF.

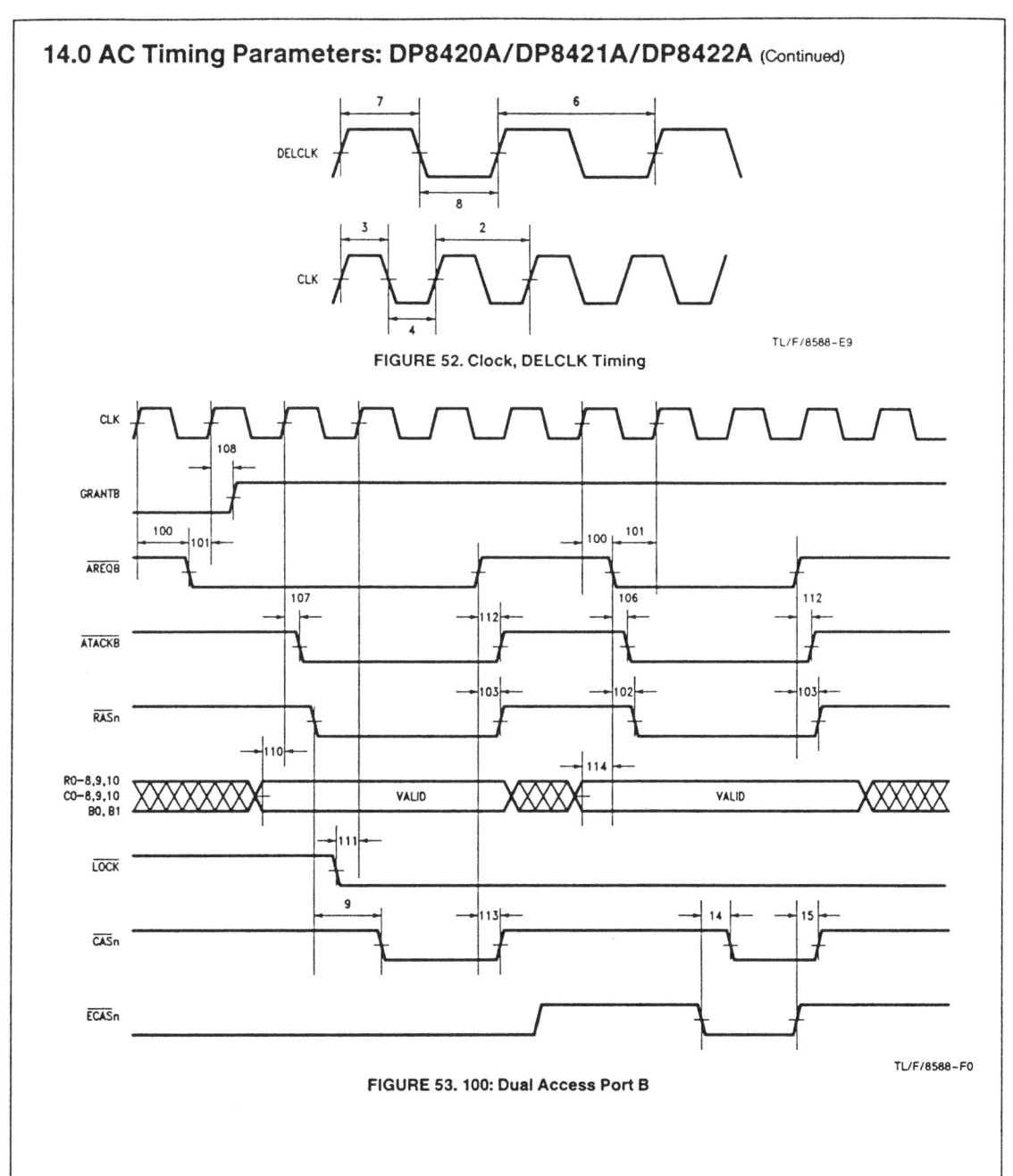

14.0 AC Timing Parameters: DP8420A/DP8421A/DP8422A (Continued)

FIGURE 52. Clock, DELCLK Timing

FIGURE 53. 100: Dual Access Port B

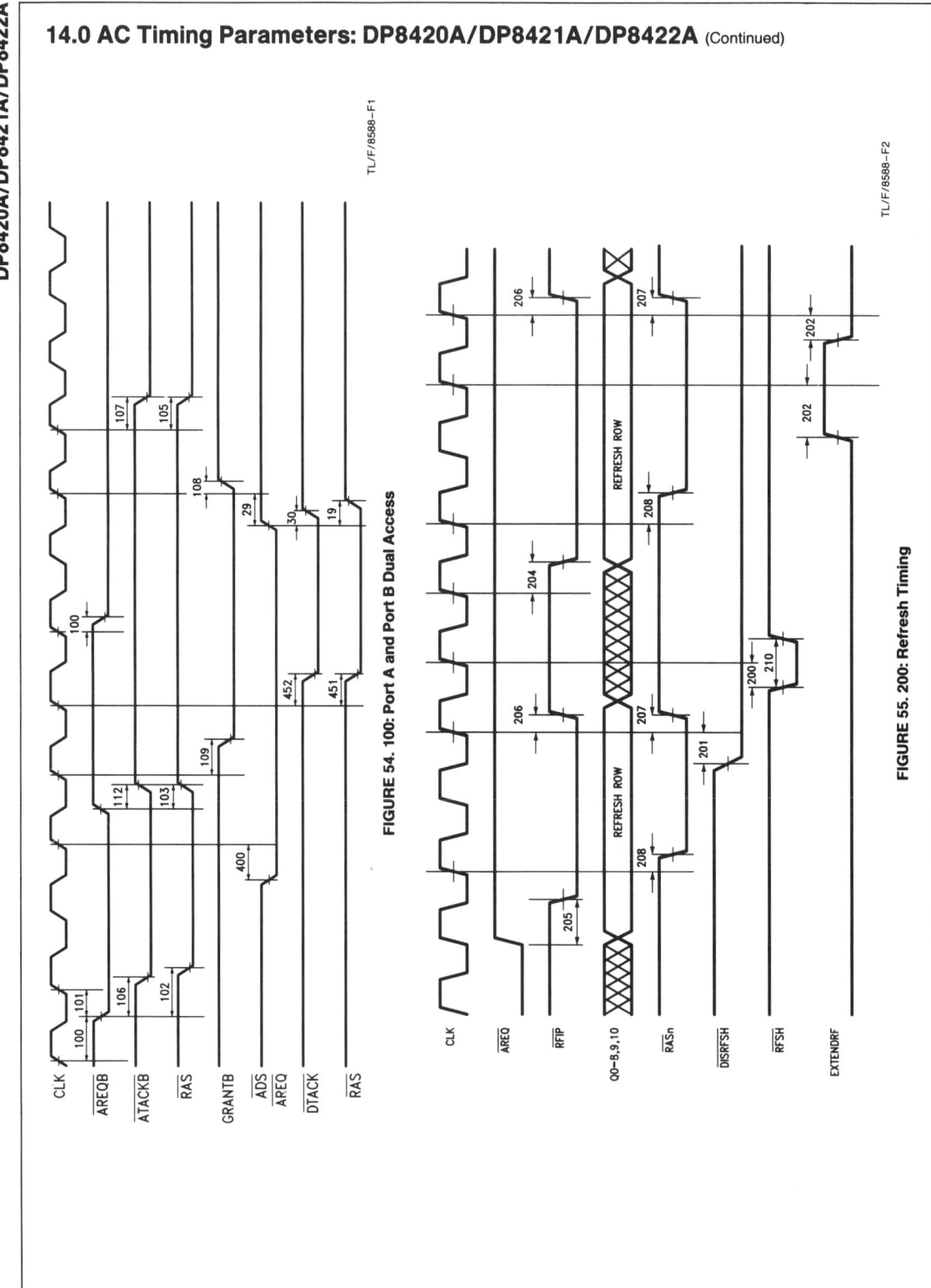

14.0 AC Timing Parameters: DP8420A/DP8421A/DP8422A (Continued)

FIGURE 54. 100: Port A and Port B Dual Access

FIGURE 55. 200: Refresh Timing

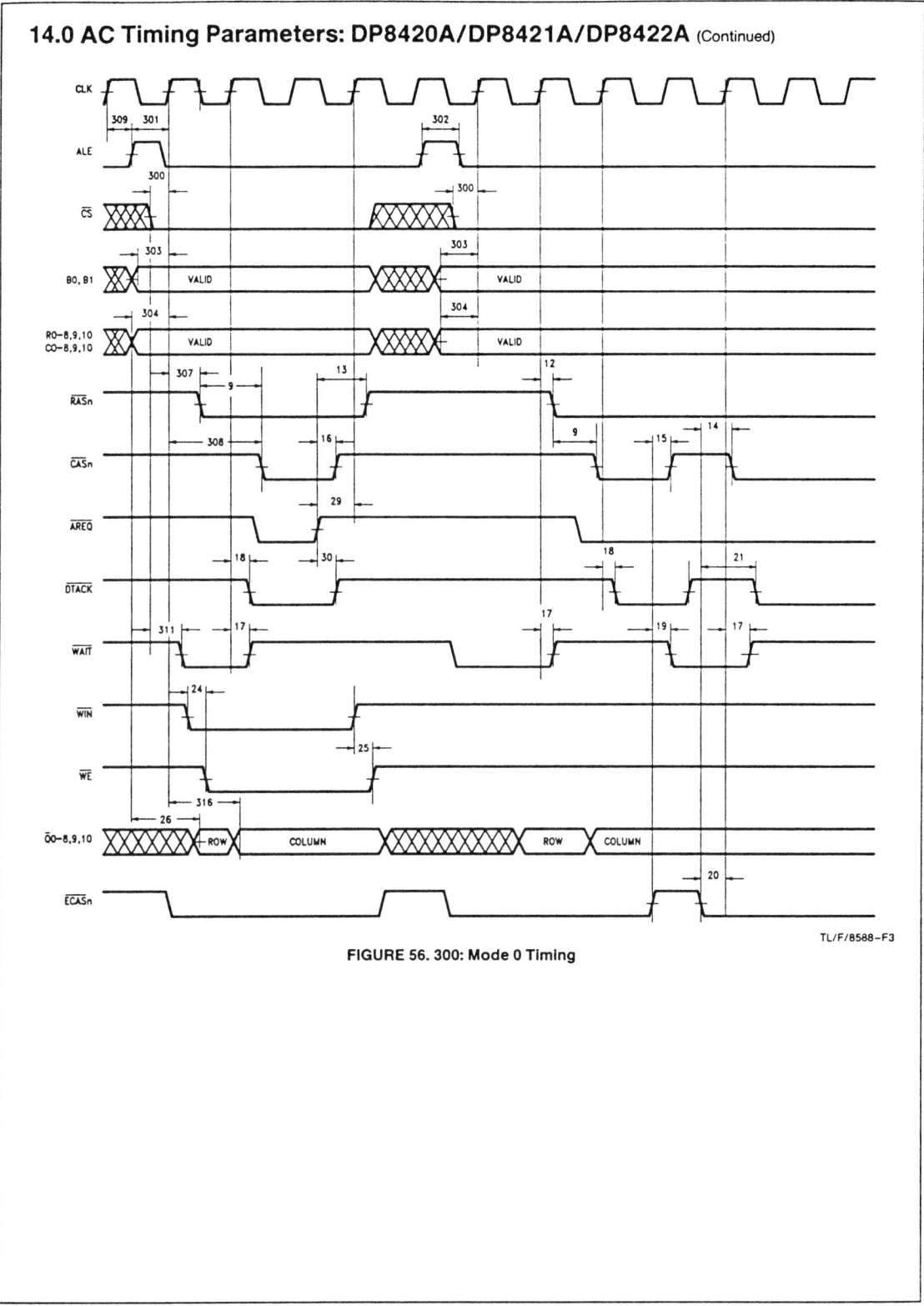

14.0 AC Timing Parameters: DP8420A/DP8421A/DP8422A (Continued)

FIGURE 56. 300: Mode 0 Timing

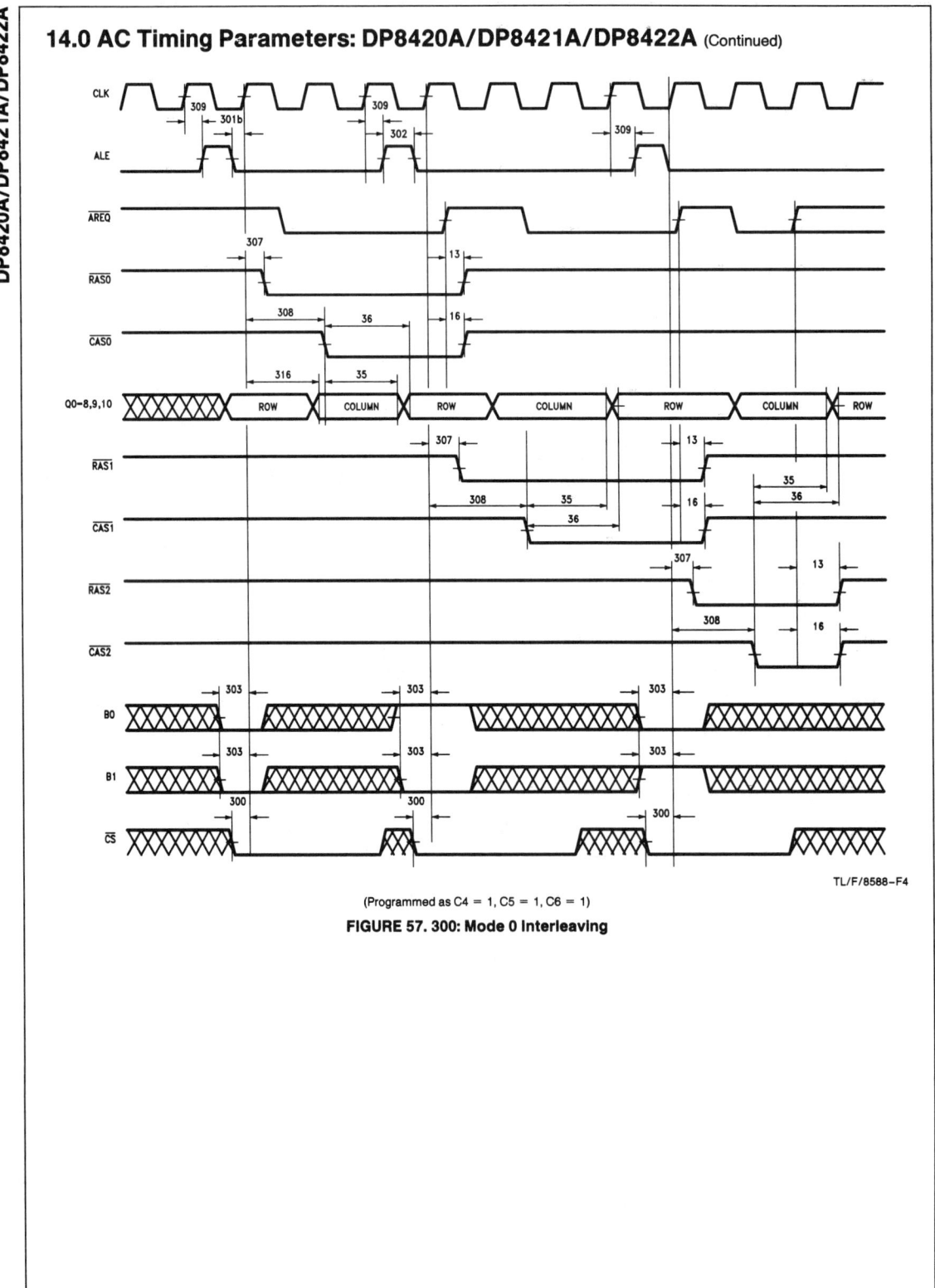

14.0 AC Timing Parameters: DP8420A/DP8421A/DP8422A (Continued)

(Programmed as C4 = 1, C5 = 1, C6 = 1)

FIGURE 57. 300: Mode 0 Interleaving

14.0 AC Timing Parameters: DP8420A/DP8421A/DP8422A (Continued)

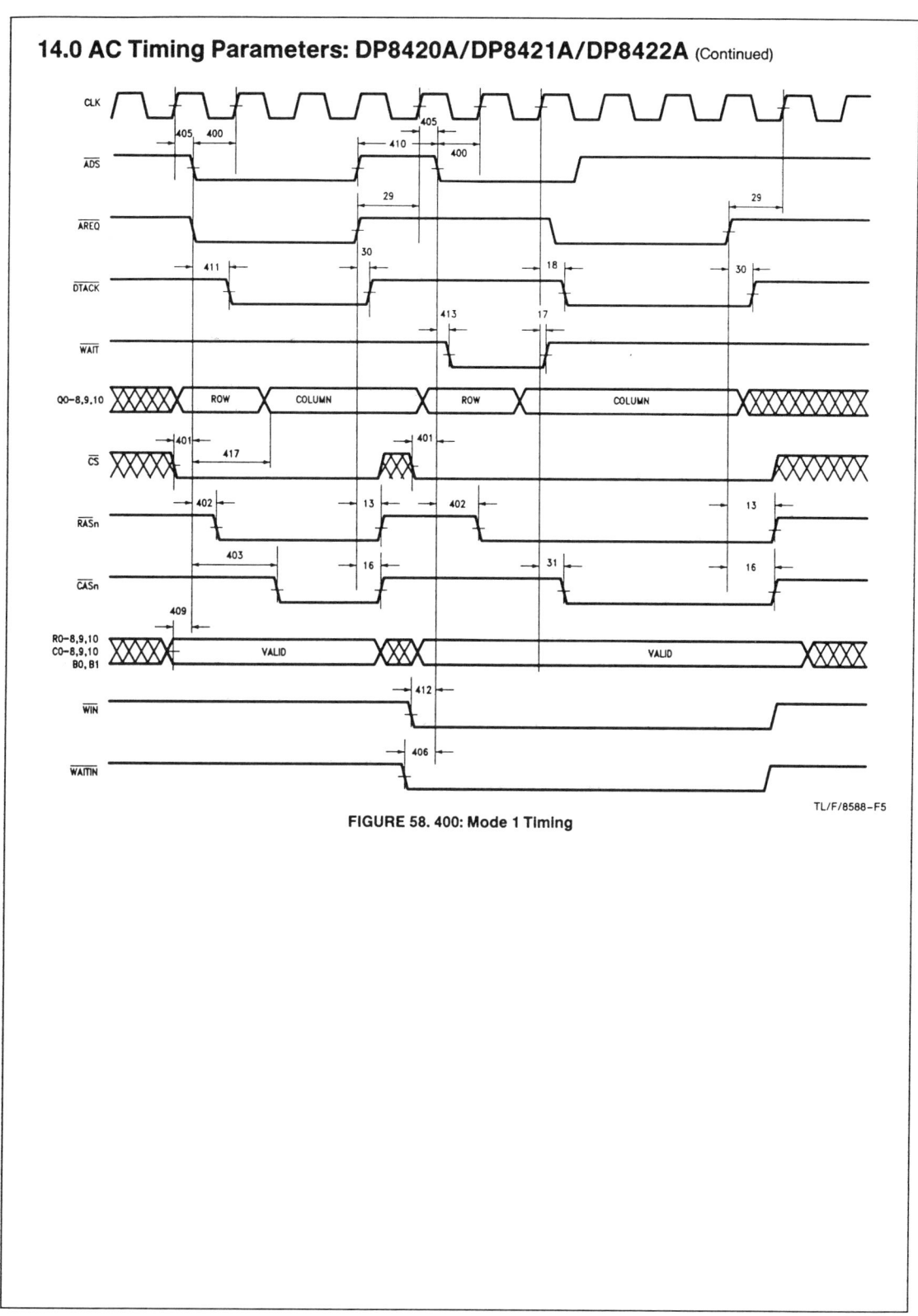

FIGURE 58. 400: Mode 1 Timing

14.0 AC Timing Parameters: DP8420A/DP8421A/DP8422A (Continued)

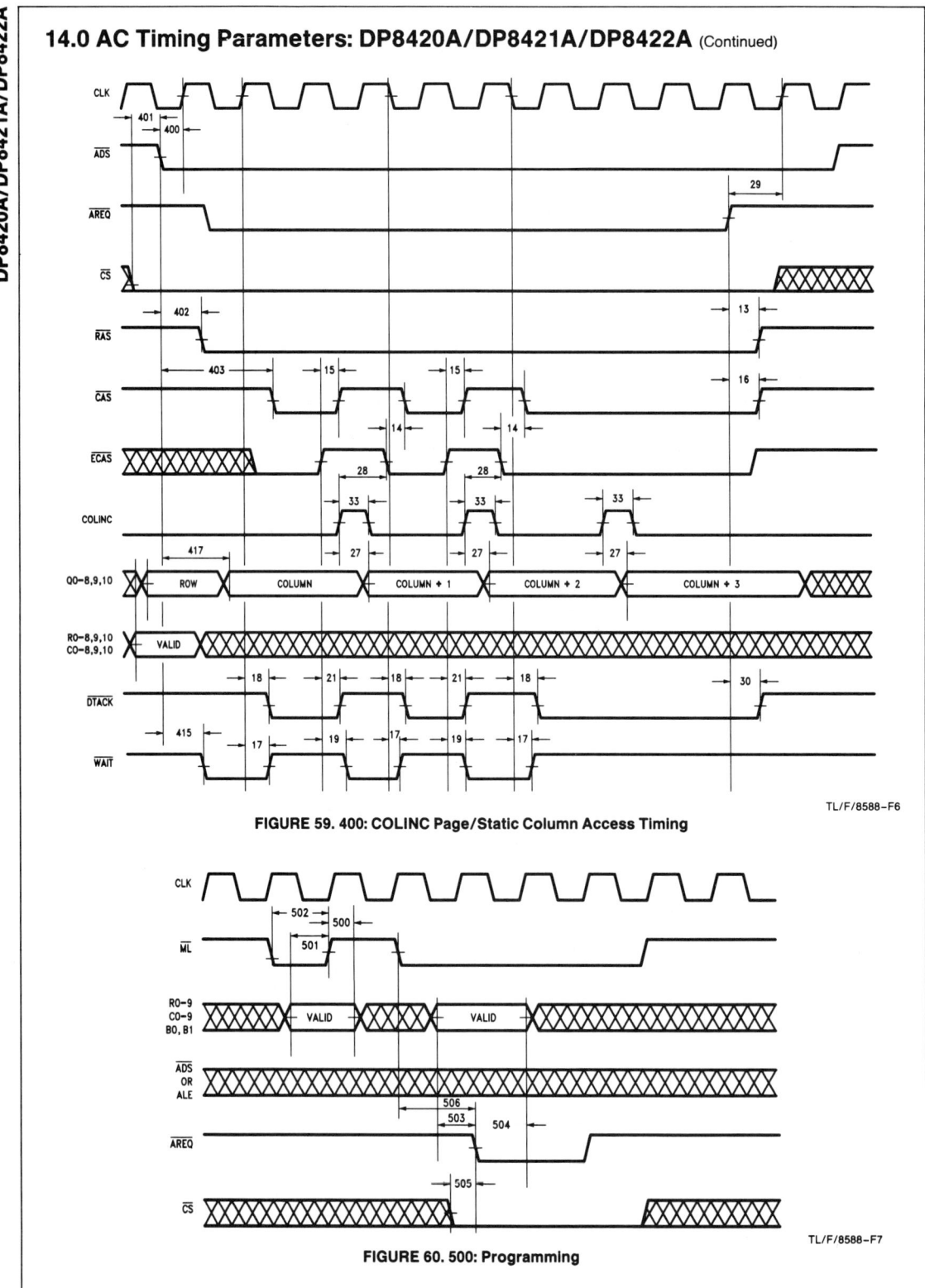

FIGURE 59. 400: COLINC Page/Static Column Access Timing

FIGURE 60. 500: Programming

15.0 Functional Differences between the DP8420A/21A/22A and the DP8420/21/22

1. **Extending the Column Address Strobe ($\overline{CAS}$) after $\overline{AREQ}$ Transitions High**

   The DP8420A/21A/22A allows $\overline{CAS}$ to be asserted for an indefinite period of time beyond $\overline{AREQ}$ (or $\overline{AREQB}$, DP8422A only. Scrubbing refreshes are not affected.) being negated by continuing to assert the appropriate $\overline{ECAS}$ inputs. This feature is allowed as long as the $\overline{ECAS}0$ input was negated during programming. The DP8420/21/22 does not allow this feature.

2. **Dual Accessing**

   The DP8420A/21A/22A asserts $\overline{RAS}$ either one or two clock periods after GRANTB has been asserted or negated depending upon how the R0 bit was programmed during the mode load operation. The DP8420/21/22 will always start $\overline{RAS}$ one clock period after GRANTB is asserted or negated. The above statements assume that $\overline{RAS}$ precharge has been completed by the time GRANTB is asserted or negated.

3. **Refresh Request Output ($\overline{RFRQ}$)**

   The DP8420A/21A/22A allows $\overline{RFRQ}$ (refresh request) to be output on the $\overline{WE}$ output pin given that $\overline{ECAS}0$ was negated during programming or the controller was programmed to function in the address pipelining (memory interleaving) mode. The DP8420/21/22 only allows $\overline{RFRQ}$ to be output during the address pipelining mode.

4. **Clearing the Refresh Request Clock Counter**

   The DP8420A/21A/22A allows the internal refresh request clock counter to be cleared by negating $\overline{DISRFSH}$ and asserting $\overline{RFSH}$ for at least 500 ns. The DP8420/21/22 clears the internal refresh request clock counter if $\overline{DISRFSH}$ remains low for at least 500 ns. Once the internal refresh request clock counter is cleared the user is guaranteed that an internally generated $\overline{RFRQ}$ will not be generated for at least 13 μs–15 μs (depending upon how programming bits C0, 1, 2, 3 were programmed).

16.0 DP8420A/21A/22A User Hints

1. All inputs to the DP8420A/21A/22A should be tied high, low or the output of some other device.

   Note: One signal is active high. COLINC (EXTNDRF) should be tied low to disable.

2. Each ground on the DP8420A/21A/22A must be decoupled to the closest on-chip supply (V_{CC}) with 0.1 μF ceramic capacitor. This is necessary because these grounds are kept separate inside the DP8420A/21A/22A. The decoupling capacitors should be placed as close as possible with short leads to the ground and supply pins of the DP8420A/21A/22A.

3. The output called "CAP" should have a 0.1 μF capacitor to ground.

4. The DP8420A/21A/22A has 20Ω series damping resistors built into the output drivers of $\overline{RAS}$, $\overline{CAS}$, address and $\overline{WE}/\overline{RFRQ}$. Space should be provided for external damping resistors on the printed circuit board (or wire-wrap board) because they may be needed. The value of these damping resistors (if needed) will vary depending upon the output, the capacitance of the load, and the characteristics of the trace as well as the routing of the trace. The value of the damping resistor also may vary between the wire-wrap board and the printed circuit board. To determine the value of the series damping resistor it is recommended to use an oscilloscope and look at the furthest DRAM from the DP8420A/21A/22A. The undershoot of $\overline{RAS}$, $\overline{CAS}$, $\overline{WE}$ and the addresses should be kept to less than 0.5V below ground by varying the value of the damping resistor. The damping resistors should be placed as close as possible with short leads to the driver outputs of the DP8420A/21A/22A.

5. The circuit board must have a good V_{CC} and ground plane connection. If the board is wire-wrapped, the V_{CC} and ground pins of the DP8420A/21A/22A, the DRAM associated logic and buffer circuitry must be soldered to the V_{CC} and ground planes.

6. The traces from the DP8420A/21A/22A to the DRAM should be as short as possible.

7. $\overline{ECAS}0$ should be held low during programming if the user wishes that the DP8420A/21A/22A be compatible with a DP8420/21/22 design.

8. **Parameter Changes due to Loading**

   All A.C. parameters are specified with the equivalent load capacitances, including traces, of 64 DRAMs organized as 4 banks of 18 DRAMs each. Maximums are based on worst-case conditions. If an output load changes then the A.C. timing parameters associated with that particular output must be changed. For example, if we changed our output load to

   C = 250 pF loads on $\overline{RAS}$0–3 and $\overline{CAS}$0–3
   C = 760 pF loads on Q0–9 and $\overline{WE}$

   we would have to modify some parameters (not all calculated here)

   $308a clock to $\overline{CAS}$ asserted

   (t_{RAH} = 15 ns, t_{ASC} = 0 ns)

   A ratio can be used to figure out the timing change per change in capacitance for a particular parameter by using the specifications and capacitances from heavy and light load timing.

   $$\text{Ratio} = \frac{\$308a \text{ w/Heavy Load} - \$308a \text{ w/Light Load}}{C_H(\overline{CAS}) - C_L(\overline{CAS})}$$

   $$= \frac{79 \text{ ns} - 72 \text{ ns}}{125 \text{ pF} - 50 \text{ pF}} = \frac{7 \text{ ns}}{75 \text{ pF}}$$

   $308a (actual) = (capacitance difference $\times$ ratio) + $308a (specified)

   $$= (250 \text{ pF} - 125 \text{ pF})\frac{7 \text{ ns}}{75 \text{ pF}} + 79 \text{ ns}$$

   $$= 11.7 \text{ ns} + 79 \text{ ns}$$

   $$= 90.7 \text{ ns} @ 250 \text{ pF load}$$

9. It is recommended that the user perform a hardware reset of the DP8420A/21A/22A before programming and using the chip. A hardware reset consists of asserting both $\overline{ML}$ and $\overline{DISRFSH}$ for a minimum of 16 positive edges of CLK, see Section 8.1.

AN-538 Interfacing the DP8420A/21A/22A to the 68000/008/010

National Semiconductor
Application Note 538
Joe Tate and Rusty Meier

INTRODUCTION

This application note explains interfacing the DP8420A/21A/22A DRAM controller to the 68000. Three different designs are shown and explained. It is assumed that the reader is familiar with the 68000 access cycles and the DP8420A/21A/22A modes of operation. This application note also applies to the 68010.

DESIGN #1 DESCRIPTION

Design #1 is a simple circuit to interface the 68000 to the DP8420A/21A/22A and up to 32 Mbytes of DRAM. The DP8420A/21A/22A is operated in Mode 1. An access cycle begins when the 68000 places a valid address on the address bus and asserts the address strobe ($\overline{AS}$). Chip select ($\overline{CS}$) is generated by a 74AS138 decoder. If a refresh or Port B access (DP8422A only) is not in progress, the DP8420A/21A/22A will assert the proper $\overline{RAS}$ depending on the bank select inputs (B0, B1). After guaranteeing the programmed value of row address hold time the DP8420A/21A/22A will switch the DRAM address (Q0–8, 9, 10) to the column address and assert $\overline{CAS}$. By this time, the 74AS245's have been enabled and the DRAMs place their data on the data bus. The DP8420A/21A/22A also asserts $\overline{DTACK}$ which is used to generate $\overline{DTACK}$ to the 68000 to complete the access.

If a refresh or Port B access had been in progress, the DP8420A/21A/22A would have delayed the 68000's access by inserting wait states into the access cycle until the refresh or Port B access was complete and the programmed amount of precharge time was met. This circuit can run up to 10 MHz with 0 wait states, with two or more banks. For 10 MHz, zero wait states with one bank, see design #2.

Timing parameters are referenced to the numbers shown in the DP8420A/21A/22A data sheet timing parameters. Numbered times starting with a "$" refer to the DP8420A/21A/22A timing parameters. Numbered times starting with "#" refer to the 68000 data sheet. Equations have been given to allow the user to calculate timing based on his frequency and application. The clock is at 10 MHz, a multiple of 2 MHz, allowing it to be tied directly to DELCLK. If DELCLK is not a multiple of 2 MHz, $\overline{ADS}$ to $\overline{CAS}$ must be recalculated.

DESIGN #1 TIMING AT 10 MHz AND 8 MHz

Clock Period = Tcp10 = 100 ns @ 10 MHz
 = Tcp8 = 125 ns @ 8 ns

$400b: $\overline{ADS}$ Asserted Setup to CLK High
 = Clock Period − CLK High to $\overline{AS}$ Asserted
 = Tcp10 − #9
 = 100 ns − 55 ns
 = **45 ns @ 10 MHz**
 = Tcp8 − #9
 = 125 ns − 60 ns
 = **65 ns @ 8 MHz**

$401: $\overline{CS}$ Setup to $\overline{ADS}$ Asserted
 = 68000 Address to $\overline{AS}$ Max
  − 74AS138 Decoder
 = #11 − Tphl Max
 = 20 ns − 9 ns
 = **11 ns @ 10 MHz**
 = #11 − Tphl
 = 30 ns − 9 ns
 = **21 ns @ 8 MHz**

$407 & $404: Address Valid Setup to $\overline{ADS}$ Asserted
 = 68000 Address to $\overline{AS}$ Max
 = #11 Max
 = **20 ns @ 10 MHz**
 = #11 Max
 = **30 ns @ 8 MHz**

$405: $\overline{ADS}$ Negated Held from CLK High
 = 68000 CLK High to $\overline{AS}$ Asserted Min
 = #10 Min
 = **0 ns @ 10 MHz**
 = #10 Min
 = **0 ns @ 8 MHz**

#47: DTACK Setup Time
 = ½ Clock Period
  − Clock to $\overline{DTACK}$ Asserted
 = ½ Tcp10 − $18
 = 50 ns − 28 ns
 = **22 ns @ 10 MHz** **Using 8420-25
 = ½ Tcp8 − $18
 = 62.5 ns − 33 ns
 = **29.5 ns @ 8 MHz** **Using 8420-25

RAS LOW DURING REFRESH

tRAS = Programmed Clock
  − [(CLK High to Refresh $\overline{RAS}$ Asserted)
  − (CLK High to Refresh $\overline{RAS}$ Negated)]
 = Tcp10 + Tcp10 − 55
 = 100 ns + 100 ns − 6 ns
 = **194 ns @ 10 MHz**
 = Tcp8 + Tcp8 − $55
 = 125 ns + 125 ns − 6 ns
 = **244 ns @ 8 MHz**

$\overline{RAS}$ PRECHARGE PARAMETERS**

tRP = (Programmed Clocks − 1)
  − [($\overline{AREQ}$ to $\overline{RAS}$ Negated)
  − (CLK to $\overline{RAS}$ Asserted)]
 = Tcp10 − $50
 = 100 ns − 16 ns
 = **84 ns @ 10 MHz**
 = Tcp8 − $50
 = 125 ns − 16 ns
 = **109 ns @ 8 MHz**

**To gain more precharge program 3t or use design #2.

tRAC AND tCAC FOR DRAMs

Timing is supplied for the system shown in *Figure 1*. *(see Figures 2, 3 and 4)*. Since systems and DRAM times vary, the user is encouraged to change the following equations to match his system requirements. Timing has been supplied for systems with 0 or 1 wait state. If DELCLK is not a multiple of 2 MHz, the timing for tRAH and tASC will increase or decrease according to the equations given in the data sheet. The $\overline{ADS}$ to $\overline{RAS}$ and $\overline{ADS}$ to $\overline{CAS}$ will also have to be changed depending on the capacitance of the DRAM array.

0 Wait States

tRAC = s2 + s3 + s4 + s5 + s6 − CLK to $\overline{AS}$ Asserted Max − $\overline{ADS}$ Asserted to $\overline{RAS}$ Asserted − 74AS245 Delay Max − 68000 Data Setup Min
 = 2½ Tcp10 − #9 − $402 − Tphl Max − #27
 = 250 ns − 55 ns − 35 ns − 7 ns − 10 ns
 = **143 ns @ 10 MHz**   Using 8420-20 w/Heavy Load
 = 2½ Tcp8 − #9 − $402 − Tphl Max − #27
 = 312.5 ns − 60 ns − 35 ns − 7 ns − 15 ns
 = **195 ns @ 8 MHz**   Using 8420-20 w/Heavy Load

1 Wait State

tRAC = s2 + s3 + s4 + sw + sw + s5 + s6 − CLK to $\overline{AS}$ Asserted Max − $\overline{ADS}$ Asserted to $\overline{RAS}$ Asserted − 74AS245 Delay Max − 68000 Data Setup Min
 = 3½ Tcp10 − #9 − $402 − Tphl Max − #27
 = 350 ns − 55 ns − 35 ns − 7 ns − 10 ns
 = **243 ns @ 10 MHz**   Using 8420-20 w/Heavy Load
 = 3½ Tcp8 − #9 − $402 − Tphl Max − #27
 = 437.5 ns − 60 ns − 35 ns − 7 ns − 15 ns
 = **320 ns @ 8 MHz**   Using 8420-20 w/Heavy Load

0 Wait States

tCAC = s2 + s3 + s4 + s5 + s6 − CLK to $\overline{AS}$ Asserted Max − $\overline{ADS}$ Asserted to $\overline{CAS}$ Asserted − 74AS245 Delay Max − 68000 Data Setup Min
 = 2½ Tcp10 − #9 − $403a − Tphl Max − #27
 = 250 ns − 55 ns − 94 ns − 7 ns − 10 ns
 = **84 ns @ 10 MHz**   Using 8420-20 w/Heavy Load
 = 2½ Tcp8 − #9 − $403a − Tphl Max − #27
 = 312.5 ns − 60 ns − 94 ns − 7 ns − 15 ns
 = **136 ns @ 8 MHz**   Using 8420-20 w/Heavy Load

1 Wait State

tCAC = s2 + s3 + s4 + sw + sw + s6 − CLK to $\overline{AS}$ Asserted Max − $\overline{ADS}$ Asserted to $\overline{CAS}$ Asserted − 74AS245 Delay Max − 68000 Data Setup Min
 = 3½ Tcp10 − #9 − $403a − Tphl Max − #27
 = 350 ns − 55 ns − 94 ns − 7 ns − 10 ns
 = **184 ns @ 10 MHz**   Using 8420-20 w/Heavy Load
 = 3½ Tcp8 − #9 − $403a − Tphl Max − #27
 = 437.5 ns − 60 ns − 94 ns − 7 ns − 15 ns
 = **261 ns @ 8 MHz**   Using 8420-20 w/Heavy Load

\multicolumn{3}{c}{Design #1 Programming Bits}
Bits	**Description**	**Value**
R0, R1	$\overline{RAS}$ Low Time During REFRESH = 2T $\overline{RAS}$ Precharge Time = 2T	R0 = 0 R1 = 1
R2, R3	$\overline{DTACK}$ Generation Modes for Non-Burst Accesses	R2 = s R3 = s
R4, R5	$\overline{DTACK}$ Generation Modes for Burst Accesses	R4 = s R5 = s
R6	Add Wait States with $\overline{WAITIN}$	R6 = s
R7	$\overline{DTACK}$ Mode Select	R7 = 1
R8	Non Interleaved Mode	R8 = 1
R9	Staggered or All RAS REFRESH	R9 = u
C0, C1, C2	Divisor for DELCLK	C0 = s C1 = s C2 = s
C3	+30 REFRESH	C3 = 0
C4, C5, C6	$\overline{RAS}$, $\overline{CAS}$ Configuration Mode *Choose All $\overline{CAS}$ Mode	C4 = u C5 = u C6 = u
C7	Select 0 ns Column Address Setup	C7 = 1
C8	Select 15 ns Row Address Setup	C8 = 1
C9	CAS is Delayed to the Next Rising CLK Edge During Writes	C9 = 1
B0	The Row/Column Bank Latches Are Fall Through Mode	B0 = 1
B1	Access Mode 1	B1 = 1
$\overline{ECAS0}$	$\overline{CAS}$ Not Extended Beyond $\overline{RAS}$	$\overline{ECAS0}$ = 0

u = user defined   **s = system dependent**

R2 = 1	R3 = 0		for 0 WAIT STATES
R2 = 1	R3 = 0	R6 = 0	for 1 WAIT STATE
C0 = 1	C1 = 0	C2 = 1	for 10 MHz
C0 = 0	C1 = 0	C2 = 1	for 8 MHz
R4 = 0	R5 = 0		for 0 WAIT STATES during write portion of test and set
R4 = 1	R5 = 1		for 1 WAIT STATE during write portion of test and set

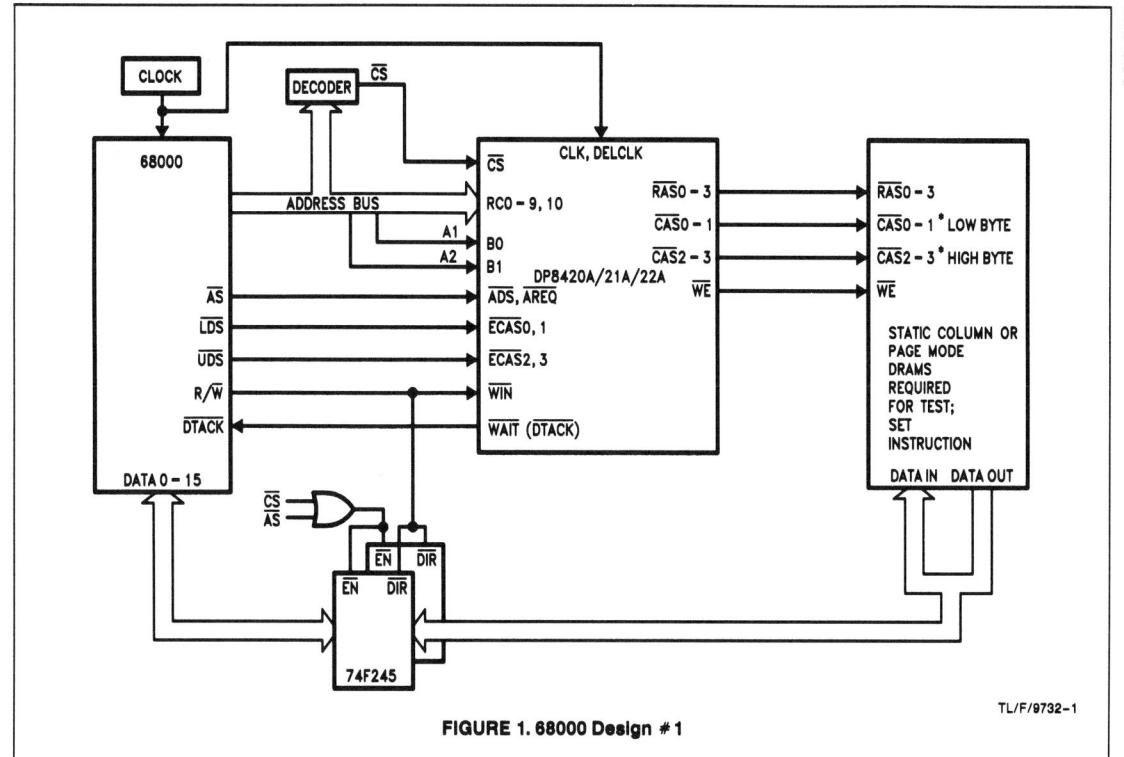

FIGURE 1. 68000 Design #1

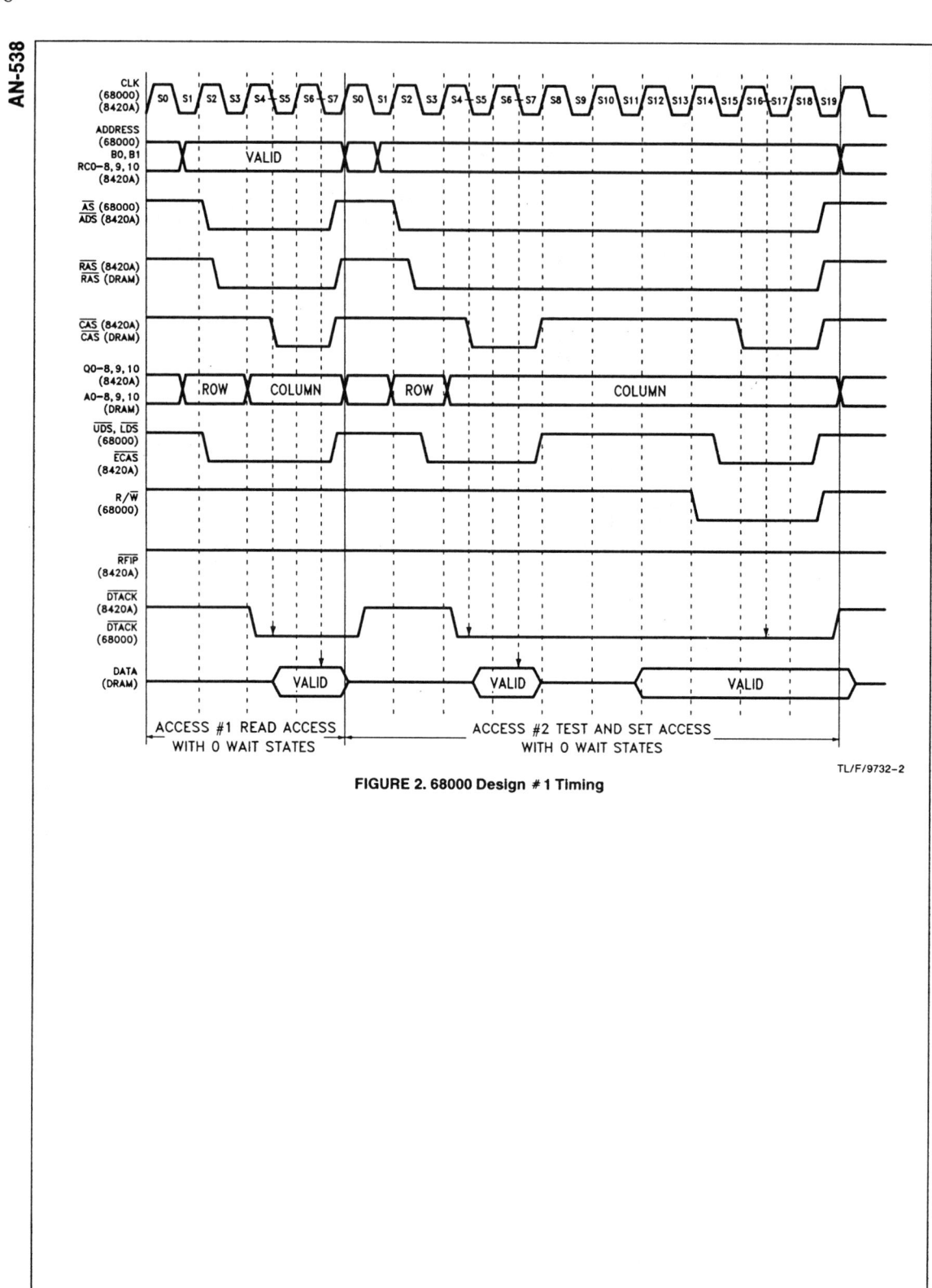

FIGURE 2. 68000 Design #1 Timing

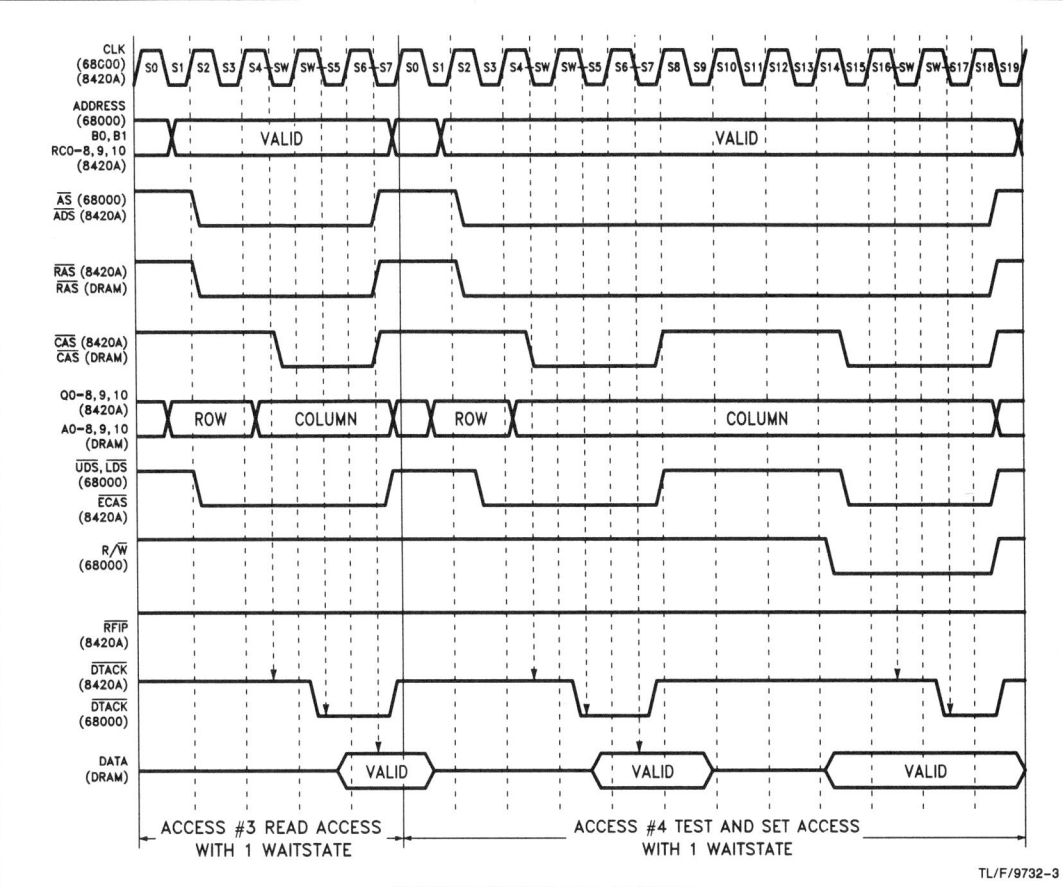

FIGURE 3. 68000 Design #1 Timing

FIGURE 4. 68000 Design #1 Timing

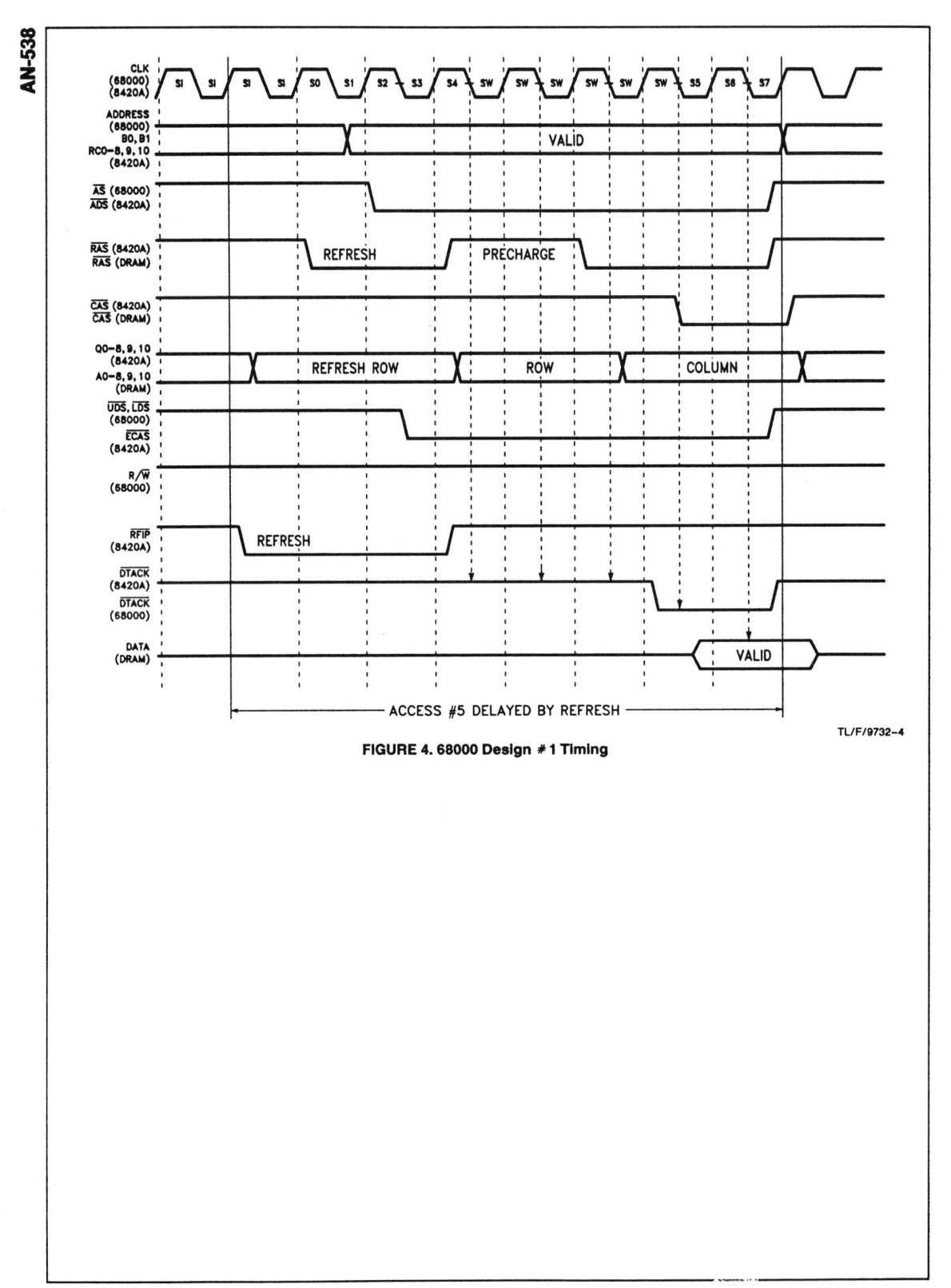

DESIGN #2 DESCRIPTION

Design #2 differs from Design #1 in that the 68000 can be run up to 12.5 MHz. This design can also run with no wait states at 10 MHz if only one bank of DRAM is being used. A latch must be used with the 68000 address strobe to guarantee the address setup to $\overline{ADS}$ asserted requirement of the DP8420A/21A/22A. Again, the DP8420A/21A/22A is operated in Mode 1.

An access cycle begins when the 68000 places a valid address on the address bus at the beginning of processor state s1. At processor state s2, the 68000 asserts the address strobe, $\overline{AS}$. This signal is qualified with CLK low to set a latch. The output of this latch produces the signal $\overline{ADS}$ to the DP8420A/21A/22A. When the signal $\overline{ADS}$ is asserted on the DP8420A/21A/22A, the chip will assert $\overline{RAS}$. After guaranteeing the row address hold time, the 8420A/21A/22A will place the column address to the DRAM address bus. After guaranteeing the column address setup time, the DP8420A/21A/22A will assert $\overline{CAS}$. After time tCAC has passed, the DRAM will place its data on the data bus. The 8420A/21A/22A will assert the $\overline{DTACK}$ output allowing the bus cycle to end.

If a refresh of a Port B access had been in progress, the access would have been delayed by inserting wait states in the Port A access cycle.

DESIGN #2 TIMING AT 12.5 MHz

Clock Period = Tcp12 80 ns @ 12.5 MHz

$400b: $\overline{ADS}$ Asserted Setup to CLK High
= Clock Period + ½ Clock Period + 74AS04 Delay Min + 74AS04 Delay Min − Clock to AS Asserted Max − 74AS04 Delay Min − 74AS02 Delay Max − 74AS02 Delay Max
= Tcp12 + ½ Tcp12 + Tphl Min + Tphl Min − #9 − Tphl Min − Tphl Max − Tphl Max
= 80 ns + 40 ns + 1 ns + 1 ns − 55 ns − 1 ns − 4.5 ns − 4.5 ns
= **57 ns @ 12.5 MHz**

$401: $\overline{CS}$ Setup to $\overline{ADS}$ Asserted
= Clock Period + 74AS04 Delay Min + 74AS04 Delay Min + 74AS02 Delay Min + 74AS02 Delay Min − 74AS04 Delay Min − Clock to ADR Max − 74AS138 Delay Max
= Tcp12 + Tphl Min + Tphl Min + Tphl Min + Tphl Min − Tphl Min − #6 − Tphl Max
= 80 ns + 1 ns + 1 ns + 1 ns + 1 ns − 1 ns − 55 ns − 9 ns
= **19 ns @ 12.5 MHz**

$407 & $404: Address Valid to $\overline{ADS}$ Asserted
= Clock Period + 74AS04 Delay Min + 74AS04 Delay Min + 74AS02 Delay Min + 74AS02 Delay Min − Clock to ADR Max − 74AS04 Min
= Tcp12 + Tphl + Tphl + Tphl + Tphl − #6 − Tphl
= 80 ns + 1 ns + 1 ns + 1 ns + 1 ns − 55 ns − 1 ns
= **28 ns @ 12.5 MHz**

$405: $\overline{ADS}$ Negated Held from CLK High
= Min 74AS04 + Min 74AS02 + Min 74AS02 + Min 74AS04 − Min 74AS04
= Tphl + Tphl + Tphl + Tphl − Tphl
= 1 ns + 1 ns + 1 ns + 1 ns − 1 ns
= **3 ns @ 12.5 MHz**

#47: DTACK Setup Time
= 1 Clock Period − CLOCK skew (74AS04) − Max Clock to DTACK
= Tcp12 − Tphl Max − $18
= 80 ns − 5 ns − 28 ns
= **47 ns @ 12.5 MHz**

$\overline{RAS}$ LOW DURING REFRESH

tRAS = Programmed Clock − [(CLK High to Refresh $\overline{RAS}$ Asserted) − (CLK High to Refresh $\overline{RAS}$ Negated)]
= Tcp12 + Tcp12 − $55
= 80 ns + 80 ns − 6 ns
= **154 ns @ 12.5 MHz**

$\overline{RAS}$ PRECHARGE PARAMETERS

tRP = Programmed Clocks − Clock to $\overline{AS}$ Negated − [($\overline{AREQ}$ to $\overline{RAS}$ Negated) − (CLK to $\overline{RAS}$ Asserted)]
= Tcp12 + Tcp12 − $50
= 80 ns + 80 ns − 16 ns
= **144 ns @ 12.5 MHz**

$29b: $\overline{AREQ}$ Negated Setup to CLK
= Clock Period + Min CLOCK Skew 74AS04 − Max 74AS02 − Max 74AS02
= Tcp12 + Tphl + Tphl − Tphl
= 80 ns + 1 ns − 4.5 ns − 4.5 ns
= **72 ns @ 12.5 MHz**

tRAC AND tCAC FOR DRAMs

Timing is supplied for the system shown in *Figure 5*. (See *Figures 6*). Since systems and DRAM times vary, the user is encouraged to change the following equations to match his system. Timing has been supplied for systems with 0 wait states and 1 bank of DRAM and 1 wait state and 4 banks of DRAM. If DELCLK is not a multiple of 2 MHz, the times of tRAH and tASC will increase or decrease according to the equations given in the data sheet. The $\overline{ADS}$ to $\overline{RAS}$ and $\overline{ADS}$ to $\overline{CAS}$ will also have to be changed depending on the capacitance of the DRAM array.

tRAC

0 wait states * does not use transceivers *

tRAC = s2 + s3 + s4 + s5 + s6 − 74AS02 Max − 74AS02 Max − Clock to AS Max − $\overline{ADS}$ to $\overline{RAS}$ − Data Setup

= 2½ Tcp12 − Tphl − Tphl − #9 − $402 − #27

= 200 − 4.5 ns − 4.5 ns − 55 ns − 25 ns − 10 ns

= **101 ns @ 12.5 MHz**   **Using 8420-25 w/Light Load

1 wait state * uses transceivers *

tRAC = s2 + s3 + s4 + sw + sw + s5 + s6 − 74AS02 Max − 7AS02 Max − Clock to $\overline{AS}$ Max − $\overline{ADS}$ to $\overline{RAS}$ − 74AS245 Delay − Data Setup

= 3½ Tcp12 − Tphl − Tphl − #9 − $402 − Tphl − #27

= 280 ns − 4.5 ns − 4.5 ns − 55 ns − 29 ns − 7 ns − 10 ns

= **170 ns @ 12.5 MHz**

tCAC

0 wait states * does not use transceivers *

tCAC = s2 + s3 + s4 + s5 + s6 − 74AS02 Max − 74AS02 Max − Clock to $\overline{AS}$ Max − $\overline{ADS}$ Asserted to $\overline{CAS}$ − Data Setup

= 2½ Tcp12 − Tphl − Tphl − #9 − $403a − #27

= 200 ns − 4.5 ns − 4.5 ns − 55 ns − 75 ns − 10 ns

= **51 ns @ 12.5 MHz**   *Using 8420-25 w/Light Load

1 wait state * uses transceivers *

tCAC = s2 + s3 + s4 + sw + sw + s5 + s6 − 74AS02 Max Delay − 74AS02 Max Delay − Clock to $\overline{AS}$ Max − $\overline{ADS}$ Asserted to $\overline{CAS}$ − 74AS245 Data Setup

= 3½ Tcp12 − Tphl − Tphl − #9 − $403a − Tphl − #27

= 280 ns − 4.5 ns − 4.5 ns − 55 ns − 75 ns − 7 ns − 10 ns

= **124 ns @ 12.5 MHz**

DESIGN #2, 0 WAIT STATES DURING WRITE ACCESS

Design #2 can be modified to allow 0 wait states during writes. To accomplish this, the chip must be programmed with the same value except that bits R2, R3 and R6 are changed to:

R2 = 0   DTACK of 0T from RAS
R3 = 0
R6 = 0   Hold off DTACK 1 extra clock period

The hardware must be modifed. The signal R/W from the 68000 is inverted and tied to the 8420 signal $\overline{WAITIN}$. This ensures that a wait state will only be asserted during read accesses (see *Figure 6*).

0 waits during write access timing

$\overline{RAS}$ Low Time

tRP = Max $\overline{AS}$ Low − ½ Clock Period − 74AS02 Delay − 74AS02 Delay + 74AS02 Delay + 74AS02 Delay − [($\overline{ADS}$ Asserted to $\overline{RAS}$) − ($\overline{AREQ}$ Negated to $\overline{RAS}$ Negated)]

= #14 − ½ Tcp12 − Tphl − Tphl + Tphl + Tphl − $52

= 160 ns − 40 ns − 0 ns

= **120 ns @ 12.5 MHz**

$\overline{CAS}$ Low Time

tCP = s2 + s3 + s4 + s5 + s6 − Max CLK to $\overline{AS}$ − 74AS02 − 74AS02 − Max $\overline{AS}$ to $\overline{CAS}$ + Min CLK to $\overline{DS}$ + Min ECAS to CAS

= 2½ Tcp12 − #9 − Tphl − Tphl − $403a + #12 + $14

= 200 ns − 55 ns − 4.5 ns − 4.5 ns − 82 ns + 0 ns + 0 ns

= **54 ns @ 12.5 MHz**

Design #2 Programming Bits

Bits	Description	Value
R0, R1	$\overline{RAS}$ Low Time = 2T $\overline{RAS}$ Precharge Time = 2T	R0 = 0 R1 = 1
R2, R3	$\overline{DTACK}$ Generation Modes for Non-Burst Accesses	R2 = 0 R3 = 1
R4, R5	$\overline{DTACK}$ Generation Modes for Burst Accesses	R4 = 0 R5 = 1
R6	Add Wait States with $\overline{WAITIN}$	R6 = 0
R7	$\overline{DTACK}$ Mode Select	R7 = 1
R8	Non Interleaved Mode	R8 = 1
R9	Staggered or All RAS REFRESH	R9 = u
C0, C1, C2	Divisor for DELCLK	C0 = u C1 = u C2 = u
C3	+30 REFRESH	C3 = 0
C4, C5, C6	$\overline{RAS}$, $\overline{CAS}$ Configuration Mode *Choose All $\overline{CAS}$ Mode	C4 = u C5 = u C6 = u
C7	Select 15 ns Column Address Setup	C7 = 1
C8	Select 15 ns Row Address Setup	C8 = 1
C9	CAS is Delayed to the Next Rising CLK Edge During Writes	C9 = 1
B0	The Row/Column Bank Latches Are Fall Through Mode	B0 = 1
B1	Access Mode 1	B1 = 1
$\overline{ECAS0}$	$\overline{CAS}$ Not Extended Beyond $\overline{RAS}$	$\overline{ECAS0}$ = 0

u = user defined
*see previous page for 0 WAIT STATES during writes

FIGURE 5. 68000 Design #2 up to 12.5 MHz

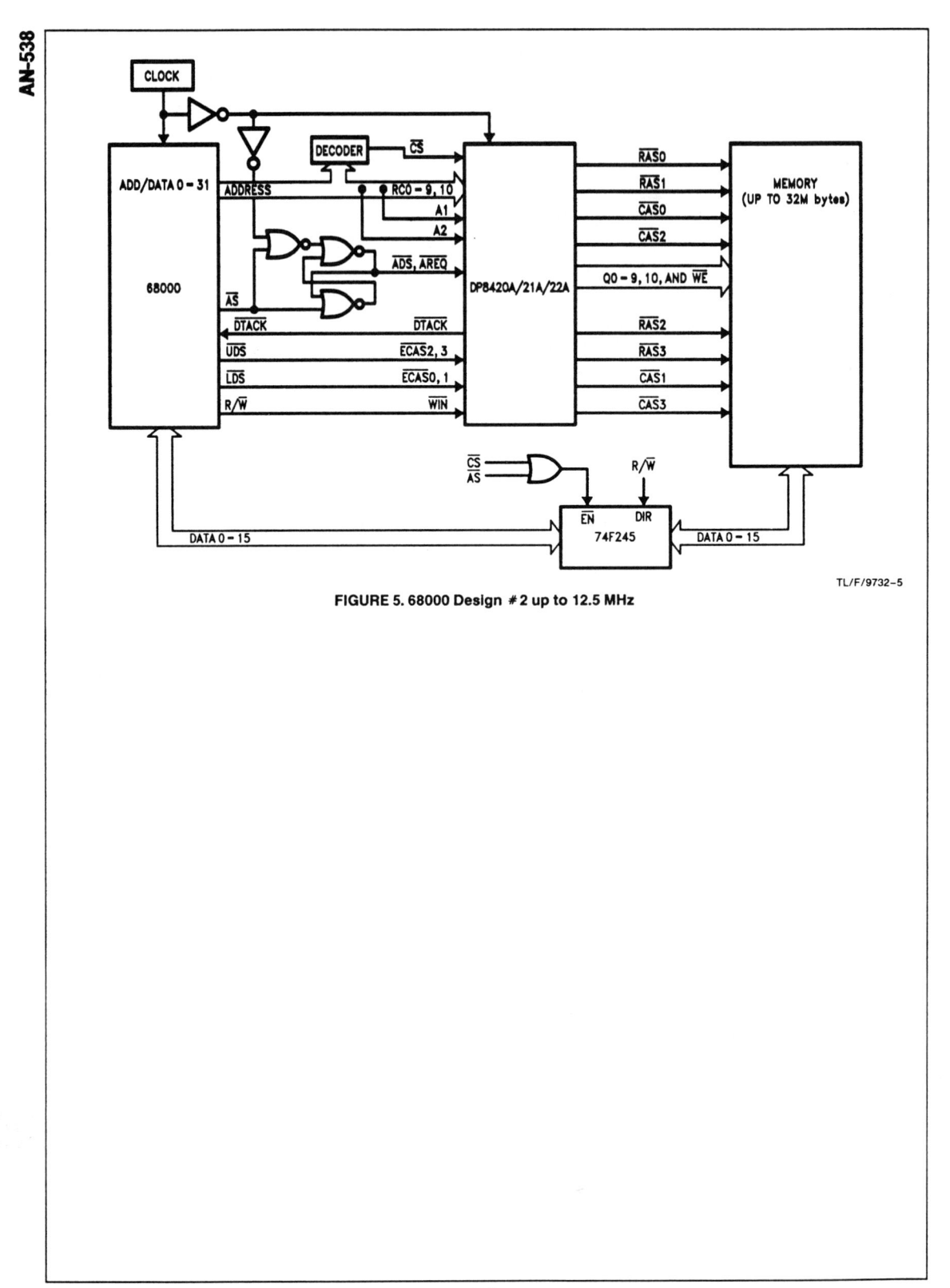

FIGURE 6. Design #2 Timing with Zero Wait States during Writes

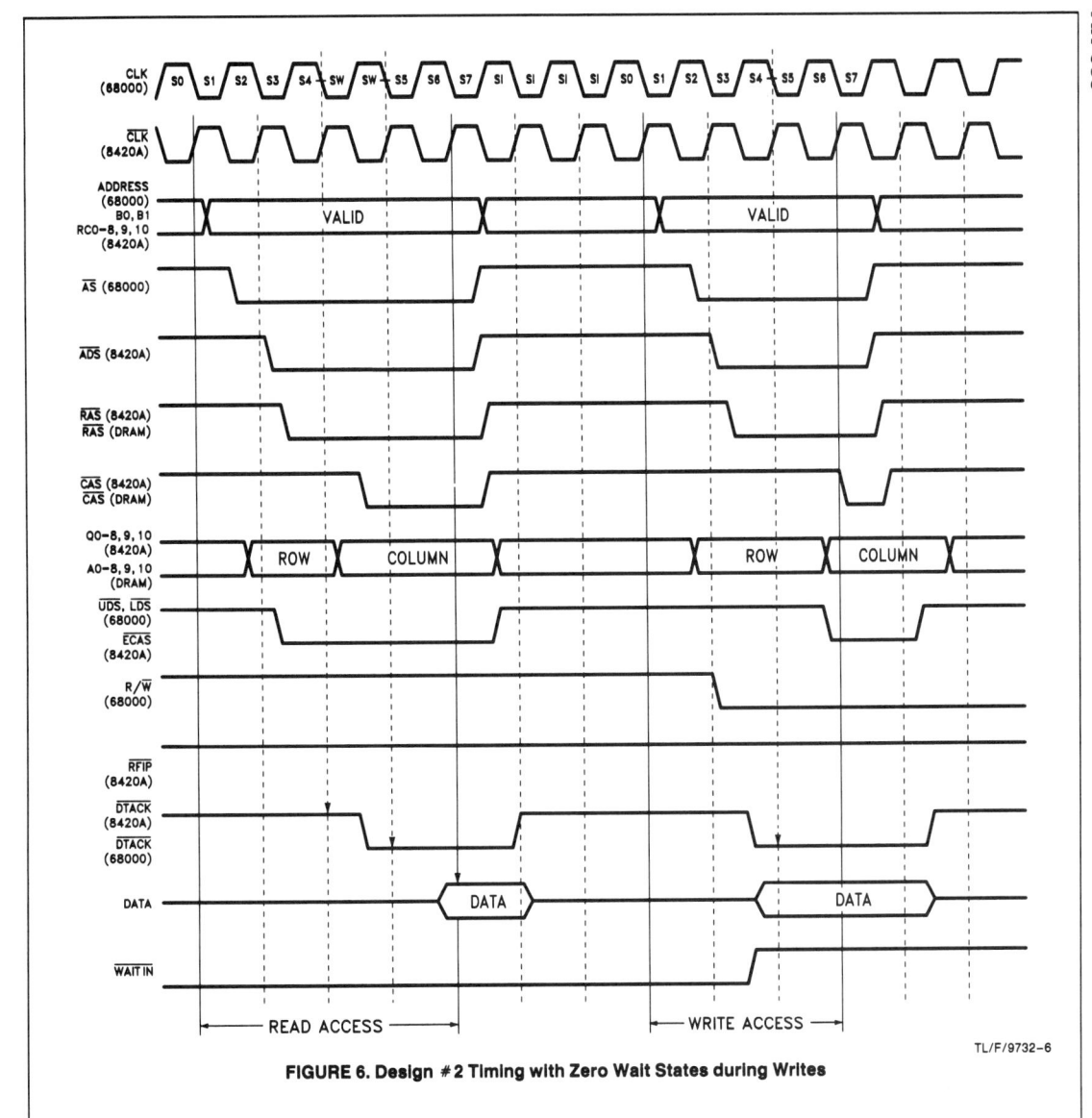

DESIGN #3 DESCRIPTION

Design #3 is a simple circuit to interface the 68000 running @ 16 MHz to the DP8420A/21A/22A and up to 32 Mbytes of DRAM. The DP8420A/21A/22A is operated in Mode 1. An access cycle begins when the 68000 places a valid address on the address bus and asserts $\overline{AS}$. $\overline{AS}$ is then clocked with a 74AS74 flip-flop. The output of the flip-flop is used to produce $\overline{ADS}$ to the DP8420A/21A/22A.

Chip Select ($\overline{CS}$) is generated by a 74AS138 decoder. If a refresh or Port B access had been in progress, the 8420A/21A/22A would hold off the access by inserting wait states in the access cycle. The DP8420A/21A/22A will place the row address on the DRAM's address bus and assert $\overline{RAS}$. After guaranteeing the row address hold time, tRAH, the DP8420A/21A/22A will place the column address on the DRAM's address bus and assert $\overline{CAS}$.

DESIGN #3 TIMING AT 16.667 MHz

Clock Period = Tcp16 = 60 ns @ 16.667 MHz

$400b: $\overline{ADS}$ Asserted Setup to CLK High
 = Clock Period − 74AS74 Delay Max
 = Tcp16 − Tphl
 = 60 ns − 9 ns
 = **51 ns @ 16.667 MHz**

$401: $\overline{CS}$ Asserted Setup to $\overline{ADS}$ Asserted
 = 1½ Clock Periods + Min 74AS74 Delay − Max Clock to Address − 74AS138 Delay
 = 1½ Tcp16 + Tphl − #6 − Tphl
 = 90 ns + 4.5 ns + 50 ns − 9 ns
 = **35.5 ns @ 16.667 MHz**

$407 & $404: Address Valid Setup to $\overline{ADS}$ Asserted
 = 1½ Clock Periods + Min 74AS74 Delay − Max Clock to Address
 = 1½ Tcp16 + Tphl − #6
 = 90 ns + 4.5 ns − 50 ns
 = **44.5 ns @ 16.667 MHz**

$405: $\overline{ADS}$ Negated Held from CLK High
 = Min 74AS74 Delay
 = **4.5 ns @ 16.667 MHz**

#47: DTACK Setup Time
 = Clock Period − 74AS74 Delay Max
 = Tcp16 − Tphl
 = 60 ns − 9 ns
 = **51 ns @ 16.667 MHz**

$\overline{RAS}$ LOW DURING REFRESH

tRAS = Programmed Clocks
 − [(CLK High to Refresh $\overline{RAS}$ Asserted)
 − (CLK High to Refresh $\overline{RAS}$ Negated)]
 = Tcp16 + Tcp16 + Tcp16 + Tcp16 − $55
 = 240 ns − 6 ns
 = **234 ns @ 16.667 MHz**

tRP = (Programmed Clocks − 1) − [($\overline{AREQ}$ to $\overline{RAS}$ Negated) − (CLK to $\overline{RAS}$ Asserted)]
 = Tcp16 + Tcp16 − $50
 = 120 ns − 16 ns
 = **104 ns @ 16.667 MHz**

$\overline{RAS}$ PRECHARGE PARAMETERS

tRP = Programmed Clocks − Clock to $\overline{AS}$ Negated − [($\overline{AREQ}$ to $\overline{RAS}$ Negated) − (CLK to $\overline{RAS}$ Asserted)]

tRAC AND tCAC FOR DRAMs

Timing is supplied for the system shown in *Figure 7*. Since system and DRAM times vary, the user is encouraged to change the following equations to match his system requirements. Timing has been supplied for systems with 2 wait states. If DELCLK is not a multiple of 2 MHz, the timing for tRAH and tASC will increase or decrease according to the times given in the data sheet. The $\overline{ADS}$ to $\overline{RAS}$ and $\overline{ADS}$ to $\overline{CAS}$ will also have to be changed depending on the capacitance of the DRAM array.

1 wait state * using 1 BANK with no transceivers

tRAC = s4 + sw + sw + s5 + s6 − 74AS74 Delay − $\overline{ADS}$ to $\overline{RAS}$ − Data Setup
 = 2½ Tcp16 − Tphl − $402 − #27
 = 150 ns − 9 ns − 25 ns − 10 ns
 = **106 ns @ 16.667 MHz** Using 8420-25 w/Light Load

2 wait states * uses 4 banks with tranceivers *

tRAC = s4 + sw + sw + sw + sw + s5 + s6 − 74AS74 Delay − $\overline{ADS}$ to $\overline{RAS}$ − Data Setup − Transceivers
 = 3½ Tcp16 − Tphl − $402 − #27 − Tphl
 = 210 ns − 9 ns − 29 ns − 10 ns − 7 ns
 = **155 ns @ 16.667 MHz**

1 wait state * using 1 BANK with no transceivers
tCAC   = s4 + sw + sw + s5 + s6
       − 74AS74 Delay − $\overline{ADS}$ to $\overline{CAS}$
       − Data Setup
       = 2½ Tcp16 − Tphl − $403a − #27
       = 150 ns − 9 ns − 75 ns − 10 ns
       = **56 ns @ 16 MHz**

2 wait states * using 4 banks with transceivers *
tCAC   = s4 + sw + sw + sw + sw + s5 + s6
       − 74AS74 Delay − $\overline{ADS}$ to $\overline{CAS}$
       − Data Setup − Transceiver
       = 3½ Tcp16 − Tphl − $403a
       − #27 − Tphl
       = 210 ns − 9 ns − 82 ns
       − 10 ns − 7 ns
       = **102 ns @ 16 MHz**

Design #3 Programming Bits

Bits	Description	Value
R0, R1	$\overline{RAS}$ Low Time = 2T $\overline{RAS}$ Precharge Time = 2T	R0 = 0 R1 = 1
R2, R3	$\overline{DTACK}$ Generation Modes for Non-Burst Accesses	R2 = 1 R3 = 0
R4, R5	$\overline{DTACK}$ Generation Modes for Burst Accesses	R4 = u R5 = u
R6	Add Wait States with $\overline{WAITIN}$	R6 = u
R7	$\overline{DTACK}$ Mode Select	R7 = 1
R8	Non Interleaved Mode	R8 = 1
R9	Staggered or All RAS REFRESH	R9 = u
C0, C1, C2	Divisor for DELCLK (÷8 for 16 MHz)	C0 = 0 C1 = 1 C2 = 0
C3	÷30 REFRESH	C3 = 0
C4, C5, C6	$\overline{RAS}$, $\overline{CAS}$ Configuration Mode *Choose All $\overline{CAS}$ Mode	C4 = u C5 = u C6 = u
C7	Select 15 ns Column Address Setup	C7 = 1
C8	Select 15 ns Row Address Setup	C8 = 1
C9	CAS is Delayed to the Next Rising CLK Edge During Writes	C9 = 1
B0	The Row/Column Bank Latches Are Fall Through Mode	B0 = 1
B1	Access Mode 1	B1 = 1
$\overline{ECAS0}$	$\overline{CAS}$ Not Extended Beyond $\overline{RAS}$	$\overline{ECAS0}$ = 0

u = user defined
*see previous page for 0 WAIT STATES during writes

AN-538

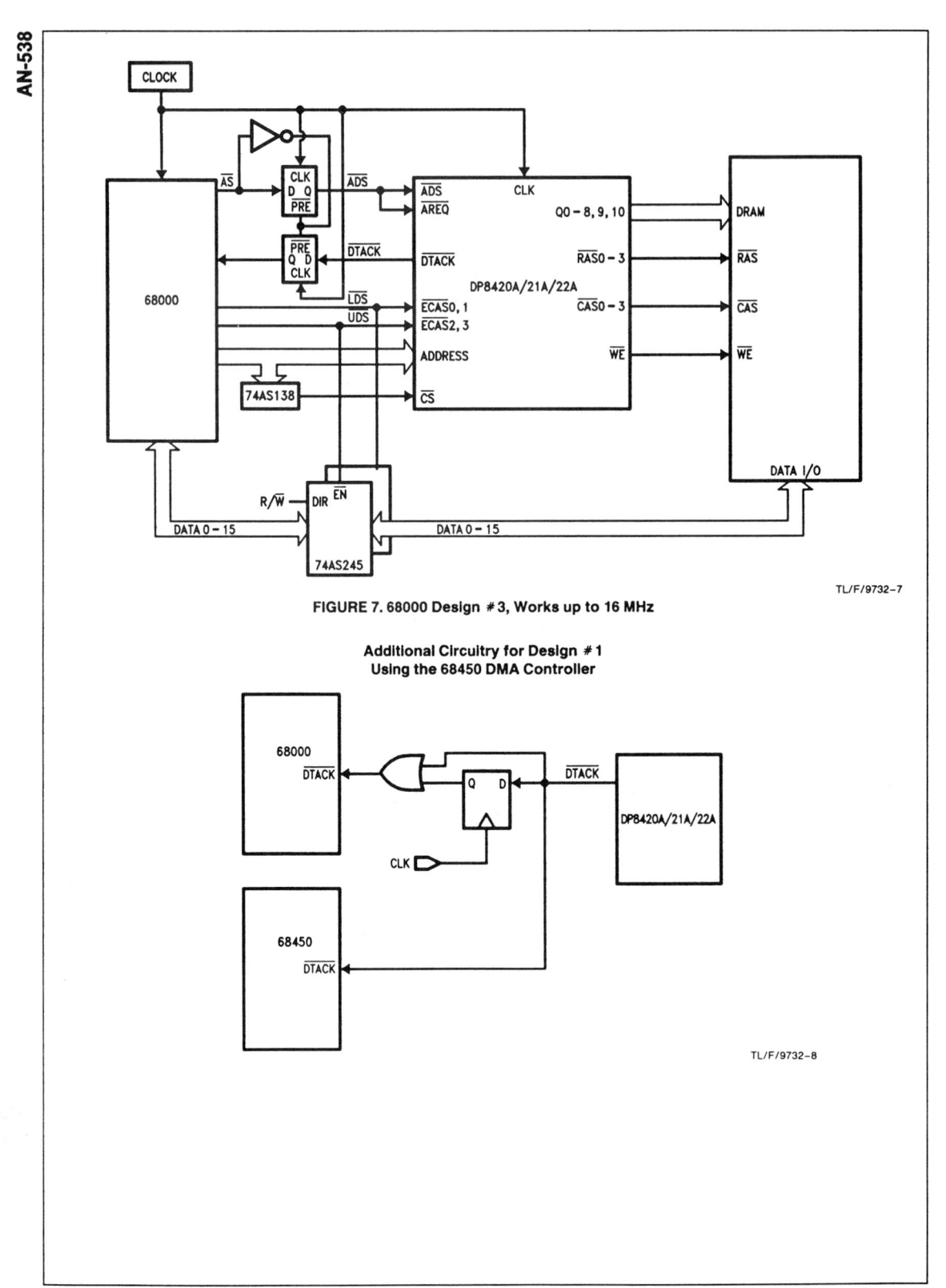

FIGURE 7. 68000 Design #3, Works up to 16 MHz

**Additional Circuitry for Design #1
Using the 68450 DMA Controller**

Because the 68450 samples $\overline{\text{DTACK}}$ on a positive edge of CLK and the 68000 samples $\overline{\text{DTACK}}$ on the negative edge, additional circuitry must be added to produce the two $\overline{\text{DTACK}}$ signals. The $\overline{\text{DTACK}}$s must be produced different to ensure $\overline{\text{RAS}}$ low time after an access delayed by a refresh. The programming bits must also be changed as follows:

For 0 WAITSTATES
R2 = 0 R3 = 1 FOR /DTACK OF 1/2

For 1 WAITSTATE
R2 = 0 R3 = 1 R6 = 0 FOR /DTACK OF 1 1/2

Tie the DP8420 signal $\overline{\text{WAITIN}}$ low for 1 waitstate and high for 0 waitstates. All timing except for the following should still apply. Times with a "#" refer to the 68000 data sheet. Times with a "!" refer to the 68450 data sheet and times with a "$" refer to the DP8420A/21A/22A data sheet.

$47: $\overline{\text{DTACK}}$ Setup Time
= ½ CLOCK Period − 74AS74 CLOCK to Q − 74AS32
= ½ Tcp10 − Tphl − Tphl
= 50 ns − 9 ns − 6 ns
= **35 ns @ 10 MHz**

= ½ Tcp8 − Tphl − Tphl
= 62.5 ns − 9 ns − 6 ns
= **47 ns @ 8 MHz**

!6: $\overline{\text{DTACK}}$ Setup Time (68450)
= ½ CLOCK Period − CLOCK to $\overline{\text{DTACK}}$
= ½ Tcp10 − $18
= 50 ns − 28 ns
= **22 ns @ 10 MHz**

= ½ Tcp8 − $18
= 62.5 ns − 33 ns
= **29 ns @ 8 MHz**

All other 68450 times are the same as the 68000.

**DRAM Speed Versus Processor Speed,
(DRAM Speed References the RAS Access Time, tRAC, of the DRAM.
Using DP8422A-25 Timing Specifications)**

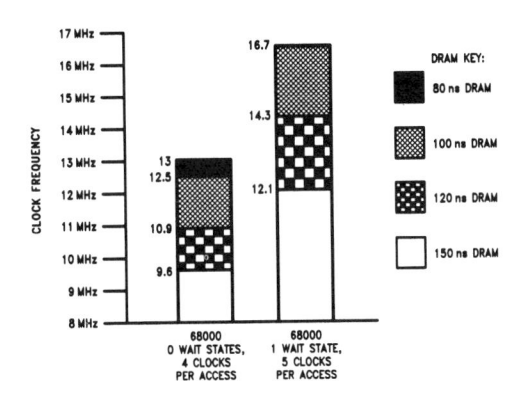

Clock Frequency	68000 0 WAIT STATES, 4 CLOCKS PER ACCESS	68000 1 WAIT STATE, 5 CLOCKS PER ACCESS
80 ns DRAM	13	16.7
100 ns DRAM	12.5	14.3
120 ns DRAM	10.9	12.1
150 ns DRAM	9.6	—

TL/F/9732-9

High-Speed Cache Directory Optimizes Throughput of New High-End Microprocessors

Nicholas Efthymiou and Loren Schiele

Texas Instruments

IMPORTANT NOTICE

Texas Instruments (TI) reserves the right to make changes to or to discontinue any semiconductor product or service identified in this publication without notice. TI advises its customers to obtain the latest version of the relevant information to verify, before placing orders, that the information being relied upon is current.

TI warrants performance of its semiconductor products to current specifications in accordance with TI's standard warranty. Testing and other quality control techniques are utilized to the extent TI deems necessary to support this warranty. Unless mandated by government requirements, specific testing of all parameters of each device is not necessarily performed.

TI assumes no liability for TI applications assistance, customer product design, software performance, or infringement of patents or services described herein. Nor does TI warrant or represent that any license, either express or implied, is granted under any patent right, copyright, mask work right, or other intellectual property right of TI covering or relating to any combination, machine, or process in which such semiconductor products or services might be or are used.

Specifications contained in this User's Guide supersede all data for these products published by TI in the United States before March 1988.

Copyright © 1988, Texas Instruments Incorporated

Contents

Title	Page
ABSTRACT	1
INTRODUCTION	1
CACHE ARCHITECTURES (WAYS OF DESIGNING A CACHE SCHEME)	1
TI PRODUCT LINE OF CACHE TAGS (SOLUTION TO ANY CACHE DESIGN)	1
CACHE DESIGNS	3
Write-Through Scheme Using Buffered Writes	3
Copy-Back Scheme Using 'ACT2158/9 Cache Tag Comparator	4
MULTIMEMORY CONSISTENCY	5
ESTABLISHING CACHE COHERENCY THROUGH BUS WATCHING	6
SUMMARY	7
REFERENCES	7

List of Illustrations

Figure	Title	Page
1	Hit Rate vs Cache Size	1
2	Functional Block Diagram	2
3	Cascading the 'ACT2159	3
4	Buffered Write-Through Cache Scheme	4
5	Copy-Back Scheme Using the 'ACT2158	5
6	Multiprocessor Architecture Utilizing Individual Processor Caches	6
7	Multiprocessor Architecture Utilizing Shared Main Memory Caches	6
8	Bus Watching Using the 'ACT2158	7

ABSTRACT

As designs change from 16-bit microprocessor based architectures to 32-bit systems, cache memories will play an essential part in optimizing system performance. In this report, advanced cache schemes (e.g., copy-back and write-through) using the Texas Instruments (TI) product line of cache tag controllers are discussed. Cache coherency in multiprocessor applications through the implementation of a bus-watching scheme using the TI SN74ACT2158/9 (8K × 9) cache tag comparator is also addressed.

INTRODUCTION

With the advent of powerful general-purpose 32-bit microprocessors, CPU speed can dramatically outpace the capabilities of Dynamic RAM memory arrays. This makes memory the constraining factor in determining system speed and performance.

Designers must adopt certain architectural refinements to fully capitalize on the performance of these 32-bit computing engines. One technique is to employ a hierarchical memory scheme in which high-speed cache or buffer memory is inserted between the processor and main memory to hold the most frequently used portions of main memory. Maximizing the cache hit rate (probability of finding referenced data in the cache) and minimizing the time needed to access data from the cache are the key requirements for optimizing the design of a cache memory. The hit rate, which measures the cache effectiveness, increases dramatically as the cache size becomes larger (see Figure 1). However, there is always an upper limit. A cache with the same size as the main memory is board-space and cost prohibitive. In addition, it results in a degradation of cache speed performance.

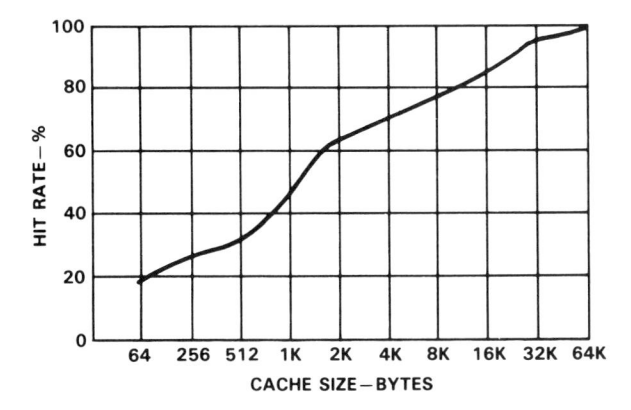

Figure 1. Hit Rate vs Cache Size

CACHE ARCHITECTURES
(WAYS OF DESIGNING A CACHE SCHEME)

Designing a cache memory is a matter of balancing complexity and cost against system performance. Cache schemes vary according to their degree of associativity, or the number of possible locations where the memory contents can be mapped to the cache. Caches can be direct-mapped, set associative, or fully associative. A direct-mapped cache has a set associativity of one (i.e., only one potential location exists for each memory address). A set-associative cache provides two or more cache locations for each memory address. In a fully-associative cache, the cache can store contents of any main-memory location to any cache location.

Of the possible schemes, the direct-mapped cache is the easiest to implement and requires less board space and cost than the associative cache methods. If high hit rates are desired, a two-way set associative cache scheme offers significant performance improvement over direct-mapped caches, but there is little additional improvement beyond four-way set associative designs. For many applications (PCs, superminis, 3-D graphics workstations), the direct-mapped approach is the easiest method for optimizing a 32-bit microprocessor performance at a reasonable cost.

TI PRODUCT LINE OF CACHE TAGS
(THE SOLUTION TO ANY CACHE DESIGN)

The 74ACT2151-4 series of cache tag RAMs from TI offers the versatility and speed needed to provide memory-access times compatible with the operating speed of microprocessors such as the 68030 and 80386. Organized in memory sizes of 2K × 8 and 1K × 11, with access speeds in the 20-ns to 25-ns range, these cache controllers simplify the design of high-speed cache memories. TI devices provide the building blocks necessary to implement direct-map cache, set-associative caches and even bus-watching schemes. These cache-address comparators are easily cascadable for wider tag addresses and deeper tag memories. In addition, significant reductions in cache memory component count, board area, and power dissipation can be achieved by implementing these devices.

To address higher density cache schemes, TI is also offering an 8K × 9 cache tag comparator 74ACT2158/9 with access times of 25 ns (see Figure 2).

The 74ACT2158 has a totem-pole match output while the 74ACT2159 has an open-drain match output for wire and tying. This simplifies cascading in both depth and width. The device incorporates on-chip parity generation and checking, as well as separate outputs (Q0-Q8) that can be effectively used in a copy-back cache scheme or can serve as fast static

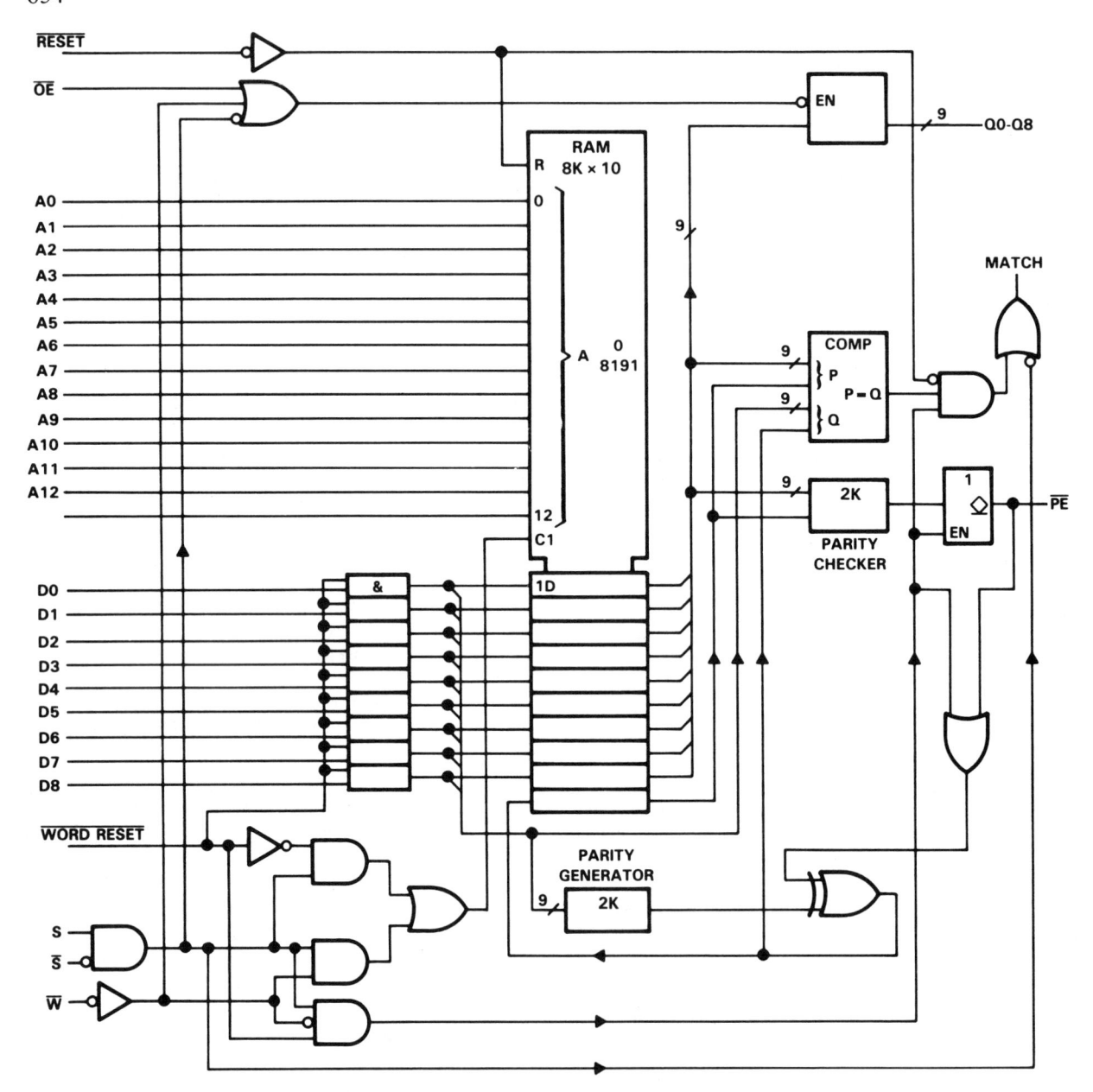

Figure 2. Functional Block Diagram (Positive Logic)

RAM buffers for storage of data from the main memory. A master reset input is also provided for initialization. When RESET is taken low, all 8K × 10 locations are cleared to zero (with valid parity).

The 'ACT2158 and 'ACT2159 are easily cascadable in both the width and depth directions. When cascading in the width direction, the same address bits (A0-A12) are tied to the address pins of each chip. The remaining address bits are tied to the individual data inputs (D0-D8) of each device. Two devices cascaded in width will provide comparison for 30 bits of address with one data pin tied high for a valid bit.

When cascading in the depth direction, two devices are required for 16K depth. Address bit A13 is tied to the $\overline{S}$ pin of the first device and to the $\overline{S}$ pin of the second device. The rest of the address bits (A14-A22) for one device wide are tied to the D0-D8 pins. When cascading more than two devices in depth, a fast decoder must be used to determine which device is selected. (The TIBPAD16N8-7 Programmable Address Decoder is the fastest PLD available. It supports a 7-ns access time and it is pin compatible with the popular 16L8 architecture.) When a device is deselected, the MATCH output of that device is forced high to allow for proper gating. When cascading in depth and/or width, a composite MATCH output is needed. When using the 'ACT2158, a high-speed gate such as the SN74AS20 or the SN74AS30 should be used to achieve fast MATCH times. The 'ACT2159 has an open-drain MATCH output, which provides for wired-AND typing. Figure 3 shows an example of cascading the 'ACT2159 in both width and depth.

CACHE DESIGNS

Write-Through Scheme Using Buffered Writes

Since the cache serves as only a temporary buffer that actually reserves and stores the most recently accessed information, the burden of updating the main memory with new modified data still lies within the CPU. There are now two general approaches used in updating main memory; write through and copy back. In a write-through scheme, the CPU updates the cache and main memory on every write operation. When using a write-through cache, the time required for the processor to write each piece of data to main memory can substantially impact the overall performance. For example, if a processor accesses memory in two clock cycles, 10 accesses would require 20 clock cycles for all 10 accesses to be serviced out of cache. With six wait states required each time main memory is written into, 15% (or 3) of the 10 accesses increase from 20 clock cycles to 38 clock cycles. As seen by this example, even if the cache hit rate is 100%, the time required to service a write causes the processor to run at slightly more than 50% of its potential.

The solution to this problem is to buffer each processor write. By loading the address and data into a shallow bank of FIFOs, performance degradation due to writes can essentially be eliminated. Programmable logic can be used to send the buffered address and data on to memory. TI FIFOs SN74ALS2232 (16 × 8) or SN74ALS2233 (16 × 9) are deep enough and fast enough to optimize performance in most systems. For more elaborate buffering schemes, the TI bidirectional FIFO, SN74ALS2238, is very useful. Figure 4 shows a typical block diagram of a write-through cache with FIFO buffers.

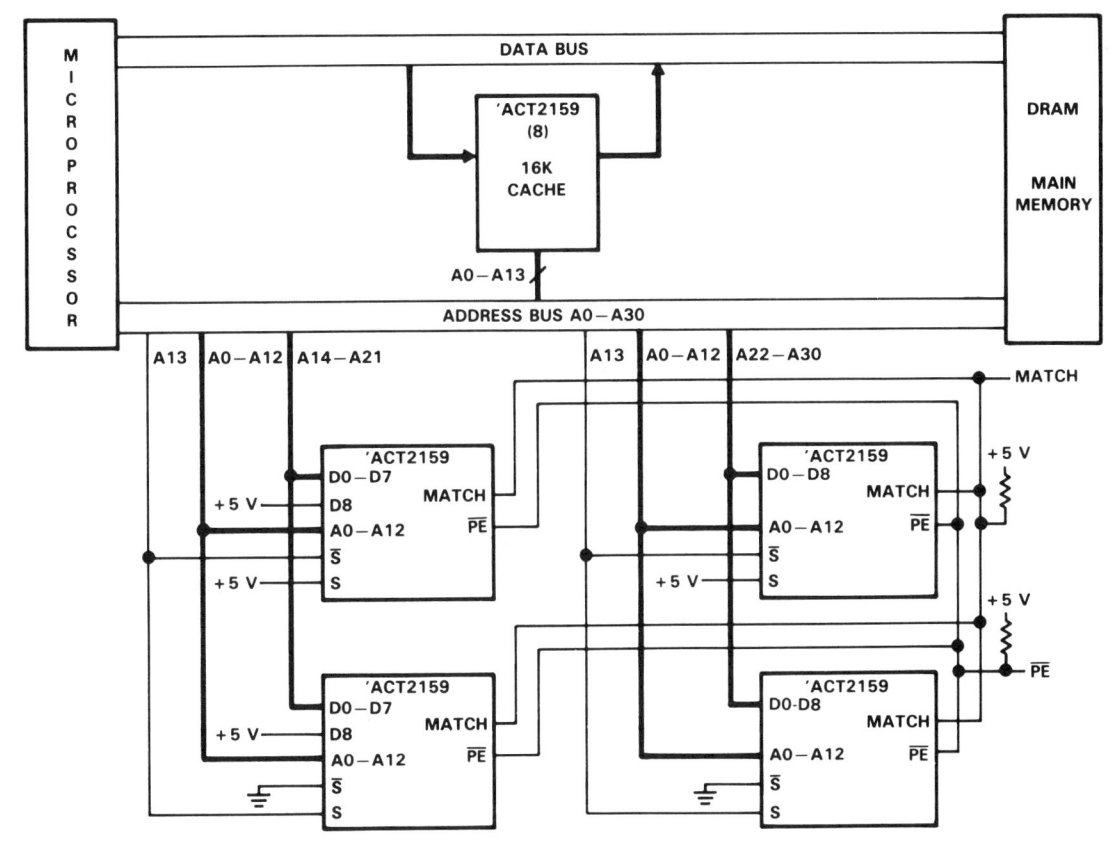

Figure 3. Cascading the 'ACT2159

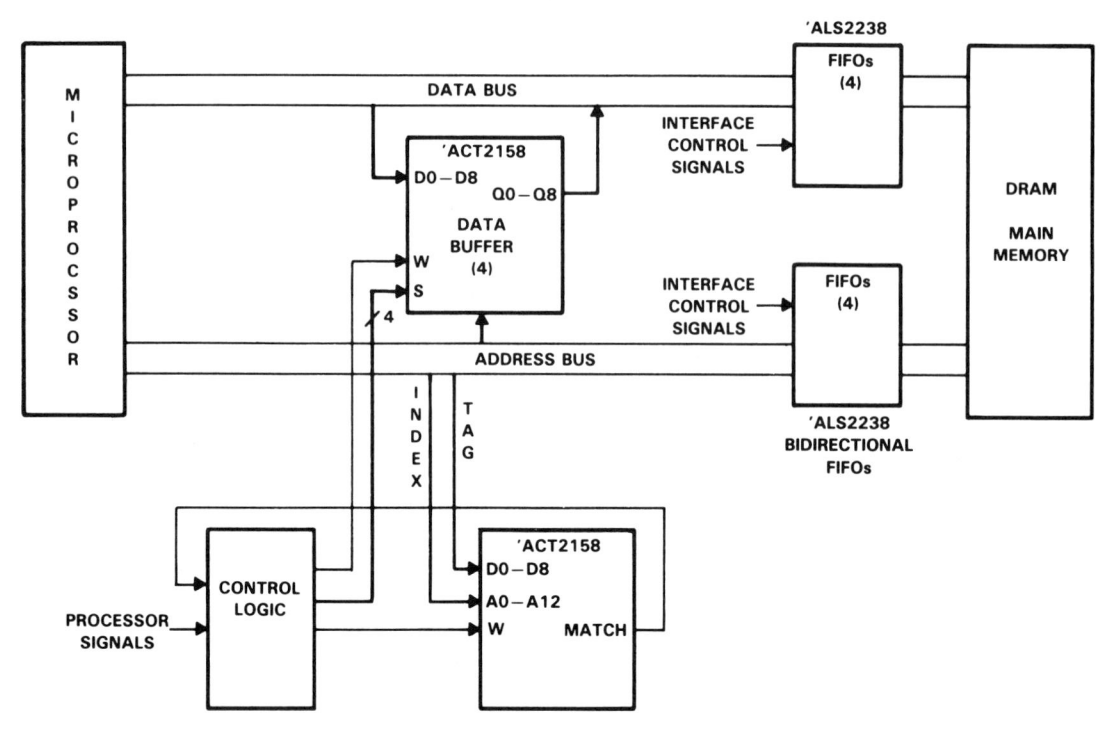

Figure 4. Buffered Write-Through Cache Scheme

Copy-Back Scheme Using 'ACT2158/9 Cache Tag Comparator

In a copy-back scheme, the processor initially modifies only the cache. Later, the modified data can be written into main memory. A major advantage in using the copy-back scheme is that most processor writes are accomplished without a performance penalty. The 'ACT2158 or 'ACT2159 can address the copy-back mode in a versatile manner and provide a reduction in main memory-write cycles by at least a factor of 2. Figure 5 shows a direct implementation of a copy-back scheme using the TI 'ACT2158 or 'ACT2159 Cache Tag Comparator.

When the CPU performs a read operation from the main memory, a copy of that data is stored in the cache, based on the principle of locality. A dirty bit is assigned to that data and is set to a zero value to indicate that the cache location contains the same copy of data as the main memory. Upon modification of the data by the microprocessor, the newly modified word is written to the cache (Index and Tag encompass the same values), and the dirty bit is modified to a value of 1. Later, when the microprocessor again tries to write to the same cache location and a cache write miss occurs (same index but different tag value), the dirty bit is monitored to determine whether or not existing data and address needs to be copied back to main memory. If the data has been modified (dirty bit 1), the referenced tag and data along with the current index needs to be stored in an intermediate buffer in order to perform copy-back.

A method of optimizing the copy-back scheme is to use spare-cache or main-memory cycles to do the copy-back of the dirty (modified) lines. The same procedure should also be followed when the CPU performs a read-miss on the cache and has to go out to main memory to retrieve the new data. If the cache location where the miss occurred has modified data (dirty bit value of 1), then this data and address must be buffered before data from main memory is written in the cache. Copy-back schemes reduce the number of main-memory writes. This improves processor performance and reduces bus traffic. Processor performance can be further improved by using one level of buffering between the processor and main memory. This level of buffering allows most of the processor writes to be serviced without the addition of wait states.

Since the 'ACT2158 has separate tag outputs (Q0-Q8), the tag can be stored in the buffer at the same time that the new tag is applied to the 'ACT2158. This avoids delays incurred when using cache tag chips with only one data port.

Several issues define the trade-off between write-through and copy-back schemes. Since a write to main memory only occurs if the data in the cache has been modified (when a miss occurs), copy-back provides significant reduction in main-memory traffic. However, since a dirty bit is required to determine whether or not to copy a line back to memory, this complicates the cache logic.

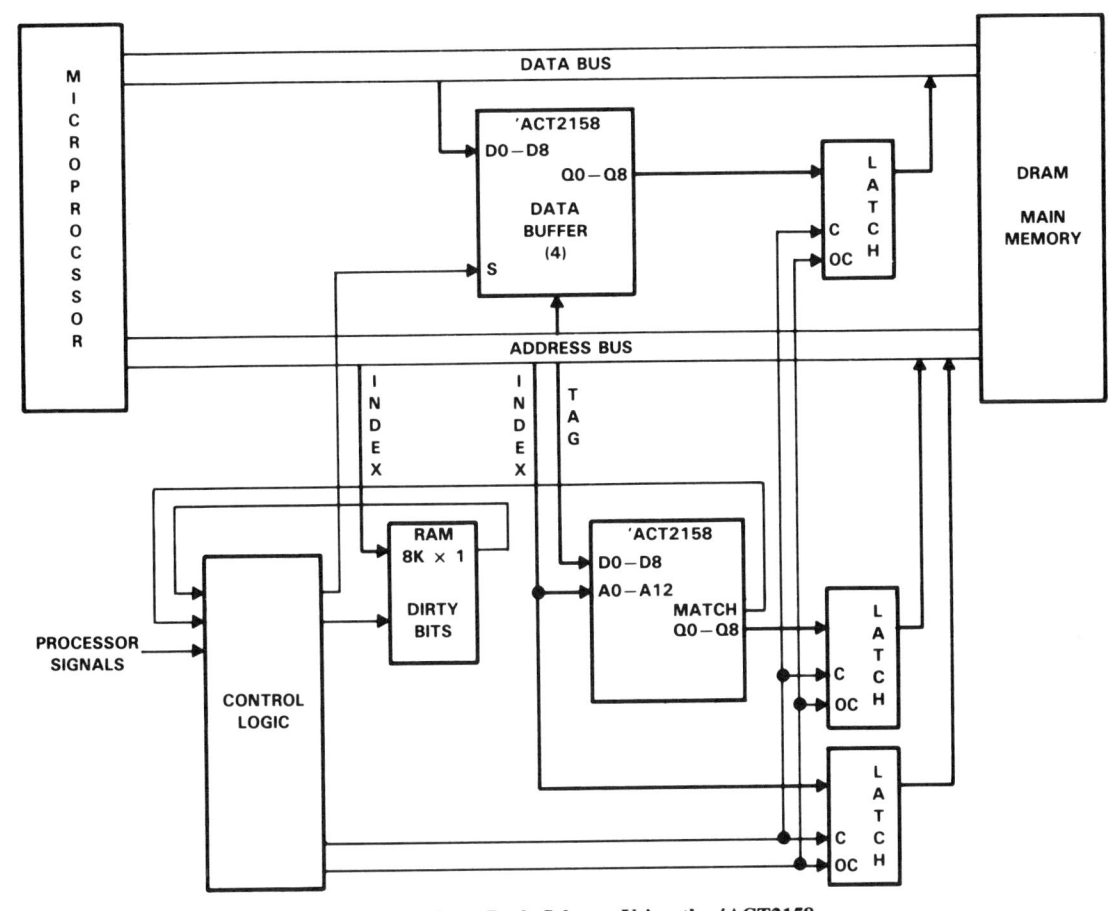

Figure 5. Copy-Back Scheme Using the 'ACT2158

In a write-through scheme, the main-memory traffic bottleneck is clearly evident because a main-memory access is required on every store. However, the scheme does not require dirty bits and keeps the cache control logic simple. In addition, the main memory always reflects the same data as the cache at any given time. This enhances the system reliability because a store-through system can often be restored more easily in the event of CPU or cache memory failure.

MULTIMEMORY CONSISTENCY

When implementing memory systems using multiple memory architectures, the problem of data coherency is always a concern. The concern is that the data will not remain coherent or consistent in all memories. The problem is usually found in multiprocessor systems where several processors can perform different tasks using the same data. However, coherency also becomes a concern in single-processor systems where direct memory accesses are performed or when data is temporarily stored in buffers as described in the buffered write-through scheme.

In a single-processor system that uses the buffered write-through cache scheme the simplest and cheapest method of ensuring data coherency with respect to the write buffer is to unload the buffer each time the processor goes to main memory for data. This ensures that the requested data does not reside in the buffer. In systems where direct memory accesses can alter main memory data which also exists in cache, a more elaborate coherency scheme must be implemented. This also applies to multiprocessor systems.

In the multiprocessor application shown in Figure 6, both of the processors caches contain the same data. Processor A writes new data into location X of its cache and the main memory as part of a write-through operation. If processor B attempts to read that data from its own cache it will continue to use the old value of X, unless it is informed of the new data residing in the processor A cache and the main memory. If the system is to perform correctly, the possibility of having several processors using different copies of the same data must be avoided.

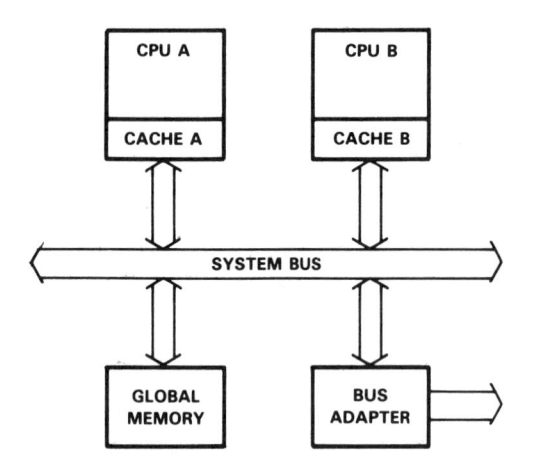

Figure 6. Multiprocessor Architecture Utilizing Individual Processor Caches

Currently, several solutions exist or have been proposed to solve this problem. One solution is for all processors to share a cache that is associated with only the main memory of the multiprocessor architecture (see Figure 7).

Although this technique eliminates the cache coherency problem, it is practically not feasible because the bandwidth of a single cache can not support more than one processor without forcing the CPU into wait states. In addition, a degradation in speed performance is associated with the transmission delays through the interconnection network and with the conflicts at the caches.

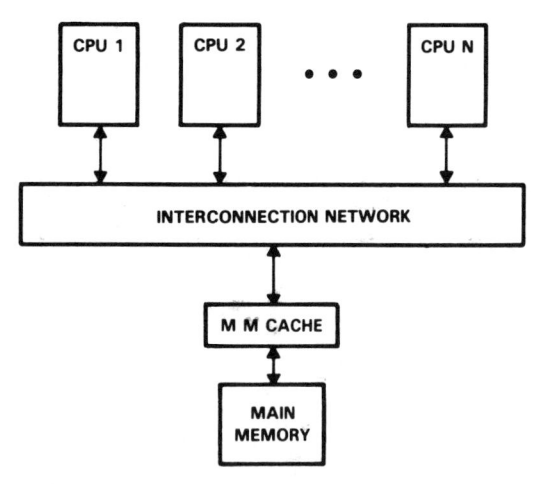

Figure 7. Multiprocessor Architecture Utilizing Shared Main Memory Caches

In a different approach, each time the CPU performs a write operation to its cache, it also sends that write to all other caches in the system. If the other caches have a copy of that store, the data can be either updated or invalidated. The major difficulty in the broadcasting of store addresses is that every cache memory in the system has to lend itself to an extra cycle for the invalidation process anytime the processor performs a write operation. Although the memory interference that occurs is usually acceptable for two processors, it significantly reduces system performance when the architecture calls for more than two processors.

ESTABLISHING CACHE COHERENCY THROUGH BUS WATCHING

Another method of achieving cache coherency is by implementing bus watching. The SN74ACT2158/9 used in a bus-watching scheme preserves cache coherency by monitoring the system bus address lines at no cost to the system or local throughput. As shown in Figure 8, each time that a DMA writes to a location in main memory which is also found in cache A, Cache A must be notified of that new data. The easiest and most efficient way to achieve this is to invalidate the cache location so that the next time the processor requests that data, it is forced to go out to main memory to obtain it.

A bus-watching scheme is used to determine whether or not invalidation is necessary. The bus watcher consists of 'ACT2158s that contain the exact same addresses as the cache tag of processor A. When a DMA write cycle occurs, the bus-watcher logic determines if a write operation takes place and if the address that the DMA is issuing also resides in the cache. Through the use of latches, the appropriate index is then fed to the cache tag of the CPU to invalidate its contents. If necessary to keep the processor from slowing down during invalidation operations, a FIFO can be used to store several index values. The invalidation then occurs when the cache controller is not checking for addressed data in the cache. Using a latch, the address between the CPU and the cache tag can be three-stated, allowing the bus-watcher index to be applied to the tag RAM for invalidation.

The word-reset function, a special feature of the 'ACT2158/9, allows any addressed memory location to be cleared to zero (with valid parity) by taking the word-reset input low. By tying one of the data inputs high, that particular bit can be used as an internal valid bit. Whenever data is compared to a memory location that has been invalidated (cleared) by word reset, a miss will occur. If a data input is not high, a false match will occur whenever a data word of zero is compared to a reset location. With the word invalidated, the next time the CPU applies the same address, it will be forced to go out to main memory to retrieve that data. The 'ACT2158/9 is also used as a high-speed SRAM to provide parity generation and checking.

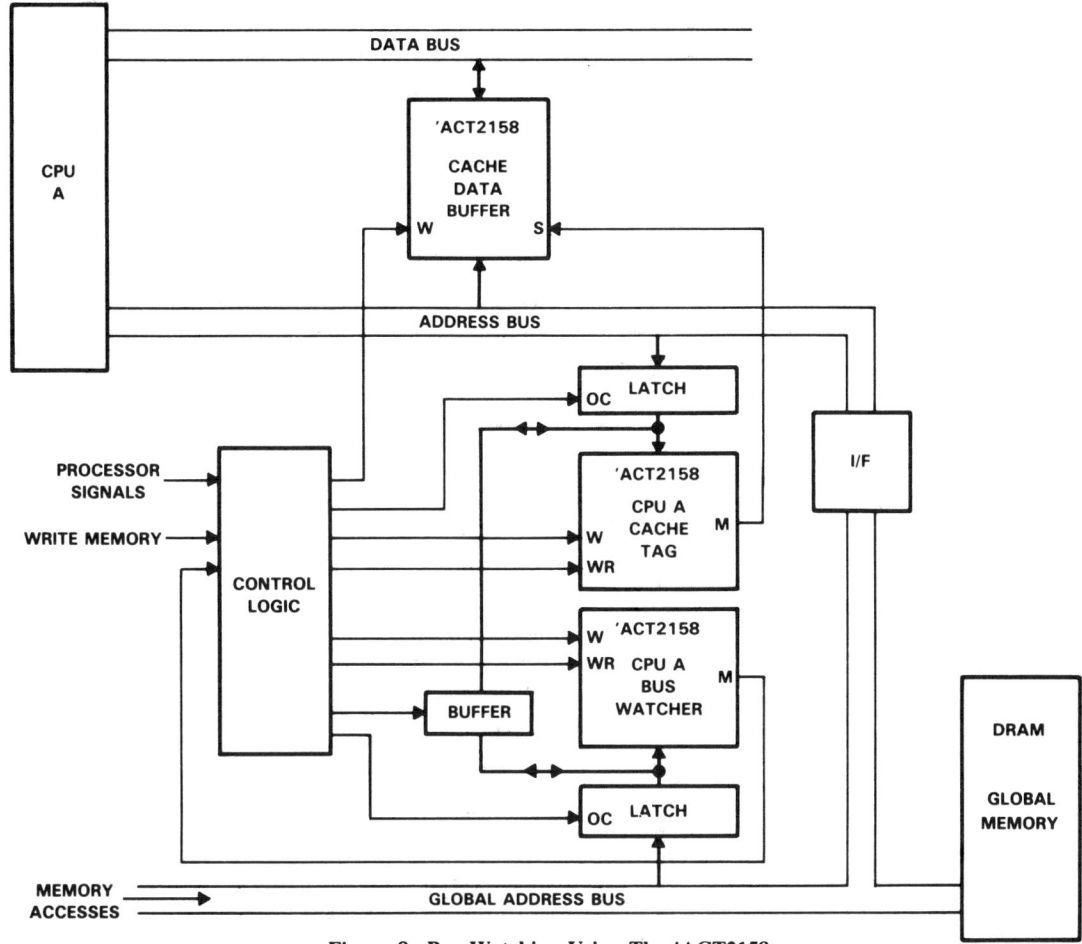

Figure 8. Bus Watching Using The 'ACT2158

SUMMARY

Cache memories play an essential role in any 32-bit computer design architecture by providing more power and capability to the system at a reasonable cost. Large cache sizes and fast access times are key requirements in obtaining high hit ratios and optimizing the microprocessor overall performance. The TI product line of caches using high-speed CMOS technology provides the flexibility, speed, and density required to configure a high-performance system.

The new generation of TI high-density cache tags (SN74ACT2158/9) are ideally suited for new advanced cache schemes such as copy-back and bus-watching for cache coherency.

REFERENCES

Alan Jay Smith, "Cache Memories," *Computing Surveys*, 14, 3, September 1982 pp 500-503.

Kai Hwang, Faye' A. Briggs, *Computer Architecture and Parallel Processing*, 1984, pp 518-519.

TI Sales Offices

ALABAMA: Huntsville (205) 837-7530.
ARIZONA: Phoenix (602) 995-1007; Tucson (602) 624-3276.
CALIFORNIA: Irvine (714) 660-1200; Sacramento (916) 929-0197; San Diego (619) 278-9600; Santa Clara (408) 980-9000; Torrance (213) 217-7000; Woodland Hills (818) 704-7759.
COLORADO: Aurora (303) 368-8000.
CONNECTICUT: Wallingford (203) 269-0074.
FLORIDA: Altamonte Springs (305) 260-2116; Ft. Lauderdale (305) 973-8502; Tampa (813) 286-0420.
GEORGIA: Norcross (404) 662-7900.
ILLINOIS: Arlington Heights (312) 640-3000.
INDIANA: Carmel (317) 573-6400; Ft. Wayne (219) 424-5174.
IOWA: Cedar Rapids (319) 395-9550.
KANSAS: Overland Park (913) 451-4511.
MARYLAND: Baltimore (301) 944-8600.
MASSACHUSETTS: Waltham (617) 895-9100.
MICHIGAN: Farmington Hills (313) 553-1500; Grand Rapids (616) 957-4200.
MINNESOTA: Eden Prairie (612) 828-9300.
MISSOURI: St. Louis (314) 569-7600.
NEW JERSEY: Iselin (201) 750-1050.
NEW MEXICO: Albuquerque (505) 345-2555.
NEW YORK: East Syracuse (315) 463-9291; Melville (516) 454-6600; Pittsford (716) 385-6770; Poughkeepsie (914) 473-2900.
NORTH CAROLINA: Charlotte (704) 527-0930; Raleigh (919) 876-2725.
OHIO: Beachwood (216) 464-6100; Dayton (513) 258-3877.
OREGON: Beaverton (503) 643-6758.
PENNSYLVANIA: Blue Bell (215) 825-9500.
PUERTO RICO: Hato Rey (809) 753-8700.
TENNESSEE: Johnson City (615) 461-2192.
TEXAS: Austin (512) 250-6769; Houston (713) 778-6592; Richardson (214) 680-5082; San Antonio (512) 496-1779.
UTAH: Murray (801) 266-8972.
VIRGINIA: Fairfax (703) 849-1400.
WASHINGTON: Redmond (206) 881-3080.
WISCONSIN: Brookfield (414) 782-2899.
CANADA: Nepean, Ontario (613) 726-1970; Richmond Hill, Ontario (416) 884-9181; St. Laurent, Quebec (514) 336-1860.

TI Regional Technology Centers

CALIFORNIA: Irvine (714) 660-8140; Santa Clara (408) 748-2220; Torrance (213) 217-7019.
COLORADO: Aurora (303) 368-8000.
GEORGIA: Norcross (404) 662-7945.
ILLINOIS: Arlington Heights (312) 640-2909.
MASSACHUSETTS: Waltham (617) 895-9196.
TEXAS: Richardson (214) 680-5066.
CANADA: Nepean, Ontario (613) 726-1970.

TI Distributors

TI AUTHORIZED DISTRIBUTORS
Arrow/Kierulff Electronics Group
Arrow Canada (Canada)
Future Electronics (Canada)
GRS Electronics Co., Inc.
Hall-Mark Electronics
Marshall Industries
Newark Electronics
Schweber Electronics
Time Electronics
Wyle Laboratories
Zeus Components

—OBSOLETE PRODUCT ONLY—
Rochester Electronics, Inc.
Newburyport, Massachusetts
(617) 462-9332

ALABAMA: Arrow/Kierulff (205) 837-6955; Hall-Mark (205) 837-8700; Marshall (205) 881-9235; Schweber (205) 895-0480.
ARIZONA: Arrow/Kierulff (602) 437-0750; Hall-Mark (602) 437-1200; Marshall (602) 496-0290; Schweber (602) 997-4874; Wyle (602) 866-2888.
CALIFORNIA: Los Angeles/Orange County: Arrow/Kierulff (818) 701-7500, (714) 838-5422; Hall-Mark (818) 716-7300, (714) 669-4100, (213) 217-8400; Marshall (818) 407-0101, (818) 459-5500, (714) 458-5395; Schweber (818) 999-4702; (714) 863-0200, (213) 320-8090; Wyle (213) 322-9953, (818) 880-9000, (714) 863-9953; Zeus (714) 921-9000; Sacramento: Arrow/Kierulff (916) 565-4800; Marshall (916) 635-9700; Schweber (916) 929-9732; Wyle (916) 638-5282; San Diego: Arrow/Kierulff (619) 565-4800; Hall-Mark (619) 268-1201; Marshall (619) 578-9600; Schweber (619) 450-0454; Wyle (619) 565-9171; San Francisco Bay Area: Arrow/Kierulff (408) 745-6600; Hall-Mark (408) 432-0900; Marshall (408) 942-4600; Schweber (408) 432-7171; Wyle (408) 727-2500; Zeus (408) 998-5121.
COLORADO: Arrow/Kierulff (303) 790-4444; Hall-Mark (303) 790-1662; Marshall (303) 451-8383; Schweber (303) 799-0258; Wyle (303) 457-9953.
CONNETICUT: Arrow/Kierulff (203) 265-7741; Hall-Mark (203) 269-0100; Marshall (203) 265-3822; Schweber (203) 748-7080.
FLORIDA: Ft. Lauderdale: Arrow/Kierulff (305) 429-8200; Hall-Mark (305) 971-9280; Marshall (305) 977-4880; Schweber (305) 977-7511; Orlando: Arrow/Kierulff (305) 725-1480, (305) 682-6923; Hall-Mark (305) 855-4020; Marshall (305) 767-8585; Schweber (305) 331-7555; Zeus (305) 365-3000; Tampa: Hall-Mark (813) 530-4543; Marshall (813) 576-1399.
GEORGIA: Arrow/Kierulff (404) 449-8252; Hall-Mark (404) 447-8000; Marshall (404) 923-5750; Schweber (404) 449-9170.
ILLINOIS: Arrow/Kierulff (312) 250-0500; Hall-Mark (312) 860-3800; Marshall (312) 490-0155; Newark (312) 784-5100; Schweber (312) 364-3750.
INDIANA: Indianapolis: Arrow/Kierulff (317) 243-9353; Hall-Mark (317) 872-8875; Marshall (317) 297-0483.
IOWA: Arrow/Kierulff (319) 395-7230; Schweber (319) 373-1417.
KANSAS: Kansas City: Arrow/Kierulff (913) 541-9542; Hall-Mark (913) 888-4747; Marshall (913) 492-3121; Schweber (913) 492-2922.
MARYLAND: Arrow/Kierulff (301) 995-6002; Hall-Mark (301) 988-9800; Marshall (301) 840-945C; Schweber (301) 840-5900; Zeus (301) 997-1118.

MASSACHUSETTS: Arrow/Kierulff (617) 935-5134; Hall-Mark (617) 667-0902; Marshall (617) 658-0810; Schweber (617) 275-5100, (617) 657-0760; Time (617) 532-6200; Zeus (617) 863-8800.
MICHIGAN: Detroit: Arrow/Kierulff (313) 971-8220; Marshall (313) 525-5850; Newark (313) 967-0600; Schweber (313) 525-8100; Grand Rapids: Arrow/Kierulff (616) 243-0912.
MINNESOTA: Arrow/Kierulff (612) 830-1800; Hall-Mark (612) 941-2600; Marshall (612) 559-2211; Schweber (612) 941-5280.
MISSOURI: St. Louis: Arrow/Kierulff (314) 567-6888; Hall-Mark (314) 291-5350; Marshall (314) 291-4650; Schweber (314) 739-0526.
NEW HAMPSHIRE: Arrow/Kierulff (603) 668-6968; Schweber (603) 625-2250.
NEW JERSEY: Arrow/Kierulff (201) 538-0900, (609) 596-8000; GRS Electronics (609) 964-8560; Hall-Mark (201) 575-4415, (609) 235-1900; Marshall (201) 882-0320, (609) 234-9100; Schweber (201) 227-7880.
NEW MEXICO: Arrow/Kierulff (505) 243-4566.
NEW YORK: Long Island: Arrow/Kierulff (516) 231-1000; Hall-Mark (516) 737-0600; Marshall (516) 273-2424; Schweber (516) 334-7555; Zeus (914) 937-7400; Rochester: Arrow/Kierulff (716) 427-0300; Hall-Mark (716) 244-9290; Marshall (716) 235-7620; Schweber (716) 424-2222; Syracuse: Marshall (607) 798-1611.
NORTH CAROLINA: Arrow/Kierulff (919) 876-3132, (919) 725-8711; Hall-Mark (919) 872-0712; Marshall (919) 878-9882; Schweber (919) 876-0000.
OHIO: Cleveland: Arrow/Kierulff (216) 248-3990; Hall-Mark (216) 349-4632; Marshall (216) 248-1788; Schweber (216) 464-2970; Columbus: Arrow/Kierulff (614) 436-0928; Hall-Mark (614) 888-3313; Dayton: Arrow/Kierulff (513) 435-5563; Marshall (513) 898-4480; Schweber (513) 439-1800.
OKLAHOMA: Arrow/Kierulff (918) 252-7537; Schweber (918) 622-8003.
OREGON: Arrow/Kierulff (503) 645-6456; Marshall (503) 644-5050; Wyle (503) 640-6000.
PENNSYLVANIA: Arrow/Kierulff (412) 856-7000, (215) 928-1800; GRS Electronics (215) 922-7037; Schweber (215) 441-0600, (412) 963-6804.
TEXAS: Austin: Arrow/Kierulff (512) 835-4180; Hall-Mark (512) 258-8848; Marshall (512) 837-1991; Schweber (512) 339-0088; Wyle (512) 834-9957; Dallas: Arrow/Kierulff (214) 380-6464; Hall-Mark (214) 553-4300; Marshall (214) 233-5200; Schweber (214) 661-5010; Wyle (214) 235-9953; Zeus (214) 783-7010; Houston: Arrow/Kierulff (713) 530-4700; Hall-Mark (713) 781-6100; Marshall (713) 895-9200; Schweber (713) 784-3600; Wyle (713) 879-9953.
UTAH: Arrow/Kierulff (801) 973-6913; Hall-Mark (801) 972-1008; Marshall (801) 485-1551; Wyle (801) 974-9953.
WASHINGTON: Arrow/Kierulff (206) 575-4420; Marshall (206) 747-9100; Wyle (206) 453-8300.
WISCONSIN: Arrow/Kierulff (414) 792-0150; Hall-Mark (414) 797-7844; Marshall (414) 797-8400; Schweber (414) 784-9020.
CANADA: Calgary: Future (403) 235-5325; Edmonton: Future (403) 438-2858; Montreal: Arrow Canada (514) 735-5511; Future (514) 694-7710; Ottawa: Arrow Canada (613) 226-6903; Future (613) 820-8313; Quebec City: Arrow Canada (418) 687-4231; Toronto: Arrow Canada (416) 672-7769; Future (416) 638-4771; Vancouver: Future (604) 294-1166; Winnipeg: Future (204) 339-0554.

Customer Response Center

TOLL FREE: (800) 232-3200
OUTSIDE USA: (214) 995-6611
(8:00 a.m. – 5:00 p.m. CST)

TEXAS INSTRUMENTS

SN74ACT2152, SN74ACT2154
2K × 8 CACHE ADDRESS COMPARATORS

D3050, SEPTEMBER 1987–REVISED NOVEMBER 1987

- Fast Address to Match Delay 25 or 35 ns Max
- Common I/O with Read Feature
- On-Chip Address/Data Comparator
- On-Chip Parity Generator and Checking
- Parity Error Output, Force Parity Error Input
- Easily Expandable
- Choice of Open-Drain or Totem-Pole MATCH Output
- EPIC™ (Enhanced Performance Implanted CMOS) 1-μm Process
- Fully TTL-Compatible

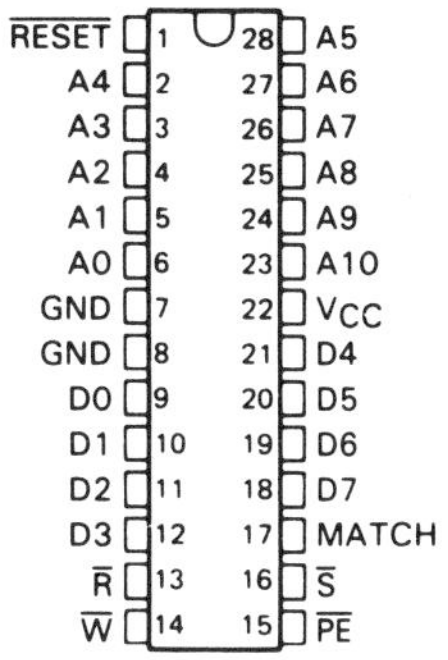

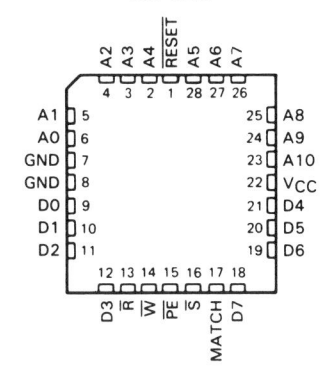

description

The 'ACT2152 and 'ACT2154 cache address comparators consist of a high-speed 2K × 9 static RAM array, parity generator, parity checker, and 9-bit high-speed comparator. They are fabricated using advanced silicon-gate CMOS technology for high speed and simple interface with bipolar TTL circuits. These cache address comparators are easily cascadable for wider tag addresses or deeper tag memories. Significant reductions in cache memory component count, board area, and power dissipation can be achieved with these devices. The 'ACT2152 has a totem-pole MATCH output while the 'ACT2154 has an open-drain MATCH output for easy AND-tying.

If $\overline{S}$ is low and $\overline{W}$ and $\overline{R}$ are high, the cache address comparator compares the contents of the memory location addressed by A0-A10 with the data D0-D7 plus generated parity. An equality is indicated by a high level on the MATCH output. A low-level output on $\overline{PE}$ signifies a parity error in the internal RAM data. $\overline{PE}$ is an N-channel open-drain output for easy OR-tying. During a write cycle ($\overline{S}$ and $\overline{W}$ low), data on D0-D7 plus generated odd parity are written in the 9-bit memory location addressed by A0-A10. Also during write, a parity error may be forced by holding $\overline{PE}$ low.

EPIC is a trademark of Texas Instruments Incorporated.

SN74ACT2152, SN74ACT2154
2K × 8 CACHE ADDRESS COMPARATORS

A read mode is provided with the 'ACT2152 and 'ACT2154, which allows the contents of RAM to be read at the D0-D7 pins. The read mode is selected when $\overline{R}$ and $\overline{S}$ are low, and $\overline{W}$ is high.

A reset input is provided for initialization. When $\overline{RESET}$ is taken low, all 2K × 9 RAM locations are cleared to zero (with valid parity) and the MATCH output is forced high. If an input data word of zero is compared to any memory location that has not been written into since reset, MATCH will be high indicating that input data, plus generated parity, is equal to the reset memory location. $\overline{PE}$ will be high after reset for every addressed memory location, indicating no parity error in the RAM data. By tying a single data input pin high, this bit will function as a valid bit and a match will not occur unless data has been written into the addressed memory location. When cascading in the width direction, only one bit must be tied high regardless of the address width.

These cache address comparators operate from a single +5-V supply and are offered in 28-pin 600-mil ceramic side-brazed, plastic dual-in-line, or PLCC packages.

The 'ACT2152 and 'ACT2154 are characterized for operation from 0°C to 70°C.

MATCH OUTPUT DESCRIPTION

MATCH = V_{OH} if: [A0-A10] = D0-D7 + parity,
or: $\overline{RESET}$ = V_{IL},
or: $\overline{S}$ = V_{IH},
or: $\overline{W}$ = V_{IL}.

MATCH = V_{OL} if: [A0-A10] ≠ D0-D7 + parity,
with $\overline{RESET}$ = V_{IH},
$\overline{S}$ = V_{IL}, and $\overline{W}$ = V_{IH}

FUNCTION TABLE

| INPUTS | | | | OUTPUTS | | I/O | FUNCTION |
$\overline{W}$	$\overline{R}$	$\overline{S}$	$\overline{RESET}$	MATCH	$\overline{PE}$	D0-D7	
H	L	L	H	L	H	Output	Read
H	H	L	H	L	L	Input	Parity error
				L	H		Not equal
				H	L		Undefined error
				H	H		Equal
L	X	L	H	H	IN	Input	Write
X	X	H	H	H	H	Hi-Z	Device disabled
X	X	X	L	H	†	†	Memory reset

†The state of these pins is dependent on inputs $\overline{W}$, $\overline{R}$, and $\overline{S}$.

TEXAS INSTRUMENTS

SN74ACT2152, SN74ACT2154
2K × 8 CACHE ADDRESS COMPARATORS

logic symbol†

Pin	'ALS2152	'ALS2154
RESET	(1)	(1)
$\overline{R}$ READ	(13)	(13)
$\overline{W}$ WRITE	(14)	(14)
$\overline{S}$ SELECT	(16)	(16)
A0	(6)	(6)
A1	(5)	(5)
A2	(4)	(4)
A3	(3)	(3)
A4	(2)	(2)
A5	(28)	(28)
A6	(27)	(27)
A7	(26)	(26)
A8	(25)	(25)
A9	(24)	(24)
A10	(23)	(23)
D0	(9)	(9)
D1	(10)	(10)
D2	(11)	(11)
D3	(12)	(12)
D4	(21)	(21)
D5	(20)	(20)
D6	(19)	(19)
D7	(18)	(18)
MATCH	(17)	(17)
PE / $\overline{PE}$	(15)	(15)

†These symbols are in accordance with ANSI/IEEE Std 91-1984.

Texas Instruments

SN74ACT2152, SN74ACT2154
2K × 8 CACHE ADDRESS COMPARATORS

functional block diagram (positive logic)

SN74ACT2152, SN74ACT2154
2K × 8 CACHE ADDRESS COMPARATORS

PIN NAME	PIN NO.	DESCRIPTION
A0	6	Address inputs. Addresses 1 of 2048 random access memory locations. Must be stable for the duration of the write cycle.
A1	5	
A2	4	
A3	3	
A4	2	
A5	28	
A6	27	
A7	26	
A8	25	
A9	24	
A10	23	
D0	9	Data inputs/outputs. D0-D7 are data inputs during the compare and write modes. D0-D7 are data outputs during the read mode.
D1	10	
D2	11	
D3	12	
D4	21	
D5	20	
D6	19	
D7	18	
GND	7,8	Ground
MATCH	17	When MATCH output is at V_{OH} during a compare cycle, D0-D7 plus generated parity equals the contents of the 9-bit memory location addressed by A0-A10. MATCH is also driven high during deselect, reset, and read. Since the 'ACT2154 features an open-drain MATCH output, an external pull-up resistor of 220 Ω minimum is required.
$\overline{PE}$	15	Parity error input/output. During compare cycles, $\overline{PE}$ at V_{OL} indicates a parity error in the stored data. During write cycles, $\overline{PE}$ can force a parity error into the 9th-bit location specified by A0-A10 when $\overline{PE}$ is taken to V_{IL}. $\overline{PE}$ is an open-drain output so an external pull-up resistor of 220 Ω minimum is required.
$\overline{R}$	13	Read input. When $\overline{R}$ and $\overline{S}$ are at V_{IL} and $\overline{W}$ is at V_{IH}, addressed data is output to the D0-D7 pins and the MATCH and $\overline{PE}$ outputs are forced high.
$\overline{RESET}$	1	Reset input. Asynchronously clears entire RAM array to zero and forces MATCH high when $\overline{RESET}$ is at V_{IL}.
$\overline{S}$	16	Chip select input. Enables device when $\overline{S}$ is at V_{IL}. Deselects device and forces MATCH and $\overline{PE}$ high when $\overline{S}$ is at V_{IH}.
V_{CC}	22	Supply voltage
$\overline{W}$	14	Write control input. Writes D0-D7 and generated parity into RAM and forces MATCH high when $\overline{W}$ and $\overline{S}$ are at V_{IL}. Places selected device in compare mode when $\overline{W}$ and $\overline{R}$ are at V_{IH}.

Texas Instruments

SN74ACT2152, SN74ACT2154
2K × 8 CACHE ADDRESS COMPARATORS

absolute maximum ratings over operating free-air temperature range (unless otherwise noted)

Supply voltage, V_{CC} (see Note 1) .... −1.5 to 7 V
Input voltage, any input .... −1.5 to 7 V
Operating free-air temperature range .... 0°C to 70°C
Storage temperature range .... −65°C to 150°C

NOTE 1: All voltage values are with respect to GND.

recommended operating conditions

		MIN	NOM	MAX	UNIT
V_{CC}	Supply voltage	4.5	5	5.5	V
V_{IH}	High-level input voltage, write or compare cycles	2.2		V_{CC}+0.5	V
V_{IH}	High-level input voltage, read cycle	2.6		V_{CC}+0.5	
V_{IL}	Low-level input voltage (See Note 2)	−0.5		0.8	V
V_{OH}	High-level output voltage, MATCH ('ACT2154) and $\overline{PE}$ outputs only			5.5	V
I_{OH}	High-level output current, MATCH ('ACT2152) and D0-D7			−8	mA
I_{OL}	Low-level output current — MATCH − 'ACT2152			8	mA
	MATCH − 'ACT2154			24	mA
	$\overline{PE}$			24	mA
	D0-D7			8	mA
T_A	Operating free-air temperature	0		70	°C

NOTE 2: The algebraic convention, in which the more negative (less positive) limit is designated as minimum, is used in this data sheet for logic voltage levels only.

electrical characteristics over recommended operating free-air temperature range (unless otherwise noted)

PARAMETER		TEST CONDITIONS	SN74ACT2152-25 SN74ACT2154-25			SN74ACT2152-35 SN74ACT2154-35			UNIT
			MIN	TYP†	MAX	MIN	TYP†	MAX	
I_{OH}	High-level output current MATCH ('ACT2154) and $\overline{PE}$	V_{OH} = 5.5 V, V_{CC} = 5.5 V			5			5	µA
V_{OH}	High-level output voltage MATCH ('ACT2152) and D0-D7	I_{OH} = −8 mA, V_{CC} = 4.5 V	3.7			3.7			V
V_{OL}	Low-level output voltage	MATCH - 'ACT2154 I_{OL} = 24 mA, V_{CC} = 4.5 V			0.4			0.4	V
		MATCH - 'ACT2152 I_{OL} = 8 mA, V_{CC} = 4.5 V			0.4			0.4	
		$\overline{PE}$ I_{OL} = 24 mA, V_{CC} = 4.5 V			0.4			0.4	
		D0-D7 I_{OL} = 8 mA, V_{CC} = 4.5 V			0.4			0.4	
I_I	Input current	V_I = 0−V_{CC}, V_{CC} = 5.5 V			±1			±1	µA
I_{OZ}	Off-state output current	V_O = 0−V_{CC}, V_{CC} = 5.5 V, $\overline{S}$ at V_{IH}			±5			±5	µA
I_{OS}‡	Short-circuit output current MATCH ('ACT2152) and D0-D7	V_O = 0, V_{CC} = 5.5 V	50		150	50		150	mA
I_{CC1}	Supply current (operative)	$\overline{RESET}$ at 3 V, V_{CC} = 5.5 V, $\overline{S}$ at 0 V		85	125		85	125	mA
I_{CC2}	Supply current (reset)	$\overline{RESET}$ at 0 V, V_{CC} = 5.5 V, $\overline{S}$ at 0 V		5	25		5	25	mA
I_{CC3}	Supply current (deselected)	$\overline{RESET}$ at 3 V, V_{CC} = 5.5 V, $\overline{S}$ at 3 V		75	105		75	105	mA
C_i	Input capacitance	f = 1 MHz		5			5		pF
C_o	Output capacitance	f = 1 MHz		6			6		pF

† All typical values are at V_{CC} = 5 V, T_A = 25°C.
‡ Not more than one output should be shorted at a time, and duration of the short-circuit should not exceed one second.

Texas Instruments

SN74ACT2152, SN74ACT2154
2K × 8 CACHE ADDRESS COMPARATORS

switching characteristics over recommended ranges of supply voltage and operating free-air temperature (unless otherwise noted), see Figures 1, 2, and 3

compare cycle

PARAMETER		SN74ACT2152-25 SN74ACT2154-25 MIN / TYP† / MAX			SN74ACT2152-35 SN74ACT2154-35 MIN / TYP† / MAX			UNIT
$t_{a(A-M)}$	Access time from address to MATCH		19	25		20	35	ns
$t_{a(A-P)}$	Access time from address to $\overline{PE}$ high or low		21	28		27	38	ns
$t_{a(S-M)}$	Access time from $\overline{S}$ to MATCH		12	15		13	18	ns
$t_{p(D-M)}$	Propagation time, data inputs to MATCH		11	16		12	18	ns
$t_{p(RST-MH)}$	Propagation time, $\overline{RESET}$ low to MATCH high		12	18		13	20	ns
$t_{p(S-MH)}$	Propagation time, $\overline{S}$ high to MATCH high		6	12		8	15	ns
$t_{p(W-MH)}$	Propagation time, $\overline{W}$ low to MATCH high		9	14		9	15	ns
$t_{p(W-PH)}$	Propagation time, $\overline{W}$ low to $\overline{PE}$ high		7	14		8	18	ns
$t_{v(A-M)}$	MATCH valid time after change of address	4			4			ns
$t_{v(D-M)}$	MATCH valid time after change of data	2			2			ns
$t_{v(S-M)}$	MATCH valid time (low) after $\overline{S}$ high	2			2			ns
$t_{v(A-P)}$	$\overline{PE}$ valid time after change of address	6			6			ns

read cycle

PARAMETER		SN74ACT2152-25 SN74ACT2154-25 MIN / TYP† / MAX			SN74ACT2152-35 SN74ACT2154-35 MIN / TYP† / MAX			UNIT
$t_{a(A-D)}$	Read Access time from address to D0-D7		25	30		26	35	ns
$t_{en(S-D)}$	Enable time, $\overline{S}$ low to D0-D7		12	20		14	20	ns
$t_{en(R-D)}$	Enable time, $\overline{R}$ low to valid D0-D7 output		16	20		18	20	ns
$t_{p(R-MH)}$	Propagation time, $\overline{R}$ low to MATCH high		7	12		8	15	ns
$t_{p(R-PH)}$	Propagation time, $\overline{R}$ low to $\overline{PE}$ high		7	15		8	18	ns
t_{dis}	D0-D7 output disable time from high or low level From $\overline{R}$, $\overline{S}$, $\overline{W}$		15	20		16	20	ns

† All typical values are at V_{CC} = 5 V, T_A = 25°C.

Texas Instruments

SN74ACT2152, SN74ACT2154
2K × 8 CACHE ADDRESS COMPARATORS

timing requirements over recommended ranges of supply voltage and operating free-air temperature (unless otherwise noted)

PARAMETER		SN74ACT2152-25 SN74ACT2154-25 MIN NOM MAX	SN74ACT2152-35 SN74ACT2154-35 MIN NOM MAX	UNIT
$t_{w(RSTL)}$	Pulse duration, $\overline{RESET}$ low	30	35	ns
$t_{w(WL)}$	Pulse duration, $\overline{W}$ low, without writing $\overline{PE}$	15	20	ns
$t_{w(WL)PE}$	Pulse duration, $\overline{W}$ low, writing $\overline{PE}$ (see Note 3)	15	20	ns
$t_{su(A)}$	Address setup time before $\overline{W}$ low	0	0	ns
$t_{su(D)}$	Data setup time before $\overline{W}$ high	10	15	ns
$t_{su(P)}$	$\overline{PE}$ setup time before $\overline{W}$ high (see Note 3)	10	15	ns
$t_{su(S)}$	Chip select setup time before $\overline{W}$ high	10	15	ns
$t_{su(RST)}$	$\overline{RESET}$ inactive setup time before $\overline{W}$ high	15	20	ns
$t_{h(A)}$	Address hold time after $\overline{W}$ high	0	0	ns
$t_{h(WH-D)}$	Data hold time after $\overline{W}$ high	5	5	ns
$t_{h(WL-D)}$	Data hold time after $\overline{W}$ low with MATCH high, (see Note 4)	10	10	ns
$t_{h(P)}$	$\overline{PE}$ hold time after $\overline{W}$ high	5	5	ns
$t_{h(S)}$	Chip select hold time after $\overline{W}$ high	0	0	ns
t_{AVWH}	Address valid to write enable high	15	20	ns

NOTES: 3. Parameters $t_{wPE(WL)}$ and $t_{su(P)}$ apply only during the write cycle timing when writing a parity error.
4. $t_{h(WL-D)}$ guarantees that when $\overline{W}$ is taken low during a compare cycle with MATCH high, match will remain high without a glitch low. (As shown in the function table, $\overline{W}$ low forces MATCH high). $t_{h(WL-D)}$ is guaranteed indirectly by $t_{v(D-M)}$ and $t_{p(W-MH)}$.

SN74ACT2152, SN74ACT2154
2K × 8 CACHE ADDRESS COMPARATORS

PARAMETER MEASUREMENT INFORMATION

FIGURE 1. TACT2152 MATCH OUTPUT

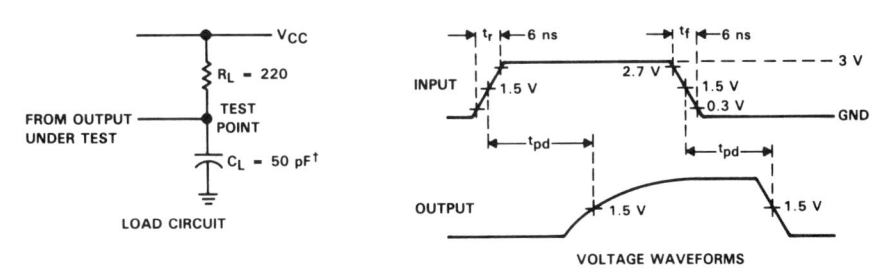

FIGURE 2. OPEN-DRAIN MATCH AND $\overline{PE}$ OUTPUTS

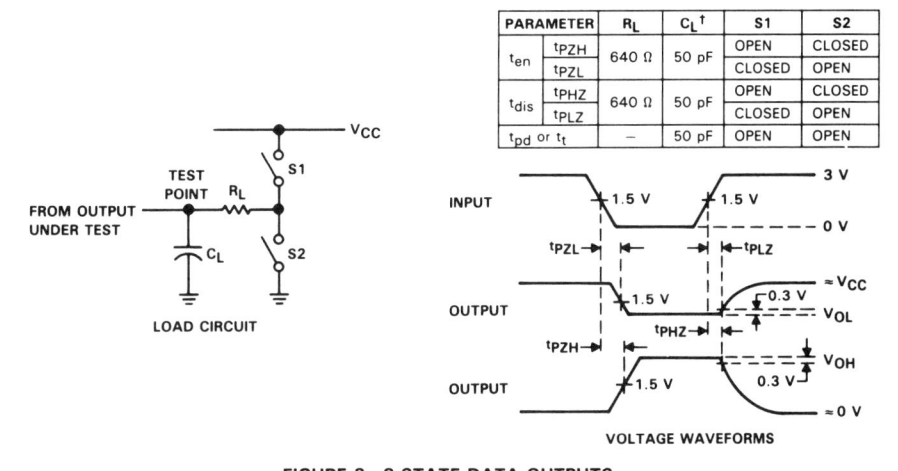

PARAMETER		R_L	C_L†	S1	S2
t_{en}	t_{PZH}	640 Ω	50 pF	OPEN	CLOSED
	t_{PZL}			CLOSED	OPEN
t_{dis}	t_{PHZ}	640 Ω	50 pF	OPEN	CLOSED
	t_{PLZ}			CLOSED	OPEN
t_{pd} or t_t		—	50 pF	OPEN	OPEN

FIGURE 3. 3-STATE DATA OUTPUTS

†C_L includes probe and test fixture capacitance.

Texas Instruments

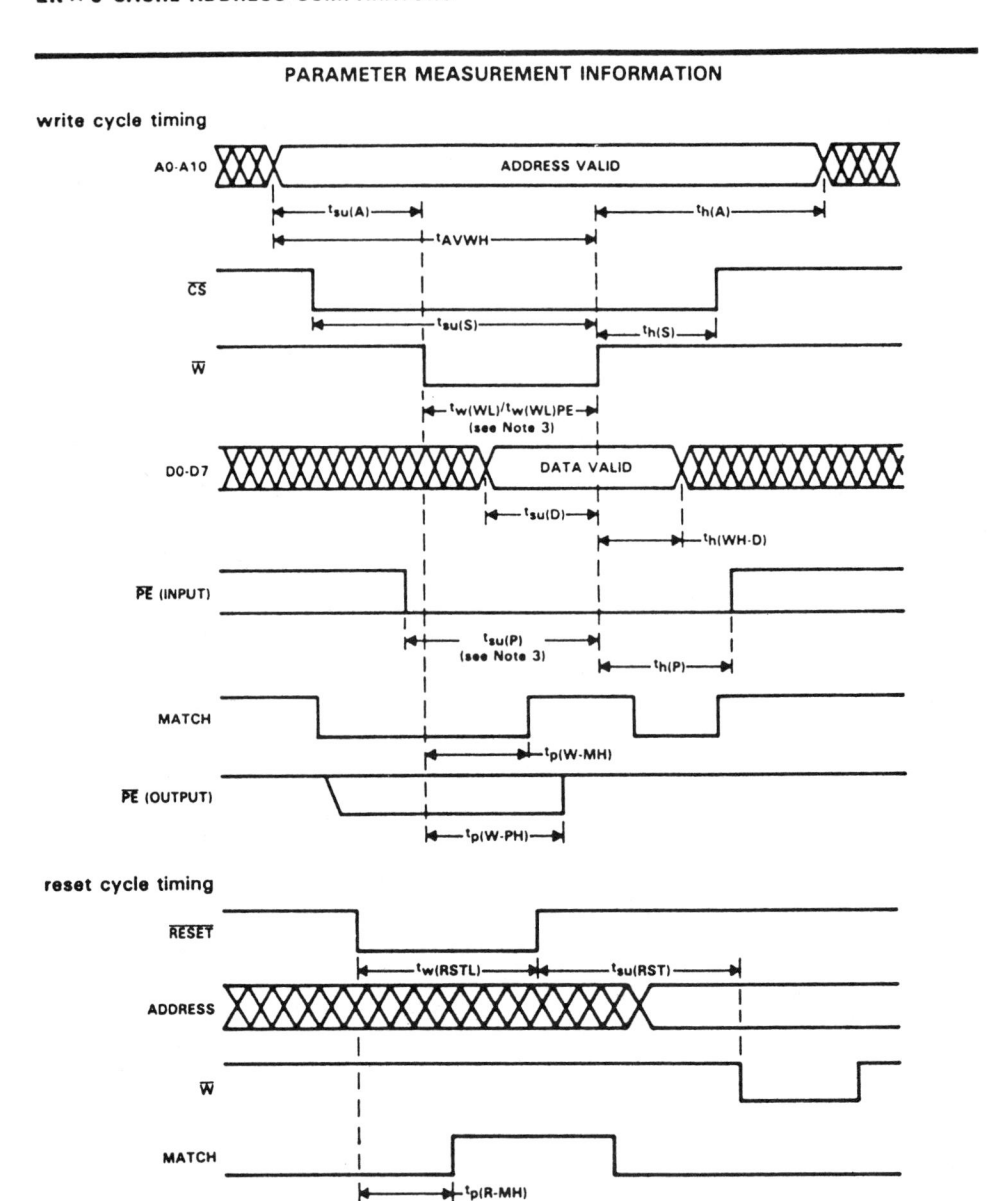

SN74ACT2152, SN74ACT2154
2K × 8 CACHE ADDRESS COMPARATORS

PARAMETER MEASUREMENT INFORMATION

write cycle timing

reset cycle timing

NOTE 3: Parameters $t_{w(WL)PE}$ and $t_{su(P)}$ apply only during the write cycle when writing a parity error.

Texas Instruments

SN74ACT2152, SN74ACT2154
2K × 8 CACHE ADDRESS COMPARATORS

PARAMETER MEASUREMENT INFORMATION

compare cycle timing

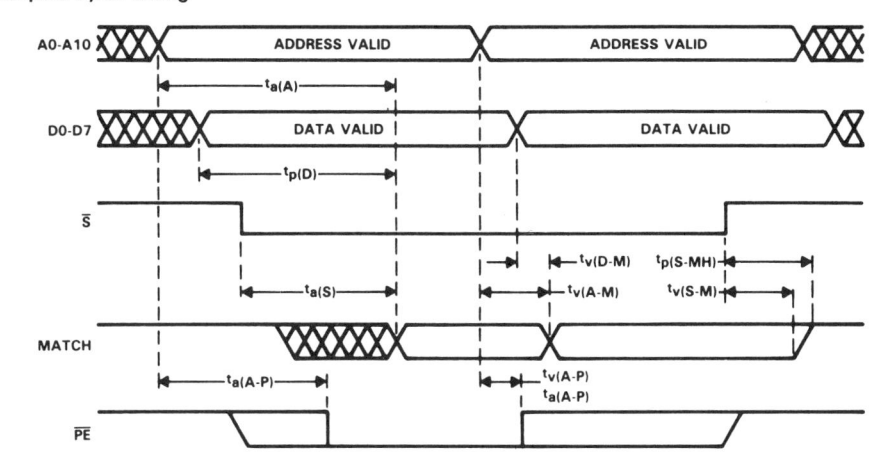

read cycle timing

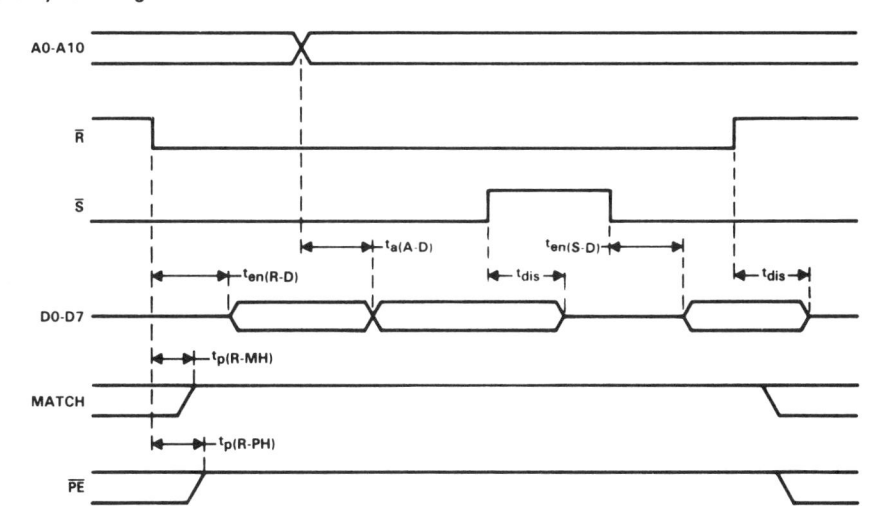

TEXAS INSTRUMENTS

SN74ACT2152, SN74ACT2154
2K × 8 CACHE ADDRESS COMPARATORS

TYPICAL APPLICATION INFORMATION

cascading the 'ACT2152 and 'ACT2154

The 'ACT2152 and 'ACT2154 are easily cascaded in width and depth. Wider addresses can be compared by driving the A0-A10 inputs of each device with the same index and applying the additional address bits to the D0-D7 inputs. The select ($\overline{S}$) input allows these devices to be cascaded in depth. When a device is deselected, the MATCH output is driven high. It should be noted that a decoder can be used to drive the select inputs since the propagation delay from select to match is much faster than from address to match. MATCH on the 'ACT2154 is an open-drain output for easy AND-tying. Figure 4 shows the 'ACT2154 cascaded.

cache coherency through bus watching

When cache designs are implemented, the problem of cache coherency is always a concern. One solution to this problem is to implement bus-watching using the 'ACT2152 or 'ACT2154. By storing the same tags in the bus-watcher RAM as are stored in the cache tag RAM, the bus-watcher will indicate a hit every time a cache address passes down the main address bus. If data is being modified in main memory, the index can be passed to the cache tag RAM for invalidation. Figure 5 shows a possible bus-watcher implementation.

application

Due to the high-performance switching characteristics of the 'ACT2152 and 'ACT2154, it is necessary that the address inputs not be allowed to float in the three-state condition. Proper termination techniques should be employed. It is recommended that the RC time constant associated with the address inputs (63.2% of rise time at A0-A10) not exceed 60 ns.

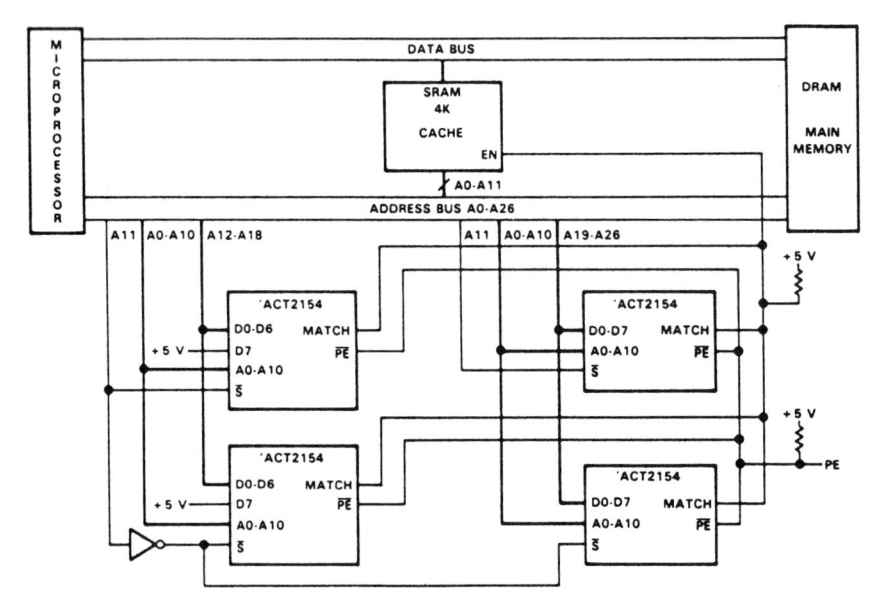

FIGURE 4. CASCADING THE 'ACT2154

Texas Instruments

SN74ACT2152, SN74ACT2154
2K × 8 CACHE ADDRESS COMPARATORS

TYPICAL APPLICATION INFORMATION

FIGURE 5. BUS WATCHING USING THE 'ACT2152

CACHE TAG RAM CHIPS SIMPLIFY CACHE MEMORY DESIGN

APPLICATION NOTE AN-07

By David C. Wyland

ABSTRACT

Cache memories are a widely used tool for increasing the throughput of computer systems. The IDT7174 Cache Tag RAM is a new component designed to support direct mapped cache designs by providing the tag comparison on-chip. This allows relatively large cache memories to be designed with low chip count. The application of the IDT7174 to cache memory design is explored by designing a simple cache memory, reviewing its operation and performance, discussing methods of extending the design, and then reviewing the theory behind the design of cache memories in general.

INTRODUCTION

Cache memories are an important design tool for increasing computer performance by increasing the effective speed of the memory. Computer memories are usually implemented with slow, inexpensive devices such as dynamic RAMs. A cache memory is a small, high-speed memory that fits between the CPU and the main memory in a computer system. It increases the effective speed of the main memory by responding quickly with a copy of the most frequently used main memory data. When the CPU tries to read data from the main memory, the high-speed cache memory will respond first if it has a copy of the requested data. Otherwise, a normal main memory cycle will take place. In typical systems, the read data will be supplied by the cache memory over 90% of the time. The result is that the large main memory appears to the CPU to have the high speed of the cache memory.

The IDT7174 Cache Tag RAM introduced by IDT simplifies the design of high-speed cache memories. It can be used to make a high-performance cache memory with a low part count. The IDT7174 Cache Tag RAM consists of a 64K-bit static RAM organized as 8K x 8 and an 8-bit comparator, as shown in Figure 1. The comparator is used in direct mapped cache memories to perform the address tag comparison, and allows a 16K byte cache for a 68000 microprocessor to be built with four memory chips. The IDT7174 also provides a single pin RAM clear control which clears all words in the internal RAM to zero when activated. This control is used to clear the tag bits for all locations at power-on or system-reset when the cache is empty of data. This allows one of the comparison bits to be used as a cache data valid bit.

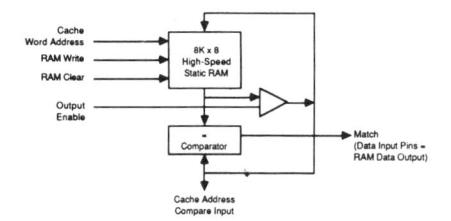

Figure 1: IDT7174 Cache Tag RAM Block Diagram

DESIGN OF A CACHE MEMORY

To understand the application of the IDT7174 to cache memories, we will begin by designing one. A block diagram of a cache memory system using IDT7174 Cache Tag memory chips is shown in Figure 2. The cache memory serves a 16-bit microprocessor with a 24-bit address bus and a main memory. In this system, the 13 least significant bits of the address bus are connected to the address inputs of both the cache tag and the cache data RAM chips. The upper 11 bits of the address bus are connected to the data I/O pins of the cache tag RAMs. The remaining five I/O pins of the cache tag RAMs are connected to a logic 1 (+5).

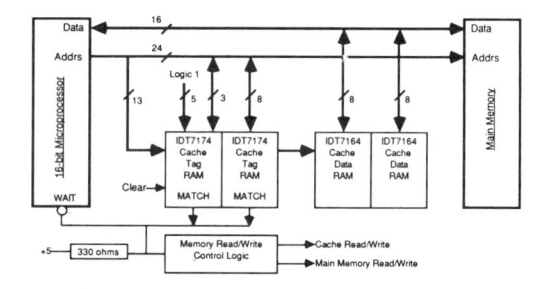

Figure 2: Cache Memory System Block Diagram

The MATCH outputs of the cache tag rams are tied together and connected to the WAIT input of the microprocessor. A 330 ohm pull-up resistor is used because the MATCH outputs are open-drain type. The MATCH outputs are positive-active. The MATCH output goes high when the contents of the internal RAM are equal to the data on the I/O pins. When several cache tag RAMs have their MATCH outputs connected together, a wire-AND function results: all of the comparators must each register a match before the common MATCH signal can go high.

In the system shown, the state of the WAIT input to the microprocessor determines whether the memory data is to come from the cache or the main memory. If the WAIT input to the microprocessor is high, the microprocessor will accept data immediately from the cache data RAMs; if the WAIT input is low, the microprocessor will wait for the slower main memory to respond with the data.

To understand how the cache memory operates, we will follow its operation from start-up in an initially empty state. When the system is powered-up, the cache tag RAMs are cleared to zero by a pulse to the initialize pins of the IDT7174 RAMs. This causes all cells in the RAM to be simultaneously cleared to logic zero. When the microprocessor begins its first read cycle, the 13 least significant bits of the address bus select a location in the cache tag RAMs. The location in the cache tag RAMs is compared against the upper bits of the address bus and against five bits of logic one.

CACHE TAG RAM CHIPS SIMPLIFY CACHE MEMORY DESIGN

APPLICATION NOTE AN-07

The MATCH output of the cache tag RAMs will be low because all cache tag RAM cells were reset to zero, and the zeros from the selected cell are being compared against the five bits of logic one. In this case, the microprocessor waits for the slower main memory to respond. This is called a cache miss.

When the main memory responds with read data for the microprocessor, this data is also written into the cache data memory at the address defined by the 13 least significant bits of the address bus. At the same time, the upper 11 bits of the address bus and the five bits of logic one are written into the cache tag memory. This 11-bit address tag, in combination with the 13 bits of RAM address select, uniquely identify the copy of the main memory data that was stored. The five logic one bits serve as a data valid bits which indicate that the data in the cell is a valid copy of main memory data.

When the microprocessor requests data from the same location that has been written into the cache, the upper address bits on the address bus will be the same as the bits which were previously written into the cache tag RAM and the MATCH signal will go high. This is called a cache hit. In this case, the cache data is gated onto the data bus and the memory cycle is complete.

If the microprocessor requests data from an address with the same 13 least significant bits as a word in the cache, but with different upper address bits, a cache miss will result and the current (more recent) data will be written into the cache. In this manner, the cache is continuously updated with the most recently used data.

Memory write cycles are treated differently from read cycles. On write cycles, data is written directly into main memory and into the cache. This is called the write-through method of cache updating. Since all data is written immediately into main memory, it always contains current information. Data is written into the cache on full word writes or on byte (i.e. partial word) writes if a match occurred. Writing bytes into the cache only if a cache match occurs ensures that the full word in the cache is valid. For example, this ensures valid data for a byte write followed by a word read.

The design in Figure 2 uses unbuffered writes. In unbuffered writes, all write cycles occur at main memory speeds. This slows down the system for all write cycles at the expense of simple memory controls; however, this may be acceptable since only 15% of all memory cycles are write cycles in typical programs. Buffered write is a slightly more complicated method which improves performance. In buffered write cycles, the write data and address are loaded into registers, and the main memory write cycle proceeds in overlap with other processor operations. Since the next few cycles will probably be read cycles and their data will come from the cache, the result is that buffered write cycles are as short as cache read cycles.

CACHE MEMORY DESIGN: PERFORMANCE

Even a simple cache memory can improve system performance. For a simple, 16-bit cache system such as described above, a hit rate (percentage of read cycles that are from the cache) of 68% can be expected. If IDT7174 Cache Tag RAMs and IDT7164 cache data RAMs are used, an access time at the chip level of 35ns results and a corresponding system cache read or write cycle time of 50ns is practical. Assuming a system cache access time of 50ns and a main memory system access time of 250ns, the average access time of an unbuffered cache would be 134ns and the average access time of a buffered cache would be 104ns. This corresponds to an improvement in access time of 1.9:1 and 2.4:1, respectively.

CACHE DESIGN DETAILS: CONTROL LOGIC

Figure 3 shows a block diagram of a control logic design and a typical timing diagram for the cache memory of Figure 2. The vertical lines in the timing diagram represent 50ns timing intervals. The microprocessor is assumed to have a 50ns clock and a 100ns memory cycle time. In the timing diagram and associated logic, a Read/Write Timing signal is used to determine whether to use the cache data or to start the main memory. This timing signal is the memory read/write request signal from the CPU delayed by 37ns; the address-to-match time of the IDT7174. If main memory is used, this timing signal is used to write the main memory data into the cache RAMs on both the main memory read and write cycles. Data is written into the cache on write cycles only if there is a match or if it is a word write operation. The state of the MATCH line is latched by the Read/write Timing signal so that i remains stable during cache write operations.

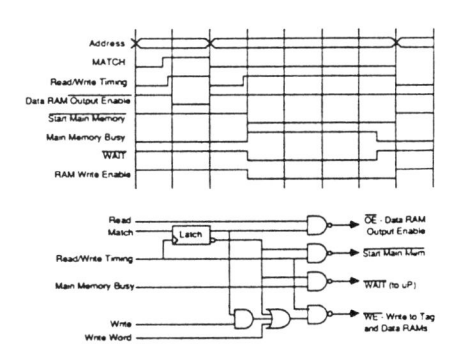

Figure 3: Cache Memory Control Timing and Logic Block Diagrams

CACHE DESIGN DETAILS: UNCACHED ADDRESSES

In the above cache design, we have assumed that all parts memory are cached; however, there are significant exceptions this assumption. Hardware I/O addresses should not be cach because they do not respond in the same way as normal memc locations. Bits in an I/O register can and must change at any tin asynchronously, with respect to the rest of the system. A cac copy of an earlier I/O state is clearly not a valid response to an I read request under these conditions. Also, an I/O regis address may be used for different functions for read and write, that what is read will not be the same as what was written. F example, write-only control bits will not appear when read, a read-only bits will not be affected by write operations. For the reasons, hardware I/O addresses must always force cac misses. This can be accomplished by adding an I/O addre decoder to the memory address bus to force a cache miss. (T decoder aleady exists in many systems to enable the subsystem.)

CACHE DESIGN DETAILS: DMA ADDRESSES

Direct Memory Access (DMA) allows I/O devices such as controllers to have direct access to main memory by tempora stopping the CPU and taking control of the memory address data busses. If DMA devices are allowed to write into n memory without updating the cache memory, cache data cc become invalidated because it would no longer be a copy of

14-48

CACHE TAG RAM CHIPS SIMPLIFY CACHE MEMORY DESIGN

contents of main memory. The simplest solution to this problem is to have the cache monitor the memory bus and be updated if an address match occurs in the same manner as CPU write-through operations. Otherwise, the I/O DMA buffer areas of memory must be forced to be uncached in the same manner as hardware I/O addresses.

CACHE DESIGN DETAILS: EXPANDING THE CACHE IN WIDTH

The cache as described above, can be expanded in both width and depth. For a 32-bit system, two additional IDT7164 cache data RAMs (for a total of 4 chips) will be required to store the 32-bit data words. A block diagram of a 32-bit cache system, with a 32-bit address bus, is shown in Figure 4. Compared with Figure 2, the number of cache data RAMs has been expanded from two to four to handle the expansion of the data bus from 16 to 32 bits, and the number of cache tag RAMs has been expanded from two to three to handle the expansion of the address bus from 24 to 32 bits.

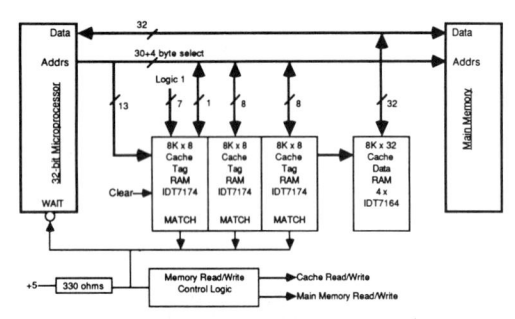

Figure 4: 32-Bit Cache Memory System

Note that the cache memory system uses the memory address lines corresponding to the 32-bit words stored in the cache. If a byte addressing memory address convention is used, the least significant bit of the address lines going to the cache RAM chips is A2, with A1 and A0 used to select the byte(s) within the word to be read or written in the cache data RAMs.

There is a benefit to expanding the cache width by adding data RAMs: the miss rate improves. The miss rate improves because of the increase in width, as well as in the amount of data stored. The miss rate for a 8K x 32-bit cache is estimated at 12.4%, as compared to 32% for a 8K x 16-bit cache. Doubling the cache width by adding RAM chips doubles the amount of data stored. We would expect an improvement in miss rate due to the increased probability of finding the data in the cache.

There is an additional improvement in miss rate, however, specifically due to the increase in width. This is because there is a high probability that the next word the CPU wants is the next word after the current one. If the cache width is doubled, there is a 50% probability that the next word is already in the cache, fetched from main memory along with the current word.

Studies have shown that the miss rate is cut almost in half for each doubling of the cache data word width — called line size in cache theory — up to 16 bytes and larger (Smith 85). The disadvantage of very wide cache data word width is either a wide main memory data bus or complex logic to transfer the word to the cache in a high-speed serial burst. Simply doubling the number of main memory cycles does not work well because you have doubled the effective access time of the main memory but have cut the miss rate by less than half, yielding a net decrease in performance.

CACHE DESIGN DETAILS: EXPANDING THE CACHE IN DEPTH

The cache memory can be expanded in depth by adding copies of the cache tag and data chips and using upper bits of the address bus for chip enable selection. An example of an expanded cache is shown in Figure 5. The primary reason for increasing the size of the cache memory is to decrease the miss rate percentage. For example, increasing the cache size from 8K x 16 to 16K x 16 decreases the estimated miss rate from 32% to 22%.

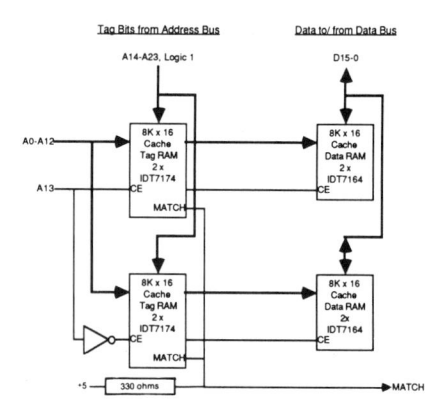

Figure 5: Depth Expanded Cache Memory System

CACHE DESIGN DETAILS: SET ASSOCIATIVE EXPANSION

A better way to expand the cache memory in depth is called set associative expansion (shown in Figure 6), and its control logic (shown in Figure 7). In this example, we have two independent cache memories which results in a two-way set associative cache. If a match is found in one of the memories, its data is gated to the data bus. If no match is found, one of the two memories is selected and updated. Selection of one of the two memories for cache write update is done by using an additional 8K x 1 memory to hold a flag for each cache word, indicating which memory was read last. This way, the least recently used cache word of the pair is updated.

The cache system described above attacks the problem of having two frequently used words mapped to the same cache word. For example, if a program loop included an instruction at 200B2 (hexadecimal) and called a subroutine at 800B2, the cache word 00B2 would be alternately registered as a cache miss and updated with memory data from each of these two addresses. The above design solves this problem by having two independent memories. One would cache the instruction at 200B2 and the other would cache 800B2.

Two way set associative expansion, while more complex in control logic, achieves a better miss rate. For example, the estimated miss rate for a 16K x 16 set associative cache is 18% versus 22% for a simple 16K x 16 cache.

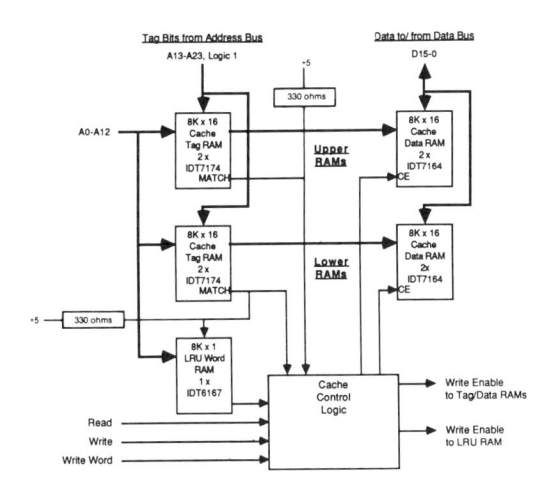

CACHE TAG RAM CHIPS SIMPLIFY CACHE MEMORY DESIGN

APPLICATION NOTE AN-07

Figure 6: 2-Way Set Associative Cache Memory System

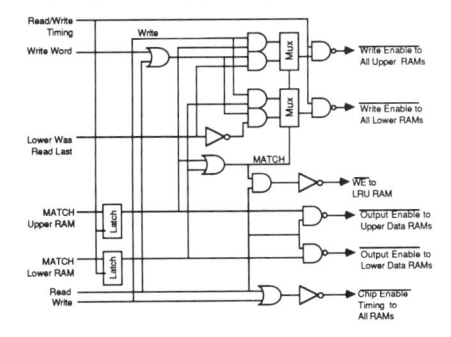

Figure 7: 2-Way Set Associative Cache Control Logic Block Diagram

CACHE THEORY: HOW IT WORKS

A cache memory cell holds a copy of one word of data corresponding to a particular address in main memory. It will respond with this word if the address on the main memory address bus matches the address of the word stored. A cache memory cell therefore has three components. These components are an address memory cell, an address comparator, and a data memory cell, as shown in Figure 8. The data and address memory cells record the cached data and its corresponding address in main memory. The address comparator checks the address cell contents against the address on the memory address bus. If they match, the contents of the data cell are placed on the data bus.

An ideal cache memory would have a large number of cache memory cells with each of them holding a copy of the most frequently used main memory data. This type of cache memory is called fully associative because access to the data in each memory cell is through its associated, stored address. This type of memory is expensive to build because the address cell and address comparator are generally several times larger, in terms of chip area or part count, than the data cell. Also, the address comparator required for each associative memory cell makes the design of the cell different from that of standard RAM memory cells. This makes a fully associative memory a custom design, precluding the use of efficient standard RAM designs.

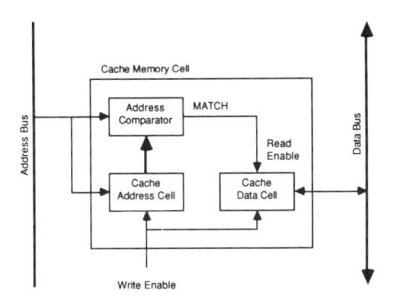

Figure 8: Cache Memory Cell Block Diagram

CACHE THEORY: WHY IT WORKS

Cache memories work because computer programs spend most of their memory cycles accessing a very small part of the memory. This is because most of the time the computer is executing instructions in program loops and using local variables for calculation. Because of this observation, a 64K byte cache can have a 90+% hit rate on programs that are megabytes in size.

HOW THE DIRECT MAPPED CACHE WORKS

The direct mapped cache memory is an alternative to the associative cache memory which uses a single address comparator for the cache memory system and standard RAM cells for the address and data cells. The direct mapped cache is based on an idea borrowed from software called hash coding which is a method for simulating an associative memory. In a hash coding approach, the memory address space is divided into a number of sets of words with the goal of each set having no more than one word of most-frequently-used data. In our case, there are 8K sets of 2048 words each.

Each set is assigned an index number derived from the main memory address by a calculation which is called the hashing algorithm. This algorithm is chosen to maximize the probability that each set has no more than one word of most-frequently-used data. In the direct mapped cache, the hashing algorithm uses the least significant bits of the memory address as the set number. This uses the concept of locality, which assumes that the most often used instructions and data are clustered in memory. If locality holds, the least significant bits of the address should be able to divide this cluster into individual words and assign each one to a separate set.

A memory map of a direct mapped cache of Figure 2 is shown in Figure 9 as an example of how the main memory words are related to the cache words. The 16M Word main memory is divided into 8K word pages, a total of 2048 pages. Each word within each 8K page is mapped to its corresponding word in the 8K words of the cache; i.e., word 0 of the cache corresponds to word 0 in each of the 2048 pages (8K sets at 2048 words/set).

14–50

CACHE TAG RAM CHIPS SIMPLIFY CACHE MEMORY DESIGN

APPLICATION NOTE AN-07

Each word in the cache stores one word out of its set of 2048 corresponding to one of the 2048 possible pages. Both the data word and the page number (i.e. upper address bits), are stored.

Since only one word in each set (one of 2048 words in our case) is assumed to be one of the most-frequently-used words, each set has a single cache memory cell associated with it. This cache cell consists of an address cell and a data cell, but no comparator. One comparator is used for the cache memory system since only one set can be selected for a given memory cycle and only one comparison need be made. In a memory cycle, one set is selected, and the single cache address cell for that set is read and compared against the memory address, and the data from the cache data cell is placed on the bus if there is a match. The advantage of this scheme is that a single comparator is used, allowing standard RAM memories to be used to store the cache address and data for each set.

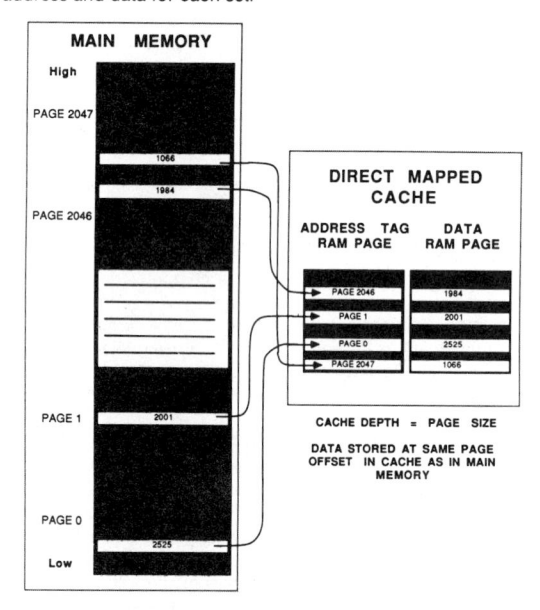

Figure 9: Cache System Memory Map

The cache cell for each set should hold the data that was most frequently used. However, since we do not know which data was the most frequently used until after the program is run, we approximate it by storing the most recently used data and replacing the least recently used (oldest) data. In the direct mapped cache, this is done by replacing the cache cell contents with the newer main memory data in the case of a cache miss.

CACHE PERFORMANCE

A cache memory improves a system by making data available from a small, high-speed memory sooner than would otherwise be possible from a larger, slower main memory. The performance of a cache memory system depends upon the speed of the cache memory relative to the speed of the main memory and on the hit rate or percentage of memory cycles that are serviced by the cache.

The cache performance equations below express the idea that the average speed of the cache memory is the weighted average of the cycle times for cache hits plus the main memory time for cache misses, with memory writes dealt with as a special case of 100% cache miss or 100% cache hit for the unbuffered and buffered cases, respectively.

CACHE SYSTEM PERFORMANCE: MISS RATE

One of the key parameters in a cache memory system is the miss rate. Miss rate figures are estimates derived from statistical studies of cache memory systems. The miss rate is an estimate because it varies, often significantly, with the program being run. Miss rate estimates for various cache memory configurations are given in Table 1. Miss rates for one example of two-way set associative expansion are also shown in this table.

Size: Words/Tag RAM	Miss Rate for Cache Data Word Width - Bits				Notes
	16	32	64	128	
2K	0.57	0.23	0.10	0.04	
4K	0.40	0.18	0.07	<0.04	
8K	0.32	0.12	0.05	<0.04	
16K	0.22	0.09	<0.04	<0.04	
16K (8K + 8K)	0.18	0.07	<0.04	<0.04	2-way Set Assoc

Table 1.

The miss rate estimates given in Table 1 are derived from simulation studies. (See references.) These studies covered cache sizes of up to 32K bytes and cache data word widths (called line sizes in cache terminology) from 4 bytes through 64 bytes. In the case of 16-bit word width caches, the figures given are extrapolations from the 32-bit data. Also, the figures for cache sizes above 32K bytes (i.e., 16K x 32, etc.) are extrapolations from 32K byte data.

CACHE SYSTEM PERFORMANCE FOR READ CYCLES

Cache memory system performance is determined by the access time of the main memory, the access time of the cache, the miss rate (the percentage of memory cycles that are not serviced by the cache) and the write time. The effective access time of a cache memory system can be expressed as a fraction of the main memory access time. This dimensionless number, Ps, is a measure of cache performance. If we consider read cycles only, the access time of a cache memory system is:

$$T_s = (1 - M)T_c + MT_m = (1 - M)T_c + MT_m$$
$$P_s = T_s/T_m = (1 - M)(T_c/T_m) + M = (1 - M)P_c + M$$

Where:
- T_s = Cache average system cycle time, averaged over read and write
- M = Miss rate of cache
- T_c = Cache cycle time, read or write (assumed to be equal)
- T_m = Main memory cycle time, read or write (assumed to be equal)
- P_c = Cache memory access time as a fraction of main memory cycle time
- P_s = Cache system access time as a fraction of main memory access time

If the miss rate of a cache memory is 100%, P_c = 1.00. If the cache memory is infinitely fast corresponding to a cache access time of zero, P_c will be equal to the miss rate, M. For real cache memories, the access time of the cache is finite. This means that the cache system access time will approach the cache access time as the miss rate approaches zero. This is shown in Figure 10.

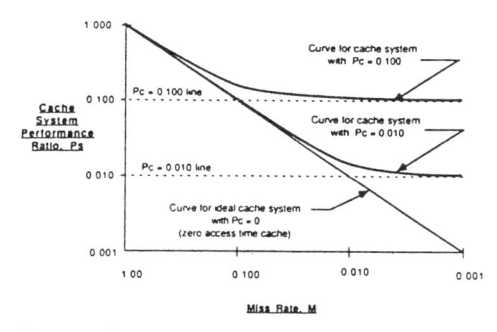

CACHE TAG RAM CHIPS SIMPLIFY CACHE MEMORY DESIGN

Figure 10: Cache Access Time vs Miss Rate for Read Cycles

CACHE SYSTEM PERFORMANCE FOR READ AND WRITE CYCLES

Memory write cycles affect the average access time of the cache system. In a write-through design, unbuffered write cycles are equivalent to cache misses, while buffered write cycles are equivalent to cache hits. Unbuffered write cycles take a main memory cycle to write data for every write. If the main memory write cycle time is the same as the read cycle time, this is equivalent to a cache miss. In buffered write, data is written into the cache and into a register for later off-line write into the memory. Thus, the write cycle in the buffered write case is equivalent to a cache cycle. Each write cycle in the buffered case is, therefore, equivalent to a cache hit. The performance equations for this case are:

$$Ps = R((1 - M)Pc + M) + W(Tw/Tm)$$

For unbuffered writes:

$$Ps = R((1 - M)Pc + M) + W$$

For buffered writes:

$$Ps = R((1 - M)Pc + M) + WPc$$

Where:

- R = Fraction of total memory cycles that are read cycles
- W = Fraction of total memory cycles that are write cycles
- Tw = Write time = Tm for unbuffered, Tc for buffered writes

The effect of unbuffered write cycles is to limit the maximum performance of the cache system. For the average case where write cycles are approximately 15% of the total number of memory cycles, this is approximately equivalent to a cache memory performance of 0.15, as shown in Figure 11.

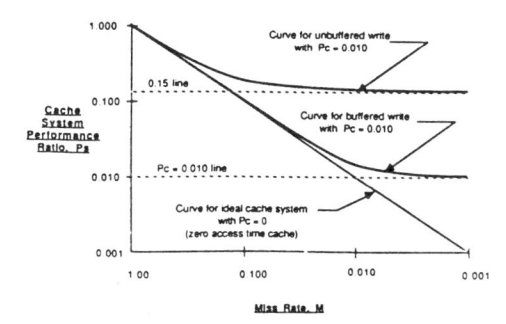

Figure 11: Cache Access Time vs Miss Rate for Buffered and Unbuffered Write Cycles

CACHE SYSTEM PERFORMANCE IN TERMS OF AVERAGE MEMORY ACCESS TIME

Although cache memory systems can be evaluated in terms of the dimensionless performance parameter, Ps, you often need to calculate the actual access time for a specific system. This is expressed by:

$$Ts = R((1 - M) Tcr + MTmr) + WTw$$

Where:

- Ts = Cache average system cycle time, averaged over read and write
- R = Percentage of memory cycles which are read cycles = 85% typical
- W = Percentage of memory cycles which are write cycles = 15% typical
- M = Miss rate of cache = 10+% typical
- Tcr = Cache read cycle time
- Tmr = Main memory read cycle time
- Tw = Write cycle time: main memory for unbuffered write, cache for buffered

For typical values:

$$Ts = 0.85(0.9Tcr + 0.1\%mr) + 0.15Tw$$
$$= \underline{0.765Tcr + 0.085Tmr + 0.15Tw}$$

For unbuffered write and Tcr = 50ns, Tmr = Tw = 250ns:

$$Ts = 0.765(50) + 0.085(250) + 0.15(250) = \underline{97.0ns}$$

For buffered write and Tcr = Tw = 50ns, Tmr = 250ns:

$$Ts = 0.765(50) + 0.085(250) + 0.15(50) = \underline{67.0ns}$$

CACHE SYSTEM PERFORMANCE IN TERMS OF CPU WAIT STATES

In many computer and microprocessor systems, the purpose of the cache memory system is to eliminate CPU wait states, clock periods where the processor is stopped waiting for the memory. The cache performance calculations for this condition are more properly expressed in terms of processor wait states as follows:

$$Ncw = R((1 - M) Ncr + (1 - H)Nmr) + WNw$$
$$= RMNmr + WNw \quad \text{If: Ncr = 0 (no wait states for cache)}$$

Where:

- Ncw = CPU average number of wait states, averaged over read and write
- R = Percentage of memory cycles which are read cycles = 85% typical
- W = Percentage of memory cycles which are write cycles = 15% typical
- M = Miss rate of cache = 10+% typical
- Ncr = Cache read cycle time wait states (typically 0)
- Nmr = Main memory read cycle wait states
- Nw = Write cycle wait states: main memory wait states for unbuffered write, cache wait states for buffered

For unbuffered write and Ncr = 0 wait states, Nmr = 3 wait states:

$$Ncw = 0.085(3) \text{ 1m1 }.15(3) = \underline{0.535 \text{ wait states}}$$

For buffered write and Ncr = Nw = 0 wait states, Nmr = 3 wait states:

$$Ncw = 0.085(3) + .15(0) = \underline{0.255 \text{ wait states}}$$

CACHE SYSTEM PERFORMANCE IN TERMS OF CPU THROUGHPUT

The reason for adding a cache to a CPU is to improve throughput by eliminating wait states. CPU throughput improvement, as a result of adding a cache, can be expressed as the ratio of the speeds before and after adding the cache. For our purposes, CPU throughput improvement can be equated to memory throughput improvement. CPU throughput for this case can be defined as the CPU clock frequency divided by the number of clock states per memory cycle. The speed improvement provided by the cache can therefore be expressed as the ratio of the throughput with the reduced number of wait states provided by the cache to the throughput with full wait states:

$$Fc = \frac{fclk/(No + Ncw)}{fclk/(No + Nm)}$$

$$= (No + Nm)/(No + Ncw)$$

Where:

- fclk = Frequency of processor clock
- N = Number of clock cycles per memory cycle
- Ncw = Number of wait states for cache system (average)
- Nm = Number of wait states for main memory
- No = Number of processor states per memory cycle with no wait states
- Fc = Processor throughput relative to throughput without cache

A 68010 microprocessor requires four clock states per memory cycle, i.e. No = 4. Assuming a 12.5MHz clock and 250ns main memory access time, Nm = 2 wait states. If we use the unbuffered write case from the clock state analysis above, Ncw = 0.535. The throughput improvement provided by the cache is therefore:

Fc = (4 + 2)/(4 + 0.535) = 6/4.535 =
1.32 = 32% throughput increase

This is equivalent to increasing the CPU clock speed from 12.5MHz to 16.5MHz.

CACHE MEMORY PERFORMANCE: HOW MUCH DO YOU NEED?

A simple, direct mapped cache memory system, as described above, is often the most cost effective design. In many cases, the effort to decrease the miss rate beyond that of a simple design may not be worth the increase in system performance.

For example, if Pc is greater than 0.20 corresponding to a cache access time greater than 20% of the main memory access time, it may not be cost effective to improve the hit rate above 90%. This is because there is a knee in the curve of performance improvement versus miss rate at the point where Pc = miss rate, as shown in Figure 10. In some cases, even the added expense of buffered write may not be justified. To examine the relationship between CPU throughput and miss rate, CPU thorughput improvement versus miss rate for various microprocessors is shown in Table 2.

Miss Rate	Throughput Relative to Uncached System			
	68010 Unbuffered	68010 Buffered	68020 Buffered	RISC Buffered
1.00	1.00	1.00	1.00	1.00
0.80	1.06	1.12	1.19	1.27
0.60	1.13	1.20	1.32	1.49
0.40	1.20	1.28	1.49	1.79
0.20	1.29	1.38	1.71	2.24
0.10	1.34	1.44	1.84	2.56
0.05	1.37	1.47	1.92	2.76
0.00	1.40	1.50	2.00	3.00

Table 2.

The data shown is for three CPU/cache systems. The 68010 microprocessor system has a 12.5MHz clock and a cache with unbuffered write. The 68020 system has a 16MHz clock and a buffered write cache. The RISC CPU assumes a 10MHz RISC computer with a 10MHz clock and a buffered write cache, and assumes one clock per memory cycle with wait states equal to an integral number of clock cycles.

Using the data in Table 2, we can make an interesting comparison between chip count and performance gained over an uncached system. Table 3 gives this comparison, showing the chip counts, miss ratios, and performance improvement gain for simple, depth expanded, and two-way set associative expanded caches. The chip counts given are for the cache tag and data RAM chips required, but do not include chip counts for the control logic. One RAM chip is added for the two-way set associative case for the least-recently-used cache flag RAM.

Tag RAM Size	68010 Unbuffered			68020 Buffered			RISC Buffered		
	Chips	Miss	Perf	Chips	Miss	Perf	Chips	Miss	Perf
8K	4	0.32	1.24	7	0.12	1.81	7	0.12	2.49
16K	8	0.22	1.28	14	0.09	1.86	14	0.09	2.60
8K+8K S.A.	9	0.17	1.31	15	0.07	1.89	15	0.07	2.68

Table 3.

Table 3 shows that the throughput improvement created by expanding the cache above a minimum chip count design is small. This table can be interpreted in two ways. In small systems where the goal is to achieve high-performance at minimum chip count, the table indicates that a mimum chip count cache is best since it buys the most performance improvement per chip; doubling the cache chip count purchased less than 10% further increase in performance in all cases. In larger systems where the goal is to achieve maximum performance at moderate chip count, the table indicates that a further increase in performance of 5-8% can be obtained by adding fewer than ten chips.

CACHE DESIGNS: DIFFERENT WAYS TO MAKE ONE

The cache memory described above is a direct mapped cache. It is a simple, commonly used design with respectable performance. Further investigation into the technology of cache memories will reveal a wealth of other approaches to cache design. Much of the variety comes from attempts to maximize the performance of relatively small cache memories typical of earlier technology. Fortunately, there exists some data to help sort out the relative value of the various approaches. This data is in the form of studies on cache memory performance as a function of cache size, organization, word width, etc., such as the excellent work done by Prof. Alan Jay Smith of the University of California

CACHE TAG RAM CHIPS SIMPLIFY CACHE MEMORY DESIGN   APPLICATION NOTE AN-07

at Berkeley (see references). These studies provide background and insight on how to achieve the highest performance out of cache memory systems, as well as documentation of a wide variety of cache schemes which do and do not work. The following comments are intended to provide a simplified guide to, and summary of, some of this data. The following comments are, in large part, judgments and opinions derived from the data in various reports and do not necessarily reflect the opinions of the original authors of the data.

WHAT WE HAVE LEARNED ABOUT CACHE MEMORY DESIGN

A simple, direct mapped cache as discussed above will give good performance if it is large enough. The ultimate measure of cache memory performance is its effect on system cycle time, which is a function of cache cycle time relative to main memory cycle time and the hit rate of the cache. Given a cache cycle time, miss rate becomes the measure of cache performance. Improving cache peformrance, therefore, means improving the hit rate. However, a simple design with a moderate miss rate may be sufficient for many applications, giving most of the performance improvement that could be achieved by a more sophisticated design.

Much of the work that has been done on cache architecture and design was aimed at maximizing the performance of relatively small caches, consistent with the capabilities of earlier technologies. With today's technology, in the form of chips such as the IDT7174, we can easily make large cache memories at low chip counts that are at the upper limit of the earlier technologies. As a result, much of the sophistication required in smaller cache designs, in order to achieve an acceptable hit rate, is not required in today's large cache designs.

CACHE ARCHITECTURE: DIRECT MAPPED vs SET ASSOCIATIVE

A pure cache memory should be an associative memory, where the cache contains all of the most recently used data words. The direct mapped and set associative designs are approximations to this which sometimes exclude recently used words when there is more than one frequently used word per set. Fortunately, the difference between associative, set associative and direct mapped can be quantified. The ratios of miss rates for set associative and fully associative, relative to the direct mapped case, are shown in Table 3A. For example, if the miss rate for a direct mapped design is estimated at 0.20, the miss rate for a two-way set associative design of the same size would be (0.78)(0.20) = 0.156.

What this chart tells us is that two-way set associative caches have a significant performance improvement over simple direct mapped caches, but there is little additional improvement beyond four-way set associative designs. As was noted earlier, the set associative method can often be included in depth expanded cache designs where the two (or more) sets of cache hardware required for the expansion can be arranged to work in a set associative manner.

Cache Type	Ratio of Miss Rate to Direct Mapped
Direct Mapped	1.00
2-Way Set Assoc	0.78
4-Way Set Assoc	0.70
8-Way Set Assoc	0.67
Fully Associative	0.66

Table 3a.

CACHE SIZE

Cache sizes on commercial systems have tended to range from 16K to 64K bytes. Caches smaller than 16K can have significantly higher miss rates, while caches larger than 64K may not significantly improve the miss rate. This is shown above in Table 1. Much work has been done on the relationship between cache size and miss rate; however, most of this work is concerned with small caches, 32K bytes and under. The IDT7164/IDT7174 combination allows 16K byte cache memory design for 16-bit systems and a 32K-byte design for 32-bit systems using a minimum number of chips, and can be easily expanded to 64K and larger if desired.

WRITE THROUGH vs COPY BACK

There are two general approaches to handling the memory write problem: write through and copy back. In the write through approach, memory data is written into main memory as it is received from the CPU. In the copy back mode, memory data is written into the cache and flagged with a "dirty write" bit which indicates that the word has been written into the cache but not into the main memory. The cache data is copied into main memory as a separate operation at some later time, and the dirty write bit is cleared. There appears to be little performance difference between the write through and copy back approaches. Since the write through approach is simpler in concept and easier to implement, it is the most often used method.

WRITE BUFFERING

A significant performance increase can be achieved with a single level of write buffering. Complete write buffering requires more than one level of buffering to cover the case of two write cycles closer together than the main memory write cycle time. A FIFO can be used to buffer more than one word of write data; however, the FIFO need be no deeper than four words, since no further performance results from making it deeper.

SPLITTING THE CACHE: INSTRUCTION/DATA, SUPERVISOR/USER

Splitting the cache into two smaller caches, one for instructions and one for data, seems like it would improve the hit rate; however, it doesn't. In theory, the CPU spends most of its instruction cycles in a small part of the program. By caching these separately from the more random data memory, the hit rate on the instruction portion could be improved. Alas, the studies show that splitting the cache into two pieces typically does no better — and in some cases does a lot worse — than leaving the cache in one piece. This is, perhaps, because the miss rate for data is degraded by more than the hit rate for instructions is improved.

LINE SIZE: MAIN MEMORY WORD WIDTH vs CACHE WORD WIDTH

We have considered cache sizes where the CPU word width, memory word width and cache data word width are the same size. Performance improvement can result if the main memory and cache words are wider than the CPU word. If the cache word width (called the line size) is doubled the miss rate is cut almost in half. This is because the next word the CPU wants from memory is often the word adjacent to the one it just used. Increasing the line size by a factor of two will lower the miss rate by almost a factor of two up to line sizes of 16 bytes and beyond. This is shown in Table 4.

Cache Size in Bytes	Miss Ratio Reduction for Increasing Line Size			
	Line Size (Size of Block From Main Mem to Cache)			
	4 bytes	8 bytes	16 bytes	32 bytes
4K	1.00	0.586	0.364	0.262
8K	1.00	0.581	0.345	0.222
16K	1.00	0.569	0.330	0.203
32K	1.00	0.564	0.324	0.194

Table 4.

There are two approaches to increasing line size in order to reduce miss rate: by increasing the memory data bus width, and by fetching a block rather than a word of data from memory. Increasing the data bus width (from 16 to 32 bits, for example) may be practical in some systems where additional performance is desired.

The other alternative is to transfer a block of bytes to the cache instead of a single word. This becomes significant in systems where there is a delay before data transfer from main memory, but where several words can be transferred quickly after the initial delay. An example of this concept is the page mode in dynamic RAM designs. In such a system, there may be an initial latency of 200ns to begin a memory read cycle but, once started, the memory may be able to transfer words at 100ns per word for blocks of up to 256 words. In this case, a line (block) size of 2-4 words may be used to significantly reduce the miss rate with moderate increase in the main memory cycle time.

SUMMARY

Cache memories have been extensively used in large computer systems to improve performance. Cache tag RAM chips allow this technology to be adapted to the small-to-medium system design at reasonable cost. Simple, direct mapped cache designs with low chip counts can be used to achieve significant performance improvements. High-performance and low miss rates are possible with simple designs due to the high speed and relatively large cache sizes possible with high-speed CMOS technology.

REFERENCES

[Smith82] Alan Jay Smith, "Cache Memories," Computing Surveys, 14, 3, September 1982, pp. 473-530.

[Smith84] Alan Jay Smith, "CPU Cache Memories," April 1986. To appear in *Handbook for Computer Designers*, ed. Flynn and Rossman.

[Smith85] Alan Jay Smith, "Line (Block) Size Selection in CPU Cache Memories," June 1985. Available as UC Berkeley CS Report UCB/CSD 85/239.

[Smith86] Alan Jay Smith, "The Memory Architecture and the Cache and Memory Management Unit for the Fairchild CLIPPER™ Processor," April 1986. Available as UC Berkeley CS Report UCB/CSD 86/289.

[Smith86b] Alan Jay Smith, "Bibliography and Readings on CPU Cache Memories and Related Topics," 1986. from Computer Science Division, EECS Department, University of California, Berkeley, CA 94720.

Error Detection and Correction using SN54/74LS630 or SN54/74LS631

Authors
Dale Hunt, Thomas J. Tyson

Contributors
David Mondeel, Larry Moriarty
Low Power Schottky Applications Engineering

INTRODUCTION

Error Detection Schemes

Simple Parity
The most common error detection scheme is simple parity. In 16-bit machines, the dataword is divided into two 8-bit bytes and the designer arbitrarily uses ODD or EVEN parity and generates a ninth parity bit to be stored along with the 8-bit dataword. Upon retrieval, the parity is checked and the data is accepted if parity sense is correct. If the sense is in error, an attempt is made to rewrite the bad data. Rewriting is the only recourse since this scheme cannot pinpoint the bad bit. An error in a single bit of either byte will result in parity sense inversion. Two errors in the same byte will not be detected since no parity inversion results. Obviously this method is valid only where the probability of a dual-bit error in a single byte is insignificant.

Checksum Generation/Checking
Memory errors can be detected by arithmetically summing a block of N memory words as they are stored, then repeating the summation during retrieval and comparing the "checksums." Obviously, this is not a real-time error detection scheme and requires considerable CPU overhead for the summations. No error correction is possible.

Increasing Need for Error Correction
With the advent of the 64K dynamic RAM, the complexity level of memory boards has reached the point where one can no longer afford to ignore the need for error correction. The memory system manufacturer must provide reasonable MTBF to the end user for large memory boards (64K bytes or larger). This desired level of quality is impossible to achieve without error correction.

In the past, the implementation of error correction with discrete MSI logic was expensive both in terms of board space and package count. The SN54/74LS630 or SN54/74LS631 EDAC chip provides a simple solution to this problem for 16-bit machines.

Semiconductor DRAMs tend to fail at the package level rather than the failure of a memory location within the chip. Obviously, 4K x 1, 16K x 1, or 64K x 1 is the preferred DRAM organization for this error correction implementation.

Real-time single-bit correction in conjunction with a reasonable maintenance schedule can control DRAM MTBF to the point where system reliability will depend almost entirely on peripheral, support ICs and not on the DRAM itself.

In addition to single-bit error correction, the 'LS630/'LS631 EDAC detects all possible dual-bit errors. This reduces the chance that the CPU will use invalid data by several orders of magnitude.

WHY ERROR DETECTION AND CORRECTION USING THE SN54/74LS630 OR SN54/74LS631?

Advantages
1. Improves system reliability
2. Virtual elimination of system downtime due to memory
3. Increases practical size of semiconductor memory systems
4. Decreases PC board complexity

THEORY OF OPERATION

Figure 1 is an overall functional block diagram and function table for the SN54/74LS630 and SN54/74LS631. The operation of the 'LS630 can be broken into nine basic functions — the description and circuit of each is also shown.

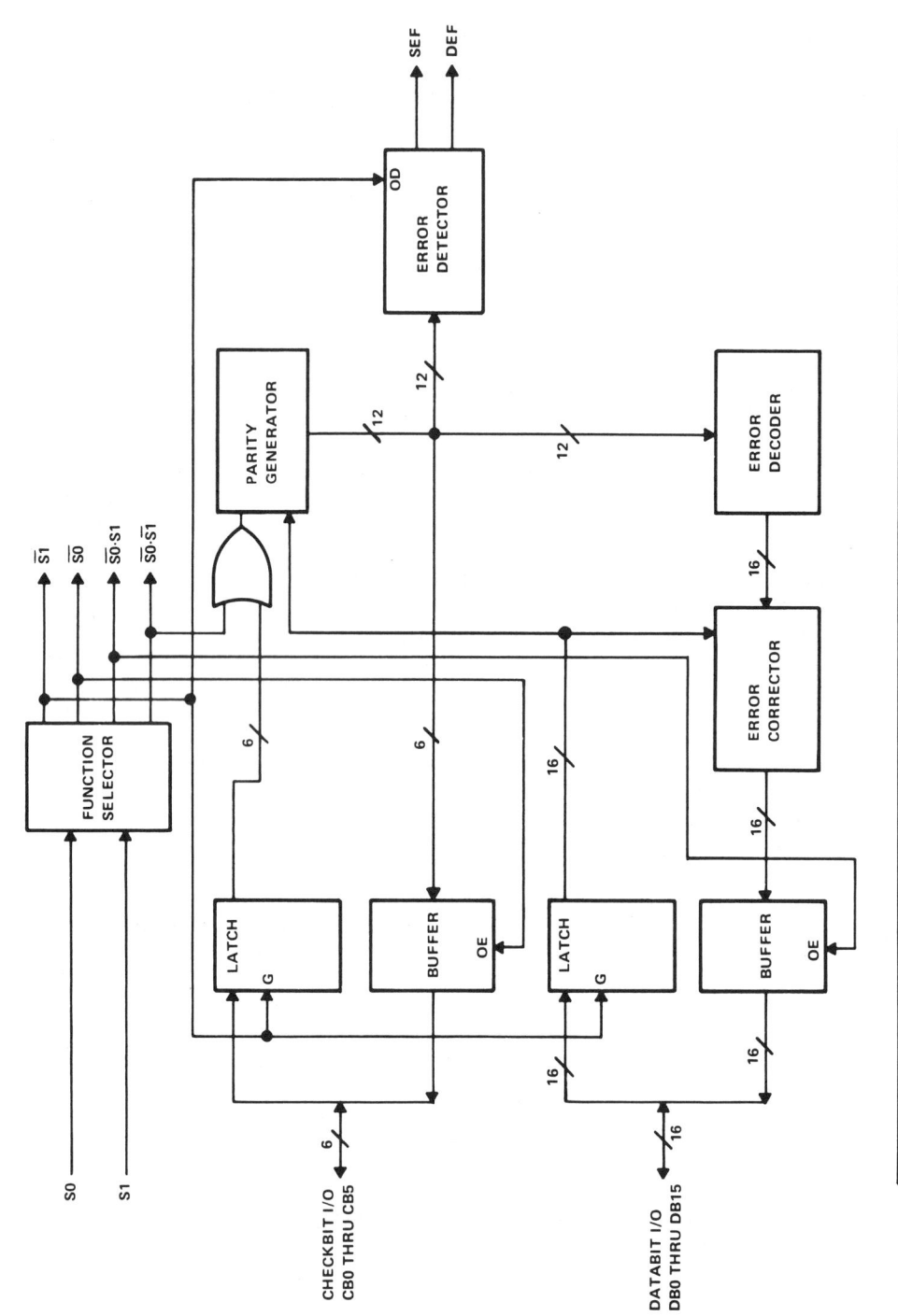

Figure 1. *SN54/74LS630 or SN54/74LS631 Functional Block Diagram and Function Table*

Operational Description

Function Selector

Ideally, S0 and S1 should be automatically generated by WRITE and READ signals from the CPU so that the EDAC/memory will appear transparent to the CPU. (See Figure 2.)

6-Bit Check Latch

During the READ cycle as S1 goes from L to H, the 6-bit checkword is latched for parity checking against the 16-bit dataword from memory. (See Figure 3.)

6-Bit Check Bit 3-State Buffers

During the WRITE cycle, these buffers present the 6-bit parity checkword for storage into memory.

During the correction portion of READ, these buffers transmit the error syndrome code to be used in locating the bad memory chip. Many systems use a syndrome latch and LED display to assist R&M. (See Figure 4.)

Parity Generator

During the WRITE cycle, the check bits from the internal latches are disregarded and the incoming dataword from the CPU is used to encode the 6-bit checkword. During the READ cycle, the internally stored check bits from memory are parity checked against a newly formed checkword generated by the parity tree using the stored data bits from memory. The result at the outputs of the parity generator is the error syndrome word in true and complement form. If no error occurs, the syndrome is all ones. (See Figure 5.)

Error Detector

The 6-input AND gate detects any error (any inverted syndrome bit), while the 6-input parity tree detects dual-bit errors. (See Figure 6.)

Error Decoder

The internal ERROR DECODER uses the same syndrome error code to pinpoint the "correctable error" located within the 16-bit dataword from memory. The proper NAND gate output goes low to provide the inverting signal for the correct cell within the ERROR CORRECTOR. (See Figure 7.)

Error Corrector

The "correctable error" signal (CORR0 through CORR15) is inputted to the proper Exclusive-Nor Gate to invert the erroneous bit. (See Figure 8.)

16-Bit Data Latch

During the READ cycle, the 16-bit dataword from memory is latched as S1 goes from L to H so that the

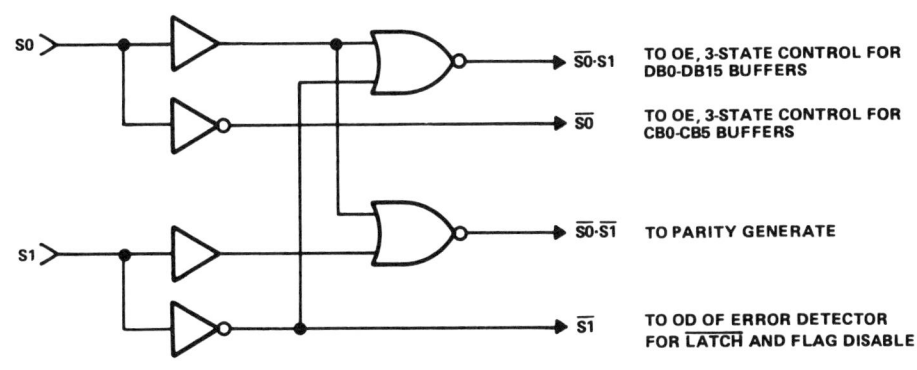

Figure 2. Function Selector

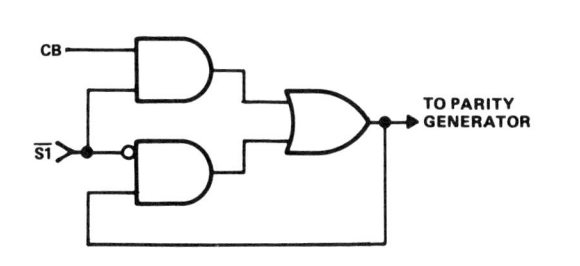

Figure 3. 6-Bit Check Latch

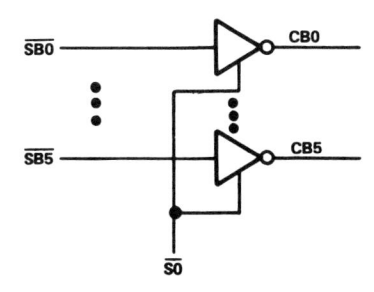

Figure 4. 6-Bit Check Bit 3-State Buffers

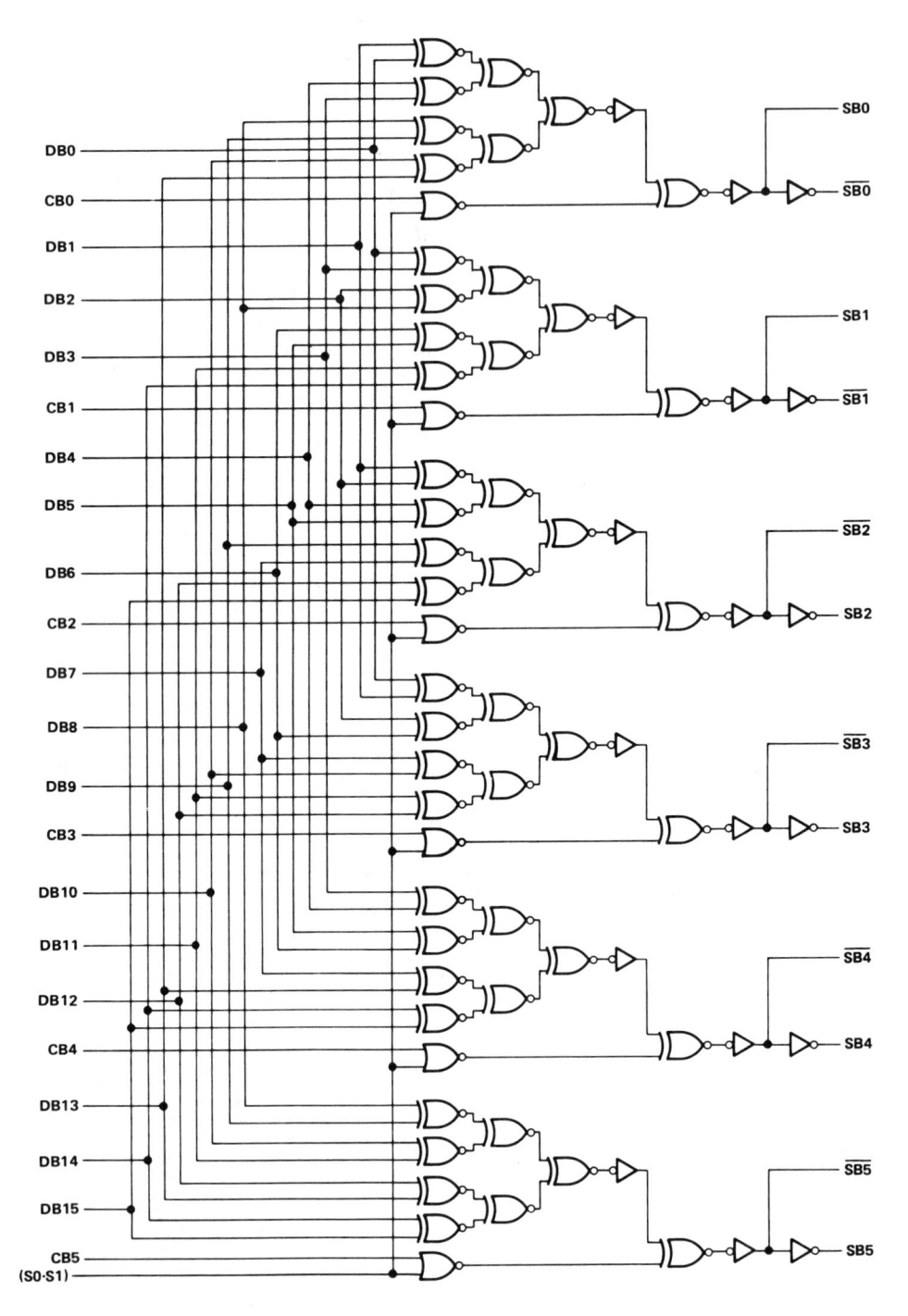

Figure 5. Parity Generator

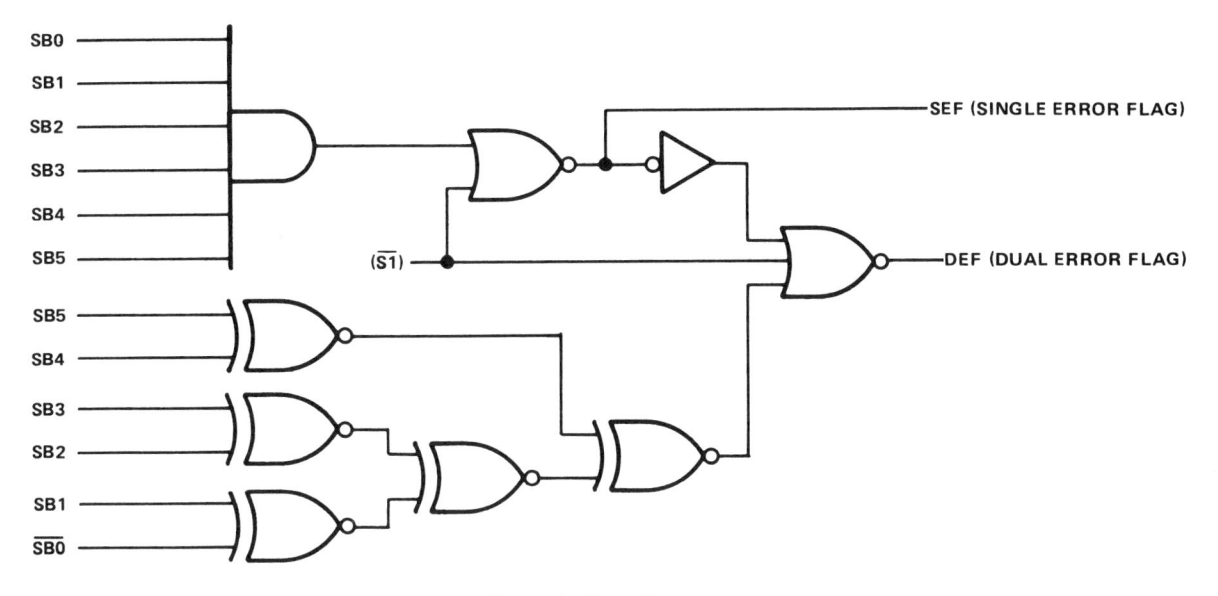

Figure 6. Error Detector

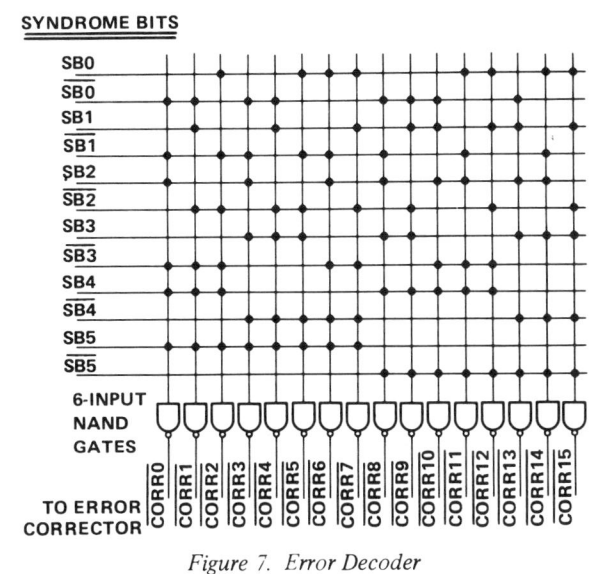

Figure 7. Error Decoder

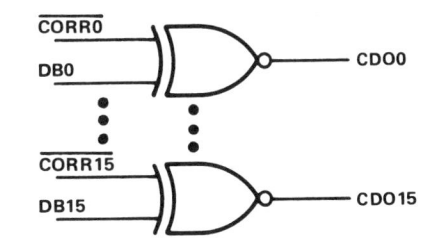

Figure 8. Error Corrector

memory word can be compared to the checkword for error decoding and correction. (See Figure 9.)

16-Bit Data 3-State Buffers

These buffers are active only during the correction cycle to present the corrected word to the CPU. (See Figure 10.)

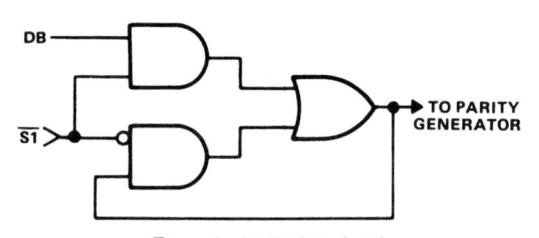

Figure 9. 16-Bit Data Latch

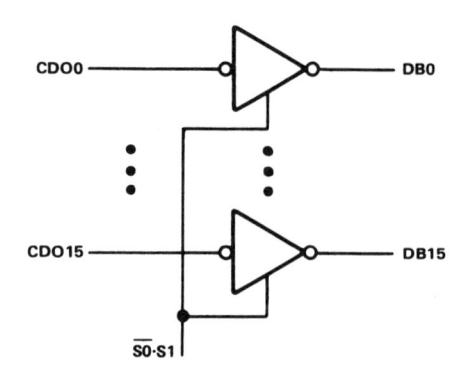

Figure 10. 16-Bit Data 3-State Buffers

Flow Charts of READ/WRITE Cycles

WRITE Cycle

$$S0 = S1 = L$$

① Present data to memory bus and EDAC

② Allow delay time for check bit formation & memory set-up time requirements

③ Write the entire 22-bit word into memory

$$CB0 = \overline{DB0 \oplus DB1 \oplus DB3 \oplus DB4 \oplus DB8 \oplus DB9 \oplus DB10 \oplus DB13 \oplus 1}$$
$$CB1 = \overline{DB0 \oplus DB2 \oplus DB3 \oplus DB5 \oplus DB6 \oplus DB8 \oplus DB11 \oplus DB14 \oplus 1}$$
$$\Big\} \text{Even Parity}$$

$$CB2 = DB1 \oplus DB2 \oplus DB4 \oplus DB5 \oplus DB7 \oplus DB9 \oplus DB12 \oplus DB15 \oplus 1$$
$$CB3 = DB0 \oplus DB1 \oplus DB2 \oplus DB6 \oplus DB7 \oplus DB10 \oplus DB11 \oplus DB12 \oplus 1$$
$$CB4 = DB3 \oplus DB4 \oplus DB5 \oplus DB6 \oplus DB7 \oplus DB13 \oplus DB14 \oplus DB15 \oplus 1$$
$$CB5 = DB8 \oplus DB9 \oplus DB10 \oplus DB11 \oplus DB12 \oplus DB13 \oplus DB14 \oplus DB15 \oplus 1$$
$$\Big\} \text{Odd Parity}$$

(With both S0 and S1 low, the check bits from the latches are excluded from the parity evaluation.)

READ Cycle (No Error)

S1	S0
L	H
H	H

① Activate memory outputs while S0=H, S1=L

② Switch S1 from L to H to latch in the 22-bit word from memory and to enable the error flags

③ Interrogate SEF and DEF to see if both are low (both will indeed be low in this case)

④ Accept the 16-bit dataword as valid

Please note that this parity generator is identical to WRITE previously discussed except that the check bits from the latch storage are used to help generate parity. (Old and new check bits are actually parity checked against each other.)

Since no error exists, all internal syndrome bits are true and no flag will occur.

READ Cycle (Single-Bit Error)

S1	S0
L	H
H	H
H	L

Read Memory
Latch and Flag
Correct and Syndrome

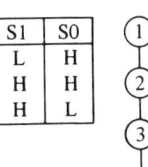

① Activate memory outputs with S0=H, S1=L

② Switch S1 from L to H to latch and enable flags

③ Interrogate SEF and DEF (in this case SEF will be high and DEF low)

④ Place memory outputs into Hiz

⑤ Switch S0 from H to L to output corrected data and syndrome bits

Parity generation is identical to READ Cycle with no error. Syndrome bits are generated during error correction and may be used to pinpoint the bad IC memory chip.

Because the error flags are identical for either a data bit or check bit single error, the EDAC must be sent through the correction cycle and corrected data outputted to the bus. In the case of the single error in one of the six check bits, the corrected data will be identical to the data from memory.

As repeated syndrome words point to hard errors, hardware can be added on the memory board to latch and display the syndrome word for R&M.

READ Cycle (Dual Bit Error)

S1	S0
L	H
H	H

① Activate memory outputs while S0=H, S1=L

② Switch S1 from L to H to latch and flag

③ Interrogate SEF and DEF (in this case both flags will be high)

④ Interrupt CPU

Syndrome bits are useless in the case of dual-bit errors.

READ Cycle (Gross Errors, All Zeros or All Ones)

S1	S0
L	H
H	H

① Activate memory outputs while S0=H, S1=L

② Switch S1 from L to H to latch and flag

③ Interrogate SEF and DEF (in this case both flags will be high)

④ Interrupt CPU

Syndrome bits are useless in the case of gross errors.

Syndrome Bit Generation Code

$$CB0 = \overline{DB0 \oplus DB1 \oplus DB3 \oplus DB4 \oplus DB8 \oplus DB9 \oplus DB10 \oplus DB13 \oplus CB0}$$

$$CB1 = \overline{DB0 \oplus DB2 \oplus DB3 \oplus DB5 \oplus DB6 \oplus DB8 \oplus DB11 \oplus DB14 \oplus CB1}$$

$$CB2 = DB1 \oplus DB2 \oplus DB4 \oplus DB5 \oplus DB7 \oplus DB9 \oplus DB12 \oplus DB15 \oplus CB2$$

$$CB3 = DB0 \oplus DB1 \oplus DB2 \oplus DB6 \oplus DB7 \oplus DB10 \oplus DB11 \oplus DB12 \oplus CB3$$

$$CB4 = DB3 \oplus DB4 \oplus DB5 \oplus DB6 \oplus DB7 \oplus DB13 \oplus DB14 \oplus DB15 \oplus CB4$$

$$CB5 = DB8 \oplus DB9 \oplus DB10 \oplus DB11 \oplus DB12 \oplus DB13 \oplus DB14 \oplus DB15 \oplus CB5$$

Syndrome For Check Bit Error

$$CBERR0 = \overline{CB0} \cdot CB1 \cdot CB2 \cdot CB3 \cdot CB4 \cdot CB5$$

$$CBERR1 = CB0 \cdot \overline{CB1} \cdot CB2 \cdot CB3 \cdot CB4 \cdot CB5$$

$$CBERR2 = CB0 \cdot CB1 \cdot \overline{CB2} \cdot CB3 \cdot CB4 \cdot CB5$$

$$CBERR3 = CB0 \cdot CB1 \cdot CB2 \cdot \overline{CB3} \cdot CB4 \cdot CB5$$

$$CBERR4 = CB0 \cdot CB1 \cdot CB2 \cdot CB3 \cdot \overline{CB4} \cdot CB5$$

$$CBERR5 = CB0 \cdot CB1 \cdot CB2 \cdot CB3 \cdot CB4 \cdot \overline{CB5}$$

IMPLEMENTATION OF THE TI EDAC (SN54/74LS630 OR SN54/74LS631) IN A TI TMS9900 MICROPROCESSOR-BASED SYSTEM.

The operation of the EDAC can be made to seem transparent to the normal operation of the TMS9900. In other words, the EDAC can generate a 6-bit checkword and place this checkword on the memory bus during the normal memory WRITE cycle and can also read, latch, flag errors (single or dual-bit), and correct a single-bit error during a normal memory READ cycle.

Application

In order to use the EDAC in a memory board, some control logic was necessary to control the operation and timing of the EDAC and to modify the operation and timing of the memory. This logic has been called the EDAC controller. Figure 11 shows the block diagram of a memory board employing the EDAC device. In this system, the following assumptions were made.

1. The memory system is 22-bit wide (16-bits for the dataword, 6-bits for the checkword) with separate inputs and outputs.
2. The memory controller supplies the EDAC controller with a signal which is labeled MRDY. This signal will be active only during a memory cycle and when the memory is ready to read or write.
3. The memory output bus is driven by 3-state drivers (in this case the 'LS244 and 'LS367A) whose outputs are controlled by the EDAC control signal MBE and the memory controller.
4. The microprocessor supplies the EDAC controller with the necessary memory control signals (i.e., $\overline{MEMEN}$, $\overline{WE}$, DBIN).

Memory WRITE cycle

The EDAC controller (Figure 12) employs the $\overline{MEMEN}$ and DBIN signals from the microprocessor to set the EDAC control signals (S1 and S0) to a low state. In this mode, the EDAC reads the 16-bit dataword outputted by the microprocessor from the data bus. It then generates a 6-bit checkword and outputs this checkword to the 6-bit checkword bus. To ensure the memory reads a valid dataword and a valid checkword, the write enable signal $\overline{WE}$ is delayed by approximately 60 ns. (This is the worst case time it takes the EDAC to generate a checkword.) The resultant signal $\overline{WED}$ is the memory write enable signal. If the memory system needs an extended write cycle because of a slow access time, the EDAC is still kept in the write mode until the memory WRITE cycle is finished. If the memory system requires an extended memory WRITE cycle, the write enable signal $\overline{WE}$ from the microprocessor need not be delayed.

Memory READ cycle

When the microprocessor enters the memory READ cycle, the EDAC input controls are set in the following states, S1=L, S0=H. In this mode, the EDAC reads the 22-bit word (16-bit dataword and 6-bit checkword) from memory. When the memory controller indicates that valid data is on the bus by setting MRDY high, the EDAC controller puts the EDAC in the latch mode (S1=H, S0=H). To avoid glitches on the EDAC error flags (i.e., DEF and SEF), valid data should be on the data bus approximately 30 ns before the MRDY goes active. In other words, valid data should be set up about 30 ns before the EDAC is put in the latch mode. If there is no error (single or dual), the EDAC will remain in this mode until the end of the memory READ cycle. If there is a single error, the EDAC controller will put the EDAC into the error correct mode, and it will also send a signal MBE to the memory

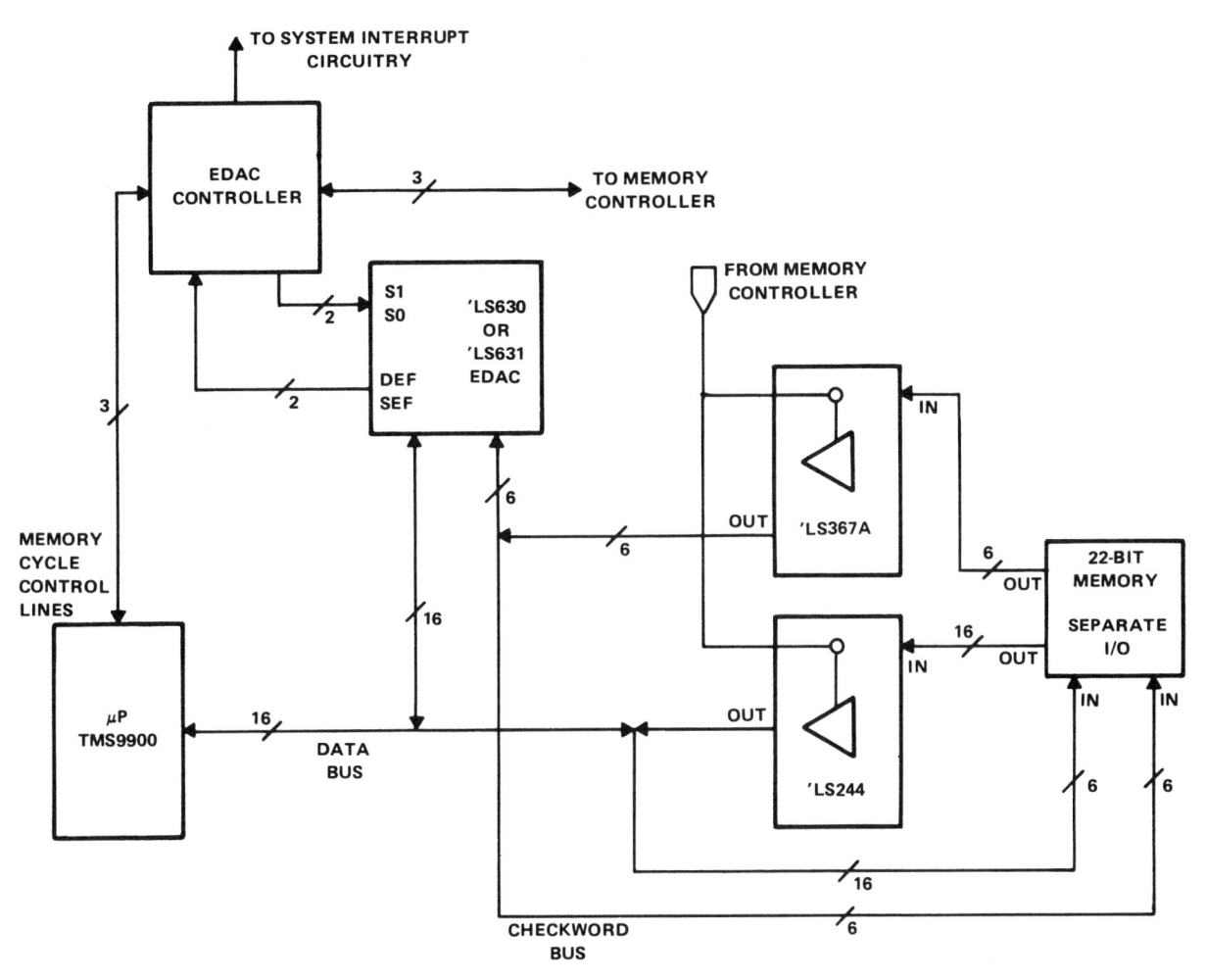

Figure 11. Block Diagram of Memory Board Using TI's SN54/74LS630 or SN54/74LS631 EDAC

controller to disable the memory output bus. In the error correct mode S1=H, S0=L, the EDAC will correct the data bit that was in error and it will also place six syndrome bits on the 6-bit bus which is an error code to indicate the bit that was in error.

If there is a dual error, the EDAC will remain in the latch mode S1=H, S0=H, until the end of the memory read cycle. The EDAC controller will set SYSIN high. This is an interrupt signal and should be an input to the system interrupt circuitry.

Figure 13 gives the timing waveforms for a memory WRITE Cycle with no wait state and for a memory READ Cycle with no errors and no wait states.

Figure 14 gives the timing waveform for a memory READ Cycle with a single-bit error and no wait states.

In order to simplify maintenance operations, a hardware indication of the error syndrome and address bank might be used. Figure 15 shows a simple circuit to perform this function. Once a single-bit error is flagged, the 'LS374 will latch in the syndrome bits and the address bank decode bits. This will inform the R&M technician what chip could be faulty, since soft errors might occur this indicated the device should still be tested. The switch S1 will clear (set low) all the LEDs. A limitation of this circuit is that only one error can be indicated, even though more than one single-bit error might occur between service checks. If there is a high probability the many single-bit errors might occur, a stack register might be employed to store the syndrome and address bank bits.

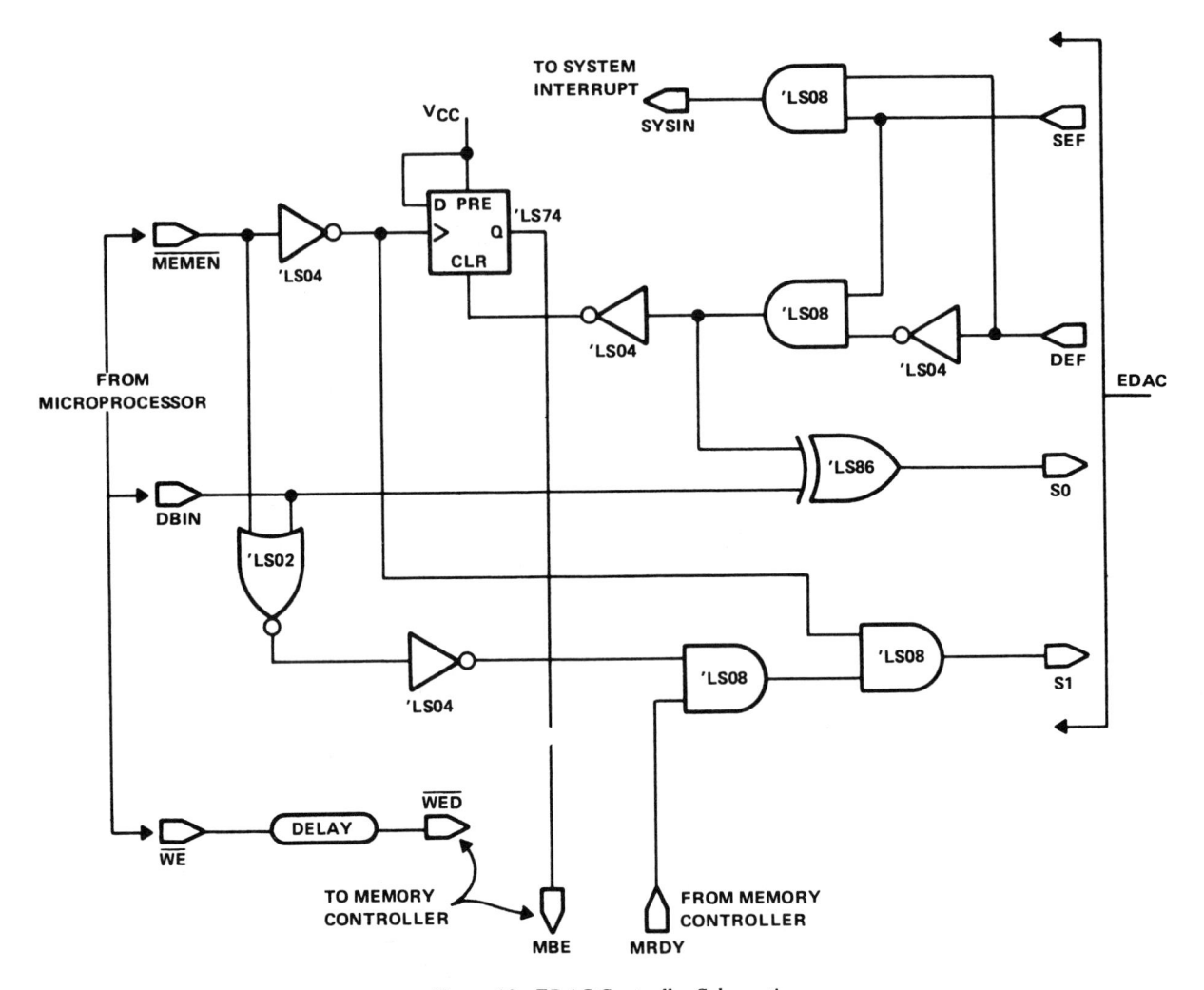

Figure 12. EDAC Controller Schematic

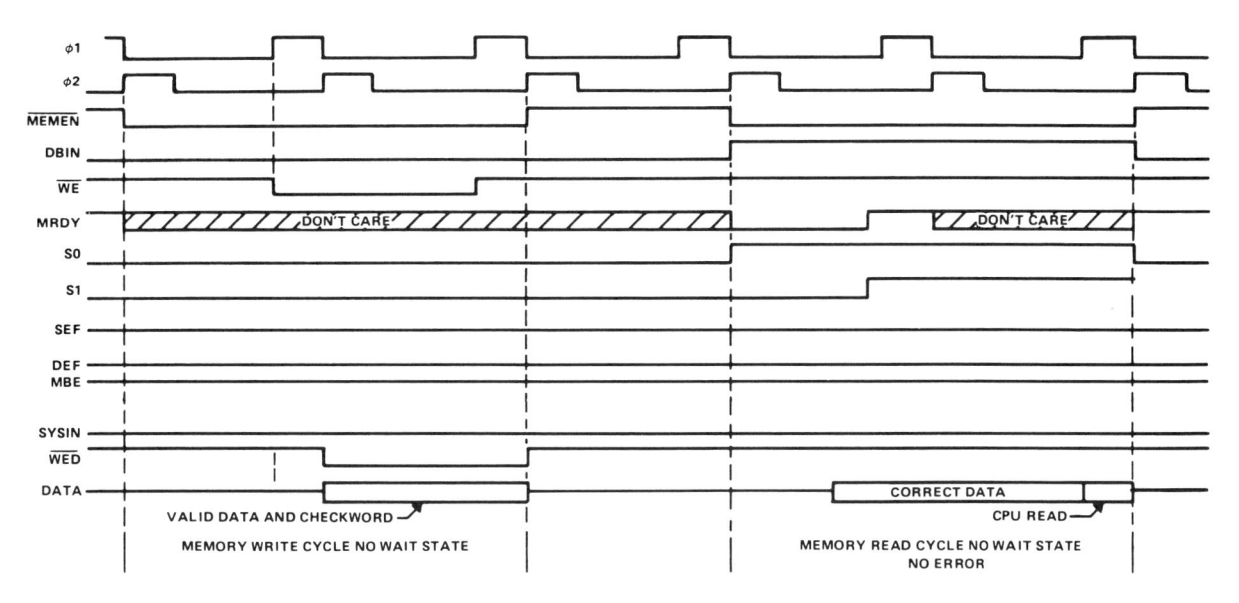

Figure 13. Memory Write and Read Cycle Timing with No Wait State

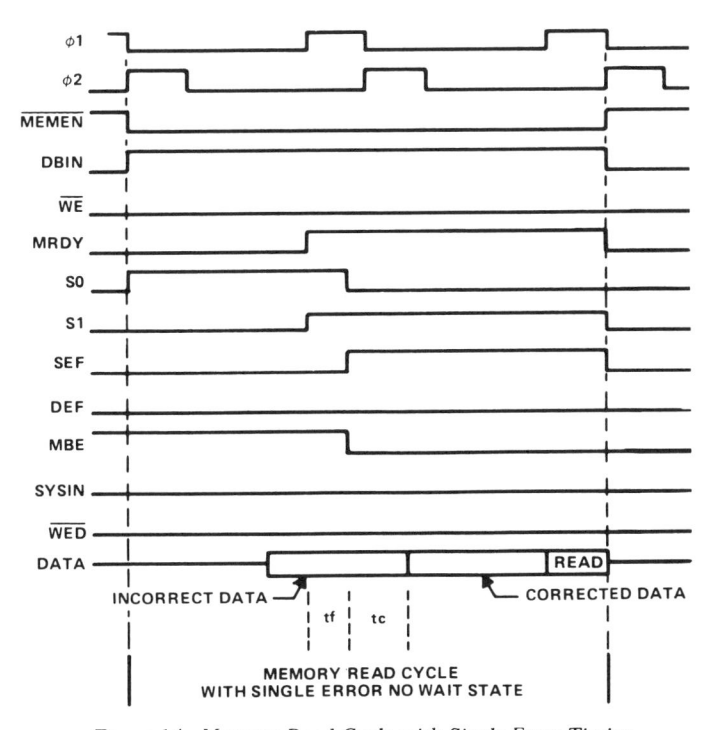

Figure 14. Memory Read Cycle with Single Error Timing

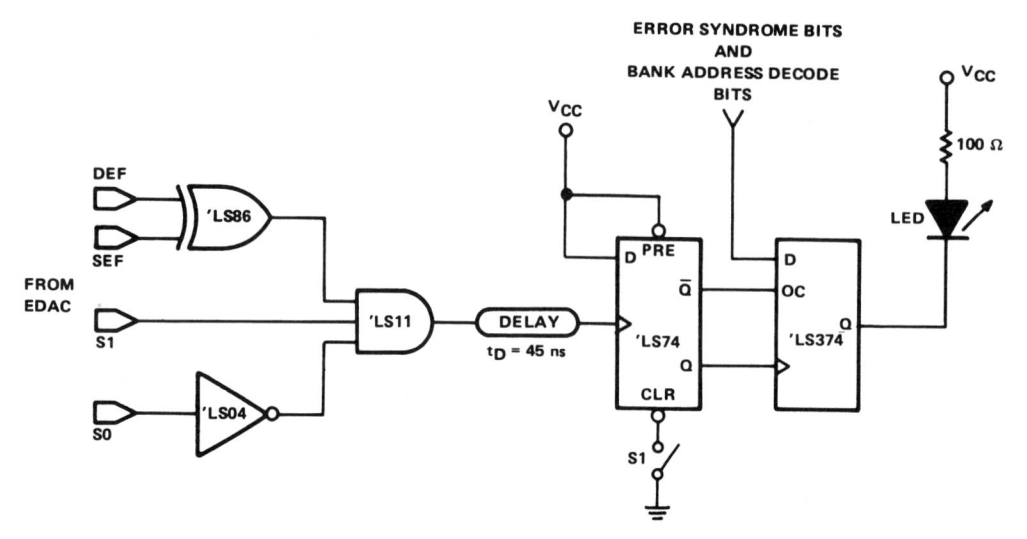

Figure 15. Hardware Single-Bit Error Mapping

TTL LSI

TYPES SN54LS630, SN54LS631, SN74LS630, SN74LS631
16-BIT PARALLEL ERROR DETECTION AND CORRECTION CIRCUITS

BULLETIN NO. DL-S 12747, MARCH 1980

(TIM99630, TIM99631)

- Detects and Corrects Single-Bit Errors
- Detects and Flags Dual-Bit Errors
- Fast Processing Times:
  - Write Cycle: Generates Check Word in 45 ns Typical
  - Read Cycle: Flags Errors in 27 ns Typical
- Power Dissipation 600 mW Typical
- Choice of Output Configurations:
  - 'LS630 . . . 3-State
  - 'LS631 . . . Open-Collector

SN54LS' . . . J PACKAGE
SN74LS' . . . N PACKAGE
(TOP VIEW)

```
         DEF  [ 1    28 ]  Vcc
         DB0  [ 2    27 ]  SEF
         DB1  [ 3    26 ]  S1     } CONTROL
         DB2  [ 4    25 ]  S0
         DB3  [ 5    24 ]  CB0
         DB4  [ 6    23 ]  CB1
DATA     DB5  [ 7    22 ]  CB2    CHECK
BITS     DB6  [ 8    21 ]  CB3    BITS
         DB7  [ 9    20 ]  CB4
         DB8  [10    19 ]  CB5
         DB9  [11    18 ]  DB15
         DB10 [12    17 ]  DB14   DATA
         DB11 [13    16 ]  DB13   BITS
         GND  [14    15 ]  DB12
```

description

The 'LS630 and 'LS631 devices are 16-bit parallel error detection and correction circuits (EDACs) in 28-pin, 600-mil packages. They use a modified Hamming code to generate a 6-bit check word from a 16-bit data word. This check word is stored along with the data word during the memory write cycle. During the memory read cycle, the 22-bit words from memory are processed by the EDACs to determine if errors have occurred in memory.

Single-bit errors in the 16-bit data word are flagged and corrected.

Single-bit errors in the 6-bit check word are flagged, and the CPU sends the EDAC through the correction cycle even though the 16-bit word is not in error. The correction cycle will simply pass along the original 16-bit word in this case and produce error syndrome bits to pinpoint the error-generating location.

Dual-bit errors are flagged but not corrected. These dual errors may occur in any two bits of the 22-bit word from memory (two errors in the 16-bit data word, two errors in the 6-bit check word, or one error in each word).

The gross-error condition of all lows or all highs from memory will be detected. Otherwise, errors in three or more bits of the 22-bit word are beyond the capabilities of these devices to detect.

CONTROL FUNCTION TABLE

Memory Cycle	Control S1	Control S0	EDAC Function	Data I/O	Check Word I/O	Error Flags SEF	Error Flags DEF
WRITE	L	L	Generate Check Word	Input Data	Output Check Word	L	L
READ	L	H	Read Data & Check Word	Input Data	Input Check Word	L	L
READ	H	H	Latch & Flag Errors	Latch Data	Latch Check Word	Enabled	Enabled
READ	H	L	Correct Data Word & Generate Syndrome Bits	Output Corrected Data	Output Syndrome Bits	Enabled	Enabled

Copyright © 1979 by Texas Instruments Incorporated

TEXAS INSTRUMENTS

TYPES SN54LS630, SN54LS631, SN74LS630, SN74LS631
16-BIT PARALLEL ERROR DETECTION AND CORRECTION CIRCUITS

functional block diagram

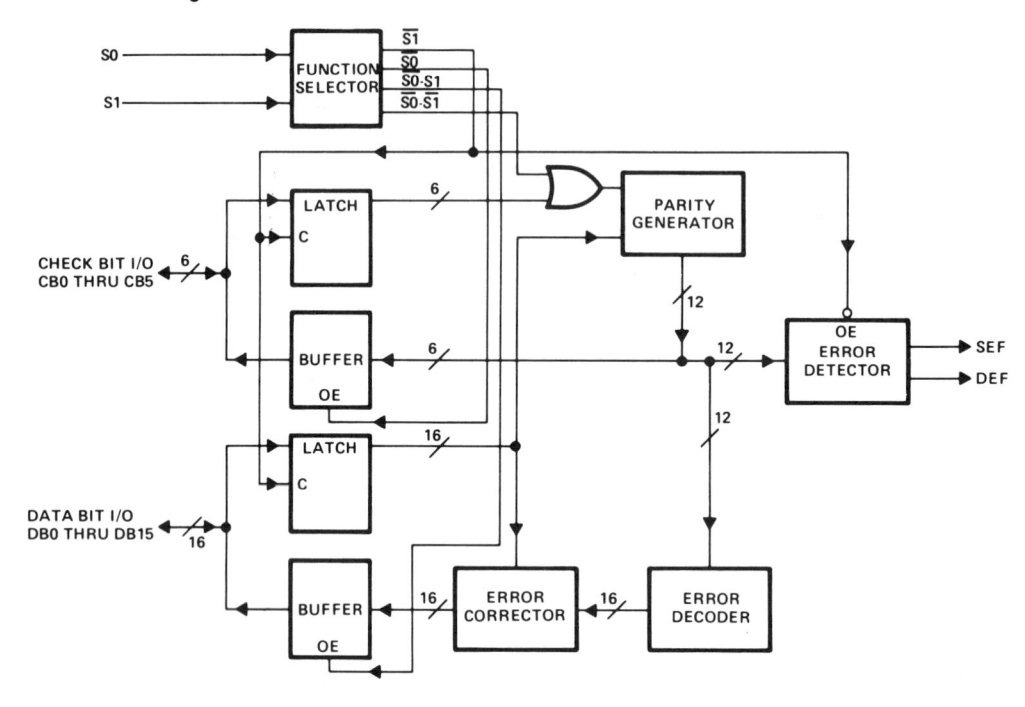

ERROR FUNCTION TABLE

| Total Number of Errors || Error Flags || Data Correction |
16-Bit Data	6-Bit Checkword	SEF	DEF	
0	0	L	L	Not Applicable
1	0	H	L	Correction
0	1	H	L	Correction
1	1	H	H	Interrupt
2	0	H	H	Interrupt
0	2	H	H	Interrupt

In order to be able to determine whether the data from the memory is acceptable to use as presented to the bus, the EDAC must be strobed to enable the error flags and the flags will have to be tested for the zero condition.

The first case in the error function table represents the normal, no-error condition. The CPU sees lows on both flags. The next two cases of single-bit errors require data correction. Although the EDAC can discern the single check bit error and ignore it, the error flags are identical to the single error in the 16-bit data word. The CPU will ask for data correction in both cases. An interrupt condition to the CPU results in each of the last three cases, where dual errors occur.

error detection and correction details

During a memory write cycle, six check bits (CB0-CB5) are generated by eight-input parity generators using the data bits as defined below. During a memory read cycle, the 6-bit check word is retrieved along with the actual data.

TEXAS INSTRUMENTS

TYPES SN54LS630, SN54LS631, SN74LS630, SN74LS631
16-BIT PARALLEL ERROR DETECTION AND CORRECTION CIRCUITS

CHECKWORD BIT	16-BIT DATA WORD															
	0	1	2	3	4	5	6	7	8	9	10	11	12	13	14	15
CB0	x	x		x	x				x	x	x		x			
CB1	x		x	x		x	x		x			x			x	
CB2		x	x		x	x		x		x			x			x
CB3	x	x	x				x	x			x	x	x			
CB4					x	x	x	x	x					x	x	x
CB5									x	x	x	x	x	x	x	x

The six check bits are parity bits derived from the matrix of data bits as indicated by "x" for each bit.

Error detection is accomplished as the 6-bit check word and the 16-bit data word from memory are applied to internal parity generators/checkers. If the parity of all six groupings of data and check bits are correct, it is assumed that no error has occurred and both error flags will be low. (It should be noted that the sense of two of the check bits, bits CB0 and CB1, is inverted to ensure that the gross-error condition of all lows and all highs is detected.)

If the parity of one or more of the check groups is incorrect, an error has occurred and the proper error flag or flags will be set high. Any single error in the 16-bit data word will change the sense of exactly three bits of the 6-bit check word. Any single error in the 6-bit check word changes the sense of only that one bit. In either case, the single error flag will be set high while the dual error flag will remain low.

Any two-bit error will change the sense of an even number of check bits. The two-bit error is not correctable since the parity tree can only identify single-bit errors. Both error flags are set high when any two-bit error is detected.

Three or more simultaneous bit errors can fool the EDAC into believing that no error, a correctable error, or an uncorrectable error has occurred and produce erroneous results in all three cases.

Error correction is accomplished by identifying the bad bit and inverting it. Identification of the erroneous bit is achieved by comparing the 16-bit data word and 6-bit check word from memory with the new check word with one (check word error) or three (data word error) inverted bits.

As the corrected word is made available on the data word I/O port, the check word I/O port presents a 6-bit syndrome error code. This syndrome code can be used to identify the bad memory chip.

ERROR SYNDROME TABLE

ERROR LOCATION	SYNDROME ERROR CODE					
	CB0	CB1	CB2	CB3	CB4	CB5
DB0	L	L	H	L	H	H
DB1	L	H	L	L	H	H
DB2	H	L	L	L	H	H
DB3	L	L	H	H	L	H
DB4	L	H	L	H	L	H
DB5	H	L	H	H	L	H
DB6	H	L	H	L	L	H
DB7	H	H	L	L	L	H
DB8	L	L	H	H	H	L
DB9	L	H	L	H	H	L
DB10	L	H	H	L	H	L
DB11	H	L	H	L	H	L
DB12	H	H	L	L	H	L
DB13	L	H	H	H	L	L
DB14	H	L	H	H	L	L
DB15	H	H	L	H	L	L
CB0	L	H	H	H	H	H
CB1	H	L	H	H	H	H
CB2	H	H	L	H	H	H
CB3	H	H	H	L	H	H
CB4	H	H	H	H	L	H
CB5	H	H	H	H	H	L
NO ERROR	H	H	H	H	H	H

Texas Instruments

TYPES SN54LS630, SN54LS631, SN74LS630, SN74LS631
16-BIT PARALLEL ERROR DETECTION AND CORRECTION CIRCUITS

schematics of inputs and outputs

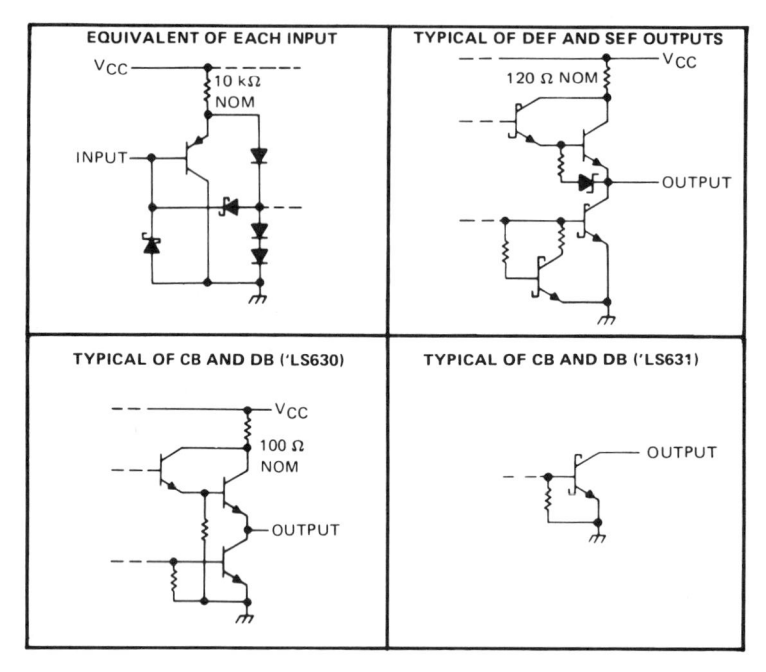

absolute maximum ratings over operating free-air temperature range (unless otherwise noted)

Supply voltage, V_{CC} (see Note 1) .... 7 V
Input voltage: S0 and S1 .... 7 V
               CB and DB .... 5.5 V
Off-state output voltage .... 5.5 V
Operating free-air temperature range: SN54LS630, SN54LS631 .... −55°C to 125°C
                                    SN74LS630, SN74LS631 .... 0°C to 70°C
Storage temperature range .... −65°C to 150°C

NOTE 1: Voltage Values are with respect to network ground terminal.

recommended operating conditions

		SN54LS630 SN54LS631			SN74LS630 SN74LS631			UNIT
		MIN	NOM	MAX	MIN	NOM	MAX	
Supply voltage, V_{CC}		4.5	5	5.5	4.75	5	5.25	V
High-level output current, I_{OH}	CB or DB, 'LS630 only			−1			−1	mA
	DEF or SEF			−0.4			−0.4	
High-level output voltage, V_{OH}	CB or DB, 'LS631 only			5.5			5.5	V
Low-level output current, I_{OL}	CB or DB			12			24	mA
	DEF or SEF			4			8	
Setup time, t_{su}	CB or DB to S1↑	10			10			ns
Hold time, t_h	CB or DB after S1↑	15			15			ns
Operating free-air temperature, T_A		−55		125	0		70	°C

† The upward-pointing arrow indicates a transition from low to high.

Texas Instruments

TYPES SN54LS630, SN54LS631, SN74LS630, SN74LS631
16-BIT PARALLEL ERROR DETECTION AND CORRECTION CIRCUITS

electrical characteristics over recommended operating free-air temperature range (unless otherwise noted)

	PARAMETERS		TEST CONDITIONS†		SN54LS630 MIN TYP‡ MAX	SN74LS630 MIN TYP‡ MAX	UNIT
V_{IH}	High-level input voltage				2	2	V
V_{IL}	Low-level input voltage				0.7	0.8	V
V_{IK}	Input clamp voltage		V_{CC} = MIN,	I_I = −18 mA	−1.5	−1.5	V
V_{OH}	High-level output voltage	CB or DB	V_{CC} = MIN, V_{IH} = 2 V, V_{IL} = V_{IL} min	I_{OH} = MAX	2.4 3.3	2.4 3.2	V
		DEF or SEF		I_{OH} = −400 µA	2.5 3.4	2.7 3.4	
V_{OL}	Low-level output voltage	CB or DB	V_{CC} = MIN, V_{IH} = 2 V, V_{IL} = V_{IL} max	I_{OL} = 12 mA	0.25 0.4	0.25 0.4	V
				I_{OL} = 24 mA		0.35 0.5	
		DEF or SEF		I_{OL} = 4 mA	0.25 0.4	0.25 0.4	
				I_{OL} = 8 mA		0.35 0.5	
I_{OZH}	Off-state output current, high-level voltage applied	CB or DB	V_{CC} = MAX, S0 and S1 at 2 V	V_O = 0.4 V,	20	20	µA
I_{OZL}	Off-state output current, low-level voltage applied	CB or DB	V_{CC} = MAX, S0 and S1 at 2 V	V_O = 0.4 V,	−20	−20	µA
I_I	Input current at maximum input voltage	CB or DB	V_{CC} = MAX, V_{IH} = 4.5 V	V_I = 5.5 V	0.1	0.1	mA
		S0 or S1		V_I = 7 V	0.1	0.1	
I_{IH}	High-level input current		V_{CC} = MAX,	V_I = 2.7 V	20	20	µA
I_{IL}	Low-level input current		V_{CC} = MAX,	V_I = 0.4 V	−0.2	−0.2	mA
I_{OS}	Short-circuit output current¶	CB or DB	V_{CC} = MAX,		−30 −130	−30 −130	mA
		DEF or SEF			−20 −100	−20 −100	
I_{CC}	Supply current		V_{CC} = MAX, S0 and S1 at 4.5 V, All CB and DB pins grounded, DEF and SEF open		143 230	143 230	mA

electrical characteristics over recommended operating free-air temperature range (unless otherwise noted)

	PARAMETER		TEST CONDITIONS†		SN54LS631 MIN TYP‡ MAX	SN74LS631 MIN TYP‡ MAX	UNIT
V_{IH}	High-level input voltage				2	2	V
V_{IL}	Low-level input voltage				0.7	0.8	V
V_{IK}	Input clamp voltage		V_{CC} = MIN,	I_I = −18 mA	−1.5	−1.5	V
V_{OH}	High-level output voltage	DEF or SEF	V_{CC} = MIN, V_{IH} = 2 V,	I_{OH} = −400 µA, V_{IL} = V_{IL} max	2.5 3.4	2.7 3.4	V
I_{OH}	High-level output current	CB or DB	V_{CC} = MIN, V_{IH} = 2 V,	V_{OH} = 5.5 V, V_{IL} = V_{IL} max	100	100	µA
V_{OL}	Low-level output voltage	CB or DB	V_{CC} = MIN, V_{IH} = 2 V, V_{IL} = V_{IL} max	I_{OL} = 12 mA	0.25 0.4	0.25 0.4	V
				I_{OL} = 24 mA		0.35 0.5	
		DEF or SEF		I_{OL} = 4 mA	0.25 0.4	0.25 0.4	
				I_{OL} = 8 mA		0.35 0.5	
I_I	Input current at maximum input voltage	CB or DB	V_{CC} = MAX, V_{IH} = 4.5 V,	V_I = 5.5 V	100	100	µA
		S0 or S1		V_I = 7 V	100	100	
I_{IH}	High-level input current		V_{CC} = MAX	V_I = 2.7 V	20	20	µA
I_{IL}	Low-level input current		V_{CC} = MAX,	V_I = 0.4 V	−0.2	−0.2	mA
I_{OS}	Short-circuit output current¶	DEF or SEF	V_{CC} = MAX,		−20 −100	−20 −100	mA
I_{CC}	Supply current		V_{CC} = MAX, S0 and S1 at 4.5 V, All CB and DB grounded, SEF and DEF open		113 180	113 180	mA

†For conditions shown as MIN or MAX, use the appropriate value specified under recommended operating conditions.
‡All typical values are at V_{CC} = 5 V, T_A = 25°C.
¶Not more than one output should be shorted at a time, and duration of the short circuit should not exceed one second.

TEXAS INSTRUMENTS

TYPES SN54LS630, SN54LS631, SN74LS630, SN74LS631
16-BIT PARALLEL ERROR DETECTION AND CORRECTION CIRCUITS

switching characteristics, V_{CC} = 5 V, T_A = 25°C, C_L = 45 pF

PARAMETER	FROM (INPUT)	TO (OUTPUT)	TEST CONDITIONS	'LS630 MAX	TYP	MAX	UNIT
t_{PLH} Propagation delay time, low-to-high-level output◊	DB	CB	S0 at 0 V, S1 at 0 V, R_L = 667 Ω, See Figure 1		31	45	ns
t_{PHL} Propagation delay time, high-to-low-level output◊	DB	CB			45	65	ns
t_{PLH} Propagation delay time, low-to-high-level output*	S1↑	DEF	S0 at 3 V, R_L = 2 kΩ, See Figure 1		27	40	ns
		SEF			20	30	
t_{PZH} Output enable time to high level#	S0↓	CB, DB	S1 at 3 V, R_L = 667 Ω, See Figure 2		24	40	ns
t_{PZL} Output enable time to low level#	S0↓	CB, DB	S1 at 3 V, R_L = 667 Ω, See Figure 1		30	45	ns
t_{PHZ} Output disable time from high level▲	S0↑	CB, DB	S1 at 3 V, R_L = 667 Ω, See Figure 2		43	65	ns
t_{PLZ} Output disable time from low level▲	S0↑	CB, DB	S1 at 3 V, R_L = 667 Ω, See Figure 1		31	45	ns

switching characteristics, V_{CC} = 5 V, T_A = 25°C, C_L = 45 pF, see Figure 1

PARAMETER	FROM (INPUT)	TO (OUTPUT)	TEST CONDITIONS	'LS631 MIN	TYP	MAX	UNIT
t_{PLH} Propagation delay time, low-to-high level output◊	DB	CB	S0 at 0 V, S1 at 0V, R_L = 667 Ω		38	55	ns
t_{PHL} Propagation delay time, high-to-low-level output◊	DB	CB			25	40	ns
t_{PLH} Propagation delay time, low-to-high-level output*	S1↑	DEF	S0 at 3 V, R_L = 2 kΩ		27	40	ns
		SEF			20	30	ns
t_{PHL} Propagation delay time, high-to-low-level output#	S0↓	CB, DB	S1 at 3 V, R_L = 667 kΩ		28	45	ns
t_{PLH} Propagation delay time, low-to-high-level output▲	S0↑	CB, DB	S1 at 3 V, R_L = 667 kΩ		33	50	ns

◊These parameters describe the time intervals taken to generate the check word during the memory write cycle.
*These parameters describe the time intervals taken to flag errors during the memory read cycle.
#These parameters describe the time intervals taken to correct and output the data word and to generate and output the syndrome error code during the memory read cycle.
▲These parameters describe the time intervals taken to disable the CB and DB buses in preparation for a new data word during the memory read cycle.

PARAMETER MEASUREMENT INFORMATION

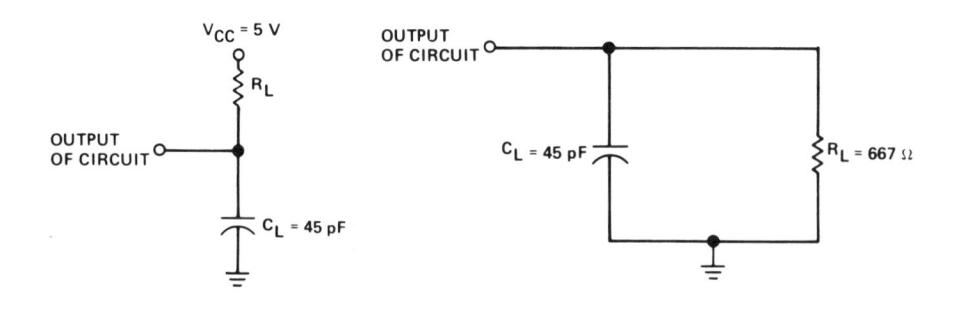

FIGURE 1—OUTPUT LOAD CIRCUIT

FIGURE 2—OUTPUT LOAD CIRCUIT

Texas Instruments

TI cannot assume any responsibility for any circuits shown or represent that they are free from patent infringement.

TEXAS INSTRUMENTS RESERVES THE RIGHT TO MAKE CHANGES AT ANY TIME IN ORDER TO IMPROVE DESIGN AND TO SUPPLY THE BEST PRODUCT POSSIBLE.

THE IDT FourPort™ RAM FACILITATES MULTIPROCESSOR DESIGNS

APPLICATION NOTE AN–43

By Robert Stodieck

THE IDT FourPort RAM

Serving as both a complex four bus interconnect network and fast "parallel" memory, the IDT FourPort RAM can greatly facilitate the creation of multiprocessor and multi–ALU systems to accelerate DSP, graphics, control and other tasks that involve large vector processing tasks.

Memory architectures based on single–port RAM allow only one device to access a memory array at one time. Hardware designed to accelerate computing processes by utilizing parallelism, or pipelining with single–port memory tend to require architectures that are either complex, specialized, or both. The advent of a fast FourPort single chip RAM greatly simplifies the task of creating generalized small multiprocessor or multi–ALU systems to accelerate a variety of vector algorithms.

Potential applications include dedicated real–time multiprocessor systems for control, graphics, and DSP systems, as well as general purpose vector co–processors to assist general purpose computers. Vector processing means any computing operation with a large number of operations that may be executed in parallel by multiple processors. In these applications the FourPort RAM serves both as a fast static RAM and as the interconnect network between processors working on a common data set.

Imagine a static RAM that allows four processors to randomly and asynchronously read or write four locations at a time in the same RAM array. For processes that can be executed in parallel, four processors can be programmed to operate simultaneously on different parts of a data set stored in the FourPort RAM. If data is being generated at different rates than it is being used, software controlled buffers can be created at will, temporarily storing data passing from one processor to the next. The buffering minimizes the time lost in handshaking between processors. Four way fully random accessibility avoids hardware imposed algorithmic constraints.

The IDT FourPort RAM has precisely these characteristics. There are only two constraints on the access patterns allowed in the FourPort. Two devices cannot write to the same address location in the RAM at the same time, since simultaneous multiple writes to any one multiport memory location may corrupt the data in that RAM location. Also, a device cannot read an address location that is being written, to avoid having the read occur when the output data is changing. There are no other restrictions on access patterns.

As it turns out address collisions are usually prohibited by the logical sequencing requirements of software, and the time lost in avoiding address collisions is often minimal. Most of the time all processors have essentially free read and write access to the memory.

FourPort RAM BASED MULTIPROCESSOR ARRAYS FOR VECTOR OPERATIONS

The FourPort RAM is both a storage and communications media. As a communications media it has little, and in some cases zero handshaking or arbitration overhead. A processing device may be able to store results in a multiport memory and spend little or no time signaling the next device to receive the results. As a communications media it also has very high bandwidth. These characteristics make the IDT7050 and the IDT7052 a ideal memory for connecting multi–element and multiprocessor computer architectures (see Figure 2).

A multiprocessor system can be created using almost any existing microprocessor system. Since the hardware interface of the FourPort RAM to the processors is that of a simple static RAM, it can be connected transparently to almost any existing system. Control signals as well as data can be handled via the RAM. Thus, microprocessor boards that were designed for entirely different applications can be used in a multiprocessor array.

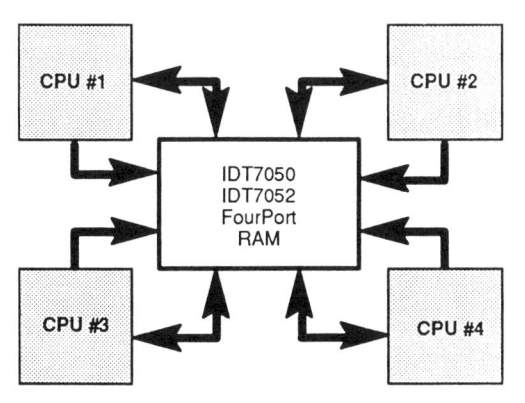

Figure 1. The IDT7052 FourPort RAM allows four simultaneous memory accesses to independent addresses a 2K or 1K x 8–bit memory array. It serves both as a interconnect network and as fast static RAM

FourPort is a trademark of Integrated Device Technology, Inc.

© 1989 Integrated Device Technology, Inc.

THE IDT FourPort RAM FACILITATES MULTIPROCESSOR DESIGNS

APPLICATION NOTE AN-43

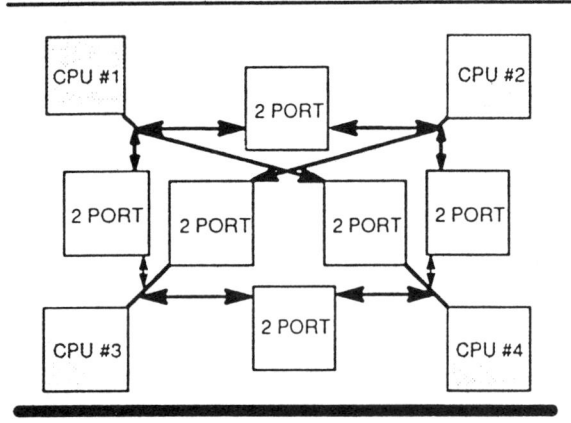

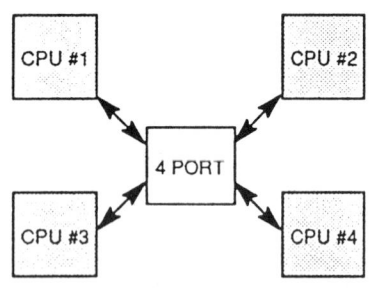

Figure 2. FourPort RAM interconnection advantages over Dual-Port RAM. The processors in both figures are inter connected with a latency of one memory access. This efficiency requires 6 separate Dual-Port RAMs but only 1 FourPort RAM

AN OVERVIEW OF THE OPERATION OF A MULTIPORT RAM BASED MULTI-PROCESSOR WITH A MULTIPORT RAM BASED CONTROL SYSTEM

Processors sharing multiport memory must avoid writing into memory locations that are simultaneously being read or written from another port. This is usually accomplished by address range segregation. That is, at any one moment processor "A" is prevented from writing to multiport memory locations that processor "B" is accessing from another port. Hardware interrupts, hardware semaphores and stalling processors with hardware busy logic are hardware based methods of controlling the accesses of processors to multiport RAM. It is also possible to control the processors in a multiprocessor array via the common RAM interface. This results in an essentially software-only control system. The control algorithms for a multiport RAM based multiprocessor array are different than those for a multiprocessor array based on single port RAM. This section describes an example of a control protocol for a multiport RAM based multiprocessor array.

Access coordination in multiport RAM based multiprocessors, overlaps with the more familiar task of process coordination in a multiprocessor and uses the same control schemes. In a single master system, the master determines the address ranges being used by all processors. This avoids the problems of arbitrating for resources. In small embedded systems, running algorithms of limited complexity, the software can be tuned so that software-only control approaches have little or no detrimental effect on overall performance. Such systems are more easily debugged if a simple single master control arrangement is used. In this section of this application note we will discuss a single master example.

In a master/slave array, the master controls all the actions of the slaves. The slaves must either have local program store in RAM or ROM or be operating out of the FourPort RAM. Each processor must have a unique ID code to be able to identify the unique command location where it is to receive its commands from the master. This can be achieved, for example, by supplying a unique firmware ID code via individual PROMs, PALs or readable DIP switches for each processor. A number of other approaches are possible.

Each slave command has a corresponding op-code. The slaves poll their command locations looking for new command opcodes. For example finding a "0" in a command location may imply no operation is requested from the slave etc. The commands can be anything that the slave processors have been programmed to do. Appropriate commands might be, multiply data values at locations 000H to 7FFH with the corresponding coefficients at locations 800H to FFFH, or multiply data values at locations 000H to 7FFH with the value at location 800H, etc. Thus, with a few memory accesses, the master processor can trigger and control lengthy slave processor operations.

MASTER/SLAVE CONTROL PROTOCOL FOR A MULTIPORT RAM BASED PROCESSOR ARRAY

A command protocol is the set of rules for passing commands from the master to the slaves. In a software-only control system, all processors must be aware that writes to certain command locations are forbidden, or forbidden without "permission" from the current owner (see Figure 4). In general, a process is given a variable address range to operate in. The command protocol, on the other hand, uses fixed address locations.

The master of an array of processors can tell slave processor #1 to execute a command "n", by writing the command opcode corresponding to command "n" to the slave processor's command location. Parameters for the process, such as constants, or the assigned address range, are placed in reserved locations prior to starting the process that will use them.

There are four problems that a multiport RAM based command protocol must solve:

1. Write-write conflicts must be avoided in the control locations.
2. Read-write synchronization problems must be avoided in the control locations.
3. The master must not issue a new command out of sequence, i.e. the slave has to acknowledge readiness to execute a new command.
4. A slave must execute each command only one time.

Figure 3 shows flow charts for a protocol that allows a master to control slaves and slaves to receive commands without risk of violating these four rules.

All slaves have unique command locations in RAM. If the reads and writes to the command locations are asynchronous, command locations must always be read at least twice. The two read results are then compared and discarded if they do not match. In this way, command data that may have been changing during the read operation, and therefore may have been read incorrectly, is discarded.

Before issuing any command, the master first reads a slave's "command" location. If the value read indicates that the slave is ready, the master places the slave's command op-code in that same command location. The slave must signal readiness for new commands by placing a "no-op/ready" value in the command location. The "no-op/ready" flag value is interpreted as a "ready-for new-command" flag by the master, and a "no-operation" command by the slave.

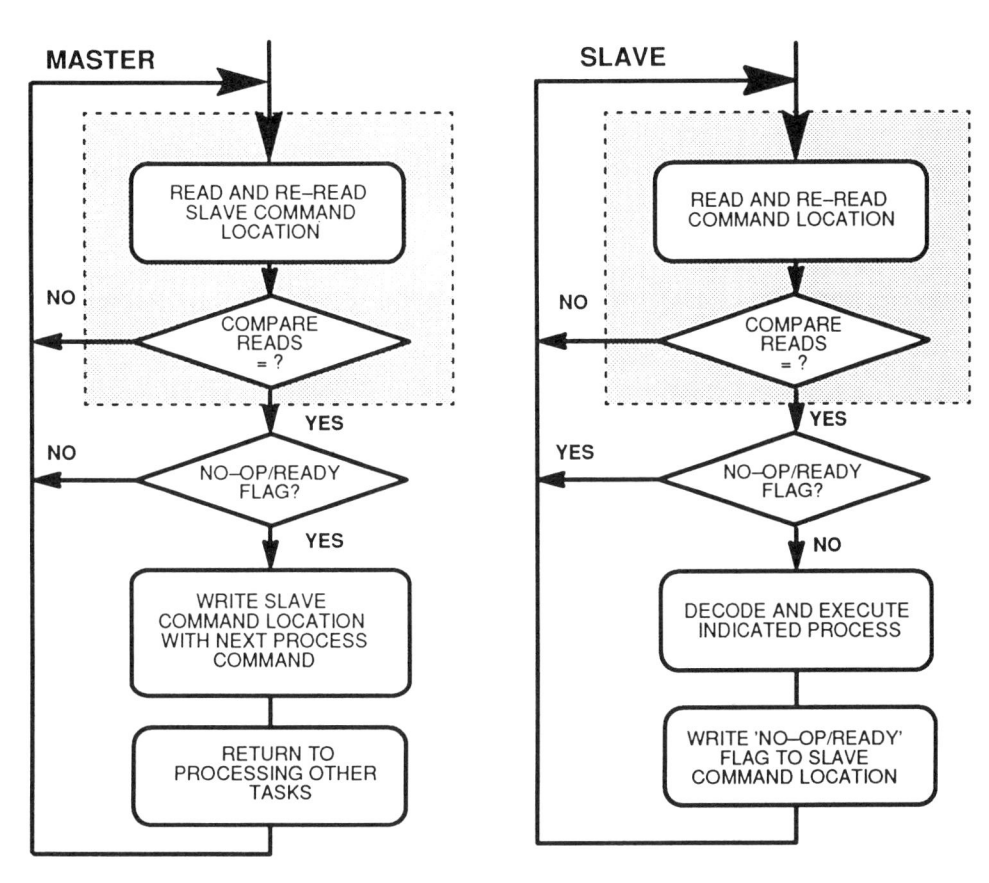

Figure 3. Flow charts for a master–slave software–only command protocol for a multiport RAM based multiprocessor. In unsynchronized systems (see shaded boxes) all commands must be read at least twice with the same result before the command keyword can be assumed to be valid

THE IDT FourPort RAM FACILITATES MULTIPROCESSOR DESIGNS
APPLICATION NOTE AN-43

```
┌─────────────────────────────┐  ┐
│ SLAVE COMMAND LOCATION #1   │  │
├─────────────────────────────┤  │
│ SLAVE COMMAND LOCATION #2   │  │
├─────────────────────────────┤  │
│ SLAVE COMMAND LOCATION #3   │  │
├─────────────────────────────┤  │  MASTER AND
│                             │  │  LOCAL SLAVE
│ SLAVE#1 PARAMETER LOCATIONS │  ├  CAN WRITE,
│                             │  │  PERMISSION
├─────────────────────────────┤  │  REQUIRED
│ SLAVE#2 PARAMETER LOCATIONS │  │
├─────────────────────────────┤  │
│ SLAVE #3 PARAMETER LOCATIONS│  │
├─────────────────────────────┤  ┘
│ SLAVE#1 STATUS LOCATIONS    │  ┐
├─────────────────────────────┤  │  ONLY WRITTEN
│ SLAVE#2 STATUS LOCATIONS    │  ├  BY LOCAL SLAVE
├─────────────────────────────┤  │
│ SLAVE #3 STATUS LOCATIONS   │  │
├─────────────────────────────┤  ┘
│                             │  ┐
│                             │  │  ACCESS
│                             │  ├  RIGHTS
│                             │  │  ALLOCATED
│                             │  │  BY MASTER
└─────────────────────────────┘  ┘
```

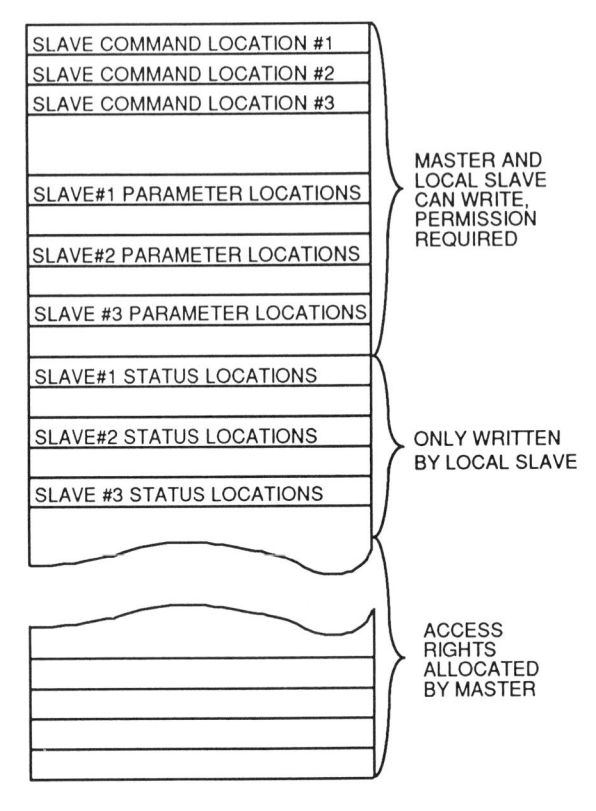

Figure 4. Write Access Allocations

Having to wait for a "ready" signal from the slave prevents the master from issuing new commands out of sequence. Conversely, by signaling "ready" in this way, the slave is also clearing the old commands from the command location. This prevents the slave from later accidently re-reading and re-executing an old command. The command locations are written alternately by the master and the designated slave, but the protocol prevents simultaneous writes that might destroy the data in the RAM location. By using the same location for both the master's command and the slave's ready indication, synchronization problems caused by differences in the memory cycle rates of different processors can also be avoided.

All slaves should also have unique slave status locations in RAM where the master looks for slave status information. The status locations are writable by the slave only. Copious use of reserved slave status locations is essential for the benefit of the programmer trying to debug untested software.

The slave may also signal "done" by writing a "done" flag to a slave status location. The meaning of "done" is that the results of the last operation are ready for use. Keep in mind that "done" is a different signal than "ready". "Ready" implies that the master can post the next command and return to executing other tasks. Depending on the overall algorithm the master is controlling, the master may write a new command as soon as "ready" is signaled, or it may need to wait until "done" is signaled also. It cannot merely check for a "done" signal before issuing a new command. To do so would make it possible to issue new commands out of sequence, based on stale "done" signals.

INITIALIZATION

Since multiport RAM is the control interface, the RAM command locations must be initialized prior to starting the execution of the slaves. One way this can be handled is by delaying the reset pulses to the slaves while the master initializes RAM. Alternatively, after reset, the master can issue a known sequence of commands that frees the slaves from a special start up routine.

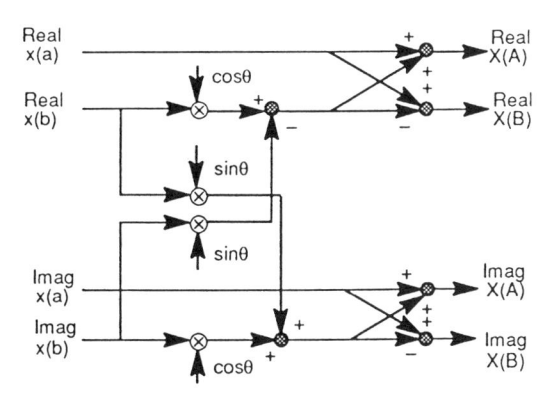

THE IDT FourPort RAM
FACILITATES MULTIPROCESSOR DESIGNS

APPLICATION NOTE AN-43

Real X(A) = Real x(a) + (Real x(b) • cos θ – Imag x(b) • sin θ)
Real X(B) = Real x(a) – (Real x(b) • cos θ – Imag x(b) • sin θ)
Imag X(A) = Imag x(a) + (Imag x(b) • cos θ + Real x(b) • sin θ)
Imag X(B) = Imag x(a) – (Imag x(b) • cos θ + Real x(b) • sin θ)

Figure 5. Flow Diagram for Calculating One FFT Butterfly. Each 'X' pattern shown in Figure 6 represents one such "Butterfly". Each end point in Figure 6 represents a complex pair of numbers input or output to or from a "Butterfly". The sine and cosine factors are sometimes called "Twiddle Factors". The angles used for calculating the Twiddle Factors for each Butterfly are shown in Figure 6

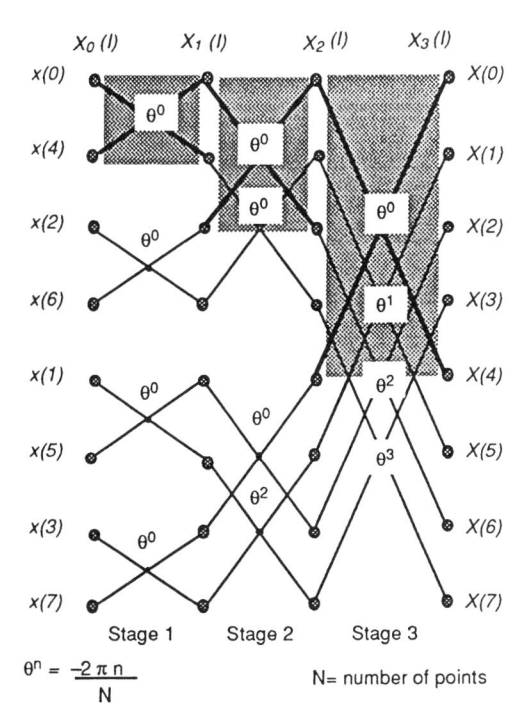

$$\theta^n = \frac{-2\pi n}{N}$$

N = number of points

Figure 6. Overview of the "Butterfly" calculations for an 8 Point FFT. To complete this 8 Point FFT requires 3 stages of Butterflies (2^3 = 8). A 1K FFT has 10 stages or levels (2^{10} = 1K). The indexes of x(n), the input sequence, are shown out of sequence for graphical clarity. Each stage in this figure has 4 Butterflies

OUTLINE OF A DIGITAL SIGNAL PROCESSING EXAMPLE

Basic DSP algorithms such as the FFT can utilize high degrees of parallelism and provide good examples of vector algorithms. The access patterns of the processor doing such an algorithm are complex and the data sets are usually small enough to fit comfortably in a multiport RAM array. Analyzing how the FFT will be processed provides a good example of the advantages of a multiport RAM based multiprocessing environment.

The objective of our example task is to translate a time series of data values into their frequency domain representation: i.e. execute a fast Fourier transform, as quickly as possible. This is a common process step in a number of systems for interpreting data from things as diverse as military radar to medical CAT scans. It is also a relatively well known algorithm among many contemporary electrical engineers, and so makes a good example for our system.

Our objective algorithm could be run on a single processor. The object of the FourPort RAM based multiprocessor arrangement is to multiply the speed of our computational process without resorting to a specialized and more expensive architecture.

The generality of this architecture implies that it can be applied to a variety of computationally involved tasks. The generality of this architecture also means that there are often a number of ways a programmer can attack a specific problem. The intent of this example is merely to illustrate one approach, not to fully optimize an algorithm.

LOAD BALANCING

The FFT calculations can be flow graphed. When they are, they appear as a repetitive array of calculations (Figure 6) of a particular set of four equations. This set of four equations is called a 'butterfly' for the appearance of its flow graph (Figure 5). The inputs and outputs are series of complex numbers.

A common bench mark of processor performance is a 1K FFT. A quick glance at the equations to be calculated shows why multiple processors are desirable for such a task. If we assume that 1024 real and 1024 imaginary data values have been loaded in the four port memory, there are now 2048 multiplications to be done as a first step. All these multiplications could be done simultaneously. Next there are 1024 additions followed by another 2048 additions to complete the first stage of FFT butterflies. Again, all of operations at any one of these three steps could be done simultaneously. For a 1K FFT there are 10 stages of butterflies.

Processing on one stage of the FFT must be completed before processing on the next stage can begin. Each processor is given an address range of FFT butterfly input data to process for each stage of FFT butterflies. A sine table is required for calculation of the FFT "twiddle" factors. This can be stored in the four port memory and, therefore, will always be available to all processors. Calculation of the "twiddle" factors is a matter of calculating the addresses used in the sine look up table. (See Figure 6 for the angle calculations).

For efficiency, the computational load between processors must be balanced. Since there are hundreds or thousands of operations that may be done in parallel at each stage of the FFT, task partitioning is a matter of assigning each processor an appropriate number of "butterflies" to work on to achieve an equity of loading.

Since the minimal FFT tasks are easily divided between the processors, and the FourPort RAM all but prevents inter–processor data transfer conflicts, the four processors in this example can be kept busy most of time.

Since there are so many tasks that can be done in parallel, other types of tasks can be included without seriously upsetting the balance. For example, if one processor is being used to handle I/O and input conditioning tasks, then it can be assigned to do fewer

THE IDT FourPort RAM
FACILITATES MULTIPROCESSOR DESIGNS
APPLICATION NOTE AN-43

butterfly calculations than the other processors. If the work load of all the processors can be balanced, the net speed advantage of this four processor array, can then in fact be close to 4 times that of a single processor.

A TMS320C2x HARDWARE INTERFACE EXAMPLE

TI's TMS320 single chip DSP processors are particularly well suited for embedded numerical processing. An example interface is shown in Figure 7. The internal RAM and ROM of the TMS320 can be used for temporary data storage and program memory, making the FourPort RAM the only external RAM required in a processor array. The FourPort RAM also cascades in depth and width as easily as a standard single port RAM. The interface shown in Figure 7 would typically require 30ns RAMs for a 20Mhz TMS320C2x type processor.

The TMS320C20 and TMS320C25 also include a "sync" pin that facilitates synchronizing the internal clock phases of multiple processors at reset. In synchronous TMS320C2x arrays, this guarantees that memory accesses are in phase with each other and there are no partial clock phase memory access collisions. This form of processor synchronization is accommodated entirely in hardware.

SUMMARY

The FourPort RAM combines features of fast static memory and a complex multiple bus interconnect network. The FourPort RAM all but eliminates stalls when transferring data between processors or to memory. This prevents bus conflicts from being a bottleneck in multiprocessor systems.

The flexibility of the FourPort RAM allows multiprocessor designs to remain generalized while achieving high speeds on critical vector processes. The fact that the RAM itself is also the interconnect network between processors eliminates the complexity of a conventional multiprocessor bus system. The straightforward static RAM interface of the FourPort RAM, allows almost any processor to be used in an array for embedded systems. These factors conspire to make practical a variety of new vector processing architectures centered around the world's first "large" truly four ported single-chip RAM.

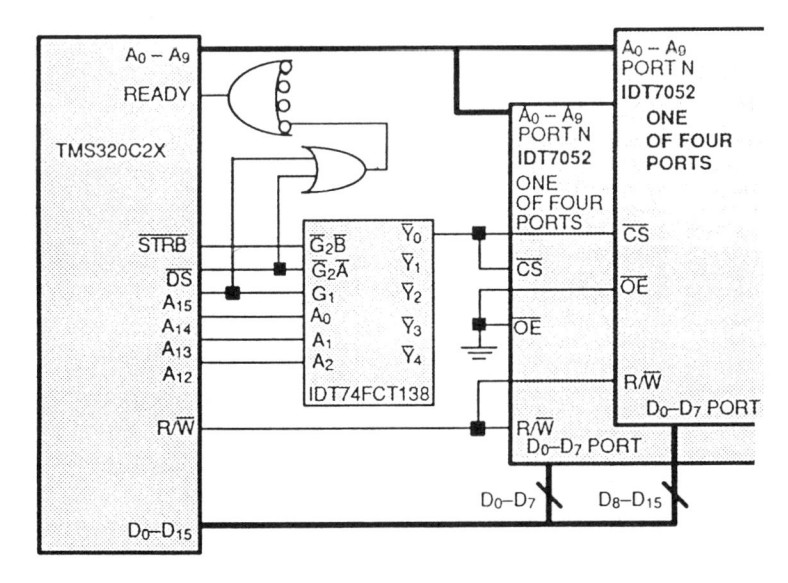

Figure 7. A 16-bit TMS320C2X to IDT7052 Interface Example

Integrated Device Technology, Inc. reserves the right to make changes to the specifications in this application note in order to improve design or performance and to supply the best possible product.

Integrated Device Technology, Inc.

3236 Scott Blvd., Santa Clara, CA 95054-3090   Telephone: (408) 727-6116   TWX 9103382070

INTRODUCTION TO IDT'S FourPort™ RAM

APPLICATION NOTE AN-45

By John R. Mick

INTRODUCTION

Integrated Device Technology is continuing to pioneer higher speed and higher density static RAMs. As IDT has improved its CEMOS™ technology, new RAM architectures and additional features have become feasible. The end result is that design engineers now have powerful new integrated circuits available for demanding applications. One such circuit is the IDT7052 FourPort RAM. This device is a Four–port 2K by 8–bit Static RAM built using a 12–transistor four ported static RAM cell. Each of the four ports is independent in terms of the byte that it can read or write.

This new IDT7052 FourPort RAM provides the system architect with better ways to look at computer system design. For example, the IDT7052 can be used in a multiprocessor environment to provide a common memory among several processors. An example of such an architecture is shown in Figure 1. Here we see each of the four RAM ports connected to a high performance microprocessor. These processors could also be intelligent controllers, DSP engines, or a combination of the two. The FourPort RAM can be used in such computer architectures as hypercubes and parallel processing machines for storage and movement of data. It offers unheard of opportunities in digital signal processing (DSP) where new architectures for Fast–Fourier–Transforms (FFTs), recursive and non–recursive digital filters, windowing functions, and special purpose algorithms can take advantage of multiple ports into a shared memory. The IDT7052 FourPort RAM can increase system performance and reduce parts count by providing simultaneous access to the data by more than one processor at a time.

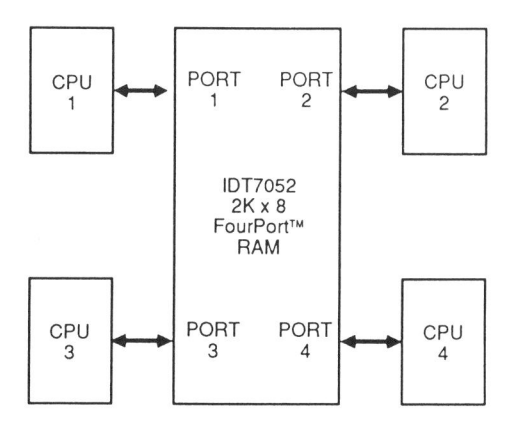

Figure 1. Four–Port RAM Providing Common Memory to Four CPUs

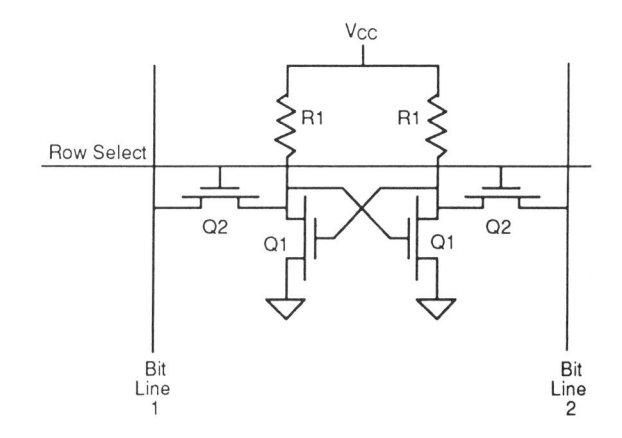

Figure 2. Typical Four–Transistor SRAM Cell

UNDERSTANDING THE FourPort RAM

In order to effectively design with the IDT7052 FourPort RAM, it is important for the design engineer to understand its construction and architectural features. This is most easily accomplished by starting with a simple single port RAM cell and evolving its architecture into the FourPort structure. Figure 2 shows a typical single port static RAM built using a four–transistor cell. This architecture is commonly used by most static RAM manufacturers to build static RAMs because it offers high density, good speed and low power.

In its simplest description, the device consists of two N–channel transistors (Q1) and two resistors (R1) that are connected so as to form two simple cross–coupled inverters. This gives a regenerative action such that one N–channel transistor is ON and the other N–channel transistor is OFF. Thereby, a single bit of memory is formed. In order to interface to this cross–coupled pair of inverters, two additional N–channel transistors (Q2) are connected between the inverter outputs and the bit–lines. The gates of these two N– channel transistors are connected to a line called the row select. These Q2 transistors connected between the cell and the bit lines are usually called transmission gates. The result is that when a particular row of cells in the RAM is addressed, these two transistors are turned on and one bit–line will reflect a HIGH and the other bit–line will reflect a LOW as determined by the current state of the static RAM cell.

An expanded example of this RAM architecture is shown in Figure 3. Here we see a 16–bit RAM, organized four–rows by four–columns internally, in a more complete form. The bit–lines of the cells are connected to the inputs of a sense amplifier by means of N–channel switches. These switches are controlled by the column address decoder. The sense amplifier will detect whether the state of the bit is a logic one or a logic zero depending on the relative polarity of the two bit–lines going into the differential sense amplifier. We usually call the transistors connected between the sense amplifier and bit–lines a data multiplexer or data selector.

FourPort is a trademark of Integrated Device Technology, Inc.

© 1989 Integrated Device Technology, Inc.

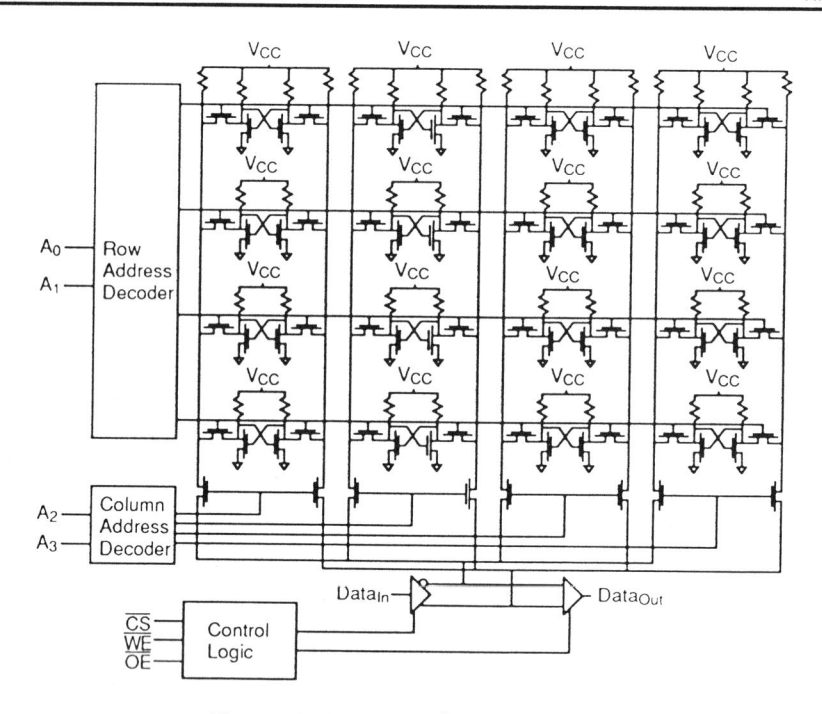

INTRODUCTION TO IDT'S FourPort RAM   APPLICATION NOTE AN–45

Figure 3. An Example Four-Transistor Cell for a 16-bit SRAM

When we wish to write the simple RAM as shown in Figure 3, a row address line is selected by the row address decoder and the N–channel pass transistors (Q2 of Figure 2) connected to the bit lines are turned on by pulling their gates high. Now however, the write amplifier driven by the Data–In line (Figure 3) is turned on by the write enable signal via the control logic. The write amplifier will drive one bit–line HIGH and the other bit–line LOW as determined by the logic state of the data input. The output of the write amplifier is more powerful than the inverter transistors (Q1 in Figure 2) in the RAM cell and it easily overpowers these inverter transistors if it is necessary to flip the static RAM bit. In its simplest form; this is all there is to the circuitry of a static RAM. Functions such as chip enable are used to simply enable or disable the entire operation of the RAM. Output enable on a static RAM is used to turn "on" the outputs during a read cycle and turn "off" the outputs during write cycles. It can be used to solve timing problems in high speed applications.

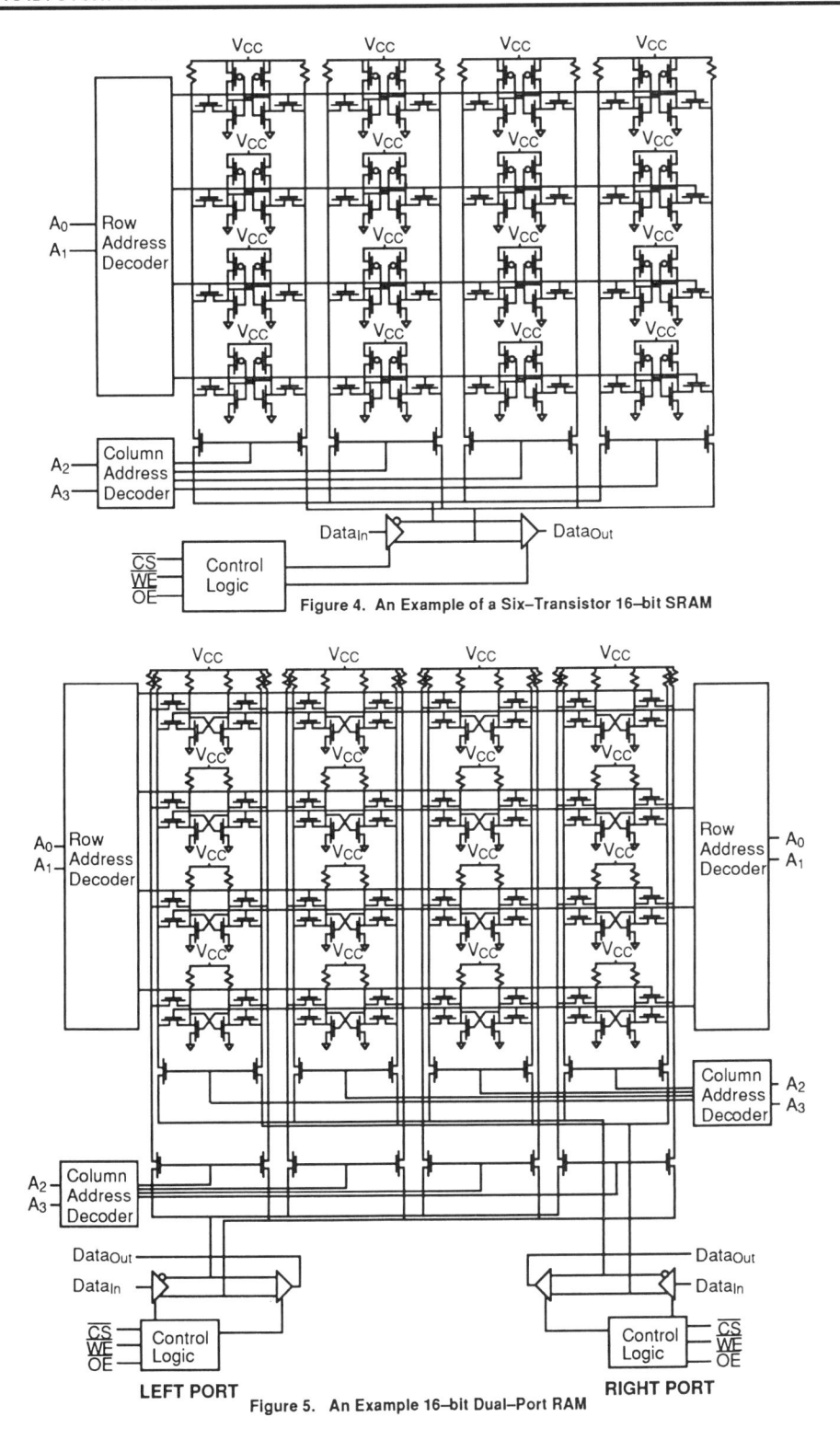

INTRODUCTION TO IDT'S FourPort RAM   APPLICATION NOTE AN–45

Figure 4. An Example of a Six–Transistor 16–bit SRAM

Figure 5. An Example 16–bit Dual–Port RAM

INTRODUCTION TO IDT'S FourPort RAM

A variation on the standard four-transistor static RAM cell is the six-transistor static RAM cell as shown in Figure 4. In this Figure we see that the two pull-up resistors (R1 of Figure 2) have been replaced by two P-channel transistors. The operation of such a six-transistor cell is identical to the four-transistor cell previously described. The difference between the two approaches is that the physical size of the cell with the P-channel transistors is larger than the cell with the resistors. The standby power can be lower for the six-transistor cell because there is no power being dissipated. In a four-transistor cell, one of the the pull-up resistors is always dissipating power since one transistor of the cell is always ON. The six-transistor cell can have higher radiation hardened characteristics than the four-transistor cell because the voltage swings in the cell are larger. This is because the internal node in the cell that is high is pulled to the +5V rail by the P-channel transistor. In addition, the six-transistor cell provides higher internal noise margins in the circuit for this same reason. Most manufacturers of static RAMs use the four-transistor cell because it allows static RAMs of higher density to be fabricated with smaller die sizes.

Next, let's look at a typical dual-port RAM such as the IDT7134, a 4K by 8-bit device. An example schematic diagram showing a sixteen-bit two-port RAM is shown in Figure 5. Here we see our standard cross-coupled inverter pairs using two N-channel transistors with resistor pull-ups (Q1 and R1 of Figure 2) to form the sixteen memory bits. Notice however, now there are two pairs of [?] nel transmission gates connected to each RAM cell's t[?] compliment outputs and two pairs of bit-lines associated w[?] cell. Each pair of bit-lines is a read/write port into the d[?] RAM. Each pair of transmission gates has its own row [?] control so that Port A can select any memory cell in the R[?] Port B can select any memory cell in the RAM. This is th[?] nique used in IDT dual-port RAMs to provide total indepen[?] cess to individual bytes. Each pair of bit-lines is connect[?] sense amplifier and a write buffer via a data multiplexer [?] each port on the 2-port RAM can read or write data at its s[?] address.

Now for the FourPort RAM operation. Figure 6 shows a [?] schematic diagram for the IDT7052 12-transistor FourPo[?] cell. The two inverters making up the basic memory cell ar[?] cated using two N-channel pulldown transistors and two P[?] nel pullup transistors. They are connected in the normal [?] coupled inverter fashion to make a single memory cell. Fo[?] vidual memory ports are achieved by using four pairs of N-c[?] pass transistors to connect to four pairs of bit-lines. Four i[?] ual row addresses are used to select each pair of transm[?] gates connected between the RAM cell outputs and the b[?] pairs. Four sense-amplifier/write-buffers are used to provic[?] vidual read/write paths from each port to all the cells in the [?]

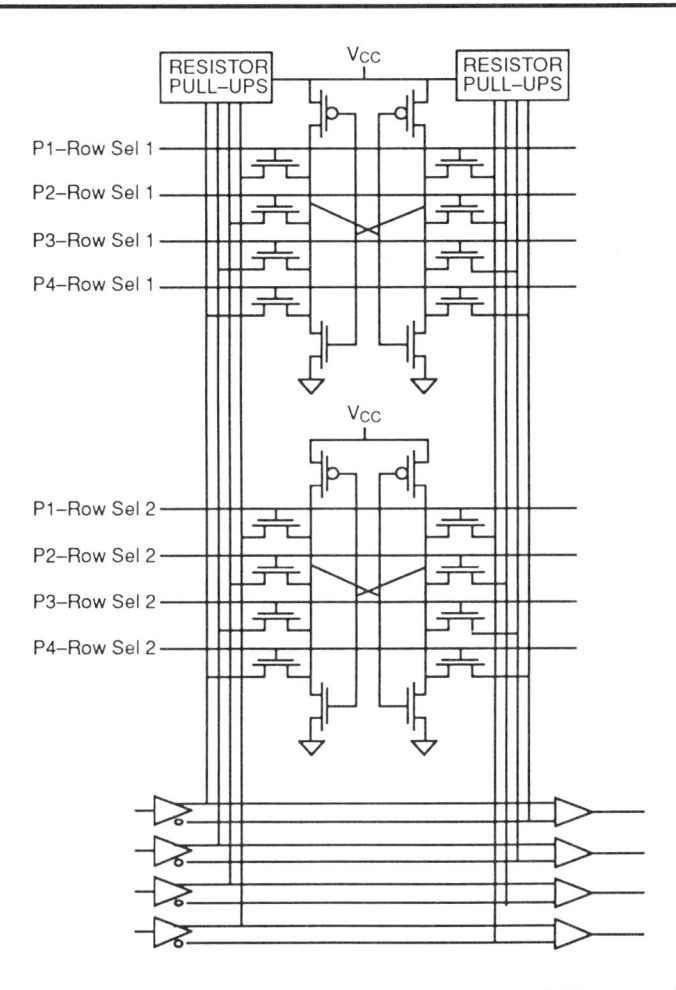

INTRODUCTION TO IDT'S FourPort RAM　　　　　　　　　　　　　　　　　　　　　　　　　　　　APPLICATION NOTE AN–45

Figure 6. A Simple Example of a Twelve Transistor FourPort RAM Configuration

From this discussion, the design engineer should understand the mechanism used to implement a FourPort RAM. As described, we can see how we can make each port of the FourPort RAM totally independent from the other ports. Do not confuse this statement to mean that independent reads and writes can always be performed without data corruption. If two ports write to the same byte at the same time, one or both values may be lost. Likewise if one port writes to a byte at the same time another port is reading the byte, the read may be corrupted even though the byte write is completed correctly. This application note does not discuss issues of data integrity in the case of multiple accesses to the same location, when one of the asynchronous accesses is a write cycle. These problems are discussed in detail in Application Note 2 and will not be further discussed here. Suffice it to say that the IDT7050 and IDT7052 FourPort RAMs have a $\overline{\text{BUSY}}$ input to allow external hardware or software arbitration schemes to be implemented to meet the specific needs of the designer's system. The $\overline{\text{BUSY}}$ input serves only to block write cycles from the port to which this signal is applied. It has no effect on a read cycle. Note that in the following applications we are not using the $\overline{\text{BUSY}}$ input of the FourPort RAM so it should be tied HIGH. We probably will not always mention this, so do not forget it or you will not be able to write into the FourPort RAM.

Once the rules are understood however, only engineering creativity is needed to visualize new architectural opportunities for FourPort RAMs. This powerful new memory technology will provide increased performance in future electronic processing systems.

CASCADING THE FourPort RAM

Perhaps the most easily understood techniques in designing with static RAMs are width and depth expansion. Width expansion of any port of the FourPort RAM is straightforward. No additional parts are needed to build 16, 24 or 32 bit wide or wider memories. Any port of the FourPort RAM can be viewed the same as a simple single port static RAM. All the same rules apply and they can be applied individually to each port of the FourPort RAM.

INTRODUCTION TO IDT'S FourPort RAM

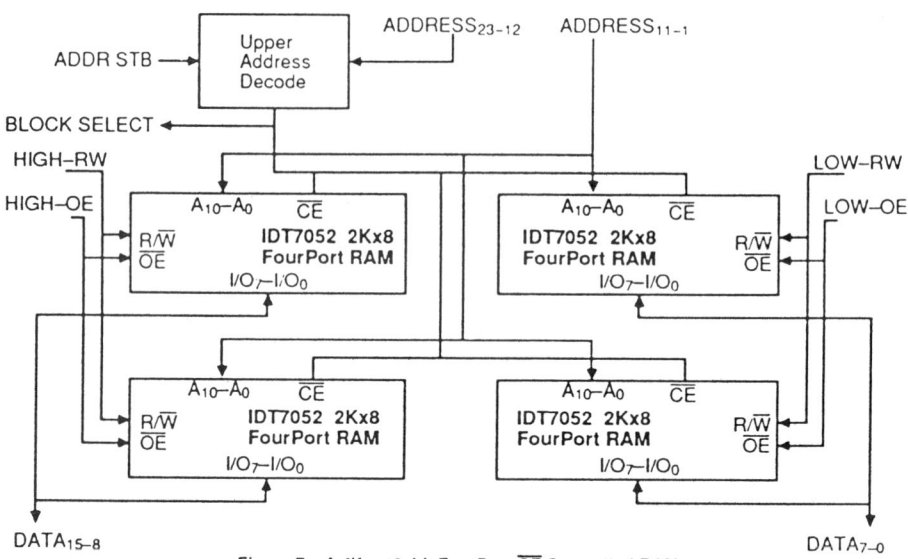

Figure 7. A 4K x 16-bit FourPort $\overline{CE}$ Controlled RAM

Depth expansion of the FourPort RAM is also quite simple. If one port is viewed as a static RAM, it is expanded similar to a single port device. Lower addresses are connected between devices and upper addresses are decoded by means of a standard decoder such as an IDT74FCT138 or IDT74FCT139. The outputs of the decoders can be used either to control the chip selects or control the write-enable and output-enable individually. Simple examples of expansion of one port of a FourPort RAM to a 4K-word by 16-bit configuration are shown in Figure 7 and Figure 8. Figure 7 shows the Chip Enable expansion method while Figure 8 shows wr enable, output-enable expansion. The two schemes are sim but, sometimes one can have a timing advantage over the ot! This is usually a function of the actual timing signals that are av able or have already been generated.

Once the depth expansion is understood, we can view the C interconnect schemes by simply looking at a one deep FourF RAM. We recognize that deeper versions can be realized as described.

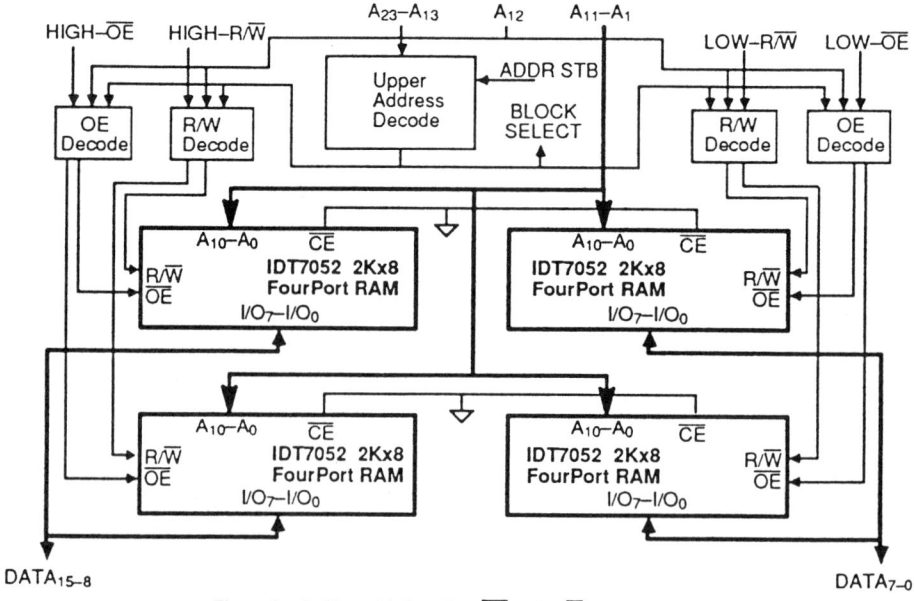

Figure 8. A 4Kx16-bit FourPort $\overline{OE}$ and R/$\overline{W}$ Controlled RAM

INTRODUCTION TO IDT'S FourPort RAM　　　　　　　　　　　　　　　　　　　　　　　　　　　　　　**APPLICATION NOTE AN–45**

CONNECTING THE FourPort RAM TO CPUs

A Z80A Example

Probably the easiest interface of the IDT7052 FourPort RAM is to a Z80A. This processor still provides a great price–performance tradeoff! By using four Z80As with the IDT7052 FourPort RAM, significant performance advantages can result. For example, no time need be lost due to DMA channels. The data placed in memory by one Z80A on one port is instantly available to another Z80A on another port. In a similar fashion, parallel processing can be performed by multiple processors working on the data in shared memory.

The typical connection scheme for the IDT7052 (or IDT7050 1Kx8 FourPort RAM) to a Z80A is shown in Figure 9. Here we see the eleven address lines, A_{10}–A_0, of the FourPort RAM are connected to the A_{10}–A_0 lines of the Z80A. This places the FourPort RAM in a contiguous 2K address space of the Z80A. The 2K byte segment actually used is determined by upper address decode circuit. A PAL or an IDT74FCT521 could be used to perform this function. The data lines are connected between the processor and the RAM. The Z80A has a $\overline{RD}$ line that can be connected to the FourPort $\overline{OE}$ and a $\overline{WR}$ line that can be connected to the FourPort R/$\overline{W}$ input. This works along the lines of the a Chip Enable expansion method just described. When the Z80A addresses the FourPort RAM, either a read or write will be performed depending on the instruction being executed. If $\overline{RD}$ goes LOW, the FourPort RAM will output data from the addressed byte. If $\overline{WR}$ goes LOW, the FourPort RAM will write data into the addressed byte.

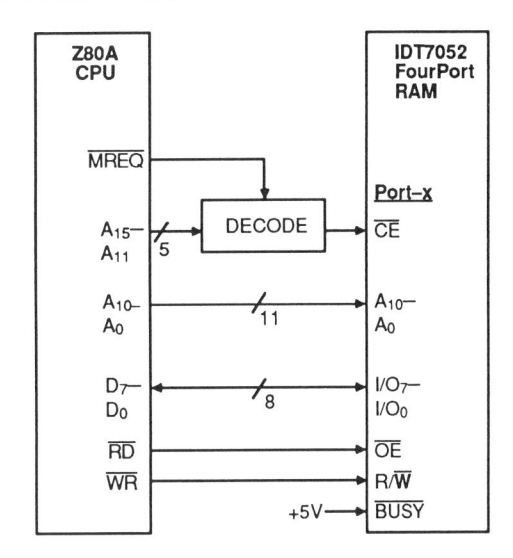

Figure 9. Interfacing the Z80A to One Port of the FourPort RAM

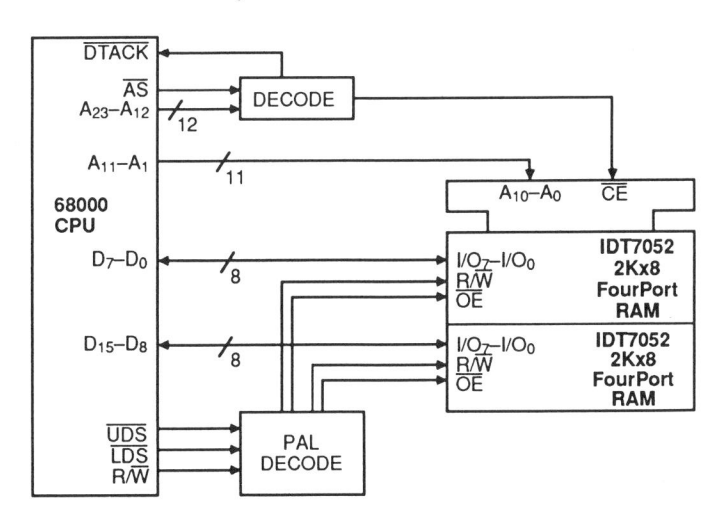

Figure 10. A 16–bit FourPort RAM with the 68000 CPU

A 68000 CONNECTION EXAMPLE

If we wish to build a 16–bit microprocessor interface to one port of the IDT7052 FourPort RAM, a typically interface might be as shown in Figure 10. Here we see two IDT7052s used in a 16–bit configuration. One FourPort RAM is connected to the lower eight data bits (D_7–D_0) and the other FourPort RAM is connected to the upper eight data bits (D_{15}–D_8). This completes a 16–bit data bus. Address lines A_{10}–A_0 of the FourPort RAM are connected between RAMs and also connected to address lines A_{11}–A_1 respectively of the 68000. Remember, the 68000 does not have an A_0 address line but uses Upper–Data–Strobe ($\overline{UDS}$) and Lower–Data–Strobe ($\overline{LDS}$) to control the upper and lower byte selection. These two signals in conjunction with the R/$\overline{W}$ signal are decoded in a PAL to generate the individual FourPort RAM R/$\overline{W}$ and $\overline{OE}$ control signals. Figure 11 shows the truth table needed for the PAL. It has been my experience when working with the 68000, that once these signals are generated, they are useful throughout the design to control other peripherals, etc. Basically, however, in this exam-

INTRODUCTION TO IDT'S FourPort RAM

ple we simply have a lower byte FourPort RAM and an upper byte FourPort RAM.

INPUTS			OUTPUTS			
R/$\overline{W}$	$\overline{UDS}$	$\overline{LDS}$	$\overline{URW}$	$\overline{LRW}$	$\overline{UOE}$	$\overline{LOE}$
X	1	1	1	1	1	1
1	0	0	1	1	0	0
0	0	0	0	0	1	1
1	1	0	1	1	1	0
0	1	0	1	0	1	1
1	0	1	1	1	0	1
0	0	1	0	1	1	1

Figure 11. 68000 16-bit Control PAL Truth Table

The upper address lines of the 68000, A_{23}–A_{12} in this case, are used to position the 2K bytes of FourPort RAM in continuous address space of the 68000. The actual location can be anywhere from 0x000000 to 0xFFFFFF as long as the overall range is on 2K byte boundaries. Usually we include address strobe ($\overline{AS}$) in the decoding as it can solve some timing problems. A timing review will show if it is needed. An output of the decode circuit can be used to generate the data acknowledge ($\overline{DTACK}$) if it is needed. Usually design engineers have an overall plan for generating the memory $\overline{CE}$s and $\overline{DTACK}$, so what is shown here is only to remind you of solving the overall problem.

Address of Bytes in a Big-Endian 32-bit Word

Word Address	Bits 31–24	Bits 23–16	Bits 15–8	Bits 7–0
0x0024	0x0024	– etc –	– etc –	– etc –
0x0020	0x0020	0x0021	0x0022	0x0023
0x001C	0x001C	0x001D	0x001E	0x001F
0x0018	0x0018	0x0019	0x001A	0x001B
0x0014	0x0014	0x0015	0x0016	0x0017
0x0010	0x0010	0x0011	0x0012	0x0013
0x000C	0x000C	0x000D	0x000E	0x000F
0x0008	0x0008	0x0009	0x000A	0x000B
0x0004	0x0004	0x0005	0x0006	0x0007
0x0000	0x0000	0x0001	0x0002	0x0003

Address of Bytes in a Big-Endian 16-bit Word

Word Address	Bits 15–8	Bits 7–0
0x0014	0x0014	– etc –
0x0012	0x0012	0x0013
0x0010	0x0010	0x0011
0x000E	0x000E	0x000F
0x000C	0x000C	0x000D
0x000A	0x000A	0x000B
0x0008	0x0008	0x0009
0x0006	0x0006	0x0007
0x0004	0x0004	0x0005
0x0002	0x0002	0x0003
0x0000	0x0000	0x0001

Address of Bytes in an 8-bit Word

Word Address	Bits 7–0
– etc –	– etc –
0x0010	0x0010
0x000F	0x000f
0x000E	0x000E
0x000D	0x000D
0x000C	0x000C
0x000B	0x000B
0x000A	0x000A
0x0009	0x0009
0x0008	0x0008
0x0007	0x0007
0x0006	0x0006
0x0005	0x0005
0x0004	0x0004
0x0003	0x0003
0x0002	0x0002
0x0001	0x0001
0x0000	0x0000

Figure 12. Memory Map for 8, 16, and 32-bit Byte Ordering

INTRODUCTION TO IDT'S FourPort RAM   APPLICATION NOTE AN–45

HOW ABOUT 8–BITS, 16–BITS AND 32–BITS IN THE SAME SYSTEM!!!

This is perhaps the most interesting example to talk about. We will use an 8-bit Z80A, a 16-bit 68000 and a 32-bit R3000 RISC microprocessor to discuss the design techniques. We have chosen the three processors because they are typical, they are fun to work with and they have had broad acceptance in the microprocessor world. First, let's look at Figure 12 to understand memory addressing and "memory space". All three of our selected microprocessors are "byte" addressable machines. That means they can address bytes as well as words in the case of the 68000 and R3000. The 68000 is a Big-Endian machine and the R3000 will be operated in Big-Endian mode to keep things simple. (DEC and Intel fans can make the appropriate transformation. In fact, the FourPort RAM might make a really exciting byte-ordering problem solver between machines by connecting one port as Big-Endian and another port as Little-Endian to the same microprocessor and similarly for the second processor.)

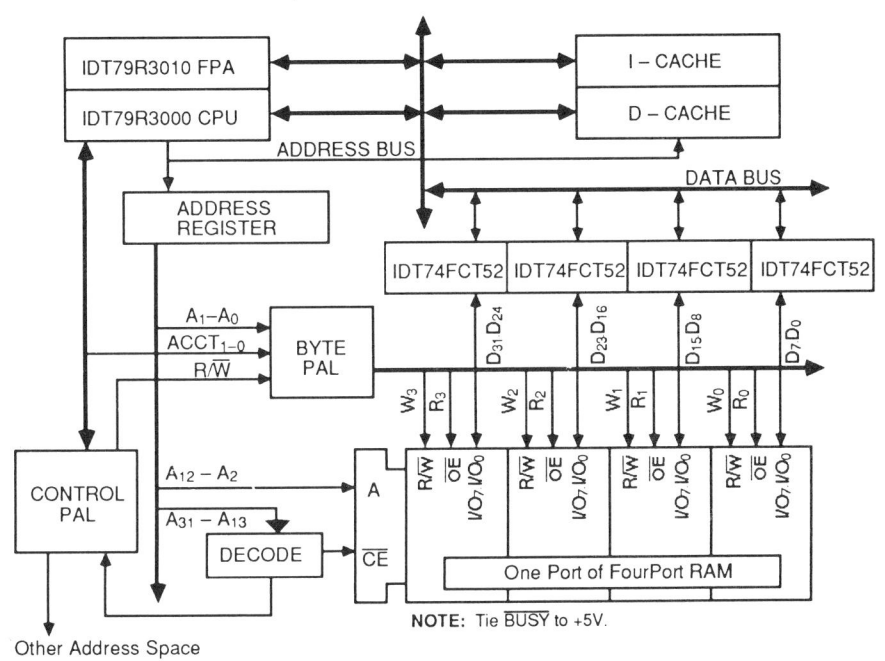

Figure 13. Using a 32–bit Wide FourPort Memory with the R3000

Since we are talking about byte addressable machines, Figure 12 shows the byte addresses of an 8-bit machine, the byte addresses of a 16-bit machine and the byte addresses of a 32-bit machine. Likewise word addresses of 16-bit and 32-bit machines are shown. What is intended here is to point out that we want the consecutive byte ordering of all of the machines to remain constant. By doing this, we keep the ability to do indexing into an array of bytes from any of the processors as a simple task. For example, a 40 byte index from any byte address is the same in all processors talking to each other through the FourPort RAM. We can look at Figure 12 as representing the Z80A, 68000 and R3000 respectively.

Next, let's look at the interface needed for each of our three processors. We will build on our previous examples in this application note but there are differences needed to allow proper addressing. Let's begin by looking at the R3000. The reader should refer to IDT's wealth of information on the IDT79R3000 RISC microprocessor if you are not familiar with the standard CPU, FPA, Cache and I/O interface. We will use four of the IDT7052 FourPort RAMs to give a 32-bit wide memory for this example. We assume the first port is connected to the R3000, the second port is connected to the 68000 and the third port is connected to the Z80A. The fourth port could be connected to a second one of any of these processors or a wide selection of other things.

A typical R3000 interface is shown in Figure 13. The key element here is to understand that we are interfacing to a 32-bit data bus, 32-bit address bus with byte encoded control signals and to an R3000 interface. We are able to implement the required byte control per Figure 12 by using the "BYTE PAL" shown in Figure 13. The truth table for this PAL is shown in Figure 14. Using this decoding, we are able to do all of the required operations. This includes 32-bit word operations, 24-bit three-byte operations, 16-bit half-word operations and 8-bit byte operations. The signals available are A_0, A_1, AccessType0, and AccessType1, all from the R3000 address register, and we assume a R/$\overline{W}$ input from the "Control PAL" shown in Figure 13. There may be other options here, but this R/$\overline{W}$ signal must be realized in some fashion. The upper address bits from the R3000 are decoded in the fashion previously discussed to locate the total 8K bytes of FourPort RAM in the R3000 address space.

Working out the timing of the R3000 interface is most of the work. Remember that at this interface point there are several flexibilities in the final timing. With an R3000 running at 16 MHz, a data transfer cycle is in multiples of 67 nanoseconds, 20 MHz gives 50 nanoseconds, and 25 MHz allows 40 nanoseconds. Thus depending on the processor speed and the FourPort RAM speed selected, block refill may or may not be desired. In any case, we

INTRODUCTION TO IDT'S FourPort RAM

APPLICATION NOTE AN-45

should be able to run with zero, one or two stall cycles. As mentioned before, the design engineer usually has a plan for address decoding and control handshake which is more closely tied to the overall system design. From this standpoint, interfacing to one port of the FourPort RAM is no different than interfacing to an EPROM, DRAM, SRAM, or peripheral.

INPUTS						OUTPUTS								COMMENTS
R/$\overline{W}$	ACCT1	ACCT0	A1	A0	W3	W2	W1	W0	R3	R2	R1	R0	
1	1	1	0	0	1	1	1	1	0	0	0	0	Word Read
0	1	1	0	0	0	0	0	0	1	1	1	1	Word Write
1	1	0	0	0	1	1	1	1	0	0	0	1	Tri-Byte Read
1	1	0	0	0	1	1	1	1	1	0	0	0	Tri-Byte Read
0	1	0	0	1	0	0	0	1	1	1	1	1	Tri-Byte Write
0	1	0	0	1	1	0	0	0	1	1	1	1	Tri-Byte Write
1	0	1	0	0	1	1	1	1	0	0	1	1	Half-Word Read
1	0	1	0	0	1	1	1	1	1	1	0	0	Half-Word Read
0	0	1	1	0	0	0	1	1	1	1	1	1	Half-Word Write
0	0	1	1	0	1	1	0	0	1	1	1	1	Half-Word Write
1	0	0	0	0	1	1	1	1	0	1	1	1	Read Byte 0
1	0	0	0	1	1	1	1	1	1	0	1	1	Read Byte 1
1	0	0	1	0	1	1	1	1	1	1	0	1	Read Byte 2
1	0	0	1	1	1	1	1	1	1	1	1	0	Read Byte 3
0	0	0	0	0	0	1	1	1	1	1	1	1	Write Byte 0
0	0	0	0	1	1	0	1	1	1	1	1	1	Write Byte 1
0	0	0	1	0	1	1	0	1	1	1	1	1	Write Byte 2
0	0	0	1	1	1	1	1	0	1	1	1	1	Write Byte 3

Figure 14. 32-bit R3000 Control PAL Truth Table (Big-Endian)

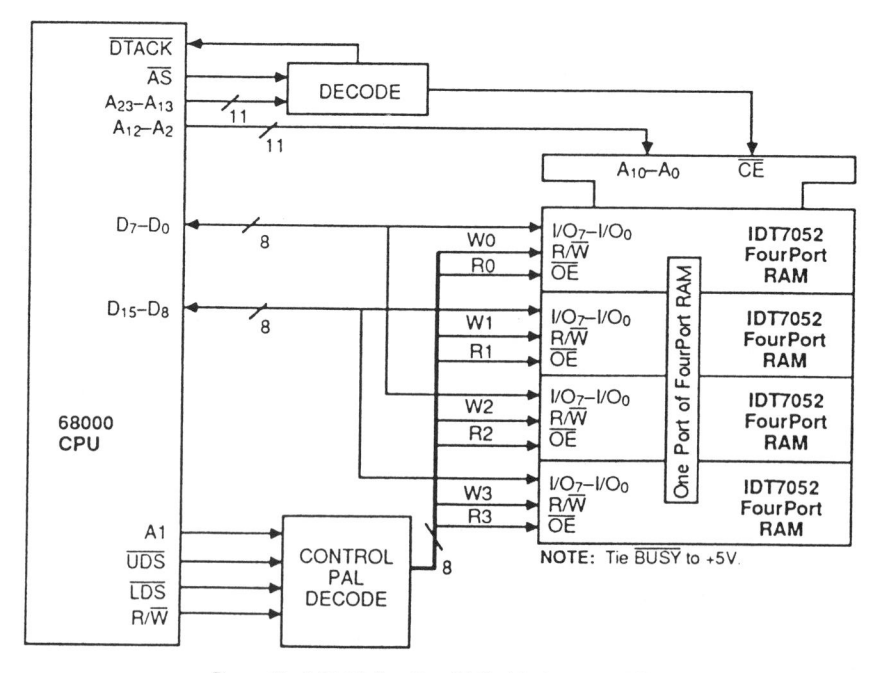

Figure 15. A 32-bit FourPort RAM with the 68000 CPU

INTRODUCTION TO IDT'S FourPort RAM   APPLICATION NOTE AN-45

INPUTS				OUTPUTS								COMMENTS
R/$\overline{W}$	A1	$\overline{LDS}$	$\overline{UDS}$	W3	W2	W1	W0	R3	R2	R1	R0	
1	0	0	0	1	1	1	1	0	0	1	1	Word Read
1	1	0	0	1	1	1	1	1	1	0	0	Word Read
0	0	0	0	0	0	1	1	1	1	1	1	Word Write
0	1	0	0	1	1	0	0	1	1	1	1	Word Write
1	0	1	0	1	1	1	1	0	1	1	1	Read Byte 0
1	0	0	1	1	1	1	1	1	0	1	1	Read Byte 1
1	0	1	0	1	1	1	1	1	1	0	1	Read Byte 2
1	0	0	1	1	1	1	1	1	1	1	0	Read Byte 3
0	1	1	0	0	1	1	1	1	1	1	1	Write Byte 0
0	1	0	1	1	0	1	1	1	1	1	1	Write Byte 1
0	1	1	0	1	1	0	1	1	1	1	1	Write Byte 2
0	1	0	1	1	1	1	0	1	1	1	1	Write Byte 3

Figure 16. 32-Bit 68000 Configuration Control PAL Truth Table

Next, let's look at the 16-bit 68000 interface in a 32-bit memory system. A detailed block diagram is shown in Figure 15. The key thing to notice here is that four of the IDT7052 FourPort RAMs are used. Notice that two of the devices are connected to the D_7–D_0 data bus and two of the devices are connected to the D_{15}–D_8 data bus on the 68000. Address line A_1 will be used to select which pair of FourPort RAMs that the processor will read or write.

For example, when A_1 is LOW, control signals W3, W2, R3 and R2 will be enabled. When A_1 is HIGH, control signals W1, W0, R1 and R0 will be enabled. This is shown in complete detail in the truth table of Figure 16. If we study this truth table, we see how we accomplish both 16-bit word (half-word) reads and writes as well as 8-bit byte reads and writes. All of this is consistent with the memory map shown in Figure 12. The technique here is actually to use A_1 to select either the lower half-word or the upper half-word in a 32-bit FourPort RAM memory system. Every thing else about the design is the same as the previous 68000 example.

Lastly, let's look at the 8-bit interface to the Z80A microprocessor. It also should be viewed as being hooked into a 32-bit memory system. A detailed block diagram is shown in Figure 17. Notice that all four of the IDT7052 FourPort RAMs are connected to the D_7–D_0 data bus. Address lines A_1 and A_0 will be used to select the device to which the Z80A processor will talk. In fact, A_1, A_0, $\overline{RD}$ and $\overline{WR}$ are inputs to the control PAL decode. The truth table to be implemented is detailed in Figure 18. This processor is only capable of performing byte reads or writes so the decoding is straightforward. A_1 and A_0 are used to do byte selection. Thus, the FourPort RAM A_{10}–A_0 address inputs are connect to the A_{12}–A_2 address lines of the Z80A. This keeps the byte addressing as desired in the memory map of Figure 12. Again the remaining part of the design is as shown in the previous Z80A example.

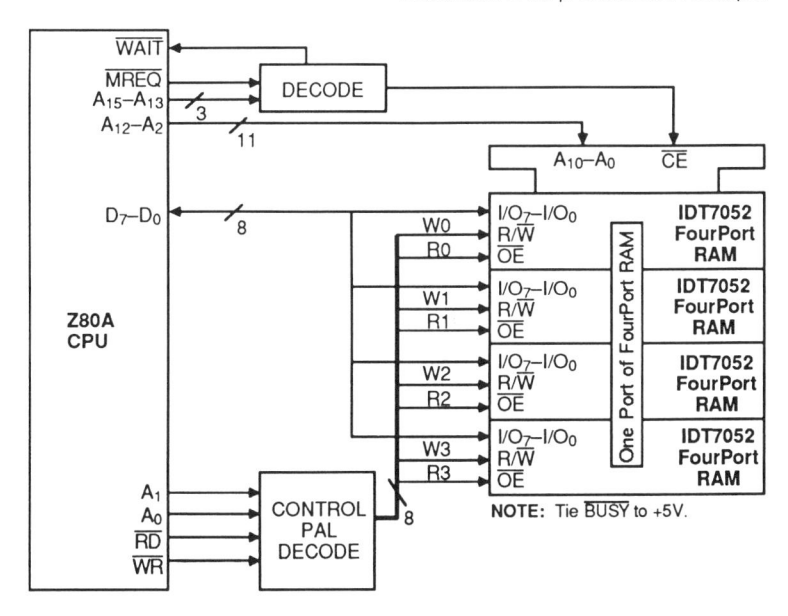

Figure 17. A 32-bit FourPort RAM with the Z80A CPU

INTRODUCTION TO IDT'S FourPort RAM　　　　　　　　　　　　　　　　　　　　　　　　　APPLICATION NOTE AN–45

INPUTS				OUTPUTS							COMMENTS
$\overline{WR}$	$\overline{RD}$	A1	A0	W3	W2	W1	W0	R3	R2	R1	R0	
1	0	0	0	1	1	1	1	0	1	1	1	Read Byte 0
1	0	0	1	1	1	1	1	1	0	1	1	Read Byte 1
1	0	1	0	1	1	1	1	1	1	0	1	Read Byte 2
1	0	1	1	1	1	1	1	1	1	1	0	Read Byte 3
0	1	0	0	0	1	1	1	1	1	1	1	Write Byte 0
0	1	0	1	1	0	1	1	1	1	1	1	Write Byte 1
0	1	1	0	1	1	0	1	1	1	1	1	Write Byte 2
0	1	1	1	1	1	1	0	1	1	1	1	Write Byte 3

Figure 18. 32–bit Z80A Control PAL Truth Table

The key in all of this discussion is to keep track of the data bus width being used in the design. Similarly, the decoding and processor address connections must take this into account. This is one point that the design engineer usually does not have to deal with when working with single port memories.

The purpose of this three processor example is to show a few interconnect schemes to typical microprocessors. From this discussion, the design engineer should be able to extend the concepts presented here to other 8–bit, 16–bit and 32–bit microprocessors. Just keep the techniques in mind and work out the desired memory mapping and timing.

SYSTEM DESIGN IDEAS

Now that we have discussed how the the FourPort RAM is built and we have a good idea of how to connect it to many processors, let's look at some system level uses for this type of FourPort RAM.

DIGITAL SIGNAL PROCESSING (DSP)

Digital signal processing applications have been expanding as new developments in semiconductor technology provide increased packing density and new architectures in integrated circuits. The IDT7052 FourPort RAM is another in the continuing growth of integrated circuits that allow design engineers to realize new system designs.

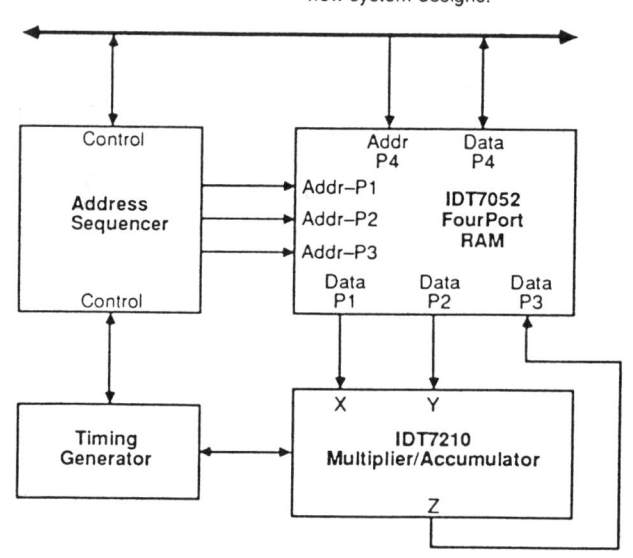

Figure 19. A Simple DSP Engine Using a FourPort RAM

One of the simplest DSP algorithms that can be implemented is the finite–impulse–response (FIR) filter. In this type of algorithm, the impulse response of the filter has nonzero values only for a finite duration. These types of filters are easily implemented using only multiplication and summation. Figure 19 shows a block diagram of a DSP machine that takes advantage of the FourPort RAM to interface to a multiplier–accumulator (MAC) such as the IDT7210. In this example, two of the four ports of the FourPort RAM are used to feed data to the MAC inputs and a third port of FourPort RAM is used to receive completed results from the MAC output. The fourth port of the FourPort RAM is connected to a local data–address bus to interface to the remainder of the system.

In the actual operation of such a processor as shown in Figure 19, data is loaded into the FourPort RAM via Port 4. The algorithm usually needs coefficients and these are also loaded into the FourPort RAM using Port 4. An address sequencer has the responsibility of providing the correct sequence of addresses to Ports 1, 2 and 3. This unit operates in conjunction with the timing generator to execute the algorithm. Let's look at an example. Suppose our algorithm is:

$$y(n) = A_0 * x(n) + A_1 * x(n-1) + A_2 * x(n-2) + A_3 * x(n-3)$$

We could read this as the current processed value is equal to the current sample times A_0, plus the first past sample times A_1, plus the second past sample times A_2, plus the third past sample times A_3. This already shows its potential as a FourPort RAM application.

Now, we initialize our system at power–up by putting the values A_0, A_1, A_2, and A_3 into the FourPort RAM. We would most likely clear the four locations for the data. Then we start taking data. Each time we receive a data value, we can overwrite the fourth past sample. For each new sample, we will compute a new $y(n)$ and put it into the FourPort RAM. At some point we will extract the sequence of values for $y(n)$ that we have computed. As can be seen, we can have several operations happening on the same clock cycle. RAM ports 1 and 2 could be outputting data, RAM port 3 could be inputting data, and RAM port 4 could be inputting or outputting data, all simultaneously. The speed implications are obvious. Needless to say, this is a simple example for the purpose of demonstration. But if we wanted to work on 1024 samples with 128 coefficients and a more complex algorithm, all we have to do is follow the same methodology. See IDT application note 42 for a more detailed example of how to use this method to implement a matrix multiplication.

CPU to CPU to CPU to CPU

Referring to Figure 1 where we started this application note, we see that we have a FourPort RAM connected between four CPUs. How do we efficiently communicate each processor's status to the other processors? You will need to work an acceptable software semaphore scheme or do hardware handshaking using external circuitry. The most obvious software scheme is token passing. After each processor has determined its order in the token passing scheme, the token passing protocol boils down to each processor taking its turn. This can be achieved by reading one memory location to see who is master. Usually multiple reads and compares are performed to avoid any data corruption problems. A good example of this mechanism is detailed in IDT application note 43.

Let's take an example. The byte at address zero contains the token. The current value is one. CPU 1 is master of the FourPort RAM and can read or write data. When finished, it writes a two at address zero. Every so often CPU 2 checks address zero and when it sees a two, it knows it is master. It performs any needed data reads or writes. When it is finished, it writes a three at address zero. CPU 3 writes a four and CPU 4 writes a one. Thus, a simple token passing scheme. "Fail safe" mechanisms can be implemented to keep the token moving if there is any failure.

Another obvious scheme is to set up a simple software semaphore path between each pair of processors. This technique can be used to pass data between processor pairs. The semaphore for each processor can use a different byte (or word) address for each semaphore in each direction. By using this method, many different software handshake techniques can be implemented. Rather than use a test and set instruction for semaphores in this application, another interlocking mechanism, like separate locations dedicated to the status of each processor, should be used to guarantee clean communications between tasks.

In addition, several hardware approaches are usually available in most multiprocessor environments. These include individual interrupts between processors as well as broadcast interrupt approaches. In either case, after the data has been set up in a private buffer, processor A can interrupt processor B to notify it of the pending message. The data structures used in such an environment can include pointer passing and linkage conventions consistent with modern day software techniques.

SUMMARY

The FourPort RAM is a truly new innovative integrated circuit memory that offers new communications methods for computing machines. It provides exceptional speeds because of its opportunity for parallelism. The IDT7052 2K x 8–bit FourPort RAM and the IDT7050 1K x 8–bit FourPort RAM are the first in a series of memories that will pioneer these new architectural frontiers. At speeds as fast as some of the fastest standard static RAMs, they bring new performance dimensions to parallel communication between tasks of a computing machine. These devices utilize the latest in IDT's CEMOS technology to provide the design engineer with an economical high performance, low power, small size and highly reliable "Speciality Memory" for todays performance–driven designs.

UNDERSTANDING THE IDT7201/7202 FIFO

APPLICATION NOTE AN-01

By Michael J. Miller

INTRODUCTION

This article discusses several different types of FIFO queues, their implementation, their performance and their use. Data, or information in computers, is processed as words or bytes in a predominantly serial fashion. There are producers and consumers of information that are connected by busses. Often there is a mismatch in the rate at which data is produced and the rate at which it can be accepted. The data is therefore buffered in serial lists until it can be used. The serial lists are stored in memory and require overhead to maintain them. These First-In-First-Out (FIFO) structures can be implemented at many levels from all software to all hardware. The software implementations are often the most flexible but yield the lowest performance. The hardware implementations, while less flexible, give the highest performance.

QUEUES

The elements of any computer or controller can be divided into three categories in relation to information: transformation, storage and transfer. Logic gates transform and combine information, memory elements store information and wires transfer information between the other elements.

Memory can be viewed as an element which transfers information with respect to time. The simplest of memory elements are latches and registers. RAMs are dense arrays of latches. While RAMs allow for dense information storage, they require an address to access individual pieces of information in the array. Therefore, addresses (information) must be generated and stored in order to access the desired information. The addresses are stored in programs and data structures such as linked lists.

Queues are a special organization of dense arrays of latches. Queues are a linear organization of groups of latches. Access to the linear string is restricted to either end. While RAMs allow for random access of any data in the array at any point in time, they require address inputs. Queues on the other hand, don't have an address thus avoiding the address generation and storage overhead. Queues can be divided into two categories: FIFOs and LIFOs.

Queues can be observed in the world about us. FIFO is an acronym for "First-In-First-Out". They can be observed in a bank line-up where customers enter at the end of a line and, after some wait, are serviced at the other end. The FIFO queue provides a mechanism by which customers, which arrive at an erratic rate, can wait until a teller can accommodate them.

LIFO is an acronym for "Last-In-First-Out". We can observe this phenomenon in the work place. As a person is working at a desk, interrupts occur. A higher priority interrupt such as a phone call or a request from people higher in management will cause the person to drop the work on the desk and start a new task. When the higher priority task is accomplished, the interrupted task on the desk is resumed. Depending on how many interrupts of sequentially higher priority tasks come in during the day, the stack of tasks on the desk grows. Another time honored example is the stacks of trays at the cafeteria. As trays are washed they are placed on a spring loaded elevator which sinks down to accommodate the new trays. When new customers enter the food line, trays are removed from the stack.

As can be seen in the above examples, queues are used to buffer between the flows of consumers and distributors of services. Groups of computing elements can be divided into consumers and producers of information with rates that must be matched. For example, a rotating Winchester disk is a source of information that must be serviced at a rate that may not be easily matched by the CPU which is consuming the information through the use of a data bus.

SIMPLE FIFO

The implementation of FIFOs is varied and presents many trade-offs. The simplest design treats the FIFO as a fixed number of memory elements in a linear array. When data is written (pushed) in at one end, all of the rest of the elements shift their data over to their neighbor at the same time. One can visualize (Figure 1) the structure as a shift register. The same structure can be implemented in software where the program manages an array of memory locations in RAM. To push data into the queue the program must first start by moving the contents of the next to the last location into the last location. The algorithm continues from the last to the first location. When all of the data has been rippled down, the first location in the queue will be vacated. The data to be pushed into the queue is written into that vacated location.

An improvement in the software solution could be made with the introduction of a pointer. A pointer is a variable which contains an address. The pointer would identify a location from which to read the output of the FIFO. When a new piece of

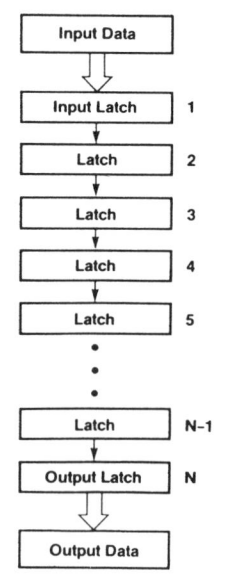

Figure 1. Hardware implementation of a fixed length FIFO.

©1985 Integrated Device Technology, Inc. Printed in U.S.A.

UNDERSTANDING THE IDT7201/7202 FIFO   APPLICATION NOTE

information is written, it would go into the location identified by the pointer after which the pointer would be incremented. The pointer now points at the new output data. When the pointer reaches the end of the array, the next increment would be replaced by setting the pointer to the beginning of the array. The obvious advantage is that the program does less work and therefore is faster. This software technique is called a circular queue with one pointer. (See Figure 2.)

FIXED LENGTH FIFO: NO FALL-THROUGH

The FIFO described previously is called a Fixed Length FIFO and has the characteristic that it takes N cycles for a piece of information that was placed into it to emerge out of it. The number N is the number of locations in the FIFO. This implementation also has the characteristic that, when first started after power up, it will produce unknown data for N cycles until the first valid data arrives at the output. The latency is therefore N read/write cycles. The fixed length FIFO does not allow for differences between the rate of input and output rates. This type of FIFO is used where the arrival of data at the output is delayed to match parallel paths in a pipelined system.

VARIABLE LENGTH FIFO

The variable length FIFO solves the rate mismatch problem but requires more overhead to implement. Where the fixed length FIFO is like a steel pipe which information is fed through and has a fixed number of locations, the variable length FIFO is like a rubber hose that can stretch, holding from one to many items. The items are removed at will instead of being required to at write time. Every variable length system has a limit and therefore must signal when it is at capacity and must be serviced before bursting.

FALL-THROUGH FIFOs

In the real world of silicon and aluminum there is no such thing as rubber. Variable length FIFOs must therefore be implemented using fixed length queues. This fact creates some limitations which translate into trade-offs. The traditional hardware implementation uses two sets of shift registers. One set is used to hold the data in much the same way as in the fixed length FIFO. Data that is placed in the top emerges at the bottom. There is a second shift register that functions in parallel. The second shift register contains flags that indicate whether the associated data element at the same chronological position in the data queue is valid data or not. When data is written into the top location of the data queue, a true flag is placed into the "valid bit" queue. The variable length quality is achieved by allowing the data and its associated valid bit to "sink down" into the next location below it if there is no valid data in that location (see Figure 3). In this way valid data "sinks" to the bottom of the queue and stacks up in much the same way as pearls being dropped into a narrow tube filled with oil. The clocking of data down through the queue is controlled by an internal self-generated clock. The maximum latency or fall-through time is a product of the number of cells in the queue and the internal clock cycle length. This approach meets the requirement that differing rates may be accommodated. The valid bit data is brought out in parallel with the queue data. The valid bit data tells the consumer when valid data is present, thus avoiding the start-up period of invalid data as in the previous implementation of the fixed length FIFO. Examples of this approach are the shorter FIFOs such as the MMI 67401. Fall-through FIFOs tend to have very long undesirable fall-through times if the FIFO is deep.

The software approach could be designed to mirror the typical hardware approach by working with two arrays. One for the data and one of the valid bits. That approach uses too much memory. An alternate could use a wider array which carried the valid bit with the data. The algorithm would then start at the end of the array and pass to the front, advancing all elements which were valid to the end of the array until all valid data was collected at the end of the array. This approach would be very costly in terms of CPU cycles for what is achieved. There is a fall-through latency which is a product of the time to execute the updated software loop times the number of locations in the queue.

TWO-POINTER FIFO

A more economical approach would utilize two pointers and one array that was as wide as the data. One pointer would point to

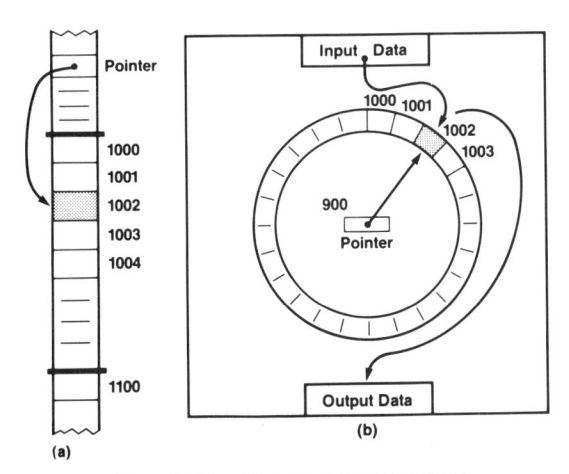

Figure 2. Circular queue with one pointer
a) As it is in memory.
b) Logical view.

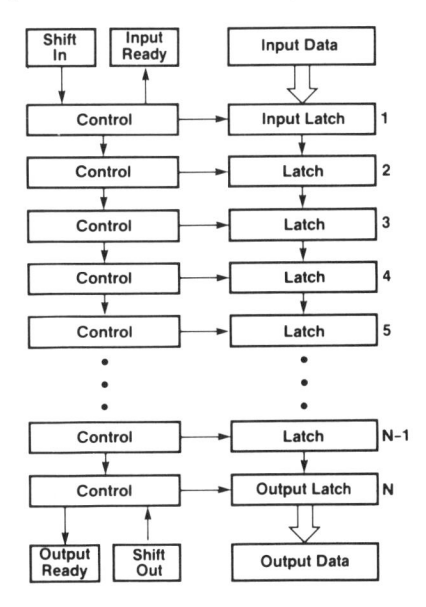

Figure 3. Classical FIFO architecture.

UNDERSTANDING THE IDT7201/7202 FIFO　　　　　　　　　　　　　　　　　　　　　　　　　　　　　　　　　　　　　**APPLICATION NOTE**

the location at which new data is written into. The second pointer identifies where data is to be read from for output from the queue (see Figure 4). When either pointer is used to access a location, it is incremented. When a pointer is incremented to the last location in the array, the next increment will be substituted with a reset of the pointer to the beginning of the array. The logical view of this structure is a circular queue with a read and a write pointer. This approach results in a much shorter fall-through time while still achieving the variable length feature. The fall-through time is the time that it takes to invoke the software to write the data into the queue, plus the time that it takes to invoke the software to read the data out of the queue. While this is much better than the previous approach, it still requires a reasonable amount of time to accomplish.

TODAY'S HIGH SPEED FIFOs

The hardware approach, which is used by the IDT7201 and IDT7202 devices, utilizes the software concepts demonstrated in the previous approach but at very fast hardware speeds (50ns typical military). The block diagram in Figure 5 shows the two pointers which locate where reading and writing is to take place in the queue (RAM Array). There is added logic which provides status about the queue: empty ($\overline{EF}$), half full ($\overline{HF}$) and full ($\overline{FF}$) ($^-$ means an active LOW signal). Two pins, one input ($\overline{XI}$) and one output ($\overline{XO}$), provide for unlimited expansion while still maintaining the 50ns fall-through time. This part functions identically to the software approach utilizing the two pointers. When either pointer reaches the last location, it is reset to the first location thus achieving a circular queue via a wraparound approach. The status flags reflect the count of how many valid pieces of data are in the queue. After the device is reset, the empty flag ($\overline{EF}$) is asserted. As soon as a datum is written into the queue, the empty flag is deasserted. The empty flag is not asserted again until all pieces of data have been read from the queue. When the count of data elements reaches one-half the number of locations in the RAM array, the half full flag ($\overline{HF}$) is asserted. If a read is performed which reduces the count to just below the half way count, then the ($\overline{HF}$) is deactivated. The full flag is asserted when the count of data elements is exactly equal to the number of locations in the RAM array, thus flagging that there are no more empty locations in the queue.

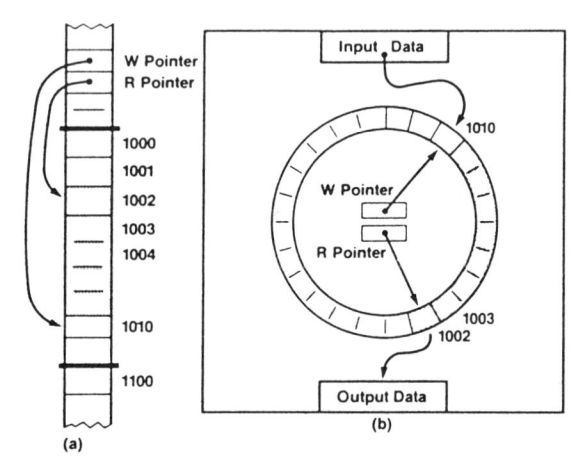

Figure 4. Circular queue with two pointers
  a) As it is in memory.
  b) Logical view.

WIDER FIFOs

Applications may vary widely as to the width and depth of the FIFO required. If an application's maximum requirement is 1024 locations or less and 9 bits in width or less, then the IDT7202 will fit. Wider word widths can be achieved by connecting two or more devices in parallel (control signals). The status flags can be detected from any one device because each device is working in lock step parallel. Figure 6 shows an example of an 18 bit-word composed of two IDT7201/7202 devices. The older classical architecture would require more external circuitry to match the Input Ready and Output Ready signals to account for differences in the internal self-generated clock frequencies. RAM-based FIFOs, such as the IDT7201/7202, do not have this problem.

DEEPER FIFOs

Some applications require deeper FIFOs. In the older architecture, deeper FIFOs mean longer fall-through times because they are connected end to end. The time increases in direct proportion to the number of devices. For example two devices yield a maximum fall-through time of twice that of one device. This can make some applications of FIFOs impractical or totally unusable.

With the two pointer approach used in the IDT7201/7202, the data input busses are connected together and the data output busses are common. This produces a parallel architecture (see Figure 7) as opposed to the serial approach above. The parallel structure is analogous to cascading standard RAM devices to achieve deeper memories.

Since FIFOs do not have chip selects and external decoding mechanisms, the task of choosing which device is selected must be provided for internally. The control (in the IDT7201/7202) is achieved through a unique serial structure. The first (or master)

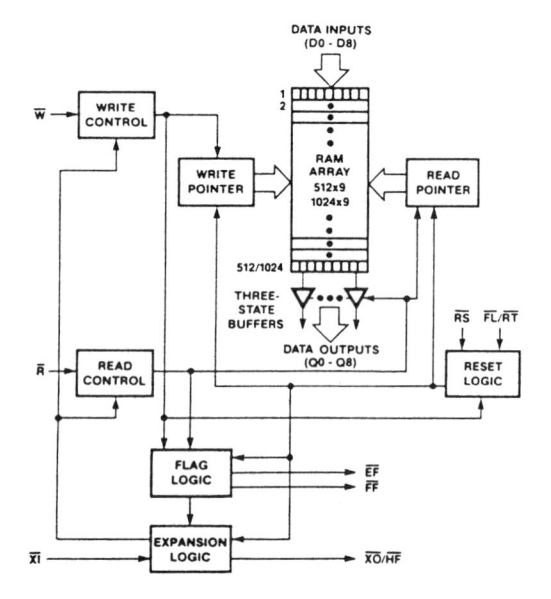

Figure 5. Functional Block diagram of IDT7201/7202 FIFO.

14-3

UNDERSTANDING THE IDT7201/7202 FIFO

APPLICATION NOTE

FIFO is identified by grounding the $\overline{FL}$ input. All other FIFOs in the structure must have the $\overline{FL}$ input pulled up to V_{CC}. The $\overline{XO}$ output of the first FIFO is connected to the $\overline{XI}$ input on the next FIFO in the queue. The $\overline{XO}$ output of that FIFO is connected to the $\overline{XI}$ input of the next and so on until the $\overline{XO}$ output of the last FIFO is connected to the $\overline{XI}$ input of the first FIFO (see Figure 7).

After reset, the active read and write pointers are in the first device. When the write pointer has progressed to the end of the first FIFO device, it outputs a pulse on $\overline{XO}$ which activates the write pointer at the beginning of the next device and simultaneously deactivates the write pointer in the first device. Thus, write enable control is passed to the second device. When the

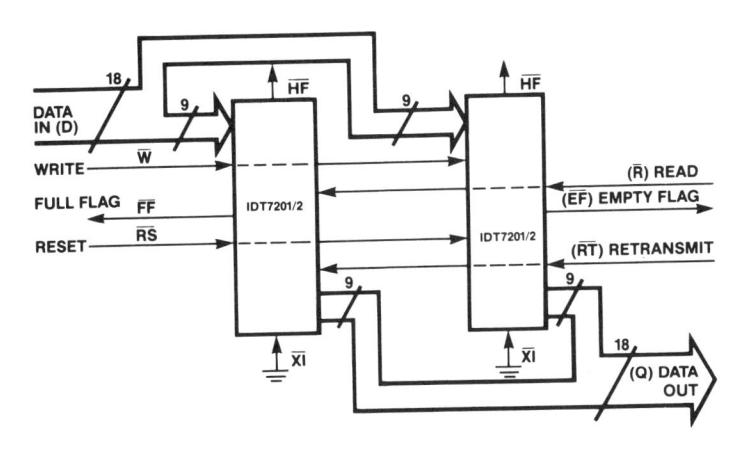

Figure 6. IDT7201/7202 FIFO Word-Width Expansion.

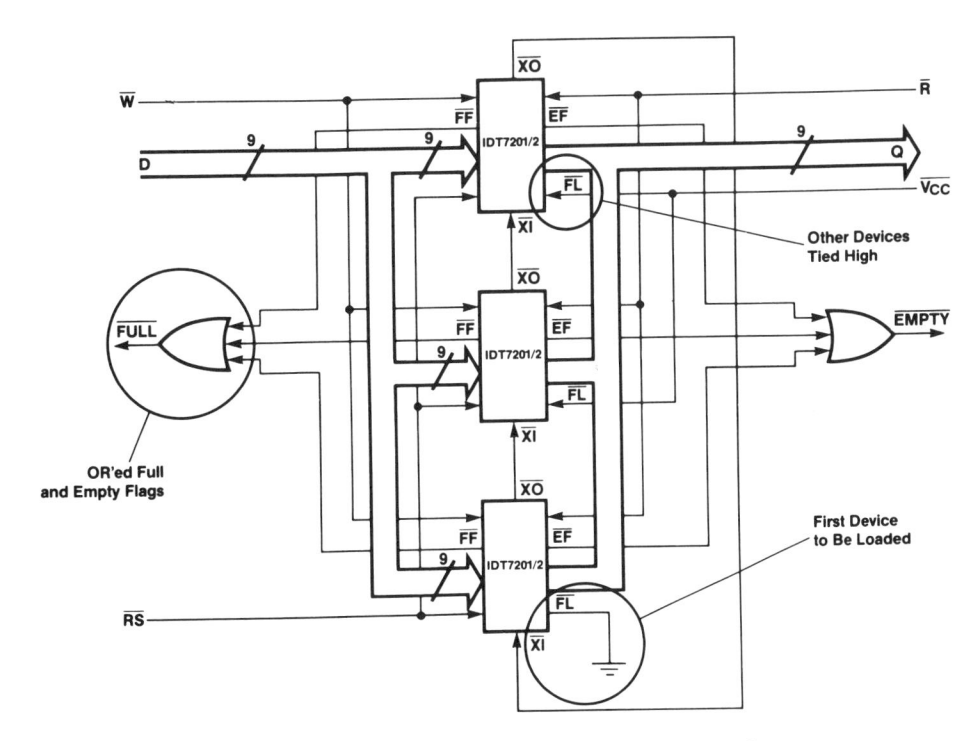

Figure 7. IDT7201/7202 FIFO Word-Depth Expansion.

UNDERSTANDING THE IDT7201/7202 FIFO　　　　　　　　　　　　　　　　　　　　　　　　　**APPLICATION NOTE**

active read pointer reaches the end of the first device, it terminates and activates the read pointer in the next device with another pulse on the $\overline{XO}$ output of the first device. Figure 8 shows the progression of read and write pointers across two devices. In this ring structure, the read pointer is always chasing the write pointer. The pointer enable crosses the device boundaries via sending an $\overline{XO}$ pulse onto the next device. This continues in a circular queue fashion.

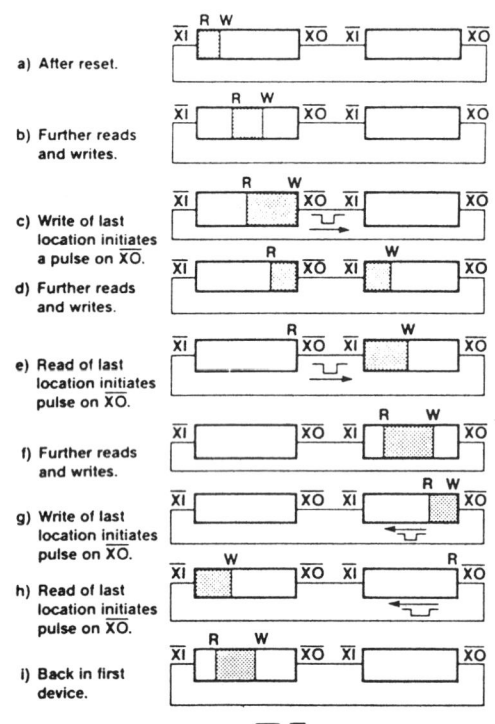

a) After reset.
b) Further reads and writes.
c) Write of last location initiates a pulse on $\overline{XO}$.
d) Further reads and writes.
e) Read of last location initiates pulse on $\overline{XO}$.
f) Further reads and writes.
g) Write of last location initiates pulse on $\overline{XO}$.
h) Read of last location initiates pulse on $\overline{XO}$.
i) Back in first device.

Figure 8. Example on $\overline{XO}/\overline{XI}$ expansion scheme.

The IDT7201/7202 has been designed such that the read and write pointer can never cross over each other even in the cascade mode. The $\overline{XO}$ pulse is synchronous with read and write. When the last location is read or written, the $\overline{XO}$ output goes low with the read or write enable input and back high with the read or write enable. To see why there is no conflict even though reads and writes are asynchronous, the usage must be examined. The case of concern is when the FIFO is empty and the read and write pointers are at the last location. It must be realized that the consumer will not read until the empty flag is deasserted. The empty flag output will go high after the write pulse has gone high again thus ensuring that the $\overline{XO}$ pulse, indicating the write pointer, has been passed on to the next device. The consumer will then read the last location causing another pulse on $\overline{XO}$ which will transfer the read pointer (see Figure 9).

There is one special case regarding read flow-through mode (discussed below). In this mode the consumer can anticipate the write, by producer, by lowering the read enable input. In this case the $\overline{XO}$ input does not go low with read enable. When write enable is lowered, $\overline{XO}$ goes low. $\overline{XO}$ goes high with write enable. At this point the empty flag is cleared, thus signaling to the consumer to terminate the read after the appropriate period specified in the data sheet. During this period the $\overline{XO}$ output, which went high at the end of the write enable pulse, has lowered again. When the read enable is raised by the consumer, the $\overline{XO}$ output goes high. In this way two pulses on $\overline{XO}$ are assured (see Figure 9).

a) Regular case.

b) The read-flow through case.

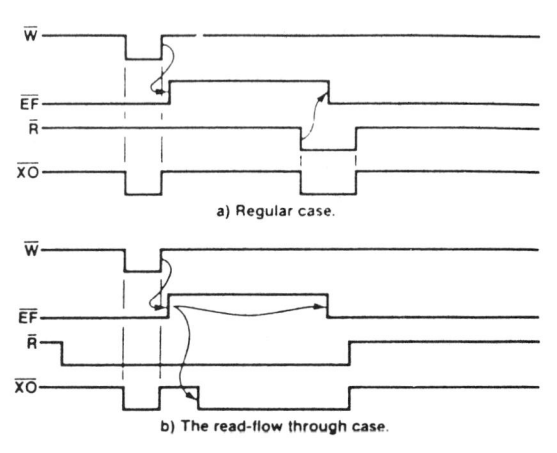

Figure 9. Generation on XO output when the FIFO is empty.
a) Regular case.   b) The read-flow through case.

Two examples of the IDT7201/7202 in expanded depth configuration are available from IDT commercially. The IDT7M203/204 are Subsystems modules which incorporate onto one ceramic substrate four FIFO LCCs and the $\overline{EF}$ & $\overline{FF}$ "OR" gating to produce 2Kx9 and 4Kx9 FIFOs. The Subsystem module has a lead frame which pins out like the 28-pin 0.6 inch IDT7201/7202. This allows for a plug compatible 4Kx9 FIFO in one socket.

SPECIAL FEATURES OF IDT7201/7202

The architecture used in the IDT7201/7202 provides some features that distinguish it from FIFOs with other architectures. One outstanding feature is the dual port implementation of the RAM array. The RAM is designed in such a way that the read and write ports are separate, allowing for simultaneous asynchronous reads and writes with no hand shaking or arbitration. In the classical architecture the consumer and producer circuits must monitor ready flags for each access.

The IDT7201/7202 support a retransmit function. In the single device solution, the $\overline{FL}/\overline{RT}$ input may be pulsed low signaling a retransmit.

A retransmit operation will set the internal read pointer to the first location and will not affect the write pointer. READ ENABLE ($\overline{R}$) and WRITE ENABLE ($\overline{W}$) must be in the high state during retransmit. This feature is useful when less than 512/1024 writes are performed between resets. The retransmit feature is not compatible with Depth Expansion Mode and will affect HALF FULL FLAG ($\overline{HF}$) depending on the relative locations of the read and write pointers. For example in a communications application, during transmission of a message, the receiver may request a retransmit of the message. This can be accomplished by always starting new messages at the beginning of the queue via a pulse on the reset input. If and when the retransmit request arrives, the $\overline{FL}/\overline{RT}$ line is pulsed. The read pointer is repositioned at the beginning of the queue. The message producer may continue to write more of the same message into the queue as the retransmit

UNDERSTANDING THE IDT7201/7202 FIFO **APPLICATION NOTE**

of the message continues. The retransmit can happen as many times as desired. At the start of the next complete message, the reset line ($\overline{RS}$) must be pulsed after the successful acknowledge by the receiver. The reset ensures that the new message will be placed in the FIFO at the start of internal queue. It should be noted that, when retransmit is possible, messages cannot be bigger than the maximum size of the queue. If the message is longer than the queue, even though the read pointer has progressed far enough to accommodate the extra data, resetting the read pointer back to the beginning with retransmit will produce data from the end of message instead of the beginning.

This architecture supports flow-through modes. In the read flow-through mode, when the buffer is empty, the consumer can anticipate the write, by the producer at the other end, by lowering the read input. When the empty flag ($\overline{EF}$) goes false, the consumer circuitry can terminate the early read cycle by reading the data and deasserting the read signal. The read input must go high for a brief period in order to clock the read pointer. The read flow-through mode avoids the standard sequence of monitoring flag going high before hitting a read cycle.

The write flow-through mode is a mode that is employed when the FIFO is full. The producer can anticipate a read by the consumer by lowering the write input before the read. When the full flag ($\overline{FF}$) raises, the producer knows that the consumer has read a location, thus freeing up a location that can receive the new data. The producer then raises the write input which actually writes the data into the RAM array. This flow-through mode avoids the overhead of monitoring the full flag before initiating a write cycle.

The IDT7201 is pin and functionally compatible with the Mostek MK4501, thus serving as an alternate source. The IDT7202 gives the same functionality as the IDT7201 but is twice as deep (1024x9). The IDT7202 is the largest FIFO made with the zero fall-through time architecture making it the logical choice for FIFO applications.

SOFTWARE VERSUS HARDWARE SOLUTIONS

With every application involving a computer or programmed controller, the designer can trade off between performing certain functions in software or hardware. In general, the software solution is a more flexible design (easily changed) but performs the task more slowly. The hardware solution is less flexible but performs the task very fast.

To clarify these concepts, a discussion of an application and how it could be solved at the various levels from software to hardware is beneficial. A good example is a file server. The server could be connected to a Local Area Network (LAN) and, on the other side, to a Winchester disk drive. Both I/O connections demand attention at unpredictable intervals and must be serviced on demand or data is lost.

If the data rate of both interfaces is sufficiently low, a total software solution might be considered. The data rate would have to be low enough such that the software code could poll the status of either I/O port. As data arrives it could be placed into software FIFO queues. When a full record is buffered, then processing would commence. During the processing, the I/O ports must still be monitored as another user on the LAN might make a request (see Figure 10). It is doubtful that a total software solution could be designed for the server application that would have acceptable system performance.

The next approach to consider might be to include hardware interrupts. Interrupts allow for one task to be running and almost immediately switching to an I/O service routine. Interrupts are something like a hardware subroutine call. This scheme would use the interrupt mechanism to call routines to move data to and from the I/O ports and the software FIFO queues. The overhead of constantly polling the I/O port status flags would be eliminated, thus allowing for higher system performance. An asynchronous-type problem is introduced with interrupts. To use interrupts properly, the I/O service routines may be called at any instance. Therefore, the interrupt routines must be designed in such a way that they do not destroy data that the interrupted task might be using. Usually, the routines must be careful to save the state of the machine, perform their task and restore the state of the machine. The extra code to maintain the state of the machine is an overhead that is not in the polled solution. Worse yet, saving the state of the machine may be too much overhead to allow for an interrupt during a time-critical piece of code. Because interrupts may not be acceptable at certain points in the code, the programmer must insert code to disable and re-enable interrupts around the critical sections.

Where the polling scheme provides a solution which has a more easily definable sequence of execution, the interrupt solution is indefinite. The programmer must spend a lot more time proving that all possible sequences caused by random interrupts will produce desirable results. Because interrupts may not be acceptable at certain points in the code, the programmer must insert code to disable and re-enable interrupts around the critical sections. The interrupt disable solution not only cuts performance by not accepting I/O during some periods, but also adds more overhead with the maintenance of the interrupt enable mechanism. In some sense, interrupts can be to software what the meta-stable flip-flop problem is to hardware.

The interrupt solution can be moved out of the software and more into the hardware realm through the use of a technique called Direct Memory Access (DMA). The DMA solution is provided by a block of circuitry which monitors the I/O ports. When the port requires attention, the DMA logic interrupts the current task at the bus transfer level and steals a memory cycle to transfer the data to or from the port and the FIFO queue in memory. The task that is running on the processor misses only a few memory cycles now and again which is much less than in the interrupt scheme where a whole subroutine of many memory cycles was executed to transfer each element of data. The DMA solution is not for free. DMA controllers are complex devices which must be programmed as well as designed into the bus structure. The DMA mechanism can only serve one source at any given instance in time thus still being a bottleneck in throughput.

So far, each solution proposed has moved the mechanism that feeds data to or from FIFOs in program memory away from the software and closer to the I/O port. The memory bus still remains the bottleneck because both FIFO queues are in memory. To simplify and improve performance, hardware FIFOs such as the IDT7201/7202 can be used. The processor would interface to the FIFO through an I/O port as before, but the FIFO would now be between the I/O port and the rest of the hardware. The software could then service the data at a steady rate and be sure that data was not lost without the problems or overhead of more complicated schemes such as interrupts or DMA.

Because the queues are between the controller and the peripheral, the peripheral can load or read the queue without interrupting the controller. Since the controller is not involved

UNDERSTANDING THE IDT7201/7202 FIFO   APPLICATION NOTE

with maintaining both queues, there is no possibility of lost data because one queue was being serviced while data for the other queue arrived. For these reasons the hardware FIFO represents the highest performance solution.

If the designer uses large FIFOs like the IDT7202, there is a minimum of device count. Assuming 2 FIFOs (transmit and receive) for each I/O port gives a count of four 28-pin devices for the FIFO solution. The DMA solution would at least be one 40-pin device and several bus buffer/control devices. The interrupt solution would require a similar parts count to the DMA solution. Therefore, the FIFO solution is not only the highest performance solution but usually has the lowest part count of the hardware solutions.

a) The total software solution.

b) The interrupt solution.

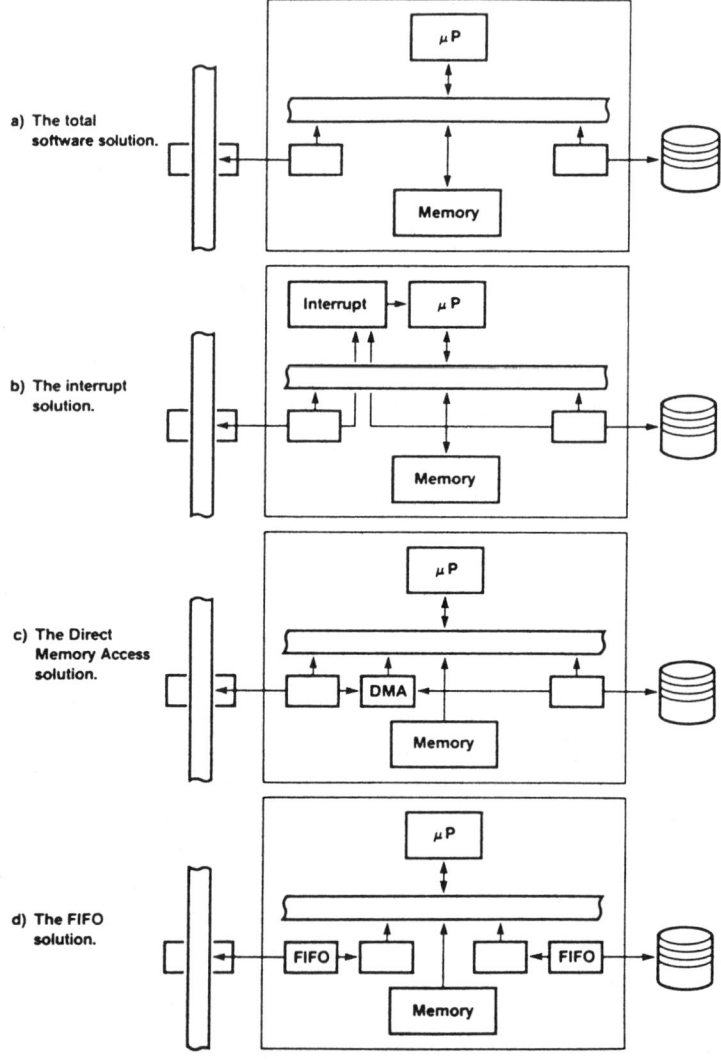

c) The Direct Memory Access solution.

d) The FIFO solution.

Figure 10. Example solutions for File Servers.

UNDERSTANDING THE IDT7201/7202 FIFO APPLICATION NOTE

COMMUNICATIONS-MULTIPLEXOR APPLICATION

Another example of a rate mismatch problem is shown in a CRT terminal and CPU interface. In order to not load the CPU with the burden of monitoring the UARTs of multiple CRTs and printers, a communications controller is employed. The controller can serve as a communications multiplexor and data concentrator (see Figure 11).

As the controller receives characters it must buffer them such that if multiple characters are received close together from several terminals, they will not be lost as more characters come in. The natural structure to store them in is a queue of the FIFO type. The CPU will then need to respond to the characters. If the controller is inputing other characters, the CPU should not have to wait until the controller is done. Therefore, a FIFO can be employed on the transmit side as well as the receive side. To make the design simple, two sets of FIFOs could be placed between the CPU and controller. When characters are received they are placed in one end of a FIFO and read from the other end by the CPU. As the CPU prepares characters for transmission, it places them in a FIFO going the other direction. The controller then reads them from the other end of the transmit FIFO and sends them out through the UART.

Conceivably, there could be a pair of FIFOs for each UART. That way it would be easy for both the controller and the CPU to keep straight which characters correspond with which UART. While this provides for a large total of buffer space for characters, it is more than needed when using a part like the IDT7201/7202. For eight UARTs, this scheme would require a minimum of sixteen FIFO devices. A better solution would be to use one FIFO device in either direction. If an IDT7202 were used, it could provide a maximum of up to 128 characters per UART if all the UARTs input at the same time and rate. While the two FIFO techniques would most likely provide plenty of buffering at a minimal device count, it presents the problem of which character belongs to which UART. The solution is to make a wider FIFO which is 18 bits wide; thus using 4 devices instead of 16 devices for 8 UARTs. This would allow for a UART number to be placed in the FIFO along side each character. The remainder of the word could be used for flag, status and command information between the CPU and the controller. For example, several of the bits in the FIFO word could indicate whether the character information was a character to send or BAUD change rate information.

The empty and full flags of the IDT7201/7202 FIFO would be used as status flags. For example, the transmit buffer must be monitored from both sides. As the CPU prepares a character to transmit, it would first examine the full flag ($\overline{FF}$) to see if the FIFO is full. If the FIFO was full, it would delay outputting the character. If the buffer is not full then it would place the character in the FIFO. The empty flag ($\overline{EF}$) would be monitored by the controller. As soon as the CPU places a character into an empty FIFO, the empty flag would change to not true. At this point the controller would know there was a character in the buffer which could be transmitted. The controller would read characters from the buffer as long as the empty flag was not true (buffer contains more than one character).

CONCLUSION

Hardware FIFOs are an economical memory organization to use when lists of data items are to be buffered. Because they do not require an address to access items in the list, there is less overhead in terms of circuitry and access time. The FIFO buffer is most often used as a "system rubber band" to stretch between the differing and fluctuating rates of different elements in a system. The IDT7201/7202 FIFO device features the newest RAM-based architecture and provides the latest in technology in terms of access time, fall-through time and size, thus providing the most economical solution for today's design needs.

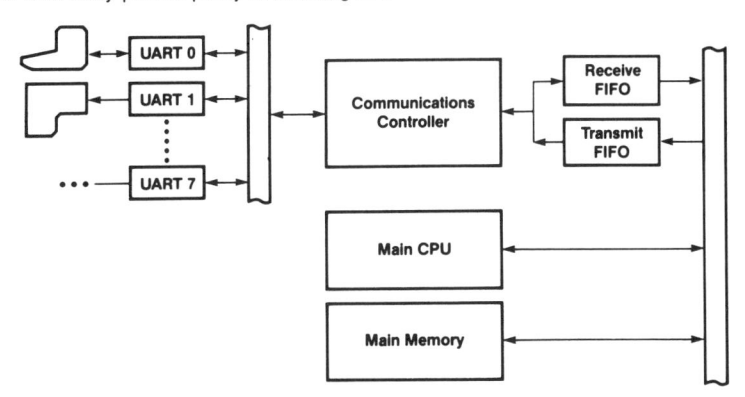

Figure 11. Communications Controller example.

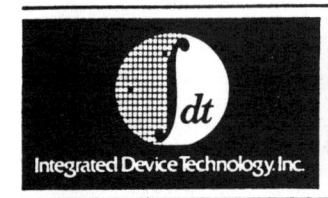

DUAL-PORT RAMS SIMPLIFY COMMUNICATION IN COMPUTER SYSTEMS

APPLICATION NOTE AN-02

By David C. Wyland

INTRODUCTION

Dual-port RAMs allow two independent devices to have simultaneous read and write access to the same memory. This allows the two devices to communicate with each other by passing data through the common memory. These devices might be a CPU and a disc controller or two CPUs working on different but related tasks. The dual-port memory approach is useful and popular because it allows the same memory to be used for both working storage and communication by both devices and avoids the need for any special data communication hardware between the devices. The latest development in dual-port RAMs has been the appearance of high speed dual-port RAM chips. These chips allow high speed access by both devices with the minimum amount of interference and delay. Integrated Device Technology offers a family of these devices as shown in Table 1.

Width	Size	Part	Interrupt	Busy Logic MASTER	Busy Logic SLAVE	Semaphore	Comments
X8	1K	IDT7130	X	X			
		IDT7140	X		X		
	2K	IDT7132		X			
		IDT7142			X		
		IDT71321	X	X			52-pin
		IDT71421	X		X		52-pin
		IDT71322				X	
	4K	IDT7134		X			
		IDT71342				X	52-pin
X16	2K	IDT7133		X			
		IDT7143			X		

Table 1. Dual-Port RAMs Available from Integrated Device Technology

DUAL-PORT RAMS: SIMULTANEOUS ACCESS

A dual-port memory has two sets of address, data and read/write control signals, each of which access the same set of memory cells. This is shown in Figure 1. Each set of memory controls can independently and simultaneously access any word in the memory including the case where both sides are accessing the same memory location at the same time. Up to this time, there have been very few true dual-port memories available. Memories have a single set of controls for address, data and read/write logic and are single-port RAMs. If you wanted a dual-port RAM function, you had to design special logic to make the single-port RAM simulate a dual-port RAM in operation.

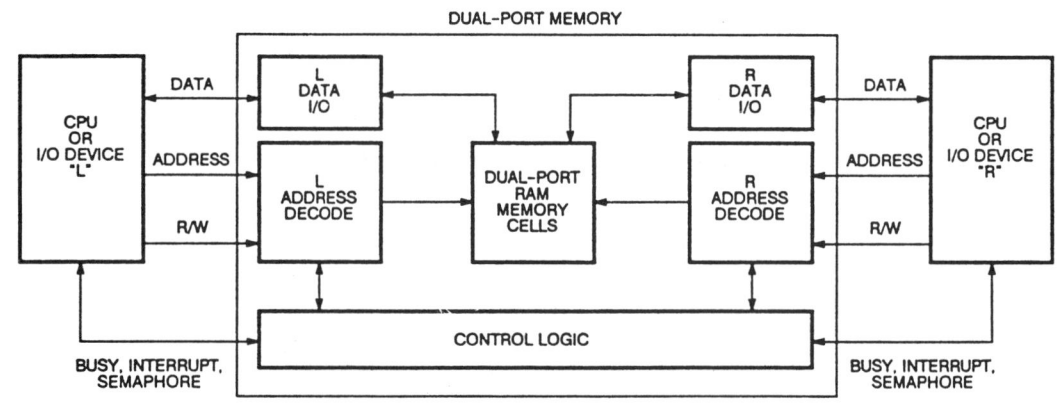

Figure 1. Dual-Port Memory Block Diagram

**DUAL-PORT RAMS SIMPLIFY
COMMUNICATION IN COMPUTER SYSTEMS**

APPLICATION NOTE AN-02

Direct Memory Access (DMA) as a Dual-Port Memory Simulation

The concept of using a conventional memory to simulate a dual-port RAM has been common in computer systems almost from the beginning. It is known under the name Direct Memory Access, or DMA. In the DMA concept, a single memory is shared between the CPU and one or more I/O devices as shown in Figure 2.

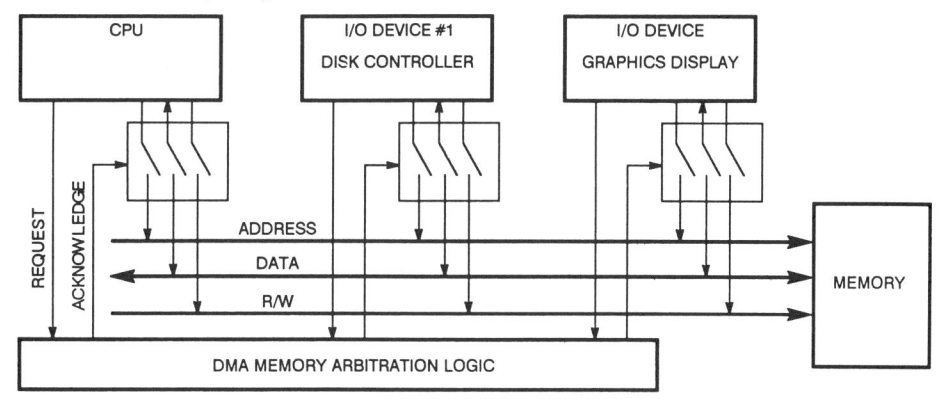

Figure 2. DMA Memory System Block Diagram

Each device wishing to use memory submits a request to the arbitration logic. The arbitration logic responds by connecting the memory address, data and control lines to one of the requesters and tells any other requesting devices to wait by issuing a busy signal. The busy signal causes the memory access logic in the device to wait until busy has gone away before performing a memory transfer.

DMA Limitations: Waiting for the Bus

In a computer system with DMA, the CPU must stop and wait while an I/O device is doing DMA transfers to memory. This works well in typical systems where the I/O devices are transferring data only a small percentage of the time and the impact on CPU processing time is minimal. These assumptions do not hold where you have two CPUs trying to use the same memory. In this case, one CPU must wait while the other uses the memory. As a result, the average speed of the CPUs will typically be cut in half.

There are two solutions to this problem: 1) You can provide local memory for both CPUs and limit use of the common memory to CPU/CPU communication only, in an attempt to reduce the time impact of DMA waiting, or 2) you can provide true hardware dual-port memory between the CPUs and all simultaneous high-speed access by both CPUs to the same memory without waiting. The introduction of high-speed dual-port RAM chips now makes the second option practical.

Dual-Port RAM Chips: How They Work

A true dual-port memory allows independent and simultaneous access of the same memory cells by both devices. This means two complete and independent sets of address, data and read/write logic and memory cells that are capable of being read and written by two different sources. An example of the dual-port memory cell is shown in Figure 3. In this cell both the left and right hand select lines can independently and simultaneously select the cell for read out. In addition, either side can write data into the cell independent of the other side. The only problem would be when both sides try to write into the same cell at the same time. We will discuss this in a moment.

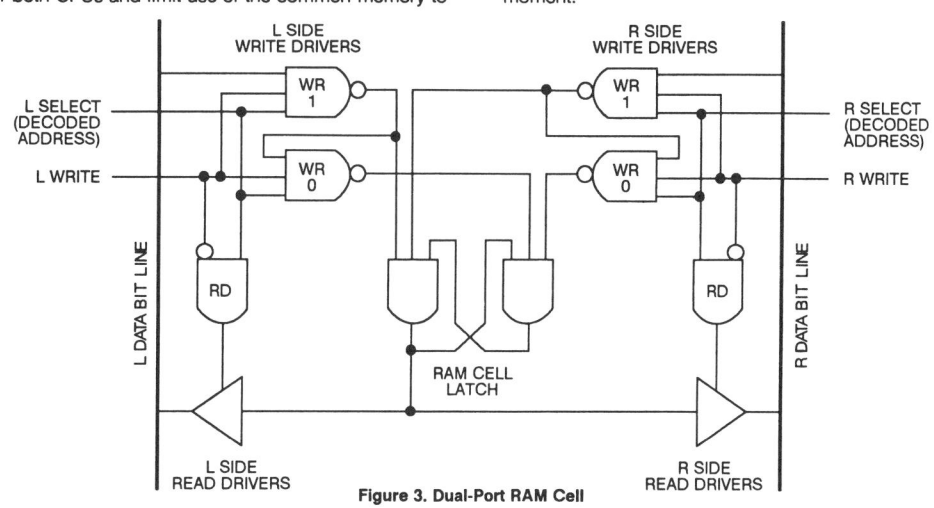

Figure 3. Dual-Port RAM Cell

DUAL-PORT RAMS SIMPLIFY
COMMUNICATION IN COMPUTER SYSTEMS

DUAL-PORT RAM CONTROL LOGIC

Dual-port RAM chips include control logic to solve three common application problems: signaling between processors, timing interactions when both are using the same location and hardware support for temporary assignment (called allocation) of a block of memory to one side only.

Interrupt Logic For Signaling

A common problem in dual-processor systems is signaling between the processors. For example, processor A needs to signal processor B to request a task to be performed, as defined by data in the common memory. When processor B has completed the task, it needs to signal processor A that the task is done. Note that the signaling must occur in both directions. A common form of signaling is for one processor to cause an interrupt on the other processor. This allows the receiving processor to be informed of a communication without having to constantly check for it.

Hardware support for this signaling function is provided by interrupt logic, available on certain IDT dual-port RAM chips. A block diagram of this logic is shown in Figure 4. In these chips, the top two addresses of the memory chip also serve as interrupt generators for each of the ports. If the left side CPU writes into the even address of this pair (3FF in a 1K RAM) an interrupt latch is set and the interrupt line to the right hand port is activated. This interrupt latch is cleared when the right hand CPU reads from the even address. A similar set of logic is provided to allow the right hand CPU to interrupt the left hand one. This logic is associated with the odd address of the pair (3FE in a 1K memory). Providing this logic on chip saves the system designer from having to design in extra logic to allow one CPU to interrupt the other.

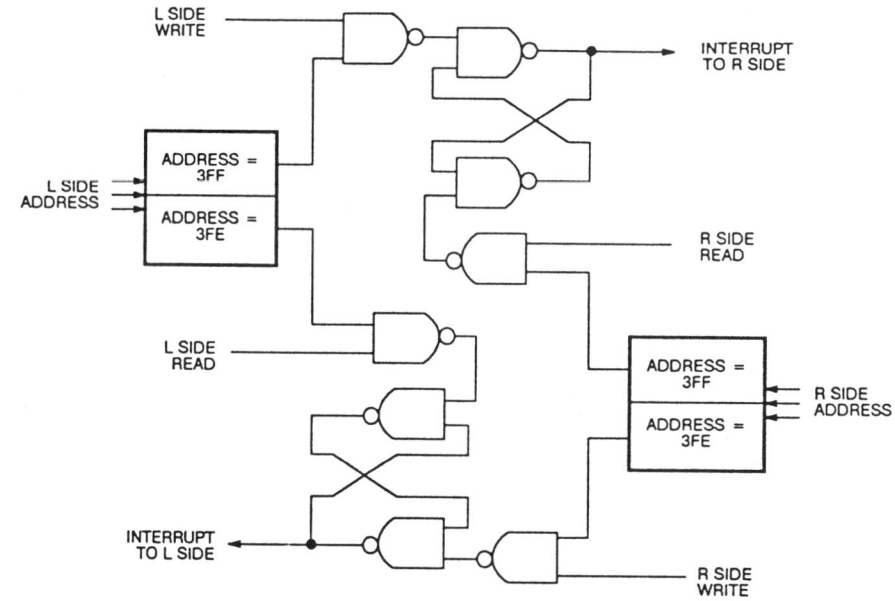

Figure 4. IDT7130 Interrupt Logic

Busy Logic Solves Interaction Problems

A problem can occur with dual-port memories when both ports attempt to access the same address at the same time. There are two significant cases: when one port is trying to read the same data that the other port is writing and when both ports attempt to write to the same word at the same time. If one port is reading while the other port is writing, the data on the read side will be changing during the read and a read error can be caused. If both ports attempt to write at the same time, the memory cell is being driven by both sides and the result can be a random combination of both data words rather than the data word from one side or the other. Busy logic solves this problem by detecting when both sides are using the same location at the same time and causing one side to wait until the other side is done.

Note that although one or the other processor may have to wait occasionally, the throughput loss is minimal, typically less than 0.1%. This is because the probability of both processors using the same location at the same time is small. For example, if there are a thousand words in memory with a relatively uniform and random access of these locations by either side, the probability of a given location being accessed by one side is of the order of one part of a thousand. The probability of both sides accessing the same location at the same time is, therefore, of the order of one part in a million. As a result, the average throughput of the system is reduced by only one part per million due to dual-port RAM access contention (again, assuming uniform random address access by both sides).

Busy Logic Design

Busy logic is called hardware address arbitration logic because it consists of hardware that decides which side will receive a busy signal if the addresses are equal. It consists of common address detection logic and a cross coupled arbitration latch. A logic diagram of the type of busy logic used in the IDT dual-port RAM chips is shown in Figure 5. The purpose of this logic is to provide a busy signal for the address that arrived last, to inhibit writing to the busy port and to make a decision in favor of one side or the other when both addresses arrive at the same time. This logic consists of a pair

DUAL-PORT RAMS SIMPLIFY
COMMUNICATION IN COMPUTER SYSTEMS

APPLICATION NOTE AN-02

of address comparators, a pair of delay buffers, a cross-coupled latch and a set of busy output drivers. The address comparator output goes true when the addresses at its inputs are equal.

In the logic shown in Figure 5, the ability to detect which address arrived last is provided by the time delay buffers between address lines and the comparators. If we assume that the L address is stable and the R address changes to match the L address, the R address comparator will go true immediately while the L address comparator will become active some time later as determined by the time delay gates.

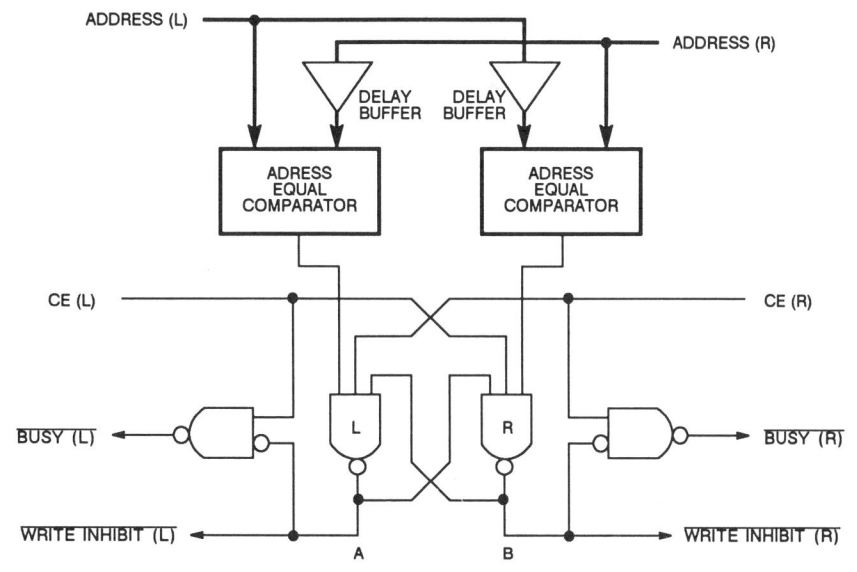

Figure 5. Dual-Port Busy Logic Design

The arbitration latch formed by the L and R gates reflects the address comparator output timing. This latch has three stable states, both latch outputs A and B high, A low/B high and A high/B low. Initially, both A and B are high because the outputs of both address comparators are low. We start with the L address stable and the R address arriving later. When the R comparator becomes active its output will go high and B will go low. The A output will remain high because its address comparator input will go high sometime later and the L gate input from B output will go low before this occurs. The result is that the R gate B output will be active inhibiting writing to the R side of the dual-port RAM and activating the busy signal to the R port.

The extreme case of busy logic decision making is when both addresses arrive at exactly the same time. In this case, the outputs of both address comparators go high at the same time activating both sides of the arbitration latch. The latch will settle into one of two states with either the A or the B latch output being active. The latch design ensures that a decision will be made in favor of one side or the other.

The chip enable lines come directly into the arbitration latch, although they could have been brought into the address comparators along with the other address lines. This is because if the chip enable for one side is inactive, both reading and writing for that side is automatically inhibited and/or arbitration is not needed. If the addresses are equal, the chip enable that arrives last will lose the arbitration. If both chip enables are active then arbitration will be determined by the settling of the address lines.

Temporary Assignment of Memory to One Side

A common problem in dual-port RAM application is the need to temporarily assign a block of memory to one side. For example, sometimes you need to update a data table as whole and you cannot allow the other processor to use the table until you are done. This is called block allocation of the memory.

Block allocation can also be used to avoid the address arbitration problem since it is a way of ensuring that both sides do not use the same address at the same time. This method is also called software arbitration because the software on both sides decides and agrees as to who has permission to use a given portion of the memory. Software allocation has the advantage of not requiring busy logic, which is useful in systems which cannot accommodate a busy signal.

The design problem with block allocation is communication of the assignments between the CPUs. A simple but time consuming method is to pass messages between the CPUs, perhaps aided by interrupt logic. In the message method, processor A requests use of a block from processor B. Processor B agrees and sends permission back to processor A. When A is finished it sends a release message to B which responds with a release acknowledge to A. In this system, four messages are sent for each block assigned and released.

Semaphore Logic Support for Memory Assignment

Although block allocation is a software technique, it can benefit from hardware support. In message passing allocation, four messages must be passed to assign and release a block of memory. Semaphore logic, available in certain IDT dual-port RAMs, can be used to eliminate this message passing and its associated overhead. Semaphore logic provides a set of flags especially designed for the block assignment function. Each flag is used as a token to indicate which CPU has permission to use a block of memory.

**DUAL-PORT RAMS SIMPLIFY
COMMUNICATION IN COMPUTER SYSTEMS**

APPLICATION NOTE AN-02

Each semaphore flag can be set to one side or the other but not both. This ensures that only one side has permission to use the block of memory.

The IDT semaphore logic bits are designed to be used in a set-and-test sequence. Each bit is normally in the logic one state, indicating that it is not assigned to either side. A processor, desiring to assign a bit and, therefore, its associated block of memory, attempts to write a zero into the bit. It then reads the bit to see if it was successful. If it was, the bit will read zero, and the processor has use of the block. If it reads a one, it was unsuccessful, and the block is in use by the other side. The processor must then wait until the bit becomes zero, indicating that the other side has released it.

Semaphore flags have a particular requirement: a given flag can be assigned to only one side at a time. Specifically, you must not have a situation where both sides simultaneously think they have permission to use a block. Semaphore logic is designed to resolve this problem. If both sides attempt to set a semaphore flag at exactly the same time, only one side sees it set.

Semaphore flags consist of eight individually addressable dual-port latches. Each latch can be read and written by either side. They are selected by a separate chip enable, addressed by the three last significant bits of the address lines and are read and written through the D_0 data bit. Except for sharing the address, data and read/write pins of the RAM, the semaphore latches are completely independent, as shown in Figure 6.

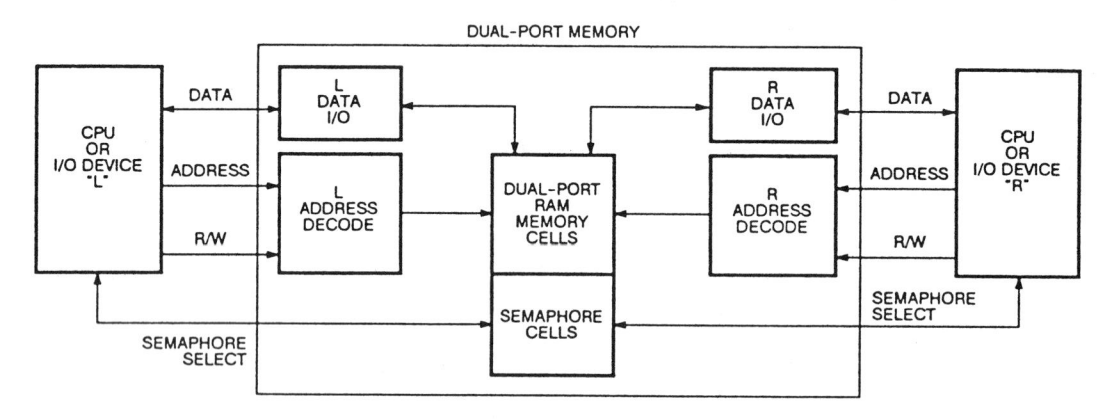

Figure 6. Dual-Port RAM Semaphore Logic

A logic diagram of a semaphore logic flag is shown in Figure 7. In this logic, both flip-flops are initially at logic one and both Grant outputs are high. If only one flip-flop is set to zero, its corresponding Grant output will go to zero. If the other flip-flop is set later, this will have no effect. If both flip-flops are set at the same time however, the latch will settle so that only one Grant output goes low, ensuring that only one side receives permission to use the resource.

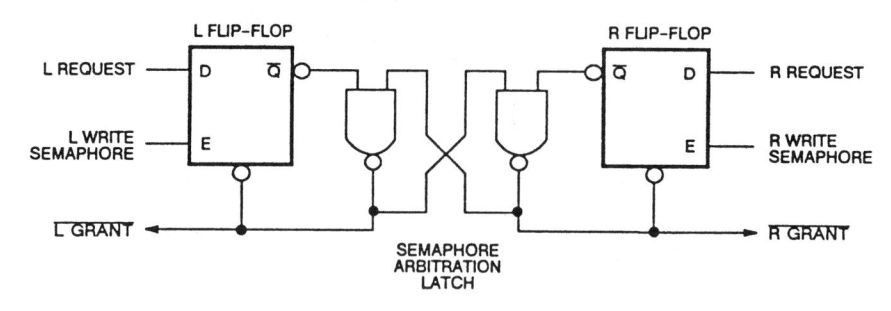

Figure 7. Semaphore Logic Design

DUAL-PORT RAM CHIP TIMING

The dual-port RAM has a simple static RAM interface and timing requirements. There are some special requirements associated with Busy, however. A timing diagram, shown in Figure 8, shows the relationships between address, data, read/write, chip select and busy signals for a dual-port RAM chip and busy logic. In this diagram, the chip select is used to enable the chip for a read or write operation after the addresses have settled. An arbitration is performed at the leading edge of the chip select.

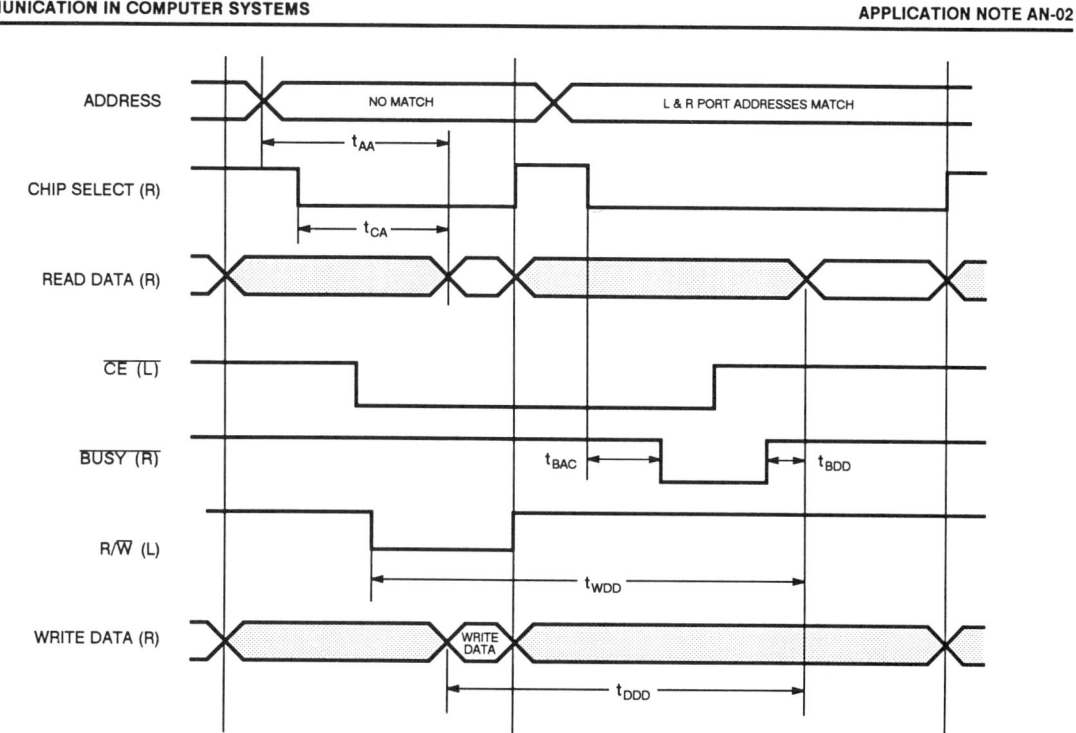

DUAL-PORT RAMS SIMPLIFY
COMMUNICATION IN COMPUTER SYSTEMS

APPLICATION NOTE AN-02

Figure 8. Dual-Port RAM Timing Diagram

Busy Logic Timing

In the case of address contention, the busy signal from the losing RAM port stabilizes some time after the leading edge of its chip select (or after its address settles, whichever comes last). If the busy signal is going to become active, it will become active during this time or not at all. If the busy signal is generated, the CPU must wait for busy to go away before completing the read or write cycle. Once the busy signal has gone high the memory read or write cycle can proceed to completion.

Note that during the arbitration time following the chip select the busy signal may be changing. Since it is possible to have a glich on the busy line during this indeterminate period, the busy line should be sensed as a level rather than as an edge.

Busy arbitration will be somewhat slower in the extreme case where both addresses arrive at exactly the same time. This is because both gates of the arbitrator latch are initially inactive and must settle into a state where only one of them is active. There will be a period of time when both gates are in transition. This is called the metastable condition and is a classic and unavoidable problem in latch and flip-flop design. As a result, the busy settling time is somewhat longer in the low probability worst case than in the commonly observed typical case. The maximum arbitration times, t_{BAA} and t_{BAC}, on the data sheet give the worst case values, including metastability setting, for these times.

Read/Write Timing with Busy

The read and write timing for either port of the dual-port RAM chip is the same as a simple static RAM in the absence of address contention. All the standard timing measures apply: read data address access time is t_{AA}, etc.

Dual-port RAMs have additional timing specifications for the case of address contention where one port is busy and waiting for access. For the most general and conservative case, the read or write cycle for the waiting side should begin after the busy signal goes away. The actual timing can be somewhat shorter than this in most cases.

For the case where the waiting side is waiting to write, the write timing requirement is that the write pulse width be measured from busy going away. For the case where both sides are reading, the data will be available at the outputs one access time after the address/chip select lines settle even though the busy line is active. In the most common case, the trailing edge of busy will occur more than one access time after the address and data for the busy side have settled. As a result, the read access time as measured from the trailing edge of busy, for this case t_{BDD}, is effectively zero.

The write/read case, waiting to read while the other side is writing to the same location, has some additional timing specifications. Since writing to a location by the L side, for example, will involve changing the data the cell being read by the R side, there is a write-to-read propagation delay time. This time is t_{WDD} for the delay for constant write data from the leading of the write pulse to the read data, and t_{DDD} for the delay for changing write data from a change of the write data to the read data.

If the writing side is running at minimum values for the write pulse or write data set-up times, the read access time, t_{BDD}, will no longer be zero. The actual t_{BDD} will be equal to t_{WDD} minus the actual write pulse width or t_{DDD} minus the actual write data set-up time, which ever is larger (and greater than zero). Note: t_{BDD} is always less than t_{AA} for the worst case of minimum write values. This is why the read or write cycle is begun from the trailing edge of busy for the most conservative case recommended above.

**DUAL-PORT RAMS SIMPLIFY
COMMUNICATION IN COMPUTER SYSTEMS**

APPLICATION NOTE AN-02

DUAL-PORT MEMORY EXPANSION: MAKING BIG ONES OUT OF LITTLE ONES

Dual-port RAM chips can be combined to form large dual-port memories. Expansion in memory depth with dual-port RAMs is similar to expansion in depth for conventional RAMs. An example of this kind of expansion is shown in Figure 9 where and 8K x 8 dual-port RAM has been made out of 2K x 8 dual-port RAM chips.

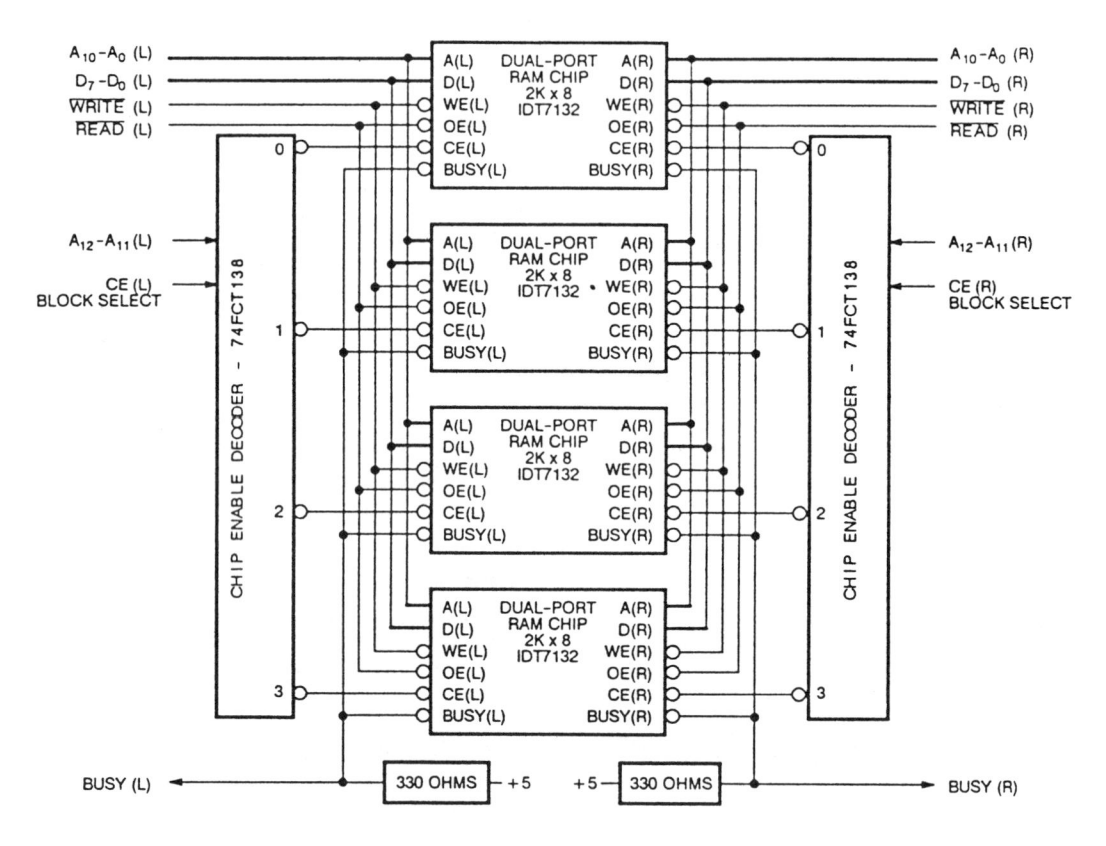

Figure 9. Depth Expansion of Dual-Port RAMs

Width Expansion: The Busy Lock-up Problem

Dual-port RAMs can also be expanded in width. However, in this case, we have a subtle problem. Expansion in width implies that several dual-port RAM chips will be active at the same time. This is a problem if several hardware arbitrators are active at the same time. If we examine the case of a 16-bit RAM made out of two 8-bit RAMs, we can better understand the problem. If the addresses for both ports arrive simultaneously at both RAMs, it is possible for one RAM arbitrator to activate its L busy signal and the other RAM to activate its R busy signal. If both busy signals are used on each side, we now have a situation where both sides are simultaneously busy. The system is now locked up since both sides will be busy and both CPUs will wait indefinitely for their port to become free.

DUAL-PORT RAMS SIMPLIFY
COMMUNICATION IN COMPUTER SYSTEMS

APPLICATION NOTE AN-02

The Busy Lock-up Solution: Use Only One Arbitrator

The solution to this busy lock-up problem is to use the arbitration logic in only one RAM and to force the other RAM to follow it. In this case, one RAM is dedicated as the arbitration MASTER and additional RAM are designated as SLAVES. Two solutions to this problem are shown in Figure 10. One solution is to add external logic to the chip-enables of additional dual-port RAM chips. The logic gates shown cause the SLAVE RAM chip select to be disabled if the MASTER RAM is busy. Since only one set of arbitration logic is controlling the system the problem of SLAVE lock-up is avoided.

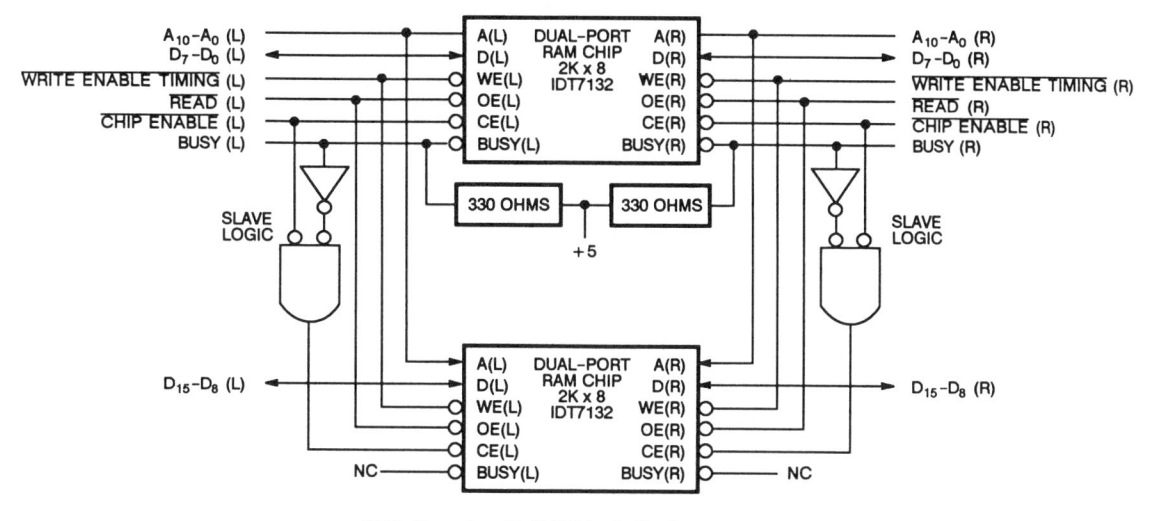

Width Expansion with SLAVE Logic (Not Recommended)

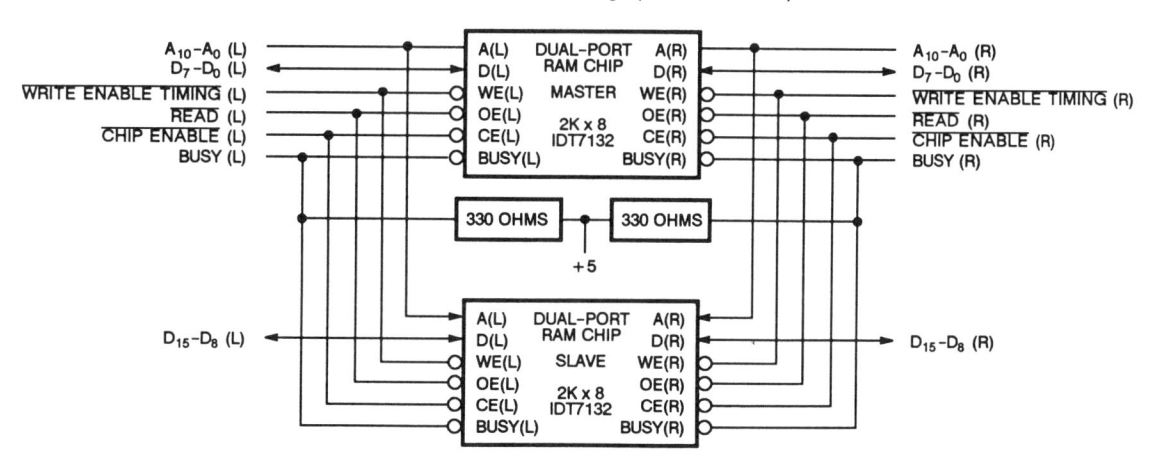

Width Expansion with SLAVE Chips (Recommended)

Figure 10. Width Expansion of Dual-Port RAMs

The second, more desirable solution, is to use specially designed dual-port RAM SLAVE chips which are part of IDT's product line. These SLAVE chips incorporate the SLAVE disable logic internally so that no additional logic is required to make a MASTER/SLAVE combination. In the SLAVE chip, the busy pin serves as an input rather than an output. If the MASTER chip activates busy, the SLAVE chip will sense this busy state and internally disable its write enable. SLAVE chips provide a speed advantage over systems which use external logic to implement the SLAVE function. Since the SLAVE logic is built into the SLAVE RAM chip, it can be designed so that there is no speed penalty when using SLAVE chips to expand the dual-port RAM width.

**DUAL-PORT RAMS SIMPLIFY
COMMUNICATION IN COMPUTER SYSTEMS**

APPLICATION NOTE AN-02

Width Expansion: Write Timing

When expanding dual-port RAMs in width, the writing of the SLAVE RAMs must be delayed until after the busy input at the SLAVE has settled. Otherwise, the SLAVE chip may begin writing while the busy signal is settling. This is true for systems using SLAVE chips and for systems using conventional dual-port RAMs with SLAVE logic. This delay can be accomplished by delaying the write enable to the SLAVE by the arbitration time of the MASTER. This is shown in Figure 11.

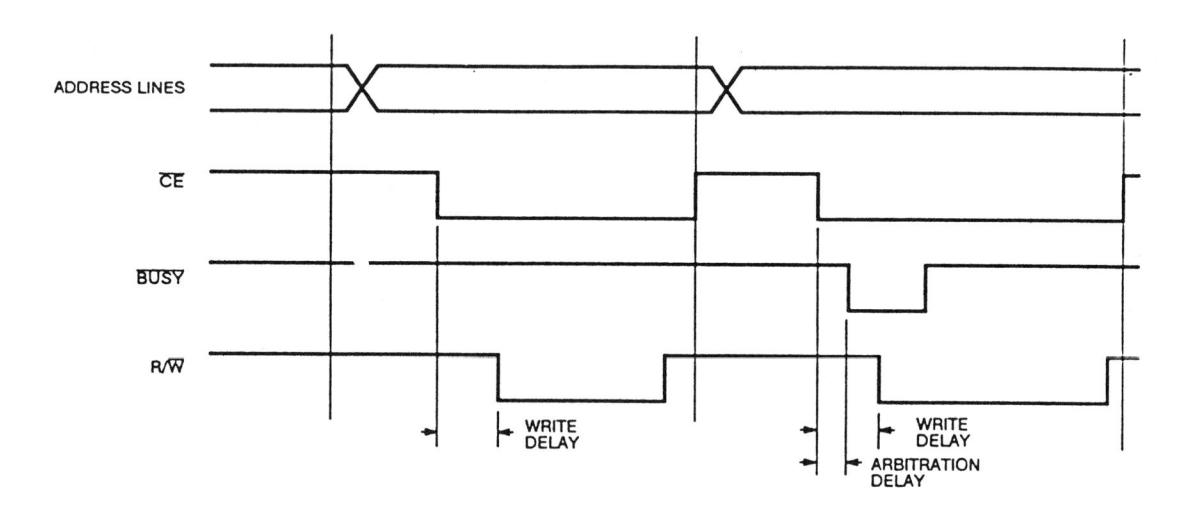

Figure 11. MASTER/SLAVE Write Timing

Note that the write delay is required only in width expanded systems which use SLAVE RAMs, not in single chip or depth expanded systems where only one chip is active at a time. This is because the individual chips have a built-in delay between the chip select and write enable inputs and the internal write enable to the RAM. Separate timing must be supplied in the SLAVE case because this internal delay time can be balanced to the arbitration time only within a chip and can vary from chip to chip. If the delay time for the SLAVE is less than the arbitration time of the MASTER, writing could begin before busy became active, as above.

**DUAL-PORT RAMS SIMPLIFY
COMMUNICATION IN COMPUTER SYSTEMS**

APPLICATION NOTE AN-02

Width and Depth Expansion: An Example

These techniques for expanding dual-port memories in width and depth are combined in the example shown in Figure 12. In this example, an 8K x 16 dual-port memory is made from 2K x 8 chips in MASTER/SLAVE combination.

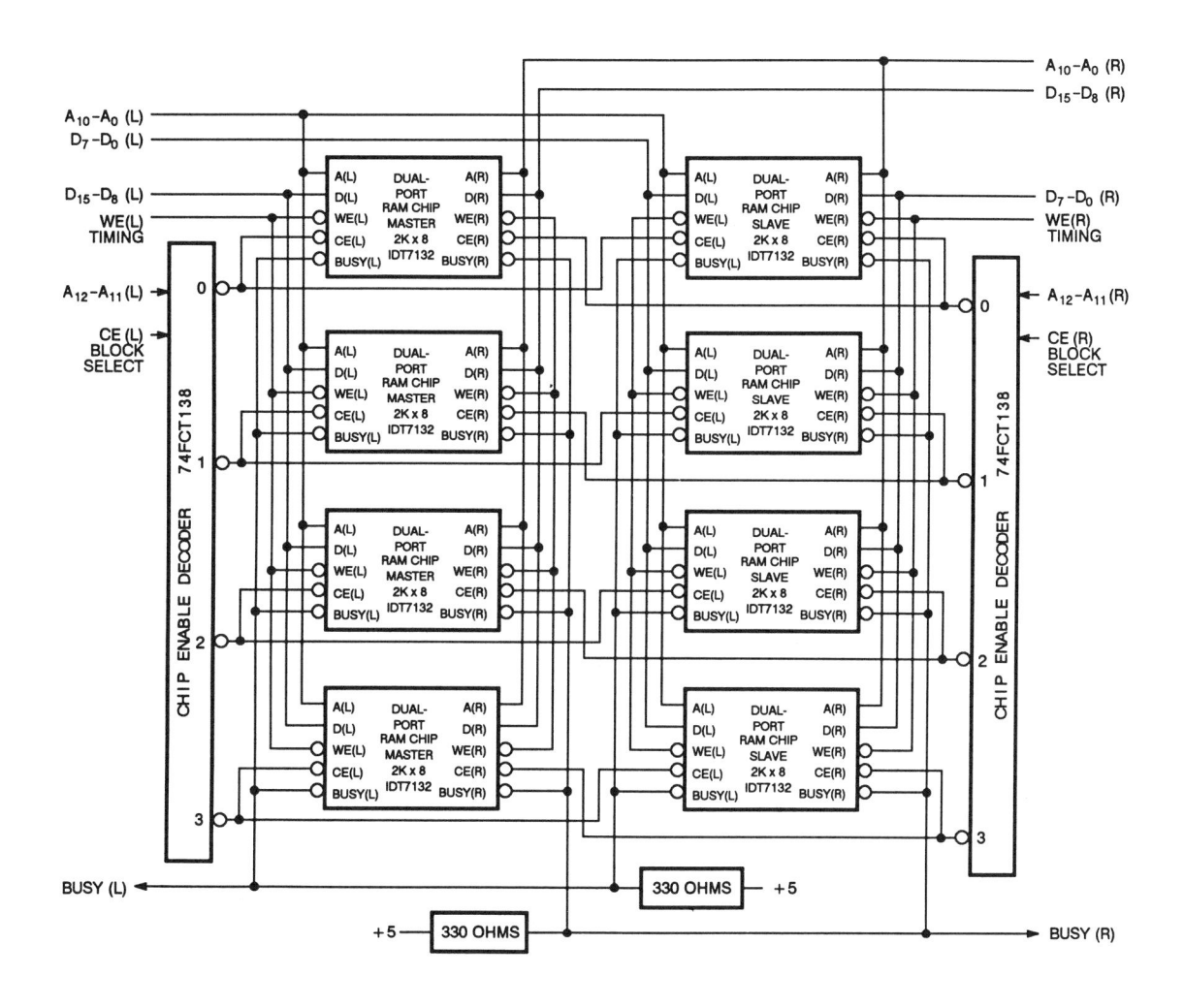

Figure 12. Width and Depth Expansion of Dual-Port RAMs

DUAL-PORT RAMS SIMPLIFY COMMUNICATION IN COMPUTER SYSTEMS

APPLICATION NOTE AN-02

USING THEM: DUAL-PORT RAM APPLICATION EXAMPLES

Examples of dual-port RAMs used for CPU-to-CPU communication are shown in Figures 13, 14 and 15. In Figure 11, a pair of 8-bit processors communicate using a single 2K x 8 dual-port RAM chip. In Figure 12, there is a similar system where a pair of 16-bit processors communicate using a pair of dual-port RAM chips and a MASTER/SLAVE configuration. Finally, in Figure 13, we have an 8-bit processor communicating with a 16-bit processor through two 2K x 8 dual-port RAMs.

In Figure 13, two Z80 microprocessors communicate using a single IDT7132 dual-port RAM chip. The IDT7132 is controlled by the chip enable. The write enable is set up in advance by the WR signal from the Z80 and the chip enable is used to write data into the RAM or to gate the read data onto the Z80 bus. The output enable (not shown) is tied to ground (continuous enable). The write enable is used to disable the output drivers.

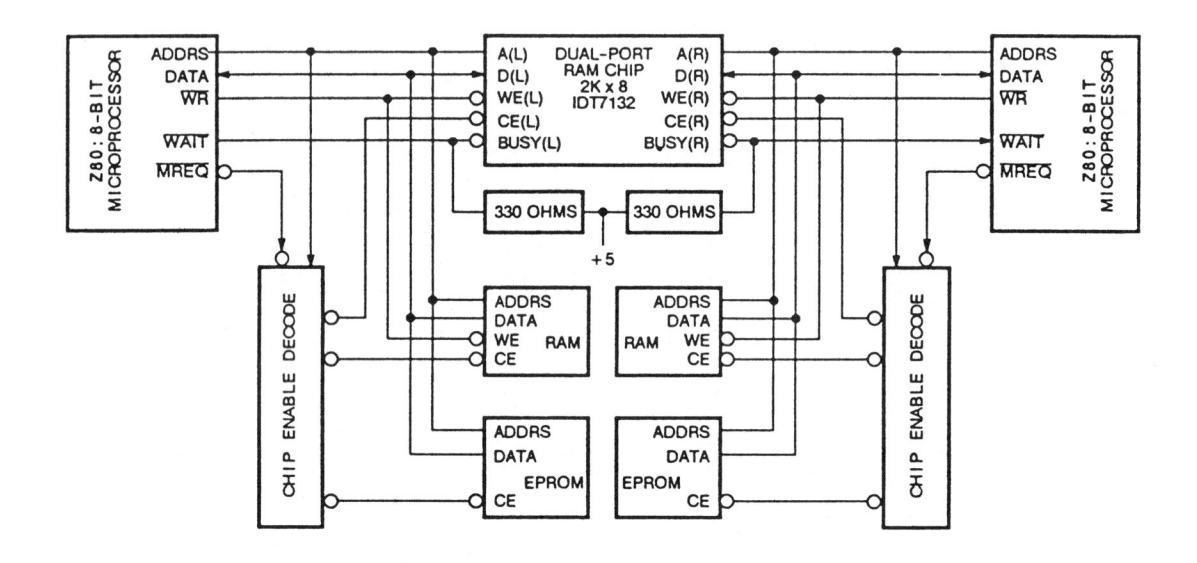

Figure 13. 8-bit to 8-bit CPU Communication

**DUAL-PORT RAMS SIMPLIFY
COMMUNICATION IN COMPUTER SYSTEMS**

APPLICATION NOTE AN-02

In Figure 14, two 68000 microprocessors communicate through a pair of dual-port RAMs. A IDT7132/7142 MASTER/SLAVE pair is used to avoid the busy lock-up problem. Note that the Address Strobe (AS) from each 68000 is used with an address decoder to enable the dual-port RAM chips. This is to maintain the address for read-modify-write cycles so that arbitration is not lost between the read and the write. This is important for test and set instructions, for example.

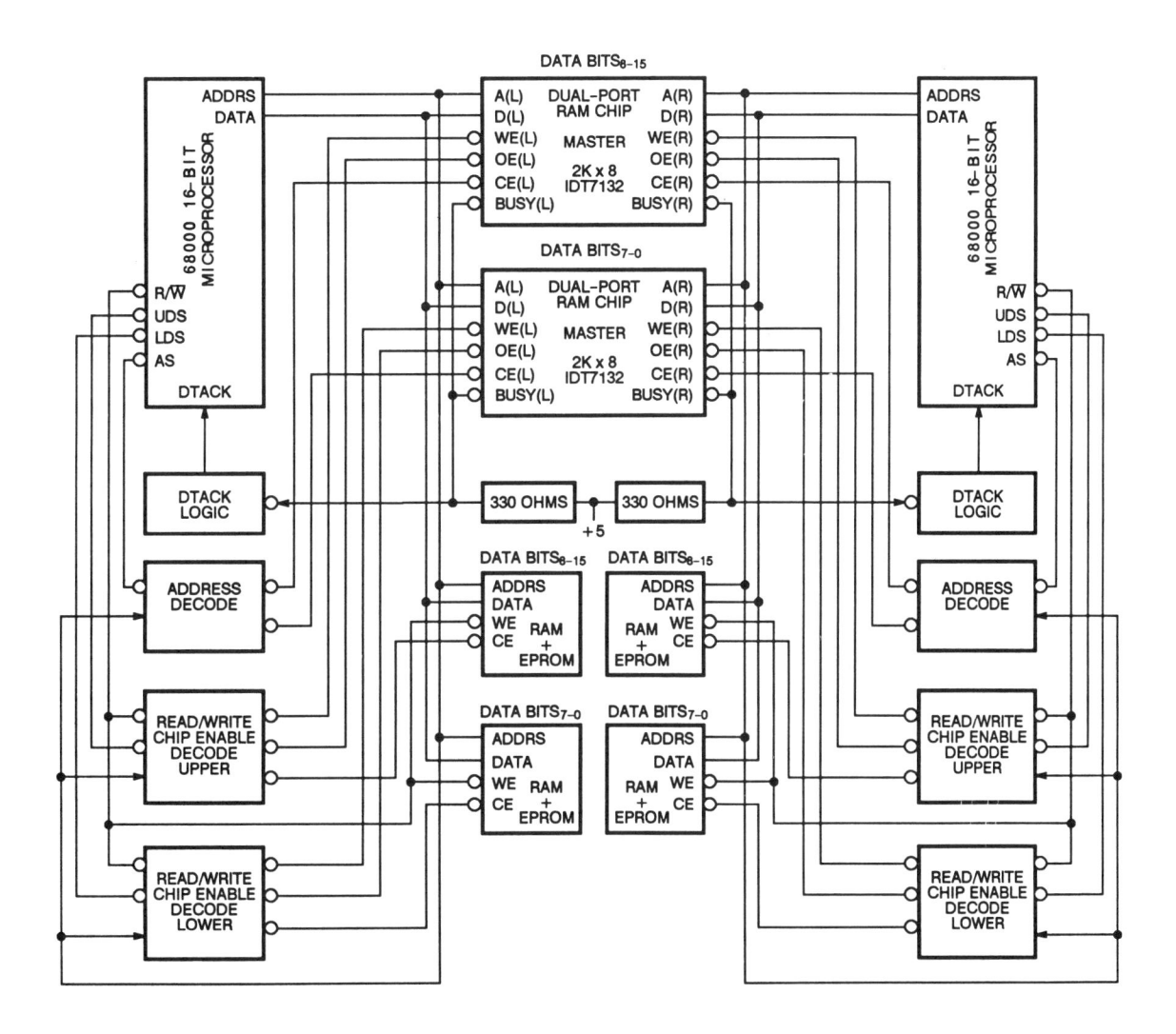

Figure 14. 16-bit to 16-bit CPU Communication

**DUAL-PORT RAMS SIMPLIFY
COMMUNICATION IN COMPUTER SYSTEMS**

APPLICATION NOTE AN-02

In Figure 15, a Z80 and a 68000 communicate using a pair of IDT7132 dual-port RAMs. No SLAVE logic is required because the Z80 side chip enable decode ensures that only one RAM chip will be enabled at a time. Otherwise, this figure is a combination of the logic from Figures 13 and 14.

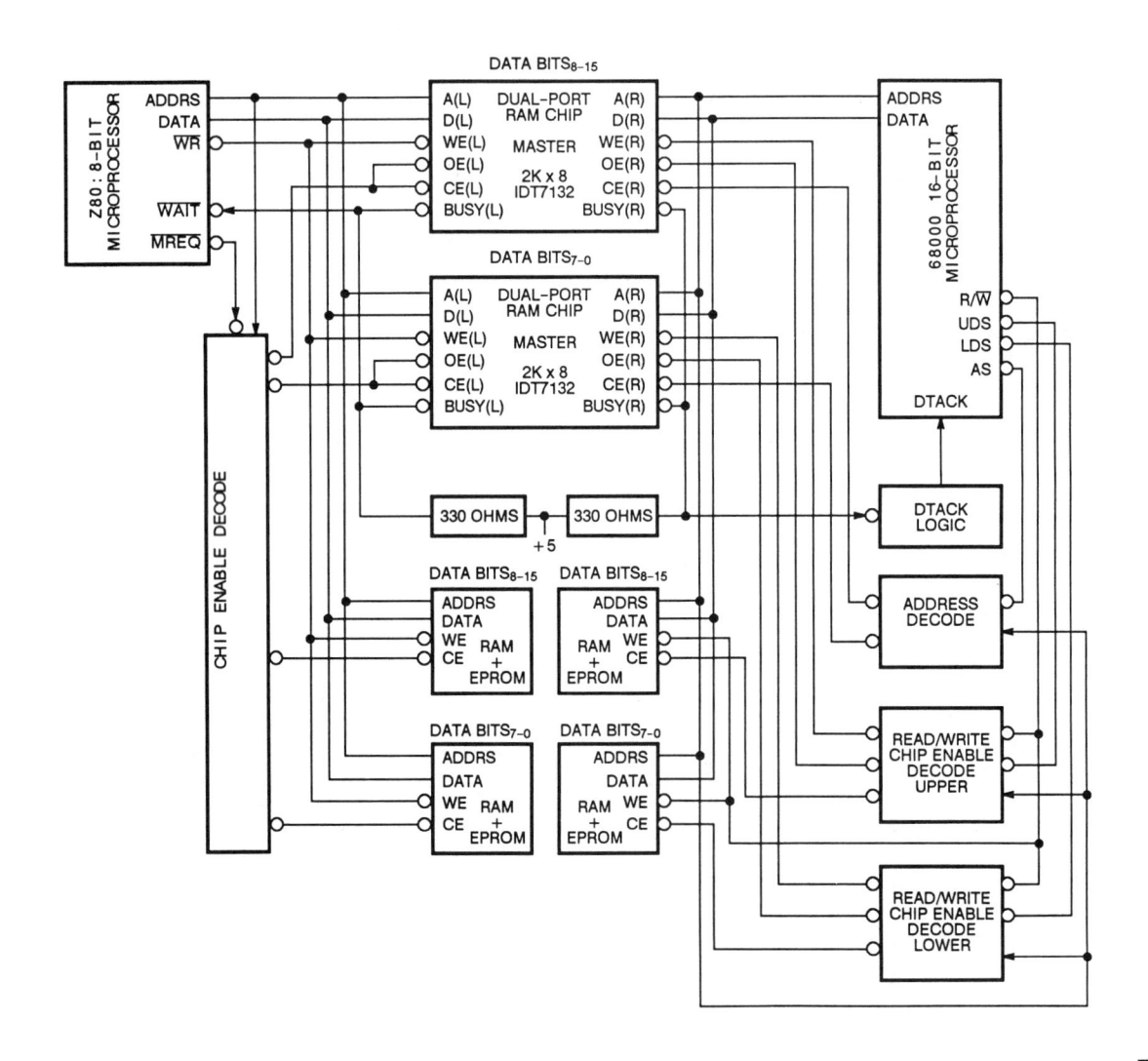

Figure 15. 8-bit to 16-bit CPU Communication

SUMMARY AND CONCLUSION

The development of true dual-port memories in integrated circuit form provides the designer with the ability to set up communication between components of a computer system while avoiding many of the problems of prior systems. While the concept of dual-port memory has been with us from the early days of computing in the form of DMA, the new dual-port ICs can provide this function at very high speeds and without the delays associated with earlier designs. Because of the utility of the dual-port memory concept these chips should come into wide spread use and become one of the standard components used by the computer designer.

CMOS PARALLEL FIRST-IN/FIRST-OUT FIFO 512 x 9-BIT & 1024 x 9-BIT

IDT7201SA/LA
IDT7202SA/LA

FEATURES:
- First-In, First-Out dual port memory
- 512 x 9 organization (IDT7201A)
- 1024 x 9 organization (IDT7202A)
- Low power consumption
- Ultra high speed — 45ns cycle time
- Asynchronous and simultaneous read and write
- Fully expandable by both word depth and/or bit width
- IDT7201A pin and functionally compatible with Mostek MK4501 but with half-full flag capability
- IDT7202A allows for deep word structure (1024) without expansion
- Half-full flag capability in single device mode
- Master/slave multiprocessing applications
- Bidirectional and rate buffer applications
- Empty and full warning flags
- Auto retransmit capability
- High-performance 1.2 micron CEMOS™ II technology
- Available in Plastic DIP, CERDIP and LCC
- Military product available 100% screened to MIL-STD-883, Class B

DESCRIPTION:
The IDT7201A/7202A is a dual port memory that utilizes a special First-In, First-Out algorithm that loads and empties data on a first-in, first-out basis. The device uses full and empty flags to prevent data overflow and underflow and expansion logic to allow for unlimited expansion capability in both word size and depth.

The reads and writes are internally sequential through the use of ring pointers, with no address information required to load and unload data. Data is toggled in and out of the device through the use of the WRITE ($\overline{W}$) and READ ($\overline{R}$) pins. The device has a read/write cycle time of 45ns (22MHz).

The device utilizes a 9-bit wide data array to allow for control and parity bits at the user's option. This feature is especially useful in data communications applications where it is necessary to use a parity bit for transmission/reception error checking. It also featues a RETRANSMIT ($\overline{RT}$) capability that allows for reset of the read pointer to its initial position when $\overline{RT}$ is pulsed low to allow for retransmission from the beginning of data. A half-full flag is available in the single device mode and width expansion modes.

The IDT7201A/7202A is fabricated using the high speed CEMOS II, 1.2 micron technology and is available in DIPs and LCCs screened to MIL-STD-883, Method 5004. It is designed for those applications requiring asynchronous and simultaneous read/writes in multiprocessing and rate buffer applications. The 1024 x 9 organization of the IDT7202A allows a 1024 deep word structure without the need for expansion.

PIN CONFIGURATIONS

DIP TOP VIEW

PLCC & LCC TOP VIEW

FUNCTIONAL BLOCK DIAGRAM

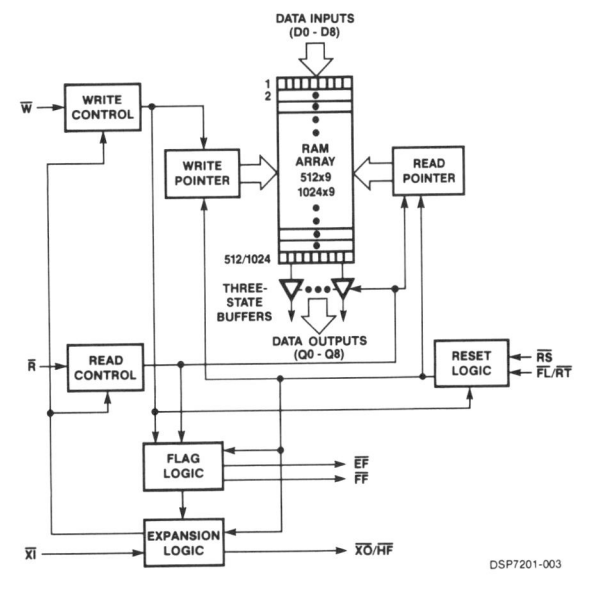

MILITARY AND COMMERCIAL TEMPERATURE RANGES

JULY 1986

©1986 Integrated Device Technology, Inc.

CEMOS is a trademark of Integrated Device Technology, Inc.

Printed in U.S.A.

IDT7201A/7202A CMOS
PARALLEL FIRST-IN/FIRST-OUT FIFO 512 x 9-BIT & 1024 x 9-BIT

MILITARY AND COMMERCIAL TEMPERATURE RANGES

ABSOLUTE MAXIMUM RATING[1]

SYMBOL	RATING	COMMERCIAL	MILITARY	UNIT
V_{TERM}	Terminal Voltage with Respect to GND	–0.5 to +7.0	–0.5 to +7.0	V
T_A	Operating Temperature	0 to +70	–55 to +125	°C
T_{BIAS}	Temperature Under Bias	–55 to +125	–65 to +135	°C
T_{STG}	Storage Temperature	–55 to +125	–65 to +155	°C
P_T	Power Dissipation	1.0	1.0	W
I_{OUT}	DC Output Current	50	50	mA

NOTE:
1. Stresses greater than those listed under ABSOLUTE MAXIMUM RATINGS may cause permanent damage to the device. This is a stress rating only and functional operation of the device at these or any other conditions above those indicated in the operational sections of this specification is not implied. Exposure to absolute maximum rating conditions for extended periods may affect reliability.

RECOMMENDED DC OPERATING CONDITIONS

SYMBOL	PARAMETER	MIN.	TYP.	MAX.	UNIT	NOTES
V_{CC}	Military Supply Voltage	4.5	5.0	5.5	V	—
V_{CC}	Commercial Supply Voltage	4.5	5.0	5.5	V	—
GND	Supply Voltage	0	0	0	V	—
V_{IH}	Input High Voltage Commercial	2.0	—	—	V	—
V_{IH}	Input High Voltage Military	2.2	—	—	V	—
V_{IL}	Input Low Voltage Commercial & Military	—	—	0.8	V	1

NOTE:
1. 1.5V undershoots are allowed for 10ns once per cycle.

DC ELECTRICAL CHARACTERISTICS
(Commercial: V_{CC} = 5V ± 10%, T_A = 0°C to +70°C; Military: V_{CC} = 5V ± 10%, T_A = –55°C to +125°C)

SYMBOL	PARAMETER	IDT7201SA/LA IDT7202SA/LA COMMERCIAL T_A = 35ns MIN. TYP. MAX.	IDT7201SA/LA IDT7202SA/LA MILITARY T_A = 40ns MIN. TYP. MAX.	IDT7201SA/LA IDT7202SA/LA COMMERCIAL T_A = 50, 65, 80, 120ns MIN. TYP. MAX.	IDT7201SA/LA IDT7202SA/LA MILITARY T_A = 50, 65, 80, 120ns MIN. TYP. MAX.	UNIT	NOTES
I_{LI}	Input Leakage Current (Any Input)	–1 — 1	–10 — 10	–1 — 1	–10 — 10	μA	1
I_{LO}	Output Leakage Current	–10 — 10	–10 — 10	–10 — 10	–10 — 10	μA	2
V_{OH}	Output Logic "1" Voltage I_{OH} = –2mA	2.4 — —	2.4 — —	2.4 — —	2.4 — —	V	—
V_{OL}	Output Logic "0" Voltage I_{OL} = 8mA	— — 0.4	— — 0.4	— — 0.4	— — 0.4	V	—
I_{CC1}	Average V_{CC} Power Supply Current	— — 100	— — 120	— 50 80	— 70 100	mA	3
I_{CC2}	Average Standby Current ($\overline{R} = \overline{W} = \overline{RS} = \overline{FL}/\overline{RT} = V_{IH}$)	— — 15	— — 20	— 5 8	— 8 15	mA	3
$I_{CC3}(L)$	Power Down Current (All Input = V_{CC} –0.2V)	— — 500	— — 900	— — 500	— — 900	μA	3
$I_{CC3}(S)$	Power Down Current (All Input = V_{CC} –0.2V)	— — 5	— — 9	— — 5	— — 9	mA	3

NOTES:
1. Measurements with 0.4 ≤ V_{IN} ≤ V_{CC}.
2. $\overline{R} ≥ V_{IH}$, 0.4 ≤ V_{OUT} ≤ V_{CC}.
3. I_{CC} measurements are made with outputs open.

IDT7201A/7202A CMOS
PARALLEL FIRST-IN/FIRST-OUT FIFO 512 x 9-BIT & 1024 x 9-BIT

MILITARY AND COMMERCIAL TEMPERATURE RANGES

AC ELECTRICAL CHARACTERISTICS
(Commercial: V_{CC} = 5V ± 10%, T_A = 0°C to +70°C; Military: V_{CC} = 5V ± 10%, T_A = −55°C to +125°C)

SYMBOL	PARAMETER	COM'L 7201A/2A-35 MIN. MAX.	MILITARY 7201A/2A-40 MIN. MAX.	7201A/2A-50 MIN. MAX.	7201A/2A-65 MIN. MAX.	7201A/2A-80 MIN. MAX.	7201A/2A-120 MIN. MAX.	UNITS
t_{RC}	Read Cycle Time	45 —	50 —	65 —	80 —	100 —	140 —	ns
t_A	Access Time	— 35	— 40	— 50	— 65	— 80	— 120	ns
t_{RR}	Read Recovery Time	10 —	10 —	15 —	15 —	20 —	20 —	ns
t_{RPW}	Read Pulse Width[2]	35 —	40 —	50 —	65 —	80 —	120 —	ns
t_{RLZ}	Read Pulse Low to Data Bus at Low Z[3]	5 —	5 —	10 —	10 —	10 —	10 —	ns
t_{WLZ}	Write Pulse High to Data Bus at Low Z[3,4]	10 —	10 —	15 —	15 —	20 —	20 —	ns
t_{DV}	Data Valid from Read Pulse High	5 —	5 —	5 —	5 —	5 —	5 —	ns
t_{RHZ}	Read Pulse High to Data Bus at High Z[3]	— 20	— 25	— 30	— 30	— 30	— 35	ns
t_{WC}	Write Cycle Time	45 —	50 —	65 —	80 —	100 —	140 —	ns
t_{WPW}	Write Pulse Width[2]	35 —	40 —	50 —	65 —	80 —	120 —	ns
t_{WR}	Write Recovery Time	10 —	10 —	15 —	15 —	20 —	20 —	ns
t_{DS}	Data Setup Time	18 —	20 —	30 —	30 —	40 —	40 —	ns
t_{DH}	Data Hold Time	0 —	0 —	5 —	10 —	10 —	10 —	ns
t_{RSC}	Reset Cycle Time	45 —	50 —	65 —	80 —	100 —	140 —	ns
t_{RS}	Reset Pulse Width[2]	35 —	40 —	50 —	65 —	80 —	120 —	ns
t_{RSR}	Reset Recovery Time	10 —	10 —	15 —	15 —	20 —	20 —	ns
t_{RTC}	Retransmit Cycle Time	45 —	50 —	65 —	80 —	100 —	140 —	ns
t_{RT}	Retransmit Pulse Width[2]	35 —	40 —	50 —	65 —	80 —	120 —	ns
t_{RTR}	Retransmit Recovery Time	10 —	10 —	15 —	15 —	20 —	20 —	ns
t_{EFL}	Reset to Empty Flag Low	— 45	— 50	— 65	— 80	— 100	— 140	ns
$t_{HFH,FFH}$	Reset to Half & Full Flag High	— 45	— 50	— 65	— 80	— 100	— 140	ns
t_{REF}	Read Low to Empty Flag Low	— 30	— 35	— 45	— 60	— 60	— 60	ns
t_{RFF}	Read High to Full Flag High	— 30	— 35	— 45	— 60	— 60	— 60	ns
t_{WEF}	Write High to Empty Flag High	— 30	— 35	— 45	— 60	— 60	— 60	ns
t_{WFF}	Write Low to Full Flag Low	— 30	— 35	— 45	— 60	— 60	— 60	ns
t_{WHF}	Write Low to Half-Full Flag Low	— 45	— 50	— 65	— 80	— 100	— 140	ns
t_{RHF}	Read High to Half-Full Flag High	— 45	— 50	— 65	— 80	— 100	— 140	ns

NOTES:
1. Timings referenced as in AC Test Conditions.
2. Pulse widths less than minimum value are not allowed.
3. Values guaranteed by design, not currently tested.
4. Only applies to read data flow through mode.

AC TEST CONDITIONS

Input Pulse Levels	GND to 3.0V
Input Rise and Fall Times	5ns
Input Timing Reference Levels	1.5V
Output Reference Levels	1.5V
Output Load	See Figure 1

CAPACITANCE (T_A = +25°C, f = 1.0MHz)

SYMBOL	PARAMETER[1]	CONDITIONS	TYP.	UNIT
C_{IN}	Input Capacitance	V_{IN} = 0V	5	pF
C_{OUT}	Output Capacitance	V_{OUT} = 0V	7	pF

NOTE:
1. This parameter is sampled and not 100% tested.

*Includes jig and scope capacitances.

Figure 1. Output Load.

NOTE:
Generating $\overline{R}/\overline{W}$ Signals — When using these high-speed FIFO devices, it is necessary to have clean inputs on the $\overline{R}$ and $\overline{W}$ signals. It is important to not have glitches, spikes or ringing on the $\overline{R}$, $\overline{W}$ (that violate the V_{IL}, V_{IH} requirements); although the minimum pulse width low for the $\overline{R}$ and $\overline{W}$ are specified in tens of nanosecond, a glitch of 5ns can affect the read or write pointer and cause it to increment.

IDT7201A/7202A CMOS
PARALLEL FIRST-IN/FIRST-OUT FIFO 512 x 9-BIT & 1024 x 9-BIT

SIGNAL DESCRIPTIONS:

INPUTS:

DATA IN (D0-D8)
Data inputs for 9-bit wide data.

CONTROLS:

RESET ($\overline{RS}$)
Reset is accomplished whenever the RESET ($\overline{RS}$) input is taken to a low state. During reset, both internal read and write pointers are set to the first location. A reset is required after power up before a write operation can take place. Both the READ ENABLE ($\overline{R}$) and WRITE ENABLE ($\overline{W}$) inputs must be in the high state during the window shown in Figure 2; i.e., t_{RPW} or t_{WPW} before the rising edge of $\overline{RS}$, and should not change until t_{RSR} after the rising edge of $\overline{RS}$. HALF-FULL FLAG ($\overline{HF}$) will be reset to high after master RESET ($\overline{RS}$).

WRITE ENABLE ($\overline{W}$)
A write cycle is initiated on the falling edge of this input if the FULL FLAG ($\overline{FF}$) is not set. Data setup and hold times must be adhered to with respect to the rising edge of the WRITE ENABLE ($\overline{W}$). Data is stored in the RAM array sequentially and independently of any ongoing read operation.

After half of the memory is filled, and at the falling edge of the next write operation, the HALF-FULL FLAG ($\overline{HF}$) will be set to low and will remain set until the difference between the write pointer and read pointer is less than or equal to one half of the total memory of the device. The HALF-FULL FLAG ($\overline{HF}$) is then reset by the rising edge of the read operation.

To prevent data overflow, the FULL FLAG ($\overline{FF}$) will go low, inhibiting further write operations. Upon the completion of a valid read operation, the FULL FLAG ($\overline{FF}$) will go high after t_{RFF}, allowing a valid write to begin. When the FIFO is full, the internal write pointer is blocked from $\overline{W}$, so external changes in $\overline{W}$ will not affect the FIFO when it is full.

READ ENABLE ($\overline{R}$)
A read cycle is initiated on the falling edge of the READ ENABLE ($\overline{R}$) provided the EMPTY FLAG ($\overline{EF}$) is not set. The data is accessed on a First-In, First-Out basis independent of any ongoing write operations. After READ ENABLE ($\overline{R}$) goes high, the Data Outputs (Q0 through Q8) will return to a high impedance condition until the next READ operation. When all the data has been read from the FIFO, the EMPTY FLAG ($\overline{EF}$) will go low, allowing the "final" read cycle but inhibiting further read operations with the data outputs remaining in a high impedance state. Once a valid write operation has been accomplished, the EMPTY FLAG ($\overline{EF}$) will go high after t_{WEF}, and a valid READ can then begin. When the FIFO is empty, the internal read pointer is blocked from $\overline{R}$, so external changes in $\overline{R}$ will not affect the FIFO when it is empty.

FIRST LOAD/RETRANSMIT ($\overline{FL}/\overline{RT}$)
This is a dual purpose input. In the Depth Expansion Mode, this pin is grounded to indicate that it is the first loaded. (See Operating Modes.) In the Single Device Mode, this pin acts as the retransmit input. The Single Device Mode is initiated by grounding the EXPANSION IN ($\overline{XI}$).

The IDT7201A/2A can be made to retransmit data when the RETRANSMIT ENABLE CONTROL ($\overline{RT}$) input is pulsed low. A retransmit operation will set the internal read pointer to the first location and will not affect the write pointer. READ ENABLE ($\overline{R}$) and WRITE ENABLE ($\overline{W}$) must be in the high state during retransmit. This feature is useful when less than 512/1024 writes are performed between resets. The retransmit feature is not compatible with Depth Expansion Mode and will affect HALF-FULL FLAG ($\overline{HF}$) depending on the relative locations of the read and write pointers.

EXPANSION IN ($\overline{XI}$)
This input is a dual purpose pin. EXPANSION IN ($\overline{XI}$) is grounded to indicate an operation in the single device mode. EXPANSION IN ($\overline{XI}$) is connected to EXPANSION OUT ($\overline{XO}$) of the previous device in the Depth Expansion or Daisy Chain Mode.

OUTPUTS:

FULL FLAG ($\overline{FF}$)
The FULL FLAT ($\overline{FF}$) will go low, inhibiting further write operation, when the write pointer is one location from the read pointer, indicating that the device is full. if the read pointer is not moved after RESET ($\overline{RS}$), the FULL FLAG ($\overline{FF}$) will go low after 512 writes for the IDT7201A and 1024 writes for the IDT7202A.

EXPANSION OUT/HALF-FULL FLAG ($\overline{XO}/\overline{HF}$)
This is a dual purpose output. In the single device mode, when EXPANSION IN ($\overline{XI}$) is grounded, this output acts as an indication of a half-full memory.

After half of the memory is filled, and at the falling edge of the next write operation, the HALF-FULL FLAG ($\overline{HF}$) will be set to low and will remain set until the difference between the write pointer and read pointer is less than or equal to one half of the total memory of the device. The HALF-FULL FLAG ($\overline{HF}$) is then reset by the rising edge of the read operation.

In the Depth Expansion Mode, EXPANSION IN ($\overline{XI}$) is connected to EXPANSION OUT ($\overline{XO}$) of the previous device. This output acts as a signal to the next device in the Daisy Chain by providing a pulse to the next device when the previous device reaches the last location of memory.

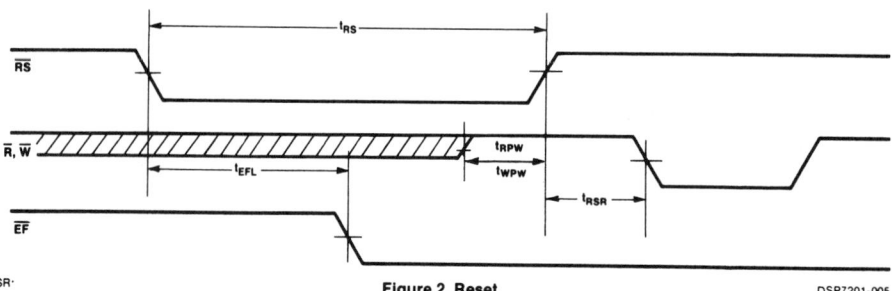

Figure 2. Reset

NOTES:
1. $t_{RSC} = t_{RS} + t_{RSR}$.
2. $\overline{W}$ and $\overline{R} = V_{IH}$ around the rising edge of $\overline{RS}$.

IDT7201A/7202A CMOS
PARALLEL FIRST-IN/FIRST-OUT FIFO 512 x 9-BIT & 1024 x 9-BIT

MILITARY AND COMMERCIAL TEMPERATURE RANGES

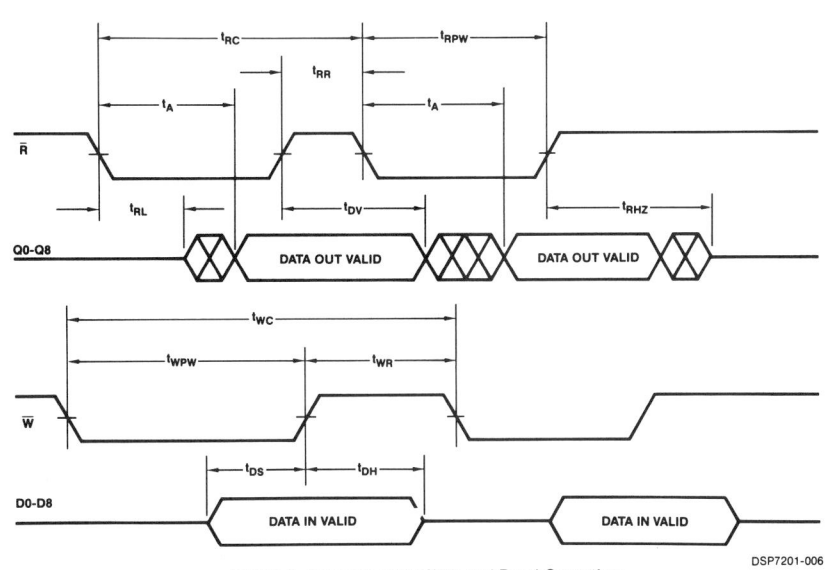

Figure 3. Asynchronous Write and Read Operation

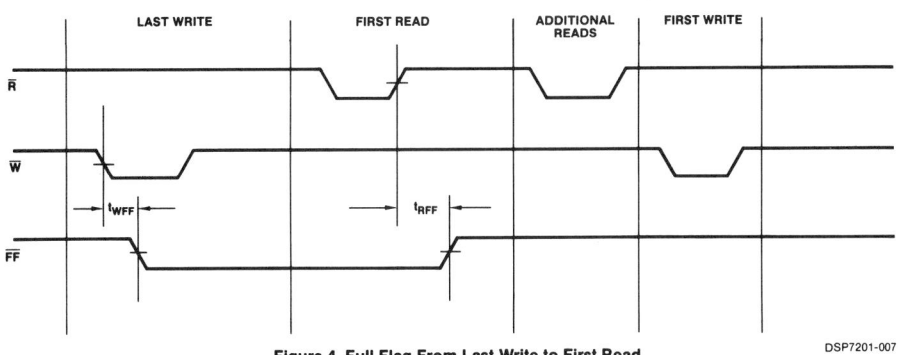

Figure 4. Full Flag From Last Write to First Read

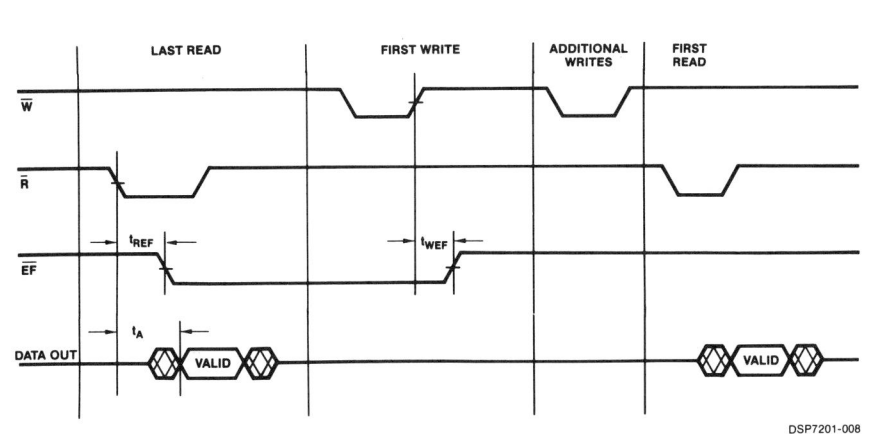

Figure 5. Empty Flag From Last Read to First Write

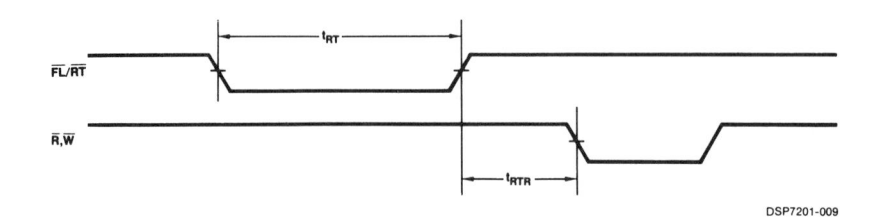

IDT7201A/7202A CMOS
PARALLEL FIRST-IN/FIRST-OUT FIFO 512 x 9-BIT & 1024 x 9-BIT

MILITARY AND COMMERCIAL TEMPERATURE RANGES

NOTES:
1. $t_{RTC} = t_{RT} + t_{RTR}$.
2. $\overline{EF}$, $\overline{HF}$ and $\overline{FF}$ may change state during retransmit as a result of the offset of the read and write pointers, but flags will be valid at t_{RTC}.

Figure 6. Retransmit

t_{RPE}: EFFECTIVE READ PULSE WIDTH AFTER EMPTY FLAG HIGH

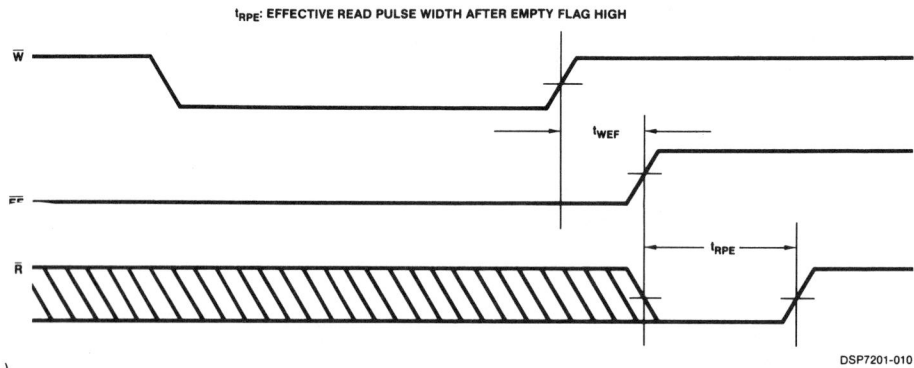

NOTE:
1. ($t_{RPE} = t_{RPW}$).

Figure 7. Empty Flag Timing

t_{WPF}: EFFECTIVE WRITE PULSE WIDTH AFTER FULL FLAG HIGH

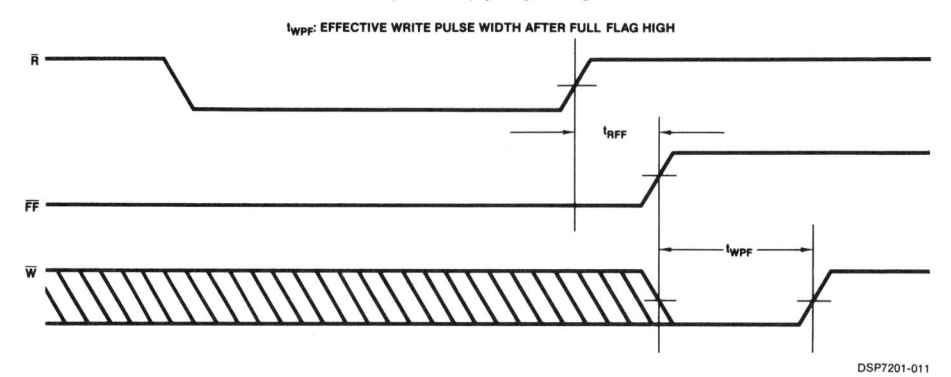

NOTE:
1. ($t_{WPF} = t_{WPW}$).

Figure 8. Full Flag Timing

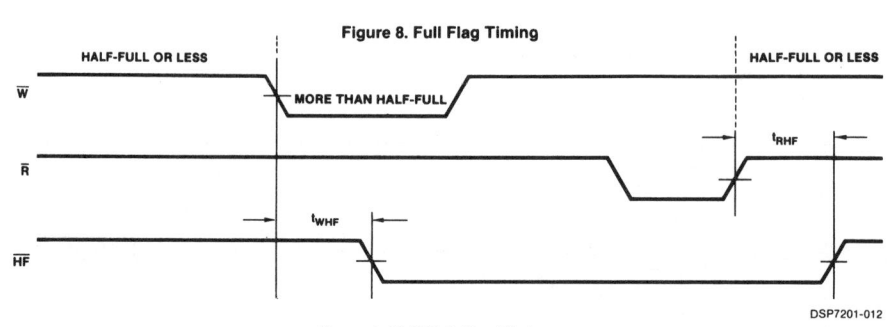

Figure 9. Half-Full Flag Timing

IDT7201A/7202A CMOS
PARALLEL FIRST-IN/FIRST-OUT FIFO 512 x 9-BIT & 1024 x 9-BIT

MILITARY AND COMMERCIAL TEMPERATURE RANGES

DATA OUTPUTS (Q0-Q8)
Data outputs for 9-bit wide data. This output is in a high impedance condition whenever READ ($\overline{R}$) is in a high state.

OPERATING MODES:

SINGLE DEVICE MODE
A single IDT7201A/2A may be used when the application requirements are for 512/1024 words or less. The IDT7201A/2A is in a Single Device Configuration when the EXPANSION IN ($\overline{XI}$) control input is grounded. (See Figure 10.) In this mode the HALF-FULL FLAG ($\overline{HF}$), which is an active low output, is shared with EXPANSION OUT ($\overline{XO}$).

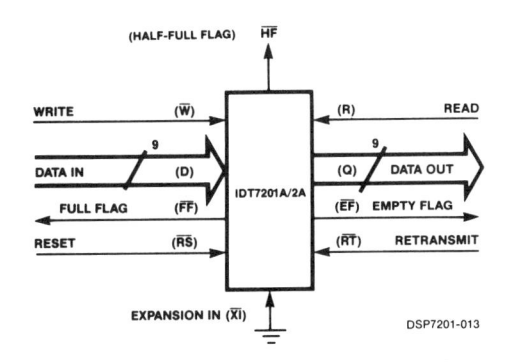

Figure 10. Block Diagram of Single 512x9/1024x9 FIFO

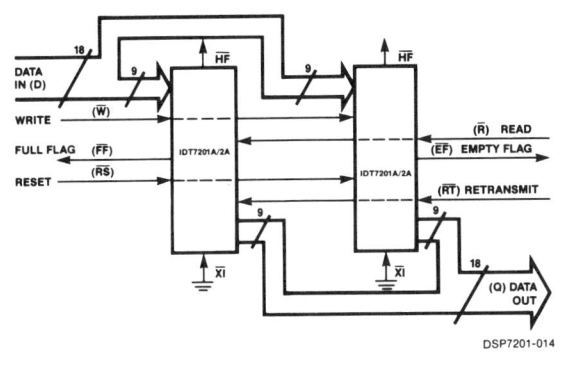

NOTES:
Flag detection is accomplished by monitoring the $\overline{FF}$, $\overline{EF}$, and the $\overline{HF}$ signals on either (any) device used in the width expansion configuration. Do not connect any output control signals together.

Figure 11. Block Diagram of 512x18/1024x18 FIFO Memory Used in Width Expansion Mode

WIDTH EXPANSION MODE
Word width may be increased simply by connecting the corresponding input control signals of multiple devices. Status flags ($\overline{EF}$, $\overline{FF}$ and $\overline{HF}$) can be detected from any one device. Figure 11 demonstrates an 18-bit word width by using two IDT7201A/2As. Any word width can be attained by adding additional IDT7201A/2As.

DEPTH EXPANSION (DAISY CHAIN) MODE
The IDT7201A/2A can easily be adapted to applications when the requirements are for greater than 512/1024 words. Figure 12 demonstrates Depth Expansion using three IDT7201A/2As. Any depth can be attained by adding additional IDT7201A/2As. The IDT7201A/2A operates in the Depth Expansion configuration when the following conditions are met:

1. The first device must be designed by grounding the FIRST LOAD ($\overline{FL}$) control input.
2. All other devices must have $\overline{FL}$ in the high state.
3. The EXPANSION OUT ($\overline{XO}$) pin of each device must be tied to the EXPANSION IN ($\overline{XI}$) pin of the next device. See Figure 12.
4. External logic is needed to generate a composite FULL FLAG ($\overline{FF}$) and EMPTY FLAG ($\overline{EF}$). This requires the ORing of all $\overline{EF}$s and ORing of all $\overline{FF}$s (i.e. all must be set to generate the correct composite $\overline{FF}$ or $\overline{EF}$). See Figure 12.
5. The RETRANSMIT ($\overline{RT}$) function and HALF-FULL FLAG ($\overline{HF}$) are not available in the Depth Expansion Mode.

COMPOUND EXPANSION MODE
The two expansion techniques described above can be applied together in a straightforward manner to achieve large FIFO arrays. (See Figure 13.)

BIDIRECTIONAL MODE
Applications which require data buffering between two systems (each system capable of READ and WRITE operations) can be achieved by pairing IDT7201A/2As as is shown in Figure 14. Care must be taken to assure that the appropriate flag is monitored by each system; (i.e. $\overline{FF}$ is monitored on the device where $\overline{W}$ is used; $\overline{EF}$ is monitored on the device where $\overline{R}$ is used). Both Depth Expansion and Width Expansion may be used in this mode.

DATA FLOW-THROUGH MODES
Two types of flow-through modes are permitted, a read flow-through and write flow-through mode. For the read flow-through mode (Figure 15), the FIFO permits a reading of a single word after writing one word of data into an empty FIFO. The data is enabled on the bus in ($t_{WEF} + t_A$)ns after the rising edge of $\overline{W}$, called the first write edge, and it remains on the bus until the $\overline{R}$ line is raised from low-to-high, after which the bus would go into a three-state mode after t_{RHZ}ns. The $\overline{EF}$ line would have a pulse showing temporary de-assertion and then would be asserted. In the interval of time that $\overline{R}$ was low, more words can be written to the FIFO (the subsequent writes after the first write edge would de-assert the empty flag); however, the same word (written on the first write edge), presented to the output bus as the read pointer, would not be incremented when $\overline{R}$ is low. On toggling $\overline{R}$, the other words that were written to the FIFO will appear on the output bus as in the read cycle timings.

In the write flow-through mode (Figure 16), the FIFO permits the writing of a single word of data immediately after reading one word of data from a full FIFO. The $\overline{R}$ line causes the $\overline{FF}$ to be de-asserted but the $\overline{W}$ line being low causes it to be asserted again in anticipation of a new data word. On the rising edge of $\overline{W}$, the new word is loaded in the FIFO. The $\overline{W}$ line must be toggled when $\overline{FF}$ is not asserted to write new data in the FIFO and to increment the write pointer.

IDT7201A/7202A CMOS
PARALLEL FIRST-IN/FIRST-OUT FIFO 512 x 9-BIT & 1024 x 9-BIT

MILITARY AND COMMERCIAL TEMPERATURE RANGES

TRUTH TABLES

TABLE I — RESET AND RETRANSMIT — SINGLE DEVICE CONFIGURATION/WIDTH EXPANSION MODE

MODE	INPUTS			INTERNAL STATUS		OUTPUTS		
	$\overline{RS}$	$\overline{RT}$	$\overline{XI}$	Read Pointer	Write Pointer	$\overline{EF}$	$\overline{FF}$	$\overline{HF}$
Reset	0	X	0	Location Zero	Location Zero	0	1	1
Retransmit	1	0	0	Location Zero	Unchanged	X	X	X
Read/Write	1	1	0	Increment[1]	Increment[1]	X	X	X

NOTE:
1. Pointer will increment if flag is high.

TABLE II — RESET AND FIRST LOAD TRUTH TABLE — DEPTH EXPANSION/COMPOUND EXPANSION MODE

MODE	INPUTS			INTERNAL STATUS		OUTPUTS	
	$\overline{RS}$	$\overline{FL}$	$\overline{XI}$	Read Pointer	Write Pointer	$\overline{EF}$	$\overline{FF}$
Reset-First Device	0	0	(1)	Location Zero	Location Zero	0	1
Reset all Other Devices	0	1	(1)	Location Zero	Location Zero	0	1
Read/Write	1	X	(1)	X	X	X	X

NOTES:
1. $\overline{XI}$ is connected to $\overline{XO}$ of previous device. See Figure 12.
$\overline{RS}$ = Reset Input, $\overline{FL}/\overline{RT}$ = First Load/Retransmit, $\overline{EF}$ = Empty Flag Output. $\overline{FF}$ = Full Flag Output, $\overline{XI}$ = Expansion Input, $\overline{HF}$ = Half-Full Flag Output.

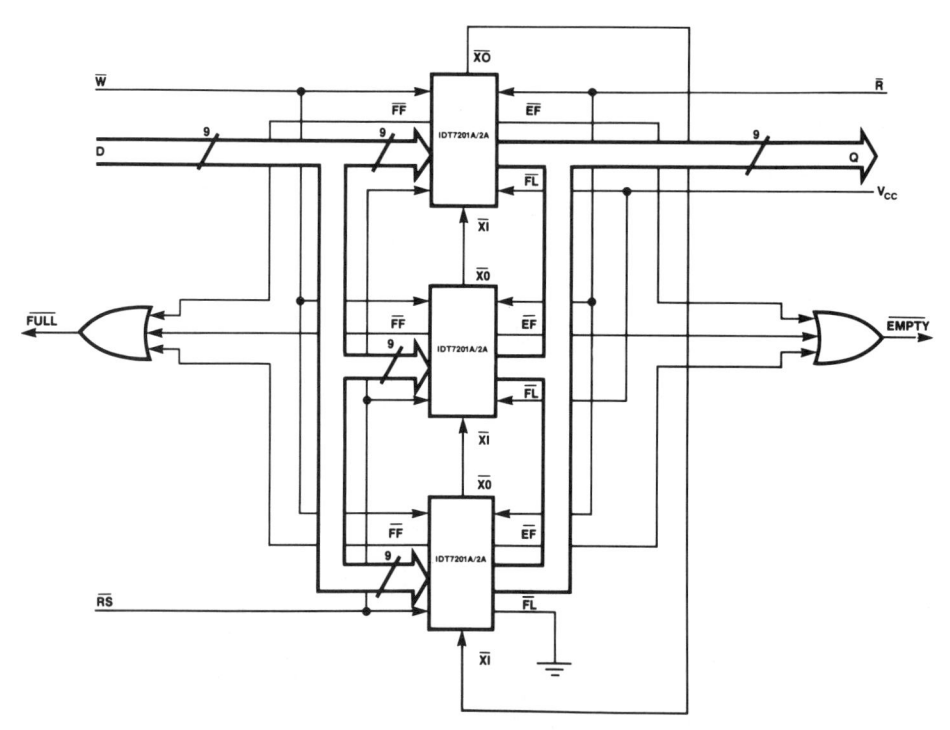

Figure 12. Block Diagram of 1536x9/3072x9 FIFO Memory (Depth Expansion)

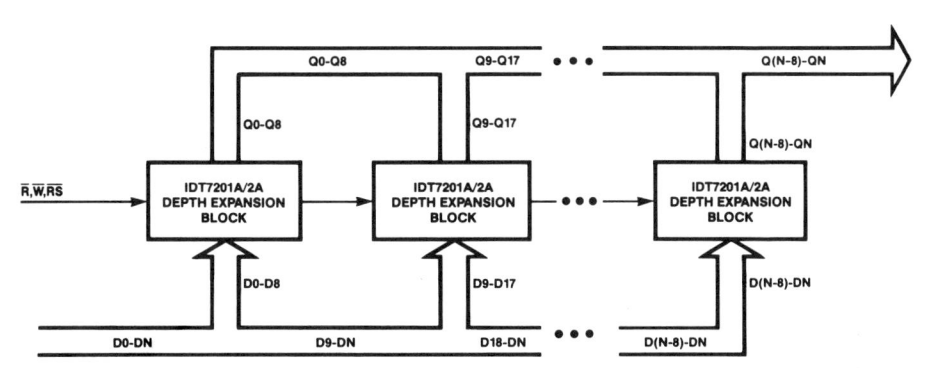

IDT7201A/7202A CMOS
PARALLEL FIRST-IN/FIRST-OUT FIFO 512 x 9-BIT & 1024 x 9-BIT

MILITARY AND COMMERCIAL TEMPERATURE RANGES

NOTES:
1. For depth expansion block see DEPTH EXPANSION Section and Figure 12.
2. For Flag detection see WIDTH EXPANSION Section and Figure 11.

Figure 13. Compound FIFO Expansion

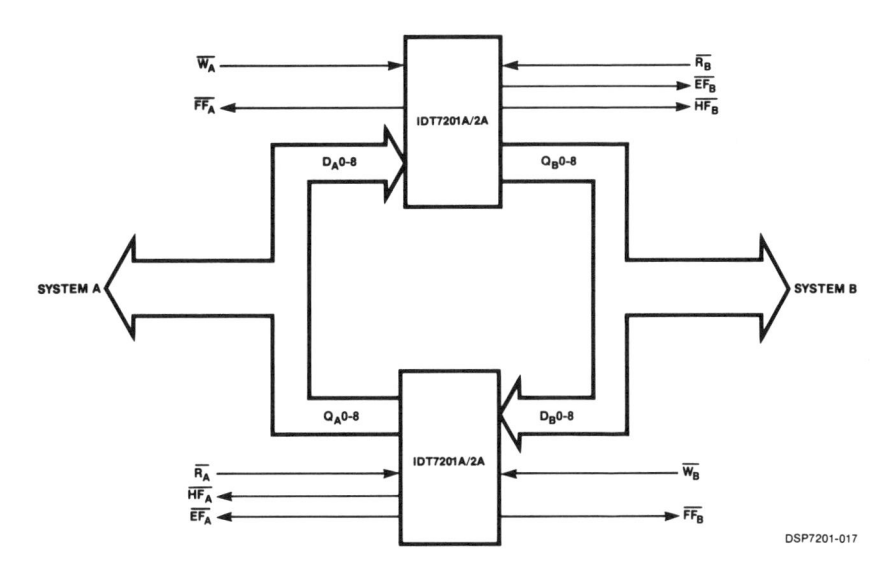

Figure 14. Bidirectional FIFO Mode

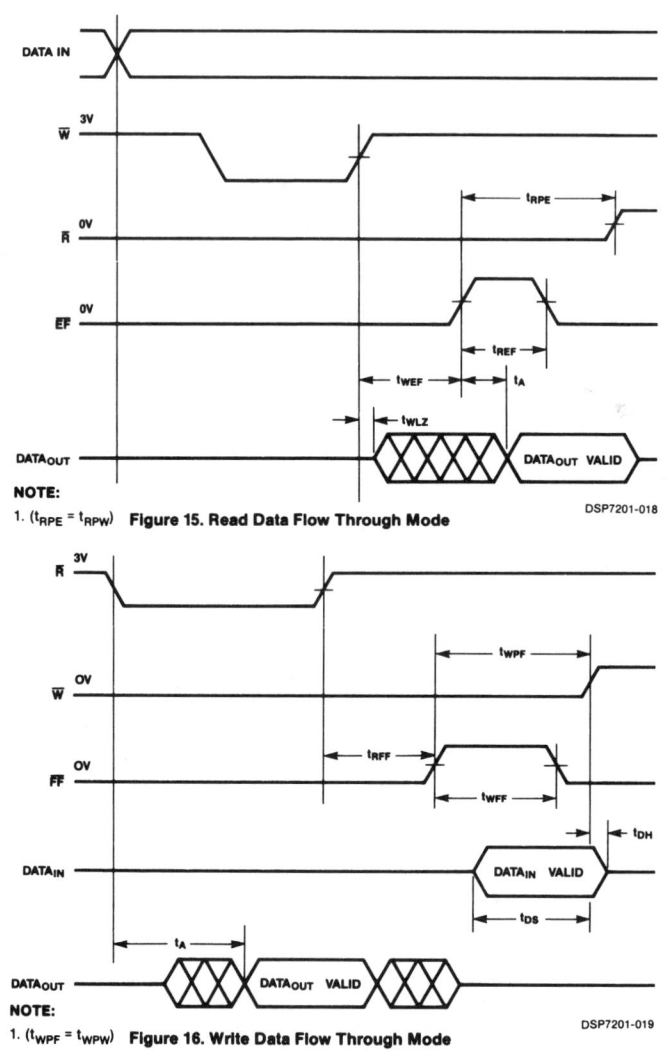

IDT7201A/7202A CMOS
PARALLEL FIRST-IN/FIRST-OUT FIFO 512 x 9-BIT & 1024 x 9-BIT   MILITARY AND COMMERCIAL TEMPERATURE RANGES

Figure 15. Read Data Flow Through Mode

NOTE:
1. ($t_{RPE} = t_{RPW}$)

Figure 16. Write Data Flow Through Mode

NOTE:
1. ($t_{WPF} = t_{WPW}$)

PACKAGE DIMENSIONS
28-PIN PLASTIC DIP

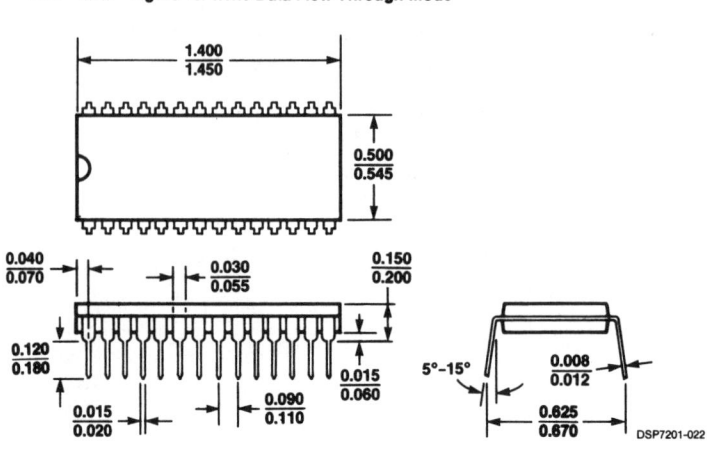

IDT7201A/7202A CMOS
PARALLEL FIRST-IN/FIRST-OUT FIFO 512 x 9-BIT & 1024 x 9-BIT

MILITARY AND COMMERCIAL TEMPERATURE RANGES

PACKAGE DIMENSIONS (continued)

28-PIN CERDIP

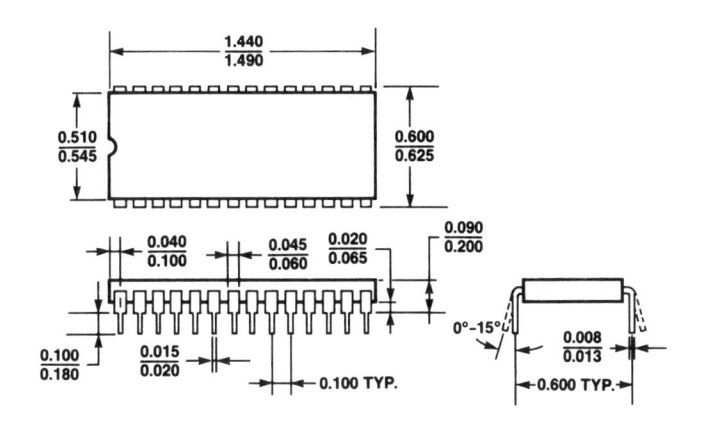

32-PIN LEADLESS CHIP CARRIER

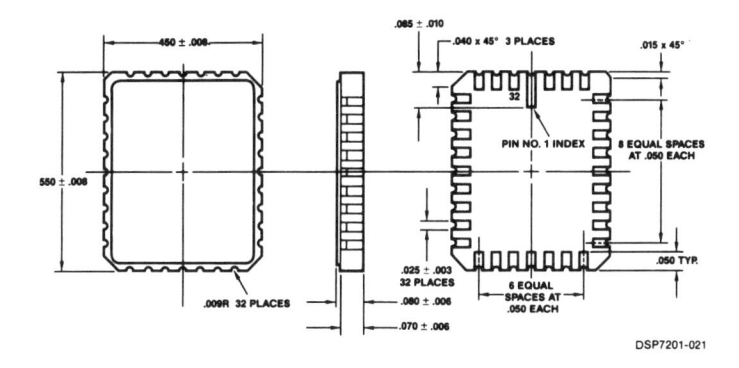

32-PIN PLASTIC LEADED CHIP CARRIER

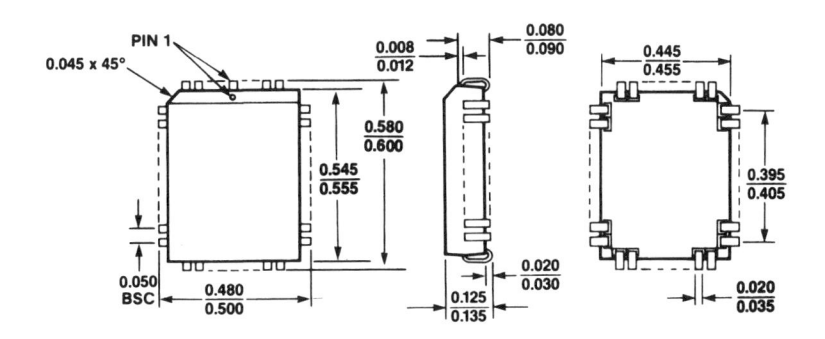

IDT7201A/7202A CMOS
PARALLEL FIRST-IN/FIRST-OUT FIFO 512 x 9-BIT & 1024 x 9-BIT

MILITARY AND COMMERCIAL TEMPERATURE RANGES

ORDERING INFORMATION

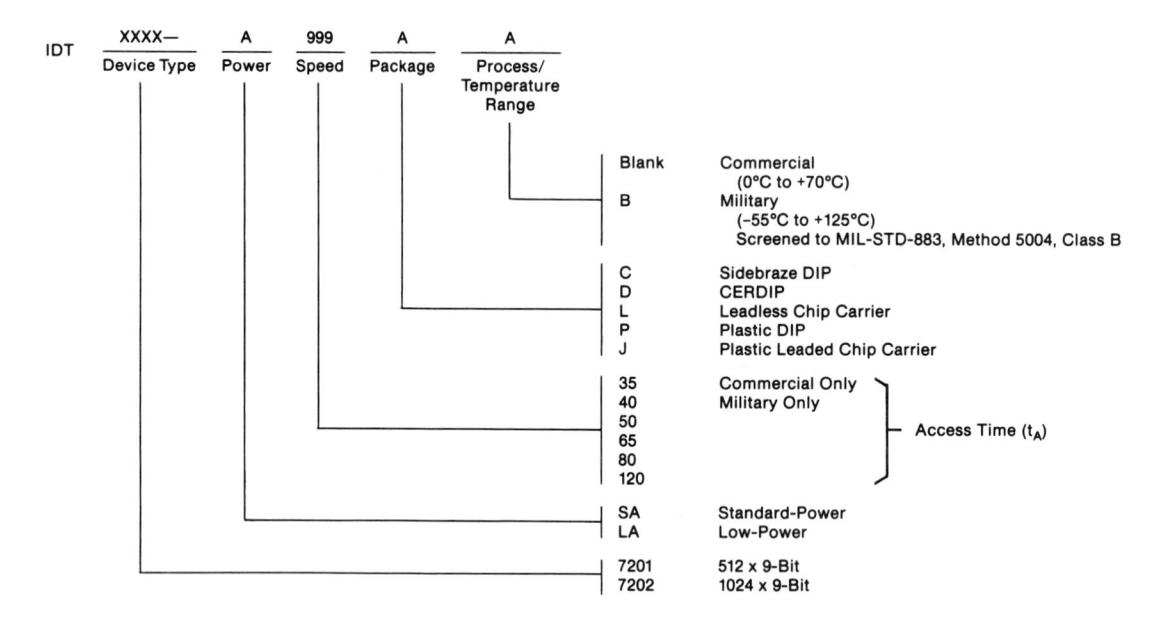

```
IDT  XXXX—    A       999     A         A
     Device   Power   Speed   Package   Process/
     Type                               Temperature
                                        Range
```

	Blank	Commercial (0°C to +70°C)
	B	Military (−55°C to +125°C) Screened to MIL-STD-883, Method 5004, Class B
	C	Sidebraze DIP
	D	CERDIP
	L	Leadless Chip Carrier
	P	Plastic DIP
	J	Plastic Leaded Chip Carrier
	35	Commercial Only
	40	Military Only
	50	
	65	⎬ Access Time (t_A)
	80	
	120	
	SA	Standard-Power
	LA	Low-Power
	7201	512 x 9-Bit
	7202	1024 x 9-Bit

IDT PROCESSING SUMMARY:

Maintaining the highest standards of quality in our monolithic hermetic products is the basis of IDT's standard manufacturing systems and procedures. IDT products begin with stringent design rules derived for use in high reliability programs. This is followed by a dedicated commitment to reliable workmanship as well as rigid controls throughout wafer fab, device assembly and electrical test, all of which are designed to produce products that are inherently reliable.

SCREENING FLOW — METHOD 5004

SCREEN	TEST METHOD	LEVEL
Visual and Mechanical		
Internal visual	2010 Condition B	100%
High-temperature storage	1008 Condition C	100%
Temperature cycle	1010 Condition C	100%
Constant acceleration	2001 Condition E (Y_1 only)	100%
Hermeticity	1014	
Fine	Condition A or B	100%
Gross	Condition C	100%
Burn-In		
Pre-burn-in electrical	Per applicable device specifications at $T_A = +25°C$	100%
Burn-in	MIL-STD-883 Grade Product Method 1015, Condition D	100%
	Standard Grade Product 48 hours minimum to same burn-in conditions as Military Grade Product	100%
Final Electrical Tests[1]	5004	
Static (dc)	a) @ $T_A = +25°C$ and power supply extremes	100%
	b) @ temperature and power supply extremes	100%
Functional	a) @ $T_A = +25°C$ and power supply extremes	100%
	b) @ temperature and power supply extremes	100%
Switching (ac)	a) @ $T_A = +25°C$ nominal power supply	100%
Percent Defective Allowable (PDA)	Method 5004	
Calculated at post burn-in @ $T_A = +25°C$	Military Grade	5%
	Standard Grade	10%
Quality Conformance	5005 Sample section as applicable	Sample
External Visual	2009 Per IDT or customer specification	100%

NOTE:
1. Commercial Grade Products are sample tested to the applicable temperature extremes.

All military grade products are manufactured and screened to the demanding requirements of MIL-STD-883, Method 5004, Class B. Commercial and industrial grade products differ from military grade only in burn-in time and electrical test temperature

IDT supplies a full line of military grade products completely screened to the Class B criteria of Method 5004, with inspection lots tested to the Quality Conformance Requirements of Method 5005 as shown. This includes 100% 160-hour burn-in at T_A = +125°C (or equivalent) per Method 1015. Cond. D, followed by 100% temperature testing of all DC, AC and functional characteristics over the full –55°C to +125°C temperature range.

Samples of the military grade products which have been processed to Method 5004 100% screening requirements are submitted to the Quality Conformance inspection requirements of MIL-STD-883. These Quality Conformance Inspections are performed to the specific requirements of Method 5005 Group A (electrical), Group B (mechanical), Group C (chip integrity), and Group D (package environmental integrity).

Documentation, design, processing and assembly workmanship guidelines are patterned after MIL-M-38510 specifications.

Specification of IDT monolithic hermetic products will ensure the user not only of high performance products, but also components tested to meet stringent reliability requirements.

QUALITY CONFORMANCE TESTING
Per MIL-STD-883, Method 5005, Class B

SCREEN	TEST METHOD	LEVEL
Quality Conformance Sample Tests	Group A (Electrical Tests)	Sample
	Group B (Mechanical Tests)	Sample
	Group C (Chip Integrity Tests)	Sample
	Group D (Module Integrity Tests)	Sample

For special customer specifications or Quality requirements beyond Class B levels of MIL-STD-883, such as SEM analysis, X-Ray, or other screening flows to meet specific user needs, contact your local IDT sales office.

Integrated Device Technology, Inc.
3236 Scott Blvd., Santa Clara, CA 95054-3090 • Telephone: (408) 727-6116 • TWX 9103382070

Chapter 5   Peripherals

In the mid 1970's the range of peripherals used in microprocessor systems was rather small: the 8-bit parallel interface, the serial interface and the timer. Today, there are very many more peripherals. For example, we have sophisticated multi-protocol parallel interfaces, disk interfaces, counter-timers, IEEE 488 bus interfaces, SCSI interfaces, display controllers, and so on. Here we have selected some typical interfaces. Instead of providing a little information about many devices, we have provided full details on a few devices.

We begin this chapter by presenting the data sheets of two important devices, the analog to digital converter and the digital to analog converter. Numerous ADCs and DACs are available to the systems designer, each with their operating parameters. Some are fast and some exhibit high precision. Some are low-cost and some are easy to interface to microprocessor. Here we have simply selected two typical devices.

Our next peripheral is the dual universal serial communications controller (or DUSCC). This device has all the normal features you would expect of a serial interface (e.g., modem control and programmable data formats) and a lot more. In particular, it supports several communications protocols (both bit-oriented and byte-oriented).

The next peripheral is the VL82C106 PC/AT combo I/O chip. The first generation of IBM PCs (and their clones) used individual logic elements to perform control and I/O functions. Semiconductor manufacturers have now put groups of functional elements on a chip to simplify the construction of today's PCs. Here we provide the data sheet of a peripheral that includes two serial ports, a parallel port, a keyboard port, and a real-time clock. Each of these peripherals behaves like its corresponding discrete-component equivalent.

Until relatively recently, the SCSI parallel interface was associated only with high-performance hard-disk interfaces. Its rapidly expanding use means that it is being found in more and more basic applications. For example, before long it will probably replace the popular Centronics interface used to connect printers to microcomputers. The reason for the SCSI's increasing popularity is its performance and flexibility. It is as much a local area network as a parallel bus. We provide both the data sheet of an SCSI interface controller and an application note showing how it is used.

Modern computers invariably have a battery-backed, real-time clock that maintains the time-of-day and the date. The real-time clock makes it possible to date-stamp files and to support activities ranging from multitasking to industrial process control. We provide the data sheet of a typical real-time clock and an application note showing how it can be interfaced to a processor.

The popular microprocessor revolution would never have occurred without low-cost mass storage systems. We therefore provide the data sheet of a typical high-performance floppy disk controller. Since the reliability of a floppy disk controller is very much dependent on the way in which it separates data and clock pulses from the disk drive, we include an application note on the design of data separators. The last part of this section provides an application note on the interface of the floppy disk controller to a PC-AT.

The final peripheral in this chapter is hardly the least important. The growth of the high-performance microcomputer has created a demand for low-cost computer networks, to permit all the computers in a building to share resources. Here we provide the data sheets and application reports for the components required to implement the popular ethernet local area network.

Mullard

SCN68562
Dual Universal Serial Communications Controller (DUSCC)

Preliminary Specification

DESCRIPTION

The Signetics SCN68562 Dual Universal Serial Communications Controller (DUSCC) is a single-chip MOS-LSI communications device that provides two independent, multi-protocol, full-duplex receiver/transmitter channels in a single package. It supports bit-oriented and character-oriented (byte count and byte control) synchronous data link controls as well as asynchronous protocols. The SCN68562 interfaces to the 68000 MPU via asynchronous bus control signals and is capable of program-polled, interrupt-driven, block-move or DMA data transfers.

The operating mode and data format of each channel can be programmed independently. Each channel consists of a receiver, a transmitter, a 16-bit multifunction counter/timer, a digital phase-locked loop (DPLL), a parity/CRC generator and checker, and associated control circuits. The two channels share a common bit rate generator (BRG), operating directly from a crystal or an external clock, which provides sixteen common bit rates simultaneously. The operating rate for the receiver and transmitter of each channel can be independently selected from the BRG, the DPLL, the counter/timer, or from an external 1× or 16× clock, making the DUSCC well suited for dual-speed channel applications. Data rates up to 4Mbits are supported.

The transmitter and receiver each contain a four-deep FIFO with appended transmitter command and receiver status bits and a shift register. This permits reading and writing of up to four characters at a time, minimizing the potential of receiver overrun or transmitter underrun, and reducing interrupt or DMA overhead. In addition, a flow control capability is provided to disable a remote transmitter when the FIFO of the local receiving device is full.

Two modem control inputs (DCD and CTS) and three modem control outputs (RTS and two general purpose) are provided. Because the modem control inputs and outputs are general purpose in nature, they can be optionally programmed for other functions.

PIN CONFIGURATIONS

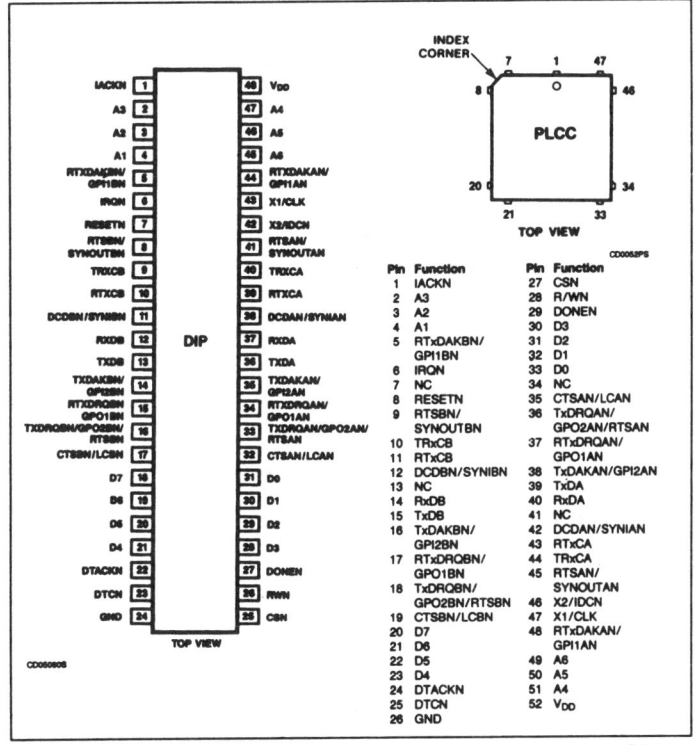

DIP TOP VIEW

Pin	Signal	Pin	Signal
1	IACKN	48	V_DD
2	A3	47	A4
3	A2	46	A5
4	A1	45	A6
5	RTxDAKBN/GPI1BN	44	RTxDAKAN/GPI1AN
6	IRQN	43	X1/CLK
7	RESETN	42	X2/IDCN
8	RTSBN/SYNOUTBN	41	RTSAN/SYNOUTAN
9	TRxCB	40	TRxCA
10	RTxCB	39	RTxCA
11	DCDBN/SYNIBN	38	DCDAN/SYNIAN
12	RxDB	37	RxDA
13	TxDB	36	TxDA
14	TxDAKBN/GPI2BN	35	TxDAKAN/GPI2AN
15	RTxDRQBN/GPO1BN	34	RTxDRQAN/GPO1AN
16	TxDRQBN/GPO2BN/RTSBN	33	TxDRQAN/GPO2AN/RTSAN
17	CTSBN/LCBN	32	CTSAN/LCAN
18	D7	31	D6
19	D6	30	D1
20	D5	29	D2
21	D4	28	D3
22	DTACKN	27	DONEN
23	DTCN	26	R/WN
24	GND	25	CSN

PLCC TOP VIEW

Pin	Function	Pin	Function
1	IACKN	27	CSN
2	A3	28	R/WN
3	A2	29	DONEN
4	A1	30	D3
5	RTxDAKBN/GPI1BN	31	D2
6	IRQN	32	D1
7	NC	33	D0
8	RESETN	34	NC
9	RTSBN/SYNOUTBN	35	CTSAN/LCAN
10	TRxCB	36	TxDRQAN/GPO2AN/RTSAN
11	RTxCB	37	RTxDRQAN/GPO1AN
12	DCDBN/SYNIBN	38	TxDAKAN/GPI2AN
13	NC	39	TxDA
14	RxDB	40	RxDA
15	TxDB	41	NC
16	TxDAKBN/GPI2BN	42	DCDAN/SYNIAN
17	RTxDRQBN/GPO1BN	43	RTxCA
18	TxDRQBN/GPO2BN/RTSBN	44	TRxCA
19	CTSBN/LCBN	45	RTSAN/SYNOUTAN
20	D7	46	X2/IDCN
21	D6	47	X1/CLK
22	D5	48	RTxDAKAN/GPI1AN
23	D4	49	A6
24	DTACKN	50	A5
25	DTCN	51	A4
26	GND	52	V_DD

Jermyn Distribution

Vestry Estate,
Sevenoaks, Kent.
Tel: (0732) 450144
Telex: 95142
Facsimile: 451251

Jermyn

December 1986

Dual Universal Serial Communications Controller (DUSCC) SCN68562

Preliminary Specification

FEATURES
General Features
- Dual full-duplex synchronous/asynchronous receiver and transmitter
- Multi-protocol operation
  - BOP: HDLC/ADCCP, SDLC, SDLC loop, X.25 or X.75 link level, e tc.
  - COP: BISYNC, DDCMP, X.21
  - ASYNC: 5 – 8 bits plus optional parity
- Four character receiver and transmitter FIFOs
- 0 to 4MHz data rate
- Programmable bit rate for each receiver and transmitter selectable from:
  - 16 fixed rates: 50 to 38.4k baud
  - One user-defined rate derived from programmable counter/timer
  - External 1× or 16× clock
  - Digital phase-locked loop
- Parity and FCS (frame check sequence LRC or CRC) generation and checking
- Programmable data encoding/decoding: NRZ, NRZI, FM0, FM1, Manchester
- Programmable channel mode: full-half-duplex, auto-echo, or local loopback
- Programmable data transfer mode: polled, interrupt, DMA, wait
- DMA interface
  - Compatible with Signetics' SCB68430 Direct Memory Access Interface (DMAI) and other DMA controllers
  - Half- or full-duplex operation
  - Single or dual address data transfers
  - Automatic frame termination on counter/timer terminal count or DMA DONE
- Interrupt capabilities
  - Daisy chain option
  - Vector output (fixed or modified by status)
  - Programmable internal priorities
  - Maskable interrupt conditions
  - 68000 compatible
- Multi-function programmable 16-bit counter/timer
  - Bit rate generator
  - Event counter
  - Count received or transmitted characters
  - Delay generator
  - Automatic bit length measurement
- Modem controls
  - RTS, CTS, DCD, and up to four general purpose I/O pins per channel
  - CTS and DCD programmable auto-enables for Tx and Rx
  - Programmable interrupt on change of CTS or DCD
- On-chip oscillator for crystal
- TTL compatible
- Single +5V power supply

Asynchronous Mode Features
- Character length: 5 to 8 bits
- Odd or even parity, no parity, or force parity
- Up to two stop bits programmable in 1/16- bit increments
- 1× or 16× Rx and Tx clock factors
- Parity, overrun, and framing error detection
- False start bit detection
- Start bit search ½ bit time after framing error detection
- Break generation with handshake for counting break characters
- Detection of start and end of received break
- Character compare with optional interrupt on match

Character-Oriented Protocol Features
- Character length: 5 to 8 bits
- Odd or even parity, no parity, or force parity
- LRC or CRC generation and checking
- Optional opening PAD transmission
- One or two SYN characters
- External sync capability
- SYN detection and optional stripping
- SYN or MARK linefill on underrun
- Idle in MARK or SYNs
- Parity, FCS, overrun, and underrun error detection
- BISYNC Features
  - EBCDIC or ASCII header, text and control messages
  - SYN, DLE stripping
  - EOM (end of message) detection and transmission
  - Auto transparency mode switching
  - Auto hunt after receipt of EOM sequence (with closing PAD check after EOT or NAK)
  - Control character sequence detection for both transparent and normal text

Bit-Oriented Protocol Features
- Character length: 5 to 8 bits
- Detection and transmission of residual character: 0 – 7 bits
- Automatic switch to programmed character length for I field
- Zero insertion and deletion
- Optional opening PAD transmission
- Detection and generation of FLAG, ABORT, and IDLE bit patterns
- Detection and generation of shared (single) FLAG between frames
- Detection of overlapping (shared zero) FLAGs
- ABORT, ABORT-FLAGs, or FCS-FLAGs line fill on underrun
- Idle in MARK or FLAGs
- Secondary address recognition including group and global address
- Single- or dual-octet secondary address
- Extended address and control fields
- Short frame rejection for receiver
- Detection and notification of received end of message
- CRC generation and checking
- SDLC loop mode capability

ORDERING INFORMATION

PACKAGES	V_{CC} = +5V ±5%, T_A = 0 to +70°C
Ceramic DIP	SCN68562C4I48
Plastic DIP	SCN68562C4N48
Plastic LCC	SCN68562C4A52

December 1986

Preliminary Specification

Dual Universal Serial Communications Controller (DUSCC)   SCN68562

BLOCK DIAGRAM

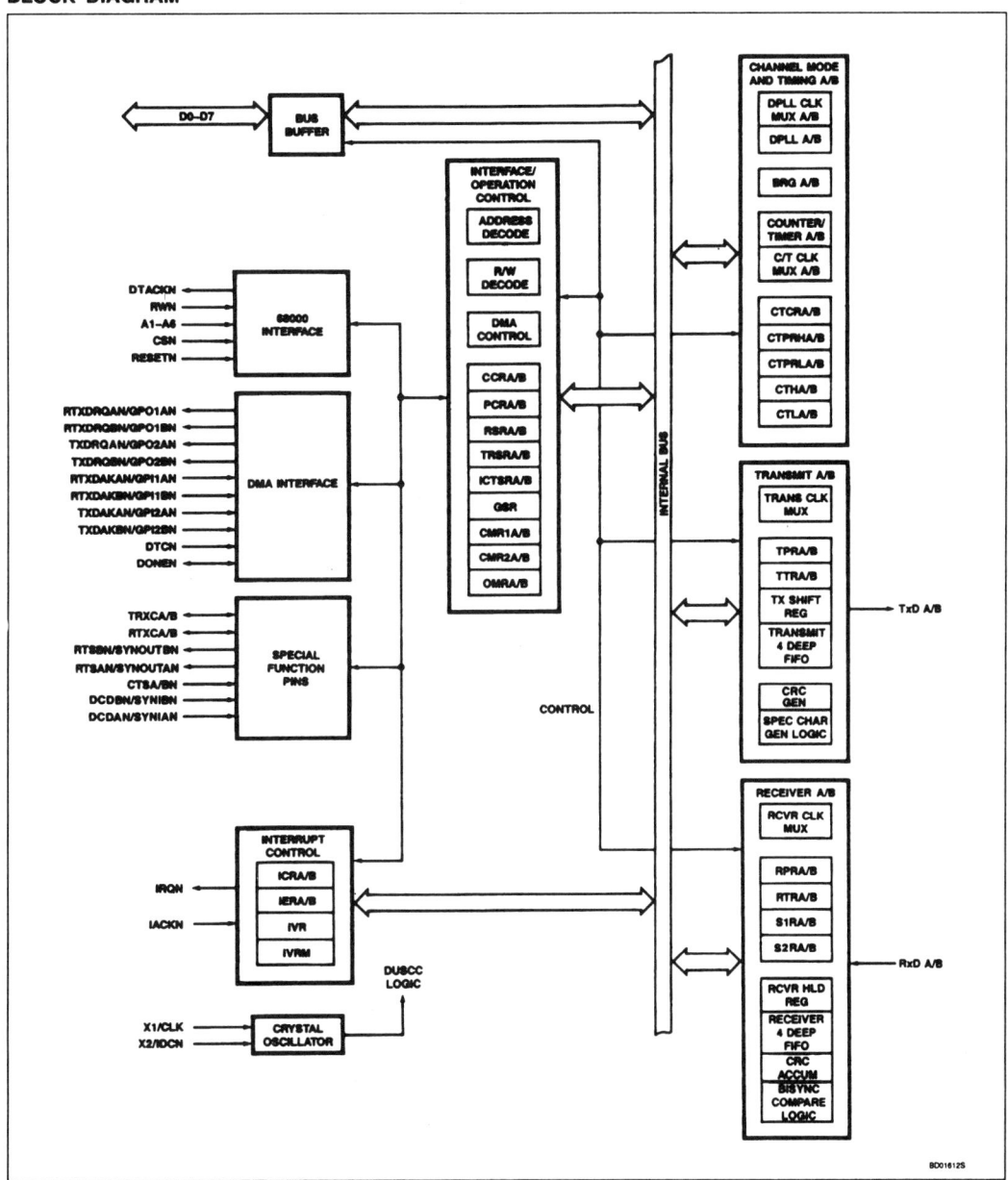

Preliminary Specification

Dual Universal Serial Communications Controller (DUSCC) SCN68562

PIN DESCRIPTION

In this data sheet, signals are discussed using the terms 'active' and 'inactive' or 'asserted' and 'negated' independent of whether the signal is active in the high (logic 1) or low (logic 0) state. N at the end of a pin name signifies the signal associated with the pin is active low (see individual pin description for the definition of the active level of each signal.) Pins which are provided for both channels are designated by either an underline (_) or by A/B after the name of the pin and the active low state indicator, N, if applicable. A similar method is used for registers provided for both channels; these are designated by either an underline or by A/B after the name.

MNEMONIC	PIN NO. DIP	PIN NO. PLCC	TYPE	NAME AND FUNCTION
A1 – A6	4 – 2, 45 – 47	5 – 3, 51 – 49	I	**Address Lines:** Active high. Address inputs which specify which of the internal registers is accessed for read/write operations.
D0 – D7	31 – 28, 21 – 18	34 – 31, 23 – 20	I/O	**Bidirectional Data Bus:** Active high, three state. Bit 0 is the LSB and bit 7 is the MSB. All data, command, and status transfers between the CPU and the DUSCC take place over this bus. The data bus is enabled when CSN is low, during interrupt acknowledge cycles and single-address DMA acknowledge cycles.
R/WN	26	29	I	**Read/Write:** A high input indicates a read cycle and a low input indicates a write cycle when a cycle is initiated by assertion of the CSN input.
CSN	25	28	I	**Chip Select:** Active low input. When low, data transfers between the CPU and the DUSCC are enabled on D0 – D7 as controlled by the R/WN and A1 – A6 inputs. When CSN is high, the DUSCC is isolated from the data bus (except during interrupt acknowledge cycles and single-address DMA transfers) and D0 – D7 are placed in the tri-state condition.
DTACKN	22	24	O	**Data Transfer Acknowledge:** Active low, 3-State. DTACKN is asserted on a write cycle to indicate that the data on the bus has been latched, and on a read cycle or interrupt acknowledge cycle to indicate valid data is on the bus. The signal is negated when completion of the cycle is indicated by negation of the CSN or IACKN input, and returns to the inactive state (3-State) a short period after it is negated. In single address DMA mode, the operation of this pin is similar to the description above. The exception is that it is negated when completion of the cycle is indicated by the assertion of DTCN or negation of DMA acknowledge inputs (whichever occurs first), and returns to the inactive state (3-State) a short period after it is negated. When negated, DTACKN becomes an open drain output and requires an external pull-up resistor.
IRQN	6	7	O	**Interrupt Request:** Active low, open drain. This output is asserted upon occurrence of any enabled interrupting condition. The CPU can read the general status register to determine the interrupting condition(s), or can respond with an interrupt acknowledge cycle to cause the DUSCC to output an interrupt vector on the data bus.
IACKN	1	2	I	**Interrupt Acknowledge:** Active low. When IACKN is asserted, the DUSCC responds by placing the contents of the interrupt vector register (modified or unmodified by status) on the data bus and asserting DTACKN. If no active interrupt is pending, DTACKN is not asserted.
X1/CLK	43	47	I	**Crystal or External Clock:** When using the crystal oscillator, the crystal is connected between pins X1 and X2. If a crystal is not used, an external clock is supplied at this input. This clock is used to drive the internal bit rate generator, as an optional input to the counter/timer or DPLL, and to provide other required clocking signals.
X2/IDCN	42	46	O	**Crystal or Interrupt Daisy Chain:** Active low. When a crystal is used as the timing source, the crystal is connected between pins X1 and X2. This pin can be programmed to provide an interrupt daisy chain output which propagates the IACKN signal to lower priority devices, if no active interrupt is pending. This pin should be grounded when an external clock is used on X1 and X2, is not used as an interrupt daisy chain output.
RESETN	7	8	I	**Master Reset:** Active low. A low on this pin resets the transmitters and receivers and resets the registers shown in Table 1. Reset is asynchronous, i.e., no clock is required.
RxDA, RxDB	37, 12	41, 13	I	**Channel A (B) Receiver Serial Data Input:** The least significant bit is received first. If external receiver clock is specified for the channel, the input is sampled on the rising edge of the clock.
TxDA, TxDB	36, 13	39, 15	O	**Channel A (B) Transmitter Serial Data Output:** The least significant bit is transmitted first. This output is held in the marking (high) condition when the transmitter is disabled or when the channel is operating in local loopback mode. If external transmitter clock is specified for the channel, the data is shifted on the falling edge of the clock.
RTxCA, RTxCB	39, 10	43, 11	I/O	**Channel A (B) Receiver/Transmitter Clock:** As an input, it can be programmed to supply the receiver, transmitter, counter/timer, or DPLL clock. As an output, can supply the counter/timer output, the transmitter shift clock (1×), or the receiver sampling clock (1×). The maximum external receiver/transmitter clock frequency is 4MHz.

December 1986

Dual Universal Serial Communications Controller (DUSCC) SCN68562

Preliminary Specification

PIN DESCRIPTION (Continued)

MNEMONIC	PIN NO. DIP	PIN NO. PLCC	TYPE	NAME AND FUNCTION
TRxCA, TRxCB	40, 9	44, 10	I/O	**Channel A (B) Transmitter/Receiver Clock:** As an input, it can supply the receiver, transmitter, counter/timer, or DPLL clock. As an output, it can supply the counter/timer output, the DPLL output, the transmitter shift clock ($1\times$), the receiver sampling clock ($1\times$,), the transmitter BRG clock ($16\times$), the receiver BRG clock ($16\times$), or the internal system clock ($\times 1/2$). The maximum external receiver/transmitter clock frequency is 4MHz.
CTSA/BN, LCA/BN	32, 17	35, 19	I/O	**Channel A (B) Clear-To-Send Input or Loop Control Output:** Active low. The signal can be programmed to act as an enable for the transmitter when not in loop mode. The DUSCC detects logic level transitions on this input and can be programmed to generate an interrupt when a transition occurs. When operating in the BOP loop mode, this pin becomes a loop control output which is asserted and negated by DUSCC commands. This output provides the means of controlling external loop interface hardware to go on-line and off-line without disturbing operation of the loop.
DCDA/BN, SYNIA/BN	38, 11	42, 12	I	**Channel A (B) Data Carrier Detected or External Sync Input:** The function of this pin is programmable. As a DCD active low input, it acts as an enable for the receiver or can be used as a general purpose input. For the DCD function, the DUSCC detects logic level transitions on this input and can be programmed to generate an interrupt when a transition occurs. As an active low external sync input, it is used in COP modes to obtain character synchronization without receipt of a SYN or FLAG character. This mode can be used in disc or tape controller applications or for the optional byte timing lead in X.21.
RTxDRQA/BN, GPO1A/BN	34, 15	37, 17	O	**Channel A (B) Receiver/Transmitter DMA Service Request or General Purpose Output:** Active low. For half-duplex DMA operation, this output indicates to the DMA controller that one or more characters are available in the receiver FIFO (when the receiver is enabled) or that the transmit FIFO is not full (when the transmitter is enabled). For full-duplex DMA operation, this output indicates to the DMA controller that data is available in the receiver FIFO. In non-DMA mode, this pin is a general purpose output that can be asserted and negated under program control.
TxDRQA/BN, GPO2A/BN, RTSA/BN	33, 16	36, 18	O	**Channel A (B) Transmitter DMA Service Request, General Purpose Output, or Request-to-Send:** Active low. For full-duplex DMA operation, this output indicates to the DMA controller that the transmit FIFO is not full and can accept more data. When not in full-duplex DMA mode, this pin can be programmed as a general purpose or a Request-to-Send output, which can be asserted and negated under program control (see Detailed Operation).
RTxDAKA/BN, GPI1A/BN	44, 5	48, 6	I	**Channel A (B) Receiver/Transmitter DMA Acknowledge or General Purpose Input:** Active low. For half-duplex single address DMA operation, this input indicates to the DUSCC that the DMA controller has acquired the bus and that the requested bus cycle (read receiver FIFO or load transmitter FIFO) is beginning. For full-duplex single address DMA operation, this input indicates to the DUSCC that the DMA controller has acquired the bus and that the requested read receiver FIFO bus cycle is beginning. Because the state of this input can be read under program control, it can be used as a general purpose input when not in single address DMA mode.
TxDAKA/BN, GPI2A/BN	35, 14	38, 16	I	**Channel A (B) Transmitter DMA Acknowledge or General Purpose Input:** Active low. When the channel is programmed for full-duplex single address DMA operation, this input is asserted to indicate to the DUSCC that the DMA controller has acquired the bus and that the requested load transmitter FIFO bus cycle is beginning. Because the state of this input can be read under program control, it can be used as a general purpose input when not in full-duplex single address DMA mode.
DTCN	23	25	I	**Device Transfer Complete:** Active low. DTCN is asserted by the DMA controller to indicate that the requested data transfer is complete.
DONEN	27	30	I/O	**Done:** Active low, open drain. See Detailed Operation for a description of the function of this pin.
RTSA/BN, SYNOUTA/BN	41, 8	45, 9	O	**Channel A (B) Sync Detect or Request-to-Send:** Active low. If programmed as a sync output, it is asserted one bit time after the specified sync character (COP or BISYNC modes) or a FLAG (BOP modes) is detected by the receiver. As a Request-to-Send modem control signal, it functions as described previously for the TxDRQ_N/RTS_N pin.
V_{DD}	48	52	I	+5V ±5% power input.
GND	24	26	I	Signal and power ground input.

December 1986

Preliminary Specification

Dual Universal Serial Communications Controller (DUSCC) SCN68562

REGISTERS

The addressable registers of the DUSCC are shown in Table 1. The following rules apply to all registers:

1. A read from a reserved location in the map results in a read from the 'null register'. The null register returns all ones for data and results in a normal bus cycle. A write to one of these locations results in a normal bus cycle without a write being performed.
2. Unused bits of a defined register are read as zeros, unless ones have been loaded after master reset.
3. Bits that are unused in the chosen mode but are used in others are readable and writable but their contents are ignored in the chosen mode.

All registers are addressable as 8-bit quantities. To facilitate operation with the 68000 MOVEP instruction, addresses are ordered such that certain sets of registers may also be accessed as words or long words.

The operation of the DUSCC is programmed by writing control words into the appropriate registers. Operational feedback is provided via status registers which can be read by the CPU. The contents of certain control registers are initialized on RESET (set to zero). Care should be exercised if the contents of a register are changed during operation, since certain changes may cause operational problems, e.g., changing the channel mode at an inappropriate time may cause the reception or transmission of an incorrect character. In general, the contents of registers which control transmitter or receiver operation, or the counter/timer, should be changed only when they are not enabled.

The DUSCC registers can be separated into five groups to facilitate their usage:

1. Channel mode configuration and pin description registers
2. Transmitter and receiver parameter and timing registers
3. Counter/timer control and value registers
4. Interrupt control and status registers
5. Command register

This arrangement is used in the following description of the DUSCC registers.

Channel Mode Configuration and Pin Description Registers

There are five registers in this group for each channel. The bit format for each of these registers is contained in Table 2. The primary function of these registers is to define configuration of the channels and the function of the programmable pins.

Channel Mode Register 1 (CMR1A, CMR1B)

[7:6] Data Encoding — These bits select the data encoding for the received and transmitted data:

00   If the DPLL is set to NRZI mode (see DPLL commands), it selects positive logic (1 = high, 0 = low). If the DPLL is set to FM mode (see DPLL commands), Manchester (bi-phase level) encoding is selected.

01   NRZI. Non-return-to-zero inverted.

10   FM0. Bi-phase space.

11   FM1. Bi-phase mark.

[5] Extended Control (BOP) —

0   No. A one-octet control field follows the address field.

1   Yes. A two-octet control field follows the address field.

[5] Parity (COP/ASYNC), Code Select (BISYNC) —

0   Even parity if with parity is selected by [4:3] or a 0 in the parity bit position if force parity is selected by [4:3]. In BISYNC protocol mode, internal character comparisons are made using EBCDIC coding.

1   Odd parity if with parity is selected by [4:3] or a 1 in the parity bit position if force parity is selected by [4:3]. In BISYNC protocol mode, internal character comparisons are made using 8-bit odd parity ASCII coding. (Note: The receiver should be programmed for 8-bit characters, RPR[1:0] = 11, with no parity, CMR1[4:3] = 00.)

[4:3] Address Mode (BOP) — This field controls whether a single octet or multiple octets follow the opening FLAG(s) for both the receiver and the transmitter. This field is activated by selection of BOP secondary mode through the channel protocol mode bits CMR1_[2:0] (see Detailed Operation).

00   Single-octet address.
01   Extended address.
10   Dual-octet address.
11   Dual-octet address with group.

[4:3] Parity Mode (COP/ASYNC) — This field selects the parity mode for both the receiver and the transmitter. A parity bit is added to the programmed character length if with parity or force parity is selected:

00   No parity. Required when BISYNC protocol mode is programmed.
01   Reserved.
10   With parity. Odd or even parity is selected by [5].
11   Force parity. The parity bit is forced to the state selected by [5].

[2:0] Channel Protocol Mode — This field selects the operational protocol and sub-mode for both the receiver and transmitter:

000   BOP Primary. No address comparison is performed. For receive, all characters received after the opening FLAG(s) are transferred to the FIFO.

001   BOP Secondary. This mode activates the address modes selected by [4:3]. Except in the case of extended address ([4:3]=01), an address comparison is performed to determine if a frame should be received. Refer to Detailed Operation for details of the various addressing modes. If a valid comparison occurs, the receiver is activated and the address octets and all subsequent received characters of the frame are transferred to the receive FIFO.

010   BOP Loop. The DUSCC acts as a secondary station in a loop. The GO-ON-LOOP and GO-OFF-LOOP commands are used to cause the DUSCC to go on and off the loop. Normally, the TxD output echoes the RxD input with a two-bit time delay. If the transmitter is enabled and the 'go active on poll' command has been asserted, the transmitter will begin sending when an EOP sequence consisting of a zero followed by seven ones is detected. The DUSCC changes the last one of the EOP to zero, making it another FLAG, and then operates as described in the detailed operation section. The loop sending status bit (TRSR[6]) is asserted concurrent with the beginning of transmission. The frame should normally be terminated with an EOM followed by an echo of the marking RxD line so that secondary stations further down the loop can append their messages to the messages from up-loop stations by the same process. If the 'go active on poll' command is not asserted, the transmitter remains inactive (other than echoing the received data) even when the EOP sequence is received.

011   BOP Loop without address comparison. Same as normal loop mode except that address field comparisons are disabled. All received frames are transmitted to the CPU.

100   COP Dual SYN. Character sync is achieved upon receipt of a bit sequence matching the contents of the appropriate bits of S1R and S2R (SYN1-SYN2), including parity bits if any.

101   COP Dual SYN (BISYNC). Character sync is achieved upon receipt of a bit

December 1986

Preliminary Specification

Dual Universal Serial Communications Controller (DUSCC) SCN68562

Table 1. DUSCC Register Address Map

ADDRESS BITS[1] 6 5 4 3 2 1	ACRONYM	REGISTER NAME	MODE	AFFECTED BY RESET
c 0 0 0 0 0	CMR1	Channel Mode Register 1	R/W	Yes — 00
c 0 0 0 0 1	CMR2	Channel Mode Register 2	R/W	Yes — 00
c 0 0 0 1 0	S1R	SYN 1/Secondary Address 1 Register	R/W	No
c 0 0 0 1 1	S2R	SYN 2/Secondary Address 2 Register	R/W	No
c 0 0 1 0 0	TPR	Transmitter Parameter Register	R/W	Yes — 00
c 0 0 1 0 1	TTR	Transmitter Timing Register	R/W	No
c 0 0 1 1 0	RPR	Receiver Parameter Register	R/W	Yes — 00
c 0 0 1 1 1	RTR	Receiver Timing Register	R/W	No
c 0 1 0 0 0	CTPRH	Counter/Timer Preset Register High	R/W	No
c 0 1 0 0 1	CTPRL	Counter/Timer Preset Register Low	R/W	No
c 0 1 0 1 0	CTCR	Counter/Timer Control Register	R/W	Yes — 00
c 0 1 0 1 1	OMR	Output and Miscellaneous Register	R/W	Yes — 00
c 0 1 1 0 0	CTH	Counter/Timer High	R	No
c 0 1 1 0 1	CTL	Counter/Timer Low	R	No
c 0 1 1 1 0	PCR	Pin Configuration Register	R/W	Yes — 00
c 0 1 1 1 1	CCR	Channel Command Register	R/W	No
c 1 0 0 X X	TxFIFO	Transmitter FIFO	W	No
c 1 0 1 X X	RxFIFO	Receiver FIFO	R	No
c 1 1 0 0 0	RSR	Receiver Status Register	R/W[2]	Yes — 00
c 1 1 0 0 1	TRSR	Transmitter and Receiver Status Register	R/W[2]	Yes — 00
c 1 1 0 1 0	ICTSR	Input and Counter/Timer Status Register	R/W[2]	Yes
d 1 1 0 1 1	GSR	General Status Register	R/W[2]	Yes — 00
c 1 1 1 0 0	IER	Interrupt Enable Register	R/W	Yes — 00
c 1 1 1 0 1		Not used		
0 1 1 1 1 0	IVR	Interrupt Vector Register — Unmodified	R/W	Yes — 0F
1 1 1 1 1 0	IVRM	Interrupt Vector Register — Modified	R	Yes — 0F
0 1 1 1 1 1	ICR	Interrupt Control Register	R/W	Yes — 00
1 1 1 1 1 1		Not used		

NOTES:
1. c = 0 for channel A, c = 1 for channel B.
   d = don't care — register may be accessed as either channel.
   x = don't care — FIFOs are addressable at any of four adjacent addresses to allow them to be addressed as byte/word/long word with the 68000 MOVEP instruction.
2. A write to this register may perform a status resetting operation.

sequence matching the contents of the appropriate bits of S1R and S2R (SYN1-SYN2). In this mode, special transmitter and receive logic is activated. Transmitter and receiver character length must be programmed to 8 bits and no parity (see Detailed Operation).

110 COP Single SYN. Character sync is achieved upon receipt of a bit sequence matching the contents of the appropriate bits of S1R (SYN1), including parity bit if any. This mode is required when the external sync mode is selected (see description of RPR[4], BOP/COP).

111 Asynchronous. Start/stop format.

Channel Mode Register 2 (CMR2A, CMR2B)

[7:6] Channel Connection — This field selects the mode of operation of the channel. The user must exercise care when switching into and out of the various modes. The selected mode will be activated immediately upon mode selection, even if this occurs in the middle of a received or transmitted character.

00 Normal mode. The transmitter and receiver operate independently in either half- or full-duplex, controlled by the respective enable commands.

01 Automatic echo mode. Automatically retransmits the received data with a half-bit time delay (ASYNC, 16× clock mode) or a two-bit time delay (all other modes). The following conditions are true while in automatic echo mode:

1. Received data is reclocked and retransmitted on the TxD output.
2. The receiver clock is used for the transmitter for Async 16X clock mode. For other modes the transmitter clock must be supplied.
3. The receiver must be enabled, but the transmitter need not be enabled.
4. The TxRDY and underrun status bits are inactive.
5. The received parity and/or FCS are checked if required, but are not regenerated for transmission, i.e., transmitted parity and/or FCS are as received.
6. In ASYNC mode, character framing is checked, but the stop bits are retransmitted as received. A received break is echoed as received.
7. CPU to receiver communication continues normally, but the CPU to transmitter link is disabled.

10 Local loopback mode. In this mode:

1. The transmitter output is internally connected to the receiver input.
2. The transmit clock is used for the receiver if NRZI or NRZ encoding is used. For FM or Manchester encoding because the receiver clock is derived from the DPLL, the DPLL source clock must be maintained.
3. The TxD output is held high.

December 1986   7

Preliminary Specification

Dual Universal Serial Communications Controller (DUSCC) — SCN68562

4. The RxD input is ignored.
5. The receiver and transmitter must be enabled.
6. CPU to transmitter and receiver communications continue normally.

11 Reserved.

[5:3] Data Transfer Interface — This field specifies the type of data transfer between the DUSCC's Rx and TxFIFOs and the CPU. All interrupt and status functions operate normally regardless of the data transfer interface programmed. Refer to Detailed Operation for details of the various DMA transfer interfaces.

000 Half-duplex single address DMA.
001 Half-duplex dual address DMA.
010 Full-duplex single address DMA.
011 Full-duplex dual address DMA.
100 Wait on receive only. In this mode a read of a non-empty receive FIFO results in a normal bus cycle. However, if the receive FIFO of the channel is empty when a read Rx FIFO cycle is initiated, the DTACKN output remains negated until a character is received and loaded into the FIFO. DTACKN is then asserted and the cycle is completed normally.
101 Wait on transmit only. In this mode a write to a non-full transmit FIFO results in a normal bus cycle. However, if the transmit FIFO of the channel is full when a write TxFIFO cycle is initiated, the DTACKN output remains negated until a FIFO position becomes available for the new character. DTACKN is then asserted and the cycle is completed normally.
110 Wait on transmit and receive. As above for both wait on receive and transmit operations.
111 Polled or interrupt. DMA and wait functions of the channel are not activated. Data transfers to the Rx and TxFIFOs are via normal bus read and write cycles in response to polling of the status registers and/or interrupts.

[2:0] Frame Check Sequence Select — This field selects the optional frame check sequence (FCS) to be appended at the end of a transmitted frame. When CRC is selected in COP, then no parity and 8-bit character length must be used. The selected FCS is transmitted as follows:
1. Following the transmission of a FIFOed character tagged with the 'send EOM' command.
2. If underrun control (TPR[7:6]) is programmed for TEOM, upon occurrence of an underrun.
3. If TEOM on zero count or done (TPR[4]) is asserted and the counter/timer is counting transmitted characters, after transmission of the character which causes the counter to reach zero count.
4. In DMA mode with TEOM on zero count or done (TPR[4]) set, after transmission of a character if DONEN is asserted when that character was loaded into the TxFIFO by the DMA controller.

000 No frame check sequence.
001 Reserved
010 LRC8: Divisor = x^8+1, dividend preset to zeros. The Tx sends the calculated LRC non-inverted. The Rx indicates an error if the computed LRC is not equal to 0. Valid for COP modes only.
011 LRC8: Divisor = x^8+1, dividend preset to ones. The Tx sends the calculated LRC non-inverted. The Rx indicates an error if the computed LRC is not equal to 0. Valid for COP modes only.
100 CRC16: Divisor = $x^{16}+x^{15}+x^2+1$, dividend preset to zeros. The Tx sends the calculated CRC non-inverted. The Rx indicates an error if the computed CRC is not equal to 0. Not valid for ASYNC mode.
101 CRC16: Divisor = $x^{16}+x^{15}+x^2+1$, dividend preset to ones. The Tx sends the calculated CRC non-inverted. The Rx indicates an error if the computed CRC is not equal to 0. Not valid for ASYNC mode.
110 CRC-CCITT: Divisor = $x^{16}+x^{12}+x^5+1$, dividend preset to zeros. The Tx sends the calculated CRC non-inverted. The Rx indicates an error if the computed CRC is not equal to 0. Not valid for ASYNC mode.
111 CRC-CCITT: Divisor = $x^{16}+x^{12}+x^5+1$, dividend preset to ones. The Tx sends the calculated CRC inverted. The Rx indicates an error if the computed CRC is not equal to H'F0B8'. Not valid for ASYNC mode.

SYN1/Secondary Address 1 Register (S1RA, S1RB)

[7:0] Character Compare — In ASYNC mode this register holds a 5- to 8-bit long bit pattern which is compared with received characters. If a match occurs, the character compare status bit (RSR[7]) is set. This field is ignored if the receiver is in a break condition.

In COP modes, this register contains the 5- to 8-bit SYN1 bit pattern, right justified. Parity bit need not be included in the value placed in the register even if parity is specified in CMR1[4:3]. However, a character received with parity error, when parity is specified, will not match. In ASYNC or COP modes, if parity is specified, then any unused bits in this register must be programmed to zeros. In BOP secondary mode it contains the address used to compare the first received address octet. The register is not used in BOP primary mode or secondary modes where address comparisons are not made, such as when extended addressing is specified.

SYN2/Secondary Address 2 Register (S2RA, S2RB)

[7:0] — This register is not used in ASYNC, COP single SYN, BOP primary modes, BOP secondary modes with single address field, and BOP secondary modes where address comparisons are not made, such as when extended addressing is specified.

In COP dual SYN modes, it contains the 5- to 8-bit SYN2 bit pattern, right justified. Parity bit need not be included in the value placed in the register even if parity is specified in CMR1[4:3]. However, a character received with parity error, when parity is specified, will not match. If parity is specified, then any unused bits in this register must be programmed to zeros. In BOP secondary mode using two address octets, it contains the partial address used to compare the second received address octet.

Pin Configuration Register (PCRA, PCRB)
This register selects the functions for multi-purpose I/O pins.

[7] X2/IDC — This bit is defined only for PCRA. It is not used in PCRB.
0 The X2/IDCN pin is used as a crystal connection.
1 The X2/IDCN pin is the interrupt daisy chain output.

[6] GPO2/RTS — The function of this pin is programmable only when not operating in full-duplex DMA mode.
0 The TxDRQ_N/GPO2_N/RTS_N pin is a general purpose output. It is low when OMR[2] is a 1 and high when OMR[2] is 0.
1 The pin is a request-to-send output (see Detailed Operation).

December 1986

8

Dual Universal Serial Communications Controller (DUSCC) SCN68562

Preliminary Specification

Table 2. Channel Configuration/Pin Definition Registers Bit Formats

CHANNEL MODE REG 1

(CMR1A, CMR1B)

BIT 7	BIT 6	BIT 5	BIT 4	BIT 3	BIT 2	BIT 1	BIT 0
\multicolumn{2}{c}{Data Encoding}	Extended Control	\multicolumn{2}{c}{Address Mode (BOP)}	\multicolumn{3}{c}{Channel Protocol Mode}

Data Encoding:
- 00 — NRZ/Manchester
- 01 — NRZI
- 10 — FM0
- 11 — FM1

Extended Control:
- BOP only
- 0 — no
- 1 — yes

Address Mode (BOP):
- 00 — 8-bit
- 01 — extended address
- 10 — 16-bit
- 11 — 16-bit w/group

#Parity:
- 0 — even
- 1 — odd

Parity Mode (COP/ASYNC):
- 00 — no parity
- 01 — reserved
- 10 — with parity
- 11 — force parity

Channel Protocol Mode:
- 000 — BOP primary
- 001 — BOP secondary
- 010 — BOP loop
- 011 — BOP loop – no adr. comp.
- 100 — COP dual SYN
- 101 — COP dual SYN (BISYNC)
- 110 — COP single SYN
- 111 — asynchronous

NOTE:
#In BISYNC protocol mode, 0 = EBCDIC, 1 = ASCII coding.

CHANNEL MODE REG 2

(CMR2A, CMR2B)

BIT 7	BIT 6	BIT 5	BIT 4	BIT 3	BIT 2	BIT 1	BIT 0

Channel Connection:
- 00 — normal
- 01 — auto echo
- 10 — local loop
- 11 — reserved

Data Transfer Interface:
- 000 — half-duplex single address DMA
- 001 — half-duplex dual address DMA
- 010 — full-duplex single address DMA
- 011 — full-duplex dual address DMA
- 100 — wait on Rx only
- 101 — wait on Tx only
- 110 — wait on Rx or Tx
- 111 — polled or interrupt

Frame Check Sequence Select:
- 000 — none
- 001 — reserved
- 010 — LRC8 preset 0s
- 011 — LRC8 preset 1s
- 100 — CRC 16 preset 0s
- 101 — CRC 16 preset 1s
- 110 — CRC CCITT preset 0s
- 111 — CRC CCITT preset 1s

SYN1/SECONDARY ADDRESS REG 1

(S1RA, S1RB)

BIT 7	BIT 6	BIT 5	BIT 4	BIT 3	BIT 2	BIT 1	BIT 0

ASYNC — Character compare (5–8 bits)
COP — SYN1 (5–8 bits)
BOP — First address octet

SYN2/SECONDARY ADDRESS REG 2

(S2RA, S2RB)

BIT 7	BIT 6	BIT 5	BIT 4	BIT 3	BIT 2	BIT 1	BIT 0

ASYNC — not used
COP — SYN2 (5–8 bits)
BOP — Second address octet

PIN CONFIGURATION REG

(PCRA, PCRB)

BIT 7	BIT 6	BIT 5	BIT 4	BIT 3	BIT 2	BIT 1	BIT 0
X2/IDC	GPO2/RTS	SYNOUT/RTS	\multicolumn{2}{c}{RTxC Pin}	\multicolumn{3}{c}{TRxC Pin}

X2/IDC:
- *
- 0 — X2
- 1 — IDC

GPO2/RTS:
- 0 — GPO2
- 1 — RTS

SYNOUT/RTS:
- 0 — SYNOUT
- 1 — RTS

RTxC Pin:
- 00 — input
- 01 — C/T
- 10 — TxCLK 1×
- 11 — RxCLK 1×

TRxC Pin:
- 000 — input
- 001 — XTAL/2
- 010 — DPLL
- 011 — C/T
- 100 — TxCLK 16×
- 101 — RxCLK 16×
- 110 — TxCLK 1×
- 111 — RxCLK 1×

NOTE:
*PCRA only. Not used in PCRB.

Dual Universal Serial Communications Controller (DUSCC)   SCN68562

Preliminary Specification

[5] SYNOUT/RTS —
0   The SYNOUT_N/RTS_N pin is an active low output which is asserted one bit time after a SYN pattern (COP modes) in HSRH/HSRL or FLAG (BOP modes) is detected in CCSR. The output remains asserted for one receiver clock period. See Figure 1 for receiver data path.

1   The pin is a request-to-send output (see Detailed Operation). The logical state of the pin is controlled by OMR[0], when set the output is zero.

[4:3] RTxC —
00   The pin is an input. It must be programmed for input when used as the input for the receiver or transmitter clock, the DPLL, or the C/T.

01   The pin is an output from the counter/timer. Refer to CTCRA/B description.

10   The pin is an output from the transmitter shift register clock.

11   The pin is an output from the receiver shift register clock.

Table 3. Transmitter and Receiver Parameter and Timing Register Bit Format

TRANSMITTER PARAMETER REG (TPRA, TPRB)

	BIT 7	BIT 6	BIT 5	BIT 4	BIT 3	BIT 2	BIT 1	BIT 7
	\multicolumn{2}{c}{Underrun Control}	Idle	TEOM On Zero Cnt Or Done	Tx RTS Control	CTS Enable Tx	\multicolumn{2}{c}{Tx Character Length}
COP	00 — FCS-idle 01 — reserved 10 — MARKs 11 — SYNs		0 — MARKs 1 — SYNs	0 — no 1 — yes	0 — no 1 — yes	0 — no 1 — yes	00 — 5 bits 01 — 6 bits 10 — 7 bits 11 — 8 bits	
	Underrun Control		Idle	TEOM On Zero Cnt Or Done				
BOP	00 — FCS-FLAG-idle 01 — reserved 10 — ABORT-MARKs 11 — ABORT-FLAGs		0 — MARKs 1 — FLAGs	0 — no 1 — yes				
ASYNC	\multicolumn{8}{l}{Stop Bits Per Character: $9/16$ to 1, $17/16$ to 1.5, $25/16$ to 2 programmable in $1/16$-bit increments}

TRANSMITTER TIMING REG (TTRA, TTRB)

	BIT 7	BIT 6	BIT 5	BIT 4	BIT 3	BIT 2	BIT 1	BIT 0
	External Source	\multicolumn{3}{c}{Transmitter Clock Select}	\multicolumn{4}{c}{Bit Rate Select}
	0 — RTxC 1 — TRxC	000 — 1× external 001 — 16× external 010 — DPLL 011 — BRG 100 — 2× other channel C/T 101 — 32× other channel C/T 110 — 2× own channel C/T 111 — 32× own channel C/T			\multicolumn{4}{c}{one of sixteen rates from BRG}

RECEIVER PARAMETER REG (RPRA, RPRB)

	BIT 7	BIT 6	BIT 5	BIT 4	BIT 3	BIT 2	BIT 1	BIT 0
	not used	not used	not used	Rx RTS Control	Strip* Parity	DCD Enable Rx	\multicolumn{2}{c}{Rx Character Length}
ASYNC				0 — no 1 — yes	0 — no 1 — yes	0 — no 1 — yes	00 — 5 bits 01 — 6 bits 10 — 7 bits 11 — 8 bits	
COP	SYN Strip	FCS to FIFO	Auto Hunt & Pad Chk	Ext Sync	Strip* Parity			
	0 — no 1 — yes	0 — no 1 — yes	0 — no 1 — yes	0 — no 1 — yes	0 — no 1 — yes			
BOP	not used	FCS to FIFO	Overrun Mode	not used	All Party Address			
		0 — no 1 — yes	0 — hunt 1 — cont		0 — no 1 — yes			

NOTE:
*If the receiver character length is 8-bits and parity is programmed, this bit must be set.

December 1986

Dual Universal Serial Communications Controller (DUSCC)   SCN68562

Table 3. Transmitter and Receiver Parameter and Timing Register Bit Format (Continued)

RECEIVER TIMING REG

(RTRA, RTRB)

BIT 7	BIT 6	BIT 5	BIT 4	BIT 3	BIT 2	BIT 1	BIT 0
External Source	\multicolumn{3}{l}{Receiver Clock Select}		\multicolumn{4}{l}{Bit Rate Select}			
0 — RTxC 1 — TRxC	\multicolumn{3}{l}{000 — 1× external 001 — 16× external ASYNC 010 — BRG protocol 011 — C/T of channel mode only 100 — DPLL, source = 64× X1/CLK 101 — DPLL, source = 32× External 110 — DPLL, source = 32× BRG 111 — DPLL, source = 32× C/T}			\multicolumn{4}{l}{one of sixteen rates from BRG}			

OUTPUT AND MISC REG

(OMRA, OMRB)

BIT 7	BIT 6	BIT 5	BIT 4	BIT 3	BIT 2	BIT 1	BIT 0
\multicolumn{3}{l}{Tx Residual Character Length}			TxRDY Activate	RxRDY Activate	OUT 2	OUT 1	RTS
\multicolumn{3}{l}{000 — 1 bit 001 — 2 bits 010 — 3 bits 011 — 4 bits 100 — 5 bits 101 — 6 bits 110 — 7 bits 111 — same as TPR[1:0]}			0 — FIFO not full 1 — FIFO empty	0 — FIFO not empty 1 — FIFO full	0 – 0 1 – 1	0 – 0 1 – 1	0 – 0 1 – 1

[2:0] TRxC —

000   The pin is an input. It must be programmed for input when used as the input for the receiver or transmitter clock, the DPLL, or the C/T.

001   The pin is an output from the crystal oscillator. (XTAL/2)

010   The pin is an output from the DPLL output clock.

011   The pin is an output from the counter/timer. Refer to CTCRA/B description.

100   The pin is an output from the transmitter BRG at 16X the rate selected by TTR [3:0].

101   The pin is an output from the receiver BRG at 16X the rate selected by RTR [3:0].

110   The pin is an output from the transmitter shift register clock.

111   The pin is an output from the receiver shift register clock.

Transmitter and Receiver Parameter and Timing Registers

This set of five registers contains the information which controls the operation of the transmitter and receiver for each channel. Table 3 shows the bit map format for each of these registers. The registers of this group are:

1. Transmitter parameter and timing registers (TPRA/B and TTRA/B)
2. Receiver parameter and timing registers (RPRA/B and RTRA/B)
3. Output and miscellaneous register (OMRA/B)

The first and second group of registers define the transmitter and receiver parameters and timing. Included in the receiver timing registers are the programming parameters for the DPLL. The last register of the group, OMR contains additional transmitter and receiver information and controls the logical state of the output pins when they are not used as a part of the channel configuration.

Transmitter Parameter Register (TPRA, TPRB)

[7:6] Underrun Control — In BOP and COP modes, this field selects the transmitter response in the event of an underrun (i.e., the TxFIFO is empty).

00   Normal end of message termination. In BOP, the transmitter sends the FCS (if selected by CMR2[2:0]) followed by a FLAG and then either MARKs or FLAGs, as specified by [5]. In COP, the transmitter sends the FCS (if selected by CMR2[2:0]) and then either MARKs or SYNs, as specified by [5].

01   Reserved.

10   In BOP, the transmitter sends an ABORT (11111111) and then places the TxD output in a marking condition until receipt of further instructions. In COP, the transmitter places the TxD output in a marking condition until receipt of further instructions.

11   In BOP, the transmitter sends an ABORT (11111111) and then sends FLAGs until receipt of further instructions. In COP, the transmitter sends SYNs until receipt of further instructions.

[5] Idle — In BOP and COP modes, this bit selects the transmitter output during idle. Idle is defined as the state following a normal end of message until receipt of the next transmitter command.

0   Idle in marking condition.

1   Idle sending SYNs (COP) or FLAGs (BOP).

[4] Transmit EOM on Zero Count or Done — In BOP and COP modes, the assertion of this bit causes the end of message (FCS in COP, FCS-FLAG in BOP) to be transmitted upon the following events:

1. If the counter/timer is counting transmitted characters, after transmission of the character which causes the counter to reach zero count. (DONEN is also asserted as an output if the channel is in a DMA operation.)
2. If the channel is operating in DMA mode, after transmission of a character if DONEN was asserted when that character was loaded into the TxFIFO by the DMA controller.

[7:4] Stop Bits per Character — In ASYNC mode, this field programs the length of the stop bit appended to the transmitted character as shown in Table 4.

Dual Universal Serial Communications Controller (DUSCC)   SCN68562

Table 4. Stop Bits — Transmitted Character

[7:4]	5 BITS/CHAR	6, 7, or 8 BITS/CHAR
0000	1.063	0.563
0001	1.125	0.625
0010	1.188	0.688
0011	1.250	0.750
0100	1.313	0.813
0101	1.375	0.875
0110	1.438	0.938
0111	1.500	1.000
1000	1.563	1.563
1001	1.625	1.625
1010	1.688	1.688
1011	1.750	1.750
1100	1.813	1.813
1101	1.875	1.875
1110	1.938	1.938
1111	2.000	2.000

Stop bit lengths of $9/16$ to 1 and $1\ 9/16$ to 2 bits, in increments of $1/16$- bit, can be programmed for character lengths of 6, 7, and 8 bits. For a character length of 5 bits, $1\ 1/16$ to 2 stop bits can be programmed in increments of $1/16$- bit. The receiver only checks for a 'mark' condition at the center of the first stop bit position (one bit time after the last data bit, or after the parity bit if parity is enabled) in all cases.

If an external $1\times$ clock (or a $2\times$ clock for counter/timer) is used for the transmitter, [7] = 0 selects one stop bit and [7] = 1 selects two stop bits to be transmitted. If Manchester, NRZI, or FM data encoding is selected, only integral stop bit lengths should be used.

[3] Transmitter Request-to-Send Control — This bit controls the deactivation of the RTS_N output by the transmitter (see Detailed Operation).
0   RTS_N is not affected by status of transmitter.
1   RTS_N changes state as a function of transmitter status.

[2] Clear-to-Send Enable Transmitter — The state of this bit determines if the CTS_N input controls the operation of the channel's transmitter (see Detailed Operation). The duration of CTS level change is described in the discussion of ICTSR[4].
0   CTS_N has no affect on the transmitter.
1   CTS_N affects the state of the transmitter.

[1:0] Transmitted Bits per Character — This field selects the number of data bits per character to be transmitted. The character length does not include the start, parity, and stop bits in ASYNC or the parity bit in COP. In BOP modes the character length for the address and control fields is always 8 bits, and the value of this field only applies to the information (I) field, except for the last character of the I field, whose length is specified by OMR[7:5].

Transmitter Timing Register (TTRA, TTRB)

[7] External Source — This bit selects the RTxC pin or the TRxC pin of the channel as the transmitter clock input when [6:4] specifies external. When used for input, the selected pin must be programmed as an input in the PCR [4:3] or [2:0].
0   External input from RTxC pin.
1   External input from TRxC pin.

[6:4] Transmitter Clock Select — This field selects the clock for the transmitter.
000   External clock from TRxC or RTxC at $1\times$ the shift (baud) rate.
001   External clock from TRxC or RTxC at $16\times$ the shift rate.
010   Internal clock from the phase locked loop at $1\times$ the bit rate. It should be used only in half-duplex operation since the DPLL will periodically resync itself to the received data if in full-duplex operation.
011   Internal clock from the bit rate generator at $32\times$ the shift rate. The clock signal is divided by two before use in the transmitter which operates at $16\times$ the baud rate. Rate selected by [3:0].
100   Internal clock from counter/timer of other channel. The C/T should be programmed to produce a clock at $2\times$ the shift rate.
101   Internal clock from counter/timer of other channel. The C/T should be programmed to produce a clock at $32\times$ the shift rate.
110   Internal clock from the counter/timer of own channel. The C/T should be programmed to produce a clock at $2\times$ the shift rate.
111   Internal clock from the counter/timer of own channel. The C/T should be programmed to produce a clock at $32\times$ the shift rate.

[3:0] Bit Rate Select — This field selects an output from the bit rate generator to be used by the transmitter circuits. The actual frequency output from the BRG is $32\times$ the bit rate shown in Table 5. With a crystal or external clock of 14.7456MHz the bit rates are as given in Table 5 (this input is divided by two before being applied to the oscillator circuit).

Table 5. Receiver/Transmitter Baud Rates

[3:0]	BIT RATE	[3:0]	BIT RATE
0000	50	1000	1050
0001	75	1001	1200
0010	110	1010	2000
0011	134.5	1011	2400
0100	150	1100	4800
0101	200	1101	9600
0110	300	1110	19.2k
0111	600	1111	38.4k

Receiver Parameter Register (RPRA, RPRB)

[7] SYN Stripping — This bit controls the DUSCC processing in COP modes of SYN 'character patterns' that occur after the initial character synchronization. Refer to Detailed Operation of the receiver for details and definition of SYN 'patterns', and their accumulation of FCS.
0   Strip only leading SYN 'patterns' (i.e. before a message).
1   Strip all SYN 'patterns' (including all odd DLE's in BISYNC transparent mode).

[6] Transfer Received FCS to FIFO — In BISYNC and BOP modes, the assertion of this bit causes the received FCS to be loaded into the RxFIFO. BOP mode operates correctly only if a minimum of two extra FLAGs (without shared zeros) are appended to the frame. If the FCS is specified to be transferred to the FIFO, the EOM status bit will be tagged onto the last byte of the FCS instead of to the last character of the message.
0   Do not transfer FCS to RxFIFO.
1   Transfer FCS to RxFIFO.

[5] Auto-Hunt and Pad Check (BISYNC) — In BISYNC mode, the assertion of this bit causes the receiver to go into hunt for character sync mode after detecting certain end-of-message (EOM) characters. These are defined in the Detailed Operations section for COP receiver operation. After the EOT and NAK sequences, the receiver also does a check for a closing PAD of four 1s.
0   Disable auto-hunt and PAD check.
1   Enable auto-hunt and PAD check.

[5] Overrun Mode (BOP) — The state of this control bit determines the operation of the receiver in the event of a data overrun, i.e., when a character is received while the RxFIFO and the Rx shift register are both full.
0   The receiver terminates receiving the current frame and goes into hunt phase, looking for a FLAG to be received.

December 1986

Dual Universal Serial Communications Controller (DUSCC)   SCN68562

1   The receiver continues receiving the current frame. The overrunning character is lost. (The five characters already assembled in the RxFIFO and Rx shift register are protected).

[4] Receiver Request-to-Send Control (ASYNC) — See Detailed Operation.
0   Receiver does not control RTS_N output.
1   Receiver can negate RTS_N output.

[4] External Sync (COP) — In COP single SYN mode, the assertion of this bit enables external character synchronization and receipt of SYN patterns is not required. In order to use this feature, the DUSCC must be programmed to COP single SYN mode, CMR1[2:0] = 110, which is used to set up the internal data paths. In all other respects, however, the external sync mode operation is protocol transparent. A negative signal on the DCD_N/SYNI_N pin will cause the receiver to establish synchronization on the next rising edge of the receiver clock. Character assembly will start at this edge with the RxD input pin considered to have the second bit of data. The sync signal can then be negated. Receipt of the active high external sync input causes the SYN detect status bit (RSR[2]) to be set and the SYNOUT_N pin to be asserted for one bit time. When this mode is enabled, the internal SYN (COP mode) detection and special character recognition (e.g., IDLE, STx, ETx, etc.) circuits are disabled. Character assembly begins as if in the I-field with character length as programmed in RPR[1:0]. Incoming COP frames with parity specified optionally can have it stripped by programming RPR[3]. The user must wait at least eight bit times after Rx is enabled before applying the SYNI_N signal. This time is required to flush the internal data paths. The receiver remains in this mode and further external sync pulses are ignored until the receiver is disabled and then reenabled to resynchronize or to return to normal mode. See Figure 2.
0   External sync not enabled.
1   External sync enabled.

Note that EXT SYNC and DCD ENABLE Rx cannot be asserted simultaneously since they use the same pin.

[3] Strip Parity — In COP and ASYNC modes with parity enabled, this bit controls whether the received parity bit is stripped from the data placed in the receiver FIFO. It is valid only for programmed character lengths of 5, 6, and 7 bits. If the bit is stripped, the corresponding bit in the received data is set to zero. This bit must be set to A '1' if 8-bit character length will parity is programmed.
0   Transfer parity bit as received.

1   Strip parity bit from data.

[3] All Parties Address — In BOP secondary modes, the assertion of this bit causes the receiver to 'wake up' upon receipt of the address H'FF' or H'FF, FF', for single- and dual-octet address modes, respectively, in addition to its normal station address. This feature allows all stations to receive a message.
0   Don't recognize all parties address.
1   Recognize all parties address.

[2] DCD Enable Receiver — If this bit is asserted, the DCD_N/SYNI_N input must be low in order for the receiver to operate. If the input is negated (goes high) while a character is being received, the receiver terminates receipt of the current message (this action in effect disables the receiver). If DCD is subsequently asserted, the receiver will search for the start bit, SYN pattern, or FLAG, depending on the channel protocol. (Note that the change of input can be programmed to generate an interrupt; the duration of the DCD level change is described in the discussion of the input and counter/timer status register ICTSR[5]).
0   DCD not used to enable receiver
1   DCD used to enable receiver

EXT SYNC and DCD ENABLE Rx cannot be asserted simultaneously since they use the same pin.

[1:0] Received Bits per Character — This field selects the number of data bits per character to be assembled by the receiver. The character length does not include the start, parity, and stop bits in ASYNC or the parity bit in COP. In BOP modes, the character length for the address and control fields is always 8 bits, and the value of this field only applies to the information field. If the number of bits assembled for the last character of the I field is less than the value programmed in this field, RCL not zero (RSR[0]) is asserted and the actual number of bits received is given in TRSR[2:0].

Receiver Timing Register (RTRA, RTRB)

[7] External Source — This bit selects the RTxC pin or the TRxC pin of the channel as the receiver or DPLL clock input, when [6:4] specifies external. When used for input, the selected pin must be programmed as an input in the PCR [4:3] or [2:0].
0   External input from RTxC pin.
1   External input from TRxC pin.

[6:4] Receiver Clock Select — This field selects the clock for the receiver.
000   External clock from TRxC or RTxC at 1× the shift (baud) rate.

001   External clock from TRxC or RTxC at 16× the shift rate. Used for ASYNC mode only.
010   Internal clock from the bit rate generator at 32× the shift rate. Clock is divided by two before use by the receiver logic, which operates at 16× the baud rate. Rate selected by [3:0]. Used for ASYNC mode only.
011   Internal clock from counter/timer of own channel. The C/T should be programmed to produce a clock at 32× the shift rate. Clock is divided by two before use in the receiver logic. Used for ASYNC mode only.
100   Internal clock from the digital phase locked loop. The clock for the DPLL is a 64× clock from the crystal oscillator or system clock input. (The input to the oscillator is divided by two).
101   Internal clock from the digital phase locked loop. The clock for the DPLL is an external 32× clock from the RTxC or TRxC pin, as selected by [7].
110   Internal clock from the digital phase locked loop. The clock for the DPLL is a 32× clock from the BRG. The frequency is programmed by [3:0].
111   Internal clock from the digital phase locked loop. The clock for the DPLL is a 32× clock from the counter/timer of the channel.

[3:0] Bit Rate Select — This field selects an output from the bit rate generator to be used by the receiver circuits. The actual frequency output from the BRG is 32× the bit rate shown in Table 5.

Output and Miscellaneous Register (OMRA, OMRB)

[7:5] Transmitted Residual Character Length — In BOP modes, this field determines the number of bits transmitted for the last character in the information field. This length applies to:
- the character in the transmit FIFO accompanied by the FIFOed TEOM command.
- the character loaded into the FIFO by the DMA controller if DONEN is simultaneously asserted and TPR[4] is asserted.
- the character loaded into the FIFO which causes the counter to reach zero count when TPR[4] is asserted.

The length of all other characters in the frame's information field is selected by TPR[1:0]. If this field is 111, the number of bits in the last character is the same as programmed in TPR[1:0].

December 1986

Dual Universal Serial Communications Controller (DUSCC)   SCN68562

[4] TxRDY Activate Mode —

0   FIFO not full. The channel's TxRDY status bit is asserted each time a character is transferred from the transmit FIFO to the transmit shift register. If not reset by the CPU, TxRDY remains asserted until the FIFO is full, at which time it is automatically negated.

1   FIFO empty. The channel's TxRDY status bit is asserted when a character transfer from the transmit FIFO to the transmit shift register causes the FIFO to become empty. If not reset by the CPU, TxRDY remains asserted until the FIFO is full, at which time it is negated.

If the TxRDY status bit is reset by the CPU, it will remain negated regardless of the current state of the transmit FIFO, until it is asserted again due to the occurrence of one of the above conditions.

[3] RxRDY Activate Mode —

0   FIFO not empty. The channel's RxRDY status bit is asserted each time a character is transferred from the receive shift register to the receive FIFO. If not reset by the CPU, RxRDY remains asserted until the receive FIFO is empty, at which time it is automatically negated.

1   FIFO full. The channel's RxRDY status bit is asserted when a character transfer from the receive shift register to the receive FIFO causes the FIFO to become full. If not reset by the CPU, RxRDY remains asserted until the FIFO is empty, at which time it is negated.

The RxRDY status bit will also be asserted, regardless of the receiver FIFO full condition, when an end-of-message character is loaded in the RxFIFO (BOP/BISYNC), when a BREAK condition (ASYNC mode) is detected in RSR[2], or when the counter/timer is programmed to count received characters and the character which causes it to reach zero is loaded in the FIFO (all modes). (Refer to the detailed operation of the receiver.)

If reset by the CPU, the RxRDY status bit will remain negated, regardless of the current state of the receiver FIFO, until it is asserted again due to one of the above conditions.

[2] General Purpose Output 2 — This general purpose bit is used to control the TxDRQ_/GP02_/RTS_N pin, when it is used as an output. The output is high when the bit is a 0 and is low when the bit is a 1.

[1] General Purpose Output 1 — This bit is used to control the RTxDRQ_N/GPO1_N output, which is a general purpose output when the channel is not in DMA mode. The output is high when the bit is a 0 and is low when the bit is a 1.

[0] Request-to-Send Output — This bit controls the TxDRQ_N/GPO2_N/RTS_N and SYNOUT_N/RTS_N pin, when either is used as a RTS output. The output is high when the bit is a 0 and is low when the bit is a 1.

Counter/Timer Control and Value Registers

There are five registers in this set consisting of the following:

1. Counter/timer control register (CTCRA/B)
2. Counter/timer preset high and low registers (CTPRHA/B, CTPRLA/B)
3. Counter/timer (current value) high and low registers (CTHA/B, CTLA/B)

The format of each of the registers of this set is contained in Table 6. The control register contains the operational information for the counter/timer. The preset registers contain the count which is loaded into the counter/timer circuits. The third group contains the current value of the counter/timer as it operates.

Counter/Timer Control Register (CTCRA, CTCRB)

[7] Zero Detect Interrupt — This bit determines whether the assertion of the C/T ZERO COUNT status bit (ICTSR[6]) causes an interrupt to be generated.

0   Interrupt disabled.

1   Interrupt enabled if master interrupt enable (ICR[1] or ICR[0]) is asserted.

[6] Zero Detect Control — This bit determines the action of the counter upon reaching zero count.

0   The counter/timer is preset to the value contained in the counter/timer preset registers (CTPRL, CTPRH) at the next clock edge.

1   The counter/timer continues counting without preset. The value at the next clock edge will be H'FFFF'.

[5] Counter/Timer Output Control — This bit selects the output waveform when the counter/timer is selected to be output on TRxC or RTxC.

1   The output is a single clock positive width pulse each time the C/T reaches zero count. (The duration of this pulse is one clock period.)

0   The output toggles each time the C/T reaches zero count. The output is cleared to low by either of the preset counter/timer commands.

[4:3] Clock Select — This field selects whether the clock selected by [2:0] is prescaled prior to being applied to the input of the C/T.

00   No prescaling.
01   Divide clock by 16.
10   Divide clock by 32.
11   Divide clock by 64.

[2:0] Clock Source — This field selects the clock source for the counter timer.

000   RTxC pin. Pin must be programmed as input.

001   TRxC pin. Pin must be programmed as input.

010   Source is the crystal oscillator or system clock input divided by four.

011   This selects a special mode of operation. In this mode the counter, after receiving the 'start C/T' command, delays the start of counting until the RxD input goes low. It continues counting until the RxD input goes high, then stops and sets the C/T zero count status bit. The CPU can use the value in the C/T to determine the bit rate of the incoming data. The clock is the crystal oscillator or system clock input divided by four.

100   Source is the 32× BRG output selected by RTR[3:0] of own channel.

101   Source is the 32× BRG output selected by TTR[3:0] of own channel.

110   Source is the internal signal which loads received characters from the receive shift register into the receiver FIFO. When operating in this mode, the FIFOed EOM status bit (RSR[7]) shall be set when the character which causes the count to go to zero is loaded into the receive FIFO.

111   Source is the internal signal which transfers characters from the data bus into the transmit FIFO. When operating in this mode, and if the TEOM on zero count or done control bit (TPR[4]) is asserted, the FIFOed Send EOM command will be automatically asserted when the character which causes the count to go to zero is loaded into the transmit FIFO.

Counter/Timer Preset High Register (CTPRHA, CTPRHB)

[7:0] MSB — This register contains the eight most significant bits of the value loaded into

December 1986

Preliminary Specification

Dual Universal Serial Communications Controller (DUSCC)   SCN68562

Table 6. Counter/Timer Control and Value Register Bit Formats

COUNTER/TIMER CONTROL REG

	BIT 7	BIT 6	BIT 5	BIT 4	BIT 3	BIT 2	BIT 1	BIT 0
(CTCRA, CTCRB)	Zero Detect Interrupt	Zero Detect Control	Output Control	Prescaler		Clock Source		
	0 — disable 1 — enabled	0 — preset 1 — continue	0 — square 1 — pulse	00 — 1 01 — 16 10 — 32 11 — 64		000 — RTxC pin 001 — TRxC pin 010 — X1/CLK divided by 4 011 — X1/CLK divided by 4 gated by RxD 100 — Rx BRG 101 — Tx BRG 110 — Rx characters 111 — Tx characters		

COUNTER/TIMER PRESET HIGH REG

	BIT 7	BIT 6	BIT 5	BIT 4	BIT 3	BIT 2	BIT 1	BIT 0
(CTPRHA, CTPRHB)	Most significant bits of counter/timer preset value							

COUNTER/TIMER PRESET REGISTER LOW

	BIT 7	BIT 6	BIT 5	BIT 4	BIT 3	BIT 2	BIT 1	BIT 0
(CTPRLA, CTPRLB)	Least significant bits of counter/timer preset value							

COUNTER/TIMER HIGH

	BIT 7	BIT 6	BIT 5	BIT 4	BIT 3	BIT 2	BIT 1	BIT 0
(CTHA, CTHB)	Most significant bits of counter/timer							

COUNTER/TIMER LOW

	BIT 7	BIT 6	BIT 5	BIT 4	BIT 3	BIT 2	BIT 1	BIT 0
(CTLA, CTLB)	Least significant bits of counter/timer							

the counter/timer upon receipt of the load C/T from preset register command or when the counter/timer reaches zero count and the zero detect control bit (CTCR[6]) is negated. The minimum 16-bit counter/timer preset value is H'0002'.

Counter/Timer Preset Low Register (CTPRLA, CTPRLB)

[7:0] LSB — This register contains the eight least significant bits of the value loaded into the counter/timer upon receipt of the load C/T from preset register command or when the counter/timer reaches zero count and the zero detect control bit (CTCR[6]) is negated. The minimum 16-bit counter/timer preset value is H'0002'.

Counter/Timer High Register (CTHA, CTHB)

[7:0] MSB — A read of this 'register' provides the eight most significant bits of the current value of the counter/timer. It is recommended that the C/T be stopped via a stop counter command before it is read in order to prevent errors which may occur due to the read being performed while the C/T is changing. This count is continued after the register is read.

Counter/Timer Low Register (CTLA, CTLB)

[7:0] LSB — A read of this 'register' provides the eight least significant bits of the current value of the counter/timer. It is recommended that the C/T be stopped via a stop counter command before it is read, in order to prevent errors which may occur due to the read being performed while the C/T is changing. This count is continued after the register is read.

Interrupt Control and Status Registers

This group of registers define mechanisms for communications between the DUSCC and the processor and contain the device status information. Four registers, available for each channel, and four common device registers comprise this group which consists of the following:

1. Interrupt enable register (IERA/B)
2. Receiver status register (RSRA/B)
3. Transmitter and receiver status register (TRSRA/B)

December 1986

Dual Universal Serial Communications Controller (DUSCC) SCN68562

4. Input and counter/timer status register (ICTSRA/B)
5. Interrupt vector register (IVR) and modified interrupt vector register (IVRM)
6. Interrupt control register (ICR)
7. General status register (GSR)

See Table 7 for bit formats and Figure 3 for table relationships.

Interrupt Enable Register (IERA, IERB)
This register controls whether the assertion of bits in the channel's status registers causes an interrupt to be generated. An additional condition for an interrupt to be generated is that the channel's master interrupt enable bit, ICR[0] or ICR[1], be asserted.

[7] DCD/CTS —
0 Interrupt not enabled.
1 Interrupt generated if ICTSR[4] or ICTSR[5] are asserted.

[6] TxRDY —
0 Interrupt not enabled.
1 Interrupt generated if TxRDY (GSR[1] or GSR[5] for channels A and B respectively) is asserted.

[5] TRSR 73 —
0 Interrupt not enabled.
1 Interrupt generated if bits 7, 6, 5, 4 or 3 of the TRSR are asserted.

[4] RxRDY —
0 Interrupt not enabled.
1 Interrupt generated if RxRDY (GSR[0] or GSR[4] for channels A and B respectively) is asserted.

[3] RSR 76 —
0 Interrupt not enabled.
1 Interrupt generated if bits 7 or 6 of the RSR are asserted.

[2] RSR 54 —
0 Interrupt not enabled.
1 Interrupt generated if bits 5 or 4 of the RSR are asserted.

[1] RSR 32 —
0 Interrupt not enabled.
1 Interrupt generated if bits 3 or 2 of the RSR are asserted.

Table 7. Interrupt Control and Status Register Bit Format

RECEIVER STATUS REG

	*BIT 7	BIT 6	BIT 5	BIT 4	BIT 3	BIT 2	BIT 1	BIT 0
(RSRA, RSRB) ASYNC	# Char compare	RTS negated	Overrun error	not used	BRK end detect	BRK start detect	# Framing error	# Parity error
COP	# EOM detect +	PAD error +	Overrun error	not used	not used	Syn detect	# CRC error	# Parity error
BOP	# EOM detect	Abort detect	Overrun error	Short frame detect	Idle detect	Flag detect	# CRC error	# RCL not zero
LOOP	# EOM detect	Abort/EOP detect	Overrun error	Short frame detect	Turn-around detect	Flag detect	# CRC error	# RCL not zero

NOTES:
\# Status bit is FIFOed.
\+ COP BISYNC mode only.
* All modes indicate character count complete.

TRANSMITTER AND RECEIVER STATUS REG

	BIT 7	BIT 6	BIT 5	BIT 4	BIT 3	BIT 2	BIT 1	BIT 0
(TRSRA, TRSRB) ASYNC	Transmitter empty	CTS underrun	not used	Send break ack	DPLL error	not used	not used	not used
COP	Transmitter empty	CTS underrun	Frame complete	Send SOM ack	DPLL error	not used	Rx hunt mode	Rx xpnt mode
BOP	Transmitter empty	CTS underrun / Loop sending*	Frame complete	Send SOM/ abort ack	DPLL error	\<td colspan=3\>Rx Residual Character Length: 000 — 0 bit, 001 — 1 bits, 010 — 2 bits, 011 — 3 bits, 100 — 4 bits, 101 — 5 bits, 110 — 6 bits, 111 — 7 bits

NOTE:
*Loop mode only

INPUT AND COUNTER/TIMER STATUS REGISTER

	BIT 7	BIT 6	BIT 5	BIT 4	BIT 3	BIT 2	BIT 1	BIT 0
(ICTSRA, ICTSRB)	C/T running	C/T zero count	Delta DCD	Delta CTS/LC	DCD	CTS/LC	GPI2	GPI1

December 1986

Preliminary Specification

Dual Universal Serial Communications Controller (DUSCC)   SCN68562

Table 7. Interrupt Control and Status Register Bit Format (Continued)

INTERRUPT ENABLE REG

	BIT 7	BIT 6	BIT 5	BIT 4	BIT 3	BIT 2	BIT 1	BIT 0
(IERA, IERB)	DCD/CTS	TxRDY	TRSR [7:3]	RxRDY	RSR [7:6]	RSR [5:4]	RSR [3:2]	RSR [1:0]
	0 — no 1 — yes	0 — no 1 — yes	0 — no 1 — yes	0 — no 1 — yes	0 — no 1 — yes	0 — no 1 — yes	0 — no 1 — yes	0 — no 1 — yes

INTERRUPT VECTOR REG AND INTERRUPT VECTOR MODIFIED REG

	BIT 7	BIT 6	BIT 5	BIT 4	BIT 3	BIT 2	BIT 1	BIT 0
(IVR, IVRM)	8-bit interrupt vector							

GENERAL STATUS REG

	BIT 7	BIT 6	BIT 5	BIT 4	BIT 3	BIT 2	BIT 1	BIT 0
	Channel B				Channel A			
(GSR)	External or C/T status	Rx/Tx status	TxRDY	RxRDY	External or C/T status	Rx/Tx status	TxRDY	RxRDY

INTERRUPT CONTROL REG

	BIT 7	BIT 6	BIT 5	BIT 4	BIT 3	BIT 2	BIT 1	BIT 0
(ICR)	Channel A/B Interrupt Priority		Vector Mode		Bits to Modify	Vector Includes Status	Channel A Master Int Enable	Channel B Master Int Enable
	00 — channel A 01 — channel B 10 — interleaved A 11 — interleaved B		00 — vectored 01 — vectored 10 — vectored 11 — non vectored		0 — 2:0 1 — 4:2	0 — no 1 — yes	0 — no 1 — yes	0 — no 1 — yes

[0] RSR 10 —
0   Interrupt not enabled.
1   Interrupt generated if bits 1 or 0 of the RSR are asserted.

Receiver Status Register (RSRA, RSRB)
This register informs the CPU of receiver status. Bits indicated as 'not used' in a particular mode will read as zero. The logical OR of these bits is presented in GSR[2] or GSR[6] (ORed with the bits of TRSR) for channels A and B, respectively. Unless otherwise indicated, asserted status bits are reset only by performing a write operation to the status register with the bits to be reset being ones in the accompanying data word, or when the RESETN input is asserted, or when a 'reset receiver' command is issued.

Certain status bits are specified as being FIFOed. This means that they occupy positions in a status FIFO that correspond to the data FIFO. As the data is brought to the top of the FIFO (the position read when the RxFIFO is read), the FIFOed status bits are logically ORed with the previous contents of the corresponding bits in the status register. This permits the user to obtain status either character by character or on a block basis. For character by character status, the SR bits should be read and then cleared before reading the character data from RxFIFO. For block status, the status register is initially cleared and then read after the message is received. Asserted status bits can be programmed to generate an interrupt (see Interrupt Enable Register).

[7] Character Count Complete (All Modes), Character Compare (ASYNC), EOM (BI-SYNC/BOP/LOOP) — If the counter/timer is programmed to count received characters, this bit is asserted when the character which causes the count to go to zero is loaded into the receive FIFO. It is also asserted to indicate the following conditions:

ASYNC   The character currently at the top of RxFIFO matched the contents of S1R. A character will not compare if it is received with parity error even if the data portion matches.

BISYNC   The character currently at the top of the FIFO was either a text message terminator or a control sequence received outside of a text or header field. See Detailed Operation of COP Receiver. If transfer FCS to FIFO (RPR[6]) is set, the EOM will instead be tagged onto the last byte of the FCS. Note that if an overrun occurs during receipt of a message, the EOM character may be lost, but this status bit will still be asserted to indicate that an EOM was received. For two-byte EOM comparisons, only the second byte is tagged (assuming the CRC is not transferred to the FIFO).

BOP, LOOP   The character currently at the top of the FIFO was the last character of the frame. If transfer FCS to FIFO (RPR[6]) is asserted, the EOM will be tagged instead onto the last byte of the FCS. Note that if an overrun occurs, the EOM character may be lost, but this status bit will still be asserted to indicate that an EOM was received. This bit will not be set when an abort is received.

December 1986

Dual Universal Serial Communications Controller (DUSCC) SCN68562

[6] RTS Negated (ASYNC), PAD Error (BISYNC), ABORT (BOP) —

ASYNC The RTSN output was negated due to receiving the start bit of a new character while the RxFIFO was full (see RPR[4]).

BISYNC PAD error detected (see RPR[5]).

BOP An ABORT sequence consisting of a zero followed by seven ones was received after receipt of the first address octet but before receipt of the closing FLAG. The user should read RxFIFO until it is empty and determine if any valid characters from a previous frame are in the FIFO. If no character with a tagged EOM detect ([7]) is found, all characters are from the current frame and should be discarded along with any previously read by the CPU. An ABORT detect causes the receiver to automatically go into search for FLAG state. An abort during a valid frame does not cause the CRC to reset; this will occur when the next frame begins.

LOOP Performs the ABORT detect function as described for BOP without the restriction that the pattern be detected during an active frame. A zero followed by seven ones is the end-of-poll sequence which allows the transmitter to go active if the 'go active on poll' command has been invoked.

[5] Overrun Error (All Modes) — A new character was received while the receive FIFO was full and a character was already waiting in the receive shift register to be transferred to the FIFO. The DUSCC protects the five characters previously assembled (four in RxFIFO, one in the Rx shift register) and discards the overrunning character(s). After the CPU reads the FIFO, the character waiting in the RxSR will be loaded into the available FIFO position. This releases the RxSR and a new character assembly will start at the next character boundary. In this way, only valid characters will be assembled, i.e. no partial character assembly will occur regardless of when the RxSR became available during the incoming data stream.

[4] Short Frame (BOP/LOOP) —

ASYNC Not used
COP Not used
BOP, LOOP A closing flag was received with missing fields in the frame. See detailed operation for BOP receiver.

[3] BREAK End Detect (ASYNC), IDLE (BOP), Turnaround (LOOP) —

ASYNC 1× clock mode: The RxD input has returned to the marking state for at least one period of the 1× receiver clock after detecting a BREAK.

16× clock mode: The RxD input has returned to the marking (high) state for at least one-half bit time after detecting a BREAK. A half-bit time is defined as eight clock cycles of the 16× receiver clock.

COP Not used

BOP An IDLE sequence consisting of a zero followed by fifteen ones was received. During a valid frame, an abort must precede an idle. However, outside of a valid frame, an idle is recognized and abort is not.

LOOP A turnaround sequence consisting of eight contiguous zeros was detected outside of an active frame. This should normally be used to terminate transmitter operation and return the system to the 'echoing RxD' mode.

[2] BREAK Start Detect (ASYNC), SYN Detect (COP), FLAG Detect (BOP/LOOP)

ASYNC An all zero character, including parity (if specified) and first stop bit, was received. The receiver shall be capable of detecting breaks which begin in the middle of a previous character. Only a single all-zero character shall be put into the FIFO when a break is detected. Additional entries to the FIFO are inhibited until the end of break has been detected (see above) and a new character is received.

COP A SYN pattern was received. Refer to Detailed Operation for definition of SYN patterns. Set one bit time after detection of SYN pattern in HSRH, HSRL. See Figure 1 for receiver data path.

BOP, LOOP A FLAG sequence (01111110) was received. Set one bit time after FLAG is detected in CCSR. See Figure 1 for receiver data path.

[1] Framing Error (ASYNC), CRC Error (COP/BOP/LOOP) —

ASYNC At the first stop bit position the RxD input was in the low (space) state. The receiver only checks for framing error at the nominal center of the first stop bit regardless of the number of stop bits programmed in TPR[7:4]. This bit is not set for BREAKS.

COP In BISYNC COP mode, this bit is set upon receipt of the BCC byte(s), if any, to indicate that the received BCC was in error. The bit is normally FIFOed with the last byte of the frame (the character preceding the first BCC byte). However, if transfer FCS to FIFO (RPR[6]) is asserted, this bit is FIFOed with the last BCC byte. The value of this bit should be ignored for non-text messages or if the received frame was aborted via an ENQ. In non-BISYNC COP modes, the bit is set with each received character if the current value of the CRC checker is not equal to the non-error value (see CMR2[2:0]).

BOP, LOOP This bit is set upon receipt of the FCS byte(s), if any, to indicate that the received FCS was in error. The bit is normally FIFOed with the last byte of the I field (the character preceding the first FCS byte). However, if transfer FCS to FIFO (RPR[6]) is asserted, this bit is FIFOed with the last FCS byte.

[0] Parity Error (ASYNC/COP), RCL Not Zero (BOP/LOOP) —

ASYNC The parity bit of the received character was not as expected. A parity error does not affect the parity bit put into the FIFO as part of the character when strip parity (RPR[3]) is negated.

COP The parity bit of the received character was not as expected. A parity error does not affect the parity bit put into the FIFO as part of the character when strip parity (RPR[3]) is negated. A SYN or other character received with parity error is treated as a data character. Thus, a SYN with parity error received while in SYN search state will not establish character sync. Characters received with parity error while in the SYN search state will not set the error bit.

BOP, LOOP The last character of the I field did not have the character length specified in RPR[1:0]. The actual received character length of this byte can be read in TRSR[2:0]. This bit is FIFOed with the EOM character but TRSR[2:0] is not. An exception occurs if the command to transfer the FCS to the FIFO is active. In this case, the bit will be FIFOed with the last byte of the FCS, i.e., with REOM. In the event

Preliminary Specification

Dual Universal Serial Communications Controller (DUSCC)   SCN68562

Figure 1. Receiver Data Path

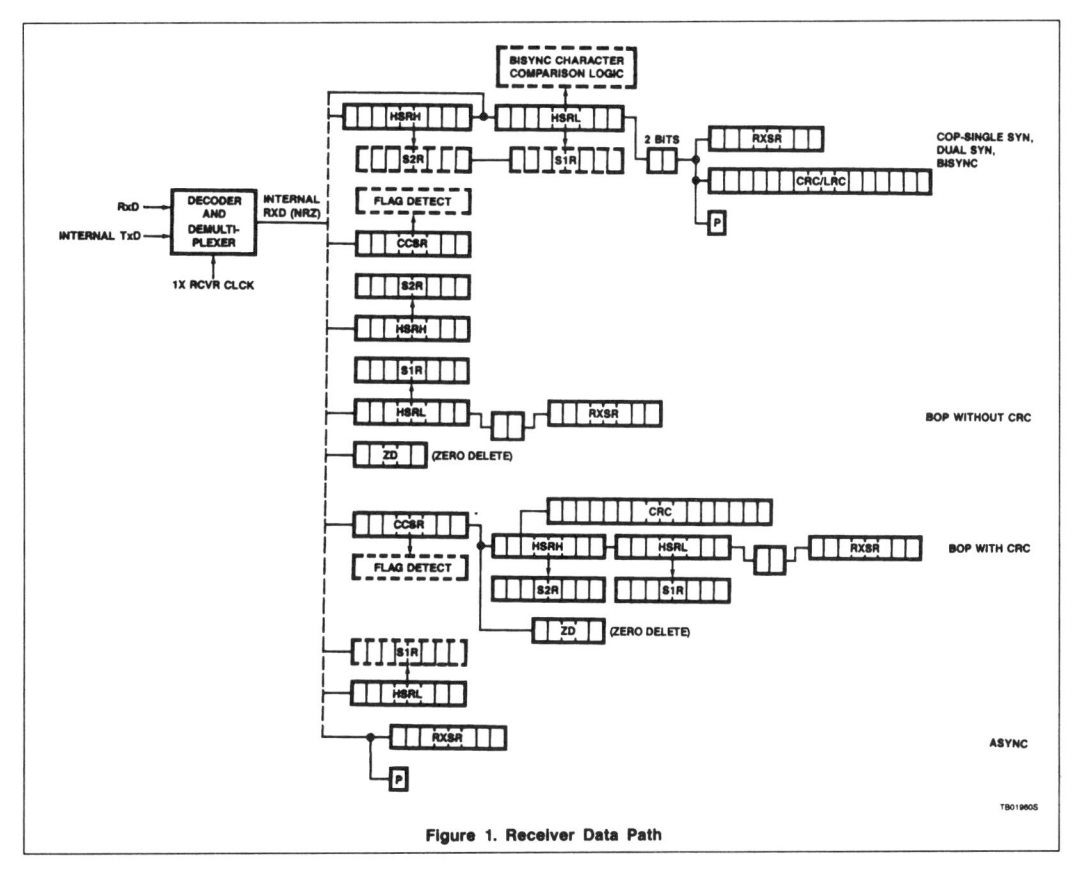

that residual characters from two consecutive frames are received and are both in the FIFO, the length in TRSR[2:0] applies to the last received residual character.

Transmitter/Receiver Status Register (TRSRA, TRSRB)
This register informs the CPU of transmitter and receiver status. Bits indicated as not used in a particular mode will read as zero, except for bits [2:0], which may not be zero. The logical-OR of bits [7:3] is presented in GSR[2] or GSR[6] (ORed with the bits of RSR) for channels A and B, respectively. Unless otherwise indicated, asserted status bits are reset only:

1. By performing a write operation to the status register with the bits to be reset being ones in the accompanying data word [7:3].
2. When the RESETN input is asserted.
3. For [7:4], when a 'reset transmitter' command is issued.
4. For [3:0], when a 'reset receiver' command is issued.
5. For [2:0], see description in BOP mode.

Asserted status bits in [7:3] can be programmed to generate an interrupt. See IER.

[7] Transmitter Empty — Indicates that the transmit shift register has completed serializing a character and found no other character to serialize in the TxFIFO. The bit is not set until at least one character from the transmit FIFO (not including PAD characters in synchronous modes) has been serialized. The transmitter action after transmitter empty depends on operating mode:

ASYNC   The TxD output is held in the MARK state until another character is loaded into the TxFIFO. Normal operation then continues.

COP   Action is specified by TPR[7:6].
BOP,   Action is specified by TPR[7:6].
LOOP

[6] CTS Underrun (ASYNC/COP/BOP), Loop sending (LOOP) —

ASYNC,   This bit is set only if CTS enable
COP,   Tx (TPR [2]) is asserted. It indi-
BOP   cates that the transmit shift register was ready to begin serializing a character and found the CTS_N input negated. In ASYNC mode, this bit will be reasserted if cleared by the CPU while the CTS_N input is negated.

LOOP   Asserted when the go active on poll command has been invoked and an EOP sequence has been detected, causing the transmitter to go active by changing the EOP to a FLAG (see detailed operation of transmitter).

December 1986   19

Dual Universal Serial Communications Controller (DUSCC) — SCN68562

[5] Frame Complete (COP/BOP) —
ASYNC Not used.
COP Asserted at the beginning of transmission of the end of message sequence invoked by which is either a TEOM command, or when TPR[4], or TPR[7:6] = 00. The CPU can invoke the TSOM command after this bit is set to control the number of SYNs between transmitted frames.
BOP Asserted at the beginning of transmission of the end of message sequence which is invoked by either a TEOM command, or when TPR[4] = 1, or TPR[7:6] = 00. The CPU can invoke the TSOM command after this bit is set to control the number of FLAGs between transmitted frames.
In COP/BOP modes, the frame complete status bit is set during the next-to-last bit (on TxD pin) of the last character in the data/information field. In BOP mode, if a 1-bit residual character is selected through OMR[7:5], then this bit is set during the next-to-last bit (on TxD pin) of the last full length character of the information field.

[4] Send Break Ack (ASYNC)/Send SOM ACK (COP)/Send SOM-Abort Ack (BOP) —
ASYNC Set when the transmitter begins transmission of a break in response to the send break command. If the command is reinvoked, the bit will be set again at the beginning of the next character time. The user can control the length of the break by counting character times through this mechanism.
COP Set when the transmitter begins transmission of a SYN pattern in response to the TSOM or TSOMP command. If the command is reinvoked, the bit will be set again at the beginning of the next transmitted SYN pattern. The user can control the number of SYNs which are sent through this mechanism.
BOP Set when the transmitter begins transmission of a FLAG/ABORT in response to the TSOM or TSOMP or TABRK command. If the command is reinvoked, the bit will be set again at the beginning of the next transmitted FLAG/ABORT. The user can control the number of FLAGs/ABORTs which are sent through this mechanism.

[3] DPLL Error — Set while the DPLL is operating in FM mode to indicate that a data transition was not detected within the detection window for two consecutive bits and that the DPLL was forced into search mode. This feature is disabled when the DPLL is specified as the clock source for the transmitter via TTR[6:4].

[2:0] Received Residual Character Length (BOP) —
BOP This field should be examined to determine the length of the last character of the I field (character tagged with REOM status bit) if RSR[0] is set to indicate that the length was not equal to the character length specified in RPR[1:0]. This field is negated when a reset receiver or disable receiver command is issued, or when the first control character for the next frame of data is in HSRL (see Figure 1). Care must be taken to read TRSR[2:0] before these bits are cleared.

[1] Receiver in Hunt Mode (COP) —
COP This bit is asserted after the receiver is reset or disabled. It indicates that the receiver is in the hunt mode, searching the data stream for a SYN sequence to establish character synchronization. The bit is negated automatically when character sync is achieved.

[0] Receiver in Transparent Mode (BISYNC) —
COP Indicates that a DLE-STx sequence was received and the receiver is operating in BISYNC transparent mode. Set two bit times after detection of STx in HSRL. See Figure 1 for receiver data path. Transparent mode operation is terminated and the bit is negated automatically when one of the terminators for transparent text mode is received (DLE-ETx/ETB/ITB/ENQ).

Input and Counter/Timer Status Register (ICTSRA, ICTSRB)

This register informs the CPU of status of the counter/timer and inputs. The logical-OR of bits [6:4] is presented in GSR[3] or GSR[7] for channels A and B, respectively. Unless otherwise specified, bits of this register are reset only:
1. By performing a write operation to the status register with the bits to be reset (ones in the accompanying data word for bits [6:4] only).
2. When the RESETN input is asserted (bits [7:4] only).

[7] Counter/Timing Running — Set when the C/T is started by start C/T command and reset when it is stopped by a stop C/T command.

[6] Counter/Timer Zero Detect — Set when the counter/timer reaches zero count, or when the bit length measurement is enabled (CTCR [2:0] = 011) and the RXD input has returned high. The assertion of this bit causes an interrupt to be generated if ICTCR[7] and the channel's master interrupt enable (ICR[1] or ICR[0]) are asserted.

[5] Delta DCD — The DCD input is sampled approximately every 6.8µs using the 32×, 4800 baud output from the BRG. After synchronizing with the sampling clock, at least two consecutive samples at the same level are required to establish the level. As a consequence, a change of state at the DCD input, lasting at least 17µs, will set this bit. The reset circuitry initializes the sampling circuits so that a change is not falsely indicated at power on time. The assertion of this bit causes an interrupt to be generated if IER[7] and the channel's master interrupt enable (ICR[1] or ICR[0]) are asserted.

[4] Delta CTS/LC — When not in loop mode, the CTS input is sampled approximately every 6.8µs using the 32×, 4800 baud output from the BRG. After synchronizing with the sampling clock, at least two consecutive samples at the same level are required to establish the level. As a consequence, a change of state at the CTS input, lasting at least 17µs, will set this bit. The reset circuitry initializes the sampling circuits so that a change is not falsely indicated at power on time. The assertion of this bit causes an interrupt to be generated if IER[7] and the channel's master interrupt enable (ICR[1] or ICR[0]) are asserted.

In SDLC loop mode, this bit is set upon transitions of the LC output. LC is asserted in response to the 'go on-loop' command when the receiver detects a zero followed by seven ones, and negated in response to the 'go off-loop' command when the receiver detects a sequence of eight ones.

[3:0] Current State of DCD, CTS, GPI2, and GPI1 Inputs — This field provides the current state of the channel's input pins. The bit's value is latched at the beginning of the read cycle.

Interrupt Vector Register (IVR) and Modified Vector Register (IVRM)

[7:0] Register Content — If ICR[2] = 0, the content of IVR register is output on the data bus when the DUSCC has issued an interrupt request and the responding interrupt acknowledge (IACKN) is received. The value in the IVR is initialized to H'0F' on master reset. If 'vector includes status' is specified by

Dual Universal Serial Communications Controller (DUSCC)   SCN68562

ICR[2] = 1, bits [2:0] or [4:2] (depending on ICR[3]), of the vector are modified as shown in Table 8 to indicate the highest priority interrupt currently active. The priority is programmable through the ICR. This modified vector is stored in the IVRM. When ICR[2] = 1, the content of the IVRM is output on to the data bus on the interrupt acknowledge. The vector is not modified, regardless of the value of ICR[2], if the CPU has not written an initial vector into this register.

Either the modified or unmodified vector can also be read by the CPU via a normal bus read cycle (see Table 1). The vector value is locked at the beginning of the IACK or read cycle until the cycle is completed.

Interrupt Control Register (ICR)

[7:6] Channel A/B Interrupt Priority — Selects the relative priority between channels A and B. The state of this bit determines the value of the interrupt vector (see Interrupt Vector Register). The priority within each channel, from highest to lowest, is as follows:

0	Receiver ready
1	Transmitter ready
2	Rx/Tx status
3	External or C/T status

00	Channel A has the highest priority. The DUSCC interrupt priorities from highest to lowest are as follows: A(0), A(1), A(2), A(3), B(0), B(1), B(2), B(3)
01	Channel B has the highest priority. The DUSCC interrupt priorities from highest to lowest are as follows: B(0), B(1), B(2), B(3), A(0), A(1), A(2), A(3)
10	Priorities are interleaved between channels, but channel A has the highest priority between events of equal channel priority. The DUSCC interrupt priorities from highest to lowest are as follows: A(0), B(0), A(1), B(1), A(2), B(2), A(3), B(3)

Table 8. Interrupt Status Encoding

IVRM [2:0]/ [4:2]	HIGHEST PRIORITY INTERRUPT CONDITION
000	Channel A receiver ready
001	Channel A transmitter ready
010	Channel A Rx/Tx status
011	Channel A external or C/T status
100	Channel B receiver ready
101	Channel B transmitter ready
110	Channel B Rx/Tx status
111	Channel B external or C/T status

11   Priorities are interleaved between channels, but channel B has the highest priority between events of equal channel priority. The DUSCC interrupt priorities from highest to lowest are as follows: B(0), A(0), B(1), A(1), B(2), A(2), B(3), A(3)

[5:4] Vector Mode — The value of this field determines the response of the DUSCC when the interrupt acknowledge (IACKN) is received from the CPU.

00 or 01 or 10   Vectored mode. Upon interrupt acknowledge, the DUSCC locks its current interrupt status until the end of the acknowledge cycle. If it has an active interrupt pending, it responds with the appropriate vector and then asserts DTACKN. If it does not have an interrupt, it propagates the acknowledge through its X2/IDCN output if this function is programmed in PCRA[7]. Otherwise, the IACKN is ignored. Locking the interrupt status at the leading edge of IACKN prevents a device at a high position in the interrupt daisy chain from responding to an IACK issued for a lower priority device while the acknowledge is being propagated to that device.

11   Non-vectored mode. The DUSCC ignores an IACK if one is received; the interrupt vector is not placed on the data bus. The internal interrupt status is locked when a read of the IVR or IVRM is performed. Except for the absence of the vector on the bus, the DUSCC performs as it does in vectored mode — the vector is prioritized and modified if programmed.

[3] Vector Bits to Modify — Selects which bits of the vector stored in the IVR are to be modified to indicate the highest priority interrupt pending in the DUSCC. See Interrupt Vector Register.

0	Modify bits 2:0 of the vector.
1	Modify bits 4:2 of the vector.

[2] Vector Includes Status — Selects whether the modified (includes status) (IVRM) or unmodified vector (IVR) is output in response to an interrupt acknowledge (see Interrupt Vector Register).

0	Unmodified vector.
1	Modified vector.

[1] Channel A Master Interrupt Enable —

0	Channel A interrupts are disabled.
1	Channel A interrupts are enabled.

[0] Channel B Master Interrupt Enable —

0	Channel B interrupts are disabled.
1	Channel B interrupts are enabled.

General Status Register (GSR)

This register provides a 'quick look' at the overall status of both channels of the DUSCC. A write to this register with 1s at the corresponding bit positions causes TxRDY (bits 5 and 1) and/or RxRDY (bits 4 and 0) to be reset. The other status bits can be reset only by resetting the individual status bits that they point to.

[7] Channel B External or Counter/Timer Status — This bit indicates that one of the following status bits is asserted: ICTSRB[6:4].

[6] Channel B Receiver or Transmitter Status — This bit indicates that one of the following status bits is asserted: RSRB[7:0], TRSRB[7:3].

[5] Channel B Transmitter Ready — The assertion of this bit indicates that one or more characters may be loaded into the channel B transmitter FIFO to be serialized by the transmit shift register. See description of OMR[4]. This bit can be asserted only when the transmitter is enabled. Disabling or resetting the transmitter negates TxRDY.

[4] Channel B Receiver Ready — The assertion of this bit indicates that one or more characters are available in the channel B receiver FIFO to be read by the CPU. See description of OMR[3]. RxRDY is initially reset (negated) by a chip reset or when a 'reset channel B receiver' command is invoked.

[3] Channel A External or Counter/Timer Status — This bit indicates that one of the following status bits is asserted: ICTSRA[6:4].

[2] Channel A Receiver or Transmitter Status — This bit indicates that one of the following status bits is asserted: RSRA[7:0], TRSRA[7:3].

[1] Channel A Transmitter Ready — The assertion of this bit indicates that one or more characters may be loaded into the channel A transmitter FIFO to be serialized by the transmit shift register. See description of OMR[4]. This bit can be asserted only when the transmitter is enabled. Disabling or resetting the transmitter negates TxRDY.

[0] Channel A Receiver Ready — The assertion of this bit indicates that one or more characters are available in the channel A receiver FIFO to be read by the CPU. See description of OMR[3]. RxRDY is initially reset (negated) by a chip reset or when a 'reset channel A receiver' command is invoked.

Channel Command Register (CCRA, CCRB) —

Commands to the DUSCC are entered through the channel command register. The format of that register is shown in Table 9. A

December 1986

Dual Universal Serial Communications Controller (DUSCC) SCN68562

Preliminary Specification

Table 9. Command Register Bit Format

CHANNEL COMMAND REG

(CCRA, CCRB)	BIT 7	BIT 6	BIT 5	BIT 4	BIT 3	BIT 2	BIT 1	BIT 0
					\multicolumn{4}{c}{Transmitter Command}
	00 = Transmitter CMD		don't care	don't care	\multicolumn{4}{l}{0000 — reset Tx 0001 — reset TxCRC* 0010 — enable Tx 0011 — disable Tx 0100 — transmit SOM (TSOM) 0101 — transmit SOM with PAD (TSOMP) 0110 — transmit EOM (TEOM)* 0111 — transmit ABORT/BREAK (TABRK) 1000 — transmit DLE (TDLE)* 1001 — go active on poll 1010 — reset go active on poll 1011 — go on-loop 1100 — go off-loop 1101 — exclude from CRC*}
					\multicolumn{4}{c}{Receiver Command}
	01 = Receiver CMD		don't care	don't care	\multicolumn{4}{l}{0000 — reset Rx 0001 — reserved 0010 — enable Rx 0011 — disable Rx}
					\multicolumn{4}{c}{Counter/Timer Command}
	10 = C/T CMD		don't care	don't care	\multicolumn{4}{l}{0000 — start 0001 — stop 0010 — preset to FFFF 0011 — preset from CTPRH/CTPRL}
					\multicolumn{4}{c}{DPLL Command}
	11 = DPLL CMD		don't care	don't care	\multicolumn{4}{l}{0000 — enter search mode 0001 — disable DPLL 0010 — set FM mode 0011 — set NRZI mode 0100 — reserved for test 0101 — reserved for test}

NOTE:
*FIFOed commands

read of this register returns the last invoked command (with bits 4 and 5 set to 1).

Transmitter Commands

0000 Reset transmitter. Causes the transmitter to cease operation immediately. The transmit FIFO is cleared and the TxD output goes into the marking state. Also clears the transmitter status bits (TRSR[7:4]) and resets the TxRDY status bit (GSR[1] or GSR[5] for channels A and B, respectively). The counter/timer and other registers are not affected.

0001 Reset transmit CRC. This command is appended to and FIFOed along with the next character loaded into the transmit FIFO. It causes the transmitter CRC generator to be reset to its initial state prior to beginning transmission of the appended character.

0010 Enable transmitter. Enables transmitter operation, conditioned by the state of the CTS ENABLE Tx bit, TPR[2]. Has no effect if invoked when the transmitter has previously been enabled.

0011 Disable transmitter. Terminates transmitter operation and places the TxD output in the marking state at the next occurrence of a transmit FIFO empty condition. All characters currently in the FIFO, or any loaded subsequently prior to attaining an empty condition, will be transmitted.

0100 Transmit start of message. Used in COP and BOP modes to initiate transmission of a frame after the transmitter is first enabled, prior to sending the contents of the FIFO. Can also be used to precisely control the number of SYN/FLAGs at the beginning of transmission or in between frames.

When the transmitter is first enabled, transmission will not begin until this command (or the transmit SOM with PAD command, see below) is issued. The command causes the SYN (COP) or FLAG (BOP) pattern to be transmitted. SEND SOM ACK (TRSR[4]) is set when transmission of the SYN/FLAG begins. The CPU may then reinvoke the command if multiple SYN/FLAGs are to be transmitted. Transmission of the FIFO characters begins when the command is no longer reinvoked. If the FIFO is empty, SYN/FLAGs continue to be transmitted until a character is loaded into the FIFO, but the status bit (TSR[4]) is not set. Insertion of SYN/FLAGs between frames can be accomplished by invoking this command after the frame complete status bit (TRSR[5]) has been asserted in response to transmission of the end-of-message sequence.

December 1986   22

Dual Universal Serial Communications Controller (DUSCC) SCN68562

0101   Transmit start of message with opening PAD. Used in COP and BOP modes after the transmitter is first enabled to send a bit pattern for DPLL synchronization prior to transmitting the opening SYN (COP) or FLAG (BOP). The SYN/FLAG is sent at the next occurrence of a transmit FIFO empty condition. All characters currently in the FIFO, or any loaded subsequently prior to attaining an empty condition, will be transmitted. While the PAD characters are transmitted, the character length is set to 8 bits, (regardless of the programmed length), and parity generation (COP), zero insertion (BOP), and LRC/CRC accumulation are disabled. SEND SOM ACK (TRSR[4]) is set when transmission of the SYN/FLAG begins. The CPU may then invoke the transmit SOM command if multiple SYN/FLAGs are to be transmitted.

The TSOM/TSOMP commands, described above, are sampled by the controller in alternate bit times of the transmitter clock. As a consequence, the first bit time of a COP/BOP frame will be transmitted on the TxD pin, after a maximum of three bit times, after the command is issued. (The additional 1-bit delay in the data path is due to the data encoding logic.)

0110   Transmit end-of-message. This command is appended to the next character loaded into the transmit FIFO. It causes the transmitter to send the end-of-message sequence (selected FCS in COP modes, FCS – FLAG in BOP modes) after the appended character is transmitted. Frame complete (TRSR[5]) is set when transmission of the FCS begins. This command is also asserted automatically if the TEOM on zero count or done control bit (TPR[4]) is asserted, and the counter/timer is programmed to count transmitted characters when the character which causes the count to go to zero is loaded into the transmit FIFO.

0111   Transmit Abort BOP/Transmit Break ASYNC. In BOP modes, causes an abort (eight ones) to be transmitted after transmission of the character currently in the shift register is completed. The transmitter then sends MARKs or FLAGs depending on the state of underrun control (TPR[7:6]). Send SOM/abort ack (TRSR[4]) is set when the transmission of the abort begins. If the command is reasserted before transmission of the previous ABORT is completed, the process will be repeated. This can be used to send the idle sequence. The 'transmit SOM' command must be used to initiate transmission of a new message. In either mode, invoking this command causes the transmit FIFO to be flushed (characters are not transmitted).

In ASYNC mode, causes a break (space) to be transmitted after transmission of the character currently in the shift register is completed. Send break ack (TRSR[4]) is set when the transmission of the break begins. The transmitter keeps track of character times. If the command is reasserted, send break ack will be set again at the beginning of the next character time. The user can use this mechanism to control the length of the break in character time multiples. Transmission of the break is terminated by issuing a 'reset Tx' or 'disable Tx' command.

1000   Transmit DLE. Used in COP modes only. This command is appended to and FIFOed with the next character loaded into the transmitter FIFO. It causes the transmitter to send a DLE, (EBCDIC H'10', ASCII H'10') prior to transmitting the appended character. If the transmitter is operating in BI-SYNC transparent mode, the transmitter control logic automatically causes a second DLE to be transmitted whenever a DLE is detected at the top of the FIFO. In this case, the TDLE command should not be invoked. An extra (third) DLE, however, will not be sent if the transmit DLE command is invoked.

1001   Go active on poll. Used in BOP loop mode only. Causes the transmitter, if it is enabled, to begin sending when an EOP sequence consisting of a zero followed by seven ones is detected. The last one of the EOP is changed to zero, making it another FLAG, and then the transmitter operates as described in the detailed operation section. The loop sending status bit (TRSR[6]) is asserted concurrent with the beginning of transmission.

1010   Reset go active on poll. Clears the stored 'go active on poll' command.

1011   Go on-loop. Used in BOP loop mode to control the assertion of the LC_N output. This output provides the means of controlling external loop interface hardware to go on-loop and off-loop. When the command is asserted, the DUSCC will look for the receipt of seven contiguous ones, at which time it will assert the LC_N output and set the delta DCD/LC status bit (ICTSR[4]). This allows the DUSCC to break into the loop without affecting loop operation. This command must be used to initiate loop mode operation.

1100   Go off-loop. Used in BOP loop mode to control the negation of the LC_N output. This output provides the means of controlling external loop interface hardware to go on-loop and off-loop. When the command is asserted, the DUSCC will look for the receipt of eight contiguous ones, at which time it will negate the LC_N output and set the delta DCD/LC status bit (ICTSR[4]). This allows the DUSCC to get off the loop without affecting loop operation. This command is normally used to terminate loop mode operation.

1101   Exclude from CRC. This command is appended to and FIFOed along with the next character loaded into the transmit FIFO. It causes the transmitter CRC generator to be disabled while the appended character is being transmitted. Thus, that character is not included in the CRC accumulation.

Receiver Commands

0000   Reset Receiver. Causes the receiver to cease operation, clears the receiver FIFO, and clears the receiver status (RSR[7:0], TRSR[3:0], and either GSR[0] or GSR[4] for channels A and B, respectively). The counter/timer and other registers are not affected.

0001   Reserved.

0010   Enable receiver. Causes receiver operation to begin, conditioned by the state of the DCD ENABLE Rx bit, RPR[2]. Receiver goes into START, SYN, or FLAG search mode depending on channel protocol mode. Has no effect if invoked when the receiver has previously been enabled.

0011   Disable receiver. Terminates operation of the receiver. Any character currently being assembled will be lost. Does not affect FIFO or any status.

Counter/Timer Commands

0000   Start. Starts the counter/timer and prescaler.

0001   Stop. Stops the counter/timer and prescaler. Since the command may be asynchronous with the selected clock source, the counter/timer and/

Dual Universal Serial Communications Controller (DUSCC) SCN68562

or prescaler may count one or more additional cycles before stopping.

0010 Preset to FFFF. Presets the counter timer to H'FFFF' and the prescaler to its initial value. This command causes the C/T output to go low.

0011 Preset from CTPRH/CTPRL. Transfers the current value in the counter/timer preset registers to the counter/timer and presets the prescaler to its initial value. This command causes the C/T output to go low.

Digital Phase-Locked Loop Commands

0000 Enter Search Mode. This command causes the DPLL counter to be set to the value 16 and the clock output will be forced high. The counter will be disabled until a transition on the data line is detected, at which point it will start incrementing and the clock output will go from high to low. After the counter reaches a count of 31, it will reset to zero and cause the clock output to go from low to high. The DPLL will then continue normal operation. This allows the DPLL to be locked onto the data without preframe transitions. This command should not be used if the DPLL is programmed to supply the clock for the transmitter and the transmitter is active.

0001 Disable DPLL. Disables operation of the DPLL.

0010 Set FM Mode. Sets the DPLL to the FM mode of operation, used when FM0, FM1, or Manchester (NRZ) is selected by CMR1[7:6].

0011 Set NRZI Mode. Sets the DPLL to the NRZI mode of operation, used when NRZ or NRZI is selected by CMR1[7:6].

0100 Reserved for test

0101 Reserved for test

DETAILED OPERATION

Interrupt Control

A single interrupt output (IRQN) is provided which is activated upon the occurrence of any of the following conditions:

Channel A external or C/T special condition

Channel B external or C/T special condition

Channel A Rx/Tx error or special condition

Channel B Rx/Tx error or special condition

Channel A TxRDY

Channel B TxRDY

Channel A RxRDY

Channel B RxRDY

Each of the above conditions occupies a bit in the general status register (GSR). If ICR[2] is set, the eight conditions are encoded into three bits which are inserted into bits [2:0] or [4:2] of the interrupt vector register. This forms the content of the IVRM during an interrupt acknowledge cycle. Unmodified and modified vectors can be read directly through specified registers. Two of the conditions are the inclusive OR of several other maskable conditions:

- Ext or C/T special condition: Delta DCD, delta CTS or C/T zero count (ICTSR[6:4]).

- Rx/Tx error or special condition: Any condition in the receiver status register (RSR[7:0]) or a transmitter or DPLL condition in the transmitter and receiver status register (TRSR[7:3]).

The TxRDY and RxRDY conditions are defined by OMR[4] and OMR[3], respectively. Also associated with the interrupt system are the interrupt enable register (IER), one bit in the counter/timer control register (CTCR), and the interrupt control register (ICR).

The IER is programmed to enable specified conditions or groups of conditions to cause an interrupt by asserting the corresponding bit. A negated bit prevents an interrupt from occurring when the condition is active and hence masks the interrupt. In addition to the IER, CTCR[7] could be programmed to enable or disable an interrupt upon the C/T zero count condition. The interrupt priorities within a channel are fixed. Priority between channels is controlled by ICR[7:6]. Refer to Table 8 and ICR[7:6].

The ICR contains the master interrupt enables for each channel (ICR[1] and ICR[0]) which must be set if the corresponding channel is to cause an interrupt. The CPU vector mode is specified by ICR[5:4] which selects either vectored or non-vectored operation. If vectored mode is selected, the content of the IVR or IVRM is placed on the data bus when IACK is activated. If ICR[2] is set, the content of IVRM is output which contains the content of IVR and the encoded status of the interrupting condition.

Upon receiving an interrupt acknowledge, the DUSCC locks its current interrupt status until the end of the acknowledge cycle. If it has an active interrupt pending, it responds with the appropriate vector and then asserts DTACKN. If it does not have an interrupt, it propagates the acknowledge through its X2/IDCN output if this function is programmed in PCRA[7]; otherwise, the IACKN is ignored. Locking the interrupt status at the leading edge of IACKN prevents a device at a high position in the interrupt daisy chain from responding to an IACK issued for a lower priority device while the acknowledge is being propagated to that device.

DMA Control

The DMA control section provides the interface to allow the DUSCC to operate with an external DMA controller. One of four modes of DMA can be programmed for each channel independently via CMR2[5:3]:

- Half-duplex single address. In this mode, a single pin provides both DMA read and write requests. Acknowledgement of the requests is via a single DMA acknowledge pin. The data transfer is accomplished in a single bus cycle — the DMA controller places the memory address of the source or destination of the data on the address bus and then issues the acknowledge signal, which causes the DUSCC to either write the data into its transmit FIFO (write request) or to output the contents of the top of the receive FIFO (read request). The cycle is completed when the DTCN input is asserted by the DMA controller. This mode can be used when channel operation is half-duplex (e.g., BISYNC). It allows a single DMA channel to service the receiver and transmitter.

- Half-duplex dual address. In this mode, a single pin provides both DMA read and write requests. Acknowledgement of the requests is via normal bus read and write cycles. The data transfer requires two bus cycles — the DMA controller acquires the data from the source (memory for a Tx DMA or DUSCC for a Rx DMA) on the first cycle and deposits it at the destination (DUSCC for a Tx DMA or memory for a Rx DMA) on the second bus cycle. This mode is used when channel operation is half-duplex (e.g., BISYNC) and allows a single DMA channel to service the receiver and transmitter.

- Full-duplex single address. This mode is similar to half-duplex single address mode but provides separate request and acknowledge pins for the receiver and transmitter.

- Full-duplex dual address. This mode is similar to half-duplex dual address mode but provides duplex dual address mode but provides separate request pins for the receiver and transmitter.

Figures 4 through 7 describe operation of the DUSCC in the various DMA environments. Table 10 summarizes pins used for the DMA request and acknowledge function for the transmitter and receiver for the different DMA modes.

December 1986

Dual Universal Serial Communications Controller (DUSCC)   SCN68562

Table 10. DMA REQ and ACK Pins for Operational Modes

FUNCTION	HALF DUPLEX SINGLE ADDR DMA	HALF DUPLEX DUAL ADDR DMA	FULL DUPLEX SINGLE ADDR DMA	FULL DUPLEX DUAL ADDR DMA
RCVR REQ	RTxDRQ_N	RTxDRQ_N	RTxDRQ_N	RTxDRQ_N
TRAN REQ	Same as RCVR REQ	Same as RCVR REQ	TxDRQ_N	TxDRQ_N
RCVR ACK	RTxDAK_N	Normal read RCVR FIFO	RTxDAK_N	Normal read RCVR FIFO
TRAN ACK	Same as RCVR ACK	Normal write TRAN FIFO	TxDAK_N	Normal write TRAN FIFO

The DMA request signals are functionally identical to the TxRDY and RxRDY status signals for each serial channel except that in DMA the signals are negated on the leading edge of the acknowledge signal when the subsequent transfer causes the FIFO to become full (transmitter request) or empty (receiver request). In non DMA operation TxRDY and RxRDY signals are negated only after the transfer is completed. The DMA read request can be programmed through OMR[3] be asserted either when any character is in the receive FIFO or only when the receive FIFO is full. Likewise, the DMA write request can be programmed through OMR[4] to be asserted either when the transmit FIFO is not full or only when the transmit FIFO is empty (the transmitter must be enabled for a DMA request to be asserted). The request signals are negated when the respective data transfer cycle is completed. When the serial channel is not operating in DMA mode, the request and acknowledge pins for the channel can be programmed for other functions (see pin descriptions).

DMA DONEN Operation
As an input, DONEN is asserted by the DMA controller concurrent with the corresponding DMA acknowledge to indicate to the DUSCC that the character being transferred into the TxFIFO is the last character of the transmission frame. In synchronous modes, the DUSCC can be programmed through TPR[4] to automatically transmit the frame termination sequence (e.g., FCS – FLAG in BOP mode) upon receipt of this signal.

As an output, DONEN is asserted by the DUSCC under the following conditions:
a. In response to the DMA acknowledge for a receiver DMA request if the FIFOed RECEIVED EOM status bit (RSR[7]) is set for the character being transferred.
b. In response to the DMA acknowledge for a transmitter DMA request if the counter/timer has been programmed to count transmitted characters and the terminal count has occurred.

Block Transfers Using DTACK
The DTACKN line may be used to synchronize data transfers to and from the DUSCC utilizing a 'wait' state. Either the receiver or the transmitter or both may be programmed for this mode of operation, independently for each channel, via CMR2[5:3].

In this mode, if the CPU attempts a write to the transmit FIFO and an empty FIFO position is not available, the DTACK line will remain negated until a position empties. The data will then be written into the FIFO and DTACKN will be asserted to signify that the transfer is complete.

Similarly, a read of an empty receive FIFO will be held off until data is available to be transferred. Potentially, this mode can cause the microcomputer system to hang up if, for example, a read request was made and no further data was available.

Timing Circuits
The timing block for each channel consists of a crystal oscillator, a bit rate generator (BRG), a digital phase locked loop (DPLL) and a 16-bit counter/timer (C/T) (see Figure 8).

Crystal Oscillator
The crystal oscillator operates directly from a crystal (normally 14.7456MHz if the internal BRG is to be used) connected across the X1/CLK and X2/IDCN pins with a minimum of external components. If an external clock of the appropriate frequency is available, it may be connected to the X1/CLK pin. This signal is divided by two to provide the internal system clock (a maximum of 16MHz input is allowed).

Bit Rate Generator
The BRG operates from the oscillator or external clock and is capable of generating 16 bit rates. These are available to the receiver, transmitter, DPLL, and C/T. The BRG output is at 32× the base bit rate. Since all sixteen rates are generated simultaneously, each receiver and transmitter may select its bit rate independently. The transmitter and receiver timing registers include a 4-bit field for this purpose (TTR[3:0], RTR[3:0]).

Digital Phase-Locked Loop
Each channel of the DUSCC includes a DPLL used in synchronous modes to recover clock information from a received data stream. The DPLL is driven by a clock at nominally 32 times the data rate. This clock can be programmed, via RTR (7:4), to be supplied from an external input, from the receiver BRG, from the C/T, or directly from the crystal oscillator.

The DPLL uses this clock, along with the data stream to construct a data clock which may then be used as the DUSCC receive clock, transmit clock, or both. The output of the DPLL is a square wave at 1× the data rate. The derived clock can also be programmed to be output on a DUSCC pin; only the DPLL receiver output clock is available on the TRxC pin. Four commands are associated with DPLL operation: Enter search mode, set FM mode, set NRZI mode, and disable DPLL. The commands are described in the command register description. Waveforms associated with the DPLL are illustrated in Figure 9.

DPLL NRZI Mode Operation
This mode is used with NRZ and NRZI data encoding. With this type of encoding, the transitions of the data stream occur at the beginning of the bit cell. The DPLL has a six-bit counter which is incremented by a 32× clock. The first edge detected during search mode sets the counter to 16 and begins operation. The DPLL output clock then rises at a count of 0 and falls at 16. Data is sampled on the rising edge of the clock. When a transition in the data stream is detected, the count length is adjusted by one or two counts, depending on the counter value when the transition occurs (see Table 11). A transition detection at the rollover point (third column in Figure 11) is treated as a transition occurring at zero count.

The count length adjustments cause the rising edge of the DPLL output clock to converge to the nominal center of the bit cell. In the worst case, which occurs when a DPLL pulse is coincident with the data edge, the DPLL converges after 12 data transitions.

For NRZ encoded data, a stream of alternating ones and zeros should be used as a synchronizing pattern. For NRZI encoded data, a stream of zeros should be used.

Table 11. NRZI Mode Count Length

COUNT WHEN TRANSITION DETECTED	COUNT LENGTH ADJUST- MENT	COUNTER RESET AFTER COUNT REACHES
0 – 7	–2	29
8 – 15	–1	30
16 – 23	+1	32
24 – 30	+2	33
None detected	0	31

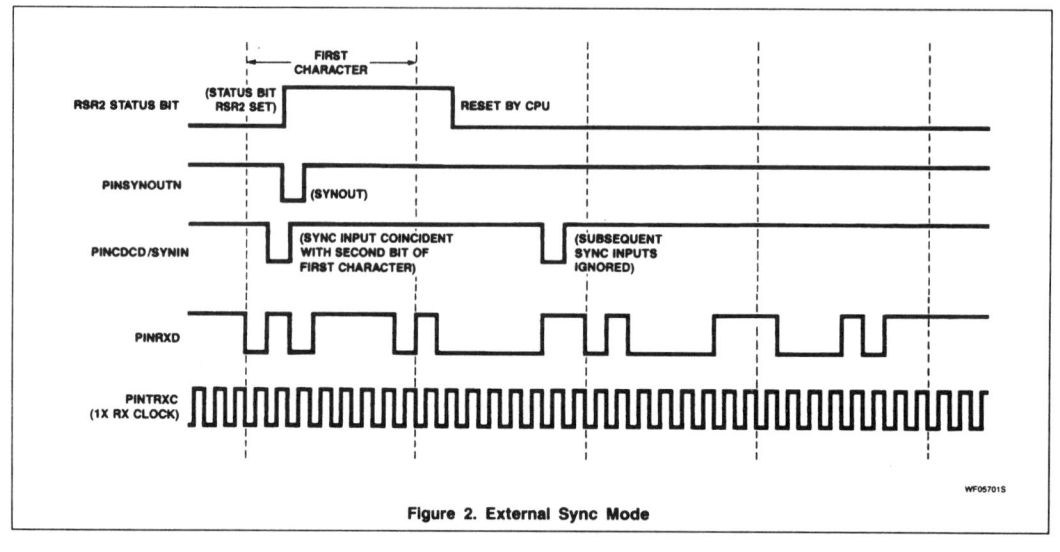

Dual Universal Serial Communications Controller (DUSCC) SCN68562

Figure 2. External Sync Mode

DPLL FM Mode Operation

FM operation is used with FM0, FM1, and Manchester data encoding. With this type of encoding, transitions in the data stream always occur at the beginning of the bit cell for FM0 and FM1, or at the center of the bit cell for Manchester. The DPLL 6-bit counter is incremented by a 32× clock. The first edge detected during search mode sets the counter to 16 and begins operation. The DPLL receiver clock then rises on a count of 8 and falls on 24. (The DPLL transmitter clock output falls on a count of 16. It rises on a count of 0 if a transition has been detected between counts of 16 and 23. For other cases, it rises 1/2 count of the 32× input clock sooner.) This provides a 1× clock with edges positioned at the nominal centers of the two halves of the bit cell. The transition detection circuit is enabled between counts of 8 and 23, inclusive. When a transition is detected, the count length is adjusted by one, depending on when the transition occurs (see Table 12).

Table 12. FM Mode Count Length

COUNT WHEN TRANSITION DETECTED	COUNT LENGTH ADJUST-MENT	COUNTER RESET AFTER COUNT REACHES
8 – 15	–1	30
16 – 23	+1	32
24 – 7	Disabled	
None detected	0	31

If a transition is not detected for two consecutive data bits, the DPLL is forced into search mode and the DPLL error status bit (TRSR [3]) is asserted. This feature is disabled when the DPLL output is used only as the transmitter clock.

To prevent the DPLL from locking on the wrong edges of the data stream, an opening PAD sequence should be transmitted. For FM0, a stream of at least 16 ones should be sent initially. For FM1, a minimum stream of 16 zeros should be sent and for Manchester encoding the initial data stream should consist of alternating ones and zeros.

Counter/Timer

Each channel of the DUSCC contains a counter/timer (C/T) consisting of a 16-bit down counter, a 16-bit preset register, and associated control circuits. Operation of the counter/timer is programmed via the counter/timer control register (CTCR). There are also four commands associated with C/T operation, as described in the Command Description section.

The C/T clock source, clock prescaling, and operating mode are programmed via CTCR[2:0], CTCR[4:3], and CTCR[6] respectively. The preset register is loaded with a minimum of 2 by the CPU and its contents can be transferred into the down counter by a command, or automatically upon reaching terminal count if CTCR[6] is negated. Commands are also available to stop and start the C/T and to preset it to an initial value of FFFF. Counting is triggered by the falling edge of the clocking input. The C/T zero count status bit, ICTSR[6], is set when the C/T reaches the terminal count of zero and ICTSR[7] indicates whether the counter is currently enabled or not.

An interrupt is generated upon reaching zero count if CTCR[7] and the channel's master interrupt enable are asserted. The output of the C/T can be programmed to be output on the channel's RTxC or TRxC pin (via PCR[4:0]) as either a single pulse or a square wave, as programmed in CTCR[5]. The contents of the C/T can be read at any time by the CPU, but the C/T should normally be stopped before this is done. Several C/T operating modes can be selected by programming of the counter/timer control register. Typical applications include:

1. Programmable divider. The selected clock source, optionally prescaled, is divided by the contents of the preset register. The counter automatically reloads itself each time the terminal count is reached. In this mode, the C/T may be programmed to be used as the Rx or Tx bit rate generator, as the input to the DPLL, or it may be output on a pin as either a pulse or a square wave. The C/T interrupt should be disabled in this mode.

2. Periodic interrupt generator. This mode is similar to the programmable divider mode, except that the C/T interrupt is enabled, resulting in a periodic interrupt to the CPU.

3. Delay timer. The counter is preset from the preset register and a clock source, optionally prescaled, is selected. An interrupt is generated upon reaching terminal count. The C/T continues counting

December 1986

Preliminary Specification

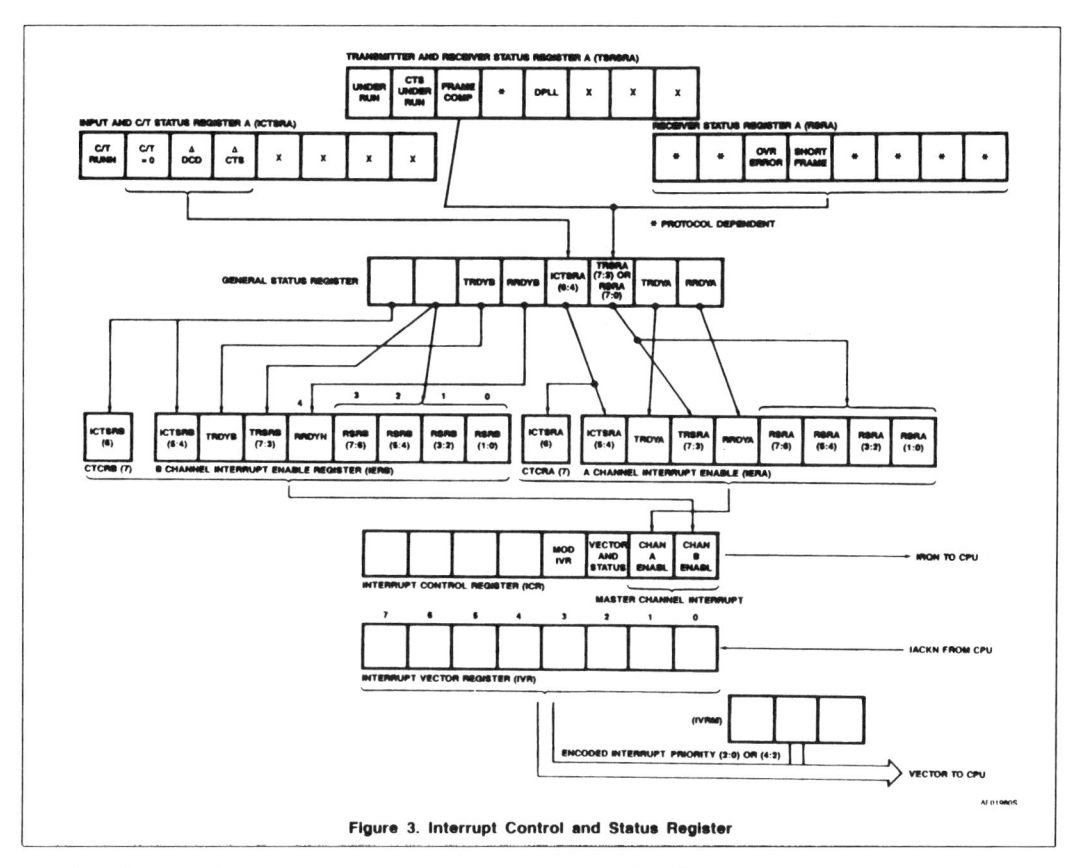

Dual Universal Serial Communications Controller (DUSCC)   SCN68562

Figure 3. Interrupt Control and Status Register

without reloading itself and its contents may be read by the CPU to allow additional delay past the zero count to be determined.

4. Character counter. The counter is preset to FFFF by command and the clock source becomes the internal signal used to control loading of the Rx or Tx characters. This operation is selected by CTCR [2:0]. The C/T counts characters loaded into the RxFIFO by the receiver or loaded into the transmit FIFO by the CPU respectively. The current character count can be determined by the CPU by reading the contents of the C/T and taking its ones complement. Optionally, a preset number may be loaded into the counter and an interrupt generated when the count is exhausted. When counting Tx characters, the terminal count condition can be programmed through TPR_ [4] to cause an end of message sequence to be transmitted. When counting received characters, the FIFOed EOM status bit is asserted when the character which causes the count to go to zero is loaded into the receive FIFO. The channel's 'reset Tx' or 'reset Rx' commands have no effect on the operation of the C/T.

5. External event counter. The counter is preset to FFFF by command and an external clock source is selected. The current count can be determined by the CPU by reading the contents of the C/T and taking its ones complement. Optionally, a preset number may be loaded into the counter and an interrupt generated when the count is exhausted.

6. Bit length measurement. The counter is preset to FFFF by command and the X1/CLK/4 clock input gated by RxD mode (optionally prescaled) is programmed. The C/T starts counting when RxD goes low and stops counting when RxD goes high. At this time, ICTSR[6] = 1 is set and an interrupt (if enabled) is generated. The resulting count in the counter can be read by the CPU to determine the bit rate of the input data. Normally this function is used for asynchronous operation.

Communication Channels A and B

Each communication channel of the DUSCC is a full-duplex receiver and transmitter that supports ASYNC, COP, and BOP transmission formats. The bit rate clock for each receiver and transmitter can be selected independently to come from the bit rate generator, C/T, DPLL, or an external input (such as a modem generated clock).

December 1986

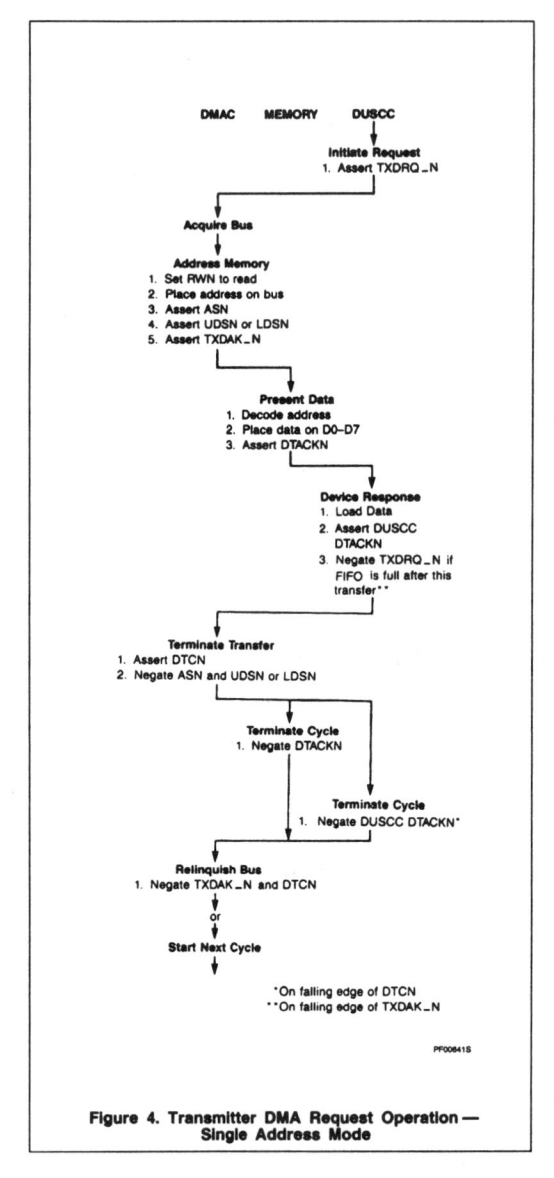

Preliminary Specification

Dual Universal Serial Communications Controller (DUSCC)   SCN68562

Figure 4. Transmitter DMA Request Operation — Single Address Mode

DMAC   MEMORY   DUSCC

Initiate Request
1. Assert TXDRQ_N

Acquire Bus

Address Memory
1. Set RWN to read
2. Place address on bus
3. Assert ASN
4. Assert UDSN or LDSN
5. Assert TXDAK_N

Present Data
1. Decode address
2. Place data on D0–D7
3. Assert DTACKN

Device Response
1. Load Data
2. Assert DUSCC DTACKN
3. Negate TXDRQ_N if FIFO is full after this transfer**

Terminate Transfer
1. Assert DTCN
2. Negate ASN and UDSN or LDSN

Terminate Cycle
1. Negate DTACKN

Terminate Cycle
1. Negate DUSCC DTACKN*

Relinquish Bus
1. Negate TXDAK_N and DTCN

or

Start Next Cycle

*On falling edge of DTCN
**On falling edge of TXDAK_N

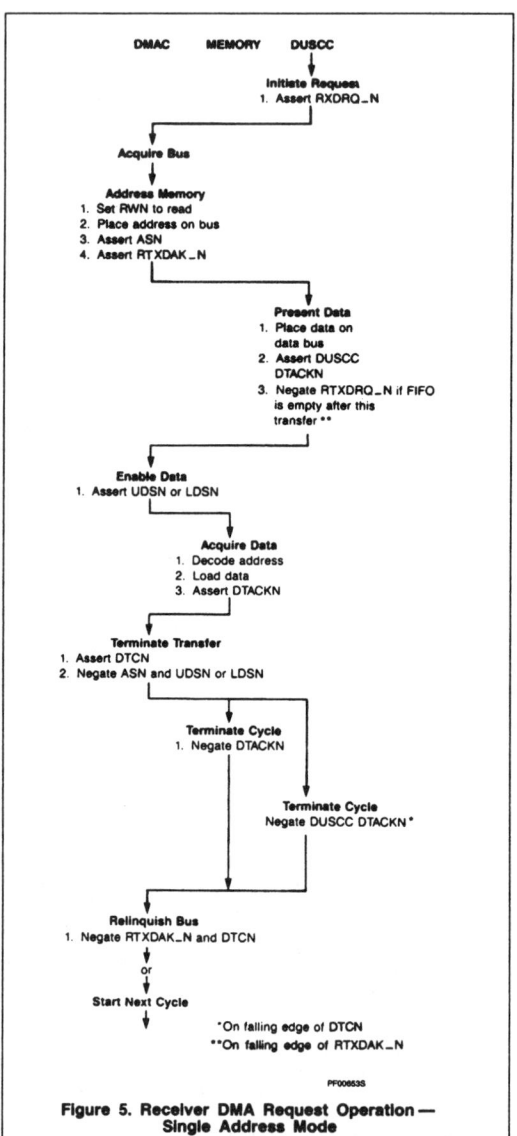

Figure 5. Receiver DMA Request Operation — Single Address Mode

DMAC   MEMORY   DUSCC

Initiate Request
1. Assert RXDRQ_N

Acquire Bus

Address Memory
1. Set RWN to read
2. Place address on bus
3. Assert ASN
4. Assert RTXDAK_N

Present Data
1. Place data on data bus
2. Assert DUSCC DTACKN
3. Negate RTXDRQ_N if FIFO is empty after this transfer**

Enable Data
1. Assert UDSN or LDSN

Acquire Data
1. Decode address
2. Load data
3. Assert DTACKN

Terminate Transfer
1. Assert DTCN
2. Negate ASN and UDSN or LDSN

Terminate Cycle
1. Negate DTACKN

Terminate Cycle
Negate DUSCC DTACKN*

Relinquish Bus
1. Negate RTXDAK_N and DTCN

or

Start Next Cycle

*On falling edge of DTCN
**On falling edge of RTXDAK_N

December 1986

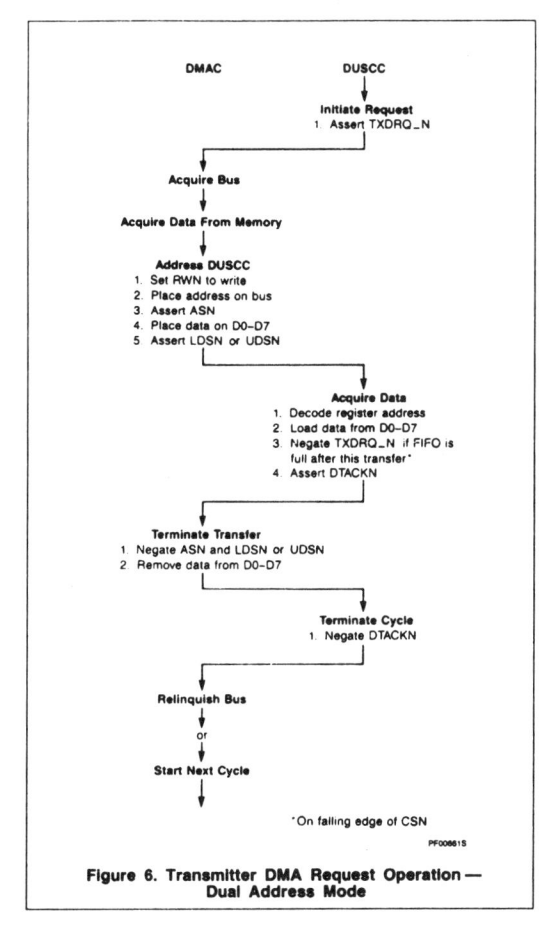

Preliminary Specification

Dual Universal Serial Communications Controller (DUSCC)　　SCN68562

Figure 6. Transmitter DMA Request Operation — Dual Address Mode

DMAC / DUSCC

- **Initiate Request** (DUSCC)
  1. Assert TXDRQ_N
- **Acquire Bus** (DMAC)
- **Acquire Data From Memory** (DMAC)
- **Address DUSCC** (DMAC)
  1. Set RWN to write
  2. Place address on bus
  3. Assert ASN
  4. Place data on D0-D7
  5. Assert LDSN or UDSN
- **Acquire Data** (DUSCC)
  1. Decode register address
  2. Load data from D0-D7
  3. Negate TXDRQ_N if FIFO is full after this transfer*
  4. Assert DTACKN
- **Terminate Transfer** (DMAC)
  1. Negate ASN and LDSN or UDSN
  2. Remove data from D0-D7
- **Terminate Cycle** (DUSCC)
  1. Negate DTACKN
- **Relinquish Bus** or **Start Next Cycle** (DMAC)

*On falling edge of CSN

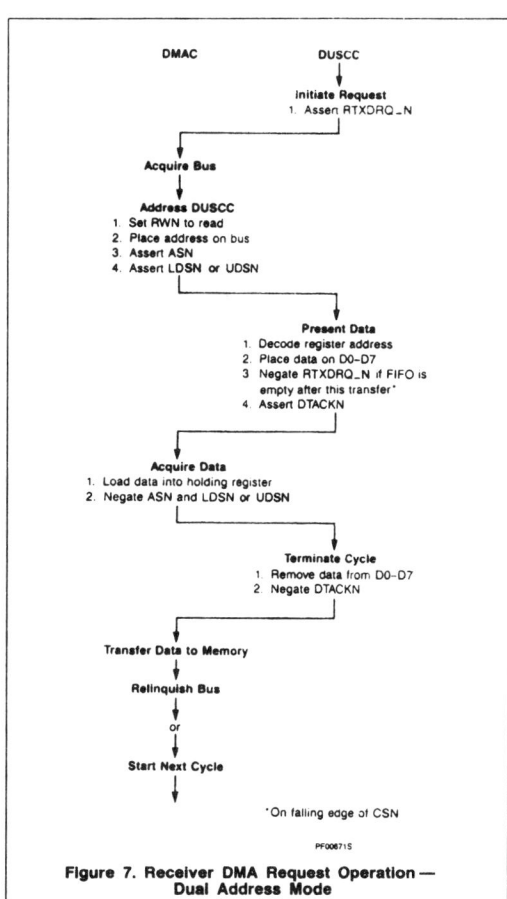

Figure 7. Receiver DMA Request Operation — Dual Address Mode

DMAC / DUSCC

- **Initiate Request** (DUSCC)
  1. Assert RTXDRQ_N
- **Acquire Bus** (DMAC)
- **Address DUSCC** (DMAC)
  1. Set RWN to read
  2. Place address on bus
  3. Assert ASN
  4. Assert LDSN or UDSN
- **Present Data** (DUSCC)
  1. Decode register address
  2. Place data on D0-D7
  3. Negate RTXDRQ_N if FIFO is empty after this transfer*
  4. Assert DTACKN
- **Acquire Data** (DMAC)
  1. Load data into holding register
  2. Negate ASN and LDSN or UDSN
- **Terminate Cycle** (DUSCC)
  1. Remove data from D0-D7
  2. Negate DTACKN
- **Transfer Data to Memory** (DMAC)
- **Relinquish Bus** or **Start Next Cycle** (DMAC)

*On falling edge of CSN

December 1986

Dual Universal Serial Communications Controller (DUSCC)   SCN68562

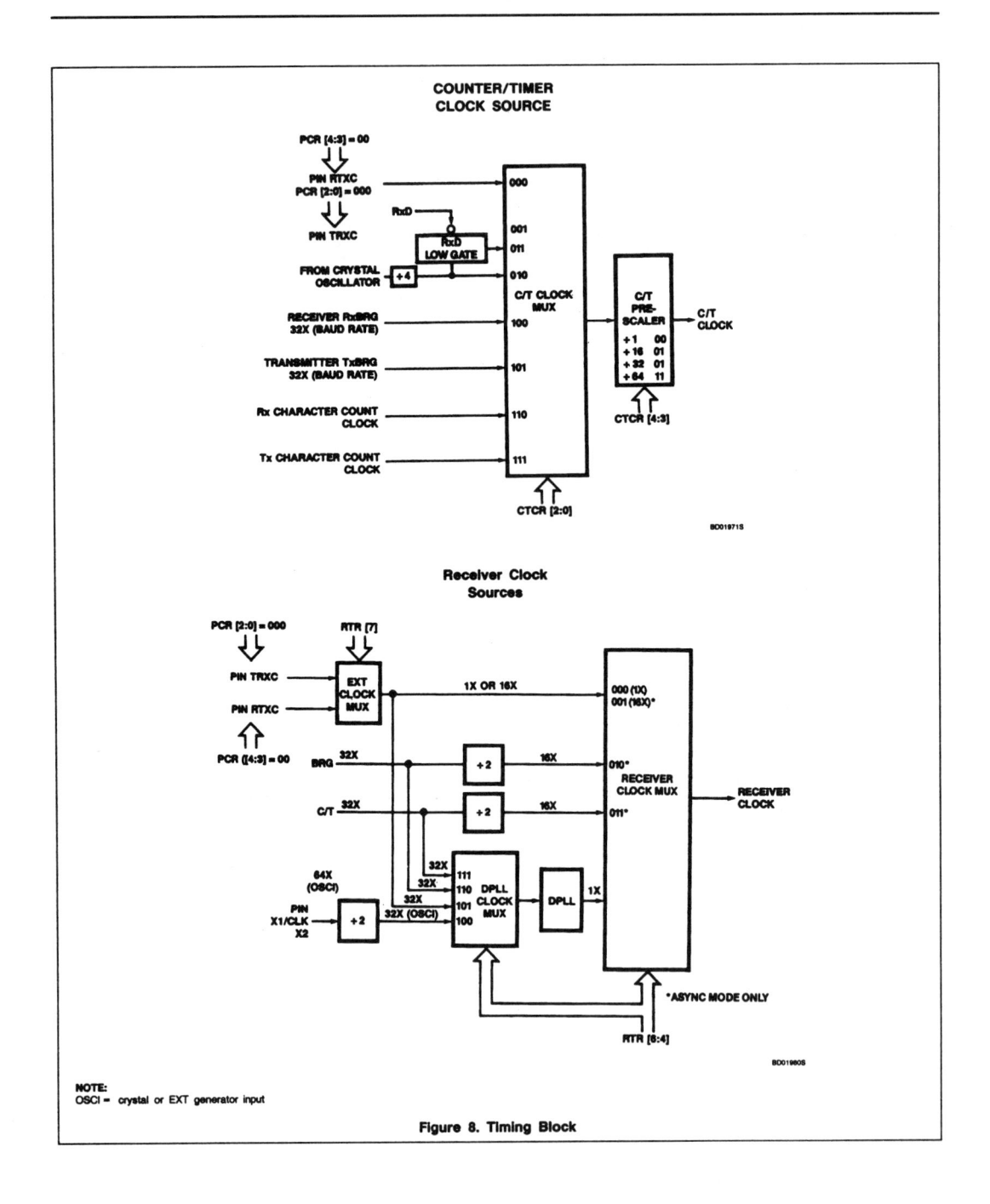

Figure 8. Timing Block

Preliminary Specification

Dual Universal Serial Communications Controller (DUSCC)   SCN68562

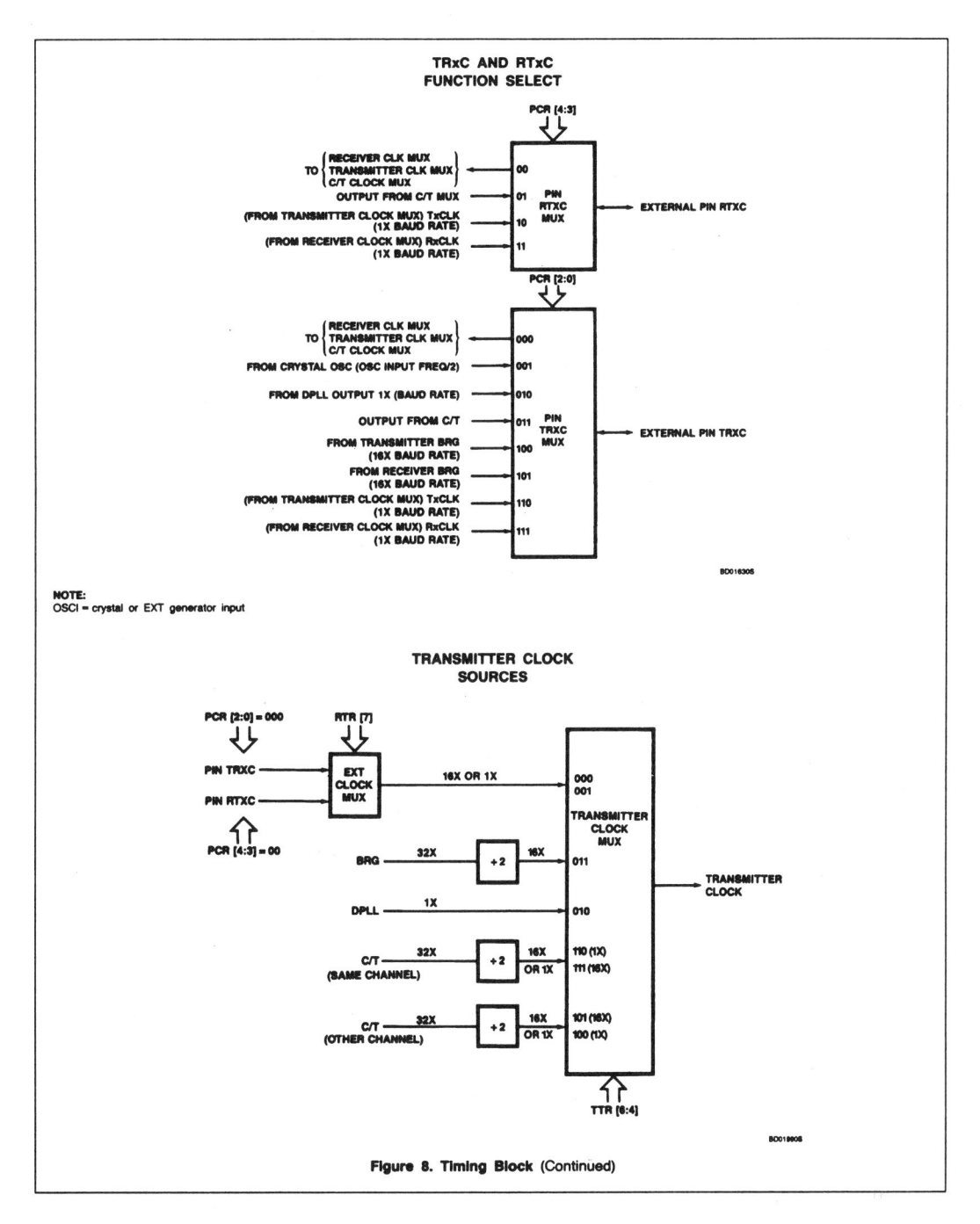

Figure 8. Timing Block (Continued)

Preliminary Specification

Dual Universal Serial Communications Controller (DUSCC) SCN68562

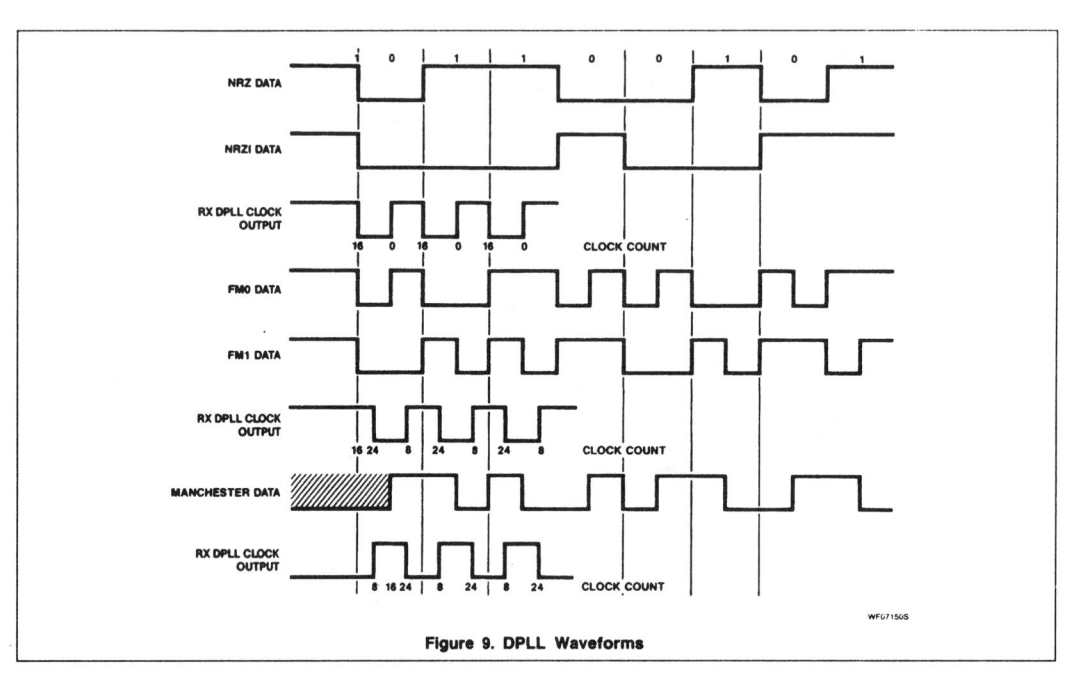

Figure 9. DPLL Waveforms

TRANSMITTER

Transmitter TxFIFO and TxRDY

The transmitter accepts parallel data from the data bus and loads it into the TxFIFO, which consists of four 8-bit holding registers. This data is then moved to the transmitter shift register, TxSR, which serializes the data according to the transmission format programmed. The TxSR is loaded from the TxFIFO, from special character logic, or from the CRC/LRC generator. The LSB is transmitted first, which requires right justification of characters by the CPU. TxRDY (GSR[5] or GSR[1]) and underrun (TRSR[7]) indicate the state of the TxFIFO. The TxFIFO may be addressed at any of four consecutive locations (see Table 1) to allow use of multiple byte word instructions. A write to any valid address always writes data to the next empty FIFO location.

TxRDY is set when the transmitter is enabled and there is an empty position in the TxFIFO (OMR[4] = 0) or when the TxFIFO becomes empty (OMR[4] = 1). The CPU may reset TxRDY through a status reset write cycle. If this is done, it will not be reasserted until a character is transferred to the TxSR (OMR[4] = 0) or when the TxFIFO becomes empty again (OMR[4] = 1). The assertion of TxRDY, enabling of the IER [6] and the enabling of the channel master interrupt ICR [0] or [1] allow an interrupt to be generated.

If DMA operation is programmed, either RTxDRN (half-duplex) or TxDRQN (full-duplex) follows the state of TxRDY if the transmitter is enabled. These operations differ from normal ready in that the request signal is negated on the leading edge of the DMA acknowledge signal when the subsequent transfer causes the transmit FIFO to become full, while the TxRDY signal is negated only after the transfer is completed. Underrun status TRS[7] set indicates that one or more data characters (not PAD characters) have been transmitted and the TxFIFO and TxSR are both empty.

In 'wait on Tx', a write to a full FIFO causes the write cycle to be extended until a FIFO position is available. DTACKN is asserted to acknowledge acceptance of the data. In nonwait modes, if an attempt is made to load data into a full TxFIFO, the TxFIFO data is preserved and the overrun data character(s) is lost. A normal DTACKN will be issued, and no indication of this occurrence is provided. The transmitter is enabled by the enable transmitter command. When the disable transmitter command is issued, the transmitter continues to operate until the TxFIFO becomes empty. The TxRDY does not become valid until the transmitter is enabled. Characters can be loaded into the FIFO while disabled. However, if the FIFO is full when the transmitter is enabled, TxRDY is not asserted.

TxRTS Control

If TxRTS CONTROL, TPR[3], is programmed, the channel's RTS output is negated five bit times after the last bit (stop bit in ASYNC mode) of the last character is transmitted. RTS is normally asserted and negated by writing to OMR_[0]. The assertion of TPR[3] causes RTS to be reset automatically (if the transmitter is not enabled) after all characters in the transmitter FIFO (if any) are transmitted and five bit times after the 'last character' is shifted out. This feature can be used to automatically terminate the transmission of a message as follows:

– Program auto-reset mode: TPR_[3] = 1.

– Enable transmitter.

– Assert RTS_N: OMR_[0] = 1.

– Send message.

– Disable transmitter after the last character is loaded into the TxFIFO.

– The last character will be transmitted and OMR_ [0] will be reset five bit times after the last bit, causing RTS_N to be negated. The TxD output will remain in the marking state until the transmitter is enabled again.

The 'last bit' in ASYNC is simply the last stop bit of the character. In BOP and COP, the last character is defined either explicitly by either appending it with TEOM or implicitly through the selection of the frame underrun control

December 1986   32

Dual Universal Serial Communications Controller (DUSCC)   SCN68562

sequence, TPR[7:6] (transmitter parameter register). Table 13 summarizes the relationship of the selected underrun sequence and the protocol mode.

Tx CTS Operation

If CTS enable Tx, TPR[2], is set, the CTSN input must be asserted for the transmitter to operate. Changes in CTSN while a character is being transmitted do not affect transmission of that character. However, if the CTS input becomes negated when TPR[2] is set and the transmitter is enabled and ready to start sending a new character, CTS underrun, TRSR[6], is asserted and the TxD output is placed in the marking (high) state. In ASYNC mode, operation resumes when CTSN is asserted again. In COP and BOP modes, the transmission of the message is terminated and operation of the transmitter will not resume until CTS is asserted and a TSOM or TSOMP command is invoked. Prior to issuing the command and retransmitting the message, the transmitter must be reset. After a change-of-state CTS is established by the input sampling circuits (refer to the description of ICTSR[4]), it is sampled by the Tx controller 1½ bit times before each new character is serialized out of the Tx shift register. (This is 2½ bits before the LSB of the new character appears on the TxD pin; there is an additional 1-bit delay in the transmitter data path due to the data encoding logic.)

Tx Special Bit Pattern Transmission

The DUSCC provides features to transmit special bit patterns (see Table 14).

The TxD pin is held marking after a hardware reset, a reset Tx command, when the transmitter is not enabled, and during underrun/idle, if this feature is selected through TPR[7:5]. The TxD pin is also held marking if the transmitter is enabled, and the TxFIFO is empty (ASYNC), or if a TSOM or TSOMP command has not been issued (SYNC modes).

The following command bits can be appended to characters in the TxFIFO: TEOM, TDLE, exclude from CRC, and reset TxCRC. An invoked command(s) is appended to the next character loaded into the TxFIFO and follows the character through the FIFO until that character is ready to be loaded into the TxSR. The transmitter data path is shown in Figure 10. The following describes the operation of the transmitter for the various protocols.

Tx ASYNC Mode

Serialization begins when the TxFIFO data is loaded into the TxSR. The transmitter first sends a start bit, then the programmed number of bits/character (TPR[1,0]), a parity bit (if specified), and the programmed number of stop bits. Following the transmission of the

Table 13. Abort Sequence — Protocol Mode

TPRA [7:6]	PROTOCOL	LAST CHARACTER
00	BOP	FLAG following either FCS (if selected) or last data character
	COP	Last byte of FCS before line begins SYN or MARKing
10	BOP	Abort sequence (11111111) prior to MARKing
	COP	Last byte of FCS before line begins SYN or MARKing
11	BOP	Abort sequence (11111111) prior to FLAG
	COP	First SYN of SYN sequence

Table 14. Special Bit Patterns

PROTOCOL	BIT PATTERN
ASYNC-BREAK	An all 0's character including parity bit (if specified) and stop bits. Used for send break command.
COP-SYN	Contained in S1R (single SYN mode) or in S1R/S2R (dual SYN modes). Used for TSOM and TSOMP commands and for non-transparent mode linefill and IDLE.
COP-DLE	Used for TDLE command and for BISYNC transparent mode linefill and to generate BISYNC control sequences.
COP-CRC	16/8 bits from the CRC/LRC accumulator used for TEOM command or for auto-EOM modes.
BOP-FLAG	01111110. Used for TSOM, TSOMP, and TEOM commands, for auto-EOM modes, and as an IDLE line fill.
BOP-ABORT	11111111. Used for send ABORT command or during TxFIFO underrun.
BOP-CRC	16 bits from the CRC accumulator used for TEOM command or for auto-EOM modes.
BOP/COP MARK	All 1's pattern on data line.

stop bits, if a new character is not available in the TxFIFO, the TxD output goes to marking and the underrun condition (TRSR[7]) is set.

Transmission resumes when the CPU loads a new character into the TxFIFO or issues a send break command. The send break command clears the TxFIFO and forces a continuous space (low) on the TxD output after the character in TxSR (if any) is serialized. A send break acknowledge (TRSR[4]) is returned to the CPU to facilitate reassertion of the send break command in order to send an integral number of break characters. The send break condition is cleared when the reset Tx or disable Tx command is issued.

Tx COP Modes

Transmitter commands associated with all COP modes are: transmit SOM (TSOM, transmit start of message), transmit SOM with PAD (TSOMP), transmit EO' (TEOM, transmit end of message), reset TxCRC, exclude from CRC, and transmit DLE.

A TSOM or send TSOMP command must be issued to start COP transmission. TSOM (without PAD) causes the TxCRC/LRC generator to be initialized and one or two SYN characters from S1R/S2R to be loaded into the TxSR and shifted out on the TxD output. A parity bit, if specified, is appended to each SYN character after the MSB. Send SOM acknowledge (TRSR[4]) is asserted when the SYN output begins. The user may reinvoke the command to cause multiple SYNs to be transmitted. If the command is not reinvoked and the TxFIFO is empty, SYN patterns continue to be transmitted until the TxFIFO is loaded. If data is present in the FIFO, the first character is loaded into the TxSR and serialization of the data begins. Note that the TxFIFO may be preloaded with data before the TSOM is issued.

The TSOMP command causes all characters in the TxFIFO (PAD characters) to be loaded into the TxSR and serialized if the Tx is enabled. Unlike the transmit SOM without PAD command, data (non-PAD characters) cannot be preloaded into the TxFIFO. While the PAD is transmitted, parity is disabled and character length is automatically set to 8 bits regardless of the value in TPR[1:0]. When the TxFIFO becomes empty after the PAD, the TxCRC/LRC generator is initialized, the SYN character(s) are transmitted with optional parity appended, and send SOM acknowledge asserted. Operation then proceeds in the same manner as the TSOM command; the user has the option to invoke the TSOM command to cause multiple SYNs to be transmitted.

After the TSOM/TSOMP command is executed, characters in the TxFIFO are loaded into the TxSR and shifted out with a parity bit, if specified, appended after the MSB. If, after the opening SYN(s) and at least one data has

December 1986   33

Dual Universal Serial Communications Controller (DUSCC)   SCN68562

Figure 10. Transmitter Data Path

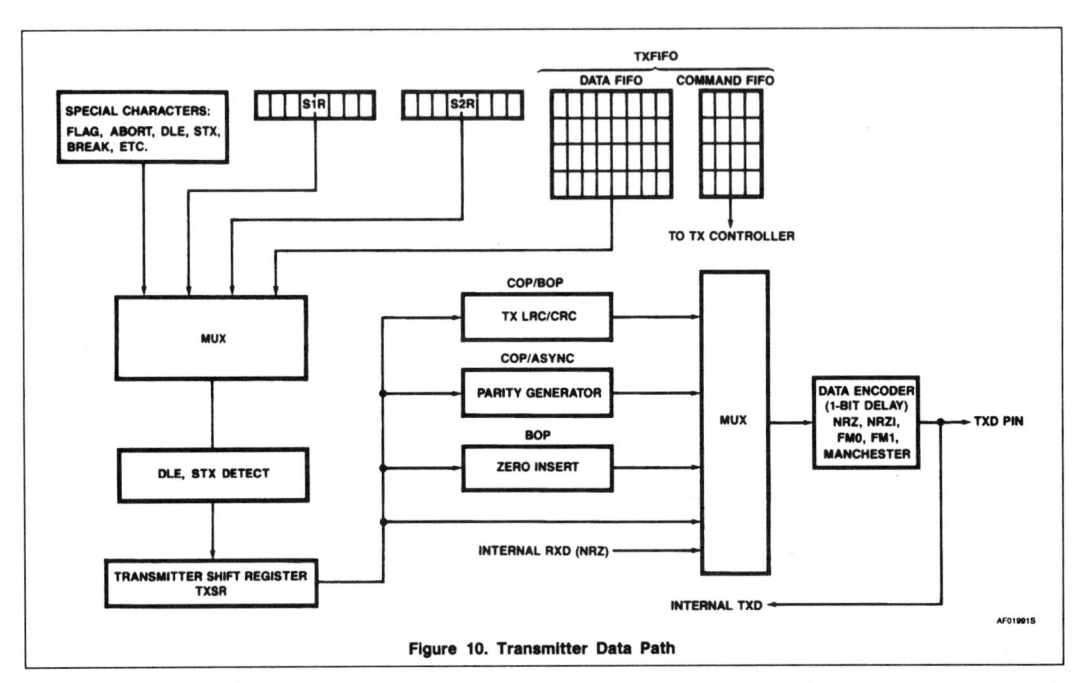

been transmitted, the TxFIFO is empty, a data underrun condition results and TRSR[7] is asserted. The transmitter's action on data underrun is determined by TPR[7:6] and the COP protocol. If TPR_[7:6] = '10', the transmitter line fills with MARK characters until a character is loaded into the FIFO. If TPR_[7:6] = '11' is selected, the transmitter line fills with SYN, SYN1-SYN2, or DLE-SYN1 for mono-sync, dual sync, and BISYNC transparent modes, respectively. If TPR[7:6] = '00', the BCC characters are transmitted and frame complete (TRSR[6]) is set. TxD then assumes the programmed idle state (TPR[5]) of MARKs or SYN1/SYN1-SYN2.

Operation resumes with the transmission of a SYN sequence when a TSOM command is invoked. A TSOMP command is ignored unless the transmitter is disabled and then reenabled.

An appended TEOM command also terminates the frame as described above. It occurs after transmission of the character to which the TEOM is appended. The TEOM command can be explicitly asserted through the channel command register. If TPR[4] = '1', the TEOM is automatically appended to a character in DMA mode, if the DONEN input is asserted when that character is loaded into the TxFIFO, or if the counter/timer is counting transmitted characters when the character which causes the counter to reach zero count is loaded.

The TDLE command when appended to a character in the TxFIFO, causes the DLE character to be loaded into the TxSR and serialized before the TxFIFO character is loaded into the TxSR and serialized. This feature is particularly useful for BISYNC operation. The DLE character will be excluded from the CRC accumulation in BISYNC transparent mode (see below), but will be included in all other COP modes.

In BISYNC mode, transmission of a DLE-STx character sequence (either via a send TDLE command appended to the STx character, or via DLE and STx loaded into the TxFIFO) puts the transmitter into the transparent text mode of operation. In this mode, normally restricted character sequences can be transmitted as 'normal' bit sequences. The switch occurs after transmission of the two characters, so that the DLE and STx are included in the BCC accumulation. If the DLE-STx is to be excluded from the CRC, the user should issue a 'reset CRC' command prior to loading the next character.

Another method of excluding the two characters from the CRC is to invoke the 'exclude from CRC' command prior to loading the character(s) into the FIFO. While in transparent mode, the transmitter line fills with DLE-SYN1 and automatically transmits an extra DLE if it finds a DLE in the TxFIFO ('DLE stuffing'). The transmitter reverts to non-transparent mode when the frame complete status is set in TRSR[5].

CRC/LRC accumulation can be specified in all COP modes; the type of CRC is specified via CMR2[2:0]. The TSOM/TSOMP commands set the CRC/LRC accumulator to its initial state and accumulation begins with the first non-SYN character after the initial SYN(s) are transmitted. PAD characters are not subject to CRC accumulation. In non-BISYNC or BISYNC normal modes, all transmitted characters except linefill characters (SYNs or MARKs) are subject to accumulation. In BISYNC transparent mode, odd (stuffed) DLEs and the DLE-SYN1 linefill are excluded from the accumulation. Characters can be selectively excluded from the accumulation by invoking the 'exclude from CRC' command prior to loading the character into the FIFO.

Accumulation stops when transmission of the first character of the BCC begins. The CPU can set the accumulator to its initial state prior to the transmission of any character by using the appended reset CRC command. The CRC generator is also automatically initialized after the EOM is sent.

TxBOP Modes

Transmitter commands associated with BOP modes are TSOM, TSOMP, TEOM, and trans-

December 1986

Dual Universal Serial Communications Controller (DUSCC)   SCN68562

Preliminary Specification

mit ABORT (TABRK). The TSOM and TSOMP commands are identical to COP modes except that a FLAG character (01111110) is used as the start of message sequence instead of the SYNs, and FLAG(s) that continue to be sent until the TxFIFO is loaded. There is no zero insertion (see below) during transmission of the PAD characters, and they are not preceded by a FLAG or accumulated in the CRC. Character length is automatically set to 8 bits regardless of TPR[1:0].

The first characters loaded into the TxSR from the TxFIFO are the address and control fields, which have fixed character lengths of eight bits. The number of address field bytes is determined by CMR1[4:3]. If extended address field is specified, the field is terminated if the first address octet is H'00' or if the LSB of the octet is a 1. The number of control field bytes is selected by CMR1_[5]. If any information field characters follow the control field (forming an I field), they are transmitted with the number of bits per character programmed in TPR[1:0]. The TEOM command can be appended to the last character either explicitly or automatically as described for COP mode. When the character with the appended TEOM is loaded from the TxFIFO, it is transmitted with the character length specified by OMR[7:5]. In this way, a residual character of 1–8 bits is transmitted without requiring the CPU to change the Tx character length for this last character.

After the opening FLAG and first address octet have been transmitted, an underrun occurs (TRSR[7] = 1) if the TxFIFO is empty when the transmitter requires a new character. The underrun control bits (TPR[7:6]) determine whether the transmitter fill ends with either ABORT-MARKs, ABORT-FLAGs (see below), or ends transmission with the 'normal' end of message sequence.

EOM on underrun is functionally similar to EOM due to an appended TEOM command. If the EOM is due to underrun, the normal character length applies to the last data character. After the last character is transmitted, the FCS (inverted CRC) and closing FLAG are sent, frame complete (TRSR[5]) is set, and the TxCRC is initialized. If the TxFIFO is empty after the closing FLAG has been sent, TxD will assume the programmed idle state of FLAGs or MARKs (TPR[5]) and wait for a character to be loaded into the FIFO or for a TSOM command to be issued. If the TxFIFO is not empty at that time, the TxFIFO data will be loaded into the TxSR and serialized. In that case, the closing FLAG is the opening FLAG of the next frame.

The user can control the number of FLAGs between frames by invoking the TSOM command after frame complete is asserted. The DUSCC then operates in the same manner as for transmission of multiple FLAGs at the beginning of a frame. When the command is no longer reinvoked, transmission of the TxFIFO data will begin. If the FIFO is empty, FLAGs continue to be transmitted.

The DUSCC provides automatic zero insertion in the data stream to prevent erroneous transmission of the FLAG sequence. All data characters loaded into the TxSR from the TxFIFO and characters transmitted from the CRC generator are subject to zero insertion. For this feature a zero is inserted in the serial data stream each time five consecutive ones (regardless of character boundaries) have been transmitted.

A send ABORT command clears the TxFIFO and inserts an ABORT character of eight ones (not subject to zero insertion) into the TxSR for transmission after the current character has been serialized. A send abort ack (TRSR[4]) facilitates reassertion of send abort by the user to guarantee transmission of multiple abort characters. This feature can be used to send the 15-ones idle sequence. The transmitter sends either marks or FLAGs after the abort character(s) has been transmitted, depending on TPR[7:6]. Operation resumes with the transmission of a FLAG when a TSOM command is invoked. A TSOMP command is ignored unless the transmitter is disabled and then reenabled.

CRC accumulation can be specified in all BOP modes. The type of CRC is specified via CMR2[2:0], and is normally selected as CRC-CCITT preset to ones, although any option is valid. Note that LRC8 option is not allowed in BOP modes.

The TSOM/TSOMP command sets the CRC accumulator to its initial state and accumulation begins with the first address octet after the initial FLAG(s). Accumulation stops when transmission of the first character of the FCS begins. The CPU can set the accumulator to its initial state prior to the transmission of any character by using the appended reset CRC command and can exclude any character from the accumulation by use of the exclude from CRC command, but these features would not normally be used in BOP modes. The CRC generator is also automatically initialized after the EOM or an ABORT are sent.

TxBOP Loop Mode
The loop modes are used by secondary stations on the loop, while the primary station operates in the BOP primary mode. Both the transmitter and receiver must be enabled. Loop operation is initiated by issuing the 'go on-loop' command. The receiver looks for the receipt of seven contiguous ones and then asserts the LC_N output to cause external loop control hardware to put the DUSCC into the loop, with the TxD output echoing the RxD input with a 2-bit time delay. The echoing process continues until a go active on poll (GAP) command is invoked. The DUSCC then looks for receipt of an EOP bit pattern (a zero followed by seven ones, 11111110) and changes the last one of the EOP into a zero making it an opening FLAG. Loop sending (TRSR[6]) is asserted at that same time. The action of the transmitter after sending the initial FLAG depends on the status of the transmit FIFO.

If the transmit FIFO is not empty, a normal frame transmission begins. The operation is then similar to normal BOP operation with the following differences:

1. An ABORT command, an underrun, or receipt of the turnaround sequence (H'01') or a FLAG cause the transmitter to cease operation and to revert to echoing the RxD input with a 2-bit time delay. A new transmission cannot begin until the GAP command is reinvoked and a new EOP sequence is received.

2. Subsequent to sending the EOM sequence of FCS-FLAG, the DUSCC examines the internal GAP flip-flop. If it is not set (having been reset by its 'reset GAP' command), the DUSCC reverts to echoing the received data. If the internal GAP flip-flop is still set, transmission of a new frame begins, with the user having control of sending multiple FLAGs between frames by use of the 'send SOM' command. If the FIFO is empty at this time, the DUSCC continues to send FLAGs until the data is loaded into the FIFO or until GAP is reset. If the latter occurs, it reverts to echoing RxD.

When the DUSCC reverts to echoing RxD in any of the above cases, the last transmitted zero and seven ones will form an EOP for the next station down the loop.

If the TxFIFO is empty when the EOP is recognized, the transmitter continues to send FLAGs until there is data in the FIFO. If a turnaround sequence or the reset GAP command is received before the FIFO is loaded, the transmitter switches to echoing RxD without any data transmission. Otherwise a frame transmission begins as above when a character is loaded into the FIFO. The mechanism provides time for the CPU to examine the received frame (the frame preceding the EOP) to determine if it should respond or not, while holding its option to initiate a transmission.

Termination of operation in the loop mode should be accomplished by use of the 'go off-loop' command. When the

December 1986   35

Preliminary Specification

Dual Universal Serial Communications Controller (DUSCC)   SCN68562

command is invoked, the DUSCC looks for the receipt of eight contiguous ones. It then negates the LC_N output to cause the external loop control hardware to remove the DUSCC from the loop without affecting operation of other units remaining on the loop.

RECEIVER

The receiver data path includes two 9-bit holding registers, HSRH and HSRL, an 8-bit character comparison register, CCSR, two synchronizing flip flops, a receiver shift register, RxSR, the programmable SYN comparison registers, S1R and S2R, and BISYNC character comparison logic. The DUSCC configures this circuitry and utilizes it according to the operational mode selected for the channel through the two mode registers CMR1 and CMR2. For all data paths, character data is assembled according to the character bit count, in the RxSR, and is moved to the RxFIFO with any appended statuses when assembly is completed. Figure 1 depicts the four data paths created in the DUSCC for the previous protocols.

Receiver RxFIFO, RxRDY

The receiver converts received serial data on RxD (LSB first) into parallel data according to the transmission format programmed. Data is shifted through a synchronizing flip flop and one or more shift registers, the last of which is the 8-bit receiver shift register (RxSR). Bits are shifted into the RxSR on the rising edge of each 1× receive clock until the LSB is in RxSR[0]. Hence, the received character is right justified, with all unused bits in the RxSR cleared to zero. A receive character length counter generates a character boundary signal for synchronization of character assembly, character comparisons, break detection (ASYNC), and RxSR to RxFIFO transfers (except for BOP residual characters). During COP and BOP hunt phases, the SYN/FLAG comparison is made each receive bit time, as abort, and idle comparisons in BOP modes.

An internal clock from the BRG, the DPLL or the counter/timer, or an external 1× or 16× clock may be used as the receiver clock in ASYNC mode. The BRG or counter/timer cannot be used directly for the receiver clock in synchronous modes, since these modes require a 1× receive clock that is in phase with the received data. This clock may come externally from the RTxC or TRxC pins, or it may be derived internally from the DPLL. Encoded data is internally converted to NRZ format for the receiver circuits by using clock pulses generated by the DPLL.

When a complete character has been assembled in the RxSR, it is loaded into the receive FIFO with appended status bits. The most significant data bits of the character are set to zero if the character length is less than eight bits. In ASYNC and COP modes the user may select, via RPR[3], whether the data transferred to the FIFO includes the received parity bit or not. The receiver indicates to the CPU or DMA controller that it has data in the FIFO by asserting the channel's RxRDY status bit (GSR[4] or GSR[0]) and, if in DMA mode, the corresponding receiver DMA request pin.

The RxFIFO consists of four 8-bit holding registers with appended status bits for character count complete indications (all modes), character compare indication (ASYNC), EOM indication (BISYNC/BOP), and parity, framing, and CRC errors. Data is loaded into the RxFIFO from the RxSR and extracted (read) by the CPU or DMA controller via the data bus. An RxFIFO read creates an empty RxFIFO position for new data from the RxSR.

RxRDY assertion depends on the state of OMR[3]:

1. If OMR[3] is 0 (FIFO not empty), RxRDY is asserted each time a character is transferred from the receive shift register to the receive FIFO. If it is not reset by the CPU, RxRDY remains asserted until the receive FIFO becomes empty, at which time it is automatically negated. If it is reset by the CPU, it will remain negated, regardless of the current state of the receive FIFO, until a new character is transferred from the RxSR to the RxFIFO.
2. If OMR[3] is 1 (FIFO full), RxRDY is asserted:
   a. when a character transfer from the receive shift register to the receive FIFO causes it to become full
   b. when a character with a tagged EOM status bit is loaded into the FIFO (BISYNC or BOP) regardless of RxFIFO full condition.
   c. when the counter/timer is programmed to count received characters and the character which causes it to reach zero count is loaded into the FIFO (ICTSR [6]).
   d. when the beginning of a break is detected in ASYNC mode regardless of the RxFIFO full condition.

If it is not reset by the CPU, RxRDY remains asserted until the FIFO becomes empty, at which time it is automatically negated. If it is reset by the CPU, it will remain negated regardless of the current state of the receive FIFO, until it is asserted again due to one of the above conditions.

The assertion of RxRDY causes an interrupt to be generated if IER[4] and the channel's master interrupt enable (ICR[0] or ICR[1]) are asserted.

When DMA operation is programmed, the RxRDY status bit is routed to the DMA control circuitry for use as the channel receiver DMA request. Assertion of RxRDY results in assertion of RTxDRQN output.

Several status bits are appended to each character in the RxFIFO. When the FIFO is read, causing it to be 'popped', the status bits associated with the new character at the top of the RxFIFO are logically ORed into the RSR. Therefore, the user should read R SR before reading the RxFIFO in response to RxRDY activation. If character-by-character status is desired, the RSR should be read and cleared each time a new character is received. The user may elect to accumulate status over several characters or over a frame by clearing RSR at appropriate times. This mode would normally also be used when operating in DMA mode. If the RxFIFO is empty when a read is attempted, and wait mode as specified in CMR2[5:3], is not being used, a H'FF' is output on the data bus.

In all modes, the DUSCC protects the contents of the FIFO and the RxSR from overrun. If a character is received while the FIFO is full and a character is already in the RxSR waiting to be transferred into the FIFO, the overrunning character is discarded and the OVERRUN status bit (RSR[5]) is asserted. If the overrunning character is an end-of-message character, the character is lost but the FIFOed EOM status bit will be asserted when the character in the RxSR is loaded into the FIFO.

Operation of the receiver is controlled by the enable receiver command. When this command is issued, the DUSCC goes into the search for start bit state (ASYNC), search for SYN state (COP modes), or search for FLAG state (BOP modes). When the disable receiver command is issued, the receiver ceases operation immediately. The RxFIFO is cleared on master reset, or by a reset receiver command. However, disabling the receiver does not affect the RxFIFO, RxRDY, or DMA request operation.

Receiver DCD and RTS Controls

If DCD enable Rx, RPR[2], is asserted, the DCD input must be asserted for the receiver to operate. If RPR[2] is asserted and the sampling circuit detects that the DCD input has been negated, the receiver ceases operation immediately. Operation resumes when the sampled DCD is asserted again. A change of state detector is provided on the DCD input of each channel. The required duration of the DCD level change is described in the discussion of ICTSR[5]. The user may program a change of state to cause an interrupt to be generated (master interrupt enable ICR[0] or [1] and IER [7] must be set) so that appropriate action can be taken.

December 1986

Preliminary Specification

Dual Universal Serial Communications Controller (DUSCC)   SCN68562

In ASYNC mode, RPR[4] can be programmed to control the deactivation of the RTS_N output by the receiver. RTS_N can be manually asserted and negated by writing to OMR_[0]. However, the assertion of RPR[4] causes RTS to be negated automatically upon receipt of a valid start bit if the channel's receive FIFO is already full. When this occurs, the RTSN negated status bit, RSR[6], is set. This may be used as a flow control feature to prevent overrun in the receiver by using the RTS_N output signal to control the CTS_N input of the remote transmitter. The new character will be assembled in the RxSR, but its transfer to the FIFO will be delayed until the CPU reads the FIFO, making the FIFO position available for the new character.

Once enabled, receiver operation depends on channel protocol mode. The following describes the receiver operation for the various protocols:

RxASYNC Mode

When first enabled, the receiver goes into the search for start bit state, looking for a high-to-low (mark-to-space) transition of the start bit on the RxD input. If a transition is detected, the state of the RxD pin is sampled again each 16× clock for 7½ clocks (16× clock mode) or at the next rising edge of the bit time clock (1× clock mode). If RxD is sampled high, the start bit is invalid and the search for a valid start bit begins again.

If RxD is still low, a valid start bit is assumed and the receiver continues to sample the input at one bit time intervals (16 periods of the 16× Rx clock; one period of the 1× Rx clock) at the theoretical center of the bit, until the proper number of data bits and the parity bit (if specified) have been assembled, and the first stop bit has been detected.

The assembled character is then transferred to the RxFIFO with appended parity error (if parity is specified) and framing error status bits. The DUSCC can be programmed to compare this character to the contents of S1R. The appended character compare status bit, RSR[7], is set if the data matches and there is no parity error.

After the stop bit is sampled, the receiver will immediately look for the next start bit. However, if a non-zero character was received without a stop bit (i.e. framing error) and RxD remains low for one-half of the bit period after the stop bit was sampled, then the receiver operates as if a new start bit transition had been detected at that point (one-half bit time after the stop bit was sampled).

If a break condition is detected (RxD low for entire character time including optional parity and first stop bit), only one character consisting of all zeros will be loaded into the RxFIFO and break start detect, RSR[2], will be set.

The RxD input must return to a high condition for at least one half of a bit time (16× clock mode) or for one bit time (1× clock mode) before the break condition is terminated and the search for the next start bit begins. At that time, the break end detect condition, RSR[3], is set.

Rx COP Modes

When the receiver is enabled in COP modes, it first goes into the SYN hunt phase, testing the received data each bit time for receipt of the appropriate SYN bit pattern, plus parity if specified, to establish character boundaries. Receipt of the SYN bit pattern terminates hunt phase and places the receiver in the data phase, in which all leading SYNs are stripped and the RxFIFO begins to load starting with the first non-SYN character. In COP single SYN protocol mode, S1R contains the SYN character required to establish character synchronization. In COP dual SYN and BISYNC protocol modes, S1R and S2R contain the first and second SYN characters, respectively, required to establish character synchronization. The SYN character length is the same as the character length programmed in RPR[1:0], plus the parity bit if parity is specified. SYN characters received with a parity error, when parity is specified, are considered invalid and will not cause synchronization to be achieved.

If external synchronization is programmed (RPR[4] = 1), the internal SYN detection and special character recognition logic are disabled and receipt of SYN characters is not required. A pulse on the SYNI input pin will establish character synchronization and terminate hunt phase. The SYNI pin is ignored after the first input on the SYNIN pin is received. The receiver must be disabled and then reenabled to resynchronize or to return to normal mode. This must be programmed in conjunction with CMR1[2:0] = 110. Refer to the description of RPR[4] for further details.

The SYN detect status bit, RSR[2], is set whenever SYN1, SYN1-SYN2, or DLE-SYN1 is detected for single SYN, dual SYN/BI-SYNC normal, and BISYNC transparent modes, respectively, and the SYNOUT pin will go active for one receive clock period one bit time after SYN detection in HSRH/HSRL. After character sync has been attained, the receiver enters the data phase and assembles characters in the RxSR, beginning with the first non-SYN character, with the least significant bit received first. It computes the BCC if specified, checks parity if specified, and checks for overrun errors.

The operation of the BCC (CRC/LRC) logic depends on the particular COP mode in use. The BCC is initialized upon first entering the data phase. For non-BISYNC modes, all received characters after entering data phase are included in the BCC computation, except for leading SYNs and SYNs which are specified to be stripped by RPR[7]. As each received character is transferred from the RxSR to the FIFO, the current value of the BCC characters is checked and the CRC ERROR status bit (RSR[1]) is set if the value of the CRC remainder is not the expected value. The EOM status bit, RSR[7], is not set since there is no defined end-of-message character. The receiver computes the BCC for text messages automatically when operating in BISYNC protocol mode.

BISYNC Features

The DUSCC provides support for both BI-SYNC normal and transparent operations. The following summarizes the features provided. Both EBCDIC and ASCII text messages can be handled by the DUSCC as selected by CMR1[5]. The receiver has the capability of recognizing special characters for the BISYNC protocol mode (see Table 15).

All sequences in Table 15, except SOH and STx, when detected explicitly cause a status to be affected. The following describes the conditions when this occurs.

The first character received when entering data phase for a header or text message should be an SOH, an STx, or a DLE-STx two-character sequence. Receipt of any of these initializes the CRC generator and starts the CRC accumulation. The SOH places the receiver in header mode, receipt of the STx places it in text mode, and receipt of the DLE-STx sequence (at any time) automatically places the receiver in transparent mode and sets the XPNT mode status bit, TRSR[0]. There is no explicit status associated with SOH and STx. If any other characters are received when entering the data phase, the message is treated as a control message and will not be accumulated in CRC.

After the data phase is established, the receiver searches the data stream for an end of message control character(s):

Header field: ENQ, ETB, or ITB

Normal text field: ENQ, ETx, ETB, or ITB

Transparent text field: DLE-ENQ, DLE-ETx, DLE-ETB, or DLE-ITB

Control message field: EOT, NAK, ACK0, ACK1, WACK, RVI or TTD

Detection of any one of these sequences causes the EOM status bit, RSR[7], to be set. Also if RPR[5] is set and the receiver does not detect a closing PAD (four 1's) after the 'EOT' or 'NAK', the PAD error status bit, RSR[6], is set. When the abort sequence ENQ or DLE-ENQ is detected, the character is tagged with an EOM status and transferred to the FIFO, but the appended CRC error

December 1986   37

Dual Universal Serial Communications Controller (DUSCC) SCN68562

Preliminary Specification

status bit should be ignored. For the other EOM control sequences, the receiver waits for the next two bytes (the CRC bytes) to be received, checks the value of the CRC generator, and tags the transferred character with a CRC error, RSR[1], if the CRC remainder is not correct. See Figure 11 for an example of BCC accumulation in various BISYNC messages.

The CRC bytes are normally not transferred to the FIFO, unless the transfer FCS to FIFO control bit, RPR[6], is asserted. In this case the EOM and CRC error status bits will be tagged onto the last byte of the last FCS byte instead of to the last character of the message. After detecting one of the end-of-message (EOM) character sequences and setting RSR[7], the receiver automatically goes into auto hunt mode for the SYNC characters and PAD check if RRP[5] is set.

SYN Pattern Stripping

Leading SYNs (before a message) are always stripped and excluded from the FCS, but SYN patterns within a message are treated by the receiver according to the RPR[7] bit. SYN character patterns are defined for the various COP modes as follows:

COP single SYN mode — SYN1

COP dual SYN mode — SYN1, and SYN2 when immediately preceded by SYN1.

BISYNC normal mode — SYN1, and SYN2 when immediately preceded by SYN1. SYN1 is always stripped, even if it is not followed by SYN2 when stripping is selected.

BISYNC transparent mode — DLE - SYN1, where the DLE is the last of an odd number of consecutive DLEs.

0   Strip only RPR[7] leading the SYN and do not accumulate in FCS.

1   Strip all SYNs. Additionally, strip odd DLEs when operating in BISYNC transparent mode. Do not accumulate stripped characters in FCS.

Processing of the SYN patterns is determined by the RPR[7] bit, the COP mode, and the position of the pattern in the frame. This is summarized in Table 16.

The value of the RPR[7] field does not affect the setting of the SYN DETECT status bit, RSR[2], and the generation of a SYNOUT pulse when a SYN pattern is received.

RxBOP Mode

In BOP protocol mode, the receiver may be in any one of four phases: hunt phase, address field (A) phase, control field (C) phase, or information field (I) phase. The character length for the A and C phases is always 8 bits. The I field character length is specified in RPR[1:0].

DESIGN NOTE: If the residual character length is not zero, the unused most significant bits in the receiver FIFO are not necessarily zero. The unused bits should be ignored, this will not cause a CRC error.

After an enable receiver command is executed, the receiver enters hunt phase, in which a comparison for the string (01111110) is done every Rx bit time. The FLAG delineates the beginning (and end) of a received frame and establishes the character boundary. Each FLAG match in CCSR causes the FLAG detect status bit (RSR[2]) to be set and SYNOUT N pin to be activated one bit time later for one receive clock period. FLAGs with an overlapping zero will be detected. All FLAGs are deleted from the data stream.

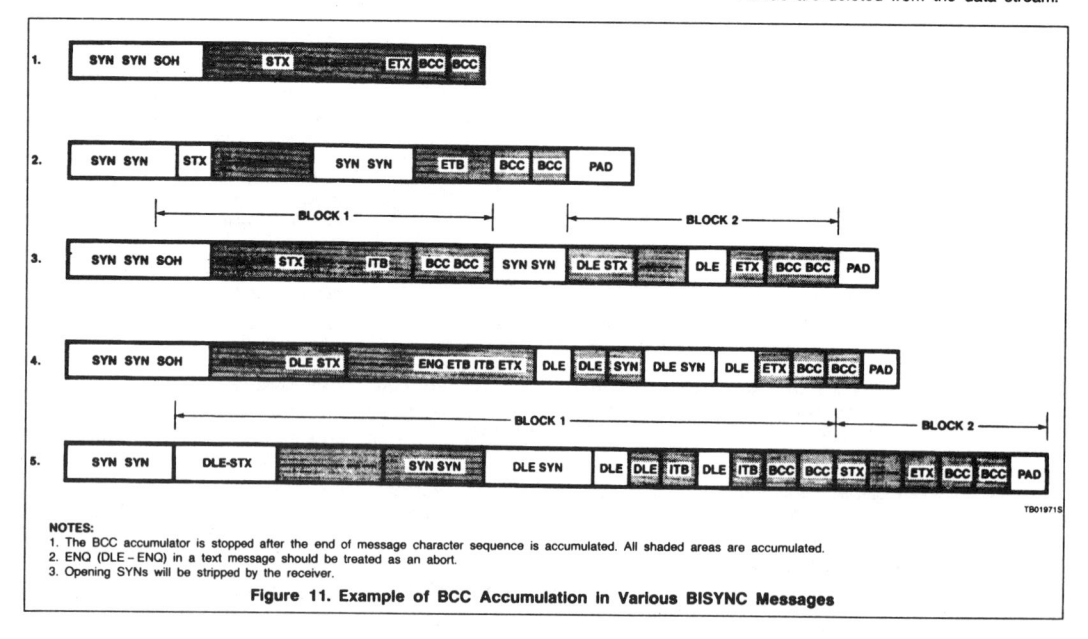

NOTES:
1. The BCC accumulator is stopped after the end of message character sequence is accumulated. All shaded areas are accumulated.
2. ENQ (DLE – ENQ) in a text message should be treated as an abort.
3. Opening SYNs will be stripped by the receiver.

Figure 11. Example of BCC Accumulation in Various BISYNC Messages

December 1986

Preliminary Specification

Dual Universal Serial Communications Controller (DUSCC)   SCN68562

Table 15. BISYNC Features

BISYNC — Single-Character Sequences

Sequence	ASCII	EBCDIC	Description
SOH	H'01	H'01	Start of header
STx	H'02	H'02	Start of text
ETx	H'83	H'03	End of text
EOT	H'04	H'37	End of transmission
ENQ	H'85	H'2D	Enquiry
DLE	H'10	H'10	Data link escape
NAK	H'15	H'3D	Negative ack
ETB	H'97	H'26	End of transmission block
ITB	H'1F	H'1F	End of intermediate transmission block

BISYNC — Two-Character Sequences

Sequence	ASCII	EBCDIC	Description
ACK0	H'10,B0	H'10,70	Acknowledge 0
ACK1	H'10,31	H'10,61	Acknowledge 1
WACK	H'10,3B	H'10,6B	Wait before transmit positive ack
RVI	H'10,BC	H'10,7C	Reverse interrupt
TTD	H'02,85	H'02,2D	Temporary text delay

BISYNC — (Transparent Text Mode) — Two-Character Sequences

Sequence	ASCII	EBCDIC	Description
DLE-ENQ	H'10,85	H'10,2D	Enquiry
DLE-ITB	H'10,1F	H'10,1F	End of intermediate transmission block
DLE-ETB	H'10,97	H'10,26	End of transmission block
DLE-ETx	H'10,83	H'10,03	End of text
DLE-STx	H'10,02	H'10,02	Start of transparent text mode

Table 16. SYN Pattern Processing

MODE	RPR [7]	LEADING SYNs	WITHIN A MESSAGE
BISYNC	0	no FCS no FIFO	no FCS Pattern into FIFO
	1	no FCS no FIFO	no FCS no FIFO
COP	0	no FCS no FIFO	Accumulate in FCS Pattern into FIFO
	1	no FCS no FIFO	no FCS no FIFO

Once a FLAG has been detected, the receiver will exit hunt phase and enter address phase. The handling of the address field is determined by the values programmed in CMR1[2:0], which selects one of the BOP modes. The BOP secondary address modes are selected by CMR1[4:3] and function as in the description that follows.

Single-Octet Address
For receive, the address comparison for a secondary station is made on the first octet following the opening FLAG. A match occurs if the first octet after the FLAG matches the contents of S1R, or if all parties address (RPR[3]) is asserted and the first octet is equal to H'FF'.

Dual-Octet Address
For receive, the address comparison for a secondary station is made on the first two octets following the opening FLAG.

A match occurs if the first two octets after the FLAG match the contents of S1R and S2R respectively, or if all parties address (RPR[3]) is asserted and the first two octets are equal to H'FF, FF'.

Dual Address with Group Mode
For receive, the address comparison for a secondary station is made on the first two octets following the opening FLAG. A match occurs for one of three possible conditions. If the first two octets after the FLAG match the contents of S1R and S2R, respectively, or if the first octet is H'FF' and the second matches the contents of S2R (group mode), or when all parties address (RPR[3]) is asserted and the first two octets are equal to H'FF, FF'. The second condition (group mode) allows a selected group of stations to receive a message.

Extended Address Mode
Extend address field to the next octet if the LSB of the current address octet is zero. Address field is terminated if the LSB of the address is a one. The address field will be terminated after the first octet if the null address H'00' is received/transmitted as the first address octet. For this mode the receiver does not perform an address comparison (all received characters after the opening FLAG are transferred to the FIFO) but does determine when the address field is terminated.

The length of the A field may be a single octet, a dual octet, or more octets, as described above. A primary station or an extended address secondary station does not perform an address comparison, and all characters in the A, C, and I fields after the flag are transferred to the FIFO. Although address field comparisons are not performed, the length of the address field is still determined by CMR1[4:3]. For the other secondary address modes, if there is a match, or the received character(s) match either of the other enabling conditions (group or all-parties address), all characters in the A, C, and I fields are transferred to the FIFO. If there is no match, the receiver returns to the FLAG hunt phase.

C phase begins after A phase is terminated. The receiver receives one or two control characters, CMR1[5]. After this phase is terminated, the character length is switched automatically from 8 bits to the number of bits specified in RPR[1:0] and the information field phase is entered.

The frame is terminated when a closing FLAG is detected. The same FLAG can also serve as the opening FLAG of the next frame. The 16 bits received prior to the closing FLAG form the frame check sequence (if an FCS is specified in CMR2[2:0]). All non-FLAG characters of the frame are accumulated in the CRC checker and the result is compared to the expected remainder. Failure to match will cause a CRC error. EOM detect RSR[7], RCL not zero RSR[0], and CRC error RSR[1] are normally FIFOed with the last character of the I field. RCL not zero RSR[0] is set if the length of the last character of the I field does not have the length programmed in RPR[1:0]. The residual character length in TRSR[2:0] is also valid at that time. The CRC characters themselves are normally not passed to the RxFIFO. However, if the transfer FCS to FIFO control bit RPR[6] is asserted, the FCS bytes will be transferred to the FIFO. In this case the EOM, CRC error, and RCL not zero status bits will be tagged onto the last byte of the CRC sequence instead of to the last character of the message.

If the closing FLAG is received prior to receipt of the appropriate number of A field, C field

December 1986   39

Dual Universal Serial Communications Controller (DUSCC)   SCN68562

as programmed in CMR1[5:3], and FCS field octets, a short frame will be detected and RSR[4] will be set. The I field need not be present in a valid frame. An abort (11111110) comparison is done after an opening FLAG has been received and up to receipt of the closing FLAG. A match causes the abort detect status bit (RSR[6]) to be set. The receiver then enters FLAG search mode. The abort is stripped from the received data stream.

If a zero followed by 15 contiguous ones is detected, the idle detect status bit RSR[3] is set. This comparison is done whenever the receiver is enabled. Therefore, it can occur before or after a received frame.

Zero deletion is performed during BOP receive. A zero after 5 contiguous ones is deleted from the data stream regardless of character boundaries. Deleted zeroes are not subject to CRC accumulation. FLAG, ABORT, and IDLE comparisons are done prior to zero deletion.

If external synchronization is programmed (RPR[4] = 1), the internal FLAG detection and address comparison logic is disabled and receipt of FLAGs is not required. In this arrangement, a pulse on the SYNI-N input pin will establish synchronization and terminate hunt phase. The receiver will then go immediately into the I-field mode with zero deletion disabled, assembling and transferring characters into the FIFO with the character length specified in RPR[1:0]. The SYNI-N pin is ignored after the first input on the SYNI-N pin is received. The receiver must be disabled and then reenabled to resynchronize or to return to normal operating mode.

This mode must be programmed in conjunction with CMR1[2:0] = 110. Refer to the description of RPR[4] for further details.

BOP Loop Mode

Operation of the receiver in BOP loop protocol mode is similar to operation in other BOP modes, except that only certain frame formats are supported. Several character detection functions that interact with the operation of the transmitter or transmitter commands are added:

1. When the 'go on-loop' command is invoked, the receiver looks for the receipt of a zero followed by seven ones and then asserts the LC_N output.
2. When the 'go off-loop' command is invoked, the receiver looks for the receipt of eight contiguous ones and then negates the LC_N output.
3. The TxD output normally echoes the receive input with a two bit time delay. When the 'go active on poll' command is asserted, the receiver looks for an EOP (a zero followed by seven ones) and then switches the TxD output line to the normal transmitter output. Receipt of an EOP or an ABORT sets RSR[6].
4. Receipt of a turnaround sequence (eight contiguous zeros) terminates the transmitter operation, if any, and returns the TxD output to echoing the RxD input. RSR[3] is set if a turnaround is received.

See transmitter operation for additional details.

SUMMARY OF COP FEATURES

COP Dual SYN Mode	
SYN detect	SYN1-SYN2
Linefill	SYN1-SYN2
SYN stripping	SYN1-SYN2 used to establish character sync, i.e. leading SYNs. Subsequent to this (after receiving first non-SYN character), SYN1 and SYN1-SYN2 if stripping is specified by RPR[7].
Excluded from FCS**	SYN1 and SYN1-SYN2 before beginning of message, i.e. leading SYNs and, if SYN stripping is specified by RPR[7] anywhere else in the message for the Rx; linefill SYN1-SYN2 for Tx regardless of RPR[7]. (If SYN stripping is not specified, then SYNs within a message will be included in FCS by Rx.)

BISYNC normal mode	
SYN detect	SYN1-SYN2
Linefill	SYN1-SYN2
SYN stripping	SYN1-SYN2 used to establish character sync, i.e. leading SYNs. Subsequent to this (after receiving first non-SYN character), SYN1 and SYN1-SYN2 if stripping is specified by RPR[7].
Excluded from FCS	All SYNs either before or within a message, regardless of RPR[7], plus additional characters as required by the protocol.

BISYNC transparent mode	
SYN detect	*DLE-SYN1
Linefill	DLE-SYN1
SYN/DLE stripping	*DLE-SYN1 and odd DLEs if stripping is specified by RPR[7].
Excluded from FCS	*DLE-SYN1 and odd DLEs, regardless of RPR[7] plus additional characters as required by the protocol.

COP single SYN mode	
SYN detect	SYN1
Linefill	SYN1
SYN stripping	SYN1 used to establish character sync, i.e. leading SYNs. Subsequent to this, SYN1 if stripping is specified by RPR[7].
Excluded from FCS**	SYN1 before beginning of message, i.e. leading SYNs, and if SYN stripping is specified by RPR[7], anywhere else in the message for the Rx; linefill SYN1 for Tx regardless of RPR[7]. (If SYN stripping is not specified, then SYNs within a message will be included in FCS by Rx.)

NOTES:
*DLE indicates last DLE of an odd number of consecutive DLEs.
**In non-BISYNC COP modes (single or dual SYN case), if SYN stripping is off, i.e. RPR[7] = 0, then SYNs within a message will be included in FCS by receiver. Therefore, the remote DUSCC transmitter should be careful not to let the TxFIFO underrun since the linefill SYN characters are not accumulated in FCS by the transmitter regardless of RPR[7]. Letting the TxFIFO underrun will result in a CRC error in the receiver.

December 1986

Preliminary Specification

Dual Universal Serial Communications Controller (DUSCC) SCN68562

ABSOLUTE MAXIMUM RATINGS[1]

SYMBOL	PARAMETER	RATING	UNIT
T_A	Operating ambient temperature[2] range	0 to +70	°C
T_{STG}	Storage temperature range	−55 to +150	°C
	All voltages with respect to ground[3]	−0.5 to +6.0	V

DC ELECTRICAL CHARACTERISTICS T_A = 0°C to +70°C, V_{CC} = 5.0V ±5%[4, 5, 6]

SYMBOL	PARAMETER	TEST CONDITIONS	Min	Typ	Max	UNIT
V_{IL}	Input low voltage All except X1/CLK X1/CLK				0.8 .04	V V
V_{IH}	Input high voltage All except X1/CLK X1/CLK		2 2.4			V V
V_{OL}	Output low voltage except DONEN, IRQN DONEN, IRQN	I_{OL} = 2.4mA I_{OL} = 8.8mA			0.5 0.5	V V
V_{OH}	Output high voltage (except o.c. outputs)	I_{OH} = −400μA	2.4			V
I_{XIL}	X1/CLK low input current	V_{IN} = 0, X2 grounded V_{IN} = 0, X2 floated	−4.5 −4.5			mA mA
I_{XIH}	X1/CLK high input current	V_{IN} = V_{CC}, X2 grounded V_{IN} = V_{CC}, X2 floated			1 30	mA mA
I_{X2L}	X2 low input current	V_{IN} = 0, X1/CLK floated	−100			μA
I_{X2H}	X2 high input current	V_{IN} = V_{CC}, X1/CLK floated			100	μA
I_{IL}	Input leakage current	V_{IN} = 0 to V_{CC}	−10		10	μA
I_{LL}	Data bus 3-State leakage current	V_O = 0 to V_{CC}	−10		10	μA
I_{OC}	Open-collector output leakage current	V_O = 0 to V_{CC}	−10		10	μA
I_{CC}	Power supply current				200	mA

NOTES:
1. Stresses above those listed under Absolute Maximum Ratings may cause permanent damage to the device. This is a stress rating only and functional operation of the device at these or at any other conditions above those indicated in the operation section of this specification is not implied.
2. For operating at elevated temperatures, the device must be derated based on +150°C maximum junction temperature and thermal resistance of 35°C/W junction to ambient for ceramic DIP, 35°C/W for plastic DIP, and 41°C/W for PLCC.
3. This product includes circuitry specifically designed for the protection of its internal devices from damaging effects of excessive static charge. Nonetheless, it is suggested that conventional precautions be taken to avoid applying any voltages larger than the rated maxima.
4. Parameters are valid over specified temperature range.
5. All voltage measurements are referenced to ground (GND). For testing, all inputs except X1/CLK swing between 0.4V and 2.4V with a transition time of 20ns maximum. For X1/CLK, this swing is between 0.4V and 4.4V. All time measurements are referenced at input voltages of 0.8V and 2V and output voltages of 0.8V and 2.0V as appropriate.
6. Typical values are at +25°C, typical supply voltages, and typical processing parameters.
7. Test condition for outputs: C_L = 150pF, except interrupt outputs. Test conditions for interrupt outputs: C_L = 50pF, R_L = 2.7kΩ to V_{CC}.
8. This specification will impose maximum 68000 CPU CLK to 6MHz. Higher CPU CLK can be used if repeating bus reads are not performed.
9. Execution of the valid command (after it is latched) requires 3-4 falling edges of X1 (see Figure 14).
10. Tests for open drain outputs are intended to guarantee switching of the output transistor. Measurement of this response is referenced from the midpoint of the switching signal to a point 0.5V above V_{OL}. This point represents noise margin that assures true switching has occurred. Beyond this level, the effects of external circuitry and test environment are pronounced and can greatly affect the resultant measurement.

December 1986

Dual Universal Serial Communications Controller (DUSCC) SCN68562

Preliminary Specification

AC ELECTRICAL CHARACTERISTICS $T_A = 0°C$ to $+70°C$, $V_{CC} = 5V \pm 5\%$ [4, 5, 6, 7]

NO.	FIGURE	PARAMETER	Min	Typ	Max	UNIT
1	12	RESETN pulse width	3			µs
2	13, 15	A0–A6 setup time to CSN low	10			ns
3	13, 15	A0–A6 hold time from CSN high	0			ns
4	13, 15	RWN setup time to CSN low	0			ns
5	13, 15	RWN hold time to CSN high	0			ns
6	13, 15	CSN high pulse width[8]	160			ns
7	13, 15, 16	CSN or IACKN high from DTACKN low	30			ns
8	13, 16	Data valid from CSN low			300	ns
9	13	Data bus floating from CSN high			270	ns
10	15	Data hold time from CSN high[9, 10, 11]	0			ns
11	13, 16	DTACKN low from read data ready	0			ns
12	13, 15	DTACKN low from CSN low			480	ns
13	13, 15	DTACKN high from CSN high			235	ns
14	13, 15	DTACKN high impedance from CSN high			360	ns
15	16	DTACKN low from IACKN low			650	ns
16	17	Port input setup time to CSN low	20			ns
17	17	Port input hold time from CSN low	120			ns
18	17	Port output valid from DTACK low			560	ns
19	18	IRQN high from:[12] Read RxFIFO (RxRDY interrupt)			300	ns
		Write TxFIFO (TxRDY interrupt)			300	ns
		Write RSR (receiver condition interrupt)			650	ns
		Write TRSR (receiver/transmitter interrupt)			650	ns
		Write ICTSR (port change and timer/counter interrupt)			650	ns
20	19	X1/CLK high or low time	25			ns
		X1/CLK frequency	2	14.7456	16	MHz
		CTCLK high or low time	100			ns
		CTCLK frequency	0		4	MHz
		RxC high or low time	110			ns
		RxC frequency (16×)	0		4	MHz
		(1×)	0		4	MHz
		TxC high or low time	110			ns
		TxC frequency (16×)	0		4	MHz
		(1×)	0		4	MHz
21	20	TxD output from 1X TxC input low			385	ns
22	20	TxD output from TxC output low	0		50	ns
23	21	RxD data setup time to RxC high	50			ns
24	21	RxD data hold time from RxC high	0			ns
25	22	IACKN low to daisy chain low			350	ns
26	24	Data valid from receive DMA ACKN			320	ns
27	23, 24	DTCN width	100			ns
28	23, 24	RDYN low to DTCN low	80			ns
29	24	Data bus float from DTCN low			300	ns
30	23, 24	DMA ACKN low to RDYN (DTACKN) low			600	ns
31	23, 24	RDYN high from DTCN low			420	ns
32	23, 24	RDYN high impedance from DTCN low			530	ns
33	24	Receive DMA REQN high from DMA ACKN low			585	ns
34	24	Receive DMA ACKN width	150			ns
35	23, 24	Receive DMA ACKN low to DONEN low			300	ns
36	23	Data setup to DTCN low	300			ns
37	23	Data hold from DTCN low	230			ns
38	23	Transmit DMA REQN high from ACKN low			550	ns
39	23	Transmit DMA ACKN deasserted width	150			ns
40	23	Transmit DMA ACKN low to DONEN low output			550	ns
41	25	CSN low to transmit DONEN low output			400	ns
42	25	CSN low to transmit DMA REQ negated			620	ns
43	25	CSN low to receive DONEN low			400	ns
44	25	CSN low to receive DMA REQ negated			620	ns

December 1986

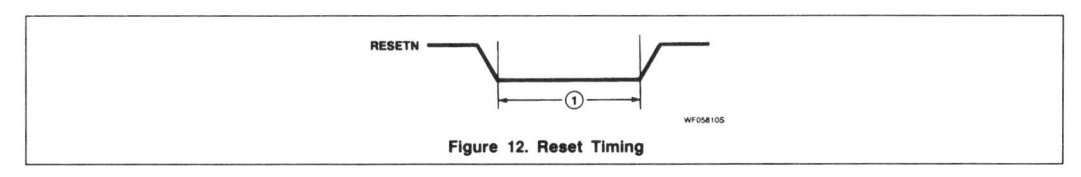

Preliminary Specification

Dual Universal Serial Communications Controller (DUSCC) SCN68562

Figure 12. Reset Timing

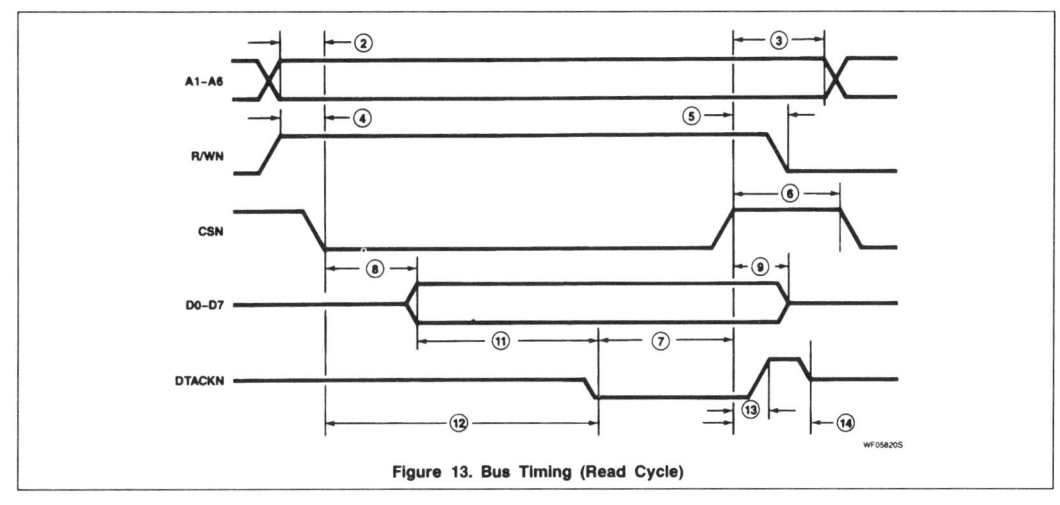

Figure 13. Bus Timing (Read Cycle)

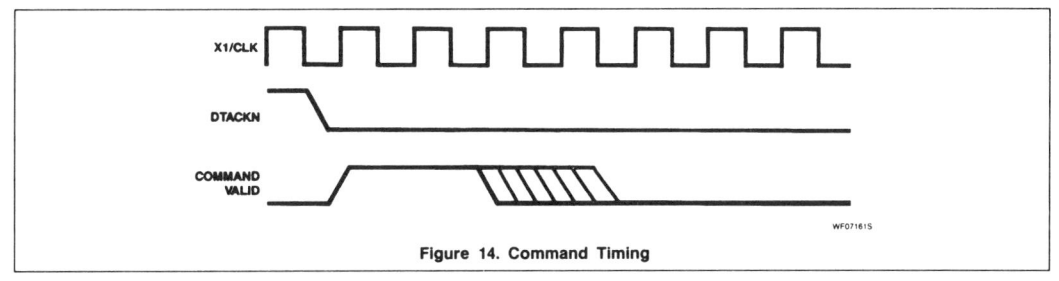

Figure 14. Command Timing

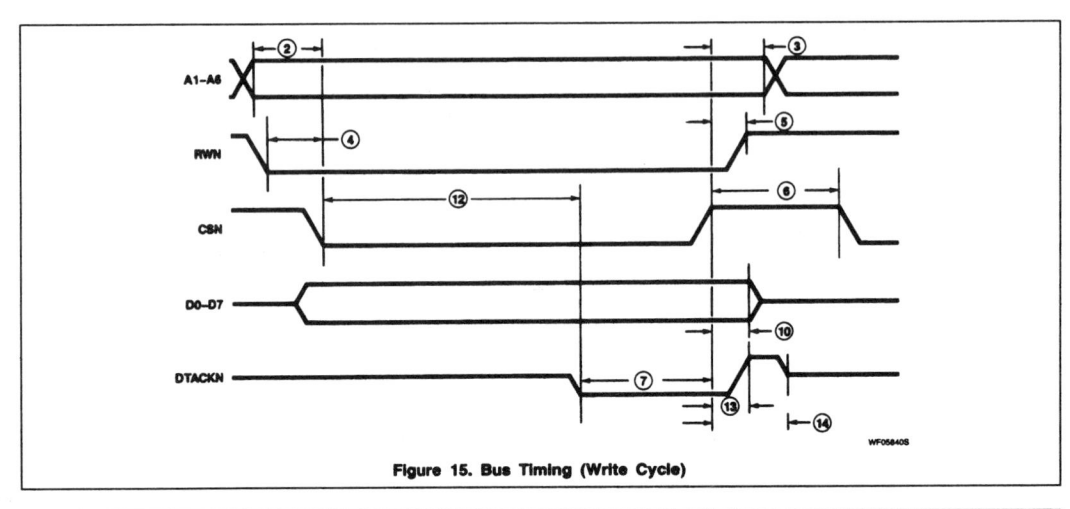

Dual Universal Serial Communications Controller (DUSCC)   SCN68562

Figure 15. Bus Timing (Write Cycle)

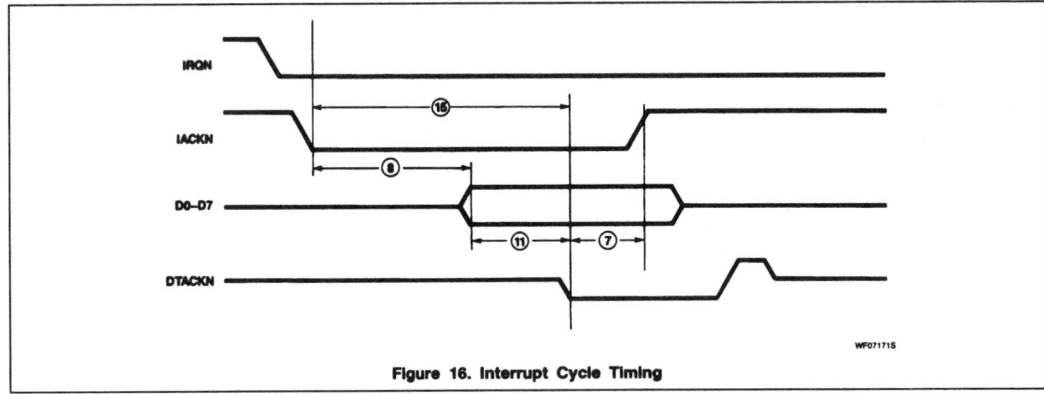

Figure 16. Interrupt Cycle Timing

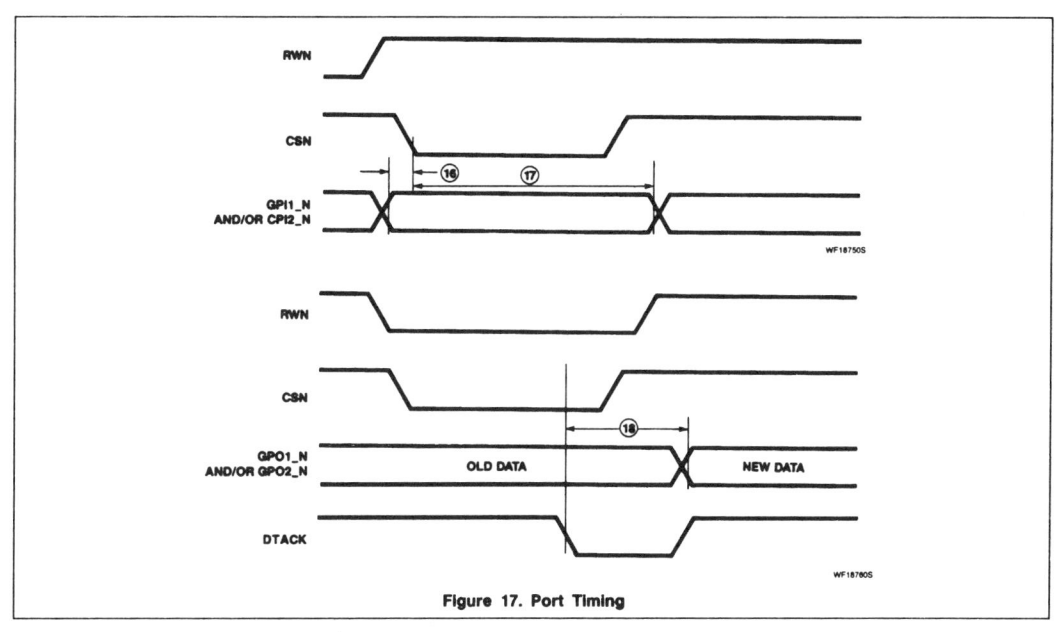

Dual Universal Serial Communications Controller (DUSCC)   SCN68562

Figure 17. Port Timing

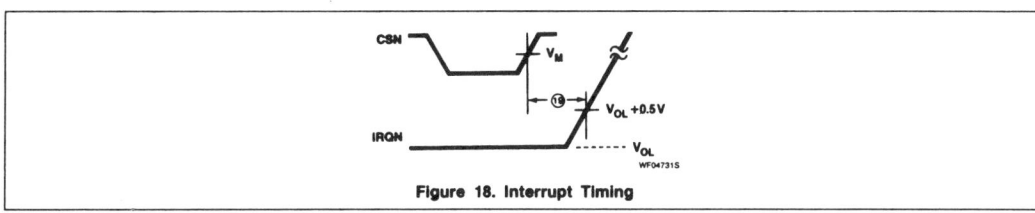

Figure 18. Interrupt Timing

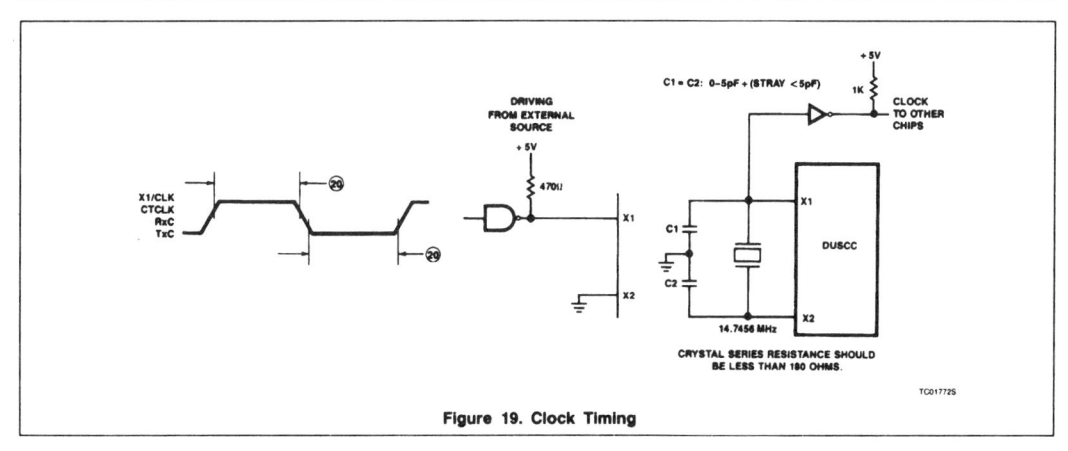

Figure 19. Clock Timing

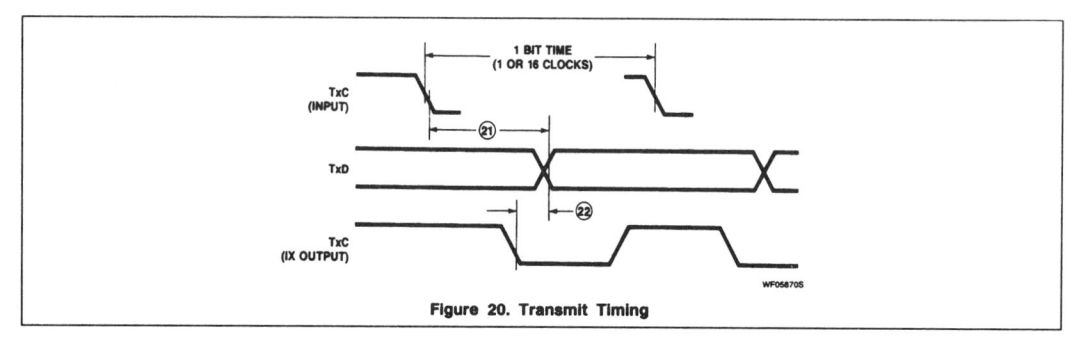

Dual Universal Serial Communications Controller (DUSCC)   SCN68562

Figure 20. Transmit Timing

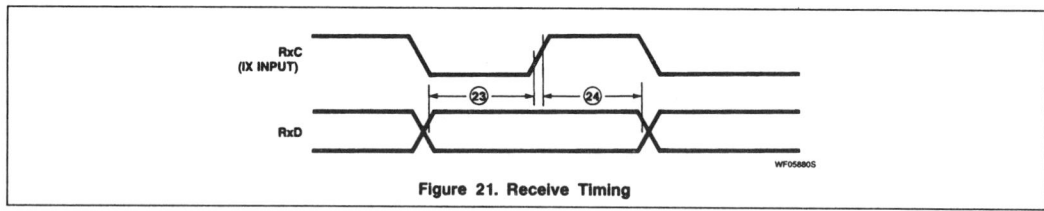

Figure 21. Receive Timing

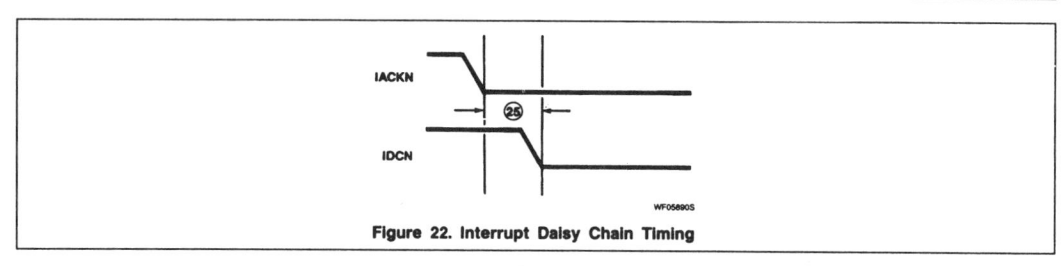

Figure 22. Interrupt Daisy Chain Timing

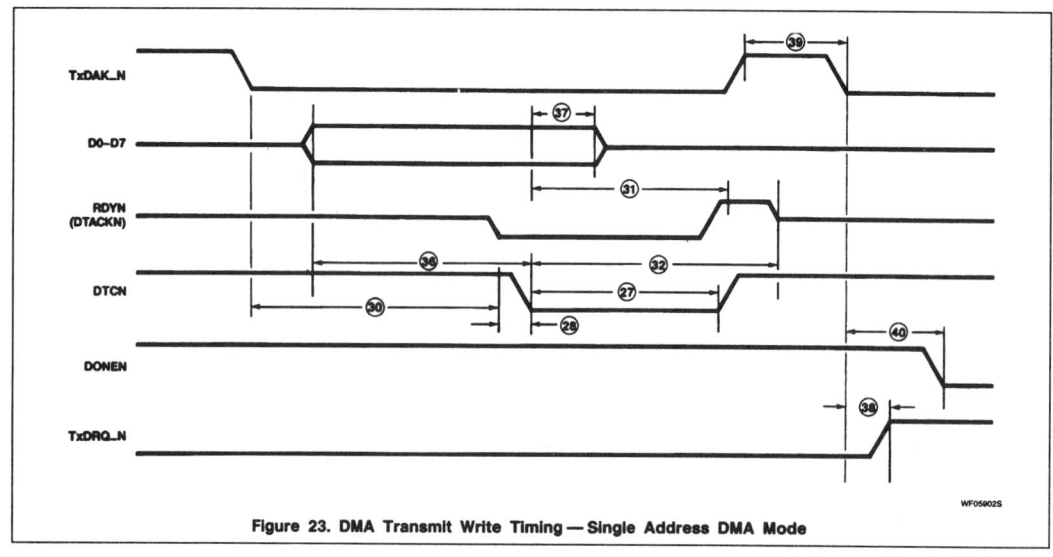

Figure 23. DMA Transmit Write Timing — Single Address DMA Mode

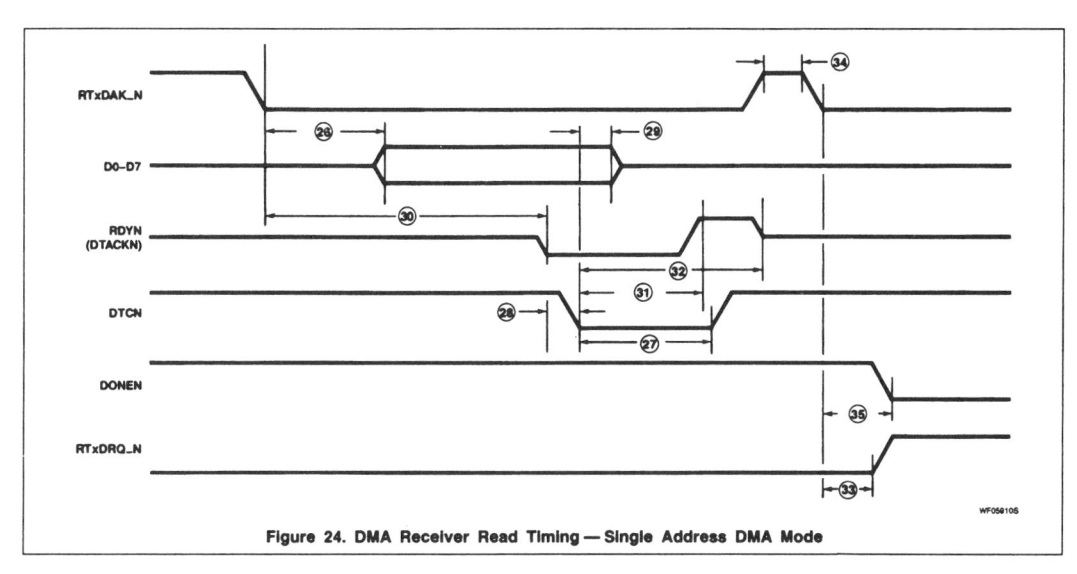

Dual Universal Serial Communications Controller (DUSCC)   SCN68562

Figure 24. DMA Receiver Read Timing — Single Address DMA Mode

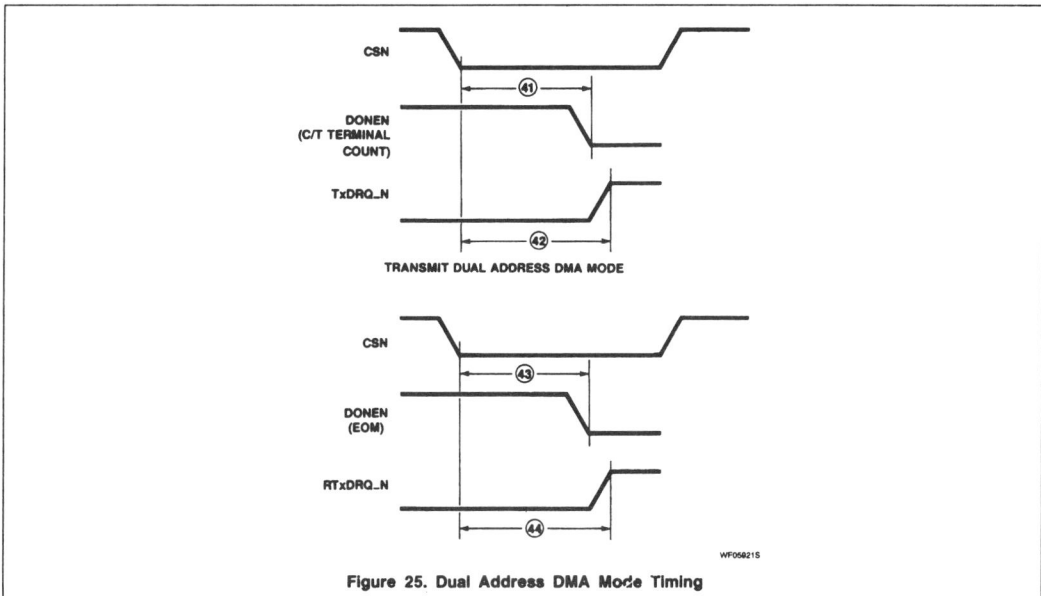

Figure 25. Dual Address DMA Mode Timing

DEFINITIONS

Data Sheet Identification	Product Status	Definition
Objective Specification	Formative or In Design	This data sheet contains the design target or goal specifications for product development. Specifications may change in any manner without notice.
Preliminary Specification	Preproduction Product	This data sheet contains preliminary data and supplementary data will be published at a later date. Signetics reserves the right to make changes at any time without notice in order to improve design and supply the best possible product.
Product Specification	Full Production	This data sheet contains Final Specifications. Signetics reserves the right to make changes at any time without notice in order to improve design and supply the best possible product.

Signetics reserves the right to make changes, without notice, in the products, including circuits, standard cells, and/or software, described or contained herein in order to improve design and/or performance. Signetics assumes no responsibility or liability for the use of any of these products, conveys no license or title under any patent, copyright, or mask work right to these products, and makes no representations or warranties that these products are free from patent, copyright, or mask work right infringement, unless otherwise specified. Applications that are described herein for any of these products are for illustrative purposes only. Signetics makes no representation or warranty that such applications will be suitable for the specified use without further testing or modification.

Mullard

Mullard Limited,
Mullard House, Torrington Place, London WC1E 7HD.
Tel: 01-580 6633 Telex: 264341

Mullard manufacture and market electronic components under their own name, and those of associated companies

Printed in England

M87-1365/Y

VLSI TECHNOLOGY, INC.

ADVANCE INFORMATION

VL82C106
PC/AT COMBO I/O CHIP

FEATURES
- Combines the following PC/AT® Peripheral Chips:
  - VL16C450 UART - COM1:
  - VL16C450 UART - COM2:
  - Parallel Printer Port - LPT1:
  - Keyboard/Mouse Ctrl. - KBD
  - Real Time Clock
- Serial ports fully 16C450 compatible
- Bidirectional line printer port
- Software control of PS/2®-compatible enhancements (LPT Port, Mouse)
- CMOS direct drive of Centronics-type parallel interface
- PC/AT- or PS/2-compatible keyboard and mouse controller
- 146818A-compatible Real Time Clock (RTC)
- 16 bytes of additional standby RAM (66 bytes total)
- IDE bus control signals included (two external 74LS245 and one 74ALS244 - or equivalent - buffers are required)
- Seven battery-backed programmable chip select registers for auto configuration
- Preprogrammed default chip selects
- Programmable wait state generation
- 5 µA standby current for RTC, RAM, and chip select registers
- Single 128-pin plastic quad flatpack

DESCRIPTION

The VL82C106 Combo chip replaces with a single 128-pin chip, several of the commonly used peripherals found in PC/AT-compatible computers. This chip when used with the VLSI PC/AT-compatible chip set allows designers to implement a very cost effective, minimum chip count motherboard containing functions that are common to virtually all PCs.

The on-chip UARTs are completely software compatible with the VL16C450 ACE.

The bidirectional parallel port provides a PS/2 software compatible interface between a Centronics-type printer and the VL82C106. Direct drive is provided so that all that is necessary to interface to the line printer port is a resistor - capacitor network. The bidirectional feature (option) is software programmable for backwards PC/AT-compatibility.

The keyboard/mouse controller is selectable as PC/AT- or PS/2-compatible.

The Real Time Clock is 146818A-compatible and offers a standby current drain of 5 µA at 3.0 V.

Included is the control logic necessary for the support of the Integrated Drive Electronics (IDE) hard disk bus interface.

The Combo I/O chip also includes seven programmable chip selects, three internal and four external. Each chip select has a programmable 16-bit base address and a mask register that allows the number of bytes corresponding to each chip select to be programmed (e.g. 3F8H-3FFH has a base address of 3F8H and a range of 8 bytes). Each chip select can be programmed for number of wait states (0-7) and 8- or 16-bit operation. 16-bit decoding is used for all I/O addresses. A default fixed decode is provided on reset for the on-chip serial ports, printer port, and off-chip floppy and hard disk controllers, which may be changed to battery-backed programmable chip selects via a control bit.

ORDER INFORMATION

Part Number	Package
VL82C106-FC	Plastic Quad Flatpack

Note: Operating temperature range is 0°C to +70°C.

INTERNAL FUNCTIONAL DIAGRAM

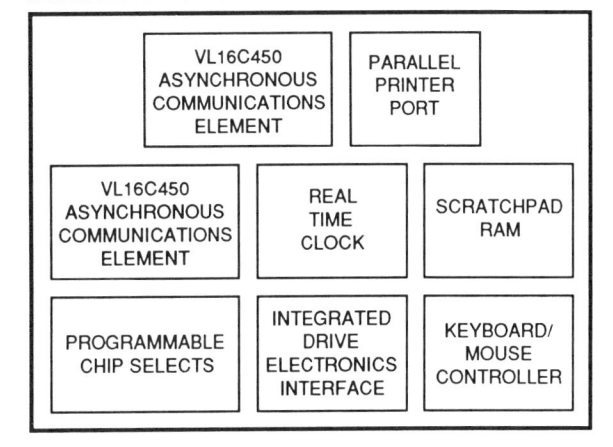

- VL16C450 ASYNCHRONOUS COMMUNICATIONS ELEMENT
- PARALLEL PRINTER PORT
- VL16C450 ASYNCHRONOUS COMMUNICATIONS ELEMENT
- REAL TIME CLOCK
- SCRATCHPAD RAM
- PROGRAMMABLE CHIP SELECTS
- INTEGRATED DRIVE ELECTRONICS INTERFACE
- KEYBOARD/ MOUSE CONTROLLER

PC/AT® and PS/2® is the registered trademark of IBM Corporation.

VLSI Technology, Inc. • 8375 South River Parkway • Tempe, AZ 85284 • 602-752-8574

… VLSI TECHNOLOGY, INC.

ADVANCE INFORMATION

VL82C106

BLOCK DIAGRAM

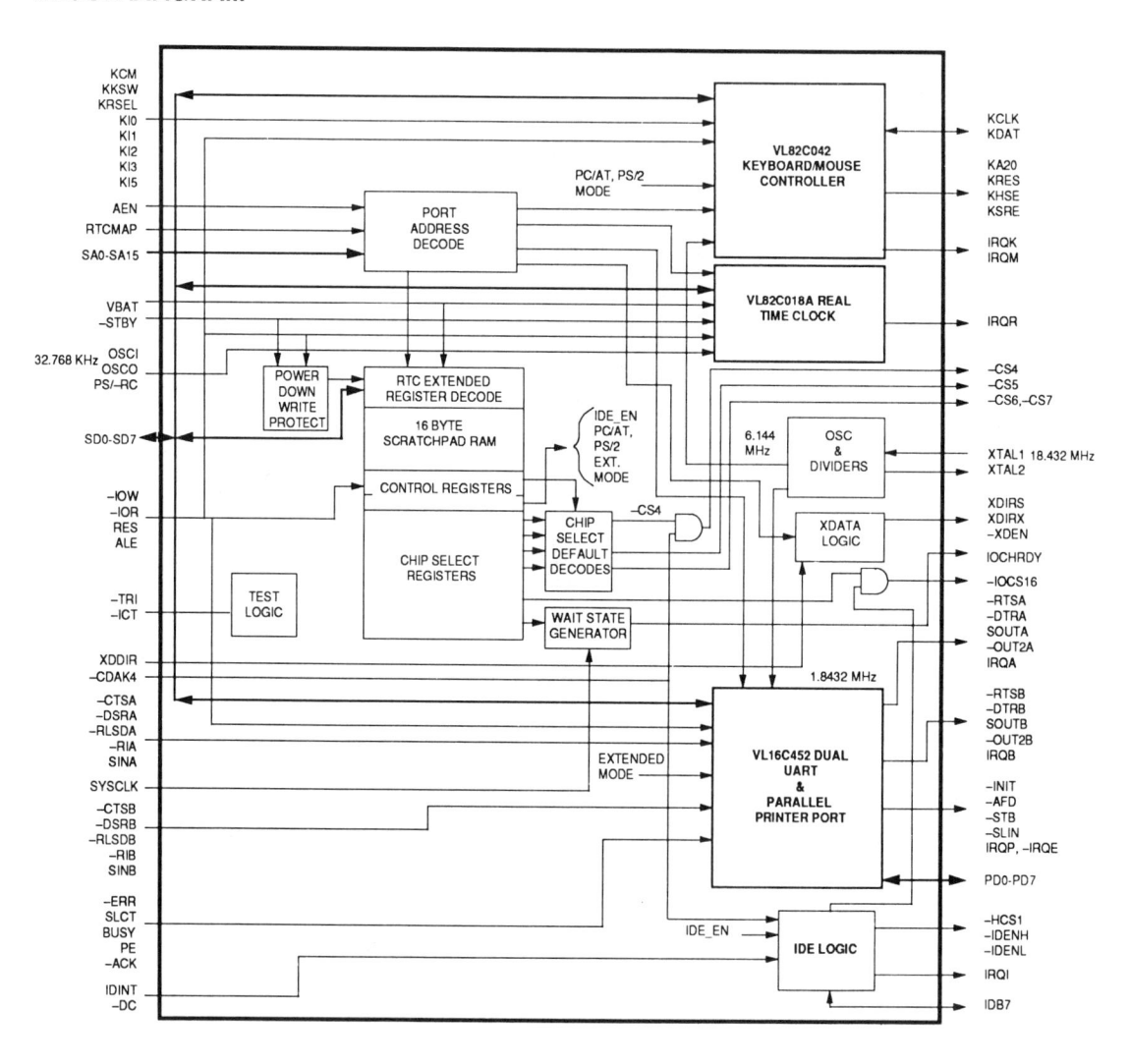

805

VLSI TECHNOLOGY, INC.

ADVANCE INFORMATION

VL82C106

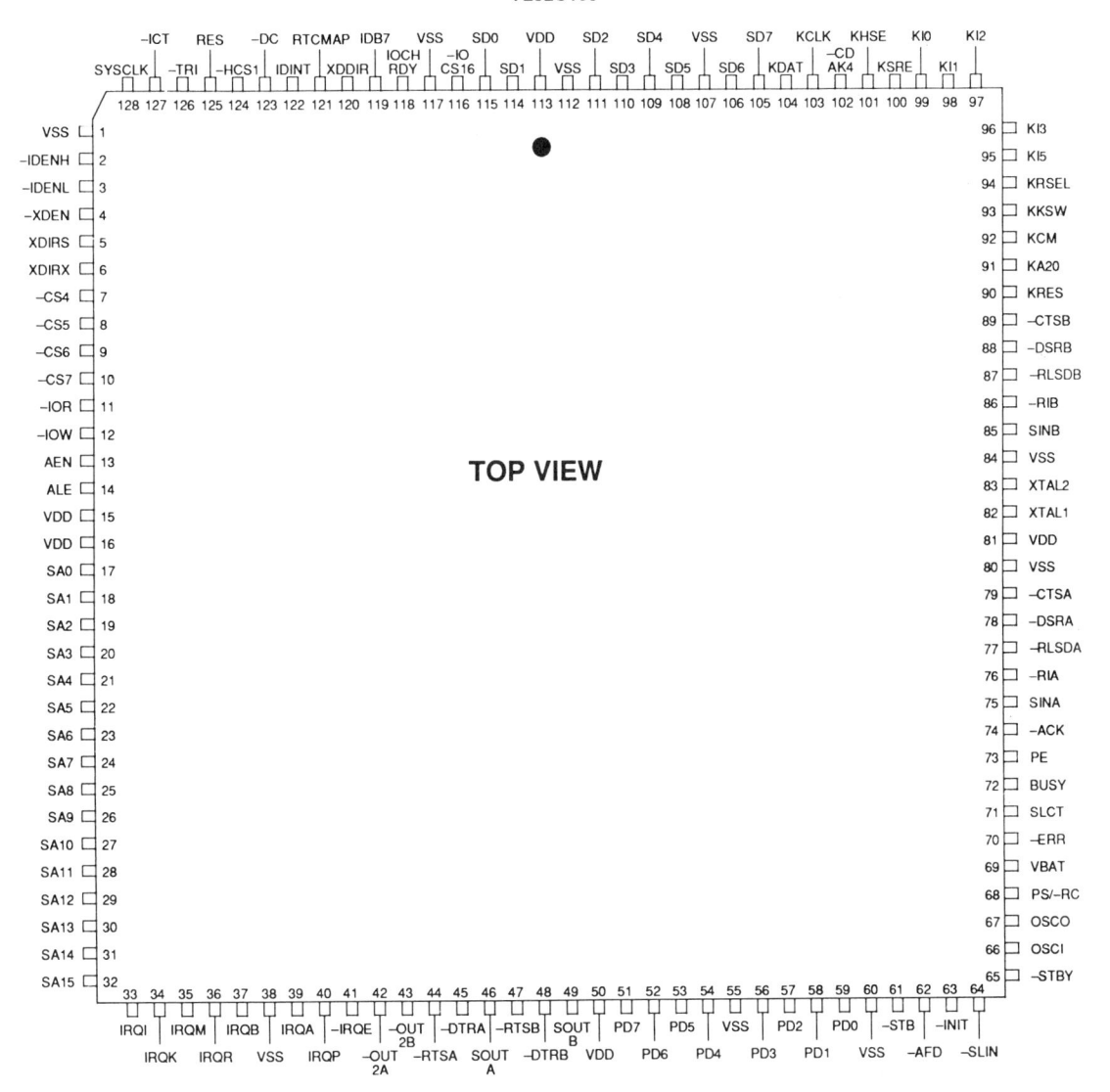

PIN DIAGRAM

VL82C106 — TOP VIEW

Top edge (pins 128–97, left to right):
SYSCLK (128), -ICT (127), -TRI (126), RES (125), -HCS1 (124), -DC (123), IDINT (122), RTCMAP (121), XDDIR (120), IDB7 (119), IOCH RDY (118), VSS (117), -IO CS16 (116), SD0 (115), SD1 (114), VDD (113), VSS (112), SD2 (111), SD3 (110), SD4 (109), SD5 (108), VSS (107), SD6 (106), SD7 (105), KDAT (104), KCLK (103), -CD AK4 (102), KHSE (101), KSRE (100), KI0 (99), KI1 (98), KI2 (97)

Left edge (pins 1–32, top to bottom):
1 VSS
2 -IDENH
3 -IDENL
4 -XDEN
5 XDIRS
6 XDIRX
7 -CS4
8 -CS5
9 -CS6
10 -CS7
11 -IOR
12 -IOW
13 AEN
14 ALE
15 VDD
16 VDD
17 SA0
18 SA1
19 SA2
20 SA3
21 SA4
22 SA5
23 SA6
24 SA7
25 SA8
26 SA9
27 SA10
28 SA11
29 SA12
30 SA13
31 SA14
32 SA15

Right edge (pins 96–65, top to bottom):
96 KI3
95 KI5
94 KRSEL
93 KKSW
92 KCM
91 KA20
90 KRES
89 -CTSB
88 -DSRB
87 -RLSDB
86 -RIB
85 SINB
84 VSS
83 XTAL2
82 XTAL1
81 VDD
80 VSS
79 -CTSA
78 -DSRA
77 -RLSDA
76 -RIA
75 SINA
74 -ACK
73 PE
72 BUSY
71 SLCT
70 -ERR
69 VBAT
68 PS/-RC
67 OSCO
66 OSCI
65 -STBY

Bottom edge (pins 33–64, left to right):
33 IRQI, 34 IRQK, 35 IRQM, 36 IRQR, 37 IRQB, 38 VSS, 39 IRQA, 40 IRQP, 41 -IRQE, 42 -OUT 2A, 43 -OUT 2B, 44 -RTSA, 45 -DTRA, 46 SOUT A, 47 -RTSB, 48 -DTRB, 49 SOUT B, 50 VDD, 51 PD7, 52 PD6, 53 PD5, 54 PD4, 55 VSS, 56 PD3, 57 PD2, 58 PD1, 59 PD0, 60 VSS, 61 -STB, 62 -AFD, 63 -INIT, 64 -SLIN

3

VLSI TECHNOLOGY, INC

ADVANCE INFORMATION

VL82C106

SIGNAL DESCRIPTIONS

Signal Name	Pin Number	Signal Type	Signal Description
COMMUNICATIONS PORT A			
–RTSA	44	O1	Request to Send, Port A
–DTRA	45	O1	Data Terminal Ready, Port A
SOUTA	46	O1	Serial Data Output, Port A
–CTSA	79	I4	Clear to Send, Port A
–DSRA	78	I4	Data Set Ready, Port A
–RLSDA	77	I4	Receive Line Signal Detect, Port A
–RIA	76	I4	Ring Indicator, Port A
SINA	75	I4	Serial Input, Port A
IRQA	39	O6	Interrupt Request, Port A
–OUT2A	42	O1	Output 2, Port A
COMMUNICATIONS PORT B			
–RTSB	47	O1	Request to Send, Port B
–DTRB	48	O1	Data Terminal Ready, Port B
SOUTB	49	O1	Serial Data Output, Port B
–CTSB	89	I4	Clear to Send, Port B
–DSRB	88	I4	Data Set Ready, Port B
–RLSDB	87	I4	Receive Line Signal Detect, Port B
–RIB	86	I4	Ring Indicator, Port B
SINB	85	I4	Serial Input, Port B
IRQB	37	O6	Interrupt Request, Port B
–OUT2B	43	O1	Output 2, Port B
PARALLEL PRINTER PORT			
PD0	59	I05	Printer Data Port, Bit 0
PD1	58	I05	Printer Data Port, Bit 1
PD2	57	I05	Printer Data Port, Bit 2
PD3	56	I05	Printer Data Port, Bit 3
PD4	54	I05	Printer Data Port, Bit 4
PD5	53	I05	Printer Data Port, Bit 5
PD6	52	I05	Printer Data Port, Bit 6
PD7	51	I05	Printer Data Port, Bit 7
–INIT	63	O4	Initialize Printer Signal
–AFD	62	O4	Autofeed Printer Signal
–STB	61	O4	Data Strobe to Printer
–SLIN	64	O4	Select Signal to Printer
–ERR	70	I4	Error Signal from Printer
SLCT	71	I4	Select Signal from Printer

VLSI TECHNOLOGY, INC.

ADVANCE INFORMATION

VL82C106

SIGNAL DESCRIPTIONS

Signal Name	Pin Number	Signal Type	Signal Description
BUSY	72	I4	Busy Signal from Printer
PE	73	I4	Paper Error Signal from Printer
−ACK	74	I4	Acknowledge Signal from Printer
IRQP	40	O6	Printer Interrupt Request Output
−IRQE	41	O1	Printer Interrupt Request Enabled Signal
REAL TIME CLOCK PORT			
VBAT	69	NA	Standby Power - Normally 3 V to 5 V, battery backed.
−STBY	65	I5	Power Down Control
OSCI	66	NA	Crystal Connection Input - 32 KHz
OSCO	67	NA	Crystal Connection Output - 32 KHz
PS/−RC	68	I5	Power Sense/RAM Clear Input
IRQR	36	O1	Real Time Clock Interrupt Request Output
RTCMAP	121	I4	High - RTC is mapped to 70H and 71H, Low - RTC is mapped to 170H and 171H.
KEYBOARD CONTROLLER PORT			
KCLK	103	IO4	Keyboard Clock
KDAT	104	IO4	Keyboard Data
KCM	92	I4	General purpose input, normally color/monochrome.
KKSW	93	I4	General purpose input, normally keyboard switch.
KA20	91	O1	General purpose output, normally A20 Gate.
KRES	90	O1	General purpose output, normally reset.
KHSE	101	O1/IO4	General purpose input, normally speed select.
KSRE	100	O1/IO4	General purpose output, normally shadow RAM enable.
IRQK	34	O1	Keyboard Interrupt Request
IRQM	35	O1	Mouse Interrupt Request
KRSEL	94	I4	General purpose input, normally RAM select.
KI0	99	I4	General purpose input, bit 0.
KI1	98	I4	General purpose input, bit 1.
KI2	97	I4	General purpose input, bit 2.
KI3	96	I4	General purpose input, bit 3.
KI5	95	I4	General purpose input, bit 5.
IDE BUS I/O			
−IDENH	2	O1	IDE Bus Transceiver High Byte Enable
−IDENL	3	O1	IDE Bus Transceiver Low Byte Enable
IDINT	122	I4	IDE Bus Interrupt Request Input
IDB7	119	IO6	IDE Bus Data Bit 7
−DC	123	I4	Floppy Disk Change Signal
−HCS1	124	O1	IDE Host Chip Select 1
−IRQI	33	O6	IDE Interrupt Request Output

VLSI TECHNOLOGY, INC.

ADVANCE INFORMATION

VL82C106

SIGNAL DESCRIPTIONS

Signal Name	Pin Number	Signal Type	Signal Description
COMMON BUS I/O			
SD0	115	IO2	System Bus Data, Bit 0
SD1	114	IO2	System Bus Data, Bit 1
SD2	111	IO2	System Bus Data, Bit 2
SD3	110	IO2	System Bus Data, Bit 3
SD4	109	IO2	System Bus Data, Bit 4
SD5	108	IO2	System Bus Data, Bit 5
SD6	106	IO2	System Bus Data, Bit 6
SD7	105	IO2	System Bus Data, Bit 7
SA0	17	I1	SystemBus Address, Bit 0
SA1	18	I1	System Bus Address, Bit 1
SA2	19	I1	System Bus Address, Bit 2
SA3	20	I1	System Bus Address, Bit 3
SA4	21	I1	System Bus Address, Bit 4
SA5	22	I1	System Bus Address, Bit 5
SA6	23	I1	System Bus Address, Bit 6
SA7	24	I1	System Bus Address, Bit 7
SA8	25	I1	System Bus Address, Bit 8
SA9	26	I1	System Bus Address, Bit 9
SA10	27	I1	System Bus Address, Bit 10
SA11	28	I1	System Bus Address, Bit 11
SA12	29	I1	System Bus Address, Bit 12
SA13	30	I1	System Bus Address, Bit 13
SA14	31	I1	System Bus Address, Bit 14
SA15	32	I1	System Bus Address, Bit 15
XTAL1	82	NA	Crystal/Clock Input - 18.432 MHz
XTAL2	83	NA	Cystal/Clock Output - 18.432 MHz
−IOR	11	I1	System Bus I/O Read
−IOW	12	I1	System Bus I/O Write
RES	125	I1	System Reset
AEN	13	I1	System Bus Address Enable
ALE	14	I1	System Bus Address Latch Enable
−IOCS16	116	O8	System Bus I/O Chip Select 16
IOCHRDY	118	O8	System Bus I/O Channel Ready
SYSCLK	128	I1	System Clock - Processor clock divide by 2.
−CS4	7	O1	Chip Select 4 - Normally for external floppy disk controller.
−CS5	8	O1	Chip Select 5 - Normally −HCS0 for IDE.

VLSI TECHNOLOGY, INC.

ADVANCE INFORMATION

VL82C106

SIGNAL DESCRIPTIONS

Signal Name	Pin Number	Signal Type	Signal Description
–CS6	9	O1	Chip Select 6 - Normally for external floppy disk controller.
–CS7	10	O1	Chip Select 7 - Normally for external floppy disk controller.
–CDAK4	102	I1	DMA Acknowledge forces –CS4 active.
XDDIR	120	I1	X Data Bus Transceiver Direction
XDIRS	5	O1	Modified X Data Bus Transceiver Direction Control Signal - Excludes real time clock and keyboard controller decodes.
XDIRX	6	O1	X Data Bus Transceiver Control Signal - Includes all CS decodes generated on chip.
–XDEN	4	O1	X Data Bus Transceiver Enable
–TRI	126	I4	Three-state Control Input - For all outputs to isolate chip for board tests.
–ICT	127	I4	In Circuit Test Mode Control
POWER, GROUND, & UNCOMMITTED			
VDD	15, 16, 50, 81, 113		System Power: +5 V
VSS	1, 38, 55, 60, 80, 84, 107, 112, 117		System Ground

I/O LEGEND

	mA	Type	Comment
O1	2	TTL	
O4	12	TTL-OD	Open drain, weak pull-up, no VDD diode
O6	4	TTL-TS	Three-State
O8	24	TTL-OD	Open drain, fast active pull-up
I1	–	TTL	
I2	–	CMOS	
I4	–	TTL	30k Ω pull-up
I5	–	TTL	Schmitt-trigger
IO2	24	TTL-TS	Three-State
IO4	12	TTL-OD	Open drain, slow turn-on
IO5	12	TTL-TS	Three-State
IO6	24	TTL-TS	Three-State, 30k Ω pull-up

VLSI TECHNOLOGY, INC.

ADVANCE INFORMATION

VL82C106

FUNCTIONAL DESCRIPTION

Below is a detailed explanation of each of the major building blocks of the VL82C106 Combo chip. The following functional blocks are covered:

- 16C450 Serial Ports
- Parallel Printer Port
- 146818A-Compatible Real Time Clock
- Keyboard Controller
- Control and Chip Selects
- IDE Interface

SERIAL COMMUNICATIONS PORTS

The chip contains two UARTs, based on the VL16C450 Megacell core. Each of these UARTs share a common baud-rate clock, which is the XTAL1 input (18.432 MHz) divided by ten. The 18.432 MHz signal is shared with the keyboard controller, which divides it by three to get an approximate 6 MHz reference clock. Please refer to the VL16C450 data sheet for the register descriptions for the UARTs.

COMA is accessed via internally generated CS1, while COMB uses internally generated CS2.

LINE PRINTER PORT

The Line Printer Port contains the functionality of the port included in the VL16C452, but offers a software programmable Extended Mode. This enhancement is the addition of a Direction Control Bit, and Interrupt Status Bit. These features are disabled on initial power-up, but may be turned on by clearing the –EMODE bit of Control Register 0 (RTC Register 69H in AT or PS/2 mode or I/O PORT 102H in PS/2 mode). When the –EMODE bit is set, the part functions exactly as a PC/AT-compatible printer port.

The Line Printer Port is accessed via internally generated programmable chip select CS3.

Register 0 - Line Printer Port Data

The Line Printer (LPT) Port is either uni- or bidirectional, depending on the state of the Extended Mode and Data Direction Control bits.

Compatibility Mode (–EMODE bit = 1) - Read operations differ according to the state of the EMODE Control bit. When in compatibility mode (–EMODE = 1), reads to this register return the last data that was written to the LPT Port. Write operations immediately output data to the LPT Port.

(–EMODE bit = 0) - Read operations return either the data last written to the LPT Data Register if the Direction Bit is set to output ("0") or the data that is present on the pins of the LPT Port if the direction is set to input ("1"). Write operations latch data into the output register, but only drive the LPT Port when the Direction Bit is set to output.

In either case, the bits of the LPT Data Register are defined as follows:

Bit	Description
0	Data Bit 0
1	Data Bit 1
2	Data Bit 2
3	Data Bit 3
4	Data Bit 4
5	Data Bit 5
6	Data Bit 6
7	Data Bit 7

Register 1 - LPT Port Status

The LPT Status Register is a read-only register that contains interrupt status and real time status of the LPT connector pins. The bits are described as follows:

Bit	Description
0	Reserved
1	Reserved
2	–IRQ
3	–ERROR
4	SLCT
5	PE
6	–ACK
7	–BUSY

Bits 0 and 1 - Reserved, read as "1's".

Bit 2 - Interrupt Status bit, a "0" indicates that the printer has acknowledged the previous transfer with a ACK handshake (bit 4 of the control register must be set to "1"). The bit is changed to "0" on the active to inactive transition of the –ACK signal. This bit is changed to a "1" after a read from the status port. The default value for this bit is "1".

When in AT mode, bit 1 RTC Register 6AH = 1, the IRQP output follows the –ACK input if enabled. When in PS/2 mode, IRQP is set during the inactive transition of the –ACK signal, and cleared following a read of the LPT status register.

Bit 3 - Error Status bit, a "0" indicates that the printer has had an error. A "1" indicates normal operation. This bit follows the state of the –ERR pin.

Bit 4 - Select Status bit, indicates the current status of the SLCT signal from the printer. A "0" indicates the printer is currently not selected (off-line). A "1" means the printer is currently selected.

Bit 5 - Paper Empty Status bit, a "0" indicates normal operation. A "1" indicates that the printer is currently out of paper. This bit follows the state of the PE pin.

Bit 6 - Acknowledge Status bit, a "0" indicates that the printer has received a character and is ready to accept another. A "1" indicates that the last operation to the printer has not been completed yet. This bit follows the state of the –ACK pin.

Bit 7 - Busy Status bit, a "0" indicates that the printer is busy and cannot receive data. A "1" indicates that the printer is ready to accept data. This bit is the inversion of the BUSY pin.

Register 2 - LPT Port Control

This port is a read/write port that is used to control the LPT direction as well as the Printer Control lines driven from the port. Write operations set or reset these bits, while read operations return the status of the last write operation to this register (except for bit 5 which is write only and is always read back as a "1"). The bits in this register are defined as follows:

VLSI TECHNOLOGY, INC.

ADVANCE INFORMATION

VL82C106

Bit	Description
0	STROBE
1	AUTO FD XT
2	–INIT
3	SLCT IN
4	IRQ EN
5	DIR (Write Only)
6	Reserved
7	Reserved

Bit 0 - Printer Strobe Control bit, when set ("1") the STROBE signal is asserted on the LPT interface, causing the printer to latch the current data. When reset ("0") the signal is negated.

Bit 1 - Auto Feed Control bit, when set ("1") the AUTO FD XT signal will be asserted on the LPT interface, causing the printer to automatically generate a line feed at the end of each line. When reset ("0") the signal is negated.

Bit 2 - Initialize Printer Control bit, when set ("1") the INIT signal is negated. When reset ("0") the INIT signal is asserted to the printer, forcing a reset.

Bit 3 - Select Input Control bit, when set ("1") the SLCT IN signal is asserted, causing the printer to go "on-line". When reset ("0") the signal is negated.

Bit 4 - Interrupt Request Enable Control bit, when set ("1") enables interrupts from the LPT Port whenever the –ACK signal is asserted by the printer. When reset ("0") interrupts are disabled.

Bit 5 - When EMODE = 1, Direction (DIR) Control bit, when set ("1") the output buffers in the LPT Port are disabled, allowing data driven from external sources to be read from the LPT Port. When reset ("0"), the output buffers are enabled, forcing the LPT pins to drive the LPT pins. The power-on-reset value of this is cleared ("0"). When –EMODE = 1, this write only bit has no effect and should be read as "1".

Bits 6 and 7 - Reserved, read as "1's".

REAL TIME CLOCK

The Real Time Clock (RTC) is the equivalent of the Motorola MC146818A Real Time Clock component. It is also compatible with the Dallas Semiconductor DS1287A RTC when an external battery and crystal are provided. Clock functions include the following:

- Time of Day Clock
- Alarm Function
- 100 Year Calendar Function
- Programmable Periodic Interrupt Output
- Programmable Square Wave Output
- 50 Bytes of User RAM
- User RAM Preset Feature

RTC PROGRAMMERS MODEL

The RTC memory consists of ten RAM bytes which contain the time, calendar, and alarm data, four control and status bytes, and 50 general purpose RAM bytes. The address map of the real time clock is shown below.

Add.	Function	Range
00-09	Time Regs.	0-99
0A	RTC Register A	(R/W)
0B	RTC Register B	(R/W)
0C	RTC Register C	(R O)
0D	RTC Register D	(R O)
0E-3F	User RAM (Standby)	

All 64 bytes are directly readable and writable by the processor program except for the following:

1) Registers C and D are read only.
2) Bit 7 of Register A is read only.

The RTC is normally accessed via internally decoded PORT 070H (RTC register address) and PORT 071H (RTC data read/write).

The RTC address and data ports can be moved to Port 170H, Port 171H by pulling the RTCMAP pin (121) to ground. This pin can be left Not connected or tied high for normal port addressing.

The RTC address map also includes additional standby RAM, plus control registers for Combo chip configuration and chip select control. The RAM and Chip Select control registers are powered via the VBAT power supply for battery-backed operation.

The total address map is shown below:

Add. (HEX)	Function
00-0D	Time Portion of RTC
0E-3F	RAM Portion of RTC
40-4F	Additional Standby RAM
50-68	Reserved
69-7F	Chip Select Control Registers

The processor program obtains time and calendar information by reading the appropriate locations. The program may initialize the time, calendar, and alarm by writing to these RAM locations. The contents of the ten time, calendar, and alarm bytes may be either binary or binary-coded decimal (BCD).

Time of Day Register

The contents of the Time of Day registers can be either in Binary or BCD format. They are relatively straightforward, but are detailed here for completeness. The address map of these registers is shown next:

Add.	Function	Range
0	Seconds (Time)	0-59
1	Seconds (Alarm)	0-59
2	Minutes (Time)	0-59
3	Minutes (Alarm)	0-59
4	Hours (Time)	1-12, 12 Hr Mode
4	Hours (Time)	0-23, 24 Hr Mode
5	Hours (Alarm)	0-23
6	Day of Week	1-7
7	Date of Month	1-31
8	Month	1-12
9	Year	0-99

VLSI TECHNOLOGY, INC.

ADVANCE INFORMATION

VL82C106

Address 0 - Seconds (Time): The range of this register is 0-59 in BCD mode, and 0-3BH in Binary mode.

Address 1 - Seconds (Alarm): The range of this register is 0-59 in BCD mode, and 0-3BH in Binary mode.

Address 2 - Minutes (Time): The range of this register is 0-59 in BCD mode, and 0-3BH in Binary mode.

Address 3 - Minutes (Alarm): The range of this register is 0-59 in BCD mode, and 0-3BH in Binary mode.

Address 4 - Hours (Time): The range of this register is:

Range	Mode	Time
1-12	BCD	AM
81-92	BCD	PM
01H-0CH	Binary	AM
81H-8CH	Binary	PM

Address 5 - Hours (Alarm): The range of this register is:

Range	Mode	Time
1-12	BCD	AM
81-92	BCD	PM
01H-0CH	Binary	AM
81H-8CH	Binary	PM

Address 6 - Day of Week: The range of this register is 1-7 in BCD mode, and 1-7H in Binary mode.

Address 7 - Date: The range of this register is 1-31 in BCD mode, and 1-1FH in Binary mode.

Address 8 - Month: The range of this register is 1-12 in BCD mode, and 1-0CH in Binary mode.

Address 9 - Year: The range of this register is 0-99 in BCD mode, and 0-63H in Binary mode.

RTC CONTROL REGISTER

The RTC has four registers which are accessible to the processor program. The four registers are also fully accessible during the update cycle.

Add.	Function	Type
0A	RTC Register A	R/W
0B	RTC Register B	R/W
0C	RTC Register C	R O
0D	RTC Register D	R O
0E-3F	User RAM (Standby)	R/W

Register A

This register contains control bits for the selection of Periodic Interrupt, Input Divisor, and the Update In Progress Status bit. The bits in the register are defined as follows:

Bit	Description	Abbr.
0	Rate Select Bit 0	RS0
1	Rate Select Bit 1	RS1
2	Rate Select Bit 2	RS2
3	Rate Select Bit 3	RS3
4	Divisor Bit 0	DV0
5	Divisor Bit 1	DV1
6	Divisor Bit 2	DV2
7	Update In Progress	UIP

Bits 0 to 3 - The four rate selection bits (RS0 to RS3) select one of 15 taps on the 22-stage divider, or disable the divider output. The tap selected may be used to generate a periodic interrupt. These four bits are read/write bits which are not affected by RESET. The Periodic Interrupt Rate that results from the selection of various tap values is as follows:

RS Value	Periodic Interrupt Rate	
0	None	
1	3.90625	ms
2	7.8125	ms
3	122.070	µs
4	244.141	µs
5	488.281	µs
6	976.562	µs
7	1.953125	ms
8	3.90625	ms
9	7.8125	ms
0AH	15.625	ms
0BH	31.25	ms
0CH	62.5	ms
0DH	125	ms
0EH	250	ms
0FH	500	ms

Bits 4 to 6 - The three Divisor Selection bits (DV0 to DV2) are fixed to provide for only a five-state divider chain, which would be used with a 32 KHz external crystal. Only bit 6 of this register can be changed allowing control of the reset for the divisor chain. When the divider reset is removed the first update cycle begins one-half second later. These bits are not affected by power-on reset (external pin).

DV Value	Condition
2	Operation Mode, Divider Running
6	Reset Mode, Divider in Reset State

Bit 7 - The Update In Progress (UIP) bit is a status flag that may be monitored by the program. When UIP is a "1" the update cycle is in progress or will soon begin. When UIP is a "0" the update cycle is not in progress and will not be for at least 244 µs. The time, calendar, and alarm information in RAM is fully available to the program when the UIP

VLSI TECHNOLOGY, INC.

ADVANCE INFORMATION

VL82C106

bit is "0". The UIP bit is a read-only bit, and is not affected by reset. Writing the SET bit in Register B to a "1" will inhibit any update cycle and then clear the UIP status bit.

Register B
Register B contains command bits to control various modes of operations and interrupt enables for the RTC. The bits in this register are defined as follows:

Bit	Description	Abbr.
0	Daylight Savings Enable	DSE
1	24/12 Mode	24/12
2	Data Mode (Binary or BCD)	DM
3	Not Used	
4	Update End Interrupt Enable	UIE
5	Alarm Interrupt Enable	AIE
6	Periodic Interrupt Enable	PIE
7	Set Command	SET

Bit 0 - The Daylight Savings Enable (DSE) bit is a read/write bit which allows the program to enable two special updates (when DSE is "1"). On the first Sunday in April the time increments from 1:59:59 AM to 3:00:00 AM. On the last Sunday in October when the time first reaches 1:59:59 AM it changes to 1:00:00 AM. These special updates do not occur when the DSE bit is a "0". DSE is not changed by any internal operations or reset.

Bit 1 - The 24/12 control bit establishes the format of the hours bytes as either the 24-hour mode ("1") or the 12-hour mode ("0"). This is a read/write bit, which is affected only by software.

Bit 2 - The Data Mode (DM) bit indicates whether time and calendar updates are to use binary or BCD formats. The DM bit is written by the processor program and may be read by the program, but is not modified by any internal functions or reset. A "1" in DM signifies binary data, while a "0" in DM specifies binary-coded-decimal (BCD) data.

Bit 3 - This bit is unused in this version of the RTC, but is used for Square Wave Enable in the Motorola MC146818.

Bit 4 - The UIE (Update End Interrupt Enable) bit is a read/write bit which enables the Update End Interrupt Flag (UF) bit in Register C to assert an IRQ. The reset pin being asserted or the SET bit going high clears the UIE bit.

Bit 5 - The Alarm Interrupt Enable (AIE) bit is a read/write bit which when set to a "1" permits the Alarm Interrupt Flag (AF) bit in Register C to assert an IRQ. An alarm interrupt occurs for each second that the three time bytes equal the three alarm bytes (including a "don't care" alarm code of 11XXXXXXb). When the AIE bit is a "0", the AF bit does not initiate an IRQ signal. The reset pin clears AIE to "0". The internal functions do not affect the AIE bit.

Bit 6 - The Periodic Interrupt Enable (PIE) bit is a read/write bit which allows the Periodic Interrupt Flag (PF) bit in Register C to cause the IRQ pin to be driven low. A program writes a "1" to the PIE bit in order to receive periodic interrupts at the rate specified by the RS3, RS2, RS1, and RS0 bits in Register A. A "0" in PIE blocks IRQ from being initiated by a periodic interrupt, but the Periodic Interrupt Flag (PF) bit is still set at the periodic rate. PIE is not modified by any internal functions, but is cleared to "0" by a reset.

Bit 7 - When the SET bit is a "0", the update cycle functions normally by advancing the counts once-per-second. When the SET bit is written to a "1", any update cycle in progress is aborted and the program may initialize the time and calendar bytes without an update occurring in the midst of initializing. SET is a read/write bit which is not modified by reset or internal functions.

Register C
Register C contains status information about interrupts and internal operation of the RTC. The bits in this register are defined as follows:

Bit	Description	Abbr.
0	Not Used, Read as 0	
1	Not Used, Read as 0	
2	Not Used, Read as 0	
3	Not Used, Read as 0	
4	Update End Interrupt Flag	UF
5	Alarm Interrupt Flag	AF
6	Periodic Interrupt Flag	PF
7	IRQ Pending Flag	IRQF

Bits 0 to 3 - The unused bits of Status Register 1 are read as "0's", and cannot be written.

Bit 4 - The Update Ended Interrupt Flag (UF) bit is set after each update cycle. When the UIE bit is a "1", the "1" in UF causes the IRQF bit to be a "1", asserting IRQ. UF is cleared by a Register C read or a reset.

Bit 5 - A "1" in the AF (Alarm Interrupt Flag) bit indicates that the current time has matched the alarm time. A "1" in the AF causes the IRQ pin to go low, and a "1" to appear in the IRQF bit, when the AIE bit also is a "1". A reset or a read of Register C clears AF.

Bit 6 - The Periodic Interrupt Flag (PF) is a read only bit which is set to a "1" when a particular edge is detected on the selected tap of the divider chain. The RS3 to RS0 bits establish the periodic rate. PF is set to a "1" independent of the state of the PIE bit. PF being a "1" initiates an IRQ signal and sets the IRQF bit when PIE is also a "1". The PF bit is cleared by a reset or a software read of Register C.

Bit 7 - The Interrupt Request Pending Flag (IRQF) is set to a "1" when one or more of the following are true:

PF = PIE = 1
AF = AIE = 1
UF = UIE = 1

The logic can be expressed in equation form as:

IRQF = PF • PIE + AF • AIE + UF • UIE

VLSI TECHNOLOGY, INC.

ADVANCE INFORMATION

VL82C106

Any time the IRQF bit is a "1", the IRQ pin is asserted. All flag bits are cleared after Register C is read by the program or when the reset pin is asserted.

Register D
This register contains a bit that indicates the status of the on-chip standby RAM. The contents of the registers are described as the following:

Bit	Description	Abbr.
0	Not Used, Read as 0	
1	Not Used, Read as 0	
2	Not Used, Read as 0	
3	Not Used, Read as 0	
4	Not Used, Read as 0	
5	Not Used, Read as 0	
6	Not Used, Read as 0	
7	Vaild RAM Data and Time	VRT

Bits 0 to 6 - The remaining bits of Register D are unused. They cannot be written, but are always read as "0's".

Bit 7 - The Valid RAM Data and Time (VRT) bit indicates the condition of the contents of the RAM, provided the power sense (PS) pin is satisfactorily connected. A "0" appears in the VRT bit when the power-sense pin is low. The processor program can set the VRT bit when the time and calendar are initialized to indicate that the RAM and time are valid. The VRT is a read only bit which is not modified by the reset pin. The VRT bit can only be set by reading Register D.

Pulling the PS/-RC pin low for a minimum of 2 μs also sets all RAM bytes from address 0E through 3F to all ones.

CMOS STANDBY RAM
The 66 general purpose RAM bytes are not dedicated within the RTC. They can be used by the processor program, and are fully available during the update cycle.

GENERAL RTC NOTES
Set Operation
Before initializing the internal registers, the SET bit in Register B should be set to a "1" to prevent time/calendar updates from occurring. The program initializes the ten locations in the selected format (binary or BCD), then indicates the format in the Data Mode (DM) bit of Register B. All ten time, calendar, and alarm bytes must use the same Data Mode, either binary or BCD. The SET bit may now be cleared to allow updates. Once initialized the RTC makes all updates in the selected Data Mode. The Data Mode cannot be changed without reinitializing the ten data bytes.

BCD vs Binary Format
The 24/12 bit in Register B establishes whether the hour locations represent 1-to-12 or 0-to-23. The 24/12 bit cannot be changed without reinitializing the hour locations. When the 12-hour format is selected the high-order bit of the hours byte represents PM when it is a "1".

Update Operation
The time, calendar, and alarm bytes are not always accessible by the processor program. Once-per-second the ten bytes are switched to the update logic to be advanced by one second and to check for an alarm condition. If any of the ten bytes are read at this time, the data outputs are undefined. The update lockout time is 1948 μs for the 32.768 KHz time base. The Update Cycle section shows how to accommodate the Update Cycle in the processor program.

Alarm Operation
The three alarm bytes may be used in two ways. First, when the program inserts an alarm time in the appropriate hours, minutes, and seconds alarm locations, the Alarm Interrupt is initiated at the specified time each day if the alarm enable bit is high. The second usage is to insert a "don't care" state in one or more of three alarm bytes. The "don't care" code is any byte from 0C0H to 0FFH. An Alarm Interrupt each hour is created with a "don't care" code in the hours alarm location. Similarly, an alarm is generated every minute with "don't care" codes in the hours and minutes alarm bytes. The "don't care" codes in all three alarm bytes create an interrupt every second.

Interrupts
The RTC plus RAM includes three separate fully automatic sources of interrupts to the processor. The Alarm Interrupt may be programmed to occur at rates from one-per-second to one-a-day. The Periodic Interrupt may be selected for rates from half-a-second to 30.517 μs. The Update Ended Interrupt may be used to indicate to the program that an update cycle is completed.

The processor program selects which interrupts, if any, it wishes to receive. Three bits in Register B enable the three interrupts. Writing a "1" to an interrupt-enable bit permits that interrupt to be initiated when the event occurs. A "0" in the interrupt-enable bit prohibits the IRQ pin from being asserted due to the interrupt cause.

If an interrupt flag is already set when the interrupt becomes enabled, the IRQ pin is immediately activated, though the interrupt initiating the event may have occurred much earlier. Thus, there are cases where the program should clear such earlier initiated interrupts before first enabling new interrupts.

When an interrupt event occurs, a flag bit is set to a "1" in Register C. Each of the three interrupt sources have separate flag bits in Register C, which are set independent of the state of the corresponding enable bits in Register B. The flag bit may be used with or without enabling the corresponding enable bits.

Divider Control
The Divider Control bits are fixed for only 32.768 KHz operation. The divider chain may be held in reset, which allows precision setting of the time. When the divider is changed from reset to an operating time base, the first update cycle is one-half a second later. The Divider Control bits are also used to facilitate testing the RTC.

Periodic Interrupt Selection
The Periodic Interrupt allows the IRQ pin to be triggered from once every 500 ms to once every 30.517 μs. The Periodic Interrupt is separate from the Alarm Interrupt which may be output from once-per-second to once-per-day.

VLSI TECHNOLOGY, INC.

ADVANCE INFORMATION

VL82C106

KEYBOARD CONTROLLER
The keyboard controller on-chip ROM contains the code that is required to support the PC/AT and PS/2 command sets and 128 bytes of conversion code.

Keyboard serial I/O is handled with hardware implementations of the receiver and transmitter. Both functions depend on an 8-bit timer for time-out detection. Enhanced status reporting is provided in hardware to simplify error handling in software. This logic is duplicated for the mouse interface.

User RAM support is provided. The program writes a command 20-3FH (read) or 60-7FH (write) with the lower five bits representing the RAM address. Data from a read or for a write are accessed through port 60H DBB.

Parallel Port 1 (input) is provided and Parallel Port 2 (output) has defined functions depending on whether the controller is in PC/AT or PS/2 mode.

Support for PORT 60H DBB (reads and writes) and Status Register (reads and writes) is provided in hardware for interface to the PC host.

KEYBOARD CONTROLLER INTERFACE TO PC/AT
The interface to the PC/AT consists of one register pair (PORT 60H/64H) for the keyboard and mouse.

The PORT 60H read operations output the contents of the Output Buffer to D0-D7 and clears the status of the Output Buffer Full (OBF/Status Register bit 0) bit.

Status read operations output the contents of the Status Register to D0-D7. No status is changed as a result of the read operation.

The PORT 60H write operations cause the Input Buffer DBB to be changed. The state of the C/D bit is cleared (Status Register bit 3, "0" indicates data) and the Input Buffer Full (IBF/Status Register bit 1) bit is set ("1").

Command write operations are to PORT 64H. The C/D bit will be set to ("1") when a valid command has been written to PORT 64H.

KEYBOARD PORT INTERFACE PROTOCOL
Data transmission between the controller, the keyboard, and mouse consist of a synchronous bit stream over the data and clock lines. The bits are defined as follows:

Bit	Function
1	Start Bit (Always 0)
2	Data Bit 0 (LSB)
3-8	Data Bits 1-6
9	Data Bit 7 (MSB)
10	Parity Bit (Odd)
11	Stop Bit (Always 1)

PROGRAMMER INTERFACE
The programmer interface to the keyboard controller is quite simple, consisting of four registers:

Register	R/W	I/O
Status	R	64H
Command	W	64H
Output Buffer	R	60H
Input Buffer	W	60H

The behavior of these registers differ according to the mode of operation (PC/AT or PS/2). There exists only one mode register and one Status Register with different bit definitions for PC/AT mode and PS/2 mode. The bit definitions for each register in each mode follows.

VLSI TECHNOLOGY, INC.

ADVANCE INFORMATION

VL82C106

FIGURE 1. PC/AT MODE REGISTER (READ PORT 60H AFTER WRITE COMMAND 20H TO PORT 64H)

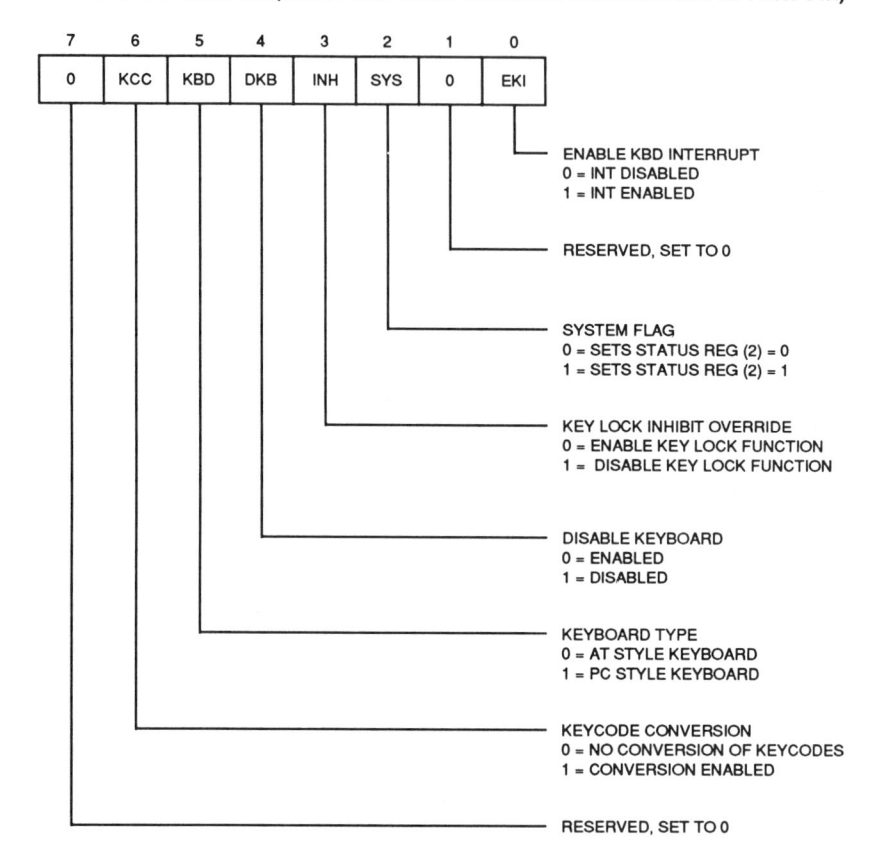

7	6	5	4	3	2	1	0
0	KCC	KBD	DKB	INH	SYS	0	EKI

ENABLE KBD INTERRUPT
0 = INT DISABLED
1 = INT ENABLED

RESERVED, SET TO 0

SYSTEM FLAG
0 = SETS STATUS REG (2) = 0
1 = SETS STATUS REG (2) = 1

KEY LOCK INHIBIT OVERRIDE
0 = ENABLE KEY LOCK FUNCTION
1 = DISABLE KEY LOCK FUNCTION

DISABLE KEYBOARD
0 = ENABLED
1 = DISABLED

KEYBOARD TYPE
0 = AT STYLE KEYBOARD
1 = PC STYLE KEYBOARD

KEYCODE CONVERSION
0 = NO CONVERSION OF KEYCODES
1 = CONVERSION ENABLED

RESERVED, SET TO 0

PC/AT MODE REGISTER

Bit 0 - Enable Keyboard Interrupt (EKI), when set ("1") causes the controller to generate a keyboard interrupt whenever data (keyboard or controller) is written into the output buffer.

Bit 1 - Reserved, should be written as "0".

Bit 2 - System Flag (SYS), when set ("1") writes the System Flag bit of the Status Register to "1". This bit is used to indicate a switch from virtual to real mode when set.

Bit 3 - Inhibit Override (INH), when set ("1") disables the keyboard lock function (KKSW Input).

Bit 4 - Disable Keyboard (DKB), when set ("1") disables the keyboard by holding the –KCKOUT line low.

Bit 5 - Keyboard Type (KBD), when set ("1") allows for compatibility with PC-style keyboards. In this mode, parity is not checked and scan codes are not converted.

Bit 6 - Keycode Conversion (KCC), when set ("1") causes the controller to convert the scan codes to PC format. When reset, the codes (AT keyboard) are passed along unconverted.

Bit 7 - Reserved, should be written as "0".

14

VLSI TECHNOLOGY, INC.

ADVANCE INFORMATION
VL82C106

FIGURE 2. PS/2 MODE REGISTER (READ PORT 60H AFTER WRITE COMMAND 20H TO PORT 64H)

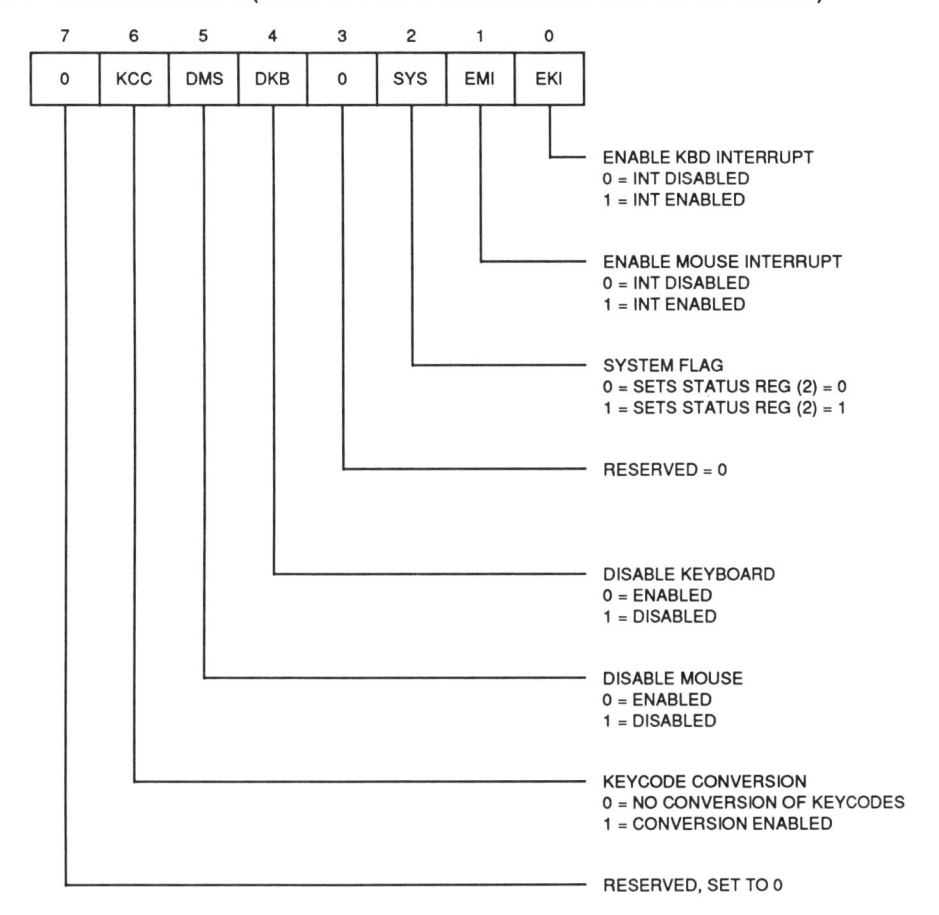

7	6	5	4	3	2	1	0
0	KCC	DMS	DKB	0	SYS	EMI	EKI

- ENABLE KBD INTERRUPT
  0 = INT DISABLED
  1 = INT ENABLED

- ENABLE MOUSE INTERRUPT
  0 = INT DISABLED
  1 = INT ENABLED

- SYSTEM FLAG
  0 = SETS STATUS REG (2) = 0
  1 = SETS STATUS REG (2) = 1

- RESERVED = 0

- DISABLE KEYBOARD
  0 = ENABLED
  1 = DISABLED

- DISABLE MOUSE
  0 = ENABLED
  1 = DISABLED

- KEYCODE CONVERSION
  0 = NO CONVERSION OF KEYCODES
  1 = CONVERSION ENABLED

- RESERVED, SET TO 0

PS/2 MODE REGISTER

Bit 0 - Enable Keyboard Interrupt (EKI), when set ("1") causes the controller to generate a keyboard interrupt whenever data (keyboard or command) is written into the output buffer.

Bit 1 - Enable Mouse Interrupt (EMI), when set ("1") allows the controller to generate a mouse interrupt when mouse data is available in the output register.

Bit 2 - System Flag (SYS), when set ("1") writes the System Flag bit of the Status Register to "1". This bit is used to indicate a switch from virtual to real mode when set.

Bit 3 - Reserved, "0".

Bit 4 - Disable Keyboard (DKB), when set ("1") disables the keyboard by holding the –KCKOUT low.

Bit 5 - Disable Mouse (DMS), when set ("1") disables the mouse by holding the –MCKOUT low.

Bit 6 - Keycode Conversion (KCC), when set ("1") causes the controller to convert the scan codes to PC format. When reset, the codes (PS/2 keyboard) are passed along unconverted.

Bit 7 - Reserved, "0".

15

VLSI TECHNOLOGY, INC.

ADVANCE INFORMATION

VL82C106

FIGURE 3. PC/AT STATUS REGISTER (READ ONLY - PORT 64H)

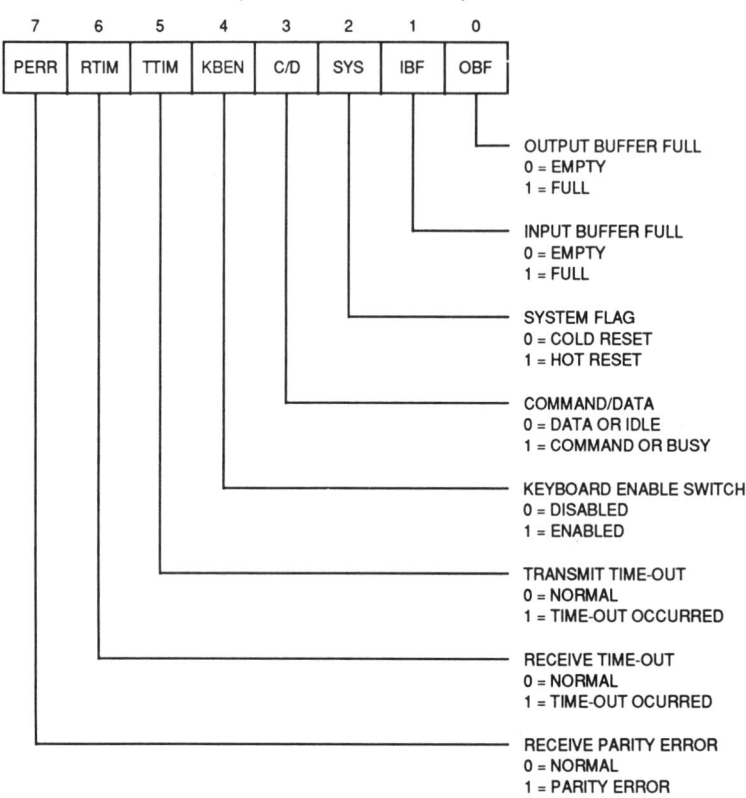

7	6	5	4	3	2	1	0
PERR	RTIM	TTIM	KBEN	C/D	SYS	IBF	OBF

- **OUTPUT BUFFER FULL**
  - 0 = EMPTY
  - 1 = FULL
- **INPUT BUFFER FULL**
  - 0 = EMPTY
  - 1 = FULL
- **SYSTEM FLAG**
  - 0 = COLD RESET
  - 1 = HOT RESET
- **COMMAND/DATA**
  - 0 = DATA OR IDLE
  - 1 = COMMAND OR BUSY
- **KEYBOARD ENABLE SWITCH**
  - 0 = DISABLED
  - 1 = ENABLED
- **TRANSMIT TIME-OUT**
  - 0 = NORMAL
  - 1 = TIME-OUT OCCURRED
- **RECEIVE TIME-OUT**
  - 0 = NORMAL
  - 1 = TIME-OUT OCURRED
- **RECEIVE PARITY ERROR**
  - 0 = NORMAL
  - 1 = PARITY ERROR

PC/AT Status Register

Bit 0 - Output Buffer Full (OBF), when set ("1") indicates that data is available in the controller Data Bus Buffer, and that the CPU has not read the data yet. CPU reads to PORT 60H to reset the state of this bit.

Bit 1 - Input Buffer Full (IBF), when set ("1") indicates thatdata has been written to PORT 60H or 64H, and the controller has not read the data.

Bit 2 - System Flag (SYS), when set ("1") indicates that the CPU has changed from virtual to real mode.

Bit 3 - Command/Data (CD), when set ("1") indicates that a command has been placed into the Input Data Buffer of the controller. The controller uses this bit to determine if the byte written is a command to be executed.

Bit 4 - Keyboard Enable (KBEN), indicates the state of the "keyboard inhibit" switch input (KKSW). "0" indicates the keyboard is inhibited.

Bit 5 - Transmit Time-out (TTIM), when set ("1") indicates that a transmission to the keyboard was not completed before the controller's internal timer timed-out.

Bit 6 - Receive Time-out (RTIM), when set ("1") indicates that a transmission from the keyboard was not completed before the controller's internal timer timed-out.

Bit 7 - Parity Error (PERR), when set ("1") indicates that a parity error (even parity = error) occurred during the last transmission (received scan code) from the keyboard. When a parity error is detected, the output buffer is loaded with FFH, the OBF Status bit is set and the KIRQ pin is set ["1" if the EKI bit/ Mode Register bit 0 is set ("1")].

VLSI TECHNOLOGY, INC.

ADVANCE INFORMATION

VL82C106

FIGURE 4. PS/2 STATUS REGISTER (READ ONLY - PORT 64H)

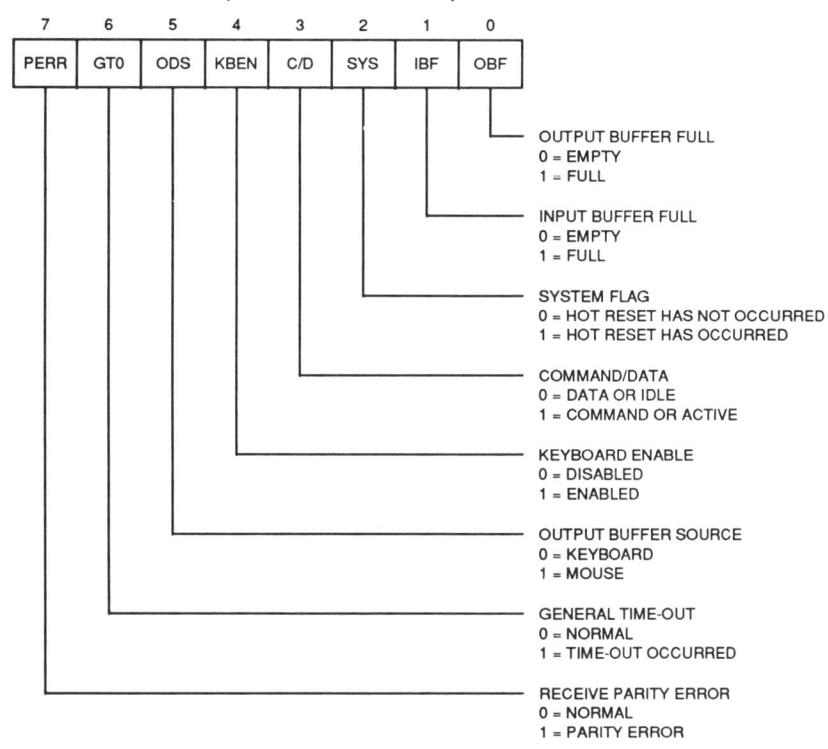

7	6	5	4	3	2	1	0
PERR	GT0	ODS	KBEN	C/D	SYS	IBF	OBF

- Bit 0 — OUTPUT BUFFER FULL
  - 0 = EMPTY
  - 1 = FULL
- Bit 1 — INPUT BUFFER FULL
  - 0 = EMPTY
  - 1 = FULL
- Bit 2 — SYSTEM FLAG
  - 0 = HOT RESET HAS NOT OCCURRED
  - 1 = HOT RESET HAS OCCURRED
- Bit 3 — COMMAND/DATA
  - 0 = DATA OR IDLE
  - 1 = COMMAND OR ACTIVE
- Bit 4 — KEYBOARD ENABLE
  - 0 = DISABLED
  - 1 = ENABLED
- Bit 5 — OUTPUT BUFFER SOURCE
  - 0 = KEYBOARD
  - 1 = MOUSE
- Bit 6 — GENERAL TIME-OUT
  - 0 = NORMAL
  - 1 = TIME-OUT OCCURRED
- Bit 7 — RECEIVE PARITY ERROR
  - 0 = NORMAL
  - 1 = PARITY ERROR

PS/2 Status Register

Bit 0 - Output Buffer Full (OBF), when set ("1") indicates that data is available in the controller Data Bus Buffer, and that the CPU has not read the data yet. The CPU reads to PORT 60H reset the state of this bit.

Bit 1 - Input Buffer Full (IBF), when set ("1") indicates that data has been written to PORT 60H or 64H, and the controller has not read the data.

Bit 2 - System Flag (SYS), when set ("1") indicates that the CPU has changed from virtual to real mode.

Bit 3 - Command/Data (CD), when set ("1") indicates that a command has been placed into the Input Data Buffer of the controller. The controller uses this bit to determine if the byte written is a command to be executed. This bit is not reset until the command has completed its operation.

Bit 4 - Keyboard Enable (KBEN) indicates the state of the "keyboard inhibit" switch input (KKSW). "0" indicates the keyboard is inhibited.

Bit 5 - Output Buffer Data Source (ODS), when set ("1") indicates that the data in the output buffer is mouse data. When reset, it indicates the data is from the keyboard.

Bit 6 - Time-out Error (TERR), when set ("1") indicates that a transmission was started and that it did not complete within the normal time taken (approximately 11 KCKIN cycles). If the transmission originated from the controller a FEH is placed in the output buffer. If the transmission originated from the keyboard a FFH is placed in the output buffer.

Bit 7 - Parity Error (PERR), when set ("1") indicates that a parity error (even parity = error) occurred during the last transmission from the keyboard. When a parity error is detected, the output buffer is loaded with FFH, the OBF Status bit is set and the KIRQ pin is set ["1" if the EKI bit/Mode Register bit 0 is set ("1")].

VLSI TECHNOLOGY, INC.

ADVANCE INFORMATION

VL82C106

COMMAND SET

The command set supported by the keyboard controller supports two modes of operation, and a set of extensions to the AT command set for the PS/2. In both modes, the command is implemented by writing the command byte to PORT 64H. Any subsequent data is read from PORT 60H (see description of command 20) or written to PORT 60H (see description of command PORT 60H). The commands for each mode are shown in the table below:

PC/AT Mode:

Comm.	Description
20	Read Mode Register
21-3F	Read Keyboard Controller RAM (Byte 1-31)
60	Write Mode Register
61-7F	Write Keyboard Controller RAM (Byte 1-31)
AA	Self Test
AB	KBD Interface Test
AC	Diagnostic Dump
AD	Disable Keyboard
AE	Enable Keyboard
C0	Read Input Port (P10-P17)
D0	Read Output Port (P20-P27)
D1	Write Output Port
E0	Read Test Inputs (T0, T1)
F0-FF	Pulse Output Port (P20-P27)

Note: If data is written to the data buffer (PORT 60H) and the command preceding it did not expect data from the port (PORT 60H) the data will be transmitted to the keyboard.

Added PS/2 Commands:

Comm.	Description
A4	Test Password
A5	Load Password
A6	Enable Password
A7	Disable Mouse
A8	Enable Mouse
A9	Mouse Interface Test
C1	Poll in Port Low (P10-P13 -> S4-S7)
C2	Poll in Port High (P14-P17 -> S4-S7)
D1	Write Output Port
D2	Write Keyboard Output Buffer
D3	Write Mouse Output Buffer
D4	Write to Mouse

The following is a description of each command:

20 Read the keyboard controller's Mode Register (PC/AT and PS/2) - The keyboard controller sends its current mode byte to the output buffer (accessed by a read of PORT 60H).

21-3F Read the keyboard controller's RAM (PC/AT and PS/2) - Bits D4-D0 specify the address.

60 Write the keyboard controller's Mode Register (PC/AT and PS/2) - The next byte of data written to the keyboard data port (PORT 60H) is placed in the controller's mode register.

61-7F Write the keyboard controller's RAM (PC/AT and PS/2) - This command writes to the internal keyboard controller RAM with the address specified in bits D4-D0.

A4 Test Password Installed (PS/2 only) - This command checks if there is currently a password installed in the controller. The test result is placed in the output buffer (the OBF bit is set) and KIRQ is asserted (if the EKI bit is set). Test result - FAH means that the password is installed, and F1H means that it is not.

A5 Load Password (PS/2 only) - This command initiates the password load procedure. Following this command the controller will take data from the input buffer port (PORT 60H) until a 00H is detected or a full eight byte password including a delimiter (e.g. <cr>) is loaded into the password latches. Note: this means that during password validation the password can be a maximum of seven bytes with a delimiter such as <cr>.

A6 Enable Password (PS/2 only) - This command enables the security feature. The command is valid only when a password pattern is written into the controller (see A5 command). No other commands will be "honored" until the security sequence is completed and command A6 is cleared.

A7 Disable Mouse (PS/2 only) - This command sets bit 5 of the Mode Register which disables the mouse by driving the –MCKOUT line low.

A8 Enable Mouse (PS/2 only) - This command resets bit 5 of the Mode Register, thus enabling the mouse.

A9 Mouse Interface Test (PS/2 only) - This command causes the controller to test the mouse clock and data lines. The results are placed in the output buffer (the OBF bit is set) and the KIRQ line is asserted (if the EKI bit is set). The results are as follows:

Data	Meaning
00	No Error
01	Mouse Clock Line Stuck Low
02	Mouse Clock Line Stuck High
03	Mouse Data Line Stuck Low
04	Mouse Data Line Stuck High

VLSI TECHNOLOGY, INC.

ADVANCE INFORMATION

VL82C106

AA Self Test command (PC/AT and PS/2) - This commands the controller to perform internal diagnostic tests. A 55H is placed in the output buffer if no errors were detected. The OBF bit is set and KIRQ is asserted (if the EKI bit is set).

AB Keyboard Interface Test (PC/AT and PS/2) - This command causes the controller to test the keyboard clock and data lines. The test result is placed in the output buffer (the OBF bit is set) and the KIRQ line is asserted (if the EKI bit is set). The results are as follows:

Data	Meaning
00	No Error
01	Keyboard Clock Line Stuck Low
02	Keyboard Clock Line Stuck High
03	Keyboard Data Line Stuck Low
04	Keyboard Data Line Stuck High

AC Diagnostic Dump (PC/AT only, Reserved on PS/2) - Sends 16 bytes of the controller's RAM, the current state of the input port, and current state of the output port to the system.

AD Keyboard Disable (PC/AT and PS/2) - This command sets bit 4 of the Mode Register to a "1". This disables the keyboard by driving the clock line (–KCKOUT) high. Data will not be sent or received.

AE Keyboard Enable (PC/AT and PS/2) - This command resets bit 4 of the mode byte to a "0". This enables the keyboard again by allowing the keyboard clock to free-run.

C0 Read P1 Input Port (PC/AT and PS/2) - This command reads the keyboard input port and places it in the output buffer. This command overwrites the data in the buffer.

C1 Poll Input Port low (PS/2 only) - P1 bits 0-3 are written into Status Register bits 4-7 until a new command is issued to the keyboard controller.

C2 Poll Input Port high (PS/2 only) - P1 bits 4-7 are written into Status Register bits 4-7 until a new command is issued to the keyboard controller.

D0 Read Output Port (PC/AT and PS/2) - This command causes the controller to read the P2 output port and place the data in its output buffer. The definitions of the bits are as follows:

Bit	Pin	PC/AT Mode	PS/2 Mode
0	P20	–RC	–RC
1	P21	A20 Gate	A20 Gate
2	P22	Speed Sel	–MDOUT
3	P23	Shadow Enable	–MCKOUT
4	P24	Output Buffer Full	KIRQ
5	P25		MIRQ
6	P26	–KCKOUT	–KCKOUT
7	P27	KDOUT	–KDOUT

Note: P22 (bit 2) is the speed control pin used by Award BIOS, and this is different from what is used by Phoenix and AMI.

D1 Write Output Port (PC/AT and PS/2) - The next byte of data written to the keyboard data port (PORT 60H) will be written to the controller's output port. The definitions of the bits are as defined above. In PC/AT mode, P26 and P27 are not modified. In PS/2 mode, P22, P23, P26 and P27 are not modified.

D2 Write Keyboard Output Buffer (PS/2 only) - The next byte written to the data buffer (PORT 60H) is written to the output buffer (60H) as if initiated by the keyboard [the OBF bit is set ("1") and KIRQ will be set if the EKI bit is set ("1")].

D3 Write Mouse Output Buffer (PS/2 only) - The next byte written to the data buffer (PORT 60H) is written to the output buffer as if initiated by the mouse [the OBF bit is set ("1") and MIRQ will be set if the EMI bit is set ("1")].

D4 Write to Mouse (PS/2 only) - The next byte written to the data buffer (PORT 60H) is transmitted to the mouse.

E0 Read Test Inputs (PC/AT and PS/2) - This command causes the controller to read the T0 and T1 input bits. The data is placed in the output buffer with the following meanings:

Bit	PC/AT Mode	PS/2 Mode
0	Keyboard Data	Keyboard Clock
1	Keyboard Clock	Mouse Clock
3-7	Read as 0's	Read as 0's

F0-FF Pulse Output Port (PC/AT and PS/2) - Bits 0-3 of the controller's output port may be pulsed low for approximately 6 µs. Bits 0-3 of the command specify which bit will be pulsed. A "0" indicates that the bit should be pulsed; a "1" indicates that the bit should not be modified. FF is treated as a special case (Pulse Null Port). In PC/AT mode, bits P26 and P27 are not pulsed. In PS/2 mode, bits P26, P27, P22 and P23 are not pulsed.

ADVANCE INFORMATION

VL82C106

IDE Bus Interface Control

Integrated Drive Electronics bus interface control signals are provided by the VL82C106 Combo chip. The timing and drive for these lines are consistent with the Conner Peripherals CP342 Integrated Hard Disk Manual.

A set of signals are used for this interface when the VL82C106 Combo chip is configured to support the IDE interface via IDE_EN, bit 5 of Control Register 1 (RTC Register 6AH).

The Combo chip has duplicated bit 1 of the "Fixed Disk Register" (I/O 3F6H) to enable IRQI.

Input Signals:

IDINT — This signal indicates an interrupt request to the system. It is used to generate IRQI.

Output Signals:

−CS4 — Chip Select 4. This signal is used as the floppy disk chip select. The default decode is 03F4H–03F5H, but may be redefined as described in the section on Combo Chip Control Registers. The IDE_EN control bit of Control Register 1 has no effect on this signal. −CS4 is also active when −CDAK4 is active.

−CS5 — Chip Select 5. This signal is used as the −HOST CS0 of the IDE bus. The default decode is 01F0H–01F7H, but may be redefined as described in the section on Combo Chip Control Registers. The IDE_EN control bit of Control Register 1 has no effect on this signal.

−HCS1 — This signal is active (low) for address 03F6H–03F7H and is used as −HOST CS1 of the IDE bus.

−IDENH — This signal is used to drive the −OE pin of an external 74LS245 buffering bits 8-15 of the IDE data bus to the SD bus. It is active (low) when:

−CS5 is active AND SA2-SA0 = 000.

−IDENL — This signal is used to drive the −OE pin of an external 74LS245 buffering bits 0-6 of the IDE data bus. It is active (low) when:

−CS5 is active OR SA0-SA9 = 3F6 OR 3F7.

This allows a simple implementation for an IDE bus that includes both the hard disk controller and the floppy disk controller.

IRQI — This is the three-state interrupt request to the CPU. It is normally tied directly to the IRQ14 signal of the system. It reflects the state of the IDINT input and is enabled by writing bit 1 = 0 of I/O 3F6H as long as IDE_EN=1. Reset or disabling the IDE system three-states IRQI.

−IOCS16 — The VL82C106 Combo chip has multiple sources for this signal. It is driven active (low) when:

(−CS5 is active AND SA0-SA2 = 000 AND IDE_EN = 1 AND {(CS_MODE = 0) OR (CS_MODE = 1 AND 16-bit operation selected for CS5)}) OR (any other CS is active with 16-bit operation selected AND CS_MODE = 1) OR (IDE_EN = 0 AND CS5 is active AND 16-bit operation is selected AND CS_MODE = 1).

Bidirectional Signals:

IDB7, −DC — The control for the transceiver between IDB7, −DC, and SD7 is as follows:

IDB7 −> SD7 when:

(−CS5 is active OR SA0-SA9 = 3F6) AND −IOR is active AND IDE_EN = 1 AND NOT SA0-SA9 = 3F7H.

−DC −> SD7 when SA0-SA9 = 3F7H AND −IOR is active AND IDE_EN = 1.

SD7 −> IDB7 at all other times when IDE_EN=1, three-stated (with internal pull-up) if IDE_EN=0.

Combo Chip Control Ports:

Contained in the VL82C106 are a set of 26 registers used for programming peripheral chip select base addresses, chip select address ranges, and enabling options. Each base address register is a 16-bit register with bits corresponding to address bits A15-A0. In addition to base address registers, there is an address range register that can be used to "don't-care" bits (A0-A4) used in the address range comparison, effectively controlling the address space occupied by the chip select from 1 to 32 bytes. There are also programmable bits to selectively generate wait states, and assert −IOCS16 whenever the corresponding address range is present. These registers are used in groups of three per chip select, and are defined as shown below:

Base Address Register (LSB):

Bit	Description
0	Base Address, Bit A0
1	Base Address, Bit A1
2	Base Address, Bit A2
3	Base Address, Bit A3
4	Base Address, Bit A4
5	Base Address, Bit A5
6	Base Address, Bit A6
7	Base Address, Bit A7

Base Address Register (MSB):

Bit	Description
0	Base Address, Bit A8
1	Base Address, Bit A9
2	Base Address, Bit A10
3	Base Address, Bit A11
4	Base Address, Bit A12
5	Base Address, Bit A13
6	Base Address, Bit A14
7	Base Address, Bit A15

VLSI TECHNOLOGY, INC.

ADVANCE INFORMATION

VL82C106

Range Register:

Bit	Description
0	Don't Care, Bit A0
1	Dont' Care, Bit A1
2	Don't Care, Bit A2
3	Don't Care, Bit A3
4	Don't Care, Bit A4
5	Wait State 0
6	Wait State 1
7	8/16 Bit I/O

The only bits that need detailed descriptions are those contained in the Range Register. These bits are defined as follows:

Bits 0 to 4 - Don't Care Bits, when set ("1") causes that corresponding bit to be ignored during the chip select generation, effectively allowing the chip select signals to correspond to a range or ranges of addresses in the space from Base Address + 0 to Base Address + 31.

Bits 5 & 6 - Wait State 0 and 1, these bits determine the number of wait states that will be generated whenever the corresponding chip select signal is generated. They generate wait states according to the following table:

WS1	WS0	Wait States*
0	0	0
0	1	1
1	0	3
1	1	7

Note: Programmed wait states can only extend the I/O cycle set by the system architecture.

Bit 7 - 8/16 Bit I/O, this bit is used to selectively assert –IOCS16 whenever the corresponding chip select signal is generated. When set ("1") the access is defined as an 8-bit access, and –IOCS16 is not asserted.

* Number of wait states = number of SYSCLK cycles IOCHRDY is forced inactive (low) by the Combo chip.

Default Chip Selects

The VL82C106 Combo chip also has several hard-wired default chip selects for the serial ports, line printer port, floppy disk chip select and hard disk chip select. These default chip selects are used after a reset until the battery-backed programmable values are enabled via bit 3 of the second control register (RTC register 6AH). The wait state and non IDE –IOCS16 values are also disabled in this mode. This allows the Combo chip to function normally without the need for programming. The default chip selects are:

Select/ Device	Address
COMA	3F8H-3FFH (Bit 3 of RTC Reg 69H = 1) 2F8H-2FFH (Bit 3 of RTC Reg 69H = 0)
COMB	2F8H-2FFH (Bit 3 of RTC Reg 69H = 1) 3F8H-3FFH (Bit 3 of RTC Reg 69H = 0)
LPT	03BCH-03BFH (Bit 5, 6 of RTC Reg 69H = 0, 0) 0378H-037BH (Bit 5, 6 of RTC Reg 69H = 1, 0) 0278H-027BH (Bit 5, 6 of RTC Reg 69H = 0,1)
–CS4	03F4H-03F5H
–CS5	01F0H-01F7H
–CS6	03F2H AND –IOW is Active
–CS7	03F7H AND –IOW is Active

Note that on reset, COMA, COMB, LPT, and –CS4 through –CS7 are enabled and set to the hard-wired values.

Combo Chip Control Register

The VL82C106 Combo chip contains a number of programmable options, including peripheral base address and chip select "hole" size. The registers used to provide this control are located in the upper bytes of the RTC address space. They are defined as follows:

Addr	Usage
69	Control Register 0*
6A	Control Register 1*
6B	CS1 COMA Base Add LSB
6C	CS1 COMA Base Add MSB
6D	CS1 COMA Range
6E	CS2 COMB Base Add LSB
6F	CS2 COMB Base Add MSB
70	CS2 COMB Range
71	CS3 LPT Base Add LSB
72	CS3 LPT Base Add MSB
73	CS3 LPT Range
74	CS4 FDC Base Add LSB
75	CS4 FDC Base Add MSB
76	CS4 FDC Range
77	CS5 HDC Base Add LSB
78	CS5 HDC Base Add MSB
79	CS5 HDC Range
7A	CS6 Base Add LSB
7B	CS6 Base Add MSB
7C	CS6 Range
7D	CS7 Base Add LSB
7E	CS7 Base Add MSB
7F	CS7 Range

* Note: Control Register 0 and 1 are not battery-backed via the VBAT supply.

Control Register 0 (RTC Register 69H or I/O PORT 102H) Bits:

This register contains bits that enable or disable functionality of the internal components of the Combo chip. The bits of this register are defined to be consistent with definitions used in the PS/2-50 family.

VLSI TECHNOLOGY, INC.

ADVANCE INFORMATION

VL82C106

This register can also be accessed at address 102H, for PS/2 compatibility. The contents of the register are detailed below:

Bit	Usage	Value After Reset
0	SYS BD EN	Enabled (1)
1	FDCS EN (CS4)	Enabled (1)
2	COMA EN (CS1)	Enabled (1)
3	COMA DEF	COM1 (1)
4	LPT EN (CS3)	Enabled (1)
5	LPT DEF 0	Paralled Port 1 (0)
6	LPT DEF 1	Disabled (0)
7	–EMODE	Compat. Mode (1)

Bit 0 - System Board Enable (SYS BD EN) Control bit, when set ("1") allows bits 1, 2, and 4 to enable and disable their respective devices. When reset ("0") the floppy disk chip select (CS4), COMA (CS1), and the LPT port (CS3) are disabled regardless of the contents of bits 1, 2, and 4.

Bit 1 - Floppy Disk CS Enable (FDCS EN) Control bit, when set ("1") allows the FD CS signal (CS4) to be asserted to an external floppy disk controller chip. When reset ("0") prevents the assertion of this chip select.

Bit 2 - Communications Port A Enable (COMA EN) Control bit, when set ("1") allows the internal COMA (CS1) port to be accessed. When reset ("0") COMA is disabled.

Bit 3 - Communications Port A Default Address (COMA DEF) Control bit, when set ("1") forces the hard-wired default base address to COMA to correspond to (3F8H-3FFH) and COMB to (2F8H-2FFH). When reset ("0") forces the COMA hard-wired address to (2F8H-2FFH) and COMB to (3F8H-3FFH). The base address will be the programmed values if bit 3 of control register 1 (RTC register 6AH) is set.

Bit 4 - Line Printer Port Enable (LPT EN) Control bit, when set ("1") enables the LPT port (CS3). When reset ("0") disables the LPT port.

Bit 5 & 6 - Line Printer Default bits 0 and 1 (LPT DEF 0 and 1) Control bits, set the Line Printer Base hard-wired address defaults as shown below:

Bit 6	Bit 5	Address Range
0	0	03BCH-03BFH
0	1	0378H-037BH
1	0	0278H-027BH
1	1	Reserved

Setting bit 3 of RTC register 6AH changes the base address to that set in the program address registers for LPT (CS3).

Bit 7 - Line Printer Extended Mode (EMODE) Control bit, when set ("1") disables the Extended Mode and forces PC/AT compatibility. When reset ("0") the Extended Mode is enabled, allowing the printer port direction to be controlled.

Control Register 1 (RTC Register 6AH) Bits

This register is used to control peripheral chip selects that are not included in Control Register 0. The bits in this register are defined as follows:

Bit	Usage	Value After Reset
0	COMB EN	Enabled (1)
1	AT/PS2 KBD	AT (1)
2	PRIV EN	Enabled (1)
3	CS MODE	Hard-wire (0)
4	HDCS EN	Enabled (1)
5	IDE EN	Enabled (1)
6	CS6 EN	Enabled (1)
7	CS7 EN	Enabled (1)

Bit 0 - Communication Port B Enable. A "1" enables COMB (CS2). A zero ("0") disables COMB.

Bit 1 - AT or PS/2 Compatible Keyboard. A "1" selects PC/AT type keyboard controller functions, while a "0" places the keyboard controller in PS/2 mode.

Bit 2 - Private Controls Enable. When in AT mode (AT/PS2_KBD = 1), this bit is used to latch the values of the keyboard controller's output signals KHSE, KSRE, and IRQM to the VL82C106 output pins. When "1" these outputs follow the keyboard controller's outputs. When "0" these outputs held at that value regardless of the keyboard controller's outputs.

When in PS/2 mode (AT/PS2_KBD = 0), this bit has no effect on the KHSE, KSRE, and IRQM output pins. The Combo chip outputs follow the keyboard controller's outputs.

Bit 3 - Chip Select Decode Mode. When "0", CS1-CS7 decodes revert to the hard-wired address decoding and non IDE –IOCS16 and IOCHRDY generation is disabled. A "1" enables the address decoding, wait state generation and 8/16-bit operation as programmed into the RTC registers 69H-7FH. (See sections on Default Chip Selects and Combo Chip Control Register.)

Bit 4 - Hard Disk Chip Select Enable. A "1" enables the Hard Disk Chip Select signal (–CS5), while a "0" disables the chip select.

Bit 5 - Integrated Drive Electronics Enable. A "1" enables the IDE functions of outputs –IDENH, –IDENL, IRQI, –IOCS16, and IDB7 as described in IDE Bus Interface Control section.

Bit 6 - Chip Select 6 Enable. When "0", the –CS6 output is disabled. A "1" enables the address decoding, wait state generation and 8/16-bit operation as programmed into the RTC registers 7A-7CH. (See sections on Default Chip Selects and Combo Chip Control Registers.)

Bit 7 - Chip Select 7 Enable. When "0", the –CS7 output is disabled. A "1" enables the address decoding, wait state generation and 8/16-bit operation as programmed into the RTC registers 7D-7FH. (See sections Default Chip Selects and Combo Chip Control Register.)

VLSI Technology, INC.

ADVANCE INFORMATION

VL82C106

Miscellaneous Control Signals

XDDIR — This input signal is generated by the VL82C101/VL82C201. It is inactive (low) when data is transferred from the XD bus to the SD bus, i.e., interrupt acknowledge cycles and I/O read accesses to addresses 000H-0FFH.

XDIRS — This output signal is to control the direction pin of a transceiver between the XD bus and the SD bus when the Combo chip is on the SD bus. Since the architecture assumes the RTC and Keyboard Controller are on the XD bus, this signal is set active (high) when XDDIR is high or either the RTC or the Keyboard Controller is selected.

XDIRX — This output signal is to control the direction pin of the transceiver between the XD bus and the SD bus when the Combo chip is on the XD bus. Since the architecture assumes the peripherals other than the RTC and Keyboard Controller are on the SD bus, this signal is inactive (low) when the XDDIR is low or when –IOR is low and any chip select (CS1-CS7) is generated.

–XDEN — This output signal is used to enable the XD bus transceiver when the VL82C106 Combo chip is placed on the XD bus and DMA's are desired for peripherals controlled by the Combo chip selects. It is the AND of –IOR and –IOW (active low when either –IOR or –IOW are active).

–CDAK4 — This input will directly produce an active low on –CS4 when active low itself and is used by the IDE logic.

–IOCS16 — This output signal is used to indicate to the system that the peripheral being accessed is a 16-bit device. It is set active (low) when a programmed chip select, which specifies 16-bit I/O, is decoded or for certain IDE functions. (See sections on Combo Chip Control Ports and IDE Bus Interface Control.)

When 16-bit programmed chip select operation is selected, –IOCS16 becomes active on the leading edge of ALE and inactive on the trailing edge of –IOW or –IOR. For 8-bit operation or default chip select operation, –IOCS16 is inactive during –IOW or –IOR active.

IOCHRDY — This output signal is used to the lengthen I/O cycle to the peripheral being accessed. It is set inactive (low) for the programmed number of wait states when a programmed chip select, which specifies one, three, or seven wait states, is decoded. (See the section IDE Bus Interface Control.) IOCHRDY transitions inactive at the falling edge of –IOW or –IOW, if enabled, and returns high at the falling edge of SYSCLK after the appropriate number of wait states (SYSCLK cycles).

Note: Programmed wait states can only extend the I/O cycle, i.e., if the system architecture provides four wait states for 8-bit I/O, programming 1 or 3 has no effect.

XTAL1 — This pin is the input to the on-board 18.432 MHz crystal oscillator. This pin may also be driven by an external CMOS clock signal at 18.432 MHz.

XTAL2 — This pin is the output pin of the internal crystal oscillator and should be left open and unloaded if an external clock signal is applied to the XTAL1 pin. This pin is not capable of driving external loads other than the crystal.

–TRI — This pin is used for in-circuit testing. When low, all outputs and I/O pins are placed in the high impedance state.

–ICT — This pin, when strobed low, places the VL82C106 into test mode, determined by the data on the SD0 through SD3 pins. The chip will remain in this mode until RES is asserted. Test mode may be changed by strobing this pin low again with different data on the SD0-SD3 pins.

VLSI TECHNOLOGY, INC.

ADVANCE INFORMATION

VL82C106

AC CHARACTERISTICS: TA = 0°C to +70°C, VCC = 5 V ±5%, GND = 0 V

Symbol	Parameter	Min	Max	Unit	Conditions
I/O Read/Write Figures 5, 6					
tSU1	Address Setup Time	55		ns	
tH2	Address Hold Time	20		ns	
tSU3	AEN Setup Time	55		ns	
tH4	AEN Hold Time	20		ns	
t5	Command Pulse Width	125		ns	
tSU6	Write Data Setup	60		ns	
tH7	Write Data Hold	20		ns	
tD8	Read Data Delay	0	130	ns	CL=200 pF
tH9	Read Data Hold	5	60	ns	CL=50 pF
Chip Select Timing (Hard-wired) Figures 7, 9					
tD11	Chip Select Delay from Address		35	ns	CL=50 pF
tD12	–CS6, –CS7 Delay from –IOW		30	ns	CL=50 pF
tD13	–IOCS16 Active from Address		60	ns	CL=200 pF
tD14	–CS4 Delay from –CDAK4		25	ns	CL=50 pF
Chip Select Timing (Programmable) Figures 7, 9					
tD11	Chip Select Delay from Address		45	ns	CL=50 pF
tD13	–IOCS16 Active from Address		70	ns	CL=200 pF
tD14	–CS4 Delay from –CDAK4		25	ns	CL=50 pF
–IOCS16/IOCHRDY Timing Figures 8, 9					
tD15	IOCHRDY Inactive from Command		50	ns	CL=200 pF
tD16	IOCHRDY Active from SYSCLK		55	ns	CL=200 pF
tD17	–IOCS16 Inactive from Command		55	ns	CL=200 pF
SYSCLK/ALE Timing Figures 8, 9					
t18	SYSCLK Period	84		ns	
t19	SYSCLK Pulse Width Low	35		ns	
t20	SYSCLK Pulse Width High	35		ns	
t21	ALE Pulse Width High	40		ns	

Note: –IOCS16, IOCHRDY are open-drain outputs with an active pull-up for approximately 10 ns. These parameters are measured at VOH = 1.5 V with a 300 ohm pull-up. Actual performance will vary depending on system configuration.

VLSI TECHNOLOGY, INC.

ADVANCE INFORMATION

VL82C106

AC CHARACTERISTICS (Cont.): TA = 0°C to +70°C, VCC = 5 V ±5%, VSS = 0 V

Symbol	Parameter	Min	Max	Unit	Conditions
IDE Interface Timing Figure 10					
tD18	IRQI Delay from IDINT		40	ns	CL=100 pF
tD19	IDENH/IDENL Delay from Address		60	ns	CL=50 pF
tD20	IDB7 Delay from SD7 Input		40	ns	CL=200 pF
tD21	SD7 Delay from IDB7 Input		40	ns	CL=200 pF
tD22	SD7 Delay from –DC Input		40	ns	CL=200 pF
tD23	SD7 Delay from –IOR During IDE Access	0	85	ns	CL=200 pF
tH24	SD7 Hold from –IOR Inactive	5	60	ns	CL=50 pF
tD25	IDB7 Delay from –IOR Inactive	0	85	ns	CL=200 pF
tH26	IDB7 Hold from –IOR Active	5	60	ns	CL=50 pF
XDATA Control Timing Figure 11					
tD27	–XDIRS/–XDIRX Delay from –XDDIR		30	ns	CL=50 pF
tD28	–XDIRX Delay from –IOR		30	ns	CL=50 pF
tD29	–XDEN Delay from Command		30	ns	CL=50 pF
Real Time Clock Timing Figure 17					
tPSPW	Power Sense Pulse Width	2		µs	
tPSD	Power Sense Delay	2		µs	
tVRTD	VRT Bit Delay		2	µs	
tSBPW	–STBY Pulse Width	2		µs	

VLSI TECHNOLOGY, INC.

ADVANCE INFORMATION

VL82C106

AC CHARACTERISTICS (Cont.): TA = 0°C to +70°C, VCC = 5 V ±5%, GND = 0 V

Symbol	Parameter	Min	Max	Units	Conditions
SERIAL, PRINTER					
Transmitter Figure 13					
tHR1	Delay from Rising Edge of –IOW (WR THR) To Reset Interrupt		175	ns	100 pF Load
tIRS	Delay from THRE Reset to Transmit Start		16	CLK Cycles	Note 2
tSI	Delay from Write to THRE	8	24	CLK Cycles	Note 2
tSTI	Delay from Stop to Interrupt (THRE)		8	CLK Cycles	Note 2
tIR	Delay from –IOR (RD IIR) to Reset Interrupt (THRE)		250	ns	100 pF Load
Modem Control Figure 14					
tMDO	Delay from –IOW (WR MCR) to Output		250	ns	100 pF Load
tSIM	Delay to Set Interrupt from MODEM Input		250	ns	100 pF Load
tRIM	Delay to Reset Interrupt from –IOR (RS MSR)		250	ns	100 pF Load
Receiver Figure 12					
tSINT	Delay from Stop to Set Interrupt		1	CLK Cycles	Note 2
tRINT	Delay from –IOR (RD RBR/RDLSR) to Reset Interrupt		1	µs	100 pF Load
Parallel Port Figure 15					
tDT	Data Time	1		µs	Software Controller
tSB	Strobe Time	1	500	µs	Software Controller
tAD	Acknowledge Delay (Busy Start to Acknowledge)			µs	Defined by Printer
tAKD	Acknowledge Delay (Busy End to Acknowledge)			µs	Defined by Printer
tAK	Acknowledge Duration Time			µs	Defined by Printer
tBSY	Busy Duration Time			µs	Defined by Printer
tBSD	Busy Delay Time			µs	Defined by Printer

Notes:
1. All timing specifications apply to pins on both serial channels (e.g. RI refers to both RI0 and RI1).
2. CLK cycle refers to external 18.432 MHz clock divided by 10, e.g. 1.8432 MHz.

VLSI TECHNOLOGY, INC.

ADVANCE INFORMATION

VL82C106

BUS TIMING

FIGURE 5. WRITE CYCLE

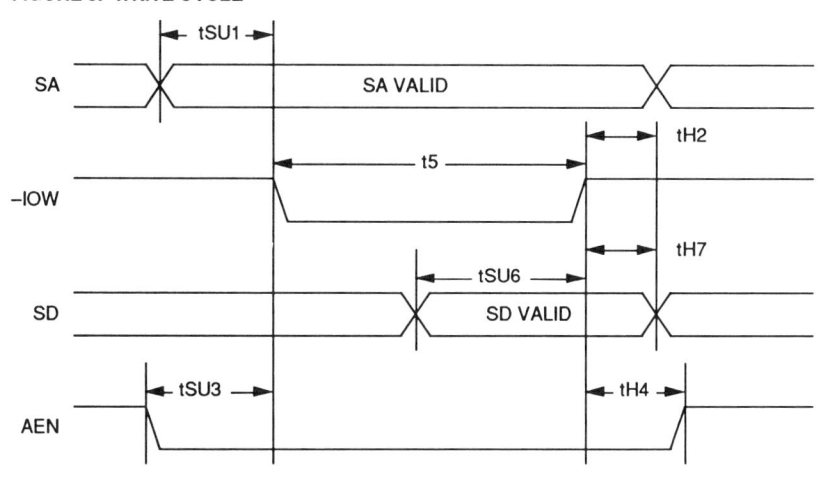

FIGURE 6. READ CYCLE

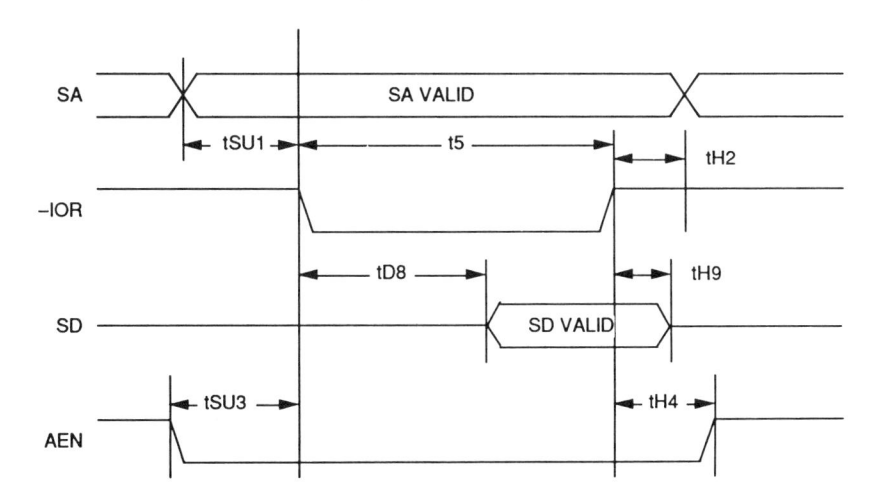

VLSI TECHNOLOGY, INC.

ADVANCE INFORMATION

VL82C106

CHIP SELECT TIMING

FIGURE 7.

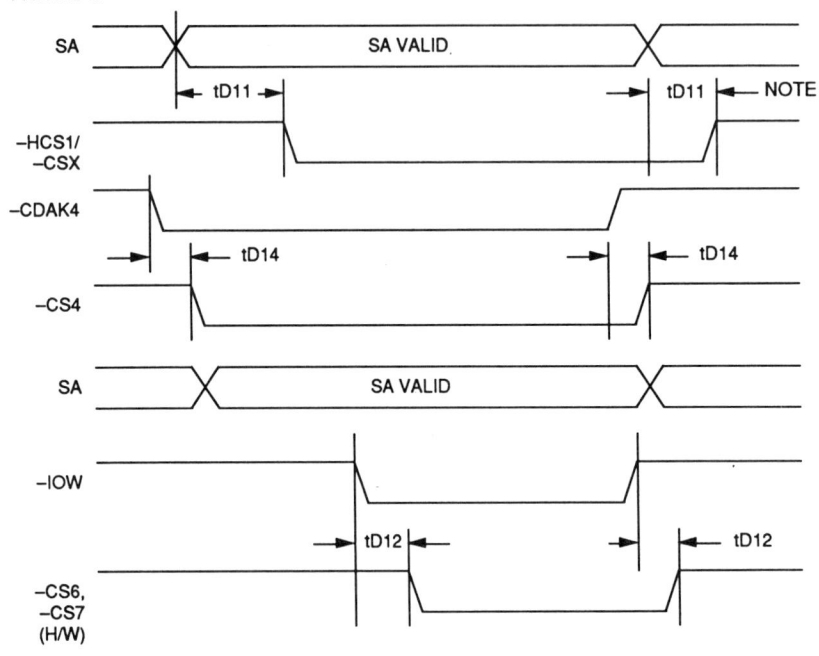

Note: Except −CS6, −CS7 hard-wired.

VLSI TECHNOLOGY, INC.

ADVANCE INFORMATION

VL82C106

IOCHRDY TIMING
FIGURE 8.

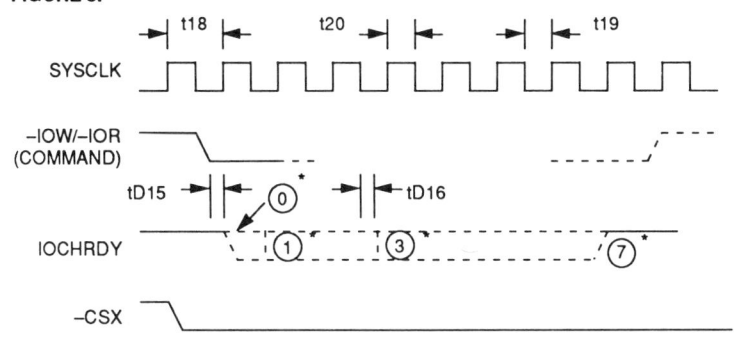

* Programmed number of wait states. 0 = 0 wait state, 1 = 1 wait states, etc.

IOCS16 TIMING
FIGURE 9.

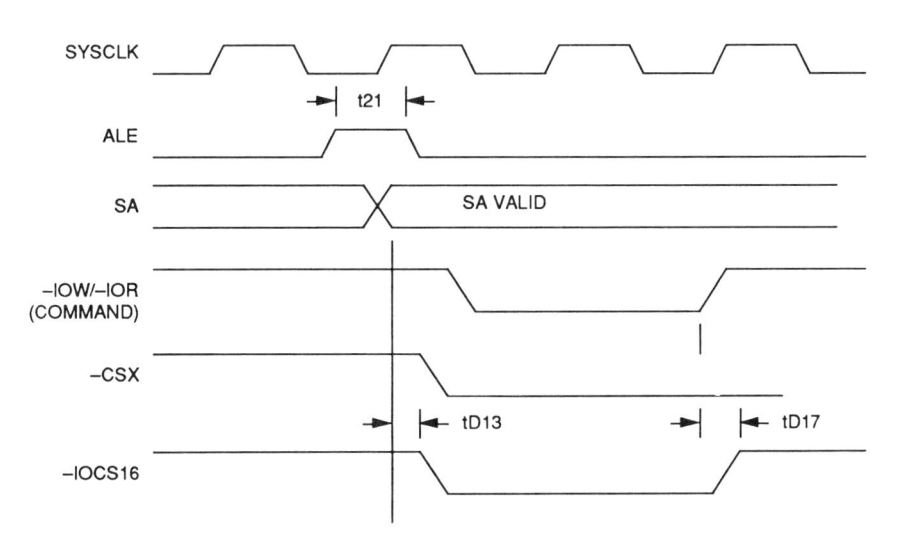

VLSI TECHNOLOGY, INC.

ADVANCE INFORMATION

VL82C106

IDE INTERFACE TIMING

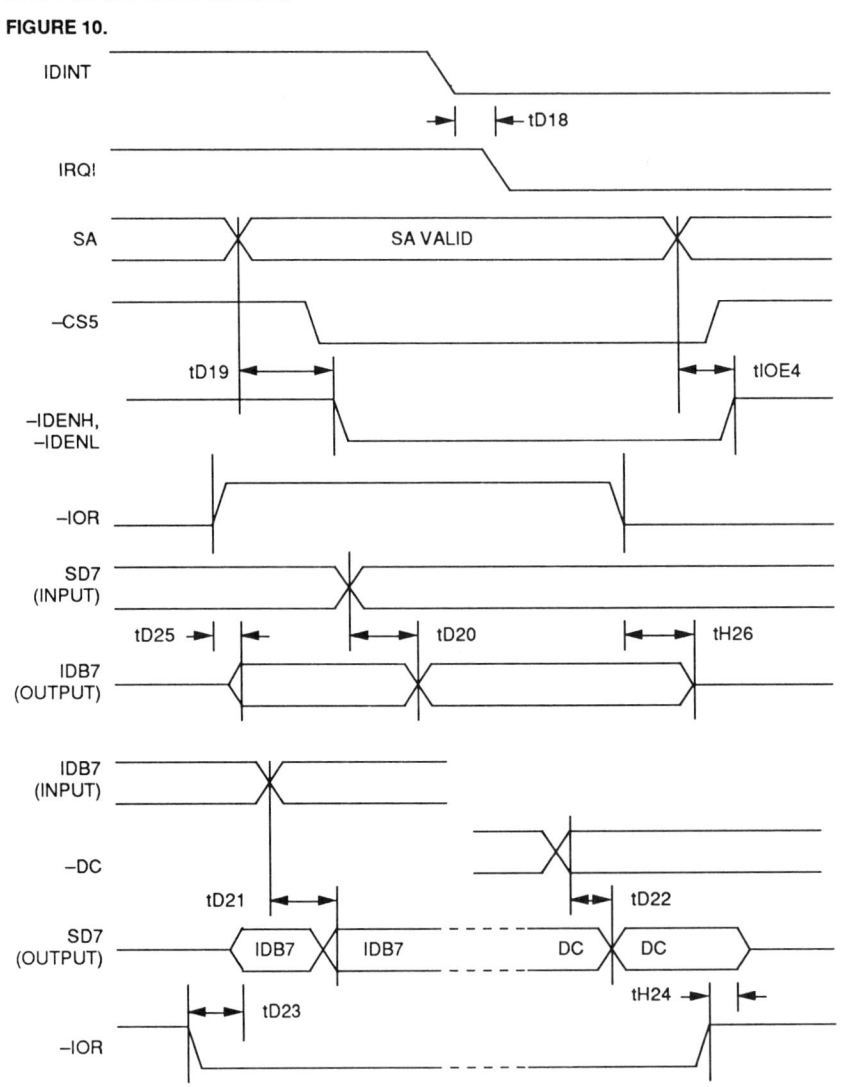

FIGURE 10.

VLSI TECHNOLOGY, INC.

ADVANCE INFORMATION

VL82C106

XDATA CONTROL TIMING

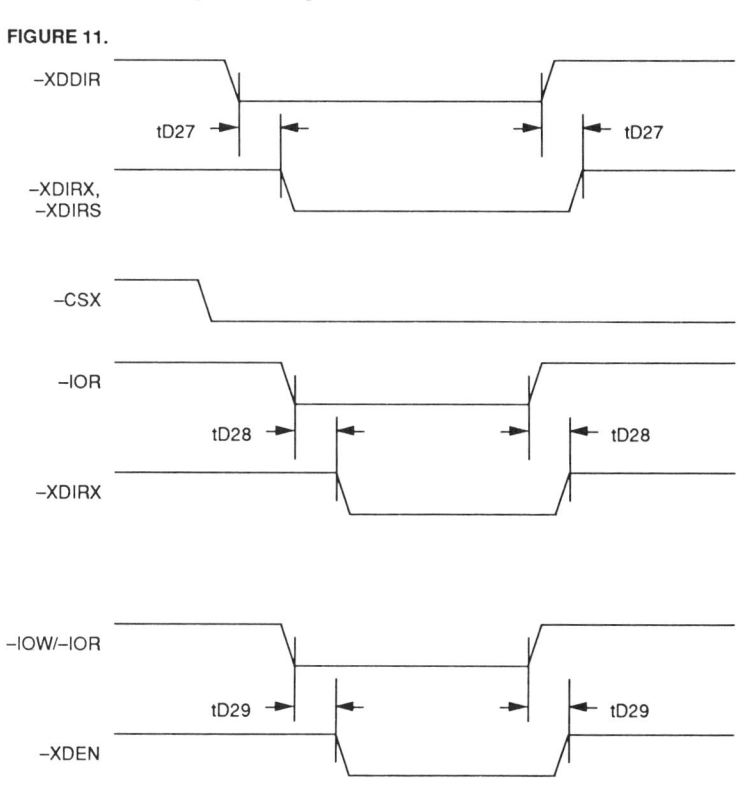

FIGURE 11.

VLSI TECHNOLOGY, INC.

ADVANCE INFORMATION

VL82C106

RECEIVER TIMING
FIGURE 12.

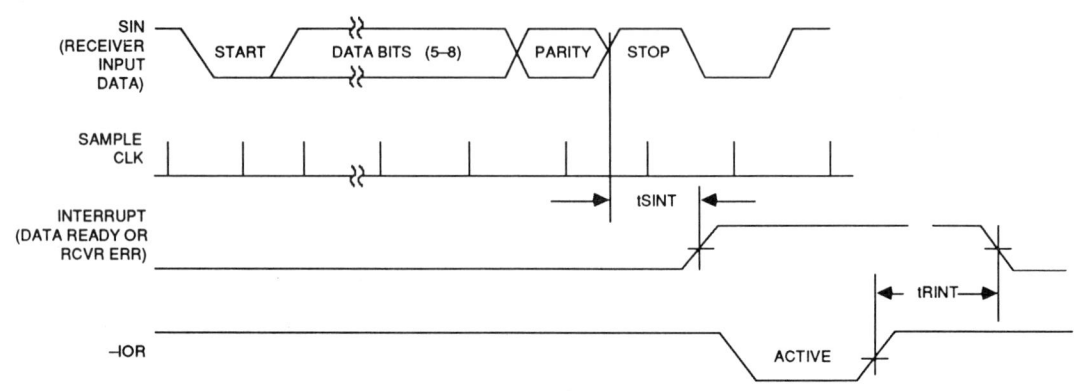

TRANSMITTER TIMING
FIGURE 13.

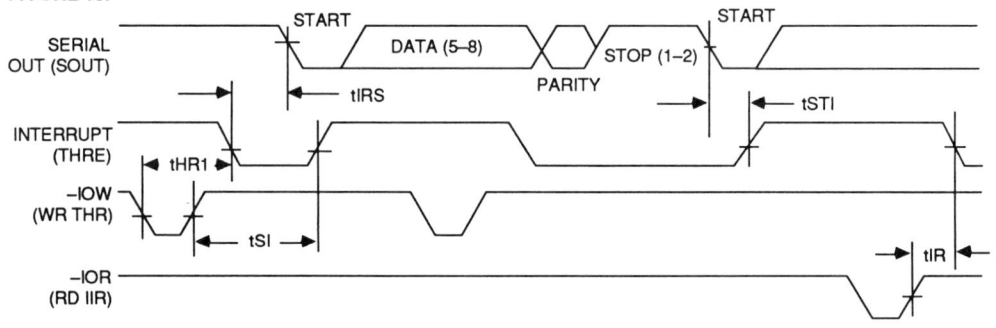

MODEM TIMING
FIGURE 14.

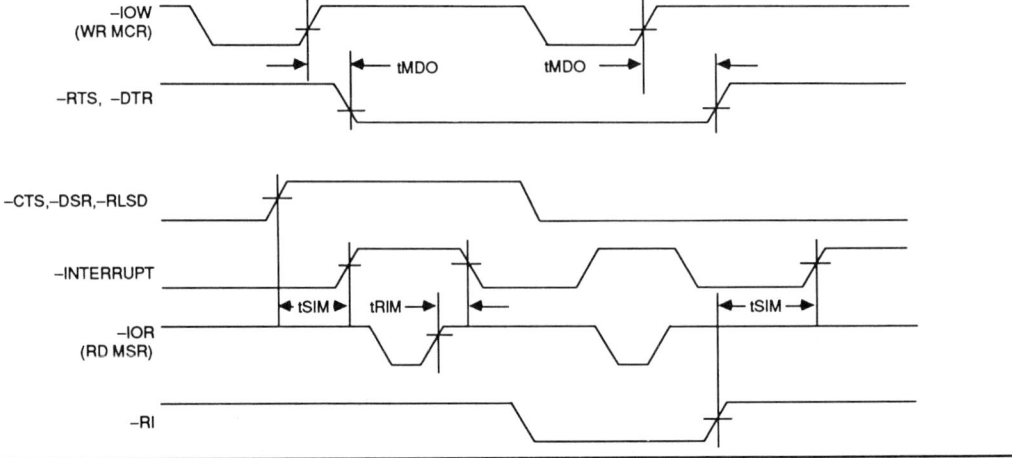

VLSI TECHNOLOGY, INC.

ADVANCE INFORMATION

VL82C106

PARALLEL PORT TIMING
FIGURE 15.

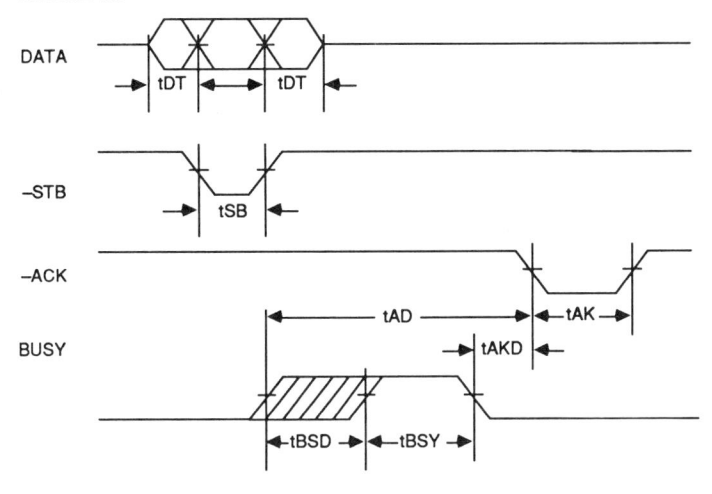

KEYBOARD CONTROLLER TIMING
FIGURE 16.

RECEIVE

* (KSRE (pin 100)) - In PS/2 Mode
KCLK (pin 103)

* (KHSE (pin 101)) - In PS/2 Mode
KDAT (pin 104)

tHOLD = 52 Periods of XTAL1 Input
(2.8 µs @ XTAL1 = 18.432 MHz)

tSU = 0 ns

TRANSMIT

* (KSRE - PS/2)
KCLK

* (KHSE - PS/2)
KDAT

tPD = 18 Periods of XTAL1 Input   Min.
(976 ns @ XTAL1 = 18.432 MHz)

52 Periods of XTAL1 Input   Max.
(2.8 µs @ XTAL1 = 18.432 MHz)

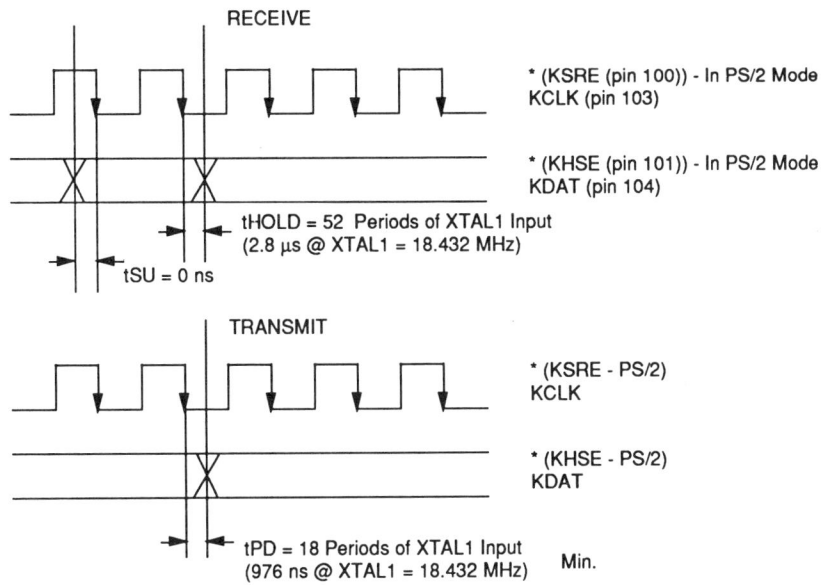

* **Note:** Specifications are identical for KHSE (pin 101) with respect to KSRE (pin 100) in PS/2 Mode.

VL82C106

ADVANCE INFORMATION

REAL TIME CLOCK TIMING
FIGURE 17.

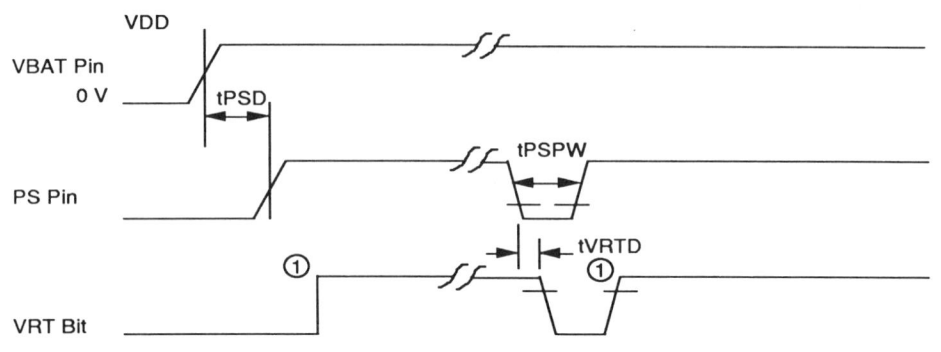

① The VRT bit is set to a "1" by reading Register D. The VRT bit can only be cleared by pulling the PS pin low (see REGISTER D ($0D)).

CRYSTAL OSCILLATOR CONFIGURATIONS
FIGURE 18.

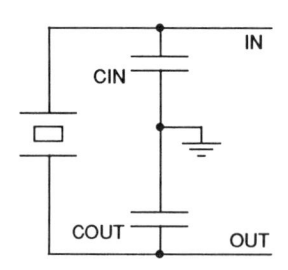

32.768 KHz
CIN = COUT = 10-22 pF
CIN may be a trimmer for precision timekeeping applications.

18.432 MHz
CIN = 10 pF
COUT = 30 pF

RECOMMENDED CRYSTAL PARAMETERS

Rs (max) ≤ 40k Ω
Co (max) ≤ 1.7 pF
Cl (max) ≤ 12.5 pF
Parallel Resonance

Rs ≤ 50 Ω
Co ≤ 7 pF
Cl ≤ 20 pF
Parallel Resonance

VLSI TECHNOLOGY, INC.

ADVANCE INFORMATION

VL82C106

ABSOLUTE MAXIMUM RATING

Ambient Temperature	–10°C to +70°C
Storage Temperature	–65°C to 150°C
Supply Voltage to Ground Potential	–0.5 V to VCC +0.3 V
Aplied Output Voltage	–0.5 V to VCC +0.3 V
Aplied Input Voltage	–0.5 V to +7.0 V
Power Dissipation	500 mW

Stresses above those listed may cause permanent damage to the device. These are stress ratings only. Functional operation of this device at these or any other conditions above those indicated in this data sheet is not implied. Exposure to absolute maximum rating conditions for extended periods may affect device reliability.

DC CHARACTERISTICS: TA = 0°C to +70°C, VDD = 5 V ±5%, VSS = 0 V

Symbol	Parameter	Min	Max	Units	Conditions
VIL	Input Low Voltage				
	Input Types (All except I2)	–0.5	0.8	V	
	Input Type I2	–0.5	VDD*0.2	V	
VIH	Input High Voltage				
	Input Types I1, I3, I4, IO2, IO4, IO5, IO6	2.0	VDD+0.5	V	
	Input Type I2	VDD*0.7	VDD+0.5	V	
	Input Type I5	2.4	VDD+0.5	V	
VOL	Output Low Voltage				
	Output Type 01		0.4	V	IOL = 2.0 mA
	Output Type 06		0.4	V	IOL = 4.0 mA
	Output Type 04, I04, I05		0.4	V	IOL = 12.0 mA
	Output Type 02, 07, 08, IO2, I06		0.4	V	IOL = 24.0 mA
VOH	Output High Voltage				
	Output Type 01, 06	2.4		V	IOH = –0.8 mA
	Output Type I05	2.4		V	IOH = –2.0 mA
	Output Type 02, IO2, I06	2.4		V	IOH = –2.4 mA
IIH	Input High Current				
	Input Types I1, I3, I4, I5		10	µA	VIN = VDD
IIL	Input Low Current				
	Input Types I1, I5	–10		µA	VIN = VSS + 0.2
	Input Types I4, IO6	–500	–50	µA	VIN = 0.8 V All other pins floating.
ILOL	Three-State Leakage Current				
	I/O Output Types 06, 07,	–50		µA	VSS+0.2
	IO2, IO4, IO5		50	µA	VDD
IODL	Open-Drain Off Current				
	I/O Output Type 04	–5.0	–1.0	mA	V = 0.8 V
CO	Output Capacitance		8	pF	
CI	Input		8	pF	
CIO	Input/Output Capacitance		16	pF	
ICC	Operating Supply Current		40	mA	
IBAT	VBAT Supply Current, Standby Mode		5.0	µA	VBAT = 3.0 V
			50.0	µA	VBAT = 5.0 V

Note: For pin types, refer to the Legend and Pin Descriptions on pages 4-7 of this data sheet.

VLSI TECHNOLOGY, INC.

ADVANCE INFORMATION

VL82C106

PACKAGE OUTLINE
128-PIN PLASTIC QUAD FLATPACK

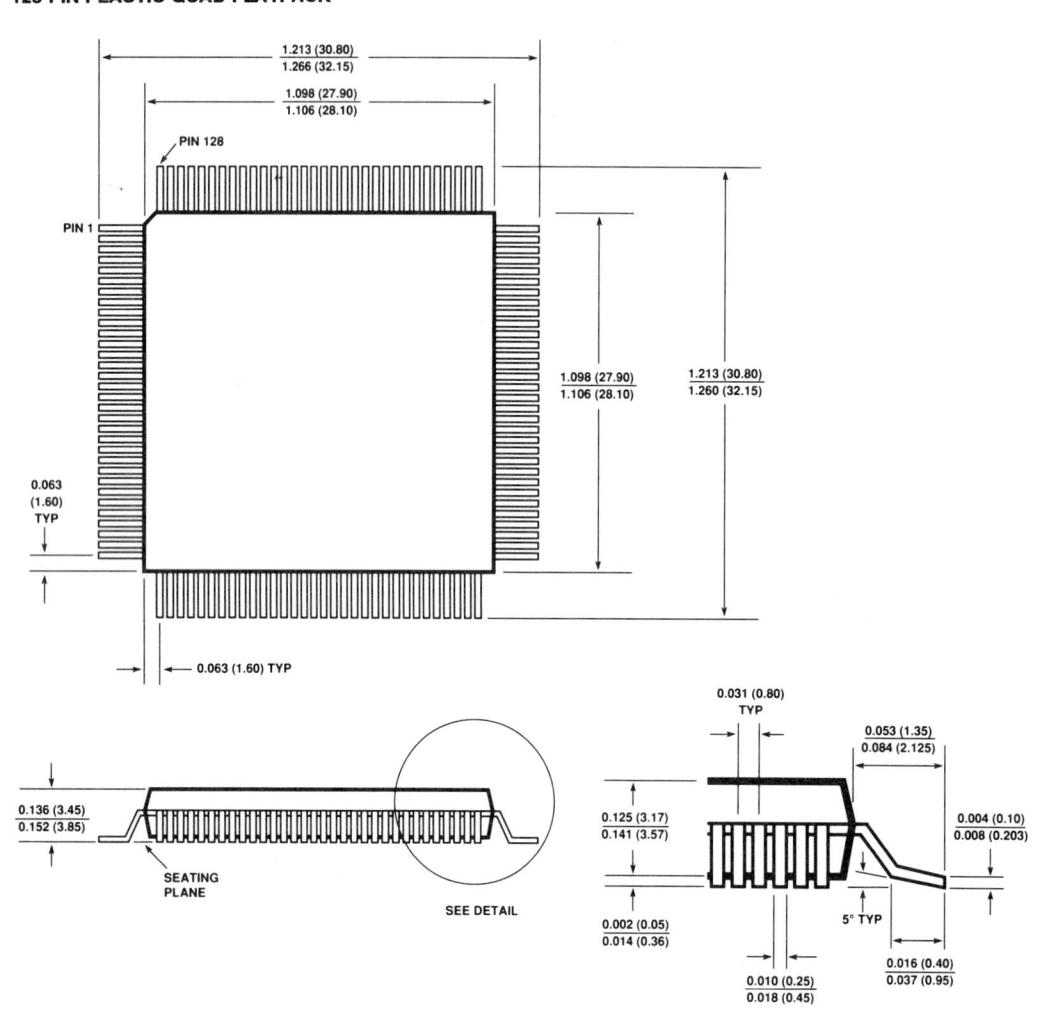

NOTES: UNLESS OTHERWISE SPECIFIED
1. THE CJQFP ARE CURRENTLY USED ONLY FOR PROTOTYPE BUILDS.
2. ALL DIMENSIONS ARE IN INCHES (MM). CONTROLLING DIMENSION IS METRIC.
3. CJQFP ARE EPOXY DIE ATTACHED AND EPOXY SEALED.
4. LEAD COPLANARITY SHALL BE WITHIN .004" (0.102 MM) MAXIMUM.
5. PJQFP SHALL BE MANUFACTURED WITH ONE OF THE FOLLOWING DIRECT MATERIALS:
   LEADFRAME: ALLOY 42, COPPER OLIN194
   DIE ATTACH MATERIAL: HITACHI CHE EN-4000, KASEI EPINAR 4110
   MOLD COMPOUND: SUMITOMO 6300, KASEI CEL 4000

25-90003

DP8490 Enhanced Asynchronous SCSI Interface (EASI)

General Description

The DP8490 EASI is a CMOS device designed to provide a low cost, high performance Small Computer Systems Interface. It complies with the ANS X3.131-1986 SCSI standard as defined by the ANSI X3T9.2 committee. It can act as both INITIATOR and TARGET, making it suitable for any application. The EASI supports selection, reselection, arbitration and all other bus phases. High-current open-drain drivers on chip reduce application chip count by interfacing direct to the SCSI bus. An on-chip oscillator provides all timing delays.

The DP8490 is pin and program compatible with the NMOS NCR5380 and CMOS DP5380 devices. NCR5380, DP5380 or AM5380 applications should be able to use it with no changes to hardware or software. The DP8490 includes new features which make this part more attractive for new designs and performance upgrades. These new features include μP data bus parity, programmable parity for both SCSI and μP busses, loopback test mode, improved arbitration support, faster timing and extended interrupt control logic. The DP8490 is available in a 40-pin DIP or a 44-pin PCC.

The EASI is intended to be used in a microprocessor based application, and achieves maximum performance with a DMA controller. The device is controlled by reading and writing several internal registers. A standard non-multiplexed address and data bus easily fits any μP environment.

Data transfers can be performed by programmed-I/O, pseudo-DMA or via a DMA controller. The EASI easily interfaces to a DMA controller using normal or Block Mode. The EASI can be used in either a polled or interrupt-driven environment. The EASI includes enhanced features for interrupt control.

Features

SCSI Interface
- Supports TARGET and INITIATOR roles
- Parity generation with optional checking
- Programmable parity polarity (ODD/EVEN)
- Arbitration support—can interrupt when done
- Direct control/monitoring of all SCSI signals
- High current outputs drive SCSI bus directly
- Faster and improved timing
- Very low SCSI bus loading

μP Interface
- Memory or I/O-mapped control transfers
- Programmed-I/O or DMA data transfers
- Normal or Block-mode DMA
- Fast DMA handshake timing
- Individually maskable interrupts
- Active interrupts identified in one register
- Optional data bus parity generation/checking
- Programmable parity polarity (ODD/EVEN)
- Loopback test mode

Connection Diagram

MPU BUS: D0..7,P, A0, A1, A2, CS, RD, WR, RESET, INT, DRQ, DACK, READY, EOP, GND

DP8490 Enhanced Asynchronous SCSI Interface (EASI)

SCSI BUS: DB0..7, DBP, RST, BSY, SEL, ACK, ATN, REQ, I/O, C/D, MSG, VCC

TL/F/9387–1

Table of Contents

1.0 FUNCTIONAL DESCRIPTION
2.0 PIN DESCRIPTION
3.0 REGISTER DESCRIPTION
4.0 DEVICE OPERATION
5.0 INTERRUPTS
6.0 RESET CONDITIONS
7.0 LOOPBACK TESTING
8.0 EXTRA FEATURES/COMPATIBILITY
9.0 APPLICATION GUIDE
10.0 ABSOLUTE MAXIMUM RATINGS
11.0 DC ELECTRICAL CHARACTERISTICS
12.0 AC ELECTRICAL CHARACTERISTICS
A1 FLOWCHARTS
A2 REGISTER CHART

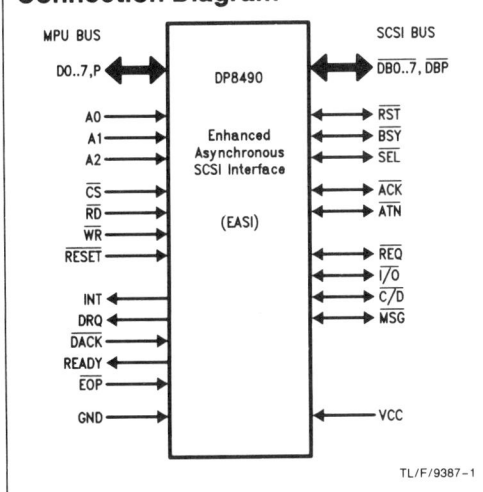

TRI-STATE® is a registered trademark of National Semiconductor Corporation.
PAL® is a registered trademark of and used under license from Monolithic Memories, Inc.

©1988 National Semiconductor Corporation   TL/F/9387

1.0 Functional Description

1.1 OVERVIEW

The EASI is designed to be used as a peripheral device in a μP-based application and appears as a number of read/write registers. Write registers are programmed to select desired functions. Status registers provide indication of operating conditions. In an application extensive use of interrupts is desirable. The EASI incorporates an improved interrupt structure which enables fully interrupt-driven operation. In the enhanced mode interrupts can be individually masked or enabled, and a status register identifies all active interrupt requests.

For best performance a DMA controller can be easily interfaced directly to the EASI. The EASI provides request/acknowledge and wait-state signals for the DMA interface.

The SCSI bus is easily controlled via the EASI registers. Any bus signal may be asserted or deasserted via a bit in the appropriate register, and the state of every signal is available by reading registers. This direct control over SCSI signals allows the user to implement all or part of the protocol in firmware. The EASI provides hardware support for much of the protocol, and all speed-critical steps are handled by the EASI.

The EASI provides the following SCSI support:

- Programmed-I/O transfers for all eight information transfer types, with or without parity.
- Data transfers via DMA, in either block or non-block mode. The DMA interface supports most devices.
- Individual setting/resetting and monitoring of every SCSI bus signal.
- Automatic release of the bus for BSY loss from a TARGET, SCSI RST, and lost arbitration.
- Automatic bus arbitration with an optional interrupt upon completion—the μP has only to check for highest priority. The 2.2 μs arbitration delay can be optionally performed by the EASI.
- Selection or Reselection of any bus device. The EASI will respond to both Selection and Reselection.
- Optional automatic monitoring of the $\overline{BSY}$ signal from a TARGET with an interrupt after releasing control of the bus.
- Optional parity polarity selection. Default after reset is ODD, but EVEN generation and checking can be programmed for diagnostic purposes and to determine whether a device supports parity when first making a connection.

Figure 1 shows an EASI in a typical application, a low cost embedded SCSI disk controller. In this application the 8051 single-chip μP acts as the controller and the dual DMA channels in the DP8475 allow one for the disk data and the other for SCSI data. The PAL provides chip selection as well as determining who has control of the bus. The advantage of using a μP with on-board ROM is that there is more free time on the external bus.

1.2 μP INTERFACE

Figure 2 shows a block diagram of the EASI. Key blocks within the EASI are Read/Write registers with associated decode and control logic, interrupt and DMA logic, SCSI bus arbitration logic, SCSI drivers/receivers with parity and the SCSI data input and output registers. The EASI has three interfaces, one to SCSI, one to a DMA controller and the third to a μP. The internal registers control all operation of the EASI.

The μP interface consists of non-multiplexed address and data busses with associated control signals. The data bus can be programmed to use either ODD or EVEN parity. Address decode logic selects a register for reading or writing. The address lines A0–A2 select the register for μP accesses while for DMA accesses the address lines are ignored. The decode logic also selects different registers or functions to be mapped into address 7, according to the programmed mode (see Section 8).

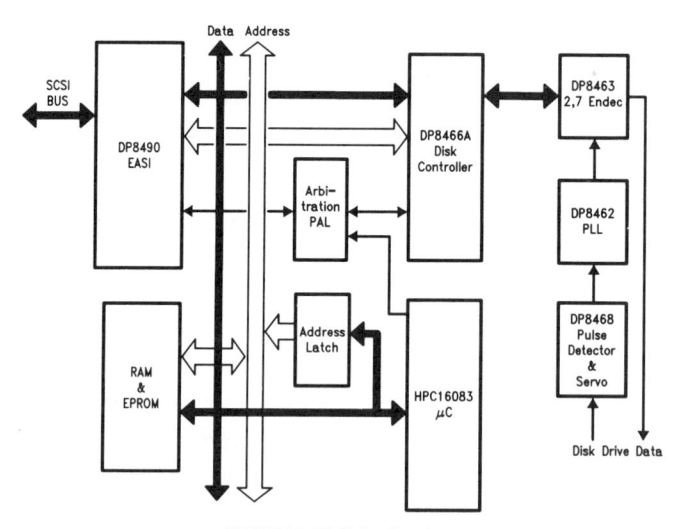

FIGURE 1. EASI Application

1.0 Functional Description (Continued)

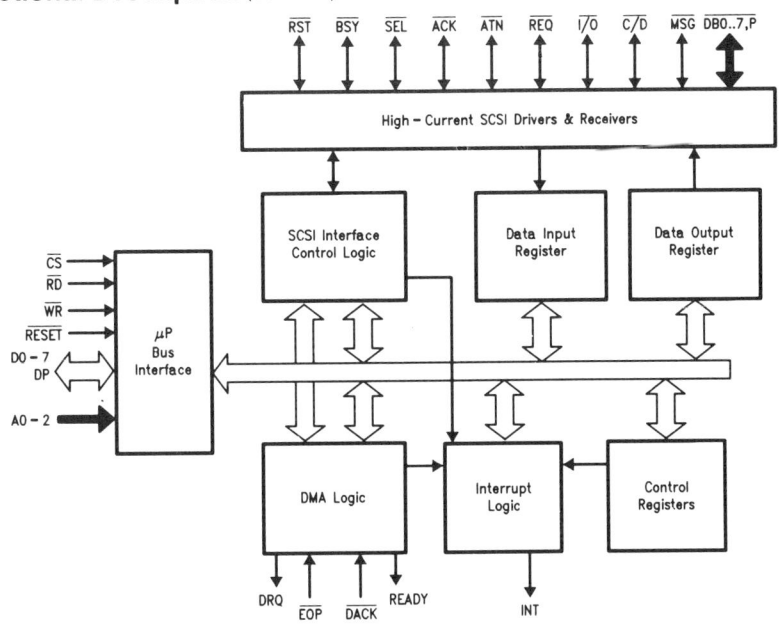

FIGURE 2. EASI Block Diagram

The register bank consists of twelve registers mapped into an address space of eight locations. Upon an external chip reset the registers are cleared (all zeroes)—the same as the NCR5380. Once the ENHANCED MODE bit in the INITIATOR COMMAND REGISTER is set, three new registers can be accessed to utilize the extra features of the DP8490 EASI.

1.3 DMA INTERFACE

The DMA logic interfaces to single-cycle, block mode, flow-through or fly-by controllers. Single byte transfers are accomplished via the DRQ/$\overline{DACK}$ handshake signals. Block mode transfers use the READY output to control the speed (insert wait-states). An End Of Process ($\overline{EOP}$) input from the DMA controller signals the EASI to halt DMA transfers. An interrupt can be generated for DMA completion or an error (see Section 5). All DMA data passes through the SCSI data input and output registers, automatically selected during DMA cycles.

1.4 SCSI INTERFACE

The EASI contains all logic required to interface directly to the SCSI bus. Direct control and monitoring of all SCSI signals is provided. The state of each SCSI signal may be determined by reading a register which continuously reflects the state of the bus. Each signal may be asserted by writing a ONE to the appropriate bit.

The EASI includes logic to automatically handle SCSI timing sequences too fast for µP control. In particular there is hardware support for DMA transfers, bus arbitration, selection/reselection, bus phase monitoring, $\overline{BSY}$ monitoring for bus disconnection, bus reset and parity generation and checking.

The EASI arbitration logic controls arbitration for use of the SCSI bus. The µP programs the SCSI device ID into the EASI, then sets the ARBITRATE bit. The EASI will interrupt the µP when one of three events occurs: arbitration is lost; arbitration has completed and the ID priorities need to be checked; or arbitration is complete and the 2.2 µs SCSI Arbitration delay has expired. Arbitration can be invoked with the enhanced feature of an interrupt on completion or the expiration of the SCSI Arbitration delay. These extra steps are programmed via the EXTRA MODE REGISTER (EMR). The INITIATOR COMMAND REGISTER (ICR) is read to determine whether arbitration has been won or lost.

The $\overline{BSY}$ signal is continuously monitored to detect bus disconnection and bus free phases. The EASI incorporates an on-board oscillator to determine Bus Settle, Bus Free and Arbitration Delays. The oscillator tolerance guarantees all timing to be within the SCSI specification.

The EASI incorporates high-current drivers and SCHMITT trigger receivers for interfacing directly to the SCSI bus. This feature reduces the chip count of any SCSI application. The driver/receivers also incorporate loopback logic which is enabled by an EMR bit. The Loopback mode enables testing of all EASI functions without interfering with the SCSI bus.

1.5 PARITY

The EASI provides for parity protection on both the µP and SCSI interfaces. Each data bus has eight data bits and one parity bit (only the PCC part provides µP parity, both the DIP and PCC provide SCSI parity). In each case the parity may

1.0 Functional Description (Continued)

be enabled via a register bit. A parity error can be programmed to cause an interrupt. Additionally the parity may be programmed to be either ODD or EVEN. This has a particular use on the SCSI interface where programming EVEN parity allows diagnostics, or determining whether a device supports parity. The inclusion of μP parity allows development of controllers that maintain data integrity right from the media to the host system.

1.6 INTERRUPTS

The EASI is intended to be used in an interrupt-driven environment. Each function can be programmed to cause an interrupt. In ENHANCED MODE two registers are used to control interrupts—the INT STATUS REGISTER (ISR) and the INT MASK REGISTER (IMR). Each interrupt can be masked from interrupting via the IMR. When an interrupt is recognized by the μP, reading the ISR will display all active interrupt sources. The ISR contents remain unchanged until an interrupt reset is programmed. A shadow register behind the ISR guarantees that interrupts occurring while others are serviced will not be lost.

2.0 Pin Descriptions

Symbol	DIP	PCC	Type	Function
CS	21	24	I	**Chip Select:** an active low enable for read or write operations, accessing the register selected by A0–A2.
A0–A2	30, 32, 33	33, 36, 37	I	**Address 0–2:** these three signals are used with CS, RD, and WR to address a register for read or write.
RD	24	27	I	**Read:** an active low enable for reading an internal register selected by A0–A2 and enabled by CS. It also selects the Input Data Register when used with DACK.
WR	29	32	I	**Write:** an active low enable for writing an internal register selected by A0–A2 and enabled by CS. It also selects the Output Data Register when used with DACK.
RESET	28	31	I	**Reset:** an active low input with a Schmitt trigger. Clears all internal registers. (SCSI RST unaffected.)
D0–D7, P	1, 40–34	2, 44–38, 1	I/O	**Data 0–7, P:** bidirectional TRI-STATE® signals connecting the active high μP data bus to the internal registers. The PCC part offers an optional parity on the μP data bus. If the parity option is not enabled the pin is TRI-STATE.
INT	23	26	O	**Interrupt:** an active high output to the μP when an error has occurred, an event requires service or has completed.
DRQ	22	25	O	**DMA Request:** an active high output asserted when the data register is ready to be read or written. DRQ occurs only if DMA mode is enabled. The signal is cleared by DACK.
DACK	26	29	I	**DMA Acknowledge:** an active low input that resets DRQ and addresses the data registers for input or output transfers. DACK is used instead of CS by the DMA controller.
READY	25	28	O	**Ready:** an active high output used to control the speed of block mode DMA transfers. Ready goes active when the chip is ready to send/receive data and remains inactive after the transfer until the byte is sent or until the DMA mode bit is reset.
EOP	27	30	I	**End of Process:** an active low signal that terminates a block of DMA transfers. It should be asserted during the transfer of the last byte.
DB0–DB7, DBP	9–2, 10	10–3, 11	I/O	**DB0–DB7, DBP:** SCSI data bus with parity. DB7 is the MSB and is the highest priority during arbitration. Parity is default ODD but can be programmed EVEN. Parity is always generated and can be optionally checked. Parity is not valid during arbitration.
RST	16	18	I/O	**Reset:** SCSI reset, monitored and can be set by EASI.
BSY	13	15	I/O	**Busy:** indicates the SCSI bus is being used. Can be driven by TARGET or INITIATOR.

2.0 Pin Descriptions (Continued)

Symbol	DIP	PCC	Type	Function
$\overline{SEL}$	12	14	I/O	**Select:** used by the INITIATOR to select a TARGET or by the TARGET to reselect an INITIATOR.
$\overline{ACK}$	14	16	I/O	**Acknowledge:** driven by the INITIATOR and received by the TARGET as part of the $\overline{REQ}/\overline{ACK}$ handshake.
$\overline{ATN}$	15	17	I/O	**Attention:** driven by the INITIATOR to indicate an attention condition to the TARGET.
$\overline{REQ}$	20	22	I/O	**Request:** driven by the TARGET and received by the INITIATOR as part of the $\overline{REQ}/\overline{ACK}$ handshake.
$\overline{I/O}$	17	19	I/O	**Input/Output:** driven by the TARGET to control the direction of transfers on the SCSI bus. This signal also distinguishes between selection and reselection.
$\overline{C/D}$	18	20	I/O	**Command/Data:** driven by the TARGET to indicate whether command or data bytes are being transferred.
$\overline{MSG}$	19	21	I/O	**Message:** driven by the TARGET during message phase to identify message bytes on the bus.
V_{CC} GND	31 11	35 12, 13	—	**V_{CC}, GND:** +5 V_{DC} is required. Because of very large switching currents, good decoupling and power distribution is mandatory.

2.1 Connection Diagrams

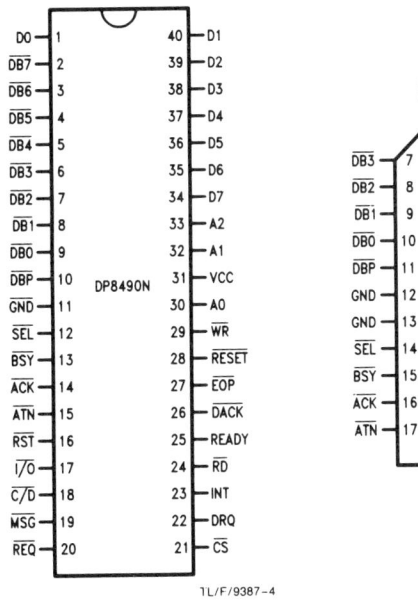

DP8490N (DIP pinout):
- Pin 1: D0, Pin 40: D1
- Pin 2: $\overline{DB7}$, Pin 39: D2
- Pin 3: $\overline{DB6}$, Pin 38: D3
- Pin 4: $\overline{DB5}$, Pin 37: D4
- Pin 5: $\overline{DB4}$, Pin 36: D5
- Pin 6: $\overline{DB3}$, Pin 35: D6
- Pin 7: $\overline{DB2}$, Pin 34: D7
- Pin 8: $\overline{DB1}$, Pin 33: A2
- Pin 9: $\overline{DB0}$, Pin 32: A1
- Pin 10: $\overline{DBP}$, Pin 31: VCC
- Pin 11: $\overline{GND}$, Pin 30: A0
- Pin 12: $\overline{SEL}$, Pin 29: $\overline{WR}$
- Pin 13: $\overline{BSY}$, Pin 28: $\overline{RESET}$
- Pin 14: $\overline{ACK}$, Pin 27: $\overline{EOP}$
- Pin 15: $\overline{ATN}$, Pin 26: $\overline{DACK}$
- Pin 16: $\overline{RST}$, Pin 25: READY
- Pin 17: $\overline{I/O}$, Pin 24: $\overline{RD}$
- Pin 18: $\overline{C/D}$, Pin 23: INT
- Pin 19: $\overline{MSG}$, Pin 22: DRQ
- Pin 20: $\overline{REQ}$, Pin 21: $\overline{CS}$

TL/F/9387–4
Order Number DP8490N
See NS Package Number N40A

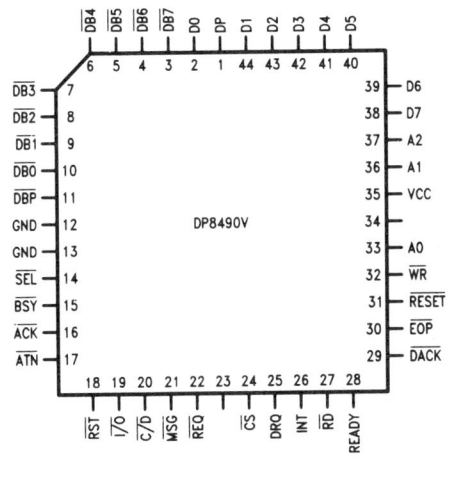

DP8490V (PCC pinout):
- Top (pins 6–1, 44–40): $\overline{DB4}$, $\overline{DB5}$, $\overline{DB6}$, $\overline{DB7}$, D0, DP, D1, D2, D3, D4, D5
- Left side: Pin 7: $\overline{DB3}$, Pin 8: $\overline{DB2}$, Pin 9: $\overline{DB1}$, Pin 10: $\overline{DB0}$, Pin 11: $\overline{DBP}$, Pin 12: GND, Pin 13: GND, Pin 14: $\overline{SEL}$, Pin 15: $\overline{BSY}$, Pin 16: $\overline{ACK}$, Pin 17: $\overline{ATN}$
- Bottom (pins 18–28): $\overline{RST}$, $\overline{I/O}$, $\overline{C/D}$, $\overline{MSG}$, $\overline{REQ}$, $\overline{CS}$, DRQ, INT, $\overline{RD}$, READY
- Right side: Pin 39: D6, Pin 38: D7, Pin 37: A2, Pin 36: A1, Pin 35: VCC, Pin 34, Pin 33: A0, Pin 32: $\overline{WR}$, Pin 31: $\overline{RESET}$, Pin 30: $\overline{EOP}$, Pin 29: $\overline{DACK}$

TL/F/9387–5
Order Number DP8490V
See NS Package Number V44A

3.0 Register Description

3.1 GENERAL

The DP8490 EASI is a register-based device with eight addressable locations used to access twelve registers. Some addresses have dual functions depending upon whether they are being read from or written to. Two basic operating "modes" result in differences in the registers accessed through address 7. Device operation is described in Section 4 but mode differences are highlighted in this section and Section 8.

The EASI operates in one of two modes—NORMAL (MODE N) and ENHANCED (MODE E). Switching between the modes is performed by setting or resetting bit 6 in the Initiator Command Register.

In MODE N, EASI registers appear the same as the DP5380. In MODE E, address 7 accesses enhanced logic features. To help identify these differences the register description is split into two subsections. The first describes the registers in MODE N. The second describes the register differences when in MODE E.

Figure 3.1 summarizes the register map in MODE N. Note that for registers reading or writing SCSI signals the SCSI name is used for each bit. Although the SCSI bus is active low, the registers invert the SCSI bus. This means an active SCSI signal is represented by a ONE in a register and an inactive signal by a ZERO.

Hex Adr	Register	Mnemonic	Bits	R/W
0	Output Data Register	ODR	8	WO
0	Current SCSI Data	CSD	8	RO
1	Initiator Command Register	ICR	8	RW
2	Mode Register 2	MR2	8	RW
3	Target Command Register	TCR	4	RW
4	Select Enable Register	SER	8	WO
4	Current SCSI Bus Status	CSB	8	RO
5	Bus and Status	BSR	8	RO
5	Start DMA Send	SDS	0	WO
6	Start DMA Target Receive	SDT	0	WO
6	Input Data Register	IDR	8	RO
7	Start DMA Initiator Receive	SDI	0	WO
7	Reset Parity/Interrupts	RPI	0	RO

FIGURE 3.1. Normal Mode Registers

3.2 NORMAL MODE REGISTERS

OUTPUT DATA REGISTER (ODR)
8 Bits   HA 0   Write Only

This is a transparent latch used to send data to the SCSI bus. The register can be written by μP cycles or via DMA. DMA writes automatically select the ODR at Hex Address 0 (HA 0). This register is also written with the ID bits required during arbitration and selection/reselection phases. Data is latched at the end of the write cycle.

Bit 7							Bit 0
DB7	DB6	DB5	DB4	DB3	DB2	DB1	DB0

Output Data Register

CURRENT SCSI DATA (CSD)
8 Bits   HA 0   Read Only

This register enables reading of the current SCSI data bus. If SCSI parity checking is enabled it will be checked at the beginning of the read cycle. The register is also used for μP accesses of SCSI data during programmed-I/O or ID checking during arbitration. Parity is not valid during arbitration. DMA transfers select the IDR (HA 6) instead of the CSD register.

Bit 7							Bit 0
DB7	DB6	DB5	DB4	DB3	DB2	DB1	DB0

Current SCSI Data

INITIATOR COMMAND REGISTER (ICR)
8 Bits   HA 1   Read/Write

This register is used to control the INITIATOR and some other SCSI signals, and to monitor the progress of bus arbitration. Most of the SCSI signals may also be asserted in TARGET mode. Bits 5 to 0 are reset when $\overline{BSY}$ is lost (see MR2 description).

Bit 7							Bit 0
$\overline{RST}$	AIP/ MODE	LA/ DIFF	$\overline{ACK}$	$\overline{BSY}$	$\overline{SEL}$	$\overline{ATN}$	DBUS

Initiator Command Register

DBUS: Assert Data Bus   Bit 0

0 Disable SCSI data bus driving.

1 Enable contents of Output Data Register onto the SCSI data bus. SCSI parity is also generated and driven on DBP.

This bit should be set when transferring data out of the EASI in either TARGET or INITIATOR mode, for both DMA or programmed-I/O. In INITIATOR mode the drivers are only enabled if: Mode Register 2 TARGET MODE bit is 0, and $\overline{I/O}$ is false, and $\overline{C/D}$, $\overline{I/O}$, $\overline{MSG}$ match the contents of the Target Command Register (phasematch is true). In TARGET mode only the MR2 bit needs to be set with this bit.

Reading the ICR reflects the state of this bit.

$\overline{ATN}$: Assert Attention   Bit 1

0 Deassert $\overline{ATN}$

1 Assert SCSI $\overline{ATN}$ signal. The MR2 TARGET MODE bit must also be false to assert the signal.

Reading the ICR reflects the state of this bit.

$\overline{SEL}$: Assert Select   Bit 2

0 Deassert $\overline{SEL}$

1 Assert SCSI $\overline{SEL}$ signal. Can be used in INITIATOR or TARGET mode.

Reading the ICR reflects the state of this bit.

$\overline{BSY}$: Assert Busy   Bit 3

0 Deassert $\overline{BSY}$.

1 Assert SCSI $\overline{BSY}$ signal. Can be used in INITIATOR or TARGET mode.

Reading the ICR reflects the state of this bit.

3.0 Register Description (Continued)

ACK: Assert Acknowledge   Bit 4
0   Deassert $\overline{ACK}$.
1   Assert SCSI $\overline{ACK}$ signal. The MR2 TARGET MODE bit must also be false to assert the signal.
Reading the ICR reflects the state of this bit.

DIFF: Differential Enable   Bit 5   Write
0   This bit must be reset to 0.

LA: Lost Arbitration   Bit 5   Read
0   Normally reset to 0 to show arbitration not lost or not enabled.
1   Will be set when the EASI loses arbitration, i.e. when SEL is true during arbitration AND the Assert $\overline{SEL}$ bit of this register is false.

A 1 in this bit means the EASI has arbitrated for the bus, asserted $\overline{BSY}$ and its ID on the data bus and another device has asserted $\overline{SEL}$. The ARBITRATE bit in MR2 or the EMR must be set to enable arbitration.

MODE: Operating Mode   Bit 6   Write
0   Normal Mode (MODE N) is selected.
1   Enhanced Mode (MODE E) is selected.

AIP: Arbitration In Progress   Bit 6   Read
0   Normally 0 to show no arbitration in progress.
1   Set when the EASI has detected BUS FREE phase and asserted $\overline{BSY}$ and the Output Data Register contents onto the SCSI data bus. This bit remains set until arbitration is disabled.

$\overline{RST}$: Assert $\overline{RST}$ Bit 7
0   Deassert $\overline{RST}$.
1   Assert SCSI $\overline{RST}$ signal. $\overline{RST}$ is asserted as long as this bit is 1, or until a µP Reset ($\overline{RESET}$).

After this bit is set the INT pin goes active and internal registers are reset (except for the interrupt latch, MR2 TARGET MODE bit, and this bit). Reading the ICR reflects the state of this bit.

MODE REGISTER 2 (MR2)

8 Bits   HA 2   Read/Write

This register is used to program basic operating conditions in the EASI. Operation as TARGET or INITIATOR, DMA mode and type as well as some interrupt controls are set via this register. This is a read/write register and when read the value reflects the state of each bit.

Bit 7							Bit 0
BLK	TARG	PCHK	PINT	EOP	BSY	DMA	ARB

Mode Register 2

ARB: Arbitrate   Bit 0
0   Disable arbitration
1   Enable arbitration. The EASI will wait for a BUS FREE phase then arbitrate for the bus. Before setting this bit the Output Data Register should contain the SCSI device ID—a single bit set only. The status of the arbitration process is given in the AIP and LA bits (6,5) in the Initiator Command Register.

DMA: DMA Mode   Bit 1
0   Disable DMA mode
1   Enable DMA operation. This bit should be set then one of address 5 to 7 written to start DMA. The TARGET MODE bit in the ICR and the phase lines in the TCR should have been set appropriately. The DBUS bit in the ICR must be set for DMA send operations. $\overline{BSY}$ must be active in order to set this bit. The phase lines must match the contents of the TCR during the actual transfers. In DMA mode EASI logic automatically controls the $\overline{REQ}/\overline{ACK}$ handshakes.

This bit should be reset by a µP write to stop any DMA transfer. An $\overline{EOP}$ signal will not reset this bit. During DMA, $\overline{CS}$ and $\overline{DACK}$ should not be active simultaneously.

This bit will be reset if $\overline{BSY}$ is lost during DMA mode.

BSY: Monitor Busy   Bit 2
0   Disable $\overline{BSY}$ monitor.
1   Monitor SCSI $\overline{BSY}$ signal and interrupt when $\overline{BSY}$ goes inactive. When this bit goes active the lower 6 bits of the ICR are reset and all signals removed from the SCSI bus. This is used to check for valid TARGET connection.

EOP: Enable $\overline{EOP}$ Interrupt   Bit 3
0   No interrupt for $\overline{EOP}$.
1   Interrupt after valid $\overline{EOP}$ condition.

PINT: Enable SCSI Parity Interrupt   Bit 4
0   No Interrupt on SCSI parity error.
1   When SCSI parity is enabled via the PCHK bit, setting this bit enables an interrupt upon a SCSI parity error.

PCHK: Enable SCSI Parity Checking   Bit 5
0   No SCSI parity checking.
1   Enable checking of SCSI parity during read operations. This applies to either programmed I/O or DMA mode.

TARG: Target Mode   Bit 6
0   Initiator Mode
1   Target Mode

BLK: Block Mode DMA   Bit 7
0   Non-block DMA
1   When set along with DMA bit (1), enables block mode DMA transfers. In block mode the READY line is used to handshake each byte with the DMA controller instead of the DRQ/$\overline{DACK}$ handshake used in non-block mode.

TARGET COMMAND REGISTER (TCR)

4 Bits   HA 3   Read/Write

This register is used to control TARGET SCSI signals and to program the desired phase during INITIATOR mode. During DMA phase transfers the SCSI phase lines ($\overline{C/D}$, $\overline{MSG}$, $\overline{I/O}$) must match the contents of the TCR for transfers to occur. A phase mismatch halts DMA transfers and generates an interrupt.

Bit 7							Bit 0
X	X	X	X	$\overline{REQ}$	$\overline{MSG}$	$\overline{C/D}$	$\overline{I/O}$

Target Command Register

3.0 Register Description (Continued)

This is a read/write register and the value read reflects the state of each bit, except bits 4–7 which always read 0.

I/O: Assert I/O   Bit 0
0   Deassert I/O.
1   Assert SCSI I/O signal. The MR2 TARGET MODE bit must also be active.

C/D: Assert C/D   Bit 1
0   Deassert C/D.
1   Assert SCSI C/D signal. The MR2 TARGET MODE bit must also be active.

MSG: Assert MSG   Bit 2
0   Deassert MSG.
1   Assert SCSI MSG signal. The MR2 TARGET MODE bit must also be active.

REQ: Assert REQ   Bit 3
0   Deassert REQ.
1   Assert SCSI REQ signal. The MR2 TARGET MODE bit must also be active. This bit is used to handshake SCSI data via programmed-I/O.

SELECT ENABLE REGISTER (SER)
8 Bits   HA 4   Write Only

This write-only register is used to program the SCSI device ID for the EASI to respond to during Selection or Reselection phases. Only one bit in the register should be set. When SEL is true, BSY false and the SER ID bit active an interrupt will occur.

This interrupt is reset or can be disabled by writing zero to this register. Parity will also be checked during Selection or Reselection if the PCHK bit in MR2 is set.

Bit 7							Bit 0
DB7	DB6	DB5	DB4	DB3	DB2	DB1	DB0

Select Enable Register

CURRENT SCSI BUS STATUS (CSB)
8 Bits   HA 4   Read Only

This read-only register is used to monitor SCSI control signals and the SCSI parity bit. The SCSI lines are monitored during programmed-I/O transfers and after an interrupt in order to determine the cause. A bit is 1 if the corresponding SCSI signal is active.

Bit 7							Bit 0
RST	BSY	REQ	MSG	C/D	I/O	SEL	DBP

Current SCSI Bus Status

BUS AND STATUS REGISTER (BSR)
8 Bits   HA 5   Read Only

This read-only register is used to monitor SCSI signals not included in the CSB, and internal status bits. This register is read after an interrupt (in MODE N) to determine the cause of an interrupt. Bit 0 or 1 are set to 1 if the SCSI signal is active.

Bit 7							Bit 0
EDMA	DRQ	SPER	INT	PHSM	BSY	ATN	ACK

Bus and Status Register

ACK: Acknowledge   Bit 0
This bit reflects the state of the SCSI ACK Signal.

ATN: Attention   Bit 1
This bit reflects the state of the SCSI ATN Signal.

BSY: Busy Error   Bit 2
0   No error
1   The SCSI BSY signal has become inactive while the MR2 BSY (Monitor BSY) bit is set. This will cause an interrupt, remove all EASI signals from the SCSI bus and reset the DMA MODE bit in MR2.

PHSM: Phase Match   Bit 3
0   Phase Match. The SCSI C/D, I/O and MSG phase lines are continuously compared with the corresponding bits in the TCR. The result of this comparison is reflected in this bit. This bit must be 1 (phase matches) for DMA transfers. A phase mismatch will stop DMA transfers and cause an interrupt.

INT: Interrupt Request   Bit 4
0   No interrupt
1   Interrupt request active. Set when an enabled interrupt condition occurs. This bit reflects the state of the INT pin. INT may be reset by performing a Reset Parity/Interrupt (RPI) function.

SPER: SCSI Parity Error   Bit 5
0   No SCSI parity error
1   SCSI parity error occurred. This bit remains set once an error occurs until the RPI function clears it. The PCHK bit in MR2 must be set for a parity error to be checked and registered.

DRQ: DMA Request   Bit 6
0   No DMA request
1   DMA request active. This bit reflects the state of the DRQ pin. DRQ is reset by asserting DACK during a DMA cycle or by resetting the DMA bit in MR2. A Busy error will reset the MR2 DMA bit and thus will also clear DRQ. A phase mismatch will not reset DRQ.

EDMA: End of DMA   Bit 7
0   Not end of DMA
1   Set when DACK, EOP and either RD or WR are active simultaneously. Normally occurs when the last byte is transferred by the DMA. During DMA send operations the last byte transferred by the DMA may not have been transferred on SCSI so REQ and ACK should be monitored to verify when the last SCSI transfer is complete. This bit is reset when the MR2 DMA bit is reset.

Note: In MODE E the EASI presents a true EDMA bit in bit 7 of the TCR. This feature removes the need to poll the REQ and ACK signals.

START DMA SEND (SDS)
0 Bits   HA 5   Write Only

This write-only register is used to start a DMA send operation. A write of don't care data should be the last thing done by the µP. The MR2 DMA, BLK and TARG bits must have been programmed previously.

Bit 7							Bit 0
X	X	X	X	X	X	X	X

Start DMA Send

3.0 Register Description (Continued)

START DMA TARGET RECEIVE
0 Bits   HA 6   Write Only

This write-only register is used to start a DMA Target Receive operation. Same comments as SDS apply.

INPUT DATA REGISTER (IDR)
8 Bits   HA 6   Read Only

This read-only register contains the SCSI data last latched during a DMA receive. Each byte from SCSI is latched into this register automatically by the EASI DMA logic. A DMA read ($\overline{DACK}$ and $\overline{RD}$) automatically selects this register. Programmed I/O SCSI data reads should use the CSD (HA 8).

START DMA INITIATOR RECEIVE (SDI)
0 Bits   HA 7   Write Only

This write-only register is used to start DMA INITIATOR Receive Operation. Same comments as SDS apply. An alternative method of performing the SDI function is available in the Enhanced Mode Register.

RESET PARITY/INTERRUPT (RPI)
0 Bits   HA 7   Read Only

This read-only register is used to reset the parity and interrupt latches. Reading this register resets the SCSI parity, μP parity, Busy Loss and Interrupt Request latches. It also resets the interrupt latches presented in the Interrupt Status Register (available in MODE E).

An alternative method of performing the Reset Parity Interrupt function is available in the Enhanced Mode Register. In MODE E writing a value of 01 to bits 2 and 1 of the EMR will reset the same bits as a read from HA 7 in MODE N. The EMR RPI will also reset enhanced logic that has bits set in the Interrupt Status Register.

3.3 ENHANCED MODE REGISTERS

Addresses 0 to 6 remain the same as MODE N except for bit 7 of TCR, as described below. Address 7 is the SDI and RPI functions in MODE N, but in MODE E it directly accesses the Enhanced Mode Register (EMR) and indirectly accesses the Interrupt Mask Register (IMR) and the Interrupt Status Register (ISR).

When bit 6 in the ICR (HA 1) is set, HA 7 accesses the read/write EMR. The SDI and RPI functions performed when writing/reading HA 7 in MODE N are disabled. To perform these functions the EMR is used instead.

Note that EMR functions are intended to be used in an interrupt-driven environment. Reading this register reflects the state of each bit.

TARGET COMMAND REGISTER (TCR)
5 Bits   HA 3   Read/Write

This is the same as MODE N except for bit 7 which is described below. Note bits 4–6 always read 0.

Bit 7							Bit 0
(true) EDMA	X	X	X	$\overline{REQ}$	$\overline{MSG}$	$\overline{C/D}$	$\overline{I/O}$

Target Command Register

EDMA: True End of DMA   Bit 7
0   Not End of DMA
1   Set when the last byte of data has been transferred. This bit is not set until $\overline{REQ}$ and $\overline{ACK}$ both go inactive following the DMA cycle during which $\overline{EOP}$ was asserted. Note that unlike the BSR EDMA bit, this bit reflects the true completion of DMA transfers.

ENHANCED MODE REGISTER (EMR)
8 Bits   HA 7   Read/Write

This register is accessed at HA 7 when the ICR MODE bit (6) is set. The register controls operation of enhanced logic and timing. Normally the application will leave the EASI permanently in MODE N or MODE E.

Bit 7							Bit 0
APHS	MPEN	MPOL	SPOL	LOOP	EFN1	EFN0	ARB

Enhanced Mode Register

ARB: Extended Arbitration   Bit 0
0   Disable extended arbitration
1   Enable extended arbitration. This is an alternative bit to the MR2 ARB function. The EASI waits for a BUS FREE phase then arbitrates for the bus, asserting the contents of the ODR onto the SCSI data bus and asserting $\overline{BSY}$. The EASI will then wait the 2.2 μs SCSI Arbitration Delay, and set the Arbitration Complete bit in the ISR and cause an interrupt. As for the MR2 ARB function, the ICR LA and AIP bit can be examined to determine arbitration status.

EFN1,0: Enhanced Function   Bits 2,1
00   No operation—these bits ALWAYS read 00.
01   When this pattern is written the Parity and Interrupt latches are reset. This pattern should be followed by any other pattern to remove the reset (usually 00). This function replaces the RPI function performed when reading HA 7 in MODE N. Only the latches with ISR bits set to 1 will be reset.
10   Start DMA Initiator Receive. At the end of the write cycle the SDI function is performed. There is no need to follow with another pattern as required with the RPI value (01). This function replaces the SDI function performed when writing HA 7 in MODE N.
11   Read/Write ISR/IMR. When this pattern is written the NEXT read of HA 7 will access the ISR, or the NEXT write to HA 7 will access the IMR. This state ONLY lasts for the ONE following read OR write cycle. Further cycles will then access the EMR.

LOOP: Loopback Mode   Bit 3
0   Normal operation
1   When set: SCSI drivers are disabled and the SCSI I/O's looped back inside the EASI, and both the TARGET and INITIATOR signals may be driven simultaneously. This enables the μP to check EASI operation without affecting the SCSI bus.

SPOL: SCSI Parity Polarity   Bit 4
0   SCSI parity is ODD (as per SCSI specification).
1   SCSI parity is EVEN. This allows diagnostics to be performed.

3.0 Register Description (Continued)

MPOL: μP Parity Polarity   Bit 5

0   μP parity is ODD.

1   μP parity is EVEN.

MPEN: μP Parity Enable   Bit 6

0   μP parity checking and generation disabled.

1   Enable checking of μP data bus parity during μP and DMA writes. Generate parity during μP and DMA reads. Parity errors will cause an interrupt and set the ISR MPE bit if not masked.

APHS: Any Phase Mismatch   Bit 7

0   Disable phase mismatch detection

1   Detect SCSI requests with a phase mismatch present. Is set when $\overline{REQ}$ goes active AND the SCSI phase lines do not match the contents of the TCR. Can be used in INITIATOR mode to interrupt on TARGET phase changes.

INTERRUPT STATUS REGISTER (ISR)

8 Bits   HA 7 (EFN = 11)   Read Only

This register is accessed during the first read cycle after the EFN bits in the EMR have both been set to 1. Once read, successive accesses of HA 7 go to the EMR. This register provides all interrupt status within one register. This is intended to make determination of interrupt sources easier in MODE E. In MODE N two registers must be read—the BSR and CSB. In MODE E only the ISR needs to be read. Additionally each interrupt (except SCSI $\overline{RST}$) has a corresponding status bit in the ISR.

When the ISR is read all unmasked, enabled, active interrupt sources set their corresponding ISR bits to a ONE. The interrupt status is sampled at the beginning of the ISR read cycle. When an RPI function is performed via the EMR, only each interrupt latch with an ISR bit set will be reset. This means interrupts occurring since the last ISR read will not be lost.

ISR bits may be individually masked via their corresponding bits in the IMR.

Bit 7							Bit 0
SPE	MPE	EDMA	DPHS	APHS	BSY	SEL	ARB

Interrupt Status Register

ARB: Arbitration Complete   Bit 0

This bit is set when arbitration enabled by EMR ARB bit (0) has completed. Completion occurs in two ways: when the EASI loses arbitration and has asserted the LA bit in the ICR; or when the EASI has asserted the ID contained in the ODR onto the data bus, asserted $\overline{BSY}$, then waited for the 2.2 μs SCSI Arbitration Delay.

SEL: Selection/Reselection   Bit 0

This bit is set when $\overline{BSY}$ is false, $\overline{SEL}$ is active, and any SER bit set to 1 has an active corresponding SCSI data bus bits. This situation occurs during Selection or Reselection phases.

BSY: Busy Loss   Bit 2

This bit is the same as BSR bit 2. Set when the SCSI $\overline{BSY}$ signal becomes inactive while the MR2 BSY (monitor $\overline{BSY}$) bit is set.

APHS: Any Phase Mismatch   Bit 3

Set when a $\overline{REQ}$ occurs while the SCSI phase lines do not match. Bit is set to enable detection of the contents of the TCR. EMR APHS (bit 7) must be 1 to allow this bit to be set.

DPHS: DMA Phase Mismatch   Bit 4

Set when a SCSI DMA mode operation occurs with a phase mismatch. Similar to the APHS condition but restricted to DMA mode only. In MODE N this condition is not as easily determined.

EDMA: End of DMA   Bit 5

Set when $\overline{REQ}$ and $\overline{ACK}$ are both false following the DMA cycle during which $\overline{EOP}$ was asserted. This represents the true end of DMA operation.

MPE: μP Parity Error   Bit 6

Set when μP parity error is detected at the end of a μP or DMA write cycle. The MPEN bit in the EMR must be set for μP parity to be checked.

SPE: SCSI Parity Error   Bit 7

Set when a SCSI parity error is detected when reading the CSD, during selection/reselection, or when the IDR is loaded during DMA operation. The MR2 PCHK bit must be set for SCSI parity to be checked, and the MR2 PINT bit must be set to enable the interrupt.

INTERRUPT MASK REGISTER (IMR)

8 Bits   HA 7 (EFN = 11)   Write Only

This register is accessed during the first write cycle after the EFN bits in the EMR have been set to 11. Once written, successive reads or writes at HA 7 access the EMR.

This register has the same bit definition as the ISR. If a bit in the IMR is set to 1 that interrupt will be masked. The interrupt will be captured internally (if enabled) but will not cause an active INT signal. A bit reset to 0 will enable that interrupt to occur (if enabled) and will enable the ISR bit to be set to 1.

Bit 7							Bit 0
SPE	MPE	EDMA	DPHS	APHS	BSY	SEL	ARB

Interrupt Mask Register

4.0 Device Operation

4.1 GENERAL

This section describes overall operation of the EASI. More detailed information of data transfers, interrupts and reset conditions are covered in later sections. The operation description covers μP accesses, SCSI bus monitoring, arbitration, selection, reselection, programmed-I/O, DMA interrupts. Programming and timing details are covered.

For information regarding interfacing to μPs and DMA controllers refer to Section 9.

In the descriptions following, program examples are given in pseudo-C. This processor-independent approach should be clearest. These are backed up by flow charts in Appendix A1.

For each section where appropriate the description is split into two, MODE N and MODE E.

4.0 Device Operation (Continued)

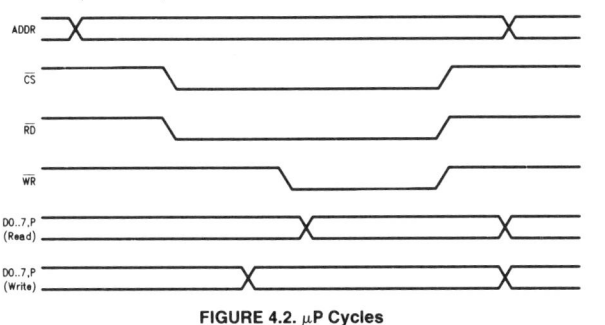

FIGURE 4.2. μP Cycles

4.2 μP ACCESSES

The μP accesses the EASI via the $\overline{CS}$, $\overline{RD}$, $\overline{WR}$ and address and data lines in order to read/write the registers. *Figure 4.1* shows typical timing. Note the use of non-multiplexed address and data lines.

4.3 SCSI BUS MONITORING/DRIVING

The SCSI bus may be monitored or driven at any time. Each bus signal is buffered and inverted by the EASI and can be read via the CSB, BSR and CSD registers. An active SCSI signal reads a 1 in the status registers.

Each SCSI signal may be asserted by setting a bit in the TCR or ICR. Setting the bit to 1 asserts the SCSI signal.

The following code demonstrates a byte transferred via programmed-I/O in INITIATOR mode.

```
|   /* Transfer one byte as Initiator */
    while (NOT (TCR: REQ));
       /* wait till TARGET asserts REQ */
    data = input (CSD);
       /* parity is checked if enabled */
    output (ICR, Assert_ACK);
    while (TCR:REQ);
       /* wait till TARGET deasserts REQ */
    output (ICR, 0);
       /* deassert ACK, ready for next byte */
|
```

4.4 ARBITRATION

This sub-section describes the arbitration support provided by the EASI and how to program it.

4.4.1 MODE N Arbitration

Since the SCSI arbitration process requires signal sequencing too fast for μPs, hardware support is provided by the EASI. The arbitration process is enabled by bit 0 MR2 (ARB). Prior to setting this bit the ODR should be programmed with the device's SCSI ID—a single bit.

The EASI will monitor the bus for a BUS FREE phase. The $\overline{BSY}$ signal is continuously monitored. If continuously inactive for at least a SCSI Bus Settle Delay (400 ns) and $\overline{SEL}$ is inactive, a valid Bus Free Phase exists. After a period of SCSI Bus Free Delay (800 ns) the EASI asserts $\overline{BSY}$ and the ODR onto the SCSI data bus. The μP should poll the ICR to determine when arbitration has started. The AIP bit in the ICR is set when the Bus Free Phase is detected and the EASI is beginning the Bus Free Delay. Following the Bus Free Delay a 2.2 μs SCSI Arbitration Delay is required before examining the data bus to resolve the priorities of the ID bits. This delay must be implemented in firmware. The ICR Lost Arbitration (LA) bit must be examined to determine whether arbitration is lost. The LA bit is set if another device asserts $\overline{SEL}$ during arbitration. If the LA bit is 0 the data bus is read via the CSD register. The data is examined to resolve ID priorities. If this device is the highest ID, assert $\overline{SEL}$ by setting ICR bit 2 to a 1. After waiting Bus Clear + Bus Settle Delays (1200 ns), the Selection Phase begins. These 2 delays must be implemented in firmware.

4.4.2 MODE E Arbitration

The extended arbitration in MODE E is enabled by bit 0 of the EMR (ARB). This alternative offers two significant advantages over MODE N. First the 2.2 μs SCSI Arbitration Delay is implemented by the EASI. Second the arbitration process may be interrupt driven.

In MODE N the EASI must be polled to see when the ICR AIP bit is set. After the appropriate delays, the LA bit and data bus are examined. If arbitration is lost or this device is not the highest priority the MR2 ARB bit must be reset to 0, then set to 1 again and the whole process restarted. This means the EASI MUST be polled until arbitration is won—potentially many SECONDS, typically ms. This ties up the host μP.

In MODE E when the EMR ARB bit is set the EASI will: wait for BUS FREE phase; delay SCSI Bus Free Delay; assert $\overline{BSY}$ and the ODR onto $\overline{DB0}$–$\overline{DB7}$; delay SCSI Arbitration Delay; interrupt the μP.

The μP should read the ISR and if the ISR bit 0 (ARB complete) is set examine ICR bits 5 and 6 (LA and AIP) to determine whether arbitration is lost. If not lost the data bus is examined to resolve ID priorities. As for MODE N if arbitration has failed the EMR ARB bit should be reset and then set again after first resetting the ARB interrupt. Note that the EMR ARB bit allows the μP to carry on with other tasks while the EASI arbitrates. This means there is NO NEED to poll the EASI.

4.5 SELECTION/RESELECTION

The EASI can be used to select or reselect a device. The EASI will also respond to selection or reselection. Selecting or reselecting a device is the same in MODE N or MODE E. Response to selection or reselection can differ between the bus modes.

4.0 Device Operation (Continued)

4.5.1 Selecting/Reselecting

Selection requires programming the ODR with the desired and own device IDs; the data bus via ICR DBUS (bit 0); asserting $\overline{\text{ATN}}$ if required via ICR bit 1; asserting $\overline{\text{SEL}}$ via ICR bit 2; then resetting the MR2/EMR ARB bit.

The SER should have been cleared to zero before Selection/Reselection to ensure the EASI does not respond. If Reselection is desired the $\overline{\text{I/O}}$ line should also be asserted before $\overline{\text{SEL}}$ via TCR bit 0.

Resetting the ARB bit causes the EASI to remove $\overline{\text{BSY}}$ and the ODR from the data bus. Thus the ICR Assert data bus bit is required to assert the bits for desired and own device IDs.

$\overline{\text{BSY}}$ is then monitored to determine when the device has responded to (re)selection. If the device fails to respond an error handler should sequence the EASI off the bus. If the device responds the ICR DBUS and $\overline{\text{SEL}}$ bits should be reset to remove these signals. If this is a Reselection the ICR $\overline{\text{BSY}}$ bit (3) should be set before removing the other signals.

The bus is now ready to handle Information Transfer Phases.

4.5.2 MODE N (Re)Selection Response

The EASI responds to Selection or Reselection when the SER is non-zero. A (re)selected interrupt is generated when $\overline{\text{BSY}}$ is false for at least a Bus Settle Delay (400 ns); and $\overline{\text{SEL}}$ is true AND any non-zero bit in the SER has its corresponding SCSI data bus bit active. A Selection is disabled by zeroing the SER. If parity is supported it should be valid during (re)selection so it must be checked via the SPE bit (5) in the BSR. SCSI specification states that (re)selection is not valid if more than 2 data bits are active. This condition is checked by reading the CSD.

When the selection interrupt occurs it is determined by reading the BSR and CSB registers. There is no dedicated status bit for (re)selection in MODE N, it must be determined by the absence of other interrupts, and the active state of the $\overline{\text{SEL}}$ signal. Reselection occurs when $\overline{\text{I/O}}$ is also active. See Section 6.

4.5.3 MODE E (Re)Selection Response

The same conditions for valid (re)selection apply as for MODE N. A (re)selection interrupt is generated as per MODE N. The difference is that the interrupt sets the SEL bit (1) in the ISR. In MODE E, reading the ISR enables exact determination of interrupt sources. This interrupt can be masked via bit 1 of the IMR. The interrupt is reset by programming the RPI function to the EMR. See Section 6 for further details.

4.6 MONITORING BSY

While an INITIATOR is connected to a TARGET the TARGET must maintain an active $\overline{\text{BSY}}$ signal. During DMA operations the $\overline{\text{BSY}}$ signal is monitored by the EASI and will halt operations if it goes inactive. To enable $\overline{\text{BSY}}$ to be monitored at other times the MR2 BSY bit (2) should be set. An interrupt will be generated if $\overline{\text{BSY}}$ goes inactive while MR2 BSY is set.

In MODE N this interrupt sets bit 2 in the BSR. In MODE E this interrupt sets bit 2 in the BSR and bit 2 in the ISR. The interrupt may be masked via bit 2 of the IMR.

4.7 COMMAND/MESSAGE/STATUS TRANSFERS

Command, message and status bytes are transferred using programmed-I/O. The SCSI $\overline{\text{REQ}}/\overline{\text{ACK}}$ handshake is accomplished by monitoring and setting lines individually. Data is output via the ODR and read in via the CSD register.

The following code shows INITIATOR and TARGET programming for two of these cases. See Appendix A1 for flowcharts.

Initiator Command Send

```
{
MR2 = monitor BSY
TCR = Command Phase   /* 02h */
while (bytes) to do) {
      while (REQ) inactive)
            idle; /* CSB bit 5 = 0 */
      if (BSR: phase_match==0)
            phase error;
      else {
            ODR = data byte;
            ICR = Assert_ACK;
            while (REQ active)
            idle; /* CSB bit 5==1 */
            ICR = deassert_ACK
            /* byte transfer complete */
            byte count - -;
      }
}
goto data phase;
}
```

Target Message Receive

```
{
/* assumed Assert_BSY already set in ICR */
MR2 = TARG MODE OR PARITY CHECK
      OR PARITY INTERRUPT;
TCR = Message_Out phase; /* 06h */
delay (Bus Settle);
TCR = Assert_REQ;
while (ACK inactive)
      idle; /* BSR bit 0 */
data = CSD; /* parity is latched */
if (BSR: parity_error)
      error routine;
else {
      TCR = deassert_REQ;
      while (ACK active)
            idle;
}
   /* message done, can change to next
      phase*/
}
```

4.8 NON-BLOCK DMA TRANSFERS

Data transfers may be effected by DMA. This method should be used for optimum performance. Two methods of DMA are available—block and non-block mode. This section describes non-block mode transfers. MODE N operation is covered first followed by MODE E.

The interface to the DMA controller uses the DRQ, $\overline{\text{DACK}}$, $\overline{\text{EOP}}$ lines in non-block mode. Each byte is requested (DRQ) and ack'd ($\overline{\text{DACK}}$). Representative timing for a DMA read is shown in *Figure 4.8.1*.

4.0 Device Operation (Continued)

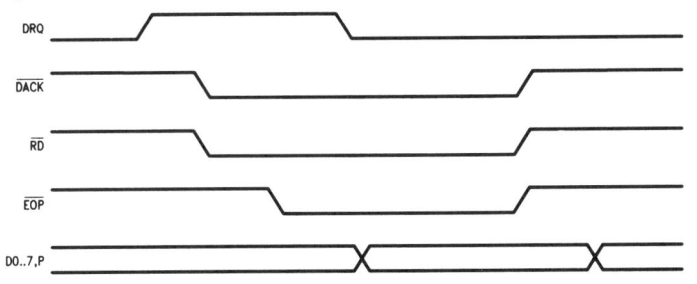

FIGURE 4.8.1. Non-Block DMA Timing

4.8.1 MODE N Non-Block DMA

DMA operation involves programming the EASI with the setup parameters, initiating the DMA cycles and checking for correct operation when the completion interrupt is received. The DMA controller should be programmed with the data byte count and the memory start address. Methods of halting DMA operation are covered in Section 4.11.

Setting up the EASI requires enabling or disabling the following: Data bus driving, DMA mode enable, $\overline{BSY}$ monitoring, $\overline{EOP}$ interrupt, parity checking, parity interrupt, TARGET Mode, bus phase.

Once set up, DMA should be initiated by writing to address 5, 6 or 7 as appropriate. The DMA controller should assert $\overline{EOP}$ during the transfer of the last byte, although this may be done by the μP if the DMA transfers (n−1) bytes and the μP transfers the last byte. See the application guide for more details (Section 9).

Upon completion the μP should check the following as required: End of DMA, Parity Error, Phase Match, Busy Error. In MODE N the end of DMA occurs as a response to $\overline{EOP}$. SCSI transfers may still be underway so $\overline{REQ}$ and $\overline{ACK}$ must still be checked to establish when the final byte is finished.

The code below shows programming of the EASI in each of the four DMA cases. One of these cases is shown in a flow diagram in Appendix A.

```
Initiator Send    /* DATA OUT PHASE */
{
  Program DMA Controller;
  TCR = 00h;     /* phase */
  ICR = 01h;     /* Assert_DBUS */
  MR2 = 0Eh;
  SDS = 00;      /* Start DMA Send */
  while (NOT interrupt)
        idle;
  while (CSD:REQ)
        idle;   /* wait for last SCSI byte
                   transfer so phase
                   is checked */
  if(BSR:Busy_error OR NOT(BSR:End_of_DMA))
        error routine;
  else {          /* DMA END */
        MR2 = 04h;  /* reset DMA bit */
        ICR = 0;
  }
}
```

```
Initiator Receive  /* DATA IN PHASE */
{
  Program DMA Controller;
  TCR = 01h;     /* phase */
  MR2 = 3Eh;
  SDI = 0;       /* Start DMA Init Rx */
  while (NOT interrupt)
        idle;
  /* no need to wait for last SCSI handshake
     done since DMA done implies it is
     checked */
  if (BSR:parity_error OR BSR:busy_error
      or NOT (BSR End of DMA))
        do error routines;
  else {          /* End of DMA */
        while (CSD:REQ)
              idle; /* wait for REQ inactive
                       to deassert ACK */
        MR2 = 04h;
  }
}

Target Receive   /* DATA OUT PHASE */
{
  Program DMA controller;
  TCR = 0;       /*phase*/
  ICR = 08h;
  MR2 = 7Ah;     /*check parity*/
  SDT = 0;       /*Start DMA Targ Rx*/
  while (not interrupt)
        idle;
  /*when End of DMA occurs the last byte
    has been read and checked*/
  if(BSR:parity_error OR NOT(BSR:End_of_DMA)
        error routine;
  else {          /* End of DMA */
        while (BSR:ACK)
              idle;
  /*Not True End of DMA in MODE N, so wait
    until SCSI bus inactive before
    changing phase */
        MR2 = 40h;
        change phase as required;
  }
}
```

4.0 Device Operation (Continued)

```
Target Send    /* DATA IN PHASE */
{
    Program DMA Controller;
    TCR = 01h;    /* phase */
    ICR = 09h;
    MR2 = 4Ah;
    SDS = 0;      /* Start DMA Send */
    while (NOT interrupt)
          idle;
    if (NOT(BSR:END_of_DMA))
        error;
    else {        /* DMA end */
        repeat {
                while (CSB:REQ OR BSR:ACK)
                loop count = 3;
                loop count - -;}
                /* decrement */
        until (loop count == 0);
        MR2 = 40h;
    Change phase as required;
    }
}
```

Some explanation of the final part of Target Send is required. In this type of DMA operation it is very difficult to exactly determine the True End of DMA in MODE N. Simply detecting REQ and ACK simultaneously inactive is not enough.

Reference to *Figure 4.8.2* will help to understand the following text.

As shown in *Figure 4.8.2*, ACK going active causes the DRQ for the next byte and also REQ to go inactive. ACK going inactive allows REQ to go active for the next byte. If the INITIATOR is slow removing ACK the µP may sample the SCSI bus after the EOP interrupt at point A. Here both REQ and ACK will be inactive, but there is one more byte to transfer on SCSI. Due to chip timing delays this condition will not last more than 200 ns. A safe way to determine the True End of DMA is to sample REQ and ACK and ONLY when both are inactive in three successive samples will the µP be at point B in the figure.

In MODE E, True End of DMA is correctly decoded and the End of DMA interrupt occurs as a result of this True condition, thus there is no need to sample REQ and ACK. For this and other reasons operation in MODE E is strongly recommended.

4.8.2 MODE E Non-Block DMA

Operation for the Non-Block DMA in MODE E is essentially the same as MODE N. A primary difference is that in MODE E True End of DMA is decoded internally and this causes an interrupt. This feature removes the need to check for DMA End before moving to the next phase. Additionally, the use of the ISR allows easier determination of the end condition. For more information on interrupts see Section 6.

Examples of code are given below.

```
Initiator Receive   /* DATA IN PHASE */
{
    Program DMA controller;
    TCR = 01h;     /* phase */
    ICR = 40h;     /* MODE E */
    MR2 = BEh;
    EMR = 04h;     /* Start DMA INIT Rx */
    while (NOT interrupt)
          idle;
    if (ISR != 20h)
        error;
    else           /* DMA end */
    MR2 = 04h;
}


Target Send    /* DATA IN PHASE */
{
    Program DMA controller;
    TCR = 01h;     /* phase */
    ICR = 49h;     /* MODE E */
    MR2 = CAh;
    SDS = 0;       /* Start DMA Send */
    while (NOT interrupt)
          idle;
    if (ISR != 20h)
        error;
    else {         /* DMA end */
    MR2 = 40h;
    Change phase as required;
    }
}
```

4.9 BLOCK MODE DMA TRANSFERS

In Block Mode the DMA interface uses the DRQ, DACK, EOP and READY lines, DRQ is asserted once at the begin-

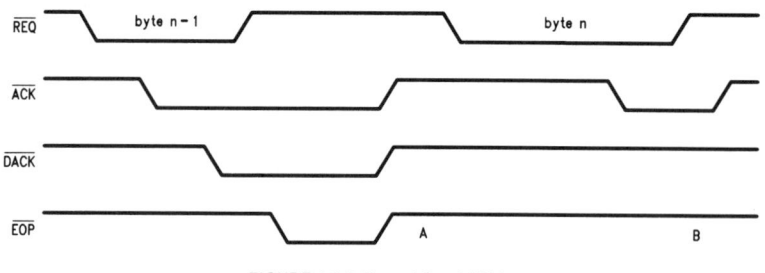

FIGURE 4.8.2. Target Send DMA

4.0 Device Operation (Continued)

ning of transfers and deasserted once $\overline{DACK}$ is received. $\overline{DACK}$ should be asserted continuously for the duration of all the transfer. $\overline{EOP}$ should be asserted during the last DMA byte signal when the next DMA byte transfers. The EASI asserts the READY signal when the next DMA byte should be transferred. As for non-block mode the End of DMA interrupt is just $\overline{EOP}$ in MODE N, but is a True End of DMA in MODE E.

The block mode is intended for systems where the overhead of handing the system busses to and from the μP and DMA controller is too great. The block mode handshake is not necessarily faster than non-block (it may be) but the overall transfer rate is improved once the bus exchange overhead is removed. Of course the μP is prevented from executing for the whole DMA operation.

If a phase mismatch occurs the READY signal is left in the inactive state. The DMA controller must hand back the bus to the μP and the inactive READY signal may need to be gated off. For more detail see Section 9.

When performing DMA as an INITIATOR the $\overline{EOP}$ signal does not deassert $\overline{ACK}$ on the SCSI bus in MODE N. Firmware must determine when $\overline{REQ}$ is inactive after the last SCSI transfer then reset the MR2 DMA bit to deassert $\overline{ACK}$. In MODE E the EASI correctly handles the deassertion of $\overline{ACK}$ after True End of DMA.

Programming the EASI in block mode is the same as non-block mode except bit 7 in MR2 should also be set.

4.10 PSEUDO DMA

The system design can utilize EASI DMA logic for non data transfers. This removes the need to poll $\overline{REQ}/\overline{ACK}$ and program the assertion/deassertion of the handshake signal. The μP can emulate a DMA controller by asserting $\overline{DACK}$ and $\overline{EOP}$ signals. DRQ may be sampled by reading the BSR. In most cases the chip decode logic can be adapted to this use for little or no cost. See Section 9 for further details.

4.11 HALTING A DMA OPERATION

There are three ways to halt a DMA operation apart from a chip or SCSI reset. These methods are: $\overline{EOP}$, phase mismatch and resetting the DMA MODE bit in MR2.

4.11.1 End of Process

$\overline{EOP}$ is asserted for a minimum period during the last DMA cycle. The $\overline{EOP}$ signal generates the End of DMA interrupt in MODE N, and enables the True End of DMA to occur later in MODE E. $\overline{EOP}$ does not cause the MR2 DMA mode bit to be reset.

4.11.2 DMA Phase Mismatch

If a $\overline{REQ}$ goes active while there is a phase mismatch the DMA will be halted and an interrupt generated. The EASI will stop driving the SCSI bus when the mismatch occurs. A phase mismatch is when the TCR phase bits do not match the SCSI bus values.

4.11.3 DMA Mode Bit

If $\overline{EOP}$ is not used, the best method is to reset the MR2 DMA Mode bit. This bit may be reset at any time, and should be reset after an End of DMA interrupt or a phase mismatch. Resetting the bit disables all DMA logic and thus should only be reset at the True End of DMA condition. Additionally, all DMA logic is reset, so this bit must be reset then set again to carry out the next DMA phase.

5.0 Interrupts

5.1 OVERVIEW

The EASI is intended to be used in an interrupt driven environment. MODE E has greatly enhanced the use of interrupts to relieve firmware overhead. This section describes the conditions for and use of each interrupt. Each description explains MODE N then MODE E operation.

Before individually describing each interrupt, an explanation of the use of interrupts is required.

5.2 USING INTERRUPTS

MODE N: In this mode interrupts are controlled by bits in MR2 if control is provided. Not all interrupts can be disabled under software control. When an interrupt occurs, both the BSR and CSD register must be read and analyzed to determine the source of interrupt. Since status is NOT provided for each interrupt, great care should be exercised when determining the interrupt source.

MODE E: In this mode every interrupt can be individually masked and enabled or disabled. In addition, when an interrupt occurs a single register (ISR) can be read to determine the source(s) of interrupt. An associated register (IMR) allows a mask bit to be set for each interrupt. Finally, the design of the logic prevents loss of interrupts occurring after the INT signal goes Active and before the Reset Parity/Interrupt function. In MODE N loss of interrupts can occur.

5.3 SCSI PARITY ERROR

MODE N: If SCSI parity checking is enabled via MR2 bit 5, an interrupt can occur as a result of a read from CSD, a selection/(re)selection, or a DMA receive operation. The parity error bit (bit 5) in the BSR will be set if checking is enabled. An interrupt will occur if Enable Parity Interrupt (bit 4) of MR2 is set. The interrupt is reset by reading HA 7. Following an interrupt the BSR and CSB should contain the values shown below.

Bit 7							Bit 0
X	X	1	1	X	X	X	X
EDMA	DRQ	SPER	INT	PHSM	BSY	$\overline{ATN}$	$\overline{ACK}$

BSR

Bit 7							Bit 0
0	1	X	X	X	X	0	X
$\overline{RST}$	$\overline{BSY}$	$\overline{REQ}$	$\overline{MSG}$	$\overline{C/D}$	$\overline{I/O}$	$\overline{SEL}$	$\overline{DBP}$

CSB

5.4 μP PARITY ERROR

MODE N: μP parity is not available under this mode.

MODE E: If bit 6 of the EMR is 1, μP parity will be checked on a write to the EASI via the μP or DMA controller. The parity polarity is determined by bit 5 of the EMR. If bit 6 of the IMR is zero, a μP parity error will set bit 6 of the ISR and cause an interrupt. The interrupt may be reset by writing 01 to bits 2 and 1 of the EMR followed by any other pattern.

5.5 END OF DMA

MODE N: If $\overline{EOP}$ is asserted during a DMA transfer, bit 7 of the BSR will be set and an interrupt generated if bit 3 of MR2 is 1. $\overline{EOP}$ is recognized when $\overline{EOP}$, $\overline{DACK}$ and either $\overline{IOR}$ or $\overline{IOW}$ are all simultaneously active for a minimum period. The interrupt may be reset by reading HA 7. Following an interrupt the BSR and CSB should contain the values shown below.

5.0 Interrupts (Continued)

Bit 7							Bit 0
1	X	X	1	X	X	0	X
EDMA	DRQ	SPER	INT	PHSM	BSY	$\overline{ATN}$	$\overline{ACK}$

BSR

Bit 7							Bit 0
0	1	X	X	X	X	0	X
$\overline{RST}$	$\overline{BSY}$	$\overline{REQ}$	$\overline{MSG}$	$\overline{C/D}$	$\overline{I/O}$	$\overline{SEL}$	$\overline{DBP}$

CSB

5.6 DMA PHASE MISMATCH

MODE N: When the SCSI $\overline{REQ}$ goes active during a DMA operation the contents of the TCR are compared with the SCSI phase lines $\overline{C/D}$, $\overline{MSG}$ and $\overline{I/O}$. If the two do not match an interrupt is generated. This interrupt will occur as long as the MR2 DMA bit is set (bit 1), i.e., it cannot be masked in MODE N. The mismatch removes the EASI from driving the SCSI data bus. The interrupt may reset by reading HA 7. Following an interrupt the BSR and CSB should contain the values shown below.

Bit 7							Bit 0
X	0	X	1	0	X	X	X
EDMA	DRQ	SPER	INT	PHSM	BSY	$\overline{ATN}$	$\overline{ACK}$

BSR

Bit 7							Bit 0
0	X	X	X	X	X	0	X
$\overline{RST}$	$\overline{BSY}$	$\overline{REQ}$	$\overline{MSG}$	$\overline{C/D}$	$\overline{I/O}$	$\overline{SEL}$	$\overline{DBP}$

CSB

MODE E: Bit 4 of the ISR will be set by a DMA phase mismatch—the same conditions as MODE N. The interrupt and setting if ISR bit 4 may be masked by setting bit 4 of the IMR. The interrupt may be reset by writing 01 to bits 2 and 1 of the EMR followed by any other pattern.

5.7 ANY PHASE MISMATCH

MODE N: This feature is not available under this mode.

MODE E: This condition is similar to DMA Phase Mismatch except that it applies to all operations—not just DMA. If the TCR contents do not match the SCSI phase lines when $\overline{REQ}$ goes active an interrupt is generated and bit 3 set in the ISR. The ISR bit and the interrupt may be masked by setting bit 3 in the IMR. The interrupt may be reset by writing 01 to bits 2 and 1 of the EMR followed by any other pattern.

5.8 BUSY LOSS

MODE N: If bit 2 in MR2 is set the SCSI $\overline{BSY}$ signal is monitored and an interrupt is generated if $\overline{BSY}$ is continuously inactive for at least a BUS SETTLE DELAY (400 ns). This interrupt may be reset by reading HA 7. Following an interrupt the BSR and CSB should contain the values shown below, where usually CSB = 00.

Bit 7							Bit 0
X	X	X	1	X	1	X	X
EDMA	DRQ	SPER	INT	PHSM	BSY	$\overline{ATN}$	$\overline{ACK}$

BSR

Bit 7							Bit 0
0	0	X	X	X	X	X	X
$\overline{RST}$	$\overline{BSY}$	$\overline{REQ}$	$\overline{MSG}$	$\overline{C/D}$	$\overline{I/O}$	$\overline{SEL}$	$\overline{DBP}$

CSB

MODE E: Bit 2 in MR2 performs the same function as in MODE N. When an interrupt is generated bit 2 in the ISR will be set. The ISR bit and the interrupt may be masked by setting bit 2 in the IMR. The interrupt may be reset by writing 01 to bits 2 and 1 of the EMR followed by any other pattern.

5.9 (RE)SELECTION

MODE N: An interrupt will be generated when: $\overline{SEL}$ is active, $\overline{BSY}$ is inactive, and the device ID is true. The device ID is determined by the value in the SER. If ANY non-zero bit in the SER has its corresponding SCSI data bit active during selection, the device ID is true. If $\overline{I/O}$ is active this is a reselection. The interrupt is disabled by writing all zeroes to the SER, and reset by reading HA 7.

If SCSI parity checking is enabled it will be checked and should be valid. Following an interrupt the BSR and CSB should contain the values shown below.

Bit 7							Bit 0
0	0	0	1	X	0	X	0
EDMA	DRQ	SPER	INT	PHSM	BSY	$\overline{ATN}$	$\overline{ACK}$

BSR

Bit 7							Bit 0
0	0	0	0	0	0	1	X
$\overline{RST}$	$\overline{BSY}$	$\overline{REQ}$	$\overline{MSG}$	$\overline{C/D}$	$\overline{I/O}$	$\overline{SEL}$	$\overline{DBP}$

CSB

MODE E: The functioning of the (re)selection logic is the same as for MODE N. Additionally, bit 1 in the ISR will be set upon a (re)selection interrupt. The interrupt and setting of bit 1 in the ISR may be masked by setting bit 1 in the IMR. The interrupt may be reset by writing all zeroes to the SER then writing 01 to bits 2 and 1 of the EMR followed by any other pattern.

5.10 ARBITRATION COMPLETE

MODE N: No interrupt is generated in MODE N.

MODE E: When bit 0 of the EMR is set to 1 the EASI monitors $\overline{BSY}$ and $\overline{SEL}$ to determine a BUS FREE Phase. The EASI then carries out all steps required for bus arbitration. When either arbitration is lost or the arbitration process has completed, bit 0 in the ISR is set and an interrupt generated. The interrupt and bit 0 in the ISR can be masked by setting bit 0 in the IMR. Note that arbitration is not affected by the IMR, just the interrupt.

6.0 Reset Conditions

6.1 GENERAL

There are three ways to reset the EASI; µP chip $\overline{RESET}$, SCSI bus reset applied externally, SCSI bus reset issued by the EASI.

6.2 CHIP RESET

When the $\overline{RESET}$ signal is asserted for the required duration the EASI clears ALL internal registers and therefore resets all logic. This action does not create an interrupt or generate a SCSI reset. Since all registers contain zeroes,

6.0 Reset Conditions (Continued)

the EASI is in MODE N under any of the three reset conditions.

6.3 EXTERNAL SCSI RESET

When an SCSI RST is applied externally the EASI resets all registers and logic and issues an interrupt. The only register bits not affected are the Assert RST bit (bit 7) in the ICR and the TARGET Mode bit (bit 6) in MR2. Note that the ISR will contain all zeroes.

6.4 SCSI RESET ISSUED

When the µP sets the Assert RST bit in the ICR, the RST signal goes active. Since the EASI monitors RST also, the same reset actions as in 6.2 apply. The SCSI RST signal will remain active as long as bit 7 in the ICR is set—i.e., until programmed 0 or a chip RESET occurs.

7.0 Loopback Testing

7.1 GENERAL

The DP8490 EASI features loopback testing, enabled by bit 3 of the EMR. When the LOOP bit is set in the EMR all SCSI drivers are disabled at the pads and the signals looped back internally. Additionally, both TARGET and INITIATOR signals may be simultaneously asserted—this is not possible in normal operation. All this enables testing of EASI operation without affecting the SCSI bus.

7.2 SCSI SIGNAL DRIVING/MONITORING

Since each SCSI signal is looped back, testing is accomplished by asserting a signal via the ICR or TCR, then reading back via CSB and BSR to check its state. The code below provides examples. The first example tests SCSI control signals and the second the SCSI data bus. Note that loopback mode must be enabled prior to asserting any signals if they are not to drive the SCSI bus.

```
SCSI_signal_test
{
  ICR = 40h;    /* MODE E */
  EMR = 08h;    /* LOOPback */
  ICR = 5Eh;    /* drive INIT controls */
  TCR = 0Fh;    /* drive TARG controls */
  if((CSB==7Eh) AND (BSR==0Fh))
       ok;
  else error;
}

SCSI_data_test
{
  ICR = 40h;    /* MODE E */
  EMR = 08h;    /* LOOPback */
  ICR = 41h;    /* assert data bus */
  MR2 = 30h;    /* parity check & interrupt */
  ODR = 0AAh;
  if(CSD==0AAh)  ok;
  else  error;
  ODR = 55h;
  if(CSD==55h)  ok;
  else  error;
  if(interrupt)error;  /* no parity errors */
  else ok;
}
```

7.3 (RE)SELECTION AND ARBITRATION

Both these features may be tested in loopback. Note that when checking BSY via the CSB register the "debounced" version of BSY is presented and will be active for 400 ns–800 ns after BSY goes inactive. Logic within the EASI continuously monitors BSY when it becomes inactive to detect a valid Bus Free Phase. One of the outputs of this logic is a clean version of BSY which is accessed through the CSB.

7.4 DATA TRANSFERS

Both programmed-I/O and DMA transfers may be performed. When doing DMA transfers the MR2 TARGET MODE bit (6) must be programmed according to the type of DMA—i.e., set to 1 for TARGET Send or Receive, reset to 0 for INITIATOR Send or Receive. Additionally, the actions of the other SCSI device must be programmed. For example, when testing INITIATOR operations the BSY signal must be set via ICR to simulate a TARGET connected, and the REQ signal must be programmed active and inactive to perform the handshake. The code below shows a single byte transfer as INITIATOR Send.

```
DMA_test
{
  program DMA controller;
  ICR = 40h;    /* MODE E */
  EMR = 08h;    /* LOOPback */
  ICR = 49h;    /* BSY & data bus on */
  MR2 = 0Ah;    /* DMA & EOP interrupt */
  TCR = 08h;    /* assert REQ */
  SDS = 0;      /* Start DMA Send */
  /* DMA cycle with EOP is done here */
  TCR = 0;      /* deassert REQ */
  if(IDR==data byte)  ok;
  else  error;
  if(BSR==90h)  ok;
  else  error;
  /* also should be an EOP interrupt */
}
```

8.0 Extra Features/Compatibility

8.1 OPERATING MODES

This section is intended to clearly identify the differences between the DP8490 and DP5380. The description covers registers, signals, timing, "bugs" and enhancements. For a more detailed description of register programming and bit functions refer to Sections 3–7.

Before discussing differences a review of DP8490 operation is required. The EASI can be operated in two modes—Normal Mode (MODE N) and Enhanced Mode (MODE E). The EASI is in one or the other mode at any time. MODE E is selected when bit 6 in the INITIATOR COMMAND REGISTER (ICR) is set to 1. A SCSI or external chip reset clears all registers (except MODE REGISTER 2 bit 6) so MODE N is selected as the default. MODE N may also be selected at any time by clearing bit 6 of the ICR to zero. If MODE E

8.0 Extra Features/Compatibility (Continued)

functions have been invoked, selecting MODE N does not in general affect their operation. This means MODE E is used to enable selection of enhanced features which can be used in either mode. Normally an application will choose to remain in MODE E always since all functions are accessible in this mode. MODE N is only required for downward compatibility with a standard 5380 device. The DP5380 operates ONLY in MODE N.

The MODE selection via bit 6 of the ICR enables existing 5380 firmware to run unaltered with the EASI. On the 5380 this is a TEST MODE bit which disables ALL output drivers—NOTHING ELSE! No logic is "tested" and no changes are made to internal logic. Since the µP data bus drivers are also disabled the internal state of the 5380 is inaccessible. This means the TEST MODE bit has very limited application for the end user. By comparison the LOOPBACK bit in the EASI enables thorough testing of device operation.

In summary, current operating firmware should not be using bit 6 of the ICR so the DP8490 uses this bit to enable enhanced operation. *As long as this bit is not set in the EASI the DP8490 appears the same as the NCR, AMD or DP5380. Programming, device operation, and timing sequences are all the same as a 5380.*

8.2 INTERNAL REGISTERS

Figure 8.1 shows the register map of the EASI. Note that in MODE N hex address 7 (HA 7) causes a Parity/Interrupt Reset when read and a Start DMA Initiator Receive when written. In MODE E the read/write EXTRA MODE REGISTER (EMR) is mapped into HA 7. Since the two original functions at HA 7 would conflict, their functions are reproduced by writing to bits 1 and 2 of the EMR. This removes any need to switch between modes. The INTERRUPT MASK REGISTER (IMR) and the INTERRUPT STATUS REGISTER (ISR) are accessed indirectly via the EMR. Setting both bits 1 and 2 of the EMR to 1 enables the next read of address 7 to access the ISR or the next write to access the IMR. Once the access is made subsequent uses of address 7 use the EMR.

ables the extra functions. Setting an EMR bit to 1 enables the corresponding function while programming a 0 disables it. Note on power-up or reset the EMR is zeroed so all extra functions are disabled. The INTERRUPT MASK REGISTER (IMR) provides the ability to individually mask out interrupts. The INTERRUPT STATUS REGISTER (ISR) shows the status of the interrupt system at any time. When an interrupt occurs the ISR will contain 1's for interrupts that are active, enabled and not masked. *Figure 8.2* shows the format of the three registers.

8.4 EMR FEATURES

8.4.1 Arbitration

In MODE N, arbitration is enabled via bit 0 of MR2. However this requires polling the device for potentially many milliseconds. The µP polls the ICR for the ARBITRATION IN PROGRESS bit (6) which is set once the ASI/EASI detects a bus free phase and enters arbitration. The µP must then wait for the 2.2 µs ARBITRATION DELAY before checking IDs on the bus. No interrupt is given for any of these events.

In MODE E, EASI arbitration is enabled by bit 0 of either MR2 or EMR. The EMR bit enables extended arbitration. The EASI will wait for bus free, arbitrate, wait the 2.2 µs ARBITRATION DELAY and the INTERRUPT the µP if not masked in the IMR. An interrupt occurs if arbitration is complete or has been lost. This feature removes the need to poll the EASI.

Hex Addr	Read Register	Write Register
00	Current SCSI Data	Output Data Register
01	Initiator Command	Initiator Command
02	Mode Register 2	Mode Register 2
03	Target Command	Target Command
04	Current SCSI Bus Status	Select Enable Register
05	Bus and Status	Start DMA Send
06	Input Data Register	Start DMA Targ Rx
07	Reset Parity/Interrupt (MODE N)	Start DMA Init Rx (MODE N)
07	Extra Mode Register (MODE E)	Extra Mode Register (MODE E)
07	Interrupt Status Reg (MODE E and bits 1 and 2 of EMR = 1)	Interrupt Mask Reg (MODE E and bits 1 and 2 of EMR = 1)

FIGURE 8.1. EASI Register Map

8.3 ENHANCEMENT MODE REGISTERS

MODE E provides three new registers to provide control for the extra features of the EASI. The EXTRA MODE REGISTER (EMR) is a read/write register which enables or dis-

ENHANCED MODE REGISTER

- Extended Arbitration
  - 00 = NOP
  - 01 = Reset Parity/Int
  - 10 = Start DMA Init Rx
  - 11 = Rd/Wr ISR/IMR
- Loopback Mode
- SCSI Parity (ODD/EVEN)
- µP Parity (ODD/EVEN)
- µP Parity Enable
- Interrupt on any Phase Mismatch

INTERRUPT MASK/STATUS REGISTER

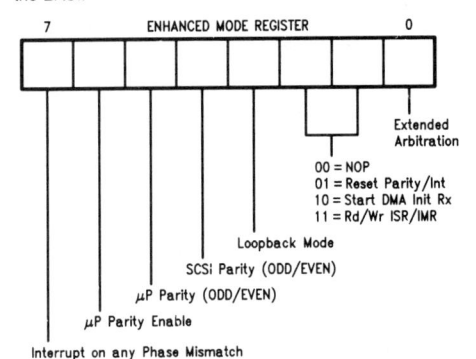

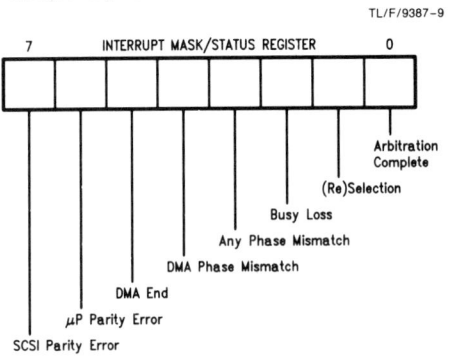

- Arbitration Complete
- (Re)Selection
- Busy Loss
- Any Phase Mismatch
- DMA Phase Mismatch
- DMA End
- µP Parity Error
- SCSI Parity Error

FIGURE 8.2. MODE E Registers

8.0 Extra Features/Compatibility (Continued)

8.4.2 Loopback Mode

When bit 3 is set in the EMR the EASI disables all SCSI drivers and loops back the signals internally. SCSI Data Out is linked to SCSI Data In. $\overline{BSY}$, $\overline{SEL}$, $\overline{ATN}$, $\overline{RST}$, $\overline{I/O}$, $\overline{C/D}$, $\overline{MSG}$, $\overline{REQ}$ and $\overline{ACK}$ outputs are fed back to their own inputs. This enables testing of EASI operations, including DMA.

8.4.3 SCSI Parity

SCSI parity may be enabled and also cause an interrupt (or be masked). Bit 4 of the EMR allows the polarity of SCSI parity to be set to either ODD or EVEN, with a default of ODD. This feature allows checking of SCSI devices and cable. It also allows an INITIATOR to automatically detect whether a device supports parity. (Send a parity error and check it is reported).

8.4.4 µP Parity

The EASI includes parity for the µP data bus. This enables controllers to validate data while it passes through their data buffers. In common with SCSI parity the µP parity may be: enabled/disabled, interrupt or be masked, be EVEN or ODD polarity.

8.4.5 Phase Mismatch

In MODE N the EASI will only give a Phase Mismatch interrupt during DMA (normally data) transfers. In MODE E an interrupt can also be programmed for a phase mismatch during any phase. The mismatch is detected if $\overline{REQ}$ is active and the phase lines do not match the phase expected as specified in the TCR bits 0 to 2. This feature allows completely interrupt-driven operation of the EASI.

8.5 INTERRUPTS

The EASI has an improved interrupt structure to ease programming. In MODE N the structure is the same as a standard 5380. In this mode interrupts can occur which may only be determined by the lack of other interrupts (e.g., Selection and Phase Mismatch). Interrupts can also be missed if they occur while servicing others. To determine the cause of an interrupt requires reading the BSR and CSB registers then interpreting the results, knowing what the 5380 was currently doing.

In MODE E interrupts can be individually masked via the IMR and active ones determined by reading the ISR. Any unmasked interrupts sets an ISR bit to 1. After the interrupt service routine the µP should perform a Reset Parity/Interrupt function via the EMR—this will ONLY RESET interrupts which had a 1 in the ISR when it was last read. This feature means interrupts will not be lost.

8.6 TIMING

The NCR5380 timing has some aspects which have been improved for MODE E of the EASI. In MODE N timing details are the same as the 5380. Of course the DP5380 ASI remains the same as the NCR5380. The timing improvements are listed below.

1. **True End of DMA:** In MODE N the end of DMA is when the last byte is transferred by the DMA controller, not when the final SCSI transfer is done. In MODE E bit 7 of the TCR shows a true end of DMA status—i.e., the last byte transferred by the later of the two events. This bit is compatible with the NCR CMOS 53C80.

2. **SCSI Handshake after EOP:** in MODE N, INITIATOR receive when a $\overline{REQ}$ is received after $\overline{EOP}$ has been given an $\overline{ACK}$ will be generated although no valid data exists since no DRQ was issued. The EASI will NOT generate this invalid $\overline{ACK}$ while in MODE E.

µP and DMA accesses have relaxed timing on both the EASI and ASI. Data setup/hold and read access times are reduced. Faster handshaking on the SCSI bus, along with faster response to the DMA signals means that higher transfer rates are possible with both devices—typically over 3 MBytes/s.

9.0 Application Guide

This section is intended to show the interface between the µP, EASI and DMA controller (DMAC). *Figure 9.1* shows a general interface when the EASI and DMAC are I/O-mapped devices. This configuration will implement a 2 to 2.5 MBytes/s SCSI port using 2 cycle compressed timing from the 5 MHz DMAC.

Using a faster DMAC and memory may allow the EASI to operate at a higher rate—but of course any system will be limited by the available DMA rate from the SCSI device currently connected to. The interface shown has several features that are examined more closely in the following text.

All the interface signal requirements are satisfied by a PAL device. The memory interface is not shown, only the relevant DMAC and µP lines are included.

The EASI data and address lines connect directly to the µP/DMAC busses. The DRQ output from the EASI goes direct to the DMAC. The $\overline{EOP}$ output from the DMAC goes to the EASI input, via the PAL, but can also be asserted via the PAL since the DMAC output is open-drain.

The PAL is programmed so that the µP can access the EASI in three ways. The three access types are: Register R/W, DMA R/W, DMA with $\overline{EOP}$. Examination of the PAL equations below shows how the µP may perform any of the three basic access types simply by accessing the EASI at different I/O address slots. This enables the µP to simulate a DMAC (pseudo-DMA). DMA mode may then be used for all information transfer phases.

In DMA mode the EASI generates all SCSI handshakes. At all other times the µP is responsible for $\overline{REQ}/\overline{ACK}$ handshakes. Using pseudo-DMA may reduce µP overhead.

When doing DMA transfers via BLOCK MODE and an error occurs, the EASI may not deassert the READY signal. For some DMA controllers this may lock the bus, so the PAL asserts READY and $\overline{EOP}$ to the DMA if an interrupt occurs while READY is false. This completes the current DMA cycle and prevents further DMA for the rest of the block thus allowing the bus to be handed back to the µP for servicing.

The PAL generates $\overline{IOR}$ and $\overline{IOW}$ strobes while the µP is bus master, but the DMAC provides the strobes while it is bus master so the PAL outputs are TRI-STATE.

The PAL details are shown in *Figure 9.2* with the signal definitions and equations following.

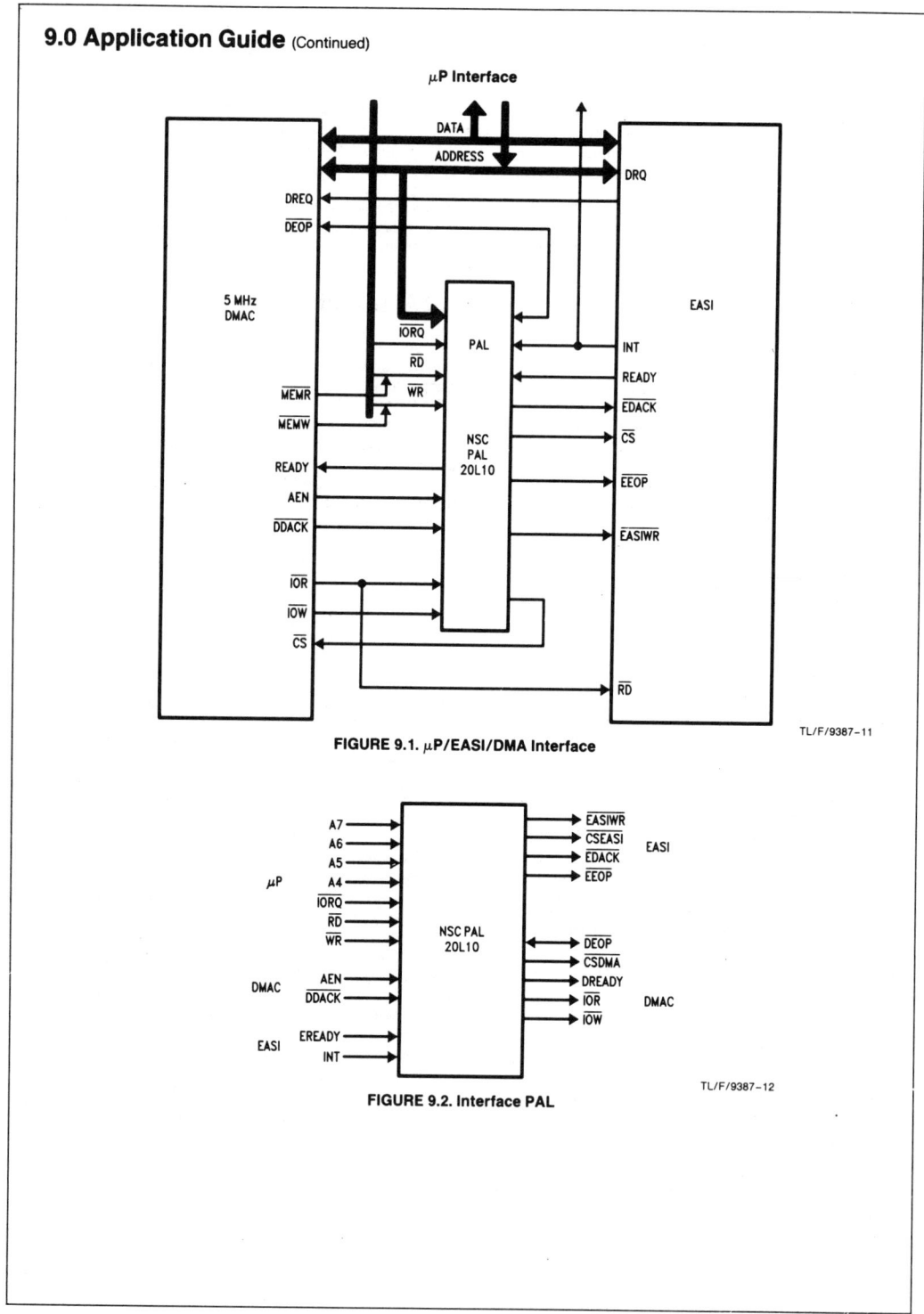

9.0 Application Guide (Continued)

FIGURE 9.1. μP/EASI/DMA Interface

FIGURE 9.2. Interface PAL

9.0 Application Guide (Continued)

```
/CSEASI = /IORQ*/A7*/A6*/A5*/A4*/AEN        ; EASI reg R/W chip select
/EDACK  = /IORQ*/A7*/A6*/A5* A4*/RD         ; µP pseudo-DMA cycle
         /IORQ*/A7*/A6*/A5* A4*/WR
         +/IORQ*/A7*/A6* A5*/A4*/RD         ; µP pseudo-DMA with EOP
         +/IORQ*/A7*/A6* A5* A4*/WR
         +/DDACK                            ; DMAC DMA cycle
/EEOP   = /IORQ*/A7*/A6* A5*/A4*/RD*/AEN    ; µP pseudo-DMA with EOP
         +/IORQ*/A7*/A6* A5*/A4*/WR*/AEN
         +/DEOP*EREADY                      ; Prevents EASI from seeing
                                            ; EOP until READY goes high.
IF(/DDACK*/EREADY*INT)/DEOP = /DDACK*/EREADY*INT; on error this will terminate the DMA
transfer.
/CSDMA  = /IORQ*/A7*/A6*A5*A4               ; DMAC register R/W
/DREADY = /EREADY*/INT                      ; EASI not READY and not INT
         +/EREADY*/DDACK                    ; EASI not READY and DMA cycle active
IF(/AEN) /IOR = /IORQ*/RD                   ; µP I/O Read cycle
IF(/AEN) /IOW = /IORQ*/WR                   ; µP I/O Write cycle
/EASIWR = /IORQ* WR*/AEN+/IOW*EREADY*AEN    ; Prevents SCSI data being
                                            ; changed before EASI is
                                            ; READY for next byte.
```

FIGURE 9.3. PAL Equations

The µP and DMA signals are defined below

A7–A4	Address bus
IORQ	Memory I/O cycle select
RD	Read Strobe
WR	Write Strobe
AEN	High DMA address enable asserted by DMAC
DDACK	DMAC DMA Acknowledge
CSDMA	DMA Chip Select
DREADY	Ready signal to DMAC—inserts wait-states when low
IOR, IOW	I/O data strobes to/from DMAC
EASIWR	EASI write strobe.

10.0 Absolute Maximum Ratings*

If Military/Aerospace specified devices are required, please contact the National Semiconductor Sales Office/Distributors for availability and specifications.

Supply Voltage (V_{CC})	−0.5V to +7.0V
DC Input Voltage (V_{IN})	−0.5V to V_{CC} + 0.5V
DC Output Voltage (V_{OUT})	−0.5V to V_{CC} + 0.5V
Storage Temperature Range (T_{STG})	−65°C to +150°C
Power Dissipation (P_D)	500 mW
Lead Temperature (T_L) (Soldering, 10 seconds)	260°C
Electro-Static Discharge Rating	2 kV

*Absolute maximum ratings are those values beyond which damage to the device may occur.

11.0 DC Electrical Characteristics

T_A = 0°C to +70°C, V_{CC} = 5.0V ±5% unless otherwise specified

Symbol	Parameter	Conditions	Typ	Limit	Units
V_{IH}	Minimum High Level Input Voltage			2.0	V
V_{IL}	Maximum Low Level Input Voltage			0.8	V
V_{OH1} V_{OH2}	Minimum High Level Output Voltage	$\|I_{OUT}\|$ = 20 µA $\|I_{OUT}\|$ = 4.0 mA		V_{CC} − 0.1 2.4	V V
V_{OL1} V_{OL2} V_{OL3}	Maximum Low Level Output Voltage	SCSI Bus Pins: $\|I_{OL}\|$ = 48 mA Other Pins: $\|I_{OL}\|$ = 20 µA $\|I_{OL}\|$ = 8.0 mA		0.5 0.1 0.4	V V V
I_{IN}	Maximum Input Current	V_{IN} = V_{CC} or GND		±1	µA
I_{OZ}	Maximum TRI-STATE Output Leakage Current	V_{OUT} = V_{CC} or GND		±10	µA
I_{CC}	Supply Current	V_{IN} = V_{CC} or GND SCSI Inputs = 3V	2.5	4	mA

Capacitance T_A = 25°C, f = 1 MHz

Symbol	Parameter (Note 3)	Typ	Units
C_{IN}	Input Capacitance	5	pF
C_{OUT}	Output Capacitance	7	pF

AC Test Conditions

Input Pulse Levels	GND to 3.0V
Input Rise and Fall Time	6 ns
Input/Output Reference Levels	1.3V
TRI-STATE Reference Levels (Note 2)	active low + 0.5V active high − 0.5V

TL/F/9387–13

Note 1: C_L = 50 pF including jig and scope capacitance.
Note 2: S1 = Open for push-pull outputs.
S1 = V_{CC} for active low to TRI-STATE.
S1 = GND for active high to TRI-STATE.
Note 3: This parameter is not 100% tested.

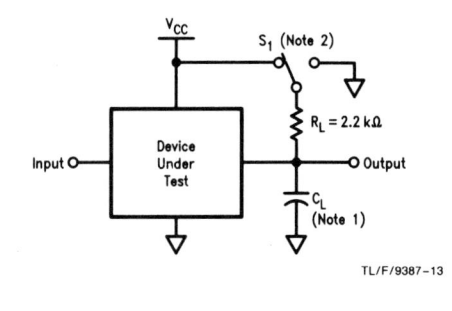

12.0 AC Electrical Characteristics
All parameters are preliminary and subject to change without notice

Symbol	Parameter	DP8490 Min	DP8490 Typ	DP8490 Max	Units
bfas	BSY false to arbitrate start	1200		2200	ns
bfbc	BSY false to bus clear			800	ns
stbc	SEL true to bus clear			500	ns
rst	RESET pulse width	100			ns

12.1 ARBITRATION

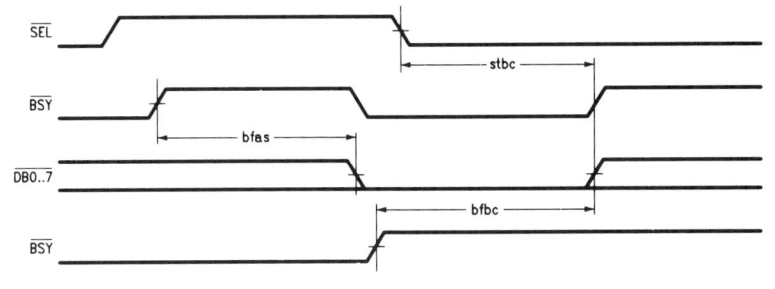

12.2 µP RESET

12.0 Electrical Characteristics

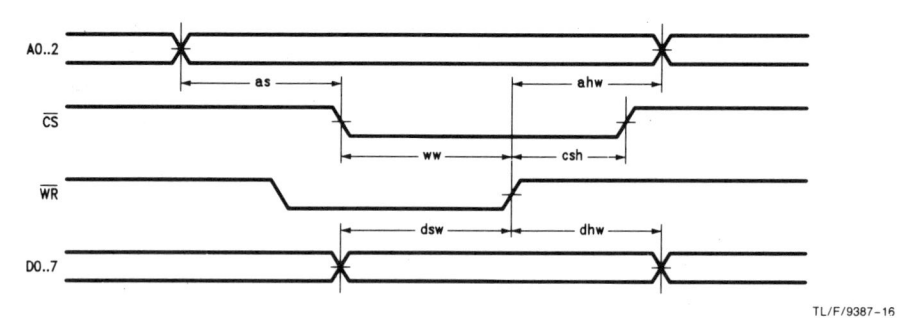

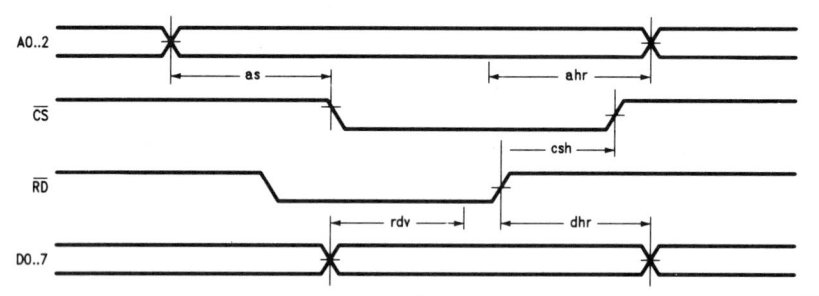

12.3 µP WRITE

12.4 µP READ

Symbol	Parameter	DP8490 Min	DP8490 Typ	DP8490 Max	Units
ahr	Address Hold from and of Read Enable (Note 1)	5			ns
ahw	Address Hold from End of Write Enable (Note 2)	5			ns
as	Address Setup to Read or Write Enable (Notes 1,2)	5			ns
csh	CS Hold from End of RD or WR	0			ns
dhr	Data Hold from End of Read Enable (Notes 1,3)	20		60	ns
dhw	µP Data Hold Time from End of WR	10			ns
dsw	Data Setup to End of Write Enable (no µP Parity) (Note 2) (with µP Parity)	35 35			ns ns
rdv	Data Valid from Read Enable (Note 1)			50	ns
ww	Write Enable Width (Note 2)	40			ns

Note 1: Read enable (µP) is CS and RD active.
Note 2: Write enable (µP) is CS and WR active.
Note 3: This includes the RC delay inherent in the test's method. These signals typically turn off after 25 ns, enabling other devices to drive these lines with no contention.

12.0 AC Electrical Characteristics (Continued)

12.5 DMA WRITE (NON-BLOCK MODE) TARGET SEND

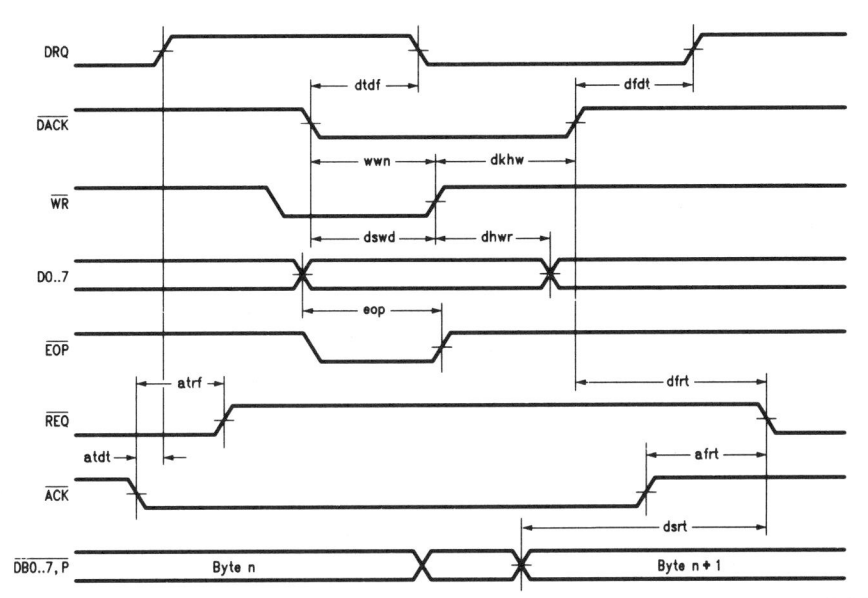

TL/F/9387-18

Symbol	Parameter	DP8490 Min	DP8490 Typ	DP8490 Max	Units
afrt	ACK False to REQ True (DACK or WR False)			75	ns
atdt	ACK True to DRQ True			55	ns
atrf	ACK True to REQ False			100	ns
dfdt	DACK False to DRQ True	30	90		ns
dfrt	DACK False to REQ True (ACK False)			75	ns
dhwr	DMA Data Hold Time from End of WR	10			ns
dkhw	DACK Hold from End of WR	0			ns
dsrt	SCSI Data Setup to REQ True	25			ns
dswd	Data Setup to End of DMA Write Enable (no µP Parity) (Note 1)	35			ns
	(with µP Parity)	35			ns
dtdf	DACK True to DRQ False			45	ns
eop	Width of EOP Pulse (Note 2)	25			ns
wwn	DMA Non-block Mode Write Enable Width (Note 2)	40			ns

Note 1: Write enable (DMA) is DACK and WR active.
Note 2: EOP, DACK, RD/WR must all be true for recognition of EOP.

12.0 AC Electrical Characteristics (Continued)

12.6 DMA WRITE (NON-BLOCK MODE) INITIATOR SEND

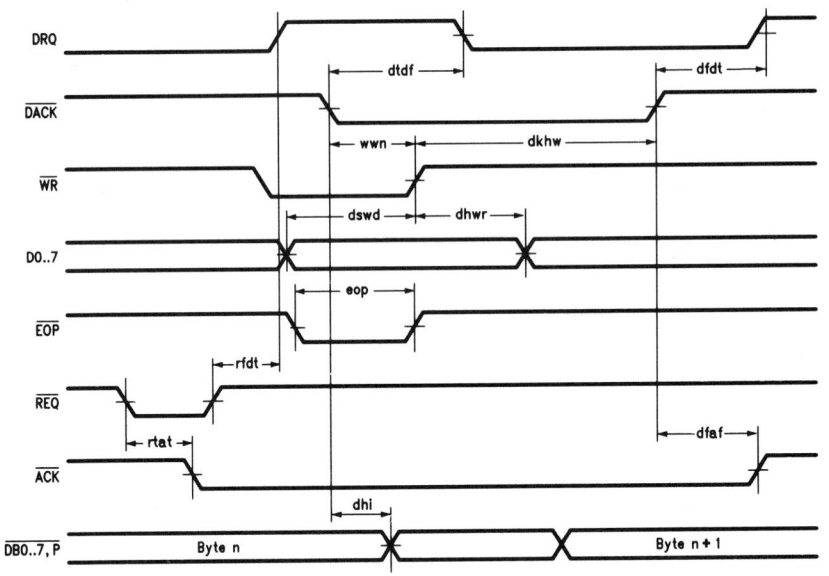

Symbol	Parameter	DP8490 Min	DP8490 Typ	DP8490 Max	Units
dfaf	DACK False to ACK False			90	ns
dfdt	DACK False to DRQ True	30	90		ns
dhi	SCSI Data Hold from Write Enable-Initiator	15			ns
dhwr	DMA Data Hold Time from End of WR	10			ns
dkhw	DACK Hold from End of WR	0			ns
dswd	Data Setup to End of DMA Write Enable (no μP Parity) (Note 1) (with μP Parity)	35 35			ns ns
dtdf	DACK True to DRQ False			45	ns
eop	Width of EOP Pulse (Note 2)	25			ns
rfdt	REQ False to DRQ True			60	ns
rtat	REQ True to ACK True			80	ns
wwn	DMA Write Enable Width (Note 1)	40			ns

Note 1: Write enable (DMA) is DACK and WR active.
Note 2: EOP, DACK, RD/WR must all be true for recognition of EOP.

12.0 AC Electrical Characteristics (Continued)

12.7 DMA READ (NON-BLOCK MODE) TARGET RECEIVE

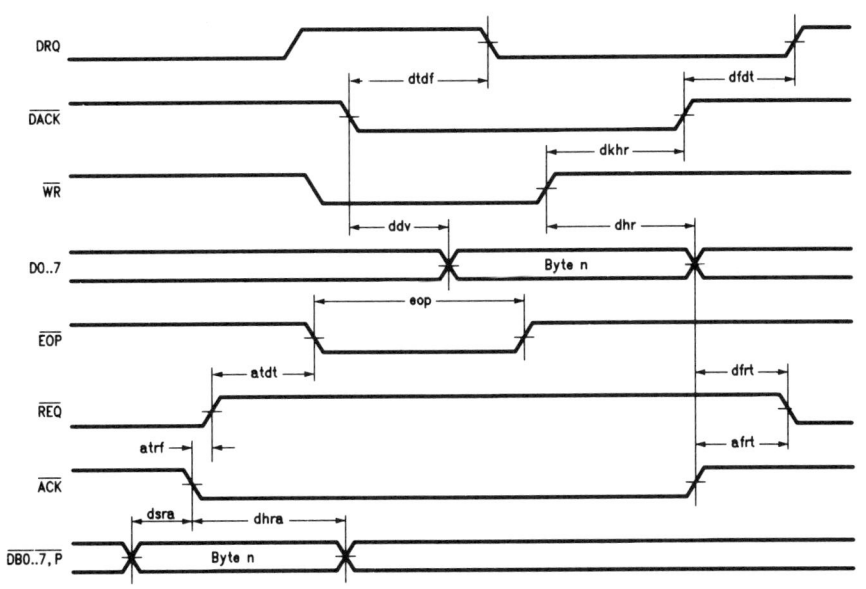

Symbol	Parameter	DP8490 Min	DP8490 Typ	DP8490 Max	Units
afrt	ACK False to REQ True (DACK or WR False)			75	ns
atdt	ACK True to DRQ True			55	ns
atrf	ACK True to REQ False			100	ns
ddv	DMA Data Valid from Read Enable (Note 1)			40	ns
dfdt	DACK False to DRQ True	30	90		ns
dfrt	DACK False to REQ True (ACK False)			75	ns
dhr	Data Hold from End of Read Enable (Notes 1, 2)	20		60	ns
dhra	SCSI Data Hold from ACK True	15			ns
dkhr	DACK Hold from End of RD	0			ns
dsra	SCSI Data Setup Time to ACK True	10			ns
dtdf	DACK True to DRQ False			45	ns
eop	Width of EOP Pulse (Note 3)	25			ns

Note 1: Read enable (DMA) is DACK and RD active.

Note 2: This includes the RC delay inherent in the test's method. These signals typically turn off after 25 ns enabling other devices to drive these lines with no contention.

Note 3: EOP, DACK, RD/WR must all be true for recognition of EOP.

12.0 AC Electrical Characteristics (Continued)

12.8 DMA READ (NON-BLOCK MODE) INITIATOR RECEIVE

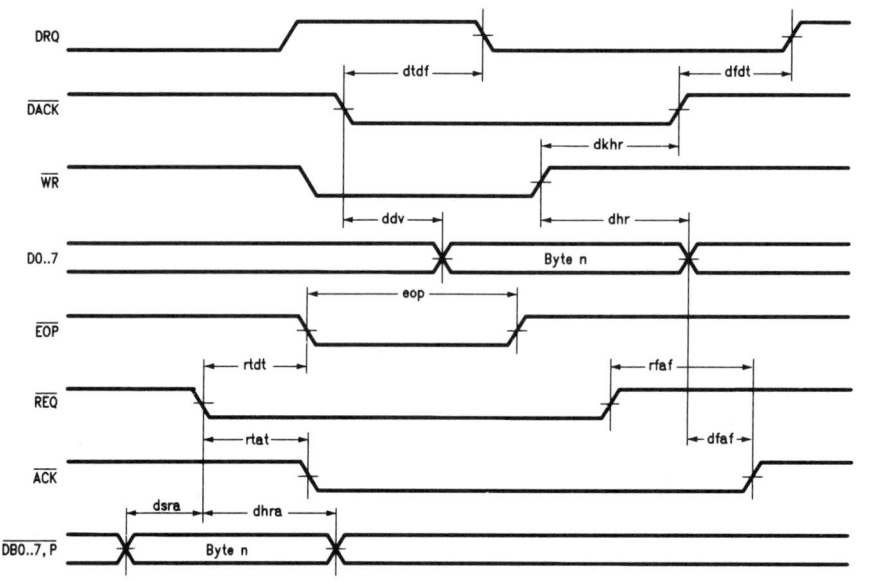

Symbol	Parameter	DP8490 Min	DP8490 Typ	DP8490 Max	Units
ddv	DMA Data Valid from Read Enable (Note 1)			40	ns
dfaf	DACK False to ACK False (REQ False)			90	ns
dfdt	DACK False to DRQ True	30	90		ns
dhr	Data Hold from End of Read Enable (Notes 1, 2)	20		60	ns
dhra	SCSI Data Hold from REQ True	15			ns
dkhr	DACK Hold from End of RD	0			ns
dsra	SCSI Data Setup Time to REQ True	10			ns
dtdf	DACK True to DRQ False			45	ns
eop	Width of EOP Pulse (Note 3)	25			ns
rfaf	REQ False to ACK False (DACK False)			90	ns
rtat	REQ True to ACK True			80	ns
rtdt	REQ True to DRQ True			70	ns

Note 1: Read enable (DMA) is DACK and RD active.

Note 2: This includes the RC delay inherent in the test's method. These signals typically turn off after 25 ns enabling other devices to drive these lines with no contention.

Note 3: EOP, DACK, RD/WR must all be true for recognition of EOP.

12.0 AC Electrical Characteristics (Continued)

12.9 DMA WRITE (BLOCK MODE) TARGET SEND

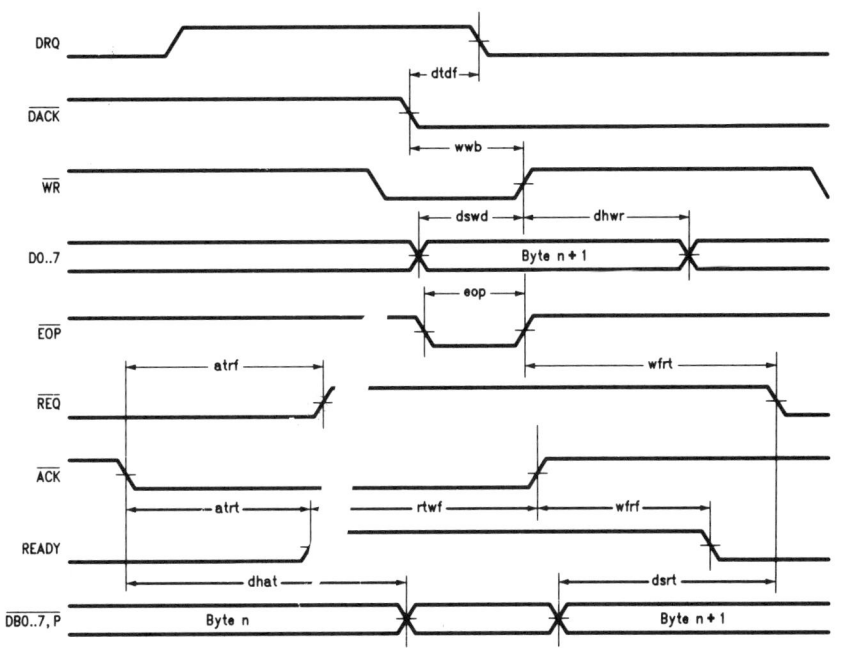

TL/F/9387-22

Symbol	Parameter	DP8490 Min	DP8490 Typ	DP8490 Max	Units
afrt	ACK False to REQ True (DACK or WR False)			75	ns
atrf	ACK True to REQ False			100	ns
atrt	ACK True to READY True			50	ns
dhat	SCSI Data Hold from ACK True	40			ns
dhwr	DMA Data Hold Time from End of WR	10			ns
dsrt	SCSI Data Setup to REQ True	35			ns
dswd	Data Setup to End of DMA Write Enable (no µP Parity) (Note 1) (with µP Parity)	35 35			ns ns
dtdf	DACK True to DRQ False			45	ns
eop	Width of EOP Pulse (Note 2)	25			ns
rtwf	READY True to WR False	40			ns
wfrf	WR False to READY False			50	ns
wfrt	WR False to REQ True (ACK False)			80	ns
wwb	DMA Write Enable Width (Note 1)	40			ns

Note 1: Write enable (DMA) is DACK and WR active.
Note 2: EOP, DACK, RD/WR must all be true for recognition of EOP.

12.0 AC Electrical Characteristics (Continued)

12.10 DMA WRITE (BLOCK MODE) INITIATOR SEND

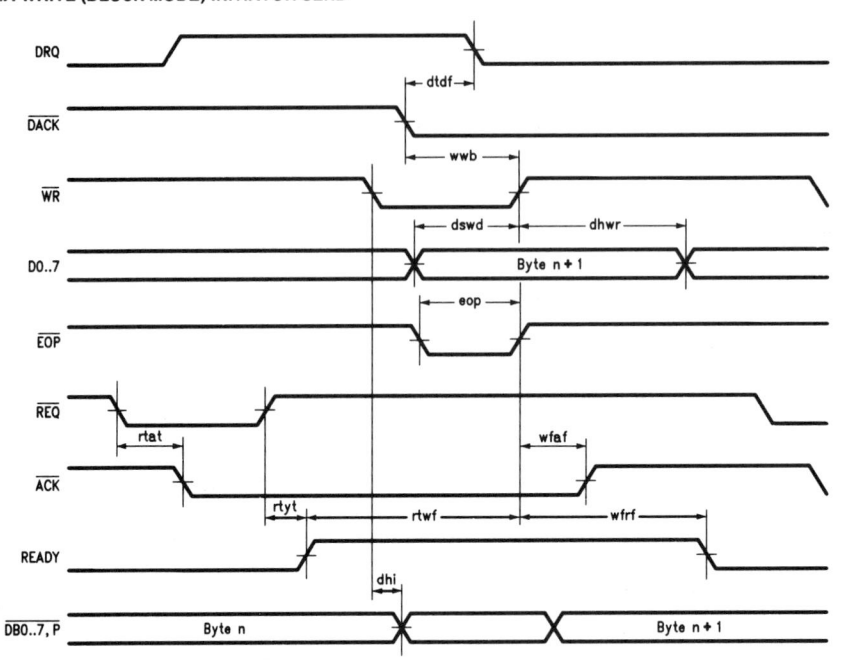

Symbol	Parameter	DP8490 Min	DP8490 Typ	DP8490 Max	Units
dhi	SCSI Data Hold from Write Enable	15			ns
dhwr	DMA Data Hold Time from End of WR	10			ns
dswd	Data Setup to End of DMA Write Enable (no μP Parity) (Note 1) (with μP Parity)	35 35			ns ns
dtdf	DACK True to DRQ False			45	ns
eop	Width of EOP Pulse (Note 2)	25			ns
rfyt	REQ False to READY True			60	ns
rtat	REQ True to ACK True			80	ns
rtwf	READY True to WR False	40			ns
wfaf	WR False to ACK False (REQ False)			95	ns
wfrf	WR False to READY False			50	ns
wwb	DMA Write Enable Width (Note 1)	40			ns

Note 1: Write enable (DMA) is DACK and WR active.
Note 2: EOP, DACK, RD/WR must all be true for recognition of EOP.

12.0 AC Electrical Characteristics (Continued)

12.11 DMA WRITE (BLOCK MODE) TARGET RECEIVE

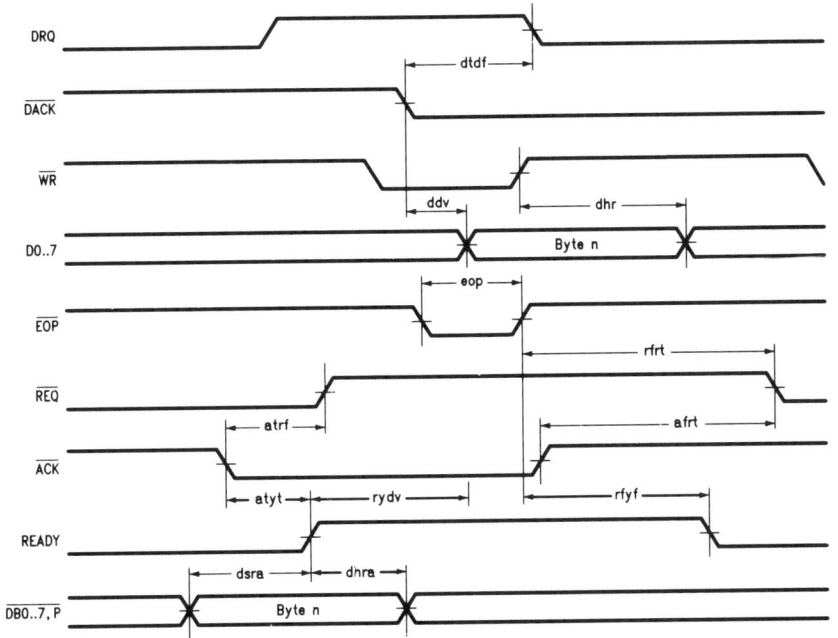

Symbol	Parameter	DP8490 Min	DP8490 Typ	DP8490 Max	Units
afrt	ACK False to REQ True (DACK or WR False)			75	ns
atrf	ACK True to REQ False			100	ns
atyt	ACK True to READY True			50	ns
ddv	DMA Data Valid from Read Enable (Note 1)			40	ns
dhr	Data Hold from End of Read Enable (Notes 1,2)	20		60	ns
dhra	SCSI Data Hold from ACK True	15			ns
dsra	SCSI Data Setup Time to ACK True	10			ns
dtdf	DACK True to DRQ False			45	ns
eop	Width of EOP Pulse (Note 3)	25			ns
rfrt	RD False to REQ True (ACK False)			75	ns
rfyf	RD False to READY False			45	ns
rydv	READY True to Data Valid			20	ns

Note 1: Read enable (DMA) is DACK and RD active.
Note 2: This includes the RC delay inherent in the test's method. These signals typically turn off after 25 ns enabling other devices to drive these lines with no contention.
Note 3: EOP, DACK, RD/WR must all be active for recognition of EOP.

12.0 AC Electrical Characteristics (Continued)

12.12 DMA READ (BLOCK MODE) INITIATOR RECEIVE

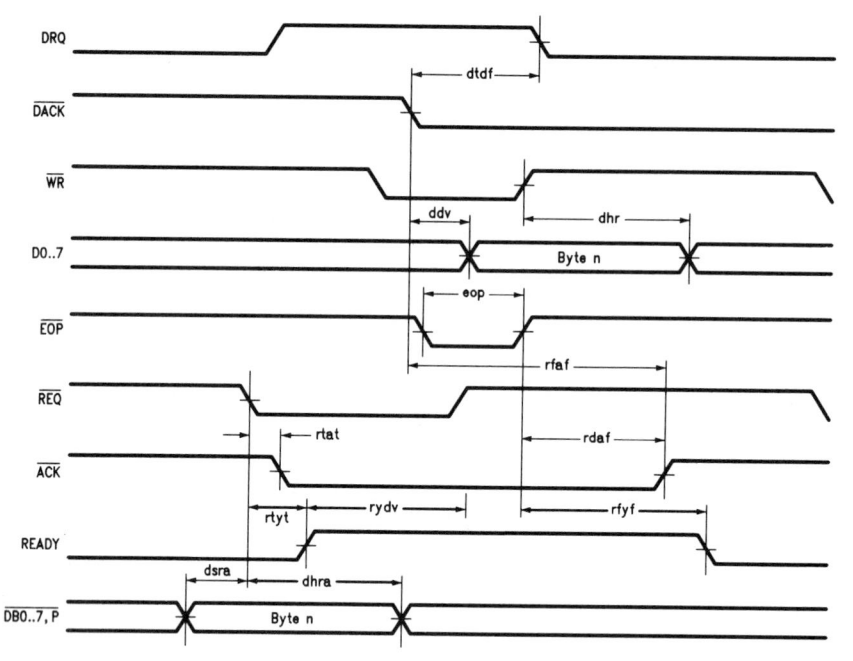

TL/F/9387-25

Symbol	Parameter	DP8490 Min	DP8490 Typ	DP8490 Max	Units
ddv	DMA Data Valid from Read Enable (Note 1)			40	ns
dhr	Data Hold from End of Read Enable (Notes 1,2)	20		60	ns
dhra	SCSI Data Hold from REQ True	15			ns
dsra	SCSI Data Setup Time to REQ True	10			ns
dtdf	DACK True to DRQ False			45	ns
eop	Width of EOP Pulse (Note 3)	25			ns
rdaf	RD False to ACK False (REQ False)			120	ns
rfaf	REQ False to ACK False (DACK False)			90	ns
rfyf	RD False to READY False			45	ns
rtat	REQ True to ACK True			80	ns
rtyt	REQ True to READY True			65	ns
rydv	READY True to Data Valid			20	ns

Note 1: Read enable (DMA) is DACK and RD active.

Note 2: This includes the RC delay inherent in the test's method. These signals typically turn off after 25 ns enabling other devices to drive these lines with no contention.

Note 3: EOP, DACK, RD/WR must all be active for recognition of EOP.

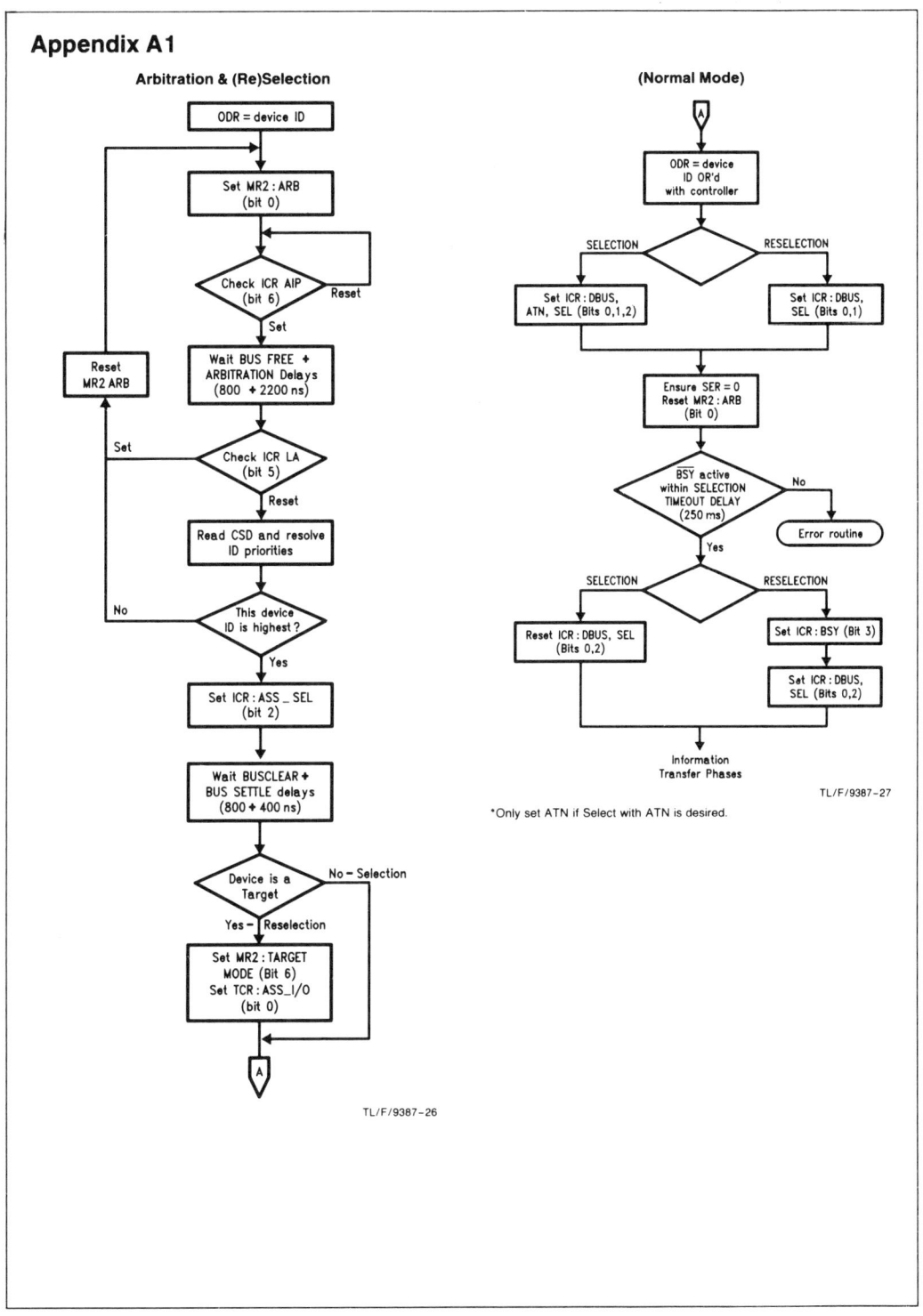

Appendix A1

Arbitration & (Re)Selection

ODR = device ID
↓
Set MR2 : ARB (bit 0)
↓
Check ICR AIP (bit 6) — Reset → (loop back)
↓ Set
Wait BUS FREE + ARBITRATION Delays (800 + 2200 ns)
↓
Check ICR LA (bit 5) — Set → Reset MR2 ARB → (loop back to ODR = device ID)
↓ Reset
Read CSD and resolve ID priorities
↓
This device ID is highest? — No → (back to Check ICR AIP)
↓ Yes
Set ICR : ASS_SEL (bit 2)
↓
Wait BUSCLEAR + BUS SETTLE delays (800 + 400 ns)
↓
Device is a Target — No = Selection →
↓ Yes = Reselection
Set MR2 : TARGET MODE (Bit 6)
Set TCR : ASS_I/O (bit 0)
↓
(A)

TL/F/9387-26

(Normal Mode)

(A)
↓
ODR = device ID OR'd with controller
↓
SELECTION / RESELECTION
- SELECTION: Set ICR : DBUS, ATN, SEL (Bits 0,1,2)
- RESELECTION: Set ICR : DBUS, SEL (Bits 0,1)
↓
Ensure SER = 0
Reset MR2 : ARB (Bit 0)
↓
BSY active within SELECTION TIMEOUT DELAY (250 ms) — No → Error routine
↓ Yes
SELECTION / RESELECTION
- SELECTION: Reset ICR : DBUS, SEL (Bits 0,2)
- RESELECTION: Set ICR : BSY (Bit 3) → Set ICR : DBUS, SEL (Bits 0,2)
↓
Information Transfer Phases

TL/F/9387-27

*Only set ATN if Select with ATN is desired.

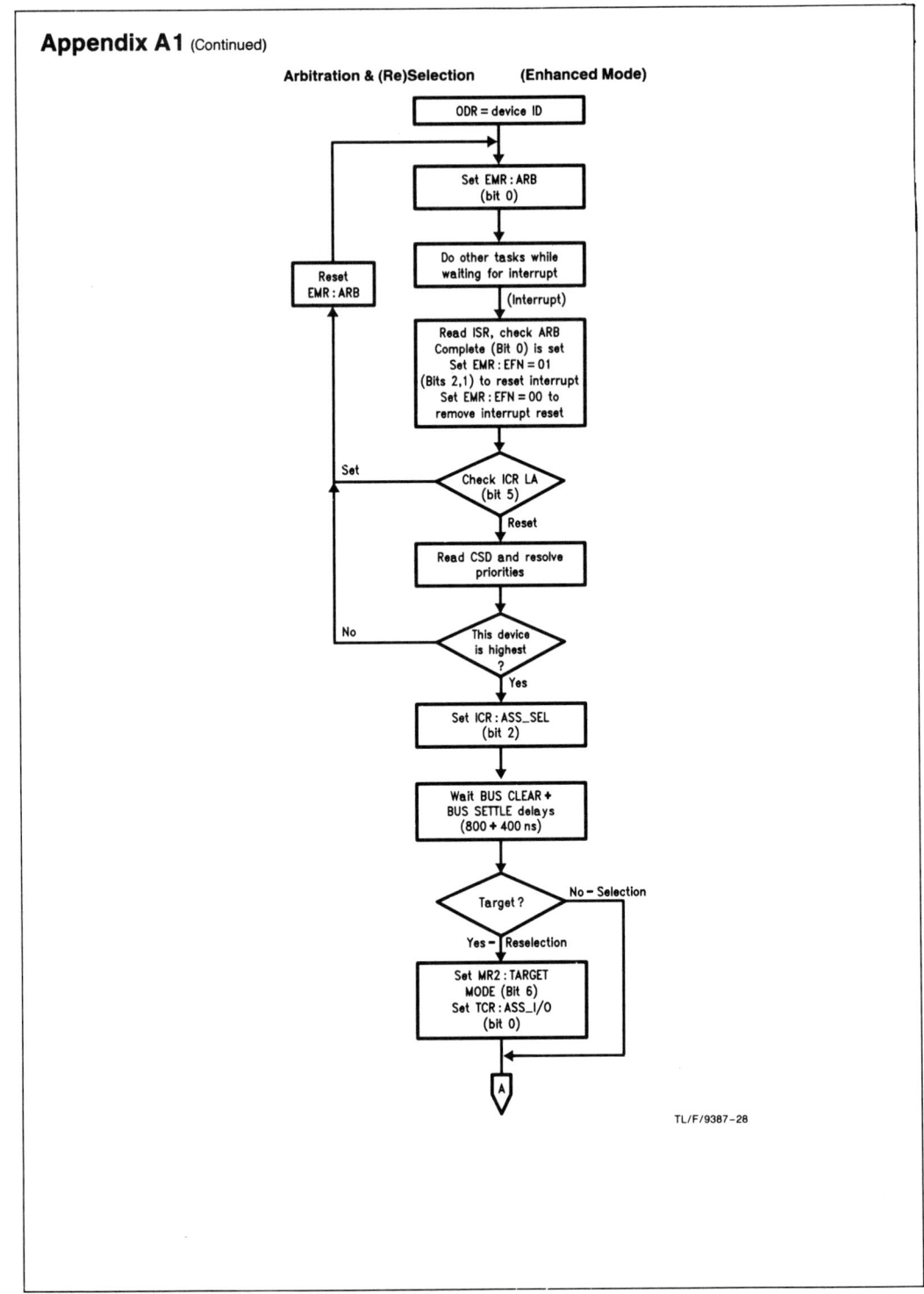

Appendix A1 (Continued)

Arbitration & (Re)Selection   (Enhanced Mode)

```
ODR = device ID
      │
      ▼
Set EMR : ARB
   (bit 0)
      │
      ▼
Do other tasks while
waiting for interrupt
      │
   (Interrupt)
      ▼
Read ISR, check ARB
Complete (Bit 0) is set
Set EMR : EFN = 01
(Bits 2,1) to reset interrupt
Set EMR : EFN = 00 to
remove interrupt reset
      │
      ▼
  Check ICR LA    ──Set──▶ Reset EMR : ARB ──▶ (back to ODR = device ID)
    (bit 5)
      │ Reset
      ▼
Read CSD and resolve
    priorities
      │
      ▼
  This device       ──No──▶ (back to Reset EMR : ARB)
  is highest
      ?
      │ Yes
      ▼
Set ICR : ASS_SEL
    (bit 2)
      │
      ▼
Wait BUS CLEAR +
BUS SETTLE delays
  (800 + 400 ns)
      │
      ▼
   Target ?  ──No – Selection──┐
      │                        │
  Yes – Reselection            │
      ▼                        │
Set MR2 : TARGET               │
  MODE (Bit 6)                 │
Set TCR : ASS_I/O              │
    (bit 0)                    │
      │                        │
      ▼◀───────────────────────┘
     (A)
```

TL/F/9387–28

Appendix A1 (Continued)

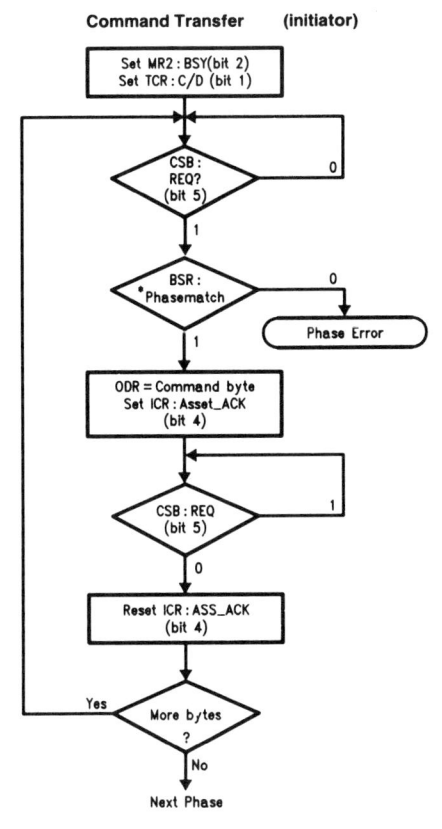

Command Transfer (initiator)

```
        Set MR2 : BSY (bit 2)
        Set TCR : C/D (bit 1)
                 │
                 ▼
            ┌─────────┐
            │  CSB :  │   0
            │  REQ?   │──────┐
            │ (bit 5) │      │
            └─────────┘      │
                 │ 1         │
                 ▼           │
            ┌─────────┐      │
            │  BSR :  │   0  │      ╭──────────────╮
            │*Phase-  │─────────────│ Phase Error  │
            │ match   │             ╰──────────────╯
            └─────────┘
                 │ 1
                 ▼
        ODR = Command byte
        Set ICR : Asset_ACK
             (bit 4)
                 │
                 ▼
            ┌─────────┐
            │ CSB:REQ │   1
            │ (bit 5) │──────┐
            └─────────┘      │
                 │ 0
                 ▼
        Reset ICR : ASS_ACK
             (bit 4)
                 │
                 ▼
            ┌─────────┐
      Yes   │  More   │
     ◄──────│  bytes  │
            │    ?    │
            └─────────┘
                 │ No
                 ▼
            Next Phase
```

TL/F/9387-29

*This step unnecessary in MODE E if the EMR : APHS (bit 7) is enabled. Logic automatically checks the phase on any transfer and interrupts on an error.

35

Appendix A1 (Continued)

Command Transfer (Target)

```
         ┌──────────────┐
         │ Message_out  │
         │    Phase     │
         └──────▲───────┘
                │
         ╱ATN active?╲
         ╲BSR bit 1  ╱
                │
         ┌──────▼───────┐
         │Set TCR: C/D (bit 1)│
         └──────┬───────┘
         ┌──────▼───────────┐
         │Wait Bus Settle Delay│
         │Set TCR: ASS_REQ (bit 3)│
         └──────┬───────────┘
          0 ╱BSR:ACK?╲
            ╲(bit 0) ╱
                │1
         ┌──────▼───────────┐
         │  Command = CSD   │
         │(Parity checked if enabled)│
         └──────┬───────────┘
         ┌──────▼───────────┐
         │Verify Operation Code│
         │  setup CDB length │
         └──────┬───────────┘
         ┌──────▼───────────┐
         │ Reset TCR: ASS_REQ│
         └──────┬───────────┘
          1 ╱BSR:ACK?╲
            ╲(bit 0) ╱
                │0
         ┌──────▼───────────┐
         │Transfer remaining │
         │    CDB bytes     │
         └──────────────────┘
```

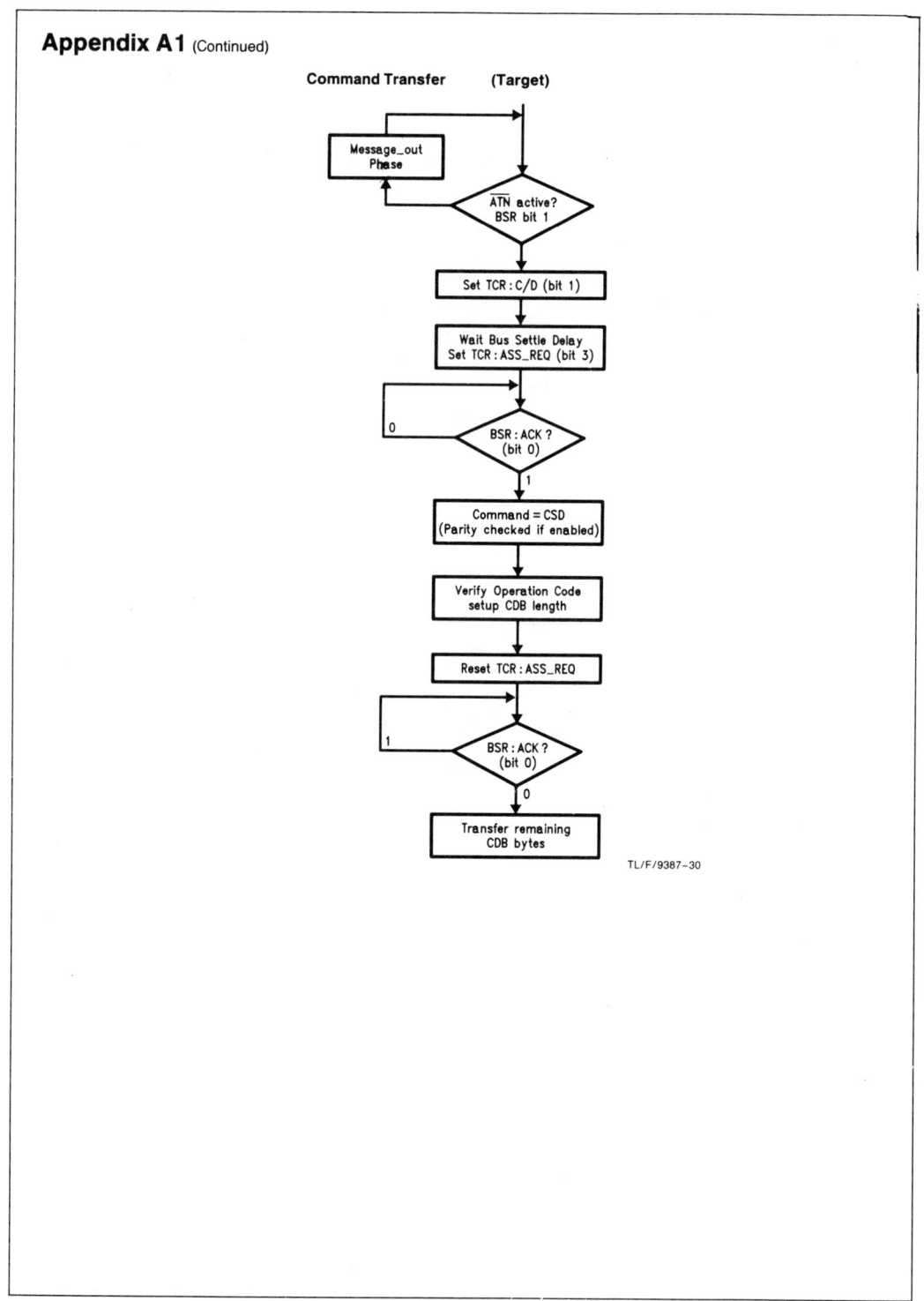

TL/F/9387–30

Appendix A1 (Continued)

**Block Mode DMA Transfer
Initiator Receive (MODE E)**

- Program DMA Controller with block length, start address and memory write operation
- TCR = 01h; ICR = 40h; MR2 = BEh
- EMR = 04h to Start DMA
- Interrupt? No → loop; Yes →
- Read ISR for interrupt status
- True End of DMA only? No → Error routine; Yes →
- Reset DMA Mode (MR2 = 04h)

TL/F/9387-31

**Block Mode DMA Transfer
Target Send (MODE E)**

- Program DMA Controller with block length, start address and memory write operation
- TCR = 01h; ICR = 39h; MR2 = CAh
- SDS = XX to Start DMA Send
- Interrupt? No → loop; Yes →
- Read ISR for interrupt status
- True End of DMA only? No → Error routine; Yes →
- Reset DMA Mode (MR2 = 04h)

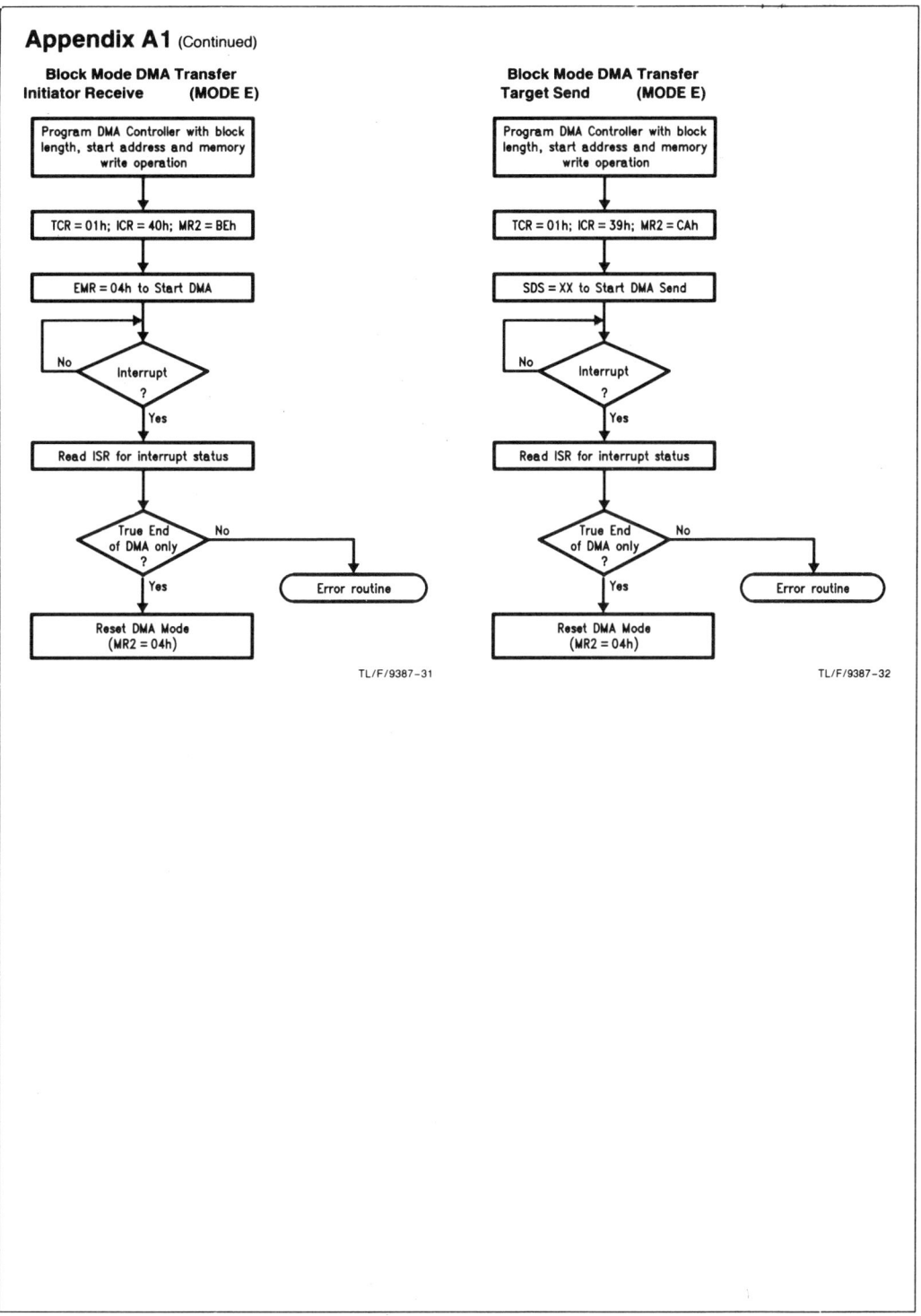

TL/F/9387-32

Appendix A2

Register Chart

Read

Current SCSI Data (CSD) — Bit 7 to Bit 0

DB7	DB6	DB5	DB4	DB3	DB2	DB1	DB0

Initiator Command Register (ICR) — Bit 7 to Bit 0

RST	AIP	LA	ACK	BSY	SEL	ATN	DBUS

Mode Register 2 (MR2) — Bit 7 to Bit 0

BLK	TARG	PCHK	PINT	EOP	BSY	DMA	ARB

Target Command Register (TCR) — Bit 7 to Bit 0

0	0	0	0	REQ	MSG	C/D	I/O

Current SCSI Bus Status (CSB) — Bit 7 to Bit 0

RST	BSY	REQ	MSG	C/D	I/O	SEL	DBP

Bus and Status Register (BSR) — Bit 7 to Bit 0

EDMA	DRQ	SPER	INT	PHSM	BSY	ATN	ACK

Input Data Register (IDR) — Bit 7 to Bit 0

DB7	DB6	DB5	DB4	DB3	DB2	DB1	DB0

Reset Parity/Interrupt (RPI) — MODE N — Bit 7 to Bit 0

X	X	X	X	X	X	X	X

Enhanced Mode Register (EMR) — Bit 7 to Bit 0

APHS	MPEN	MPOL	SPOL	LOOP	EFN1	EFN0	ARB

Interrupt Status Register (ISR) — Bit 7 to Bit 0

SPE	MPE	EDMA	DPHS	APHS	BSY	SEL	ARB

Target Command Register (TCR) — MODE E — Bit 7 to Bit 0

(true) EDMA	0	0	0	REQ	MSG	C/D	I/O

X = Unknown

Write

Output Data Register (ODR) — Bit 7 to Bit 0

DB7	DB6	DB5	DB4	DB3	DB2	DB1	DB0

Initiator Command Register (ICR) — Bit 7 to Bit 0

RST	MODE E	DIFF EN	ACK	BSY	SEL	ATN	DBUS

Mode Register 2 (MR2) — Bit 7 to Bit 0

BLK	TARG	PCHK	PINT	EOP	BSY	DMA	ARB

Target Command Register (TCR) — Bit 7 to Bit 0

X	X	X	X	REQ	MSG	C/D	I/O

Select Enable Register (SER) — Bit 7 to Bit 0

DB7	DB6	DB5	DB4	DB3	DB2	DB1	DB0

Start DMA Send (SDS) — Bit 7 to Bit 0

X	X	X	X	X	X	X	X

Start DMA Target Receive (SDT) — Bit 7 to Bit 0

X	X	X	X	X	X	X	X

Start DMA Initiator Receive (SDI) — MODE N — Bit 7 to Bit 0

X	X	X	X	X	X	X	X

Enhanced Mode Register (EMR) — Bit 7 to Bit 0

APHS	MPEN	MPOL	SPOL	LOOP	EFN1	EFN0	ARB

Interrupt Mask Register (IMR) — Bit 7 to Bit 0

SPE	MPE	EDMA	DPHS	APHS	BSY	SEL	ARB

Target Command Register (TCR) — MODE E — Bit 7 to Bit 0

X	X	X	X	REQ	MSG	C/D	I/O

X = Don't Care

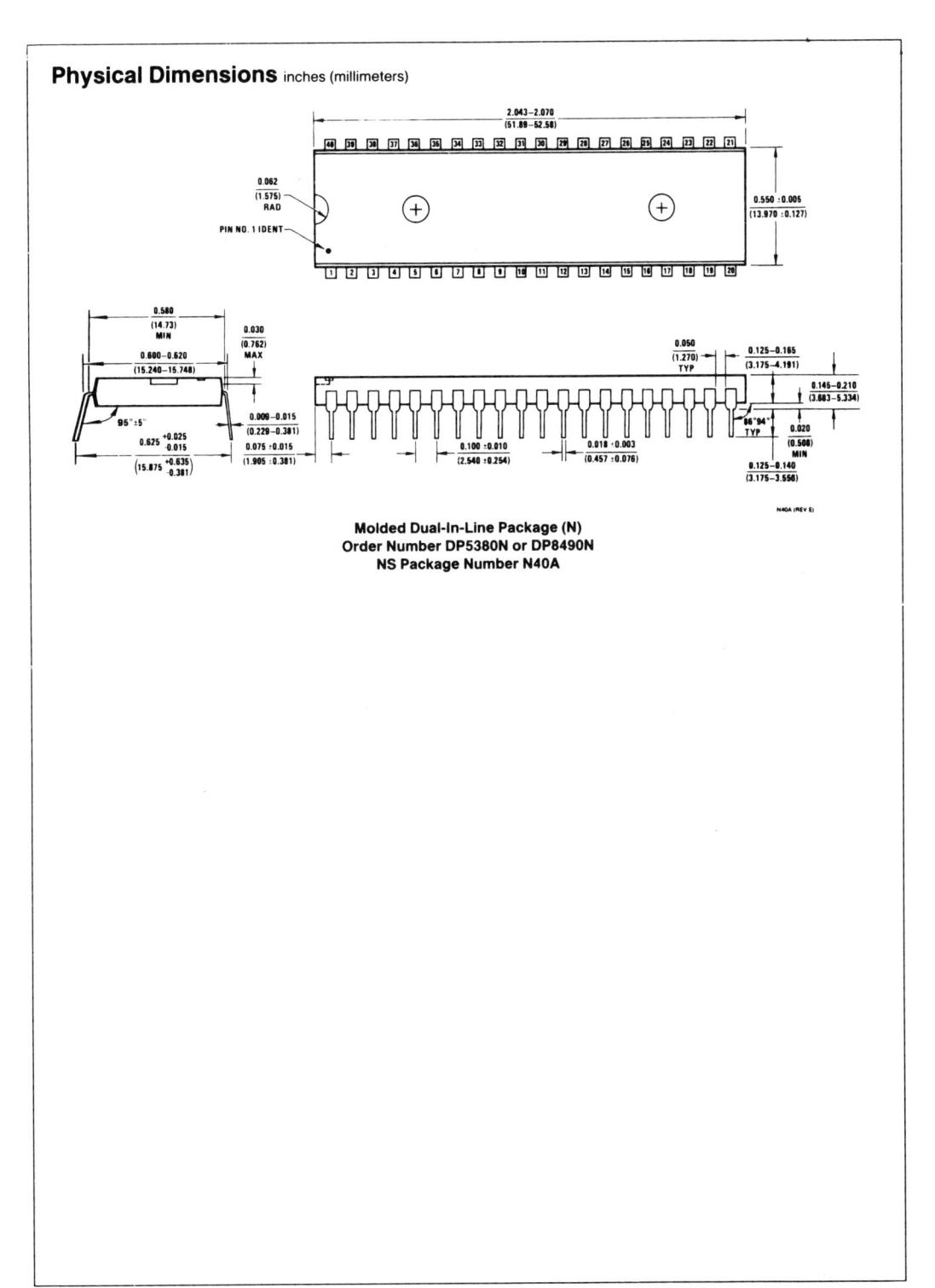

Physical Dimensions inches (millimeters)

Molded Dual-In-Line Package (N)
Order Number DP5380N or DP8490N
NS Package Number N40A

DP8490 Enhanced Asynchronous SCSI Interface (EASI)

Physical Dimensions inches (millimeters) (Continued) Lit. # 103201

Plastic Chip Carrier
Order Number DP5380V or DP8490V
NS Package Number V44A

V44A (REV H)

LIFE SUPPORT POLICY

NATIONAL'S PRODUCTS ARE NOT AUTHORIZED FOR USE AS CRITICAL COMPONENTS IN LIFE SUPPORT DEVICES OR SYSTEMS WITHOUT THE EXPRESS WRITTEN APPROVAL OF THE PRESIDENT OF NATIONAL SEMICONDUCTOR CORPORATION. As used herein:

1. Life support devices or systems are devices or systems which, (a) are intended for surgical implant into the body, or (b) support or sustain life, and whose failure to perform, when properly used in accordance with instructions for use provided in the labeling, can be reasonably expected to result in a significant injury to the user.

2. A critical component is any component of a life support device or system whose failure to perform can be reasonably expected to cause the failure of the life support device or system, or to affect its safety or effectiveness.

National Semiconductor GmbH	National Semiconductor S.A.	National Semiconductor S.p.A.	National Semiconductor AB	National Semiconductor	National Semiconductor (UK) Ltd.
Industriestraße 10 D-8080 Fürstenfeldbruck Tel. (0 81 41) 103-0 Telex 527 649 Fax (0 81 41) 10 35 54	Centre d'Affaires «La Boursidière» Bâtiment Champagne, B.P. 90 Route Nationale 186 F-92357 Le Plessis Robinson Tel. (1) 40 94 88 88 Telex 631 065 Fax (1) 40 94 88 11	Strada 7 - Palazzo R/3 I-20089 Rozzano - Milanofiori Tel. (02) 8 24 20 46/7/8/9 Telex 352 647 Fax (02) 8 25 47 58	Box 2016 Stensträvägen 13 S-12702 Skärholmen Tel. (08) 97 01 90 Telex 10 731 Fax (08) 97 68 12	Postbus 90 1380 AB Weesp Tel. (0 29 40) 3 04 48 Telex 10 956 Fax (0 29 40) 3 04 30	The Maple, Kembrey Park Swindon, Wiltshire SN2 8UT Tel. (07 93) 61 41 41 Telex 444 674 Fax (07 93) 69 75 22

National does not assume any responsibility for use of any circuitry described, no circuit patent licenses are implied and National reserves the right at any time without notice to change said circuitry and specifications.

A SCSI Printer Controller Using Either the DP8490 EASI or DP5380 ASI and Users Guide

National Semiconductor
Application Note 563
Andrew M. Davidson
December 1988

The DP8490 Enhanced Asynchronous SCSI Interface and DP5380 Asynchronous SCSI Interface are CMOS devices, which offer a low cost high performance Small Computer Systems Interface. These devices are pin compatible, and software compatible until an enhanced mode bit is set in the DP8490. This enhanced mode offers many new features which can yield increases in system performance through software, in addition to the improvements in speed and power shown by both devices over existing NMOS devices.

This application note shows how the hardware and software can be designed for a SCSI Printer Controller (SPC) so that it can incorporate either the DP5380 or DP8490. Since the software automatically detects which device is inserted either can be used, although the enhanced mode of the DP8490 offers a better system in terms of throughput and error tolerance. All of the software discussed is available on a floppy disk.

1.0 Introduction

The SCSI Printer Controller (Figure 1.1) consists of five main parts:

1. Microprocessor NSC800™
2. Printer Interface NSC831 PIO (Parallel Input/Output)
3. SCSI Interface DP8490 or DP5380
4. Memory
   CMOS EPROM NMC27C256
   CMOS Static RAM 62256
5. DMA Controller 9517 or 8237

The NSC800 is an eight bit CMOS microprocessor which is the central processing unit of the National Semiconductor

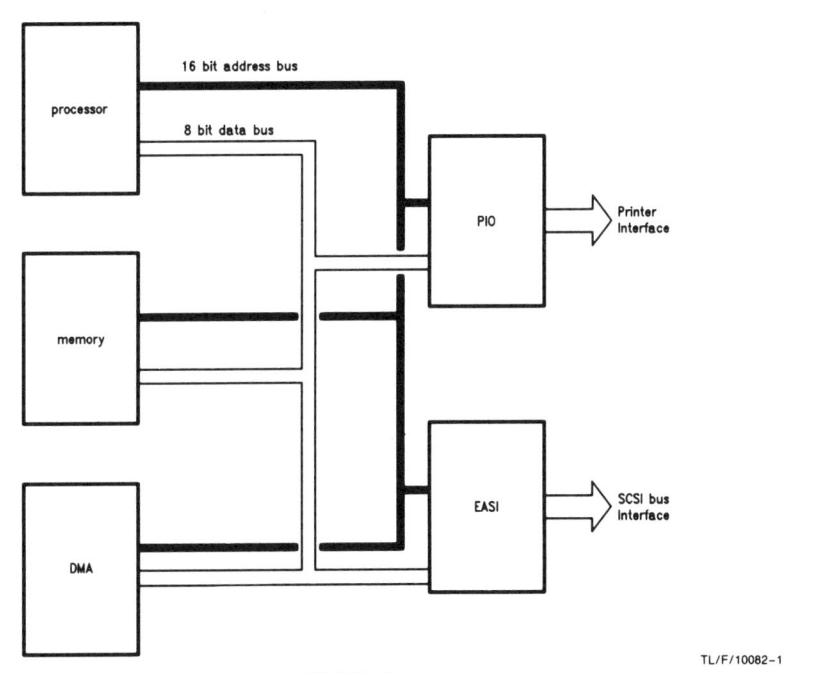

FIGURE 1.1

TRI-STATE® is a registered trademark of National Semiconductor Corporation.
NSC800™ is a trademark of National Semiconductor Corporation.
IBM® is a registered trademark of International Business Machines Corp.
ASC-88™, SCSI Manager II™, SCSI-BIOS™ and SCSI-PRO™ are trademarks of Advanced Storage Concepts.
Microsoft® and MS-DOS® are registered trademarks of Microsoft Corporation.
PAL® is a registered trademark of and used under license from Monolithic Memories, Inc.
Z80® is a registered trademark of Zilog Corp.

1.0 Introduction (Continued)

NSC800 microcomputer family. It is capable of addressing 64 kbytes of memory and 256 I/O devices using a multiplexed address and data bus. The instruction set is fully compatible with that of the Z80.

The NSC831 is the parallel input/output (PIO) device of the NSC800 family. This provides 20 I/O bits which can be individually programmed to be inputs or outputs. These are configured as two eight bit ports and one four bit port. The PIO is used as the interface between the microprocessor and the printer.

The DP5380 and DP8490 comply with the ANS X3.131-1986 SCSI standard as defined by the ANSI X3T9.2 committee. They can act as both Initiator and Target supporting all bus phases. Due to the on-chip high-current open-drain drivers the devices interface directly to the SCSI bus.

The 64 kbyte memory space is split between a 32 kbyte EPROM, containing the run-time software, and a 32 kbyte static RAM.

Data transfers between memory and the EASI can be controlled by the DMA controller, which supplies all read and write strobes for a transfer in either direction. This is a synchronous device which uses the 4 MHz clock output from the microprocessor.

The microprocessor is normally in control of the internal busses, giving it the ability to read or write memory or I/O devices. During a DMA transfer the DMA takes control of these busses and can pass data between the EASI and memory. As the microprocessor is the only device with 'intelligence' it must control these transfers. It commences and controls operations by setting registers in the DMA and EASI.

2.0 Hardware

2.1 DEVICES

2.1.1 Introduction

This section is intended to explain the hardware of the SPC referring to the circuit diagram given in Appendix A. This will describe the devices used and the signals generated but not the way in which they are programmed. What will be discussed is how the devices relate to each other, their relative timings, and any extra hardware required. A more exact description of the internal registers of the EASI, DMA and PIO, and their operation, will be given in the diagnostics section.

2.1.2 Microprocessor

The NSC800 is an eight bit microprocessor which multiplexes the eight bit data bus with the lower half of the address bus to create a sixteen bit address bus. An octal latch, MM74HCT373, is required to hold the address on the bus during a memory read or write, with the ALE (Address Latch Enable) output from the NSC800 strobing the latch.

Since the bus has devices which use both TTL and CMOS levels pull-up resistors are required. The $\overline{\text{IOM}}$ output signifies whether the processor is on a memory or an I/O cycle and is used along with the $\overline{\text{RD}}$ and $\overline{\text{WR}}$ outputs to strobe bus data.

This processor allows control of the bus to be passed to an I/O device, using the $\overline{\text{BRQ}}$ (bus request) input and $\overline{\text{BACK}}$ (bus acknowledge) output. On this board the DMA is the only device which may request control of the bus.

When the EASI issues a DMA request the DMA signals that it needs control of the bus by issuing a $\overline{\text{BRQ}}$. The processor acknowledges this by issuing $\overline{\text{BACK}}$, and then drives the bus and all related control signals to their high impedance state. The MM74HCT373 must TRI-STATE® its outputs; the output enable is driven by the DMA signal AEN (address enable), which it asserts when it has control of the bus. At the end of its operations the DMA releases AEN (enabling the latch outputs) and negates $\overline{\text{BRQ}}$. The processor de-asserts $\overline{\text{BACK}}$ and retakes control of the bus by enabling its outputs. To prevent spurious signals being generated when the bus is not driven the control signals must have pull-up resistors.

The NSC800 provides five hardware interrupts of which only $\overline{\text{RSTA}}$ is used. The others are tied inactive. $\overline{\text{RSTA}}$ is driven by the EASI interrupt output, with an inversion between them to allow for the difference in active levels. An interrupt on this pin causes a software reset to a particular address in memory (more details of interrupt servicing will be given in the diagnostic software section).

The system clock is generated using an 8 MHz crystal, with the frequency divided by two for use by the processor and the DMA. There is also power-on reset circuitry, which resets the processor, and causes a reset output to the other devices when power is first applied.

2.1.3 DMA Controller

The 9517 and 8237 are compatible direct memory access devices, controlled by setting internal registers. These registers are selected using a chip select, with the particular register selected by the lower nibble of the address bus, and data strobed by $\overline{\text{IOR}}$ (I/O read) or $\overline{\text{IOW}}$ (I/O write). These registers control the type of data transfer, the number of bytes transferred and the memory address the transfer is to/from.

The two modes of transfer used are single mode and block mode. In single mode the DRQ (DMA request) input and $\overline{\text{DACK}}$ (DMA acknowledge) output are used to "handshake" every byte of data transferred. In block mode, after an initial DRQ, $\overline{\text{DACK}}$ must be active until after the last byte has been transferred. The rate of transfer is controlled by the READY input.

Another device required in conjunction with the DMA is an MM74HCT74 'D' type flip-flop. This is used to overcome the metastability problem introduced by the asynchronous READY signal driving a synchronous device. The flip-flop synchronizes the input with the system clock.

The DMA supplies all the necessary control signals to move data from memory to EASI or vice-versa. When all data has been transferred the DMA drives $\overline{\text{EOP}}$ (end of process) active. $\overline{\text{EOP}}$ can also be used as an input to the DMA, prematurely terminating any transfer. It is configured as an open-drain driver, so requires a pull-up resistor, of an advised value of 4.7 kΩ.

As in the processor the DMA has a multiplexed address and data bus, also requiring an MM74HCT373 to latch the address. In this case it is the upper byte of the address bus which is multiplexed with the data bus. The strobe signal is similar to ALE but is generated in the DMA and called ADSTB (address strobe). The TRI-STATE for this latch is driven by the inverse of AEN, since the device must drive the bus only when the DMA has control.

2.0 Hardware (Continued)

The 9517 and 8237 have four DMA channels, of which only one is used. The DRQ inputs for the other three channels are tied inactive. The RESET input causes the control registers to be cleared and the DRQ inputs to be masked.

2.1.4 EASI

The EASI is controlled by setting internal registers, written by an $\overline{IOW}$ and chip select, read by an $\overline{IOR}$ and chip select. The particular register selected is determined by the lowest three bytes of the address bus. The EASI has an eight bit microprocessor data bus, and on the PLCC part a microprocessor bus parity pin. However the NSC800 does not support parity checking so this bit is tied low. The other microprocessor controlled pin is the $\overline{RESET}$. This causes the EASI to clear all registers and therefore reset all logic.

The DRQ output and $\overline{DACK}$ input "handshake" single mode DMA, while the READY output is used to control the speed of a block mode transfer. $\overline{EOP}$ is an input which terminates DMA, and can be used to cause an interrupt. This interrupt output INT can interrupt the microprocessor if the EASI detects an error, or has completed some task.

SCSI uses an eight bit data bus with a parity bit; which must support odd parity during all bus transactions except arbitration. Other SCSI signals are the bus selection control signals $\overline{SEL}$ and $\overline{BSY}$, phase control signals $\overline{MSG}$, $\overline{C/D}$ and $\overline{I/O}$, data transfer handshakes $\overline{REQ}$ and $\overline{ACK}$ and the message flag $\overline{ATN}$. Further details on the use of these signals will be given in the software sections of this document. The SCSI reset $\overline{RST}$ is similar to a chip reset, but generates an interrupt.

The EASI interfaces directly to the SCSI bus using high-current open-drain drivers. This bus is a 50-way ribbon cable (maximum length 6.0 meters) on which all SCSI devices are daisy-chained. The devices at the end of this cable must terminate the SCSI signals i.e., a 220Ω resistor to power, and a 330Ω to ground. This power can be V_{CC} or TERMPWR (terminator power).

TERMPWR is V_{CC} fed through a Schottky barrier diode. This can be fed to pin 26 of the connector allowing the SCSI device at one end of the bus to be powered down without affecting the bus. Since the terminators are receiving power the bus can operate effectively. To make this optional TERMPWR is fed to the connector through a jumper link. The Schottky diode must be included to prevent backflow of power into the printer controller board.

Other manufacturers CMOS devices can not be used in a configuration like this, as they pull the bus low when not powered. The DP5380 and DP8490 have a special input protection to overcome this problem.

All other pins on the connector should be tied to ground, except pin 25 which must be left floating.

2.1.5 PIO

The printer controller uses a NSC831 PIO to interface between the processor and the printer. The NSC831 has 20 individually programmable I/O bits which are arranged as three ports; A, B and C. As it is part of the NSC800 family the NSC831 has a compatible multiplexed address and data bus with an ALE input, to strobe the address into an internal latch. This eight bit address can then be used to access an internal register, in conjunction with an $\overline{IOR}$ or $\overline{IOW}$ and chip select. The device has two chip selects, one of which is tied active (low), with the other coming from the address decode PAL®. The RESET input causes the three ports to become all inputs.

In this application the three ports are set up so that each port is all input or all output. Port A is an output, which drives the printer data bus through an MM74HCT241 octal buffer. Port B is an input, with the lower nibble coming from the printer outputs, driven by half an MM74HCT240 octal inverting buffer. The pull-down resistors on the $\overline{ERROR}$ and PE (paper error) inputs give a proper error signal if the printer is unconnected i.e., the same as when the printer is off-line. The upper three bytes of Port B are connected to switches, which are used to set the board's SCSI I.D. This will be further explained in the SCSI diagnostics section. Only three bytes of Port C are used, all as outputs, driven by the same device used by Port B. Bits 0 and 2 are printer outputs, while bit 3 drives a LED. This displays the 'health' of the board i.e., when the board is operational the LED is on, when non-operational it is off and when there is a known hardware error a message is 'flashed'.

The printer connector is a standard IBM® 25 way 'D' range, with all unused pins grounded, except 13.

2.2 PAL EQUATIONS

2.2.1 Introduction

PAL's are used in this board to generate chip selects, overcome potential timing problems and to invert signals for devices with different active levels. The PAL's used are National Semiconductor's PAL16L8. The equations are written in a form compatible with PLAN (Programmable Logic Analysis by National) Ver. 2.00.

2.2.2 Decode PAL

The decode PAL supplies the five main devices with their chip selects.

Memory is split into two 32k byte blocks, each of which represents one device. These devices are an EPROM, at the base of memory, and RAM in the upper half. Thus the memory map and chip select equations are as follows:

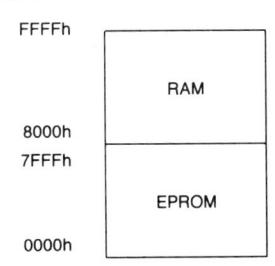

```
FFFFh
          +----------+
          |          |
          |   RAM    |
          |          |
8000h     +----------+
7FFFh     +----------+
          |          |
          |  EPROM   |
          |          |
0000h     +----------+
```

/EPROMZ = /A15*BACKZ*/IOMZ

/RAMZ = A15*BACKZ*/IOMZ + A15 * AEN

EPROMZ (the Z at the end of the name shows it is active low) can only be selected by the processor, while RAMZ is available to the DMA.

2.0 Hardware (Continued)

The DMA, PIO and EASI fit into the I/O map shown, which results in the equations following. These equations show that the devices can only be selected by the microprocessor on an I/O cycle.

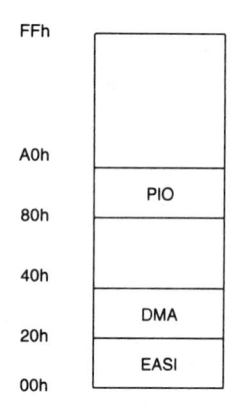

```
FFh  ┌─────────┐
     │         │
     │         │
 A0h │         │
     │   PIO   │
 80h │         │
     │         │
 40h │         │
     │   DMA   │
 20h │         │
     │  EASI   │
 00h └─────────┘
```

/EASIZ = /A7 * /A6 * /A5 * IOMZ * BACKZ
/DMAZ = /A7 * /A6 * A5 * IOMZ * BACKZ
/PIOZ = A7 * /A6 * /A5 * IOMZ * BACKZ

The other three outputs are used as inverters:

/BRQZ = HRQ
/AENOZ = AEN
/HLDA = BACKZ

The two spare inputs are tied low.

2.2.3 Control PAL

The NSC800 uses $\overline{IOM}$ to distinguish between memory and I/O cycles. This allows $\overline{IOR}$ and $\overline{IOW}$, for processor cycles to be generated.

/IORZ = /RDZ * IOMZ
IORZ.TRST = BACKZ
/IOWZ = /WRZ * IOMZ
IOWZ.TRST = BACKZ

The second line of these equations shows that the output is TRI-STATE when the processor relinquishes control of the bus.

Before examining the next five equations it is important to fully understand the relative signal sequencing involved in a DMA transfer. When the transfer is initiated the EASI issues a DRQ, causing the DMA to respond with $\overline{DACK}$, and take control of the bus by asserting $\overline{BRQ}$. The processor asserts $\overline{BACK}$ and allows the bus and relevant control signals to go TRI-STATE. The DMA then asserts AEN, to show it is in control of the bus, causing the microprocessor address latch to TRI-STATE its outputs and the DMA latch to enable its outputs.

For single mode each byte transferred must have a DRQ and a $\overline{DACK}$. For block mode DRQ returns inactive after the DMA responds with a $\overline{DACK}$, but the $\overline{DACK}$ must remain active until the transfer is complete. The READY line controls the rate of block mode DMA transfer i.e., when READY is low the byte transfer is 'frozen', until READY returns high (Figures 2.1a and 2.1b).

The READY signal is asynchronous but the DMA is synchronous, testing the READY signal on the negative edge of each clock. For this reason the READY input must be synchronized, using a flip-flop which updates its output on the positive edge of the clock.

When READY is detected as being inactive on the negative edge of the clock an extra cycle is inserted to the read and write.

While data is being transferred a register keeps track of how many bytes are left. When this reaches zero an $\overline{EOP}$ signal is generated, concurrent with the last read and write, causing the EASI to set an EOP flag.

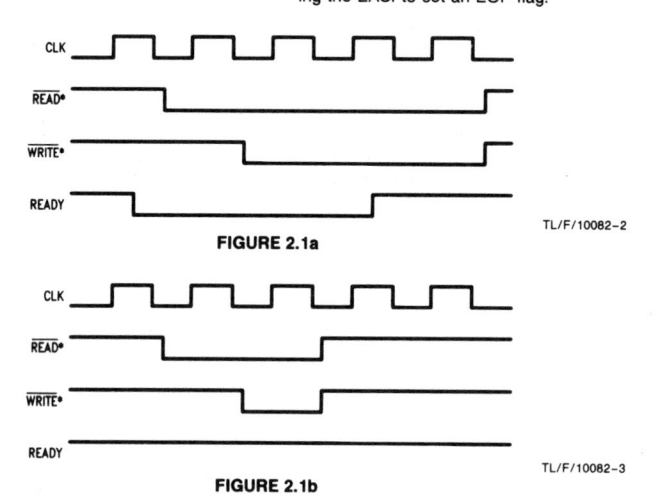

FIGURE 2.1a

FIGURE 2.1b

*Note:
Of these read and write signals, one will refer to memory the other I/O, depending on the direction of data transfer.

2.0 Hardware (Continued)

The tranfer is now complete so $\overline{BRQ}$, AEN and $\overline{DACK}$ are all driven inactive. AEN going inactive causes the address latches to swop back over; the microprocessor latch outputs are driven, the DMA latch outputs are TRI-STATE. When the microprocessor detects that $\overline{BRQ}$ is inactive it deasserts $\overline{BACK}$ and retakes control of the bus. The microprocessor then clocks flags in the EASI to determine the success, or otherwise, of the transfer.

The other form of DMA available is pseudo-DMA; that is the EASI 'thinks' it is a proper DMA transfer but the processor is still in control of the bus. To do this the microprocessor must set the DMA to mask off any DRQ, initialize the EASI for single mode DMA and monitor the EASI flags for DRQ going active. At this point the processor must generate an $\overline{IOR}$ or an $\overline{IOW}$ and a $\overline{DACK}$, to properly simulate DMA. This can be done by causing an I/O read or write at a certain address, to generate a $\overline{DACK}$, and if wanted an $\overline{EOP}$. The particular address does not matter since during DMA the EASI ignores the address bus.

The following equations generate pseudo-DMA signals, while allowing the true DMA signals to operate normally:

/DACKOZ = /IORZ * BACKZ * /A7 * A6
 + /IOWZ * BACKZ * /A7 * A6
 + /DACKIZ

/EOP2 = /IORZ * BACKZ * /A7 * A6 * /A5
 + /IOWZ * BACKZ * /A7 * A6 * /A5
 + /EOP1 * EREADY

DACKIZ is the $\overline{DACK}$ from the DMA, which the output normally follows. The rest of the equation generates the pseudo-DMA $\overline{DACK}$.

$\overline{EOP2}$ is the input to the EASI, $\overline{EOP1}$ is the output from the DMA. The first two lines of this equation generate an $\overline{EOP}$ for the pseudo-DMA cycle, on the lower half of the address space that also generates the pseudo DACK. This means the final I/O map now is as follows:

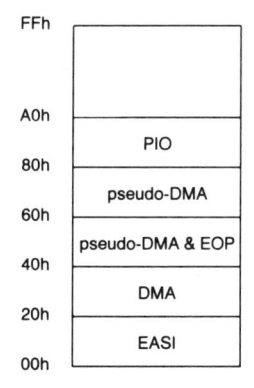

```
FFh
      +-----------------+
      |                 |
A0h   |                 |
      |      PIO        |
80h   |                 |
      | pseudo-DMA      |
60h   |                 |
      | pseudo-DMA & EOP|
40h   |                 |
      |      DMA        |
20h   |                 |
      |     EASI        |
00h   +-----------------+
```

The third part of the $\overline{EOP1}$ equation is to overcome a problem that occurs during a block mode DMA transfer. On the last byte the READY signal from the EASI (this will be called EREADY to distinguish it from the READY into the DMA, DREADY) may be driven low to freeze the transfer. However $\overline{DACK}$, $\overline{EOP}$ and $\overline{IOR}$ or $\overline{IOW}$ will all be active causing a valid EOP condition. This is not a problem until we consider the next equation.

/DREADY = /EREADY * /INT * /DACKIZ

This equation causes the DMA READY input to go high if any of EREADY, SCSI interrupt INT or DACKIZ go high. This overcomes a potential bus lockup, caused by an error occurring during a block mode DMA transfer. On a phase or parity error the EASI will stop the transfer and generate an interrupt. If DREADY is left inactive the DMA will keep control of the microprocessor bus. This equation holds DREADY high on interrupt, allowing the DMA to pass control of the bus back to the microprocessor.

Although this equation prevents an error it also introduces a problem in the $\overline{EOP2}$ equation. As explained previously, during the last byte of a block mode transfer although READY is low a valid EOP condition may exist, which will cause an interrupt. This interrupt will then cause DREADY to be driven high, allowing the DMA to finish the transfer, but lose the last byte since the EASI is not ready. To overcome this the $\overline{EOP2}$ equation gates $\overline{EOP1}$ with EREADY, so the EASI does not see a valid EOP condition until it is able to transfer the last byte. **This problem does not occur in MODE E since it generates a true end of DMA interrupt.**

Another problem is introduced by the DREADY equation. If an error occurs during a DMA transfer the EASI will generate an interrupt and DREADY will be forced high. This allows the DMA to 'run free', writing garbage into MEMORY, and wasting SCSI bus time. To prevent this an external $\overline{EOP}$ must be applied to the DMA on error. The next equation does this:

/EOP1 = /DACKIZ * /EREADY * INT
EOP1.TRST = /DACKIZ * /EREADY * INT

$\overline{EOP1}$ is an I/O pin which normally acts as input from the DMA, so the output is TRI-STATE unless DACKIZ and EREADY are low, with INT active. When this error condition occurs an $\overline{EOP}$ is output to the DMA, terminating the data transfer. A prerequisite of this equation working properly is the existence of the DREADY equation, since an externally applied $\overline{EOP}$ will have no effect on the DMA if READY is held low.

The next equation is also required to prevent a fault occurring during a block mode DMA transfer:

/EASIWRZ = /IOWZ * BACKZ
 + /IOWZ * EREADY * /BACKZ

When the processor is in control of the bus a straightforward $\overline{IOW}$ can write to the EASI, but if the DMA is in control EREADY must be high to allow the write. The inclusion of EREADY is required because during DMA the EASI is a flowthrough latch; an $\overline{IOW}$ passes the data from the processor bus onto the SCSI bus. Therefore if the DMA reaches its next byte before the data on the SCSI bus has been transferred, the DMA will overwrite the data. To prevent this $\overline{EASIWR}$ can only be allowed when EREADY is high.

The final PAL output is used to invert the SCSI interrupt signal, to make this active high output compatible with the processor's active low input.

/SCSIINTZ = INT

3.0 Diagnostic Software

3.1 INTRODUCTION

The diagnostic software resides at the base of RAM, with the purpose of checking and initializing the printer controller board after power up and every hard reset. It must be at location zero, since after any hardware reset the program counter is cleared. The interrupt service routine is included here, since it must exist at an exact location in memory.

3.0 Diagnostic Software (Continued)

The software uses the NSC800 instruction set (fully Z80® compatible) which makes the required low level board operations faster, and allows the exact positioning of code in memory. The assembler and linking loader are from Microtec Research.

3.2 DEVICE CHECKING AND INITIALIZATION

This section will refer to the diagnostic software PRINTER.SRC and the files PRNSYM.SRC and EASISYM.SRC which contain the constants used. A full listing of all programs described in this document are available on floppy disk.

3.2.1 Memory Check and Interrupt Servicing

An 'org' statement can be used to position this program at the base of ROM, address 0. After a jump to the ROM check, the next byte stores a label defining which version of the software is installed. On an EPROM the version number can be read from address 0002h. The upper and lower nibbles should be considered as two numbers i.e., '10' defines version 1.0.

The interrupts are disabled, since the service routines are not initialized, and the EPROM is checked; simply consisting of reading the same address twice and ensuring the same value returns both times. If this test fails the system halts since a drastic error must have occurred. If the EPROM passes the test the program jumps to the RAM check.

On interrupt the NSC800 stops before its next instruction, pushes the program counter onto the stack and loads it with a new address, depending on the highest priority interrupt channel active. Interrupt A, the only channel used, causes a jump to location 003Ch. The interrupt service routine at this location pushes all of the processor registers onto the stack, and tests for a SCSI reset. SCSI reset must be checked, as this is a special condition, causing the board to be reset by a general restart routine (this routine is in EASIO.SRC and explained in section 3.5).

If the interrupt is not a SCSI reset the software makes a call to the bottom of RAM, where a jump command should be followed by a public variable RESETA. This variable can be loaded with the starting address of the routine to service the interrupt. On return from this routine the processor registers are popped back off the stack, and program control returns from the interrupt. Since RESETA is public any external software may use it, and therefore control which routine services an interrupt. RAM must be verified before this jump table is set up.

The RAM test simply checks each BIT can contain a zero and a one, then clears every byte. Once RAM has been verified the stack pointer may be initialized, allowing call statements to be included. It should be remembered that although the interrupt routine is between the ROM and RAM tests, when the program runs it will jump straight from the ROM test to the RAM test. Since the interrupt jump table is in volatile RAM the code for a jump instruction must be written into RAM after every reset.

3.2.2 PIO Initialization

The file PRNSYM.SRC contains the port addresses for all of the I/O devices. The upper nibble contains the location the device takes within the I/O map, the lower nibble selects the register within the device to be accessed.

The PIO has five types of registers which control data transfers through the device:

1. **Data Register**—Each port has an eight bit data register containing the data passed between the PIO and the processor. These are either read or write registers, depending on whether the port bits are inputs or outputs.
2. **Data Direction Registers**—Each port also has a data direction register which controls whether each of the 20 bits is an input or an output (an input is defined by a zero, an output by a one).
3. **Mode Register**—There is one three bit mode register which selects which of the four modes the device is in. This board will always require the PIO to be in Mode 0 which is the basic I/O mode. The alternatives use Port C for handshaking.
4. **Bit Clear Register**—Each port has an eight bit, bit clear register which clears any output bit whose corresponding bit in this register is high.
5. **Bit Set Register**—These are similar to the above, but allow the output bits to be set.

The PIO is set for Mode 0, with Ports A and C outputs and Port B an input. Port A drives the printer data bus, while Ports B and C read and write the printer control signals. The four printer outputs, to the Port B input, are $\overline{ACK}$, BUSY, PE and $\overline{ERROR}$. $\overline{ERROR}$ goes low when there is no paper, the printer is off-line or an error occurs; PE goes high when the printer is out of paper; BUSY goes high to show the printer can accept data; $\overline{ACK}$ pulses low to acknowledge data after a byte has been transferred. Since BUSY and $\overline{ACK}$ are both handshake signals the user must decide which to use. This software will use BUSY.

The top three bytes of Port B are used to read in the SCSIID from a block of switches. The SCSIID is the boards identifier on the SCSI bus, used in selection and arbitration, consisting of a single active bit. This is read in as a three bit number, converted to the correct bit pattern and stored in a public variable called SCSIID.

Port C is an output, driving the printer signals $\overline{INIT}$ and $\overline{STROBE}$. A low pulse on $\overline{INIT}$ of 50 μs causes the printer to be initialized; a 500 ns low pulse on $\overline{STROBE}$ causes the printer to read the data on the bus. The highest bit of Port C drives a LED, which is on during board operation, and can display an error message by occulting a fixed code. Errors are displayed by the routine *ERROR*, a public function which displays the error number as a four bit code. It does this by occulting the LED for 1 second to show a 1, ½ a second for a 0.

Since none of the PIO registers are true read/write the device cannot be tested, only initialized.

3.2.3 DMA Test and Initialization

The DMA consists of address, word-count and control registers for four channels, of which only one is used. There follows a description of the registers used, with the addresses shown in PRNSYM.SRC.

1. **Word-Count Registers**—There are two word-count registers; a sixteen bit write only base word-count register and a sixteen bit read only current word-count register. As with all sixteen bit registers in this device the two bytes of data are accessed by two successive selections of the same address. An internal flip-flop determines which byte is read or written, with the lower byte selected first after a reset. To transfer the correct number of bytes the word-count register must be written with the number of bytes to be transferred minus one.

3.0 Diagnostic Software (Continued)

2. **Address Registers**—The address registers are also sixteen bit, and called base and current. The base address register is written with the address the transfer must start at, while the current address register contains the next address to be written or read. *The DMA is tested by writing a value to the base address and reading it back from the current address.*

3. **Control Registers**—There are three types of control register which will be discussed only in the way they are used by this software. The master clear register (DMAMCL) is the software equivalent of a hardware reset; all registers are cleared and DRQ is masked off. Masking of DRQ is controlled by the parallel mask register (DMAMSK), in this case always used to allow DRQ on Channel 0 and mask off the others. The final type of register used is the mode register (DMAMOD) which controls whether a transfer is block or single mode, memory read or memory write. Each of the four channels has a six bit mode register, all at the same address, with the bottom two bits of the data bus determining which is selected.

The DMA test program initializes the device for a one byte, block mode, memory write, to an address TSTBYT. The DMA is initialized in this particular manner for use in the EASI loopback DMA test routine.

An error in DMA test will cause error signal 0.

3.3 EASI TEST AND INITIALIZATION

3.3.1 Introduction

This document will show how the registers within the DP5380 and DP8490 are used in an application. For a full register description refer to the DP5380 and DP8490 datasheets. Register names and their locations are given in PRNSYM.SRC.

3.3.2 DP8490 and DP5380 Test

The DP5380 and DP8490 are completely pin and hardware compatible, and also software compatible until the enhanced mode bit is set in the DP8490. The DP8490 powers up in normal mode (MODE N which is software compatible with the 5380) and is in MODE N after any chip reset. The initial test is therefore the same for both devices, writing to the Mode Register 2 (EASIMR2) and reading back the data. The EASIMR2 is selected since it has the most read/write bits that do not directly affect the bus.

An error in EASI test will cause error signal 1.

For the DP5380 testing is now complete, but the DP8490 enhanced mode offers a loopback test facility, where the SCSI drivers are disabled and the SCSI I/O's looped back inside the EASI. **Using this feature the user can fully check the device, and by doing a DMA transfer fully check the board.**

At this stage it is therefore necessary to determine which device is inserted. In the DP5380 bit 6 of the Initiator Command Register (EASIICR) selects the 'test mode', which disables all output drivers on the device, making it invisible to the system. Although the device can still be written to no data can be read back, making applications very limited. In the DP8490 this bit selects the enhanced mode (MODE E) of the device.

In MODE E addresses 0-6 access the same registers as in MODE N, but address 7 is different. Instead of accessing the Reset Parity/Interrupt (EASIRPI) and Start DMA Initiator receive (EASISDI) registers, address 7 directly accesses the Enhanced Mode Register (EASIEMR) and indirectly accesses the Interrupt Mask Register (EASIIMR) and Interrupt Status Register (EASIISR). The only other difference to registers occurs in the Target Command Register (EASITCR) where one of the previously unused bits becomes a flag (explained in section 3.3.4).

To test which device is inserted EASI test sets the enhanced/test mode bit, writes data to address 7 and reads back from the same address. If the device is a DP5380 it will be in 'test mode' and the data read will be 0FFh due to the pull-up resistors on the data bus. If it is a DP8490 the data read back will be the data written, providing the only bits set are read/write. For a DP5380 the program jumps to initialization for selection; if it is a DP8490 loopback testing follows.

3.3.3 DP5380 Initialization for Selection

The DP5380 initialization involves programming the device to respond to a selection. First, the public variable DP8490 (which other programs should treat as a constant) is set FALSE to indicate that a DP8490 is not inserted. For the DP5380 to be selected it must have the SCSIID of the device in its Select Enable Register (EASISER) and parity must be enabled. The processor must enable its interrupts and set up the jump table; *intA* is an external routine which sets the processor's interrupt mask to only allow interrupts on A, and enables the interrupts (see EASIO.SRC); *main_* is an external program which responds to a SCSI selection (further explained in the Run-time software section).

If BSY is inactive for a bus settle delay (400 ns), SEL is true and the bit on the data bus corresponding to the SCSIID is active, a SCSI selection interrupt is generated. This causes a processor interrupt, taking the program from the continuous 'jr Now' instruction to the interrupt service routine, and through the jump table to *main_*. When *main_* has finished servicing the interrupt it will return to the interrupt servicing routine and then return from the interrupt. Thus the board's 'idle' state is to continually execute the 'jr Now' instruction.

3.3.4 DP8490 Loopback Test

The loopback test mode of the DP8490 allows all signals to be fully tested, including a DMA transfer, without affecting the SCSI bus. To allow this DMA transfer loopback allows the user to drive both initiator and target signals simultaneously.

After setting the DP8490 flag true all initiator signals, and then target signals, are asserted and checked. This includes a check on the data bus by writing a test value to it and reading it back. The data bus test value has an odd number of bits, and since the specified SCSI parity is ODD, the parity bit must be inactive. One of the new features offered in MODE E is programmable SCSI parity, which is tested by ensuring the parity bit becomes active when EVEN parity is enabled, and inactive when a test value with an even number of active bits is written. Parity is then returned ODD, and the parity bit should become active.

An error in Loopback testing causes error signal 2.

The next test in loopback mode is a DMA target receive transfer to a location in memory, TSTBYT. This not only tests the EASI, but also the interrupt servicing, the DMA and memory. Although the device is in loopback the software must carry out the transfer as it would normally i.e., the bus phase must be correct, $\overline{BSY}$ must be active and the SCSI bus must be asserted. The EASIMR2 must be properly set

3.0 Diagnostic Software (Continued)

up for block mode DMA, with interrupt on $\overline{EOP}$ or parity error. Since this test is interrupt driven the interrupt jump table must be loaded with the address of the routine which will service the DMA loopback test, and the interrupts must be enabled. The DMA has already been initialized for this.

Before enabling interrupts the SCSI interrupts should be reset, which in MODE N would involve reading address 7. Since the EASIEMR is now at this address the resetting of interrupts, and the start of DMA initiator receive are initiated by writing to function bits of the EASIEMR.

When the Start DMA Target receive (EASISDT) register is written the EASI will issue a $\overline{REQ}$ to show it is ready to receive data on the SCSI bus. At this point the device at the other end of the bus would normally assert $\overline{ACK}$ to show it has data available. In this case there is no other device so the user must wait for $\overline{REQ}$ to go active and then assert $\overline{ACK}$. Thus both initiator and target signals must be asserted simultaneously. The user waits for $\overline{REQ}$ to go inactive, and deasserts $\overline{ACK}$ to show the bus transfer is complete.

The program then goes into a continuous loop awaiting a SCSI interrupt. This interrupt will occur because the EASI will have issued a DRQ, when $\overline{ACK}$ went active, and the DMA wil have transferred the last test byte written to the EASI to TSTBYT, finishing with an $\overline{EOP}$. On interrupt the program will jump to address 003Ch, where all the processor registers are pushed onto the stack, a call is made to the base of RAM, and from there it will jump to the subroutine *DIAGA*.

One of the problems with DMA in a DP5380 is that end of DMA is flagged when $\overline{DACK}$, $\overline{EOP}$ and $\overline{IOR}$ or $\overline{IOW}$ are simultaneously active; although the data may not yet have been transferred on the SCSI bus. To overcome this the software must examine the SCSI handshake signals $\overline{REQ}$ and $\overline{ACK}$, both of which must be inactive on three successive samples for a true end of DMA. This is more fully explained in the DP5380 and DP8490 datasheets. **MODE E of the DP8490 detects true end of DMA, after $\overline{ACK}$ goes inactive, before generating an interrupt.**

DIAGA responds to the interrupt after the loopback DMA by checking all the correct flags have been set. The DP5380 only uses four bits of EASITCR so MODE E uses the free bit 7 as a flag, to show true end of DMA. This is the first flag checked. Address 7 not only directly addresses the EASIEMR it also indirectly addresses the EASIIMR and EASIISR. After writing the correct code to the function bits of the EASIEMR the next access of address 7 will be to the EASIISR, if it is a read, or the EASIIMR, if it is a write. **The advantage of the EASIISR over the DP5380 registers is that all interrupt information is available in one register, and every interrupt is flagged.** To check DMA the user need only read the EASIISR and ensure that the only flag active is end of DMA. The user should note that SCSI reset causes the device to revert to MODE N, from which the EASIISR can not be read, so is not flagged.

The final EASI flag test is the 'conventional' end of DMA flag in the Bus and Status Register (EASIBSR). This flag, the interrupt flag and the flag to show no phase mismatch has occurred must be the only bits active. The final loopback DMA test is to ensure that memory location TSTBYT contains the correct data.

An error in Loopback DMA test causes error signal 3.

3.3.5 DP8490 Initialization for Selection

The DP8490 initialization is very similar to that of the 5380, even calling the same routine, *main__*, to respond to selection. However, this device stays in MODE E and uses these enhancements to only allow interrupts on selection and parity by setting the EASIIMR. Selection and parity are the only valid interrupts at this point.

At the end of the DIAGA routine program control returns to the 'jr Here' instruction, which it will continually enact until a parity or selection interrupt causes a jump to *main__*.

3.4 ERROR HANDLING

Throughout all software for this board the error handling is the same for errors considered non-recoverable, which includes all errors in diagnostics. On error the board continually displays an error number, as a four bit binary code, using the LED. The subroutine *ERROR* carries out this function.

On error register 'I' should be loaded with the error number and routine *ERROR* called. This routine then occults the LED for ½ second to display a zero, 1 second to display a one, with the LED on for ½ second between flashes. The four bits, most significant bit first, are repeatedly displayed between 2 second intervals, during which time the LED is on. The timing delays are generated using a routine *DELAY*, which gives a number of ¼ second delays, the number of delays being determined by the value in the 'I' register.

The following list shows the possible errors in diagnostics. The run-time software section contains a similar listing of its error codes.

Error 0
  The DMA can not be accessed.
Error 1
  The EASI can not be accessed.

All other diagnostic errors concern a DP8490

Error 2
  An error has occurred in the assertion of SCSI signals in loopback test mode.
Error 3
  An error has occurred during the loopback DMA transfer.

3.5 EASIO.SRC

This file contains public assembly language routines, which can be called by either the diagnostics or run-time software, to implement low level commands. As these routines may need to be called by routines written in 'C' any variables passed to the routines are passed in the 'hl' register pair, then in 'de', then 'bc' and then on the stack. *ERROR* and *DELAY* both use the 'I' register to pass a variable. Data passed back to the calling routine is returned in the accumulator.

Functions *'read'* and *'write'* implement general purpose I/O register accesses, while *'dmaread'* and *'dmawrite'* handle the special case of the 16-bit DMA registers. These require two accesses of the same address. *'intA'* initializes the processor interrupt mask, with a 'pseudo' I/O write, which selects a register internal to the processor, setting the mask to only allow interrupts on RSTA. This function also enables the interrupts, which can be done by *'eni'*, with *'dsi'* disabling interrupts.

'IDtest' is used by the run-time software, during selection, to read the number of bits active on the bus. This routine checks the number of high bits in the byte passed in the 'I' register, returning the value in the accumulator.

3.0 Diagnostic Software (Continued)

Function *'restrt'* is a general purpose reset routine, which can be called on an error condition. This causes every device on the board to be reset, and the diagnostics to be rerun, thus clearing all memory and reinitializing the stack.

4.0 Run-Time Software

4.1 INTRODUCTION

Following the diagnostic software must be a program which will control all SCSI bus transfers, beginning with selection. This program is written in 'C', using a PARAGON 'C' cross compiler for an NSC800. The relevant files for this section are SCSI.C, COMMAND.LIB, PROCESS.LIB, DMA.LIB, PRINT.LIB, ARBITRAT.LIB, SYM.H, CONST.H and COMMANDS.H, all of which are on the supplied floppy disk.

SCSI.C contains the main program, called *'main'* by 'C' and *'main_'* by assembler. The '.LIB' files contain the functions called by *main* and the '.H' files contain the constant values used by all of these files.

4.2 MANDATORY PHASES

This section outlines all the bus phases and transfers which a target must support. It would be perfectly legal for a target to respond to a selection, fetch a command block from the initiator, and then returns status and message bytes before releasing the bus. Although no process would be actuated this would be a legal sucession of events.

4.2.1 Selection Response

main( ) is the program jumped to when an interrupt is generated as the program circles the continuous loop at the end of diagnostics. This routine should only be entered after a selection interrupt, if any other interrupt is active it is considered an error. The function *select( )* checks an interrupt to ensure it is valid.

The type of device installed is checked by reading the 'DP8490' flag, and this determines how the interrupt is verified. In a DP5380 a selection interrupt is determined by the absence of any other interrupt, with $\overline{SEL}$ active. The user must check the EASIBSR to ensure that no error flags are active, only the INT and PHSM (phase match) bits.

Any other flag will cause error number 4 to be displayed.

After reading the EASIBSR and finding no interrupt flags the user knows the interrupt should be a selection. The only other unflagged interrupt is a reset, which would have been handled by the low level interrupt service routine. Therefore the EASICSB register must also be read, and if $\overline{SEL}$ is inactive an error condition exists.

This causes error number 5 to be displayed.

In MODE E the user need only read the EASIISR to determine which interrupt is active, including selection. All interrupts, other than parity and selection, should be masked off, so any error concerns only these two flags.

A SCSI parity error is displayed as error number 8, if the selection flag is not active error number 9 is displayed.

The common tests for MODES E and N both concern the EASICSD; error 6 shows that the correct SCSIID bit was not active on the bus; error 7 shows there was more than two bits active on the bus. During selection an initiator is only allowed to assert two bits on the bus, its own ID and the target ID.

The target must assert $\overline{BSY}$ to show it has recognized the selection, then when the initiator deasserts $\overline{SEL}$ the selection phase is complete.

4.2.2 Command Phase

A selection phase is followed by a command phase, where the target reads a command block from the initiator. This command block specifies the actions the initiator requires the target to execute, plus the length of any data transfers requested. This board only allows six byte command blocks, which are transferred into the target by function *fetch_cmd( )*.

The command block is transferred using programmed I/O; that is each byte is individually handshaken under processor control. The bus phase must be Command Out (out and in always refer to the initiator, so Command Out is a command block sent by the initiator), bus phase being set in the EASITCR. To transfer a byte (see *Figure 4.1*) the user must assert $\overline{REQ}$, then wait for $\overline{ACK}$ to go active to show data is available. On $\overline{ACK}$ the target can read the data on the bus, then deassert $\overline{REQ}$ to show it has received the data. When the initiator deasserts $\overline{ACK}$ the byte transfer is complete.

4.2.3 Status Phase

The target must send a status byte to the initiator during the status phase, at the termination of each command. A list of status codes is given in COMMANDS.H. *status( )* is the function which enters a Status In phase, and transfers the code.

The EASITCR must be written with the Status In phase and the EASIODR with the status code. This code should be asserted onto the bus, remembering to keep $\overline{BSY}$ asserted, and if neccessary the MODE E bit. $\overline{REQ}$ is then asserted to show data is available. The initiator should assert $\overline{ACK}$ when it has read the data, allowing the target to deassert $\overline{REQ}$ and take the data off the bus. The transfer is complete when the initiator deasserts $\overline{ACK}$ (See *Figure 4.2*).

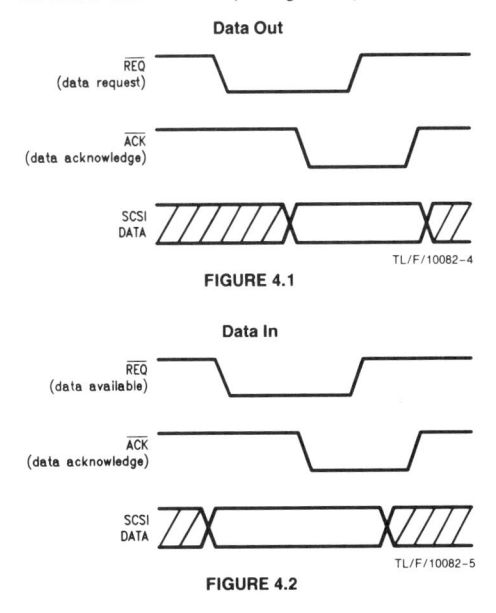

FIGURE 4.1

FIGURE 4.2

4.0 Run-Time Software (Continued)

4.2.4 Message Phases

A selection must be terminated by the target entering the Message In phase, and sending a relevant message code (as listed in COMMANDS.H). The target follows the Status phase with the Message In phase, usually to send the COMMAND COMPLETE message. This indicates valid status exists, so the target can release $\overline{BSY}$ to free the bus. *messin( )* enters the Message In phase and sends the message using programmed I/O, as in *status( )*.

The only message which must be supported is COMMAND COMPLETE. If a device on the bus supports other messages it indicates this by responding to or asserting $\overline{ATN}$. An initiator asserts $\overline{ATN}$ if it has a message for the target. It is common for an initiator to assert $\overline{ATN}$ during selection and, if the target responds by entering a Message Out phase, it sends the IDENTIFY message. This establishes whether the target can respond to a greater set of messages, and whether the initiator supports disconnection. It shows this by setting bit 6 of the message. In a disk controller IDENTIFY would also establish the path by including a Logical Unit Number (LUN), but the printer controller uses all LUN's at this SCSIID.

getmes( ) enters a Message Out phase and fetches a message using programmed I/O. Although this software only supports single byte messages it must be prepared to accept messages from the initiator of up to the maximum length, 256 bytes. If an initiator wishes to send further bytes after the first it must keep $\overline{ATN}$ asserted, only deasserting after the final byte has been transferred. *getmes( )* will handshake up to 256 bytes, after which if $\overline{ATN}$ is asserted it will be considered an error. Only the first byte is used in determining the message sent.

The use of messages during disconnection, arbitration, reselection and in error handling will be discussed in those sections of this document.

4.2.5 Command Termination

After sending status and message codes to the initiator, the target should then release $\overline{BSY}$ to free the SCSI bus. However, before allowing a bus free phase the target must initialize itself for the next selection, as in function *reset( )*.

The interrupt jump table is loaded with *main( )* and the EASISER with the SCSIID. For a MODE E device the interrupt mask is set. For both types of device the interrupts must be reset, and $\overline{BSY}$ deasserted. Interrupts are disabled at the start of this function, so must be enabled after calling *reset( )*.

4.3 COMMAND PROCESSING

After the command block has been fetched the command has to be determined and executed. The printer command set is shown in COMMANDS.H.

process_cmd( ) reads the first byte of the command block which contains the operation code. This determines what actions are taken. If a command has been sent which this software does not recognize, a status of CHECK CONDITION is returned, with sense set to ILLEGAL REQUEST.

4.3.1 Test Unit Ready

TEST UNIT READY will be sent by an initiator before a print to ensure the printer is on, on-line, has paper and is not in an error condition. On this command the function *ck_printer( )* is called to read the printer signals through Port B of the PIO and check for errors.

If the error line is not high the printer is operational, so the status is GOOD, and the sense is set to NO SENSE (no error). If there is an error the paper error line must be checked. If this is low (active low input) the printer is out of paper and status of CHECK CONDITION is returned, with sense set to MEDIUM ERROR. If it is not a paper error the printer is assumed to be off or off-line, so status is CHECK CONDITION with sense UNIT ATTENTION.

4.3.2 Request Sense

An initiator will send a REQUEST SENSE command after the target has returned a status of CHECK CONDITION. The sense data is sent to the initiator in an effort to understand an error condition, and if possible recover from it. Byte 4 of this command block contains the length of an extended sense message, which must not be greater than 4, or sense is set to ILLEGAL REQUEST and CHECK CONDITION status returned. This software only supports four bytes of sense data.

Sense data is sent to the initiator using single mode DMA. The DMA is initialized by *send_sense( )*; it is reset by a master clear, the mode set, the DMA mask written and the address and word-count loaded. The function *single_dma_in( )* (explained in section 4.4.1) enters the correct phase (Data In) and transfers the data.

4.3.3 Reserve Unit

In a multi-initiator system a printer controller must be reserved before a print commences, or the data from two different initiators may be mixed. This command stores the initiator's ID in a variable called 'reserved', and on subsequent selections only this initiator may execute commands.

Any initiator wishing to reserve the unit must put its own ID on the bus during selection, along with the target's ID, so this software knows which initiator is reserving the unit. If the initiator's ID is not on the bus it can not reserve the unit, and since this is a prerequisite of printing, it cannot use the printer.

If an initiator attempts to reserve this unit without making its ID available sense is set to ILLEGAL REQUEST and status of CHECK CONDITION returned. If another device attempts to reserve the board when it is already reserved status returned is RESERVATION CONFLICT.

4.3.4 Release Unit

This is the reciprocal command to the previous, freeing the printer for other initiators after a print has been completed. If an initiator other than the reserver attempts this command a status of RESERVATION CONFLICT is returned; if this is attempted when there is no reservation current the returned status is CHECK CONDITION with sense set to ILLEGAL REQUEST.

Run-Time Software (Continued)

4.3.5 Print

Since transferring data to a printer is a very slow process, typical Epson Fx range 80 cps–160 cps, the print command transfers the data to a buffer, leaving it to be printed later. Bytes 2, 3 and 4 of the command block contain the data length, byte 2 the most significant. The use of three bytes to define the size means that in theory block transfers of up to 16 MB are allowed. This software could support blocks of up to 64 kB, bytes 3 and 4 are read, but blocks are limited to a size BUFFLIM, which will be determined in section 4.4.3.

Possible errors in a print occur when the unit is not reserved, the data length is set to zero, or the block size is too large. The block size is too large if byte 2 of the command block is active or if the data length is greater than BUFFLIM. In response to an error the transfer is cancelled and status of CHECK CONDITION is returned with sense set to ILLEGAL REQUEST.

The data is not transferred unless sense is set to NO SENSE. If an error has occurred in the printer sense will have been set to indicate the error source. The error condition must be rectified and TEST UNIT READY sent to reset the sense data.

print_cmd( ) is the function which transfers data into a buffer using block mode DMA. The print buffer is a circular queue, allowing the user to take data off the front and put data on the rear (see *Figure 4.3*).

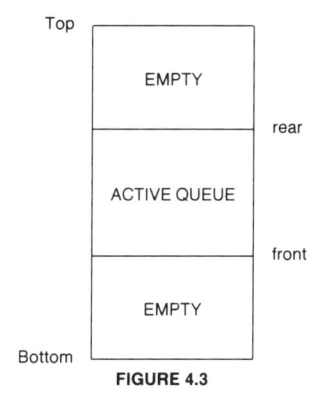

FIGURE 4.3

front points to the next byte to be printed, rear points to the next available byte for entering data. When front equals rear the queue is empty, when rear is one less than front the queue is full.

The buffer limits are called top and bottom. When either the front or rear pointers reach the top the next increment takes them to the bottom. Thus the queue may look like *Figure 4.4*.

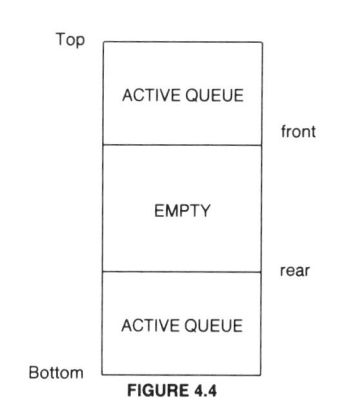

FIGURE 4.4

When *print_cmd( )* is first called after a reset it must set up the queue, defining top and bottom, and setting front equal to rear equal to bottom. Before any transfer the data length must be checked to ensure that the DMA will not take rear past top. If it would the data must be transferred in two blocks; one to the top of the queue, and one starting at the bottom of the queue. This routine should not be called unless there is sufficient free space for the size of transfer. Function *dma_data( )* sets up the EASI and DMA for a block mode transfer, enters the Data Out phase and calls *dma( )* to handle the transfer. This will be more fully explained in section 4.4.

4.3.6 Flush Buffer

The purpose of this command is to allow an initiator to terminate an unwanted print. An initialization pulse is sent to the printer, in case it has its own buffer, and the SPC buffer is cleared. This is done by setting front equal to rear equal to bottom.

4.4 DATA TRANSFERS

This section is concerted with the way in which software controls DMA transfers, both block and single mode, and how the print data block length is determined.

4.4.1 Single Mode DMA

The four bytes of sense data are sent to the initiator using single mode DMA. *send_sense( )* sets up the DMA to transfer four bytes of data from the sense buffer to the EASI, leaving function *single_dma_in( )* to complete the transfer.

The phase, which in this case is Data In, is determined by the calling function, and the EASIMR2 initialized for a single mode transfer. Parity checking is enabled with interrupt on $\overline{EOP}$ or parity error. The data bus is asserted and the routine *dma( )* is called to handle the transfer.

Since both single and block mode DMA transfers, either in or out of the initiator, have large portions of code that are common, a function can be written for general purpose DMA handling. This function, *dma( )*, sets up the interrupt response and makes a write to the register which initiates the required type of transfer, which could be target receive or target send etc. *dma( )* checks the success of the transfer when the transfer is complete.

Run-Time Software (Continued)

In MODE E the interrupt mask can be set to only allow parity, $\overline{EOP}$ and DMA phase mismatch interrupts. The interrupt jump table is loaded with the starting address of a service routine, which sets a flag to show an interrupt has occurred. The program can then sit in a loop waiting for this flag to go active.

After the interrupt the cause has to be checked to ensure the transfer was successful, but this checking depends on the device installed. For a DP8490 the user can read the EASIISR and if any flag other than end of DMA is active an error has occurred. On such an error, status is set to CHECK CONDITION with a sense of ABORTED COMMAND. The interrupt mask can be reset, masking off end of DMA and DMA phase mismatch, and the DMA mode in the EASIMR2 disabled. The interrupt service routine is set to jump to the general interrupt handler *gen_int( )*. The interrupt must then be reset and the processor's interrupts enabled.

In MODE E true end of DMA is detected, but in MODE N the end of DMA interrupt is generated when $\overline{EOP}$, $\overline{DACK}$ and $\overline{IOR}$ or $\overline{IOW}$ are active concurrently. True end of DMA, after the transfer is complete, must be detected by software. $\overline{REQ}$ and $\overline{ACK}$ must both be inactive on three successive samples for true end of DMA. **This additional code makes the MODE N DMA routine slower**. After detecting this the user can check the end of DMA flag is active in the EASIBSR, giving status CHECK CONDITION and setting sense ABORTED COMMAND bit if it is not. The DMA bit is then reset and *gen_int( )* called to ensure there was no parity or phase errors, and to reset the interrupt service routine.

On return to the function *single_dma_in( )* it will deassert the SCSI bus and return to the calling routine.

4.4.2 Block Mode DMA

blk_dma_in( ) is the equivalent routine to *single_dma_in( )* initiating a block mode transfer to the initiator. *send_sense( )* could call this routine, after setting the DMA for block mode, and transfer the data using block mode, with no difference to the other software. The only difference between these two routines is the setting of the EASIMR2 BLK mode bit.

print_cmd( ) is used to transfer the data to be printed from the initiator. This function checks that the transfer will not exceed the constraints of the queue and calls *dma_data( )* to initiate a block mode transfer. *dma_data( )* sets the EASI and DMA registers for a block mode target receive, with the data size set by *print_cmd( )* and the destination starting address equal to rear. *dma( )* is again used to handle the interrupts.

4.4.3 Determination of Data Block Length

The block of data to be transferred for a print has been specified as a maximum length of BUFFLIM, but the value of this constant has to be evaluated. The two main considerations in this are the latencies of the bus and the SPC.

Since this board is a printer controller it will be of a very low priority. Thus the device should not hold the bus for too great a time per transfer, as this would slow down initiator's accesses of a high priority peripheral. The second consideration is the time that the target takes from selection to entering a data phase, the data block must take longer than this to transfer, or the command is inefficient. Measurements taken on various models of personal computers showed that, with an Advanced Storage Concepts ASC-88™ SCSI Host Adaptor, the time taken from the initiator asserting $\overline{SEL}$ to the target commencing the data transfer is approximately 3 ms.

The block mode DMA rate was measured as between 200 kB/s and 500 kB/s limited by the DMA in the PC. A block length of 2 kB was selected, since this would take between 4 ms and 10 ms to transfer. Thus the data transfer time is greater than the selection to data phase time, but the overall time on the bus in not too long.

4.5 PRINTING DATA

Due to the inherent slowness of printers data can not be printed while the controller is in command of the SCSI bus. To prevent tying up the bus data is stored in a buffer and printed after $\overline{BSY}$ has been released. The function *printit( )* handles the transfer of data to the printer.

printit( ) checks the printer is not in an error condition, and transfers the byte of data at address front to the printer. The data is transferred through Port A of the PIO with BUSY and $\overline{STROBE}$ used to handshake the data. BUSY must be high and $\overline{STROBE}$ pulsed low a minimum of 0.5 μs for the printer to accept data. front is incremented, to point to the next byte to be transferred. To print out the queue *printit( )* is called until front equals rear.

If the function does detect an error it sets sense to the appropriate value, and implements any outstanding reconnection. The printer is continually polled to determine when it comes back on-line, at which time the print continues.

While the board is printing it must still be available for selection. An initiator can reset every device on the SCSI bus if a target does not respond to a selection within a Selection Timeout, normally 250 ms. The interrupt service routine has been set to jump to *main( )*, but the program is currently in *main( )*, so this function must be re-entrant.

main( ) is written in such a way that it will only process commands if it is unreserved and not printing, or reserved by the current initiator but not printing. If the board is reserved, but not by the current initiator, a status of RESERVATION CONFLICT is returned. If the board is printing, and the reserving initiator attempts to send it another command a status of BUSY can be returned.

However, the controller has a 30 kB print buffer, of which only 2 kB would be used at a time. It would be much more efficient to continue executing commands until the buffer is full. One method of achieving this can be seen in *outbuf( )*. In this function a flag, called next, is set if the free space in the queue is greater than BUFFLIM. If a selection occurs while a print is on, this flag is checked and the command processed if it is active. If it is inactive a status of BUSY is returned. This method has the disadvantage that the initiator must continually poll the SPC to determine when it is ready to accept data. This 'loads' the SCSI bus and slows down the print, since the SPC will be responding to selection. It also restricts other device's use of the bus. The alternative method of utilizing the buffer space is to use disconnection and reconnection.

Run-Time Software (Continued)

4.6 DISCONNECTION AND RECONNECTION

4.6.1 Disconnection

If an initiator sends the IDENTIFY message to the target, it can indicate that it supports disconnection by setting bit 6 of this message. If this bit is not set, or the message not sent, the initiator is assumed not to support disconnection, and no disconnection is attempted. This software uses disconnection in two places; 1) the board is reserved, currently printing and selected by the reserving initiator; 2) the board has been released, but not finished printing.

If the board is busy printing it can be advantageous to disconnect from the reserved initiator after reading in the command block, and then reconnect having ensured that the command can be implemented i.e., there is enough free space in the queue. For any other initiator the response is the same as before, the status of RESERVATION CONFLICT is returned until the unit is released. After this the controller will disconnect from any initiator attempting to select it, until it has enough free space in its buffer.

disconnect( ) is the routine which carries out the disconnection procedure. This can only take place after the command phase and before the data phase, with an initiator that sent the IDENTIFY message with the disconnection bit set. The target follows the Command Out phase with a Message In phase and sends the DISCONNECT message, after which it can drop $\overline{BSY}$, to release the bus. If the DISCONNECT message is not sent the initiator will treat the deasserting of $\overline{BSY}$ as an illegal termination of command. Before releasing $\overline{BSY}$ *exchange( )* is called to store the ID of the initiator and the command it wants to execute. *reset( )* is again used to initialize the interrupts and release $\overline{BSY}$.

disconnect( ) stores in a variable, recon__data, the amount of print buffer which will be required to execute the command. For a print, this depends on the length of data to be transferred, for any other command the amount is zero. recon__data is used to determine when the target can reconnect to the initiator i.e., once there is enough space free in the buffer.

After disconnection the program will return to printing, at which point it may be selected by another initiator. As it can only disconnect from one initiator at a time it must return a status of BUSY.

In *outbuf( )* it can be seen that reconnection takes place when the free space in the print buffer is greater than the number of bytes to be transferred. *reconnect( )* calls the functions which action the correct reconnection procedure, beginning with arbitration.

4.6.2 Arbitration

Arbitration requires two functions, one for MODE E another for MODE N, due to the fundamental difference in their methods of arbitration i.e., **MODE E is interrupt driven, MODE N is polled**. These two functions, *Narbitrate( )* and *Earbitrate( )*, carry out the same basic operation, arbitrate for the bus until successful, but do so in distinctly different ways.

To arbitrate for the bus a device must wait for a Bus Free Phase, when $\overline{BSY}$ is continuously inactive for 400 ns with $\overline{SEL}$ inactive. After a Bus Free Delay of 800 ns the SCSI device should assert $\overline{BSY}$, and assert its SCSIID bit on the bus. After a further 2.2 μs Arbitration Delay the data bus should be examined, and the device with the highest priority SCSI ID bit asserted wins arbitration.

In MODE N the EASIODR should be written with the SCSIID and the arbitration bit in the EASIMR2 set. The interrupts must be initialized to cause a jump to routine *servn( )*. The EASI will wait for a Bus Free Phase then, after a Bus Free Delay, it will assert $\overline{BSY}$, and put the contents of the EASIODR onto the bus. The user must poll the AIP bit of the EASIICR to determine when arbitration has begun and check the LA bit to ensure that arbitration has not been lost. The LA bit is set if another initiator asserts $\overline{SEL}$ during arbitration. If AIP is active and LA inactive the user must examine the EASICSD to determine whether it is the highest priority device arbitrating. If arbitration is lost the arbitration bit in the EASIMR2 must be reset and the whole procedure begun again. The device shows it has won arbitration by asserting $\overline{SEL}$. An interrupt during MODE N arbitration is treated as a selection, so *servn( )* enables parity, since there is no parity checking during arbitration, and calls *main( )*.

In MODE E arbitration is interrupt driven, allowing the board to continue printing while waiting to arbitrate. For a busy SCSI bus typically many milliseconds, and potentially many seconds, can be taken up arbitrating. In MODE E this time can be utilized, thus increasing the system throughput. For this application the time gained is used to continue printing, in other applications, such as a disk controller, it could be used in data cacheing, overlapped seeks, etc.

In MODE E arbitration the EASIODR must be written with the SCSIID and parity checking cancelled in the EASIMR2. The interrupts are set to allow selection and arbitration interrupts, with the jump table loaded with *serva( )*. This sets a flag to show an interrupt has occurred. The enhanced arbitration is initiated, with a write to the arbitration bit of the EASIEMR. This causes the EASI to wait for a Bus Free Phase; delay a Bus Free Delay; assert $\overline{BSY}$ and the EASIODR; delay an Arbitration Delay then interrupt the processor. The interrupt causes the flag to be set that shows the state of arbitration should be examined. While waiting for this interrupt the printer carries on printing out data, until front equals rear.

After the interrupt is detected the EASIISR can be read to determine the cause. If it is not an arbitration interrupt parity checking is enabled and the interrupt treated as a selection, by calling *main( )*. If the interrupt is signalling the commencement of arbitration, the procedure is the same as in MODE N, with the LA bit of the EASIICR being examined, and priority determined by reading the EASICSD. If arbitration is lost, interrupts must be reset, the arbitration bit in the EASIEMR reset, and the whole procedure begun again. If arbitration is successful the user asserts $\overline{SEL}$.

Function *printarb( )* is used to print data during arbitration in the same way as *printit( )* does normally. The difference here is that the arbitration interrupt should be serviced as quickly as possible.

This routine must be left if an interrupt occurs. Instead of waiting for the printer to come on-line if an error occurs this routine simply does not send the character.

4.0 Run-Time Software (Continued)

4.6.3 Reselection

When arbitration has been won the disconnected initiators ID and command block must be restored, and the initiator reselected. This reselection is carried out by function *reselect( )*.

The EASISER is cleared to stop it responding to the reselection and the initiator ID bit written into the EASIODR along with the SCSIID. I/O must be asserted to show this is a reselection. The EASIODR is asserted onto the bus and the relevant arbitration bit, depending on MODE reset. This deasserts $\overline{BSY}$. Parity checking is enabled, and the board waits a selection timeout delay of 250 ms for the initiator to respond, by asserting $\overline{BSY}$. If it does not, $\overline{RST}$ is asserted, resetting the whole SCSI bus.

If the initiator does respond, the target must assert $\overline{BSY}$, then deassert $\overline{SEL}$ and take the EASIODR off the bus. For MODE E the interrupt mask can be set, and the arbitration interrupt reset. Reselection is now complete.

4.6.4 Reconnection

After *reselect( )* the IDENTIFY message is sent by the target with the disconnect bit set. The command that was previously sent is processed, if this message is received successfully. After processing the command status and the COMMAND COMPLETE message are returned. *reset( )* is again used to set the interrupts and release $\overline{BSY}$.

If a print was not current at the time of reconnection, and the reconnected command was PRINT, this is handled within *reconnect( )*. As long as the status is GOOD the print is carried out. However, *reconnect( )* will be mostly called from *outbuf( )*, when the queue has enough free space to process the outstanding command. Any reconnection missed by *outbuf( )* is captured in *main( )*.

4.7 ERROR HANDLING

Throughout this document the handling of errors has been discussed as the errors arose, but there are some remaining to be discussed. These are phase or parity errors during command transfers and message errors.

4.7.1 General Error Handling

While this board is connected to the SCSI bus, with $\overline{BSY}$ active, *gen_int( )* is generally used to handle interrupts. This routine responds to an unexpected interrupt by checking the parity and phase flags, in the EASIBSR for MODE N, the EASIISR for MODE E. It then resets the interrupts, before setting appropriate flags. Sense is set to ABORTED COMMAND with a status of CHECK CONDITION if an error has occurred.

After calling *select( )*, to respond to selection, *main( )* then calls function *set_up( )*, which sets the mask and jump table to respond to an interrupt. This is also the routine which checks to see if $\overline{ATN}$ is asserted. If it is *getmes( )* is called to enter the Message Out phase and fetch the IDENTIFY message. *messout( )* processes the message. This is where the target determines if the initiator supports disconnection.

If an error occurs during *select( )* or *set-up( )*, detected by *gen_int( )*, the board sends relevant status and a message of COMMAND COMPLETE, before releasing the bus. No attempt is made to recover from the error. Similarly, if a phase error occurs during the command phase, status and sense are returned. However, if a parity error occurs during a command phase, the target can attempt a recovery by sending the RESTORE POINTERS message. This message instructs the initiator to reset the command pointer to the beginning of the command block, allowing the target to re-enter the Command Out phase and re-transfer the command block. If there is a parity error again, status is returned, along with the COMMAND COMPLETE message. If the second transfer is successful, the command execution continues as normal.

4.7.2 Message Errors

If an initiator wishes to respond to a message sent by the target it indicates this by asserting $\overline{ATN}$, before releasing $\overline{ACK}$ to finish the transfer. The target should then enter the Message In phase, and transfer message bytes until $\overline{ATN}$ goes inactive, up to 256 bytes. If the initiator attempts to send more than this the target will send the MESSAGE REJECT message and terminate the command with status of CHECK CONDITION and sense set to ABORTED COMMAND. If a parity error occurs during the message transfer, the target must wait until the transfer is complete, then instruct the initiator to resend all previous message bytes, by asserting $\overline{REQ}$ before changing phase. If the parity error occurs again the command will be terminated with status of CHECK CONDITION and sense ABORTED COMMAND.

The relevant functions for message phases are *messin( )*, *getmes( )* and *messout( )*. *messin( )* is used to send a message to the initiator, while *getmes( )* fetches a message from the initiator. *messout( )* processes the message sent by the initiator. The supported messages are now explained.

IDENTIFY: establishes the use of a greater message set, indicates the ability to support, or not support, disconnection and gives a LUN number, if necessary.

ABORT: if the reserving initiator sends this message to the target it causes the board to be reset.

BUS DEVICE RESET: This is similar to ABORT, except any initiator can implement it, resetting the board.

MESSAGE PARITY ERROR: On receiving this message the target attempts to resend the last message sent, and if this fails, terminates the command with status of CHECK CONDITION and sense set to HARDWARE ERROR.

MESSAGE REJECT: the targets response to this message is determined by the message being rejected. If the last message sent was COMMAND COMPLETE, the MESSAGE REJECT is ignored, since the initiator must support the mandatory message. A MESSAGE REJECT in reply to a DISCONNECT causes the disconnection to be cancelled. If it was in reply to a MESSAGE REJECT sent, the command is terminated with a status of CHECK CONDITION and sense set to HARDWARE ERROR. If the message sent was IDENTIFY (during reconnection) the target immediately goes to the Bus Free Phase and aborts the command. No Status or Message In phases are attempted, though sense is set to HARDWARE ERROR.

4.0 Run-Time Software (Continued)

It should be noted that *messin( )* has been written to only allow *getmes( )* to be called, from inside *messin( )*, once. Alternatively on parity error the target and initiator could eternally cycle, sending each other the MESSAGE PARITY ERROR message.

4.7.3 Non-Recoverable Errors

The following are errors which occur **during selection**, so severe that system operation is terminated, with an error code displayed on the LED.

The first two concern a DP5380.

Error 4
  Wrong interrupt flags active.

Error 5
  SEL inactive.

These errors concern either a DP5380 or a DP8490.

Error 6
  SCSIID bit not active on bus.

Error 7
  More than two bits active on bus.

The final errors are for a DP8490 only.

Error 8
  SCSI parity error.

Error 9
  Select flag inactive.

5.0 User Guide

The SCSI Printer Controller (SPC) is a Small Computer System Interface target board, which can use either the DP5380 Asynchronous SCSI interface (ASI) or DP8490 Enhanced Asynchronous SCSI Interface (EASI). It can be selected and used by an initiator as described in the ANSX3.131-1986 SCSI standard as defined by the ANSI X3T9.2 committee. This document will explain the installation of SPC, and show how it can be used in a system. By way of example a description is given of how the SPC can be used with an Advanced Storage Concepts ASC-88 IBM PC-SCSI Manager II™ host adaptor.

5.1 INSTALLATION

This section explains how the board must be set up before use, and the type of connectors required to interface to it. Any reference made to an EASI also applies for an ASI.

5.1.1 POWER

The SPC can be installed in a Personal Computer (PC) where it takes power from the backplane. There are two connections to +5V and two to ground. These are the only connections made to the backplane. Alternatively power can be taken from the available connector block *(Figure 5.1)*.

When the board receives power the LED will come on, and stay on if the board passes its self diagnostics. If it fails an error message will be displayed by the LED, indicating the source of error. This will be further explained in section 5.1.5. If the LED does not come on some fatal error has occurred.

5.1.2 SCSI CONNECTOR

The SCSI bus should consist of a 50 way flat ribbon cable a maximum of 6.0 meters long. Both ends of this cable should have all SCSI lines terminated, with a 330Ω resistor to ground and a 220Ω resistor to power. SCSI devices are daisy chained along this cable, with SCSI signals common to all devices.

The SPC contains sockets for two DP8490 or DP5380 devices, one a PLCC socket the other DIL. **Only one of these sockets should contain a device**.

To comply with the regulations on terminating resistors six SIL resistor packs are available between the SCSI connector and the EASI DIL package *(Figure 5.1)*. If this board is not to be used at the end of the SCSI cable all of these six packs must be removed. The resistors have been socketed for this purpose.

An option available in the SCSI standard is to supply terminator power on the cable, so terminators at either end of the bus can use the same power. If the device at one end of the bus is unpowered its terminators can receive power from the cable, and the bus will operate correctly. It should be noted that other manufacturer's CMOS devices will not work in this configuration since if not powered they pull the SCSI lines low. National Semiconductor's DP5380 and DP8490 have special input protection to prevent this.

Pin 26 on the SCSI connector is the terminator power pin, which can be left floating by leaving the jumper in position 2, see *Figure 5.2*. By moving the jumper to position 1 the terminator power is supplied to this pin. *Figure 5.3* shows the two possible configurations. Terminator power is fed through a Schottky barrier diode to prevent a backflow of power into the board.

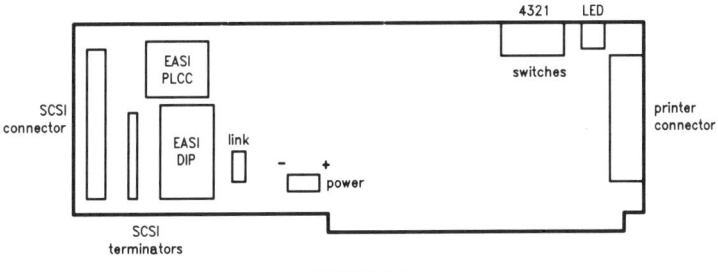

FIGURE 5.1

5.0 User Guide (Continued)

LINK

```
jumper  [X]       [X]
        [X]       [X]    jumper
        [X]       [X]
       Position 1  Position 2
```

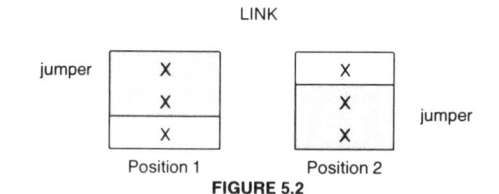

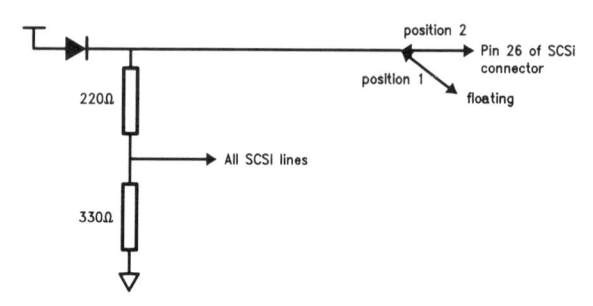

FIGURE 5.2

```
          position 2
   ─▷├─────•──────◀── Pin 26 of SCSI
           │   position 1    connector
          [220Ω]       •── floating
           │
           •──────▶ All SCSI lines
           │
          [330Ω]
           │
           ▽
```

FIGURE 5.3

TL/F/10082-7

5.1.3 SWITCH BLOCK SETTINGS

The switch block is included to allow the user to select the SCSI ID of the SPC. This is used to identify the board during selection phases and determine its priority in arbitration. The SCSI ID is read in as a three bit binary number and converted in software to an eight bit pattern, with one bit active. For an ID of 0 the least significant bit is active, for an ID of 7 the most significant bit is active.

The three bit number is taken from the switch block (see *Figure 5.1*), using switches 1, 2 and 3. 1 is the most significant bit, 4 is unused. These switches should be set to give a unique ID for this bus. An ID of 0 is suggested, making the printer the lowest priority device on the bus. These switches must be set up before power is applied, as they are only checked during the board diagnostics.

5.1.4 PRINTER CONTROLLER

The printer connector is a standard IBM 25 way 'D' type connector. The SPC software controls data transfers to the printer using the BUSY and $\overline{\text{STROBE}}$ signals.

5.1.5 ERROR MESSAGES

Non-recoverable errors cause a four bit binary number to be displayed by the LED. This indicates the source of error.

The LED displays the number by occulting for 1 second to show a 1, ½ second to show a 0. The code is displayed most significant bit first, with the LED on for 1 second between bits. The error number is repeatedly displayed, between breaks of 2 seconds when the LED is on.

The error numbers, and their cause, are shown below. The first four errors are generated during the board diagnostics, the others occur when a selection fails. The SCSI controller will be referred to as an EASI, unless the error is specific to either a DP5380 or DP8490. These possible differences are due to the DP8490's more extensive testing, and because it handles a selection differently.

Error 0
 The DMA could not be accessed. This is possibly a damaged device.

Error 1
 EASI can not be accessed. Device could be damaged, or not properly terminated.

Error 2
 DP8490 has failed loopback test. Device could be damaged.

Error 3
 DP8490 has failed the loopback DMA test. In this test a DMA transfer to memory from the DP8490 is attempted. Failure here could indicate an error in memory, processor, DMA, DP8490 or PALs.

These first four errors concern problems internal to the board. The following errors indicate a bus problem that occurred while the board was waiting for selection.

Error 4
 DP5380 error flag active.

Error 5
 DP5380 found select line inactive. Possible selection timeout or bus error.

Error 6
 EASI ID bit not active on bus. Possible selection timeout or bus error.

Error 7
 EASI detected more than two bits active on bus. During a selection phase the initiator can only assert its own ID and the targets ID on the bus. This indicates a bus error.

Error 8
 DP8490 SCSI parity error. Error on bus.

Error 9
 DP8490 select flag inactive. This is board hardware error, possibly in DP8490 or PAL, possibly interrupt line.

For any of these problems the user should try switching the device on and off again. For bus errors the cable and terminators should be checked along with any other devices on the bus.

5.0 User's Guide (Continued)

5.2 DRIVER SOFTWARE

By way of example it will be shown how software can be written to drive the SPC from an ASC-88 host adaptor, installed in a PC. The ASC-88, like many commercially available host adaptors, handles all low level SCSI signal controls. The user controls the command to be implemented by means of a Job Control Block, JCB, which is passed to it.

The software required to use an ASC-88 is on the supplied floppy disk. This disk contains both the source code, explained later, and the executable code. This is called **printout.exe** and can be implemented in the form:

   printout *filename*

The *filename* supports the MS-DOS use of directories and paths:

   e.g., printout a: /ASC /ASC.C

The second executable command, called **stoprint.exe**, can be used to terminate a print. When this command is executed the SPC flushes the print queue and initializes the printer.

5.2..1 SEQUENCE OF COMMANDS

Although the SPC requires no prescribed sequence of SCSI commands **an initiator must reserve the unit before a print can take place**. Otherwise the sequence of commands specified here indicate how the SPC could be used.

Any print should begin with a TEST UNIT READY command. On receiving this command the SPC tests the printer to ensure it has paper and is not in an error condition. If the status of GOOD is returned the user should then attempt RESERVE UNIT. If the command is successful the SPC will only execute commands from this initiator. This prevents print data from two or more sources being mixed.

The user then sends PRINT commands with a maximum length of data per transfer of 2 kB. If the file to be printed is larger than 2 kB the data must be sent in several blocks. This limit, set to minimize bus latency, is more fully explained in Section 4.4.3.

These PRINT commands transfer the data to the SPC print buffer. To check if an error occurred when the SPC attempted to transfer the data to the printer the user can send a REQUEST SENSE command. If the returned data is NO SENSE the printer is still operational. Any other sense indicates an error.

The final command sent is RELEASE UNIT. This allows the SPC to be used by other initiators.

5.2.2 ASC-88 SOFTWARE

All ASC-88 software is written for a Microsoft® C compiler. File ASC.C contains the main run-time software. ASC STRUC.LIB sets up the structure which is used as a Job Control Block, while ASCCOM.LIB defines the JCBs for particular SCSI commands. The routines used in ASC.C, other than SCSI command calls, are in ASCROT.LIB and UTIL.LIB. CONSTANT.H contains the constants used throughout these files. If the user wishes the SCSI ID of the target board to be anything other than zero they should change the value of *TARGET_ID* in this file, and recompile the code for both commands.

STDIO.H is a Microsoft library containing constants, macro definitions and function declarations for I/O stream operations. This controls opening files, reading files, detecting file ends and closing files. DOS.H, also a Microsoft library, handles the interface to MS-DOS®. This allows the user to set up registers and execute interrupts. The final Microsoft library CONIO.H allows the user to fetch information from the keyboard.

The SCSI-BIOS™ EPROM in the ASC-88 is accessed by generating interrupt number forty. The value in the 'AH' register determines what SCSI-BIOS software is used. SCSI-PRO™ implements the SCSI command specified in the JCB whose starting address is passed in the 'BX' and 'ES' registers.

ASC.C must be compiled and loaded to produce a file printout.exe. When executed this code opens the file specified in the command line and instigates the sequence of commands, outlined in the previous section, to print it out. After these commands the file is closed.

If the printer enters an error state while communicating with the initiator a message will be displayed on screen and the user is given the option of terminating the print. If the print is to be continued the user must correct the printer error and press a key to restart the print. If an error occurs when the SPC has already received all the data from the PC the board will wait until the print error is corrected and continue printing.

STOPRINT.C contains the run-time software required to terminate a print. This uses the library CHECK.LIB, which checks the success of a FLUSH BUFFER command. To terminate an unwanted long print the user can: take the printer off-line; press escape to leave *printout*; send the *stoprint* command.

Appendix A

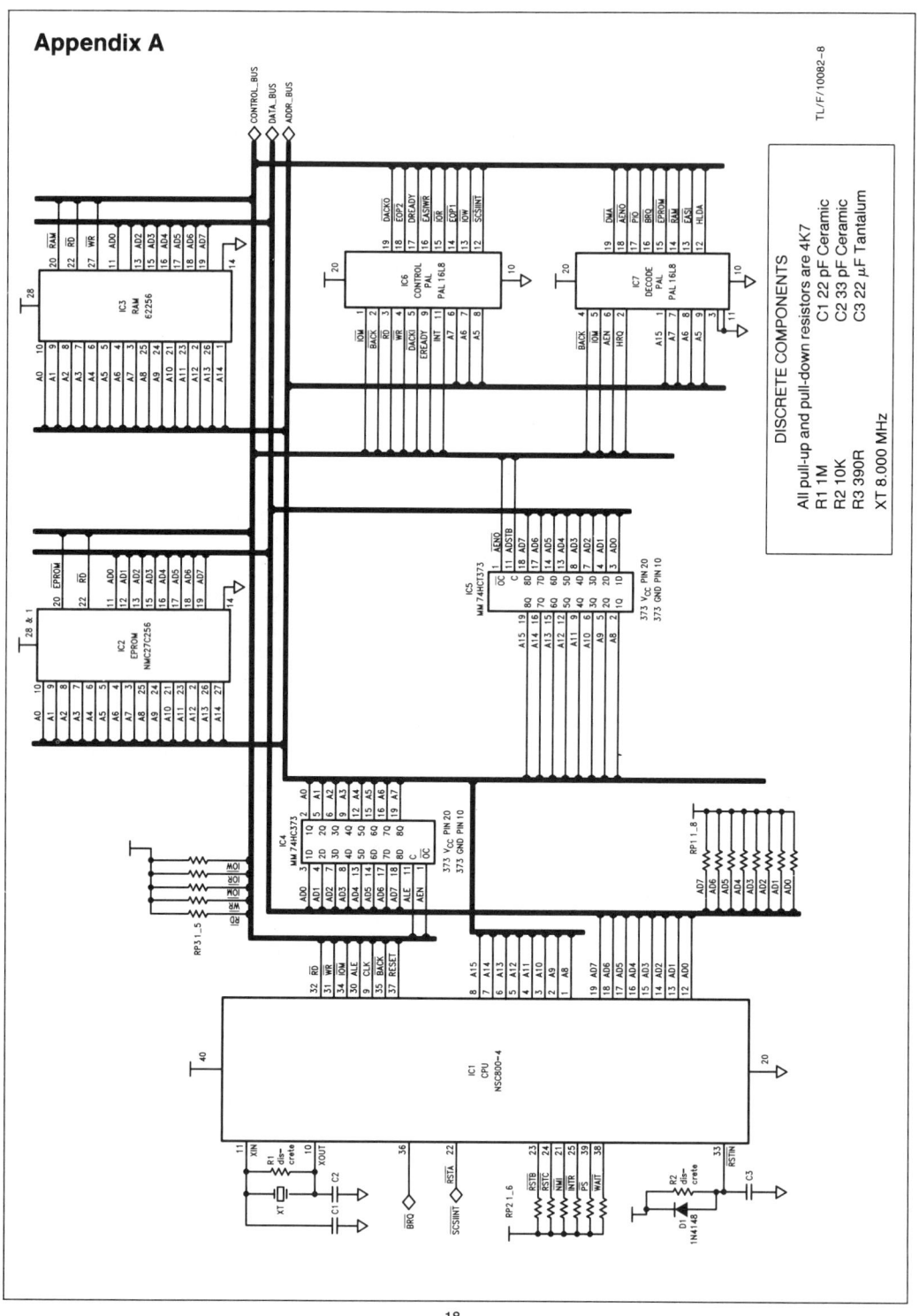

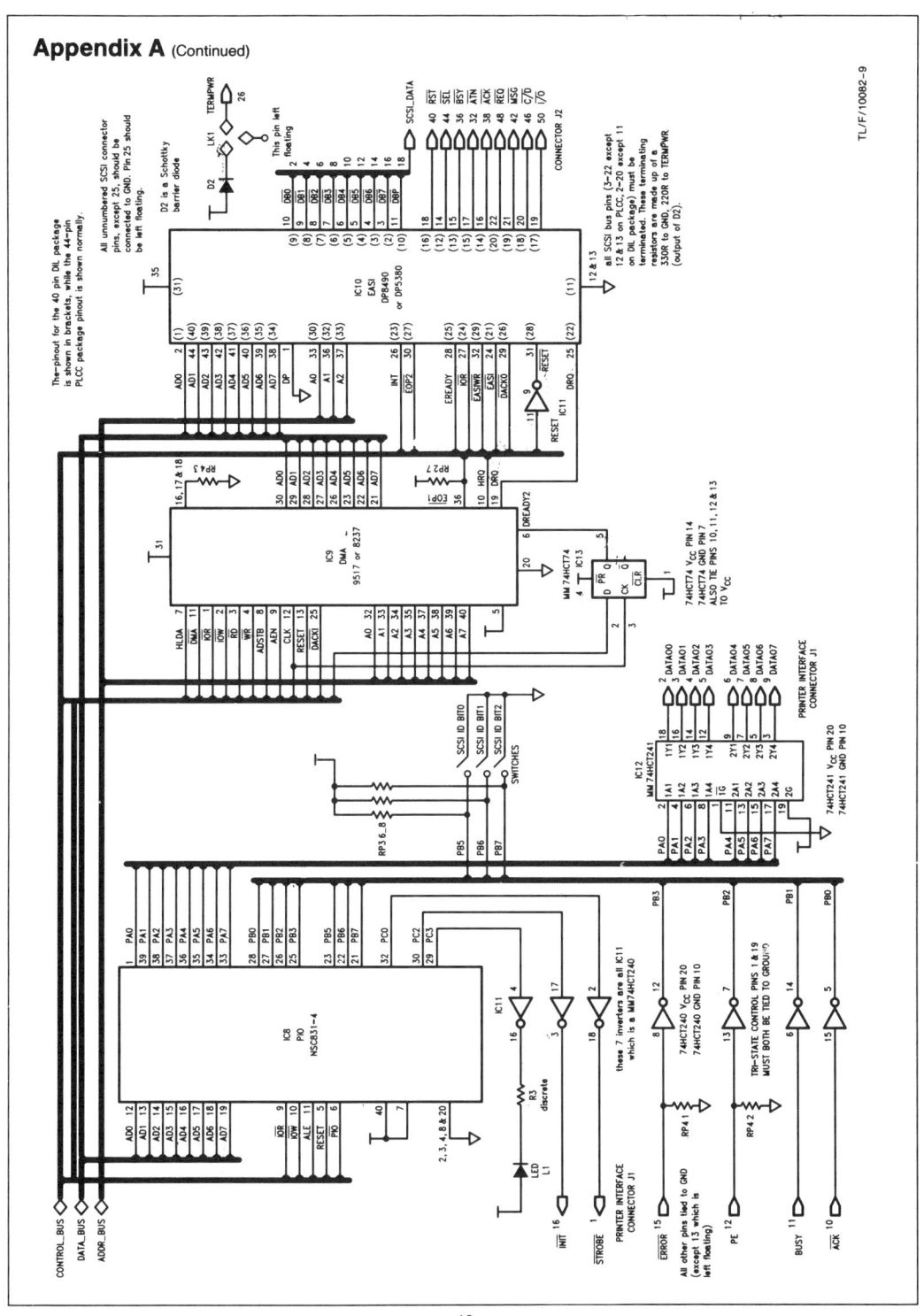

Appendix A (Continued)

A SCSI Printer Controller Using Either the DP8490 EASI or DP5380 ASI and Users Guide

LIFE SUPPORT POLICY

NATIONAL'S PRODUCTS ARE NOT AUTHORIZED FOR USE AS CRITICAL COMPONENTS IN LIFE SUPPORT DEVICES OR SYSTEMS WITHOUT THE EXPRESS WRITTEN APPROVAL OF THE PRESIDENT OF NATIONAL SEMICONDUCTOR CORPORATION. As used herein:

1. Life support devices or systems are devices or systems which, (a) are intended for surgical implant into the body, or (b) support or sustain life, and whose failure to perform, when properly used in accordance with instructions for use provided in the labeling, can be reasonably expected to result in a significant injury to the user.

2. A critical component is any component of a life support device or system whose failure to perform can be reasonably expected to cause the failure of the life support device or system, or to affect its safety or effectiveness.

National Semiconductor GmbH	National Semiconductor S.A.	National Semiconductor S.p.A.	National Semiconductor AB	National Semiconductor	National Semiconductor (UK) Ltd.
Industriestraße 10	Centre d'Affaires	Strada 7 - Palazzo R/3	Box 2016	Postbus 90	The Maple, Kembrey Park
D-8080 Fürstenfeldbruck	«La Boursidiere»	I-20089 Rozzano - Milanofiori	S-12702 Skärholmen	1380 AB Weesp	Swindon, Wiltshire SN2 6UT
Tel. (0 81 41) 103-0	Route Nationale 186	Tel. (02) 8 24 20 46/7/8/9	Tel. (0 29 40) 3 04 48	Tel. (08) 97 01 90	Tel. (07 93) 61 41 41
Telex 527 649	F-92357 Le Plessis Robinson	Telex 352 647	Telex 10 731	Telex 10 956	Telex 444 674
Fax (0 81 41) 10 35 54	Tel. (1) 40 94 88 88	Fax (02) 8 25 47 58	Fax (08) 97 68 12	Fax (0 29 40) 3 04 30	Fax (07 93) 69 75 22
	Telex 631 065				
	Fax (1) 40 94 88 11				

National does not assume any responsibility for use of any circuitry described, no circuit patent licenses are implied and National reserves the right at any time without notice to change said circuitry and specifications.

Lit. # 100563

AN-563

MM58274C Microprocessor Compatible Real Time Clock

General Description

The MM58274C is fabricated using low threshold metal gate CMOS technology and is designed to operate in bus oriented microprocessor systems where a real time clock and calendar function are required. The on-chip 32.768 kHz crystal controlled oscillator will maintain timekeeping down to 2.2V to allow low power standby battery operation. This device is pin compatible with the MM58174B but continues timekeeping up to tens of years. The MM58274C is a direct replacement for the MM58274 offering improved Bus access cycle times.

Applications

- Point of sale terminals
- Teller terminals
- Word processors
- Data logging
- Industrial process control

Features

- Same pin-out as MM58174A and MM58274B
- Timekeeping from tenths of seconds to tens of years in independently accessible registers
- Leap year register
- Hours counter programmable for 12 or 24-hour operation
- Buffered crystal frequency output in test mode for easy oscillator setting
- Data-changed flag allows simple testing for time rollover
- Independent interrupting time with open drain output
- Fully TTL compatible
- Low power standby operation (10 μA at 2.2V)
- Low cost 16-pin DIP and 20-pin PCC

Block Diagram

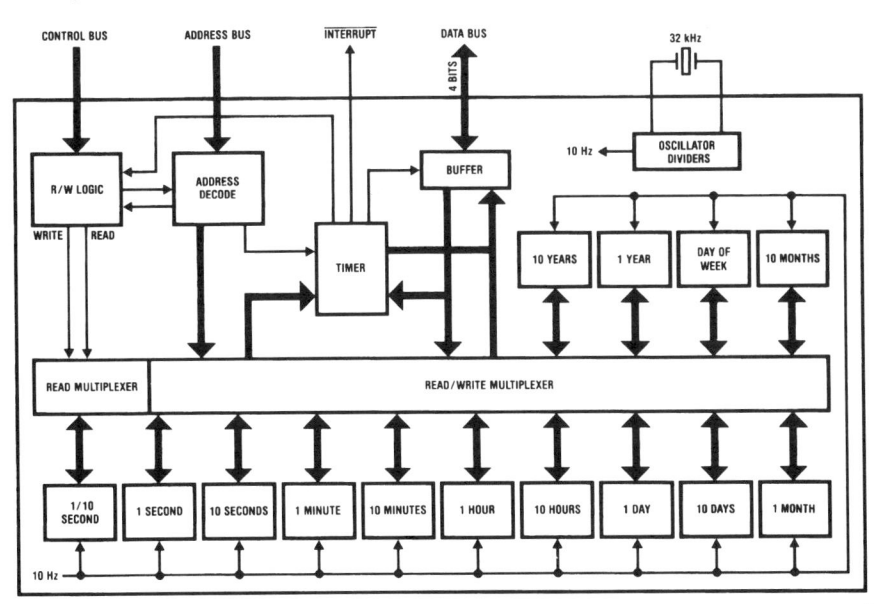

FIGURE 1

Absolute Maximum Ratings (Note 1)

If Military/Aerospace specified devices are required, please contact the National Semiconductor Sales Office/Distributors for availability and specifications.

DC Input or Output Voltage	-0.3V to $V_{DD} + 0.3$V
DC Input or Output Diode Current	± 5.0 mA
Storage Temperature, T_{STG}	$-65°C$ to $+150°C$
Supply Voltage, V_{DD}	6.5V
Power Dissipation, P_D	500 mW
Lead Temperature (Soldering, 10 seconds)	260°

Operating Conditions

	Min	Max	Units
Operating Supply Voltage	4.5	5.5	V
Standby Mode Supply Voltage	2.2	5.5	V
DC Input or Output Voltage	0	V_{DD}	V
Operating Temperature Range	-40	85	°C

Electrical Characteristics

$V_{DD} = 5$V $\pm 10\%$, T $= -40°C$ to $+85°C$ unless otherwise stated.

Symbol	Parameter	Conditions	Min	Typ	Max	Units
V_{IH}	High Level Input Voltage (except XTAL IN)		2.0			V
V_{IL}	Low Level Input Voltage (except XTAL IN)				0.8	V
V_{OH}	High Level Output Voltage (DB0–DB3)	$I_{OH} = -20\ \mu A$ $I_{OH} = -1.6$ mA	$V_{DD} - 0.1$ 3.7			V V
V_{OH}	High Level Output Voltage (INT)	$I_{OH} = -20\ \mu A$ (In Test Mode)	$V_{DD} - 0.1$			V
V_{OL}	Low Level Input Voltage (DB0–DB3, $\overline{INT}$)	$I_{OL} = 20\ \mu A$ $i_{OL} = 1.6$ mA			0.1 0.4	V V
I_{IL}	Low Level Input Current (AD0–AD3, DB0–DB3)	$V_{IN} = V_{SS}$ (Note 2)	-5		-80	μA
I_{IL}	Low Level Input Current ($\overline{WR}$, $\overline{RD}$)	$V_{IN} = V_{SS}$ (Note 2)	-5		-190	μA
I_{IL}	Low Level Input Current ($\overline{CS}$)	$V_{IN} = V_{SS}$ (Note 2)	-5		-550	μA
I_{OZH}	Ouput High Level Leakage Current ($\overline{INT}$)	$V_{OUT} = V_{DD}$			2.0	μA
I_{DD}	Average Supply Current	All $V_{IN} = V_{CC}$ or Open Circuit $V_{DD} = 2.2$V (Standby Mode) $V_{DD} = 5.0$V (Active Mode)		4	10 1	μA mA
C_{IN}	Input Capacitance			5	10	pF
C_{OUT}	Output Capacitance	(Outputs Disabled)		10		pF

Note 1: Absolute Maximum Ratings are those values beyond which damage to the device may occur. All voltages referenced to ground unless otherwise noted.

Note 2: The DB0–DB3 and AD0–AD3 lines all have active P-channel pull-up transistors which will source current. The $\overline{CS}$, $\overline{RD}$, and $\overline{WR}$ lines have internal pull-up resistors to V_{DD}.

AC Switching Characteristics

READ TIMING: DATA FROM PERIPHERAL TO MICROPROCESSOR V_{DD} = 5V ±0.5V, C_L = 100 pF

Symbol	Parameter	Commercial Specification T_A = −40°C to +85°C			Units
		Min	Typ	Max	
t_{AD}	Address Bus Valid to Data Valid		390	650	ns
t_{CSD}	Chip Select On to Data Valid		140	300	ns
t_{RD}	Read Strobe On to Data Valid		140	300	ns
t_{RW}	Read Strobe Width (Note 3, Note 7)			DC	
t_{RA}	Address Bus Hold Time from Trailing Edge of Read Strobe	0			ns
t_{CSH}	Chip Select Hold Time from Trailing Edge of Read Strobe	0			ns
t_{RH}	Data Hold Time from Trailing Edge of Read Strobe	70	160		ns
t_{HZ}	Time from Trailing Edge of Read Strobe Until O/P Drivers are TRI-STATE®			250	ns

WRITE TIMING: DATA FROM MICROPROCESSOR TO PERIPHERAL V_{DD} = 5V ±0.5V

Symbol	Parameter	Commercial Specification T_A = −40°C to +85°C			Units
		Min	Typ	Max	
t_{AW}	Address Bus Valid to Write Strobe ⌐ (Note 4, Note 6)	400	125		ns
t_{CSW}	Chip Select On to Write Strobe ⌐	250	100		ns
t_{DW}	Data Bus Valid to Write Strobe ⌐	400	220		ns
t_{WW}	Write Strobe Width (Note 6)	250	95		ns
t_{WCS}	Chip Select Hold Time Following Write Strobe ⌐	0			ns
t_{WA}	Address Bus Hold Time Following Write Strobe ⌐	0			ns
t_{WD}	Data Bus Hold Time Following Write Strobe ⌐	100	35		ns
t_{AWS}	Address Bus Valid Before Start of Write Strobe	70	20		ns

Note 3: Except for special case restriction: with interrupts programmed, max read strobe width of control register (ADDR 0) is 30 ms. See section on Interrupt Programming.

Note 4: All timings measured to the trailing edge of write strobe (data latched by the trailing edge of $\overline{WR}$).

Note 5: Input test waveform peak voltages are 2.4V and 0.4V. Output signals are measured to their 2.4V and 0.4V levels.

Note 6: Write strobe as used in the Write Timing Table is defined as the period when both chip select and write inputs are low, ie., $\overline{WS}$ = $\overline{CS}$ + $\overline{WR}$. Hence write strobe commences when both signals are low, and terminates when the first signal returns high.

Note 7: Read strobe as used in the Read Timing Table is defined as the period when both chip select and read inputs are low, ie., $\overline{RS}$ = $\overline{CS}$ + $\overline{RD}$.

Note 8: Typical numbers are at V_{CC} = 5.0V and T_A = 25°C.

Switching Time Waveforms

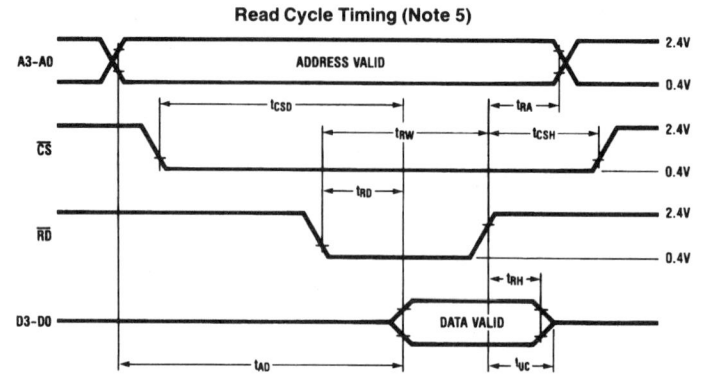

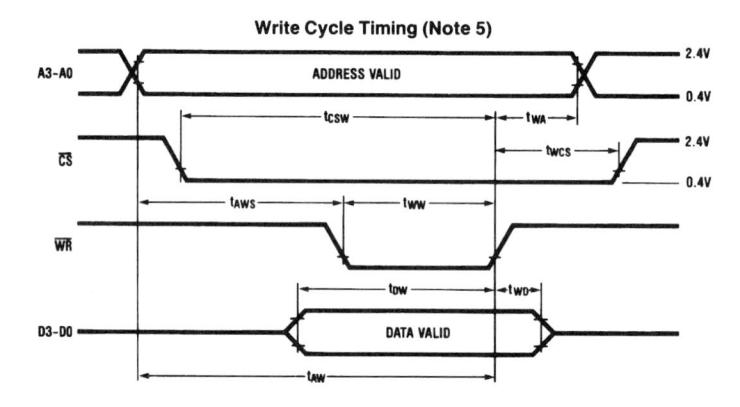

Read Cycle Timing (Note 5)

Write Cycle Timing (Note 5)

Connection Diagrams

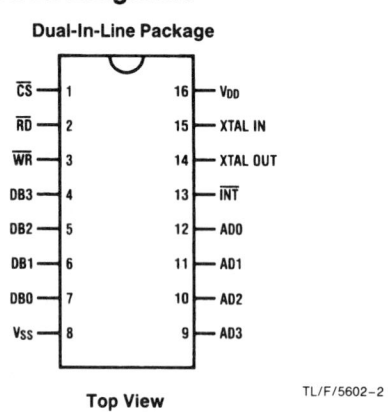

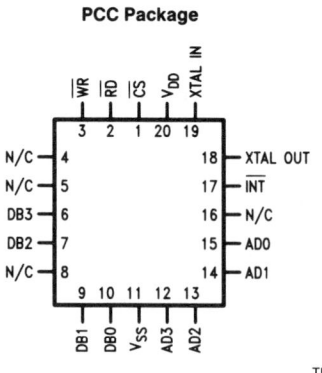

Dual-In-Line Package — Top View

PCC Package

FIGURE 2

Order Number MM58274CJ, MM58274CN or MM58274CV
See NS Package J16A, N16A, or V20A

Functional Description

The MM58274C is a bus oriented microprocessor real time clock. It has the same pin-out as the MM58174A while offering extended timekeeping up to units and tens of years. To enhance the device further, a number of other features have been added including: 12 or 24 hours counting, a testable data-changed flag giving easy error-free time reading and simplified interrupt control.

A buffered oscillator signal appears on the interrupt output when the device is in test mode. This allows for easy oscillator setting when the device is initially powered up in a system.

The counters are arranged as 4-bit words and can be randomly accessed for time reading and setting. The counters output in BCD (binary coded decimal) 4-bit numbers. Any register which has less than 4 bits (e.g., days of week uses only 3 bits) will return to a logic 0 on any unused bits. When written to, the unused inputs will be ignored.

Writing a logic 1 to the clock start/stop control bit resets the internal oscillator divider chain and the tenths of seconds counter. Writing a logic 0 will start the clock timing from the nearest second. The time then updates every 100 ms with all counters changing synchronously. Time changing during a read is detected by testing the data-changed bit of the control register after completing a string of clock register reads.

Interrupt delay times of 0.1s, 0.5s, 1s, 5s, 10s, 30s or 60s can be selected with single or repeated interrupt outputs. The open drain output is pulled low whenever the interrupt timer times out and is cleared by reading the control register.

CIRCUIT DESCRIPTION

The block diagram in *Figure 1* shows the internal structure of the chip. The 16-pin package outline is shown in *Figure 2*.

Crystal Oscillator

This consists of a CMOS inverter/amplifier with an on-chip bias resistor. Externally a 20 pF capacitor, a 6 pF–36 pF trimmer capacitor and a crystal are required to complete the 32.768 kHz timekeeping oscillator circuit.

The 6 pF–36 pF trimmer fine tunes the crystal load impedance, optimizing the oscillator stability. When properly adjusted (i.e., to the crystal frequency of 32.768 kHz), the circuit will display a frequency variation with voltage of less than 3 ppm/V. When an external oscillator is used, connect to oscillator input and float (no connection) the oscillator output.

When the chip is enabled into test mode, the oscillator is gated onto the interrupt output pin giving a buffered oscillator output that can be used to set the crystal frequency when the device is installed in a system. For further information see the section on Test Mode.

Divider Chain

The crystal oscillator is divided down in three stages to produce a 10 Hz frequency setting pulse. The first stage is a non-integer divider which reduces the 32.768 kHz input to 30.720 kHz. This is further divided by a 9-stage binary ripple counter giving an output frequency of 60 Hz. A 3-stage Johnson counter divides this by six, generating a 10 Hz output. The 10 Hz clock is gated with the 32.768 kHz crystal frequency to provide clock setting pulses of 15.26 μs duration. The setting pulse drives all the time registers on the

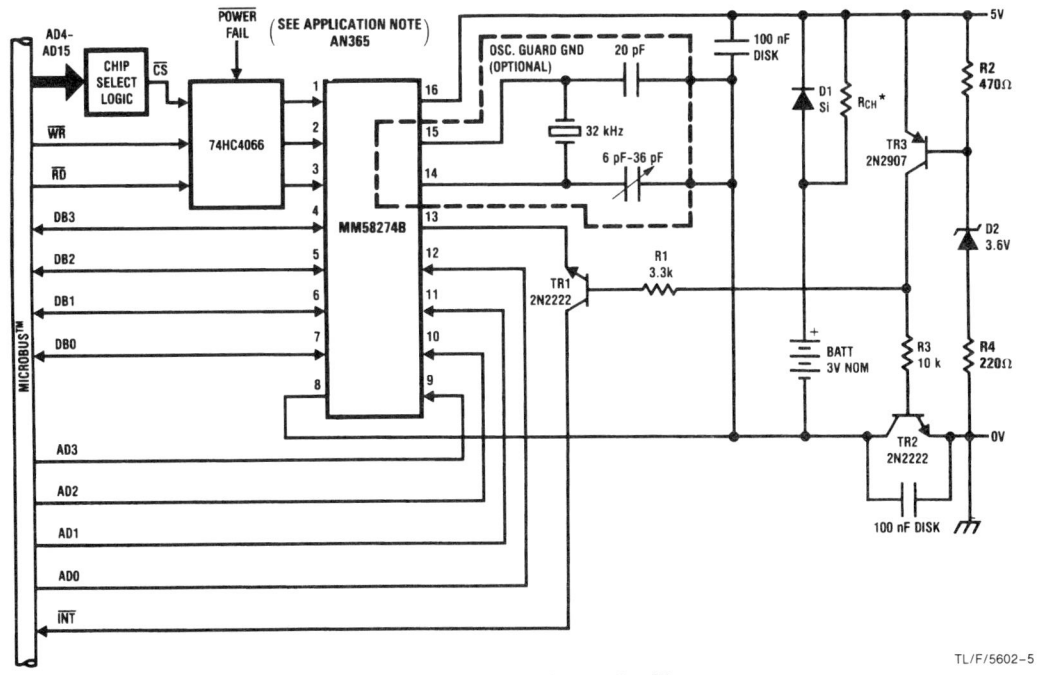

FIGURE 3. Typical System Connection Diagram

Functional Description (Continued)

device which are synchronously clocked by this signal. All time data and data-changed flag change on the falling edge of the clock setting pulse.

Data-Changed Flag

The data-changed flag is set by the clock setting pulse to indicate that the time data has been altered since the clock was last read. This flag occupies bit 3 of the control register where it can be tested by the processor to sense data-changed. It will be reset by a read of the control register. See the section, "Methods of Device Operation", for suggested clock reading techniques using this flag.

Seconds Counters

There are three counters for seconds:
a) tenths of seconds
b) units of seconds
c) tens of seconds.

The registers are accessed at the addresses shown in Table I. The tenths of seconds register is reset to 0 when the clock start/stop bit (bit 2 of the control register) is set to logic 1. The units and tens of seconds are set up by the processor, giving time setting to the nearest second. All three registers can be read by the processor for time output.

Minutes Counters

There are two minutes counters:
a) units of minutes
b) tens of minutes.

Both registers may be read to or written from as required.

Hours Counters

There are two hours counters:
a) units of hours
b) tens of hours.

Both counters may be accessed for read or write operations as desired.

In 12-hour mode, the tens of hours register has only one active bit and the top three bits are set to logic 0. Data bit 1 of the clock setting register is the AM/PM indicator; logic 0 indicating AM, logic 1 for PM.

When 24-hour mode is programmed, the tens of hours register reads out two bits of data and the two most significant bits are set to logic 0. There is no AM/PM indication and bit 1 of the clock setting register will read out a logic 0.

In both 12/24-hour modes, the units of hours will read out four active data bits. 12 or 24-hour mode is selected by bit 0 of the clock setting register, logic 0 for 12-hour mode, logic 1 for the 24-hour mode.

Days Counters

There are two days counters:
a) units of days
b) tens of days.

The days counters will count up to 28, 29, 30 or 31 depending on the state of the months counters and the leap year counter. The microprocessor has full read/write access to these registers.

Months Counters

There are two months counters:
a) units of months
b) tens of months.

Both these counters have full read/write access.

Years Counters

There are two years counters:
a) units of years
b) tens of years.

Both these counters have full read/write access. The years will count up to 99 and roll over to 00.

TABLE I. Address Decoding of Real-Time Clock Internal Registers

Register Selected		AD3	AD2	AD1	AD0	(Hex)	Access
0	Control Register	0	0	0	0	0	Split Read and Write
1	Tenths of Seconds	0	0	0	1	1	Read Only
2	Units Seconds	0	0	1	0	2	R/W
3	Tens Seconds	0	0	1	1	3	R/W
4	Units Minutes	0	1	0	0	4	R/W
5	Tens Minutes	0	1	0	1	5	R/W
6	Unit Hours	0	1	1	0	6	R/W
7	Tens Hours	0	1	1	1	7	R/W
8	Units Days	1	0	0	0	8	R/W
9	Tens Days	1	0	0	1	9	R/W
10	Units Months	1	0	1	0	A	R/W
11	Tens Months	1	0	1	1	B	R/W
12	Units Years	1	1	0	0	C	R/W
13	Tens Years	1	1	0	1	D	R/W
14	Day of Week	1	1	1	0	E	R/W
15	Clock Setting/ Interrupt Registers	1	1	1	1	F	R/W

Functional Description (Continued)

Day of Week Counter

The day of week counter increments as the time rolls from 23:59 to 00:00 (11:59 PM to 12:00 AM in 12-hour mode). It counts from 1 to 7 and rolls back to 1. Any day of the week may be specified as day 1.

Clock Setting Register/Interrupt Register

The interrupt select bit in the control register determines which of these two registers is accessible to the processor at address 15. Normal clock and interrupt timing operations will always continue regardless of which register is selected onto the bus. The layout of these registers is shown in Table II.

The clock setting register is comprised of three separate functions:

a) leap year counter: bits 2 and 3
b) AM/PM indicator: bit 1
c) 12/24-hour mode set: bit 0 (see Table IIA).

The leap year counter is a 2-stage binary counter which is clocked by the months counter. It changes state as the time rolls over from 11:59 on December 31 to 00:00 on January 1.

The counter should be loaded with the 'number of years since last leap year' e.g., if 1980 was the last leap year, a clock programmed in 1983 should have 3 stored in the leap year counter. If the clock is programmed during a leap year, then the leap year counter should be set to 0. The contents of the leap year counter can be read by the μP.

The AM/PM indicator returns a logic 0 for AM and a logic 1 for PM. It is clocked when the hours counter rolls from 11:59 to 12:00 in 12-hour mode. In 24-hour mode this bit is set to logic 0.

The 12/24-hour mode set determines whether the hours counter counts from 1 to 12 or from 0 to 23. It also controls the AM/PM indicator, enabling it for 12-hour mode and forcing it to logic 0 for the 24-hour mode. The 12/24-hour mode bit is set to logic 0 for 12-hour mode and it is set to logic 1 for 24-hour mode.

IMPORTANT NOTE: *Hours mode and AM/PM bits cannot be set in the same write operation. See the section on Initialization (Methods of Device Operation) for a suggested setting routine.*

All bits in the clock setting register may be read by the processor.

The interrupt register controls the operation of the timer for interrupt output. The processor programs this register for single or repeated interrupts at the selected time intervals.

The lower three bits of this register set the time delay period that will occur between interrupts. The time delays that can be programmed and the data words that select these are outlined in Table IIB.

Data bit 3 of the interrupt register sets for either single or repeated interrupts; logic 0 gives single mode, logic 1 sets for repeated mode.

Using the interrupt is described in the Device Operation section.

TABLE IIA. Clock Setting Register Layout

Function	DB3	DB2	DB1	DB0	Comments	Access
Leap Year Counter	X	X			0 Indicates a Leap Year	R/W
AM/PM Indicator (12-Hour Mode)			X		0 = AM 1 = PM 0 in 24-Hour Mode	R/W
12/24-Hour Select Bit				X	0 = 12-Hour Mode 1 = 24-Hour Mode	R/W

TABLE IIB. Interrupt Control Register

Function	Comments	DB3	DB2	DB1	DB0
No Interrupt	Interrupt output cleared, start/stop bit set to 1.	X	0	0	0
0.1 Second		0/1	0	0	1
0.5 Second		0/1	0	1	0
1 Second	DB3 = 0 for single interrupt	0/1	0	1	1
5 Seconds	DB3 = 1 for repeated interrupt	0/1	1	0	0
10 Seconds		0/1	1	0	1
30 Seconds		0/1	1	1	0
60 Seconds		0/1	1	1	1

Timing Accuracy: single interrupt mode (all time delays): ±1 ms
Repeated Mode: ±1 ms on initial timeout, thereafter synchronous with first interrupt (i.e., timing errors do not accumulate).

Functional Description (Continued)

Control Register

There are three registers which control different operations of the clock:

a) the clock setting register
b) the interrupt register
c) the control register.

The clock setting and interrupt registers both reside at address 15, access to one or the other being controlled by the interrupt select bit; data bit 1 of the control register.

The clock setting register programs the timekeeping of the clock. The 12/24-hour mode select and the AM/PM indicator for 12-hour mode occupy bits 0 and 1, respectively. Data bits 2 and 3 set the leap year counter.

The interrupt register controls the operation of the interrupt timer, selecting the required delay period and either single or repeated interrupt.

The control register is responsible for controlling the operations of the clock and supplying status information to the processor. It appears as two different registers; one with write only access and one with read only access.

The write only register consists of a bank of four latches which control the internal processes of the clock.

The read only register contains two output data latches which will supply status information for the processor. Table III shows the mapping of the various control latches and status flags in the control register. The control register is located at address 0.

The write only portion of the control register contains four latches:

A logic 1 written into the test bit puts the device into test mode. This allows setting of the oscillator frequency as well as rapid testing of the device registers, if required. A more complete description is given in the Test Mode section. For normal operation the test bit is loaded with logic 0.

The clock start/stop bit stops the timekeeping of the clock and resets to 0 the tenths of seconds counter. The time of day may then be written into the various clock registers and the clock restarted synchronously with an external time source. Timekeeping is maintained thereafter.

A logic 1 written to the start/stop bit halts clock timing. Timing is restarted when the start/stop bit is written with a logic 0.

The interrupt select bit determines which of the two registers mapped onto address 15 will be accessed when this address is selected.

A logic 0 in the interrupt select bit makes the clock setting register available to the processor. A logic 1 selects the interrupt register.

The interrupt start/stop bit controls the running of the interrupt timer. It is programmed in the same way as the clock start/stop bit; logic 1 to halt the interrupt and reset the timer, logic 0 to start interrupt timing.

When no interrupt is programmed (interrupt control register set to 0), the interrupt start/stop bit is automatically set to a logic 1. When any new interrupt is subsequently programmed, timing will not commence until the start/stop bit is loaded with 0.

In the single interrupt mode, interrupt timing stops when a timeout occurs. The processor restarts timing by writing logic 0 into the start/stop bit.

In repeated interrupt mode the interrupt timer continues to count with no intervention by the processor necessary.

Interrupt timing may be stopped in either mode by writing a logic 1 into the interrupt start/stop bit. The timer is reset and can be restarted in the normal way, giving a full time delay period before the next interrupt.

In general, the control register is set up such that writing 0's into it will start anything that is stopped, pull the clock out of test mode and select the clock setting register onto the bus. In other words, writing 0 will maintain normal clock operation and restart interrupt timing, etc.

The read only portion of the control register has two status outputs:

Since the MM58274C keeps real time, the time data changes asynchronously with the processor and this may occur while the processor is reading time data out of the clock.

Some method of warning the processor when the time data has changed must thus be included. This is provided for by the data-changed flag located in bit 3 of the control register. This flag is set by the clock setting pulse which also clocks the time registers. Testing this bit can tell the processor whether or not the time has changed. The flag is cleared by a read of the control register but not by any write operations. No other register read has any effect on the state of the data-changed flag.

Data bit 0 is the interrupt flag. This flag is set whenever the interrupt timer times out, pulling the interrupt output low. In a polled interrupt routine the processor can test this flag to determine if the MM58274C was the interrupting device. This interrupt flag and the interrupt output are both cleared by a read of the control register.

TABLE III. The Control Register Layout

Access (addr0)	DB3	DB2	DB1	DB0
Read From:	Data-Changed Flag	0	0	Interrupt Flag
Write To:	Test 0 = Normal 1 = Test Mode	Clock Start/Stop 0 = Clock Run 1 = Clock Stop	Interrupt Select 0 = Clock Setting Register 1 = Interrupt Register	Interrupt Start/Stop 0 = Interrupt Run 1 = Interrupt Stop

Functional Description (Continued)

Both of the flags and the interrupt output are reset by the trailing edge of the read strobe. The flag information is held latched during a control register read, guaranteeing that stable status information will always be read out by the processor.

Interrupt timeout is detected and stored internally if it occurs during a read of the control register, the interrupt output will then go low only after the read has been completed.

A clock setting pulse occurring during a control register read will *not* affect the data-changed flag since time data read out before or after the control read will not be affected by the time change.

METHODS OF DEVICE OPERATION

Test Mode

National Semiconductor uses test mode for functionally testing the MM58274C after fabrication and again after packaging. Test mode can also be used to set up the oscillator frequency when the part is first commissioned.

Figure 4 shows the internal clock connections when the device is written into test mode. The 32.768 kHz oscillator is gated onto the interrupt output to provide a buffered output for initial frequency setting. This signal is driven from a TRI-STATE output buffer, enabling easy oscillator setting in systems where interrupt is not normally used and there is no external resistor on the pin.

If an interrupt is programmed, the 32.768 kHz output is switched off to allow high speed testing of the interrupt timer. The interrupt output will then function as normal.

The clock start/stop bit can be used to control the fast clocking of the time registers as shown in *Figure 4*.

Initialization

When it is first installed and power is applied, the device will need to be properly initialized. The following operation steps are recommended when the device is set up (all numbers are decimal):

1) Disable interrupt on the processor to allow oscillator setting. Write 15 into the control register: *The clock and interrupt start/stop bits are set to 1, ensuring that the clock and interrupt timers are both halted. Test mode and the interrupt register are selected.*

2) Write 0 to the interrupt register: *Ensure that there are no interrupts programmed and that the oscillator will be gated onto the interrupt output.*

3) Set oscillator frequency: *All timing has been halted and the oscillator is buffered out onto the interrupt line.*

4) Write 5 to the control register: *The clock is now out of test mode but is still halted. The clock setting register is now selected by the interrupt select bit.*

5) Write 0001 to all registers. This ensures starting with a valid BCD value in each register.

6) Set 12/24 Hours Mode: *Write to the clock setting register to select the hours counting mode required.*

7) Load Real-Time Registers: *All time registers (including Leap Years and AM/PM bit) may now be loaded in any order. Note that when writing to the clock setting register to set up Leap Years and AM/PM, the Hours Mode bit must not be altered from the value programmed in step 5.*

8) Write 0 to the control register: *This operation finishes the clock initialization by starting the time. The final control register write should be synchronized with an external time source.*

In general, timekeeping should be halted before the time data is altered in the clock. The data can, however, be altered at any time if so desired. Such may be the case if the user wishes to keep the clock corrected without having to stop and restart it; i.e., winter/summer time changing can be accomplished without halting the clock. This can be done in software by sensing the state of the data-changed flag and only altering time data just after the time has rolled over (data-changed flag set).

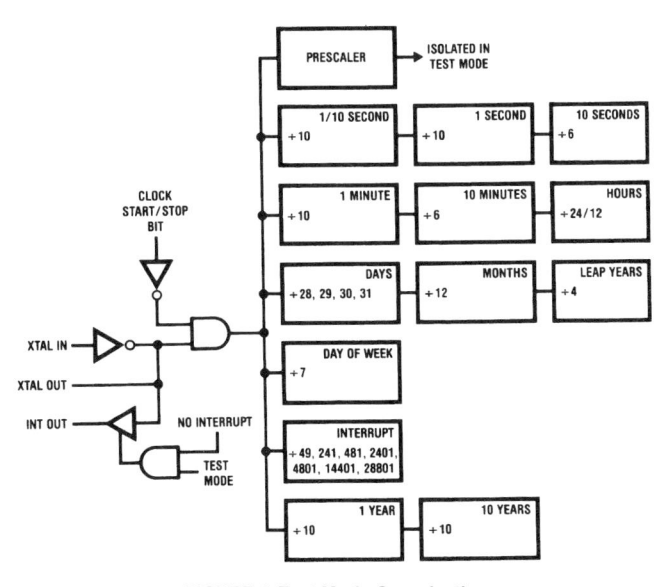

FIGURE 4. Test Mode Organization

Functional Description (Continued)

Reading the Time Registers

Using the data-changed flag technique supports microprocessors with block move facilities, as all the necessary time data may be read sequentially and then tested for validity as shown below.

1) Read the control register, address 0: *This is a dummy read to reset the data-changed flag (DCF) prior to reading the time registers.*

2) Read time registers: *All desired time registers are read out in a block.*

3) Read the control register and test DCF: *If DCF is cleared (logic 0), then no clock setting pulses have after occurred since step 1. All time data is guaranteed good and time reading is complete.*

If DCF is set (logic 1), then a time change has occurred since step 1 and time data may not be consistent. Repeat steps 2 and 3 until DCF is clear. The control read of step 3 will have reset DCF, automatically repeating the step 1 action.

Interrupt Programming

The interrupt timer generates interrupts at time intervals which are programmed into the interrupt register. A single interrupt after delay or repeated interrupts may be programmed. Table IIB lists the different time delays and the data words that select them in the interrupt register.

Once the interrupt register has been used to set up the delay time and to select for single or repeat, it takes no further part in the workings of the interrupt system. All activity by the processor then takes place in the control register.

Initializing:

1) Write 3 to the control register (AD0): *Clock timing continues, interrupt register selected and interrupt timing stopped.*

2) Write interrupt control word to address 15: *The interrupt register is loaded with the correct word (chosen from Table IIB) for the time delay required and for single or repeated interrupts.*

3) Write 0 or 2 to the control register: *Interrupt timing commences. Writing 0 selects the clock setting register onto the data bus; writing 2 leaves the interrupt register selected. Normal timekeeping remains unaffected.*

On Interrupt:

Read the control register and test for Interrupt Flag (bit 0).

If the flag is cleared (logic 0), then the device is not the source of the interrupt.

If the flag is set (logic 1), then the clock did generate an interrupt. The flag is reset and the interrupt output is cleared by the control register read that was used to test for interrupt.

Single Interrupt Mode:

When appropriate, write 0 or 2 to the control register to restart the interrupt timer.

Repeated Interrupt Mode:

Timing continues, synchronized with the control register write which originally started interrupt timing. No further intervention is necessary from the processor to maintain timing.

In either mode interrupt timing can be stopped by writing 1 into the control register (interrupt start/stop set to 1). Timing for the full delay period recommences when the interrupt start/stop bit is again loaded with 0 as normal.

IMPORTANT NOTE: Using the interrupt timer places a constraint on the maximum Read Strobe width which may be applied to the clock. Normally all registers may be read from with a t_{RW} down to DC (i.e., $\overline{CS}$ and $\overline{RD}$ held continuously low). When the interrupt timer is active however, the maximum read strobe width that can be applied to the control register (Addr 0) is 30 ms.

This restriction is to allow the interrupt timer to properly reset when it times out. Note that it only affects reading of the control register—all other addresses in the clock may be accessed with DC read strobes, regardless of the state of the interrupt timer. Writes to any address are unaffected.

NOTES ON AC TIMING REQUIREMENTS

Although the Switching Time Waveforms show Microbus control signals used for clock access, this does not preclude the use of the MM58274C in other non-Microbus systems. *Figure 5* is a simplified logic diagram showing how the control signals are gated internally to control access to the clock registers. From this diagram it is clear that $\overline{CS}$ could be used to generate the internal data transfer strobes, with $\overline{RD}$ and $\overline{WR}$ inputs set up first. This situation is illustrated in *Figure 6*.

The internal data busses of the MM58274C are fully CMOS, contributing to the flexibility of the control inputs. When determining the suitability of any given control signal pattern for the MM58274B the timing specifications in AC Switching Characteristics should be examined. As long as these timings are met (or exceeded) the MM58274C will function correctly.

When the MM58274C is connected to the system via a peripheral port, the freedom from timing constraints allows for very simple control signal generation, as in *Figure 7*. For reading (*Figure 7a*), Address, $\overline{CS}$ and $\overline{RD}$ may be activated simultaneously and the data will be available at the port after t_{AD}-max (650 ns). For writing (*Figure 7b*), the address and data may be applied simultaneously; 70 ns later $\overline{CS}$ and $\overline{WR}$ may be strobed together.

Functional Description (Continued)

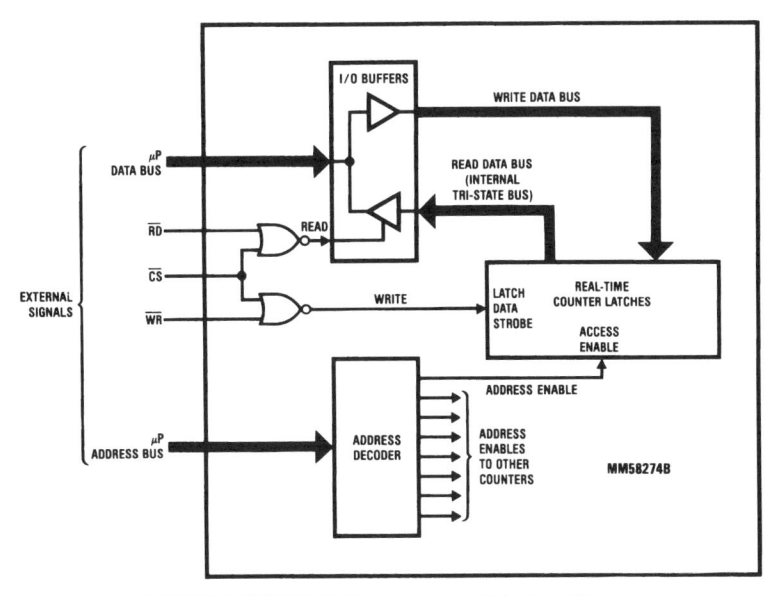

FIGURE 5. MM58274C Microprocessor Interface Diagram

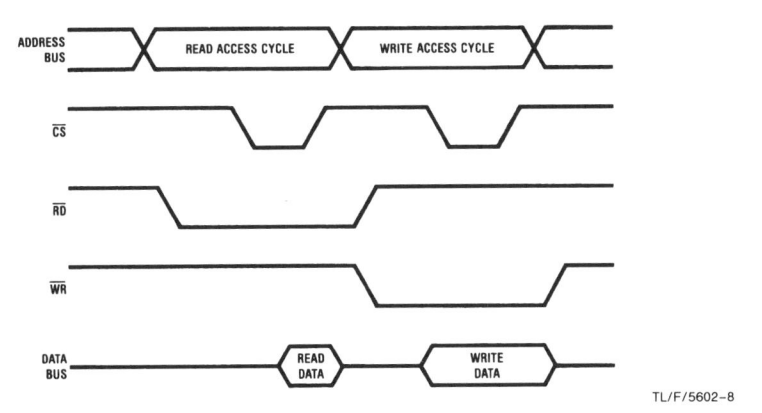

FIGURE 6. Valid MM58274C Control Signals Using Chip Select Generated Access Strobes

Functional Description (Continued)

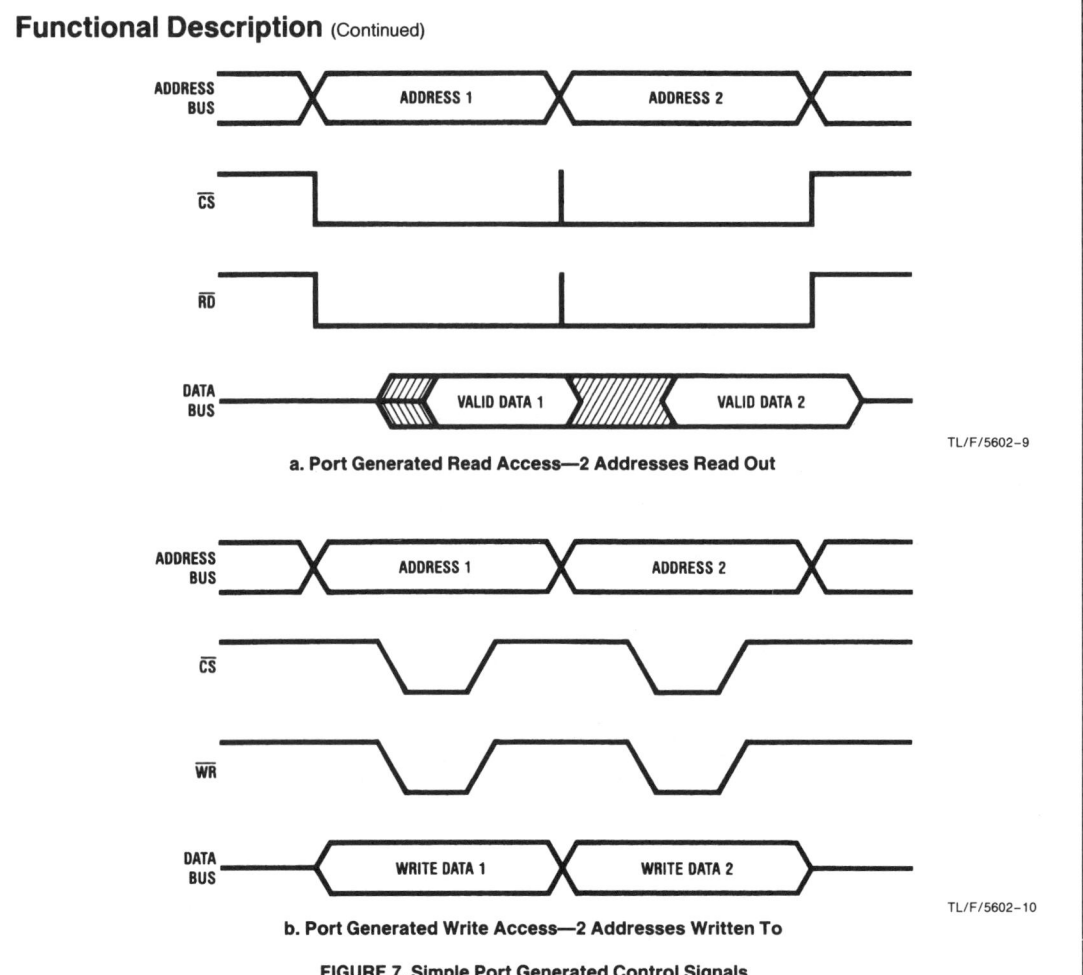

a. Port Generated Read Access—2 Addresses Read Out

b. Port Generated Write Access—2 Addresses Written To

FIGURE 7. Simple Port Generated Control Signals

Functional Description (Continued)

APPLICATION HINTS

Time Reading Using Interrupt

In systems such as point of sale terminals and data loggers, time reading is usually only required on a random demand basis. Using the data-changed flag as outlined in the section on methods of operation is ideal for this type of system. Some systems, however, need to sense a change in real time; e.g., industrial timers/process controllers, TV/VCR clocks, any system where real time is displayed.

The interrupt timer on the MM58274C can generate interrupts synchronously with the time registers changing, using software to provide the initial synchronization.

In single interrupt mode the processor is responsible for initiating each timing cycle and the timed period is accurate to ±1 ms.

In repeated interrupt mode the period from the initial processor start to the first timeout is also only accurate to ±1 ms. The following interrupts maintain accurate delay periods relative to the first timeout. Thus, to utilize interrupt to control time reading, we will use repeated interrupt mode.

In repeated mode the time period between interrupts is exact, which means that timeouts will always occur at the same point relative to the internal clock setting pulses. The case for 0.1s interrupts is shown in *Figure A-1*. The same is true for other delay periods, only there will be more clock setting pulses between each interrupt timeout. If we set up the interrupt timer so that interrupt always times out just after the clock setting pulse occurs (*Figure A-2*), then there is no need to test the data-changed flag as we know that the time data has just changed and will not alter again for another 100 ms.

This can be achieved as outlined below:

1) Follow steps 1 and 2 of the section on interrupt programming. In step 2 set up for repeated interrupt.

2) Read control register AD0: *This is a dummy read to reset the data-changed flag.*

3) Read control register AD0 until data-changed flag is set.

4) Write 0 or 2 to control register. Interrupt timing commences.

Time Reading with Very Slow Read Cycles

If a system takes longer than 100 ms to complete reading of all the necessary time registers (e.g., when CMOS processors are used) or where high level interpreted language routines are used, then the data-changed flag will always be set when tested and is of no value. In this case, the time registers themselves must be tested to ensure data accuracy.

The technique below will detect both time changing *between* read strobes (i.e., between reading tens of minutes and units of hours) and also time changing *during* read, which can produce invalid data.

1) Read and store the value of the *lowest* order time register required.

2) Read out all the time registers required. The registers may be read out in any order, simplifying software requirements.

3) Read the lowest order register and compare it with the value stored previously in step 1. If it is still the same, then all time data is good. If it has changed, then store the new value and go back to step 2.

In general, the rule is that the first and last reads *must* both be of the lowest order time register. These two values can then be compared to ensure that no change has occurred. This technique works because for any higher order time register to change, all the lower order registers must also change. If the lowest order register does not change, then no higher order register has changed either.

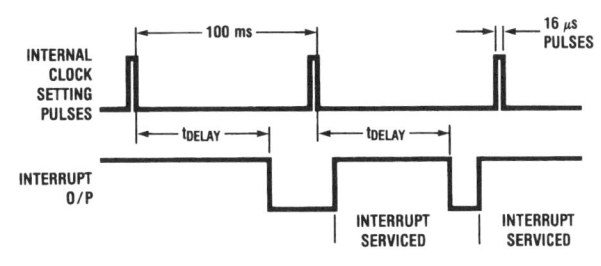

FIGURE A-1. Time Delay from Clock Setting Pulses to Interrupt is Constant

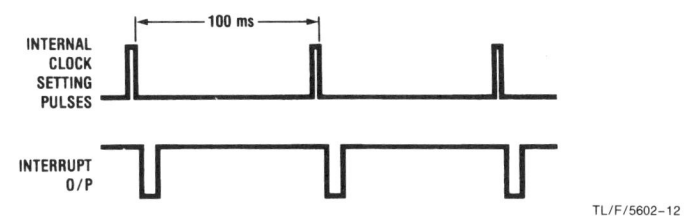

FIGURE A-2. Interrupt Timer Synchronized with Clock Setting Pulses

The MM58274C Adds Reliable Real-Time Keeping to Any Microprocessor System

National Semiconductor
Application Note 365
Peter K. Thomson

INTRODUCTION

When a Real-Time Clock (RTC) is to be added into a digital system, the designer will face a number of design constraints and problems that do not usually occur in normal systems. Attention to detail in both hardware and software design is necessary to ensure that a reliable and trouble free product is implemented.

The extra circuitry required for an RTC falls into three main groups: a precise oscillator to control real-time couting; a backup power source to maintain time-keeping when the main system power is removed; power failure detection and write protection circuitry. The MM58274C in common with most RTC devices uses an on-chip oscillator circuit and an external watch crystal (frequency 32.768 kHz) as the time reference. A battery is the usual source of backup power, along with circuitry to isolate the battery-backed clock from the rest of the system. Like any CMOS component, the RTC must be protected against data corruption when the main system power fails; a problem that is very often not fully appreciated.

Rather than dealing strictly with any one particular application, this applications note discusses all of the aspects involved in adding a reliable RTC function to a microprocessor system, with descriptions of suitable circuitry to achieve this. Hardware problems, component selection, and physical board layout are examined. The software examples given in the data sheet are explained and clarified, and some other software suggestions are presented. Finally a number of otherwise unrelated topics are lumped together under "Miscellany"; including a discussion on how the MM58274C may be used directly to upgrade an existing MM58174A installation.

CONTENTS

1.0 HARDWARE
    1.1 COMPONENT SELECTION
        1.1.1 Crystal
        1.1.2 Loading Capacitors
        1.1.3 Backup Battery:
            Capacitors
            Nickel-Cadmium Cells
            Alkaline
            Lithium
            Other Cells and Notes
            Temperature Range
    1.2 BOARD LAYOUT
        1.2.1 Oscillator Connection
        1.2.2 Battery Placement
        1.2.3 Other Components
    1.3 POWER SUPPLY ISOLATION SCHEMES
        1.3.1 The Need for Isolation
        1.3.2 Isolation Techniques I—5V Supply Only
        1.3.3 Isolation Techniques II—Negative Supply Switched
        1.3.4 Other Methods
    1.4 POWER FAIL PROTECTION
        1.4.1 Write Protect Switch
        1.4.2 5V Sensing
        1.4.3 Supply Pre-Sense
        1.4.4 Switching Power Supplies
        1.4.5 Summary

2.0 SOFTWARE
    2.1 DATA VALIDATION
        2.1.1 Post-Read Synchronization
        2.1.2 Pre-Read Synchronization
    2.2 INTERRUPT AS A "DATA-CHANGED" FLAG
    2.3 WRITING WITHOUT HALTING TIME-KEEPING
    2.4 THE CLOCK AS A μP WATCHDOG
    2.5 THE JAPANESE CALENDAR

3.0 MISCELLANY
    3.1 CONNECTION TO NON-MICROBUS SYSTEMS
    3.2 TEST MODE
    3.3 TEST MODE AND OSCILLATOR SETTING
    3.4 UPGRADING AN MM58174A SYSTEM WITH THE MM58274C
    3.5 WAIT STATE GENERATION FOR FAST μPs

APPENDIX A-1 Reading Valid Real-Time Data (Reprinted from the MM58274C Data Sheet)

APPENDIX A-2 MM58274C Functional Truth Tables

1.0 HARDWARE

Selecting the correct components for the job and implementing a good board layout is crucial to developing an accurate and reliable Real-Time Clock function. The range of component choices available is large and the suitability of different types depends on the demands of the system.

1.1 COMPONENT SELECTION

With reference to *Figure 1*, the oscillator components and the battery are examined and the suitability of different types is discussed.

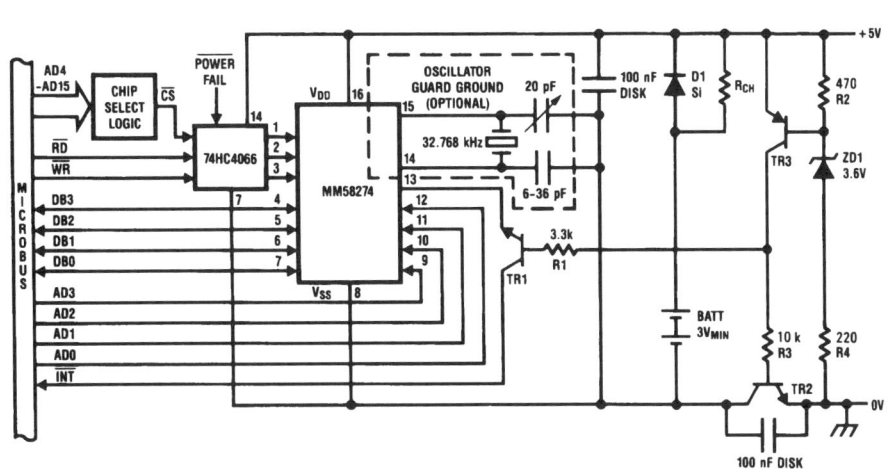

FIGURE 1. MM58274C System Installation

1.1.1 Crystal

The oscillator is designed to work with a standard low power NT cut or XY Bar clock crystal of 32.768 kHz frequency. The circuit is a Pierce oscillator and is shown complete in *Figure 2*. The 20 MΩ resistor biases the oscillator into its linear region and ensures oscillator start-up. The 200 kΩ resistor prevents the oscillator amplifier from overdriving the crystal. If very low power crystals are used (i.e., less than 1 µW) an external resistor of around 200 kΩ may have to be added to reduce the drive to the crystal.

The oscillator will drive most normal watch crystals, with up to 20 µW drive available from the on-chip oscillator.

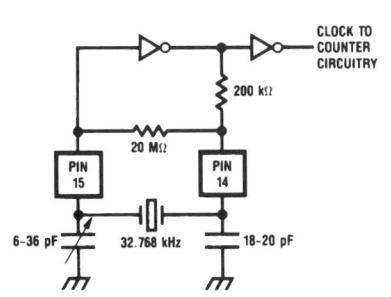

FIGURE 2. Complete Oscillator Diagram

1.1.2 Loading Capacitors

Two capacitors are used to provide the correct output loading for the crystal. One is a fixed value capacitor in the range 18 pF–20 pF and the other is a variable 6 pF–36 pF trimmer capacitor. Adjusting the trimmer allows the crystal loading (and hence the oscillator frequency) to be fine tuned for optimal results.

The capacitors are the components most likely to affect the overall accuracy of the oscillator and care must be exercised in selection. Ceramic capacitors offer good operating temperature range with close tolerance and low temperature coefficients (typically ±3 ppm/K, for good quality examples). If trimming is undesirable a pair of close tolerance (±5% or better) capacitors in the range 18 pF–20 pF may be used. The average time-keeping accuracy for this configuration is within ±20 seconds per month.

1.1.3 Backup Battery

There are a number of different cell types available that can be used for time-keeping retention. Some cells are more suitable than others, and the way in which the system is used also influences the choice of cell. Ideally the standby voltage of the RTC should be kept as low as possible, as the supply current increases with increasing voltage *(Figure 3)*. Four different power sources are discussed: capacitors, nickel-cadmium rechargeable cells, alkaline and lithium primary cells.

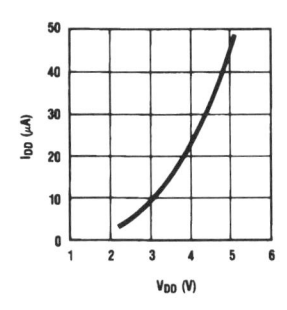

FIGURE 3. Typical I_{DD} (µA) vs V_{DD} (V) for MM58274C in Standby Mode ($T_A = 25°C$)

Capacitors

When the system is permanently powered, and any long term removal of system power (i.e., more than a few hours) requires complete restarting, then a 1–2 Farad capacitor may be sufficient to run the clock during the power down. This can keep the clock running for 48–72 hours.

Nickel-Cadmium Cells

Nickel-cadmium (Ni-Cad) cells can be trickle-charged from the system power supply using a resistor as shown in *Figure 1*. The exact value of resistor used depends on the capacity and number of cells in the battery. Consult the manufacturers data for information on charging rates and times.

A 3- or 4-cell battery should be used to power the clock (the nominal battery voltages are 3.6V for 3 cells in series and 4.8V for 4 cells), with 3 cells preferable. PCB mounting batteries of 100 mAh capacity are available and these will give around 6 months data retention (at normal room temperature). For this cell type to be used the system must spend a large proportion of its time turned on to keep the battery charged (i.e., used daily).

Alkaline

Alkaline cells are among the least expensive primary cells which are suitable for use in real-time clock applications. They are available in a large range of capacities and shapes and have a very good storage (shelf) life.

Two cells in series will provide a nominal 3V, which is adequate to power the clock (via the isolating diode). The main problem with the alkaline system is that the cell terminal voltage drops slowly over the life of the cell. When the voltage at the clock supply pin drops to 2.2V, the cells must be replaced (battery voltage around 2.6V–2.7V). With present alkaline cells, this point is usually reached when the cells are only ½ to ⅔ discharged.

Provisions must be made either to check the battery voltage at regular intervals or to replace the cells regularly enough to avoid the danger of using discharged cells. Once again the manufacturers data regarding capacity and cell voltage against time must be examined to determine a suitable cell selection. A good alkaline system will supply 1–2 years continuous time-keeping.

Lithium

Lithium cells are the most suitable for real-time clock applications. A single cell with 3V potential is sufficient to power the system. The cell potential is very stable over use and the storage life is excellent. The energy density of lithium cells is very high, giving enough capacity in a physically small cell to power the clock continuously for at least 5 years (at room temperature using a 1,000 mAh cell).

Several cells which are recommended for RTC use are D2/3A*, D2A*, and 1/6DEL/P*. Each have 1,000 mAh capacity. These cells are available with solder pin connections for PCB mounting, giving a reliable backup supply.

Other Cells and Notes

There are many other types of cells, both primary and secondary, which may be adapted for RTC use. When selecting a cell type, attention must be paid to:

a. Cell capacity and physical size.
b. Storage (shelf) life.
c. Voltage variation over use.
d. Operating temperature range.
e. The method of battery connection and mounting.

In general, soldered cells are preferable to connector mounted cells. With replaceable batteries, the battery and connector contacts must be kept thoroughly clean. Dirty or corroded contacts can cause the clock to be starved of power, giving erratic and unreliable performance. The ease of operator access for cell replacement should also be considered.

Temperature Range

The performance of any cell will be satisfactory for most office or domestic environments. When "ruggedized" equipment is to be used (i.e., field portable equipment, automotive, etc.) the temperature specification of different cell types should be taken into account when selecting a cell. Lithium cells offer good performance over 0°C–70°C with little loss in capacity. Once again, the manufacturer's data should be examined to determine suitability, especially since different cells of the same type can have markedly different characteristics.

Few types of cells will offer any useful capacity at temperatures in or below the range 0°C–10°C, and fewer still will operate over the full military temperature range (−55°C to +125°C). Solid lithium cells and mercury-cadmium cells are two systems which can cover this range.

1.2 BOARD LAYOUT

1.2.1 Oscillator Connection

The oscillator components must be built as close to the pins of the clock chip as is physically possible. The ideal configuration is shown in *Figure 4*. From *Figure 2*, the oscillator circuit, it can be seen that both Osc In and Osc Out are high impedance nodes, susceptible to noise coupling from adjacent lines. Hence the oscillator should, as far as is practicable, be surrounded by a guard ground. The absolute maximum length of PCB tracking on either pin is 2.5 cm (1 inch). Longer tracks increase the parasitic track to track capacitances, increasing the risk of noise coupling and hence reducing the overall oscillator stability.

Where the system operates in humid or very cold environments (below 5°C), condensation or ice may form on the PCB. This has the effect of adding parasitic resistances and capacitances between pins 14 and 15, and also to ground. This variation in loading adversely affects the stability of the oscillator and in extreme cases may cause the oscillator to stop.

*Duracell Trade Number.
**Tadiran Trade Number.

Keeping the PCB tracks as short as possible will help to minimize the problem, and on its own this may be sufficient. Where the operating conditions are particularly severe, the PCB and oscillator components should be coated with a suitable water repellent material, such as lacquer or silicon grease (suitability being determined by the electrical properties of the materials—high impedance and low dielectric constant).

Figures 2 and *4* show the trimmer placed on Osc Out. The placement of the trimmer capacitor on either Osc In or Osc Out is not critical. Placing the trimmer on Osc Out yields a smaller trim range, but less susceptibility to changes in trimmer capacitance. Placement of the trimmer capacitor on Osc In gives a wider trim span, but slightly greater susceptibility to capacitance changes.

1.2.2 Battery Placement

For the battery, placement is less critical than with the oscillator components. Practical considerations are of greater importance now; i.e., accessibility. The battery should be placed where it is unlikely to be accidentally shorted or disconnected during routine operation and servicing of the equipment.

When replaceable cells are used, connecting a 100 μF capacitor across the RTC supply lines will keep the clock operating for 30–40 seconds with the battery disconnected *(Figure 5)*. This allows the battery to be replaced regardless of whether or not the main supply is active.

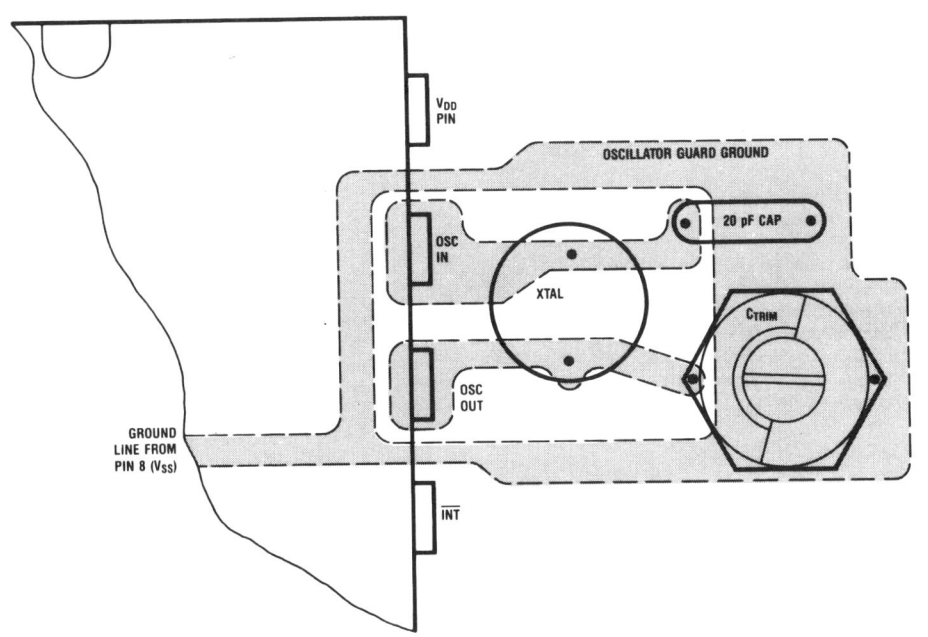

FIGURE 4. Oscillator Board Layout

FIGURE 5. Simplified Power Supply Diagram with 100 μF Capacitor Added

1.2.3 Other Components

The placement of the other RTC dedicated components (e.g., supply disconnection and power failure protection components) is not particularly critical. However, the same guidelines as applied to the battery should be followed when the PCB layout is designed.

1.3 POWER SUPPLY ISOLATION SCHEMES

1.3.1 The Need for Isolation

There are two reasons for disconnecting the clock circuit from the rest of the system:

1. To prevent the backup battery from trying to power the whole system when the main power fails.
2. To minimize the battery current (and extend battery life) by preventing current leakage out of the RTC input pins.

The MM58274C inputs have internal pull-up devices which pull the inputs to V_{DD} in power down mode. This turns off the internal TTL input buffers and causes the μP interface functions of the clock to go to full CMOS logic levels, drawing no supply current (except for the unavoidable leakage current of the internal MOS transistors). For the MM58274C this is achieved by isolating the ground (V_{SS}) supply line from the rest of the system.

Figures 6a and *6b* show the two cases where first V_{DD} *(6a)* and then V_{SS} *(6b)* are open-circuited. The line out from the MM58274C represents any of the Control, Address, or Data lines on the RTC, with the internal pull-up resistor shown. The two diodes and resistor R_S represent the logic device connected to the RTC input and the resistance of the rest of the system with no power applied.

When V_{DD} is open-circuit as in *Figure 6a*, there is a complete current path, shown by the arrows, out of the RTC input and through the external circuitry. This battery current is a complete waste and serves only to reduce the cell life. Depending on the value of R_S, the voltage level at the pin may fall low enough to turn on the internal TTL level buffer, wasting further current as the buffer is no longer fully CMOS.

With V_{SS} disconnected *(Figure 6b)*, there is no return path to the battery and the pin is pulled completely up to V_{DD}. The TTL buffer is switched off and no power is lost.

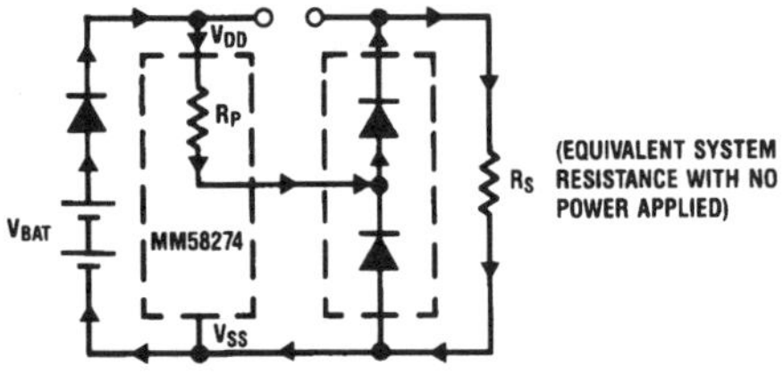

1.3.2 Isolation Techniques I—5V Supply Only

Figure 7 shows the isolation circuit suggested in the MM58274C data sheet. This circuit provides complete disconnection where only the system +5V is available for switching control.

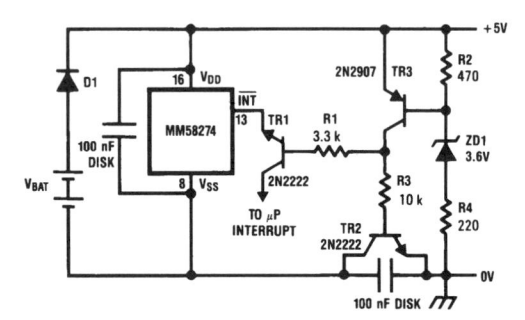

FIGURE 7. 5V Isolation Circuit

TR2 is the disconnecting device, which is controlled by TR3 and its associated circuitry. TR3 is turned on by its bias chain R2, ZD1, R4 as the system supply rises up to 4.2V. TR3 and R3 then turn on TR2 to connect the clock to the system supply. D1 isolates the backup battery when the system supply is active. The 100 nF disk capacitors decouple the supply during R/W operations and should be included in any disconnection scheme.

TR3 is necessary to prevent R3 and TR2 from leaking battery current in the power down condition. The circuit without TR3 is shown in *Figure 8* where TR2 has been replaced by equivalent diodes to clearly show the problem. The circuitry could be simplified by replacing TR3 with a Zener diode *(Figure 9)*. There will be a small loss of current down through TR2 however, as the Zener will pass a small leakage current at below its "knee" voltage. Thus the Zener should be selected for its low current capability.

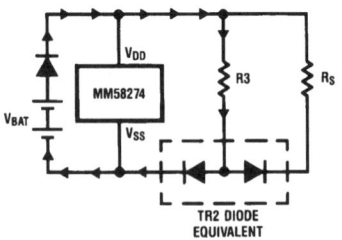

a) V_{DD} Disconnection

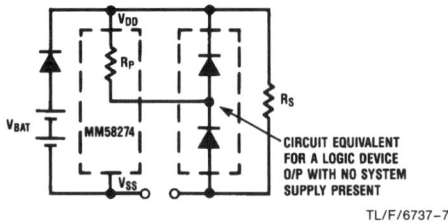

b) V_{SS} Disconnection

FIGURE 6. Current Leakage Prevention by Proper Supply Disconnection

R_S = System Resistance with No Power Applied

FIGURE 8. Current Leakage in Simplified Disconnection Schemes

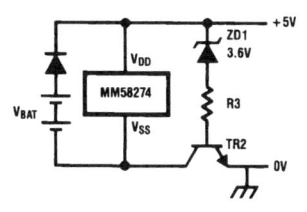

FIGURE 9. Alternative Supply Disconnection Scheme Sensing 5V (Decoupling Capacitors Omitted for Clarity)

Finally TR1 and R1 *(Figure 7)* are optional components which are only required when the interrupt output is used. If interrupts are left programmed when the power fails, the interrupt timer will still time-out setting the interrupt output. Since this is an active low pull-down transistor it effectively shorts directly across TR2, destroying the RTC isolation and discharging the battery into the rest of the system *(Figure 10)*. In order to prevent this from occurring, TR1 and R1 are added.

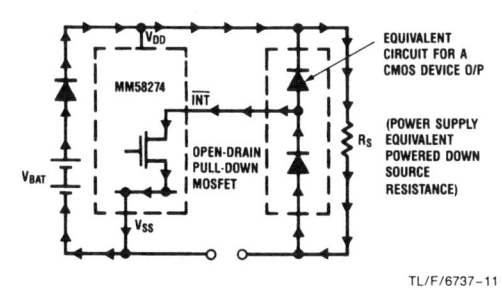

FIGURE 10. Battery Discharge Path via Unisolated Interrupt Output

None of the disconnection components are at all critical, with general purpose transistors being completely adequate for the task. D1 should be a small-signal silicon or germanium diode.

1.3.3 Isolation Techniques II—Negative Supply Switched

Where a negative voltage supply is available (either regulated or unregulated) the circuit of *Figure 11* may be used. This is similar in operation to its diode equivalent shown in *Figure 12*, where the voltage drops across the diodes provide the correct potential to the clock. *Figure 11* has the advantage, however, that the clock power is supplied from the ground line by transistor action, rather than via the resistor as in *Figure 12*. Less lower is dissipated in the resistor as only transistor bias current need be drawn.

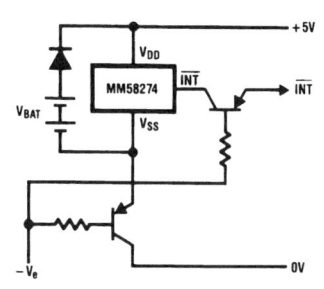

FIGURE 11. Negative Voltage Driven Supply Disconnection Scheme (Decoupling Capacitors Omitted for Clarity)

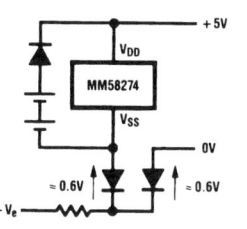

FIGURE 12. Diode Equivalent Circuit of *Figure 11*

1.3.4 Other Methods

There are many other possibilities for supply disconnection schemes, i.e., relay disconnection. When designing a disconnection scheme, the performance must be analyzed both with the system power applied and with system power absent. Check for leakage paths and undue voltage drops and try to set up so that disconnection and reconnection will take place as near to the backup voltage as possible.

1.4 POWER FAIL PROTECTION

One of the major causes of unreliability in RTC designs is due to inadequate power failure protection. As the system is powered up and down, the μP and surrounding logic can produce numerous spurious signals, including spurious writes and illegal control signals (i.e., RD and WR both active together).

Bipolar logic devices can produce spikes and glitches as the internal biasing switches off around 3V–3.5V, and the transistors operate in their linear region for a short time. Any such spurious signals, if applied to the RTC, could cause the time data to be corrupted. Systems using 74HC logic and CMOS processors are less stringent in their power failure requirements as the devices tend to work right down to around 2V. Some form of write protection is still required, however.

In order to protect the time data, the system must be physically prevented from writing to the clock when the power supply is not stable. The ideal situation is to ban Write access to the clock before the system +5V starts to fail, and then keep the chip "locked-out" until the power is restored and stabilized. This ideal access control signal is illustrated in *Figure 13*.

Three methods of power fail protection are discussed, although there are also many other possibilities.

1.4.1 Write Protect Switch

By far the simplest and potentially the most hazard-free method is to use a switch on the WR control line to the clock *(Figure 14)*. This is completely adequate, but requires the intervention of an operator to alter time data or program interrupts.

Some thought must be given to ensuring that the operator cannot accidentally leave the WR line switched in. This may be achieved by the physical access method used (i.e., the machine is impossible to operate or switch off when in the time setting mode, because of the placement of access hatches, etc.) or with software. The switch state could be sensed by trying to alter the data in the Tens of Years counter or Interrupt register just prior to leaving the clock setting routine, and refusing to leave the routine until the WR switch has been opened. The switch condition should similarly be checked whenever the system is initialized or reset.

The physical location of the switch should also be considered for ease of accessibility. How easy the switch is to reach will depend on the system; i.e., in some cases a "tamper proof" clock may be required.

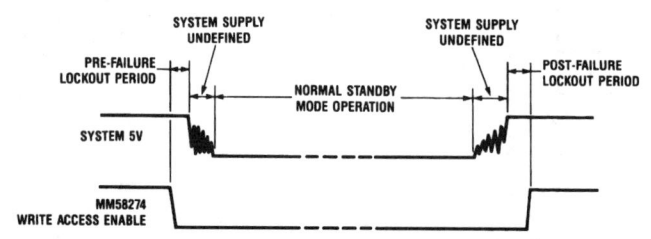

FIGURE 13. RTC Access Lockout Definition

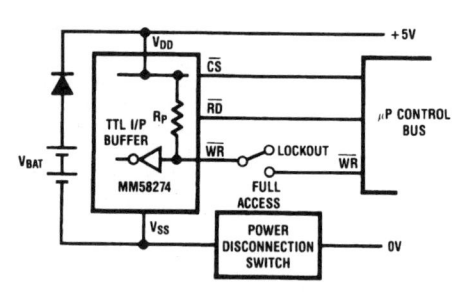

FIGURE 14. Write Protection by Manually Switching WR

1.4.2 5V Sensing

The circuit of *Figure 15* senses the system 5V supply and prevents access to the clock if the supply falls below 4.2V–4.3V. This circuit should be used where only the system 5V is available for reference. The LM139 comparator and associated components sense the 5V supply and generate the power fail signal (P.Fail). The 74HC75 and components disconnect the WR line.

R3 and ZD1 provide a reference voltage of 2V–3V for the comparator. R4 and VR1 form a potential divider chain sensing the 5V line, and VR1 is adjusted to switch the comparator output at 4.2V–4.3V. An alternative to VR1 would be to use a pair of close tolerance resistors ($\pm 2\%$) with values selected to suit the Zener diode reference used. The combination of R4, D3 and C2 provide an RC time constant to delay the comparator when sensing the return of 5V (to provide the post-failure delay in *Figure 13*). The LM139 has an open-collector output which is held low when 5V is present and is switched off when 5V fails. This line is pulled high by R5 to flag power failure (P.Fail). Since the comparator is a linear device drawing a bias current, it is powered by the system 5V supply to avoid consuming battery power.

One 74HC75 package contains four latches, of which two are used. These are transparent latches controlled by the "G" input. With G high, the latch is transparent and the Q and $\overline{Q}$ outputs follow the Data input. When G is low, the state of Q and $\overline{Q}$ on the falling edge is latched. In this way, F2 prevents P.Fail from locking out the clock if there is a Write cycle in progress. F1 isolates the WR input on the clock when F2 passes the P.Fail signal. C1, R2 and D1 do not slow the advent of P.Fail, but they cause a delay in the release of the function to mask any comparator noise or oscillation as the comparator switches off or on (i.e., during the undefined supply periods).

D2, C3 and R6 smooth the comparator supply and help it to function effectively. The time constants of the RC networks should be selected to suit the power supply of the system that is used. Comparing the functioning of this circuit with the ideal case of *Figure 13* shows that most of the conditions can be satisfied, except that there is no real pre-failure lock-out period. This cannot be achieved without some form of look ahead power failure.

As an alternative to F1 a permanently powered 74HC4066 analog switch could be used as the isolating component *(Figure 16)*. The 74HC4066 does not require pull-up resistors on its inputs as there are no internal CMOS buffers inside this device which must be controlled. The resistor on the WR line is for the benefit of the 74HC75.

Note that both of the devices mentioned must be permanently powered from the battery to be useful in this way. Unused gates in any such device must *NOT* be used in combinational logic that is not permanently powered. All unused inputs should be tied to V_{DD} or V_{SS} to render them inactive.

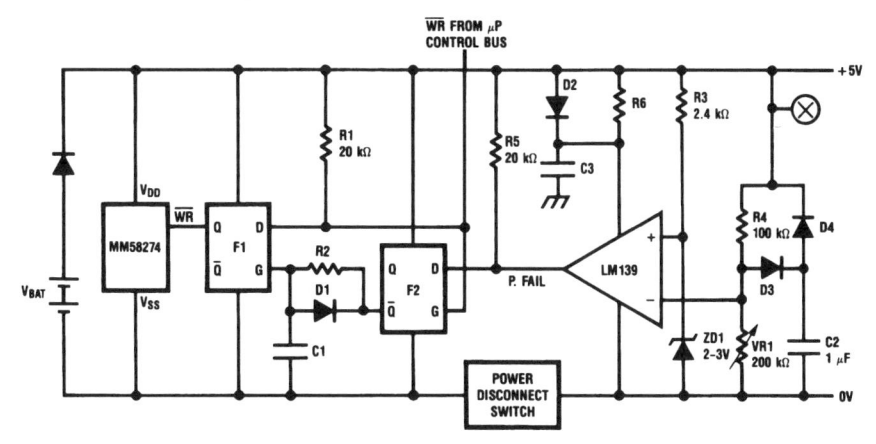

FIGURE 15. Power Supply Failure Detection and Write Protection Circuitry

1.4.3 Supply Pre-Sense

The same circuit of *Figure 15* can be used with unregulated supplies or other voltage lines which will fail before the 5V line. To achieve this, point X is connected to the sensed voltage instead of 5V, and the R4/VR1 ratio is adjusted to suit. The major benefit here is that advance warning of an impending 5V failure can be detected, allowing a pre-failure lockout signal to be generated.

Less precision is required to sense the unregulated supply than the system 5V supply. Consequently less complex circuitry can be used to do the detection and this is reflected in the circuit of *Figure 17*. Most 5V regulators will operate with an input voltage from 7V to 25V. Typically the input voltage is around 9V to 12V, giving some headroom. In *Figure 17* this voltage is high enough to drive a current through the Zener diode and turn on transistor TR1, holding P.Fail low. R_{LIM} limits the Zener current. The Zener voltage is selected to switch off before the regulator fails, around 7.5V–8.5V depending on the time constant of the supply. With no current, TR1 switches off and R_P pulls P.Fail high.

When power is re-applied the 5V supply will stabilize before the Zener switches on, removing P.Fail. To provide a longer post-failure lockout period R_{LIM} could be replaced with two resistors and a diode/capacitor delay as in *Figure 15*.

Figure 18 is another extension of the same basic idea to provide an advance interrupt signal to allow µP housekeeping before the RTC (and CMOS RAM) is locked out. The extra rectifying components D1, C_t and R_t keep $\overline{NMI}$ off as long as input power is present. Time constant τ_2 is selected to be at 2–3 times faster than τ_1, the supply time constant. The interrupt signal is thus asserted before P.Fail.

1.4.4 Switching Power Supplies

Switching power supplies are available which generate power failure signals. This signal may be adequate for direct use as a P.Fail line, but the manufacturer's information should be consulted to determine the suitability of a given power unit. P.Fail must still be gated with the Write signal for the clock, regardless of the actual detection method employed.

1.4.5 Summary

The general guidelines for power fail protection are:

1. Physically isolate the WR input to the clock. The µP cannot be relied upon to logically operate the isolation mechanism.
2. The clock should be isolated before the 5V power line starts to fail, and stay isolated until after it has reestablished.
3. Consider the action of the sensing and protection circuitry if the supplies oscillate or if a momentary glitch occurs.
4. The Power Fail signal must be gated with Write strobes to the RTC. A foreshortened Write may also cause data corruption.
5. Logic components (and ICs in general) should be avoided when designing power failure schemes. Discrete components are far more predictable in their performance when the power supplies are not well defined. The exception to this general rule is when using permanently powered HCMOS logic devices. They will function in a reliable manner down to 2V.

System-powered logic devices cannot be relied on for power failure or Write isolation (not even CMOS).

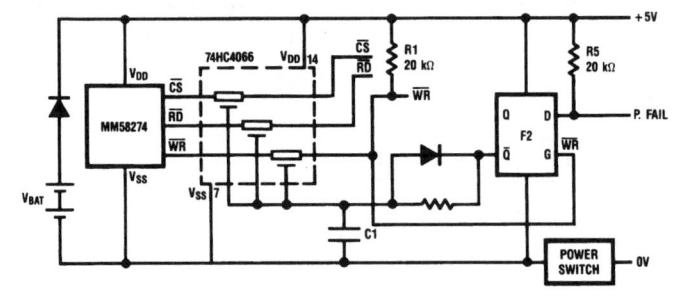

FIGURE 16. F1 Replaced by a 74HC4066 Analog Switch (Pull-Up Resistors Not Required on $\overline{CS}$ or $\overline{RD}$ Inputs)

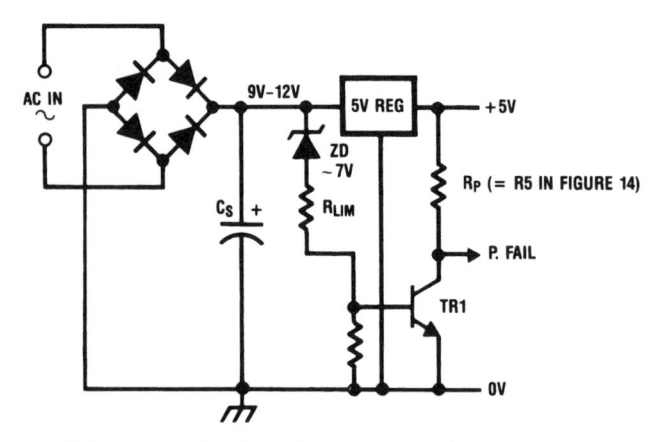

FIGURE 17. Power Fail Signal Generation from Unregulated Supplies

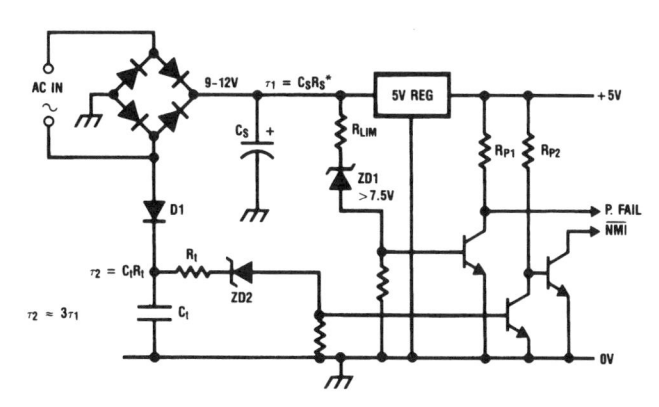

FIGURE 18. Power Fail Circuit with μP Housekeeping Interrupt

*R_S = Equivalent Resistance of the System.

2.0 SOFTWARE

2.1 DATA VALIDATION

The MM58274C data sheet describes in some detail three different methods of reading the clock and validating the real-time data. These techniques are reproduced in Appendix A-1. Rather than repeating the data sheet examples, this applications note examines the principles that lie behind the techniques suggested.

The basic problem is that the μP must somehow be synchronized with the changes in real-time in order to read valid data. This synchronization can either be done prior to reading the time data (pre-read), or after reading the data (post-read synchronization).

2.1.1 Post-Read Synchronization

Using the Data-Changed Flag (DCF) or the lowest order time register as outlined in the appendix: Time Reading using DCF and Time Reading with very slow Read cycles; are both examples of post-read synchronization.

What this means is that the data is read out first, and then verified. This is achieved by defining a random time-slot, started by the first DCF or low order register read, and ended by the second such read. If DCF has not been set during the time-slot or the lowest order register has not changed, then no real-time change occurred during that time-slot. All real-time reads during the time-slot are thus guaranteed.

2.1.2 Pre-Read Synchronization

The Interrupt Timer technique uses pre-read synchronization. Once it has been initialized as described, the interrupt timer times out just after the real-time data has changed. Thus the μP is guaranteed a full 100 ms period in which to read the time counters before the next change occurs.

The interrupt timer has to be synchronized with the real-time counters because it is an independent unit which may be started and stopped at any time by the μP. This software synchronization is achieved by using another pre-read technique. The timer is set up and ready to go, but then the μP waits for DCF to occur before issuing the start command. The same technique could be used to actually read the time-data, but post-read synchronization is faster.

2.2 INTERRUPT AS A "DATA-CHANGED" FLAG

DCF is set every 100 ms when the 1/10ths of seconds counter is changed. When the time is only being read to the nearest second or minute, it would be useful to have a flag which is only set by a change in the lowest order counter being used.

If the interrupt output from the clock is not being used, the timer can be used as a programmable data-changed flag. To achieve this, the timer is set up and started in exactly the same way as described for interrupt time reading (Appendix A-1). The interrupt output, however, should be left unconnected. When reading the real-time data, the technique used is the same as for the normal Data-Changed Flag except that the Interrupt Flag is tested instead of DCF.

Note that the lowest order real-time register which is to be read out should be used to initially synchronize the counter. The interrupt timer is started when the real-time counter value is seen to change.

2.3 WRITING WITHOUT HALTING TIME-KEEPING

For most purposes the RTC should be halted when the time is being set, especially if large numbers of counters are being updated. The clock can also then be re-started in synchronism with an external time reference. If only a few counters are to be altered and the clock is already synchronized, then this can be done without stopping the clock. An example of a minor change which may be undertaken in this way is daylight savings (winter/summer change of hour).

The problem to be overcome when writing in this way is that the write strobe may coincide with a time change pulse. As the time counters are synchronous, the 100 ms clock pulse is fed to each one. Writing to one counter may cause a spurious carry to be generated from that counter, causing the next one up the chain to be incremented.

Since a spurious carry will only affect the next counter if it coincides with a time update pulse, the solution is once again to synchronize clock access with the real-time change. The most suitable method for this is pre-read synchronization. In other words, the μP must wait for DCF to be set before starting to write data to the clock, giving a guaranteed 100 ms period for writing.

2.4 THE CLOCK AS A µP WATCHDOG

The interrupt timer can be used as a µP watchdog circuit, operating on a non-maskable interrupt input to the µP. The timer is set up in either single or repeat interrupt mode for the watchdog period required: 0.1s, 0.5s or 1 second are probably the most useful times for this. Synchronization with real-time is not required.

In the main program loop the µP writes to the clock, stopping and then re-starting the interrupt timer. The timer period selected will depend on how long the main loop takes to execute. As long as the µP continues to execute the loop, no time-outs occur and no interrupts are generated. If the µP fails for some reason to reset the timer, it eventually times out, generating the initializing interrupt to restore operations.

2.5 THE JAPANESE CALENDAR

Because the MM58274C has a programmable leap year counter, this allows the possibility of programming for the Japanese Showa calendar. The Japanese calender counts years from the time that the present Japanese Emperor comes to power.

The normal procedure for the MM58274C is to program "the number of years since last leap year." This remains the same whether the clock is loaded with the Gregorian or Showa year. When software is used to calculate the leap year count value from the year, then the formula used must be modified.

The formula for the Gregorian year is:
  Leap Year Value = [Gregorian Year/4] REMAINDER

Whereas for the Showa year the formula is:
  Leap Year Value = [(Showa Year + 1)/4] REMAINDER

Leap Year Value is the number from 0 to 3 which is written into the leap year counter, and is the REMAINDER of the integer calculations shown above.

3.0 MISCELLANY

3.1 CONNECTION TO NON Microbus™ SYSTEMS

Adding the MM58274C to non Microbus processors is made fairly straightforward because of the flexibility of the control signal timing. *Figure 19* shows two examples of logic to connect to clock to a 6502/6800 microprocessor bus.

Figure 19a the $\overline{RD}$ and $\overline{WR}$ inputs are strobed, generating reasonably typical Microbus type control signals. In *Figure 19b*, CS is used as the strobe signal. There is no particular advantage to either circuit, they are just variations on the same theme. This circuit flexibility may be used to advantage to save SSI packages in the board design.

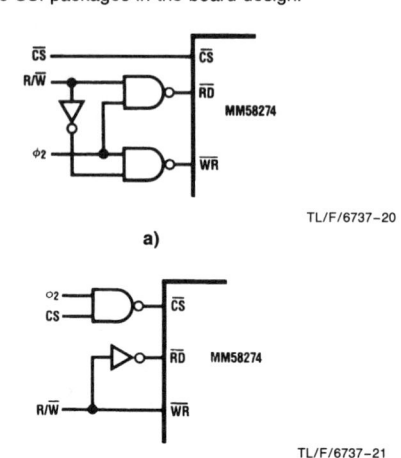

FIGURE 19. 6800/6502 µP Bus Interface

3.2 TEST MODE

Test Mode is used by National Semiconductor when the MM58274C is tested during manufacture. It enables the real-time counters to be clocked rapidly through their full count sequence.

The MM58274C counters are clocked synchronously to simplify µP access, with ripple carry signals from each counter to the next. In Test Mode some of these carries are intercepted and permanently asserted causing the counters to count each clock pulse. The prescaler is also bypassed so that the counters count every clock applied to the Osc In pin. The Test Mode counter connection is shown in *Figure 20*.

If Test Mode is to be used for incoming inspection or device verification, then the clock waveform of *Figure 21* should be applied to the oscillator input (Osc In, pin 15). The MM58274C uses semi-dynamic flip-flops in the counters which are only fully static when the oscillator input is high. Thus *Figure 21* shows that the oscillator waveform is normally high, pulsing low to clock the real-time counters. The time data in the counters changes on the rising edge of Osc In.

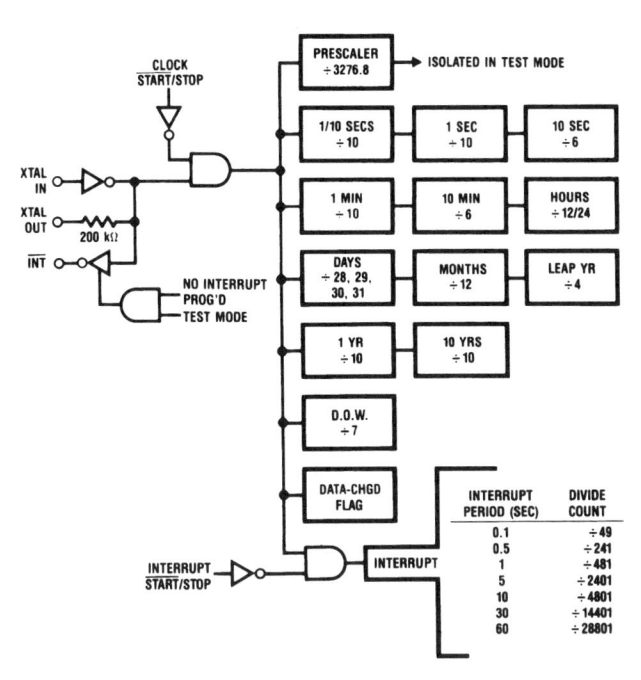

FIGURE 20. Test Mode Interconnection Diagram of Internal Counter Stages

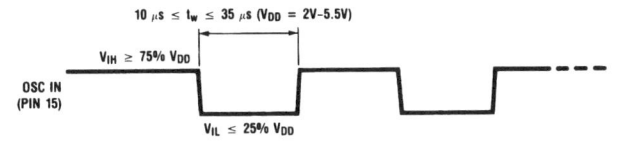

FIGURE 21. Oscillator Waveform for Counter Clocking in Test Mode

The pulse width limits for reliable clocking are shown on the diagram. When running with a 32 kHz crystal, the normal pulse width is 15.26 μs. With no forcing input, the oscillator will self bias to around 2.5V (V_{DD} = 5V). While a few hundred mV swing above and below this level is sufficient to drive the oscillator, for guaranteed test clocking the input should swing between $V_{IH} \geq 75\%$ V_{DD} and $V_{IL} \leq 25\%$ V_{DD}.

3.3 TEST MODE AND OSCILLATOR SETTING

When Test Mode is used to set the oscillator frequency, the interrupt timer must be disabled (interrupt register programmed with all 0s) for the oscillator frequency to appear on the interrupt output. No test equipment should be connected directly to either oscillator pin, as the added loading will alter the characteristics of the oscillator making precise tuning impossible.

Note that oscillator frequency will vary slightly as the supply varies between operating and standby voltages. Typically this variation will be around ±6 seconds per month ($V_{STANDBY}$ = 2.4V), slowing at standby voltage. When the clock will spend the greater part of its working life in standby mode, it may prove worthwhile to correct for this in the tuning. This can be done by tuning at standby voltage (by writing the RTC into test mode, then disconnecting it from the system to tune on battery backup). Alternatively, the clock can be slightly overtuned at operational voltage, tuning to 32.7681 kHz.

In a similar way, where the RTC spends equal amounts of time in both operational and standby modes (i.e., powered by day, standby at night), the oscillator may be tuned somewhere between the two conditions. Following these tuning suggestions will not eliminate time-keeping errors, but they will help in minimizing them.

Time-keeping accuracy cannot be exactly specified. It depends on the quality of the components used in the oscillator circuit and their physical layout, also the stability of the supply voltage, the variations in ambient temperture, etc. With good components and a reasonably stable environment however, time-keeping accuracy to within 4 seconds/month can be achieved, although 8 seconds/month is somewhat more typical in practical systems.

3.4 UPGRADING AN MM58174A SYSTEM WITH THE MM58274C

The MM58274C has the same pin-out as the MM58174A and can be used as a direct replacement, with certain reservations. The two devices are not quite the same in their external circuit appearances, and this is reflected in their applications circuits. In addition, the MM58274C is not software compatible with the MM58174A, requiring a change in the operating system to use the MM58274C.

Figure 22 shows the circuit diagram for the MM58174A system connection. There are two major differences between this and the MM58274C diagram *(Figure 1);* a) the oscillator circuit and b) the supply disconnection scheme.

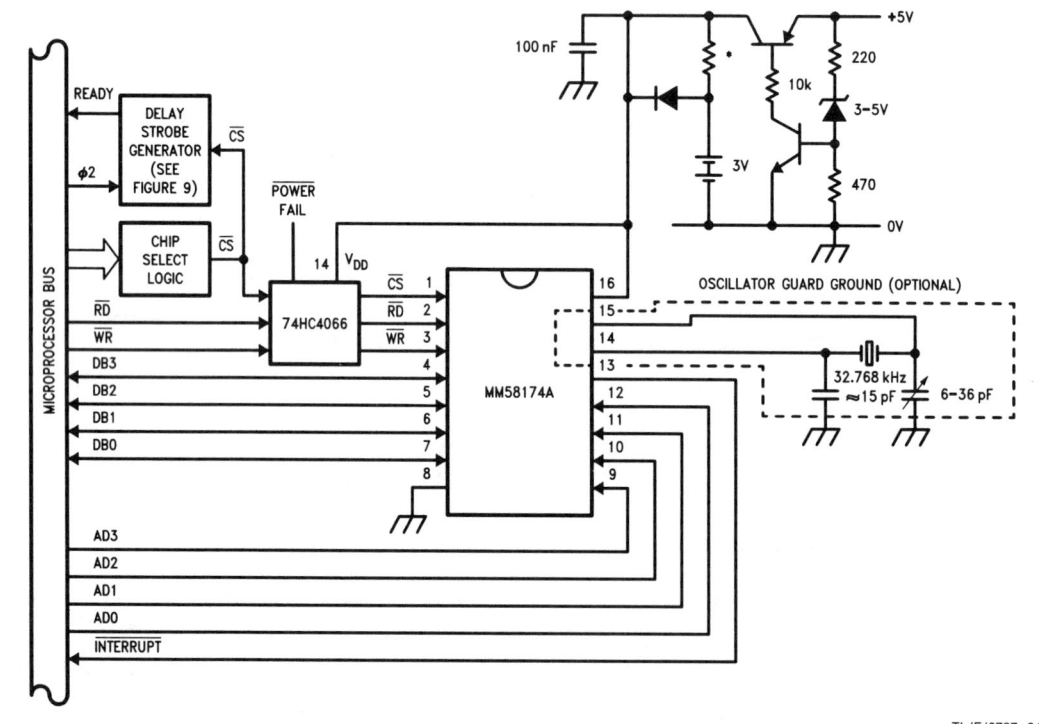

*Use resistor with Ni-Cad cells only.

FIGURE 22. MM58174A System Installation

a) The Oscillator Circuit

The MM58274C normally operates with an 18 pF–20 pF fixed loading capacitor as opposed to the 15 pF of the MM58174A. This is a reflection of the greater internal capacitance of the MM58174A, rather than any change in the characteristics of the oscillator itself. The MM58274C will operate using a 15 pF capacitor, but the oscillator will probably need to be retrimmed.

Operating with a 15 pF capacitor will make the oscillator more sensitive to change in the environment, i.e., temperature, voltage, moisture, etc. This will result in lower accuracy in time-keeping. The oscillator is more prone to stopping at low voltage. Oscillation would normally be maintained down to 1.8V–1.9V (although not guaranteed); with a 15 pF load it may only oscillate down to 2.0V–2.1V. It is thus important to check the battery regularly and replace it before the RTC voltage falls below 2.2V.

Where possible the 15 pF capacitor should be replaced by an 18 pF–20 pF capacitor (anywhere in the range 18 pF–20 pF is adequate), or a second 3 pF–5 pF capacitor may be added in parallel with the 15 pF.

Note: When components have been soldered into the oscillator circuit, allow the circuit to cool to room temperature before attempting to retune the oscillator.

The change of pin of the tuning capacitor (from OSC Out to Osc In) is not critical.

b) The Supply Disconnection Scheme

The MM58174A uses mostly pull-down devices on its μP inputs to pull the inputs to CMOS levels, and so the 5V power line is disconnected on this device. The two exceptions to this are the $\overline{CS}$ and $\overline{WR}$ inputs which have pull-up resistors to inactivate the internal write strobe. As *Figure 5a* shows, there is a leakage path through these pins, which in most MM58174A installations are individually isolated.

The largest penalty in inserting an MM58274C into an MM58174A circuit is the battery current that is lost through the pull-up devices. This will increase the typical supply current from 4 μA to 50–100 μA and it is up to the individual user to decide whether or not this drain is tolerable in a particular application.

The most important requirement is that the $\overline{WR}$ input should be electrically isolated or current leakage through pin inputs may force the inputs low enough to cause spurious writes to occur. Since it is already customary to isolate these inputs for the MM58174A, this may not be a problem. Where this has not been done, either the circuit will have to be modified or the $\overline{WR}$ PCB track can be cut and a switch or some extra circuitry added to allow isolation.

Note that power fail disconnection and input isolation may be achieved using the same components. In *Figure 22* the MM74HC4066 analog switch will do both jobs.

The current drained by the input pull-ups may be minimized with some attention to the data/address driving devices. It is often possible to replace LSTTL devices with standard 7400 series devices and reduce the leakage (at the cost of some increase in operating current). Many 7400 series device outputs lack diodes in the right places to pass leakage currents. LSTTL devices will, for the main part, have these diodes. CMOS devices will always have diodes to both power rails on inputs and outputs.

There is no hard and fast rule for this. Where devices from one manufacturer work, the same part from a different one may not. Some trial and error experimentation may prove worthwhile in selected devices.

3.5 WAIT STATE GENERATION FOR FAST μPs

Although the MM58274C has faster access times than the MM58174A, in many cases, the μP will be too fast to directly access the RTC. *Figure 23* shows a circuit which will produce wait states of any length required to enable the RTC to be accessed, using the 74HC74 dual D-type flip-flop.

The RTC $\overline{CS}$ signal clocks up a logic 1 on the Q output of the first F/F, removing the Preset from all the other F/Fs and pulling the μP WAIT line low, via the transistor. The other F/Fs 1 to n, form a shift register clocked by the ϕ_2 system clock.

After n ϕ_2 clocks (where n is the number of flip-flops in the shift register) a logic 0 shifts out from the nth F/F, resetting the main flip-flop. The main F/F then presets the shift register and clears the WAIT signal, ready for the next $\overline{CS}$ edge to repeat the cycle. On power-up the delay generator will initialize itself after a maximum of n system clocks have occurred so no reset signal is required. Some μPs demand that a $\overline{WAIT}$/READY input is synchronized with ϕ_2 of the system clock. This can readily be achieved by selecting the correct ϕ_2 edge as the clock signal for the shift register chain.

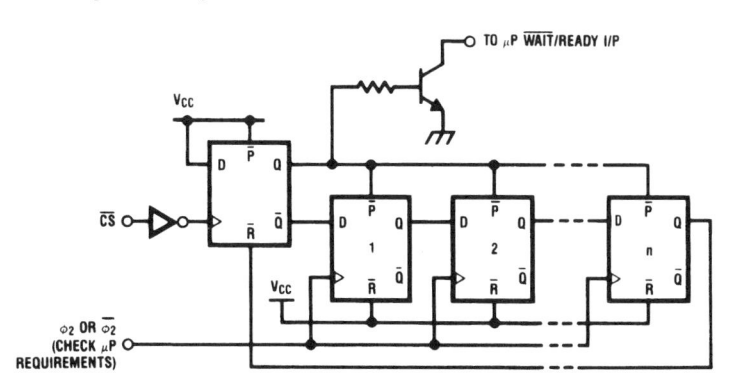

Flip-Flop—MM74HC74 D-Type Latch

FIGURE 23. Access Delay Generator (Clocked Wait State Generator)

APPENDIX A-1. READING VALID REAL-TIME DATA

TIME READING USING DCF

Using the Data-Changed Flag (DCF) technique supports microprocessors with block move facilities, as all the necessary time data may be read sequentially and then tested for validity as shown below.

1. Read the control register, address 0: *This is a dummy read to reset the data-changed flag (DCF) prior to reading the time registers.*
2. Read time registers: *All desired time registers are read out in a block.*
3. Read the control register and test DCF: *If DCF is still clear (logic 0), then no clock setting pulses have occurred since step 1. All time data is guaranteed good and time reading is complete.*

If DCF is set (logic 1), then a time change has occurred since step 1 and time data may not be consistent. Repeat steps 2 and 3 until DCF is clear. The control read of step 3 will have reset DCF, automatically repeating the step 1 action.

TIME READING USING AN INTERRUPT

In systems such as point-of-sale terminals and data loggers, time reading is usually only required on a random demand basis. Using the data-changed flag as outlined above is ideal for this type of system. Where the µP must respond to any change in real-time (e.g., industrial timers/process controllers, TV/VCR clocks or any system where real-time is displayed) then the interrupt timer may be for time reading. Software is used to synchronize the interrupt timer with the time changing as outlined below:

1. Select the interrupt register (write 2 or 3 to ADDR0).
2. Program for repeated interrupts of the desired time interval (see Table IIb in Appendix A-2): *Do not start the timer yet.*
3. Read control register AD0: *This is a dummy read to reset the data-changed flag.*
4. Read control register AD0 repeatedly until data-changed flag is set.
5. Write 0 or 2 to control register. Interrupt timing commences.

When interrupt occurs, read out all required time data. There is no need to test DCF as the interrupt "pre-synchronizes" the time reading already. The interrupt flag is automatically reset by reading from ADDR0 to test it. In repeat interrupt mode, the timer continues to run with no further µP intervention necessary.

TIME READING WITH VERY SLOW READ CYCLES

If a system takes longer than 100 ms to complete reading of all the necessary time registers (e.g., when CMOS processors are used or where high level interpreted language routines are used) then the data-changed flag will always be set when tested and is of no value. In this case, the time registers themselves must be tested to ensure data accuracy.

The technique below will detect both time changing *between* read strobes (i.e., between reading tens of minutes and units of hours) and also time changing *during* read, which can produce invalid data.

1. Read and store the value of the *lowest* order time register required.
2. Read out all the time registers required. The registers may be read out in any order, simplifying software requirements.
3. Re-read the lowest order register and compare it with the value stored previously in step 1. If it is still the same, then all time data is good. If it has changed, then store the new value and go back to step 2.

In general, the rule is that the first and last reads *must* both be of the lowest order time register. These two values can then be compared to ensure that no change has occurred. This technique works because for any higher order time register to change, all the lower order registers must also change. If the lowest order register does not change, then no other register has changed either.

APPENDIX A-2. FUNCTIONAL TRUTH TABLES FOR MM58274C

TABLE I. Address Decoding for Internal Registers

Register Selected		AD3	AD2	AD1	AD0	Access
0	Control Register	0	0	0	0	Split Read and Write
1	Tenths of Secs	0	0	0	1	Read Only
2	Units Seconds	0	0	1	0	R/W
3	Tens Seconds	0	0	1	1	R/W
4	Units Minutes	0	1	0	0	R/W
5	Tens Minutes	0	1	0	1	R/W
6	Units Hours	0	1	1	0	R/W
7	Tens Hours	0	1	1	1	R/W
8	Units Days	1	0	0	0	R/W
9	Tens Days	1	0	0	1	R/W
10	Units Months	1	0	1	0	R/W
11	Tens Months	1	0	1	1	R/W
12	Units Years	1	1	0	0	R/W
13	Tens Years	1	1	0	1	R/W
14	Day of Week	1	1	1	0	R/W
15	Clock Setting/Interrupt Registers	1	1	1	1	R/W

TABLE IIa. Clock Setting Register Layout

Function	Data Bits Used				Comments	Access
	DB3	DB2	DB1	DB0		
Leap Year Counter	X	X			0 Indicates a Leap Year	R/W
AM/PM Indicator (12 Hour Mode)			X		0 = AM 1 = PM 0 in 24 Hour Mode	R/W
12–24 Hour Select Bit				X	0 = 12 Hour Mode 1 = 24 Hour Mode	R/W

TABLE IIb. Interrupt Control Register

Function	Comments	Control Word			
		DB3	DB2	DB1	DB0
No Interrupt	Interrupt Output Cleared, Start/Stop Bit Set to 1.	X	0	0	0
0.1 Second		0/1	0	0	1
0.5 Second		0/1	0	1	0
1 Second		0/1	0	1	1
5 Seconds		0/1	1	0	0
10 Seconds		0/1	1	0	1
30 Seconds		0/1	1	1	0
60 Seconds		0/1	1	1	1

Timing Accuracy:
Single Interrupt Mode (all time delays): ±1 ms
Repeated Mode: ±1 ms on initial timeout, thereafter synchronous with first interrupt (i.e., timing errors do not accumulate).

DB3 = 0 for Single Interrupt   DB3 = 1 for Repeated Interrupt

TABLE III. The Control Register Layout

Access (ADDR0)	DB3	DB2	DB1	DB0
Reaf From:	Data Changed Flag	0	0	Interrupt Flag
Write To:	Test 0 = Normal 1 = Test Mode	Clock Start/Stop 0 = Clock Run 1 = Clock Stop	Interrupt Select 0 = Clk. Set Reg. 1 = Int. Reg.	Interrupt Start/Stop 0 = Int. Run 1 = Int. Stop

DP8473 Floppy Disk Controller PLUS-2™

PRELIMINARY
July 1988

General Description

This device is a derivative of the DP8472/4 Floppy Disk Controller which incorporates additional logic specifically required for an IBM® PC, PC-XT®, PC-AT®, or PS/2® design. This controller is a full featured floppy disk controller that is software compatible with the µPD765A, but also includes many additional hardware and software enhancements.

This controller incorporates a precision analog data separator, that includes a self trimming delay line and VCO. Up to three external filters are switched automatically depending on the data rate selected. This provides optimal performance at the standard PC data rates of 250/300 kb/s, and 500 kb/s. It also enables optimum performance at 1 Mb/s (MFM). These features combine to provide the lowest possible PLL bandwidth, with the greatest lock range, and hence the widest window margin.

This controller includes write precompensation circuitry. A shift register is used to provide a fixed 125 ns early-late precompensation for all tracks at 500k/300k/250 kb/s (83 ns for 1 MB/s), or a precompensation value that scales with
(Continued)

Features

- Fully µPD765A and IBM-BIOS compatible
- Integrates all PCXT®, PCAT®, and most PS/2® Logic
  — On chip 24 MHz Crystal Oscillator
  — DMA enable logic
  — IBM compatible address decode of A0–A2
  — 12 mA µP bus interface buffers
  — 40 mA floppy drive interface buffers
  — Data rate and drive control registers
- Precision analog data separator
  — Self-calibrating PLL and delay line
  — Automatically chooses one of three filters
  — Intelligent read algorithm
- Two pin programmable precompensation modes
- DP8472/4 core with its enhancements
  — up to 1 Mb/s data rate
  — Implied seek up to 4000 tracks
  — IBM or ISO formatting
- Low power CMOS, with power down mode

Connection Diagrams

Plastic Leaded Chip Carrier

Pin assignments (Top View):
- 8 – MTR0
- 9 – HD SEL
- 10 – TRK0
- 11 – INDEX
- 12 – WRT PRT
- 13 – V_CCA
- 14 – V_CC
- 15 – RESET
- 16 – WR
- 17 – RD
- 18 – CS
- 19 – A0
- 20 – A1
- 21 – A2
- 22 – D0
- 23 – D1
- 24 – D2
- 25 – D3
- 26 – D4
- 27 – GNDB
- 28 – D5
- 29 – D6
- 30 – D7
- 31 – DRQ
- 32 – DAK
- 33 – TC
- 34 – INT
- 35 – DSKCHG/RG
- 36 – GNDC
- 37 – OSC2/CLOCK
- 38 – OSC1
- 39 – GNDA
- 40 – FILTER
- 41 – FGND500
- 42 – FGND250
- 43 – DR3
- 44 – RDATA
- 45 – DR2
- 46 – PUMP/PREN
- 47 – DRVTYP
- 48 – SETCUR
- 49 – WGATE
- 50 – STEP
- 51 – RPM/LC
- 52 – MTR2
- 1 – MTR3
- 2 – GND0
- 3 – WDATA
- 4 – DIR
- 5 – DR1
- 6 – DR0
- 7 – MTR1

TL/F/9384-1

Top View

Order Number DP8473V
See NS Package Number V52A

Dual-In-Line Package

- WDATA – 1 | 48 – GNDD
- DIR – 2 | 47 – RPM/LC
- DR1 – 3 | 46 – STEP
- DR0 – 4 | 45 – WGATE
- MTR1 – 5 | 44 – SETCUR
- MTR0 – 6 | 43 – DRVTYP
- HD SEL – 7 | 42 – PUMP/PREN
- TRK0 – 8 | 41 – RDATA
- INDEX – 9 | 40 – FGND250
- WRT PRT – 10 | 39 – FGND500
- V_CCA – 11 | 38 – FILTER
- V_CC – 12 | 37 – GNDA
- RESET – 13 | 36 – OSC1
- WR – 14 | 35 – OSC2/CLOCK
- RD – 15 | 34 – GNDC
- CS – 16 | 33 – DSKCHG/RG
- A0 – 17 | 32 – INT
- A1 – 18 | 31 – TC
- A2 – 19 | 30 – DAK
- D0 – 20 | 29 – DRQ
- D1 – 21 | 28 – D7
- D2 – 22 | 27 – D6
- D3 – 23 | 26 – D5
- D4 – 24 | 25 – GNDB

TL/F/9384-2

Top View

Order Number DP8473N
See NS Package Number N48A

TRI-STATE® is a registered trademark of National Semiconductor Corporation.
IBM®, PCXT®, PCAT® are registered trademarks of International Business Machines Corporation.

©1988 National Semiconductor Corporation   TL/F/9384   DS08288S/10M78   Printed in W. Germany

General Description (Continued)

the data rate, 83 ns/125 ns/208 ns/ 250 ns for data rates of 1.0M/500k/300k/250 kb/s respectively.

Specifically to support the PC-AT and PC-XT design, the Floppy Disk Controller PLUS-2 includes address decode for the A0–A2 address lines, the motor/drive select register, data rate register for selecting 250/300/500 kb/s, Disk Changed status, dual speed spindle motor control, low write current and DMA/interrupt sharing logic. The controller also supports direct connection to the µP bus via internal 12 mA buffers. The controller also can be connected directly to the disk drive via internal open drain high drive outputs, and Schmitt inputs.

In addition to this logic the DP8473 includes many features to ease design of higher performance drives and future controller upgrades. These include 1.0 Mb/s data rate, extended track range to 4096, Implied seeking, working Scan Commands, motor control timing, both standard IBM formats as well as Sony 3.5″ (ISO) formats, and other enhancements.

This device is available in a 52 pin Plastic Chip Carrier, and in a 48 pin Dual-In-Line package.

Table of Contents

General Description
Pin Description
Functional Description
Register Description
Result Phase Register Description
Processor Software Interface
Command Description Table
Command Descriptions
DC and AC Characteristics

Block Diagram

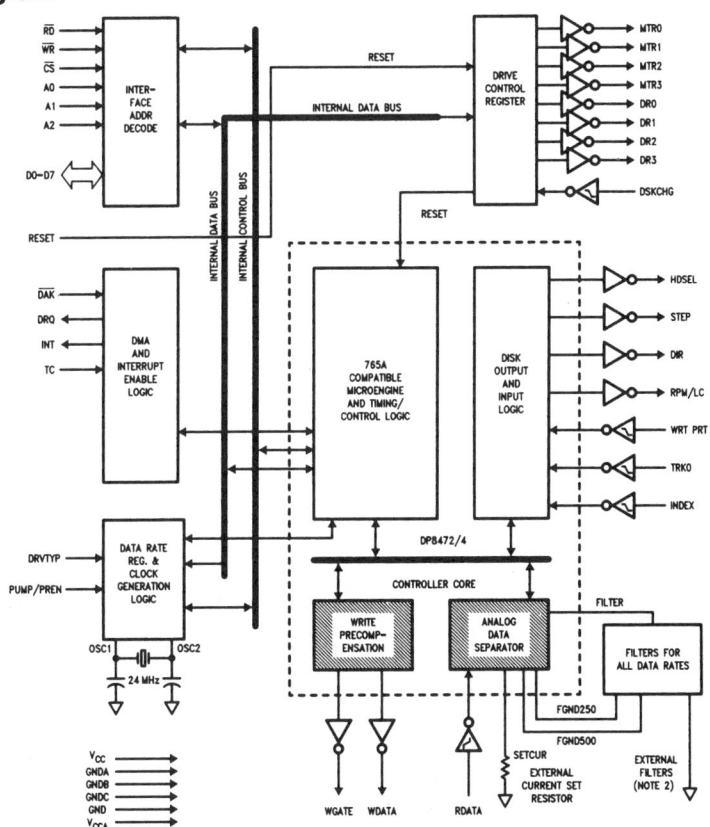

Note 1: The MTR2, MTR3, DR2, and DR3 are not available on the 48 pin DIP (DP8473N, J) versions.
Note 2: See *Figure 4* for filter description.
Note 3: Total transistor count is 29,700 (approx).

FIGURE 1. DP8473 Functional Block Diagram

Pin Descriptions

Symbol	DP8473 PCC	DP8473 DIP	Function
MTR2	1	—	This is an active low motor enable line for drive 2, which is controlled by the Drive Control register. This is a high drive open drain output.
GNDD	2	48	This pin is the digital ground for the disk interface output drivers.
WDATA	3	1	This is the active low open drain write precompensated serial data to be written onto the selected disk drive. This is a high drive open drain output.
DIR	4	2	This output determines the direction of the head movement (low = step in, high = step out). When in the write or read modes, this output will be high. This is a high drive open drain output.
DR1	5	3	This is an active low drive select line for drive 1 that is controlled by the Drive Control register bits D0, D1. The Drive Select bit is ANDed with the Motor Enable of the same number. This is a high drive open drain output.
DR0	6	4	This is an active low drive select similar to DR1 line except for drive 0.
MTR1	7	5	This is an active low motor enable line for drive 1. Similar to MTR2.
MTR0	8	6	This is an active low motor enable line for drive 0. Similar to MTR2.
HD SEL	9	7	This output determines which disk drive head is active. Low = Head 1, Open (high) = Head 0. This is a high drive open drain output.
TRK0	10	8	This active low Schmitt input tells the controller that the head is at track zero of the selected disk drive.
INDEX	11	9	This active low Schmitt input signals the beginning of a track.
WRT PRT	12	10	This active low Schmitt input indicates that the disk is write protected. Any command that writes to that disk drive is inhibited when a disk is write protected.
V_{CCA}	13	11	This pin is the 5V supply for the analog data separator circuitry.
V_{CC}	14	12	This pin is the 5V supply for the digital circuitry.
RESET	15	13	Active high input that resets the controller to the idle state, and resets all the output lines to the disk drive to their disabled state. The Drive Control register is reset to 00. The Data Rate register is set to 250 kb/s. The Specify command registers are not affected. The Mode Command registers are set to the default values. Reset should be held active during power up. To prevent glitches activating the rest sequence, a small capacitor (1000 pF) should be attached to this pin.
$\overline{WR}$	16	14	Active low input to signal a write from the microprocessor to the controller.
$\overline{RD}$	17	15	Active low input to signal a read from the controller to the microprocessor.
$\overline{CS}$	18	16	Active low input to enable the $\overline{RD}$ and $\overline{WR}$ inputs. Not required during DMA transfers. This should be held high during DMA transfers.
A0, A1, A2	19–21	17–19	Address lines from the microprocessor. This determines which registers the microprocessor is accessing as shown in Table IV in the Register Description Section. Don't care during DMA transfers.
D0–D4	22–26	20–24	Bi-directional data lines to the microprocessor. These are the lower 5 bits and have buffered 12 mA outputs.
GNDB	27	25	This pin is the digital ground for the 12 mA microprocessor interface buffers. This includes D0–D7, INT, and DRQ.
D5–D7	28–30	26–28	Bi-directional data lines to the microprocessor. These upper 3 bits have buffered 12 mA outputs.
DRQ	31	29	Active high output to signal the DMA controller that a data transfer is needed. This signal is enabled when D3 of the Drive Control Register is set.
$\overline{DAK}$	32	30	Active low input to acknowledge the DMA request and enable the $\overline{RD}$ and $\overline{WR}$ inputs. This signal is enabled when D3 of the Drive Control Register is set.
TC	33	31	Active high input to indicate the termination of a DMA transfer. This signal is enabled when the DMA Acknowledge pin is active.
INT	34	32	Active high output to signal that an operation requires the attention of the microprocessor. The action required depends on the current function of the controller. This signal is enabled when D3 of the Drive Control Register is set.

Pin Descriptions (Continued)

Symbol	DP8473 PCC	DP8473 DIP	Function
DSKCHG/RG	35	33	This latched Schmitt input signal is inverted and routed to D7 of the data bus and is read when address xx7H is enabled. When the RG bit in the Mode Command is set, this pin functions as a Read Gate signal that when low forces the data separator to lock to the crystal, and when high it locks to data for diagnostic purposes.
GNDC	36	34	This pin is the digital ground for the controller's digital logic, including all internal registers, micro-engine, etc.
OSC2/CLOCK	37	35	One side of the external 24 MHz crystal is attached here. If a crystal is not used, a TTL or CMOS compatible clock is connected to this pin.
OSC1	38	36	One side of an external 24 MHz crystal is attached here. This pin is tied low if an external clock is used.
GNDA	39	37	This pin is the analog ground for the data separator, including all the PLLs, and delay lines.
FILTER	40	38	This pin is the output of the charge pump and the input to the VCO. One or more filters are attached between this pin and the GNDA, FGND250 and FGND500 pins.
FGND500	41	39	This pin connects the PLL filter for 500k(MFM)/250k(FM) b/s to ground. This is a low impedance open drain output.
FGND250	42	40	This pin connects the PLL filter for 250k(MFM)/125k(FM) b/s or 300k(MFM)/150k(FM) b/s to ground. This is a low impedance open drain output.
DR3	43	—	This is the same as DR0 except for drive 3.
RDATA	44	41	The active low raw data read from the disk is connected here. This is a Schmitt input.
DR2	45	—	This is the same as DR0 except for drive 2.
PUMP/PREN	46	42	When the PU bit is set in Mode Command this pin is an output that indicates when the charge pump is making a correction. Otherwise this pin is an input that sets the precomp mode as shown in Table VI. If pin is configured as PUMP, PREN is assumed high.
DRV TYP	47	43	This is an input used by the controller to enable the 300 kb/s mode. This enables the use of floppy drives with either dual or single speed spindle motors. For dual speed spindle motors, this pin is tied low. When low, and 300 kb/s data rate is selected in the data rate register, the PLL actually uses 250 kb/s. This pin is tied high for single speed spindle motor drives (standard AT drive). When this pin is high and 300 kb/s is selected 300 kb/s is used. (See also RPM/LC pin).
SETCUR	48	44	An external resistor connected from this pin to analog ground programs the amount of charge pump current that drives the external filters. The PLL Filter Design section shows how to determine the values.
WGATE	49	45	This active low open drain high drive output enables the write circuitry of the selected disk drive. This output has been designed to prevent glitches during power up and power down. This prevents writing to the disk when power is cycled.
STEP	50	46	This active low open drain high drive output will produce a pulse at a software programmable rate to move the head during a seek operation.
RPM/LC	51	47	This high drive open drain output pin has two functions based on the selection of the DRVTYP pin. 1. When using a dual speed spindle motor floppy drive (DRVTYP pin low), this output is used to select the spindle motor speed, either 300 RPM or 360 RPM. In this mode this output goes low when 250/300 kb/data rate is chosen in the data rate register, and high when 500 kb/s is chosen. 2. When using a single speed spindle motor floppy drive (DRVTYP pin high), this pin indicates when to reduce the write current to the drive. This output is high for high density media (when 500 kb/s is chosen).
MTR3	52	—	This is an active low motor enable line for drive 3.

Typical Application

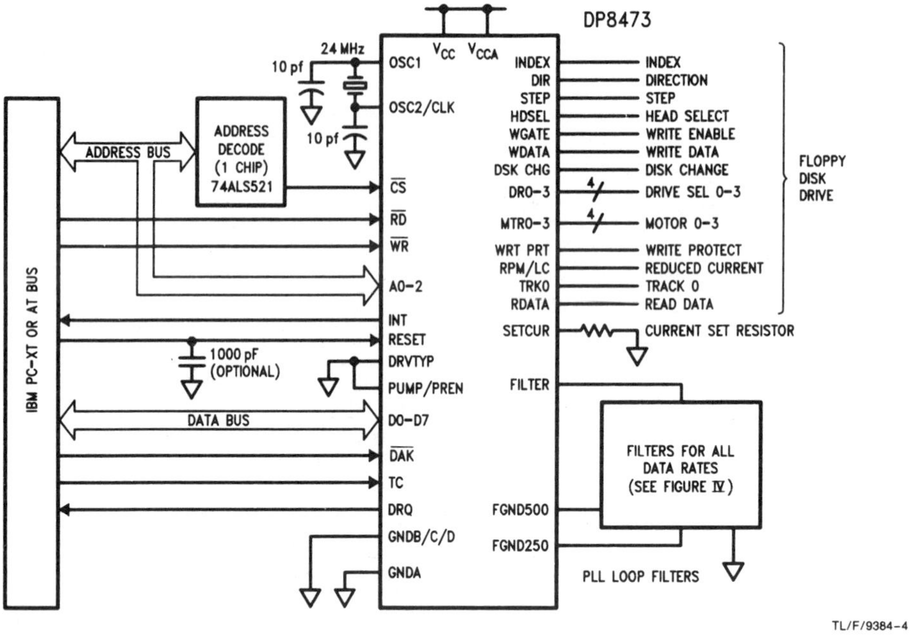

FIGURE 2. DP8473 Typical Application

Recommended Plastic Chip Carrier Socket:
AMP P/N 821551-1 or equivalent.

Functional Description

This section describes the basic architectural features of the DP8473, and many of the enhancements provided. Refer to *Figure 1*.

765A COMPATIBLE MICRO-ENGINE

The core of the DP8473 is the same µPD765A compatible microcoded engine that is used in the DP8472/4. This engine consists of a sequencer, program ROM, and disk/misc registers. This core is clocked by either a 4 MHz, 4.8 MHz or 8 MHz clock selected in the Data Rate Register. Upon this core is added all the glue logic used to implement a PC-XT or AT, or PS/2 floppy controller, as well as the data separator and write precompensation logic.

The controller consists of a microcoded engine that controls the entire operation of the chip including coordination of data transfer with the CPU, controlling the drive controls, and actually performing the algorithms associated with reading and writing data to/from the disk. This includes the read algorithm for the data separator.

Like the µPD765A, this controller takes commands and returns data and status through the Data Register in a byte serial fashion. Handshake for command/status I/O is provided via the Main Status Register. All of the µPD765A commands are supported, as are many of the DP8472 superset commands.

DATA SEPARATOR

The internal data separator consists of an analog PLL and its associated circuitry. The PLL synchronizes the raw data signal read from the disk drive. The synchronized signal is used to separate the encoded clock and data pulses. The data pulses are de-serialized into bytes and then sent to the µP by the controller.

The main PLL consists of four main components, a phase comparator, a filter, a voltage controlled oscillator (VCO), and a programmable divider. The phase comparator detects the difference between the phase of the divider's output and the phase of the raw data being read from the disk. This phase difference is converted to a current which either charges or discharges one of the three external filters. The resulting voltage on the filter changes the frequency of the VCO and the divider output to reduce the phase difference between the input data and the divider's output. The PLL is "locked" when the frequency of the divider is exactly the same as the average frequency of the data read from the disk. A block diagram of the data separator is shown in *Figure 3*.

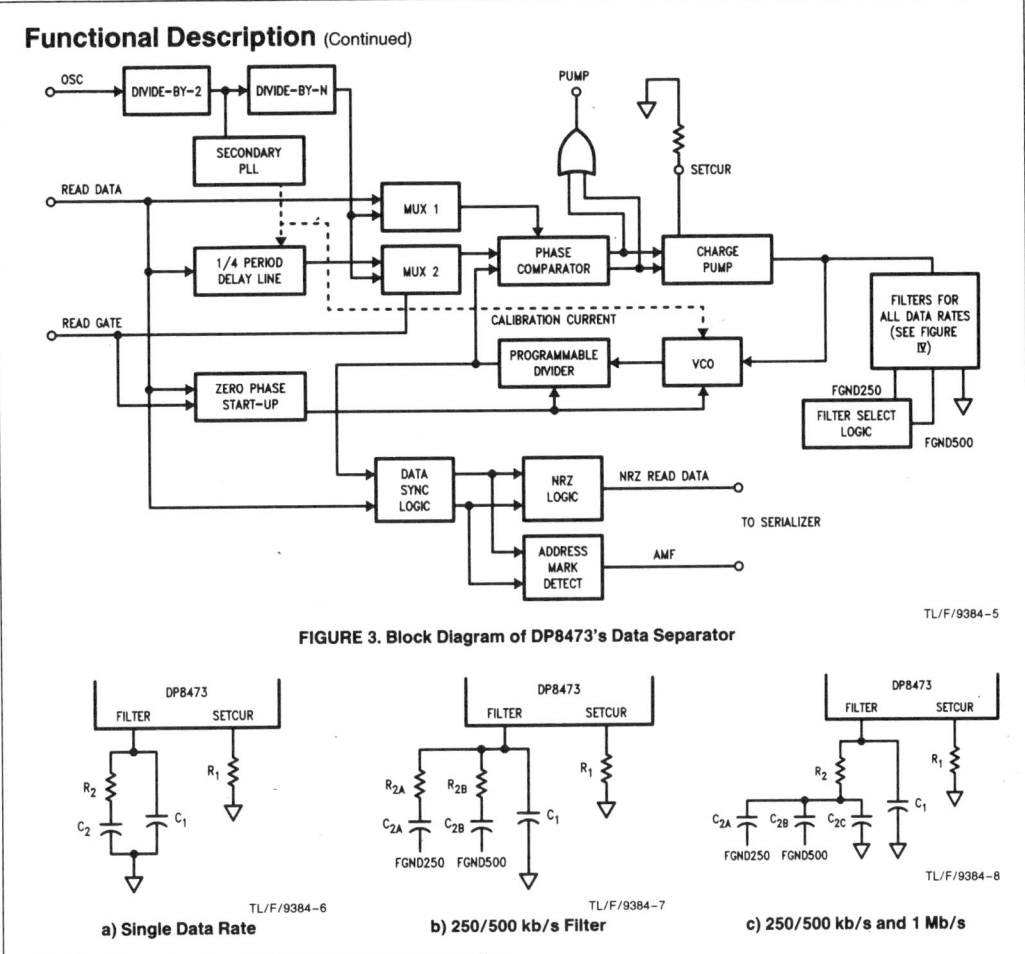

Functional Description (Continued)

FIGURE 3. Block Diagram of DP8473's Data Separator

a) Single Data Rate

b) 250/500 kb/s Filter

c) 250/500 kb/s and 1 Mb/s

Note: For all filter configurations, 250kb/s and 300 kb/s share the same filter.

FIGURE 4. Typical Configuration for Loop Filters for the DP8473 Showing Component Labels

To ensure optimal performance, the data separator incorporates several additional circuits. The quarter period delay line is used to determine the center of each bit cell. A secondary PLL is used to automatically calibrate the quarter period delay line. The secondary PLL also calibrates the center frequency of the VCO.

To eliminate the logic associated with controlling multiple data rates the DP8473 supports the connection of three filters to the chip via the FGND250, and FGND500 pins (filter ground switches). The controller chooses which filter components to use based on the value loaded in the Data Rate Register. If 500 k(MFM) is being used then the FGND500 is enabled (FGND250 is disabled). If 250 k(MFM) or 300 k(MFM) is being used the FGND250 pin is enabled, and FGND500 is disabled. For 1 Mb/s (MFM) both FGND pins are disabled.

Figure 4 shows several possible filter configurations. For a filter to cover all data rates (*Figure 4c*), the DP8473 has a 1 Mb/s filter always connected and other capacitor filter components for the other data rates are switched in parallel to this filter. The actual loop filter for 500 kb/s is the parallel combination of the two capacitors, C_{2C} and C_{2B}, attached to the FGND500 pin and to ground. The 250/300 kb/s filter is the parallel combination of the capacitors, C_{2C} and C_{2A}, attached to the FGND250, and ground. If 1 Mb/s need not be supported then the filter configuration of *Figure 4b* can be used. This configuration allows more optimal performance for both 500k and 250/300 kb/s. *Figure 4a* is a simple filter configuration primarily for a single data rate (or multiple data rates with a performance compromise). Table II shows some typical filter values. Other filter configurations and values are possible, these result in good general performance.

While the controller and data separator support both FM and MFM encoding, the filter switch circuitry only supports

Functional Description (Continued)

the IBM standard MFM data rates. To provide both FM and MFM filters external logic may be necessary.

The controller takes best advantage of the internal data separator by implementing a sophisticated ID search algorithm. This algorithm, shown in *Figure 5*, enhances the PLL's lock characteristics by forcing the PLL to relock to the crystal any time the data separator attempts to lock to a non-preamble pattern. This algorithm ensures that the PLL is not thrown way out of lock by write splices or bad data fields.

TABLE II. Typical Filter Values for the Various Data Rates (Assuming ±6% Capture Range)

Data Rate (MFM b/s)	C_2	R_2	C_1	R_1
Filter Values when Using All 3 Data Rates				
1.0M	C_{2C} = 0.012 µF	560Ω	510 pF	5.6 kΩ
500k	C_{2B} = 0.015 µF			
250/300k	C_{2A} = 0.033 µF			
Filter Values when Using 250/300 and 500 kb/s				
500k	C_{2B} = 0.027 µF	560Ω	1000 pF	5.6 kΩ
250/300k	C_{2A} = 0.047 µF	560Ω		
Filter Using Only One Data Rate				
1.0M	C_2 = 0.012 µF	560Ω	510 pF	5.6 kΩ
500k	C_2 = 0.027 µF	560Ω	1000 pF	5.6 kΩ
300/250k	C_2 = 0.047 µF	560Ω	2000 pF	5.6 kΩ

(These values are preliminary and thus are subject to change.)

TABLE III. Data Rates (MFM) versus VCO Divide-By Factor

Data Rate	N
1 Mb/s	4
500 kb/s	8
300 kb/s	16
250 kb/s	16

PLL DIAGNOSTIC MODES

In addition, the DP8473 has two diagnostic modes to enable filter optimization, 1) enabling the Charge Pump output signal onto the PUMP/PREN pin, and 2) providing external control of the Read Gate signal to the data separator. Both modes are enabled in the last byte of the Mode Command.

The Pump output signal indicates when the charge pump is making a phase correction, and hence whether the loop is locked or not.

The Read Gate function, when enabled, allows the designer to manually force the data separator to lock to the incoming data or back to the reference clock. This enables easy verification of the lock characteristics of the PLL, by monitoring the FILTER pin, and the Pump signal.

PLL FILTER DESIGN

This section provides information to enable design of the data separator's external filter and charge pump set resistor. This discussion is for a single data rate filter, and can be easily extrapolated to the other filters of *Figure 4*. Table II shows some typical filter component values, but if a custom filter is desired, the following parameters must be considered:

R_1: Charge pump current setting resistor. The current set by this resistor is multiplied by the charge pump gain, K_P which is ~2.5. Thus the charge pump current is:

I_{PUMP} = (2.5) 1.2V/R_1. R_1 should be set to between 3–12 kΩ. This resistor determines the gain of the phase detector, which is $K_D = I_{PUMP}/2\pi$.

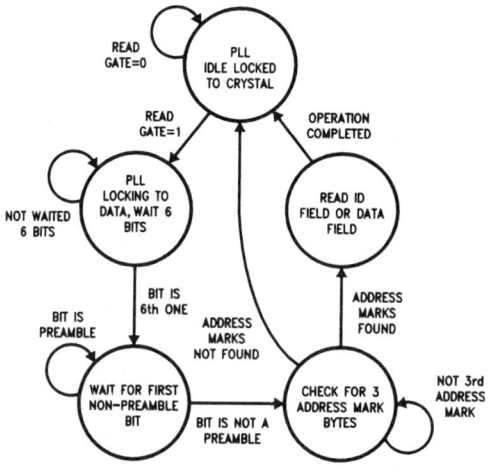

FIGURE 5. Read Algorithm-State Diagram for Data

Functional Description (Continued)

C_2: Filter capacitor in series with R_2. With pump current this determines loop bandwidth.

R_2: Filter resistor. Determines the PLL damping factor.

C_1: This filter capacitor improves the performance of the PLL by providing additional filtering of bit jitter and noise.

K_{VCO}: The ratio of the change in the frequency of the VCO output due to a voltage change at the VCO input. $K_{VCO} \approx 25$ Mrad/s/V. The VCO is followed by a divider to achieve the desired frequency for each data rate. VCO center frequency is 4 MHz for data rates of 1 Mb/s, 500 kb/s, and 250 kb/s (MFM), and is 4.8 MHz for 300 kb/s (MFM).

K_{PLL}: This is the gain of the internal PLL circuitry, and is the product of $V_{REF} \times K_{VCO} \times K_P$. This value is specified in the Phase Locked Loop Characteristics table.

ω_n: This is the bandwidth of the PLL, and is given by,

$$\omega_n = \sqrt{\frac{K_{PLL}}{2\pi C_2 N R_1}}$$

where N is the number of VCO cycles between two phase comparisons. The value of N for the various data rates are shown in Table III.

ζ: The damping factor is set to 0.7 to 1.2 and is given by,

$$\zeta = \frac{\omega_n R_2 C_2}{2}$$

The trade off, when choosing filter components is between acquisition time while the PLL is locking and jitter immunity while reading data. To select the proper components for a standard floppy disk application the following procedure can be used:

1. Choose FM or MFM, and data rate. Determine N from Table III. Determine preamble length (MFM = 12). The PLL should lock within ½ the preamble time.

2. Determine loop bandwidth (ω_n) required, and set the charge pump resistor R_1.

3. Calculate C_2 using:

$$C_2 = \frac{K_{PLL}}{2\pi R_1 N \omega_n^2}$$

4. Choose R_2 using:

$$R_2 = \frac{2\zeta}{\omega_n C_2}$$

6. Select C_1 to be about 1/20th of C_2.

The above procedure will yield adequate loop performance. If optimum loop performance is required, or if the nature of the loop performance is very critical, then some additional consideration must be given to choosing ω_n and the damping factor. (For a detailed description on how to choose ω_n and ζ, see: **AN-505 Floppy Disk Data Separator Design Guide for the DP8472, DP8473, and DP8474**).

WRITE PRECOMPENSATION

The DP8473 incorporates a single fixed 3-bit shift register. This shift register outputs are tapped and multiplexed onto the write data output. The taps are selected by a standard precompensation algorithm. This precompensation value can be selected from the PUMP/PREN pin. When this pin is low 125 ns precomp is used for all data rates except 1 Mb/s which uses 83 ns. When PREN is tied high, the precompensation-value scales with data rate at 250 kb/s its 250 ns, for 300 kb/s its 208 ns, at 500 kb/s its 125 ns, and at 1.0 Mb/s its 83 ns. These values are shown in Table VI.

PC-AT AND PC-XT LOGIC BLOCKS

This section describes the major functional blocks of the PC logic that have been integrated on the controller. Refer back to *Figure 1*, the block diagram.

DMA Enable Logic: This is gating logic that disables the DMA lines and the Interrupt output, under the control of the DMA Enable bit in the Drive control register. When the DMA Enable bit is 0 then the INT, and DRQ are held TRI-STATE, and $\overline{DAK}$ is disabled.

Drive Output Buffers/Input Receivers: The drive interface output pins can drive 150Ω ±10% termination resistors. This enables connection to a standard floppy drive. All drive interface inputs are TTL compatible schmitt trigger inputs with typically 250 mV of hysteresis. *The only functional differences between the 52 pin PLCC and the 48 pin DIP version are that the MTR2 and 3, and DR2 and 3 pins have been removed in order to accommodate the 48 pin package.*

Bus Interface-Address Decode: The address decode circuit allows software access to the controller, Drive Control Register, and Data Rate Register (see Table IV for the memory map) using the same address map as is used in the XT, AT, or PS/2. The decoding is provided for A0–A2, so only a single address decoder connected to the chip select is needed to complete the decode. The bus interface logic includes the 8-bit data bus and DRQ/INT signals. The output drive for these pins is 12 mA.

TABLE IV. Address Memory Map for DP8473

A2	A1	A0	R/W	Register
0	0	0	X	None (Bus TRI-STATE)
0	0	1	X	None (Bus TRI-STATE)
0	1	0	W	Drive Control Register
0	1	1	X	None (Bus TRI-STATE)
1	0	0	R	Main Status Register
1	0	1	R/W	Data Register
1	1	0	X	None (Bus TRI-STATE)
1	1	1	W	Data Rate Register
1	1	1	R	Disk Changed Bit*

*When this location is accessed only bit D7 is driving, all others are held TRI-STATE.

Drive Control Register: This 8-bit write only register controls the drive selects, motor enables, DMA enable, and Reset. See Register Description.

Reset Logic: The reset input pin is active high, and directly feeds the Drive Control Register and the Data Rate Register. After a hardware reset the Drive Control Register is reset to all zeros, and the Data Rate Register is set to 250 kb/s data rate. The controller is held reset until the software sets the Drive Control Register reset bit, after which the controller may be initialized. A software reset to the controller core can be issued by resetting then setting this bit. A software reset does not reset the Drive Control Register, or the Data Rate Register.

Functional Description (Continued)

Data Rate Register and Clock Logic: This is a two bit register that controls the data rate that the controller uses. See Register Description. This register feeds logic that selects the data rates by programming a prescaler that divides the crystal or clock input by either 3, 5, or 6. This causes either 4 MHz, 4.8 MHz and 8 MHz to be input as the master clock for the controller core. If the Drive Type pin is high and a 300 kb/s data rate is chosen, 4.8 MHz is used to generate 300 kb/s, but when the DRVTYP pin is low and 300 kb/s is selected, 4 MHz is used, and the actual data rate is 250 kb/s. See Table VI.

Low Power Mode Logic: This logic is an enhancement over the standard XT, AT, PS/2 design. In the Low Power Mode the crystal oscillator, controller and all linear circuitry are turned off. When the oscillator is turned off the controller will typically draw about 100 μA. The internal circuitry is disabled while the oscillator is off because the internal circuitry is driven from this clock. The oscillator will turn back on automatically after it detects a read or a write to the Main Status or Data Registers. It may take a few milli-seconds for the oscillator to stabilize and the μP will be prevented from trying to access the Data Register during this time through the normal Main Status Register protocol. (The Request for Master bit in the Main Status Register will be inactive.) There are two ways to go into the low power mode. One is to command the controller to switch to low power immediately. The other method is to set the controller to automatically go into the low power mode 500 ms after the beginning of the idle state (based on a 500 kb/s (MFM) data rate). This would be invisible to the software. The low power mode is programmed through the Mode Command.

The Data Rate Register and the Drive Control Register are unaffected by the power down mode. They will remain active. It is up to the user to ensure that the Motor and Drive select signal are turned off.

TABLE V. Truth Table for Drive Control Register

D7	D6	D5	D4	D1	D0	Function
X	X	X	1	0	0	Drive 0 Selected (DR0 = 0)
X	X	1	X	0	1	Drive 1 Selected (DR1 = 0)
X	1	X	X	1	0	Drive 2 Selected (DR2 = 0)
1	X	X	X	1	1	Drive 3 Selected (DR3 = 0)

Crystal Oscillator: The DP8473 is clocked by a single 24 MHz signal. An on-chip oscillator is provided, to enable the attachment of a crystal, or a clock. If a crystal is used, a 24 MHz fundamental mode, parallel resonant crystal should be used. This crystal should be specified to have less than 150Ω series resistance, and shunt capacitance of less than 7 pF. Typically a series resonant crystal can be used, it will just oscillate in parallel mode 30–300 ppm from its ideal frequency.

If an external oscillator circuit is used, it must have a duty cycle of at least 40–60%, and minimum input levels of 2.4V and 0.4V. The controller should be configured so that the clock is input into the OSC2 pin, and OSC1 is tied to ground.

Crystals: Staytek: CX1-SM1-24 MHz(B)
SaRonix: SRX 3164

Register Description

This section describes the register bits for all the registers that are directly accessible to the μP. Table IV (previous page) shows the memory map for these registers. Note that in the PC some of the registers are partially decoded, this is not the case here. All registers occupy only their documented addresses.

MAIN STATUS REGISTER (Read Only)

The read only Main Status Register indicates the current status of the disk controller. The Main Status Register is always available to be read. One of its functions is to control the flow of data to and from the Data Register. The Main Status Register indicates when the disk contoller is ready to send or receive data. It should be read before each byte is transferred to or from the Data Register except during a DMA transfer. No delay is required when reading this register after a data transfer.

D7 Request for Master: Indicates that the Data Register is ready to send or receive data from the μP. This bit is cleared immediately after a byte transfer and will become set again as soon as the disk controller is ready for the next byte.

D6 Data Direction: Indicates whether the controller is expecting a byte to be written to (0) or read from (1) the Data Register.

D5 Non-DMA Execution: Bit is set only during the Execution Phase of a command if it is in the non-DMA mode. In other words, if this bit is set, the multiple byte data transfer (in the Execution Phase) must be monitored by the μP either through interrupts, or software polling as described in the Processor Software Interface section.

D4 Command in Progress: Bit is set after the first byte of the Command Phase is written. Bit is cleared after the last byte of the Result Phase is read. If there is no result phase in a command, the bit is cleared after the last byte of the Command Phase is written.

D3 Drive 3 Seeking: Set after the last byte of the Command Phase of a Seek or Recalibrate command is issued for drive 3. Cleared after reading the first byte in the Result Phase of the Sense Interrupt Command for this drive.

D2 Drive 2 Seeking: Same as above for drive 2.

D1 Drive 1 Seeking: Same as above for drive 1.

D0 Drive 0 Seeking: Same as above for drive 0.

DATA REGISTER (Read/Write)

This is the location through which all commands, data and status flow between the CPU and the DP8473. During the Command Phase the μP loads the controller's commands into this register based on the Status Register Request for Master and Data Direction bits. The Result Phase transfers the Status Registers and header information to the μP in the same fashion.

Register Description (Continued)

TABLE VI. Data Rate and Precompensation Programming Values

D1	D0**	DRVTYP Pin	Data Rate MFM (kb/s)	Normal Precomp* (ns)	Alternate Precomp* (ns)	FGND Pin Enabled	RPM/LC Pin Level
0	0	X	500	125	125	FGND500	High
0	1	0	250	125	250	FGND250	Low
0	1	1	300	125	208	FGND250	Low
1	0	0	250	125	250	FGND250	Low
1	0	1	250	125	250	FGND250	Low
1	1	0	1000	83	83	None	High
1	1	1	1000	83	83	None	Low

*Normal values when PUMP/PREN pin set low; Alternate values when PUMP/PREN pin set high.
**D0 and D1 are Data Rate Control Bits.

DRIVE CONTROL REGISTER (Write Only)

D7 Motor Enable 3: This controls the Motor for drive 3, MTR3. When 0 the output is high, when 1 the output is low. (Note this signal is not output to a pin on 48 pin DIP version.)

D6 Motor Enable 2: Same function as D7 except for drive 2's motor. (Note this signal is not brought out to a pin on DIP.)

D5 Motor Enable 1: This bit controls the Motor for drive 1's motor. When this bit is 0 the MTR1 output is high.

D4 Motor Enable 0: Same as D5 except for drive 0's motor.

D3 DMA Enable: When set to a 1 this enables the DRQ, DAK, INT pins. A zero disables these signals.

D2 Reset Controller: This bit when set to a 0 resets the controller, and when a 1 enables normal operation. It does not affect the Drive Control or Data Rate Registers which are reset only by a hardware reset.

D1-D0 Drive Select: These two pins are encoded for the four drive selects, and are gated with the motor enable lines, so that only one drive is selected when it's Motor Enable is active. (See Table V.)

DATA RATE REGISTER (Write Only)

D7-D2: Not used.

D1, D0 Data Rate Select: These bits set the data rate and the write precompensation values for the disk controller. After a hardware reset these bits are set to 10 (250 kb/s). They are encoded as shown in Table VI.

DISK CHANGED REGISTER (Read Only)

D7 Disk Changed: This bit is the latched complement of the Disk Changed input pin. If the DSKCHG input is low this bit is high.

D6-D0: These bits are reserved for use by the hard disk controller, thus during a read of this register, these bits are TRI-STATE.

Result Phase Status Registers

The Result Phase of a command contains bytes that hold status information. The format of these bytes are described below. Do not confuse these register bytes with the Main Status Register which is a read only register that is always available. The Result Phase status registers are read from the Data Register only during the Result Phase.

STATUS REGISTER 0 (ST0)

D7-D6 Interrupt Code:

  00 = Normal Termination of Command.

  01 = Abnormal Termination of Command. Execution of Command was started, but was not successfully completed.

  10 = Invalid Command Issue. Command Issued was not recognized as a valid command.

  11 = Ready changed state during the polling mode.

D5 Seek End: Seek or Recalibrate Command completed by the Controller. (Used during Sense Interrupt command.)

D4 Equipment Check: After a Recalibrate Command, Track 0 signal failed to occur. (Used during Sense Interrupt command.)

D3 Not Used: 0

D2 Head Address (at end of Execution Phase).

D1, D0 Drive Select (at end of Execution Phase).

  00 = Drive 0 selected. 01 = Drive 1 selected.

  10 = Drive 2 selected. 11 = Drive 3 selected.

STATUS REGISTER 1 (ST1)

D7 End of Track: Controller transferred the last byte of the last sector without the TC pin becoming active. The last sector is the End Of Track sector number programmed in the Command Phase.

D6 Not Used: 0

D5 CRC Error: If this bit is set and bit 5 of ST2 is clear, then there was a CRC error in the Address Field of the correct sector. If bit 5 of ST2 is set, then there was a CRC error in the Data Field.

D4 Over Run: Controller was not serviced by the μP soon enough during a data transfer in the Execution Phase.

Result Phase Status Registers (Continued)

TABLE VII. Maximum Time Allowed to Service an Interrupt or Acknowledge a DMA Request in Execution Phase

Data Rate	Time to Service
125	62.0 μs
250	30.0 μs
500	14.0 μs
1000	6.0 μs

Time from rising edge of DRQ or INT to trailing edge of $\overline{DAK}$ or $\overline{RD}$ or $\overline{WR}$.

D3 Not Used: 0

D2 No Data: Three possible problems: 1) Controller cannot find the sector specified in the Command Phase during the execution of a Read, Write, or Scan command. An address mark was found however so it is not a blank disk. 2) Controller cannot read any Address Fields without a CRC error during Read ID command. 3) Controller cannot find starting sector during execution of Read A Track command.

D1 Not Writable: Write Protect pin is active when a Write or Format command is issued.

D0 Missing Address Mark: If bit 0 of ST2 is clear then the disk controller cannot detect any Address Field Address Mark after two disk revolutions. If bit 0 of ST2 is set then the disk controller cannot detect the Data Field Address Mark.

STATUS REGISTER 2 (ST2)

D7 Not Used: 0

D6 Control Mark: Controller tried to read a sector which contained a deleted data address mark during execution of Read Data or Scan commands. Or, if a Read Deleted Data command was executed, a regular address mark was detected.

D5 CRC Error in Data Field: Controller detected a CRC error in the Data Field. Bit 5 of ST1 is also set.

D4 Wrong Track: Only set if desired sector not found, and the track number recorded on any sector of the current track is different from that stored in the Track Register.

D3 Scan Equal Hit: "Equal" condition satisfied during any Scan Command.

D2 Scan Not Satisfied: Controller cannot find a sector on the track which meets the desired condition during Scan Command.

D1 Bad Track: Only set if the desired sector is not found, and the track number recorded on any sector on the track is different from that stored in the Track Register and the recorded track number is FF.

D0 Missing Address Mark in Data Field: Controller cannot find the Data Field Address Mark during Read/Scan command. Bit 0 of ST1 is also set.

STATUS REGISTER 3 (ST3)

D7 Not Used: 0

D6 Write Protect Status

D5 Not Used: 1

D4 Track 0 Status

D3 Not Used: 0

D2 Head Select Status

D1, D0 Drive Selected:
00 = Drive 0 selected. 01 = Drive 1 selected.
10 = Drive 2 selected. 11 = Drive 3 selected.

Processor Software Interface

Bytes are transferred to and from the disk controller in different ways for the different phases in a command.

COMMAND SEQUENCE

The disk controller can perform various disk transfer, and head movement commands. Most commands involve three separate phases.

Command Phase: The μP writes a series of bytes to the Data Register. These bytes indicate the command desired and the particular parameters required for the command. All the bytes must be written in the order specified in the Command Description Table. The Execution Phase starts immediately after the last byte in the Command Phase is written. Prior to performing the Command Phase, the Drive Control and Data Rate Registers should be set.

Execution Phase: The disk controller performs the desired command. Some commands require the μP to read or write data to or from the Data Register during this time. Reading data from a disk is an example of this.

Result Phase: The μP reads a series of bytes from the data register. These bytes indicate whether the command executed properly and other pertinent information. The bytes are read in the order specified in the Command Description Table.

A new command may be initiated by writing the Command Phase bytes after the last bytes required from the Result Phase have been read. If the next command requires selecting a different drive or changing the data rate the Drive Control and Data Rate Registers should be updated. If the command is the last command, then the software should deselect the drive. *(Note as a general rule the operation of the controller core is independent of how the μP updates the Drive Control and Data Rate Registers. The software must ensure that manipulation of these registers is coordinated with the controller operation.)*

During the Command Phase and the Result Phase, bytes are transferred to and from the Data Register. The Main Status Register is monitored by the software to determine when a data transfer can take place. Bit 6 of the Main Status Register must be clear and bit 7 must be set before a byte can be written to the Data Register during the Command Phase. Bits 6 and 7 of the Main Status Register must both be set before a byte can be read from the Data Register during the Result Phase.

If there is information to be transferred during the Execution Phase, there are three methods that can be used. The DMA mode is used if the system has a DMA controller. This allows the μP to do other things during the Execution Phase data transfer. If DMA is not used, an interrupt can be issued for each byte transferred during the Execution Phase. If interrupts are not used, the Main Status Register can be polled to indicate when a byte transfer is required.

Processor Software Interface (Continued)

DMA MODE

If the DMA mode is selected, a DMA request will be generated in the Execution Phase when each byte is ready to be transferred. To enable DMA operations during the Execution Phase, the DMA mode bit in the Specify Command must be enabled, and the DMA signals must be enabled in the Drive Control Register. The DMA controller should respond to the DMA request with a DMA acknowledge and a read or write strobe. The DMA request will be cleared by the active edge of the DMA acknowledge. After the last byte is transferred, an interrupt is generated, indicating the beginning of the Result Phase. During DMA operations the Chip Select input must be held high. TC is asserted to terminate an operation. Due to the internal gating TC is only recognized when the $\overline{DAK}$ input is low.

INTERRUPT MODE

If the non-DMA mode is selected, an interrupt will be generated in the Execution Phase when each byte is ready to be transferred. The Main Status Register should be read to verify that the interrupt is for a data transfer. Bits 5 and 7 of the Main Status Register will be set. The interrupt will be cleared when the byte is transferred to or from the Data Register. The µP should transfer the byte within the time allotted by Table VII. If the byte is not transferred within the time allotted, an Overrun Error will be indicated in the Result Phase when the command terminates at the end of the current sector.

An interrupt will also be generated after the last byte is transferred. This indicates the beginning of the Result Phase. Bits 7 and 6 of the Main Status Register will be set and bit 5 will be clear. This interrupt will be cleared by reading the first byte in the Result Phase.

SOFTWARE POLLING

If the non-DMA mode is selected and interrupts are not suitable, the µP can poll the Main Status Register during the Execution Phase to determine when a byte is ready to be transferred. In the non-DMA mode, bit 7 of the Main Status Register reflects the state of the interrupt pin. Otherwise, the data transfer is similar to the Interrupt Mode described above.

Command Description Table

READ DATA

Command Phase

MT	MFM	SK	0	0	1	1	0
IPS	X	X	X	X	HD	DR1	DR0
Track Number							
Drive Head Number							
Sector Number							
Number of Bytes per Sector							
End of Track Sector Number							
Intersector Gap Length							
Data Length							

Note 1

Result Phase

Status Register 0
Status Register 1
Status Register 2
Track Number
Head Number
Sector Number
Bytes/Sector

READ ID

Command Phase

0	MFM	0	0	1	0	1	0
X	X	X	X	X	HD	DR1	DR0

Result Phase

Status Register 0
Status Register 1
Status Register 2
Track Number
Head Number
Sector Number
Bytes/Sector

FORMAT A TRACK

Command Phase

0	MFM	0	0	1	1	0	1
X	X	X	X	X	HD	DR1	DR0
Number of Bytes per Sector							
Number of Sectors per Track							
Intersector Gap Length							
Data Pattern							

Result Phase

Status Register 0
Status Register 1
Status Register 2
Track Number
Head Number
Sector Number
Bytes/Sector

Command Description Table (Continued)

READ DELETED DATA
Command Phase

MT	MFM	SK	0	1	1	0	0
IPS	X	X	X	X	HD	DR1	DR0
Track Number								
Drive Head Number								
Sector Number								
Number of Bytes per Sector								
End of Track Sector Number								
Intersector Gap Length								
Data Length								

Result Phase

Status Register 0
Status Register 1
Status Register 2
Track Number
Head Number
Sector Number
Bytes/Sector

WRITE DATA
Command Phase

MT	MFM	0	0	0	1	0	1
IPS	X	X	X	X	HD	DR1	DR0
Track Number								
Drive Head Number								
Sector Number								
Number of Bytes per Sector								
End of Track Sector Number								
Intersector Gap Length								
Data Length								

Result Phase

Status Register 0
Status Register 1
Status Register 2
Track Number
Head Number
Sector Number
Bytes/Sector

SCAN EQUAL
Command Phase

MT	MFM	SK	1	0	0	0	1
IPS	X	X	X	X	HD	DR1	DR0
Track Number								
Drive Head Number								
Sector Number								
Number of Bytes per Sector								
End of Track Sector Number								
Intersector Gap Length								
Sector Step Size								

Result Phase

Status Register 0
Status Register 1
Status Register 2
Track Number
Head Number
Sector Number
Bytes/Sector

READ A TRACK
Command Phase

0	MFM	SK	0	0	0	1	0
IPS	X	X	X	X	HD	DR1	DR0
Track Number								
Drive Head Number								
Sector Number								
Number of Bytes per Sector								
End of Track Sector Number								
Intersector Gap Length								
Data Length								

Result Phase

Status Register 0
Status Register 1
Status Register 2
Track Number
Head Number
Sector Number
Bytes/Sector

WRITE DELETED DATA
Command Phase

MT	MFM	0	0	1	0	0	1
IPS	X	X	X	X	HD	DR1	DR0
Track Number								
Drive Head Number								
Sector Number								
Number of Bytes per Sector								
End of Track Sector Number								
Intersector Gap Length								
Data Length								

Result Phase

Status Register 0
Status Register 1
Status Register 2
Track Number
Head Number
Sector Number
Bytes/Sector

SCAN LOW OR EQUAL
Command Phase

MT	MFM	SK	1	1	0	0	1
IPS	X	X	X	X	HD	DR1	DR0
Track Number								
Drive Head Number								
Sector Number								
Number of Bytes per Sector								
End of Track Sector Number								
Intersector Gap Length								
Sector Step Size								

Result Phase

Status Register 0
Status Register 1
Status Register 2
Track Number
Head Number
Sector Number
Bytes/Sector

Command Description Table (Continued)

SCAN HIGH OR EQUAL
Command Phase

MT	MFM	SK	1	1	1	0	1
IPS	X	X	X	X	HD	DR1	DR0
Track Number
Drive Head Number
Sector Number
Number of Bytes per Sector
End of Track Sector Number
Intersector Gap Length
Sector Step Size

Result Phase

Status Register 0
Status Register 1
Status Register 2
Track Number
Head Number
Sector Number
Bytes/Sector

SEEK
Command Phase

0	0	0	0	1	1	1	1
X	X	X	X	X	X	DR1	DR0
New Track Number
MSB of Track	0	0	0	0

Note 2

RECALIBRATE
Command Phase

0	0	0	0	0	1	1	1
0	0	0	0	0	0	DR1	DR0

SENSE INTERRUPT
Command Phase

0	0	0	0	1	0	0	0

Result Phase

Status Register 0
Present Track Number (PTN)
MSN PTN	0	0	0	0

Note 2

SENSE DRIVE STATUS
Command Phase

0	0	0	0	0	1	0	0
X	X	X	X	X	HD	DR1	DR0

Result Phase

Status Register 3

SPECIFY
Command Phase

0	0	0	0	0	0	1	1
Step Rate Time				Motor Off Time			
Motor On Time							DMA

MODE
Command Phase

0	0	0	0	0	0	0	1
TMR	IAF	IPS	0	LW	PR	1	ETR
0	0	0	0	0	0	0	0
1	1	0	WLD	Head Settle			
0	0	0	0	0	RG	0	PU

Note 3

SET TRACK
Command Phase

0	R/W	1	0	0	0	0	1
0	0	1	1	0	MSB	DR1	DR0
New Track Number

Result Phase

Value

Note 3

INVALID COMMAND
Command Phase

Invalid Op Codes

Result Phase

Status Register 0

Note 1: The IPS bit is only enabled if the IPS bit in the mode command is set. Otherwise this bit is a don't care.

Note 2: Shaded byte only written or read if the extended track range mode is enabled in the Mode Command (ET) = 1.

Note 3: These commands are additional enhanced commands.

Note: Mnemonic Definitions

X = DON'T CARE
MFM = Data Encoding Scheme
MSN PTN = Most Significant Nibble Present Track Number
MT = Multi-Track
IPS = Implied Seek (In individual commands this bit is a don't care unless the IPS bit in the mode command is set.)
SK = Skip Sector
HD = Head Number
DRn = Drive to Select (encoded)
TMR = Motor/Head Timer Mode
IAF = Index Address Field
LW PR = Low Power Mode
ETR = Extended Track Range
WLD = Wildcard in Scan
RG = Enables the Read Gate Input on the DSKCHG pin for the Data Separator.
PU = Enables Charge Pump PUMP signal to be output on the PUMP/PREN pin.
MSB = Selects whether the most significant or least significant byte of the track is read. 1 = MSB.
R/W = Selects whether the track is written or read (Read = 0, Write = 1).

Command Description

READ DATA

The Read Data op-code is written to the data register followed by 8 bytes as specified in the Command Description Table. After the last byte is written, the controller starts looking for the correct sector header. Once the sector is found the controller sends the data to the μP. After one sector is finished, the Sector Number is incremented by one and this new sector is searched for. If MT (Multi-Track) is set, both sides of one track can be read. Starting on side zero, the sectors are read until the sector number specified by End of Track Sector Number is reached. Then, side one is read starting with sector number one.

In DMA mode the Read Data command continues to read until the TC pin is set. This means that the DMA controller should be programmed to transfer the correct number of bytes. TC could be controlled by the μP and be asserted when enough bytes are received. An alternative to these methods of stopping the Read Data command is to program the End of Track Sector Number to be the last sector number that needs to be read. The controller will stop reading the disk with an error indicating that it tried to access a sector number beyond the end of the track.

The Number of Data Bytes per Sector parameter is defined in Table VIII. If this is set to zero then the Data Length parameter determines the number of bytes that the controller transfers to the μP. If the data length specified is smaller than 128 the controller still reads the entire 128 byte sector and checks the CRC, though only the number of bytes specified by the Data Length parameter are transferred to the μP. Data Length should not be set to zero. If the Number of Bytes per Sector parameter is not zero, the Data Length parameter has no meaning and should be set to FF (hex).

If the Implied Seek Mode is enabled by both the Mode command and the IPS bit in this command, a Seek will be performed to the track number specified in the Command Phase. The controller will also wait the Head Settle time if the implied seek is enabled.

After all these conditions are met, the controller searches for the specified sector by comparing the track number, head number, sector number, and number bytes/sector given in the Command Phase with the appropriate bytes read off the disk in the Address Fields.

If the correct sector is found, but there is a CRC error in the Address Field, bit 5 of ST1 (CRC Error) is set and an abnormal termination is indicated. If the correct sector is not found, bit 2 of ST1 (No Data) is set and an abnormal termination is indicated. In addition to this, if any Address Field track number is FF, bit 1 of ST2 (Bad Track) is set or if any Address Field track number is different from that specified in the Command Phase, bit 4 of ST2 (Wrong Track) is set.

After finding the correct sector, the controller reads that Data Field. If a Deleted Data Mark is found and the SK bit is set, the sector is not read, bit 6 of ST2 (Control Mark) is set, and the next sector is searched for. If a deleted data mark is found and the SK bit is not set, the sector is read, bit 6 of ST2 (Control Mark) is set, and the read terminates with a normal termination. If a CRC error is detected in the Data Field, bit 5 is set in both ST1 and ST2 (CRC Error) and an abnormal termination is indicated.

If no problems occur in the read command, the read will continue from one sector to the next in logical order (not physical order) until either TC is set or an error occurs.

If a disk has not been inserted into the disk drive, there are many opportunities for the controller to appear to hang up. It does this if it is waiting for a certain number of disk revolutions for something. If this occurs, the controller can be forced to abort the command by writing a byte to the Data register. This will place the controller into the Result Phase.

TABLE VIII. Sector Size Selection

Bytes/Sector Code	Number of Bytes in Data Field
0	128
1	256
2	512
3	1024
4	2048
5	4096
6	8192

An interrupt will be generated when the Execution Phase of the Read Data command terminates. The values that will be read back in the Result Phase are shown in Table IX. If an error occurs, the result bytes will indicate the sector being read when the error occurred.

READ DELETED DATA

This command is the same as the Read Data command except for its treatment of a Deleted Data Mark. If a Deleted

TABLE IX. Result Phase Termination Values with No Error

MT	HD	Last Sector	Track	Head	Sector	B/S
0	0	< EOT	NC	NC	S+1	NC
0	0	= EOT	T+1	NC	1	NC
0	1	< EOT	NC	NC	S+1	NC
0	1	= EOT	T+1	NC	1	NC
1	0	< EOT	NC	NC	S+1	NC
1	0	= EOT	NC	1	1	NC
1	1	< EOT	NC	NC	S+1	NC
1	1	= EOT	T+1	0	1	NC

EOT = End of Track Sector Number from Command Phase
NC = No Change in Value
S = Sector Number last operated on by controller
T = Track Number programmed in Command Phase

Command Description (Continued)

Data Mark is read, the sector is read normally. If a Regular Data Mark is found and the SK bit is set, the sector is not read, bit 6 of ST2 (Control Mark) is set, and the next sector is searched for. If a Regular Data Mark is found and the SK bit is not set, the sector is read, bit 6 of ST2 (Control Mark) is set, and the read terminates with a normal termination.

WRITE DATA

The Write Data command is very similar to the Read Data command except that data is transferred from the μP to the disk rather than the other way around. If the controller detects the Write Protect signal, bit 1 of ST1 (Not Writable) is set and an abnormal termination is indicated.

WRITE DELETED DATA

This command is the same as the Write Data Command except a Deleted Data Mark is written at the beginning of the Data Field instead of the normal Data Mark.

READ A TRACK

This command is similar to the Read Data command except for the following. The controller starts at the index hole and reads the sectors in their physical order, not their logical order.

Even though the controller is reading sectors in their physical order, it will still perform a comparison of the header ID bytes with the Data programmed in the Command Phase. The exception to this is the sector number. Internally, this is initialized to a one, and then incremented for each successive sector read. Whether or not the programmed Address Field matches that read from the disk, the sectors are still read in their physical order. If a header ID comparison fails, bit 2 of ST1 (No Data) is set, but the operation will continue. If there is a CRC error in the Address Field or the Data Field, the read will also continue.

The command will terminate when it has read the number of sectors programmed in the EOT parameter.

READ ID

This command will cause the controller to read the first Address Field that it finds. The Result Phase will contain the header bytes that are read. There is no data transfer during the Execution Phase of this command. An interrupt will be generated when the Execution Phase is completed.

FORMAT A TRACK

This command will format one track on the disk. After the index hole is detected, data patterns are written on the disk including all gaps, address marks, Address Fields, and Data Fields. The exact details of the number of bytes for each field is controlled by the parameters given in the Format A Track command, and the IAF (Index Address Field) bit in the Mode command. The Data Field consists of the Fill Byte specified in the command, repeated to fill the entire sector.

To allow for flexible formatting, the μP must supply the four Address Field bytes (track, head, sector, number of bytes) for each sector formatted during the Execution Phase. In other words, as the controller formats each sector, it will request four bytes through either DMA requests or interrupts. This allows for non-sequential sector interleaving. Some typical values for the programmable GAP size are shown in Table X.

The Format Command terminates when the index hole is detected a second time, at which point an interrupt is generated. Only the first three status bytes in the Result Phase are significant.

TABLE X. Gap Length for Various Sector Sizes and Disk Types

Mode	Sector Size	Sector Code	EOT	Gap	Format Gap
\multicolumn{6}{c}{8″ Drives (360 RPM, 500 kb/s)}					
FM	128	00	1A	07	1B
	256	01	0F	0E	2A
	512	02	08	1B	3A
	1024	03	04	47	8A
	2048	04	02	C8	FF
	4096	05	01	C8	FF
MFM	256	01	1A	0E	36
	512	02	0F	1B	54
	1024	03	08	35	74
	2048	04	04	99	FF
	4096	05	02	C8	FF
	8192	06	01	C8	FF
\multicolumn{6}{c}{5.25″ Drives (300 RPM, 250 kb/s)}					
FM	128	00	12	07	09
	128	00	10	10	19
	256	01	08	18	30
	512	02	04	46	87
	1024	03	02	C8	FF
	2048	04	01	C8	FF
MFM	256	01	12	0A	0C
	256	01	10	20	32
	512	02	08	2A	50
	1024	03	04	80	F0
	2048	04	02	C8	FF
	4096	05	01	C8	FF
\multicolumn{6}{c}{3.5″ Drives (300 RPM, 250 kb/s)}					
FM	128	00	0F	07	1B
	256	01	09	0E	2A
	512	02	05	1B	3A
MFM	256	01	0F	0E	36
	512	02	09	1B	54
	1024	03	05	35	74

Note: Format Gap is the gap length used only for the Format command.

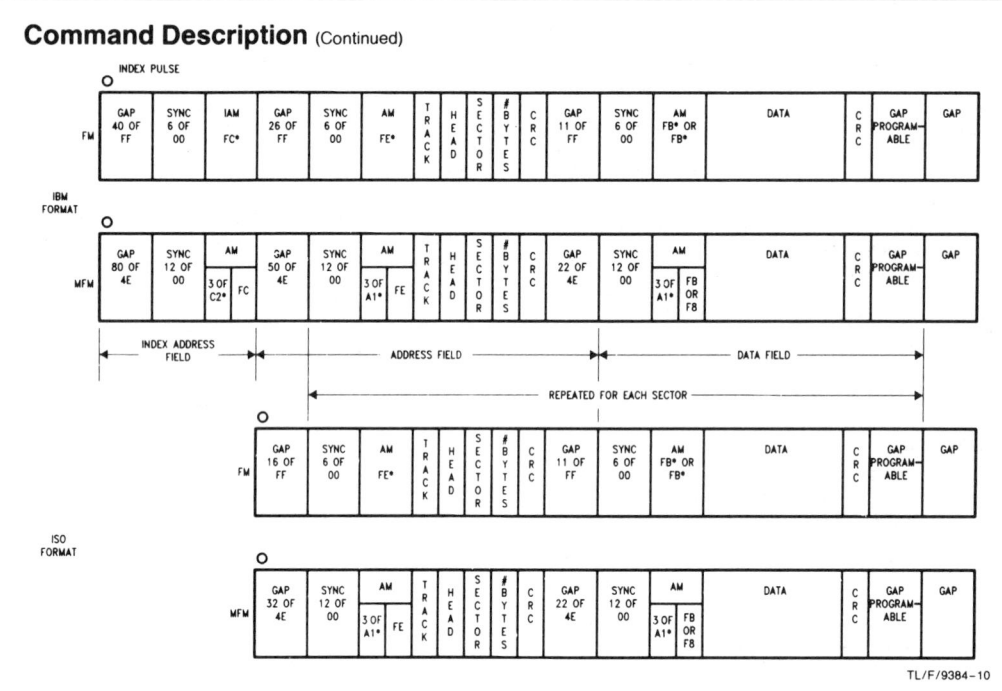

Command Description (Continued)

FIGURE 6. IBM and ISO Formats Supported by the Format Command

Notes:

FE* = Data pattern of FE, Clock pattern of C7
FC* = Data pattern of FC, Clock pattern of D7
FB* = Data pattern of FB, Clock pattern of C7
F8* = Data pattern of F8, Clock pattern of C7
A1* = Data pattern of A1, Clock pattern of 0A
C2* = Data pattern of C2, Clock pattern of 14

All byte counts in decimal.
All byte values in hex.
CRC uses standard polynomial $x^{16} + x^{12} + x^5 + 1$.

SCAN COMMANDS

The Scan Commands allow data read from the disk to be compared against data sent from the μP. There are three Scan Commands to choose from:

Scan Equal	Disk Data $=$ μP Data
Scan Less Than or Equal	Disk Data $\leq$ μP Data
Scan Greater Than or Equal	Disk Data $\geq$ μP Data

Each sector is interpreted with the most significant bytes first. If the Wildcard mode is enabled from the Mode command, an FF(hex) from either the disk or the μP is used as a don't care byte that will always match equal. After each sector is read, if the desired condition has not been met, the next sector is read. The next sector is defined as the current sector number plus the Sector Step Size specified. The Scan command will continue until the scan condition has been met, or the End of Track Sector Number has been reached, or if TC is asserted.

If the SK bit is set, sectors with deleted data marks are ignored. If all sectors read are skipped, the command will terminate with D3 of ST2 set (Scan Equal Hit). The result phase of the command is shown in Table XI.

TABLE XI. Scan Command Termination Values

Command	Status Register 2 D2	Status Register 2 D3	Conditions
Scan Equal	0	1	Disk $=$ μP
	1	0	Disk $\neq$ μP
Scan Low or Equal	0	1	Disk $=$ μP
	0	0	Disk $<$ μP
	1	0	Disk $>$ μP
Scan High or Equal	0	1	Disk $=$ μP
	0	0	Disk $>$ μP
	1	0	Disk $<$ μP

Command Description (Continued)

SEEK

There are two ways to move the disk drive head to the desired track number. Method One is to enable the Implied Seek Mode. This way each individual Read or Write command will automatically move the head to the track specified in the command.

Method Two is using the Seek Command. During the Execution Phase of the Seek Command, the track number to seek to is compared with the present track number and a step pulse is produced to move the head one track closer to the desired track number. This is repeated at the rate specified by the Specify Command until the head reaches the correct track. At this point an interrupt is generated and a Sense Interrupt Command is required to clear the interrupt.

During the Execution Phase of the Seek Command the only indication via software that a Seek Command is in progress is bits 0-3 (Drive Busy) of the Main Status Register. Bit 4 of the Main Register (Controller Busy) is not set. While the internal microengine is capable of multiple seeks on 2 or more drives at the same time since the drives are selected via the Drive Control Register in software, software should ensure that only one drive is seeking at one time. No other command except the Sense Interrupt Command should be issued while a Seek Command is in progress.

If the extended track range mode is enabled, a fourth byte should be written in the Command Phase to indicate the four most significant bits of the desired track number. Otherwise, only three bytes should be written.

RECALIBRATE

The Recalibrate Command is very similar to the Seek Command. It is used to step a drive head out to track zero. Step pulses will be produced until the track zero signal from the drive becomes true. If the track zero signal does not go true before 77 step pulses are issued, an error is generated. If the extended track range mode is enabled, an error is not generated until 3917 pulses are issued.

Recalibrations on more than one drive at a time should not be issued for the same reason as explained in the Seek Command. No other command except the Sense Interrupt Command should be issued while a Recalibrate Command is in progress.

SENSE INTERRUPT STATUS

An interrupt is generated by the controller when any of the following conditions occur:

1. Upon entering the Result Phase of:
   a. Read Data Command
   b. Read Deleted Data Command
   c. Write Data Command
   d. Write Deleted Data Command
   e. Read a Track Command
   f. Read ID Command
   g. Format Command
   h. Scan Commands
2. During data transfers in the Execution Phase while in the Non-DMA mode
3. Internal Ready signal changes state (only occurs immediately after a hardware or software reset).
4. Seek or Recalibrate Command termination

An interrupt generated for reasons 1 and 2 above occurs during normal command operations and are easily discernible by the μP. During an execution phase in Non-DMA Mode, bit 5 (Execution Mode) in the Main Status Register is set to 1. Upon entering Result Phase this bit is set to 0. Reasons 1 and 2 do not require the Sense Interrupt Status command. The interrupt is cleared by reading or writing information to the data register.

Interrupts caused by reasons 3 and 4 are identified with the aid of the Sense Interrupt Status Command. This command resets the interrupt when the command byte is written. Use bits 5, 6 and 7 of ST0 to identify the cause of the interrupt as shown in Table XII.

TABLE XII. Status Register 0 Termination Codes

Status Register 0			Cause
Interrupt Code		Seek End	
D7	D6	D5	
1	1	0	Internal Ready Went True
0	0	1	Normal Seek Termination
0	1	1	Abnormal Seek Termination

TABLE XIII. Step, Head Load and Unload Timer Definitions (500 kb/s MFM)

Timer	Mode 1		Mode 2		Unit
	Value	Range	Value	Range	
Step Rate	(16 − N)	1−16	(16 − N)	1−16	ms
Head Unload	N × 16	0−240	N × 512	0−7680	ms
Head Load	N × 2	0−254	N × 32	0−4064	ms

Issuing a Sense Interrupt Status Command without an interrupt pending is treated as an invalid command.

If the extended track range mode is enabled, a third byte should be read in the Result Phase which will indicate the four most significant bits of the Present Track Number. Otherwise, only two bytes should be read.

SPECIFY

The Specify Command sets the initial values for each of the three internal timers. The timer programming values are shown in Table XIII.

The Head Load and Head Unload timers are artifacts of the μPD765A. These timers determine the delay from loading the head until a read or write command is started, and unloading the head sometime after the command was completed. Since the DP8473's head load signal is now the software controlled Motor lines in the Drive Control Register, these timers only provide some delay from the initiation of a command until it is actually started. Like the DP8474 these times can be extended by setting the TMR bit in the Mode Command.

The Step Rate Time defines the time interval between adjacent step pulses during a Seek, Implied Seek, or Recalibrate Command.

The times stated in the table are affected by the Data Rate. The values in the table are for 500 kb/s MFM (250 kb/s FM) and 1 Mb/s MFM (500 kb/s FM). For a 300 kb/s MFM data rate (150 kb/s FM) these values should be multiplied by 1.6667, and for 250 kb/s MFM (125 kb/s FM) these values should be doubled.

The choice of DMA or Non-DMA operation is made by the NON-DMA bit. When this bit is 1 then Non-DMA mode is selected, and when this bit is 0, the DMA mode is selected.

This command does not generate an interrupt.

Command Description (Continued)

LOW PWR (LOW PoWeR mode)

SENSE DRIVE STATUS

This two byte command obtains the status of a disk drive. Status Register 3 is returned in the result phase and contains the drive status. This command does not generate an interrupt.

MODE

This command is used to select the special features of the controller. The bits for the command phase bytes are shown in the command description table, and their function is described below. The defaults after a hardware or software reset are shown by the "bullets" to the left of each item.

- **TMR = 0 (motor TiMeR):** Timers for motor on and motor off are defined for Mode 1. (See Specify Command)
  TMR = 1: Timers for motor on and motor off are defined for Mode 2. (See Specify Command)

- 00 Completely disable the low power mode. (default)
  01 Go into low power mode 500 ms after the head unload timer times out.
  10 Go into low power mode now.
  11 Not Used.

- **IAF = 0 (Index Address Format):** The controller will format tracks with the Index Address Field included. (IBM Format)
  IAF = 1: The controller will format tracks without including the Index Address Mark Field. (ISO Format)

- **IPS = 0 (ImPlied Seek):** The implied seek bit in the command is ignored.
  IPS = 1: The implied seek bit in the command is enabled so that if the bit is set in the command, a Seek will be performed automatically.

- **ETR = 0 (Extended Track Range):** Header format is the IBM System 34 (double density) or System 3740 (single density).
  ETR = 1: Header format is the same as above but there are 12 bits of track number. The MSB's of the track number are in the upper four bits of the head number byte.

- **WLD = 0 (scan WiLD card):** An FF(hex) from either the μP or the disk during a Scan Command is interpreted as a wildcard character that will always match true.
  WLD = 1: The Scan commands do not recognize FF(hex) as a wildcard character.

  Head Settle: Time allowed for head to settle after an Implied Seek. Time = N × 4 ms, (0 ms–60 ms). (Based on 500 kb/s and 1 Mb/s MFM data rates. Double for 250 kb/s.)

PU (PUMP Pulse Output): When set enables a signal that indicates when the Data Separator's charge pump is making a phase correction. This is a series of pulses. This signal is output on the PUMP/PREN pin when this bit is set.

This is intended as a test mode to aid in evaluation of the Data Separator. (Default mode is off)

RG (Read Gate): Like the PUMP output, when this bit is set it enables a pin (the DSKCHG pin) to act as an external Read Gate signal for the Data Separator. This is intended as a test mode to aid in evaluation of the Data Separator. (Default mode is off)

SET TRACK

This command is used to inspect or change the value of the internal Present Track Register. This could be useful for recovery from disk mis-tracking errors, where the real current track could be read through the Read ID command and then the Set Track Command can set the internal present track register to the correct value.

The first byte of the command contains the command opcode and the R/W bit. If the R/W bit is low, a track register is to be read. In this case, the result phase contains the value in the internal register specified, and the third byte of the command is a dummy byte.

If the R/W bit is high, data is written to a track register. In this case the 3rd byte of the command phase is forced into the specified internal register, and the result phase contains the new byte value written.

The particular track register chosen to operate on is determined by the least significant 3 bits of the second byte of the command. The two LSB's select the drive (DR1, DR0), and the next bit (MSB) determines whether the least significant byte (MSB = 0) or the most significant byte (MSB = 1) of the track register is to be read/written. When not in the extended track range mode, only the LSB track register need be updated. In this instance, the MSB bit is set to 0.

This command does not generate an interrupt.

INVALID COMMAND

If an invalid command (i.e., a command not defined) is received by the controller, the controller will respond with ST0 in the Result Phase. The Controller does not generate an interrupt during this condition. Bits 6 and 7 in the Main Status Register are both set to one's indicating to the processor that the Controller is in the Result Phase and the contents of ST0 must be read. When the system reads ST0 it will find an 80(hex) indicating an invalid command was received.

Typical Performance Characteristics

Typical Window Margin Performance Characteristics at 250 kb/s MFM

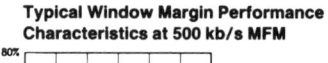

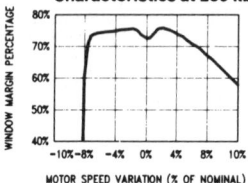

Typical Window Margin Performance Characteristics at 500 kb/s MFM

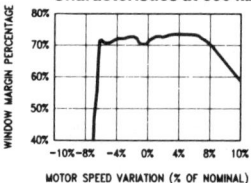

Absolute Maximum Ratings (Notes 1 and 2)

If Military/Aerospace specified devices are required, contact the National Semiconductor Sales Office/Distributors for availability and specifications.

Supply Voltage (V_{CC})	−0.5V to +7V
DC Input Voltage (V_{IN})	−0.5V to V_{CC} + 0.5V
DC Output Voltage (V_{OUT})	−0.5V to V_{CC} + 0.5V
Storage Temperature Range (T_{STG})	−65°C to +165°C
Package Power Dissipation (P_D)	750 mW
Lead Temperature (T_L) (Soldering, 10 seconds)	260°C
$|V_{CC} - V_{CCA}|$	0.6V

Operating Conditions

	Min	Max	Units
Supply Voltage (V_{CC})	4.5	5.5	V
Operating Temperature (T_A)	0	+70	°C
ESD Tolerance: C_{ZAP} = 100 pF, R_{ZAP} = 1.5 kΩ	2000		V

DC Electrical Characteristics V_{CC} = 5V ± 10% unless otherwise specified (Note 3)

Symbol	Parameter	Conditions	Min	Max	Units
V_{IH}	High Level Input Voltage	(except OSC2/CLK)	2.0		V
V_{IL}	Low Level Input Voltage	(except OSC2/CLK)		0.8	V
I_{IN}	Input Current (except OSC pins)	V_{IN} = V_{CC} or GND		±10.0	µA
I_{CCA}	Average V_{CCA} Supply Current	V_{IN} = 2.4V or 0.5V, I_O = 0 mA (Note 4)		20.0	mA
	Quiescent V_{CCA} Supply Current in Low Power Mode	V_{IN} = V_{CC} or GND, I_O = 0 mA (Note 4)		100	µA
I_{CC}	Average V_{CC} Supply Current	V_{IN} = 2.4V or 0.5V, I_O = 0 mA (Note 4)		20.0	mA
	Quiescent V_{CC} Supply Current in Low Power Mode	V_{IN} = V_{CC} or GND, I_O = 0 mA (Note 4)		100	µA

OSCILLATOR PINS (OSC2/CLK)

Symbol	Parameter	Conditions	Min	Max	Units
I_{IL}	OSC2 Input Current (OSC1 = GND)	V_{IN} = V_{CC} or GND	−1.6		mA
V_{IH}	OSC2 High Level Input Voltage	OSC1 = GND	2.4		V
V_{IL}	OSC2 Low Level Input Voltage	OSC1 = GND		0.4	V

MICROPROCESSOR INTERFACE PINS (D0–D7, INT, DAK, TC, DRQ, RD, WR, CS, A0–A3)

Symbol	Parameter	Conditions	Min	Max	Units
V_{OH}	High Level Output Voltage	I_{OUT} = −20 µA	V_{CC} − 0.1		V
		I_{OUT} = −4.0 mA	3.5		V
V_{OL}	Low Level Output Voltage	I_{OUT} = 20 µA		0.1	V
		I_{OUT} = 12 mA		0.4	V
I_{OZ}	Output TRI-STATE® Leakage Current	V_{OUT} = V_{CC} or GND		±10.0	µA

DISK DRIVE INTERFACE PINS
(MTR0–3, DR0–3, WDATA, WGATE, RDATA, DIR, HDSEL, TRK0, WRTPRT, RPM, STEP, DSKCHG, INDEX)

Symbol	Parameter	Conditions	Min	Max	Units
V_H	Input Hysteresis		250 Typical		mV
V_{OL}	Low Level Output Voltage	I_{OUT} = 40 mA		0.4	V
I_{LKG}	Output High Leakage Current	V_{OUT} = V_{CC} or GND		±10.0	µA

Note 1: Absolute Maximum Ratings are those values beyond which damage to the device may occur.
Note 2: Unless otherwise specified all voltages are referenced to ground.
Note 3: These DC Electrical Characteristics are measured statically, and not under dynamic conditions.
Note 4: I_{CC} is measured with a 0.1 µF supply decoupling capacitor to ground.

PRELIMINARY

Phase Locked Loop Characteristics $V_{CC} = 5V \pm 10\%$, $F_{XTAL} = 24$ MHz unless otherwise specified

Symbol	Parameter	Conditions	Min	Typ	Max	Units
V_{REF}	SETCUR Pin Reference Voltage	$R_1 = 5.6$ kΩ, $V_{CC} = 5V$		1.1		V
K_{VCO}	VCO Gain (Note 5)	$t_{DATA} = 1$ μs $\pm 10\%$		25		Mrad/s/V
R_1	Recommended Pump Resistor Range			3–12		kΩ
$K_{P(UP)}$	Charge Pump Up Current Gain ($I_{REF}/I_{P(UP)}$) (Note 6)	$R_1 = 5.6$ kΩ		2.50		(none)
$K_{P(DWN)}$	Charge Pump Down Current Gain ($I_{REF}/I_{P(DWN)}$) (Note 6)	$R_1 = 5.6$ kΩ		2.25		(none)
K_{PLL}	Internal Phase Locked Loop Gain (Note 7)	DP8473-1 ($R_1 = 5.6$ kΩ) Pump Up Pump Down	 58 54	 75 70	 90 82	 Mrad Mrad
		DP8473-2 ($R_1 = 5.6$ kΩ) Pump Up Pump Down	 48 44	 75 68	 100 92	 Mrad Mrad
T_{SW}	Static Window (Note 8)		Max Early		Max Late	
		DP8473-1 ($R_1 = 5.6$ kΩ) 250 kb/s 300 kb/s 500 kb/s 1.0 Mb/s	 1050 840 530 265		 850 710 425 215	 ns ns ns ns
		DP8473-2 ($R_1 = 5.6$ kΩ) 250 kb/s 300 kb/s 500 kb/s 1.0 Mb/s	 1100 920 560 275		 800 670 400 200	 ns ns ns ns

Note 5: The VCO gain is measured at the 1.0 Mb/s data rate by forcing the data period over a range from 900 ns to 1100 ns, and measuring the resulting voltage on the filter pin. The best straight line gain is fit to the measured points.

Note 6: This is the current gain of the charge pump, which is defined as the output current divided by the current through R_1.

Note 7: This is the product of: $V_{REF} \cdot K_P \cdot K_{VCO}$. The total variation in this specification indicates the total loop gain variation contributed by the internal circuitry. The K_{VCO} portion of this specification is measured at the 1.0 Mb/s data rate by forcing the data period over a range of 900 ns to 1100 ns, and measuring the resultant K_{VCO}. K_P is measured by forcing the Filter pin to 2.1V and measuring the ratio of the charge pump current over the input current.

Note 8: The DP8473 is guaranteed to correctly decode a single shifted clock pulse at the end of a long series of non-shifted preamble bits as long as the single shifted pulse is shifted less than the amount specified in T_{SW}. The length of the preamble is long enough for the PLL to lock. The filter components used are those in Table II.

AC Electrical Characteristics

MICROPROCESSOR READ TIMING

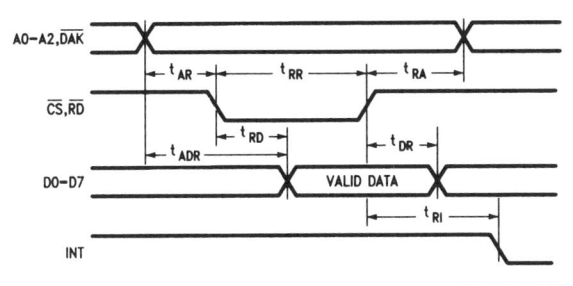

Symbol	Parameter	Min	Max	Units
t_{AR}	Address Valid prior to Read Strobe	10		ns
t_{RA}	Address Hold from Read Strobe	0		ns
t_{RR}	Read Strobe Width	90		ns
t_{RD}	Read Strobe and Chip Select to Data Valid		90	ns
t_{ADR}	Address Valid to Read Data		100	ns
t_{DR}	Data Hold from Read Strobe to High Impedance	5	30	ns
t_{RI}	Clear INT from Read Strobe		50	ns

MICROPROCESSOR WRITE TIMING

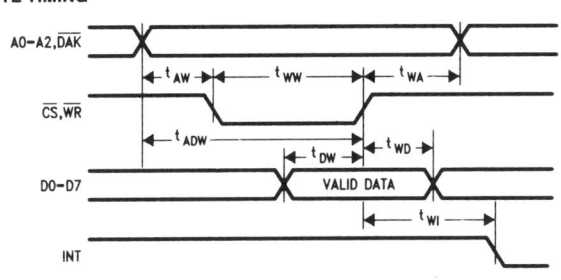

Symbol	Parameter	Min	Max	Units
t_{AW}	Address Valid to Leading Edge of Write Strobe	10		ns
t_{WA}	Address Hold from Write Strobe	0		ns
t_{WW}	Write Strobe Width	80		ns
t_{ADW}	Address Valid to Trailing Edge of Write Strobe	110		ns
t_{DW}	Data Setup to End of Write Strobe or Chip Select	65		ns
t_{WD}	Data Hold from Write Strobe	0		ns
t_{WI}	Clear INT from Write Strobe		50	ns

AC Electrical Characteristics (Continued)

OSC2/CLOCK AND RESET TIMING

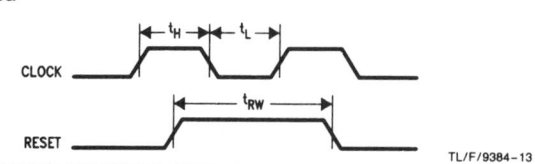

Symbol	Parameter	Min	Max	Units
t_H	Clock High Time	16		ns
t_L	Clock Low Time	16		ns
t_{RW}	Reset Pulse Width	100		ns

DMA TIMING (Note 9)

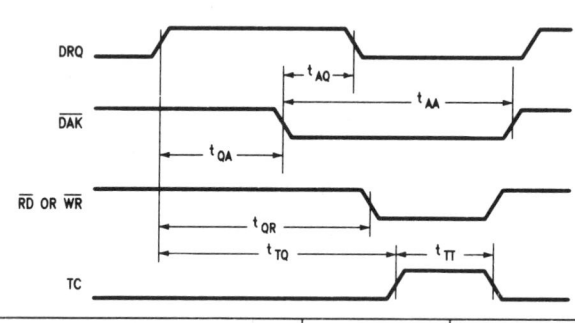

Symbol	Parameter	Min	Max	Units
t_{AQ}	End of DRQ from DAK		75	ns
t_{QA}	DAK Assertion from DRQ	0		ns
t_{AA}	DAK Pulse Width	90		ns
t_{QR}	DRQ to Read or Write Strobe	0		ns
t_{TT}	TC Strobe Width	50		ns
t_{TQ}	Time after Last DRQ That TC Must Be Asserted By		(Note 10)	ns

Note 9: DMA Acknowledge is sufficient to acknowledge a data transfer. Read or Write Strobes are neccessary only if data is to be presented to the data bus. If Read/Write Strobes are applied, then they and the Acnowledge must be removed within 1 μs of each other.

Note 10: TC is the terminal count pin which terminates the data transfer operation. There are several constraints placed on the timing of TC. 1) TC is enabled by $\overline{DAK}$, so TC must be pulsed while $\overline{DAK}$ is low. 2) TC must occur before ((1/data rate × 8) − 1 μs). Data rate is the exact data transfer rate being used.

DRIVE READ TIMING

Symbol	Parameter	Min	Max	Units
t_{RDW}	Read Data Pulse Width	50		ns

AC Electrical Characteristics (Continued)

DRIVE WRITE TIMING

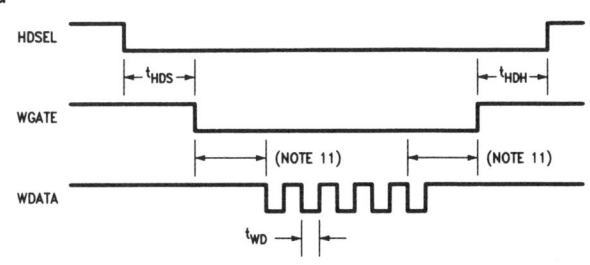

Symbol	Parameter	Conditions	Min	Max	Units
t_{WD}	Write Data Pulse Width	250 kb/s (MFM)	500		ns
		300 kb/s (MFM)	416		ns
		1.0 Mb/s 500 kb/s(MFM)	250		ns
t_{HDS}	Head Select Setup to Write Gate Assertion		10		µs
t_{HDH}	Head Select Hold from Write Gate		4		µs

Note 11: Whenever WGATE is asserted the WDATA line is active. At the end of each write one dummy byte is written before WGATE is deasserted.

DRIVE TRACK ACCESS TIMING

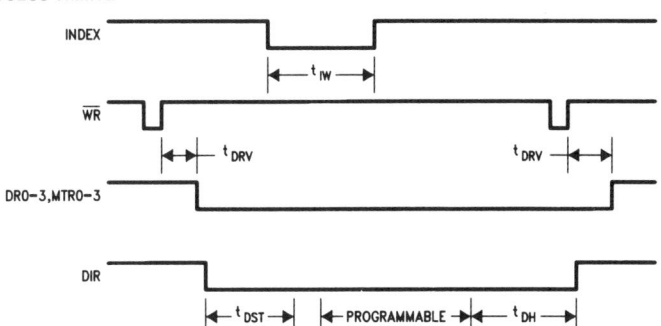

Symbol	Parameter	Min	Max	Units
t_{DST}	Direction Setup prior to Step	6		µs
t_{DH}	Direction Hold from End of Step	1 step time		
t_{STP}	Step Pulse Width	8		µs
t_{IW}	Index Pulse Width	100		ns
t_{DRV}	Drive Select or Motor Time from Write Strobe		100	ns

AC Test Conditions (Notes 11, 12, 13)

Input Pulse Levels	GND to 3V
Input Rise and Fall Times	6 ns
Input and Output Reference Levels	1.3V
TRI-STATE Reference Levels	Active High − 0.5V
	Active Low + 0.5V

Note 11: C_L = 100 pF, includes jig and scope capacitance.

Note 12: S1 = open for push-pull outputs. S1 = V_{CC} for high impedance to active low and active low to high impedance measurements. S1 = GND for high impedance to active high and active high to high impedance measurements. R_L = 1.0 kΩ for µP interface pins.

Note 13: For the Open Drain Drive Interface Pins S1 = V_{CC} and R_L = 150Ω.

Capacitance T_A = 25°C, f = 1 MHz (Note 14)

Symbol	Parameter	Typ	Units
C_{IN}	Input Capacitance	5	pF
C_{OUT}	Output Capacitance	8	pF

Note 14: This parameter is not 100% tested.

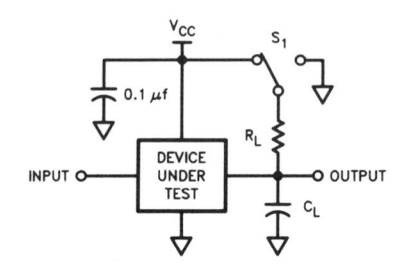

Physical Dimensions inches (millimeters)

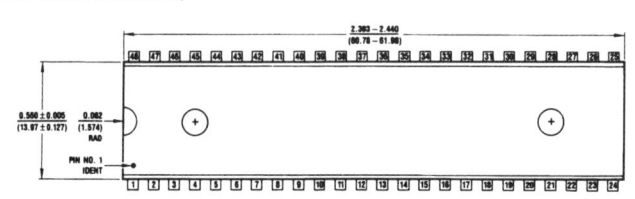

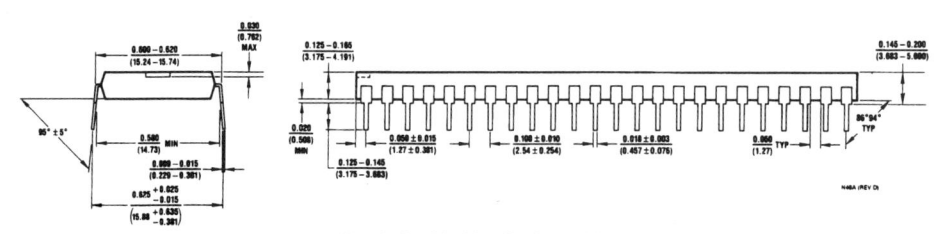

Plastic Dual-In-Line Package (N)
Order Number DP8473N
NS Package Number N48A

DP8473 Floppy Disk Controller PLUS-2

Physical Dimensions inches (millimeters) (Continued)

Lit. # 103175

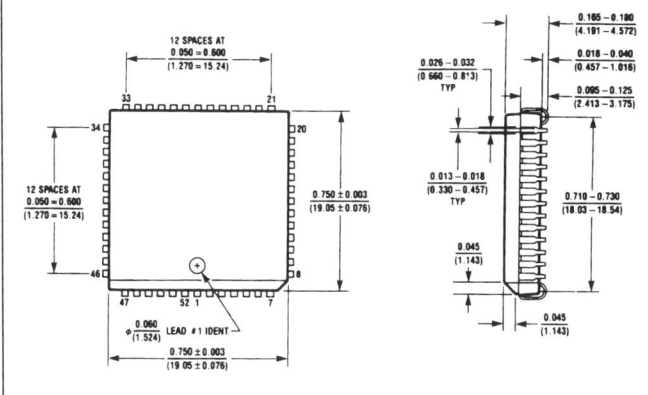

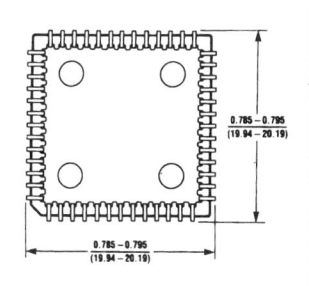

Plastic Leaded Chip Carrier (V)
Order Number DP8473V
NS Package Number V52A

LIFE SUPPORT POLICY

NATIONAL'S PRODUCTS ARE NOT AUTHORIZED FOR USE AS CRITICAL COMPONENTS IN LIFE SUPPORT DEVICES OR SYSTEMS WITHOUT THE EXPRESS WRITTEN APPROVAL OF THE PRESIDENT OF NATIONAL SEMICONDUCTOR CORPORATION. As used herein:

1. Life support devices or systems are devices or systems which, (a) are intended for surgical implant into the body, or (b) support or sustain life, and whose failure to perform, when properly used in accordance with instructions for use provided in the labeling, can be reasonably expected to result in a significant injury to the user.

2. A critical component is any component of a life support device or system whose failure to perform can be reasonably expected to cause the failure of the life support device or system, or to affect its safety or effectiveness.

National Semiconductor GmbH	National Semiconductor France S.A.	National Semiconductor S.p.A.	National Semiconductor AB	National Semiconductor	National Semiconductor (UK) Ltd.
Industriestraße 10 D-8080 Fürstenfeldbruck Tel. (0 81 41) 103-0 Telex 527 649 Fax (0 81 41) 10 35 54	Expansion 10000 28, rue de la Redoute F-92260 Fontenay-aux-Roses Tel. (1) 46 60 81 40 Telex 250 956 Fax (1) 46 60 81 40	Strada 7 - Palazzo R/3 I-20089 Rozzano - Milanofiori Tel. (02) 8 24 20 46/7/8/9 Telex 352 647 Fax (02) 8 25 47 58	Box 2016 Stensätravägen 13 S-12702 Skärholmen Tel. (08) 97 01 90 Telex 10 731 Fax (08) 97 68 12	Postbus 90 1380 AB Weesp Tel. (0 29 40) 3 04 48 Telex 10 956 Fax (0 29 40) 3 04 30	The Maple, Kembrey Park Swindon, Wiltshire SN2 6UT Tel. (07 93) 61 41 41 Telex 444 674 Fax (07 93) 69 75 22

National does not assume any responsibility for use of any circuitry described, no circuit patent licenses are implied and National reserves the right at any time without notice to change said circuitry and specifications.

Floppy Disk Data Separator Design Guide for the DP8473

National Semiconductor
Application Note 505
Bob Lutz, Paolo Melloni
and Larry Wakeman

Table of Contents

1.0 INTRODUCTION

2.0 THE FUNCTION OF A DATA SEPARATOR
  2.1 Encoding Techniques
  2.2 Typical Floppy Format
  2.3 Obstacles in Reading Data
  2.4 Performance Measures of a Data Separator
  2.5 Analog Data Separator Basics
  2.6 Operation of an Analog Data Separator
  2.7 Digital Data Separators

3.0 DP8473 DATA SEPARATOR FUNCTIONAL DESCRIPTION
  3.1 Block Diagram Description
  3.2 Self Calibration
  3.3 Data Separator Read Algorithm

4.0 DESIGNING WITH THE DP8473 DATA SEPARATORS
  4.1 Basic Phase Lock Loop Theory
    Initially Locked Model
    The Charge Pump
    The VCO and Programmable Divider
    The PLL Loop Filter
  4.2 System Performance and Filter Design
    Acquisition to the Data Stream
    Theoretical Dynamic Window Margin Determination
    Open Loop Bode Plots and the Second Capacitor
    Choosing Component Tolerances and Types

5.0 ADVANCED TOPICS
  5.1 Design and Performance Testing
  5.2 Understanding the Window Margin Curves
  5.3 DP8473 Filter Switching Design Considerations
    Designing with a Single Filter
    Designing for 250K/300K/500K MFM
    Designing for 1.0 Mb/s and a 2nd Data Rate
    Designing for All Possible Data Rates
  5.4 DP847x Oscillator Design
  5.5 Trimming for Perfection
    Trimming the Loop Gain
    Trimming the Quarter Period Delay Line
  5.6 Initially Unlocked Model
    Acquisition to the Crystal

Bibliography

1.0 INTRODUCTION

Due to the increase in CMOS processing capabilities it is now possible to integrate both the analog and digital circuitry to achieve a high performance monolithic data separator. The choice of CMOS technology also enables the integration of an analog data separator function with good performance onto a floppy disk controller, resulting in National's DP8473 integrated floppy data separator/controller.

This paper discusses the functionality of the DP8473 data separator blocks, after a brief introduction to floppy disk data separator theory. It then delves into the detail of PLL design theory, providing design equations and considerations that enables the user to optimize the performance of the PLL for various applications.

2.0 THE FUNCTION OF A DATA SEPARATOR

2.1 Encoding Techniques

The floppy disk controller writes data to the floppy disk drive in a bit serial fashion as a series of encoded pulses. These pulses are then converted by the drive into magnetic flux reversals on the floppy disk media. The pulses can be later read by the drive and converted back to encoded pulses which can be decoded by the controller into the original data.

Since data is one serial set of bits, and because "real world" imperfections in the writing/reading process can cause the serial information to vary and jitter, the clocking information is embedded into the data stream, which enables synchronization to the data by the circuitry in charge of reading the data.

It is the purpose of the Data Separator circuit to take the encoded data from the disk, and to recover and separate out the clock signal. The separated clock and data signals are then sent to the controller's deserializer which converts the data to bytes of data suitable for microprocessor manipulation.

The two most popular encoding schemes used on floppy disks are: FM (Frequency Modulation), and MFM (Modified Frequency Modulation). FM defines a bit cell for each bit of data. Each cell contains a position for a clock pulse and a position for a data pulse. Each of these positions are referred to as windows. The clock pulse is present in every cell and a data pulse is present only if the data bit for that cell is a one. When this data is read back from a disk, a read clock can be generated from the clock pulses of the signal. An example of FM encoded data is shown in *Figure 1*.

FM encoding was the first method used for recording data on a floppy disk. It is still used in some low cost systems where storage capacity is not a critical issue. This method works very well and requires relatively simple circuitry to separate the clock pulses from the data pulses when the data is read back. However, only 50% of the useful disk space is used for recording data. The other 50% is used to record clock pulses.

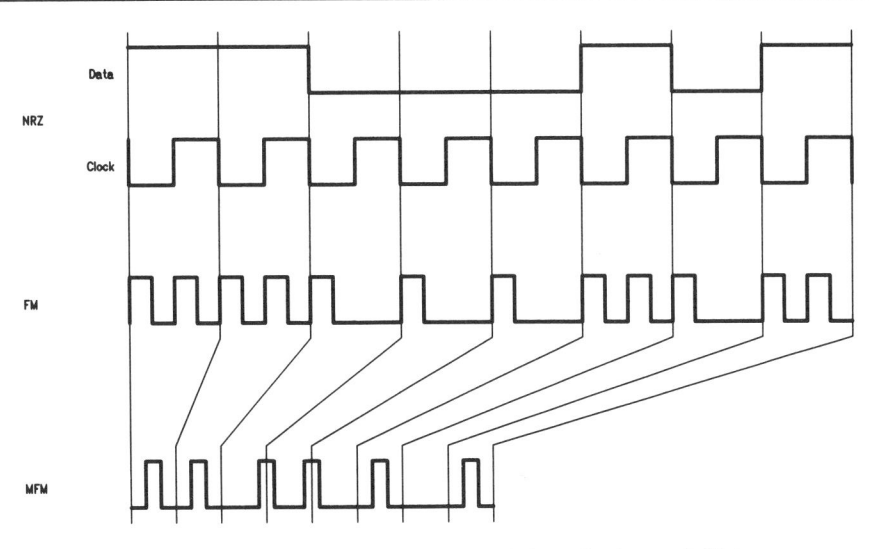

FIGURE 1. Examples of how clock and data information is encoded into
FM and MFM formats. Notice the increased density of MFM over FM.

MFM encoding allows 100% of the useful disk space for storing data, and is currently the most widely used recording format used for floppy disks. MFM defines a bit cell for each bit of data, similar to FM. Again, each cell contains a position for a clock pulse (clock window) and a position for a data pulse (data window). A data pulse is present if the data bit is a one. A clock pulse is present only if the data bit is a zero and the data bit in the previous bit window was a zero.

A comparison of FM and MFM can be seen in *Figure 1*. Because MFM requires fewer pulses to encode the same amount of data, the information can be stored in half the area required for FM encoded data. The only drawback of MFM is that it requires better read/write head and accompanying electronics. It also requires a higher precision data separator than FM requires. This is to resolve the location of each pulse more precisely than with FM.

2.2 Typical Floppy Format

A disk consists of many separate tracks. These tracks are configured as a set of concentric circles. Each track contains a set of many sectors. Each sector contains one Address Field and one Data Field. *Figure 2* shows the most common track format used on floppy disks today. It is the IBM double density standard.

The Address Field within a sector is used to identify what sector the following data field belongs to. It begins with a synchronization field or preamble to allow the data separator (which will be described soon) to synchronize to the speed at which the data is being read from the disk.

Next an address mark uniquely identifies this field as an Address Field. The address marks are encoded with a unique illegal pattern that is an MFM encoding rule violation. The violation is a missing clock pulse from a particular location within the byte. This illegal pattern guarantees that this is an address mark field and not a data pattern from some other area of the disk. The following four bytes identify the sector being read. This is followed by a CRC (Cyclic Redundancy Check). The CRC allows the controller to verify that the information read is free of errors. This is followed by a gap which is simply a series of bytes that physically separates the Address Field from the Data Field.

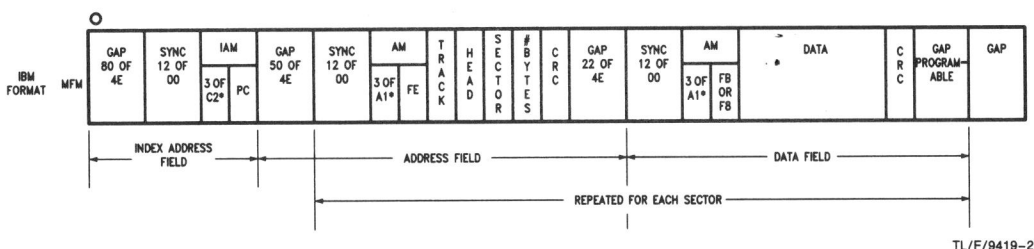

FIGURE 2. Typical format of a floppy disk. This is the IBM MFM standard.

Notes:
C2* = Data Pattern of C2, Clock Pattern of 14
A1* = Data Pattern of A1, Clock Pattern of 0A

The Data Field contains the data that the sector represents. It begins with a preamble and data field address mark, similar to the beginning of the Address Field. The actual sector data follows this. The data is followed by a CRC and then a gap, that separates this sector from the next.

2.3 Obstacles in Reading Data

Since MFM is the most popular floppy disk data encoding method the following discussion will refer specifically to MFM.

The floppy controller must be able to decode the data read from a disk drive. Theoretically, this could be a fairly easy process. The controller must first synchronize to the clock pulses in the preamble field of a sector. After that, it is just a matter of checking when the next encoded data pulse arrives. The pulse can arrive, 1, 1.5 or 2 bit periods later. Using this information the controller can decode this and all subsequent bits, reconstructing the original data from this information.

Unfortunately, this simple method is not so simple. The data pulses read back from a disk drive will generally be somewhat different from the data originally written.

There are three major sources of data degradation.

1. **Bit Shift**—As data is written, magnetic interaction of adjacent bits cause the data to be shifted in time from its nominal position. When these flux transitions are recorded close to each other, the superposition of their magnetic fields tends to move their apparent position. Thus when they are read, the floppy drive's peak detector moves the peak of these flux transitions apart from each other. This is the major cause of instantaneous bit shift.

   (This type of data degradation is mostly predictable and can be partially compensated for by shifting the data as it is written to the disk in the opposite direction that the bit is predicted to shift. This is called Write Precompensation. The DP8473 contains circuitry required to perform this function. The write precompensation circuit intercepts the serial data being written to the disk and shifts the data early, late, or none, based on the data pattern.)

   Several other factors can contribute to jitter. The drive's peak detector may be unbalanced, resulting in a bit shift similar to that described above. This could cause positive going peaks to appear earlier than negative going peaks or vice-versa. Also, a long cable between the disk drive and the disk controller may contribute to bit shift.

2. **Motor Speed Variation (MSV)**—This is an error in the spindle motor speed from the nominal, and causes the data rate to vary typically 1–2% for each drive. For design purposes this value is doubled since drive media is interchangeable. Thus a slow drive can record data that is read on a fast drive.

3. **Instantaneous Speed Variation (ISV)**—This is an additional speed error that is a constantly changing affect due to disk-jacket friction, and mechanical resonances. Usually this variation has a frequency component less than 1 kHz and causes the data rate to vary an additional 1–2% (again doubled).

While bit shift is the primary cause of decoding problems, the speed variations create difficulty in locking to the frequency of the data stream, and degrade jitter tolerance. The data separator must be able to synthesize the average frequency of the data coming in, and if the disk data rate differs from the nominal value then synchronization is more difficult.

2.4 Performance Measures of a Data Separator

There are several measures of data separator performance. The most universal one is window margin. Window margin measurements themselves can be subdivided into two categories, Dynamic, and Static (described shortly). Window margin is defined as the amount of bit shift that can be tolerated by a data separator without mis-decoding the data.

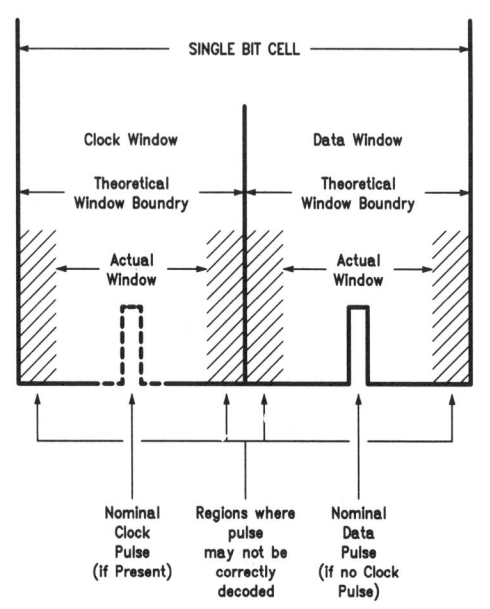

FIGURE 3. Window Margin Timing Definition

Figure 3 shows a timing diagram of a typical bit cell, and its composite data and clock windows. Theoretically a pulse (either clock or data) could be shifted in time either early or late relative to its nominal position by up to $\frac{1}{4}$ of a bit period and still be decoded correctly. This is the theoretical window boundary region in *Figure 3*. If the pulse is shifted more than this, then it would fall into another pulse's window. In reality, due to the limitations of practical data separator implementations, the actual window boundary in which data will be directly decoded is less than the full $\frac{1}{4}$ period. This is shown in *Figure 3* as the actual window region. Typically this actual window size is measured as either a percentage of the theoretical maximum or in terms of nanoseconds. The former is the more popular method and will be used here as well. In equation form:

$$WM\% = \frac{(\text{Actual Window Size})}{(\text{Nominal Theoretical Window Size})} \times 100\%$$

Window margin should be measured with a specified amount of ISV and MSV, at a specific data rate, with a specified data field and format pattern, and a known bit shift algorithm. If any of these are unspecified, then a true comparison of data separator performance is more difficult. Window margin is usually specified as a percentage of the nominal frequency window (as opposed to the nominal frequency plus the MSV).

There are two basic types of window margin tests. One test is where the data separator and controller must read a sector of data, in which all the bits are shifted until the data separator cannot read the sector correctly. This is called dynamic window margin. A second test is to present a long sequence of perfectly centered MFM clock pulses except for one bit. This bit is shifted until an error occurs. This second test is called static window margin.

Do Not Confuse the Two Measurements. The first one more correctly reflects the "Real World". The second one is an indicator of the accuracy of the circuits that compose the PLL, but does not include most of the errors due to the response of the PLL.

As an example of a dynamic window margin test: A data separator has a 70% window margin at 500 Kb/s, with a total ±1.5% MSV, and ±1% ISV. The encoding is MFM, and the data pattern is a repeating DB6DB6 ... (HEX) pattern. A reverse write precompensation algorithm is used for pattern dependent bit jitter (all bits are jittered). These conditions are one of the worst case conditions for analog data separators. This means that the data separator will correctly decode a pulse so long as it is shifted no more than ±350 ns from its nominal position (the theoretical window at 500 Kb/s is ±500 ns) over the full MSV and ISV range.

Another data separator performance measurement is Bit Error Rate (BER). This is a measure that is defined as ratio of the number of bit errors during long term reading divided by the total number of bits read. A small bit error rate is desirable. BER is better for evaluation of total system performance, since the performance of the whole system effects on this figure. Thus as a final system checkout the manufacturer can specify the media, drives, data separator, and determine the error rate of this system. It is relatively difficult to isolate the bit errors due solely to the data separator. This specification is a less exact performance measurement than window margin for a data separator.

Why is Window Margin Important?

The greater the window margin the lower the error rate. For example if a bit is read with too much bit jitter for the data separator, then that data cannot be read and the whole sector (or file) is lost. This is especially fatal since floppies do not have error correction capability.

Another area where window margin is important, is manufacturing yields. A larger window margin ensures that when intermixing best/worst case drives and controllers there is minimum fallout. Thus larger volume vendors tend to try to optimize window margin to improve yields as well as data integrity.

Each designer needs to decide for himself what margin requirements are necessary. In general, many high quality analog designs typically achieve 60–65% window margin under worst case conditions, with the best designs approaching 70%.

Later in sections 4.2 and 5.1 theoretical calculation and practical measurements of window margin are described.

2.5 Analog Data Separator Basics

The job of a data separator is to produce a read clock that follows the slow data rate change caused by the drive motor variation (MSV and ISV), but not track the instantaneous bit jitter. This read clock then is used to clock in the serial data into some type of deserializer. Generating a read clock for MFM encoded data is potentially difficult. Because of the MFM encoding rules, many clock pulses are missing from clock or data windows. As a matter of fact, in a long string of one's, there are no clock pulses at all. A data separator must use both clock pulses and data pulses to synchronize to the encoded signal. The most popular method to do this is with a Phase Locked Loop (PLL).

A PLL consists of three main components, a phase detector, a filter, and a voltage controlled oscillator (VCO), as shown in *Figure 4*. Also in most cases a divider is used to divide the VCO frequency as needed. The basic operation of a PLL is fairly straight-forward. The phase detector detects the difference between the phase of the VCO (or divider) output and the phase of a periodic input signal. This phase difference is converted to a current which either charges or discharges a filter. The resulting filter voltage changes the frequency of the VCO (and also the divider) output in an attempt to reduce the phase difference between the two phase detector input signals. A PLL is "locked" when the frequency of the two phase detector input signals is the same.

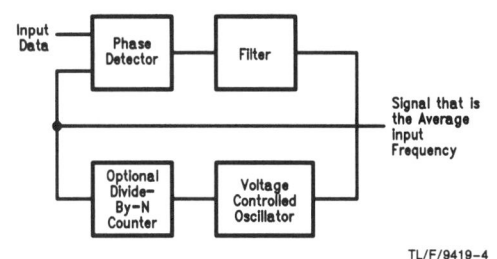

FIGURE 4. Simplified Block Diagram of Phase Locked Loop

With a slight modified version of the basic PLL, an MFM data separator can be made. A modification is required because MFM encoded data is not a periodic signal. A phase comparison can only be made when a pulse arrives from the disk. When there is no clock or data pulse, the PLL should continue generating the frequency it was generating before the missing pulse. This is called a phase only comparison, and it is the usual method of tracking an MFM signal.

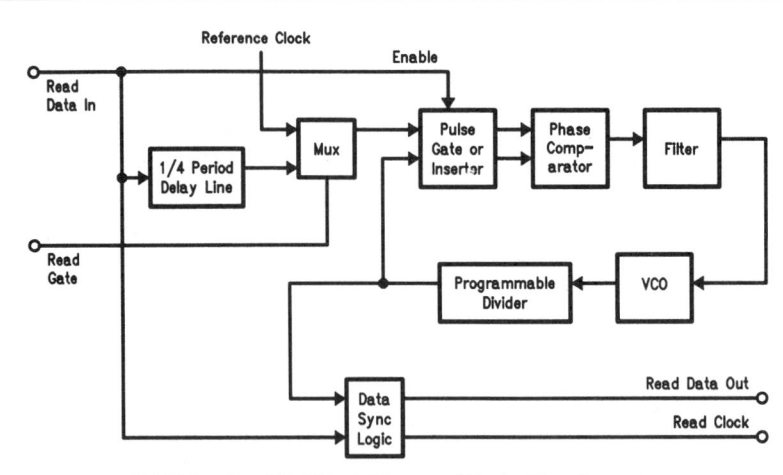

FIGURE 5. Simplified Block Diagram of Typical Data Separator

A typical data separator is shown in *Figure 5*. In addition to the components of a typical PLL, it includes a quarter period delay line and either a pulse gate or pulse inserter (not both). Both of these blocks enable the pulse gate or pulse inserter to decide when the phase comparison should be made. The quarter period delay line delays the incoming data pulses a quarter of a bit cell, and this feeds the pulse gate or pulse inserter. The pulse gate will disable phase comparisons when a VCO pulse occurs but read data pulses are missing. The pulse inserter will insert fake read data pulses into the phase detector when there is a VCO pulse but no read data pulse. These components are required to determine the proper timing of the phase comparisons for MFM encoded data. The need for these blocks can be demonstrated by referring to *Figure 6*. Only the use of the pulse gate is described since this is what is implemented in the DP8473 data separator.

Figure 6a shows two MFM bit cells, each with a clock pulse. The VCO output provides two clocks per cell since an MFM pulse can appear in either of the two windows that compose the bit cell. (Note for simplicity the divider block is ignored.) To achieve lock, the data separator tries to line up the rising edge of the input pulses with the rising edges of the VCO output cycles. MFM encoded data is not periodic, that is some of the cells are missing pulses. The data separator must decide when to make a valid phase comparison. This can be seen from *Figure 6a* where the phase detector first makes a comparison to an early pulse, which is correct, but then on the next VCO cycle the phase detector now compares this VCO edge even though no input pulses are present. Hence, there must be a mechanism for fooling the phase detector into not making a comparison. The method chosen in the DP8473 is to use a pulse gate to eliminate the unwanted VCO edge.

However, disabling the phase detector's input does not completely solve the problem, as shown in *Figure 6b*. Here the first early pulse is compared correctly. At the beginning the next VCO cycle the data separator does not know whether to do a phase comparison, since it does not know whether the pulse is missing or just late. Thus by the time the next pulse does arrive the PLL is lost.

Therefore, the DP8473 data separator uses a $\frac{1}{4}$ period ($\frac{1}{2}$ bit window) long delay line with the pulse gate. Now with this delay line, all phase comparisons are made to the delayed data. Thus the PLL is operating $\frac{1}{4}$ of a period behind the data coming from the disk, but this allows the phase comparison enable logic to determine whether a pulse will occur in a bit cell or not, and make the proper comparison.

Figure 6c shows how this works. For an early bit, the data input enables a phase comparison, and the phase detector compares the delayed data bit to the VCO edge. In the case of this early bit, the proper pump up is generated. On the next VCO cycle, the quarter period delay has detected no pulse, and so no comparison is made. For the late bit, the comparison is enabled prior to the VCO clock, so a pump down is generated until the delayed data bit is seen by the phase detector.

At nominal frequency, a delay of $\frac{1}{4}$ of a bit ensures that the phase detector will be properly enabled even if the data bit is late all the way to the edge of its clock or data window. (Remember one bit cell contains a clock and a data window. A data pulse will appear within either (but not both) of these windows. Therefore the theoretical maximum amount of bit shift is $\frac{1}{4}$ of a bit cell.)

The quarter period delay line solves this problem of forecasting the future. It causes the MFM encoded data pulses to be delayed by a quarter of a bit period. This allows the pulse gate to determine when data pulses exist ahead of time and thus enable the phase detector only at the appropriate times.

It is important that the quarter period delay line be accurate. If the quarter period delay line is not accurate (ie. it's too long or too short), then the window margin performance of the data separator will be reduced. This performance reduction is due to the PLL's inability to correctly resolve bit shift near the edge of a bit window. For example, if at 500 Kb/s the delay line were shorter than it should be, say 400 ns long instead of 500 ns, then a bit shifted 450 ns from its nominal position is incorrectly decoded. The window margin in this case is immediately reduced 20% from it's ideal. The same degradation occurs when the delay line is too long.

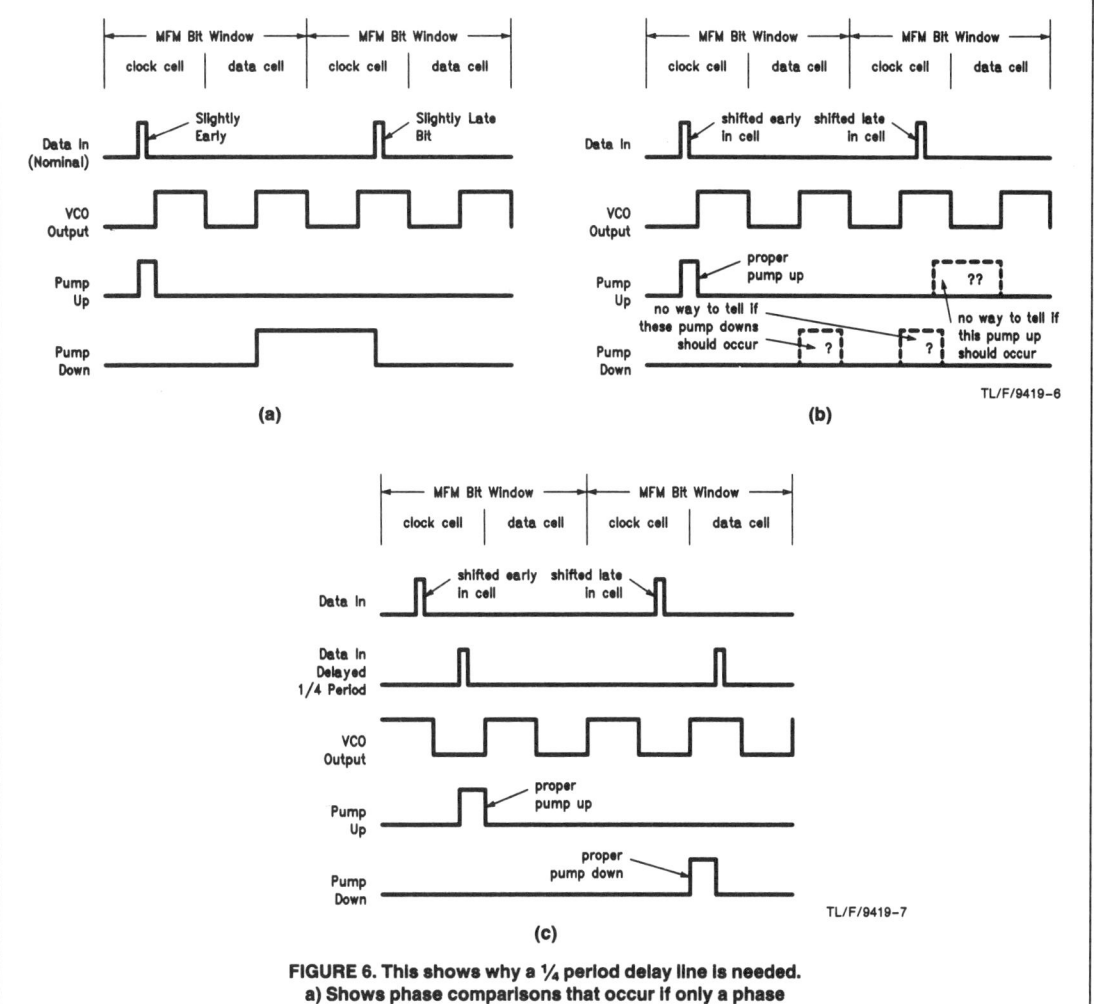

FIGURE 6. This shows why a ¼ period delay line is needed.
a) Shows phase comparisons that occur if only a phase
detector is used; b) Shows the data separator's need to predict
the arrival of a pulse; and c) Shows how the ¼ period delay fixes this.

2.6 Operation of an Analog Data Separator

A data separator can be described as operating in one of three phases during each read cycle: Idle Phase, Initial Locking Phase, and the Tracking Phase.

Initially, when the data separator is not being used to read data from the disk, it is in the Idle Phase. While in the Idle Phase, the PLL is both phase and frequency locked to a reference frequency. (Frequency comparison is implemented by forcing a phase comparison every VCO clock.) The PLL must eventually lock to both clock and data pulses of the encoded data when it is read from the disk, so the reference frequency is generally two times the data rate frequency.

When data is to be read from the disk, the PLL switches from the reference frequency to the incoming data stream. Because the encoded data read from the disk is not a periodic signal, only phase comparisons are made. Since the PLL was initially locked to a frequency very close to twice the actual data rate, the time required for the PLL to lock onto the data read from the disk is minimized.

To further minimize this locking time, the beginning of each Address Field and Data Field starts with a preamble (or synchronization field). The preamble is a series of bytes with a zero data pattern (all clock pulses and no data pulses). When read, the preamble will produce a periodic signal with little bit jitter. The data separator can lock to this signal with the least chance of an error. It would be ideal for the floppy controller to switch the data separator from the Idle Phase to the Initial Locking Phase at the beginning of a preamble to enable the maximum amount of lock time.

Once the PLL is locked to the average frequency of the data being read from the disk, it should simply track the data frequency. This means tracking the slow data rate speed variations caused by the drive motor, yet ignoring instantaneous bit jitter. This is the Tracking Phase. The data separator then allows the controller's deserializer to start decoding the incoming data.

2.7 Digital Data Separators

A second method of separating clock and data information is to use a digital data separator. While the circuits for the analog solution has evolved significantly, digital data separators have also improved somewhat, in a (less than successful) attempt to match the performance of the analog approach. These circuits are described below.

First Generation Digital Data Separator (DDS)—This circuit is a very convenient all digital data separator. Its primary advantages are simplicity, and low external parts count. This circuit usually consists of a set of counter timing circuits and some control logic to count times between individual pulses, and thus determine whether a pulse is clock or data. The SMC9216 is representative of this technology. Its major disadvantage is performance, and the inability to optimize the window margin for various lock ranges. The dynamic window margin for these types of circuits is usually around 55% with no MSV and as low as 30% with a ±3% total MSV.

Second Generation DDS—A sophisticated digital data separator can be compared functionally to an analog data separator. The ideal digital separator consists of a sampler (phase detector), a ROM look up table, with memory (filter), and a programmable counter (VCO). The pulse gate can be implemented as an extension of the ROM look up table. These circuits are typically much better than 1st generation circuits, but still far short of the analog approach. These circuits have dynamic window margins of 50–55% over a ±6% lock range (total MSV variation).

3.0 DP8473 DATA SEPARATOR FUNCTIONAL DESCRIPTION

The integrated floppy disk data separator from National Semiconductor combine the performance of an analog PLL and the ease of use of a digital data separator. It does not require any external trimmed components, and it has a data rate range from 125 Kbits/sec up through 1.0 Mbits/sec. It is built using CMOS technology to achieve good linear performance as well as low power operation. A block diagram for the data separator is shown in *Figure 7*.

3.1 Block Diagram Description

The heart of the DP8473 data separator is the main PLL. The main PLL consists of the VCO, programmable divider, phase detector, and the charge pump. The entire operation of the PLL and data separator logic is based on the Reference Clock which should be an accurate reference frequency. The Reference Clock is divided by two, then feeds the Secondary PLL, and the divide-by-N counter. As discussed later the Secondary PLL is used to calibrate the operation of the quarter period delay and Primary VCO. The Reference Clock's Divide-By-N counter and the Programmable Divider are both programmable counters whose divide by factor is determined by the data rate selected. The output of the divide-by-N and the Programmable divider is always twice the data rate. The output of the divide-by-N is used as a reference frequency for the PLL to lock to when the PLL is idle. The output of the Programmable Divider is the separated clock that is used to strobe the incoming pulses into the controller's deserializer.

Note: Throughout this discussion, the Reference Clock as shown in *Figure 7* is the master clock for the data separator block. This Reference Clock also generates several other clock frequencies that are used by the data separator sub-sections. In the following discussions the term Reference Clock refers only to the signal in *Figure 7* that feeds the divide-by-2 and divide-by-N blocks. Also the term Divide-By-N counter is used for the counter driven by the Reference Clock, whereas the Programmable Divider refers to the counter driven by the VCO.

In the DP8473, the Reference Clock of *Figure 7* is derived from a prescaler circuit that is operating at 24 MHz. The output of this prescaler circuit is 8 MHz for all data rates except 300 Kb/s. At 300 Kb/s the equivalent prescaler output is 9.6 MHz. The 24 MHz is intended to be a fixed frequency, but could be scaled lower for unique applications if desired.

Under normal operation the Secondary PLL and the Primary VCO run at one half the Reference Clock frequency. Thus the Primary VCO's output is 4 MHz (except at 300 Kb/s where the VCO output is 4.8 MHz.) Operation at different data rates is accomplished by changing the Divide-By-N and Programmable Divider.

The basic operation of the Phase Locked Loop is fairly standard although there are several added features in the DP8473 PLL. The phase detector determines the phase

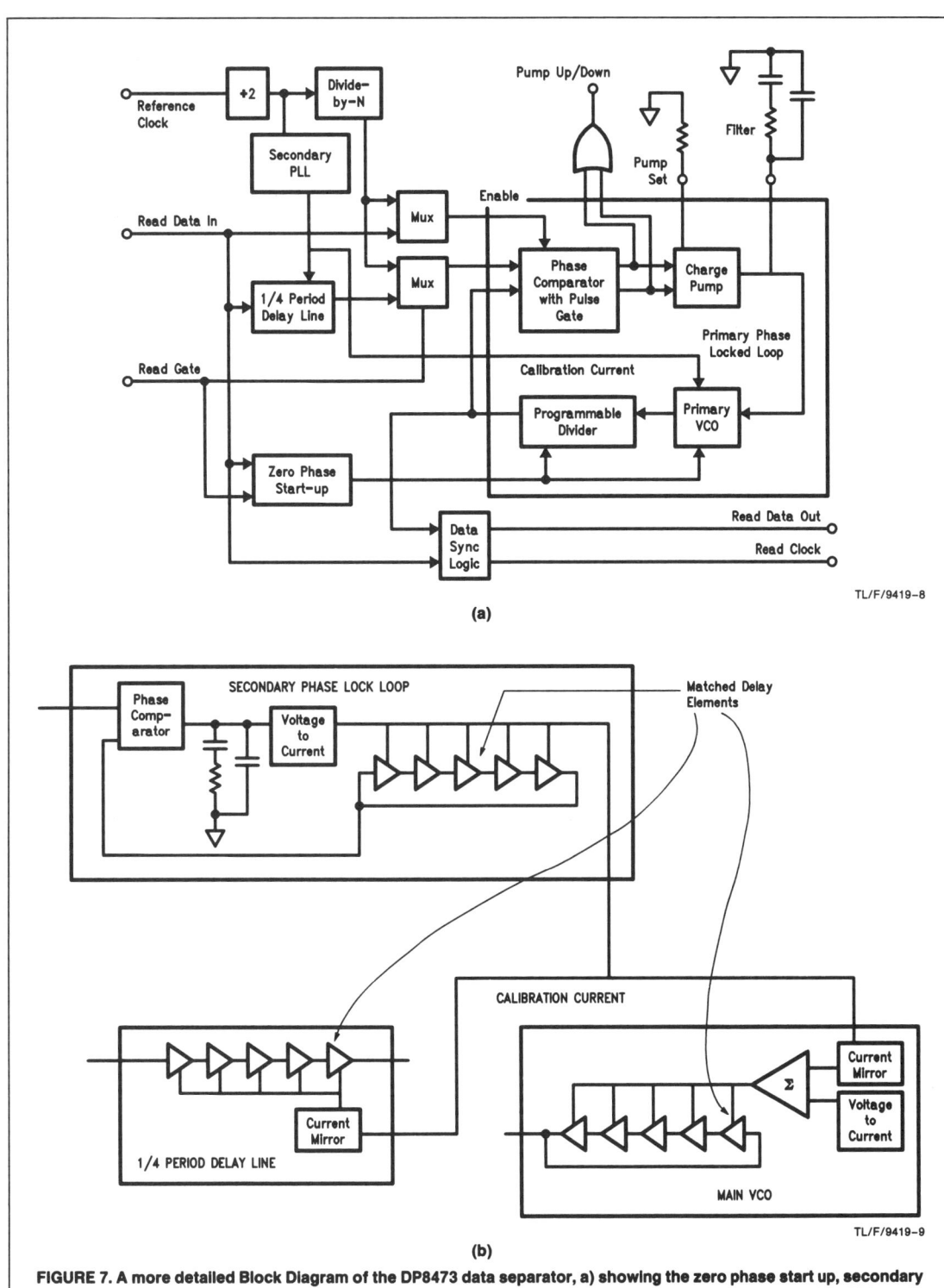

FIGURE 7. A more detailed Block Diagram of the DP8473 data separator, a) showing the zero phase start up, secondary PLL, control logic block and external R's and C's. b) Also showing a more detailed diagram of the secondary PLL, main VCO and delay line. This highlights the self calibration technique.

difference between two inputs. One input is always the divided output of the Programmable Divider.

The other input is either a reference frequency (derived from the Reference Clock) of twice the data rate while the data separator is in the idle mode, or when reading the disk the encoded data from the drive after it has passed through the quarter-period delay line.

While in the idle mode, the phase detector determines both phase and frequency information. This is accomplished by forcing the pulse gate to enable all phase comparisions. This state is set by the top two multiplexers of *Figure 7* which in idle mode select clocks generated by the Reference Clock to enable the pulse gate on every clock edge (hence all clocks are compared by the phase detector). By locking to the Reference Clock generated frequencies in both phase and frequency, the PLL is preventing from locking back to a harmonic of this frequency.

When the data separator is told to read the incoming data pulses, Read Gate is asserted. This changes the signals selected by the multiplexers. The top multiplexer switches to inputting the "raw" read data pulses into the pulse gate, and the bottom multiplexer sends read data delayed by a quarter bit period to the phase detector input. When this switch occurs, the Zero Phase Start-Up logic synchronizes the Programmable Divider's output to be in phase with the very next arriving data pulse. This causes the PLL to acquire lock to the data quicker since it is starting with a "near zero" phase error between the Programmable Divider and the encoded data.

The quarter-period delay line consists of a series of voltage controlled delay elements. The encoded data from the disk drive enters the beginning of the delay line. The output is derived from the output of one of the delay elements. The delay element used for the output depends upon the data rate used.

When locked to either the read data pulses or the reference, the phase detector issues either a pump up signal or a pump down signal depending upon whether the VCO should increase or decrease its frequency. The length of this pump signal is proportional to the amount of phase difference between the two input signals. When locked, the phase difference between the VCO and the delayed data will be small. In this case the width of the pump signal could be so small that the rise time may prevent the signal from ever being recognized by the charge pump. To ensure that the charge pump can recognize even the smallest pump signal, both pump up and pump down signals are asserted at each phase comparison and the appropriate signal is extended by an amount proportional to the phase difference between the two input signals. The pump signals are then subtracted from each other by the charge pump. Therefore, the rise time of the pump up or down signals will not degrade the performance of the charge pump.

The charge pump simply adds or removes an amount of charge proportional to the length of the pump signal to or from an external filter. The voltage of the external filter determines the frequency produced by the VCO.

Finally a synchronized clock and serial data signal is sent to the controller's deserializer by the Data Synch Logic Block. This circuit takes the output of the Programmable Divider and the Read Data pulses, and synchronizes these two signals, by centering the read data pulse in the appropriate Programmable Divider's clock cycle. This allows the controller to easily deserialize and decode the data pulses properly.

An additional block not shown here, but used on the DP8473 is the filter selection logic. This logic is used to select different filters for different data rates. The description and use of this circuitry is described in section 5.3.

3.2 Self Calibration

Normally, most VCO implementations would need an external precision capacitor (maybe trimmed) to set its center frequency. Also, the quarter period delay line would require an external trimming resistor to set the delay to exactly a quarter of the data rate. The actual delay of the delay elements used in these functions would normally vary from one part to another due to normal process variations. However, the DP8473 has been designed to eliminate the need for these external trims. There are actually two PLLs in the DP8473; the Primary PLL, and a Secondary PLL. The Primary PLL is used for data separation. The Secondary PLL is used to calibrate all of the delay elements used in the chip. This includes the quarter-period delay line and the main VCO.

The VCO of the Secondary PLL is a ring oscillator of delay elements. The amount of delay that each inverter produces is regulated by a control voltage which is the internally connected output of the Secondary PLL. The Secondary PLL is locked to the same frequency as the Primary VCO, half of the reference clock frequency. The delay elements used in the secondary VCO are identical to those used in the Primary VCO and are regulated by the same control voltage. There are also the same number of delay elements in each. Therefore, the center frequency of the Primary VCO is internally trimmed to exactly half of the reference frequency. Since the delay elements in the Secondary PLL have a known delay, any number of identical elements that are set by the calibration voltage will have a very accurate delay. Thus the quarter period delay line is just a chain of these delay elements that have the desired total length.

The delay elements used in the quarter-period delay line are also of the same type used in the secondary VCO. Because the delay of each element is accurately set by the Secondary PLL, there is no need for any trimmed tuning components for any of these circuits. Essentially, the only external passive components required for the DP8473 are for the filter(s) (two capacitors and a resistor per data rate), and a resistor to set the gain of the charge pump (ie. the amount of charge pump current).

3.3 Data Separator Read Algorithm

Since the DP8473 floppy disk controller incorporates the analog data separator, it takes advantage of close proximity of the controller and data separator blocks to implement a read algorithm that is much more sophisticated than previous floppy controller integrated circuits.

Before describing the details of the algorithm, a brief discussion of a disk read (or write) is necessary. When the controller is issued a read (or write) command, it is asked for a specified sector. The controller starts to look at the incoming MFM information. It scans this information, trying to locate the proper address field. In order to do this, the data separator is first told to lock to the disk data, and once locked the controller looks at the incoming information. In most controllers, the data separator is told to look at the data continuously until the correct sector address is found. This makes the data separator susceptible to being thrown out of lock, since the controller is not "watching" the data separator to see if it has maintained lock through the search process (which it can easily lose). Once the correct address is found the controller must then ensure that the associated data area is read, by re-locking to the incoming signal and then reading (or writing) the data.

Figure 8 shows in state diagram form the algorithm used by the DP8473 controller to ensure that disk data is correctly read (or written). This algorithm is much more sophisticated than previous generation controllers. The following describes the operation. When the controller is idle, the data separator is locked to the crystal/clock reference in a frequency comparison mode. (Frequency comparisons are made by disabling the pulse gate, and ensuring that all reference and VCO cycles are compared.)

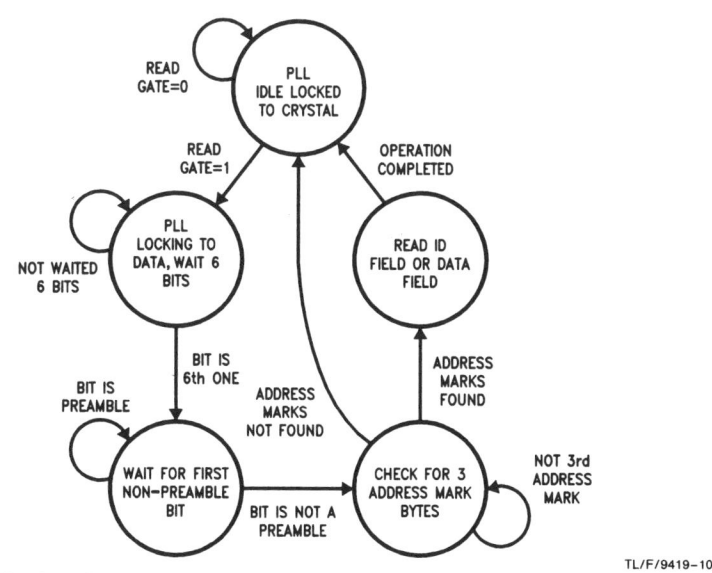

FIGURE 8. State Diagram for a Data Separator Synchronization and Read Operation

When a read command is issued to the controller the controller asserts an internal Read Gate signal to the data separator. This causes the PLL to switch from locking to the reference to locking to the data with the pulse gate enabled, and the primary VCO/divider started in phase with the next incoming pulse. The PLL waits 6-bit times to lock to the data. When the seventh bit arrives the data separator assumes that this bit is a preamble bit (thus an MFM clock bit). The controller/data separator then continues looking at the data until a non-preamble pattern is detected (ie. an MFM data pulse). It then checks to see if it has now encountered an address mark with the proper rule violation. If it has not, read gate is deasserted. The data separator returns to the idle state for 6 byte times, and then starts all over again.

If three address mark bytes are found then the data separator remains locked to the data while the controller looks to see if it has found the right address field. If the controller discovers that this field is not the correct address field then it deasserts read gate for 6 bytes, and tries again.

If the correct address field is encountered, the controller deasserts read gate during the gap between the address and data fields. It then re-asserts read gate, and follows the state diagram to read the data field (ie. looking for preamble, address marks etc.).

This comparison is done on a bit-by-bit basis, therefore ensuring that the PLL never tries to lock on an unwanted field for more than one bit time. In other words, the PLL will never loose lock. This algorithm provides a very fast lock to the data stream, and ensures that the data separator never falls out of lock while reading the data. Both of these features reduce the need to do retries of operations to ensure correct execution.

4.0 DESIGNING WITH THE DP8473 DATA SEPARATOR

The following section is a fairly in-depth description of the design characteristics of the PLL in the DP8473 controller. *(National Semiconductor cannot be responsible for the sanity of any one who ventures into this section. Hence we recommend using the filter values supplied in the datasheets.)*

Two elements determine the overall performance of a Phase Locked Loop: the loop gain and the loop filter design. When using the DP8473 both of these elements are controlled by the user. The amount of current in the charge pump circuit can be set with an external resistor. This will set the overall gain of the PLL. The filter is external to the DP8473 and is user definable. This gives the user the possibility of tailoring the data separator performance to his own application requirements and design criteria. The following information will present some tradeoffs that apply in choosing the external components for typical applications.

4.1 Basic Phase Lock Loop Theory

This section will first start with the basic control systems model for a second order PLL, and then apply these basic equations to the individual blocks that compose the data separator in the DP8473 Controller.

Initially Locked Model

In order to understand the behavior of the data separator and to discuss the tradeoffs of the different design parameters, some background in the theory of PLLs will be presented.

Figure 9 shows a diagram of a PLL which is assumed to be in a locked state. Each box contains the phase transfer function of the corresponding block and each node has the relative phase signal. The PLL is locked, which means that the VCO output and the input signal are at the same frequency and in-phase with each other (the phase error is a constant). This model is useful for understanding the phase locking process when the PLL is switched from the reference frequency to the incoming data.

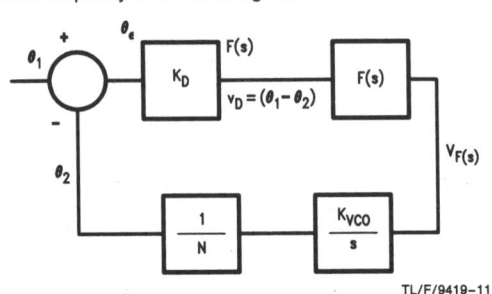

FIGURE 9. The Block Diagram for an Initially Locked PLL Control System, Showing the Transfer Functions for Each Block

The frequency variations that this model take into account are assumed small enough so that the loop stays locked in frequency. Therefore, only the effect on the phase difference is considered. Also, it is easier to refer to bit shift tolerance in terms of phase. Hence the phase transfer functions yield the most appropriate information.

The concept of phase is very much related with the concept of time. The advantage of phase information is that it is independent of frequency of the signal and it is measured as a pure number (radians).

A simple RC filter model will be used to simplify the math. This is actually a good model because the effect of a second capacitor is only seen at high frequencies. This simplification allows the use of second order PLL theory that is easily available in literature. (See for example: R.E. Best, Phase Locked Loops, McGraw-Hill, 1984)

From *Figure 9* the closed loop phase transfer function for the loop can be derived using standard control system theory techniques, and reduces to:

$$H(s) = \frac{\theta_2(s)}{\theta_1(s)} = \frac{K_D K'_{VCO} F(s)}{s + K_D K'_{VCO} F(s)} \quad (1)$$

where $K'_{VCO} = K_{VCO}/N$, N being the number of VCO cycles between phase comparisons due to the divider that typically is inserted between the VCO and phase detector. The closed loop phase error function can be written as:

$$H_\epsilon(s) = \frac{\theta_\epsilon(s)}{\theta_1(s)} = 1 - H(s) =$$

$$\frac{\theta_1 - \theta_2}{\theta_1} = \frac{s}{s + K_D K'_{VCO} F(s)} \quad (2)$$

Now we'll evaluate the variables that appear in expressions (1) and (2).

The Charge Pump

In the phase detector/charge pump circuit a current is generated in the correct direction (positive or negative) every time the edges of the VCO/divider output and the incoming data pulses are not coincident. The current is a pulse with amplitude equal to I_{PUMP} and length equal to the phase error between the two signals. The pump current is zero for the rest of the period, so the average current is:

$$I_Z = \frac{I_{PUMP} \theta_\epsilon}{2\pi} \quad (3)$$

where θ_ϵ is the phase error between the VCO/divider and input pulses. The phase detector and charge pump gain is:

$$K_D = \frac{I_Z}{\theta_\epsilon} = \frac{I_{PUMP}}{2\pi} \quad (4)$$

where $I_{PUMP} = K_P \times I_R$. I_R is the current set by an external resistor at SETCUR pin. The current at this pin is: $I_R = V_{REF}/R_1$, $K_P = 2.5$, and $V_{REF} \cong 1.2V$. Thus combining equations:

$$I_{PUMP} = \frac{(2.5)(1.2)}{R_1} \quad (5)$$

Note that the maximum current that can flow to or from the charge pump is 500 μA, which corresponds to a resistor value of 3 kΩ. The minimum current is limited to 125 μA by stability and leakage constraints on the internal reference circuits. So R_1 must be smaller than 12 kΩ. In conclusion, the charge pump current and resistor can be set in the following range:

$$125 \, \mu A \leq I_{PUMP} \leq 500 \, \mu A$$
$$3 \, k\Omega \leq R_1 \leq 12 \, k\Omega$$

Any value in this range can be chosen, and usually the choice is dependent on the PLL filter capacitor's mechanical size and cost. We have chosen in the datasheet a value of 5.6 kΩ since it represents a good compromise of all these considerations.

The VCO and Programmable Divider

The VCO gain is defined as the ratio between a frequency change at the output vs. a voltage change at the input (FILTER pin). This value cannot be set by the user and has been designed to be immune to process, temperature and voltage variation. There is a variation of less than ±20% between different parts, and the typical value of K_{VCO} is:

$$K_{VCO} = 25 \frac{Mrad/sec}{volt} \quad (6)$$

The actual value K'_{VCO} for expressions (1) and (2) differs by a factor of N. N is the ratio between the frequency of the internal VCO and the "instantaneous" frequency of the data. This takes into account both the factor due to the way the data is encoded, and the factor due to the internal programmable divider used for the data rate selection. The following table gives the value of N for different codes, data rates, and data patterns. K'_{VCO} can be derived from these values of N.

TABLE I. VCO Gain Reduction Factor for the DP8473 with a 24 MHz Crystal/Clock

Data Rate	Code	Data Patterns	N
1 Mb/s	MFM	all 0's, all 1's	4
		010101...	8
500 Kb/s	MFM	all 0's, all 1's	8
		010101...	16
	FM	all 0's	8
		all 1's	4
300 Kb/s	MFM	all 0's, all 1's	16
		010101...	32
250 Kb/s	MFM	all 0's, all 1's	16
		010101...	32
	FM	all 0's	16
		all 1's	8
125 Kb/s	FM	all 0's	32
		all 1's	16

So for a 250 Kb/s MFM data rate N = 16 and K'_{VCO} = 1.56 Mrad/set/volt.

The loop filter calculation is made assuming lock and acquisition during a preamble (all 0's pattern), so these values of N are used in the bandwidth and damping calculations shown later.

The PLL Loop Filter

Inside the data separator, the charge pump output is connected directly to the VCO input. A filter is attached externally to this point. The typical configuration of this filter is shown in *Figure 10*. The output of the phase detector/charge pump circuit is basically a current generator with a very high output impedance (hundreds of kΩ). This high impedance combined with the external capacitor, C_2, of the filter provide a small (close to 0) steady phase error after a frequency step in the input signal. The charge pump setting along with C_2 sets the bandwidth on the PLL. The DP8473's charge pump circuit eliminates the need for an external active filter. The resistor R_2 is the damping resistor and it controls the stability of the loop.

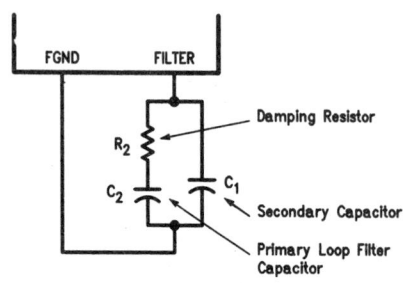

AN-505

FIGURE 10. Simple Schematic of Typical DP8473 Data Separator Filter Configuration

The filter design is usually improved by adding another capacitor in parallel, C_1. This second capacitor is intended to improve the low-pass filtering action of the PLL. In our subsequent filter discussions, C_1 is ignored initially since its value will be much smaller than C_2.

In the DP8473, the input of the filter is a current from the phase detector/charge pump, the output is a voltage to the VCO. Therefore, the transfer function of the filter of Figure 9 is simply its impedance:

$$F(s) = Z(s) = \frac{1 + sR_2C_2}{sC_2} \quad (7)$$

(As mentioned, we are ignoring the effect of C_1 for now.) Substituting these equations into (1) and (2) produces:

$$H(s) = \frac{\dfrac{K'_{VCO}K_D(sR_2C_2 + 1)}{C_2}}{\left(s^2 + s(K'_{VCO}R_2K_D) + \dfrac{K'_{VCO}K_D}{C_2}\right)} \quad (8)$$

This then reduces to a standard second order equation of the form:

$$H(s) = \frac{(2s\zeta\omega_n + \omega_n^2)}{(s^2 + 2\zeta\omega_n s + \omega_n^2)}$$

and similarly the error function has the form:

$$H_\epsilon(s) = \frac{s^2}{(s^2 + 2\zeta\omega_n s + \omega_n^2)} \quad (9)$$

In this discussion we won't be making use of the error function equation, however it is the basis of much of the acquisition equations which are used to derive Figures 11 and 12. The complete analysis is beyond the scope of this paper, but is discussed in detail by most of the references. From equation 8, and the standard second order equation we can solve for the bandwidth, ω_n and damping, ζ. The natural frequency is:

$$\omega_n^2 = \frac{K'_{VCO}K_D}{C_2}$$

$$\omega_n = \sqrt{\frac{I_{PUMP}K'_{VCO}}{2\pi C_2}} \quad (10)$$

By combining K_{VCO} and I_{PUMP} constants, the design equations for the PLL can be simplified, by introducing K_{PLL}, which is defined as the product of $V_{REF} \times K_P \times K_{VCO}$. By using the K_{PLL} constant the above equation becomes:

$$\omega_n = \sqrt{\frac{K_{PLL}}{2\pi NR_1C_2}} \quad (11)$$

where N is the VCO divider as defined in Table I. The damping factor is given by:

$$2\zeta\omega_n = K'_{VCO}R_2K_D \frac{K'_{VCO}R_2K_DC_2}{C_2}$$

$$2\zeta\omega_n = \omega_n^2 R_2 C_2$$

$$\zeta = \frac{\omega_n R_2 C_2}{2} \quad (12)$$

These two parameters (equations 11 and 12) will allow us to calculate the PLL filter components based on bandwidth, and damping. The closed loop phase transfer function shows that the PLL behaves like a low-pass filter. It passes signals for input phase signals whose frequency spectrum is between 0 and ω_n. This means that a second-order PLL is able to track a phase and frequency modulation of the input signal up to a frequency equal to $\omega_n/2\pi$, and it will not follow input variations of higher frequencies.

4.2 System Performance and Filter Design

The system performance of a data separator can be described by three main criteria.

1) Acquisition Time—ability to guarantee lock during a preamble.
2) Window Margin—ability to recognize data shifted in time from its ideal position (bit jitter) without incorrectly decoding it.
3) Tracking of Disk Data—ability to follow slow (<1 kHz) disk data speed variations.

The filter design must meet the requirements set by these performance characteristics. The two conflicting requirements are that the bandwidth must be large enough to ensure proper locking to the data stream, but as small as possible while tracking the data to maximize bit shift rejection and window margin. Primarily the filter sets the bandwidth, and it is determined by the required acquisition time as shown later.

To illustrate this, a numerical example (for 500 Kb/s data rate) is presented following the design considerations step by step. After we have completed the paper design a discussion of performace measurements is provided. Once the initially calculated paper values are chosen, then real measurements must be made, and adjustments to these values are decided upon.

Acquisition to the Data Stream

Acquisition means to achieve phase lock and to bring the phase error of the VCO to zero, or close to it. This includes acquisition of phase lock to either the data input or to the reference frequency. The lock mechanisms for these two cases may be different.

At the moment just before Read Gate is asserted, it will be assumed that the VCO is locked to the reference frequency. It has been locked for a relatively long period of time, therefore the phase error between the VCO output and the reference frequency is nearly zero. Because the system is initially locked, the initially locked model can be used.

When Read Gate is asserted by the controller, the input to the phase detector is switched from the reference frequency to the data stream. This is an instantaneous change of both phase and frequency to the input of the phase detector. The loop must be designed to assure that it can achieve both phase and frequency lock to the incoming data. Phase and Frequency Lock implies that the steady state phase and frequency error at the phase detector input is near zero.

The goal is to lock to the data stream within the length of the preamble, very often half of the preamble to increase the probability of locking successfully. In fact, during a preamble the data pulses are relatively free of bit shift and the frequency is constant. There are two basic requirements to ensure that the data separator correctly locks to the data stream in the required amount of time.

1. The loop bandwidth must be large enough to ensure that phase and frequency error of the VCO goes to zero within the required time (usually within ½ the length of the preamble). This implies that the shorter the preamble the larger ω_n.

2. The filter must also be designed to guarantee that a data pulse will never fall out of the data window during the lock process. The peak phase error during acquisition must be less than ½ of a data or clock window (i.e. or $< \pi/2$). If the filter is incorrectly designed and the data pulse falls outside the window (called cycle slipping) during acquisition, the loop may never lock within the desired acquisition time and the encoded data will not be decoded correctly. This requires that to guarantee lock over a wider variation of data rate, a larger ω_n is required.

Both of these requirements can be approximately derived from *Figures 11* and *12*. These curves plot relative phase error normalized to ω_n versus time in units normalized to ω_n. This period of time, and the amount of phase error present is dependent upon ω_n and damping.

For the first requirement, the phase error settles close to 0 in about:

$$t_{acq} = \frac{5}{\omega_n} \quad (13)$$

This equation yields a minimum starting bandwidth, and is valid for systems where the speed variation of the incoming data is small (± 1–2%).

For the second requirement, the peak phase error, $\theta_{e(PEEK)}$, during acquisition can be determined from *Figures 11* and *12*. The design must ensure that the sum of this peak phase error due to a phase step, $\theta_{e(PHASE)}$, the peak phase error due to a frequency step, $\theta_{e(FREQ)}$, and the phase error due to PLL noise and non-linearities, $\theta_{e(PLL)}$, must all be less than $\pi/2$. In equation form:

$$\theta_{e(PEAK)} = \theta_{e(FREQ)} + \theta_{e(PHASE)}$$
$$+ \theta_{e(PLL)} < \frac{\pi}{2} \quad (14)$$

Thus the sum of the peak phase errors for a phase step and a frequency step must be calculated.

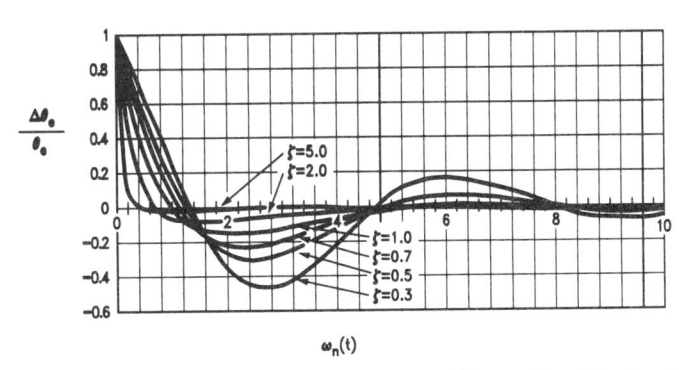

FIGURE 11. A Plot of Normalized Phase Error of a PLL to a Phase Step Input

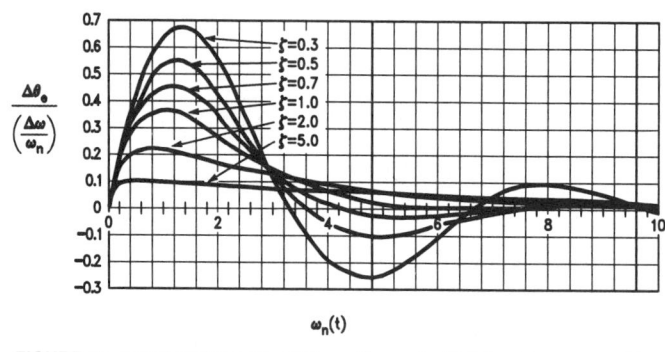

FIGURE 12. A Plot of Normalized Phase Error of a PLL to a Frequency Step

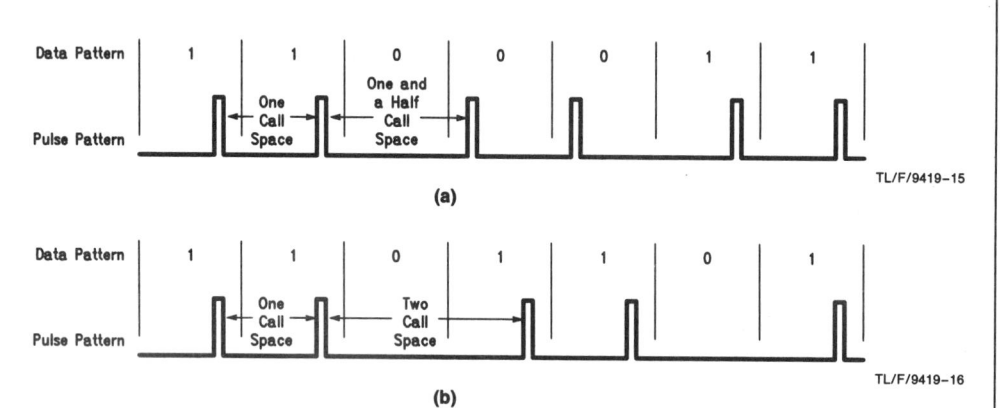

FIGURE 13. Two "Worst Case" Data Field Patterns for Measurement of PLL Bit Shift Tolerance a) "11000" Pattern with 2/3 μs Pulse Spacing (at 500 Kb/s), and b) A "DB6" ("110") Pattern with 2/4 μs Pulse Spacing (at 500 Kb/s)

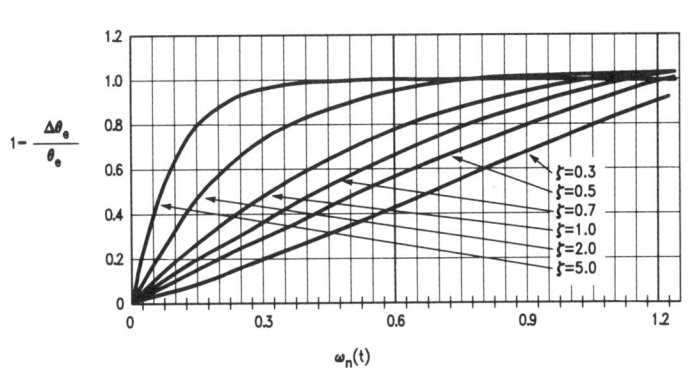

FIGURE 14. A Plot of Normalized Bit Shift Resistance Versus Time for a PLL

For determining the peak phase step error, the value of $\theta_{e(PHASE)}$ is the maximum Y-axis value of the chosen curve multiplied by the input phase step, and the result is in radians. For example with a damping of 0.7 a maximum error of −0.20 occurs at about 3 ω_n. If the maximum phase step is $\pi/2$ when switching to the data, then the peak phase error is 0.1 π radians. The choice of which curve to use depends on damping factor desired.

Since a frequency step is also present at the transition of READ GATE, the peak phase error of the frequency step given by the normalized plot in *Figure 12* must also be derived. The phase error can be calculated by reading the peak Y-axis value from the desired curve and using:

$$\theta_{e(FREQ)} = Y \frac{\Delta \omega}{\omega_n} \qquad (15)$$

to determine $\theta_{e(FREQ)}$. Where Y is the value read from the Y axis, and $\Delta \omega$ is the maximum frequency step times 2π. The maximum frequency step is the worst case frequency variation of the data being read from the disk drive, which is the sum of the MSV and the ISV.

With the equations for loop bandwidth, damping factor, and the relationship between acquisition time and bandwidth, the following example demonstrates the first steps at arriving at the loop filter components.

Example 1: Design a data separator using the DP8473. Determine the loop bandwidth, dampening factor, and C_2, R_1, R_2 component values for a data separator that decodes MFM data at a data rate of 500 kbits/sec. The preamble is 12 bytes long, and the total MSV/ISV is ±6%.

Select a value of pump current resistor. For example 5.6 kΩ.

Find out the minimum acquisition time required. Generally, half preamble which is 6 bytes. Thus:

$$t_{acq} = ((6 \times 8) \text{ bits}) \times 2 \text{ }\mu\text{s/bit} = 96 \text{ }\mu\text{s}$$

Next calculate ω_n based on the larger of the two acquisition requirements. The first requirement for ω_n is: $5/\omega_n < t_{acq}$; Thus:

$$\omega_n > 52.5 \text{ Krad/sec.}$$

Calculate ω_n for the second acquisition requirement, i.e. ensuring the maximum phase error is less than $\pi/2$. Due to the Zero Phase start-up block within the Data Separator, the maximum phase step when switching to the data is $\pi/8$. We'll choose a damping of 0.7. From *Figure 11* the maximum overshoot is 0.2 ($\pi/8$) = 0.08 rad (Note 1.0 rad = 314 ns at 500 kb/s). Assume that the data separator contributes a total 0.1 rad noise error. This is for charge pump asymmetry, delay line variation, and VCO jitter.

Using equation 18:

$\theta_{e(PEAK)} = \theta_{e(FREQ)} + 0.08$ rad + 0.1 read < $\pi/2$ or ...

500 ns > 25 ns + 31 ns + $\theta_{e(FREQ)}$ which results in:

$\theta_{e(FREQ)} < 443$ ns or 1.41 rad

This yields the maximum tolerable phase error due to a frequency step (which is dependent on ω_n).

First find the maximum frequency step in radians that the data separator must undergo. The design requirement was 6%, but in order to account for gain variations in the data separator some margin on top of this is required, so we will design to 8% total speed variation. Thus:

$$\Delta\omega = 0.08 (500k) 2\pi = 251 \text{ Krad}$$

The relative phase step error from *Figure 12* using a damping factor of 0.7, is Y = 0.45, plugging into equation 15:

$$\omega_n \geq \frac{\Delta\omega}{\theta_e} = \frac{0.45 (251 \text{ Krad})}{1.41} = 80 \text{ Krad/s}.$$

The larger of the two calculated ω_n's is 80 Krad/sec, so that is the chosen bandwidth.

$$C_2 = \frac{K_{PLL}}{2\pi R_1 N \omega_n^2} \quad (16)$$

$$= \frac{75 \text{ Mrad}}{2\pi (8)(5.6 \text{ k}\Omega)(80 \text{ k}^2)} \approx 0.041 \text{ }\mu F$$

We will round down to the next lowest standard value of 0.039 μF.

$$R_2 = \frac{2\zeta}{\omega_n C_2} \quad (17)$$

$$R_2 = \frac{2\zeta}{C_2 \omega_n} \approx \frac{2(0.7)}{(0.039)(80K)} \approx 450\Omega$$

In the above example we have calculated a set of component values for the acceptable bandwidth based on acquisition. This calculation yields a good starting point from which experimental measurements can be made, but may not be the optimum values. Depending on other considerations we may decide to chose a value of ω_n that is slightly different depending on window margin, or bit shift performance; as will be shown.

Theoretical Dynamic Window Margin Determination

Previously in section 2.4 Window Margin was discussed in terms of distortions of the data that degrade the window margin. Here, using a model for these distortions we will arrive at a calculation of the expected Dynamic Window Margin for the DP8473 analog PLL. The effects of Window error, VCO jitter, and PLL response to a previous or present bit shift all cause a reduction of the available window margin available. (Remember the goal is to maximize the total available bit window.) The following analysis provides a feel for the amount of degradation due to various parameters, and serves to provide an indicator of the expected PLL performance. The following is a list of parameters that cause loss of window:

Internal Window Error (or static phase error): The window error is related to the accuracy of the internal delay line in the data path before the phase detector. As explained previously, this delay line is automatically trimmed using the crystal frequency as reference. The static phase error is the sum of two factors. One of these factors is the difference between the data stream frequency and the nominal and unavoidable internal mismatches. Another contributing factor is charge pump leakage. This factor causes a perceived phase error that is equivalent to varying the delay line length. This is usually > 2%–4%.

VCO Jitter: The VCO jitter is caused by the modulation of the VCO frequency with secondary VCO frequency, crystal oscillator and other noise. This can account for another 2–5% percent of error.

PLL Response: The PLL response to a data bit shifted from its nominal position because of noise or jitter is directly translated in a margin loss for the bits following any shifted bit. For a highly accurate PLL circuit this is the primary source of error, and it typically results in a window loss of up to 20%, depending on data pattern and frequency variation constraints.

All of these degradations are summed into the Window Margin specification.

$$\theta_{wm} = (\tfrac{1}{2} \text{ Bit Window}) - \theta_{e(PLL)} - \theta_{e(SWL)} \quad (18)$$

This yields the margin loss, where θ_{wm} is the total window margin or the total amount of a half bit cell in which a data pulse will be properly recognized, $\theta_{e(PLL)}$ is the error due to PLL response, and $\theta_{e(SWL)}$ is the total error contributed by imperfections of the PLL circuitry (including delay line accuracy, leakage, and noise).

The window margin loss contributed by the device accuracies are relatively straightforward, and are supplied by National. The more difficult task is to determine the window margin loss due to the PLL response.

Tracking of the Disk Data

The bit shift produced by an average disk depends on the pattern of encoded pulses recorded on the media. Pulses that are placed close together appear to push each other apart when they are read back. This is the primary cause of bit shift.

The PLL tolerance to bit shift is dependent on the amount of data bits that are shifted, and the data pattern. It can tolerate more bit shift by individual bits if only a few of the bits within a pulse stream are shifted. However, if most of the data read by the PLL is shifted (both early and late), the loop is constantly correcting itself and is never really phase locked. Because it is not phase locked the maximum tolerable bit shift is less.

Some data patterns are better at determining PLL performance than others. These are patterns that are particularly difficult for a PLL remain locked to while the data pulses are shifted early and late. For example a bit pattern of "11000" is difficult to decode since it has pulses alternately spaced by 1 and 1.5 bit cells. See *Figure 13a*. This is difficult to decode in some cases because under maximum bit shift conditions all pulses are equally spaced.

Another bit pattern that is difficult to decode is a repeated bit pattern triplet of "110" (or "101" also referred to as a "DB6" pattern), which has a pair of pulses one bit cell apart, and the next pair two bit cells apart. This pattern is particularly difficult to decode because it contains two pulses of minimum spacing, followed by two maximum spaced pulses. This pulse pattern is shown in *Figure 13b*.

Unlike the acquisition process described earlier, the PLL must largely ignore individual bit shifts during the tracking phase. The PLL should only follow the longer term average data rate. The desired PLL response during the tracking phase is somewhat different from the response required during the acquisition phase. Instead of a high bandwidth to decrease the lock time, a low bandwidth is preferred to prevent the PLL from following individual bit shifts. When choosing the filter bandwidth, the lowest possible value should be used that still satisfies the acquisition time requirement.

Figure 14 can be used to determine the theoretical window margin. This curve of phase bit shift resistance plots the amount of error introduced in the PLL's VCO by a phase error. The following example will show the steps involved in calculating expected window margin, and illustrate some concepts of tracking data.

Example 2: Determine the total dynamic window margin for a loop with a bandwidth of 90 Krad/sec, a damping of 0.7, and at a 500 Kb/s data rate. The intrinsic circuit errors amount to 0.1 radians of the window. Calculate the margin for both the 110 and 11000 pattern.

First step is to calculate the bandwidth of the PLL while tracking these data patterns. The bandwidth is the square root of the ratio of the pulses per bit cell relative to preamble data, multiplied by the bandwidth. Preamble data has 1 pulse/bit cell, and a 110 pattern has 2 pulses per 3 bit cells, while a 1100 pattern has 4 pulses for every 5 cells. Thus:

$$\omega_{n(110)} = (90 \text{ Krad/sec}) \times \sqrt{0.66}$$
$$= 72 \text{ Krad/sec}$$
$$\omega_{n(11000)} = (90 \text{ Krad/sec}) \times \sqrt{0.8}$$
$$= 81 \text{ Krad/sec}$$

To analyze the problem, we need to look at two bits. The present bit which has an early phase step, and the next bit which has an equal late phase step. We must determine how large a shift in the first bit can be tolerated such that the same amount of shift in the second bit will still fall within the proper window. In equation form:

$$\theta_{e(2)} + K_{WM}\Delta\theta_{e(1)} + \theta_{PLL} \geq \theta_T \quad (19)$$

where $\Delta\theta_{e(1)}$, the shift of the first bit, is multiplied by K_{WM}, which is the affect the first bit has on the VCO when then next bit arrives. θ_T is the total window available (in this case ±500 ns). θ_{PLL} is the static error degradation due to the DP847x and is 10% of 500 ns or 50 ns. $\Delta\theta_{e(2)}$ is the maximum phase error of the second bit. We can solve for $\Delta\theta_{e(2)}$ assuming that $\Delta\theta_{e(1)} = \Delta\theta_{e(2)}$:

$$\Delta\theta_{e(2)} = \frac{\theta_T - \theta_{PLL}}{1} + K_{WM} \quad (20)$$

The value of K_{WM} is the value read off the y-axis of Figure 14, at a time normalized to ω_n. This time is the time between two bits or:

For a "110" pattern $\omega_n(t) = (4 \mu s)(72 \text{ Krad/s}) = 0.29$, resulting in a value from Figure 14 of $K_{WM} = 0.37$. And for a "11000" pattern $\omega_n(t) = (3 \mu s)(81 \text{ Krad/s}) = 0.24$, which results in $K_{WM} = 0.28$. Using 0.37, the window margin is:

$$\Delta\theta_{e(2)} = \frac{500 \text{ ns} - 31 \text{ ns}}{1.37} \doteq 342 \text{ ns}$$

In terms of percentage this is 100*(342/500)% = 68%. Note that this is the window margin at nominal frequency.

Open Loop Bode Plots and the Second Capacitor

Figure 15 shows the open loop gain Bode plot for the second order PLL. This plot is useful as a double check to make sure that we have a stable design and is important to show the affect of the second capacitor in the filter. In this plot, it is assumed that the charge pump output impedance is infinity.

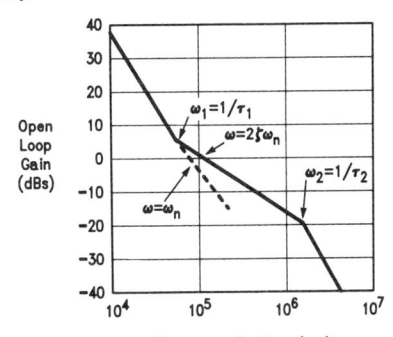

FIGURE 15. Bode Plot of Open Loop Gain of DP8472/3/4 Using a Typical Filter at 500 Kb/s (from Example 1)

As can be seen, the gain starts off with a slope of 40 dB per decade due to the two poles of the filter and VCO. The phase angle starts at −180°C. The stabilizing zero is introduced at $\omega_1 = 1/R_2C_2$, and causes the slope to change to 20 dB per decade. ω_n is the extrapolation of 40 dB/dec line to 0 gain, and the actual crossing point is $\omega = 2\zeta\omega_n$. Example 4 discusses how to plot this curve.

The further reduce the effect of unwanted changes in the VCO phase the second capacitor can be added to the filter. This capacitor, C_1, introduces a pole, Figure 15, between the loop natural frequency and the data rate frequency. The pole due to the second capacitor occurs at $\omega_2 = 1/\tau_2 = 1/R_2C_1$. This capacitor provides further attenuation of bit shift caused frequency components, and the pump pulse noise, both of which have frequency components that are around the data frequency. The only considerations in choosing the value of this capacitor are related to the stability of the loop, and inadvertently affecting ω_n. A good criterion for stability is that the Bode plot of the open loop gain, Figure 15 must cross the 0 dB gain with a slope of 20 dB/dec, ie. before the break caused by C_1's pole.

To determine a simple method for deriving C_1, we must look at the open loop gain of the PLL along with the transfer function of the loop filter. The open loop gain is:

$$\frac{\theta_2}{\theta_1} = \frac{K_D K'_{VCO}}{s} F(s) \quad (21)$$

Where $F(s)$ is the filters transfer function:

$$F(s) = Z(s) = \frac{\left(\frac{1 + sR_2C_2}{sC_1}\right)}{\left(\frac{sR_2C_1C_2}{(C_1+C_2)}\right)} \quad (22)$$

Combining these two equations and manipulating we find that the second pole occurs at:

$$\omega_P = \frac{1}{R_2C_1} \text{ (confirming Figure 15)} \quad (23)$$

This is assuming $C_2 >> C_1$ (which we ought to assume to maintain the validity of previous filter assumptions).

The zero introduced by C_2 and R_2 should be designed to be close to the 0 dB gain crossing. Its frequency is $\omega_z = 1/R_2C_2$. The frequency of the pole due to R_2 and C_1 is approximately $\omega_P = 1/R_2C_1$. This pole must not significantly change the slope around the 0 dB line. If we choose $C_1 = C_2/20$ the effect on the slope of the transfer function is less than 1 dB/decade at the frequencies around the 0 dB gain line crossing. Thus as a guide:

$$C_1 \leq \frac{C_2}{20} \quad (24)$$

Example 3: For example 1, determine C_1.

Very simply for example 1a:

$$C_1 \leq \frac{0.039 \, \mu F}{20} \cong 2000 \, pF$$

The 1/20 factor provides the approximate value for C_1.

Example 4: Plot the Bode diagram of the open loop gain for the DP8472 with $C_2 = 0.027$, $C_1 = 1000$ pF, $R_1 = 5.6$ kΩ, $R_2 = 545\Omega$.

There are two easy methods of doing this. One method involves determining the open loop gain at a low frequency, where the poles and zeros don't have any affect, and then using this gain point at a start drawing the properly sloped lines to the break point frequencies. A second method is to calculate the ω_n, and $2\zeta\omega_n$ frequencies.

For the first method, first calculate the locations of τ_1 and τ_2 in Figure 15. Thus

$$\omega_1 = \frac{1}{\tau_1} = \frac{1}{R_2C_2}$$

$$= \frac{1}{(545\Omega)(0.027 \, \mu F)} = 6.8 \times 10^4$$

$$\omega_2 = \frac{1}{\tau_2} = \frac{1}{R_2C_1}$$

$$= \frac{1}{(545\Omega)(1000 \, pF)} = 1.8 \times 10^6$$

Now pick a point that is below ω_1, and calculate the open loop gain, which is:

$$K_{LOOP} = 20 \log \left[\left(\frac{K_{VCO}K_P V_{REF}}{2\pi R_1 N \omega}\right)\left(\frac{1}{\omega C_2}\right) \right]$$

At $\omega = 10^4$, the K_{LOOP} is:

$$K_{LOOP} = 20 \log \left(\frac{12 \times 10^6}{R_1 C_2 \omega^2 N}\right) = 39 \, dB$$

Now draw a line from $\omega = 10^4$ to ω_1 with a 40 dB/decade slope. Then at ω_1 draw a line to ω_2 with a 20 dB/decade slope, and finally draw a line from ω_2 with a 40 dB/decade slope.

To understand the affect of C_1, the additional attenuation introduced can be determined as using:

$$A_P(w) = \frac{1}{1 + \frac{\omega}{\omega_P}} \quad (25)$$

Using our previous example 1:

$$\omega_P = \frac{1}{1000 \, pF \; 545\Omega} = 1834 \, \text{Krad, and}$$

$$\omega = 2\pi \, (500 \, kHz) = 3142 \, \text{Krad/sec.}$$

Thus

$$A_P(w) = \frac{1}{1 + \frac{3142}{1834}} = 0.37$$

which yields an additional 9 dB of attenuation.

Choosing Component Tolerances and Types

One of the most often asked questions is how accurate should the filter and charge pump resistors and capacitors be? The answer depends on how accurate a data separator is required. For a good performance design, the following criteria can be followed for each component:

R_1: The pump set resistor's tolerance affects the loop bandwidth, ω_n. The loop bandwidth directly affects window margin. Due to the square root relationship, a 5% change in this resistor changes ω_n by 2.5% which in turn affects the window margin by 1–2%. It is thus recommended that R_1 be a 1% resistor. A standard carbon or metal film resistor with a low series inductance should be chosen.

C₂: The main filter capacitor also affects ω_n in the same way as R_1 so it too should be relatively accurate. 5% is recommended. This capacitor should be a very good quality capacitor, with good high frequency response and low dielectric absorption. Mica is a good choice although maybe too expensive. Polypropalene and metal film are good as well. Avoid Mylar or Polystyrene.

R₂: This resistor has a much lower affect on window margin, and thus standard 5% resistors can be used.

C₁: The second capacitor's accuracy in not critical 10%–20%, but its high frequency characteristics should be quite good, similar to a good high frequency power supply decoupling capacitor.

5.0 ADVANCED TOPICS

The following sections discuss several specialized areas of evaluation and design of the PLL for the DP8473 controller. Also a short discussion of the crystal oscillator design considerations is given.

5.1 Design and Performance Testing

Testing data separators can get rather complicated. Once the PLL circuitry, gain, bandwidth, and damping are set, then data separator lock range testing, window margin testing, and finaly bit error rate testing may be undertaken. This testing can require some fairly sophisticated setups, and is time consuming. To help the designer get started, a rigorous approach to floppy disk PLL design verification is described with reference to desired performance and available equipment. If the designer is not concerned about optimum performance for custom applications, then the values for the PLL filter and pump resistor provided in the DP8473 datasheet should prove more than adequate.

Step 1—Calculating the Filter/Pump Resistor: Following the examples above, the optimum "paper design" filter components should be calculated (or use the values provided in the datasheets and tweak them).

Step 2—Testing Lock Range and Damping: For characterization of the PLL's acquisition, a fairly simple setup can be used, which utilizes a pulse generator to provide a pulse train that simulates a preamble to be input into the data separator's read data input. A second synchronous squarewave that is 20–50 times the period of the data pulses is applied to read gate. The frequency of the read data pulses should be varied from the nominal data rate to the limits of the desired lock range, and the lock range requirement should be verified by monitoring the Filter pin, and the Pump outputs. *Figure 16* shows a typical setup, and some typical waveforms on the pump and filter pins during acquisition. *Figure 16b* shows a proper locking PLL which does not exhibit "cycle slipping". Cycle slipping is denoted by the saw tooth waveform on the filter pin, which can be seen by the locking shown in *Figure 16c*.

In both *Figure 16b* and *c* the total amplitude, ΔV, of the filter pin waveform is typically less than 200 mV.

The object is to adjust the PLL bandwidth to ensure that when locking over the desired range of data rates (for example 500 Kb/s ±6%) that no cycle slipping occurs. If slipping does occur then the bandwidth should be increased.

A second piece of information that these curves provide is verification of the damping of the PLL. As in *Figure 16b* the filter pin waveform should slightly overshoot, and then (maybe) slightly undershoot the eventual locked voltage. If there is very little overshoot then the loop may be overdamped, and if the filter pin voltage "rings" for a few cycles the loop is probably underdamped. For example, *Figure 16c* not only cycle slips, but does not overshoot, therefore the loop bandwidth may be fine, and the loop is just too heavily damped.

Step 3—Window Margin Evaluation: Usually to perform this test a disk simulator should be used to simulate the worst case drive read data conditions, and measure the error performance of the data separator. This simulator can be used to vary the following parameters:

1. Motor Speed Variation
2. Instantaneous Speed Variation
3. Instantaneous bit shift
   (to determine the window edge)
4. Data pattern.

The test setup shown in *Figure 17* can accomplish some of this testing. The disk simulator looks like a formatted disk to the controller/data separator. To test the data separator, software in the host computer performs a repeated series of read operations over a period of time, while the designer programs the disk simulator to vary the data rate from one end of the lock range to the other. At each data rate the bit shift is increased until the error rate increases above a minimal threshold.

This testing should measure window margin over the entire lock range, and under conditions as described in section 2.4. Typically this process is a trial-and-error process. The bandwidth and damping can be adjusted based on the results of these tests.

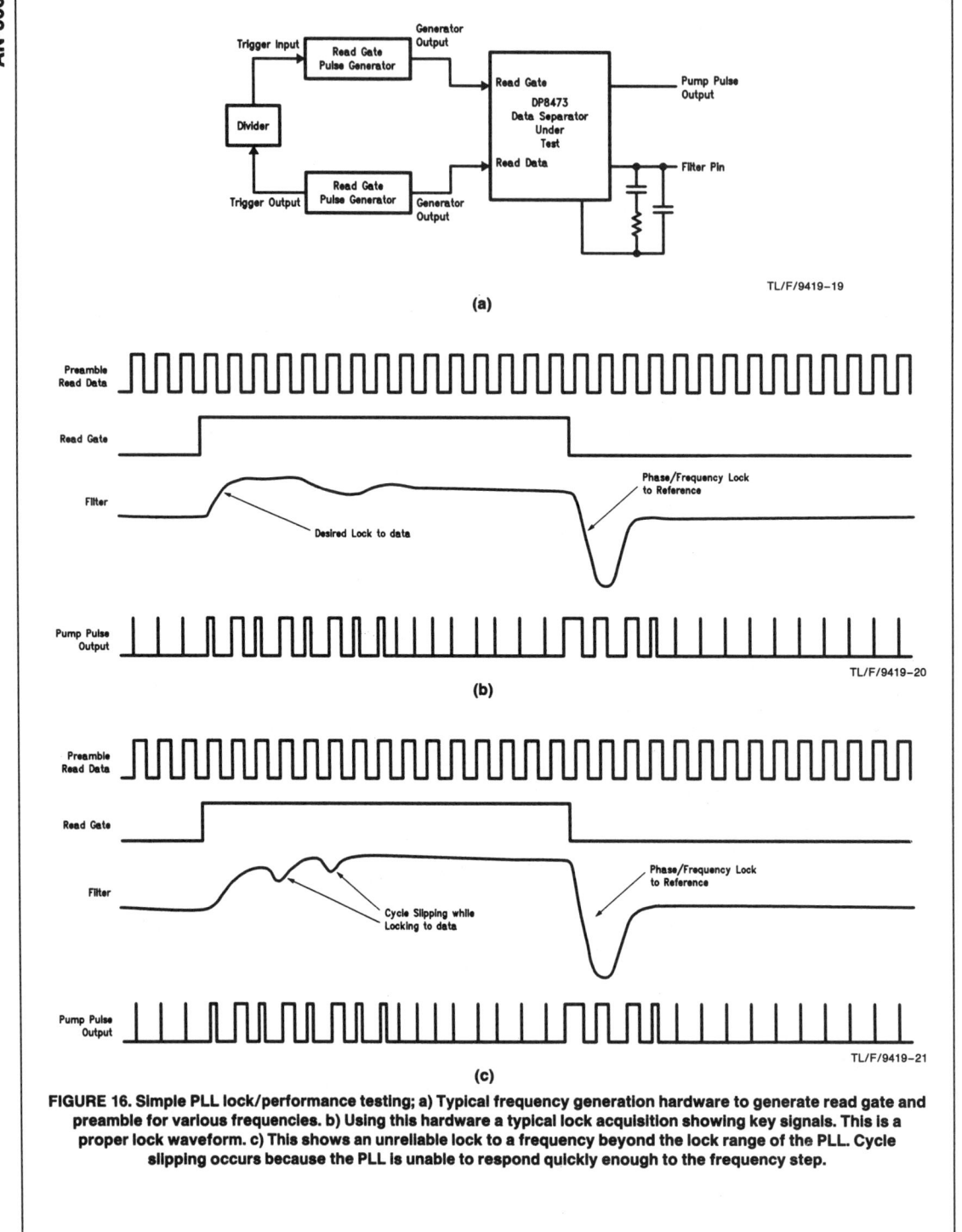

FIGURE 16. Simple PLL lock/performance testing; a) Typical frequency generation hardware to generate read gate and preamble for various frequencies. b) Using this hardware a typical lock acquisition showing key signals. This is a proper lock waveform. c) This shows an unreliable lock to a frequency beyond the lock range of the PLL. Cycle slipping occurs because the PLL is unable to respond quickly enough to the frequency step.

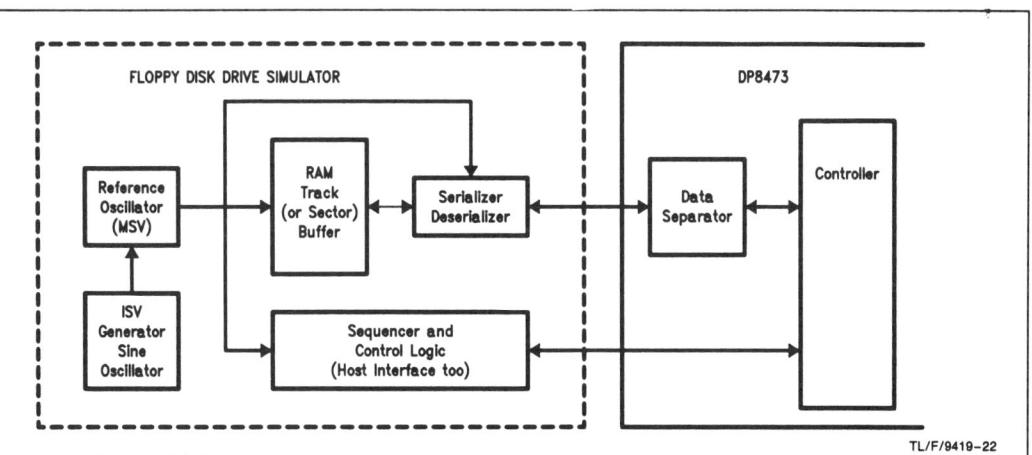

FIGURE 17. Block Diagram of Connection of Disk Simulator to Data Separator to be Tested

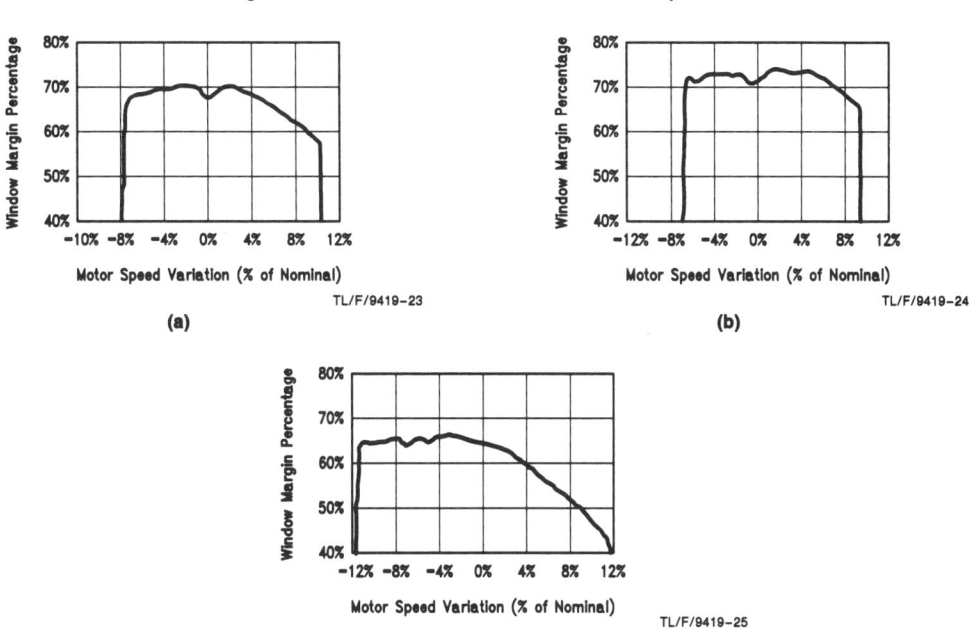

FIGURE 18. Various window margin graphs. a) Data separator with slightly long delay line and fairly normal lock range; b) data separator with short delay line; c) a relatively poor data separator with a long delay line, and wide bandwidth.

ISV may also be simulated. Unfortunately most disk simulators do not easily test for this. Therefore it is very likely that the window margin measurement will be made with only MSV variation initially. This may also be desirable since MSV-only testing will yield a better understanding of the PLL's lock range. If MSV-only testing is done initially then the design of the PLL and total frequency range for evaluation should be the sum of the desired MSV and ISV. For example, a design requirement of ±3% ISV and ±3% MSV can be approximated by ±6% MSV.

If an MSV-only test is done, it may be useful to then follow this with a simple ISV test just to double check that the loop will follow the ISV properly with little degradation in window margin. It is also sometimes useful to vary the ISV frequency until the window margin degrades, just to give an indication of how high the ISV frequency can go. A typical PLL's ISV performance should not degrade until the frequency is greater than about 800 Hz.

Step 4—Considering Temperature, Supply and Device Variations: The next step is to take the above "optimum" filter and simulate variations in gain induced by temperature, supply and processing. This is accomplished with the same disk simulator setup as in step 3. To simulate these variations, the designer need only vary the pump resistor's value. This will affect the open loop gain identically to device variations. The above tests should be re-run with a minimum and maximum resistor. If the performance "falls-off-a-cliff" then some compromises and adjustments to the filter may be required.

For relatively small temperature ranges a resistor variation of ±10–15% should be adequate. For full 0°–70° simulation, resistor variation of ±20% is a more accurate reflection of the DP8473 performance. If for some reason the performance of the data separator cannot be maintained at the desired window margin, then trimming the pump resistor may be needed to meet performance over the full process spread.

The final simulation involves varying the delay line length. It is possible to simulate variations in the delay line by forcing a leakage current onto the filter pin (at the VCO input) using a pull up or pull down resistor. This leakage current will cause a phase error within the loop and will cause the loop to act as if the delay line length were changed. Before actually running dynamic window margin tests, the designer must determine the actual length of the delay line, and then adjust this length to the limits specified in the datasheet. Then at each limit the dynamic window margin test can be performed.

In order to measure the length of the delay line a static window margin test can be performed. This test uses a disk simulator with a format that has all 0's or 1's data fields. Within the data field one bit is shifted until the PLL mis-decodes the data. Using the maximum tolerable bit shift number, the delay line length can be extrapolated. This same measurement can be done with various filter pull up/down resistors. By proper resistor selection the limits of the delay line length can be simulated. Using these simulated delay line techniques a dynamic window margin test can be run, and the resultant PLL performance can be characterized.

Step 5a—Bit Error Rate Measurements: This test is performed as a verification of the final total system. It consists of putting together a complete floppy drive, floppy media, separator, and controller system then running long term read/write tests randomly across the disk media. For a known number of read/writes and the resultant value of read errors, a number can be derived that is the ratio of bit errors to total number of bits read. This test proves the integrity of the entire system, and should be performed over some manufacturing spread of products. While this test is useful to verify the complete system integration, the data separator is only one small part of the total contribution to the total system error rate.

Step 5b—Real World Worst-Case Tests: As a final analysis and proof that the data separator is solid it is often useful to test the data separator on a known "worst case" drive and disk. Generally the evaluation is done with a disk that is recorded off speed and off track. This ensures that a maximum amount of bit shift and speed variation is present. The main problem is to ensure that the disk still has acceptable data. If excessive errors are encountered, evaluation of the types of error that are occurring can be used to determine whether the bandwidth of the PLL needs to be increased (acquisition related errors) or decreased (bit shift related errors).

Alternate Simple In-System Data Separator Evaluation

Provided here are some quick tips on evaluating the data separator without any test equipment, but just by running long term in-system tests.

1. If after some initial testing a lot of sector or ID address mark or data address mark not found errors are given by the controller, then in all likelihood the data separator is not locking properly. The bandwidth of the loop should be increased, or possibly the loop is too heavily overdamped.

2. If the disk controller is experiencing a large amount of address or data field CRC errors, then probably the loop is being sent out of lock by bit shift noise. Thus, the gain of the loop may be too high or the loop is underdamped.

5.2 Understanding the Window Margin Curves

Careful analysis of a window margin curve can yield quite a bit of information about the data separator characteristics. *Figures 18a, b,* and *c* show some typical window margin plots. These curves were made using a disk simulator that outputs a "reverse write precompensated" bit shift pattern to the data separator.

These curves plot the maximum bit shift tolerance (vertical axis) versus motor speed variation (MSV-only) (horizontal axis). The controller was programmed to perform a repetitive single sector read, while the simulator outputs a formatted track at the programmed MSV and bit shift amount. All the data read with MSV and bit shift amounts that fall under and within the curve (shaded area) can be consistently read correctly. All data read with MSV and bit shift amounts outside and above the curve either could not be correctly located by the controller, or had errors in it.

The first thing to note is that as the MSV variation from the nominal frequency increases, there is a point at which the PLL performance drops to zero bit shift (the vertical lines of the curve). This is an indication of the lock range of the PLL. Thus for *Figure 18a* the lock range is −8% to +10%. This asymmetry in the lock range is typical of the DP847x series PLLs and is due to a slight skew in the charge pump. This skew adds some bit shift rejection improvement. *Figure 18b* has a lock range of −7% to +9%, while *Figure 18c* has a much wider lock range of nearly ±12%. This is an indication of the loop bandwidth. Here *Figure 18a* has the lowest bandwidth, then *Figure 18b*, and finally *Figure 18c* has the highest bandwidth.

Another characteristic of each of these curves is the downward slope in the positive MSV direction. The start of this slope is the point at which the delay line becomes longer than the $1/4$ period of the data rate. For example, in *Figure 18a* the slope starts at about 0 MSV (ignoring the dip at zero MSV for the moment which is due to some internal noise). This indicates that the delay line is a quarter period of the nominal data rate. Unfortunately, this causes a degradation in performance at higher MSV. In *Figure 18b* the slope starts at about +3%–4% MSV, and so the delay line is about 3%–4% short at a nominal data rate. This is an optimal delay line length to maximize performance over a ±6%–8% lock range. In *Figure 18c* the delay line is about 3%–4% too long and so there is quite a bit of degradation in window margin in the +MSV portion of the curve.

A final observation is that the wider the bandwidth the lower the window margin, assuming other things like data rate remain constant.

Another interesting fact to note is the type of errors that occur when the bit shift/MSV exceeds the PLL performance. Generally to the left or right of the curve (extreme MSV) the controller will give "Address Mark not Found" errors which is an indication of the PLLs' inability to lock properly. Errors for bit shift that exceeds the curve but for MSV within the PLL's lock range are generally CRC errors, although at very high bit shift a mixture of CRC and Address Mark Errors are expected.

5.3 DP8473 Filter Switching Design Considerations

Due to the desire to handle multiple data rates, the DP8473 incorporates on-chip data rate selection logic, and also filter switching logic. In this section we will discuss how this logic works and how to design a set of filters to maximize performance at various combinations of data rates.

Designing with a Single Filter

Previous design examples have dealt with optimization of a single filter at a single data rate. If the DP8473 is to be used at one data rate, the circuit connection is shown in *Figure 19*, and its design is straight forward. It is possible to use a single filter and obtain a reasonable performance at two data rates (i.e., 250 Kb/s and 500 Kb/s or 500 Kb/s and 1 Mb/s). This can be accomplished since the loop bandwidth is scaled by the PLL's divider. Since the divider value increases by a factor of two when going from the high to the low data rate, the bandwidth scales by:

$$\omega_{n250} = \frac{\omega_{n500}}{\sqrt{2}}$$

This scaling is probably not enough to optimize the lower data rate and the damping is affected too, but the single filter approach can still provide acceptable performance in many instances.

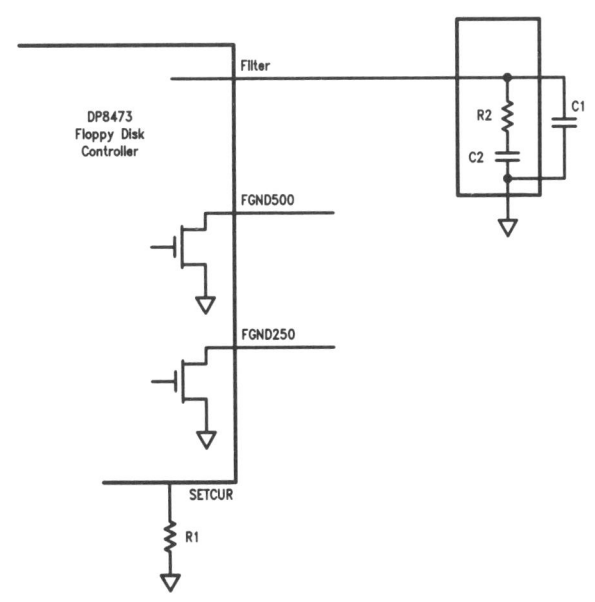

FIGURE 19. DP8473 Filter Configuration for a Single Data Rate Filter, May Be Used as Compromise Filter for Any Two Data Rates

AN-505

Designing for 250K/300K/500K MFM

This next filter configuration allows fewer compromises (and hence better performance) when using all 3 PCAT data rates. This configuration is shown in *Figure 20*. This circuit uses two pairs of R's and C's that compose two independent filters. One filter is connected to the FGND250, and the other to FGND500.

To implement the design of *Figure 20*, first design single optimum filters at 250 Kb/s, 300 Kb/s, and 500 Kb/s. Use a single pump resistor for all data rates. Verify and tweak the performance of these filters individually. The 500 Kb/s filter can be directly used. The 250 Kb/s and 300 Kb/s individual filter values need to be compromised into a single R and C filter. C_1 should be 1/20th of C_2, and if desired C_3 can be added with a value that is 1/20th of C_4.

Designing for 1.0 Mb/s and a 2nd Data Rate

By adding an additional capacitor to *Figure 19* 1.0 Mb/s data rate can be supported, as shown in *Figure 21*. In this figure a single damping resistor is used, but depending on the chosen data rate, one or both capacitors is selected. This configuration allows the bandwidth to be adjusted more flexibly when designing for the various data rates. The design process is, however, a little more complex.

To implement the design of *Figure 21*, first design single optimum filters at 1.0 Mb/s and 500 Kb/s (or 250 Kb/s). Use a single pump resistor for all data rates. Verify and tweak the performance of these filters individually. Using these values, choose a value for R_2 (damping resistor) that is a good compromise for all data rates. Next choose C_2 to be the optimum value from the initial individual design verification of the 1.0 Mb/s design. Next choose C_3 such that the sum of C_2 and C_3 equals the value for the optimum 500 Kb/s (or 250 Kb/s) design. Finally, choose C_1 to be 1/20th of C_3.

Designing for All Possible Data Rates

To support all possible data rates the simplest circuit configuration is one similar to *Figure 20*, but with an additional capacitor selected for the 1 Mb/s data rate.

To implement the design of *Figure 22*, first design single optimum filters at 250 Kb/s, 500 Kb/s, (300 Kb/s also if needed), and 1 Mb/s. Use a single pump resistor for all data rates. Verify and tweak the performance of these filters individually. Using all of these values, choose a value for R_2 (damping resistor) that is a good compromise for all data rates. Next choose C_2 to be the optimum value from the initial individual design verification for 1 Mb/s data rate filter. Next choose C_3 from the 500 Kb/s initial design. C_3 should be chosen such that $C_2 + C_3$ equals the optimum 500 Kb/s filter value. Then the 250 Kb/s (and 300 Kb/s if used) filter capacitor, C_4, must be chosen in a similar manner. C_4 should be chosen such that $C_4 + C_2$ equals the optimum 250 Kb/s filter capacitor value, or if using 300 Kb/s it must equal the best compromise 250/300 Kb/s filter capacitor. C_1 should be chosen to be 1/20th of C_2.

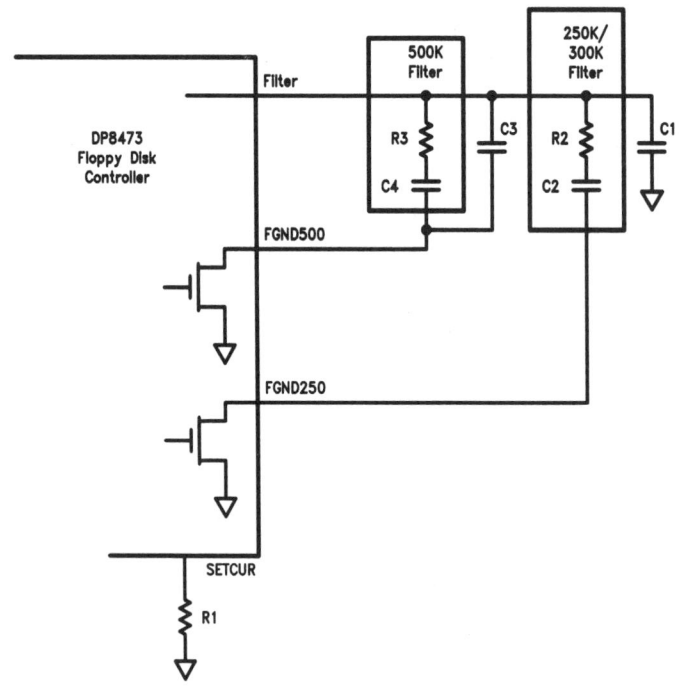

FIGURE 20. DP8473 Filter Configuration for Optimum 250/300/500 Kb/s (2 Filter) Design

FIGURE 21. DP8473 Filter Configuration for a Slight Tradeoff Filter Design at 1 Mb and 500 Kb/s

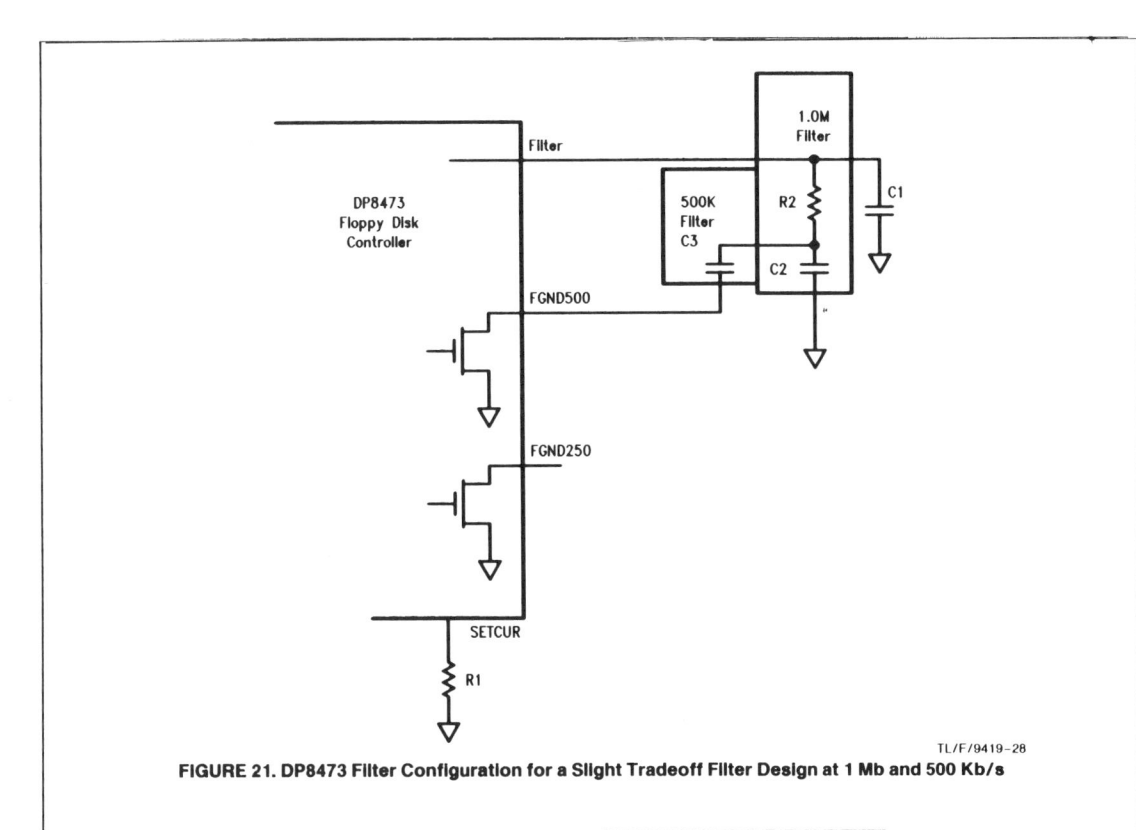

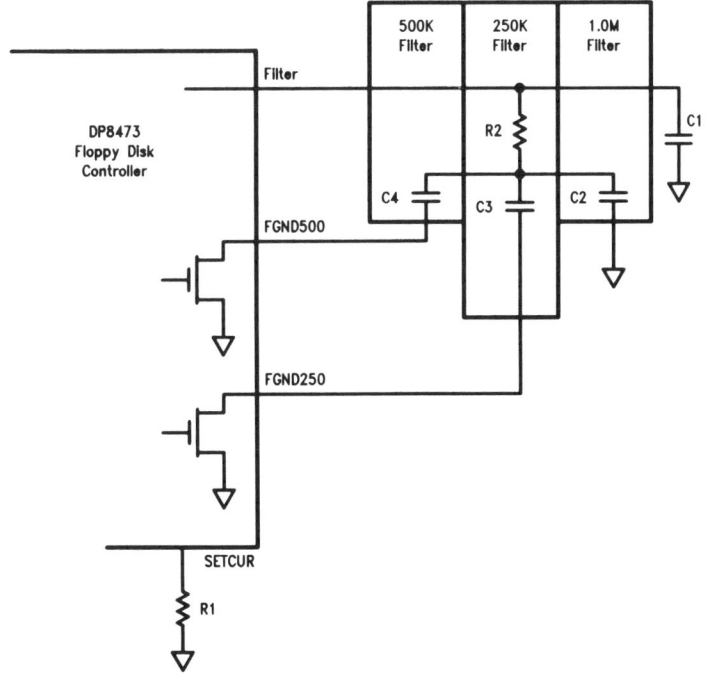

FIGURE 22. DP8473 Filter Configuration for All Data Rates

The previous filter does, however, have some minor performance tradeoffs if all data rates are to be implemented. If the best performance is desired, then the configuration shown in *Figure 23* should be used. In this figure, 3 individual filters are used for each of the data rates. The 1.0 Mb/s filter must be switched via some external circuitry, labeled "ground switch" in the figure. The circuit should enable this filter only when 1 Mb/s data is used, and this filter should be disabled when any other data rate is needed. The circuit to accomplish this could be as simple as an open collector gate derived from the RPM/LC pin, or an alternative that uses no additional hardware is to use an unused drive select output. If the latter option is chosen, then software will have to select the 1 Mb/s filter prior to using this data rate by enabling this bit in the Drive Control Register.

The design of this filter network is very straightforward. Simply design and optimize each filter individually, and use these filter values directly. (Again if 300 Kb/s is also used the 250 Kb/s filter used will be a compromise between the optimal 250 Kb/s and 300 Kb/s filters derived individually.)

5.4 DP8473 Oscillator Design

Figure 24 shows the schematic of the crystal oscillator used on the floppy disk controller. This circuit consists of a simple inverter whose impedance has been optimized for use as an oscillator. The inverter is biased into its linear operating region by a high value (>1 MΩ) resistor that is in parallel with the crystal. This biasing allows the inverter to operate as a simple inverting linear gain element.

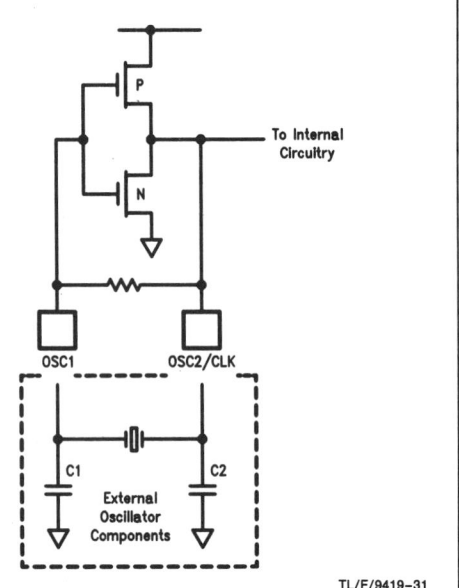

FIGURE 24. Simplified Schematic of Oscillator Circuit for DP8473

The DP8473 oscillator is intended to be used with a fundamental mode parallel resonant crystal. The only external components required are the crystal and two external capacitors. These capacitors are usually very small (picofarads).

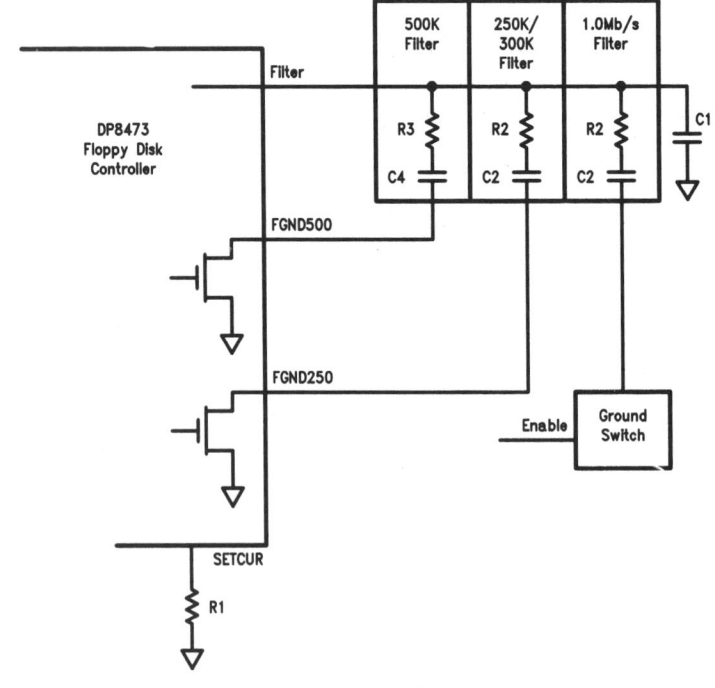

FIGURE 23. Filter Configuration for Optimal Filter Design at All Data Rates

TABLE II. Important Parameters for Crystal Selection

Parameter	DP8473
Crystal Frequency	24 MHz
Oscillatory Mode	Fundamental
Oscillator Resonance	Parallel
Accuracy	<0.5%
Series Resistance	<100Ω
Shunt Capacitance	<7 pF
External Parallel Capacitors (include parasitics)	10 pF

Table II shows the important parameters to check for when selecting a crystal to use with the DP8473. While the recommended resonance mode is parallel, a series resonant crystal can be used. It will just oscillate in parallel mode 30 ppm–300 ppm from its ideal frequency.

If an external oscillator circuit is used, it must have a duty cycle of at least 40%–60%, and minimum input levels of 2.4V and 0.4V. The controller should be configured so that the clock is input into the OSC2 pin, and OSC1 is tied to ground.

5.5 Trimming for Perfection

The integrated data separator was designed to achieve excellent performance. However, product, temperature, and power supply variations can degrade performance somewhat. This can lead up to a 10% variation in window margin performance. While this is still exceptional for any analog design, it is possible to trim out this variability.

The two major factors that contribute to data separator performance degradation are: 1) Loop Gain variation; 2) ¼ period delay line length variation.

Trimming the Loop Gain

The loop gain variation can be trimmed by replacing the pump resistor, R_1, with a variable resistor. This resistor should be trimmed based on the ideal lock range of the PLL desired. For example, if a ±6% lock range is desired, then during final board product test, a tester can be used to measure the total lock range and R_1 can be adjusted larger to reduce lock range, or smaller to increase it.

Trimming the Quarter Period Delay Line

The perceived length of the quarter period delay line can be modified by causing a static phase error in the loop to compensate for the quarter period delay line's error. This is accomplished by placing a pull up or pull down resistor on the filter pin. This resistor can be adjusted by measuring the Static Window margin for both an early and late single bit shift. Based on these measurements the delay line can effectively be adjusted by changing the value of the filter pull up/down resistor.

5.6 Initially Unlocked Model

This section is provided in order to complete the full discussion of the theoretical operation of a data separator. It is useful to discuss how the controller locks back to the crystal/clock reference when it needs to. This operation is taken care of by the controller so that the user need not concern himself with the design aspects of this section. However, if the user desires a more complete understanding of the entire lock process that the data separator goes through, this section is presented.

Another model is used to analyze the behavior of the PLL in an unlocked state. It is assumed in this model that the loop is not locked, and the VCO frequency is different from the input frequency. This model can be used to evaluate how the PLL re-locks to the reference clock after reading bad data and being thrown off frequency.

The first operation of the PLL is to frequency lock, so for this model each block is described as a function in terms of frequency, not phase. An equation can be derived for the frequency error that is similar to equation (2) for the phase error:

$$K_e = \frac{\Omega_e}{\Omega_1} = \frac{1}{[1 + K_{DF} K'_{VCO} F(s)]} \quad (26)$$

In the initially unlocked model, the phase detector has a key role. The phase detector compares the VCO output with the input signal. If the VCO output rising edge is leading the input signal, a pump-down signal is generated from this edge of the VCO to the next rising edge of the input signal. If the VCO is lagging the input signal, a pump-up signal is generated from the edge of the input signal to the rising edge of the VCO.

There is no overshoot of the VCO frequency. Only one type of pump signal is generated, up or down, to bring the VCO frequency toward the input frequency. For example, if the VCO frequency is higher than the input frequency, only pump-down signals are generated. It can also be seen that the larger the frequency difference between the VCO and the input, the longer the pump pulses become. The average current flowing from the charge pump is roughly proportional to the frequency difference of the signals at the input of the phase (and now also frequency) detector. The phase detector gain is:

$$K_{DF} = \frac{I_{PUMP}}{\Delta \omega} \quad (27)$$

for $\omega_2 < 2\omega_1$, where $\Delta\omega = \omega_2 - \omega_1$, and

$$K_{DF} = -I_{PUMP} \quad (28)$$

for $\omega_2 > 2\omega_1$. Therefore, if ω_2 is not too far from ω_1, the expression (8) can be written for our PLL as:

$$K_e(s) = \frac{\dfrac{s\Delta\omega C_2}{K'_{VCO} I_{PUMP}}}{1 + s\left(R_2 C_2 + \dfrac{\Delta\omega C_2}{K'_{VCO} I_{PUMP}}\right)} \quad (29)$$

This expression will allow the calculation of the time that the loop requires to lock back to the reference after a read operation goes through a bad data field or write splice.

Acquisition to the Crystal

After the completion of a read attempt, it is important to ensure that under the worst case conditions the PLL will properly re-lock itself to the reference clock. In order to achieve the required performance during the acquisition to the data stream, the PLL must have reached the lock to the crystal before it is allowed to lock back to the data.

If the PLL attempts to lock to a write splice the VCO may be pulled way off frequency. To prevent this, read gate should be deasserted as soon as a wrong or bad data field is detected. This will prevent the VCO frequency from being pulled too far away. The DP8473 read algorithm has been optimized to prevent this.

If the PLL is locked to the data stream, when read gate is deasserted, the lock mechanism to lock back to the reference frequency is quite similar to locking to the data. If the frequency of the VCO has been swept way off frequency because of a bad data field, noise, write splice, or missing data, the unlocked PLL model must be used. Since when locking to the crystal the phase-frequency comparison is always enabled, the phase detector acts as a frequency discriminator. Switching to the crystal imposes a frequency step to the PLL. It can be demonstrated that the frequency error generated is an exponential function of time, going to 0 with the time constant of:

$$T_P = R_2 C_2 + \frac{C_2 \Delta \omega R_2 N}{K_{PLL}} \qquad (30)$$

Thus T_P can be assumed to be the worst case acquisition time to the reference clock.

Example 3: From the previous example 1, we have chosen the values for the components of: $C_2 = 0.039\ \mu F$, $R_2 = 535 \Omega$. We would like to find the worst case time required to re-lock to the crystal. Assume that the maximum frequency range of the VCO is $\pm 30\%$.

If the VCO is pulled 30% off center, then:

$\Delta \omega = 2\pi (500\ \text{kHz})(0.30) = 942\ \text{Krad/sec}$

Using equation (10) for example 1, we can obtain:

$T_P = (0.039\ \mu F)(545 \Omega)$
$\quad + \dfrac{(0.039\mu)(942K)(5.6K\Omega)(8)}{(75\ \text{Mrad})}$

$T_P = 41\ \mu s$

This is about 4 byte times. Thus, read gate must be deasserted for this length of time before re-asserted to assure that the PLL has re-locked to the reference. Most floppy controllers de-assert read gate much longer than this, and the DP8473 deasserts its internal read gate for 6 bytes.

Bibliography

Gardner, Floyd M. PhD., *Phase Locked Techniques*, 2nd edition. New York: John Wiley and Sons, 1979

Best, Roland E., *Phase Locked Loops*, New York: McGraw-Hill, 1984

DP8470 Phase Locked Loop Data Sheet, National Semiconductor Corp. 1986,7

DP8472 Floppy Disk Controller PLUS Data Sheet, National Semiconductor Corp. 1986,7

DP8473 Floppy Disk Controller PLUS/2 Data Sheet, National Semiconductor Corp. 1987

Design Guide for DP8473 in a PC-AT

National Semiconductor
Application Note 631
Robert Lutz

OVERVIEW

When designing a floppy interface circuit for a PC-AT in the past, there was very little flexibility given to the design engineer. The NEC765, which was designed into the original IBM PC, was the only floppy controller available that would guarantee compatibility. Compatibility is extremely important in a PC design.

There were many design issues that had to be resolved when using the NEC765 to produce a fully functional floppy controller interface. A data separator had to be selected or designed. A write precompensation circuit needed to be included. A whole score of miscellaneous logic had to be designed to handle all of the unique PC functions that the NEC765 does not handle itself.

The DP8473 from National Semiconductor was developed to eliminate all of these design problems. All of the extra functions and logic normally required for an XT or an AT design were included inside the chip. Even an analog data separator, which is classically the hardest function to design, has been integrated into the DP8473.

Compatibility has been completely retained. The DP8473 is software compatible with the NEC765A. We have not found any current software available for the PC that fails to work properly with the DP8473. This includes software running under DOS and OS/2.

This application note will discuss some of the issues involved in a floppy controller design with the DP8473. Even though on the surface, there may not seem to be many options when designing a floppy controller for a PC, there really are quite a few. Some of these options include: signal swapping in the floppy cable, different types of floppy drives, and data separator filter selection.

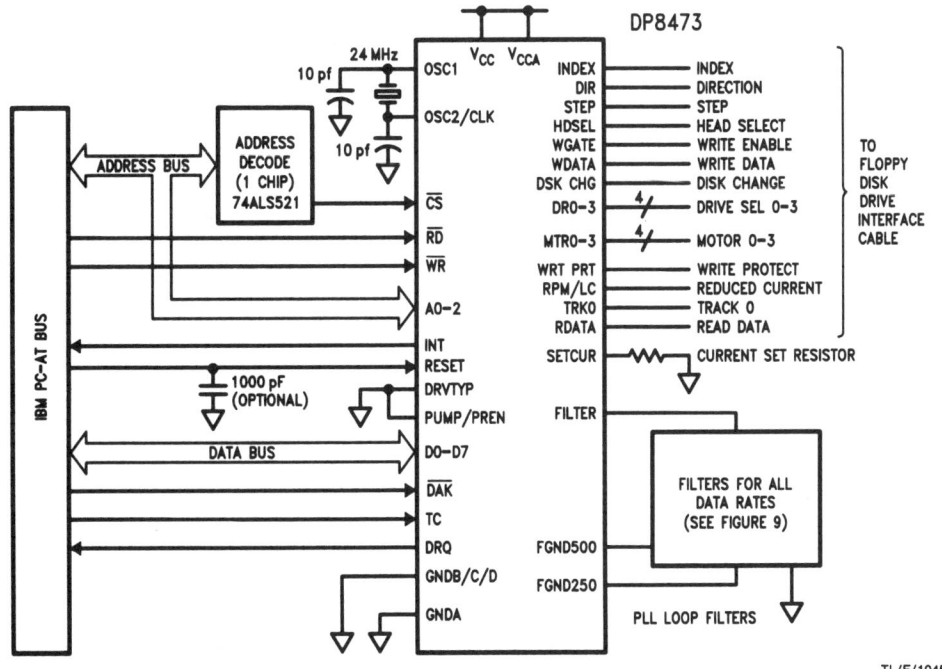

FIGURE 1. Schematic of Typical DP8473 Floppy Controller Design

HARDWARE ENVIRONMENT

A typical floppy controller design with the DP8473 will look something like the schematic shown in *Figure 1*. You may be surprised that the entire schematic for the floppy controller design fits on less than one page. Especially if you consider that the schematic for a similar function in the IBM PC-XT technical reference manual takes four full pages.

The heart of the design is, of course, the DP8473. Most of the interface pins to and from the DP8473 go directly to the peripheral bus or the disk drive cable without additional logic or buffering.

DRIVE CABLE INTERFACE

The DP8473 disk interface signals connect directly to the drive cable. Most disk drives terminate the drive cable with resistors. Termination is required because the output buffers of the floppy controller are open-collector.

Terminated signals are used because historically, relatively long cables have been used to connect the floppy controller to the disk drives. The cable termination will decrease the amount of crosstalk and noise on the drive cable signals.

A typical disk interface circuit consists of an open-collector output buffer at the signal source, and a termination resistor and a Schmitt input buffer at the signal destination. For example, the STEP output pin on the DP8473 is an open-collector output buffer that is capable of sinking up to 48 mA (See Note 1). If the output is off, the buffer is disabled, and the termination resistor on the disk drive will pull the signal high. If the STEP output is on, the DP8473 buffer will pull the signal low. An example of the cable termination logic is shown in *Figure 2*.

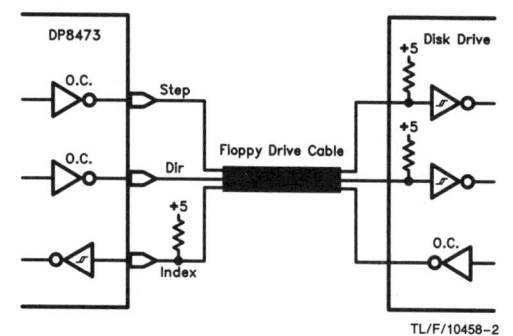

Note 1: The DP8473 actually contains open-drain output buffers due to its CMOS design. The end result is the same as TTL open-collector buffers.

FIGURE 2. Example of Buffers and Terminators Used for Floppy Drive Interface

The termination resistors used with 5.25″ drives or 8″ drives typically have a value of 150Ω. With this resistor value, the output buffers must be capable of sinking about 35 mA each in order to pull the drive signal to a logic low level. The DP8473 is able to sink this current without external buffers.

The termination resistors used with 3.5″ drives are often 1 kΩ. 1 kΩ termination resistors are also sometimes found on low power 5.25″ drives. Drive manufacturers have recognized that the floppy interface cable used in a PC is relatively short. Also, the drives are installed in the same grounded enclosure as the PC and the floppy controller. This reduces the amount of noise introduced on the floppy interface cable.

If both 3.5″ drives and 5.25″ drives are to be used in an application, the termination resistors used with the DP8473 must be chosen carefully. The termination resistor value used with the DP8473 must be the larger of the termination resistors used on the drives. For example, if the 5.25″ drive has 150Ω termination resistors and the 3.5″ drive has 1 kΩ termination resistors, the termination resistors used on the inputs to the DP8473 should be 1 kΩ.

The termination resistors for the inputs to the DP8473 should be placed near the DP8473. The termination resistors for the outputs for the DP8473 to the disk drive are contained in the disk drive itself.

Additional disk interface buffering is not normally required when using the DP8473. It can sink up to 48 mA for each disk output signal. This is more than enough capacity for a typical floppy design.

DRIVE CONFIGURATION

A PC-XT can typically interface to up to four disk drives. A PC-AT, however, is usually limited to two disk drives. The two drive limitation is due to the ROM BIOS used in most AT's. More than two drives can be used if special software drivers are written.

The connection between the floppy drives and the floppy controller in a PC is slightly different than the SA450 standard used in non-PC's. The advantage to the method used in a PC is that each disk drive installed in the PC can be configured identically. Even the Drive Select strap is the same. Each drive is configured as drive 1 (or B). Even if four drives are all connected, they are each strapped as drive 1.

The trick used to accomplish this feat is cable wire swapping. The drive cable is cut up and wires are moved around. The cable swapping re-routes the four DRIVE select signals (DR0, DR1, DR2, DR3) to the DR1 signal of each individual drive. For example, DR0 is routed to drive A's DR1 input. DR1 is routed to Drive B's DR1 input, and so on. In a similar manner, the cable swapping also re-routes the four MOTOR signals (MTR0, MTR1, MTR2, MTR3) to the MOTOR signal of each drive. *Figure 3a* demonstrates how the cable is configured for two floppy disk drives. A second cable would be used if more than two drives are required as shown in *Figure 3b*.

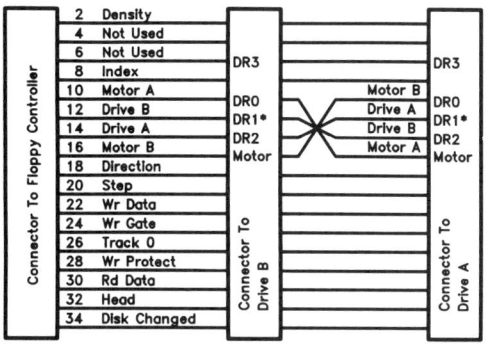

FIGURE 3a. Cable Swapping Used for Drives A and B

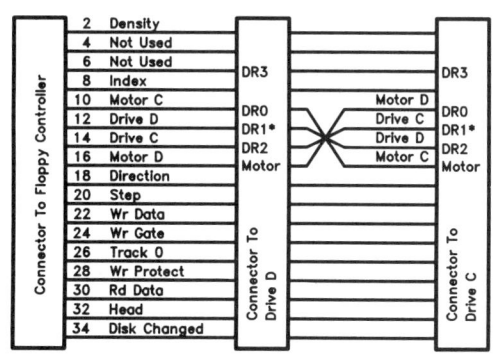

FIGURE 3b. Cable Swapping Used for Drives C and D

Note: The asterisk (*) next to DR1 indicates that this drive is strapped as Drive 1. In the PC, all drives are strapped as Drive 1.

If more than one disk drive is attached to the floppy controller, there may be more than one drive terminated. This may cause a current overloading problem. The controller may not be able to drive an active signal low.

It is easy to prevent current overloading in a two disk drive PC. Simply make sure that only one drive is terminated. Ideally the terminated drive should be the drive at the end of the drive cable, although it could be either drive.

If both drives are terminated, the output buffers will be driving too much current. The system will be out of specification. This is a common situation and is largely ignored by PC manufacturers. The output buffers can usually handle the additional load.

It three or four drives are to be used, things become more complex. For example, if four 150Ω terminated drives are all attached to the DP8473, the DP8473 will have to sink 139 mA for each drive interface signal. The DP8473 is guaranteed to sink up to 48 mA. Therefore, this configuration would not work without additional buffering. Please refer to the How to Calculate the Maximum Current Required for Output Buffers section for a description on current calculation.

There are three techniques that can be used to prevent overloading the output buffers:

Technique 1:

Using larger termination resistors can reduce the load on the DP8473 to acceptable levels. If 1 kΩ resistors can be used instead of 150Ω resistors. In the worst case where all four disk drives are terminated, only 21 mA will be generated instead of 140 mA. This is well within the specification of the DP8473. 1 kΩ resistors are commonly used with 3.5" drives.

Technique 2:

The most direct technique that can be used is simply adding additional buffers for the extra disk drives. Drives A and B can be driven directly by the DP8473. The outputs to drives C and D can pass through an open-collector buffer such as the 7407. This is shown in *Figure 4*. 1k pullup resistors are required for some of the DP8473 outputs because they are not terminated by drives A or B.

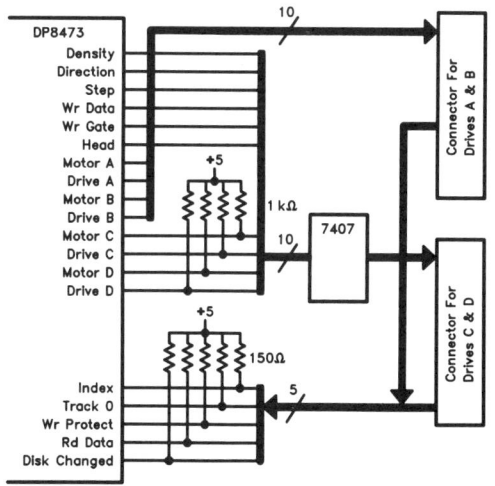

FIGURE 4. Extra Buffers Required for Four Drive System

Technique 3:

Daisy chain the floppy drives with the controller on one end of the drive interface cable, and one terminated drive at the other end. One to three additional non-terminated drives can be added in the middle as shown in *Figure 6*. With this technique, the four Motor signals from the DP8473 should be wire-ORed together as shown in *Figure 5*. Each drive must be strapped for the proper drive select (0–3).

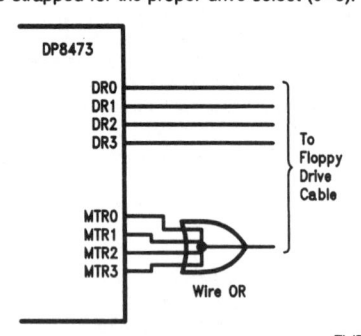

FIGURE 5. Wire OR Required for Daisy Chain Connection

How to Calculate the Maximum Current Required for Output Buffers.

Since a floppy controller design may not work correctly due to current overloading, it is important to understand exactly how to calculate the maximum current required by the floppy controller for each output signal to the disk drive. This is largely determined by the termination resistors used by the disk drives. A formula that can be used is:

$$\frac{(V_{CC}) + (\text{Max } V_{CC} \text{ Variation}) - (V_{OL(max)})}{(\text{Termination Resistor}) \bullet \frac{1}{N} \bullet (1 - \text{Resistor Accuracy})}$$

$V_{CC} = 5.0$

Max V_{CC} Variation = Power supply variation (0.5V)

$V_{OL(Max)}$ = Maximum active low output voltage of buffer (0.8V)

Termination Resistor = Termination on disk drive

N = Number of terminated disk drives

Resistor Accuracy = Accuracy of termination resistors (10% or 0.10)

Example 1:
One terminated drive with 150 termination resistors.

$$\frac{5.0 + 0.5 - 0.8}{150 \bullet 1 \bullet 0.9} = 34.8 \text{ mA}$$

Example 2:
Four terminated drives with 1k termination resistors.

$$\frac{5.0 + 0.5 - 0.8}{1000 \bullet \frac{1}{4} \bullet 0.9} = 20.9 \text{ mA}$$

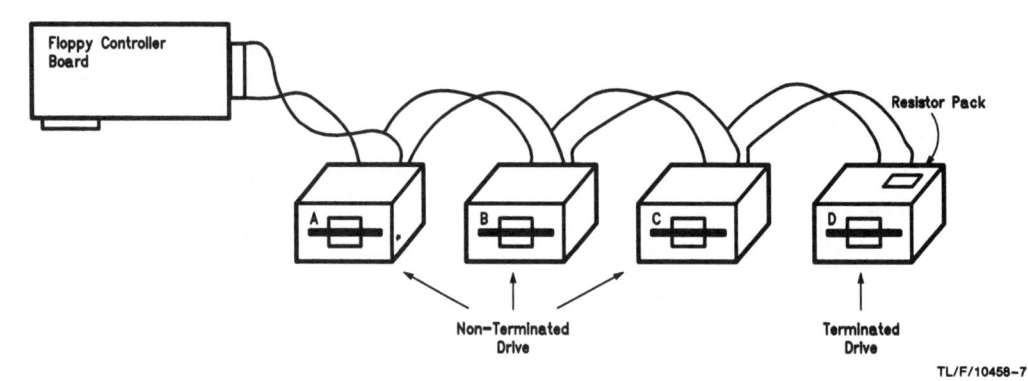

FIGURE 6. Daisy Chain of Four Drives

DRIVE TYPES

There are many types of disk drives that can be connected to the DP8473 floppy disk controller. The DP8473 is compatible with 8", 5.25", and 3.5" floppy disk drives, although 8" drives are rarely used today.

Other types of peripherals may be connected to the floppy controller as well. A streaming tape drive that is used to back up the hard disk is often connected to the floppy controller. A tape drive of this type is very specialized. It has been designed to look like a floppy disk drive from an electrical interface point of view. It does not perform exactly like a disk drive, however. Special software is usually required to make it work correctly. The STEP signal is often used to issue commands to the tape drive. For example, to rewind the tape, four step pulses may need to be issued. To start a read, six step pulses might be issued.

All of these different drive types have one thing in common, a similar electrical interface. This allows them all to be connected to a common drive interface cable. For example, the READ DATA signal on pin 30 of the floppy interface cable is the MFM encoded serial stream of data from the disk. The INDEX signal on pin 8 of the cable identifies the beginning of a track.

However, there are some minor differences between drive tapes that must be considered. The DENSITY signal is a good example of a difference. This signal is active for *high* density transfers on a dual density 3.5" drives. But, this signal is active for *low* density transfers on a dual density 5.25" drive. This difference makes the floppy system design more complex when 5.25" dual density and 3.5" dual density drives are both used in the same PC.

The DP8473 has a signal called RPM/LC that normally connects to the Density or Low Current input on a dual density 5.25" drive. If a 3.5" drive is used, the RPM/LC output should be inverted.

A design such as the one shown in *Figure 7* could be used to create the proper DENSITY signal for both 5.25" and 3.5" drives. The jumpers and logic allow the user to select between drive types for each individual drive.

Another solution is simply to use a 3.5" drive that contains a built-in jumper to vary the polarity of the DENSITY signal directly on the drive. This option eliminates the need for external logic and jumpers.

Another signal that is drive type dependent is DISK CHANGED. This signal exists on low and dual density 3.5" drives and also on dual density 5.25" drives. It does not exist on low density 5.25" drives. If a low density 5.25" drive is to be used, the DISK CHANGED signal should be held active (low level). It may be held active by the drive by itself. If not, a pull-down resistor could be used to activate the non-driven signal.

One thing to consider while choosing drive types is media compatibility. Of course, you can't put a 3.5" disk in a 5.25" drive. But, there are more subtle incompatibilities even within similar media types. For example, a low density 5.25" disk written to in low density mode by a dual density drive cannot be read reliably by a low density drive. Table I lists the compatabilities between different drives and media types.

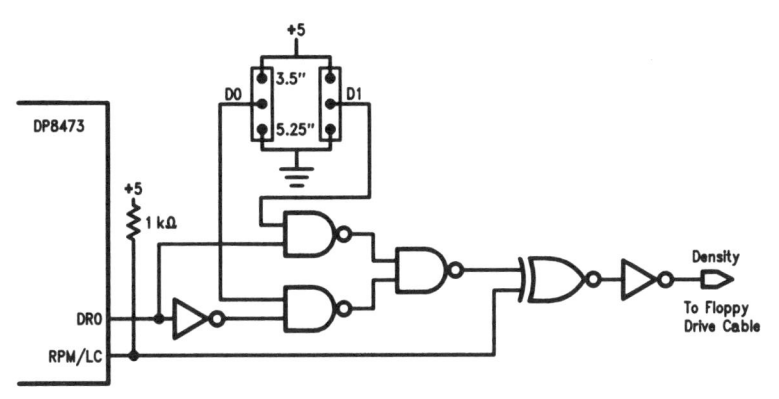

FIGURE 7. Density Select Logic for 5.25" or 3.5" Drives

TABLE I. Drive and Media Compatibility

Drive Type	Mode	3.5"	3.5" HD	5.25" DD	5.25" HD
3.5" LD	720k	R/W			
3.5" LD,HD	720k 1.44M	R/W	R/W		
5.25" LD	360k			R/W	
5.25" LD,HD	360k 1.2M			R Only	R/W

Notes:
LD = Low Density
HD = High Density
R/W = Readable & Writable
R Only = Readable Only

DATA RATES

Different drive types may use different data rates. The data rate is specified by the number of bits that are transferred in a second. For example, 250 kb/s is translated to 250 thousand bits per second.

The data rate used by a disk drive is determined by the electronics of the drive and the specifications of the media. Low Density media require data to be transferred at 250 kb/s. High Density media is twice as fast at 500 kb/s.

There is a complication with Low Density 5.25" media in a Dual Density drive. The 5.25" Dual Density drive spins the disk faster (360 RPM) than a Low Density drive (300 RPM). When a Low Density disk is read from a Dual Density drive, the data rate will be 300 kb/s instead of 250 kb/s because of the rotational speed difference.

The DP8473 can operate at all the data rates required for a PC. This includes 250, 300, and 500 kb/s. In addition, the DP8473 can operate at 1 Mb/s. This high data rate is starting to appear in both floppy disks and streaming tape drives.

μP INTERFACE

It is hard to imagine how the interface between the DP8473 and the μP bus could be made any simpler than it is. Only one interface function is not integrated in the DP8473. That function is address decoding. The typical μP connections are shown in *Figure 8*.

The DP8473 requires a CS (chip select) enable signal that is generated elsewhere. Typically, this could be generated by an ALS521 8-bit comparator or a similar circuit. Address decoding of the three least significant bits of the address is performed by the DP8473. The AEN (address enable) signal from the μP bus should be included in the CS logic. This will prevent DMA transfers from generating a CS.

The eight bit data bus from the DP8473 connects directly to the data bus of the μP. Bus transceivers are not required because 12 mA buffers are built into the DP8473.

The IOR (I/O read), IOW (I/O write), RESET, TC (terminal count), DRQ (DMA request), and IRQ (interrupt request) signals can be connected directly to the DP8473. No additional buffering or gating is required. The logic required to TRI-STATE® the DRQ and INT pins is integrated in the DP8473.

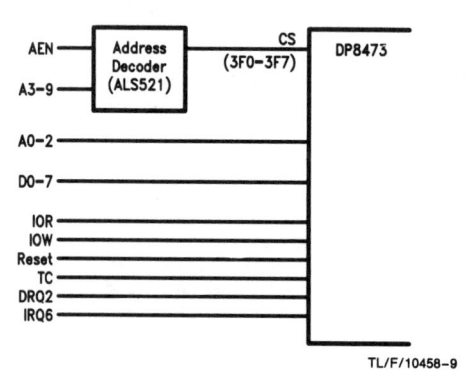

FIGURE 8. μP Interface to DP8473

FILTER SELECTION

The internal data separator in the DP8473 requires an external filter to operate correctly. This filter is part of an analog Phase-Locked Loop (PLL) that is used by the data separator integrated into the DP8473. This is commonly referred to as an analog data separator due to the analog nature of the PLL's filter and Voltage Controlled Oscillator (V_{CO}).

As the floppy controller changes from one data rate to another, the filter used by the PLL must change also. This is done automatically in the DP8473. Two or three filters are connected externally to different pins of the DP8473. The correct filter is selected and activated by the DP8473 itself.

The filter selection is performed by grounding or forcing TRI-STATE the appropriate filters. For example, a two-filter arrangement is shown in *Figure 9*. If 500 kb/s is selected, the FGND500 pin is grounded and the FGND250 pin is at TRI-STATE. From the filter pin's point of view, this appears as filter F2 in parallel with capacitor C1. F1 is electrically out of the picture. When 250 kb/s is selected, it is the opposite. The FGND500 pin is at TRI-STATE and the FGND250 pin is grounded. Since 300 kb/s data rate is close to 250 kb/s, the same filter is used for both data rates.

The two-filter arrangement shown in *Figure 9* would be used in most PC applications. It supports 250, 300, and 500 kb/s data rate. Other filter combinations may be used for specialized applications. These filter combinations are described in the DP8473 data sheet.

DATA SEPARATOR PERFORMANCE

One of the most important features of the DP8473 is the high level of data separator performance. This performance translates directly to reduced disk I/O errors. There is quite a bit of information that can be discussed on data separator performance. A lot of this information is presented in an application note titled "Floppy Disk Data Separator Design Guide for the DP8473, AN-505".

It might seem desirable to specify the maximum error rate of the DP8473. This could be expressed in terms of the number of bits that can be read correctly before an error occurs (10^{12}, for example). This is not a practical parameter to specify, however. One problem is the amount of time this type of measurement would take. A test of 10^{12} could take well over 400 days to measure.

Another problem is defining the test conditions. Is the test performed under ideal disk conditions? Are bits jittered or is motor speed variation simulated? There are so many variables in this type of a test that it would be difficult for two different people to produce the same results. In addition, the error rate is related to other factors beside the DP8473 such as the disk drive or the floppy media.

There are many different types of measurements that can be made with a data separator. Most of these measurements indicate how well only one particular section of the data separator performs. For example, the gain of the VCO (Voltage Controlled Oscillator) or the accuracy of the $\frac{1}{4}$ period delay line. They don't, however, give a good indication of how well the data separator will perform under real-life conditions.

There is one measurement that produces meaningful data. This measurement is called "Dynamic Window Margin". Dynamic window margin attempts to indicate total data separator performance under real-life conditions. It measures how much bit jitter the data separator can handle while reading a worst case data pattern (DB6 hex) with a drive that has a motor speed of varying accuracy. The data is jittered in a manner similar to real world jitter with a reverse write pre-compensation pattern. The measurement is taken at the worst point over a motor speed variation of ±6%. An example of a dynamic window margin measurement is shown in *Figure 10*.

The typical dynamic window margin is specified in the DP8473 data sheet. National Semiconductor has made measurements of the dynamic window margin of many data separators, and the results show that no other data separator performs as well as the DP8473s.

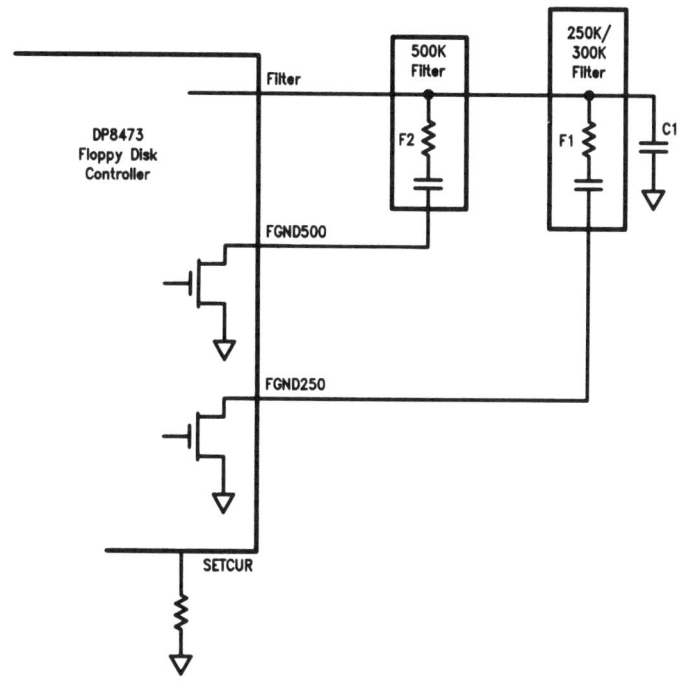

FIGURE 9. Typical Filter Connection for 250, 300, and 500 kb/s

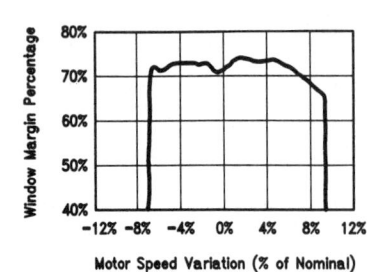

FIGURE 10. Typical Dynamic Window Margin for DP8473

LAYOUT AND COMPONENT CONSIDERATIONS

The DP8473 contains a combination of digital and analog circuitry. Because of the analog nature of the data separator, some precautions should be taken when designing a board with the DP8473.

A good DP8473 board layout design will distinguish between analog and digital, V_{CC} and Ground. The supply pins used by the digital circuitry are labeled V_{CC}, GNDB, GNDC, and GNDD. The supply pins used for the analog data separator are labeled $V_{CC}A$ and GNDA. Standard digital decoupling techniques should be used with the digital supply pins. This typically involves 0.1 μF capacitors connected between V_{CC} and GND.

The analog supply pins require a bit more consideration than the digital supply pins. Any noise or ripple on the analog supplies will degrade the data separator performance. It is recommended to minimize this noise as much as possible. Less than 50 mV noise would be good.

There are many methods that can be used to minimize noise on the analog supply pins. One of the best methods is a 5.0V voltage regulator dedicated to the analog section. This guarantees a very clean signal. The voltage regulator can be driven by the 12V power supply.

Another method is to place the DP8473 on the board close to the entry point of the power supply. At the very least, separate supply lines should be dedicated from the power supply entry point to the DP8473 $V_{CC}A$ and GNDA.

In addition to the analog supply, any noise or crosstalk introduced to the external filters will adversely effect the performance of the data separator. The data separator filters should be positioned as close as possible to the DP8473. A ground plane surrounding the filters would also be advisable. The resistor attached to the SETCUR pin should also be close to the DP8473.

If a 24 MHz crystal is used, it should be placed close to the DP8473 as well as the 10 pF capacitor attached to both sides of the crystal.

The component types and tolerances may also effect the data separator performance. The accuracy of the resistor attached to the SETCUR pin is important. It should have a 1% tolerance rating. A 5% could be used, but the accuracy may effect the data separator performance.

The capacitor in series with the resistor attached to the FILTER pin is also critical. It should have a 5% tolerance. The series resistor in the same network is not as critical as the SETCUR resistor. Therefore, a normal 5% tolerance can be used here.

Finally, the capacitor in parallel with the filter network can be rated as low as 10%–20%. It does not effect the data separator very much.

The component tolerances mentioned here are only recommendations. The DP8473 will work properly with a wide range of filter values. These recommendations should be followed if data separator performance is an important issue in a particular design (which is normally true).

TROUBLE-SHOOTING

If the floppy controller does not appear to operate correctly, there are some key areas that can be looked at for the source of the problem.

Drive's in Use light remains on at all times.
- Drive Interface cable plugged in backward.
- Drive Interface signals not properly routed.

DOS returns a "Not ready error reading drive X".
This error can be caused by many different problems. At this point it would be best to run the "Floppy Demo Program" as described later in this section.

DOS Directory command returns an old directory.
- Disk Changed signal improperly routed.

POST produces a "601" error while booting.
- Drive Interface cable unplugged.
- Hardware problem with μP interface.

PC locks up while booting.
- Hardware problem with μP interface.

Parity Error.
- DMA interface problem.

Many advanced diagnostics may be used while running the "Floppy Demo Program" which is available from National Semiconductor. This program allows you to issue individual floppy commands such as Read Data, Format Track, and Seek. The Result Phase of these commands can be analyzed to help determine where a problem exists. The following list describes some likely sources of problems based on what is observed with the "Floppy Demo Program".

"Missing Address Mark in Address Field" error.
This indicates that the floppy controller could not find any valid data on the track being read. No sectors could be found.
- Track has not been formatted. Could be a blank disk. The controller does recognize Index Pulses, however. This indicates that the drive cable interface is at least partially intact.
- Read Data drive interface signal not properly routed.
- General data separator problem. See the data separator discussion later.

"Did not receive interrupt" error.
This is usually accompanied with all FF's in the Result Phase of the command. This indicates that the controller could not read anything from the disk drive and Index pulses were not seen. Be sure to reset the floppy controller after this error.
- Disk not inserted in drive or drive door open.
- Wrong drive selected in command.
- Drive Interface cable unplugged.

MSR (Main Status Register) does not return to 80 (hex) after a reset.
- Floppy controller not inserted in socket properly.
- Hardware problem with address decode (CS) or µP interface signals.
- Crystal or external clock not operating properly.
- Floppy controller is bad.

"No Data" error or "CRC error".
- Bad disk media.
- General data separator problem.

Fewer bytes read than requested or error in Result Phase.
- Noise on Reset pin. Insert capacitor between Reset and digital GND as specified in the data sheet.

Long term read produces some errors.
- General data separator problem.

Many of the problems described above refer to a general data separator problem. There are many things that can be looked at for data separator problems.
- Filter wired incorrectly.
- Incorrect filter component values.
- Filter layed-out poorly.
- Too much noise on analog V_{CCA} or GNDA.

FLOPPY CONTROLLER DESIGN WITH NEC765A

In order to appreciate the amount of circuitry integrated into the DP8473, it may be useful to analyze a typical floppy controller design for the PC-AT using the standard NEC765A floppy disk controller. To simplify the analysis, only the block diagram will be presented. The actual schematic would require many pages. The block diagram is shown in *Figure 11*.

The NEC765A was developed many years ago originally for 8" floppy disk drives. It became very popular because it was designed into the original IBM PC. The NEC765A performs many functions, but there are also many functions that it does not perform.

A data separator isolates the individual pulses read from the disk drive and allows the floppy controller to distinguish between MFM clock pulses and data pulses. It typically incorporates an analog PLL. An analog data separator design could require as many as 10 chips to design. A single chip discrete digital data separator could also be used, although the performance will not be as good.

The write precompensation circuit shifts the MFM encoded data as it is being written to the disk. This shifting compensates for known bit shifts that will occur due to the magnetic influences of the individual bits recorded on the disk. This is typically designed with a shift register and a multiplexer.

The NEC765 cannot interface directly to the Drive Interface cable. Separate 48 mA buffers are required for each output signal. This requires three 7406's. Also, the inputs from the disk drive required Schmitt inverters.

Additional buffering is also required for the µP data bus.

The PC-AT requires that the Disk Changed signal be read from a particular port. This involves address decoding and buffering.

The only method available to vary the data rate used by the NEC765A is by altering the input clock frequency to the chip. This must be done with a complex divider circuit that generates the different frequencies required for 250, 300, and 500 kb/s data rates.

Due to timing incompatabilities in the NEC765A, the Drive Select and Motor On signals must be generated by an external port that is controlled by software. This port also controls the software reset and DMA and INT enable circuitry.

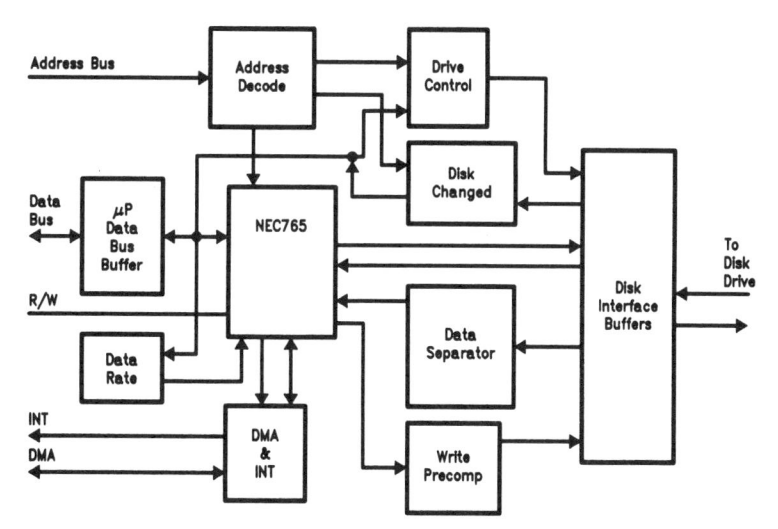

FIGURE 11. Block Diagram of Floppy Controller Design with NEC765

DMA transfers must be slowed down due to handshaking problems with the NEC765A. This delay is performed with an external shift register.

This entire design easily requires at least a couple of dozen chips. The amount of board space used is quite large. A considerable amount of current is also consumed. This compares to the DP8473 solution which only requires two chips. It is easy to see that the DP8473 solution is much more economical.

CONCLUSION

This design guide was created to answer the most common questions encountered while designing with the DP8473. Any new design can be based on the information given in this guide. A two drive system can be created or more drives can be added if required. A variety of disk drives may be used including 3.5" drives.

The address decoding is the only function not integrated into the DP8473. However, the integrated data separator requires external filtering which should be carefully layed-out on the board.

If problems arise, there are many items that can be looked at to help identify where the problem exists. It may be useful to obtain a floppy controller diagnostic program similar to the "Floppy Demo Program" available from a National Semiconductor sales office.

If more information is desired concerning the performance of the data separator or the trade-offs of designing a custom filter for the PLL, please read the application note titled "Floppy Disk Data Separator Design Guide for the DP8473", AN-505.

seeq

8005
Advanced Ethernet Data Link Controller

PRELIMINARY — December 1989

Features

- **Conforms to IEEE 802.3 Standard**
  - Ethernet (10BASE5) Cheapernet (10BASE2) and Twisted Pair (10BASE-T)
- **Recognizes One to Six Selectable Station or Multicast Addresses**
- **Advanced Error Detection/Handling:**
  - Automatic Re-Transmit after collision
  - Auto discard of bad packets
- **Software Selection of 2 Byte or 6 Byte Station Addresses**
- **Optional Preamble and Cyclic Redundancy Code (CRC) Generation/Checking**
- **Manages 64K Bytes of Local Packet Buffer**
  - Connects to RAS/CAS/Data/Control of 64K x 4 Dynamic RAMS
  - Automatic DRAM Refresh
  - Automatic Posting of Status Packet in Buffer
- **Flexible System Bus Interface**
  - 8 or 16 Bit Data Transfers with Byte Swap Capability
  - Programmable DMA Burst Length
  - Selectable for Intel or Motorola Compatible Bus Signals
- **Connects Directly to 8020 Manchester Code Converter**
- **Uses Fewer Support Chips**
  - Lower System Costs
  - Higher Reliability
- **68 Pin Surface Mount Plastic Leaded Chip Carrier Package**

Pin Configuration

8005 TOP VIEW

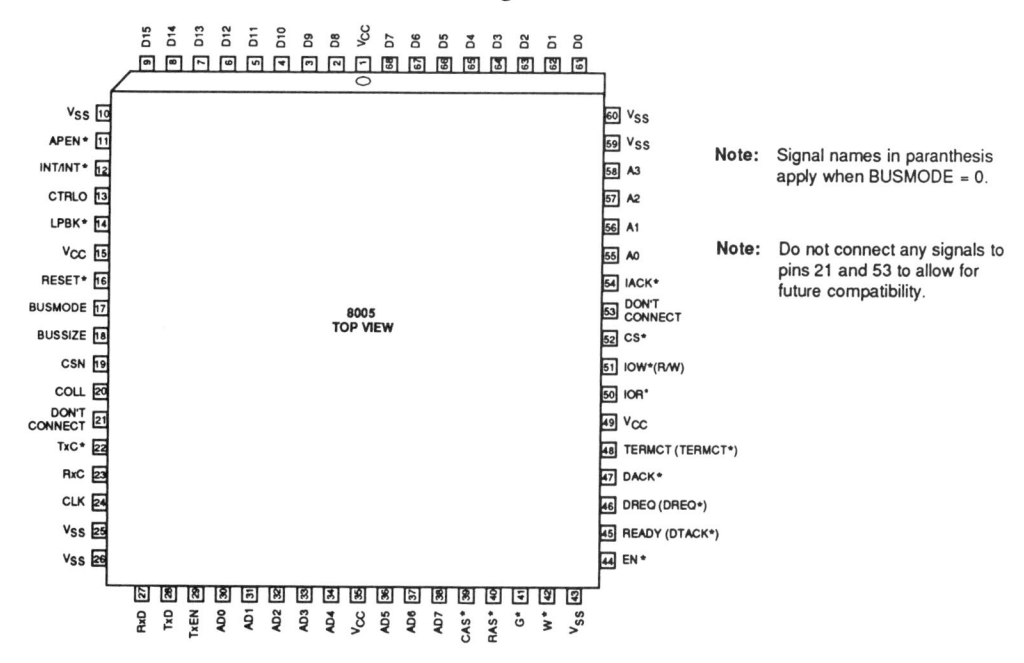

Note: Signal names in paranthesis apply when BUSMODE = 0.

Note: Do not connect any signals to pins 21 and 53 to allow for future compatibility.

seeq Technology, Incorporated

MD400031/C

PRELIMINARY

8005

Pin Description

(An asterisk after a signal name signifies an active low signal)

D0-D15: A 16 bit bidirectional system data bus. If BUSSIZE = 0, the bus is configured as 8 bits and D8-D15 are not used for data transfer. Byte order for local buffer data transfers on a 16 bit bus is software configured. D8-D15 are used to provide address information to the optional external address PROM in both 8 and 16 bit modes.

EN*: An output which can be used to control the three-state control pin of external bi-directional drivers such as the 74LS245.

APEN*: Active low address PROM enable output.

IOW*(R/W): If BUSMODE = 1, this input defines the current bus cycle as a write. If BUSMODE = 0, this input defines the bus cycle as a read if a 1 or a write if a 0.

IOR*: If BUSMODE = 1, this input defines the current bus cycle as a read. If BUSMODE = 0, this input is not used.

CS*: The chip select input, used to access internal registers and the packet buffer.

A0-A3: Address select inputs used to select internal registers for reading or writing. A0 is not used in 16-bit mode.

DACK*: An input used to acknowledge granting of the system bus for external DMA transfers. When DREQ is active, DACK* functions as a chip select for reads and writes.

DREQ(DREQ*): An output to an external DMA controller used to signal that a DMA request is being made. This signal is high active when BUSMODE = 1, low active when BUSMODE = 0. A three-state output.

TERMCT(TERMCT*): An input which signals that the last byte or word of a DMA access is on the bus. When BUSMODE = 1, this input is high active; when BUSMODE = 0, it is low active.

READY(DTACK*): A three-state output. When BUSMODE = 1, this output functions as a READY pin (Intel compatible); when BUSMODE = 0, this output is DTACK* (Motorola compatible).

INT/INT*: When BUSMODE = 1, this is a high active interrupt output; when BUSMODE = 0 this output is low active.

IACK*: Active low interrupt acknowledge input. When this input is asserted and INT is also asserted, the contents of the Interrupt Vector register are placed on D0-D7.

RESET*: The low active reset input. Asserting RESET* clears all configuration and pointer to 00. Following reset, a wait of 4 µs is necessary before accessing the part.

BUSMODE: An input which selects Intel-compatible bus signals when high or Motorola-compatible bus signals when low.

BUSSIZE: An input that selects the 8-bit system bus when low or 16-bit system bus when high.

AD0-AD7: A multiplexed address and data bus used to provide row and column address and read/write data to the packet buffer dynamic RAM.

RAS*: Row Address Strobe to the packet buffer memory.

CAS*: Column Address Strobe to the packet buffer memory. Page mode addressing is used when possible to speed access to the buffer.

W*: An output to the dynamic RAM buffer that indicates the current cycle is a write.

G*: An output to the dynamic RAM buffer that enables read data onto the AD bus.

TxEN: An output to the Manchester Code Converter that indicates a transmission is in progress.

TxC*: An input from the Manchester Code Converter that is used to synchronize transmitted data.

TxD: The transmit data output to the Manchester Code Converter.

RxC: An input from the Manchester Code Converter used to synchronize received data.

RxD: The receive data input from the Manchester Code Converter.

COLL: The collision input from the Manchester Code Converter.

CSN: The carrier sense input from the Manchester Code Converter.

CTRLO: Control Output, a general purpose control pin, level follows bit 12 of Configuration Register #2.

LPBK*: The loopback control output.

CLK: The master 20 MHz input clock.

Block Diagram Description

The 8005 has three major blocks: the AEDLC Ethernet Data Link Controller, Buffer Controller and Bus Interface.

The 8005 supports the link layer (layer 2) of the IEEE 802.3 standard. It performs serialization/deserialization, preamble generation/stripping, CRC generation/stripping, transmission deferral, collision handling and address recognition of up to 6 station addresses as well as multicast/broadcast addresses. CTRLO and LPBK* are general purpose outputs that can be used to control, for example,

seeq Technology, Incorporated
MD400031/C

PRELIMINARY *8005*

the loopback function of the 8020/8023A MCC. For non-IEEE 802.3 applications such as serial backplane buses, support is also provided for 2 byte address recognition, reduced slot time and reduced preamble length.

The Buffer Controller provides management for a 64K byte local packet buffer consisting of two 64K x 4 dynamic RAMS. This block provides arbitration and control for four different memory ports: the 8005 Transmitter, for network transmit packets; the 8005 Receiver, for received frames; the Bus Interface, for system data and control; and an internal DRAM refresh generator. To minimize pin count, dynamic RAM addresses and data are time multiplexed on a single 8 bit bus. A control line and an 8 bit address is also provided to permit reading from a locally attached EEPROM or PROM. This permits configuring a P.C. board with its station address(es) and configuration data independent of the network layer software used.

The Buffer Controller interfaces to the system bus and provides access to internal configuration/status registers, the local packet buffer and a control signal interface to permit DMA or programmed I/O transfer of packet data. The data path between the system bus and the local DRAM buffer is buffered by a 16 byte FIFO called DMA FIFO. This permits high speed data transfers to occur even when the Buffer Controller is busy servicing the Transmitter or Receiver or refreshing the DRAM. Both 8 and 16 bit transfers are supported, and byte ordering on a 16 bit bus is under software control. The 8005 supports both Intel-compatible and Motorola-compatible buses.

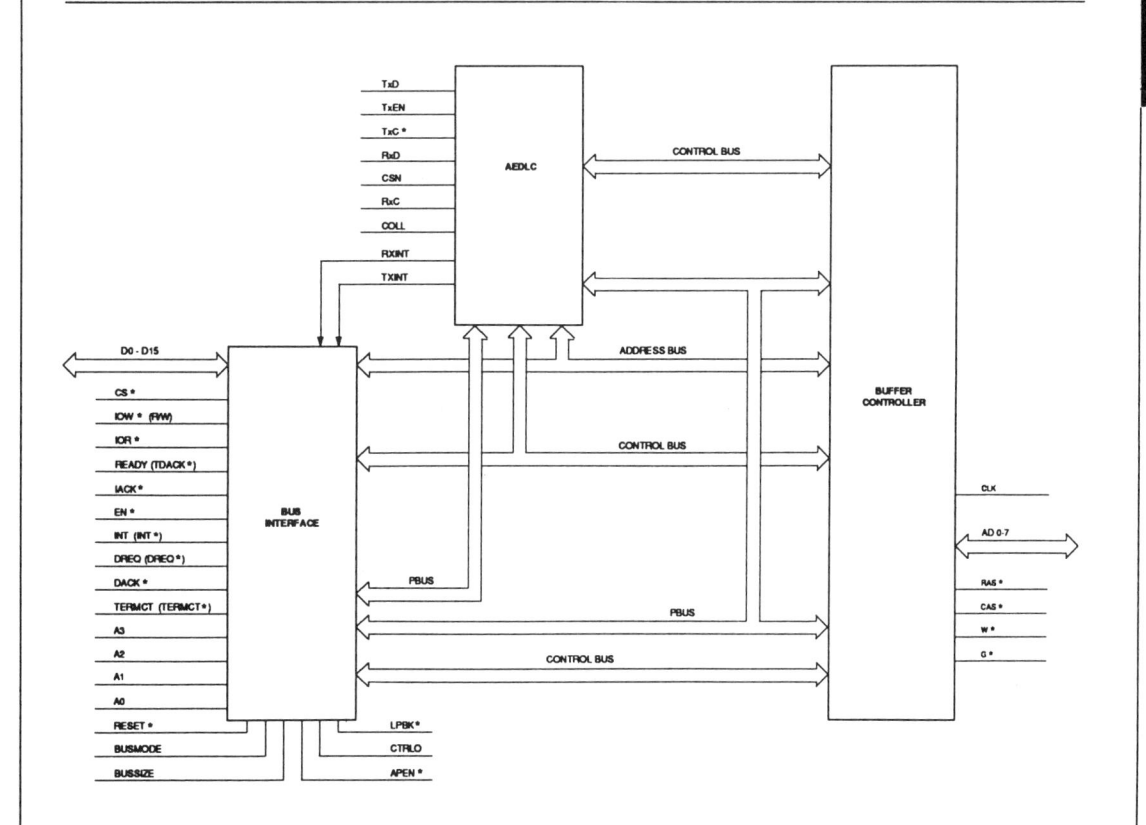

Block Diagram

seeq *Technology, Incorporated*
MD400031/C

PRELIMINARY

8005

The 8005 Interconnect Diagram

The interconnect diagram shows the 8005 in a typical system configuration, connecting to the LAN via an 8020 Manchester Code Converter. The Attachment Unit Interface connects to an Ethernet (10BASE5); Cheapernet (10BASE2); or a twisted pair (10BASE-T) network.

Separate TMS 4464-120 64K DRAMs store received packets, or packets waiting for transmission. AD_0-AD_7 address both RAMs. Data is exchanged on the AD leads, DQ_0 - DQ_3 to one RAM, and DQ_4 - DQ_7 to the other RAM.

The System Bus exchanges data with the Buffer Controller in the 8005. Two bi-directional data buffers (74LS45) interface 16-bit data, only one buffer is used for 8-bit data. The 2804 PROM stores the node address. The 8005 has six 6-byte address fields.

Buffer Management

The Buffer Controller manages a 64K byte packet buffer into which packets that are received are temporarily stored until the system either reads or disposes of them and packets placed there by the system are held for transmission over the link. The buffer is logically divided into separate receive and transmit areas of selectable size. The transmit area always originates at address 0. Each packet in the buffer is prefixed by a header of 4 bytes that contains command and status information and a 16 bit pointer to the start of the next packet in the buffer.

To transmit packets, the system loads one or more packets of data, complete with header information, into the transmit area of the buffer and commands the 8005 to begin transmission, starting from the address contained in the Transmit Pointer. When transmission is complete, the 8005 updates the status byte in the header and interrupts the system if so programmed. The Transmit Pointer automatically wraps to location 0 when the Transmit End Area is reached.

The Buffer Controller manages the buffer area as a circular buffer with automatic wraparound. As data is received from the 8005 it is stored in the buffer beginning at the location specified by the Receive Pointer. The Receive Pointer will wrap from FF,FF to Transmit End Area + 1,00. For example, if TEA = 80 the Receive Pointer wraps to 81,00. If the Receive Pointer reaches Receive End Area,00 an overflow has occurred. The Receiver is turned off and an interrupt is issued. Restarting the Receiver is accomplished by freeing up buffer space and turning the Receiver back on.

Transmit Packet Format

Each Packet to be transmitted consists of a four byte header and up to 65,532 bytes of data which are placed into the local buffer via the Bus Interface. The header contains the following information in the indicated order:

1. *Most significant byte of the address of the next packet header.*
2. *Least significant byte of the address of the next packet header.*
3. *A transmit command byte.*
4. *A transmit status byte which should be initialized to zero by the system and will contain status for this packet when transmission is complete.*

Bytes 1 and 2, *called the Next Packet Pointer, point to the location immediately following the last byte of the packet, which is the first byte of the next packet header, if it exists. In 16 bit mode, the user should note the order of these bytes to be sure it is compatible with the MSB-LSB storage convention of the processor/bus being used. Byte 1 is the more significant byte.*

Byte 3 *is the Transmit Command byte. It contains information to guide the controller in processing the packet associated with this block.*

Bit 0: Xmit Babble Int. Enable. *The 8005 will transmit packets as large as the Transmit buffer can hold but will abort packets and interrupt if this bit is set to a one. This condition is caused by an attempt to transmit a packet larger than the allowed 1514 bytes, excluding preamble and CRC. If babble occurs with bit 0-Xmit Babble Int. Enable set to a 1 on byte 3-the Transmit Command byte, the Transmitter will abort transmission and turn itself off. When the bit is set to 0, no interrupt is generated, and the Transmitter is not turned off, but a status bit is set in the Status Header.*

Bit 1: Xmit Collision Interrupt Enable. *When set to a one, a Transmit Interrupt will be generated if a collision occurs during a transmit attempt.*

Bit 2: 16 Collisions Enable. *When set to a one, a Transmit Interrupt will be generated if 16 collisions occur during a transmit attempt, and the transmitter will be turned off. When set to 0 no interrupt is generated, and the transmitter will not be turned off, but a status bit is set in the Status Header.*

Bit 3: Xmit Success Interrupt Enable. *When set to a one, a Transmit Interrupt will be generated if the transmission is successful, that is, fewer than 16 collisions occurred.*

seeq *Technology, Incorporated*

PRELIMINARY

8005

Bit 4: Not used.

Bit 5: Data Follows: If this bit is cleared to a zero, the transmitter will process this header as a pointer only, with no data associated with it. This provides a means to redirect the Transmit Pointer.

Bit 6: Chain Continue. If set to a one, there are more headers in the chain to be processed. If this bit is a zero, the header is the last one in the chain.

Bit 7: Xmit/Receive. If this bit is a one, the current header is for a packet to be transmitted. If this bit is a zero, the packet header will be processed as a header only and no data follows (bit 5).

Byte 4 is the Transmit Status byte, which is written by the Buffer Controller upon conclusion of each packet transmission or retransmission attempt. It provides for reporting of both normal and error termination conditions of each transmission.

Bit 0: Xmit Babble. If set to a one, transmit babble occurred during the transmission attempt. This is caused by an attempt to transmit a packet larger than

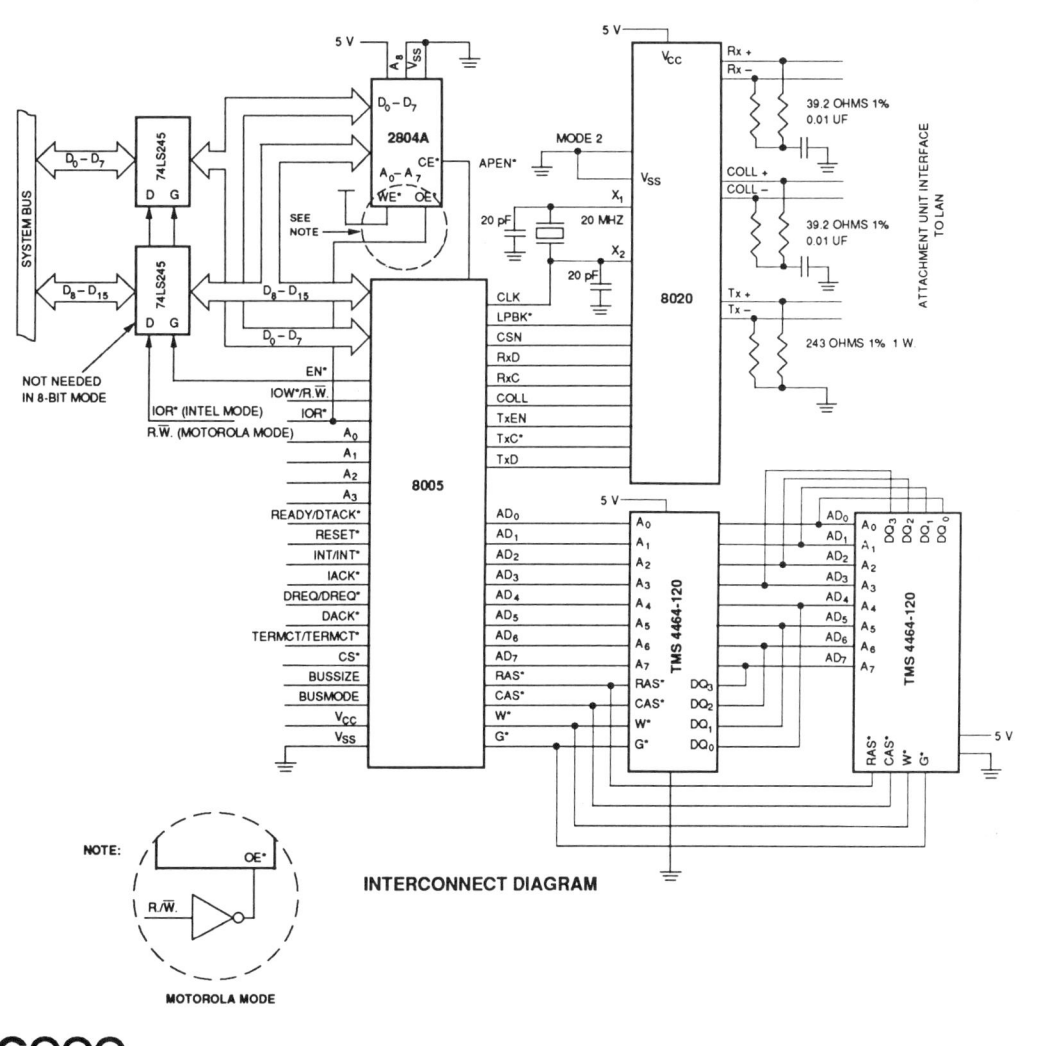

INTERCONNECT DIAGRAM

NOTE: MOTOROLA MODE

seeq *Technology, Incorporated*
MD400031/C

4-47

PRELIMINARY

8005

the allowed 1514 bytes, excluding preamble and CRC. If babble occurs with bit 0-Xmit Babble Int. Enable set to a 1 on byte 3, the Transmit Command byte, the transmitter will abort transmission and turn itself off.

Bit 1: Xmit Collision. If set to a one, a collision occurred during the transmission attempt.

Bit 2: 16 Collisions. If set to 1, 16 collisions occurred during the transmission attempt.

Bit 3, 4, 5 and 6: Reserved.

Bit 7: Done. If set to a one, the controller has completed all processing of the packet associated with this header (either the packet has been sent successfully or 16 collisions occurred) and there is now valid status in the Status byte.

The data field follows the fourth byte.

Receive Packet Format

Each Packet received is preceded by a four byte header and is placed into the local buffer via the Buffer Controller. The header contains the following information in the indicated order:

1. Most significant byte of the address of the next packet header.
2. Least significant byte of the address of the next packet header.
3. Header Status byte.
4. Packet Status byte.

Bytes 1 and 2, called the Next Packet Pointer, point to the first byte of the next receive packet header. The next packet header starts immediately after the end of the current packet. The packet length is equal to the difference between the starting addresses of the two packet headers minus 4. If the value of the Next Packet Pointer is less than the current one, the pointer has wrapped around from the end of the buffer to the Receive Start Area (the Receive Start Area equals the Transmit End Area address + 1). When in 16 bit mode, the user should note the order of these bytes to be sure it is compatible with the MSB-LSB storage convention of the processor/bus being used.

The third byte of the header contains header information associated with this packet.

Bits 0 through 5: Not Used.

Bit 6: Chain Continue. If this bit is set to a one, there are more packets in this chain to be processed. If this bit is a zero, this packets is the last one in the chain and this header space will be used for the next packet that is received.

Bit 7: Xmit/Receive. This bit is always set to 0 by the controller to indicate a receive packet header.

The fourth byte of the header, called the Packet Status byte, contains status information resulting from processing the packet associated with this block.

Bit 0: Oversize Packet. If this bit is a one, the packet was larger than 1514 bytes, excluding the Preamble and CRC fields.

Bit 1: CRC Error. If this bit is a one, a CRC Error occurred in this frame. CRC status is captured on byte boundaries, so that 7 or less dribble bits will not cause a CRC error.

Bit 2: Dribble Error. Packets are integral multiples of octets (bytes). If this bit is a one, the received packet did not end on an octet (byte) boundary.

Bit 3: Short Packet. If this bit is a one, the packet contained less than 64 bytes including CRC. Short packets are properly received as long as they are at least 6 bytes long; packets with less than 6 bytes will only be received if the match mode bits in Configuration Register #1 specify promiscuous mode, multicast/broadcast is selected and the first bit of the destination address is a 1, or the 2-byte address mode has been selected.

Bits 4, 5 and 6: Not used.

Bit 7: Done. If this bit is a one, the controller has completed all processing of this packet and there are now valid pointers and status in this header. The user may now move this packet out of the local buffer, if desired, and reuse this buffer space.

The data field follows this byte, unless this is a header only packet.

Registers

There are nine directly accessible 16 bit registers in the 8005, one of which is used as a "window" into indirectly accessed registers as well as the local buffer memory. Access is controlled by chip select, I/O read, I/O write and four address inputs, A0-A3. The following description assumes a 16 bit wide system interface; as such, the low order address input, A0, is shown as "X," a don't care. In 8 bit mode, input pin A0 selects bits 0 through 7 of the register when a zero, and bits 8 through 15 when a one. Note that the byte swap bit does not affect the byte order of these registers.

All "not used" bits should be set to 0 to maintain future compatibility. When read, "not used" bits read as '1'.

seeq Technology, Incorporated
MD400031/C
4-48

PRELIMINARY

8005

Command Register, A3-0 = 000X (Write only)

Bit 0: DMA Interrupt Enable. When set to a 1, completion of a DMA operation, as signaled by Terminal Count, will generate an interrupt.

Bit 1: Rx Interrupt Enable. When set to a 1, this bit enables interrupts whenever a packet becomes available in the packet buffer.

Bit 2: Tx Interrupt Enable. When set to a 1, this bit enables interrupts for completion of transmit operations. See the Transmit Header Command byte description for conditions that can cause an interrupt.

Bit 3: Buffer Window Interrupt Enable. Setting this bit to a one enables interrupts for Buffer Window register reads from the packet buffer.

Bit 4: DMA Interrupt Acknowledge. Setting this bit to a one causes a pending DMA interrupt to be cleared.

Bit 5: Rx Interrupt Acknowledge. Setting this bit to a one causes a pending Receive interrupt to be cleared.

Bit 6: Tx Interrupt Acknowledge. Setting this bit to a one causes a pending Transmit interrupt to be cleared.

Bit 7: Buffer Window Interrupt Acknowledge. Setting this bit to a one causes a pending Buffer Window interrupt to be cleared.

Bit 8: Set DMA On. Setting this bit to a one enables the DMA request logic. If the DMA FIFO is set to the read direction, a DMA Request will be asserted when the DMA FIFO has enough bytes to satisfy the burst size. If the DMA FIFO is in the write direction the DMA Request will be asserted immediately. Clearing this bit has no effect. Setting this bit with bit 11 set will force a DMA Interrupt, provided the DMA Interrupt Enable bit is set, which permits testing the interrupt without actually performing DMA operations.

Bit 9: Set Rx On. Setting this bit to a one enables the Receiver. Clearing this bit to a 0 has no effect. Setting this bit with bit 12 set will force an interrupt, provided the Receive Interrupt Enable bit is set, which permits testing the interrupt without receiving packet data.

Bit 10: Set Tx On. Setting this bit to a 1 enables the Transmitter. The Buffer Controller will read the header information pointed to by the Transmit pointer and process the packet accordingly (see transmit packet header description). The conditions for interrupting upon completing packet processing are specified in the Transmit Header Command byte, which is stored in the buffer memory. Setting this bit with bit 13 set will force a transmit interrupt for test purposes.

Bit 11: Set DMA Off. Setting this bit to a one disables the DMA Request logic.

Bit 12: Set Rx Off. Setting this bit to a one disables the receive logic. If the 8005 is actively receiving a packet when bit 12 is set, the Receiver will be disabled after completing reception of the packet. Bit 9 Rx On will be '1' until the receiver is disabled.

Bit 13: Set Tx Off. Setting this bit to a one disables the transmitter. If a packet is being transmitted when this bit is set, the packet will be aborted.

Bit 14: FIFO Read. When set to a one, the DMA FIFO direction is set to read from the packet buffer. The FIFO direction should not be changed from a write to a read until it is empty (see FIFO status bits).

Bit 15: FIFO Write. When set to a one, the DMA FIFO direction is set to write to the packet buffer. Changing the DMA FIFO direction clears the DMA FIFO.

Status Register, A3-0=000X (Read only)

Bit 0: DMA Interrupt Enable. When set, this bit indicates that interrupts are enabled for terminal count during a DMA operation.

Bit 1: Rx Interrupt Enable. When set, this bit indicates that interrupts are enabled for receive events.

Bit 2: Tx Interrupt Enable. When set, this bit indicates that interrupts are enabled for transmit events.

Bit 3: Buffer Window Interrupt Enable. When set, this bit indicates that interrupts are enabled for Buffer Window reads from the packet buffer.

Bit 4: DMA Interrupt. When set, this bit indicates that DMA has been terminated, either due to terminal count or the DMA On bit being written off. If the associated Interrupt Enable bit is set, an interrupt will also be asserted.

Bit 5: Rx Interrupt. When set, this bit indicates that a Receive packet chain is available. If the associated Interrupt Enable bit is set, an interrupt is also asserted.

Bit 6: Tx Interrupt. When set, this bit indicates that a Transmit interrupt condition has occurred. The following are valid Tx Interrupt conditions: Xmit Babble, Xmit Collisions, Xmit 16 Collisions and Xmit Success. If the Tx Interrupt enable bit is set, an interrupt is also asserted.

Bit 7: Buffer Window Interrupt. When set, this bit indicates that data has been read from the local buffer into the DMA FIFO and is ready to be read via the Bus Interface. If the associated interrupt enable bit has been set, an interrupt is asserted.

Bit 8: DMA On. When set, this bit indicates that the DMA logic is enabled. When Terminal Count is asserted during a DMA transfer, this bit will be reset to

SeeQ Technology, Incorporated

MD400031/C

4-49

DATA COM

PRELIMINARY

8005

indicate that the DMA activity has been completed. When reset, this bit three-states the DREQ pin.

Bit 9: Rx On. When set, this bit indicates that the Receiver is enabled. This bit remains set during active reception of a packet and turns 'off' at the end of reception if bit 12 Rx off is set.

Bit 10: Tx On. When set, this bit indicates that the Transmitter is enabled.

Bits 11 & 12: Not used.

Bit 13: DMA FIFO Full. When set, this bit indicates that the DMA FIFO is full.

Bit 14: DMA FIFO Empty. When set, this bit indicates that the DMA FIFO is empty.

Bit 15: FIFO Direction. When set, this bit indicates that the DMA FIFO is in the read direction; when cleared, it indicates that the DMA FIFO is in the write direction. After hardware or software reset, this bit is cleared.

Configuration Register 1, A3-0=001X

Bits 0-3: Buffer Code. These four bits are the Buffer Window Code bits, which determine the source of Buffer Window register reads and the destination of buffer window register writes. Buffer code bits 3-0 should be set to '1000' by pointing to local buffer memory before turning FIFO to read direction to perform reads.

Buffer Code Selection Table

Buffer Code Bits				Buffer Window Reg. Contents
3	2	1	0	
0	0	0	0	Station addr. reg. 0
0	0	0	1	Station addr. reg. 1
0	0	1	0	Station addr. reg. 2
0	0	1	1	Station addr. reg. 3
0	1	0	0	Station addr. reg. 4
0	1	0	1	Station addr. reg. 5
0	1	1	0	Address PROM
0	1	1	1	Transmit end area
1	0	0	0	Local buffer memory
1	0	0	1	Interrupt vector
1	0	1	X	Reserved — do not use
1	1	X	X	Reserved — do not use

Bits 4-5: DmaBurstInterval. These two bits specify the interval between DMA requests.

If configured for continous mode, the DMA request will persist until Terminal Count is asserted.

5	4	Burst Interval
0	0	Continuous
0	1	800 nanoseconds
1	0	1600 nanoseconds
1	1	3200 nanoseconds

DMA Burst Size Selection

Bits 6-7: DmaBurstSize. These two bits specify the DMA Burst Transfer count.

7	6	# of DMA Tranfers/Burst
0	0	1
0	1	4
1	0	8
1	1	16 (Illegal in word mode)

Bits 8-13: These six bits select which of the station address register sets (each register set contains 6 bytes) will be used to compare incoming destination addresses. Bit 8 corresponds to station address register set 0, bit 9 to register set 1, ... bit 13 to register set 5. A '1' in any bit enables that Station Address register set for reception. These bits are both read and write.

Bits 14-15: These two bits define the match modes for the Receiver logic.

15	14	Matchmode Description
0	0	Specific addresses only
0	1	Specific + broadcast addresses
1	0	Above + multicast addresses
1	1	All frames (promiscuous mode)

Configuration Register 2, A0-A3=010X

Bit 0: ByteSwap. The normal order for packing packet bytes into a 16 bit word is low byte first, i.e., the first byte of a packet is contained in bits 0 through 7, the second byte in bits 8 through 15. Setting this bit to a 1 causes the high and low order bytes to be swapped for data reads and writes to the Buffer Window Register when the 8005 is in 16 bit mode. Control registers are not affected. This bit has no effect when the 8005 is in 8 bit mode. It should not be changed when a DMA is in progress. Changing this bit will not affect the sequence of receive data bytes in the local buffer memory since the swap occurs on the system (Bus Interface) side of the buffer memory. This bit is both read and write.

Bit 1: AutoUpdREA. If this bit is set to 1, the Receive End Area register will be updated with the most significant byte of the DMA pointer whenever the Buffer

Seeq Technology, Incorporated
MD400031/C

4-50

PRELIMINARY

8005

Controller crosses a packet buffer page while reading DMA data. In this way, as buffer memory space is released by reading from it, free buffer space is automatically allocated to the Receive logic. Turn Auto Upd REA off before enabling reads from transmit space.

Bit 2: Not Used. This bit should be written to '0' for future compatibility.

Bit 3: CRC Error Enable. When set, the receiver will accept packets with CRC errors, place them in the local buffer and indicate that a packet is available via the Rx Interrupt Status bit.

Bit 4: Dribble Error. When set, the receiver will accept packets with a byte alignment error.

Bit 5: Short Frame Enable. When set, packets of less than 512 bits (64 bytes) exclusive of preamble and start packet delimiter bits, will be received and placed in the local buffer. Packets shorter than 6 bytes (2 bytes if bit 8 = 1) will always be rejected unless the Receiver is in promiscuous mode (all addresses match) or multicast/broadcast mode and the packet is a multicast/broadcast packet.

Bit 6: SlotSelect. This bit selects the slot time used to calculate backoff time following a collision. When a 0, which is the state after reset, the slot time is 512 bits and meets the IEEE 802.3 standard; when a 1, the slot time is 128 bits, the interframe spacing is 24 bits and the collision jam is 2 bytes long, which is useful for smaller networks such as serial backplane buses.

Bit 7: PreamSelect. When this bit is a 0, which is the state after reset, the 8005 automatically transmits an IEEE 802.3 compatible 64 bit preamble; when set to 1, the user must supply the preamble as part of the packet data. the preamble must still follow the 802.3 form in order to be recognized by other 8005's, but may have arbitrary length. Note that a minimum of 16 preamble bits are required by the 8005 on reception.

Bit 8: AddrLength. This bit selects the length of address to be used in address matching. When a 0, which is the state after reset, the length is 6 bytes, which conforms with the IEEE 802.3 standard; when set to 1 the length is 2 bytes, which is useful in limited networks such as serial backplane buses.

Bit 9: RecCrc. If set to a 1, received packets will include the CRC. If set to a 0, which is the state after reset, the 4 byte CRC will be stripped when received.

Bit 10: XmitNoCrc. If set to a 1, the Transmitter will not append the 4 byte frame check sequence to each packet transmitted. This is useful in local loopback to perform diagnostic checks, since it allows the software to provide its own CRC as the last four bytes of a packet to check the Receiver CRC logic. It is initialized to 0 after hardware or software reset.

Bit 11: Loopback. This bit controls the External Loopback pin. When set to a 1, the loopback output pin is at Vol; after reset or when cleared to a 0, the External Loopback output pin is at Voh.

Bit 12: CTRLO This bit controls the Control Output pin. When set to a 1, the CTRLO pin is at Voh; when cleared to 0 or after reset, this pin is at Vol.

Bits 13-14: Not used. Reserved for future use.

Bit 15: Reset. Writing a 1 to this bit is the same as asserting the hardware reset input. Reset should be followed by a 4 µs wait before attempting another access. Reads as a 0.

Receive End Area Register, A3-0 = 0110

Bits 0-7: ReaPtr. The Receive End Area pointer contains the high order byte of the local buffer address at which the Receive logic must stop to prevent writing over previously received packets. If the Receive logic reaches this address it will stop; the Receiver will be turned off and an interrupt will be issued. The Receiver can be re-started by freeing up buffer space and turning the Receiver back 'ON' again. This register can be updated automatically by setting bit 1 in Configuration Register #2, which causes ReaPtr to be updated each time the high byte of the DMA_Ptr is updated by the Buffer Controller during reads via the DMA FIFO. It is both read and write.

Buffer Window Register, A3-0 = 100X

This register provides access to the area specified by the Buffer Code bits (bits 0-3) in Configuration Register #1. When the Buffer Code points to either the buffer memory (Buffer Code = 1000_2), or the address PROM (Buffer Code = 0110_2), the address of the data transferred through this register is determined by the DMA pointer. All Buffer Code registers are byte wide except data.

Receive Pointer Register, A3-0 = 101X

The Receive pointer provides a 16 bit address that points to the next buffer memory location into which data or header information will be placed by the Receive logic. The low order 8 bits contain the least significant byte of the address. Prior to enabling the Receiver, this register should be set to point to the beginning of the Receive Area in the local buffer. This initial value should be remembered by system software since it will be the address of the first byte of the header block of the first packet received. While receiving, the Receive pointer will be incremented for each byte stored into the local buffer. When the Receive pointer

PRELIMINARY

8005

increments past hex FFFF the most significant byte will be set equal to the value of the Transmit End Area + 1 and the least significant byte will be set to 00. Reading this register may be done at any time. It should be written only when the receiver is idle.

Transmit Pointer Register, A3-0 = 110X
The Transmit pointer points to the current location being accessed by the Transmit logic. Before starting the Transmitter, software loads this register with the address of the beginning of a transmit packet chain.

DMA Address Register, A3-0 = 111X
The DMA address register provides 16 bits of address information to the local buffer memory and 8 bits of address to the address PROM, depending on the buffer code written into Configuration Register 1. Its normal use is to provide an auto-incremented address to the local buffer so that the packet data can be moved via the Bus Interface. **When the DMA Address register is loaded, the DMA FIFO is cleared.** Therefore it is important to insure that the DMA FIFO is empty if it is in the write direction before loading the DMA register. When writing a packet to be transmitted, the DMA Address register automatically wraps around to 0000 when the Transmit End Area (contained in an indirect register, Buffer Code 0111) has been reached. When reading receive packets, the DMA Address register automatically wraps around to the Receive Start Area (Transmit End Area + 1,00) when address hex FFFF has been read.

Indirectly Accessed Registers
Infrequently used registers, such as, those normally loaded only when initially configuring the 8005, are accessed indirectly by first loading the Buffer Code bits in Configuration Register #1 with a code that points to the desired register. Reads and writes occur through the Buffer Window register. All indirect registers (a total of 38) are 8 bits wide, thus only D0-D7 are used.

Station Address Registers
The 8005 contains six 48-bit Station Address registers, which permits one network connection to provide up to 6 different server functions. Each of these Station Address registers is comprised of six 8-bit registers which must be loaded through the Buffer Window Register. Only those Station Address registers to be enabled for address matching need to be loaded.

To load a Station Address register, first turn the Receiver off. Select the desired station number (0-5) by writing the Buffer Code bits in Configuration Register #1. Next do 6 sequential **byte** writes to the Buffer Window register as follows: Write the least significant byte of the 6 byte Station Address; its low order bit, bit 0, will be the first bit received. Next write the remaining 5 bytes in ascending order. To read a Station Address register, first turn the receiver off by setting 'bit 12' Rx off on the Command register and verifying that the Receiver is off. Then select the desired station number by writing the Buffer Code bits in Configuration Register #1. Do 6 sequential reads to the Buffer Window Register; the first byte read will be the least significant byte. If the 8005 is configured to match 2 byte instead of 6 byte addresses, only the first 2 station address bytes are significant, although all 6 will read and write properly.

Transmit End Area Pointer
The 8-bit value of this pointer defines, with 256 location granularity, the end of the Transmit Packet Buffer area by specifying the highest value permitted in the most significant byte of the Transmit Pointer Register and, when loading a packet to be transmitted, the DMA Address register. It also indirectly defines the Receive Start Area address, since the Buffer Controller automatically calculates the high order byte of the address by adding 1 to the Transmit End Area pointer. To read or write this value, set Buffer Code = 0111, and do a read or write to the Buffer Window Register.

Interrupt Vector Register
This Read/Write register is accessed through the Buffer Window register when the Buffer Code in Configuration Register #1 is 9. It contains an 8 bit vector which is placed on data bits D0-D7 during an Interrupt Acknowledge cycle. If BUSMODE = 0, an Interrupt Acknowledge cycle is defined by INT* = 0, IACK* = 0, and READ/WRITE = 1. When BUSMODE = 1, an Interrupt Acknowledge cycle is defined by INT = 1, IACK* = 0, and IOR* = 0.

Other Buffer Window Register Uses
Address PROM Access

The 8005 supports access to up to 256 bytes of configuration data contained in a PROM or EEPROM. This can be used for any purpose, such as storing station addresses, register configurations, network connection data, etc. The address to the PROM is supplied by the DMA register through data bus bits D8-D15; the data lines from the PROM are connected to D0-D7. Chip select for the PROM is provided by output APEN*. Before accessing this PROM, insure that Transmit, Receive and DMA sections of the 8005 are disabled. Next load the PROM starting address which you wish to access into both the low byte and the high byte of DMA register. Set the Buffer Code bits in Configuration Register #1 to point to the address PROM. Each access to the Buffer Window register will chip enable the PROM, permitting reads. Successive accesses will in-

seeq Technology, Incorporated
MD400031/C

4-52

PRELIMINARY **8005**

Example of Chained Receive Frames

Bit #	7	6	5	4	3	2	1	0
Addr. ptr 1	\multicolumn{8}{c	}{Upper byte of next packet pointer}
Addr. ptr 2	\multicolumn{8}{c	}{Lower byte of next packet pointer}
Header status	0	1	1	X	X	X	X	X
Packet status	1	0	0	1	0	0	0	0
Data								
"								
Addr. ptr 1	\multicolumn{8}{c	}{Upper byte of next packet pointer}
Addr. ptr 2	\multicolumn{8}{c	}{Lower byte of next packet pointer}
Header status	0	1	1	X	X	X	X	X
Packet status	1	0	0	1	0	0	0	0
Data								
"								
Addr. ptr 1	\multicolumn{8}{c	}{Upper byte of next packet pointer}
Addr. ptr 2	\multicolumn{8}{c	}{Lower byte of next packet pointer}
Header status	0	1	1	X	X	X	X	X
Packet status	1	0	0	1	0	0	0	0
Data								
"								
Addr. ptr 1	0	0	0	0	0	0	0	0
Addr. ptr2	0	0	0	0	0	0	0	0
Header status	0	0	0	0	0	0	0	0
Packet status	0	0	0	0	0	0	0	0

Next receive packet header goes here.

Last header in chain.

Packet Header Bytes

Transmit Header Command Byte (Byte #3)

7	6	5	4	3	2	1	0
1	Chain Continue	Data Follows	Not Used	Xmit Success Enable	16 Coll. Enable	Coll. Int. Enable	Babble Int. Enable

Receive Header Status Byte (Byte #3)

7	6	5	4	3	2	1	0
0	Chain Continue	Not Used	Not Used	Not Used	Not Used	Not Used	Not Used

Transmit Packet Status Byte (Byte #4)

7	6	5	4	3	2	1	0
Done	←	Reserved	→		16 Coll.	Collision	Babble

Receive Packet Status Byte (Byte #4)

7	6	5	4	3	2	1	0
Done	Not Used	Not Used	Not Used	Short Frame	Drib. Error	CRC Error	Oversize

seeq Technology, Incorporated
MD400031/C
4-53

PRELIMINARY

8005

8005 Configuration and Pointer Registers

Command (write only) (A3-0 = 000X)

15	14	13	12	11	10	9	8	7	6	5	4	3	2	1	0
FIFO Write	FIFO Read	Set Tx Off	Set Rx Off	Set DMA Off	Set Tx On	Set Rx On	Set DMA On	Buffer Window Ack	Tx Int Ack	Rx Int Ack	DMA Int Ack	Buffer Window Enable	Tx Int Enable	Rx Int Enable	DMA Int Enable

Status (read only) (A3-0 = 000X)

15	14	13	12	11	10	9	8	7	6	5	4	3	2	1	0
FIFO Dir	FIFO Empty	FIFO Full	Not Used	Not Used	Tx On	Rx On	DMA On	Buffer Window Int	Tx Int	Rx Int	DMA Int	Buffer Window Enable	Tx Int Enable	Rx Int Enable	DMA Int Enable

Configuration Register #1 (A3-0 = 001X)

15	14	13	12	11	10	9	8	7	6	5	4	3	2	1	0
Addr Match Mode	Addr Match Mode	Sta. 5 Enable	Sta. 4 Enable	Sta. 3 Enable	Sta. 2 Enable	Sta. 1 Enable	Sta. 0 Enable	DMA Burst Lngth	DMA Burst Lngth	DMA Burst Intvl	DMA Burst Intvl	Buffer Code 3	Buffer Code 2	Buffer Code 1	Buffer Code 0

Configuration Rgister #2 (A3-0 = 010X)

15	14	13	12	11	10	9	8	7	6	5	4	3	2	1	0
Reset	Not Used	Not Used	Control Output	Loop-Back	Xmit No CRC	Recv. CRC	Addr Leng.	Xmit No Pream	Slot Time Sel.	Short Frame Enable	Drib. Error Enable	CRC Error Enable	Not Used	Auto Update REA	Byte Swap

Receive End Area Register (A3-0 = 0110 [2])

15	14	13	12	11	10	9	8	7	6	5	4	3	2	1	0
X	X	X	X	X	X	X	X	\multicolumn{8}{c}{Receive End Area Pointer}							

Receive Pointer Register (A3-0 = 101X)

15	14	13	12	11	10	9	8	7	6	5	4	3	2	1	0
\multicolumn{16}{c}{LOCAL BUFFER ADDRESS FOR NEXT RECEIVE BYTE}															

Transmit Pointer Register (A3-0 = 110X)

15	14	13	12	11	10	9	8	7	6	5	4	3	2	1	0
\multicolumn{16}{c}{LOCAL BUFFER ADDRESS FOR NEXT TRANSMIT BYTE}															

DMA Address Register (A3-0 = 111X)

15	14	13	12	11	10	9	8	7	6	5	4	3	2	1	0
\multicolumn{16}{c}{LOCAL BUFFER ADDRESS FOR SYTSEM READS OR WRITES}															

Buffer Window Register (A3-0 = 100X [2])

15	14	13	12	11	10	9	8	7	6	5	4	3	2	1	0
\multicolumn{16}{c}{BUFFER CODE BITS DETERMINE SOURCE/DESTINATION FOR READS AND WRITES}															

NOTES: 1. In 16 bit mode address A0 is a don't care for all registers except REA.
2. Both 8 and 16 bit modes.

seeq Technology, Incorporated

PRELIMINARY

8005

Station Address Register Format
2 of 6 Station Address Registers Shown

7 6 5 4 3 2 1 0
LEAST SIGNIFICANT BYTE STATION ADDRESS REGISTER 0 BYTE 0 BUFFER CODE = 0000
STATION ADDRESS REGISTER 0 BYTE 1 BUFFER CODE = 0000
STATION ADDRESS REGISTER 0 BYTE 2 BUFFER CODE = 0000
STATION ADDRESS REGISTER 0 BYTE 3 BUFFER CODE = 0000
STATION ADDRESS REGISTER 0 BYTE 4 BUFFER CODE = 0000
STATION ADDRESS REGISTER 0 BYTE 5 BUFFER CODE = 0000 MOST SIGNIFICANT BYTE

7 6 5 4 3 2 1 0
LEAST SIGNIFICANT BYTE STATION ADDRESS REGISTER 1 BYTE 0 BUFFER CODE = 0001
STATION ADDRESS REGISTER 1 BYTE 1 BUFFER CODE = 0001
STATION ADDRESS REGISTER 1 BYTE 2 BUFFER CODE = 0001
STATION ADDRESS REGISTER 1 BYTE 3 BUFFER CODE = 0001
STATION ADDRESS REGISTER 1 BYTE 4 BUFFER CODE = 0001
STATION ADDRESS REGISTER 1 BYTE 5 BUFFER CODE = 0001 MOST SIGNIFICANT BYTE

crement the DMA register to point to the next byte in the PROM. If a 16 bit wide bus is used, the address supplied to the PROM will also be read on D8-D15.

Buffer Memory Access

The normal state of the Buffer Code bits, once the 8005 has been initialized with station addresses and buffer areas have been allocated, is with Buffer Memory selected. Access to the local buffer memory is provided by the DMA register, which automatically increments after each byte or word transfer. To write to the local buffer, set the buffer code to select the buffer memory, set the FIFO direction to write (Command Register bits 14 and 15), load a starting address into the DMA register and write to the Buffer Window register. To read from the local buffer, the same steps as above must be followed except that the FIFO direction should be changed to the read direction **after** the DMA register has been written. This is the simplest way to access the local buffer as it requires no system DMA activity. It also permits network layer software to read network control data at the beginning of a received packet to determine if it is necessary to move the packet into global memory for further processing or simply reuse the area occupied by the packet by updating the Receive End Area register. For fastest transfer speed, e.g., to move packet data, an external system DMA Controller is supported via the DMA Request output, DMA Acknowledge input and Terminal Count input signals.

Asynchronous Bus Control

The 8005 supports asynchronous bus control via the READY/DTACK* pin. By using READY/DTACK*, the cycle time minimums listed in the tables A thru' J need not be observed. READY/DTACK* takes care of these cycle times. This greatly simplifies the task of interfacing to the 8005 and also results in a higher overall data rate. To acheive the highest possible data rate, all data transfers should terminate within 100 ns of READY/DTACK* being asserted. This permits a sustained system bus transfer rate of 3.33 Mbytes/sec in 16 bit mode or 2.5 Mbytes/sec in 8 bit mode.

SeeQ Technology, Incorporated

MD400031/C

PRELIMINARY

8005

Absolute Maximum Stress Ratings

Temperature:
  Storage .... −65°C to +150°C
  Under Bias .... −10°C to +80°C
All Inputs and Outputs with
  Respect to V_{ss} .... +6 V to −0.3 V

*COMMENT: Stresses above those listed under "Absolute Maximum Ratings" may cause permanent damage to the device. This is a stress rating only and functional operation of the device at these or any other conditions above those indicated in the operational sections of this specification is not implied. Exposure to absolute maximum rating conditions for extended periods may affect device reliability.

Recommended Operating Conditions

V_{CC} Supply Voltage	5V ± 5%
Ambient Temperature	0°C to 70°C

DC Operating Charateristics
(Over operating temperature and V_{cc} range, unless otherwise specified)

Symbol	Parameter	Min.	Max.	Unit	Test Condition
I_{IL}	Input/Output Leakage		10 −10	µA µA	$V_{IN} = V_{CC}$ $V_{IN} = 0.1\,V$
I_{CC}	Active I_{cc} Current @ $T_A = 0°C$		350	mA	CS* = V_{IL}, Outputs Open $T_A = 0°C$
	Active I_{cc} Current @ $T_A = 70°C$		280	mA	CS* = V_{IL}, Outputs Open $T_A = 70°C$
V_{IL1}	Input Low Voltage (except TXC*, RXC, CLK)	−0.3	0.8	V	
V_{IL2}	Input Low Voltage (TXC*, RXC, CLK)	−0.3	0.4	V	
V_{IH1}	Input High Voltage (except TXC*, RXC, CLK)	2.0	$V_{cc} + 1$	V	
V_{IH2}	Input High Voltage (TXC*, RXC, CLK)	3.5	$V_{cc} + 1$	V	
V_{OL1}	Output Low Voltage (except AD_{0-7})		0.40	V	$I_{OL} = 2.1\,mA$
V_{OL2}	Output Low Voltage (AD_{0-7})		0.40	V	$I_{OL} = 200\,µA$
V_{OH1}	Output High Voltage (except AD_{0-7})	2.4		V	$I_{OH} = -400\,µA$
V_{OH2}	Output High Voltage (AD_{0-7})	2.4		V	$I_{OH} = -200\,µA$

seeQ *Technology, Incorporated*
MD400031/C

PRELIMINARY **8005**

A.C. Test Conditions

Output Load:
 AD0-AD7, I(load) = ± 200 µA
 C(load) = 50 pF.
 All Other Outputs: 1 TTL Gate and C(load) = 100 pF.
Input Rise and Fall Times (except TXC, RXC, CLK):
 10 ns maximum.

Input Rise and Fall Times (TXC, RXC, CLK):
 5 ns maximum.
Input Pulse Levels: 0.45 V to 2.4 V
Timing Measurement Reference Level:
 Inputs: 1 V and 2 V
 Outputs: 0.8 V and 2 V

Capacitance [1]   Ambient Temperature = 25°C, F = 1 MHz

Symbol	Parameter	Limits Min.	Limits Max.	Unit	Test Condtion
C_{IN}	Input Capacitance		15	pF	$V_{IN} = 0$
C_{OUT}	Output Capacitance		15	pF	$V_{OUT} = 0$

Electrostatic Discharge Characteristics

Symbol	Parameter	Value	Test Condtion
V_{ZAP} [2]	E.S.D. Tolerance	> 2000 V	Mil-STD 883 Meth. 3015

NOTES: 1. This parameter is measured only for the initial qualification and after process or design changes which may affect capacitance.
2. Characterized. Not tested.

seeq Technology, Incorporated

PRELIMINARY

8005

A.C. Characteristics (Assuming 20 MHz Input Master Clock)
(Over operating temperature and V_{cc} range, unless otherwise specified)

Table A. Bus Write Cycle — BUSMODE = 0

Ref. #	Symbol	Description	Min.	Max.	Units
1	TAVCSL	Address Setup Time	30		ns
2	TRWLCSL	R/W* Setup Time	30		ns
3	TCSLCSH	CS* Pulse Width	100		ns
4	TDVCSH	Data Setup Time	70		ns
5	TCSHDX	Data Hold Time	20		ns
6	TCSLDTL	DTACK* Assertion Delay[4]		60	ns
7	TCSHDTH	DTACK* Deassertion Delay		60	ns
8	TDTHDTZ	DTACK* Hi-Z Delay		50	ns
9	TCSHAX	Address Hold Time	20		ns
10	TCSHRWX	R/W* Hold Time	20		ns
11	TCSHCSL	CS* High Time	200		ns
12	TCSHDTL	Write Recovery Time: a. FIFO Data Write[1] b. Configuration Regs [1,2] c. Pointer Regs. [3]		800 800 1800	ns ns ns ns
13	TCSLENL	EN* Assert Delay		50	ns
14	TCSHENH	EN* Deassert Delay		50	ns
15	TCSLDTV	CS* Assert to DTACK* Valid		50	ns

NOTES:
1. Write Recovery Time is for 16 bit writes. If BUSSIZE = 0 (8 bit writes), subtract 200 ns.
2. Configuration Registers are: Command/Status Register, Configuration Register #1 & 2, Interrupt Vector Register, and Station Address Registers.
3. Pointer Registers are: Receive End Area Pointer, Receive Pointer Register, Transmit Pointer Register, Transmit End Area Register, and DMA register. If BUSSIZE = 0, subtract 600 ns.
4. The trailing edge of CS* initiates an internal write sequence. Should another CS* occur during this time, the assertion of DTACK* will be delayed until the internal write sequence has finished (Ref. # 12, TCSHDTL).
5. After changing the Buffer Code (Config. Reg. #1 bits 0-3), Ref. #11 must be increased to 800 ns before a Buffer Window access is done in order to allow time for the new Buffer Code to propagate internally.

seeq Technology, Incorporated
MD400031/C

PRELIMINARY

8005

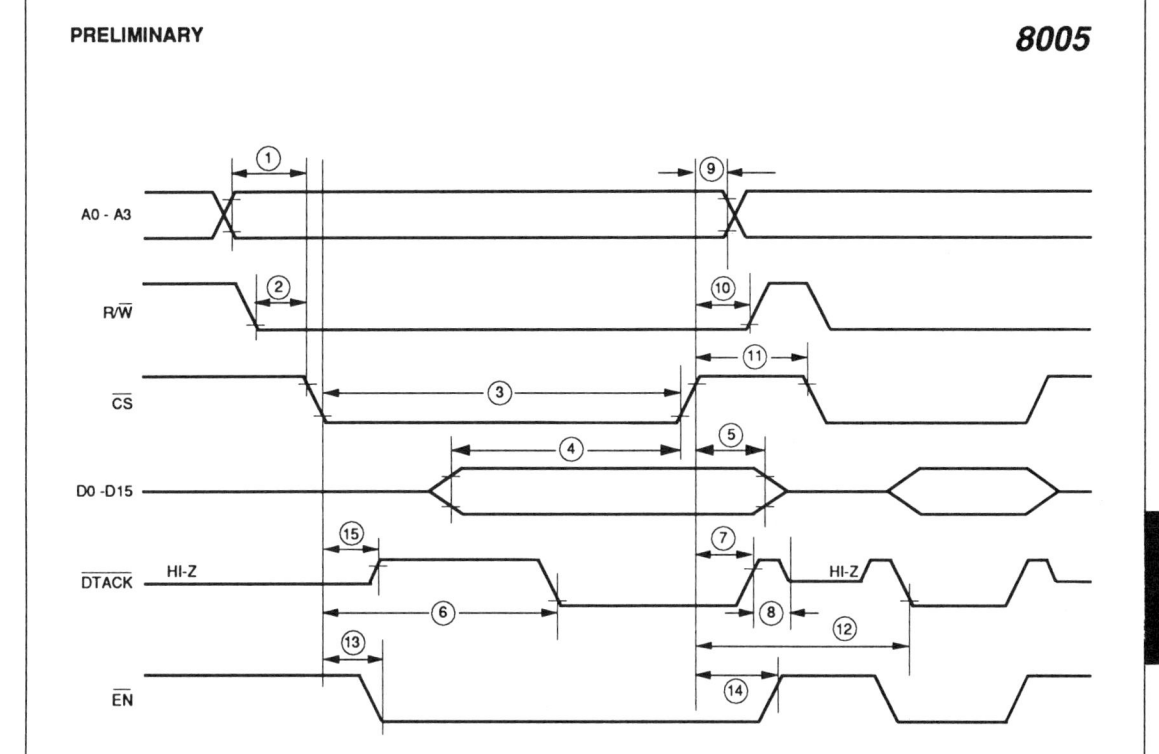

Figure A. Bus Write Cycle Timing Diagram — BUSMODE = 0

PRELIMINARY

8005

A.C. Characteristics (Assuming 20 MHz Input Master Clock)
(Over operating temperature and V_{cc} range, unless otherwise specified)

Table B. Bus Read Cycle — BUSMODE = 0

Ref. #	Symbol	Description	Min.	Max.	Units
1	TAVCSL	Address Setup Time	30		ns
2	TRWHCSL	R/W* Setup Time	30		ns
3	TCSLDTL	DTACK* Assert Delay			ns
		a. FIFO Data[1]		60	ns
		b. Configuration Regs.[2]		800	ns
		c. Other Pointer Regs.[3]		1800	ns
4	TDTLDV	Time from DTACK* Asserted to Data Valid		50	ns
5	TCSLCSH	CS* Pulse Width	100		ns
6	TCSHDTH	DTACK* Deassertion Delay		60	ns
7	TDTHDTZ	DTACK* Hi-Z Delay		50	ns
8	TCSHDZ	Data Hi-Z Delay		100	ns
9	TCSHDX	Data Hold Time	20		ns
10	TCSHRWX	R/W* Hold Time	20		ns
11	TCSHAX	Address Hold Time	20		ns
12	TCSHCSL	CS* High Time	200		ns
13	TCSLAPL	APEN* Assert Delay		400	ns
14	TCSHAPH	APEN* Deassert Delay		50	ns
15	TCSLENL	EN* Assert Delay		50	ns
16	TCSHENH	EN* Deassert Delay		50	ns
17	TCSLDTV	CS* Assert to DTACK* Valid		50	ns

NOTES:
1. The BIU prefetches one word (byte) of FIFO data. Thus, data is generally available immediately and DTACK* will assert within 50 ns. Following the read, the BIU will fetch the next word (byte) of data. Should another data read occur before the BIU has completed the prefetch, DTACK* will be delayed until the prefetch is completed. The assert delay in this case is 650 ns max (450 ns in 8 bit mode).
2. Configuration Registers are: Command/Status Register, Configuration Register # 1 & 2, Interrupt Vector Register, DMA Pointer Register, and Station Address Registers. If BUSSIZE = 0 (8 bit reads), subtract 200 ns.
3. Pointer Registers are: Receive End Area Pointer, Receive Pointer Register, Transmit Pointer Register, and Transmit End Area Register. If BUSSIZE = 0, subtract 600 ns.

seeq *Technology, Incorporated*

PRELIMINARY

8005

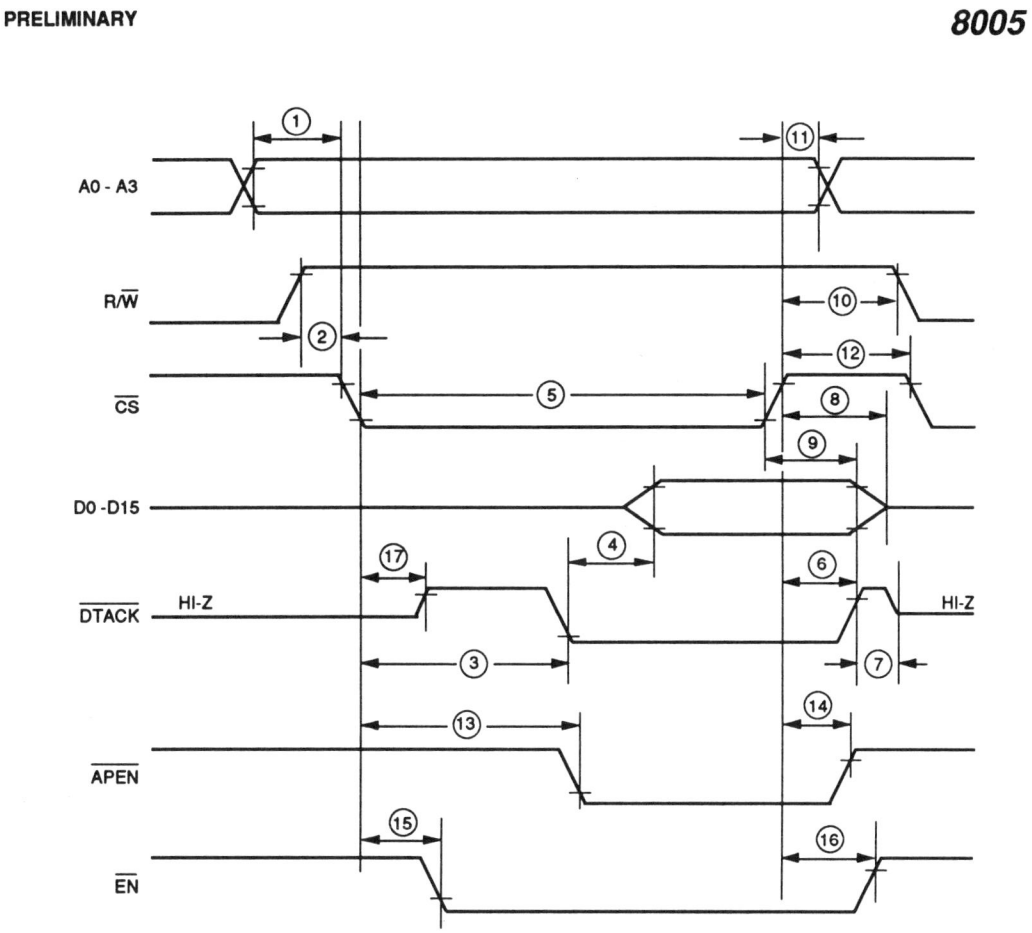

Figure B. Bus Read Cycle Timing Diagram — BUSMODE = 0

PRELIMINARY

8005

A.C. Characteristics (Assuming 20 MHz Input Master Clock)
(Over operating temperature and V_{cc} range, unless otherwise specified)

Table C. Interrupt Cycle — BUSMODE = 0

Ref. #	Symbol	Description	Min.	Max.	Units
1	TDTLDV	Time from DTACK* Assert to Data Valid		50	ns
2	TIALDTV	DTACK* Assert Delay		600	ns
3	TIAHDX	Data Hold from IACK* Deassert	20		ns
4	TIAHDZ	Data Hi-Z from IACK* Deassert		100	ns
5	TIAHDTH	DTACK* Deassert Delay		60	ns
6	TDTHDTZ	DTACK* Hi-Z Delay		50	ns
7	TRWHIAL	R/W* Setup Time	30		ns
8	TIAHRWX	R/W* Hold Time from IACK*	20		ns
9	TIALENL	EN* Assert Delay		50	ns
10	TIAHENH	EN* Deassert Delay		50	ns
11	TIALDTV	IACK* Assert to DTACK* Valid		50	ns

seeq *Technology, Incorporated*
MD400031/C

4-62

PRELIMINARY

8005

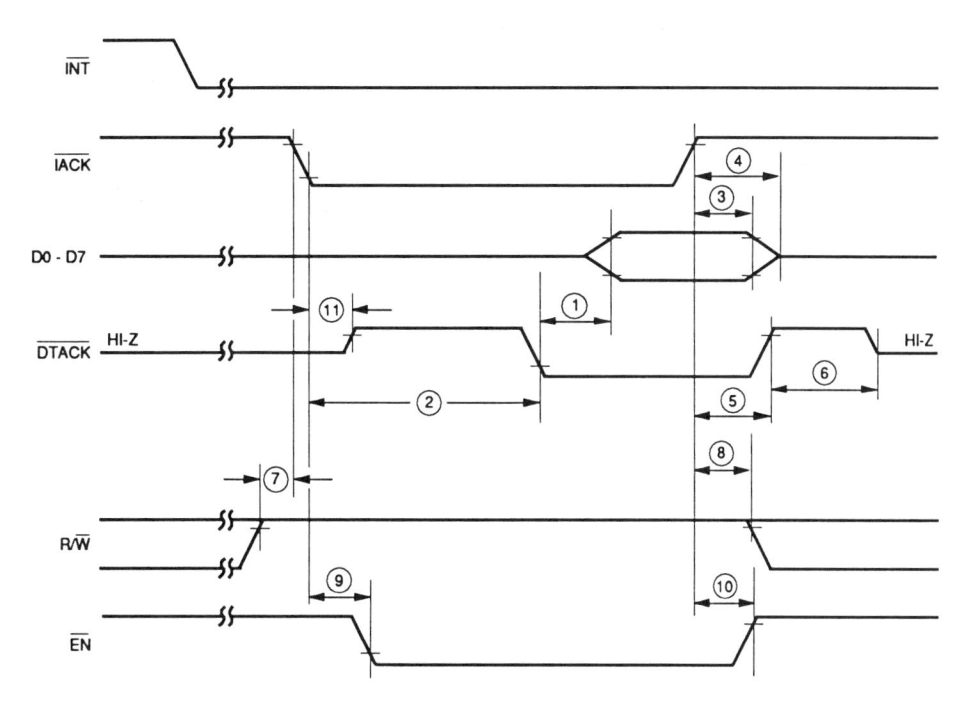

Figure C. Interrupt Cycle Timing Diagram — BUSMODE = 0

PRELIMINARY

8005

A.C. Characteristics (Assuming 20 MHz Input Master Clock)
(Over operating temperature and V_{cc} range, unless otherwise specified)

Table D. DMA Read Cycle — BUSMODE = 0

Ref. #	Symbol	Description	Min.	Max.	Units
1	TRWHDAL	R/W* Setup Time	30		ns
2	TDALDAH	DACK* Pulse Width[1]	100		ns
3	TDTLDV	Time from DTACK* Asserted to Data Valid		50	ns
4	TDAHDX	Data Hold Time	20		ns
5	TDAHDZ	Data Hi-Z Delay		100	ns
6	TDAHDAL	DACK* High Time	200		ns
7	TDAHRWX	R/W* Hold Time	20		ns
8	TDALTCL	TERMCT* Asserted While DACK* Asserted	125		ns
9	TTCLDRH	DREQ* Delay[5]		175	ns
10	TDALDRH	DREQ Delay After End of DMA Burst[6]		100	ns
11	TDALDTL1	DTACK* Assertion Delay[2]		60	ns
12	TDAHDTH	DTACK* Deassertion Delay		60	ns
13	TDTHDTZ	DTACK* Hi-Z Delay		50	ns
14	TDALENL	EN* Assert Delay		50	ns
15	TDAHENH	EN* Deassert Delay		50	ns
16	TDALDTV	DACK* Assert to DTACK* Valid		50	ns
17	TDALDTL2	Read Recovery Time[3,4]		800	ns

NOTE:
1. DACK* must be asserted until DTACK* is asserted and for a minimum of 100 ns.
2. This delay applies only if the 8005 is "ready" when DACK* is asserted i.e. the first read of a burst, or a read that occurs after the Ref. #17 TDALDTL2 period has elapsed.
3. The BIU pre-fetches FIFO data. Thus, data is available immediately for the first read of any burst. Once the BIU detects a read operation, it begins fetching the next byte or word of data. This occurs during the Ref. #17 TDALDTL2 period. If a subsequent DACK* occurs within the Ref. #17 TDALDTL2 period, DTACK* will stay de-asserted until the FIFO data has been fetched. If the subsequent DACK* does not occur until after the Ref. #17 TDALDTL2 period has elapsed, then the 8005 is "ready" and Ref. #11 TDALDTL1 applies.
4. Subtract 200 ns if BUSSIZE = 0 (8 bit mode).
5. DACK* and TERMCT* must both be active at the same time and for a minimum of 125 ns. The de-assertion of DREQ* is timed from the last one to assert.
6. Ref. #10 TDALDRH applies for normal DMA burst terminations — not those due to TERMCT.

All the timing in this table also apply when reading data with programmed I/O; CS* replaces DACK* and the DREQ* and TERMCT* signals do not apply. A0-A3 setup times are the same as R/W*.

seeq *Technology, Incorporated*
MD400031/C

PRELIMINARY *8005*

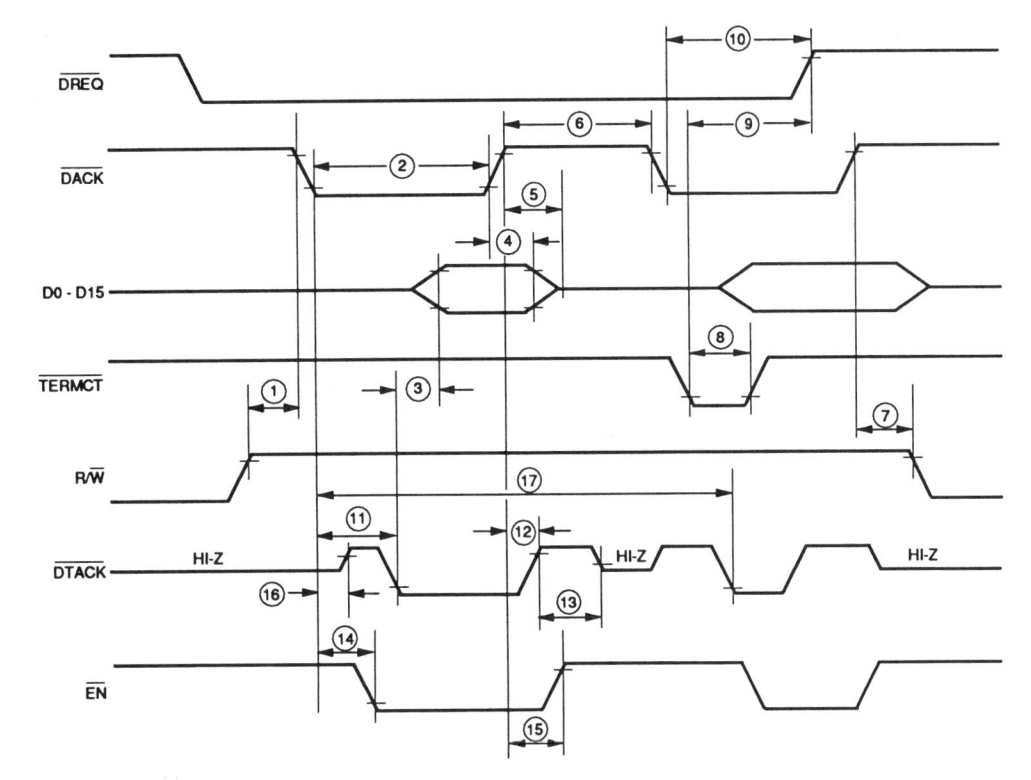

Figure D. DMA Read Cycle Timing Diagram — BUSMODE = 0

PRELIMINARY   *8005*

A.C. Characteristics (Assuming 20 MHz Input Master Clock)
(Over operating temperature and V_{CC} range, unless otherwise specified)

Table E. DMA Write Cycle — BUSMODE = 0

Ref. #	Symbol	Description	Min.	Max.	Units
1	TRWLDAL	R/W* Setup Time	30		ns
2	TDALDAH	DACK* Pulse Width[1]	100		ns
3	TDVDAH	Data Setup Time	70		ns
4	TDAHDX	Data Hold Time	20		ns
5	TDAHDAL	DACK* High Time	200		ns
6	TDALTCL	TERMCT* Asserted While DACK* Asserted	125		ns
7	TDAHRWX	R/W* Hold Time	20		ns
8	TTCLDRH	DREQ* Delay[5]		175	ns
9	TDALDRH	DREQ Delay After End of DMA Burst[6]		100	ns
10	TDALDTL	DTACK* Assertion Delay[2]		60	ns
11	TDAHDTH	DTACK* Deassertion Delay		60	ns
12	TDTHDTZ	DTACK* Hi-Z Delay		50	ns
13	TDALENL	EN* Assert Delay		50	ns
14	TDAHENH	EN* Deassert Delay		50	ns
15	TDALDTV	DACK* Assert to DTACK* Valid		50	ns
16	TDAHDTL	Write Recovery Time[3,4]		800	ns

NOTES:
1. DACK* must be asserted until DTACK* is asserted and for a minimum of 100 ns.
2. This delay applies only if the 8005 is "ready" when DACK* is asserted i.e. the first write of a burst, or a write that occurs after Ref. # 16 TDAHDTL period has elapsed.
3. The trailing edge of DACK* initiates an internal write sequence that lasts a maximum of 800 ns in 16 bit mode. Should another DACK* occur during this period, DTACK* will remain de-asserted until Ref. #16 TDAHDTL period has elapsed. If the subsequent DACK* does not occur until after the internal write sequence has ended, then the 8005 is "ready" and Ref. # 10 TDALDTL applies.
4. Subtract 200 ns when BUSSIZE = 0 (8 bit mode).
5. DACK* and TERMCT* must both be active at the same time and for a minimum of 125 ns. The de-assertion of DREQ* is timed from the last one to assert.
6. Ref. #9 TDALDRH applies for normal DMA burst terminations — not those due to TERMCT.

All the timing in this table also apply when writing data with programmed I/O; CS* replaces DACK* and the DREQ*, TERMCT* signals do not apply. A0-A3 times are the same as R/W*.

seeq *Technology, Incorporated*

PRELIMINARY

8005

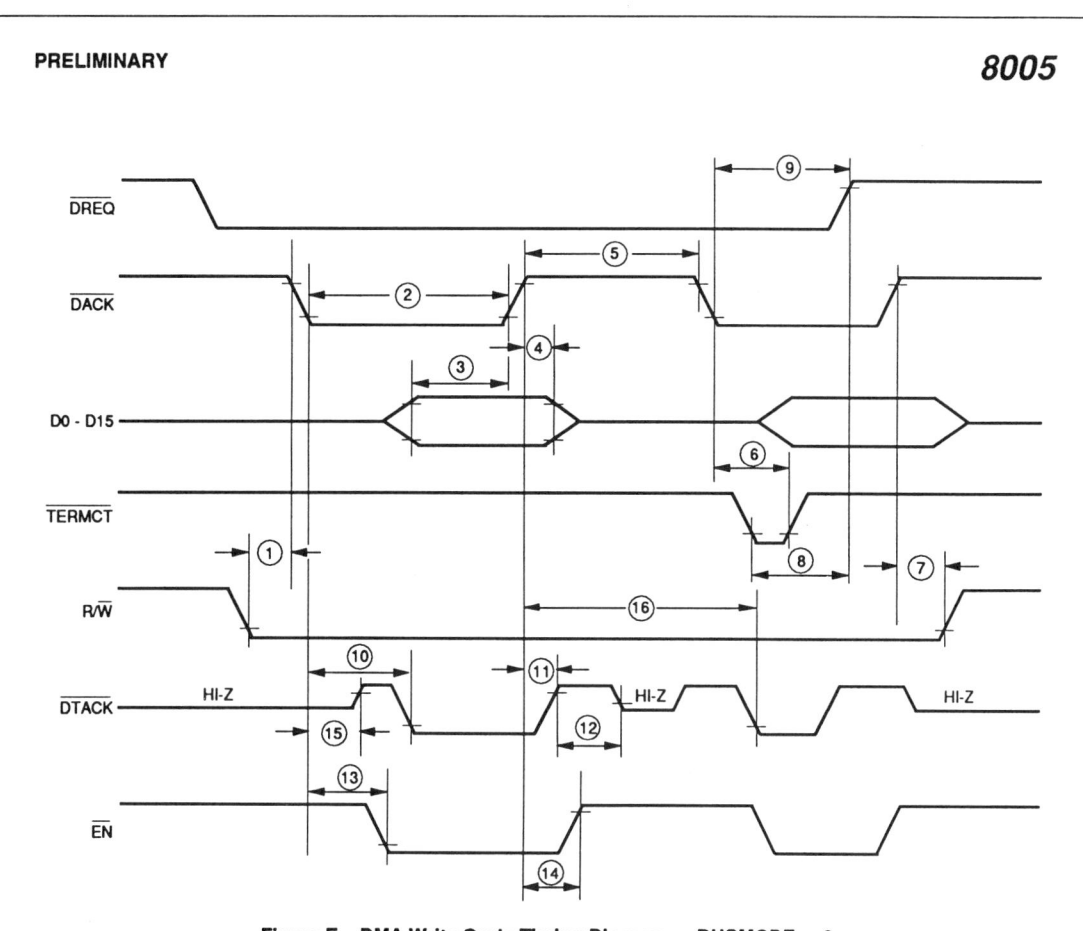

Figure E. DMA Write Cycle Timing Diagram — BUSMODE = 0

PRELIMINARY

8005

A.C. Characteristics (Assuming 20 MHz Input Master Clock)
(Over operating temperature and V_{cc} range, unless otherwise specified)

Table F. Bus Write Cycle — BUSMODE = 1

Ref. #	Symbol	Description	Min.	Max.	Units
1	TAVWL	Address Setup Time	30		ns
2	TCSLWL	CS* Setup Time	30		ns
3	TWLWH	IOW* Pulse Width	100		ns
4	TDVWH	Data Setup Time	70		ns
5	TWHDX	Data Hold Time	20		ns
6	TWLRYL	READY Deassert Delay		35	ns
7	TCSLRYV	CS* Asserted to READY Valid[4]		50	ns
8	TCSHRYZ	READY Delay to Hi-Z		50	ns
9	TWHAX	Address Hold Time	20		ns
10	TWHCSH	CS* Hold Time	20		ns
11	TWHWL	IOW* High Time[5]	200		ns
12	TWHRYH	Write Recovery Time: a. FIFO Data Write[1] b. Configuration Regs. [1,2] c. Pointer Registers. [3]		800 800 1800	ns ns ns ns
13	TCSLENL	EN* Assert Delay		50	ns
14	TCSHENH	EN* Deassert Delay		50	ns

NOTES:
1. Recovery time is for 16 bit writes. If BUSSIZE = 0 (8 bit writes), subtract 200 ns.
2. Configuration Registers are: Command/Status Register, Configuration Register #1, & 2, Interrupt Vector Register, and Station Address Registers.
3. Pointer Registers are: Receive End Area Pointer, Receive Pointer Register, Transmit Pointer Register, Transmit End Area Register, and DMA Register. If BUSSIZE = 0, subtract 600 ns.
4. The trailing edge of IOW* initiates an internal write sequence. Should another IOW* occur during this sequence, READY de-asserts (Ref. # 6 TWLRYL) and then asserts after the internal write sequence has finished (Ref. #12 TWHRYH). If the subsequent IOW* does not occur until after the internal write sequence has ended, then Ref. # 6 TWLRYL has no meaning since READY does not de-assert under this condition.
5. After changing the Buffer Code (Config. Reg. #1 bits 0-3), Ref. #11 must be increased to 800 ns before a Buffer Window access is done in order to allow time for the new Buffer ZCode to propagate internally.

seeq *Technology, Incorporated*

PRELIMINARY *8005*

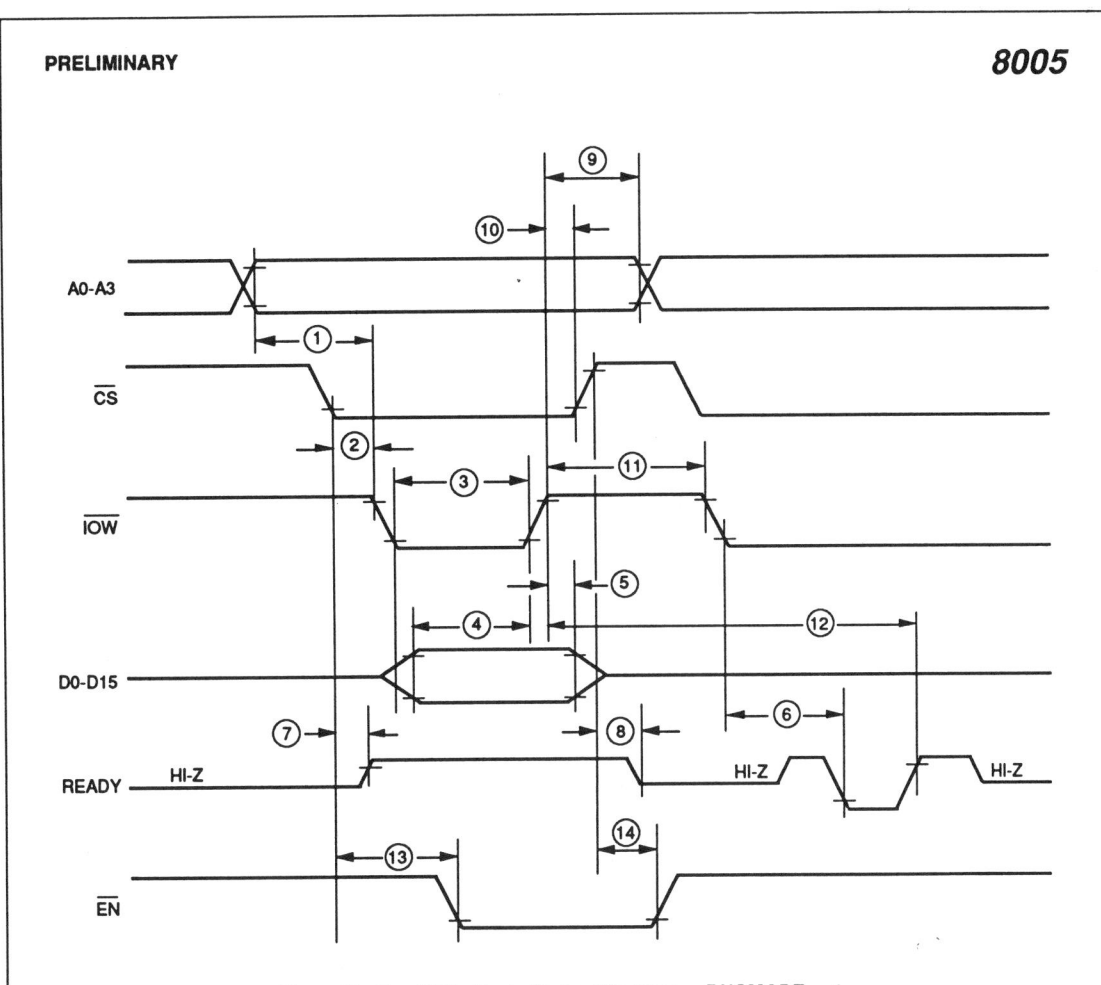

Figure F. Bus Write Cycle Timing Diagram — BUSMODE = 1

PRELIMINARY

8005

A.C. Characteristics (Assuming 20 MHz Input Master Clock)
(Over operating temperature and V_{cc} range, unless otherwise specified)

Table G. Bus Read Cycle — BUSMODE = 1

Ref. #	Symbol	Description	Min.	Max.	Units
1	TAVRL	Address Setup Time	30		ns
1a	TCSLRL	CS* Setup Time	30		ns
2	TRHRL	IOR* High Time	200		ns
3	TRLRYH	READY Assert Delay			ns
		a. FIFO Data [1]		35	ns
		b. Configuration Regs. [2]		800	ns
		c. Pointer Registers. [3]		1800	ns
4	TRLRYL	READY Deassertion Delay		35	ns
5	TRYHDV	READY Assert to Data Valid		50	ns
6	TCSHRYZ	READY Delay to Hi-Z		50	ns
7	TRHDX	Data Hold Time	20		ns
8	TRHDZ	Data Delay to Hi-Z		100	ns
9	TRHAX	Address Hold Time	20		ns
10	TRHCSH	CS* Hold Time	20		ns
11	TRLRH	IOR* Pulse Width	100		ns
12	TRLAPL	APEN* Assert Delay		400	ns
13	TRHAPH	APEN* Deassert Delay		50	ns
14	TCSLENL	EN* Assert Delay		50	ns
15	TCSHENH	EN* Deassert Delay		50	ns
16	TCSLRYV	CS* Assert to READY Valid		50	ns

NOTES:
1. The BIU prefetches one word (byte) of FIFO data. Thus, data is generally available immediately and READY will not de-assert during a data read. Following the read, the BIU will fetch the next word (byte) of data. Should another data read occur before the BIU has completed the prefetch, READY will first de-assert and then assert after the prefetch is completed. The assert delay in this case is 800 ns max (600 ns in 8 bit mode).
2. Configuration Registers are: Command/Status Register, Configuration Register # 1, & 2, Interrupt Vector Register, DMA Pointer Register, and Station Address Registers. If BUSSIZE = 0 (8 bit reads), subtract 200 ns.
3. Pointer Registers are: Receive End Area Pointer, Receive Pointer Register, Transmit Pointer Register, and Transmit End Area Register. If BUSSIZE = 0, subtract 600 ns.

seeq *Technology, Incorporated*

PRELIMINARY

8005

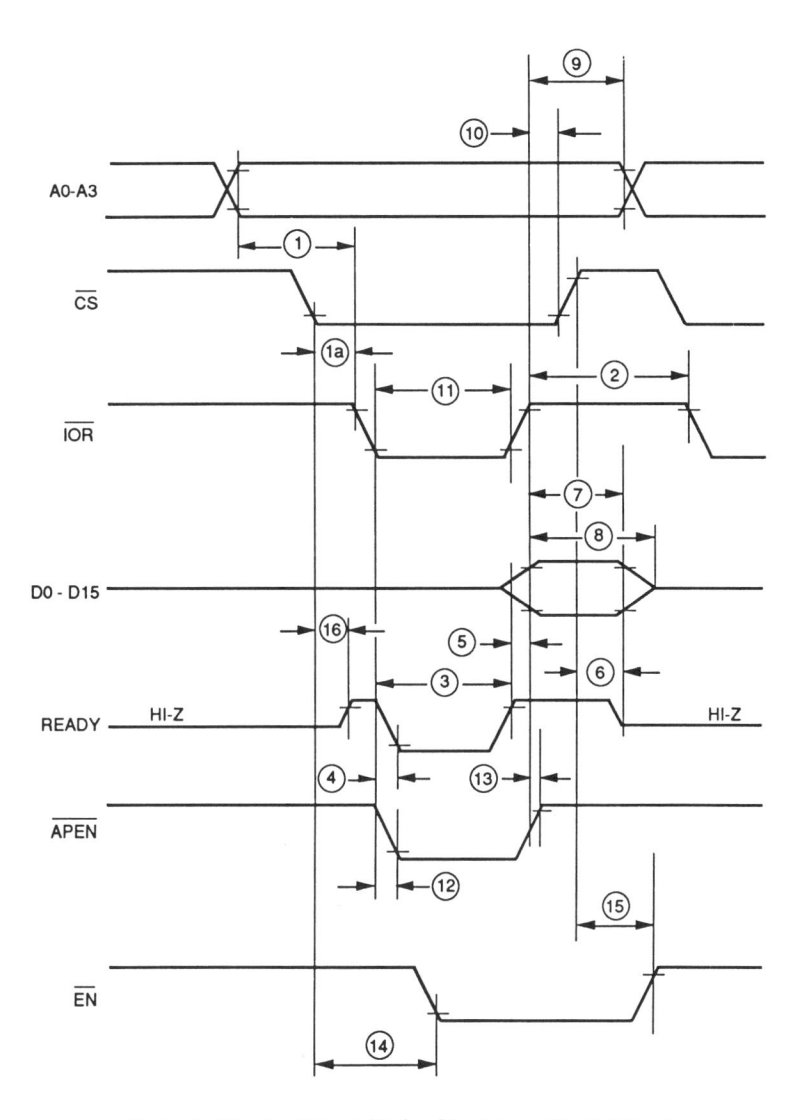

Figure G. Bus Read Cycle Timing Diagram — BUSMODE = 1

PRELIMINARY

8005

A.C. Characteristics (Assuming 20 MHz Input Master Clock)
(Over operating temperature and V_{cc} range, unless otherwise specified)

Table H. Interrupt Cycle — BUSMODE = 1

Ref. #	Symbol	Description	Min.	Max.	Units
1	TRYHDV	READY Assert to Data Valid		50	ns
2	TRLRYL	READY Deassertion Delay		35	ns
3	TRLRYH	READY Assert Delay		600	ns
4	TRHDZ	Data Delay to Hi-Z		100	ns
5	TIAHRYZ	READY Delay to Hi-Z		50	ns
6	TRHDX	Data Hold from IOR*	20		ns
7	TIALRL	IACK* Setup Time	30		ns
8	TIALENL	EN* Assert Delay		50	ns
9	TIAHENH	EN* Deassert Delay		50	ns
10	TIALRYV	IACK* Assert to READY Valid		50	ns
11	TRHIAH	IACK* Hold Time from IOR*	20		ns

seeq *Technology, Incorporated*
MD400031/C

PRELIMINARY

8005

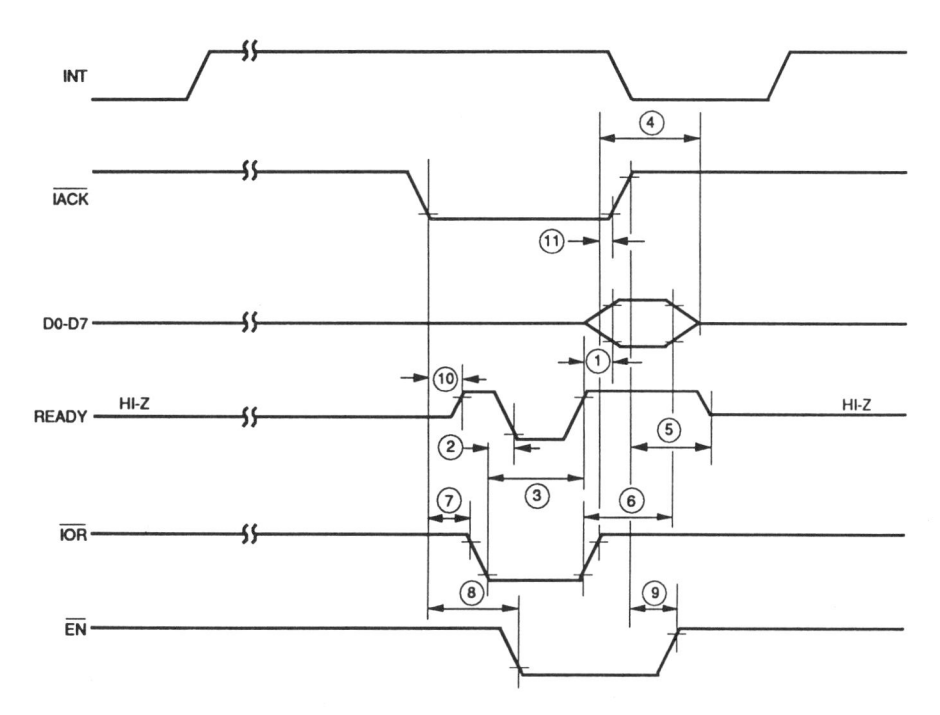

Figure H. Interrupt Cycle Timing Diagram — BUSMODE = 1

PRELIMINARY

8005

A.C. Characteristics (Assuming 20 MHz Input Master Clock)
(Over operating temperature and V_{cc} range, unless otherwise specified)

Table I. DMA Write Cycle — BUSMODE = 1

Ref. #	Symbol	Description	Min.	Max.	Units
1	TDALWL	DACK* Setup Time	30		ns
2	TWLWH	IOW* Pulse Width[1]	100		ns
3	TDVWH	Data Setup Time	70		ns
4	TWHDX	Data Hold Time	20		ns
5	TWHWL	IOW* High Time	200		ns
6	TTCHTCL	TERMCT Asserted While DACK* Asserted	125		ns
7	TTCHDRL	DREQ Delay from TERMCT[4]		175	ns
8	TWLDRL	DREQ Delay from IOW*[5]		100	ns
9	TWHDAH	DACK* Hold Time	20		ns
10	TDALENL	EN* Assert Delay		50	ns
11	TDAHENH	EN* Deassert Delay		50	ns
12	TWLRYL	READY Deassert Delay[2]		35	ns
13	TWHRYH	Write Recovery Time[3]		800	ns
14	TDAHRYZ	READY Delay to Hi-Z		50	ns
15	TDALRYV	DACK* Asserted to READY Valid		50	ns

NOTES:
1. IOW* must be asserted until READY is asserted and for a minimum of 100 ns.
2. The trailing edge of IOW* initiates an internal write sequence that lasts a maximum of 800 ns in 16 bit mode. Should another IOW* occur during this period, READY de-asserts (Ref. #12 TWLRYL) and then asserts after the internal write sequence has finished (Ref. #13 TWHRYH). If the subsequent IOW* does not occur until after the internal write sequence has ended, then Ref. #12 TWLRYL has no meaning since READY does not de-assert under this condition.
3. Subtract 200 ns when BUSSIZE = 0 (8 bit mode).
4. DACK* and TERMCT must both be asserted at the same time and for a minimum of 125 ns. The de-assertion of DREQ is timed from the last one to assert.
5. Ref. #8 TWLDRL applies for normal DMA burst terminations — not those due to TERMCT.

All the timing in this table also apply when writing data with programmed I/O; CS* replaces DACK* and the DREQ*, TERMCT signals do not apply. A0-A3 times are the same as CS*.

seeq *Technology, Incorporated*

PRELIMINARY

8005

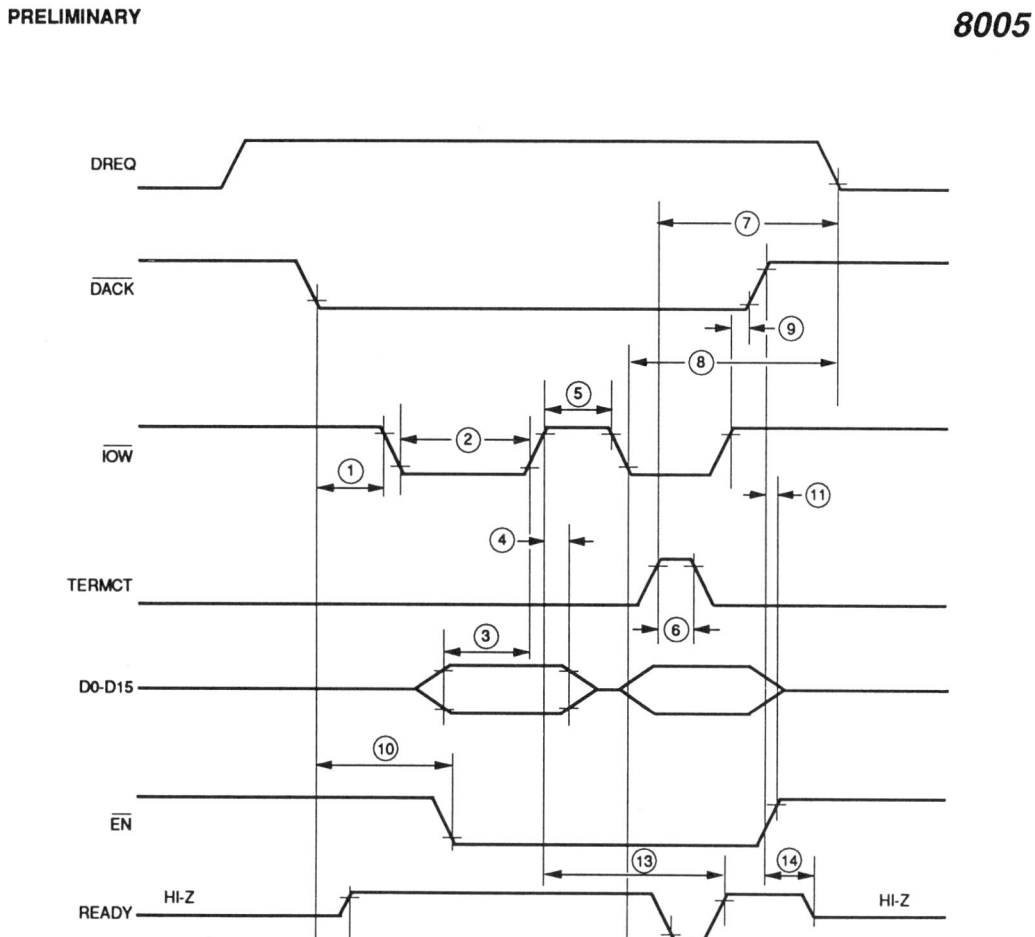

Figure I. DMA Write Cycle Timing Diagram — BUSMODE = 1

PRELIMINARY *8005*

A.C. Characteristics (Assuming 20 MHz Input Master Clock)
(Over operating temperature and V_{cc} range, unless otherwise specified)

Table J. DMA Read Cycle — BUSMODE = 1

Ref. #	Symbol	Description	Min.	Max.	Units
1	TDALRL	DACK* Setup Time	30		ns
2	TRLRH	IOR* Pulse Width[1]	100		ns
3	TDVRYH	READY Asserted to Data Valid		50	ns
4	TRHDX	Data Hold Time	20		ns
5	TRHRL	IOR* High Time	200		ns
6	TTCHTCL	TERMCT Asserted While DACK* Asserted	125		ns
7	TTCHDRL	DREQ Delay from TERMCT[4]		175	ns
8	TRLDRL	DREQ Delay from IOR*[5]		100	ns
9	TRHDAH	DACK* Hold Time	20		ns
10	TRHDZ	Data Hi-Z Delay		100	ns
11	TDALENL	EN* Assert Delay		50	ns
12	TDAHENH	EN* Deassert Delay		50	ns
13	TRLRYL	READY Deassert Delay[2]		35	ns
14	TRLRYH	Read Recovery Time[3]		800	ns
15	TDAHRYZ	READY Delay to Hi-Z		50	ns
16	TDALRYV	DACK* Assert to READY Valid		50	ns

NOTES:
1. IOR* must be asserted until READY is asserted and for a minimum of 100 ns.
2. The BIU pre-fetches FIFO data. Thus, data is available immediately for the first read of any burst. Once the BIU detects a read operation, it begins fetching the next byte or word of data. This occurs during the Ref. #14 TRLRYH period. If a subsequent IOR* occurs within the Ref. #14 TRLRYH period, READY will de-assert (Ref. #13 TRLRYL) and then assert after the FIFO data has been fetched. If the subsequent IOR* does not begin until Ref. #14 has ended, then Ref. #13 has no meaning since READY does not de-assert under this condition.
3. Subtract 200 ns if BUSSIZE = 0 (8 bit mode).
4. DACK* and TERMCT must be asserted at the same time and for a minimum of 125 ns. The de-assertion of DREQ is timed from the last one to assert.
5. Ref. #8 TRLDRL applies for normal DMA burst terminations — not those due to TERMCT.

All the timing in this table also apply when reading data with programmed I/O: CS* replaces DACK* and the DREQ, TERMCT signals do not apply. A0-A3 times are the same as CS*.

Seeq *Technology, Incorporated*
MD400031/C

PRELIMINARY

8005

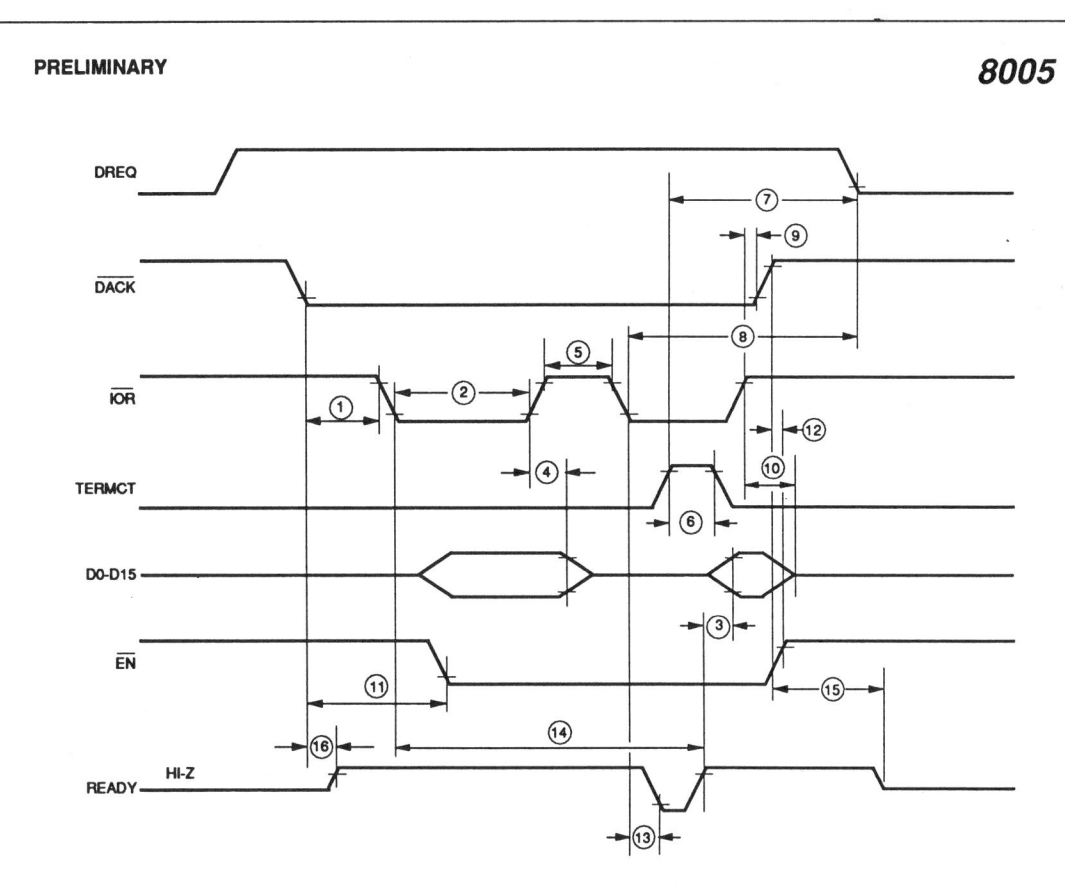

Figure J. DMA Read Cycle Timing Diagram — BUSMODE = 1

PRELIMINARY *8005*

A.C. Characteristics (Assuming 20 MHz Input Master Clock)
(Over operating temperature and V_{cc} range, unless otherwise specified)

Table K. Local Buffer Read or Write Cycle

Ref. #	Symbol	Description	Min.	Max.	Units
1	TRSLAX	Row Address Hold Time	100		ns
2	TAVRSL	Row Address Setup Time	25		ns
3	TRSHRSL	RAS* Pulse Width High	200		ns
4	TCSLAX	Column Address Hold Time	45		ns
5	TAVCSL	Column Address Setup Time	10		ns
6	TCSHCSL	CAS* Pulse Width — High	60		ns
7	TCSLCSH	CAS* Pulse Width — Low	110		ns
8	TAZGL	Address Hi-Z to G* Low Time	0		ns
9	TGLCSH	G* Setup Time to CAS*	70		ns
10	TGLDV	G* to Data Valid		40	ns
11	TCSHDX	Data Hold from CAS Deassert	0		ns
12	TCSHDZ	Data Hi-Z from CAS Deassert		40	ns
13	TAVAV	Read or Write Cycle Time a. Single Cycle b. Page Mode	 600 200		 ns ns
14	TDVWL	Data Setup Time	5		ns
15	TWLDX	Data Hold Time	60		ns
16	TWLWH	Write Pulse Width	60		ns
17	TCSLWL	CAS* Setup to W*	60		ns
18	TWLCSH	Write Setup Time	40		ns
19	TRSLRSL	RAS* Cycle Time	600		ns

NOTE: TMS 4464-10, −12 or equivalent satisfies the above timing.

seeq *Technology, Incorporated*
MD400031/C

4-78

PRELIMINARY *8005*

A.C. Characteristics (Assuming 20 MHz Input Master Clock)
(Over operating temperature and V_{CC} range, unless otherwise specified)

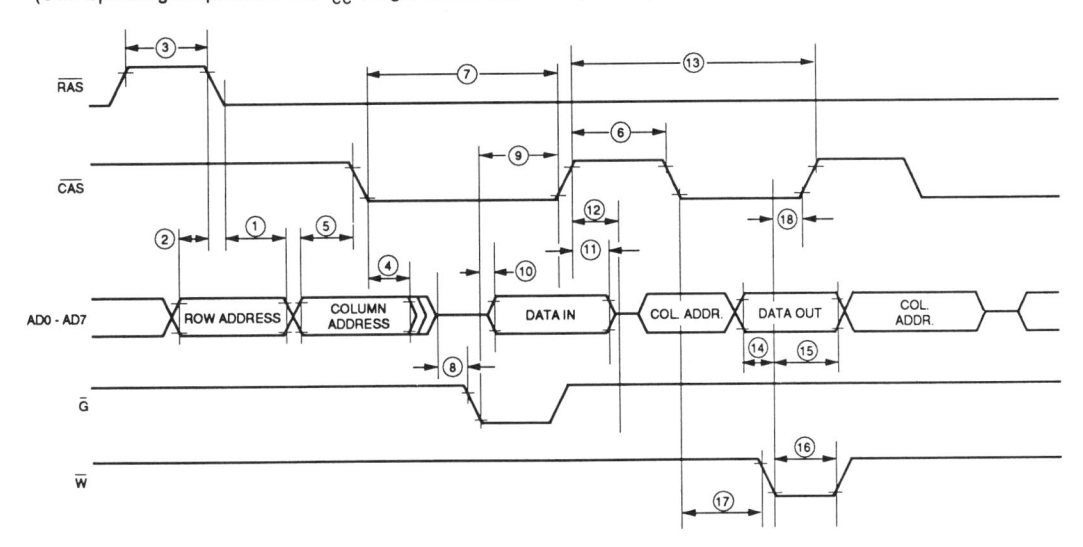

Figure K1. Local Dram Buffer Page-Mode Read and Write Cycle Timing Diagram

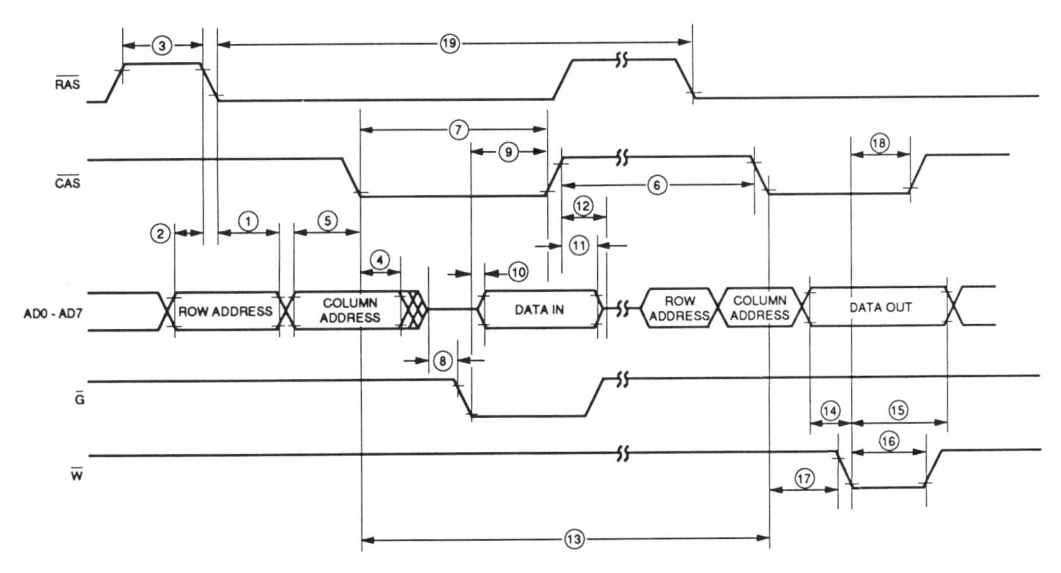

Figure K2. Local Dram Buffer Single Cycle Read and Write Cycle Timing Diagram

seeq Technology, Incorporated

PRELIMINARY

8005

A.C. Characteristics (Assuming 20 MHz Input Master Clock)
(Over operating temperature and V_{cc} range, unless otherwise specified)

Table L. Local Buffer Refresh Cycle

Ref. #	Symbol	Description	Min.	Max.	Units
1	TAVRSL	Address Setup Time to RAS*	25		ns
2	TRSLAX	Address Hold Time from RAS*	100		ns
3	TRSLRSH	RAS* Pulse Width	200		ns
4	TRSLRSL	RAS* Cycle Time	400		ns

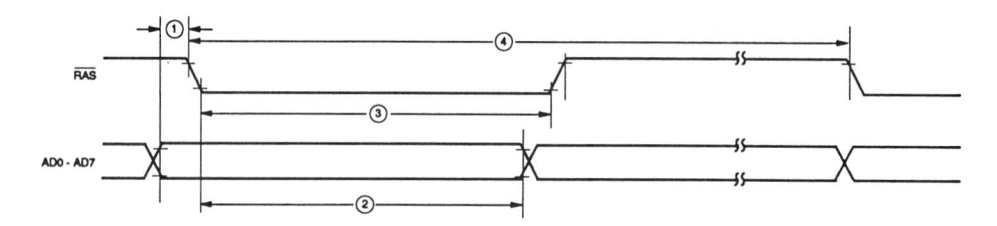

Figure L. Local Dram Buffer Refresh Cycle Timing Diagram

PRELIMINARY

8005

A.C. Characteristics (Assuming 20 MHz Input Master Clock)
(Over operating temperature and V_{cc} range, unless otherwise specified)

Table M. Serial Interface Timing

Ref. #	Symbol	Description	Min.	Max.	Units
1	TCKHCKH	TXC*/RXC Cycle Time	100		ns
2	TCKHCKL	TXC*/RXC High Width	45		ns
3	TCKLCKH	TXC*/RXC Low Width	45		ns
4	TCKLDV	TXD Delay from TXC*		60	ns
5	TDVCKH	RXD Setup to RXC	30		ns
6	TCKHDX	RXD Hold Time from RXC	20		ns
7	TCKLTEH	TXEN Delay from TXC*		60	ns
8	TCKLTEL	TXEN Hold Time from TXC*	20		ns
9	TCSHCKH	CSN Setup to RXC	20		ns
10	TCKHCSL	CSN Hold Time from RXC	20		ns
11	TCHCL	COLL Pulse Width	200		ns

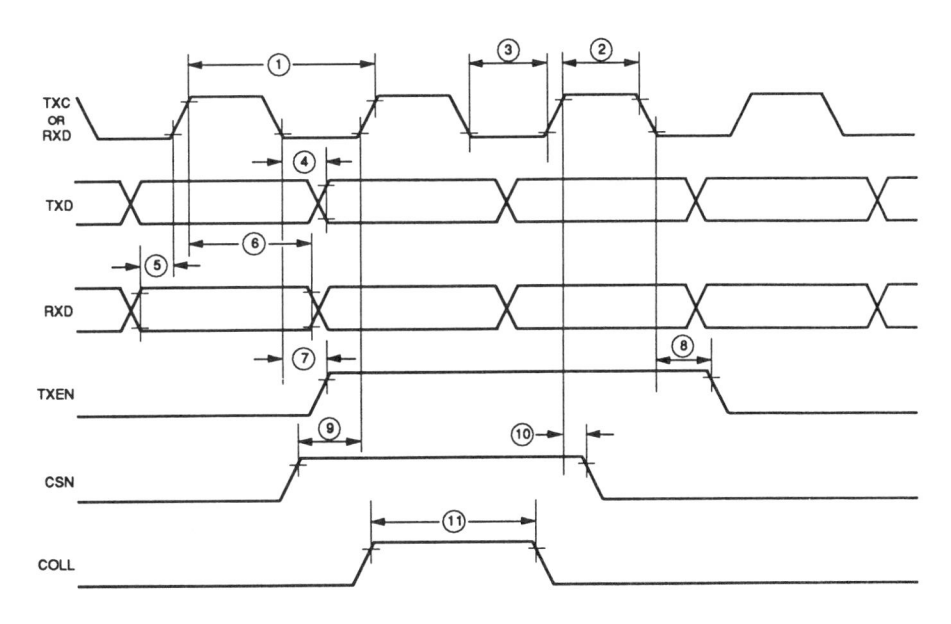

Figure M. Serial Transmit & Receive Interface Timing

seeq Technology, Incorporated

MD400031/C

4-81

PRELIMINARY

8005

A.C. Characteristics (Assuming 20 MHz Input Master Clock)
(Over operating temperature and V_{cc} range, unless otherwise specified)

Table N. Master Clock and Reset Timing

Ref. #	Symbol	Description	Min.	Max.	Units
1	TCKHCKL	CLK Pulse Width High	15	25	ns
2	TCKLCKH	CLK Pulse Width Low	15	25	ns
3	TCKHCKH	CLK Cycle Time	49.9	50.1	ns
4	TRSLRSH	Reset Pulse Width	1		µs

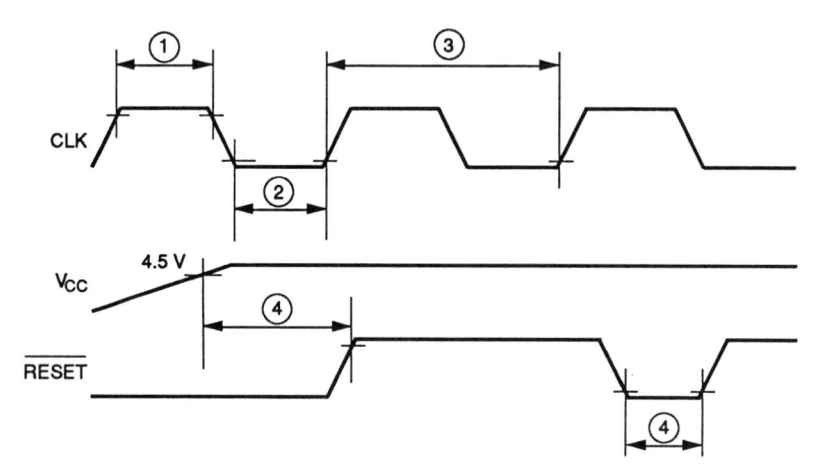

Figure N. Master Clock and Reset Timing

Ordering Information

PART NUMBER

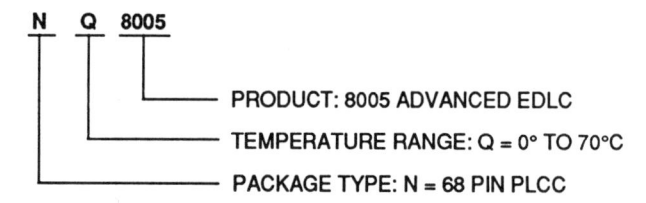

N Q 8005

- PRODUCT: 8005 ADVANCED EDLC
- TEMPERATURE RANGE: Q = 0° TO 70°C
- PACKAGE TYPE: N = 68 PIN PLCC

seeq *Technology, Incorporated*

seeq

8020 MCC™ Manchester Code Converter

November 1989

Features

- Compatible with IEEE 802.3 /Ethernet (10BASE5), IEEE802.3/Cheapernet (10BASE2) and Ethernet Rev. 1 Specifications
- Compatible with 8003 ELDC®, 8005 Advanced EDLC
- Manchester Data Encoding/Decoding and Receiver Clock Recovery with Phase Locked Loop (PLL)
- Receiver and Collision Squelch Circuit and Noise Rejection Filter
- Differential TRANSMIT Cable Driver
- Loopback Capability for Diagnostics and Isolation
- Fail-Safe Watchdog Timer Circuit to Prevent Continuous Transmission
- 20 MHz Crystal Oscillator
- Transceiver Interface High Voltage (16 V) Short Circuit Protection
- Low Power CMOS Technology with Single 5V Supply
- 20 pin DIP & PLCC Packages

Description

The SEEQ 8020 Manchester Code Converter chip provides the Manchester data encoding and decoding functions of the Ethernet Local Area Network physical layer. It interfaces to the SEEQ 8003 and 8005 Controllers and any standard Ethernet transceiver as defined by IEEE 802.3 and Ethernet Revision 1.

The SEEQ 8020 MCC is a functionally complete Encoder/Decoder including ECL level balanced driver and receivers, on board oscillator, analog phase locked loop for clock

Pin Configuration

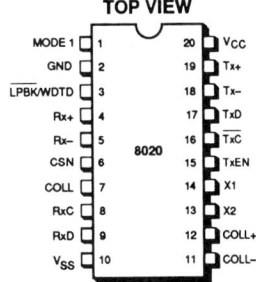

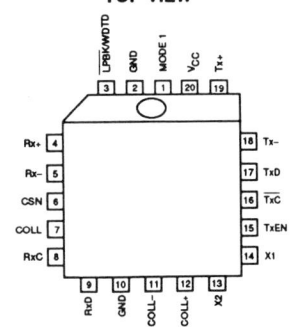

Functional Block Diagram

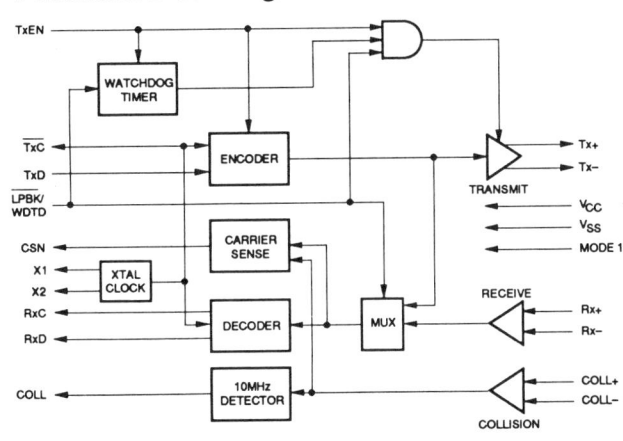

Figure 1. 8020 MCC Manchester Code Converter Block Diagram.

MCC is a trademark of SEEQ Technology Inc.
EDLC is a registered trademark of SEEQ Technology Inc.

seeq Technology, Incorporated
MD400023/C

8020

recovery and collision detection circuitry. In addition, the 8020 includes a watchdog timer, a 4.5 microsecond window generator, and a loopback mode for diagnostic operation.

Together with the 8003 or 8005 and a transceiver, the 8020 Manchester Code Converter provides a high performance minimum cost interface for any system to Ethernet.

Functional Description

The 8020 Manchester Code Converter chip has two portions, transmitter and receiver. The transmitter uses Manchester encoding to combine the clock and data into a serial stream. It also differentially drives up to 50 meters of twisted pair transmission line. The receiver detects the presence of data and collisions. The 8020 MCC recovers the Manchester encoded data stream and decodes it into clock and data outputs. Manchester Encoding is the process of combining the clock and data stream so that they may be transmitted on a single twisted pair of wires, and the clock and data may be recovered accurately upon reception. Manchester encoding has the unique property of a transition at the center of each bit cell, a positive going transition for a "1", and a negative going transition for a "0" (See Figure 2). The encoding is accomplished by exclusive-ORing the clock and data prior to transmission, and the decoding by deriving the clock from the data with a phase locked loop.

Clock Generator

The internal oscillator is controlled by a 20 MHz parallel resonant crystal or by an external clock on X1. The 20 MHz clock is then divided by 2 to generate a 10 MHz ±0.01% transmitter clock. Both 10 MHz and 20 MHz clocks are used in Manchester data encoding.

Manchester Encoder and Differential Output Driver

The encoder combines clock and data information for the transceiver. In Manchester encoding, the first half of the bit cell contains the complement of the data and the second half contains the true data. Thus a transition is always guaranteed in the middle of a bit cell.

Data encoding and transmission begin with TxEN going active; the first transition is always positive for Tx(-) and negative for Tx(+). In IEEE mode, at the termination of a transmission, TxEN goes inactive and transmit pair approach to zero differential. In Ethernet mode, at the end of the transmission, TxEN goes inactive and the transmit pair stay differentially high. The transmit termination can occur at bit cell center if the last bit is a one or at a bit boundary if the last bit is a zero. To eliminate DC current in the transformer during idle, Tx± is brought to 100 mV differential in 600 ns after the last transition (IEEE mode). The back swing voltage is guaranteed to be less than .1 V.

Watchdog Timer

A watchdog timer is built on chip. It can be enabled or disabled by the $\overline{LPBK}$/WDTD signal. The timer starts counting at the beginning of the transmission. If TxEN goes inactive before the timer expires, the timer is reset and ready for the next transmission. If the timer expires before the transmission ends, transmission is aborted by disabling the differential transmitter. This is done by idling the differential output drivers (differential output voltage becomes zero) and deasserting CSN.

Differential Input Circuit (Rx+ and Rx–, COLL+ and COLL–).

As shown in Figure 3, the differential input for Rx+ and Rx- and COLL+ and COLL- are externally terminated by a pair of 39.2 Ω ± 1% resistors in series for proper impedance matching.

The center tap has a 0.01μF capacitor, tied to ground, to provide the AC common mode impedance termination for the transceiver cable.

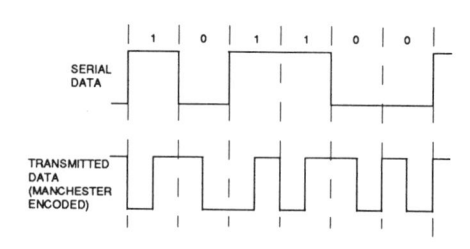

Figure 2. Manchester Coding

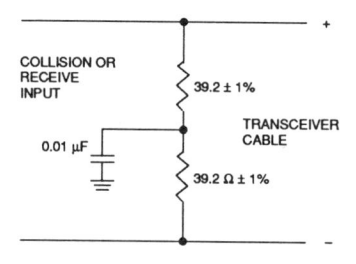

Figure 3. Differential Input Terminator

8020

Both collision and receiver input circuits provide a static noise margin of -140 mV to -300mV (peak value). Noise rejection filters are provided at both input pairs to prevent spurious signals. For the receiver pair, the range is 15 ns to 30 ns. For the collision pair, the range is 10 ns to 18 ns. The D.C. threshold and noise rejection filter assure that differential receiver data signals less than -140 mV in amplitude or narrower than 15 ns (10 ns for collision pair) are always rejected, signals greater than -300 mV and wider than 30 ns (18 ns for collision pair) are always accepted.

Manchester Decoder and Clock Recovery Circuit

The filtered data is processed by the data and clock recovery circuit using a phase-locked loop technique. The PLL is designed to lock onto the preamble of the incoming signal with a transition width asymmetry not greater than +8.25 ns to -8.25 ns within 12 bit cell times worst case and can sample the incoming data with a transition width asymmetry of up to +8.25 ns to -8.25 ns. The RxC high or low time will always be greater than 40 ns. RxC follows $\overline{TxC}$ for the first 1.2 μs and then switches to the recovered clock. In addition, the Encoder/Decoder asserts the CSN signal while it is receiving data from the cable to indicate the receiver data and clock are valid and available. At the end of the frame, after the node has finished transmitting, CSN is deasserted and will not be asserted again for a period of 4.5 μs regardless of the state of the state of the receiver pair or collision pair. This is called inhibit period. There is no inhibit period after packet reception. During clock switching, RxC may stay high for 200ns maximum.

Collision Circuit

A collision on the Ethernet cable is sensed by the transceiver. It generates a 10 MHz ±15% differential square wave to indicate the presence of the collision. During the collision period, CSN is asserted asynchronously with RxC. However, if a collision arrives during inhibit period 4.5 μs from the time CSN was deasserted, CSN will not be reasserted.

Loopback

In loopback mode, encoded data is switched to the PLL instead of Tx+/Tx- signals. The recovered data and clock are returned to the Ethernet Controller. All the transmit and receive circuits, including noise rejection filter, are tested except the differential output driver and the differential input receiver circuits which are disabled during loopback. At the end of frame transmission, the 8020 also generates a 650 ns long COLL signal 550 ns after CSN was deasserted to simulate the IEEE 802.3 SQE test. The watchdog timer remains enabled in this mode.

Pin Description

The MCC chip signals are grouped into four categories:

- Power Supply and Clock
- Controller Interface
- Transceiver Interface
- Miscellaneous

Power Supply

V_{CC} ....+5V
V_{SS} ....Ground

X1 and X2 clock (Inputs): *Clock Crystal: 20 MHz crystal oscillator input. Alternately, pin X1 may be used at a TTL level input for external timing by floating pint X2,*

Controller Interface

RxC Receive Clock (Output): *This signal is the recovered clock from the phase decoder circuit. It is switched to $\overline{TxC}$ when no incoming data is present from which a true receive clock is derived. 10 MHz nominal and TTL compatible.*

RxD Receive Data (Output): *The RxD signal is the recovered data from the phase decoder. During idle periods, the RxD pin is LOW under normal conditions. TTL and MOS level compatible. Active HIGH.*

CSN Carrier Sense (Output): *The Carrier Sense Signal indicates to the controller that there is activity on the coaxial cable. It is asserted when receive data is present or when a collision signal is present. It is deasserted at the end of frame or at the end of collision, whichever occurs later. It is asserted or deasserted synchronously with RxC. TTL compatible.*

$\overline{TxC}$ Transmit Clock (Output): *A 10 MHz signal derived from the internal oscillator. This clock is always active. TTL and MOS level compatible.*

TxD Transmit Data (Input): *TxD is the NRZ serial input data to be transmitted. The data is clocked into the MCC by $\overline{TxC}$. Active HIGH, TTL compatible.*

TxEN Transmit Enable (Input): *Transmit Enable, when asserted, enables data to be sent to the cable. It is asserted synchronously with $\overline{TxC}$. TxEN goes active with the first bit of transmission. TTL compatible.*

seeq Technology, Incorporated
MD400023/C

8020

COLL Collision (Output): When asserted, indicates to the controller the simultaneous transmission of two or more stations on network cable. TTL Compatible.

Transceiver Interface

Rx+ and Rx− Differential Receiver Input Pair (Input): Differential receiver input pair which brings the encoded receive data to the 8020. The last transition is always positive-going to indicate the end of the frame.

COLL+ and COLL− Differential Collision Input Pair (Input): This is a 10 MHz± 15% differential signal from the transceiver indicating collision. The duty cycle should not be worse than 60%/40% — 40%/60%. The last transition is positive-going. This signal will respond to signals in the range of 5 MHz to 11.5 MHz. Collision signal may be asserted if 'MAU not available' signal is present.

Tx+ and Tx- Differential Transmit Output Pair (Output): Differential transmit pair which sends the encoded data to the transceiver. The cable driver buffers are source follower and require external 243 Ω resistors to ground as loading. These resistors must be rated at 1 watt to withstand the fault conditions specified by IEEE 802.3. If MODE 1=1, after 200 ns following the last transition, the differential voltage is slowly reduced to zero volts in 8 μs to limit the back swing of the coupling transformer to less than 0.1 V.

Miscellaneous

MODE 1 (Input): This pin is used to select between AC or DC coupling. When it is tied high or left floating, the output drivers provide differential zero signal during idle (IEEE 802.3 specification). When pin 1 is tied low, then the output is differentially high when idle (Ethernet Rev.1 specification).

$\overline{LPBK}$/WDTD $\overline{Loopback}$ /Watchdog Timer Disable (Input):

Normal Operation: For normal operation this pin should be HIGH or tied to V_{cc}. In normal operation the watchdog timer is enabled.

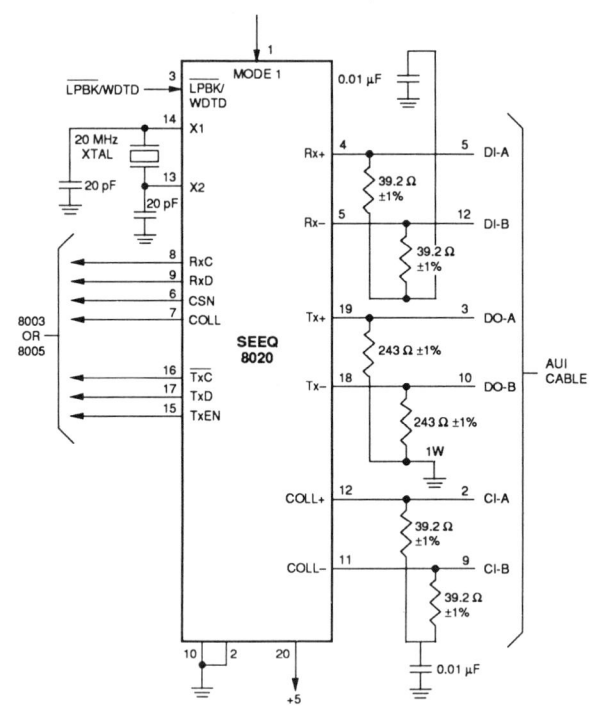

Figure 4. 8020 Interface

8020

Loopback: When this pin is brought low, the Manchester encoded transmit data from TxD and $\overline{TxC}$ is routed through the receiver circuit and sent back onto the RxD and RxC Pins. During loopback, Collision and Receive data inputs are ignored. The transmit pair is idled. At the end of transmission, the signal quality error test (SQET) will be simulated by asserting collision during the inhibit window. During loopback, the watchdog timer is enabled.

Watchdog Timer Disable: When this pin is between 10 V (Min.) and 16 V (Max.), the on chip 25 ms Watchdog Timer will be disabled. The watchdog timer is used to monitor the transmit enable pin. If TxEN is asserted for too long, then the watchdog timer (if enabled) will automatically deassert CSN and inhibit any further transmissions on the Tx+ and Tx- lines. The watchdog timer is automatically reset each time TxEN is deasserted.

Interconnection to a Data Link Controller

Figure 5 shows the interconnections between the 8020 MCC and SEEQ's 8003 or 8005. There are three connections for each of the two transmission channels, transmit and receive, plus the Collision Signal line (COLL).

Transmitter connections are:

    Transmit Data, TxD
    Transmit Clock, $\overline{TxC}$
    Transmit Enable, TxEN
    Collision, COLL

Receiver connections are:

    Receive Data, RxD
    Receive Clock, RxC
    Carrier Sense, CSN

D.C. and A.C. Characteristics and Timing

Crystal Specification

Resonant Frequency (C_L = 20 pF) .... 20 MHz
                                                     ± 0.005% 0-70° C
                                           and ± 0.003% at 25° C
Type .... Fundamental Mode
Circuit .... Parallel Resonance
Load Capacitance (C_L) 20pF
Shunt Capacitance (C_0) .... 7pF Max.
Equivalent Series Resistance (R1) .... 25Ω Max.
Motional Capacitance (C1) .... 0.02 pF Max.
Drive Level 2mW

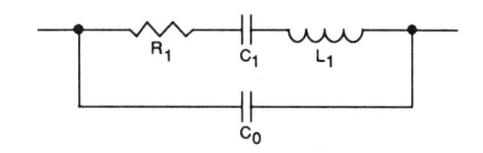

EQUIVALENT CIRCUIT OF CRYSTAL

Figure 6.

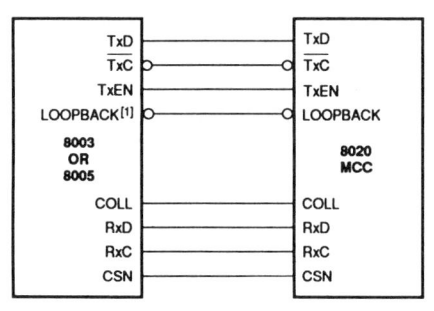

Figure 5. Interconnection of 8020 and 8003/8005

NOTE
1. Loopback output on 8005 only.

8020

Absolute Maximum Rating*

Storage Temperature .... −65°C to +150°C
All Input or Output Voltage .... −0.3 to V_{CC} +0.3
V_{CC} .... −0.3 to 7V
(Rx±, Tx±, COLL±) High Voltage
 Short Circuit Immunity .... −0.3 to 16V

*COMMENT: Stresses above those listed under "Absolute Maximum Ratings" may cause permanent damage to the device. This is a stress rating only and functional operation of the device at these or any other conditions above those indicated in the operational sections of this specification is not implied. Exposure to absolute maximum rating conditions for extended periods may affect device reliability.

DC Characteristics T_A = 0°C to 70°C; V_{CC} = 5 V ± 5%

Symbol	Parameter	Min.	Max.	Unit	Conditions
I_{IL}	Input Leakage Current (except MODE 1, Receive and Collision Pairs)		10	μA	0 ≤ V_{IN} ≤ V_{CC}
	MODE 1 Input Leakage Current		200	μA	0 ≤ V_{IN} ≤ V_{CC}
	Receive and Collision Pairs (Rx±, COLL±) Input Leakage Current		2	mA	V_{IN} = 0
I_{CC}	V_{CC} Current		75	mA	All Inputs, Outputs Open
V_{IL}	TTL Input Low Voltage	−0.3	0.8	V	
V_{IH}	TTL Input High Voltage (except X1)	2.0	V_{CC} + 0.3	V	
	X1 Input High Voltage	3.5	V_{CC} + 0.3	V	
V_{OL}	TTL Output Low Voltage except TxC		0.4	V	I_{OL} = 2.1 mA
	$\overline{TxC}$ Output Low Voltage		0.4	V	I_{OL} = 4.2 mA
V_{OH}	TTL Output High Voltage (except RxC, $\overline{TxC}$, RxD)	2.4		V	t_{OH} = −400 μA
	RxC, $\overline{TxC}$, RxD Output High Voltage	3.9		V	t_{OH} = −400 μA
V_{ODF}	Differential Output Swing	±0.55	±1.2	V	78Ω Termination Resistor and 243Ω Load Resistors
V_{OCM}	Common Mode Output Voltage	V_{CC} −2.5	V_{CC} −1	V	78Ω Termination Resistor and 243Ω Load Resistors
V_{BKSV}	Tx± Backswing Voltage During Idle		0.1	V	Shunt Inductive Load ≤ 27 μH
V_{IDF}	Input Differential Voltage (measured differentially)	±0.3	±1.2	V	
V_{ICM}	Input Common Mode Voltage	0	V_{CC}	V	
C_{IN}[1]	Input Capacitance		15	pF	
C_{OUT}[1]	Output Capacitance		15	pF	

NOTE:
1. Characterized. Not tested

seeq Technology, Incorporated

8020

AC Test Conditions

Output Loading TTL Output:	50% point of swing
Differential Output:	20% to 80% points
Differential Signal Delay Time Reference Level:	
Differential Output Rise and Fall Time:	High time measured at 3.0V
RxC, $\overline{TxC}$, X1 High and Low Time:	Low time measured at 0.6V
RxD, RxC, $\overline{TxC}$, X1 Rise and Fall Time:	Measured between 0.6V and 3.0V points
TTL Input Voltage (except X1):	0.8V to 2.0V with 10 ns rise and fall time
X1 Input Voltage:	0.8V to 3.5V with 5 ns rise and fall time
Differential Input Voltage:	At least ± 300 mV with rise and fall time of 10 ns measured between −0.2V and +0.2V

1 TTL gate and 20 pF capacitor.

243Ω resistor and 10 pF capacitor from each pin to V_{ss} and a termination 78Ω resistor load resistor in parallel with a 27 μH inductor between the two differential output pins

20 MHz TTL Clock Input Timing T_A = 0°C to 70°C; V_{CC} = 5 V ± 5%

Symbol	Parameter	Min.	Max.	Unit
t_1	X1 Cycle Time	49.995	50.005	ns
t_2	X1 High Time	15		ns
t_3	X1 Low Time	15		ns
t_4	X1 Rise Time		5	ns
t_5	X1 Fall Time		5	ns
t_{5A}	X1 to $\overline{TxC}$ Delay Time	10	45	ns

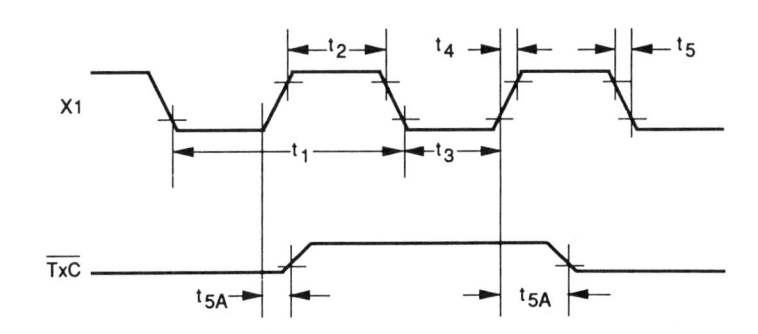

Figure 12. 20 MHz TTL Clock Timing

seeq Technology, Incorporated

8020

Transmit Timing $T_A = 0°C$ to $70°C$; $V_{CC} = 5V \pm 5\%$

Symbol	Parameter	Min.	Max.	Unit
t_6[1]	$\overline{TxC}$ Cycle Time	99.99	100.01	ns
t_7	$\overline{TxC}$ High Time	40		ns
t_8	$\overline{TxC}$ Low Time	40		ns
t_9[1]	$\overline{TxC}$ Rise Time		5	ns
t_{10}[1]	$\overline{TxC}$ Fall Time		5	ns
t_{11}	TxEN Setup Time	40		ns
t_{12}	TxD Setup Time	40		ns
t_{13}[1]	Bit Center to Bit Center Time	99.5	100.5	ns
t_{14}[1]	Bit Center to Bit Boundary Time	49.5	50.5	ns
t_{15}[1]	Tx+ and Tx− Rise Time		5	ns
t_{16}[1]	Tx+ and Tx− Fall Time		5	ns
t_{17}	Transmit Active Time From The Last Positive Transition	200		ns
t_{17A}[1]	From Last Positive Transition of the Transmit Pair to Differential Output Approaches within 100 mV of 0 V	400	600	ns
t_{17B}[1]	From Last Positive Transition of the Transmit Pair to Differential Output Approaches within 40 mV of 0 V		7000	ns
t_{18}	Tx+ and Tx− Output Delay Time		70	ns
t_{19}	TxD Hold Time	15		ns
t_{20}	TxEN Hold Time	15		ns

NOTE:

1. Characterized. Not tested.

seeq *Technology, Incorporated*

8020

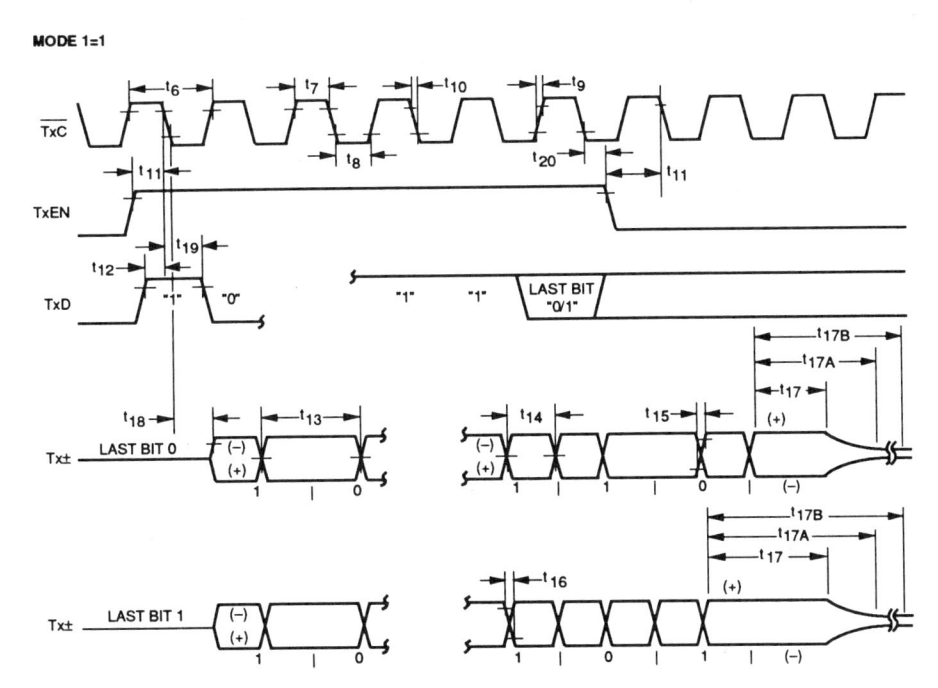

Figure 7. Transmit Timing

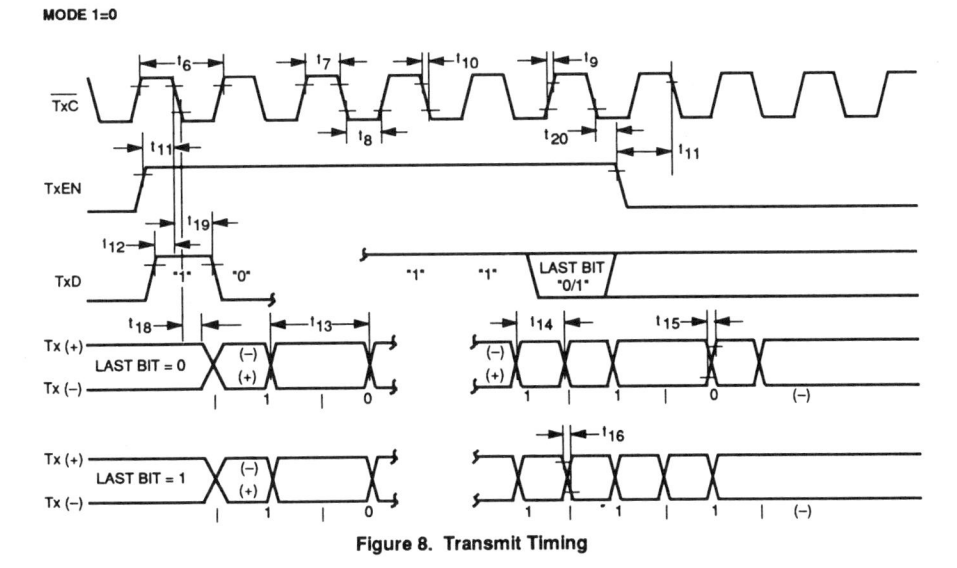

Figure 8. Transmit Timing

8020

Receive Timing $T_A = 0°C$ to $70°C$; $V_{CC} = 5 V \pm 5\%$

Symbol	Parameter	Min.	Max.	Unit
t_{21}	CSN Assert Delay Time		240	ns
t_{22}	CSN Deasserts Delay Time (measured from Last Bit Boundary)		240	ns
t_{23A}	CSN Hold Time	30		ns
t_{23B}	CSN Set up Time	30		ns
t_{24}	RxD Hold Time	30		ns
t_{25}	RxD Set up Time	30		ns
t_{26}[1]	RxC Rise and Fall Time		5	ns
t_{27}[1]	During Clock Switch RxC Keeps High Time	40	200	ns
t_{28}	RxC High and Low Time	40		ns
t_{29}[1]	RxC Clock Cycle Time (during data period)	95	105	ns
t_{30}	CSN Inhibit Time (on Transmission Node only)	4.3	4.6	µs
t_{31}	Rx+/Rx− Rise and Fall Time		10	ns
t_{32}[1]	Rx+/Rx− Begin Return to Zero from Last Positive-Going Transition	160		ns
t_{33}[1]	RxD Rise Time		10	ns
t_{34}[1]	RxD Fall Time		10	ns

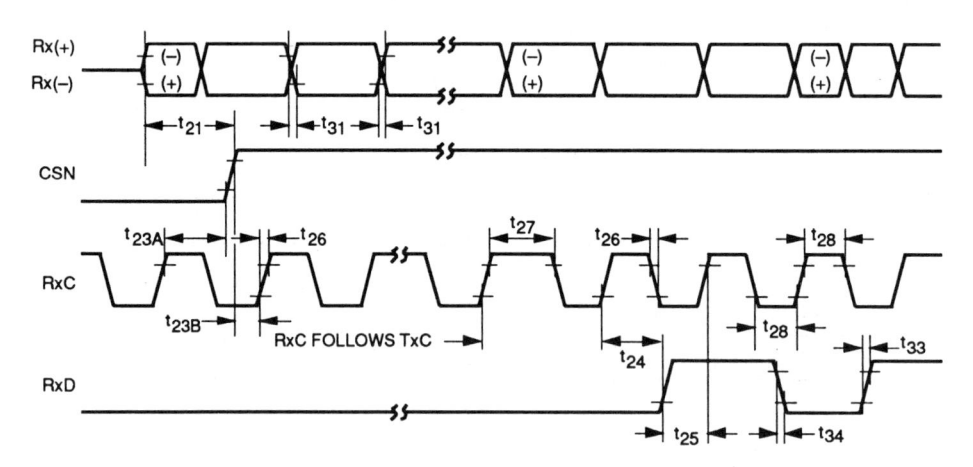

Figure 9. Receive Timing-Start of Packet

seeq Technology, Incorporated

8020

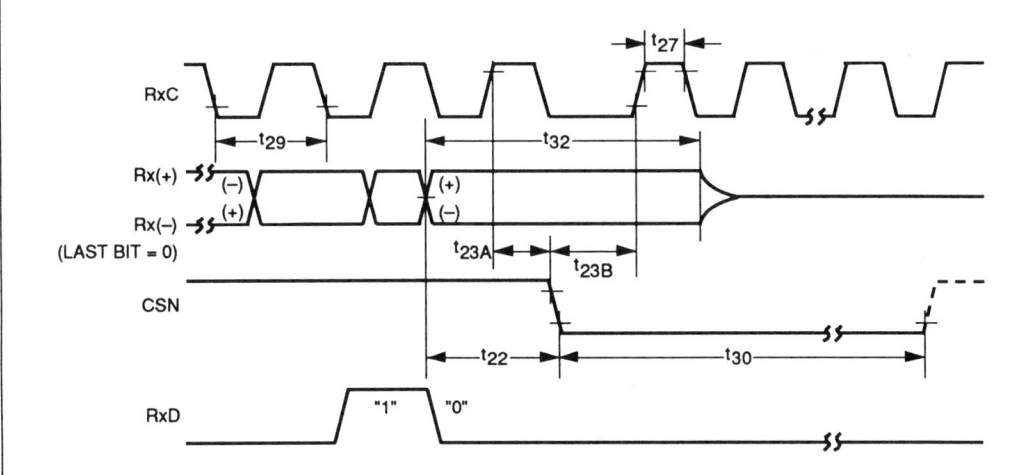

Figure 10. Receive Timing-End of Packet

8020

Collision Timing $T_A = 0°C$ to $70°C$; $V_{CC} = 5V \pm 5\%$

Symbol	Parameter	Min.	Max.	Unit
t_{51}	COLL+ /COLL — Cycle Time	86	118	ns
t_{52}	COLL+/COLL — Rise and Fall Time		10	ns
t_{53}	COLL+/COLL — High and Low Time	35	70	ns
t_{54}	COLL+/COLL — Width (measured at –0.3 V)	26		ns
t_{55}	COLL Asserts Delay Time		300	ns
t_{56}	COLL Deasserts Delay Time		500	ns
t_{57}	CSN Asserts Delay Time		400	ns
t_{58}	CSN Deasserts Delay Time		600	ns

NOTES:
1. COLL + and COLL – asserts and deasserts COLL, asynchronously, and asserts and deasserts CSN synchronously with RxC.
2. If COLL + and COLL – arrives within 4.5µs from the time CSN was deasserted; CSN will not be reasserted (on transmission node only).
3. When COLL + and COLL – terminates, CSN will not be deasserted if Rx+ and Rx– are still active.
4. When the node finishes transmitting and CSN deasserted, it cannot be asserted again for 4.5 µs.

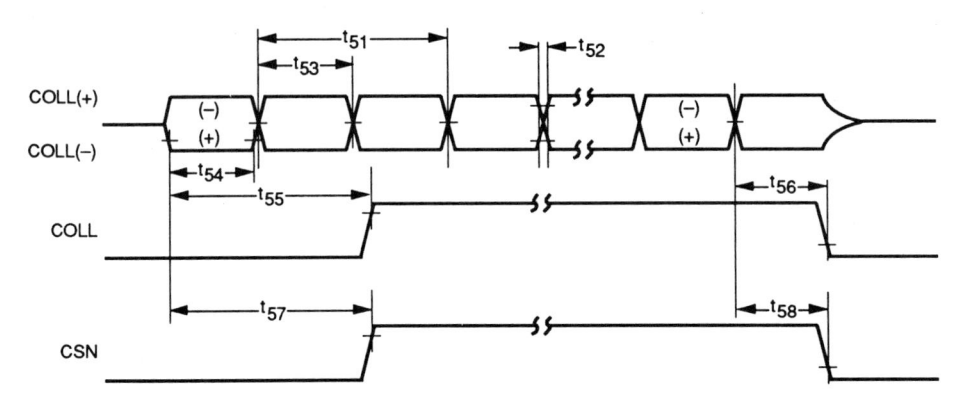

Figure 11. Collision Timing

seeq Technology, Incorporated

8020

Loopback Timing $T_A = 0°C$ to $70°C$; $V_{CC} = 5V \pm 5\%$

Symbol	Parameter	Min.	Max.	Unit
t_{61}	LPBK Setup Time	500		ns
t_{62}	LPBK Hold Time	5		µs
t_{63}	In Collision Simulation, COLL Signal Delay Time	475	625	ns
t_{64}	COLL Duration Time	600	750	ns

NOTES:

1. PLL needs 12-bit cell times to acquire lock, RxD is invalid during this period.

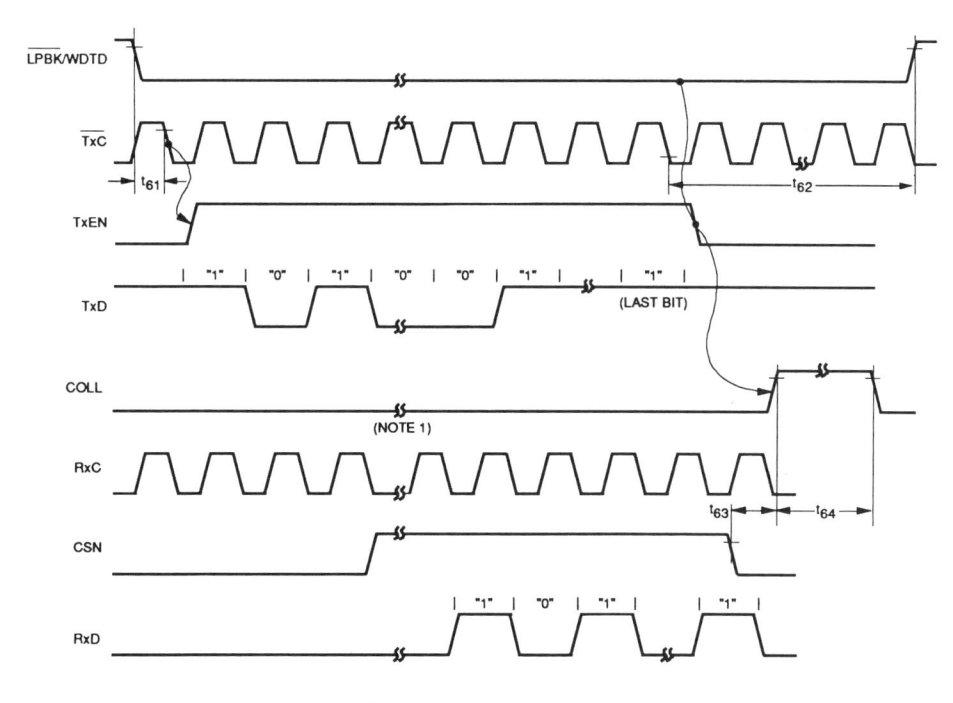

Figure 13. Loopback Timing

8020

Ordering Information

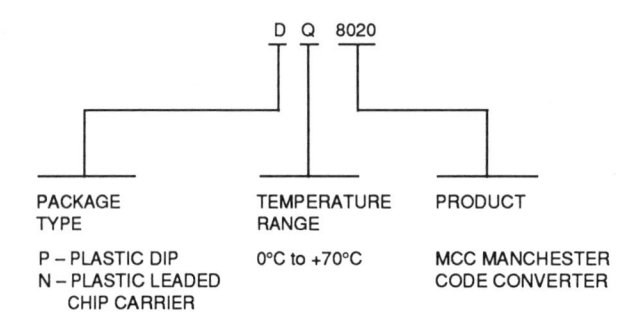

```
            D  Q  8020
            │  │   │
            │  │   │
PACKAGE     TEMPERATURE   PRODUCT
TYPE        RANGE

P – PLASTIC DIP        0°C to +70°C     MCC MANCHESTER
N – PLASTIC LEADED                      CODE CONVERTER
    CHIP CARRIER
```

seeq Technology, Incorporated
MD400023/C

**Communications Products
Application Brief**

8005 ADVANCED EDLC® USER'S GUIDE

September 1987

**SEEQ
Technology, Incorporated**

EDLC is a registered trademark of SEEQ Technology, Inc.

8005
Advanced EDLC
User's Guide

Introduction

Ethernet was developed by the Palo Alto Research Center (PARC) of the Xerox Corporation. The first network was implemented in 1975, as a result of a joint effort by Stanford University and PARC. Over the years, it was proven to be reliable and efficient in a wide variety of network applications. As a result of that success, it became the first industry standard protocol for LANs, supported internationally by computer manufacturers in the United States and Europe.

The network allows equal access by all nodes, can support upwards of 1000 nodes, and can operate with a coaxial cable length in excess of 500 meters. Ethernet is easy to realize, due in large part to currently available LSI chips which implement it.

In 1980 the Institute of Electrical and Electronics Engineers (IEEE) sponsored a committee to review, document, and publish this protocol as an international industry standard. After three years of review and refinement, this specification has been published by the IEEE press under the title, "ANSI/IEEE 802. 3-1985 CSMA/CD Local Area Network Standard Protocol". The medium access method is described by the abbreviation CSMA/ CD, or Carrier Sense, Multiple Access with Collision Detection.

CSMA/CD: Carrier Sense, Multiple Access with Collision Detection

Carrier Sense

All nodes on the network can detect all signals transmitted from any source. A node is any connection to the coaxial cable via transceiver, shown in Figure 1.

The transceiver makes a connection to the cable via connectors or has barbs to pierce the cable and establish an electrical connection when a screw or bolt is tightened.

The transceiver provides collision detection, electrical isolation and voltage level translation between the system at the node and the cable carrying data.

Multiple Access

All nodes have equal access to the network. There is no priority assigned to any node. Also, there is no central control, nor is there any token passing. Any given node may transmit if the net work is not already busy. If two or more nodes transmit at the same time, a collision occurs.

Collision Detection

All nodes can detect a collision by monitoring the medium. When a collision occurs, the transmitting nodes jointly decide which node will retransmit first by a technique known as truncated binary exponential backoff, which provides for a random timeout at each node before each retransmit attempt.

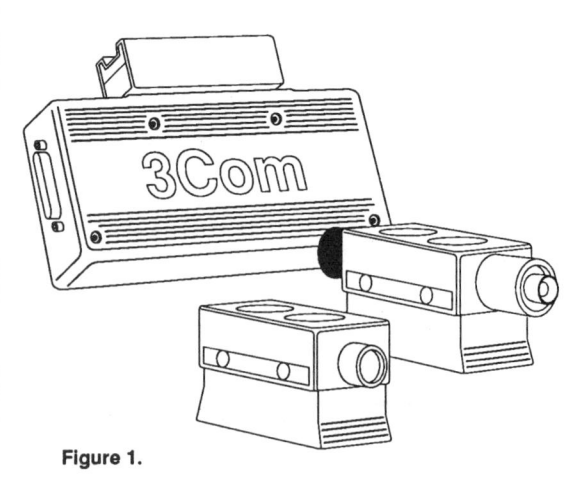

Figure 1.

seeq Technology, Incorporated

Ethernet Data Format

Data is formatted and transmitted in "packets" or "frames", as shown in Figure 2. These frames begin with a preamble for synchronization, and end with a CRC field for error detection. In between, the frame has destination and source addresses, a byte count field, and a data field. This data field contains from 46 to 1500 bytes of information which is passed to a higher layer of software for processing. It is transparent to the media access layer of Ethernet, and may contain any arbitrary sequence of bytes.

Total frame length is 72 to 1526 bytes, including preamble (8 bytes), and frame check sequence (4 bytes).

The signaling method used in Ethernet is base-band Manchester code, transmitted at 10 Megabits per second. Manchester code is such that each bit is defined by a transition at its mid-bit point: a ONE is encoded as a high going signal and a ZERO is a low going signal. Thus, the data is said to be self clocked. This technique provides a continuous supply of bit framing information for the receiver, since the transmitted signal is never static for more than one bit time.

Addressing Scheme

An Ethernet address contains six bytes to define a station address. This allows for over 140 trillion unique addresses. The 48th bit in the address is reserved to indicate a broadcast or multicast address. Xerox Corporation controls issuing addresses for Ethernet. As a system manufacturer, you receive your block of addresses when you receive a license. It is necessary to assign a unique address for each product that communicates on Ethernet.

Direct Memory Access System Interface

There are two basic DMA techniques for interfacing the network to the system bus. The first, in Figure 3a, uses DMA to transfer data directly between the Ethernet controller and the system memory. In Figure 3b, a temporary buffer memory intervenes between the system memory and the controller chip. This buffer eliminates the need to service LAN traffic in real-time.

Why a Local Buffer?

Consider the first approach, where no local buffer is used at the node. Since the LAN data rate is 10 Megabits per second, the DMA controller must be capable of handling system data at a minimum of 1.25 Megabytes per second. If the controller can not operate at this rate continuously, LAN data will be lost. Additionally, if the system is to support loopback diagnostics, both transmit and receive must operate simultaneously, together transferring 2.5 Megabytes per second. Clearly, a garden variety DMA controller will not get the job done. Particular attention

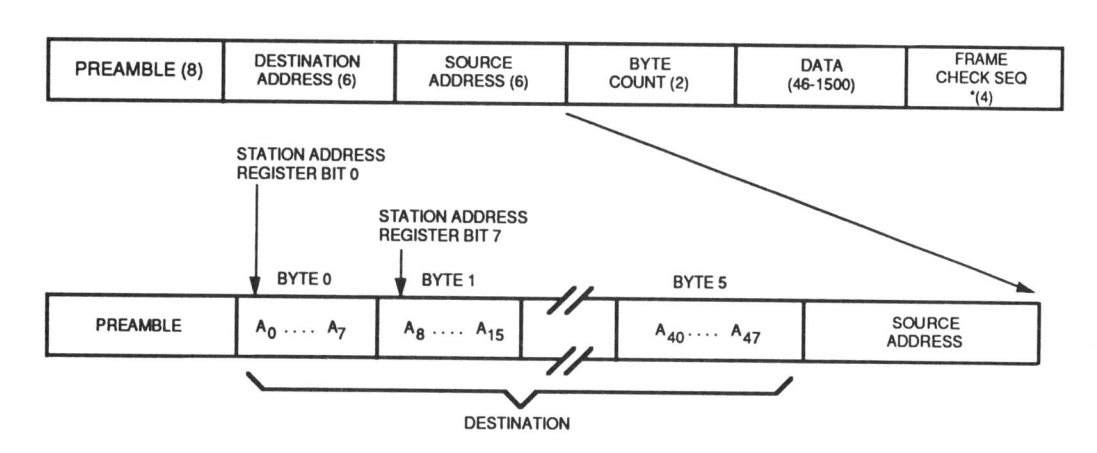

Figure 2. Ethernet frame format. Numbers in parentheses indicate the length of each field. Bits within a byte are transmitted and received LBS first and MSB last.

seeq Technology, Incorporated

must be paid to how long it takes the controller to acquire the system bus. If too long, Ethernet data will be lost.

Collision Effects

Collisions normally occur during transmission of the first 64 bytes of data. If packets are retrieved via DMA from system memory, when a collision occurs these 64 bytes must be retransmitted. This is an inefficient use of bus bandwidth.

An Ethernet Controller is a True Asynchronous Peripheral

Prudent system design calls for buffering any peripherals which are asynchronous in nature. Buffering makes the resource much more manageable at the system level.

Implementing a Local Buffer

Most currently available Ethernet controllers have a modest buffer built in, usually on the order of 16 bytes. This is sometimes adequate to handle system bus acquisition delay, but it does not make efficient use of bus bandwidth in three important areas:

1. Collisions during transmit. As network traffic increases, the probability of a collision increases. Each time a collision occurs the Ethernet controller must retransmit from the beginning of the packet. The time spent retransmitting due to collision uses bus bandwidth unnecessarily.

2. Frame check sequence (CRC) errors after receive. Since errors are not detected until after a packet has been received, bus bandwidth will be wasted when receiving packets with errors.

3. A significant number of receive packets are minimum size (64 bytes) yet contain much less than 64 bytes of information. For example, packet acknowledgments contain less than 20 bytes of information and are padded to the 64 byte minimum required. Transfer of these pad bytes over the system bus cannot be avoided without some large local buffer.

Supplementing the Controller Buffer

RAM can be added to the Ethernet board to add to the modest buffer already on the controller chip. Figures 4a and 4b show two possible ways.

The buffer should be at least 1514 bytes long. Static RAMs were chosen in Figure 4a to avoid having to include refresh control circuits in the dual port memory control logic.

The memory control must regulate access to the buffer by two buses: the system bus, and the data bus from the controller. The SRAMs are costly.

If DRAMs are used as in Figure 4b the cost is lower but they do require refresh circuitry in the memory controller.

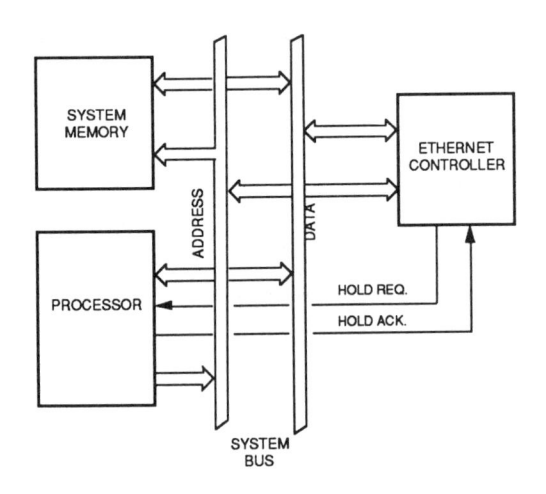

Figure 3a.

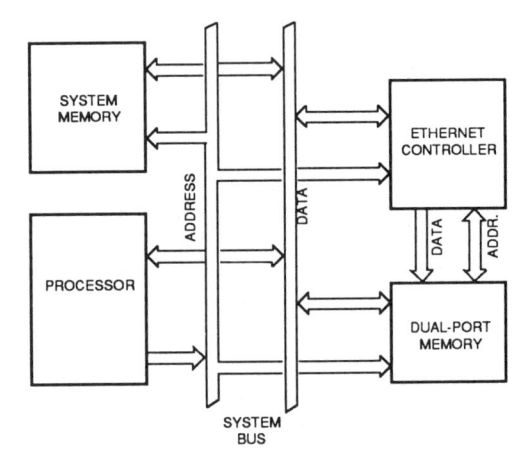

Figure 3b.

Local Buffering with the 8005

The 8005 Advanced Ethernet Data Link Controller combines several unique approaches to the problem of implementing an Ethernet connection. Look at the design in Figure 5.

First consider the local buffer: the 8005 is designed to work with 64K x 4 DRAMs which are readily available, and inexpensive. It has on board refresh circuitry, and just two DRAM chips provide 64 Kbytes of local buffer storage.

The 8005 treats the DRAM in a unique fashion: it multiplexes both address and data over eight lines. This saves on circuit board traces: only 12 lines are required to interface with the DRAMs, compared with 26 lines if static RAMs are used.

The 8005 also directly supports an address (EE) PROM, which allows for storage of the 8005's Ethernet address and configuration data.

The 8005 supports six unique station addresses. Thus, one physical connection on the Ethernet suffices for six logical connections. You could make effective use of this feature by, for example, connecting six devices to one Ethernet node, and controlling access to each device.

Figure 6 illustrates a cluster controller which services three printers and three PCs or terminals, and provides access to the Ethernet for the devices.

The printer controller services the cluster of three printers, and a low cost, low speed LAN provides coverage for the PCs. This LAN coverage may represent a relatively small geographic area, like a single corporate department. Note, however, that each device has access to the Ethernet, and each has a specific Ethernet address.

Design Examples

In this section, we'll briefly examine the way in which the 8005 can put two popular microprocessor bus formats on Ethernet, by way of using the Intel and the Motorola bus modes built into the 8005. Then we'll look in detail at a intelligent Ethernet controller which could realistically reside on a PC board, and usurp a minimal amount of resources from the system in which it is installed.

The Intel Mode

Figure 7 shows an implementation of the 8005 in an environment using an Intel processor. Note that BUSMODE is pulled up, indicating that the 8005 will produce Intel-compatible output signals, and accept inputs from an Intel bus. Also, in this example, we have selected a 16 bit bus, since BUSSIZE is high.

The Motorola Mode

In Figure 8, the 8005 is configured for use with Motorola processors, and the interface fits that processor family. BUSMODE is a ZERO, and we have specified a 16 bit bus, as before with the Intel mode.

A Board Level Ethernet Controller

Figure 9 illustrates a design using the Intel 80186 as a co-processor with the 8005, on the same PC board, to implement Ethernet. The 80186 is a particularly good choice for this application, because it has an on-chip DMA controller.

The 80186 has multiplexed address and data lines, here shown being demultiplexed by the latch. The data bus is buffered by the 74LS245s, but these may not be required, depending on the fanout required by the specific application.

The important signals between the two chips are the following; refer also to Intel 80186 and SEEQ 8005 data sheets.

Use DREQ from the 8005 into DRQ0 of the 80186. This is the highest priority DMA request on the 80186. Since the 80186 has no explicit DMA acknowledgment signal, you need to use the peripheral chip select signal: PCS1 is used as the DMA acknowledge, and PCS0 is the 8005 CS (chip select). The 8005 INTerrupt is connected to the 80186 INT0, and IACK of the 8005 is pulled up, since the 80186 does not provide for its use.

The RDY line of the 8005 is connected to the ARDY (asynchronous ready), since the two chips are each running off their own clocks. At the 80186, pull up SRDY (synchronous ready).

The 80186 does not provide a terminal count output, as do many other DMA controllers, to indicate to the 8005 to drop its DMA request. Therefore, when the 80186 Terminal Count Interrupt occurs, software must disable the DMA request in the 8005 by setting bit 11 in the command register.

Other Support Circuits

The 8005 supports a PROM, shown here as a 2804A E²PROM. The PROM is used primarily to store its Ethernet address and configuration data, but other convenient data may be stored there too.

The 8005 supports the TI TMS 4464 DRAMs (or equivalent) with a minimum of PC board circuit traces by multiplexing both address and data lines to the DRAMs. Two

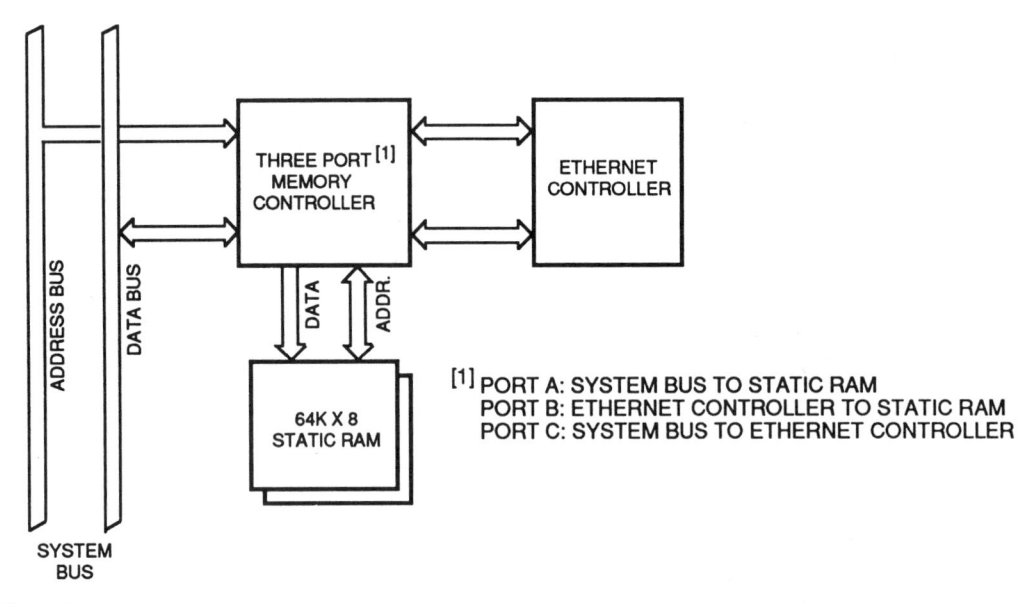

Figure 4a.

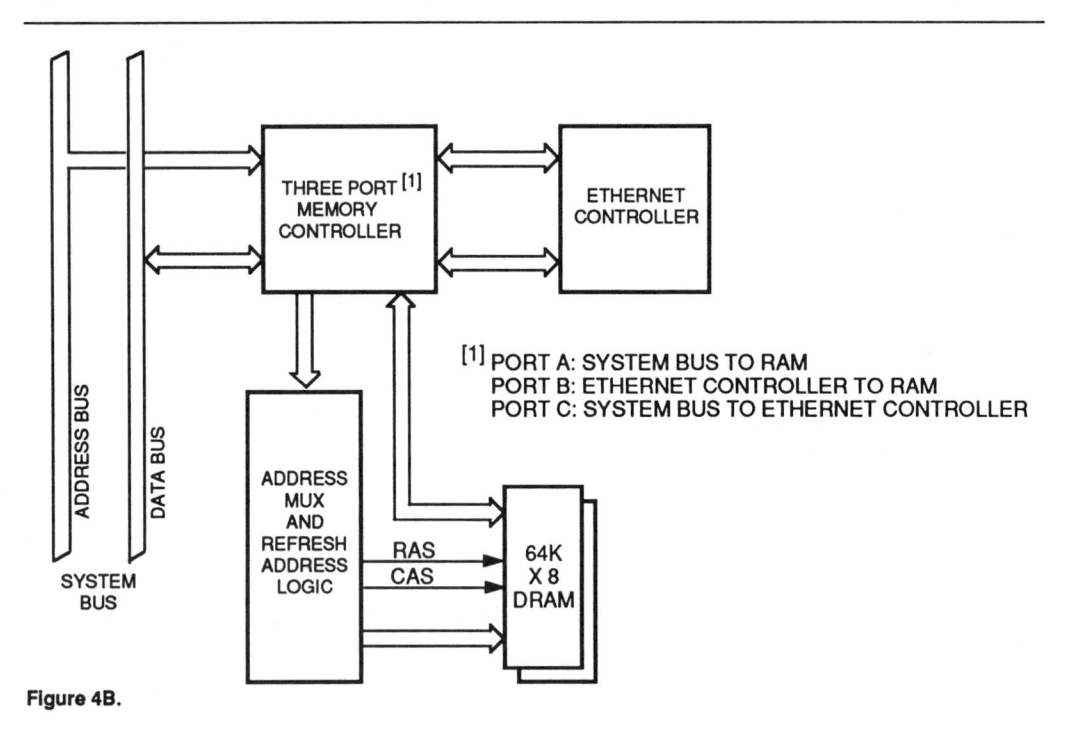

[1] PORT A: SYSTEM BUS TO STATIC RAM
PORT B: ETHERNET CONTROLLER TO STATIC RAM
PORT C: SYSTEM BUS TO ETHERNET CONTROLLER

Figure 4B.

[1] PORT A: SYSTEM BUS TO RAM
PORT B: ETHERNET CONTROLLER TO RAM
PORT C: SYSTEM BUS TO ETHERNET CONTROLLER

Figure 4. Implementing a local bufer for Ethernet traffic, using static RAM(a), and dynamic RAM (b). DRAMs are lower in cost, but require refresh circuitry.

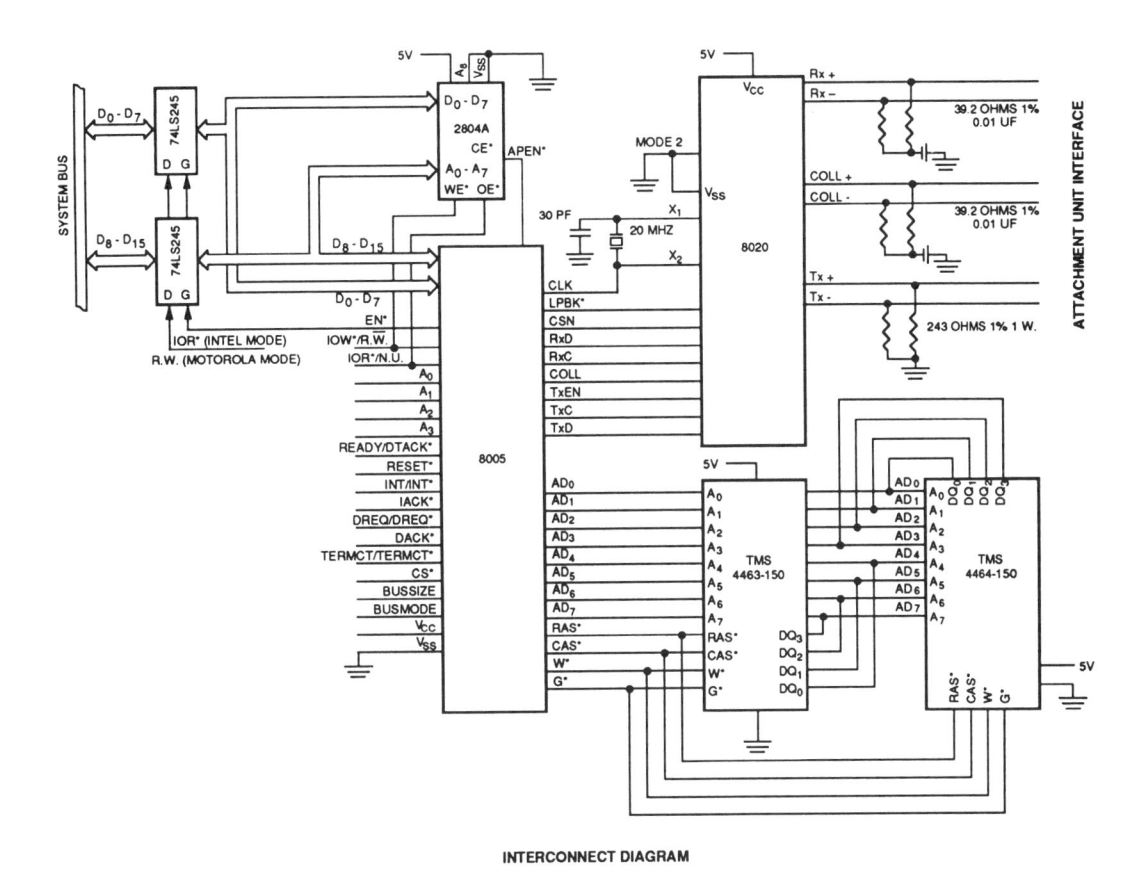

Figure 5. The 8005 Advanced Ethernet Datalink Controller: it supports a local buffer via DRAM, keeps its Ethernet address and configuration data in its own on-board PROM, and provides a very flexible and sophisticated link between your system and Ethernet.

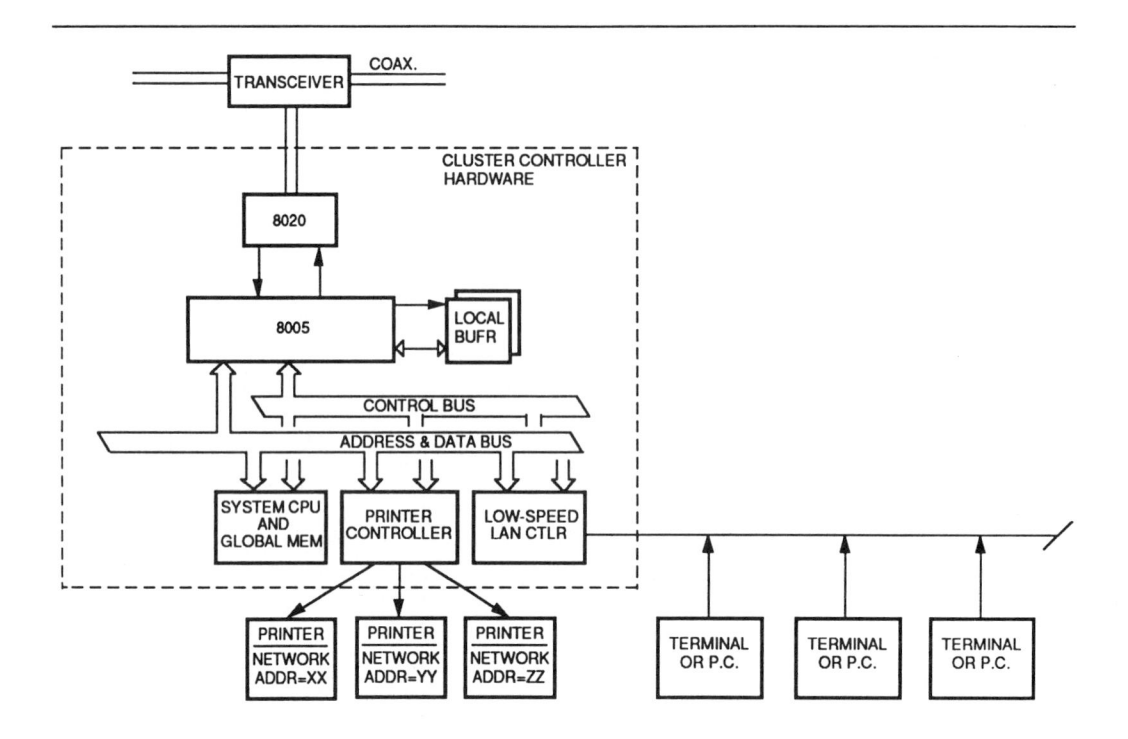

Figure 6. You can connect up to six devices to one Ethernet node using the capability of the 8005 to decode up to six station addresses. In this example, three printers and three PCs or terminals are connected to one Ethernet Node. The 8005 and its system CPU controls Ethernet access to and from the devices.

DRAM chips provide an ample 64 Kbytes of Packet Buffer storage. The 8005 allows you to partition this buffer into receive and transmit areas of your own choice.

Finally, the diode RC network provides a power on reset pulse (minimum 10 microseconds wide) for both the 8005 and 80186.

The 8005 in Non Ethernet Applications

The Ethernet, because of its simplicity and high speed, is often used in smaller physical configurations than those for which it was originally intended. Applications include communications between processors in a large parallel processing engine.

The 8005, because of its configurability, can be "trimmed down" for use in networks which need not strictly follow the Ethernet format.

The Ethernet address is six bytes long. The 8005 may be configured to accept just a 2 byte address, saving four bytes per address in a packet. Since there are two address fields per packet (destination and source), eight bytes are saved.

Ethernet specifies a minimum "slot time" of 51.2 microseconds. This represents the time required for one round trip of a packet on a maximum length cable, and is required for reliable collision detection. The 8005 may be configured for a slot time of 12 microseconds, which shortens waiting time after a collision. Additionally, when you select the shorter slot time, the 8005 automatically reduces the Collision Jam Pattern from 8 to two bytes, and reduces the interframe spacing from 9.6 to 2.4 microseconds.

Refer to the 8005 data sheet for more detail on selecting these optional parameters.

Configuring the 8005

This step is required following hardware reset or software reset. Note that a hardware reset must be provided following power on. Following reset, allow 10 microseconds after the reset before attempting access to the part.

1055

INTERCONNECT DIAGRAM
8005, 16 BIT BUS, INTEL MODE

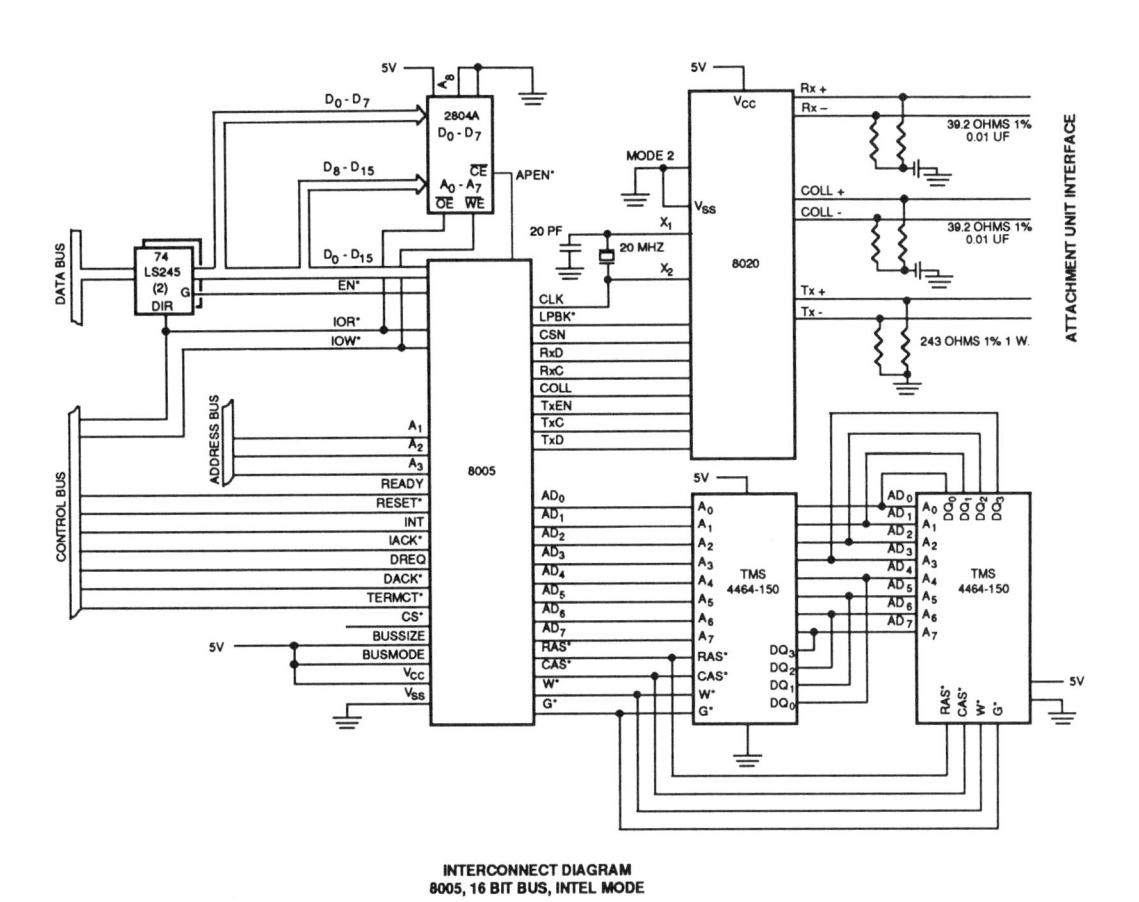

Figure 7. The 8005 interfaced with an Intel processor. This example illustrates the use of a 16 bit bus, since BUSSIZE is a ONE.

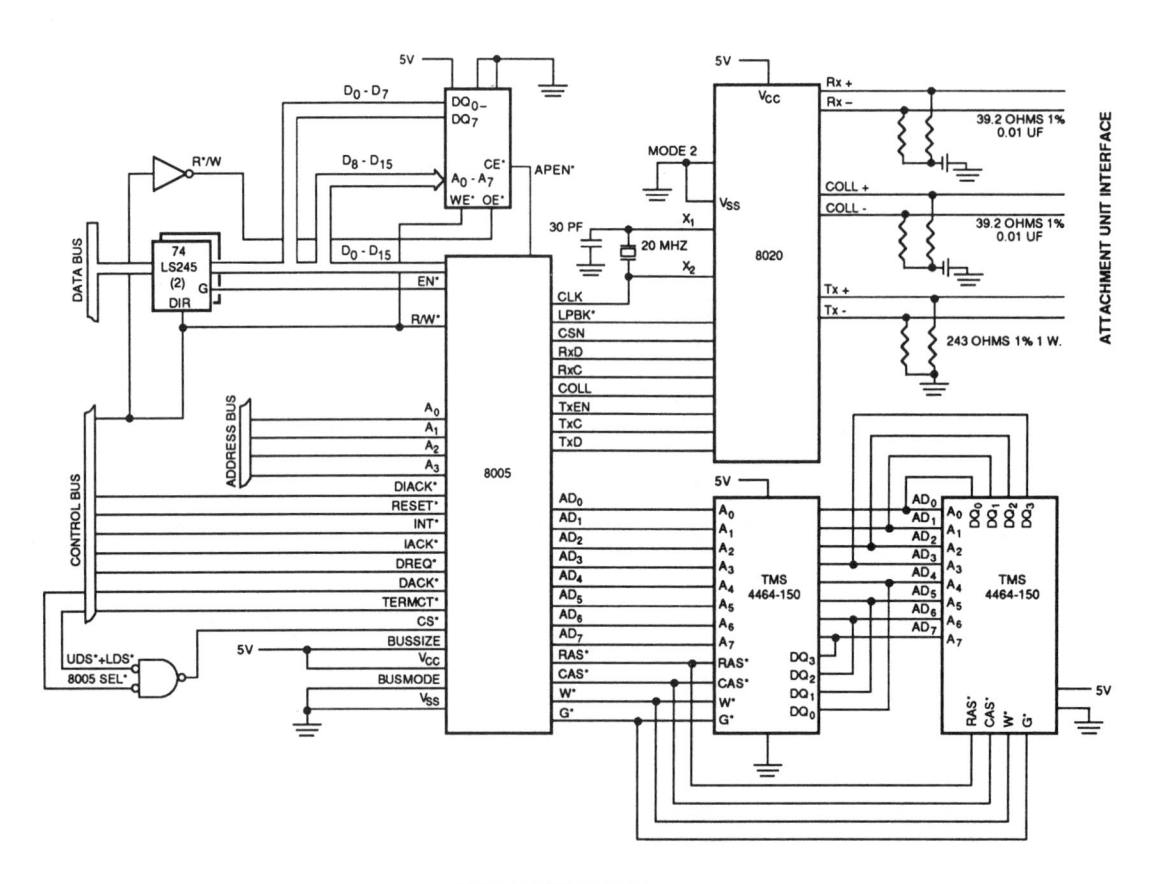

Figure 8. The 8005 in a Motorola environment, and with a 16 bit bus size.

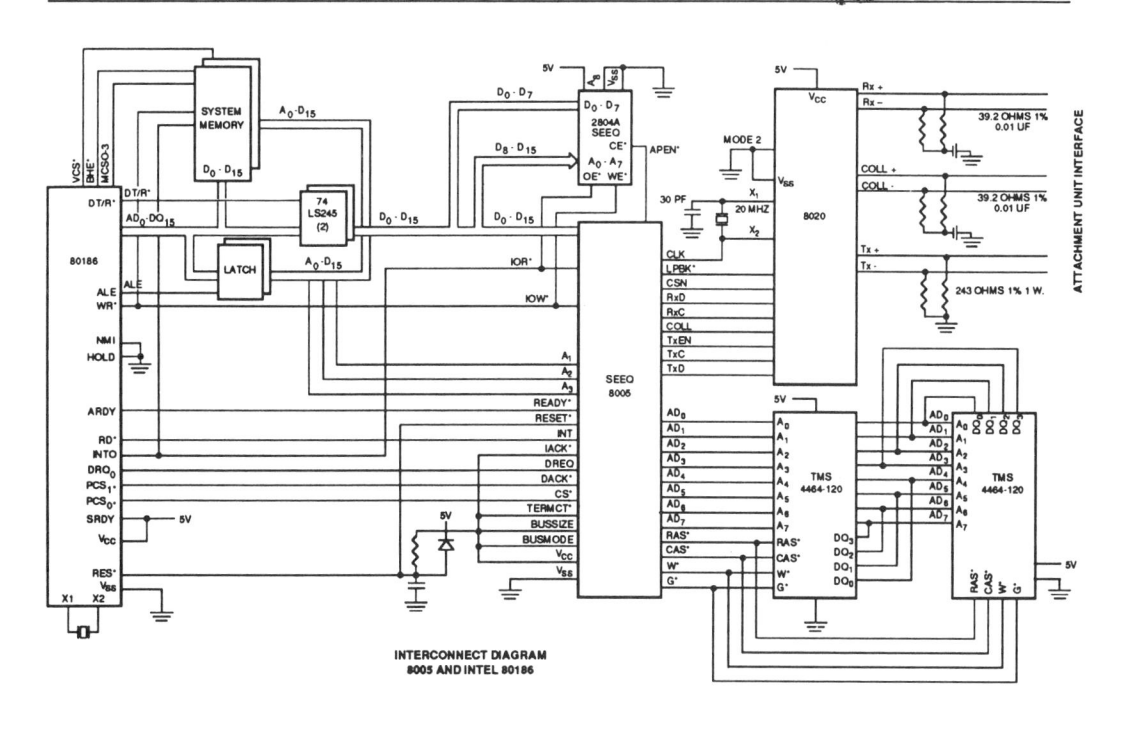

Figure 9. The use of the 8005 and the Intel 80186 to implement a board level intelligent Ethernet data link. The 80186 is a good companion for the 8005, since it has an on-chip DMA controller. The 8005 supports an address PROM, and 64 Kbytes of DRAM to serve as a local Packet Buffer.

Configuring includes loading the Ethernet station address(es), selecting transmit and receive packet buffer size and defining interrupt conditions and an optional interrupt vector.

All this information may be stored in a PROM on the same PC board as the 8005. This allows the assigned Ethernet station address(es) to travel with the board.

Register Architecture

The general approach to initializing the 8005 consists of reading information from the PROM into system RAM and writing it back into several registers inside the chip. See Figure 10, which depicts the Register Model of the 8005.

There are nine 16-bit registers which are directly accessible by using the signals Chip Select, I/O read, I/O write and A_1 through A_3. There are also four registers which are selected by the buffer window code bits and accessed indirectly through the buffer window register.

In the discussion below, note that the 8005 has been configured for a 16 bit bus. Input A_0 (pin 54) is ignored when in 16 bit mode, and is shown as a "Don't Care" (X). In 8 bit mode, A_0 selects the low order byte when a ZERO, and the high order byte when a ONE.

Reading the Address (EE) PROM

After reset, if you are using a local Address PROM, write that location to the DMA Address Register which points to the first configuration byte in the PROM. Select access to the Address PROM by writing 0006 to the Buffer Code Bits in Configuration Register #1. The 8005 will then drive the chip enable line of the PROM via APEN (pin 10) for each Read or Write to the Buffer Window Register. When all configuration and station address bytes have been moved into system RAM, the next step is to write them into the 8005.

Loading Indirect Registers

Indirect registers are selected by the buffer code in Configuration Register #1 and accessed through the buffer window register. All indirect registers are 8 bits wide and therefore only use data bits D_0D_7.

Station Address Registers

To load the station address registers, select the desired station address register set by writing a value from 0000 to 0005 to Configuration Register #1. Then write the appropriate 6 byte address to the buffer window register, one byte at a time, with the most significant byte first, and the least significant byte last. Each write automatically increments an internal pointer register to the next byte of the station address. Repeat this process until you have loaded all desired station address registers.

Specify Transmit Buffer Size

Write a 0007 to Configuration Register #1 to select the Transmit End Area register. Write an 8 bit value to the Buffer Window register which specifies the most significant byte of the last address in the Transmit Buffer space.

For example, to define space for four packets, each 1514 bytes long:

```
1514 X 4 =   6056  bytes for data
   4 X 4 =     16  bytes for header
              ────
              6072  bytes required;
6072/256 =       23+, or hex 0017
```

Thus, we would write hex 0017 to the transmit end area register. This also sets the receive buffer area, by default, to start at hex 1800, which leaves 58 Kbytes (hex FFFF minus hex 1800) for receive packets.

If interrupts will be enabled and an interrupt vector is required, write a 9 into Configuration Register #1 to select the Interrupt Vector Register, and then write the 8 bit interrupt vector into the Buffer Window Register.

Specify Receive Buffer Size

Write an 8 bit value into the least significant byte of the Receive End Area Register to specify the most significant byte of the last buffer address for receive packets. This would normally be hex FF if the rest of the local buffer is to be used for received frames.

Loading Direct Access Registers

Initialize Transmit Pointer Register

Write 0000 to this register.

Configuration Register #1

Loading this register defines receiver match modes, enables station address register sets and sets up DMA burst interval and size. Access this register by setting $A_3 - A_0$ to 001X.

Configuration Register #2

Following reset, this register is configured for IEEE 802.3 compatible network interface. It contains bits to select non-IEEE 802.3 network operation, diagnostic modes (CRC enable/disable for both receive and transmit), enable receiving packets with errors (short frames, dribble errors, CRC errors, overflow errors), select byte order for 16 bit bus and enable automatic receive end area update.

Initialize Receive Pointer Register

Load this register with the same value as the Receive Start Area (16 bit Transmit End Area address plus hex 0100). Save this value, since it points to the first byte of the next packet header, and you will need it to find the next received packet.

In the example above, the Transmit End Area address was hex 17FF. Therefore, the Receive Pointer Register should be loaded with hex 1800.

Initialize DMA Address Register

If no packets are to be loaded into the transmit area, load this register with the contents of the Receive Pointer Register.

Command/Status Register

Set RxON (bit 9), and, if desired, RxINT Enabl (bit 1) to ONEs. If you are not using interrupts, you may poll RxINT (bit 5) to see if a frame has been received.

Transmitting a Frame

This discussion assumes that the system is connected to an IEEE 802.3 compatible network. The contents of a Transmit frame have no meaning to the Packet Buffer Controller and the Ethernet Data Link Controller circuitry, and can be arbitrary in length and content. As discussed above, transmission of the Preamble and CRC (frame check sequence) can be suppressed under software control for specialized network requirements or diagnostic tests.

After you have gone through the configuring as outlined above, the 8005 is ready to receive or transmit frames. Refer to Figure 2 and recall that a frame consists of from 64 to 1514 bytes, which includes a 6 byte destination address, a 6 byte source address, and an area for data all of which is supplied by your system software. The entire frame has a prefix containing a 62 bit preamble (which synchronizes the phase-locked-loop in the Manchester Code Converter with respect to the received packet), and a 2 bit start frame delimiter. Following the data field there is a 4 byte frame check sequence. All of the components of the prefix and the CRC are supplied by the 8005.

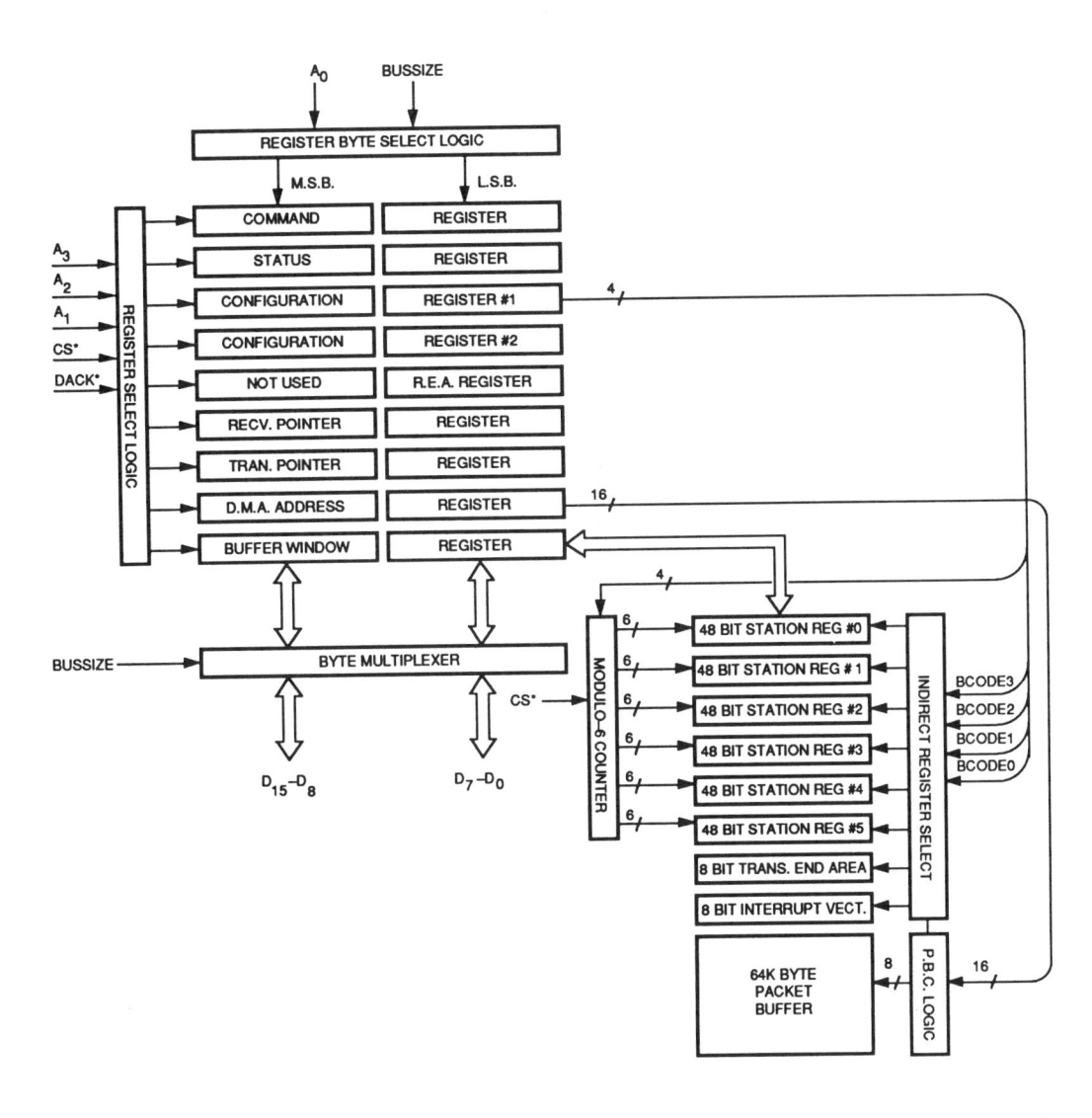

Figure 10. Register Model, which illustrates the register architecture inside the 8005. Using both directly and indirectly accessable registers lowers pin count. All access to indirect registers and the Packet Buffer is through the Buffer Window Register.

A packet is prepared for transmission by writing into the Transmit Buffer Area a 4 byte header, followed by the destination address, the source address, and finally the data field. Refer to Figure 11. You may choose to do this via programmed I/O, or via an external DMA controller. Frames may be chained together up to the capacity of the available Transmit Buffer Area by using the Next Packet Pointer (first two bytes) and the Chain Continue bit (bit 6) in the Transmit Header Command byte.

Refer to Figure 12. Read the Status Register to see if the DMA FIFO direction is set to write to the Packet Buffer (bit 15 cleared). If it is and the DMA register is not going to be loaded with a new value then data can be written immediately. If the DMA register is to be changed, then check to ensure that the FIFO is empty (Status Register bit 4 set). If the FIFO is not empty, continue testing bit 14 until the FIFO is empty. If you change the FIFO direction or write to the DMA Register, FIFO contents will be cleared.

If necessary, load the DMA Register with the address for the first byte of the Packet Header, and write Packet Header and data into the FIFO. The first Packet Header address is normally 0000.

Figure 13 depicts the same operation, only under DMA control, After you set up the system DMA controller, set DMA ON (Command Register, bit 8), and DMA Interrupt Enable (Command Register, bit 0), if desired. The former enables the DMA request logic, and the latter causes an interrupt to be generated at the completion of a DMA operation i.e., when terminal count has been input.

After all of the packets in a given chain have been written into the Transmit Buffer Area, load the Transmit Pointer Register with the address of the first byte of the first transmit packet header, set TxON (bit 10) and, optionally, TxINT Enabl (bit 2) to ONEs in the Command/Status Register.

The 8005 will then read the first header, which is pointed to by the Transmit Pointer Register, and process that packet, and all additional packets in the packet chain in turn. Any retransmission of a packet due to a collision will be automatically handled by the 8005, thus relieving your system from having to transfer that packet of data more than once.

When a packet has been successfully transmit ted (or 16 collisions occur), the Done bit (bit 7) in the transmit header status byte will be set to a ONE. The Transmit Buffer Area occupied by that packet is now available for another packet, and may be written to at the same time as subsequent packets are being transmitted. The 8005 will move to the next packet in the chain.

When all packets in a chain have been completed (transmitted successfully or collided 16 times), the 8005 resets TxON (bit 10) in the status register to indicate that it is

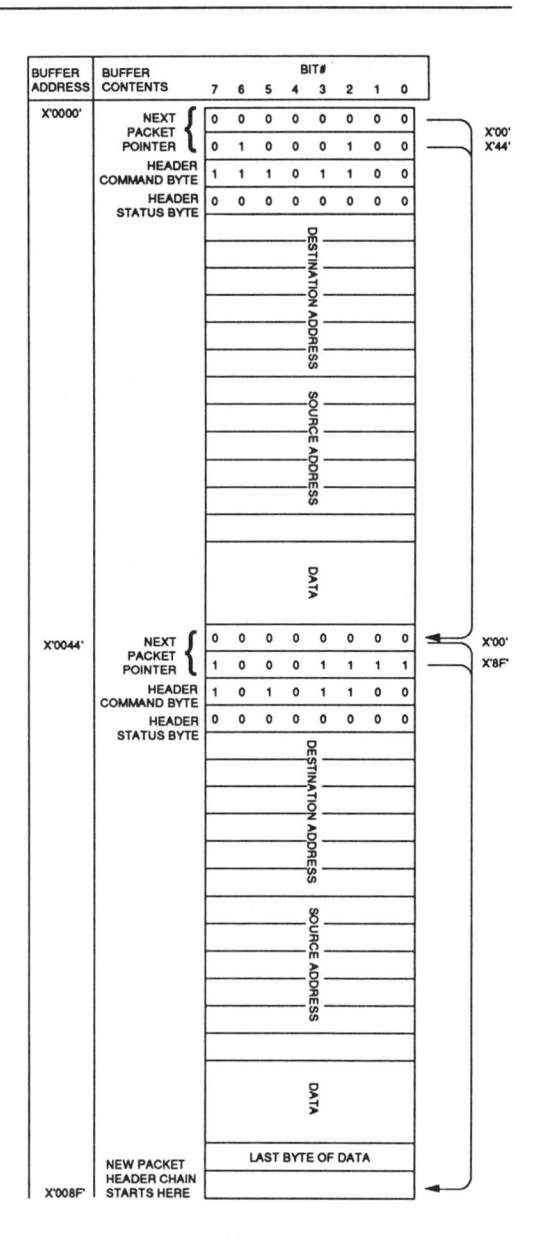

Figure 11. Transmit Packet Chain, residing in the Packet Buffer, and ready to be transmitted. Two packets are in this chain. Note that the Packet Buffer is nondestructively read, and the packets are still in the buffer after they have been transmitted. after transmission, the 8005 updates the Header Status Byte (byte 4). The first two bytes of the Packet Header point to the address of the first byte of the second Packet Header.

ready to transmit another packet chain. If 16 collisions occur on a packet, the 8005 stops transmission attempts for that packet only and moves to the next packet in the chain, if one exists. In the example in Figure 11, bits 2 and 3 are ON in the transmit header command byte which will cause the 8005 to set the transmit interrupt bit in the status register and, if enabled, interrupt the processor when 16 collisions occur or the transmission is successful.

The last packet in the chain is denoted by having the Chain Continue bit cleared to a ZERO. The Next Packet Pointer points to the address follow ing the last byte of the last packet.

You may treat the transmit packet buffer in one of two ways:

1. As a circular buffer with wraparound, where you remember the address to load new packet headers and packet data. The DMA register automatically wraps around to address 0 when the transmit end area has been reached.

2. As a linear buffer, where you reset the transmit pointer to 0000 after each packet chain transmission.

Receiving Frames

Once the 8005 has been configured and the receiver enabled, frames which meet the match mode and station address requirements specified in Configuration Register #1 and the enable bits 2 - 5 in Configuration Register #2 will be moved into the Receive Buffer Area beginning at the address contained in the Receive Pointer Register.

When one or more packets are available in the receive area, the 8005 sets Rx Interrupt (bit 5) in the Command/Status Register to a ONE. If receive interrupts are enabled. (Command Register bit 1 set), then the external interrupt (pin 11) is asserted. Frame header and data can now be read by loading the DMA Register with the start ing address of the Packet Header and executing successive

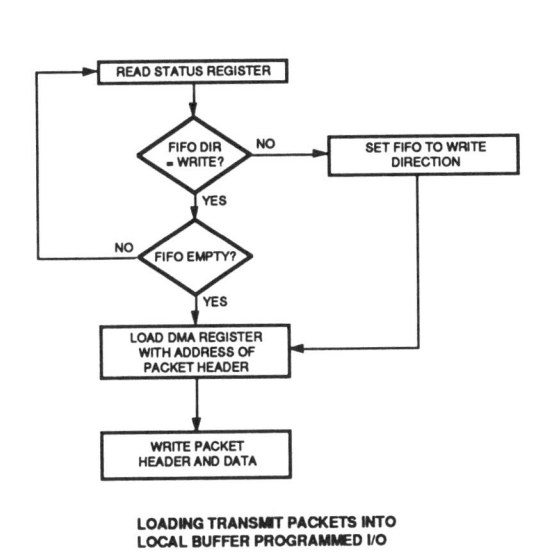

Figure 12. Loading Transmit Packets into the local Buffer, under Programmed I/O conditions. Note that, if you change the direction of the DMA FIFO, or load the DMA Pointer Register, you will lose any data stored in the FIFO.

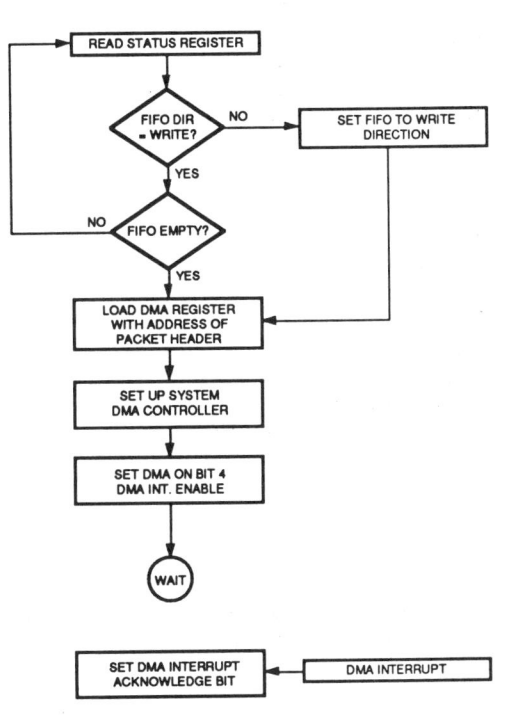

Figure 13. Loading Transmit Packets into the Local Buffer under DMA transfer, using Interrupt.

reads. If Auto Update REA (bit 1 of Configuration Register #2) is set, the Receive End Area Register will be updated with the upper byte of the DMA register each time a DMA read occurs. This releases buffer space as its contents are read, and allows for the receipt of more data at the same time as data is being read out.

The action taken on a receive packet depends on the status of the packet and its contents. If the packet status is bad, it may be skipped entirely without transferring any of its data to system memory by loading the Receive End Area Register with the most significant byte of the next packet pointer. This will release the buffer space of the previous packet for future packets. In like fashion, if the packet data shows it to be an "overhead" packet (such as a Packet Acknowledgement), this can be so noted in network software and the packet skipped. Thus, unnecessary transfer of the packet over the system bus can be avoided, and system bandwidth preserved. If the packet data must be processed, just the information portion of a packet (exclusive of any bytes used to pad the packet to a minimum size) can be read to system memory by programmed I/O or by an external DMA controller.

Receive Packet Chaining

The 8005 automatically chains together receive packets using a circular FIFO buffer structure. Each packet is prefaced by a 4 byte header whose first two bytes form a 16 bit address that points to the next header. A chain of packets always ends with a header-only packet whose 4 bytes equal 00. The address of this header-only packet should be saved, since it will contain the header of the next packet received. It is a simple matter to follow the packet chain from header to header until the chain Continue/End bit is read as a ZERO, calculate the length of the chain and set up the DMA Register and an external DMA controller to transfer the entire chain of packets to system memory if desired. This is advisable in applications where high average receive data rates are expected and data must be moved quickly from the local buffer to the system memory at the expense of bus bandwidth. To minimize system bus utilization, packets can be moved one at a time; this permits moving only the information content of a packet.

Calculating Packet Chain Length

In order to perform a DMA transfer, you need to give the DMA controller the "count"; i.e., how many bytes (or words, in a 16 bit system) will be transferred. To do that, you need to calculate how many bytes are available in the Packet Buffer as a result of receive activity.

Refer to Figure 17. This flow chart illustrates the steps required to calculate the length of the packet chain.

The first step requires that you know the Packet Buffer address of the last packet header read in the most previous receipt of Ethernet data. If the 8005 has just been initialized, the address is the beginning of the Receive Packet Buffer which was determined earlier in this note (hex 1800). If packets have been previously been read this address will be the location of the header last read that had the chain continue/end bit reset.

The next step, referring to Figure 17, is to turn off the Auto Updat REA (Configuration Register #2, bit 1). This insures that the 8005 will not use the area occupied by this packet chain for new receive data.

Read each Packet Pointer in turn, and then read the Header Status byte immediately after the Pointer, which is Byte #3. Bit 6 of Byte #3 is the Chain Continue bit. Continue reading this bit in each packet header until this bit goes to ZERO. This signals the end of the chain. Save the local buffer address of the first byte of this last header as this is the address of the header for the next packet received. Subtract the address of the first header in the chain from this address. If the result is a positive number, you have the chain length directly.

If the result is negative it denotes that the Receive Pointer Register has wrapped around past the beginning address of the receive area. The chain length will be equal to the sum of the receive buffer size plus the value (including sign) of this result. You already know the buffer size, since you defined it during configuration of the 8005: hex FFFF minus the receive start address (defined during configuration) plus 1. For the previous example, the buffer length is hex E800 (FFFF - 1800 + 1). Load the chain length into the DMA controller, and set Auto Updat REA. You are now ready to read data out of the receive buffer and into system memory.

There are two ways to read Packets out of the Local Buffer:

1. Via programmed I/O.
2. Via DMA transfer.

The front end portion of each procedure is the same: first, check to see if the FIFO is empty; then set it to Read. If the FIFO is not empty, check to see if it is in the Write direction. If not, load the DMA Register with the address of the next Packet Header. If this is the first Packet to be read, this address will be that which was derived when you defined the Transmit Buffer size during configuration of the 8005.

Reading Packets Using Programmed I/O

The data path between the local buffer and the host bus is buffered by a 16 byte FIFO called the DMA FIFO. It serves as a rate buffer between the host and the local buffer, especially for 16-bit data transfers. Because the local buffer is a shared resource (there are 4 ports including the DRAM refresh port), the initial read from the buffer window which follows loading the DMA register may take eight microseconds worst case. The 8005 signals this delay by deasserting Ready (if Busmode = 1) or delaying Dtack (if Busmode = 0). If this initial read wait state is unacceptable, then the buffer window interrupt feature can be used. The buffer window interrupt is asserted for programmed I/O reads (not DMA reads) when the DMA FIFO has data available.

Under Programmed I/O control (see Figure 14), after you load the DMA Register, read Status Register bit 7, Buffer Window Interrupt or wait for a hardware Buffer Window Interrupt if it is enabled. When the interrupt is asserted, read Packet Header and data out of the receive FIFO, via the Buffer Window, until all bytes have been transferred.

Reading Packets Using DMA

The second approach is by DMA transfer. See Figure 15. After loading the DMA Register, load the system DMA controller with the destination address in system memory, and the previously calculated packet chain length. Then set DMA ON (bit 8 in the Command Register). This enables the DMA Request logic inside the 8005. Optionally, set DMA Interrupt Enable, which will cause an Interrupt to be generated when the DMA controller has asserted Terminal Count. The DMA Request output signal will be asserted when there are a sufficient number of bytes in the DMA FIFO to satisfy the DMA Burst Size (2, 4, 8, or 16 bytes) which you selected earlier when configuring the 8005.

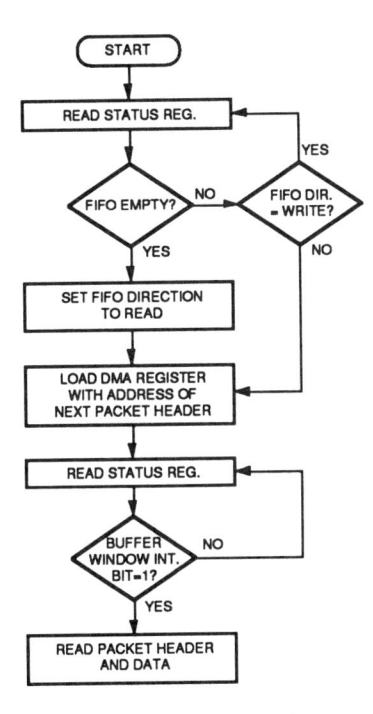

READING A PACKET FROM LOCAL BUFFER PROGRAMMED I/O

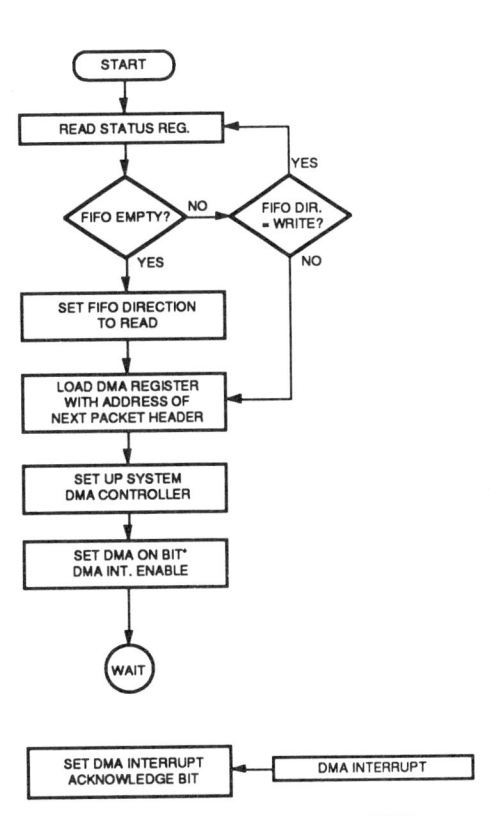

READING A PACKET FROM LOCAL BUFFER DMA TRANSFER WITH INTERRUPT

Figure 14. Reading Packets out of the FIFO using the Programmed I/O procedure.

Fgure 15. Reading the Local Buffer under DMA control.

Interrupts

There are several interrupt sources in the 8005. This section describes these interrupts and how to service them. For this discussion, refer to Figure 18.

Transmit Interrupts

There are four transmit interrupt sources in the 8005; Babble, Collision, 16 Collisions, and Transmit Success. Each of these can set the transmit interrupt bit in the status register if so programmed in the transmit header command byte. If T_x Interrupt Enable (Command Register bit 2) is set, the 8005 will also assert an interrupt on pin 11. The transmit interrupt is cleared by setting TxINTACK (bit 6) in the command register.

Babble Interrupt

The 8005 will transmit packets as large as will fit in the transmit buffer. The IEEE 802.3 standard specifies a maximum packet size of 1514 bytes. The babble interrupt indicates that a packet larger than 1514 bytes was transmitted.

Collision Interrupt

When a packet collision occurs, the 8005 packet buffer controller automatically restores its transmit pointer to the beginning of the packet and schedules retransmission following the back off time. In some applications it may be desirable to record the number of collisions that occur. This bit enables setting the TxINT bit in the status register for each collision.

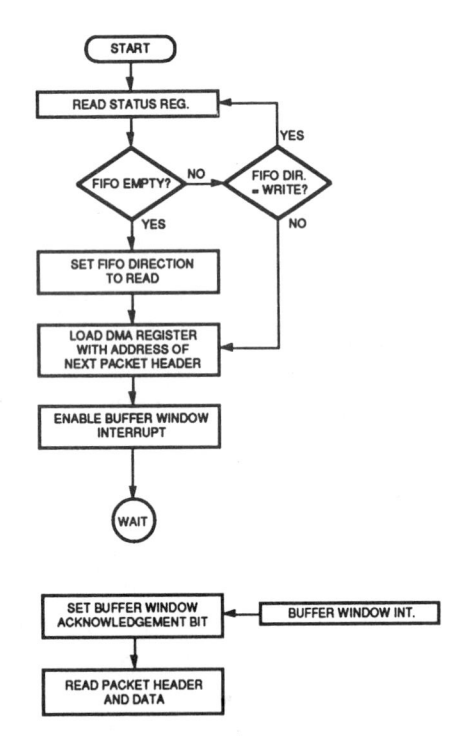

Figure 16. Reading a Packet from the local Packet Buffer using the Buffer Window Interrupt approach.

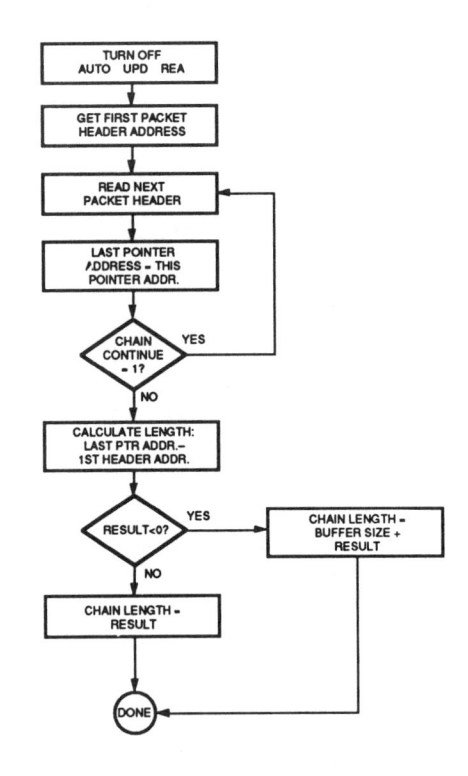

Figure 17. The steps necessary to calculate the length of a Packet Chain. You need to save address of the last header in the last the packet read, in order to perform the calculation.

seeq Technology, Incorporated

8-44

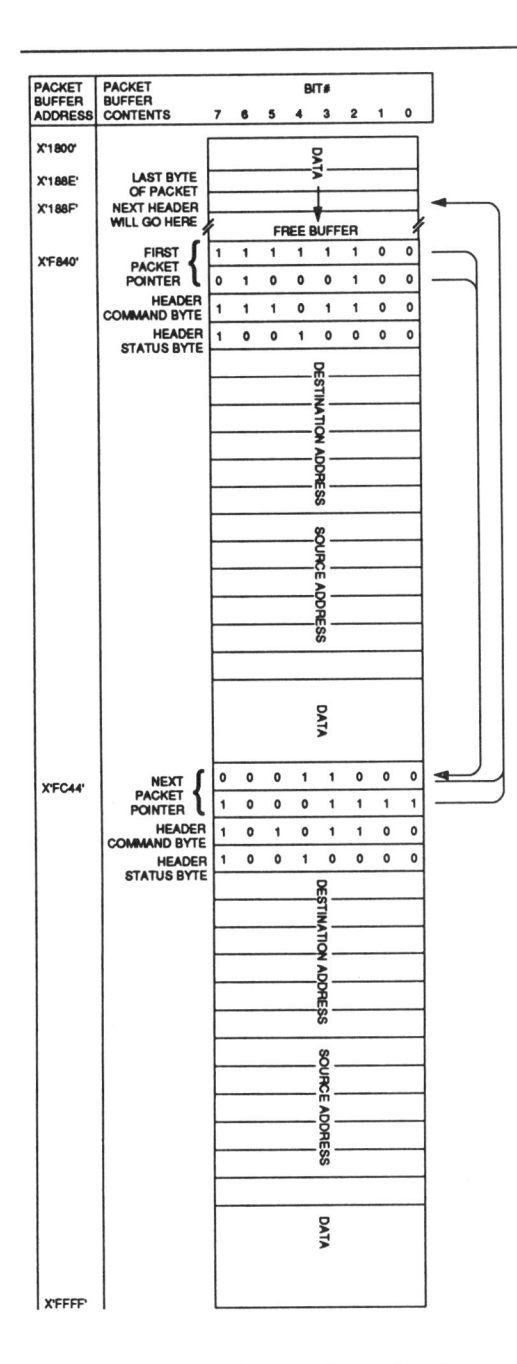

PACKET BUFFER ADDRESS	PACKET BUFFER CONTENTS	BIT# 7 6 5 4 3 2 1 0
X'1800'		DATA
X'188E'	LAST BYTE OF PACKET	
X'188F'	NEXT HEADER WILL GO HERE	
		FREE BUFFER
X'F840'	FIRST PACKET POINTER	1 1 1 1 1 1 0 0 0 1 0 0 0 1 0 0
	HEADER COMMAND BYTE	1 1 1 0 1 1 0 0
	HEADER STATUS BYTE	1 0 0 1 0 0 0 0
		DESTINATION ADDRESS
		SOURCE ADDRESS
		DATA
X'FC44'	NEXT PACKET POINTER	0 0 0 1 1 0 0 0 1 0 0 0 1 1 1 1
	HEADER COMMAND BYTE	1 0 1 0 1 1 0 0
	HEADER STATUS BYTE	1 0 0 1 0 0 0 0
		DESTINATION ADDRESS
		SOURCE ADDRESS
		DATA
X'FFFF'		

Figure 17a. Example of two receive packets in a packet chain with wraparound.

16 Collisions Interrupt

The 8005 counts the number of collisions that occur on each packet. If a packet has collided 16 times, the usual cause is a network fault such as an unterminated coaxial cable or an open in the cable. This interrupt notifies the host that a packet has collided 16 times, and the packet buffer controller will now abandon transmit attempts for that packet and move on to the next packet in the chain if one exists.

Transmit Successful Interrupt

This interrupt indicates that a packet was successfully transmitted with less than 16 collisions.

Receive Interrupts

The 8005 sets the receive interrupt bit (status register bit 5) whenever a packet that meets the criteria in bits 2 - 5 of Configuration Register #2 has been placed in the local buffer. It will remain set and, if the receive interrupt enable bit is also set, the external interrupt will remain asserted until the receive interrupt acknowledge bit is set. If a separate interrupt for each packet is desired, the receive interrupt should be acknowledged within 70 microseconds, which is the minimum time for receipt of a subsequent 64 byte packet. If more than 70 microseconds elapses before acknowledging a receive interrupt, it is possible for additional packets to be added to the packet chain.

The 8005 protects the receive interrupt condition such that if a new interrupt is being generated while the host is setting the receive interrupt acknowledge, the receive interrupt will persist. If, however, a new frame is received after the interrupt acknowledge and before the calculation of the packet chain length, the packet chain which is read will include the new packet associated with the new interrupt. The new interrupt, when serviced, will now be associated with an empty packet since it was part of the previous chain.

DMA Interrupts

The DMA interrupt bit in the status register is set following receipt of terminal count from the external DMA controller. If the DMA interrupt enable bit (command register bit #0) is also set, an external interrupt will be asserted. The interrupt is cleared by writing a 1 to the DMA interrupt acknowledge bit.

Self-Test and Network Diagnostics

The 8005 contains a number of special features for self-test and network diagnostic support.

Loopback

Two forms of loopback are possible with the 8005. Local loopback is accomplished when the 8005 is connected to an 8020 Manchester Code Converter. When bit 11 of Configuration Register #2 is set, the loopback pin of the 8020 will be brought low. This causes transmitted data to be looped back to the receiver of the 8020. If the packet transmitted meets the match mode and is addressed to one of the 8005's enabled station addresses, it will be received and placed in the local buffer. Using diagnostic control bits 9 and 10 in Configuration Register #2, it is possible to transmit packets with CRC errors to check the receive CRC logic, and to include the CRC in a receive packet to check the transmit CRC logic. Loopback can also be accomplished by connecting the 8020 to an Ethernet transceiver. Because the network is half-duplex, any data transmitted will also be received. Thus the same loopback test as above can be performed while the network is active by simply sending a packet to oneself.

Interrupts

The 8005 has separate control bits for turning on an off the receive logic, transmit logic and DMA logic. The interrupts for these functions can be tested without actually performing the function by setting both the on and off control bits simultaneously. For example, if the receive interrupt logic is to be tested set both RxON and RxOFF bits in the command register. This will cause the receive interrupt bit in the status register to be set and, if the receive interrupt enable bit is also set, will cause an external interrupt. This mode has no effect on any logic other than the interrupt logic and associated status register bit, i.e., packets can be transmitted and received while this diagnostic mode is set.

Detecting Network Cable Faults

It is possible to make a gross determination of cable faults by taking advantage of the full-duplex nature of the 8005: although it will not transmit while receiving (that would violate the Ethernet specification), it does receive while transmitting, as long as the packet destination address fits the receiver match mode.

Cable Opens/Missing Terminator

An open coaxial cable or a missing cable terminator results in the transmission line being terminated in an infinite impedance. Thus, any data transmitted will be reflected back from the impedance mismatch some time delay after it is transmitted. This time delay depends on the physical distance to the impedance mismatch, so the length of the packet must be large enough to insure that data are still being transmitted after one round trip propagation delay to the mis match. A 256 byte packet should be an adequate size. The reflected signal will partially cancel the transmitted signal and cause a collision to be detected by the transceiver. Thus an open is indicated by repeated collisions when transmitting a packet or, if the network is known to be quiet (no other nodes active), a single collision when transmitting. It is also possible to make a rough determination of where the fault is by enabling receipt of packets with errors (Configuration Register #2 bits 3 - 5) and then counting the number of bytes correctly received. Note that if the cable open is very close to the transmitting node, the collision may occur during the preamble and the 8005 would unconditionally reject the receive packet.

Cable Shorts

A shorted coaxial cable causes premature loss of carrier sense to the receiver of the 8005 while it is transmitting. It is therefore possible to send a packet of at least 256 bytes to oneself with the receiver enabled to accept frames with errors. A cable short results in a truncated receive packet; the size of the receive packet indicates the rough distance to the cable short.

seeq Technology, Incorporated

8-46

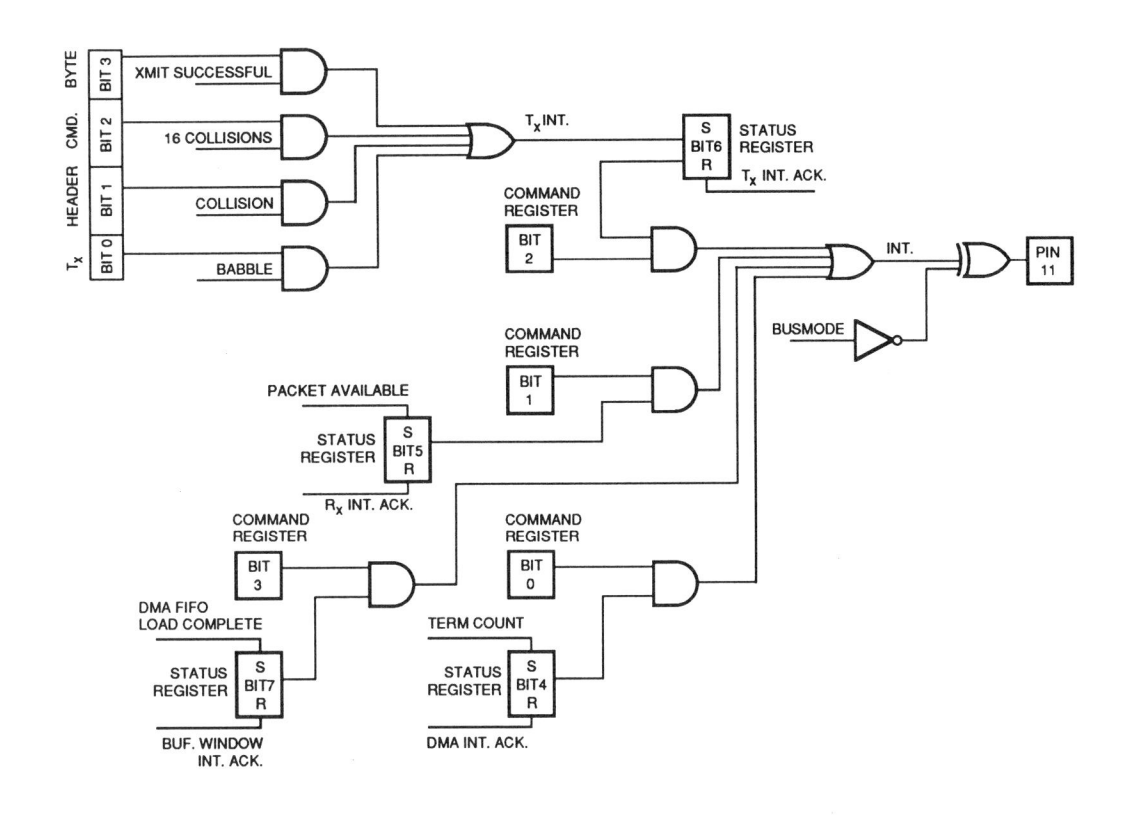

Figure 18. Functional diagram of interrupt logic.

Index

27210 word-wide EPROM, 378–387
27C203 fast-pipelined EPROM, 388–403
28F020 256 K × 8 CMOS flash memory, 404–431
28HC256 high speed EEPROM, 493–502
74ACT2152 cache address comparator, 642–654
74F786 4-input asynchronous arbiter, 288
74LS logic, 268
74LS630 16-bit error detection/correction circuit, 675–680
8005 Ethernet controller, 993–1032
8005 User's Guide, 1047–1067
8020 Manchester code converter, 1033–1046
8086 EEPROM interface, 515
8088 EPROM interface, 513

AC characteristics of Schottky logic, 276
ACL (advanced CMOS logic), 172
Acquisition of data in PLL, 967
Address comparator cache, 642–654
Alkaline manganese batteries, 345
ALS logic, 161
Am27H010 1 Megabit CMOS EPROM, 360–377
Am29PL141 fuse programmable controller, 237
Analog data separator, 956–959
Arbiter, asynchronous, 288
Arbitration, 293
Arbitration, dual-port RAM, 728
Arbitration in shared resource systems, 286–290
Arbitration logic, 296
Arrhenius relation, 477–478
Asynchronous arbiter (74F786), 288
Asynchronous transmission (DUSC), 755

Balanced interface circuit, 126
Bandgap reference, 65
Batteries, 345–350
  energy density, 349
  primary, 345
  secondary, 345
  service life, 349
Battery backup design, 342
Battery backup for real-time clocks, 913
Battery backup, selection of cell, 350
Battery charging circuits, 355–357
Battery isolation, 353
BiSync DUSC, 791
Bit-oriented protocol, 756
Bit shift, in MFM, 456
Buses, 72–156

Cache address comparator, 661–673
Cache coherency, 658
Cache director application, 650–660
Cache memory, copy-back scheme, 657
Cache memory design tutorial, 674–682
Cache system performance, 678
Cache tag comparator, 656
Cache tag RAM, applications, 674–682
Cache, write-through scheme, 656
Capacitor
  ceramic, 119
  in decoupling, 117–125

Carbon zinc batteries, 345
Carrier sense multiple access, 1048
CAS-before-RAS, 532
Cascading four port RAM, 711
Ceramic capacitor, 119–125
Ceramic capacitor, voltage effects, 122
Chained receive frames, Ethernet controller, 1003
Characteristic impedance, 75
Charge pump, 965
CMOS circuits, operating voltages, 210
CMOS devices, TTL compatible, 205–216
CMOS dynamic RAM, 520–537
CMOS EPROM, Am27H010, 360–377
CMOS flash memory, 404–431
CMOS noise margins, 211–213
Code converter, Manchester, 1033–1046
Coherency, cache memory, 658
Combo I/O chip for PC/AT, 803–838
Communications controller. 755–802
Communications via multi-port memory, 292
Connecting four port RAMS to CPUs, 713
Controller, 562–633
  DRAM, 542–613
  floppy disk, 928–953
Correction, errors in memory, 683–694
Counter/timer (DUSC), 768
Coupling, ground trace, 98
Crosstalk in logic circuits, 100, 194–198
Crosstalk, ribbon cable, 108
Crystal oscillators, 913, 925
CSMA/CD. 1048

Data acquisition, in PLL, 967
Data controller, Ethernet, 993–1032
Data retention design, 341
Data separator
  analog, 957
  designing with the DP8473, 964–982
  digital, 960–964
  floppy disk, 954–982
  performance, 991
DC characteristics, of Schottky logic, 272
DC noise margin of Schottky logic, 274
Decoder, Manchester, 1035
Decoupling capacitor, 117–125
Decoupling, transmission lines, 95
Design considerations, transmission lines, 97–116
Design for data retention, 341
Designing for testability with the PROSE device, 261–265
Designing testable state machines, 256–260
Detection, errors in memory, 683–694
Detector
  low voltage, 355
  power fail, 351
Digital data separator, 960–964
Digital phase-locked loop, DUSC, 778
Direct mapped cache, 677
Disk drives, types, 989
Distributed refresh, in power minimization, 344
DMA and dual-port RAMs, 728

INDEX

DMA control, DUSC, 778
DMA interface, in SCSI chip, 841
DP8420A DRAM controller, 562–633
DP8473 floppy disk controller, 928–953
DP8490 application in printer controller, 879–898
DP8490 SCSI interface, 839–878
DRAM controller, 562–633
   application, 634–649
   dual port accessing, 614–617
     RAS and CAS configuration, 602–606
   refresh, 573–579
DRAM
   nibble mode, 538
   page mode, 531, 538
   read cycle, 531
   refresh, 532
   static column mode, 538
   write cycle, 531
Drive configuration, disk interface, 984
Drive types, disk drives, 987
Dual-port access, DRAM controller, 614–617
Dual universal serial communications controller, 755–802
Dual voltage switch, 352
Dual-port RAM
   applications, 738–740
   control logic, 730
   memory arbitration, 735
   memory expansion, 734
   timing, 732
Dual-port RAMs in computer systems, 728–740
DUSC (dual universal serial communications controller), 755–802
   BiSync, 790, 791
   DMA control, 778
   interrupts, 770
   pins, 758
   receiver commands, 776
   registers, 760
   transmitter commands, 776
Dynamic RAM, $4 M \times 1$ CMOS 520–537

EEPROM, 28HC256, 493–502
EEPROM interfacing, 503–519
EEPROM microprocessor interfaces, 506–560
EEPROM, special functions, 496
EEPROM, write mode, 494
Encoding techniques, 954–957
Energy density, batteries, 350
Enhanced SCSI interface, 839–878
EPROM, CMOS Am27H010. 361–378
EPROM
   erasing, 366
   family pin-out, 380
   programming, quick-pulse, 386
   read mode, 366, 384
Error detection/correction using 74LS630, 682–693
Error detection/correction chip, 695–700
Ethernet addressing, 1049
Ethernet controller registers, 998
Ethernet controller system design, 1051–1057
Ethernet controller user's guide, 1047–1068
Ethernet data controller, 993–1032
Ethernet data format, 1049
Ethernet packet format, 996

Failure rate circulations, 462, 476–480
Fall-through in FIFO, 721

FAST ICs, 291–303
Fast pipelined EPROM, 389–404
Fast static RAM, 313–318, 319–328
FIFO
   flow-through mode, 747
   IDT7201, 741–753
   software versus hardware, 725
   width expansion, 723, 747
   understanding, 720–727
Flash memory, 28F020, 404–431
Flash memory
   applications of, 407
   erase algorithm, 422–447
   principles of, 408, 435–439
   programming guide, 432–455
   reliability, 456–492
Flip-flops, PAL, 227
Floppy controller design, 991
Floppy disk controller
   register, 936
Floppy disk controller in PC-AT, 983–992
Floppy disk cable interface to PC, 984
Floppy disk controller
   commands, 939
   DP8473, 928–953
   functional description, 931
Floppy disk data separator, 954–982
Format a track, floppy disk controller, 943
Format, of track, 955
Four-port RAM to 68000 connection, 712
Four-port RAM
   introduction, 706–718
   in multiprocessor design, 700–705
   cascading, 710
   in DSP, 704
Fuse programmable controller (Am29PL141), 237

Ground trace couping, 98
Guide to flash memory reprogramming, 432–455

Hazards, in latches, 253
HCMOS devices, 169
Hidden refresh, 532
High speed 28HC256 EEPROM, 493–502
High speed cache directory application, 650–660
Hit rate, cache, 679
HM628128 $128 K \times 8$ static RAM, 319–328

I/O combo chip for PC/AT, 803–838
IDT7201 FIFO introduction, 720–727
IDT7201 FIFO data, 741–753
Importance of minimization, in testable circuits, 250
Input characteristics, AC/HC logic, 174
Interface TTL/CMOS, 208
Interfacing, DRAM controller to 68000, 634–649
Interfacing, multiple microprocessors, 291–303
Interfacing the 68000 to the Multibus, 298–303
Interrupts, DUSC, 770

Latch hazards, 253
Latched EEPROM, 503
Lattice diagram, 78
Lead acid batteries, 348
Line driving, 90, 93
Line driving and system design, 74–125
Line printer port, 810
Lines, mismatched, 84

INDEX

Lithium batteries, 345
LM2984C microprocessor power supply system, 51–64
Logic circuits, comparison of, 177
Logic circuits, power consumption, 177
Logic families, 158–180
Logic families, design considerations, 181–204
Logic families, power supply consideration, 184–189
Logic hazards, 250
Loop filter, 965
Low voltage detector, 355
Low-power memory systems design, 329–359
Low-power Schottky logic, 268–281

Manchester
    code converter, 1033–1046
    decoder, 1035
MC34063 switching regulator, 2–38
MC34164 micropower undervoltage sensing circuit, 65–71
MCM514100 4M × 1 CMOS dynamic RAM, 520–535
MCM6206 32 K × 8 fast static RAM, 313–318
Mealy and Moore state machines, 221–223
Memory expansion, dual-port RAM, 734, 737
Memory, flash, 404–431
Memory systems, 312–753
Memory systems design, low power, 329–359
Memory systems design using CHMOS, 335–345
Mercuric oxide batteries, 345
Metastability
    and MTBF, 283
    characteristics, 282
    in shared resource systems, 287
    primer, 282–285
Metastable characteristics of logic elements, 199–204
MFM, 954–956
Micropower undervoltage sensing circuit, 65–71
Microprocessor
    interfaces to EEPROM, 506–561
    power supply system, 61–64
    power supply controller, 39–50
    reset, by LM2984C, 62
Mismatched lines, 84
Miss rate, cache, 679
MM58274C real time clock, 899–911
MM58274C real time clock application, 912–927
Modulation rate, in RS422/423 circuits, 127
MTBF (and metastability), 283
Multiple microprocessor interfacing, 291–303
Multiple reflections, 77
Multiprocessor design using fourport RAM, 701–706

Nickel cadmium batteries, 348
Noise decoupling, 99
Noise margins, CMOS, 211–213
Non-volatile (low power) storage, 330

Operating voltages, of CMOS, 210
Oscillator design, floppy disk controller, 980–982
Oscillators, 255
Output characteristics, AC/HC logic, 175
Oxide breakdown in flash memory, 461

PAL devices as sequencers, 227
Parallel
    I/O communications, 292
    printer port, 810
    priority resolution, arbitration, 293
    termination, 94

Partial power-down in CMOS systems, 186
PC/AT combo I/O chip VL82C106, 803–838
Phase lock loop theory, 964–973
Phase locked loop, 957
    floppy disk controller, 934
    DUSC, 778
PHD16N8-5 programmable decoder, 304–311
Pipelined bus appplication of EPROM, 395
Pipelined EPROM, 3880403
PMS14R31 (PROSE sequencer), 236
Power analysis
    im memory systems, 339
    in low power design, 344
Power consumption
    im HCT circuits, 214
    logic circuits, 177
    minimization, 343
Power fail
    detector, 351
    protection, 918
Power sensing, 351
Power supplies and real-time clocks, 916–920
Power supply considerations, in logic families, 184–189
Power supply controller (TCA5600), applications, 47–49
Power supply controller, 39–50
Power supply isolation, 916
Power supply
    power-down override, 62
    standby output. 61
Power switching circuits, 351–355
Power up/down protection in flash memory, 416
Power-down override, 62
Primary batteries, 345
Printer controller using SCSI controller, 879–898
Printer port, 810
Programmable decoder logic, 304–311
Programmable logic sequencers (PLS), 233–238
Programmable voltage regulator, 48
Propagation delay, logic circuits, 180
Propagation velocity, 75
PROSE sequencer (PMS14R21), 236
Protection circuitry
    ACL, 173
    HCMOS, 170

Queues, in FIFOs, 720
Quick-erase, flash memory, 413, 451
Quick-pulse EPROM programming algorithm, 385
Quick-pulse programming, flash memory, 413

RAM, 4 M × 1 CMOS dynamic, 520–537
RAS and CAS configuration DRAM controller, 602–606
Real time clock
    application, 912–927
    application hints, 911
    functional description, 903
    in I/O combo chip, 811
    MM58274C, 899–911
    power supplies, 916–920
Receiver, DUSC, 790
Reconvergent fanout, 248
Reflections, 75
    multiple, 77
Refresh
    DRAM, 532
    DRAM controller, 573–579

INDEX

Reliability
   of flash memory, 456–492
   of PLCC packages, 472
Reset circuits, 353
Reset control, by microprocessor controller, 39–50
Ribbon cable, crosstalk, 108
Ringing, 88
Rise time versus line delay, 86
RS232C, 140
RS422, 139, 140, 149
RS422/423 drivers and receivers, 126–135
RS423, 140
RS485, 139, 141, 149, 151

Scan commands, floppy disk controller, 944
Schottky devices, characteristics of, 269
Schottky diodes in LS logic, 268
Schottky logic, design guidelines, 279–281
SCN6862 dual universal serial communications controller, 755–802
SCSI controller
   applications, 857–859
   facilities, 856
   operation, 849
   registers, 844
SCSI interface DP8490, 839–878
SCSI printer controller, 879–898
SCSI run-time software, 887–893
Secondary batteries, 345
Seek command, floppy disk controller, 945
Semaphore arbitration, 298
Sequencers, programmable logic, 233–238
Serial communication port, 810
Serial communications controller, 755–802
Serial interface standards, 136–148
Series termination, 79
Service life, batteries, 349
Shared bus architecture, 293
Shorted line, 78
Silver oxide batteries, 345
Simulating single stuck-at faults, 247
Simultaneous switching in logic systems, 189–193
SRAMs, description of, 332
State machine
   applications, 218
   appplication of pipelined EPROM. 396
   definition, 217
   design, 217–245
   device selection, 223
   representation, 240
   syntax, 241

   theory, 219–221
   tutorial, 238–245
Step-down switching regulator operation, 5–7
Step-up switching regulator operation, 9–11
Stripline, 90
Supply voltage ripple, 184
Switching regulator
   design examples, 7, 11, 17
   principles of operation, 2
   step-down, 5–7
   uA78S40, 2–38
   voltage inverting, 13–16
SYN pattern stripping, 792

TCA5600 power supply controller, 39–50
Termination, 75
   parallel, 94
   series, 79
Test vectors, 266
Testability, 246–267
Testability, defining, 246
Testable combinatorial circuits, designing, 247–251
Testable sequential circuits, 252–255
Testable state machines, designing, 256–260
Timer EEPROM, 503
Two-pointer FIFO, 721
Transmission line concepts, 75–96
Transmission line drivers and receivers for RS422/RS423, 126–135
Transmission lines, design considerations, 97–116
Transmitter, DUSC, 786
TTL/CMOS interface, 208
TTL compatible CMOS devices, 205–216

uA78S40 switching regulator, 2–38
Unbalanced interface circuit, 126
Undervoltage sensing circuit, 65–71

Variable length FIFOs, 721
VL82C106 PC/AT combo I/O chip, 803–838
Voltage-inverting switching regulator operation, 13–16

Watchdog circuit, microprocessor controller, 48
Watchdog timer, in flash EPROM programming, 447, 452
Watchdog, using real-time clock, 922
Word-wide EPROM, 378–387
Write precompensation, 935
Write protection, EEPROM, 495
Write-through, cache memory, 658

Z80 EEPROM interface, 509